Endocrine system
  Acts by means of hormones secreted into
  the blood to control processes that require
  duration rather than speed – e. g., metabolic
  activities and water and electrolyte balance.
  See chapters 18 and 19.

Immune system
  Defends against foreign invaders and cancer
  cells; paves way for tissue repair.
  See chapter 12.

Integumentary system
  Serves as protective barrier between external
  environment and remainder of body; sweat
  glands and adjustments in skin blood flow
  important in temperature regulation.
  See chapters 12 and 17.

Muscular and skeletal systems
  Support and protect body parts and allow body
  movement; heat-generating muscle contractions
  important in temperature regulation; calcium
  stored in bone.
  See chapters 8, 17, and 19.

Exchanges with all other systems

Protects against
foreign invaders

Keeps internal
fluids in

Keeps foreign
material out

Enables body
to interact
with external
environment

Body systems
maintain
homeostasis

HOMEOSTASIS
  A dynamic steady state of the
  constituents in the internal fluid
  environment that surrounds and
  exchanges materials with cells.
  See chapter 1.
  Factors homeostatically maintained are:
  • Concentration of nutrient molecules
    See chapters 16, 17, 18, and 19.
  • Concentration of O_2 and CO_2
    See chapter 13.
  • Concentration of waste products
    See chapter 14.
  • pH   See chapter 15.
  • Concentration of water, salt, and other
    electrolytes
    See chapters 14, 15, 18, and 19.
  • Temperature   See chapter 17.
  • Volume and pressure
    See chapters 10, 14, and 15.

Homeostasis is
essential for
survival of cells

CELLS
  Need homeostasis for their own
  survival and for performing
  specialized functions essential for
  survival of the whole body.
  See chapters 1, 2, and 3.
  Need a continual supply of nutrients
  and O_2 and ongoing elimination of
  acid-forming CO_2 to generate the
  energy needed to power life-sustaining
  cellular activities as follows:
  Food + $O_2 \rightarrow CO_2 + H_2O$ + Energy
  See chapter 17.

Cells make up
body systems

HUMAN PHYSIOLOGY

From Cells to Systems

SECOND EDITION

HUMAN PHYSIOLOGY

From Cells to Systems

SECOND EDITION

LAURALEE SHERWOOD

Department of Physiology
School of Medicine
West Virginia University

WEST PUBLISHING COMPANY

Minneapolis/St. Paul
New York
Los Angeles
San Francisco

PRODUCTION CREDITS

COPYEDITING   Patricia Lewis
INTERIOR AND COVER DESIGN   Diane Beasley
COMPOSITION   G&S Typesetters, Inc.
ARTWORK   Wayne Clark, Darwen and Vally Hennings, Carlyn
Iverson, Sandra McMahon, Elizabeth Morales-
Denney, Precision Graphics, Publication Services,
Rolin Graphics, John and Judy Waller, and Cyndie
C.H.-Wooley. Individual illustration credits follow
Index.
INDEX   E. Virginia Hobbs
COVER IMAGE   Custom Medical Stock Photo, Inc.
PRODUCTION, PREPRESS, PRINTING, AND BINDING   West Publishing
Company

WEST'S COMMITMENT TO THE ENVIRONMENT

In 1906, West Publishing Company began recycling materials
left over from the production of books. This began a tradition
of efficient and responsible use of resources. Today, up to
95 percent of our legal books and 70 percent of our college
and school texts are printed on recycled, acid-free stock. West
also recycles nearly 22 million pounds of scrap paper annually—the equivalent of 181,717 trees. Since the 1960s, West
has devised ways to capture and recycle waste inks, solvents,
oils, and vapors created in the printing process. We also recycle plastics of all kinds, wood, glass, corrugated cardboard,
and batteries, and have eliminated the use of styrofoam book
packaging. We at West are proud of the longevity and the
scope of our commitment to the environment.

Library of Congress Cataloging-in-Publication Data

Sherwood, Lauralee.
    Human physiology : from cells to systems / Lauralee
Sherwood. —
    2nd ed.
        p.   cm.
    Includes indexes.
    ISBN 0-314-01225-7 (hard)
    1. Human physiology.   I. Title.
    [DNLM: 1. Anatomy.   2. Physiology.   QT 104 S554h]
QP34.5.S48   1993
612—dc20
DNLM/DLC
for Library of Congress   92-48672
                                                            CIP

To my family, with love and gratitude
for their support, encouragement, and understanding:

my husband, Peter Marshall
my parents, Larry and Lee Sherwood
my daughter and son-in-law, Melinda and Mark Marple
my daughter, Allison Marshall
my mother-in-law, Sarah Marshall
and in memory of my father-in-law, Harold Marshall

BRIEF CONTENTS

Preface   xiii

CHAPTER 1   Homeostasis: The Foundation of Physiology   1

CHAPTER 2   Cellular Physiology   15

CHAPTER 3   Plasma Membrane and Membrane Potential   45

CHAPTER 4   Neuronal Physiology   78

CHAPTER 5   Central Nervous System   104

CHAPTER 6   Peripheral Nervous System: Afferent Division; Special Senses   150

CHAPTER 7   Peripheral Nervous System: Efferent Division   197

CHAPTER 8   Muscle Physiology   213

CHAPTER 9   Cardiac Physiology   258

CHAPTER 10   Blood Vessels and Blood Pressure   299

CHAPTER 11   Blood   347

CHAPTER 12   Body Defenses   368

CHAPTER 13   Respiratory System   412

CHAPTER 14   Urinary System   464

CHAPTER 15   Fluid and Acid-Base Balance   510

CHAPTER 16   Digestive System   541

CHAPTER 17   Energy Balance and Temperature Regulation   595

CHAPTER 18   Principles of Endocrinology; Central Endocrine Organs   614

CHAPTER 19   Peripheral Endocrine Organs   649

CHAPTER 20   Reproductive System   695

APPENDIXES

A   A Review of Chemical Principles   A–1

B   Storage, Replication, and Expression of Genetic Information   B–1

C   Control, Functions, and Interactions of Immune-Response Effectors   C–1

D   Answers to End-of-Chapter Objective Questions   D–1

E   Supplemental Reading List   E–1

Glossary   G–1

Index   I–1

Credits   CR–1

CONTENTS

Preface   xiii

 CHAPTER 1

Homeostasis: The Foundation of Physiology   1

Introduction   2
Levels of Organization in the Body   2
● CONCEPTS, CHALLENGES, AND CONTROVERSIES: A First: Brain
    Cells Growing in Laboratory Dishes   3
Concept of Homeostasis   5
● A CLOSER LOOK AT EXERCISE PHYSIOLOGY: What Is Exercise
    Physiology?   7
Chapter in Perspective: Focus on Homeostasis   12
Chapter Summary   12

 CHAPTER 2

Cellular Physiology   15

Introduction   16
● CONCEPTS, CHALLENGES, AND CONTROVERSIES: HeLa Cells:
    Problems in a "Growing" Industry   17
Organelles   18
● A CLOSER LOOK AT EXERCISE PHYSIOLOGY: Aerobic Exercise:
    What For and How Much?   34
Cytosol and Cytoskeleton   34
Chapter in Perspective: Focus on Homeostasis   42
Chapter Summary   42

 CHAPTER 3

Plasma Membrane and Membrane Potential   45

Membrane Structure and Composition   46
● CONCEPTS, CHALLENGES, AND CONTROVERSIES: Cystic Fibrosis:
    A Fatal Defect in Membrane Transport   49
Membrane Receptors and Postreceptor Events   50
Cell-to-Cell Adhesions   54
Membrane Transport   56
● A CLOSER LOOK AT EXERCISE PHYSIOLOGY: Exercising Muscles
    Have a "Sweet Tooth"   64
Membrane Potential   68
Chapter in Perspective: Focus on Homeostasis   74
Chapter Summary   75

 CHAPTER 4

Neuronal Physiology   78

Electrical Signals: Graded Potentials and Action
    Potentials   79
Synapses and Neuronal Integration   91
● A CLOSER LOOK AT EXERCISE PHYSIOLOGY: The Yells and
    Grunts of an Athlete May Serve a Physiological
    Purpose   97
● CONCEPTS, CHALLENGES, AND CONTROVERSIES: Parkinson's
    Disease, Pollution, and Ethical Problems   100
Chapter in Perspective: Focus on Homeostasis   101
Chapter Summary   101

 CHAPTER 5

Central Nervous System   104

Introduction   105
Comparison of the Endocrine and Nervous Systems   105
Organization of the Nervous System   107
Protection and Nourishment of the Brain   109
● CONCEPTS, CHALLENGES, AND CONTROVERSIES: The Human
    Brain: In Search of Self-Understanding   110
Cerebral Cortex   117
Subcortical Structures and Their Relationship with the
    Cortex in Higher Brain Functions   125
● CONCEPTS, CHALLENGES, AND CONTROVERSIES: Alzheimer's
    Disease: A Tale of Beta Amyloid and Dementia   132
Cerebellum   135
Brain Stem   136
Spinal Cord   139
● A CLOSER LOOK AT EXERCISE PHYSIOLOGY: Swan Dive or Belly
    Flop: It's a Matter of CNS Control   144
Chapter in Perspective: Focus on Homeostasis   146
Chapter Summary   147

 CHAPTER 6

Peripheral Nervous System: Afferent Division; Special Senses   150

Introduction   151
Receptor Physiology   152

● A CLOSER LOOK AT EXERCISE PHYSIOLOGY: Back Swings and Pre-Jump Crouches: What Do They Have in Common? 155
Pain 157
● CONCEPTS, CHALLENGES, AND CONTROVERSIES: Acupuncture: Is It for Real? 160
Eye: Vision 161
Ear: Hearing and Equilibrium 177
Chemical Senses: Taste and Smell 190
Chapter in Perspective: Focus on Homeostasis 193
Chapter Summary 194

CHAPTER 7

Peripheral Nervous System: Efferent Division 197

Introduction 198
Autonomic Nervous System 198
● CONCEPTS, CHALLENGES, AND CONTROVERSIES: The Autonomic Nervous System and Aging: A Fortuitous Find 204
Somatic Nervous System 204
Neuromuscular Junction 205
● A CLOSER LOOK AT EXERCISE PHYSIOLOGY: Loss of Muscle Mass: A Plight of Space Flight 209
Chapter in Perspective: Focus on Homeostasis 210
Chapter Summary 210

CHAPTER 8

Muscle Physiology 213

Introduction 214
Structure of Skeletal Muscle 215
Molecular Basis of Skeletal-Muscle Contraction 218
Skeletal-Muscle Mechanics 226
Skeletal-Muscle Metabolism and Fiber Types 234
● A CLOSER LOOK AT EXERCISE PHYSIOLOGY: Are Athletes Who Use Steroids to Gain Competitive Advantage Really Winners or Losers? 239
Control of Motor Movement 240
● CONCEPTS, CHALLENGES, AND CONTROVERSIES: Muscular Dystrophy: When One Small Step is a Big Deal 241
Smooth and Cardiac Muscle 246
Chapter in Perspective: Focus on Homeostasis 254
Chapter Summary 255

CHAPTER 9

Cardiac Physiology 258

Introduction 259
Anatomical Considerations 260
● CONCEPTS, CHALLENGES, AND CONTROVERSIES: Fetal Circulation: A Fetus Has It All Mixed Up 262
Electrical Activity of the Heart 268
● A CLOSER LOOK AT EXERCISE PHYSIOLOGY: The What, Who, and When of Stress Testing 278

Mechanical Events of the Cardiac Cycle 278
Cardiac Output and Its Control 283
Nourishing the Heart Muscle 290
Chapter in Perspective: Focus on Homeostasis 295
Chapter Summary 296

CHAPTER 10

Blood Vessels and Blood Pressure 299

Introduction 300
● CONCEPTS, CHALLENGES, AND CONTROVERSIES: From Humors to Harvey: Historical Highlights in Circulation 302
Arteries 305
Arterioles 308
Capillaries 317
Veins 328
Blood Pressure 332
● A CLOSER LOOK AT EXERCISE PHYSIOLOGY: The Ups and Downs of Hypertension and Exercise 340
Chapter in Perspective: Focus on Homeostasis 343
Chapter Summary 344

CHAPTER 11

Blood 347

Introduction 348
Plasma 348
Erythrocytes 349
● A CLOSER LOOK AT EXERCISE PHYSIOLOGY: Blood Doping: Is More of a Good Thing Better? 352
● CONCEPTS, CHALLENGES, AND CONTROVERSIES: In Search of a Blood Substitute 353
Leukocytes 355
Platelets and Hemostasis 358
Chapter in Perspective: Focus on Homeostasis 364
Chapter Summary 365

CHAPTER 12

Body Defenses 368

Introduction 369
Nonspecific Immune Responses 371
Specific Immune Responses: General Concepts 380
B Lymphocytes: Antibody-Mediated Immunity 381
● CONCEPTS, CHALLENGES, AND CONTROVERSIES: Vaccination: A Victory over Many Dreaded Diseases 388
T Lymphocytes: Cell-Mediated Immunity 392
● A CLOSER LOOK AT EXERCISE PHYSIOLOGY: A Help or Hindrance to Immune Defense: 401
Immune Diseases 401
External Defenses 405
Chapter in Perspective: Focus on Homeostasis 408
Chapter Summary 408

CHAPTER 13

Respiratory System 412

Introduction 413
Respiratory Mechanics 417
Gas Exchange 437
Gas Transport 442
Control of Respiration 450
● CONCEPTS, CHALLENGES, AND CONTROVERSIES: Effects of
  Heights and Depths on the Body 451
● A CLOSER LOOK AT EXERCISE PHYSIOLOGY: How to Find Out
  How Much Work You're Capable of Doing 459
Chapter in Perspective: Focus on Homeostasis 460
Chapter Summary 460

CHAPTER 14

Urinary System 464

Introduction 465
Glomerular Filtration 471
Tubular Reabsorption 479
Tubular Secretion 488
Urine Excretion and Plasma Clearance 490
● A CLOSER LOOK AT EXERCISE PHYSIOLOGY: When Protein in the
  Urine Does Not Mean Kidney Disease 503
● CONCEPTS, CHALLENGES, AND CONTROVERSIES: Dialysis:
  Cellophane Tubing or Abdominal Lining as an Artificial
  Kidney 504
Chapter in Perspective: Focus on Homeostasis 506
Chapter Summary 506

CHAPTER 15

Fluid and Acid-Base Balance 510

Fluid Balance 511
● A CLOSER LOOK AT EXERCISE PHYSIOLOGY: A Potentially Fatal
  Clash: When Exercising Muscles and Cooling Mechanisms
  Compete for an Inadequate Plasma Volume 517
● CONCEPTS, CHALLENGES, AND CONTROVERSIES: Breaching the
  Blood-Brain Barrier 519
Acid-Base Balance 524
Chapter in Perspective: Focus on Homeostasis 537
Chapter Summary 538

CHAPTER 16

Digestive System 541

Introduction 542
Mouth 550
Pharynx and Esophagus 553
Stomach 555
● A CLOSER LOOK AT EXERCISE PHYSIOLOGY: Pregame Meal:
  What's In and What's Out? 559

● CONCEPTS, CHALLENGES, AND CONTROVERSIES: Beaumont: The
  Beginnings of Physiology in America 561
Pancreatic and Biliary Secretions 567
Small Intestine 575
Large Intestine 587
Overview of the Gastrointestinal Hormones 590
Chapter in Perspective: Focus on Homeostasis 591
Chapter Summary 592

CHAPTER 17

Energy Balance and Temperature Regulation 595

Energy Balance 596
Temperature Regulation 601
● A CLOSER LOOK AT EXERCISE PHYSIOLOGY: What the Scales
  Don't Tell You 602
● CONCEPTS, CHALLENGES, AND CONTROVERSIES: The Extremes
  of Heat and Cold Can Be Fatal 610
Chapter in Perspective: Focus on Homeostasis 611
Chapter Summary 611

CHAPTER 18

**Principles of Endocrinology; Central Endocrine
Organs 614**

Introduction 615
General Principles of Endocrinology 616
● CONCEPTS, CHALLENGES, AND CONTROVERSIES: Tinkering With
  Our Biological Clocks 629
Hypothalamus and Pituitary 631
● A CLOSER LOOK AT EXERCISE PHYSIOLOGY: The Endocrine
  Response to the Challenge of Combined Heat and
  Marching Feet 633
Endocrine Control of Growth 639
● CONCEPTS, CHALLENGES, AND CONTROVERSIES: Growth and
  Youth in a Bottle? 644
Chapter in Perspective: Focus on Homeostasis 646
Chapter Summary 647

CHAPTER 19

Peripheral Endocrine Organs 649

Thyroid Gland 650
Adrenal Glands 656
Endocrine Control of Fuel Metabolism 667
● CONCEPTS, CHALLENGES, AND CONTROVERSIES: Diabetics and
  Insulin: Some Have It and Some Don't 676
Endocrine Control of Calcium Metabolism 681
● A CLOSER LOOK AT EXERCISE PHYSIOLOGY: Osteoporosis: The
  Bane of Brittle Bones 691
Chapter in Perspective: Focus on Homeostasis 691
Chapter Summary 692

≡● CHAPTER 20

Reproductive System   695

Introduction   696
Male Reproductive Physiology   702
Sexual Intercourse between Males and Females   711
Female Reproductive Physiology   714

● A CLOSER LOOK AT EXERCISE PHYSIOLOGY: Menstrual Irregularities: When Cyclists and Other Athletes Do Not Cycle   724
● CONCEPTS, CHALLENGES, AND CONTROVERSIES: Reproductive-Related Technology: Taking the Making of Life into Our Own Hands   726
● A CLOSER LOOK AT EXERCISE PHYSIOLOGY: Exercise and Pregnancy: A Compatible or Contraindicated Combination?   735

Chapter in Perspective: Focus on Homeostasis   742
Chapter Summary   742

Appendix A: A Review of Chemical Principles   A–1
Appendix B: Storage, Replication, and Expression of Genetic Information   B–1
Appendix C: Control, Functions, and Interactions of Immune-Response Effectors   C–1
Appendix D: Answers to End-of-Chapter Objective Questions   D–1
Appendix E: Supplemental Reading List   E–1
Glossary   G–1
Index   I–1
Credits   CR–1

PREFACE

PHILOSOPHY, GOALS, AND THEME

A baby first discovering that it can control its own hands will spend many hours, fascinated, as it manipulates them in front of its face. Most of us, even infants, have a natural curiosity about how our bodies work. Our bodies are quite miraculous. No machine has been constructed that can take over even a portion of a natural body function as effectively. By capitalizing on students' natural curiosity about themselves, I have strived to make physiology a subject that they can enjoy learning.

Even the most tantalizing subject matter, however, can be drudgery to study and difficult to comprehend if not effectively presented. Therefore, this book has a logical, understandable format that is unencumbered by unnecessary details and emphasizes how each concept is an integral part of the whole subject matter. Too often, students view isolated sections of a physiology course as separate entities; by understanding how each component of the body depends on other components, a student can appreciate the integrated functioning of the human body. The text focuses on the mechanisms of body function from cells to systems and is organized around the central theme of homeostasis—how the body meets changing demands while maintaining the internal constancy necessary for all cells and organs to function.

The text is written with its primary target audience—undergraduate students preparing for health-related careers—in mind. Its approach and depth are appropriate, however, for other undergraduate student populations. Because it is intended to serve as an introductory text and, for most students, may be their only exposure to a formal physiology text, there is broad coverage of all aspects of physiology, yet depth, where needed, is not sacrificed. The scope of this text has been limited by judicious selection of pertinent content that a student can reasonably be expected to assimilate in a one-semester physiology course.

This edition has been revised with the goal of offering instructors two choices in length and depth of physiology textbooks while retaining all of the features that made the first edition successful. Now available is a carefully condensed version of the parent book for those students who do not have a math and physics background, *Fundamentals of Physiology: A Human Perspective* (L. Sherwood, West Publishing Company, 1991). In preparing the abbreviated text, the parent book was selectively pruned to a length and depth more suitable for lower-level undergraduate courses of shorter duration and intensity. This second edition of the parent book is designed primarily for upper-level undergraduate students who have more math and science background and need more breadth and depth of coverage in physiology. Accordingly, this edition is slightly more quantitative than the first edition while largely retaining the scope, length, and depth of the original book. Chemistry and genetic appendixes are provided for those who need a review or a quick reference source, and a particular effort has been made to clarify all discussions of math and physics. Also, because anatomy is not a prerequisite course, enough relevant anatomy is integrated within the text to make the inseparable relation between structure and function meaningful.

To keep pace with today's rapid advances in the health sciences, students in the health professions must be able to draw on their conceptual understanding of physiology instead of merely recalling isolated facts that soon may be outdated. Therefore, this text is designed to promote understanding of the basic principles and concepts of physiology rather than memorization of details. The text is written in simple, straightforward language, and every effort has been made to assure smooth reading through good transitions, logical reasoning, and integration of ideas throughout the text.

In consideration of the clinical orientation of the target group, research methodologies and data are not emphasized, although the material is based on up-to-date evidence. New information based on recent discoveries has been added to all chapters. The first edition was carefully pared of redundant and extraneous points so that addition of cutting-edge content and helpful new pedagogy has not increased the length of the book for the second edition. Some controversial ideas and hypotheses are presented to illustrate that physiology is a dynamic, changing discipline.

CHANGES IN THIS EDITION

Organizational Changes   This edition retains the same number and sequence of chapters as the first edition. However, the content within some chapters has been slightly reorganized as the following examples illustrate:

- The genetics section in chapter 2 has been moved to an appendix because most physiology instructors do not cover this topic. Placement in an appendix frees up valuable text space but still enables students to review the concepts as necessary.
- The comparison of the nervous and endocrine systems has been expanded in chapter 5 (Central Nervous System) to provide a more extensive, early foundation in endocrinology. This expansion, coupled with the introduction to endocrine control and hormones in chapter 1 and the weaving in of the roles of specific hormones in appropriate chapters (such as coverage of vasopressin in the kidney chapter), exposes students to endocrine function throughout the text. A thorough, integrated coverage of the endocrine system is presented late in the text (chapters 18 and 19) after the students have the knowledge base for the in-depth discussion of this topic that its importance merits.
- Several sections in chapter 8 are reorganized so that all of the material related to skeletal-muscle mechanics is presented together.
- The discussion of fluid balance in chapter 15 is reorganized so that control of salt balance and regulation of ECF volume are grouped together and control of water balance and regulation of ECF osmolarity are together. The material flows better now than in the first edition where the regulation of ECF volume and osmolarity were grouped together as were control of salt and water balance.

Increased Quantitative Coverage   As mentioned in the preceding section, this edition is a little more quantitatively oriented than the first but still does not rely heavily on formulas and graphs. The following examples illustrate the types of quantitative treatments added:

- The Nernst equation is now included in the discussion of membrane potential.
- The lever action of muscles is presented in the section on skeletal-muscle mechanics.
- Coverage of the law of LaPlace in relation to alveolar surface tension and the role of pulmonary surfactant is expanded.
- A new graph depicts the magnitude of increase in muscle tension during motor-unit recruitment in muscles with differing sizes of motor units.

Updated Content   All chapters have been updated based on current research findings, as evidenced by the following new or expanded topics:

- Introduction of vaults, a newly discovered organelle.
- Expanded discussion of neuropeptides and neuromodulation.
- Presentation of new findings related to memory.
- New section on alternative theories of how T-tubule potential is coupled with Ca^{++} release from the sarcoplasmic reticulum.

- More extensive coverage of vasoactive substances released from endothelial cells; introduction of nitric oxide and its recently identified multiple roles in the body.
- Updated discussion of steroid-hormone receptors, incorporating new information on hormone response elements (HREs).
- Coverage of insulin's role in glucose transporter recruitment.
- Incorporation of new information on factors determining the onset of parturition.

Revised Coverage   Other topics that do not involve newly discovered information were added, expanded, or rewritten for the sake of clarity, accuracy, and thoroughness. Reviewers' comments were especially helpful in determining which topics needed to be revised. The following are representative of these changes:

- Reworking of the discussion of negative feedback to include the terms controlled variable, sensor, set point, integrator, and effector; addition of the concept of feedforward mechanisms.
- Expansion of the discussion of the electron-transport chain to include the concept of chemiosmosis.
- Revised section on the mechanism of summation and tetanus at the cellular level.
- Modified section on heart failure that omits discussion of the descending limb of the Frank-Starling curve and describes instead the lowering of the curve downward and to the right.
- New discussion, analogy, and figure to clarify flow rate compared to velocity of flow.
- New coverage of the role of urea recycling in the renal medulla.

New Features and Pedagogical Aids   In addition to these organizational and content changes, new features and pedagogical aids have been added to complement those in the first edition, as will be described in the next section.

FEATURES AND PEDAGOGICAL AIDS NEW TO THIS EDITION

Unique Homeostatic Model   A new, unique, easy-to-follow, pictorial homeostatic model depicting the relationship among cells, systems, and homeostasis is developed in the introductory chapter and presented on the inside front cover as a quick reference. Unlike the homeostatic model in the first edition, this new model features a realistic human figure and is less intimidating. A specialized version of this model begins each of the following chapters to emphasize how each body system functionally fits in with the body as a whole.

Chapter Opening   Each chapter begins with (1) the pictorial homeostatic model highlighting the features of that chapter; (2) a didactic introduction that emphasizes how the system discussed in that chapter fits in with the body as a whole; and (3) a **Chapter Contents at a Glance**, which lists the principal topics to be covered in that chapter. These opening features are designed to orient the student and help put the material that follows in perspective.

Chapter Closing Focusing on Homeostasis   Each chapter concludes with a narrative, **Chapter in Perspective: Focus on Homeostasis**, which helps the students put into perspective how the part of the body just discussed contributes to homeostasis. This pedagogical feature, the opening homeostatic model, and the introductory comments are designed to work together to facilitate the students' comprehension of the interactions and interdependency of body systems, even though each system is discussed separately.

Narrative Chapter Summaries   A more detailed, section-by-section, narrative **Chapter Summary** at each chapter's end enables students to focus on the main concepts before moving on. Furthermore, an instructor can assign the summary (or part of it) in lieu of the full chapter (or section) for a topic that needs only a concise overview in a particular course. Thus, the instructor has maximum flexibility to cover topics in depth or superficially.

Expanded End-of-Chapter Learning Activities   The **Review Exercises** at the end of each chapter now include objective questions, with answers provided in an appendix, as well as essay questions.

The first edition had one "A Point to Ponder" question at the end of the essay questions. Because reviewers especially liked this feature, it has been expanded into a separate **Points to Ponder** section. An average of five thought-provoking questions not specifically covered in the text are provided for each chapter. These questions could be assigned or used as a basis for class discussions. Explanations of the Points to Ponder questions are provided in the *Instructor's Manual*.

New Boxed Feature   In addition to the exercise physiology boxes in the first edition, each chapter in this edition has a new boxed feature entitled **Concepts, Challenges, and Controversies**. These boxes expose students to high-interest, tangentially relevant information on such diverse topics as environmental impact on the body, aging, ethical issues, new discoveries regarding common diseases, historical perspectives, and body responses to new environments such as those encountered in space flight and deep-sea diving.

Improved Illustrations and Captions   The illustrations use color and three-dimensional effects even more extensively than in the first edition to enhance understanding and be more aesthetically pleasing. For example, more color has been added to the flow diagrams to help distinguish between organs and actions.

New figures, photographs, and tables have been added as relevant to further complement the written material.

The figure captions have been expanded as needed to improve understanding of the figures.

A colored bullet (●) precedes each figure number and title and also precedes the first reference to the figure in the text. This feature enables students to easily find the text description of a figure and enables them to return quickly to the text they were reading before they referred to the figure.

Full Glossary   In lieu of the glossary index in the first edition, a full glossary has been added in this edition to enable students to quickly review key terms when they occur later in the book.

Supplemental Reading List   An appendix of suggested current readings has been added as a guide for students who wish to explore specific topics in greater detail.

RETAINED FEATURES AND PEDAGOGICAL AIDS

Analogies   Many analogies and frequent references to everyday experiences are included to help students relate to the physiology concepts presented.

Pathophysiology   Another effective way to keep students' interest is to help them realize that they are learning worthwhile and applicable material. Because most students using this text will have health-related careers, frequent references to pathophysiology and clinical physiology demonstrate the contents' relevance to their professional goals.

Full-Color Illustrations   A full-color art program is used as a functional tool to learning. Anatomical illustrations, schematic representations, photographs, tables, and graphs are designed to complement and reinforce the written material. Flow diagrams are used extensively to help the students integrate the didactic information presented.

Integrated Color-Coded Figure/Table Combinations   Figure/table combinations enable students to better visualize what part of the body is responsible for what activities. For example, an anatomical depiction of the brain is integrated with a table of the functions of the major brain components, with each component shown in the same color in the figure and the table.

Diversity of Human Models   A unique feature of this book is that the people depicted in the various illustrations are realistic representatives of a cross section of humanity (they were drawn from photographs of real people). Sensitivity to the various races, sexes, and ages of undergraduate students should enable all students to identify with the material being presented.

Boxed Feature on Exercise Physiology   A relevant box entitled **A Closer Look at Exercise Physiology** is featured in each chapter. Current concepts related to exercise physiology are included for three reasons: increasing national awareness of the importance of physical fitness; increasing recognition of the value of prescribed therapeutic exercise programs for a variety of conditions; and growing career opportunities related to fitness and exercise.

Feedforward Statements as Subsection Titles   Instead of traditional topic titles for each subsection (for example, **Heart valves**), feedforward statements alert the student to the main point of the subsection to come (for example, **Heart valves ensure the proper direction of blood flow through the heart**).

Cross-References   Cross-references to related material in earlier chapters enable students to quickly refresh their memories and also give instructors more flexibility in organizing the presentation of materials.

Key Terms   Key terms are defined as they appear in the text. Word derivations are provided as necessary to enhance understanding of new words.

Cellular Approach   Even though the text is primarily organized according to body systems, it provides extensive coverage of cellular physiology at the beginning and incorporates explanations of function at the cellular and molecular level throughout as a basis for understanding organ function. The cellular/molecular approach is slighted in most general undergraduate physiology textbooks, yet this approach is at the cutting edge of much of physiology research.

Chemistry Appendix   In addition to the new genetics appendix, a chemistry appendix is provided. Most undergraduate physiology texts have a chapter on chemistry, yet physiology instructors rarely teach basic chemistry concepts. The decision was made, therefore, to reserve valuable text space for physiological concepts and to provide instead a chemistry appendix as a handy reference for students who need a review of basic chemistry concepts that are essential to understanding physiology.

ORGANIZATION

There is no ideal organization of physiological processes into a logical sequence. With the sequence chosen, most chapters build on material presented in immediately preceding chapters, yet each chapter is designed to stand on its own to allow the instructor flexibility in curriculum design. The general flow is from introductory background information to cells to excitable tissue to organ systems. Every attempt has been made to provide logical transitions from one chapter to the next. For example, chapter 8, *Muscle Physiology,* ends with a discussion of cardiac muscle, which is carried forward into chapter 9, *Cardiac Physiology.* Even topics that seem unrelated in sequence, such as chapter 12, *Defense Mechanisms of the Body,* and chapter 13, *Repiratory System,* are linked together, in this case by ending chapter 12 with a discussion of respiratory defense mechanisms.

Several organizational features warrant specific mention. The most difficult decision in organizing this text was the placement of the chapters on the endocrine system. Intermediary metabolism of absorbed nutrient molecules is largely under endocrine control, providing a link from digestion (chapter 16) and energy balance (chapter 17) to the endocrine system (chapters 18 and 19). There is merit in placing the chapters on the nervous and endocrine systems in close proximity because of these systems' roles as the body's major control systems. Placing the endocrine system chapters earlier, immediately after the discussion of the nervous system (chapters 4 through 7), however, would have created two problems. First, it would have disrupted the logical flow of material related to excitable tissue. Second, the endocrine system could not have been covered at the level of depth its importance warrants if it had been discussed before the students were provided the background essential to understanding this system's roles in maintaining homeostasis. Placing the endocrine system chapters late in the book does not mean, however, that students are not exposed to endocrine function or hormones until near the book's completion. Endocrine control and hormones are defined in chapter 1 and are compared with nervous control in chapter 5. Specific hormones are introduced in appropriate chapters, such as vasopressin and aldosterone in the chapters on kidney and fluid balance. Chapters 18 and 19 explore the basic characteristics of endocrine glands and hormones as well as the control and functions of specific endocrine secretions.

Unique to this book, the skin is covered in the chapter on defense mechanisms of the body in consideration of the skin's newly recognized immune functions. Bone is also covered more extensively in the endocrine chapters than in most undergraduate physiology texts, especially with regard to hormonal control of bone growth and bone's dynamic role in calcium metabolism.

Departure from traditional groupings of material in several important instances has permitted more independent and more extensive coverage of topics that are frequently omitted or buried within chapters concerned with other subject matter. For example, a separate chapter is devoted to fluid balance and acid-base regulation, topics often tucked within the kidney chapter. The grouping of the autonomic nervous system, motor neurons, and the neuromuscular junction in an independent chapter on the efferent division of the peripheral nervous system, which serves as a link between the nervous system chapters and the muscle chapter, is another example.

Although there is a rationale for covering the various aspects of physiology in the order given here, it is by no means the only logical way of presenting the topics. Each chapter is able to stand on its own, especially with the cross-references provided, so that the sequence of presentation can be varied at the instructor's discretion. Some chapters may even be omitted, depending on the students' needs and interests and the time constraints of the course. For example, a cursory explanation of the defense role of the leukocytes is covered in the chapter on blood, so an instructor could choose to omit the more detailed explanations of immune defense in chapter 12. Similarly, the in-depth coverage of topics in chapters 2,

6, 15, 17, and 19 could selectively be omitted without sacrificing a student's general appreciation of systems-approach physiology.

As an alternative to total omission of certain chapters, the **Chapter Summary** could be used as a supplement for less comprehensive coverage of these topics.

ANCILLARIES

Learning Resource Manual   This companion guide is a vehicle for student review. Various self-study learning aids are provided to accompany each chapter.

Instructor's Manual   This teaching aid offers lecture suggestions and a list of pertinent films and software that may be ordered to complement the lectures. Explanations of **Points to Ponder** questions in the textbook's review exercises are provided. For the instructor's convenience, an extensive assortment of author-generated test questions of various objective formats and degrees of difficulty is also included.

Computerized Test Service   Computerized tests composed of questions of the instructor's choice from the *Instructor's Manual* are available from West Publishing Company. Contact your West sales representative for details.

Colored Transparency Acetates   Ready-to-use colored transparency acetates are available for selected illustrations in the text. A complete list of those available (by figure number and title) is included in the *Instructor's Manual.*

Videotapes   Free selections from the **West Life Science Video Library** are available to qualified adopters. Contact your West sales representative for more information.

Software   A **Human Physiology Software** series to be used with Hypercard on Macintosh computers is available from West Publishing Company. This series is a self-paced learning environment that can be used by students as an introduction, a study guide, a progress monitor, an evaluation tool, or a review of material already presented in the text or lecture. The programs follow a learning path that branches off to provide increasing depth with each level. Graphics with animation and three-dimensional effects help the students visualize the material presented. A free annual site license is available to adopters of the text. The software and workbook can be purchased by nonadopters and students. Contact your West sales representative for further detail.

ACKNOWLEDGMENTS

I gratefully acknowledge the many people who helped with the first edition or this edition of the textbook. A special thank-you goes to two individuals who contributed substantially to the content of the book: Rachel Yeater (professor, Sports Exercise Program, and director, Human Performance Laboratory, School of Physical Education, West Virginia University), who contributed the material for the boxed features **A Closer Look at Exercise Physiology**; and Spencer Seager (chairman, Chemistry Department, Weber State College) who prepared Appendix A, **A Review of Chemical Principles**.

During the book's creation and revision, many colleagues provided assistance. George Hedge and Robert Goodman, Department of Physiology, West Virginia University, deserve a special note of gratitude for their willingness to share materials used for their publication, **Clinical Endocrine Physiology** (Saunders, 1987), and for their thoughtful reviews of this book's chapters on endocrinology and reproduction. Appreciation is also extended to Elizabeth Walker and Dennis Overman, Department of Anatomy, West Virginia University, who provided many custom-made light and electron micrographs for the book.

Others at West Virginia University deserving of recognition for providing resource materials or countless clarifications include James Culberson and William Beresford, Department of Anatomy; Marta Henderson, Division of Medical Technology; Ronald Gaskins, Department of Medicine; Sidney Schochet, Jr., and Val Vallyathan, Department of Pathology; Karen Curto and Robert Stitzel, Department of Pharmacology; and Christine Baylis, Paul Brown, John Connors, Gunter Franz, Wilbert Gladfelter, Linda Huffman, Michael Johnson, Ping Lee, Philip Miles, Ronald Millechia, and William Stauber, Department of Physiology.

I express sincere appreciation to the following reviewers for their conscientious reading and assessment of the manuscript and for offering valuable advice on ways to improve the text.

Milton Fingerman
Tulane University

Gregory Fitch
Kansas State University

Jack M. Goldberg
University of California—Davis

Donald E. Grogan
University of Missouri—St. Louis

Anne Hedlin
University of Toronto

Robert E. Leipheimer
Youngstown State University

Paul F. Mercer
University of Western Ontario

Steven D. Mercurio
Mankato State University

Scott K. Powers
University of Florida

Martin Roeder
Florida State University

Mary Ann Rokitka
SUNY at Buffalo

Dee U. Silverthorn
University of Texas at Austin

A special note of gratitude is extended to the two reviewers who meticulously examined the final draft of this edition: Gregory Fitch and Dee Silverthorn. Their invaluable suggestions and attention to every detail helped assure the clarity and accuracy of this edition.

It was my personal and professional pleasure and good fortune to work with the same two highly competent editors at West Publishing Company who oversaw the development and production of my two previous textbooks. My deepest appreciation goes to Jerry Westby, executive editor, an excellent facilitator who guided me through the development of this revised edition with unfailing support, insightful suggestions, and helpful decision making; and Barbara Fuller, a premier production editor whose efficiency, resourcefulness, painstaking attention to detail, and unflagging dedication brought this edition smoothly and punctually through the production process. A warm thank-you is extended to Jerry and Barb for the extra tasks they did "above and beyond the call of duty" to help make this the best book it could possibly be. Also at West Publishing Company are Denis Ralling, the developmental editor who helped with the review process and, together with Dean DeChambeau, another developmental editor, oversaw the development of the ancillary materials; Ellen Stanton, promotion manager for the book; Patricia Lewis, copyeditor; and Diane Beasley, interior designer. I further wish to thank the text's artists for their excellent contributions: John and Judy Waller, Darwen and Vally Hennings, Cyndie C. H.-Wooley, Wayne Clark, Sandy McMahon, Carlyn Iverson, Rolin Graphics, Publication Services, and Precision Graphics.

Finally, my love and gratitude go to my family for another year of sacrifices in family life as this second edition was being developed and produced. I want to thank them for caring and for being there for me. I could not have completed the project without their patience, understanding, and support.

1 Homeostasis: The Foundation of Physiology

During the minute that it will take you to read this page:

Your eyes will convert the image from this page into electrical signals (nerve impulses) that will transmit the information to your brain for processing.

Besides receiving and processing information such as visual input, your brain also will provide output to your muscles to help maintain your posture, move your eyes across the page as you read, and turn the page as needed. Chemical messengers will carry signals between your nerves and muscles to trigger appropriate muscle contraction.

Your heart will beat seventy times, pumping 5 liters (about 5 quarts) of blood to your lungs and another 5 liters to the rest of your body.

You will breathe in and out about twelve times, exchanging 6 liters of air between the atmosphere and your lungs.

More than 1 liter of blood will flow through your kidneys, which will act on the blood to conserve the "wanted" materials and eliminate the "unwanted" materials in the urine. Your kidneys will produce 1 ml (about a thimbleful) of urine during this minute.

Your cells will consume 250 ml (about a cup) of oxygen and produce 200 ml of carbon dioxide.

Your digestive system will be processing your last meal for transfer into your bloodstream for delivery to your cells.

You will use about 2 calories of energy derived from food to support your body's "cost of living" plus your contracting muscles will burn additional calories.

≡ **CHAPTER CONTENTS AT A GLANCE** ≡

INTRODUCTION
LEVELS OF ORGANIZATION IN THE BODY
Cells as the basic units of life
Organizational levels of tissues, organs, and systems
● CONCEPTS, CHALLENGES, AND CONTROVERSIES A First: Brain Cells Growing in Laboratory Dishes
CONCEPT OF HOMEOSTASIS
Significance of the internal environment
Necessity of homeostasis
Factors that are homeostatically maintained
● A CLOSER LOOK AT EXERCISE PHYSIOLOGY What Is Exercise Physiology?
Contributions of various systems to homeostasis
Negative feedback
Illness and death related to loss of ability to maintain homeostasis

≡ INTRODUCTION

The activities described on the preceding page are a sampling of the various processes that occur in our bodies all the time just to keep us alive. We usually take these life-sustaining activities of the body for granted and don't really think about "what makes us tick," but that's what physiology is all about. **Physiology** is the study of the functions of the body, or how the body works.

Physiologists view the body as a machine whose mechanisms of action can be explained in terms of cause-and-effect sequences of physical and chemical processes—the same types of processes that occur in other components of the universe. It is important to distinguish between the mechanistic and teleological approaches to explaining the various events that occur in the body. With the **mechanistic approach**, which is employed by physiologists, the *mechanisms* of action are emphasized. That is, physiologists explain the "how" of events that occur in the body. With a **teleological approach**, phenomena that occur in the body are explained in terms of their particular purpose in fulfilling a bodily *need*, without considering how this outcome is accomplished. That is, the teleological approach emphasizes the "why" of body processes.

A simple example will help you to distinguish between these approaches. A teleological explanation of why you shiver when you are cold is "to keep warm," because shivering generates heat. A physiologist's mechanistic explanation of why you shiver is that when temperature-sensitive nerve cells detect a fall in body temperature, they signal the hypothalamus, the part of the brain responsible for temperature regulation. In response, the hypothalamus activates nerve pathways that ultimately bring about involuntary, oscillating muscle contractions (that is, shivering).

Because those mechanisms of action that are most beneficial to survival have prevailed throughout evolutionary history, it is helpful when studying physiology to predict what mechanistic process would be teleologically useful to the body under a particular circumstance. Chances are such a mechanism will exist. This does not mean that the discipline of physiology is an elaborate guessing game. Predictions must be verified by facts ascertained by careful scientific investigation. However, the fact that most bodily mechanisms do serve a useful purpose (having been naturally selected throughout evolutionary time) allows you to apply a certain amount of logical reasoning to each new situation you encounter in your study of physiology. If you always try to find the thread of logic in what you are studying, you can avoid a good deal of pure memorization, and, more importantly, you will better understand and appreciate the concepts being presented.

Physiology is closely interrelated with **anatomy**, the study of the structure of the body. Just as the functioning of an automobile depends on the shapes, organization, and interactions of its various parts, the same is true of the human body. Structure and function are inseparable. Therefore, as we tell the story of how the body works, we will provide sufficient anatomical background for you to understand the function of the body part being discussed.

≡ LEVELS OF ORGANIZATION IN THE BODY

Cells are the basic units of life.

The basic unit of both structure and function is the **cell**, the foundation of all living organisms. The cell is the smallest unit capable of carrying out the processes associated with life. In fact, simple life-forms include **unicellular** (single-celled) **organisms** such as bacteria and amoebae. Humans are **multicellular** (many-celled) **organisms**—the adult human body is an aggregate of more than 75 trillion cells.

All cells, whether they exist as solitary cells or as part of a multicellular organism, perform certain basic functions essential for survival of the cell and, in turn, survival of the organism. These basic cell functions include:

1. obtaining food (nutrients) and oxygen (O_2) from the environment surrounding the cell;
2. performing various chemical reactions that use nutrients and O_2 to provide energy for the cell;
3. eliminating to the cell's surrounding environment carbon dioxide (CO_2) and other by-products or wastes produced during these chemical reactions;
4. synthesizing proteins and other components needed for cellular structure, for growth, and for carrying out particular cell functions;
5. being sensitive and responsive to changes in the environment surrounding the cell;
6. controlling to a large extent the exchange of materials between the cell and its surrounding environment;
7. moving materials from one part of the cell to another in carrying out cellular activities, with some cells even being able to move in entirety through their surrounding environment; and
8. in the case of most cells, reproducing. Some body cells, such as nerve cells and muscle cells, have lost the ability

A FIRST: BRAIN CELLS GROWING IN LABORATORY DISHES

The immature nerve cells present during fetal life have the ability to divide and reproduce more nerve cells as the brain and other parts of the nervous system develop. Once the nervous system is formed, the nerve cells for reasons unclear lose the ability to divide any further, so humans are usually born with all of the brain cells that they will ever have.

Most human cells can be *cultured;* that is, when removed from the body, they will continue to thrive and reproduce in laboratory dishes when supplied with appropriate nutrients and other supportive materials. Cell cultures have been invaluable in studying cell functions and drug actions.

Until recently, human brain cells have never been maintained in cell culture because of the inability of mature brain cells to reproduce themselves. In 1990, however, the science world was startled by the announcement that, for the first time, brain cells removed from an 18-month-old infant for the treatment of a rare brain disorder had been coaxed into a continuous culture of living, dividing human brain cells. The infant suffered seizures because one side of her brain grew more than the other as a result of

the unusual proliferation of immature brain cells. Why the immature brain cells underwent too many cell divisions and overproduced before they matured is unclear. The condition was treated by surgically removing excess brain tissue. Some of the tissue was placed in culture dishes and treated with nerve growth factors. Most of the cells died, as had occurred in all previous attempts at maintaining human-nerve-cell cultures, but two small clusters survived. These few surviving cells were nurtured into a stable cell line that was proven to be human brain cells and that could be expanded at will.

Successful culturing of human brain cells offers new opportunities for research on brain function and brain diseases as well as for the development and testing of drugs for various brain disorders. Furthermore, laboratory-grown brain cells may eventually be used to replace brain cells that have been destroyed as a result of head injuries, strokes, or degenerative brain disorders such as Alzheimer's disease.

In even more recent research, scientists have prompted brain cells taken from adult mice to grow and divide in culture by treating them with *epidermal growth factor*

(EGF). EGF is a naturally occurring protein that promotes development of nervous tissue and skin in embryos. Apparently, the brains of adult mice contain previously undiscovered undifferentiated "stem cells" that can give rise to the two different types of brain cells: *neurons,* which generate and transmit electrical signals, and *glial cells,* which provide physical and biochemical support for the neurons. EGF has the ability to "turn on" these stem cells, resulting in the development and multiplication of both neurons and glial cells.

Researchers suggest that adult human brains may contain similar stem cells that never get turned on because EGF may not be able to penetrate the formidable barrier that exists between the blood and the brain. If this is true, these scientists speculate that injecting EGF into injured regions of the brain might induce the dormant stem cells to proliferate and replace dead or damaged nerve cells. They warn, however, that prompting growth of new brain cells carries the risk of muddling normal brain activities. Therefore, clinical applications of these new discoveries must await further careful investigation.

---

to reproduce. When these cells are destroyed through trauma or disease processes, they cannot be replaced. (See the accompanying boxed feature, Concepts, Challenges, and Controversies.) With other body cells, repair is possible.

Cells are remarkable in the similarity with which they carry out these functions. Thus, all cells share many common characteristics. In multicellular organisms, each cell also performs a specialized function, which is usually a modification or elaboration of one of the basic cell functions. The following are a few examples:

- By taking special advantage of their protein-synthesizing ability, the gland cells of the digestive system secrete digestive enzymes, which are all proteins.
- Capitalizing on the basic ability of cells to respond to changes in their surrounding environment, nerve cells gen-

erate and transmit to other regions of the body electrical impulses that relay information about changes to which the nerve cells are responsive.

- The ability of kidney cells to selectively retain the substances needed by the body while eliminating unwanted substances in the urine depends on these cells' highly specialized ability to control exchange of materials between the cell and its environment.
- Muscle contraction, which involves selective movement of internal structures to bring about shortening of muscle cells, is an elaboration of the inherent capability of these cells to produce intracellular ("within the cell") movement.

It is important to recognize that each cell performs these specialized activities in addition to carrying on the unceasing, fundamental activities required of all cells. The fundamental cellular activities are essential for survival of each individual cell, whereas the specialized contributions and interactions among the cells of a multicellular organism are essential for the survival of the whole organism.

Cells are progressively organized into tissues, organs, systems, and, finally, the whole body.

Just as a machine does not function unless all its various parts are properly assembled, the cells of the body must be specifically organized to carry out the life-sustaining processes of the body as a whole, such as digestion, respiration, and circulation. The body is made up of four levels of organization: cells, tissues, organs, and systems.

Cells of similar structure and function are organized into **tissues**, of which there are four primary types: *muscle, nervous, epithelial,* and *connective tissue.* Each tissue consists of cells of a single specialized type, along with varying amounts of extracellular ("outside of the cell") material.

- **Muscle tissue** is composed of cells specialized for contraction and force generation. There are three types of muscle tissue: *skeletal muscle,* which accomplishes movement of the skeleton; *cardiac muscle,* which is responsible for pumping blood out of the heart; and *smooth muscle,* which encloses and controls movement of contents through hollow tubes and organs, such as movement of food through the digestive tract.

- **Nervous tissue** consists of cells specialized for initiation and transmission of electrical impulses, sometimes over long distances. These electrical impulses serve as signals for relaying information from one part of the body to another. Nervous tissue is found in (1) the brain; (2) the spinal cord; (3) the nerves that signal information about the external environment and about the status of various internal factors in the body that are subject to regulation, such as blood pressure; and (4) the nerves that influence muscle contraction or gland secretion.

- **Epithelial tissue** is made up of cells specialized in the exchange of materials between the cell and its environment. This tissue is organized into two general types of structures: epithelial sheets and secretory glands. Epithelial cells are joined together very tightly to form sheets of tissue that cover and line various parts of the body. For example, the outer layer of the skin is epithelial tissue, as is the lining of the digestive tract. In general, these epithelial sheets serve as boundaries that separate the body from the external environment and from the contents of cavities that communicate with the external environment, such as the digestive-tract lumen. (A **lumen** is the cavity within the interior of hollow organs and tubes.) Only selected transfer of materials is permitted between the regions separated by an epithelial barrier. The type and extent of controlled exchange vary, depending on the location and function of the epithelial tissue. For example, very little can be exchanged between the body and external environment across the skin, whereas the epithelial cells lining the digestive tract are specialized for absorption of nutrients.

● **FIGURE 1–1   Exocrine- and Endocrine-Gland Formation**   (a) Glands arise during development from the formation of pocketlike invaginations of surface epithelial cells. (b) If the cells at the deepest part of the invagination become secretory and release their product through the connecting duct to the surface, an exocrine gland is formed. (c) If the connecting cells are lost and the deepest secretory cells release their product into the blood, an endocrine gland is formed.

Glands are epithelial-tissue derivatives that are specialized for secretion. **Secretion** is the release from a cell, in response to appropriate stimulation, of specific products that have in large part been synthesized by the cell. Glands are formed during embryonic development by pockets of epithelial tissue that dip inward from the surface. There are two categories of glands—*exocrine* and *endocrine* (● Fig. 1–1). If during development the connecting cells between the epithelial surface and the secretory-gland cells within the depths of the invagination remain intact as a duct between the gland and the surface, an exocrine gland is formed. **Exocrine glands** (*exo* means "external"; *crine* means "secretion") secrete through ducts to the outside of the body (or into a cavity that communicates with the outside). Examples are sweat glands and glands that secrete digestive juices. If, on the other hand, the connecting cells disappear during development and the secretory-gland cells are isolated from the surface, an endocrine gland is formed. **Endocrine glands** (*endo* means "internal") lack ducts and release their secretory products, known as *hormones,* internally into the blood within the body. For example, the parathyroid gland secretes parathyroid hormone into the blood, which transports this hormone to its sites of action at the bones and kidneys.

■ **Connective tissue** is distinguished by having relatively few cells dispersed within an abundance of extracellular material. As its name implies, connective tissue serves to connect, support, and anchor various body parts. It includes such diverse structures as the loose connective tissue that attaches epithelial tissue to underlying structures; tendons that attach skeletal muscles to bones; bone, which gives the body shape, support, and protection; and blood, which transports materials from one part of the body to another. Except for blood, the cells within connective tissue produce specific molecules that they release into the extracellular spaces between the cells. One such molecule is the rubber-band–like protein fiber **elastin**, whose presence facilitates the stretching and recoiling of structures such as the lungs, which alternately inflate and deflate during breathing.

Muscle, nervous, epithelial, and connective tissue are the primary tissues in a classical sense; that is, each is an integrated collection of cells of the same specialized structure and function. The term *tissue* is also frequently used in another sense to refer to the aggregate of various cellular and extracellular components that make up a particular organ (for example, lung tissue or liver tissue).

Organs are composed of two or more types of primary tissue organized to perform a particular function or functions. The stomach is an example of an organ made up of all four primary tissue types. The tissues that compose the stomach function collectively to store the ingested food and move it forward into the remainder of the digestive tract as well as to begin the digestion of protein. The stomach is lined with epithelial tissue that restricts the transfer of harsh digestive chemicals and undigested food from the stomach lumen into the blood. Epithelially derived gland cells in the stomach include exocrine cells, which secrete protein-digesting juices into the lumen, and endocrine cells, which secrete a hormone that helps regulate the stomach's exocrine secretion and muscle contraction. The walls of the stomach contain smooth-muscle tissue whose contraction mixes ingested food with the digestive juices and propels the mixture forward into the intestine. Also within the walls is nervous tissue, which, along with hormones, controls muscle contraction and gland secretion. These various tissues are all bound together by connective tissue.

Organs are further organized into **body systems**, each of which is a collection of organs that perform related functions and interact to accomplish a common activity that is essential for survival of the whole body. For example, the digestive system consists (in addition to the stomach) of the mouth, pharynx (throat), esophagus, small intestine, large intestine, salivary glands, pancreas, liver, and gallbladder. These digestive organs cooperate to accomplish the breakdown of dietary food into small nutrient molecules that can be absorbed into the blood.

The **total body**—a single, independently living individual—is composed of the various organ systems structurally and functionally linked together as an entity that is separate from the **external** (outside of the body) **environment**. Thus, the body is made up of living cells organized into life-sustaining systems.

CONCEPT OF HOMEOSTASIS

Body cells are in contact with a privately maintained internal environment instead of with the external environment that surrounds the body.

If each cell possesses basic survival skills, why can't the body cells live without performing specialized tasks and being organized according to specialization into systems that accomplish functions essential for the whole body's survival? The cells in a multicellular organism must contribute to the survival of the organism as a whole and cannot live and function without contributions from the other body cells because the vast majority of the cells are not in direct contact with the external environment in which the organism lives. A unicellular organism such as an amoeba can directly obtain nutrients and O_2 from and eliminate wastes into its immediate external surroundings. A muscle cell or any other cell in a multicellular organism has the same need for life-supporting nutrient and O_2 uptake and waste elimination, yet the muscle cell cannot directly make these exchanges with the environment surrounding the body because the cell is isolated from this external environment.

How is it possible for a muscle cell to make vital exchanges with the external environment with which it has no contact? The key is the presence of an aqueous **internal environment** with which the body cells have direct contact. This internal environment is *outside* the cells but *inside* the body. It consists of the **extracellular** (*extra* means "outside of") **fluid**, which is made up of **plasma**, the fluid portion of the blood, and **interstitial fluid**, which surrounds and bathes the cells (● Fig. 1–2). Various body systems accomplish exchanges between the external environment and the internal environment. For example, the digestive system transfers from the external environment into the plasma the nutrients required by all body cells. Likewise, the respiratory system transfers O_2 from the external environment into the plasma. The circulatory system distributes these nutrients and O_2 throughout the body. Thorough mixing and exchange of materials take place between the plasma and the interstitial fluid across the thin, pore-lined walls of the capillaries, the smallest of the blood vessels. As a result, the nutrients and O_2 originally obtained from the external environment are delivered to the

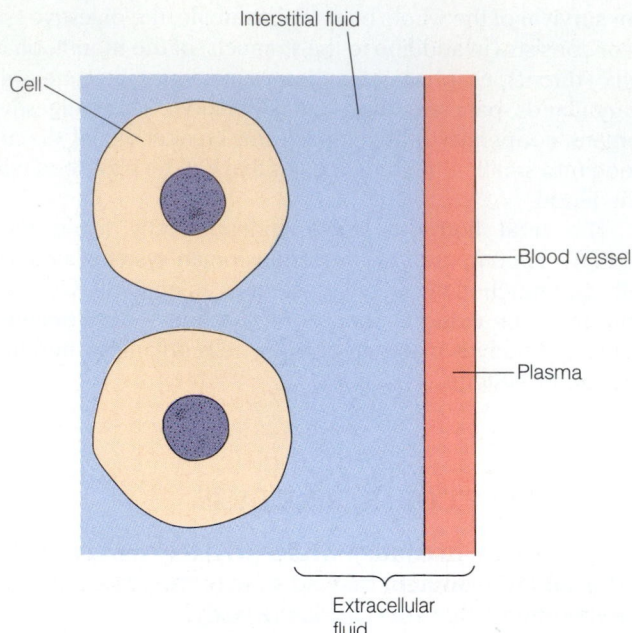

● **FIGURE 1–2 Components of the Extracellular Fluid (Internal Environment)**

interstitial fluid that surrounds the cells. The body cells, in turn, make life-sustaining exchanges with the internal environment. No matter how remote a cell is from the body surface, it is able to take in from the internal environment the nutrients and O_2 needed to support its own existence. Similarly, wastes produced by the cells are extruded into the interstitial fluid, picked up by the plasma, and transported to the organs that specialize in eliminating these wastes from the internal environment to the external environment. The lungs remove CO_2 from the plasma, and the kidneys remove other wastes for elimination in the urine.

Thus, a body cell takes in essential nutrients from and eliminates wastes into its watery surroundings, just as an amoeba does. The major difference is that the internal environment must constantly be maintained at the composition required to support the existence of all the cells by the same cells that it supports. In contrast, an amoeba does nothing to regulate it surroundings.

Homeostasis is essential for cell survival, and each cell, as part of an organized system, contributes to homeostasis.

The body cells can live and function only when they are bathed by extracellular fluid that is compatible with their survival; thus, the chemical composition and physical state of the internal environment can be allowed to deviate only within narrow limits. As cells remove nutrients and O_2 from the internal environment, these essential materials must constantly be replenished in order for the cells' ongoing maintenance of life processes to continue. Likewise, wastes must constantly be removed from the internal environment so that

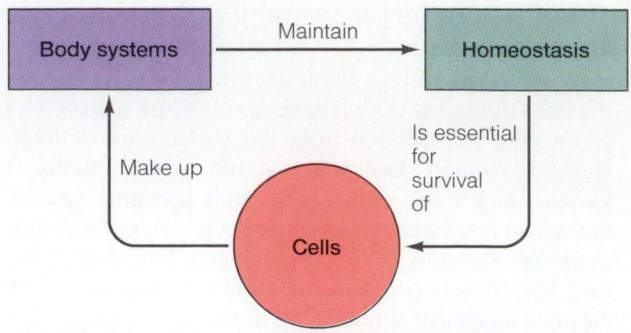

● **FIGURE 1–3 Interdependent Relationship of Cells, Body Systems, and Homeostasis** The depicted interdependent relationship serves as the foundation for modern-day physiology: *homeostasis is essential for survival of cells, body systems maintain homeostasis, and cells make up body systems.*

they do not reach toxic levels. Other elements in the internal environment that are important for the maintenance of life also must be kept relatively constant. Maintenance of a relatively stable internal environment is termed **homeostasis** (*homeo* means "the same"; *stasis* means "to stand or stay").

The functions performed by each body system contribute to maintaining homeostasis, thereby maintaining within the body the environment required for the survival and function of all the cells of which the body is composed. This is the central theme of physiology and of this book: *homeostasis is essential for the survival of each cell, and in turn, each cell, through its specialized activities, contributes as part of a body system to the maintenance of the internal environment shared by all cells* (● Fig. 1–3).

The fact that the internal environment must be kept relatively stable does not mean that its composition, temperature, and so on are absolutely unchanging. External and internal factors continuously threaten to disrupt homeostasis. For example, exposure to a cold environmental temperature tends to reduce the body's internal temperature. Likewise, addition of CO_2 into the internal environment as a result of energy-generating chemical reactions tends to raise the concentration of this gas within the body. When any factor starts to move the internal environment away from optimal conditions, appropriate counterreactions are initiated to restore these conditions. In our examples, when the body temperature starts to fall on a cold day, compensatory shivering is initiated. This shivering internally generates heat that restores the body temperature to normal. Similarly, a rise in the CO_2 levels within the internal environment triggers an increase in breathing. The extra CO_2 is blown off to the external environment, restoring the CO_2 concentration in the extracellular fluid to normal. Thus, homeostasis should be viewed not as a fixed state but as a dynamic steady state in which the changes that do occur are minimized by compensatory physiological responses. For each factor in the internal environment, the small fluctuations around the optimal level are normally kept within the narrow limits compatible with life by carefully regulated mechanisms. (See the accompanying boxed feature, A Closer Look at Exercise Physiology.)

WHAT IS EXERCISE PHYSIOLOGY?

Exercise physiology is the study of both the functional changes that occur in response to a single session of exercise and the adaptations that occur as a result of regular, repeated exercise sessions. Exercise initially disrupts homeostasis. The changes that occur in response to exercise are the body's attempt to meet the challenge of maintaining homeostasis when increased demands are placed on the body.

Heart rate is one of the easiest factors to monitor that shows both an immediate response to exercise and long-term adaptation to a regular exercise program. When a person begins to exercise, the active muscle cells use more O_2 to support their increased energy demands. Heart rate increases to deliver more oxygenated blood to the exercising muscles. The heart adapts to regular exercise of sufficient intensity and duration by increasing its strength and efficiency so that it pumps more blood per beat. Because of increased pumping ability, the heart does not have to beat as rapidly to pump a given quantity of blood as it did before physical training.

Exercise physiologists study the mechanisms responsible for the changes that occur as a result of exercise. Much of the knowledge gained from the study of exercise is used to develop appropriate exercise programs to increase the functional capacities of people ranging from athletes to the infirm.

The factors of the internal environment that must be homeostatically maintained include the following (● Fig. 1–4):

1. *Its concentration of nutrient molecules.* Cells need a constant supply of nutrient molecules to serve as a metabolic fuel for energy production. Energy, in turn, is needed to support life-sustaining and specialized cellular activities.

2. *Its concentration of O_2 and CO_2.* Cells need O_2 to perform chemical reactions that extract from nutrient molecules the most energy possible for use by the cell. The CO_2 produced during these chemical reactions must be balanced by CO_2 removal from the lungs so that acid-forming CO_2 does not increase the acidity of the internal environment.

3. *Its concentration of waste products.* Various chemical reactions produce end products that exert a toxic effect on the body's cells if these wastes are allowed to accumulate beyond a certain limit.

4. *Its pH.* Among the most pronounced effects of changes in the acidity of the internal fluid environment are alterations in the electrical signaling mechanism of nerve cells and alterations in enzyme activity of all cells.

5. *Its concentration of water, salt, and other electrolytes.* Because the relative concentrations of salt (NaCl) and water in the extracellular fluid (internal environment) influence how much water enters or leaves the cells, these concentrations are carefully regulated to maintain the proper volume of the cells. Cells do not function normally when they are swollen or shrunken. Other electrolytes perform a variety of vital functions. For example, the rhythmic beating of the heart depends on a relatively constant concentration of potassium (K^+) in the extracellular fluid.

6. *Its temperature.* Body cells function optimally within a narrow temperature range. Cells slow down too much if they are too cold and, worse yet, their structural and enzymatic proteins are impaired if they get too hot.

7. *Its volume and pressure.* The circulating component of the internal environment, the plasma, must be maintained at adequate volume and blood pressure to assure bodywide distribution of this important link between the external environment and the cells.

There are eleven major body systems (Table 1–1 and Fig. 1–4); their most important contributions to homeostasis are listed here:

1. The *circulatory system* is the transport system that carries materials such as nutrients, O_2, CO_2, wastes, electrolytes, and hormones from one part of the body to another.

2. The *digestive system* breaks down dietary food into small nutrient molecules that can be absorbed into the plasma for distribution to the body cells. It also transfers water and electrolytes from the external environment into the internal environment. It eliminates undigested food residues to the external environment in the feces.

3. The *respiratory system* obtains O_2 from and eliminates CO_2 to the external environment. By adjusting the rate of removal of acid-forming CO_2, the respiratory system is also important in maintaining the proper pH of the internal environment.

4. The *urinary system* removes excess water, salt, acid, and other electrolytes from the plasma and eliminates them in the urine, along with waste products other than CO_2.

5. The *skeletal system* provides support and protection for the soft tissues and organs. It also serves as a storage reservoir for calcium (Ca^{++}), an electrolyte whose plasma concentration must be maintained within very narrow limits. Together with the muscular system, the skeletal system also enables movement of the body and its parts.

6. The *muscular system* moves the bones to which the skeletal muscles are attached. From a purely homeo-

BODY SYSTEMS
Made up of cells organized according to specialization to maintain homeostasis.
See chapter 1.

Nervous system
Acts through electrical signals to control rapid responses of the body; also responsible for higher functions—e.g., conciousness, memory, and creativity.
See chapters 4, 5, 6, and 7.

Information from external environment relayed through nervous system

Control

O_2

CO_2

Respiratory system
Obtains O_2 from and eliminates CO_2 to external environment; helps regulate pH by adjusting the rate of removal of acid-forming CO_2
See chapters 13 and 15.

Urinary system
Important in regulating the volume, electrolyte composition, and pH of the internal environment. Removes wastes and excess water, salt, acid and other electrolytes from the plasma and eliminates them in the urine.
See chapters 14 and 15.

Urine containing wastes and excess water and electrolytes

Nutrients, water, electrolytes

Feces containing undigested food residues

Digestive system
Obtains nutrients, water, and electrolytes from the external environment and transfers them into the plasma; eliminates undigested food residues to external environment.
See chapter 16.

Sperm leave male

Sperm enter female

Reproductive system
Not essential for homeostasis, but essential for perpetuation of species.
See chapter 20.

Exchanges with all other systems

External environment

Circulatory system
Transports nutrients, O_2, CO_2, wastes, electrolytes, and hormones throughout body.
See chapters 9, 10, and 11.

● **FIGURE 1–4   Role of the Body Systems in Maintaining Homeostasis**

static view, this system enables an individual to move toward food or away from harm. Furthermore, the heat generated by muscle contraction is important in temperature regulation. In addition, because skeletal muscles are under voluntary control, a person is able to use them to accomplish myriad other movements of his or her own choice. These movements, which range from the fine motor skills required for delicate needlework to the powerful movements involved in weight lifting, are not necessarily directed toward maintaining homeostasis.

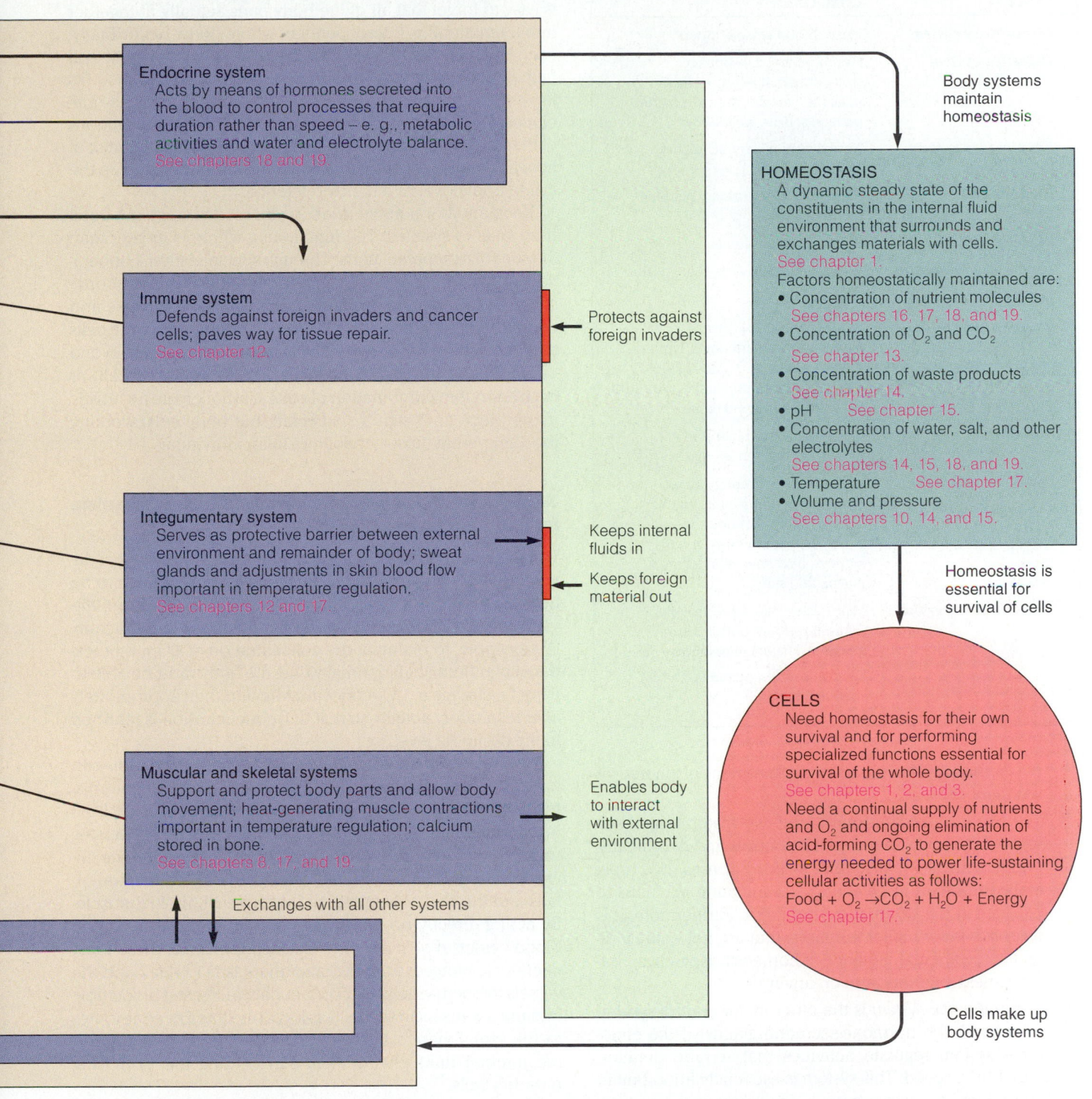

7. The *integumentary system* serves as an outer protective barrier that prevents internal fluid from being lost from the body and foreign microorganisms from entering the body. This system is also important in the regulation of body temperature. The amount of heat lost from the body surface to the external environment can be adjusted by controlling sweat production and by regulating the flow of warm blood through the skin.

8. The *immune system* defends against foreign invaders and body cells that have become cancerous. It also

TABLE 1–1 Components of Body Systems

SYSTEM	COMPONENTS
Circulatory system	Heart, blood vessels, blood
Digestive system	Mouth, pharynx, esophagus, stomach, small intestine, large intestine, salivary glands, exocrine pancreas, liver, gallbladder
Respiratory system	Nose, pharynx, larynx, trachea, bronchi, lungs
Urinary system	Kidneys, ureters, urinary bladder, urethra
Skeletal system	Bones, cartilage, joints
Muscular system	Skeletal muscles
Integumentary system	Skin, hair, nails
Immune system	White blood cells, thymus, bone marrow, tonsils, adenoids, lymph nodes, spleen, appendix, gut-associated lymphoid tissue, skin-associated lymphoid tissue
Nervous system	Brain, spinal cord, peripheral nerves, special sense organs
Endocrine system	All hormone-secreting tissues, including hypothalamus, pituitary, thyroid, adrenals, endocrine pancreas, parathyroids, gonads, kidneys, intestine, heart, thymus, pineal, and skin
Reproductive system	Male: testes, penis, prostate gland, seminal vesicles, bulbourethral glands, and associated ducts
	Female: ovaries, oviducts, uterus, vagina, breasts

paves the way for repair or replacement of injured or worn-out cells.

9. The *nervous system* is one of the two major control systems of the body. In general, it controls and coordinates bodily activities that require swift responses. It is especially important in detecting and initiating reactions to changes in the external environment. Furthermore, it is responsible for higher functions that are not entirely directed toward maintenance of homeostasis, such as consciousness, memory, and creativity.

10. The *endocrine system* is the other major control system. In general, the hormone-secreting glands of the endocrine system regulate activities that require duration rather than speed. This system is especially important in controlling the concentration of nutrients and, by adjusting kidney function, controlling the internal environment's volume and electrolyte composition.

11. The *reproductive system* is not essential for homeostasis and therefore is not essential for survival of the individual. It is essential, however, for perpetuation of the species.

As we examine each of these systems in greater detail, always keep in mind that the body is a coordinated whole even though each system provides its own special contributions. It is easy to forget that all of the body parts actually fit together into a functioning, interdependent whole body. Accordingly, each chapter begins with a figure and discussion that will help you focus on how the body system to be described fits in with the body as a whole. In addition, each chapter concludes with a brief overview of the homeostatic contributions of the body system. As a further tool to help you keep track of how all the pieces fit together, Figure 1–4 is duplicated on the inside front cover as a handy reference.

Keep another point in mind as you read through the book from cells to systems: The functioning whole is greater than the sum of its separate parts. Through specialization, cooperation, and interdependence, cells combine to form a coordinated, unique, single living organism with more diverse and complex capabilities than is possessed by any one of the cells that make it up. For humans, these capabilities go far beyond the processes needed to maintain life. Obviously a cell, or even a random combination of cells, cannot create an artistic masterpiece or design a spacecraft, but body cells working together permit those capabilities in an individual.

Negative feedback is a common regulatory mechanism for maintaining homeostasis.

To maintain homeostasis, the body must be able to detect deviations in the internal environmental factors that need to be held within narrow limits, and it must be able to control the various body systems responsible for adjusting these factors. For example, to maintain the concentration of CO_2 in the extracellular fluid at an optimal value, the body must be able to detect a change in CO_2 concentration and then appropriately alter respiratory activity so that CO_2 concentration is returned to the desirable level.

Control systems that operate to maintain homeostasis can be grouped into two classes—intrinsic and extrinsic controls. **Intrinsic (local) controls** (*intrinsic* means "within") are built in or inherent to an organ. For example, as an exercising muscle rapidly uses up O_2 and produces CO_2 to generate energy to support its contractile activity, the O_2 concentration falls and the CO_2 concentration increases within the muscle. By acting directly on the smooth muscle in the walls of the blood vessels that supply the exercising muscle, these local chemical changes cause the smooth muscle to relax and the vessels to open widely to accommodate increased blood flow into the exercising muscle. This local mechanism contributes to the maintenance of an optimal level of O_2 and CO_2 in the internal fluid environment surrounding the exercising muscle's cells.

Most factors in the internal environment are maintained, however, by **extrinsic controls** (*extrinsic* means "outside of"), which are regulatory mechanisms initiated outside of an organ to alter the activity of the organ. Extrinsic control of the various organs and systems is accomplished by the nervous and endocrine systems, the two major control systems of the body. Extrinsic control permits coordinated regulation of several organs toward a common goal; in contrast, intrinsic controls are self-serving for the organ in which they occur. Coordinated, overall regulatory mechanisms are critical for maintaining the dynamic steady state in the internal environment as a whole. For example, to restore blood pressure to the proper level when it falls too low, the nervous system simultaneously acts on the heart and the blood vessels throughout the body to increase the blood pressure to normal.

The body's homeostatic control mechanisms primarily operate on the principle of negative feedback. **Negative feedback** occurs when a change in a controlled variable triggers a response that opposes the change, driving the variable in the opposite direction of the initial change.

A common example of negative feedback is control of room temperature (the **controlled variable**) by a **control system** that includes a furnace, a thermostatic device, and all their electrical connections. The room temperature is determined by the activity of the furnace, a heat source that can be turned on or off. In order to switch on or off appropriately, the control system as a whole must "know" what the *actual* room temperature is, "compare" it with the *desired* room temperature, and "adjust" the output of the heat source to bring the actual temperature to the desired level. Information about the actual room temperature is provided by a thermometer in the thermostat; the thermometer is the **sensor** that monitors the magnitude of the controlled variable. The desired temperature level or **set point** is provided by the thermostat setting. The thermostat acts as an **integrator**: it compares the sensor's input with the set point and adjusts the heat output of the **effector**, the furnace, to bring about the appropriate effect or response to oppose the deviation from the set point.

For example, if the room temperature has fallen below the set-point level (● Fig. 1–5a), the thermostat, through connecting circuitry, activates the furnace, which produces heat to increase the room temperature. Once the room temperature reaches the set point, the activating mechanism in the thermostat and consequently the furnace are switched off. Thus, the heat produced by the furnace counteracts or is "negative" to the original fall in temperature. If heat production were to continue unabated, the room temperature would be increased above the set point. Overshooting beyond the set point does not occur because the heat "feeds back" to shut off the thermostat that triggered its output, thereby limiting its own production by controlling the signal that initiated the heat production. Thus, the control system takes corrective actions to prevent the controlled variable from drifting too far below or too far above the set point.

A homeostatic negative-feedback system operates in the same way to maintain a controlled factor in a relatively steady state. For example, when the pressure-monitoring nerve cells detect a *decrease* in blood pressure below the desired level, they bring about a sequence of events that culminate in nerve-controlled changes in the circulatory system to *increase* blood pressure to the proper level (Fig. 1–5b). When the blood pressure increases to the set point, the stimulatory signal to the heart and blood vessels arising from the pressure-monitoring nerve cells is turned off. As a result, the blood pressure does not continue to increase above the set point.

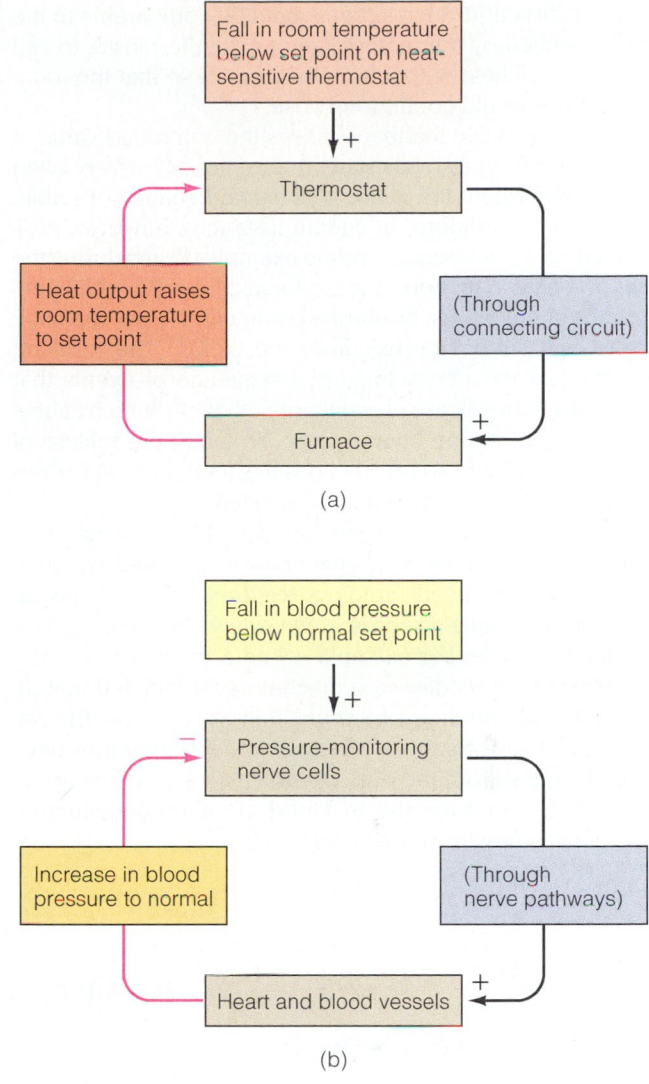

(a)

(b)

+ = Stimulates or activates
− = Inhibits or turns off

● **FIGURE 1–5   Negative Feedback** (a) Negative-feedback control of home heating system. (b) Negative-feedback control of blood pressure.

The opposite events occur when the original change is an elevation in blood pressure above normal. The pressure-monitoring nerve cells bring about a reduction in blood pressure to normal by triggering compensatory responses in the circulatory system. The blood pressure does not fall too low, because the pressure-monitoring nerve cells cease triggering the pressure-reducing responses when the blood pressure reaches the right level.

Positive feedback occurs less frequently in the body. In negative feedback, a control system's output is regulated to resist change, so that the regulated variable is maintained at a relatively steady set point. With positive feedback, however, the output is continually enhanced so that the controlled variable continues to be moved in the direction of the initial change. Instead of bringing about a response that counteracts the initial change, positive feedback reinforces the change in the same direction. Such action would be comparable to the heat generated by a furnace triggering the thermostat to call for even more heat output from the furnace so that the room temperature would continuously rise.

Because positive feedback moves the controlled variable even farther from a steady state, it does not occur very often in the body, where the major goal is maintenance of stable, homeostatic conditions. In certain instances, however, positive feedback does occur; a prime example occurs during the birth of a baby. The hormone oxytocin causes powerful contractions of the uterus. As uterine contractions push the baby against the cervix (the exit from the uterus), the resultant stretching of the cervix triggers a sequence of events that brings about the release of even more oxytocin, which causes even stronger uterine contractions, triggering the release of more oxytocin, and so on. This positive-feedback cycle does not cease until the baby is finally expelled.

In addition to feedback mechanisms, which bring about a response in *reaction to* a change in a regulated variable, the body less frequently employs **feedforward** mechanisms, which bring about a response in *anticipation of* a change in a regulated variable. For example, when a meal is still in the digestive tract, a feedforward mechanism is initiated that increases the secretion of a hormone that will promote the cellular uptake and storage of ingested nutrients after they have been absorbed from the digestive tract. This anticipatory response helps limit the rise in blood nutrient concentration that occurs following nutrient absorption.

Disruptions in homeostasis can lead to illness and death.

When one or more of the body's systems fail to function properly, homeostasis is disrupted and all of the cells suffer because they no longer have an optimal environment in which to live and function. Various pathophysiological states ensue, depending on the type and extent of homeostatic disruption. **Pathophysiology** refers to the abnormal functioning of the body (altered physiology) associated with disease. When a homeostatic disruption becomes so severe that it is no longer compatible with survival, death results.

 ## CHAPTER IN PERSPECTIVE: FOCUS ON HOMEOSTASIS

In this chapter, you learned what homeostasis is: a dynamic steady state of the constituents in the internal fluid environment (the extracellular fluid) that surrounds and exchanges materials with the cells. Maintenance of homeostasis is essential for the survival and normal functioning of cells, and each cell, through its specialized activities, contributes as part of a body system to the maintenance of homeostasis.

We have already described how cells are organized according to specialization into body systems. How homeostasis is essential for cell survival and how body systems maintain this internal constancy are the topics of the remainder of this book.

 CHAPTER SUMMARY

Levels of Organization in the Body

The human body is composed of an interactive society of cells, which are the basic units of both structure and function. Each cell performs basic functions essential for its own survival, such as obtaining O_2 and nutrients, which the cell uses to acquire energy; eliminating wastes; synthesizing needed cellular components; reacting to changes in the surrounding environment; controlling movement of materials within the cell and between the cell and its environment; and reproducing.

In multicellular organisms, each cell performs, in addition to these fundamental cell functions, a specialized activity that is usually an elaboration of one of the basic cell functions. The body's cells are highly organized into functional groupings, with cells of similar structure and specialized activity organized into tissues. There are four primary types of tissue: (1) muscle tissue, which is specialized for contraction and force generation; (2) nervous tissue, which is specialized for initiation and transmission of electrical impulses; (3) epithelial tissue, which lines and covers various body surfaces and cavities and also forms secretory glands; and (4) connective tissue, which connects, supports, and anchors various body parts. Tissues are further organized into organs, which are structures composed of several types of primary tissue that act together to perform a particular function or functions. Organs make up body systems, which are

collections of organs that perform related functions and interact to accomplish a common activity essential for survival of the whole body. Organ systems, in turn, compose the whole body.

Concept of Homeostasis

Homeostasis refers to the maintenance of a dynamic steady state within the internal fluid environment that bathes all of the body's cells. Because the body's cells are not in direct contact with the external environment, cell survival depends on maintenance of a stable internal fluid environment with which the cells directly make exchanges. For example, O_2 and nutrients must constantly be replenished in the internal environment to keep pace with the rate at which the cells use these materials for energy production. The factors of the internal environment that must be homeostatically maintained are its (1) concentration of nutrient molecules, (2) concentration of O_2 and CO_2, (3) concentration of waste products, (4) pH, (5) concentration of water, salt, and other electrolytes, (6) temperature, and (7) volume and pressure.

The functions performed by each of the eleven body systems are directed toward maintaining homeostasis. The body systems' functions ultimately depend on the specialized activities of the cells composing each system. Thus, homeostasis is essential for each cell's survival, and each cell contributes to homeostasis.

Control systems that regulate the body systems' various activities to maintain homeostasis can be classified as (1) intrinsic controls, which are inherent compensatory responses of an organ to a change, and (2) extrinsic controls, which are responses of an organ that are triggered by factors external to the organ, namely, by the nervous and endocrine systems. Both intrinsic and extrinsic control systems generally operate on the principle of negative feedback: a change in a regulated variable triggers a response that drives the variable in the opposite direction of the initial change, thus opposing the change.

Pathophysiological states ensue when one or more of the body systems fail to function properly so that an optimal internal environment can no longer be maintained. Serious homeostatic disruption leads to death.

REVIEW EXERCISES

Objective Questions (Answers on p. D–1.)

1. Which of the following activities is *not* carried out by every cell in the body?
   a. obtaining O_2 and nutrients
   b. performing chemical reactions to acquire energy for the cell's use
   c. eliminating wastes
   d. controlling to a large extent exchange of materials between the cell and its external environment
   e. reproducing

2. Which of the following is the proper progression of the levels of organization in the body?
   a. cells, organs, tissues, body systems, whole body
   b. cells, tissues, organs, body systems, whole body
   c. cells, tissues, organs, whole body, body systems
   d. cells, organs, tissues, whole body, body systems
   e. cells, tissues, body systems, organs, whole body

3. Which of the following is *not* a type of connective tissue?
   a. bone
   b. blood
   c. the spinal cord
   d. tendons
   e. the tissue that attaches epithelial tissue to underlying structures

4. The term *tissue* can apply either to one of the four primary tissue types or to a particular organ's aggregate of cellular and extracellular components. (True or false?)

5. Cells in a multicellular organism have specialized to such an extent that they have little in common with single-celled organisms. (True or false?)

6. Cellular specializations are usually a modification or elaboration of one of the basic cell functions (True or false?)

7. The four primary types of tissue are _____, _____, _____, and _____.

8. _____ refers to the release from a cell, in response to appropriate stimulation, of specific products that have in large part been synthesized by the cell.

9. _____ glands secrete through ducts to the outside of the body, whereas _____ glands release their secretory products, known as _____, internally into the blood.

10. _____ controls are inherent to an organ, whereas _____ controls are regulatory mechanisms initiated outside of an organ that alter the activity of the organ.

11. Match the following:
    ____ 1. circulatory system
    ____ 2. digestive system
    ____ 3. respiratory system
    ____ 4. urinary system
    ____ 5. muscular and skeletal systems
    ____ 6. integumentary system
    ____ 7. immune system
    ____ 8. nervous system
    ____ 9. endocrine system
    ____ 10. reproductive system

    a. obtains O_2 and eliminates CO_2
    b. support and protect body parts and allow movement
    c. controls, via hormones it secretes, processes that require duration
    d. transport system
    e. removes wastes and excess water, salt, and other electrolytes
    f. essential for perpetuation of species
    g. obtains nutrients, water, and electrolytes
    h. defends against foreign invaders and cancer
    i. acts through electrical signals to control body's rapid responses
    j. serves as protective barrier between body and external environment

Essay Questions

1. Define physiology.
2. What are the basic cell functions?
3. Distinguish between the external environment and internal environment.
4. Of what fluid compartments is the internal environment composed?
5. Define homeostasis.
6. Describe the interrelationship among cells, body systems, and homeostasis.
7. What factors of the internal environment must be homeostatically maintained?
8. Compare negative and positive feedback.

1. Considering the nature of negative-feedback control and the function of the respiratory system, what effect do you predict that a decrease in CO_2 in the internal environment would have on how rapidly and deeply a person breathes?

2. Would the O_2 levels in the blood be (a) normal, (b) below normal, or (c) elevated in a patient with severe pneumonia in whom exchange of O_2 and CO_2 between the air and blood in the lungs is impaired? Would the CO_2 levels in the same patient's blood be (a) normal, (b) below normal, or (c) elevated? Because CO_2 reacts with H_2O to form carbonic acid (H_2CO_3), would the patient's blood (a) have a normal pH, (b) be too acidic, or (c) not be acidic enough (that is, be too alkaline), assuming that other compensatory measures have not yet had time to act?

3. Given that most AIDS victims die from overwhelming infections or rare types of cancer, what body system do you think is impaired by the AIDS virus?

4. The hormone insulin enhances the transport of glucose (sugar) from the blood into most of the body's cells. Its secretion is controlled by a negative-feedback system between the concentration of glucose in the blood and the insulin-secreting cells. Therefore, which of the following statements is correct?

   a. A decrease in blood glucose concentration stimulates insulin secretion, which in turn further lowers the blood glucose concentration.

   b. An increase in blood glucose concentration stimulates insulin secretion, which in turn lowers the blood glucose concentration.

   c. A decrease in blood glucose concentration stimulates insulin secretion, which in turn increases the blood glucose concentration.

   d. An increase in blood glucose concentration stimulates insulin secretion, which in turn further increases the blood glucose concentration.

   e. None of the above are correct.

2 Cellular Physiology

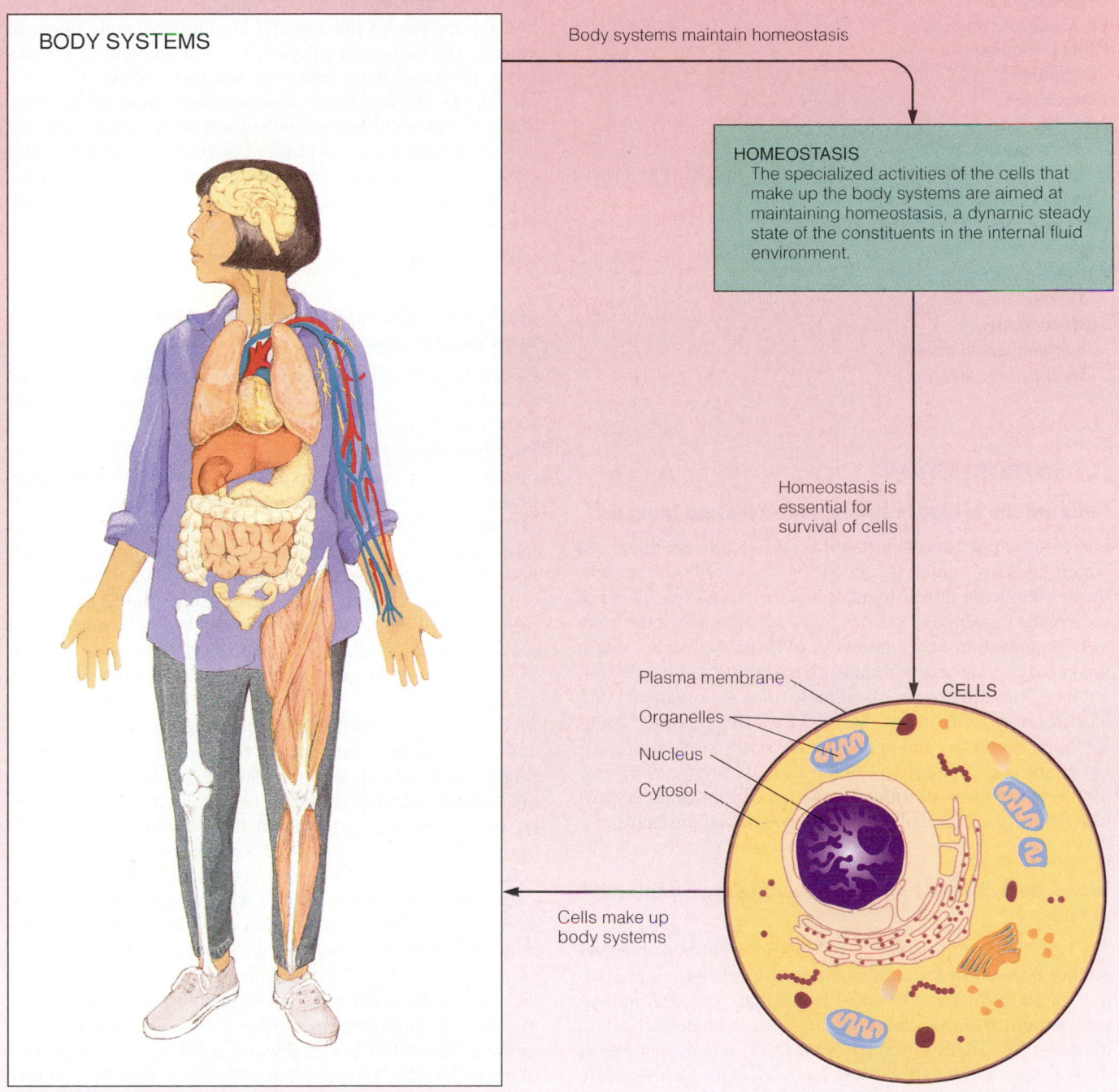

BODY SYSTEMS

Body systems maintain homeostasis

HOMEOSTASIS
The specialized activities of the cells that make up the body systems are aimed at maintaining homeostasis, a dynamic steady state of the constituents in the internal fluid environment.

Homeostasis is essential for survival of cells

Plasma membrane
Organelles
Nucleus
Cytosol

CELLS

Cells make up body systems

Cells are the body's living building blocks. Just as the body as a whole is highly organized, so too is a cell's interior. A cell is made up of three major parts: a **plasma membrane** that encloses the cell; the **nucleus**, which houses the cell's genetic material; and the **cytoplasm**, which is organized into discrete, highly specialized *organelles* dispersed throughout a gelatin-like liquid, the *cytosol*. The cytosol is pervaded by a protein scaffolding, the *cytoskeleton,* that serves as the "bone and muscle" of the cell.

Through the coordinated action of each of these cellular components, every cell is capable of performing certain basic functions essential to its own survival and a specialized task that contributes to the maintenance of homeostasis. Cells are organized according to their specialization into body systems that maintain the stable internal environment essential for the whole body's survival. All body functions ultimately depend on the activities of the individual cells that compose the body.

≡ **CHAPTER CONTENTS AT A GLANCE** ≡

INTRODUCTION
Cells as the bridge between molecules and humans
Observations of cells
General cell structure
● CONCEPTS, CHALLENGES, AND CONTROVERSIES HeLa Cells:
    Problems in a "Growing" Industry
ORGANELLES
Endoplasmic reticulum
Golgi complex
Lysosomes
Peroxisomes
Mitochondria
    ATP production
● A CLOSER LOOK AT EXERCISE PHYSIOLOGY Aerobic Exercise:
    What For and How Much?
CYTOSOL AND CYTOSKELETON
Cytosol
Cytoskeleton
    Microtubules
    Microfilaments
    Intermediate filaments
    Microtrabecular lattice

 INTRODUCTION

Cells are the bridge between molecules and humans.

Even though researchers have analyzed the chemicals of which cells are made, it has not been possible to organize these chemicals into a living cell in a laboratory. It is not merely the presence of various molecules but rather the complex organization and interaction of these molecules within the cell that confers the unique characteristics of life. Within each cell, inanimate chemical molecules are organized into a living entity. Cells, in turn, serve as the living building blocks for the immensely complicated whole body. Modern physiologists are unraveling many of the broader mysteries of how the body works by probing deeper into the molecular structure and organization of the cells that make up the body.

Increasingly better tools are revealing the complexity of cells.

The cells that compose the human body are so small that they cannot be seen by the unaided eye. The smallest visible particle is about five to ten times larger than a typical human cell, which averages about 10 to 20 micrometers (μm) in diameter. (1 μm is equal to 1/1,000,000 meter; a meter is 39.37 inches. Refer to the inside back cover for a comparison of metric units and their English equivalents.) About one hundred average-sized cells lined up side by side would stretch a distance of only 1 mm.

Not until the microscope was invented in the middle of the seventeenth century was the existence of cells revealed. In the early part of the nineteenth century, with the development of better light microscopes, researchers learned that all plant and animal tissues are composed of individual cells. Another important discovery was that cells are filled with a fluid, which, with the microscopic capabilities of the time, appeared to be a rather homogeneous, soupy mixture believed to be the elusive "stuff of life." Not until the 1940s, when the technique of electron microscopy was first employed to observe living matter, did an understanding of the great diversity and complexity of the internal structure of cells begin to emerge. (Electron microscopes are about one hundred times more powerful than light microscopes.) Now, with the availability of even more sophisticated microscopic tools, biochemical techniques, cell-culture technology (see the accompanying boxed feature, Concepts, Challenges, and Controversies), and genetic engineering, the concept of the cell as a microscopic bag of amorphous fluid has given way to our present-day knowledge of the cell as a complex, highly organized, compartmentalized structure.

Most cells are subdivided into the plasma membrane, nucleus, and cytoplasm.

Even though there is no such thing as a "typical" cell because of diverse structural and functional specializations, different cells share many common features. Most cells have three major subdivisions: the *plasma membrane,* the *nucleus,* and the *cytoplasm* (Table 2–1). The **plasma membrane** or **cell membrane** is a very thin membranous structure that encloses each cell, separating the cell's contents from its surroundings. The fluid contained within all of the cells of the body is known collectively as **intracellular fluid** (**ICF**), and the fluid outside of the cells is referred to as **extracellular fluid** (**ECF**). The plasma membrane does not merely serve as a mechanical barrier to hold in the contents of the cells; it has the ability to selectively control movement of molecules between the ICF and ECF.

The two major parts of the cell's interior are the nucleus and the cytoplasm. The **nucleus**, which is typically the largest single organized cellular component, can be seen as a distinct spherical or oval structure, usually located near the center of the cell. It is surrounded by a double-layered membrane, which separates the nucleus from the remainder of the cell. Sequestered within the nucleus is the cell's genetic material, **deoxyribonucleic acid** (**DNA**), which has two important functions. First, DNA provides codes or "instructions" for directing synthesis of specific structural and enzymatic proteins within the cell. By directing the kinds and amounts of various enzymes and other proteins that are produced, the nucleus indirectly governs most cellular activities and serves as the cell's control center. Second, DNA serves as a genetic blueprint during cell replication to ensure that the cell produces additional cells just like itself, thus continuing the identical type of cell line within the body. (See Appendix B for further details.) Furthermore, in the reproductive cells, the DNA blueprint serves to pass on genetic characteristics to future generations.

HeLa CELLS: PROBLEMS IN A "GROWING" INDUSTRY

Many basic advances in cell physiology and related fields such as genetics and cancer research have come about through the use of cells grown outside the body in what is known as *in vitro* culture (*in vitro* is a Latin phrase that means "in glass"). In the early 1950s, many attempts were made to culture human cells using tissues obtained from biopsies or surgical procedures. These early attempts usually met with failure; the cells died after a few days or weeks in culture, mostly without undergoing mitosis. These difficulties continued until February 1951, when a researcher at Johns Hopkins University received a sample of a cervical cancer from a woman named Henrietta Lacks. Following convention, the culture was named HeLa by combining the first two letters of the donor's first and last names. This cell line not only grew but prospered under culture conditions and represented one of the earliest cell lines successfully grown outside the body.

Researchers were eager to have human cells available on demand to study the effects of drugs, toxic chemicals, radiation, and viruses on human tissue. For example, it was found that polio virus reproduced well in HeLa cells, providing a breakthrough in the development of a polio vaccine. As cell-culture techniques improved, human cell lines were started from other cancers and normal tissues, including heart, kidney, and liver tissues. By the early 1960s, a central collection of cell lines had been established in Washington, D.C., and cultured human cells were an important tool in many areas of biological research.

The first clouds in this happy picture began to gather in 1966, when Stanley Gartler, a geneticist at the University of Washington, discovered that eighteen different human cell lines he had analyzed had all been contaminated and taken over by HeLa cells. Over the next two years, it was confirmed that twenty-four of the thirty-four cell lines in the central repository were actually HeLa cells. Researchers who had spent years studying what they thought were heart or kidney cells had in reality been working with a cervical cancer cell instead. Gartler's discovery meant that hundreds of thousands of experiments performed in laboratories around the world were invalid.

As painful as this lesson was, scientists started over, preparing new cell lines and using new, stricter rules to prevent contamination with HeLa cells. Unfortunately, the problem did not end. In 1974, Walter Nelson-Rees published a paper demonstrating that five cell lines extensively used in cancer research were in fact HeLa cells. In 1976, another paper announced that eleven additional cell lines, each widely used in research, were also HeLa cells; and in 1981, Nelson-Rees listed twenty-two more cell lines that were contaminated with HeLa. In all, one-third of all cell lines used in cancer research were apparently really HeLa cells. The result was an enormous waste of dollars and resources. Clearly, not enough care was taken to prevent the spread of HeLa cells to other cultured lines. This spread can be attributed to poor record keeping, mislabeled cultures, and sloppy laboratory techniques. The response of the scientific community has been interesting. Some scientists quickly stepped forward to acknowledge the problem and retract their conclusions. Others engaged in denials and steadfastly refused to acknowledge that their work was invalid. Henrietta Lacks provided the scientific world with two valuable gifts: a cell line and proof that science is a human endeavor, subject to human traits that include not only honesty, candor, and veracity, but also ego, fear, and denial.

The **cytoplasm** is that portion of the cell interior not occupied by the nucleus. It contains a number of distinct, highly organized, membrane-bound structures—the **organelles**—dispersed within a complex, gel-like mass called the **cytosol**. Nearly all cells contain five main types of organelles—the endoplasmic reticulum, Golgi complex, lysosomes, peroxisomes, and mitochondria (● Fig. 2–1). These organelles are similar in all cells, although there are some variations depending on the specialized capabilities of each cell type. Organelles are like intracellular "specialty shops." Each is a separate internal compartment that contains a specific set of chemicals for carrying out a particular cellular function. This compartmentalization is advantageous because it permits chemical activities that would not be compatible with each other to occur simultaneously within the cell. For example, the enzymes that destroy unwanted proteins in the cell do so within the protective confines of the lysosomes without the risk of destroying essential cellular proteins. About half of the total cell volume is occupied by organelles.

The remainder of the cytoplasm (the part not occupied by organelles) consists of cytosol, a semiliquid mass laced with an elaborate protein network that constitutes the cytoskeleton. Many of the chemical reactions that are compatible with each other are carried on in the cytosol. The cytoskeletal network gives the cell its shape, provides for its internal organization, and regulates its various movements. In this chapter, we will examine each of the cytoplasmic components in more detail, concentrating first on the organelles.

TABLE 2–1 Summary of Cell Structures and Functions

CELL PART	NUMBER PER CELL	STRUCTURE	FUNCTION
Plasma Membrane	1	Lipid bilayer studded with proteins and small amounts of carbohydrate	Selective barrier between cellular contents and extracellular fluid; control of traffic in and out of the cell
Nucleus	1	DNA and specialized proteins enclosed by a double-layered membrane	Control center of the cell, providing storage of genetic information
			Provision of codes for the synthesis of structural and enzymatic proteins that determine the cell's specific nature
			Blueprint for cell replication
Cytoplasm			
Organelles			
Endoplasmic reticulum	1	Extensive, continuous membranous network of fluid-filled tubules and flattened sacs, partially studded with ribosomes	Formation of new cell membrane and other cell components and manufacture of products for secretion
Golgi complex	1 to several hundred	Sets of stacked, flattened membranous sacs	Modification, packaging, and distribution center for newly synthesized proteins
Lysosomes	300	Membranous sacs containing hydrolytic enzymes	Digestive system of the cell, destroying unwanted material such as foreign substances and cellular debris
Peroxisomes	200	Membranous sacs containing oxidative enzymes	Detoxification activities
Mitochondria	100–2,000	Rod- or oval-shaped bodies enclosed by two membranes, with the inner membrane folded into cristae that project into the interior matrix	Energy organelles; major site of ATP production; contain enzymes for citric-acid cycle and electron-transport chain
Vaults	Thousands	Shaped like octagonal barrels	Unclear; may transport messenger RNA from nucleus to cytoplasm; may be important in cellular contractile systems
Cytosol			
Intermediary metabolism enzymes	Many	Sequential arrangement within the cytoskeleton	Intracellular reactions involving the degradation, synthesis, and transformation of small organic molecules
Ribosomes	Many	Granules of RNA and proteins—some attached to rough endoplasmic reticulum, some free in the cytoplasm	Protein synthesis
Secretory vesicles	Varies	Membrane-bound packages of secretory products	Storage of secretory products until signaled to empty to the outside
Inclusions	Varies	Glycogen granules, fat droplets	Storage of excess nutrients

ORGANELLES

The endoplasmic reticulum is a synthesizing factory.

The **endoplasmic reticulum** (**ER**) is an elaborate fluid-filled membranous system distributed extensively throughout the cytosol. Two distinct types of endoplasmic reticulum—the smooth ER and the rough ER—can be distinguished. The **smooth ER** is a meshwork of tiny interconnected tubules, whereas the **rough ER** projects outward from the smooth ER as stacks of relatively flattened sacs (• Fig. 2–2). Even though these two regions differ considerably in appearance and function, they are thought to be continuous with each other. In other words, the ER is one continuous organelle with many interconnected channels. The outer surface of the rough-ER membrane is studded with small, dark-staining particles that give it a "rough" or granular appearance. These particles are

TABLE 2-1 Summary of Cell Structures and Functions (continued)

CELL PART	NUMBER PER CELL	STRUCTURE	FUNCTION
Cytoskeleton			
Microtubules	Many	Long, slender, hollow tubes composed of tubulin molecules	Maintain asymmetrical cell shapes
			Coordinate complex cell movements
			Facilitate transport of secretory vesicles within cell, such as axonal transport
			Serve as dominant structural and functional component of cilia and flagella
			Form mitotic spindle during cell division
Microfilaments	Many	Helically intertwined chains of actin molecules; microfilaments composed of myosin molecules also present in muscle cells	Play a vital role in various cellular contractile systems
			Play a dominant role in muscle contraction
			Form nonmuscle contractile assemblies, such as in white blood cells during amoeboid movement
			Serve as a mechanical stiffener for microvilli
			Increase the surface area available for absorption in intestines and kidneys
			Are specialized to detect sound and positional changes in ear
Intermediate filaments	Many	Irregular, threadlike proteins	Play a structural role in parts of the cell subject to mechanical stress
Microtrabecular lattice	1	Meshwork of exceedingly fine interlinked filaments	Suspends and functionally links larger cytoskeletal elements and various organelles
			Organizes cytosolic enzymes
			Integrated Functions of Entire Cytoskeleton:
			Responsible for the shape, rigidity, and spatial geometry of each type of cell; the "bone" of the cell
			Responsible for directing intracellular transport and for regulating cellular movements; the "muscle" of the cell
			Appears to play a role in regulating growth and division of cells

ribosomes, which are special ribonucleic acid (RNA)-protein complexes that synthesize proteins under the direction of nuclear DNA. The genetic code for protein synthesis is transcribed into a messenger molecule that carries the genetic message from the nucleus to the ribosome "workbench," where protein synthesis takes place (see p. B-7). Not all ribosomes in the cell are attached to the rough ER. Unattached or "free" ribosomes are dispersed throughout the cytosol. The relative amount of smooth and rough ER varies between cells, depending on the activity of the cell.

Rough Endoplasmic Reticulum The rough ER, in association with its ribosomes, synthesizes and releases a variety of new proteins into the ER lumen, the fluid-filled space enclosed by the ER membrane. These proteins serve one of two purposes: (1) Some proteins are destined for export to

● **FIGURE 2–1   Schematic Three-Dimensional Illustration of Cell Structures Visible under Electron Microscope**

● **FIGURE 2–2   Endoplasmic Reticulum (ER)**   (a) Schematic three-dimensional representation of the relationship between the rough and smooth ER. The smooth ER is a meshwork of tiny interconnected tubules. The rough ER, which is studded with ribosomes, projects outward from the smooth ER as stacks of relatively flattened sacs. (b) Electron micrograph of rough ER. Note the layers of flattened sacs studded with small, dark-staining ribosomes.

the cell's exterior as secretory products, such as proteinaceous hormones or enzymes, all of which are proteins (see p. A–4). (2) Other proteins are transported to sites within the cell for use in the construction of new cellular membrane (either new plasma membrane or new organelle membrane) as well as other protein components of organelles. Cellular membranes consist predominantly of lipids (fats) and proteins. The membranous wall of the ER also contains enzymes essential for the synthesis of nearly all the lipids needed for the production of new membranes. These newly synthesized lipids enter the ER lumen along with the proteins. Predictably, the rough ER is most abundant in cells specialized for protein secretion (for example, cells that secrete digestive enzymes) or in cells that require extensive membrane synthesis (for example, rapidly growing cells such as immature egg cells).

Once assembled, each ribosome participates in the synthesis of only one type of protein. The free and ER-bound ribosomes differ only in the proteins that they help to synthesize. In contrast to the rough-ER ribosomes, the free ribosomes synthesize enzymatic proteins that are used intracellularly within the cytosol. All ribosomes are produced in the cell's nucleus under the direction of DNA, with each being "programmed" at any given time to facilitate the synthesis of only one specific protein needed by that particular cell.

Currently, it is thought that newly formed ribosomes destined to become attached to the rough ER start to synthesize a **leader sequence** that acts as a signal, much like an address on a letter (● Fig. 2–3). Another protein present in the cytosol, the **signal-recognition protein**, binds to one of these ribosomes upon recognizing the newly synthesized leader sequence. Subsequently, the signal-recognition protein, acting much like a mail carrier, delivers the ribosome to the proper "address" on the ER membrane. How does the signal-recognition protein know the proper address? Special proteins called **ribophorins** found exclusively in the rough regions of the ER membrane act as the "house address" on this membrane. The ribophorins serve as binding sites for preferential ribosomal attachment. The signal-recognition protein (the "mail carrier") can recognize both the leader-sequence signal on the ribosome (the "address on the envelope") and the ribophorin binding site on the ER (the "house address") and therefore delivers the proper ribosome to the proper site on the rough ER for binding.

Once a ribosome is attached to the ER membrane and starts directing synthesis of a specific protein, the developing protein threads its way across the ER membrane into the lumen. Here the leader sequence is removed and the protein chain is folded into its final conformation. The protein may also be modified in other ways, such as being "pruned" (● Fig. 2–4) or having sugar molecules attached to it. After this processing within the ER lumen, a new protein is unable to pass through the ER membrane and therefore becomes permanently separated from the cytosol as soon as it has been synthesized. In this way, the ER provides cells with a mecha-

---

● **FIGURE 2–3   Delivery of Ribosomes to Rough Endoplasmic Reticulum**   (a) The signal-recognition protein "recognizes" the correct leader sequence on the developing protein chain of a cytosolic ribosome destined to become attached to the rough ER. (b) The signal-recognition protein carries the ribosome to the correct ribophorin "address" on the ER membrane. (c) The signal-recognition protein departs, and the protein chain threads through the ER membrane into the lumen; the proper ribosome is now attached. Other ribosomes synthesizing cytosolic proteins are not recognized by the signal-recognition protein.

(a)   (b)   (c)

• **FIGURE 2–4 Formation of Insulin Molecule within Endoplasmic Reticulum** Insulin is originally formed as a single long polypeptide chain that is folded back on itself, with the two overlapping ends being joined by two disulfide (sulfur-to-sulfur) bonds. The connecting piece (in gray) between the two overlapping ends is then pruned away, leaving the insulin molecule (in green) consisting of an A chain and a B chain connected by the disulfide bonds.

Pruned connecting section

A chain

—Disulfide bonds—

B chain

Each circle represents a specific amino acid.

nism for separating the newly produced molecules that are destined for export out of the cell or for synthesis of new cellular components (those synthesized by the ER) from those that belong in the cytosol (those produced by the free ribosomes).

How do the newly synthesized molecules within the ER lumen get to their destinations at other intracellular sites or to the exterior of the cell if these molecules cannot pass out through the ER membrane? The smooth ER is important in accomplishing this feat.

Smooth Endoplasmic Reticulum The smooth ER does not contain ribosomes; hence it is "smooth." Lacking ribosomes, it is not involved in protein synthesis. Instead it serves a variety of other purposes that vary in different cell types.

In the majority of cells, the smooth ER is rather sparse and serves primarily as a central packaging and discharge site for molecules that are to be transported from the ER. Newly synthesized proteins and lipids pass from the rough ER to gather in the smooth ER. Portions of the smooth ER then "bud off" (that is, are pinched off), giving rise to **transport vesicles** that contain the new molecules enclosed in a membrane layer derived from the smooth-ER membrane (• Fig. 2–5). Newly synthesized membrane components are rapidly incorporated into the ER membrane itself to replace the membrane that was used to "wrap" the transport vesicle. Transport vesicles move to the Golgi complex for further processing of their cargo.

In contrast to the sparseness of smooth ER in most cells, some specialized types of cells have an extensive smooth ER, which has additional responsibilities as follows:

■ The smooth ER is abundant in cells that specialize in lipid metabolism—for example, cells that secrete steroid hormones. (A steroid is a special type of lipid derived from cholesterol.) The membranous wall of the smooth ER, like that of the rough ER, contains enzymes for synthesis of lip-

ids. The lipid-producing enzymes in the membranous wall of the rough ER alone are insufficient to carry out the extensive lipid synthesis necessary to maintain adequate steroid-hormone secretion levels. These cells have an expanded smooth-ER compartment to house the additional enzymes necessary to keep pace with demands for hormone secretion.

■ In liver cells, the smooth ER has a special capability. It contains enzymes that are involved in detoxifying harmful substances produced within the body by metabolism or sub-

• **FIGURE 2–5 Vesicular Transport between the Endoplasmic Reticulum and the Golgi Complex and Release of Secretory Vesicles from the Golgi Complex after Further Processing**

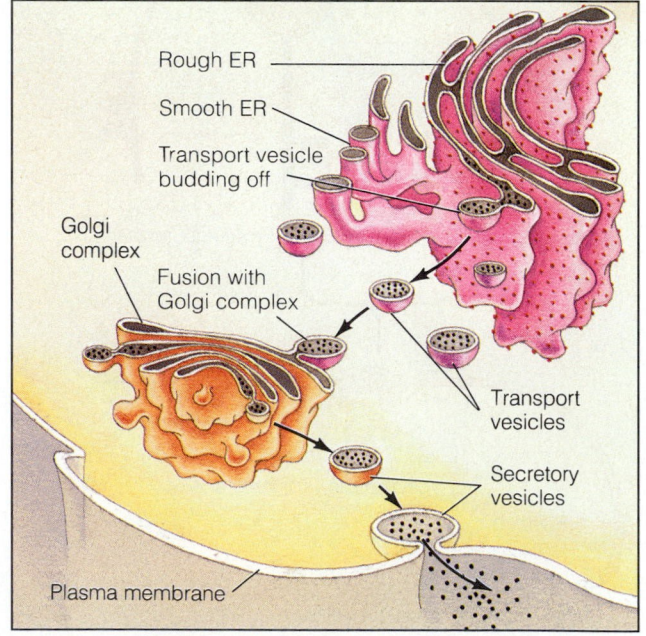

Rough ER

Smooth ER

Transport vesicle budding off

Golgi complex

Fusion with Golgi complex

Transport vesicles

Secretory vesicles

Plasma membrane

stances that enter the body from the outside in the form of drugs or other foreign compounds. These detoxification enzymes alter toxic substances so that the latter can be eliminated more readily in the urine. The amount of smooth ER available in liver cells for the task of detoxification can vary dramatically, depending on the need. For example, if phenobarbital (a barbiturate drug used as a sedative) is administered in large quantities, the amount of smooth ER with its associated detoxification enzymes doubles within a few days, only to return to normal within five days after drug administration ceases. The mechanisms involved in regulating these changes are not well understood. Unfortunately, in some cases the same enzymes transform otherwise harmless substances into carcinogenic substances that play a role in cancer development.

■ Muscle cells have developed another specialized use for the smooth ER. They have an elaborate, modified smooth ER known as the *sarcoplasmic reticulum,* which stores calcium and plays an important role in the process of muscle contraction.

The Golgi complex is a refining plant and directs molecular traffic.

Closely associated with the endoplasmic reticulum is the **Golgi complex**, which consists of sets of flattened, slightly curved, membrane-bounded sacs, or cisternae, stacked in layers (● Fig. 2–6). Note that the flattened sacs are thin in the middle but have dilated edges. The number of Golgi stacks varies, depending on the cell type. Some cells have only one stack, whereas cells highly specialized for protein secretion may have hundreds of stacks.

The majority of the newly synthesized molecules that have just budded off from the smooth ER enter a Golgi stack. When a transport vesicle carrying its newly synthesized cargo reaches a Golgi stack, the vesicle membrane fuses with the membrane of the sac closest to the center of the cell. The vesicle membrane opens up and becomes a new part of the Golgi membrane, and the contents of the vesicle are released to the interior of the sac (Fig. 2–5).

These newly synthesized raw materials from the ER travel by means of vesicle formation through the layers of the Golgi stack, where two important interrelated functions take place:

1. *Processing the raw materials into finished products.* Within the Golgi complex, the "raw" proteins from the ER are modified into their final form, largely by adjustments made in the sugars attached to the protein. The biochemical pathways that the proteins undergo during their passage through the Golgi complex are elaborate, precisely programmed, and specific for each final product.

2. *Sorting and directing the finished products to their final destinations.* The Golgi complex is responsible for sorting and segregating different types of products according to their function and destination, namely, molecules that are destined for secretion to the exterior, molecules that will become part of the plasma membrane, and molecules that will become incorporated into other organelles.

The means by which the Golgi complex directs this molecular traffic is complex and still is not clearly understood. It is known, however, that products destined for intracellular transport are packaged differently than those to be extruded from the cell. Those products targeted for intracellular sites are packaged in *coated vesicles* whereas those to be secreted are packaged in *secretory vesicles.*

Coated vesicles originate from the dilated rims of the Golgi sacs, which pinch off to form clusters of small vesicles that can be seen along the edge of a Golgi stack (Fig. 2–6). These

---

● **FIGURE 2–6   Golgi Complex**   (a) Schematic three-dimensional representation of a Golgi complex. Newly synthesized proteins arriving from the ER are progressively modified into their final form as they move by vesicular transport through the layers of the Golgi complex. (b) Electron micrograph of a Golgi stack. Note the vesicles at the dilated edges of the sacs. These vesicles contain finished protein products packaged for distribution to their final destination.

Transport vesicles from endoplasmic reticulum

Golgi sacs

Vesicles containing finished product

(a)

Golgi stack

Nucleus

(b)

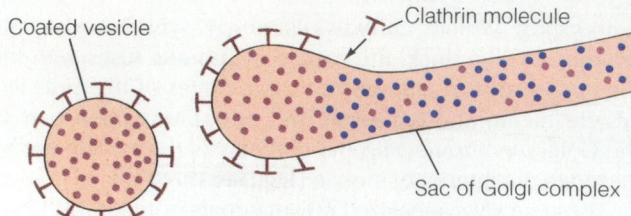

Coated vesicle

Clathrin molecule

Sac of Golgi complex

● **FIGURE 2–7** **Formation of a Coated Vesicle** A coated vesicle, coated with clathrin molecules and specific accessory proteins, captures specific cargo from the lumen of the Golgi complex and then buds off the dilated rim of a Golgi sac.

Secretory vesicle

Fusion of secretory vesicle with plasma membrane

Secretion of vesicle contents

● **FIGURE 2–8** **Exocytosis of Secretory Product**

vesicles are "coated" or enclosed in a cagelike protein framework consisting of **clathrin**, a structural protein that forms the latticework of the coat, plus accessory proteins derived from the Golgi membrane (● Fig. 2–7). The accessory proteins serve two important functions. First, specific accessory proteins on the interior surface of the coat act as "recognition markers" for the recognition and attachment of specific molecules that have been processed in the Golgi lumen. These markers ensure that the proper cargo is captured and packaged for the appropriate destination. Second, other specific accessory proteins on the outer surface of each vesicle act as "docking markers" that are able to fuse only with the membrane of the designated intracellular site. These docking 517markers ensure that the coated transport vesicle docks and unloads its cargo only at the appropriate intracellular "address." Thus, coated vesicles appear to act as miniature sorting machines: specific molecules are selectively trapped in a vesicle before it buds off, with its destiny being fusion with only one designated type of intracellular structure. There are different classes of coated vesicles, each of which contains specific captured cargo that is bound to a specified intracellular compartment.

Cells specialized for secretion, such as endocrine cells that secrete proteinaceous hormones, handle molecules destined for export out of the cell somewhat differently than intracellular cargo is handled. In secretory cells, numerous large **secretory vesicles** (about two hundred times larger than coated vesicles) can be seen in association with a Golgi stack (Fig. 2–5). The molecules to be secreted, such as proteinaceous hormones, are trapped within these large secretory vesicles, which pinch off the dilated rims of the outermost sac closest to the plasma membrane. The secretory proteins trapped within these structures are in a dilute solution, but the contents become concentrated over time as the secretory vesicle gradually loses water. The concentrated secretory proteins remain stored within the secretory vesicles until the cell is stimulated by a specific signal that indicates a need for release of that particular secretory product. On appropriate stimulation, the vesicles move to the cell's periphery. Vesicular contents are quickly released to the cell's exterior as the vesicle fuses with the plasma membrane, opens, and empties its contents to the outside (Figs. 2–5 and ● 2–8). Release of the contents of the secretory vesicle constitutes the process of **secretion**. This mechanism—extrusion to the exterior of substances originating within the cell—is referred to as **exocyto-**

sis (*exo* means "out of"; *cyto* means "cell"). Secretory vesicles fuse only with the plasma membrane and not with any of the internal membranes that bound the organelles, thereby preventing fruitless or even dangerous discharge of secretory products into the organelles.

Lysosomes serve as the intracellular digestive system.

Lysosomes are membrane-bound sacs containing powerful hydrolytic enzymes (see *hydrolysis* p. A–11) capable of digesting and thereby removing various unwanted cellular debris and foreign material, such as bacteria that have been internalized within the cell. Thus, lysosomes serve as the intracellular "digestive system."

On the average, a cell contains about three hundred lysosomes. Instead of having a uniform structure as is characteristic of all other organelles, lysosomes vary in size and shape, depending on the contents they are digesting. Most commonly, lysosomes are small (0.2–0.5 μm in diameter) oval or spherical bodies that appear granular during inactivity. These granules are protein aggregates of the powerful digestive enzymes contained within. The surrounding membrane that confines these enzymes normally prevents them from destroying the cell that houses them. Both the membrane and the enzymes are derived from the Golgi complex. A specific collection of newly synthesized hydrolytic enzymes is captured in a coated vesicle that is pinched off from the Golgi complex to become a new lysosome.

Extracellular material to be attacked by lysosomal enzymes is brought into the interior of the cell through the process of **endocytosis** (*endo* means "within"). Endocytosis can be accomplished in one of two ways. In most cases, the plasma membrane invaginates (dips inward), forming a pouch that contains a small bit of extracellular fluid, usually with a specific particle that has bound to a surface receptor (● Fig. 2–9a). The plasma membrane then seals at the surface of the pouch, forming a small, intracellular, membrane-bound vesicle with the contents of the pouch trapped inside. A few cell types, most notably white blood cells, perform a special form of endocytosis known as **phagocytosis** (Fig. 2–9b and c). When a white blood cell encounters a large multimolecular particle, such as a bacterium or tissue debris, it extends surface projections that completely surround or engulf the particle, forming an internalized vesicle that traps the particle within. A lysosome fuses with the membrane of the internalized vesicle and releases its hydrolytic enzymes into the vesicle. These potent enzymes safely attack

FIGURE 2-9 Endocytosis (a) Receptor-mediated endocytosis. When a particle attaches to a specific surface receptor site, the membrane dips inward to form a pouch, then seals the surface to internalize the particle within an intracellular vesicle. (b) Phagocytosis. White blood cells perform a special form of endocytosis known as phagocytosis. They internalize multimolecular particles such as bacteria by extending surface projections that seal in the targeted material. A lysosome fuses with the internalized vesicle, releasing enzymes that attack the engulfed material within the confines of the vesicle. (c) Scanning-electron-micrograph series of a white blood cell phagocytizing an old, worn-out red blood cell.

the bacterium or other trapped material within the enclosed confines of the vesicle without damaging the remainder of the cell.

The hydrolytic enzymes largely break down the engulfed material into products such as amino acids, glucose, and fatty acids that can be used by the cell. These small products can readily pass through the lysosomal membrane to enter the cytoplasm for future use. Lysosomes that have completed their digestive activities are known as **residual bodies**. Usually, undigestible substances left within the residual body are eventually eliminated from the cell by exocytosis. Less commonly, they are retained and accumulate within the cell as it ages.

Lysosomes also can fuse with aged or damaged organelles to remove these useless parts of the cell. This selective self-digestion makes way for new replacement parts. All organelles are renewable. If the whole cell is severely damaged or dies, the lysosomes rupture and release their destructive enzymes into the cytosol so that the cell digests itself entirely. In most tissues, elimination of a nonfunctional cell clears the way for its replacement with a healthy new one through cell division. However, in tissues in which cell reproduction is im-

possible, such as the heart and brain, scar tissue replaces the self-destructed dead cells.

In specific instances, lysosomes cause intentional self-destruction of healthy cells. This happens as a normal part of embryonic development when certain unwanted tissues that form are programmed for destruction. For example, embryonic ducts capable of forming a male reproductive tract are deliberately destroyed during the development of a female fetus. Lysosomes also play an important role in tissue regression, such as during the normal reduction in the uterine lining following pregnancy. The mechanisms that control lysosomal activity under such circumstances are still unknown.

An inherent danger even in a healthy, intact cell is that lysosomal membranes may inadvertently rupture. Because these organelles have the potential ability to self-destruct the cell, their discoverer named them "suicide bags." However, two factors make this threat less severe than imagined. First, the lysosomal hydrolytic enzymes function best in an acid environment. The lysosomal membrane transports hydrogen ions (acid formers) into the lysosome, making it considerably more acidic than the remainder of the cell. Should the enzymes accidentally leak into the cytosol, they would be less potent than they are in their acidic home. Second, in most instances, cells could tolerate the limited damage that would occur if only one or two lysosomes inadvertently ruptured because most parts of the cell are renewable. The biggest danger is accidental digestion of part of the irreplaceable DNA molecule within the nucleus. Such nuclear damage would alter the cell's genetic properties, and the defect would be perpetuated to all of the cell's progeny.

Occasionally, individuals lack the ability to synthesize one or more of the lysosomal enzymes. The result is massive accumulation within the lysosomes of the specific compound that is normally digested by the missing enzyme. Clinical manifestations often accompany such disorders because the engorged lysosomes interfere with normal cell activity. The nature and severity of the symptoms depend on the type of substance that is accumulating, which in turn depends on what lysosomal enzyme is missing. Among these so-called storage diseases is **Tay-Sachs disease**. It is characterized by abnormal accumulation of gangliosides, which are complex molecules found in nerve cells. Profound symptoms of progressive nervous-system degeneration result as the accumulation process continues.

Peroxisomes house oxidative enzymes that detoxify various wastes.

Typically, several hundred small **peroxisomes** about one-third to one-half the average size of lysosomes are present in a cell. Peroxisomes are similar to lysosomes in that they are membrane-bound sacs containing enzymes, but unlike the lysosomes, which contain hydrolytic enzymes, peroxisomes house several powerful oxidative enzymes and contain most of the cell's catalase. **Oxidative enzymes**, as the name implies, use oxygen (O_2), in this case to strip hydrogen from specific molecules. Such a reaction is important in detoxifying various wastes produced within the cell or foreign compounds that have entered the cell, such as the ethanol consumed in alcoholic beverages. The major product generated in the peroxisome is **hydrogen peroxide (H_2O_2)**, which is formed by molecular oxygen and the hydrogen atoms stripped from the waste.

Hydrogen peroxide, itself a powerful oxidant, is potentially destructive if it is allowed to accumulate or escape from the confines of the peroxisome. However, peroxisomes also contain an abundance of **catalase**, an antioxidant enzyme that decomposes potent H_2O_2 into harmless H_2O and O_2. This latter reaction is an important safety mechanism that destroys the potentially deadly peroxide at its site of production, thereby preventing its possible devastating escape into the cytosol.

Mitochondria are the energy organelles.

Mitochondria are the energy organelles or "powerhouses" of the cell; they extract energy from the nutrients in food and transform it into a usable form to energize cellular activities.

The number of mitochondria per cell varies greatly, depending on the energy needs of each particular cell type. A single cell may contain as few as one hundred or as many as several thousand mitochondria. In some cell types, the mitochondria are densely compacted in cellular regions that use most of the cell's energy. For example, mitochondria are packed between the contractile units in the muscle cells of the heart.

Mitochondria are rod- or oval-shaped structures about the size of bacteria. In fact, considerable evidence indicates that mitochondria are descendants of bacteria that invaded or were engulfed by primitive cells early in evolutionary history. It is generally believed that a symbiotic ("living together") relationship subsequently evolved, with the bacteria eventually becoming permanent organelles.

Each mitochondrion is enclosed by a double membrane—a smooth outer membrane that surrounds the mitochondrion itself and an inner membrane that forms a series of infoldings or shelves called **cristae**, which project into an inner cavity filled with a gel-like solution known as the **matrix** (● Fig. 2–10). These cristae contain crucial proteins that ultimately are responsible for converting much of the energy in food into a usable form (the electron-transport proteins, to be described shortly). The generous folds of the inner mem-

● **FIGURE 2–10 Mitochondrion** (a) Schematic representation of a mitochondrion. The electron-transport proteins embedded in the cristae folds of the mitochondrial inner membrane are ultimately responsible for converting much of the energy of food into a usable form. (b) Electron micrograph of a mitochondrion.

(a)

(b)

brane greatly increase the surface area available for housing these important proteins. The matrix consists of a concentrated mixture of hundreds of different dissolved enzymes (the citric-acid cycle enzymes, soon to be described) that are important in preparing nutrient molecules for the final extraction of usable energy by the cristae proteins.

Energy derived from food is stored in ATP.

The source of energy for the body is the chemical energy stored in the carbon-hydrogen bonds in ingested food. Body cells, however, are not equipped to use this energy directly. Instead they must extract energy from food nutrients and convert it into an energy form that they can use—namely, the high-energy phosphate bonds of **adenosine triphosphate**, (**ATP**), which consists of adenosine with three phosphate groups attached (see p. A–12). When a high-energy bond such as that binding the terminal phosphate to adenosine is split, a substantial amount of energy is released. Adenosine triphosphate is the universal energy carrier or common energy "currency" of the body. Cells can "cash in" ATP to pay the energy "price" for running the cellular machinery. To obtain immediately usable energy, cells split the terminal phosphate bond of ATP, which yields **adenosine diphosphate** (**ADP**, adenosine with two phosphate groups attached) plus inorganic phosphate (P_i) plus energy:

$$\text{ATP} \quad \xrightarrow{\text{splitting}} \quad \text{ADP} + P_i + \text{energy for use by the cell}$$

In this energy scheme, food might be thought of as the "crude fuel," whereas ATP is the "refined fuel" for operating the body's machinery. Let us elaborate on this fuel conversion process (Table 2–2). Dietary food is digested or broken down by the digestive system into smaller absorbable units that can be transferred from the digestive-tract lumen into the circulatory system. For example, dietary carbohydrates are broken down primarily into glucose, which can be absorbed into the blood. No usable energy is released during the digestion of food. When delivered to the cells by the blood, the nutrient molecules are transported across the plasma membrane into the cytosol. Among the thousands of enzymes within the cytosol are those responsible for **glycolysis**, a chemical process involving nine separate sequential reactions that break down the simple six-carbon sugar molecule, glucose, into two pyruvic-acid molecules, each of which contains three carbons. During this process, some of the energy stored in the chemical bonds of glucose is used to convert ADP into ATP (● Fig. 2–11). However, glycolysis is not very efficient in terms of energy extraction; one molecule of glucose has a net yield of only two molecules of ATP. Much of the energy originally contained in the glucose molecule is still locked in the chemical bonds of the pyruvic-acid molecules. The low-

TABLE 2–2	Overview of Cellular Energy Production from Glucose				
REACTION	SUBSTANCE PROCESSED	LOCATION	ENERGY YIELD (per glucose molecule processed)	END PRODUCTS AVAILABLE FOR FURTHER ENERGY EXTRACTION (per glucose molecule processed)	NEED FOR OXYGEN
Glycolysis	Glucose	Cytosol	Two molecules of ATP	Two pyruvic-acid molecules	No; anaerobic
Citric-acid cycle	Acetyl CoA, which is derived from pyruvic acid, the end product of glycolysis. Two acetyl CoA molecules result from the processing of one glucose molecule	Mitochondrial matrix	Two molecules of ATP	Eight NADH and two $FADH_2$ hydrogen carrier molecules	Yes, derived from molecules involved in citric-acid cycle reactions
Electron-transport chain	High-energy electrons stored in hydrogen atoms in the hydrogen carrier molecules NADH and $FADH_2$ derived from citric-acid cycle reactions	Mitochondrial inner-membrane cristae	Thirty-two molecules of ATP	None	Yes, derived from molecular oxygen acquired from breathing

• FIGURE 2–11   A Simplified Summary of Glycolysis
Glycolysis involves the breakdown of glucose into pyruvic acid, with a net yield of two molecules of ATP for every glucose molecule processed.

energy yield of glycolysis is insufficient to support the body's demand for ATP. This is where the mitochondria come into play.

The pyruvic acid produced by glycolysis in the cytosol can be selectively transported into the mitochondrial matrix. Here it is further broken down into a two-carbon molecule, acetic acid, by enzymatic removal of one of the carbons in the form of carbon dioxide (CO_2), which eventually is eliminated from the body as an end product, or waste (• Fig. 2–12). During this process, a carbon-hydrogen bond is disrupted, so a hydrogen atom is also released. This hydrogen atom is held by a hydrogen carrier molecule, the function of which will be discussed shortly. The acetic acid thus formed combines with coenzyme A, a derivative of pantothenic acid (a B vitamin), producing the compound acetyl coenzyme A (acetyl CoA).

Acetyl CoA then enters the **citric-acid cycle**, which consists of a cyclical series of eight separate biochemical reactions that are directed by the enzymes of the mitochondrial matrix. This cycle of reactions can be compared to one revolution around a ferris wheel. (Keep in mind that Figure 2–12 is highly schematic. It depicts a cyclical series of biochemical reactions. The molecules themselves are not physically moved around in a cycle.) On the top of the ferris wheel, acetyl CoA, a two-carbon molecule, enters a seat already occupied by oxaloacetic acid, a four-carbon molecule. These two molecules link together to form a six-carbon citric-acid molecule, and the trip around the citric-acid cycle begins. (This cycle is alternatively known as the **Krebs cycle** in honor of its principal discoverer, Sir Hans Krebs, or the **tricarboxylic-acid cycle**, because citric acid contains three carboxylic acid groups.) As the seat moves around the cycle, at each new position, matrix enzymes modify the passenger molecule to form a slightly different molecule. These molecular alternations have the following important consequences:

1. Two carbons are sequentially "kicked off the ride" as they are removed from the six-carbon citric-acid molecule, converting it back into the four-carbon oxaloacetic acid, which is now available at the top of the cycle to pick up another acetyl CoA for another revolution through the cycle.

2. The released carbon atoms, which were originally present in the acetyl CoA that entered the cycle, are converted into two molecules of CO_2. This CO_2, as well as the CO_2 produced during the formation of acetic acid from pyruvic acid, passes out of the mitochondrial matrix and subsequently out of the cell to enter the blood. In turn, the

blood carries it to the lungs, where it is finally eliminated into the atmosphere through the process of breathing. The oxygen used to make CO_2 from these released carbon atoms is derived from the molecules that were involved in the reactions, not from free molecular oxygen supplied by breathing.

3. Hydrogen atoms are also "bumped off" during the cycle at four of the chemical conversion steps. These hydrogens are "caught" by two other compounds that act as hydrogen carrier molecules—**nicotinamide adenine dinucleotide** (**NAD**), a derivative of the B vitamin niacin, and **flavine adenine dinucleotide** (**FAD**), a derivative of the B vitamin riboflavin. These compounds are converted by the transfer of hydrogen to NADH and $FADH_2$, respectively.

4. One more molecule of ATP is produced for each molecule of acetyl CoA processed. Actually, ATP is not directly produced by the citric-acid cycle. The released energy is used to directly link inorganic phosphate to **guanosine diphosphate** (**GDP**) to form **guanosine triphosphate** (**GTP**), a high-energy molecule similar to ATP. The energy from GTP can then be transferred to ATP as follows:

$$ADP + GTP \rightleftharpoons ATP + GDP$$

Because each glucose molecule is converted into two acetic-acid molecules, thus permitting two turns of the citric-acid cycle, two more ATP molecules are produced from each glucose molecule.

These few additional ATPs are still not much of an energy profit. However, the citric-acid cycle is important in preparing the hydrogen carrier molecules for their entry into the **electron-transport chain**, which produces far more energy than the sparse amount of ATP produced by the cycle itself. Considerable untapped energy is still stored in the released hydrogen atoms, which contain electrons at high energy levels. The "big payoff" comes when NADH and $FADH_2$ enter the electron-transport chain, which consists of electron carrier molecules located in the inner mitochondrial membrane lining the cristae (• Fig. 2–13a). The high-energy electrons are extracted from the hydrogens held in NADH and $FADH_2$ and are transferred sequentially to the electron carrier molecules, freeing NAD and FAD to pick up more hydrogen atoms. The electron-transport molecules are arranged in a specifically ordered fashion on the inner membrane so that the high-energy electrons are progressively transferred through a chain of

In cytosol

Pyruvic acid (3C)
(from glycolysis)

Mitochondrial outer and inner membranes

Crista

CO_2 (1C)

NAD

H

NADH

Acetic acid (2C)

Coenzyme A

Acetyl CoA (2C)

(joins)

In mitochondrial matrix

Oxaloacetic acid (4C)

Malic acid (4C)

Citric acid (6C)

NADH

NAD

H

*

Fumaric acid (4C)

Isocitric acid (6C)

NAD

H

NADH

CO_2 (1C)

FADH₂

FAD

H

Succinic acid (4C)

α ketoglutaric acid (5C)

NAD

GTP

*

Succinyl CoA (4C)

H

Pᵢ

CO_2 (1C)

NADH

ADP

GDP

ATP

C = carbon atom.

* H_2O enters the cycle at the steps marked with an asterisk.

● **FIGURE 2–12   Citric-Acid Cycle**   A simplified version of the citric-acid cycle, showing how the two carbons entering the cycle by means of acetyl CoA are eventually converted to CO_2, with oxaloacetic acid, which accepts the acetyl CoA, being regenerated at the end of the cyclic pathway. Also denoted is the release of hydrogen atoms at specific points along the pathway, with these hydrogens binding to the hydrogen carrier molecules NAD and FAD for further processing. One molecule of ATP is generated for each molecule of acetyl CoA that enters the citric-acid cycle, for a total of two molecules of ATP for each molecule of processed glucose.

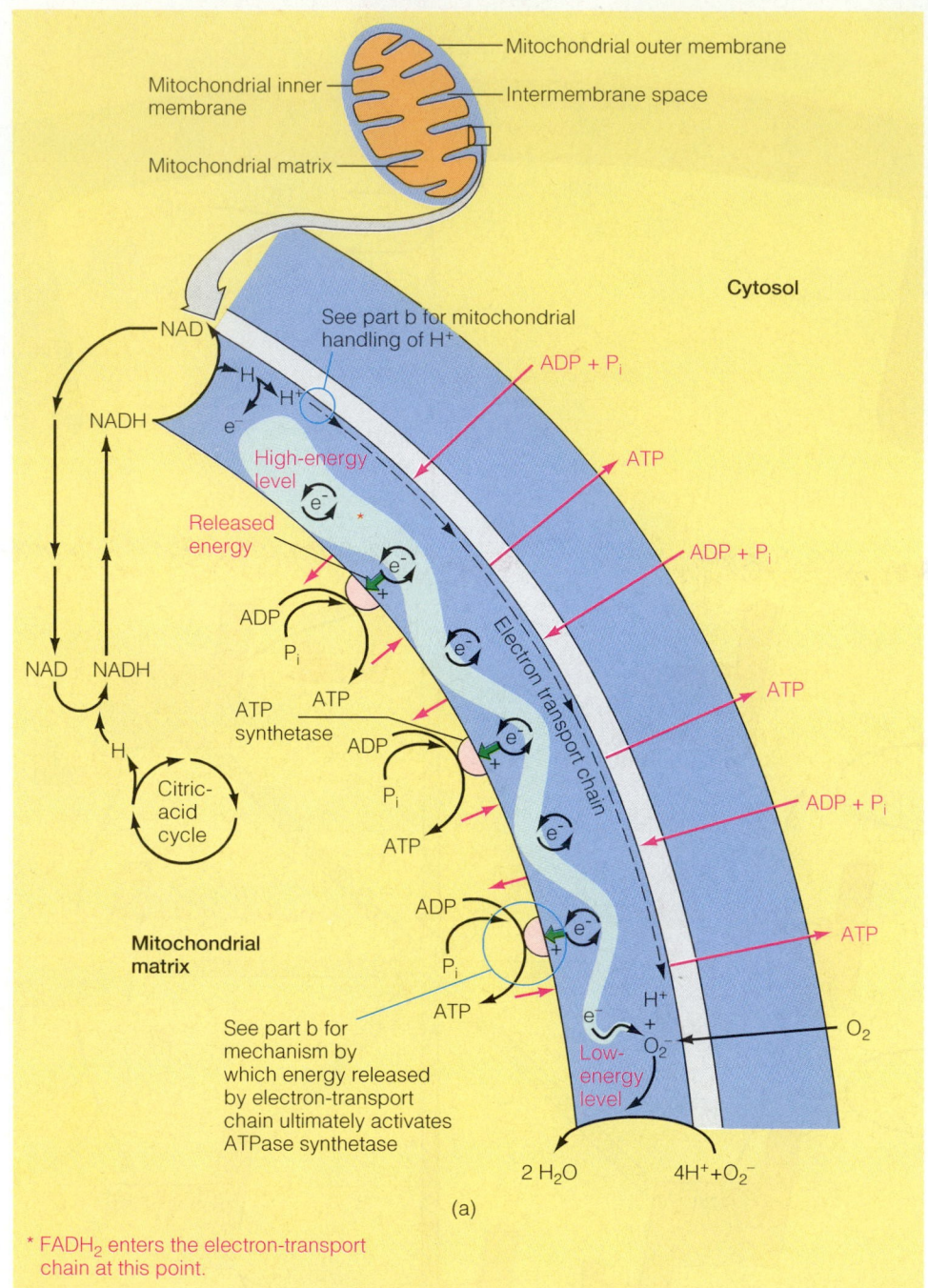

Mitochondrial outer membrane

Mitochondrial inner membrane

Intermembrane space

Mitochondrial matrix

Cytosol

NAD

See part b for mitochondrial handling of H^+

ADP + P_i

NADH

H
H^+
e^-

ATP

High-energy level

e^-

Released energy

ADP + P_i

NAD   NADH

e^-

ADP

P_i

e^-

Electron transport chain

ATP

H

ATP synthetase

ATP

e^-

ADP

ATP

ADP + P_i

Citric-acid cycle

P_i

e^-

ATP

Mitochondrial matrix

ADP

e^-

ATP

P_i

See part b for mechanism by which energy released by electron-transport chain ultimately activates ATPase synthetase

ATP

e^- H^+
      +
     O_2^-

Low-energy level

O_2

2 H_2O $4H^+ + O_2^-$

(a)

* $FADH_2$ enters the electron-transport chain at this point.

● **FIGURE 2–13   ATP Synthesis by Mitochondrial Inner Membrane**   (a) ATP synthesis resulting from the passage of high-energy electrons through the mitochondrial electron-transport chain. Hydrogen (H) that is released during the degradation of carbon-containing nutrient molecules by the citric-acid cycle in the mitochondrial matrix is carried to the mitochondrial inner membrane by hydrogen carriers such as NADH. After releasing hydrogen at the inner membrane, NAD shuttles back to pick up more hydrogen generated by the citric-acid cycle in the matrix. Meanwhile, high-energy electrons extracted from the hydrogen are passed through the electron-transport chain located on the mitochondrial inner membrane. Energy is gradually released as the electrons fall to successively lower energy levels by moving through the electron-transport chain of reactions. The released energy triggers a sequence of steps (shown in part b) that ultimately results in activation of the enzyme ATP synthetase within

reactions, with the electrons falling to successively lower energy levels with each step.

This electron-transport chain is also called the **respiratory chain** because it is critical to cellular respiration, which refers to the intracellular oxidation of nutrient derivatives. Ultimately, the electrons are passed to molecular oxygen (O_2)

derived from the air we breathe. Electrons bound to O_2 are in their lowest energy state. Oxygen breathed in from the atmosphere enters the mitochondria to serve as the final electron acceptor of the electron-transport chain. This negatively charged oxygen (negative because it has acquired additional electrons) then combines with the positively charged hydro-

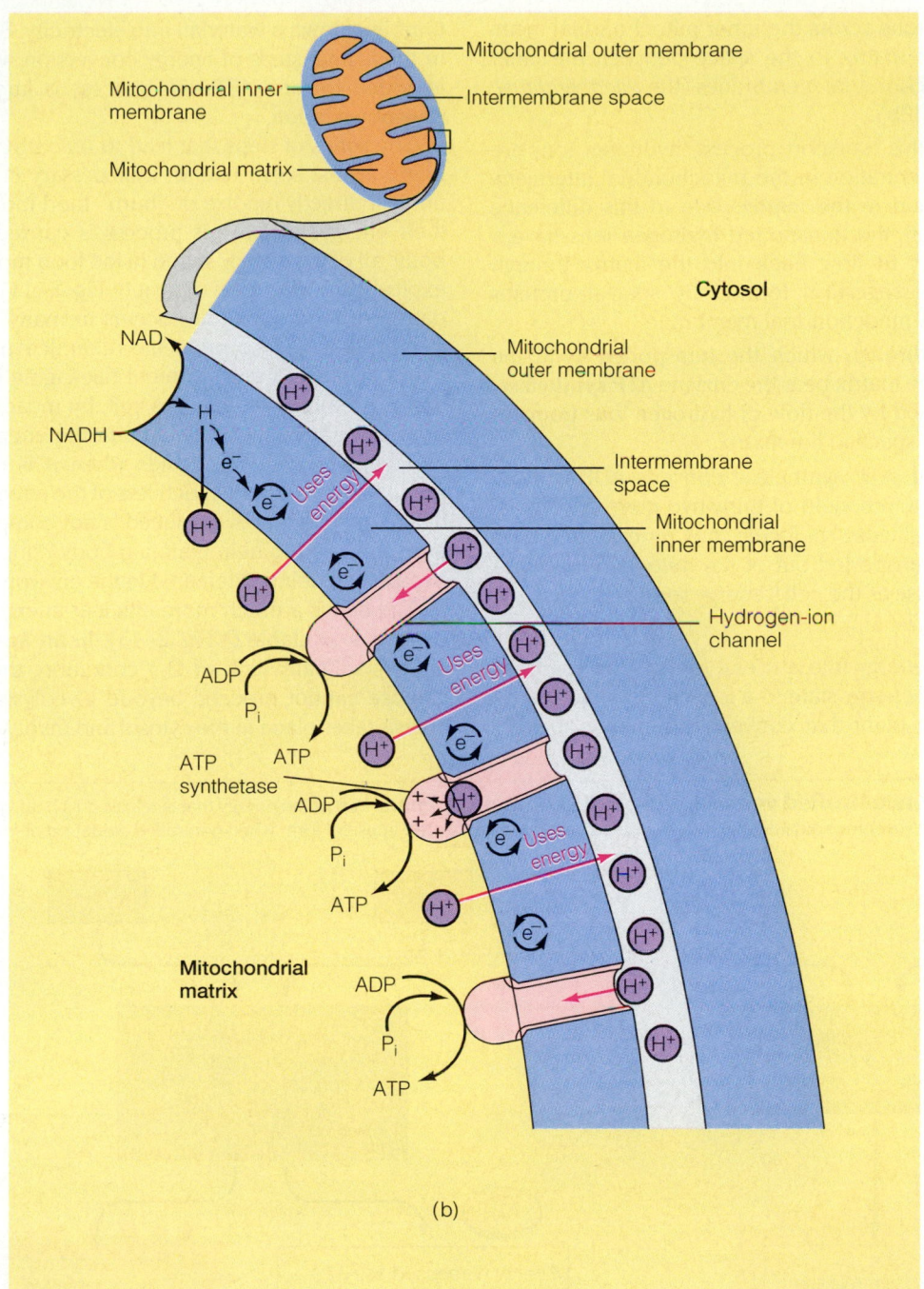

Mitochondrial outer membrane

Mitochondrial inner membrane

Intermembrane space

Mitochondrial matrix

Cytosol

NAD

NADH

Mitochondrial outer membrane

Intermembrane space

Mitochondrial inner membrane

Hydrogen-ion channel

Uses energy

ADP

P_i

ATP synthetase

ATP

ADP

P_i

ATP

Uses energy

Uses energy

Mitochondrial matrix

ADP

P_i

ATP

(b)

the mitochondrial inner membrane. ATP synthetase synthesizes ATP from ADP and P_i. Molecular oxygen, after serving as the final electron acceptor, combines with the hydrogen ions (H^+) generated from hydrogen upon extraction of high-energy electrons to produce water. (b) Activation of ATP synthetase by movement of H^+. Energy released during the transfer of electrons by the electron-transport chain is used to transport hydrogen ions from the matrix to the intermembrane space. The resultant buildup of hydrogen ions in the intermembrane space brings about the flow of hydrogen ions from the intermembrane space to the matrix through special channels in the mitochondrial inner membrane. The flow of hydrogen ions through the channel activates ATP synthetase, which is located at the matrix end of the channel. Activation of ATP synthetase brings about synthesis of ATP.

gen ions (positive because they have donated the electrons at the beginning of the electron-transport chain) to form water.

As the electrons move through this chain of reactions to ever-lower energy levels, they release energy. Part of the released energy is lost as heat, but some of it is harnessed by

the mitochondrion to synthesize ATP through the following steps, which are collectively known as the **chemiosmotic mechanism**:

1. At three sites in the electron-transport chain, the energy released during the transfer of electrons is used to trans-

port hydrogen ions across the inner mitochondrial membrane from the matrix to the space between the inner and outer mitochondrial membranes (the *intermembrane space*) (Fig. 2–13b).

2. As a result of this transport process, hydrogen ions are in greater concentration in the mitochondrial intermembrane space than in the matrix. Due to this difference in concentration, the transported hydrogen ions have a strong tendency to flow back into the matrix through channels or passageways formed by special proteins within the inner mitochondrial membrane.

3. The channels through which the transported hydrogen ions return to the matrix bear the enzyme **ATP synthetase**, which is activated by the flow of hydrogen ions from the intermembrane space to the matrix.

4. Upon activation, ATP synthetase converts ADP + P_i to ATP, providing a rich yield of thirty-two more ATP molecules for each glucose molecule thus processed. ATP is subsequently transported out of the mitochondrion into the cytosol for use as the cell's energy source.

The harnessing of energy into a useful form as the electrons tumble from a high-energy state to a low-energy state can be likened to a power plant that converts the energy of water tumbling down a waterfall into electricity. Because O_2 is used in these final steps of energy conversion when a phosphate is added to form ATP, this process is known as **oxidative phosphorylation**.

The series of steps that lead to oxidative phosphorylation might at first seem like an unnecessary complication. Why not just directly oxidize or "burn" food molecules to release their energy? When this process is carried out outside the body, all of the energy stored in the food molecule is released explosively in the form of heat (• Fig. 2–14). In the body, oxidation of food molecules occurs in many small, controlled steps so that the food molecule's chemical energy is gradually made available for convenient packaging in a storage form that is useful to the cell. The cell, by means of its mitochondria, can more efficiently capture the energy from the food molecules within ATP bonds when it is released in small quantities. In this way, much less of the energy is converted to heat. The heat that is produced is not completely wasted energy; it is used to help maintain body temperature, with any excess heat being eliminated to the environment.

The cell is a much more efficient energy converter when oxygen is available (• Fig. 2–15). In an **anaerobic** ("lack of air," specifically lack of O_2) condition, the degradation of glucose cannot proceed beyond glycolysis. Recall that glycolysis takes place in the cytosol and involves the breakdown

• **FIGURE 2–14   Uncontrolled versus Controlled Oxidation of Food** Part of the energy that is released as heat when food undergoes uncontrolled oxidation (burning) outside the body is instead harnessed and stored in useful form when controlled oxidation of food occurs inside the body.

• **FIGURE 2–15** **Comparison of Energy Yield and Products under Anaerobic and Aerobic Conditions** In anaerobic conditions, only two ATPs are produced for every glucose molecule processed, but in aerobic conditions, a total of thirty-six ATPs are produced per glucose molecule.

of glucose into pyruvic acid, producing a low yield of two molecules of ATP per molecule of glucose. The untapped energy of the glucose molecule remains locked in the bonds of the pryuvic-acid molecules, which are eventually converted to lactic acid if they do not enter the pathway that ultimately leads to oxidative phosphorylation. When sufficient O_2 is present—an **aerobic** ("with air" or "with O_2") condition— mitochondrial processing (that is, the citric-acid cycle in the matrix and the electron-transport chain on the cristae) harnesses sufficient energy to generate thirty-four more molecules of ATP, for a total net yield of thirty-six ATPs per molecule of glucose processed. (For a description of aerobic exercise, see the accompanying boxed feature, A Closer Look at Exercise Physiology.) The overall reaction for the oxidation of food molecules to yield energy is as follows:

food +	O_2	→	CO_2	+	H_2O	+	ATP
	(necessary for oxidative phos- phoryla- tion)		(produced primarily by the citric-acid cycle)		(produced by the electron- transport chain)		(produced primarily by the electron- transport chain)

Glucose, the principal nutrient derived from dietary carbohydrates, is the fuel preference of most cells. However, nutrient molecules derived from fats (fatty acids) and, if necessary, from protein (amino acids) can also participate in these chemical reactions at specific points to eventually produce energy. Amino acids are usually used for protein synthesis instead of energy production, but they can be used as fuel if insufficient glucose and fat are available (chapter 17).

Note that the oxidative reactions within the mitochondria generate energy, unlike the oxidative reactions controlled by the peroxisome enzymes. Both organelles use O_2 but for different purposes.

The energy stored within ATP is used for synthesis, transport, and mechanical work.

Once formed, ATP is transported out of the mitochondria and is then available as an energy source as needed within the cell. Cellular activities that require energy expenditure fall into three main categories:

1. *Synthesis of new chemical compounds,* such as protein synthesis by the endoplasmic reticulum. Some cells, especially cells with a high rate of secretion and cells in the growth phase, use up to 75% of the ATP they generate just to synthesize new chemical compounds.

2. *Membrane transport,* such as the selective transport of molecules across the kidney tubules during the process of urine formation. Kidney cells can expend as much as 80% of their ATP currency to operate their selective membrane-transport mechanisms.

3. *Mechanical work,* such as contraction of the heart muscle to pump blood or contraction of skeletal muscles to lift an object. These activities require tremendous quantities of ATP.

As a result of cellular energy expenditure to support these various activities, large quantities of ADP are produced. These energy-depleted ADP molecules enter the mitochondria for "recharging" and then cycle back into the cytosol as energy-rich ATP molecules after participating in oxidative phosphorylation. A single ADP/ATP molecule may shuttle back and forth between the mitochondria and cytosol for this recharging/expenditure cycle thousands of times per day.

The high demands for ATP render glycolysis alone an insufficient as well as inefficient supplier of power for most cells. If it were not for the mitochondria, which house the metabolic machinery for oxidative phosphorylation, our energy capability would be very limited. However, glycolysis

AEROBIC EXERCISE: WHAT FOR AND HOW MUCH?

Aerobic ("with O_2") **exercise** involves large muscle groups and is performed at a low-enough intensity and for a long-enough period of time that fuel sources can be converted to ATP by using the citric-acid cycle as the predominant metabolic pathway. Aerobic exercise can be sustained for from fifteen to twenty minutes to several hours at a time. Short-duration, high-intensity activities, such as weight training and the 100 meter dash, which last for a matter of seconds and rely solely on energy stored in the muscles and on glycolysis, are forms of **anaerobic** ("without O_2") **exercise**.

Inactivity is associated with increased risk of developing both hypertension (high blood pressure) and coronary artery disease (blockage of the arteries that supply the heart). To reduce the risk of hypertension and coronary artery disease and to improve physical work capacity, the American College of Sports Medicine recommends that an individual participate in aerobic exercise a minimum of three times per week for twenty to sixty minutes. The intensity of the exercise should be based on a percentage of the individual's maximal capacity to work. The easiest way to establish the proper intensity of exercise and to monitor intensity levels is by check-

ing heart rate. The estimated maximal heart rate is determined by subtracting the person's age from 220. Significant benefits can be derived from aerobic exercise performed between 70% and 80% of maximal heart rate. For example, the estimated maximal heart rate for a twenty year old is 200 beats per minute. If this person exercised three times per week for twenty to sixty minutes at an intensity that increased the heart rate to 140 to 160 beats per minute, the participant should significantly improve his or her aerobic work capacity and reduce the risk of cardiovascular disease.

---

does provide cells with a sustenance mechanism to produce at least some ATP under anaerobic conditions. Skeletal-muscle cells in particular take advantage of this ability during short bursts of strenuous exercise, when energy demands for contractile activity outstrip the body's ability to bring adequate O_2 to the exercising muscles to support oxidative phosphorylation. Also, red blood cells, which are the only cells that do not contain any mitochondria, rely solely on glycolysis for their limited energy production. However, the energy needs of red blood cells are low because they also lack a nucleus and therefore are not capable of synthesizing new substances, the biggest energy expenditure for most noncontractile cells.

Vaults are a newly discovered organelle.

In addition to the five well-documented organelles, researchers have recently identified a sixth type of organelle —**vaults**. Vaults, which are three times as large as ribosomes, are shaped like octagonal barrels. Their name comes from the fact that their multiple arches reminded their discoverers of vaulted or cathedral ceilings. A cell may contain thousands of vaults. Why would the presence of such relatively large, numerous organelles have been elusive until recently? The reason is that they do not show up with ordinary staining techniques. One clue as to the function of vaults may be their octagonal shape. Interestingly, the pores in the membrane surrounding the nucleus are also octagonal shaped and the same size as vaults, leading to speculation that vaults may be cellular "trucks." According to this proposal, vaults would

dock at nuclear pores, pick up molecules synthesized in the nucleus, and deliver their cargo elsewhere in the cell. One possibility is that vaults may be carrying messenger RNA from the nucleus to sites of protein synthesis within the cytoplasm (see p. B–7). Also, because vaults are especially abundant in regions of the cell where actin is being assembled, they may somehow be involved with cellular contractile systems.

CYTOSOL AND CYTOSKELETON

Occupying about 55% of the total cell volume, the cytosol is the semiliquid portion of the cytoplasm that surrounds the organelles. Its amorphous appearance under an electron microscope belies the fact that the cytosol is not a uniform liquid mixture but is actually more like a highly organized, gelatinous mass with differences in composition and consistency between various regions of the cell.

The cytosol is important in intermediary metabolism, ribosomal protein synthesis, and storage of fat and glycogen.

Three general categories of activities are associated with the cytosol: (1) enzymatic regulation of intermediary metabolism; (2) ribosomal protein synthesis; and (3) storage of fat, carbohydrate, and secretory vesicles. Dispersed throughout the cytosol is a cytoskeleton that gives shape to the cell, provides an intracellular organizational framework, and is responsible for various cell movements.

Enzymatic Regulation of Intermediary Metabolism

Intermediary metabolism refers collectively to the large set of intracellular chemical reactions that involve the degradation, synthesis, and transformation of small organic molecules such as simple sugars, amino acids, and fatty acids. These reactions are critical for ultimately capturing energy to be used for cellular activities and for providing the raw materials needed for maintenance of the cell's structure and function and for the cell's growth. All intermediary metabolism occurs in the cytoplasm, with most of it being accomplished in the cytosol. Thousands of enzymes involved in glycolysis and other intermediary biochemical reactions are found in the cytosol.

Ribosome Protein Synthesis

Also dispersed throughout the cytosol are the free ribosomes, which synthesize proteins for use in the cytosol itself. In contrast, rough-ER ribosomes synthesize proteins for secretion and for construction of new cellular components. Often, cytosolic ribosomes that are synthesizing identical proteins are clustered together in "assembly lines" known as **polyribosomes**.

Storage of Fat and Glycogen

Excess nutrients not immediately used for ATP production are converted in the cytosol into storage forms that are readily visible even under a light microscope. Such nonpermanent masses of stored material are known as **inclusions**. The largest and most important storage product is fat. Small fat droplets can be seen within the cytosol in various cells. In **adipose tissue**, the tissue specialized for fat storage, the stored fat molecules can occupy almost the entire cytosol, coalescing to form one large fat droplet (Fig. 2–16a). The other visible storage product is **glycogen**, the storage form of glucose, which appears as aggregates or clusters dispersed throughout the cell (Fig. 2–16b). Cells vary in their ability to store glycogen, with liver and muscle cells having the greatest stores. When food

is not available to provide fuel for the citric-acid cycle and electron-transport chain, stored glycogen and fat are broken down to release glucose and fatty acids, respectively, which can feed the mitochondrial energy-producing machinery. An average adult has enough glycogen stored to provide sufficient energy for about a day of normal activities, and typically enough fat is stored to provide energy for two months.

Secretory vesicles that have been processed and packaged by the endoplasmic reticulum/Golgi complex also remain in the cytosol until signaled to empty their contents to the outside.

Presence of Cytoskeleton

Permeating the cytosol is the **cytoskeleton**, a complex protein network that acts as the "bone and muscle" of the cell. The distinct shape, size, complexity, and intracellular specialization of the various body cells necessitate intracellular scaffolding to support and organize the cellular components into an appropriate arrangement and to control their movements. These functions are performed by the cytoskeleton. There are at least four distinct elements in this elaborate network: (1) *microtubules*, (2) *microfilaments*, (3) *intermediate filaments*, and (4) the *microtrabecular lattice* (Table 2–1). The different parts of the cytoskeleton are structurally linked and functionally coordinated to provide certain integrated functions for the cell. Because of the complexity of this network and the variety of functions it serves, its elements will be addressed separately.

Microtubules are essential for maintaining asymmetrical cell shapes and are important in complex cell movements.

The **microtubules** are the largest of the cytoskeletal elements. They are very slender [22 nanometer (nm) diameter; 1 nm = 1 billionth of a meter], long, hollow, unbranched tubes composed primarily of **tubulin**, a small, globular protein molecule

---

FIGURE 2–16   Inclusions (a) Light micrograph depicting fat storage in an adipose cell. Note that the fat droplet occupies almost the entire cytosol. (b) Light micrograph depicting glycogen storage in a liver cell. The red-staining granules throughout the liver cell's cytosol are glycogen deposits.

(a)   Fat droplet   Nucleus of adipose cell

Glycogen deposits   Liver cell

(b)

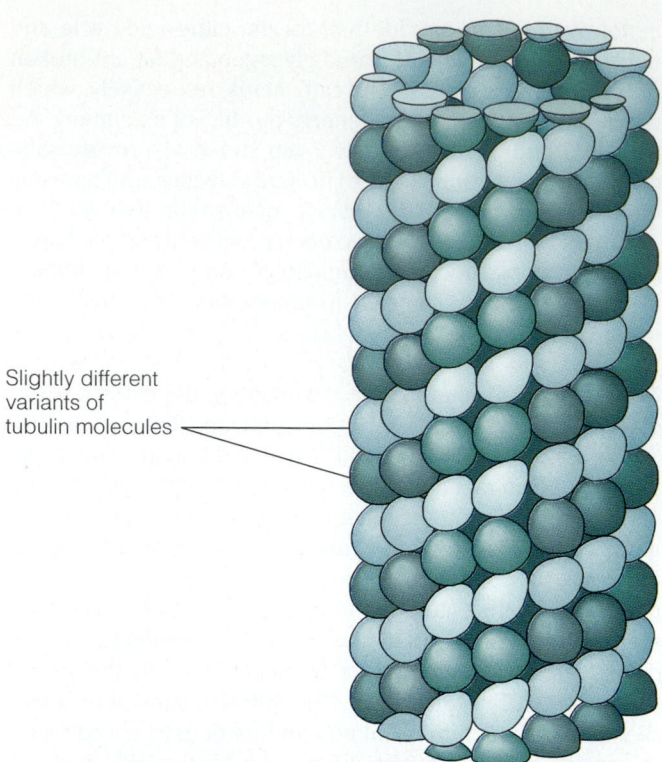

Slightly different variants of tubulin molecules

• **FIGURE 2–17   Arrangement of Tubulin Molecules in a Microtubule** Microtubules, the largest of the cytoskeletal elements, are long, hollow tubes formed by globular-shaped tubulin molecules.

(6 nm diameter) (• Fig. 2–17). Microtubules are essential for maintaining an asymmetrical cell shape, such as that of a nerve cell, whose elongated axon may extend up to a meter in length from the origin of the cell body in the spinal cord to the termination of the axon at a muscle (• Fig. 2–18). Microtubules, along with specialized intermediate filaments, stabilize this asymmetrical axonal extension. In cells whose mature shape and organization do not depend on the presence of microtubules, these cytoskeletal structures are present at

an earlier developmental stage of the cell. Here they function as temporary scaffolding, creating a shape and organization that are later maintained by other cellular elements. For example, spiral rings of microtubules wrap around the base of the tail of a developing sperm and subsequently disappear. Before their disappearance, the microtubules serve as a transient "mold" for arranging the mitochondria spirally around the tail, where these energy organelles will be readily available to meet the cell's energy needs for motility.

Microtubules also play an important role in coordinating numerous complex cell movements, including (1) transport of secretory vesicles from one region of the cell to another; (2) movement of specialized cell projections such as cilia and flagella; and (3) distribution of chromosomes during cell division through formation of a mitotic spindle.

Transport of Secretory Vesicles   Axonal transport provides a good example of the importance of an organized system for moving secretory vesicles. In a nerve cell, specific chemicals are released from the terminal end of the elongated axon to influence a muscle or another structure that the nerve cell controls. These chemicals are largely produced within the cell body, where the nuclear DNA blueprint, endoplasmic reticular factory, and Golgi packaging and distribution outlet are located. Yet these chemicals ultimately function at the end of the axon, which may be a meter away. If these chemicals had to diffuse on their own from the cell body to a distant axon terminal, it would take them about fifty years to get there—obviously an impractical solution. The microtubules provide a "highway" for vesicular traffic along the axon, with the driving force depending on ATP (Fig. 2–18). Specific protein molecules attach to the particle to be transported and "walk" along the microtubule, powered by ATP. **Kinesin** is one such transport molecule. It consists of a long tail and two globular heads, much like a double-ended golf club. Kinesin's tail binds to the particle to be moved while its globular heads act like little feet, alternately attaching to the microtubule, bending and pushing forward, letting

• **FIGURE 2–18   Two-Way Vesicular Axonal Transport Facilitated by the Microtubular "Highway" in a Nerve Cell**

Endoplasmic reticulum

Golgi complex

Microtubular "highway"

Secretory vesicle

Axon

Debris

Axon terminal

Lysosome

Cell body

go, then recocking and reattaching to a site farther down the microtubule. The process is repeated over and over as kinesin and its cargo move to the end of the axon. Reverse vesicular traffic also occurs along these microtubular highways. Vesicles that contain debris are transported by different protein motors from the axon terminal to the cell body for degradation by lysosomes, which are confined within the cell body. Coincidentally, these microtubule highways may also serve as pathways for the movement of such infectious agents as herpes virus, poliomyelitis virus, and rabies virus, which travel retrograde (backwards) along nerves from their surface site of contamination to the central nervous system (brain and spinal cord).

Movement of Cilia and Flagella Microtubules are also the dominant structural and functional components of cilia and flagella. These specialized protrusions from the cell surface allow a cell to move materials across its surface (in the case of a stationary cell) or to propel itself through its environment (in the case of a motile cell). **Cilia** are numerous tiny, hairlike protrusions, whereas a **flagellum** is a single, long, whiplike appendage. Even though they project from the surface of the cell, cilia and flagella are both intracellular structures covered by the plasma membrane.

Cilia beat or stroke in unison, much like the coordinated efforts of a rowing team. Each cilium exerts a rapid active stroke, which moves material on the cell surface forward. This stroke is followed by a recovery phase in which the cilium more slowly returns to its original position with a kind of unrolling, backward movement that does not exert much force. In this way, the material just pushed forward is not futilely pushed backward but instead is moved only in the direction of the forward stroke.

In humans, ciliated cells are found in the stationary cells that line the respiratory tract and the oviduct of the female reproductive tract. Respiratory cilia help keep foreign particles out of the lungs (• Fig. 2–19). The thousands of cilia lining the respiratory airways project into a layer of sticky mucus that traps dust and other inspired particles. The coordinated stroking action of these cilia sweeps this dust-laden mucus up to the throat, where it can be expectorated (spit out) or swallowed and eventually eliminated in the feces. In the female reproductive tract, the sweeping action of the cilia that line the oviduct draws the egg (ovum) released from the ovary during ovulation into the oviduct and then guides it toward the uterus (womb).

The only human cells that bear flagella are sperm. The whiplike motion of the flagellum or "tail" enables a sperm to move through its environment. This ability is particularly useful when the sperm maneuvers for final penetration of the ovum during fertilization.

Cilia and flagella have the same basic internal structure. Both consist of nine fused pairs of microtubules (doublets) arranged in an outer ring around two single unfused microtubules in the center (• Fig. 2–20). This characteristic "nine plus two" array of microtubules extends throughout the length of the motile appendage. A cilium or flagellum originates from a specialized cytoplasmic structure, the **basal**

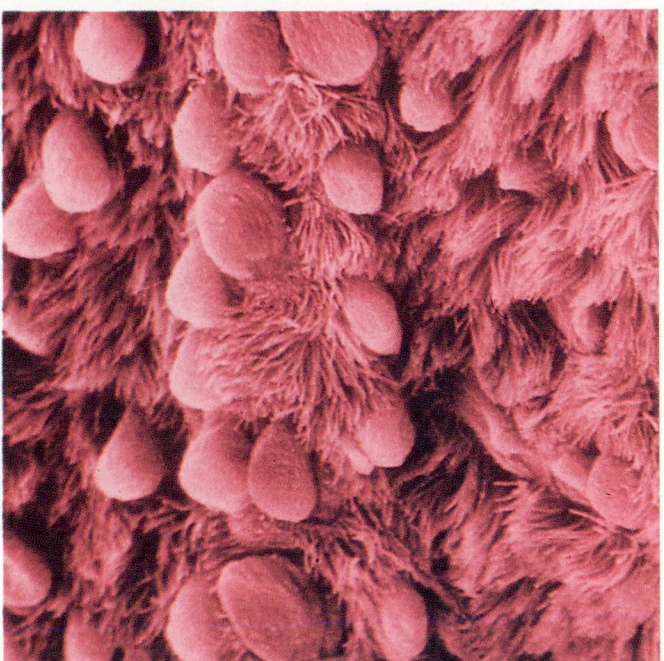

• **FIGURE 2–19   Scanning Electron Micrograph of Cilia on Cells Lining the Respiratory Tract in Humans** The respiratory airways are lined by goblet cells, which secrete a sticky mucus that traps inspired particles, and epithelial cells that bear numerous hairlike cilia. The cilia beat in unison to sweep inspired particles up and out of the airways.

body, which is located inside the main part of the cell and provides a base from which the microtubules grow to form the appendage. Each basal body is a short cylinder composed of a parallel microtubular symmetry similar to that of the cilium or flagellum.

Associated with these microtubules are accessory proteins that maintain the microtubules' organization and play an essential part in the microtubular movement that causes the entire structure to bend. The most important of these accessory proteins is **dynein**, which forms a set of armlike projections from each doublet of microtubules (Fig. 2–20a). The bending movements of cilia and flagella are produced by the sliding of adjacent microtubule doublets past each other. Sliding is accomplished by the dynein arms, which have the capability of splitting ATP and then using the released energy to "crawl" along the neighboring microtubule doublet (similar to kinesin's movement) to cause relative displacement of the doublets (• Fig. 2–21). Groups of cilia working together are oriented to beat in the same direction and contract in a synchronized manner through controlling mechanisms that are poorly understood. These mechanisms apppear to involve the single microtubules at the cilium's center and their surrounding accessory proteins, the **inner sheath** (Fig. 2–20a).

Defects in flagellate and ciliary function have been observed in humans. Some hereditary forms of male sterility involving nonmotile sperm have been traced to defects in the dynein arms or inner sheath. These same individuals usually also have long histories of recurrent respiratory-tract disease because the same types of defects are present in their respi-

Inner
sheath

Microtubule
doublet

Central singlet
microtubule

Dynein arms

Plasma
membrane

100 nm

(a)

(b)

• **FIGURE 2–20   Internal Structure of Cilia and Flagella** (a) Schematic diagram of a cilium in cross section showing the characteristic "nine plus two" arrangement of microtubules along with the dynein arms and other accessory proteins. (b) Electron micrograph of numerous cilia in cross section. Note the "nine plus two" arrangement of microtubules.

ratory cilia, which are unable to clear mucus and inhaled particles from the respiratory system.

Formation of the Mitotic Spindle   Cell division involves two discrete but related activities: *mitosis* (nuclear division) and *cytokinesis* (cytoplasmic division). **Mitosis** involves rep-

• **FIGURE 2–21   Bending of Cilium Accomplished by Relative Sliding of Anchored Microtubule Doublets** (a) Dynein arms extend between the microtubule doublets within a cilium. Relative sliding of adjacent microtubule doublets is believed to be accomplished by ATP-powered dynein "crawling." (b) If the microtubule doublets were not anchored at their base, they would freely slide past each other as a result of this dynein activity. (c) Because the doublets are anchored at their base, bending occurs when the doublets slide with respect to each other as a result of dynein activity.

(a)   (b)   (c)

lication of the DNA-containing chromosomes, which are then evenly distributed in the two halves of the cell. In **cytokinesis**, the plasma membrane is constricted in the middle of the cell, and the two halves separate into two new daughter cells, each with a full complement of chromosomes.

Microtubules are transiently assembled to form a **mitotic spindle**, which organizes and directs the movement of the replicated chromosomes away from each other toward opposite ends of the cell so that the genetic material is evenly distributed when the cell divides. The mitotic spindle is formed by the **centrioles**, a pair of short cylindrical structures that lie at right angles to each other near the nucleus (Fig. 2–1). The centrioles also duplicate during cell division. After self-replication, the centriole pairs move toward opposite ends of the cell and form the spindle apparatus between them through a precisely organized assemblage of microtubules.

Besides their role in mitotic-spindle formation, the centrioles and surrounding complex of densely staining proteinaceous material together assemble the many microtubules that normally radiate throughout the cytoskeleton. The centrioles are identical in structure to basal bodies. In fact, under some circumstances the centrioles and basal bodies are interconvertible. During development of ciliated human cells, the centriole pair migrates to the region of the cell where the cilia will be formed and duplicates itself to produce the many basal bodies that will form the cilia.

Microfilaments are important in cellular contractile systems and as mechanical stiffeners.

The **microfilaments** are the smallest (6 nm diameter) elements of the cytoskeleton visible with a conventional elec-

Actin molecule

● **FIGURE 2–22   Arrangement of Actin Molecules in a Microfilament**

tron microscope. The most obvious microfilaments in most cells are those composed of **actin,** a protein molecule that has a globular shape similar to tubulin. Unlike tubulin, which forms a hollow tube, actin is assembled into two twisted strands, much like two strings of pearls twisted into a helix (spiral) to form a microfilament (● Fig. 2–22). In muscle cells, another protein called **myosin** forms a different kind of microfilament. In most cells, myosin is not as abundant and does not form such distinct filaments.

Microfilaments serve at least two functions: (1) they play a vital role in various cellular contractile systems, and (2) they act as mechanical stiffeners for several specific cellular projections.

Microfilaments in Cellular Contractile Systems   Actin-based assemblies are involved in muscle contraction, cell division, and cell locomotion. The most obvious, best organized, and most clearly understood cellular contractile system is that found in muscle. Muscle contains an abundance of actin and myosin filaments, which are organized so that when they are triggered by electrically induced ATP splitting to slide past each other, they generate a contractile force (chapter 9).

Surprisingly, nonmuscle cells may also contain "muscle-like" assemblies. Some of these microfilament contractile systems are transiently assembled to perform a specific function when needed. A good example is the contractile ring that forms during cytokinesis to split apart the duplicate cell halves. The ring consists of a beltlike bundle of actin filaments located just beneath the plasma membrane in the middle of the cell. When this ring of fibers contracts, it pinches the cell in two (● Fig. 2–23).

Complex actin-based assemblies are also responsible for most cell locomotion. Four types of human cells are capable of moving on their own—sperm, white blood cells, fibroblasts, and skin cells. Sperm move by the flagellar mechanism already described. Motility for the other cells is accomplished by **amoeboid movement**, a process that depends on actin filaments and is similar to the mechanism used by amoebae to

maneuver through their environment. White blood cells leave the circulatory system and travel by amoeboid movement to areas of infection or inflammation, where they engulf and destroy microorganisms and cellular debris. Fibroblasts ("fiber formers") move into a wound from adjacent connective tissue to help repair the damage and are responsible for scar formation. Skin cells, which are ordinarily stationary, can become modestly mobile and move by amoeboid motion toward a cut area to restore the skin surface.

Two regions of cytosol can be distinguished in a cell undergoing amoeboid movement: an outer layer of cytosol that is quite gelatinous, the **plasma gel** or **ectoplasm** ("outer cytoplasm"), and an inner core of a more fluid cytosol, the **plasma sol** or **endoplasm** ("inner cytoplasm"). The more gelatinous nature of the ectoplasm appears to be caused by a greater abundance of actin filaments just beneath the plasma membrane compared to the remainder of the cell.

Although the explanation is still controversial, amoeboid movement apparently involves local transitions between the gel and sol states within the cell as a result of alternate assembly and disassembly of actin filaments (● Fig. 2–24). The trigger for initiating amoeboid movement might be the proximity of a food particle (in the case of an amoeba) or the presence of a bacterium (in the case of a white blood cell). Movement is accomplished by *solation* at the end of the cell closest to the prey, causing thinning of the ectoplasm at the "front," accompanied by simultaneous *gelation* of endoplasm at the opposite end, causing thickening of the ectoplasm in the "rear." The thickened ectoplasm in the rear (thicker presumably because of the assembly of more actin filaments during gelation) contracts, squeezing out the more fluid endoplasm, which streams toward the leading end of the cell, much as toothpaste moves to the end when the tube is squeezed. This streaming endoplasm causes the cell to bulge out into a fingerlike extension, a **pseudopod** ("false foot"), at the thinned front end of the cell (thinned because of the disassembly of actin filaments during solation). The streaming of endoplasm from rear to front into the pseudopod moves the cell forward. After reaching the tip of the pseudopod, the streaming endoplasm turns toward the sides of the cell like a fountain spray and gels to stop the flow and restore ectoplasm in this region. The cycle then repeats itself so that the cell progressively moves forward.

Microfilaments as Mechanical Stiffeners   Besides their role in cellular contractile systems, the actin filaments' second major function is to serve as mechanical supports or stiffeners for several cellular extensions, of which the most common are **microvilli**. Microvilli are microscopic, nonmotile, hairlike projections from the surface of epithelial cells lining

● **FIGURE 2–23   Dividing Cell**

Nucleus

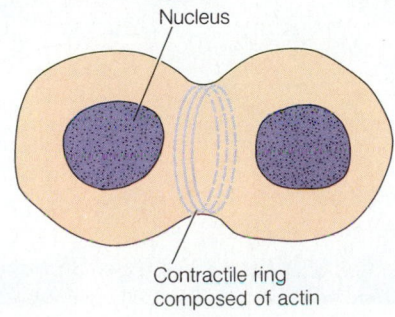

Contractile ring
composed of actin

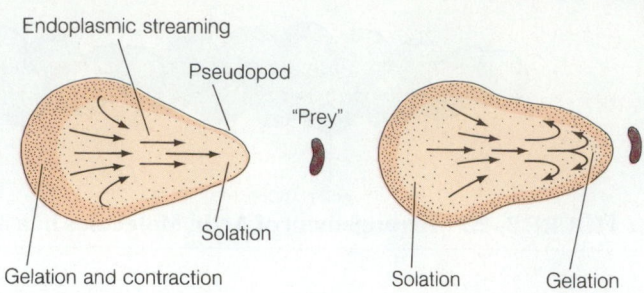

● **FIGURE 2–24 Amoeboid Movement** During amoeboid movement, a motile cell undergoes simultaneous (1) solation, or thinning, of the ectoplasm at the end of the cell closest to a "prey" (a particle to be engulfed) as a result of disassembly of actin filaments and (2) gelation, or thickening, of the endoplasm at the opposite end as a result of assembly of more actin filaments. Contraction of the actin filaments in the "rear" squeezes the more fluid endoplasm forward, causing the front of the cell to bulge out into a fingerlike pseudopod. The cell crawls forward by repeated cycles of pseudopod formation, driven by cyclical assembly and disassembly of actin filaments.

Sol-like endoplasm

Gel-like ectoplasm

Endoplasmic streaming

Pseudopod

"Prey"

Solation

Gelation and contraction

Solation

Gelation

the small intestine and kidney tubules. A single small-intestinal cell may have several thousand microvilli, which are packed together like the bristles of a brush, projecting from the cell's free surface (● Fig. 2–25). This bristly appearance has given these microvilli the alternative name of **brush border**. Their presence greatly increases the surface area available for transferring material across the plasma membrane. In the case of the small intestine, the microvilli increase the area available for absorbing digested nutrients. In the kidney tubules, brush borders enlarge the absorptive surface that salvages useful substances passing through the kidney so that these materials are saved for the body instead of being eliminated in the urine. Within each microvillus, a core consisting of parallel actin filaments linked together forms a rigid mechanical stiffener that keeps these valuable surface projections intact.

A remarkable specialization of microvilli is found in the hair cells of the inner ear. In the portion of the inner ear responsible for hearing, the actin-stiffened projections on the surface of the hair cells are exquisitely sensitive to vibrations produced by incoming sound. In the part of the inner ear that

plays a key role in equilibrium and balance, the specialized microvilli are responsive to changes in head movement and position.

Both microtubules and microfilaments form *stable* structures, such as cilia, flagella, muscle contractile units, and microvilli, and also form *labile* structures, such as mitotic spindles and contractile rings, as the need arises. Pools of unassembled tubulin and actin subunits in the cytosol can be rapidly assembled into organized structures to perform specific activities and then can be disassembled when they are no longer needed.

Intermediate filaments are important in regions of the cell subject to mechanical stress.

In contrast to the other cytoskeletal elements, the **intermediate filaments** are highly stable structures. There is no evidence for a reversible pool between unassembled and assembled intermediate-filament proteins. The intermediate filaments are intermediate in size between the microtubules

● **FIGURE 2–25 Scanning Electron Micrograph of Intestinal Microvilli** [From *Tissues and Organs: A Text-Atlas of Scanning Electron Microscopy* by Richard G. Kessel and Randy H. Kardon. Copyright © 1979 by W. H. Freeman and Company. Reprinted with permission.]

Microvilli

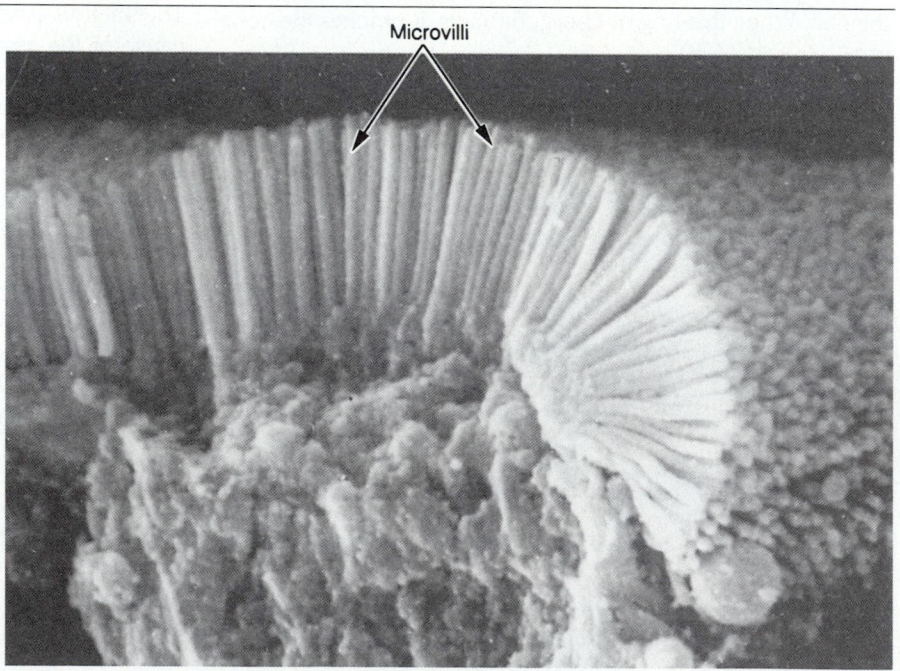

and the microfilaments (7–11 nm diameter)—hence their name. The proteins that compose the intermediate filaments vary between cell types, but in general they appear as irregular, threadlike molecules. These proteins form tough, durable fibers that are structurally important in the parts of the cell subject to mechanical stress. There are different types of intermediate filaments, depending on their structural or tension-bearing role in specific cell types. In general, only one class of intermediate filament is found in a particular cell type. Several important examples follow:

- Neurofilaments are intermediate filaments found in nerve-cell axons. Together with microtubules, neurofilaments strengthen and stabilize these elongated cellular extensions.
- Intermediate filaments in skeletal-muscle cells hold the actin-myosin contractile units in proper alignment.
- Skin cells contain irregular networks of keratin-containing intermediate filaments. These intracellular filaments interconnect with extracellular filaments that tie adjacaent cells together, thereby creating a continuous filamentous network that extends throughout the skin and gives it strength. When the surface skin cells die, their tough keratin skele-tons persist to form a protective, waterproof outer layer. Hair and nails are also keratin structures.

The microtrabecular lattice provides a dynamic structural framework.

The **microtrabecular lattice** is the most recently discovered element of the cytoskeleton, having first been made visible by high-voltage electron microscopy in the early 1970s. Using this technique, which provides a three-dimensional view of the internal organization of the cell, the microtrabecular lattice is visible as a meshwork of exceedingly fine (< 2 nm diameter), interlinked filaments that pervade the cytoplasm and are connected to the inner layer of the plasma membrane. Some cell biologists believe that the microtrabecular lattice is not a separate entity but consitutes intricate interconnections between the three other types of cytoskeletal structures. This latticework appears to suspend the microtubules and microfilaments, as well as various organelles. The free ribosomes are not freely floating in the cytosol as originally thought; they are entrapped as polyribosomal clusters at junctions of the microtrabecular lattice and thus have the appearance of flies caught in a spider web (● Fig. 2–26).

● **FIGURE 2–26   Microtrabecular Lattice in Relation to Other Cytoskeletal Structures and Organelles**

Endoplasmic reticulum

Ribosome on rough endoplasmic reticulum

Free ribosomes

Microtrabecular lattice

Plasma membrane

Microtubule

Microfilament

Mitochondrion

By attaching to the inner surface of the plasma membrane and forming an internal scaffolding, the microtrabecular lattice, in association with the other cytoskeletal elements, supports the plasma membrane and is responsible for the particular shape, rigidity, and spatial geometry of each different cell type. Thus, this internal framework acts as the cell's "skeleton." In addition, by lacing throughout the entire cell, the microtrabecular lattice appears to link the different components of the cytoplasm into a functional unit. For example, the coordinated action of the cytoskeletal elements is responsible for directing intracellular transport and for regulating numerous cellular movements and thereby also serves as the cell's "musculature."

The lattice apparently also plays a role in organizing the cytosolic enzymes. Reactions that take place in the cytosol, such as glycolysis, are too well choreographed and too rapid to occur by random contacts between enzymes and their substrates (the substances acted on by the enzymes). There is convincing evidence that these enzymes are somehow incorporated into the lattice, probably in some sort of sequential alignment that guides glucose through the steps of the glycolytic pathway. Furthermore, the lattice, by sequestering the free ribosomes at its intersections, plays an important role in governing the dispensation of newly synthesized cytosolic proteins. Once proteins have been synthesized, they are available in a controlled, nonrandom fashion for enzymatic activities and for the assembly of microtubules, microfilaments, and other cytosolic components.

The microtrabecular lattice is not a rigid, static structure. Its structure has been observed to vary reversibly when the cell changes shape or when intracellular movements are occurring. The lattice changes constantly through local contractions, expansions, and deformations so that the organelles and cytoskeletal fibers are continually redistributed and reoriented as the cell carries on its various activities. Recent observations suggest that signals received via an organized cytoskeleton normally play a role in regulating cell growth and division, a factor that might be disrupted in cancer cells. Conspicuous cytoskeletal changes accompanied by abnormal growth behavior are often present in cancer cells. Very little is known about the interactions between the components of the cytoskeleton in carrying out these and other integrated activities.

CHAPTER IN PERSPECTIVE: FOCUS ON HOMEOSTASIS

The ability of cells to perform functions essential for their own survival as well as specialized tasks that contribute to the maintenance of homeostasis ultimately depends on the successful, cooperative operation of the intracellular components. For example, to support life-sustaining activities, all cells must generate energy in a usable form from nutrient molecules. Energy is generated intracellularly by chemical reactions that take place within the cytosol and mitochondria.

In addition to being essential for basic cell survival, the organelles and cytoskeleton also participate in many cells' specialized tasks that contribute to homeostasis. Following are several examples:

- Nerve and endocrine cells both release proteinaceous chemical messengers that are important in regulatory activities aimed toward maintaining homeostasis—for example, chemical messengers released from nerve cells stimulate the respiratory muscles, which accomplish life-sustaining exchanges of O_2 and CO_2 between the body and atmosphere through breathing. These proteinaceous chemical messengers (neurotransmitters in nerve cells and hormones in endocrine cells) are all produced by the endoplasmic reticulum/Golgi complex and released by exocytosis from the cell when needed.

- Muscle cells' ability to contract depends on their highly developed cytoskeletal microfilaments sliding past each other. Muscle contraction is responsible for many homeostatic activities, including (1) contraction of the heart muscle, which pumps life-supporting blood throughout the body; (2) contraction of the muscles attached to bones, which enables the body to procure food; and (3) contraction of the muscle in the walls of the stomach and intestine, which moves the food along the digestive tract so that ingested nutrients can be progressively broken down into a form that can be absorbed into the blood for delivery to the cells.

- White blood cells help the body resist infection by making extensive use of lysosomal destruction of engulfed particles as they police the body for microbial invaders.

As we begin to examine the various organs and systems, keep in mind that proper cellular functioning is the foundation of all organ activities.

 CHAPTER SUMMARY

Introduction

The complex organization and interaction of the various chemicals within a cell confer the unique characteristics of life. Cells, in turn, are the living building blocks of the body.

Body cells, which are too small to be seen by the unaided eye, have been shown by microscopic techniques to consist of three major subdivisions: (1) the plasma membrane, which encloses the cell and separates the intracellular and extracellular fluid; (2) the nucleus, which contains deoxyribonucleic acid (DNA), the cell's genetic material; and (3) the cytoplasm, the portion of the cell's interior not occupied by the nucleus. The cytoplasm consists of cytosol, a complex gelatinlike mass, and organelles, which are highly organized, membrane-bound structures dispersed within the cytosol. Compartmentalization of specific sets of chemicals within the organelles permits chemical activities that would not be compatible with each other to occur simultaneously within separate organelle compartments.

Organelles

Five established types of organelles are found in most cells: endoplasmic reticulum, Golgi complex, lysosomes, peroxisomes, and mitochondria. A sixth organelle type, vaults, has recently been discovered. The endoplasmic reticulum (ER) is a single, complex membranous network that encloses a fluid-filled lumen. The primary function of the ER is to serve as a factory for synthesizing proteins and lipids to be used for (1) the production of new cellular components, particularly cell membranes; and (2) secretion of special products such as enzymes and hormones to the exterior of the cell. There are two types of endoplasmic reticulum: rough endoplasmic reticulum, which is studded with ribosomes, and smooth endoplasmic reticulum, which lacks ribosomes. The rough endoplasmic reticular ribosomes synthesize proteins, which are released into the ER lumen so that they are separated from the cytosol. Also entering the lumen are lipids produced within the membranous walls of the ER. Synthesized products move from the rough ER to the smooth ER, where they are packaged and discharged as transport vesicles. Transport vesicles are formed as a portion of the smooth ER "buds off," containing a collection of newly synthesized proteins and lipids wrapped in smooth-ER membrane.

The Golgi complex, which consists of stacks of flattened, membrane-bound sacs, serves a twofold function: (1) it acts as a refining plant for modifying into a finished product the newly synthesized molecules delivered to it in crude form from the endoplasmic reticular factory; and (2) it sorts, packages, and directs molecular traffic to appropriate intracellular and extracellular destinations.

Each cell contains several hundred lysosomes, which are membrane-bound sacs that contain powerful hydrolytic (digestive) enzymes. Serving as the intracellular digestive system, lysosomes destroy phagocytized foreign material such as bacteria,

demolish worn-out cell parts to make way for new replacement parts, and eliminate the entire cell if it is severely damaged or dead.

Peroxisomes, small membrane-bound sacs containing powerful oxidative enzymes, are specialized for carrying out particular oxidative reactions, including certain detoxification activities.

The rod-shaped mitochondria are the energy organelles of the cell. They house the enzymes of the citric-acid cycle and electron-transport chain, which efficiently convert the energy in food molecules to the usable energy stored in ATP molecules. During this process, which is known as oxidative phosphorylation, the mitochondria utilize molecular oxygen and produce carbon dioxide and water as by-products. The body's cells use ATP as an energy source for synthesis of new chemical compounds, for membrane transport, and for mechanical work.

Vaults, which are shaped like octagonal barrels, are the same shape and size as the nuclear pores. It is speculated that vaults may transport messenger RNA from the nucleus to the cytoplasmic sites of protein synthesis. They are especially abundant at sites of actin assembly.

Cytosol and Cytoskeleton

The cytosol contains the enzymes involved in intermediary metabolism and the ribosomal machinery essential for synthesis of these enzymes as well as other cytosolic proteins. Furthermore, many cells store unused nutrients within the cytosol in the form of glycogen granules or fat droplets. Also present in the cytosol are secretory vesicles containing products that are to be discharged from the cell on appropriate stimulation. Pervading the cytosol is the cytoskeleton, which serves as the "bone and muscle" of the cell. The four types of cytoskeletal elements—microtubules, microfilaments, intermediate filaments, and microtrabecular lattice—are each composed of different proteins and perform different roles. Collectively, the cytoskeletal elements give the cell shape and support, enable it to organize and move its internal structures as needed, and, in some cells, allow movement between the cell and its environment.

REVIEW EXERCISES

Objective Questions (Answers on p. D–1.)

1. The barrier that separates and controls movement between the cellular contents and extracellular fluid is the _____.

2. The chemical that directs protein synthesis and serves as a genetic blueprint is _____, which is found in the _____ of the cell.

3. The cytoplasm consists of _____, which are specialized, membrane-bound intracellular compartments, and a gel-like mass known as _____, which contains an elaborate protein network called the _____.

4. Transport vesicles from the _____ fuse with and enter the _____ for modification and sorting.

5. The (what kind of) _____ enzymes within the peroxisomes primarily detoxify various wastes produced within the cell, generating _____ in the process.

6. The universal energy carrier of the body is _____.

7. The largest cells in the human body can be seen by the unaided eye. (True or false?)

8. Inadvertent rupture of lysosomal membranes inevitably leads to complete self-destruction of the cell. (True or false?)

9. Amoeboid movement is accomplished by transitions of the cytosol between a gel and a sol state as a result of alternate assembly and disassembly, respectively, of actin filaments. (True or false?)

10. Using the answer code at the right, indicate which type of ribosome is being described:

  ____ 1. synthesizes proteins used to construct new cell membrane
  ____ 2. synthesizes proteins used intracellularly within the cytosol
  ____ 3. synthesizes secretory proteins such as enzymes or hormones

  a. free ribosome
  b. rough-ER–bound ribosome

11. Using the answer code at the right, indicate which form of energy production is being described:

  ____ 1. takes place in the mitochondrial matrix
  ____ 2. produces H_2O as a by-product
  ____ 3. rich yield of ATP
  ____ 4. takes place in the cytosol
  ____ 5. processes acetyl CoA
  ____ 6. located in the mitochondrial inner-membrane cristae
  ____ 7. converts glucose into two pyruvic-acid molecules
  ____ 8. utilizes molecular oxygen

  a. glycolysis
  b. citric-acid cycle
  c. electron-transport chain

Essay Questions

1. What are a cell's three major subdivisions?
2. Distinguish between intracellular and extracellular fluid.
3. State an advantage of organelle compartmentalization.
4. List the types of organelles.
5. Describe the structure of the endoplasmic reticulum, distinguishing between the rough and smooth ER. What is the function of each?
6. Compare exocytosis and endocytosis. Define secretion and phagocytosis.
7. Which organelles serve as the intracellular digestive system? What type of enzymes do they contain? What functions do they serve?
8. Compare lysosomes with peroxisomes.
9. Describe the structure of mitochondria and explain their role in oxidative phosphorylation.
10. Distinguish between the oxidative enzymes found in peroxisomes and those found in mitochondria.
11. What three categories of cellular activities require energy expenditure?
12. List and describe the functions of each of the components of the cytoskeleton.

POINTS TO PONDER

1. Let's consider how much ATP you synthesize in a day. Assume that you consume one mole of O_2 per hour or 24 moles/day (a mole is the number of grams of a chemical equal to its molecular weight). About 6 moles of ATP are produced per mole of O_2 consumed. The molecular weight of ATP is 507.

   $$24 \text{ moles } O_2/\text{day} \times 6 \text{ moles ATP/mole } O_2$$
   $$= 144 \text{ moles ATP/day}$$

   $$144 \text{ moles ATP/day} \times 507 \text{ gm ATP/mole}$$
   $$= 73,000 \text{ gm ATP/day}$$

   Given that 1,000 gm equal 2.2 pounds, how many pounds of ATP do you produce per day at this rate? (This is under relatively inactive conditions!)

2. After a mother stops breast-feeding her infant, the highly developed milk-secreting glands and supportive structures in the breasts gradually diminish. What organelle do you think is responsible for diminution of the breast tissue upon cessation of lactation (milk production)?

3. Hydrogen peroxide, which belongs to a class of very unstable compounds known as *free radicals*, can bring about drastic, detrimental changes in a cell's structure and function by reacting with almost any molecule with which it comes in contact, including DNA. The resultant cellular changes can lead to genetic mutations, cancer, or other serious consequences. Furthermore, some researchers speculate that the cumulative effects of more subtle cellular damage resulting from free radical reactions over a period of time might contribute to the gradual deterioration associated with aging. Related to this speculation, studies have shown that longevity decreases in fruit flies in direct proportion to a decrease in a specific chemical found in one of the cellular organelles. Based on your knowledge of how the body rids itself of dangerous hydrogen peroxide, what do you think this chemical in the organelle is?

4. Why do you think a person is able to perform anaerobic exercise (such as lifting and holding a heavy weight) only briefly but can sustain aerobic exercise (such as walking or swimming) for long periods? (Hint: muscles have limited energy stores.)

5. One type of the affliction *epidermolysis bullosa* is caused by a genetic defect that results in production of abnormally weak keratin. Based on your knowledge of the role of keratin, what part of the body do you think would be affected by this condition?

3 Plasma Membrane and Membrane Potential

BODY SYSTEMS

Body systems maintain homeostasis

HOMEOSTASIS
The plasma membranes of the cells that make up the body systems play a dynamic role in exchanges and interactions between constituents in the intracellular and extracellular fluid. Many of these plasma–membrane activities, including controlled changes in membrane potential, are important in maintaining homeostasis.

Homeostasis is essential for survival of cells

Plasma membrane

Membrane potential

CELLS

Cells make up body systems

All cells are enveloped by a **plasma membrane**, a thin, flexible, lipid barrier that separates the contents of the cell from its surroundings. To carry on life-sustaining and specialized activities, each cell must exchange materials across its plasma membrane with the homeostatically maintained internal fluid environment that surrounds it. This discriminating barrier contains specific proteins, some of which enable selective passage of materials through the membrane. Other proteins in the plasma membrane serve as receptor sites for interaction with specific chemical messengers in the cell's environment. These chemical messengers control many cellular activities critical to the maintenance of homeostasis.

Cells have a **membrane potential**, which refers to a slight excess of negative charges lined up along the inside of the membrane and a slight excess of positive charges on the outside. The specialization of nerve and muscle cells depends on the ability of these cells to alter their potential on appropriate stimulation.

≡ **CHAPTER CONTENTS AT A GLANCE** ≡

MEMBRANE STRUCTURE AND COMPOSITION
Trilaminar appearance
Membrane composition; fluid mosaic model
● CONCEPTS, CHALLENGES, AND CONTROVERSIES Cystic Fibrosis:
   A Fatal Defect in Membrane Transport
MEMBRANE RECEPTORS AND POSTRECEPTOR EVENTS
Receptor-activated channels
Second messenger systems
CELL-TO-CELL ADHESIONS
Extracellular matrix
Specialized cell junctions
MEMBRANE TRANSPORT
Influences of lipid solubility and size of molecule
Passive versus active forces
Passive diffusion down electrochemical gradient
Osmosis
Carrier-mediated transport
● A CLOSER LOOK AT EXERCISE PHYSIOLOGY Exercising Muscles
   Have a "Sweet" Tooth
Endocytosis; exocytosis
MEMBRANE POTENTIAL
Definition of potential
Ionic basis of resting membrane potential

● FIGURE 3–1   Trilaminar Appearance of a Plasma Membrane in an Electron Micrograph Depicted are the plasma membranes of two adjacent cells. Note the trilaminar structure (that is, two dark layers separated by a light middle layer) of each membrane.

≡● **MEMBRANE STRUCTURE AND COMPOSITION**

The plasma membrane separates the intracellular and extracellular fluid.

The survival of every cell depends on the maintenance of intracellular contents unique for that cell type despite the remarkably different composition of the extracellular fluid surrounding it. This difference in fluid composition inside and outside of a cell is maintained by the **plasma membrane**, an extremely thin layer of lipids and proteins that forms the outer boundary of every cell and encloses the intracellular contents. In addition to serving as a mechanical barrier that traps needed molecules within the cell, the plasma membrane plays an active role in determining the composition of the cell by selectively permitting specific substances to pass between the cell and its environment. Many of the functional differences between cell types are due to subtle variations in the composition of their plasma membranes, which in turn enable the cells to interact in different ways with essentially the same extracellular fluid environment.

The plasma membrane is a fluid lipid bilayer embedded with proteins.

The plasma membrane is too thin to be seen under an ordinary light microscope, but with an electron microscope it appears as a **trilaminar** (three-layered) **structure** consisting of two dark layers separated by a light middle layer (● Fig. 3–1). The specific arrangement of the molecules that make up the plasma membrane is believed to be responsible for this three-layered "sandwich" appearance.

All plasma membranes consist mostly of lipids (fats) and proteins plus small amounts of carbohydrate. The most abundant membrane lipids are phospholipids, with lesser amounts of cholesterol. **Phospholipids** have a polar (electrically charged) head containing a negatively charged phosphate group and two nonpolar (electrically neutral) fatty acid tails (● Fig. 3–2a). The polar end is hydrophilic ("water-loving") because it can interact with water molecules, which are also polar; the nonpolar end is hydrophobic ("water-fearing") and will not mix with water. Such two-sided molecules self-assemble into a **lipid bilayer**, a double layer of lipid molecules, when in contact with water (Fig. 3–2b). The hydrophobic tails bury themselves in the center away from the water, while the hydrophilic heads line up on both sides in contact with the water. The outer surface of the layer is exposed to extracellular fluid (ECF), whereas the inner surface is in contact with the intracellular fluid (ICF).

This lipid bilayer is not a rigid structure but instead is fluid in nature, with a consistency more like liquid cooking oil than solid shortening. The phospholipids, which are not held together by chemical bonds, are able to twirl around rapidly as

Choline

Phosphate

Glycerol

Head (polar, hydrophilic)

Fatty acid

Tails (nonpolar, hydrophobic)

(a)

ECF (water)

Polar heads (hydrophilic)

Nonpolar tails (hydrophobic)

Lipid bilayer

Polar heads (hydrophilic)

ICF (water)

(b)

• FIGURE 3–2   **Structure and Organization of Phospholipid Molecules in a Lipid Bilayer** (a) Phospholipid molecule. (b) When in contact with water, phospholipid molecules organize themselves into a lipid bilayer with the polar heads interacting with the polar water molecules at each surface and the nonpolar tails all facing the interior of the bilayer.

well as move about within their own half of the layer, much like skaters on a crowded skating rink.

Also contributing to the fluidity as well as the stability of the membrane is **cholesterol**. By being tucked in between the phospholipid molecules, the cholesterol molecules prevent the fatty acid chains from packing together and crystallizing, a process that would drastically reduce membrane fluidity.

The fluid nature of the membrane permits it to be flexible so that the cell can change its shape. Red blood cells, for example, must change shape considerably as they squeeze their way single file through the capillaries, the tiniest of blood vessels. It is suspected that other essential membrane functions, such as transport processes, are also dependent on the fluidity of the lipid bilayer.

Attached to or inserted within the lipid bilayer are the **membrane proteins** (• Fig. 3–3). Some of these proteins, having polar regions at both ends joined by a nonpolar central portion, extend entirely through the thickness of the membrane. Other proteins stud only the outer or inner surface; they are anchored by interactions with a protein that spans

the membrane or by attachment to the lipid bilayer. The fluidity of the lipid bilayer enables most membrane proteins to float freely like "icebergs" in a moving "sea" of lipid, although the mobility of proteins that perform a specialized function in a specific area of the cell is restricted. This view of membrane structure is known as the **fluid mosaic model**, in reference to the membrane fluidity and the ever-changing mosaic pattern of the proteins embedded within the lipid bilayer.

The small amount of **membrane carbohydrate** is located only at the outer surface. Short-chain carbohydrates, which protrude from the outer surface, are bound primarily to membrane proteins and to a lesser extent to lipids, forming *glycoproteins* and *glycolipids*, respectively (Fig. 3–3).

This proposed structure can account for the trilaminar appearance of the plasma membrane. The two dark lines are believed to be caused by the preferential staining of the hydrophilic polar regions of the lipid and protein molecules, whereas the light space between corresponds to the hydrophobic core formed by the nonpolar regions of these molecules.

Even though the outer and inner layers have the same appearance when viewed with an electron microscope, it is obvious from our description that the plasma membrane actually is asymmetrical; that is, the surface of the membrane facing the extracellular fluid is strikingly different from the surface facing the cytoplasm. Carbohydrate is located only on the outer surface; the outer and inner surfaces bear different amounts and types of protein; and even the lipid composition of the outer half of the bilayer varies somewhat from the inner half. This distinct sidedness of the membrane is related to the different functions carried out at the outer and inner surfaces.

The lipid bilayer forms the primary barrier to diffusion, whereas proteins perform most of the specific membrane functions.

The various components of the plasma membrane are responsible for carrying out different functions as discussed in the following paragraphs.

Lipid Bilayer   The lipid bilayer serves at least three important functions:

1. It forms the basic structure of the membrane (the "fence" around the cell).
2. Its hydrophobic interior serves as a barrier to passage of water-soluble substances between the ICF and ECF. Water-soluble substances cannot dissolve in and pass through the lipid bilayer. (However, water molecules themselves are small enough to pass between the molecules that form this barrier.)
3. It is responsible for the fluidity of the membrane.

Membrane Proteins   A variety of different proteins within the plasma membrane serve the following specialized functions:

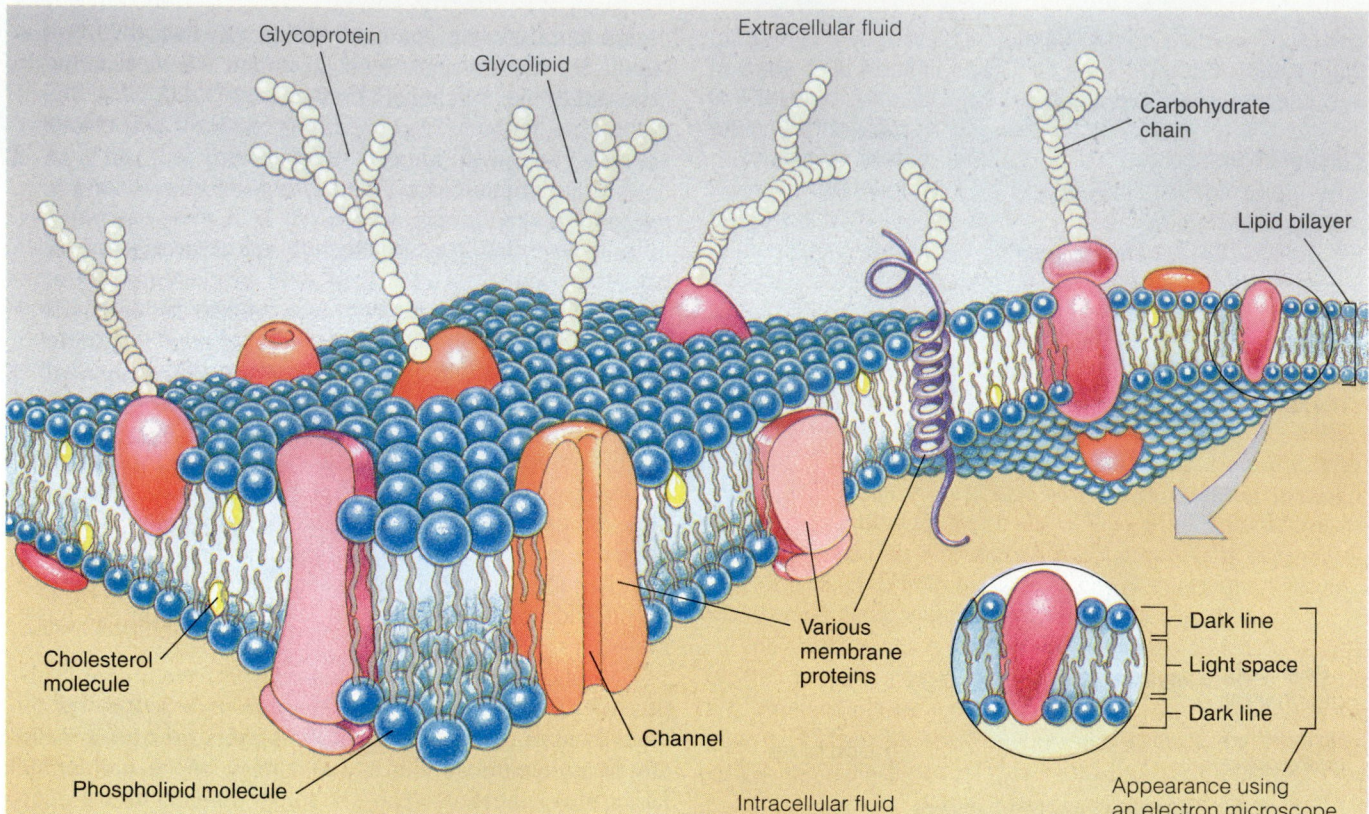

FIGURE 3–3 **Fluid Mosaic Model of Plasma Membrane Structure** The plasma membrane is composed of a lipid bilayer embedded with proteins that penetrate the thickness of the membrane, are partially submerged in the membrane, or are loosely attached to the surface of the membrane. Short carbohydrate chains are attached to proteins or lipids on the outer surface only.

1. Some proteins that span the membrane form water-filled pathways or **channels** across the lipid bilayer. Their presence enables water-soluble substances that are small enough to enter a channel, such as ions, to pass through the membrane without coming into direct contact with the hydrophobic lipid interior (Fig. 3–3). The channels are highly selective. Not only does their small diameter preclude passage of particles greater than 0.8 nanometer (nm) in diameter (a nm is a billionth of a meter or 40 billionths of an inch), but a given channel can also selectively attract or repel particular ions. For example, sodium (Na^+) channels and potassium (K^+) channels can accommodate the passage of only Na^+ and K^+, respectively. This selectivity is believed to be due to specific arrangements of charged amino acid groups on the interior surfaces of the proteins that form the channel walls. Cells vary in the number, kind, and activity of channels they possess. It is even possible for a given channel to be *open* or *closed* to its specific ion as a result of changes in channel shape in response to a controlling mechanism. (See the accompanying boxed feature, Concepts, Challenges, and Controversies.)

2. Other proteins serve as **carrier molecules** that transfer substances unable to cross the membrane on their own. Thus, channels and carrier molecules are both important in the transport of substances between the ECF and ICF. Each carrier can transport only a particular molecule or closely related molecules. Variability in the kinds of carriers different cells possess permits them to selectively transport different substances across their membranes. For example, the thyroid gland requires iodine for the synthesis of thyroid hormone. Accordingly, the plasma membranes of thyroid gland cells uniquely possess carriers for iodine, enabling this essential element to be transported from the blood into thyroid gland cells, a capability not present in other body cells.

3. Many of the proteins on the outer surface serve as **receptor sites** that "recognize" and bind with specific molecules in the environment of the cell. This binding initiates a series of membrane and intracellular events that alter the activity of the particular cell. (The postreceptor pathways involved in altering cell function are discussed in the next section.) In this way, chemical messengers in the blood, such as hormones, are able to influence only the specific cells that possess receptors for the messenger while having no effect on other cells, even though every cell is exposed to the same messenger via its widespread distribution by the blood. To illustrate, the anterior pituitary gland secretes into the blood thyroid-stimulating hormone (TSH), which can attach only to the surface of thyroid gland cells to stimulate secretion of thyroid hormone. No other cells have receptor sites for TSH, so only thyroid

CYSTIC FIBROSIS: A FATAL DEFECT IN MEMBRANE TRANSPORT

Cystic fibrosis (CF), the most common fatal genetic disease in the United States, strikes 1 in every 2,000 Caucasian children. With cystic fibrosis, the body's exocrine glands (see p. 4) secrete an abnormally thick, sticky mucus. Researchers have recently found that cystic fibrosis is caused by any one of several different genetic defects that lead to production of a flawed version of a protein known as *cystic fibrosis transmembrane conductance regulator* (*CFTR*). CFTR normally helps regulate the chloride (Cl⁻) channels in the plasma membrane of exocrine-gland cells. With CF, the defective CFTR "gets stuck" in the endoplasmic reticulum/Golgi system, which normally manufactures and processes this product and ships it to the plasma membrane (see p. 23); that is, in CF patients, the mutated version of CFTR is only partially processed and never makes it to the cell surface. The resultant absence of CFTR protein in the plasma membrane's Cl⁻ channels leads to membrane impermeability to Cl⁻. Because Cl⁻ transport across the membrane is closely linked to Na⁺ transport, and, in turn, water transport across the membrane is closely linked to salt (NaCl) transport, CF patients are unable to secrete sufficient salt and water to dilute their mucus secretions to a normal consistency.

Most dramatically affected are the respiratory airways and the pancreas. The presence of the thick, sticky mucus in the respiratory airways makes it difficult to get adequate air in and out of the lungs. Also, because bacteria thrive in the accumulated mucus, CF patients suffer from repeated respiratory infections. Gradually, the involved lung tissue becomes scarred (fibrotic), losing its elasticity, so that the lungs become harder to inflate. This complication increases the work of breathing beyond the extra work required to move air through the clogged airways.

Similarly, the pancreatic duct, which carries secretions from the pancreas to the small intestine, becomes plugged with thick mucus in CF patients. Because the pancreas produces enzymes important in the digestion of food, malnourishment eventually results. Furthermore, as the pancreatic digestive secretions accumulate behind the blocked pancreatic duct, fluid-filled cysts form in the pancreas, with the affected pancreatic tissue gradually degenerating and becoming fibrotic. The name "cystic fibrosis" aptly describes long-term changes that occur in the pancreas and lungs as the result of a single genetic flaw in the plasma-membrane Cl⁻ channels.

Treatment consists of physical therapy to help clear the airways of the excess mucus and antibiotic therapy to combat respiratory infections, plus special diets and administration of supplemental pancreatic enzymes to maintain adequate nutrition. Despite this supportive treatment, most CF victims do not survive beyond their early twenties, with most dying from lung complications.

With the recent discovery of the genetic defect responsible for the majority of CF cases, investigators are hopeful of developing a means to correct or compensate for the defective gene. For example, one group of researchers recently reported success in inserting a healthy human CFTR gene into an altered cold virus, which, when inhaled by experimental animals, penetrated the cells lining the respiratory airways. Once carried into the cells, the stowaway gene produced human CFTR. If successful in humans, this strategy could be a potential cure for CF. Another potential cure being studied is development of a drug that induces the mutated CFTR to be "finished off" and inserted in the plasma membrane. Furthermore, several promising new drug therapy approaches, such as a mucus-thinning aerosol drug that can be inhaled, offer hope of reducing the number of lung infections and extending the life span of CF victims until a cure can be found.

cells are influenced by TSH despite its widespread distribution.

4. Another group of proteins function as **membrane-bound enzymes** that control specific chemical reactions at either the inner or outer cell surface. Cells display specialization in the types of enzymes embedded within their plasma membranes. For example, the outer layer of the plasma membrane of skeletal-muscle cells contains an enzyme that destroys the chemical messenger that triggers muscle contraction, thus enabling the muscle to relax.

5. Some proteins are arranged in a **filamentous meshwork** on the inner surface of the membrane and are secured to certain internal protein elements of the cytoskeleton. These membrane proteins appear to be structurally important in the maintenance of cell shape and probably participate in surface changes accompanying cell movements.

6. Other proteins serve as **cell adhesion molecules (CAMs)**. These molecules protrude from the membrane surface and form loops or other appendages that the cells use to

grip ahold of each other and to the connective-tissue fibers that interlace between cells. Thus, these molecules help hold tissues and organs together.

7. Finally, still other proteins, especially in conjunction with carbohydrates, are important in the cells' ability to recognize "self" (that is, cells of the same type) and in cell-to-cell interactions.

Membrane Carbohydrates The function of carbohydrates on the outer-membrane surface remains obscure. The following activities are among the leading suggestions for possible roles of these short sugar chains:

1. They may orient and anchor membrane proteins.
2. The complexity and diversity of these carbohydrate chains as well as their location on the external surface suggest that they play an important role in recognition of "self" and in cell-to-cell interactions. Cells are able to recognize other cells of the same type and join together to form tissues. If cultures of embryonic cells of two different types, such as nerve cells and muscle cells, are mixed together, the cells will sort themselves into separate aggregates of nerve cells and muscle cells. Apparently, the unique combination of sugar chains projecting from the surface membrane proteins serves as the "trademark" of a particular cell type, enabling a cell to recognize others of its own kind in tissue formation.
3. Carbohydrate-containing surface markers also appear to be involved in tissue growth, which is normally held within certain limits of cell density. Cells do not "trespass" across the boundaries of neighboring tissues; that is, they do not overgrow their own territory. Abnormal surface carbohydrate markers have been identified in certain tumor cells, suggesting that this abnormality might underlie the uncontrolled growth of tumor cells.

≡● MEMBRANE RECEPTORS AND POSTRECEPTOR EVENTS

Binding of chemical messengers to membrane receptors brings about a wide range of responses in different cells though only a few remarkably similar pathways are used.

Dispersed within the outer surface of the plasma membrane are specialized protein receptors that bind with the selected chemical messengers that come into contact with the cell—for example, hormones delivered by the blood or chemicals released from nerve endings. This combination of messenger with receptor triggers a sequence of cellular events that ultimately controls a particular cellular activity important in the maintenance of homeostasis, such as membrane transport, secretion, contraction, or metabolism.

In spite of the wide range of possible responses, there are only two general means by which binding of the receptor with the extracellular chemical messenger (the **first messenger**) brings about the desired intracellular response: (1) by opening or closing specific channels in the membrane to regulate the movement of particular ions into or out of the cell or (2) by transferring the signal to an intracellular chemical messenger (the **second messenger**), which in turn triggers a preprogrammed series of biochemical events within the cell. Because of the universal nature of these postreceptor events, let us examine each more closely.

Channel Regulation Mechanism The first mechanism, that of altering channels, regulates the flow of specific ions across the membrane. This ionic movement can be responsible for two different cellular events:

1. A small, short-lived movement of Na^+, K^+, or both across the membrane (Pathway 1 in ● Fig. 3–4) alters the electrical activity of cells that are capable of generating electrical signals (or impulses), such as nerve and muscle cells.
2. A transient flow of calcium (Ca^{++}) into the cell through opened Ca^{++} channels (Pathway 2) triggers an alteration in shape and function of specific intracellular proteins, which leads to the cell's response. Illustrative is the increase in cytosolic Ca^{++} responsible for triggering the release of secretory product from many gland cells.

Upon completion of the response, the ions that moved across the membrane through opened channels to trigger the response are quickly returned to their original location by special carrier mechanisms in the membrane.

In some instances, the chemical messenger (or another stimulus) acts indirectly to open the Ca^{++} channels by altering Na^+ and K^+ channels to induce an electrical impulse in the cell. The electrical impulse, in turn, is directly responsible for opening the Ca^{++} channels (Pathway 3). Release of chemicals from nerve cells in response to a nerve impulse is one such example.

In some cells, a rise in cytosolic Ca^{++} can be brought about by release of Ca^{++} from intracellular stores instead of Ca^{++} entry through membrane channels. For example, large amounts of Ca^{++} are stored within a modified endoplasmic reticulum (the sarcoplasmic reticulum) in skeletal-muscle cells. An electrical impulse in these cells triggers the release of Ca^{++} from this organelle into the cytosol (Pathway 4). This increased cytosolic Ca^{++} then alters a specific protein within the skeletal-muscle cell to initiate the events leading to contraction.

Cytosolic Ca^{++} may be increased in yet another way; intracellular Ca^{++} can be released in a second messenger pathway independent of any electrical events, as will be described next.

Second Messenger Mechanism The intracellular pathways activated by a second messenger in response to binding of the first messenger to a surface receptor are remarkably similar among different cells despite the diversity of ultimate responses to that signal. The variability in response depends on the specialization of the cell, not on the mechanism utilitized.

Two major second messenger pathways are now known: one utilizes **cyclic adenosine monophosphate** (**cyclic AMP** or

• FIGURE 3–4   Postreceptor Event: Channel Regulation   Binding of an extracellular chemical messenger with a surface membrane receptor can regulate ionic movement through channels to bring about the desired cellular response in the following ways: a short-lived opening or closing of membrane Na^+ and K^+ channels, which generates electrical impulses ①; a transient opening of membrane Ca^{++} channels, either directly by the binding of an extracellular messenger to a surface receptor ② or indirectly in response to electrical impulses ③; and release of Ca^{++} from intracellular stores resulting from the opening of Ca^{++} channels in organelles in response to electrical impulses ④. An increase in cytosolic Ca^{++} arising from pathways ②, ③, or ④ causes changes in the shape and function of specific intracellular proteins to produce the desired cellular response.

cAMP) as a second messenger, and the other employs Ca^{++} in this role. The two pathways have much in common. The initial stages are similar in that binding of a chemical messenger to a receptor on the outer surface of the membrane brings about almost identical events within the membrane. In both cases, this binding leads, via a series of biochemical steps, to activation of an enzyme on the cytoplasmic side of the membrane. This enzyme, in turn, activates intracellular second messengers that diffuse throughout the cell to trigger the appropriate cellular response. In both pathways, the cellular response is accomplished by altering the structure and subsequent function of particular cell proteins. For example, a particular enzymatic protein regulating a specific metabolic event may be modified so that its activity is increased or decreased.

In the cyclic AMP pathway, binding of an appropriate extracellular messenger to a special surface receptor eventually activates (or in some instances inhibits) the enzyme **adenylate cyclase** (Step 1 in • Fig. 3–5) on the inner surface of the membrane. A membrane-bound "middleman," a **G protein**, acts as an intermediary between the receptor and adenylate cyclase. G proteins are so named because they are bound to guanine nucleotides. Binding of an extracellular messenger to a surface receptor activates a specific G protein, which in turn activates adenylate cyclase. Adenylate cyclase then induces the conversion of intracellular ATP to cyclic adenosine monophosphate by cleaving off two of the phosphates (Step 2). (This is the same ATP used as the common energy currency in the body.) The extracellular messenger cannot gain entry into the cell to "personally" deliver its message to the proteins that carry out the desired response. Instead, it initiates membrane events that "arouse" an intracellular messenger, cAMP, which triggers a sequence of intracellular events to bring about the response dictated by the extracellular mes-

● **FIGURE 3–5 Postreceptor Event: Cyclic AMP Second Messenger System** Binding of an extracellular chemical messenger, the *first messenger,* to a surface membrane receptor activates through a series of steps the membrane-bound enzyme adenylate cyclase ①, which in turn converts intracellular ATP into cyclic AMP ②. Cyclic AMP acts an intracellular *second messenger,* triggering the desired cellular response by activating cAMP-dependent protein kinase ③, which in turn phosphorylates ④ and thereby modifies ⑤ a particular intracellular protein. The altered protein then accomplishes the cellular response dictated by the extracellular messenger.

senger. To fulfill this function, cAMP activates a specific intracellular enzyme, **cAMP-dependent protein kinase** (Step 3). Protein kinase in turn phosphorylates (Step 4) a specific intracellular protein, such as an enzyme important in a particular metabolic pathway. **Phosphorylation** refers to the transfer of a phosphate group from ATP to the protein at the expense of degrading ATP to ADP. Attachment of a phosphate group to the protein induces the protein to change its shape and function (either activating or inhibiting it) to bring about the desired response (Step 5). After the response is accomplished, intracellular enzymes inactivate the participating chemicals so that the response can be terminated. Otherwise, once triggered, the response would go on indefinitely until the cell ran out of necessary supplies.

It is important to recognize that different types of cells have different proteins available for phosphorylation and modification by protein kinase. Therefore, a *common second messenger, cAMP, can induce widely differing responses in different cells,* depending on what proteins are modified. Cyclic AMP can be thought of as a commonly used molecular "switch" that can "turn on" (or "turn off") different cellular events de-

pending on the unique specialization of a particular cell type. The variable responsiveness once the switch is turned on is due to the genetically programmed differences in the sets of proteins within different cells. For example, activation of the cAMP system brings about modification of heart rate in the heart, stimulation of the formation of female sex hormones in the ovaries, and control of water conservation during urine formation in the kidneys.

Instead of cAMP, some cells use Ca^{++} as a second messenger. In such cases, binding of the first messenger to the surface receptor eventually leads to activation of the enzyme **phospholipase C**, which is bound to the inner side of the membrane (Step 1 in ● Fig. 3–6). This enzyme breaks down a special type of phospholipid within the membrane itself (Step 2). The products of this breakdown mobilize intracellular Ca^{++} stores to increase cytosolic Ca^{++} (Step 3). Many of the Ca^{++}-dependent cellular events are triggered by activation of **calmodulin**, an intracellular Ca^{++}-binding protein (Step 4). Activation of calmodulin by Ca^{++} is similar to activation of protein kinase by cAMP. From here the patterns of the two pathways are similar. Once bound to Ca^{++}, activated calmo-

Extracellular chemical messenger

Surface membrane receptor

ECF

① (Activates through a series of steps)
PIP₂ (a component of the phospholipid tails)

Phospholipase C

Plasma membrane ICF

② (PIP₂ converted by phospholipase C to DG and IP₃)

DG and IP₃

③ (Mobilizes)

Intracellular Ca^{++} stored within organelles

Increase in cytosolic Ca^{++}

④ (Activates)

Calmodulin

⑤ (Induces protein to take shape)

Altered protein shape and function

Brings about cellular response

PIP_2 = Phosphatidylinositol 4,5-biphosphate

DG = Diacylglycerol

IP_3 = Inositol triphosphate

• **FIGURE 3–6 Postreceptor Event: Calcium Second Messenger System**
Binding of an extracellular chemical messenger to a surface membrane receptor activates through a series of steps the membrane-bound enzyme phospholipase C ①. Phospholipase C converts PIP_2, a membrane component, into DG and IP_3 ②, which in turn mobilize Ca^{++} stored within organelles ③. Ca^{++}, acting as a second messenger, activates calmodulin ④, which induces a change in the shape and function of a particular intracellular protein to produce the desired cellular response ⑤.

dulin alters other cellular proteins (Step 5), either activating or inhibiting them, to bring about the ultimate desired cellular response.

The cAMP and Ca^{++} pathways frequently overlap in bringing about a particular cellular activity. For example, cAMP and Ca^{++} can influence each other. Calcium-activated calmodulin can regulate adenylate cyclase and thus influence cAMP, whereas cAMP-dependent kinase may phosphorylate and thereby change the activity of Ca^{++} channels or carriers. Furthermore, in some instances, both Ca^{++} and cAMP regulate the same intracellular protein.

Even though the Ca^{++} and cAMP effects can become complexly intertwined, it is still notable that a great many diverse cellular events can be traced to a suprisingly small number of pathways: channel effects, one of the two major second messenger pathways, or some combination of these. It would not be correct, however, to leave the impression that these are the only possible pathways. For example, **cyclic guanosine monophosphate** (**cyclic GMP**) is suspected of playing a role in postreceptor events in some cells.

Several remaining points about receptor activation and the ensuing events merit attention. First, a multiplying (cascading) effect of these pathways greatly amplifies the initial signal (• Fig. 3–7). Binding of one chemical messenger molecule to a receptor activates a number of adenylate cyclase molecules (let us arbitrarily say ten), each of which activates many (in

• **FIGURE 3–7   Amplification of Initial Signal by Second Messenger System** Through amplification, very low concentrations of extracellular chemical messengers, such as hormones, can trigger pronounced cellular responses.

Extracellular chemical messenger bound to membrane receptor

Total number of molecules

1

Amplification

Activated adenylate cyclase (10)   10

Amplification

Cyclic AMP (100)   1,000

Activated protein kinase 1,000

Amplification

Phosphorylated (activated) protein (e.g., an enzyme) (100)   100,000

Amplification

Products of activated enzyme (100)   10,000,000

our example one hundred) cAMP molecules. Each cAMP molecule then acts on a single protein kinase, which phosphorylates and thereby influences many (again, let us say one hundred) specific proteins, such as enzymes. Each enzyme, in turn, is responsible for producing many (perhaps one hundred) molecules of a particular product, such as a secretory product. The result of this cascade of events, with one event triggering the next event in sequence, is a tremendous amplification of the initial signal; in other words, the magnitude of the output is much greater than the input. In the hypothetical example in Figure 3–7, one chemical messenger molecule has been responsible for inducing a yield of ten million molecules of a secretory product. In this way, very low concentrations of hormones and other chemical messengers can trigger pronounced cellular responses.

While membrane receptors serve as links between extracellular first messengers and intracellular second messengers in the regulation of specific cellular activities, the receptors themselves are also frequently subject to regulation. In many instances, the number and affinity (attraction of a receptor for its chemical messenger) can be altered, depending on the circumstances.

Many disease processes can be linked to malfunctioning receptors or defects in one of the components of the ensuing pathways. As an example, defective receptors are responsible for the extreme muscular weakness that characterizes *myasthenia gravis.* With this disease, affected skeletal-muscle receptors are unable to respond to the chemical messenger released by nerves that normally triggers muscle contraction. On the other hand, in the intestinal disease *cholera,* the recep-

tors are all functional, but the toxin produced by the cholera pathogen prevents the normal inactivation of cAMP in intestinal cells. In these cells, cAMP induces fluid secretion into the lumen. The severe diarrhea characteristic of cholera is caused by the continued secretion of this fluid into the gut, triggered by the continual presence of cAMP.

CELL-TO-CELL ADHESIONS

In multicellular organisms such as humans, plasma membranes not only serve as the outer boundaries of all cells but also participate in cell-to-cell adhesions that allow groups of cells to bind together into tissues and to be packaged further into organs. Organization of cells into appropriate groupings may be at least partially attributable to the carbohydrate chains on the membrane surface. Once arranged, cells are held together by three different means: (1) cell adhesion molecules in the cells' plasma membranes, (2) the extracellular matrix, and (3) specialized cell junctions.

The extracellular matrix serves as the biological "glue."

The cells within a tissue for the most part are not in direct physical contact with neighboring cells. Instead, they are held together by the **extracellular matrix**, an intricate meshwork of fibrous proteins embedded in a watery, gel-like substance composed of complex carbohydrates. The watery gel provides a pathway for diffusion of nutrients, wastes, and other water-soluble traffic between the blood and tissue cells. Interwoven within this gel are three major types of protein fibers: collagen, elastin, and fibronectin.

1. **Collagen** forms cablelike fibers or sheets that provide tensile strength (resistance to longitudinal stress). In *scurvy*, a condition caused by vitamin C deficiency, these fibers are not properly formed. As a result, the tissues, especially those of the skin and blood vessels, become very fragile. This leads to bleeding in the skin and mucous membranes, which is especially noticeable in the gums.

2. **Elastin** is a rubberlike protein fiber most abundant in tissues that must be capable of easily stretching and then recoiling after the stretching force is removed. It is found, for example, in the lungs, which stretch and recoil as air moves in and out.

3. **Fibronectin** promotes cell adhesion and holds cells in position. Reduced amounts of this protein have been found within certain types of cancerous tissue, possibly accounting for the fact that cancerous tissue cells do not adhere well to each other but tend to break loose and metastasize (spread elsewhere in the body).

The extracellular matrix is secreted by local cells, most commonly by **fibroblasts** ("fiber formers") present in the matrix. Often the matrix plus the cells within it are known collectively as *connective tissue* because they connect cells together into tissues and tissues into organs. In some tissues, the matrix becomes highly specialized to form such structures as cartilage or tendons or, upon appropriate calcification, the hardened structures of bones and teeth.

Some cells are directly linked together by specialized cell junctions.

In addition to the tissue cohesion provided by the extracellular matrix, some cells are directly linked together by one of three types of specialized cell junctions: (1) *desmosomes* (adhering junctions); (2) *tight junctions* (impermeable junctions); and (3) *gap junctions* (communicating junctions).

At a **desmosome**, filaments of unknown composition extend between the plasma membranes of two closely adjacent but nontouching cells, acting as "spot rivets" to anchor the cells together (• Fig. 3–8). Desmosomes are distributed widely throughout the body, being most abundant in tissues that are subject to considerable stretching, such as the skin, heart, muscle, and uterus.

Sheets of epithelial tissue are joined by **tight junctions**. Epithelial tissue covers the surface of the body and lines all of its internal cavities. All of these epithelial sheets serve as highly selective barriers between two compartments that have considerably different chemical compositions. For example, the epithelial sheet lining the digestive tract separates the food and potent digestive juices within the inner cavity (lumen) from the blood vessels that lie on the other side. It is important that only completely digested food particles and not undigested food particles or digestive juices move across the epithelial sheet from the lumen to the blood. Accordingly, the lateral (side) edges of adjacent cells in the epithelial sheet are

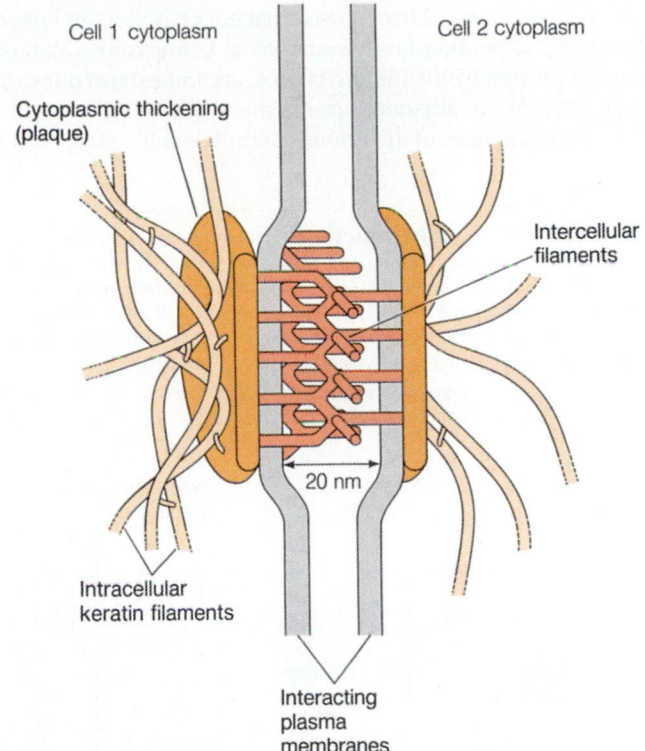

• **FIGURE 3–8   Spot Desmosome** Spot desmosomes are adhering junctions that anchor cells together in tissues subject to considerable stretching.

joined together in a tight seal near their luminal border by direct fusion of proteins on the outer surfaces of the two interacting plasma membranes (• Fig. 3–9). These tight junctions are impermeable and thus prevent materials from passing between the cells. Passage across the epithelial barrier, therefore, must take place *through* the cells, not *between* them. This transcellular (across the cell) traffic is regulated by means of the channels and carriers present. If the cells were not joined by tight junctions, uncontrolled exchange of molecules could take place between the compartments by unpoliced traffic through the spaces between adjacent cells.

Tight junctions also perform another function besides preventing undesirable leaks within epithelial sheets. They restrict the movement of membrane proteins to one of two cell surfaces: the luminal border (facing the lumen) or basolateral border (facing the blood) (Fig. 3–9). Proteins are free to wander within their own surface but cannot move to the other surface due to the presence of the tight junction. The separation of proteins between the two parts of the surface membrane is functionally important. These proteins include carriers specific for transporting materials either from the lumen into the epithelial cell (those located on the luminal border) or from the epithelial cell into the blood (basolateral carriers). The tight junctions restrict these different carriers to the regions of the membrane where they can properly carry out their specific job.

The third type of junction between cells is a communicating junction known as a **gap junction**. As implied by the name, a gap exists between two adjacent cells that are linked by small connecting tunnels known as **connexons**. Connexons are formed by the joining of proteins that extend outward from each of the adjacent plasma membranes (● Fig. 3–10). The small diameter of the tunnels permits small water-soluble

● **FIGURE 3–9   Tight Junction** Tight junctions are impermeable junctions that join the lateral edges of epithelial cells near their luminal borders, thus preventing passage of materials *between* the cells. Only regulated passage of materials can occur *through* these cells, which form highly selective barriers that separate two compartments of highly different chemical composition.

● **FIGURE 3–10   Gap Junction** Gap junctions are communicating junctions consisting of connexons, small connecting tunnels that permit movement of charge-carrying ions between two adjacent cells.

particles such as ions (electrically charged particles) to pass between the connected cells but precludes passage of large molecules such as proteins. Gap junctions are found in heart muscle and smooth muscle. Movement of ions between cells through gap junctions plays an important role in transmitting electrical activity throughout an entire muscle mass. Because this electrical activity brings about contraction, the presence of gap junctions enables synchronized contraction of a whole muscle mass, such as the heart. Many tissues that are not electrically active also are connected by gap junctions, but their significance in nonmuscle tissue is still unclear.

≡● MEMBRANE TRANSPORT

Lipid-soluble substances and small ions can passively diffuse through the plasma membrane down their electrochemical gradient.

Anything that passes between a cell and the surrounding extracellular fluid must be able to penetrate the plasma membrane. If a substance can cross the membrane, the membrane is said to be **permeable** to that substance; if a substance is unable to pass, the membrane is **impermeable** to it. The plasma membrane is **selectively permeable** in that it permits some particles to pass through while excluding others.

Two properties of particles influence whether they can permeate the plasma membrane without any assistance: (1) the relative solubility of the particle in lipid and (2) the size of the particle. Highly lipid-soluble particles are able to dissolve in the lipid bilayer and pass through the membrane. Uncharged or nonpolar molecules (such as O_2, CO_2, and fatty acids) are highly lipid-soluble and readily permeate the membrane. On the other hand, charged particles (ions such as Na^+ and K^+) and polar molecules (such as glucose and proteins) have low lipid solubility but are very soluble in water. The lipid bilayer serves as an impermeable barrier to particles poorly soluble in lipid. For water-soluble ions less than 0.8 nm in diameter, the protein channels serve as an alternate route for passage across the membrane. Only ions for which specific channels are available can permeate the membrane. Particles that have low lipid solubility and are too large for channels cannot permeate the membrane on their own.

Even though a particle might be capable of permeating the membrane by virtue of its lipid solubility or its ability to fit through a channel, some force is needed to produce its movement across the membrane (a particle does not just decide it wants to be on the other side). Two general types of forces are involved: (1) forces that do not require the cell to expend energy to produce movement (**passive forces**) and (2) forces requiring expenditure of cellular energy (ATP) in the transport of a substance across the membrane (**active forces**). We will now examine the various methods of membrane transport, indicating whether each uses a passive or active mechanism.

Diffusion down a Concentration Gradient All molecules are in continuous random motion at temperatures above absolute zero as a result of heat (thermal) energy. This motion is most evident in liquids and gases, where the individual molecules have more room to move before colliding with another molecule. Each molecule moves separately and randomly in any direction. As a consequence of this haphazard movement, the molecules frequently collide, bouncing off each other in different directions like billiard balls striking each other. Such random intermingling is known as **diffusion**. Obviously, the greater the molecular concentration of a substance in a solution, the greater the likelihood of collisions. For example, in ● Figure 3–11a, the concentration differs between area A and area B in a solution. Such a difference in concentration between two adjacent areas is referred to as a **concentration gradient** (or **chemical gradient**). Random molecular collisions will occur more frequently in area A because of its greater concentration of molecules. As a result, more molecules will bounce from area A into area B than in the opposite direction. In both areas, the individual molecules will move randomly and in all directions, but the net movement of molecules by diffusion will be from the area of higher concentration to the area of lower concentration.

Net diffusion refers to the difference between two oppos-

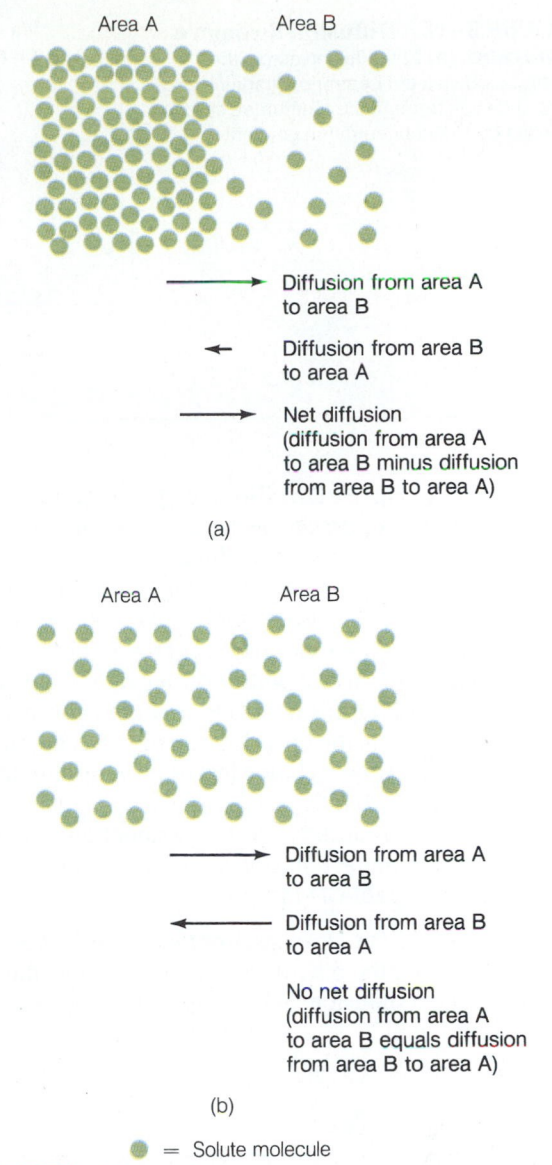

(a)

= Solute molecule

● **FIGURE 3–11 Diffusion** (a) Diffusion down concentration gradient. (b) State of equilibrium.

ing movements. If ten molecules move from A to B while two molecules simultaneously move from B to A, the net diffusion is eight molecules moving from A to B. Molecules will spread in this way until the substance is uniformly distributed between the two areas and a concentration gradient no longer exists (Fig. 3–11b). At this point, there will be no net diffusion. Movement of molecules from A to B will be exactly matched by movement of molecules from B to A. This situation is known as a **state of equilibrium** or **steady state**. Even though movement is still taking place, no net diffusion is occurring.

What happens if different concentrations of a substance are separated by a plasma membrane? If the substance can permeate the membrane, net diffusion of the substance will occur through the membrane down its concentration gradient from the area of high concentration to the area of low

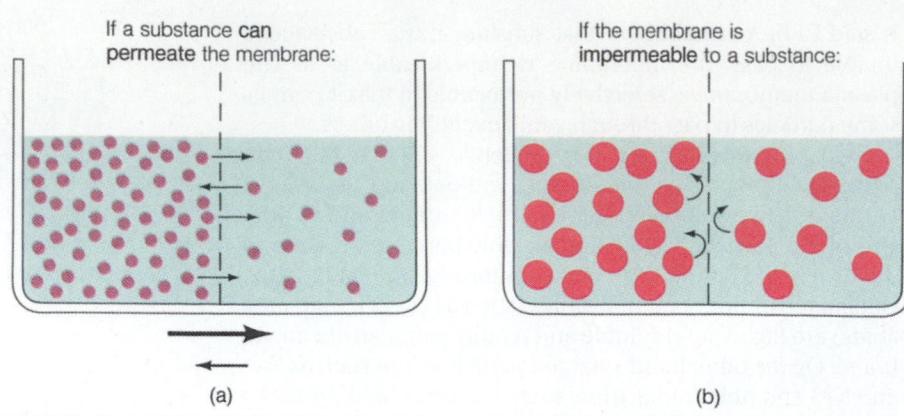

● **FIGURE 3–12 Diffusion through a Membrane** (a) Net diffusion across the membrane down a concentration gradient. (b) No diffusion through the membrane despite the presence of a concentration gradient.

If a substance can permeate the membrane:

If the membrane is impermeable to a substance:

(a)

(b)

concentration (● Fig. 3–12a). No energy is required for this movement, so it is a *passive* mechanism of membrane transport. The transfer of O_2 across the lung membrane occurs by this means. The blood carried to the lungs is low in O_2, having given up O_2 to the body tissues for cellular metabolism. The air in the lungs, on the other hand, is high in O_2 because it is continuously exchanged with fresh air by the process of breathing. Because of this concentration gradient, net diffusion of O_2 occurs from the lungs into the blood as blood flows through the lungs. Thus, as blood leaves the lungs for delivery to the tissues, it is high in O_2.

Several factors in addition to the concentration gradient influence the rate of net diffusion across a membrane (**Fick's law of diffusion**) (Table 3–1).

1. *The magnitude* (or *steepness*) *of the concentration gradient.* The greater the difference in concentration, the faster the rate of net diffusion. For example, if the tissues are using O_2 more rapidly as a result of exercise, the blood delivered to the lungs will be lower in O_2 than normal, creating

a larger than normal O_2 difference between the air in the lungs and the blood entering the lungs. Accordingly, more O_2 will diffuse from the lungs into the blood before equilibrium is achieved. This greater transfer of O_2 provides more O_2 for exercise.

2. *The permeability of the membrane to the substance.* The more permeable the membrane is to a substance, the more rapidly the substance can diffuse down its concentration gradient. Note that if the membrane is impermeable to the substance, no diffusion can take place across the membrane, even though a concentration gradient may exist (Fig. 3–12b). For example, because the plasma membrane is impermeable to the vital intracellular proteins, they are unable to escape from the cell, even though they are in much greater concentration in the ICF than in the ECF.

3. *The surface area of the membrane across which diffusion is taking place.* Obviously, the larger the surface area available, the greater the rate of diffusion it can accommodate. For example, absorption of nutrients across the membranes in the small intestine is enhanced by surface foldings and projections that greatly increase the available absorptive surface.

4. *The molecular weight of the substance.* Lighter molecules such as O_2 and CO_2 bounce farther upon collision than do heavier molecules. Consequently, O_2 and CO_2 diffuse rapidly, permitting rapid exchanges of these gases across the lung membranes.

5. *The distance through which diffusion must take place.* The greater the distance, the slower the rate of diffusion. Accordingly, membranes across which diffusing particles must travel are normally relatively thin, such as the membranes separating the air and blood in the lungs. Thickening of this air-blood interface (as in pneumonia, for example) slows down exchange of O_2 and CO_2. Furthermore, diffusion is efficient only for short distances between cells and their surroundings. It becomes an inappropriately slow process for distances of more than a few centimeters. To illustrate, it would take months to years for O_2 to diffuse from the surface of the body to the cells in the interior. This makes it imperative that the circulatory system provide a network of tiny vessels that deliver and

TABLE 3–1 Factors Influencing Rate of Net Diffusion of a Substance across a Membrane (Fick's Law of Diffusion)

FACTOR	EFFECT ON RATE OF NET DIFFUSION
↑ Concentration gradient of substance (ΔC)	↑
↑ Permeability of membrane to substance (P)	↑
↑ Surface area of membrane (A)	↑
↑ Molecular weight of substance (MW)	↓
↑ Distance (thickness) (ΔX)	↓

Modified Fick's equation:

Net rate of diffusion $(Q) = \dfrac{\Delta C \cdot P \cdot A}{MW \cdot \Delta X}$

$\left[\dfrac{P}{\sqrt{MW}} = \text{diffusion coefficient (D)} \right]$

Restated: $Q \propto \dfrac{\Delta C \cdot A \cdot D}{\Delta X}$

Positively charged area

Negatively charged area

Cations (positively charged ions) attracted toward negative area

Anions (negatively charged ions) attracted toward positive area

• **FIGURE 3–13   Movement along Electrical Gradient**

pick up materials at every "block" of a few cells, with diffusion accomplishing short local exchanges between the blood and surrounding cells.

Movement along an Electrical Gradient   Movement of ions (electrically charged particles that have either lost or gained an electron) is also affected by their electrical charge. Like charges (those with the same kind of charge) repel each other whereas opposite charges attract each other. If a relative difference in charge exists between two adjacent areas (• Fig. 3–13), the positively charged ions (**cations**) tend to move toward the more negatively charged area, whereas the negatively charged ions (**anions**) tend to move toward the more positively charged area. A difference in charge between two adjacent areas thus produces an **electrical gradient** that passively induces the movement of ions. When an electrical gradient exists between the ICF and ECF, only ions that can permeate the plasma membrane are able to move along this gradient. The simultaneous existence of an electrical gradient and concentration (chemical) gradient for a particular ion is referred to as an **electrochemical gradient**. Later in this chapter you will learn how electrochemical gradients are important in the determination of the electrical properties of the plasma membrane.

Osmosis is the net diffusion of water down its own concentration gradient.

Water can readily permeate the plasma membrane. The driving force for diffusion of water across the membrane is the same as for any other diffusing molecule, namely, its concentration gradient. Usually, the term *concentration* refers to the density of the solute (dissolved substance) in a given volume of water. It is important to recognize, however, that the addition of a solute to pure water in essence decreases the water concentration. As a generalization, one molecule of a solute displaces one molecule of water.

Compare the water and solute concentrations in the two containers in • Figure 3–14. Container A is full of pure water, so the water concentration is 100% while the solute concen-

tration is 0%. In container B, 10% of the water molecules have been replaced by solute. The water concentration is now 90% (a lower water concentration than A), and the solute concentration is 10% (a higher solute concentration compared to A). Note that as the solute concentration increases, the water concentration decreases correspondingly.

If solutions of unequal solute concentration (and hence unequal water concentration) are separated by a membrane that permits passage of water, such as the plasma membrane (• Fig. 3–15), water will diffuse down its own concentration gradient from the area of higher water concentration (lower solute concentration) to the area of lower water concentration (higher solute concentration). This net diffusion of water is known as **osmosis**. Because solutions are always referred to in terms of concentration of solute, *water moves by osmosis to the area of higher solute concentration.* Very loosely, then, the solute can be thought of as "drawing" or attracting water, but in reality, osmosis is nothing more than the diffusion of water down its own concentration gradient.

Thus far in our discussion of osmosis, we have ignored any solute movement. Let us compare the results of osmosis in the

• **FIGURE 3–14   Relationship between Solute and Water Concentration in a Solution**   (a) Pure water. (b) Solution.

100% water concentration
0% solute concentration

(a)

90% water concentration
10% solute concentration

(b)

● = Water molecule   ● = Solute molecule

Higher H₂O
concentration,
lower solute
concentration

Lower H₂O
concentration,
higher solute
concentration

● = Water molecule   ● = Solute molecule

● **FIGURE 3–15   Osmosis** Osmosis is the net diffusion of water down its own concentration gradient (to the area of higher solute concentration).

two cases when the solute (1) can and (2) cannot permeate the membrane:

1. If the membrane is permeable to the solute as well as to water, the solute is able to move down its own concentration gradient in the opposite direction of the net water movement (● Fig. 3–16). This movement continues until both the solute and water are evenly distributed across the membrane. Even though some transitory differences in volume between the two compartments may occur because of differences in the speed with which water and the solute diffuse through the membrane, the final volume of the compartments when equilibrium is achieved is the same as at the onset. Water and solute molecules have merely exchanged places between the two compartments until their distributions have equalized; that is, an equal number of water molecules have moved from side 1 to side 2 as solute molecules have moved from side 2 to side 1. With all concentration gradients abolished, osmosis ceases.

2. If the membrane is impermeable to the solute, the solute is not able to cross the membrane down its concentration gradient (● Fig. 3–17). At the onset, the concentration gradients are similar to the previous example. However, even though net diffusion of water takes place from side 1 to side 2, the solute cannot move. As a result of water movement alone, side 2's volume increases while the volume of side 1 correspondingly decreases. Loss of water from side 1 increases the concentration of solutes on side 1, whereas addition of water to side 2 reduces solute concentration on that side. Eventually, the concentrations of water and solute on the two sides of the membrane become equal, and net diffusion of water ceases. Unlike the situation in which the solute can also permeate, diffusion of water alone has resulted in a change in final volumes of the two compartments. The volume of the side originally containing the greater solute concentration is larger, having gained water.

What will happen if a nonpenetrating solute is present on side 2 but pure water is present on side 1 (● Fig. 3–18)? Osmosis occurs from side 1 to side 2, but the concentrations between the two compartments can never become equal. No matter how dilute side 2 becomes because of water diffusing into it, it can never become pure water, nor can side 1 ever acquire any solute. Since equilibrium is impossible to achieve, does net diffusion of water (osmosis) continue unabated until all of the water has left side 1? No. As the volume expands in compartment 2, a difference in **fluid** (**hydrostatic**) **pressure** between the two compartments is created, and it opposes osmosis. The magnitude of opposing pressure necessary to completely stop osmosis is equal to the **osmotic**

● **FIGURE 3–16   Movement of Water and a Permeable Solute Unequally Distributed across a Membrane**

Membrane (permeable to both water and solute)

Side 1   Side 2

Higher H₂O
concentration,
lower solute
concentration

Lower H₂O
concentration,
higher solute
concentration

H₂O moves from side 1 to side 2 down its concentration gradient

Solute moves from side 2 to side 1 down its concentration gradient

Side 1   Side 2

• Water concentrations equal
• Solute concentrations equal
• No further net diffusion
• State of equilibrium exists

● = Water molecule   ⟶ = Direction of net diffusion
○ = Solute molecule   ⟵

Membrane (permeable to H₂O but impermeable to solute)

Side 1   Side 2

H₂O

Solute

Higher H₂O concentration, lower solute concentration

Lower H₂O concentration, higher solute concentration

H₂O moves from side 1 to side 2 down its concentration gradient

Solute unable to move from side 2 to side 1 down its concentration gradient

Side 1   Side 2

Original level of solutions

· Water concentrations equal
· Solute concentrations equal
· No further net diffusion
· State of equilibrium exists

= Water molecule   ⟶ = Direction of net diffusion

= Solute molecule

• FIGURE 3–17 Osmosis in the Presence of an Unequally Distributed Nonpenetrating Solute

Membrane (permeable to H₂O but impermeable to solute)

Side 1   Side 2

H₂O

Solute

Pure water

Lower H₂O concentration, higher solute concentration

H₂O moves from side 1 to side 2 down its concentration gradient

Solute unable to move from side 2 to side 1 down its concentration gradient

Side 1   Side 2

Hydrostatic (fluid) pressure difference

Original level of solutions

Osmosis

Hydrostatic pressure

· Water concentrations not equal
· Solute concentrations not equal
· Tendency for water to diffuse by osmosis into side 2 is exactly balanced by opposing tendency for hydrostatic pressure difference to push water into side 1
· Osmosis ceases
· Opposing pressure necessary to completely stop osmosis is equal to osmotic pressure of solution

= Water molecule   ⟶ = Direction of net diffusion

= Solute molecule

• FIGURE 3–18 Osmosis When Pure Water Is Separated from a Solution Containing a Nonpenetrating Solute

pressure of the solution on side 2. The osmotic pressure can be related directly to the concentration of nonpenetrating solute. The greater the concentration of nonpenetrating solute → the lower the concentration of water → the greater the drive for water to move by osmosis from pure water into the solution → the greater the opposing pressure required to stop the osmotic flow → the greater the osmotic pressure of the solution. Therefore, a solution with a high solute concentration exerts greater osmotic pressure than does a solution with a lower solute concentration.

Approximately one hundred times the volume of water in a cell crosses the plasma membrane every second. Even though water is rapidly entering and leaving cells, they nor-

mally do not experience any net gain (swelling) or loss (shrinking) of volume. This is because the concentration of solutes in the ECF is normally carefully regulated (primarily by the kidneys) to maintain the same level of osmotic activity as is present within the cells; thus, no net diffusion of water occurs. For example, the plasma in which the red blood cells are suspended normally has the same osmotic activity as the fluid inside these cells, enabling the cells to maintain a constant volume. If red blood cells are placed in a solution with a lower concentration of solutes (and, therefore, a higher concentration of water), water enters the cells by osmosis, causing them to swell, perhaps to the point of rupturing or *lysing*.

Special mechanisms are used to transport selected molecules unable to cross the plasma membrane on their own.

All of the kinds of transport we have discussed thus far—diffusion down concentration gradients, diffusion along electrical gradients, and osmosis—produce net movement of molecules capable of permeating the plasma membrane by virtue of their lipid solubility or small size. Large, poorly lipid-soluble molecules such as proteins, glucose, and amino acids cannot cross the plasma membrane on their own no matter what forces are acting on them. This impermeability ensures that the large polar intracellular proteins cannot escape from

the cell. It also means, however, that the cell must provide mechanisms for transporting into the cell essential nutrients, such as glucose for energy and amino acids for the synthesis of proteins, as well as for transporting out of the cell metabolic wastes and secretory products, such as proteinaceous hormones and enzymes. Furthermore, passive diffusion alone cannot always account for the movement of small ions. Cells utilize two different mechanisms to accomplish these selective transport processes: carrier-mediated transport and vesicular transport.

Carrier-Mediated Transport All carrier proteins span the thickness of the plasma membrane and are able to undergo reversible changes in shape so that specific binding sites can alternately be exposed at either side of the membrane. The details of the conformational changes that carriers undergo are unknown, but • Figure 3–19 is a schematic representation

• **FIGURE 3–19   Carrier-Mediated Transport: Facilitated Diffusion**

Step 1
Conformation X of carrier (binding sites exposed to outside of membrane)
Molecule to be transported binds to carrier

Step 2
Upon binding with molecules to be transported, carrier changes its conformation

Step 3
Conformation Y of carrier (binding sites exposed to inside of membrane)
Transported molecule detaches from carrier

Molecule to be transported

Concentration gradient

(High)

ECF

Plasma membrane

ICF

(Low)

Carrier molecule

Conformation X of carrier

Conformation Y of carrier

ECF

ICF

of how this transport process might take place. As the molecule to be transported attaches to a binding site on the carrier on one side of the membrane (Step 1), it triggers a change in the carrier's shape to expose the same site to the other side of the membrane (Step 2). Then, having been moved in this way from one side of the membrane to the other, the bound molecule detaches from the carrier (Step 3).

Carrier-mediated transport systems display three important characteristics that determine the kind and amount of material that can be transferred across the membrane: specificity, saturation, and competition.

1. **Specificity**. Each carrier protein is specialized to transport a specific substance, or at most a few closely related chemical compounds. For example, amino acids cannot bind to glucose carriers, although several similar amino acids may be able to utilize the same carrier. Cells vary in the types of carriers they possess, thus permitting transport selectivity among cells. A number of inherited diseases involve defects in transport systems for a particular substance. *Cysteinuria* (cysteine in the urine) is such a disease involving defective cysteine carriers in the kidney membranes. This transport system normally removes cysteine from the fluid destined to become urine and returns this essential amino acid to the blood. When this carrier is malfunctional, large quantities of cysteine remain in the urine, where it is relatively insoluble and tends to precipitate. This is one cause of urinary stones.

2. **Saturation**. There is a limit to the amount of a substance that can be transported across the membrane via a carrier in a given time; that is, a limited number of carrier binding sites are available within a particular plasma membrane for a specific substance. This limit is known as the **transport maximum (T_m)**. Until the T_m is reached, the number of carrier binding sites occupied by a substance and, accordingly, the substance's rate of transport across the membrane are directly related to its concentration. The more of a substance available to be transported, the more will be transported. When the T_m is reached, the carrier is saturated (all binding sites are occupied), and the rate of the substance's transport across the membrane is maximal. Further increases in concentration of the substance are not accompanied by corresponding increases in rate of transport (• Fig. 3–20).

   As an analogy, consider a ferry boat that can maximally carry one hundred people across a river in an hour. If twenty-five people board the ferry, twenty-five will be transported that hour. Doubling the number of people boarding to fifty will double the rate of transport to fifty people per hour. Such a direct relationship will exist between the number of people waiting to board (the concentration) and the rate of transport until the ferry is fully occupied (its T_m is reached). The ferry can maximally transport one hundred people per hour. Even if one hundred and fifty people are waiting to board, still only one hundred will be transported per hour.

   Saturation of carriers is a critical rate-limiting factor in the transport of selected substances across the kidney membranes during urine formation and across the intesti-

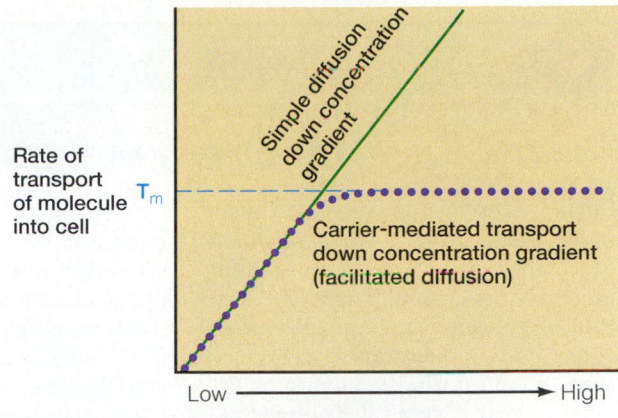

Rate of transport of molecule into cell

T_m

Concentration of transported molecules in ECF

Low ————————————→ High

• **FIGURE 3–20   Comparison of Carrier-Mediated Transport and Simple Diffusion down Concentration Gradient**   With simple diffusion of a molecule down its concentration gradient, the rate of transport of the molecule into the cell is directly proportional to the extracellular concentration of the molecule. With carrier-mediated transport of a molecule down its concentration gradient, the rate of transport of the molecule into the cell is directly proportional to the extracellular concentration of the molecule until the carrier is saturated, at which time the rate of transport reaches a maximal value (transport maximum, or T_m). The rate of transport does not increase with further increases in the ECF concentration of the molecule.

nal membranes during absorption of digested foods. Furthermore, in some important instances, it is possible to regulate (for example, by hormones) the rate of carrier-mediated transport by varying the affinity (attraction) of the binding site for its passenger or by varying the number of binding sites. For example, the hormone insulin greatly increases the carrier-mediated transport of glucose into most cells of the body by promoting an increase in the number of glucose carriers in the cells' plasma membranes. Deficiency of insulin (*diabetes mellitus*) drastically impairs the body's ability to utilize glucose as the primary energy source.

3. **Competition**. Several closely related compounds may compete for a ride across the membrane on the same carrier. If a given binding site can be occupied by more than one type of molecule, the rate of transport of each substance is less when both molecules are present than when either is present by itself. To illustrate, assume the ferry has one hundred seats (binding sites) that can be occupied by either men or women. If only men are waiting to board, up to one hundred men can be transported during each trip; the same holds true if only women are waiting to board. If, however, both men and women are waiting to board, they will compete for the available seats so that fewer men and fewer women will be transported than when either group is present alone. Fifty of each might make the trip, although the total number of people transported will still be the same, one hundred people. In other words, when a carrier is able to transport two closely related substances, such as the amino acids gly-

EXERCISING MUSCLES HAVE A "SWEET TOOTH"

During exercise, muscle cells use more glucose and other nutrient fuels than usual to power their increased contractile activity. The rate of glucose transport into exercising muscle may increase more than tenfold during moderate or intense physical activity. The mechanisms responsible for increased glucose uptake by exercising muscle are still unclear. In many cells, including resting muscle cells, facilitated diffusion of glucose into the cells depends on the hormone insulin. Because plasma-insulin levels fall during exercise, however, insulin is probably not responsible for the increased transport of glucose into exercising muscles. One possibility currently being studied is that exercise increases the availability of glucose

carriers in the plasma membrane of muscle cells. This has been demonstrated in rats that have undergone physical training.

Exercise influences glucose transport into cells in yet another way. Regular aerobic exercise (see p. 34) has been shown to increase both the affinity (degree of attraction) and number of plasma-membrane receptor sites that bind specifically with insulin. This adaptation results in an increase in insulin sensitivity; that is, the cells are more responsive than normal to a given level of circulating insulin.

Because insulin enhances the facilitated diffusion of glucose into most cells, an exercise-induced increase in insulin sensitivity is one of the factors that makes exercise

a beneficial therapy for controlling diabetes mellitus, a disorder characterized by insulin deficiency (see chapter 19). As a result of inadequate insulin action, glucose entry into most cells is impaired. Plasma levels of glucose become elevated because glucose remains in the plasma instead of being transported into the cells. In the type II form of the disease, insulin is being produced, but not in sufficient quantities to meet the body's need for glucose uptake. By increasing the cells' responsiveness to the limited amount of insulin available, regular aerobic exercise helps drive glucose into the cells, where it can be used for energy production, instead of remaining in the plasma, where it leads to detrimental consequences for the body.

cine and alanine, the presence of both diminishes the rate of transfer of either.

Carrier-mediated transport takes two forms, depending on whether energy must be supplied to complete the process: facilitated diffusion (does not need energy) and active transport (needs energy). **Facilitated diffusion** uses a carrier to facilitate (assist) the transfer of a particular substance across the membrane "downhill" from high to low concentration. This process is passive and does not require energy because movement occurs naturally down a concentration gradient. **Active transport**, on the other hand, requires energy expenditure by the carrier to transfer its passenger "uphill" against a concentration gradient, from an area of lower concentration to an area of higher concentration. An analogous situation is a car on a hill. To move the car downhill requires no energy; it will coast from the top down. Driving the car uphill, however, requires the utilization of energy (gasoline).

The most notable example of *facilitated diffusion* is the transport of glucose into cells. Glucose is in higher concentration in the blood than in the tissues. Fresh supplies of this nutrient are regularly added to the blood by eating and from reserve energy stores in the body. Simultaneously, the cells metabolize glucose almost as rapidly as it enters the cells from the blood. As a result, there is a continuous gradient for net diffusion of glucose into the cells. Being a polar molecule, however, glucose cannot cross cell membranes on its own. Without the glucose carrier molecules to facilitate membrane transport of glucose, the cells would be deprived of their pre-

ferred source of fuel. (See the accompanying boxed feature, A Closer Look at Exercise Physiology.)

The carrier binding sites involved in facilitated diffusion can bind with their passenger molecules when exposed to either side of the membrane (Fig. 3–19). Because more passengers are waiting to board on the high-concentration side than on the low-concentration side, the net movement always proceeds down the concentration gradient from higher to lower concentration. As is characteristic of mediated transport, the rate of facilitated diffusion is limited by saturation of the carrier binding sites, unlike the rate of simple diffusion, which is always directly proportional to the concentration gradient (Fig. 3–20).

Active transport also involves the use of a protein carrier to transfer a specific substance across the membrane, but in this case the carrier transports the substance *against* its concentration gradient. For example, the uptake of iodine by thyroid gland cells necessitates active transport because 99% of the iodine in the body is concentrated in the thyroid. To move iodine from the blood, where its concentration is low, into the thyroid, where its concentration is high, requires expenditure of energy to drive the carrier. Specifically, energy in the form of ATP is required in active transport to vary the affinity of the binding site when exposed on opposite sides of the plasma membrane. In contrast, the affinity of the binding site in facilitated diffusion is the same when exposed to either the outside or inside of the cell.

With active transport, the binding site has a greater affinity for its passenger on the low-concentration side as a result of

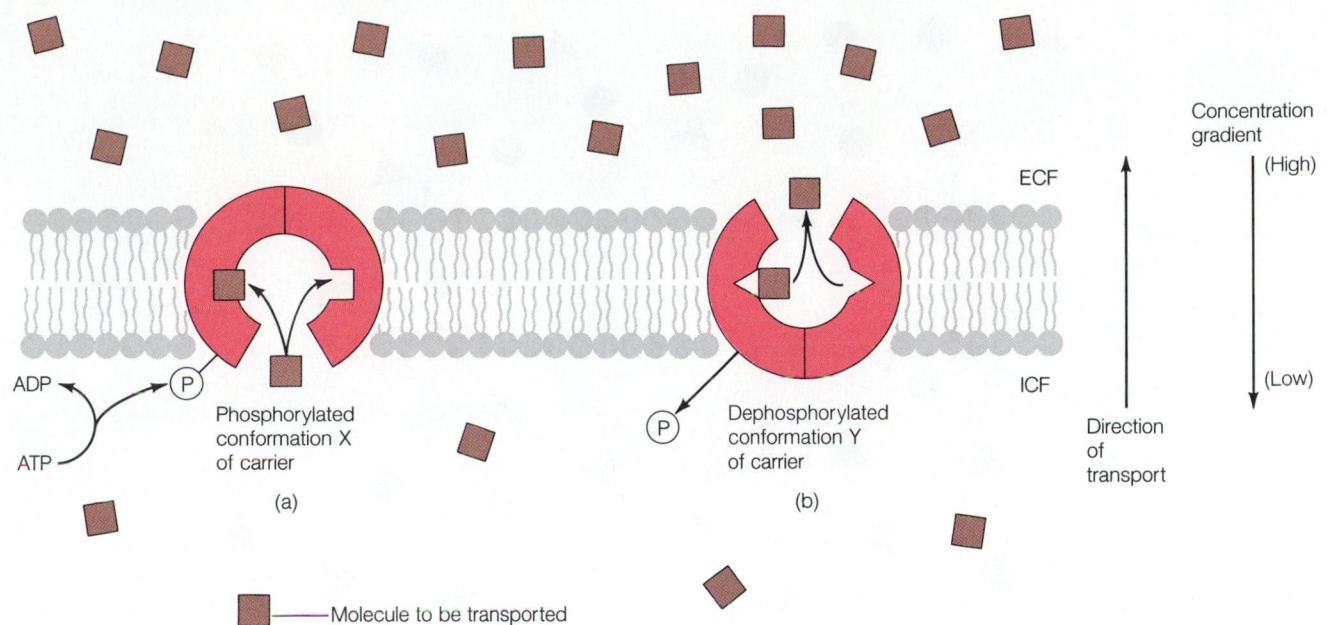

ADP
ATP
(P)
Phosphorylated
conformation X
of carrier
(a)

(P)
Dephosphorylated
conformation Y
of carrier
(b)

ECF

ICF

Concentration
gradient
(High)

(Low)

Direction
of
transport

☐——Molecule to be transported

• **FIGURE 3–21   Active Transport**   The energy of ATP is required in the phosphorylation-dephosphorylation cycle of the carrier to transport the molecule uphill from a region of low concentration to a region of high concentration. (a) On the low-concentration side, the phosphorylated conformation of the carrier has high-affinity binding sites for the molecule to be transported so the molecule binds to the carrier. (b) On the high-concentration side, the dephosphorylated conformation of the carrier has low-affinity binding sites for the molecule being transported so the molecule is released from the carrier.

phosphorylation of the carrier on this side (• Fig. 3–21a). The carrier exhibits ATPase activity in that it splits the terminal phosphate from an ATP molecule to yield ADP plus a free inorganic phosphate (see p. 27). Phosphorylation involves the binding of this phosphate group to the carrier. Phosphorylation and the binding of the passenger on the low-concentration side induce a conformational change in the carrier protein so that the passenger is now exposed to the high-concentration side of the membrane (Fig. 3–21b). The change in carrier shape is accompanied by **dephosphorylation**; that is, the phosphate group detaches from the carrier. Removal of phosphate reduces the affinity of the binding site for the passenger, so the passenger is released on the high-concentration side. The carrier then returns to its original conformation. Thus, ATP energy is used in the phosphorylation-dephosphorylation cycle of the carrier. It alters the affinity of the carrier's binding sites on opposite sides of the membrane so that transported particles are moved uphill from an area of low concentration to an area of higher concentration. These active transport mechanisms are frequently called **pumps**, analogous to water pumps that require energy to raise water up against the downward pull of gravity.

The simplest active transport systems pump a single type of passenger. An example is the **hydrogen-ion (H^+) pump** used by specialized stomach cells to transport H^+ into the stomach lumen in association with the secretion of hydrochloric acid during digestion of a meal. This pump moves H^+ against a gradient three to four million times more concentrated.

Other more complicated active transport mechanisms involve the transfer of two different passengers, either simultaneously in the same direction or sequentially in opposite directions. For example, the plasma membrane of all cells contains a sequentially active **Na^+-K^+ ATPase pump** (**Na^+-K^+ pump** for short). This carrier transports Na^+ out of the cell, concentrating it in the ECF, and picks up K^+ from the outside, concentrating it in the ICF (• Fig. 3–22). Splitting of ATP through ATPase activity and the subsequent phosphorylation of the carrier on the intracellular side increases the carrier's affinity for Na^+ and induces a change in carrier shape, leading to deposition of Na^+ to the exterior. The subsequent dephosphorylation of the carrier increases its affinity for K^+ on the extracellular side and restores the original carrier conformation, thereby transferring K^+ into the cytoplasm. There is not a direct exchange of Na^+ for K^+, however. The Na^+-K^+ pump moves three Na^+ out of the cell for every two K^+ it pumps in. (To appreciate the magnitude of active Na^+-K^+ pumping that takes place, consider that a single nerve-cell membrane contains perhaps one million Na^+-K^+ pumps capable of transporting about 200 million ions/second.) The Na^+-K^+ pump plays three important roles:

1. It establishes Na^+ and K^+ concentration gradients across the plasma membrane of all cells; these gradients are critically important in the ability of nerve and muscle cells to generate electrical impulses essential to their functioning (a topic that will soon be discussed more thoroughly).

2. It helps regulate cell volume by controlling the concentrations of solutes inside the cell and thus minimizing osmotic effects that would induce swelling or shrinking of the cell.

3. The energy used to run the Na^+-K^+ pump also indirectly serves as the energy source for the cotransport of glucose and amino acids across intestinal and kidney cells.

Phosphorylated
conformation X of
Na⁺-K⁺ pump has
high affinity for Na⁺
when exposed to ICF

Dephosphorylated
conformation Y of
Na⁺-K⁺ pump has
high affinity for K⁺
when exposed to ECF

ECF

ICF

ADP

ATP

= Sodium (Na⁺)

= Potassium (K⁺)

● **FIGURE 3–22 Na⁺-K⁺ ATPase Pump** The plasma membrane of all cells contains an active transport carrier, the Na⁺-K⁺ ATPase pump, which uses energy in the carrier's phosphyorylation/dephosphorylation cycle to sequentially transport Na⁺ out of the cell and K⁺ into the cell against these ions' concentration gradients.

Unlike most cells of the body, the intestinal and kidney cells actively transport glucose and amino acids by moving them uphill from low to high concentration. The intestinal cells transport these nutrients from inside the intestinal lumen into the blood, concentrating them in the blood until none of these molecules are left in the lumen to be lost in the feces. The kidney cells save these nutrient molecules for the body by transporting them out of the fluid that is to become urine, moving them against a concentration gradient into the blood. However, energy is not directly supplied to the carrier in these instances. The carriers that transport glucose against its concentration gradient from the lumen in the intestine and kidneys are distinct from the glucose facilitated-diffusion carriers. The luminal carriers in intestinal and kidney cells are *cotransport carriers* in that they have two binding sites, one for Na⁺ and one for the nutrient molecule. The Na⁺-K⁺ pumps in these cells are located in the basolateral membrane. Since more Na⁺ is present in the lumen than inside the cells because of the energy-requiring Na⁺-K⁺ pump that transports Na⁺ out of the cell, more Na⁺ binds to the luminal cotransport carrier when it is exposed to the outside (● Fig. 3–23). Binding of Na⁺ to the cotransport carrier increases the carrier's affinity for its other passenger (for example, glucose), so the carrier has a high affinity for glucose when exposed to the outside. When both Na⁺ and glucose are bound to the carrier, it undergoes a conformational change and opens to the inside of the cell. Both Na⁺ and glucose are released to the interior, Na⁺ because of the lower intracellular Na⁺ concentration, and glucose because of the reduced affinity of the binding site upon release of Na⁺.

The movement of Na⁺ into the cell by this cotransport carrier is downhill because the intracellular Na⁺ concentration is low, but the movement of glucose is uphill because glucose becomes concentrated in the cell. The released Na⁺ is quickly pumped out by the active Na⁺-K⁺ transport mechanism, keeping the level of intracellular Na⁺ low. The energy expended in this process is not directly utilized to run the cotransport carrier, because phosphorylation is not required to alter the affinity of the binding site to glucose. Instead, it is the establishment of a Na⁺ concentration gradient by a primary active transport mechanism (the Na⁺-K⁺ pump) that drives this secondary active transport mechanism (Na⁺-glucose cotransport carrier) to move glucose against its concentration gradient. With **primary active transport**, energy is directly required to move a substance uphill. With **secondary active transport**, energy is required in the entire process, but it is not directly required to run the pump. Rather it uses "second-hand" energy stored in the form of an **ion concentration gradient** (for example, a Na⁺ gradient) to move the cotransported molecule uphill. This is a very efficient interaction. The cotransported molecule is essentially getting a free ride, because Na⁺ must be pumped out anyway to maintain the electrical and osmotic integrity of the cell.

The glucose that is carried into the cell across the luminal border by secondary active transport subsequently moves out of the cell across the basolateral border passively down its concentration gradient and enters the blood. This movement is accomplished by facilitated diffusion, mediated by another carrier in the plasma membrane. This passive carrier is identical to the one that transports glucose into other cells, but in intestinal and kidney cells, it transports glucose out of the cell. The difference depends on the direction of the glucose concentration gradient. In the case of intestinal and kidney cells, glucose is in higher concentration inside the cells.

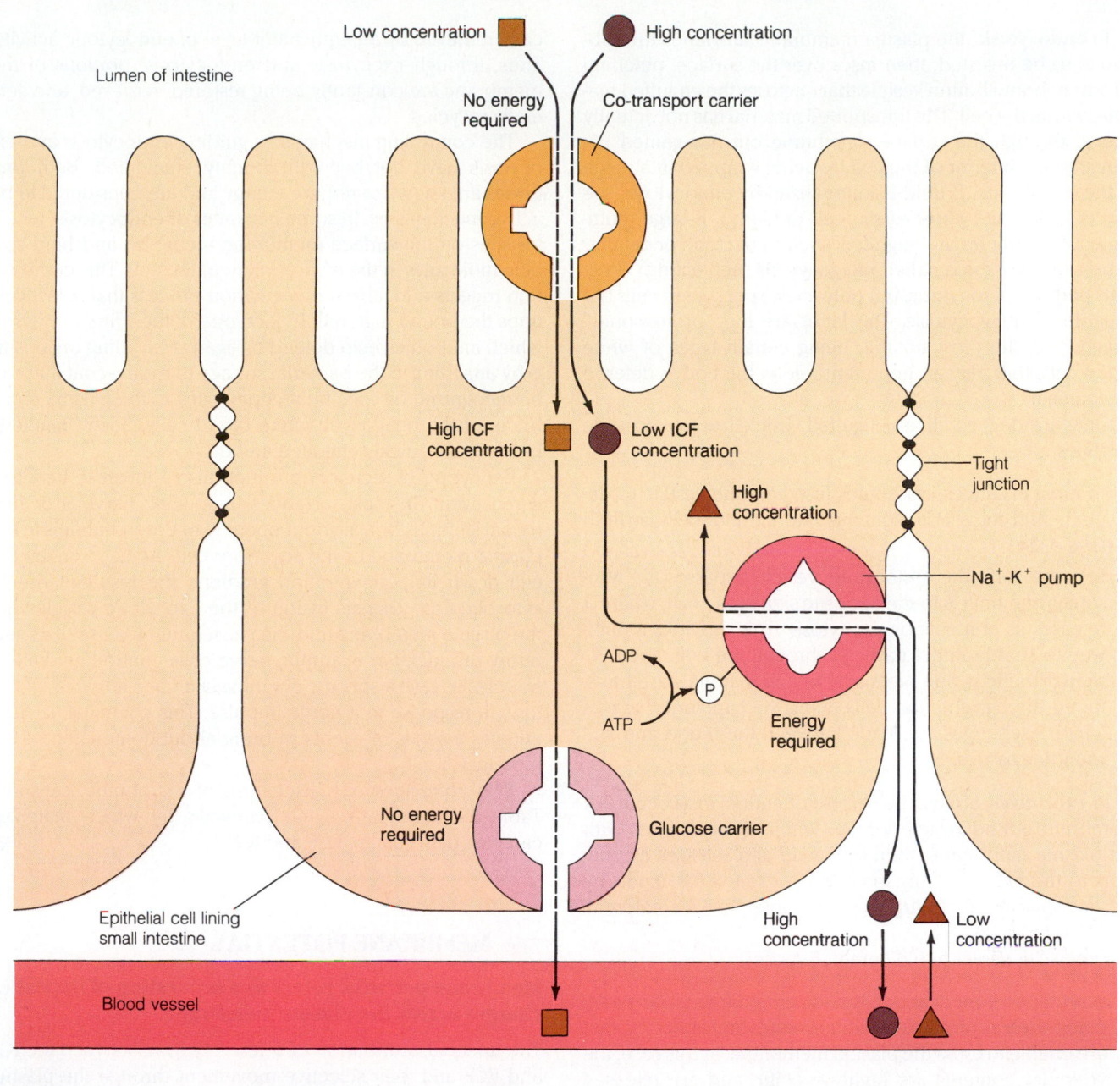

Low concentration █   High concentration ●

Lumen of intestine

No energy required   Co-transport carrier

High ICF concentration   Low ICF concentration

Tight junction

High concentration ▲

Na⁺-K⁺ pump

ADP

ATP   (P)

Energy required

No energy required   Glucose carrier

High concentration   Low concentration

Epithelial cell lining small intestine

Blood vessel

● = Sodium (Na⁺)   ▲ = Potassium (K⁺)   ■ = Glucose

● **FIGURE 3–23   Secondary Active Transport** Glucose (as well as amino acids) is transported across intestinal and kidney cells against its concentration gradient by means of secondary active transport. A cotransport carrier at the luminal border simultaneously transfers glucose against a concentration gradient and Na⁺ down a concentration gradient from the lumen into the cell. No energy is directly used by the cotransport carrier to move glucose uphill. Instead, operation of the cotransport carrier is driven by the Na⁺ concentration gradient established by the energy-using Na⁺-K⁺ pump, which actively transports Na⁺ out of the cell at the basolateral border, keeping the ICF Na⁺ concentration lower than the luminal Na⁺ concentration. After entering the cell by secondary active transport, glucose is transported down its concentration gradient from the cell into the blood by facilitated diffusion, mediated by a passive glucose carrier.

Vesicular Transport: Endocytosis/Exocytosis The special carrier-mediated transport systems embedded in the plasma membrane can selectively transport ions and small polar molecules. But what about large polar molecules or even multimolecular materials that must leave or enter the cell, such as during secretion of protein hormones by endocrine cells or during ingestion of invading bacteria by white blood cells? To be transferred between the ICF and ECF, these large particles are enclosed in a membrane-bound vesicle. Transport into the cell in this manner is termed *endocytosis,* whereas transport out of the cell is called *exocytosis* (see p. 24).

In **endocytosis**, the plasma membrane surrounds the substance to be ingested, then fuses over the surface, pinching off a membrane-bound vesicle that encloses the engulfed material within the cell. The transported material has not actually passed through the surface membrane but has gained entrance to the interior of the cell by being wrapped in a piece of the membrane. If fluid is internalized by endocytosis, the process is termed **pinocytosis** (cell drinking). If large multimolecular particles are engulfed, such as bacteria or cellular debris, the process is called **phagocytosis** (cell eating). Most cells perform pinocytosis, but only a few specialized cells are capable of phagocytosis. The latter are the "professional" phagocytes, the most notable being certain types of white blood cells that play an important role in the body's defense mechanisms.

Once inside the cell, an engulfed vesicle has two possible destinies:

1. In most instances, lysosomes fuse with the vesicle to degrade and release its contents into the intracellular fluid (see p. 24).
2. In some cells, the endocytotic vesicle bypasses the lysosomes and travels to the opposite side of the cell, where it releases its contents by exocytosis. This provides a pathway to shuttle intact particles through the cell. Such vesicular traffic is one means by which materials are transferred through the thin cells lining the capillaries, across which exchanges are made between the blood and surrounding tissues.

In **exocytosis**, almost the reverse of endocytosis occurs. A membrane-bound vesicle formed within the cell fuses with the plasma membrane, then opens up and releases its contents to the exterior. Only materials packaged for export by the endoplasmic reticulum/Golgi complex can be externalized by exocytosis.

Exocytosis serves two different purposes:

1. It provides a mechanism for secreting large polar molecules, such as proteinaceous hormones and enzymes, that are unable to cross the plasma membrane. In this case, the vesicular contents are highly specific and are released only upon receipt of appropriate signals.
2. It enables the cell to add specific components to the membrane, such as selected carriers, channels, or receptors, depending on the cell's needs. In such cases, the composition of the membrane surrounding the vesicle is important, and the contents may be merely a sampling of ICF.

Furthermore, the rates of endocytosis and exocytosis must be kept in balance to maintain a constant membrane surface area and cell volume. More than 100% of the plasma membrane may be used in an hour to wrap internalized vesicles in a cell actively involved in endocytosis, necessitating rapid replacement of surface membrane by exocytosis. On the other hand, when a secretory cell is stimulated to secrete, it may temporarily insert up to thirty times its surface membrane through exocytosis. This added membrane must be specifically retrieved by an equivalent level of endocytotic activity. Thus, through exocytosis and endocytosis, portions of the membrane are constantly being restored, retrieved, and generally recycled.

The controlling mechanisms guiding endocytosis and exocytosis have not been thoroughly elucidated. Both processes are known to require energy and are considered to be active mechanisms. In some instances of endocytosis, receptor sites on the surface membrane recognize and bind specific molecules in the environment of the cell. This combination triggers a localized invagination process that selectively traps the bound material. In fact, one of the primary ways in which antibodies help defend us against invading organisms is by attaching to the bacterial surface to form a coat that can be recognized by specific receptor sites in the plasma membrane of the phagocytic white blood cells. Such "marked" bacteria are quickly engulfed and destroyed.

Exocytosis of secretory products also appears to be a triggered event. In most instances, a specific nervous or hormonal stimulus brings about opening of Ca^{++} channels in the plasma membrane of the secretory cell. As Ca^{++} enters the cell down its concentration gradient, the resultant rise in cytosolic Ca^{++} triggers fusion of the exocytotic vesicle with the plasma membrane and the subsequent release of its secretory product. For example, nerve cells controlling skeletal muscles release a specific chemical via Ca^{++}-induced exocytosis in response to a nerve impulse. This chemical, in turn, initiates a series of events to bring about contraction of the muscle.

This completes our discussion of membrane transport. Table 3–2 summarizes the pathways by which materials can pass between the ECF and ICF.

▬ MEMBRANE POTENTIAL

Membrane potential refers to a separation of opposite charges across the plasma membrane.

The unequal distribution of a few key ions between the ICF and ECF and their selective movement through the plasma membrane are responsible for the electrical properties of the membrane. All plasma membranes have a membrane potential or are polarized electrically. **Membrane potential** refers to a separation of charges across the membrane or to a difference in the relative number of cations and anions in the ICF and ECF. Recall that opposite charges tend to attract each other and like charges tend to repel each other. Work must be performed (energy expended) to separate opposite charges after they have come together. Conversely, when oppositely charged particles have been separated, the electrical force of attraction between them can be harnessed to perform work when the charges are permitted to come together again. This is the basic principle underlying electrically powered devices. Because separated charges have the "potential" to do work, a separation of charges across the membrane is referred to as a *membrane potential*. Potential is measured in units of volts (the same unit used for the voltage in electrical

TABLE 3–2 Characteristics of the Methods of Membrane Transport

METHODS OF TRANSPORT	SUBSTANCES INVOLVED	ENERGY REQUIREMENTS AND FORCE-PRODUCING MOVEMENT	LIMIT TO TRANSPORT
Diffusion			
Through lipid bilayer	Nonpolar molecules of any size (e.g., O_2, CO_2, fatty acids)	Passive; moves down concentration gradient (from high to low concentration)	Continues until the gradient is abolished (state of equilibrium with no net diffusion)
Through protein channel	Specific small ions (e.g., Na^+, K^+, Ca^{++}, Cl^-)	Passive; moves down electrochemical gradient (from high to low concentration and attraction of ion to area of opposite charge)	Continues until there is no net movement and a state of equilibrium is established
Special case of osmosis	Water only	Passive; water moves down its own concentration gradient (water moves to area of lower water concentration, i.e., higher solute concentration)	Continues until concentration difference is abolished or until stopped by an opposing hydrostatic pressure or until cell is destroyed
Carrier-Mediated Transport			
Facilitated diffusion	Specific polar molecules for which a carrier is available (e.g., glucose)	Passive; moves down concentration gradient (from high to low concentration)	Displays a transport maximum (T_m); carrier can become saturated
Primary active transport	Specific ions or polar molecules for which carriers are available (e.g., Na^+, K^+, amino acids)	Active; moves against concentration gradient (from low to high concentration); requires ATP	Displays a transport maximum; carrier can become saturated
Secondary active transport	Specific polar molecules and ions for which carriers are available (e.g., glucose, amino acids, some ions, including Ca^{++})	Active; moves against concentration gradient (from low to high concentration); driven directly by ion gradient (usually Na^+) established by ATP-requiring primary pump	Displays a transport maximum; carrier can become saturated
Vesicular Transport			
Endocytosis			
Pinocytosis	Small volume of ECF fluid, perhaps with specific bound molecules (e.g., protein); also important in membrane recycling	ATP required	Control poorly understood
Phagocytosis	Multimolecular particles (e.g., bacteria and cellular debris)	ATP required	Necessitates binding to specific receptor site on membrane surface
Exocytosis	Secretory products (e.g., hormones and enzymes) as well as large molecules passing through cell intact; also important in membrane recycling	ATP required; increase in cytosolic Ca^{++} induces fusion of vesicle with plasma membrane	Secretion triggered by specific neural or hormonal stimuli; other controls involved in transcellular traffic and membrane recycling not known

devices), but because the membrane potential is relatively low, the unit used is **millivolts** (mV) (1 mV = 1/1,000 volt).

Since the concept of potential is fundamental to understanding nerve and muscle physiology, it is important to understand clearly what this term means. The membrane in • Figure 3–24a is electrically neutral. An equal number of positive (+) and negative (−) charges are on each side of the membrane, so no membrane potential exists. In Figure 3–24b, some of the + charges from the right side have been moved to the left. Now the left has an excess of + charges, leaving behind an excess of − charges on the right. In other words, there is a separation of charges across the membrane,

● FIGURE 3–24 **Determination of Membrane Potential by Unequal Distribution of Positive and Negative Charges across the Membrane** (a) When the positive and negative charges are equally balanced on each side of the membrane, no membrane potential exists. (b) When there is a separation of charges across the membrane, membrane potential exists. (c) The unbalanced charges responsible for the potential accumulate in a thin layer along opposite surfaces of the membrane. (d) The vast majority of the fluid in the ECF and ICF is electrically neutral. The unbalanced charges accumulate in a thin layer along the plasma membrane. (e) Membrane B has more potential than membrane A and less potential than membrane C.

or a difference in the relative number of + and − charges between the two sides (i.e., a membrane potential exists). The attractive force between these separated charges will cause them to accumulate in a thin layer along the outer and inner surfaces of the plasma membrane (Fig. 3–24c). These separated charges represent only a small fraction of the total number of charged particles (ions) present in the ICF and ECF. The vast majority of the fluid inside and outside the cells is electrically neutral (Fig. 3–24d). The electrically balanced ions can be ignored, because they do not contribute to membrane potential. Thus, an almost insignificant fraction of the total number of charged particles present in the body fluids is responsible for the membrane potential.

The magnitude of the potential depends on the degree of separation of the opposite charges: the greater the number of charges separated, the larger the potential. Therefore, in Figure 3–24e, membrane B has more potential than A and less potential than C.

Membrane potential is primarily due to differences in distribution and membrane permeability of sodium, potassium, and large intracellular anions.

All living cells have a membrane potential characterized by a slight excess of positive charges outside and a correspondingly slight excess of negative charges on the inside. In the body, electric charges are carried by ions. The ions primarily responsible for the generation of membrane potential are Na^+, K^+, and A^-. The latter refers to the large, negatively charged (anionic) intracellular proteins. Other ions (calcium, magnesium, chloride, bicarbonate, and phosphate, to name a few) do not make a direct contribution to the electrical properties of the plasma membrane in most cells, even though they play other important roles in the body.

The concentrations and relative permeabilities of the ions critical to membrane electrical activity are compared in Table 3-3. Note that *Na$^+$ is in greater concentration in the extracellular fluid and K$^+$ is in much higher concentration in the intracellular fluid.* These concentration differences are maintained by the Na^+-K^+ pump at the expense of energy. Because the plasma membrane is virtually impermeable to A^-, these large, negatively charged proteins are found *only inside* the cell. After they have been synthesized from amino acids transported into the cell, they remain trapped within the cell. In addition to the active carrier mechanism, Na^+ and K^+ can passively cross the membrane through protein channels specific for them. It is usually much easier for K^+ than Na^+ to get through the membrane because typically more K^+ channels than Na^+ channels are open. In a nerve cell at rest (i.e., when it is not conducting a nerve impulse), the membrane is about fifty to seventy-five times more permeable to K^+ than to Na^+.

Armed with a knowledge of the relative concentrations and permeabilities of these ions, we can now analyze the forces acting across the plasma membrane. This analysis will be broken down as follows: first, we will consider the direct contribution of the Na^+-K^+ pump to membrane potential; second, the effect the movement of K^+ alone would have on membrane potential; third, the effect of Na^+ alone; and finally the situation that exists in the cells when both K^+ and Na^+ effects are taking place concurrently.

Effect of the Sodium-Potassium Pump on Membrane Potential
About 20% of the membrane potential is directly generated by the Na^+-K^+ pump. This active transport mechanism pumps three Na^+ out for every two K^+ it transports in. Because Na^+ and K^+ are both positive ions, this unequal transport generates a membrane potential, with the outside becoming relatively more positive than the inside as more positive ions are transported out than in. However, most of the membrane potential—the remaining 80%—is caused by the passive diffusion of K^+ and Na^+ down concentration gradients. Thus, most of the Na^+-K^+ pump's role in producing membrane potential is indirect through its critical contribution to maintaining the concentration gradients directly responsible for the ion movements that generate most of the potential.

Effect of the Movement of Potassium Alone on Membrane Potential
Let's consider a hypothetical situation characterized by (1) the concentrations that exist for K^+ and A^- across the plasma membrane, (2) free permeability of the membrane to K^+ only, and (3) no potential as yet present. The concentration gradient for K^+ would tend to move this ion out of the cell (• Fig. 3-25). Because the membrane is permeable

• **FIGURE 3-25   Equilibrium Potential for K$^+$**   At the equilibrium potential for K^+ (E_{K^+}), the outward concentration gradient is exactly counterbalanced by the inward electrical gradient. The membrane potential at this point is -90 mV.

to K^+, this ion would readily pass through. As potassium ions moved to the outside, they would carry their positive charge with them, so more positive charges would be on the outside whereas negative charges in the form of A^- would be left behind on the inside, similar to the situation shown in Figure 3-24b. (Remember that the large protein anions cannot diffuse out, despite a tremendous concentration gradient.) A membrane potential would now exist. Because an electrical gradient also would be present, K^+, being a positively charged ion, would be attracted toward the negatively charged interior and repelled by the positively charged exterior. Thus, two opposing forces would now be acting upon K^+: the concentration gradient tending to move K^+ out of the cell and the electrical gradient tending to move these same ions into the cell.

Initially, the concentration gradient would be stronger than the electrical gradient, so net diffusion of K^+ out of the cell would continue and the membrane potential would increase. As more and more K^+ moved down its concentration

| | CONCENTRATION (MILLIMOLES/LITER) | | RELATIVE |
ION	Extracellular	Intracellular	PERMEABILITY
Na$^+$	150	15	1
K$^+$	5	150	50–75
A$^-$	0	65	0

TABLE 3-3   Concentration and Permeability of Ions Responsible for Membrane Potential in a Resting Nerve Cell

gradient and out of the cell, however, the opposing electrical gradient would also become greater as the outside became increasingly more positive and the inside more negative. Net outward diffusion would gradually be reduced as the strength of the electrical gradient approached that of the concentration gradient. Finally, when these two forces exactly balanced each other, no further net movement of K^+ would occur. The potential that would exist at this equilibrium is known as the **equilibrium potential** for **K^+ (E_{K^+})**. At this point, a large concentration gradient for K^+ would still exist, but no more K^+ would move out down this concentration gradient because of the exactly equal opposing electrical gradient (Fig. 3–25).

The membrane potential at E_{K^+} is -90 mV. It is not really a negative potential. By convention, *the sign always designates the polarity of the excess charge on the inside of the membrane.* A membrane potential of -90 mV means that the potential is of a magnitude of 90 mV, with the inside being negative relative to the outside. A potential of $+90$ mV would have the same strength, but in this case the inside would be more positive than the outside.

The equilibrium potential for a given ion of differing concentrations across a membrane can be calculated by means of the **Nernst equation** as follows:

$$E = 61 \log \frac{C_o}{C_i}$$

where

$E =$ equilibrium potential for the ion in mV
$61 =$ a constant that incorporates the universal gas constant (R), absolute temperature (T), the ion's valence (z), and an electrical constant known as Faraday (F); $61 = RT/zF$
$C_o =$ concentration of the ion outside the cell in millimoles(mM)/liter
$C_i =$ concentration of the ion inside the cell in mM/liter

Given that the ECF concentration of K^+ is 5 mM/liter and the ICF concentration is 150 mM/liter,

$$E_{K^+} = 61 \log \frac{5 \text{ mM/liter}}{150 \text{ mM/liter}}$$
$$= 61 \log \frac{1}{30}$$

Since the log of $\frac{1}{30} = -1.477$,

$$E_{K^+} = 61(-1.477) = -90 \text{ mV}$$

Since 61 is a constant, the equilibrium potential is essentially a measure of the membrane potential (that is, the magnitude of the electrical gradient) that exactly counterbalances the concentration gradient that exists for the ion (that is, the ratio between the ion's concentration outside and inside the cell). Note that the larger the concentration gradient for an ion, the greater the ion's equilibrium potential. A comparably greater opposing electrical gradient would be required to counterbalance the larger concentration gradient.

Plasma membrane

$E_{Na^+} = +60$ mV

● **FIGURE 3–26   Equilibrium Potential for Na^+** At the equilibrium potential for Na^+ (E_{Na^+}), the inward concentration gradient is exactly counterbalanced by the outward electrical gradient. The membrane potential at this point is $+60$ mV.

Effect of the Movement of Sodium Alone on Membrane Potential A similar hypothetical situation could be developed for Na^+ alone (● Fig. 3–26). The concentration gradient for Na^+ would move this ion into the cell, producing a buildup of positive charges on the interior of the membrane and leaving negative charges unbalanced outside [primarily in the form of chloride (Cl^-); Na^+ and Cl^- are the predominant ECF ions]. Net diffusion inward would continue until equilibrium was established by the development of an opposing electrical gradient that exactly counterbalanced the concentration gradient. At this point, given the concentrations for Na^+, the **Na^+ equilibrium potential (E_{Na^+})** would be $+60$ mV. In this case the inside of the cell would be positive, in contrast to the equilibrium potential for K^+. The magnitude of E_{Na^+} is somewhat less than for E_{K^+} (60 mV compared to 90 mV) because the concentration gradient for Na^+ is not as large (Table 3–3); thus, the opposing electrical gradient (membrane potential) is not as great at equilibrium.

Concurrent Potassium and Sodium Effects on Membrane Potential In the cell, the effects of both K^+ and Na^+ must be taken into account. Because the membrane at rest is fifty to seventy-five times more permeable to K^+ than to Na^+, K^+ passes through more readily than Na^+; thus, K^+ influences the resting membrane potential to a much greater extent than does Na^+. Recall that K^+ acting alone would establish an equilibrium potential of -90 mV. The membrane is somewhat permeable to Na^+, however, so some Na^+ enters the cell in a limited attempt to reach its equilibrium potential. This Na^+ influx neutralizes or cancels some of the potential produced by K^+ alone.

To facilitate an understanding of this concept, assume that each separated pair of charges in ● Figure 3–27 represents 10 mV of potential. (This is not technically correct because in

Plasma membrane

ECF

ICF

K+

K$^+$ — Relatively large net diffusion of K$^+$ outward establishes an E$_{K+}$ of −90 mV

A$^-$ — No diffusion of A$^-$ across membrane

Na$^+$ and associated Cl$^-$

Na$^+$ — Relatively small net diffusion of Na$^+$ inward neutralizes some of the potential created by K$^+$ alone

Resting membrane potential = −70 mV

(A$^-$ = large intracellular anionic proteins)

● **FIGURE 3–27 Effect of Concurrent K$^+$ and Na$^+$ Movement on Establishing the Resting Membrane Potential** Given the concentration gradients that exist across the plasma membrane, K$^+$ tends to drive the membrane potential to K$^+$'s equilibrium potential (−90 mV), whereas Na$^+$ tends to drive the membrane potential to Na$^+$'s equilibrium potential (+60 mV). However, K$^+$ exerts the dominant effect on the resting membrane potential because the membrane is more permeable to K$^+$. As a result, the resting potential (−70 mV) is much closer to E$_{K+}$ than to E$_{Na+}$. During the establishment of resting potential, the relatively large net diffusion of K$^+$ outward does not produce a potential of −90 mV because the resting membrane is slightly permeable to Na$^+$ and the relatively small net diffusion of Na$^+$ inward neutralizes some of the potential that would be created by K$^+$ alone, bringing the resting potential to −70 mV, slightly less than E$_{K+}$.

reality, many separated charges must be present to account for a potential of 10 mV.) In this simplified example, nine separated pluses and minuses, with the minuses on the inside, would represent the E$_{K+}$ of −90 mV. Superimposing the slight influence of Na$^+$ on this K$^+$-dominated membrane, assume that two sodium ions enter the cell down the Na$^+$ concentration and electrical gradients. (Because more Na$^+$ is outside the cell and the inside is negative, both gradients favor the inward movement of Na$^+$.) The inward movement of these two positively charged sodium ions neutralizes some of the potential established by K$^+$, so now only seven pairs of charges are separated, and the potential is −70 mV. This is the **resting membrane potential** of a typical nerve cell. The resting potential is much closer to E$_{K+}$ than to E$_{Na+}$ because of the greater permeability of the membrane to K$^+$, but it is slightly less than E$_{K+}$ (−70 mV is a lower potential than −90 mV) because of the weak influence of Na$^+$.

At resting potential, neither K$^+$ nor Na$^+$ is at equilibrium. A potential of −70 mV does not exactly counterbalance the concentration gradient for K$^+$; it takes a potential of −90 mV to do that. Thus, there is a continual tendency for K$^+$ to passively *leak* out through its channels. In the case of Na$^+$, the concentration and electrical gradients do not even oppose each other; they both favor the inward movement of Na$^+$. Therefore, Na$^+$ continually leaks inward down its electrochemical gradient, but only slowly because of its low permeability.

Since such leaking goes on all the time, why doesn't the intracellular concentration of K$^+$ continue to fall and the concentration of Na$^+$ inside the cell progressively increase? This does not happen because of the Na$^+$-K$^+$ pump. This active transport mechanism counterbalances the rate of leakage (● Fig. 3–28). At resting potential, the pump transports back into the cell essentially the same number of potassium ions that have leaked out and simultaneously transports to the outside the sodium ions that have leaked in. Through this balance between leak and pump, the concentration gradients for K$^+$ and Na$^+$ remain constant across the membrane. Thus, not only is the Na$^+$-K$^+$ pump initially responsible for the Na$^+$ and K$^+$ concentration differences across the membrane, but it also maintains these differences.

As just discussed, it is the presence of these concentration gradients, together with the difference in permeability of the membrane to these ions, that accounts for the resting membrane potential. In this resting state, the potential remains constant. There is no net movement of any ions. All passive forces are exactly balanced by active forces. A steady state exists, even though there is still a strong concentration gradient for both K$^+$ and Na$^+$ in opposite directions, as well as a slight excess of positive charges in the ECF accompanied by a corresponding slight excess of negative charges in the ICF (enough to account for a potential of the magnitude of 70 mV). At this point, although movement across the membrane is taking place by means of passive leaks and active

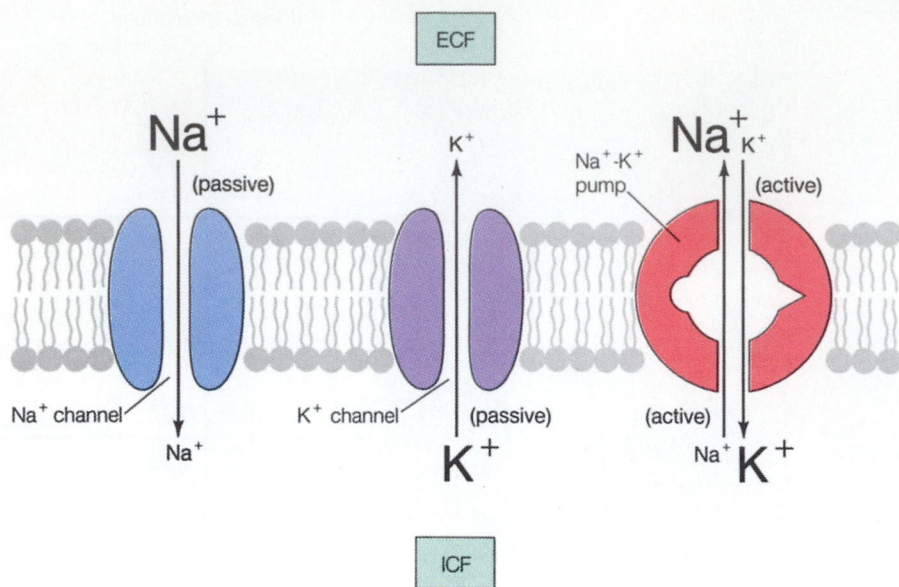

• **FIGURE 3–28   Counterbalance between Passive Na⁺ and K⁺ Leaks and the Active Na⁺-K⁺ Pump** At resting membrane potential, the passive leaks of Na⁺ and K⁺ down their electrochemical gradients are exactly counterbalanced by the active Na⁺-K⁺ pump, so that there is no net movement of Na⁺ and K⁺ and the membrane potential remains constant.

pumping, the exchange of charges between the ICF and ECF is exactly balanced, with the potential that has been established by these forces remaining constant.

Thus far, we have largely ignored one other ion present in high concentration in the ECF—Cl^-. Chloride is the principal ECF anion. Its equilibrium potential is -70 mV, exactly the same as the resting membrane potential. Movement alone of negatively charged Cl^- into the cell down its concentration gradient would produce an opposing electrical gradient, with the inside negative compared to the outside. When physiologists were first examining the ionic effects that could account for the membrane potential, it was tempting to think that Cl^- movements and establishment of the Cl^- equilibrium potential could be solely responsible for producing the identical resting membrane potential. Actually, the reverse is the case. The membrane potential is responsible for driving the distribution of Cl^- across the membrane.

Most cells are highly permeable to Cl^- but have no active transport mechanisms for Cl^-. With no active forces acting on it, Cl^- passively distributes itself to achieve an individual state of equilibrium. In this case, the established electrical gradient (the membrane potential, which is negative on the inside) drives the negative Cl^- out of the cell until an exactly counterbalanced opposing concentration gradient is established, with more Cl^- outside than inside the cells. Thus, the concentration difference for Cl^- between the ECF and ICF is passively brought about by the presence of the membrane potential, rather than being maintained by an active pump, as is the case for K⁺ and Na⁺. Therefore, in most cells Cl^- does not influence membrane potential; instead, membrane potential passively influences Cl^- distribution. (Some specialized cells have an active Cl^- pump, with subsequent movement of Cl^- accounting for part of the potential.)

Nerve and muscle cells have developed a specialized use for membrane potential. They are able to rapidly and transiently alter their membrane permeabilities to the involved ions in response to appropriate stimulation, thereby bringing about fluctuations in membrane potential. The rapid fluctuations in potential are responsible for producing nerve impulses in nerve cells and for triggering contraction in muscle cells. These activities will be the focus of the next five chapters. Even though all cells display a membrane potential, its significance in other cells is uncertain, although the potential of some secretory cells may somehow be linked to their level of secretory activity.

 CHAPTER IN PERSPECTIVE: FOCUS ON HOMEOSTASIS

All cells of the body must obtain vital materials such as nutrients and O_2 from the surrounding ECF and must transfer to the ECF wastes to be eliminated as well as secretory products such as chemical messengers and digestive enzymes. Thus, transport of materials across the plasma membrane between the ECF and ICF is essential for cell survival, and the constituents in the ECF must be homeostatically maintained in order to support these life-sustaining exchanges.

Many cell types utilize membrane transport to carry out their specialized activities geared toward maintaining homeostasis. Following are several examples:

1. Absorption of nutrients from the digestive-tract lumen involves transport of these energy-giving molecules across the membranes of the cells lining the tract.
2. Exchange of O_2 and CO_2 between the air and blood in the lungs involves the transport of these gases across the membranes of the cells lining the lungs' air sacs and blood vessels.
3. Urine formation is accomplished by the selective transfer of materials between the blood and the fluid within the

kidney tubules across the membranes of the cells lining the tubules.

4. The beating of the heart is triggered by cyclical changes in the transport of Na^+, K^+, and Ca^{++} across the heart cells' membranes.

5. Secretion of chemical messengers such as neurotransmitters from nerve cells and hormones from endocrine cells involves the transport of these regulatory products to the ECF on appropriate stimulation.

In addition to providing selective transport of materials between the ECF and ICF, the plasma membrane contains receptor sites for binding with specific chemical messengers that regulate various cell activities, many of which are specialized activities aimed toward maintaining homeostasis. For example, the hormone vasopressin, which is secreted in response to a water deficit in the body, binds with receptor sites in the plasma membrane of a specific type of kidney cell. This binding triggers these cells to conserve water during urine for-

mation, thus helping alleviate the water deficit that initiated the response.

All living cells have a membrane potential, with the cell's interior being slightly more negative than the fluid surrounding the cell when the cell is electrically at rest. The specialized activities of nerve and muscle cells depend on these cells' ability to change their membrane potential rapidly on appropriate stimulation. These transient, rapid changes in potential in nerve cells serve as electrical signals or nerve impulses, which provide a means to transmit information along nerve pathways. This information is used to accomplish homeostatic adjustments, such as restoring blood pressure to normal when signaled that it has fallen too low.

Rapid changes in membrane potential in muscle cells trigger muscle contraction, the specialized activity of muscle. Muscle contraction contributes to homeostasis in many ways, including the pumping of blood by the heart and moving food through the digestive tract.

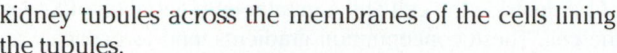

CHAPTER SUMMARY

Membrane Structure and Composition

All cells are bounded by a plasma membrane, a thin lipid bilayer in which proteins are interspersed and to which carbohydrates are attached on the outer surface. The electron microscopic appearance of the plasma membrane as a trilaminar structure (two dark lines separated by a light interspace) is believed to be caused by the arrangement of the molecules composing it. The phospholipids orient themselves to form a bilayer with a hydrophobic interior (light interspace) sandwiched between the hydrophilic outer and inner surfaces (dark lines). This lipid bilayer forms the structural boundary of the cell, serving as a barrier for water-soluble substances and being responsible for the fluid nature of the membrane. Cholesterol molecules tucked between the phospholipids contribute to the fluidity and stability of the membrane.

Membrane proteins, the type and distribution of which vary among cells, serve as channels for passage of small ions across the membrane; carriers for transport of specific substances in or out of the cell; receptor sites for detecting and responding to chemical messengers that alter cell function; membrane-bound enzymes that govern specific chemical reactions; and a support meshwork on the inner surface to help maintain cell shape in association with the cytoskeleton.

The role of membrane carbohydrates is largely unknown, but it is speculated that the sugar chains projecting from glycoproteins are important in recognition of "self" in cell-to-cell interactions such as tissue formation and immune processes.

Membrane Receptors and Postreceptor Events

Attachment of the first messenger (an extracellular chemical messenger such as a hormone) to a membrane receptor initiates one of several related intracellular pathways to bring about the desired response. Chemical messengers trigger cellular responses by two major methods: (1) opening or closing specific channels or (2) activating an intracellular messenger (the second messenger). Two commonly employed second messengers

are cyclic AMP and Ca^{++}. Once activated, these second messengers initiate a similar cascade of intracellular events that ultimately lead to a change in the shape and function of particular proteins to cause the appropriate cellular response. Despite the often widespread distribution of a single chemical messenger and the similarity of intracellular pathways employed, cells vary in their response because (1) different cell types are equipped with different sets of receptors that can bind with only selected types of messengers from among the many that might come into contact with each cell; and (2) various cell types contain different intracellular proteins, each of which responds uniquely to an identical second messenger.

Membrane Transport

Materials can pass between the ECF and ICF by the following pathways. Nonpolar (lipid-soluble) molecules of any size can dissolve in and pass through the lipid bilayer. Small ions traverse through protein channels specific for them. Particles move passively through these pathways down electrochemical gradients. Osmosis is a special case of water moving down its own concentration gradient.

Other substances can be selectively transferred across the membrane by specific carrier proteins, being moved either down a concentration gradient without the need for energy expenditure (facilitated diffusion) or moved against a concentration gradient at the expense of cellular energy (active transport). Carriers can move a single substance in one direction, two substances in opposite directions, or two substances in the same direction. Primary active transport requires the direct utilization of ATP to drive the pump, whereas secondary active transport is driven by an ion concentration gradient established by a primary active transport system. Carrier mechanisms are important for transfer of small polar molecules and for selected movement of ions.

Large polar molecules and multimolecular particles can leave or enter the cell by being wrapped in a piece of membrane to form vesicles that can be internalized (endocytosis) or externalized (exocytosis). Cells are differentially selective in what enters or leaves because they vary in the number and kind of channels, carriers, and mechanisms for vesicular transport they possess. Large polar molecules (too large for channels and not lipid-soluble) for which there are no special transport mechanisms are unable to permeate.

Membrane Potential

All cells have a membrane potential, which is a separation of opposite charges across the plasma membrane. The Na^+-K^+ pump makes a small direct contribution to membrane potential through its unequal transport of positive ions; it transports more Na^+ ions out than K^+ ions in. The primary role of the Na^+-K^+ pump, however, is to actively maintain a greater concentration of Na^+ outside the cell and a greater concentration of K^+ inside the cell. These concentration gradients tend to passively move K^+ out of the cell and Na^+ into the cell. Because the resting membrane is much more permeable to K^+ than to Na^+, substantially more K^+ leaves the cell than Na^+ enters. This results in an excess of positive charges outside the cell and leaves an unbalanced excess of negative charges inside in the form of large protein anions (A^-) that are trapped within the cell. When the resting membrane potential of -70 mV is achieved, no further net movement of K^+ and Na^+ takes place because any further leaking of these ions down their concentration gradients is quickly reversed by the Na^+-K^+ pump. The distribution of Cl^- across the membrane is passively driven by the established membrane potential so that Cl^- is concentrated in the ECF.

REVIEW EXERCISES

Objective Questions (Answers on p. D–1.)

1. The nonpolar tails of the phospholipid molecules bury themselves in the interior of the plasma membrane. (True or false?)

2. The hydrophobic regions of the molecules composing the plasma membrane correspond to the two dark layers of this structure visible under an electron microscope. (True or false?)

3. Second messenger systems ultimately bring about the desired cellular response by inducing a change in the shape and function of particular intracellular proteins. (True or false?)

4. The Na^+-K^+ pump is directly responsible for separating sufficient charges through its unequal pumping to establish a resting membrane potential of -70 mV. (True or false?)

5. The two general ways in which interaction of a chemical messenger with a membrane receptor can bring about the desired intracellular responses are _____ and _____.

6. Engulfment of a small volume of fluid by a cell is known as _____, whereas cellular ingestion of a multimolecular particle is referred to as _____. Collectively, these processes are termed _____.

7. At resting membrane potential, there is a slight excess of _____ (positive/negative) charges on the inside of the membrane, with a corresponding slight excess of _____ charges on the outside.

8. Using the answer code at the right, indicate which membrane component is responsible for the function in question:

   ___ 1. channel formation
   ___ 2. barrier to passage of water-soluble substances
   ___ 3. receptor sites
   ___ 4. membrane fluidity
   ___ 5. recognition of "self"
   ___ 6. membrane-bound enzymes
   ___ 7. structural boundary
   ___ 8. carriers

   a. lipid bilayer
   b. proteins
   c. carbohydrates

9. Using the answer code at the right, indicate the direction of net movement in each case:

   ___ 1. simple passive diffusion
   ___ 2. facilitated diffusion
   ___ 3. primary active transport
   ___ 4. Na^+ during secondary active transport
   ___ 5. cotransported molecule during secondary active transport
   ___ 6. water with regard to the water concentration gradient during osmosis
   ___ 7. water with regard to the solute concentration gradient during osmosis

   a. movement from high to low concentration
   b. movement from low to high concentration

10. Using the answer code at the right, indicate the type of cell junction described:

___ 1. adhering junction
___ 2. impermeable junction
___ 3. communicating junction
___ 4. consists of connexons, which permit passage of ions and small molecules between cells
___ 5. consists of interconnecting fibers, which spot rivet adjacent cells
___ 6. consists of an actual fusion of proteins on the outer surfaces of two interacting cells
___ 7. important in tissues subject to mechanical stretching
___ 8. important in synchronizing contractions within heart and smooth muscle by allowing spread of electrical activity between the cells composing the muscle mass
___ 9. important in preventing passage between cells in epithelial sheets that separate compartments of two different chemical compositions

a. gap junction
b. tight junction
c. desmosome

Essay Questions

1. Describe the fluid mosaic model of membrane structure.

2. What are the functions of the three major types of protein fibers in the extracellular matrix?

3. Compare the cAMP and Ca^{++} second messenger pathways.

4. What two properties of a particle influence whether it can permeate the plasma membrane?

5. List and describe the methods of membrane transport. Indicate what types of substances are transported by each method, and state whether each is a passive or active means of transport.

6. As stated by Fick's law of diffusion, what factors influence the rate of net diffusion across a membrane?

7. State three important roles of the Na^+-K^+ pump.

8. Describe the contribution of each of the following to the establishment and maintenance of membrane potential: (a) the Na^+-K^+ pump; (b) passive movement of K^+ across the membrane; (c) passive movement of Na^+ across the membrane; (d) the large intracellular anions.

POINTS TO PONDER

1. An increase in cAMP in the intestinal cells opens the Cl^- channels in the luminal membranes of these cells, thereby increasing the secretion of Cl^- from the cells into the intestinal-tract lumen. By what mechanisms would Na^+ and water be secreted into the lumen in accompaniment with Cl^- secretion? How does this secretory response account for the severe diarrhea that is characteristic of cholera? (Hint: see p. 54 for the cause of cholera.)

2. Assume that a membrane permeable to Na^+ but not to Cl^- separates two solutions. The concentration of sodium chloride on side 1 is much higher than on side 2. Which of the following ionic movements would occur?

   a. Na^+ would move until its concentration gradient is dissipated (i.e., until the concentration of Na^+ on side 2 is the same as the concentration of Na^+ on side 1).
   b. Cl^- would move down its concentration gradient from side 1 to side 2.
   c. A membrane potential, negative on side 1, would develop.
   d. A membrane potential, positive on side 1, would develop.
   e. None of the above are correct.

3. Compared to resting potential, would the membrane potential become more negative or more positive if the membrane were more permeable to Na^+ than to K^+?

4. Which of the following methods of transport is being utilized to transfer the substance into the cell in the accompanying graph?

   a. diffusion down a concentration gradient
   b. osmosis
   c. facilitated diffusion
   d. active transport
   e. vesicular transport
   f. It is impossible to tell with the information provided.

Concentration of substance in the extracellular fluid (moles $\times 10^4$)

5. Calculate the equilibrium potential for Na^+ using the Nernst equation. (Hint: the ECF and ICF concentrations for Na^+ are provided in Table 3–3 on p. 71.)

4 Neuronal Physiology

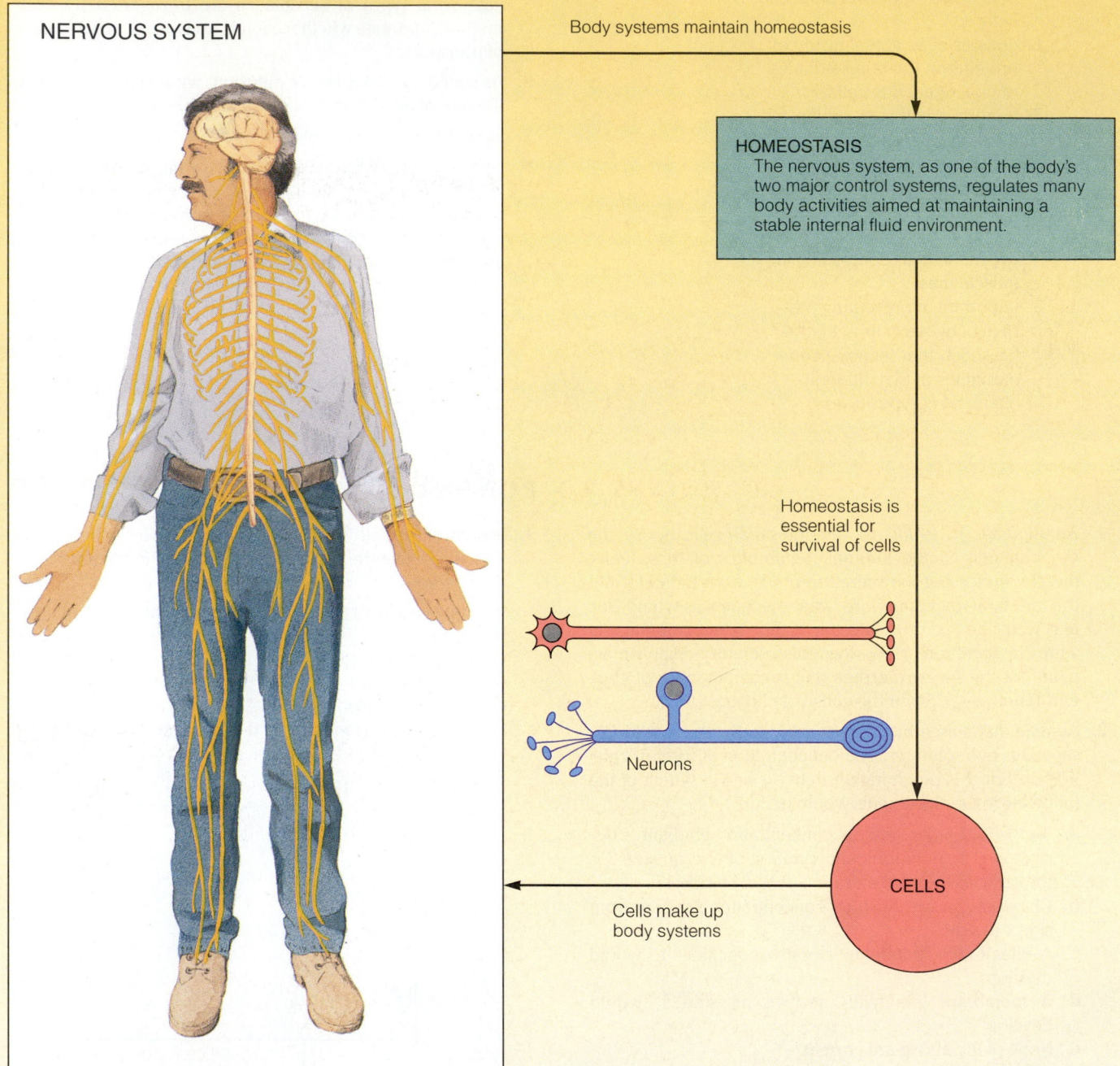

NERVOUS SYSTEM

Body systems maintain homeostasis

HOMEOSTASIS
The nervous system, as one of the body's two major control systems, regulates many body activities aimed at maintaining a stable internal fluid environment.

Homeostasis is essential for survival of cells

Neurons

CELLS

Cells make up body systems

Nerve cells, or **neurons**, make up the nervous system, one of the two major control systems of the body. The nervous system exerts control over much of the body's muscular and glandular activities, most of which are directed toward maintaining homeostasis.

Neurons are specialized for rapid electrical and chemical signaling. They are able to process, initiate, code, and conduct changes in their membrane potential as a means of rapidly transmitting a message throughout their length. Furthermore, neurons have developed chemical means of passing this information through intricate nerve pathways from neuron to neuron as well as to muscles and glands.

ELECTRICAL SIGNALS: GRADED POTENTIALS AND
  ACTION POTENTIALS
Graded potentials—short-distance signals
Action potentials—long-distance signals
  Permeability changes; ionic fluxes
  Potential changes
  Propagation of action potentials; role of myelin
  Refractory period
  All-or-none law
SYNAPSES AND NEURONAL INTEGRATION
Chemical transmission of electrical signal
Excitatory and inhibitory synapses
Neurotransmitters
Grand postsynaptic potential; summation
● A CLOSER LOOK AT EXERCISE PHYSIOLOGY The Yells and
  Grunts of an Athlete May Serve a Physiological Purpose
Action potential initiation at axon hillock
Modification of synaptic effectiveness
● CONCEPTS, CHALLENGES, AND CONTROVERSIES Parkinson's
  Disease, Pollution, and Ethical Problems
Convergence; divergence

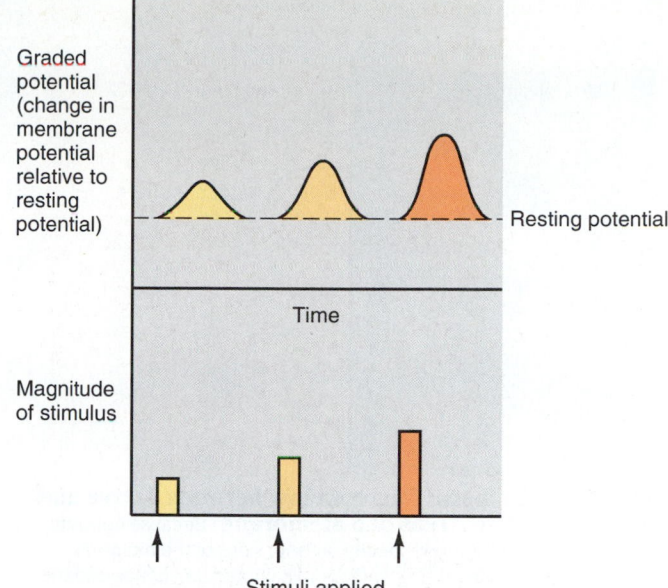

● **FIGURE 4–1 Graded Potentials** The greater the magnitude of a triggering event such as a stimulus, the larger the graded potential.

ELECTRICAL SIGNALS: GRADED POTENTIALS AND ACTION POTENTIALS

Nerve and muscle are excitable tissues.

All cells of the body possess a membrane potential related to the nonuniform distribution of and differential permeability to Na^+, K^+, and large intracellular anions. Two types of cells, *nerve cells* and *muscle cells,* have developed a specialized use for this membrane potential. Specifically, these cells are able to undergo transient, rapid changes in their membrane potentials. These fluctuations in potential, which serve as electrical signals, take two basic forms: (1) *graded potentials,* which serve as short-distance signals, and (2) *action potentials,* which signal over long distances.

Nerve and muscle are considered to be **excitable tissues** because they are capable of producing electrical signals when excited. The constant membrane potential that exists when an excitable tissue cell is not displaying rapid changes in potential is referred to as the *resting membrane potential* (although the membrane is far from "resting" because of the balanced leak-pump activity constantly going on). Let us examine these types of signals in more detail.

Graded potentials die out over short distances.

Graded potentials are local changes in membrane potential that occur in varying grades or degrees of magnitude or strength. For example, membrane potential could change from $-70mV$ to $-60mV$ (a 10 mV graded potential change) or from $-70mV$ to $-50mV$ (a 20 mV graded potential change). The magnitude of a graded potential is related to the magnitude of the triggering event that brings about the potential change; that is, *the stronger the triggering event, the larger*

the graded potential (● Fig. 4–1). Depending on the location or function of the graded potential, a triggering event might be (1) a stimulus, such as light stimulating specialized nerve cells in the eye; (2) an interaction of a chemical messenger with a surface receptor on a nerve- or muscle-cell membrane; or (3) a spontaneous change of potential caused by imbalances in the leak-pump cycle.

When a graded potential occurs locally in a nerve- or muscle-cell membrane, a different potential exists in this area than in the remainder of the membrane, which is still at resting potential. Because opposite charges attract each other, current (movement of charges) passively flows between the involved area and the adjacent resting regions on both the inside and outside of the membrane. For example, assume that a triggering event has temporarily reversed the charges in a particular region of an excitable tissue membrane, a region now called an *active area* (● Fig. 4–2). Current will flow on both sides of the membrane between the active and neighboring *inactive* (still at resting potential) *areas.* By convention, the direction of current flow is always designated by the movement of the positive charges, but keep in mind that negative charges are moving simultaneously in the opposite direction.

The charges in the active area do not actually have to reverse for local current flow to occur. A reduction in potential in the active area compared with that in the remainder of the membrane (the usual case with graded potentials) will also initiate current flow between the active and neighboring inactive areas. Current flow is easier to visualize, however, if we assume a reversal of charges. This local flow of current alters the potential in the previously inactive area, so this area's potential now differs from that of the region immediately next to it on the other side, inducing further current flow at this next site, and so on.

Plasma membrane

Extracellular fluid (ECF) — | + Intracellular fluid (ICF)

Active area — | +

Local current flow

Inactive area (at resting potential) + | —

• FIGURE 4–2 Local Current Flow between Active and Adjacent Inactive Areas of a Membrane Because opposite charges attract, current flows locally on both sides of the membrane between the active area that is undergoing a potential change and the adjacent inactive area that is still at resting potential.

This passive current flow is similar to the means by which current is carried through electrical wires. We know from experience that current leaks out of an electrical wire unless the wire is covered with an insulating material such as rubber. The regions on excitable tissues at which graded potentials take place are not wrapped in insulating material. Because current from the activated region of the membrane leaks into the surrounding extracellular fluid (ECF), the magnitude of the graded potential continues to decrease the farther it moves away from the initial active area; that is, the spread of a graded potential is *decremental* (• Fig. 4–3). In fact, these local currents die out within a few millimeters from the initial site of potential change and consequently can function as signals only for very short distances. This situation is similar to two people speaking directly to each other. Over short dis-

tances, the voices are clearly understood, but as the two people move farther apart, the sounds progressively diminish until further communication by this means becomes impossible. Just as there are methods, such as telephones and telegraphs, for long-distance communication between people, action potentials provide a means of transmitting an electrical signal long distances through the body. The limited signaling distance of graded potentials does not mean that they are of no value, however. The following graded potentials are critically important to the body: (1) postsynaptic potentials, (2) receptor potentials, (3) end-plate potentials, and (4) pacemaker potentials. These terms are unfamiliar to you now, but you will become well acquainted with them as we continue discussing nerve and muscle in the next several chapters.

Action potentials are brief reversals of membrane potential brought about by rapid changes in membrane permeability.

Because the passive current flow accompanying a graded potential fades very quickly as it moves away from its site of initiation, there must be another mechanism by which an electrical signal can be transmitted over long distances, with the strength of the signal being maintained as it travels away from its site of initiation. When appropriately triggered, nerve and muscle cell membranes undergo brief, rapid reversals of membrane potential; these reversals, known as **action potentials**, are able to spread throughout the membrane in nondecremental fashion. To understand the processes that occur during an action potential, it is necessary to be familiar with the following terms (• Fig. 4–4):

1. **Polarization**. The membrane has potential; there is a separation of opposite charges.

2. **Depolarization**. The membrane potential is reduced from resting potential; it has decreased or moved toward 0 mV; fewer charges are separated than at resting potential.

3. **Hyperpolarization**. The potential is greater than resting potential; it has increased or become even more negative; more charges are separated than at resting potential.

4. **Repolarization**. The membrane returns to resting potential after having been depolarized.

• FIGURE 4–3 Decremental Spread of Graded Potentials Due to leakage of current, the magnitude of a graded potential continues to decrease as it passively spreads from the initial active area. The potential dies out altogether within a few mm of its site of initiation.

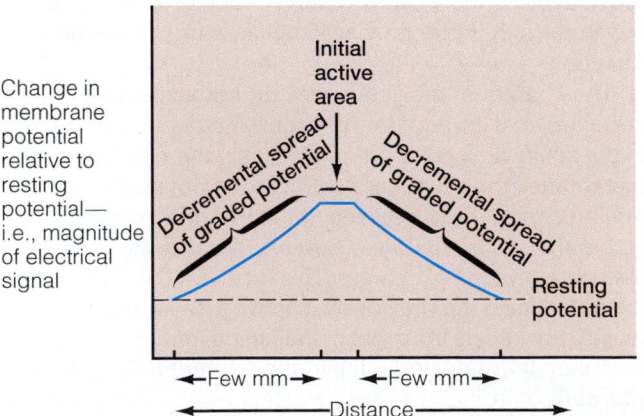

Change in membrane potential relative to resting potential—i.e., magnitude of electrical signal

Initial active area

Decremental spread of graded potential

Decremental spread of graded potential

Resting potential

←Few mm→ ←Few mm→

←————Distance————→

• FIGURE 4–4 Types of Changes in Membrane Potential

Upward deflection = Decrease in potential
Downward deflection = Increase in potential

Membrane potential (mV)

+20
+10
0
−10
−20
−30
−40
−50
−60
−70
−80
−90

Repolarization

Hyperpolarization

Depolarization

Resting potential

Time (msec)

FIGURE 4–5 Changes in Membrane Potential during an Action Potential

One possibly confusing point should be clarified. On the device used for recording rapid changes in potential, a *decrease* in potential is represented as an *upward* deflection whereas an *increase* in potential is represented as a *downward* deflection.

With these terms in mind, we will consider the changes that occur in the membrane potential during an action potential (● Fig. 4–5). Recall that the resting potential of a typical nerve cell is −70 mV. To initiate an action potential, a triggering event causes the membrane to depolarize. Depolarization proceeds slowly at first until it reaches a critical level known as **threshold potential**, typically between −50 and −55 mV. At threshold potential, an explosive depolarization takes place. A recording of the potential at this time shows a sharp upward deflection to +30 mV as the potential rapidly decreases toward 0 mV, then reverses itself so that the inside of the cell becomes positive compared to the outside. Just as rapidly, the potential drops back to resting potential as the membrane repolarizes. Often the forces responsible for driving the membrane back to resting potential push it too far,

causing a transient hyperpolarization (the **after hyperpolarization**), during which the inside of the membrane becomes even more negative than normal (for example −80 mV). The entire rapid change in potential from threshold to peak reversal and then back to resting is called the *action potential.* In a nerve cell, an action potential lasts for only 1 msec (0.001 sec). It lasts longer in muscle, with the duration varying depending on the muscle type. Often an action potential is referred to as a **spike** because of its spikelike recorded appearance. Alternatively, when an excitable membrane is triggered to undergo an action potential, it is said to **fire**. Thus, the terms *action potential, spike,* and *firing* all refer to the same phenomenon of rapid potential reversal.

How is the membrane potential, which is usually maintained at a constant resting level by the counterbalancing leak and pump activities described in the last chapter, thrown out of balance to such an extent as to produce an action potential? Recall that K^+ makes the greatest contribution to the establishment of the resting potential because the membrane at rest is considerably more permeable to K^+ than to Na^+. During an action potential, marked changes in membrane permeability to Na^+ and K^+ take place, permitting rapid fluxes of these ions down their electrochemical gradients. These ion movements carry the current responsible for the potential changes that occur during an action potential.

The membrane channels behave as if they have "gates" that can be open or closed, depending on the circumstances. The channels are formed by proteins that span the thickness of the membrane. Changes in conformation (shape) of these proteins are believed to alternatively block the channel or permit passage through it. Channels apparently are able to exist in at least three different conformations (● Fig. 4–6): (1) gates closed but capable of opening; (2) gates open (activated); and (3) gates closed and not capable of opening (inactivated).

There are three kinds of channels, depending on the factor that induces the change in channel conformation: (1) **voltage-gated channels**, which open or close in response to changes in membrane potential; (2) **chemical messenger–gated channels**, which change conformation in response to the binding of a specific chemical messenger with a membrane receptor that is in close association with the channel; and (3) **mechanically gated channels**, which

Extracellular fluid (ECF)

Plasma membrane

Closed but capable of opening

Open (activated)

Closed and not capable of opening (inactivated)

Intracellular fluid (ICF)

● **FIGURE 4–6 Conformations of Voltage-Gated Na⁺ Channel**

respond to stretching or other mechanical deformation. Voltage-gated channels are the ones involved in action potentials. (The roles of the other types of channels will be described later.)

Because proteins that compose the volgage-gated channels contain a number of charged groups, the electric field (potential) surrounding them can exert a distorting force on the channel structure; that is, various charged portions of the channel protein are electrically attracted or repelled by an unequal distribution of cations and anions in the fluids surrounding the membrane. Unlike the majority of membrane proteins, which remain stable in spite of fluctuations in membrane potential, the voltage-gated channel proteins are especially sensitive to voltage changes. Small distortions in shape induced by potential changes can cause them to flip to another conformation.

At resting potential (-70 mV), many K^+ channels are open but most of the Na^+ channels are closed; thus, the resting membrane is fifty to seventy-five times more permeable to K^+. When a membrane starts to depolarize toward threshold as a result of a triggering event, some of its voltage-gated Na^+ channels open. Since both the concentration and electrical gradients for Na^+ favor its movement into the cell, Na^+ starts to move in, carrying its positive charge with it. This depolarizes the membrane further, thereby opening more voltage-gated Na^+ channels and allowing more Na^+ to enter. As a result, still further depolarization occurs, opening more Na^+ channels, and so on, in a positive-feedback cycle (● Fig. 4–7).

At threshold potential, there is an explosive increase in Na^+ permeability (**P Na^+**) as the membrane becomes six hundred times more permeable to Na^+ than to K^+. Each individual channel is either closed or open and cannot be

partially open. However, the delicately poised gating mechanisms of the various Na^+ channels are jolted open by slightly different voltage changes. During the early depolarizing phase, Na^+ channels are gradually opened as the potential progressively decreases. By threshold the gates of all the Na^+ channels have swung open so that Na^+ permeability now dominates the membrane, in contrast to the K^+ domination of the resting potential. Thus, at threshold Na^+ rushes into the cell, rapidly eliminating the internal negativity and even making the inside of the cell more positive than the outside. The potential reaches $+30$ mV, close to the Na^+ equilibrium potential. The potential does not become any more positive because, at the peak of the action potential, the Na^+ channels close to the inactivated state, and P Na^+ falls to its low resting value. What causes the Na^+ channels to close? When the membrane potential reaches threshold, two simultaneous events are believed to take place in the gates of each Na^+ channel. First, the outer gates adjacent to the ECF are triggered to *open rapidly,* converting the channel to its open (activated) conformation (Fig. 4–6). Simultaneously, the same potential charge triggers the inner gates adjacent to the intracellular fluid (ICF) to *close slowly.* Consequently, after the channel opens, there is a time delay before it is converted to its inactivated conformation. Meanwhile, the channel has remained open for about 0.5 msec, and Na^+ has rushed into the cell, bringing the action potential to its peak of $+30$ mV. Then the inner gates slam closed and remain shut in an inactivated state until the membrane potential has been restored to its negative resting value.

Simultaneous with inactivation of the Na^+ channels, K^+ permeability (**P K^+**) greatly increases to about three hundred times the resting P Na^+. This opening of even more K^+ channels is also a delayed voltage-gated response triggered by the initial depolarization to threshold. The marked increase in P K^+ causes K^+ to rush out of the cell down its concentration and electrical gradients, carrying positive charges back to the outside. Note that at the peak of the action potential, the internal positivity of the cell tends to repel the positive K^+ ions, so the electrical gradient for K^+ is outward, unlike at resting potential. The outward movement of K^+ rapidly restores the internal negativity and returns the potential to resting.

To review (● Fig. 4–8), *the rising phase of the action potential (depolarization) is due to Na^+ influx* (Na^+ entering the cell) induced by an explosive increase in P Na^+ at threshold. *The falling phase (repolarization) is brought about by K^+ efflux* (K^+ leaving the cell) caused by the marked increase in P K^+ occurring simultaneously with the inactivation of the Na^+ channels at the peak of the action potential. As the potential returns to resting, the changing voltage shifts the Na^+ channels to their "closed but able to be opened" conformation. The newly opened K^+ channels also close, so the membrane returns to the resting number of open K^+ channels. Sometimes more K^+ leaves than is necessary to bring the potential to resting because the K^+ channels do not close quickly enough. This slight excessive K^+ efflux makes the interior of the cell transiently even more negative than resting potential, causing the after hyperpolarization.

● **FIGURE 4–7   Positive-Feedback Cycle Responsible for Opening Sodium Channels at Threshold**

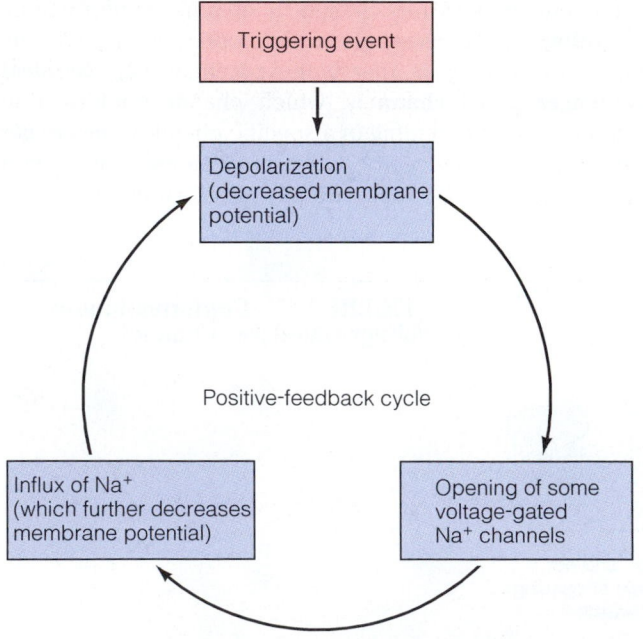

Triggering event

Depolarization (decreased membrane potential)

Positive-feedback cycle

Influx of Na^+ (which further decreases membrane potential)

Opening of some voltage-gated Na^+ channels

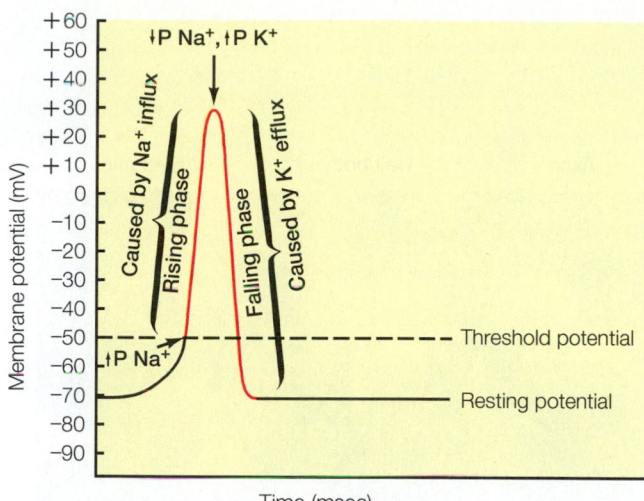

● **FIGURE 4–8   Permeability Changes and Ionic Fluxes during an Action Potential**

The Na⁺-K⁺ pump gradually restores the concentration gradients disrupted by action potentials.

At the completion of an action potential, the membrane potential has been restored to its resting condition, but the ion distribution has been altered slightly. Sodium has entered the cell during the rising phase, and a comparable amount of K^+ has left during the falling phase. It is the task of the Na^+-K^+ pump to restore these ions to their original locations in the long run, but not after each action potential.

The active pumping process takes much longer to restore Na^+ and K^+ to their original locations than it takes for the passive fluxes of these ions during an action potential. However, the membrane does not need to wait until the Na^+-K^+ pump slowly restores the concentration gradients before it can undergo another action potential. Actually, the movement of only relatively few of the total number of Na^+ and K^+ ions present is responsible for the dramatic swings in potential that occur during an action potential. Only about 1 out of 100,000 K^+ ions present in the cell leaves during an action potential, while a comparable number of Na^+ ions enter from the ECF. The shifting locations of these extremely small percentages of the total Na^+ and K^+ during a single action potential produce only infinitesimal changes in their ICF and ECF concentrations. There is still much more K^+ inside the cell than outside, and Na^+ is still predominantly an extracellular cation. Consequently, the Na^+ and K^+ concentration gradients still exist, so repeated action potentials can occur without the necessity of the pump keeping pace to restore the gradients.

Of course, were it not for the pump, even tiny fluxes accompanying repeated action potentials would eventually "run down" the concentration gradients so that further action potentials would be impossible. If the concentrations of Na^+ and K^+ were equal between the ECF and ICF, changes in permeability to these ions would not bring about ionic fluxes, so no change in potential would occur. Thus, the Na^+-K^+ pump

is critical in the long run to maintaining the concentration gradients. However, it does not have to perform its role between every action potential, nor is it directly involved in the ion fluxes or potential changes that occur during an action potential.

Once initiated, action potentials are propagated throughout an excitable cell.

A single action potential involves only a small patch of the total surface membrane of an excitable cell. If action potentials are to serve as long-distance signals, obviously they cannot be merely isolated events occurring in a limited area of a nerve or muscle cell membrane. Mechanisms must exist to conduct or spread the action potential throughout the entire cell membrane. Furthermore, the signal must be transmitted from one cell to the next cell (for example, along specific nerve pathways). Let us first examine how an action potential (nerve impulse) is conducted throughout a nerve cell before turning our attention to how the impulse is passed to another cell.

A single nerve cell or **neuron** typically consists of three basic parts: the cell body, the dendrites, and the axon, although there are variations in structure, depending on the location and function of the neuron. (The distinctions of other specialized neurons will be described later.) The nucleus and organelles are housed in the **cell body** (● Fig. 4–9), from which numerous extensions known as **dendrites** typically project like antennae to increase the surface area available for receiving signals from other nerve cells. Dendrites carry signals *toward* the cell body. In most neurons the plasma membrane of the cell body and dendrites contains protein receptors for binding chemical messengers from other neurons. The **axon** or **nerve fiber** is a single, elongated tubular extension that conducts action potentials *away from* the cell body and eventually terminates at other cells. The axon frequently gives off side branches or **collaterals** along its course. The first portion of the axon plus the region of the cell body from which the axon leaves is known as the **axon hillock**. It is the site where action potentials are initiated in a neuron (with the exception of neurons specialized to carry sensory information, a topic described in a later chapter). The impulses are then propagated along the axon to its typically highly branched ending at the **axon terminals**. These terminals release chemical messengers that simultaneously influence numerous other cells with which they come into close association.

Axons vary in length from less than a millimeter in neurons that communicate only with neighboring cells to longer than a meter in neurons that communicate with distant parts of the nervous system or with peripheral organs. For example, the axon of the nerve cell innervating your big toe must traverse the distance between the origin of its cell body within the spinal cord in the lower region of your back all the way down your leg to your toe.

Once an action potential is initiated at the axon hillock, no further triggering event is necessary to activate the remainder

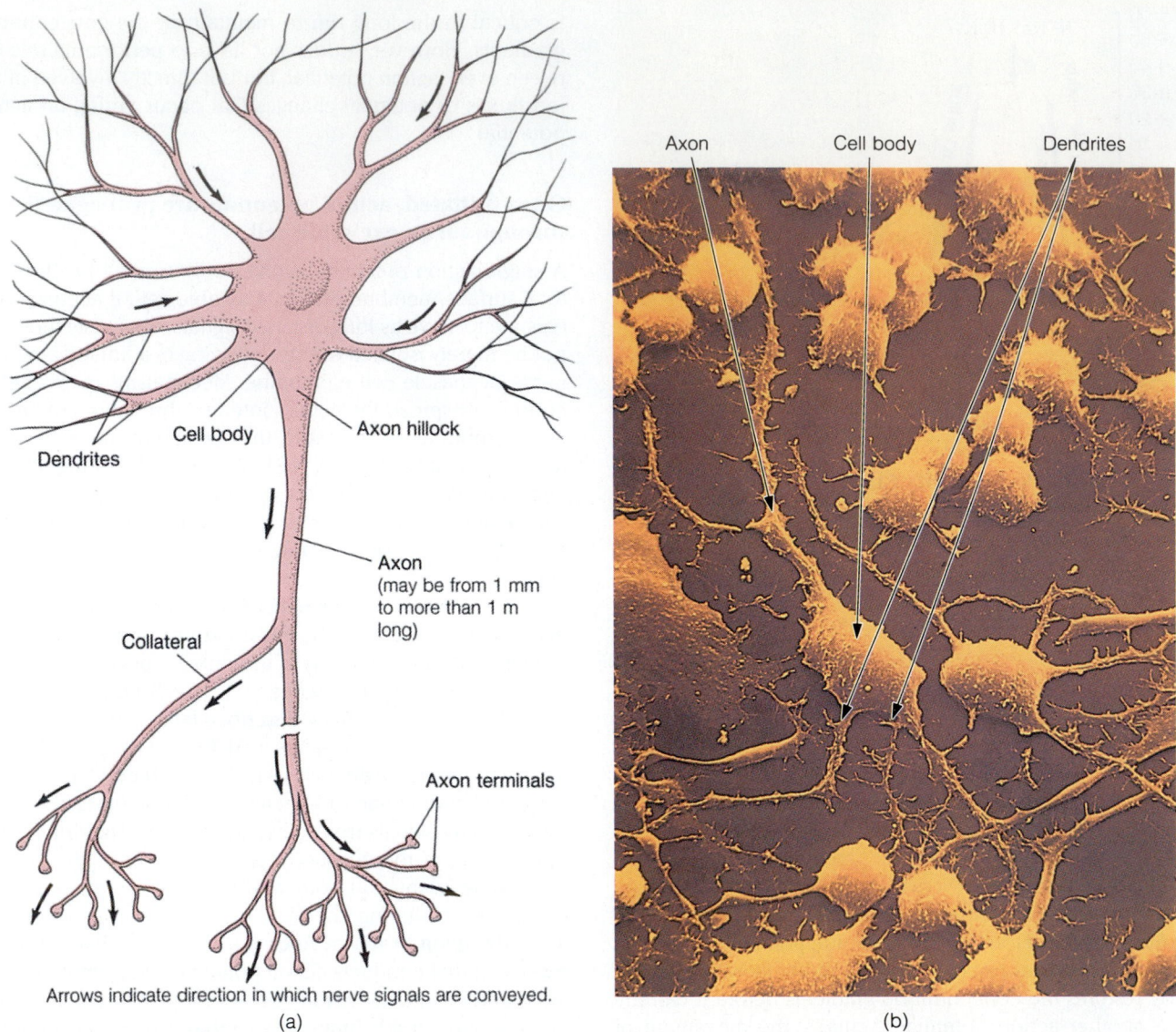

Axon   Cell body   Dendrites

Dendrites

Cell body

Axon hillock

Axon
(may be from 1 mm
to more than 1 m
long)

Collateral

Axon terminals

Arrows indicate direction in which nerve signals are conveyed.

(a)   (b)

● **FIGURE 4–9** **Anatomy of a Neuron (Nerve Cell)** (a) Most but not all neurons consist of the basic parts schematically represented in the figure. (b) An electron micrograph of the cell body, dendrites, and part of the axon of a neuron within the central nervous system.

of the nerve fiber. The impulse is automatically conducted throughout the neuron without further stimulation by one of two methods of propagation: *conduction by local current flow* or *saltatory conduction.*

● Figure 4–10 illustrates **conduction by local current flow**. You are viewing a schematic representation of a longitudinal section of the axon hillock and the portion of the axon immediately beyond it. The axon hillock is at the peak of an action potential. The inside of the cell is positive in this active area because Na^+ has already entered the nerve cell at this point. The remainder of the axon, still at resting potential and negative inside, is considered to be inactive. For the action potential to spread from the active to the inactive areas, the inactive areas must somehow be depolarized to threshold before they can undergo an action potential. This depolarization is accomplished by local current flow between the area already undergoing an action potential (positive inside, nega-

tive outside) and the adjacent inactive area (negative inside, positive outside), similar to the current flow responsible for the spread of graded potentials. Because opposite charges attract, current is able to flow locally between the active area and the neighboring inactive area on both the inside and the outside of the membrane. This local current flow in effect neutralizes or eliminates some of the unbalanced charges in the inactive area; that is, it reduces the number of opposite charges separated across the membrane or reduces the potential in this area. This depolarizing effect quickly brings the involved inactive area to threshold, at which time the voltage-gated Na^+ channels in this region of the membrane are all thrown open, leading to an action potential in this previously inactive area.

Meanwhile, the original active area returns to resting potential as a result of K^+ efflux. (Local current flow does *not* make a significant contribution to returning the original ac-

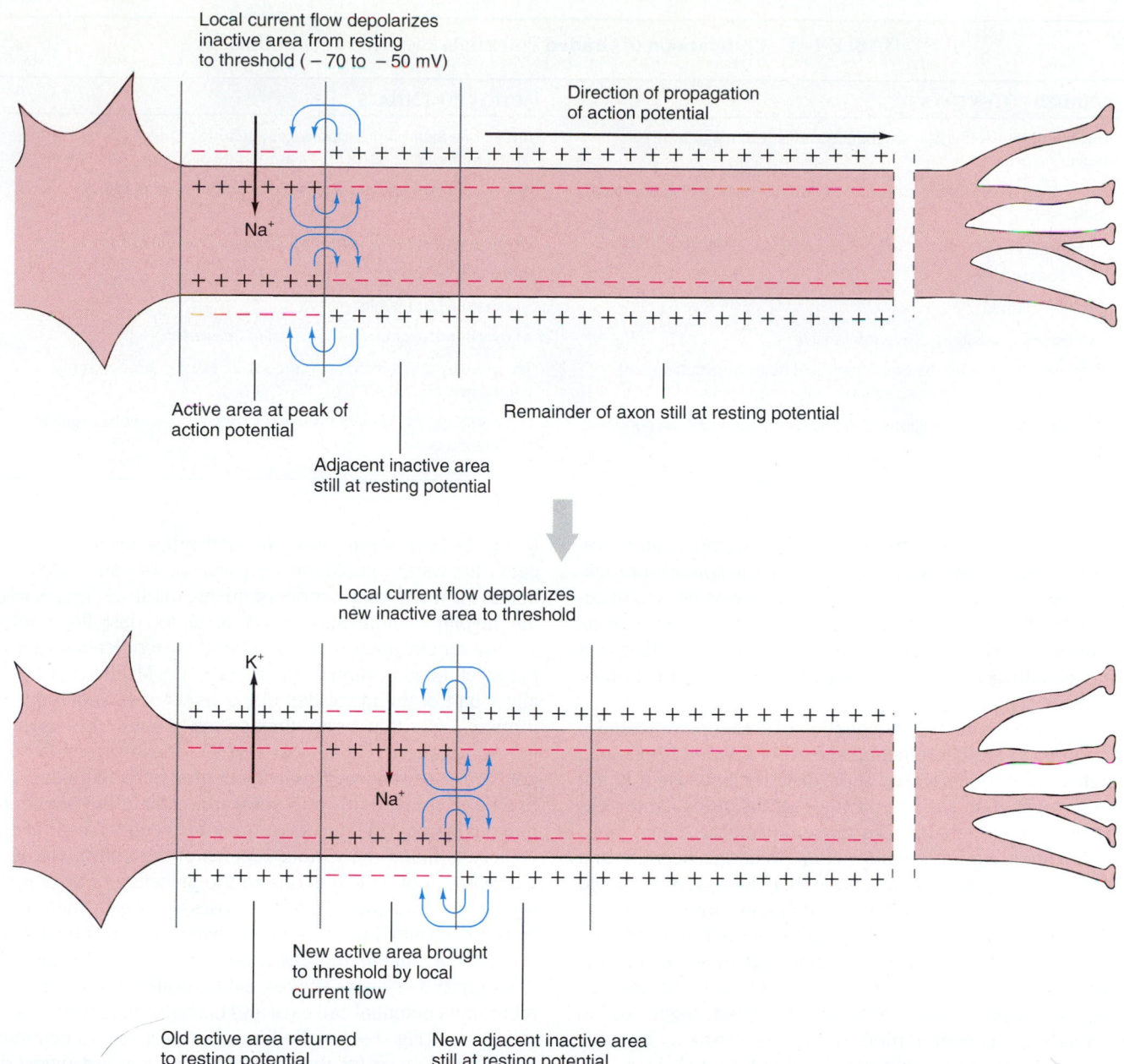

Local current flow depolarizes
inactive area from resting
to threshold (−70 to −50 mV)

Direction of propagation
of action potential

Na⁺

Active area at peak of
action potential

Adjacent inactive area
still at resting potential

Remainder of axon still at resting potential

Local current flow depolarizes
new inactive area to threshold

K⁺

Na⁺

New active area brought
to threshold by local
current flow

Old active area returned
to resting potential

New adjacent inactive area
still at resting potential

● **FIGURE 4–10** **Conduction by Local Current Flow** Local current flow between the active area at the peak of an action potential and the adjacent inactive area still at resting potential reduces the potential in the inactive area to threshold, which triggers an action potential in the previously inactive area. The original active area returns to resting potential, and the new active area induces an action potential in the next adjacent inactive area by local current flow as the cycle repeats itself down the length of the axon.

tive area to resting. Repolarization is accomplished primarily by a concurrent drop in P Na⁺ and rise in P K⁺, as described earlier.) In turn, beyond the new active area is another inactive area, so the same thing happens again. Local current flow brings this next inactive area to threshold, causing it to fire and become a new active area. This cycle repeats itself until the action potential has spread to the end of the axon. *Once an action potential is initiated in one part of a nerve cell membrane, a self-perpetuating cycle is initiated so that the action potential is propagated throughout the rest of the fiber automatically.* In this way, the axon is similar to a firecracker fuse

that needs to be lit at only one end. Once ignited, the fire spreads down the fuse; it is not necessary to hold a match to every separate section of the fuse.

Note that the original action potential does not travel along the membrane. Instead, it triggers an identical new action potential in the adjacent area of the membrane, with this process being repeated along the axon's length. Since each new action potential in the conduction process is a fresh event dependent on the induced permeability changes and electrochemical gradients, which are virtually identical down the length of the axon, the last action potential at the end

TABLE 4–1 Comparison of Graded Potentials and Action Potentials

GRADED POTENTIALS	ACTION POTENTIALS
Graded potential change; magnitude varies with magnitude of triggering event	All-or-none membrane response; magnitude of triggering event coded in frequency rather than amplitude of action potentials
Decremental conduction; magnitude diminishes with distance from initial site	Propagated throughout membrane in undiminishing fashion
Passive spread to neighboring inactive areas of membrane	Self-regeneration in neighboring inactive areas of membrane
No refractory period	Refractory period
Can be summed	Summation impossible
Can be depolarization or hyperpolarization	Always depolarization and reversal of charges
Triggered by stimulus, by combination of neurotransmitter with receptor or by spontaneous shifts in leak-pump cycle	Triggered by depolarization to threshold, usually through spread of graded potential
Occurs in specialized regions of membrane designed to respond to triggering event	Occurs in regions of membrane with abundance of voltage-gated Na^+ channels

of the axon is identical to the original one, no matter how long the axon. Thus, an action potential is spread throughout the axon in undiminished fashion. It always goes to maximal amplitude, rather than getting progressively smaller as it moves down the axon. In this way, action potentials can serve as faithful long-distance signals without attenuation or distortion.

The nondecremental propagation of an action potential is in contrast to the decremental spread of a graded potential, which dies out over a very short distance because it is not able to regenerate itself. Typically, regions of excitable cells where graded potentials take place do not have an achievable threshold for undergoing action potentials because of a sparsity of voltage-gated Na^+ channels. (There is no threshold phenomenon for the occurrence of graded potentials themselves.) Therefore, sites specialized for graded potentials do not undergo action potentials, even though they might be depolarized considerably. However, as will be elaborated on later, graded potentials can, before dying out, trigger action potentials in adjacent portions of the membrane by bringing these more sensitive regions to threshold through local current flow spreading from the site of the graded potential. Table 4–1 summarizes the differences between graded potentials and action potentials, some of which are yet to be discussed.

Myelination increases the speed of conduction of action potentials and conserves energy in the process.

The velocity, or speed, with which an action potential travels down the axon depends on two factors: (1) whether the fiber is myelinated and (2) the diameter of the fiber. Conduction by local current flow occurs in unmyelinated fibers. A faster method of propagation, *saltatory conduction*, takes place in myelinated fibers.

Myelinated fibers, as the name implies, are covered with **myelin** at regular intervals along the length of the axon

(● Fig. 4–11a). Myelin is composed primarily of lipids. Because the water-soluble ions responsible for carrying current across the membrane cannot permeate this thick lipid barrier, the myelin coating acts as an insulator, just like rubber around an electrical wire, to prevent current leakage across the myelinated portion of the membrane. Myelin is not actually a part of the nerve cell but consists of separate myelin-forming cells that wrap themselves around the axon in jelly-roll fashion (Fig. 4–11b and c). These myelin-forming cells are **oligodendrocytes** in the central nervous system (the brain and spinal cord) and **Schwann cells** in the peripheral nervous system (the nerves running between the central nervous system and the various regions of the body). The lipid composition of myelin is due to the presence of layer upon layer of the lipid bilayer that composes the plasma membrane of these myelin-forming cells. Between the myelinated regions, the axonal membrane is bare and exposed to the ECF. It is only at these bare spaces, called **nodes of Ranvier**, that membrane potential can exist and current can flow across the membrane (Fig. 4–11d). Sodium channels are concentrated at the nodal areas; the myelin-covered regions are almost devoid of these special passageways.

The nodes are usually about 1 mm apart, a distance short enough that local current from an active node can reach an adjacent node before dying off. When an action potential occurs at one of the nodes, opposite charges attract from the adjacent inactive node, reducing its potential to threshold so that it undergoes an action potential, and so on. Consequently, in a myelinated fiber, the impulse "jumps" from node to node, skipping over the myelinated sections of the axon (● Fig. 4–12); this process is called **saltatory conduction** (*saltere* means "to jump or leap"). Saltatory conduction propagates action potentials more rapidly than does conduction by local current flow, because the action potential leaps over myelinated sections but must be regenerated within every section of an unmyelinated axonal membrane from beginning to end. Myelinated fibers conduct impulses about fifty times faster than nonmyelinated fibers of comparable

FIGURE 4–11 Myelinated Fibers

(a) A myelinated fiber is surrounded by myelin at regular intervals. The intervening unmyelinated regions are known as nodes of Ranvier. (b) In the peripheral nervous system, each patch of myelin is formed by a separate Schwann cell that wraps itself jelly-roll fashion around the nerve fiber. In the central nervous system, each of the several processes of a myelin-forming oligodendrocyte forms a patch of myelin around a separate nerve fiber. (c) An electron micrograph of a myelinated fiber in cross section. (d) Membrane potential exists only at the nodes of Ranvier where the bare axon is exposed to the ECF. No charges exist across the insulated myelinated regions.

Myelin

Axon

Nodes of Ranvier

1 mm

(a)

Schwann cell

Nucleus

Cytoplasm

Myelin sheath

Peripheral axon

Central axon

Central axon

Central axon

Oligodendrocyte

(b)

Myelin sheath

Axon

Plasma membrane

(c)

+ + + +   + + + +   + + + +

– – – –   – – – –   – – – –

Myelinated axon (nerve fiber)

– – – –   – – – –   – – – –

+ + + +   + + + +   + + + +

Schwann cell

Node of Ranvier

(d)

• **FIGURE 4–12 Saltatory Conduction** The impulse "jumps" from node to node in a myelinated fiber.

Local current flow brings adjacent node to threshold

Direction of propagation of action potential

Na^+

Active node at peak of action potential

Adjacent inactive node still at resting potential

Remainder of nodes still at resting potential

K^+

Na^+

Old active node returned to resting

New active node

New adjacent inactive node

size. As a general rule, the most urgent types of information are transmitted via myelinated fibers, whereas the nervous pathways carrying less urgent information are unmyelinated.

In addition to permitting action potentials to travel faster, a second advantage of myelination is that it conserves energy. Since the ion fluxes associated with action potentials are confined to the nodal regions, the energy-consuming Na^+-K^+ pump must restore fewer ions to their respective sides of the membrane following propagation of an action potential.

Fiber diameter also influences the velocity of action potential propagation.

Besides the effect of myelination, the speed with which an axon can conduct action potentials is also influenced by the *diameter* of the fiber. The magnitude of current flow (that is, the amount of charge that moves) depends not only on the difference in potential between two adjacent electrically charged regions but also on the *resistance* or hindrance to electrical charge movement between the two regions. When fiber diameter increases, the resistance decreases. Thus, the larger the diameter of the nerve fiber, the faster it can propagate action potentials.

Large myelinated fibers, such as those supplying skeletal

muscles, can conduct action potentials at a speed of up to 120 meters (m)/sec (360 miles/hr), compared with a conduction velocity of 0.7 m/sec (2 miles/hr) in small unmyelinated fibers such as those supplying the digestive tract. This variability in speed of propagation of action potentials is related to the urgency of the information being conveyed. A signal to skeletal muscles to execute a particular movement (for example, to prevent you from falling as you trip on something) must be transmitted more rapidly than a signal to modify a slow-acting digestive process. Were it not for myelination, axon diameters within urgent nerve pathways would have to be very large and cumbersome to achieve the necessary conduction velocities. Indeed, this is the case in many invertebrates. Vertebrates have escaped the necessity of very large fibers by wrapping the axons in myelin to permit economic, rapid long-distance signaling.

Myelin plays a central role in several disorders.

Multiple sclerosis (MS) is a pathophysiological condition in which demyelination of nerve fibers occurs in various locations throughout the nervous system. MS is an autoimmune (*auto* means "self"; *immune* means "defense against") disease in which the body's defense system erroneously attacks the myelin sheath surrounding myelinated nerve fibers. The symptoms vary considerably, depending on the extent and location of the myelin damage. Loss of myelin slows transmission of impulses in the affected neurons. Also, scarring associated with myelin damage can injure the underlying axons, further interfering with action potential propagation.

Myelin defects may be the culprit in MS, but the presence of myelinating cells can be of tremendous benefit when an axon is cut. In the case of a cut axon in a peripheral nerve, the portion of axon farthest from the cell body degenerates and the surrounding Schwann cells phagocytize the debris. The Schwann cells themselves remain and form a **regeneration tube** to guide the regenerating nerve fiber to its proper destination (● Fig. 4–13). The remaining portion of the axon connected to the cell body starts to grow and move forward within the Schwann cell column by ameoboid movement (see p. 39). It is believed that the growing axon tip "sniffs" its way forward in the proper direction, guided by a chemical secreted into the regeneration tube by the Schwann cells. Successful fiber regeneration is responsible for the return of sensation and movement after a period of time following traumatic peripheral nerve injuries, although regeneration is not always successful.

Fibers in the central nervous system, which are myelinated by oligodendrocytes rather than Schwann cells, do not have this same regenerative ability. Therefore, damaged neuronal

fibers in the brain and spinal cord never regenerate. In the future, however, with the help of exciting new findings, it may be possible to induce significant regeneration of damaged fibers in the central nervous system by chemical means. Nerve growth in the brain and spinal cord appears to be controlled by a delicate balance between *nerve-growth enhancing* and *nerve-growth inhibiting proteins*. During fetal development, nerve growth in the central nervous system is possible as the brain and spinal cord are being formed. Researchers speculate that the nerve-growth inhibitors, which appear late in fetal development in the myelin sheaths surrounding central nerve fibers, may normally serve as "guardrails" to keep new nerve endings from "straying outside their proper paths." Scientists have been able for the first time to induce significant nerve regeneration in rats with severed spinal cords by chemically blocking these nerve-growth inhibitors, thereby allowing the nerve-growth enhancers to promote abundant sprouting of new nerve fibers at the site of injury.

The refractory period assures unidirectional propagation of the action potential and limits the frequency of action potentials.

What assures the one-way propagation of an action potential away from the initial site of activation? Note in ● Figure 4–14

(a)

(b)

(c)

Axon severed

Axon degenerates

Skeletal muscle

Regeneration tube

Axon regenerates

● **FIGURE 4–13   Regeneration of a Peripheral Nerve Fiber** (a) A severed axon can regenerate in the peripheral nervous system. First, the segment from the cut to the terminal end degenerates. (b) The myelin sheath remains and forms a regeneration tube to guide the regenerating axon to its proper destination. (c) The remaining portion of the axon, which is connected to the cell body, regenerates the missing portion, often reestablishing previous contacts and restoring the neuron's function.

● FIGURE 4–14   Value of Refractory Period "Backward" current flow is prevented by the refractory period. During and slightly beyond the time when a particular patch of membrane is undergoing an action potential, that area cannot be restimulated to undergo another action potential as a result of current flow.

"Backward" current flow does not reexcite old active area because this area is in its refractory period

"Forward" current flow excites new inactive area

Direction of propagation of action potential

Old active area returned to resting

New active area

New adjacent inactive area

---

that once the action potential has been regenerated at a new neighboring site (now positive inside) and the original active area has returned to resting (once again negative inside), the close proximity of opposite charges between these two areas is conducive to local current flow taking place in the backward direction as well as in the forward direction (into as yet unexcited portions of the membrane). If such backward current flow were able to bring the just inactivated area to threshold, another action potential would be initiated here, which would spread both forward and backward, initiating another action potential, and so forth. The situation would be chaotic, with numerous action potentials bouncing back and forth along the axon until the nerve cell eventually fatigued. Fortunately, neurons are saved from this fate of oscillating action potentials by the existence of the **refractory period** which has two components: the absolute refractory period and the relative refractory period. During the time that a particular patch of axonal membrane is undergoing an action potential, it is incapable of initiating another action potential, no matter how strongly it is stimulated. This time period when a recently activated patch of membrane is completely refractory (unresponsive) to further stimulation is known as the *absolute refractory period* (● Fig. 4–15). It corresponds to the time period during which the Na⁺ gates are first opened and then closed and inactivated. Not until the potential has returned to resting and the voltage-dependent Na⁺ channels are restored to their "closed but capable of opening" conformation can they respond to another depolarization with an explosive increase in P Na⁺ to initiate another action potential.

Following the absolute refractory period is a *relative refractory period* during which a second action potential can be produced only by a stimulus considerably stronger than is usually necessary. During this time, the K⁺ gates (those which had opened at the peak of the action potential to bring about repolarization) are still closing. Only when all channels have been restored to their resting conformation is the patch of membrane that has just undergone an action potential ready to respond again in a normal fashion. Meanwhile, the impulse has continued to be rapidly propagated in the forward direction only. By the time the original site has recovered from its

refractory period and is capable of being restimulated by normal current flow, the action potential is so far away that it can no longer influence the original site. Thus, *the refractory period assures the unidirectional propagation of the action potential down the axon away from the initial site of activation.*

The refractory period is also responsible for setting an upper limit on the frequency of action potentials; that is, it determines the maximum number of new action potentials that can be initiated and propagated along the fiber in a given period of time. The original site must recover from its refractory period before a new impulse can be triggered to follow the first impulse. The length of the refractory period varies for different types of neurons. The longer the refractory period, the greater the delay before the new action potential can be

---

● FIGURE 4–15   Refractory Period During the absolute refractory period, the portion of the membrane that has just undergone an action potential cannot be restimulated. It corresponds to the time during which the Na⁺ gates are not in their resting conformation. During the relative refractory period, the membrane can be restimulated only by a stronger stimulus than is usually necessary. It corresponds to the time during which the K⁺ gates that were opened during the action potential are still closing.

initiated and the lower the frequency with which a nerve cell can respond to repeated or ongoing stimulation.

Action potentials occur in all-or-none fashion.

If any portion of the neuronal membrane is depolarized to threshold, an action potential is initiated and relayed throughout the membrane in undiminished fashion. Furthermore, once threshold has been reached, the resultant action potential always goes to maximal height, because the changes in voltage during an action potential are due to ion movements down concentration and electrical gradients, which are not affected by stimulus strength. A stimulus stronger than one necessary to bring the membrane to threshold does not produce a larger action potential. On the other hand, a stimulus that fails to depolarize the membrane to threshold does not trigger an action potential at all. Thus, *an excitable membrane either responds to a stimulus with a maximal action potential that spreads nondecrementally throughout the membrane, or it does not respond with an action potential at all.* This is called the **all-or-none law**.

This all-or-none concept is analogous to firing a gun. Either the trigger is not pulled sufficiently to fire the bullet at all (threshold is not reached), or it is pulled hard enough to elicit the full firing response of the gun (threshold is reached). Squeezing the trigger harder does not produce a greater explosion. Just as it is not possible to fire a gun halfway, it is not possible to have a halfway action potential.

The threshold phenomenon is a means by which some discrimination can take place between important and unimportant stimuli. Stimuli too weak to bring the membrane to threshold do not initiate an action potential and therefore do not clutter up the nervous system with transmission of insignificant signals. How is it possible, however, to differentiate between two stimuli of varying strengths if both bring the membrane to threshold and generate action potentials of the same magnitude? For example, how can one distinguish between touching a warm object or a very hot object if both trigger identical action potentials in a nerve fiber relaying information about skin temperature to the central nervous system? The answer lies in the *frequency* with which the action potentials are generated. A stronger stimulus does not produce a larger action potential, but it does trigger a greater number of action potentials per second to be propagated along the fiber. In addition, a stronger stimulus in a region will cause the thresholds of more neurons to be reached, increasing the total information sent to the central nervous system.

SYNAPSES AND NEURONAL INTEGRATION

A neurotransmitter carries the signal across a synapse.

What happens once an action potential reaches the end of an axon? A neuron may terminate at one of three structures: a muscle, a gland, or another neuron. The junctions between nerves and the muscles and glands that they innervate will be described later. For now we will concentrate on the junction between two neurons—a **synapse**.

Typically, a neuron-to-neuron synapse involves a junction between an axon terminal of one neuron and the dendrites or cell body of a second neuron. Less frequently, axon-to-axon and dendrite-to-dendrite connections occur. Most neuronal cell bodies and associated dendrites receive thousands of synaptic inputs, which are axon terminals from many other neurons. It has been estimated that some neurons within the central nervous system receive as many as 100,000 synaptic inputs (• Fig. 4–16a and b).

The anatomy of one of these thousands of synapses is shown in Figure 4–16c. The axon terminal of the **presynaptic neuron**, which conducts its action potentials *toward* the synapse, ends in a slight swelling, the **synaptic knob**. The synaptic knob contains **synaptic vesicles**, which store a specific chemical messenger, a **neurotransmitter**, that has been synthesized and packaged by the presynaptic neuron. The synaptic knob comes into close proximity to, but does not actually directly contact, the **postsynaptic neuron**, the neuron whose action potentials are propagated *away* from the synapse. The space between the presynaptic and postsynaptic neurons, the **synaptic cleft**, is too wide for the direct spread of current from one cell to the other and therefore prevents action potentials from electrically passing between the neurons. The portion of the postsynaptic membrane immediately underlying the synaptic knob is referred to as the **subsynaptic membrane**.

Synapses operate in one direction only; that is, the presynaptic neuron influences the postsynaptic neuron, but the postsynaptic neuron does not influence the presynaptic neuron. The reason for this becomes readily apparent when one examines the events that occur at a synapse.

When an action potential in a presynaptic neuron has been propagated to the axon terminal, this change in potential triggers the opening of voltage-gated Ca^{++} channels in the synaptic knob (Step 1 in Fig. 4–16c). Because Ca^{++} is in much higher concentration in the ECF, this ion flows into the synaptic knob (Step 2). Here it induces the release by exocytosis (see p. 24) of a neurotransmitter from some of the synaptic vesicles into the synaptic cleft (Step 3). The released neurotransmitter diffuses across the cleft and combines with specific protein receptor sites on the subsynaptic membrane (Step 4). This binding triggers the opening of specific ion channels in the subsynaptic membrane, thereby altering the permeability of the postsynaptic neuron (Step 5). This is an example of chemical messenger–gated channels, in contrast to the voltage-gated channels responsible for the action potential and for the Ca^{++} influx into the synaptic knob. Because only the presynaptic terminal can release a neurotransmitter and only the subsynaptic membrane of the postsynaptic neuron has receptor sites for the neurotransmitter, the synapse can operate only in the direction from presynaptic to postsynaptic neuron.

Some synapses excite the postsynaptic neuron whereas others inhibit it.

There are two types of synapses, depending on the permeability changes induced in the postsynaptic neuron by the combination of transmitter substance with receptor sites:

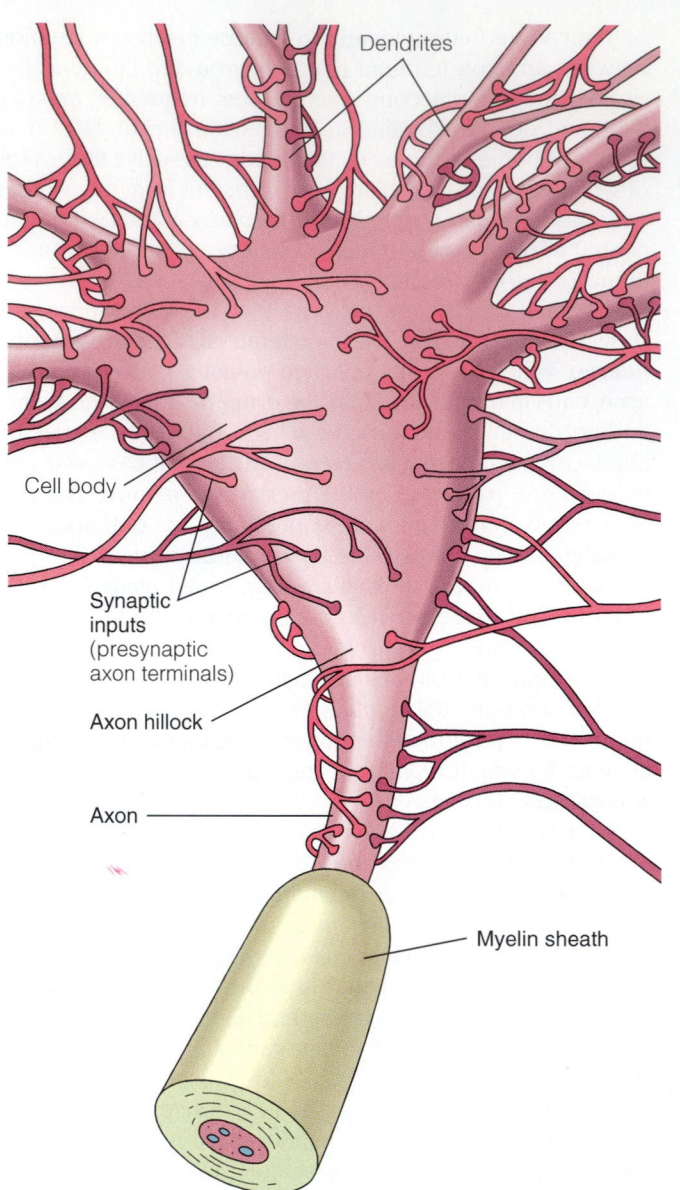

Dendrites

Cell body

Synaptic
inputs
(presynaptic
axon terminals)

Axon hillock

Axon

Myelin sheath

(a)

● **FIGURE 4–16   Synaptic Structure and Function**
(a) Schematic representation of presynaptic inputs to a single neuron.
(b) Electron micrograph showing multiple presynaptic inputs to a single
postsynaptic cell body. (c) Schematic representation of the structure of
and events that occur at a single synapse. Propagation of an action
potential to the axon terminal of a presynaptic neuron ① triggers the
opening of voltage-gated Ca^{++} channels and the subsequent entry of
Ca^{++} into the synaptic knob ②. Calcium induces the release by exocytosis of neurotransmitter from synaptic vesicles into the synaptic
cleft ③. After diffusing across the cleft, the neurotransmitter binds with
its receptor sites on the subsynaptic membrane ④. This binding triggers
the opening of specific ion channels in the subsynaptic membrane ⑤,
which alters the permeability of the postsynaptic neuron.

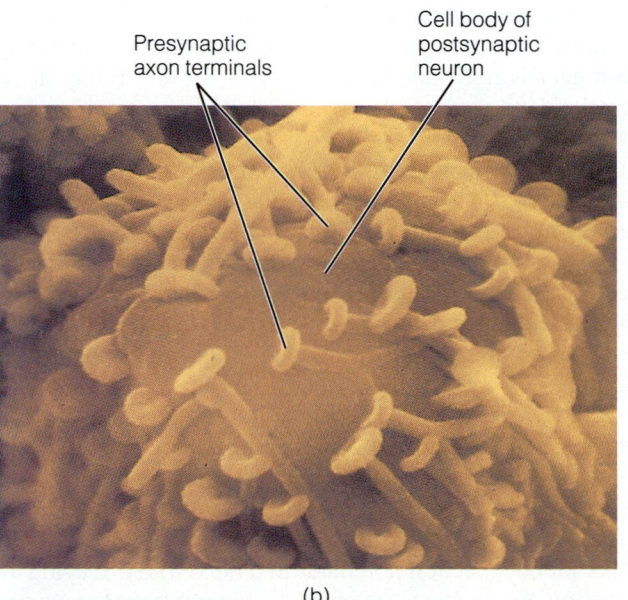

Presynaptic
axon terminals

Cell body of
postsynaptic
neuron

(b)

---

excitatory synapses and *inhibitory synapses*. At an **excitatory synapse**, the response to the receptor-neurotransmitter combination is an opening of Na^+ and K^+ channels within the subsynaptic membrane, thus increasing permeability to both of these ions. Both the concentration and electrical gradients for Na^+ favor its movement into the postsynaptic neuron at resting potential, whereas only the concentration gradient for K^+ favors its movement outward. Therefore, the permeability change induced at an excitatory synapse results in the simultaneous movement of a few K^+ ions out of the postsynaptic neuron while a relatively larger number of Na^+ ions enter this neuron. The result is a net movement of positive ions into the cell. This makes the inside of the membrane less negative than at resting potential and produces a *small depolarization* of the postsynaptic neuron. Activation of one excitatory synapse can rarely depolarize the postsynaptic membrane suffi-

ciently to bring it to threshold. Too few channels are involved at a single subsynaptic membrane to permit adequate depolarizing fluxes to reduce the potential to threshold. This slight depolarization, however, does bring the membrane of the postsynaptic neuron closer to threshold, increasing the likelihood that threshold will be reached and an action potential will occur. Accordingly, such a postsynaptic potential change occurring at an excitatory synapse is called an **excitatory postsynaptic potential**, or **EPSP** (● Fig. 4–17a).

At an **inhibitory synapse**, the combination of the released chemical messenger with its receptor sites increases the permeability of the subsynaptic membrane to either K^+ or Cl^- by altering these ions' respective channel conformations. In either case, the resulting ion movements bring about a *small hyperpolarization* of the postsynaptic neuron (greater internal negativity). In the case of increased P K^+, more positive

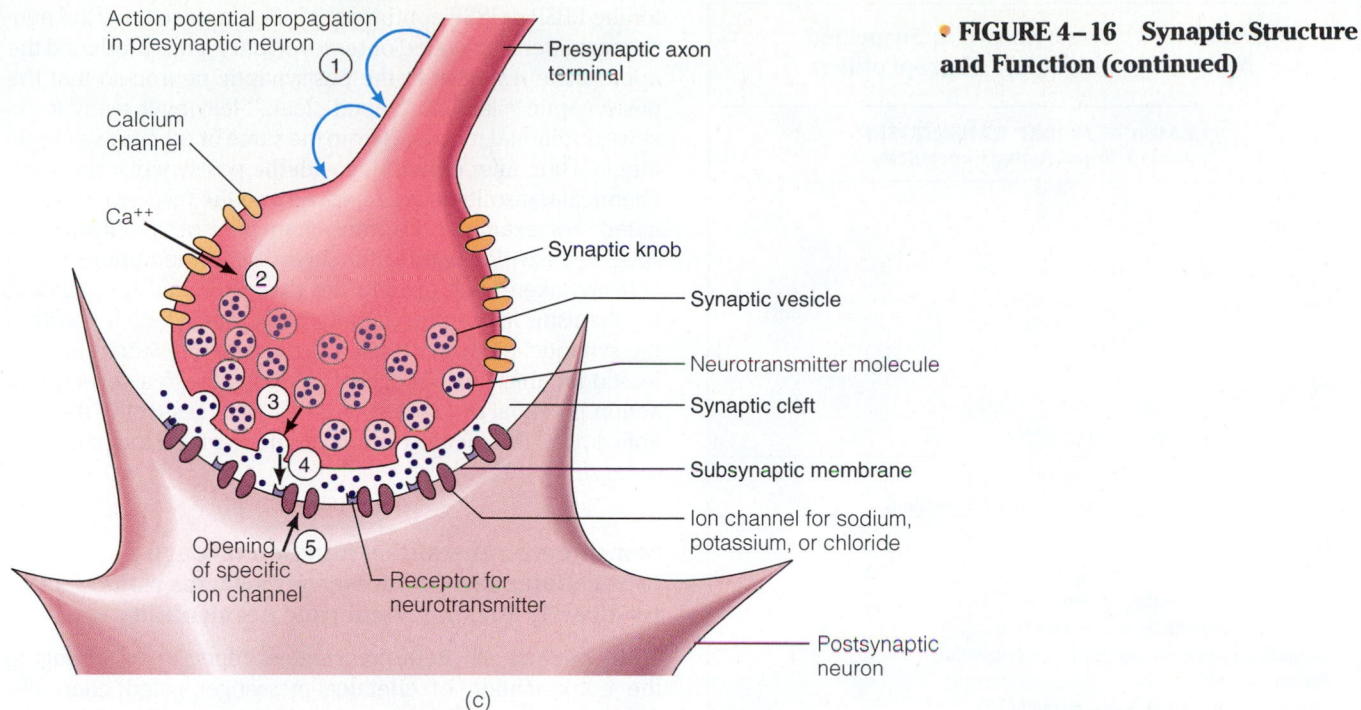

Action potential propagation in presynaptic neuron ①

Presynaptic axon terminal

Calcium channel

Ca++

②

Synaptic knob

Synaptic vesicle

Neurotransmitter molecule

Synaptic cleft

③

Subsynaptic membrane

④

Ion channel for sodium, potassium, or chloride

Opening of specific ion channel ⑤

Receptor for neurotransmitter

Postsynaptic neuron

(c)

• FIGURE 4–16 Synaptic Structure and Function (continued)

charges leave the cell via K⁺ efflux, leaving more negative charges behind on the inside; in the case of increased P Cl⁻, negative charges enter the cell in the form of Cl⁻ ions because Cl⁻ concentration is higher outside of the cell. This slight hyperpolarization moves the membrane potential even farther away from threshold (Fig. 4–17b), lessening the likelihood that the postsynaptic neuron will reach threshold and undergo an action potential. The membrane is said to be inhibited under these circumstances, and the small hyperpolarization of the postsynaptic cell is called an **inhibitory postsynaptic potential**, or **IPSP**. (In cells where the equilibrium potential for Cl⁻ exactly equals the resting potential, an increased P Cl⁻ does not result in a hyperpolarization because there is no driving force to produce Cl⁻ movement.)

Opening of Cl⁻ channels in these cells tends to hold the membrane at resting potential, thus reducing the likelihood that threshold will be reached.)

This conversion of the electrical signal in the presynaptic neuron (an action potential) to an electrical signal in the postsynaptic neuron (either an EPSP or IPSP) by chemical means (via the neurotransmitter-receptor combination) takes time. This **synaptic delay** is usually about 0.5 to 1 msec. Chains of neurons often must be traversed along a specific neural pathway. The more complex the pathway, the more synaptic delays, and the longer the *total reaction time* (the time required to respond to a particular event).

Many different chemicals are known or suspected to serve as neurotransmitters (Table 4–2). Even though transmitter

• **FIGURE 4–17 Postsynaptic Potentials** (a) Excitatory synapse. An excitatory postsynaptic potential (EPSP) brought about by activation of an excitatory presynaptic input brings the postsynaptic neuron closer to threshold potential. (b) Inhibitory synapse. An inhibitory postsynaptic potential (IPSP) brought about by activation of an inhibitory presynaptic input moves the postsynaptic neuron farther from threshold potential.

(a)

(b)

TABLE 4–2   Some Known or Suspected Neurotransmitters and Neuropeptides

CLASSICAL NEUROTRANSMITTERS
(Small, Rapid-Acting Molecules)

Acetylcholine
Dopamine
Norepinephrine
Epinephrine
Serotonin
Histamine
Glycine
Glutamate
Aspartate
Gamma-aminobutyric acid (GABA)

NEUROPEPTIDES
(Large, Slow-Acting Molecules)

β-endorphin
Adrenocorticotropic hormone (ACTH)
α-melanocyte-stimulating hormone (MSH)
Thyrotropin-releasing hormone (TRH)
Gonadotropin-releasing hormone (GnRH)
Somatostatin
Vasoactive intestinal polypeptide (VIP)
Cholecystokinin (CCK)
Gastrin
Substance P
Neurotensin
Leucine enkephalin
Methionine enkephalin
Motilin
Insulin
Glucagon
Angiotensin II
Bradykinin
Vasopressin
Oxytocin
Carnosine
Bombesin

for the EPSP or IPSP continues. It is desirable to have the neurotransmitter inactivated or removed after it has produced the appropriate response in the postsynaptic neuron so that the postsynaptic "slate" is "wiped clean," leaving it ready to receive additional messages from the same or other presynaptic inputs. Thus, after combining with the postsynaptic receptor, chemical transmitters are removed and the response is terminated. For example, the transmitter may be inactivated by specific enzymes within the subsynaptic membrane or be actively taken back up into the axon terminal by transport mechanisms in the presynaptic membrane. Once it is within the synaptic knob, the transmitter can be (1) stored and released another time (recycled) in response to a subsequent action potential or (2) destroyed by enzymes within the synaptic knob. The method employed depends on the particular synapse.

Some neurotransmitters function through intracellular second messenger systems rather than by directly altering membrane permeability.

Most, but not all, neurotransmitters function by changing the conformation of chemical-messenger gated channels, thereby altering membrane permeability and ionic fluxes across the postsynaptic membrane. Another mode of synaptic transmission utilized by some neurotransmitters, for example serotonin, involves the activation of second messengers (see p. 50) within the postsynaptic neuron. Activation of the intracellular messenger, cyclic AMP, can induce both short- and long-term effects. In the short term, cAMP can lead to opening of ionic gates, a task that the other neurotransmitter-receptor combinations do directly without involving the intracellular messenger. The gating effects can be either excitatory or inhibitory. In addition, cAMP may trigger more long-term changes in the postsynaptic neuron cell, even to the extent of possibly altering the cell's genetic expression. Such long-term cellular changes may play a role in learning and memory.

The grand postsynaptic potential depends on the sum of the activities of all presynaptic inputs.

The events that occur at a single synapse result in either an EPSP or an IPSP at the postsynaptic neuron. If a single EPSP is inadequate to bring the postsynaptic neuron to threshold and an IPSP moves it even farther from threshold, how is it possible to initiate an action potential in the postsynaptic neuron? Recall that a typical neuronal cell body receives thousands of presynaptic inputs from many other neurons. Some of these presynaptic inputs may be carrying sensory information brought from the environment; some may be signaling internal changes in homeostatic balance; others may be transmitting signals from control centers in the brain; and still others may arrive carrying other bits of information. At any given time, any number of these presynaptic neurons (probably hundreds) may be firing and thus influencing the postsynaptic neuron's level of activity. The total potential in the postsynaptic neuron, the **grand postsynaptic potential**

substances vary from synapse to synapse, the same transmitter is always released at a particular synapse. The importance of neurotransmitters lies not so much in their names or chemical nature but in the particular responsiveness of the postsynaptic membrane upon their binding with subsynaptic receptors. One particular neurotransmitter will always induce EPSPs whereas another will always induce IPSPs. Yet another may even produce an EPSP at one synapse and an IPSP at another synapse. What is important is that the response of a given transmitter-receptor combination is always constant. *A given synapse is either always excitatory or always inhibitory.* It does not give rise to an EPSP under one circumstance and produce an IPSP at another time.

Neurotransmitters are quickly removed from the synaptic cleft to wipe the postsynaptic slate clean.

As long as the neurotransmitter remains bound to the receptor sites, the alteration in membrane permeability responsible

(**GPSP**), is a composite of all EPSPs and IPSPs occurring at approximately the same time.

The postsynaptic neuron can be brought to threshold in two ways: (1) *temporal summation* and (2) *spatial summation*. To illustrate these methods of summation, we will examine the possible interactions of three presynaptic inputs—two excitatory (E1 and E2) and one inhibitory (I1)—on a hypothetical postsynaptic neuron (● Fig. 4–18). The recording represents the potential in the postsynaptic cell. Bear in mind during our discussion of this simplified version that many thousands of synapses are actually interacting in the same way on a single cell body.

Now suppose that E1 has an action potential that causes the release of neurotransmitter from a few of its synaptic vesicles to bring about an EPSP in the postsynaptic neuron. Because EPSPs (as well as IPSPs) are graded potentials, the EPSP spreads only a short distance before dying off. If another action potential subsequently occurs in E1, an EPSP of the same

magnitude takes place (Panel A in Fig. 4–18). Next assume that E1 has two action potentials in close succession to each other (Panel B). The first action potential in E1 produces an EPSP in the postsynaptic neuron. While the postsynaptic membrane is still partially depolarized from this first EPSP (before it has returned to resting), the second presynaptic action potential produces a second EPSP in the postsynaptic neuron. The second EPSP will add on to the first EPSP, bringing the membrane to threshold so that an action potential can occur in the postsynaptic neuron. Graded potentials do not have a refractory period, so this additive effect is possible. The summing of several EPSPs occurring very close together in time because of successive firing of a single presynaptic neuron is known as **temporal summation** (*tempus* means "time"). The actual situation is much more complex than the one just described. Up to fifty EPSPs might have to sum to bring the postsynaptic membrane to threshold. Each action potential in a presynaptic neuron triggers the emptying of a

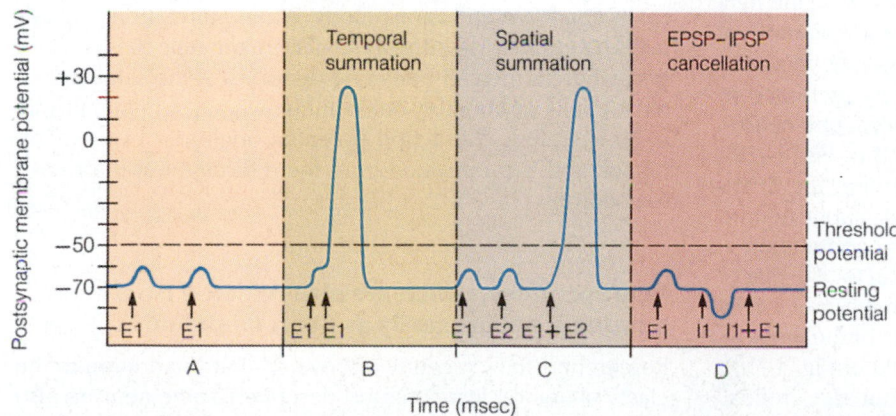

● **FIGURE 4–18 Determination of Grand Postsynaptic Potential by Sum of Activity in Presynaptic Inputs** Two excitatory (E1 and E2) and one inhibitory (I1) presynaptic inputs terminate on this hypothetical postsynaptic neuron. The potential of the postsynaptic neuron is being recorded. If an excitatory presynaptic input (E1) is stimulated a second time after the first EPSP in the postsynaptic cell has died off, a second EPSP of the same magnitude will occur (Panel A). If, however, E1 is stimulated a second time before the first EPSP has died off, the second EPSP will add onto or sum with the first EPSP, resulting in temporal summation, which may bring the postsynaptic cell to threshold (Panel B). The postsynaptic cell may also be brought to threshold by spatial summation of EPSPs that are initiated by simultaneous activation of two (E1 and E2) or more excitatory presynaptic inputs (Panel C). Simultaneous activation of an excitatory (E1) and inhibitory (I1) presynaptic input does not change the postsynaptic potential because the resultant EPSP and IPSP cancel each other out (Panel D).

certain number of synaptic vesicles. The amount of neurotransmitter released and the resultant magnitude of the change in postsynaptic potential are thus directly related to the frequency of presynaptic action potentials. One way, then, in which the postsynaptic membrane can be brought to threshold is through rapid, repetitive excitation from a single persistent input.

Let us now see what will happen in the postsynaptic neuron if both excitatory inputs are stimulated simultaneously (Panel C). An action potential in either E1 or E2 will produce an EPSP in the postsynaptic neuron; however, neither of these alone will bring the membrane to threshold to elicit a postsynaptic action potential. Simultaneous action potentials in E1 and E2, however, will each produce EPSPs that add to each other, bringing the postsynaptic membrane to threshold so that an action potential occurs. Such summation of EPSPs originating simultaneously from several different presynaptic inputs (i.e., from different points in "space") is known as **spatial summation**. A second way, therefore, to elicit an action potential in a postsynaptic cell is through concurrent activation of several excitatory inputs. Again, in reality up to fifty EPSPs arriving simultaneously on the postsynaptic membrane are required to bring it to threshold. Similarly, IPSPs can undergo temporal and spatial summation. As IPSPs add together, however, they progressively move the potential farther from threshold.

If an excitatory and an inhibitory input are simultaneously activated, the concurrent EPSP and IPSP more or less cancel each other out (the extent of cancellation depends on their respective magnitudes). In most cases, the postsynaptic membrane potential remains close to resting (Panel D).

Thus, the grand postsynaptic potential depends on the sum of activity in all presynaptic inputs. The following oversimplified real-life example demonstrates the benefits of this neuronal integration. The explanation is not completely accurate technically, but the principles of summation are accurate. Assume for simplicity's sake that urination is controlled by a postsynaptic neuron supplying the urinary bladder. (Actually, voluntary control of urination is accomplished by postsynaptic integration at the neuron controlling the external urethral sphincter rather than the bladder itself.) As the bladder starts to fill with urine and becomes stretched, a reflex is initiated that ultimately produces EPSPs in the postsynaptic neuron responsible for causing bladder contraction. Partial filling of the bladder does not cause sufficient excitation to bring the neuron to threshold, so urination does not take place (Panel A of Fig. 4-18). As the bladder becomes progressively filled, the frequency of action potentials is progressively increased in the presynaptic neuron that signals the postsynaptic neuron of the extent of bladder filling (E1 in Panel B of Fig. 4-18). When the frequency becomes great enough that the EPSPs are temporally summed to threshold, the postsynaptic neuron has an action potential that stimulates bladder contraction.

What if the time is inopportune for urination to take place? IPSPs can be produced at the bladder postsynaptic neuron by presynaptic inputs originating in higher levels of the brain responsible for voluntary control (I1 in Panel D of Fig. 4-18). These "voluntary" IPSPs in effect cancel out the "reflex" EPSPs triggered by stretching of the bladder. Thus, the postsynaptic neuron remains at resting potential and does not have an action potential, so the bladder is prevented from contracting and emptying even though it is full.

What if the bladder is only partially filled, so that the presynaptic input originating from this source is insufficient to bring the postsynaptic neuron to threshold to cause bladder contraction, yet the person needs to supply a urine specimen for laboratory analysis? The person can voluntarily activate an excitatory presynaptic neuron (E2 in Panel C of Fig. 4-18), which spatially summates with the reflex-activated presynaptic neuron (E1) to bring the postsynaptic neuron to threshold. This achieves the action potential necessary to stimulate bladder contraction, even though the bladder is not full.

This example illustrates the importance of postsynaptic neuronal integration. Each postsynaptic neuron in a sense "computes" all the input it receives and makes a "decision" about whether to pass the information on (i.e., whether or not to reach threshold). Each postsynaptic neuron filters out and does not pass on information it receives that is not significant enough to bring it to threshold. If every action potential in every presynaptic neuron that impinges upon a particular postsynaptic neuron were to cause an action potential in the postsynaptic neuron, the neuronal pathways would be overwhelmed with trivia. Only if an excitatory presynaptic signal is reinforced by other supporting signals through summation will the information be passed on. Furthermore, interaction of postsynaptic potentials provides a way for one set of signals to offset another set (IPSPs negating EPSPs). This allows a fine degree of discrimination and control in determining what information will be passed on. (See the accompanying boxed feature, A Closer Look at Exercise Physiology.)

Action potentials are initiated at the axon hillock because it has the lowest threshold.

Threshold potential is not uniform throughout the postsynaptic neuron. The lowest threshold is present at the axon hillock because this region has an abundance of voltage-gated Na^+ channels, making it considerably more sensitive to changes in potential than the remainder of the cell body and dendrites. The latter regions have a significantly higher threshold than the axon hillock. Because of local current flow, changes in membrane potential occurring anywhere on the cell body or dendrites (EPSPs or IPSPs) spread throughout the cell body, dendrites, and axon hillock. When summation of EPSPs takes place, the lower threshold of the axon hillock is reached first, whereas the cell body and dendrites at the same potential are still considerably below their own much higher thresholds. Therefore, the action potential originates in the axon hillock and is propagated from there throughout the rest of the neuron.

Neuropeptides, which are also released from axon terminals, act primarily as neuromodulators.

Researchers have recently discovered that in addition to the classical neurotransmitters just described, some neurons also release **neuropeptides** (Table 4-2). Neuropeptides differ

THE YELLS AND GRUNTS OF AN ATHLETE
MAY SERVE A PHYSIOLOGICAL PURPOSE

Skeletal muscles are innervated by a special type of neuron called motor neurons (see chapter 7). Action potentials in motor neurons initiate the contractile process in the muscle on which they terminate. Both excitatory and inhibitory presynaptic inputs from a variety of sources impinge on the cell body and dendrites of the motor neurons. These neurons fire, thereby initiating muscle contraction, when their membrane is brought to threshold by summation of EPSPs. A person can voluntarily contract a given skeletal muscle by consciously activating nerve pathways that generate EPSPs at the appropriate motor neurons.

Among the inhibitory inputs to motor neurons are those arising from the Golgi tendon organ, a sensory device located within the tendon that attaches a muscle to the skeleton (see chapter 8). The Golgi tendon organ is sensitive to stretch imposed on it by the muscle's contraction. When stretched, the Golgi tendon organ initiates a nerve pathway that generates IPSPs on the motor neurons that innervate the very same contracting muscle in which the sensory device is located. If the grand postsynaptic potential of the involved motor neurons is reduced below threshold by the inhibitory input from the Golgi tendon organ, the muscle is no longer stimulated and it relaxes. This reflex relaxation is protective in nature; its purpose is to prevent the muscle from contracting hard enough to rupture the muscle or tendon or break a bone.

Weight training requires the voluntary control of this natural muscular inhibitory reflex. When persons begin to weight train, their ability to lift dramatically increases in the first two weeks even though no physiologic change has yet occurred in the muscle. This initial improvement in strength is thought to be brought about by the body learning to voluntarily produce more EPSPs, thus voluntarily overriding the muscles' inhibition by their stretched Golgi tendon organs. As weight training progresses, muscles get physiologically stronger as the muscle cells enlarge, the tendons get thicker, and the Golgi tendon organs decrease their sensitivity.

Power lifters sometimes yell as they lift to try to further increase the excitatory stimuli during heavy lifting. Similarly, tennis players and other athletes often grunt during the execution of powerful maneuvers. In sports such as arm wrestling, the inhibitory reflex is overridden by voluntary control to the point that participants have ruptured muscles or tendons and have even broken the bones in the arm during the excitement of competition.

from classical neurotransmitters in several important ways (Table 4–3). Classical neurotransmitters are small, rapid-acting molecules that typically bring about a response in the postsynaptic neuron within a few milliseconds or less, usually a change in postsynaptic potential (an EPSP or IPSP) as a result of triggering the opening of specific ion channels. Classical neurotransmitters are synthesized and packaged locally in vesicles in the cytosol of the synaptic knob. These chemical messengers are primarily amino acids or closely related compounds.

TABLE 4–3 Comparison of Classical Neurotransmitters and Neuropeptides

CHARACTERISTIC	CLASSICAL NEUROTRANSMITTERS	NEUROPEPTIDES
Size	Small; one amino acid or similar chemical	Large; 2 to 40 amino acids in length
Site of synthesis	Cytosol of synaptic knob	Endoplasmic reticulum/Golgi complex in cell body; travel to synaptic knob by axonal transport
Site of release	Axon terminal	Axon terminal; may be cosecreted with neurotransmitter
Speed and duration of action	Rapid, brief response	Slow, prolonged response
Site of action	Subsynaptic membrane of postsynaptic cell	Nonsynaptic sites on either presynaptic or postsynaptic cell at much lower concentrations than classical neurotransmitters
Effect	Usually alter potential of postsynaptic cell by opening specific ion channels	Usually enhance or suppress synaptic effectiveness by long-term changes in neurotransmitter synthesis or postsynaptic receptor sites

Neuropeptides are larger molecules made up of anywhere from two to about forty amino acids. They are synthesized in the neuronal cell body in the endoplasmic reticulum/Golgi complex (see p. 18) and are subsequently moved by axonal transport along the microtubular highways to the axon terminal (see p. 36). A given axon terminal releases only a single classical neurotransmitter, but the same terminal may also contain one or more neuropeptides that may be secreted along with the classical neurotransmitter. It is unclear whether the neuropeptides are packaged in the same or separate synaptic vesicles within the axon terminal.

Even though neuropeptides are currently the subject of intense investigation, our knowledge about their functions and control is still sketchy. They are known to diffuse locally and act on other adjacent neurons at much lower concentrations than classical neurotransmitters, and they bring about slower, more prolonged responses. Some neuropeptides released at synapses may function as true neurotransmitters, but most are believed to function as **neuromodulators**. Neuromodulators are chemical messengers that bind to neuronal receptors at nonsynaptic sites (that is, not at the subsynaptic membrane). In so doing, they bring about long-term changes that subtly depress or enhance synaptic effectiveness, often by activating second messenger systems. Neuromodulators may act at either presynaptic or postsynaptic sites. For example, a neuromodulator may influence the level of an enzyme critical in the synthesis of a neurotransmitter by a presynaptic neuron, or it may alter the sensitivity of the postsynaptic neuron to a particular neurotransmitter by causing long-term changes in the number of postsynaptic receptor sites for the neurotransmitter. Thus, neuromodulators delicately fine-tune the synaptic response. The effect may last for days or even months or years.

Interestingly, the synaptically released neuromodulators include many substances that also have distinctly different roles in other body regions as hormones released into the blood from neural and nonneural tissues. For example, *cholecystokinin (CCK)* is a well-known hormone released from the small intestine that causes the gallbladder to contract and release bile into the intestine, among other digestive activities. CCK has also been found in synaptic vesicles in diffuse areas of the brain, where it is believed to mediate satiety (the feeling of no longer being hungry) and perhaps decrease the perception of painful stimuli, among other unknown activities. In fact, in many instances neuropeptides are named for their first-discovered role as hormones, as is the case with cholecystokinin (*chole* means "bile"; *cysto* means "bladder"; *kinin* means "contraction"). It appears that a number of chemical messengers are quite versatile and can assume different roles, depending on their source, distribution, and interaction with a particular type of cell.

Presynaptic inhibition or facilitation can selectively alter the effectiveness of a given presynaptic input.

Besides neuromodulation, presynaptic inhibition or facilitation is another means of depressing or enhancing synaptic effectiveness. Sometimes a third neuron influences activity

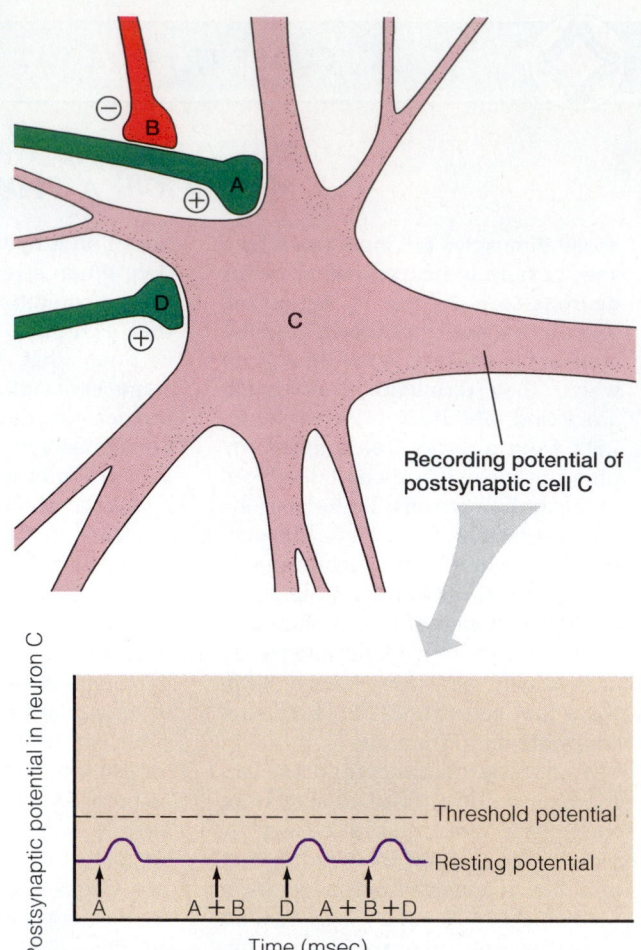

Recording potential of postsynaptic cell C

• **FIGURE 4–19   Presynaptic Inhibition** A, an excitatory terminal ending on postsynaptic cell C, is itself innervated by inhibitory terminal B. Stimulation of terminal A alone produces an EPSP in cell C, but simultaneous stimulation of terminal B prevents the release of excitatory transmitter from terminal A. Consequently, no EPSP is produced in cell C despite the fact that terminal A has been stimulated. Such presynaptic inhibition selectively depresses activity from terminal A without suppressing any other excitatory input to cell C. Stimulation of excitatory terminal D produces an EPSP in cell C even though inhibitory terminal B is simultaneously stimulated because terminal B only inhibits terminal A.

between a presynaptic ending and a postsynaptic neuron. A presynaptic axon terminal (labeled A in • Fig. 4–19) may itself be innervated by another axon terminal (labeled B) that can modify the amount of transmitter released from presynaptic terminal A. When the neurotransmitter released from terminal B binds with receptor sites on terminal A, the amount of transmitter released from terminal A in response to action potentials is altered. If the amount of transmitter released from A is reduced, the phenomenon is known as **presynaptic inhibition**. If the release of transmitter is enhanced, the effect is called **presynaptic facilitation**. The mechanism for these effects is unclear, but it most likely involves Ca^{++} because the amount of cytosolic Ca^{++} in the synaptic knob determines the extent to which synaptic vesicles release neurotransmitter.

What is the importance of this presynaptic modulation? The amount of transmitter released from presynaptic terminal A influences the potential in the postsynaptic neuron with which it is interacting (labeled C on Fig. 4–19). For example, if A, an excitatory input to C, is prevented from releasing its transmitter by presynaptic inhibition via B, the formation of EPSPs on postsynaptic membrane C will specifically be prevented from input A. As a result, no change in the potential of the postsynaptic neuron will occur in spite of action potentials in A. Could the same thing be accomplished by a simultaneous production of an IPSP through activation of an inhibitory input to negate an EPSP produced by activation of A? Not quite. The entire postsynaptic membrane is hyperpolarized by IPSPs, thereby negating excitatory information fed into any part of the cell from any presynaptic input. Presynaptic inhibition (or presynaptic facilitation), on the other hand, provides a means by which certain inputs to the postsynaptic neuron can be selectively altered without affecting the contributions of any other inputs. For example, firing of B does not have any influence on another excitatory presynaptic input (labeled D on Fig. 4–19). This type of neuronal integration is another means by which electrical signaling between nerve cells can be carefully fine-tuned.

The effectiveness of synaptic transmission can be modified by drugs and diseases.

Numerous opportunities exist for influencing synaptic transmission by drugs. In fact, the vast majority of drugs that influence the nervous system perform their function by altering synaptic mechanisms. Synaptic drugs may act to block an undesirable effect or to enhance a desirable effect. Possible drug actions include (1) altering the synthesis, axonal transport, storage, or release of a neurotransmitter; (2) modifying neurotransmitter interaction with the postsynaptic receptor; (3) influencing neurotransmitter reuptake or destruction; or (4) replacing a deficient neurotransmitter with a substitute transmitter.

For example, the illegal drug *cocaine* blocks the reuptake of the neurotransmitter *dopamine* at presynaptic terminals by competitively binding with the dopamine reuptake transporter. The transporter is a protein molecule that serves the ferrying function of picking up released dopamine from the synaptic cleft and shuttling it back to the axon terminal for reuptake. With cocaine occupying the dopamine transporter, dopamine remains in the synaptic cleft longer than usual and continues to interact with its postsynaptic receptor sites. The result is prolonged activation of neural pathways that use this chemical as a neurotransmitter. Among these pathways are those believed to play a role in emotional responses, especially feelings of pleasure.

Whereas cocaine abuse leads to excessive dopamine activity, **Parkinson's disease** is attributable to a deficiency of dopamine in a particular region of the brain involved in controlling complex movements. (See the accompanying boxed feature, Concepts, Challenges, and Controversies.) Treatment of Parkinson's disease is an example of a deficient neurotransmitter being replaced with a substitute transmitter.

When patients with this disease are given *levodopa* (*L-dopa*), a compound closely related to dopamine, the L-dopa can gain access to the brain and be taken up by the dopamine-deficient synaptic knobs, thereby substituting for the lacking "home-grown variety" of this neurotransmitter. This greatly alleviates the symptoms associated with the deficit in most patients. Dopamine itself cannot be administered because it is unable to cross the blood-brain barrier (see p. 113), whereas L-dopa can enter the brain from the blood.

Synaptic transmission is also vulnerable to a number of disease processes, including defects at both presynaptic and postsynaptic sites. For example, two different neural poisons, strychnine and tetanus toxin, act at different synaptic sites to block inhibitory impulses while leaving the excitatory inputs unchecked. *Strychnine* competes with one of the inhibitory neurotransmitters, *glycine*, at the postsynaptic receptor site. This poison combines with the receptor but does not directly alter the potential of the postsynaptic cell in any way. However, strychnine blocks the receptor so that it is not available for interaction with glycine when the latter is released from the inhibitory presynaptic ending. Thus, postsynaptic inhibition (formation of IPSPs) is abolished in nerve pathways that use glycine as an inhibitory transmitter. Unchecked excitatory pathways lead to convulsions, muscle spasticity, and death. *Tetanus toxin*, on the other hand, prevents the release of another inhibitory transmitter, *gamma-aminobutyric acid* (*GABA*) from presynaptic inputs terminating on neurons that supply skeletal muscles. Unchecked excitatory inputs to these neurons result in uncontrolled muscle spasms. These spasms occur especially in the jaw muscles early in the disease, giving rise to the common name of lockjaw for this condition. Later they progress to the muscles responsible for breathing, at which point the disease proves to be fatal. The outcomes of both strychnine and tetanus toxin poisoning are similar, but one (strychnine) blocks specific postsynaptic inhibitory receptors, whereas the other (tetanus toxin) prevents the presynaptic release of a specific inhibitory neurotransmitter. Other drugs and diseases that influence synaptic transmission are too numerous to mention, but as these examples illustrate, any site along the synaptic pathway is vulnerable to interference, either pharmacological (drug-induced) or pathological (disease-induced).

Neurons are linked to each other through convergence and divergence to form vast and complex nerve pathways.

Two important relationships exist between neurons: convergence and divergence. Any given neuron may have many other neurons synapsing upon it. Such a relationship is known as **convergence** (• Fig. 4–20). Through this converging input, a single cell is influenced by thousands of other cells. This single cell, in turn, influences the level of activity in many other cells by divergence of output. **Divergence** refers to the branching of axon terminals so that a single cell synapses with many other cells.

Note that a particular neuron is postsynaptic to the neurons converging upon it but presynaptic to the other cells

PARKINSON'S DISEASE, POLLUTION, AND ETHICAL PROBLEMS

In 1983, Dr. J. W. Langston, a California neurologist, was confronted with a medical puzzle unlike any he had ever seen. Several young men and women were admitted to the hospital in a rigid, stuporous state as if they had been frozen in their tracks. Confined to their beds, the patients lay immobile day after day. They could neither talk nor feed themselves and could not move their limbs.

Langston began an intensive study of the victims and found they all were drug addicts. Each of them had injected *meperidine (MPPP)*, a synthetic form of heroin. However, the MPPP was contaminated with a slightly different drug, MPTP, which is converted by a naturally occurring enzyme in the body into a paralyzing chemical, MPP$^+$. Made in a basement by one of California's small-time drug pushers, the synthetic heroin is one of the "designer drugs" available on the black market today. Sloppy chemical technique resulted in contamination of the MPPP with MPTP.

Researchers have shown that MPP$^+$ destroys dopamine-secreting cells in a part of the brain called the *substantia nigra*. Axons from these cells terminate in the *basal nuclei,* another region of the brain involved in the coordination of slow, sustained movements, inhibition of muscle tone, and suppression of useless patterns of movement. Reduced dopamine activity in the basal nuclei resulting from destruction of these substantia nigra cells was responsible for the ensuing symptoms. Most prominently, loss of the basal nuclei's inhibitory influence on muscle tone led to the pronounced muscular rigidity experienced by these patients.

A gradual destruction of these same dopamine-secreting cells in the substantia nigra and the resultant loss of basal nuclei function are responsible for *Parkinson's disease.* The symptoms, which develop slowly as dopamine activity in the basal nuclei gradually diminishes, begin with involuntary tremors at rest, such as involuntary rhythmic shaking of the hands or head. As the disease worsens, patients speak in a slow monotone, become increasingly stiff, and walk with a shuffling, stooped gait. In later stages, memory and thinking become severely impaired. Ultimately, debilitating rigidity ensues, similar to that seen in the MPTP victims.

Interestingly, MPP$^+$ is very similar to *paraquat,* a commonly used agricultural pesticide. This observation led to the hypothesis that Parkinson's disease may be linked to pesticide use. Indeed, epidemiological studies have shown a remarkable correlation between the use of pesticides and the incidence of Parkinson's disease. Other chemical pollutants are also suspected of contributing to the destruction of the dopamine-secreting brain cells. Researchers note that Parkinson's disease was unheard of before the Industrial Revolution. As the Industrial Revolution and environmental pollution spread, the incidence of the disease rose sharply, reaching a plateau in the early 1900s. Currently, an estimated one million people in the United States have Parkinson's disease. Some investigators warn that increased use of paraquat and other pesticides and a rise in industrial pollution could substantially increase the incidence of the disease in years

to come, underscoring the fact that to protect our health we must also protect the quality of our environment.

A standard treatment for Parkinson's disease has been administration of levodopa (*L-dopa*), a precursor of dopamine. Dopamine itself is unable to cross the blood-brain barrier to get into the brain, but L-dopa can. By serving as a substitute neurotransmitter for dopamine, L-dopa has been an effective form of therapy for many patients. Unfortunately, however, with prolonged use, the beneficial effects of L-dopa tend to diminish and troublesome side effects develop. Therefore, researchers have been seeking other means to manage this disabling condition. One new drug that holds promise is *selegiline,* which enhances the effectiveness of dopamine at synapses and hopefully can forestall the symptoms of Parkinson's disease.

One alternative approach to drug therapy for Parkinson's disease has stirred considerable controversy— harvesting dopamine-secreting cells from aborted fetuses and transplanting them into the brains of Parkinson's patients. The legal and ethical issues encumbering fetal-cell transplants came into focus in 1988 when a group of researchers sought permission to transplant fetal brain cells into a patient with Parkinson's disease. This was the first request to perform a fetal-cell transplant in humans in the United States, although the procedure had been performed in other countries. The debate sparked by the request led to a politically heated ban on federally funded research using cells from aborted fetuses.

upon which it terminates. Thus, the terms *presynaptic* and *postsynaptic* refer only to a single synapse. Most neurons are presynaptic to one group of neurons and postsynaptic to another group.

There are an estimated 100 billion neurons in the brain alone. When you consider the vast and intricate interconnections possible between these neurons through converging and diverging pathways, you can begin to imagine how com-

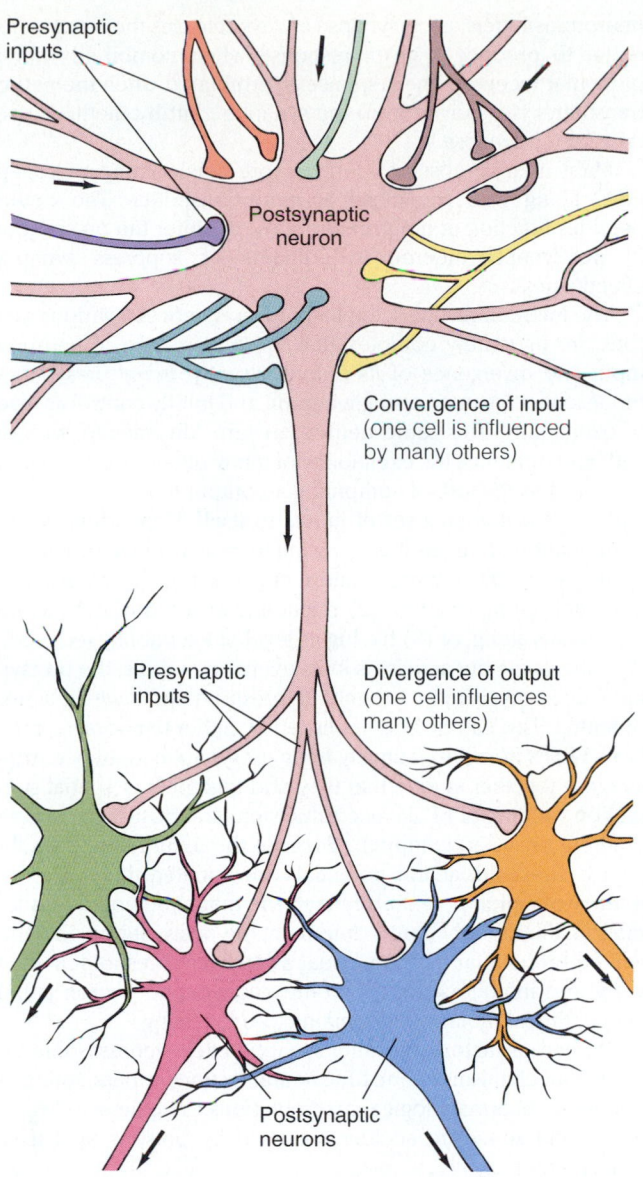

Presynaptic inputs

Postsynaptic neuron

Convergence of input
(one cell is influenced by many others)

Presynaptic inputs

Divergence of output
(one cell influences many others)

Postsynaptic neurons

Arrows indicate direction in which information is being conveyed.

• **FIGURE 4–20   Convergence and Divergence**

plex the wiring mechanism of our nervous system really is. Even the most sophisticated computers are far less complex than the human brain.

 CHAPTER IN PERSPECTIVE: FOCUS ON HOMEOSTASIS

All cells depend on maintenance of a stable internal environment for their survival, and all cells contribute to this homeostatic condition by means of their specialized activities. The specialization of nerve cells (neurons) and muscle cells depends on their ability to rapidly alter their membrane potential and thus produce electrical signals in response to appropriate triggering events.

Nerve cells are specialized to receive, process, encode, and rapidly transmit information from one part of the body to another. The information is transmitted over intricate nerve pathways by propagation of action potentials along the nerve cells' length as well as by chemical transmission of the signal from neuron to neuron and from neuron to muscles and glands through neurotransmitter-receptor interactions at synapses.

Collectively, the nerve cells comprise the nervous system, one of the two major control systems of the body. Many of the activities controlled by the nervous system are geared toward maintaining homeostasis. Some neuronal electrical signals convey information about changes to which the body must respond in order to maintain homeostasis—for example, these signals convey information about a fall in blood pressure. Other neuronal electrical signals convey messages to muscles and glands to stimulate appropriate responses to counteract these changes—for example, the nervous system through its electrical signals initiates adjustments in heart and blood-vessel activity to restore blood pressure to normal when it starts to fall.

The specialization of muscle cells, contraction, also depends on these cells' ability to undergo action potentials. Action potentials trigger muscle contractions, many of which are important in maintaining homeostasis. For example, beating of the heart, mixing of ingested food with digestive enzymes, and shivering to generate heat when the body is cold are all accomplished by action potential–induced muscle contractions.

 CHAPTER SUMMARY

**Electrical Signals:
Graded Potentials and Action Potentials**

Nerve and muscle cells are known as excitable tissues because they can rapidly alter their membrane permeabilities and thus undergo transient membrane potential changes when excited. There are two kinds of potential change: (1) graded potentials, which serve as short-distance signals that quickly die off within a close range of the small patch of membrane where they are first triggered, and (2) action potentials, the long-distance signals.

During an action potential, depolarization of the membrane to threshold potential triggers sequential changes in permeability caused by conformational changes in voltage-gated channels. These permeability changes bring about a brief reversal of membrane potential, with Na^+ influx being responsible for the rising phase (from -70 mV to $+30$ mV), followed by K^+ efflux during the falling phase (from peak back to resting potential). Before an action potential returns to resting, it regenerates an identical new action potential in the area next to it by means of current flow that brings the previously inactive area to threshold. This self-perpetuating cycle continues until the action potential has spread throughout the cell membrane in undiminished fashion. There are two types of action potential propagation: (1) conduction by local current flow in unmyelinated fibers, in which the action potential spreads along every portion of the membrane; and (2) the more rapid saltatory conduction in myelinated fibers, where the impulse jumps over the sections of the fiber covered with insulating myelin. The Na^+-K^+ pump gradually restores the ions that moved during propagation of the action potential to their original location to maintain the concentration gradients.

It is impossible to restimulate the portion of the membrane where the impulse has just passed until it has recovered from its refractory period. The refractory period assures the unidirectional propagation of action potentials away from the original site of activation.

Action potentials occur either maximally in response to stimulation or not at all. Variable strengths of stimuli are coded by varying the frequency of action potentials, not their magnitude.

Synapses and Neuronal Integration

The primary means by which one neuron directly interacts with another neuron is through a synapse. An action potential in the presynaptic neuron triggers the release of a neurotransmitter, which combines with receptor sites on the postsynaptic neuron. This combination alters the postsynaptic cell in one of two ways. (1) The most typical response is the opening of chemical messenger–gated channels. If both Na^+ and K^+ channels are opened, the resultant ionic fluxes cause an EPSP, a small depolarization that brings the postsynaptic cell closer to threshold. On the other hand, the likelihood that the postsynaptic neuron will reach threshold is diminished when an IPSP, a small hyperpolarization, is produced as a result of the opening of either K^+ or Cl^- channels, or both. (2) In an alternate synaptic mechanism, an intracellular second messenger system, such as cyclic AMP, is activated by the neurotransmitter-receptor combination. Cyclic AMP can bring about channel opening or can have more prolonged effects within the cell, including alterations of the cell's genetic expression. Even though there are a number of different neurotransmitters, each synapse always releases the same transmitter to produce a given response when combined with a particular receptor. The response is terminated when the neurotransmitter is removed from the synaptic cleft by methods specific for the synapse.

Many neurons cosecrete larger, more slowly acting neuropeptides along with the classical neurotransmitter. The neuropeptides function at nonsynaptic sites on either the presynaptic or postsynaptic neuron to enhance or suppress synaptic effectiveness.

The interconnecting synaptic pathways between various neurons are incredibly complex due to convergence of neuronal input and divergence of its output. Usually, many presynaptic inputs converge upon a single neuron and jointly control its level of excitability. This same neuron, in turn, diverges to synapse with and influence the excitability of many other cells. Each neuron thus has the task of computing an output to numerous other cells from a complex set of inputs to itself. Depending on the combination of signals it is receiving from its various presynaptic inputs, at any given time a neuron may react by (1) firing action potentials along its axon, (2) remaining at rest and not passing any signals along, or (3) having its level of excitability reduced.

If the dominant activity is in its excitatory inputs, the postsynaptic cell is likely to be brought to threshold and have an action potential. This can be accomplished by either temporal summation (EPSPs from a repetitively firing presynaptic input occurring so close together in time that they add together) or spatial summation (adding of EPSPs occurring simultaneously from several different presynaptic inputs). Because the axon hillock has the lowest threshold, action potentials are initiated there. The frequency of action potentials reflects the magnitude of EPSP summation. If inhibitory inputs dominate, the postsynaptic potential will be brought farther than usual away from threshold. If excitatory and inhibitory activity to the postsynaptic neuron is balanced, the membrane will remain close to resting.

Numerous factors may alter synaptic effectiveness: some are built-in mechanisms to fine-tune neural responsiveness, some are deliberate pharmacological manipulations to achieve a desired result, and some are accidents caused by poisons or disease processes.

REVIEW EXERCISES

Objective Questions (Answers on p. D–1.)

1. Conformational changes in channel proteins brought about by voltage changes are believed to be responsible for opening and closing Na^+ and K^+ gates during the generation of an action potential. (True or false?)

2. The Na^+-K^+ pump restores the membrane to resting potential after it reaches the peak of an action potential. (True or false?)

3. Following an action potential, there is more K^+ outside the cell than inside because of the K^+ efflux. (True or false?)

4. Synapses permit two-way transmission of signals between two neurons. (True or false?)

5. The unidirectional propagation of action potentials away from the original site of activation is assured by the _____.

6. The _____ is the site of action potential initiation in most neurons because it has the lowest threshold.

7. A junction in which electrical activity in one neuron influences the electrical activity in another neuron by means of a neurotransmitter is called a _____.

8. Summing of EPSPs occurring very close together in time as a result of repetitive firing of a single presynaptic input is known as _____.

9. Summing of EPSPs occurring simultaneously from several different presynaptic inputs is known as _____.

10. The neuronal relationship where synapses from many presynaptic inputs act upon a single postsynaptic cell is called _____, whereas the relationship in which a single presynaptic neuron synapses with and thereby influences the activity of many postsynaptic cells is known as _____.

11. Using the answer code at the right, indicate which potential is being described:

_____ 1. behaves in all-or-none fashion

_____ 2. magnitude of potential change varies with the magnitude of triggering response

_____ 3. decremental spread away from original site

_____ 4. spreads throughout the membrane in nondiminishing fashion

_____ 5. serves as long-distance signal

_____ 6. serves as short-distance signal

a. graded potential
b. action potential

Essay Questions

1. What are the two types of excitable tissue?
2. Define the following terms: polarization, depolarization, hyperpolarization, repolarization, resting membrane potential, threshold potential, action potential, refractory period, all-or-none law.
3. Compare the three kinds of channels in terms of the factor that induces the change in channel conformation.
4. Describe the permeability changes and ionic fluxes that occur during an action potential.
5. Compare conduction by local current flow and saltatory conduction.
6. Compare the events that occur at excitatory and inhibitory synapses.
7. Distinguish between classical neurotransmitters and neuropeptides. Compare neurotransmitters and neuromodulators.
8. Discuss the possible outcomes of the grand postsynaptic potential brought about by interactions between EPSPs and IPSPs.
9. Distinguish between presynaptic inhibition and an inhibitory postsynaptic potential.

POINTS TO PONDER

1. Which of the following would occur if a neuron were experimentally stimulated simultaneously at both ends?

   a. The action potentials would pass in the middle and travel to the opposite ends.

   b. The action potentials would meet in the middle and then be propagated back to their starting positions.

   c. The action potentials would stop as they met in the middle.

   d. The stronger action potential would override the weaker action potential.

   e. Summation would occur when the action potentials met in the middle, resulting in a larger action potential.

2. Compare the expected changes in membrane potential of a neuron stimulated with a *subthreshold stimulus* (a stimulus not sufficient to bring the membrane to threshold), a *threshold stimulus* (a stimulus just sufficient to bring the membrane to threshold), and a *suprathreshold stimulus* (a stimulus larger than that necessary to bring the membrane to threshold).

3. Assume you touched a hot stove with your finger. Contraction of the biceps muscle causes flexion (bending) of the elbow, whereas contraction of the triceps muscle causes extension (straightening) of the elbow. What pattern of postsynaptic potentials (EPSPs and IPSPs) would you expect to be initiated reflexly in the cell bodies of the neurons controlling these muscles to pull your hand away from the painful stimulus?

   Now assume your finger is being pricked to obtain a blood sample. The same *withdrawal reflex* would be initiated. What pattern of postsynaptic potentials would you voluntarily produce in the neurons controlling the biceps and triceps to keep your arm extended in spite of the painful stimulus?

4. Schizophrenia is believed to be caused by excessive dopamine activity in a particular region of the brain. Explain why symptoms of schizophrenia sometimes occur as a side effect in patients being treated for Parkinson's disease.

NERVOUS SYSTEM
(CENTRAL NERVOUS SYSTEM)

Body systems maintain homeostasis

HOMEOSTASIS
The nervous system, as one of the body's two major control systems, regulates many body activities aimed at maintaining a stable internal fluid environment.

Homeostasis is essential for survival of cells

CELLS

Cells make up body systems

The **nervous system** is one of the two major control systems of the body, the other being the endocrine system. A complex interactive network of three basic types of nerve cells—afferent neurons, efferent neurons, and interneurons—constitutes the nervous system. The **central nervous system (CNS)** is composed of the brain and spinal cord, which receive input about the external and internal environment from the afferent neurons. The CNS sorts and processes this input, then initiates appropriate directions in the efferent neurons, which carry the instructions to glands or muscles to bring about the desired response—some type of secretion or movement. Many of these neurally controlled activities are directed toward maintaining homeostasis. In general, the nervous system acts by means of its electrical signals (action potentials) to control the rapid responses of the body.

CHAPTER CONTENTS AT A GLANCE

INTRODUCTION
COMPARISON OF NERVOUS AND ENDOCRINE SYSTEMS
ORGANIZATION OF THE NERVOUS SYSTEM
● CONCEPTS, CHALLENGES, AND CONTROVERSIES The Human
   Brain: In Search of Self-Understanding
PROTECTION AND NOURISHMENT OF THE BRAIN
Neuroglia; meninges; cerebrospinal fluid
Blood-brain barrier
CEREBRAL CORTEX
Cortical structure
Sensory perception
Motor control
Language ability
Association areas
SUBCORTICAL STRUCTURES AND THEIR RELATIONSHIP
   WITH THE CORTEX IN HIGHER BRAIN FUNCTIONS
Basal nuclei
Thalamus
Hypothalamus
Limbic system
Learning; memory
● CONCEPTS, CHALLENGES, AND CONTROVERSIES Alzheimer's
   Disease: A Tale of Beta Amyloid and Dementia
CEREBELLUM
BRAIN STEM
Components; functions
Consciousness; sleep-wake cycle
SPINAL CORD
Anatomy of spinal cord
Spinal reflexes
● A CLOSER LOOK AT EXERCISE PHYSIOLOGY Swan Dive or Belly
   Flop: It's a Matter of CNS Control

 # INTRODUCTION

The way humans act and react depends on complex, organized, discrete neuronal processing. Many of the basic life-supporting neuronal patterns, such as those controlling respiration and circulation, are similar in all individuals. However, there must be subtle differences in neuronal integration between someone who is a talented composer versus someone who cannot carry a tune, or someone who is a math wizard versus someone who struggles with long division. Some differences in the nervous systems between individuals are genetically endowed. The rest, however, are due to environmental encounters and experiences. When the immature nervous system develops according to its genetic plan, an overabundance of neurons and synapses are formed. Depending on external stimuli and the extent these pathways are used, some are retained, firmly established, and even enhanced, whereas others are eliminated. A case in point is **amblyopia** (**lazy eye**), in which the weaker of the two eyes in not used for vision. A lazy eye that does not get appropriate visual stimulation during a critical developmental period will almost completely and permanently lose the power of vision. The blind eye itself is completely normal; the defect lies in the lost neuronal connections in the brain's visual pathways. If, however, the weak eye is forced to work by covering the stronger eye with a patch during the sensitive developmental period, its vision will be retained. The maturation of the nervous system truly does involve instances of "use it or lose it." Once matured, ongoing modifications still occur as we continue to learn from our unique set of experiences. For example, the act of reading this page is somehow altering the neuronal activity of your brain as you (it is hoped) tuck the information away in your memory.

 # COMPARISON OF NERVOUS AND ENDOCRINE SYSTEMS

The nervous system is a "wired" system, and the endocrine system is a "wireless" system, even though both systems influence their target cells by means of chemical messengers.

The nervous and endocrine systems are the two main control systems of the body. The **nervous system**, through its swift transmission of electrical impulses, generally coordinates the rapid activities of the body, such as muscle movements. The **endocrine system**, which secretes *hormones* into the blood for delivery to distant sites of action, primarily controls metabolic and other activities that require duration rather than speed, such as maintenance of blood glucose levels. Although these two systems differ in many respects, they have much in common (Table 5–1). They both ultimately alter their **target cells** (their sites of action) by releasing chemical messengers (neurotransmitters in the case of nerve cells, hormones in the case of endocrine cells), which interact in particular ways with specific receptors (particular plasma-membrane proteins) of the target cells.

Anatomically, the nervous and endocrine systems are quite different. In the nervous system, each nerve cell terminates directly on its specific target cells; that is, the nervous system is "wired" in a very specific way into highly organized, distinct anatomical pathways for transmission of signals from one part of the body to another. Information is carried along chains of neurons to the desired destination through action potential propagation coupled with synaptic transmission (chapter 4). In contrast, the endocrine system is a "wireless" system in that the endocrine glands are not anatomically linked with their target cells. Instead, the endocrine chemical messengers are secreted into the blood and delivered to distant target sites. In fact, the components of the endocrine system itself are not anatomically interconnected; the endocrine glands are scattered throughout the body (see Fig. 18–2, p. 617). These glands constitute a system in a functional sense, however, because they all secrete hormones and many interactions take place between various endocrine glands.

As a result of these anatomical differences, the two systems

TABLE 5–1   Comparison of the Nervous System and the Endocrine System

PROPERTY	NERVOUS SYSTEM	ENDOCRINE SYSTEM
Anatomical arrangement	A "wired" system; specific structural arrangement between neurons and their target cells; structural continuity in the system	A "wireless" system; endocrine organs widely dispersed and not structurally related to one another or to their target cells
Type of chemical messenger	Neurotransmitters released into synaptic cleft	Hormones released into blood
Distance of action of chemical messenger	Very short distance (diffuses across synaptic cleft)	Long distance (carried by blood)
Means of specificity of action on target cell	Dependent on close anatomical relationship between nerve cells and their target cells	Dependent on specificity of target cell binding and responsiveness to a particular hormone
Speed of response	Rapid (milliseconds)	Slow (minutes to hours)
Duration of action	Brief (milliseconds)	Long (minutes to days or longer)
Major functions	Coordinates rapid, precise responses	Controls activities that require long duration rather than speed
Influence on other major control system?	Yes	Yes

accomplish specificity of action by distinctly different means. Specificity of neural communication depends on nerve cells and their target cells having a close anatomical relationship so that each neuron has a very narrow range of influence. A neurotransmitter is released for restricted distribution only to specific adjacent target cells, then is swiftly inactivated by enzymes at the nerve–target cell juncture or is taken back up by the nerve terminal before it is able to gain access to the blood. The target cells for a particular neuron have receptors for the neurotransmitter, but so do many other cells in other locations, and they could respond to this same mediator if it were delivered to them. For example, the entire system of nerve cells supplying all of your body's skeletal muscles (motor neurons) use the same neurotransmitter, *acetylcholine (ACh)*, and all of your skeletal muscles bear complementary ACh receptors (chapter 8). Yet specific muscles are able to be discretely activated to contract because of precise structural arrangements, or wiring patterns, between motor neurons and muscle cells. You are able to specifically wiggle your big toe without influencing any of your other muscles because ACh can be discretely released from the motor neurons that are specifically wired to the muscles controlling your toe. If ACh were indiscriminately released into the blood, as are the hormones of the wireless endocrine system, all of the skeletal muscles would simultaneously respond by contracting because they all have identical receptors for ACh. This does not happen, of course, due to the direct lines of communication between neurons and their target cells.

This specificity is in sharp contrast to the way specificity of communication is built into the endocrine system. Because hormones travel in the blood, they are able to reach virtually all tissues. Yet despite this ubiquitous distribution, only specific target cells are able to respond to each hormone. Specificity of hormonal action depends on specialization of target-cell receptors. For a hormone to exert its effect, the essential

first step is the binding of the hormone with receptors specific for it that are located only on or in the hormone's target cells. Target-cell receptors are highly discerning in their binding function. They will recognize and bind only a certain hormone, even though they are exposed simultaneously to many other blood-borne hormones, some of which are structurally very similar to the one that they discriminately bind. A receptor recognizes a specific hormone because the conformation of a portion of the receptor molecule matches a unique portion of its binding hormone in "lock-and-key" fashion. Binding of a hormone with target-cell receptors initiates a reaction (or series of reactions) that culminates in the hormone's final effect. The hormone cannot influence any other cells because they lack the right binding receptors. This lock-and-key type of interaction between a hormone and its target cells assures that the regulation of a particular tissue will be governed by the appropriate hormone.

Note that in contrast to neural communication, specificity of endocrine communication is built into the receiving end, that is, at the target cells. At the transmitting end of endocrine control systems, the original hormonal messenger is widely dispersed instead of being directed specifically toward the target cells. Radio transmission and reception provide an analogy. Radio waves (a specific hormone) are nondiscriminately dispersed in all directions from their point of origin (an endocrine gland). Even though radio waves are present in the air surrounding all of your appliances (hormone circulating throughout the body in the blood), only your radio (the target cell) is able to respond to them because it alone has receiving components specific for radio waves (receptors). Thus, even though your radio and coffee pot are being equally bombarded by radio waves, only your radio can respond. Similarly, even though all tissues are equally exposed to a given hormone through the blood, only the target cells with the appropriate receptors for that hormone are able to respond to it.

Even though the nervous and endocrine systems have their own realms of authority, they are functionally interconnected.

The nervous and endocrine systems are specialized for controlling different types of activities. In general, the nervous system is responsible for coordinating rapid, precise responses. Neural signals in the form of action potentials are rapidly propagated along nerve-cell fibers, resulting in the release at the nerve terminal of a neurotransmitter that has to diffuse only a microscopic distance to its target cell before a response is effected. Not only is a neurally mediated response rapid, but it is also brief; the action is quickly brought to a halt as the neurotransmitter is swiftly removed from the target site. This permits either termination of the response, almost immediate repetition of the response, or rapid initiation of an alternate response, depending on the circumstances (for example, the swift changes in commands to muscle groups needed to coordinate walking). This mode of action makes neural communication extremely rapid and precise. The target tissues of the nervous system are the muscles and glands, especially exocrine glands, of the body.

The supremacy of the endocrine system lies in its ability to control activities that require duration rather than speed, such as regulating organic metabolism and H_2O and electrolyte balance, promoting smooth, sequential growth and development, controlling reproduction, and regulating red blood cell production.

The endocrine system responds more slowly to its triggering stimuli than the nervous system does for several reasons. First, the endocrine system must depend on blood flow to convey its hormonal messengers over long distances. Second, hormones' mechanism of action at their target cells is more complex than that of neurotransmitters and thus requires more time before a response occurs. The ultimate effect of some hormones cannot be detected until a few hours after they bind with target-cell receptors. Also, due to the receptors' high affinity for their respective hormone, hormones often remain bound to receptors for some time, thus prolonging their biological effectiveness. Furthermore, unlike the brief, neurally induced responses that come to a halt almost immediately after the neurotransmitter is removed, endocrine effects usually last for some time after the hormone's withdrawal. Neural responses to a single burst of neurotransmitter release usually last only milliseconds to seconds, whereas the alterations in target cells induced by hormones range from minutes to days or, in the case of growth-promoting effects, even a lifetime. Thus, hormonal action is relatively slow and prolonged, making endocrine control particularly suitable for the regulation of metabolic activities that require long-term stability.

Although the endocrine and nervous systems have their own areas of specialization, they are intimately interconnected functionally. Some nerve cells do not release neurotransmitters at synapses but instead terminate at blood vessels and release their chemical messengers into the blood, where these chemicals act as hormones. A given messenger may even be a neurotransmitter when released from a nerve end-

ing and a hormone when secreted by an endocrine cell. The nervous system directly or indirectly controls the secretion of many hormones. At the same time, many hormones act as neuromodulators (see p. 98), altering synaptic effectiveness and thereby influencing the excitability of the nervous system. The presence of certain key hormones is even essential for the proper development and maturation of the brain during fetal life. Furthermore, in many instances both the nervous and endocrine systems influence the same target cells in supplementary fashion. For example, these two major control systems both contribute to the regulation of the circulatory and digestive systems. Thus, many important regulatory interfaces exist between the nervous and endocrine systems; the study of these relationships is known as **neuroendocrinology**.

For now we will concentrate on the nervous system and will examine the endocrine system in more detail in later chapters. Throughout the text we will continue to point out the numerous ways in which these two control systems interact so that the body is a coordinated whole, even though each system has its own realm of authority.

ORGANIZATION OF THE NERVOUS SYSTEM

The nervous system is organized into the central nervous system and the peripheral nervous system.

The nervous system is organized into the **central nervous system (CNS)**, consisting of the **brain** and **spinal cord**, and the **peripheral nervous system (PNS)**, consisting of nerve fibers that carry information between the CNS and other parts of the body (the periphery) (● Fig. 5–1). The PNS is further subdivided into afferent and efferent divisions. The **afferent division** (*afferent* means "carrying toward") carries information *to* the CNS, apprising it of the external environment and providing status reports on internal activities being regulated by the nervous system. Instructions *from* the CNS are transmitted via the **efferent division** (*efferent* means "carrying from") to **effector organs**—the muscles or glands that carry out the orders to bring about the desired effect. The efferent nervous system is divided into the **somatic nervous system**, which consists of the fibers of the **motor neurons** that supply the skeletal muscles, and the **autonomic nervous system** fibers, which innervate smooth muscle, cardiac muscle, and glands. The latter system is further subdivided into the **sympathetic nervous system** and the **parasympathetic nervous system**, both of which innervate most of the organs supplied by the autonomic system.

It is important to recognize that all of these "nervous systems" are really subdivisions of a single, integrated nervous system. They are arbitrary divisions based on differences in the structure, location, and functions of the various diverse parts of the whole nervous system.

There are three classes of neurons.

Three classes of neurons make up the nervous system: *afferent neurons, efferent neurons,* and *interneurons.* The afferent ner-

● **FIGURE 5–1   Organization of Nervous System**

vous system is composed of **afferent neurons**, which are shaped differently than efferent neurons and interneurons (● Fig. 5–2). At its peripheral ending, an afferent neuron has a **sensory receptor** that generates action potentials in response to a particular type of stimulus. (This stimulus-sensitive receptor should not be confused with the plasma membrane's special protein receptors, which bind chemical messengers.) The afferent neuron cell body, which is devoid of dendrites and presynaptic inputs, is located adjacent to the spinal cord. A long *peripheral axon,* commonly called the *afferent fiber,* extends from the receptor to the cell body, and a short *central axon* passes from the cell body into the spinal cord. Action potentials are initiated at the receptor end of the peripheral axon in response to a stimulus and are propagated along the peripheral axon and central axon toward the spinal cord. The terminals of the central axon diverge and synapse with other neurons within the spinal cord; in this way they disseminate information about the stimulus. Afferent neurons thus lie primarily within the peripheral nervous system. Only a small portion of their central axon endings project into the spinal cord to relay peripheral signals.

Efferent neurons also lie primarily in the peripheral nervous system (Fig 5–2). The cell bodies of efferent neurons originate in the CNS, where many centrally located presynaptic inputs converge upon them to influence their outputs to the effector organs. Efferent axons (*efferent fibers*) leave the CNS to course their way to the muscles or glands they innervate, conveying their integrated output for the effector organs to put into effect. (An autonomic nerve pathway actually consists of a two-neuron chain between the CNS and effector organ.)

Interneurons lie entirely within the CNS. About 99% of all neurons belong to this category. They serve two main roles. First, as their name implies, they lie between (*inter* means "between") the afferent and efferent neurons and are important in the integration of peripheral responses to peripheral information. For example, upon receiving information through afferent neurons that you are touching a hot object, appropriate interneurons signal efferent neurons that transmit to your hand and arm muscles the message, "Pull the hand away from the hot object!" The more complex the required action, the greater the number of interneurons interposed

Central nervous system (spinal cord)

Peripheral nervous system

Axon terminals

Cell body

Afferent neuron

Central axon

Peripheral axon (afferent fiber)

Receptor

Interneuron—

Efferent neuron*

Cell body

Axon (efferent fiber)

Effector organ (muscle or gland)

Axon terminals

* Efferent autonomic nerve pathways consist of a two-neuron chain between the CNS and effector organ.

between the afferent message and efferent response. Second, interconnections between interneurons themselves are responsible for the abstract phenomena associated with the "mind," such as thoughts, emotions, memory, creativity, intellect, and motivation. These activities are the least well understood functions of the nervous system. (See the accompanying boxed feature, Concepts, Challenges, and Controversies.)

With this brief introduction to the types of neurons and their location in the various divisions of the nervous system, we will now turn our attention to the central nervous system, followed in the next two chapters by a discussion of the two divisions of the peripheral nervous system.

PROTECTION AND NOURISHMENT OF THE BRAIN

Neuroglia physically support the interneurons and help sustain them metabolically.

About 90% of the cells within the CNS are not neurons but **glial cells** or **neuroglia**. Despite their large numbers, the glial cells occupy only about half the volume of the brain because they do not branch as extensively as the neurons do. Unlike neurons, glial cells do not initiate or conduct nerve impulses. They are important, however, in the viability of the CNS. For much of the time since their discovery in the nineteenth century, the glial cells were thought to be passive "mortar" that physically supported the functionally important neurons. In the last decade, however, the varied and important roles of these dynamic cells have become apparent. The glial cells serve as the connective tissue of the CNS and as such help support the neurons both physically and metabolically. The four major types of glial cells in the CNs are *astrocytes, oligodendrocytes, ependymal cells,* and *microglia* (Table 5–2).

Named for their starlike shape (*astro* means "star," *cyte* means "cell") (● Fig. 5–3), **astrocytes** provide a number of critical functions. First, as the main "glue" (*glia* means "glue") of the CNS, they hold the neurons together in proper spatial relationships. Second, astrocytes serve as a scaffold to guide neurons to their proper final destination during fetal brain development. Third, astrocytes induce the small blood vessels of the brain to undergo the anatomical and functional changes that are responsible for the establishment of the blood-brain barrier, a highly selective barricade between the blood and brain that will soon be described in greater detail. Fourth, astrocytes are important in the repair of brain injuries and in neural scar formation. Fifth, they help support the neurons metabolically. Astrocytes take up glutamate and gamma-aminobutyric acid (GABA), excitatory and inhibitory neurotransmitters, respectively, and then degrade these chemical messengers into raw materials for the neurons to use in making more of these neurotransmitters. Sixth, the astrocytes take

THE HUMAN BRAIN: IN SEARCH OF SELF-UNDERSTANDING

The responsibilities and mechanisms of function of much of the human brain are still very poorly understood. Part of our lack of understanding derives from the complexity of the brain. It is the most complicated, mysterious, awesome organ on earth. Considering the myriad interconnections possible between the estimated 100 billion neurons in the brain, it is a wonder that scientists have been able to unravel even the bits of information that they have.

Furthermore, unlike other organs of the body, the human brain is so unique that no experimental animal models are available for study of its most complex and sophisticated functions, such as language and creativity. In contrast, because human hearts are essentially the same as dog hearts, investigations probing canine heart function provide valuable insights into human heart function. Many aspects of brain function involve subjective feelings that cannot be measured or even communicated by nonverbal animals. We do not know to what extent, if any, animals experience what we know of as happiness, sadness, love, fear, jealousy, and so on. We can only observe in them behaviors that we associate with certain emotional states (e.g., a dog wagging its tail when it is "happy"). Therefore, studies on experimental animals have been limited to examining less advanced, objectively measurable brain activities, such as control of movement, that are shared by humans and other species.

In the past, knowledge about the brain was gleaned from (1) stimulation during brain surgery; (2) analysis of changes in electrical activity reaching the surface of the skull when a person is engaged in various activities; (3) observation of deficits in function associated with diseases or destruction of particular areas of the brain; (4) detailed anatomical studies in cadavers; and (5) inferences from animal studies. The latter, of course, are limited to characteristics shared by these species and hu-

mans, leaving most of the very characteristics that confer upon us our uniqueness as humans largely unexplored.

Now, however, several exciting new noninvasive techniques of imaging the living human brain, as well as new methods of probing the cellular and molecular functions of neurons, are contributing pieces of knowledge to our efforts to bridge the gap between our understanding of the biology of nerve cells and the behavior and intellect of the human brain. In fact, it is the human brain itself that is developing techniques for unraveling the secrets of its own complex nature.

These new imaging techniques, which are also becoming useful diagnostic tools, include **computerized axial tomography (CAT) scans**, **magnetic resonance imaging (MRI) scans**, and **positron emission tomography (PET) scans**. All three of these techniques depend on a computerized reconstruction of an image built from a series of "snapshots" that reveal specific differences in the imaged tissue. In CAT scans, images are reconstructed from differences in X-ray absorption by various areas in the brain or other body parts being examined. With MRI scans, differences in the vi-

bration of hydrogen ions (protons) in various areas of the brain are detected through the use of an externally applied magnetic field and reconstructed into a computerized image. Whereas CAT and MRI scans are especially useful in detecting tumors, intracranial hemorrhages, and other physical abnormalities, PET scans are particularly helpful in detecting blood flow in the brain. PET scans depend on the injection of short-lived radioactively labeled O_2. As this tracer circulates in the blood, it emits positrons, which collide with electrons and produce a tiny burst of gamma-ray energy. The PET equipment detects this gamma radiation. Increased gamma activity means that more blood is flowing through that area. A computer then reconstructs an image of blood flow in the brain or other body part from a rapid series of PET snapshots. Because more blood flows into a particular region of the brain when it is more active, neuroscientists are able to "take pictures" of the brain at work. For example, different areas of the brain "light up" on PET scans as a person performs different tasks, indicative of increased activity in these areas (see the accompanying figure).

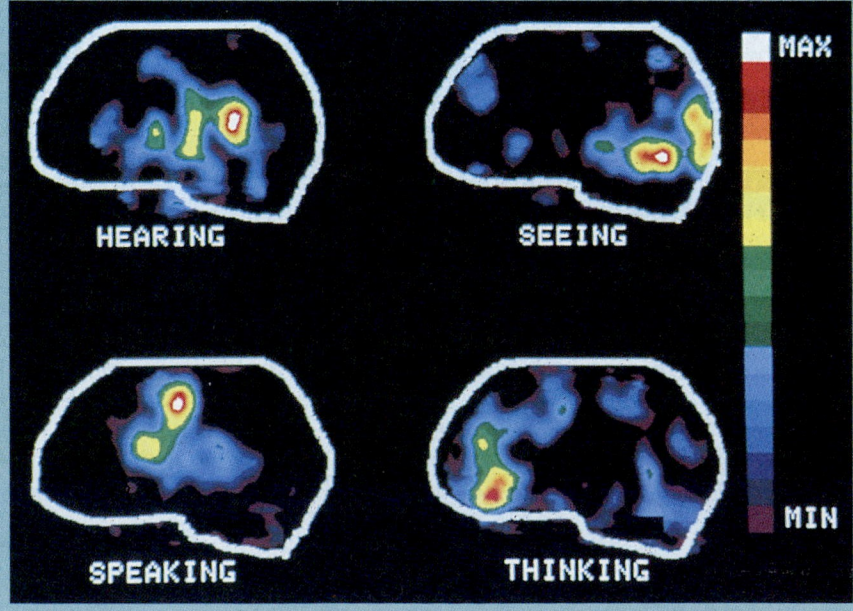

TABLE 5–2 Functions of Glial Cells

TYPE OF GLIAL CELL	FUNCTIONS
Astrocytes	Physically support neurons in proper spatial relationships
	Serve as scaffold during fetal brain development
	Induce formation of blood-brain barrier
	Form neural scar tissue
	Take up and degrade released neurotransmitters into raw materials for synthesis of more neurotransmitters by neurons
	Take up excess K^+ to help maintain proper brain ECF ion concentration and normal neural excitability
	Possess receptors for neurotransmitters, which may be important in a chemical signaling system
Oligodendrocytes	Form myelin sheaths in CNS
Ependymal cells	Line internal cavities of brain and spinal cord
	Contribute to formation of cerebrospinal fluid
Microglia	Play a role in defense of brain as phagocytic scavengers

up excess K^+ from the brain ECF when high action potential activity outpaces the ability of the Na^+-K^+ pump to return the effluxed K^+ to the neurons. By doing so, the astrocytes help maintain the proper brain ECF ion concentration to sustain normal neural excitability. If brain ECF K^+ levels were allowed to rise, the resultant lower K^+ concentration gradient between the neuronal ICF and surrounding ECF would reduce the neuronal membrane closer to threshold, even at rest. This would increase the excitability of the brain. In fact, an elevation in brain ECF K^+ concentration may be one of the factors responsible for the brain cells' explosive convulsive discharge that occurs during epileptic seizures. Finally, astro-

• **FIGURE 5–3** **Astrocytes** Note the starlike shape of these two astrocytes that have been grown in tissue culture.

cytes possess receptors for the same neurotransmitters as neurons do. Researchers have shown that binding of the common neurotransmitter glutamate to astrocyte receptors causes these cells to release stored calcium ions. The purpose of this calcium unleashing is unclear, but it may represent a type of chemical signaling system between cells in the brain; that is, a way in which these glial cells "talk" to each other or to neurons.

Oligodendrocytes form the insulative myelin sheaths around axons in the CNS. An oligodendrocyte has several elongated projections, each of which is wrapped jelly-roll fashion around a section of an interneuronal axon to form a patch of myelin (see Fig. 4–11b).

Ependymal cells line the internal cavities of the CNS. As the nervous system develops embryonically from a hollow neural tube, the original central cavity of this tube is maintained and modified to form the **ventricles** of the brain and the **central canal** of the spinal cord. The ependymal cell lining of the ventricles contributes to the formation of cerebrospinal fluid.

The **microglia** are the scavengers of the CNS. They are phagocytic cells delivered by the blood to the central nervous tissue, where they remain stationary until activated by an infection or injury. They then migrate to the affected area to remove any foreign invaders or tissue debris.

Unlike neurons, glial cells do not lose the ability to undergo cell division, so most brain tumors of neural origin consist of glial cells (**gliomas**). Neurons themselves do not form tumors because they are unable to divide and multiply. Brain tumors of nonneural origin are of two types: (1) those that metastasize (spread) to the brain from other sites and (2) **meningiomas**, which originate from the meninges, the protective membranes covering the central nervous system.

The delicate central nervous tissue is well protected.

Central nervous tissue is very delicate. This characteristic, coupled with the fact that damaged nerve cells cannot be replaced because neurons are unable to divide, makes it imperative that this fragile, irreplaceable tissue be well protected. Four major features help protect the CNS from injury:

1. It is enclosed by hard, bony structures. The **cranium** (**skull**) encases the brain, and the **vertebral column** surrounds the spinal cord.

2. Three protective and nourishing membranes, the **meninges**, lie between the bony covering and the nervous tissue.

3. The brain "floats" in a special cushioning fluid, the **cerebrospinal fluid** (**CSF**).

4. A highly selective **blood-brain barrier** limits access of blood-borne materials into the vulnerable brain tissue.

The role of the first of these protective devices, the bony covering, is self-evident. However, the latter three protective mechanisms warrant further discussion.

Three meningeal "mothers" provide protection and nourishment for the central nervous system.

The meninges, the three membranes that wrap the central nervous system are, from the outermost to the innermost layer, the dura mater, the arachnoid mater, and the pia mater (• Fig. 5–4). (*Mater* means "mother", indicative of the protective and supportive role played by these membranes.)

The **dura mater** (*dura* means "tough") is a tough, inelastic covering consisting of two layers. Usually, these layers adhere

---

• **FIGURE 5–4 Relationship of Meninges and Cerebrospinal Fluid to Brain and Spinal Cord** (a) Brain, spinal cord, and meninges in sagittal section. The arrows depict the direction of flow of cerebrospinal fluid (in yellow), which is produced by the choroid plexuses ①, circulates throughout the ventricles ②, exits the fourth ventricle at the base of the brain ③, flows in the subarachnoid space between the meningeal layers ④, and is finally reabsorbed from the subarachnoid space into the venous blood across the arachnoid villi ⑤. (b) Frontal section in the region between the two cerebral hemispheres of the brain, depicting the meninges in greater detail.

closely, but in some regions they are separated to form blood-filled cavities, **dural sinuses**, or in the case of the larger cavities, **venous sinuses**. Venous blood draining from the brain empties into these sinuses to be returned to the heart. Cerebrospinal fluid also reenters the blood at these sinus sites.

The **arachnoid mater** (*arachnoid* means "spiderlike") is a delicate, richly vascularized layer with a "cobwebby" appearance. The space between the arachnoid layer and the underlying pia mater, the **subarachnoid space**, is filled with CSF. Protrusions of arachnoid tissue, the **arachnoid villi**, penetrate through gaps in the overlying dura and project into the dural sinuses. It is across the surfaces of these villi that CSF is reabsorbed into the blood circulating within the sinuses.

The innermost meningeal layer, the **pia mater** (*pia* means "gentle"), is the most fragile. It is highly vascular and closely adheres to the surfaces of the brain and spinal cord, following every ridge and valley. In certain areas it dips deeply into the brain to bring a rich blood supply into close contact with the ependymal cells lining the ventricles. This relationship is important in the formation of CSF, a topic to which we now turn our attention.

The central nervous system is suspended in its own special cerebrospinal fluid.

Cerebrospinal fluid is formed primarily by the **choroid plexuses** found in particular regions of the ventricle cavities of the brain. Choroid plexuses consist of richly vascularized, cauliflower-like masses of pia mater tissue that dip into pockets formed by ependymal cells. Once CSF is formed, it flows through the four interconnected ventricles within the interior of the brain and through the spinal cord's narrow central canal, which is continuous with the last ventricle. Cerebrospinal fluid escapes through small openings from the fourth ventricle at the base of the brain to enter the subarachnoid space and subsequently flows between the meningeal layers over the entire surface of the brain and spinal cord (Fig 5–4). When the CSF reaches the upper regions of the brain, it is reabsorbed from the subarachnoid space into the venous blood through the arachnoid villi.

Flow of CSF through this system is facilitated by circulatory and postural factors that result in a CSF pressure of about 10 mm Hg. Reduction of this pressure by removal of even a few milliliters (ml) of CSF during a spinal tap for laboratory analysis may produce severe headaches.

Through the ongoing processes of formation, circulation, and reabsorption, the entire volume of CSF of about 125 to 150 ml is replaced more than three times a day. **Hydrocephalus** ("water on the brain") occurs if any one of these processes is defective so that excess CSF accumulates. The resulting increase in CSF pressure can lead to brain damage and mental retardation if untreated. Treatment consists of surgically shunting the excess CSF to veins elsewhere in the body.

The density of the CSF is about the same as that of the brain itself, so the brain is essentially suspended in its special fluid environment. The major function of CSF is to serve as a shock-absorbing (cushioning) fluid to prevent the brain from bumping against the interior of the hard skull when the head is subjected to sudden, jarring movements.

In addition to protecting the delicate brain from mechanical trauma, the CSF performs an important role related to the exchange of materials between the body fluids and the brain. In all body tissues, the spaces between the cells are filled with a type of extracellular fluid known as interstitial fluid. It is the interstitial fluid of the brain and not the blood plasma or the CSF that comes into direct contact with the neuronal and glial cells (● Fig. 5–5). Since the brain interstitial fluid directly bathes the neural cells, its composition is critical. The composition of brain interstitial fluid is influenced more by changes in the composition of CSF than by alterations in blood plasma because there is fairly free exchange of materials between the CSF and brain interstitial fluid but only limited exchange between the blood and the brain interstitial fluid. Accordingly, it is essential that the composition of the CSF be regulated within very narrow limits.

Cerebrospinal fluid is formed as a result of selective transport mechanisms across the membranes of the choroid plexuses. The composition of CSF differs from that of plasma. For example, CSF is lower in K^+ and higher in Na^+, making it an ideal environment for the movement of these ions down concentration gradients, a process essential for conduction of nerve impulses (see pp. 83–88).

A highly selective blood-brain barrier carefully regulates exchanges between the blood and brain.

The brain is carefully shielded from harmful changes in the blood by the blood-brain barrier. Exchange of materials between the blood and surrounding interstitial fluid can take place only across the walls of capillaries, the smallest of blood vessels. Unlike the rather free exchange across capillaries elsewhere in the body, there are strict limitations on permissible exchanges across brain capillaries. Changes in most plasma constituents do not easily influence the composition of brain interstitial fluid because only selected exchanges can be made. For example, even if the K^+ level in the blood is doubled, little change occurs in the K^+ concentration of the fluid bathing the central neurons. This is beneficial because alterations in interstitial fluid K^+ would be detrimental to neuronal function.

The blood-brain barrier consists of both anatomical and physiological factors. Capillary walls throughout the body are formed by a single layer of cells. Usually, there are holes or pores between the cells making up the capillary wall that permit free exchange of all plasma components with the surrounding interstitial fluid, with the exception of the large plasma proteins. In the brain capillaries, however, the cells are joined by **tight junctions**, which completely seal the capillary wall so that nothing can be exchanged across the wall by passing *between* the cells (● Fig. 5–6). The only possible exchanges are *through* the capillary cells themselves. Lipid-soluble substances such as O_2, CO_2, alcohol, and steroid hormones penetrate these cells easily by dissolving in their lipid plasma membrane. Small water molecules also diffuse through readily, apparently by passing between the phospholipid molecules that compose the plasma membrane. All other substances exchanged between the blood and brain interstitial fluid, including such essential materials as glucose,

• **FIGURE 5–5 Exchanges between Blood and Brain Cells** The brain cells (neurons and neuroglia) exchange materials with the surrounding brain interstitial fluid. The composition of the brain interstitial fluid is influenced primarily by the composition of the cerebrospinal fluid (CSF), not by the composition of the blood, because there is free exchange between the CSF and brain interstitial fluid whereas exchange between the blood and brain interstitial fluid is restricted by the highly discriminating blood-brain barrier. The composition of CSF, in turn, is determined by selective transport processes from the blood to the CSF across the choroid plexuses of the brain.

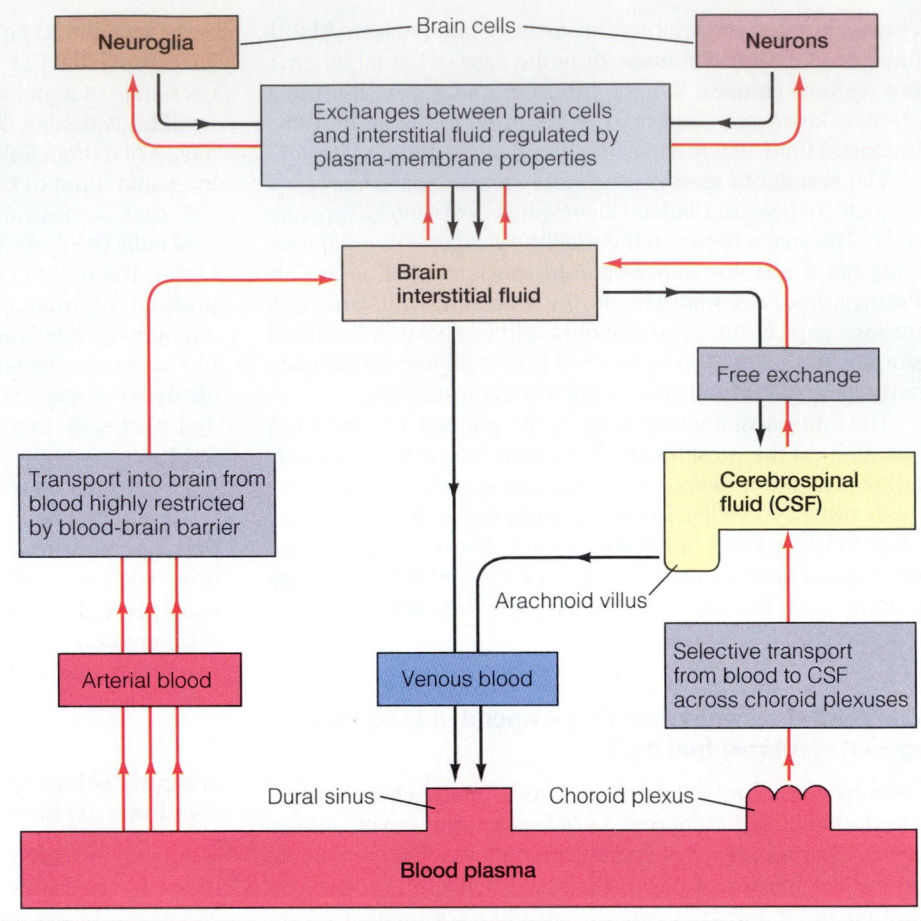

• **FIGURE 5–6 Blood-Brain Barrier** Unlike most capillaries in the body, the cells forming the walls of brain capillaries are joined by tight junctions that prevent materials from passing between the cells. The only passage across brain capillaries is through the cells that form the capillary walls. With the exception of lipid-soluble substances and water, passage of all other materials through these cells is physiologically regulated by carrier-mediated systems, which are not present in capillaries elsewhere.

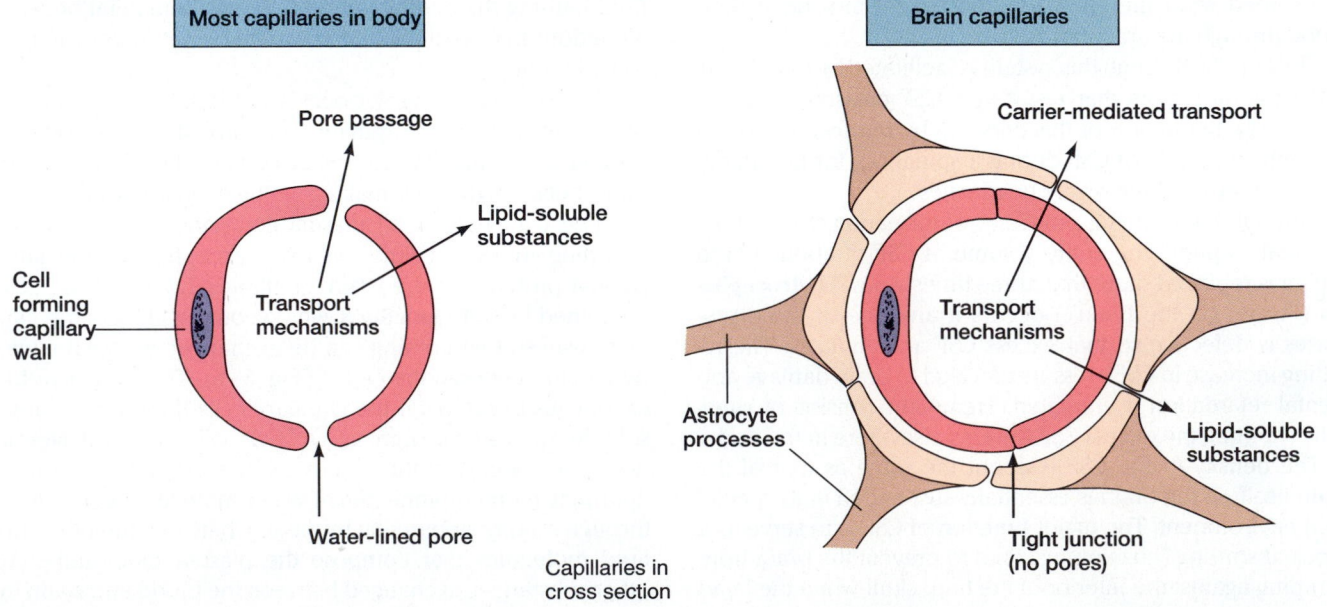

amino acids, and ions, are transported by highly selective membrane-bound carriers. Accordingly, transport across the capillary walls between the cells is anatomically prevented, and transport through the cells is physiologically restricted.

The brain capillaries are surrounded by astrocyte processes, which at one time were thought to be physically responsible for the blood-brain barrier. Evidence now indicates that the astrocytes have two roles regarding the blood-brain barrier: (1) They appear to signal the cells forming the brain capillaries to "get tight." Capillary cells do not have an inherent ability to form tight junctions; they do so only at the command of a signal within their neural environment. (2) Astrocytes are believed to participate in the cross-cellular transport of some substances, such as K^+.

The blood-brain barrier protects the delicate brain and spinal cord from chemical fluctuations in the blood and minimizes the possibility that potentially harmful blood-borne substances might reach the central neural tissue. It further prevents certain circulating hormones that could also act as neurotransmitters from reaching the brain, where they could produce uncontrolled nervous activity. On the negative side, the blood-brain barrier limits the use of drugs for the treatment of brain and spinal cord disorders because many drugs are unable to penetrate this barrier.

Certain areas of the brain are not subject to the blood-brain barrier, most notably a portion of the hypothalamus. Functioning of the hypothalamus depends on its "sampling" the blood and adjusting its controlling output accordingly in order to maintain homeostasis. Part of this output is in the form of hormones that must enter hypothalamic capillaries to be transported to their sites of action. Appropriately, these hypothalamic capillaries are not sealed by tight junctions.

The brain depends on constant delivery of oxygen and glucose by the blood.

Even though many substances in the blood never actually come in contact with the brain tissue, the brain, more than any other tissue, is highly dependent on a constant blood supply. Unlike most tissues that can resort to anaerobic metabolism to produce ATP in the absence of O_2 for at least short periods (see p. 33), the brain cannot produce ATP in the absence of O_2. Furthermore, in contrast to most tissues, which can use other sources of fuel for energy production in lieu of glucose, the brain normally uses only glucose but does not store any of this nutrient. Therefore, the brain is absolutely dependent on a continuous, adequate blood supply of O_2 and glucose. Accordingly, brain damage results if this organ is deprived of its critical O_2 supply for more than four to five minutes or if its glucose supply is cut off for more than ten to fifteen minutes.

Brain damage may occur in spite of protective mechanisms.

Even though the brain is carefully protected in its cushiony vault, traumatic head injuries may damage the delicate brain tissue. Direct brain damage occurs if the brain is violently shaken or jarred by a forceful impact, such as a fall or a blow, or if crushed cranial bones are pushed against the underlying neural tissue. Further indirect brain damage may occur following a head injury as a consequence of swelling or hemorrhaging within the enclosed confines of the cranium. The resultant increase in intracranial pressure may cause compression and damage of brain tissue.

The most common cause of brain damage is not traumatic head injuries, however, but **cerebrovascular accidents (strokes)**. When a brain (cerebral) blood vessel ruptures or is blocked by a clot, the brain tissue being supplied by that vessel is deprived of its vital O_2 and glucose supply. The result is damage and usually death of the deprived tissue. Recently, researchers have learned that neural damage (and the subsequent loss of neural function) extends well beyond the blood-deprived area as a result of a toxic release of glutamate, a common excitatory neurotransmitter, from the O_2-starved neurons. Glutamate or other neurotransmitters are normally released in small amounts from neurons as a means of chemical communication between brain cells (see p. 91). Damaged brain cells release excessive amounts of glutamate, which binds with and overexcites surrounding neurons. The excitatory overdose of glutamate subsequently destroys these surrounding cells by triggering damaging chemical reactions within them. These doomed neighbors then spew out more glutamate, which in turn kills even more neuronal victims as the damaging cascade of chemical reactions spreads from the initial site of O_2 deprivation. This glutamate cascade is believed to be responsible for the majority of neuron deaths following a stroke.

Armed with this new knowledge, scientists are currently investigating new drugs, such as glutamate-receptor blockers, that will halt this domino effect. The goal, of course, is to limit the extent of neuronal damage and thus minimize or even prevent clinical symptoms such as paralysis. This new therapeutic approach, coupled with recent successes in restoring blood flow through blocked cerebral vessels by administration of clot-dissolving drugs, holds much promise in the near future for treating strokes, which are the most prevalent cause of adult disability and the third leading cause of death in the United States. Until now, treatment of strokes has been limited to rehabilitative therapy after the damage was already complete.

Brain damage may also occur as a result of infectious or degenerative neural disorders or brain tumors, all of which can lead to destruction of brain tissue. No matter what the cause, the nature of the ensuing loss of neurological function depends on the area of the brain involved and the extent of permanent damage. The following exemplify the range of possible outcomes: (1) death if an area responsible for maintenance of a vital function is destroyed (for example, the brain center controlling respiration); (2) severe or mild loss of specific sensory awareness; (3) motor (movement) disorders of varying severity; (4) language impairment; (5) impaired mental abilities; or (6) full functional recovery.

Because mature neurons are unable to divide, destroyed CNS neurons cannot be replaced through cell division. Furthermore, damaged axons cannot regenerate within the CNS as they can in the peripheral nervous system due to the release of nerve-growth inhibiting proteins by the myelin-forming oligodendrocytes of the CNS. However, the brain displays a degree of **plasticity**, that is, an ability to change or be functionally remolded in response to the demands placed on it. This ability is more pronounced in the early developmental years, but even adults retain some plasticity. When an area of the brain associated with a particular activity is destroyed, in some instances other areas of the brain gradually assume some or all of the responsibilities of the damaged region. The underlying molecular mechanisms responsible for the brain's plasticity are only beginning to be unraveled. Current evidence suggests that the formation of new neural pathways (not new neurons, but new connections between existing neurons) in response to changes in experience are mediated in part by alterations in dendritic shape resulting from modifications in certain cytoskeletal elements (see p. 35). As its dendrites become more branched and elongated, a neuron

is able to receive and integrate more signals from other neurons.

Headaches are seldom due to brain damage.

Headaches are the most common form of pain. Almost everyone experiences headaches at least occasionally. Fortunately, most headaches are not associated with brain damage. They usually result from tension or increased pressure within pain-sensitive structures inside the cranium. The following are among the causes of headaches:

1. *Tension* associated with sustained tightening of muscles in the neck, scalp, and forehead in conjunction with anxiety, stress, or fatigue.
2. *Swelling of the mucous membranes* lining the sinuses in response to respiratory infections or allergies.
3. *Eye disorders* accompanied by straining of eye muscles.

1. Brain stem
2. Cerebellum
3. Forebrain
   a. Diencephalon
      1. Hypothalamus
      2. Thalamus
   b. Cerebrum
      1. Basal nuclei
      2. Cerebral cortex

MAJOR FUNCTIONS

1. Sensory perception
2. Voluntary control of movement
3. Language
4. Personality traits
5. Sophisticated mental events, such as thinking, memory, decision making, creativity, and self-consciousness

1. Inhibition of muscle tone
2. Coordination of slow, sustained movements
3. Suppression of useless patterns of movement

1. Relay station for all synaptic input
2. Crude awareness of sensation
3. Some degree of consciousness
4. Role in motor control

1. Regulation of many homeostatic functions, such as temperature control, thirst, urine output, and food intake
2. Important link between nervous and endocrine systems
3. Extensive involvement with emotion and basic behavioral patterns

1. Maintenance of balance
2. Enhancement of muscle tone
3. Coordination and planning of skilled voluntary muscle activity

1. Origin of majority of peripheral cranial nerves
2. Cardiovascular, respiratory, and digestive control centers
3. Regulation of muscle reflexes involved with equilibrium and posture
4. Reception and integration of all synaptic input from spinal cord; arousal and activation of cerebral cortex
5. Sleep centers

The order in which these components are listed generally represents both their anatomical location and their complexity and sophistication of function, from the least specialized, oldest level to the newest, most specialized level.

A primitive nervous system consists of comparatively few interneurons interspersed between afferent and efferent neurons. During evolutionary development, the interneuronal component progressively expanded, formed more complex interconnections, and became localized at the head end of the nervous system, forming the brain. Newer, more sophisticated layers of the brain were added on to the older, more primitive layers. The human brain represents the present peak of development.

The *brain stem*, the oldest and smallest region of the brain, is continuous with the spinal cord. It controls many of the life-sustaining processes, such as breathing, circulation, and digestion, that are common to many of the lower vertebrate forms. These processes are often referred to as "vegetative" functions because, with the loss of higher brain functions, these lower brain levels, in accompaniment with appropriate supportive therapy such as intravenous feeding, can still sustain the functions essential for survival. However, because the person has no awareness or control of that life, the condition is sometimes referred to as "being a vegetable."

Attached at the top rear portion of the brain stem is the *cerebellum*, which is concerned with maintaining proper position of the body in space and subconscious coordination of motor activity (movement). On top of the brain stem, tucked within the interior of the cerebrum is the *diencephalon*. It houses two brain components: the *hypothalamus*, which controls many homeostatic functions important in maintaining stability of the internal environment, and the *thalamus*, which performs some primitive sensory processing. On top of this "cone" of lower brain regions is the *cerebrum*, whose "scoop" gets progressively larger and more highly convoluted (that is, has tortuous ridges delineated by deep grooves or folds) the more advanced the vertebrate species is. The cerebrum is most highly developed in humans, where it constitutes about 80% of the total brain weight. The outer layer of the cerebrum is the highly convoluted *cerebral cortex*, which caps an inner core that houses the *basal nuclei*. The cerebral cortex plays a key role in the most sophisticated neural functions, such as voluntary initiation of movement, final sensory perception, conscious thought, language, personality traits, and other factors we associate with the mind or intellect. It is the highest, most complex integrating area of the brain. Each

4. *Dilation of cerebral blood vessels* in association with high blood pressure, hangovers, or migraine headaches.
5. *Increased intracranial pressure* accompanying brain tumors or intracranial hemorrhaging.
6. *Inflammation and swelling* in association with meningeal infections (**meningitis**) or infection of the brain itself (**encephalitis**).

CEREBRAL CORTEX

Newer, more sophisticated regions of the brain are piled on top of older, more primitive regions.

Although the brain is a functional whole, it is organized into several different regions. The parts of the brain can be arbitrarily grouped in various ways based on anatomical distinctions, functional specialization, and evolutionary development. We will use the following grouping (Table 5–3):

of these regions of the brain will be discussed in turn, starting with the cerebral cortex.

The cerebral cortex is an outer shell of gray matter covering an inner core of white matter.

The **cerebrum**, by far the largest portion of the human brain, is divided into two halves, the right and left **cerebral hemispheres**. They are connected to each other by the **corpus callosum**, a thick band consisting of an estimated 300 million neuronal axons traversing between the two hemispheres.

Each hemisphere is composed of a thin outer shell of *gray matter*, the **cerebral cortex**, covering a thick central core of *white matter* (see Fig. 5–14). Located deep within the white matter is another region of gray matter, the basal nuclei. Throughout the entire CNS, **gray matter** consists predominantly of densely packaged cell bodies and their dendrites as well as glial cells. Bundles or tracts of myelinated nerve fibers (axons) constitute the **white matter**; its white appearance is due to the lipid (fat) composition of the myelin. The fiber tracts in the white matter transmit signals from one part of the cerebral cortex to another or between the cortex and other regions of the CNS. Such communication enables integration between different areas of the cortex and elsewhere. This integration is essential for even a relatively simple task such as picking a flower. Vision of the flower is received by one area of the cortex, reception of its fragrance takes place in another area, and movement is initiated by still another area. More subtle neuronal responses, such as appreciation of the flower's beauty and the urge to pick it, are poorly understood but undoubtedly extensively involve interconnecting fibers between different cortical regions.

The cerebral cortex is organized into layers and functional columns.

The cerebral cortex is organized into six well-defined layers based on varying distributions of the cell bodies and locally associated fibers of several distinctive cell types. These layers are organized into functional vertical columns that extend perpendicularly from the surface down through the depths of the cortex to the underlying white matter. The neurons within a given column are believed to function as a "team," with each cell being involved in different aspects of the same specific activity—for example, perceptual processing of the same stimulus from the same location.

The functional differences between various areas of the cortex result from different layering patterns within the columns and from different input-output connections, not from the presence of unique cell types or different neuronal mechanisms. For example, those regions of the cortex responsible for perception of senses have an expanded layer 4, a layer rich in **stellate cells**, which are responsible for initial processing of sensory input to the cortex. In contrast, the cortical areas that control output to skeletal muscles have a thickened layer 5, which contains an abundance of large **pyramidal cells**. These cells send fibers down the spinal cord from the cortex to terminate on the efferent motor neurons that innervate the skeletal muscles.

The four pairs of lobes in the cerebral cortex are specialized for different activities.

It is important to recognize that even though a discrete activity is ultimately attributed to a particular region of the brain, no part of the brain functions in isolation. Each part depends on complex interplay among numerous other regions for both incoming and outgoing messages. With this in mind, let us now consider the locations of the major functional areas of the brain.

The anatomical landmarks used in cortical mapping are certain deep folds that divide each half of the cortex into four major lobes: the *occipital, temporal, parietal,* and *frontal lobes* (● Fig. 5–7). Refer to the basic functional map of the cortex in ● Figure 5–8 during the following discussion of the major activities attributed to various regions of these lobes.

● **FIGURE 5–7   Cortical Lobes** Each half of the cerebral cortex is divided into the occipital, temporal, parietal, and frontal lobes, as depicted in this schematic lateral view of the brain.

Supplementary motor area
(on inner surface—not visible;
programming of complex movements)

Premotor cortex (coordination
of complex movements)

Prefrontal association cortex
(planning for voluntary
activity; decision making;
personality traits)

Frontal lobe

Broca's area
(speech formation)

Primary auditory cortex
surrounded by higher-order
auditory cortex (hearing)

Limbic association cortex
(mostly on inner and bottom
surface of temporal lobe;
motivation and emotion; memory)

Temporal lobe

Brain stem

Spinal cord

Primary motor cortex
(voluntary movement)

Somatosensory cortex
(somesthetic sensation
and proprioception)

Posterior parietal cortex
(integration of somato-
sensory and visual input;
important for complex
movements)

Wernicke's area
(speech understanding)

Parietal lobe

Parietal-temporal-
occipital association cortex
(integration of all
sensory input; important
in language)

Occipital lobe

Primary visual cortex
surrounded by higher-
order visual cortex (sight)

Cerebellum

● **FIGURE 5–8   Functional Areas of the Cerebral Cortex** Various regions of the cerebral cortex are primarily responsible for various aspects of neural processing, as indicated in this schematic lateral view of the brain.

Occipital and Temporal Lobes The **occipital lobes**, which are located posteriorly (at the back of the head), are responsible for initially processing visual input. Sound sensation is initially received by the **temporal lobes**, located laterally (on the sides of the head).

Parietal Lobes The parietal lobes and frontal lobes, located on the top of the head, are separated by a deep infolding, the **central sulcus**, which runs roughly down the middle of the lateral surface of each hemisphere. The parietal lobes lie to the rear of the central sulcus on each side, and the frontal lobes lie in front of it.

The **parietal lobes** are primarily responsible for receiving and processing sensory input such as touch, pressure, heat, cold, and pain from the surface of the body. These sensations are collectively known as **somesthetic sensations** (*somesthetic* means "body feelings"). The parietal lobes also perceive awareness of body position, a phenomenon referred to as **proprioception**. The **somatosensory cortex**, the site for initial cortical processing of this somesthetic and proprioceptive input, is located at the front of each parietal lobe immediately behind the central sulcus (● Fig. 5–9a). Each region within the somatosensory cortex receives sensory input from a specific area of the body. This distribution of cortical sensory processing is depicted in Figure 5–9b. Note that on this so-called

sensory homunculus (*homunculus* means "little man") the body is represented upside down on the somatosensory cortex and, more importantly, *different parts of the body are not equally represented*. The size of each body part in this homunculus is indicative of the relative proportion of the somatosensory cortex devoted to that area. The exaggerated size of the face, tongue, hands, and genitalia is indicative of the high degree of sensory perception associated with these body parts.

The somatosensory cortex on each side of the brain for the most part receives sensory input from the opposite side of the body, because most of the ascending pathways carrying sensory information up the spinal cord cross over to the opposite side before eventually terminating in the cortex. Thus, damage to the left half of the somatosensory cortex produces sensory deficits on the right side of the body, whereas sensory losses on the left side are associated with damage to the right half of the cortex.

Simple awareness of touch, pressure, or temperature is detected by the thalamus, a lower level of the brain, but the somatosensory cortex goes beyond pure recognition of sensations to fuller sensory perception. The thalamus makes you aware that something hot versus something cold is touching your body but it does not tell you where or of what intensity. The somatosensory cortex localizes the source of sensory input and perceives the level of intensity of the stimulus. It also

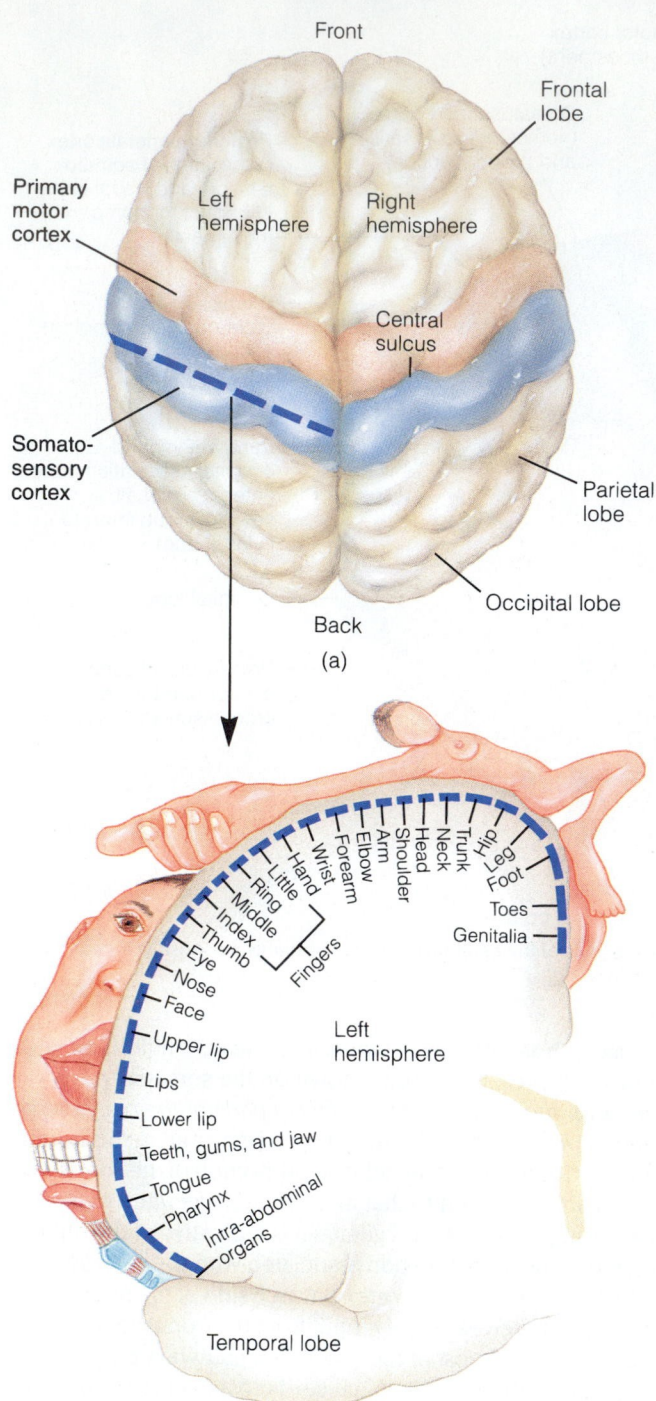

FIGURE 5–9 Somatotopic Map of Somatosensory Cortex (a) Top view of cerebral hemispheres. (b) Sensory homunculus showing the distribution of sensory input to the somatosensory cortex from different parts of the body. The distorted graphic representation of the body parts is indicative of the relative proportion of the somatosensory cortex devoted to reception of sensory input from each area.

Labels in figure:
- Front
- Primary motor cortex
- Left hemisphere
- Right hemisphere
- Frontal lobe
- Central sulcus
- Somato-sensory cortex
- Parietal lobe
- Occipital lobe
- Back
- (a)
- Toes
- Genitalia
- Fingers
- Little, Ring, Middle, Index, Thumb, Hand, Wrist, Forearm, Elbow, Arm, Shoulder, Head, Neck, Trunk, Hip, Leg, Foot
- Eye, Nose, Face, Upper lip, Lips, Lower lip, Teeth, gums, and jaw, Tongue, Pharynx, Intra-abdominal organs
- Left hemisphere
- Temporal lobe
- (b)

is capable of spatial discrimination, so it can discern shapes of objects being held and can distinguish subtle differences in similar objects that come into contact with the skin.

The somatosensory cortex, in turn, projects this sensory input via white matter fibers to adjacent higher sensory areas for even further elaboration, analysis, and integration of sensory information. These higher areas are important in the perception of complex patterns of somatosensory stimulation—for example, simultaneous appreciation of the texture, firmness, temperature, shape, position, and location of an object you are holding.

Frontal Lobes The **frontal lobes**, lying at the front of the cortex, are responsible for three main functions: (1) voluntary motor activity, (2) speaking ability, and (3) elaboration of thought. The area at the rear of the frontal lobe immediately in front of the central sulcus and adjacent to the somatosensory cortex is the **primary motor cortex** (● Fig. 5–10a). It confers voluntary control over movement produced by skeletal muscles. As in sensory processing, the motor cortex on each side of the brain primarily controls muscles on the opposite side of the body. Neuronal tracts originating in the motor cortex of the left hemisphere cross over before passing down the spinal cord to terminate on efferent motor neurons that trigger skeletal-muscle contraction on the right side of the body. Accordingly, damage to the motor cortex on the left side of the brain produces paralysis on the right side of the body and the converse is also true.

Stimulation of different areas of the primary motor cortex brings about movement in different regions of the body. Like the sensory homunculus for the somatosensory cortex, the **motor homunculus**, which depicts the location and relative amount of motor cortex devoted to output to the muscles of each body part, is upside down and distorted (Fig. 5–10b). The fingers, thumbs, and muscles important in speech, especially those of the lips and tongue, are grossly exaggerated, indicative of the fine degree of motor control with which these body parts are endowed. Compare this to how little brain tissue is devoted to the trunk, arms, and lower extremities, which are not capable of such complex movements. Thus, the extent of representation in the motor cortex is proportional to the precision and complexity of motor skills required of the respective part.

Other regions of the nervous system besides the primary motor cortex are important in motor control.

Even though signals from the primary motor cortex terminate on the efferent neurons that trigger voluntary skeletal-muscle contraction, the motor cortex is not the only region of the brain involved with motor control. First, lower brain regions and the spinal cord control involuntary skeletal-muscle activity, such as the maintenance of posture. Some of these same regions also play an important role in monitoring and coordinating voluntary motor activity that has been set in motion by the primary motor cortex. Second, although fibers originating from the motor cortex can activate motor neurons to bring about muscle contraction, the motor cortex itself does not initiate voluntary movement. The motor cortex is activated by a widespread pattern of neuronal discharge, the **readiness potential**, which occurs about 750 msec before specific electrical activity is detectable in the motor cortex. The higher motor areas of the brain believed to be involved in this voluntary decision-making period include the *supplementary* *motor area*, the *premotor cortex*, and the *posterior parietal cortex* (Fig. 5–8). These higher areas all command the primary motor cortex. Furthermore, a subcortical region of the brain, the *cerebellum*, appears to play an important role in the planning, initiation, and timing of certain kinds of movement by sending input to the motor areas of the cortex.

These four regions of the brain are important in programming and coordinating complex movements involving simultaneous contraction of many muscles. Even though electrical stimulation of the primary motor cortex brings about contraction of particular muscles, no purposeful coordinated movement can be elicited, just as pulling on isolated strings of a puppet does not produce any meaningful movement. A puppet displays purposeful movements only through coordinated manipulation of the strings by a skilled puppeteer. In the same way, these four regions (and perhaps other areas as yet undetermined) develop a **motor program** for the specific voluntary task and then "pull" the appropriate pattern of "strings" in the primary motor cortex to bring about the sequenced contraction of appropriate muscles to accomplish the desired complex movement.

The **supplementary motor area** lies on the medial (inner) surface of each hemisphere anterior to (in front of) the primary motor cortex. It appears to play a preparatory role in programming complex sequences of movement. Stimulation of various regions of this motor area brings about complex patterns of movement, such as opening or closing of the hand. Lesions here do not result in paralysis, but they do interfere with performance of more complex, useful integrated movements.

The **premotor cortex**, located on the lateral surface of each hemisphere in front of the primary motor cortex, is believed to be important in orienting the body and arms toward a specific target. The premotor cortex is guided by sensory input processed by the **posterior parietal cortex**, a region that lies posterior to (in back of) the primary somatosensory cortex. These two higher motor areas have many anatomical interconnections and appear to be closely interrelated functionally. When either of these areas is damaged, the individual cannot process complex sensory information to accomplish purposeful movement in a spatial context. These patients, for example, cannot successfully manipulate silverware when eating.

Even though these higher motor areas command the primary motor cortex and are important in preparing for the execution of deliberate, meaningful movement, we cannot say that voluntary movement is actually *initiated* by these areas. This pushes the question of how and where voluntary activity is initiated one step further. Probably, no single area is responsible; undoubtedly, numerous pathways can ultimately bring about deliberate movement.

For example, think about the neural systems called into play during the simple act of picking up an apple to eat. You know that the fruit is located in a bowl on the kitchen counter because of your memory. Sensory systems, coupled with your knowledge based on past experience, enable you to identify the apple from the other varieties of fruit in the bowl. Motor systems, upon receiving this integrated sensory information, issue commands to the exact muscles of the body in the

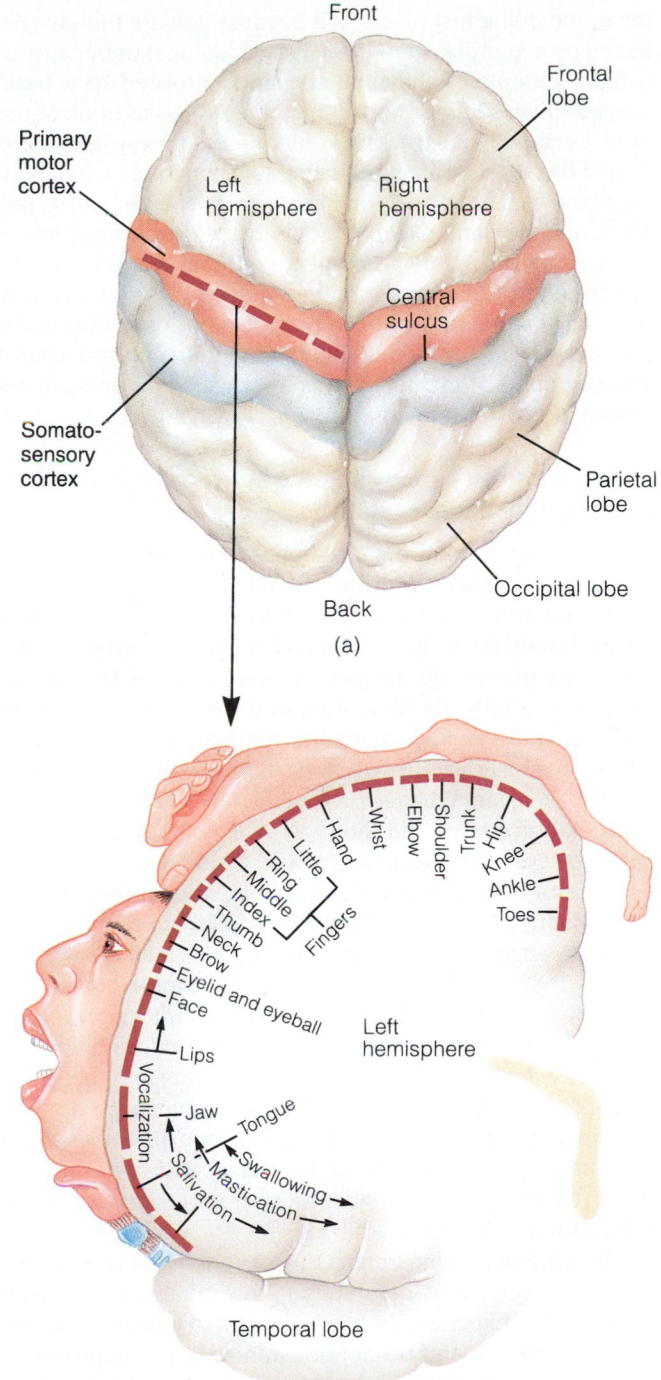

(a)

(b)

• FIGURE 5–10 Somatotopic Map of Primary Motor Cortex (a) Top view of cerebral hemispheres. (b) Motor homunculus showing the distribution of motor output from the primary motor cortex to different parts of the body. The distorted graphic representation of the body parts is indicative of the relative proportion of the primary motor cortex devoted to controlling skeletal muscles in each area.

proper sequence to enable you to move to the fruit bowl and pick up the targeted apple. During execution of this act, minor adjustments in the motor command are made as needed, based on continual updating provided by sensory input about the position of the body relative to the goal. Then there is the issue of motivation and behavior. Why are you reaching for the apple in the first place? Is it because you are hungry (detected by a neural system in the hypothalamus) or because of a more complex behavioral scenario unrelated to a basic hunger drive, such as the fact that you started to think about food because you just saw someone eating on television? Why did you select an apple rather than a banana when both are in the fruit bowl and you like the taste of both, and so on? Thus, initiation and execution of purposeful voluntary movement actually include a complex neuronal interplay involving output from the motor regions guided by integrated sensory information and ultimately dependent on motivational systems and elaboration of thought. All this is played against a background of memory stores from which meaningful decisions about desirable movements can be made.

Somatotopic maps vary slightly between individuals and are dynamic, not static.

Although the general organizational pattern of sensory and motor somatotopic ("body representation") maps of the cortex is similar in all people, the precise distribution is unique for each individual. Just as each of us has two eyes, a nose, and a mouth and yet no two faces have these features arranged in exactly the same way, so it is with brains. Furthermore, an individual's somatotopic mapping is not "carved in stone" but is subject to constant subtle modifications based on use. The general pattern is governed by genetic and developmental processes, but the individual cortical architecture appears capable of being influenced by **use-dependent competition** for cortical space. For example, when monkeys were encouraged to use their middle fingers instead of their other fingers to press a bar for food, after only several thousand bar presses the "middle finger area" of the motor cortex was greatly expanded and encroached upon territory previously devoted to the other fingers. This competition for cortical territory based on use probably accounts for the fact that some individuals born without arms have been able to accomplish amazing "dexterity" with their feet. They are able to use their feet almost like hands to feed themselves and perform other skills requiring a much finer degree of control over toe muscles than most people have.

Thus, the particular architecture of your own rather plastic brain has been and continues to be influenced by your unique set of experiences. It is important to realize, however, that what you do and do not do cannot totally shape the organization of your cortex. Some limits are genetically established. Also, there are developmental limits on the extent to which modeling can be influenced by patterns of use. Whereas some cortical activities maintain their plasticity throughout life, especially the ability to add new memory stores and learn, other cortical regions can be modified by use for only a specified time after birth before becoming permanently fixed. The length of this critical developmental period varies for different cortical regions.

Language ability has several discrete components controlled by different regions of the cortex.

Language ability is an excellent example of early cortical plasticity coupled with later permanence. Unlike the sensory and motor regions of the cortex, which are present in both hemispheres, the areas of the brain responsible for language ability are found in only one hemisphere—the left hemisphere in the vast majority of the population. However, if a child under the age of two accidentally suffers damage to the left hemisphere, language functions are transferred to the right hemisphere with no delay in language development but at the expense of less obvious nonverbal abilities for which the right hemisphere is normally responsible. Up to about the age of ten, after damage to the left hemisphere, language ability can usually be reestablished in the right hemisphere following a temporary period of loss. If damage occurs beyond the early teens, however, language ability is permanently impaired, even though some limited restoration may be possible. The regions of the brain involved in comprehending and expressing language apparently are permanently assigned before adolescence.

Even in normal individuals, there is evidence for early plasticity and later permanence in language development. Infants are able to distinguish between and articulate the entire range of speech sounds, but each language utilizes only a portion of these sounds. As children mature, they often lose the ability to distinguish between or express speech sounds that are not important in their native language. For example, Japanese children can distinguish between the sounds of "r" and "l," but many Japanese adults cannot perceive the difference between them.

Language is a complex form of communication in which written or spoken words symbolize objects and convey ideas. It involves the integration of two distinct capabilities—namely, *expression* and *comprehension*—each of which is related to a specific area of the cortex. The primary areas of cortical specialization for language are Broca's area and Wernicke's area. **Broca's area**, which is responsible for speaking ability, is located in the left frontal lobe in close association with the motor areas of the cortex that control the muscles necessary for articulation (Figs. 5–8 and ● 5–11). **Wernicke's area**, located in the left cortex at the juncture of the parietal, temporal, and occipital lobes, is concerned with language comprehension. It plays a critical role in understanding both spoken and written messages. Furthermore, it is responsible for formulating coherent patterns of speech that are transferred via a bundle of fibers to Broca's area, which in turn controls articulation of this speech. Wernicke's area receives input from the visual cortex in the occipital lobe, a pathway important in reading comprehension and in describing objects seen, as well as from the auditory cortex in the temporal lobe, a pathway essential for understanding spoken words. According to the leading model of language, precise interconnecting pathways between these localized cortical areas are involved in the various aspects of speech (Fig. 5–11).

Because various aspects of language are localized in different regions of the cortex, damage to specific regions of the brain can result in selective disturbances of language. Damage to Broca's area results in a failure of word formation, although the patient can still understand the spoken and written word. Such individuals know what they want to say but are unable to express themselves. Even though they can

Facial area of
motor cortex

Angular gyrus of
parietal-temporal-occipital
association cortex

Broca's
area

3

2

4

Wernicke's
area

1

Bundle of
interconnecting fibers

Visual cortex

● **FIGURE 5–11 Cortical Pathway for Speaking a Written Word or Naming a Visual Object** To describe an object seen, the brain first transfers the visual information from the visual cortex to a specific area (the angular gyrus) of the parietal-temporal-occipital association cortex, a region concerned with integrating such sensory inputs as sight, sound, and touch (path 1). From here the information is transferred to Wernicke's area (path 2), where the choice and sequence of words to be spoken are formulated. The language command is then transmitted to Broca's area (path 3), which, in turn, translates the message from Wernicke's area into a programmed sound pattern that is conveyed to the precise areas of the primary motor cortex (path 4) to activate the appropriate facial muscles that will cause the desired words to be spoken. Similarly, appropriate muscles of the hand can be commanded to write the desired words.

move their lips and tongue, they cannot establish the proper motor command to articulate the desired words. In contrast, patients with a lesion in Wernicke's area cannot understand words they see or hear. They are able to speak fluently even though their perfectly articulated words make no sense. They cannot attach meaning to words or choose appropriate words to convey their thoughts. Such language disorders caused by damage to specific cortical areas are known as **aphasias**, most of which result from strokes. Aphasias should not be confused with **speech impediments**, which are caused by a defect in the mechanical aspect of speech, such as weakness or incoordination of the muscles controlling the vocal apparatus. It is probable that **dyslexia**, a difficulty in learning to read because of inappropriate interpretation of letters or words as a reverse image (for example, *bad* is "seen" as *dab*), arises from developmental abnormalities in connections between the visual and language areas of the cortex or within the language areas themselves.

The association areas of the cerebral cortex are involved in many higher functions.

The motor, sensory, and language areas account for only about half of the total cerebral cortex. The remaining areas, called **association areas**, are involved in higher functions. During brain surgery, a neurosurgeon carefully examines the portion of the patient's brain that has been exposed, looking for appropriate landmarks on the neural map. Before cutting into this precious, nonregenerative tissue, the surgeon explores with a tiny stimulating electrode, asking the patient to

describe what happens with each stimulation—the flick of a finger . . . a prickly feeling on the bottom of the foot . . . nothing? Because the brain itself is insensitive to pain, all of this is done while the patient is awake with only local anesthesia used along the cut scalp, allowing the surgeon to be absolutely certain of location before penetrating the brain. At one time the association areas were called "silent" areas because stimulation does not produce any observable motor response or sensory perception. There are three association areas (Fig. 5–8): (1) the prefrontal association cortex, (2) the parietal-temporal-occipital association cortex, and (3) the limbic association cortex.

The **prefrontal association cortex** is the front portion of the frontal lobe just anterior to the premotor cortex. The roles attributed to this region are (1) planning for voluntary activity; (2) weighing consequences of future actions and choosing between different options for various social or physical situations; and (3) personality traits. Stimulation of this area does not produce any observable effects, but deficits in this area result in changes in personality and social behavior.

The **parietal-temporal-occipital association cortex** is found at the interface of the three lobes for which it is named. In this strategic location, it pools and integrates somatic, auditory, and visual sensations projected from these three lobes for complex perceptual processing. It enables us to "get the complete picture" of the relationship of various parts of our bodies with the external world. For example, it integrates visual information with proprioceptive input to enable you to place what you are seeing in proper perspective, such as realizing that a bottle is in an upright position in spite of the angle from which you view it (that is, whether you are standing up, lying down, or hanging upside down from a tree branch). This region is also involved in the language pathway connecting Wernicke's area to the visual and auditory cortices.

The **limbic association cortex** is located mostly on the bottom and adjoining inner portion of each temporal lobe. This area is concerned primarily with motivation and emotion and is extensively involved in memory.

Indeed, all association areas appear to be involved in sophisticated mental events such as memory, thinking, decision making, creativity, and self-consciousness. None of these higher brain functions are controlled by a specific cortical region. All are believed to depend on complex interrelated pathways involving several different regions. The cortical association areas are all interconnected by bundles of fibers within the cerebral white matter. Collectively, the association areas integrate diverse information for purposeful action. An oversimplified basic sequence of linkage between the various functional areas of the cortex is schematically represented in ● Figure 5–12.

The cerebral hemispheres have some degree of specialization.

The cortical areas described thus far appear to be equally distributed in both the right and left hemispheres, except for the language areas, which are found only on one side, usually the left. The left side is also most commonly the dominant hemi-

Sensory input	Relayed from afferent neuronal receptors
Primary sensory areas (somatosensory, 1° visual, 1° auditory cortices)	Initial cortical processing of specific sensory input
Higher sensory areas	Further elaboration and processing of specific sensory input
Association areas	Integration, storage, and use of diverse sensory input for planning of purposeful action
Higher motor areas	Programming sequences of movement in context of diverse information provided
Primary motor cortex	Commanding of efferent motor neurons to initiate voluntary movement
Motor output	Relayed through efferent motor neurons to appropriate skeletal muscles, which carry out desired action

For simplicity, a number of interconnections have been omitted.

● **FIGURE 5–12   Schematic Linking of Various Regions of Cortex**

sphere for fine motor control. Thus, most people are right-handed, because the left side of the brain controls the right side of the body. Furthermore, each hemisphere is somewhat specialized in the types of mental activities it carries out best. The **left cerebral hemisphere** excels in the performance of logical, analytical, sequential, and verbal tasks, such as math, language forms, and philosophy. In contrast, the **right cerebral hemisphere** excels in nonlanguage skills, especially spatial perception and artistic and musical endeavors. Whereas the left hemisphere tends to process information in a fragmentary way, the right hemisphere views the world holistically. Normally, there is much sharing of information between the two hemispheres so that they complement each other, but in many individuals the skills associated with one hemisphere appear to be more strongly developed. Left cerebral hemisphere dominance tends to be associated with "thinkers," whereas the right hemispheric skills dominate in "creators."

An electroencephalogram is a record of postsynaptic activity in cortical neurons.

Extracellular current flow arising from electrical activity within the cerebral cortex can be detected by placing recording electrodes on the scalp to produce a graphic record known as an **electroencephalogram** or **EEG**. These "brain waves" for the most part are not due to action potentials but instead represent the momentary collective postsynaptic potential activity (that is, EPSPs and IPSPs; see p. 92) in the cell bodies and dendrites located in the cortical layers under the recording electrode.

Electrical activity can always be recorded from the living brain, even during sleep and unconscious states, but the wave forms vary depending on the degree of activity of the cerebral cortex. Often the wave forms appear irregular, but sometimes distinct patterns can be observed on the basis of the wave's amplitude and frequency. A dramatic example of this is illustrated in ● Figure 5–13, in which the EEG wave form recorded over the occipital (visual) cortex changes markedly in response to simply opening and closing the eyes.

The EEG has three major uses:

1. It is often used as a *clinical tool in the diagnosis of cerebral dysfunction.* Diseased or damaged cortical tissue often gives rise to altered EEG patterns. One of the most common neurological diseases accompanied by a distinctively abnormal EEG is **epilepsy**. Epileptic seizures occur when a large collection of neurons abnormally undergo synchronous action potentials that produce stereotypical, involuntary spasms and alterations in behavior. The seizures may be partial or generalized, depending on the location and extent of the abnormal neuronal discharge. Each type of seizure displays different EEG features.

2. The EEG is also used to *distinguish various stages of sleep,* as will be described later in this chapter.

3. The EEG finds further use in the *legal determination of brain death.* Even though a person may have stopped breathing and the heart may have stopped pumping blood, it is often possible to restore and maintain circulatory and respiratory activity if resuscitative measures are instituted soon enough. Yet, because of the susceptibility of the brain to O_2 deprivation, irreversible brain damage may have already occurred before heart and lung function have been reestablished, resulting in the paradoxical situation of a dead brain in a living body. There are important medical, legal, and social implications in determining whether a comatose patient being maintained by artificial respiration and other supportive measures is alive or dead. The need for viable organs for modern transplant surgery has made the timeliness of such life/death determinations of utmost importance. Physicians, lawyers, and the public in general have accepted the notion of brain

● **FIGURE 5–13   Replacement of Alpha Rhythm on EEG with Beta Rhythm by Eyes' Opening**

Eyes closed	Eyes open	Eyes closed
Alpha waves	Beta waves	Alpha waves

death—that is, a brain that is not functioning, with no possibility of recovery—as the determinant of death under such circumstances. The most widely accepted indication of brain death is *electrocerebral silence*—an essentially flat EEG. This, however, must be coupled with other stringent criteria, such as absence of eye reflexes, to guard against a false terminal diagnosis in individuals with a flat EEG that is due to causes that can be reversed, as in certain kinds of drug intoxication.

SUBCORTICAL STRUCTURES AND THEIR RELATIONSHIP WITH THE CORTEX IN HIGHER BRAIN FUNCTIONS

The **subcortical** ("under the cortex") **regions** of the brain interact extensively with the cortex in the performance of their functions. These regions include the *basal nuclei* located in the cerebrum and the *thalamus* and *hypothalamus* located in the diencephalon.

The basal nuclei play an important inhibitory role in motor control.

The **basal nuclei** (also known as **basal ganglia**) consist of several masses of gray matter located deep within the cerebral white matter (Table 5–3 and ● Fig. 5–14). In the nervous system, a **nucleus** (plural, **nuclei**) refers to a functional aggregation of neuronal cell bodies. The basal nuclei play a complex role in the control of movement in addition to having nonmotor functions that are less well understood. In particular, the basal nuclei are important in (1) inhibiting muscle tone throughout the body (proper muscle tone is normally maintained by a balance of excitatory and inhibitory inputs to the neurons that innervate skeletal muscles); (2) selecting and maintaining purposeful motor activity while suppressing useless or unwanted patterns of movement; and (3) helping monitor and coordinate slow sustained contractions, especially those related to posture and support. The basal nuclei do not directly influence the efferent motor neurons that bring about muscle contraction but act instead by modifying ongoing activity in motor pathways.

● **FIGURE 5–14   Frontal Section of Brain** The cerebral cortex, an outer shell of gray matter, surrounds a core of white matter. Deep within the cerebral white matter are several masses of gray matter, the basal nuclei. The ventricles are cavities in the brain through which the cerebrospinal fluid flows. The thalamus forms the walls of the third ventricle.

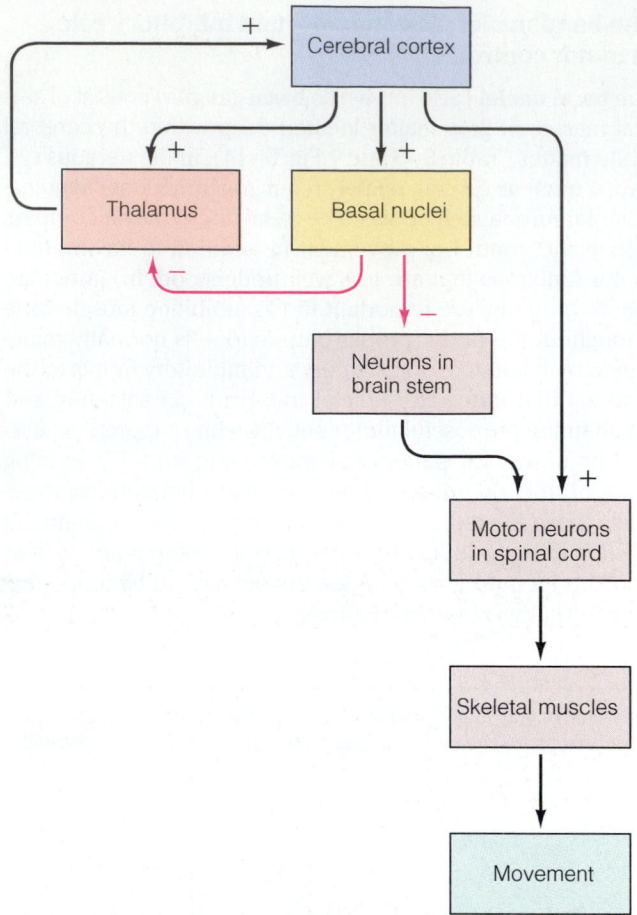

FIGURE 5-15   Relationship of Thalamus, Basal Nuclei and Cortex in Control of Movement The thalamus positively reinforces voluntary motor activity initiated by the cerebral cortex. The basal nuclei modulate voluntary motor activity by exerting an inhibitory effect on the thalamus and by inhibiting brain stem neurons that influence the motor neurons supplying the skeletal muscles.

To accomplish these complex integrative roles, the basal nuclei receive and send out much information, as is indicated by the tremendous number of fibers linking them to other regions of the brain. One important pathway consists of strategic interconnections that form a complex feedback loop linking the cerebral cortex (especially its motor regions), the basal nuclei, and the thalamus (• Fig. 5-15). It is speculated that the thalamus positively reinforces voluntary motor behavior initiated by the cortex, whereas the basal nuclei modulate this activity by exerting an inhibitory effect on the thalamus to eliminate antagonistic or unnecessary movements. The basal nuclei also exert an inhibitory effect on motor activity by acting through neurons in the brain stem.

The importance of the basal nuclei in motor control is evident in diseases involving this region, the most common of which is **Parkinson's disease**. This condition is associated with a deficiency of dopamine, an important neurotransmitter in the basal nuclei (see p. 100). As a result of the basal nuclei lacking sufficient dopamine to exert their normal roles, three types of motor disturbances characterize Parkinson's disease: (1) increased muscle tone or rigidity; (2) involuntary,

useless, or unwanted movements, such as *resting tremors* (for example, hands rhythmically shaking making it difficult or impossible to hold a cup of coffee); and (3) slowness in initiating and carrying out different motor behaviors. It is difficult for those who suffer from Parkinson's disease to stop ongoing activities. If sitting down, they tend to remain seated, and if they get up, they do so very slowly.

The thalamus is a sensory relay station and is important in motor control.

Deep within the brain near the basal nuclei is the **diencephalon**, a midline structure that forms the walls of the third ventricular cavity, one of the spaces through which cerebrospinal fluid flows. The diencephalon consists of two main parts, the *thalamus* and the *hypothalamus* (Table 5-3 and Figs. 5-14 and • 5-16).

The **thalamus** serves as a "relay station" and synaptic integrating center for preliminary processing of all sensory input on its way to the cortex. It screens out insignificant signals and routes the important sensory impulses to appropriate areas of the somatosensory cortex, as well as to other regions of the brain. The thalamus, along with the brain stem and cortical association areas, is important in our ability to direct attention to stimuli of interest. For example, parents can sleep soundly through the noise of outdoor traffic but be instantly aware of their baby's slightest whimper. The thalamus is also capable of crude awareness of various types of sensation but cannot distinguish their location or intensity. Some degree of consciousness resides here as well. As described in the preceding section, the thalamus also plays an important role in motor control by positively reinforcing voluntary motor behavior initiated by the cortex.

The hypothalamus regulates many homeostatic functions.

The **hypothalamus** is a collection of specific nuclei and associated fibers that lie beneath the thalamus. It is an integrating center for many important homeostatic functions and serves as an important link between the autonomic nervous system and the endocrine system. Specifically, the hypothalamus (1) controls body temperature; (2) controls thirst and urine output; (3) controls food intake; (4) controls anterior pituitary hormone secretion; (5) produces posterior pituitary hormones; (6) controls uterine contractions and milk ejection; (7) serves as a major autonomic nervous system coordinating center, which in turn affects all smooth muscle, cardiac muscle, and exocrine glands; and (8) plays a role in emotional and behavioral patterns.

The hypothalamus is the area of the brain most notably involved in the direct regulation of the internal environment. For example, when the body is cold, the hypothalamus initiates internal responses to increase heat production (such as shivering) and to decrease heat loss (such as constricting the skin blood vessels to reduce the flow of warm blood to the body surface, where heat could be lost to the external environment). Other areas of the brain, such as the cerebral cortex, act more indirectly to regulate the internal environment. For example, a person who feels cold is motivated to voluntarily put on warmer clothing, close the window, turn up the thermostat, and so on. Even these voluntary behavioral activities are strongly influenced by the hypothalamus, which, as a part of the limbic system, functions in conjunction with the cortex in controlling emotions and motivated behavior.

The limbic system plays a key role in emotion and behavior.

The **limbic system** is not a separate structure but refers to a ring of forebrain structures that surround the brain stem and are interconnected by intricate neuronal pathways (● Fig.

5–17). It includes portions of each of the following: the lobes of the cerebral cortex, the basal nuclei, the thalamus, and the hypothalamus. This complex interacting network is associated with emotions, basic survival and sociosexual behavioral patterns, motivation, and learning.

The concept of **emotion** encompasses subjective emotional feelings and moods (such as anger, fear, and happiness) plus the overt physical responses that occur in association with these feelings. These responses include specific behavioral patterns (for example, preparation for attack or defense when angered by an adversary) and observable emotional expressions (for example, laughing, crying, or blushing). Evidence points to a central role for the limbic system in all aspects of emotion. Stimulation of specific regions within the limbic system of humans during brain surgery produces

● **FIGURE 5–17   Limbic System** This partially transparent view of the brain reveals the structures composing the limbic system.

various vague subjective sensations described by the patient as joy, satisfaction, or pleasure in one region and discouragement, fear, or anxiety in another.

Behavioral patterns controlled at least in part by the limbic system include those aimed at survival of the individual (attack, searching for food) and those directed toward perpetuation of the species (sociosexual behaviors conducive to mating). In experimental animals, stimulation of the limbic system brings about complex and even bizarre behaviors. For example, stimulation in one area can elicit responses of anger and rage in a normally docile animal, whereas stimulation in another area results in placidity and tameness, even in an otherwise vicious animal. Stimulation in yet another limbic area can induce sexual behaviors such as copulatory movements.

The relationships among the hypothalamus, limbic system, and higher cortical regions regarding emotions and behavior are still not well understood. It appears that the extensive involvement of the hypothalamus in the limbic system is responsible for the involuntary internal responses of various body systems in preparation for appropriate action to accompany a particular emotional state. For example, the increased heart rate and respiratory rate, elevation of blood pressure, and diversion of blood to skeletal muscles that occur in anticipation of attack when angered are controlled by the hypothalamus. These preparatory changes in internal state require no conscious control.

In executing complex behavioral activities such as attack, flight, or mating, the individual (animal or human) must interact with the external environment. Higher cortical mechanisms are called into play to connect the limbic system and hypothalamus with the outer world so that appropriate overt behaviors are manifested. At the simplest level, the cortex provides the neural mechanisms necessary for implementation of the appropriate skeletal-muscle activity required to approach or avoid an adversary, participate in sexual activity, or display emotional expression. For example, the stereotypical sequence of movement for the universal human emotional expression of smiling is apparently preprogrammed in the cortex and can be called forth by the limbic system. One can also voluntarily call forth the smile program, as when posing for a picture. Even individuals blind from birth have normal facial expressions; that is, they do not learn to smile by observation. Smiling means the same thing in every culture, despite widely differing environmental experiences. Such behavior patterns shared by all members of a species are believed to be more abundant in lower animals.

In humans and to an undetermined extent in other species, the cortex is additionally crucial for conscious awareness of emotional feelings. It also is capable of reinforcing, modifying, or suppressing basic behavioral responses so that actions can be guided by planning, strategy, and judgment based on an understanding of the situation. For example, even if you were angry at someone and your body was internally preparing for attack, you probably would judge that an attack would be inappropriate and could consciously suppress the external manifestation of this basic emotional behavior. Thus, the cortex, particularly the prefrontal and limbic association areas, is important in conscious learned control of innate behavioral patterns.

An individual tends to reinforce behaviors that have proved to be gratifying and suppress behaviors that have been associated with unpleasant experiences. Certain regions of the limbic system have been designated as **"reward"** and **"punishment" centers** because stimulation in these respective areas gives rise to pleasant or unpleasant sensations. An experimental animal with an implanted self-stimulating device will self-deliver up to 5,000 stimulations per hour when the device is implanted in a reward center, even shunning food when starving in preference for the pleasure derived from self-stimulation. On the other hand, animals will avoid stimulation at all costs when the device is implanted in a punishment center. Reward centers are found most abundantly in regions involved in mediating the highly motivated behavioral activities of eating, drinking, and sexual activity.

Motivation is the ability to direct behavior toward specific goals. Some goal-directed behaviors are aimed at satisfying specific identifiable physical needs related to homeostasis. **Homeostatic drives** represent the subjective urges associated with specific bodily needs that motivate appropriate behavior to satisfy those needs. As an example, the sensation of thirst accompanying a water deficit in the body drives an individual to drink to satisfy the homeostatic need for water. However, whether water, a soft drink, or another beverage is chosen as the thirst-quencher is unrelated to homeostasis. Much human behavior is not dependent on purely homeostatic drives related to simple tissue deficits such as thirst, but is influenced by experience, learning, and habit. Thus, human behavior is shaped in a complex framework of unique personal gratifications blended with cultural expectations. To what extent, if any, motivational drives unrelated to homeostasis, such as the drive to pursue a particular career or win a certain race, are involved with the reinforcing effects of the reward and punishment centers is unknown. Indeed, some individuals motivated toward a particular goal may even deliberately "punish" themselves in the short term to achieve their long-range gratification (for example, the temporary pain of training in preparation for winning a competitive athletic event).

Norepinephrine, dopamine, and serotonin serve as neurotransmitters in motivated behavioral and emotional pathways.

The underlying neurophysiological mechanisms responsible for the psychological observations of motivated behavior and emotions largely remain a mystery, although the neurotransmitters **norepinephrine**, **dopamine**, and **serotonin** all have been implicated. Norepinephrine and dopamine, both chemically classified as *catecholamines*, are known to be transmitters in the regions that elicit the highest rates of self-stimulation in animals equipped with do-it-yourself devices. A number of drugs affect moods in humans, and some of these drugs have also been shown to influence self-stimulation in experimental animals. For example, increased self-stimulation is observed after the administration of drugs that increase catecholamine synaptic activity, such as amphetamine, an "upper" drug. Although most of these drugs are used therapeutically to treat various mental disorders, others, unfortunately, are abused.

In some cases, the effectiveness of various drugs in treating a specific disorder has provided an important clue to the underlying biochemical defect responsible for the condition. For example, two different lines of biochemical evidence suggest that **schizophrenia**, a mental disorder characterized by delusions and hallucinations, probably results from excess dopamine transmission. First, all drugs effective in the treatment of schizophrenia interfere with dopamine transmission in one way or another. Second, drugs that enhance dopamine activity can induce symptoms resembling those of schizophrenia. An unanswered question has been the anatomic site of the abnormal dopamine transmission. Studies identifying the regions of the brain that extensively use dopamine as a neurotransmitter have narrowed down the possible sites. Among these are neural networks linked to the limbic system. Since the limbic system is extensively involved in emotions and in triggering behavior appropriate to environmental circumstances—functions that are abnormal in schizophrenic patients—it is speculated that this might be the defective site in schizophrenia.

The molecular mechanisms of other mental disorders are also beginning to be unraveled through similar pharmacological evidence. For example, a functional deficiency of serotonin or norepinephrine or both is implicated in **depression**, a disorder characterized by a pervasive unpleasant mood accompanied by a generalized loss of interests and inability to experience pleasure. All effective antidepressant drugs increase the available concentration of these neurotransmitters in the CNS. The fact that serotonin and norepinephrine are synaptic messengers in the regions of the brain involved in pleasure and motivation suggests that the pervasive sadness and lack of interest (no motivation) in depressed patients is related at least in part to disruption of these regions due to deficiencies or decreased effectiveness of these neurotransmitters. Researchers are optimistic that as our understanding of the molecular mechanisms of mental disorders is expanded in the future, many psychiatric problems can be corrected or managed through drug intervention, a hope of great medical significance.

Learning is the acquisition of knowledge as a result of experiences.

Learning is the acquisition of knowledge or skills as a consequence of experience, instruction, or both. It is widely believed that rewards and punishments are integral parts of many types of learning. If an animal is rewarded upon responding in a particular way to a stimulus, the likelihood increases that it will respond in the same way again to the same stimulus as a consequence of this experience. Conversely, if a particular response is accompanied by punishment, the animal is less likely to repeat the same response to the same stimulus. When behavioral responses that give rise to pleasure

are reinforced or those accompanied by punishment are avoided, learning has taken place. Housebreaking a puppy is an example. If the puppy is praised when it urinates outdoors but scolded when it wets the carpet, it will soon learn the acceptable place to empty its bladder. Wild animals, in contrast, have no such training experience, so they do not learn to confine bladder voiding to particular locations. Thus, learning is a change in behavior that occurs as a result of experiences. It is highly dependent on the organism's interaction with its environment. The only limits to the effects that environmental influences can have on learning are the biological constraints imposed by species-specific and individual genetic endowments.

Memory is laid down in stages.

Memory is the storage of acquired knowledge for later recall. Learning and memory form the basis by which individuals adapt their behavior to their particular external circumstances. Without these mechanisms, planning for successful interactions and intentional avoidance of predictably disagreeable circumstances would be impossible.

The neural change responsible for retention or storage of knowledge is known as the **memory trace**. Concepts, not verbatim information, are generally stored. As you read this page, you are storing the concept discussed, not the specific words. Later, when you retrieve the concept from memory, you will convert it into your own words. It is possible, however, to memorize bits of information word by word.

Storage of acquired information is believed to be accomplished in at least two stages: short-term memory and long-term memory (● Fig. 5–18 and Table 5–4). **Short-term memory** lasts for seconds to hours, whereas **long-term memory** is retained for days to years. The process of transferring and fixing short-term memory traces into long-term memory stores is known as **consolidation**. Stored knowledge is of no use unless it can be retrieved and used to influence current or future behavior. A recently developed concept is that of a **working memory**, or what has been coined "the blackboard of the mind." Working memory involves comparing current sensory data with relevant stored knowledge and manipulating that information, as in being able to carry on a conversation or knowing to put on warm clothing if you see snow outside. Working memory enables people to string thoughts together in a logical sequence and plan for future action.

Newly acquired information is initially deposited in short-term memory, which has a limited capacity for storage. Information in short-term memory has one of two eventual fates. Either it is soon forgotten (for example, forgetting a telephone number after you have looked it up and finished dialing), or it is transferred into the more permanent long-term memory mode through *active practice* or *rehearsal*. The recycling of newly acquired information through short-term memory increases the likelihood of long-term memory consolidation. (Therefore, when you cram for an exam, your long-term retention of the information is poor!) This relationship can be

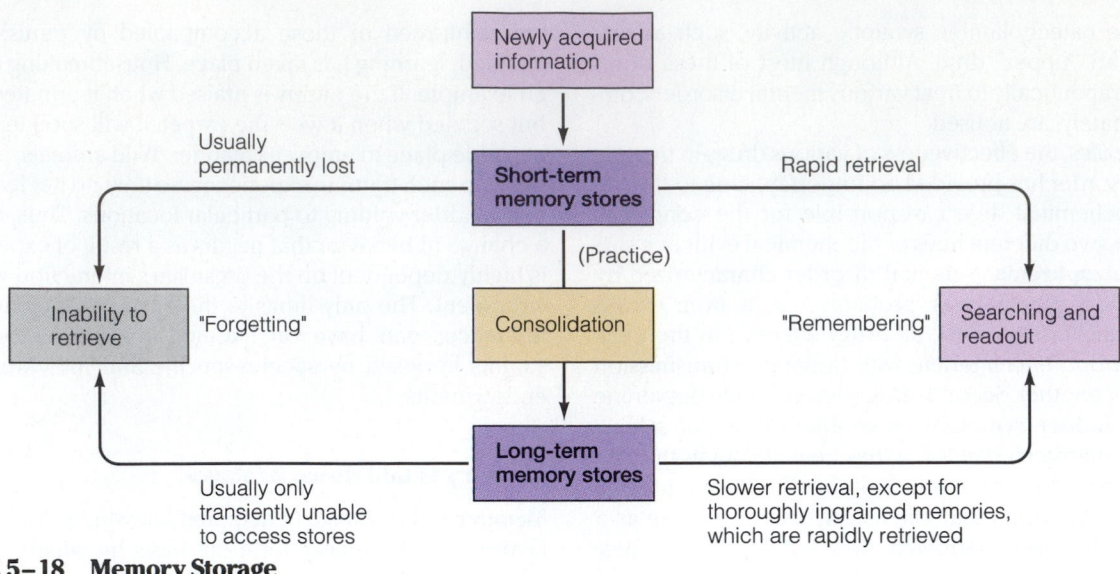

• **FIGURE 5–18   Memory Storage**

likened to developing photographic film. The originally developed image (short-term memory) will rapidly fade unless it is chemically fixed (consolidation) to provide a more enduring image (long-term memory). Sometimes only parts of memories are fixed while others fade away. Information of interest or importance to the individual is more likely to be recycled and fixed in long-term stores, whereas less important information is quickly erased.

The storage capacity of the long-term memory bank is much larger than the capacity of short-term memory. Different informational aspects of long-term memory traces seem to be processed and codified, then stored in conjunction with other memories of the same type; for example, visual memories are stored separately from auditory memories. This organization facilitates future searching of memory stores to retrieve desired information. For example, in remembering a woman you once met, you may use various recall cues from different

storage pools, such as her name, her appearance, the fragrance she wore, the song playing in the background, and so on.

Because long-term memory stores are larger, it takes longer to retrieve information from long-term memory than from short-term memory. *Remembering* is the process of retrieving specific information from memory stores; *forgetting* is the inability to retrieve stored information. Information lost from short-term memory is permanently forgotten, but information in long-term storage is frequently forgotten only transiently. Often you are only temporarily unable to access the information—for example, being unable to remember an acquaintance's name, then having it suddenly "come to you" later.

Some forms of long-term memory involving information or skills used on a daily basis are essentially never forgotten and are rapidly accessible, such as knowing your own name or

TABLE 5–4 Comparison of Short-Term and Long-Term Memory		
CHARACTERISTIC	**SHORT-TERM MEMORY**	**LONG-TERM MEMORY**
Time of storage after acquisition of new information	Immediate	Later; must be transferred from short-term to long-term memory through consolidation; enhanced by practice or recycling of information through short-term mode
Capacity of storage	Limited	Very large
Retrieval time (remembering)	Rapid retrieval	Slower retrieval, except for thoroughly ingrained memories, which are rapidly retrieved
Inability to retrieve (forgetting)	Permanently forgotten; memory fades quickly unless consolidated into long-term memory	Usually only transiently unable to access; relatively stable memory trace
Mechanism of storage	Involves transient modifications in functions of preexisting synapses, such as altering amount of neurotransmitter released	Involves relatively permanent functional or structural changes between existing neurons, such as formation of new synapses; synthesis of new proteins plays a key role

being able to write. Even though long-term memories are relatively stable, stored information may be gradually lost or modified over time unless it is thoroughly ingrained as a result of years of practice.

Occasionally, individuals suffer from a lack of memory that involves whole portions of time rather than isolated bits of information. This condition, known as **amnesia**, occurs in two forms. The most common form, *retrograde* ("going backwards") *amnesia,* is the inability to recall recent past events. It usually follows a traumatic event that interferes with electrical activity of the brain, such as a concussion or stroke. If a person is knocked unconscious, the content of short-term memory is essentially erased, resulting in loss of memory of activities that occurred within about the last half hour before the event. Severe trauma may interfere with access to recently acquired information in long-term stores as well.

Anterograde ("going forward") *amnesia,* on the other hand, is the inability to store memory in long-term storage for later retrieval. It is usually associated with lesions of the medial portions of the temporal lobes, which are generally considered to be critical regions for memory consolidation. Individuals suffering from this condition may be able to recall things they learned before the onset of their problem, but they are unable to establish new permanent memories. New information is lost as quickly as it fades from short-term memory. In one case study, the individual could not remember where the bathroom was in his new home but still had total recall of his old home.

Memory traces are present in multiple regions of the brain.

These observations raise the question, What parts of the brain are responsible for memory? Apparently, the neurons involved in memory traces are widely distributed throughout the subcortical and cortical regions of the brain, because some traces of memory remain even after extensive damage. However, there is evidence that certain learning tasks are profoundly affected by damage to specific regions of the brain. The regions of the brain implicated in memory include the temporal lobes, the prefrontal cortex, other regions of the cerebral cortex, the limbic system, and the cerebellum.

The temporal lobes and limbic system are essential for transferring new memories into long-term storage. The **hippocampus**, the elongated, medial portion of the temporal lobe that is part of the limbic system (Fig. 5–17), plays a vital role in short-term memory involving the integration of various related stimuli and is also crucial for consolidation into long-term memory. The hippocampus is believed to only temporarily store new long-term memories and then transfer them to other cortical sites for more permanent storage. Accessing and manipulating these long-term stores through operation of the working memory appears to be carried out by the prefrontal region of the cerebral cortex. Furthermore, the hippocampus and surrounding regions play an important role in **declarative memories**—memories of facts that often result after only one experience and that can be declared in a statement such as, "I saw the Statue of Liberty last summer." Interestingly, extensive damage in the hippocampus region is evident in Alzheimer's patients during autopsy. (See the accompanying boxed feature, Concepts, Challenges, and Controversies.) In contrast to the role of the hippocampus and surrounding regions of the temporal lobes and limbic system, the cerebellum seems to play an essential role in **procedural memories** involving motor skills gained through repetitive training, such as memorizing a particular dance routine. The distinct localization of these two types of memory is apparent in individuals who have temporal/limbic lesions. They are able to perform a skill, such as playing a piano, but the next day they have no recollection that they played it.

Short-term and long-term memory appear to involve different molecular mechanisms.

Another question besides the "where" of memory is the "how" of memory. Despite a vast amount of psychological data, only a few tantalizing scraps of physiological evidence concerning the cellular basis of memory traces are available. Obviously, some change must take place within the neural circuitry of the brain to account for the altered behavior that follows learning. A single memory resides in changes in the pattern of signals transmitted across synapses within a vast neuronal network, not in a single neuron. Different mechanisms appear to be responsible for short-term and long-term memory. Evidence suggests that short-term memory involves transient modifications in the function of preexisting synapses, such as a temporary alteration in the amount of neurotransmitter released in response to stimulation within affected nerve pathways. Long-term memory, in contrast, is believed to involve relatively permanent functional or structural changes between existing neurons in the brain.

Short-term memory Ingenious experiments in the sea snail, *Aplysia,* have shown that two forms of short-term memory—habituation and sensitization—are due to modification of different channel proteins in presynaptic terminals of specific afferent neurons. This modification in turn brings about changes in transmitter release. **Habituation** is a decreased responsiveness to repetitive presentations of an indifferent stimulus—that is, one that is neither rewarding nor punishing. **Sensitization** refers to increased responsiveness to mild stimuli following a strong or noxious stimulus. *Aplysia* reflexly withdraws its gill when its siphon, a breathing organ at the top of its gill, is touched. Afferent neurons responding to touch of the siphon (presynaptic neurons) directly synapse on efferent motor neurons (postsynaptic neurons) controlling gill withdrawal. The snail becomes habituated when its siphon is repeatedly touched; that is, it learns to ignore the stimulus and no longer withdraws its gill in response. Sensitization, a more complex form of learning, takes place in *Aplysia* when it is given a hard bang on the siphon. Subsequently, the snail withdraws its gill more vigorously in response to even mild touch. Interestingly, these different forms of learning affect the same site—the synapse between a siphon affer-

ALZHEIMER'S DISEASE: A TALE OF BETA AMYLOID AND DEMENTIA

"I can't remember where I put my keys. I must be getting Alzheimer's." The incidence and awareness of **Alzheimer's disease**, which is characterized in its early stages by loss of recent memories, have become so commonplace that people sometimes jest about having it when they can't remember something. Alzheimer's disease is no joking matter, however.

About four million Americans currently have Alzheimer's disease, and because it is an age-related condition and the population is aging, the incidence is expected to climb. About 0.1% of those between 60 and 65 years of age are afflicted with the disease, but the incidence rises to 30% to 47% among those over age 85. According to the *National Institute of Aging Report to Congress,* the most rapidly growing segment of our population by percentage is the over-85 age group.

Alzheimer's disease accounts for about two-thirds of the cases of *senile dementia,* which refers to a generalized age-related diminution of mental abilities. In the earliest stages, only short-term memory is impaired, but as the disease progresses, even firmly entrenched long-term memories, such as recognition of family members, are lost. Confusion, disorientation, and personality changes characterized by irritability and emotional outbursts are common. Higher mental abilities gradually deteriorate as the patient inexorably loses the ability to read, write, and calculate. Language ability and speech are often impaired. In later stages, Alzheimer's victims become childlike and are unable to feed, dress, and groom themselves. Patients usually die in a severely debilitated state four to twelve years after the onset of the disease.

First described eighty-five years ago by Alois Alzheimer, a German neurologist, the condition can only be confirmed at autopsy upon finding the characteristic brain lesions associated with the disease. Prior to death, Alzheimer's is diagnosed by a process of elimination; that is, all other disorders that could produce dementia, such as a stroke or brain tumor, must be ruled out. The characteristic brain lesions, *neuritic (senile) plaques* and *neurofibrillary tangles,* are dispersed throughout the cerebral cortex and are especially abundant in the hippocampus. A neuritic plaque consists of a central core of extracellular, sticky, fibrous protein known as **beta amyloid** surrounded by degenerating dendritic and axonal nerve endings. Neurofibrillary tangles are dense bundles of abnormal, paired helical filaments that accumulate in the cell bodies of affected neurons. Alzheimer's disease is also characterized by degeneration of cell bodies of certain neurons in the basal forebrain. The acetylcholine-releasing axons of these neurons normally terminate in the cerebral cortex and hippocampus, so the loss of these neurons results in a deficiency of acetylcholine in these areas. Neuron death and loss of synaptic communication are

---

ent and a gill efferent—in opposite ways. Habituation depresses this synaptic activity, whereas sensitization enhances it (● Fig. 5–19).

In habituation, closing of the Ca^{++} channels reduces Ca^{++} entry into the presynaptic terminal, which leads to a decrease in neurotransmitter release. As a consequence, the postsynaptic potential is reduced compared to normal, resulting in a decrease or absence of the behavioral response controlled by the postsynaptic efferent neuron (gill withdrawal). Thus, the memory for habituation in *Aplysia* is stored in the form of modification of specific Ca^{++} channels. With no further training, this reduced responsiveness lasts for several hours. A similar process is responsible for short-term habituation in other invertebrates studied. This suggests that Ca^{++} channel modification might be a general mechanism of habituation, although in higher species the involvement of intervening interneurons makes the process somewhat more complicated. Habituation is probably the most common form of learning and is believed to be the first learning process to take place in human infants. By learning to ignore indifferent stimuli, the animal or person is free to attend to other more important stimuli.

Sensitization in *Aplysia* has also been shown to involve channel modification, but a different channel and mechanism are involved. In contrast to habituation, Ca^{++} entry into the presynaptic terminal is enhanced in sensitization. The subsequent increase in neurotransmitter release produces a larger postsynaptic potential, resulting in a more vigorous gill withdrawal response. Sensitization does not have a direct effect on the presynaptic Ca^{++} channels. Instead, it indirectly enhances Ca^{++} entry via presynaptic facilitation (see p. 98). The neurotransmitter serotonin is released from a facilitating interneuron that synapses on the presynaptic terminal to bring about increased release of presynaptic neurotransmitter in response to an action potential. It does so by triggering activation of a cyclic AMP second messenger (see p. 50) within the presynaptic terminal, which ultimately brings about blocking of K^+ channels. This blockage prolongs the action potential in the presynaptic terminal since K^+ efflux through opened K^+ channels is required for the action potential to re-

ALZHEIMER'S DISEASE: A TALE OF BETA AMYLOID AND DEMENTIA (continued)

responsible for the ensuing dementia.

If the brain of an individual who did not have Alzheimer's disease and lived beyond the seventies is examined after death, some neuritic plaques and neurofibrilllary tangles can be found, especially in the hippocampus and other areas important in memory. Apparently, Alzheimer's disease is an exaggeration or acceleration of processes normally associated with aging.

The cause of Alzheimer's disease is unknown in the majority of cases. Many investigators believe that the condition probably has many underlying causes. Genetic defects appear to contribute to 15% to 20% of the cases. Environmental factors have been implicated, but the evidence has not been conclusive. One unsettled controversy concerns the abnormally high concentration of aluminum found in the degenerating neurons. This finding led to speculation that aluminum cooking utensils might be the culprit, although studies have not confirmed this link.

Even though the initial triggering factors are unclear except for the few known genetic defects, much progress has been made in the last few years in understanding the pathology underlying the condition. The plasma membranes of neurons contain **amyloid precursor protein (APP)**, whose function is unknown. APP is normally degraded within the cell after a portion of the membrane is internalized by endocytosis. There are two intracellular pathways for breaking down APP: one involves lysosomes (see p. 24), which produce small beta amyloid fragments from APP, and the other involves cytosolic enzymes that break APP down into larger, harmless protein fragments. According to the most recent proposal, Alzheimer's disease develops when for some reason more of the APP is steered into the lysosomal beta amyloid–yielding pathway. Even though beta amyloid itself is not toxic, it has been shown to increase the influx of Ca^{++} into the nerve cells in response to the excitatory neurotransmitter glutamate. The resultant flooding of nerve cells with Ca^{++} triggers a chain of biochemical events that kills the cells. Thus, beta amyloid indirectly destroys neurons, with resultant memory loss and dementia.

No treatment is currently available for Alzheimer's disease, but as researchers continue to unravel the underlying factors, the likelihood of finding a means to block the gradual, relentless progression of the disease increases. Prevention or treatment of Alzheimer's disease cannot come too soon in view of the tragic toll the condition is taking on its victims, their families, and society. The cost of custodial care for Alzheimer's patients is currently estimated at $80 billion annually and will continue to rise as a greater percentage of our population ages and becomes afflicted with the condition.

---

turn to resting (repolarization). Since the presence of a local action potential is responsible for opening of Ca^{++} channels in the terminal, a prolonged action potential permits the greater Ca^{++} influx associated with sensitization.

Thus, existing synaptic pathways may be functionally interrupted (habituated) or enhanced (sensitized) during simple learning. It is speculated that much of short-term memory is similarly a temporary modification of already existing processes. Several lines of research suggest that the cyclic AMP cascade, especially activation of protein kinase, plays an important role, at least in elementary forms of learning and memory.

Further studies have revealed that declarative memories, which involve conscious awareness and are more complex than habituation and sensitization, are initially stored by means of more persistent changes in activity of existing synapses. Specifically, initial storage of declarative information appears to be accomplished by means of **long-term potentiation (LTP)**, which refers to an increase in strength of existing synaptic connections in involved pathways following brief periods of stimulation. LTP has been shown to last for days or even weeks—long enough for this short-term memory to be consolidated into more permanent long-term memory. LTP is especially prevalent in the hippocampus. The development of LTP involves changes in both the presynaptic and postsynaptic neurons at a given synapse. Both of these neurons must be active simultaneously, which restricts the development of LTP to the pathway that is stimulated. Postsynaptic depolarization leads to an influx of Ca^{++}, which activates Ca^{++}-dependent second-messenger pathways in the postsynaptic neuron. As a result, the postsynaptic neuron is believed to release a retrograde ("going backwards") factor which diffuses to the presynaptic neuron. If the presynaptic neuron is active (assuring that only the specific active pathway and not neighboring pathways is being potentiated), the retrograde factor activates a second-messenger system in the presynaptic neuron to enhance transmitter release from this neuron. This enhanced transmitter release mantains LTP. Note that the development of LTP involves a chemical factor from the postsynaptic neuron influencing the presynaptic neuron, just

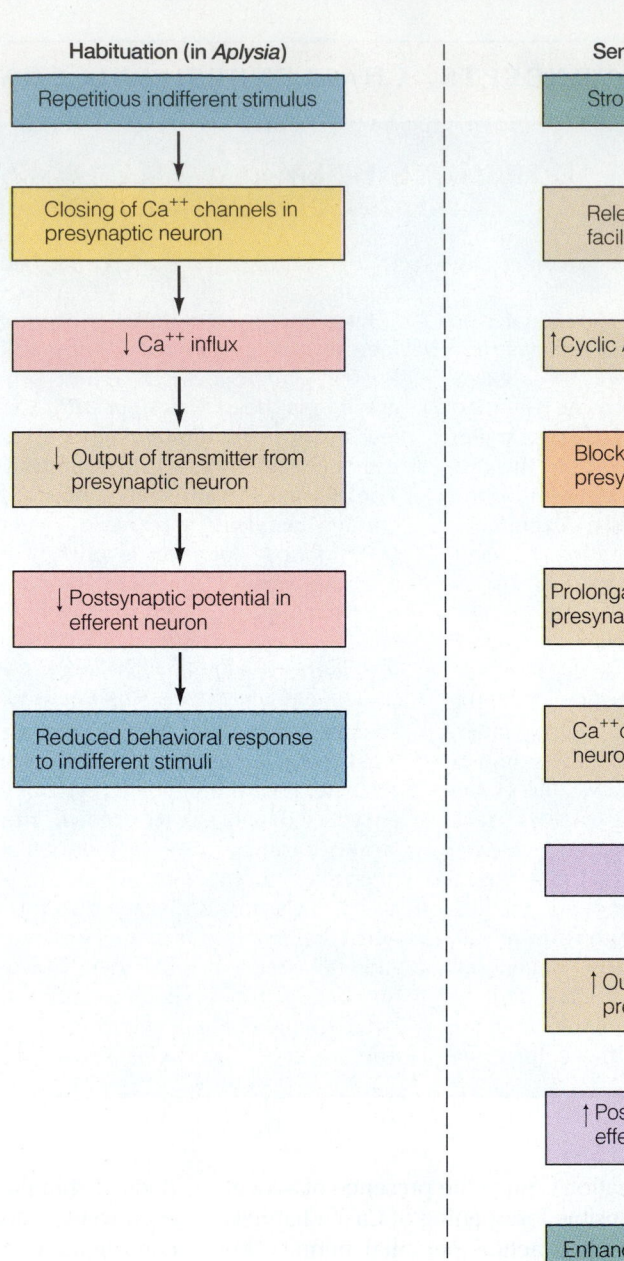

● **FIGURE 5–19 Habituation and Sensitization in *Aplysia*** Researchers have shown that in the sea snail *Aplysia* two forms of short-term memory, habituation and sensitization, result from opposite changes in neurotransmitter release from the same presynaptic neuron due to different transient channel modifications.

Habituation (in *Aplysia*)

Repetitious indifferent stimulus

↓

Closing of Ca^{++} channels in presynaptic neuron

↓

↓ Ca^{++} influx

↓

↓ Output of transmitter from presynaptic neuron

↓

↓ Postsynaptic potential in efferent neuron

↓

Reduced behavioral response to indifferent stimuli

Sensitization (in *Aplysia*)

Strong or noxious stimulus

↓

Release of serotonin from facilitating interneuron

↓

↑ Cyclic AMP in presynaptic neuron

↓

Blockage of K$^+$ channels in presynaptic neuron

↓

Prolongation of action potential in presynaptic neuron

↓

Ca^{++} channels in presynaptic neuron kept open longer

↓

↑ Ca^{++} influx

↓

↑ Output of transmitter from presynaptic neuron

↓

↑ Postsynaptic potential in efferent neuron

↓

Enhanced behavorial response to mild stimuli

the opposite direction of neurotransmitter activity at a synapse. Synapses traditionally operate unidirectionally, with the presynaptic neuron releasing a neurotransmitter that influences the postsynaptic neuron. The cell body and dendrites of the postsynaptic neuron do not contain any transmitter vesicles. Thus, the retrograde factor is distinct from classical neurotransmitters or neuropeptides. The retrograde messenger is believed to be **nitric oxide**, a chemical that recently has been found to exert a bewildering array of other functions in the body. These other functions range from dilation of blood vessels in the penis during erection to destruction of foreign invaders by the immune system.

Long-term memory   Long-term memory storage appears to involve rather permanent physical changes in the brain,

changes that endure the ongoing turnover of cellular materials over the years. No new neurons are formed, but permanent structural or functional alterations between existing neurons can take place. Examples of such alterations include the formation of new synaptic connections, permanent changes in pre- or postsynaptic membranes, or an increase or decrease in transmitter synthesis.

Studies comparing the brains of experimental animals reared in a sensory-deprived environment with those raised in a sensory-rich environment demonstrate readily observable microscopic differences. The animals afforded more environmental interactions—and supposedly, therefore, more opportunity to learn—displayed greater branching and elongation of dendrites in nerve cells in regions of the brain suspected to be involved with memory storage. Greater dendritic surface area presumably provides more sites for synapses. Thus, long-term memory may be stored at least in part by a particular pattern of dendritic branching and synaptic contacts.

Other studies suggest that protein synthesis plays a critical role in long-term memory storage. Drugs that block protein synthesis have been shown to interfere with long-term retention of learned behavior. Some researchers believe the consolidation of a memory into a long-term storage mode involves the switching on of genes, which results in the production of new proteins. One recently discovered group of genes, so-called **immediate early genes (IEGs)**, may play a critical role in memory consolidation. These genes have been shown to be activated in response to brief bursts of action-potential activity and may be the critical link between short-term memory and the synthesis of proteins that encode long-term memory. The exact role that these critical new long-term memory proteins might play remains speculative. They may be needed for structural changes in dendrites or used for synthesis of more neurotransmitters or additional receptor sites. Alternatively, they may accomplish long-term modification of neurotransmitter release by sustaining biochemical events initially activated by short-term memory processes.

To complicate the issue even further, numerous hormones and neuropeptides are known to affect learning and memory processes.

CEREBELLUM

The cerebellum is important in balance as well as in planning and execution of voluntary movement.

The **cerebellum**, which is attached to the back of the upper portion of the brain stem, lies underneath the occipital lobe of the cortex (Table 5–3 and Fig 5–16). It consists of three functionally distinct parts, which are believed to have developed successively during evolution (● Fig. 5–20). These parts have different sets of inputs and outputs, and, accordingly, each has different functions.

1. The **vestibulocerebellum** is important for the maintenance of balance and controls eye movement.

2. The **spinocerebellum** regulates muscle tone and coordinates skilled, voluntary movements. When cortical motor areas send messages to muscles for the execution of a particular movement, the spinocerebellum is also informed of the intended motor command. In addition, this region receives input from peripheral receptors that apprise it of what is actually taking place regarding body movement and position. The spinocerebellum essentially acts as "middle management," comparing the "intentions" or "orders" of the higher centers with the "performance" of the muscles and then correcting any "errors" or deviations from the intended movement (● Fig. 5–21). The spi-

● **FIGURE 5–20   Cerebellum**   (a) Gross structure of cerebellum. (b) Unfolded cerebellum, revealing its three functionally distinct parts. (c) Internal structure of cerebellum.

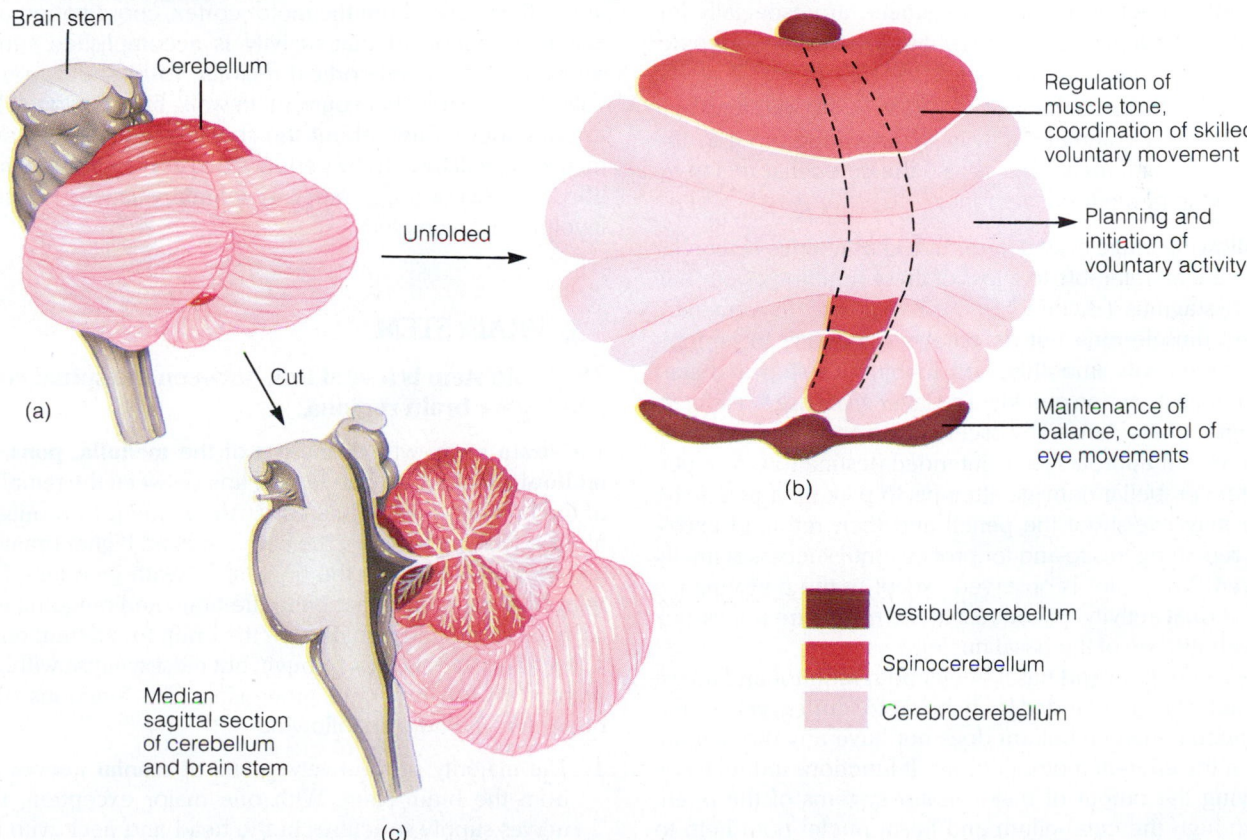

(a)

Brain stem
Cerebellum

Unfolded

Cut

Median sagittal section of cerebellum and brain stem

(c)

(b)

Regulation of muscle tone, coordination of skilled voluntary movement

Planning and initiation of voluntary activity

Maintenance of balance, control of eye movements

■ Vestibulocerebellum

■ Spinocerebellum

■ Cerebrocerebellum

● **FIGURE 5-21   Role of the Spinocerebellum** In coordinating rapid, phasic motor activity, the spinocerebellum compares the "intentions" of higher motor centers with the "performance" of the muscles and corrects any "errors" by making the necessary adjustments to accomplish the intended movement.

nocerebellum even appears to be able to predict the position of a body part in the next fraction of a second and to make adjustments accordingly. If you are reaching for a pencil, for example, this region "puts on the brakes" soon enough to stop the forward movement of your hand at the intended location rather than allowing you to overshoot your target. These ongoing adjustments, which assure smooth, precise, directed movement, are especially important for rapidly changing (phasic) activities like typing, playing the piano, or running.

3. The **cerebrocerebellum** plays a role in the planning and initiation of voluntary activity by providing input to the cortical motor areas. This is also the cerebellar region involved in procedural memories.

The following range of symptoms that characterize cerebellar disease are all referable to a loss of these functions: poor balance, nystagmus (rhythmical, oscillating eye movements), reduced muscle tone but no paralysis, inability to perform rapid movements smoothly, and inability to stop and start skeletal-muscle action quickly. The latter gives rise to an *intention tremor* characterized by oscillating to-and-for movements of a limb as it approaches its intended destination. As a person with cerebellar damage attempts to pick up a pencil, he or she may overshoot the pencil and then rebound excessively, repeating this to-and-for process until success is finally achieved. No tremor is observed except in the performance of intentional activity, in contrast to the resting tremor associated with disease of the basal nuclei.

The cerebellum and basal nuclei both monitor and adjust motor activity commanded from the motor cortex, yet like the basal nuclei, the cerebellum does not have any direct influence on the efferent motor neurons. It functions indirectly by modifying the output of major motor systems of the brain. Even though the cerebellum and basal nuclei both help to

subconsciously coordinate voluntary motor activity, they play different roles. The cerebellum helps maintain balance; smooths out fast, phasic motor activity; and enhances muscle tone. The basal nuclei are important in coordinating slow, sustained movement related to posture and support, and they also function in inhibiting muscle tone.

Even though the motor command for a particular voluntary activity arises from the motor cortex, coordination of the actual execution of that activity is accomplished subconsciously by these subcortical regions. To illustrate, you can voluntarily decide that you want to walk, but you do not have to consciously think about the specific sequence of movements that will have to be performed to accomplish this intentional act. Accordingly, much of voluntary activity is actually involuntarily regulated.

⚏● BRAIN STEM

The brain stem is a vital link between the spinal cord and higher brain regions.

The **brain stem**, which consists of the **medulla**, **pons**, and **midbrain**, is a critical connecting link between the remainder of the brain and the spinal cord. All incoming and outgoing fibers traversing between the periphery and higher brain centers must pass through the brain stem, with incoming fibers relaying sensory information to the brain and outgoing fibers carrying command signals from the brain for efferent output. A few fibers merely pass through, but most synapse within the brain stem for important processing. The functions of the brain stem include the following:

1. The majority of the twelve pairs of **cranial nerves** arise from the brain stem. With one major exception, these nerves supply structures in the head and neck with both

sensory and motor fibers. They are important in sight, hearing, taste, smell, sensation of the face and scalp, eye movement, chewing, swallowing, facial expressions, and salivation. The major exception is cranial nerve X, the **vagus nerve**. Instead of innervating regions in the head, most of the branches of the vagus nerve supply organs in the thoracic and abdominal cavities. The vagus is the major nerve of the parasympathetic nervous system.

2. Collected within the brain stem are neuronal clusters or "centers" that control heart and blood vessel function, respiration, and many digestive activities.

3. This region also plays a role in modulating the sense of pain (see chapter 6).

4. The brain stem plays a role in the regulation of muscle reflexes involved with equilibrium and posture.

5. Running throughout the entire brain stem and into the thalamus is a widespread network of interconnected neurons called the **reticular formation**. This network receives and integrates all synaptic input. Ascending fibers originating in the reticular formation carry signals upward to arouse and activate the cerebral cortex (● Fig. 5–22). These fibers compose the **reticular activating system (RAS)**, which controls the overall degree of cortical alertness and is important in the ability to direct attention.

6. The centers responsible for sleep are also housed within the brain stem.

Sleep is an active process consisting of alternating periods of slow-wave and paradoxical sleep.

Consciousness refers to subjective awareness of the external world and self, including awareness of the private inner world of one's own mind; that is, awareness of thoughts, perceptions, dreams, and so on. Even though the final level of awareness resides in the cerebral cortex and a crude sense of awareness is detected by the thalamus, conscious experience depends on the integrated functioning of many parts of the nervous system.

The following states of consciousness are listed in decreasing order of level of arousal, based on the extent of interaction between peripheral stimuli and the brain:

- Maximum alertness
- Wakefulness
- Sleep (several different types)
- Coma

Maximum alertness depends on attention-getting sensory input that "energizes" the RAS and subsequently the level of activity of the CNS as a whole. At the other extreme, **coma** refers to the total unresponsiveness of a living person to external stimuli, caused either by brain stem damage that interferes with the RAS or by widespread depression of the cerebral cortex, such as that accompanying O_2 deprivation.

The **sleep-wake cycle** is a normal cyclical variation in awareness of surroundings. In contrast to being awake, sleeping individuals are not consciously aware of the external world, but they do have inward conscious experiences such as dreams. Furthermore, they can be aroused by external stimuli, such as an alarm going off.

Sleep is an *active* process, not just the absence of wakefulness. The brain's overall level of activity is not reduced during sleep. During certain stages of sleep, O_2 uptake by the brain is even increased above normal waking levels.

There are two types of sleep, characterized by different EEG patterns and different behaviors: **slow-wave sleep** and

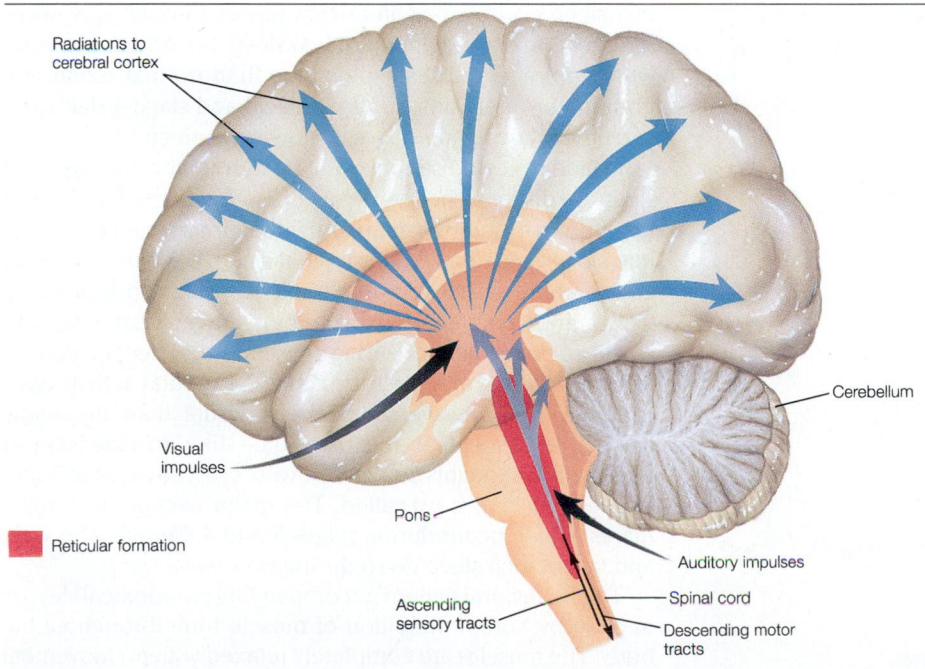

Radiations to cerebral cortex

Visual impulses

Reticular formation

Ascending sensory tracts

Pons

Cerebellum

Auditory impulses

Spinal cord

Descending motor tracts

● **FIGURE 5–22 The Reticular Activating System** The reticular formation, a widespread network of neurons within the brain stem, receives and integrates all synaptic input. The reticular activating system, which promotes cortical alertness and helps direct attention toward specific events, consists of ascending fibers (in blue) originating in the reticular formation that carry signals upward to arouse and activate the cerebral cortex.

TABLE 5–5   Comparison of Slow-Wave and Paradoxical Sleep

CHARACTERISTIC	TYPE OF SLEEP	
	Slow-Wave Sleep	Paradoxical Sleep
EEG	Displays slow waves	Similar to EEG of alert, awake person
Motor activity	Considerable muscle tone; frequent shifting	Abrupt inhibition of muscle tone; no movement
Heart rate, respiratory rate, blood pressure	Minor reductions	Irregular
Dreaming	Rare (mental activity is extension of waking-time thoughts)	Common
Arousal	Sleeper easily awakened	Sleeper hard to arouse but apt to wake up spontaneously
Percentage of sleeping time	80%	20%
Other important characteristics	Has 4 stages; sleeper must pass through this type of sleep first	Rapid eye movements

paradoxical or **REM sleep** (Table 5–5). Slow-wave sleep occurs in four stages, each displaying progressively slower EEG waves of higher amplitude (hence "slow-wave" sleep) (● Fig. 5–23). At the onset of sleep, an individual moves from the light sleep of stage 1 to the deep sleep of stage 4 during a period of thirty to forty-five minutes, then reverses through the same stages in the same amount of time. A ten- to fifteen-minute episode of paradoxical sleep punctuates the end of each slow-wave sleep cycle. Paradoxically, the EEG pattern during this time abruptly becomes similar to that of a wide-awake, alert individual, even though the person is still asleep (Fig. 5–23). Following the paradoxical episode, the stages of slow-wave sleep are repeated once again. A person cyclically alternates between the two types of sleep throughout the night. On the average, paradoxical sleep occupies 20% of total sleeping time throughout adolescence and most of adulthood. Infants spend considerably more time in paradoxical sleep. In contrast, paradoxical as well as stage 4 slow-wave sleep decline in the elderly.

Paradoxical sleep can be considered either the deepest sleep, since it is hardest to arouse sleepers from this stage, or the lightest sleep, since sleepers are most apt to awaken on their own during this stage. Several other observations support the contention that it is the deepest sleep. First, in a normal sleep cycle, a person always passes through slow-wave sleep before entering paradoxical sleep. Second, individuals who require less total sleeping time than normal spend proportionately more time in paradoxical and stage 4 sleep and less time in the lighter stages of slow-wave sleep.

In addition to distinctive EEG patterns, the two types of sleep are distinguished by behavioral differences. It is difficult to pinpoint exactly when an individual drifts from drowsiness into slow-wave sleep. In this type of sleep, the person still has considerable muscle tone and frequently shifts body position. Only minor reductions in respiratory rate, heart rate, and blood pressure occur. During this time the sleeper can be easily awakened and rarely dreams. The mental activity associated with slow-wave sleep is less visual than dreaming. It is more conceptual and plausible—like an extension of waking-time thoughts concerned with everyday events—and it is less likely to be recalled. The major exception is nightmares, which occur during stages 3 and 4. People who walk and talk in their sleep do so during slow-wave sleep.

The behavioral pattern accompanying paradoxical sleep is marked by abrupt inhibition of muscle tone throughout the body. The muscles are completely relaxed with no movement

● **FIGURE 5–23   EEG Patterns during Different Types of Sleep** Note that the EEG pattern during paradoxical sleep is similar to that of an alert, awake person, whereas the pattern during slow-wave sleep displays distinctly different waves.

Slow-wave sleep, stage 4 (delta waves)

Paradoxical sleep (beta waves)

Awake, eyes open (beta waves)

taking place. Paradoxical sleep is further characterized by *rapid eye movements*, which give it the name REM sleep. Heart rate and respiratory rate become irregular and blood pressure may fluctuate. Another characteristic of REM sleep is *dreaming*. There is little evidence that the rapid eye movements are related to "watching" the dream imagery. The eye movements seem to be driven in a locked oscillating pattern uninfluenced by dream content.

The sleep-wake cycle is probably controlled by interactions among three brain stem regions.

The sleep-wake cycle as well as the various stages of sleep are believed to be due to the cyclical interplay of three different neural systems in the brain stem: (1) an *arousal system,* which is part of the reticular activating system, (2) a *slow-wave sleep center,* and (3) a *paradoxical sleep center*. The pattern of interaction among the three identified neural regions, which brings about the fairly predictable cyclical sequence between being awake and passing alternatively between the two stages of sleep, is the subject of intense investigation. Nevertheless, the molecular mechanisms controlling the sleep-wake cycle remain poorly understood.

The normal cycle can easily be interrupted, with the arousal system more readily overriding the sleep systems than vice versa; that is, it is easier to stay awake when you are sleepy than to fall asleep when you are wide-awake. The arousal system can be activated by afferent sensory input (for example, a person has difficulty falling asleep when it is noisy) or by input descending to the brain stem from higher brain regions. Intense concentration or strong emotional states, such as anxiety or excitement, can keep a person from falling asleep, just as motor activity, such as getting up and walking around, can arouse a drowsy person.

An unusual sleep disturbance is **narcolepsy**. It is characterized by brief (five- to thirty-minute) irresistable **sleep attacks** during the day. A person suffering from this condition suddenly falls asleep during any ongoing activity, often without warning. Narcolepsy is suspected of being linked to abnormal activation of neurons in the paradoxical sleep center during the waking state. Narcoleptic patients, in fact, can enter into paradoxical sleep directly without the normal prerequisite passage through slow-wave sleep.

Even though humans spend about a third of their lives sleeping, the reason sleep is needed largely remains a mystery. Sleep is not accompanied, as once was suspected, by a *reduction* in neural activity (that is, the brain cells are not "resting"), but rather by a profound *change* in activity. It is speculated that sleep is necessary to allow the brain to "shift gears" in order to accomplish the long-term structural and chemical adjustments necessary for learning and memory. This theory might explain why infants require so much sleep. Their highly plastic brains are rapidly undergoing profound synaptic modifications in response to environmental stimulation. In contrast, mature individuals in whom neural changes are less dramatic sleep less. Not much is known about the brain's need for the two types of sleep, although a specified amount of paradoxical sleep appears to be required. Individ-

uals experimentally deprived of paradoxical sleep for a night or two by being aroused every time the paradoxical EEG pattern appeared suffered hallucinations and spent proportionally more time in paradoxical sleep during subsequent undisturbed nights, as if to "make up for lost time."

SPINAL CORD

The spinal cord extends through the vertebral canal and is connected to the spinal nerves.

Extending from the brain stem is a long slender cylinder of nerve tissue, the **spinal cord**. It is about 45 cm long (18 inches) and 2 cm in diameter (about the size of your little finger). Exiting through a large hole in the base of the skull, the spinal cord is enclosed by the protective vertebral column as it descends through the vertebral canal. Paired spinal nerves emerge from the spinal cord through spaces formed between the bony, winglike arches of adjacent vertebrae. The spinal nerves are named according to the region of the vertebral column from which they emerge (● Fig. 5–24): there are eight pairs of *cervical* nerves (that is, C1–C8), twelve *thoracic* nerves, five *lumbar* nerves, five *sacral* nerves, and one *coccygeal* nerve.

During development, the vertebral column grows about 25 cm longer than the spinal cord. Because of this differential growth, segments of the spinal cord that give rise to various spinal nerves are not aligned with the corresponding intervertebral spaces. Most of the spinal nerve roots must descend along the cord before emerging from the vertebral column at the corresponding space. The spinal cord itself extends only to the level of the first or second lumbar vertebra (about waist level), so the nerve roots of the remaining nerves are greatly elongated in order to exit the vertebral column at their appropriate space. The thick bundle of elongated nerve roots within the lower vertebral canal is known as the **cauda equina** ("horse's tail") because of its appearance (Fig. 5–24b). This region is the site where "spinal taps" are performed to obtain a sample of CSF. Insertion of a needle into the canal below the level of the second lumbar vertebra does not run the risk of penetrating the spinal cord. The needle pushes aside the nerve roots of the cauda equina so that the sample of the surrounding fluid can be withdrawn safely.

Although there are some slight regional variations, the cross-sectional anatomy of the spinal cord is generally the same throughout its length (● Fig. 5–25). In contrast to the gray matter in the brain, the gray matter in the spinal cord forms a butterfly-shaped region on the inside and is surrounded by the outer white matter. As in the brain, the cord gray matter consists primarily of neuronal cell bodies and their dendrites, short interneurons, and glial cells. The white matter is organized into **tracts**, which are bundles of nerve fibers (axons of long interneurons) with a similar function. The bundles are grouped into columns that extend the length of the cord. Each of these tracts begins or ends within a particular area of the brain, and each is specific in the type of information that it transmits. Some are **ascending** (cord to

(a)

Cervical
cord

Thoracic
cord

Lumbar
cord

Sacral
cord

(b)

Cervical
nerves

1
2
3
4
5
6
7
8

Vertebra

Thoracic
nerves

1
2
3
4
5
6
7
8
9
10
11
12

Lumbar
nerves

1
2
3
4
5

Cauda
equina

Sacral
nerves

1
2
3
4
5

Coccygeal
nerve

1

● **FIGURE 5–24 Spinal Nerves** There are thirty-one pairs of spinal nerves named according to the region of the vertebral column from which they emerge. Because the spinal cord is shorter than the vertebral column, spinal nerve roots must descend along the cord before emerging from the vertebral column at the corresponding intervertebral space, especially those beyond the level of the first lumbar vertebra (L1). Collectively these rootlets are called the cauda equina, literally "horse's tail." (a) Posterior view of brain, spinal cord, and spinal nerves (on the right side only). (b) Lateral view of spinal cord and spinal nerves emerging from vertebral column.

brain) **tracts** that transmit to the brain signals derived from afferent input. Others are **descending** (brain to cord) **tracts** that relay messages from the brain to efferent neurons (● Fig. 5–26).

The tracts are generally named for their origin and termination. For example, the **corticospinal tract** is a descending pathway; its cell bodies originate primarily in the motor region of the cerebral cortex, and its axons travel down to terminate in the spinal cord on the cell bodies of efferent motor neurons supplying skeletal muscles. In contrast, the **lateral spinothalamic tract** is an ascending pathway that originates in the spinal cord and runs laterally through the cord until it synapses in the thalamus. It carries sensory information about pain and temperature that has been delivered from various regions of the body through the spinal cord to the thalamus, which in turn sorts and relays the information to the somato-

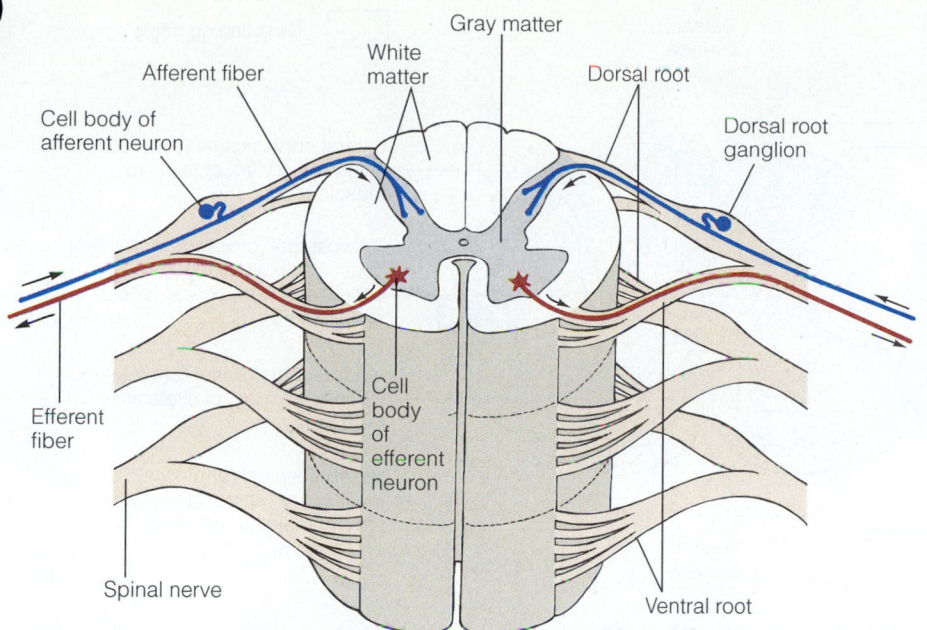

● **FIGURE 5–25 Spinal Cord in Cross Section** The afferent fibers enter through the dorsal root, and the efferent fibers exit through the ventral root. Afferent and efferent fibers are enclosed together within a spinal nerve.

Labels on figure:
Afferent fiber
Gray matter
White matter
Dorsal root
Cell body of afferent neuron
Dorsal root ganglion
Efferent fiber
Cell body of efferent neuron
Spinal nerve
Ventral root

sensory cortex. It is important to appreciate that various types of signals are segregated within the spinal cord and, accordingly, that damage to particular areas of the cord can interfere with some functions while others remain intact.

The centrally located gray matter is also functionally organized (● Fig. 5–27). The central canal, which is filled with CSF, lies in the center of the gray matter. Each half of the gray matter is arbitrarily divided into a **dorsal (posterior) horn**, a **ventral (anterior) horn**, and a **lateral horn**. The dorsal horn contains cell bodies of interneurons upon which afferent neurons terminate. The ventral horn contains cell bodies of the efferent motor neurons supplying skeletal muscles. Autonomic nerve fibers supplying cardiac and smooth muscle and exocrine glands originate at cell bodies found in the lateral horn.

Spinal nerves connect with each side of the spinal cord by a **dorsal root** and a **ventral root** (Fig. 5–25). Afferent fibers carrying incoming signals enter the spinal cord through the dorsal root; efferent fibers carrying outgoing signals leave through the ventral root. The cell bodies for the afferent neurons at each level are clustered together in a **dorsal root ganglion**. (A collection of neuronal cell bodies located outside of the CNS is called a *ganglion,* whereas a functional collection of cell bodies within the CNS is referred to as a *center* or a *nucleus*). The cell bodies for the efferent neurons originate in the gray matter and send axons out through the ventral root.

The dorsal and ventral roots at each level join to form a **spinal nerve** that emerges from the vertebral column. A spinal nerve contains both afferent and efferent fibers traversing between a particular region of the body and the spinal cord. Note the relationship between a *nerve* and a *neuron.* A **nerve** is a bundle of peripheral neuronal axons, some afferent and some efferent, which are enclosed by a connective-tissue covering and follow the same pathway. A nerve does not contain a complete nerve cell, only the axonal portions of many neurons. (By this definition, there are no nerves in the CNS! Bundles of axons in the CNS are called tracts.) The individual fibers within a nerve generally do not have any direct influence on each other. They travel together for convenience, just as many individual telephone lines are carried within a telephone cable, yet any particular connection can be private without interference or influence from other lines in the cable.

The thirty-one pairs of spinal nerves, along with the twelve pairs of cranial nerves that arise from the brain, constitute the *peripheral nervous system.* After they emerge, the spinal nerves progressively branch to form a vast network of peripheral nerves that supply the tissues. Each segment of the spinal cord gives rise to a pair of spinal nerves that ultimately supply a particular region of the body with both afferent and efferent fibers. Thus, the location and extent of sensory and motor deficits associated with spinal cord injuries can be clinically important in determining the level and extent of the cord injury.

With reference to sensory input, each specific region of the body surface supplied by a particular spinal nerve is called a **dermatome**. These same spinal nerves also carry fibers that branch off to supply internal organs, and sometimes pain originating from one of these organs is "referred" to the corresponding dermatome supplied by the same spinal nerve. **Referred pain** originating in the heart, for example, may appear to come from the left shoulder and arm. The mechanism responsible for referred pain is not completely understood. Presumably, inputs arising from the heart share a pathway to the brain in common with inputs from the left upper extremity. The higher perception levels, being more accustomed to receiving sensory input from the left arm than from the heart, may interpret the input from the heart as having arisen from the left arm.

Ascending tracts   Dorsal surface   Descending tracts

Dorsal columns:
1. Fasciculus gracilis
2. Fasciculus cuneatus
(conscious muscle sense concerned
with awareness of body position;
crossed touch, pressure, vibration)

Dorsal spinocerebellar (uncrossed;
unconscious muscle sense—
important in control of muscle
tone and posture)

Ventral spinocerebellar (crossed;
unconscious muscle sense)

Lateral spinothalamic (crossed;
pain and temperature)

Ventral spinothalamic (crossed; touch)

Gray   matter

Lateral corticospinal (crossed;
voluntary control of skeletal
muscles)

Rubrospinal (crossed; involuntary
control of skeletal muscle
concerned with muscle tone
and posture)

Ventral corticospinal (uncrossed;
voluntary control of skeletal
muscles)

Vestibulospinal (uncrossed;
involuntary control of muscle
tone to maintain balance and
equilibrium)

Ventral surface

(a)

• **FIGURE 5–26   Ascending and Descending Tracts in the White Matter of the Spinal Cord** (a) Location and functions of major ascending and descending tracts in the spinal cord in cross section. (b) Cord-to-brain pathways of several ascending tracts (fasciculus cuneatus and ventral spinocerebellar tract). (c) Brain-to-cord pathways of several descending tracts (lateral corticospinal and ventral corticospinal tracts).

The spinal cord is responsible for the integration of many basic reflexes.

The spinal cord is strategically located between the brain and afferent and efferent fibers of the peripheral nervous system; this location enables the spinal cord to fulfill its two primary functions: (1) serving as a link for transmission of information between the brain and the remainder of the body; and (2) integrating reflex activity between afferent input and efferent output without involving the brain. This type of reflex activity is known as a **spinal reflex**.

A **reflex** is any response that occurs automatically without conscious effort. There are two types of reflexes: **simple** or **basic reflexes**, which are built-in, unlearned responses, such as closing the eyes when an object moves toward them; and **acquired** or **conditioned reflexes**, which are a result of practice and learning, such as a pianist striking a particular key on seeing a given note on the music staff. The musician does this automatically, but only after considerable conscious training effort. (See the accompanying boxed feature, A Closer Look at Exercise Physiology.)

The neural pathway involved in accomplishing reflex activity is known as a **reflex arc**, which typically includes the five following basic components:

1. Receptor
2. Afferent pathway
3. Integrating center
4. Efferent pathway
5. Effector

The **receptor** responds to a **stimulus**, which is a detectable physical or chemical change in the environment of the receptor. In response to the stimulus, the receptor produces an action potential that is relayed by the **afferent pathway** to the integrating center for processing. Usually, the **integrating center** is the CNS. The spinal cord and brain stem are responsible

• **FIGURE 5–27   Regions of the Gray Matter**

Central canal

Dorsal horn (cell bodies of interneurons upon which afferent neurons terminate)

Lateral horn (cell bodies of autonomic efferent nerve fibers)

Ventral horn (cell bodies of somatic efferent neurons)

Somatosensory area of cerebral cortex

Thalamus

Cerebral cortex

Primary motor cortex

Midbrain

Cerebellum

Pons

Ventral spinocerebellar tract

Medulla

Muscle stretch receptor

Fasciculus cuneatus

Ventral corticospinal tract

Lateral corticospinal tract

Spinal cord

(b)

Spinal cord

Pressure receptor in skin

Skeletal muscle cell

Spinal cord

(c)

• **FIGURE 5–26   Ascending and Descending Tracts in the White Matter of the Spinal Cord (continued)**

SWAN DIVE OR BELLY FLOP: IT'S A MATTER OF CNS CONTROL

Sport skills must be learned. Much of the time, strong basic reflexes must be overridden in order to perform the skill. Learning to dive into water, for example, is very difficult initially. Strong head-righting reflexes controlled by sensory organs in the neck and ears initiate a straightening of the neck and head before the beginning diver enters the water, causing what is commonly known as a "belly flop." In a backward dive, the head-righting reflex causes the beginner to land on his or her back or even in a sitting position. To perform any motor skill that involves body inversions, somersaults, back flips, or other abnormal postural movements, the person must learn to consciously inhibit basic postural reflexes. This is accomplished by having the person concentrate on specific body positions during the movement. For example, in the somersault, the person must concentrate on keeping the chin tucked and grabbing the knees in order to perform the skill. After the skill is performed repeatedly, new synaptic patterns are formed in the CNS, and the new or conditioned response substitutes for the natural reflex responses. Sport skills must be practiced until the movement becomes automatic; then the athlete is free during competition to think about strategy or the next move to be performed in a routine.

---

for the integration of the basic reflexes, whereas higher brain levels usually process acquired reflexes. The integrating center processes all information available to it from this receptor as well as from all other inputs, then "makes a decision" about the appropriate response. The instructions from the integrating center are transmitted via the **efferent pathway** to the **effector**—a muscle or gland—which carries out the desired response. Unlike conscious behavior, in which any one of a number of responses is possible, a reflex response is predictable because the pathway between the receptor and effector is always the same.

A basic spinal reflex is one integrated by the spinal cord; that is, all components necessary for linking afferent input to efferent response are present within the spinal cord. The **withdrawal reflex** can serve to illustrate a basic spinal reflex (• Fig. 5–28). When a person touches a hot stove (or receives another painful stimulus), a reflex is initiated to pull the hand away from the stove (to withdraw from the painful stimulus). The skin has different receptors for warmth, cold, light touch, pressure, and pain. Even though all information is sent to the CNS by way of action potentials, the CNS can distinguish between various stimuli because different receptors and consequently different afferent pathways are activated by different stimuli. When a receptor is stimulated sufficiently to reach threshold, an action potential is generated in the afferent neuron. The stronger the stimulus, the greater the frequency of action potentials generated and propagated to the CNS. Once the afferent neuron enters the spinal cord, it diverges to synapse with the following different interneurons (the numbers correspond to those on Fig. 5–28).

1. An excited afferent neuron stimulates excitatory interneurons that in turn stimulate the efferent motor neurons supplying the biceps, the muscle in the arm that flexes (bends) the elbow joint. The resultant contraction of the biceps pulls the hand away from the hot stove.

2. It also stimulates inhibitory interneurons that in turn inhibit the efferent neurons supplying the triceps to prevent it from contracting. The triceps is the muscle in the arm that extends (straightens out) the elbow joint. When the biceps is contracting to flex the elbow, it would be counterproductive for the triceps to be contracting. Therefore, built into the withdrawal reflex is inhibition of the muscle that antagonizes (opposes) the desired response. This type of neuronal connection involving stimulation of the nerve supply to one muscle and simultaneous inhibition of the nerves to its antagonistic muscle is known as **reciprocal innervation**.

3. It stimulates still other interneurons that carry the signal up the spinal cord to the brain via an ascending pathway. Only when the impulse reaches the sensory area of the cortex is the person aware of the pain, its location, and the type of stimulus. Also, when the impulse reaches the brain, the information can be stored as memory, and the person can start thinking about the situation—how it happened, what to do about it, and so on. All this activity at the conscious level is above and beyond the basic reflex.

As is characteristic of all spinal reflexes, the brain can modify the withdrawal reflex. Impulses may be sent down descending pathways to the efferent neurons supplying the involved muscles to override the input from the receptors, actually preventing the biceps from contracting in spite of the painful stimulus. When your finger is being pricked to obtain a blood sample, pain receptors are stimulated, initiating the withdrawal reflex. However, knowing that you must be brave and not pull your hand away, you can consciously override the reflex by sending IPSPs via descending pathways to the

Thermal pain receptor in finger

Afferent pathway

Stimulus

Effector organs

Biceps (flexor)

Triceps (extensor)

Efferent pathway

Ascending pathway to brain

Integrating center (spinal cord)

Components of a reflex arc

Receptor
Afferent pathway
Integrating center
Efferent pathway
Effector organs

+ = Stimulates
− = Inhibits
⤙ = Synapse
○⤙ = Excitatory interneuron
○⤙ = Inhibitory interneuron

Response

• **FIGURE 5–28 Withdrawal Reflex** When a painful stimulus activates a receptor in the finger, action potentials are generated in the corresponding afferent pathway, which propagates the electrical signals to the CNS. Once the afferent neuron enters the spinal cord, it diverges and terminates on three different types of interneurons (only one of each type is depicted): (1) excitatory interneurons, which in turn stimulate the efferent motor neurons to the biceps, causing the arm to flex and pull the hand away from the painful stimulus; (2) inhibitory interneurons, which inhibit the efferent motor neurons to the triceps, thus preventing counterproductive contraction of this antagonistic muscle; and (3) interneurons that carry the signal up the spinal cord via an ascending pathway to the brain for awareness of pain, memory storage, and so on.

motor neurons supplying the biceps and EPSPs to those supplying the triceps. The activity in these efferent neurons depends on the sum of activity of all their synaptic inputs. Because the neurons supplying the biceps are now receiving more IPSPs from the brain (voluntary) than EPSPs from the afferent pain pathway (reflex), these neurons are inhibited and do not reach threshold. Therefore, the biceps is not stimulated to contract and withdraw the hand. Simultaneously, the neurons to the triceps are receiving more EPSPs from the brain than IPSPs via the reflex arc, so they reach threshold, fire, and consequently stimulate the triceps to contract. Thus, the arm is kept extended in spite of the painful stimulus. In this way, the withdrawal reflex has been voluntarily overridden.

Only one reflex is simpler than the withdrawal reflex—the **stretch reflex**, in which an afferent neuron originating at a stretch-detecting receptor in a skeletal muscle terminates directly on the efferent neuron supplying the same skeletal muscle to cause it to contract and counteract the stretch. This is a **monosynaptic** ("one synapse") **reflex**, because the only synapse in the reflex arc is the one between the afferent and efferent neuron. The withdrawal reflex and all other reflexes are **polysynaptic** ("many synapses"), because interneurons are interposed in the reflex pathway and, therefore, a number of synapses are involved.

Spinal reflex action is not necessarily limited to motor responses on the side of the body to which the stimulus is applied. Assume that a person steps on a tack instead of burning a finger. A reflex arc is initiated to withdraw the injured foot from the painful stimulus, while the opposite leg simultaneously prepares to suddenly bear all of the weight so that the person does not lose balance or fall (• Fig. 5–29). Unimpeded bending of the injured extremity's knee is accomplished by concurrent reflex stimulation of the muscles that flex the knee and inhibition of the muscles that extend the knee. At the same time, unimpeded extension of the opposite limb's knee is accomplished by activation of pathways that cross over to the opposite side of the spinal cord to reflexly stimulate this knee's extensors and inhibit its flexors. This **crossed extensor reflex** assures that the opposite limb will be in a position to bear the weight of the body as the injured limb is withdrawn from the stimulus.

Besides protective reflexes such as the withdrawal reflex and simple postural reflexes such as the crossed extensor reflex, basic spinal reflexes also mediate emptying of pelvic organs (for example, urination, defecation, and expulsion of semen). All spinal reflexes can be voluntarily overridden at least temporarily by higher brain centers.

Not all reflex activity involves a clear-cut reflex arc, although the basic principles of an automatic response to a de-

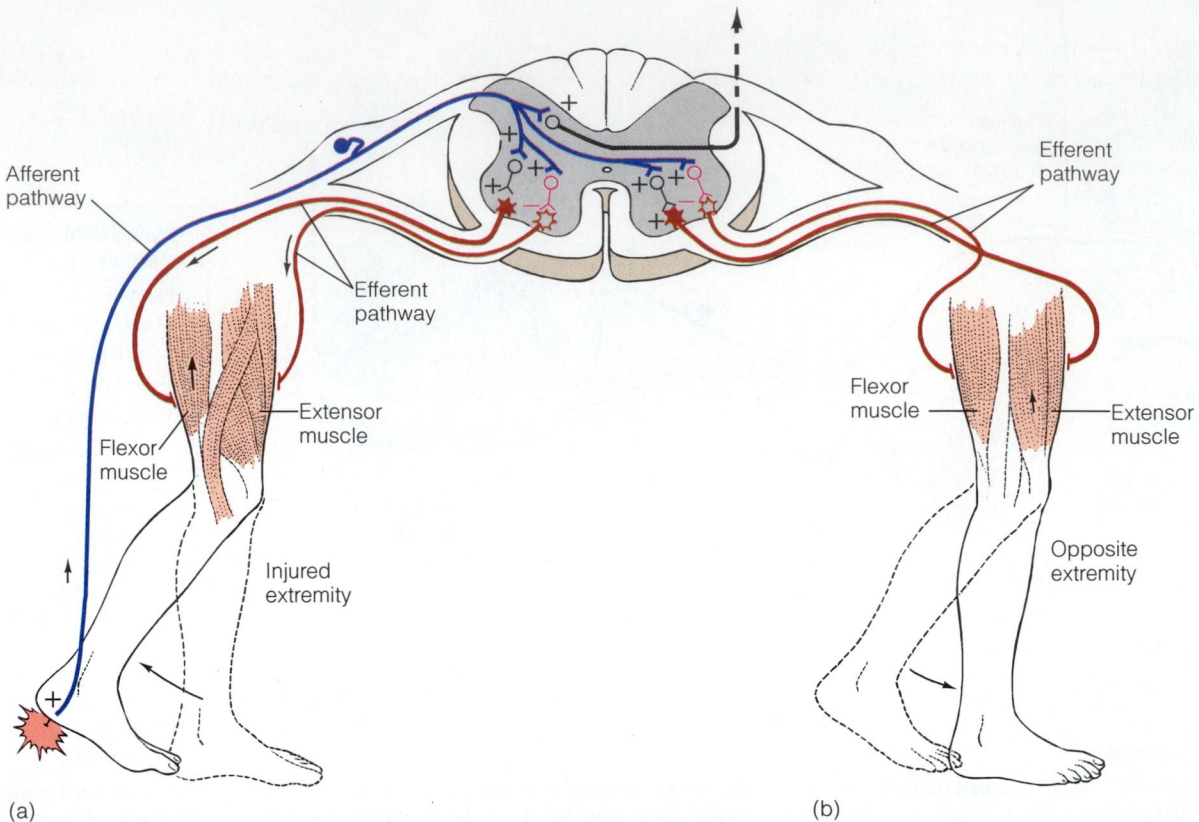

(a)   (b)

● **FIGURE 5–29   Crossed Extensor Reflex Coupled with Withdrawal Reflex** (a) Withdrawal reflex, which causes flexion of injured extremity to withdraw from painful stimulus. (b) Crossed extensor reflex, which extends opposite limb to support full weight of body.

tectable change are still present. Pathways for unconscious responsiveness digress from the typical reflex arc in two general ways:

1. *Responses mediated at least in part by hormones.* A particular reflex may be mediated solely by either neurons or hormones or may involve a pathway utilizing both.

2. *Local responses that do not involve either nerves or hormones.* For example, the blood vessels in an exercising muscle dilate because of local metabolic changes, thereby increasing blood flow to match the active muscle's metabolic needs.

CHAPTER IN PERSPECTIVE:
FOCUS ON HOMEOSTASIS

In order to interact in appropriate ways with the external environment to sustain the body's viability, such as in the acquisition of food, and to make the internal adjustments necessary to maintain homeostasis, the body must be informed about any changes taking place in the external and internal environment and must be able to process this information and send messages to various muscles and glands to accomplish the desired results. The nervous system, one of the body's two major control systems, plays a central role in this life-sustaining communication. The central nervous system

(CNS), which consists of the brain and spinal cord, receives information about the external and internal environment by means of the afferent peripheral nerves. After sorting, processing, and integrating this input, the CNS sends directions by means of efferent peripheral nerves to bring about appropriate muscular contractions and glandular secretions.

With its swift electrical signaling system, the nervous system is especially important in controlling the rapid responses of the body. Many neurally controlled muscular and glandular activities are aimed toward maintaining homeostasis. The CNS is the main site of integration between the afferent input and efferent output. It is responsible for linking the appropriate response to a particular input so that conditions compatible with life are maintained in the body. For example, when informed by the afferent nervous system that blood pressure has fallen, the CNS sends appropriate commands to the heart and blood vessels to increase the blood pressure to normal. Were it not for this processing and integrative ability of the CNS, maintenance of homeostasis in an organism as complex as a human would be impossible.

At the simplest level, the spinal cord integrates many basic protective and evacuative reflexes that do not require conscious participation, such as withdrawal from a painful stimulus and emptying of the urinary bladder. The brain, in addition to serving as a more complex integrating link between afferent input and efferent output, is also responsible for the initiation of all voluntary movement, for complex perceptual awareness of the external environment and self, for language, and for abstract neural phenomena such as thinking, learning, remembering, consciousness, emotions, and personality traits. All neural activity—from the most private thoughts to commands for motor activity, from enjoying a concert to retrieving memories from the distant past—is ultimately attributable to propagation of action potentials along individual nerve cells and chemical transmission between cells.

The nervous system has become progressively more complex during evolutionary development. Newer, more complicated, and more sophisticated layers of the brain have been piled on top of older, more primitive regions. Many of the basic activities necessary for survival are built into the older parts of the brain. The newer, higher levels progressively modify, enhance, or nullify actions coordinated by lower levels in a hierarchy of command, and they also add new capabilities. Many of these higher neural activities are not aimed toward maintaining life, but they add immeasurably to the quality of being alive.

CHAPTER SUMMARY

Introduction

The nervous system is one of the two control systems of the body, the other being the endocrine system. In general, the nervous system coordinates rapid responses, whereas the endocrine system regulates activities that require duration rather than speed.

The nervous system consists of the central nervous system (CNS), which includes the brain and spinal cord, and the peripheral nervous system, which includes the nerve fibers carrying information to (afferent division) and from (efferent division) the CNS. Three classes of neurons—afferent neurons, efferent neurons, and interneurons—compose the excitable cells of the nervous system. Afferent neurons apprise the CNS of conditions in both the external and internal environment. Efferent neurons carry instructions from the CNS to effector organs, namely, muscles and glands. Interneurons are responsible for integrating afferent information and formulating an efferent response, as well as for all higher mental functions associated with the "mind."

Protection and Nourishment of the Brain

Glial cells form the connective tissue within the CNS and physically and metabolically support the neurons. The brain is provided with several protective devices, which are important because neurons cannot divide to replace damaged cells. The brain is wrapped in three layers of protective membranes—the meninges—and is further surrounded by a hard bony covering. Cerebrospinal fluid flows within and around the brain to cushion it against physical jarring. Protection against chemical injury is conferred by a blood-brain barrier that limits access of blood-borne substances to the brain.

The brain depends on a constant blood supply for delivery of O_2 and glucose because it is unable to generate ATP in the absence of either of these substances.

Cerebral Cortex

The cerebral cortex is the outer shell of gray matter that caps an underlying core of white matter; the white matter consists of bundles of nerve fibers that interconnect various cortical regions with other areas. The cortex itself consists primarily of neuronal cell bodies and dendrites.

Ultimate responsibility for many discrete functions is known to be localized in particular regions of the cortex as follows: (1) the occipital lobes house the visual cortex; (2) the auditory cortex is found in the temporal lobes; (3) the parietal lobes are responsible for reception and perceptual processing of somato-sensory input; and (4) voluntary motor movement is set into motion by frontal lobe activity. Language ability depends on the integrated activity of two primary language areas located in the left hemisphere only.

The association areas are areas of the cortex not specifically assigned to processing sensory input or commanding motor output or language ability. These areas provide an integrative link between diverse sensory information and purposeful action; they also play a key role in higher brain functions such as memory and decision making.

Subcortical Structures and Their Relationship with the Cortex in Higher Brain Functions

The subcortical brain structures, which include the basal nuclei, thalamus, and hypothalamus, interact extensively with the cortex in the performance of their functions. The basal nuclei inhibit muscle tone; coordinate slow, sustained postural contractions; and suppress useless patterns of movement. The thalamus serves as a relay station for preliminary processing of sensory input on its way to the cortex. It also accomplishes a crude awareness of sensation and some degree of consciousness. The hypothalamus regulates many homeostatic functions, in part through its extensive control of the autonomic nervous system and endocrine system. The limbic system, which includes portions of the hypothalamus and other forebrain structures, is responsible for emotion as well as basic, inborn behavior patterns related to survival. It also plays an important role in motivation and learning.

There are two types of memory: (1) a short-term memory with limited capacity and brief retention, coded at least in part by temporary modifications in transmitter release; and (2) a long-term memory with large storage capacity and enduring memory traces, presumably involving relatively permanent structural or functional changes between already existing neurons.

Cerebellum and Brain Stem

The cerebellum helps to maintain balance, enhances muscle tone, and helps coordinate voluntary movement. It is especially important in smoothing out fast, phasic motor activities.

The brain stem is an important link between the spinal cord and higher brain levels. It is the origin of the cranial nerves; contains centers that control cardiovascular, respiratory, and digestive function; regulates postural muscle reflexes; controls the overall degree of cortical alertness; and establishes the sleep-wake cycle. The prevailing state of consciousness depends on the cyclical interplay between an arousal system (the reticular activating system), a slow-wave sleep center, and a paradoxical sleep center, all located in the brain stem.

Spinal Cord

The spinal cord has two vital functions. First, it serves as the neuronal link between the brain and the peripheral nervous system. All communication up and down the spinal cord is located in well-defined, independent ascending and descending tracts in the cord's outer white matter. Second, it is the integrating center for spinal reflexes, including some of the basic protective and postural reflexes and those involved with the emptying of the pelvic organs. The components of a basic reflex arc include a receptor, an afferent pathway, an integrating center, an efferent pathway, and an effector. The centrally located gray matter of the spinal cord contains the interneurons interposed between the afferent input and efferent output as well as the cell bodies of efferent neurons. The afferent and efferent fibers, which carry signals to and from the spinal cord, respectively, are bundled together into spinal nerves. These nerves are attached to the spinal cord in paired fashion throughout its length. They supply specific regions of the body.

REVIEW EXERCISES

Objective Questions (Answers on p. D–1)

1. The major function of the CSF is to nourish the brain. (True or false?)

2. The brain can perform anaerobic metabolism in emergencies when O_2 supplies are low. (True or false?)

3. Damage to the left cerebral hemisphere brings about paralysis and loss of sensation on the left side of the body. (True or false?)

4. The hands and structures associated with the mouth have a disproportionately large share of representation in both the sensory and motor cortices. (True or false?)

5. The left cerebral hemisphere specializes in artistic and musical ability, whereas the right side excels in verbal and analytical skills. (True or false?)

6. The specific function a particular cortical region will carry out is permanently determined during embryonic development. (True or false?)

7. _____ is a decreased responsiveness to an indifferent stimulus that is repeatedly presented.

8. The process of transferring and fixing short-term memory traces into long-term memory stores is known as _____.

9. Afferent fibers enter through the _____ root of the spinal cord, whereas efferent fibers leave through the _____ root.

10. List the five components of a basic reflex arc: 1. _____ 2. _____ 3. _____ 4. _____ 5. _____ .

11. Using the answer code at the right, indicate which neurons are being described (a characteristic may apply to more than one class of neurons):
    ___ 1. have receptor at peripheral endings
    ___ 2. lie entirely within the CNS
    ___ 3. lie primarily within the peripheral nervous system
    ___ 4. innervate muscles and glands
    ___ 5. cell body is devoid of presynaptic inputs
    ___ 6. predominant type of neuron
    ___ 7. responsible for thoughts, emotions, memory, etc.

    a. afferent neurons
    b. efferent neurons
    c. interneurons

12. Match the following:
    ___ 1. coordinates slow, sustained movements; inhibits muscle tone; suppresses unwanted patterns of movement
    ___ 2. enhances muscle tone; smoothes out fast, phasic motor activities; compares "intentions" of higher motor centers with the "performance" by muscles and corrects any "errors"
    ___ 3. relay station for synaptic input; reinforces voluntary movement initiated by the motor cortex
    ___ 4. regulates the internal environment; links the autonomic nervous system and endocrine system
    ___ 5. contains centers that control cardiovascular, respiratory, and digestive functions; controls overall degree of cortical alertness and establishes the sleep-wake cycle; regulates postural reflexes
    ___ 6. an interconnected ring of forebrain structures surrounding the brain stem; extensively involved with emotions, motivated behavior, and learning; contains "reward" and "punishment" centers

    a. thalamus
    b. hypothalamus
    c. basal nuclei
    d. limbic system
    e. cerebellum
    f. brain stem

13. Match the following:

___ 1. consists of nerves carrying information between the periphery and the CNS

___ 2. consists of the brain and spinal cord

___ 3. division of the peripheral nervous system that transmits signals to the CNS

___ 4. division of the peripheral nervous system that transmits signals from the CNS

___ 5. supplies skeletal muscles

___ 6. supplies smooth muscle, cardiac muscle, and glands

a. somatic nervous system
b. autonomic nervous system
c. central nervous system
d. peripheral nervous system
e. efferent division
f. afferent division

Essay Questions

1. Compare the general responsibilities of the nervous system and the endocrine system.
2. Discuss the function of each of the following: glial cells, the cranium, the vertebral column, the meninges, the cerebrospinal fluid, and the blood-brain barrier.
3. Compare the composition of white and gray matter.
4. Draw and label the major functional areas of the cerebral cortex, indicating the functions attributable to each area.
5. Define somesthetic sensations and proprioception.
6. What is an electroencephalogram?
7. Discuss the roles of Broca's area and Wernicke's area in language ability.
8. Compare short-term and long-term memory.
9. What is the reticular activating system?
10. Compare slow-wave and paradoxical sleep.
11. Draw and label a cross section of the spinal cord.
12. Distinguish between a monosynaptic and polysynaptic reflex.

 POINTS TO PONDER

1. Special studies designed to assess the specialized capacities of each cerebral hemisphere have been performed on "split-brain" patients. These are individuals in whom the corpus callosum—the bundle of fibers that links the two halves of the brain together—has been surgically cut to prevent the spread of epileptic seizures from one hemisphere to the other. Even though no overt changes in behavior, intellect, or personality occur in these patients because both hemispheres individually receive the same information, deficits are observable with tests designed to restrict information to one brain hemisphere at a time. One such test involves limiting a visual stimulus to only one-half of the brain. Because of a cross-over in the nerve pathways from the eyes to the occipital cortex, the visual information to the right of a midline point is transmitted to only the left half of the brain, whereas visual information to the left of this point is received by only the right half of the brain. A split-brain patient presented with a visual stimulus that reaches only the left hemisphere accurately describes the object seen, but when a visual stimulus is presented to only the right hemisphere, the patient denies having seen anything. The right hemisphere does receive the visual input, however, as demonstrated by nonverbal tests. Even though a split-brain patient denies having seen anything after an object is presented to the right hemisphere, he or she can correctly match the object by picking it out from among a number of objects, usually to the patient's surprise. What is your explanation of this finding?

2. The hormone insulin enhances the carrier-mediated transport of glucose into most of the body's cells but not into brain cells. The uptake of glucose from the blood by neurons is not dependent on insulin. Knowing the brain's need for a continuous supply of blood-borne glucose, what effect do you predict that insulin excess would have on the brain?

3. Which of the following symptoms are most likely to occur as the result of a severe blow to the back of the head?
   a. paralysis
   b. hearing impairment
   c. visual disturbances
   d. burning sensations
   e. personality disorders

4. Give examples of conditioned reflexes you have acquired.

5. Under what circumstances might it be inadvisable to administer a clot-dissolving drug to a stroke victim?

6 Peripheral Nervous System: Afferent Division; Special Senses

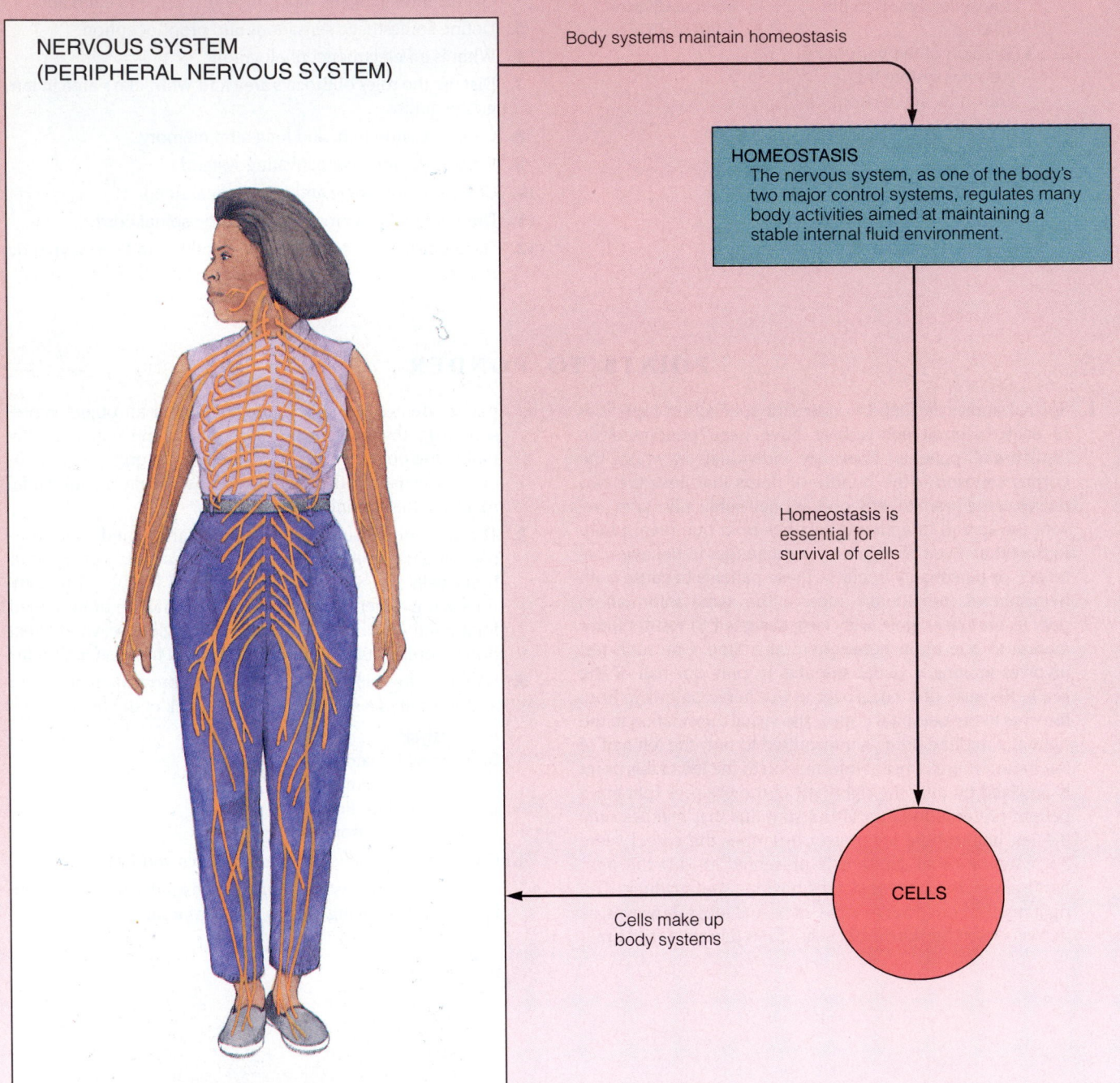

NERVOUS SYSTEM
(PERIPHERAL NERVOUS SYSTEM)

Body systems maintain homeostasis

HOMEOSTASIS
The nervous system, as one of the body's two major control systems, regulates many body activities aimed at maintaining a stable internal fluid environment.

Homeostasis is essential for survival of cells

Cells make up body systems

CELLS

The nervous system, one of the two major control systems of the body, consists of the central nervous system (CNS), composed of the brain and spinal cord, and the **peripheral nervous system**, composed of the afferent and efferent fibers that relay signals between the CNS and periphery (other parts of the body).

The **afferent division** of the peripheral nervous system de-

tects, encodes, and transmits peripheral signals *to* the CNS for processing. It is the communication link by which the CNS is informed about the internal and external environment. This input to the controlling centers of the CNS is essential in the maintenance of homeostasis. The CNS has to "know" what is going on in order to make appropriate adjustments.

INTRODUCTION
RECEPTOR PHYSIOLOGY
Receptor types
Receptor potential; receptor adaptation
Labeled lines for afferent input
● A CLOSER LOOK AT EXERCISE PHYSIOLOGY Back Swings and
   Pre-Jump Crouches: What Do They Have in Common?
Receptive field
PAIN
Receptors and mechanisms of pain
Built-in analgesic system
● CONCEPTS, CHALLENGES, AND CONTROVERSIES Acupuncture:
   Is It for Real?
EYE: VISION
Light
Refractive structures; accommodation
Phototransduction
Comparison of rod and cone vision
Visual pathways; visual processing
EAR: HEARING AND EQUILIBRIUM
Sound waves
Roles of external and middle ear
Sound transduction by organ of Corti
Auditory pathway
Vestibular apparatus
CHEMICAL SENSES: TASTE AND SMELL

● **FIGURE 6–1   Do You "See" a White Square That Is Not
Really There?**

INTRODUCTION

The peripheral nervous system consists of nerve fibers that carry information between the CNS and other parts of the body. The afferent division of the peripheral nervous system sends information about the internal and external environment to the CNS. Afferent information about the internal environment, such as the concentration of CO_2 in the body fluids, never reaches the level of conscious awareness, but this input is essential for determining the appropriate efferent output to maintain homeostasis. Afferent input that does reach the level of conscious awareness is known as **sensory input** and includes **somatic sensation** [*somesthetic* and *proprioceptive sensation* (see p. 119)] and **special senses** (*vision, hearing, taste,* and *smell*). Final processing of sensory input by the CNS not only is essential for interaction with the environment for basic survival (for example, food procurement and defense from danger) but also adds immeasurably to the richness of life.

   Perception is our conscious interpretation of the external world as created by the brain from a pattern of nerve impulses delivered to it from sensory receptors. Is the world as we perceive it reality? The answer is a resounding no. Our perception is different from what is really "out there" for several reasons. First, humans have receptors to detect only a limited number of existing energy forms. We perceive sounds, colors, shapes, textures, smells, tastes, and temperature but are not informed of magnetic forces, polarized light waves, radio waves, or X rays because we do not have receptors to respond to the latter energy forms. Our response range is limited even for the energy forms for which we do have receptors. For example, dogs can hear a dog whistle whose pitch is above our level of detection. Second, the information channels to our brains are not high-fidelity recorders. During precortical processing of sensory input, some features of stimuli are accentuated and others are suppressed or ignored. Third, the cerebral cortex further manipulates the data, comparing it with other incoming information as well as with memories of past experiences to extract the significant features—for example, sifting out a friend's words from the hubbub of sound in a school cafeteria. In the process, the cortex often fills in or distorts the information to abstract a logical perception; that is, it "completes the picture." As a simple example, you "see" a white square in ● Figure 6–1 even though there is no white square but merely right-angle wedges taken out of four purple circles. Optical illusions illustrate how the brain interprets reality according to its own rules. Do you see two faces in profile or a wineglass in ● Figure 6–2? You can alternately see one or the other out of identical visual input. Thus, our perceptions do not replicate reality. Other species equipped with different receptor types and sensitivities and with different neural processing perceive a markedly different world than we do.

● **FIGURE 6–2   Variable Perceptions from the Same
Visual Input** Do you see two faces in profile or a wineglass?

RECEPTOR PHYSIOLOGY

Receptors have differential sensitivities to various stimuli.

At their peripheral endings, afferent neurons have **receptors** that apprise the CNS of detectable changes, or **stimuli**, in both the external world and the internal environment by generating action potentials in response to the stimuli. These action potentials are transmitted via the afferent fibers to the CNS. Stimuli exist in a variety of energy forms, or **modalities**, such as heat, light, sound, pressure, and chemical changes. Because the only way that afferent neurons can transmit information to the CNS is via action potential propagation, receptors must convert these other forms of energy into electrical energy (action potentials). This energy conversion process is known as **transduction**.

Each type of receptor is specialized to respond more readily to one type of stimulus, its **adequate stimulus**, than to other stimuli. For example, receptors in the eye are most sensitive to light, receptors in the ear to sound waves, and warmth receptors in the skin to heat energy. We cannot "see" with our ears or "hear" with our eyes because of this differential sensitivity of receptors, a principle known as the **law of specific nerve energies**. Some receptors can respond weakly to stimuli other than their adequate stimulus, but even when activated by a different stimulus, a receptor still gives rise to the sensation usually detected by that receptor type. As an example, the adequate stimulus for eye receptors (photoreceptors) is light, to which they are exquisitely sensitive, but these receptors can also be activated to a lesser degree by mechanical stimulation. When hit in the eye, a person often "sees stars" because the mechanical pressure stimulates the photoreceptors. Thus, the sensation perceived depends on the type of receptor stimulated rather than on the type of stimulus. However, because receptors typically are activated by their adequate stimulus, the sensation usually corresponds to the stimulus modality.

Depending on the type of energy to which they ordinarily respond, receptors are categorized as follows:

- **Photoreceptors** are responsive to light.
- **Mechanoreceptors** are sensitive to mechanical energy. Examples include skeletal-muscle receptors sensitive to stretch, the receptors in the ear containing fine hair cells that are bent as a result of sound waves, and blood-pressure–monitoring baroreceptors.
- **Thermoreceptors** are sensitive to heat and cold.
- **Osmoreceptors** detect changes in the concentration of solutes in the body fluids and the resultant changes in osmotic activity (see p. 59).
- **Chemoreceptors** are sensitive to specific chemicals. Chemoreceptors include the receptors for smell and taste, as well as those located deeper within the body that detect O_2 and CO_2 concentrations in the blood or the chemical content of the digestive tract.
- **Nociceptors** or **pain receptors** are sensitive to tissue damage such as pinching or burning or to distortion of tissue.

Intense stimulation of any receptor is also perceived as painful.

Some sensations are compound sensations in that their perception arises from central integration of several simultaneously activated primary sensory inputs. For example, the perception of wetness comes from touch, pressure, and thermal receptor input; there is no such thing as a "wet receptor."

The information detected by receptors is conveyed via afferent neurons to the CNS, where it is used for a variety of purposes. First, afferent input is essential for the control of efferent output, both for regulation of motor behavior in accordance with external circumstances and for coordination of internal activities directed toward maintenance of homeostasis. At the most basic level, afferent input provides information (of which the individual may or may not be consciously aware) for the CNS to use in directing activities necessary for survival. Second, processing of sensory input by the reticular activating system in the brain stem is critical for cortical arousal and consciousness (see p. 137). Third, central processing of sensory information gives rise to our perceptions of the world around us. Finally, selected information delivered to the CNS may be stored for future reference.

Altered membrane permeability of receptors in response to a stimulus produces a graded receptor potential.

A receptor may be either a specialized ending of the afferent neuron or a separate cell closely associated with the peripheral ending of the neuron. Stimulation of a receptor alters its membrane permeability, usually by causing a nonselective opening of all small ion channels. The means by which this permeability change takes place is individualized for each receptor type. Because the electrochemical driving force is greater for Na^+ than for other small ions at resting potential, the predominant effect is an inward flux of Na^+, which depolarizes the receptor membrane. (There are exceptions; for example, photoreceptors are hyperpolarized upon stimulation.)

This local depolarizing change in potential is known as a **receptor potential** in the case of a separate receptor or as a **generator potential** if the receptor is a specialized ending of an afferent neuron. The receptor (or generator) potential is a graded potential (see p. 79) whose amplitude and duration can vary, depending on the strength and the rate of application or removal of the stimulus. The stronger the stimulus, the greater the permeability change and the larger the receptor potential. As is true of all graded potentials, receptor potentials have no refractory period, so summation in response to rapidly successive stimuli is possible. Because the receptor region has a very high threshold, action potentials do not take place at the receptor itself. For long-distance transmission, the receptor potential must be converted into action potentials that can be propagated along the afferent fiber. This conversion is accomplished by the opening of Na^+ channels in the afferent neuron membrane adjacent to the receptor in response to the presence of a receptor (or generator) potential. If the resulting Na^+ influx is sufficient to bring this region adjacent to the receptor to threshold, an action potential is initiated that is self-propagated along the afferent fiber to the CNS.

The means by which the Na+ channels in the adjacent afferent membrane are opened differ depending on whether the receptor is a separate cell or a specialized afferent ending. In the case of a separate receptor, a receptor potential triggers the release of a chemical messenger that diffuses across the small space separating the receptor from the ending of the afferent neuron, similar to a synapse (● Fig. 6–3a). Binding of the chemical messenger with specific protein receptor sites on the afferent neuron opens chemical messenger–gated Na+ channels (see p. 81). If the magnitude of the resulting ionic flux is sufficient to bring the adjacent membrane to threshold, a self-propagating action potential is initiated in the afferent neuron. In the case of a specialized afferent ending (Fig. 6–3b), local current flow between the activated receptor ending undergoing a generator potential and the cell membrane adjacent to the receptor brings about opening of voltage-gated Na+ channels in this adjacent region. Once threshold is reached, an action potential is initiated that is propagated along the afferent fiber. (For convenience, from here on we will refer to both receptor potentials and generator potentials as receptor potentials.)

The intensity of the stimulus is reflected by the magnitude of the receptor potential, which in turn is encoded by the frequency of action potentials generated in the afferent neuron. A larger receptor potential cannot bring about a larger action potential (all-or-none law), but it can induce more rapid firing of action potentials. This is one way stimulus strength is coded; the stronger the stimulus, the greater the frequency of action potentials. Stimulus strength is also reflected by the size of the area stimulated. Stronger stimuli usually affect larger areas, so a correspondingly greater population of receptors responds. For example, a light touch does not activate as many pressure receptors in the skin as a more forceful touch applied to the same area. Stimulus intensity is therefore distinguished both by the frequency of action potentials generated in the afferent neuron (**frequency code**) and by the number of receptors activated within the area (**population code**).

Receptors may adapt slowly or rapidly to sustained stimulation.

Because of **adaptation**, stimuli of the same intensity do not always elicit receptor potentials of the same magnitude from the same receptor. Some receptors have the ability to diminish the extent of their depolarization in spite of a sustained stimulus strength, with a subsequent decrease in the frequency of action potentials generated in the afferent neuron. The receptor "adapts" to the stimulus by no longer responding to it to the same degree.

There are two types of receptors—*tonic receptors* and *phasic receptors*—based on their speed of adaptation. **Tonic receptors** do not adapt at all or adapt slowly. These receptors are important in situations where maintained information about a stimulus is valuable. Examples are muscle stretch receptors, which monitor muscle length, and joint proprioceptors, which measure the degree of joint flexion. The CNS must be continually apprised of the degree of muscle length and joint position to maintain posture and balance. It is important, therefore, that these receptors do not adapt to a stimulus but continue to generate action potentials to relay this information to the CNS (● Fig. 6–4a).

Phasic receptors, on the other hand, are rapidly adapting receptors. These receptors often exhibit an **off response** (Fig. 6–4b). The receptor rapidly adapts by no longer responding to a maintained stimulus, but when the stimulus is removed, the receptor responds with a slight depolarization. Phasic receptors are useful in situations where it is important to signal a *change* in stimulus intensity rather than to relay status quo information. Rapidly adapting receptors include *tactile (touch)* receptors in the skin that signal changes in

---

● **FIGURE 6–3 Conversion of Receptor and Generator Potentials into Action Potentials** (a) Receptor potential. The chemical messenger released from a separate receptor initiates an action potential in the fiber by opening chemical messenger–gated Na+ channels. (b) Generator potential. The local current flow between the depolarized receptor ending and the afferent fiber initiates an action potential in the fiber by opening voltage-gated Na+ channels.

(a)

(b)

• FIGURE 6–4 Tonic and Phasic Receptors (a) Tonic receptor. This receptor type does not adapt at all or adapts slowly to a sustained stimulus and thus provides continuous information about the stimulus. (b) Phasic receptor. This receptor type adapts rapidly to a sustained stimulus and frequently exhibits an off response when the stimulus is removed. Thus, the receptor signals changes in stimulus intensity rather than relaying status quo information.

pressure on the skin surface. Because these receptors adapt rapidly, you are not continually conscious of wearing your watch, rings, and clothing. When you put something on, you soon become accustomed to it because of these receptors' rapid adaptation. When you take the item off, you are aware of its removal because of the off response.

The mechanism by which adaptation is accomplished varies for different receptors and has not been fully elucidated for all receptor types. The mechanism of adaptation for a **Pacinian corpuscle**, a skin receptor that detects pressure and vibration, is known to be due to its physical properties. A Pacinian corpuscle is a specialized receptor ending that consists of concentric layers of connective tissue resembling layers of an onion wrapped around the peripheral terminal of an afferent neuron (• Fig. 6–5). When pressure is first applied to the Pacinian corpuscle, the underlying terminal responds with a receptor potential of a magnitude that reflects the intensity of the stimulus. As the stimulus continues, the pressure energy is dissipated because it causes the receptor layers to slip (just as steady pressure on a peeled onion causes its layers to slip). Because this physical effect filters out the steady component of the applied pressure, the underlying neuronal ending no longer responds with a receptor potential; that is, adaptation has occurred. Adaptation in this case is due entirely to the physical features of the receptor. Experimentally stripping a

Pacinian corpuscle of its connective-tissue layers converts the naked terminal into a slowly adapting receptor.

Adaptation should not be confused with habituation (see p. 131). Although both these phenomena involve decreased neural responsiveness to repetitive stimuli, they operate at different points in the neural pathway. Adaptation is a receptor adjustment in the PNS, whereas habituation involves a modification in synaptic effectiveness in the CNS.

Each somatosensory pathway is "labeled" according to modality and location.

On reaching the spinal cord, afferent information has two possible destinies: (1) it may become part of a reflex arc, bringing about an appropriate effector response, or (2) it may be relayed upward to the brain via ascending pathways for further processing and possible conscious awareness. Although some afferent information concerns the status of the internal environment, such as blood pressure and concentration of CO_2 in the body fluids, one is not made aware of this information, so it is not considered to be sensory information. The incoming pathway for subconscious information derived from the internal viscera is called a **visceral afferent**. If the information detected by the receptors is propagated to the conscious levels of the brain, it is called **sensory information**, and the incoming pathway is considered to be a **sensory afferent**.

Sensory information is categorized as either (1) **somatic sensation** arising from the body surface, including *somesthetic sensation* arising from the skin and *proprioception* from muscles, joints, skin, and inner ear, or (2) **special sensation**, including *seeing, hearing, tasting,* and *smelling*. (See the accompanying boxed feature, A Closer Look at Exercise Physiology.)

Pathways conveying somatic sensation, the **somatosensory pathways**, consist of discrete chains of neurons synaptically interconnected in a particular sequence to accomplish progressively more sophisticated processing of the sensory information. The afferent neuron with its peripheral receptor that first detects the stimulus is known as a **first-order sensory neuron**. It synapses on a **second-order sensory neuron**, either in the spinal cord or the medulla, depending on which sen-

• FIGURE 6–5 Pacinian Corpuscle The Pacinian corpuscle, a skin receptor that signals a change in pressure and detects vibration, consists of concentric layers of connective tissue wrapped around the peripheral terminal of an afferent neuron.

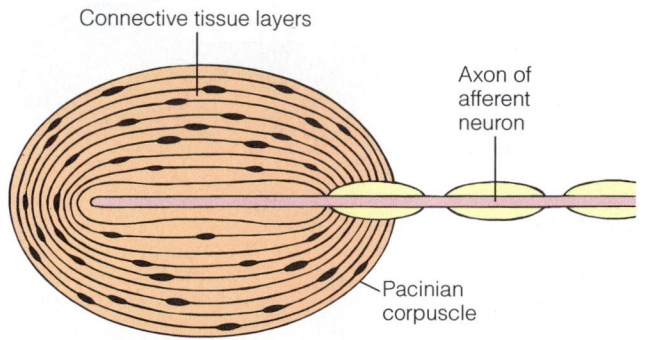

Connective tissue layers

Axon of afferent neuron

Pacinian corpuscle

BACK SWINGS AND PRE-JUMP CROUCHES: WHAT DO THEY SHARE IN COMMON?

Proprioception, the sense of the body's position in space, is critical to any movement and is especially important in athletic performance, whether it be a figure skater performing triple jumps on ice, a gymnast performing a difficult floor routine, or the football quarterback throwing perfectly to a spot 60 yards down field. In order to control skeletal-muscle contraction to achieve the desired movement, the CNS must be continuously apprised of the results of its actions by means of sensory feedback information.

A number of receptors provide proprioceptive input. Muscle proprioceptors provide feedback information on muscle tension and length. Joint proprioceptors provide feedback on joint acceleration, angle, and direction of movement. Skin proprioceptors inform the CNS of weight-bearing pressure on the skin. Proprioceptors in the inner ear, along with those in neck muscles, provide information about head and neck position so that the CNS can orient the head correctly. For example, neck reflexes facilitate essential trunk and limb movements during

somersaults, and divers and tumblers use strong movements of the head to maintain spins.

The most complex and probably one of the most imporant proprioceptors is the muscle spindle (see p. 243). Muscle spindles are found throughout a muscle but tend to be concentrated in its center. Each spindle lies parallel to the muscle fibers within the muscle. The spindle is sensitive to both the muscle's rate of change in length and the final length achieved. If a muscle is stretched, each muscle spindle within the muscle is also stretched, and the afferent neuron whose peripheral axon terminates on the muscle spindle is stimulated. The afferent fiber passes into the spinal cord and synapses directly on the motor neurons that supply the same muscle. Stimulation of the stretched muscle as a result of this stretch reflex causes the muscle to contract sufficiently to relieve the stretch.

Older persons or those with weak quadriceps (thigh) muscles unknowingly take advantage of the muscle spindle by pushing on the center of the thighs when they get up from a

sitting position. Contraction of the quadriceps muscle extends the knee joint, thus straightening the leg. The act of pushing on the center of the thighs when getting up slightly stretches the quadriceps muscle in both limbs, stimulating the muscle spindles. The resultant stretch reflex aids in contraction of the quadriceps muscles and helps the person to assume a standing position.

In sports, people use the muscle spindle to advantage all the time. To jump high, as in basketball jumpball, an athlete starts by crouching down. This action stretches the quadriceps muscles and increases the firing rate of their spindles, thus triggering the stretch reflex that reinforces the quadriceps muscles' contractile response so that these extensor muscles of the legs gain additional power. The same is true for crouch starts in running events. The back swing in tennis, golf, and baseball similarly provides increased muscular excitation through reflex activity initiated by stretched muscle spindles.

sory pathway is involved. This neuron then synapses on a **third-order sensory neuron** in the thalamus, and so on. With each step, further processing of the input occurs. A particular sensory modality detected by a specialized receptor type is sent over a specific afferent and ascending pathway (a neural pathway committed to that modality) to excite a defined area in the somatosensory cortex. This process is known as **projection**; that is, a particular sensory input is "projected" to a specific region of the cortex. Thus, information is kept separated within specific **labeled lines** between the periphery and the cortex. In this way, even though all information is propagated to the CNS via the same type of signal (action potentials), the brain can decode the type and location of the stimulus. Table 6–1 summarizes how the CNS is informed of the type (what?), location (where?), and intensity (how much?) of a stimulus.

Activation of a sensory pathway at any point gives rise to the same sensation that would be produced by stimulation of the receptors in the body part itself. This is the basis for **phan-**

tom pain (for example, pain perceived as originating in the foot by a person whose leg has been amputated at the knee). Irritation of the severed endings of the afferent pathways in the stump triggers action potentials that, upon reaching the foot region of the somatosensory cortex, are interpreted as pain in the missing foot.

Acuity is influenced by receptive field size and lateral inhibition.

Each sensory neuron responds to stimulus information only within a circumscribed region of the skin surface surrounding it; this region is known as its **receptive field**. The size of a receptive field varies inversely with the density of receptors in the region; the more closely receptors of a particular type are spaced, the smaller the area of skin each monitors. The smaller the receptive field in a region, the greater its **acuity** or **discriminative ability**. Compare the tactile (touch) discrimination in your fingertips with that in your elbow by "feeling"

TABLE 6–1	Coding of Sensory Information
STIMULUS PROPERTY	**MECHANISM OF CODING**
Type of stimulus (stimulus modality)	Distinguished by the type of receptor activated and the specific pathway over which this information is transmitted to a particular area of the cerebral cortex
Location of stimulus	Distinguished by the location of the activated receptor field and the pathway that is subsequently activated to transmit this information to the area of the somatosensory cortex representing that particular location
Intensity of stimulus (stimulus strength)	Distinguished by the frequency of action potentials initiated in an activated afferent neuron and the number of receptors (and afferent neurons) activated

the same object with both. You are able to discern more precise information about the object with your richly innervated fingertips because the receptive fields there are small; as a result, each neuron signals information about small, discrete portions of the object's surface. In contrast, the skin over the elbow is served by relatively few sensory endings with larger

receptive fields. Subtle differences within each large receptive field cannot be detected (● Fig. 6–6). The distorted cortical representation of various body parts in the sensory homunculus (see p. 119) corresponds precisely with the innervation density; more cortical space is allotted for sensory reception from areas with smaller receptive fields and, accordingly, greater tactile discriminative ability.

Besides receptor density, a second factor influencing acuity is **lateral inhibition**. You can appreciate the importance of this phenomenon by slightly indenting the surface of your skin with the point of a pencil (● Fig. 6–7a). The receptive field is excited immediately under the center of the pencil point where the stimulus is most intense, but the surrounding receptive fields are also stimulated, only to a lesser extent because they are less distorted. If information from these marginally excited afferent fibers in the fringe of the stimulus area were to reach the cortex, localization of the pencil point would be blurred. To facilitate localization and sharpen contrast, lateral inhibition occurs within the CNS (Fig. 6–7b). The most strongly activated signal pathway originating from the center of the stimulus area inhibits the less excited pathways from the fringe areas. This occurs via inhibitory interneurons that pass laterally between ascending fibers serving neighboring receptive fields. Blockage of further transmission in the weaker inputs increases the contrast between wanted and unwanted information so that the pencil point can be precisely localized. The extent of lateral inhibitory connections within sensory pathways varies for different modalities. Those with

● **FIGURE 6–6 Comparison of Discriminative Ability of Regions with Small versus Large Receptive Fields**
The relative tactile acuity of a given region can be determined by the **two-point threshold of discrimination test**. If the two points of a pair of calipers applied to the surface of the skin stimulate two different receptive fields, two separate points will be felt. If the two points touch the same receptive field, they will be perceived as only one point. By adjusting the distance between the caliper points, one can determine the minimal distance at which the two points can be recognized as two rather than one, which is a reflection of the size of the receptive fields in the region. With this technique, it is possible to plot the discriminative ability of the body surface. The two-point threshold ranges from 2 mm in the fingertip (enabling one to read Braille, where the raised dots are spaced 2.5 mm apart) to 48 mm in the poorly discriminative skin of the calf. (a) Region with small receptive fields. (b) Region with large receptive fields.

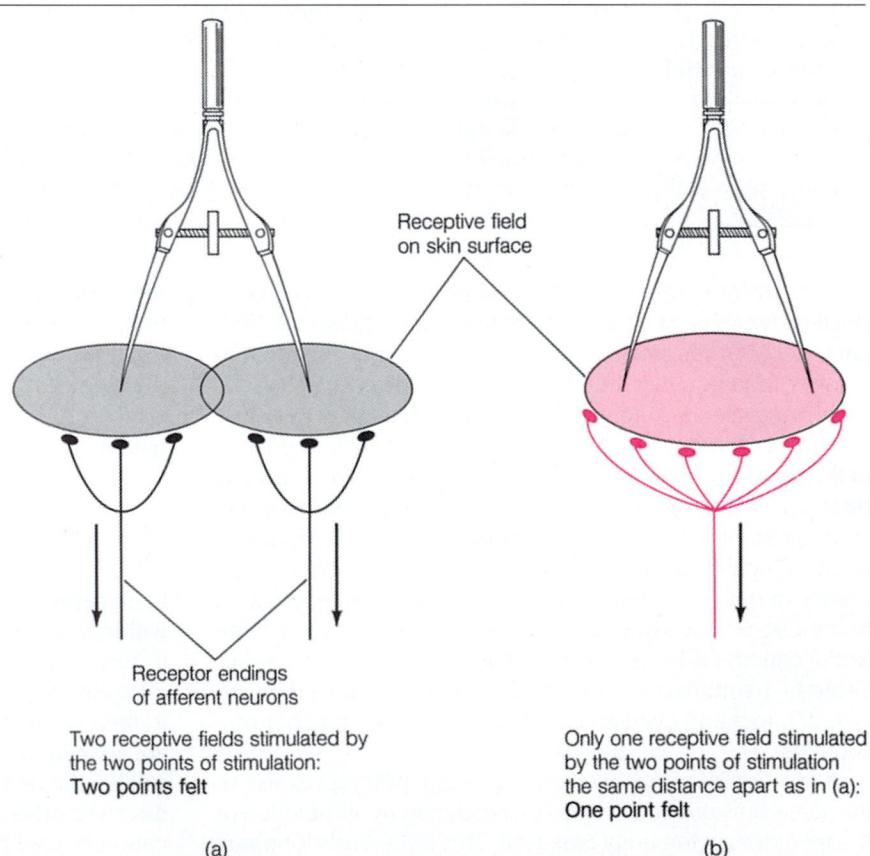

Receptive field on skin surface

Receptor endings of afferent neurons

Two receptive fields stimulated by the two points of stimulation: **Two points felt**

Only one receptive field stimulated by the two points of stimulation the same distance apart as in (a): **One point felt**

(a)

(b)

• **FIGURE 6–7 Lateral Inhibition** (a) The receptor at the site of most intense stimulation is activated to the greatest extent. Surrounding receptors are also stimulated but to a lesser degree. (b) The most intensely activated receptor pathway halts transmission of impulses in the less intensely stimulated pathways through lateral inhibition. This process facilitates localization of the site of stimulation.

Skin surface

Receptor pathways

Stimulated less

Stimulated less

Stimulated most

(a)

Lateral inhibition

Transmission stopped

Transmission stopped

Transmission continues

(b)

the most lateral inhibition—touch and vision—bring about the most accurate localization.

PAIN

Stimulation of nociceptors elicits the perception of pain plus motivational and emotional responses.

Pain is primarily a protective mechanism meant to bring to conscious awareness the fact that tissue damage is occurring or is about to occur. Unlike other somatosensory modalities, it is accompanied by motivated behavioral responses (such as withdrawal or defense) as well as emotional reactions (such as crying or fear). Also, unlike other sensations, the subjective perception of pain can be influenced by other past or present experiences (for example, heightened pain perception accompanying fear of the dentist or lowered pain perception in an injured athlete during a competitive event).

There are three categories of pain receptors: **mechanical nociceptors** respond to mechanical damage such as cutting, crushing, or pinching; **thermal nociceptors** respond to temperature extremes, especially heat; and **polymodal nociceptors** respond equally to all kinds of damaging stimuli, including irritating chemicals released from injured tissues. None of the nociceptors have specialized receptor structures; they are all naked nerve endings. Because of their value to survival,

TABLE 6-2   Characteristics of Pain

FAST PAIN	SLOW PAIN
Carried by A-delta fibers	Carried by C fibers
Sharp, prickling sensation	Dull, aching, burning sensation
Easily localized	Poorly localized
Occurs first	Occurs second; persists for longer time; more unpleasant
Occurs upon stimulation of mechanical and thermal nociceptors	Occurs upon stimulation of polymodal nociceptors

nociceptors do not adapt to sustained or repetitive stimulation. On the other hand, all nociceptors can be sensitized by the presence of *prostaglandins,* which greatly enhance the receptor response to noxious stimuli (that is, it hurts more when prostaglandins are present). Prostaglandins are a special group of fatty acid derivatives that act locally on being released. Aspirinlike drugs inhibit the synthesis of prostaglandins, accounting at least in part for the drugs' **analgesic** (pain-relieving) properties.

Pain impulses originating at nociceptors are transmitted to the CNS via one of two types of afferent fibers (Table 6–2). Signals arising from mechanical and thermal nociceptors are transmitted over large myelinated **A-delta fibers** at rates of up to 30 meters/sec (the **fast pain pathway**). Impulses from polymodal nociceptors are carried by small unmyelinated **C fibers** at a much slower rate of 12 meter/sec (the **slow pain pathway**). Think about the last time you cut or burned your finger. You undoubtedly felt a sharp twinge of pain at first, with a more diffuse, disagreeable pain commencing shortly thereafter. Pain typically is perceived initially as a brief, sharp, prickling sensation that is easily localized (fast pain pathway originating from specific mechanical or heat nociceptors). This feeling is followed by a dull, aching, poorly localized sensation that persists for a longer time and is more unpleasant (slow pain pathway activated by chemicals, especially **bradykinin,** a normally inactive substance that is activated by enzymes released into the ECF from damaged tissue). Bradykinin and related compounds not only provoke pain, presumably by stimulating the polymodal nociceptors, but they also contribute to the inflammatory response to tissue injury (Chapter 12). The persistence of these chemicals might explain the long-lasting, aching pain that continues after removal of the mechanical or thermal stimulus that caused the tissue damage.

The primary afferent fibers synapse with specific second-order interneurons in the dorsal horn of the spinal cord (● Fig. 6–8a). One of the neurotransmitters released from these afferent pain terminals is **substance P**, which is believed to be unique to pain fibers. Ascending pain pathways have poorly understood destinations in the *somatosensory cortex,* the *thalamus,* and the *reticular formation.* The role of the cortex in pain perception is not clear, although it is probably important at least in localizing the pain. Pain can still be perceived in the absence of the cortex, presumably at the level

of the thalamus. The reticular formation increases the level of alertness associated with the noxious encounter. Interconnections from the thalamus and reticular formation to the hypothalamus and limbic system elicit the behavioral and emotional responses accompanying the painful experience.

In contrast to the pain accompanying peripheral injury, which serves as a normal protective mechanism to warn of impending or actual damage to the body, abnormal chronic pain states are speculated to result from damage within the pain pathways in the peripheral nerves or the CNS. That is, pain is perceived because of abnormal signaling within the pain pathways in the absence of peripheral injury or typical painful stimuli. For example, strokes that damage ascending pathways can lead to an abnormal, persistent sensation of pain.

The brain has a built-in analgesic system.

In addition to the chain of neurons connecting peripheral nociceptors with higher CNS structures for pain perception, the CNS also contains a neuronal system that suppresses pain. Our knowledge about this built-in **analgesic system** is still fragmentary. It appears that neural mechanisms exist to suppress transmission in the pain pathways as they enter the spinal cord. Electrical stimulation of the **periaqueductal gray matter** (gray matter surrounding the cerebral aqueduct, a narrow canal that connects the third and fourth ventricular cavities) results in profound analgesia, as does stimulation of the reticular formation within the brain stem. These two regions are thought to be part of a descending analgesic pathway (Fig. 6–8b) that blocks by presynaptic inhibition (see p. 98) the release of substance P from afferent pain fiber terminals.

The built-in analgesic system is dependent on the presence of **opiate receptors**. It has long been known that **morphine,** a derivative of the opium poppy, is a powerful analgesic. It seemed highly unlikely that the body would be endowed with opiate receptors only to interact with chemicals derived from a flower! A search was therefore undertaken to discover the substances that normally bind with these opiate receptors. The result was the discovery of **endogenous opiates** (morphinelike substances), the **endorphins, enkephalins,** and **dynorphin,** which are important in the body's natural analgesic system. According to a proposed model for the analgesic system, these endogenous opiates serve as analgesic neurotransmitters; they are released from a descending analgesic pathway and bind with opiate receptors on the afferent presynaptic terminal. This binding suppresses the release of substance P, thereby blocking further transmission of the pain signal. Morphine binds to these same opiate receptors; this accounts for its analgesic properties.

It is unclear how the natural pain-suppressing mechanisms are activated. Factors known to modulate pain include exercise (endorphins are believed to be released during prolonged exercise and presumably are responsible for the "runner's high"), acupuncture, hypnosis, and stress. (See the accompanying boxed feature, Concepts Challenges, and Controversies.) There is evidence that some types of stress induce analgesia via the opiate pathway and other less well under-

(a)

(b)

● FIGURE 6–8 Pain Pathway and Proposed Analgesic Pathway (a) Pain pathway. When activated by a noxious stimulus, afferent pain pathways release substance P, which is unique to pain fibers. Substance P activates ascending pain pathways that provide various brain regions with input for processing different aspects of the painful experience. (b) Proposed analgesic pathway. Endogenous opiates released from descending analgesic (pain-relieving) pathways are believed to bind with opiate receptors at the synaptic knob of the afferent pain fiber. This binding inhibits the release of substance P, thereby blocking transmission of pain impulses along the ascending pain pathways.

stood nonopiate mechanisms. It is sometimes disadvantageous for a stressed organism to display the normal reaction to pain. For example, when two males are fighting for dominance of the herd, withdrawing, escaping, or resting when injured would mean certain defeat.

Pain can frequently be managed by drugs that suppress transmitter activity at some point along the pain pathway. Surgical intervention is occasionally necessary for relief of intractable pain, as may occur in terminally ill cancer patients. Surgical relief from pain entails either interrupting the ascending pain pathways within the spinal cord or severing certain pathways in the brain to modify emotional response to the pain.

ACUPUNCTURE: IS IT FOR REAL?

It sounds like science fiction. How can a needle inserted in the hand relieve a toothache? **Acupuncture analgesia (AA)**, the technique of relieving pain by the insertion and manipulation of threadlike needles at key points, has been practiced in China for over 2,000 years but is relatively new to Western medicine and still remains controversial in our country. Many Western scientists were skeptical because, until recently, the phenomenon could not be explained on the basis of any known, logical, physiological principles, although a tremendous body of anecdotal evidence in support of the effectiveness of AA existed in China.

According to a leading expert in acupuncture, the technique was not embraced in Western culture because of a clash in philosophies between West and East:

> Western medical science is quick to reject a phenomenon if it does not fit the current scientific theories. Chinese Taoism had a distaste for explanatory theories and chose instead merely to observe phenomena in order to be in harmony with mother nature. If a needle in the hand cured a toothache, that was sufficient for Chinese Taoism. For Western medicine acupuncture was impossible and hence was relegated to the wastebasket of placebo effects.*

The *placebo effect* refers to a chemical or technique that brings about a desired response through the power of suggestion or distraction rather than through any direct action. The placebo effect was first documented in 1945 when a physician injected patients with what they thought was morphine for pain relief, but some received sugar (a placebo) instead. Pain was relieved in 70% of those who actually received morphine, but, surprisingly, 35% of those who received sugar but believed they were receiving morphine, also reported pain relief.

Because the Chinese were content with anecdotal evidence for the success of AA, this phenomenon did not come under close scientific scrutiny until the last two decades when European and American scientists started studying it. As a result of these efforts, an impressive body of rigorous scientific investigation supports the contention that AA really works (that is, by a physiological rather than a placebo/psychological effect). Furthermore, its mechanisms of action have become apparent. Indeed, more is known about the underlying physiological mechanisms of AA than of many conventional medical techniques, such as gas anesthesia.

AA has been proven to be effective in treating chronic pain and to exert a real physical effect in that it is more effective than placebo controls. In fact, AA compares favorably with morphine for treating chronic pain. In controlled clinical studies, 55% to 85% of patients were helped by AA. (By comparison, 70% of patients benefit from morphine therapy.) Pain relief was reported by only 30% to 35% of placebo controls (individuals who thought they were receiving proper AA treatment, but in whom needles were inserted in the wrong places or not deep enough).

The overwhelming body of evidence supports the acupuncture endorphin hypothesis as the primary mechanism of AA's action. According to this hypothesis, acupuncture needles activate specific afferent nerve fibers, which send impulses to the central nervous system. Here the incoming impulses activate three centers (a spinal cord center, a midbrain center, and a hormonal center, the hypothalamus/anterior pituitary unit) to cause analgesia. All three centers have been shown to block pain transmission through use of endorphins and closely related compounds. Several other neurotransmitters, such as serotonin and norepinephrine, as well as cortisol, the major hormone released during stress, are implicated as well.

In the United States, AA is not used in mainstream medicine, even by physicians who have been convinced by scientific evidence that the technique is valid. AA methodology is not taught in our medical colleges, and the techniques take time to learn. Also, AA is much more time-consuming than using drugs. Western physicians who have been trained to use drugs to solve most pain problems are generally reluctant to scrap their known methods for an unfamiliar, time-consuming technique. In a limited way, however, acupuncture is gaining favor as an alternative treatment for relief of chronic pain, especially since analgesic drugs can have troublesome side effects. AA has caught on in Europe more extensively than in the United States. For example, in a recent survey of pain clinics in Germany, over 90% of the physicians reported using acupuncture.

Because acupuncture is relatively new in the United States, the laws governing its use vary from state to state. Some states permit only trained physicians to perform AA, whereas others register nonphysician acupuncturists. Eleven schools in the United States are currently offering four-year training programs in AA for nonphysicians, an indication that AA will come to be more commonly used for pain relief in our country, if not by physicians then by others trained in this now scientifically legitimate technique.

*Gabriel Stux and Bruce Pomeranz, *Basics of Acupuncture* (Springer Verlag, 1991), p. 1.

Recall that a painful experience includes the sensation of pain plus an emotional and behavioral reaction to it. These two components can be dissociated. The person still feels pain but does not mind as much.

⇛ ● EYE: VISION

Somatic sensation is detected by widely distributed receptors that provide information about the body's interactions with the environment in general. In contrast, each of the special senses has highly localized, extensively specialized receptors that respond to unique environmental stimuli. The special senses include **vision, hearing, taste** and **smell**, to which we now turn our attention, starting with vision.

The eye is a fluid-filled sphere enclosed by three specialized tissue layers.

The eyes capture the patterns of illumination in the environment as an "optical picture" on a layer of light-sensitive cells, the retina, much as a camera captures an image on film. Just as film can be developed into a visual likeness of the original image, the coded image on the retina is transmitted through a series of progressively more complex steps of visual processing until it is finally consciously perceived as a visual likeness of the original image.

The **eye** (Table 6–3) is a spherical, fluid-filled structure enclosed by three layers. From outermost to innermost, these are (1) the sclera/cornea; (2) the choroid/ciliary body/iris; and (3) the retina (● Fig. 6–9a). Most of the eyeball is covered

TABLE 6–3 Functions of Major Components of the Eye		
STRUCTURE	**LOCATION**	**FUNCTION**
(in alphabetical order)		
Aqueous humor	Anterior cavity between cornea and lens	Clear watery fluid that is continually formed and carries nutrients to the cornea and lens
Bipolar neurons	Middle layer of nerve cells in retina	Important in retinal processing of light stimulus
Blind spot	Point slightly off-center on retina that is devoid of photoreceptors (also known as optic disc)	Route for passage of optic nerve and blood vessels
Choroid	Middle layer of eye	Pigmented to prevent scattering of light rays in eye; contains blood vessels that nourish retina; anteriorly specialized to form ciliary body and iris
Ciliary body	Specialized anterior derivative of the choroid layer; forms a ring around the outer edge of the lens	Produces aqueous humor and contains ciliary muscle
Ciliary muscle	Circular muscular component of ciliary body; attaches to lens by means of suspensory ligaments	Important in accommodation
Cones	Photoreceptors in outermost layer of retina	Responsible for high acuity, color, and day vision
Cornea	Anterior clear outermost layer of eye	Contributes most extensively to eye's refractive ability
Fovea	Exact center of retina	Region with greatest acuity
Ganglion cells	Inner layer of retina	Important in retinal processing of light stimulus; form optic nerve
Iris	Visible pigmented ring of muscle within aqueous humor	Varies size of pupil by variable contraction; responsible for eye color
Lens	Between aqueous humor and vitreous humor; attaches to ciliary muscle by suspensory ligaments	Provides variable refractive ability during accommodation
Macula lutea	Area immediately surrounding the fovea	Has high acuity because of abundance of cones
Optic disc	(*see* blind spot)	
Optic nerve	Leaves each eye at optic disc (blind spot)	First part of visual pathway to the brain
Pupil	Anterior round opening in middle of iris	Permits variable amounts of light to enter eye
Retina	Innermost layer of eye	Contains the photoreceptors (rods and cones)
Rods	Photoreceptors in outermost layer of retina	Responsible for high sensitivity, black-and-white, and night vision
Sclera	Tough outer layer of eye	Protective connective-tissue coat; forms visible white part of eye; anteriorly specialized to form cornea
Suspensory ligaments	Suspended between ciliary muscle and lens	Important in accommodation
Vitreous humor	Between lens and retina	Semifluid, jellylike substance that helps maintain spherical shape of eye

External eye muscle

Suspensory ligament

Choroid

Retina

Ciliary body

Sclera

Conjunctiva—

Iris —

Fovea

Pupil

Lens

Cornea

Aqueous humor

Canal for tear drainage

Sclera

Iris

Pupil

(b)

Optic nerve

Vitreous humor

Optic disc

Blood vessels

(a)

• **FIGURE 6–9**   **Structure of the Eye**   (a) Internal sagittal view. (b) External front view.

by a tough outer layer of connective tissue, the **sclera**, which forms the visible white part of the eye (Fig. 6–9b). Anteriorly (toward the front), the outer layer consists of the transparent **cornea** through which light rays pass into the interior of the eye. The middle layer underneath the sclera is the highly pigmented **choroid**, which contains many blood vessels that nourish the retina. The choroid layer becomes specialized anteriorly to form the *ciliary body* and *iris*. The innermost coat under the choroid is the **retina**, which consists of an outer pigmented layer and an inner nervous-tissue layer. The latter contains the **rods** and **cones**, the photoreceptors that convert light energy into nerve impulses. Like the black walls of a photographic studio, the pigment in the choroid and retina absorbs light after it strikes the retina to prevent reflection or scattering of light within the eye.

The interior of the eye consists of two fluid-filled cavities, separated by a **lens**, all of which are transparent to permit light to pass through the eye from the cornea to the retina. The anterior (front) cavity between the cornea and lens contains a clear watery fluid, the **aqueous humor**, and the larger posterior (rear) cavity between the lens and retina contains a semifluid, jellylike substance, the **vitreous humor**.

The vitreous humor is important in maintaining the spherical shape of the eyeball. The aqueous humor carries nutrients for the cornea and lens, both of which lack a blood supply. Blood vessels in these structures would impede the passage of light to the photoreceptors. Aqueous humor is produced at a rate of about 5 ml/day by a capillary network within the **ciliary body**, a specialized anterior derivative of the choroid layer. This fluid drains into a canal at the edge of the cornea

and eventually enters the blood (• Fig. 6–10). If aqueous humor is not drained as rapidly as it forms (for example, because of a blockage in the drainage canal), the excess will accumulate in the anterior cavity, causing the intraocular ("within the eye") pressure to rise. This condition is known as **glaucoma**. The excess aqueous humor pushes the lens backward into the vitreous humor, which in turn is pushed against the inner neural layer of the retina. This compression causes retinal and optic nerve damage that can lead to blindness if it is not treated.

The amount of light entering the eye is controlled by the iris.

Not all of the light passing through the cornea reaches the light-sensitive photoreceptors because of the presence of the **iris**, a thin, pigmented smooth muscle that forms a visible ringlike structure within the aqueous humor (Fig. 6–9a and b). The pigment in the iris is responsible for eye color. The round opening in the center of the iris through which light enters the interior portions of the eye is the **pupil**. The size of this opening can be adjusted by variable contraction of the iris muscles to admit more or less light as needed, much as the shutter controls the amount of light entering a camera. The iris contains two sets of smooth-muscle networks, one *circular* (the muscle fibers run in a ringlike fashion within the iris) and the other *radial* (the fibers project outward from the pupillary margin like bicycle spokes) (• Fig. 6–11). Because muscle fibers shorten when they contract, the pupil gets smaller when the **circular** (or **constrictor**) **muscle** contracts

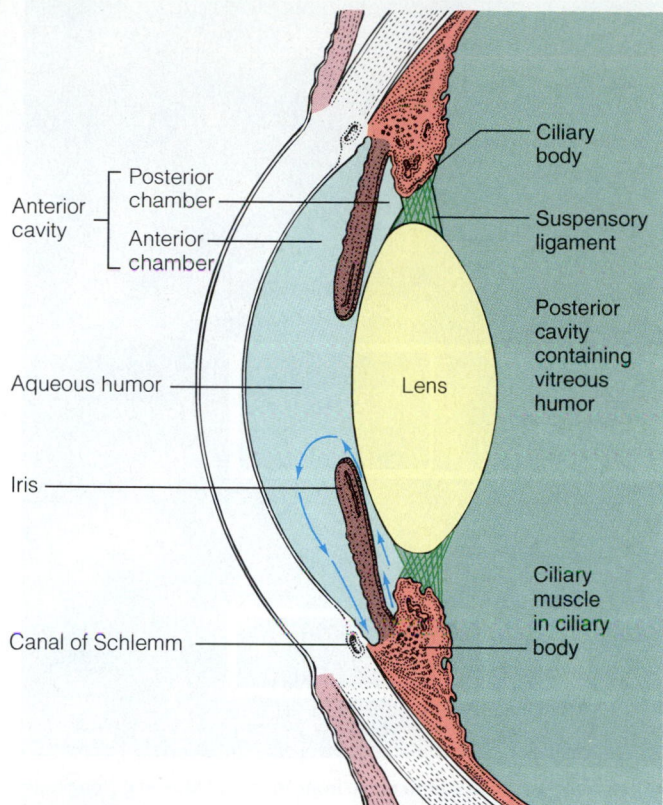

● FIGURE 6–10 Formation and Drainage of Aqueous Humor Aqueous humor is formed by a capillary network in the ciliary body, then drains into the canal of Schlemm, and eventually enters the blood.

and forms a smaller ring. This reflex pupillary constriction occurs in bright light to decrease the amount of light entering the eye. When the **radial** (or **dilator**) **muscle** shortens, the size of the pupil increases. Such pupillary dilation occurs in dim light to allow the entrance of more light.

Iris muscles are controlled by the autonomic nervous system. Parasympathetic nerve fibers innervate the circular mus-

cle, and sympathetic fibers supply the radial muscle. Acting via the autonomic nervous system, conditions other than light can induce changes in pupillary size. For example, dilation of the pupils accompanies generalized discharge of the sympathetic nervous system in response to actual or impending danger.

The eye refracts the entering light to focus the image on the retina.

Light is a form of electromagnetic radiation composed of particlelike individual packets of energy called **photons** that travel in wavelike fashion. The distance between two wave peaks is known as the **wavelength** (● Fig. 6–12). The photoreceptors in the eye are sensitive only to wavelengths between 400 and 700 nanometers (nm; billionths of a meter between peaks). This **visible light** is only a small portion of the total electromagnetic spectrum (● Fig. 6–13). Light of different wavelengths in this visible band is perceived as different color sensations. Short wavelengths are sensed as violet and blue; long wavelengths are interpreted as orange and red.

In addition to having variable wavelengths, light energy also varies in **intensity**, that is, the amplitude or height of the wave (Fig. 6–12). Dimming a bright red light does not change its color; it just becomes less intense or less bright.

Light waves *diverge* (radiate outward) in all directions from every point of a light source. The forward movement of a light wave in a particular direction is known as a **light ray**. Divergent light rays reaching the eye must be bent inward to be focused back into a point on the light-sensitive retina to provide an accurate image of the light source (● Fig. 6–14).

The bending of a light ray (**refraction**) occurs when the ray passes from a medium of one density into a medium of a different density (● Fig. 6–15). Light travels faster through air than through other transparent media such as water and glass. When a light ray enters a medium of greater density, it is slowed down (the converse is also true). The ray changes its course of direction if it strikes the surface of the new medium at any angle other than perpendicular.

● FIGURE 6–11 Control of Pupillary Size

Parasympathetic stimulation +

Iris

Sympathetic stimulation +

Circular (constrictor) muscle runs circularly

Pupil

Radial (dilator) muscle runs radially

Pupillary constriction

Pupillary dilation

● FIGURE 6–12 Properties of an Electromagnetic Wave A wavelength is the distance between two wave peaks. Intensity refers to the amplitude of the wave.

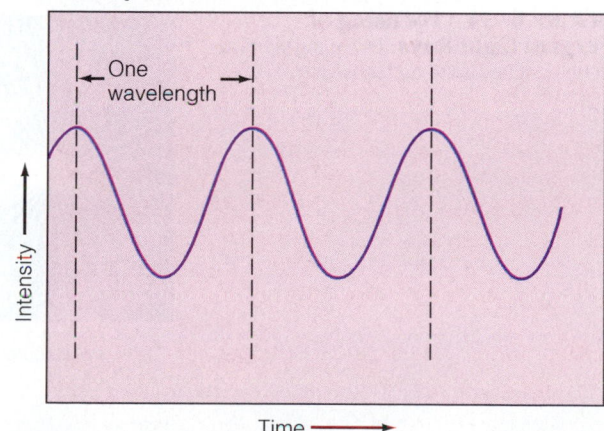

One wavelength

Intensity

Time ⟶

INVISIBLE LONG WAVES | VISIBLE LIGHT SPECTRUM | INVISIBLE SHORT WAVES

Infrared rays (beyond red)

Ultraviolet rays (beyond violet)

1500 1000 700 600 500 400 300

| Radio | TV | Microwaves | Infra-red | | U-V | X-rays | Gamma rays | Cosmic rays |

• **FIGURE 6–13 Electromagnetic Spectrum** The wavelengths in the electromagnetic spectrum range from 10^4 m (10 km—e.g., long radio waves) to less than 10^{-14} m (quadrillionths of a meter—e.g., gamma and cosmic rays). The visible spectrum includes wavelengths ranging from 400 to 700 nanometers (nm; billionths of a meter). Numbers on the figure refer to wavelengths in nanometers.

Two factors contribute to the degree of refraction: the comparative densities of the two media (the greater the difference in density, the greater the degree of bending) and the angle at which the light strikes the second medium (the greater the angle, the greater the refraction).

With a curved surface such as a lens, the greater the curvature, the greater the degree of bending and the stronger the lens. When a light ray strikes the curved surface of any object of greater density, the direction of refraction depends on the angle of the curvature (• Fig. 6–16). A lens with **convex** surfaces converges light rays, bringing them closer together, a requirement for bringing an image to a focal point. Refractive surfaces of the eye are therefore convex. A lens with **concave** surfaces diverges light rays (spreads them farther apart); a concave lens is useful for correcting certain refractive errors of the eye, such as nearsightedness.

The two structures most important in the eye's refractive ability are the *cornea* and the *lens*. The curved corneal surface, the first structure light passes through as it enters the eye, contributes most extensively to the eye's total refractive ability because the difference in density at the air/corneal interface is much greater than the differences in density be-

• **FIGURE 6–14 Focusing of Divergent Light Rays** Diverging light rays must be bent inward to be focused.

Point source of light   Light rays   Eye structures that bend light rays   Light rays focused on retina

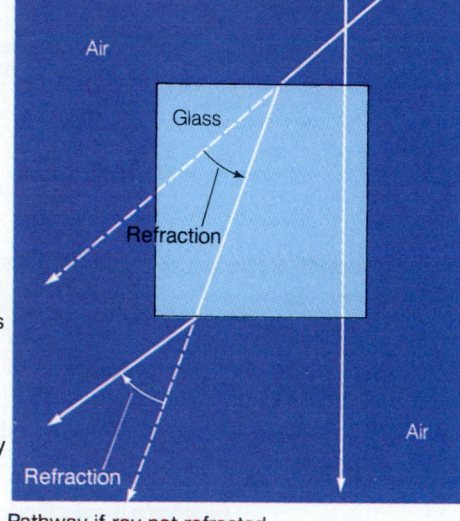

Light ray striking
a glass surface
perpendicularly

Air

Glass

Refraction

Pathway if ray not refracted
by passing from air into glass

Actual
refracted pathway

Refraction

Air

Pathway if ray not refracted
by passing from glass into air

(a)

(b)

● **FIGURE 6–15** **Refraction** (a) A light ray is bent (refracted) when it strikes the surface of a medium of different density than the one in which it had been traveling (for example, moving from air into glass) at any angle other than perpendicular to the new medium's surface. (b) The pencil in the glass of water appears to bend. What is happening, though, is that the light rays coming to the camera (or your eyes if you see this in person) are bent as they pass through the water, then the glass, and then the air. Consequently, the pencil appears distorted.

● **FIGURE 6–16** **Refraction by Convex and Concave Lenses** (a) Convex lens, which converges the rays (brings them closer together). (b) Concave lens, which diverges the rays (spreads them farther apart).

Focal point

Light from
distant source

Focal length

(a)

Light from
distant source

(b)

= Points of stimulation of the retina

tween the lens and the fluids surrounding it. In **astigmatism**, the curvature of the cornea is uneven, so light rays are unequally refracted. The refractive ability of a person's cornea remains constant because the curvature of the cornea never changes. In contrast, the refractive ability of the lens can be adjusted by changing its curvature as needed for near or far vision.

The refractive structures of the eye must bring light images into focus on the retina for clear vision. If an image is focused before it reaches the retina or is not yet focused when it reaches the retina, it will be blurred (• Fig. 6–17). Light rays originating from near objects are more divergent when they reach the eye than are rays from distant sources. Rays from light sources greater than twenty feet are considered to be parallel by the time they reach the eye. For a given refractive ability of the eye, a near source of light requires a greater distance behind the lens for focusing than a far source does, because the near source rays are still diverging when they reach the eye (• Fig. 6–18a and b).

In a particular eye, the distance between the lens and the retina always remains the same. To bring both near and far light sources into focus on the retina (that is, in the same distance), a stronger lens must be used for the near source (Fig. 6–18c). The strength of the lens can be adjusted through the process of accommodation, to which we now turn our attention.

Accommodation increases the strength of the lens for near vision.

The ability to adjust the strength of the lens so that both near and far sources can be focused on the retina is known as **accommodation**. The strength of the lens depends on its shape, which in turn is regulated by the ciliary muscle.

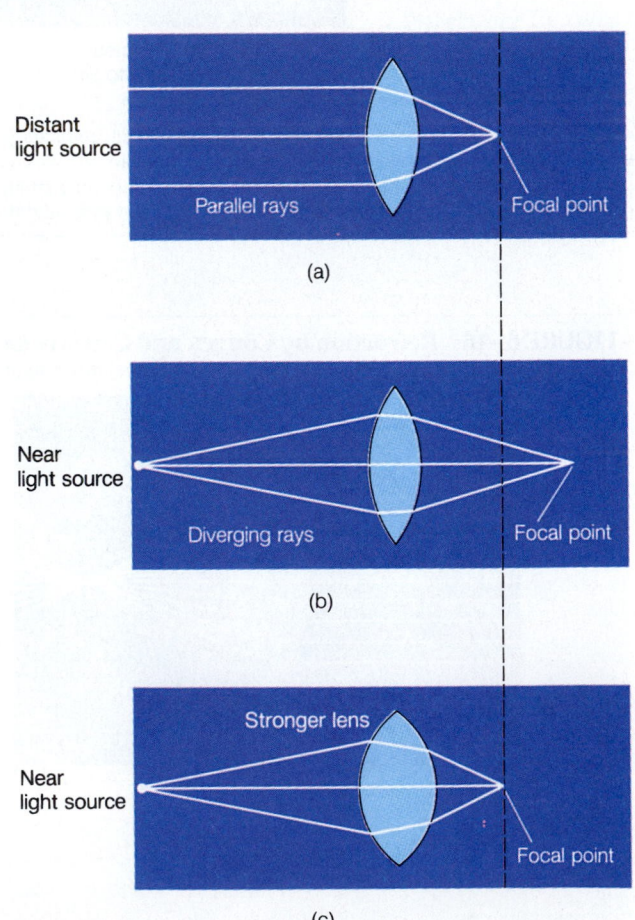

• **FIGURE 6–18   Focusing of Distant and Near Sources of Light**   (a) A distant (far) light source is a light source greater than 20 feet from the eye. By the time the rays reach the eye from a distant source, they are considered to be parallel. (b) The rays from a near light source (a light source less than 20 feet from the eye) are still diverging when they reach the eye. A longer distance is required for a lens of a given strength to bend the diverging rays from a near light source into focus (compared to the parallel rays from a distant light source). (c) To focus both a distant and a near light source in the same distance (the distance between the lens and retina), a stronger lens must be used for the near source. A stronger lens is able to focus a near image in the same distance as a weaker lens focuses a distant image.

Ciliary muscle
Lens
Suspensory ligaments
Pupil
Suspensory ligaments
(a)

Suspensory ligaments
Lens
(b)

Sympathetic stimulation
+
Relaxed ciliary muscle
Cornea
Iris
Flattened, weak lens
Taut suspensory ligaments
(c)

Parasympathetic stimulation
+
Contracted ciliary muscle
Rounded, strong lens
Slackened suspensory ligaments
(d)

● **FIGURE 6–19   Mechanism of Accommodation**   (a) Schematic representation of suspensory ligaments extending from the ciliary muscle to the outer edge of the lens. (b) Scanning electron micrograph showing the suspensory ligaments attached to the lens. (c) When the ciliary muscle is relaxed, the suspensory ligaments are taut, putting tension on the lens so that it is flat and weak. (d) When the ciliary muscle is contracted, the suspensory ligaments become slack, reducing the tension on the lens. The lens can then assume a stronger, rounder shape because of its elasticity.

The **ciliary muscle** is part of the ciliary body, an anterior specialization of the choroid layer. The ciliary body has two major components: the ciliary muscle and the capillary network that produces the aqueous humor. The ciliary muscle is a circular ring of smooth muscle attached to the lens by **suspensory ligaments** (● Fig. 6–19a and b).

When the ciliary muscle is relaxed, the suspensory ligaments are taut and pull the lens into a flattened, weakly refractive shape (Fig. 6–19c). As the muscle contracts, its circumference decreases, slackening the tension in the suspensory ligaments (Fig. 6–19d). When the lens is subjected to less tension by the suspensory ligaments, it assumes a more spherical shape because of its inherent elasticity. The greater curvature of the more rounded lens increases its strength, causing greater bending of light rays.

In the normal eye, the ciliary muscle is relaxed and the lens is flat for far vision, but the muscle contracts to allow the lens to become more convex and stronger for near vision. The ciliary muscle is controlled by the autonomic nervous system. Sympathetic nerve fibers induce relaxation of the ciliary muscle for far vision, whereas the parasympathetic nervous system causes the muscle's contraction for near vision.

The lens is an elastic structure consisting of transparent fibers. Occasionally, these fibers become opaque so that light rays cannot pass through, a condition known as a **cataract**. The defective lens can usually be surgically removed and vision restored by an implanted artificial lens or compensating eyeglasses.

Throughout life, only cells at the outer edges of the lens are replaced. Cells in the center of the lens are in double jeopardy. Not only are they the oldest, but they also are the farthest away from the aqueous humor, the lens' nutrient source. With advancing age, these nonrenewable central cells die and become stiff. With loss of elasticity, the lens is no longer able to assume the spherical shape required to accommodate for near vision. This age-related reduction in accommodative ability, **presbyopia**, affects most people by middle age (45 to 50), requiring them to resort to corrective lens for near vision (reading).

Other common vision disorders are nearsightedness (myopia) and farsightedness (hyperopia). In a normal eye (● Fig. 6–20a), a far light source is focused on the retina without accommodation, whereas the strength of the lens is increased by accommodation to bring a near source into focus. In **myo-** pia (Fig. 6–20b1), because the eyeball is too long or the lens is too strong, a near light source is brought into focus on the retina without accommodation (even though accommodation is normally used for near vision), whereas a far light source is focused in front of the retina and is blurry. Thus, a myopic individual has better near vision than far vision, a condition that can be corrected by a concave lens (Fig. 6–20b2). With **hyperopia** (Fig. 6–20c1), either the eye-

● **FIGURE 6–20   Myopia and Hyperopia**

(a)

Far source

Near source

No accommodation

Accommodation

Normal eye (Emmetropia)

Far source focused on retina without accommodation

Near source focused on retina with accommodation

(b)

1.

Image out of focus

Focus

No accommodation

No accommodation

Nearsightedness (Myopia)— Eyeball too long or lens too strong

1. Uncorrected

Far source focused in front of retina (where retina would be in eye of normal length)

Near source focused on retina without accommodation

2.

No accommodation

Accommodation

2. Corrected with concave lens, which diverges light rays before they reach the eye

Far source focused on retina without accommodation

Near source focused on retina with accommodation

(c)

1.

Image out of focus

Focus

Accommodation

Accommodation

Farsightedness (Hyperopia)— Eyeball too short or lens too weak

1. Uncorrected

Far source focused on retina with accommodation

Near source focused behind retina even with accommodation

2.

No accommodation

Accommodation

2. Corrected with convex lens, which converges light rays before they reach the eye

Far source focused on retina without accommodation

Near source focused on retina with accommodation

ball is too short or the lens is too weak. Far objects are focused on the retina only with accommodation, whereas near objects are focused behind the retina even with accommodation and, accordingly, are blurry. Thus, a hyperopic individual has better far vision than near vision, a condition that can be corrected by a convex lens (Fig. 6–20c2). Such vision tends to get worse as the person gets older because of loss of accommodative ability with the onset of presbyopia.

Light must pass through several retinal layers before reaching the photoreceptors.

The major function of the eye is to focus light rays from the environment on the rods and cones, the photoreceptor cells of the retina. The photoreceptors then transform the light energy into electrical signals for transmission to the CNS.

The receptor-containing portion of the retina is actually an extension of the CNS and not a separate peripheral organ. During embryonic development, the retinal cells "back out" of the nervous system, so the retinal layers, surprisingly, are facing backward! The neural portion of the retina consists of three layers (• Fig. 6–21): (1) the outermost layer (closest to the choroid) containing the **rods** and **cones**, whose light-sensitive ends face the choroid (i.e., away from the incoming

light); (2) a middle layer of **bipolar neurons**; and (3) an inner layer of **ganglion cells**. Axons of the ganglion cells join together to form the **optic nerve**, which leaves the retina slightly off-center. The point on the retina at which the optic nerve leaves and through which blood vessels pass is the **optic disc** (Figs. 6–9a and • 6–22). This region is often referred to as the **blind spot**; no image can be detected in this area because it is devoid of rods and cones. We are normally not aware of the blind spot because central processing somehow "fills in" the missing spot. You can discover the existence of your own blind spot by a simple demonstration (• Fig. 6–23).

Light must pass through the ganglion and bipolar layers before reaching the photoreceptors in all areas of the retina except the **fovea**. In the fovea, which is a pinhead-sized depression located in the exact center of the retina (Fig. 6–9a), the bipolar and ganglion cell layers are pulled aside so that light directly strikes the photoreceptors. This feature, coupled with the fact that only cones (which have greater acuity or discriminative ability than the rods) are found here, makes the fovea the point of most distinct vision. Thus, we turn our eyes so that the object at which we are looking is focused on the fovea. The area immediately surrounding the fovea, the **macula lutea**, also has a high concentration of cones and fairly high acuity (Fig. 6–22). Macular acuity, however, is less

---

• **FIGURE 6–21** **Retinal Layers** The retinal visual pathway extends from the photoreceptor cells (rods and cones) to the bipolar cells to the ganglion cells. The horizontal and amacrine cells act locally for retinal processing of visual input.

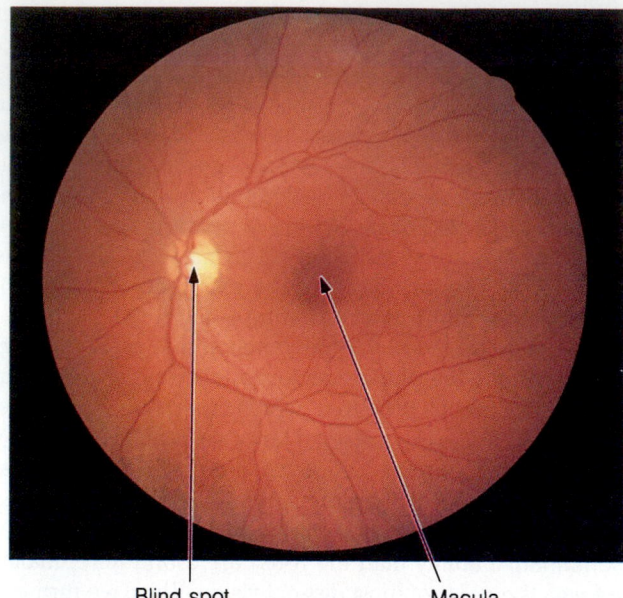

Blind spot   Macula

● **FIGURE 6–22** **View of the Retina Seen through an Ophthalmoscope** With an ophthalmoscope, a lighted viewing instrument, it is possible to view the optic disc (blind spot) and macula within the retina at the rear of the eye.

than that of the fovea because of the overlying ganglion and bipolar cells in the macula.

Phototransduction by retinal cells converts light stimuli into neural signals.

Photoreceptors consist of three parts (● Fig. 6–24a): (1) an *outer segment,* which lies closest to the eye's exterior, facing the choroid, and detects the light stimulus; (2) an *inner segment,* which lies in the middle of the photoreceptor's length and contains the metabolic machinery of the cell; and (3) a *synaptic terminal,* which lies closest to the eye's interior, facing the bipolar neurons, and transmits the signal generated in the photoreceptor upon light stimulation to these next cells in the visual pathway. The outer segment, which is rod-shaped

● **FIGURE 6–23** **Demonstration of Blind Spot** Discover the blind spot in your right eye by closing your left eye and holding the book about four inches from your face. While focusing on the circle, gradually move the book away from you until the cross vanishes from view. At this time, the image of the cross is striking the blind spot of your right eye. You can similarly discover the blind spot in your left eye by closing your right eye and focusing on the cross. The circle will disappear when its image strikes the blind spot of your left eye.

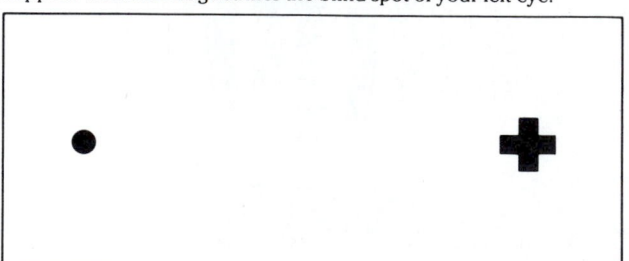

in rods and cone-shaped in cones (Fig. 6–24b and c), is composed of stacked, flattened membranous discs containing an abundance of photopigment molecules. Over a billion of these molecules may be packed into the outer segment of each photoreceptor. Photopigments undergo chemical alterations when activated by light. A photopigment consists of an enzymatic protein called **opsin** combined with **retinene,** a derivative of vitamin A. There are four different photopigments, one in the rods and one in each of three types of cones. Retinene is identical in all four photopigments, but the photoreceptors' opsins vary slightly, enabling the photopigments to differentially absorb various wavelengths of light. **Rhodopsin,** the rod photopigment, cannot discriminate between various wavelengths in the visible spectrum; it absorbs all visible wavelengths. Therefore, rods provide vision only in shades of gray by detecting different intensities, not different colors. The photopigments in the three types of cones—**red, green,** and **blue cones**—respond selectively to various wavelengths of light, making **color vision** possible (● Fig. 6–25).

Phototransduction, the mechanism of excitation, is basically the same for all photoreceptors (● Fig. 6–26b). When a photopigment molecule absorbs light, it dissociates into its retinene and opsin components, and its retinene portion changes shape, triggering the enzymatic activity of opsin. Through a series of steps, this light-induced biochemical change in the photopigment brings about a hyperpolarizing receptor potential that influences transmitter release from the synaptic terminal of the photoreceptor.

Specifically, the light-induced biochemical change in the photopigment leads to closing of chemical messenger–gated Na^+ channels in the outer segment's membrane. Unlike other chemical-gated channels that respond to external chemical messengers, these channels respond to an internal second messenger, **cyclic GMP** (cyclic guanosine monophosphate), which links photopigment light absorption with Na^+ channel closing. Intracellular cyclic GMP, which is in high concentration in the dark, keeps the Na^+ channels in the outer segment's plasma membrane open (Fig. 6–26a). Therefore, unlike most receptors, the Na^+ channels of a photoreceptor are open in the absence of stimulation, that is, in the dark. The resultant passive inward Na^+ leak depolarizes the photoreceptor and keeps the Ca^{++} channels in its synaptic terminal open, thereby permitting release of transmitter while in the dark.

Upon exposure to light, the concentration of cyclic GMP is decreased through a series of biochemical steps triggered by retinene-opsin dissociation (Fig. 6–26b). During the light-excitation process, the reduction in cyclic GMP brings about closure of Na^+ channels, which stops the depolarizing Na^+ leak and causes membrane hyperpolarization. This hyperpolarization, which is the receptor potential, passively spreads from the outer segment to the synaptic terminal of the photoreceptor. Here the potential change leads to a reduction in transmitter release from the synaptic terminal. Thus, photoreceptors are inhibited by their adequate stimulus (hyperpolarized by light) and excited in the absence of stimulation (depolarized by darkness). The hyperpolarizing potential and subsequent decrease in transmitter release are graded ac-

Rod

- Synaptic terminal
- Inner segment
- Outer segment

Nucleus

Mitochondria

Connecting cilium

Plasma membrane

Cytoplasmic space

Discs

Cone

- Synaptic terminal
- Inner segment
- Outer segment

(a)

Rod   Cone

(b)

Outer segment of rod

Outer segment of cone

(c)

• FIGURE 6–24 **Photoreceptors** (a) Schematic representation of the three parts of the rods and cones, the eye's photoreceptors (oriented in the same direction as in Fig. 6–21). (b) Schematic three-dimensional representation of the outer segments of the rods and cones. Note the stacked, flattened membranous discs, which contain an abundance of photopigment molecules. (c) Scanning electron micrograph of the outer segments of the rods and cones. Note the rod shape in rods and the cone shape in cones.

Bipolar cells display graded potentials similar to the photoreceptors. Action potentials do not originate until the ganglion cells, the first neurons in the chain that must propagate the visual message over long distances to the brain.

The altered photopigments are restored to their original

• FIGURE 6–25 **Sensitivity of the Three Types of Cones to Different Wavelengths** The ratios of stimulation of the three cone types are shown for three sample colors.

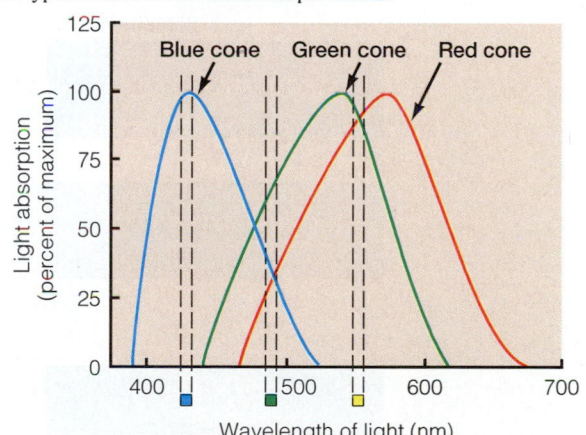

Color perceived	Percent of maximum stimulation		
	Red cones	Green cones	Blue cones
🟦	0	0	100
🟩	31	67	36
🟨	83	83	0

cording to the intensity of light. The brighter the light, the greater the hyperpolarizing response and the greater the reduction in transmitter release.

How does the retina signal the brain about light stimulation through such an inhibitory response? The photoreceptors synapse with bipolar cells. These cells in turn terminate on the ganglion cells, whose axons from the optic nerve for transmission of signals to the brain. The answer to our seeming paradox lies in the fact that the transmitter released from the photoreceptors' synaptic terminal has an *inhibitory* action on many of the bipolar cells. The reduction in transmitter release that accompanies light-induced receptor hyperpolarization decreases this inhibitory action on the bipolar cells. Removal of inhibition has the same effect as direct excitation of the bipolar cells. The greater the illumination on the receptor cells, the greater the removal of inhibition from the bipolar cells and the greater in effect the excitation of these next cells in the visual pathway to the brain.

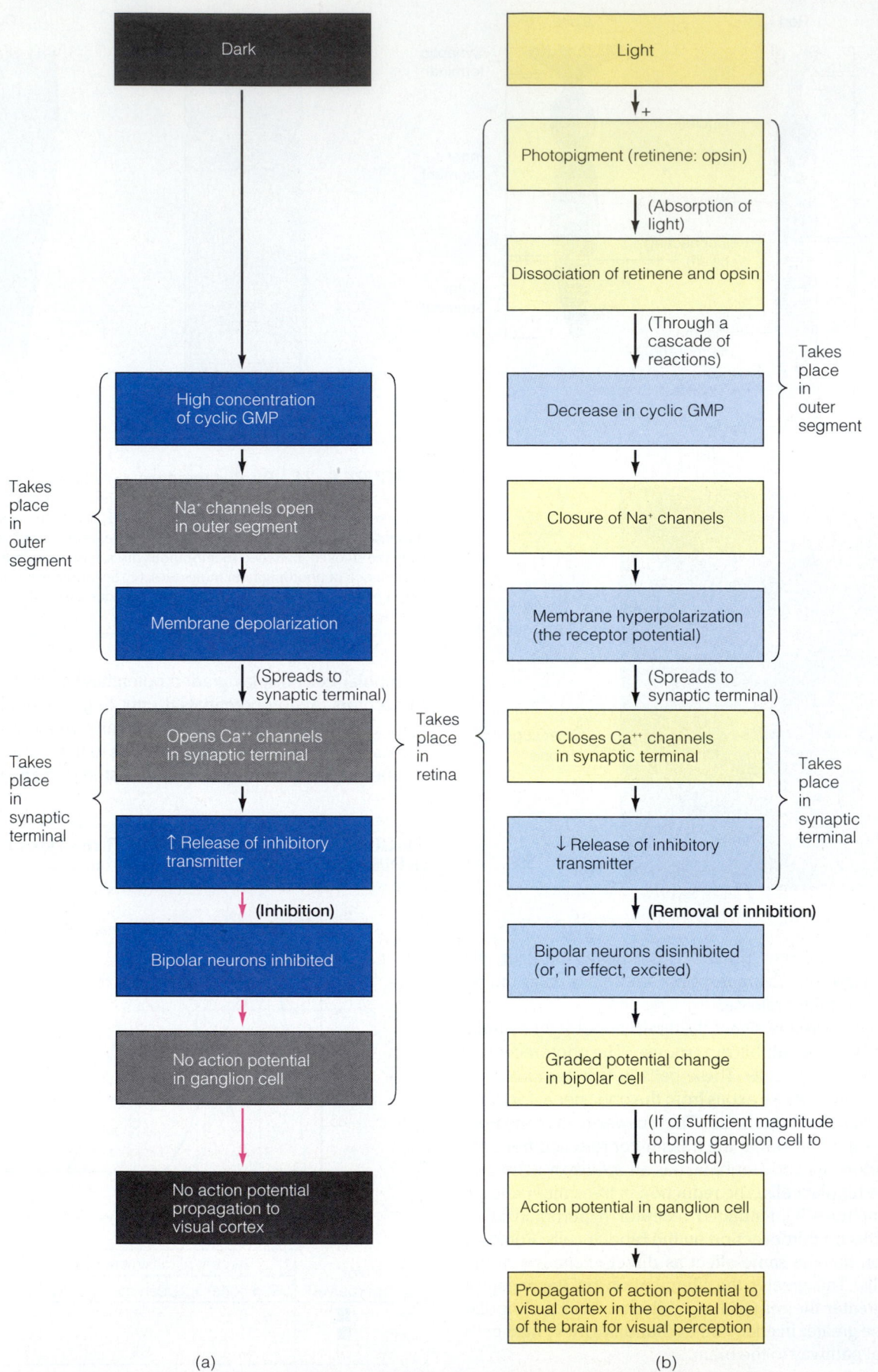

● **FIGURE 6–26   Phototransduction and Initiation of Action Potential in the Visual Pathway** (a) Events occurring in a photoreceptor in response to the dark that prevent action potentials from being initiated in the visual pathway. (b) Events occurring in a photoreceptor in response to a light stimulus that initiate an action potential in the visual pathway (phototransduction).

conformation in the dark by enzyme-mediated mechanisms. Thereupon the membrane potential and rate of transmitter release of the photoreceptor are returned to their unexcited state, and no action potentials are transmitted to the visual cortex.

Rods provide indistinct gray vision at night whereas cones provide sharp color vision during the day.

The retina contains more than thirty times more rods than cones (100 million rods compared to 3 million cones per eye). Cones are most abundant in the center of the retina in the macula. From this point outward, the concentration of cones decreases and the concentration of rods increases. Rods are most abundant in the periphery. Because of differential absorption of various wavelengths of light, cones provide color vision whereas rods provide vision only in shades of gray. The capabilities of the rods and cones also differ in other respects due to a difference in the "wiring patterns" between these photoreceptor types and other retinal neuronal layers (Table 6–4). Cones have low sensitivity to light, being "turned on" only by bright daylight, but they have high acuity (sharpness; ability to distinguish between two nearby points). Thus, cones provide sharp vision with high resolution for fine detail. Humans use cones for day vision, which is in color and distinct. Rods, on the other hand, have low acuity but high sensitivity, so they respond to the dim light of night. We are able to see at night with our rods but at the expense of color and distinctness. Let us see how wiring patterns influence sensitivity and acuity.

There is little convergence of neurons (see p. 99) in the retinal pathways for cone output. Each cone generally has a private line connecting it to a particular ganglion cell. In contrast, there is much convergence in rod pathways. Output from more than one hundred rods may converge via bipolar cells on a single ganglion cell.

Before a ganglion cell can have an action potential that will be propagated to the CNS, the cell must be brought to threshold through influence of the graded potentials in the receptors to which it is wired. Because a single cone ganglion cell is often influenced by only one cone, only bright daylight is intense enough to induce a sufficient receptor potential in the cone to ultimately bring the ganglion cell to threshold. The abundant convergence in the rod visual pathways, in contrast, offers good opportunities for summation of subthreshold events in a rod ganglion cell (see p. 95). Whereas a small receptor potential induced by dim light in a single cone would not be sufficient to bring its ganglion cell to threshold, similar small receptor potentials induced by the same dim light in multiple rods converging on a single ganglion cell would have an additive effect to bring the rod ganglion cell to threshold. Because rods can bring about action potentials in response to small amounts of light, they are much more sensitive than cones. However, because cones have private lines into the optic nerve, each cone transmits information about an extremely small receptive field on the retinal surface. Cones are thus able to provide highly detailed vision at the expense of sensitivity. With rod vision, acuity is sacrificed for sensitivity. Since many rods share a single ganglion cell, once an action potential is initiated, it is impossible to discern which of the multiple rod inputs were activated to bring the ganglion cell to threshold. Objects appear fuzzy when rod vision is used because of this poor ability to distinguish between two nearby points.

The sensitivity of the eyes can vary markedly through dark and light adaptation.

The eyes' sensitivity to light depends on the amount of photopigment present in the rods and cones. When you go from bright sunlight into darkened surroundings, you cannot see anything at first, but gradually you begin to distinguish objects as a result of the process of **dark adaptation**. Breakdown of photopigments during exposure to sunlight tremendously decreases photoreceptor sensitivity. For example, a reduction in rhodopsin content of only 0.6% from its maximum value decreases rod sensitivity approximately three thousand times. In the dark, the photopigments broken down during light exposure are gradually regenerated. As a result, the sensitivity of your eyes gradually increases so that you can begin to see in the darkened surroundings. However, only the highly sensitive, rejuvenated rods are "turned on" by the dim light.

Conversely, when you move from the dark to the light (for example, leaving a movie theater and entering the bright sunlight), your eyes are very sensitive to the dazzling light at first. With little contrast between lighter and darker parts, the entire image appears bleached. As some of the photopigments are rapidly broken down by the intense light, the sensitivity of the eyes decreases and normal contrasts can once again be detected, a process known as **light adaptation**. The rods are so sensitive to light that sufficient rhodopsin is broken down to essentially "burn out" the rods in bright light; that is, the rod photopigments, having already been broken down by the bright light, are no longer able to respond to the light. Furthermore, a central neural adaptive mechanism switches the eye from the rod system to the cone system on exposure to bright light. Therefore, only the less-sensitive cones are used for day vision.

TABLE 6–4 Properties of Rod and Cone Vision	
RODS	**CONES**
100 million per retina	3 million per retina
Vision in shades of gray	Color vision
High sensitivity	Low sensitivity
Low acuity	High acuity
Night vision	Day vision
Much convergence in retinal pathways	Little convergence in retinal pathways
More numerous in periphery	Concentrated in fovea

It is estimated that our eyes' sensitivity can change as much as one million times as they adjust to various levels of illumination through dark and light adaptation. These adaptive measures are also enhanced by pupillary reflexes that adjust the amount of available light permitted to enter the eye.

Since retinene, one of the photopigment components, is a derivative of vitamin A, adequate amounts of this nutrient must be available for the ongoing resynthesis of photopigments. **Night blindness** occurs because of dietary deficiencies of vitamin A. Although photopigment concentrations in both rods and cones are reduced in this condition, there is still sufficient cone photopigment to respond to the intense stimulation of bright light, except in the most severe cases. However, even modest reductions in rhodopsin content can decrease the sensitivity of rods so much that they are unable to respond to dim light. The person can see in the day using cones but cannot see at night because the rods are no longer functional. Thus, carrots are "good for your eyes" because they are rich in vitamin A.

Color vision depends on the ratios of stimulation of the three cone types.

Vision depends on stimulation of retinal photoreceptors by light. Certain objects in the environment such as the sun, fire, and light bulbs, emit light. But how do we see objects like chairs, trees, and people, which do not emit light? The pigments in various objects selectively absorb particular wavelengths of light transmitted to them from light-emitting sources, and the unabsorbed wavelengths are reflected from the objects' surfaces. These reflected light rays enable us to see the objects. An object perceived as blue absorbs the longer red and green wavelengths of light and reflects the shorter blue wavelengths, which can be absorbed by the photopigment in the eye's blue cones, thereby activating them.

Each cone type is most effectively activated by a particular wavelength of light in the range of color indicated by its name—blue, green, or red. However, cones also respond in varying degrees to other wavelengths (Fig. 6–25). Our perception of the many colors of the world depends on the three cone types' various *ratios of stimulation* in response to different wavelengths. A wavelength perceived as blue does not stimulate red or green cones at all but excites blue cones maximally (the percentage of maximal stimulation for red, green, and blue cones, respectively, is 0:0:100). The sensation of yellow, in comparison, arises from a stimulation ratio of 83:83:0, red and green cones each being stimulated 83% of maximum while blue cones are not excited at all. The ratio for green is 31:67:36, and so on, with various combinations giving rise to the sensation of all the different colors. White is a mixture of all wavelengths of light, whereas black is the absence of light.

The extent each of the cone types is excited is coded and transmitted in separate parallel pathways to the brain. A distinct color vision center in the primary visual cortex has recently been identified. This center combines and processes these inputs to generate the perception of color, taking into consideration the object in comparison with its background.

• FIGURE 6–27   Color Blindness Chart   People with red-green color blindness cannot detect the number 29 in this chart.

The concept of color is therefore in the mind of the beholder. Most of us agree on what color we see because we have the same types of cones and use similar neural pathways for comparing their output. Occasionally, however, individuals lack a particular cone type, so their color vision is a product of the differential sensitivity of only two types of cones, a condition known as **color blindness**. Not only do color-defective individuals perceive certain colors differently, but they are also unable to distinguish as many varieties of colors (• Fig. 6–27). For example, people with certain color defects are unable to distinguish between red and green. At a traffic light they can tell which light is "on" by its intensity, but they must rely on the position of the bright light to know whether to stop or go.

Visual information is separated and modified within the visual pathway before it is integrated into a perceptual image of the visual field by the cortex.

The field of view that can be seen without moving the head is known as the **visual field**. The information that reaches the visual cortex in the occipital lobe is not a replica of the visual field for several reasons:

1. The image detected on the retina at the onset of visual processing is upside down and backwards because of bending of the light rays (• Fig. 6–28). Once it is projected to the brain, the inverted image is interpreted as being in its correct orientation.

2. The information transmitted from the retina to the brain is not merely a point-to-point record of photoreceptor activation. Before the information reaches the brain, the retinal neuronal layers beyond the rods and cones reinforce selected information and suppress other information to enhance contrast. One mechanism of retinal processing is

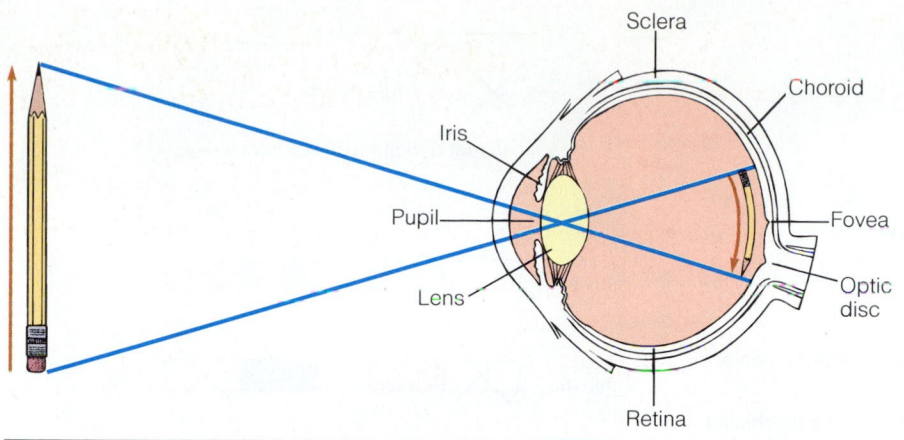

• **FIGURE 6–28   Inversion of Image on Retina**

Labels on figure: Sclera, Choroid, Iris, Pupil, Fovea, Lens, Optic disc, Retina

---

lateral inhibition, by which strongly excited cone pathways suppress activity in surrounding pathways of weakly stimulated cones. This increases the dark-bright contrast to enhance the sharpness of boundaries.

Another mechanism of retinal processing involves differential activation of two types of ganglion cells, **on-center** and **off-center ganglion cells**. The receptive field of a cone ganglion cell is determined by the field of light detection by the cone with which it is linked. On-center and off-center ganglion cells respond in opposite ways, depending on the relative comparison of illumination between the center and periphery of their receptive fields. Think of the receptive field as a donut. An on-center ganglion cell increases its rate of firing when light is most intense at the center of its receptive field (that is, when the donut hole is lit up). In contrast, an off-center cell increases its firing rate when the periphery of its receptive field is most intensely illuminated (that is, when the donut itself is lit up). This is useful for enhancing the difference in light level between one small area at the center of a receptive field and the illumination immediately around it. By emphasizing differences in relative brightness, this mechanism helps define contours of images, but in so doing, information about absolute brightness is sacrificed (● Fig. 6–29).

● **FIGURE 6–29   Example of Outcome of Retinal Processing by On-Center and Off-Center Ganglion Cells** Note that the gray circle surrounded by black appears brighter than the one surrounded by white, even though the two circles are identical (same shade and size). Retinal processing by on-center and off-center ganglion cells is largely responsible for enhancing differences in relative (rather than absolute) brightness, which helps to define contours.

 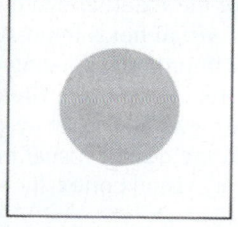

3. Various aspects of visual information such as form, color, depth, and movement are separated and projected in parallel pathways to different regions of the cortex. Only when these separate bits of processed information are integrated by higher visual regions is a reassembled picture of the visual scene perceived. This is similar to the blobs of paint on an artist's palette versus the finished portrait; the separate pigments do not represent a portrait of a face until they are appropriately integrated on a canvas.

Patients with lesions in specific visual processing regions of the brain may be unable to completely combine components of a visual impression. For example, a person may be unable to discern movement of an object but have reasonably good vision for shape, pattern, and color. Sometimes the defect can be remarkably specific, like being unable to recognize familiar faces while retaining the ability to recognize inanimate objects.

4. Because of the pattern of wiring between the eyes and the visual cortex, the left half of the cortex receives information only from the right half of the visual field as detected by both eyes, and the right half receives input only from the left half of the visual field of both eyes.

As a result of refraction (● Fig. 6–30a), light rays from the left half of the visual field fall on the right half of the retina of both eyes [the medial (inner) half of the left retina and the lateral (outer) half of the right retina]. Similarly, rays from the right half of the visual field reach the left half of each retina (the lateral half of the left retina and the medial half of the right retina). Each optic nerve exiting the retina carries information from both halves of the retina it serves. This information is separated as the optic nerves meet at the **optic chiasm** (*chiasm* means "cross") located underneath the hypothalamus. Within the optic chiasm, the fibers from the medial half of each retina cross to the opposite side, but those from the lateral half remain on the original side. The reorganized bundles of fibers leaving the optic chiasm are known as **optic tracts**. Each optic tract carries information from the lateral half of one retina and the medial half of the other retina. Therefore, this partial crossover brings together from the two eyes fibers that carry information from the same half of the visual field. Each optic tract, in turn, delivers to the half of the

• FIGURE 6–30   Visual Pathway and Visual Deficits Associated with Lesions in the Pathway (a) Visual pathway. Note that the left half of the visual cortex in the occipital lobe receives information from the right half of the visual field of both eyes (in blue), and the right half of the cortex receives information from the left half of the visual field of both eyes (in red). (b) Visual deficits with specific lesions in visual pathway. Each visual deficit illustrated is associated with a lesion at the corresponding numbered point in the visual pathway in part (a).

brain on its same side information about the opposite half of the visual field. A knowledge of these pathways can facilitate diagnosis of visual defects arising from interruption of the visual pathway at various points (Fig. 6–30b).

The thalamus and visual cortices elaborate the visual message.

The first stop in the brain for information in the visual pathway is the **lateral geniculate nucleus** in the thalamus (Fig. 6–30a). It separates information received from the eyes and relays it via fiber bundles known as **optic radiations** to different zones in the cortex, each of which processes different aspects of the visual stimulus (for example, color, form, depth, movement). This sorting process is no small task, because each optic nerve contains more than one million fibers carrying information from the photoreceptors in one retina. This is more than all the afferent fibers carrying somatosensory input from all the regions of the body! Yet the connections are precise. The lateral geniculate nucleus and each of the zones in the cortex that processes visual information have a topographical map representing the retina point-for-point. As with the somatosensory cortex, the neural maps of the retina are distorted. The fovea, the retinal region capable of greatest acuity, has much greater representation in the neural map than do the more peripheral regions of the retina.

Although each half of the visual cortex receives information simultaneously from the same part of the visual field as

received by both eyes, the messages from the two eyes are not identical. Each eye views an object from a slightly different vantage point, even though there is a tremendous area of overlap (• Fig. 6–31). The overlapping area seen by both eyes at the same time is known as the **binocular** ("two-eyed") field of vision, which is important for **depth perception**. Like other areas of the cortex, the primary visual cortex is organized into functional columns, each processing information from a small region of the retina. Independent alternating columns are devoted to information about the same point in the visual field from the right and left eyes. The brain uses the slight disparity in the information received from the two eyes to estimate distance, allowing us to perceive three-dimensional objects in spatial depth. Some depth perception is possible using only one eye, based on experience and comparison with other cues. For example, if your one-eyed view includes a car and a building and the car is much larger, you correctly interpret that the car must be closer to you than the building.

Occasionally, the two views are not successfully merged. This condition may occur for two reasons: (1) the eyes are not both focused on the same object simultaneously due to defects of the external eye muscles that make fusion of the two eyes' visual fields impossible, or (2) the binocular information is improperly integrated during visual processing. The result is double vision or **diplopia**, a condition in which the disparate views from both eyes are seen simultaneously.

Within the cortex, visual information is first processed in the primary visual cortex, then is projected to higher-order vi-

= Binocular overlap

= Monocular field of vision—left eye

= Monocular field of vision—right eye

• **FIGURE 6–31   Visual Fields**

sual areas for even more complex processing and abstraction. The cortex contains a hierarchy of visual cells that respond to increasingly complex stimuli. Three types of visual cortical neurons have been identified based on the complexity of stimulus requirements needed for the cell to respond; these are called **simple, complex**, and **hypercomplex cells**. Simple and complex cells are stacked on top of each other within the cortical columns of the primary visual cortex, whereas hypercomplex cells are found in the higher visual processing areas. Unlike a retinal cell, which responds to the amount of light, a cortical cell fires only when it receives a particular pattern of illumination for which it is programmed. These patterns are built up by converging connections that originate from closely aligned photoreceptor cells in the retina. For example, some simple cells fire only when a bar is viewed vertically in a specific location, others when a bar is horizontal, and others at various oblique orientations. Movement of a critical axis of orientation becomes important for response by some of the complex cells. Hypercomplex cells add a new dimension to visual processing by responding only to particular edges, corners, and curves. Each level of cortical visual neurons has increasingly greater capacity for abstraction of information built up from the increasing convergence of input from lower-order neurons. In this way, the dotlike pattern of photoreceptors stimulated to varying degrees by varying light intensities in the retinal image is transformed in the cortex into information about depth, position, orientation, movement, contour, and length. Other aspects of this information, such as color

perception, are processed simultaneously through a similar hierarchical organization. How and where the entire image is finally put together is still unresolved.

Visual input goes to other areas of the brain not involved in vision perception.

Not all fibers in the visual pathway terminate in the visual cortices. Some are projected to other regions of the brain for purposes other than direct vision perception. Following are examples of nonsight activities dependent on input from the retina:

1. Control of pupil size.
2. Synchronization of biological clocks to cyclical variations in light intensity (for example, the sleep-wake cycle synchronized to the night-day cycle).
3. Contribution to cortical alertness and attention.
4. Control of eye movements.

In the latter regard, each eye is equipped with a set of six **external eye muscles** that position and move the eye so that it can better locate, see, and track objects. Eye movements are among the fastest, most discretely controlled movements of the body.

Protective mechanisms help prevent eye injuries.

Several mechanisms help protect the eyes from injury. Except for its anterior portion, the eyeball is sheltered by the bony socket in which it is positioned. The **eyelids** act like shutters to protect the anterior portion of the eye from environmental insults. They close reflexly to cover the eye under threatening circumstances such as rapidly approaching objects, dazzling light, and instances when the cornea or eyelashes are touched. Frequent spontaneous blinking of the eyelids helps disperse the lubricating, cleansing, bactericidal ("germ-killing") tears. **Tears** are produced continuously by the **lacrimal gland** in the upper lateral corner under the eyelid. This eye-washing fluid flows across the surface of the cornea and drains into tiny canals in the corner of each eye (Fig. 6–9b), eventually emptying into the back of the nasal passageway. This drainage system cannot handle the profuse tear production during crying, so the tears overflow from the eyes. The eyes are also equipped with protective **eyelashes**, which trap fine airborne debris such as dust before it can fall into the eye.

EAR: HEARING AND EQUILIBRIUM

The **ear** consists of three parts: the *external*, the *middle*, and the *inner* ear (• Fig. 6–32 and Table 6–5). The external and middle portion of the ear transmit airborne sound waves to the fluid-filled inner ear, amplifying the sound energy in the process. The inner ear houses two different sensory systems: the *cochlea*, which contains the receptors for conversion of sound waves into nerve impulses, making hearing possible;

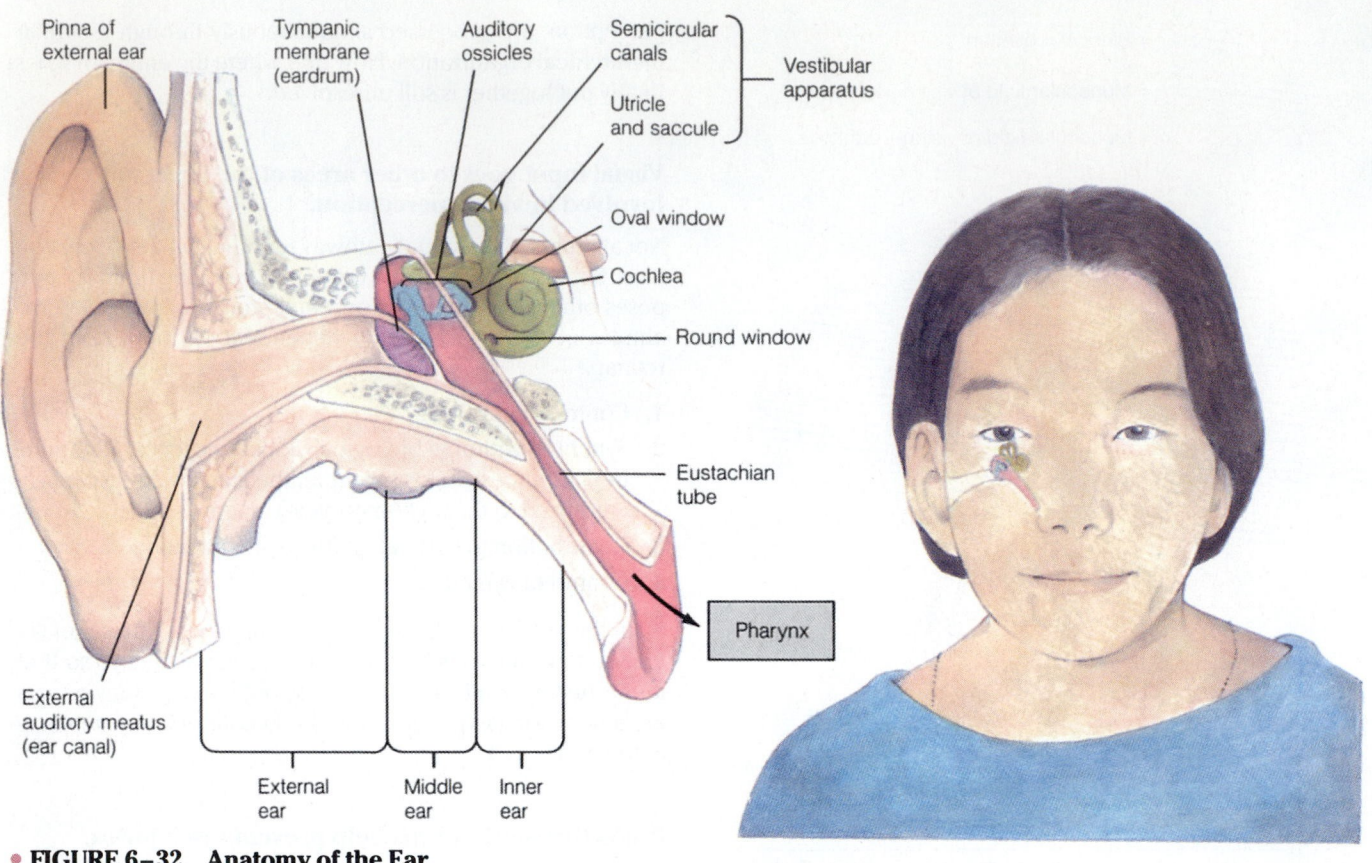

Pinna of
external ear

Tympanic
membrane
(eardrum)

Auditory
ossicles

Semicircular
canals

Utricle
and saccule

Vestibular
apparatus

Oval window

Cochlea

Round window

Eustachian
tube

Pharynx

External
auditory meatus
(ear canal)

External
ear

Middle
ear

Inner
ear

• **FIGURE 6–32   Anatomy of the Ear**

and the *vestibular apparatus,* which is necessary for the sense of equilibrium.

Sound waves consist of alternate regions of compression and rarefaction of air molecules.

Hearing is the neural perception of sound energy. **Sound waves** are traveling vibrations of air that consist of regions of high pressure caused by compression of air molecules alternating with regions of low pressure caused by rarefaction of the molecules (• Fig. 6–33a). Any device capable of producing such a disturbance pattern in air molecules is a source of sound. A simple example is a tuning fork. When a tuning fork is struck, its prongs vibrate. As a prong of the fork moves in one direction (Fig. 6–33b), air molecules ahead of it are pushed closer together, or compressed, increasing the pressure in this area. Simultaneously, the air molecules behind the prong spread out or are rarefied as the prong moves forward, lowering the pressure in that region. As the prong moves in the opposite direction, an opposite wave of compression and rarefaction is created. Even though individual molecules are moved only short distances as the tuning fork vibrates, alternating waves of compression and rarefaction spread out considerable distances in a rippling fashion. Disturbed air molecules disturb other molecules in adjacent regions, setting up new regions of compression and rarefaction, and so on (Fig. 6–33c). Sound energy is gradually dissipated as sound

waves travel farther from the original sound source. The intensity of the sound decreases, until it finally dies out when the last sound wave is too weak to disturb the air molecules around it.

Sound waves can also travel through media other than air, such as water. They do so less efficiently, however; greater pressures are required to cause movements of fluid than movements of air because of the fluid's greater inertia (resistance to change).

Sound is characterized by its pitch (tone), intensity (loudness), and timbre (quality) (• Fig. 6–34):

■ The **pitch** or **tone** of a sound (for example, whether it is a C or a G note) is determined by the *frequency* of vibrations. The greater the frequency of vibration, the higher the pitch. Human ears can detect sound waves with frequencies from 20 to 20,000 cycles per second but are most sensitive to frequencies between 1,000 and 4,000 cycles per second.

■ The **intensity**, or **loudness**, of a sound depends on the *amplitude* of the sound waves, or the pressure difference between a high-pressure region of compression and a low-pressure region of rarefaction. Within the hearing range, the greater the amplitude, the louder the sound. Human ears can detect a wide range of sound intensities, from the slightest whisper to the painfully loud takeoff of a jet. Loudness is expressed in **decibels (dB)**, which is a logarithmic measure of intensity compared with the faintest sound that

TABLE 6–5   Functions of Major Components of the Ear

STRUCTURE	LOCATION	FUNCTION
External ear		**Collects and transfers sound waves to middle ear**
Pinna (ear)	Skin-covered flap of cartilage located on each side of the head	Collects sound waves and channels them down the ear canal; contributes to sound localization
External auditory meatus (ear canal)	Tunnels from the exterior through the temporal bone to the tympanic membrane	Directs sound waves to tympanic membrane; contains filtering hairs and secretes earwax to trap foreign particles
Tympanic membrane (eardrum)	Thin membrane that separates the external and middle ear	Vibrates in synchrony with sound waves that strike it, setting middle ear bones in motion
Middle ear		**Transfers vibrations of tympanic membrane to the fluid in the cochlea, amplifying the sound energy in the process**
Malleus, incus, stapes	Movable chain of bones that extends across middle ear cavity; malleus attaches to the tympanic membrane and stapes attaches to the oval window	Oscillate in synchrony with tympanic-membrane vibrations and set up wavelike movements in the cochlear perilymph at the same frequency
Inner ear: cochlea		**Houses sensory system for hearing**
Oval window	Thin membrane at entrance to cochlea; separates the middle ear from the scala vestibuli	Vibrates in unison with movement of stapes, to which it is attached; oval-window movement sets cochlear perilymph in motion
Scala vestibuli *Scala tympani*	Upper compartment of cochlea Lower compartment of cochlea	Contain perilymph that is set in motion by oval-window movement driven by oscillation of middle ear bones
Cochlear duct (scala media)	Middle compartment of cochlea	Contains endolymph; houses the basilar membrane
Basilar membrane	Forms the floor of the cochlear duct	Vibrates in unison with perilymph movements; bears the organ of Corti, the sense organ for hearing
Organ of Corti	Rests on top of basilar membrane throughout its length	Contains hair cells, the receptors for sound, which undergo receptor potentials when bent as a result of fluid movement in cochlea
Tectorial membrane	Stationary membrane that overhangs the organ of Corti and within which the surface hairs of the receptor hair cells are embedded	Site at which the embedded hairs of the receptor cells are bent and undergo receptor potentials as the vibrating basilar membrane moves in relation to the stationary tectorial membrane
Round window	Thin membrane that separates the scala tympani from the middle ear	Vibrates in unison with fluid movements in perilymph to dissipate pressure in cochlea; does not contribute to sound reception
Inner ear: vestibular apparatus		**Houses sensory systems for equilibrium, and provides input essential for maintenance of posture and balance**
Semicircular canals	Three semicircular canals arranged three-dimensionally in planes at right angles to each other near the cochlea deep within the temporal bone	Detect rotational or angular acceleration or deceleration
Utricle	Saclike structure in a bony chamber between the cochlea and semicircular canals	Detects (1) changes in head position away from vertical and (2) horizontally directed linear acceleration and deceleration
Saccule	Lies next to utricle	Detects (1) changes in head position away from horizontal and (2) vertically directed linear acceleration and deceleration

● **FIGURE 6–33 Formation of Sound Waves** (a) Sound waves are alternating regions of compression and rarefaction of air molecules. (b) A vibrating tuning fork sets up sound waves as the air molecules ahead of the advancing arm of the tuning fork are compressed while the molecules behind the arm are rarefied. (c) Disturbed air molecules bump into molecules beyond them, setting up new regions of air disturbance more distant from the original source of sound. In this way, sound waves travel progressively farther from the source, even though each individual air molecule travels only a short distance when it is disturbed. The sound wave dies out when the last region of air disturbance is too weak to disturb the region beyond it.

Region of compression   Region of rarefaction

Pressure

Time

(a)

Normal density of air molecules when tuning fork is at rest

Region of rarefaction   Region of compression

(b)

(c)

● **FIGURE 6–34 Properties of Sound Waves**

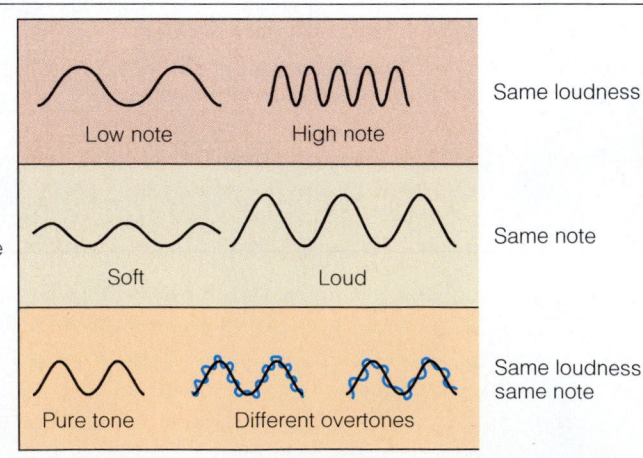

Pitch (tone) depends on frequency

Low note   High note   Same loudness

Intensity (loudness) depends on amplitude

Soft   Loud   Same note

Timbre (quality) depends on overtones

Pure tone   Different overtones   Same loudness, same note

TABLE 6-6	Relative Magnitude of Common Sounds	
SOUND	LOUDNESS IN DECIBELS (dB)	COMPARISON TO FAINTEST AUDIBLE SOUND (HEARING THRESHOLD)
Rustle of leaves	10 dB	10 times louder
Ticking of watch	20 dB	100 times louder
Normal conversation	60 dB	1 million times louder
Shouting	90 dB	1 billion times louder
Loud rock concert	120 dB	1 trillion times louder
Takeoff of jet plane	150 dB	1 quadrillion times louder

can be heard—the **hearing threshold**. Because of the logarithmic relationship, every ten decibels indicates a tenfold increase in loudness. A few examples of common sounds illustrate the magnitude of these increases (Table 6–6). Note that the rustle of leaves at 10 dB is ten times louder than hearing threshold, but the sound of a jet taking off at 150 dB is a quadrillion (a million billion) times, not 150 times, louder than the faintest audible sound. Sounds greater than 100 dB can permanently damage the sensitive sensory apparatus in the cochlea.

The **timbre**, or **quality**, of a sound depends on its *overtones,* which are additional frequencies superimposed on top of the fundamental pitch or tone. A tuning fork has a pure tone, but most sounds lack purity. For example, complex mixtures of overtones impart different sounds to different instruments playing the same note (a C note sounds different on a trumpet than on a piano). Overtones are likewise responsible for the characteristic differences in voices. Timbre enables the listener to distinguish the source of sound waves, because each source produces a different pattern of overtones. Thanks to timbre, you can tell whether it is your mother or girlfriend calling on the telephone before you say the wrong thing.

The external and middle ear convert airborne sound waves into fluid vibrations in the inner ear.

The specialized receptors for sound are located in the fluid-filled inner ear. Airborne sound waves must therefore be channeled toward and transferred into the inner ear, compensating in the process for the loss in sound energy that naturally occurs as sound waves pass from air into water. This function is performed by the external ear and the middle ear.

The **external ear** (Fig. 6–32) consists of the *pinna (ear), external auditory meatus (ear canal),* and *tympanic membrane (eardrum).* The **pinna**, a prominent skin-covered flap of carti-

lage, collects sound waves and channels them down the external ear canal. Many species (dogs, for example) can cock their ears in the direction of sound to collect more sound waves, but human ears are relatively immobile. Because of its shape, the pinna partially shields sound waves that approach the ear from the rear and thus helps a person distinguish whether a sound is coming from directly in front or behind.

Sound localization for sounds approaching from the right or left is determined by two cues. First, the sound wave reaches the ear closer to the sound source slightly before it arrives at the farther ear. Second, the sound is less intense as it reaches the farther ear, because the head acts as a sound barrier that partially disrupts the propagation of sound waves. The auditory cortex integrates all of these cues to determine the location of the sound source. It is difficult to localize sound with only one ear.

The entrance to the **ear canal** is guarded by fine hairs. The skin lining the canal contains modified sweat glands that produce **cerumen** (earwax), a sticky secretion that traps fine foreign particles. Together the hairs and earwax help prevent airborne particles from reaching the inner portions of the ear canal, where they could accumulate or injure the tympanic membrane and interfere with hearing.

The **tympanic membrane**, which is stretched across the entrance to the **middle ear**, vibrates when struck by sound waves. The alternating higher- and lower-pressure regions of a sound wave cause the exquisitely sensitive eardrum to bow inward and outward in unison with the wave's frequency.

The resting air pressure on both sides of the tympanic membrane must be equal for the membrane to be free to move as sound waves strike it. The outside of the eardrum is exposed to atmospheric pressure that reaches it through the ear canal. The inside of the eardrum facing the middle ear cavity is also exposed to atmospheric pressure via the **eustachian (auditory) tube**, which connects the middle ear to the **pharynx** (back of the throat) (Fig. 6–32). The eustachian tube is normally closed, but it can be pulled open by yawning, chewing, and swallowing. Such opening permits air pressure within the middle ear to equilibrate with atmospheric pressure so that pressures on both sides of the tympanic membrane are equal. During rapid external pressure changes (for example, during air flight), each eardrum bulges painfully as the pressure outside the ear changes while the pressure in the middle ear remains unchanged. Opening the eustachian tube by yawning allows the pressures on both sides of the tympanic membrane to equalize, relieving the pressure distortion as the eardrum "pops" back into place. Infections originating in the throat sometimes spread through the eustachian tube to the middle ear. The resultant fluid accumulation in the middle ear not only is painful but also interferes with conduction of sound across the middle ear.

The middle ear transfers the vibratory movements of the tympanic membrane to the fluid of the inner ear. This transfer is facilitated by a moveable chain of three small bones or **ossicles** (the **malleus, incus**, and **stapes**) that extend across the middle ear (● Fig. 6–35a). The first bone, the malleus, is attached to the tympanic membrane, and the last bone, the stapes, is attached to the **oval window**, the entrance into the

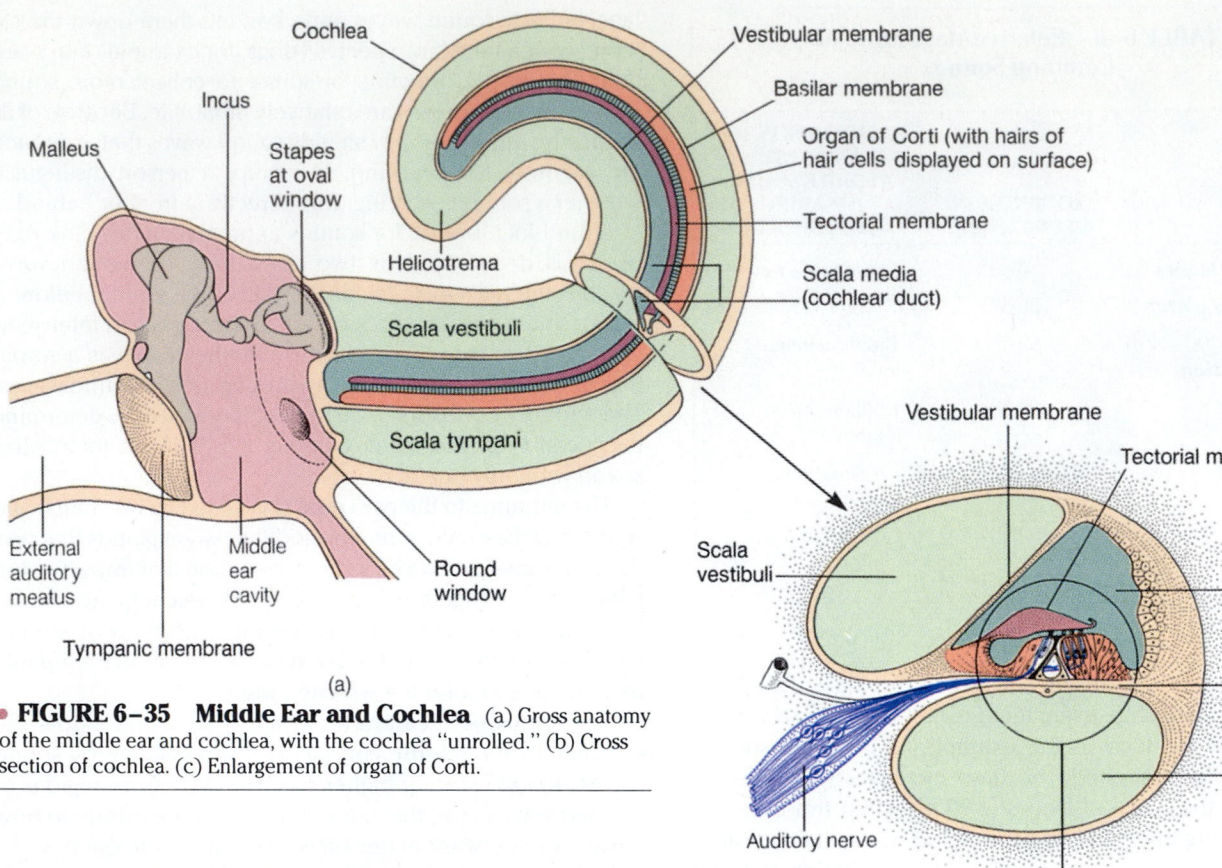

(a)

• **FIGURE 6–35   Middle Ear and Cochlea** (a) Gross anatomy of the middle ear and cochlea, with the cochlea "unrolled." (b) Cross section of cochlea. (c) Enlargement of organ of Corti.

(b)

(c)

fluid-filled cochlea. As the tympanic membrane vibrates in response to sound waves, the chain of bones is set into motion at the same frequency, transmitting this frequency of movement from the tympanic membrane to the oval window. The resultant pressure on the oval window with each vibration produces wavelike movements in the inner ear fluid at the same frequency as the original sound waves. However, as noted earlier, greater pressure is required to set fluid in motion. Two mechanisms related to the ossicular system amplify the pressure of the airborne sound waves to set up fluid vibrations in the cochlea. First, because the surface area of the tympanic membrane is much larger than that of the oval window, pressure is increased as force exerted on the tympanic membrane is conveyed to the oval window (pressure = force/unit area). Second, the lever action of the ossicles provides an additional mechanical advantage. Together, these mechanisms increase the force exerted on the oval window by twenty times what it would be if the sound wave struck the oval window directly. This additional pressure is sufficient to set the cochlear fluid in motion.

Several tiny muscles in the middle ear contract reflexly in response to loud sounds (over 70 dB), causing the tympanic membrane to tighten and limiting movement of the ossicular chain. This reduced movement of middle ear structures diminishes the transmission of loud sound waves to the inner ear to protect the delicate sensory apparatus from damage. This reflex response is relatively slow, however, happening at least 40 msec after exposure to a loud sound. It thus provides protection only from prolonged loud sounds, not from sudden loud sounds like an explosion.

Hair cells in the organ of Corti transduce fluid movements into neural signals.

The snail-shaped cochlear portion of the **inner ear** is a coiled tubular system lying deep within the temporal bone (Fig. 6–32). It is easier to understand the functional components of the **cochlea** by "unrolling" it, as shown in Figure 6–35a. The cochlea is divided throughout most of its length into three fluid-filled longitudinal compartments. A blind-ended **cochlear duct**, which is also known as the **scala media**, constitutes the middle compartment. It tunnels lengthwise through the center of the cochlea, almost but not quite reaching its end. The upper compartment, the **scala vestibuli** (Figs. 6–35a and b), follows the inner contours of the spiral, and the **scala tympani**, the lower compartment, follows the outer contours. The fluid within the cochlear duct is called **endolymph** (● Fig. 6–36a). The scala vestibuli and scala tympani both contain a slightly different fluid, the **perilymph**. The region beyond the tip of the cochlear duct where the fluid in the

● **FIGURE 6–36   Transmission of Sound Waves** (a) Fluid movement within the perilymph set up by vibration of the oval window follows two pathways: (1) through the scala vestibuli, around the helicotrema, and through the scala tympani, causing the round window to vibrate; and (2) a "shortcut" from the scala vestibuli through the basilar membrane to the scala tympani. The first pathway just dissipates sound energy, but the second pathway triggers activation of the receptors for sound by bending the hairs of the hair cells as the organ of Corti on top of the vibrating basilar membrane is displaced in relation to the overlying tectorial membrane. (b) Different regions of the basilar membrane vibrate maximally at different frequencies. (c) The narrow, stiff end of the basilar membrane nearest the oval window vibrates best with high-frequency pitches. The wide, flexible end of the basilar membrane by the helicotrema vibrates best with low-frequency pitches.

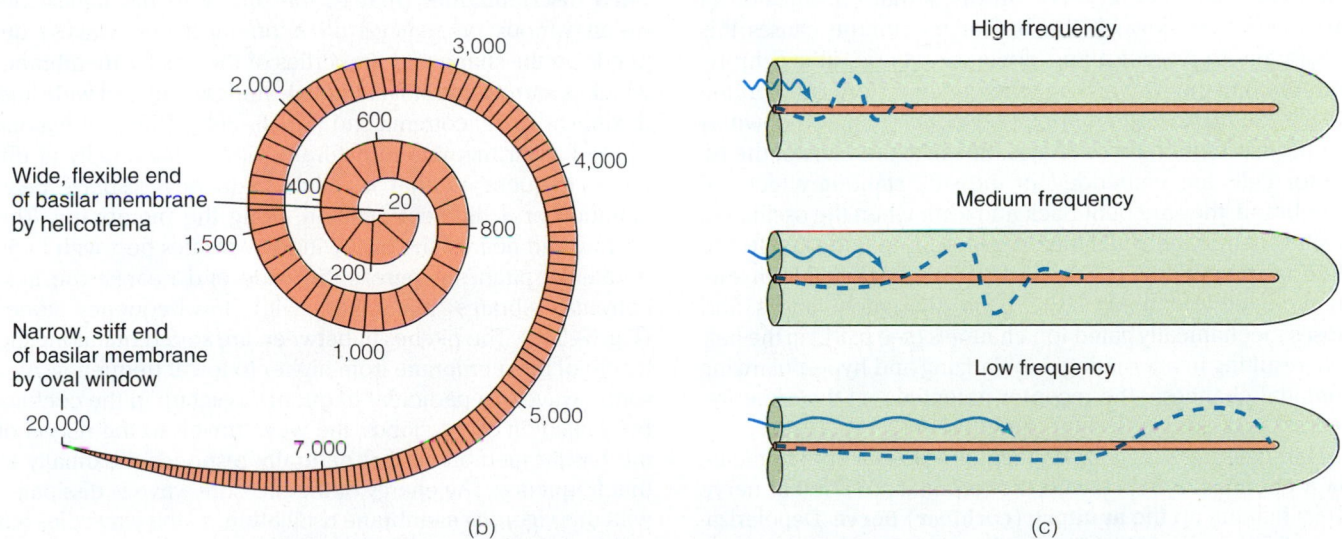

The numbers indicate the frequencies with which different regions of the basilar membrane maximally vibrate.

upper and lower compartments is continuous is called the **helicotrema**. The scala vestibuli is sealed from the middle ear cavity by the oval window, to which the stapes is attached. Another small membrane-covered opening, the **round window**, seals the scala tympani from the middle ear. The thin **vestibular membrane** separates the cochlear duct from the scala vestibuli. The **basilar membrane** forms the floor of the cochlear duct, separating it from the scala tympani. The basilar membrane is especially important because it bears the **organ of Corti**, the sense organ for hearing.

The organ of Corti, which rests on top of the basilar membrane throughout its full length, contains **hair cells** that are the receptors for sound. Hair cells generate neural signals when their surface hairs are mechanically deformed in association with fluid movements in the inner ear. These hairs are mechanically embedded in the **tectorial membrane**, an awninglike projection overhanging the organ of Corti throughout its length (Fig. 6–35b and c).

The pistonlike action of the stapes against the oval window sets up pressure waves in the upper compartment. Because fluid is incompressible, pressure is dissipated in two ways as the stapes causes the oval window to bulge inward: (1) displacement of the round window and (2) deflection of the basilar membrane (Fig. 6–36a). In the first of these pathways, the pressure wave pushes the perilymph forward in the upper compartment, then around the helicotrema, and into the lower compartment, where it causes the round window to bulge outward into the inner ear cavity to compensate for the pressure increase. As the stapes rocks backward and pulls the oval window outward toward the middle ear, the perilymph shifts in the opposite direction, displacing the round window inward. This pathway does not result in sound reception; it just dissipates pressure.

Pressure waves of frequencies associated with sound reception take a "shortcut." Pressure waves in the upper compartment are transferred through the thin vestibular membrane, into the cochlear duct, and then through the basilar membrane into the lower compartment, where they cause the round window to alternately bulge outward and inward. The main difference in this pathway is that transmission of pressure waves through the basilar membrane causes this membrane to move up and down, or vibrate, in synchrony with the pressure wave. Since the organ of Corti rides on the basilar membrane, the hair cells also move up and down as the basilar membrane oscillates. Because the hairs of the receptor cells are embedded in the stiff, stationary tectorial membrane, they are bent back and forth when the oscillating basilar membrane shifts their position in relationship to the tectorial membrane (● Fig. 6–37). This back-and-forth mechanical deformation of the hairs alternately opens and closes mechanically gated ion channels (see p. 81) in the hair cell, resulting in alternating depolarizing and hyperpolarizing potential changes—the receptor potential—at the same frequency as the original sound stimulus.

Hair cells are specialized receptor cells that communicate via a chemical synapse with the terminals of afferent nerve fibers making up the **auditory (cochlear) nerve**. Depolarization of hair cells (when the basilar membrane is deflected

Bending of hairs, which produces a receptor potential

Hair cells

Deflection of basilar membrane caused by fluid movements in the cochlea

● **FIGURE 6–37** **Bending of Hair Cells upon Deflection of Basilar Membrane**

upward) increases their rate of transmitter release, which steps up the rate of firing in the afferent fibers. Conversely, the firing rate decreases as the hair cells release less transmitter when they are hyperpolarized upon displacement in the opposite direction.

Thus, the ear converts sound waves in the air into oscillating movements of the basilar membrane that bend the hairs of the receptor cells back and forth. This shifting mechanical deformation of the hairs alternately opens and closes the receptor cells' channels, which bring about graded potential changes in the receptor that lead to changes in the rate of action potentials propagated to the brain. In this way, sound waves are translated into neural signals that can be perceived by the brain as sound sensations (● Fig. 6–38).

Pitch discrimination depends on the region of the basilar membrane that vibrates; loudness discrimination depends on the amplitude of the vibration.

Pitch discrimination (that is, the ability to distinguish between various frequencies of incoming sound waves) depends on the shape and properties of the basilar membrane, which is narrow and stiff at its oval-window end and wide and flexible at its helicotrema end (Fig. 6–36b). Different regions of the basilar membrane naturally vibrate maximally at different frequencies; that is, each frequency displays peak vibration at a different position along the membrane. The narrow end nearest the oval window vibrates best with high-frequency pitches, whereas the wide end nearest the helicotrema vibrates maximally with low-frequency tones (Fig. 6–36c). The pitches in between are sorted out along the length of the membrane from higher to lower frequency. As a sound wave of a particular frequency is set up in the cochlea by oscillation of the stapes, the wave travels to the region of the basilar membrane that naturally responds maximally to that frequency. The energy of the pressure wave is dissipated with this vigorous membrane oscillation, so the wave dies out at the region of maximal displacement.

The hair cells in the region of peak vibration of the basilar membrane undergo the most mechanical deformation and, accordingly, are the most excited. This information is propagated to the CNS, which interprets the pattern of hair cell stimulation as a sound of a particular frequency. Modern techniques have determined that the basilar membrane is so fine-tuned that the peak membrane response to a single pitch extends probably no more than the width of a few hair cells.

Overtones of varying frequencies cause many points along the basilar membrane to vibrate simultaneously but less intensely than the fundamental tone, enabling the CNS to distinguish the timbre of the sound (**timbre discrimination**).

Intensity (**loudness**) **discrimination** depends on the amplitude of vibration. As sound waves originating from louder sound sources strike the eardrum, they cause it to vibrate more vigorously (that is, bulge in and out to a greater extent) but at the same frequency as a softer sound of the same pitch. This greater tympanic-membrane deflection is converted into

• **FIGURE 6–39   Loss of Hair Cells Caused by Loud Noises** Effect of intense sound on the inner ear. (a) Organ of Corti of a normal guinea pig, showing three rows of outer hair cells and one row of inner hair cells. (b) Organ of Corti of a guinea pig after 24-hour exposure to intense noise at a sound level approached by loud rock music (2,000 Hz octave band at 120 decibels SPL).

greater amplitude of basilar-membrane movement in the region of peak responsiveness. The CNS interprets greater basilar-membrane oscillation as a louder sound.

The auditory system is so sensitive and can detect sounds so faint that the distance of basilar-membrane deflection is comparable to only a fraction of the diameter of a hydrogen atom, the smallest of atoms. It is no wonder that very loud sounds, which cannot be sufficiently attenuated by protective middle ear reflexes (for example, the sounds of a typical rock concert), can set up such violent vibrations of the basilar membrane that irreplaceable hair cells are actually sheared off or permanently distorted, leading to partial hearing loss (• Fig. 6–39).

The auditory cortex is mapped according to tone.

Just as various regions of the basilar membrane are associated with particular tones, the auditory cortex is also *tonotopically* organized. Each region of the basilar membrane is linked to a specific region of the auditory cortex in the temporal lobe. Accordingly, specific cortical neurons are activated only by particular tones; each region of the auditory cortex becomes excited only in response to a specific tone detected by a selected portion of the basilar membrane.

The afferent neurons that pick up the auditory signals from the hair cells exit the cochlea via the auditory nerve. The neural pathway between the organ of Corti and the auditory cortex involves several synapses en route, the most notable of which are in the brain stem and medial geniculate nucleus of the thalamus. The brain stem uses the auditory input for alert-ness and arousal. The thalamus sorts and relays the signals upward. Unlike the visual pathways, auditory signals from each ear are transmitted to both temporal lobes because the fibers partially cross over in the brain stem. For this reason, a disruption of the auditory pathway on one side beyond the brain stem does not affect hearing in either ear to any extent.

Like other regions of the cortex, the **primary auditory cortex** is organized into columns. In contrast to the visual system, however, no hierarchy of cells within the columns that respond to increasingly complex auditory signals has been identified. The primary auditory cortex appears to perceive discrete sounds while the surrounding higher-order auditory cortex integrates the separate sounds into a coherent, meaningful pattern. Think about the complexity of the task accomplished by your auditory system. When you are at a concert, your organ of Corti responds to the simultaneous mixture of the instruments, the applause and hushed talking of the audience, and the background noises in the theater. You are able to distinguish these separate parts of the many sound waves reaching your ears and pay attention to those of importance to you.

Deafness is caused by defects either in conduction or neural processing of sound waves.

Loss of hearing, or **deafness**, may be temporary or permanent, partial or complete. Deafness is classified into two types—*conductive deafness* and *sensorineural deafness*—depending on the part of the hearing mechanism that fails to function adequately. **Conductive deafness** occurs when

(a)

(b)

Hair cell depolarized when stereocilia are bent toward kinocilium

Hair cell hyperpolarized when stereocilia are bent away from kinocilium

(c)

(d)

• **FIGURE 6–40 Vestibular Apparatus** (a) Gross anatomy of vestibular apparatus. (b) Receptor cell unit in ampulla of semicircular canals. (c) Schematic representation of "hairs" on sensory hair cells of semicircular canals. (d) Scanning electron micrograph of kinocilium and stereocilia on the hair cells within the vestibular apparatus.

can lie in the organ of Corti, in the auditory nerves or ascending auditory pathways, or, rarely, in the auditory cortex.

Hearing aids are helpful in noncurable cases of conductive deafness but are less beneficial for sensorineural deafness. These devices bypass a defective conduction system by amplifying and transmitting the sound waves through the bone to the inner ear to set up fluid motion. However, the receptor-cell–neural-pathway system must still be intact for the sound to be perceived.

The vestibular apparatus detects position and motion of the head and is important for equilibrium and coordination of head, eye, and body movements.

In addition to its cochlear-dependent role in hearing, the inner ear has another specialized component, the **vestibular apparatus**, which provides information essential for the sense of equilibrium and for coordinating head movements with eye and postural movements. The vestibular apparatus consists of two sets of structures lying within a tunneled-out region of the temporal bone near the cochlea—the *semicircular canals* and the *otolith organs,* namely, the *utricle* and *saccule* (• Fig. 6–40a).

sound waves are not adequately conducted through the external and middle portions of the ear to set the fluids in the inner ear in motion. Possible causes include physical blockage of the ear canal with earwax, rupture of the eardrum, middle-ear infections with accompanying fluid accumulation, or restriction of the ossicular movement because of bony adhesions between the stapes and the oval window. In **sensorineural deafness**, the sound waves are transmitted to the inner ear, but they are not translated into nerve signals that are interpreted by the brain as sound sensations. The defect

The vestibular apparatus detects changes in position and motion of the head. As in the cochlea, all components of the vestibular apparatus contain endolymph and are surrounded by perilymph. Also, similar to the organ of Corti, the vestibular components each contain hair cells that respond to mechanical deformation triggered by specific movements of the endolymph. Also, like the auditory hair cells, the vestibular receptors may be either depolarized or hyperpolarized, depending on the direction of the fluid movement. Unlike the auditory system, however, much of the information provided by the vestibular apparatus does not reach the level of conscious awareness.

The **semicircular canals** detect rotational or angular acceleration or deceleration of the head, such as when starting or stopping spinning, somersaulting, or turning the head. Each ear contains three semicircular canals arranged three-dimensionally in planes that lie at right angles to each other. The receptive hair cells of each semicircular canal are situated on top of a ridge located in the **ampulla**, a swelling at the base of the canal (Figs. 6–40a and b). The hairs are embedded in an overlying caplike, gelatinous layer, the **cupula**, which protrudes into the endolymph within the ampulla. The cupula sways in the direction of fluid movement, much like seaweed leaning in the direction of the prevailing tide.

Acceleration or deceleration during rotation of the head in any direction causes endolymph movement in at least one of the semicircular canals because of their three-dimensional arrangement. As the head starts to move, the bony canal and the ridge of hair cells embedded in the cupula move with the head. However, the fluid within the canal, not being attached to the skull, does not initially move in the direction of the rotation but lags behind because of its inertia. (Because of inertia, a resting object remains at rest, and a moving object continues to move in the same direction unless the object is acted on by some external force that induces change.) When the endolymph is left behind as the head starts to rotate, the endolymph that is in the same plane as the head movement is in effect shifted in the opposite direction from the movement (similar to your body tilting to the right as the car in which you are riding suddenly turns to the left) (• Fig. 6–41). This fluid movement causes the cupula to lean in the opposite direction from the head movement, bending the sensory hairs embedded in it. If the head movement continues at the same rate in the same direction, the endolymph catches up and moves in unison with the head so that the hairs return to their unbent position. When the head slows down and stops, the reverse situation occurs. The endolymph briefly continues to move in the direction of the rotation while the head decelerates to a stop. As a result, the cupula and its hairs are transiently bent in the direction of the preceding spin, which is opposite to the way they were bent during acceleration. When the endolymph gradually comes to a halt, the hairs straighten again. Thus, the semicircular canals detect changes in the rate of rotational movement of the head. They do not respond when the head is motionless or during circular motion at a constant speed.

The hairs of a vestibular hair cell consist of twenty to fifty **stereocilia**, which are actin-stiffened microvilli (see p. 39),

Direction of head rotation

Left horizontal semicircular canal

Direction of fluid movement in semicircular canals

Direction of bending of hair cells

Right horizontal semicircular canal

• **FIGURE 6–41   Activation of Hair Cells in Semicircular Canals**

and one cilium (see p. 37), the **kinocilium** (Fig. 6–40c and d). Each hair cell is oriented so that it depolarizes when its stereocilia are bent toward the kinocilium; bending in the opposite direction hyperpolarizes the cell. The hair cells form a chemically mediated synapse with terminal endings of afferent neurons whose axons join with those of the other vestibular structures to form the **vestibular nerve**. This nerve unites with the auditory nerve from the cochlea to form the **vestibulocochlear nerve**. Depolarization of the hair cells increases the rate of firing in the afferent fibers; conversely, when the hair cells are hyperpolarized, the frequency of action potentials in the afferent fibers is reduced.

Whereas the semicircular canals provide the CNS with information about rotational changes in head movement, the **otolith organs** provide information about the position of the head relative to gravity and also detect changes in rate of linear motion (moving in a straight line regardless of direction). The **utricle** and **saccule** are saclike structures housed within a bony chamber situated between the semicircular canals and the cochlea (Fig. 6–40a). The hairs of the receptive hair cells in these sense organs also protrude into an overlying gelatinous sheet, whose movement displaces the hairs and results in changes in hair cell potential. Many tiny crystals of calcium carbonate—the **otoliths** ("ear stones")—are suspended within the gelatinous layer, making it heavier and giving it more inertia than the surrounding fluid (• Fig. 6–42a). When a person is in an upright position, the hairs within the utricles are oriented vertically and the saccule hairs are lined up horizontally.

Let's look at the utricle as an example. Its otolith-embedded, gelatinous mass shifts positions and bends the hairs in two ways:

1. When the head is tilted in any direction other than vertical (that is, other than straight up and down), the hairs are bent in the direction of the tilt because of the gravita-

(a)

tional force exerted on the top-heavy gelatinous layer (Fig. 6–42b). Within the utricles on each side of the head, some of the hair cell bundles are oriented to depolarize and others to hyperpolarize when the head is in any position other than upright. The CNS thus receives different patterns of neural activity depending on head position with respect to gravity.

2. The utricle hairs are also displaced by any change in horizontal linear motion (such as moving straight forward, backward, or to the side). As a person starts to walk forward (Fig. 6–42c), the top-heavy otolith membrane at first lags behind the endolymph and hair cells because of its greater inertia. The hairs are thus bent to the rear, in the opposite direction of the forward movement of the head. If the walking pace is maintained, the gelatinous layer soon catches up and moves at the same rate as the head so that the hairs are no longer bent. When the person stops walking, the otolith sheet continues to move forward briefly as the head slows and stops, bending the hairs toward the front. Thus, the hair cells of the utricle detect horizontally directed linear acceleration and deceleration, but they do not provide information about movement in a straight line at constant speed.

The saccule functions similarly to the utricle, except that it responds selectively to the tilting of the head away from a horizontal position (such as getting up from bed) and to vertically directed linear acceleration and deceleration (such as jumping up and down or riding in an elevator).

• **FIGURE 6–42 Utricle** (a) Receptor unit in utricle. (b) Activation of utricle by change in head position. (c) Activation of utricle by horizontal linear acceleration.

(b)

(c)

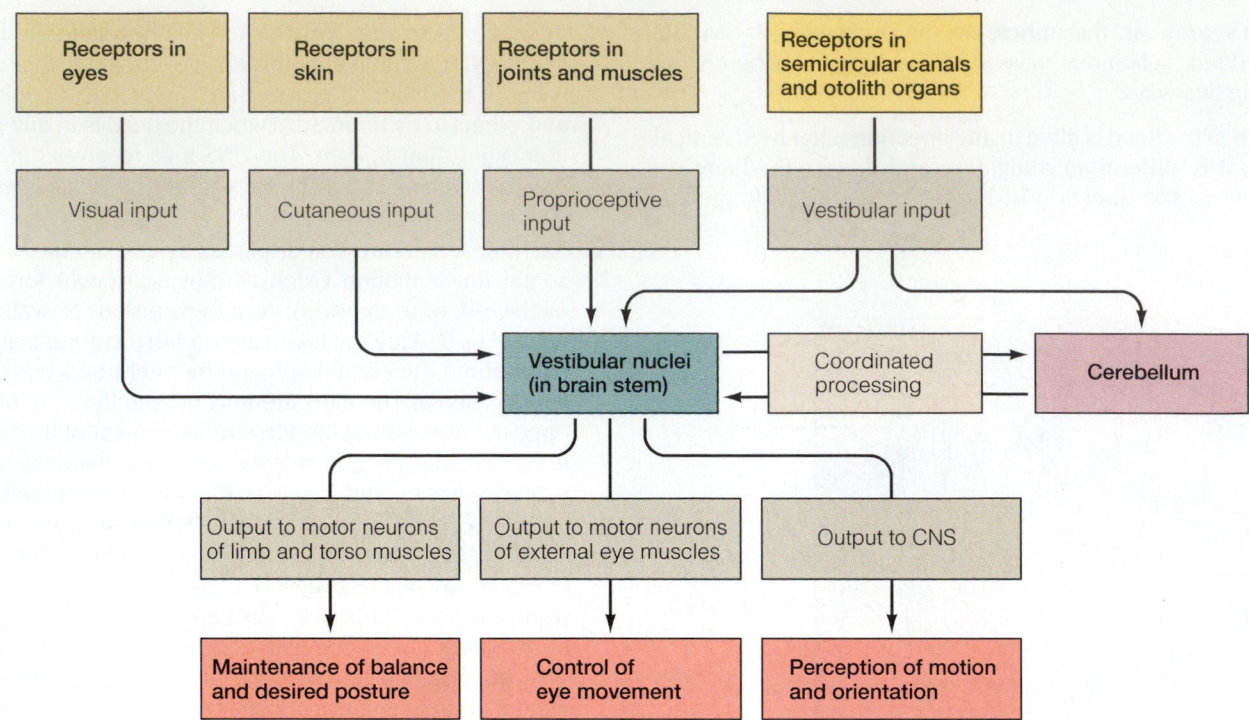

• **FIGURE 6–43   Input and Output of Vestibular Nuclei**

Signals arising from the various components of the vestibular apparatus are carried through the vestibulocochlear nerve to the **vestibular nuclei**, a cluster of neuronal cell bodies in the brain stem, and to the cerebellum. Here the vestibular information is integrated with input from the skin surface, eyes, joints, and muscles for: (1) maintaining balance and desired posture; (2) controlling the external eye muscles so that the eyes remain fixed on the same point, despite movement of the head; and (3) perceiving motion and orientation (• Fig. 6–43).

Some individuals, for poorly understood reasons, are especially sensitive to particular motions that activate the vestibular apparatus and cause symptoms of dizziness and nausea; this sensitivity is called **motion sickness**. Occasionally, fluid imbalances within the inner ear lead to **Meniere's disease**. Not surprisingly, because both the vestibular apparatus and cochlea contain the same inner ear fluids, both vestibular and auditory symptoms occur. An afflicted individual suffers transient attacks of severe **vertigo** (dizziness) accompanied by pronounced ringing in the ears and some loss of hearing. During these episodes, the person cannot stand upright and reports feeling as though self or surrounding objects in the room are spinning around.

━━ ● **CHEMICAL SENSES: TASTE AND SMELL**

Unlike the photoreceptors of the eye and mechanoreceptors of the ear, the receptors for taste and smell are chemoreceptors, which generate neural signals upon binding with particular chemicals in their environment. The sensations of taste and smell in association with food intake influence the flow of digestive juices and affect appetite. Furthermore, stimulation of taste or smell receptors induces pleasurable or objectionable sensations and signals the presence of something to seek (a desirable food) or to avoid (a toxic substance). In lower animals, smell also plays a major role in finding direction, in seeking prey or avoiding predators, and in sexual attraction to a mate. The sense of smell is less sensitive in humans and much less important in influencing our behavior (although millions of dollars are spent annually on perfumes and deodorants to make us "smell" better and thereby be more socially attractive). We will first examine the mechanism of taste (**gustation**) and then turn our attention to smell (**olfaction**).

Taste sensation is coded by patterns of activity in various taste bud receptors.

The chemoreceptors for taste sensation are packaged in **taste buds** (• Fig. 6–44), about ten thousand of which are present in the oral cavity and throat with the greatest percentage on the upper surface of the tongue. A taste bud consists of about fifty receptor cells packaged with supporting cells in an arrangement like slices of an orange. Each taste bud has a small opening, the **taste pore**, through which fluids in the mouth come into contact with the surface of its receptor cells.

Taste receptor cells are modified epithelial cells with many surface folds, or microvilli, that protrude slightly through the taste pore, greatly increasing the surface area exposed to the oral contents. The plasma membrane of the microvilli contains receptor sites that bind selectively with chemical molecules in the environment. Only chemicals in

Taste bud

Papilla

Sensory nerve fibers

Surface of tongue

Taste pore

Taste receptor cell

Supporting cell

Tongue

● **FIGURE 6–44   Location and Structure of Taste Buds**   Taste buds are located primarily along the edges of moundlike papillae on the upper surface of the tongue. The receptor cells and supporting cells of a taste bud are arranged like slices of an orange.

solution—either ingested liquids or solids that have been dissolved in saliva—can attach to receptor cells. Binding of a taste-provoking chemical with a receptor cell alters its ionic channels to produce a depolarizing receptor potential. This receptor potential, in turn, initiates action potentials within terminal endings of afferent nerve fibers with which the receptor cell synapses.

Most receptors are carefully sheltered from direct exposure to the environment, but the taste receptor cells, by virtue of their task, frequently come into contact with potent chemicals. Unlike the eye or ear receptors, which are irreplaceable, taste receptors have a life span of about ten days. Epithelial cells surrounding the taste bud differentiate first into supporting cells and then into receptor cells to constantly renew the taste bud components.

Terminal afferent endings of several cranial nerves synapse with taste buds in various regions of the mouth. Signals in these sensory inputs are conveyed via synaptic stops in the brain stem and thalamus to the **cortical gustatory area**, a region in the parietal lobe adjacent to the "tongue" area of the somatosensory cortex. Unlike most sensory input, the gustatory pathways are primarily uncrossed. The brain stem also projects fibers to the hypothalamus and limbic system, presumably to add affective dimensions, such as whether the taste is pleasant or unpleasant, and to process behavioral aspects associated with taste and smell.

We can discriminate among thousands of different taste sensations, yet all tastes are varying combinations of four **pri-**

mary tastes: *salty, sour, sweet,* and *bitter.* Salt taste is stimulated by chemical salts, especially NaCl (table salt). Acids cause a sour taste. The citric-acid content of lemons, for example, accounts for their distinctly sour taste. The sensation of sweetness is evoked by the particular configuration of glucose. Other organic molecules with similar structure can also interact with "sweet" receptor binding sites. Alkaloids (such as caffeine, nicotine, strychnine, morphine, and other toxic plant derivatives) or poisonous substances elicit a bitter taste, presumably as a protective mechanism to discourage ingestion of these potentially dangerous compounds.

Each receptor cell responds in varying degrees to all four primary tastes but is generally preferentially responsive to one of the taste modalities (● Fig. 6–45). The richness of fine taste discrimination beyond the four primary tastes depends on subtle differences in the stimulation patterns of all the taste buds in response to various substances, similar to the variable stimulation of the three cone types that gives rise to the range of color sensations. Taste perception is also influenced by information derived from other receptors, especially odor. When you temporarily lose your sense of smell because of swollen nasal passageways during a cold, your sense of taste is also markedly reduced, even though your taste receptors are unaffected by the cold. Other factors affecting taste include temperature and texture of the food as well as psychological factors associated with past experiences with the food. How the cortex accomplishes the complex perceptual processing of taste sensation is presently not known.

Smell is the least understood of the special senses.

The **olfactory** (**smell**) **mucosa**, located in the ceiling of the nasal cavity, contains three cell types: *olfactory receptors, supporting cells,* and *basal cells* (● Fig. 6–46). The supporting cells secrete **mucus**, which coats the nasal passages. The basal cells are precursors for new olfactory receptor cells, which are replaced about every two months. This is remarkable because, unlike the other special sense receptors, olfactory receptors are specialized endings of afferent neurons, not separate cells. The entire neuron, including its afferent axon projecting into the brain, is replaced. These are the only neurons that undergo cell division. The axons of the receptor cells collectively form the **olfactory nerve**. The receptor portion of the olfactory receptor cells consists of an enlarged knob bearing several long cilia that extend to the surface of the mucosa. These cilia contain the binding sites for attachment of odoriferous molecules. During quiet breathing, odorants typically reach the sensitive receptors only by diffusion because the olfactory mucosa is above the normal path of air flow. The act of sniffing enhances this process by drawing the air currents upward within the nasal cavity so that a greater percentage of the odoriferous molecules in the air come into contact with the olfactory mucosa.

To be smelled, a substance must be (1) sufficiently volatile (easily vaporized) that some of its molecules can enter the nose in the inspired air and (2) sufficiently water-soluble that it can dissolve in the mucus layer coating the olfactory mucosa. As with taste receptors, molecules must be dissolved in order to be detected by olfactory receptors. Binding of an odoriferous molecule to a specialized attachment site on the cilia leads to opening of both Na^+ and K^+ channels. The resultant ion movement brings about a depolarizing receptor potential that generates action potentials in the afferent fiber. The frequency of the action potentials depends on the concentration of the stimulating chemical molecules.

The afferent fibers pass through tiny holes in the flat bone plate separating the olfactory mucosa from the overlying brain tissue (Fig. 6–46). They immediately synapse in the **olfactory bulb**, a complex neural structure containing several different layers of cells that are functionally similar to the reti-

● **FIGURE 6–46** **Location and Structure of the Olfactory Receptors**

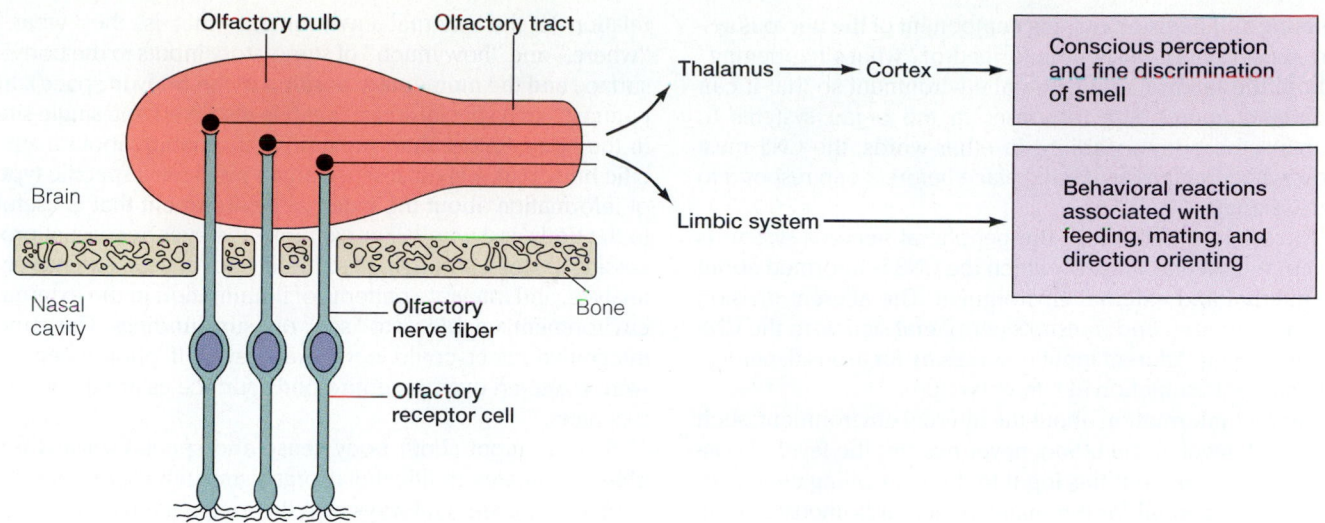

Olfactory bulb   Olfactory tract

Thalamus ⟶ Cortex ⟶ Conscious perception and fine discrimination of smell

Brain

Nasal cavity

Olfactory nerve fiber

Bone

Olfactory receptor cell

Limbic system ⟶ Behavioral reactions associated with feeding, mating, and direction orienting

• **FIGURE 6–47   Olfactory Pathways** See Figure 6–46 for a location reference point.

nal layers of the eye. Fibers leaving the olfactory bulb travel in two different routes (• Fig. 6–47): (1) a *subcortical route* going primarily to regions of the limbic system, especially the lower medial sides of the temporal lobes (considered to be the **primary olfactory cortex**), and (2) a *thalamic-cortical route*. Until recently, the subcortical route was thought to be the only olfactory pathway. This route, which includes hypothalamic involvement, permits close coordination between smell and behavioral reactions associated with feeding, mating, and direction orienting. The thalamic-cortical route, as it is for other senses, is important for conscious perception and fine discrimination of smell.

The physiological mechanism of smell discrimination is far from being understood. Humans can distinguish tens of thousands of different odors. Researchers generally believe that perception of these various odors depends on combinations of **primary odors**, similar to color vision and taste discrimination. However, there is no general agreement on how many primary odors there are or what they are. One researcher recently found genes for well over one hundred different types of smell receptors in the olfactory mucosa, and she believes there may be as many as one thousand. This large number of distinct receptors for smell discrimination is presumably needed to respond differentially to the variety of shapes and sizes of odoriferous molecules. According to the leading theory of odor, molecules of similar odor share a particular configuration in common, not a similar chemical composition. Accordingly, each type of receptor binding site is speculated to have a distinct shape and size (a lock) that matches the configuration of a particular primary odor (the key).

Whatever the mechanism for sorting out and distinguishing different odors, it is very effective, even in humans, who have a poor sense of smell compared to other species. A noteworthy example is our ability to detect methyl mercaptan (garlic odor) at a concentration of 1 molecule per 50,000 million molecules of air! This substance is added to odorless natural gas to enable us to detect potentially lethal gas leaks.

Even though the olfactory system is sensitive and highly discriminating, it is also quickly adaptive. Our sensitivity to a new odor rapidly diminishes after a short period of exposure to it, even though the odor source continues to be present. This reduced sensitivity does not involve receptor adaptation, as has been thought for years; actually, the olfactory receptors themselves adapt slowly. It apparently involves some sort of adaptation process in the CNS. Adaptation is specific for a particular odor, and responsiveness to other odors remains unchanged.

What clears the odoriferous molecules away from their binding sites on the olfactory receptors so that the sensation of smell doesn't "linger" after the source of the odor is removed? Several "odor-eating" enzymes have recently been discovered in the olfactory mucosa that could serve as molecular janitors, clearing away the odoriferous molecules so that they do not continue to stimulate the olfactory receptors. Interestingly, these odorant-clearing enzymes are very similar chemically to detoxification enzymes found in the liver (see p. 22). This resemblance may not be coincidental. Researchers speculate that these enzymes may serve the dual purpose of clearing the olfactory mucosa of old odorants and transforming potentially harmful chemicals into harmless molecules. Such detoxification would serve a very useful purpose, considering the extremely close proximity of the olfactory mucosa to the brain.

CHAPTER IN PERSPECTIVE: FOCUS ON HOMEOSTASIS

To maintain a life-sustaining stable internal environment, adjustments must constantly be made to compensate for the myriad external and internal factors that continuously threaten to disrupt homeostasis, such as external exposure to cold or internal acid production. Many of these adjustments are directed by the nervous system, one of the body's two major control systems. The central nervous system (CNS), the integrating and decision-making component of the nervous system, must continuously be informed of "what's happening" in both the internal and external environment so that it can command appropriate responses in the organ systems to maintain the body's viability. In other words, the CNS must know what changes are taking place before it can respond to these changes.

The afferent division of the peripheral nervous system is the communication link by which the CNS is informed about the internal and external environment. The afferent division detects, encodes, and transmits peripheral signals to the CNS for processing. Afferent input is necessary for arousal, perception, and determination of efferent output.

Afferent information about the internal environment, such as the CO_2 level in the blood, never reaches the level of conscious awareness, but this input to the controlling centers of the CNS is essential for the maintenance of homeostasis. Afferent input reaching the level of conscious awareness is known as sensory information and includes somesthetic and proprioceptive sensation (body sense) and special senses (vision, hearing, taste, and smell).

The body-sense receptors are distributed over the entire body surface as well as throughout the joints and muscles. Afferent signals from these receptors provide information about what's happening directly to each specific body part in relation to the external environment (that is, the "what," "where," and "how much" of stimulatory inputs to the body's surface and the momentary position of the body in space). In contrast, each special-sense organ is restricted to a single site in the body. Rather than providing information about a specific body part, a special-sense organ provides a specific type of information about the external environment that is useful to the body as a whole. For example, the eyes and visual processing system, through their ability to detect, extensively analyze, and integrate patterns of illumination in the external environment, enable us to "see" our surroundings. The same integrative effect could not be achieved if photoreceptors were scattered over the entire body surface as are the touch receptors.

Sensory input (both body sense and special senses) enables a complex multicellular organism such as a human to interact in meaningful ways with the external environment in the procurement of food, defense against danger, and other behavioral actions geared toward maintaining homeostasis. In addition to providing information essential for interactions with the external environment for basic survival, the perceptual processing of sensory input adds immeasurably to the richness of life, such as enjoyment of a good book, concert, or meal.

 CHAPTER SUMMARY

Receptor Physiology

Receptors are specialized peripheral endings of afferent neurons; they respond to particular stimuli, translating the energy forms of the stimuli into electrical signals, the language of the nervous system. There are discrete labeled-line pathways from the receptors to the CNS so that information about the type and location of the stimuli can be deciphered by the CNS, even though all the information arrives in the form of action potentials. What the brain perceives from its input, however, is an abstraction and not reality. The only stimuli that can be detected are those for which receptors are present. Furthermore, as sensory signals ascend through progressively more complex processing, some of the information may be suppressed, whereas other parts of it may be enhanced.

Stimulation of a receptor produces a graded receptor potential. The strength and rate of change of the stimulus are reflected in the magnitude of the receptor potential, which in turn determines the frequency of action potentials generated in the afferent neuron. The magnitude of the receptor potential is also influenced by the extent of receptor adaptation, which refers to a reduction in receptor potential in spite of sustained stimulation. Tonic receptors adapt slowly or not at all and thus provide continuous information about the stimuli they monitor. Phasic receptors adapt rapidly and frequently exhibit off responses, thereby providing information about changes in the energy form they monitor.

Pain

Painful experiences are elicited by noxious mechanical, thermal, or chemical stimuli and consist of two components: the perception of pain coupled with emotional and behavioral responses to it. Pain signals are transmitted over two afferent pathways: a fast pathway that carries sharp, prickling pain signals, and a slow pathway that carries dull, aching, persistent pain signals. Afferent pain fibers terminate in the spinal cord on ascending pathways that transmit the signals to the brain for processing. Descending pathways from the brain use endogenous opiates to suppress the release of substance P, the neurotransmitter from the afferent pain fiber terminal. Thus, these descending pathways block further transmission of the pain signal and serve as a built-in analgesic system.

Eye: Vision

The eye is a specialized structure housing the light-sensitive receptors essential for vision perception—namely, the rods and cones found in its retinal layer. The iris controls the size of the pupil, thereby adjusting the amount of light permitted to enter the eye. The cornea and lens are the primary refractive structures that bend the incoming light rays to focus the image on the retina. The cornea contributes most to the total refractive ability of the eye. The strength of the lens can be adjusted through action of the ciliary muscle to accommodate for differences in near and far vision.

Rods and cones are activated when the photopigments they contain differentially absorb various wavelengths of light. Light absorption causes a biochemical change in the photopigment that is ultimately converted into a change in the rate of action potential propagation in the visual pathway leaving the retina. The visual message is transmitted to the visual cortex in the brain for perceptual processing.

Cones display high acuity but can be used only for day vision because of their low sensitivity to light. Different ratios of stimulation of three cone types by varying wavelengths of light lead to color vision. Rods provide only indistinct vision in shades of gray, but because they are very sensitive to light, they can be used for night vision.

Ear: Hearing and Equilibrium

The ear performs two unrelated functions: (1) hearing, which involves the external ear, middle ear, and cochlea of the inner ear; and (2) sense of equilibrium, which involves the vestibular apparatus of the inner ear. In contrast to the photoreceptors of the eye, the ear receptors located in the inner ear—the hair cells in the cochlea and vestibular apparatus—are mechanoreceptors. Hearing depends on the ear's ability to convert airborne sound waves into mechanical deformations of receptive hair cells, thereby initiating neural signals. Sound waves consist of high-pressure regions of compression alternating with low-pressure regions of rarefaction of air molecules. The pitch (tone) of a sound is determined by the frequency of its waves and the loudness (intensity) by the amplitude of the waves. Sound waves are funneled through the external ear canal to the tympanic membrane, which vibrates in synchrony with the waves. Middle ear bones bridging the gap between the tympanic membrane and the inner ear amplify the tympanic movements and transmit them to the oval window, whose movement sets up traveling waves in the cochlear fluid. These waves, which are at the same frequency as the original sound waves, set the basilar membrane in motion. Various regions of this membrane selectively vibrate more vigorously in response to different frequencies of sound. On top of the basilar membrane are the receptive hair cells of the organ of Corti, whose hairs are bent as the basilar membrane is deflected up and down in relation to the overhanging stationary tectorial membrane in which the hairs are embedded. This mechanical deformation of specific hair cells in the region of maximal basilar-membrane vibration is transduced into neural signals that are transmitted to the auditory cortex in the brain for sound perception.

The vestibular apparatus in the inner ear consists of (1) the semicircular canals, which detect rotational acceleration or deceleration in any direction, and (2) the utricle and saccule, which detect changes in the rate of linear movement in any direction and provide information important for determining head position in relation to gravity. Neural signals are generated in response to mechanical deformation of hair cells caused by specific movement of fluid and related structures within these sense organs. This information is important for the sense of equilibrium and for maintaining posture.

Chemical Senses: Taste and Smell

Taste and smell are chemical senses. In both cases, attachment of specific dissolved molecules to binding sites on the receptor membrane causes receptor potentials that, in turn, set up neural impulses that signal the presence of the chemical. Taste receptors are housed in taste buds on the tongue; olfactory receptors are located in the mucosa in the upper part of the nasal cavity. Both sensory pathways include two routes: one to the limbic system for emotional and behavioral processing and one through the thalamus to the cortex for conscious perception and fine discrimination. Taste and olfactory receptors are continuously renewed, unlike visual and hearing receptors, which are irreplaceable.

REVIEW EXERCISES

Objective Questions (Answers on p. D–2.)

1. Conversion of the energy forms of stimuli into electrical energy by the receptors is known as _____.

2. The type of stimulus to which a particular receptor is most responsive is called its _____.

3. The sensation perceived depends on the type of receptor stimulated rather than on the type of stimulus. (True or false?)

4. All afferent information is sensory information. (True or false?)

5. Off-center ganglion cells increase their rate of firing when a beam of light strikes the periphery of their receptive field. (True or false?)

6. During dark adaptation, rhodopsin is gradually regenerated to increase the sensitivity of the eyes. (True or false?)

7. An optic nerve carries information from the lateral and medial halves of the same eye, whereas an optic tract carries information from the lateral half of one eye and the medial half of the other. (True or false?)

8. Displacement of the round window generates neural impulses that are perceived as sound sensations. (True or false?)

9. Hair cells in different regions of the organ of Corti and neurons in different regions of the auditory cortex are activated by different tones. (True or false?)

10. Each taste receptor responds to just one of the four primary tastes. (True or false?)

11. Rapid adaptation to odors results from adaptation of the olfactory receptors. (True or false?)

12. Using the answer code at the right, indicate which properties apply to taste and/or smell:
    ___ 1. Receptors are separate cells that synapse with terminal endings of afferent neurons.
    ___ 2. Receptors are specialized endings of afferent neurons.
    ___ 3. Receptors are regularly replaced.
    ___ 4. Specific chemicals in the environment attach to special binding sites on receptor surface, leading to a depolarizing receptor potential.
    ___ 5. Has two processing pathways: a limbic system route and a thalamic-cortical route.

    a. applies to taste
    b. applies to smell
    c. applies to both taste and smell

13. Match the following:

____ 1. layer that contains the photoreceptors

____ 2. point from which the optic nerve leaves the retina

____ 3. forms the white part of the eye

____ 4. thalamic structure that processes visual input

____ 5. colored diaphragm of muscle that controls amount of light entering the eye

____ 6. contributes most to refractive ability of the eye

____ 7. supplies nutrients to the lens and cornea

____ 8. produces aqueous humor

____ 9. contains vascular supply for the retina and pigment that minimizes scattering of light within the eye

____ 10. has adjustable refractive ability

____ 11. portion of retina with greatest acuity

____ 12. point at which fibers from the medial half of each retina cross to the opposite side

a. choroid
b. aqueous humor
c. fovea
d. lateral geniculate nucleus
e. cornea
f. retina
g. lens
h. optic disc; blind spot
i. iris
j. ciliary body
k. optic chiasm
l. sclera

Essay Questions

1. List and describe the receptor types according to their adequate stimulus.

2. Compare tonic and phasic receptors.

3. Explain how acuity is influenced by receptive field size and by lateral inhibition.

4. Compare the fast and slow pain pathways.

5. Describe the built-in analgesic system of the brain.

6. Describe the process of phototransduction.

7. Compare the functional characteristics of rods and cones.

8. What are sound waves? What is responsible for the pitch, intensity, and timbre of a sound?

9. Describe the function of each of the following parts of the ear: pinna, ear canal, tympanic membrane, ossicles, oval window, and the various parts of the cochlea. Include a discussion of how sound waves are transduced into action potentials.

10. Discuss the functions of the semicircular canals, the utricle, and the saccule.

11. Describe the location, structure, and activation of the receptors for taste and smell.

POINTS TO PONDER

1. Patients with certain nerve disorders are unable to feel pain. Why is this disadvantageous?

2. Ophthamologists frequently instill eye drops in their patients' eyes to bring about pupillary dilation, which makes it easier for the physician to view the eye's interior. In what way would the drug in the eye drops affect autonomic-nervous-system activity in the eye in order to cause the pupils to dilate?

3. A patient complains of not being able to see the right half of the visual field with either eye. At what point in the patient's visual pathway is the defect?

4. Explain how middle ear infections interfere with hearing. Of what value are the "tubes" that are sometimes surgically placed in the eardrums of patients with a history of repeated middle ear infections accompanied by chronic fluid accumulation?

5. Explain why your sense of smell is reduced when you have a cold, even though the cold virus does not directly adversely affect the olfactory receptor cells.

Peripheral Nervous System: Efferent Division

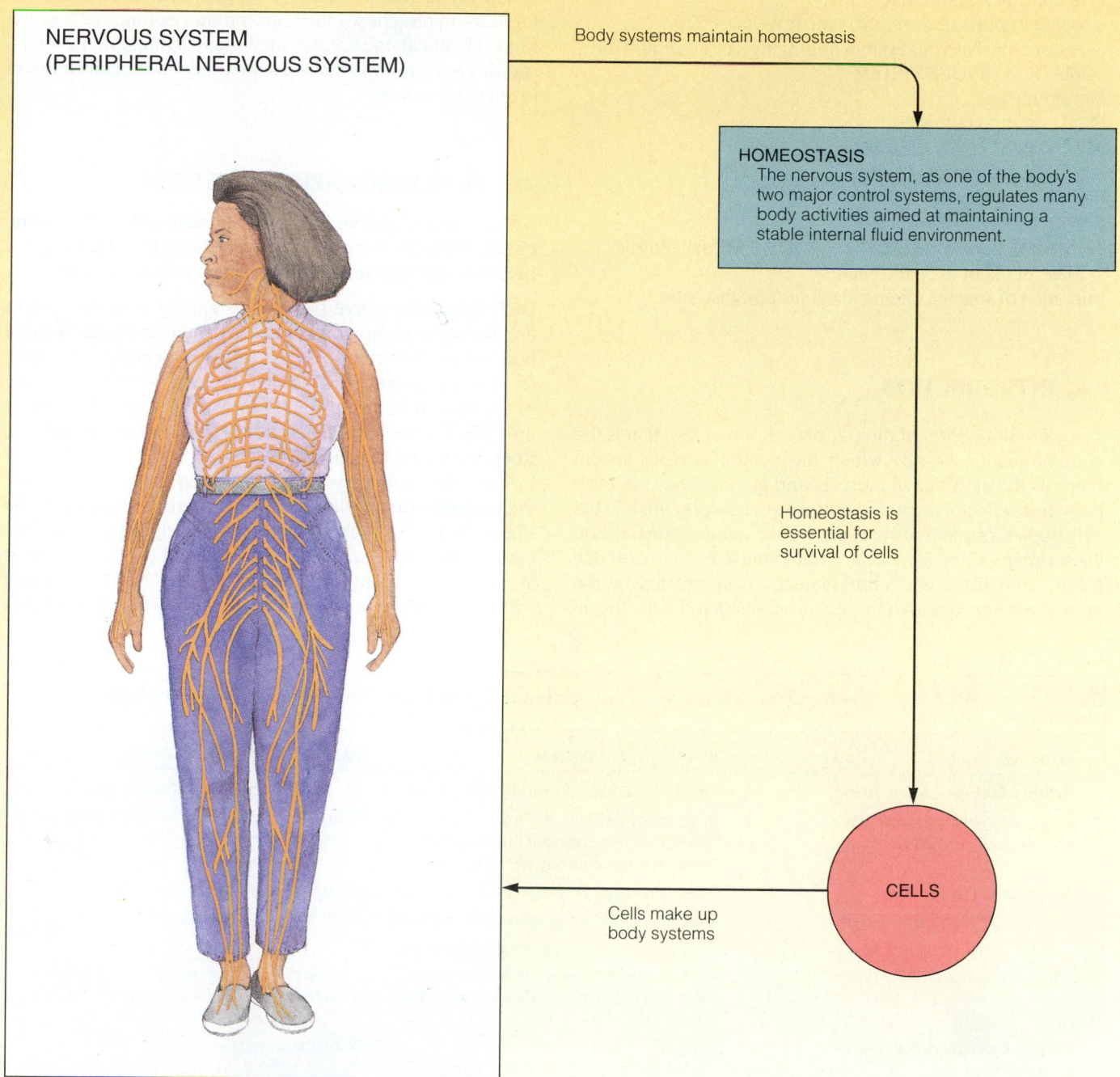

NERVOUS SYSTEM
(PERIPHERAL NERVOUS SYSTEM)

Body systems maintain homeostasis

HOMEOSTASIS
The nervous system, as one of the body's two major control systems, regulates many body activities aimed at maintaining a stable internal fluid environment.

Homeostasis is essential for survival of cells

CELLS

Cells make up body systems

The nervous system, one of the two major control systems of the body, consists of the central nervous system (CNS), composed of the brain and spinal cord, and the **peripheral nervous system**, composed of the afferent and efferent fibers that relay signals between the CNS and periphery (other parts of the body).

The CNS controls effector organs (muscles and glands) by transmitting signals *from* the CNS to these organs through the **efferent division** of the peripheral nervous system. Much of this efferent output is directed toward maintaining homeostasis.

CHAPTER CONTENTS AT A GLANCE

INTRODUCTION
AUTONOMIC NERVOUS SYSTEM
Anatomy and neurotransmitters of autonomic fibers
Dominance patterns of sympathetic and parasympathetic systems
Autonomic receptor types
CNS control of autonomic activity
● CONCEPTS, CHALLENGES, AND CONTROVERSIES The Autonomic Nervous System and Aging: A Fortuitous Find
SOMATIC NERVOUS SYSTEM
Motor neurons
Final common pathway
NEUROMUSCULAR JUNCTION
Anatomy; sequence of events
End-plate potentials
Role of acetylcholinesterase
● A CLOSER LOOK AT EXERCISE PHYSIOLOGY Loss of Muscle Mass: A Plight of Space Flight
Influence of specific chemical agents and diseases

INTRODUCTION

The efferent division of the peripheral nervous system is the communication link by which the central nervous system controls the activities of muscles and glands. The CNS regulates these effector organs by initiating action potentials in the cell bodies of efferent neurons whose axons terminate on these organs. Cardiac muscle, smooth muscle, most exocrine glands, and some endocrine glands are innervated by the **autonomic nervous system**, which is considered to be the in-

voluntary branch of the peripheral efferent division. Skeletal muscle is innervated by the **somatic nervous system**, which is the voluntary branch of the efferent division.

How many different neurotransmitters would you guess are released from these efferent neuronal terminals to elicit essentially all the neurally controlled effector organ responses? Only two—acetylcholine and norepinephrine! Acting independently, they bring about such wide-ranging effects as salivary secretion, bladder contraction, and voluntary motor movements. These effects are a prime example of how the same chemical messenger may elicit a multiplicity of responses from various tissues, depending on specialization of the effector organs.

AUTONOMIC NERVOUS SYSTEM

An autonomic nerve pathway consists of a two-neuron chain, with the terminal neurotransmitter differing between sympathetic and parasympathetic nerves.

Each autonomic nerve pathway extending from the CNS to an innervated organ consists of a two-neuron chain. The cell body of the first neuron in the series is located in the CNS. Its axon, the **preganglionic fiber**, synapses with the cell body of the second neuron, which lies within a ganglion outside the CNS. The axon of the second neuron, the **postganglionic fiber**, innervates the effector organ.

The autonomic nervous system consists of two divisions—the **sympathetic** and the **parasympathetic nervous systems** (Table 7–1). Sympathetic nerve fibers originate in the thoracic and lumbar regions of the spinal cord (● Fig. 7–1). Most sympathetic preganglionic fibers are very short, synapsing with cell bodies of postganglionic neurons within ganglia that

TABLE 7–1   Distinguishing Features of the Sympathetic and Parasympathetic Nervous Systems

FEATURE	SYMPATHETIC SYSTEM	PARASYMPATHETIC SYSTEM
Origin of preganglionic fiber	Thoracic and lumbar regions of spinal cord	Brain and sacral region of spinal cord
Origin of postganglionic fiber (location of ganglion)	Sympathetic ganglion chain (near spinal cord) or collateral ganglia (about halfway between spinal cord and effector organs)	Terminal ganglia (in or near effector organs)
Length and type of fiber	Short cholinergic preganglionic fibers	Long cholinergic preganglionic fibers
	Long adrenergic postganglionic fibers (most)	Short cholinergic postganglionic fibers
	Long cholinergic postganglionic fibers (few)	
Effector organs innervated	Cardiac muscle, almost all smooth muscle, most exocrine glands, and some endocrine glands	Cardiac muscle, most smooth muscle, most exocrine glands, and some endocrine glands
Types of receptors for neurotransmitters	α, β_1, β_2	Nicotinic, muscarinic
Dominance	Dominates in emergency "fight-or-flight" situations; prepares body for strenuous physical activity	Dominates in quiet, relaxed situations; promotes "general housekeeping" activities such as digestion
Types of discharge	Frequently mass discharge of whole system; may involve only discrete organs	Normally involves discrete organs rather than mass discharge

——— = Sympathetic system	≡≡≡ = Preganglionic fiber	**ACh** = Acetylcholine	O = Cell body
——— = Parasympathetic system	≡≡≡ = Postganglionic fiber	**NE** = Norepinephrine	——< = Axon terminal
		E = Epinephrine	

• **FIGURE 7–1 Autonomic Nervous System** The sympathetic nervous system, which originates in the thoracolumbar regions of the spinal cord, has short cholinergic (acetylcholine-releasing) preganglionic fibers and long adrenergic (norepinephrine-releasing) postganglionic fibers. The parasympathetic nervous system, which originates in the brain and sacral region of the spinal cord, has long cholinergic preganglionic fibers and short cholinergic postganglionic fibers. In most instances, sympathetic and parasympathetic postganglionic fibers both innervate the same effector organs. The adrenal medulla is a modified sympathetic ganglion, which releases epinephrine and norepinephrine into the blood.

lie in a **sympathetic ganglion chain** (the **sympathetic trunk**) located along either side of the spinal cord. Long postganglionic fibers originating in the ganglion chain terminate on the effector organs. Some preganglionic fibers pass through the ganglion chain without synapsing and terminate later in sympathetic **collateral ganglia** located about halfway between the CNS and the innervated organs, with postganglionic fibers traveling the remainder of the distance.

Parasympathetic preganglionic fibers arise from the cranial and sacral areas of the CNS. (Some cranial nerves contain parasympathetic fibers.) These fibers are long in com-

parison to sympathetic preganglionic fibers because they do not end until they reach **terminal ganglia** that lie in or near the effector organs. Very short postganglionic fibers terminate on the cells of an organ itself.

Sympathetic and parasympathetic preganglionic fibers release the same neurotransmitter, **acetylcholine (ACh)**, but the postganglionic endings of these two systems release different neurotransmitters (the neurotransmitters that influence the effector organs). Parasympathetic postganglionic fibers release acetylcholine. Accordingly, they, along with all autonomic preganglionic fibers, are called **cholinergic fibers**.

TABLE 7–2 Sites of Release for Acetylcholine and Norepinephrine	
ACETYLCHOLINE	**NOREPINEPHRINE**
All preganglionic terminals of the autonomic nervous system	Most sympathetic postganglionic terminals
All parasympathetic postganglionic terminals	Adrenal medulla
Sympathetic postganglionic terminals at sweat glands and some blood vessels in skeletal muscle	Central nervous system
Terminals of efferent neurons supplying skeletal muscle (motor neurons)	
Central nervous system	

Most sympathetic postganglionic fibers, in contrast, are called **adrenergic fibers** because they release **noradrenaline**, commonly known as **norepinephrine**. Both acetylcholine and norepinephrine also serve as chemical messengers elsewhere in the body (Table 7–2).

Postganglionic autonomic fibers do not end in a single terminal swelling like a synaptic knob. Instead, the terminal branches of autonomic fibers contain numerous swellings, or **varicosities**, that simultaneously release neurotransmitter over a large area of the innervated organ rather than on single cells. This diffuse release of neurotransmitter, coupled with the fact that any resulting change in electrical activity is spread throughout a smooth- or cardiac-muscle mass via gap junctions (see p. 55), means that whole organs instead of discrete cells are typically influenced by autonomic activity.

The autonomic nervous system controls involuntary visceral-organ activities.

The autonomic nervous system regulates visceral activities normally outside the realm of consciousness and voluntary control, such as circulation, digestion, sweating, and pupillary size. Accordingly, it is considered to be the *involuntary* branch of the efferent division, in contrast to the somatic *voluntary* branch, which supplies skeletal muscles that can be consciously controlled. It is not entirely true, however, that an individual has no control over activities governed by the autonomic system. Visceral afferent information usually does not reach the conscious level, so individuals have no way of consciously controlling the resultant efferent output. With the technique of **biofeedback**, however, people are provided with a conscious signal regarding visceral afferent information, such as a sound, a light, or a graphic display on a computer screen. This signal enables them to exert some voluntary control over events that are normally considered to be subconscious activities. For example, individuals have learned to consciously lower their blood pressure when they "hear" that it is elevated via special devices that convert blood-pressure levels into sound signals. Such biofeedback techniques are gaining wider acceptance and usage.

The sympathetic and parasympathetic nervous systems dually innervate most visceral organs.

Most visceral organs are innervated by both sympathetic and parasympathetic nerve fibers. Table 7–3 summarizes the major effects of these autonomic branches. Although the details of this wide array of autonomic responses are more fully described in later chapters that discuss the individual organs involved, several general concepts can be derived now. As can be seen from the table, the sympathetic and parasympathetic nervous systems generally exert opposite effects in a particular organ. Sympathetic stimulation increases the heart rate, whereas parasympathetic stimulation decreases it; sympathetic stimulation slows down movement within the digestive tract, whereas parasympathetic stimulation enhances digestive motility. Note that one system is not always excitatory and the other always inhibitory. Both systems increase the activity of some organs and reduce the activity of others.

Rather than memorizing a list such as that presented in the table, it is better to logically deduce the actions of the two systems based on an understanding of the circumstances under which each system dominates. Usually, both systems are partially active; that is, normally some level of action-potential activity exists in both the sympathetic and the parasympathetic fibers supplying a particular organ. This ongoing activity is called **sympathetic** or **parasympathetic tone** or **tonic activity**. Under given circumstances, activity of one division can dominate the other. *Sympathetic dominance* to a particular organ exists when the sympathetic fibers' rate of firing to that organ increases above tonic level, coupled with a simultaneous decrease below tonic level in the parasympathetic fibers' frequency of action potentials to the same organ. The reverse situation is true for *parasympathetic dominance*. Shifts in balance between sympathetic and parasympathetic activity can be accomplished discretely for individual organs to meet specific demands, or a more generalized, widespread discharge of one system in favor of the other can be elicited to control bodywide functions. Massive widespread discharges take place more frequently in the sympathetic system. The value of this potential for massive sympathetic discharge is evident considering the circumstances during which this system usually dominates.

The sympathetic system promotes responses that prepare the body for strenuous physical activity in the face of emergency or stressful situations, such as a physical threat from the outside environment. This response is typically referred to as a **fight-or-flight response** because the sympathetic system readies the body to fight against or flee from the threat. Think about the body resources needed in such circumstances. The heart beats more rapidly and more forcefully; blood pressure is elevated because of generalized constriction of the blood vessels; the respiratory airways open wide to permit maximal air flow; glycogen (stored sugar) and fat stores are broken down to release extra fuel in the blood; and blood vessels

TABLE 7–3 Effects of the Autonomic Nervous System on Various Organs

ORGAN	TYPE OF SYMPATHETIC RECEPTOR	EFFECT OF SYMPATHETIC STIMULATION	EFFECT OF PARASYMPATHETIC STIMULATION
Heart	β_1	Increased rate, increased force of contraction (of whole heart)	Decreased rate, decreased force of contraction (of atria only)
Blood vessels	α (most organs)	Constriction	Dilation of vessels supplying the penis and clitoris only
	β_2 (heart and skeletal-muscle vessels)	Dilation	
	*Cholinergic (skeletal-muscle vessels)	Dilation	
Lungs	β_2 (airways)	Dilation of bronchioles (airways)	Constriction of bronchioles
	? (gland cells)	Inhibition (?) of mucus secretion	Stimulation of mucus secretion
Digestive tract	α, β_2 (organs)	Decreased motility (movement)	Increased motility
	α (sphincters)	Contraction of sphincters (to prevent forward movement of contents)	Relaxation of sphincters (to permit forward movement of contents)
	? (gland cells)	Inhibition (?) of digestive secretions	Stimulation of digestive secretions
Gallbladder	?	Relaxation	Contraction (emptying)
Urinary bladder	β_2	Relaxation	Contraction (emptying)
Eye	α (iris)	Dilation of pupil	Constriction of pupil
	β_2 (ciliary muscle)	Adjustment of eye for far vision	Adjustment of eye for near vision
Liver (glycogen stores)	β_2	Glycogenolysis (glucose released)	None
Adipose cells (fat stores)	β_2	Lipolysis (fatty acids released)	None
Exocrine glands			
Exocrine pancreas	α	Inhibition of pancreatic exocrine secretion	Stimulation of pancreatic exocrine secretion (important for digestion)
Sweat glands	α, cholinergic	Stimulation of secretion by sweat glands	None
Salivary glands	α	Stimulation of small volume of thick saliva rich in mucus	Stimulation of large volume of watery saliva rich in enzymes
Endocrine glands			
Adrenal medulla	Cholinergic	Stimulation of epinephrine and norepinephrine secretion	None
Endocrine pancreas	α	Inhibition of insulin secretion; stimulation of glucagon secretion	Stimulation of insulin and glucagon secretion
Genitals	α	Ejaculation and orgasmic contractions (males); orgasmic contractions (females)	Erection (caused by dilation of blood vessels in penis [male] and clitoris [female])
Brain activity	?	Increased alertness	None

*Uncertain as to whether cholinergic sympathetic receptors are present in skeletal-muscle vessels in humans, although they have been demonstrated in other species.

?Receptor type unknown.

supplying skeletal muscles dilate (open more widely). All of these responses are aimed at providing increased flow of oxygenated, nutrient-rich blood to the skeletal muscles in anticipation of strenuous physical activity. Furthermore, the pupils dilate and the eyes adjust for far vision, enabling the person to make a quick visual assessment of the entire threatening scene. Sweating is promoted in anticipation of excess heat production by the physical exertion. Because digestive and urinary activities are unessential in meeting the threat, the sympathetic system inhibits these activities.

The parasympathetic system, on the other hand, domi-nates in quiet, relaxed situations. Under such nonthreatening circumstances, the body can be concerned with its own "general housekeeping" activities, such as digestion and emptying of the urinary bladder. The parasympathetic system promotes these types of bodily functions while slowing down those activities that are enhanced by the sympathetic system. There is no need, for example, to have the heart beating rapidly and forcefully when the person is in a tranquil setting.

What is the advantage of dual innervation of organs with nerve fibers whose actions oppose each other? It enables precise control over an organ's activity, similar to having both an accelerator and a brake to control the speed of a car. If an animal suddenly starts across the road as you are driving, you could eventually stop if you simply took your foot off the accelerator. However, you can come to a more rapid, controlled stop by simultaneously applying the brake as you lift up on the accelerator. In a similar manner, a sympathetically accelerated heart rate could gradually be reduced to normal following a stressful situation by decreasing the rate of firing in the cardiac sympathetic nerve (letting up on the accelerator), but the heart rate can be reduced more rapidly by simultaneously increasing activity in the parasympathetic supply to the heart (applying the brake). Indeed, the two divisions of the autonomic nervous system are usually reciprocally controlled; increased activity in one division is accompanied by a corresponding decrease in the other.

Inhibition of the parasympathetic nervous system by *cocaine,* an illegal addictive drug, may be a major contributing factor to sudden death in cocaine overdose. If cocaine blocks the protective parasympathetic brakes, as it appears to do, the sympathetic nervous system could proceed unchecked in accelerating the heart beat. Sudden death results if the heart beat becomes too rapid and irregular to adequately pump blood.

There are several exceptions to the general rule of dual reciprocal innervation by the two branches of the autonomic nervous system; the most notable are the following:

■ *Innervated blood vessels* (most arterioles and veins are innervated, arteries and capillaries are not) receive only sympathetic nerve fibers. Regulation is accomplished by increasing or decreasing the firing rate above or below the tonic level in these sympathetic fibers. The only blood vessels to receive parasympathetic fibers are those supplying the penis and clitoris. The precise vascular control this dual innervation affords these organs is important in accomplishing erection.

■ *Sweat glands* are innervated only by sympathetic nerves. The postganglionic fibers of these nerves are unusual because they secrete acetylcholine rather than norepinephrine.

■ *Salivary glands* are innervated by both autonomic divisions, but unlike elsewhere, sympathetic and parasympathetic activity is not antagonistic. Both stimulate salivary secretion, but the saliva's volume and composition differ, depending on which autonomic branch is dominant.

The adrenal medulla, an endocrine gland, is a modified part of the sympathetic nervous system.

The **adrenal medulla**, the inner portion of an endocrine gland known as the *adrenal gland,* is considered to be a modified sympathetic ganglion that does not give rise to postganglionic fibers. Instead, it secretes hormones into the blood upon stimulation by the preganglionic fiber that originates in the CNS (Figs. 7–1 and ● 7–2). Not surprisingly, the hormones are identical or similar to postganglionic sympathetic neurotrans-

● = Acetylcholine   ● = Norepinephrine   ● = Epinephrine

● **FIGURE 7–2   Comparison of Release and Binding to Receptors of Epinephrine and Norepinephrine**   Norepinephrine is released both as a neurotransmitter from sympathetic postganglionic fibers and as a hormone from the adrenal medulla. Alpha (α) and beta₁ (β₁) receptors bind with both norepinephrine and epinephrine, whereas beta₂ (β₂) receptors bind primarily with epinephrine.

mitters. About 20% of the adrenal medullary hormone output is norepinephrine, and the remaining 80% is the closely related substance, **epinephrine (adrenaline)**. These hormones, in general, act to reinforce activity of the sympathetic nervous system.

There are several different types of membrane receptor proteins for each autonomic neurotransmitter.

Because each autonomic neurotransmitter and medullary hormone stimulate activity in some tissues but inhibit activity in others, the particular responses must depend on specializa-

TABLE 7-4 Location of Nicotinic and Muscarinic Cholinergic Receptors

TYPE OF RECEPTOR	SITE OF RECEPTOR	RESPOND TO ACETYLCHOLINE RELEASED FROM:
Nicotinic receptors	All autonomic ganglia	Sympathetic and parasympathetic preganglionic fibers
	Motor end plates of skeletal-muscle fibers	Motor neurons
	Some CNS cell bodies and dendrites	Some CNS presynaptic terminals
Muscarinic receptors	Effector cells (cardiac muscle, smooth muscle, glands)	Parasympathetic postganglionic fibers
	Some CNS cell bodies and dendrites	Some CNS presynaptic terminals

tion of the tissue cells rather than on properties of the chemicals themselves. Responsive tissue cells possess one or more of several different types of plasma-membrane receptor proteins for these chemical messengers. Binding of a neurotransmitter to a receptor induces the tissue-specific response by means of a second messenger system within the cell (see p. 50).

Two types of acetylcholine (cholinergic) receptors—nicotinic and muscarinic receptors—have been identified on the basis of their response to particular drugs (Table 7-4). **Nicotinic receptors** (activated by the tobacco plant derivative, nicotine) are found in all autonomic ganglia. They respond to acetylcholine released from both sympathetic and parasympathetic preganglionic fibers. **Muscarinic receptors** (activated by the mushroom poison, muscarine) are found on effector cell membranes (smooth muscle, cardiac muscle, and exocrine glands). They bind with acetylcholine released from parasympathetic postganglionic fibers.

There are two major classes of adrenergic receptors for norepinephrine and epinephrine based on the ability of various drugs to either initiate or prevent responses in the effector organ. These receptors are designated as **alpha** (α) and **beta** (β) **receptors**, with the latter further subclassified into β_1 and β_2 **receptors**. These various receptor types are distinctly distributed among the effector organs. Receptors of the β_2 type bind primarily with epinephrine, whereas β_1 and α receptors have about equal affinities for norepinephrine and epinephrine (Fig. 7-2). Activation of α receptors usually brings about an excitatory response in the effector organ—for example, arteriolar constriction caused by increased contraction of the smooth muscle in the walls of these blood vessels. Stimulation of β_1 receptors, which are found primarily in the heart, also causes an excitatory reponse, namely, increased rate and force of cardiac contraction. The response to β_2 receptor activation is generally inhibitory, such as arteriolar or bronchio-

lar (respiratory airway) dilation caused by relaxation of the smooth muscle in the walls of these tubular structures.

Because activation of various receptor types brings about different responses to the same autonomic messenger, these receptors can be manipulated fairly selectively by drugs. Drugs are available that selectively enhance or mimic (**agonists**) or block (**antagonists**) autonomic responses at each of the receptor types. Some are only of experimental interest, but others are very important therapeutically. For example, *atropine* blocks the effect of acetylcholine at muscarinic receptors but does not affect nicotinic receptors. Since the acetylcholine released at both parasympathetic and sympathetic preganglionic fibers combines with nicotinic receptors, blockage at nicotinic synapses would knock out both of these autonomic branches. By acting selectively to interfere with acetylcholine action only at muscarinic junctions, which are the sites of parasympathetic postganglionic action, atropine effectively blocks parasympathetic effects but does not influence sympathetic activity at all. This principle is used to suppress salivary and bronchial secretions before surgery to reduce the risk of a patient inhaling these secretions into the lungs.

Likewise, drugs that act selectively at α and β adrenergic receptor sites to either activate or block specific sympathetic effects are widely used. *Salbutamol* is an excellent example. It selectively activates β_2 adrenergic receptors at low doses, making it possible to dilate the bronchioles in the treatment of asthma without undesirably stimulating the heart (the heart has β_1 receptors). Other drugs that act selectively at α and β receptors are beneficial in manipulating blood pressure and heart rate in the treatment of hypertension and cardiac arrhythmias.

Many regions of the central nervous system are involved in the control of autonomic activities.

Messages from the CNS are delivered to cardiac muscle, smooth muscle, and glands via the autonomic nerves, but what regions of the CNS regulate autonomic output?

■ Some autonomic reflexes, such as urination, defecation, and erection, are integrated at the spinal cord level, but all of these spinal reflexes are subject to control by higher levels of consciousness.

■ The medulla within the brain stem is the region most directly responsible for autonomic output. Centers for controlling cardiovascular, respiratory, and digestive activity via the autonomic system are located there.

■ The hypothalamus plays an important role in integrating the autonomic, somatic, and endocrine responses that automatically accompany various emotional and behavioral states. For example, the increased heart rate, blood pressure, and respiratory activity associated with anger or fear are brought about by the hypothalamus acting through the medulla.

THE AUTONOMIC NERVOUS SYSTEM AND AGING:
A FORTUITOUS FIND

Often in scientific research, an investigator will unexpectedly come across an important finding while studying a totally different phenomenon. This happened with Robert Schmidt, a neuropathologist, who accidentally stumbled across the first evidence of the effects of aging on the autonomic nervous system while investigating a different problem—what might be responsible for the intestinal disorders that often accompany diabetes mellitus. Since nerve damage is known to be one of the long-term complications of diabetes, Schmidt suspected that damage to the autonomic nerves that help regulate the intestine might be the cause of the intestinal symptoms. Upon microscopic examination of the sympathetic ganglia supplying the intestine in diabetic rats, Schmidt was astonished to see that some of the preganglionic fibers (axons) near their terminal endings were swollen to more than thirty times their normal size. Thinking this dramatic pathology must somehow be linked to diabetes, he examined the same ganglia in nondiabetic control rats for comparison, fully expecting to find normal preterminal axons. To his surprise, some of the axon endings in these normal rats were as swollen as those in the diabetic rats.

Puzzling over his results, Schmidt finally thought of one thing that the two groups had in common. They were both old. Most studies on rats involve young specimens, but Schmidt's animals were old because he was investigating a long-term effect of diabetes.

In 1990, Schmidt published results in the *American Journal of Pathology* demonstrating comparable evidence of this autonomic aging in humans. In autopsy examinations of sympathetic ganglia in 56 people who had died between the ages of 15 and 93 from various causes, Schmidt found very few enlarged preterminal axons in persons younger than 60, whereas the number of swollen axons increased dramatically in those over 60. In fact, abnormal preterminal axons were significantly more frequent in older humans than in aging rats.

Further study revealed that the abnormal sympathetic axons contain an overabundance of disoriented neurofilaments, cytoskeletal components that normally serve as internal scaffolding to help maintain a nerve cell's shape (see p. 41). Neurofilaments are normally degraded in the axon terminals for recycling, but this process is apparently impaired in connection with the age-related autonomic damage.

Not all of the preganglionic fibers in a sympathetic ganglion are swollen and packed with neurofilaments. Only terminals that contain a recently discovered neuromodulator (see p. 98), **neuropeptide Y (NPY)**, are affected. The role that NPY plays in sympathetic ganglia is unknown, but other investigators have found evidence that NPY is a cotransmitter in sympathetic control of vascular (blood-vessel) smooth muscle.

The autonomic nervous system helps control many body functions such as regulation of blood pressure and movement of food through the digestive tract via its influence on smooth muscle. Aging is often accompanied by autonomic malfunctions, such as becoming lightheaded when standing up too quickly because of inadequate vascular control or constipation due to sluggish bowel motility. These age-related autonomic dysfunctions are assuming greater significance as life expectancy and consequently the number of elderly are increasing.

Normal function of the smooth muscle of the intestine and blood vessels depends on complex, incompletely understood control mechanisms involving the interaction of multiple neural elements and hormones. One can speculate that an abnormality in NPY neuromodulation of these activities might contribute to age-related autonomic malfunctions, although this remains to be confirmed. According to Schmidt, "We predict that when the function of NPY-containing processes in human prevertebral (sympathetic) ganglia is understood, that function will be compromised in the elderly."

■ Autonomic activity can also be influenced by the frontal cortex through its involvement with emotional expression characteristic of the individual's personality. An example is blushing when embarrassed, which is caused by dilation of blood vessels supplying the skin of the cheeks. Such responses are mediated through hypothalamic-medullary pathways.

A wide variety of autonomic malfunctions accompany aging. Preliminary studies are providing clues to the age-related decline in autonomic control. (See the accompanying boxed feature, Concepts, Challenges, and Controversies.)

SOMATIC NERVOUS SYSTEM

Motor neurons supply skeletal muscle.

Skeletal muscle is innervated by **motor neurons**, the axons of which constitute the **somatic nervous system**. The cell bodies of these motor neurons are located within the ventral horn of the spinal cord. Unlike the two-neuron chain of autonomic nerve fibers, the axon of a motor neuron is continuous from its origin in the spinal cord to its termination on skeletal muscle. Motor-neuron axon terminals release acetylcholine, which brings about excitation and contraction of the innervated muscle fibers. Motor neurons can only stimulate skeletal muscles, in contrast to autonomic fibers, which can either stimulate or inhibit their effector organs. Inhibition of skeletal-muscle activity can be accomplished only within the CNS through activation of inhibitory synaptic input to the cell bodies and dendrites of the motor neurons supplying that particular muscle.

Motor neurons are the final common pathway.

Motor neurons are influenced by many converging presynaptic inputs, both excitatory and inhibitory. Some of these inputs are part of spinal-reflex pathways originating with peripheral sensory receptors. Others are part of descending pathways originating within the brain. Areas of the brain that exert control over skeletal-muscle movements include the motor regions of the cortex, the basal nuclei, the cerebellum, and the brain stem.

Motor neurons are considered to be the **final common pathway** since the only way any other parts of the nervous system can influence skeletal-muscle activity is by acting on these motor neurons. The level of activity in a motor neuron and its subsequent output to the skeletal-muscle fibers it innervates depend on the relative balance of EPSPs and IPSPs brought about by its presynaptic inputs originating from these diverse sites in the brain.

The somatic system is considered to be under voluntary control, but much of the skeletal-muscle activity involving posture, balance, and stereotypical movements is subconsciously controlled. You may decide you want to start walking, but you do not have to consciously bring about alternate contraction and relaxation of the involved muscles because these movements are involuntarily coordinated by lower brain centers.

The cell bodies of the crucial motor neurons may be selectively destroyed by *polio* virus. The result is paralysis of the muscles innervated by the affected neurons.

Table 7–5 summarizes the features of the two divisions of the efferent nervous system discussed in this chapter. Table 7–6 compares the three types of neurons that have been examined in the last three chapters.

NEUROMUSCULAR JUNCTION

Acetylcholine chemically links electrical activity in motor neurons with electrical activity in skeletal-muscle cells.

An action potential in a motor neuron is rapidly propagated from the CNS to the skeletal muscle along the large myelinated fiber (axon) of the neuron. As the axon approaches a muscle, it divides into many terminal branches and loses its myelin sheath. Each of these axon terminals forms a special junction, a **neuromuscular junction**, with one of the many muscle cells that compose the whole muscle (● Fig. 7–3). A single muscle cell, referred to as a **muscle fiber**, is long and cylindrical in shape. The axon terminal is enlarged into

TABLE 7–5 Comparison of the Autonomic Nervous System and the Somatic Nervous System		
FEATURE	**AUTONOMIC NERVOUS SYSTEM**	**SOMATIC NERVOUS SYSTEM**
Site of origin	Brain or lateral horn of spinal cord	Ventral horn of spinal cord
Number of neurons from origin in CNS to effector organ	Two-neuron chain (preganglionic and post-ganglionic)	Single neuron (motor neuron)
Organs innervated	Cardiac muscle, smooth muscle, exocrine and some endocrine glands	Skeletal muscle
Type of innervation	Most effector organs dually innervated by the two antagonistic branches of this system (sympathetic and parasympathetic)	Effector organs innervated only by motor neurons
Neurotransmitter at effector organs	May be acetylcholine (parasympathetic terminals) or norepinephrine (sympathetic terminals)	Only acetylcholine
Effects on effector organs	Either stimulation or inhibition (antagonistic actions of two branches)	Stimulation only (inhibition possible only centrally through IPSPs on cell body of motor neuron)
Types of control	Under involuntary control; may be voluntarily controlled with biofeedback techniques and training	Subject to voluntary control; much activity subconsciously coordinated
Higher brain centers involved in control	Spinal cord, medulla, hypothalamus, frontal cortex	Spinal cord, motor cortex, basal nuclei, cerebellum, brain stem

TABLE 7–6 Comparison of Types of Neurons

FEATURE	AFFERENT NEURON	EFFERENT NEURON Autonomic Nervous System	EFFERENT NEURON Somatic Nervous System	INTERNEURON
Origin, structure, location	Receptor at peripheral ending; elongated peripheral axon, which travels in peripheral nerve; cell body located in dorsal root ganglion; short central axon entering spinal cord	Two-neuron chain; first neuron (preganglionic fiber) originating in CNS and terminating on a ganglion; second neuron (postganglionic fiber) originating in the ganglion	Cell body of motor neuron lying in spinal cord; long axon traveling in peripheral nerve	Various shapes; lying entirely within CNS; some cell bodies originating in brain, with long axons traveling down the spinal cord in descending pathways; some originating in spinal cord, with long axons traveling up the cord to the brain in ascending pathways; others forming short local connections
Termination	Interneurons*	Effector organs (cardiac muscle, smooth muscle, exocrine and some endocrine glands)	Effector organs (skeletal muscle)	Other interneurons and efferent neurons
Function	Carries information about the external and internal environment to CNS	Carries instructions from CNS to effector organs	Carries instructions from CNS to effector organs	Processes and integrates afferent input; initiates and coordinates efferent output; responsible for thought and other higher mental functions
Convergence of input on cell body	No (only input is through receptor)	Yes	Yes	Yes
Effect of input on neuron	Can only be excited (through receptor potential induced by stimulus; must reach threshold for action potential)	Can be excited or inhibited (through EPSPs and IPSPs at first neuron; must reach threshold for action potential)	Can be excited or inhibited (through EPSPs and IPSPs; must reach threshold for action potential)	Can be excited or inhibited (through EPSPs and IPSPs; must reach threshold for action potential)
Site of action potential initiation	First excitable portion of membrane adjacent to receptor	Axon hillock	Axon hillock	Axon hillock
Divergence of output	Yes	Yes	Yes	Yes
Effect of output on structure on which it terminates	Only excites	Postganglionic fiber either excites or inhibits	Only excites	Either excites or inhibits

*Afferent neuron terminates directly on the alpha motor neuron in the case of the monosynaptic stretch reflex (see p. 145).

a knoblike structure, the **terminal button**, which fits into a shallow depression or groove in the underlying muscle fiber (● Fig. 7–4). The specialized portion of the muscle cell membrane immediately under the terminal button is known as the **motor end plate**.

Nerve and muscle cells do not actually come into direct contact at a neuromuscular junction. The space, or cleft, between these two structures is too large to permit electrical transmission of an impulse between them (that is, the action potential cannot "jump" that far). Therefore, just as at a neuronal synapse (see p. 91), a chemical messenger is used to carry the signal between the neuron terminal and the muscle

fiber. Each terminal button contains thousands of vesicles that store the chemical transmitter acetylcholine (ACh). Propagation of an action potential to the axon terminal (Step 1, Fig. 7–4) triggers the opening of voltage-gated Ca^{++} channels (see p. 81) in the terminal button. Opening of Ca^{++} channels permits Ca^{++} to diffuse into the terminal button from its higher extracellular concentration (Step 2), which in turn causes the release of ACh from several hundred of the vesicles into the cleft (Step 3).

The released ACh diffuses across the cleft and binds with specific receptor sites, which are specialized membrane proteins unique to the motor end-plate portion of the muscle-fi-

• **FIGURE 7–3   Motor Neuron Innervating Skeletal-Muscle Cells** When a motor neuron reaches a skeletal muscle, it divides into many terminal branches, each of which forms a neuromuscular junction with a single muscle cell.

motor end plate. This permits a greater net influx of positive ions and a larger depolarization. As with an EPSP, an EPP is a graded potential, whose magnitude depends on the amount and duration of ACh at the end plate.

The motor end-plate region itself does not have a threshold potential, so an action potential cannot be initiated at this site. However, an EPP brings about an action potential in the rest of the muscle fiber, as follows. The neuromuscular junction is usually located in the middle of the long cylindrical muscle fiber. When an EPP takes place at the motor end plate, local current flow occurs between the depolarized end plate and the adjacent, resting cell membrane in both directions, reducing the potential to threshold in the adjacent areas (Step 6). The subsequent action potential initiated at these sites is propagated throughout the muscle-fiber membrane by conduction by local current flow (see p. 84). The spread occurs in both directions, away from the motor end plate toward both ends of the fiber. This electrical activity triggers contraction of the muscle fiber. Thus, by means of ACh, an action potential in a motor neuron brings about an action potential and subsequent contraction in the muscle fiber. (See the accompanying boxed feature, A Closer Look at Exercise Physiology.)

Unlike synaptic transmission, the magnitude of an EPP is normally sufficient to cause an action potential in the muscle cell. Typically, therefore, one-to-one transmission of an action potential occurs at a neuromuscular junction; one action potential in a nerve cell triggers one action potential in a muscle cell that it innervates. Other comparisons of neuromuscular junctions with synapses can be found in Table 7–7.

Acetylcholinesterase terminates acetylcholine activity at the neuromuscular junction.

If ACh continued to remain in contact with the motor end plate, the channels would remain open and the EPP would continue to exist. Because this persistent end-plate depolarization would continue to initiate action potentials in the remainder of the muscle cell membrane, the muscle fiber would remain contracted until fatigued, even in the absence of further action potentials in the motor neuron. This situation would not be desirable, because there could be no controlled, purposeful alterations in movement. To control muscle contraction, electrical activity in the muscle fiber must be switched off promptly when there is no longer a signal from its motor neuron. The muscle's electrical response is turned off by an enzyme present in the motor end-plate membrane, **acetylcholinesterase** (AChE), which inactivates ACh.

Acetylcholine binds very briefly (for about 1 billionth of a second) with a receptor site, then detaches. Some of the ACh molecules quickly rebind with receptor sites, keeping the channels open, but some diffuse deeper into the folds of the motor end plate, where AChE is located (Step 7). As this process is repeated, more and more ACh is inactivated until it has been virtually removed from the cleft within a few milliseconds after its release. Removal of ACh terminates the EPP so that no more action potentials are initiated.

ber membrane (Step 4). (These cholinergic receptors are of the nicotinic type.) Binding of ACh with these receptor sites induces the opening of chemical messenger–gated channels in the motor end plate. These channels permit a small amount of cation traffic through them (both Na^+ and K^+) but no anions (Step 5). Because the permeability of the end-plate membrane to Na^+ and K^+ on opening of these channels is essentially equal, the relative movement of these ions through the channels depends on their electrochemical driving forces. Recall that at resting potential, the net driving force for Na^+ is much greater than that for K^+, because the resting potential is much closer to the K^+ equilibrium potential. Both the concentration and electrical gradients for Na^+ are inward, whereas the outward concentration gradient for K^+ is almost, but not quite, balanced by the opposing inward electrical gradient. As a result, when ACh triggers the opening of these channels, considerably more Na^+ moves inward than K^+ outward, bringing about a depolarization of the motor end plate. This potential change is known as the **end-plate potential** (**EPP**). It is similar to an EPSP (excitatory postsynaptic potential; see p. 92), except that the magnitude of an EPP is much larger because (1) more transmitter is released from a terminal button than from a presynaptic knob in response to an action potential; (2) the motor end plate has a larger surface area and, accordingly, more transmitter receptor sites than a subsynaptic membrane; and (3) many more ion channels are opened in response to the transmitter-receptor complex in the

• **FIGURE 7–4   Events at a Neuromuscular Junction** Propagation of an action potential to the terminal button of a motor neuron ① triggers the opening of voltage-gated Ca^{++} channels and the subsequent entry of Ca^{++} into the terminal button ②. The Ca^{++} triggers the exocytosis of vesicles of acetylcholine ③. Acetylcholine diffuses across the space separating the nerve and muscle cells and binds with receptor sites specific for it on the motor end plate of the muscle-cell membrane ④. This binding brings about the opening of cation channels, leading to a relatively large movement of Na^+ into the muscle cell compared to a smaller movement of K^+ outward. The result is an end-plate potential ⑤. Local current flow between the depolarized end plate and adjacent membrane initiates an action potential, which is propagated throughout the muscle fiber ⑥. Acetylcholine is subsequently destroyed by acetylcholinesterase, an enzyme located in the muscle cell membrane, terminating the muscle cell's response ⑦.

TABLE 7–7 Comparison of a Synapse and a Neuromuscular Junction	
SIMILARITIES	**DIFFERENCES**
Both consist of two excitable cells separated by a narrow cleft that prevents direct transmission of electrical activity between them.	A synapse is a junction between two neurons. A neuromuscular junction exists between a motor neuron and a skeletal-muscle fiber.
The axon terminals of both store chemical messengers (neurotransmitters) that are released by the Ca^{++}-induced exocytosis of storage vesicles when an action potential reaches the terminal.	There is a one-to-one transmission of action potentials at a neuromuscular junction, whereas one action potential in a presynaptic neuron cannot by itself bring about an action potential in a postsynaptic neuron. An action potential in a postsynaptic neuron occurs only when the summation of EPSPs brings the membrane to threshold.
In both, binding of the neurotransmitter with receptor sites in the membrane of the cell underlying the axon terminal opens specific channels in the membrane, permitting ionic movements that alter the membrane potential of the cell.	A neuromuscular junction is always excitatory (an EPP); a synapse may be either excitatory (an EPSP) or inhibitory (an IPSP). The inhibition of skeletal muscles cannot be accomplished at the neuromuscular junction; it can take place only in the CNS through IPSPs at the cell body of the motor neuron.
The resultant change in membrane potential in both cases is a graded potential.	

LOSS OF MUSCLE MASS: A PLIGHT OF SPACE FLIGHT

Skeletal muscles are a case of "use it or lose it." Stimulation of skeletal muscles by motor neurons is essential not only to induce the muscles to contract but also to maintain their size and strength. Muscles that are not routinely stimulated gradually atrophy, or diminish in size and strength.

When humans entered the weightlessness of space, it became apparent that the muscular system required the stress of work or gravity to maintain its size and strength. In 1991, the space shuttle Columbia was launched for a nine-day mission dedicated among other things to comprehensive research on physiological changes brought on by weightlessness. The three female and four male astonauts aboard suffered a dramatic and significant 25% reduction of mass in their weight-bearing muscles. The effort required to move the body is remarkably less in space than on earth, and there is no need for active muscular opposition to gravity. The result is what some refer to as *functional atrophy*.

The muscles most affected are those in the lower extemities, the gluteal (buttocks) muscles, the extensor muscles of the neck and back, and the muscles of the trunk. Changes include a decrease in muscle volume and mass, decrease in strength and endurance, increased breakdown of muscle protein, and loss of muscle nitrogen (an important component of muscle protein). The exact biologic mechanisms that induce muscle atrophy are unknown, but a majority of scientists believe that the lack of customary forcefulness of contraction plays a major role.

Space programs in the United States and the former Soviet Union have employed intervention techniques that emphasize both diet and exercise in an attempt to prevent muscle atrophy. Faithful performance of vigorous, carefully designed physical exercise has helped reduce the severity of functional atrophy. Studies of nitrogen and mineral balances, however, suggest that muscle atrophy continues to progress during exposure to weightlessness despite efforts to prevent it.

Furthermore, only half of the muscle mass was restored in the Columbia crew after the astronauts had been back on the ground an equal length of time to their flight. These and other findings suggest that extended stays in space will be difficult.

The neuromuscular junction is vulnerable to several chemical agents and diseases.

Several chemical agents and diseases are known to affect the neuromuscular junction by acting at different sites in the transmission process (Table 7–8). Two well-known toxins— *black widow spider venom* and *botulinum toxin*—alter the release of ACh, but in opposite directions. The venom of black widow spiders exerts its deadly effect by causing an explosive release of ACh from the storage vesicles, not only at neuromuscular junctions but at all cholinergic sites. All cholinergic sites undergo prolonged depolarization, the most detrimental consequence of which is respiratory failure. Breathing is accomplished by alternate contraction and relaxation of skeletal muscles, particularly the diaphragm. Inability to relax the diaphragm in the presence of a prolonged EPP caused by excessive amounts of ACh results in respiratory paralysis.

Botulinum toxin, on the other hand, exerts its lethal blow by blocking the release of ACh from the terminal button in response to an action potential in the motor neuron. *Clostridium botulinum* toxin is responsible for **botulism**, a form of food poisoning. It most frequently results from improperly canned foods contaminated with clostridial bacteria that survive and multiply, producing their toxin in the process. When this toxin is consumed, it prevents muscles from responding

TABLE 7–8   Examples of Chemical Agents and Diseases That Affect the Neuromuscular Junction

MECHANISM	CHEMICAL AGENT OR DISEASE
Alters Release of Acetylcholine	
Causes explosive release of acetylcholine	Black widow spider venom
Blocks release of acetylcholine	*Clostridium botulinum* toxin
Blocks Acetylcholine Receptor Sites	
Reversibly binds with acetylcholine receptor sites	Curare
Self-produced antibodies inactivate acetylcholine receptor sites	Myasthenia gravis
Prevents Inactivation of Acetylcholine	
Irreversibly inhibits acetylcholinesterase	Organophosphates (certain pesticides and military nerve gases)

to nerve impulses. Death is due to respiratory failure caused by the inability to contract the diaphragm. Botulinum toxin is one of the most lethal poisons known; ingestion of less than 0.0001 milligram can kill an adult.

Other chemicals interfere with neuromuscular junction activity by blocking the effect of released ACh. The best known example is **curare**, which reversibly binds to the ACh receptor sites on the motor end plate. Unlike ACh, however, curare does not alter membrane permeability nor is it inactivated by AChE. When ACh receptor sites are occupied by curare, ACh cannot combine with these sites to open the channels that would permit the ionic movement responsible for an EPP. Consequently, because muscle action potentials cannot occur in response to nerve impulses to these muscles, paralysis ensues. When sufficient curare is present to effectively block a significant number of ACh receptor sites, the person dies from respiratory paralysis caused by an inability to contract the diaphragm. Curare was used in the past as a deadly arrowhead poison. Curare and related drugs have also been used medically during surgery to help achieve more complete skeletal-muscle relaxation with less anesthetic. Under these circumstances, the amount of the agent is carefully administered, and facilities are available to maintain respiration artificially, if necessary, until the effects of the drug wear off.

Organophosphates are a group of chemicals that modify neuromuscular junction activity in yet another way—namely, by irreversibly inhibiting AChE. Inhibition of AChE prevents the inactivation of released ACh, so excited muscles remain in a contracted state, unable to repolarize and return to resting conditions. Death from organophosphates is also due to respiratory paralysis, but in this case the diaphragm is unable to relax and then contract again to bring in a fresh breath of air. These toxic agents are found in some pesticides and military nerve gases.

One disease known to involve the neuromuscular junction is **myasthenia gravis** (*myasthenia* means "muscle weakness," *gravis* means "severe"), a condition characterized by extreme muscular weakness. It is an autoimmune condition (*autoimmune* means "immunity against self") in which the body erroneously produces antibodies against its own motor end-plate ACh receptors. Consequently, not all of the released ACh molecules are able to find a functioning receptor site with which to bind. As a result, much of the ACh is destroyed by AChE without ever having an opportunity to interact with a receptor site and contribute to the EPP. Treatment consists of administration of a drug that temporarily inhibits AChE (in

contrast to the toxic organophosphates, which irreversibly block this enzyme). *Neostigmine* is such a short-term antiacetylcholinesterase drug. In myasthenia gravis patients, this drug prolongs the action of ACh at the neuromsucular junction by permitting it to build up. The resultant EPP is of sufficient magnitude to initiate action potentials and subsequent contraction in the muscle fiber, as it normally would.

 CHAPTER IN PERSPECTIVE: FOCUS ON HOMEOSTASIS

The nervous system, along with the other major control system, the endocrine system, is responsible for controlling most of the body's muscular and glandular activities. Whereas the afferent division of the peripheral nervous system detects and carries information to the central nervous system (CNS) for processing and decision making, the efferent division of the peripheral nervous system carries directives from the CNS to the effector organs (muscles and glands), which carry out the intended response. Much of this efferent output is directed toward maintaining homeostasis.

The autonomic nervous system, which is the efferent branch that innervates smooth muscle, cardiac muscle, and glands, plays a major role in the following range of homeostatic activities:

■ Regulation of blood pressure.

■ Control of digestive-juice secretion and of digestive-tract contractions that mix ingested food with the digestive juices.

■ Control of sweating to help maintain body temperature.

The somatic nervous system, the efferent branch that innervates skeletal muscle, brings about muscular contractions that enable the body to move in relation to the external environment. These contractions contribute to homeostasis by moving the body toward food or away from harm. Skeletal-muscle contraction also accomplishes breathing to maintain appropriate levels of O_2 and CO_2 in the body. Additionally, efferent output to skeletal muscles accomplishes many movements that are not aimed at maintaining a stable internal environment but nevertheless enrich our lives and enable us to contribute to society, such as dancing, building bridges, or performing surgery.

 CHAPTER SUMMARY

Introduction

The CNS controls muscles and glands by transmitting signals to these effector organs through the efferent division of the peripheral nervous system. There are two types of efferent output: the autonomic nervous system, which is under involuntary control and supplies cardiac and smooth muscle as well as most exocrine and some endocrine glands, and the somatic nervous sys-

tem, which is subject to voluntary control and supplies skeletal muscle.

Autonomic Nervous System

The autonomic nervous system consists of two subdivisions—the sympathetic and parasympathetic nervous systems. An autonomic nerve pathway consists of a two-neuron chain. The preganglionic fiber originates in the CNS and synapses with the cell body of the postganglionic fiber in a ganglion outside the CNS. The postganglionic fiber terminates on the effector organ. All preganglionic fibers and parasympathetic postganglionic fibers release acetylcholine. Sympathetic postganglionic fibers release norepinephrine. The same neurotransmitter elicits various responses from different tissues. Thus, the response depends on specialization of the tissue cells, not on the properties of the messenger. Tissues innervated by the autonomic nervous system possess one or more of several different receptor types for the postganglionic chemical messengers.

A given autonomic fiber either excites or inhibits activity in the organ it innervates. Most visceral organs are innervated by both sympathetic and parasympathetic nerve fibers, which in general produce opposite effects in a particular organ. Dual innervation of visceral organs by both branches of the autonomic nervous system permits precise control over an organ's activity. The sympathetic system dominates in emergency or stressful situations and promotes responses that prepare the body for strenuous physical activity. The parasympathetic system dominates in quiet, relaxed situations and promotes body maintenance activities such as digestion.

Somatic Nervous System

The somatic nervous system consists of the axons of motor neurons, which originate in the spinal cord and terminate on skeletal muscle. Acetylcholine, the neurotransmitter released from a motor neuron, stimulates muscle contraction. Motor neurons are the final common pathway by which various regions of the CNS exert control over skeletal-muscle activity.

Neuromuscular Junction

Each axon terminal of a motor neuron forms a neuromuscular junction with a single muscle cell (fiber). Because these structures do not make direct contact, signals are passed between the nerve terminal and muscle fiber by means of the chemical messenger acetylcholine (ACh). An action potential in the axon terminal causes the release of ACh from its storage vesicles. The released ACh diffuses across the space separating the nerve and muscle cell and binds to special receptor sites on the underlying motor end plate of the muscle cell membrane. This combination of ACh with the receptor sites triggers the opening of specific channels in the motor end plate. The subsequent ion movements depolarize the motor end plate, producing the end plate potential (EPP). Local current flow between the depolarized end plate and adjacent muscle cell membrane brings these adjacent areas to threshold, initiating an action potential that is propagated throughout the muscle fiber. This muscle action potential triggers muscle contraction. Acetylcholinesterase inactivates ACh, terminating the EPP and, subsequently, the action potential.

REVIEW EXERCISES

Objective Questions (Answers on p. D–2.)

1. Sympathetic preganglionic fibers originate in the thoracic and lumbar segments of the spinal cord. (True or false?)

2. Action potentials are transmitted on a one-to-one basis at both a neuromuscular junction and a synapse. (True or false?)

3. The sympathetic nervous system:
   a. is always excitatory.
   b. innervates only tissues concerned with protecting the body against challenges from the outside environment.
   c. has short preganglionic and long postganglionic fibers.
   d. is part of the afferent division of the peripheral nervous system.
   e. is part of the somatic nervous system.

4. Acetylcholinesterase:
   a. is stored in vesicles in the terminal button.
   b. combines with receptor sites on the motor end plate to bring about an end-plate potential.
   c. is inhibited by organophosphates.
   d. is the chemical transmitter at the neuromuscular junction.
   e. paralyzes skeletal muscle by strongly binding with acetylcholine receptor sites.

5. The two divisions of the autonomic nervous system are the _____ nervous system, which dominates in fight-or-flight situations, and the _____ nervous system, which dominates in quiet, relaxed situations.

6. The _____ is a modified sympathetic ganglion that does not give rise to postganglionic fibers but instead secretes hormones similar or identical to sympathetic postganglionic neurotransmitters into the blood.

7. Using the answer code at the right, identify the autonomic transmitter being described:
   ___ 1. secreted by all preganglionic fibers
   ___ 2. secreted by sympathetic postganglionic fibers
   ___ 3. secreted by parasympathetic postganglionic fibers
   ___ 4. secreted by the adrenal medulla
   ___ 5. secreted by motor neurons
   ___ 6. binds to muscarinic or nicotinic receptors
   ___ 7. binds to α or β receptors

   a. acetylcholine
   b. norepinephrine

8. Using the answer code at the right, indicate which type of efferent output is being described:

____ 1. composed of two-neuron chains
____ 2. innervates cardiac muscle, smooth muscle, and glands
____ 3. innervates skeletal muscle
____ 4. consists of the axons of motor neurons
____ 5. exerts either an excitatory or inhibitory effect on its effector organs
____ 6. dually innervates its effector organs
____ 7. exerts only an excitatory effect on its effector organs

a. characteristic of the somatic nervous system
b. characteristic of the autonomic nervous system

Essay Questions

1. Distinguish between preganglionic and postganglionic fibers.
2. Compare the origin, preganglionic- and postganglionic-fiber length, and neurotransmitters of the sympathetic and parasympathetic nervous systems.
3. What is the advantage of dual innervation of many organs by both branches of the autonomic nervous system?
4. Distinguish among the following types of receptors: nicotinic receptors, muscarinic receptors, α receptors, β_1 receptors, and β_2 receptors.
5. What regions of the CNS regulate autonomic output?
6. Why are motor neurons called the final common pathway?
7. Describe the sequence of events that occurs at a neuromuscular junction.
8. Discuss the effect each of the following has at the neuromuscular junction: black widow spider venom, botulinum toxin, curare, organophosphates, myasthenia gravis, and neostigmine.

POINTS TO PONDER

1. Explain why epinephrine, which causes arteriolar constriction (narrowing) in most tissues, is frequently administered in conjunction with local anesthetics.
2. Would skeletal-muscle activity be affected by atropine? Why or why not?
3. Heart pain, or *angina pectoris,* occurs whenever the blood supply to the heart muscle cannot meet the muscle's need for oxygen delivery, usually because the blood vessels supplying the heart are narrowed by cholesterol-containing deposits. Most persons with this condition do not have any pain at rest but experience bouts of pain whenever the heart's need for oxygen increases, such as during exercise or emotionally stressful situations that increase sympathetic nervous activity. Immediate relief of angina attacks can usually be obtained by prompt administration of a vasodilator drug such as *nitroglycerin,* which relaxes the smooth muscle in the walls of the narrowed vessels. Consequently, the vessels open more widely and more blood can flow through them. For prolonged treatment, patients experience fewer and less severe angina attacks when they take a β_1-blocker drug, such as *metoprolol,* on a regular basis. Explain why.
4. Considering that you can voluntarily control the emptying of your urinary bladder by contracting (preventing emptying) or relaxing (permitting emptying) your external urethral sphincter, a ring of muscle that guards the exit from the bladder, of what type of muscle is this sphincter composed and what branch of the nervous system supplies it?
5. The venom of certain poisonous snakes contains α bungarotoxin, which binds tenaciously to acetylcholine receptor sites on the motor end-plate membrane. What would the resultant symptoms be?

8 Muscle Physiology

MUSCULAR SYSTEM

Body systems maintain homeostasis

HOMEOSTASIS
Skeletal muscles contribute to homeostasis by playing a major role in the procurement of food, breathing, heat generation for maintenance of body temperature, and movement away from harm.

Homeostasis is essential for survival of cells

CELLS

Cells make up body systems

Muscles are the contraction specialists of the body. **Skeletal muscle** attaches to the skeleton. Contraction of skeletal muscles causes the bones to which they are attached to move, allowing the body to perform a variety of motor activities. Skeletal muscles that support homeostasis include those important in the acquisition, chewing, and swallowing of food and those essential for breathing. Skeletal-muscle contraction is also used to move the body away from harm. Heat-generating muscle contractions are important in temperature regulation. Skeletal muscles are also used for nonhomeostatic activities, such as dancing or operating a computer. **Smooth muscle** is found in the walls of hollow organs and tubes. Controlled contraction of smooth muscle is responsible for regulating movement of blood through the blood vessels, food through the digestive tract, air through the respiratory airways, and urine to the exterior. **Cardiac muscle** is found only in the walls of the heart, whose contraction pumps life-sustaining blood throughout the body.

CHAPTER CONTENTS AT A GLANCE

INTRODUCTION
STRUCTURE OF SKELETAL MUSCLE
Levels of organization in muscle
Thick- and thin-filament composition
MOLECULAR BASIS OF SKELETAL-MUSCLE CONTRACTION
Sliding filament mechanism
Excitation-contraction coupling
SKELETAL-MUSCLE MECHANICS
Gradation of contraction
  Motor-unit recruitment
  Summation; tetanus
  Length-tension relationship
Isotonic and isometric contractions
Load-velocity relationship
Muscle efficiency; heat production
Lever systems of muscles and bones
SKELETAL-MUSCLE METABOLISM AND FIBER TYPES
Pathways supplying ATP
Muscle fatigue
Oxygen debt
Muscle-fiber types
  ● A CLOSER LOOK AT EXERCISE PHYSIOLOGY Are Athletes Who
    Use Steroids to Gain Competitive Advantage Really
    Winners or Losers?
  ● CONCEPTS, CHALLENGES, AND CONTROVERSIES Muscular
    Dystrophy: When One Small Step Is a Big Deal
CONTROL OF MOTOR MOVEMENT
Inputs influencing motor-unit output
Local reflexes; muscle spindles; Golgi tendon organs
SMOOTH AND CARDIAC MUSCLE
Multiunit and single-unit smooth muscle
Myogenic activity
Factors modifying smooth-muscle activity; gradation of
  smooth-muscle contraction
Cardiac muscle

 INTRODUCTION

Almost all living cells possess rudimentary intracellular machinery for producing such movement as redistributing various components of the cell during cell division. White blood cells use intracellular contractile proteins to propel themselves through their environment. The contraction specialists of the body, however, are the muscle cells. Through their highly developed ability to contract, muscle cells are capable of shortening and developing tension, which enables them to produce movement and to do work. In contrast to sensory systems, which transform other forms of energy in the environment into electrical signals, muscles in response to electrical signals convert the chemical energy of ATP into mechanical energy that can act on the environment. Controlled contraction of muscles allows (1) purposeful movement of the whole body or parts of the body in relation to the

environment (such as walking or waving your hand); (2) manipulation of external objects (such as driving a car or moving a piece of furniture); (3) propulsion of contents through various hollow internal organs (such as circulation of blood or movement of materials through the digestive tract); and (4) emptying the contents of certain organs to the external environment (such as urination or giving birth).

Muscle comprises the largest group of tissues in the body, accounting for approximately half of the body's weight. Skeletal muscle alone makes up about 40% of body weight in men and 32% in women, with smooth and cardiac muscle making up another 10% of the total weight. Although the three muscle types are structurally and functionally distinct, they can be classified in several different ways according to their common characteristics (● Fig. 8–1). First, muscles are categorized as *striated* (skeletal and cardiac muscle) or *unstriated* (smooth muscle), depending on whether alternating dark and light bands or striations can be seen when the muscle is viewed under a light microscope. Second, muscles are categorized as *voluntary* (skeletal muscle) or *involuntary* (cardiac and smooth muscle), depending respectively on whether they are innervated by the somatic nervous system and are subject to voluntary control or are innervated by the autonomic nervous system and are not subject to voluntary control (see p. 198).

Most of this chapter will be devoted to a detailed examination of how the most abundant and best understood muscle, skeletal muscle, works. The chapter will conclude with a discussion of the unique properties of smooth and cardiac muscle in comparison to skeletal muscle.

● **FIGURE 8–1   Categorization of Muscle**

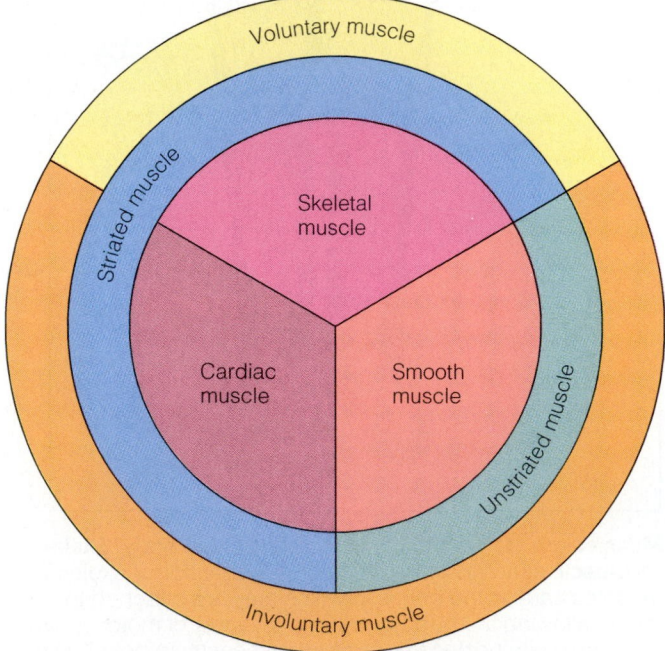

STRUCTURE OF SKELETAL MUSCLE

Skeletal-muscle fibers have a highly organized internal arrangement that creates a striated appearance.

Skeletal muscles are stimulated to contract via release of acetylcholine (ACh) at neuromuscular junctions between motorneuron terminals and muscle cells (see p. 205). A basic understanding of the structural components of a skeletal-muscle fiber is essential to understanding how the muscle action potential initiated by ACh brings about contraction.

A single skeletal-muscle cell, known as a **muscle fiber**, is relatively large, elongated, and cylinder-shaped, measuring from 10 to 100 micrometers (1 μ = 1 millionth of a meter) in diameter and up to 750,000 μ, or 2.5 feet, in length. A skeletal muscle consists of a number of muscle fibers lying parallel to each other and bundled together by connective tissue (• Fig. 8–2a). The fibers extend the entire length of the

• FIGURE 8–2 Levels of Organization in Skeletal Muscle (a) Enlargement of a cross section of whole muscle. (b) Enlargement of a myofibril within a muscle fiber. (c) Cytoskeletal components of myofibril. (d) Protein components of thick and thin filaments.

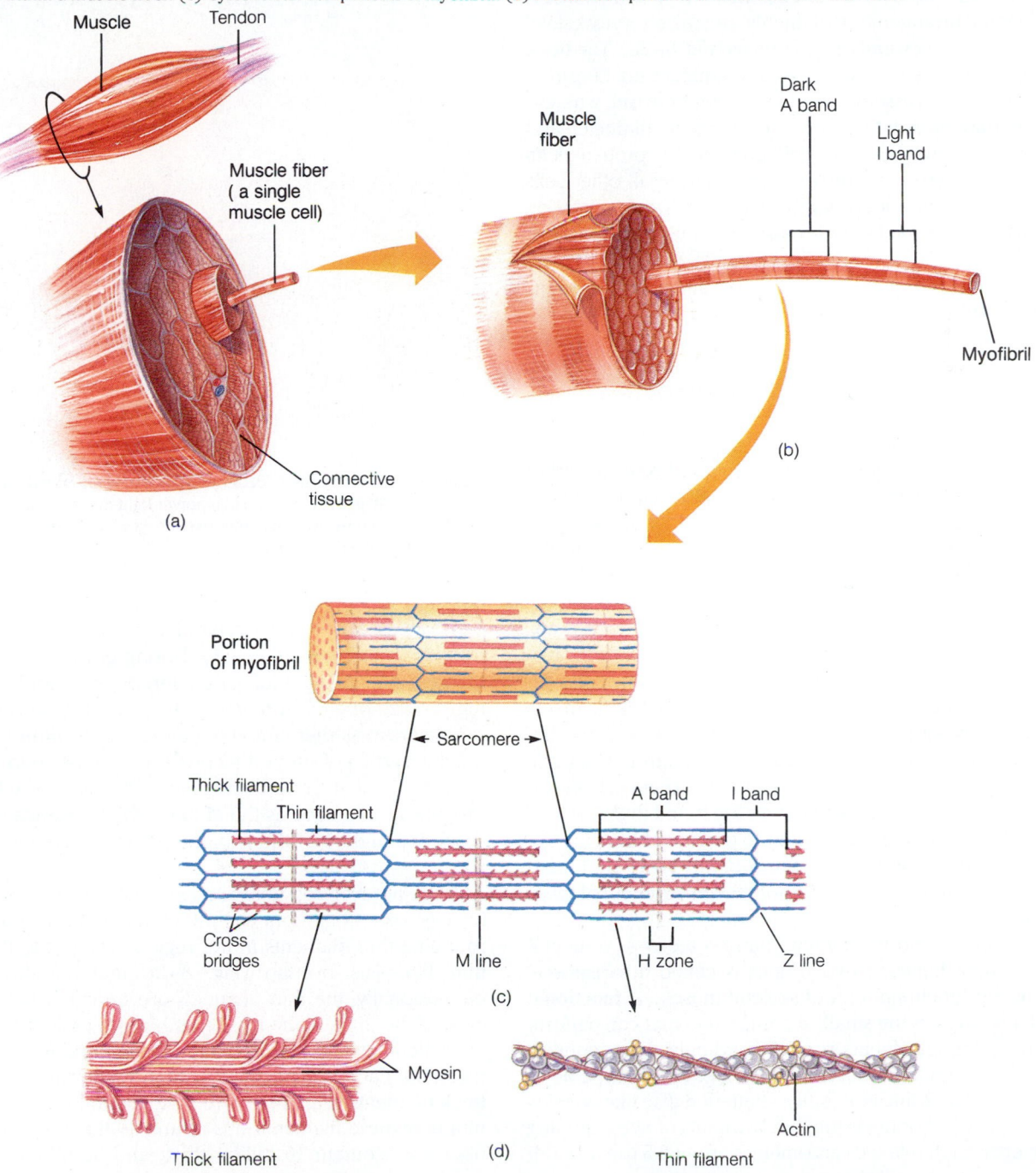

muscle. During embryonic development, the huge skeletal-muscle fibers are formed by the fusion of many smaller cells; thus, one striking feature is the presence of multiple nuclei in a single muscle cell. Another feature is the abundance of mitochondria, the energy-generating organelles, as would be expected with the high energy demands of a tissue as active as skeletal muscle.

The most predominant structural feature of a skeletal-muscle fiber is the presence of numerous **myofibrils**. These specialized contractile elements, which constitute 80% of the volume of the muscle fiber, are cylinder-shaped intracellular structures 1 μm in diameter that extend the entire length of the muscle fiber (Fig. 8–2b). Each myofibril consists of a regular arrangement of highly organized cytoskeletal elements—the thick and thin filaments (Fig. 8–2c). The **thick filaments**, which are 12 to 18 nm in diameter and 1.6 μm in length, are special assemblies of the protein *myosin*, whereas the **thin filaments**, which are 5 to 8 nm in diameter and 1.0 μm long, are made up primarily of the protein *actin* (Fig. 8–2d). These same proteins are found in all other cells of the body but in a less-organized fashion. The levels of organization in a skeletal muscle can be summarized as follows:

whole muscle	muscle fiber	myofibril	thick and thin filaments	myosin and actin
(an organ)	(a cell)	(a specialized intracellular structure)	(cytoskeletal elements)	(proteins)

Viewed with a light microscope, a relaxed myofibril (● Fig. 8–3a) displays alternating dark bands (the A bands) and light bands (the I bands). The bands of all of the myofibrils lined up parallel to each other collectively lead to the *striated* appearance of a skeletal-muscle fiber (Fig. 8–3b). Alternate stacked sets of thick and thin filaments that slightly overlap each other are responsible for the A and I bands (Fig. 8–2c). An **A band** consists of a stacked set of thick filaments along with the portions of the thin filaments that overlap on both ends of the thick filaments. The thick filaments are found only within the A band and extend its entire width. The lighter area within the middle of the A band, where the thin filaments do not reach, is known as the **H zone**. Only the central portions of the thick filaments are found in this region. The **I band** consists of the remaining portion of the thin filaments that do not project into the A band. Thus, the I band contains only thin filaments but not the entire length of these filaments.

Visible in the middle of each I band is a dense, vertical **Z line**. The area between two Z lines is called a **sarcomere**, which is the functional unit of skeletal muscle. A **functional unit** of any organ is the smallest component that can perform all of the functions of that organ. Accordingly, a sarcomere is the smallest component of a muscle fiber that is capable of contraction. The Z line is actually a flattened disc-like cytoskeletal protein that connects the thin filaments of two adjoining sarcomeres. Each relaxed sarcomere is about 2.5 μm in width

(a)

(b)

● **FIGURE 8–3   Light-Microscope View of Skeletal-Muscle Components** (a) High-power light-microscope view of myofibril. (b) Low-power light-microscope view of skeletal-muscle fibers. Note striated appearance.

and consists of one whole A band and half of each of the two I bands located on either side. During growth, a muscle increases in length by adding new sarcomeres, not by increasing the size of each sarcomere. Just as the Z lines hold the sarcomeres together in a chain along the myofibril's length, another system of supporting proteins is believed to hold the thick filaments together vertically within each stack. These proteins can be seen as the **M line**, which extends vertically down the middle of the A band within the center of the H zone.

With an electron microscope, fine **cross bridges** can be seen extending from each thick filament toward the surrounding thin filaments in the regions where the thick and thin filaments overlap (Fig. 8–2c and ● 8–4a). Three-dimensionally, the thin filaments are arranged hexagonally around the thick filaments. Cross bridges project from each thick filament in all six directions toward the surrounding thin filaments. Each thin filament, in turn, is surrounded by three thick filaments (Fig. 8–4b). To give you an idea of the magnitude of these filaments, it is estimated that a single muscle fiber may contain 16 billion thick and 32 billion thin fila-

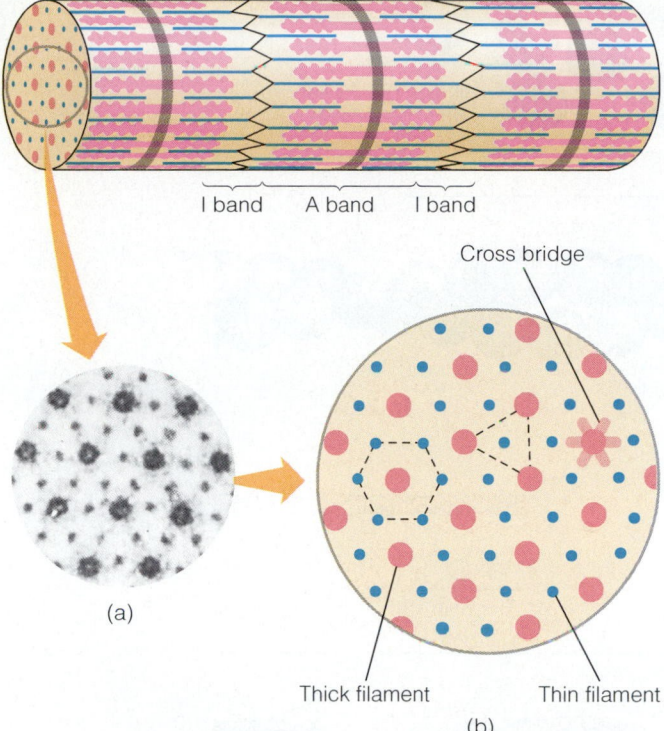

(a)

(b)

● FIGURE 8–4 Cross-Sectional Arrangement of Thick and Thin Filaments (a) Electron-micrograph cross section through the A band in the region of thick and thin filament overlap. Note the fine cross bridges extending from the thick filaments. (b) Schematic representation of the geometric relation among thick and thin filaments and cross bridges.

ments, all arranged in this very precise pattern within the myofibrils.

Myosin forms the thick filaments, whereas actin is the main structural component of the thin filaments.

Each thick filament is composed of several hundred myosin molecules packed together in a specific arrangement. A **myosin** molecule is a protein consisting of two identical subunits, each shaped somewhat like a golf club (● Fig. 8–5a). The protein's tail ends are intertwined around each other, with the two globular heads projecting out at one end. The two halves of each thick filament are mirror images made up of myosin molecules lying lengthwise in a regular staggered array, with their tails oriented toward the center of the filament and their globular heads protruding outward at regularly spaced intervals (Fig. 8–5b). These heads form the cross bridges between the thick and thin filaments. Each cross bridge has two important sites crucial to the contractile process: an **actin binding site** and a **myosin ATPase site**.

Thin filaments are composed of three proteins—actin, tropomyosin, and troponin (● Fig. 8–6). **Actin** molecules, the primary structural proteins of the thin filament, are spherical in shape. The backbone of a thin filament is formed by actin molecules joined into two strands and twisted together, like two chains of pearls wrapped around each other. Each actin molecule has a special binding site for attachment with a myosin cross bridge. By a mechanism to be described shortly, binding of actin and myosin molecules at the cross bridges results in energy-consuming contraction of the muscle fiber.

● FIGURE 8–5 Structure of Myosin Molecules and Their Organization within a Thick Filament (a) Myosin molecule. Each myosin molecule consists of two identical golf-club–shaped subunits with their tails intertwined and their globular heads, each of which contains an actin binding site and a myosin ATPase site, projecting out at one end. (b) Thick filament. A thick filament is made up of myosin molecules lying lengthwise parallel to each other. Half are oriented in one direction and half in the opposite direction so that the tails from the two halves line up end-to-end in the middle of the filament. The globular heads, which protrude at regular intervals along the thick filament, form the cross bridges.

(a)

(b)

Actin molecules

Binding site for attachment
with myosin cross bridge

Actin helix

+

Tropomyosin   Troponin

Thin filament

● FIGURE 8–6   Composition of Thin Filament The main structural component of a thin filament is two chains of spherical-shaped actin molecules that are twisted together. Troponin molecules, which consist of three small spherical subunits, and threadlike tropomyosin molecules are arranged to form a ribbon that lies alongside the groove of the actin helix and physically covers the binding sites on actin molecules for attachment with myosin cross bridges. (The thin filaments shown here are not drawn in proportion to the thick filaments in Figure 8–5. Thick filaments are two to three times larger in diameter than thin filaments.)

Accordingly, actin and myosin are often referred to as **contractile proteins**, even though, as we will see, neither myosin nor actin actually contracts.

In a relaxed muscle fiber, contraction does not take place; actin is not able to bind with cross bridges because of the position of the two other types of protein within the thin filament—tropomyosin and troponin. **Tropomyosin** molecules are threadlike proteins that lie end to end alongside the groove of the actin spiral. In this position, tropomyosin covers the actin sites that bind with the cross bridges, thus blocking the interaction that leads to muscle contraction. Tropomyosin is stabilized in this blocking position by **troponin** molecules, which fasten down the ends of each tropomyosin molecule. Troponin is a protein complex consisting of three polypeptide units: one that binds to tropomyosin, one that binds to actin, and a third that can bind with Ca^{++}. When Ca^{++} binds to troponin, the shape of this protein is changed in such a way that tropomyosin is allowed to slide away from its blocking position (● Fig. 8–7). With tropomyosin out of the way, actin and myosin can bind and interact at the cross bridges, resulting in muscle contraction. Tropomyosin and troponin are often referred to as **regulatory proteins** because of their role in covering (preventing contraction) or exposing (permitting contraction) the binding sites for cross-bridge interaction between actin and myosin.

Several important links in the contractile process remain to be discussed. How does cross-bridge interaction between actin and myosin bring about muscle contraction? How does a muscle action potential trigger this contractile process? What is the source of the Ca^{++} that physically repositions troponin and tropomyosin to permit cross-bridge binding? We will turn our attention to these topics in the next section.

MOLECULAR BASIS OF SKELETAL-MUSCLE CONTRACTION

Cycles of cross-bridge binding and bending pull the thin filaments closer together between the thick filaments during contraction.

The thin filaments on each side of a sarcomere slide inward toward the A band's center during contraction (● Fig. 8–8).

Thin filament

Tropomyosin

Actin

Cross-bridge binding sites

Myosin cross bridge

Troponin

Cross-sectional view

Actin

Cross-bridge binding site

Troponin

Tropomyosin

Actin binding site

Myosin cross bridge

Muscle fiber relaxed;
no cross-bridge binding
because the cross-bridge
binding site on actin is
physically covered by the
troponin-tropomyosin complex

Relaxed

Excited

Ca++

Muscle fiber excited;
released Ca++ binds
with troponin, pulling
troponin-tropomyosin
complex aside to
expose cross-bridge
binding site; cross-bridge
binding occurs

Longitudinal view

Ca++   Ca++   Ca++

Binding of actin and myosin cross
bridge triggers power stroke that
pulls thin filament inward during contraction

• **FIGURE 8–7   Schematic Representation of Role of Calcium in Turning on Cross Bridges**

As they slide inward, the thin filaments pull the Z lines to which they are attached closer together, so the sarcomere shortens. As all the sarcomeres throughout the muscle fiber's length shorten simultaneously, the entire fiber becomes shorter. This is known as the **sliding-filament mechanism** of muscle contraction. The H zone, the region in the center of the A band where the thin filaments do not reach, becomes smaller as the thin filaments approach each other when they slide more deeply inward. The H zone may even disappear if the thin filaments meet in the middle of the A band. The I band, which consists of the portions of the thin filaments that do not overlap with the thick filaments, decreases in width as the thin filaments further overlap the thick filaments during their inward slide. The thin filaments themselves do not change length during muscle-fiber shortening. The width of the A band remains unchanged during contraction, because

its width is determined by the length of the thick filaments, and the thick filaments do not change length during the shortening process. Note that neither the thick nor thin filaments decrease in length to shorten the sarcomere. Instead, contraction is accomplished by the thin filaments sliding closer together between the thick filaments.

The thin filaments are pulled inward relative to the stationary thick filaments by cross-bridge activity. During contraction, with the tropomyosin and troponin "chaperones" pulled out of the way by Ca++, the myosin cross bridges from a thick filament are able to bind with the actin molecules in the surrounding thin filaments. Let's concentrate on a single cross-bridge interaction (Fig. 8–9a). When myosin and actin make contact at a cross bridge, the conformation of the bridge is altered so that it bends inward as if it were on a hinge, "stroking" toward the center of the thick filament, simi-

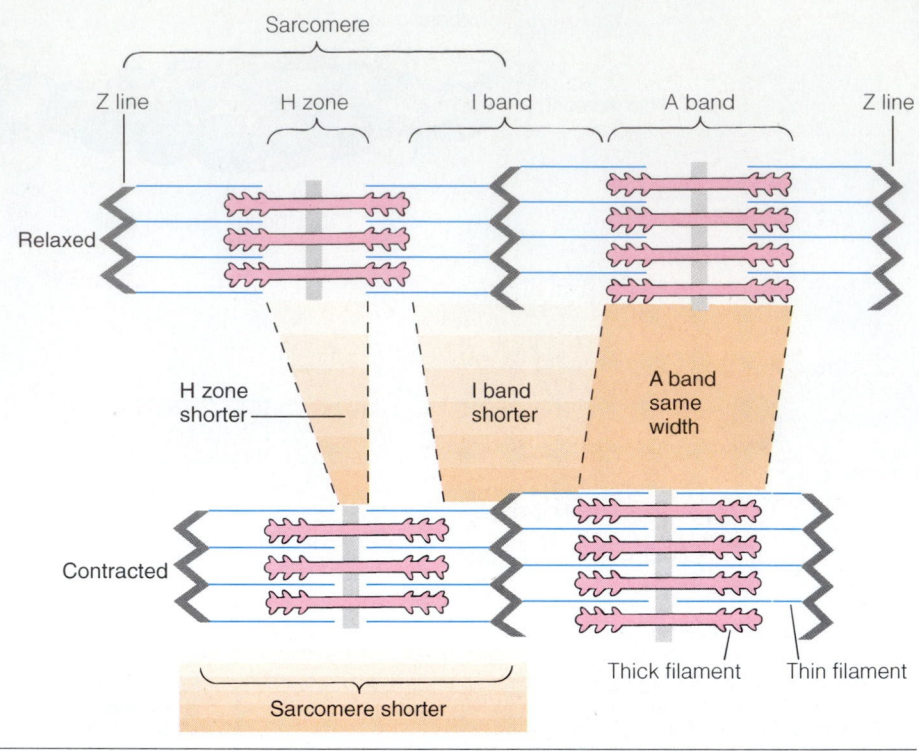

● **FIGURE 8–8   Changes in Banding Pattern during Shortening** During muscle contraction, each sarcomere shortens as the thin filaments slide closer together between the thick filaments so that the Z lines are pulled closer together. The width of the A bands does not change as a muscle fiber shortens, but the I bands and H zones become shorter.

lar to the stroking of a boat oar. This so-called **power stroke** of a cross bridge pulls the thin filament to which it is attached inward. A single power stroke pulls the thin filament inward only a small percentage of the total shortening distance. Complete shortening is accomplished by repeated cycles of cross-bridge binding and bending. At the end of one cross-bridge cycle, the link between the myosin cross bridge and actin molecule is broken. The cross bridge returns to its original conformation and binds to the next actin molecule positioned behind its previous actin partner. The cross bridge bends once again to pull the thin filament in further, then detaches and repeats the cycle. Repeated cycles of cross-bridge binding and bending successively pull in the thin filaments, much like pulling in a rope hand over hand. Because of the orientation of the myosin molecules within a thick filament (Fig. 8–9b), all of the cross bridges' power strokes are directed toward the center so that all six of the surrounding thin filaments are pulled inward simultaneously (Fig. 8–9c). The cross bridges aligned with given thin filaments do not all stroke in unison, however. At any time during contraction, part of the cross bridges are attached to the thin filaments and are stroking while others are returning to their original conformation in preparation for binding with another actin molecule. Thus, some cross bridges are "holding on" to the thin filaments, while others "let go" to bind with new actin. If it were not for this asynchronous cycling of the cross bridges, the thin filaments would be able to slip back toward their resting position between strokes.

Calcium is the link between excitation and contraction.

How is this cross-bridge cycling switched on by muscle excitation? **Excitation-contraction coupling** refers to the series of events linking muscle excitation (the presence of an action potential in a muscle fiber) to muscle contraction (cross-bridge activity that causes the thin filaments to slide closer together to produce sarcomere shortening) (Table 8–1).

Skeletal muscles are stimulated to contract by release of acetylcholine (ACh) at neuromuscular junctions between motor-neuron terminals and muscle fibers. Recall that the binding of ACh with the motor end plate of a muscle fiber brings about permeability changes in the muscle fiber that result in an action potential that is conducted over the entire surface of the muscle cell membrane.

At each junction of an A band and an I band, the surface membrane dips into the muscle fiber to form a **transverse tubule** (**T tubule**), which runs perpendicularly from the surface of the muscle cell membrane into the central portions of the muscle fiber (● Fig. 8–10). Because the T-tubule membrane is continuous with the surface membrane, an action potential on the surface membrane also spreads down into the T tubule, providing a means of rapidly transmitting the surface electric activity into the central portions of the fiber. The presence of a local action potential in the T tubules induces permeability changes in a separate membranous network within the muscle fiber, the sarcoplasmic reticulum.

Actin molecules in thin filament

Myosin cross bridge

Z line

Binding of cross bridge to actin molecule

Power stroke; bending of cross bridge pulls thin filament inward

Detachment of cross bridge at end of power stroke; return to original conformation

Binding of cross bridge to more distal actin molecule; cycle repeated

(a)

(b)

Thin filament

Thick filament

(c)

FIGURE 8–9 Cross-Bridge Activity (a) During each cross-bridge cycle, the cross bridge binds with an actin molecule, bends to pull the thin filament inward during the power stroke, then detaches and returns to its resting conformation, ready to repeat the cycle. (b) The power strokes of all cross bridges extending from a thick filament are directed toward the center of the thick filament. (c) Each thick filament is surrounded by six thin filaments, all of which are pulled inward simultaneously through cross-bridge cycling during muscle contraction.

The **sarcoplasmic reticulum** is a modified endoplasmic reticulum (see p. 18) that consists of a fine network of interconnected tubules surrounding each myofibril like a mesh sleeve (Fig. 8–10). This membranous network runs longitudinally down the myofibril (that is, encircles the myofibril throughout its length), but is not continuous. Separate segments of sarcoplasmic reticulum are wrapped around each A and I band. The ends of each segment expand to form saclike regions, the **lateral sacs**, which lie in close proximity to but do not directly contact, the adjacent T tubules (Figs. 8–10

TABLE 8–1   Steps of Excitation-Contraction Coupling and Relaxation

Acetylcholine released from the terminal of a motor neuron initiates an action potential in the muscle cell that is propagated over the entire surface of the membrane.

The surface electric activity is carried into the central portions of the muscle fiber by the T tubules.

Spread of the action potential down the T tubules triggers the release of stored Ca^{++} from the adjacent lateral sacs of the sarcoplasmic reticulum.

Released Ca^{++} binds with troponin and changes its shape so that the troponin-tropomyosin complex is physically pulled aside, uncovering actin's cross-bridge binding sites.

Exposed actin sites bind with myosin cross bridges, which have previously been energized by the splitting of ATP into ADP + P_i + energy by the myosin ATPase site on the cross bridges.

Binding of actin and myosin at a cross bridge causes the cross bridge to bend, producing a power stroke that pulls the thin filament inward. Inward sliding of all the thin filaments surrounding a thick filament shortens the sarcomere (i.e., muscle contraction).

ADP and P_i are released from the cross bridge during the power stroke.

Attachment of a new molecule of ATP permits detachment of the cross bridge, which returns to its original conformation.

Splitting of the fresh ATP molecule by myosin ATPase energizes the cross bridge once again.

If Ca^{++} is still present so that the troponin-tropomyosin complex remains pulled aside, the cross bridges go through another cycle of binding and bending, pulling the thin filament in even further.

When there is no longer a local action potential and Ca^{++} has been actively returned to its storage site in the sarcoplasmic reticulum's lateral sacs, the troponin-tropomyosin complex slips back into its blocking position, actin and myosin no longer bind at the cross bridges, and the thin filaments slide back to their resting position as relaxation takes place.

and ● 8–11). The sarcoplasmic reticulum's lateral sacs store Ca^{++}. Spread of an action potential down a T tubule triggers release of Ca^{++} from the sarcoplasmic reticulum into the cytosol.

Several mechanisms have been proposed to explain how a change in T-tubule potential is linked with the release of Ca^{++} from the sarcoplasmic reticulum's lateral sacs. Recent evidence suggests that T-tubule current activates a voltage-dependent enzyme that leads to the formation of a chemical messenger from a component of the T-tubule membrane. In

this proposed mechanism, the messenger diffuses to the adjacent lateral sacs, where it binds to membrane receptors to trigger the release of Ca^{++} from these intracellular stores. Alternatively, other investigators have discovered a mechanical connection between the T-tubule membrane and adjacent lateral sacs. So-called *foot proteins* have been observed to extend from the sarcoplasmic reticulum and span the gap between the lateral sac and T tubule. It is speculated that these foot proteins not only bridge the gap but serve as Ca^{++} channels. According to this proposed mechanism, voltage-

● **FIGURE 8–10   T Tubules and Sarcoplasmic Reticulum in Relationship to Myofibrils** The transverse (T) tubules are membranous perpendicular extensions of the surface membrane that dip deep into the muscle fiber at the junctions between the A and I bands of the myofibrils. The sarcoplasmic reticulum is a fine membranous network that runs longitudinally and surrounds each myofibril, with separate segments encircling each A and I band. The ends of each segment are expanded to form lateral sacs that lie next to, but do not directly contact, the adjacent T tubules.

Propagation of action potential across surface membrane and down T tubules of muscle fiber   T tubule

Lateral sacs of sarcoplasmic reticulum

Ca⁺⁺

Z line   Thin filament   Thick filament

(a)

Calcium ions released from lateral sac
bind to troponin on actin filaments;
tropomyosin physically moved to uncover
cross-bridge binding sites

(from lateral sac)   Tropomyosin

Ca⁺⁺

Troponin
Actin molecule
Myosin cross bridge

Myosin cross bridges attach to
actin and bend, pulling actin filaments
toward the center of the sarcomere;
powered by energy provided by ATP

(b)

• **FIGURE 8–11   Calcium Release in Excitation-Contraction Coupling** (a) Calcium (Ca^{++}) is released from the lateral sacs of the sarcoplasmic reticulum in response to a local action potential in the adjacent T tubule. (b) Calcium binds with troponin on the thin filaments to permit cross-bridge binding and power stroking.

sensitive sites that have been observed in the T-tubule membrane are activated when an action potential is propagated down the T tubule. These activated T-tubule sites, which line up with the foot proteins that protrude from the adjacent sarcoplasmic reticulum, trigger a mechanical change that opens the directly abutting foot proteins, alias Ca^{++} channels. Consequently, Ca^{++} is released from the lateral sacs. Whichever proposal, or combination thereof, turns out to be correct, this released Ca^{++}, by slightly repositioning the troponin and tropomyosin molecules, exposes the binding sites on the actin molecules so that they can link with the myosin cross bridges at their complementary binding sites.

Recall that a myosin cross bridge has two special sites, an actin binding site and an ATPase site. The latter is an enzymatic site that can bind the energy carrier, *adenosine triphosphate* (*ATP*), and split it into *adenosine diphosphate* (*ADP*) and *inorganic phosphate* (*P_i*), yielding energy in the process.

In skeletal muscle, magnesium (Mg^{++}) must be attached to ATP before myosin ATPase can split the ATP. The breakdown of ATP occurs on the myosin cross bridge before the bridge ever links with an actin molecule (Step 1 in • Fig. 8–12). The ADP and P_i remain tightly bound to the myosin, and the generated energy is stored within the cross bridge to produce a high-energy form of myosin. To use an analogy, the cross bridge is "cocked" like a gun, ready to be fired when the trigger is pulled. When the muscle fiber is excited, Ca^{++} pulls the troponin-tropomyosin complex out of its blocking position so that the energized (cocked) myosin cross bridge can bind with an actin molecule (Step 2a). This contact between myosin and actin "pulls the trigger." The energy stored within the myosin cross bridge is released to cause the cross-bridge bending responsible for the power stroke that pulls the thin actin filament inward (Step 3). The mechanism by which the chemical energy released from ATP is stored within the myo-

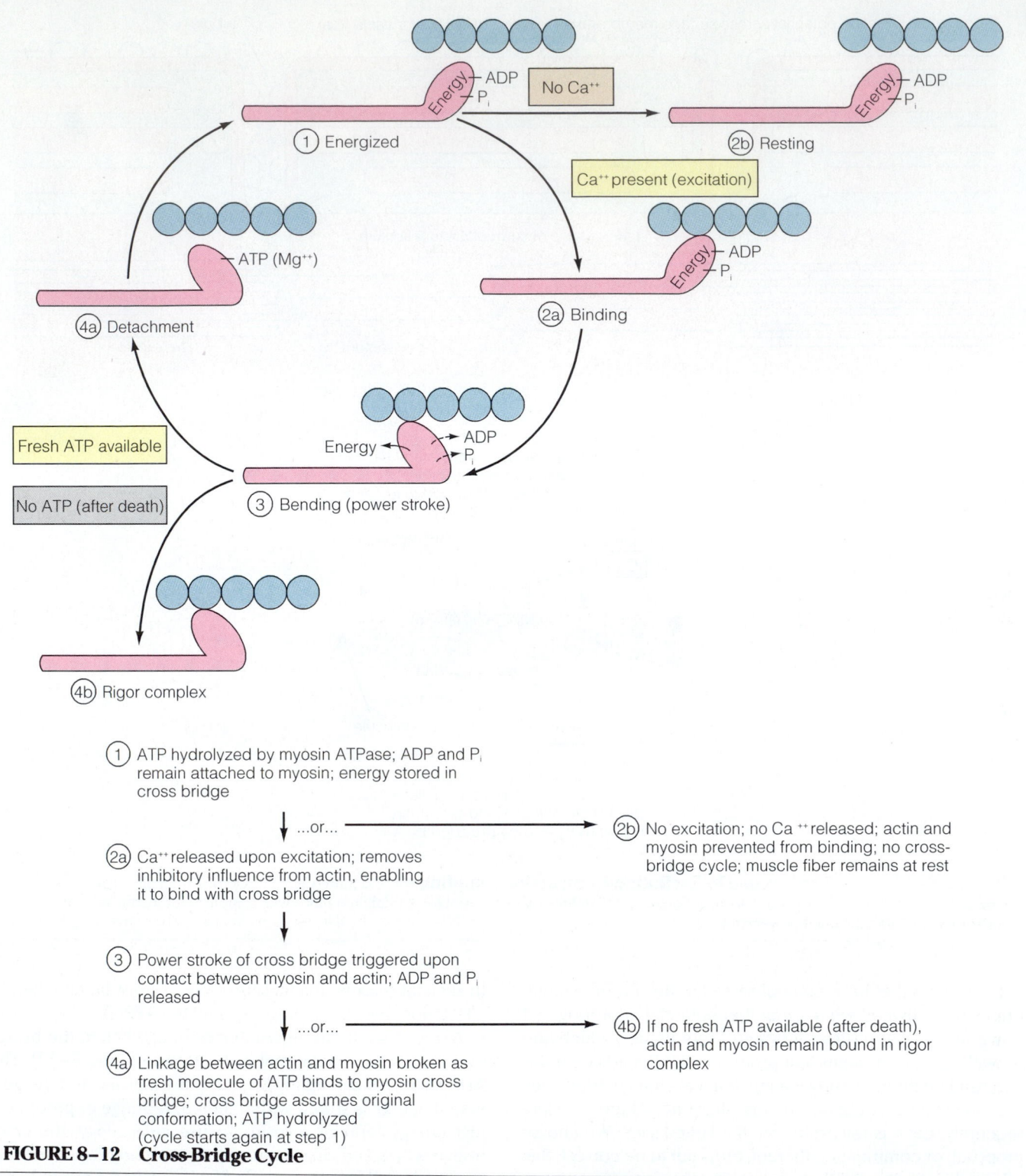

① ATP hydrolyzed by myosin ATPase; ADP and P_i remain attached to myosin; energy stored in cross bridge

⎯→ ...or... ⎯⎯⎯⎯⎯⎯⎯⎯⎯⎯⎯→ ②b No excitation; no Ca^{++} released; actin and myosin prevented from binding; no cross-bridge cycle; muscle fiber remains at rest

②a Ca^{++} released upon excitation; removes inhibitory influence from actin, enabling it to bind with cross bridge

③ Power stroke of cross bridge triggered upon contact between myosin and actin; ADP and P_i released

⎯→ ...or... ⎯⎯⎯⎯⎯⎯⎯⎯⎯⎯⎯→ ④b If no fresh ATP available (after death), actin and myosin remain bound in rigor complex

④a Linkage between actin and myosin broken as fresh molecule of ATP binds to myosin cross bridge; cross bridge assumes original conformation; ATP hydrolyzed (cycle starts again at step 1)

• **FIGURE 8–12   Cross-Bridge Cycle**

sin cross bridge and then translated into the mechanical energy of the power stroke is not known. When the muscle is not excited and Ca^{++} is not released, troponin and tropomyosin remain in their blocking position so that actin and the myosin cross bridges do not bind and no power stroking takes place (Step 2b).

Adenosine diphosphate and inorganic phosphate are also rapidly released from myosin upon contact with actin during a power stroke (Step 3). This frees the myosin ATPase site for attachment of another ATP molecule. The actin and myosin remain linked together at the cross bridge until a fresh molecule of ATP attaches to myosin at the end of the power stroke.

Attachment of the new ATP molecule permits detachment of the cross bridge, which returns to its original conformation, ready to start another cycle (Step 4a). The newly attached ATP is then split by myosin ATPase, energizing the myosin cross bridge once again (Step 1). Upon binding with another actin molecule, the energized cross bridge again bends, and so on, successively pulling the thin filament inward to accomplish contraction.

Note that fresh ATP must attach to myosin to permit the cross-bridge link between myosin and actin to be broken at the end of a cycle, even though the ATP is not split during this dissociation process. The necessity for ATP in the separation of myosin and actin is amply demonstrated by the phenomenon of **rigor mortis**. This "stiffness of death" is a generalized locking in place of the skeletal muscles that begins three to four hours after death and becomes complete in about twelve hours. Following death, the cytosolic concentration of Ca^{++} begins to rise, most likely because the inactive muscle cell membrane is unable to keep out extracellular Ca^{++} and perhaps also because Ca^{++} leaks out of the lateral sacs. This Ca^{++} moves the regulatory proteins aside, permitting actin to bind with the myosin cross bridges, which were already charged with ATP before death. Since dead cells cannot produce any more ATP, actin and myosin, once bound, are unable to detach because of the absence of fresh ATP. The thick and thin filaments thus remain linked together by the immobilized cross bridges, resulting in the stiffened condition of dead muscles (Step 4b). During the next several days, rigor mortis gradually subsides as the proteins involved in the rigor complex begin to degrade.

How is **relaxation** normally accomplished in a living muscle? Just as an action potential in a muscle fiber turns on the contractile process by triggering the release of Ca^{++} from the lateral sacs into the cytosol, the contractile process is turned off when Ca^{++} is returned to the lateral sacs upon cessation of local electrical activity. The sarcoplasmic reticulum possesses an energy-consuming carrier, a Ca^{++}-ATPase pump, which actively transports Ca^{++} from the cytosol and concentrates it in the lateral sacs. When acetylcholinesterase removes ACh from the neuromuscular junction, the muscle-fiber action potential ceases. When there is no longer a local action potential in the T tubules to trigger the release of Ca^{++}, the ongoing activity of the sarcoplasmic reticulum's Ca^{++} pump returns the released Ca^{++} back into its lateral sacs. Removal of cytosolic Ca^{++} allows the troponin-tropomyosin complex to slip back into its blocking position so that actin and myosin are no longer able to bind at the cross bridges. The thin filaments, freed from cycles of cross-bridge attachment and pulling, are able to return to their resting position. Relaxation has occurred.

Contractile activity far outlasts the electrical activity that initiated it.

A single action potential in a skeletal-muscle fiber lasts only 1 to 2 msec. The onset of the resultant contractile response lags

The duration of the action potential is not drawn to scale but is exaggerated.

● **FIGURE 8–13   Relationship of Action Potential to Resultant Muscle Twitch**

behind the action potential because the entire excitation-contraction coupling process must take place before cross-bridge activity begins. In fact, the action potential is completed before the contractile apparatus even becomes operational. This time delay of a few milliseconds between stimulation and the onset of contraction is known as the **latent period** (● Fig. 8–13). Time is also required for the generation of tension within the muscle fiber produced by means of the sliding interactions between the thick and thin filaments through cross-bridge activity. The time from the onset of contraction until peak tension is developed—the **contraction time**—averages about 50 msec, although this time varies depending on the type of muscle fiber. The contractile response does not cease until the lateral sacs have taken up all of the Ca^{++} released in response to the action potential. This reuptake of Ca^{++} is also time-consuming. Even after Ca^{++} is removed, it takes time for the filaments to return to their resting positions. The time from peak tension until relaxation is complete, the **relaxation time**, usually lasts slightly longer than contraction time, another 50 msec or more. Consequently, the entire contractile response to a single action potential may last up to 100 msec or more; this is considerably longer than the duration of the action potential that initiated it (100 msec compared to 1 to 2 msec). This fact is important in the body's ability to produce muscle contractions of variable strength, as you will discover in the next section.

SKELETAL-MUSCLE MECHANICS

Whole muscles are groups of muscle fibers bundled together by connective tissue and attached to bones by tendons.

Thus far we have described the contractile response in a single muscle fiber. In the body, groups of muscle fibers are organized into whole muscles. We will now turn our attention to contraction of whole muscles. Each person has about six hundred skeletal muscles, which range in size from the delicate external eye muscles with only a few hundred fibers to the large, powerful leg muscles containing several hundred thousand fibers.

Each muscle is covered by a sheath of connective tissue that penetrates from the surface into the muscle to envelop each individual fiber and divide the muscle into columns or bundles. The connective tissue further extends beyond the ends of the muscle to form tough, collagenous **tendons** that attach the muscle to bones. A tendon may be quite long, attaching to a bone some distance from the fleshy portion of the muscle. For example, some of the muscles involved in movement of the fingers are found in the forearm, with long tendons extending down to attach to the bones of the fingers. (You can readily observe the movement of these tendons on the top of your hand when you wiggle your fingers.) This arrangement permits greater dexterity; the fingers would be much thicker and more awkward if all the muscles involved in finger movement were actually located in the fingers.

Contractions of a whole muscle can be of varying strength.

A single action potential in a muscle fiber produces a brief, weak contraction known as a **twitch**, which is too short and too weak to be useful and normally does not take place in the body. Muscle fibers are arranged into whole muscles where they can function cooperatively to produce contractions of variable grades of strength stronger than a twitch. In other words, the force exerted by the same muscle can be made to vary, depending on whether the person is picking up a piece of paper, a book, or a fifty-pound weight. Two primary factors can be adjusted to accomplish gradation of whole-muscle tension: (1) *the number of muscle fibers contracting within a muscle* and (2) *the tension produced by each contracting fiber* (Table 8–2). We will discuss each of these factors in turn.

The number of fibers contracting within a muscle depends on the extent of motor-unit recruitment.

Because the greater the number of fibers contracting, the greater the total muscle tension, larger muscles consisting of more muscle fibers are obviously capable of generating more tension than are smaller muscles with fewer fibers.

Each whole muscle is innervated by a number of different motor neurons. When a motor neuron enters a muscle, it branches, with each axon terminal supplying a single muscle fiber (• Fig. 8–14). One motor neuron innervates a number of muscle fibers, but each muscle fiber is supplied by only one motor neuron. When a motor neuron is activated, all of the muscle fibers it supplies are stimulated to contract simultaneously. This functional unit—one motor neuron plus all of the muscle fibers it innervates—is called a **motor unit**. The muscle fibers that compose a motor unit are dispersed throughout the whole muscle; thus, their simultaneous contraction results in an evenly distributed, although weak, contraction of the whole muscle. Each muscle consists of a number of intermingled motor units. For a weak contraction of the whole muscle, only one or a few of its motor units are activated. For stronger and stronger contractions, more and more motor units are *recruited*, or stimulated to contract, a phenomenon known as **motor-unit recruitment**.

How much stronger the contraction will be with the recruitment of each additional motor unit depends on the size of the motor units (that is, the number of muscle fibers controlled by a single motor neuron) (• Fig. 8–15). The number of muscle fibers per motor unit and the number of motor units

---

TABLE 8–2   Determinants of Whole-Muscle Tension in Skeletal Muscle

NUMBER OF FIBERS CONTRACTING	TENSION DEVELOPED BY EACH CONTRACTING FIBER
Number of motor units recruited*	Frequency of stimulation (twitch summation and tetanus)*
Number of muscle fibers per motor unit	Length of fiber at onset of contraction (length-tension relationship)
Number of muscle fibers available to contract	Extent of fatigue
Size of muscle (i.e., number of muscle fibers in muscle)	Duration of activity
Presence of disease (e.g., muscular dystrophy)	Amount of asynchronous recruitment of motor units
Extent of recovery from traumatic losses	Type of fiber (fatigue-resistant oxidative or fatigue-prone glycolytic)
	Thickness of fiber
	Type of fiber (small-diameter oxidative or large-diameter glycolytic)
	Pattern of neural activity (hypertrophy, atrophy)
	Amount of testosterone (larger fibers in males than females)

*Factors controlled to accomplish gradation of contraction.

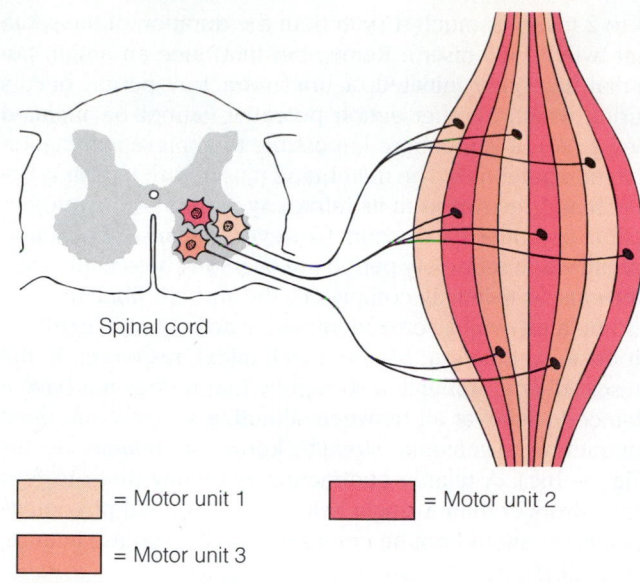

Spinal cord

☐ = Motor unit 1   ■ = Motor unit 2

■ = Motor unit 3

● **FIGURE 8–14   Schematic Representation of Motor Units in a Skeletal Muscle**

per muscle vary widely, depending on the specific function of the muscle. For muscles that produce precise, delicate movements, such as the muscles of the external eye and the hand, a single motor unit may contain as few as a dozen muscle fibers. Recruitment of each additional motor unit results in only a small additional increment in the whole muscle's strength of contraction, because so few muscle fibers are involved with each motor unit. These small motor units allow a very fine degree of control over muscle tension. However, in muscles designed for powerful, coarsely controlled movement, such as those of the legs, a single motor unit may contain 1,500 to 2,000 muscle fibers. Recruitment of motor units in these muscles results in large incremental increases in whole muscle tension. More powerful contractions occur at the expense of less precisely controlled gradations. Thus, the number of muscle fibers participating in the whole muscle's

total contractile effort depends on the number of motor units recruited and the number of muscle fibers per motor unit in that muscle.

To delay or prevent **fatigue** (inability to maintain muscle tension at a given level) during a sustained contraction involving only a portion of a muscle's motor units, as is necessary in muscles supporting the weight of the body against the force of gravity, **asynchronous recruitment of motor units** takes place. The body alternates motor-unit activity, like shifts at a factory, to give motor units that have been active an opportunity to rest while others take over. Changing of the shifts is carefully coordinated so that the sustained contraction is smooth rather than jerky. Asynchronous motor-unit recruitment is possible only for submaximal contractions during which only a portion of the motor units are required to maintain the desired level of tension. During maximal contractions, when participation of all of the muscle fibers is essential, it is impossible to alternate motor-unit activity to prevent fatigue. This is one reason why you cannot support a heavy object as long as one that is light.

Furthermore, the type of muscle fiber that is activated varies with the extent of gradation. Most muscles consist of a mixture of fiber types that differ metabolically, some being more resistant to fatigue than others. During weak or moderate endurance-type activities (aerobic exercise), the motor units most resistant to fatigue are recruited first. The last fibers to be called into play in the face of demands for further increases in tension are those that fatigue rapidly. An individual can therefore engage in endurance activities for prolonged periods of time but can only briefly maintain bursts of all-out, powerful effort. Of course, even the muscle fibers most resistant to fatigue will eventually fatigue if required to maintain a certain level of sustained tension.

The frequency of stimulation can influence the tension developed by each muscle fiber.

Whole-muscle tension depends not only on the number of muscle fibers contracting but also on the tension developed

(a)   (b)

● **FIGURE 8–15   Comparison of Motor-Unit Recruitment in Muscles with Small Motor Units and Muscles with Large Motor Units** (a) Small incremental increases in strength of contraction occur during motor-unit recruitment in muscles with small motor units because only a few additional fibers are called into play as each motor unit is recruited. (b) Large incremental increases in strength of contraction occur during motor-unit recruitment in muscles with large motor units because so many additional fibers are stimulated with the recruitment of each additional motor unit.

by each contracting fiber. Various factors influence the extent to which tension can be developed. These factors include:

1. the frequency of stimulation;
2. the length of the fiber at the onset of contraction;
3. the extent of fatigue; and
4. the thickness of the fiber.

We will examine each of these factors in turn, concentrating first on the effect of frequency of stimulation.

Even though a single action potential in a muscle fiber produces only a twitch, contractions with longer duration and greater tension can be achieved by repetitive stimulation of the fiber. Let us see what happens when a second action potential occurs in a muscle fiber. If the muscle fiber has completely relaxed before the next action potential takes place, a second twitch of the same magnitude as the first occurs (• Fig. 8–16a). The same excitation-contraction events take place each time, resulting in identical twitch responses. If, however, the muscle fiber is stimulated a second time before it has completely relaxed from the first twitch, a second action potential occurs that causes a second contractile response, which is added "piggyback" on top of the first twitch (Fig. 6–16b). The two twitches resulting from the two action potentials add together, or sum, to produce greater tension in the fiber than that produced by a single action potential. This **summation** of twitches is similar to temporal summation of EPSPs at the postsynaptic neuron (see p. 95). Summation is possible only because the duration of the action potential

(1 to 2 msec) is much shorter than the duration of the resultant twitch (100 msec). Remember that once an action potential has been initiated, a brief refractory period occurs during which another action potential cannot be initiated (see p. 90). It is therefore impossible to achieve summation of action potentials. The membrane must return to resting potential and recover from its refractory period before another action potential can occur. However, because the action potential and refractory period are over long before the resultant muscle twitch is completed, the muscle fiber may be restimulated while some contractile activity still exists to produce summation of the mechanical response. If the muscle fiber is stimulated so rapidly that it does not have a chance to relax at all between stimuli, a smooth, sustained contraction of maximal strength known as **tetanus** occurs (Fig. 8–16c). A tetanic contraction is usually three to four times stronger than a single twitch. (This normal physiological tetanus should not be confused with the disease tetanus; see p. 99.)

What is the mechanism of summation and tetanus at the cellular level? The tension produced by a contracting muscle fiber increases as the thin filaments slide further inward as a result of greater cross-bridge cycling. As the frequency of action potentials increases, the extent of filament sliding and resultant tension development increase until a maximum tetanic contraction is achieved. Sufficient Ca^{++} is released in response to a single action potential to interact with all of the troponin within the cell. As a result, all of the cross bridges are free to participate in the contractile response. How then,

---

• **FIGURE 8–16   Summation and Tetanus** (a) If a muscle fiber is restimulated after it has completely relaxed, the second twitch is the same magnitude as the first twitch. (b) If a muscle fiber is restimulated before it has completely relaxed, the second twitch is added on to the first twitch resulting in summation. (c) If a muscle fiber is stimulated so rapidly that it does not have an opportunity to relax at all between stimuli, a maximal sustained contraction known as tetanus occurs.

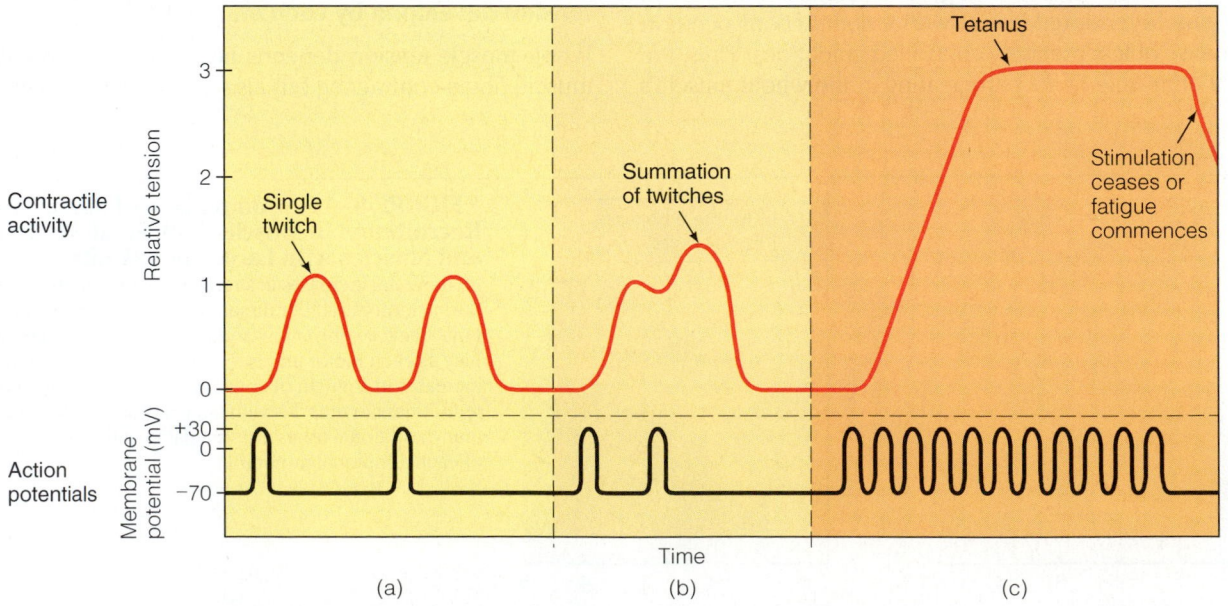

The duration of the action potentials is not drawn to scale but is exaggerated.

can repetitive action potentials bring about a greater contractile response? The difference depends on how long sufficient Ca^{++} is available.

The cross bridges will remain active and continue to cycle as long as sufficient Ca^{++} is present to keep the troponin-tropomyosin complex away from the cross-bridge binding sites on actin. Each troponin-tropomyosin complex spans a distance of seven actin molecules. Thus, binding of one Ca^{++} ion to one troponin molecule leads to the uncovering of only seven cross-bridge binding sites on the thin filament.

As soon as Ca^{++} is released in response to an action potential, the sarcoplasmic reticulum starts pumping Ca^{++} back into the lateral sacs. As the cytosolic Ca^{++} concentration declines with the reuptake of Ca^{++} by the lateral sacs, less Ca^{++} is present to bind with troponin, so some of the troponin-tropomyosin complexes slip back into their blocking positions. Consequently, not all of the cross-bridge binding sites remain available to participate in the cycling process during a single twitch induced by a single action potential.

If action potentials and twitches occur far enough apart in time for all of the released Ca^{++} from the first contractile response to be pumped back into the lateral sacs between the action potentials, an identical twitch response will occur as a result of the second action potential. With both action potentials, the same extent of cross-bridge cycling will take place. If, however, a second action potential occurs and more Ca^{++} is released while the Ca^{++} that was released in response to the first action potential is being taken back up, the cytosolic Ca^{++} concentration remains elevated. This prolonged availability of Ca^{++} in the cytosol permits more of the cross bridges to continue participating in the cycling process for a longer time.

With summation, some of the contractile activity that resulted from the first action potential is still present when the second action potential takes place (that is, the thin filaments have not been completely returned to their resting position) by the time the second action potential occurs. Because cross-bridge cycling is sustained as a result of another spurt of Ca^{++} release in response to the second action potential, the thin filaments slide in even further. As the frequency of action potentials increases, the duration of elevated cytosolic Ca^{++} concentration increases, and the magnitude of cross-bridge cycling and tension development increase correspondingly until a maximum tetanic contraction is reached. With tetanus, the maximum number of cross-bridge binding sites remain uncovered so that cross-bridge cycling, and consequently tension development, are at their peak.

Because skeletal muscle must be stimulated by motor neurons to contract, the nervous system plays a key role in regulating the strength of contraction. The two main factors subject to control to accomplish gradation of contraction are the *number of motor units stimulated* and the *frequency of their stimulation*. The areas of the brain responsible for directing motor activity use a combination of tetanic contractions and precisely timed shifts of asynchronous motor-unit recruitment to execute smooth rather than jerky contractions.

Additional factors not directly under nervous control also influence the tension developed during contraction. Among these is the length of the fiber at the onset of contraction, to which we now turn our attention.

There is an optimal muscle length at which maximal tension can be developed upon a subsequent contraction.

A relationship exists between the length of the muscle before the onset of contraction and the tetanic tension that each contracting fiber can subsequently develop at that length. For every muscle there is an **optimal length** (l_o) at which maximal force can be achieved upon a subsequent tetanic contraction. The tension that can be achieved during tetanus at the optimal muscle length is greater than the tetanic tension that can be achieved when the contraction begins with the muscle less than or greater than its optimal length. This **length-tension relationship** can be explained by the sliding-filament mechanism of muscle contraction. At l_o, when maximum tension can be developed (Point A in • Fig. 8–17), the thin filaments optimally overlap the regions of the thick filaments from which the cross bridges project. The central region of thick filaments is void of cross bridges; only myosin tails are found here. At l_o, because a maximal number of cross-bridge sites are accessible to the actin molecules for binding and bending, maximum tension can be developed. At greater lengths, as when a muscle is passively stretched (Point B), the thin filaments are pulled out from between the thick filaments, decreasing the number of actin sites available for cross-bridge binding; that is, some of the actin sites and cross bridges no longer "match up" so they "go unused." When less cross-bridge activity can occur, less tension can be developed. In fact, when the muscle is stretched to about 70% longer than its l_o (Point C), the thin filaments are completely pulled out from between the thick filaments so that no cross-bridge activity and consequently no contraction can occur. If a muscle is shorter than l_o before contraction (Point D), less tension can be developed for three reasons:

1. The thin filaments from the opposite sides of the sarcomere become overlapped, decreasing the number of actin sites exposed to the cross bridges.
2. The thick filaments become forced against the Z lines, so further shortening is impeded.
3. Besides these two mechanical factors, at muscle lengths less than 80% of l_o, not as much Ca^{++} is released during excitation-contraction coupling for reasons unknown. Consequently, fewer actin sites are uncovered for participation in cross-bridge activity.

The extremes in muscle length that prevent the development of tension occur only under experimental conditions, when a muscle is removed and stimulated at various lengths. In the body the muscles are so positioned that their relaxed length is approximately their optimal length; thus, they are capable of achieving near-maximal tetanic contraction most of the time. Because of limitations imposed by attachment to

• FIGURE 8–17 Length-Tension Relationship Maximal tetanic contraction can be achieved when a muscle fiber is at its optimal length (l_o) before the onset of contraction because of optimal overlap of thick-filament cross bridges and thin-filament cross-bridge binding sites (Point A). The percentage of maximal tetanic contraction that can be achieved decreases when the muscle fiber is longer or shorter than l_o prior to contraction. When it is longer, fewer thin-filament binding sites are accessible for binding with thick-filament cross bridges because the thin filaments are pulled out from between the thick filaments (Points B and C). When the fiber is shorter, fewer thin-filament binding sites are exposed to thick-filament cross bridges because the thin filaments overlap (Point D). Also, further shortening and tension development are impeded as the thick filaments become forced against the Z lines (Point D). In the body, the resting muscle length is at l_o. Furthermore, because of restrictions imposed by skeletal attachments, muscles cannot vary beyond 30% of their l_o in either direction (the range screened in light green). At the outer limits of this range, muscles are still able to achieve about 50% of their maximal tetanic contraction.

the skeleton, a muscle cannot be stretched or shortened more than 30% of its resting optimal length, and usually it deviates much less than 30% from normal length. Even at the outer limits (130% and 70% of l_o), the muscles are still able to generate half their maximum tension.

The factors we have discussed thus far that influence how much tension can be developed by a contracting muscle fiber—the frequency of stimulation and the muscle length at the onset of contraction—can vary from contraction to contraction. Other determinants of muscle-fiber tension—the metabolic capability of the fiber relative to resistance to fatigue and the thickness of the fiber—do not vary from contraction to contraction but depend on the fiber type and can be modified over a period of time. We will consider these other factors in the next section on skeletal-muscle metabolism and fiber types after we complete our discussion of skeletal-muscle mechanics.

The two primary types of contraction are isotonic and isometric.

Cross-bridge activity produces **tension** within the sarcomeres, the **contractile component** of the muscle. However, the sarcomeres are not directly attached to the bones. Instead, the tension generated by these contractile elements must be transmitted to the bone via the connective tissue and tendons before the bone can be moved. Connective tissue, as well as other components of the muscle such as the sarcoplasmic reticulum, exhibits a certain degree of passive elasticity. These noncontractile tissues are referred to as the **series-elastic component** of the muscle; they behave like a stretchy spring

placed between the internal tension-generating elements and the bone that is to be moved against an external load (• Fig. 8–18). Shortening of the sarcomeres stretches the series-elastic component. Muscle tension is transmitted to the bone by means of this tightening of the series-elastic component. This externally applied tension is responsible for moving the bone against a load.

Typically, a muscle is attached to at least two different bones across a joint by means of tendons that extend from each end of the muscle (• Fig. 8–19). When the muscle shortens during contraction, the position of the joint is changed as one bone is moved in relationship to the other—for example, *flexion* of the elbow joint by contraction of the biceps muscle and *extension* of the elbow by contraction of the triceps. The end of the muscle attached to the more stationary part of the skeleton is called the **origin**, whereas the end attached to the skeletal part that moves is referred to as the **insertion**.

Not all muscle contractions result in muscle shortening and movement of bones, however. For a muscle to shorten during contraction, the tension developed in the muscle must exceed the forces that oppose movement of the bone to which the muscle's insertion is attached. In the case of elbow flexion, the opposing force, or **load**, is the weight of an object being lifted. When you flex your elbow without lifting any external object, there is still a load, albeit minimal—the weight of your forearm being moved against the force of gravity.

There are two primary types of contraction, depending on whether the muscle changes length during contraction. In an **isotonic contraction**, muscle tension remains constant as the muscle changes length. In an **isometric contraction**, the muscle is prevented from shortening, so the development of

• **FIGURE 8–18   Relationship between Contractile Component and Series-Elastic Component in Transmission of Muscle Tension to Bone** Muscle tension is transmitted to the bone by means of the stretching and tightening of the muscle's elastic connective tissue and tendon as a result of sarcomere shortening brought about by cross-bridge cycling.

Contractile component
(sarcomeres)

Series-elastic
component (connective
tissue/tendon)

Load

Load

tension occurs at constant muscle length. The same internal events occur in both isotonic and isometric contractions: the tension-generating contractile process is turned on by muscle excitation; the cross bridges start cycling; and filament sliding shortens the sarcomeres, which stretches the series-elastic component to exert tension on the bone at the site of the muscle's insertion.

Considering your biceps as an example, assume that you are going to lift an object. When the tension developing in your biceps becomes great enough to overcome the weight of the object in your hand, you can lift the object, with the whole muscle shortening in the process. Because the weight of the object does not change as it is lifted, the muscle tension remains constant throughout the period of shortening. This is an *isotonic* (literally, "constant tension") contraction. Isotonic contractions are used for body movements and to do work by moving external objects. What happens if you try to lift an object too heavy for you (that is, if the tension you are capable of developing in your arm muscles is less than that

required to lift the load)? In this case, the muscle cannot shorten and lift the object but remains at constant length in spite of the development of tension, so an *isometric* ("constant length") contraction occurs. In addition to occurring when the load is too great, isometric contractions also take place when the tension developed in the muscle is deliberately less than that needed to move the load. In this case, the goal is to keep the muscle at fixed length although it is capable of developing more tension. These submaximal isometric contractions are important for maintaining posture (such as keeping the legs stiff while standing) and for supporting objects in a fixed position. During a given movement, a muscle may shift between isotonic and isometric contractions. For example, when you pick up a book to read, your biceps undergoes an isotonic contraction while the book is being lifted, but the contraction becomes isometric as you stop to hold the book in front of you.

There are actually two types of isotonic contraction—**concentric** and **eccentric**. In both cases the muscle changes

• **FIGURE 8–19   Extension and Flexion of Elbow Joint**

length at constant tension. With concentric contractions, however, the muscle shortens, whereas with eccentric contractions the muscle lengthens because it is being stretched by an external force while contracting. With an eccentric contraction, the contractile activity is resisting the stretch. An example is lowering a load to the ground. During this action, the muscle fibers in the biceps are lengthening but are still contracting in opposition to being stretched. This tension supports the weight of the object.

The body is not limited to pure isotonic and isometric contractions. Muscle length and tension frequently vary throughout a range of motion. Think about drawing a bow and arrow. The biceps' muscle tension continuously increases to overcome the progressively increasing resistance as the bow is stretched further. At the same time, the muscle progressively shortens as the bow is drawn farther back. Such a contraction occurs at neither constant tension nor constant length.

Some skeletal muscles do not attach to bones at both ends but still produce movement. For example, the muscles of the tongue are not attached at the free end. Isotonic contractions of the tongue muscles maneuver the free unattached portion of the tongue to facilitate speech and eating. The external eye muscles attach to the skull at their origin but to the eye at their insertion. Isotonic contractions of these muscles produce the eye movements that enable us to track moving objects, read, and so on. A few skeletal muscles that are completely unattached to bone actually prevent movement. These are the voluntarily controlled rings of skeletal muscles known as **sphincters** that guard the exit of urine and feces from the body by isotonically contracting.

The velocity of shortening is related to the load.

The load is also an important determinant of the **velocity** or speed of shortening (• Fig. 8–20). The greater the load, the lower the velocity at which the muscle fiber shortens during an isotonic tetanic contraction. The velocity of shortening is maximal when there is no external load, progressively decreases with an increasing load, and falls to zero (no shortening—isometric contraction) when the load cannot be overcome by maximal tetanic tension. You have frequently experienced this load-velocity relationship. You can lift light objects requiring little muscle tension quickly, whereas you

• **FIGURE 8–20   Load-Velocity Relationship** The velocity of shortening decreases as the load increases.

can lift very heavy objects only slowly, if at all. This relationship between load and shortening velocity is a fundamental property of muscle, presumably because it takes the cross bridges longer to stroke against a greater load.

Although muscles can accomplish work, much of the energy is converted to heat.

Muscle accomplishes work in a physical sense only when an object is moved. **Work** is defined as force times distance. **Force** can be equated to the muscle tension required to overcome the load (the weight of the object). The amount of work accomplished by a contracting muscle therefore depends on how much an object weighs and how far it is moved. In an isometric contraction when no object is moved, the muscle contraction's efficiency as a producer of external work is zero. All energy consumed by the muscle during the contraction is converted to heat. In an isotonic contraction, the muscle's efficiency is about 25%. Of the energy consumed by the muscle during the contraction, 25% is realized as external work whereas the remaining 75% is converted to heat. Much of this heat is not really wasted in a physiological sense because it is used in maintaining body temperature. In fact, shivering, a form of involuntarily induced skeletal-muscle contraction, is a well-known mechanism for increasing heat production on a cold day. Heavy exercise on a hot day, on the other hand, may overheat the body, because the normal

heat-loss mechanisms may be unable to compensate for this increase in heat production (chapter 17).

Interactive units of skeletal muscles, bones, and joints form lever systems.

Skeletal muscles are attached to bones across joints, forming lever systems. A **lever** is a rigid structure capable of moving around a pivot point known as a **fulcrum**. In the body, the bones function as levers, the joints serve as fulcrums, and the skeletal muscles provide the force to move the bones. The portion of a lever between the fulcrum and the point where an upward force is applied is called the **power arm**; the portion between the fulcrum and the downward force exerted by a load is known as the **load arm** (● Fig. 8–21a).

The most common type of lever system in the body is exemplified by flexion of the elbow joint. Skeletal muscles, such as the biceps whose contraction flexes the elbow joint, consist of many parallel (side-by-side) tension-generating fibers that can exert a large force at their insertion but shorten only a small distance and at relatively slow velocity. The lever system of the elbow joint amplifies the slow, short movements of the biceps to produce more rapid movements of the hand that cover a greater distance. Consider how an object weighing 5 kg is lifted by the hand (Fig. 8–21b). When the biceps contracts, it exerts an upward force at the point where it inserts on the forearm bone about 5 cm away from the elbow

● **FIGURE 8–21 Lever Systems of Muscles, Bones, and Joints** (a) Schematic representation of the most common type of lever system in the body, showing the location of the fulcrum, the upward force, the downward force, the power arm, and the load arm. (b) Flexion of the elbow joint as an example of lever action in the body. Note that the lever ratio (length of power arm to length of load arm) is 1:7 (5 cm:35 cm), which amplifies the distance and velocity of movement seven times [distance moved by muscle (extent of shortening) = 1 cm, distance moved by hand = 7 cm; velocity of muscle shortening = 1 cm/unit of time, hand velocity = 7 cm/unit of time], but at the expense of the muscle having to exert seven times the force of the load (muscle force = 35 kg, load = 5 kg).

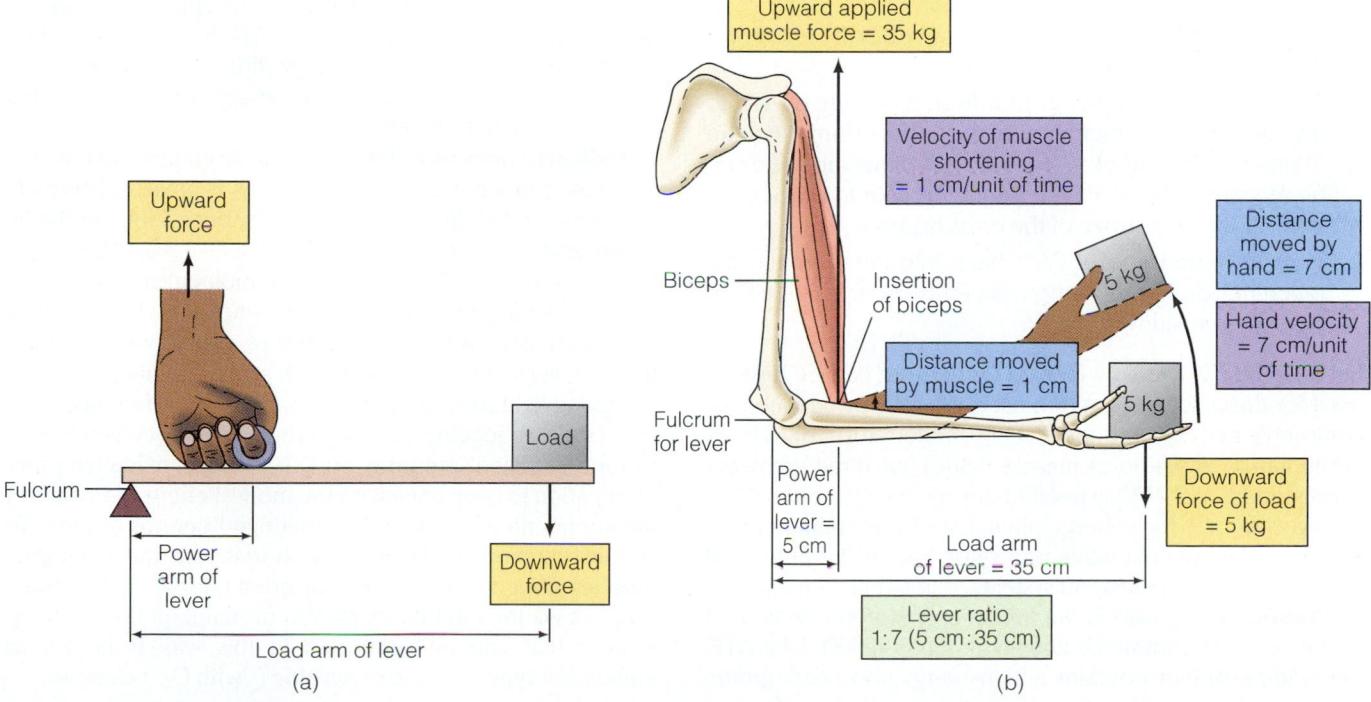

joint, the fulcrum. Thus, the power arm of this lever system is 5 cm long. The length of the load arm, the distance from the elbow joint to the hand, averages 35 cm. In this case, the load arm is seven times longer than the power arm, which enables the load to be moved a distance seven times greater than the shortening distance of the muscle (while the biceps shortens a distance of 1 cm, the hand moves the load a distance of 7 cm) and at a velocity seven times greater (the hand moves 7 cm during the same length of time the biceps shortens 1 cm). The disadvantage of this lever system is that at its insertion the muscle must exert a force seven times greater than the load. The product of the length of the power arm times the upward force applied must equal the product of the length of the load arm times the downward force exerted by the load. Because the load arm times the downward force is 35 cm × 5 kg, the power arm times the upward force must be 5 cm × 35 kg (the force that must be exerted by the muscle to be in mechanical equilibrium). Thus, skeletal muscles typically work at a mechanical disadvantage in that they must exert a considerably greater force than the actual load to be moved. Nevertheless, the amplification of velocity and distance afforded by the lever arrangement enables muscles to move loads faster over greater distances than would otherwise be possible. This amplification provides valuable maneuverability and speed.

SKELETAL-MUSCLE METABOLISM AND FIBER TYPES

Muscle fibers have alternate pathways for forming ATP.

Three different steps in the contraction-relaxation process require ATP:

1. Splitting of ATP by myosin ATPase provides the energy for the power stroke of the cross bridge.

2. Binding (but not splitting) of a fresh molecule of ATP to myosin permits detachment of the bridge from the actin filament at the end of a power stroke so that the cycle can be repeated. This ATP is subsequently split to provide energy for the next stroke of the cross bridge.

3. The active transport of Ca^{++} back into the sarcoplasmic reticulum during relaxation depends on energy derived from the breakdown of ATP.

Because ATP is the only energy source that can be directly used for these activities, ATP must constantly be supplied for contractile activity to continue. Only limited stores of ATP are immediately available in muscle tissue, but three pathways supply additional ATP as needed during muscle contraction: (1) transfer of a high-energy phosphate from creatine phosphate to ADP; (2) oxidative phosphorylation (the citric-acid cycle and electron-transport system); and (3) glycolysis.

Creatine phosphate is the first energy storehouse tapped at the onset of contractile activity (● Fig. 8–22ⓐ). Like ATP, creatine phosphate contains a high-energy phosphate group,

which can be donated directly to ADP to form ATP. Just as energy is released when the terminal phosphate bond in ATP is split, similarly energy is released when the bond between phosphate and creatine is broken. The energy released from the hydrolysis of creatine phosphate, along with the phosphate, can be donated directly to ADP to form ATP. This reaction, which is catalyzed by the muscle cell enzyme **creatine kinase**, is reversible; energy and phosphate from ATP can be transferred to creatine to form creatine phosphate:

$$\text{creatine phosphate} + \text{ADP} \underset{}{\overset{\text{creatine kinase}}{\rightleftharpoons}} \text{creatine} + \text{ATP}$$

As the energy reserves are built up in a resting muscle, the increased concentration of ATP favors the transfer of the high-energy phosphate group to creatine phosphate, in accordance with the law of mass action (see p. 443). Thus, most energy is stored in muscle in creatine phosphate pools. A rested muscle contains about five times as much creatine phosphate as ATP. At the onset of contraction, when the meager reserves of ATP are rapidly utilized, the reaction is reversed. Additional ATP is quickly formed by the transfer of energy and phosphate from creatine phosphate to ADP. Because only one enzymatic reaction is involved in this energy transfer, ATP can be formed rapidly (within a fraction of a second) by using creatine phosphate.

Thus, creatine phosphate is the first source for supplying additional ATP when exercise begins. Muscle ATP levels actually remain fairly constant early in contraction, but creatine phosphate stores become depleted. In fact, short bursts of high-intensity contractile effort, such as high jumps or sprints, are supported primarily by ATP derived at the expense of creatine phosphate. Other energy systems do not have a chance to become operable before the activity is over. If the energy-dependent contractile activity is to be continued, the muscle shifts to the alternate pathways of oxidative phosphorylation and glycolysis to form ATP. These multistepped pathways require time to pick up their rates of ATP formation to match the increased demands for energy, time that has been provided by the immediate supply of energy from the one-step creatine phosphate system.

Oxidative phosphorylation (see p. 32) takes place within the muscle mitochondria if sufficient O_2 is present. This pathway is fueled by glucose or fatty acids, depending on the intensity and duration of the activity (Fig. 8–22ⓑ). Although it provides a rich yield of thirty-six ATP molecules for each glucose molecule processed, oxidative phosphorylation is relatively slow because of the number of steps involved, and it necessitates a constant supply of O_2 and nutrient fuel.

During light exercise (such as walking) to moderate exercise (such as jogging or swimming), muscle cells are able to form sufficient amounts of ATP through oxidative phosphorylation to keep pace with the modest energy demands of the contractile machinery for prolonged periods of time. To sustain ongoing oxidative phosphorylation, the exercising muscles depend on delivery of adequate O_2 and nutrient supplies via the circulatory system to maintain their activity. Activity that can be supported in this way is known as **endurance-type exercise** or **aerobic** ("with O_2") **exercise**.

● FIGURE 8–22 Metabolic Pathways Producing ATP Utilized during Muscle Contraction and Relaxation
During muscle contraction, ATP is split by myosin ATPase to power cross-bridge stroking; during relaxation, ATP is needed to run the Ca^{++} pump that transports Ca^{++} back into the sarcoplasmic reticulum's lateral sacs. The metabolic pathways that supply the ATP needed to accomplish contraction and relaxation are ⓐ transfer of a high-energy phosphate from creatine phosphate to ADP (the immediate source); ⓑ oxidative phosphorylation (the main source when O_2 is present), fueled by glucose derived from muscle glycogen stores or by glucose and fatty acids delivered by the blood; and ⓒ glycolysis (the main source when O_2 is not present). The end product of glycolysis, pyruvic acid, is converted into lactic acid when the lack of O_2 prevents pyruvic acid from being further processed by the oxidative-phosphorylation pathway.

The O_2 required for oxidative phosphorylation is primarily delivered by the blood. Increased O_2 is made available to muscles during exercise by several mechanisms: deeper, more rapid breathing brings in more O_2; the heart contracts more rapidly and forcefully to pump more oxygenated blood to the tissues; more blood is diverted to the exercising muscles by dilation of the blood vessels supplying them; and the hemoglobin molecules that carry the O_2 in the blood release more O_2 in exercising muscles. (These mechanisms are discussed further in later chapters.) Furthermore, some types of muscle fibers have an abundance of **myoglobin**, which is similar to hemoglobin. Myoglobin can store small amounts of

O_2, but, more importantly, it increases the rate of O_2 transfer from the blood into muscle fibers.

Glucose and fatty acids, ultimately derived from ingested food, are also delivered to the muscle cells by the blood. Excess ingested nutrients not immediately utilized are stored in the liver as glycogen (chains of glucose) and in adipose tissue as fat. Additionally, muscle cells are able to store limited quantities of glucose in the form of glycogen.

There are cardiovascular limits to the amount of O_2 that can be delivered to a muscle. In near-maximal contractions, the blood vessels that course through the muscle are almost closed by the powerful contraction, severely limiting the O_2 available to the muscle fibers. Furthermore, even when O_2 is available, the relatively slow oxidative-phosphorylation system may not be able to produce ATP rapidly enough to meet the muscle's needs during intense activity. When O_2 delivery or oxidative phosphorylation cannot keep pace with the demand for ATP formation as the intensity of exercise increases, the muscle fibers rely increasingly on **glycolysis** (Fig. 8–22©) to generate ATP (see p. 27). The chemical reactions of glycolysis yield products for ultimate entry into the oxidative phosphorylation pathway, but glycolysis can also proceed alone in the absence of further processing of its products by oxidative phosphorylation. During glycolysis, a glucose molecule is broken down into two **pyruvic-acid** molecules, yielding two ATP molecules in the process. Pyruvic acid can be further degraded by oxidative phosphorylation to extract more energy. However, glycolysis alone has two advantages over the oxidative-phosphorylation pathway: (1) glycolysis can form ATP in the absence of O_2 (operating *anaerobically,* that is "without O_2"), and (2) it can proceed more rapidly than oxidative phosphorylation because it requires fewer steps. Although glycolysis extracts considerably fewer ATP molecules from each nutrient molecule processed, it can proceed so much more rapidly that it can outproduce oxidative phosphorylation over a given period of time if sufficient glucose is present.

Even though anaerobic glycolysis provides a means of performing intense exercise when the O_2 delivery/oxidative-phosphorylation capacity is exceeded, using this pathway has two consequences. First, large amounts of nutrient fuel must be processed, because glycolysis is much less efficient than oxidative phosphorylation in converting nutrient energy into the energy of ATP. (Glycolysis yields a net of two ATP molecules for each glucose molecule degraded, whereas the oxidative-phosphorylation pathway can extract thirty-six molecules of ATP from each glucose molecule.) Muscle cells are able to store limited quantities of glucose in the form of glycogen, but anaerobic glycolysis rapidly depletes the muscle's glycogen supplies. Second, the end product of anaerobic glycolysis, pyruvic acid, is converted to **lactic acid** when it cannot be further processed by the oxidative-phosphorylation pathway. Lactic-acid accumulation has been implicated in the muscle soreness that occurs during the time that intense exercise is actually taking place. (The delayed-onset pain and stiffness that begin the day after unaccustomed muscular exertion, however, are probably caused by reversible structural damage.) Furthermore, lactic acid picked up by the blood is responsible for the metabolic acidosis accompanying intense exercise. Both depletion of energy reserves and the fall in muscle pH caused by lactic-acid accumulation are believed to play a role in the onset of muscle fatigue, a topic to which we next turn our attention. Therefore, **high-intensity** or **anaerobic exercise** can be sustained for only a short duration in contrast to the body's prolonged ability to sustain aerobic activities.

Fatigue has multiple causes.

Contractile activity in a particular skeletal muscle cannot be maintained at a given level indefinitely. Eventually, the tension in the muscle declines as fatigue sets in. There are apparently three different types of fatigue: muscle fatigue, neuromuscular fatigue, and central fatigue.

Muscle fatigue occurs when an exercising muscle can no longer respond to stimulation with the same degree of contractile activity. The underlying causes of muscle fatigue are unclear. The primary implicated factors include (1) accumulation of lactic acid, which may inhibit key enzymes in the energy-producing pathways or excitation-contraction coupling process, and (2) depletion of energy reserves. The time of onset of fatigue varies with the type of muscle fiber, some fibers being more resistant to fatigue than others, and the intensity of the exercise, with more rapid onset of fatigue associated with high-intensity activities.

Evidence suggests that the limiting factor during fast, powerful activities may be at the neuromuscular junction. With **neuromuscular fatigue**, the active motor neurons are not able to synthesize acetylcholine rapidly enough to sustain chemical transmission of action potentials from the motor neurons to the muscles.

Central fatigue, also known as **psychological fatigue**, occurs when the CNS no longer adequately activates the motor neurons supplying the working muscles. The person slows down or stops exercising even though the muscles are still able to perform. During strenuous exercise, central fatigue may stem from discomfort associated with the activity; it takes strong motivation (a will to win) to deliberately persevere when in pain. In less strenuous activities, central fatigue may bring about reduced physical performance in association with boredom and monotony (such as assembly-line work) or tiredness (lack of sleep). The mechanisms involved in central fatigue are poorly understood.

Increased oxygen consumption is necessary to recover from exercise.

A person continues to breathe deeply and rapidly for a period of time after having stopped exercising. The necessity for the elevated O_2 uptake during recovery from exercise is due to a variety of factors. The best known is repayment of an **oxygen debt** that was incurred during exercise, when contractile activity was being supported by ATP derived from nonoxidative sources such as creatine phosphate and anaerobic glycolysis. Oxygen is needed for recovery of the energy systems. During exercise, the creatine phosphate stores of active muscles are reduced, lactic acid may accumulate, and glycogen stores may be tapped; the extent of these effects depends on the intensity and duration of the activity. During the recovery period, fresh supplies of ATP are formed by oxidative phosphorylation using the newly acquired O_2, which is provided by the sustained increase in respiratory activity after exercise has stopped. Most of this ATP is used to resynthesize creatine phosphate to restore its reserves. This can be accomplished in a matter of a few minutes. Any accumulated lactic acid is converted back into pyruvic acid, part of which is used by the oxidative-phosphorylation system for ATP production. The remainder of the pyruvic acid is converted back into glucose by the liver. Most of this glucose, in turn, is used to replenish the glycogen stores drained from the muscles and liver during exercise. These biochemical transformations involving pyruvic acid require O_2 and take several hours for completion. Thus, as the O_2 debt is repaid, the creatine phosphate system is restored, lactic acid is removed, and glycogen stores are at least partially replenished. Unrelated to increased O_2 uptake is the need to restore nutrients after grueling exercise, such as marathon races, in which glycogen stores are severely depleted. In such cases, long-term recovery can take a day or more because the exhausted energy stores require nutrient intake for full replenishment. Therefore, depending on the type and duration of activity, recovery can be complete within a few minutes or can require more than a day.

Part of the extra O_2 uptake during recovery is not directly related to the repayment of energy stores but instead is the result of a general metabolic disturbance following exercise. For example, the local increase in muscle temperature arising from heat-generating contractile activity speeds up the rate of all chemical reactions in the muscle tissue, including those dependent on O_2. Likewise, the secretion of epinephrine, a hormone that increases O_2 consumption by the body, is elevated during exercise. Until the circulating level of epinephrine returns to its pre-exercise state, O_2 uptake will be increased above normal.

There are three types of skeletal-muscle fibers based on differences in ATP hydrolysis and synthesis.

Based on their biochemical capacities, there are three major types of muscle fibers (Table 8–3):

1. Slow-oxidative (type I) fibers.
2. Fast-oxidative (type IIa) fibers.
3. Fast-glycolytic (type IIb) fibers.

As their names imply, the two main differences between these fiber types are their speed of contraction (slow or fast) and the type of enzymatic machinery they primarily use for ATP formation (oxidative or glycolytic). Fast fibers have higher myosin ATPase activity than slow fibers. The higher the ATPase activity, the more rapidly ATP is split and the faster the rate at which energy is made available for cross-bridge cycling. The result is a fast twitch, compared to the slower twitches of those fibers that split ATP more slowly. Thus, two factors determine the speed with which a muscle contracts:

TABLE 8–3	Characteristics of Skeletal Muscle Fibers		
	TYPE OF FIBER		
CHARACTER-ISTIC	**Slow-Oxidative (Type I)**	**Fast-Oxidative (Type IIa)**	**Fast-Glycolytic (Type IIb)**
Myosin-ATPase activity	Low	High	High
Speed of contraction	Slow	Fast	Fast
Resistance to fatigue	High	Intermediate	Low
Oxidative-phosphorylation capacity	High	High	Low
Enzymes for anaerobic glycolysis	Low	Intermediate	High
Mitochondria	Many	Many	Few
Capillaries	Many	Many	Few
Myoglobin content	High	High	Low
Color of fiber	Red	Red	White
Glycogen content	Low	Intermediate	High
Fiber diameter	Small	Intermediate	Large
Intensity of contraction	Low	Intermediate	High

the load (load-velocity relationship) and the myosin ATPase activity of the contracting fibers (fast or slow twitch).

Fiber types also differ in their ATP-synthesizing ability. Those with a greater capacity to form ATP are more resistant to fatigue. Some fibers are better equipped for oxidative phosphorylation, whereas others rely primarily on anaerobic glycolysis for synthesizing ATP. Because oxidative phosphorylation yields considerably more ATP from each nutrient molecule processed, it does not readily deplete energy stores. Furthermore, it does not result in lactic-acid accumulation. Oxidative types of muscle fibers are therefore more resistant to fatigue than are glycolytic fibers.

Other related characteristics distinguishing these three fiber types are summarized in Table 8–3. As would be expected, the oxidative fibers, both slow and fast, contain an abundance of mitochondria, the organelles that house the enzymes involved in the oxidative-phosphorylation pathway. Because adequate oxygenation is essential to support this pathway, these fibers are richly supplied with capillaries. Oxidative fibers also have a high myoglobin content. Myoglobin not only helps support oxidative fibers' O_2 dependency, but it also imparts a red color to them, just as oxygenated hemoglobin is responsible for the red color of arterial blood. Accordingly, these muscle fibers often are referred to as **red fibers**.

In contrast, the fast fibers specialized for glycolysis contain few mitochondria but have a high content of glycolytic enzymes instead. Also, to supply the large amounts of glucose necessary for glycolysis, they contain an abundance of stored glycogen. Because the glycolytic fibers need relatively less O_2 to function, they receive only a meager capillary supply compared with the oxidative fibers. The glycolytic fibers contain very little myoglobin and therefore are pale in color, so they are sometimes called **white fibers**. (The most readily observable comparison between red and white fibers is the dark and white meat in poultry.) In addition to their higher myosin ATPase content, these fibers are also larger in diameter than the slow-oxidative fibers because of a greater abundance of actin and myosin filaments. With these additional power-generating filaments, the fast-glycolytic fibers are capable of rapidly producing large amounts of tension, but only for a short duration because of their dependence on glycolysis for energy production.

Fast-oxidative fibers share characteristics with each of the other two types. They have high ATPase activity like the fast-glycolytic fibers and high oxidative capacity like the slow-oxidative fibers. They contract more rapidly than the slow-oxidative fibers and can maintain the contraction for a longer period of time than the fast-glycolytic fibers. However, because their rate of ATP production by oxidative phosphorylation cannot keep pace with the high rate of ATP splitting, they rely partially on glycolysis and are more prone to fatigue than the slow-oxidative fibers.

In humans, most of the muscles contain a mixture of all three fiber types; the percentage of each type is largely determined by the type of activity for which the muscle is specialized. Accordingly, a high proportion of slow-oxidative fibers are found in muscles specialized for maintaining low-intensity contractions for long periods of time without fatigue, such as the muscles of the back and legs that support the body's weight against the force of gravity. A preponderance of fast-glycolytic fibers are found in the arm muscles, which are adapted for performing rapid, forceful movements such as lifting heavy objects.

The percentage of these various fibers not only differs between muscles within an individual but also varies considerably among individuals. Most of us have an average of about 50% each of fast and slow fibers. Those genetically endowed with a higher percentage of the fast-glycolytic fibers are good candidates for power and sprint events, whereas those with a greater proportion of slow-oxidative fibers are more likely to be successful in endurance activities such as marathon races. Of course, success in any event will depend on many factors other than genetic endowment, such as the extent and type of training and the level of dedication.

Muscle fibers adapt considerably in response to the demands placed on them.

Not only is the nerve supply to a skeletal muscle essential for initiating contraction, but the motor neurons supplying a skeletal muscle are also important in maintaining the muscle's integrity and chemical composition. Different types of exercise produce different patterns of neuronal discharge to the muscle involved. Depending on the pattern of neural activity, long-term adaptive changes occur in the muscle fibers, enabling them to respond most efficiently to the types of demands placed on the muscle. Two types of changes can be induced in muscle fibers: changes in their ATP-synthesizing capacity and changes in their diameter. Regular endurance (aerobic) exercise, such as long-distance jogging or swimming, induces metabolic changes within the oxidative fibers, which are the ones primarily recruited during aerobic exercises. These changes enable the muscles to use O_2 more efficiently. For example, mitochondria, the organelles that house the enzymes involved in the oxidative-phosphorylation pathway, increase in number in the oxidative fibers. In addition, the number of capillaries supplying blood to these fibers increases. Muscles so adapted are better able to endure prolonged activity without fatiguing, but they do not change in size.

The actual size of the muscles can be increased by regular bouts of anaerobic, short-duration, high-intensity resistance training, such as weight lifting. The resulting muscle enlargement comes primarily from an increase in diameter (**hypertrophy**) of the fast-glycolytic fibers that are called into play during such powerful contractions. Most of the fiber thickening is a consequence of increased synthesis of myosin and actin filaments, which permits a greater opportunity for cross-bridge interaction and subsequently increases the muscle's contractile strength. The resultant bulging muscles are better adapted to activities that require intense strength for brief periods, but endurance has not been improved. Furthermore, **hyperplasia** (an increase in the number of muscle cells) is believed to contribute to a small extent to muscle enlargement. Muscle cells are unable to divide by mitotic cell division, but experimental evidence suggests that greatly enlarged fibers can split lengthwise down the middle, resulting in an increase in the number of fibers.

Men's muscle fibers are thicker and their muscles, accordingly, are larger and stronger than those of women, even without weight training, because of the actions of testosterone, a steroid hormone secreted primarily in males. Testosterone promotes the synthesis and assembly of myosin and actin and is responsible for the naturally larger muscle mass of men. This fact has led some athletes, both males and females, to the dangerous practice of taking this or closely related steroids to increase their athletic performance. (See the accompanying boxed feature, A Closer Look at Exercise Physiology.)

Apparently, all of the muscle fibers within a single motor unit are of the same fiber type. This pattern usually is established early in life, but there is good evidence that the two types of fast-twitch fibers are interconvertible, depending on training efforts. Regular endurance activities can convert fast-glycolytic fibers into fast-oxidative fibers, whereas fast-oxidative fibers can be shifted to fast-glycolytic fibers in response to power events such as weight training. Slow and fast fibers are not interconvertible, however. Although training can induce changes in muscle fibers' metabolic support systems, whether a fiber is fast or slow twitch is apparently established early in development by the fiber's nerve supply.

ARE ATHLETES WHO USE STEROIDS TO GAIN COMPETITIVE ADVANTAGE REALLY WINNERS OR LOSERS?

The testing of athletes for drugs and the much publicized exclusion from competition of those found to be using substances outlawed by sports federations have stirred considerable controversy. One such group of drugs are **anabolic androgenic steroids** (*anabolic* means "buildup of tissues"; *androgenic* means "male producing"; *steroids* are a class of hormone). These agents are closely related to testosterone, the natural male sex hormone, which is responsible for promoting the increased muscle mass characteristic of males.

Although their use is outlawed (possession of anabolic steroids without a prescription became a federal offense in 1991), these agents are taken by many athletes who specialize in power events such as weight lifting and sprinting in the hopes of increasing muscle mass and, accordingly, increasing muscle strength. Both male and female athletes have resorted to using these substances in an attempt to gain a competitive edge. Anabolic steroids are also taken by bodybuilders. Unfortunately, the use of these agents has spread into our nation's high schools. In a 1989 survey of 17,000 high school youths by the National Institute of Drug Abuse, 5% of the respondents reported current or past steroid use to "build muscles" or to "improve appearance."

Studies have confirmed that steroids can increase muscle mass when used in large amounts and coupled with heavy exercise. One reputable study demonstrated an average 8.9-pound gain of lean muscle in bodybuilders who used steroids during a ten-week period. Anecdotal evidence suggests that some steroid users have added as much as 40 pounds of muscle in a year.

The adverse effects of these drugs, however, outweigh any benefits derived. In females, who lack potent androgenic hormones, anabolic steroid drugs not only promote "male-type" muscle mass and strength but also "masculinize" them in other ways, such as by inducing growth of facial hair and by lowering the voice. More importantly, in both males and females, these agents adversely affect the reproductive and cardiovascular systems and the liver and may have an impact on behavior.

Adverse Effects on the Reproductive System

In males, testosterone secretion and sperm production by the testes are normally controlled by hormones from the anterior pituitary gland. In negative-feedback fashion, testosterone inhibits secretion of these controlling hormones so that a constant circulating level of testosterone is maintained. The anterior pituitary is similarly inhibited by androgenic steroids taken as a drug. As a result, because the testes do not receive their normal stimulatory input from the anterior pituitary, testosterone secretion and sperm production decrease and the testes atrophy.

In females, inhibition of the anterior pituitary by the androgenic drugs results in repression of the hormonal output that controls ovarian function. The result is failure to ovulate, menstrual irregularities, and decreased secretion of "femnizing" female sex hormones. Their decline results in diminution in breast size and other female characteristics.

Adverse Effects on the Cardiovascular System

Use of anabolic steroids induces several cardiovascular changes that increase the risk of developing atherosclerosis, which in turn is associated with an increased incidence of heart attacks and strokes (see p. 291). Among these adverse cardiovascular effects are (1) a reduction in high-density lipoproteins (HDL), the "good" cholesterol carriers that help remove cholesterol from the body, and (2) an elevation in blood pressure. Damage to the heart muscle itself has also been demonstrated in animal studies.

Adverse Effects on the Liver

Liver dysfunction is common with high steroid intake, because the liver, which normally inactivates steroid hormones and prepares them for urinary excretion, is overloaded by the excess steroid intake.

Adverse Effects on Behavior

Although the evidence is still controversial, anabolic steroid use appears to promote aggressive, even hostile behavior—so-called roid rages.

Addictive Effects

A troubling new concern is the addiction to anabolic steroids of some who abuse these drugs. In one study involving face-to-face interviews, 14% of steroid users were judged on the basis of their responses to be addicted. In another survey using anonymous, self-administered questionnaires, 57% of steroid users qualified as being addicted. This apparent tendency to become chemically dependent on steroids is alarming because the potential for adverse effects on health increases with long-term, heavy use, the kind of use that would be expected in someone hooked on the drug.

Thus, for health reasons, not even taking into account the legal and ethical issues, people should not use anabolic steroids. However, the problem appears to be growing worse. Federal agencies estimate that $100 million was spent for black-market steroids in the United States in 1988 and the expenditure leaped to $300 to $400 million in 1989.

Experimental switching of motor neurons supplying slow muscle fibers with those supplying fast fibers results in a gradual reversal of the speed at which these fibers contract.

Adaptative changes that take place in skeletal muscle gradually reverse to their original state over a period of months if the regular exercise program that induced these changes is discontinued.

At the other extreme, if a muscle is not used, its actin and myosin content decreases, its fibers become smaller, and the muscle accordingly decreases in mass (**atrophies**) and becomes weaker. Muscle atrophy can result in two ways. **Disuse atrophy** occurs when a muscle is not used for a long period of time even though the nerve supply is intact, as when a cast or brace must be worn or during prolonged bed confinement. **Denervation atrophy** occurs after the nerve supply to a muscle is lost. If the muscle is stimulated electrically until innervation can be reestablished, such as during regeneration of a severed peripheral nerve, atrophy can be diminished but not entirely prevented. Contractile activity itself obviously plays an important role in preventing atrophy; however, poorly understood factors released from active nerve endings, perhaps packaged with the ACh vesicles, apparently contribute to the integrity and growth of muscle tissue.

When a muscle is damaged, limited repair is possible, even though muscle cells cannot divide mitotically to replace damaged cells. A small population of the same undifferentiated cells that formed the muscle during embryonic development, **myoblasts**, remain close to the muscle surface. When a muscle fiber is damaged, it can be replaced by fusion of a group of these myoblasts to form a large multinucleated cell, which immediately begins to synthesize and assemble the intracellular machinery characteristic of the muscle. With extensive injury, this limited mechanism is not adequate to completely replace all of the lost fibers. In that case, the remaining fibers often hypertrophy to compensate.

Transplantation of myoblasts provides one of two glimmers of hope for victims of **muscular dystrophy**, a hereditary pathological condition characterized by progressive degeneration of contractile elements, which are ultimately replaced by fibrous tissue. (See the accompanying boxed feature, Concepts, Challenges, and Controversies.)

With age a person experiences a slow, continuous reduction in skeletal-muscle mass and an accompanying loss of strength. These losses are attributable to the following age-related changes:

■ A reduction in the number of skeletal-muscle fibers. Up to 30% of the fibers in skeletal muscles may be lost by age 80. Many of the lost fibers are replaced by fat tissue, with islands of fat often developing between muscle cells in older individuals. The loss of muscle fibers reduces the size of individual motor units in skeletal muscle. As a consequence, more motor units must be recruited to move a particular weight, so moving the weight requires increased effort in elderly persons.

■ A reduction in diameter of remaining skeletal-muscle fibers resulting from a loss of tension-generating myofibrils. The fast-twitch, glycolytic fibers seem to atrophy earlier than the slow-twitch, oxidative fibers do. Consequently, even though capacity for quick, powerful contractions declines with advancing age, the slow, prolonged contractions used for maintaining posture are not affected significantly, if at all, until very late in life.

■ A reduction in the ability of muscle to adapt to exercise. Some of the loss of muscle mass is nothing more than disuse atrophy—people in general become less physically active as they get older. In fact, the best defense against age-related muscle atrophy and loss of strength is regular exercise. Even modest increases in physical activity have been shown to improve muscle mass and strength in senior citizens in their eighties and beyond. However, even a regular exercise program cannot completely prevent or reverse the age-related decline in skeletal-muscle ability, because a given exercise regimen will produce less pronounced adaptive changes in the skeletal muscles of an older individual than in a younger person. The reason for this reduced adaptive ability is unclear but may result from impairment in the synthesis of new muscle proteins.

■ A reduction in nerve activity to skeletal muscles. The age-related decline in skeletal-muscle mass and strength is partially associated with age-related changes in the nervous system. There is a progressive loss of motor neurons with advancing age, as well as a reduction in the synthesis of acetylcholine by these neurons. The loss of motor-neuron innervation contributes to muscle-fiber atrophy, and lowered levels of acetylcholine reduce the efficiency of muscle stimulation. Furthermore, a reduction in release of newly discovered neuropeptide Y from the motor-neuron terminals may play a role (see p. 204).

CONTROL OF MOTOR MOVEMENT

Many inputs influence motor-unit output.

Particular patterns of motor-unit output are responsible for motor activity, ranging from maintenance of posture and balance to stereotypical locomotor movements, such as walking, to individual, highly skilled motor activity, such as gymnastics. Control of any motor movement, no matter what its level of complexity, depends on converging input to the motor neurons of specific motor units. The motor neurons, in turn, trigger contraction of the muscle fibers within their respective motor units by means of the events that occur at the neuromuscular junction. Three levels of input control motor neuron output:

1. Input from afferent neurons, usually through intervening interneurons, at the level of the spinal cord—that is, spinal reflexes (see p. 142).

2. Input from the primary motor cortex. Fibers originating from cell bodies of pyramidal cells within the primary motor cortex (see p. 118) descend directly without synaptic interruption to terminate on motor neurons (or on local interneurons that terminate on motor neurons). These

MUSCULAR DYSTROPHY:
WHEN ONE SMALL STEP IS A BIG DEAL

Hope of treatment is on the horizon for **muscular dystrophy**, a fatal muscle-wasting disease that primarily strikes boys and unrelentlessly leads to their premature death before age twenty. Muscular dystrophy encompasses a variety of hereditary pathological conditions, which have in common a progressive degeneration of contractile elements and their replacement by fibrous tissue. The gradual muscle wasting in muscular dystrophy is characterized by progressive weakness over a period of years, usually resulting in death from respiratory failure when the respiratory muscles become too weak to function adequately or from heart failure when the heart becomes too weak to beat.

The disease is caused by a recessive genetic defect on the X sex chromosome, of which males have only one copy. (Males have XY sex chromosomes; females have XX sex chromosomes.) If a male inherits from either parent an X chromosome bearing the defective dystrophic gene, he is destined to develop the disease, which affects one out of every 3,500 boys worldwide. Females, on the other hand, must inherit a dystrophic-carrying gene from both parents before acquiring the condition, a much rarer occurrence.

The defective gene responsible for *Duchenne muscular dystrophy*, the most common and most devastating form of the disease, was pinpointed in 1986. The gene normally produces **dystrophin**, a large protein found in the plasma membrane of muscle cells. Dystrophin is linked with the regulation of Ca^{++} flow into muscle cells through special Ca^{++} "leak" channels. Dystrophic muscles are characterized by a lack of dystrophin. Even though this protein represents only 0.002% of the total amount of skeletal-muscle protein, its presence is crucial. Its absence appears to permit a constant leakage

of Ca^{++} into the muscle cells through unregulated Ca^{++} channels. This Ca^{++} presumably activates proteases, protein-snipping enzymes that harm the muscle fibers. The resultant damage leads to the muscle wasting and ultimate fibrosis characterizing the disorder.

Although the disease is still untreatable and fatal, two different lines of research offer hope for the first time to the victims of muscular dystrophy—a cell-transplant and a gene-therapy approach. The cell-transplant technique involves the injection of myoblasts into the dystrophic muscles. **Myoblasts** ("muscle formers") are the undifferentiated cells that fuse to form the large, multinucleated skeletal-muscle cells during embryonic development. After development, a small group of these immature myoblasts remain close to the muscle surface. Even though muscle cells cannot divide mitotically to replace damaged cells, limited repair of a damaged muscle is possible through the activity of these myoblasts. A group of myoblasts can fuse together to form a new skeletal-muscle cell to replace a damaged cell. When the loss of muscle cells is extensive, however, as in muscular dystrophy, this limited mechanism is not adequate to replace all of the lost fibers.

A new therapeutic approach for muscular dystrophy involves the transplantation of dystrophin-producing myoblasts harvested from muscle biopsies of healthy donors into the patient's dwindling muscles. The still-experimental technique was attempted for the first time in 1990 by injection of myoblasts into one big toe of a Duchenne patient. The encouraging results: the patient could wriggle his toe more vigorously. The healthy myoblasts fused with the boy's dystrophic muscle cells, and the donated cell's normal genes produced the deficient dystro-

phin protein. As a result, the treated muscle increased in mass and strength. This early success was followed by more extensive procedures in which solutions containing billions of healthy myoblasts were injected into sixty-nine major muscle groups in the legs and buttocks of twenty-one boys with Duchenne's muscular dystrophy. Preliminary data showed a 43% increase in strength in the treated muscles, with 38% staying the same and 19% continuing to lose strength. Although the interpretation of these results is somewhat controversial, most specialists in the field remain optimistic that myoblast transplantation offers great promise for staving off the gradual weakening and certain death faced by those suffering from muscular dystrophy. Even though this technique cannot be applied to all weakened muscles, the hope is to salvage the muscles essential for breathing and walking so that these patients can lead longer, more normal lives.

An alternative therapeutic approach being sought is a possible "gene fix." One group of researchers have succeeded in inserting normal copies of the human dystrophin gene into the thigh muscles of mice. The gene was incorporated into about 1% of the thigh-muscle cells of the mice and, importantly, produced dystrophin, which was inserted in its proper location in the cells' plasma membranes. If the technique can be improved to increase the number of treated cells that take up the introduced genes, it offers hope for an eventual gene fix for muscular dystrophy.

These steps toward an eventual treatment mean that hopefully one day the afflicted boys will be able to take steps on their own instead of being destined to wheelchairs and early death.

fibers make up the **corticospinal** (or **pyramidal**) **motor system**.

3. Input from the **multineuronal** (or **extrapyramidal**) **motor system**. The pathways composing this system include a number of synapses that involve many regions of the brain. The final link in multineuronal pathways is the brain stem, especially the reticular formation, which in turn is influenced by motor regions of the cortex, the cerebellum, and the basal nuclei. In addition, the motor cortex itself is interconnected with the thalamus as well as with premotor and supplementary motor areas. Only the primary motor cortex and brain stem directly influence mo-

tor neurons; the other involved brain regions indirectly regulate motor activity by adjusting motor output from the motor cortex and brain stem. A number of complex interactions take place between these various brain regions; the most important are represented in ● Figure 8–23. (See chapter 5 for further discussion of the specific roles and interactions of these brain regions.)

The corticospinal system primarily mediates performance of fine, discrete, voluntary movements of the hands and fingers, such as those required for doing intricate needlework. Premotor and supplementary motor areas, with input from the cerebrocerebellum, plan the voluntary motor command

● **FIGURE 8–23 Motor Control** The CNS is constantly apprised of muscle length and tension and other peripheral events via pathways conveying afferent input, so it can program coordinated, purposeful skeletal-muscle activity. Motor movement is controlled by input to the motor neurons from (1) afferent neuron terminals at the level of the spinal cord; (2) the primary motor cortex, via the corticospinal motor system, and (3) brain stem nuclei, which serve as the final link in the complex multineuronal motor system involving many regions of the brain.

that is issued to the appropriate motor neurons by the primary motor cortex through this descending system. The multineuronal system, in contrast, is primarily concerned with regulation of overall body posture involving involuntary movements of large muscle groups of the trunks and limbs. Considerable complex interaction and overlapping of function exist between these two systems. For example, to voluntarily manipulate your fingers to do needlework, you subconsciously assume a particular posture of your arms that enables you to hold your work.

Some of the inputs converging on motor neurons are excitatory, whereas others are inhibitory. Coordinated movement depends on an appropriate balance of activity in these inputs. If an inhibitory system originating in the brain stem is disrupted, muscles become hyperactive (increased muscle tone; augmented limb reflexes) because of the unopposed activity in excitatory inputs to motor neurons, a condition known as **spastic paralysis**. In contrast, loss of excitatory input, such as that accompanying destruction of descending excitatory pathways exiting the primary motor cortex, brings about **flaccid paralysis** (muscles relaxed; inability to voluntarily contract muscles, although reflex activity is still present). Damage to the primary motor cortex on one side of the brain, as with a stroke, leads to flaccid paralysis on the opposite half of the body (**hemiplegia**, or paralysis of one side of the body). Disruption of all descending pathways, as in traumatic severence of the spinal cord, is accompanied by flaccid paralysis below the level of the damaged region— **quadriplegia** (paralysis of all four limbs) in upper spinal cord damage and **paraplegia** (paralysis of the legs) in lower spinal cord injury. Destruction of motor neurons—either their cell bodies or efferent fibers—causes flaccid paralysis and lack of reflex reponsiveness in the affected muscles. Damage to the cerebellum or basal nuclei does not result in paralysis but instead in uncoordinated, clumsy activity and inappropriate patterns of movement. These regions normally smooth out activity initiated voluntarily. Damage to higher cortical regions involved in planning motor activity results in the inability to establish appropriate motor commands to accomplish desired goals.

Muscle spindles and the Golgi tendon organs provide afferent information essential for controlling skeletal-muscle activity.

Coordinated, purposeful skeletal-muscle activity depends on afferent input from a variety of sources. At a simple level, afferent signals indicating your finger is touching a hot stove trigger reflex contractile activity in appropriate arm muscles to withdraw the hand from the injurious stimulus. At a more complex level, if you are going to catch a ball, the motor systems of your brain must program sequential motor commands that will move and position your body correctly for the catch, using predictions of the ball's direction and rate of movement provided by visual input. Many muscles acting simultaneously or alternately at different joints are called into play to shift your body's location and position rapidly, while

maintaining your balance in the process. It is critical to have ongoing input about your body position with respect to the surrounding environment, as well as the position of your various body parts in relationship to each other. This information is necessary for establishing a neuronal pattern of activity to perform the desired movement. Your CNS must know the starting position of your body to appropriately program muscle activity. Further, it must be constantly apprised of the progression of movement it has initiated so that it can make adjustments as needed. Your brain receives this information, which is known as *proprioceptive* input (see p. 119), from receptors in your eyes, joints, vestibular apparatus, and skin, as well as from the muscles themselves. You can demonstrate your joint and muscle proprioceptive receptors in action by closing your eyes and bringing the tips of your right and left index fingers together at any point in space. You can do so without seeing where your hands are because your brain is informed of the position of your hands and other body parts at all times by afferent input from the joint and muscle receptors.

Two types of muscle receptors—*muscle spindles* and *Golgi tendon organs*—monitor changes in muscle length and tension. This information is used in two ways: (1) to apprise motor areas of the brain of muscle length and tension and (2) to control muscle length and tension in negative-feedback fashion by means of local spinal reflexes. Muscle length is monitored by muscle spindles, whereas changes in muscle tension are detected by Golgi tendon organs. Both of these receptor types are activated by muscle stretch, but they are designed to convey different types of information. Let us see how.

Muscle spindles, which are distributed throughout the fleshy part of a skeletal muscle, consist of collections of specialized muscle fibers known as **intrafusal fibers** (*fusus* means "spindle") which lie within spindle-shaped connective-tissue capsules parallel to the "ordinary" **extrafusal fibers** (● Fig. 8–24). Unlike an ordinary skeletal-muscle fiber, which contains contractile elements (myofibrils) throughout its entire length, an intrafusal fiber has a noncontractile central portion, with the contractile elements being limited to both ends. Each muscle spindle has its own private efferent and afferent nerve supply. The efferent neuron that innervates a muscle spindle's intrafusal fibers is known as a **gamma motor neuron**, whereas the motor neurons that supply the ordinary extrafusal fibers are designated as **alpha motor neurons**. Two types of afferent sensory endings terminate on the intrafusal fibers and serve as muscle-spindle receptors, both of which are activated by stretch. The **primary (annulospiral) endings** are wrapped around the central portions of the intrafusal fibers; they detect changes in the length of the fibers during stretching as well as the speed with which it occurs. The **secondary (flower-spray) endings**, which are clustered at the end segments of many of the intrafusal fibers, are sensitive only to changes in length.

Whenever the whole muscle is passively stretched, the intrafusal fibers within its muscle spindles are likewise stretched, increasing the rate of firing in the afferent nerve fibers whose sensory endings terminate on the stretched

Alpha motor-
neuron axon

Gamma motor-
neuron axon

Secondary (flower spray)
endings of afferent
fibers

Extrafusal ("ordinary")
muscle fibers

Capsule

Intrafusal (spindle)
muscle fibers

Contractile end portions
of intrafusal fiber

Noncontractile
central portion
of intrafusal
fiber

Primary (annulospiral)
endings of afferent fibers

• **FIGURE 8–24   Muscle Spindle** A muscle spindle consists of a collection of specialized intrafusal fibers that lie within a connective-tissue capsule parallel to the ordinary extrafusal skeletal-muscle fibers. The muscle spindle is innervated by its own gamma motor neuron and is supplied by two types of afferent sensory terminals, the primary (annulospiral) endings and the secondary (flower-spray) endings, both of which are activated by stretch.

spindle fibers. The afferent neuron directly synapses on the alpha motor neuron that innervates the extrafusal fibers of the same muscle, resulting in contraction of that muscle (• Fig. 8–25a, pathway ① → ②). This **stretch reflex** serves as a local negative-feedback mechanism to resist any passive changes in muscle length so that optimal resting length can be maintained.

The classical example of the stretch reflex is the **patellar tendon,** or **knee-jerk, reflex** (• Fig. 8–26). The extensor muscle of the knee is the *quadriceps femoris,* which forms the anterior portion of the thigh and is attached just below the knee to the tibia (the shin bone) by the *patellar tendon.* Tapping this tendon with a rubber mallet passively stretches the quadriceps muscle, activating its spindle receptors. The resultant stretch reflex brings about contraction of this extensor muscle, causing the knee to extend and raise the foreleg in the well-known knee-jerk fashion. This test is routinely performed as a preliminary assessment of nervous-system function. A normal knee jerk indicates to a physician that a number of neural and muscular components—muscle spindle, afferent input, motor neurons, efferent output, neuromuscular junctions, and the muscles themselves—are functioning normally. It also indicates the presence of an appropriate bal-

ance of excitatory and inhibitory input to the motor neurons from higher brain levels. Muscle jerks may be absent or depressed with loss of higher-level excitatory inputs or may be greatly exaggerated with loss of inhibitory input to the motor neurons from higher brain levels.

The primary purpose of the stretch reflex is to resist the tendency for the passive stretch of extensor muscles caused by gravitational forces when a person is standing upright. Whenever the knee joint tends to buckle because of gravity, the quadriceps muscle is stretched. The resultant enhanced contraction of this extensor muscle brought about by the stretch reflex quickly straightens out the knee, holding the limb extended so that the person remains standing.

Gamma motor neurons initiate contraction of the muscular end regions of intrafusal fibers (Fig. 8–25a, pathway ③). This contractile response is too weak to have any influence on whole-muscle tension, but it does have an important localized effect on the muscle spindle itself. If there were no compensating mechanisms, shortening of the whole muscle by alpha motor-neuron stimulation of extrafusal fibers would cause slack in the spindle fibers so that they would be less sensitive to stretch and therefore not as effective as muscle-length detectors (Figs. 8–25b and c). **Coactivation** of the

1 = Afferent input from annulospiral endings of muscle spindle fiber

2 = Alpha motor-neuron output to regular skeletal muscle fiber

1 → 2 = Stretch reflex pathway

3 = Gamma motor-neuron output to contractile end portions of spindle fiber

4 = Descending pathways coactivating alpha and gamma motor neurons

(a)

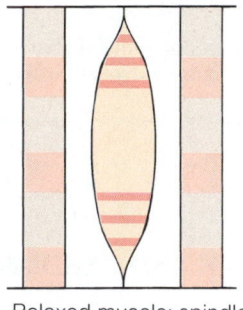

Relaxed muscle; spindle fiber sensitive to stretch of muscle

(b)

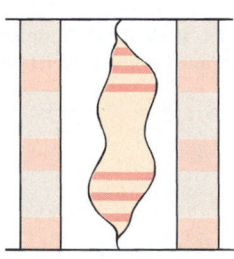

Contracted muscle; slackened spindle fiber not sensitive to stretch of muscle

(c)

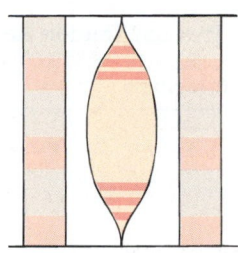

Contracted muscle; contracted spindle fiber sensitive to stretch of muscle

(d)

• **FIGURE 8–25 Muscle-Spindle Function** (a) Pathways involved in the monosynaptic stretch reflex and coactivation of alpha and gamma motor neurons. (b) Status of muscle spindle when muscle is relaxed. (c) Status of muscle spindle when muscle is contracted upon alpha motor-neuron stimulation. (d) Status of muscle spindle when both muscle and muscle spindle are contracted upon alpha and gamma motor-neuron coactivation.

gamma motor-neuron system along with the alpha motor-neuron system during reflex and voluntary contractions (Fig. 8–25a, pathway ④) takes the slack out of the spindle fibers as the whole muscle shortens, permitting these receptor structures to maintain their high sensitivity to stretch over a wide range of muscle lengths. When gamma motor-neuron stimulation triggers simultaneous contraction of both end muscular portions of an intrafusal fiber, the noncontractile central portion is pulled in opposite directions, tightening this region and taking out the slack (Fig. 8–25d). Whereas the extent of alpha motor-neuron activation depends on the in-

tended strength of the motor response, the extent of simultaneous gamma motor-neuron activity to the same muscle depends on the anticipated distance of shortening.

In contrast to muscle spindles, which lie within the belly of the muscle, **Golgi tendon organs** are located in the tendons of the muscle, where they are able to respond to changes in the muscle's externally applied tension rather than to changes in its length. Because a number of factors determine the tension developed in the whole muscle during contraction (for example, frequency of stimulation or length of muscle at the onset of contraction), it is essential that motor control systems

• FIGURE 8–26 Patellar Tendon Reflex (a Stretch Reflex) Tapping the patellar tendon with a rubber mallet stretches the muscle spindles in the quadriceps femoris muscle. The resultant monosynaptic stretch reflex results in contraction of this extensor muscle, causing the characteristic knee-jerk response.

be apprised of the tension actually achieved so that adjustments can be made if necessary.

The Golgi tendon organs consist of endings of afferent fibers entwined within bundles of connective-tissue fibers that make up the tendon. When the extrafusal muscle fibers contract, the resultant pull on the tendon tightens the connective-tissue bundles, which, in turn, increase the tension exerted on the bone to which the tendon is attached. In the process, the entwined Golgi organ afferent receptor endings are stretched, causing the afferent fibers to fire; the frequency of firing is directly related to the tension developed.

The afferent information is sent to the brain. In addition, other branches of the afferent neuron arising from the Golgi tendon organ inhibit, by means of an interneuron, the alpha motor neurons of the same muscle. This reflex is apparently protective in nature. When the tension becomes great enough, the high level of inhibitory input from the activated Golgi tendon organs counterbalances excitatory inputs to the alpha motor neurons. This inhibitory response halts further contraction and brings about sudden reflex relaxation, thus helping prevent damage to muscle or tendon from excessive, tension-developing muscle contractions.

SMOOTH AND CARDIAC MUSCLE

Smooth and cardiac muscle share some basic properties with skeletal muscle.

The two other types of muscle—smooth muscle and cardiac muscle—share some basic properties with skeletal muscle, but each also displays unique characteristics (Table 8–4). The three muscle types have several features in common. First, they all have a specialized contractile apparatus made up of thick myosin and thin actin filaments that interact in response to a rise in cytosolic Ca^{++} to accomplish contraction. Second, they all directly use ATP as the energy source for cross-bridge cycling. However, the structure and organization of fibers within these different muscle types vary, as do their mechanisms of excitation and the means by which excitation and contraction are coupled. Furthermore, there are important distinctions in the contractile response itself. We will spend the remainder of this chapter highlighting unique features of smooth and cardiac muscle as compared with skeletal muscle, reserving a more detailed discussion of their function for chapters devoted to organs containing these muscle types.

Smooth-muscle cells are small and unstriated.

The majority of smooth-muscle cells are found in the walls of hollow organs and tubes. Their contraction exerts pressure on and regulates the forward movement of the contents of these structures. Both smooth- and skeletal-muscle cells are elongated, but in contrast to their large, cylinder-shaped, multinucleated skeletal-muscle counterparts, smooth-muscle cells are spindle-shaped, have a single nucleus, and are considerably smaller (2 to 10 μm in diameter and 50 to 400 μm long). Also unlike skeletal-muscle cells, a single smooth-muscle cell does not extend the full length of a muscle. Instead, groups of smooth-muscle cells are typically arranged in sheets (● Fig. 8–27a, p. 250).

Three types of filaments are found in a smooth-muscle cell: (1) thick myosin filaments, which are longer than those found in skeletal muscle; (2) thin actin filaments, which lack troponin and tropomyosin; and (3), unique to smooth muscle, filaments of intermediate size, which do not appear to directly participate in the contractile process but probably serve as part of the cell's elastic component, as part of the cytoskeletal framework that supports the shape of the cell, or as part of both. Smooth-muscle filaments do not appear to form myofibrils and are not arranged in the sarcomere pattern found in skeletal muscle. Thus, smooth-muscle cells do not display the banding or striation found in skeletal muscle, giving rise to the term *smooth* for this muscle type. Lacking sarcomeres, smooth muscle does not have Z lines as such, but irregularly positioned **dense bodies** containing the same protein constituent found in Z lines are present (Fig. 8–27b). The actin filaments are anchored either to the dense bodies or to the internal surface of the plasma membrane. Considerably more actin is present in smooth-muscle cells than in skeletal-muscle cells, with ten to fifteen thin filaments for each thick myosin filament in smooth muscle and two thin filaments for each thick filament in skeletal muscle. The thick- and thin-filament contractile units are oriented slightly diagonally from side-to-side within the smooth-muscle cell in an elongated, diamond-shaped lattice, rather than running parallel with the long axis as myofibrils do in skeletal muscle (● Fig. 8–28a, p. 251). Relative sliding of the thin filaments past the thick filaments during contraction causes the filament lattice to reduce in length and expand from side-to-side. As a result, the whole cell shortens and bulges out between the points where the thin filaments are attached to the inner surface of the plasma membrane (Fig. 8–28b).

Smooth-muscle cells are turned on by Ca++-dependent phosphorylation of myosin.

Because the thin filaments of smooth-muscle cells do not contain the troponin and tropomyosin blocking proteins, what prevents actin and myosin from binding at the cross bridges in the resting state, and how is cross-bridge activity switched on in the excited state? Smooth-muscle myosin is able to interact with actin only when the myosin is *phosphorylated* (that is, has a phosphate group attached to it). During excita-

tion, the increased cytosolic Ca++ acts as an intracellular messenger, initiating a chain of biochemical events that results in phosphorylation of myosin (● Fig. 8–29, p. 251). Smooth-muscle Ca++ binds with **calmodulin**, an intracellular protein found in most cells that is structurally similar to troponin (see p. 52). This Ca++-calmodulin complex binds to and activates another protein, **myosin kinase**, which in turn phosphorylates myosin. Phosphorylated myosin then binds with actin so that cross-bridge cycling can begin. When Ca++ is removed, myosin is dephosphorylated (phosphate is removed) and can no longer interact with actin, so the muscle relaxes. Thus, smooth muscle is triggered to contract by a rise in cytosolic Ca++, similar to what happens in skeletal muscle. In smooth muscle, however, Ca++ ultimately turns on the cross bridges by inducing a *chemical* change in myosin in the *thick* filaments, whereas in skeletal muscle it exerts its effects by invoking a *physical* change at the *thin* filaments (● Fig. 8–30, p. 252). Recall that in skeletal muscle, Ca++ moves troponin and tropomyosin from their blocking position so that actin and myosin are free to bind with each other.

The means by which excitation brings about an increase in cytosolic Ca++ concentration in smooth-muscle cells also differs from that for skeletal muscle. A smooth-muscle cell has no T tubules and a poorly developed sarcoplasmic reticulum. The increased cytosolic Ca++ that triggers the contractile response comes from two sources: some Ca++ is released intracellularly from the meager sarcoplasmic reticulum stores, but most enters down its concentration gradient from the ECF as Ca++ channels in the plasma membrane are opened. Because smooth-muscle cells are so much smaller in diameter than skeletal-muscle fibers, this Ca++ influx from the ECF is able to influence cross-bridge activity, even in the central portions of the cell, without the necessity of an elaborate T tubule–sarcoplasmic reticulum mechanism. Relaxation is accomplished by removal of Ca++ as it is actively transported back into the sarcoplasmic reticulum and out across the plasma membrane.

Most groups of smooth-muscle tissue are capable of self-excitation.

We still have not addressed the question of how smooth muscle becomes excited to contract; that is, what opens the Ca++ channels in the plasma membrane and sarcoplasmic reticulum? Smooth muscle is grouped into two categories—multiunit and single-unit smooth muscle—based on differences in how the muscle fibers become excited. **Multiunit smooth muscle** exhibits properties partway between skeletal muscle and single-unit smooth muscle. As the name implies, a multiunit smooth muscle consists of multiple discrete units that function independently of each other and must be separately stimulated by nerves to contract, similar to skeletal-muscle motor units. Thus, skeletal muscle and multiunit smooth muscle are both **neurogenic**; that is, they depend on their nerve supply to initiate contraction. Whereas skeletal muscle is innervated by the voluntary somatic nervous system

TABLE 8–4　Comparison of Muscle Types

CHARACTERISTIC	TYPE OF MUSCLE			
	Skeletal	**Multiunit Smooth**	**Single-Unit Smooth**	**Cardiac**
Location	Attached to skeleton	Large blood vessels, eye, and hair follicles	Walls of hollow organs in digestive, reproductive, and urinary tracts and in small blood vessels	Heart only
Function	Movement of body in relation to external environment	Varies with structure involved	Movement of contents within hollow organs	Pumps blood out of heart
Mechanism of contraction	Sliding-filament mechanism	Sliding-filament mechanism	Sliding-filament mechanism	Sliding-filament mechanism
Innervation	Somatic nervous system (alpha motor neurons)	Autonomic nervous system	Autonomic nervous system	Autonomic nervous system
Level of control	Under voluntary control; also subject to subconscious regulation	Under involuntary control	Under involuntary control	Under involuntary control
Initiation of contraction	Neurogenic	Neurogenic	Myogenic (pacemaker activity and slow-wave potentials)	Myogenic (pacemaker activity)
Role of nervous stimulation	Initiates contraction; accomplishes gradation	Initiates contraction; contributes to gradation	Modifies contraction; can excite or inhibit; contributes to gradation	Modifies contraction; can excite or inhibit; contributes to gradation
Modifying effect of hormones	No	Yes	Yes	Yes
Presence of thick myosin and thin actin filaments	Yes	Yes	Yes	Yes
Striated due to orderly arrangement of filaments	Yes	No	No	Yes
Presence of troponin and tropomyosin	Yes	No	No	Yes
Presence of T tubules	Yes	No	No	Yes

(motor neurons), multiunit (as well as single-unit) smooth muscle is supplied by the involuntary autonomic nervous system. Multiunit smooth muscle is found (1) in the walls of large blood vessels; (2) in large airways to the lungs; (3) in the muscle of the eye that adjusts the lens for near or far vision; (4) in the iris of the eye, which alters the size of the pupil to adjust the amount of light entering the eye; and (5) at the base of hair follicles, contraction of which causes "goose bumps."

Most smooth muscle is of the **single-unit** variety. It is alternatively called **visceral smooth muscle** because it is found in the walls of the hollow organs or viscera (for example, the digestive, reproductive, and urinary tracts and small blood vessels). The term *single-unit smooth muscle* derives from the fact that the muscle fibers that make up this type of muscle become excited and contract as a single unit. The muscle fibers in single-unit smooth muscle are electrically linked by gap junctions (see p. 55). When an action potential occurs anywhere within a sheet of single-unit smooth muscle, it is quickly propagated via these special points of electrical contact throughout the entire group of interconnected cells, which then contract as a single coordinated unit. Such a group of interconnected muscle cells that function electrically and mechanically as a unit is known as a **functional syncytium**.

Thinking about the role of the uterus during the process of labor will help you appreciate the significance of this arrangement. Muscle cells composing the uterine wall act as a functional syncytium. They repetitively become excited and contract as a unit during labor, exerting a series of coordinated "pushes" that are eventually responsible for delivering the baby. Independent, uncoordinated contractions of individual muscle cells in the uterine wall would not exert the uniformly

TABLE 8–4 Comparison of Muscle Types (continued)

CHARACTERISTIC	TYPE OF MUSCLE			
	Skeletal	Multiunit Smooth	Single-Unit Smooth	Cardiac
Level of development of sarcoplasmic reticulum	Well developed	Poorly developed	Poorly developed	Moderately developed
Cross bridges turned on by Ca^{++}	Yes	Yes	Yes	Yes
Source of increased cytosolic Ca^{++}	Sarcoplasmic reticulum	Extracellular fluid and sarcoplasmic reticulum	Extracellular fluid and sarcoplasmic reticulum	Extracellular fluid and sarcoplasmic reticulum
Site of Ca^{++} regulation	Troponin in thin filaments	Myosin in thick filaments	Myosin in thick filaments	Troponin in thin filaments
Mechanism of Ca^{++} action	Physically repositions troponin-tropomyosin complex to uncover actin cross-bridge binding sites	Chemically brings about phosphorylation of myosin cross bridges so they can bind with actin	Chemically brings about phosphorylation of myosin cross bridges so they can bind with actin	Physically repositions troponin-tropomyosin complex
Presence of gap junctions	No	Yes (very few)	Yes	Yes
ATP used directly by contractile apparatus	Yes	Yes	Yes	Yes
Myosin ATPase activity; speed of contraction	Fast or slow, depending on type of fiber	Very slow	Very slow	Slow
Means by which gradation accomplished	Varying number of motor units contracting (motor-unit recruitment) and frequency at which they're stimulated (summation of twitches)	Varying number of muscle fibers contracting and varying cytosolic Ca^{++} concentration in each fiber by autonomic and hormonal influences	Varying cytosolic Ca^{++} concentration through myogenic activity and influences of the autonomic nervous system, hormones, mechanical stretch, and local metabolites	Varying length of fiber (depending on extent of filling) and varying cytosolic Ca^{++} concentration through autonomic, hormonal, and local metabolite influence
Presence of tone in absence of external stimulation	No	No	Yes	No
Clear-cut length-tension relationship	Yes	No	No	Yes

applied pressure needed to expel the baby. A similar situation applies for single-unit smooth muscle elsewhere in the body.

Single-unit smooth muscle is **self-excitable** rather than requiring nervous stimulation for contraction. Clusters of specialized smooth-muscle cells within a functional syncytium display spontaneous electrical activity; that is, they are able to undergo action potentials without any external stimulation. In contrast to the other excitable cells we have been discussing (such as neurons, skeletal-muscle fibers, and multiunit smooth muscle), the self-excitable cells of single-unit smooth muscle do not maintain a constant resting potential. Instead, their membrane potential inherently fluctuates without any influence by factors external to the cell.

Two major types of spontaneous depolarizations displayed by self-excitable cells are pacemaker activity and slow-wave potentials. In **pacemaker activity** (• Fig. 8–31a), the mem-brane potential gradually depolarizes on its own because of shifts in passive ionic fluxes accompanying automatic changes in channel permeability. When the membrane has depolarized to threshold, an action potential is initiated. After repolarizing, the membrane potential once again depolarizes to threshold, cyclically continuing in this manner to self-generate action potentials. **Slow-wave potentials** (Fig. 8–31b), on the other hand, are gradually alternating hyperpolarizing and depolarizing swings in potential caused by automatic cyclical changes in the rate at which sodium ions are actively transported across the membrane. The potential is moved farther from threshold during each hyperpolarizing swing and closer to threshold during each depolarizing swing. If threshold is reached, a burst of action potentials occurs at the peak of a depolarizing swing. Threshold is not always reached, however, so the oscillating slow-wave po-

● FIGURE 8–27 Microscopic View of Smooth-Muscle Cells (a) Low-power light micrograph of smooth-muscle cells. Note the spindle shape and single, centrally located nucleus. (b) Electron micrograph of smooth-muscle cells at 14,000× magnification. Note the presence of dense bodies and lack of banding.

(a)   Smooth-muscle cell   Nucleus

Smooth-muscle cell   Dense bodies

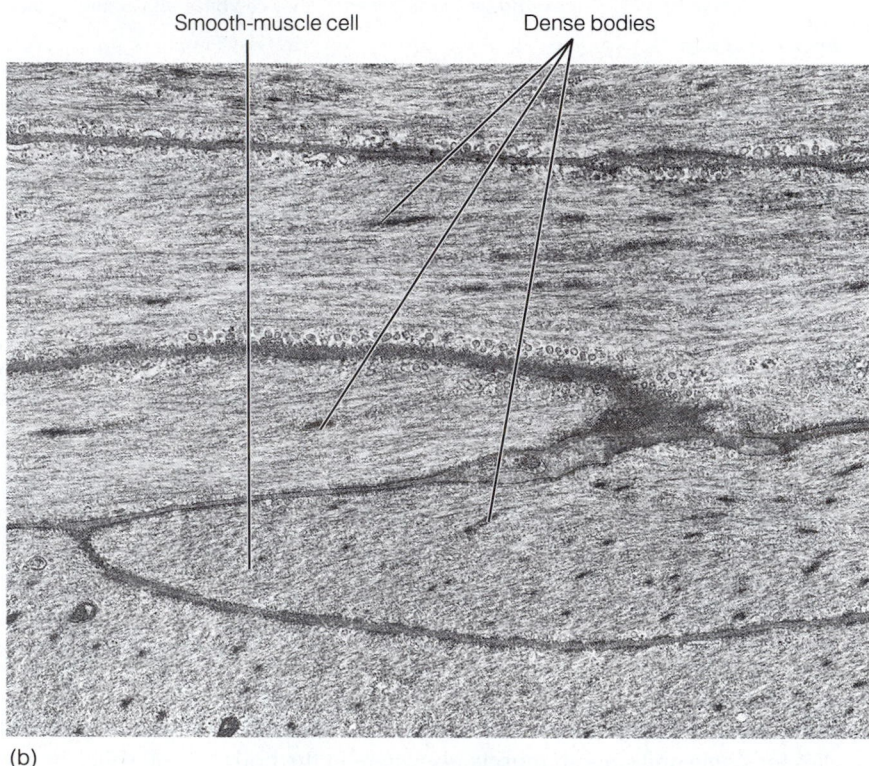

(b)

tentials can continue without generating action potentials. Whether threshold is reached depends on the starting point of the membrane potential at the onset of its depolarizing swing. The starting point, in turn, is influenced by neural and local factors.

Not all single-unit smooth-muscle cells undergo spontaneous changes in potential. However, once an action potential is initiated by a self-excitable smooth-muscle cell, it is conducted to the remaining cells of the functional syncytium via gap junctions so that the entire group of cells contracts without any nervous input (● Fig. 8–32). Such nerve-independent contractile activity initiated by the muscle itself is called **myogenic activity**, in contrast to the neurogenic activity of skeletal muscle and multiunit smooth muscle.

Gradation of single-unit smooth-muscle contraction differs considerably from skeletal muscle.

Single-unit smooth muscle differs from skeletal muscle in the way in which gradation of contraction is accomplished. Gra-

(a)

(b)

Intermediate
filament

Thick filament

Thin filament

Dense body

● **FIGURE 8–28   Schematic Representation of the
Arrangement of Thick and Thin Filaments in a Smooth-
Muscle Cell in Contracted and Relaxed States** (a) Relaxed
smooth-muscle cell. (b) Contracted smooth-muscle cell.

dation of skeletal muscle is entirely under neural control, primarily involving motor-unit recruitment and summation of twitches. In smooth muscle, the gap junctions assure that an entire smooth-muscle mass contracts as a single unit, making it impossible to vary the number of muscle fibers contracting. Only the tension of the fibers can be modified to achieve varying strengths of contraction of the whole organ. The portion of cross bridges activated and the tension subsequently developed in single-unit smooth muscle can be graded by varying the cytosolic Ca^{++} concentration. A single excitation in smooth muscle does not cause all of the cross bridges to switch on, in contrast to skeletal muscle where a single action potential triggers the release of sufficient Ca^{++} to permit all cross bridges to cycle. As Ca^{++} concentration increases in smooth muscle, more cross bridges are brought into play, and greater tension develops.

Many single-unit smooth-muscle cells have sufficient levels of cytosolic Ca^{++} to maintain a low level of tension, or **tone**, even in the absence of action potentials. A sudden drastic change in Ca^{++}, such as accompanies a myogenically induced action potential, brings about a contractile response superimposed on the ongoing tonic tension. Besides self-induced action potentials, a number of other factors can influence contractile activity and the development of tension in smooth-muscle cells by altering their cytosolic Ca^{++} concentration, including autonomic neurotransmitters. Smooth muscle is typically innervated by both branches of the autonomic nervous system. In single-unit smooth muscle, this nerve supply does not *initiate* contraction, but it can *modify* the rate and strength of contraction, either enhancing or retarding the inherent contractile activity of a given organ. Recall that the isolated motor end-plate region of a skeletal-muscle fiber interacts with ACh released from a single axon terminal of a motor neuron. In contrast, the receptor proteins that bind with autonomic transmitters are dispersed throughout the entire surface membrane of a smooth-muscle cell. Smooth-muscle cells are sensitive to varying degrees and in varying ways to autonomic transmitters, depending on their distribution of cholinergic and adrenergic receptors (see p. 203). Each terminal branch of a postganglionic autonomic fiber

● **FIGURE 8–29   Calcium
Activation of Myosin in Smooth
Muscle**

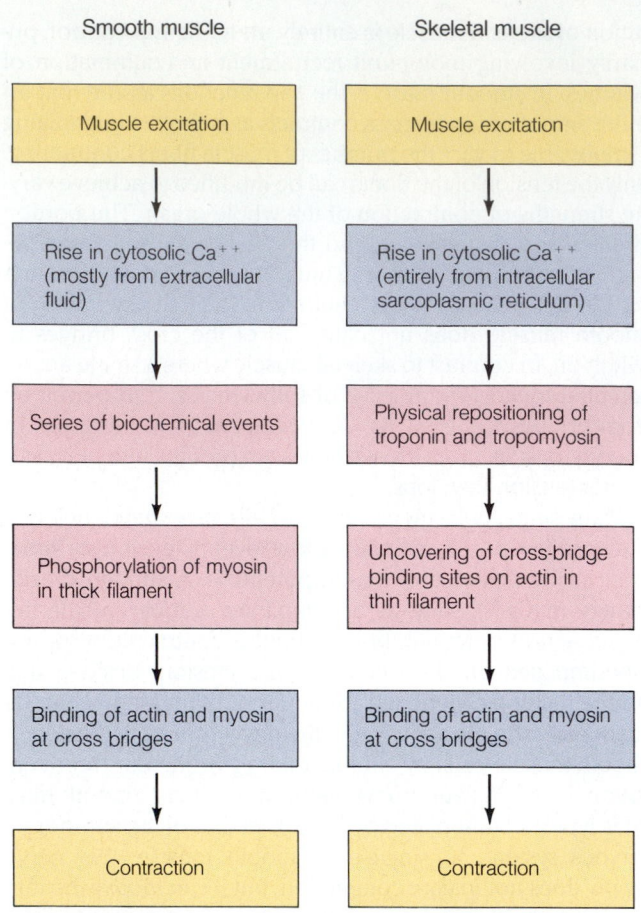

Smooth muscle	Skeletal muscle
Muscle excitation	Muscle excitation
Rise in cytosolic Ca++ (mostly from extracellular fluid)	Rise in cytosolic Ca++ (entirely from intracellular sarcoplasmic reticulum)
Series of biochemical events	Physical repositioning of troponin and tropomyosin
Phosphorylation of myosin in thick filament	Uncovering of cross-bridge binding sites on actin in thin filament
Binding of actin and myosin at cross bridges	Binding of actin and myosin at cross bridges
Contraction	Contraction

FIGURE 8-30   Comparison of the Role of Calcium in Bringing about Contraction in Smooth Muscle and Skeletal Muscle

travels across the surface of one or more smooth-muscle cells, releasing transmitter from the vesicles within its multiple bulges (varicosities) as an action potential passes along the terminal (•Fig. 8–33). The transmitter diffuses to the many receptor sites specific for it on the cells underlying the termi-

nal. Thus, in contrast to the discrete one-to-one relationship at motor end plates, a given smooth-muscle cell can be influenced by more than one type of neurotransmitter, and each autonomic terminal can influence more than one smooth-muscle cell.

Other factors (besides autonomic neurotransmitters) can influence the rate and strength of both multiunit and single-unit smooth-muscle contraction, including certain hormones, local metabolites, mechanical stretch, and specific drugs. All of these factors ultimately act by modifying the permeability of Ca++ channels in the plasma membrane, the sarcoplasmic reticulum, or both, through a variety of mechanisms. Examples of such influences accompany discussions in other chapters of the various organs that contain smooth muscle. Thus, smooth muscle is subject to more external influences than is skeletal muscle, even though smooth muscle is capable of contracting on its own, whereas skeletal muscle is not.

The relationship between the length of the muscle fibers before contraction and the tension that can be developed upon a subsequent contraction is less closely linked in smooth muscle than in skeletal muscle. The range of lengths over which a smooth-muscle fiber is able to develop near maximal tension is much greater than for skeletal muscle. Smooth muscle can still develop considerable tension even when stretched up to two and one-half times its resting length, presumably because the thin filaments still overlap the much longer thick filaments even in the stretched-out position. In contrast, the thick and thin filaments of skeletal muscle are completely pulled apart and no longer able to interact when the muscle is stretched only three-fourths longer than its resting length.

The ability of a considerably stretched smooth-muscle fiber to still develop tension is important, because the smooth-muscle fibers within the wall of a hollow organ are progressively stretched as the volume of the organ's contents increases. Consider the urinary bladder as an example. Even though the muscle fibers in the urinary bladder are stretched as the bladder gradually fills with urine, they still maintain their tone and are even capable of developing further tension

FIGURE 8-31   Self-Generated Electrical Activity in Smooth Muscle (a) In pacemaker activity, the membrane gradually depolarizes to threshold on a regular periodic basis without any nervous stimulation. These regular depolarizations cyclically trigger self-induced action potentials. (b) In slow-wave potentials, the membrane gradually undergoes self-induced hyperpolarizing and depolarizing swings in potential. A burst of action potentials occurs if a depolarizing swing brings the membrane to threshold.

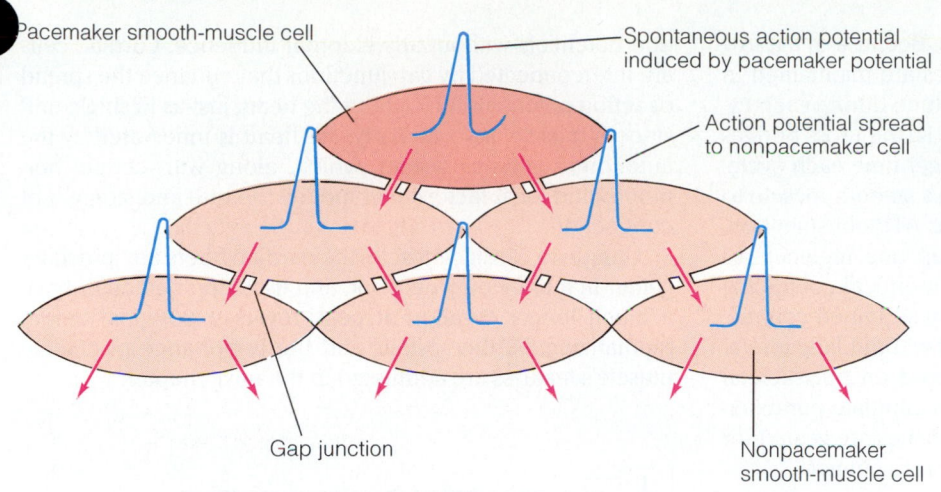

Pacemaker smooth-muscle cell

Spontaneous action potential induced by pacemaker potential

Action potential spread to nonpacemaker cell

• **FIGURE 8–32 Functional Syncytium** An action potential spontaneously initated in a pacemaker smooth-muscle cell spreads to surrounding nonpacemaker cells through gap junctions, exciting the entire sheet of connected smooth-muscle cells to contract as single coordinated unit.

Gap junction

Nonpacemaker smooth-muscle cell

Arrows indicate spread of action potentials through gap junctions in single-unit smooth muscle.

in response to inputs regulating bladder emptying. If considerable stretching prevented development of tension, as it does in skeletal muscle, a filled bladder would not be capable of contracting to empty.

In fact, when smooth muscle is suddenly stretched, it initially increases its tension, much like the tension created in a stretched rubber band. The muscle quickly adjusts to this new length, however, and inherently relaxes to the tension level prior to the stretch, probably as a consequence of rearrangement of cross-bridge attachments. Smooth-muscle cross bridges detach comparatively slowly. Upon sudden stretching, it is speculated that any attached cross bridges would strain against the stretch, contributing to a passive (not actively generated) increase in tension. As these cross bridges detach, the filaments would be permitted to slide into an unstrained stretched position, restoring the tension to its original level. This inherent property of smooth muscle is called the **stress relaxation response**.

These two responses of smooth muscle to being stretched—being able to develop tension even when considerably stretched and inherently relaxing when stretched—are highly advantageous. They enable smooth muscle to exist at a variety of lengths with little change in tension. As a result, a hollow organ enclosed by smooth muscle can accommodate variable volumes of contents with little change in the pressure exerted on the contents, except when the contents are to be

pushed out of the organ. At that time, the tension is deliberately increased by fiber shortening. It is possible for smooth-muscle fibers to contract to one-half their normal length, enabling hollow organs to dramatically empty their contents upon increased contractile activity; thus, smooth-muscled viscera can easily accommodate large volumes but can empty to practically zero volume. This length range in which smooth muscle normally functions (anywhere from one-half to two and one-half times the normal length) is considerably greater than the limited length range within which skeletal muscle remains functional.

Smooth muscle is slow and economical.

A smooth-muscle contractile response proceeds at a more leisurely pace than does a skeletal-muscle twitch. A single smooth-muscle contraction may last as long as 3 seconds (3,000 msec), compared to the maximum of 100 msec required for a single contractile response in skeletal muscle. The rate of ATP splitting by myosin ATPase is much slower in smooth muscle, so cross-bridge activity and filament sliding occur more slowly. Smooth muscle also relaxes more slowly because of a slower rate of Ca^{++} removal. However, slowness should not be equated with weakness. Smooth muscle is able to generate the same contractile tension per unit of cross-sectional area as skeletal muscle, but it does so more slowly

Sympathetic postganglionic fiber

Smooth-muscle cell

Axon varicosities

Parasympathetic postganglionic fiber

• **FIGURE 8–33 Schematic Representation of Innervation of Smooth Muscle by Autonomic Postganglionic Nerve Terminals** and at considerably less energy expense. Because of the low rate of cross-bridge cycling, cross bridges are maintained in the attached state for a longer period of time during each cycle compared with skeletal muscle; that is, the cross bridges "latch onto" the thin filaments for a longer time each cycle. This so-called **latch phenomenon** enables smooth muscle to maintain tension with comparatively less ATP consumption, because each cross-bridge cycle uses up one molecule of ATP. Smooth muscle is therefore an economical contractile tissue, making it well suited for long-term sustained contractions with little energy consumption and without fatigue. Unlike the rapidly changing demands placed on our skeletal muscles as we maneuver through and manipulate our external environment, our smooth-muscle activities are geared for long-term duration and slower adjustments to change.

Because of its slowness and the less-ordered arrangement of its filaments, smooth muscle has often been mistakenly viewed as a poorly developed version of skeletal muscle. Actually, smooth muscle is just as highly specialized for the demands placed on it; that is, being able to economically maintain tension for prolonged periods without fatigue and being able to accommodate considerable variations in the volume of contents it encloses with little change in tension. It is an extremely adaptive, efficient tissue.

Nutrient and O_2 delivery are generally adequate to support the smooth-muscle contractile process. Smooth muscle can use a wide variety of nutrient molecules for ATP production. There are no energy storage pools comparable to creatine phosphate in smooth muscle; they are not necessary. Oxygen delivery is usually adequate to keep pace with the low rate of oxidative phosphorylation needed to provide ATP for the energy-efficient smooth muscle. If necessary, anaerobic glycolysis can sustain adequate ATP production if O_2 supplies are diminished.

Cardiac muscle blends features of both skeletal and smooth muscle.

Cardiac muscle, found only in the heart, structurally and functionally shares characteristics with both skeletal and single-unit smooth muscle. Like skeletal muscle, cardiac muscle is striated because its thick and thin filaments are highly organized into a regular banding pattern. Cardiac thin filaments contain troponin and tropomyosin, which constitute the site of Ca^{++} action in turning on cross-bridge activity, as in skeletal muscle. Also similar to skeletal muscle, cardiac muscle contracts according to the sliding-filament mechanism and has a clear-cut length-tension relationship. Like the oxidative skeletal-muscle fibers, cardiac muscle cells have an abundance of mitochondria and myoglobin. They also possess T tubules and a moderately well-developed sarcoplasmic reticulum.

As in smooth muscle, Ca^{++} enters the cytosol from both the sarcoplasmic reticulum and the ECF during cardiac excitation. Like single-unit smooth muscle, the heart displays pacemaker (but not slow-wave) activity, initiating its own action potentials without any external influence. Cardiac cells are interconnected by gap junctions that enhance the spread of action potentials throughout the heart, just as in single-unit smooth muscle. Also similarly, the heart is innervated by the autonomic nervous system, which, along with certain hormones and local factors, can modify the rate and strength of contraction.

Unique to cardiac muscle, the cardiac fibers are joined together in a branching network, and its action potentials have a much longer duration at peak reversed potential before repolarizing. Further details and the importance of cardiac muscle's features are addressed in the next chapter.

CHAPTER IN PERSPECTIVE: FOCUS ON HOMEOSTASIS

The skeletal muscles comprise the muscular system itself. Cardiac and smooth muscle are part of the organs that comprise other body systems. Cardiac muscle is found only in the heart, which is part of the circulatory system. Smooth muscle is found in the walls of hollow organs and tubes, including the blood vessels in the circulatory system, airways in the respiratory system, bladder in the urinary system, stomach and intestines in the digestive system, and uterus and ductus deferens (the duct that provides a route of exit for sperm from the testes) in the reproductive system.

Contraction of skeletal muscles accomplishes movement of the body parts in relation to each other and movement of the whole body in relation to the external environment. Thus, these muscles permit us to move through and manipulate our external environment. At a very general level, some of these movements are aimed at maintaining homeostasis, such as moving the body toward food or away from harm. Examples of more specific homeostatic functions accomplished by skeletal muscles include the chewing and swallowing of food for further breakdown in the digestive system into usable energy-producing nutrient molecules (the mouth and throat muscles are all skeletal muscles) and the process of breathing to obtain O_2 and eliminate CO_2 (the respiratory muscles are all skeletal muscles). Generation of heat by contracting muscles also serves as the major source of heat production in the maintenance of body temperature. The skeletal muscles further accomplish many nonhomeostatic activities that enable us to work and play so that we may contribute to society and enjoy ourselves.

All of the other systems of the body, except the immune (defense) system, depend on their nonskeletal-muscle components to enable them to accomplish their homeostatic functions. For example, contraction of cardiac muscle in the heart pushes life-sustaining blood forward into the blood vessels, and contraction of smooth muscle in the stomach and intestines pushes the ingested food through the digestive tract at a rate appropriate for the digestive juices secreted along the route to break down the food into usable units.

Structure of Skeletal Muscle

Muscle cells are specialized for contraction. There are three types of muscle: skeletal, smooth, and cardiac. Skeletal muscles are made up of bundles of long, cylindrical muscle cells known as muscle fibers, wrapped in connective tissue. Muscle fibers are packed with myofibrils, with each myofibril consisting of alternating, slightly overlapping stacked sets of thick and thin filaments. This arrangement leads to a skeletal-muscle fiber's striated microscopic appearance. Thick filaments are composed of the protein myosin. Cross bridges made up of the myosin molecules' globular heads project from each thick filament. Thin filaments are composed primarily of the protein actin, which has the ability to bind and interact with the myosin cross bridges to bring about contraction. However, two other proteins, tropomyosin and troponin, lie across the surface of the thin filament to prevent this cross-bridge interaction in the resting state.

Molecular Basis of Skeletal-Muscle Contraction

Excitation of a skeletal-muscle fiber by its motor neuron brings about contraction through a series of events that results in the thin filaments sliding closer together between the thick filaments. This sliding-filament mechanism of muscle contraction is switched on by the release of Ca^{++} from the lateral sacs of the sarcoplasmic reticulum. Calcium release occurs in response to the spread of a muscle-fiber action potential into the central portions of the fiber by means of the T tubules. Released Ca^{++} binds to the troponin-tropomyosin complex of the thin filament, causing a slight repositioning of the complex to uncover actin's cross-bridge binding sites. After the exposed actin attaches to a myosin cross bridge, the molecular interaction between actin and myosin releases the energy within the myosin head that was stored from the prior splitting of ATP by the myosin ATPase site. This released energy powers cross-bridge stroking. During a power stroke, an activated cross bridge bends toward the center of the thick filament, "rowing" in the thin filament to which it is attached. With the addition of a fresh ATP molecule to the myosin cross bridge, myosin and actin detach, the cross bridge returns to its original shape, and the cycle is repeated. Repeated cycles of cross-bridge activity slide the thin filaments inward step-by-step. When there is no longer a local action potential, the lateral sacs actively take up the Ca^{++}, troponin and tropomyosin slip back into their blocking position, and relaxation occurs. The entire contractile response lasts about one hundred times longer than the action potential.

Skeletal-Muscle Mechanics

Gradation of whole-muscle contraction can be accomplished by (1) varying the number of muscle fibers contracting within the muscle and (2) varying the tension developed by each contracting fiber. The greater the number of active muscle fibers, the greater the whole-muscle tension. The number of fibers contracting depends on (1) the size of the muscle (the number of muscle fibers present); (2) the extent of motor-unit recruitment (how many motor neurons supplying the muscle are active); and (3) the size of each motor unit (how many muscle fibers are activated simultaneously by a single motor neuron).

Also, the greater the tension developed by each contracting fiber, the stronger the contraction of the whole muscle. Two readily variable factors having an effect on the fiber tension are (1) the frequency of stimulation, which determines the extent of summation, and (2) the length of the fiber before the onset of contraction. Summation refers to the increase in tension accompanying repetitive stimulation of the muscle fiber. After undergoing an action potential, the muscle cell membrane recovers from its refractory period and is able to be restimulated again while some contractile activity triggered by the first action potential still remains. As a result, the contractile responses induced by the two rapidly successive action potentials are able to sum, increasing the tension developed by the fiber. If the muscle fiber is stimulated so rapidly that it does not have a chance to start relaxing between stimuli, a smooth, sustained maximal (maximal for the fiber at that length) contraction known as tetanus takes place.

The tension developed upon a tetanic contraction also depends on the length of the fiber at the onset of contraction. At the optimal length (l_o), which is the resting muscle length, there is maximal opportunity for cross-bridge interaction due to optimal overlap of thick and thin filaments; thus, the greatest tension can be developed. At lengths shorter or longer than l_o, less tension can be developed upon contraction, primarily because a portion of the cross bridges are unable to participate.

The two primary types of muscle contraction—isometric (constant length) and isotonic (constant tension)—depend on the relationship between muscle tension and the load. If tension is less than the load, the muscle cannot shorten and lift the object but remains at constant length, producing an isometric contraction. In an isotonic contraction, the tension exceeds the load so the muscle can shorten and lift the object, maintaining constant tension throughout the period of shortening.

The bones, muscles, and joints form lever systems. The most common type amplifies the velocity and distance of muscle shortening to increase the speed and range of motion of the body part moved by the muscle. This increased maneuverability is accomplished at the expense of the muscle having to exert considerably more force than the load.

Skeletal-Muscle Metabolism and Fiber Types

Three biochemical pathways furnish the ATP needed for muscle contraction: (1) the transfer of high-energy phosphates from stored creatine phosphate to ADP, providing the first source of ATP at the onset of exercise; (2) oxidative phosphorylation, which efficiently extracts large amounts of ATP from nutrient molecules if sufficient O_2 is available to support this system; and (3) glycolysis, which can synthesize ATP in the absence of O_2 but uses large amounts of stored glycogen and produces lactic acid in the process.

There are three types of muscle fibers, classified by the pathways they use for ATP synthesis (oxidative or glycolytic) and the rapidity with which they split ATP and subsequently contract (slow twitch or fast twitch): slow-oxidative fibers, fast-oxidative fibers, and fast-glycolytic fibers.

Control of Motor Movement

Control of any motor movement depends on the level of activity in the presynaptic inputs that converge on the motor neurons supplying various muscles. These inputs come from three sources: (1) spinal-reflex pathways, which originate with afferent neurons; (2) the corticospinal descending system, which originates at the primary motor cortex and is concerned primarily with discrete, intricate movements of the hands; and (3) the multineuronal descending system, which originates in the brain stem and is mostly involved with postural adjustments and involuntary movements of the trunk and limbs. The final motor output from the brain stem is influenced by the cerebellum, basal nuclei, and cerebral cortex.

Establishment and adjustment of motor commands depend on continuous afferent input, especially feedback about changes in muscle length (monitored by muscle spindles) and muscle tension (monitored by Golgi tendon organs).

Smooth and Cardiac Muscle

The thick and thin filaments of smooth muscle are not arranged in an orderly pattern, so the fibers are not striated. Cytosolic Ca^{++}, which enters from the extracellular fluid as well as being released from sparse intracellular stores, activates cross-bridge cycling by initiating a series of biochemical reactions that result in phosphorylation of the myosin cross-bridges to enable them to bind with actin. Multiunit smooth muscle is neurogenic, requir-

ing stimulation of individual muscle fibers by its autonomic nerve supply to trigger contraction. Single-unit smooth muscle is myogenic; it is able to initiate its own contraction without any external influence as a result of spontaneous depolarizations to threshold potential brought about by automatic shifts in ionic fluxes. Once an action potential is initiated within a single-unit smooth muscle cell, this electrical activity spreads by means of gap junctions to the surrounding cells within the functional syncytium so that the entire sheet becomes excited and contracts as a unit. The autonomic nervous system as well as hormones and local metabolites can modify the rate and strength of the self-induced contractions. Smooth-muscle contractions are energy efficient, enabling this type of muscle to economically sustain long-term contractions without fatigue. This economy, coupled with the fact that single-unit smooth muscle is able to exist at a variety of lengths with little change in tension, makes single-unit smooth muscle ideally suited for its task of forming the walls of distensible hollow organs.

Cardiac muscle is found only in the heart. It has highly organized striated fibers like skeletal muscle. Like single-unit smooth muscle, some cardiac muscle fibers are capable of generating action potentials, which are spread throughout the heart with the aid of gap junctions.

REVIEW EXERCISES

Objective Questions (Answers on p. D–2.)

1. Upon completion of an action potential in a muscle fiber, the contractile activity initiated by the action potential ceases. (True or false?)

2. The velocity at which a muscle shortens is dependent entirely upon the ATPase activity of its fibers. (True or false?)

3. When a muscle is maximally stretched, it can develop maximal tension upon contraction because the actin filaments can slide in a maximal distance. (True or false?)

4. A pacemaker potential always initiates an action potential. (True or false?)

5. A slow-wave potential always initiates an action potential. (True or false?)

6. Smooth muscle can develop tension even when considerably stretched because the thin filaments still overlap with the long thick filaments. (True or false?)

7. A(n) _____ contraction is an isotonic contraction in which the muscle shortens, whereas the muscle lengthens in a(n) _____ isotonic contraction.

8. _____ motor neurons supply extrafusal muscle fibers whereas intrafusal fibers are innervated by _____ motor neurons.

9. The two types of atrophy are _____ and _____.

10. Which of the following is *not* involved in bringing about muscle relaxation?

   a. reuptake of Ca^{++} by the sarcoplasmic reticulum
   b. no more ATP
   c. no more action potential

   d. removal of ACh at the end plate by acetylcholinesterase
   e. filaments sliding back to their resting position

11. Which of the following provide(s) direct input to alpha motor neurons? (Indicate all correct answers.)

   a. primary motor cortex
   b. brain stem
   c. cerebellum
   d. basal nuclei
   e. spinal-reflex pathways

12. Match the following (with reference to skeletal muscle):

   ___ 1. Ca^{++}
   ___ 2. T tubule
   ___ 3. ATP
   ___ 4. lateral sac of the sarcoplasmic reticulum
   ___ 5. myosin
   ___ 6. troponin-tropomyosin complex
   ___ 7. actin

   a. cyclically binds with the myosin cross bridges during contraction
   b. has ATPase activity
   c. supplies energy for the power stroke of a cross bridge
   d. rapidly transmits the action potential to the central portion of the muscle fiber
   e. stores Ca^{++}
   f. pulls the troponin-tropomyosin complex out of its blocking position
   g. prevents actin from interacting with myosin when the muscle fiber is not excited

13. Using the answer code at the right, indicate what happens in the banding pattern during contraction:

_____ 1. thick myofilament
_____ 2. thin myofilament
_____ 3. A band
_____ 4. I band
_____ 5. H zone
_____ 6. sarcomere

a. remains the same size during contraction
b. decreases in length (shortens) during contraction

Essay Questions

1. Describe the levels of organization in a skeletal muscle.
2. What is responsible for the striated appearance of skeletal muscles? Describe the arrangement of thick and thin filaments that gives rise to the banding pattern.
3. What is the functional unit of skeletal muscle?
4. Describe the composition of thick and thin filaments.
5. Describe the sliding-filament mechanism of muscle contraction. How do cross-bridge power strokes bring about shortening of the muscle fiber?
6. Compare the excitation-contraction coupling process in skeletal muscle with that in smooth muscle.
7. By what means can gradation of skeletal-muscle contraction be accomplished?
8. What is a motor unit? Compare the size of motor units in finely controlled muscles with those specialized for coarse, powerful contractions.
9. Explain the phenomenon of summation and tetanus.
10. What effect does a skeletal-muscle fiber's length at the onset of contraction have on the strength of the subsequent contraction?
11. Compare isotonic and isometric contractions.
12. Describe the role of each of the following in powering skeletal-muscle contraction: ATP, creatine phosphate, oxidative phosphorylation, and glycolysis. Distinguish between aerobically and anaerobically supported exercise.
13. Compare the three types of skeletal-muscle fibers.
14. What are the roles of the corticospinal system and multineuronal system in the control of motor movement?
15. Describe the structure and function of muscle spindles and Golgi tendon organs.
16. Distinguish between multiunit and single-unit smooth muscle.
17. Differentiate between neurogenic and myogenic muscle activity.
18. How can smooth-muscle contraction be graded?
19. Compare the contractile speed and relative energy expenditure of skeletal muscle with that of smooth muscle.
20. In what ways is cardiac muscle functionally similar to skeletal muscle and to smooth muscle?

POINTS TO PONDER

1. Why does regular aerobic exercise provide more cardiovascular benefit than weight training does? (Hint: The heart responds to the demands placed on it in a way similar to skeletal muscle.)

2. If the biceps muscle of a child inserts 4 cm from the elbow and the length of the arm from the elbow to the hand is 28 cm, how much force must the biceps generate in order for the child to lift an 8 kg stack of books with one hand?

3. Put yourself in the position of the scientists who discovered the sliding-filament mechanism of muscle contraction by considering what molecular changes must be involved to account for the observed alterations in the banding pattern during contraction. If you were comparing a relaxed and contracted muscle fiber under a high-power light microscope (see Fig. 8–3a), how could you determine that the thin filaments do not change in length during muscle contraction? You cannot see or measure a single thin filament at this magnification. (Hint: What landmark in the banding pattern represents each end of the thin filament? If these landmarks are the same distance apart in a relaxed and contracted fiber, then the thin filaments must not change in length.)

4. What type of off-the-snow training would you recommend for a competitive downhill skier versus a competitive cross-country skier? What adaptive skeletal-muscle changes would you hope to accomplish in the athletes in each case?

5. A deadly toxin has turned out to be good news for sufferers of a number of painful, disruptive neuromuscular diseases known categorically as *dystonias*. These conditions are characterized by spasms (excessive muscle-contracting activity) that result in involuntary twisting or abnormal postures, depending on the body part affected. For example, painful neck spasms that twist the head to one side occur as a result of *spasmodic torticollis*, the most common dystonia. The Food and Drug Administration has already approved the treatment of some forms of dystonia with botulinum toxin, the toxin responsible for fatal botulism food poisoning. The therapeutic dose, however, is considerably below the amount of toxin needed to induce even mild symptoms of botulism poisoning. Explain how this toxin could be a useful therapy for dystonias. (Hint: See the effect of botulinum toxin on p. 209.)

9 Cardiac Physiology

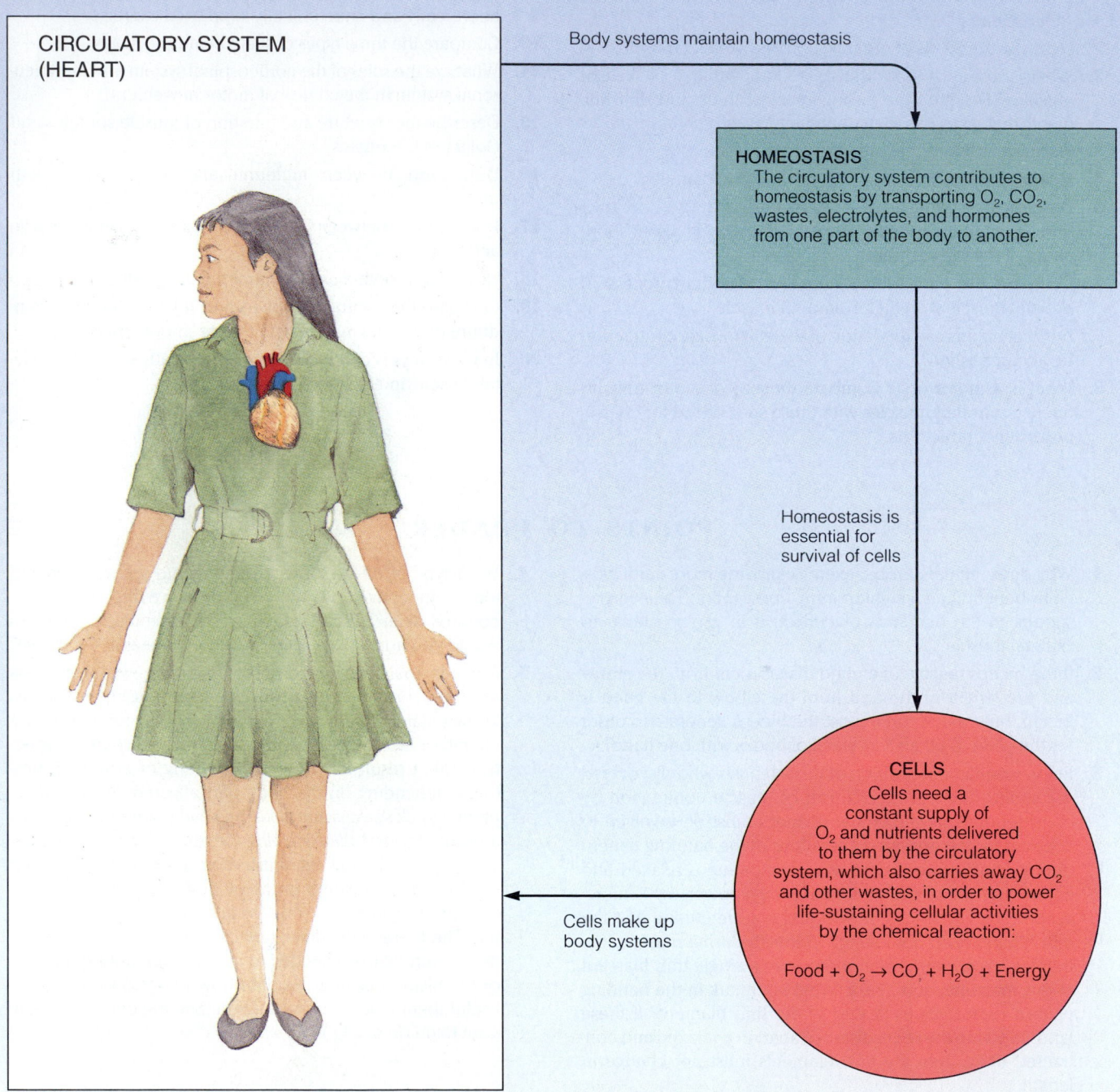

CIRCULATORY SYSTEM (HEART)

Body systems maintain homeostasis

HOMEOSTASIS
The circulatory system contributes to homeostasis by transporting O_2, CO_2, wastes, electrolytes, and hormones from one part of the body to another.

Homeostasis is essential for survival of cells

CELLS
Cells need a constant supply of O_2 and nutrients delivered to them by the circulatory system, which also carries away CO_2 and other wastes, in order to power life-sustaining cellular activities by the chemical reaction:

Food + O_2 → CO_2 + H_2O + Energy

Cells make up body systems

The maintenance of homeostasis depends on essential materials such as O_2 and nutrients being continually picked up from the external environment and delivered to the cells and on waste products being continually removed. Homeostasis also depends on the transfer of hormones, which are important regulatory chemical messengers, from their site of production to their site of action. The **circulatory system**, which contributes to homeostasis by serving as the body's transport system, consists of the heart, blood vessels, and blood.

All body tissues constantly depend on the life-supporting blood flow provided to them by the contraction or beating of the **heart**. The heart drives the blood through the blood vessels for delivery to the tissues in sufficient amounts, whether the body is at rest or engaging in vigorous exercise.

CHAPTER CONTENTS AT A GLANCE

INTRODUCTION
ANATOMICAL CONSIDERATIONS
Location of the heart
The heart as a dual pump
● CONCEPTS, CHALLENGES, AND CONTROVERSIES Fetal
   Circulation: A Fetus Has It All Mixed Up
Heart valves
Heart walls; cardiac muscle
Pericardial sac
ELECTRICAL ACTIVITY OF THE HEART
Pacemaker activity
Spread of cardiac excitation
Action potential of cardiac contractile cell
Cardiac refractory period
Electrocardiography
● A CLOSER LOOK AT EXERCISE PHYSIOLOGY The What, Who,
   and When of Stress Testing
MECHANICAL EVENTS OF THE CARDIAC CYCLE
Electrical, pressure, and volume relationships during diastole
   and systole
Heart sounds
CARDIAC OUTPUT AND ITS CONTROL
Determinants of cardiac output
Control of heart rate
Control of stroke volume
NOURISHING THE HEART MUSCLE
Coronary circulation
Coronary artery disease

INTRODUCTION

From just a matter of days following conception until death, the beat goes on. In fact, throughout an average human life span, the heart contracts about three billion times, never stopping to rest except for a fraction of a second between beats. Within about three weeks after conception, even before the mother can confirm she is pregnant, the heart of the developing embryo starts to function. It is believed to be the first organ to become functional. At this time the human embryo is only a few millimeters long, about the size of a capital letter on this page.

Why does the heart develop so early, and why is it so crucial throughout life? It is because the circulatory system is the transport system of the body. A human embryo, having very little yolk available as food, depends on the prompt establishment of a circulatory system that can interact with the maternal circulation to pick up and distribute to the developing tissues the supplies so critical for survival and growth. Thus begins the story of the circulatory system, which continues throughout life to be a vital pipeline for transporting materials on which the cells of the body are absolutely dependent.

The **circulatory system** consists of three basic components:

1. The **heart** serves as the pump that imparts pressure to the blood to establish the pressure gradient needed for blood to flow to the tissues. Blood, like all liquids, flows from an area of higher pressure to an area of lower pressure down a pressure gradient.

2. The **blood vessels** serve as the passageways through which blood is directed and distributed from the heart to all parts of the body and subsequently returned to the heart (chapter 10).

3. The **blood** serves as the transport medium within which materials being transported are dissolved or suspended (chapter 11).

The blood travels continuously through the circulatory system to and from the heart through two separate vascular (blood-vessel) loops, both originating and terminating at the heart (● Fig. 9–1). The **pulmonary** (lung) **circulation** consists

● **FIGURE 9–1 Pulmonary and Systemic Circulation in Relation to the Heart** The circulatory system consists of two separate vascular loops: the pulmonary circulation, which carries blood between the heart and lungs, and the systemic circulation, which carries blood between the heart and organ systems.

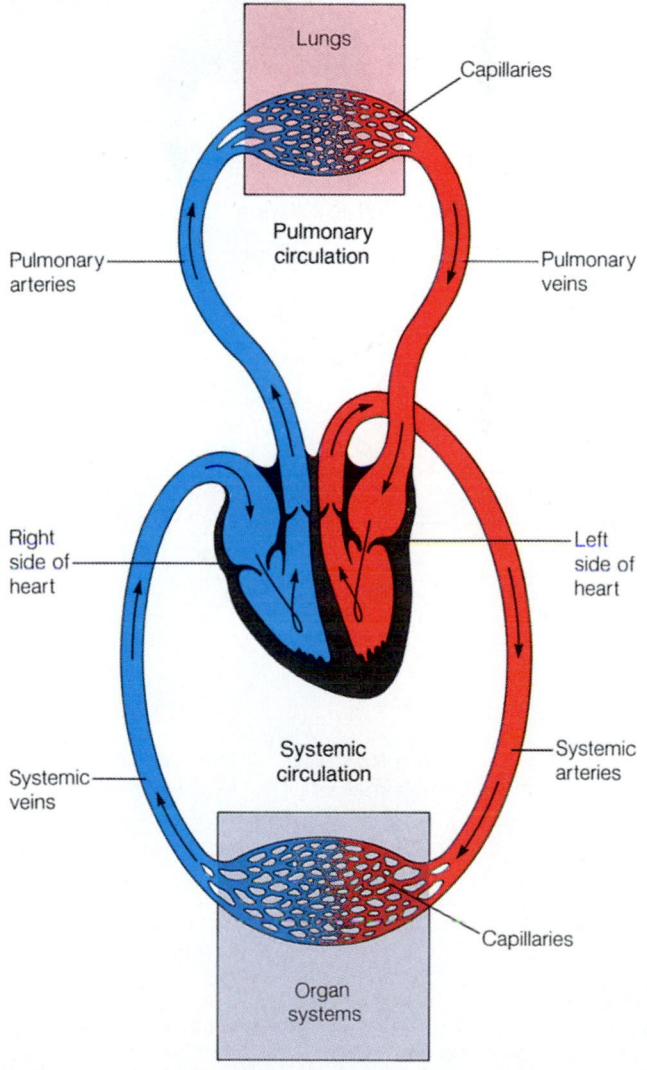

of a closed loop of vessels carrying blood between the heart and lungs, whereas the **systemic circulation** consists of a circuit of vessels carrying blood between the heart and organ systems.

ANATOMICAL CONSIDERATIONS

The heart is located in the middle of the chest cavity.

The heart is a hollow muscular organ about the size of a clenched fist. It is located in the **thoracic** (chest) **cavity** approximately midline between the **sternum** or breastbone anteriorly and the **vertebrae** (backbone) posteriorly (● Fig. 9–2a). The midline location of the heart brings up a potentially confusing point. Place your hand over your heart as if to recite the Pledge of Allegiance. Where did you place your hand? People usually place their hand on the left side of the

chest, even though the heart is actually in the middle of the chest. The heart has a broad base at the top and tapers to a pointed tip known as the **apex** at the bottom. It is situated at an angle under the sternum so that its base lies predominantly to the right and the apex to the left of the sternum. When the heart beats, especially when it contracts forcefully, the apex actually thumps against the inside of the chest wall on the left side. Because we become aware of the beating heart through the apex beat on the left side of the chest, we tend to think—erroneously—that the entire heart is on the left.

The fact that the heart is positioned between two bony structures, the sternum and vertebrae, makes it possible to manually drive blood out of the heart when it is not pumping effectively by rhythmically depressing the sternum (Fig. 9–2b). This maneuver compresses the heart between the sternum and vertebrae so that blood is squeezed out as if the heart were beating. In many instances, this *external cardiac compression*, which is part of **cardiopulmonary resuscitation**

● **FIGURE 9–2   Location and External Compression of the Heart within the Thoracic Cavity**   (a) Location of heart within the thoracic cavity. (b) External cardiac compression during cardiopulmonary resuscitation. Manual compression of the heart between the sternum anteriorly and the vertebrae posteriorly forces blood out of a nonfunctioning heart as if the heart were beating.

Sternum

Ribs

Heart

RA = Right atrium   LA = Left atrium
RV = Right ventricle   LV = Left ventricle

(a)

(b)

(CPR), serves as a lifesaving measure until appropriate therapy can be instituted to restore the heart to normal function.

Although lifesaving, the intense local pressure exerted on the sternum during manual external cardiac compression sometimes causes ribs to break. However, this traditional method of CPR may soon be replaced by a new, safer method in situations where the necessary equipment is available. Tests are currently being conducted on a new computer-controlled inflatable vest that delivers regular bursts of pressure, which is equally distributed around the circumference of the rib cage. The bursts of increased external pressure periodically compress the heart and squeeze out blood but with much less risk of damage to the rib cage because the pressure is uniformly distributed around the chest. Furthermore, the inflatable vest pushes out a larger volume of blood with greater consistency than classical chest compression. The vest's designers speculate that resuscitation success rates will increase with this new CPR technique.

The heart is a dual pump.

Even though anatomically the heart is a single organ, the right and left sides of the heart function as two separate pumps. The heart is divided into right and left halves and has four chambers, an upper and a lower chamber within each half (● Fig. 9–3a). The upper chambers, the **atria** (**atrium**, singular) receive blood returning to the heart and transfer it to the

● **FIGURE 9–3 Blood Flow through and Pump Action of the Heart** (a) Blood flow through the heart. (b) Dual pump action of the heart. The right side of the heart receives partially deoxygenated blood from the systemic circulation and pumps it into the pulmonary circulation. The left side of the heart receives highly oxygenated blood from the pulmonary circulation and pumps it into the systemic circulation. (c) Comparison of thickness of right and left ventricular walls. Note that the left ventricular wall is much thicker than the right wall.

FETAL CIRCULATION: A FETUS HAS IT ALL MIXED UP

Blood does not follow the same course in a fetus as it does after birth. The major differences between fetal circulation and circulation after birth are accommodations to the fact that the fetus is not breathing so the lungs are not functional. The fetus obtains its O_2 and eliminates its CO_2 via exchanges with the maternal blood across the placenta. Since the blood has no need to go to the fetal lungs to pick up O_2 and remove CO_2, there are two bypasses in the fetal circulation: (1) the **foramen ovale**, an opening in the septum between the right and left atrium, and (2) the **ductus arteriosus**, a vessel connecting the pulmonary artery and aorta as they both leave the heart (see the accompanying figure).

The role of these bypasses can best be visualized by tracing the flow of blood through the fetal heart. High-oxygenated blood is carried from the placenta via the umbilical vein and is emptied into the fetus's inferior vena cava. Thus, as the blood is returned to the right atrium from the systemic circulation, it is a mixture of high-oxygenated blood from the umbilical vein and low-oxygenated venous blood returning from the fetal tissues. During fetal life, because of the tremendous resistance offered by the collapsed lungs, the pressures in the right half of the heart and pulmonary circulation are higher than in the left half of the heart and systemic circulation, the opposite of the situation after

lower chambers, the **ventricles**, which pump the blood from the heart. The vessels that return blood from the tissues to the atria are **veins**, and those that carry blood away from the ventricles to the tissues are **arteries**. The two halves of the heart are separated by the **septum**, a continuous muscular partition that prevents mixture of blood from the two sides of the heart. This separation is extremely important, because the right half of the heart is receiving and pumping low-oxygenated blood while the left side of the heart receives and pumps high-oxygenated blood.

Let us examine how the heart functions as a dual pump by tracing a drop of blood through one complete circuit (Fig. 9–3a and b). Blood returning from the systemic circulation enters the right atrium via large veins known as the **venae cavae**. The drop of blood entering the right atrium has returned from the body tissues, where O_2 has been extracted from it and CO_2 has been added to it. This partially deoxygenated blood flows from the right atrium into the right ventricle, which pumps it out through the **pulmonary artery** to the lungs. Thus, the *right side of the heart pumps blood into the pulmonary circulation*. Within the lungs, the drop of blood loses its extra CO_2 and picks up a fresh supply of O_2 before being returned to the left atrium via the **pulmonary veins**. This richly oxygenated blood returning to the left atrium subsequently flows into the left ventricle, the pumping chamber that propels the blood to all body systems except the lungs; that is, *the left side of the heart pumps blood into the systemic circulation*. The large artery carrying blood away from the left

FETAL CIRCULATION (continued)

birth. Because of the pressure difference between the right and left atrium, a part of the mixed, moderately oxygenated blood returning to the right atrium is immediately shunted to the left atrium through the foramen ovale. This blood then continues into the left ventricle and is pumped back out into the systemic circulation. In addition to supplying the tissues, the systemic circulation of the fetus also includes passage of the blood through the umbilical arteries to allow exchange with the maternal blood across the placenta. The remainder of the blood in the right atrium that is not immediately shunted to the left atrium flows into the right ventricle, which pumps the blood into the pulmonary artery. Because the pressure in the pulmonary artery is greater than the pressure in the aorta, blood is shunted from the pulmonary artery into the aorta through the ductus arteriosus down the pressure gradient. Thus, most of the blood pumped out of the right ventricle destined for the pulmonary circulation is immediately shunted into the aorta and delivered to the systemic circulation instead, bypassing the nonfunctional lungs.

At birth, the foramen ovale closes and becomes a small scar known as the **fossa ovalis** in the atrial septum. The ductus arteriosus collapses and eventually degenerates into a thin, ligamentous strand referred to as the **ligamentum arteriosum.**

On occasion, these bypasses fail to close properly after birth. A *patent* (open) *foramen ovale* normally does not cause much of a problem because a valvelike flap is present on the left side of the septum. This flap closes over the opening when the left atrial pressure is greater than the right atrial pressure. Because the pressures in the right half of the heart and pulmonary circulation drop in the newborn as soon as breathing begins and the lungs are inflated, the pressures in the left half of the heart and systemic circulation are greater than the pulmonary pressures after birth. Therefore, the left atrial pressure is greater than the right atrial pressure in the neonate, the opposite of before birth. This pressure difference closes the flap over the foramen ovale, preventing any mixture between the two atrial chambers even if the foramen ovale is not completely closed.

A *patent ductus arteriosus* is a more serious situation. With the fall in lung resistance at the onset of breathing, the pressure in the pulmonary artery falls below that in the aorta, a relationship that exists throughout life. Because the aortic pressure is now higher than the pulmonary arterial pressure, some of the blood is shunted from the aorta into the pulmonary artery through the still-open ductus arteriosus, the reverse of the direction of flow through this connecting vessel during fetal life. As a result of this abnormal shunting, not all of the blood pumped out by the left ventricle goes into the systemic circulation, and excessive blood enters the pulmonary circulation. If the condition is not surgically corrected by tying off the patent ductus arteriosus, the left ventricle compensates by hypertrophying (enlarging and becoming stronger) so that it can pump out even more blood. This extra output provides adequate systemic circulation even though part of the left-ventricular output is diverted to the pulmonary circulation. The right ventricle also hypertrophies, enabling it to pump against the elevated pulmonary arterial pressure, which is increased due to the excess volume of blood shunted into the pulmonary circulation. This extra workload on the heart eventually leads to heart failure and premature death if the condition is not corrected.

---

ventricle is the **aorta**. Major arteries branch from the aorta to supply the various tissues of the body.

In contrast to the pulmonary circulation, in which all the blood flows through the lungs, the systemic circulation may be viewed as a series of parallel pathways. Part of the blood pumped out by the left ventricle goes to the muscles, part to the kidneys, part to the brain, and so on. Thus, the output of the left ventricle is distributed so that each part of the body receives a fresh blood supply; the same arterial blood does not pass from tissue to tissue. Accordingly, the drop of blood we are tracing goes to only one of the systemic tissues. Tissue cells take O_2 from the blood and use it to oxidize nutrients for energy production; in the process, the tissue cells form CO_2 as a waste product that is added to the blood. The drop of blood,

now partially depleted of O_2 content and increased in CO_2 content, returns to the right side of the heart, which once again will pump it to the lungs. One circuit is complete. (See the accompanying boxed feature, Concepts, Challenges, and Controversies.)

Both sides of the heart simultaneously pump equal amounts of blood. The volume of low-oxygenated blood being pumped to the lungs by the right side of the heart soon becomes the same volume of high-oxygenated blood being delivered to the tissues by the left side of the heart. The pulmonary circulation is a low-pressure, low-resistance system, whereas the systemic circulation is a high-pressure, high-resistance system. Therefore, even though the right and left sides of the heart pump the same amount of blood, the left side

• **FIGURE 9–4 Mechanism of Valve Action**

Valve opened

When pressure is greater behind the valve, it opens.

Valve closed; does not open in opposite direction

When pressure is greater in front of the valve, it closes. Note that, when pressure is greater in front of the valve, it does not open in the opposite direction; that is, it is a one-way valve.

performs more work, because it pumps an equal volume of blood at a higher pressure into a higher-resistance system. Accordingly, the heart muscle on the left side is much thicker than the muscle on the right side, making the left side a stronger pump (Fig. 9–3c).

Heart valves ensure that the blood flows in the proper direction through the heart.

Blood flows through the heart in one fixed direction from veins to atria to ventricles to arteries. The presence of four one-way heart valves ensures this unidirectional flow of blood. The valves are positioned so that they open and close passively because of pressure differences, similar to a one-way door (• Fig. 9–4). A forward pressure gradient forces the valve open, much as you open a door by pushing on one side of it, whereas a backward pressure gradient forces the valve closed, just as you apply pressure to the opposite side of the door to close it. Note that a backward gradient can force the valve closed but cannot force it to swing open in the opposite direction; that is, heart valves are not like swinging, saloon-type doors.

Two of the heart valves, the **right** and **left atrioventricular (AV) valves**, are positioned between the atrium and the ventricle on the right and left sides, respectively (• Fig. 9–5a). These valves allow blood to flow from the atria into the ventricles during ventricular filling (when atrial pressure exceeds ventricular pressure), but prevent the backflow of blood from the ventricles into the atria during ventricular emptying (when ventricular pressure greatly exceeds atrial pressure). If the rising ventricular pressure did not force the AV valves to close as the ventricles contracted to empty, much of the blood would inefficiently be forced back into the atria and veins instead of being pumped into the arteries. The right AV valve is also called the **tricuspid valve** because it consists of three cusps or leaflets (Fig. 9–5b). Likewise, the left AV valve, which consists of two cusps, is often called the **bicuspid valve** or, alternatively, the **mitral valve** (because of its physical resemblance to a mitre or bishop's headgear).

The edges of the AV-valve leaflets are fastened by tough, thin, fibrous cords of tendinous-type tissue, the **chordae tendineae**, which prevent the valves from being everted, that is, from being forced by the high ventricular pressure to open in

the opposite direction into the atria. These cords extend from the edges of each cusp and attach to small, nipple-shaped **papillary muscles** (*papilla* means "nipple"), which protrude from the inner surface of the ventricular walls. When the ventricles contract, the papillary muscles also contract, pulling downward on the chordae tendineae. This pulling exerts tension on the closed AV-valve cusps to hold them in position, thus helping them remain tightly sealed in the face of a strong backward pressure gradient (Fig. 9–5c).

The two remaining heart valves, the **aortic** and **pulmonary valves**, are located at the juncture where the major arteries leave the ventricles (Fig. 9–5a). They are known as **semilunar** ("half-moon") **valves** because they are composed of three cusps, each resembling a shallow half–moon shaped pocket (Fig. 9–5b). These valves are forced open when the left and right ventricular pressures exceed the pressure in the aorta and pulmonary arteries, respectively, during ventricular contraction and emptying. Closure results when the ventricles relax and ventricular pressures fall below the aortic and pulmonary artery pressures. The closed valves prevent blood from flowing from the arteries back into the ventricles from which it has just been pumped. The semilunar valves are prevented from everting by the anatomical structure and positioning of the cusps. When a backward pressure gradient is created upon ventricular relaxation, the back surge of blood fills the pocketlike cusps and sweeps them into a closed position, with their unattached upturned edges fitting together in a deep, leakproof seam (Fig. 9–5d).

Even though there are no valves between the atria and veins, backflow of blood from the atria into the veins usually is not usually a significant problem for two reasons: (1) atrial pressures usually are not much higher than venous pressures, and (2) the sites where the venae cavae enter the atria are partially compressed during atrial contraction.

Four interconnecting rings of dense connective tissue provide a firm base for attachment of the four heart valves (• Fig. 9–6). This fibrous skeleton, which separates the atria from the ventricles, also provides a fairly rigid structure for attachment of the cardiac muscle. The atrial muscle mass is anchored above the rings, and the ventricular muscle mass is attached to the bottom of the rings. It might seem rather surprising that the inlet valves to the ventricles (the AV valves) and the outlet valves from the ventricles (the semilunar valves) all lie on the same plane through the heart, as delin-

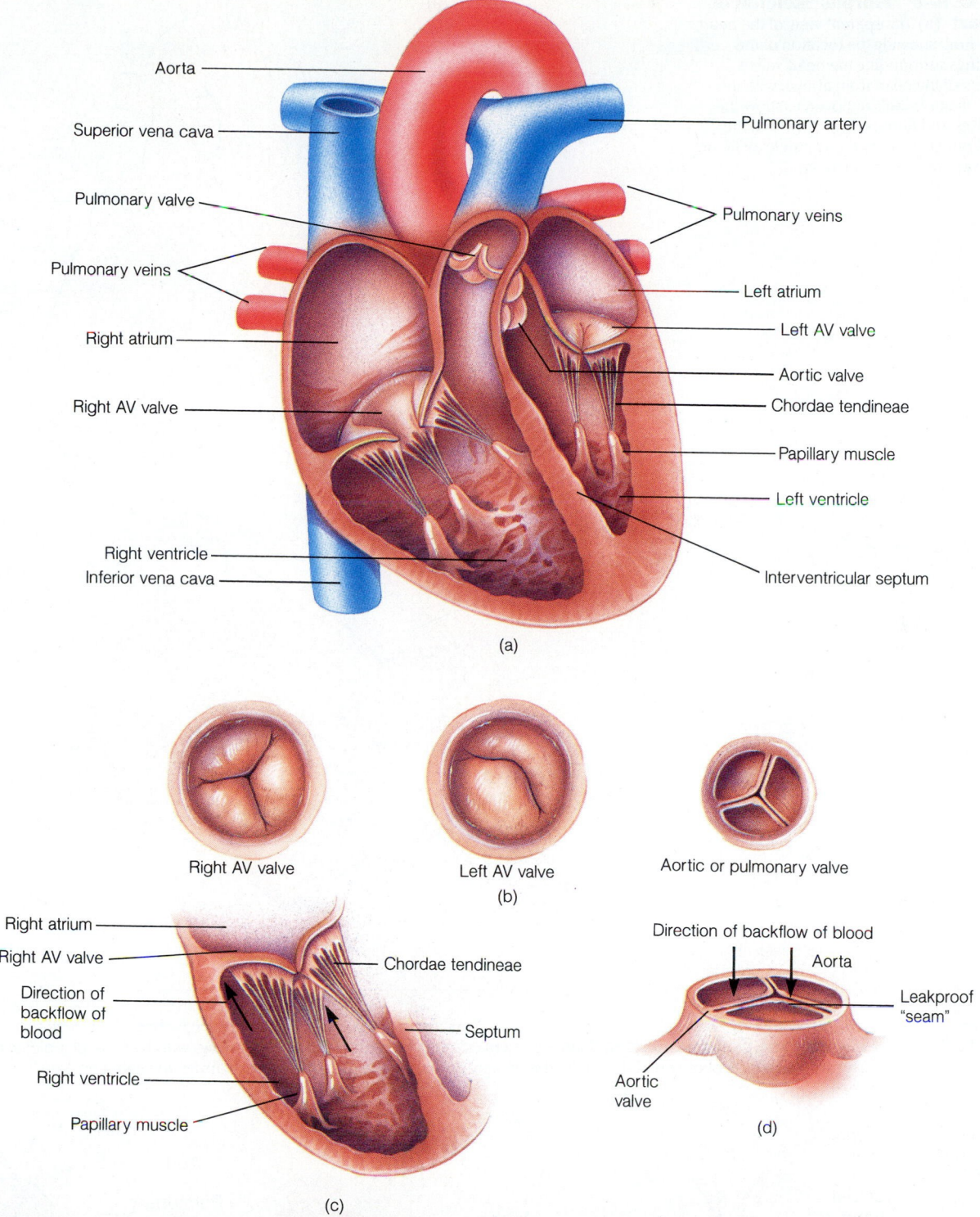

Aorta

Superior vena cava

Pulmonary valve

Pulmonary veins

Right atrium

Right AV valve

Right ventricle

Inferior vena cava

Pulmonary artery

Pulmonary veins

Left atrium

Left AV valve

Aortic valve

Chordae tendineae

Papillary muscle

Left ventricle

Interventricular septum

(a)

Right AV valve

Left AV valve

Aortic or pulmonary valve

(b)

Right atrium

Right AV valve

Direction of backflow of blood

Right ventricle

Papillary muscle

Chordae tendineae

Septum

(c)

Direction of backflow of blood

Aorta

Leakproof "seam"

Aortic valve

(d)

● **FIGURE 9–5   Heart Valves** (a) Longitudinal section of heart, depicting location of the four heart valves. (b) Top view of heart valves in closed position. (c) Prevention of eversion of AV valves. The eversion of AV valves is prevented by tension on valve leaflets exerted by chordae tendineae when the papillary muscles contract. (d) Prevention of eversion of semilunar valves. When the semilunar valves are swept closed, their upturned edges fit together in a deep, leakproof seam that prevents valve eversion.

eated by the fibrous skeleton. This relationship comes about because the heart forms from a single tube that bends upon itself and twists on its axis during embryonic development (● Fig. 9–7). Although this turning and twisting makes the study of the structural relationships of the heart more difficult,

the twisted structure has functional importance in that it helps the heart pump more efficiently. We will see how by turning our attention to the portion of the heart that actually generates the forces responsible for blood flow, the cardiac muscle within the heart walls.

(a)

• **FIGURE 9–6   Fibrous Skeleton of the Heart** (a) Transparent view of the heart from the front showing the location of the fibrous rings surrounding the heart valves. (b) A view of the heart from above, with the atria and major vessels removed to show the heart valves and fibrous rings. Note that the inlet and outlet valves to the ventricle all lie on the same plane through the heart.

Pulmonary valve

Aortic valve

Right AV valve

Fibrous skeleton

Left AV valve

(Rear)

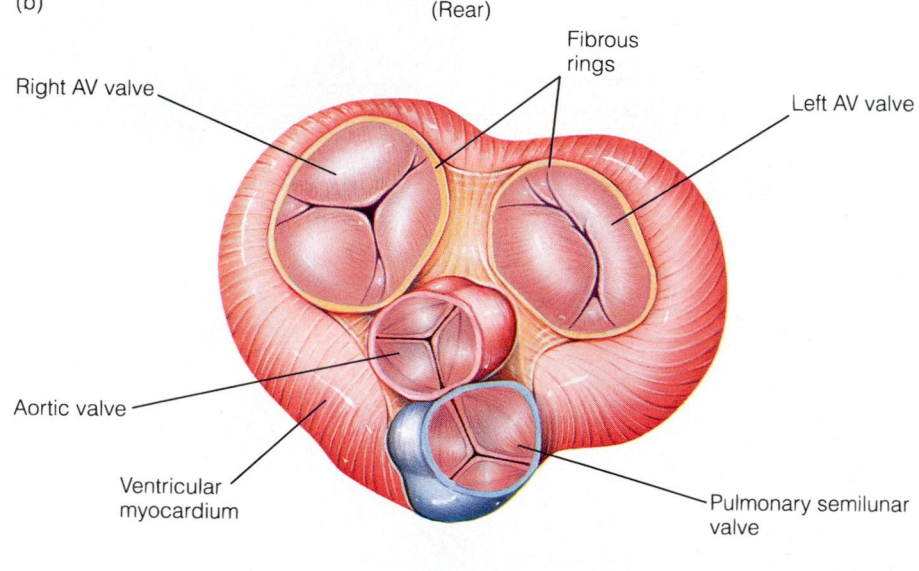

(b)

Right AV valve

Fibrous rings

Left AV valve

Aortic valve

Ventricular myocardium

Pulmonary semilunar valve

(Front)

• **FIGURE 9–7   Twisting of the Embryonic Heart on Its Axis during Development** The arrows within the heart indicate the direction of blood flow. The arrows outside of the heart depict the direction of twisting of the embryonic heart during development.

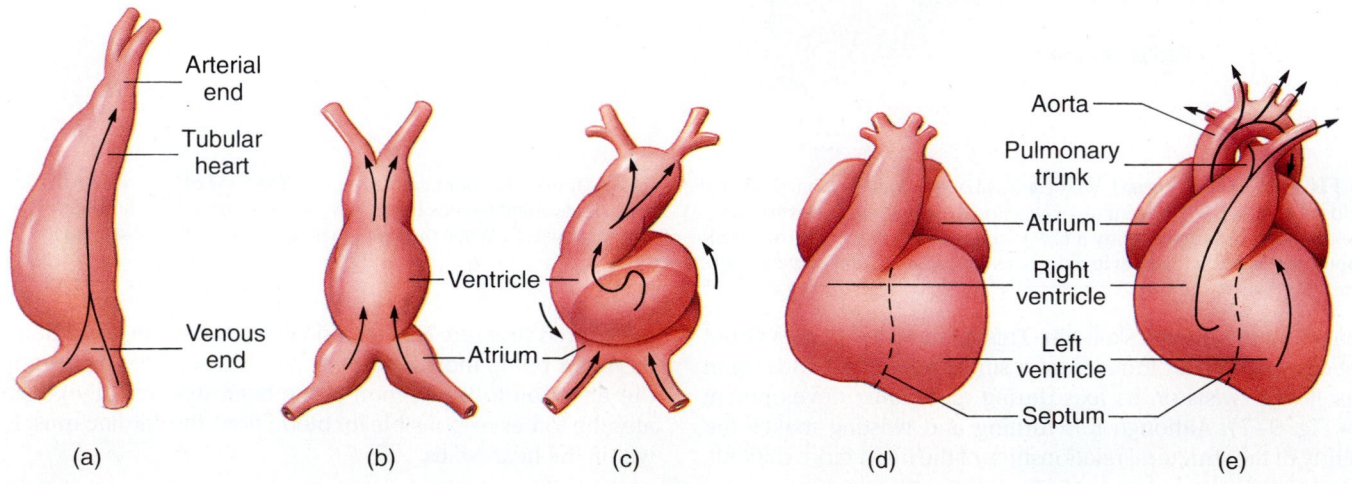

Arterial end

Tubular heart

Venous end

Ventricle

Atrium

Aorta

Pulmonary trunk

Atrium

Right ventricle

Left ventricle

Septum

(a)   (b)   (c)   (d)   (e)

The heart walls are composed primarily of spirally arranged cardiac-muscle fibers interconnected by intercalated discs.

The heart wall consists of three distinct layers:

■ The **endocardium** (*endo* means "within"; *cardia* means "heart") is a thin inner layer of **endothelium**, a unique type of epithelial tissue that lines the entire circulatory system.

■ The **myocardium** (*myo* means "muscle"), the middle layer composed of cardiac muscle, constitutes the bulk of the heart wall.

■ The **epicardium** (*epi* means "upon") is a thin external membrane covering the heart.

The myocardium consists of interlacing bundles of cardiac-muscle fibers arranged spirally around the circumference of the heart. The spiral arrangement is due to the heart's complex twisting during development. As a result of this arrangement, when the ventricular muscle contracts and shortens, the diameter of the ventricular chambers is reduced while the apex is simultaneously pulled upward toward the top of the heart in a rotating manner. This exerts a "wringing" effect, efficiently exerting pressure on the blood within the enclosed chambers and directing it upward toward the openings of the major arteries that exit at the base of the ventricles.

The individual cardiac-muscle cells are interconnected to form branching fibers, with adjacent cells joined end-to-end at specialized structures known as **intercalated discs**. Within an intercalated disc, there are two types of membrane junctions: desmosomes and gap junctions (● Fig. 9–8). A desmosome, a type of adhering junction that mechanically holds cells together, is particularly abundant in tissues, such as the heart, that are subject to considerable mechanical stress (see p. 55). At intervals along the intercalated disc, the opposing membranes approach each other very closely to form gap junctions, which are areas of low electrical resistance that allow action potentials to spread from one cardiac cell to adjacent cells (see p. 55). Cardiac muscle is capable of generating action potentials without any nervous stimulation. When one of the cardiac cells spontaneously undergoes an action potential, the electrical impulse spreads to all the other cells that are joined by gap junctions in the surrounding muscle mass so that they become excited and contract as a single functional syncytium (see p. 251). The atria and the ventricles each form a functional syncytium and contract as separate units. The synchronous contraction of the muscle cells composing the walls of each of these chambers produces the force necessary to eject the enclosed blood.

There are no gap junctions between the atrial and ventricular contractile cells, and, furthermore, these muscle masses are separated by the electrically nonconductive fibrous skeleton that surrounds the valves. However, an important specialized conducting system is present to facilitate and coordinate the transmission of electrical excitation from the atria to the ventricles to assure synchronization between atrial and ventricular pumping.

Due to both the syncytial nature of cardiac muscle and the conducting system between the atria and ventricles, an impulse spontaneously generated in one part of the heart

● **FIGURE 9–8 Organization of Cardiac-Muscle Fibers** Adjacent cardiac-muscle cells are joined end-to-end by intercalated discs, which contain two types of specialized junctions: desmosomes, which act as spot rivets mechanically holding the cells together, and gap junctions, which permit action potentials to spread from one cell to adjacent cells.

spreads throughout the entire heart. Therefore, unlike skeletal muscle, where graded contractions can be produced by varying the number of muscle cells that are contracting within the muscle (recruitment of motor units), either all the cardiac-muscle fibers contract or none of them do. A "half-hearted" contraction is not possible. Gradation of cardiac contraction is accomplished by varying the strength of contraction of all the cardiac-muscle cells.

Cardiac-muscle cells contain an abundance of mitochondria, the O_2-dependent energy organelles. In fact, up to 40% of the cell volume of cardiac-muscle cells is attributable to mitochondria, indicative of how much the heart depends on aerobic metabolism to generate the energy necessary for contraction. Cardiac muscle also has an abundance of myoglobin, which stores limited amounts of O_2 within the heart for immediate use and also facilitates the transport of O_2 arriving at the cell to the mitochondria for rapid processing.

No new cardiac-muscle cells are produced after infancy. Any cardiac enlargement that occurs is due to cardiac hypertrophy, an increase in the size of the cardiac-muscle fibers. Cardiac hypetrophy occurs in response to increased work de-

FIGURE 9–9 Pericardial Sac

Pericardial sac:
- Outer fibrous covering
- Secretory lining

Heart

Pericardial cavity filled with pericardial fluid

mands placed on the heart, such as when the heart must pump blood against increased resistance (for example, through a heart valve narrowed by disease or into blood vessels stiffened by "hardening of the arteries"), or when it must pump an increased volume of blood (as in response to a regular exercise program).

The heart is enclosed by the pericardial sac.

The heart is enclosed in the double-walled, membranous **pericardial sac** (•Fig. 9–9). The outer layer of the sac is a tough, fibrous membrane that is attached to the connective-tissue partition that separates the lungs. This attachment anchors the heart so that it remains properly positioned within the chest. The sac is lined by a membrane that secretes a thin **pericardial fluid**, which provides lubrication to prevent friction between the pericardial layers as they glide over each other with every beat of the heart. **Pericarditis**, an inflammation of the pericardial sac that results in a painful friction rub between the two pericardial layers, occurs occasionally because of viral or bacterial infection.

If blood hemorrhages into the pericardial sac as a result of a penetrating heart wound or rupture of the heart wall, the sac, because of its tough outer membrane, is unable to expand outward to accommodate the extra fluid volume. Instead, it distends inward, compressing the heart and restricting cardiac filling. This effect is similar to squeezing a soft rubber ball with your hand. The heart simply has less room to expand, so the amount of blood that can enter from the veins is limited. Because less blood is returned to the heart to be pumped out to supply the tissues, cardiac failure subsequently ensues. Such distension of the pericardial sac that impinges on cardiac filling is known as **cardiac tamponade**.

▬● ELECTRICAL ACTIVITY OF THE HEART

The sinoatrial node is the normal pacemaker of the heart.

Contraction of cardiac-muscle cells to bring about ejection of blood is triggered by action potentials sweeping across the muscle cell membranes. The heart contracts or beats rhythmically as a result of action potentials that it generates by itself, a property known as **autorhythmicity**.

There are two specialized types of cardiac-muscle cells:

1. Ninety-nine percent of the cardiac-muscle cells are **contractile cells**, which do the mechanical work of pumping. These working cells normally do not initiate their own action potentials.

2. In contrast, the small but extremely important remainder of the cardiac cells, the **autorhythmic cells**, do not contract but instead are specialized for initiating and conducting the action potentials responsible for contraction of the working cells.

Let us examine the role of the specialized autorhythmic cells in the origin and spread of the heartbeat. In contrast to nerve and skeletal-muscle cells, in which the membrane remains at constant resting potential unless the cell is stimulated (see p. 79), the cardiac autorhythmic cells do not have a resting potential. Instead they display **pacemaker activity**; that is, their membrane potential slowly depolarizes, or drifts, between action potentials until threshold is reached, at which time the membrane fires or has an action potential (•Fig. 9–10; see also p. 251). Through repeated cycles of drift and fire, these autorhythmic cells cyclically initiate action potentials, which then spread throughout the heart to trigger rhythmic beating without any nervous stimulation.

The basis for the membrane potential's slow drift to threshold is still unresolved. It is generally believed to be caused by a cyclical decrease in the passive outward flux of K^+ super-

•FIGURE 9–10 Pacemaker Activity of Cardiac Autorhythmic Cells The membrane potential does not remain constant between action potentials in the cardiac autorhythmic cells. Instead, following an action potential, the membrane slowly depolarizes or drifts to threshold as a result of inactivation of K^+ channels. At the same time that less K^+ is leaving the cell because of the decreased P_{K^+}, Na^+, whose permeability remains unaltered, continues to leak into the cell. As a result, the inside gradually becomes less negative; that is, the membrane slowly depolarizes to threshold. Once threshold is reached and Ca^{++} channels open, a rapid influx of Ca^{++} occurs, producing the rising phase of the self-induced action potential. The falling phase is brought about by a rapid K^+ efflux upon activation of K^+ channels. Inactivation of these channels after the action potential is over initiates the next slow depolarization to threshold.

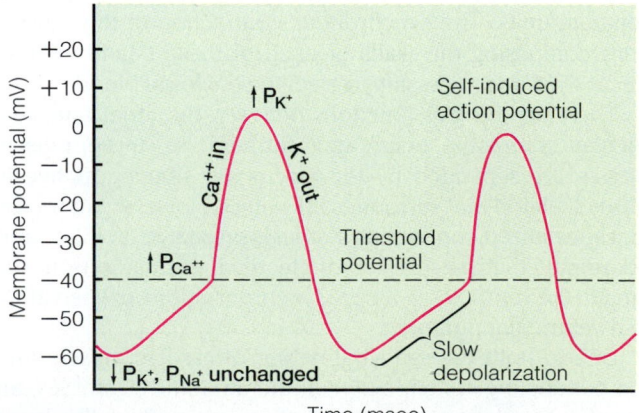

imposed on a slow, unchanging inward leak of Na^+. In cardiac autorhythmic cells, permeability to K^+ does not remain constant between action potentials as it does in nerve and skeletal-muscle cells. Instead, membrane permeability to K^+ decreases between action potentials because K^+ channels are inactivated, which diminishes the outflow of positive potassium ions down their concentration gradient. Since a small passive influx of Na^+ continues unchanged, the inside gradually becomes less negative; that is, the membrane gradually depolarizes and drifts toward threshold. Once threshold is reached, the rising phase of the action potential occurs in response to activation of Ca^{++} channels and the subsequent influx of Ca^{++}; this phase differs from skeletal muscle, where Na^+ influx rather than Ca^{++} influx swings the potential in the positive direction. The falling phase is due, as usual, to the K^+ efflux that occurs when K^+ permeability increases as a result of activation of K^+ channels. After the action potential is over, inactivation of these K^+ channels initiates the next slow depolarization to threshold.

The cardiac cells capable of autorhythmicity are found in the following specific locations (Fig. 9–11):

1. The **sinoatrial node** (**SA node**), a small specialized region in the right atrial wall near the opening of the superior vena cava.

2. The **atrioventricular node** (**AV node**), a small bundle of specialized cardiac-muscle cells located at the base of the right atrium near the septum, just above the junction of the atria and ventricles.

3. The **bundle of His** (**atrioventricular bundle**), a tract of specialized cells that originates at the AV node and enters the interventricular septum, where it divides to form the right and left bundle branches that travel down the septum, curve around the tip of the ventricular chambers, and travel back toward the atria along the outer walls.

4. **Purkinje fibers**, small terminal fibers that extend from the bundle of His and spread throughout the ventricular myocardium much like small twigs of a tree branch.

●FIGURE 9–11   Specialized Conducting System of the Heart

Atrioventricular (AV) node

Sinoatrial (SA) node

Right atrium

Right branch of bundle of His

Right ventricle

Left atrium

Left branch of bundle of His

Left ventricle

Purkinje fibers

TISSUE	ACTION POTENTIALS PER MINUTE*
SA node (normal pacemaker)	70–80
AV node	40–60
Bundle of His and Purkinje fibers	20–40

TABLE 9–1   Normal Rate of Action Potential Discharge in Autorhythmic Tissues of the Heart

*In the presence of parasympathetic tone; see p. 200.

Because of differences in their rates of slow depolarization to threshold, these various autorhythmic cells differ in the rates at which they are normally capable of generating action potentials (Table 9–1). Comparing two autorhythmic cells (Fig. 9–12), cell A has a faster rate of depolarization and thus reaches threshold more quickly and generates action potentials more rapidly than cell B. The heart cells with the fastest rate of action potential initiation are localized in the SA node. Once an action potential occurs in any cardiac-muscle cell, it is propagated throughout the rest of the myocardium via gap junctions and the specialized conducting system. Therefore, the SA node, which normally exhibits the fastest rate of autorhythmicity at 70 to 80 action potentials/minute, drives the rest of the heart at this rate and is known as the **pacemaker** of the heart. The other autorhythmic tissues are unable to assume their own naturally slower rates, because they are activated by action potentials originating in the SA node before they are able to reach threshold at their own slower rhythm.

The following analogy demonstrates how the SA node drives the remainder of the heart at its own pace. Suppose a train consists of a hundred cars, three of which are engines capable of moving on their own; the other ninety-seven cars must be pulled in order to move (Fig. 9–13a). One engine (the SA node) can travel at 70 miles/hour on its own, another engine (the AV node) at 50 miles/hour, and the last engine (the Purkinje fibers) at 30 miles/hour. If all of these cars are joined together, the engine capable of traveling at 70 miles/hour will pull the remainder of the cars at that speed. The engines that can travel at lower speeds on their own will be pulled at a faster speed by the fastest engine and will therefore be unable to assume their own slower rate as long as they are being driven by a faster engine. The other ninety-seven cars (nonautorhythmic, contractile working cells), being unable to move on their own, will likewise travel at whatever speed the fastest engine pulls them.

If for some reason the fastest engine breaks down (SA node damage), the next fastest engine (AV node) takes over and the entire train travels at 50 miles/hour; that is, if the SA node becomes nonfunctional, the AV node assumes pacemaker activity (Fig. 9–13b). The non-SA nodal autorhythmic tissues are **latent pacemakers** that can take over, although at a lower rate, should the normal pacemaker fail. If conduction of the impulse becomes blocked between the atria

(a)   (b)

● FIGURE 9–12 Different Autorhythmic Rates (a) Recording from autorhythmic cell A. (b) Recording from autorhythmic cell B. Because Cell A has a faster rate of depolarization, it reaches threshold more quickly than Cell B and therefore generates action potentials more rapidly.

● FIGURE 9–13 Analogy of Pacemaker Activity (a) Normal pacemaker activity by the SA node. (b) Takeover of pacemaker activity by the AV node when the SA node is nonfunctional. (c) Takeover of ventricular rate by the slower ventricular autorhythmic tissue in the condition of heart block even though the SA node is still functioning. (d) Takeover of pacemaker activity by an ectopic focus.

(a)   Whole train will go 70 mph
        (heart rate set by SA node, the fastest autorhythmic tissue)

(b)   Train will go 50 mph
        (the next fastest autorhythmic tissue, the AV node, will set the heart rate)

(c)   First part of train will go 70 mph; last part will go 30 mph
        (atria will be driven by SA node; ventricles will assume own, much slower rhythm)

(d)   Train will be driven by ectopic focus, which is now going faster than the SA node.
        (the whole heart will be driven more rapidly by an abnormal pacemaker)

and the ventricles, the atria continue at the typical rate of 70 beats/minute, and the ventricular tissue, not being driven by the faster SA nodal rate, assumes its own much slower autorhythmic rate of about 30 beats/minute, initiated by the ventricular autorhythmic cells (Purkinje fibers). This situation is comparable to a breakdown of the second engine (AV node) so that the lead engine (SA node) becomes disconnected from the slow third engine (Purkinje fibers) and remainder of the cars (Fig. 9–13c). The lead engine continues at 70 miles/hour while the remainder of the train proceeds at 30 miles/hour. Such a phenomenon, known as **complete heart block**, occurs when the conducting tissue between the atria and ventricles is damaged and becomes nonfunctional. A ventricular rate of 30 beats/minute will support only a very sedentary existence; in fact, the patient usually becomes comatose. In circumstances of abnormally low heart rate, as in SA-node failure or heart block, an **artificial pacemaker** can be used. Such an implanted device rhythmically generates impulses that spread throughout the heart to drive both the atria and ventricles at the typical rate of 70 beats/minute.

Occasionally, an area of the heart, such as a Purkinje fiber, becomes overly excitable and depolarizes at a more rapid rate than the SA node. (The slow engine suddenly has the capability of going faster than the lead engine; Fig. 9–13d.) This abnormally excitable area, an **ectopic focus**, initiates a premature action potential that spreads throughout the rest of the heart before a normal action potential can be initiated by the SA node. An occasional abnormal impulse from an ectopic focus produces a **premature beat** or an **extrasystole**. If the ectopic focus continues to discharge at its more rapid rate, pacemaker activity is shifted from the SA node to the ectopic focus. The heart rate abruptly becomes greatly accelerated and continues this rapid rate for a variable time period until the ectopic focus returns to normal. Such overly irritable areas may be associated with organic heart disease, but more frequently they occur in response to anxiety, lack of sleep, or excess caffeine, nicotine, or alcohol consumption.

The spread of cardiac excitation is coordinated to assure efficient pumping.

Once initiated in the SA node, an action potential spreads throughout the rest of the heart. For efficient cardiac function, the spread of excitation should satisfy three criteria:

1. *Atrial excitation and contraction should be complete before the onset of ventricular contraction.* Complete ventricular filling requires that atrial contraction precede ventricular contraction. During the period of cardiac relaxation, the AV valves are open so that venous blood entering the atria continues to flow directly into the ventricles. Almost 80% of ventricular filling occurs by this means prior to atrial contraction. When the atria do contract, additional blood is squeezed into the ventricles to complete ventricular filling. Ventricular contraction then occurs to eject blood from the heart into the arteries. If the atria and ventricles were to contract simultaneously, the AV valves would be closed immediately because ventricular pressures would greatly exceed atrial pressures. The ventricles have much

thicker walls and, accordingly, can generate more pressure. Atrial contraction would be unproductive because the atria could not squeeze blood into the ventricles through closed valves. Therefore, to assure complete filling of the ventricles—to obtain the remaining 20% of ventricular filling that occurs during atrial contraction—it is imperative that the atria become excited and contract before ventricular excitation and contraction.

2. *Excitation of cardiac-muscle fibers should be coordinated to assure that each heart chamber contracts as a unit to accomplish efficient pumping.* If the muscle fibers in a heart chamber were to become excited and contract randomly rather than contracting simultaneously in a coordinated fashion, they would be unable to eject blood. A smooth, uniform ventricular contraction is essential to squeeze out the blood. As an analogy, assume you have a basting syringe full of water. If you merely poke a finger here or there into the rubber bulb of the syringe, you will not eject much water. However, if you compress the bulb in a smooth, coordinated fashion, you can squeeze out the water. In a similar manner, contraction of isolated cardiac-muscle fibers is not successful in pumping blood. Such random, uncoordinated excitation and contraction of the cardiac cells is known as **fibrillation**. Ventricular fibrillation rapidly causes death because the heart is not able to pump blood into the arteries. This condition can often be corrected by **electrical defibrillation**, in which a very strong electrical current is applied on the chest wall. When this current reaches the heart, it essentially stimulates all parts of the heart simultaneously. Usually, the first part of the heart to recover is the SA node, which takes over pacemaker activity, once again initiating impulses that trigger the synchronized contraction of the remainder of the heart.

3. *The pair of atria and pair of ventricles should be functionally coordinated so that both members of the pair contract simultaneously.* This permits synchronized pumping of blood into the pulmonary and systemic circulation.

The normal spread of cardiac excitation is carefully orchestrated to assure that these criteria are met and the heart functions efficiently (• Fig. 9–14).

Atrial Excitation An action potential originating in the SA node first spreads throughout both atria, primarily from cell to cell via gap junctions. In addition several poorly delineated, specialized conduction pathways hasten conduction of the impulse through the atria:

■ The **interatrial pathway** extends from the SA node within the right atrium to the left atrium. Because of this pathway, a wave of excitation can spread across the gap junctions throughout the left atrium at the same time a similar spread is being accomplished throughout the right atrium. This assures that both atria become depolarized to contract more or less simultaneously.

■ The **internodal pathway** extends from the SA node to the AV node. The AV node is the only point of electrical contact between the atria and ventricles; in other words, be-

• FIGURE 9–14 Spread of Cardiac Excitation An action potential initiated at the SA node first spreads throughout both atria. Its spread is facilitated by two specialized atrial conduction pathways, the interatrial and internodal pathways. The AV node is the only point where an action potential can spread from the atria to the ventricles. From the AV node, the action potential spreads rapidly throughout the ventricles, hastened by a specialized ventricular conduction system consisting of the bundle of His and Purkinje fibers.

cause the atria and ventricles are structurally connected by electrically nonconductive fibrous tissue, the only way an action potential in the atria can spread to the ventricles is by passing through the AV node. The internodal conduction pathway directs the spread of an action potential originating at the SA node to the AV node to assure sequential contraction of the ventricles following atrial contraction.

Transmission between Atria and Ventricles The action potential is conducted relatively slowly through the AV node. This slowness is advantageous because it allows time for complete ventricular filling to occur. The impulse is delayed about 0.1 second (the **AV nodal delay**), which enables the atria to become completely depolarized and to contract, emptying their contents into the ventricles, before ventricular depolarization and contraction occur.

Ventricular Excitation Following this delay, the impulse rapidly travels down the bundle of His and throughout the ventricular myocardium via the Purkinje fibers. The network of fibers in this ventricular conduction system is specialized for rapid propagation of action potentials. Its presence hastens and coordinates the spread of ventricular excitation to assure that the ventricles contract as a unit. Although this system carries the action potential rapidly to a large number of cardiac-muscle cells, it does not terminate on every cell. The impulse quickly spreads from the excited cells to the remainder of the ventricular muscle cells by means of gap junctions.

The ventricular conduction system is more highly organized and more important than the interatrial and internodal conduction pathways. Because the ventricular mass is so much larger than the atrial mass, it is crucial that a rapid conduction system be present to hasten the spread of excitation

in the ventricles. If the entire ventricular depolarization process depended on the cell-to-cell spread of the impulse via gap junctions, the ventricular tissue immediately adjacent to the AV node would become excited and contract before the impulse had even passed to the apex of the heart. This, of course, would not allow efficient pumping. The rapid conduction of the action potential down the bundle of His and its swift, diffuse distribution throughout the Purkinje network lead to almost simultaneous activation of the ventricular myocardial cells in both ventricular chambers, which assures a single, smooth, coordinated contraction that can efficiently eject blood into both the systemic and pulmonary circulations at the same time.

The action potential of contractile cardiac-muscle cells shows a characteristic plateau.

The action potential in contractile cardiac-muscle cells, although initiated by the nodal pacemaker cells, varies considerably in ionic mechanisms and shape from the SA node potential (compare Figures 9–10 and • 9–15). Unlike autorhythmic cells, the membrane of contractile cells remains essentially at rest at about −90 mV until excited by electrical activity propagated from the pacemaker. Once the membrane of a ventricular myocardial contractile cell is excited,

• FIGURE 9–15 Action Potential in Contractile Cardiac-Muscle Cells The action potential in cardiac contractile cells differs considerably from the action potential in cardiac autorhythmic cells (compare with Fig. 9–10). The membrane potential of cardiac contractile cells remains at a resting potential of −90 mV until excited. Similar to most excitable cells, the rising phase of the action potential is caused by a fast Na^+ influx and the falling phase by a fast K^+ efflux. Unique to cardiac contractile cells, the membrane potential is maintained near the peak of the action potential for several hundred milliseconds. This plateau phase of the action potential results from a slow influx of Ca^{++} coupled with a marked decrease in K^+ permeability.

an action potential is generated by a complicated interplay of permeability changes and membrane-potential changes as follows (Fig. 9–15):

1. During the rising phase of the action potential, the membrane potential rapidly becomes reversed to a positive value of +30 mV as a result of an explosive increase in membrane permeability to Na$^+$ and a subsequent massive Na$^+$ influx. Thus far, the process is the same as in neurons and skeletal-muscle cells (see p. 81). The Na$^+$ permeability then rapidly plummets to its low resting value, but, unique to these cardiac-muscle cells, the membrane potential is maintained at this positive level for several hundred milliseconds, producing a *plateau phase* of the action potential. In contrast, the short action potential of neurons and skeletal-muscle cells lasts less than a millisecond.

2. The sudden change in voltage occurring during the rising phase of the action potential brings about two voltage-dependent permeability changes that are responsible for maintaining this plateau: activation of "slow" Ca^{++} channels and a marked decrease in K$^+$ permeability. Opening of the Ca^{++} channels results in a slow, inward diffusion of Ca^{++} because Ca^{++} is in greater concentration in the ECF. This continued influx of positively charged Ca^{++} prolongs the positivity inside the cell and is primarily responsible for the plateau portion of the action potential. This effect is enhanced by the concomitant decrease in K$^+$ permeability. The resultant reduction in outflux of positively charged K$^+$ prevents rapid repolarization of the membrane and thus contributes to prolongation of the plateau phase.

3. The rapid falling phase of the action potential results from inactivation of the Ca^{++} channels and activation of K$^+$ channels. The decrease in Ca^{++} permeability diminishes the slow, inward movement of positive Ca^{++}, whereas the sudden increase in K$^+$ permeability simultaneously promotes rapid outward diffusion of positive K$^+$. Thus, rapid repolarization at the end of the plateau is accomplished primarily by K$^+$ efflux, which once again makes the inside of the cell more negative than the outside and restores membrane potential to resting.

The mechanism by which an action potential in a cardiac-muscle fiber brings about contraction of that fiber is quite similar to the excitation-contraction coupling process of skeletal muscle (● Fig. 9–16). The presence of a local action potential within the T tubules causes Ca^{++} to be released into the cytosol from the intracellular stores in the sarcoplasmic reticulum. In contrast to skeletal-muscle cells, Ca^{++} also diffuses into the cytosol across the plasma membrane from the ECF during a cardiac action potential. This entering Ca^{++} triggers even further release of Ca^{++} from the sarcoplasmic reticulum. This extra supply of Ca^{++} is not only the major factor responsible for the prolongation of the cardiac action potential but is also responsible for the subsequent lengthening of the period of cardiac contraction, which lasts about three times longer than a single skeletal-muscle fiber contraction.

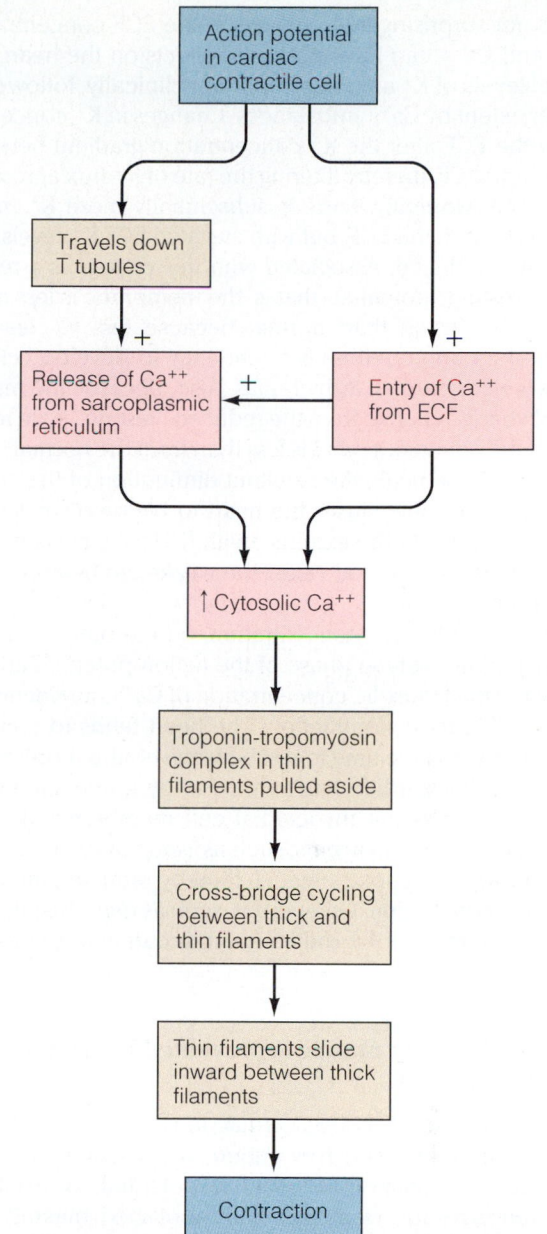

● **FIGURE 9–16   Excitation-Contraction Coupling in Cardiac Contractile Cells**

This increased contractile time ensures adequate time to eject the blood.

The role of Ca^{++} within the cytosol, as in skeletal muscle, is to bind with the troponin-tropomyosin complex and physically pull it aside so that cross-bridge cycling and contraction can take place (see p. 218). However, unlike skeletal muscle, in which sufficient Ca^{++} is always released to turn on all of the cross bridges, in cardiac muscle the extent of cross-bridge activity varies with the amount of cytosolic Ca^{++}. Removal of Ca^{++} from the cytosol by active pumps in both the plasma membrane and sarcoplasmic reticulum restores the blocking action of troponin and tropomyosin so that contraction ceases and the heart muscle relaxes.

It is not surprising that changes in the ECF concentration of K+ and Ca++ can have profound effects on the heart. Abnormal levels of K+ are most important clinically, followed to a lesser extent by Ca++ imbalances. Changes in K+ concentration in the ECF alter the K+ concentration gradient between the ICF and ECF, thereby altering the rate of K+ flux across the membrane. Normally, there is substantially more K+ inside the cells than in the ECF, but with elevated ECF K+ levels, this gradient is reduced. Associated with this change is a reduction in "resting" potential (that is, the membrane is less negative on the inside than normal because less K+ leaves). Among the consequences is a tendency to develop ectopic foci as well as cardiac arrhythmias. Also, because the magnitude of voltage change from the reduced "resting" state to the peak of the action potential is less than from the normal "resting" state to the peak, the resultant diminution of the action potentials' intensity causes the heart to become weak, flaccid, and dilated. At the extreme, with K+ levels elevated two to three times the normal value, the weakened heart may actually stop pumping.

A rise in ECF Ca++ concentration, on the other hand, by prolonging the plateau phase of the action potential and by increasing the cytosolic concentration of Ca++, augments the strength of cardiac contraction. The heart tends to contract spastically, with little time to rest between contractions. Some drugs that alter cardiac function do so by influencing Ca++ movement across the myocardial cell membranes. For example, Ca++-blocking agents, such as *verapamil*, block Ca++ influx during an action potential, thereby reducing the force of cardiac contraction. Other drugs, such as *digitalis*, increase cardiac contractility by inducing an accumulation of cytosolic Ca++.

Tetanus of cardiac muscle is prevented by a long refractory period.

Like other excitable tissues, cardiac muscle has a refractory period. During the refractory period, which occurs immediately after the initiation of an action potential, an excitable membrane's responsiveness is totally abolished, making it impossible for another action potential to be generated. In skeletal muscle, the refractory period is very short compared with the duration of the resultant contraction, so the fiber can be restimulated again before the first contraction is complete to produce summation of contractions. Rapidly repetitive stimulation that does not allow the muscle fiber to relax between stimulations results in a sustained, maximal contraction known as tetanus (see Fig. 8–16, p. 228). In contrast, cardiac muscle has a long refractory period that lasts about 250 msec because of the prolonged action potential. This is almost as long as the period of contraction initiated by the action potential; a cardiac-muscle fiber contraction averages about 300 msec in duration (● Fig. 9–17). Consequently, cardiac muscle cannot be restimulated until contraction is almost over, making summation of contractions and tetanus of cardiac muscle impossible. This is a valuable protective mechanism, because the pumping of blood requires alternate periods of contrac-

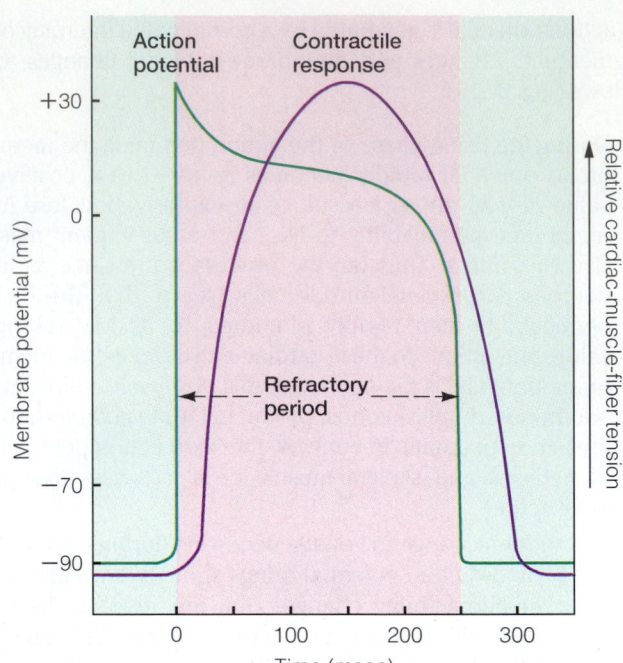

● **FIGURE 9–17   Relationship of Action Potential and Refractory Period to Duration of Contractile Response in Cardiac Muscle**

tion (emptying) and relaxation (filling). A prolonged tetanic contraction would prove fatal. The heart chambers could not be filled and emptied again.

The chief factor responsible for the long refractory period is inactivation, during the prolonged plateau phase, of the Na+ channels that were activated during the initial Na+ influx of the rising phase. Not until the membrane recovers from this inactivation process (when the membrane has already repolarized to resting), can the Na+ channels be activated once again to begin another action potential.

The ECG is a record of the overall spread of electrical activity through the heart.

The electrical currents generated by cardiac muscle during depolarization and repolarization (see p. 80) spread into the tissues surrounding the heart and are conducted through the body fluids. A small portion of this electrical activity reaches the body surface, where it can be detected using recording electrodes. The record produced is an **electrocardiogram**, or **ECG**. (Originally the term EKG was used, because this technique was developed by a German-speaking scientist, William Einthoven, and "kardia" is the word for *heart* in German.) Three important points should be remembered when considering what an ECG actually represents:

1. An ECG is a recording of that portion of the electrical activity induced in the body fluids by the cardiac impulse that reaches the surface of the body, not a direct recording of the actual electrical activity of the heart.

2. The ECG is a complex recording representing the overall spread of activity throughout the heart during depolarization and repolarization. It is not a recording of a single action potential in a single cell at a single point in time. The record at any given time represents the sum of electrical activity in all of the cardiac-muscle cells, some of which may be undergoing action potentials while others may not yet be activated. For example, immediately after firing of the SA node, the atrial cells are undergoing action potentials while the ventricular cells are still at rest. At a later point, the electrical activity will have spread to the ventricular cells while the atrial cells will be repolarizing. Therefore, the overall pattern of cardiac electrical activity varies with time as the impulse passes throughout the heart.

3. The recording represents comparisons in voltage detected by electrodes at two different points on the body surface, not the actual potential. For example, the ECG does not record a potential at all when the ventricular muscle is either completely depolarized or completely repolarized; both electrodes are "viewing" the same potential, so no difference in potential between the two electrodes is recorded.

The exact pattern of electrical activity recorded from the body surface depends on the orientation of the recording electrodes. Electrodes may be loosely thought of as "eyes" that "see" electrical activity and translate it into a visible recording, the ECG record. Whether an upward deflection or downward deflection is recorded is determined by the orientation of electrodes with respect to the current flow in the heart. For example, the spread of excitation across the heart is seen differently from the right arm than from the left foot, and both of these are seen differently than a recording directly over the heart. Even though the same electrical events are occurring in the heart, different wave forms representing the same electrical activity result when this activity is recorded by electrodes at different points on the body.

To provide standard comparisons, ECG records routinely consist of twelve conventional electrode systems, or leads. When an electrocardiograph machine is connected between recording electrodes at two points on the body, the specific arrangement of each pair of connections is called a **lead**. The twelve different leads each record electrical activity in the heart from different locations—six different electrical arrangements from the limbs and six chest leads at various sites around the heart. The same twelve leads are routinely used in all ECG recordings to provide a common basis for comparison and for recognizing deviations from normal (● Fig. 9–18).

Various components of the ECG record can be correlated to specific cardiac events.

Interpretation of the wave configurations recorded from each lead depends on a thorough knowledge of the sequence of the spread of cardiac excitation and the position of the heart relative to the placement of the electrodes. A normal ECG exhibits three distinct wave forms: the P wave, the QRS complex, and the T wave (● Fig. 9–19). (The letters do not signify anything other than the orderly sequence of the waves. Einthoven simply started in the middle of the alphabet when naming the waves.)

■ The **P wave** represents atrial depolarization.
■ The **QRS complex** represents ventricular depolarization.
■ The **T wave** represents ventricular repolarization.

The following important points about the ECG record should also be noted:

1. Firing of the SA node does not generate sufficient electrical activity to reach the surface of the body, so no wave is recorded for SA nodal depolarization. Therefore, the first recorded wave, the P wave, occurs when the impulse spreads across the atria.

2. In a normal ECG, there is no separate wave for atrial repolarization. The electrical activity associated with atrial repolarization normally occurs simultaneously with ventricular depolarization and is masked by the QRS complex.

3. The P wave is much smaller than the QRS complex because the atria have a much smaller muscle mass than the ventricles and consequently generate less electrical activity.

4. There are three times when no current is flowing in the heart musculature and the ECG remains at baseline:
   a. During the AV nodal delay. This delay is represented by the interval of time between the end of the P wave and the onset of the QRS wave; this interval is known as the **PR segment**. (It is called the PR segment rather than the PQ segment because the Q deflection is small and sometimes absent, whereas the R deflection is the dominant wave of the complex.) Current is flowing through the AV node, but the magnitude is too small to be detected by the ECG electrodes.
   b. When the ventricles are completely depolarized and the cardiac-contractile cells are undergoing the plateau phase of their action potential before they repolarize again, represented by the **ST segment**. This segment is the interval between QRS and T; it coincides with the time during which ventricular activation is complete and the ventricles are contracting and emptying.
   c. When the heart muscle is completely at rest and ventricular filling is taking place, after the T wave and before the next P wave; this time segment is called the **TP interval**.

The ECG can be useful in diagnosing abnormal heart rates, arrhythmias, and damage of heart muscle.

Because electrical activity triggers mechanical activity, abnormal electrical patterns are usually accompanied by abnormal contractile activity of the heart. Evaluation of ECG patterns can provide useful information about the status of the heart,

(a)   (b)

• **FIGURE 9–18 Electrocardiogram Leads** (a) Limb leads. The six limb leads include leads I, II, III, aVR, aVL, and aVF. Leads I, II, and III are bipolar leads, because two recording electrodes are used. The tracing records the *difference* in potential between the two electrodes. For example, lead I records the difference in potential detected at the right arm and left arm. The electrode placed on the right leg serves as a ground and is not a recording electrode. The aVR, aVL, and aVF leads are unipolar leads. Only the potential under one electrode, the exploring electrode, is recorded. For example, aVR records the potential reaching the right arm in comparison to the rest of the body. (b) Chest leads. The six chest leads, V_1 through V_6, are also unipolar leads. The exploring electrode mainly records the electrical potential of the cardiac musculature immediately beneath the electrode in six different locations surrounding the heart.

including its rate, rhythm, and the health of its musculature. The principle deviations from normal that can be ascertained through electrocardiography (Table 9–2) are discussed in the following paragraphs.

Abnormalities in Rate The distance between two consecutive QRS complexes on an ECG record is calibrated to the beat-to-beat heart rate. A rapid heart rate of more than 100 beats per minute is known as **tachycardia** (*tachy* means "fast"), whereas a slow heart rate of fewer than 60 beats per minutes is referred to as **bradycardia** (*brady* means "slow").

Abnormalities in Rhythm Rhythm refers to the regularity of the ECG waves. Any variation from the normal rhythm and sequence of excitation of the heart is termed an **arrhythmia**. It may result from the presence of ectopic foci, alterations in SA-node pacemaker activity, or interference with conduction. Oftentimes heart rate is also altered. *Extrasystoles* or *prema-*

ture beats originating from an ectopic focus are common deviations from normal rhythm. Other abnormalities in rhythm easily detected on an ECG include atrial flutter, atrial fibrillation, ventricular fibrillation, and heart block.

Atrial flutter is characterized by a rapid but regular sequence of atrial depolarizations at rates between 200 to 380 beats per minute. The ventricles rarely keep pace with the racing atria. Because the conducting tissue's refractory period is longer than that of the atrial muscle, the AV node is unable to respond to every impulse that converges on it from the atria. Maybe only one out of every two or three atrial impulses successfully passes through the AV node to the ventricles. Such a situation is referred to as a *2:1 or 3:1 rhythm*. The fact that not every atrial impulse reaches the ventricle in atrial flutter is important, because it precludes a rapid ventricular rate of more than 200 beats per minute. Such a high rate would not allow adequate time for ventricular filling between beats. In such a case, the output of the heart would be reduced to

P wave = Atrial depolarization
PR segment = AV nodal delay
QRS complex = Ventricular depolarization (atria repolarizing simultaneously)
ST segment = Time during which ventricles are contracting and emptying
T wave = Ventricular repolarization
TP interval = Time during which ventricles are relaxing and filling

• **FIGURE 9–19 Electrocardiogram Wave Forms in Lead II**

TABLE 9–2 Representative Heart Conditions Detectable through Electrocardiography

the extent that loss of consciousness or even death could result because of decreased blood flow to the brain.

Atrial fibrillation is characterized by rapid, irregular, uncoordinated depolarizations of the atria with no definite P waves. Accordingly, atrial contractions are chaotic and asynchronized. Since impulses reach the AV node erratically, the ventricular rhythm is also very irregular. The QRS complexes are normal in shape but occur sporadically. Variable lengths of time between ventricular beats are available for ventricular filling. Some ventricular beats come so close together that little filling can occur between beats. When less filling occurs, the subsequent contraction is weaker. In fact, some of the ventricular contractions may be too weak to eject enough blood to produce a palpable wrist pulse. In this situation, if the heart rate is determined directly, either by the apex beat or via the ECG, and the pulse rate is taken concurrently at the wrist, the heart rate will exceed the pulse rate. Such a difference in heart rate and pulse rate is known as a **pulse deficit**. Normally, the heart rate coincides with the pulse rate, because each cardiac contraction initiates a pulse wave as it ejects blood into the arteries.

Ventricular fibrillation is a very serious rhythmic abnormality in which the ventricular musculature exhibits uncoordinated, chaotic contractions. Multiple impulses travel erratically in all directions around the ventricles. The ECG tracing in ventricular fibrillation is very irregular with no detectable pattern or rhythm. The ventricles are ineffectual as pumps when contractions are so disorganized. If circulation is not restored in less than four minutes through external cardiac compression or electrical defibrillation, irreversible brain damage occurs and death is imminent.

Another type of arrhythmia, **heart block**, arises from defects in the cardiac conducting system. The atria still beat regularly, but the ventricles occasionally fail to be stimulated and thus do not contract following atrial contraction. Im-

pulses between the atria and ventricles can be blocked to varying degrees. In some forms of heart block, only every second or third atrial impulse is passed to the ventricles. This is known as *2:1* or *3:1 block*, which can be distinguished from the 2:1 or 3:1 rhythm associated with atrial flutter by the rates involved. In heart block, the atrial rate is normal but the ventricular rate is considerably below normal, whereas in atrial flutter, the atrial rate is very high in accompaniment with a normal or above-normal ventricular rate. *Complete heart block* is characterized by complete dissociation between atrial and ventricular activity, with impulses from the atria not being conducted to the ventricles at all. The atrial beat continues to be governed by the SA node, but the ventricles generate their own impulses at a rate much slower than the atria. On the ECG, the P waves exhibit a normal rhythm. The QRS and T waves also occur regularly but at a much slower rate than the P waves and are completely independent of P wave rhythm.

THE WHAT, WHO, AND WHEN OF STRESS TESTING

Stress tests, or **graded exercise tests,** are conducted primarily to aid in diagnosing or quantifying heart or lung disease and to evaluate the functional capacity of asymptomatic individuals. The tests are usually given on motorized treadmills or bicycle ergometers (stationary, variable-resistance bicycles). Work-load intensity (how hard the subject is working) is adjusted by progressively increasing the speed and incline of the treadmill or by progressively increasing the pedaling frequency and resistance on the bicycle. The test starts at a low intensity and continues until a prespecified work load is achieved, physiologic symptoms occur, or the subject is too fatigued to continue.

During diagnostic testing, the patient is monitored with an ECG, and blood pressure is taken each minute. A test is considered to be positive if ECG abnormalities occur (such as ST segment depression, inverted T waves, or dangerous arrhythmias) or if physical symptoms such as chest pain develop. A test that is interpreted as being positive in a person who does not have heart disease is called a false positive test. In men, false positives occur only about 10% to 20% of the time, so the diagnostic stress test for men has a *specificity* of 80% to 90%. Women have a greater frequency of false positive tests with a corresponding lower specificity of about 70%.

The *sensitivity* of a test means that those individuals with disease are correctly identified and there are few false negatives. The sensitivity of the stress test is reported to be 60% to 80%; that is, if one hundred individuals with heart disease were tested, sixty to eighty would be correctly identified, but twenty to forty would have a false negative test. Although stress testing is now an important diagnostic tool, it is just one of several tests used to determine the presence of coronary artery disease.

Stress tests are also conducted on individuals not suspected of having heart or lung disease to determine their present functional capacity. These functional tests are administered in the same way as diagnostic tests, but they are conducted by exercise physiologists and a physician need not be present. These tests are used to establish safe exercise prescriptions, to aid athletes in establishing optimal training programs, and as research tools to evaluate the effectiveness of a particular training regimen. Functional stress testing is becoming more prevalent as more people are joining hospital- or community-based wellness programs for disease prevention.

Cardiac Myopathies Abnormal ECG waves are also important in the recognition and assessment of **cardiac myopathies** (damage of the heart muscle). **Myocardial ischemia** refers to inadequate blood supply to the heart tissue. Actual death or **necrosis** of heart-muscle cells, usually caused by blockage of a blood vessel supplying that area of the heart, is termed **acute myocardial infarction**, commonly known as a **heart attack**. Abnormal QRS wave forms can be seen when a portion of the heart muscle becomes necrotic.

Interpretation of an ECG is a complex task requiring extensive knowledge and training. The foregoing discussion is not intended to make you an ECG expert by any means but to give you an appreciation of the ways in which the ECG can be used as a diagnostic tool, as well as to present an overview of some of the more common abnormalities of heart function. (For a further use of the ECG, see the accompanying boxed feature, A Closer Look at Exercise Physiology.)

⊜ MECHANICAL EVENTS OF THE CARDIAC CYCLE

The heart alternately contracts to empty and relaxes to fill.

The cardiac cycle consists of alternate periods of **systole** (contraction and emptying) and **diastole** (relaxation and filling). The atria and ventricles go through separate cycles of systole and diastole. Contraction occurs as a result of the spread of excitation across the heart, whereas relaxation follows the subsequent repolarization of the cardiac musculature. The following discussion correlates various events that occur concurrently during the cardiac cycle, including ECG features, pressure changes, volume changes, valve activity, and heart sounds. Reference to ● Figure 9–20 will facilitate this discussion. Only the events on the left side of the heart will be described, but keep in mind that identical events are

---

● **FIGURE 9–20   Cardiac Cycle**   This graph depicts various events that occur concurrently during the cardiac cycle. Follow each horizontal strip across to see the changes that take place in the electrocardiogram; aortic, ventricular, and atrial pressures; ventricular volume; and heart sounds throughout the cycle. Late diastole, one full systole and diastole (one full cardiac cycle), and another systole are shown for the left side of the heart. Follow each vertical strip downward to see what happens simultaneously with each of these factors during each phase of the cardiac cycle. See the text (pages 280–281) for a detailed explanation of the circled numbers. The figures of the heart illustrate the flow of low-oxygenated (dark blue) and high-oxygenated (bright red) blood in and out of the heart during the cardiac cycle.

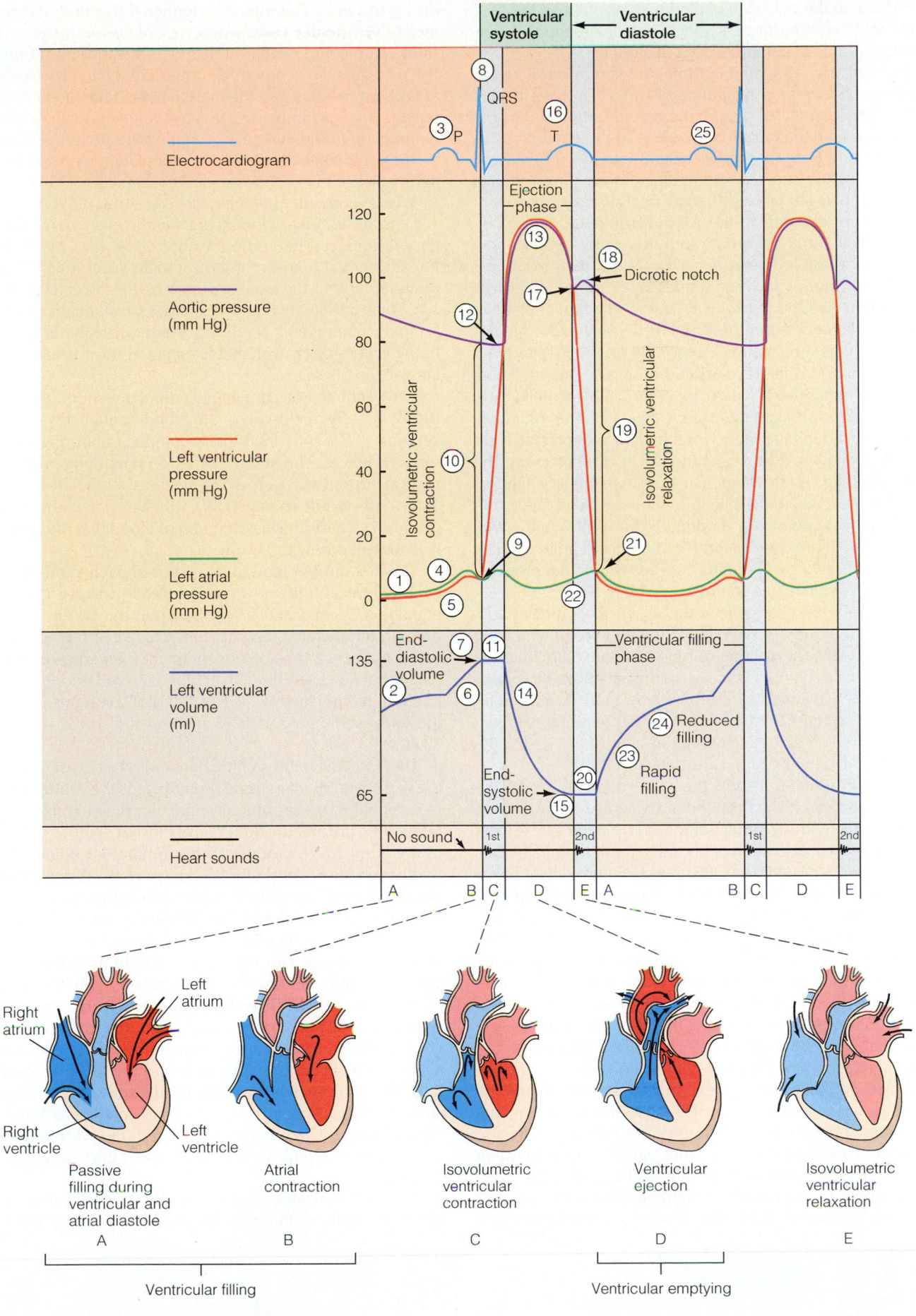

Ventricular systole

Ventricular diastole

Electrocardiogram

③ P ⑧ QRS ⑯ T ㉕

Ejection phase

120 ⑬

100 ⑱ Dicrotic notch
⑰

Aortic pressure (mm Hg)

⑫

80

60

Left ventricular pressure (mm Hg)

40 ⑩

Isovolumetric ventricular contraction

⑲ Isovolumetric ventricular relaxation

20 ⑨ ㉑

Left atrial pressure (mm Hg)

① ④ ㉒

0 ⑤

End-diastolic volume ⑦ ⑪

135 Ventricular filling phase

Left ventricular volume (ml)

② ⑥ ⑭

㉔ Reduced filling

㉓ Rapid filling

End-systolic volume ⑳

65 ⑮

Heart sounds

No sound 1st 2nd 1st 2nd

A B C D E A B C D E

Right atrium — Left atrium

Right ventricle — Left ventricle

Passive filling during ventricular and atrial diastole

Atrial contraction

Isovolumetric ventricular contraction

Ventricular ejection

Isovolumetric ventricular relaxation

A B C D E

Ventricular filling

Ventricular emptying

occurring on the right side of the heart, except that the pressures are lower. Our discussion will begin and end with ventricular diastole to complete one full cardiac cycle.

During early ventricular diastole, the atrium is still also in diastole. This stage corresponds to the TP interval on the ECG—the interval after ventricular repolarization and before another atrial depolarization. Because of the continuous inflow of blood from the venous system into the atrium, atrial pressure slightly exceeds ventricular pressure even though both chambers are relaxed (Point 1 in Fig. 9–20). Because of this pressure differential, the AV valve is open, and blood flows directly from the atrium into the ventricle throughout ventricular diastole (Heart A in Fig. 9–20). As a result, the ventricular volume slowly continues to rise even before atrial contraction takes place (Point 2). Late in ventricular diastole, the SA node reaches threshold and fires. The impulse spreads throughout the atria, which is recorded on the ECG as the P wave (Point 3). Atrial depolarization brings about atrial contraction, which squeezes more blood into the ventricle, causing a rise in the atrial pressure curve (Point 4). The excitation-contraction coupling process is taking place during the short delay between the P wave and the rise in atrial pressure. The corresponding rise in ventricular pressure (Point 5) that occurs simultaneous to the rise in atrial pressure is due to the additional volume of blood added to the ventricle by atrial contraction (Point 6 and Heart B). Throughout atrial contraction, atrial pressure still slightly exceeds ventricular pressure, so the AV valve remains open.

Ventricular diastole ends at the onset of ventricular contraction. By this time, atrial contraction and ventricular filling are completed. The volume of blood in the ventricle at the end of diastole (Point 7) is known as the **end-diastolic volume** (**EDV**), which averages about 135 ml. No more blood will be added to the ventricle during this cycle. Therefore, the end-diastolic volume is the maximum amount of blood that the ventricle will contain during this cycle.

Following atrial excitation, the impulse passes through the AV node and specialized conducting system to excite the ventricle. Simultaneously, atrial contraction is occurring. By the time ventricular activation is complete, atrial contraction is already accomplished. The QRS complex represents this ventricular excitation (Point 8), which induces ventricular contraction. The ventricular pressure curve sharply increases shortly after the QRS complex, signaling the onset of ventricular systole (Point 9). The slight delay between the QRS complex and the actual onset of ventricular systole is the time required for the excitation-contraction coupling process to occur. As ventricular contraction begins, ventricular pressure immediately exceeds atrial pressure. This backward pressure differential forces the AV valve closed (Point 9).

After ventricular pressure exceeds atrial pressure and the AV valve has closed, the ventricular pressure must continue to increase before it exceeds aortic pressure to open the aortic valve. Therefore, there is a brief period of time between closure of the AV valve and opening of the aortic valve when the ventricle remains a closed chamber (Point 10). Because all valves are closed, no blood can enter or leave the ventricle

during this time. This interval is termed the period of **isovolumetric ventricular contraction** (*isovolumetric* means "constant volume and length") (Heart C). Because no blood enters or leaves the ventricle, the ventricular chamber remains at constant volume and the muscle fibers remain at constant length. This isovolumetric condition is similar to an isometric contraction in skeletal muscle. During the period of isovolumetric ventricular contraction, ventricular pressure continues to increase as the volume remains constant (Point 11).

When ventricular pressure exceeds aortic pressure (Point 12), the aortic valve is forced open and ejection of blood begins (Heart D). The aortic pressure curve rises as blood is forced into the aorta from the ventricle faster than blood is draining off into the smaller vessels at the other end (Point 13). The ventricular volume decreases substantially as blood is rapidly pumped out (Point 14). Ventricular systole includes both the period of isovolumetric contraction and the ventricular ejection phase.

The ventricle does not empty completely during ejection. Normally, only about half of the blood contained within the ventricle at the end of diastole is pumped out during the subsequent systole. The amount of blood remaining in the ventricle at the end of systole when ejection is complete is known as the **end-systolic volume** (**ESV**), which averages about 65 ml (Point 15). This is the least amount of blood that the ventricle will contain during this cycle.

The amount of blood pumped out of each ventricle with each contraction is known as the **stroke volume** (**SV**); it is equal to the end-diastolic volume minus the end-systolic volume; in other words, the difference between the volume of blood in the ventricle before contraction and the volume after contraction is the amount of blood ejected during the contraction. In our example, the end-diastolic volume is 135 ml, the end-systolic volume is 65 ml, and the stroke volume is 70 ml.

The T wave signifies ventricular repolarization occurring at the end of ventricular systole (Point 16). As the ventricle starts to relax upon repolarization, ventricular pressure falls below aortic pressure and the aortic valve closes (Point 17). Closure of the aortic valve produces a disturbance or notch on the aortic pressure curve (Point 18) known as the **dicrotic notch**. No more blood leaves the ventricle during this cycle because the aortic valve has closed. The AV valve is not yet open, however, because ventricular pressure still exceeds atrial pressure, so no blood can enter the ventricle from the atrium. Therefore, all valves are once again closed for a brief period of time known as **isovolumetric ventricular relaxation** (Point 19 and Heart E). The muscle-fiber length and chamber volume (Point 20) remain constant. No blood leaves or enters as the ventricle continues to relax and the pressure steadily falls. When the ventricular pressure falls below the atrial pressure, the AV valve opens (Point 21) and ventricular filling occurs once again. Ventricular diastole includes both the period of isovolumetric ventricular relaxation and the ventricular filling phase.

Atrial repolarization and ventricular depolarization occur simultaneously, so the atria are in diastole throughout ventricular systole. Blood continues to flow from the pulmonary veins into the left atrium. As this incoming blood pools in the atrium, atrial pressure rises continuously (Point 22). When the AV valve opens at the end of ventricular systole, the blood that accumulated in the atrium during ventricular systole rapidly pours into the ventricle. Ventricular filling thus occurs rapidly at first (Point 23) because of the increased atrial pressure resulting from the accumulation of blood in the atria. Then ventricular filling slows down (Point 24) as the accumulated blood has already been delivered to the ventricle, and atrial pressure starts to fall. During this period of reduced filling, blood continues to flow from the pulmonary veins into the left atrium and through the open AV valve into the left ventricle. During late ventricular diastole, when ventricular filling is proceeding slowly, the SA node fires again (Point 25), and the cardiac cycle starts over.

It is significant that much of ventricular filling occurs early in diastole during the rapid-filling phase. During times of rapid heart rates, the length of diastole is reduced to a much greater extent than is the length of systole. For example, if the heart rate increases from 75 to 180 beats per minute, the duration of diastole decreases about 75%, from 500 msec to 125 msec. This greatly reduces the time available for ventricular relaxation and filling. However, because much of ventricular filling is accomplished during early diastole, filling is not seriously impaired during periods of increased heart rate, such as during exercise (• Fig. 9–21). There is a limit, however, to how rapidly the heart can beat without decreasing the period of diastole to the point that ventricular filling is severely impaired. At heart rates greater than 200 beats per minute, diastolic time is too short to allow adequate ventricular filling. With inadequate filling, the resultant cardiac output is deficient. Normally, ventricular rates do not exceed 200 beats per minute because the relatively long refractory period of the AV node will not allow impulses to be conducted to the ventricles more frequently than this.

Two heart sounds associated with valve closures can be heard during the cardiac cycle.

Two major heart sounds normally can be heard during the cardiac cycle when listening with a stethoscope. The **first heart sound** is low-pitched, soft, and relatively long—often said to sound like "lub." The **second heart sound** has a higher pitch, is shorter and sharper—often said to sound like "dup." Thus, one normally hears "lub, dup, lub, dup, lub, dup, ..." The first heart sound is associated with closure of the AV valves, whereas the second sound is associated with closure of the semilunar valves. Opening of valves does not produce any sound. The sounds are caused by vibrations set up within the walls of the ventricles and major arteries during valve closure, not by the valves snapping shut. Because closure of the AV valves occurs at the onset of ventricular contraction when ventricular pressure first exceeds atrial pressure, the first heart sound signals the onset of ventricular systole. Closure of the semilunar valves occurs at the onset of ventricular relaxation as the left and right ventricular pressures fall below the aortic and pulmonary artery pressures, respectively. The second heart sound, therefore, signals the onset of the ventricular diastole.

Turbulent blood flow produces heart murmurs.

Abnormal heart sounds, or **murmurs**, are usually (but not always) associated with cardiac disease. Murmurs not involving heart pathology, so-called **functional murmurs**, are more common in young people.

Blood normally flows in a *laminar* fashion; that is, layers of the fluid slide smoothly over each other. Laminar flow does not produce any sound. When blood flow becomes turbulent, however, a sound can be heard (• Fig. 9–22). Such an abnormal sound is due to vibrations created in the surrounding structures by the turbulent flow.

• **FIGURE 9–21 Ventricular Filling Profiles during Normal and Rapid Heart Rates** Because much of ventricular filling occurs early in diastole during the rapid-filling phase, filling is not seriously impaired when diastolic time is reduced as a result of an increase in heart rate.

Laminar flow (does not create any sound)

Turbulent flow (can be heard)

• **FIGURE 9–22   Comparison of Laminar and Turbulent Flow**

The most common cause of turbulence is valve malfunction, either a stenotic or an insufficient valve. A **stenotic valve** is a stiff, narrowed valve that does not open completely. Blood must be forced through the constricted opening at tremendous velocity, resulting in turbulence that produces an abnormal whistling sound similar to the sound produced when you force air rapidly through narrowed lips to whistle.

An **insufficient valve** is one that cannot close completely, usually because the valve edges are scarred and do not fit together properly. Turbulence is produced when blood flows backward through the insufficient valve and collides with blood moving in the opposite direction, creating a swishing or gurgling murmur. Such backflow of blood is known as **regurgitation**. Commonly, an insufficient heart valve is called a **leaky valve**, because it allows blood to leak back through at a time when the valve should be closed.

Most often, both valvular stenosis and insufficiency are caused by **rheumatic fever**, an autoimmune ("immunity against self") disease triggered by a streptococcus bacterial infection. Antibodies formed against toxins produced by these bacteria interact with many of the body's own tissues, resulting in immunological damage. The heart valves are among the most susceptible tissues in this regard. Large, hemorrhagic, fibrinous lesions form along the inflamed edges of an affected heart valve, causing the valve to become thickened, stiff, and scarred. Sometimes the leaflet edges permanently adhere to each other. Depending on the extent and specific nature of the lesions, the valve may become either stenotic or insufficient or some degree of both.

The valve involved and the type of defect can usually be detected by the *location* and *timing* of the murmur. Each heart valve may be heard best at a specific location on the chest. Noting the location at which a murmur is loudest helps the diagnostician determine which valve is involved. The timing of the murmur refers to the part of the cardiac cycle during which the murmur is heard. Recall that the first heart sound signals the onset of ventricular systole and the second heart sound signals the onset of ventricular diastole. Thus, a murmur occurring between the first and second heart sounds (lub-murmur-dup, lub-murmur-dup) signifies a **systolic murmur**. A **diastolic murmur**, on the other hand, occurs between the second and first heart sound (lub-dup-murmur, lub-dup-murmur). The sound of the murmur characterizes it as either a stenotic (whistling) murmur or an insufficient (swishy) murmur. Armed with these facts, one can determine the cause of a valvular murmur (Table 9–3). As an example, a whistling murmur (denoting a stenotic valve) occurring between the first and second heart sounds (denoting a systolic murmur) signifies the presence of stenosis in a valve that should be open during systole. It could be either the aortic or the pulmonary semilunar valve through which blood is being ejected. Identifying which of these valves is stenotic is accomplished by determining the location over which the murmur is best heard. The main concern with heart murmurs, of course, is not the murmur itself but the accompanying detrimental circulatory consequences caused by the defect.

TABLE 9–3 Timing of Murmur				
	SYSTOLIC MURMUR		**DIASTOLIC MURMUR**	
CHARACTERISTIC	First heart – murmur – heart sound sound		Second heart – murmur – heart sound sound	
TYPE OF MURMUR	Whistling murmur ↓		Whistling murmur ↓	
VALVE DISORDER	Stenotic semilunar valve		Stenotic AV valve	
TYPE OF MURMUR	Swishy murmur ↓		Swishy murmur ↓	
VALVE DISORDER	Insufficient AV valve		Insufficient semilunar valve	

CARDIAC OUTPUT AND ITS CONTROL

Cardiac output depends on the heart rate and the stroke volume.

Cardiac output (CO) is the volume of blood pumped by *each ventricle* per minute (not the total amount of blood pumped by the heart). During any period of time, the volume of blood flowing through the pulmonary circulation is equivalent to the volume flowing through the systemic circulation. Therefore, the cardiac output from each ventricle normally is identical, although on a beat-to-beat basis, minor variations may occur. The two determinants of cardiac output are *heart rate* (beats per minute) and *stroke volume* (volume of blood pumped per beat or stroke).

The average heart rate is 70 beats per minute, established by SA-node rhythmicity, whereas the average stroke volume is 70 ml per beat, producing an average cardiac output of 4,900 ml/min or close to 5 liters (l)/min:

$$\text{cardiac output} = \text{heart rate} \times \text{stroke volume}$$
$$CO = 70 \text{ beats/min} \times 70 \text{ ml/beat}$$
$$= 4,900 \text{ ml/min} \approx 5 \text{ l/min}$$

Because the body's total blood volume averages 5 to 5.5 liters, each half of the heart pumps the equivalent of the entire blood volume each minute. In other words, each minute the right ventricle normally pumps 5 liters of blood through the lungs, and the left ventricle pumps 5 liters of blood through the systemic circulation. At this rate, each half of the heart would pump about 2½ million liters of blood in just one year. Yet this is only the resting cardiac output! During exercise the cardiac output can increase to 20 to 25 liters per minute, and outputs as high as 40 liters per minute have been recorded in trained athletes during heavy exercise. The difference between the cardiac output at rest and the maximum volume of blood the heart is capable of pumping per minute is known as the **cardiac reserve**. How can cardiac output vary so tremendously, depending on the demands of the body? You can readily answer this question by thinking about how your own heart pounds rapidly (increased heart rate) and forcefully (increased stroke volume) when you engage in strenuous physical activities (need for increased cardiac output). Thus, the regulation of cardiac output depends on the control of both heart rate and stroke volume, topics that will be discussed next.

Heart rate is determined primarily by autonomic influences on the SA node.

The SA node is normally the pacemaker of the heart because it has the fastest spontaneous rate of depolarization to threshold. Recall that this automatic gradual reduction of membrane potential between beats is generally believed to be due to a reduction in K^+ permeability. When the SA node reaches threshold, an action potential is initiated that spreads throughout the heart, inducing the heart to contract or have a "heart beat." This happens about 70 times per minute, setting the average heart rate at about 70 beats per min.

The heart is innervated by both divisions of the autonomic nervous system, which can modify the rate (as well as the strength) of contraction, even though nervous stimulation is not required to initiate contraction. The parasympathetic nerve to the heart, the **vagus nerve**, primarily supplies the atrium, especially the SA and AV nodes. There is no significant parasympathetic innervation of the ventricles. The cardiac sympathetic nerves also supply the atria, including the SA and AV nodes, and richly innervate the ventricles as well.

Parasympathetic and sympathetic stimulation have the following effects on the heart (Table 9–4).

TABLE 9–4 Effects of the Autonomic Nervous System on the Heart and Structures that Influence the Heart

AREA AFFECTED	EFFECT OF PARASYMPATHETIC STIMULATION	EFFECT OF SYMPATHETIC STIMULATION
SA node	Decreases rate of depolarization to threshold; decreases heart rate	Increases rate of depolarization to threshold; increases heart rate
AV node	Decreases excitability; increases AV nodal delay	Increases excitability; decreases AV nodal delay
Ventricular conduction pathway	No effect	Increases excitability; hastens conduction through bundle of His and Purkinje cells
Atrial muscle	Decreases contractility; weakens contraction	Increases contractility; strengthens contraction
Ventricular muscle	No effect	Increases contractility; strengthens contraction
Adrenal medulla (an endocrine gland)	No effect	Promotes adrenomedullary secretion of epinephrine, a hormone that augments the sympathetic nervous system's actions on the heart
Veins	No effect	Increases venous return, which increases strength of cardiac contraction through the Frank-Starling mechanism

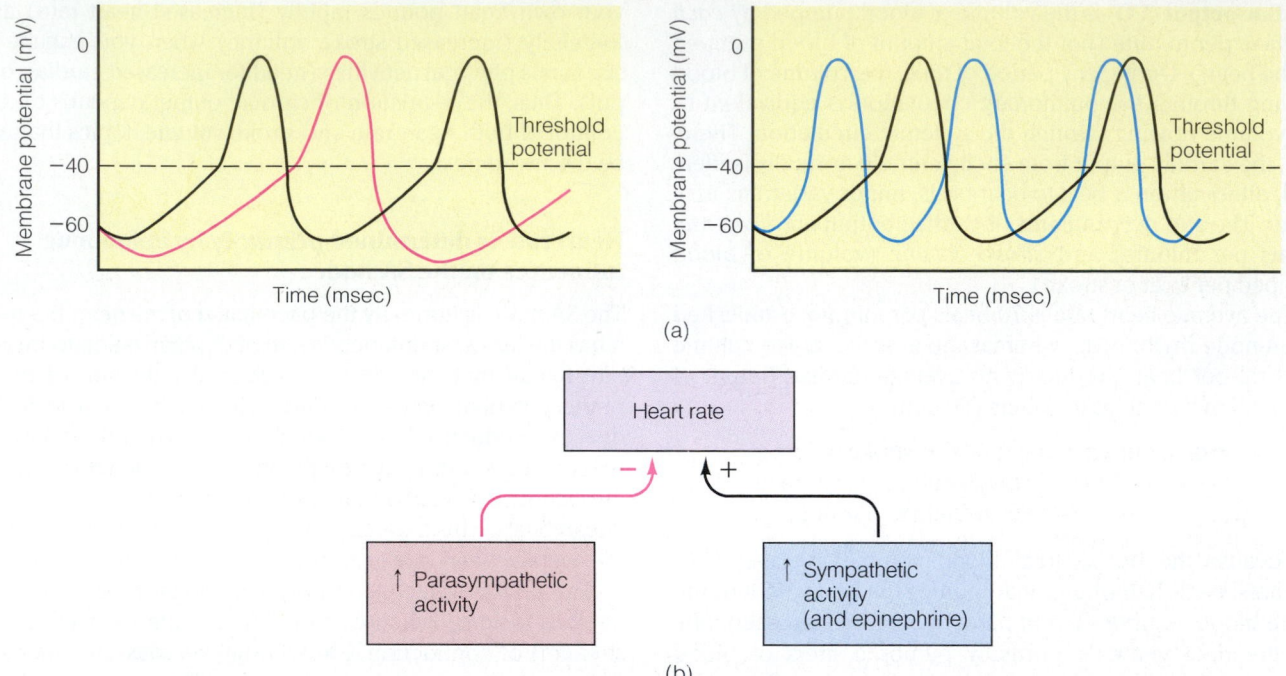

— = Inherent SA-node pacemaker activity
— = SA-node pacemaker activity on parasympathetic stimulation
— = SA-node pacemaker activity on sympathetic stimulation

(a)

Heart rate

− +

↑ Parasympathetic activity

↑ Sympathetic activity (and epinephrine)

(b)

● **FIGURE 9–23   Autonomic Control of SA-Node Activity and Heart Rate**   (a) Autonomic influence on SA-node potential. Parasympathetic stimulation decreases the rate of SA nodal depolarization so that the membrane reaches threshold more slowly and has fewer action potentials, whereas sympathetic stimulation increases the rate of depolarization of the SA node so that the membrane reaches threshold more rapidly and has more frequent action potentials. (b) Control of heart rate by the autonomic nervous system. Because each SA-node action potential ultimately leads to a heart beat, increased parasympathetic activity decreases the heart rate, whereas increased sympathetic activity increases the heart rate.

Effect of Parasympathetic Stimulation on the Heart

▪ The parasympathetic nervous system's influence on the SA node is to decrease the heart rate (● Fig. 9–23). Acetylcholine released upon increased parasympathetic activity increases the permeability of the SA node to K^+ by slowing the closure of K^+ channels. As a result, the rate at which spontaneous action potentials are initiated is reduced through a two-fold effect:

1. Enhanced K^+ permeability hyperpolarizes the SA-node membrane because more positive potassium ions leave than normal, making the inside even more negative. Because the "resting" potential starts even farther away from threshold, it takes longer to reach threshold.

2. The enhanced K^+ permeability induced by vagal stimulation also opposes the automatic reduction in K^+ permeability that is responsible for gradually depolarizing the membrane to threshold. This countering effect decreases the rate of spontaneous depolarization, prolonging the time required to drift to threshold. Therefore, the SA node reaches threshold and fires less frequently, decreasing the heart rate.

▪ Parasympathetic influence on the AV node decreases the node's excitability, prolonging transmission of impulses to the ventricles even longer than the usual AV nodal delay. This effect is brought about by increasing K^+ permeability, which hyperpolarizes the membrane, thereby retarding the initiation of excitation in the AV node.

▪ Parasympathetic stimulation of the atrial contractile cells shortens the action potential, an effect that is believed to be caused by a reduction in the slow inward current carried by Ca^{++}; that is, the plateau phase is reduced. As a result, atrial contraction is weakened.

▪ The parasympathetic system has no effect on ventricular contraction due to the lack of parasympathetic innervation to the ventricles.

Thus, the heart is more "leisurely" under parasympathetic influence—it beats less rapidly, the time between atrial and ventricular contraction is stretched out, and atrial contraction is weaker. These actions are appropriate considering that the parasympathetic system controls heart action in quiet, relaxed situations when the body is not demanding an enhanced cardiac output.

Effect of Sympathetic Stimulation on the Heart

▪ In contrast, the sympathetic nervous system, which controls heart action in emergency or exercise situations when there is a need for greater blood flow, speeds up the heart rate through its effect on the pacemaker tissue. The main effect of sympathetic stimulation on the SA node is to increase its rate of depolarization so that threshold is reached more rapidly (Fig. 9–23 and Table 9–4). Norepinephrine released from the sympathetic nerve endings appears to decrease K^+ permeability by accelerating inactivation of the K^+ channels. With fewer positive potassium ions leaving, the inside of the cell becomes less negative, creating a depolarizing effect. This swifter drift to threshold under sympathetic influence permits a greater frequency of action potentials and a correspondingly more rapid heart rate.

- Sympathetic stimulation of the AV node reduces the AV nodal delay by increasing conduction velocity, presumably by enhancing the slow, inward Ca^{++} current.
- Similarly, sympathetic stimulation speeds up the spread of the action potential throughout the specialized conducting pathway.
- In the atrial and ventricular contractile cells, both of which have an abundance of sympathetic nerve endings, sympathetic stimulation increases contractile strength so that the heart beats more forcefully and squeezes out more blood. This effect is brought about by increasing Ca^{++} permeability, which enhances the slow Ca^{++} influx and intensifies Ca^{++} participation in the excitation-contraction coupling process.

The overall effect of sympathetic stimulation on the heart, therefore, is to improve its effectiveness as a pump by increasing the heart rate, decreasing the delay between atrial and ventricular contraction, decreasing conduction time throughout the heart, and increasing the force of contraction.

Thus, as is typical of the autonomic nervous system, parasympathetic and sympathetic effects on heart rate are antagonistic (oppose each other). At any given moment, the heart rate will be determined largely by the existing balance between the inhibitory effects of the vagus nerve and the stimulatory effects of the cardiac sympathetic nerves. Under resting conditions, parasympathetic discharge is dominant. In fact, if all autonomic nerves to the heart were blocked, the resting heart rate would increase from its average value of 70 beats per minute to about 100 beats per minute, which is the inherent rate of the SA node's spontaneous discharge when not subjected to any nervous influence. (We used 70 beats per minute as the normal rate of SA-node discharge because this is the average rate under normal conditions in the body.) Alterations in the heart rate beyond this resting level in either direction can be accomplished by shifting the balance of autonomic nervous stimulation. Heart rate is increased by simultaneously increasing sympathetic and decreasing parasympathetic activity; a reduction in heart rate is brought about by a concurrent rise in parasympathetic activity and decline in sympathetic activity. The relative level of activity in these two autonomic branches to the heart, in turn, is primarily coordinated by the *cardiovascular control center* located in the brain stem.

Although autonomic innervation is the primary means by which heart rate is regulated, other factors affect it as well. The most important of these is epinephrine, a hormone that is secreted into the blood from the adrenal medulla upon sympathetic stimulation and that acts on the heart in a manner similar to norepinephrine to increase the heart rate. Epinephrine therefore reinforces the direct effect that the sympathetic nervous system has on the heart.

Stroke volume is determined by the extent of venous return and by sympathetic activity.

The other component that determines the cardiac output is stroke volume, the amount of blood pumped out by each ventricle during each beat. Two types of controls influence stroke volume: (1) *intrinsic control* related to the extent of venous return and (2) *extrinsic control* related to the extent of sympathetic stimulation of the heart. Both factors increase stroke volume by increasing the strength of contraction of the heart (• Fig. 9–24). Let us examine each of these factors in more detail to see how they influence the stroke volume.

Increased end-diastolic volume results in increased stroke volume.

As more blood is returned to the heart, the heart pumps out more blood, but the relationship is not quite as simple as it appears, because the heart does not eject all the blood it contains. The direct correlation between end-diastolic volume and stroke volume constitutes the **intrinsic control** of stroke volume, which refers to the heart's inherent ability to vary the stroke volume. This intrinsic control depends on the length-tension relationship of cardiac muscle, which is similar to that of skeletal muscle. For skeletal muscle, the resting-muscle

---

• **FIGURE 9–24   Intrinsic and Extrinsic Control of Stroke Volume**

● FIGURE 9–25 Intrinsic Control of Stroke Volume (Frank-Starling Curve) The cardiac-muscle fiber's length, which is determined by the extent of venous filling, is normally less than the optimal length for developing maximal tension. Therefore, an increase in end-diastolic volume (that is, an increase in venous return), by moving the cardiac-muscle-fiber length closer to optimal length, increases the contractile tension of the fibers on the next systole. A stronger contraction squeezes out more blood. Thus, as more blood is returned to the heart and the end-diastolic volume increases, the heart automatically pumps out a correspondingly larger stroke volume.

length is approximately the optimal length at which maximal tension can be developed during a subsequent contraction. When the skeletal muscle is longer or shorter than this optimal length, the subsequent contraction is weaker (see Fig. 8–17, p. 230). For cardiac muscle, the resting cardiac-muscle-fiber length is less than optimal length. Therefore, an increase in cardiac-muscle-fiber length, by moving closer to the optimal length, increases the contractile tension of the heart on the following systole (● Fig. 9–25).

What causes cardiac-muscle fibers to vary in length before contraction? Skeletal-muscle length can vary before contraction due to the positioning of the skeletal parts to which the muscle is attached, but cardiac muscle is not attached to any bones. The main determinant of cardiac-muscle-fiber length is the degree of diastolic filling. An analogy is a balloon filled with water—the more water you put in, the larger the balloon becomes and the more it is stretched. Likewise, the greater the extent of diastolic filling, the larger the end-diastolic volume and the more the heart is stretched. The more the heart is stretched, the longer the initial cardiac-fiber length before contraction. The increased length results in a greater force on the subsequent cardiac contraction and, consequently, a greater stroke volume. This intrinsic relationship between end-diastolic volume and stroke volume is known as the **Frank-Starling law of the heart**. Stated simply, the law says that the heart normally pumps all the blood returned to it; increased venous return results in increased stroke volume. In Figure 9–25, assume that the end-diastolic volume increases from point A to point B. You can see that this increase in end-diastolic volume is accompanied by a corresponding increase in stroke volume from point A¹ to point B¹. The extent of filling is referred to as the **preload**, because it is the work load imposed on the heart before contraction begins.

Unlike skeletal muscle, the length-tension curve of cardiac muscle normally does not have a descending limb. That is,

within physiological limits, cardiac muscle does not get stretched beyond its optimal length to the point that contractile strength diminishes with further stretching.

The built-in relationship matching stroke volume with venous return has two important advantages. First, one of the most important functions served by this intrinsic mechanism is equalization of output between the right and left sides of the heart, so that the blood pumped out by the heart is equally distributed between the pulmonary and systemic circulation. If, for example, the right side of the heart ejects a larger stroke volume, more blood enters the pulmonary circulation, so venous return to the left side of the heart is increased accordingly. The increased end-diastolic volume of the left side of the heart causes it to contract more forcefully, so it too pumps out a larger stroke volume. In this way, equality of output of the two ventricular chambers is maintained. If such equalization did not happen, excessive damming of blood would occur in the venous system preceding the ventricle with the lower output.

Second, when a larger cardiac output is needed, such as during exercise, venous return is increased through action of the sympathetic nervous system and other mechanisms to be described in the next chapter. The resultant increase in end-diastolic volume automatically increases stroke volume correspondingly. Because exercise also increases heart rate, these two factors act together to increase the cardiac output so that more blood can be delivered to the exercising muscles.

The contractility of the heart is increased by sympathetic stimulation.

In addition to intrinsic control, stroke volume is also subject to **extrinsic control** by factors originating outside of the heart, the most important of which are actions of the cardiac sym-

End-diastolic volume
135 ml

Stroke volume
70 ml

End-systolic volume
65 ml

(a)

End-diastolic volume
135 ml

Stroke volume
100 ml

End-systolic volume
35 ml

(b)

End-diastolic volume
175 ml

Stroke volume
140 ml

End-systolic volume
35 ml

(c)

• **FIGURE 9–26 Effect of Sympathetic Stimulation on Stroke Volume** (a) Normal stroke volume. (b) Stroke volume during sympathetic stimulation. (c) Stroke volume with combination of sympathetic stimulation and increased end-diastolic volume.

pathetic nerves and epinephrine (Table 9–4). Sympathetic stimulation and epinephrine enhance the heart's **contractility**, which refers to the strength of contraction at any given end-diastolic volume; in other words, the heart contracts more forcefully and squeezes out a greater percentage of the blood it contains on sympathetic stimulation, leading to more complete ejection. This increased contractility is due to the increased Ca^{++} influx triggered by norepinephrine and epinephrine. The extra cytosolic Ca^{++} allows the myocardial fibers to generate more force than they would without sympathetic influence. Normally, the end-diastolic volume is 135 ml and the end-systolic volume is 65 ml for a stroke volume of 70 ml (• Fig. 9–26a). Under sympathetic influence, for the same end-diastolic volume of 135 ml, the end-systolic volume might be 35 ml and the stroke volume 100 ml (Fig. 9–26b). In effect, sympathetic stimulation shifts the Frank-Starling curve to the left (• Fig. 9–27). The curve can be shifted to varying degrees, depending on the extent of sympathetic stimulation, up to a maximal increase in contractile strength of about 100% greater than normal.

Sympathetic stimulation increases stroke volume not only by strengthening cardiac contractility but also by enhancing venous return (Fig. 9–26c). Sympathetic stimulation constricts the veins, which squeezes more blood forward from the veins to the heart, increasing the end-diastolic volume and subsequently increasing the stroke volume even further.

The strength of cardiac-muscle contraction and, accordingly, the stroke volume can thus be graded by (1) varying the initial length of the muscle fibers, which in turn depends on the degree of ventricular filling before contraction (intrinsic control); and (2) varying the extent of sympathetic stimulation (extrinsic control). This is in contrast to gradation of skeletal muscle. In skeletal muscle, summation of twitches and recruitment of motor units are employed to produce variable strength of muscle contraction, but these mechanisms are not applicable to cardiac muscle. Twitch summation is impossible because of the long refractory period, and recruitment of motor units is not possible because the heart-muscle cells are arranged into functional syncytia instead of distinct motor units that can be discretely activated.

Increase in stroke volume at same end-diastolic volume

C

B

A

Stroke volume

End-diastolic volume

Frank-Starling curve on sympathetic stimulation

Normal Frank-Starling curve

• **FIGURE 9–27 Shift of Frank-Starling Curve to the Left by Sympathetic Stimulation** For the same end-diastolic volume (Point A), there is a larger stroke volume (from Point B to Point C) upon sympathetic stimulation as a result of increased contractility of the heart. Frank-Starling's curve is shifted to the left to variable degrees, depending on the extent of sympathetic stimulation.

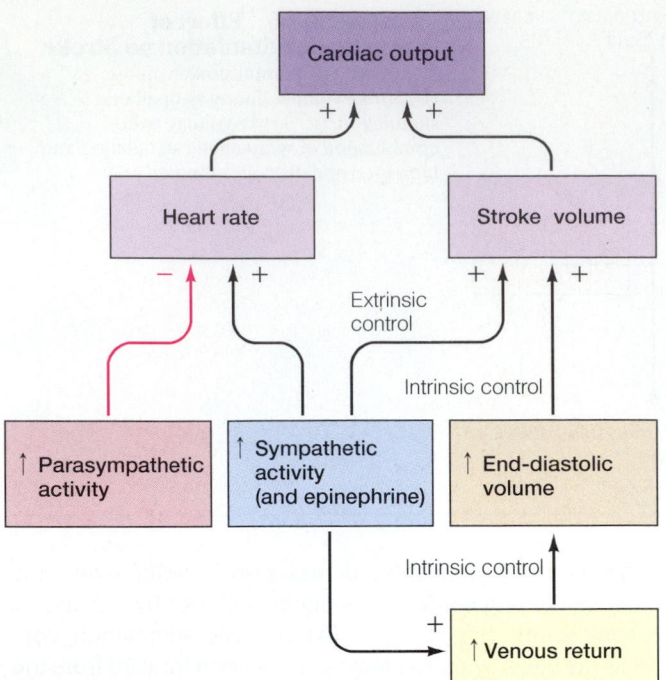

• **FIGURE 9–28 Control of Cardiac Output** Since cardiac output equals heart rate times stroke volume, this figure is a composite of Figure 9–23 (control of heart rate) and Figure 9–24 (control of stroke volume).

All the factors that determine the cardiac output by influencing the heart rate or stroke volume are summarized in • Figure 9–28. Note that sympathetic stimulation increases the cardiac output by increasing both heart rate and stroke volume. Sympathetic activity to the heart increases, for example, during exercise when the working skeletal muscles need increased delivery of O_2-laden blood to support their high rate of ATP consumption.

The ability of the heart to meet physiological demands for increased cardiac output declines as a person gets older; that is, the cardiac reserve decreases with aging. Researchers have recently discovered that this age-related decline is due in part to a reduction in norepinephrine release from the heart's sympathetic nerves. A dwindling of Ca^{++} channels in the sympathetic terminals is believed to be responsible for the diminished norepinephrine release. As with neurotransmitters elsewhere, norepinephrine release depends on the entry of Ca^{++} into the terminals through voltage-gated Ca^{++} channels. As these channels are lost in the older heart, less norepinephrine is released when the sympathetic nerve fibers to the heart are activated, resulting in a reduction in cardiac responsiveness.

High blood pressure increases the work load of the heart.

When the ventricles contract, they must generate sufficient pressure to exceed the blood pressure in the major arteries in order to force open the semilunar valves. The arterial blood pressure is referred to as the **afterload** because it is the work load imposed on the heart after the contraction has begun. If the arterial blood pressure is chronically elevated (high blood pressure) or if the exit valve is stenotic, the ventricle has to generate more pressure to eject blood. For example, instead of generating the normal pressure of 120 mm Hg, the ventricular pressure may need to go as high as 400 mm Hg to force blood through a narrowed aortic valve.

The heart may be able to compensate for a sustained increase in afterload by enlarging (through hypertrophy of the cardiac-muscle fibers; see p. 238). This enables it to contract more forcefully and maintain a normal stroke volume despite the impedance to ejection. A diseased heart or a heart weakened with age may not be able to compensate completely, however; in that case, heart failure ensues. Even if the heart is initially able to compensate for a chronic increase in afterload, the sustained extra work load placed on the heart can eventually cause pathological changes in the heart that lead to heart failure. In fact, a chronically elevated afterload is one of the two major factors that cause heart failure, a topic to which we now turn our attention.

The contractility of the heart is decreased in heart failure.

Heart failure refers to the inability of the cardiac output to keep pace with the body's demands for supplies and removal of wastes. Either one or both ventricles may fail. When a failing ventricle becomes unable to pump out all of the blood returned to it, the veins behind the failing ventricle become congested with blood. Heart failure may occur for a variety of reasons, but the two most common are (1) damage to the heart muscle as a result of a heart attack or impaired circulation to the cardiac muscle and (2) prolonged pumping against a chronically increased afterload, as with a stenotic semilunar valve or a sustained elevation in blood pressure.

The prime defect in heart failure is a decrease in cardiac contractility; that is, the intrinsic ability of the heart to develop pressure and eject a stroke volume is reduced so that the heart operates on a lower length-tension curve (• Fig. 9–29a). The Frank-Starling curve is shifted downward and to the right such that for a given end-diastolic volume, a failing heart will pump out a smaller stroke volume than a normal healthy heart.

Two major compensatory measures help restore the stroke volume to normal in the early stages of heart failure. First, sympathetic activity to the heart is reflexly increased, which increases the contractility of the heart toward normal (Fig. 9–29b). Sympathetic stimulation can only help compensate for a limited period of time, however. The heart becomes less responsive to norepinephrine after prolonged exposure, and furthermore, the norepinephrine stores in the heart become depleted. Second, when cardiac output is reduced, the kidneys, in a compensatory attempt to improve their reduced blood flow, retain extra salt and water in the body during urine formation to expand the blood volume. The increase in circulating blood volume increases the end-diastolic volume. The resultant stretching of the cardiac-muscle fibers enables

(a)

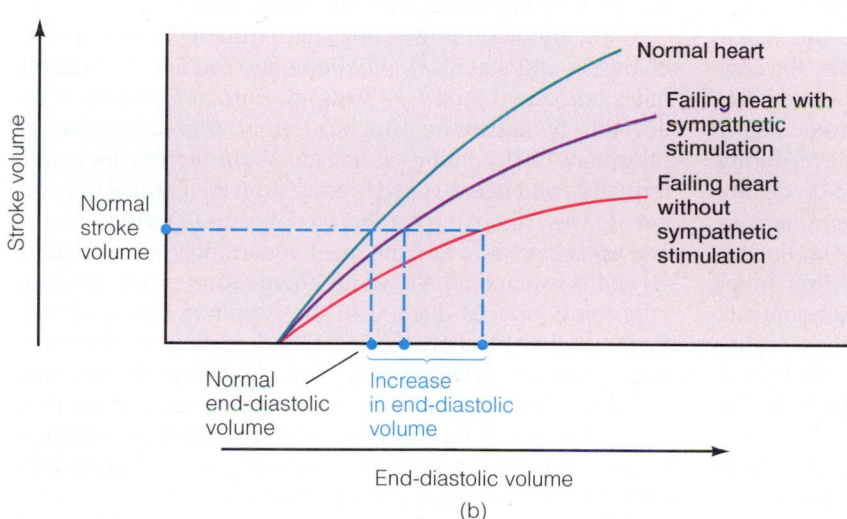

(b)

● **FIGURE 9–29 Frank-Starling Curve in Heart Failure** (a) Shift of Frank-Starling curve downward and to the right in a failing heart. Because its contractility is decreased, the failing heart pumps out a smaller stroke volume at the same end-diastolic volume than a normal heart does. (b) Compensations for heart failure. Reflex sympathetic stimulation shifts the Frank-Starling curve of a failing heart to the left, increasing the contractility of the heart toward normal. A compensatory increase in end-diastolic volume as a result of blood-volume expansion further increases the strength of contraction of the failing heart. Operating at a longer cardiac-muscle-fiber length, a compensated failing heart is able to eject a normal stroke volume.

the weakened heart to pump out a normal stroke volume (Fig. 9–29b). The heart is now pumping out the blood returned to it but is operating at a longer cardiac-muscle-fiber length.

As the disease progresses and the contractility of the heart deteriorates further, the heart reaches a point at which it is no longer able to pump out a normal stroke volume (that is, cannot pump out all of the blood returned to it) despite compensatory measures. In fact, as the end-diastolic volume continually increases, in part as a compensatory measure and in part because more blood is being returned to the heart than can be pumped out, the cardiac-muscle fibers become stretched beyond their optimal length. This overstretching decreases rather than increases the force of contraction so stroke volume is reduced even further. Thus, a vicious cycle ensues that ends in death if there is no therapeutic intervention. *Backward failure* occurs as blood that cannot enter and be pumped out by the heart continues to dam up in the venous system. *Forward failure* occurs simultaneously as the heart fails to pump an adequate amount of blood forward to the tissues because the stroke volume becomes progressively

smaller. The congestion in the venous system is the reason this condition is sometimes termed **congestive heart failure**.

Left-sided failure has more serious consequences than right-sided failure. Backward failure of the left side leads to pulmonary edema (excess tissue fluid in the lungs) because blood dams up in the lungs. This fluid accumulation in the lungs reduces exchange of O_2 and CO_2 between the air and blood in the lungs, leading to reduced arterial oxygenation and an elevation of acid-forming CO_2 in the blood. In addition, one of the more serious consequences of left-sided forward failure is an inadequate blood flow to the kidneys, which causes a twofold problem. First, vital kidney function is depressed, and second, the kidneys retain even more salt and water in the body during urine formation as they attempt to expand the plasma volume even further to improve their reduced blood flow. Excessive fluid retention further deteriorates the already existing problems of venous congestion. Treatment of congestive heart failure therefore includes measures that reduce salt and water retention and increase urinary output as well as drugs that enhance the contractile ability of the weakened heart—digitalis, for example.

NOURISHING THE HEART MUSCLE

The heart receives most of its own blood supply through the coronary circulation during diastole.

Although all the blood passes through the heart, the heart muscle is unable to extract O_2 or nutrients from the blood within its chambers for two reasons. First, the watertight endocardial lining does not permit blood to pass from the chamber into the myocardium. Second, the heart walls are too thick to permit diffusion of O_2 and other supplies from the blood in the chamber to the individual cardiac cells. Therefore, like other tissues of the body, heart muscle must receive blood through blood vessels, specifically by means of the **coronary circulation**. The coronary arteries branch from the aorta just beyond the aortic valve (see Fig. 9–34), and the coronary veins empty into the right atrium.

The heart muscle receives most of its blood supply during diastole. Blood flow to the heart-muscle cells is substantially reduced during systole for two reasons. First, the major branches of the coronary arteries are compressed by the contracting myocardium, and second, the entrance to the coronary vessels is partially blocked by the open aortic valve. Thus, most coronary arterial flow (about 70%) occurs during diastole, driven by the aortic blood pressure, with flow declining as aortic pressure drops. Only about 30% of coronary arterial flow occurs during systole (● Fig. 9–30). The limited time for coronary blood flow becomes especially important during rapid heart rates, when diastolic time is substantially reduced. Just when increased demands are placed on the heart to pump more rapidly, it has less time to provide O_2 and nourishment to its own musculature to accomplish the increased work load.

Nevertheless, under normal circumstances the heart muscle does receive adequate blood flow to support its activities, even during exercise, when the rate of coronary blood flow increases up to five times its resting rate. Increased delivery of blood to the cardiac cells is accomplished primarily by vasodilation, or enlargement, of the coronary vessels, which allows more blood to flow through them, especially during diastole. The increased coronary blood flow is imperative to meet the heart's increased O_2 requirements because, unlike most other tissues, the heart is unable to remove much additional O_2 from the blood passing through its vessels to support increased metabolic activities. Most other tissues under resting conditions extract only about 25% of the O_2 available from the blood flowing through them, leaving a considerable O_2 reserve that can be drawn on when a tissue has increased O_2 needs; that is, the tissue can immediately increase the O_2 available to it by removing a greater percentage of O_2 from the blood passing through it. In contrast, the heart, even under resting conditions, removes up to 65% of the O_2 available in the coronary vessels, far more than is withdrawn by other tissues. This leaves little O_2 in reserve in the coronary blood should cardiac O_2 demands increase. Therefore, the primary means by which more O_2 can be made available to the heart muscle is by increasing coronary blood flow.

Coronary blood flow is adjusted primarily in response to changes in the heart's O_2 requirements. The link that coordinates coronary blood flow with myocardial O_2 needs is believed to be *adenosine,* which is formed from adenosine triphosphate (ATP) during cardiac metabolic activity. Increased formation and release of adenosine from the cardiac cells occur (1) when there is a cardiac O_2 deficit or (2) when cardiac activity is increased and the heart accordingly requires more O_2 and is using more ATP as an energy source. The released adenosine induces dilation of the coronary blood vessels, thereby allowing more O_2-rich blood to flow to the more active cardiac cells to meet their increased O_2 demand (● Fig. 9–31). This matching of O_2 delivery with O_2 needs is critical because the heart muscle is dependent on oxidative processes to generate energy. The heart cannot obtain sufficient ATP through anaerobic metabolism.

Although the heart has little ability to support its energy

● **FIGURE 9–31   Matching of Coronary Blood Flow to Oxygen Need of Cardiac-Muscle Cells**

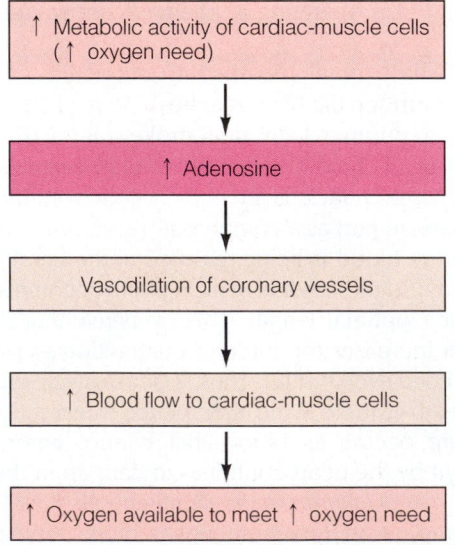

● **FIGURE 9–30   Coronary Blood Flow** Most coronary blood flow occurs during diastole because the coronary vessels are compressed almost completely closed during systole.

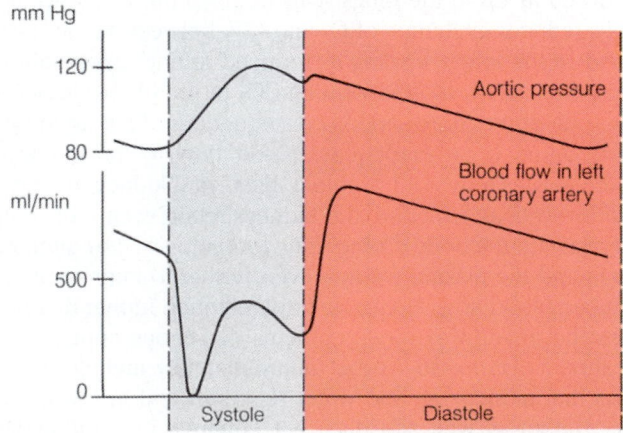

needs by means of anaerobic metabolism and must rely heavily on its O_2 supply, it can tolerate wide variations in its nutrient supply. The heart primarily uses free fatty acids and, to a lesser extent, glucose and lactate as fuel sources, depending on their availability. Because the cardiac muscle is remarkably adaptable and can shift metabolic pathways to use whatever nutrient is available, the primary danger of insufficient coronary blood flow is not fuel shortage but O_2 deficiency.

Atherosclerotic coronary artery disease can deprive the heart of essential oxygen.

Adequacy of coronary blood flow is relative to the heart's O_2 demands at any given moment. In the normal heart, coronary blood flow increases correspondingly as O_2 demands rise. With **coronary artery disease**, however, it may not be possible for coronary blood flow to keep pace with rising O_2 needs. A given rate of coronary blood flow may be adequate at rest but insufficient upon physical exertion or other stressful situations.

Complications of coronary artery disease make it the single leading cause of death in the United States. Coronary artery disease can cause *myocardial ischemia* (insufficient circulation of oxygenated blood through the coronary circulation to maintain aerobic metabolism in the heart muscle) by the three following mechanisms: (1) profound vascular spasm of the coronary arteries; (2) the formation of atherosclerotic plaques; and (3) thromboembolism. We will discuss each of these in turn.

Vascular spasm is an abnormal spastic constriction that transiently narrows the coronary vessels; it is most often triggered by exposure to cold, physical exertion, or anxiety. Vascular spasms are associated with the early stages of coronary artery disease. The condition is reversible and usually of insufficient duration to produce damage to the cardiac muscle. Recent evidence suggests that reduced O_2 availability in the coronary vessels causes the release of **platelet-activating factor** (**PAF**) from the endothelium (lining) of the vessels. PAF, which exerts a variety of biological actions, was named for its first discovered effect. Among its effects other than activating platelets, PAF, once released from the endothelium, diffuses to the underlying vascular smooth muscle and causes it to contract, bringing about vascular spasm.

Atherosclerosis is a progressive, degenerative arterial disease that leads to occlusion (gradual blockage) of affected vessels, thereby reducing blood flow through them. It is believed that atherosclerosis begins as **atheromas**, which are benign (noncancerous) tumors of smooth-muscle cells within the blood-vessel walls. These cells migrate from the muscular layer of the blood vessel to a position just beneath the endothelial lining, where they continue to divide and enlarge. Later, cholesterol and other lipids accumulate in the abnormal smooth-muscle cells to produce a **plaque**. The plaques bulge into the lumen of the vessel as they continue to develop (● Fig. 9–32). A thickening plaque interferes with nutrient exchange for the cells of the involved arterial wall, leading to

Lumen   Plaque

● **FIGURE 9–32   Severe Atherosclerotic Plaque in a Coronary Vessel** Note how the thickened plaque narrows the lumen, thereby impeding blood flow through the vessel.

degeneration of the wall in the vicinity of the plaque. The damaged area is invaded by fibroblasts (scar tissue-forming cells), and in the later stages of the disease, Ca^{++} often precipitates in the plaque. A vessel so afflicted becomes hard and poorly distensible, a condition dubbed "hardening of the arteries."

Atherosclerosis attacks arteries throughout the body, but the most serious consequences involve damage to the vessels of the brain and heart. In the brain, atherosclerosis is the prime cause of strokes, whereas in the heart, it brings about myocardial ischemia and its complications. The following are potential complications of coronary atherosclerosis:

1. Gradual enlargement of the protruding plaque continues to narrow the vessel lumen and progressively diminishes coronary blood flow, triggering increasingly frequent bouts of transient myocardial ischemia as the ability to match blood flow with cardiac O_2 needs becomes more limited. Although the heart cannot normally be "felt," pain is associated with myocardial ischemia. Such cardiac pain, known as **angina pectoris** ("pain of the chest"), can be felt beneath the sternum and is often referred to the left shoulder and down the left arm (see p. 141). The symptoms of angina pectoris recur whenever cardiac O_2 demands become too great in relation to the coronary blood flow—for example, during exertion or emotional stress. The pain is thought to result from stimulation of cardiac nerve endings by the accumulation of lactic acid when the heart shifts to its limited ability to perform anaerobic metabolism. The ischemia associated with the characteristically brief anginal attacks is usually temporary and re-

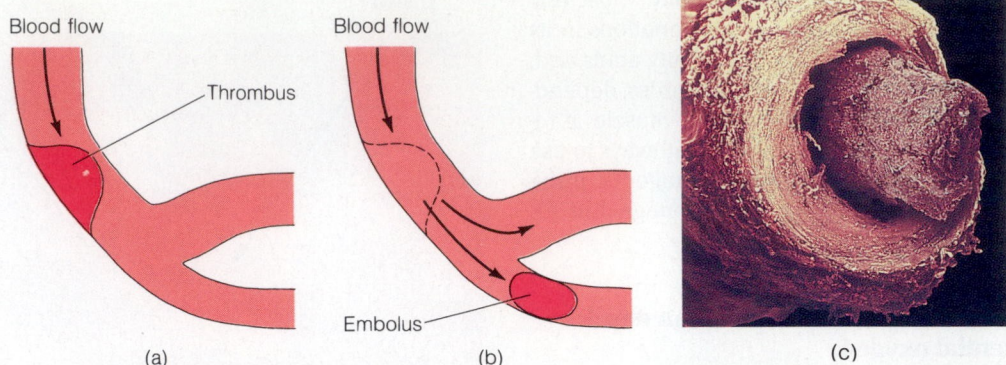

• FIGURE 9–33 Consequences of Thromboembolism (a) A thrombus may enlarge gradually until it completely occludes the vessel at that site. (b) A thrombus may break loose from its attachment, forming an embolus that may completely occlude a smaller vessel downstream. (c) Scanning electron micrograph of a vessel completely occluded by a thromboembolic lesion.

versible and can be relieved by rest, administration of vasodilator drugs such as *nitroglycerin,* or both. Nitroglycerin brings about coronary vasodilation by being metabolically converted to nitric oxide, which in turn relaxes the vascular smooth muscle.

2. The enlarging atherosclerotic plaque can break through the weakened endothelial lining that covers it, exposing blood to the underlying collagen (a component of connective tissue). Blood platelets (formed elements of the blood involved in plugging vessel defects and in clot formation) normally do not adhere to smooth, healthy vessel linings. However, when platelets come into contact with collagen at the site of vessel damage, they stick to the site and contribute to the formation of a blood clot. Such an abnormal clot attached to a vessel wall is known as a **thrombus**. The thrombus may enlarge gradually until it completely blocks the vessel at that site, or the continued flow of blood past the thrombus may break it loose from its attachment. Such a freely floating clot, or **embolus**, may completely plug a smaller vessel as it flows downstream (• Fig. 9–33). Thus, through **thromboembolism**, atherosclerosis can result in a gradual or sudden occlusion of a coronary vessel (or any other vessel).

3. When a coronary vessel is completely plugged, the cardiac tissue served by the vessel soon dies from O_2 deprivation and a heart attack occurs, unless the area can be supplied with blood from nearby vessels. Sometimes a deprived area is fortunate enough to receive blood from more than one pathway. **Collateral circulation** exists when small terminal branches from adjacent blood vessels nourish the same area. These accessory vessels cannot develop suddenly following an abrupt blockage but may be lifesaving if already developed. Such alternate vascular pathways often develop over a period of time when an atherosclerotic constriction progresses slowly, or they may be induced by sustained demands on the heart through a regular aerobic exercise program.

In the absence of collateral circulation, the extent of the damaged area during a heart attack depends on the size of the blocked vessel. The larger the vessel occluded, the greater the area deprived of its blood supply. As • Figure 9–34 illustrates, a blockage at point A in the coronary circulation would cause more extensive damage than would a blockage at point B. Because there are only two major coronary arteries, complete blockage of either one of these main branches results in extensive myocardial damage. Left coronary artery blockage is most devastating because this vessel is responsible for supplying 85% of the cardiac tissue. A heart attack has four possible outcomes: immediate death, delayed death from complications, full functional recovery, or recovery with impaired function (Table 9–5).

• FIGURE 9–34 Extent of Myocardial Damage as Function of Size of Occluded Vessel

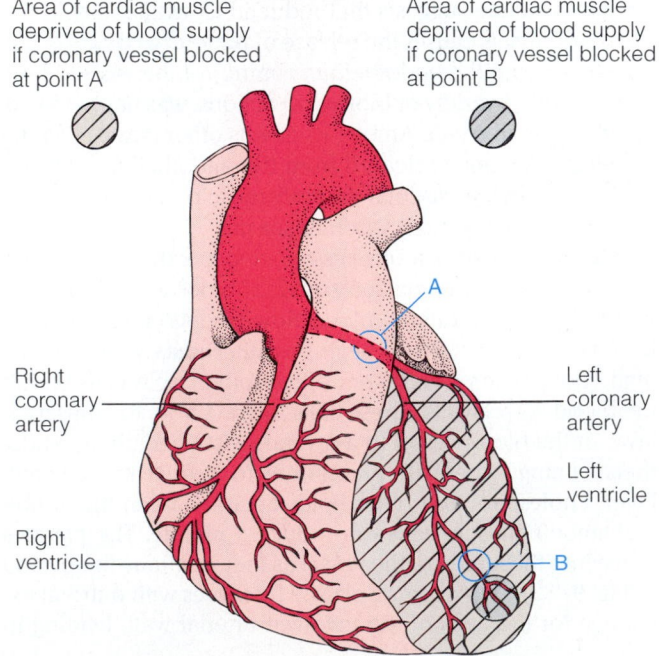

Area of cardiac muscle deprived of blood supply if coronary vessel blocked at point A

Area of cardiac muscle deprived of blood supply if coronary vessel blocked at point B

Right coronary artery

Left coronary artery

Right ventricle

Left ventricle

A

B

TABLE 9–5	Possible Outcomes of Acute Myocardial Infarction (Heart Attack)		
IMMEDIATE DEATH	**DELAYED DEATH FROM COMPLICATIONS**	**FULL FUNCTIONAL RECOVERY**	**RECOVERY WITH IMPAIRED FUNCTION**
Acute cardiac failure occurring because the heart is too weakened to pump effectively to support the body tissues Fatal ventricular fibrillation brought about by damage to the specialized conducting tissue or induced by O_2-deprivation	Fatal rupture of the dead, degenerating area of the heart wall Slowly progressing congestive heart failure occurring because the weakened heart is unable to pump out all the blood returned to it	Replacement of damaged area with a strong scar, accompanied by enlargement of remaining normal contractile tissue to compensate for the lost cardiac musculature	Persistence of permanent functional defects, such as bradycardia or conduction blocks, caused by destruction of irreplaceable autorhythmic or conductive tissues

The amount of "good" cholesterol versus "bad" cholesterol in the blood is linked to atherosclerosis.

The cause of atherosclerosis is still not entirely clear. Certain high-risk factors have been associated with an increased incidence of atherosclerosis and coronary heart disease. Included among them are genetic predisposition, obesity, advanced age, smoking, hypertension, diabetes mellitus, lack of exercise, nervous tension, and, most significantly, excess cholesterol levels in the blood.

There are two sources of cholesterol for the body: (1) dietary intake of cholesterol, with animal products such as egg yolk, red meats, and butter being especially rich in this lipid (animal fats contain cholesterol, whereas plant fats do not), and (2) manufacture of cholesterol by many organs within the body, particularly the liver. Because of the body's ability to synthesize cholesterol, there is not a direct correlation between the amount of cholesterol ingested and the cholesterol levels in the blood, although modest reductions in blood cholesterol can be accomplished by lowering the intake of animal fats. For some individuals, drugs may be necessary to satisfactorily lower blood cholesterol levels.

Actually, it is not the total blood cholesterol level but the amount of cholesterol bound to various plasma-protein carriers that appears to be most important with regard to the risk of developing atherosclerotic heart disease. Because cholesterol is a lipid, it is not very soluble in blood. Most cholesterol in the blood is attached to specific plasma-protein carriers in the form of lipoprotein complexes, which are soluble in blood. There are three major lipoproteins, named for their density of protein as compared to lipid: (1) **high-density lipoproteins** (**HDL**), which contain the most protein and least cholesterol; (2) **low-density lipoproteins** (**LDL**), which contain less protein and more cholesterol; and (3) **very-low-density lipoproteins** (**VLDL**), which contain the least protein and most lipid, but the lipid they carry is neutral fat, not cholesterol. Cholesterol carried in LDL complexes has been termed "bad" cholesterol, because cholesterol is transported *to* the cells, including those lining the blood-vessel walls, by means of LDL. In contrast, cholesterol carried in HDL complexes has been dubbed "good" cholesterol, because HDL removes cholesterol *from* the cells and transports it to the liver for partial elimination from the body.

Unlike most lipids, cholesterol is not used as metabolic fuel by cells. Instead, it serves as an essential component of plasma membranes. In addition, a few special cell types use cholesterol as a precursor for the synthesis of secretory products, such as steroid hormones and bile salts. Although most cells are capable of synthesizing some of the cholesterol needed for their own plasma membranes, they cannot manufacture sufficient amounts and therefore must rely on supplemental cholesterol being delivered by the blood. This additional cholesterol is supplied either by the diet or by cells that specialize in cholesterol synthesis, especially liver cells.

Cells accomplish cholesterol uptake from the blood by synthesizing receptor proteins specifically capable of binding LDL and inserting these receptors into the cells' plasma membranes (• Fig. 9–35). When an LDL particle binds to one of the membrane receptors, the cell engulfs the particle by endocytosis (see p. 24). Within the cell, lysosomal enzymes break down the LDL to free the cholesterol, making it available to the cell for synthesis of new cellular membrane. If too much free cholesterol accumulates in the cell, there is a shutdown of both the synthesis of LDL receptor proteins (so that less cholesterol is taken up) and the cell's own cholesterol synthesis (so that less new cholesterol is made). Faced with a cholesterol shortage, on the other hand, the cell makes more LDL receptors so that it can engulf more cholesterol from the blood.

The maintenance of a blood-borne cholesterol supply to the cells involves an interaction between dietary cholesterol and the synthesis of cholesterol by the liver. When the amount of dietary cholesterol is increased, hepatic (liver) synthesis of cholesterol is turned off because cholesterol in the blood directly inhibits a hepatic enzyme essential for cholesterol synthesis. Thus, as more cholesterol is ingested, less is produced by the liver. Conversely, when cholesterol intake from food is reduced, the liver synthesizes more of this lipid because the inhibitory effect of cholesterol on the crucial hepatic enzyme is removed. In this way, the blood concentration of cholesterol is maintained at a fairly constant level despite changes in cholesterol intake; thus, it is difficult to significantly reduce cholesterol levels in the blood by decreasing cholesterol intake.

In contrast to LDL complexes, which transport cholesterol to the cells, HDL removes cholesterol from cells and trans-

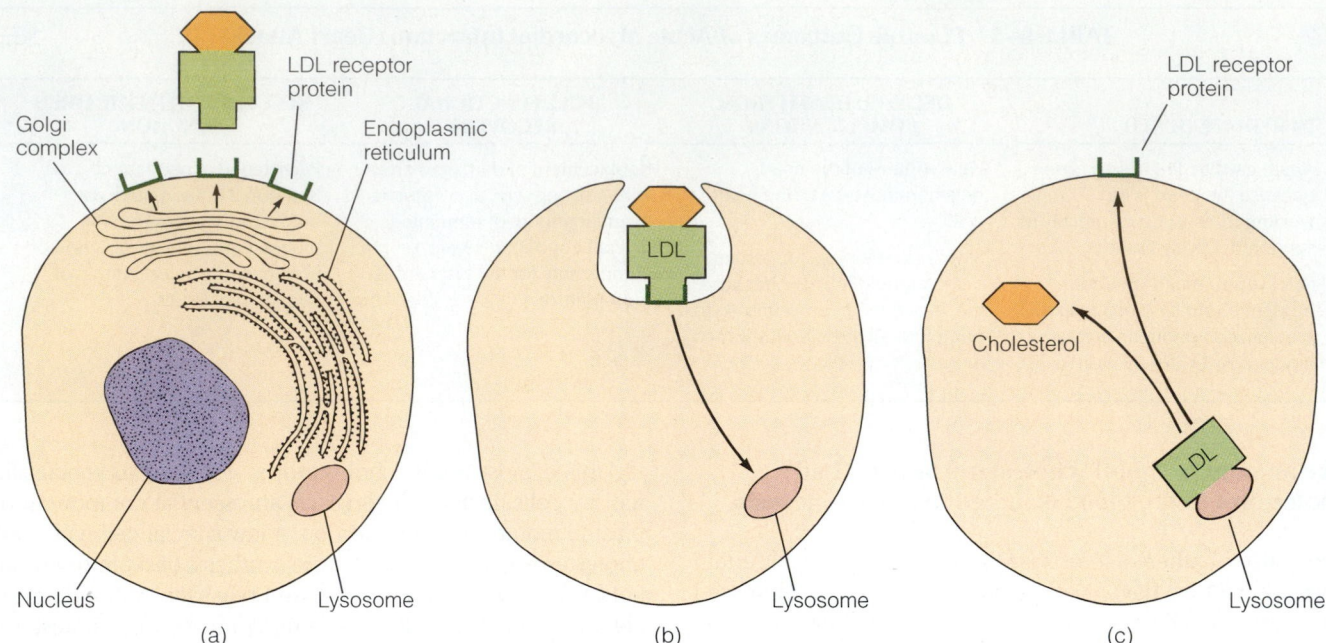

● **FIGURE 9–35   Uptake of Cholesterol by Cells** (a) A cell that needs additional cholesterol synthesizes LDL receptor proteins. (b) Receptor-bound LDL is internalized by endocytosis. (c) A lysosome frees cholesterol, and the LDL receptor is recycled back to the plasma membrane.

ports it to the liver. The liver, in turn, secretes cholesterol as well as cholesterol-derived bile salts into the bile. Bile enters the intestinal tract, where bile salts participate in the digestive process. Most of the secreted cholesterol and bile salts are subsequently reabsorbed from the intestinal tract into the blood to be recycled to the liver. Cholesterol molecules not reclaimed by absorption are eliminated in the feces.

Obviously, the liver has a central role in cholesterol me-

tabolism. At any given time, the liver may be manufacturing new cholesterol, extracting old cholesterol from the blood and secreting it into the bile, or converting old cholesterol into bile salts, which likewise are secreted into the bile (● Fig. 9–36). The first process is a mechanism for adding cholesterol to the blood to supplement dietary intake. The other two pathways lead to a net loss of cholesterol from the body. Thus, the liver has a primary role in determining

● **FIGURE 9–36   Role of the Liver in Cholesterol Metabolism** The letters in the illustration correspond to the following steps:

(a) Liver extracts cholesterol (carried by HDL) from the blood.
(b) Liver converts most of the cholesterol into bile salts.
(c) Liver secretes bile salts and cholesterol into the bile, which empties into the intestinal lumen.
(d) Part of the cholesterol and bile salts is eliminated in the feces.
(e) Some of the bile salts are reabsorbed into the blood and recycled to the liver.
(f) Liver synthesizes new cholesterol, which is carried by LDL within the blood away from the liver to other cells.

Saturated fatty acids and polyunsaturated fatty acids are known to influence these pathways at the designated points.

total blood cholesterol levels, and the interplay between LDL and HDL determines the traffic flow of cholesterol between the liver and the other cells. Whenever these mechanisms are altered, blood cholesterol levels may be affected in such a way as to influence the individual's predisposition to atherosclerosis. The following are specific examples:

■ Evidence suggests that the propensity toward developing atherosclerosis substantially increases with elevated levels of LDL. In one hereditary disease, afflicted individuals lack the genes for making LDL receptor proteins. Because their cells cannot take up LDL from the blood, the blood concentration of these cholesterol-loaded lipoproteins becomes greatly elevated. Such persons have an extraordinary tendency toward developing atherosclerosis early in life, and many die in their youth from coronary artery disease. Recent studies suggest that even in individuals with normal LDL receptors, atherosclerosis may develop if they have aberrant levels of the sole protein component of LDL, **apolipoprotein B-100 (apo-B)**. This is the portion of the LDL complex that interacts with the receptors.

■ The risk of atherosclerosis is inversely related to the concentration of HDL in the blood; that is, elevated levels of HDL are associated with a low incidence of atherosclerotic heart disease. In fact, a more accurate predictor of the risk of developing atherosclerosis than the total blood cholesterol level is the *blood HDL-cholesterol / total cholesterol ratio*. The higher the HDL-cholesterol concentration in relationship to the total blood cholesterol level, the lower the risk. Some other factors known to influence atherosclerotic risk can be related to HDL levels; for example, cigarette smoking lowers HDL, and the HDL level is higher in individuals who exercise regularly. Moreover, premenopausal women, who have a lower incidence of atherosclerotic heart disease than their male counterparts, have a higher concentration of HDL, presumably because of some influence of the female sex hormone, estrogen. After production of estrogen ceases at menopause, the incidence of coronary heart disease in women parallels that in men.

■ Varying the intake of dietary fatty acids may alter total blood cholesterol levels by influencing one or more of the mechanisms involving cholesterol balance (Fig. 9–36). The blood cholesterol level tends to be raised by ingestion of saturated fatty acids found predominantly in animal fats and tropical plant oils, such as palm oil and coconut oil. These fatty acids stimulate the synthesis of cholesterol and inhibit its conversion to bile salts. On the other hand, polyunsaturated fatty acids, the predominant fatty acids of most plants, tend to reduce blood cholesterol levels by enhancing the elimination of both cholesterol and cholesterol-derived bile salts in the feces. Furthermore, dietary soluble-fiber supplements such as oat bran and psyllium-seed husks have been shown to reduce the level of blood cholesterol, particularly LDL-cholesterol, presumably by physically interfering with its absorption from the intestine. Thus, dietary manipulations can reduce the cholesterol-related risk of atherosclerosis, but not just by reducing cholesterol intake.

In addition to LDL and HDL, researchers recently discovered another cholesterol carrier, **lipoprotein(a)**. The blood concentration of lipoprotein(a) varies nearly 1,000-fold among individuals, with the tendency for higher or lower concentrations being an inherited trait. A positive correlation has been found between higher blood concentrations of lipoprotein(a) and the incidence of atherosclerosis. Lipoprotein(a)'s normal physiological role and its possible role in the development of atherosclerosis are currently under investigation. Also under further study is a recent finding that an increased iron level in the blood is positively correlated with the incidence of atherosclerosis and coronary artery disease.

As you can see, the relationship between atherosclerosis, cholesterol, and other environmental and genetic factors is far from clear. Much research concerning this complex disease is presently in progress, because the incidence of atherosclerosis is so high and its consequences are potentially fatal.

CHAPTER IN PERSPECTIVE: FOCUS ON HOMEOSTASIS

Survival depends on continual delivery of needed supplies to all of the cells throughout the body and on ongoing removal of wastes generated by the cells. Furthermore, regulatory chemical messengers, such as hormones, must be transported from their site of production to their site of action, where they control a variety of activities, most of which are directed toward maintaining a stable internal environment.

The circulatory system contributes to homeostasis by serving as the body's transport system. It provides a means of rapidly moving materials from one part of the body to another. Without the circulatory system, materials would not get where they need to go to support life-sustaining activities nearly rapidly enough. For example, O_2 would take months to years to diffuse from the surface of the body to internal organs, yet O_2 and other substances can be picked up by the blood and delivered to all the cells in a few seconds through the heart's swift pumping action.

The heart serves as a dual pump to continuously circulate blood between the lungs, where O_2 is picked up, and the other body tissues, which use O_2 to support their energy-generating chemical reactions. As blood is pumped through the various tissues, other substances besides O_2 are also exchanged between the blood and tissues. For example, the blood picks up nutrients as it flows through the digestive organs, and other tissues remove nutrients from the blood as it flows through them.

Although all of the body tissues constantly depend on the life-supporting blood flow provided to them by the heart, the heart itself is quite an independent organ. It is able to take care of many of its own needs without any outside influence. Contraction of this magnificent muscle is self-generated through a carefully orchestrated interplay of changing ionic permeabilities. Local mechanisms within the heart assure that blood flow to the cardiac muscle normally meets the heart's need for O_2. In addition, the heart has built-in capabilities to vary its strength of contraction, depending on the amount of blood returned to it. The heart does not act entirely autonomously, however. It is innervated by the autonomic nervous system and is influenced by the hormone epinephrine, both of which can vary the rate and contractility of the heart, depending on the body's needs for blood delivery. Furthermore, as with all tissues, the cells that compose the heart depend on the other body systems for maintenance of a stable internal environment in which they can survive and function.

CHAPTER SUMMARY

Anatomical Considerations

The heart is basically a dual pump that provides the driving pressure for blood flow through the pulmonary and systemic circulations. The heart has four chambers: each half consists of an atrium, or venous input chamber, and a ventricle, or arterial output chamber. Four heart valves direct the blood in the proper direction and prevent it from flowing in the reverse direction. The heart is self-excitable, initiating its own rhythmic contractions. Contraction of the spirally arranged cardiac-muscle fibers produces a wringing effect important for efficient pumping. Also important for efficient pumping is the fact that the muscle fibers in each chamber act as a functional syncytium, contracting as a coordinated unit.

Electrical Activity of the Heart

The cardiac impulse originates at the SA node, the pacemaker of the heart, which has the fastest rate of spontaneous depolarization to threshold. Once initiated, the action potential spreads throughout the right and left atria, partially facilitated by specialized conduction pathways, but mostly by cell-to-cell spread of the impulse through gap junctions. The impulse passes from the atria into the ventricles through the AV node, the only point of electrical contact between these chambers. The action potential is delayed briefly at the AV node, assuring that atrial contraction precedes ventricular contraction to allow complete ventricular filling. The impulse then rapidly travels down the interventricular septum via the bundle of His and is rapidly dispersed throughout the myocardium by means of the Purkinje fibers. The remainder of the ventricular cells are activated by cell-to-cell spread of the impulse through gap junctions. Thus, the atria contract as a single unit, followed after a brief delay by a synchronized ventricular contraction.

The action potentials of contractile cardiac-muscle fibers exhibit a prolonged positive phase, or plateau, accompanied by a prolonged period of contraction, which assures adequate ejection time. This plateau is primarily due to activation of slow Ca^{++} channels. Because a long refractory period occurs in conjunction with this prolonged plateau phase, summation and tetanus of cardiac muscle are impossible, thereby assuring the alternate periods of contraction and relaxation essential for pumping of blood.

The spread of electrical activity throughout the heart can be recorded from the surface of the body. This record, the ECG, can provide useful information about the status of the heart.

Mechanical Events of the Cardiac Cycle

The cardiac cycle consists of three important events:

1. The generation of electrical activity as the heart autorhythmically depolarizes and repolarizes.
2. Mechanical activity consisting of alternate periods of systole (contraction and emptying) and diastole (relaxation and filling), which are initiated by the rhythmical electrical cycle.
3. Directional flow of blood through the heart chambers, guided by valvular opening and closing induced by pressure changes that are generated by mechanical activity.

Valve closing gives rise to two normal heart sounds. The first heart sound is caused by closure of the atrioventricular (AV) valves and signals the onset of ventricular systole. The second heart sound is due to closure of the aortic and pulmonary valves at the onset of diastole.

Reviewing the pressure changes associated with the cardiac cycle (see Figure 9–20), the atrial pressure curve remains low throughout the entire cardiac cycle, with only minor fluctuations (normally varying between 0 and 8 mm Hg). The aortic pressure curve remains high the entire time, with moderate fluctuations (normally varying between a systolic pressure of 120 mm Hg and a diastolic pressure of 80 mm Hg). The ventricular pressure curve fluctuates dramatically because ventricular pressure must be below the low atrial pressure during diastole to allow the AV valve to open for filling to take place, and it must be above the high aortic pressure during systole to force the aortic valve open to allow emptying to occur. Therefore, ventricular pressure normally varies from 0 mm Hg during diastole to slightly more than 120 mm Hg during systole.

Defective valve function produces turbulent blood flow, which is audible as a heart murmur. Abnormal valves may be either stenotic and not open completely or insufficient and not close completely.

Cardiac Output and Its Control

Cardiac output, the volume of blood ejected by each ventricle each minute, is determined by the heart rate times the stroke volume. Heart rate is varied by altering the balance of parasympathetic and sympathetic influence on the SA node. Parasympathetic stimulation slows the heart rate, and sympathetic stimulation speeds it up.

Stroke volume depends on (1) the extent of ventricular filling, with an increased end-diastolic volume resulting in a larger stroke volume by means of the length-tension relationship (intrinsic control), and (2) the extent of sympathetic stimulation, with increased sympathetic stimulation resulting in increased contractility of the heart, i.e., increased strength of contraction and increased stroke volume at a given end-diastolic volume (extrinsic control).

Nourishing the Heart Muscle

Cardiac muscle is supplied with oxygen and nutrients by blood delivered to it by the coronary circulation, not by blood within the heart chambers. Most coronary blood flow occurs during diastole, because the coronary vessels are compressed by the contracting heart muscle during systole. Coronary blood flow is normally varied to keep pace with cardiac oxygen needs.

Coronary blood flow may be compromised by the development of atherosclerotic plaques, which can lead to ischemic heart disease ranging in severity from mild chest pain on exertion

to fatal heart attacks. The exact cause of atherosclerosis is unclear, but apparently the ratio of cholesterol carried in the plasma in conjunction with high-density lipoproteins (HDL) compared to low-density lipoproteins (LDL) is an important factor.

REVIEW EXERCISES

Objective Questions (Answers on p. D–2.)

1. Adjacent cardiac-muscle cells are joined end-to-end at specialized structures known as _____, which contain two types of membrane junctions—_____ and _____.

2. _____ is an abnormally slow heart rate whereas _____ is a rapid heart rate.

3. The link that coordinates coronary blood flow with myocardial oxygen needs is believed to be _____.

4. The left ventricle is a stronger pump than the right ventricle because more blood is needed to supply the body tissues than to supply the lungs. (True or false?)

5. The heart lies in the left half of the thoracic cavity. (True or false?)

6. The only point of electrical contact between the atria and ventricles is the fibrous skeletal rings. (True or false?)

7. The atria and ventricles both act as a functional syncytium. (True or false?)

8. Which of the following is the proper sequence of cardiac excitation?
   a. SA node → AV node → atrial myocardium → bundle of His → Purkinje fibers → ventricular myocardium.
   b. SA node → atrial myocardium → AV node → bundle of His → ventricular myocardium → Purkinje fibers.
   c. SA node → atrial myocardium → ventricular myocardium → AV node → bundle of His → Purkinje fibers.
   d. SA node → atrial myocardium → AV node → bundle of His → Purkinje fibers → ventricular myocardium.

9. What percentage of ventricular filling is normally accomplished before atrial contraction begins?
   a. 0%
   b. 20%
   c. 50%
   d. 80%
   e. 100%

10. Sympathetic stimulation of the heart:
    a. increases the heart rate.
    b. increases the contractility of the heart muscle.
    c. shifts the Frank-Starling curve to the left.
    d. Both (a) and (b) above are correct.
    e. All of the above are correct.

11. (Circle the correct choice in each instance to complete the statements.) During ventricular filling, ventricular pressure must be (greater than/less than) atrial pressure, while during ventricular ejection, ventricular pressure must be (greater than/less than) aortic pressure. Atrial pressure is always (greater than/less than) aortic pressure. During isovolumetric ventricular contraction and relaxation, ventricular pressure is (greater than/less than) atrial pressure and (greater than/less than) aortic pressure.

12. (Circle the correct choice in each instance to complete the statement.) The first heart sound is associated with closure of the (AV/semilunar) valves and signals the onset of (systole/diastole), whereas the second heart sound is associated with closure of the (AV/semilunar) valves and signals the onset of (systole/diastole).

13. Match the following:
    ___ 1. receives low-oxygenated blood from the venae cavae
    ___ 2. prevent backflow of blood from the ventricles to the atria
    ___ 3. pumps highly oxygenated blood into the aorta
    ___ 4. prevent backflow of blood from the arteries into the ventricles
    ___ 5. pumps low-oxygenated blood into the pulmonary artery
    ___ 6. receives highly oxygenated blood from the pulmonary veins
    ___ 7. permit AV valves to function as one-way valves

    a. AV valves
    b. semilunar valves
    c. left atrium
    d. left ventricle
    e. right atrium
    f. right ventricle
    g. chordae tendineae and papillary muscles

14. Match the following:
    ___ 1. characterized by uncoordinated excitation and contraction of the cardiac cells
    ___ 2. caused by AV nodal damage
    ___ 3. an overly irritable area that takes over pacemaker activity

    a. heart block
    b. ectopic focus
    c. ventricular fibrillation

Essay Questions

1. What are the three basic components of the circulatory system?

2. Trace a drop of blood through one complete circuit of the circulatory system.

3. Describe the location and function of the four heart valves. What prevents eversion of each of these valves?

4. What are the three layers of the heart wall? Describe the distinguishing features of the structure and arrangement of cardiac-muscle cells. What are the two specialized types of cardiac-muscle cells?

5. Why is the SA node the pacemaker of the heart?

6. Describe the normal spread of cardiac excitation. What is the significance of the AV nodal delay? Why is the ventricular conduction system important?

7. Compare the changes in membrane potential associated with an action potential in a nodal pacemaker cell with those in a myocardial contractile cell. What is responsible for the plateau phase?

8. Why is tetanus of cardiac muscle impossible? Why is this advantageous?

9. Draw and label the wave forms of a normal ECG. What electrical event does each component of the ECG represent?

10. Describe the mechanical events (that is, pressure changes, volume changes, valve activity, and heart sounds) that occur during the cardiac cycle. Correlate these mechanical events with the changes that take place in electrical activity.

11. Distinguish between a stenotic and an insufficient valve.

12. Define the following: end-diastolic volume, end-systolic volume, stroke volume, heart rate, cardiac output, and cardiac reserve.

13. Discuss autonomic nervous system control of heart rate.

14. Describe the intrinsic and extrinsic control of stroke volume.

15. By what means is the heart muscle provided with blood? Why does the heart receive most of its own blood supply during diastole?

16. What are the pathological changes and consequences of coronary artery disease?

17. Discuss the sources, transport, and elimination of cholesterol in the body. Distinguish between "good" cholesterol and "bad" cholesterol.

POINTS TO PONDER

1. The stroke volume ejected on the next heart beat after a premature beat is usually larger than normal. Can you explain why? (Hint: At a given heat rate, the interval between a premature beat and the next normal beat is longer than the interval between two normal beats.)

2. Trained athletes usually have lower resting heart rates than normal (for example, 50 beats/min in an athlete compared to 70 beats/min in a sedentary individual). Considering that the resting cardiac output is 5,000 ml/min in both trained athletes and sedentary individuals, what is responsible for the bradycardia of trained athletes?

3. A characteristic murmur accompanies patent ductus arteriosus (see the boxed feature on p. 263). A harsh blowing murmur can be heard throughout the cardiac cycle, but becomes much more intense during ventricular systole and much less intense during ventricular diastole. What would account for the waxing and waning of the murmur with each beat of the heart?

4. Through what regulatory mechanisms is a transplanted heart that does not have any innvervation able to adjust the cardiac output to meet the body's changing needs?

5. There are two branches of the bundle of His, the right and left bundle branches, each of which travels down its respective side of the ventricular septum (see Fig. 9–11). Occasionally, conduction through one of these branches becomes blocked (so-called *bundle-branch block*). In this case, the wave of excitation spreads out from the terminals of the intact branch and eventually depolarizes the whole ventricle, but the normally stimulated ventricle becomes completely depolarized a considerable time before the ventricle on the side of the defective bundle branch. For example, if the left bundle branch is blocked, the right ventricle will be completely depolarized two to three times more rapidly than the left ventricle. What effect would this defect have on the heart sounds?

10 Blood Vessels and Blood Pressure

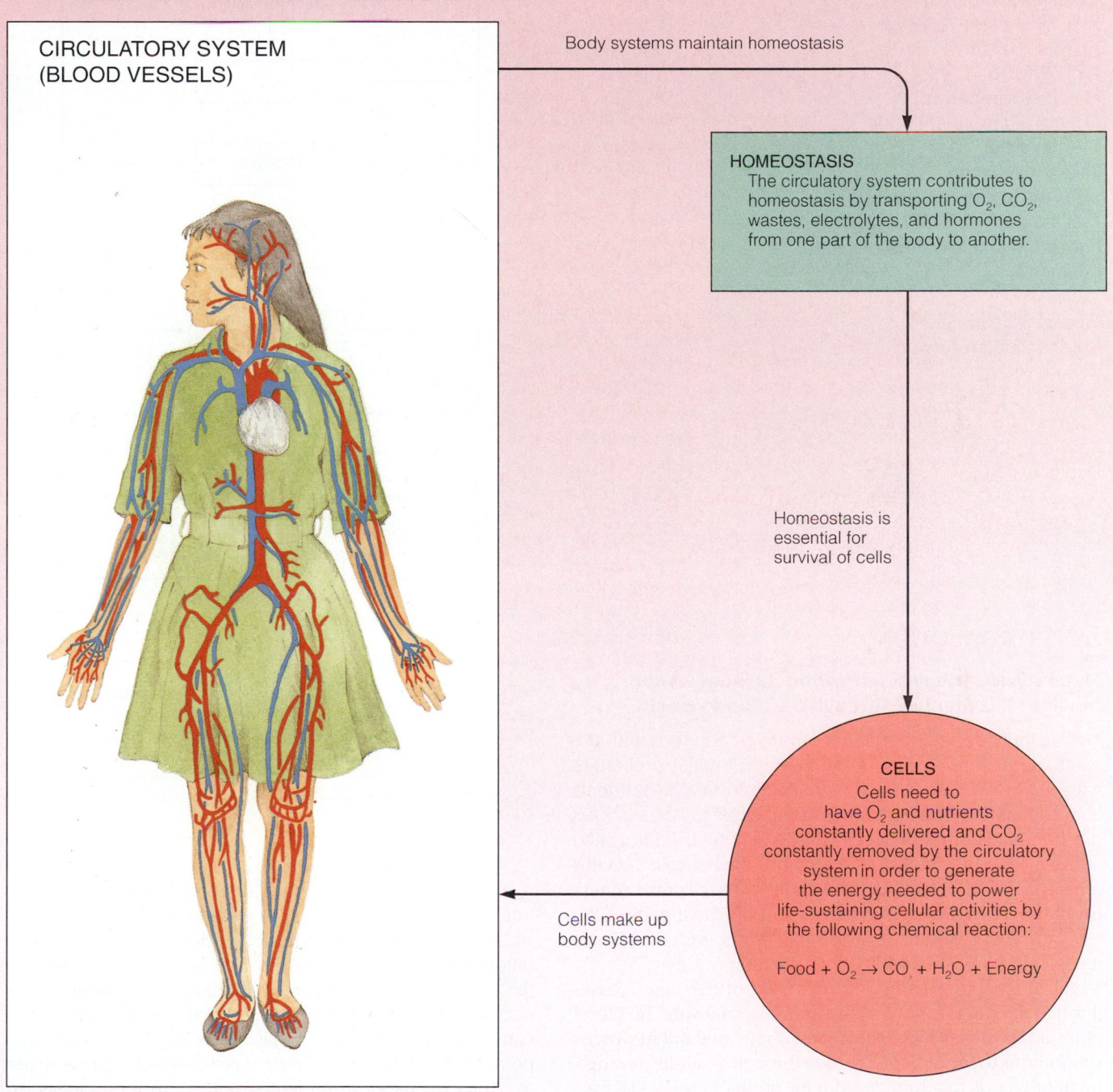

CIRCULATORY SYSTEM
(BLOOD VESSELS)

Body systems maintain homeostasis

HOMEOSTASIS
The circulatory system contributes to homeostasis by transporting O_2, CO_2, wastes, electrolytes, and hormones from one part of the body to another.

Homeostasis is essential for survival of cells

CELLS
Cells need to have O_2 and nutrients constantly delivered and CO_2 constantly removed by the circulatory system in order to generate the energy needed to power life-sustaining cellular activities by the following chemical reaction:

$$Food + O_2 \rightarrow CO_2 + H_2O + Energy$$

Cells make up body systems

The **circulatory system** contributes to homeostasis by serving as the body's transport system. The blood vessels transport and distribute blood pumped through them by the heart to meet the body's needs for O_2 and nutrient delivery, waste removal, and hormonal signaling. The highly elastic **arteries** transport blood from the heart to the tissues and serve as a pressure reservoir to continue driving blood forward when the heart is relaxing and filling. The **mean arterial blood pressure** is closely regulated to assure adequate blood delivery to the tissues. The amount of blood that flows through a given tissue depends on the caliber of the highly muscular **arterioles** that supply the tissue. Arteriolar caliber is subject to control so that the distribution of the cardiac output can be constantly readjusted to best serve the body's needs at the moment. The thin-walled, pore-lined **capillaries** are the actual site of exchange between the blood and the surrounding tissues. The highly distensible **veins** return the blood from the tissues to the heart and also serve as a blood reservoir.

≋ 🌸 **CHAPTER CONTENTS AT A GLANCE** 🌸 ≋

INTRODUCTION
Role of the blood vessels
Flow, pressure, resistance relationships
● CONCEPTS, CHALLENGES, AND CONTROVERSIES From Humors
   to Harvey: Historical Highlights in Circulation
ARTERIES
Passageways to tissues
Role as pressure reservoir
Arterial pressure
ARTERIOLES
Major resistance vessels
Control of arteriolar radius
CAPILLARIES
Sites of exchange
Diffusion across capillary wall
Bulk flow across capillary wall
Formation and functions of lymph
Edema
VEINS
Passageways to heart
Role as blood reservoir
Venous return
BLOOD PRESSURE
Factors influencing mean arterial pressure
Baroreceptor reflex
Hypertension
● A CLOSER LOOK AT EXERCISE PHYSIOLOGY The Ups and
   Downs of Hypertension and Exercise
Hypotension; circulatory shock

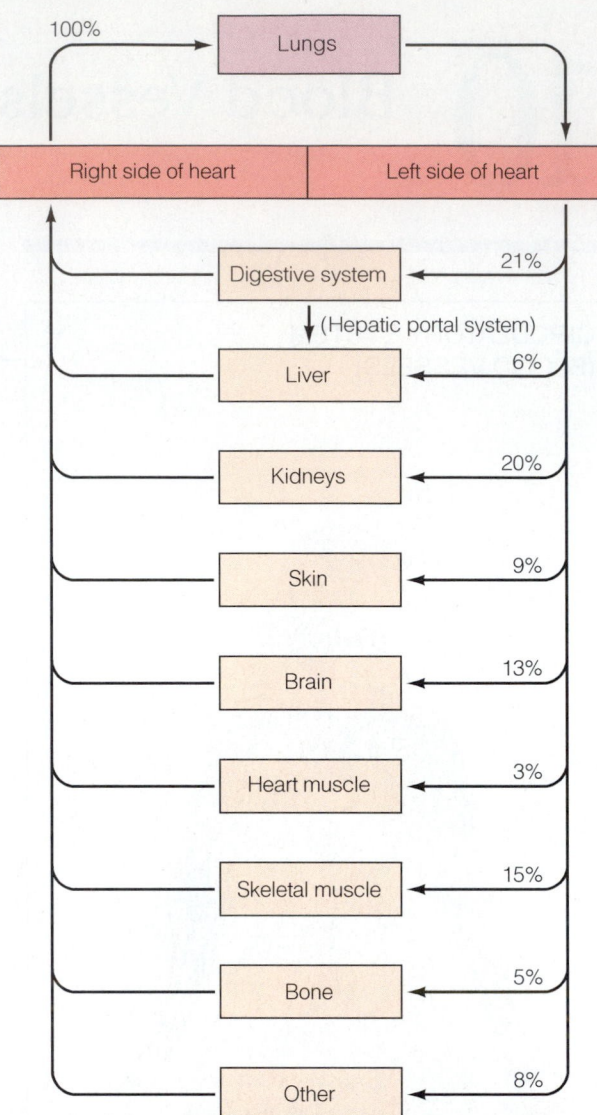

● **FIGURE 10–1   Distribution of Cardiac Output at
Rest** The lungs receive all of the blood pumped out by the right side of
the heart whereas the systemic organs each receive a portion of the
blood pumped out by the left side of the heart. The percentage of
pumped blood received by the various organs under resting conditions
is indicated. This distribution of cardiac output can be adjusted as
needed.

≋🌸 **INTRODUCTION**

**Materials are transported within the body by the
blood as it is pumped through the blood vessels.**

The majority of body cells are not in direct contact with the
external environment, yet these cells must make exchanges
with the environment, such as picking up O_2 and nutrients
and eliminating wastes. Furthermore, chemical messengers
must be transported between cells to accomplish integrated
activity. To achieve these long-distance exchanges, the cells
are linked with each other and with the external environment
by vascular highways. Blood is transported to all parts of the
body through a system of vessels that brings fresh supplies to
the vicinity of all cells while removing their wastes.

All blood pumped by the right side of the heart passes
through the lungs for O_2 pickup and CO_2 removal. The blood
pumped by the left side of the heart is parceled out in various
proportions to the systemic organs through a parallel arrangement of vessels that branch from the aorta (● Fig. 10–1). This
arrangement assures that all organs receive blood of the same
composition; that is, one organ does not receive "leftover"
blood that has passed through another organ. Because of this
parallel arrangement, the flow of blood through each systemic organ can be independently adjusted without directly
influencing blood flow through any other organ.

Blood is constantly "reconditioned" so that its composition remains relatively constant despite an ongoing drain of
supplies to support metabolic activities and the continual addition of wastes from the tissues. The organs that recondition
the blood normally receive substantially more blood than is
necessary to meet their basic metabolic needs so that they
can perform homeostatic adjustments on the blood. Large
percentages of the cardiac output are distributed to the digestive tract (to pick up nutrient supplies), to the kidneys (to
eliminate metabolic wastes and adjust water and electrolyte
composition), and to the skin (to eliminate heat). Blood flow
to the other organs—heart, skeletal muscles, and so on—is
solely for the purpose of supplying these tissues' metabolic
needs and can be adjusted according to their level of activity.
For example, during exercise, additional blood is delivered to

the active muscles to meet their increased metabolic needs.

Because reconditioning organs receive blood flow in excess of their own needs, they can withstand temporary reductions in blood flow much better than can the other organs that do not have this extra margin of blood supply. The brain in particular suffers irreparable damage when transiently deprived of blood supply. Permanent brain damage occurs after only four minutes without O_2. Therefore, a high priority in the overall operation of the circulatory system is the constant delivery of adequate blood to the brain, which can least tolerate a disruption in its blood supply. An adequate blood supply for the brain is even more imperative because this vital organ is not enzymatically equipped to support its metabolic needs anaerobically. In contrast, the digestive organs, kidneys, and skin can tolerate significant reductions in blood flow for a considerable length of time. In fact, they frequently do so. For example, during exercise, some of the blood that normally flows through the digestive organs and kidneys is diverted instead to the skeletal muscles. Likewise, blood flow through the skin is markedly restricted during exposure to cold to conserve body heat. We will see how the distribution of cardiac output is adjusted according to the body's momentary needs as we examine the roles of the various vessels that make up the vascular system.

Blood flow through vessels depends on the pressure gradient and vascular resistance.

The systemic and pulmonary circulations each consist of a closed system of vessels (•Fig. 10–2). (See the accompanying boxed feature, Concepts, Challenges, and Controversies.) **Arteries**, which carry blood from the heart to the tissues, branch into a "tree" of progressively smaller vessels, with the various branches delivering blood to different regions of the

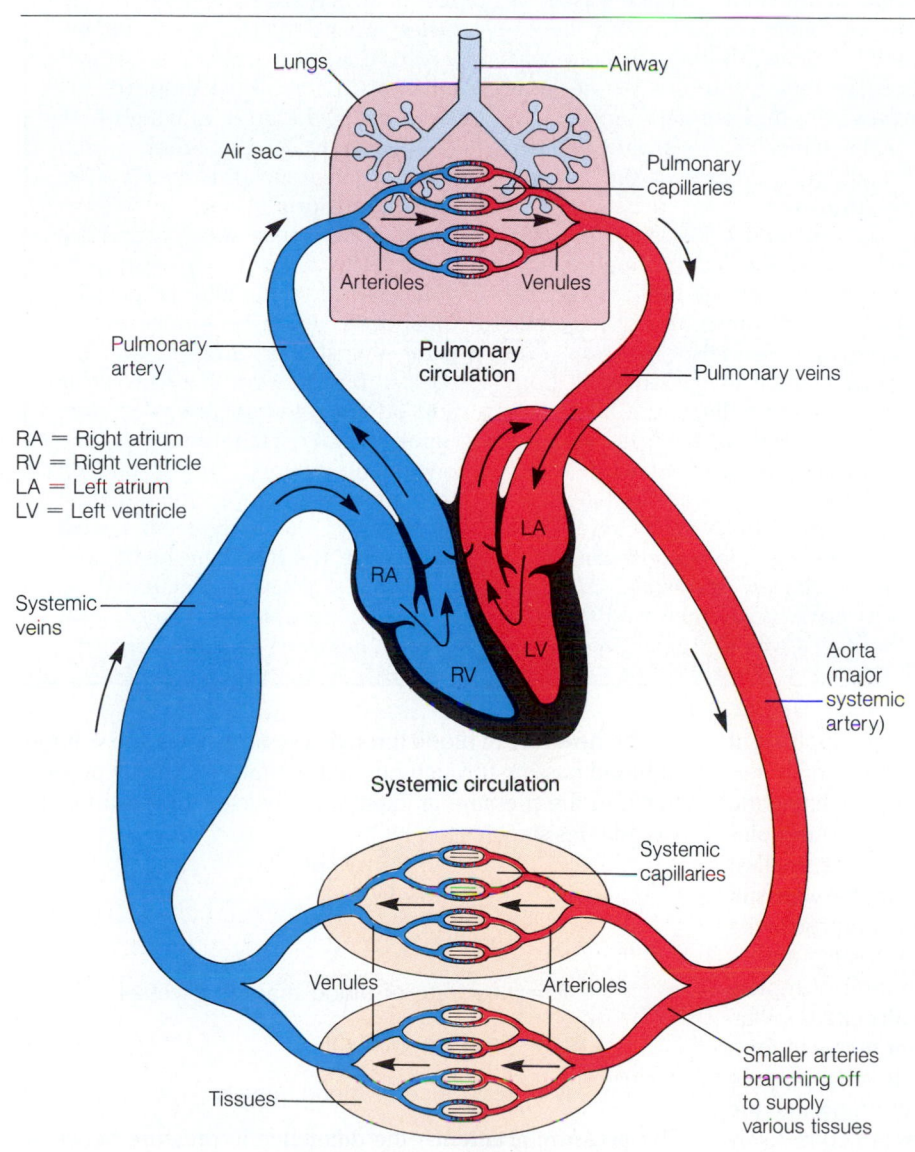

• FIGURE 10–2   Basic Organization of the Cardiovascular System Arteries progressively branch as they carry blood from the heart to the tissues. A separate small arterial branch delivers blood to each of the various organs. As a small artery enters the organ it is supplying, it branches into arterioles, which further branch into an extensive network of capillaries. The capillaries rejoin to form venules, which further unite to form small veins that leave the organ. The small veins progressively merge as they carry blood back to the heart.

RA = Right atrium
RV = Right ventricle
LA = Left atrium
LV = Left ventricle

For simplicity, only two capillary beds within two organs are illustrated.

FROM HUMORS TO HARVEY:
HISTORICAL HIGHLIGHTS IN CIRCULATION

Even grade school children today know that blood is pumped by the heart and continually circulates throughout the body in a system of blood vessels. Furthermore, it is accepted without question that the blood picks up O_2 in the lungs from the air we breathe and delivers it to the body tissues. This common knowledge was unknown for the vast majority of human history, however. Even though the function of blood was described by the intellects of the time as early as the fifth century B.C., our modern-day concept of circulation was not conceived until over 2,000 years later in 1628 when William Harvey published his now classical study, *De Motu Cordis et Sanquinis in Animalibus* ("On the Motion of the Heart and of Blood in Animals").

The ancient Greeks believed that everything material in the universe was composed of just four elements: earth, air, fire, and water. Extending this view to the human body, they thought that these four elements were present in the form of four "humors": *black bile* (representing "earth"), *blood* (representing "air"), *yellow bile* (representing "fire"), and *phlegm* (representing "water"). According to the Greeks, disease resulted when one of these

humors was out of normal balance with the rest. The "cure" for the disease was logical—drain off whichever humor was in excess to restore normal balance. Since the easiest humor to drain off was the blood, bloodletting became the standard procedure for the treatment of many illnesses—a practice that persisted well into the Renaissance.

Although the ancient Greek notion of the four humors was erroneous, their concept of the necessity of balance within the body was remarkably accurate. As we now know, life depends on homeostasis, maintenance of the proper balance of all the elements of the internal environment.

Aristotle (384–322 B.C.), a biologist as well as a philosopher, was among the first to rightly describe the heart at the center of a system of blood vessels. However, he thought that the heart was both the seat of intellect (the brain was not identified as the seat of intellect until over a century later) and a furnace that heated the blood. He considered this warmth to be the vital force of life because the body cools quickly at death. Aristotle also erroneously theorized that the "furnace" was ventilated by breath-

ing, with air serving as a cooling agent. Aristotle could grossly observe the arteries and veins in cadavers but did not have the microscopic capabilities to observe capillaries. Thus, he did not think there was a direct connection between arteries and veins.

In the third century B.C., Erasistratus, a Greek who many consider the first "physiologist," proposed that the liver used food to make blood, which was delivered to the other organs by the veins. He believed that the arteries contained air, not blood. According to his view, *pneuma* ("air"), a living force, was taken in by the lungs, which transferred it to the heart. The heart transformed the air into a "vital spirit" that was carried by the arteries to the other organs.

Galen (A.D. 130–206), a prolific, outspoken, dogmatic Roman physician, philosopher, and scholar, expanded on the work of Erasistratus and others who had preceded him. Galen further elaborated on the pneumatic theory. He proposed that there were three fundamental members in the body, from lowest to highest, the liver, the heart, and the brain. Each was dominated by a special *pneuma* or "spirit." (In Greek, *pneuma* encompassed the

---

body. When a small artery reaches the organ it is supplying, it branches into numerous **arterioles**. The volume of blood flowing through an organ can be adjusted by regulating the caliber (internal diameter) of the organ's arterioles. Arterioles branch further within the organs into **capillaries**, the smallest of vessels, across which all exchanges are made with surrounding cells. Capillary exchange is the entire purpose of the circulatory system; all other activities of the system are directed toward assuring an adequate distribution of replenished blood to capillaries for exchange with all cells. Capillaries rejoin to form small **venules**, which further merge to form small **veins** that leave the organs. The small veins progressively unite to form larger veins that eventually empty into the heart. The arterioles, capillaries, and venules are collectively referred to as the **microcirculation** because they are only visible microscopically.

The **flow rate** of blood through a vessel (that is, the volume of blood passing through per unit of time) is directly proportional to the pressure gradient and inversely proportional to vascular resistance:

$$F = \frac{\Delta P}{R}$$

where

F = flow rate of blood through a vessel

ΔP = pressure gradient

R = resistance of blood vessels

The **pressure gradient**—the difference in pressure between the beginning and end of a vessel—is the main driving force for flow through the vessel; that is, blood flows from an area

FROM HUMORS TO HARVEY:
HISTORICAL HIGHLIGHTS IN CIRCULATION (continued)

related ideas of "air," "breath," and "spirit.") Like Erasistratus, Galen believed that blood was made in the liver from food, taking on a "natural" or "physical" spirit (*pneuma physicon*) in the process. The newly formed blood then proceeded through the veins to the organs. The natural spirit, which Galen considered to be a vapor rising from the blood, controlled the functions of nutrition, growth, and reproduction. Once its spirit supply was depleted, the blood moved in the opposite direction through the same venous pathways, returning to the liver to be replenished. When the natural spirit was carried in the venous blood to the heart, it was mixed with air that was breathed in and transferred from the lungs to the heart. Contact with air in the heart transformed the natural spirit into a higher-level spirit, the "vital" spirit (*pneuma zotikon*). The vital spirit, which was carried by the arteries, conveyed heat and life throughout the body. The vital spirit was transformed further into a yet higher "animal" or "psychical" spirit (*pneuma psychicon*) in the brain. This ultimate spirit regulated the brain, nerves, feelings, and so on. Thus, according to Galenic theory, the veins and arteries were conduits for carrying different levels of pneuma, and there was no direct connection between the veins and arteries. The heart was not implicated in the movement of blood but instead was the site where blood and air were mixed. (We now know that blood and air meet in the lungs for the exchange of O_2 and CO_2.)

Galen was one of the first to understand the need for experimentation, but, unfortunately, his impatience and his craving for philosophical and literary fame led him to expound comprehensive theories that were not always based upon the time-consuming collection of evidence. Even though his assumptions about bodily structure and functions often were incorrect, his theories were convincing because they seemed to be a logical way of pulling together what was known at the time. Furthermore, the sheer quantity of his writings helped establish him as an authority. In fact, his writings remained the anatomical and physiological "truth" for nearly fifteen centuries, enduring throughout the Middle Ages and well into the Renaissance. So firmly entrenched was Galenic doctrine that those who challenged its accuracy risked their lives by being declared secular heretics.

Not until the Renaissance and the revival of classical learning did Galen's theories begin to be challenged by independent-minded European investigators. Most notably, the English physician William Harvey (1578–1657) revolutionized the view of the role of the heart, the blood vessels, and the blood. Through careful observations, experimentation, and deductive reasoning, Harvey was the first to correctly identify the heart as a pump that repeatedly moves a small volume of blood forward in one fixed direction in a circular path through a closed system of blood vessels (the *circulatory system*). He also correctly proposed that blood travels to the lungs to mix with air instead of air traveling to the heart to mix with blood. Even though he could not see physical connections between arteries and veins, he speculated their existence. Not until the discovery of the microscope later in the century was the existence of these connections, the capillaries, confirmed by Marcello Malpighi (1628–1694).

---

of higher pressure to an area of lower pressure down a pressure gradient. Contraction of the heart imparts pressure to the blood, but because of frictional losses (resistance), the pressure decreases as blood flows through a vessel. Since pressure drops throughout the vessel's length, it is higher at the beginning than at the end. This establishes a pressure gradient for forward flow of blood through the vessel. The greater the pressure gradient forcing blood through a vessel, the greater the rate of flow through that vessel (• Fig. 10–3a). Think of a garden hose attached to a faucet. If you turn on the faucet slightly, a small stream of water will flow out of the end of the hose because the pressure is slightly greater at the beginning than at the end of the hose. If you open the faucet all the way, the pressure gradient will increase tremendously so that the rate of water flow through the hose will be much greater, and water will spurt forth from the end of the hose.

Note that the *difference* in pressure between the two ends of a vessel, not the absolute pressures within the vessel, determines flow rate (Fig. 10–3b).

The other factor influencing flow rate through a vessel is the **resistance**, which is a measure of the hindrance to blood flow through a vessel caused by friction between the moving fluid and the stationary vascular walls. As resistance to flow increases, it is more difficult for blood to pass through the vessel, so flow decreases (as long as the pressure gradient remains unchanged). When resistance increases, the pressure gradient must increase correspondingly to maintain the same flow rate. Accordingly, when the vessels offer more resistance to flow, the heart must work harder to maintain adequate circulation.

Resistance to blood flow depends on three factors: (1) viscosity of the blood; (2) vessel length; and (3) vessel radius,

• FIGURE 10–3   Relationship of Flow to Pressure Gradient in a Vessel (a) As the difference in pressure (ΔP) between the two ends of a vessel increases, the flow rate increases proportionately. (b) Flow rate is determined by the difference in pressure between the two ends of a vessel, not the absolute pressures.

length remains constant in the body, it is not a variable factor in the control of vascular resistance. Therefore, the major determinant of resistance to flow is the vessel's radius. Fluid passes more readily through a large vessel than through a smaller vessel, because a given volume of blood comes into contact with much more of the surface area of a small-radius vessel than of a larger-radius one, resulting in greater resistance (• Fig. 10–4a). Furthermore, a slight change in the radius of a vessel brings about a notable change in flow, because the resistance is inversely proportional to the fourth power of the radius (multiplying the radius by itself four times):

$$R \propto \frac{1}{r^4}$$

Thus, doubling the radius (Fig. 10–4b) decreases the resistance sixteen times ($r^4 = 2 \times 2 \times 2 \times 2 = 16$; $R \propto 1/16$) and therefore increases flow through the vessel sixteenfold (at the same pressure gradient). The converse is also true. Only one-sixteenth as much blood flows through a vessel at the same driving pressure when its radius is halved. It is important to note that the radius of arterioles is subject to regulation and is the most important factor in the control of resistance to blood flow throughout the vascular tree.

• FIGURE 10–4   Relationship of Flow and Resistance to Vessel Radius (a) The same volume of blood comes into contact with a greater surface area of a small-radius vessel compared to a larger-radius vessel. Accordingly, the smaller-radius vessel offers more resistance to blood flow, because the blood "rubs" against a larger surface area. (b) Doubling the radius decreases the resistance to 1/16 and increases the flow 16 times, because the resistance is inversely proportional to the fourth power of the radius.

(a)

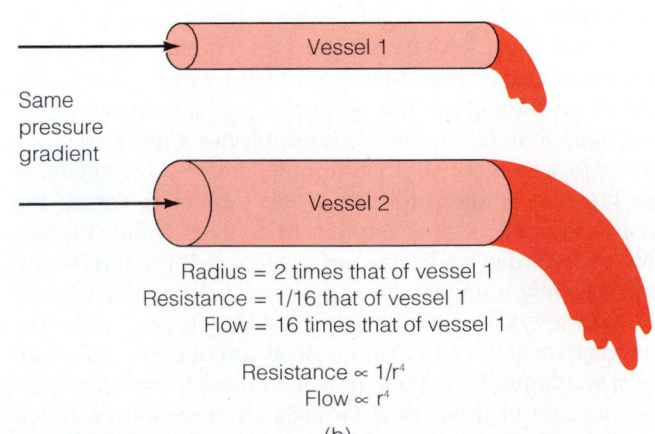

(b)

which is by far the most important. **Viscosity** refers to the friction developed between the molecules of a fluid as they slide over each other during flow of the fluid. The greater the viscosity, the greater the resistance to flow. In general, the thicker a liquid, the more viscous it is. For example, molasses flows more slowly than water because molasses has greater viscosity. Viscosity of blood is determined by two factors: the concentration of plasma proteins and, more importantly, the number of circulating red blood cells. Normally, these two factors are relatively constant and are not important in the control of resistance. Occasionally, however, blood viscosity and, accordingly, resistance to flow are altered because of an abnormal number of red blood cells. For example, blood flow is more sluggish than normal when excessive red blood cells are present.

Because blood "rubs" against the lining of the vessels as it flows past, the greater the vessel surface area in contact with the blood, the greater the resistance to flow. Surface area is determined by both the length (L) and radius (r) of the vessel. At a constant radius, the longer the vessel, the greater the surface area and the greater the resistance to flow. Since vessel

The factors that affect the flow rate through a vessel are integrated in an idealized way in **Poiseuille's law** as follows:

$$\text{flow rate} = \frac{\pi \Delta P \, r^4}{8 \eta \, L}$$

where

$$\eta = \text{viscosity}$$

The significance of the relation among flow, pressure, and resistance, as largely determined by vessel radius, will become even more apparent as we embark on a voyage through the vessels in the next section.

ARTERIES

Arteries serve as rapid-transit passageways to the tissues and as a pressure reservoir.

The consecutive segments of the vascular tree—arteries, arterioles, capillaries, and veins—are specialized to perform specific tasks (Table 10–1). Arteries are specialized to serve as rapid-transit passageways for blood from the heart to the tissues (because of their large radii, arteries offer little resistance to blood flow) and to act as a *pressure reservoir* to provide the driving force for blood when the heart is relaxing.

The heart alternately contracts to pump blood into the arteries and then relaxes to refill from the veins. No blood is pumped out when the heart is relaxing and refilling. However, capillary flow does not fluctuate between cardiac systole and diastole; blood flow is continuous through the capillaries supplying the tissues. The driving force for the continued flow of blood to the tissues during cardiac relaxation

is provided by the elastic properties of the arterial walls. All vessels are lined with a layer of smooth, flattened endothelial cells that are continuous with the endocardial lining of the heart. Surrounding the arteries' endothelial lining is a thick wall containing smooth muscle and an abundance of two types of connective-tissue fibers: collagen fibers, which provide tensile strength against the high driving pressure of blood ejected from the heart, and elastin fibers, which give the arterial walls elasticity so that they behave much like a balloon (●Fig. 10–5).

As the heart pumps blood into the arteries during ventricular systole, a greater volume of blood enters the arteries from the heart than leaves them to flow into the smaller vessels downstream, because the smaller vessels have a greater resistance to flow. The arteries' elasticity enables them to expand to temporarily hold this excess volume of ejected blood, storing some of the pressure energy imparted by cardiac contraction in their stretched walls—just as a balloon expands to accommodate the extra volume of air that you blow into it (●Fig. 10–6a). When the heart relaxes and ceases pumping blood into the arteries, the stretched arterial walls passively recoil, like an inflated balloon that is released. This recoil pushes the excess blood contained in the arteries into the vessels downstream, ensuring continued blood flow to the tissues when the heart is relaxing and not pumping blood into the system (Fig. 10–6b).

Arterial pressure fluctuates in relation to ventricular systole and diastole.

Blood pressure, the force exerted by the blood against a vessel wall, depends on the volume of blood contained within the vessel and the **compliance**, or **distensibility**, of the ves-

		TABLE 10–1	Features of Blood Vessels			
			VESSEL TYPE			
FEATURE	Aorta	Large Arterial Branches	Arterioles	Capillaries	Large Veins	Vena Cava
Number	One	Several hundred	Half a million	Ten billion	Several hundred	Two
Wall thickness	2 mm (2,000 μm)	1 mm (1,000 μm)	20 μm	1 μm	0.5 mm (500 μm)	1.5 mm (1,500 μm)
Internal radius	1.25 cm (12,500 μm)	0.2 cm (2,000 μm)	30 μm	3.5 μm	0.5 cm (5,000 μm)	3 cm (30,000 μm)
Total cross-sectional area	4.5 cm²	20 cm²	400 cm²	6,000 cm²	40 cm²	18 cm²
Special features	Thick, highly elastic, walls; large radii		Highly muscular, well-innervated walls; small radii	Thin walled; large total cross-sectional area	Thin walled; highly distensible; large radii	
Functions	Passageway from heart to tissues; serve as pressure reservoir		Primary resistance vessels; determine distribution of cardiac output	Site of exchange; determine distribution of extracellular fluid between plasma and interstitial fluid	Passageway to heart from tissues; serve as blood reservoir	

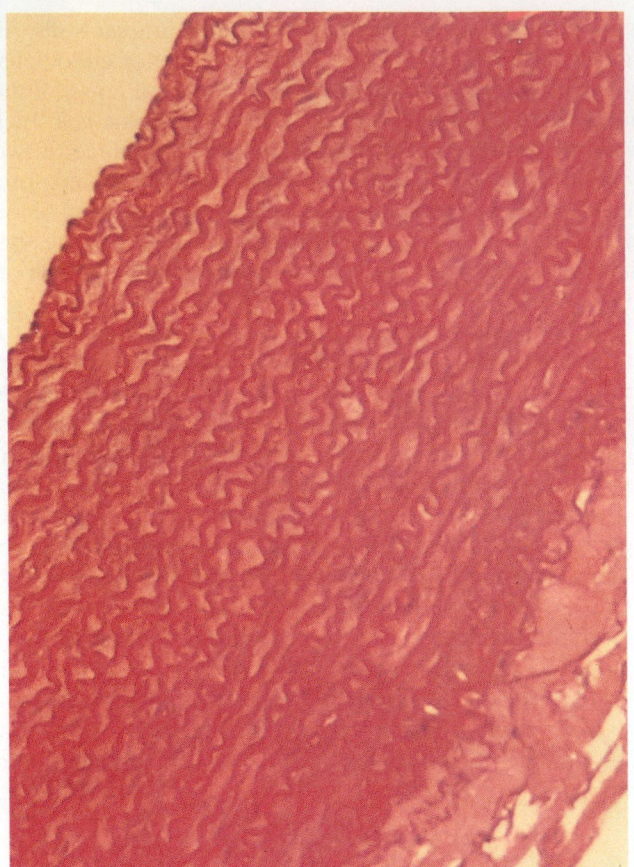

• FIGURE 10–5   Elastin Fibers in Artery Light micrograph of a portion of the wall of the aorta in cross section, showing numerous wavy elastin fibers common to all arteries.

sel walls (how easily they can be stretched). If the volume of blood that enters the arteries were equal to the volume of blood that left the arteries during the same period, arterial blood pressure would remain constant. This is not the case, however. During ventricular systole, a stroke volume of blood enters the arteries from the ventricle while only about one-third as much blood leaves the arteries to enter the arterioles. During diastole, no blood enters the arteries, while blood continues to leave, driven by elastic recoil. The maximum pressure exerted in the arteries when blood is ejected into them during systole, the **systolic pressure**, averages 120 mm Hg. The minimum pressure within the arteries when blood is draining off into the remainder of the vessels during diastole, the **diastolic pressure**, averages 80 mm Hg. The arterial pressure does not fall to 0 mm Hg, because the next cardiac contraction occurs and refills the arteries before all of the blood drains off (• Fig. 10–7; also see Fig. 9–20).

Blood pressure can be indirectly measured by using a sphygmomanometer.

The changes in arterial pressure throughout the cardiac cycle can be measured directly by connecting a pressure-measuring device to a needle inserted in an artery. However, it is more convenient and reasonably accurate to measure the pressure indirectly through use of a **sphygmomanometer**, an externally applied inflatable cuff attached to a pressure gauge. When the cuff is wrapped around the upper arm and then inflated with air, the pressure of the cuff is transmitted through the tissues to the underlying brachial artery, the main vessel carrying blood to the forearm (• Fig. 10–8). The technique involves balancing the pressure in the cuff against the

• FIGURE 10–6   Arteries as Pressure Reservoir Because of their elasticity, arteries act as a pressure reservoir. (a) The elastic arteries distend during cardiac systole as more blood is ejected into them than drains off into the narrow, high-resistance arterioles downstream. (b) The elastic recoil of arteries during cardiac diastole continues driving the blood forward when the heart is not pumping.

(a)

(b)

● FIGURE 10-7 Arterial Blood Pressure The systolic pressure is the peak pressure exerted in the arteries when blood is pumped into them during ventricular systole. The diastolic pressure is the lowest pressure exerted in the arteries when blood is draining off into the vessels downstream during ventricular diastole. The pulse pressure is the difference between systolic and diastolic pressure. The mean pressure is the average pressure throughout the cardiac cycle; it equals diastolic pressure + (1/3 × pulse pressure).

● FIGURE 10-8 Sphygmomanometry (a) Use of sphygmomanometer in determining blood pressure. The pressure in the inflatable cuff can be varied to prevent or permit blood flow in the underlying brachial artery. Turbulent blood flow can be detected through use of a stethoscope whereas smooth laminar flow and no flow are inaudible. (b) Pattern of sounds in relation to cuff pressure compared with blood pressure. The numbers on the illustration refer to the following key points during a blood-pressure determination: ① Cuff pressure exceeds blood pressure throughout cardiac cycle. No sound heard. ② First sound heard at peak systolic pressure. ③ Intermittent sounds heard as blood pressure cyclically exceeds cuff pressure. ④ Last sound heard at minimum diastolic pressure. ⑤ Blood pressure exceeds cuff pressure throughout cardiac cycle. No sound heard. (c) Blood flow through brachial artery in relation to cuff pressure and sounds.

(a)

Pressure-recording device

Inflatable cuff

Stethoscope

(b)

Cuff pressure   Blood pressure

(c) When blood pressure is 120/80:

Cuff pressure is greater than 120 mm Hg.

No blood flows through vessel.

No sound is heard.

Cuff pressure is between 120 and 80 mm Hg.

Blood flow through vessel is turbulent whenever blood pressure exceeds cuff pressure.

Intermittent sounds are heard as blood pressure fluctuates throughout cardiac cycle.

Cuff pressure is less than 80 mm Hg.

Blood flows through vessel in smooth, laminar fashion.

No sound is heard.

pressure in the artery. When cuff pressure is greater than the pressure in the vessel, the vessel is pinched closed so that no blood flows through it. When blood pressure is greater than cuff pressure, the vessel is open and blood flows through.

During the determination of blood pressure, a stethoscope is placed over the brachial artery at the inside bend of the elbow just below the cuff. No sound can be detected either when blood is not flowing through the vessel or when blood is flowing in the normal, smooth laminar flow (see p. 281). Turbulent blood flow, on the other hand, creates vibrations that can be heard. At the onset of a blood-pressure determination, the cuff is inflated to a pressure greater than systolic blood pressure so that the brachial artery collapses. Because the externally applied pressure is greater than the peak internal pressure, the artery remains completely pinched closed throughout the entire cardiac cycle; no sound can be heard, since no blood is passing through (Point 1 in Fig. 10–8b). As air in the cuff is slowly released, the pressure in the cuff is gradually reduced. When the cuff pressure falls to just below the peak systolic pressure, the artery transiently opens slightly when the blood pressure reaches this peak. Blood escapes through the partially occluded artery for a brief interval before the arterial pressure falls below the cuff pressure and the artery collapses once again. This spurt of blood is turbulent, so it can be heard. Thus, the highest cuff pressure at which the *first sound* can be heard is indicative of the *systolic blood pressure* (Point 2). As the cuff pressure continues to fall, blood intermittently spurts through the artery and produces a sound with each subsequent cardiac cycle whenever the arterial pressure exceeds the cuff pressure (Point 3). When the cuff pressure finally falls below diastolic pressure, the brachial artery is no longer pinched closed during any part of the cardiac cycle, and blood can flow uninterrupted through the vessel (Point 5). With the return of nonturbulent blood flow, no further sounds can be heard. Therefore, the highest cuff pressure at which the *last sound* can be detected is indicative of the *diastolic pressure* (Point 4). In clinical practice, arterial blood pressure is expressed as systolic pressure over diastolic pressure, with the average blood pressure being 120/80 (one hundred twenty over eighty) mm Hg.

The pulse that can be felt in an artery lying close to the surface of the skin is due to the difference between systolic and diastolic pressures. This pressure difference is known as the **pulse pressure**. When the blood pressure is 120/80, pulse pressure is 40 mm Hg (120 mm Hg − 80 mm Hg).

Mean arterial pressure is the main driving force for blood flow.

More important than the fluctuating systolic and diastolic pressures or the pulse pressure is the **mean arterial pressure**, which is the *average pressure* responsible for driving blood forward into the tissues throughout the cardiac cycle. Contrary to what you might expect, mean arterial pressure is not the halfway value between systolic and diastolic pressure (for example, with a blood pressure of 120/80, mean pressure is not 100 mm Hg), because arterial pressure remains closer to diastolic than to systolic pressure for a longer portion of each cardiac cycle. At resting heart rate, about two-thirds of the cardiac cycle is spent in diastole and only one-third in systole. As an analogy, if a race car traveled 80 miles per hour (mph) for 40 minutes and 120 mph for 20 minutes, its average speed would be 93 mph, not the halfway value of 100 mph. Similarly, a good approximation of the mean arterial pressure can be determined using the following formula:

$$\text{mean arterial pressure} =$$
$$\text{diastolic pressure} + (1/3 \times \text{pulse pressure})$$

$$\text{at 120/80, mean arterial pressure} =$$
$$80 \text{ mm Hg} + (1/3 \times 40 \text{ mm Hg}) = 93 \text{ mm Hg}$$

It is this mean arterial pressure, not the systolic or diastolic pressures, that is monitored and regulated by blood-pressure reflexes to be described later in the chapter.

Arterial pressure—whether it be systolic, diastolic, pulse, or mean—is essentially the same throughout the arterial tree. Because arteries offer little resistance to flow, only negligible frictional loss of pressure energy takes place in them.

ARTERIOLES

Arterioles are the major resistance vessels.

When an artery reaches the organ it is supplying, it branches into numerous arterioles, whose radii are small enough to offer considerable resistance to flow. In fact, the arterioles are the major resistance vessels in the vascular tree. (Even though the capillaries have smaller radii than the arterioles, we will see later how collectively the capillaries do not offer as much resistance to flow as the arteriolar level of the vascular tree does.) In contrast to the low resistance of the arteries, the high degree of arteriolar resistance causes a marked drop in mean pressure as the blood flows through these vessels. On average, the pressure falls from 93 mm Hg, the mean arterial pressure, to 37 mm Hg, the pressure at the beginning of the capillaries (• Fig. 10–9). This decline in pressure helps establish the pressure differential that encourages the flow of blood from the heart to the various organs downstream. Arteriolar resistance is also responsible for converting the pulsatile systolic-to-diastolic pressure swings in the arteries into the nonfluctuating pressure present in the capillaries.

The radii (and, accordingly, the resistances) of arterioles supplying individual organs can be adjusted independently to determine the distribution of cardiac output and to regulate arterial blood pressure. We will discuss the mechanisms involved in adjusting arteriolar resistance before considering how such adjustments are important in accomplishing these two functions.

Unlike arteries, arteriolar walls contain very little elastic connective tissue. However, they do have a thick layer of smooth muscle that is richly innervated by sympathetic nerve fibers. The smooth muscle is also sensitive to many local chemical changes and to a few circulating hormones. The smooth-muscle layer runs circularly around the arteriole (• Fig. 10–10a) so that when it contracts, the vessel's circumference (and its radius) becomes smaller, thus increasing re-

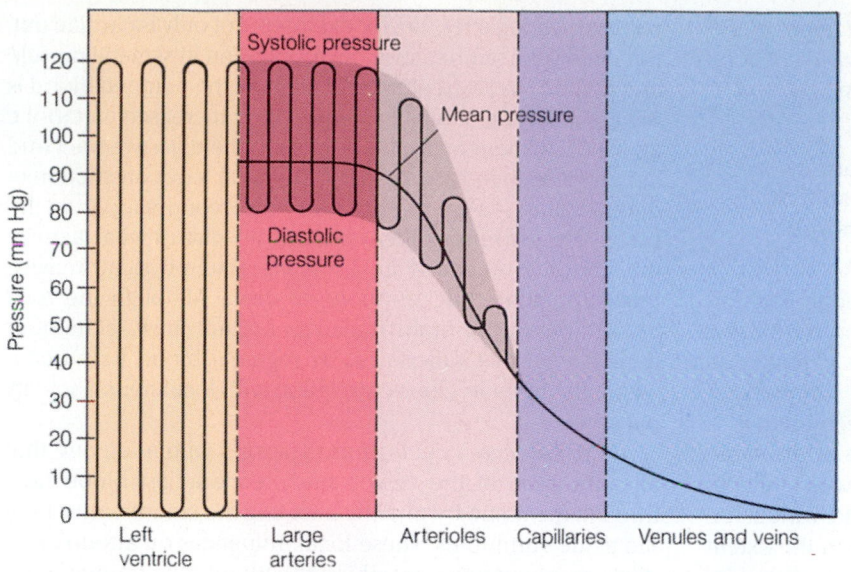

• FIGURE 10–9 **Pressures throughout Systemic Circulation**
• FIGURE 10–9 **Pressures throughout Systemic Circulation** Arterial blood pressure, which fluctuates between peak systolic pressure and low diastolic pressure each cardiac cycle, is of the same magnitude throughout the large arteries. Because of the arterioles' high resistance, the pressure drops precipitously and the systolic-to-diastolic swings in pressure are converted to a nonpulsatile pressure when blood flows through the arterioles. The pressure continues to decline but at a slower rate as blood flows through the capillaries and venous system.

sistance and decreasing the flow through that vessel. **Vasoconstriction** is the term applied to such narrowing of a vessel (Fig. 10–10c). **Vasodilation** refers to enlargement in the circumference and radius of a vessel as a result of relaxation of its smooth-muscle layer (Fig. 10–10d). Vasodilation leads to decreased resistance and increased flow through that vessel.

Arteriolar smooth muscle normally displays a state of partial constriction known as **vascular tone**, which establishes a baseline of arteriolar resistance (Fig. 10–10b). Two factors are responsible for vascular tone. First, arteriolar smooth muscle has considerable myogenic activity; that is, its membrane potential fluctuates without any external influences, leading to self-induced contractile activity (see p. 252). Sec-

ond, the sympathetic fibers supplying most arterioles continually release norepinephrine, which further enhances the vascular tone.

This ongoing tonic activity makes it possible either to increase or decrease the level of contractile activity to accomplish vasoconstriction or vasodilation, respectively. Were it not for tone, it would be impossible to reduce the tension in an arteriolar wall to accomplish vasodilation; only varying degrees of vasoconstriction would be possible.

A variety of factors can influence the level of contractile activity in arteriolar smooth muscle, thereby substantially changing resistance to flow in these vessels. These factors fall into two categories: *local (intrinsic) controls*, which are im-

(a)   Smooth muscle cells

• FIGURE 10–10 **Arteriolar Vasoconstriction and Vasodilation** (a) A scanning electron micrograph of an arteriole showing how the smooth-muscle cells run circularly around the vessel wall. (b) Schematic representation of an arteriole in cross section showing normal arteriolar tone. (c) Outcome of and factors causing arteriolar vasoconstriction. (d) Outcome of and factors causing arteriolar vasodilation.

Normal arteriolar tone

Cross section
of arteriole

(b)

Vasoconstriction
(increased contraction
of circular smooth
muscle in arteriolar
wall, which leads to
increased resistance
and decreased
flow through the vessel)

(c)

↑ Myogenic activity
↑ Oxygen (O_2)
↓ Carbon dioxide (CO_2)
  and other metabolites
↑ Sympathetic stimulation
  Vasopressin; angiotensin II
  Cold

Vasodilation
(decreased contraction
of circular smooth
muscle in arteriolar
wall, which leads to
decreased resistance
and increased flow
through the vessel)

(d)

↓ Myogenic activity
↓ O_2
↑ CO_2 and other metabolites
↓ Sympathetic stimulation
  Histamine release
  Heat

portant in matching blood flow to the metabolic needs of the specific tissues in which they occur, and *extrinsic controls,* which are important in blood-pressure regulation.

Local control of arteriolar radius is important in determining distribution of cardiac output so that blood flow is matched with the tissues' metabolic needs.

The fraction of the total cardiac output delivered to each organ is not always constant; it varies, depending on the demands for blood at the time. The amount of the cardiac output received by each organ is determined by the number and caliber of the arterioles supplying that area. Recall that $F = \Delta P/R$. Because blood is delivered to all tissues at the same mean arterial pressure, the driving force for flow is identical for each organ. Therefore, differences in flow to various organs are completely determined by differences in the extent of vascularization and by differences in resistance offered by the arterioles supplying each organ. On a moment-to-moment basis, the distribution of cardiac output can be varied by differentially adjusting arteriolar resistance in the various vascular beds.

As an analogy, consider a pipe carrying water with a number of adjustable valves located throughout its length (● Fig. 10–11). Assuming water pressure in the pipe is constant, differences in the amount of water flowing into a beaker under each valve depend entirely on which valves are open and the extent to which they are open. No water enters beakers under closed valves (high resistance), and more water flows into beakers under valves that are opened completely (low resistance) than into beakers under valves that are only partially opened (moderate resistance). Similarly, more blood flows to areas whose arterioles offer the least resistance to its

passage. During exercise, for example, not only is cardiac output increased, but because of vasodilation in skeletal muscle and in the heart, a greater percentage of the pumped blood is diverted to these organs to support their increased metabolic activity. Simultaneously, blood flow to the digestive tract and kidneys is reduced as a result of arteriolar vasoconstriction in these organs (● Fig. 10–12). Only the blood supply to the brain remains remarkably constant no matter what activity the person is engaged in, be it vigorous physical activity, intense mental concentration, or sleep. Although the *total* blood flow to the brain remains constant, new techniques demonstrate that differences in regional blood flow occur within the brain in close correlation with local neural-activity patterns.

Local (intrinsic) controls are changes within a tissue that alter the radii of the vessels and hence adjust blood flow through the tissue by directly affecting the smooth muscle of the tissue's arterioles. These local influences on arteriolar radius are important in matching the blood flow through a tissue with the tissue's metabolic needs. Local controls are especially important in skeletal muscle and in the heart, the tissues whose metabolic activity and need for blood supply normally vary most extensively, and in the brain, whose overall metabolic activity and need for blood supply remain constant. Local controls help maintain the constancy of blood flow to the brain. Local influences may be either chemical or physical in nature.

Local Chemical Influences

1. *Local metabolic changes.* Arterioles lie within the tissue they are supplying and can be acted on by local factors within the tissue. During increased metabolic activity, such as when a skeletal muscle is contracting during exercise, the local concentrations of a number of the tissue's

● **FIGURE 10–11 Flow Rate as a Function of Resistance**

Constant pressure in pipe (mean arterial pressure)

From pump (heart)

High resistance   Moderate resistance   Low resistance

No flow   Moderate flow   Large flow

Valves = Arterioles

Moderate exercise

Percent change in flow		Percent cardiac output	Flow (ml/min)
↓ 56%	Digestive tract, liver	4.8%	600
↓ 45%	Kidneys	4.4%	550
↑ 370%	Skin	13.6%	1,700
No change	Brain	5.2%	650
↑ 367%	Heart	4.4%	550
↑ 1066%	Skeletal muscle	64%	8,000
↓ 30%	Bone, other	3.6%	450
		Total cardiac output	12,500

At rest

Percent cardiac output	Flow (ml/min)	
27%	1,350	Digestive tract, liver
20%	1,000	Kidneys
9%	450	Skin
13%	650	Brain
3%	150	Heart
15%	750	Skeletal muscle
13%	650	Bone, other
Total cardiac output	5,000	

● **FIGURE 10–12   Magnitude and Distribution of Cardiac Output at Rest and during Moderate Exercise** Not only does cardiac output increase during exercise but the distribution of cardiac output is adjusted to support the heightened physical activity. The percentage of cardiac output going to the skeletal muscles and heart rises, thereby delivering extra O_2 and nutrients needed to support these muscles' stepped-up rate of ATP consumption. The percentage of cardiac output going to the skin is increased as a means of eliminating from the body surface the extra heat generated by the exercising muscles. The large percentage increases to the skeletal muscles and heart are offset by reductions in the percentage of cardiac output going to the other organs. Only the magnitude of blood flow to the brain remains unchanged as the distribution of cardiac output is readjusted during exercise.

chemicals change. For example, the local O_2 concentration decreases as the actively metabolizing cells use up more O_2 to support oxidative phosphorylation for ATP production. This and other local chemical changes produce local arteriolar dilation by triggering relaxation of the arteriolar smooth muscle without involving nerves or hormones. Local arteriolar vasodilation subsequently increases blood flow to that particular area, a response called **active hyperemia**. When cells are more active metabolically, they need more blood to bring in O_2 and nutrients and to remove metabolic wastes. The increased blood flow meets these increased local needs.

Conversely, when a tissue such as a relaxed muscle is less active metabolically and thus has reduced needs for blood delivery, the resultant local chemical changes (for example, increased local O_2 concentration) bring about local arteriolar vasoconstriction and a subsequent reduction in blood flow to the area.

Exactly which local chemical changes are responsible for these "selfish" local adjustments in arteriolar caliber that match a tissue's blood flow with its metabolic needs is still uncertain. The following local chemical factors have been shown experimentally to produce relaxation of arteriolar smooth muscles:

- *Decreased O_2*
- *Increased CO_2*—More CO_2 is generated as a by-product during the stepped-up pace of oxidative phosphorylation that accompanies increased activity.
- *Increased acid*—More carbonic acid is generated from the increased CO_2 produced as the metabolic activity of a cell increases. Also, lactic acid accumulates if the glycolytic pathway is used for ATP production.
- *Increased K^+*—Repeated action potentials that outpace the ability of the Na^+-K^+ pump to restore the resting concentration gradients result in an increase in K^+ in the tissue fluid of an actively contracting muscle or a more active region of the brain.
- *Increased osmolarity*—Osmolarity (the concentration of osmotically active solutes) increases during elevated cellular metabolism because of increased formation of osmotically active particles.
- *Adenosine release*—Especially in cardiac muscle, adenosine is released in response to increased metabolic activity or O_2 deprivation (see p. 290).
- *Prostaglandin release*—Prostaglandins are local chemical messengers derived from fatty acid chains within the plasma membrane. Their production and release is poorly understood.

The relative contributions of these various chemical changes (and possibly others) in the local metabolic control of blood flow in systemic arterioles is still being investigated.

Also under investigation is the mechanism by which local metabolic factors change the contractile state of arteriolar smooth muscle. These factors do not appear to act

TABLE 10–2   Functions of Endothelial Cells

Line blood vessels and heart chambers; serve as physical barrier between blood and remainder of vessel wall

Secrete vasoactive substances, such as endothelial-derived relaxing factor, in response to local chemical and physical changes; these substances cause relaxation (vasodilation) or contraction (vasoconstriction) of underlying smooth muscle

Secrete substances that stimulate new vessel growth and proliferation of smooth-muscle cells in vessel walls

Participate in exchange of materials between blood and surrounding tissues across capillaries through vesicular transport (see p. 67)

Influential in formation of platelet plugs, clotting, and clot dissolution (see chapter 11)

Participate in determination of capillary permeability by contracting to vary the size of the pores between adjacent endothelial cells

directly on vascular smooth muscle as once was thought. Instead, it has become increasingly clear that the **endothelial cells**, the single layer of specialized epithelial cells that line the lumen of all blood vessels, release chemical mediators that play a key role in the local regulation of arteriolar caliber. Until recently, endothelial cells were regarded as little more than a passive barrier between the blood and the remainder of the vessel wall. It is now known that endothelial cells are active participants in a variety of vessel-related activities, some of which will be described elsewhere (Table 10–2). Among these functions, endothelial cells release locally acting chemical messengers in response to chemical changes in their environment (such as a reduction in O_2) or physical changes (such as stretching of the vessel wall). These local chemical mediators act on the underlying smooth muscle to alter its state of contraction. Among the best studied of these local vasoactive ("acting on vessels") mediators is **endothelial-derived relaxing factor** (**EDRF**), which causes local arteriolar vasodilation by inducing relaxation of arteriolar smooth muscle in the vicinity. EDRF has been identified as **nitric oxide** (**NO**), a small, short-lived molecule. Recent studies have revealed an astonishing number of biological roles for NO, which is produced in numerous other tissues besides endothelial cells. In fact, it appears that NO serves as one of the body's most important messenger molecules, as exemplified by the following range of newly discovered functions of this chemical.

- As mentioned, NO (alias EDRF) causes relaxation of arteriolar smooth muscle.
- By dilating the arterioles of the penis, NO is the direct mediator of penile erection. Erection is accomplished by rapid engorgement of the penis with blood.
- Macrophages, large phagocytic cells of the immune sys-

tem, produce NO, which they use as "chemical warfare" against bacteria and cancer cells.

- NO interferes with platelet function and blood clotting at sites of vessel damage.

- NO serves as a novel type of neurotransmitter in the brain and elsewhere. Unlike classical neurotransmitters, NO is not stored in vesicles and released by exocytosis and it does not bind with receptors on its target cell. NO is synthesized on demand in the neuron terminal, then diffuses out of the terminal and into the adjacent target cell, where it brings about its effect.

- NO plays a role in the changes underlying memory (see p. 134).

The endothelial cells release other important chemicals besides EDRF/NO. Other endothelial vasoactive substances, such as **endothelin**, bring about vasoconstriction by causing arteriolar smooth-muscle contraction. Still other chemicals released from the endothelium in response to chronic changes in blood flow to an organ are believed to trigger long-term vascular changes that permanently influence blood flow to a region. Some chemicals, for example, are thought to stimulate *angiogenesis* (new vessel growth).

2. *Histamine release.* Histamine is another local chemical mediator that influences arteriolar smooth muscle, but it is not released in response to local metabolic changes and is not derived from the endothelial cells. Although histamine does not normally participate in the control of blood flow, it is important in certain pathological conditions. Histamine is synthesized and stored within special connective-tissue cells in many tissues and in certain types of circulating white blood cells. When tissues are injured or during allergic reactions, histamine is released in the damaged region. By promoting relaxation of arteriolar smooth muscle, histamine is the major cause of vasodilation in an injured area. The resultant increase in blood flow into the area is responsible for the redness and contributes to the swelling associated with inflammatory responses.

Local Physical Influences

1. *Heat or cold.* Heat application is a useful therapeutic agent for promoting increased blood flow to an area, because it causes localized arteriolar vasodilation. Conversely, applying ice packs to an inflamed area produces vasoconstriction, which reduces swelling by counteracting histamine-induced vasodilation.

2. *Myogenic responses to stretch.* Evidence indicates that arteriolar smooth muscle responds to being passively stretched by myogenically increasing its tone, thereby acting to resist the initial passive stretch. Conversely, a decrease in arteriolar stretching is accompanied by a reduction in myogenic vessel tone. Endothelial-derived vasoactive substances may also contribute to these mechanically induced responses. The extent of passive stretch varies with the volume of blood delivered to the arterioles from the arteries. An increase in mean arterial pressure drives more blood forward into the arterioles and stretches them further, whereas arterial occlusion blocks blood flow into the arterioles and reduces arteriolar stretch.

Myogenic responses, coupled with metabolically induced responses, appear to be important in reactive hyperemia and pressure autoregulation.

a. *Reactive hyperemia.* When the blood supply to a region is completely occluded, arterioles in the region dilate due to (1) myogenic relaxation, which occurs in response to the diminished stretch accompanying no blood flow, and (2) changes in local chemical composition. Many of the same chemical changes occur in a blood-deprived tissue that occur during metabolically induced active hyperemia. When a tissue's blood supply is blocked, O_2 levels decrease in the deprived tissue; the tissue continues to consume O_2, but no fresh supplies are being delivered. Meanwhile, the concentrations of CO_2, acid, and other metabolites rise. Even though their production does not increase as it does when a tissue is more active metabolically, these substances accumulate in the tissue when the normal amounts produced are not "washed away" by blood.

After the occlusion is removed, blood flow to the previously deprived tissue is transiently much higher than normal because the arterioles are widely dilated. This postocclusion increase in blood flow, called **reactive hyperemia**, can take place in any tissue. Such a response is beneficial for rapidly restoring the local chemical composition to normal. Of course, prolonged blockage of blood flow leads to irreversible tissue damage. You can demonstrate reactive hyperemia yourself by wrapping a rubber band around one of your fingers about an inch from the end for a minute. On removal, you will note that your fingertip turns red as the blood flow increases.

b. *Pressure autoregulation.* When mean arterial pressure falls (for example, because of hemorrhage or a weakened heart), the driving force is reduced, so blood flow to tissues decreases. This is a milder version of what happens during vessel occlusion. The resultant changes in local metabolites and the reduced stretch in the arterioles collectively bring about arteriolar dilation to restore tissue blood flow to normal despite the reduced driving pressure. On the negative side, widespread arteriolar dilation reduces the mean arterial pressure still further, which aggravates the problem. Conversely, in the presence of sustained elevations in mean arterial pressure (hypertension), local chemical and myogenic influences triggered by the initial increased flow of blood to tissues bring about an increase in arteriolar tone and resistance. This greater degree of vasoconstriction subsequently reduces tissue blood flow toward normal despite the elevated blood pressure (● Fig. 10–13). **Pressure autoregulation** is the term applied to these local arteriolar

• **FIGURE 10–13   Pressure Autoregulation of Tissue Blood Flow** Even though blood flow through a tissue immediately increases in response to a rise in arterial pressure, the tissue blood flow gradually is reduced as a result of pressure autoregulation within the tissue, despite a sustained increase in arterial pressure.

mechanisms that are aimed at keeping tissue blood flow fairly constant in spite of rather wide deviations in mean arterial driving pressure.

Active hyperemia, reactive hyperemia, and histamine release all deliberately increase blood flow to a particular tissue for a specific purpose by inducing local arteriolar vasodilation. In contrast, pressure autoregulation is a means by which each tissue resists alterations in its own blood flow secondary to changes in mean arterial pressure by making appropriate adjustments in arteriolar radius.

Extrinsic control of arteriolar radius is primarily important in the regulation of arterial blood pressure.

Extrinsic control of arteriolar radius includes both neural and hormonal influences, with the effects of the sympathetic nervous system being the most important. Sympathetic nerve fibers supply arteriolar smooth muscle everywhere except in the brain. A certain level of ongoing sympathetic activity contributes to vascular tone. Increased sympathetic activity produces generalized arteriolar vasoconstriction, whereas decreased sympathetic activity leads to generalized arteriolar vasodilation.

These widespread changes in arteriolar resistance bring about changes in mean arterial blood pressure. The formula $F = \Delta P/R$ applies to the entire circulation as well as to a single vessel:

- ■ F: Looking at the circulatory system as a whole, flow (F) through all of the vessels is equal to the cardiac output.
- ■ ΔP: The pressure gradient (ΔP) for the entire circulation is the mean arterial pressure. ΔP equals the difference in pres-

sure between the beginning and the end of the circulatory system. Because the beginning pressure is the mean arterial pressure as the blood leaves the left ventricle at an average of 93 mm Hg and the end pressure in the right atrium is 0 mm Hg, $\Delta P = 93$ mm Hg $- 0$ mm Hg $= 93$ mm Hg, which is equivalent to the mean pressure.

- ■ R: By far the greatest percentage of the total resistance (R) offered by all of the peripheral vessels (**total peripheral resistance**) is due to arteriolar resistance, because arterioles are the primary resistance vessels in the vascular tree.

Therefore, for the entire system, rearranging

$$F = \Delta P/R \text{ to } \Delta P = F \times R,$$

makes the equation

mean arterial pressure =
cardiac output $\times$ total peripheral resistance

Thus, the extent of total peripheral resistance offered collectively by all the arterioles influences the mean arterial blood pressure immensely. A dam provides an analogy to this relationship. At the same time it restricts the flow of water downstream, a dam increases the pressure upstream by elevating the water level in the reservoir behind the dam. Similarly, generalized, sympathetically induced vasoconstriction reflexly reduces blood flow downstream to the tissue cells while elevating the upstream mean arterial pressure, thereby increasing the main driving force for blood flow to all the organs.

These effects seem to be counterproductive. Why increase the driving force for flow to the organs by increasing arterial blood pressure while reducing flow to the organs by narrowing the vessels supplying them? In effect, the sympathetically induced arteriolar responses help to maintain the appropriate driving pressure head to all organs. The extent to which each organ actually receives blood flow is determined by local arteriolar adjustments that override the sympathetic constrictor effect. If all arterioles were dilated, blood pressure would fall substantially, so there would not be an adequate driving force for blood flow. Thus, tonic sympathetic activity constricts most vessels (with the exception of those in the brain) to help maintain a pressure head on which organs can draw as needed through local mechanisms that control arteriolar radius.

Skeletal and cardiac muscles have the most powerful local control mechanisms with which to override generalized sympathetic vasoconstriction. For example, if you are pedaling a bicycle, the increased activity in the skeletal muscles of your legs induces overriding local vasodilation in those particular muscles, despite the generalized sympathetic vasoconstriction that accompanies exercise. As a result, more blood flows through your leg muscles but not through your inactive arm muscles. In the heart, coronary vasodilation actually accompanies sympathetic stimulation. The sympathetically induced increase in heart rate and stroke volume elevates cardiac metabolic activity, triggering a profound, locally mediated vasodilation that overpowers the weaker sympathetic vasoconstrictor effect.

The norepinephrine released from sympathetic nerve endings combines with α adrenergic receptors (see p. 203) on arteriolar smooth muscle to bring about vasoconstriction. Cerebral (brain) arterioles are the only ones that do not have α receptors, so no vasoconstriction occurs in the brain. It is important that cerebral arterioles not be reflexly constricted by neural influences, since brain blood flow must remain constant despite what is going on elsewhere in the body. Cerebral vessels are almost entirely controlled by local mechanisms that maintain a constant blood flow to support a constant level of brain metabolic activity. In fact, reflex vasconstrictor activity in the remainder of the cardiovasuclar system is aimed at maintaining an adequate pressure head for blood flow to the vital brain and heart.

Thus, sympathetic activity contributes in an important way to the maintenance of mean arterial pressure, assuring an adequate driving force for blood flow to the brain at the expense of organs and tissues that can better withstand reduced blood flow. Other tissues that really need additional blood, such as active muscles (including active heart muscle), obtain it through local controls that override the sympathetic effect.

There is no significant level of parasympathetic innervation to arterioles, with the exception of the abundant parasympathetic vasodilator supply to the arterioles of the penis and clitoris. The rapid, profuse vasodilation induced by parasympathetic stimulation in these organs is largely responsible for accomplishing erection. Vasodilation elsewhere is produced by decreasing sympathetic vasoconstrictor activity below its tonic level.[1] When mean arterial pressure becomes elevated above normal, reflex reduction in sympathetic vasoconstrictor activity accomplishes generalized arteriolar vasodilation to help restore the driving pressure down toward normal.

The main region of the brain responsible for adjusting sympathetic output to the arterioles is the **cardiovascular control center** in the medulla of the brain stem. This is the integrating center for blood-pressure regulation. Several other brain regions also influence blood distribution, the most notable being the hypothalamus, which, as part of its temperature-regulating function, controls blood flow to the skin to adjust heat loss to the environment.

In addition to neural reflex activity, several hormones also extrinsically influence arteriolar radius. These hormones include the adrenal medullary hormones *epinephrine* and *norepinephrine,* which generally reinforce the sympathetic nervous system in most tissues, as well as *vasopressin* and *angiotensin II,* which are important in the control of fluid balance.

Sympathetic stimulation of the adrenal medulla causes this endocrine gland to release epinephrine and norepineph-

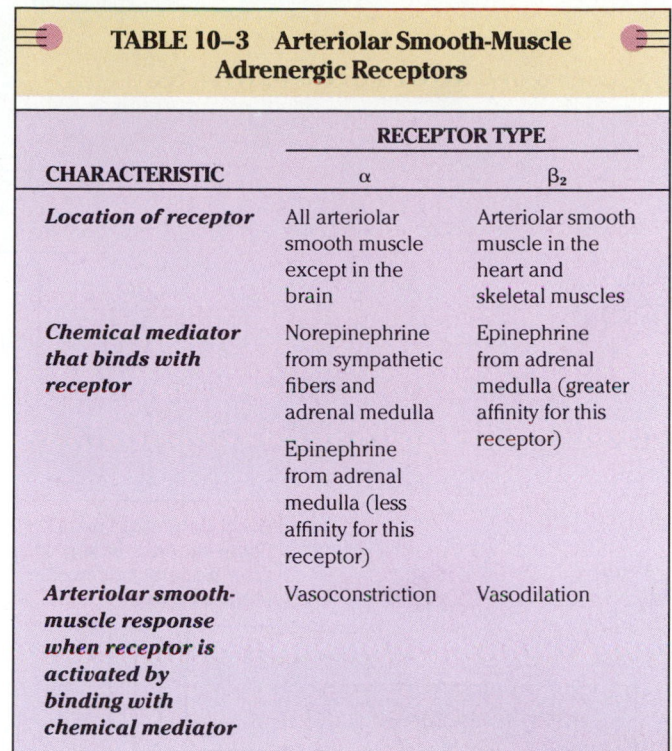

TABLE 10–3	Arteriolar Smooth-Muscle Adrenergic Receptors	
	RECEPTOR TYPE	
CHARACTERISTIC	α	β₂
Location of receptor	All arteriolar smooth muscle except in the brain	Arteriolar smooth muscle in the heart and skeletal muscles
Chemical mediator that binds with receptor	Norepinephrine from sympathetic fibers and adrenal medulla	Epinephrine from adrenal medulla (greater affinity for this receptor)
	Epinephrine from adrenal medulla (less affinity for this receptor)	
Arteriolar smooth-muscle response when receptor is activated by binding with chemical mediator	Vasoconstriction	Vasodilation

rine. Adrenal medullary norepinephrine combines with the same α receptors as sympathetically released norepinephrine to produce generalized vasoconstriction. However, epinephrine, the more abundant of the adrenal medullary hormones, combines with both β₂ and α receptors but has a greater affinity for the β₂ receptors. Activation of β₂ receptors produces vasodilation, but not all tissues have β₂ receptors; they are most abundant in the arterioles of the heart and skeletal muscles. During sympathetic discharge, the released epinephrine combines with the β₂ receptors in the heart and skeletal muscle to reinforce local vasodilatory mechanisms in these tissues. Arterioles in digestive organs and kidneys, equipped only with α receptors, do not respond to epinephrine. Therefore, the arterioles of these organs, affected only by norepinephrine during generalized sympathetic discharge, undergo more profound vasoconstriction than those in heart and skeletal muscle do (Table 10–3).

The two other hormones that extrinsically influence arteriolar tone are vasopressin and angiotensin II. Vasopressin is primarily involved in maintaining water balance by regulating the amount of water the kidneys retain for the body during urine formation. Angiotensin II is part of a hormonal pathway important in the regulation of the body's salt balance, which, because of the osmotic effect of salt in the ECF, has a secondary effect on water balance. Thus, both of these hormones play important roles in maintaining fluid balance in the body, which, in turn, is an important determinant of plasma volume and blood pressure. In addition, both vasopressin and angio-

---

[1] A portion of the skeletal-muscle fibers in some species are supplied by sympathetic cholinergic (ACh-releasing) fibers that bring about vasodilation in anticipation of exercise. However, the existence of such sympathetic vasodilator fibers in humans remains questionable.

Major factors affecting arteriolar radius

● **FIGURE 10–14 Factors Affecting Total Peripheral Resistance** The primary determinant of total peripheral resistance is the adjustable arteriolar radius. Two major categories of factors influence arteriolar radius: (1) local (intrinsic) control, which is primarily important in matching blood flow through a tissue with the tissue's metabolic needs and is mediated by local factors acting on the arteriolar smooth muscle, and (2) extrinsic control, which is important in the regulation of blood pressure and is mediated primarily by sympathetic influence on arteriolar smooth muscle.

tensin II are potent vasoconstrictors. Their role in this regard is especially important during hemorrhage. A sudden loss of blood reduces the plasma volume, which triggers increased secretion of both of these hormones to help restore plasma volume. Their vasoconstrictor effect also helps maintain blood pressure in the face of the abrupt loss of plasma vol-

ume. (The functions and control of these hormones are discussed more thoroughly in later chapters.)

This completes our discussion of the various factors that affect total peripheral resistance, the most important of which are controlled adjustments in arteriolar radius. These factors are summarized in ● Figure 10–14.

CAPILLARIES

Capillaries are ideally suited to serve as sites of exchange.

Capillaries, the sites for exchange[2] of materials between the blood and tissues, branch extensively to bring blood within the reach of every cell. There are no carrier-mediated transport systems across capillaries, with the exception of those in the brain that play a role in the blood-brain barrier (see p. 113). Exchange of materials across capillary walls is accomplished primarily by the process of diffusion. Capillaries are ideally suited to enhance diffusion in accordance with Fick's law of diffusion (see p. 58). They minimize diffusion distances while maximizing surface area and time available for exchange as follows:

1. Diffusing molecules have only a short distance to travel between the blood and surrounding cells because of the thin capillary wall and small capillary diameter, coupled with the close proximity of each and every cell to a capillary. This short distance is important because the rate of diffusion slows down as the diffusion distance increases.

    a. Capillary walls are very thin (1 μm in thickness; in comparison, the diameter of a human hair is 100 μm). Capillaries are composed of only a single layer of flattened endothelial cells—essentially, the lining of the other vessel types. No smooth muscle or connective tissue is present (• Fig. 10–15).

    b. Each capillary is so narrow (7 μm average diameter) that red blood cells (8 μm diameter) have to squeeze through single file. Consequently, plasma contents are either in direct contact with the inside of the capillary wall or are only a short diffusing distance from it.

    c. Because of extensive capillary branching, it is estimated that no cell is farther than 0.01 centimeter (4/1,000 inch) from a capillary.

2. Because capillaries are distributed in such incredible numbers (estimates range from 10 to 40 billion capillaries), a tremendous total surface area is available for exchange (an estimated 600 square meters). In spite of this large number of capillaries, at any point in time they contain only 5% of the total blood volume (250 ml out of a total of 5,000 ml). As a result, a small volume of blood is exposed to an extensive surface area. If all the capillary surfaces were stretched out in a flat sheet and the volume of blood contained within the capillaries were spread over the top, this would be roughly equivalent to spreading a half pint of paint over the floor of a high-school gymnasium. Imagine how thin the paint layer would be!

3. Blood flows more slowly in the capillaries than elsewhere in the circulatory system. The extensive capillary branch-

---

[2] Actually, some exchange takes place across the other microcirculatory vessels, especially the postcapillary venules. The entire vasculature is a continuum and does not abruptly change from one vascular type to the other. When the term *capillary exchange* is used, it tacitly refers to all exchange at the microcirculatory level, the majority of which occurs across the capillaries.

Nucleus of endothelial cell

Junction between edges of two endothelial cells

Capillary lumen

• FIGURE 10–15   Capillary Wall in Cross Section
Electron micrograph of a cross section of a capillary. Note that the capillary wall consists of a single layer of endothelial cells. The nucleus of one of the endothelial cells is shown.

---

ing is also responsible for this slow velocity of blood flow through the capillaries. Let us clarify a potentially confusing point. The term "flow" can be used in two different contexts—the *flow rate*, which refers to the *volume* of blood flowing through a given segment of the circulatory system per unit of time (this is the flow we have been talking about in relationship to the pressure gradient and resistance), and *velocity of flow*, which refers to the *speed* with which blood flows forward through a given segment of the circulatory system. Because the circulatory system is a closed system, the volume of blood flowing through any level of the system must equal the cardiac output. For example, if the heart pumps out 5 liters of blood per minute and 5 liters of blood per minute returns to the heart, then 5 liters of blood per minute must flow through the arteries, arterioles, capillaries, and veins. Therefore, the flow rate is the same at all levels of the circulatory system.

However, the velocity with which blood flows through the different segments of the vascular tree varies because velocity of flow is inversely proportional to the total cross-sectional area of all the vessels at any given level of the circulatory system. Even though the cross-sectional area of each capillary is extremely small compared to that of the large aorta, the total cross-sectional area of all the capillaries added together is about thirteen hundred times greater than the cross-sectional area of the aorta because there are so many capillaries (Table 10–1). Accordingly, blood slows considerably as it passes through the capillaries (• Fig. 10–16). This slow velocity allows adequate time for exchange of nutrients and metabolic end products be-

tween blood and tissues, which is the sole purpose of the entire circulatory system. As the capillaries rejoin to form veins, the total cross-sectional area is once again reduced, and the velocity of blood flow increases as blood returns to the heart.

As an analogy, consider a river (the arterial system) that widens into a lake (the capillaries), then narrows into a river again (the venous system) (• Fig. 10–17). The flow rate is the same throughout the length of this body of water; that is, identical volumes of water are flowing past all the points along the bank of the river and lake. However, the velocity of flow is slower in the wide lake than in the narrow river because the

• **FIGURE 10–16   Comparison of Blood Flow and Velocity of Flow in Relation to Total Cross-Sectional Area**   The blood flow rate (red curve) is identical through all levels of the circulatory system and is equal to the cardiac output (5 liters/min at rest). The velocity of flow (purple curve) varies throughout the vascular tree and is inversely proportional to the total cross-sectional area (green curve) of all the vessels at a given level. Note that the velocity of flow is slowest in the capillaries, which have the largest total cross-sectional area.

Lake
(Capillaries)

River
(Arterial system)

River
(Venous system)

A1 B1 C1

A2 B2 C2

A3 B3 C3

• **FIGURE 10–17   Relationship between Total Cross-Sectional Area and Velocity of Flow**   The three dark blue areas represent equal volumes of water. During one minute, this volume of water moves forward from points A to points C. Therefore, an identical volume of water flows past points B1, B2, and B3 during this minute; that is, the flow rate is the same at all points along the length of this body of water. However, during that minute the identical volume of water moves forward a much shorter distance in the wide lake (A2 to C2) than in the much narrower river (A1 to C1 and A3 to C3). Thus, velocity of flow is much slower in the lake than in the river. Similarly, velocity of flow is much slower in the capillaries than in the arterial and venous systems.

identical volume of water, now spread out over a larger cross-sectional area, moves forward a much shorter distance in the wide lake than in the narrow river during a given period of time. You could readily observe the forward movement of water in the swift-flowing river but the forward motion of water in the lake would be unnoticeable.

Also because of the capillaries' tremendous total cross-sectional area, the resistance offered by all of the capillaries is much lower than that offered by all of the arterioles, even though each capillary has a smaller radius than each arteriole. For this reason, the arterioles contribute more to total peripheral resistance. Furthermore, arteriolar caliber (and, accordingly resistance) is subject to control whereas capillary caliber cannot be adjusted.

Water-filled pores in the capillary wall permit passage of small, water-soluble substances that cannot cross the endothelial cells themselves.

Diffusion across capillary walls also depends on the wall's permeability to the materials being exchanged. The endothelial cells forming the capillary walls fit together in jigsaw-puzzle fashion, but the closeness of the fit varies considerably between organs. In most capillaries, narrow, water-filled clefts, or **pores,** are present at the junctions between the cells (• Fig. 10–18). These pores permit passage of water-soluble substances. Lipid-soluble substances can readily pass through the endothelial cells themselves by dissolving in the lipid bilayer barrier. The size of the capillary pores varies from organ to organ. At one extreme, the endothelial cells in brain capillaries are joined by tight junctions so that pores are nonexistent. These junctions prevent transcapillary passage of materials between the cells and thus constitute part of the

protective blood-brain barrier. In most tissues, small, water-soluble substances such as ions, glucose, and amino acids can readily pass through the water-filled clefts, but large, non-lipid-soluble materials such as plasma proteins are excluded from passage. At the other extreme, liver capillaries have such large pores that even proteins pass through readily. This is adaptive, because the liver's functions include synthesis of plasma proteins and the metabolism of protein-bound substances such as cholesterol. These proteins must all pass through the liver's capillary walls. The leakiness of various capillary beds is therefore a function of how tightly the endothelial cells are joined, which varies according to the different organs' needs.

The capillary wall has traditionally been considered a passive sieve, like a brick wall with permanent gaps in the mortar acting as pores. Recent studies, however, suggest that even under normal conditions, changes in endothelial cells are involved in the active regulation of capillary membrane permeability; that is, in response to appropriate signals, the "bricks" can readjust themselves to vary the size of the holes. Thus, the degree of leakiness does not necessarily remain constant for a given capillary bed. For example, researchers speculate that histamine increases capillary permeability by triggering contractile responses in endothelial cells to widen the intercellular gaps. This is not a muscular contraction, because no smooth-muscle cells are present in capillaries. It is due to an actin-myosin contractile apparatus in the nonmuscular capillary cells. These larger gaps make the affected capillary wall leakier so that normally retained plasma proteins escape into the surrounding tissue, where they exert an osmotic effect. The resultant additional local fluid retention contributes to inflammatory swelling, along with histamine-induced vasodilation.

Vesicular transport also plays a limited role in the passage of materials across the capillary wall. Large non-lipid-soluble molecules such as proteinaceous hormones that must be exchanged between the blood and surrounding tissues are transported from one side of the capillary wall to the other in endocytotic-exocytotic vesicles (see p. 67).

Many capillaries are not open under resting conditions.

The branching and reconverging arrangement within capillary beds varies somewhat, depending on the tissue. Capillaries typically branch either directly from an arteriole or from a thoroughfare channel known as a **metarteriole,** which runs between an arteriole and a venule. Likewise, capillaries may rejoin either at a venule or at a metarteriole (• Fig. 10–19). Unlike the true capillaries within a capillary bed, metarterioles are sparsely surrounded by wisps of spiraling smooth-muscle cells. These cells also form **precapillary sphincters,** each of which consists of a ring of smooth muscle around the entrance to a capillary as it arises from a metarteriole.[3] Pre-

---

[3] Although generally accepted, the existence of precapillary sphincters in humans has not been conclusively established.

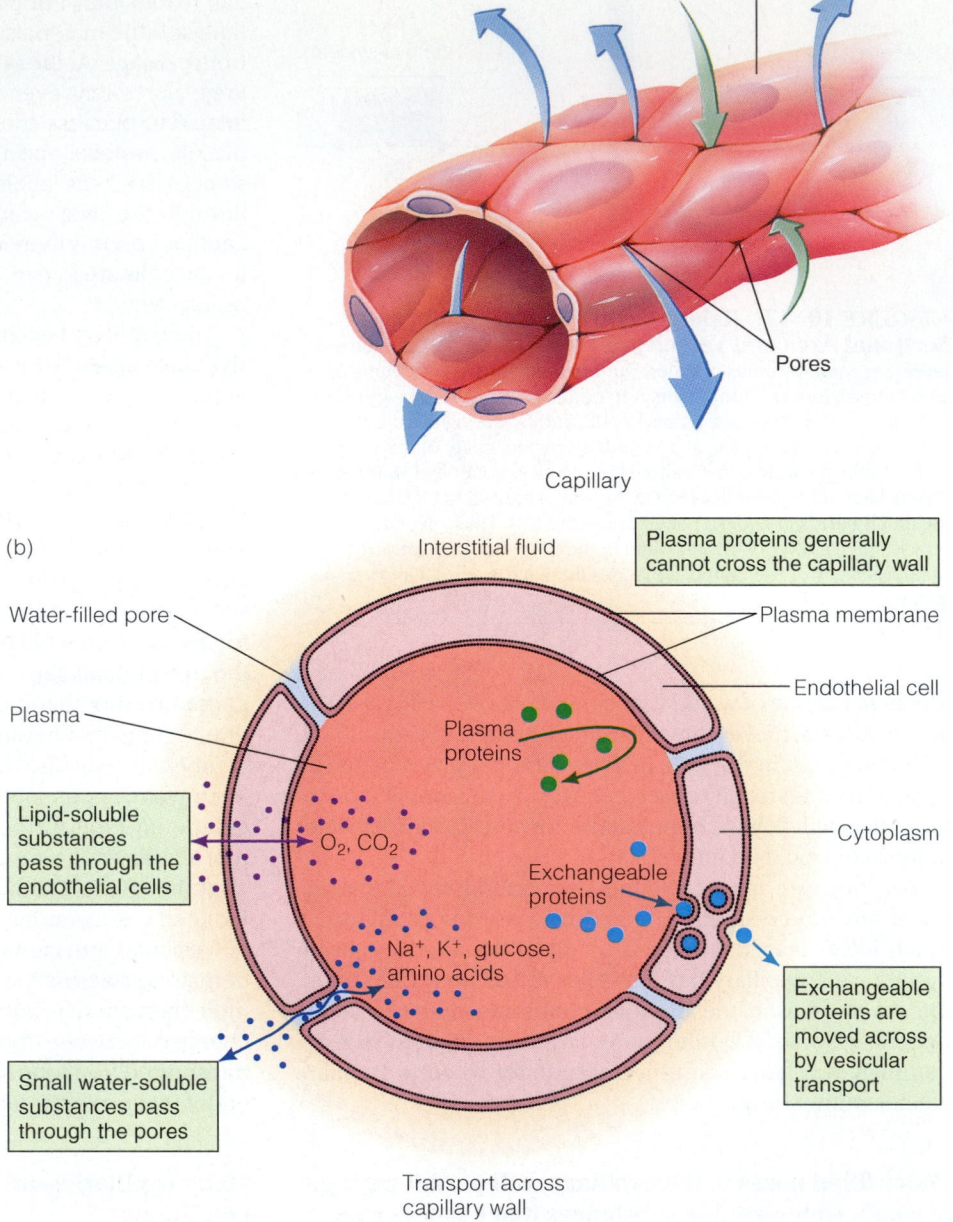

● **FIGURE 10–18   Exchanges across Capillary Wall** (a) Slitlike gaps between adjacent endothelial cells form pores within the capillary wall. (b) As depicted in this schematic representation of a cross section of a capillary wall, small water-soluble substances are exchanged between the plasma and the interstitial fluid by passing through the water-filled pores, whereas lipid-soluble substances are exchanged across the capillary wall by passing through the endothelial cells. Proteins to be moved across are exchanged by vesicular transport. The plasma proteins generally cannot escape from the plasma across the capillary wall.

(a)

Endothelial cell

Pores

Capillary

(b)   Interstitial fluid

Plasma proteins generally cannot cross the capillary wall

Water-filled pore

Plasma membrane

Plasma

Endothelial cell

Plasma proteins

Lipid-soluble substances pass through the endothelial cells

O_2, CO_2

Cytoplasm

Exchangeable proteins

Na^+, K^+, glucose, amino acids

Small water-soluble substances pass through the pores

Exchangeable proteins are moved across by vesicular transport

Transport across capillary wall

capillary sphincters are not innervated, but they have a high degree of myogenic tone and are sensitive to local metabolic changes. They act as stopcocks to control blood flow through the particular capillary that each one guards. Arterioles perform a similar function for a small group of capillaries. Capillaries themselves have no smooth muscle, so they cannot actively participate in the regulation of their own blood flow.

Generally, tissues that are more metabolically active have a greater density of capillaries. Muscles, for example, have relatively more capillaries than their tendinous attachments. Only about 10% of the precapillary sphincters in a resting muscle are open at any moment, however, so blood is flowing through only about 10% of the muscle's capillaries. As

chemical concentrations start to change in a region of the muscle tissue supplied by closed-down capillaries, the precapillary sphincters and arterioles in that region relax. Restoration of the chemical concentrations to normal as a result of increased blood flow to that region removes the impetus for vasodilation, so the precapillary sphincters close once again and the arterioles return to normal tone. In this way, blood flow through any given capillary is often intermittent as a result of arteriolar and precapillary sphincter action working in concert. When the muscle as a whole becomes more active, a greater percentage of the precapillary sphincters relax, simultaneously opening up more capillary beds, while concurrent arteriolar vasodilation increases total flow to the organ.

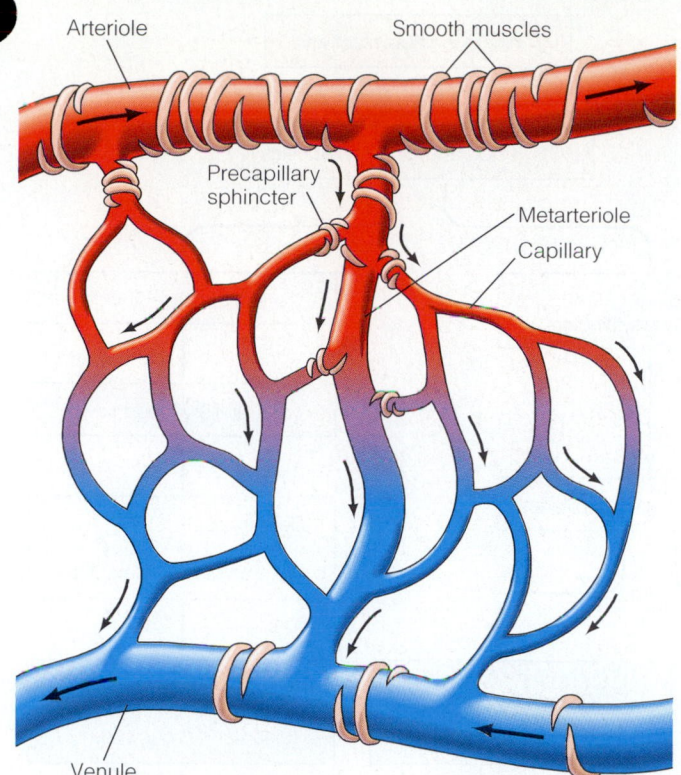

• FIGURE 10–19 Capillary Bed Capillaries branch either directly from an arteriole or from a metarteriole, a thoroughfare channel between an arteriole and venule. Capillaries rejoin at either a venule or metarteriole. Metarterioles are surrounded by smooth-muscle cells, which also form precapillary sphincters that encircle capillaries as they arise from a metarteriole.

Arteriole

Smooth muscles

Precapillary sphincter

Metarteriole

Capillary

Venule

As a result of more blood flowing through more open capillaries, the total volume and surface area available for exchange increase, and the diffusion distance between the cells and an open capillary decreases (• Fig. 10–20). Thus, blood flow through a particular tissue (assuming a constant blood pressure) is regulated by (1) the degree of resistance offered by the arterioles in the organ, controlled by sympathetic activity and local factors; and (2) the number of open capillaries, controlled by action of the same local metabolic factors on precapillary sphincters.

Diffusion across the capillary wall is important in solute exchange.

Exchanges are not made directly between blood and the tissue cells. Interstitial fluid, the true internal environment in immediate contact with the cells, acts as the go-between (• Fig. 10–21). Only 25% of the ECF circulates as plasma. The remaining 75% consists of interstitial fluid, which bathes all of the cells in the body. Cells exchange materials directly with the interstitial fluid, the type and extent of exchange being governed by the properties of the cellular plasma membranes. Movement across the plasma membrane may involve either passive diffusion or carrier-mediated transport. In contrast, only passive exchanges occur across the capillary wall between the plasma and interstitial fluid. Because of the per-

meability of the capillary walls, exchange is so thorough that the interstitial fluid takes on essentially the same composition as the incoming arterial blood, with the exception of the large plasma proteins that usually do not escape from the blood. Therefore, when we speak of exchanges between blood and tissue cells, we tacitly include interstitial fluid as a passive intermediary.

Exchanges between blood and surrounding tissues across the capillary walls are accomplished by (1) passive diffusion down concentration gradients, the primary mechanism for exchange of individual solutes; and (2) bulk flow, a process that accomplishes the totally different function of determining the distribution of the ECF volume between the vascular and interstitial-fluid compartments. We will examine each of these mechanisms in more detail, starting with diffusion.

Because there are no carrier-mediated transport systems in capillary walls, solutes cross primarily by diffusion down concentration gradients. The chemical composition of arterial blood is carefully regulated to maintain the concentrations of individual solutes at levels that will promote each solute's movement in the appropriate direction across the capillary walls. The reconditioning organs primarily contribute to this homeostatic process, continuously adding nutrients and O_2 and removing CO_2 and other wastes as blood passes through them. Meanwhile, cells are constantly using up supplies and generating metabolic wastes. Diffusion of each solute continues independently until there is no longer a concentration difference for that solute between the blood and surrounding cells (• Fig. 10–22). This process repeats itself continuously. As cells use up O_2 and glucose, the blood constantly brings in fresh supplies of these vital materials, maintaining concentration gradients that favor the net diffusion of these substances from blood to cells. Simultaneously, ongoing net diffusion of CO_2 and other metabolic wastes from cells to blood is maintained by the continual production of these wastes at the cellular level and their constant removal from the tissue level by the circulating blood.

Because the capillary wall does not limit the passage of any constituent except plasma proteins, the extent of exchanges for each solute is independently determined by the magnitude of its concentration gradient between blood and surrounding tissues. As cells increase their level of activity, they use up more O_2 and produce more CO_2, among other things. This creates larger concentration gradients for O_2 and CO_2 between cells and blood, so more O_2 diffuses out of the blood into the cells and more CO_2 proceeds in the opposite direction to help support the increased metabolic activity.

Bulk flow across the capillary wall is important in extracellular-fluid distribution.

The second means by which exchange is accomplished across capillary walls is bulk flow. A volume of protein-free plasma actually filters out of the capillary, mixes with the surrounding interstitial fluid, and is subsequently reabsorbed. This process is called **bulk flow** because the various constituents of the fluid are moving together in bulk, or as a unit, in contrast to the discrete diffusion of individual solutes down

● **FIGURE 10–20   Complementary Action of Precapillary Sphincters and Arterioles in Adjusting Blood Flow through a Tissue in Response to Changing Metabolic Needs**

↑ Tissue metabolic activity

↓ O_2, ↑ CO_2, and other metabolites

Relaxation of precapillary sphincters

Arteriolar vasodilation

↑ Capillary blood flow

↑ Number of open capillaries

↑ Delivery of O_2, more rapid removal of CO_2 and other metabolites

↑ Capillary surface area available for exchange

↓ Diffusion distance from cell to open capillary

↑ Concentration gradient for these materials between blood and tissue cells

↑ Exchange between blood and tissue to support increased metabolic activity

---

● **FIGURE 10–21   Interstitial Fluid Acting as an Intermediary between Blood and Cells**

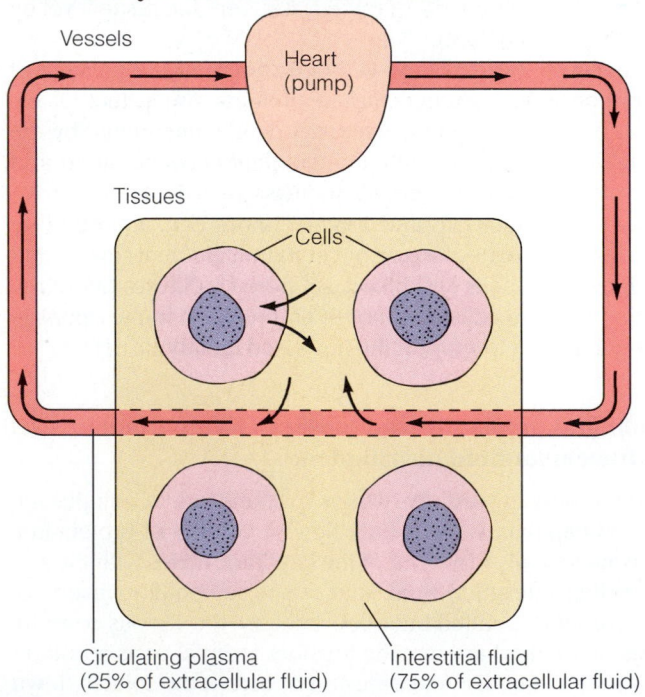

Vessels

Heart (pump)

Tissues

Cells

Circulating plasma (25% of extracellular fluid)

Interstitial fluid (75% of extracellular fluid)

● **FIGURE 10–22   Independent Exchange of Individual Solutes Down Their Own Concentration Gradients across the Capillary Wall**

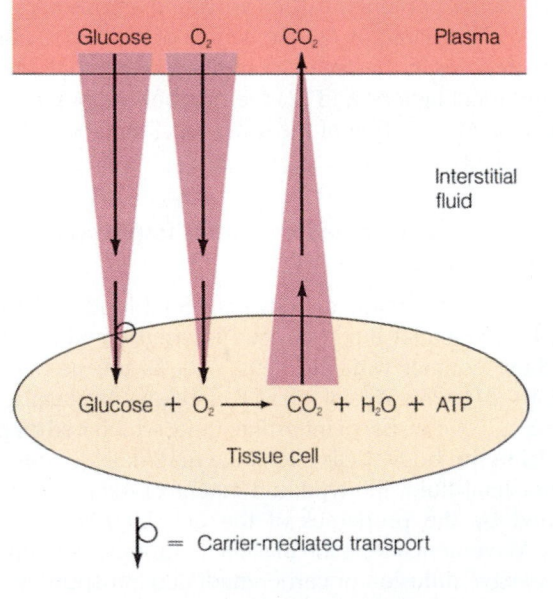

Glucose O_2 CO_2 Plasma

Interstitial fluid

Glucose + O_2 ⟶ CO_2 + H_2O + ATP

Tissue cell

= Carrier-mediated transport

concentration gradients. The capillary wall acts like a sieve, with the fluid moving through its water-filled pores. When pressure inside the capillary exceeds pressure on the outside, fluid is pushed out through the pores in a process known as **ultrafiltration**. The majority of the plasma proteins are retained on the inside during this process because of the pores' filtering effect, although a few do escape. Since all other constituents in the plasma are dragged along as a unit with the volume of fluid leaving the capillary, the filtrate is essentially a protein-free plasma. When inward-driving pressures exceed outward pressures across the capillary wall, net inward movement of fluid from the interstitial-fluid compartment into the capillaries takes place through the pores, a process known as **reabsorption**.

Bulk flow occurs because of differences in the hydrostatic and colloid osmotic pressures between the plasma and interstitial fluid. Even though pressure differences exist between plasma and surrounding fluid elsewhere in the circulatory system, only the capillaries have pores and walls thin enough

to allow fluids to pass through. Four forces influence fluid movement across the capillary wall (● Fig. 10–23):

1. **Capillary blood pressure (P_C)** is the fluid or hydrostatic pressure exerted on the inside of the capillary walls by the blood. This pressure tends to force fluid out of the capillaries into the interstitial fluid. Mean blood pressure has dropped substantially by the level of the capillaries because of frictional losses in pressure in the high-resistance arterioles upstream. On the average, the hydrostatic pressure is 37 mm Hg at the arteriolar end of a tissue capillary and has declined even further to 17 mm Hg at the venular end.

2. **Plasma-colloid osmotic pressure (π_P)**, also known as *oncotic pressure*, is a force caused by the colloidal dispersion (see p. A–6) of plasma proteins; it encourages fluid movement into the capillaries. Since the plasma proteins remain in the plasma rather than entering the interstitial fluid, a protein-concentration difference exists between

● **FIGURE 10–23 Bulk Flow across Capillary Wall** (a) Schematic representation of ultrafiltration and reabsorption as a result of imbalances in the forces acting across the capillary wall. (b) Calculation of ultrafiltration pressure at the arteriolar end of the capillary and reabsorption pressure at the venular end.

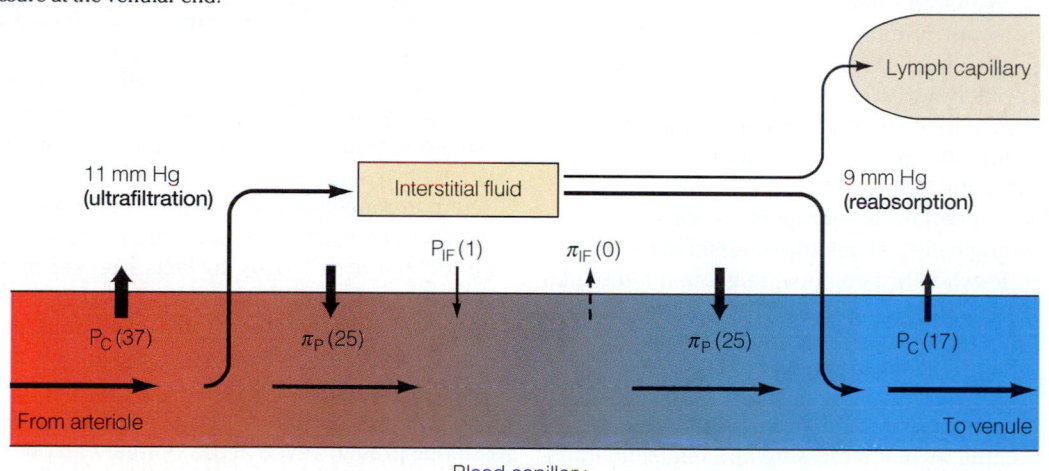

Forces at Arteriolar End
of Capillary

- Outward pressure

P_C	37
π_{IF}	0
	37

- Inward pressure

π_P	25
P_{IF}	1
	26

Net outward pressure
of 11 mm Hg =
Ultrafiltration pressure

Forces at Venular End
of Capillary

- Outward pressure

P_C	17
π_{IF}	0
	17

- Inward pressure

π_P	25
P_{IF}	1
	26

Net inward pressure
of 9 mm Hg =
Reabsorption pressure

(b)

All values are given in mm Hg.

the plasma and interstitial fluid. Accordingly, there is also a water-concentration difference between these two regions. The plasma has a higher protein concentration and a lower water concentration than does the interstitial fluid. This difference exerts an osmotic effect (see p. 59) that tends to move water from the area of higher water concentration in the interstitial fluid to the area of lower water concentration (or higher protein concentration) in the plasma. Thus, the plasma proteins may be thought of as "attracting" water, although this is not actually the underlying force involved. The other plasma constituents do not exert an osmotic effect because they readily pass through the capillary wall, so their concentrations are equal in the plasma and interstitial fluid. The plasma-colloid osmotic pressure averages 25 mm Hg.

3. **Interstitial-fluid hydrostatic pressure (P_{IF})** is the fluid pressure exerted on the outside of the capillary wall by the interstitial fluid. This pressure tends to force fluid into the capillaries. Because of the difficulties encountered in measuring interstitial-fluid hydrostatic pressure, the actual value of this pressure is a controversial issue. It is either at, slightly above, or slightly below atmospheric pressure. For purposes of illustration, we will say it is 1 mm Hg.

4. **Interstitial-fluid-colloid osmotic pressure (π_{IF})** is another force that does not normally contribute significantly to bulk flow. The small fraction of plasma proteins that leak across the capillary walls into the interstitial spaces are normally returned to the blood by means of the lymphatic system. Therefore, the protein concentration in the interstitial fluid is extremely low, and the interstitial-fluid-colloid osmotic pressure is very close to zero. If plasma proteins pathologically leak into the interstitial fluid, however, as they do when histamine widens the intercellular clefts during tissue injury, the leaked proteins exert an osmotic effect that tends to promote the movement of fluid out of the capillaries into the interstitial fluid.

Therefore, the two pressures that tend to force fluid out of the capillary are capillary blood pressure and interstitial-fluid-colloid osmotic pressure. The two opposing pressures that tend to force fluid into the capillary are plasma-colloid osmotic pressure and interstitial-fluid hydrostatic pressure. We are now prepared to analyze the fluid movement that occurs across a capillary wall because of imbalances in these opposing physical forces (Fig. 10–23). Net exchange at a given point across the capillary wall can be calculated using the following equation:

$$\text{net exchange pressure} = (P_C + \pi_{IF}) - (\pi_P + P_{IF})$$
$$\text{(outward pressure)} \quad \text{(inward pressure)}$$

A positive net exchange pressure (when the outward pressure exceeds the inward pressure) represents an ultrafiltration pressure. A negative net exchange pressure (when the inward pressure exceeds the outward pressure) represents a reabsorption pressure.

At the arteriolar end of the capillary, the outward pressure totals 37 mm Hg, whereas the inward pressure totals 26 mm Hg, for a net outward pressure of 11 mm Hg. Ultrafiltration takes place at the beginning of the capillary as this outward pressure gradient forces a protein-free filtrate through the capillary pores.

By the time the venular end of the capillary is reached, the capillary blood pressure has dropped, but the other pressures have remained essentially constant. At this point the outward pressure has fallen to a total of 17 mm Hg, whereas the total inward pressure is still 26 mm Hg, for a net inward pressure of 9 mm Hg. Reabsorption of fluid takes place as this inward pressure gradient forces fluid back into the capillary at its venular end. Ultrafiltration and reabsorption, collectively known as bulk flow, are thus due to a shift in the balance between the passive physical forces acting across the capillary wall. No active forces or local energy expenditures are involved in the bulk exchange of fluid between the plasma and surrounding interstitial fluid. With only minor contributions from the interstitial fluid, ultrafiltration occurs at the beginning of the capillary because capillary blood pressure exceeds plasma-colloid osmotic pressure, whereas by the end of the capillary, reabsorption takes place because blood pressure has fallen below osmotic pressure.

It is important to realize that we have taken "snapshots" at two points—one point at the beginning and one point at the end—in a hypothetical capillary. Actually, blood pressure gradually diminishes along the length of the capillary so that progressively diminishing quantities of fluid are filtered out in the first half of the vessel and progressively increasing quantities of fluid are reabsorbed in the last half (● Fig. 10–24). Even this situation is idealized. The pressures used in this figure are average values and controversial at that. Some capillaries have such high blood pressure that filtration actually occurs throughout their entire length, whereas others have such low hydrostatic pressure that reabsorption takes place throughout their length. In fact, a recent theory that has received considerable attention is that net filtration occurs throughout the length of all *open* capillaries whereas net reabsorption occurs throughout the length of all *closed* capillaries. According to this theory, when the precapillary sphincter is relaxed, capillary blood pressure exceeds the plasma osmotic pressure even at the venular end of the capillary, promoting filtration throughout the length. When the precapillary sphincter is closed, the reduction in blood flow through the capillary diminishes capillary blood pressure below the plasma osmotic pressure even at the beginning of the capillary, so reabsorption takes place all along the capillary. Whichever mechanism is involved, the net effect is the same. A protein-free filtrate exits the capillaries and is ultimately reabsorbed.

Bulk flow does not play an important role in the exchange of individual solutes between blood and tissues, because the quantity of solutes moved across the capillary wall by bulk flow is extremely small compared to the much larger transfer of solutes by diffusion. If ultrafiltration and reabsorption are not important in the exchange of nutrients and wastes, then of what significance is bulk fluid exchange across the capillary wall? The answer is that it plays an extremely important role in regulating the distribution of ECF between the plasma and interstitial fluid. Maintenance of proper arterial blood pressure depends in part on an appropriate volume of circulating blood. If plasma volume is reduced (for example, by

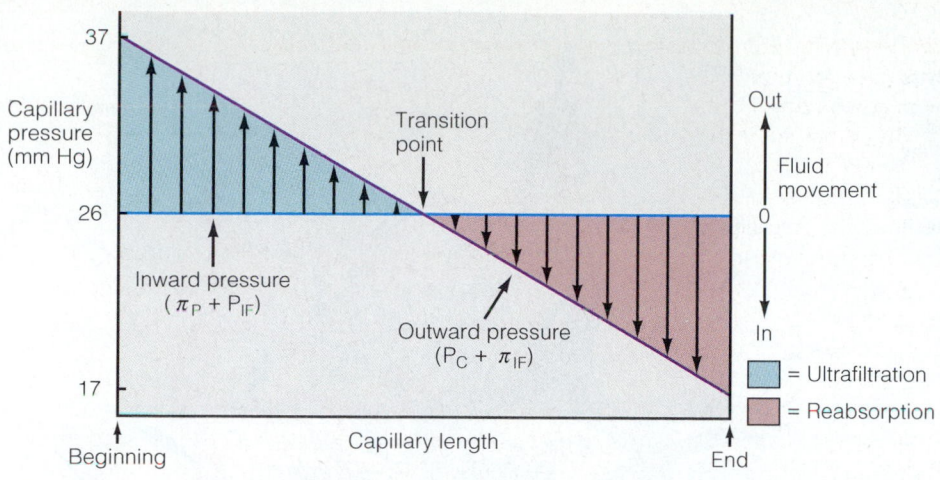

● **FIGURE 10–24 Net Filtration and Net Reabsorption along Vessel Length** The inward pressure (π_P + P_{IF}) remains constant throughout the length of the capillary whereas the outward pressure (P_C + π_{IF}) progressively declines throughout the capillary's length. In the first half of the vessel, where the declining outward pressure still exceeds the constant inward pressure, progressively diminishing quantities of fluid are filtered out (depicted by the upward arrows). In the last half of the vessel, progressively increasing quantities of fluid are reabsorbed (depicted by the downward arrows) as the declining outward pressure falls farther below the constant inward pressure.

hemorrhage), blood pressure falls. The resultant lowering of capillary blood pressure alters the balance of forces across the capillary walls. Because the net outward pressure is decreased while the net inward pressure remains unchanged, extra fluid is shifted from the interstitial compartment into the plasma as a result of reduced filtration and increased reabsorption. The extra fluid soaked up from the interstitial fluid provides additional fluid for the plasma to temporarily compensate for the loss of blood. Meanwhile, reflex mechanisms acting on the heart and blood vessels (to be described later) also come into play to help maintain blood pressure until long-term mechanisms, such as thirst and reduction of urinary output, can restore the fluid volume to completely compensate for the loss.

Conversely, if the plasma volume becomes overexpanded, as with excessive fluid intake, the resultant elevation in capillary blood pressure forces extra fluid from the capillaries into the interstitial fluid, temporarily relieving the expanded plasma volume until the excess fluid can be eliminated from the body by long-term measures.

These internal fluid shifts between the two ECF compartments occur automatically and immediately whenever the balance of forces acting across the capillary walls is changed; they provide a temporary mechanism to help keep the plasma volume fairly constant. In the process of restoring the plasma volume to an appropriate level, the interstitial-fluid volume fluctuates, but it is much more important that the plasma volume be maintained at a constant level to ensure that the circulatory system functions effectively.

The lymphatic system is an accessory route by which interstitial fluid can be returned to the blood.

Even under normal circumstances, slightly more fluid is filtered out of the capillaries into the interstitial fluid than is reabsorbed from the interstitial fluid back into the plasma. The average net ultrafiltration pressure is 11 mm Hg, whereas the net reabsorption pressure is only 9 mm Hg (Fig. 10–23). The extra fluid filtered out as a result of this filtration-reabsorption imbalance is picked up by the **lymphatic system.** This system consists of an extensive network of one-way vessels that provide an accessory route by which fluid can

be returned from the interstitial fluid to the blood. Small, blind-ended terminal lymph vessels (**lymphatic capillaries**) permeate almost every tissue of the body (● Fig. 10–25a). The endothelial cells forming the walls of lymphatic capillaries slightly overlap, with their overlapping edges being free instead of being attached to the surrounding cells. This arrangement creates valvelike openings in the vessel wall (Fig. 10–25b). Fluid pressure on the outside of the vessel pushes the edges inward, opening the valves and permitting interstitial fluid to enter. Once interstitial fluid enters a lymphatic vessel, it is called **lymph.** Fluid pressure on the inside forces the overlapping edges together, closing the valves so that lymph does not escape. These lymphatic valvelike openings are much larger than the pores in blood capillaries. Consequently, large particulates in the interstitial fluid, such as escaped plasma proteins and bacteria, can gain access to lymphatic capillaries but are excluded from blood capillaries.

Lymphatic capillaries converge to form larger and larger lymph vessels, which eventually empty into the venous system near the point where the blood enters the right atrium (● Fig. 10–26a). Because there is no "lymphatic heart" to provide driving pressure, you may wonder how lymph is directed from the tissues toward the venous system in the thoracic cavity. Lymph flow is accomplished by two mechanisms. First, since lymph vessels lie between skeletal muscles, contraction of these muscles squeezes the lymph out of the vessels. One-way valves spaced at intervals within the lymph vessels direct the flow of lymph toward its venous outlet in the chest. Second, lymph vessels beyond the lymphatic capillaries are surrounded by smooth muscle. When this muscle is stretched because the vessel is distended with lymph, the muscle inherently contracts more forcefully, thereby propelling the lymph through the vessel.

Following are the most important functions of the lymphatic system:

■ *Return of excess filtered fluid.* Normally, capillary filtration exceeds reabsorption by about 3 liters per day (20 liters filtered, 17 liters reabsorbed) (Fig. 10–26b). Yet the entire blood volume is only 5 liters, and only 2.75 liters of that is plasma. (Blood cells make up the remainder of the blood volume.) With an average cardiac output, 7,200 liters of blood pass through the capillaries daily under resting

• **FIGURE 10–25 Lymphatic Capillaries** (a) Relationship between lymphatic and blood capillaries. Blind-ended lymphatic capillaries pick up excess fluid filtered by blood capillaries and return it to the venous system in the chest. (b) Arrangement of endotheial cells in a lymphatic capillary. Note that the overlapping edges of the endothelial cells create valvelike openings in the vessel wall.

To venous system

Arteriole

Tissue cells

Interstitial fluid

Venule

Blood capillary

Lymphatic capillary

(a)

Fluid pressure on the outside of the vessel pushes the endothelial cell's free edge inward, permitting entrance of interstitial fluid (now lymph).

Interstitial fluid

Overlapping endothelial cells

Lymph

(b)

Fluid pressure on the inside of the vessel forces the overlapping edges together so that lymph cannot escape.

conditions (more when cardiac output increases). Even though only a small fraction of the filtered fluid is not reabsorbed by the blood capillaries, the cumulative effect of this process being repeated with every heartbeat results in the equivalent of more than the entire plasma volume being left behind in the interstitial fluid each day. Obviously, this fluid must be returned to the circulating plasma, and this task is accomplished by the lymph vessels. The average rate of flow through the lymph vessels is 3 liters per day, compared with 7,200 liters per day through the circulatory system.

■ *Defense against disease.* The lymph percolates through **lymph nodes** located en route within the lymphatic system. Passage of this fluid through the lymph nodes is an important aspect of the body's defense mechanism against disease. For example, bacteria picked up from the interstitial fluid are destroyed by special phagocytic cells located within the lymph nodes.

■ *Transport of absorbed fat.* The lymphatic system is important in the absorption of fat from the digestive tract. The end products of the digestion of dietary fats are packaged

Systemic circulation

Lymph node

Pulmonary circulation

Lymph vessel

Lymphatic capillaries

Valve

Blood capillaries

Veins

Arteries

Heart

Lymph node

Lymphatic capillaries

Blood capillaries

(a)

Lymph

20 l/day — 3 l/day

7,200 l/day — 17 l/day

Blood

(b)

● FIGURE 10–26 **Lymphatic System** (a) Lymph empties into the venous system near its entrance to the right atrium. (b) Lymph flow averages 3 liters per day whereas blood flow averages 7,200 liters per day.

by cells lining the digestive tract into fatty particles that are too large to gain access to the blood capillaries but can easily enter the terminal lymphatic vessels.

■ *Return of filtered protein.*   Most capillaries permit leakage of some plasma proteins during filtration. These proteins cannot readily be reabsorbed back into the blood capillaries but can easily gain access to the lymphatic capillaries. If the proteins were allowed to accumulate in the interstitial fluid rather than being returned to the circulation via the lymphatics, the interstitial-fluid-colloid osmotic pressure (an outward pressure) would progressively increase while the plasma-colloid osmotic pressure (an inward pressure) would progressively fall. As a result, filtration forces would gradually increase and reabsorption forces would gradually decrease, resulting in progressive accumulation of fluid in the interstitial spaces at the expense of loss of plasma volume.

Edema occurs when too much interstitial fluid accumulates.

Occasionally, excessive interstitial fluid does accumulate when one of the physical forces acting across the capillary walls becomes abnormal for some reason. Swelling of the tissues because of excess interstitial fluid is known as **edema**.

The causes of edema can be grouped into four general categories:

1. *A reduced concentration of plasma proteins* causes a decrease in plasma-colloid osmotic pressure. Such a drop in the major inward pressure allows excess fluid to be filtered out while less than normal amounts of fluid are reabsorbed; hence extra fluid remains in the interstitial spaces. Edema caused by decreased concentrations of plasma proteins can arise in several different ways: excessive loss of plasma proteins in the urine caused by kidney disease; reduced synthesis of plasma proteins as a result of liver disease (the liver synthesizes almost all plasma proteins); a diet deficient in protein; or significant loss of plasma proteins from large burned surfaces.

2. *Increased permeability of the capillary walls* allows more plasma proteins than usual to pass from the plasma into the surrounding interstitial fluid—for example, via histamine-induced widening of the capillary pores during tissue injury or allergic reactions. The resultant fall in plasma-colloid osmotic pressure decreases the effective inward pressure while the resultant rise in interstitial-fluid-colloid osmotic pressure caused by the excess protein in the interstitial fluid increases the effective outward force. This imbalance contributes in part to the localized edema

associated with injuries (for example, blisters) and allergic responses (for example, hives).

3. *Increased venous pressure,* as when blood dams up in the veins, is accompanied by an increased capillary blood pressure, since the capillaries drain into the veins. This elevation in outward pressure across the capillary walls is largely responsible for the edema seen with congestive heart failure (see p. 289). Regional edema can also occur because of localized restriction of venous return. An example is the swelling often occurring in the legs and feet during pregnancy. The enlarged uterus compresses the major veins that drain the lower extremities as these vessels enter the abdominal cavity. The resultant damming of blood in these veins causes a rise in blood pressure in the capillaries of the legs and feet, which promotes regional edema of the lower extremities.

4. *Blockage of lymph vessels* produces edema because the excess filtered fluid is retained in the interstitial fluid rather than being returned to the blood through the lymphatics. The protein accumulation in the interstitial fluid compounds the problem through its osmotic effect. Local lymph blockage can occur, for example, in the arms of women whose major lymphatic drainage channels from the arm have been blocked as a result of lymph-node removal during surgery for breast cancer. More widespread lymph blockage occurs with *filariasis,* a mosquito-borne parasitic disease that is found predominantly in tropical coastal regions. In this condition, small, threadlike filaria worms infect the lymph vessels, where their presence prevents proper lymph drainage. The affected body parts, particularly the scrotum and extremities, become grossly edematous. The condition is often called *elephantiasis* because of the elephantlike appearance of the swollen extremities (• Fig. 10–27).

Whatever the cause of edema, an important consequence is a reduction in exchange of materials between the blood and cells. As excess interstitial fluid accumulates, the distance between the blood and cells across which nutrients, O_2, and wastes must diffuse is increased, so the rate of diffusion decreases. Therefore, cells within edematous tissues may not be adequately supplied.

VEINS

Veins serve as a blood reservoir as well as passageways back to the heart.

The venous system completes the circulatory circuit. Blood leaving the capillary beds enters the venous system for transport back to the heart. Veins have large radii, so they offer little resistance to flow. Furthermore, because the total cross-sectional area of the venous system gradually decreases as smaller veins converge into progressively fewer but larger vessels, the velocity of blood flow increases as the blood approaches the heart.

In addition to serving as low-resistance passageways to return blood from the tissues to the heart, systemic veins also serve as a *blood reservoir.* Because of their storage capacity, veins are often referred to as **capacitance vessels**. Veins have

• **FIGURE 10–27 Elephantiasis** This tropical condition is caused by a mosquito-borne parasitic worm that invades the lymph vessels. As a result of the interference with lymph drainage, the affected body parts, usually the extremities, become grossly edematous, appearing elephantlike.

much thinner walls with less smooth muscle than do arteries. Since collagen fibers are considerably more abundant than elastin fibers in venous connective tissue, veins have very little elasticity, in contrast to arteries. Also, unlike arteriolar smooth muscle, venous smooth muscle has little inherent myogenic tone. Because of these features, veins are highly distensible, or stretchable, and have little elastic recoil. They easily distend to accommodate additional volumes of blood with only a small increase in venous pressure. Arteries stretched by an excess volume of blood recoil because of the elastic fibers in their walls, driving the blood forward. Veins containing an extra volume of blood simply stretch to accommodate the additional blood without tending to recoil. In this way veins serve as a blood reservoir; that is, when demands for blood are low, the veins can store extra blood in reserve because of their passive distensibility. Under resting conditions, the veins contain more than 60% of the total blood volume (• Fig. 10–28). When the stored blood is needed, such as during exercise, extrinsic factors (soon to be described) drive the extra blood from the veins to the heart so that it can be pumped to the tissues. Increased venous return induces an increased cardiac stroke volume in accordance with the Frank-Starling law of the heart (see p. 286). If too much blood pools in the veins instead of being returned to the heart, cardiac output is abnormally diminished. Thus, a delicate balance exists between the capacity of the veins, the extent of venous return, and the cardiac output. We will now turn our attention to the factors that affect venous capacity and contribute to venous return.

Venous return is enhanced by a number of extrinsic factors.

Venous capacity (the volume of blood that the veins can accommodate) depends on the distensibility of the vein walls

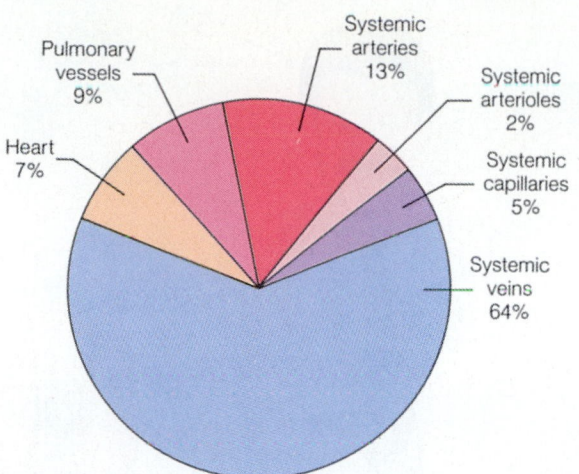

● **FIGURE 10–28   Percentage of Total Blood Volume in Different Parts of the Circulatory System**

(how much they can stretch to hold blood) and the influence of any externally applied pressure squeezing inwardly on the veins. At a constant blood volume, as venous capacity increases, more blood remains in the veins instead of being returned to the heart. Such venous storage decreases the effective circulating volume. Conversely, when venous capacity decreases, more blood is returned to the heart and continues circulating. Thus, changes in venous capacity directly influence the magnitude of venous return, which in turn is an important (although not the only) determinant of effective cir-

culating blood volume. The magnitude of the total blood volume is also influenced on a short-term basis by passive shifts in bulk flow between the vascular and interstitial-fluid compartments and on a long-term basis by factors that control total ECF volume, such as salt and water balance.

Venous return refers to the volume of blood entering each atrium per minute from the veins. Recall that the magnitude of flow through a vessel is directly proportional to the pressure gradient. Much of the driving pressure imparted to the blood by cardiac contraction has been lost by the time the blood reaches the venous system because of frictional losses along the way, especially during passage through the high-resistance arterioles. By the time the blood enters the venous system, mean pressure averages only 17 mm Hg (Fig. 10–9). However, since atrial pressure is near 0 mm Hg, a small but adequate driving pressure still exists to promote the flow of blood through the large-radius, low-resistance veins. If atrial pressure becomes pathologically elevated, as in the presence of a leaky AV valve, the venous-to-atrium pressure gradient is decreased, reducing venous return and causing blood to dam up in the venous system (congestive heart failure).

In addition to the driving pressure imparted by cardiac contraction, five other factors enhance venous return: sympathetically induced venous vasoconstriction, skeletal-muscle activity, the effect of venous valves, respiratory activity, and cardiac-suction effect (● Fig. 10–29). Most of these secondary factors affect venous return by influencing the pressure gradient between the veins and the heart. We will examine each in turn.

● **FIGURE 10–29   Factors That Facilitate Venous Return**

↑ Cardiac output

↑ Stroke volume

↑ End-diastolic volume

| Passive bulk-flow shift of fluid from interstitial fluid into plasma | Salt and water retention |

Venous valves (mechanically prevent backflow of blood) — + → ↑ Venous return ← + — ↑ Blood volume (↑ venous pressure → ↑ pressure gradient)

Cardiac-suction effect (↓ pressure in heart → ↑ pressure gradient) — + +

Respiratory pump (↓ pressure in chest veins → ↑ pressure gradient)

+

Pressure imparted to blood by cardiac contraction (↑ venous pressure → ↑ pressure gradient)

↑ Sympathetic vasoconstrictor activity (↑ venous pressure → ↑ pressure gradient)

Skeletal-muscle pump (↑ venous pressure → ↑ pressure gradient)

□ = Short-term control measures   □ = Long-term control measures

Effect of Sympathetic Activity on Venous Return Veins are not very muscular and have little inherent tone, but venous smooth muscle is abundantly supplied with sympathetic nerve fibers. Sympathetic stimulation produces venous vasoconstriction, which modestly elevates venous pressure; this, in turn, increases the pressure gradient to drive more blood from the veins into the right atrium. The veins normally have such a large diameter that the moderate vasoconstriction accompanying sympathetic stimulation has little effect on resistance to flow. Even when constricted, the veins still have a relatively large diameter and are still low-resistance vessels.

In addition to mobilizing the stored blood, venous vasoconstriction enhances venous return by decreasing venous capacity. With the filling capacity of the veins reduced, less blood draining from the capillaries remains in the veins but continues to flow instead toward the heart.

It is important to recognize the different outcomes of vasoconstriction in arterioles and veins. Arteriolar vasoconstriction *reduces* flow through these vessels because of their increased resistance (less blood can enter and flow through a narrowed arteriole), whereas venous vasoconstriction *increases* flow through these vessels because of their decreased capacity (narrowing of veins squeezes out more of the blood that is already present in the veins, thus increasing blood flow through these vessels).

Effect of Skeletal-Muscle Activity on Venous Return Many of the large veins in the extremities lie between skeletal muscles so that when the muscles contract, the veins are compressed. This external venous compression decreases venous capacity and increases venous pressure, in effect squeezing fluid contained in the veins forward toward the heart (● Fig. 10–30). This pumping action, known as the **skeletal-muscle pump**, is one way by which extra blood stored in the veins is returned to the heart during exercise. Increased muscular activity pushes more blood out of the veins and into the heart. Increased sympathetic activity and the resultant venous vasoconstriction also accompany exercise, further enhancing venous return.

The skeletal-muscle pump also counters the effect of gravity on the venous system. The average pressures provided thus far for various regions of the vascular tree are for a person in the horizontal position. When a person is lying down, the force of gravity is uniformly applied, so it does not have to be taken into consideration. When a person stands up, however, gravitational effects are not uniform. In addition to the usual pressure that results from cardiac contraction, vessels below the level of the heart are subjected to pressure caused by the weight of the column of blood extending from the heart to the level of the vessel (● Fig. 10–31). There are two important consequences of this increased pressure. First, the distensible veins yield under the increased hydrostatic pressure, further expanding so that their capacity is increased. Even though the arteries are subjected to the same gravitational effects, they do not expand like the veins because arteries are not nearly as distensible. Much of the blood entering from the capillaries tends to pool in the expanded lower-leg veins instead of returning to the heart. Because venous return is reduced, car-

Vein

Skeletal muscle relaxed

To heart

Skeletal muscle contracted

● **FIGURE 10–30 Skeletal-Muscle Pump Enhancing Venous Return**

diac output is decreased and the effective circulating volume is reduced. Second, the marked increase in capillary blood pressure resulting from the effect of gravity causes excessive fluid to filter out of capillary beds in the lower extremities, producing localized edema (that is, swollen feet and ankles).

Two compensatory measures normally counteract these gravitational effects. First, the resultant fall in mean arterial pressure that occurs when a person moves from a lying-down to upright position triggers sympathetically induced venous vasoconstriction, which drives some of the pooled blood forward. Second, the skeletal-muscle pump "interrupts" the column of blood by completely emptying given vein segments intermittently so that a particular portion of a vein is not subjected to the weight of the entire venous column from the

Pressure =
0 mm Hg

1.5 m

Pressure = 100 mm Hg

90 mm Hg caused by gravitational effect

10 mm Hg caused by pressure imparted
by cardiac contraction

Pressure =
90 mm Hg

(a)

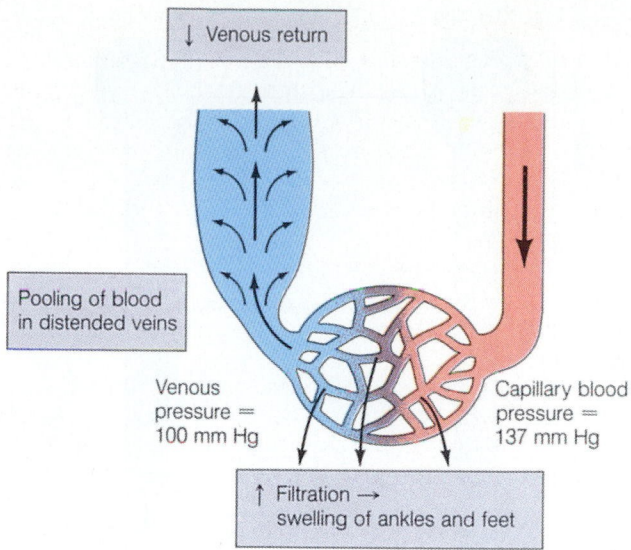

↓ Venous return

Pooling of blood
in distended veins

Venous
pressure =
100 mm Hg

Capillary blood
pressure =
137 mm Hg

↑ Filtration →
swelling of ankles and feet

• **FIGURE 10–31   Effect of Gravity on Venous Pressure**
(a) In an upright adult, the blood in the vessels extending between the
heart and foot is equivalent to a 1.5-meter column of blood. The pressure exerted by this column of blood as a result of the effect of gravity
is 90 mm Hg. The pressure imparted to the blood by the heart has
declined to about 10 mm Hg in the lower-leg veins because of frictional
losses in preceding vessels. The pressure caused by gravity (90 mm Hg)
added to the pressure imparted by the heart (10 mm Hg) produces a
venous pressure of 100 mm Hg in the ankle and foot veins. Similarly, the
capillaries in the region are subjected to these same gravitational effects.
(b) Because of the increased pressure caused by the gravitational effect,
blood pools in the distended veins, resulting in decreased venous return. Filtration also increases across the capillary walls, resulting in
swollen ankles and feet, unless compensatory measures are able to
counteract the effect of gravity.

heart to its level (Figs. 10–30 and • 10–32). Reflex venous vasoconstriction cannot completely compensate for gravitational effects without the assistance of skeletal-muscle activity. Therefore, when a person stands still for a long time,
blood flow to the brain is reduced because of the decline in
effective circulating volume, despite reflexes aimed at maintaining mean arterial pressure. Reduced flow of blood to the
brain, in turn, leads to fainting, which returns the person to a
horizontal position, thereby eliminating the gravitational effects on the vascular system and enabling effective circulation to be restored. For this reason, it is counterproductive to
try to hold someone who has fainted upright. Fainting is the
remedy to the problem, not the problem itself.

Since the skeletal-muscle pump facilitates venous return, it
is advisable to move around when you are on your feet and

to get up periodically when you are working at a desk. The
mild muscular activity "gets the blood moving." It is further
recommended that individuals who must be on their feet for
long periods of time use elastic stockings that apply a continuous gentle external compression, similar to the effect of skeletal-muscle contraction, to further counter the effect of gravitational pooling of blood in the leg veins.

Effect of Venous Valves on Venous Return   Venous vasoconstriction and external venous compression both drive
blood in the direction of the heart. Yet if you squeeze a fluid-
filled tube in the middle, fluid is pushed in both directions
from the point of constriction (• Fig. 10–33a). Why, then, isn't
blood driven backward as well as forward by venous vasoconstriction and the skeletal-muscle pump? Blood can only
be driven forward because the large veins are equipped with
one-way valves spaced at 2- to 4-cm intervals; these valves permit blood to move forward toward the heart but prevent it
from moving back toward the tissues (Fig. 10–33b). These venous valves also play a role in counteracting the gravitational
effects of upright posture by helping minimize the backflow
of blood that tends to occur when a person stands up and by
temporarily supporting portions of the column of blood when
the skeletal muscles are relaxed (Fig. 10–33c).

Standing | Walking

Heart

Thigh

150 cm

Calf

34 cm

Foot

| 100 mm Hg | Venous pressure in foot | 27 mm Hg |

Foot vein supporting column of blood 1.5 m (150 cm) in height

Foot vein supporting column of blood 34 cm in height

● **FIGURE 10–32   Effect of Contraction of the Skeletal Muscles of the Legs in Counteracting the Effects of Gravity** Contraction of skeletal muscles (as in walking) completely empties given vein segments, interrupting the column of blood that must be supported by the lower veins.

Varicose veins occur when the venous valves become incompetent and can no longer support the column of blood above them. Individuals predisposed to this condition usually have an inherited overdistensibility and weakness of their vein walls. Aggravated by frequent, prolonged standing, the veins become so distended as blood pools in them that the edges of the valves are no longer able to meet to form a seal. Varicosed superficial leg veins become visibly overdistended and tortuous. Contrary to what might be expected, the chronic pooling of blood in the pathologically distended veins does not reduce cardiac output because of a compensatory increase in total circulating blood volume. Instead, the most serious consequence of varicosed veins is the possibility of abnormal clot formation in the sluggish, pooled blood. Particularly dangerous is the risk that these clots may break loose and block small vessels elsewhere, especially the pulmonary capillaries.

Effect of Respiratory Activity on Venous Return As a result of respiratory activity, the pressure within the chest cavity averages 5 mm Hg less than atmospheric pressure. As the venous system returning blood to the heart from the lower regions of the body travels through the chest cavity, it is ex-

posed to this subatmospheric pressure. Because the venous system in the limbs and abdomen is subjected to normal atmospheric pressure, an externally applied pressure gradient exists between the lower veins (at atmospheric pressure) and the chest veins (at 5 mm Hg less than atmospheric pressure). This pressure difference squeezes blood from the lower veins to the chest veins, promoting increased venous return (● Fig. 10–34). This mechanism of facilitating venous return is known as the **respiratory pump** because it results from respiratory activity. Increased respiratory activity as well as the effects of the skeletal-muscle pump and venous vasoconstriction all enhance venous return during exercise.

Effect of Cardiac Suction on Venous Return The extent of cardiac filling does not depend entirely on factors affecting the veins. The heart plays a role in its own filling. During ventricular contraction, the AV valves are drawn downward, enlarging the atrial cavities. As a result, the atrial pressure transiently drops below 0 mm Hg, thus increasing the venous-to-atrial pressure gradient so that venous return is enhanced. In addition, the rapid expansion of the ventricular chamber during ventricular relaxation appears to create a transient negative pressure in the ventricles so that blood is "sucked in" from the atria and veins; that is, the negative ventricular pressure increases the vein-to-atria-to-ventricle pressure gradient, further enhancing venous return. Thus, the heart functions as a "suction pump" to facilitate cardiac filling.

☰ BLOOD PRESSURE

Regulation of mean arterial blood pressure is accomplished by controlling cardiac output, total peripheral resistance, and blood volume.

Mean arterial blood pressure is the main driving force for propelling blood to the tissues. This pressure must be closely regulated for two reasons. First, it must be high enough to assure sufficient driving pressure; without this pressure, the brain and other tissues will not receive adequate flow, no matter what local adjustments are made in the resistance of the arterioles supplying them. Second, the pressure must not be so high that it creates extra work for the heart and increases the risk of vascular damage and possible rupture of small blood vessels.

Elaborate mechanisms involving the integrated action of the various components of the circulatory system and other body systems are vital in the regulation of this all-important mean arterial pressure (● Fig. 10–35). Remember from an earlier discussion that the two determinants of mean arterial pressure are cardiac output and total peripheral resistance:

$$\text{mean arterial pressure} = \text{cardiac output} \times \text{total peripheral resistance}$$

A number of factors, in turn, determine cardiac output (Fig. 9–28, p. 288) and total peripheral resistance (Fig. 10–14, p. 316). Thus, one can quickly appreciate the complexity of

Vein

Open venous valve permits flow of blood toward heart

Contracted skeletal muscle

Closed venous valve prevents backflow of blood

(a)   (b)   (c)

● **FIGURE 10–33   Function of Venous Valves** (a) Fluid is pushed in both directions when a tube is squeezed in the middle. (b) Venous valves permit the flow of blood only toward the heart. (c) Venous valves play a role in counteracting gravitational effects by minimizing the backflow of blood and temporarily supporting portions of the column of blood.

blood-pressure regulation. Altering any of the pertinent factors will produce a change in blood pressure unless a compensatory change in another variable keeps the blood pressure constant. Blood flow to any given tissue depends on the driving force of the mean arterial pressure and on the degree of vasoconstriction of the tissue's arterioles. Because mean arterial pressure depends on the cardiac output and the degree of arteriolar vasoconstriction, if the arterioles in one tissue dilate, the arterioles in other tissues will have to constrict to maintain an adequate arterial blood pressure to provide a driving force to push blood not only to the vasodilated tissue but also to the brain, which depends on a constant blood supply. Thus, the cardiovascular variables must be continu-

ously juggled to maintain a constant blood pressure in spite of tissues' varying needs for blood.

Mean arterial pressure is constantly monitored by **baroreceptors** (pressure sensors) within the circulatory system. When deviations from normal are detected, multiple reflex responses are initiated to return the arterial pressure to its normal value. *Short-term* (within seconds) adjustments are accomplished by alterations in cardiac output and total peripheral resistance, mediated by means of autonomic nervous-system influences on the heart, veins, and arterioles. *Long-term* (requiring minutes to days) control involves adjusting total blood volume by restoring normal salt and water balance through mechanisms that regulate urine output and

● **FIGURE 10–34   Respiratory Pump Enhancing Venous Return** As a result of respiratory activity, the pressure surrounding the chest veins is lower than the pressure surrounding the veins in the extremities and abdomen. This establishes an externally applied pressure gradient on the veins that drives blood toward the heart.

5mm Hg less than atmospheric pressure

Atmospheric pressure

5mm Hg less than atmospheric pressure

Atmospheric pressure

• FIGURE 10–35   Determinants of Mean Arterial Blood Pressure   Note that this figure is basically a composite of Figure 9–28, Control of Cardiac Output; Figure 10–14, Factors Affecting Total Peripheral Resistance; and Figure 10–29, Factors That Facilitate Venous Return.

thirst (chapters 14 and 15). The magnitude of the total blood volume, in turn, has a profound effect on cardiac output and mean arterial pressure. Let us now turn our attention to the short-term reflex mechanisms involved in the ongoing regulation of this pressure.

The baroreceptor reflex is the most important mechanism for short-term regulation of blood pressure.

Any change in mean blood pressure triggers an autonomically mediated **baroreceptor reflex** that influences the heart and blood vessels to adjust cardiac output and total peripheral resistance in an attempt to restore blood pressure to normal. Like any reflex, the baroreceptor reflex includes a receptor, an afferent pathway, an integrating center, an efferent pathway, and effector organs.

The most important receptors involved in moment-to-moment regulation of blood pressure, the **carotid-sinus** and **aortic-arch baroreceptors**, are mechanoreceptors sensitive to

changes in both mean arterial pressure and pulse pressure. Their responsiveness to fluctuations in pulse pressure enhances their sensitivity as pressure sensors, because small changes in systolic or diastolic pressure may alter the pulse pressure without changing the mean pressure. These baroreceptors are strategically located (• Fig. 10–36) to provide critical information about arterial blood pressure in the vessels leading to the brain (the carotid-sinus baroreceptor) and in the major arterial trunk before it gives off branches that supply the rest of the body (the aortic-arch baroreceptor).

The baroreceptors constantly provide information about blood pressure; in other words, they continuously generate action potentials in response to the ongoing pressure within the arteries. When arterial pressure (either mean or pulse pressure) increases, the receptor potential of these baroreceptors increases, thus increasing the rate of firing in the corresponding afferent neurons. Conversely, when blood pressure decreases, the rate of firing generated in the afferent neurons by the baroreceptors decreases (• Fig. 10–37).

The integrating center that receives the afferent impulses

● **FIGURE 10–36   Location of Arterial Baroreceptors**

Carotid-sinus
baroreceptor

Neural signals to
cardiovascular
control center
in medulla

Common carotid arteries
Blood to the brain

Aortic arch

Aortic-arch baroreceptor

Aorta
Blood to rest of body

Left ventricle

about the status of arterial pressure is the **cardiovascular control center**,[4] located in the medulla within the brain stem. The efferent pathway is the autonomic nervous system. The cardiovascular control center alters the ratio between sympathetic and parasympathetic activity to the effector organs (the heart and blood vessels). To show how autonomic changes alter arterial blood pressure, ● Figure 10–38 provides a review of the major effects of parasympathetic and sympathetic stimulation on the heart and blood vessels.

Let us fit all the pieces of the baroreceptor reflex together now by tracing through the reflex activity occurring to compensate for an elevation or a fall in blood pressure. If for any reason arterial pressure becomes elevated above normal (● Fig. 10–39a), the carotid-sinus and aortic-arch baroreceptors increase the rate of firing in their respective afferent neurons. Upon being informed by increased afferent firing that arterial pressure has become too high, the cardiovascular control center responds by decreasing sympathetic and in-

creasing parasympathetic activity to the cardiovascular system. These efferent signals decrease heart rate, decrease stroke volume, and produce arteriolar and venous vasodilation, which in turn lead to a decrease in cardiac output and a decrease in total peripheral resistance, with a subsequent decrease in blood pressure back toward normal.

Conversely, when blood pressure falls below normal

● **FIGURE 10–37   Firing Rate in Afferent Neurons from Carotid-Sinus Baroreceptor in Relation to Magnitude of Mean Arterial Pressure**

Firing rate in afferent neuron arising
from carotid-sinus baroreceptor

[4] The cardiovascular control center is sometimes divided into cardiac and vasomotor centers, which are occasionally further classified into smaller subdivisions, such as cardioaccelleratory and cardioinhibitory centers and vasoconstrictor and vasodilator areas. Since these regions are highly interconnected and functionally interrelated, we will refer to them collectively as the cardiovascular control center.

● **FIGURE 10–38** **Summary of Effects of Parasympathetic and Sympathetic Nervous Systems on Factors That Influence Mean Arterial Blood Pressure**

● **FIGURE 10–39** **Baroreceptor Reflexes to Restore Blood Pressure to Normal** (a) Baroreceptor reflex in response to an elevation in blood pressure. (b) Baroreceptor reflex in response to a fall in blood pressure.

| When blood pressure becomes elevated above normal | → | ↑ Carotid-sinus and aortic-arch receptor potential | → | ↑ Rate of firing in afferent nerves | → | Cardiovascular center |

| ↓ Sympathetic cardiac-nerve activity and ↓ sympathetic vasoconstrictor-nerve activity and ↑ parasympathetic nerve activity | → | ↓ Heart rate and ↓ stroke volume and arteriolar and venous vasodilation | → | ↓ Cardiac output and ↓ total peripheral resistance | → | Blood pressure decreased toward normal |

(a)

| When blood pressure falls below normal | → | ↓ Carotid-sinus and aortic-arch receptor potential | → | ↓ Rate of firing in afferent nerves | → | Cardiovascular center |

| ↑ Sympathetic cardiac-nerve activity and ↑ sympathetic vasoconstrictor-nerve activity and ↓ parasympathetic nerve activity | → | ↑ Heart rate and ↑ stroke volume and arteriolar and venous vasoconstriction | → | ↑ Cardiac output and ↑ total peripheral resistance | → | Blood pressure increased toward normal |

(b)

(Fig. 10–39b), baroreceptor activity decreases, inducing the cardiovascular center to increase sympathetic cardiac- and vasoconstrictor-nerve activity while decreasing its parasympathetic output. This efferent pattern of activity leads to an increase in heart rate and stroke volume coupled with arteriolar and venous vasoconstriction. These changes result in an increase in both cardiac output and total peripheral resistance, producing an elevation in blood pressure back toward normal.

Other reflexes and responses influence blood pressure.

Besides the baroreceptor reflex, whose sole function is blood-pressure regulation, several other reflexes and responses influence the cardiovascular system even though they primarily are concerned with the regulation of other body functions. Some of these other influences deliberately move arterial pressure away from its normal value temporarily, overriding the baroreceptor reflex to accomplish a particular goal. These factors include the following:

1. Left atrial volume receptors and hypothalamic osmoreceptors are primarily important in water and salt balance in the body; thus, they affect the long-term regulation of blood pressure by controlling the plasma volume.

2. Chemoreceptors located in the carotid and aortic arteries, in close association with but distinct from the baroreceptors, are sensitive to low O_2 or high acid levels in the blood. These chemoreceptors' main function is to reflexly increase respiratory activity to bring in more O_2 or to blow off more acid-forming CO_2, but they also reflexly increase blood pressure by sending excitatory impulses to the cardiovascular center.

3. Cardiovascular responses associated with certain behaviors and emotions are mediated through the cerebral cortex-hypothalamic pathway and appear to be preprogrammed. These responses include the widespread changes in cardiovascular activity accompanying the generalized sympathetic fight-or-flight response, the characteristic marked increase in heart rate and blood pressure associated with sexual orgasm, and the localized cutaneous vasodilation characteristic of blushing.

4. Pronounced cardiovascular changes accompany exercise, including a substantial increase in skeletal-muscle blood flow; a significant increase in cardiac output; a fall in total peripheral resistance (because of widespread vasodilation in skeletal muscles despite generalized arteriolar vasoconstriction in most organs); and a modest increase in mean arterial pressure (Table 10–4). Evidence suggests that discrete exercise centers yet to be identified

TABLE 10–4 Cardiovascular Changes During Exercise		
CARDIOVASCULAR VARIABLE	**CHANGE**	**COMMENT**
Heart rate	Increases	Occurs as a result of increased sympathetic and decreased parasympathetic activity to the SA node
Venous return	Increases	Occurs as a result of sympathetically induced venous vasoconstriction and increased activity of the skeletal-muscle pump and respiratory pump
Stroke volume	Increases	Occurs both as a result of increased venous return by means of the Frank-Starling mechanism (unless diastolic filling time is significantly reduced due to a high heart rate) and as a result of a sympathetically induced increase in myocardial contractility
Cardiac output	Increases	Occurs as a result of increases in both heart rate and stroke volume
Blood flow to active skeletal muscles and heart muscle	Increases	Occurs as result of locally controlled arteriolar vasodilation, which is reinforced by the vasodilatory effects of epinephrine and overpowers the weaker sympathetic vasoconstrictor effect
Blood flow to brain	Unchanged	Occurs because sympathetic stimulation has no effect on brain arterioles; local control mechanisms maintain constant cerebral blood flow whatever the circumstances
Blood flow to skin	Increases	Occurs because the hypothalamic temperature control center induces vasodilation of skin arterioles; increased skin blood flow brings heat produced by exercising muscles to the body surface where the heat can be lost to the external environment
Blood flow to digestive system, kidneys, and other organs	Decreases	Occurs as a result of generalized sympathetically induced arteriolar vasoconstriction
Total peripheral resistance	Decreases	Occurs because resistance in skeletal muscles, heart, and skin decreases to a greater extent than resistance in the other organs increases
Mean arterial blood pressure	Increases (modest)	Occurs because cardiac output increases to a greater extent than total peripheral resistance decreases

induce the appropriate cardiac and vascular changes at the onset of exercise or even in anticipation of exercise. These effects are then reinforced by afferent inputs to the medullary cardiovascular center from chemoreceptors in exercising muscles as well as by local mechanisms important in maintaining vasodilation in active muscles. The baroreceptor reflex further modulates these cardiovascular responses.

5. Hypothalamic control over cutaneous (skin) arterioles for the purpose of temperature regulation takes precedence over control that the cardiovascular center has over these same vessels for the purpose of blood-pressure regulation. As a result, blood pressure can fall when the skin vessels are widely dilated to eliminate excess heat from the body, even though the baroreceptor responses are calling for cutaneous vasoconstriction to help maintain aequate total peripheral resistance.

6. Vasoactive substances released from the endothelial cells may play a role in the ongoing regulation of blood pressure, as exemplified by the following piece of evidence. Experimental inhibition of the enzyme that catalyzes the synthesis of EDRF/NO causes a rapid increase in blood pressure, suggestive that this chemical may normally exert an ongoing vasodilatory effect.

7. Recent studies suggest that numerous neurotransmitters from various regions of the brain may play a role in the control of blood pressure. Their importance and relations to the traditional pathways for blood-pressure regulation are currently unclear, as are any possible roles they may play in the poorly understood condition of hypertension.

Hypertension is a serious national public-health problem, but its causes are largely unknown.

Sometimes blood-pressure control mechanisms do not function properly or are unable to completely compensate for changes that have taken place. Blood pressure may be above the normal range (**hypertension** if above 140/90 mm Hg) or below normal (**hypotension** if less than 100/60 mm Hg). Hypotension in its extreme form is *circulatory shock*. We will first examine hypertension and then conclude this chapter with a discussion of hypotension and shock.

A definite cause for hypertension can be established in only 10% of the cases. Hypertension that occurs secondary to another primary problem is called **secondary hypertension**. The causes of secondary hypertension fall into four categories:

1. *Cardiovascular hypertension* is usually associated with chronically elevated total peripheral resistance caused by atherosclerosis (hardening of the arteries; see p. 291).

2. *Renal (kidney) hypertension* may occur as a result of either of two kidney defects: partial occlusion of the renal arteries or diseases of the kidney tissue itself.

   a. Atherosclerotic lesions protruding into the lumen of a renal artery or external compression of the vessel by a tumor may reduce blood flow through the kidney. The kidney responds by initiating the hormonal pathway involving angiotensin II. This pathway promotes salt and water retention during urine formation, thus increasing the blood volume to compensate for the reduced renal blood flow. Recall that angiotensin II is also a powerful vasoconstrictor. Although these two effects (increased blood volume and angiotensin-induced vasoconstriction) are compensatory mechanisms to improve blood flow through the narrowed renal artery, they also are responsible for elevating the arterial pressure as a whole.

   b. Renal hypertension also occurs if the kidneys are diseased and unable to eliminate the normal salt load. Salt retention induces water retention, which expands the plasma volume and leads to hypertension.

3. *Endocrine hypertension* arises from at least two different endocrine disorders: pheochromocytoma and Conn's syndrome.

   a. A *pheochromocytoma* is an adrenal medullary tumor that secretes excessive epinephrine and norepinephrine. Abnormally elevated levels of these hormones induce increased cardiac output and generalized peripheral vasoconstriction, both of which contribute to the hypertension characteristic of this disorder.

   b. *Conn's syndrome* is associated with increased production of aldosterone by the adrenal cortex. This hormone is part of the hormonal pathway responsible for salt and water retention by the kidneys (the renin-angiotensin-aldosterone pathway). Once again, the excessive salt and water load in the body caused by increased aldosterone levels leads to an elevation in blood pressure.

4. *Neurogenic hypertension* occurs secondary to neural lesions.

   a. The problem may be erroneous blood-pressure control caused by a defect in the cardiovascular control center or in the baroreceptors.

   b. Neurogenic hypertension may also occur as a compensatory response to a reduction in cerebral blood flow—for example, because a major cerebral vessel is compressed by a tumor. In response to a decrease in brain blood flow, reflexes are initiated that elevate the blood pressure in an attempt to provide sufficient driving pressure to adequately supply the O_2-dependent brain tissue with blood.

The underlying cause is unknown in the remaining 90% of hypertension cases. Such hypertension is known as **primary** (**essential** or **idiopathic**) **hypertension**. Primary hypertension is undoubtedly a catchall category for elevated blood pressure caused by a variety of unknown causes rather than a single disease entity. There is a strong genetic tendency to develop primary hypertension, which can be hastened or worsened by contributing factors such as obesity, stress, smoking, and excessive ingestion of salt. The following illustrate the range of potential causes for primary hypertension that are currently being investigated:

- *Defects in salt management.* Disturbances in kidney function too minor to produce outward signs of renal disease could nevertheless insidiously lead to gradual accumulation of salt and water in the body, resulting in progressive elevation in arterial pressure.

- *Plasma-membrane abnormalities such as defective Na^+-K^+ pumps.* Such defects, by altering the electrochemical gradient across plasma membranes, could change the excitability and contractility of the heart and the smooth muscle in blood-vessel walls in such a way as to lead to high blood pressure. In addition, the Na^+-K^+ pump is critical to salt management by the kidneys. A genetic defect in the Na^+-K^+ pump of hypertensive-prone laboratory rats is the first gene-hypertension link to be discovered.

- *Physical pressure on the cardiovascular control center by an overlying artery.* One neurosurgeon, in a limited number of operations, has successfully reduced high blood pressure by moving an enlarged loop of artery that pulsated against the medullary brain tissue.

- *Endogenous digitalis-like substances.* Such substances act similarly to the drug digitalis (see p. 274) to increase cardiac contractility as well as constrict blood vessels and reduce salt elimination in the urine, all of which could cause chronic hypertension.

- *Changes in the regulation of EDRF/NO or other locally acting vasoactive chemicals* could contribute to the development of hypertension.

Whatever the underlying defect, once initiated, hypertension appears to be self-perpetuating. Constant exposure to elevated blood pressure predisposes vessel walls to the development of atherosclerosis, which further elevates blood pressure.

The baroreceptors do not respond to bring the blood pressure back to normal during hypertension because they adapt or are "reset" to operate at a higher level. In the presence of chronically elevated blood pressure, the baroreceptors still function to regulate blood pressure, but they maintain it at a higher mean pressure.

Hypertension imposes stresses on both the heart and the blood vessels. The heart has an increased work load because it is pumping against an increased total peripheral resistance, whereas blood vessels may be damaged by the high internal pressure, particularly when the vessel wall is weakened by the degenerative process of atherosclerosis. Complications of hypertension include congestive heart failure caused by the heart's inability to pump continuously against a sustained elevation in arterial pressure (see p. 288), strokes caused by rupture of brain vessels, or heart attacks caused by rupture of coronary vessels. Spontaneous hemorrhage due to bursting of small vessels elsewhere in the body also may occur but with less serious consequences, such as rupture of blood vessels in the nose resulting in nosebleeds. Another serious complication of hypertension is renal failure caused by progressive impairment of blood flow through damaged renal blood vessels. Furthermore, retinal damage caused by changes in the blood vessels supplying the eyes may result in progressive loss of vision. Until complications occur, hypertension is symptomless because the tissues are adequately supplied with blood. Therefore, unless blood-pressure measurements are made on a routine basis, the condition can go undetected until a precipitous complicating event results. When one becomes aware of these potential complications of hypertension and considers that 25% of all adults in America are estimated to be afflicted with chronic elevated blood pressure, one can appreciate the magnitude of this national health problem.

Once hypertension is detected, therapeutic intervention can reduce the course and severity of the problem. Cornerstones of treatment include reduction in salt intake and administration of diuretics (drugs that enhance urine excretion) to reduce the salt and water load in the body, thereby decreasing plasma volume. In addition, other antihypertensive drugs reduce total peripheral resistance by manipulating some aspect of autonomic function to promote arteriolar vasodilation. No matter what the original cause, agents that reduce the plasma volume or total peripheral resistance (or both) will decrease the blood pressure toward normal. Furthermore, a regular aerobic exercise program can be employed to help reduce high blood pressure. (See the accompanying boxed feature, A Closer Look at Exercise Physiology.)

Inadequate sympathetic activity is responsible for dizziness or fainting accompanying transient hypotension.

Hypotension, or low blood pressure, occurs either when there is a disproportion between vascular capacity and blood volume or when the heart is too weak to impart sufficient driving pressure to the blood.

Two common situations in which hypotension occurs transiently are orthostatic hypotension and emotional fainting. Both are due to inadequate sympathetic activity. **Orthostatic (postural) hypotension** is a transient hypotensive condition resulting from insufficient compensatory responses to the gravitational shifts in blood that occur when a person moves from a horizontal to a vertical position, especially following a prolonged bed rest. The fall in blood pressure that results from blood pooling in the leg veins upon standing up is normally detected by the baroreceptors, which initiate immediate compensatory responses to restore blood pressure to its proper level. When a long-bedridden patient first starts to rise, however, these reflex compensatory adjustments are temporarily lost or reduced because of disuse. Sympathetic control of the leg veins is inadequate, so when the patient first stands up, blood pools in the lower extremities. The resultant orthostatic hypotension and decrease in brain blood flow are responsible for the dizziness or actual fainting that occurs. Because postural compensatory mechanisms are depressed during prolonged bed confinement, patients sometimes are put on a tilt table so that they can be moved gradually from a horizontal to an upright position. This allows the body to adjust slowly to the gravitational shifts in blood.

Transient hypotension because of emotional stress can

THE UPS AND DOWNS OF HYPERTENSION AND EXERCISE

When blood pressure is up, one way to bring it down is to increase the level of physical activity. An impressive epidemiologic study, which followed a group of 14,998 male graduates from Harvard for sixteen to fifty years, found that participation in collegiate sports, climbing as many as fifty stairs a day, walking five blocks, or light sports activities were not protective against the development of primary hypertension. Participation in vigorous sports such as running, swimming, handball, tennis, and cross-country skiing, on the other hand, was protective against the development of hypertension, even if other risk factors were present. Other studies have supported the finding that participation in aerobic activities is protective against the development of hypertension.

A logical question to ask is whether exercise can be used as a therapy to reduce hypertension once it has already developed. Antihypertensive medication is available to lower blood pressure in severely hypertensive patients, but sometimes undesirable side effects occur. The side effects of diuretics include electrolyte imbalances, glucose intolerance, and increased blood cholesterol levels. The side effects of drugs that manipulate total peripheral resistance, such as beta-blocker drugs, include increased blood triglyceride levels, lower HDL-cholesterol levels (the "good" form of cholesterol), weight gain, sexual dysfunction, and depression.

Patients with mild hypertension, arbitrarily defined as a diastolic blood pressure between 90 and 100 mm Hg and a systolic pressure of 160 mm Hg, pose a dilemma for physicians. The risks of taking the drugs may outweigh the benefits gained from lowering the blood pressure. Because of the drug therapy's possible side effects, nondrug treatment of mild hypertension may be most beneficial. The most common nondrug therapies are weight reduction, salt restriction, and exercise. Although losing weight will almost always reduce blood pressure, research has shown that weight reduction programs usually result in the loss of only twelve pounds, and the overall long-term success in keeping the weight off is only about 20%. Salt restriction is beneficial for many hypertensives, but adherence to a low-salt diet is difficult for many people because fast foods and foods prepared in restaurants usually contain high amounts of salt. Some recent studies employing exercise as a therapeutic tool have shown that blood pressure in cases of mild to moderate hypertension has been decreased regardless of whether salt was restricted or weight was lowered. The preponderance of evidence in the literature suggests that moderate aerobic exerciseperformed three times per week forfifteen to sixty minutes is a beneficial therapy in mild to moderate hypertension. It is wise, therefore, to include a regular aerobic exercise program in conjunction with weight loss and salt reduction to optimally reduce high blood pressure without drug intervention.

---

also cause dizziness or fainting. In this situation, higher brain centers act on the cardiovascular center to inappropriately decrease sympathetic output to the vasculature. The resultant loss of vascular tone precipitates widespread arteriolar vasodilation that leads to a decrease in total peripheral resistance. Furthermore, widespread arteriolar vasodilation causes pooling of blood in the capillaries so that venous return is decreased and cardiac output is subsequently reduced. Thus, hypotension is brought about by a decrease in both total peripheral resistance and cardiac output. As the blood pressure falls, the person becomes light-headed or faints because blood flow to the brain becomes inadequate. If the person faints or lies down, cardiac output is rapidly improved as the pooled blood quickly returns to the heart. The inappropriate autonomic nerve activity is only transient; when it is discontinued, vasoconstrictor tone is quickly restored. The person normally recovers quite rapidly with no aftereffects. Even though fainting in reaction to a highly emotional situation is considered to be an inappropriate response, there is speculation that it may actually be adaptive, analogous to the "playing dead" ploy displayed by some animals. It temporarily relieves the person from the responsibility of coping with the emotionally charged situation.

Circulatory shock can become irreversible.

When blood pressure falls so low that adequate blood flow to the tissues can no longer be maintained, the condition known as **circulatory shock** occurs. Circulatory shock is categorized into four main types (● Fig. 10–40):

1. *Hypovolemic shock* is induced by a fall in blood volume, which occurs either directly through severe hemorrhage or indirectly through loss of fluids derived from the plasma (for example, severe diarrhea, excessive urinary losses, or extensive sweating).

2. *Cardiogenic shock* is due to a weakened heart's failure to pump blood adequately.

3. *Vasogenic shock* is caused by widespread vasodilation triggered by the presence of vasodilator substances. There are two types of vasogenic shock: septic and anaphylactic. *Septic shock,* which may accompany massive infections,

• FIGURE 10–40   Causes of Circulatory Shock   Circulatory shock, which occurs when mean arterial blood pressure falls so low that adequate blood flow to the tissues can no longer be maintained, may result from (1) extensive loss of blood volume (hypovolemic shock), (2) failure of the heart to pump blood adequately (cardiogenic shock), (3) widespread arteriolar vasodilation induced by toxic or allergic vasodilator substances (vasogenic shock), or (4) neurally defective vasoconstrictor tone (neurogenic shock).

is due to vasodilator substances released from the infective agents. Similarly, extensive histamine release accompanying severe allergic reactions can cause widespread vasodilation in *anaphylactic shock.*

4. *Neurogenic shock* also involves generalized vasodilation but not by means of the release of vasodilator substances. In this case, loss of sympathetic vascular tone leads to generalized vasodilation, similar to emotional hypotension but more pronounced and prolonged. This undoubtedly is responsible for the shock accompanying crushing injuries when blood loss has not been sufficient to cause hypovolemic shock. Deep, excruciating pain apparently inhibits sympathetic vasoconstrictor activity.

We will now examine the consequences of and compensations for shock, using hemorrhage as an example (• Fig. 10–41). This figure may look intimidating, but we will work through it step by step. It is an important example that pulls together many of the principles discussed in this chapter. The circled numbers in the text correspond to the numbers in the figure and indicate the portion of the figure being discussed.

■ Following severe loss of blood, the resultant reduction in circulating blood volume leads to a decrease in venous re-

turn ① and a subsequent fall in cardiac output and arterial blood pressure. (Note the blue boxes, which indicate consequences of hemorrhage.)

■ Compensatory measures immediately attempt to maintain adequate blood flow to the brain. (Note the pink boxes, which indicate compensations for hemorrhage.)

■ The baroreceptor reflex response to the fall in blood pressure brings about increased sympathetic and decreased parasympathetic activity to the heart ②. The result is an increase in heart rate ③ to offset the reduced stroke volume ④ brought about by the loss of blood volume. With severe fluid loss, the pulse is weak because of the reduced stroke volume but rapid because of the increased heart rate.

■ As a result of increased sympathetic activity to the veins, generalized venous vasoconstriction occurs ⑤, increasing venous return and leading in turn to an increase in stroke volume by means of the Frank-Starling mechanism ⑥.

■ Simultaneously, sympathetic stimulation of the heart increases the heart's contractility ⑦ so that it beats more forcefully and ejects a greater volume of blood, likewise increasing the stroke volume.

■ The increase in heart rate and increase in stroke volume collectively lead to an increase in cardiac output ⑧.

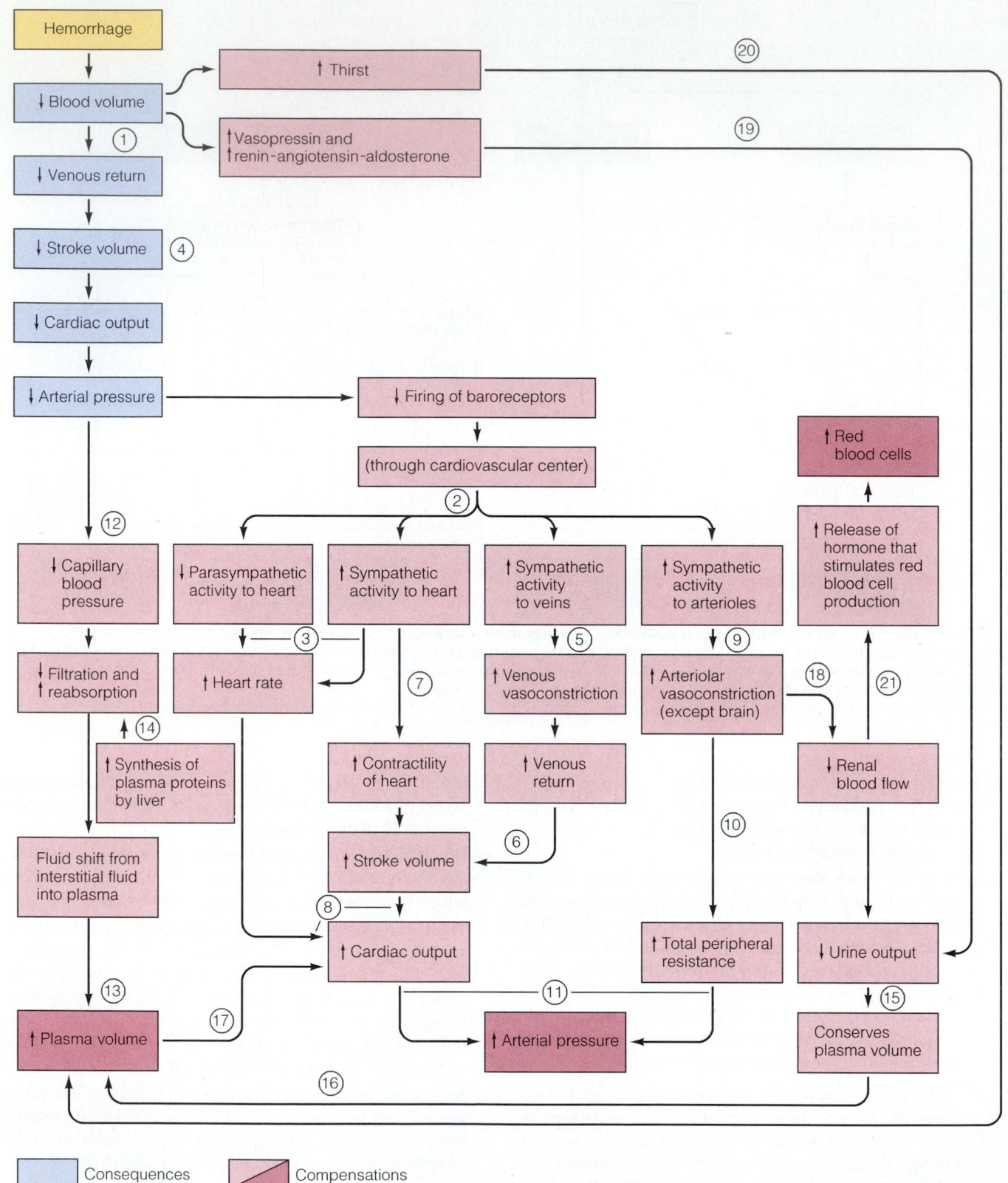

Consequences ▨ Compensations

● **FIGURE 10–41   Consequences and Compensations of Hemorrhage** The reduction in blood volume resulting from hemorrhage leads to a fall in arterial pressure. (Note the blue boxes representing consequences of hemorrhage.) A series of compensations ensue (light pink boxes) that ultimately restore plasma volume, arterial pressure, and the number of red blood cells toward normal (dark pink boxes). Refer to the text (pages 341–343) for an explanation of the circled numbers and a detailed discussion of the compensations.

- Sympathetically induced generalized arteriolar vaso-constriction ⑨ leads to an increase in total peripheral resistance ⑩.
- Together, the increase in cardiac output and total periph-eral resistance bring about a compensatory increase in ar-terial pressure ⑪.
- The original fall in arterial pressure is also accompanied by a fall in capillary blood pressure ⑫, which results in fluid shifts from the interstitial fluid into the capillaries to ex-pand the plasma volume ⑬. This response is sometimes termed **autotransfusion,** because it restores the plasma vol-ume as a transfusion does.
- This ECF fluid shift is enhanced by plasma-protein synthesis by the liver during the next few days following hemorrhage ⑭. The plasma proteins exert a colloid osmotic pressure to attract and retain extra fluid in the plasma.
- Urinary output is reduced, thereby conserving water that normally would have been lost from the body ⑮. This additional fluid retention helps to expand the reduced plasma volume ⑯. Expansion of plasma volume further augments the increase in cardiac output brought about by the baroreceptor reflex ⑰. Reduction in urinary output results from decreased renal blood flow caused by com-pensatory renal arteriolar vasoconstriction ⑱. The re-duced plasma volume also triggers increased secretion of the hormone vasopressin and activation of the salt- and water-conserving renin-angiotensin-aldosterone hormonal pathway, which bring about a further reduction in urinary output ⑲.
- Increased thirst is also stimulated by a fall in plasma vol-ume ⑳. The resultant increased fluid intake contributes to restoration of plasma volume.
- Over a longer course of time (a week or more), lost red blood cells are replaced through increased red blood cell production triggered by a reduction in O_2 delivery to the kidneys ㉑.

These compensatory mechanisms are often insufficient in the face of substantial fluid loss. Even if they are able to main-tain an adequate blood-pressure level, the short-term mea-sures cannot continue indefinitely. Ultimately, fluid volume must be replaced from the outside through drinking, transfu-sion, or a combination of both. Blood supply to the kidneys, digestive tract, skin, and other organs can be compromised to maintain blood flow to the brain only so long before organ damage begins to occur. A point may be reached at which blood pressure continues to drop rapidly because of tissue damage, despite vigorous therapy. This condition is fre-quently termed **irreversible shock**, in contrast to **reversible shock**, which can be corrected by compensatory mecha-nisms and effective therapy.

Although the exact mechanism underlying irreversibility is not currently known, there are many logical possibilities that could contribute to the unrelenting, progressive circulatory deterioration that characterizes irreversible shock. Metabolic acidosis arises when lactic-acid production increases as blood-deprived tissues resort to anaerobic metabolism. Aci-dosis deranges the enzymatic systems responsible for energy production, limiting the capability of the heart and other tis-sues to produce ATP. Prolonged depression of kidney func-tion results in electrolyte imbalances that may lead to cardiac arrhythmias. The blood-deprived pancreas releases a chemi-cal that is toxic to the heart (**myocardial toxic factor**), further weakening the heart. Vasodilator substances build up within ischemic organs, inducing local vasodilation that overrides the generalized reflex vasoconstriction. As cardiac output progressively declines because of the heart's diminishing ef-fectiveness as a pump and total peripheral resistance contin-ues to fall, hypotension becomes increasingly more severe. This causes further cardiovascular failure, which leads to a further decline in blood pressure. Thus, when shock pro-gresses to the point that the cardiovascular system itself starts to fail, a vicious positive-feedback cycle ensues that ulti-mately results in death.

 ## CHAPTER IN PERSPECTIVE: FOCUS ON HOMEOSTASIS

Homeostatically, the blood vessels serve as passageways to transport blood to and from the cells for the purpose of O_2 and nutrient delivery, waste removal, distribution of fluid and electrolytes, and hormonal signaling, among other things. Cells soon die if deprived of their blood supply, with brain cells succumbing within four minutes. Blood is constantly re-cycled and reconditioned as it travels through the various or-gans by means of the vascular highways; hence, the body needs only a very small volume of blood to maintain the ap-propriate chemical composition of the entire internal fluid en-vironment upon which the cells depend for their survival. For example, O_2 is continually picked up by the blood in the lungs and constantly delivered to all the body cells.

The smallest of blood vessels, the capillaries, are the actual site of exchange between the blood and surrounding cells. Capillaries bring homeostatically maintained blood within one-hundredth centimeter from every cell in the body; this proximity is critical because beyond a few centimeters mate-rials cannot diffuse rapidly enough to support life-sustaining activities. Oxygen that would take months to years to diffuse from the lungs to all the cells of the body is continuously de-livered at the "doorstep" of every cell, where diffusion can efficiently accomplish short local exchanges between the capillaries and surrounding cells. Likewise, hormones must be rapidly transported through the circulatory system from their sites of production in endocrine glands to their sites of action in other parts of the body because they could not dif-fuse nearly rapidly enough to effectively exert their control-ling effects, many of which are aimed toward maintaining homeostasis.

The remainder of the circulatory system is designed to transport blood to and from the capillaries. The arteries and arterioles distribute blood pumped by the heart to the capil-laries for life-sustaining exchanges to take place, and the ven-ules and veins collect blood from the capillaries and return it to the heart, where the process is repeated.

Introduction

The blood-vessel network, which connects various parts of the body with each other and with the external environment, enables materials to be exchanged. Organs that replenish nutrient supplies and remove metabolic wastes from the blood receive a greater percentage of the cardiac output than is warranted by their metabolic needs. These "reconditioning" organs can better tolerate reductions in blood supply than can organs that receive blood solely for the purpose of meeting their own metabolic needs. The brain is especially vulnerable to reductions in its blood supply. Therefore, the maintenance of adequate flow to this vulnerable organ is a high priority in circulatory function.

Blood flows in a closed loop between the heart and the tissues. The arteries transport blood from the heart throughout the body. The arterioles regulate the amount of blood that flows through each organ. The capillaries are the actual site where materials are exchanged between the blood and surrounding tissue. The veins return the blood from the tissues to the heart.

The flow rate of blood through a vessel is directly proportional to the pressure gradient and inversely proportional to the resistance. The higher pressure at the beginning of a vessel is established by the pressure imparted to the blood by cardiac contraction. The lower pressure at the end is due to frictional losses as flowing blood rubs against the vessel wall. Resistance, the hindrance to blood flow through a vessel, is influenced most by the vessel's radius. Resistance is inversely proportional to the fourth power of the radius, so small changes in radius profoundly influence flow. As the radius increases, resistance decreases and flow increases.

Arteries

Arteries are large-radius, low-resistance passageways from the heart to the tissues and also serve as a pressure reservoir. Because of their elasticity, arteries expand to accommodate the extra volume of blood pumped into them by cardiac contraction and then recoil to continue driving the blood forward when the heart is relaxing.

Systolic pressure is the peak pressure exerted by the ejected blood against the vessel walls during cardiac systole. Diastolic pressure is the minimum pressure in the arteries when blood is draining off into the vessels downstream during cardiac diastole.

The average driving pressure throughout the cardiac cycle is the mean arterial pressure, which can be estimated using the following formula: mean arterial pressure = diastolic pressure + (1/3 × pulse pressure).

Arterioles

Arterioles are the major resistance vessels. Their high resistance produces a large drop in mean pressure between the arteries and capillaries. This decline enhances blood flow by contributing to the pressure differential between the heart and the tissues. Tone, a baseline of contractile activity, is maintained in arterioles at all times. Arteriolar vasodilation, an expansion of arteriolar caliber above tonic level, decreases resistance and increases blood flow through the vessel, whereas vasoconstriction, a narrowing of the vessel, increases resistance and decreases flow.

Arteriolar caliber is subject to two types of control mechanisms: local (intrinsic) controls and extrinsic controls. Local controls involve local chemical changes associated with changes in the level of metabolic activity in a tissue; these controls act directly on the arteriolar smooth muscle in the vicinity to induce changes in the caliber of the arterioles supplying the tissue. By adjusting the resistance to blood flow in this manner, the local control mechanism adjusts blood flow to the tissue to match the momentary metabolic needs of the tissue. Adjustments in arteriolar caliber can be accomplished independently in different tissues by local control factors. Such adjustments are important in determining the distribution of cardiac output.

Extrinsic control is accomplished primarily by sympathetic nerve influence and to a lesser extent by hormonal influence over arteriolar smooth muscle. Extrinsic controls are important in maintaining mean arterial blood pressure. Arterioles are richly supplied with sympathetic nerve fibers, whose increased activity produces generalized vasoconstriction and a subsequent increase in mean arterial pressure. Decreased sympathetic activity produces generalized arteriolar vasodilation, which lowers mean arterial pressure. These extrinsically controlled adjustments of arteriolar caliber help maintain the appropriate pressure head for driving blood forward to the tissues.

Capillaries

The thin-walled, small-radius, extensively branched capillaries are ideally suited to serve as sites of exchange between the blood and surrounding tissues. Anatomically, the surface area for exchange is maximized and diffusion distance is minimized in the capillaries. Furthermore, because of the capillaries' large total cross-sectional area, the velocity of blood flow through them is relatively slow, providing adequate time for exchanges to take place.

Two types of passive exchanges—diffusion and bulk flow—take place across capillary walls. Individual solutes are exchanged primarily by diffusion down concentration gradients. Lipid-soluble substances pass directly through the single layer of endothelial cells lining a capillary, whereas water-soluble substances pass through water-filled pores between the endothelial cells. Plasma proteins generally do not escape.

Imbalances in physical pressures acting across capillary walls are responsible for bulk flow of fluid through the pores back and forth between the plasma and interstitial fluid. Fluid is forced out of the first portion of the capillary (ultrafiltration) where outward pressures (mainly capillary blood pressure) exceed inward pressures (mainly plasma-colloid osmotic pressure). Fluid is returned to the capillary along its last half when outward pressures fall below inward pressures. The reason for the shift in balance down the length of the capillary is the continuous decline in capillary blood pressure while the plasma-colloid osmotic pressure remains constant. Bulk flow is responsible for the distribution of extracellular fluid between the plasma and interstitial fluid.

Normally, slightly more fluid is filtered than is reabsorbed. The extra fluid, any leaked proteins, and tissue contaminants such as bacteria are picked up by the lymphatic system. Bacteria are destroyed as lymph passes through the lymph nodes en route to being returned to the venous system.

Veins

Veins are large-radius, low-resistance passageways for return of blood from the tissues to the heart. Additionally, they can accommodate variable volumes of blood and therefore act as a blood reservoir. The capacity of veins to hold blood can change markedly with little change in venous pressure. Veins are thin-walled, highly distensible vessels that can passively stretch to store a larger volume of blood.

The primary force responsible for venous flow is the pressure gradient between the veins and atrium (that is, what remains of the driving pressure imparted to the blood by cardiac contraction.) Venous return is enhanced by sympathetically induced venous vasoconstriction and by external compression of the veins resulting from contraction of surrounding skeletal muscles, both of which drive blood out of the veins. One-way venous valves assure that blood is driven toward the heart and prevented from flowing back toward the tissues. Venous return is also enhanced by the respiratory pump and the cardiac-suction effect. Respiratory activity produces a less-than-atmospheric pressure in the chest cavity, thus establishing an external pressure gradient that encourages flow from the lower veins that are exposed to atmospheric pressure to the chest veins that empty into the heart. In addition, slightly negative pressures created within the atria during ventricular systole and within the ventricles during ventricular diastole exert a suctioning effect that further enhances venous return and facilitates cardiac filling.

Blood Pressure

Regulation of mean arterial pressure depends on control of its two main determinants, cardiac output and total peripheral resistance. Control of cardiac output, in turn, depends on regulation of heart rate and stroke volume, whereas total peripheral resistance is determined primarily by the degree of arteriolar vasoconstriction. Short-term regulation of blood pressure is accomplished primarily by the baroreceptor reflex. Carotid-sinus and aortic-arch baroreceptors continuously monitor mean arterial pressure. When they detect a deviation from normal, they signal the medullary cardiovascular center, which responds by adjusting autonomic output to the heart and blood vessels to restore the blood pressure to normal. Long-term control of blood pressure involves maintenance of proper plasma volume through the kidneys' control of salt and water balance.

Blood pressure can be abnormally high (hypertension) or abnormally low (hypotension). Severe sustained hypotension resulting in generalized inadequate blood delivery to the tissues is known as circulatory shock.

REVIEW EXERCISES

Objective Questions (Answers on p. D–3.)

1. In general, the parallel arrangement of the vascular system enables each organ to receive its own separate arterial blood supply. (True or false?)

2. More blood flows through the capillaries during cardiac systole than during diastole. (True or false?)

3. The capillaries contain only 5% of the total blood volume at any point in time. (True or false?)

4. The same volume of blood passes through the capillaries in a minute as passes through the aorta, even though the velocity of blood flow is much slower in the capillaries. (True or false?)

5. Because there are no carrier transport systems in capillary walls, all capillaries are equally permeable. (True or false?)

6. Due to gravitational effects, venous pressure in the lower extremities is greater when a person is standing up than when lying down. (True or false?)

7. Which of the following functions is (are) attributable to arterioles? (Indicate all correct answers.)
   a. responsible for a significant decline in mean pressure that helps establish the driving pressure gradient between the heart and tissues
   b. site of exchange of materials between the blood and surrounding tissues
   c. main determinant of total peripheral resistance
   d. determine the pattern of distribution of cardiac output
   e. play a role in regulating mean arterial blood pressure
   f. convert the pulsatile nature of arterial blood pressure into a smooth, nonfluctuating pressure in the vessels further downstream
   g. act as a pressure reservoir

8. Using the answer code at the right, indicate what kind of compensatory changes occur in the factors in question to restore the blood pressure to normal in response to hypovolemic hypotension resulting from severe hemorrhage:

   ___ 1. rate of afferent firing generated by the carotid-sinus and aortic-arch baroreceptors   a. increased
      b. decreased
      c. no effect

   ___ 2. sympathetic output by the cardiovascular center

   ___ 3. parasympathetic output by the cardiovascular center

   ___ 4. heart rate

   ___ 5. stroke volume

   ___ 6. cardiac output

   ___ 7. arteriolar radii

   ___ 8. total peripheral resistance

   ___ 9. venous radii

   ___ 10. venous return

   ___ 11. urinary output

   ___ 12. fluid retention within the body

   ___ 13. fluid movement from the interstitial fluid into the plasma across the capillaries

9. Indicate whether the following factors increase or decrease venous return:

_____ 1. sympathetically induced venous vasoconstriction

_____ 2. skeletal-muscle activity

_____ 3. gravitational effects on the venous system

_____ 4. respiratory activity

_____ 5. increased atrial pressure associated with a leaky AV valve

_____ 6. ventricular pressure change associated with diastolic recoil

a. increases venous return

b. decreases venous return

c. has no effect on venous return

Essay Questions

1. Compare blood flow through reconditioning organs and through organs that do not recondition the blood.

2. Discuss the relationships among flow rate, pressure gradient, and vascular resistance. What is the major determinant of resistance to flow?

3. Describe the structure and major functions of each segment of the vascular tree.

4. How do the arteries serve as a pressure reservoir?

5. Describe the indirect technique of measuring arterial blood pressure by means of a sphygmomanometer.

6. Define vasoconstriction and vasodilation.

7. Discuss the local and extrinsic controls that regulate arteriolar resistance.

8. What is the primary means by which individual solutes are exchanged across the capillary walls? What forces are responsible for bulk flow across the capillary walls? Of what importance is bulk flow?

9. How is lymph formed? What are the functions of the lymphatic system?

10. Define edema and discuss its possible causes.

11. How do veins serve as a blood reservoir?

12. Compare the effect of vasoconstriction on the rate of blood flow in arterioles and veins.

13. Discuss the factors that determine mean arterial pressure.

14. Review the effects on the cardiovascular system of parasympathetic and sympathetic stimulation.

15. Differentiate between secondary hypertension and primary hypertension. What are the potential consequences of hypertension?

16. Define circulatory shock. What are its consequences and compensations? What is irreversible shock?

POINTS TO PONDER

1. During coronary bypass surgery, a piece of vein is removed from the patient's leg and surgically attached within the coronary circulatory system so that blood is detoured around an occluded coronary artery segment. For an extended period of time following surgery, why must the patient wear an elastic support stocking on the limb from which the vein was removed?

2. Assume a person has a blood-pressure recording of 125/77:

   a. What is the systolic pressure? _____

   b. What is the diastolic pressure? _____

   c. What is the pulse pressure? _____

   d. What is the mean arterial pressure? _____

   e. Would any sound be heard when the pressure in an external cuff around the arm was 130 mm Hg? (Yes or no?)

   f. Would any sound be heard when cuff pressure was 118 mm Hg? _____

   g. Would any sound be heard when cuff pressure was 75 mm Hg? _____

3. A classmate who has been standing still for several hours working on a laboratory experiment suddenly faints. What is the probable explanation? What would you do if the person next to him tried to get him up?

4. A drug applied to a piece of excised arteriole causes the vessel to relax, but an isolated piece of arteriolar muscle stripped from the other layers of the vessel fails to respond to the same drug. What is the probable explanation?

5. Explain how each of the following antihypertensive drugs would lower arterial blood pressure:

   a. drugs that block alpha-adrenergic receptors (ex., phentolamine)

   b. drugs that block beta-adrenergic receptors (ex., propranolol) (Hint: More important than the action of these drugs on the heart and blood vessels in reducing arterial blood pressure is their inhibitory effect on the hormonal pathway involving angiotensin II that promotes salt and water conservation.)

   c. drugs that directly relax arteriolar smooth muscle (ex., hydralazine)

   d. diuretic drugs that increase urinary output (ex., furosemide)

   e. drugs that block release of norepinephrine from sympathetic endings (ex., guanethidine)

   f. drugs that act on the brain to reduce sympathetic output (ex., clonidine)

   g. drugs that block Ca^{++} channels (ex., verapamil)

   h. drugs that interfere with the production of angiotensin II (ex., captopril)

11 Blood

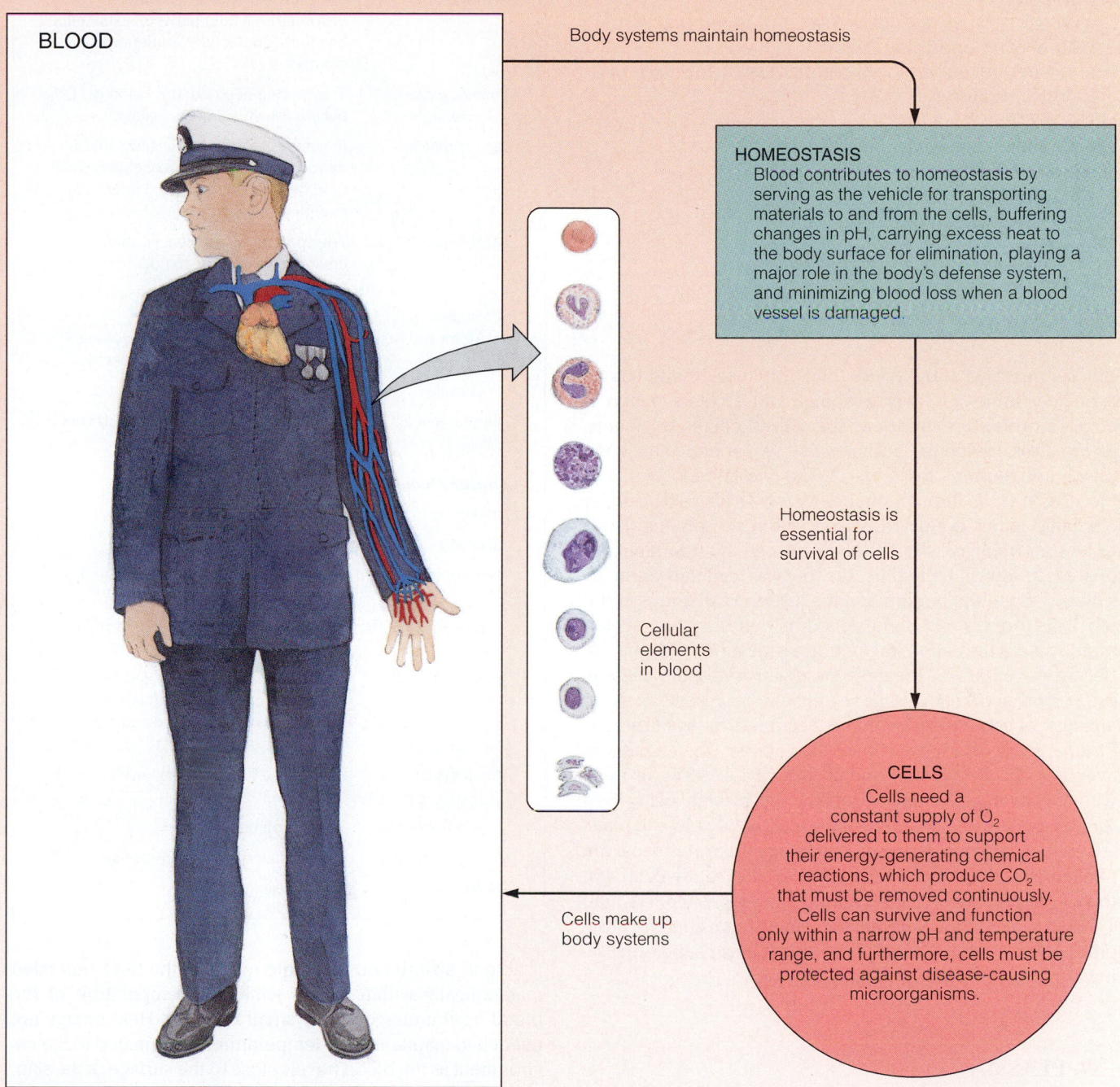

BLOOD

Body systems maintain homeostasis

HOMEOSTASIS
Blood contributes to homeostasis by serving as the vehicle for transporting materials to and from the cells, buffering changes in pH, carrying excess heat to the body surface for elimination, playing a major role in the body's defense system, and minimizing blood loss when a blood vessel is damaged.

Cellular elements in blood

Homeostasis is essential for survival of cells

Cells make up body systems

CELLS
Cells need a constant supply of O_2 delivered to them to support their energy-generating chemical reactions, which produce CO_2 that must be removed continuously. Cells can survive and function only within a narrow pH and temperature range, and furthermore, cells must be protected against disease-causing microorganisms.

Blood is the vehicle for long-distance, mass transport of materials between the cells and the external environment or between the cells themselves. Such transport is essential for maintenance of homeostasis. Blood consists of a complex liquid **plasma** in which the cellular elements—*erythrocytes, leukocytes,* and *platelets*—are suspended.

Erythrocytes (**red blood cells** or **rbcs**) are essentially plasma-membrane–enclosed bags of **hemoglobin** that trans-port O_2 and, to a lesser extent, CO_2 in the blood. **Leukocytes** (**white blood cells** or **wbcs**), the immune system's mobile defense units, are transported in the blood to sites of injury or invasion by disease-causing microorganisms.

Given the importance of blood, it is imperative that mechanisms exist to minimize blood loss when a vessel is injured. **Platelets** (**thrombocytes**) are important in **hemostasis**, the stopping of bleeding from an injured vessel.

CHAPTER CONTENTS AT A GLANCE

INTRODUCTION
PLASMA
Composition and functions of plasma
Plasma proteins
ERYTHROCYTES
Structure and function of erythrocytes
Erythropoiesis
● A CLOSER LOOK AT EXERCISE PHYSIOLOGY Blood Doping: Is More of a Good Thing Better?
● CONCEPTS, CHALLENGES, AND CONTROVERSIES In Search of a Blood Substitute
LEUKOCYTES
Types and functions of leukocytes
Leukocyte production
PLATELETS AND HEMOSTASIS
Platelet structure and production
Hemostasis

 ## INTRODUCTION

Blood represents about 8% of total body weight and has an average volume of 5 liters in women and 5.5 liters in men. It consists of the three types of specialized cellular elements, *erythrocytes, leukocytes,* and *platelets,* suspended in the complex liquid *plasma* (Table 11–1). The constant movement of blood as it flows through the blood vessels keeps its cellular elements rather evenly dispersed within the plasma. However, if a sample of whole blood is placed in a test tube and treated to prevent clotting, the heavier cellular elements slowly settle to the bottom and the lighter plasma rises to the top. This process can be hastened by centrifugation, which rapidly packs the cells in the bottom of the tube (● Fig. 11–1). Because over 99% of the cells are erythrocytes, the **hematocrit**, or **packed cell volume**, essentially represents the percentage of total blood volume occupied by erythrocytes. Plasma accounts for the remaining volume. The hematocrit averages 42% for women and slightly higher, 45%, for men, with the average volume occupied by plasma being 58% for women and 55% for men. The white blood cells and platelets, which are colorless and less dense than red cells, are packed in a thin, cream-colored layer, the "buffy coat," on top of the packed red-cell column. They represent less than 1% of the total blood volume. We will first consider the properties of the largest portion of the blood, plasma, before turning our attention to the cellular elements.

 ## PLASMA

Many of the functions of plasma are carried out by plasma proteins.

Plasma, being a liquid, is composed of 90% water, which serves as a medium for materials being carried in the blood. Also, because water has a high capacity to hold heat, plasma

CONSTITUENT	FUNCTIONS
Plasma	
Water	Transport medium; carries heat
Electrolytes	Membrane excitability; osmotic distribution of fluid between extracellular and intracellular fluid; buffering of pH changes
Nutrients, wastes, gases, hormones	Transported in blood; the blood gas CO_2 plays a role in acid-base balance
Plasma proteins	In general, exert osmotic effect that is important in distribution of extracellular fluid between vascular and interstitial compartments; buffering of pH changes
Albumins	Transport many substances; make greatest contribution to colloid osmotic pressure
Globulins	
Alpha and beta	Transport many substances; clotting factors; inactive precursor molecules
Gamma	Antibodies
Fibrinogen	Inactive precursor for fibrin meshwork of clot
Cellular Elements	
Erythrocytes	Transport O_2 and CO_2 (mainly O_2)
Leukocytes	
Neutrophils	Phagocytes that engulf bacteria and debris
Eosinophils	Attack parasitic worms; important in allergic reactions
Basophils	Release histamine, which is important in allergic reactions, and heparin, which helps clear fat from blood and may function as anticoagulant
Monocytes	In transit to become tissue macrophages
Lymphocytes	
B lymphocytes	Production of antibodies
T lymphocytes	Cell-mediated immune responses
Platelets	Hemostasis

is able to absorb and distribute much of the heat generated metabolically within tissues while the temperature of the blood itself undergoes only small changes. Heat energy not needed to maintain body temperature is eliminated to the environment as the blood travels close to the surface of the skin.

A large number of organic and inorganic substances are dissolved in the plasma. The most plentiful organic constituents by weight are the plasma proteins, which compose 6% to 8% of plasma's total weight. Inorganic constituents account for approximately 1% of plasma weight. The most abundant electrolytes (ions) in the plasma are Na^+ and Cl^- (common

Plasma
58–55%

"Buffy coat"
< 1% { Platelets

White blood cells

Packed cell
volume, or
hematocrit { Erythrocytes
(red blood cells)
42–45%

● **FIGURE 11–1   Hematocrit**

salt). There are lesser amounts of HCO_3^-, K^+, Ca^{++}, and others. The most notable functions of these extracellular-fluid (ECF) ions are their roles in membrane excitability, osmotic distribution of fluid between the ECF and cells, and buffering of pH changes; these functions are discussed elsewhere. The remaining small percentage of plasma is occupied by nutrients (for example, glucose, amino acids, lipids, and vitamins); waste products (creatinine, bilirubin, and nitrogenous substances such as urea); dissolved gases (O_2 and CO_2); and hormones. Most of these substances are merely being transported in the plasma. For example, endocrine glands secrete hormones into the plasma, which transports these chemical messengers to their sites of action.

The **plasma proteins** are the one group of plasma constituents not present just for the ride. These important components normally remain in the plasma, where they perform many valuable functions. Because they are the largest of the plasma constituents, plasma proteins usually do not exit through the narrow pores in the capillary walls. Also, unlike other plasma constituents that are dissolved in the plasma water, the plasma proteins exist in a colloidal dispersion (see p. A–6).

There are three groups of plasma proteins—*albumins, globulins,* and *fibrinogen*—which are classified according to their various physical and chemical properties. The functions of the plasma proteins are elaborated on elsewhere in the text, but the following list illustrates the wide range of these functions:

■ By virtue of their presence as a colloidal dispersion in the plasma and their absence in the interstitial fluid, plasma proteins establish an osmotic gradient between blood and interstitial fluid. This colloid osmotic pressure is the primary force responsible for preventing excessive loss of plasma from the capillaries into the interstitial fluid and thus helps maintain plasma volume.

■ Plasma proteins are partially responsible for the plasma's capacity to buffer changes in pH.

■ Plasma proteins contribute to blood viscosity, but erythrocytes are far more important in this regard.

■ Plasma proteins are not normally used as metabolic fuels, but in a state of starvation they can be degraded to provide energy for cells.

■ In addition to these general functions, each type of plasma protein performs important specific tasks:

1. **Albumins**, the most abundant of the plasma proteins, bind many substances (for example, bilirubin, bile salts, and penicillin) for transport through the plasma and contribute most extensively to the colloid osmotic pressure by virtue of their numbers.

2. There are three subclasses of **globulins: alpha (α), beta (β),** and **gamma (τ).**
   a. Specific alpha and beta globulins bind and transport a number of substances in the plasma, such as thyroid hormone, cholesterol, and iron.
   b. Many of the factors involved in the process of blood clotting are alpha or beta globulins.
   c. Inactive precursor protein molecules, which are activated as needed by specific regulatory inputs, belong to the alpha-globulin group (for example, angiotensinogen is activated to angiotensin, which plays an important role in the regulation of salt balance in the body).
   d. The gamma globulins are the immunoglobulins (antibodies), which are crucial to the body's defense mechanism.

3. **Fibrinogen** is a key factor in the blood-clotting process.

The plasma proteins generally are synthesized by the liver, with the exception of the gamma globulins, which are produced by lymphocytes, one of the types of white blood cells.

ERYTHROCYTES

The structure of erythrocytes is well suited to their primary function of oxygen transport in the blood.

Each milliliter of blood contains about 5 billion erythrocytes (red blood cells) on average, commonly reported clinically in a **red blood cell count** as 5 million cells per cubic millimeter (mm^3). Erythrocytes are flat, disc-shaped cells indented in the middle on both sides, like a doughnut with a flattened center instead of a hole (they are biconcave discs 8 μm in diameter, 2 μm thick at the outer edges, and 1 μm thick in the center) (● Fig. 11–2). This unique shape contributes in two ways to the efficiency with which erythrocytes perform their main function of O_2 transport in the blood. First, the biconcave shape provides a larger surface area for diffusion of O_2 across the membrane than a spherical cell of the same volume would provide. Second, the thinness of the cell enables O_2 to diffuse rapidly between the exterior and innermost regions of the cell.

Another feature of erythrocytes that facilitates their transport function is the flexibility of their membrane, which enables them to travel through the narrow, tortuous capillaries

● **FIGURE 11–2** **Anatomical Characteristics of Erythrocytes** Appearance of erythrocytes under scanning electron microscope. Note their biconcave shape.

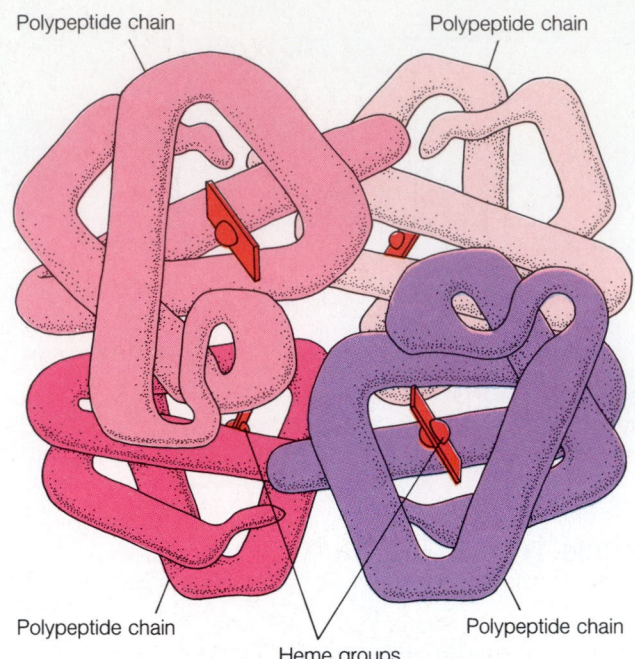

Polypeptide chain   Polypeptide chain

Polypeptide chain   Polypeptide chain

Heme groups

FIGURE 11–3 **Hemoglobin Molecule** A hemoglobin molecule consists of four highly folded polypeptide chains (the globin portion) and four iron-containing heme groups.

to deliver their O_2 cargo at the tissue level without rupturing in the process. Red blood cells, whose diameter is normally 8μm, are able to deform amazingly as they squeeze single file through capillaries as narrow as 3 μm in diameter.

Erythrocytes' most important feature that enables them to transport O_2 is the **hemoglobin** they contain. A hemoglobin molecule consists of two parts: (1) the **globin portion**, a protein made up of four highly folded polypeptide chains, and (2) four iron-containing, nonprotein nitrogenous groups known as **heme** groups, each of which is bound to one of the polypeptides (● Fig. 11–3). Each of the four iron atoms can combine reversibly with one molecule of O_2; thus, each hemoglobin molecule can pick up four O_2 passengers. Because O_2 is poorly soluble in the plasma, 98.5% of the O_2 carried in the blood is bound to hemoglobin. Hemoglobin is a pigment (that is, it is naturally colored). Because of its iron content, it appears reddish when combined with O_2 and bluish when deoxygenated. Thus, fully oxygenated arterial blood is red in color, and venous blood that has lost some of its O_2 load at the tissue level has a bluish cast.

In addition to carrying O_2, hemoglobin can also combine with the following:

1. *Carbon dioxide.* Thus, hemoglobin contributes to the transport of this gas from the tissues back to the lungs.

2. *The acidic hydrogen-ion portion (H⁺) of ionized carbonic acid,* which is generated at the tissue level from CO_2. Hemoglobin thereby buffers this acid so that it minimally alters the pH of the blood.

3. *Carbon monoxide (CO).* This gas is not normally in the blood but, if inhaled, preferentially occupies the O_2-

binding sites on hemoglobin, causing carbon-monoxide poisoning.

Therefore, hemoglobin plays the key role in O_2 transport while contributing significantly to CO_2 transport and the buffering capacity of blood.

To maximize its hemoglobin content, a single erythrocyte is stuffed with several hundred million hemoglobin molecules to the exclusion of almost everything else. Erythrocytes contain no nucleus, organelles, or ribosomes. These structures are extruded during the cell's development to make room for more hemoglobin. Thus, a red blood cell is mainly a plasma-membrane–enclosed sac full of hemoglobin.

Only a few crucial, nonrenewable enzymes remain within a mature erythrocyte: these are glycolytic enzymes and carbonic anhydrase. The **glycolytic enzymes** are necessary for generating the energy needed to fuel the active transport mechanisms involved in maintaining proper ionic concentrations within the cell. Ironically, even though erythrocytes are the vehicles for transport of O_2 to all other tissues of the body, they themselves cannot use the O_2 they are carrying for energy production. Erythrocytes, lacking the mitochondria that house the enzymes for oxidative phosphorylation, must rely entirely on glycolysis for ATP formation (see p. 27).

The other important enzyme within red blood cells, **carbonic anhydrase**, is critical in CO_2 transport. This enzyme catalyzes a key reaction that ultimately leads to the conversion of metabolically produced CO_2 into **bicarbonate ion** (**HCO_3^-**), which is the primary form in which CO_2 is transported in the blood. Thus, erythrocytes contribute to CO_2 transport in two ways—by means of its carriage on hemoglobin and its carbonic anhydrase–induced conversion to HCO_3^-.

The bone marrow continuously replaces worn-out erythrocytes.

Each of us has a total of 25 to 30 trillion red blood cells streaming through our blood vessels at any given time (100,000 times more than the entire population of the United States)! Yet, these vital gas-transport vehicles are short-lived and must be replaced at the average rate of 2 to 3 million cells per second. The price erythrocytes pay for their generous content of hemoglobin to the exclusion of the usual specialized intracellular machinery is a shortened life span. Without DNA and RNA, red blood cells cannot synthesize proteins for cellular repair, growth, and division or for renewal of enzyme supplies. Equipped only with initial supplies synthesized before extrusion of their nucleus, organelles, and ribosomes, erythrocytes are able to survive an average of only 120 days, in contrast to nerve and muscle cells, which survive during a person's entire life. During its short life span of four months, each erythrocyte travels about 700 miles as it circulates through the vasculature. As a red blood cell ages, its non-reparable plasma membrane becomes fragile and prone to rupture as the cell squeezes through tight spots in the vascular system.

Most old red blood cells meet their final demise in the **spleen**, because this organ's narrow, winding capillary network is a tight fit for these fragile cells. The spleen lies in the upper left part of the abdomen. In addition to removing most of the old erythrocytes from circulation, the spleen has a limited ability to store healthy erythrocytes in its pulpy interior; serves as a reservoir site for platelets; and contains an abundance of lymphocytes, a type of white blood cell.

Because erythrocytes cannot divide to replenish their own numbers, the old ruptured cells must be replaced by new cells produced in an erythrocyte factory—the **bone marrow**—which is the soft, highly cellular tissue that fills the internal cavities of bones. The bone marrow normally generates new red blood cells, a process known as **erythropoiesis**, at the amazing rate of 2 to 3 million per second to keep pace with the demolition of old cells.

During intrauterine development, erythrocytes are produced first by the yolk sac and then by the developing liver and spleen, until the bone marrow is formed and takes over erythrocyte production exclusively. In children most bones are filled with red bone marrow that is capable of blood-cell production. As a person matures, however, fatty yellow marrow that is incapable of erythropoiesis gradually replaces red marrow, which remains only in the sternum (breastbone), vertebrae (backbone), ribs, base of the skull, and upper ends of the long limb bones. Red marrow not only produces red blood cells but is the ultimate source for leukocytes and platelets as well. Undifferentiated **pluripotential stem cells** reside in the red marrow, where they continuously divide and differentiate to give rise to each of the types of blood cells. The different types of cells, along with the stem cells, are intermingled in the red marrow at various stages of development. The mature cells are released into the rich supply of capillaries that permeate the red marrow. Regulatory factors act on the *hemopoietic* ("blood-producing") marrow to govern the type and number of cells generated and discharged

into the blood. Of the blood cells, the mechanism for regulating red blood cell production is the best understood. We will consider it now.

Erythropoiesis is controlled by erythropoietin from the kidneys.

The number of circulating erythrocytes normally remains fairly constant, indicative that erythropoiesis must be closely regulated. Because O_2 transport in the blood is the erythrocytes' primary function, you might logically suspect that the primary stimulus for increased erythrocyte production would be reduced O_2 delivery to the tissues. You would be correct, but low O_2 levels do not stimulate erythropoiesis by acting directly on the red bone marrow. Instead, reduced O_2 delivery to the kidneys stimulates them to secrete the hormone **erythropoietin** into the blood, and this hormone in turn stimulates erythropoiesis by the bone marrow (•Fig. 11–4). The exact cells in the kidney responsible for erythropoietin secretion have not been identified, but this hormone's final site of action is known. Erythropoietin acts on derivatives of undifferentiated stem cells that are already committed to becoming red blood cells, stimulating their proliferation and maturation into mature erythrocytes. This increased erythropoietic activity elevates the number of circulating red blood cells, thereby increasing the O_2-carrying capacity of the blood and restoring O_2 delivery to the tissues to normal. Once normal O_2 delivery to the kidneys is achieved, erythropoietin secretion is turned down until needed once again. In this way, erythrocyte production is normally balanced against destruction or loss of these cells so that O_2-carrying capacity in the blood remains fairly constant. (See the accompanying boxed feature, A Closer Look at Exercise Physiology.) In response to severe loss of erythrocytes, as in hemorrhage or abnormal destruction of young circulating erythrocytes, the rate of erythropoiesis can be increased by more than six times the normal level.

The preparation of an erythrocyte for its departure from the marrow involves several steps, such as synthesis of hemoglobin and extrusion of the nucleus and organelles. Cells closest to maturity need a few days to be "finished off" and released into the blood in response to erythropoietin, and less-developed and newly proliferated cells may take up to several weeks to reach maturity. Therefore, the time required for complete replacement of lost red blood cells depends on how many are needed to return the number to normal. (When you donate blood, your circulating erythrocyte supply is replenished in less than a week.) When demands for red blood cell production are high (for example, following hemorrhage), the bone marrow may release large numbers of immature erythrocytes, known as **reticulocytes**, into the blood to quickly meet the need. These immature cells can be recognized by staining techniques that make visible the residual ribosomes and organelle remnants that have not yet been extruded. Their presence above the normal level of 0.5% to 1.5% of the total number of circulating erythrocytes is indicative of a high rate of erythropoietic activity. At very rapid rates, more than 30% of the circulating red cells may be at the immature reticulocyte stage.

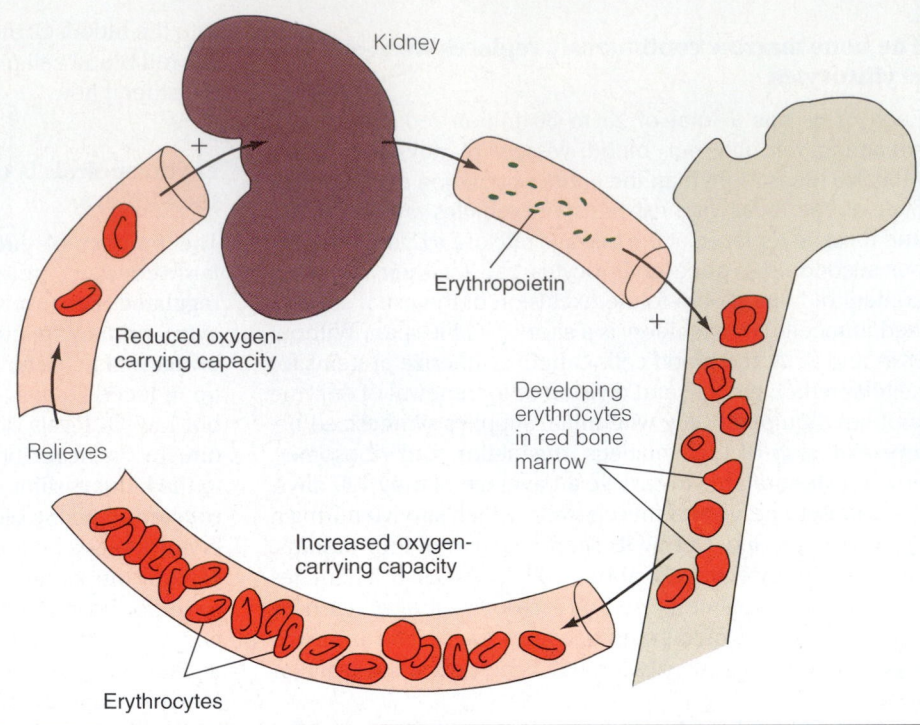

● **FIGURE 11–4 Control of Erythropoiesis** When less O_2 is delivered to the kidneys, they secrete the hormone erythropoietin into the blood. Erythropoietin in turn stimulates erythropoiesis (erythrocyte production) by the bone marrow. The additional circulating erythrocytes increase the O_2-carrying capacity of the blood, thus relieving the initial stimulus that triggered erythropoietin secretion.

Kidney

Reduced oxygen-carrying capacity

Erythropoietin

Relieves

Developing erythrocytes in red bone marrow

Increased oxygen-carrying capacity

Erythrocytes

A CLOSER LOOK AT EXERCISE PHYSIOLOGY

BLOOD DOPING: IS MORE OF A GOOD THING BETTER?

Exercising muscles require a continual supply of O_2 for generating energy to sustain endurance activities (see p. 234). Blood doping is a technique designed to temporarily increase the O_2-carrying capacity of the blood in an attempt to gain a competitive advantage. Blood doping involves removing blood from an athlete, then promptly reinfusing the plama but freezing the red blood cells for reinfusion one to seven days before a competitive event. One to four units of blood (one unit equals 450 ml) are usually withdrawn at three- to eight-week intervals before the competition. In the periods between blood withdrawals, increased erythropoietic activity restores the red blood cell count to a normal level.

Reinfusion of the stored red blood cells temporarily increases the red blood cell count and hemoglobin level above normal. Theoretically, blood doping would be beneficial to endurance athletes by improving the blood's O_2-carrying capacity. If too many red cells were infused, however, performance could suffer because the increased blood viscosity would decrease blood flow.

Early research into athletic performance after blood doping produced contradictory results. More recent research, however, indicates that, in a standard laboratory exercise test, athletes may realize a 5% to 13% increase in aerobic capacity; a reduction in heart rate during exercise compared to the rate during the same exercise in the absence of blood doping; improved performance; and reduced lactic-acid levels in the blood. (Lactic acid is produced when muscles resort to less efficient anaerobic glycolysis for energy production—see p. 236.)

Blood doping, although probably effective, is illegal in both collegiate athletics and Olympic competition for ethical and medical reasons. These prohibitive regulations are very difficult to enforce, however. Current procedures permit only the testing of urine, not blood, for the use of illegal drugs among athletes. Blood doping cannot be detected by means of a urine sample. The only way of exposing the practice of blood doping is through witnesses or self-admission.

The recent development of synthetic erythropoietin magnifies the problem of blood doping. Injection of synthetic erythropoietin would stimulate red blood cell production and thus temporarily increase the O_2-carrying capacity of the blood. Use of erythropoietin would be more convenient for the athlete and would minimize the number of witnesses involved, thereby reducing the probability of disclosure.

IN SEARCH OF A BLOOD SUBSTITUTE

One of the hottest medical contests of the 1990s is the race to develop a safe, inexpensive, disease-free, universal substitute for human blood. The benefits for society of a safe blood substitute that could be administered without regard for the recipient's blood type are great, as will be the profits for the manufacturer of the first successful product. Experts estimate that the world market for a good blood substitute may be as much as $10 billion per year.

The search for an alternative to whole-blood transfusions has been given new impetus by the rising incidence of AIDS and the concomitant concern over the safety of the nation's blood supply. Infectious diseases such as AIDS and viral hepatitis can be transmitted from infected blood donors to recipients of blood transfusions. Although careful screening of our blood supply minimizes the possibility that infectious diseases will be transmitted through transfusion, the public remains wary and would welcome a safe substitute.

Eliminating the risk of disease transmission is only one advantage of finding an alternative to whole-blood transfusion. Whole blood must be kept refrigerated, and even then it has a shelf life of only forty-two days. Also, transfusion of whole blood requires blood typing and cross matching, which cannot be done at the scene of an accident or on a battlefield.

The goal is not to find a replacement for whole blood but to duplicate its O_2-carrying capacity. The biggest need for blood transfusions is to replace acute blood loss in accident victims, surgical patients, and wounded soldiers. These individuals require short-term replenishment of blood O_2-carrying capacity un-til their own bodies synthesize replacement erythrocytes. The many other important elements in blood are not as immediately critical in sustaining life as is the hemoglobin within the red blood cells. It is the red blood cells in whole blood that require refrigeration, have a short shelf life, and bear the markers for the various blood types. Therefore, the search for a blood substitute has focused on pure hemoglobin, which can be stored at room temperature for up to six months to a year. When suspended in saline solution, hemoglobin could be injected to bolster the O_2-carrying capacity of the recipients' blood no matter what their blood type.

The following strategies are among those being pursued to develop a hemoglobin product to serve as a substitute for whole-blood transfusion:

- One problem is that hemoglobin behaves differently when it is outside red blood cells. "Naked" hemoglobin splits into halves that do not release O_2 for tissue use as readily as normal hemoglobin does. Also, these hemoglobin fragments can cause kidney damage. A cross-binding reagent has been developed that keeps hemoglobin molecules intact when they are outside the confines of red blood cells, thus surmounting one major obstacle to administering free hemoglobin.

- One plan under investigation is to extract, purify, sterilize, and chemically stabilize hemoglobin from donated human blood. This strategy relies on the continued practice of collecting human blood donations.

- Another candidate as a blood substitute is a genetically engineered, slightly modified form of human hemoglobin that holds together and releases more O_2 than normal when existing free outside red blood cells, thus precluding the need for a cross-binding reagent. The scientists working on this tactic looked for a mutant gene that produced a modified form of hemoglobin with the characteristics they sought, then inserted the gene into bacteria. The bacterial "factory" produces the desired hemoglobin product, but the problem to date has been the persistence of a toxin that the bacteria add on.

- Another group of genetic engineers have created pigs that carry the gene for producing normal human hemoglobin and retain the gene for producing swine hemoglobin. About 10% to 20% of the hemoglobin produced by these pigs is of the human variety. Without compromising the health of the pigs, blood can be drawn, the red blood cells ruptured, and the human and swine hemoglobin separated. The harvested human hemoglobin can be pasteurized and thus is free of any infectious agents. The safety and value of this product must be confirmed by clinical trials.

- Another nonhuman source of hemoglobin that has been explored is the million gallons per year of blood from cattle at slaughterhouses. Most of this blood is dried and used for fertilizers, animal feed, or landfill. Although slaughterhouses would be a low-cost source of hemoglobin, there are concerns that the human immune system will attack injected bovine hemoglobin. Early clinical trials using animal hemoglobin had to be discontinued because of side effects.

The gene that directs erythropoietin synthesis has recently been identified and cloned so this hormone can now be produced in abundance in a laboratory. This capability will have a profound impact on the need for blood transfusions. For example, administration of erythropoietin to surgical patients will stimulate their own red-cell production, thus reducing the need for transfused blood. (See the accompanying boxed feature, Concepts, Challenges, and Controversies, for an update on other alternatives to whole-blood transfusions under investigation.)

In addition to erythropoietin, which is the primary regulatory factor that adjusts the ongoing rate of red blood cell production as needed to keep erythrocyte levels constant, testosterone, the major male sex hormone, increases the basal rate of erythropoiesis. This hormone is undoubtedly responsible, at least in part, for the normally larger hematocrit in males compared to females. The additional red blood cells in males provide extra O_2-carrying capacity to meet the needs of the larger muscle mass in men, for which testosterone is also responsible.

Anemia can be caused by a variety of disorders.

In spite of control measures, O_2-carrying capacity cannot always be maintained to meet tissue needs. **Anemia** refers to a reduction below normal in the O_2-carrying capacity of the blood and is characterized by a low hematocrit (● Fig. 11–5a and b). It can be brought about by a decreased rate of erythropoiesis, excessive losses of erythrocytes, or a deficiency in the hemoglobin content of erythrocytes. The various causes of anemia can be grouped into six categories:

1. *Nutritional anemia* is caused by a dietary deficiency of a factor needed for erythropoiesis. The production of erythrocytes depends on an adequate supply of essential raw ingredients, some of which are not synthesized in the body but must be provided by dietary intake. For example, *iron-deficiency anemia* occurs when insufficient iron is available for the synthesis of hemoglobin because of either an iron-deficient diet or poor iron absorption from the digestive tract. The usual numbers of erythrocytes are produced, but they contain less hemoglobin than normal, are subsequently smaller than usual, and are able to transport less O_2. A dietary deficiency of folic acid, a member of the B-vitamin complex, can likewise produce anemia. This vitamin is essential for the formation of

DNA, which is critical in governing stem-cell division and maturation of erythrocytes. When the supply of folic acid is inadequate, fewer erythrocytes are formed, and those produced are larger and more fragile than normal. Despite their larger size, they contain the normal amount of hemoglobin. The resulting anemia is due to the production of fewer erythrocytes and their earlier demise because of their fragility. Many other tissues are affected by a deficiency of folic acid as well, but red blood cell production is especially sensitive because erythrocytes are among the most rapidly proliferating cells of the body.

2. *Pernicious anemia* is caused by an inability to absorb adequate amounts of vitamin B_{12} from the digestive tract. Like folic acid, vitamin B_{12} is essential for DNA production and its associated role in the proliferation and maturation of erythrocytes. Therefore, pernicious anemia is also characterized by a reduced number of erythrocytes that are larger and more fragile than normal. Unlike the nutritional anemias, however, pernicious anemia is not due to insufficient dietary supplies of vitamin B_{12}. This nutrient is found abundantly in a variety of common foods. With pernicious anemia, the problem is a deficiency of **intrinsic factor**, a special substance secreted by the lining of the stomach. Only when vitamin B_{12} is in combination with intrinsic factor can it be absorbed from the intestinal tract by special transport mechanisms. When intrinsic factor is deficient, insufficient amounts of ingested vitamin B_{12} are absorbed. The resulting impairment of red blood cell production and maturation leads to anemia. This condition is treated by injections of vitamin B_{12} to bypass the defective absorptive mechanism.

3. *Aplastic anemia* is caused by failure of the bone marrow to produce adequate numbers of red blood cells, even though all ingredients necessary for erythropoiesis are available. Reduced erythropoietic capability can be caused by destruction of red bone marrow by toxic chemicals, (such as benzene, arsenic, and certain drugs, es-

● **FIGURE 11–5 Hematocrit under Various Circumstances** The hematocrit is lower than normal in anemia because of too few circulating erythrocytes and above normal in polycythemia due to excess circulating erythrocytes. The hematocrit can also be elevated in dehydration when the normal number of circulating erythrocytes are concentrated in a reduced plasma volume.

Normal (hematocrit = 45%) (a)   Anemia (hematocrit = 30%) (b)   Polycythemia (hematocrit = 70%) (c)   Dehydration (hematocrit = 70%) (d)

= Plasma   = Erythrocytes

pecially chloramphenicol); heavy exposure to radiation (fallout from a nuclear-bomb explosion, for example, or excessive exposure to X rays); or invasion of the marrow by cancer cells. The destructive process may selectively reduce the marrow's output of erythrocytes, or it may reduce the productive capability for leukocytes and platelets as well. The anemia's severity depends on the extent to which erythropoietic tissue is destroyed, with severe losses being fatal.

4. *Renal anemia* may be a consequence of kidney disease. Since erythropoietin from the kidneys is the primary stimulus for promoting erythropoiesis, inadequate erythropoietin secretion as a result of kidney disease causes insufficient red blood cell production and anemia.

5. *Hemorrhagic anemia* is caused by the loss of substantial quantities of blood. The loss can either be acute, such as that occurring because of bleeding from a wound, or chronic, such as that accompanying a history of excessive menstrual flow. Anemia is present until the lost cells are replaced by transfusion or by increased erythropoietic activity.

6. *Hemolytic anemia* is caused by the rupture of excessive numbers of circulating erythrocytes. **Hemolysis**, the rupture of red blood cells, occurs either because the cells are defective, as in sickle cell anemia, or because otherwise normal cells are induced to rupture by external factors *Sickle-cell anemia* is the best-known example among various hereditary abnormalities of red blood cells that make these cells very fragile. Even as young cells, the defective, sickle-shaped erythrocytes are fragile and prone to rupture as they travel through the narrow splenic capillaries (● Fig. 11–6). Despite an accelerated rate of erythropoiesis triggered by the constant excessive loss of red blood cells, production may not be able to keep pace with the rate of destruction, and anemia may result.

Normally formed cells can also hemolyze prematurely if attacked by various external factors. An example is *malaria*, which is caused by protozoan parasites introduced into a victim's blood by the bite of a carrier mosquito.

These parasites selectively invade red blood cells, where they multiply to the point that the mass of malarial organisms ruptures the cells, releasing hundreds of new active parasites that quickly invade other red blood cells. As this cycle continues and more erythrocytes are destroyed, the anemic condition progressively worsens.

Polycythemia is an excess of circulating erythrocytes.

Polycythemia, in contrast to anemia, is characterized by an excess of circulating red blood cells and an elevated hematocrit (Fig. 11–5c). There are two general types of polycythemia, depending on the circumstances triggering the excess red blood cell production: primary polycythemia, or polycythemia vera (*vera* means "true"), and secondary polycythemia, or physiological polycythemia.

Primary polycythemia is caused by a tumorlike condition of the bone marrow in which erythropoiesis proceeds at an excessive, uncontrolled rate instead of being subject to the normal erythropoeitin regulatory mechanism. The red blood cell count may reach 11 million cells/mm³ (normal is 5 million cells/mm³), and the hematocrit may be as high as 70–80% (normal is 42–45%). No benefit is derived from the extra O_2-carrying capacity of the blood, because O_2 delivery is more than adequate with normal red blood cell numbers. Inappropriate polycythemia has adverse effects, however. The excessive number of red cells increases the blood's viscosity up to five to seven times normal, causing the blood to flow very sluggishly, which may actually reduce O_2 delivery to the tissues. The increased viscosity also increases the total peripheral resistance, which may elevate the blood pressure, thus increasing the work load of the heart, unless blood-pressure control mechanisms are able to compensate.

Secondary polycythemia, in contrast, is an appropriate erythropoietin-induced adaptive mechanism to improve the blood's O_2-carrying capacity in response to a prolonged reduction in O_2 delivery to the tissues. It occurs normally in persons living at high altitudes, where less O_2 is available in the atmospheric air, or in individuals in whom O_2 delivery to the tissues is impaired as a result of chronic lung disease or cardiac failure. The red cell count in secondary polycythemia is usually lower than that in primary polycythemia, typically averaging 6 to 8 million cells/mm³. The price paid for improved O_2 delivery is an increased viscosity of the blood.

An elevated hematocrit can occur when the body loses fluid but not erythrocytes, as in dehydration accompanying heavy sweating or profuse diarrhea (Fig. 11–5d). This is not a true polycythemia, however, because the number of circulating red blood cells is not increased. A normal number of erythrocytes are simply concentrated in a smaller plasma volume. This condition is sometimes termed *relative polycythemia*.

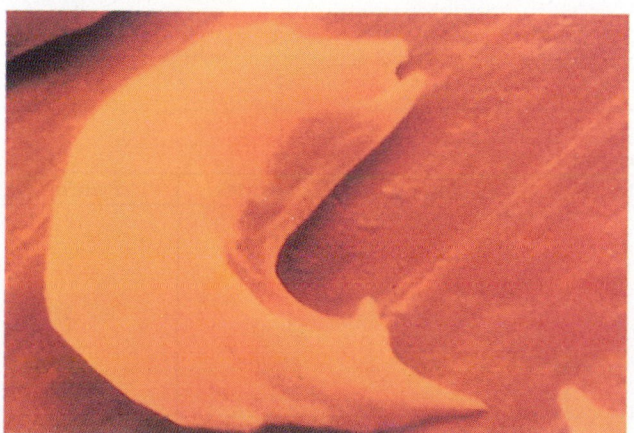

● **FIGURE 11–6   Sickle-Shaped Red Blood Cell in Sickle-Cell Anemia**

⇛ LEUKOCYTES

Leukocytes primarily function outside the blood.

Leukocytes, or white blood cells, are the mobile units of the body's immune defense system. **Immunity** refers to the body's ability to resist or eliminate potentially harmful foreign materials or abnormal cells. The leukocytes and their derivatives (1) defend against invasion by **pathogens** (disease-causing microorganisms such as bacteria and viruses) by phagocytizing (see p. 24) the foreigners or causing their destruction by more subtle means; (2) identify and destroy cancer cells that arise within the body; and (3) function as a "cleanup crew" that removes the body's "litter" by phagocytizing debris resulting from dead or injured cells. The latter is essential for wound healing and tissue repair.

To carry out their functions, the leukocytes largely employ a "seek out and attack" strategy; that is, they go to sites of invasion or tissue damage. The main reason white blood cells are present in the blood is so that they can be rapidly transported from their site of production or storage to wherever they are needed. Therefore, although the specific circulating leukocytes will be introduced here to round out our discussion of blood, we will leave a more detailed discussion of their phagocytic and immunological functions, which take place primarily in the tissues, for the next chapter, "Body Defenses."

There are five different types of leukocytes.

Leukocytes lack hemoglobin (in contrast to erythrocytes) so they are colorless (that is, "white") unless specifically stained for microscopic visibility. Unlike erythrocytes, which are of uniform structure, identical function, and constant number, leukocytes vary in structure, function, and number. There are five different types of circulating leukocytes—neutrophils, eosinophils, basophils, monocytes, and lymphocytes—each with a characteristic structure and function. They are all somewhat larger than erythrocytes.

The five types of leukocytes fall into two main categories, depending on the appearance of their nuclei and the presence or absence of granules in their cytoplasm when viewed microscopically (• Fig. 11–7). Neutrophils, eosinophils, and basophils are categorized as **polymorphonuclear** ("many-shaped nucleus") **granulocytes** ("granule-containing cells"). Their nuclei are segmented into several lobes of varying shapes, and their cytoplasm contains an abundance of membrane-bound granules. The three types of granulocytes are distinguished on the basis of the varying affinity of their granules for dyes: *eosinophils* have an affinity for the red dye

eosin, *basophils* preferentially take up a basic blue dye, and *neutrophils* are neutral, showing no dye preference. Monocytes and lymphocytes are known as **mononuclear** ("single nucleus") **agranulocytes** ("cells lacking granules"). Both have a single, large, nonsegmented nucleus and few granules. *Monocytes* are the larger of the two and have an oval or kidney-shaped nucleus. *Lymphocytes*, the smallest of the leukocytes, characteristically have a large spherical nucleus that occupies most of the cell.

Leukocytes are produced at varying rates depending on the changing defense needs of the body.

All leukocytes ultimately originate from the same undifferentiated stem cells in the red bone marrow that also give rise to erythrocytes and platelets (• Fig. 11–8). The cells destined to become leukocytes eventually differentiate into various committed cell lines and proliferate under the influence of appropriate stimulating factors. Granulocytes and monocytes are produced only in the bone marrow, which releases these mature leukocytes into the blood. Lymphocytes are originally derived from precursor cells in the bone marrow, but most new lymphocytes are actually produced by already-existing lymphocytes residing in the **lymphoid** (lymphocyte-containing) **tissues**, such as the lymph nodes and tonsils.

The total number of leukocytes normally ranges from 5 to 10 million cells per milliliter of blood, with an average of 7 million cells/ml, expressed as an average **white blood cell count** of 7,000/mm³. Leukocytes are the least numerous of the cellular elements in the blood (about 1 white blood cell for every 700 red blood cells), not because fewer are produced but because they are merely in transit while in the blood. Normally, approximately two-thirds of the circulating leukocytes are granulocytes, mostly neutrophils, whereas one-third are agranulocytes, predominantly lymphocytes (Table 11–2). However, the total number of white cells and the percentage of each type may vary considerably to meet changing defense needs. Depending on the type and extent of assault the body is combating, different types of leukocytes are selectively produced at varying rates. Chemical messengers arising from invaded or damaged tissues or from activated leukocytes themselves govern the rates of production of the various leukocytes. Specific hormones analogous to erythropoietin are required to direct the differentiation and proliferation of

• **FIGURE 11–7**   **Normal Blood Cellular Elements**

| Leukocytes | | | | | Erythrocytes | Platelets |
| Polymorphonuclear granulocytes | | | Mononuclear agranulocytes | | | |
Neutrophil	Eosinophil	Basophil	Monocyte	Lymphocyte	Erythrocytes	Platelets

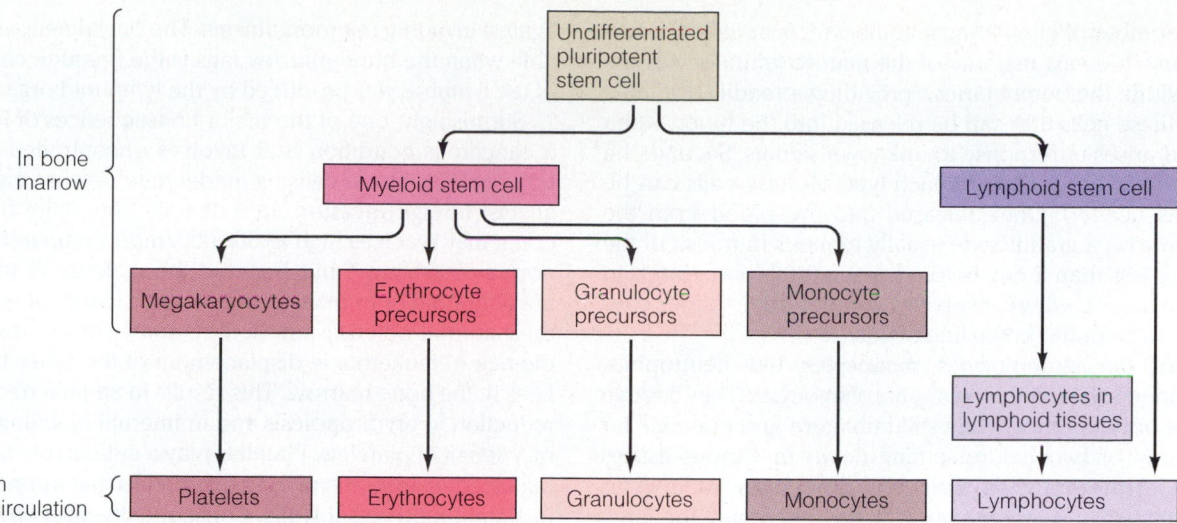

● **FIGURE 11–8   Blood-Cell Production (Hemopoiesis)** All the blood-cell types ultimately originate from the same undifferentiated pluripotent stem cells in the bone marrow. The bone marrow produces all circulating blood cells except lymphocytes, the majority of which are produced by lymphocyte colonies in lymphoid tissues. These lymphocyte colonies are originally derived from precursor cells in the bone marrow.

each cell type. Some of these hormones have been identified and can be produced in the laboratory; an example is **granulocyte colony-stimulating factor**, which stimulates increased replication and release of granulocytes, especially neutrophils, from the bone marrow. This achievement has opened up the ability to administer these hormones as a powerful new therapeutic tool to bolster a person's normal defense against infection or cancer. Examining the functions of the various types of leukocytes provides clues as to what conditions normally prompt their production.

Among the granulocytes, **neutrophils** are phagocytic specialists. They invariably are the first defenders on the scene of bacterial invasion and, accordingly, are very important in in-

flammatory responses. Furthermore, they scavenge to clean up debris. As might be expected in view of these functions, an increase in circulating neutrophils (**neutrophilia**) typically accompanies acute bacterial infections. In fact, a differential white blood cell count (a determination of the proportion of each type of leukocyte present) can be useful in making an immediate, reasonably accurate prediction of whether an infection, such as pneumonia or meningitis, is of bacterial or viral origin. Obtaining a definitive answer as to the causative agent by culturing a sample of the infected tissue's fluid takes several days. Since an elevated neutrophil count is highly indicative of bacterial infection, it is appropriate to initiate antibiotic therapy long before the true causative agent is actually known. (Bacteria generally succumb to antibiotics, whereas viruses do not.)

Eosinophils are specialists of another type. An increase in circulating eosinophils (**eosinophilia**) is associated with allergic conditions (such as asthma and hay fever) and with internal parasite infestations (for example, worms). Eosinophils obviously cannot engulf a much larger parasitic worm, but they do attach to the worm and secrete substances that kill it.

Basophils are the least numerous and most poorly understood of the leukocytes. They are quite similar structurally and functionally to **mast cells**, which never circulate in the blood but instead are dispersed in the connective tissue throughout the body. It was once believed that basophils became mast cells by migrating from the circulatory system, but researchers have shown that basophils arise from the bone marrow whereas mast cells are derived from precursor cells located in the connective tissue. Both basophils and mast cells synthesize and store *histamine* and *heparin,* powerful chemical substances that can be released upon appropriate stimulation. Histamine release is important in allergic reactions, whereas heparin hastens the removal of fat particles from the blood following a fatty meal. Heparin can also prevent blood clotting (coagulation), but whether it plays a physiological role as an anticoagulant is still being debated.

TABLE 11–2 Typical Human Blood Cell Count		
Total erythrocytes = 5,000,000,000 cells/ml blood		
Red blood cell count = 5,000,000/mm³		
Total leukocytes = 7,000,000 cells/ml blood		
White blood cell count = 7,000/mm³		
Differential white blood cell count (percentage distribution of types of leukocytes)		
Polymorphonuclear granulocytes	Mononuclear agranulocytes	
Neutrophils 60–70%	Lymphocytes	25–33%
Eosinophils 1–4%	Monocytes	2–6%
Basophils 0.25–0.5%		
Total platelets = 250,000,000/ml blood		
Platelet count = 250,000/mm³		

The number of circulating granulocytes is adjusted in two ways. First, the vast majority of the mature granulocytes are stored within the bone marrow, providing a readily available pool of these cells that can be released into the blood when the need arises in response to unknown signals. Second, the actual rate of production of each type of these cells can be varied as needed. Once released into the blood from the bone marrow, a granulocyte usually remains in transit in the blood for less than a day before leaving the blood vessels to enter the tissues, where it survives another three to four days unless it dies sooner in the line of duty.

Among the agranulocytes, **monocytes**, like neutrophils, are destined to become professional phagocytes. They emerge from the bone marrow while still immature and circulate for only a day or two before settling down in various tissues throughout the body. At their new residences, monocytes continue to mature and greatly enlarge, becoming the large tissue phagocytes known as **macrophages**. Macrophages' life span may range from months to years unless they are destroyed sooner while performing their phagocytic activity. A phagocytic cell can ingest only a limited amount of foreign material before it succumbs itself.

Lymphocytes provide immune defense against targets for which they are specifically programmed. There are two types of lymphocytes, B lymphocytes and T lymphocytes. **B lymphocytes** produce **antibodies**, which circulate in the blood. An antibody binds with and marks for destruction (by phagocytosis or other means) the specific kinds of foreign matter, such as bacteria, that induced production of the antibody. **T lymphocytes** do not produce antibodies; instead they directly destroy their specific target cells, a process known as a **cell-mediated immune response.** The target cells of T lymphocytes include body cells invaded by viruses and cancer cells. Lymphocytes have life spans estimated at 100 to 300 days. During this period, the majority of them continually recycle among the lymphoid tissues, lymph, and blood, spending only a few hours at a time in the blood. Therefore, only a small proportion of the total lymphocytes are in transit in the blood at any given moment.

By unknown mechanisms, the number of circulating lymphocytes is frequently elevated in association with a chronic infection. In **infectious mononucleosis**, not only does the number of lymphocytes in the blood increase, but many of the lymphocytes are atypical in structure. This condition, which is caused by the Epstein-Barr virus, is characterized by pronounced fatigue, a mild sore throat, and low-grade fever. Full recovery usually requires a month or more.

Even though circulating-leukocyte levels may vary, changes in these levels are normally controlled and adjusted according to the body's needs. However, abnormalities in leukocyte production can occur that are not subject to regulatory mechanisms; that is, either too few or too many leukocytes can be produced. The bone marrow can greatly slow down or even stop its production of white blood cells when it is exposed to certain toxic physical agents (such as radiation) or chemical agents (such as benzene and anticancer drugs). As you might predict, the most serious consequence is the reduction in professional phagocytes (neutrophils and macrophages), which leads to a notable reduction in the body's defense capabilities against invading microorganisms. The only defense still available when the bone marrow fails is the immune capabilities of the lymphocytes produced by the lymphoid organs.

Surprisingly, one of the major consequences of **leukemia**, a cancerous condition that involves uncontrolled proliferation of white blood cells, is inadequate defense capabilities against foreign invasion. In leukemia, the white blood cell count may reach as high as $500,000/mm^3$, compared with the normal $7,000/mm^3$, but because the majority of these cells are abnormal or immature, they are incapable of performing their normal defense functions. Another devastating consequence of leukemia is displacement of the other blood-cell lines in the bone marrow. This results in anemia because of a reduction in erythropoiesis and in internal bleeding because of a deficit of platelets. Platelets play a critical role in preventing bleeding from the myriad tiny breaks that normally occur in small blood-vessel walls. Consequently, overwhelming infections or hemorrhage are the most common causes of death in leukemic patients. The next section will examine the platelets' role in greater detail to show how they normally minimize the threat of hemorrhage.

PLATELETS AND HEMOSTASIS

Platelets are cell fragments derived from megakaryocytes.

In addition to erythrocytes and leukocytes, platelets are a third type of cellular element present in the blood. Platelets are not whole cells but small cell fragments (about 2–4 μm in diameter) that have budded off the outer edges of extraordinarily large (up to 60 μm in diameter) bone-marrow–bound cells known as **megakaryocytes**. Megakaryocytes are derived from the same undifferentiated stem cells that give rise to the erythrocytic and leukocytic cell lines. Platelets are essentially detached vesicles containing portions of megakaryocyte cytoplasm wrapped in plasma membrane.

An average of 250,000,000 platelets are normally present in each milliliter of blood (range of $150,000–350,000/mm^3$). Platelets remain functional for an average of ten days, at which time they are removed from circulation by the tissue macrophages, especially those in the spleen and liver, and are replaced by newly released platelets from the bone marrow.

Platelets do not leave the blood as white blood cells do, but at any given time about one-third of them are in storage in blood-filled spaces in the spleen. These stored platelets can be released from the spleen into the circulating blood as needed (for example, during hemorrhage) by sympathetically induced splenic contraction.

Because platelets are cell fragments, they lack nuclei. However, they are equipped with organelles and cytosolic enzyme systems for generating energy and synthesizing secretory products, which they store in numerous granules dispersed throughout the cytosol. Furthermore, platelets contain high concentrations of actin and myosin, which enable them to contract. Their secretory and contractile abilities are important in hemostasis, a topic to which we now turn our attention.

Hemostasis prevents blood loss from damaged small vessels.

Hemostasis is the arrest of bleeding from a broken blood vessel, that is, the stopping of hemorrhage. (Be sure not to confuse this with the term *homeostasis.*) For bleeding to take place from a vessel, there must be a break in the vessel wall, and the pressure inside the vessel must be greater than the pressure outside it to force the blood out through the defect. The body's inherent hemostatic mechanisms normally are adequate to seal defects and stop loss of blood through small damaged capillaries, arterioles, and venules. These small vessels are frequently ruptured by the minor traumas of everyday life; such traumas are the most common source of bleeding, although we often are not even aware that any damage has taken place. The hemostatic mechanisms normally keep blood loss from these minor vascular traumas to a minimum.

The much rarer occurrence of bleeding from medium- to large-size vessels usually cannot be stopped by the body's hemostatic mechanisms alone. Bleeding from a severed artery is more profuse and therefore more dangerous than venous bleeding because the outward driving pressure is greater in the arteries (that is, arterial blood pressure is considerably higher than venous pressure). First-aid measures for a severed artery include application over the wound of external pressure of greater magnitude than the arterial blood pressure to temporarily halt the bleeding until the torn vessel can be surgically closed. Hemorrhage from a traumatized vein can often be stopped simply by elevating the bleeding body part to reduce gravity's effects on pressure in the vein (see p. 330). If the accompanying drop in venous pressure is not sufficient to stop the bleeding, mild external compression is usually adequate.

Hemostasis involves three major steps: (1) *vascular spasm,* (2) *formation of a platelet plug,* and (3) *blood coagulation (clotting).* Platelets obviously play a major part in forming a platelet plug, but they contribute significantly to the other two steps as well.

Vascular spasm reduces blood flow through an injured vessel.

A cut or torn blood vessel immediately constricts as a result of both an inherent vascular response to injury and sympathetically induced vasoconstriction. This constriction slows blood flow through the defect and thus minimizes blood loss. Also, as the opposing endothelial (inner) surfaces of the vessel are pressed together by this initial **vascular spasm,** they become sticky and adhere to each other, further sealing off the damaged vessel. These physical measures alone cannot completely prevent further blood loss, but they are important in minimizing blood flow through the break in the vessel until the other hemostatic measures are able to actually plug up the defect.

Platelets aggregate to form a plug at a vessel defect.

Platelets normally do not adhere to the smooth endothelial surface of blood vessels, but when this lining is disrupted because of vessel injury, platelets attach to the exposed collagen, which is a fibrous protein present in the underlying connective tissue. Once platelets start aggregating at the site of the defect, they release several important chemicals from their storage granules (● Fig. 11–9). Among these chemicals is **adenosine diphosphate (ADP),** which causes the surface of

---

● **FIGURE 11–9   Formation of a Platelet Plug** Platelets aggregate at a vessel defect through several positive-feedback mechanisms involving the release of adenosine diphosphate (ADP) and thromboxane A₂ from platelets that stick to exposed collagen at the site of the injury. Platelets are prevented from aggregating at the adjacent normal vessel lining by the release of prostacyclin from the endothelial cells that form the vessel lining.

nearby circulating platelets to become sticky so that they adhere to the first layer of aggregated platelets. These newly aggregated platelets release more ADP, which causes more platelets to aggregate, and so on; thus, a plug of platelets is rapidly built up at the defect site in a positive-feedback fashion.

This aggregating process is reinforced by the formation of a chemical messenger, **thromboxane A₂**, from a component of the platelet plasma membrane upon contact with collagen. Thromboxane A₂ is closely related to the prostaglandins, a group of locally functioning chemical messengers found extensively throughout the body (see chapter 20). These local messengers are fatty acid derivatives formed from arachidonic acid, one of the fatty acids found in membrane phospholipids. Thromboxane A₂ directly promotes platelet aggregation and further enhances it indirectly by triggering the release of even more ADP from the platelet granules.

Given the self-perpetuating nature of platelet aggregation, why is the platelet plug limited to the site of vessel injury once it is initiated? (In other words, why doesn't the platelet plug continue to develop and expand over the surface of the adjacent normal vessel lining?) A key reason why this does not happen is that the normal endothelium releases **prostacyclin**, a chemical that profoundly inhibits platelet aggregation. Thus, the platelet plug is limited to the defect and does not spread to normal vascular tissue (Fig. 11–9).

The aggregated platelet plug not only physically seals the break in the vessel but also performs three other important roles. First, the actin-myosin protein complex within the aggregated platelets contracts to compact and strengthen what was originally a fairly loose plug. Second, the chemicals released from the platelet plug include several powerful vasoconstrictors (serotonin, epinephrine, and thromboxane A₂), which induce profound constriction of the affected vessel to reinforce the initial, self-induced vascular spasm. Third, the platelet plug releases other chemicals that enhance blood coagulation, the next step of hemostasis. Although the platelet-plugging mechanism alone is often sufficient to seal the myriad minute tears in capillaries and other small vessels that occur many times daily, larger holes in these vessels require the formation of a blood clot to completely stop the bleeding.

A triggered chain reaction involving clotting factors in the plasma results in blood coagulation.

Blood coagulation, or **clotting**, is the transformation of blood from a liquid into a solid gel. Formation of a clot on top of the platelet plug strengthens and supports the plug, reinforcing the seal over a break in a vessel. Furthermore, as blood in the vicinity of the vessel defect solidifies, it can no longer flow. Coagulation is the body's most powerful hemostatic mechanism, and it is required to stop bleeding from all but the most minute defects.

The ultimate step in clot formation is the conversion of **fibrinogen**, a large, soluble plasma protein produced by the liver and normally always present in the plasma, into **fibrin**, an insoluble, threadlike molecule. The conversion into fibrin is catalyzed by the enzyme **thrombin** at the site of vessel injury.

Fibrin molecules adhere to the damaged vessel surface, forming a loose, netlike meshwork that traps the cellular elements of the blood. The resultant mass, or **clot**, typically appears red because of the abundance of trapped red blood cells, but the foundation of the clot is formed from fibrin derived from the plasma (• Fig. 11–10). Except for platelets, which play an important role in ultimately bringing about the conversion of fibrinogen to fibrin, clotting can take place in the absence of all other cellular elements in the blood.

• **FIGURE 11–10 Erythrocytes Trapped in Fibrin Meshwork of a Clot**

● **FIGURE 11–11 Roles of Thrombin in Hemostasis** Thrombin, which is a component of the clotting cascade, plays multiple roles in hemostasis: (1) conversion of fibrinogen to fibrin; (2) activation of the factor that stabilizes the fibrin meshwork of the clot; (3) positive-feedback activation of more prothrombin into thrombin; and (4) enhancement of platelet aggregation. In positive-feedback fashion, aggregated platelets secrete PF3, which stimulates the clotting cascade that results in thrombin activation.

The original fibrin web is rather weak because the fibrin strands are only loosely interlaced. However, chemical linkages rapidly form between adjacent strands to strengthen and stabilize the clot meshwork. This cross-linkage process is catalyzed by a clotting factor known as **factor XIII** (**fibrin-stabilizing factor**), which normally is present in the plasma in inactive form. In addition to converting fibrinogen into fibrin, thrombin also (1) activates factor XIII to stabilize the resulting fibrin meshwork; (2) enhances platelet aggregation, which in turn is essential to the clotting process; and (3) acts in a positive-feedback fashion to facilitate its own formation (● Fig. 11–11).

Because thrombin's action converts the ever-present fibrinogen molecules in the plasma into a blood-stanching clot, thrombin must normally be absent from the plasma except in the vicinity of vessel damage. Otherwise, blood would always be coagulated, a situation incompatible with life. How can thrombin normally be absent from the plasma, yet be readily available to trigger fibrin formation when a vessel is injured? The solution lies in thrombin's existence in the plasma in the form of an inactive precursor called **prothrombin**.

The next question is, what converts prothrombin into thrombin when blood clotting is desirable? Yet another activated plasma clotting factor, **factor X**, is responsible; factor X itself is normally present in the blood in inactive form and must be converted into its active form by still another activated factor, and so on. Altogether, twelve plasma clotting factors participate in essential steps that lead to the final conversion of fibrinogen into a stabilized fibrin meshwork (● Fig. 11–12). These factors are designated by Roman numerals in the order in which they were discovered, not the order in which they participate in the clotting process.[1] Most of these clotting factors are plasma proteins synthesized by

the liver. Normally they are always present in the plasma in an inactive form, similar to fibrinogen and prothrombin. In contrast to fibrinogen, which is converted into insoluble fibrin strands, prothrombin and the other precursors, when converted to their active form, act as proteolytic (protein-splitting) enzymes, which activate another specific factor in the clotting sequence. Once the first factor in the sequence is activated, it in turn activates the next factor, and so on, in a series of sequential reactions known as a **cascade**, until thrombin catalyzes the final conversion of fibrinogen into fibrin. Several of these steps require the presence of plasma Ca^{++} and **platelet factor 3** (**PF3**), a phospholipid secreted by the aggregated platelet plug. Thus, platelets also contribute to clot formation.

The clotting cascade may be triggered by the *intrinsic pathway* or the *extrinsic pathway*.

■ The **intrinsic pathway** precipitates intravascular clotting as well as clotting of blood samples in test tubes. All elements necessary to bring about clotting by means of the intrinsic pathway are present in the blood. This pathway, which involves seven separate steps, (shown in blue in Fig. 11–12), is set off when **factor XII** (**Hageman factor**) is activated by coming into contact with either exposed collagen in an injured vessel or a foreign surface such as a glass test tube. Remember that exposed collagen also initiates platelet aggregation. Thus, formation of a platelet plug and the chain reaction leading to clot formation are simultaneously set in motion when a vessel is damaged. Furthermore, these complementary hemostatic mechanisms reinforce each other. The aggregated platelets secrete PF3, which is essential for the clotting cascade that in turn enhances further platelet aggregation (Figs. 11–11 and ● 11–13).

■ The **extrinsic pathway** takes a shortcut and requires only four steps (shown in gray in Fig. 11–12). This pathway, which requires contact with tissue factors external to the

[1] The term *factor VI* is no longer used. What once was considered to be a separate factor VI has now been determined to be an activated form of factor V.

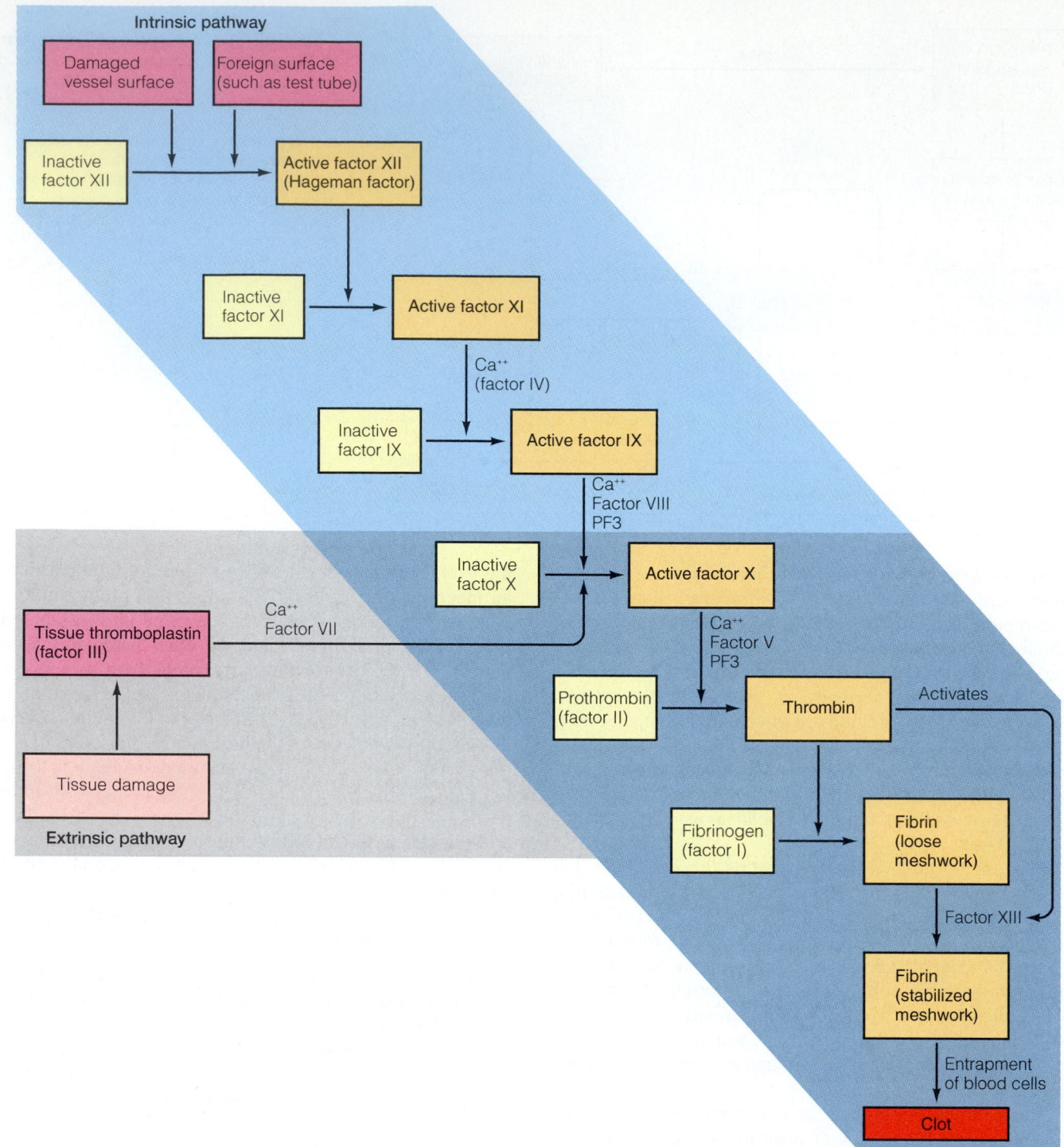

● **FIGURE 11–12 Clot Pathways** The intrinsic clotting pathway (in blue) is initiated when factor XII (Hageman factor) is activated by contact with exposed collagen at a damaged vessel surface or contact with a foreign surface. This pathway brings about clotting within damaged vessels and clotting of blood samples in test tubes. The shorter extrinsic clotting pathway (in gray) is initiated when factor X, a factor activated halfway through the intrinsic pathway, is activated by tissue thromboplastin released from damaged tissue. The extrinsic pathway causes clotting of blood that escapes from blood vessels into the surrounding tissue during injury. The steps in the two pathways are identical from factor X on (in blue-gray).

blood, initiates clotting of blood that has escaped into the tissues. When a tissue is traumatized, it releases a protein complex known as **tissue thromboplastin**. Tissue thromboplastin directly activates factor X, thereby bypassing all preceding steps of the intrinsic pathway. From this point on, the two pathways are identical.

The intrinsic and extrinsic mechanisms usually operate simultaneously. When tissue injury involves rupture of vessels, the intrinsic mechanism stops blood in the injured vessel, whereas the extrinsic mechanism clots the blood that escaped into the tissues before the vessel was sealed off. Typically, clot formation is fully developed in three to six minutes.

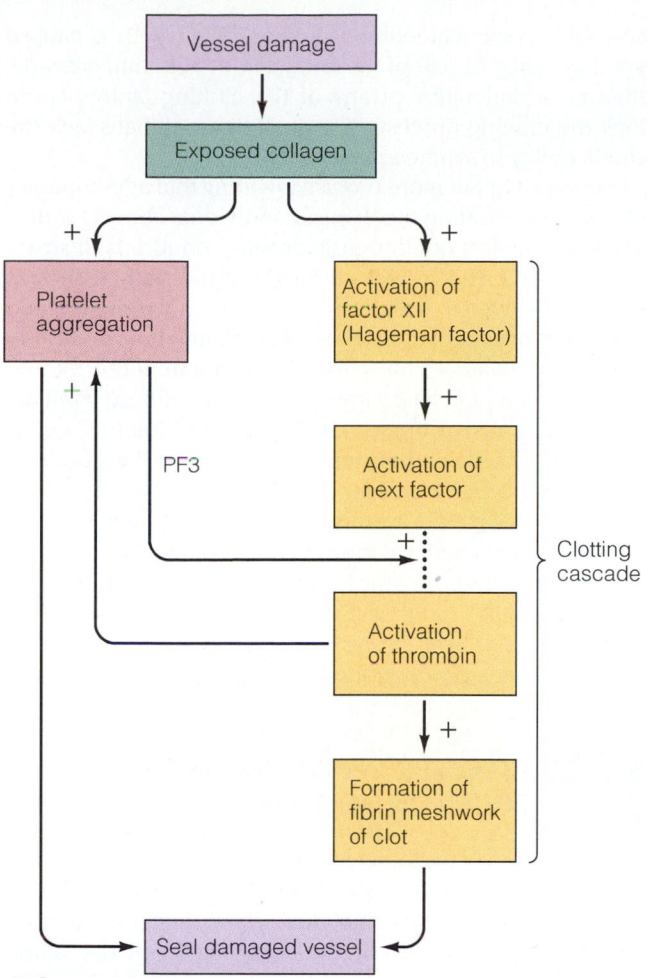

• FIGURE 11–13 Concurrent Platelet Aggregation and Clot Formation Exposed collagen at the site of vessel damage simultaneously initiates platelet aggregation and the clotting cascade. These two hemostatic mechanisms positively reinforce each other as they seal the damaged vessel.

Once a clot is formed, contraction of the platelets trapped within the clot shrinks the fibrin meshwork, pulling the edges of the damaged vessel closer together. During **clot retraction**, fluid is squeezed from the clot. This fluid, which is essentially plasma minus fibrinogen and other clotting precursors that have been removed during the clotting process, is called **serum**.

Although a clotting process that involves so many steps might seem inefficient, the advantage is the amplification accomplished during many of the steps. One molecule of an activated factor can activate perhaps a hundred molecules of the next factor in the sequence, each of which can activate many more molecules of the next factor, and so on. In this way, large numbers of the final factors involved in clotting are rapidly activated as a result of the initial activation of only a few molecules in the beginning step of the sequence. How then is the clotting process, once initiated, confined to the site of vessel injury? If the activated clotting factors were allowed to circulate, they would induce inappropriate widespread clotting that would plug up vessels throughout the body. Fortunately, after participating in the local clotting process, the massive number of factors activated in the vicinity of vessel injury are rapidly inactivated by enzymes and other factors present in the plasma or tissue.

Fibrinolytic plasmin dissolves clots and prevents inappropriate clot formation.

A clot is not meant to be a permanent solution to vessel injury. It is a transient device to stop bleeding until the vessel can be repaired. The aggregated platelets secrete a chemical that is at least partially responsible for the invasion of fibroblasts ("fiber-formers") from the surrounding connective tissue into the wounded area of the vessel. Fibroblasts form a scar at the vessel defect. Simultaneous with the healing process, the clot, which is no longer needed to prevent hemorrhage, is slowly dissolved by a fibrinolytic (fibrin-splitting) enzyme called **plasmin**. If clots were not removed after they performed their hemostatic function, the vessels, especially the small ones that endure tiny ruptures on a regular basis, would eventually become obstructed by clots.

Plasmin, like the clotting factors, is a plasma protein produced by the liver and present in the blood in an inactive precursor form, **plasminogen**. Plasmin is activated cascade-fashion by many factors, among them factor XII (Hageman factor), which also triggers the chain reaction leading to clot formation (• Fig. 11–14). When a clot is being formed, activated plasmin becomes trapped in the clot and subsequently dissolves it by slowly breaking down the fibrin meshwork. Phagocytic white blood cells gradually remove the products of clot dissolution. You have observed the slow removal of blood that has clotted after escaping into the tissue layers of your skin following an injury. The "black-and-blue marks" of bruised skin result from deoxygenated clotted blood within

• FIGURE 11–14 Role of Factor XII in Clot Formation and Dissolution Activation of factor XII (Hageman factor) simultaneously initiates a fast cascade of reactions that result in clot formation and a fast cascade of reactions that result in plasmin activation. Plasmin, which is trapped in the clot, subsequently slowly dissolves the clot. This action removes the clot when it is no longer needed after the vessel has been repaired.

the skin; this blood is eventually cleared by plasmin action, followed by the phagocytic cleanup crew.

In addition to removing clots that are no longer needed, plasmin functions continually to prevent clots from forming inappropriately. Small amounts of fibrinogen are constantly being converted into fibrin throughout the vasculature, triggered by unknown mechanisms. Clots do not develop, however, because the fibrin is quickly disposed of by plasmin that is activated by **tissue plasminogen activator** (**tPA**) derived from the tissues, especially the lungs. Normally, the low level of fibrin formation is counterbalanced by a low level of fibrinolytic activity, so inappropriate clotting does not occur. Only when a vessel is damaged do additional factors precipitate the explosive chain reaction that leads to more extensive fibrin formation and results in local clotting at the site of injury.

Naturally occurring anticoagulants that interfere with one or more steps in the clotting process are known to exist in the body. Whether they play a physiological role in clot prevention is still being debated. The best known of these is **heparin**, which is present in basophils and mast cells. It is used extensively as an anticoagulant drug as well as to prevent clotting of blood samples drawn for clinical analysis.

In recent years, genetically engineered tPA and other similar chemicals that trigger clot dissolution have been used successfully to limit damage to cardiac muscle during heart attacks. Administration of a clot-busting drug within the first hours after a clot has blocked a coronary vessel often dissolves the clot in time to restore blood flow to the cardiac muscle supplied by the blocked vessel before the muscle dies of O_2 deprivation.

Inappropriate clotting is responsible for thromboembolism.

In spite of protective measures, clots occasionally form in intact vessels. An abnormal intravascular clot attached to a vessel wall is known as a **thrombus**, and free floating clots are called **emboli**. Several factors, acting independently or simultaneously, can cause *thromboembolism:* (1) Roughened vessel surfaces associated with atherosclerosis can lead to thrombus formation (see p. 292). (2) Imbalances in the clotting-anticlotting systems can likewise trigger clot formation. (3) Slow-moving blood is more apt to clot, probably because small quantities of fibrin are formed and allowed to accumulate in the stagnant blood, for example, in blood pooled in varicosed leg veins. (4) Widespread clotting is occasionally triggered by the release of tissue thromboplastin into the blood from large amounts of traumatized tissue. Similar widespread clotting can occur in **septicemic shock**, in which bacteria or their toxins initiate the clotting cascade.

Hemophilia is the primary condition responsible for excessive bleeding.

In contrast to inappropriate clot formation in intact vessels, the opposite hemostatic disorder is failure of clots to form promptly in injured vessels, resulting in life-threatening hemorrhage from even relatively mild traumas. The most common cause of excessive bleeding is **hemophilia**, which is caused by a deficiency of one of the factors in the clotting cascade. Although a deficiency of any of the clotting factors could block the clotting process, 80% of all hemophiliacs lack the genetic ability to synthesize factor VIII.

In contrast to the more profuse bleeding that accompanies defects in the clotting mechanism, individuals having a deficiency of platelets continuously develop hundreds of small, confined hemorrhagic areas throughout the body tissues as blood is permitted to leak from tiny breaks in the small blood vessels before coagulation takes place. Platelets normally are the primary sealers of these ever-occurring minute ruptures. In the skin of a platelet-deficient person, the diffuse capillary hemorrhages are visible as small, purplish blotches, giving rise to the term **thrombocytopenia purpura** ("the purple of thrombocyte deficiency") for this condition. (Recall that the term "thrombocytes" is an alternate name for platelets.)

Vitamin K deficiency can cause a bleeding tendency. This vitamin is essential for the liver's synthesis of prothrombin and several other clotting factors. Furthermore, one of the consequences of liver disease is prolonged clotting time because of reduced production of clotting factors.

CHAPTER IN PERSPECTIVE: FOCUS ON HOMEOSTASIS

Blood contributes to homeostasis in a variety of ways. First, the composition of the interstitial fluid, the true internal environment that surrounds and directly exchanges materials with the cells, depends on the composition of the blood plasma. Because of the thorough exchange that occurs between the interstitial and vascular compartments, the interstitial fluid has the same composition as the plasma with the exception of plasma proteins, which cannot escape through the capillary walls. Thus, the blood serves as the vehicle for rapid, long-distance mass transport of materials to and from the cells, and the interstitial fluid serves as the go-between.

Homeostasis depends on the blood carrying materials such as O_2 and nutrients to the cells as rapidly as the cells consume these supplies and carrying materials such as metabolic wastes away from the cells as rapidly as the cells produce these products. It also depends on the blood carrying hormonal messengers from their site of production to their distant site of action. Once a substance enters the blood, it can be transported throughout the body within seconds, whereas diffusion of the substance over long distances in a large multicellular organism such as a human would take months to years—a situation incompatible with life. Diffusion can, however, effectively accomplish short local exchanges of materials between the blood and surrounding cells through the intervening interstitial fluid.

The blood has special transport capabilities that enable it to move its cargo efficiently throughout the body. For example, life-sustaining O_2 is poorly soluble in water, but the blood is equipped with O_2-carrying specialists, the erythrocytes (red blood cells), which are stuffed full of hemoglobin, a complex molecule that transports O_2. Likewise, homeostatically important water-insoluble hormonal messengers are shuttled in the blood by plasma-protein carriers.

Specific components of the blood perform the following additional homeostatic activities that are unrelated to blood's transport function:

■ The blood helps maintain the proper pH in the internal environment by buffering changes in the acid-base load of the body.

■ The blood helps maintain body temperature by absorbing heat produced by heat-generating tissues such as contracting skeletal muscles and distributing it throughout the body. Excess heat is carried by the blood to the body surface for elimination to the external environment.

■ The electrolytes in the plasma are important in membrane excitability, which in turn forms the basis of nerve and muscle function.

■ The electrolytes in the plasma are also important in the osmotic distribution of fluid between the extracellular and intracellular fluid, and the plasma proteins play a critical role in the distribution of extracellular fluid between the plasma and interstitial fluid.

■ Through their hemostatic functions, the platelets and clotting factors minimize the loss of life-sustaining blood following vessel injury.

■ The leukocytes (white blood cells) and certain types of plasma proteins, such as antibodies, defend the body against invading disease-causing agents, destroy cancer cells, and pave the way for wound healing and tissue repair by clearing away debris from dead or injured cells. These actions indirectly contribute to homeostasis by helping to keep the organs that directly maintain homeostasis healthy.

 CHAPTER SUMMARY

Plasma

The 5- to 5.5-liter volume of blood in an adult consists of 42–45% erythrocytes, less than 1% leukocytes and platelets, and 55–58% plasma. The percentage of whole blood volume occupied by erythrocytes is known as the hematocrit.

Plasma is a complex liquid that serves as a transport medium for substances being carried in the blood. All plasma constituents are freely diffusible across the capillary walls except the plasma proteins, which remain in the plasma and perform a variety of functions.

Erythrocytes

Erythrocytes (red blood cells) are specialized for their primary function of O_2 transport in the blood. They do not contain a nucleus, organelles, or ribosomes but instead are packed full of hemoglobin, which is an iron-containing molecule that can loosely, reversibly bind with O_2. Because O_2 is poorly soluble in blood, hemoglobin is indispensible for O_2 transport. Hemoglobin also contributes to CO_2 transport and buffering of blood by reversibly binding with CO_2 and H^+.

Unable to replace cell components, erythrocytes are destined to a short life span of about 120 days. Undifferentiated stem cells in the red bone marrow give rise to all cellular elements of the blood. Erythrocyte production (erythropoiesis) by the marrow normally keeps pace with the rate of erythrocyte loss to keep the red cell count constant. Erythropoiesis is stimulated by erythropoietin, a hormone secreted by the kidneys in response to reduced O_2 delivery.

Leukocytes

Leukocytes (white blood cells) are the defense corps of the body. They attack foreign invaders, destroy abnormal cells that arise in the body, and clean up cellular debris. There are five types of leukocytes, each with a different task: (1) Neutrophils, the phagocytic specialists, are important in engulfing bacteria and debris. (2) Eosinophils specialize in attacking parasitic worms and play a key role in allergic responses. (3) Basophils release two chem-

icals: histamine, which is also important in allergic responses, and heparin, which helps clear fat particles from the blood. (4) Monocytes, upon leaving the blood, set up residence in the tissues and greatly enlarge to become the large tissue phagocytes known as macro-phages. (5) Lymphocytes provide immune defense against bacteria, viruses, and other targets for which they are specifically programmed. Their defense tools include the production of antibodies and cell-mediated immune responses.

Leukocytes are present in the blood only while in transit from their site of production and storage in the bone marrow (and also in the lymphoid organs in the case of the lymphocytes) to their site of action in the tissues. At any given time, the majority of the leukocytes are out in the tissues on surveillance missions or performing actual combative activities. All leukocytes have a limited life span and must be replenished by ongoing differentiation and proliferation of precursor cells. The total number and percentage of each of the different types of leukocytes produced vary depending on the momentary defense needs of the body.

Platelets and Hemostasis

Platelets are cell fragments derived from large megakaryocytes in the bone marrow. Platelets play an important role in hemostasis, the arrest of bleeding from an injured vessel. The three main steps in hemostasis are (1) vascular spasm, (2) platelet plugging, and (3) clot formation. Vascular spasm reduces blood flow through an injured vessel, whereas aggregation of platelets at the site of vessel injury quickly plugs the defect. Platelets start to aggregate upon contact with exposed collagen in the damaged vessel wall.

Clot formation (blood coagulation) reinforces the platelet plug and converts blood in the vicinity of a vessel injury into a nonflowing gel. The majority of factors necessary for clotting are always present in the plasma in inactive precursor form. When a vessel is damaged, exposed collagen initiates a cascade of reactions involving successive activation of these clotting factors, ultimately converting fibrinogen into fibrin. Fibrin, an insoluble threadlike molecule, is laid down as the meshwork of the clot;

the meshwork in turn entangles blood cells to complete clot formation. Blood that has escaped into the tissues is also coagulated upon exposure to tissue thromboplastin, which likewise sets the clotting process into motion. When no longer needed, clots are dissolved by plasmin, a fibrinolytic factor also activated by exposed collagen.

REVIEW EXERCISES

Objective Questions (Answers on p. D–3.)

1. Blood can absorb metabolic heat while undergoing only small changes in temperature. (True or false?)
2. Hemoglobin can carry only O_2. (True or false?)
3. Erythrocytes originate from the same undifferentiated stem cells as leukocytes and platelets. (True or false?)
4. Erythrocytes are unable to utilize the O_2 they contain for their own ATP formation. (True or false?)
5. White blood cells spend the majority of their time in the blood. (True or false?)
6. The body's defense capabilities are reduced in leukemia even though there are an excessive number of leukocytes. (True or false?)
7. Which type of leukocyte is produced primarily in lymphoid tissue? _____
8. The majority of plasma clotting factors are synthesized by the _____.
9. Which of the following is *not* a function served by the plasma proteins?
   a. facilitate retention of fluid in the blood vessels
   b. play an important role in blood clotting
   c. bind and transport certain hormones in the blood
   d. transport O_2 in the blood
   e. serve as antibodies
   f. contribute to buffering capacity of the blood
10. Which of the following is *not* triggered by exposed collagen?
    a. vascular spasm
    b. platelet aggregation
    c. activation of the clotting cascade
    d. activation of plasminogen
11. Match the following (an answer may be used more than once):
    ___ 1. causes platelets to aggregate in positive-feedback fashion
    ___ 2. activates prothrombin
    ___ 3. fibrinolytic enzyme
    ___ 4. inhibits platelet aggregation
    ___ 5. first factor activated in intrinsic clotting pathway
    ___ 6. forms the meshwork of the clot
    ___ 7. stabilizes the clot
    ___ 8. activates fibrinogen
    ___ 9. activated by tissue thromboplastin

    a. prostacyclin
    b. plasmin
    c. ADP
    d. fibrin
    e. thrombin
    f. factor X
    g. factor XII
    h. factor XIII

12. Match the following blood abnormalities with their causes:
    ___ 1. deficiency of intrinsic factor
    ___ 2. insufficient amount of iron to synthesize adequate hemoglobin
    ___ 3. destruction of bone marrow
    ___ 4. abnormal loss of blood
    ___ 5. tumorlike condition of bone marrow
    ___ 6. inadequate erythropoietin secretion
    ___ 7. excessive rupture of circulating erythrocytes
    ___ 8. associated with living at high altitudes

    a. hemolytic anemia
    b. aplastic anemia
    c. iron-deficiency anemia
    d. hemorrhagic anemia
    e. pernicious anemia
    f. renal anemia
    g. primary polycythemia
    h. secondary polycythemia

Essay Questions

1. What is the average blood volume?
2. What is the normal percentage of blood occupied by erythrocytes and by plasma? What is the hematocrit? What is the buffy coat?
3. What is the composition of plasma?
4. List and state the functions of the three major groups of plasma proteins.
5. Describe the structure and functions of erythrocytes.
6. Why are erythrocytes able to survive for only about 120 days?
7. Describe the process and control of erythropoiesis.
8. Compare the structure and functions of the five types of leukocytes.
9. Discuss the derivation of platelets.
10. Describe the three steps of hemostasis, including a comparison of the intrinsic and extrinsic pathways by which the clotting cascade is triggered.
11. Compare plasma and serum.
12. What factors normally prevent inappropriate coagulation in the vasculature?

1. Why do you think it is important to closely monitor blood cell counts in cancer patients who are being treated with chemotherapeutic drugs designed to destroy rapidly multiplying cells, such as cancer cells?

2. A person has a hematocrit of 62. Can you conclude from this finding that the person has polycythemia? Explain.

3. A person diagnosed as having pneumonia has a white blood cell count of 7,200/mm³, with 67% of the white blood cells being neutrophils. Explain whether antibiotics would likely combat the infection.

4. Low on the list of popular animals are vampire bats, leeches, and ticks, yet these animals may someday indirectly save your life. Scientists are currently examining the "saliva" of these blood-sucking creatures in search of new chemicals that might limit cardiac-muscle damage in heart attack victims. What do you suspect the nature of these sought-after chemicals is?

5. Scientists are also seeking ways to extend the life span of red blood cells. Prolonging the life of erythrocytes even a week or two, either in blood banks or within the body of patients with acute or chronic loss of red blood cells, would have important benefits. Explain how.

12 Body Defenses

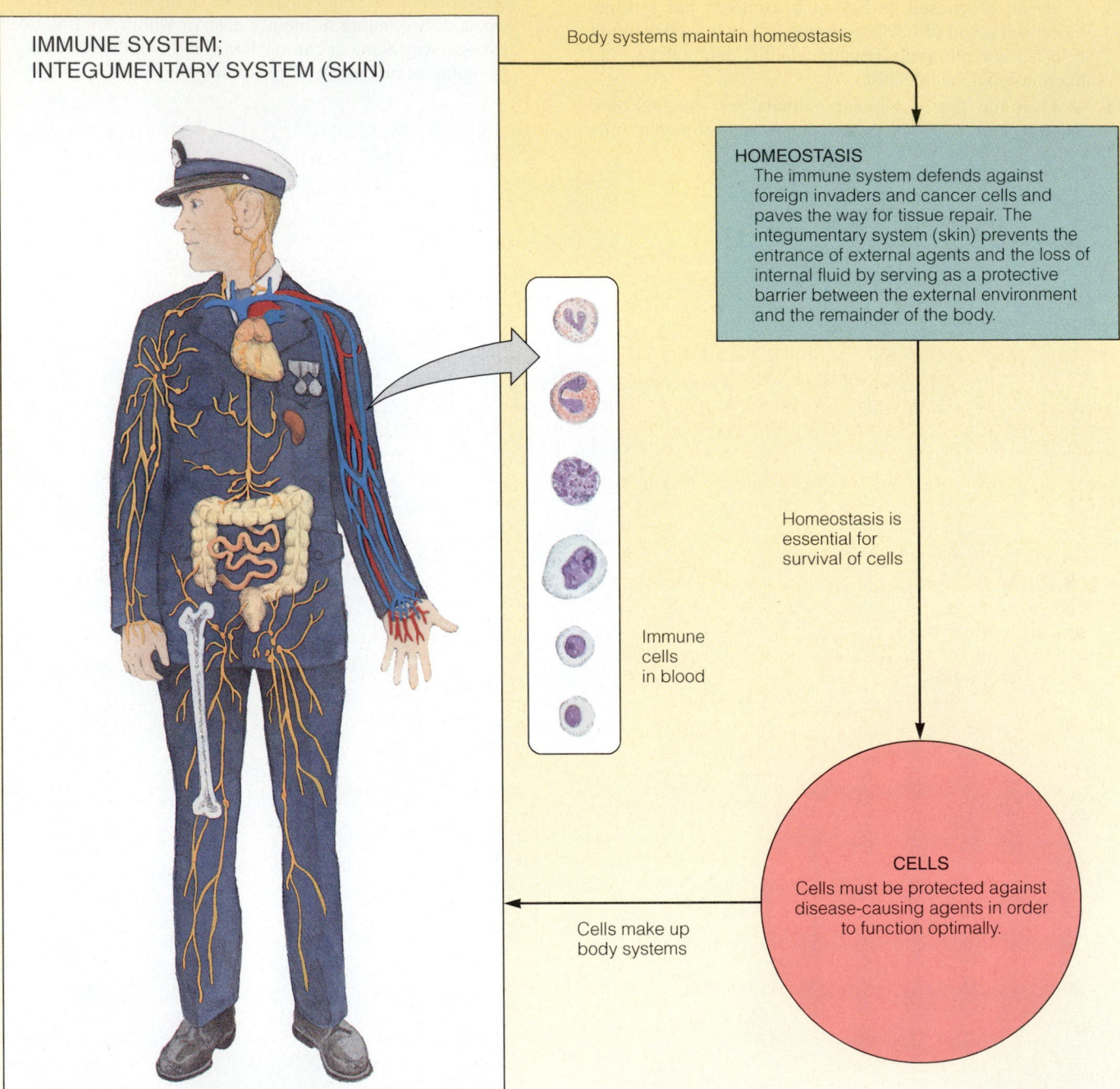

IMMUNE SYSTEM;
INTEGUMENTARY SYSTEM (SKIN)

Body systems maintain homeostasis

HOMEOSTASIS
The immune system defends against foreign invaders and cancer cells and paves the way for tissue repair. The integumentary system (skin) prevents the entrance of external agents and the loss of internal fluid by serving as a protective barrier between the external environment and the remainder of the body.

Immune cells in blood

Homeostasis is essential for survival of cells

CELLS
Cells must be protected against disease-causing agents in order to function optimally.

Cells make up body systems

Humans are constantly coming into contact with external agents that could be harmful if they gained entry into the body, the most serious of which are disease-causing microorganisms. In the event that bacteria or viruses gain entry, the body is equipped with a complex, multifaceted internal defense system—the **immune system**—which provides continual protection against invasion by foreign agents. The body surfaces that are exposed to the external environment, such as the **skin**, serve as a first line of defense to resist penetration by foreign microorganisms.

INTRODUCTION
Bacteria and viruses as targets
Leukocytes as effector cells
NONSPECIFIC IMMUNE RESPONSES
Inflammation
Interferon
Natural killer cells
Complement system
SPECIFIC IMMUNE RESPONSES: GENERAL CONCEPTS
B LYMPHOCYTES: ANTIBODY-MEDIATED IMMUNITY
Role of antibodies
Clonal selection theory
CONCEPTS, CHALLENGES, AND CONTROVERSIES Vaccination: A
    Victory over Many Dreaded Diseases
Active, passive, and natural immunity
Antigen-presenting cells
T LYMPHOCYTES: CELL-MEDIATED IMMUNITY
Cytotoxic, helper, and suppressor T cells
Major histocompatibility complex; human leukocyte-
    associated antigens
Immune surveillance against cancer
A CLOSER LOOK AT EXERCISE PHYSIOLOGY Exercise: A Help or
    Hindrance to Immune Defense?
IMMUNE DISEASES
Immune deficiency diseases
Inappropriate immune attacks
EXTERNAL DEFENSES
Skin
    Composition; cell types; functions
    Role of skin in immunity
Protective measures within body cavities that communicate
    with the external environment

=== INTRODUCTION

The immune defense system provides protection against foreign and abnormal cells and removes cellular debris.

Immunity refers to the body's ability to resist or eliminate potentially harmful foreign materials or abnormal cells. The following activities are attributable to the immune defense system, which plays a key role in recognizing and either destroying or neutralizing materials within the body that are foreign to the "normal self."

1. Defense against invading **pathogens** (disease-producing microorganisms such as viruses and bacteria).
2. Removal of "worn-out" cells (such as aged red blood cells) and tissue debris (for example, tissue damaged by trauma or disease). The latter is essential for wound healing and tissue repair.
3. Identification and destruction of abnormal or mutant cells that have originated in the body. This function, termed *im-*

mune surveillance, is the primary internal defense mechanism against cancer.

4. Inappropriate immune responses that lead either to *allergies,* which occur when the body turns against a normally harmless environmental chemical entity, or to *autoimmune diseases,* which happen when the defense system erroneously produces antibodies against itself, leading to destruction of a particular type of the body's own cells.
5. Rejection of tissue cells of foreign origin, which constitutes the major obstacle to successful organ transplantation.

Pathogenic bacteria and viruses are the major targets of the immune defense system.

The primary foreign enemies against which the immune system defends are bacteria and viruses. **Bacteria** are nonnucleated, single-celled microorganisms self-equipped with all machinery essential for their own survival and reproduction. Pathogenic bacteria that invade the body induce tissue damage and produce disease largely by releasing enzymes or toxins that physically injure or functionally disrupt affected cells and organs. The disease-producing power of a pathogen is known as its **virulence**.

Viruses, in contrast to bacteria, are not self-sustaining cellular entities. They consist only of nucleic acids (DNA or RNA) enclosed by a protein coat. Because they lack cellular machinery for energy production and protein synthesis, viruses are unable to carry out metabolism and reproduce unless they invade a **host cell** (a body cell of the infected individual) and take over the cellular biochemical facilities for their own purposes. Not only do viruses sap the host cell's energy resources, but the viral nucleic acids also direct the host cell to synthesize proteins needed for viral replication. The effect of viral invasion and replication on the host cell varies with different types of viruses.

Viruses can lead to cellular damage or death in four general ways: (1) depletion of essential cellular components by the virus; (2) cellular production of substances toxic to the cell under the dictatorship of the virus; (3) transformation of normal host cells into cancer cells; and (4) incorporation of the virus into the cell so that the body's own defense mechanisms destroy the cell, because it is no longer recognized as a "normal-self" cell.

Leukocytes are the effector cells of the immune defense system.

The cells responsible for the various immune defense strategies are the leukocytes (white blood cells) and their derivatives, briefly reviewed as follows (see pp. 356–358):

1. Neutrophils are highly mobile phagocytic specialists that engulf and destroy unwanted materials.
2. Eosinophils secrete chemicals that destroy parasitic worms and are involved in allergic manifestations.

3. Basophils release histamine and heparin and also are involved in allergic manifestations.

4. Lymphocytes
   a. B lymphocytes are transformed into plasma cells, which secrete antibodies that indirectly lead to the destruction of foreign material.
   b. T lymphocytes are responsible for cell-mediated immunity involving direct destruction of virus-invaded cells and mutant cells through nonphagocytic means.

5. Monocytes are transformed into macrophages, which are large, tissue-bound phagocytic specialists.

A given leukocyte is present in the blood only transiently. Most of the leukocytes are out in the tissues on defense missions. As a result, the immune system's effector cells are widely dispersed throughout the body and are able to defend in any location.

Almost all leukocytes originate from common precursor stem cells in the bone marrow and are subsequently released into the blood. The only exception is lymphocytes, which arise in part from lymphocyte colonies in various lymphoid tissues that were originally populated by cells derived from the bone marrow. **Lymphoid tissues** refer collectively to the tissues that store, produce, or process lymphocytes. These include the lymph nodes, spleen, thymus, tonsils, adenoids, appendix, aggregates of lymphoid tissue in the lining of the gastrointestinal tract called **Peyer's patches** or **gut-associated lymphoid tissue** (**GALT**), and the bone marrow (• Fig. 12–1). Lymphoid tissues are strategically located to intercept invading microorganisms before they have a chance to spread very far. For example, the lymphocytes populating the tonsils and adenoids are situated advantageously to respond to inhaled microbes, whereas microorganisms invading through the digestive system immediately encounter lymphocytes in the appendix and GALT. Potential pathogens that gain access to the lymph are filtered through lymph nodes, where they are exposed to lymphocytes as well as to macrophages that line the lymphatic passageways. The spleen, the largest of the lymphoid tissues, performs immune functions on the blood similar to those that the lymph nodes perform on the lymph. Through actions of its lymphocyte and macrophage population, the spleen clears the blood that passes through it of microorganisms and other foreign matter and also removes worn-out red blood cells. The thymus and bone marrow play important roles in processing T and B lymphocytes, respectively, to prepare them to carry out their specific immune strategies. Table 12–1 summarizes the major functions of the various lymphoid tissues, some of which were described in chapter 11 and others of which are yet to be discussed in this chapter.

Immune responses may be either nonspecific or specific.

Immune responses are classified as nonspecific or specific immune responses, depending on the degree of selectivity of

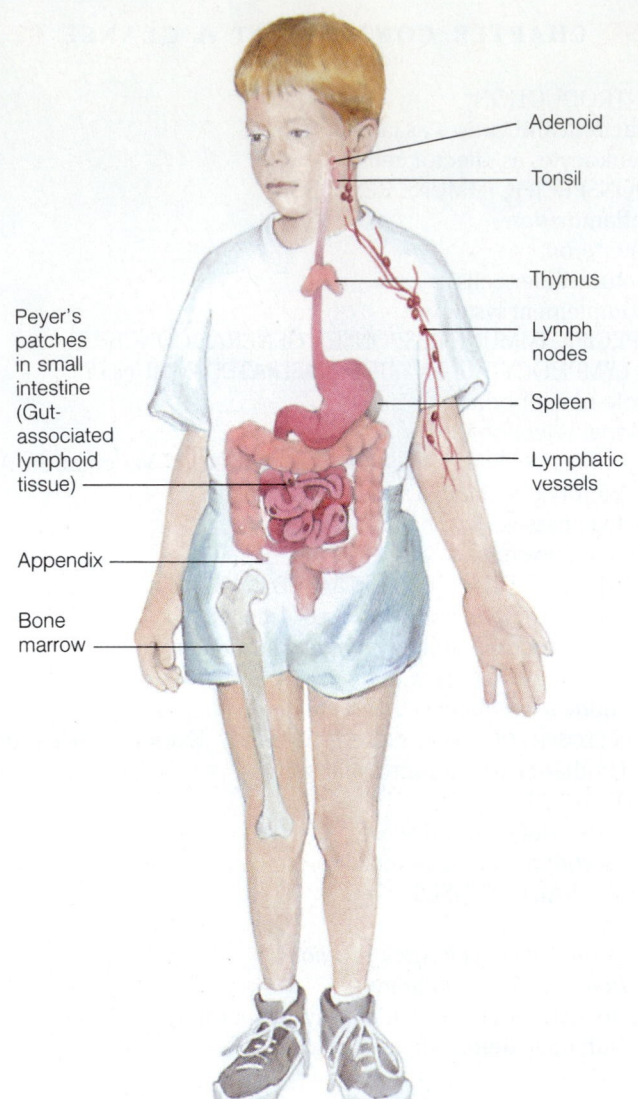

• FIGURE 12–1 Lymphoid Tissues The lymphoid tissues, which are dispersed throughout the body, store, produce, or process lymphocytes.

the defense mechanism. **Nonspecific immune responses** are inherent defense responses that nonselectively defend against foreign or abnormal material of any type, even upon initial exposure to it. Such responses provide a first line of defense against a wide range of threatening factors, including infectious agents, chemical irritants, and tissue injury accompanying mechanical trauma and burns. **Specific immune responses**, on the other hand, are selectively targeted against particular foreign material to which the body has previously been exposed. These specific responses are mediated by lymphocytes, which, upon subsequent exposure to the same offending agent, recognize and discriminately defend against it. We will first examine the nonspecific responses before turning our attention to the specific responses.

TABLE 12–1 Functions of Lymphoid Tissues

LYMPHOID TISSUE	FUNCTION
Bone marrow	Origin of all blood cells
	Site of maturational processing for B lymphocytes
Lymph nodes, tonsils, adenoids, appendix, gut-associated lymphoid tissue	Exchange lymphocytes with lymph (remove, store, produce, and add them)
	Resident lymphocytes produce antibodies and sensitized T cells, which are released into the lymph
	Resident macrophages remove microbes and other particulate debris from lymph
Spleen	Exchanges lymphocytes with blood (removes, stores, produces, and adds them)
	Resident lymphocytes produce antibodies and sensitized T cells, which are released into the blood
	Resident macrophages remove microbes and other particulate debris, most notably worn-out red blood cells, from blood
	Stores a small percentage of red blood cells, which can be added to the blood by splenic contraction as needed
Thymus	Site of maturational processing for T lymphocytes
	Secretes the hormone thymosin

NONSPECIFIC IMMUNE RESPONSES

Nonspecific defenses include inflammation, interferon, natural killer cells, and the complement system.

Nonspecific defenses that come into play whether or not there has been prior experience with the offending agent include the following:

1. *Inflammation,* a nonspecific response to tissue injury in which the phagocytic specialists—neutrophils and macrophages—play a major role, along with supportive input from other immune cell types.
2. *Interferon,* a family of proteins that nonspecifically defend against viral infection.
3. *Natural killer cells,* a special class of lymphocytelike cells that spontaneously and relatively nonspecifically lyse (rupture) and thereby destroy virus-infected host cells and cancer cells.
4. *The complement system,* a group of inactive plasma proteins that, when sequentially activated, bring about de-

struction of foreign cells by attacking their plasma membranes. The complement system can nonspecifically be called into play by the presence of any foreign invader. It also may be activated by antibodies produced as part of the specific immune response to a particular microorganism.

The fact that the complement system is involved in both nonspecific and specific defense mechanisms illustrates an important point. The various components of the immune system are highly interactive and interdependent, making the system highly sophisticated and effective but also complex and difficult to sort out. The most significant cooperative relationships known to exist among the immune effector cells will be pointed out as the various components of the immune system are discussed separately. As further reference, Appendix C summarizes the major immune functions and interactions addressed in this chapter.

Inflammation is a nonspecific response to foreign invasion or tissue damage.

Inflammation refers to an innate, nonspecific series of highly interrelated events that are set into motion in response to foreign invasion, tissue damage, or both. The ultimate goal of inflammation is to bring to the invaded or injured area phagocytes and plasma proteins that can (1) isolate, destroy, or inactivate the invaders; (2) remove debris; and (3) prepare for subsequent healing and repair. The overall inflammatory response is remarkably similar no matter what the triggering event (be it bacterial invasion, chemical injury, or mechanical trauma), although some subtle differences may be evident, depending on the injurious agent or the site of damage. The following sequence of events typically occurs during the inflammatory response. As an example, we will use bacterial entry into a break in the skin (Table 12–2).

Defense by Resident Tissue Macrophages Upon bacterial invasion through a break in the external barrier of skin, the macrophages already present in the area immediately begin phagocytizing the foreign microbes. Although the resident macrophages are usually not present in sufficient numbers to meet the challenge alone, they defend against infection during the first hour or so, before other mechanisms can be mobilized. Macrophages are usually rather stationary, gobbling debris and contaminants that come their way, but when necessary, they become mobile and migrate to sites of battle against invaders.

Localized Vasodilation Almost immediately upon microbial invasion, arterioles within the area dilate, increasing blood flow to the site of injury. This localized vasodilation is primarily induced by histamine that has been released in the area of tissue damage from mast cells, a type of connective-tissue–bound cell similar to circulating basophils. Increased local delivery of blood brings to the site more phagocytic leukocytes and plasma proteins, both of which are crucial to the defense response.

TABLE 12–2 Sequence of Events in Inflammatory Response to Bacterial Entry into a Break in the Skin

EVENT	COMMENT
Defense by resident tissue macrophages	Accomplish defense before other mechanisms can be mobilized
Localized vasodilation	Induced by histamine released from mast cells
	Increases local delivery of blood to bring more phagocytic leukocytes and crucial plasma proteins, such as those of the clotting and complement systems, to the site
	Responsible for localized redness and heat
Increased capillary permeability	Induced by histamine
	Allows plasma proteins to escape into inflamed tissue
Localized edema	Results from increased colloid osmotic pressure in interstitial fluid caused by leaked plasma proteins and increased capillary blood pressure caused by increased local blood flow
	Responsible for localized swelling and contributes to pain
Walling off of inflamed area	Accomplished by formation of clots in interstitial fluid surrounding bacteria after leaked clotting factors have been activated by exposure to tissue thromboplastin
Emigration of leukocytes, especially monocytes, which mature into tissue macrophages, and neutrophils	Accomplished by the processes of margination, diapedesis, amoeboid movement, and chemotaxis
Leukocyte proliferation	Caused by release of preformed leukocytes from bone marrow as well as increased production of new leukocytes
Leukocytic destruction of bacteria	Accomplished by neutrophils and macrophages on the scene, enhanced by opsonin action
Secretion of inflammatory mediators by phagocytes	Kill bacteria by nonphagocytic means
	Stimulate release of histamine
	Induce systemic manifestations such as fever
	Trigger clotting and anticlotting systems
	Activate kinin system, which augments many inflammatory events and activates local pain receptors
	Decrease plasma concentration of iron, which is needed for bacterial multiplication
	Stimulate release from the liver of acute-phase proteins, which exert a wide array of immune responses
	Stimulate neutrophil production
	Enhance proliferation and differentiation of B and T lymphocytes
Tissue Repair	Accomplished by replacement of lost cells through division of surrounding healthy, organ-specific cells or formation of scar tissue by connective-tissue fibroblasts

Increased Capillary Permeability Released histamine also increases the capillaries' permeability by enlarging the capillary pores (the spaces between the endothelial cells, see p. 319) so that plasma proteins that normally are prevented from leaving the blood can escape into the inflamed tissue.

Localized Edema As the leaked plasma proteins accumulate in the interstitial fluid, they exert a colloid osmotic pressure. This elevation in local osmotic pressure, coupled with the increased capillary blood pressure caused by increased local blood flow, favors enhanced filtration and reduced reabsorption of fluid across the involved capillaries. The end result of this shift in fluid balance is localized edema (see p. 327). Thus, the familiar swelling that accompanies inflammation is due to histamine-induced vascular changes. Like-

wise, the other well-known gross manifestations of inflammation such as redness and heat are largely attributable to the enhanced flow of warm arterial blood to the damaged tissue. Pain is caused both by local distention within the swollen tissue and by the direct effect of locally produced substances on the receptor endings of afferent neurons that supply the area. Keep in mind that these observable characteristics of the inflammatory process are coincidental to the primary purpose of the vascular changes in the injured area—to increase the number of leukocytic phagocytes and crucial plasma proteins in the area.

Walling Off of the Inflamed Area The leaked plasma proteins most critical to the immune response are those involved in the complement system and kinin system (to be described later) as well as clotting and anticlotting factors (see p. 360). Upon exposure to tissue thromboplastin in the injured tissue and to specific chemicals secreted by phagocytes on the scene, fibrinogen, the final factor in the clotting system, is converted into fibrin. Fibrin forms interstitial-fluid clots in the spaces around the bacterial invaders and damaged cells. This walling off of the injured region from the surrounding tissues prevents or at least delays the spread of bacterial invaders and their toxic products. Later, the more slowly activated anticlotting factors gradually dissolve the clots after they are no longer needed. A few bacteria, such as streptococci, produce enzymes that act on plasminogen, an inactive plasma-protein precursor, to convert it into plasmin, a proteolytic enzyme that dissolves fibrin clots. This action disrupts the walling-off process and allows the streptococcal organisms to spread extensively.

Emigration of Leukocytes Within an hour after the injury, the involved area is teeming with leukocytes that have exited from the vessels. Neutrophils are the first to arrive, followed during the next eight to twelve hours by the slower-moving monocytes. The latter swell and mature into macrophages during another eight-to-twelve-hour period.

Leukocyte emigration from the blood into the tissues involves the processes of margination, diapedesis, amoeboid movement, and chemotaxis. **Margination** refers to the sticking of blood-borne leukocytes, especially neutrophils and monocytes, to the inner endothelial lining of capillaries in the affected tissue. Soon the leukocytes start exiting by a mechanism known as **diapedesis**. Assuming amoebalike behavior (see p. 39), an adhered leukocyte pushes a long, narrow projection through a capillary pore; then the remainder of the cell flows forward into the projection (• Fig. 12–2). In this way, the leukocyte is able to wriggle its way through the capillary pore—even though it is much larger than the pore—and move in an amoeboid fashion toward the site of

• **FIGURE 12–2   Leukocyte Emigration from Blood**
Leukocytes emigrate from the blood into the tissues by assuming amoebalike behavior and squeezing through the capillary pores, a process known as diapedesis.

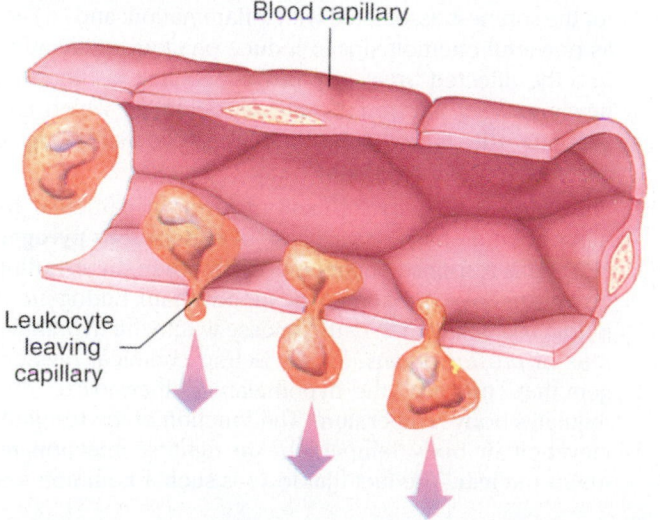

Blood capillary

Leukocyte leaving capillary

tissue damage and bacterial invasion. Neutrophils arrive on the inflammatory scene earliest because they are more mobile than monocytes.

Phagocytic cells are guided in their direction of migration by attraction to certain chemical mediators, or **chemotaxins**, released at the site of damage. This process is referred to as *chemotaxis*. Binding of chemotaxins with protein receptors on the plasma membrane of a phagocytic cell increases Ca^{++} entry into the cell. Calcium, in turn, switches on the cellular contractile apparatus that leads to amoebalike crawling. Because the concentration of chemotaxins progressively increases toward the site of injury, phagocytic cells move unerringly toward this site along a chemotaxin concentration gradient.

Leukocyte Proliferation Resident tissue macrophages as well as leukocytes that exited from the blood and migrated to the inflammatory site are soon joined by new phagocytic recruits from the bone marrow. Within a few hours after the onset of the inflammatory response, the number of neutrophils in the blood may increase up to four to five times that of normal. This increase is due partly to transfer into the blood of large numbers of preformed neutrophils stored in the bone marrow and partly to increased production of new neutrophils by the bone marrow. A slower-commencing but longer-lasting increase in monocyte production by the bone marrow also occurs, making larger numbers of these macrophage precursor cells available. In addition, multiplication of resident macrophages adds to the pool of these important immune cells. Proliferation of new neutrophils, monocytes, and macrophages and mobilization of stored neutrophils are stimulated by various chemical mediators released from the inflamed region.

Leukocytic Destruction of Bacteria Neutrophils and macrophages clear the inflamed area of infectious and toxic agents as well as tissue debris; this clearing action is the primary function of the inflammatory response. They accomplish this by both phagocytic and nonphagocytic means.

Phagocytosis involves the engulfment and intracellular degradation (breakdown) of foreign particles and tissue debris. Macrophages can engulf a bacterium in less than 0.01 second. Phagocytic cells contain an abundance of lysosomes, which are organelles filled with hydrolytic enzymes (the same type that degrade ingested food in the digestive tract). After a phagocyte has internalized targeted material, a lysosome fuses with the membrane that encloses the engulfed matter and releases its hydrolytic enzymes within the confines of the vesicle, where the enzymes begin breaking down the entrapped material (see p. 24). Many of the resultant breakdown products are low-molecular-weight organic compounds, such as amino acids and glucose, which can be used by the phagocyte itself or, in the case of macrophages, released into the extracellular fluid for use elsewhere in the body.

Nondegradable inorganic particles that have been engulfed (for example, dust, metal particles, or tattoo dyes) usually are retained indefinitely within the phagocytic cell. In fact, certain bacteria, most notably those causing tuberculosis, are likewise engulfed but not destroyed because they are resistant to lysosomal chemicals. These organisms may survive within phagocytes for many years and produce no obvious effects because they are walled off within a fibrous capsule. They cause overt disease only if allowed to escape. Such a more-or-less permanently walled-off structure in which nondestructible offending material is imprisoned is known as a **granuloma**.

Phagocytes eventually die due to accumulation of toxic by-products from foreign particle degradation or inadvertent release of destructive lysosomal chemicals into the cytosol. Neutrophils usually succumb after phagocytizing from five to twenty-five bacteria, whereas macrophages survive much longer and can engulf up to one hundred or more bacteria. Indeed, the longer-lived macrophages even clear the area of dead neutrophils in addition to other tissue debris. The **pus** that forms in an infected wound is a collection of these phagocytic cells, both living and dead; necrotic (dead) tissue liquified by lysosomal enzymes released from the phagocytes; and bacteria.

Obviously, phagocytes must be able to distinguish between normal cells and foreign or abnormal cells before accomplishing their destructive mission. Otherwise they could not selectively engulf and destroy only unwanted materials. Several different selective procedures enable phagocytes to "recognize" targets for destruction. First, dead tissue and many foreign materials have surface characteristics that differ from normal body cells. For example, the surface roughness accompanying traumatic injury to cells increases the likelihood of the cellular debris being phagocytized. Second, foreign particles are deliberately marked for phagocytic ingestion by being coated with chemical mediators generated by the immune system. Such body-produced chemicals that make bacteria more susceptible to phagocytosis are known as **opsonins**. The most important opsonins are antibodies and one of the activated proteins of the complement system.

An opsonin enhances phagocytosis by linking the foreign cell to a phagocytic cell (• Fig. 12–3). One portion of an op-

sonin molecule binds nonspecifically to the surface of an invading bacterium, whereas another portion of the opsonin molecule binds to receptor sites specific for it on the phagocytic cell's plasma membrane. This link assures that the bacterial victim does not have a chance to "get away" before the phagocyte can perform its lethal attack.

Mediation of the Inflammatory Response by Phagocyte-Secreted Chemicals Microbe-stimulated phagocytes release many chemicals, which function as mediators of the inflammatory response. These chemical mediators induce a broad range of interrelated immune activities, varying from local responses to the systemic manifestations that accompany microbe invasion. The following are among the most important functions of phagocytic secretions:

1. Some of the chemicals, which are highly destructive, directly kill microbes that have not been phagocytized. As a more subtle means of destruction, neutrophils secrete **lactoferrin**, a protein that tightly binds with iron, making it unavailable for use by invading bacteria. Bacterial multiplication depends on high concentrations of available iron. Therefore, phagocytes are able to destroy foreign invaders both by phagocytosis and by nonphagocytic means.

2. They stimulate the release of histamine from mast cells in the vicinity. Histamine, in turn, induces the local vasodilation and increased vascular permeability that accompany inflammation.

3. They trigger both the clotting and anticlotting systems to first enhance the walling-off process and then facilitate the gradual dissolution of the fibrous clot after it is no longer needed.

4. They split **kininogens**, which are inactive precursor plasma proteins synthesized by the liver, into active **kinins**. In particular, **kallikrein**, which is released by neutrophils, can accomplish this activation. Once kinins are generated by kallikrein, they augment a variety of inflammatory events. Specifically, they (a) stimulate several key steps in the complement system; (b) reinforce the vascular changes induced by histamine; (c) activate nearby pain receptors, and are thus partially responsible for the soreness associated with inflammation; and (d) act as powerful chemotaxins to induce phagocyte migration into the affected area. In positive-feedback fashion, the newly arriving neutrophils release kallikrein, which can generate still more kinins that chemotactically entice more neutrophils to join the battle (• Fig. 12–4).

5. They induce the development of fever, in particular by means of the phagocytic release of **endogenous pyrogen** (**EP**). This response occurs especially when the invading organisms have spread into the bloodstream. Endogenous pyrogen is believed to cause release within the hypothalamus of *prostaglandins,* locally acting chemical messengers that "turn up" the hypothalamic "thermostat" that regulates body temperature. The function of the resultant elevation in body temperature in fighting infection remains unclear. The fact that fever is such a common sys-

• **FIGURE 12–3 Mechanism of Opsonin Action** One of the activated complement molecules, C3b, links a foreign cell, such as a bacterium, and a phagocytic cell by nonspecifically binding with the surface of the foreign cell and specifically binding with plasma-membrane receptors on the surface of the phagocyte. This link assures that the foreign victim does not escape before it can be engulfed by the phagocyte.

Bacterium

Activated complement molecule, C3b (an opsonin)

Receptor specific for activated C3b molecule

Phagocyte

Structures are not drawn to scale.

• **FIGURE 12–4   Relationship between Kinin System and Neutrophils in Inflammatory Response** The green arrows indicate a positive-feedback loop between the kinin system and neutrophils during an inflammatory response. Neutrophils release kallikrein, which converts kininogens, inactive plasma proteins, into kinins. Active kinins augment a variety of inflammatory events, including acting as chemotaxins to attract more neutrophils to the area.

temic manifestation of inflammation suggests that the higher body temperature plays an important beneficial role in the overall inflammatory response. Recent evidence strongly supports this supposition. For example, higher temperatures augment the process of phagocytosis and increase the rate at which the many enzyme-dependent inflammatory activities proceed. One theory proposes that an elevated body temperature increases bacterial requirements for iron while reducing the plasma concentration of iron. The resultant iron deficiency, according to this theory, interferes with bacterial multiplication. Resolving the controversial issue of whether a fever can be beneficial is extremely important in view of the widespread use of drugs that suppress fever. Although a mild fever may possibly be beneficial, there is no doubt that an extremely high fever can be detrimental, particularly because of its harmful effects on the central nervous system. It is not uncommon for young children, whose temperature-regulating mechanisms are not as stable as those of more mature individuals, to have convulsions in association with high fevers.

6. They decrease plasma concentration of iron by altering iron metabolism within the liver, spleen, and other tissues, thus reducing the amount of iron available to support bacterial multiplication.

7. They stimulate **granulopoiesis**, the synthesis and release of neutrophils and other granulocytes by the bone marrow. This effect is especially prominent in response to bacterial infections.

8. They stimulate the release of **acute-phase proteins** from the liver. This collection of proteins, which have not yet been sorted out by scientists, exert a multitude of wide-ranging effects associated with the inflammatory process, tissue repair, and immune cell activities.

The latter three effects (plasma iron reduction, enhanced granulopoiesis, and release of acute-phase proteins) are all attributed to **leukocyte endogenous mediator** (**LEM**), a chemical mediator secreted by macrophages. Evidence suggests that LEM and EP are the same substance or at least very closely related.

9. They enhance the proliferation and differentiation of both B and T lymphocytes, which, in turn, are responsible for antibody production and cell-mediated immunity, respectively. Specifically, **interleukin l** (**IL-l**), a secretory product released by macrophages, is responsible for this effect on lymphocytes. Amazingly, IL-l is identical to (or closely related to) EP and LEM. Apparently, the same chemical substance is responsible for a diverse array of effects throughout the body, all of which are geared toward defending the body against infection or tissue injury. In fact, the release of EP/LEM/IL-l can be elicited by stressful situations not associated with microbe invasion (for example, during endurance exercise). Accordingly, this mediator may be part of a generalized nonspecific protective response.

This list of events that are augmented by chemicals secreted by phagocytes is not complete, but it serves to illustrate the diversity and complexity of responses elicited by these mediators. As will soon become evident, there are other important macrophage-lymphocyte interactions that are not dependent on the release of chemicals from phagocytic cells. Thus, the effect that phagocytes, especially macrophages, ultimately have on microbial invaders far exceeds their "engulf and destroy" tactics (● Fig. 12–5).

Tissue Repair   The ultimate purpose of the inflammatory process is to isolate and destroy injurious agents and to clear the area for tissue repair. In some tissues (for example, skin, bone, and liver), the healthy organ-specific cells surrounding the injured area undergo cell division to replace the lost cells, often accomplishing perfect repair. In nonregenerative tissues such as nerve and muscle, however, lost cells are replaced by **scar tissue**. Fibroblasts, a type of connective-tissue cell, start to divide rapidly in the vicinity and secrete large quantities of the protein collagen (see p. 55), which fills in the region vacated by the lost cells and results in the formation of scar tissue. Even in a tissue as readily replaceable as the skin, scar formation sometimes takes place when complex underlying structures, such as hair follicles and sweat glands, are permanently destroyed by deep wounds.

EP/LEM/IL-1 = Endogenous pyrogen/leukocyte endogenous mediator/interleukin 1

● **FIGURE 12-5 Phagocyte Responsibilities** Phagocytes not only destroy foreign particles or damaged cells by phagocytosis but also secrete chemical mediators that destroy the targeted material by nonphagocytic means and augment many aspects of inflammation and other immune processes.

Salicylates and glucocorticoid drugs suppress the inflammatory response.

Numerous drugs can suppress the inflammatory process; the most effective are the *salicylates* and related compounds (aspirin-type drugs) and *glucocorticoids* (drugs similar to the steroid hormone cortisol that is secreted by the adrenal cortex). Salicylates interfere with the inflammatory response by decreasing histamine release, resulting in a reduction in swelling, redness, and pain. Furthermore, salicylates reduce fever by inhibiting the production of prostaglandins, the local mediators of EP-induced fever.

Glucocorticoids, which are potent anti-inflammatory drugs, suppress almost every aspect of the inflammatory response. In addition, they destroy lymphocytes within lymphoid tissue and reduce antibody production. These therapeutic agents are useful for treating undesirable immune responses, such as allergic reactions (for example, poison ivy rash and asthma) and the inflammation associated with arthritis. Unfortunately, however, by suppressing inflammatory and other immune responses that localize and eliminate bacteria, such therapy also reduces the body's ability to resist infection. For this reason, glucocorticoids should be administered discriminately.

What about the normal role of the adrenal cortical hormone cortisol in this regard? Is cortisol secretion counterproductive to the immune defense system? Traditionally, cortisol has not been considered to display anti-inflammatory activity at normal blood concentrations. Instead, anti-inflammatory action has been attributed only to pharmacological levels of glucocorticoids (that is, at blood concentrations brought about by the administration of cortisol-like drugs that are higher than the normal physiological range). A recent hypothesis, however, suggests that cortisol, whose secretion is increased in response to any stressful situation, does exert anti-inflammatory activity even at normal physiological levels. According to this proposal, the anti-inflammatory effect of cortisol modulates stress-activated immune responses, preventing them from overshooting and thus protecting us against damage by potentially overreactive defense mechanisms.

Interferon transiently inhibits multiplication of viruses in most cells.

Besides the inflammatory response, another nonspecific defense mechanism is the release of **interferon** from virus-infected cells. Interferon briefly provides nonspecific resistance to viral infections by transiently interfering with replication of the same or unrelated viruses in other host cells. When a virus invades a cell, the presence of viral nucleic acid induces the cell's genetic machinery to synthesize interferon, which is secreted into the extracellular fluid.

Once released, interferon binds with receptors on the plasma membranes of neighboring cells or even distant cells that it reaches through the bloodstream, signaling these cells to prepare for the possibility of impending viral attack. Interferon does not have a direct antiviral effect; instead, it triggers the production of viral-blocking enzymes by potential host cells. Binding with interferon induces these other cells to synthesize enzymes that can break down viral messenger RNA (see p. B–7) and inhibit protein synthesis, both of which are essential for viral replication. Although viruses are still able to invade these forewarned cells, they are unable to govern cellular protein synthesis for their own replication (● Fig. 12–6).

These newly synthesized inhibitory enzymes remain inactive within the potential host cell unless it is actually invaded by virus, at which time the enzymes are activated by the presence of viral nucleic acid. This activation requirement protects the cell's own messenger RNA and protein-synthesizing machinery from unnecessary inhibition by these enzymes should viral invasion not occur. Because activation can take place only during a limited time span, this is a short-term defense mechanism.

Interferon is released nonspecifically from any cell infected by any virus and, in turn, can induce temporary self-protective activity against many different viruses in any other cells that it reaches. Thus, it provides a general, rapidly responding defense strategy against viral invasion until more specific but slower-responding immune mechanisms come into play.

● **FIGURE 12–6   Mechanism of Action of Interferon in Preventing Viral Replication**   Interferon, which is released from virus-infected cells, binds with other uninvaded host cells and triggers these cells to produce inactive enzymes capable of blocking viral replication. These inactive enzymes are activated only if a virus subsequently invades one of these prepared cells.

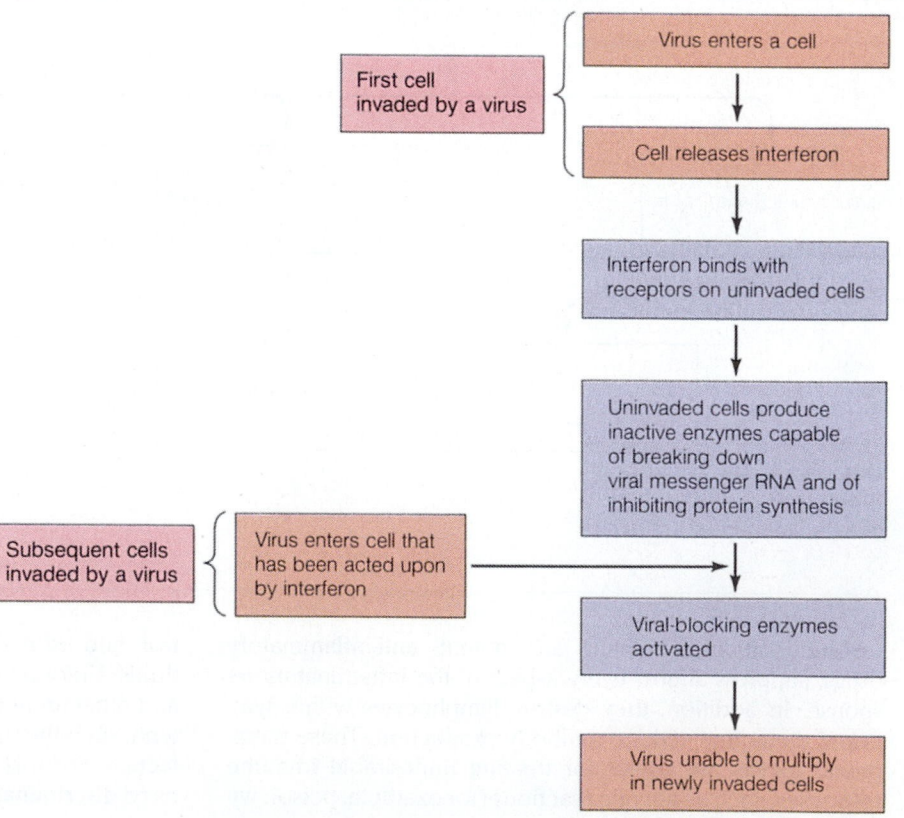

TABLE 12–3   Functions of Interferon

FUNCTIONS	ANTIVIRAL EFFECT	ANTICANCER EFFECT
Released by virus-infected cells and transiently interferes with viral replication in other host cells by inducing them to produce enzymes that destroy viral messenger RNA and inhibit virus-directed protein synthesis	√	
Enhances macrophage phagocytic activity	√	√
Stimulates production of antibodies	√	
Enhances actions of natural killer cells and cytotoxic T cells	√	√
Slows cell division and suppresses tumor growth		√

In addition to facilitating inhibition of viral replication, interferon reinforces other immune activities (Table 12–3). For example, it enhances macrophage phagocytic activity and stimulates the production of antibodies. Interferon also exerts anticancer as well as antiviral effects. Fortunately, its anticancer effects are not limited to virally induced cancers. Most types of human cancer are not caused by viruses. Interferon markedly enhances the actions of cell-killing cells—the natural killer cells and a special type of T lymphocyte, cytotoxic T cells—which attack and destroy both virus-infected cells and cancer cells. Furthermore, interferon itself slows cell division and suppresses tumor growth.

The discovery of interferon's antiviral and anticancer effects led to great excitement and anticipation about its potential role as a miracle therapeutic weapon against enemies ranging from the common cold virus to fatal invasive cancers. For a quarter of a century after its discovery in 1957, however, it was impossible to collect human interferon in sufficient quantities to make it feasible to study its therapeutic possibilities. (Interferon is species specific; each species of higher animals produces a type of interferon capable of conferring protection only to other members of the same species.) During the last decade, technological advances enabled researchers to introduce human interferon genes into bacterial genes so that these bacteria produce human interferon. Through this **recombinant DNA technology**, bacteria can be turned into interferon "factories" for large-scale commercial production of this valuable immune agent. Much research is currently underway to determine interferon's potential effectiveness in treating viral diseases and cancer. The research thrust is toward treatment rather than prevention, since it would not be economically feasible to maintain a continuous preventive level of this substance in the population because of the transient nature of interferon's defense capabilities. However, even though interferon is currently being used to successfully treat one previously fatal, rare form of leukemia, other investigative studies to date have not been as encouraging as hoped.

Natural killer cells destroy virus-infected cells and cancer cells upon first exposure to them.

Natural killer cells are naturally occurring, lymphocytelike cells that nonspecifically destroy virus-infected cells and cancer cells by directly lysing their membranes upon first exposure to them. Their mode of action and major targets are similar to cytotoxic T cells, but the latter can fatally attack only the specific types of virus-infected cells and cancer cells to which they have been previously exposed. Furthermore, following exposure, cytotoxic T cells require a maturation period before they are capable of launching their lethal assault. The natural killer cells provide an immediate, nonspecific defense against virus-invaded cells and cancer cells before the more specific and more abundant cytotoxic T cells become functional.

The complement system kills microorganisms directly both on its own and in conjunction with antibodies while augmenting the inflammatory response.

The **complement system** is another defense mechanism brought into play nonspecifically in response to invading organisms. It can also be triggered by antibodies as part of the specific immune strategy. In fact, the system derives its name from the fact that it *complements* the action of antibodies, being the primary mechanism activated by antibodies to kill foreign cells.

In the same tradition as the clotting and anticlotting systems and the kinin system, the complement system consists of plasma proteins that are produced by the liver and circulate in the blood in inactive form. Once the first component, C1, is activated, it activates the next component, C2, and so on, in a sequential cascade of activation reactions. The five final components, C5 through C9, form a large, doughnut-shaped protein complex, the **membrane-attack complex**, which attacks the surface membrane of nearby microorganisms by imbedding itself so that a large channel is created through the microbial surface membrane (● Fig. 12–7). This hole-punching technique makes the membrane extremely leaky; the resulting osmotic flux of water into the victim cell causes it to swell and burst. This complement-induced lysis is the major means of directly killing microbes without phagocytizing them.

The powerful complement cascade can be set into motion in two ways: (1) by exposure to particular carbohydrate chains present on the surfaces of microorganisms but not found on human cells (the **alternate pathway**, a nonspecific immune response) and (2) by exposure to antibodies produced against a specific foreign invader (the **classical path-**

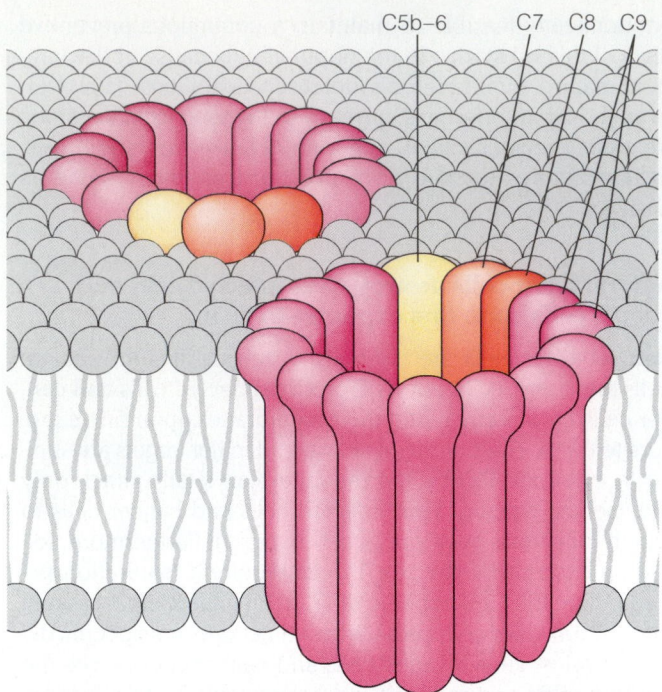

• FIGURE 12–7   Membrane-Attack Complex (MAC) of Complement System Activated complement proteins C5, C6, C7, C8, and a number of C9s aggregate to form a porelike channel in the plasma membrane of the target cell. The resultant leakage leads to destruction of the cell.

way, a specific immune response). In either case, activation of the complement system brings about direct lysis of the invader and reinforcement of other general inflammatory tactics.

Unlike the other cascade systems, in which the sole function of the various components is activation of the next precursor in the sequence, several of the activated proteins in the complement cascade perform additional important functions on their own (Table 12–4). Besides the direct destruction of foreign cells accomplished by the membrane-attack complex, various other activated complement components augment the inflammatory process by (1) serving as chemotaxins, which attract and guide professional phagocytes to the

TABLE 12–4   Functions of the Complement System
Complement components C5 through C9, when activated, form a membrane-attack complex that punches lethal holes in the target cell

Other activated complement components augment the inflammatory process by
    Serving as chemotaxins
    Acting as opsonins
    Promoting localized vasodilation and increased capillary permeability
    Stimulating histamine release from mast cells in vicinity
    Activating kinins

site of complement activation (that is, the site of microbial invasion); (2) acting as opsonins by binding with microbes and thereby enhancing their phagocytosis; (3) promoting vasodilation and increased vascular permeability to increase blood flow to the invaded area; (4) stimulating the release of histamine from mast cells in the vicinity, which in turn enhances the local vascular changes characteristic of inflammation; and (5) activating kinins, which further reinforce inflammatory reactions.

What restricts the activated complement system's destructive tactics to undesirable cells, such as invading bacteria? Several of the activated components in the cascade are very unstable. Because these unstable components are able to perpetuate the sequence only in the immediate vicinity in which they are activated before they decompose, the complement attack is confined to the surface membrane of the microbe whose presence initiated activation of the system. Nearby host cells are thus spared from lytic attack.

⊜ SPECIFIC IMMUNE RESPONSES: GENERAL CONCEPTS

Specific immune responses include antibody-mediated immunity accomplished by B lymphocyte derivatives and cell-mediated immunity accomplished by T lymphocytes.

A specific immune response is a selective attack aimed at limiting or neutralizing a particular offending target for which the body has been specially prepared following prior exposure to it. There are two classes of specific immune responses: **antibody-mediated,** or **humoral, immunity** involving the production of antibodies by B lymphocyte derivatives known as *plasma cells* and **cell-mediated immunity** involving the production of *activated T lymphocytes,* which directly attack unwanted cells.

B and T lymphocytes (B and T cells) have different life histories and, more importantly, different properties and functions. Both types of lymphocytes, like all blood cells, are derived from common stem cells in the bone marrow. Whether a lymphocyte and all of its progeny are destined to be B or T cells depends on the site of final maturation and differentiation of the original cell in the lineage (• Fig. 12–8). During fetal life and early childhood, some of the immature lymphocytes migrate through the blood to the thymus, where they undergo further processing to become T lymphocytes. The **thymus** is a lymphoid tissue located midline within the chest cavity above the heart in the space between the lungs (Fig. 12–1). Lymphocytes that mature without benefit of "thymic education" become B lymphocytes. B lymphocytes were first discovered in birds, where the maturational processing takes place in a gut-related lymphoid tissue unique to birds, the bursa of Fabricius; hence the name B cells. In humans, the site of B-cell maturation and differentiation is uncertain, although it is generally assumed to be the bone marrow.

Upon being released into the blood from either the bone marrow or the thymus, mature B and T cells take up residence and establish lymphocyte colonies in the peripheral lym-

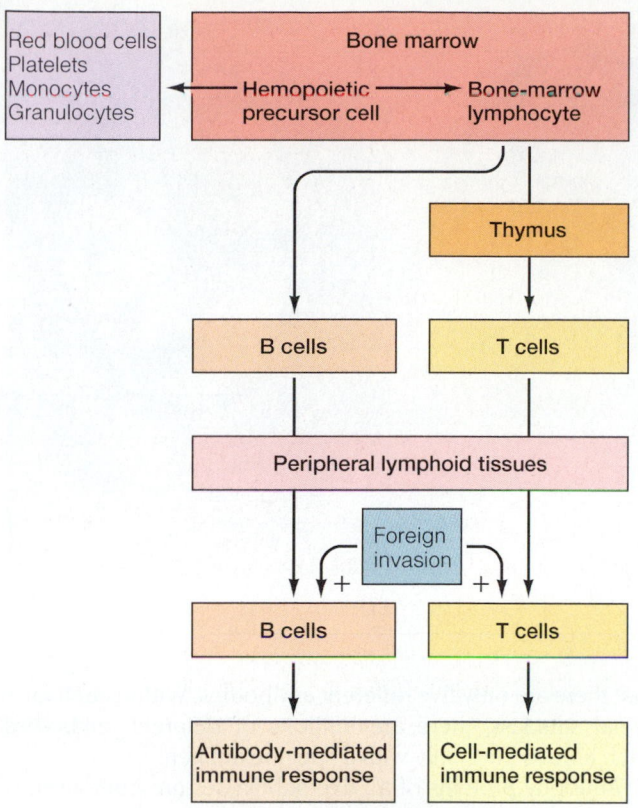

● **FIGURE 12–8 Origins of B and T Cells** B cells are derived from lymphocytes that matured and differentiated in the bone marrow whereas T cells are derived from lymphocytes that originated in the bone marrow but matured and differentiated in the thymus. New B and T cells are produced by colonies of B and T cells in peripheral lymphoid organs. The B and T cells defend against foreign invasion by producing an antibody-mediated immune response and cell-mediated immune response, respectively.

phoid tissues. Here, upon appropriate stimulation, they undergo cell division to produce new generations of either B or T cells, depending on their ancestry. After early childhood, most new lymphocytes are derived from these peripheral lymphocyte colonies rather than from the bone marrow.

The role of the thymus remained obscure until recently because its removal from an adult had no obvious effect. Because most of the migration and differentiation of T cells occurs early in development, the thymus gradually atrophies and becomes less important as the individual matures. It does, however, continue to produce **thymosin**, a hormone important in maintaining the T-cell lineage. Thymosin enhances proliferation of new T cells within the peripheral lymphoid tissues and augments the immune capabilities of existing T cells. Recent evidence indicates that secretion of thymosin decreases after about thirty to forty years of age. This decline has been implicated as being a contributing factor in aging. It is further speculated that diminishing T-cell capacity with advancing age might somehow be linked to the increased susceptibility to viral infections and cancer that occurs as a person ages. T cells play an especially important role in defense against viruses and viral-induced cancer.

Lymphocytes are able to specifically recognize and selectively respond to an almost limitless variety of foreign agents

as well as cancer cells. The recognition and response processes are different for B and T cells. In general, B cells recognize free-existing foreign invaders such as bacteria and their toxins and a few viruses, which they combat by secreting antibodies specific for the invaders. T cells specialize in recognizing and destroying body cells gone awry, including virus-infected cells and cancer cells.

Each of us has an estimated total of two trillion lymphocytes, which, if aggregated together in a mass, would be comparable to the size of the brain or liver. At any one time, the majority of these lymphocytes are concentrated in the various strategically located lymphoid tissues, but both B and T cells continually circulate among the lymph, blood, and body tissues, where they remain on constant surveillance.

An antigen induces an immune response against itself.

Both B and T cells must be able to specifically recognize unwanted cells and other material to be destroyed or neutralized as being distinct from normal-self cells. The presence of antigens enables them to make this distinction. An **antigen** is a large, complex molecule that triggers a specific immune response against itself when it gains entry into the body. In general, the more complex a molecule, the greater its antigenicity. Foreign proteins are the most common antigens because of their size and structural complexity, although other macromolecules, such as large polysaccharides, can also act as antigens. Antigens may exist as isolated molecules, such as bacterial toxins, or they may be an integral part of a multimolecular structure, such as being present on the surface of an invading foreign microbe. A complex antigen may have many **antigenic determinant sites**, each capable of stimulating the production of and interacting with different antibodies. For simplicity, we will consider each antigenic determinant site to be a separate antigen.

Many low-molecular-weight organic substances that are not antigenic by themselves can become antigenic if they attach to body proteins. Such small molecules are known as **haptens**. Antibodies developed against the hapten-protein combination can react in the future against the hapten alone should it be reintroduced to the body. Examples of haptens include poison ivy toxin, various drugs (such as penicillin), and other agents that are otherwise harmless but can elicit inappropriate immune responses known as allergies in sensitized individuals.

⊜ B LYMPHOCYTES: ANTIBODY-MEDIATED IMMUNITY

Antibodies amplify the inflammatory response to promote destruction of the antigen that stimulated their production.

Each B and T cell has receptors on its surface for binding with one particular type of the multitude of possible antigens. In the case of B cells, binding with antigen induces the cell to differentiate into a **plasma cell**, which produces antibodies that are able to combine with the specific type of antigen that

• **FIGURE 12–9** **Comparison of Unactivated B Cell and Plasma Cell** Electron micrograph of (a) an unactivated B cell, or small lymphocyte, and (b) a plasma cell. A plasma cell is an activated B cell. It is filled with an abundance of rough endoplasmic reticulum distended with antibody molecules.

(a)

(b)

stimulated the antibodies' production. During differentiation into a plasma cell, a B lymphocyte swells as the rough endoplasmic reticulum (the site for synthesis of proteins to be exported) greatly expands (• Fig. 12–9). Because antibodies are proteins, plasma cells essentially become prolific protein factories, producing up to two thousand antibody molecules per second. So great is the commitment of a plasma cell's protein-synthesizing machinery to antibody production that it is unable to maintain protein synthesis for its own viability and growth. Consequently, it dies after a brief five-to-seven-day, highly productive life span.

Antibodies are secreted into the blood or lymph, depending on the location of the activated plasma cells, but all antibodies eventually gain access to the blood, where they are known as **gamma globulins** or **immunoglobulins**. According to differences in their biological activity, antibodies are grouped into the following five subclasses:

■ **IgM** immunoglobulin serves as the B-cell surface receptor for antigen attachment and is secreted in the early stages of plasma-cell response.

■ **IgG**, the most abundant immunoglobulin in the blood, is produced copiously when the body is subsequently exposed to the same antigen.

Together, IgM and IgG antibodies are responsible for most specific immune responses against bacterial invaders and a few types of viruses.

■ **IgE** is the antibody mediator for common allergic responses such as hay fever, asthma, and hives.

■ **IgA** immunoglobulins are found in secretions of the digestive, respiratory, and genitourinary systems, as well as in milk and tears.

■ **IgD** is present on the surface of many B cells, but its function is uncertain.

It is important to note that this classification is based on different ways in which antibodies function. It does not imply

that there are only five different antibodies. Within each functional subclass, there are millions of different antibodies, each able to bind only with a specific antigen.

Antibody proteins of all five subclasses are composed of four interlinked polypeptide chains—two long, heavy chains and two short, light chains—arranged in the shape of a Y (• Fig. 12–10). Characteristics of the arm regions of the Y determine with what antigen the antibody can bind (that is, the *specificity* of the antibody). Properties of the tail portion of the

• **FIGURE 12–10** **Antibody Structure** An antibody is Y-shaped. It is able to bind only with the specific antigen that "fits" its antigen-binding sites (Fab) on the arm tips. The tail region (Fc) binds with particular mediators of antibody-induced activities.

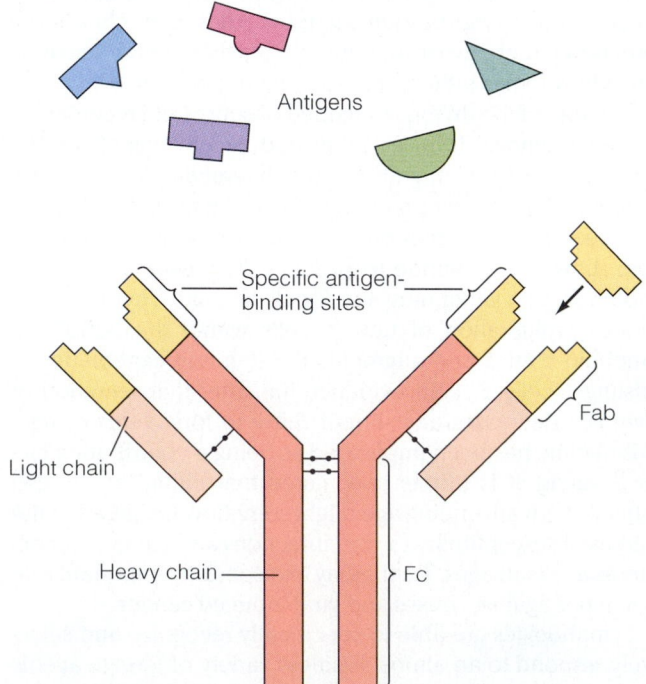

Antigens

Specific antigen-binding sites

Fab

Light chain

Heavy chain

Fc

antibody, on the other hand, determine the *functional properties* of the antibody (what the antibody does once it binds with antigen). An antibody has two identical antigen-binding sites, one at the tip of each arm. These **antigen-binding fragments** (**Fab**) are unique for each different antibody so that each antibody can interact only with an antigen that specifically matches it, much like a lock and key. The tremendous variation in the antigen-binding fragments of different antibodies is responsible for the extremely large number of unique antibodies that are capable of binding specifically with millions of different antigens.

In contrast to these variable Fab regions at the arm tips, the tail portion of every antibody within each immunoglobulin subclass is identical. The tail, the antibody's so-called **constant** (**Fc**) **region**, contains binding sites for particular mediators of antibody-induced activities, which vary among the different subclasses. In fact, differences in the constant region are the basis for distinguishing between the different immunoglobulin subclasses. For example, the constant tail region of IgG antibodies, when activated by antigen binding in the Fab region, binds with phagocytic cells and serves as an opsonin to enhance phagocytosis. In comparison, the constant tail region of IgE antibodies attaches to mast cells and basophils, even in the absence of antigen. When the appropriate antigen or hapten gains entry to the body and binds with the attached antibodies, this triggers the release of histamine from the affected mast cells and basophils. Histamine, in turn, induces the allergic manifestations that follow.

Immunoglobulins cannot directly destroy foreign organisms or other unwanted materials upon binding with antigens on their surfaces. Instead, antibodies exert their protective influence in one of two general ways: physical hindrance of antigens and amplification of nonspecific immune responses (● Fig. 12–11).

Antibodies can physically hinder some antigens from exerting their detrimental effects. For example, by combining with bacterial toxins, antibodies can prevent these harmful chemicals from interacting with susceptible cells. This process is known as **neutralization**. Similarly, antibodies are able to bind with surface antigens on some types of viruses, preventing these viruses from entering cells, where they could exert their damaging effects. Sometimes multiple antibody molecules can cross-link numerous antigen molecules into chains or lattices of antibody-antigen complexes. **Agglutination** is the term applied to the process in which foreign cells, such as bacteria or mismatched transfused red blood cells, bind together in such a clump. When linked antibody-antigen complexes involve soluble antigens, such as tetanus toxin, the lattice can become so large that it precipitates out of solution. (**Precipitation** is the process in which a substance separates from a solution.) Within the body, these physical-hindrance mechanisms play only a minor protective role against invading agents. However, the tendency for certain antigens to agglutinate or precipitate upon forming large complexes with antibodies specific for them is useful clinically and experimentally for detecting the presence of particular antigens or antibodies. Pregnancy-diagnosis tests, for example, employ this principle to detect the presence in the urine of a hormone secreted soon after conception.

Antibodies' most important function by far is to profoundly augment the nonspecific immune responses already initiated by the invaders. Antibodies mark or identify foreign material as targets for actual destruction by the complement system, phagocytes, or killer cells while enhancing the activity of these other defense systems as follows:

1. *Activation of the complement system.* When an appropriate antigen binds with an antibody, receptors on the tail portion of the antibody are able to bind with and activate C1, the first component of the complement system. This sets off the cascade of events leading to formation of the membrane-attack complex, which is specifically directed at the membrane of the invading cell that bears the antigen that initiated the activation process. In fact, antibody is the most powerful activator of the complement system. The biochemical attack subsequently unleashed against the invader's membrane is the most important mechanism by which antibodies exert their protective influence. Furthermore, various activated complement components enhance virtually every aspect of the inflammatory process. Note that the same complement system is activated by an antigen-antibody complex regardless of the type of antigen. Although the binding of antigen to antibody is highly specific, the outcome, which is determined by the antibody's constant tail region, is identical for all activated antibodies within a given subclass; for example, all IgG antibodies activate the same complement system.

2. *Enhancement of phagocytosis.* As mentioned previously, antibodies, especially IgG, act as opsonins. The tail portion of an antigen-bound IgG antibody is able to bind with a receptor on the surface of a phagocyte and subsequently promote the phagocytosis of the antigen-containing victim attached to the antibody.

3. *Stimulation of killer (K) cells.* The binding of antibody to antigen can also induce attack of the antigen-bearing cell by a **killer** (**K**) **cell**. K cells are similar to natural killer cells except that K cells require the target cell to be coated with antibodies before they can destroy it by lysing its plasma membrane. K cells have receptors for the constant tail portion of antibodies.

In these ways, antibodies, though unable to directly destroy invading bacteria or other undesirable material, bring about destruction of the antigens to which they are specifically attached by amplifying other nonspecific lethal defense mechanisms.

Occasionally, an overzealous antigen-antibody response can inadvertently cause damage to normal cells as well as to invading foreign cells. Typically, antigen-antibody complexes, formed in response to foreign invasion, are removed by phagocytic cells after having revved up nonspecific defense strategies. If large numbers of these complexes are continuously produced, however, the phagocytes are unable to clear away all of the immune complexes formed. Antigen-antibody complexes that are not removed continue to activate the complement system, among other things. Excessive amounts of activated complement and other inflammatory agents may "spill over," damaging the surrounding normal

Neutralization

Invading bacterium

Bacterial toxin

Antibody neutralizing toxin

Antibody specific to toxin

Agglutination (clumping of antigenic cells) and **precipitation** (if soluble antigen-antibody complex is too large to stay in solution)

Foreign cells (e.g., transfused mismatched red blood cells)

Antigen

Lattice

Antibodies specific to foreign cells

(a)

Activation of complement system

Inactive C1 complement molecule

Antibody

Antigenic foreign cell, such as invading bacterium

Activated by binding with antigen-attached antibody

Formation of C5–C9, the membrane-attack complex

Membrane-attack complex

Lysis (rupture) of cell

Enhancement of phagocytosis (opsonization)

Invading bacterium coated with antibodies specific to it

Phagocyte

Stimulation of killer cells

Invading bacterium coated with antibodies specific to it

Lysis induced by killer cell

Killer cell

(b)

Structures are not drawn to scale.

● **FIGURE 12–11   How Antibodies Help Eliminate Invading Microbes**   (a) Physical hindrance of antigens. (b) Amplification of nonspecific immune responses.

cells as well as the unwanted cells. Furthermore, destruction is not necessarily restricted to the initial site of inflammation. Antigen-antibody complexes may freely circulate and become trapped in the kidneys, joints, brain, small vessels of the skin, and elsewhere, causing widespread inflammation and tissue damage. Such damage produced by immune com-

plexes is referred to as an **immune-complex disease**, which can be a complicating outcome of bacterial, viral, or parasitic infection.

More insidiously, immune-complex disease can also occur as a result of overzealous inflammatory activity prompted by the presence of immune complexes formed by "self-antigens" (proteins synthesized by the person's own body) and antibodies erroneously produced against them. *Rheumatoid arthritis* is believed to be brought about in this way.

Each antigen stimulates a different clone of B lymphocytes to produce antibodies.

Consider the diversity of foreign molecules that a person can potentially encounter during a lifetime. Yet each B lymphocyte is preprogrammed to respond to only one of these millions of different antigens. Other antigens cannot combine with the same B cell and induce it to secrete different antibodies. The astonishing implication is that each of us is equipped with millions of different preformed B lymphocytes, at least one for every possible antigen that we might ever encounter—including those specific for synthetic substances that do not exist in nature.

According to early immunological theory, antibodies were believed to be "made to order" whenever a foreign antigen gained entry to the body. In contrast, the currently accepted **clonal selection theory** proposes that diverse B lymphocytes

are produced during fetal development, each capable of synthesizing antibody against a particular antigen before ever being exposed to it. All offspring of a particular ancestral B lymphocyte form a family of identical cells, or a **clone**, which is committed to producing the same specific antibody. B cells remain dormant, not actually secreting their particular antibody product until (or unless) they come into contact with the appropriate antigen. When an antigen gains entry to the body, it activates the particular clone of B cells that bear receptors on their surface uniquely specific for that antigen (● Fig. 12–12).

The first antibodies produced by a newly formed B cell are IgM immunoglobulins, which are inserted into the cell's plasma membrane rather than being secreted. Here they serve as receptor sites for binding with a specific kind of antigen, almost like "advertisements" for the kind of antibody the cell can produce. Binding of the appropriate antigen to a B cell amounts to "placing an order" for the manufacture and secretion of large quantities of that particular antibody.

Antigen binding causes the activated B-cell clone to multiply and differentiate into two cell types—*plasma cells* and

---

● **FIGURE 12–12 Clonal Selection Theory** (a) The B-cell clone specific to the antigen proliferates and differentiates. (b) Plasma cells secrete antibodies that bind with free antigen (antigen not attached to B cells). (c) Memory cells expand the specific clone and are primed and ready for subsequent exposure to the same antigen.

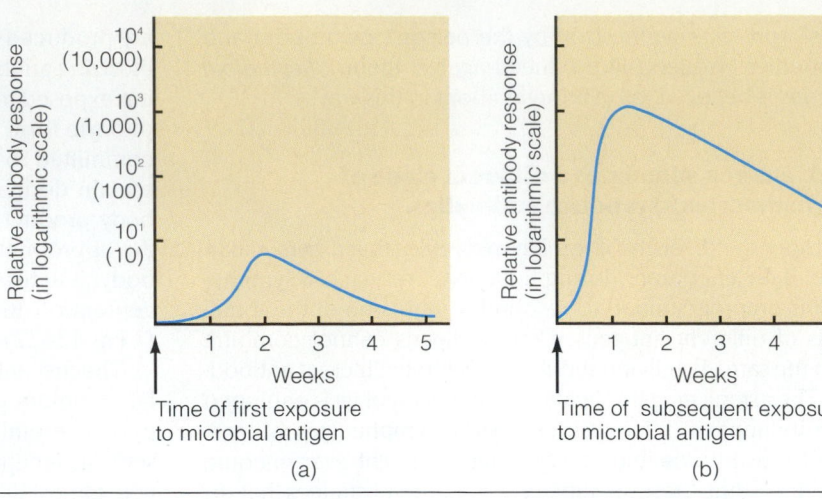

• **FIGURE 12–13   Primary and Secondary Immune Responses** (a) Primary response on first exposure to a microbial antigen. (b) Secondary response on subsequent exposure to the same microbial antigen. Note that the secondary response peaks in a week, whereas the primary response does not peak for several weeks. Also, the magnitude of the secondary response is 100 times that of the primary response. (The relative antibody response is in the logarithmic scale.)

memory cells. Most progeny are transformed into plasma cells, which are prolific producers of customized antibodies that contain the same antigen-binding sites as the surface receptors. However, plasma cells switch to the production of IgG antibodies, which are secreted rather than remaining membrane bound. In the blood, the secreted antibodies combine with invading free (not bound to lymphocytes) antigen, marking it for destruction by the complement system, phagocytic ingestion, or other means.

Not all of the new B lymphocytes produced by the specifically activated clone differentiate into antibody-secreting plasma cells. A small proportion of them become **memory cells**, which do not participate in the current immune attack against the antigen but instead remain dormant and expand the specific clone. Should the person ever be exposed to the same antigen again, these memory cells are primed and ready for even more immediate action than were the original lymphocytes in the clone.

During initial contact with a microbial antigen, the antibody response is delayed for several days until plasma cells are formed and does not reach its peak for several weeks (• Fig. 12–13). This response is known as the **primary response**. Meanwhile, symptoms characteristic of the particular microbial invasion persist until either the invader succumbs to the mounting specific immune attack against it or the infected individual dies. After reaching the peak, the antibody levels gradually decline over a period of time, although some circulating antibody from this primary response may persist for a prolonged period. Long-term protection against the same antigen, however, is primarily attributable to the memory cells. If the same antigen ever reappears, the long-lived memory cells launch a more rapid, more potent, and longer-lasting **secondary response** than occurred during the primary response. This swifter, more powerful immune attack is frequently adequate to prevent or minimize overt infection upon subsequent exposures to the same microbe, forming the basis of long-term immunity against a specific disease.

The original antigenic exposure that induces the formation of memory cells can occur through either actually having the disease or being vaccinated (• Fig. 12–14). During vaccination, the individual is deliberately exposed to a pathogen that

has been stripped of its disease-inducing capability but can still induce antibody formation against it. (See the accompanying boxed feature, Concepts, Challenges, and Controversies.)

Even though each of us has essentially the same original pool of different B-lymphocyte clones, the pool gradually becomes appropriately biased to respond most efficiently to each individual's particular antigenic environment. Those clones specific for antigens to which an individual is never exposed remain dormant for life, whereas those specific for antigens in the individual's environment typically become expanded and enhanced through the formation of highly responsive memory cells.

Memory cells are not formed for some diseases, so no lasting immunity is conferred by an initial exposure, as in the case of "strep throat." The course and severity of the disease are the same each time a person is reinfected with a microbe that the immune system does not "remember," regardless of the number of prior exposures.

Considering the millions of different antigens against which each of us has the potential to actively produce antibodies, how is it possible for an individual to have such a tremendous diversity of B lymphocytes, each capable of producing a different antibody? Antibodies are proteins that are produced in accordance with a nuclear DNA blueprint. Because all cells of the body, including the antibody-producing cells, contain the same nuclear DNA, it is hard to imagine how enough DNA could be packaged within the nuclei of every cell to code for the millions of different antibodies (a different portion of the genetic code being used by each B-cell clone), along with all of the other genetic instructions used by other cells. It is now known that only a relatively small number of gene fragments actually code for antibody synthesis, but these fragments are cut, reshuffled, and spliced in a vast number of different combinations during B-cell development. Each different combination gives rise to a unique B-cell clone. Even further diversification of antibody genes is subsequently accomplished by somatic mutation (see p. B–14). The antibody genes of already-formed B cells are highly prone to mutations in the region that codes for the variable antigen-binding sites on the antibodies. Each different mutant

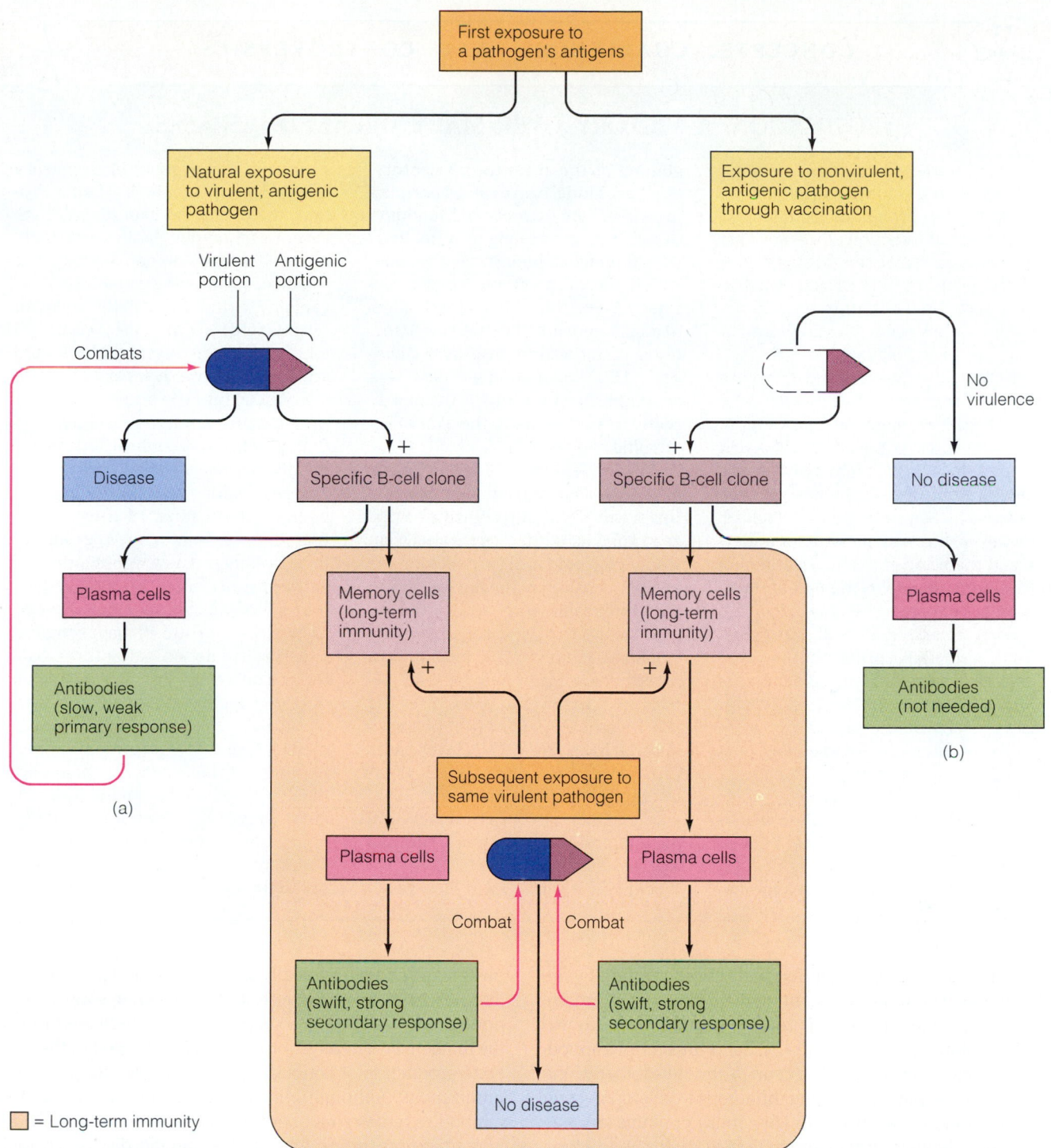

• **FIGURE 12–14   Means of Acquiring Long-Term Immunity** Long-term immunity against a pathogen can be acquired through having the disease or being vaccinated against it. (a) Exposure to virulent (disease-producing) pathogen. (b) Vaccination with modified pathogen that is no longer virulent (that is, can no longer produce disease) but is still antigenic. In both cases, long-term memory cells are produced that mount a swift, secondary response that prevents or minimizes symptoms on a subsequent natural exposure to the same virulent pathogen.

cell, in turn, gives rise to a new clone. Thus, the great diversity of antibodies is made possible by the reshuffling of a small set of gene fragments during B-cell development as well as by further somatic mutation in already-formed B cells. In this way, a huge antibody repertoire is possible using only a modest share of the genetic blueprint.

Active immunity is self-generated; passive immunity is "borrowed."

The production of antibodies as a result of exposure to an antigen is referred to as **active immunity** against that antigen. A second way in which an individual can acquire antibodies

VACCINATION: A VICTORY OVER MANY DREADED DISEASES

Modern society has come to hope and even expect that vaccines can be developed to protect us from almost any dreaded infectious disease. This expectation has been brought into sharp focus by our current frustration over the inability to date to develop a successful vaccine against HIV, the virus that causes AIDS.

Nearly twenty-five hundred years ago, our ancestors were aware of the existence of immune protection. Writing about a plague that struck Athens in 430 B.C., Thucydides observed that the same person was never attacked twice by this disease. However, the ancients did not understand the basis of this protection, so they were unable to manipulate it to their advantage.

Early attempts at deliberately acquiring lifelong protection against smallpox, a dreaded disease that was highly infectious and frequently fatal (up to 40% of the sick died), consisted of intentionally exposing oneself by coming into direct contact with a person suffering from a milder form of the disease. The hope was to protect against a future fatal bout of smallpox by deliberately inducing a mild case of the disease. By the be-ginning of the seventeenth century, this technique had evolved into using a needle to extract small amounts of pus from active smallpox pustules (the fluid-filled bumps on the skin, which leave a characteristic depressed scar or "pox" mark after healing) and introducing this infectious material into healthy individuals. This inoculation process was accomplished by applying the pus directly to slight cuts in the skin or by inhaling of dried pus.

Edward Jenner, an English physician, was the first to demonstrate that immunity against cowpox, a disease similar to but less serious than smallpox, could also protect humans against smallpox. Having observed that milkmaids who acquired cowpox seemed to be protected from smallpox, Jenner in 1796 inoculated a healthy boy with pus he had extracted from cowpox boils. After the boy recovered, Jenner (not being restricted by modern ethical standards of research on human subjects) deliberately inoculated him with what was considered to be a normally fatal dose of smallpox infectious material. The boy survived.

Jenner's results were not taken seriously, however, until a century later when, in the 1880s, Louis Pasteur, the first great experimental immunologist, extended Jenner's technique. Pasteur demonstrated that the disease-inducing capability of organisms could be greatly reduced (attenuated) so that they could no longer produce disease but would still induce antibody formation when introduced into the body—the basic principle of modern vaccines. His first vaccine was against anthrax, a deadly disease of sheep and cows. Pasteur isolated and heated anthrax bacteria, then injected these attenuated organisms into a group of healthy sheep. A few weeks later at a gathering of fellow scientists, Pasteur injected these vaccinated sheep as well as a group of unvaccinated sheep with fully potent anthrax bacteria. The result was dramatic—all of the vaccinated sheep survived while all of the unvaccinated sheep died. Pasteur's notorious public demonstrations such as this, coupled with his charismatic personality, caught the attention of physicians and scientists of the time, sparking the development of modern immunology.

---

is by the direct transfer of antibodies actively formed by another person (or animal). The immediate "borrowed" immunity conferred upon receipt of preformed antibodies is known as **passive immunity** (Table 12–5). Such transfer of antibodies of the IgG class normally occurs from the mother to the fetus across the placenta during intrauterine development. In addition, a mother's colostrum (first milk) contains IgA antibodies that provide further protection for breast-fed babies. Passively transferred antibodies are usually broken down in less than a month, but meanwhile the newborn is provided important immune protection (essentially the same as its mother's) until it can begin actively mounting its own immune responses. Antibody-synthesizing ability does not develop for about a month after birth.

Passive immunity is sometimes employed clinically to provide immediate protection or to bolster resistance against an extremely virulent infectious agent or potentially lethal toxin to which a person has been exposed (for example, rabies virus, tetanus toxin in nonimmunized individuals, and poison-ous snake venom). Typically, the administered preformed antibodies have been harvested from another source (often nonhuman) that has been exposed to an attenuated form of the antigen. Frequently, horses or sheep are used in the deliberate production of antibodies to be collected for passive immunizations. Although injection of serum containing these antibodies (**antiserum** or **antitoxin**) is beneficial in providing immediate protection against the specific disease or toxin, the recipient may mount an immune response against the injected antibodies themselves because they are foreign proteins. The result may be a severe allergic reaction to the treatment, a condition known as **serum sickness**.

Natural immunity is actually a special case of actively acquired immunity.

Certain antibodies were once thought to occur naturally in the blood. Antibodies associated with blood types are the classic example of "natural antibodies." The surface mem-

TABLE 12–5   Active versus Passive Immunity

CHARACTERISTIC	ACTIVE IMMUNITY	PASSIVE IMMUNITY
Exposure to antigen required for immunity to develop?	Yes, either through natural exposure during infection or through artificial exposure during vaccination	No
Source of circulating antibodies	Antibodies self-generated on exposure to antigen	Antibodies "borrowed"—that is, produced from another source and transferred to the individual, either naturally across the placenta or through the milk from mother to offspring, or artificially through injection of antibodies harvested from vaccinated animal
Injection required?	Injection of attenuated antigen required for artificial active immunity	Injection of borrowed antibodies required for artificial passive immunity
Time to develop resistance to antigen exposure	Several weeks (primary response to antigen) to several days (secondary response to antigen) required for antibody production	Immediate on acquisition of borrowed antibodies
Duration of resistance	Long—may be lifelong	Short—less than a month
Indication for artificial acquisition of this type of immunity	Well before potential exposure to virulent pathogen to allow enough time for appropriate B-cell clone to respond; boosters as needed to maintain active immunity	Immediately after exposure to extremely virulent virus or potentially lethal toxin when protection is needed immediately (rather than waiting for the more slowly responding active immunity to develop)

branes of human erythrocytes contain inherited antigens that vary depending on blood type. With the major blood-group system, the **ABO system**, the erythrocytes of individuals with type-A blood contain A antigens; those with type-B blood contain B antigens; those with type-AB blood have both A and B antigens; and those with type-O blood do not have any A or B red blood cell surface antigens. Antibodies against erythrocyte antigens not present on the body's own erythrocytes begin to appear in human plasma after a person is about six months of age. Accordingly, the plasma of type-A blood contains anti-B antibodies; type-B blood contains anti-A antibodies; no antibodies related to the ABO system are present in type-AB blood; and both anti-A and anti-B antibodies are present in type-O blood. Typically, one would expect antibody production against A or B antigen to be induced only if blood containing the alien antigen were injected into the body. However, high levels of these antibodies are found in the plasma of individuals who have never been exposed to a different type blood. Consequently, these were considered to be naturally occurring antibodies, that is, produced without any known exposure to the antigen. It is now known that individuals are unknowingly exposed at an early age to small amounts of A- and B-like antigens associated with common intestinal bacteria. Antibodies produced against these foreign antigens coincidentally also interact with a nearly identical foreign-blood-group antigen, even upon first exposure to it.

If a person is administered blood of an incompatible type, two different antigen-antibody interactions take place. By far the most serious consequences arise from the effect of the antibodies in the recipient's plasma on the incoming donor erythrocytes. The effect of the donor's antibodies on the recipient's erythrocyte-bound antigens is less important unless a large amount of blood is transfused, because the donor's antibodies are so diluted by the recipient's plasma that little red blood cell damage takes place in the recipient.

Antibody interaction with erythrocyte-bound antigen may result in agglutination (clumping) or hemolysis (rupture) of the attacked red blood cells. Agglutination and hemolysis of donor red blood cells by antibodies in the recipient's plasma can lead to a sometimes fatal **transfusion reaction**. Agglutinated clumps of incoming donor cells can plug small blood vessels. In addition, one of the most lethal consequences of mismatched transfusions is acute kidney failure caused by the release of large amounts of hemoglobin from damaged donor erythrocytes. If the free hemoglobin in the plasma rises above a critical level, it will precipitate in the kidneys and block the urine-forming structures, leading to acute kidney shutdown.

Because type-O individuals do not have any A or B antigens, their erythrocytes will not be attacked by either anti-A or anti-B antibodies, so they are considered to be **universal donors**. Their blood can be transfused into persons of any blood type. However, type-O individuals can receive only type-O blood because the anti-A and anti-B antibodies present in their plasma will attack either A or B antigens in incoming blood. In contrast, type-AB individuals are called **universal recipients**. Lacking both anti-A and anti-B antibodies, they can accept donor blood of any type, although they can donate blood only to other AB persons. Because their erythrocytes possess both A and B antigens, their cells would

Invading bacteria

Macrophages secrete interleukin 1, which enhances B-cell proliferation and antibody secretion

Macrophages "process and present" bacterial antigen to B and T lymphocyte clones specific to the antigen

Interleukin 1

Macrophage

Helper T cell

+

+

B cell

Antibodies enhance phagocytosis by coating the bacteria and serving as opsonins

Activated helper T cell

Helper T cells secrete B-cell growth factor that enhances B-cell proliferation and antibody secretion

+

B-cell growth factor

Plasma cell

Plasma cells secrete antibodies that bind with the antigenic bacteria

Antibodies

● **FIGURE 12–15 Synergistic Interactions among Macrophages, B Cells, and Helper T Cells** B and T cells cannot react to a newly entering foreign antigen until the antigen has been processed and presented to them by macrophages. These antigen-presenting cells also secrete interleukin 1, which stimulates proliferation of the activated B cells. These B cells are transformed into plasma cells, which produce antibodies against the antigen. Activated helper T cells secrete B-cell growth factor, which further stimulates B-cell proliferation and antibody production. The antibodies not only lead to the demise of the foreign antigen but also serve as opsonins to enhance phagocytosis by the macrophages.

be attacked if transfused into individuals with antibodies against either of these antigens.

The terms *universal donor* and *universal recipient* are somewhat misleading, however. In addition to the ABO system, numerous other erythrocyte antigens and plasma antibodies can cause transfusion reactions, the most important of which is the Rh factor. Individuals who possess the **Rh factor** (an erythrocyte antigen first observed in rhesus monkeys,

hence the designation *Rh*) are said to have *Rh-positive* blood, whereas those lacking the Rh factor are considered to be *Rh-negative*. In contrast to the ABO system, no naturally occurring antibodies develop against the Rh factor. Anti-Rh antibodies are produced only by Rh-negative individuals when (and if) they are first exposed to the foreign Rh antigen present in Rh-positive blood. A subsequent transfusion of Rh-positive blood could produce a transfusion reaction in such a sensitized Rh-negative person. Rh-positive individuals, in contrast, never produce antibodies against the Rh factor that they themselves possess. Therefore, Rh-negative individuals should be administered only Rh-negative blood, whereas Rh-positive persons can safely receive either Rh-negative or Rh-positive blood. The Rh factor is of particular medical importance when an Rh-negative mother develops antibodies against the erythrocytes of an Rh-positive fetus she is carrying, a condition known as **erythroblastosis fetalis**, or **hemolytic disease of the newborn**.

Except in extreme emergencies, it is safest to individually cross-match blood before a transfusion is undertaken even though the ABO and Rh typing is already known, because there are approximately twelve other minor human erythrocyte antigen systems. Compatibility is determined by mixing the red blood cells from the potential donor with plasma from the recipient. If no clumping occurs, the blood is considered to be adequately matched for transfusion.

In addition to being an important consideration in transfusions, the various blood-group systems are also of legal importance in disputed paternity cases, because the erythrocyte antigens are inherited. In recent years, however, DNA "fingerprinting" has become a more definitive test.

Lymphocytes respond only to antigens that have been processed and presented to them by macrophages.

B cells cannot typically perform their task of antibody production without assistance from macrophages and, in most cases, from T cells as well (● Fig. 12–15). Relevant B-cell clones are not able to recognize and produce antibodies in response to "raw" foreign antigens entering the body; a B-cell clone must be formally "introduced" to the antigen before it will react to it. Invading organisms or other antigens are first engulfed by macrophages, which cluster around the appropriate B-cell clone and handle the formal introduction. During phagocytosis, the macrophage processes the raw antigen intracellularly and then "presents" the processed antigen by exposing it on the outer surface of the macrophage's plasma membrane in such a way that the adjacent B cells can recognize and be activated by it. In addition, these antigen-presenting macrophages secrete **interleukin 1**, a multipurpose chemical mediator that enhances the differentiation and proliferation of the now-activated B-cell clone. Interleukin 1 (alias endogenous pyrogen/leukocyte endogenous mediator) is also largely responsible for the fever and malaise accompanying many infections. In collaborative fashion, activated lymphocytes secrete antibodies that, among other things, enhance further phagocytic activity.

Many antigens are similarly presented to T cells. One specialized class of T lymphocytes, called helper T cells, help B cells upon being activated by macrophage-presented antigen. The helper T cells secrete a chemical mediator, **B-cell growth factor**, which further contributes to B-cell function in concert with the interleukin 1 secreted by macrophages. Therefore, mutually supportive interactions among macrophages, B cells, and helper T cells synergistically reinforce the phagocyte-antibody immune attack against the foreign intruder. Table 12–6 summarizes the nonspecific and specific immune strategies that defend against bacterial invasion.

TABLE 12–6   Nonspecific and Specific Immune Responses to Bacterial Invasion

NONSPECIFIC IMMUNE MECHANISMS	SPECIFIC IMMUNE MECHANISMS
Inflammation	Processing and presenting of bacterial antigen by macrophages to B cells specific to the antigen
Engulfment of invading bacteria by resident tissue macrophages	Proliferation and differentiation of activated B-cell clone into plasma cells and memory cells
Histamine-induced vascular responses to enhance delivery of increased blood flow to area, bringing in additional immune-effector cells and plasma proteins	Secretion by plasma cells of customized antibodies, which specifically bind to invading bacteria
Walling off of invaded area by fibrin clot	Enhancement by interleukin 1 secreted by macrophages
Emigration of neutrophils and monocytes/macrophages to the area to engulf and destroy foreign invaders and to remove cellular debris	Enhancement by helper T cells, which have been activated by the same bacterial antigen processed and presented to them by macrophages
Secretion by phagocytic cells of chemical mediators, which enhance both nonspecific and specific immune responses and induce local and systemic symptoms associated with infection	Binding of antibodies to invading bacteria and enhancement of nonspecific mechanisms that lead to their destruction
Nonspecific Activation of the Complement System	Action as opsonins to enhance phagocytic activity
Formation of hole-punching membrane-attack complex that lyses bacterial cells	Activation of lethal complement system
Enhancement of many steps of inflammation	Stimulation of killer cells, which directly lyse bacteria
	Persistence of memory cells that are capable of responding more rapidly and more forcefully should the same bacteria be encountered again

T LYMPHOCYTES: CELL-MEDIATED IMMUNITY

The three types of T cells are specialized to kill virus-infected host cells and to help or suppress other immune cells.

As important as B lymphocytes and their antibody products are in specific defense against invading bacteria and other foreign material, they represent only half of the body's specific immune defense corps. The T lymphocytes are equally important in defense against most viral and fungal infections and also play an important regulatory role in immune mechanisms (Table 12–7).

Unlike B cells, which secrete antibodies that can attack antigen at long distances, T cells do not secrete antibodies. Instead, they must be in direct contact with their targets, a process known as *cell-mediated immunity*. Like B cells, T cells are clonal and exquisitely antigen-specific. On its plasma membrane, each T cell bears unique receptor proteins, similar although not identical to the surface receptors on B cells. Unlike B cells, T cells are activated by foreign antigen only when it is present on the surface of a cell that also carries a marker of the individual's own identity; that is, both foreign antigens and **self-antigens** must be present on a cell's surface before a T cell can bind with it (with one important exception being whole transplanted foreign cells). It is during thymic education that T cells learn to recognize foreign antigens only in combination with the individual's own tissue antigens, a lesson that is passed on to all T cells' future progeny. The importance of this dual antigen requirement and the nature of the self-antigens will be described shortly.

A delay of a few days generally follows exposure to the appropriate antigen before **sensitized**, or **activated, T cells** are prepared to launch a cell-mediated immune attack. When exposed to a specific antigen combination, cells of the complementary T-cell clone proliferate and differentiate for several days, yielding large numbers of activated T cells that carry out various cell-mediated responses. There are three subpopulations of T cells, depending on their roles when activated by antigen:

1. **Cytotoxic T cells**, which destroy host cells bearing foreign antigen, such as body cells invaded by viruses, cancer cells, and transplanted cells.
2. **Helper T cells**, which enhance the development of antigen-stimulated B cells into antibody-secreting cells, enhance activity of the appropriate cytotoxic and suppressor T cells, and activate macrophages.
3. **Suppressor T cells**, which suppress both B-cell antibody production and cytotoxic and helper T-cell activity.

The vast majority of the billions of T lymphocytes are believed to be of the helper or suppressor varieties, which do not directly participate in the immune destruction of invading pathogens. Collectively, these subpopulations are referred to as **regulatory T cells**, because they modulate the activities of B cells and cytotoxic T cells as well as their own activities and those of macrophages.

TABLE 12–7   B versus T Lymphocytes

CHARACTERISTIC	B LYMPHOCYTES	T LYMPHOCYTES
Ancestral origin	Bone marrow	Bone marrow
Site of maturational processing	Bone marrow	Thymus
Receptors for antigens	Antibodies inserted in plasma membrane serve as surface receptors; highly specific	Surface receptors present but differing from antibodies; highly specific
Bind with	Extracellular antigens such as bacteria, free viruses, and other circulating foreign material	Foreign antigen in association with self-antigen, such as virus-infected cells
Antigen must be processed and presented by macrophages	Yes	Yes
Types of active cells	Plasma cells	Cytotoxic T cells, helper T cells, suppressor T cells
Formation of memory cells	Yes	Yes
Type of immunity	Antibody-mediated immunity	Cell-mediated immunity
Secretory product	Antibodies	Lymphokines
Function	Help eliminate free foreign invaders by enhancing nonspecific immune responses against them; provide immunity against most bacteria and a few viruses	Lyse virus-infected cells and cancer cells; provide immunity against most viruses and fungi and a few bacteria; aid B cells in antibody production
Life span	Short	Long

Like B cells, not all activated T-cell progeny become effector T cells. A small proportion of them remain dormant, serving as a pool of memory T cells that are primed and ready to respond even more swiftly and vigorously should the same foreign antigen ever reappear within a body cell. T cells, even activated ones, generally have long life spans, in contrast to B cells, which rapidly work themselves to death producing antibodies once they have been converted into plasma cells upon antigen stimulation. Thus, immunity for cell-mediated responses is similar to that for antibody responses, but it is generally of longer duration. Passive cell-mediated immunity against a particular pathogen can also be conferred through administration of activated T lymphocytes that have been harvested from another individual or animal who possesses active immunity against the particular microbe.

Exposure to antigen frequently activates both the B- and T-cell mechanisms simultaneously. Just as the regulatory T cells can facilitate or suppress the secretion of antibody by B cells, antibodies may either enhance or block the ability of cytotoxic T cells to destroy a victim cell, depending on the circumstances. Most of the effects exerted by lymphocytes on other immune cells (such as other lymphocytes and macrophages) are mediated by means of the secretion of chemical messengers. All chemicals other than antibodies that are secreted by lymphocytes are collectively called **lymphokines**, the majority of which are produced by T cells. Unlike antibodies, lymphokines do not interact directly with the antigen responsible for inducing their production. The roles of specific lymphokines will be described in the following discussion of each of the T-cell subpopulations.

Cytotoxic T Cells The targets of cytotoxic T cells most frequently are host cells infected with viruses. When a virus invades a body cell, as it must to survive, the envelope of antigenic proteins surrounding the virus are incorporated into the host cell's surface membrane (● Fig. 12–16). To attack the intracellular virus, cytotoxic T cells must destroy the infected host cell in the process. Cytotoxic T cells of the clone specific for this particular virus recognize and bind to the viral antigens and self-antigens on the surface of the infected cell. Thus

sensitized by viral antigen, a cytotoxic T cell destroys the victim cell by releasing chemicals that lyse the attacked cell before viral replication can begin.

Recent studies demonstrate that one means by which cytotoxic T cells and natural killer cells destroy a targeted cell is by releasing **perforin** molecules, which penetrate into the target cell's surface membrane and join together to form pore-like channels (● Fig. 12–17). This technique of killing a cell by punching holes in its membrane is similar to the method employed by the membrane-attack complex of the complement cascade. The virus released upon destruction of the host cell is then directly destroyed in the extracellular fluid by phagocytic cells, neutralizing antibodies, and the complement system. Meanwhile, the cytotoxic T cell, which has not been harmed in the process, can move on to kill other infected host cells. The surrounding healthy cells replace the lost cells by means of cell division. Usually, not many of the host cells have to be destroyed to halt a viral infection. If the virus has had a chance to multiply, however, with replicated virus leaving the original cell and spreading to other host cells, so many of the host cells may be sacrificed by the cytotoxic T-cell defense mechanism that serious malfunction may ensue.

Recall that other nonspecific defense mechanisms also come into play to combat viral infections, most notably natural killer cells and interferon, macrophages, and the complement system. As usual, an intricate web of interplay exists among the immune defenses that are launched against viral invaders (Table 12–8).

The usual method of destroying virus-infected host cells is not appropriate for the nervous system. If cytotoxic T cells destroyed virus-infected neurons, the lost cells could not be replaced because neurons cannot reproduce. Fortunately, virus-infected neurons are spared from extermination by the immune system, but how then are neurons protected from viruses? Immunologists long thought that the only antiviral defenses for neurons were those aimed at free viruses in the extracellular fluid. Surprising new research has revealed, however, that antibodies not only target viruses for destruction in the extracellular fluid but can also eliminate viruses inside

(a)   (b)   (c)

● **FIGURE 12–16 Cytotoxic T Cell Lysing Virus-Invaded Host Cell**
(a) When a virus invades a host cell, the viral antigen is displayed on the surface of the host cell alongside the cell's self-antigen. (b) The cytotoxic T cell recognizes and binds with a specific foreign antigen (viral antigen) in association with the self-antigen. (c) The cytotoxic T cell releases chemicals that destroy the attacked cell before the virus can enter the nucleus and start to replicate.

(a)   (b)

● **FIGURE 12–17   Mechanism of Killing by Killer Cells**   (a) Details of the killing process. A rise in the lymphocyte's calcium-ion level, apparently triggered by the receptor-mediated binding of the killer cell to its target ①, brings about exocytosis, in which the granules fuse with the cell membrane ② and disgorge their perforin ③ into the small intercellular space abutting the target. Calcium there changes the conformation of the individual perforin molecules, or monomers ④, which then bind to the target-cell membrane ⑤ and insert into it ⑥. The monomers polymerize like staves of a barrel ⑦ to form pores ⑧ that admit water and salts and kill the cell. (b) Enlargement of perforin-formed pores in a target cell. Note the similarity to the membrane-attack complex formed by complement molecules (see Figure 12–7).

neurons. It is unclear whether antibodies actually enter the neurons and interfere directly with viral replication (neurons have been shown to take up antibodies near their synaptic endings) or bind with the surface of nerve cells and trigger intracellular changes that stop viral replication. The fact that some viruses, such as the herpes virus, persist for years in nerve cells, occasionally "flaring up" to produce symptoms, demonstrates that the antibodies' intraneuronal mechanism does not provide a foolproof antiviral defense for neurons.

Helper T Cells   Helper T cells enhance many aspects of the immune response, primarily by secretion of lymphokines.

The following are among the best known of these T-cell chemical messengers:

1. As noted earlier, helper T cells secrete *B-cell growth factor,* which enhances the antibody-secreting ability of the activated B-cell clone. Antibody secretion is greatly reduced in the absence of helper T cells, even though T cells themselves do not produce antibodies.

2. Helper T cells similarly secrete **T-cell growth factor**, also known as **interleukin 2 (IL-2)**, which augments the activity of cytotoxic T cells, suppressor T cells, and even other helper T cells responsive to the invading antigen. In typical interplay fashion, interleukin l secreted by macro-

TABLE 12–8 Defenses against Viral Invasion

When the virus is free in the extracellular fluid,

Macrophages:

Destroy the free virus by phagocytosis

Process and present the viral antigen to both B and T cells

Secrete interleukin 1, which activates B- and T-cell clones specific to the viral antigen

Plasma cells derived from B cells specific to the viral antigen secrete antibodies that:

Neutralize the virus to prevent its entry into a host cell

Activate the complement cascade that directly destroys the free virus and enhances phagocytosis of the virus by acting as an opsonin

When the virus has entered a host cell (which it must do to survive and multiply, with the replicated viruses leaving the original host cell to enter the extracellular fluid in search of other host cells),

Interferon:

Is secreted by virus-infected cells

Binds with and prevents viral replication in other host cells

Enhances killing power of macrophages, natural killer cells, and cytotoxic T cells

Natural killer cells:

Nonspecifically lyse virus-infected host cells

Cytotoxic T cells:

Are specifically sensitized by the viral antigen; lyse the infected host cells before the virus has a chance to replicate

Helper T cells:

Secrete lymphokines, which enhance cytotoxic T-cell activity and B-cell antibody production

When a virus-infected cell is destroyed, the free virus is released into the extracellular fluid, where it is attacked directly by macrophages, antibodies, and the activated complement components

---

the gathered macrophages. These so-called **angry macrophages** have more powerful destructive ability. They are especially important in defending against the bacteria that cause tuberculosis, because such microbes are able to survive simple phagocytosis by nonactivated macrophages.

Helper T cells are by far the most numerous of the T cells, making up 60–80% of circulating T cells. Because of the important role these cells play in "turning on" the full power of all of the other activated lymphocytes and macrophages, helper T cells may constitute the immune system's "master switch." It is for this reason that **acquired immune deficiency syndrome (AIDS)** caused by the **human immunodeficiency virus (HIV)** is so devastating to the immune defense system. The AIDS virus selectively invades helper T cells, destroying or incapacitating the cells that normally orchestrate much of the immune response (Fig. 12–18). The virus also invades macrophages, further crippling the immune system, and sometimes enters brain cells as well, leading to the dementia (severe impairment of intellectual capacity) noted in some AIDS victims.

Suppressor T Cells Much less is known about suppressor T cells than the other T-cell subpopulations. They apparently serve to limit immune reactions in a "check and balance" relationship with the other lymphocytes. Whereas B cells, cytotoxic T cells, and especially helper T cells reinforce each other's immune activities, suppressor T cells limit the responses of all the other immune cells. In negative-feedback fashion, helper T cells prod suppressor T cells into action; the suppressor T cells, in turn, inhibit the helper T cells and the other cells revved up for duty by helper T-cell influence. Because of this feedback loop, the immune response tends to be self-limiting. Such an inhibitory effect by the suppressor T cells helps prevent excessive immune reactions that might be detrimental to the body. Suppressor T cells generally increase in number more slowly in response to a viral infection than the cytotoxic and helper T cells do, so the suppressor cells help shut down the immune response after it has already served its purpose.

The immune system is normally tolerant of self-antigens.

Suppressor T cells probably also play an important role in preventing the immune system from attacking the person's own tissues, a phenomenon known as **tolerance**. Presumably, during the genetic "cut, shuffle, and paste process" that goes on during lymphocyte development, some B and T cells would by chance be formed that could react against the body's own tissue antigens. If these lymphocyte clones were allowed to function, they would destroy the individual's own body. Fortunately, the immune system normally does not produce antibodies or activated T cells against its own body antigens but instead directs its destructive tactics only at foreign

---

phages not only enhances the activity of both the appropriate B- and T-cell clones but also stimulates the secretion of interleukin 2 by the activated helper T cells.

3. Some chemicals secreted by T cells act as *chemotaxins* to lure more neutrophils and macrophages-to-be to the invaded area.

4. Once macrophages are attracted to the area, **macrophage-migration inhibition factor,** another important lymphokine released from helper T cells, keeps these large phagocytic cells in the region by inhibiting their outward migration. As a result, a great number of chemotactically attracted macrophages accumulate in the infected area. This factor also confers greater phagocytic power on

• **FIGURE 12–18 AIDS Virus** Human immunodeficiency virus (HIV) (in purple), the AIDS-causing virus, on a helper T lymphocyte, HIV's primary target.

antigens. At least four different mechanisms appear to be involved in tolerance:

1. *Clonal deletion.* In response to continuous exposure to body antigens early in development, lymphocyte clones specifically capable of attacking these self-antigens in some cases are permanently destroyed by unknown means. This is the major mechanism by which tolerance is developed.

2. *Clonal anergy.* Evidence for a probable backup to clonal deletion, clonal anergy, has recently been identified. According to this proposal, a T cell must receive two specific simultaneous signals to be activated (turned on), one from its compatible antigen and one from the antigen-presenting cell. Both signals are present for foreign antigens, which are introduced to T cells by antigen-presenting cells such as macrophages. These dual signals are not present for self-antigens, however, because these antigens are not handled by antigen-presenting cells. The first exposure to a single signal from a self-antigen turns *off* the compatible T cell, rendering the cell unresponsive to further exposure to the antigen instead of spurring the cell to proliferate. This reaction is referred to as clonal anergy (*anergy* means "lack of energy") because T cells are being inactivated (that is, "become lazy") rather than activated by their antigens.

3. *Inhibition by suppressor T cells.* Some lymphocyte clones specific for the body's own tissues that are not eliminated during early development are perhaps inhibited throughout life by suppressor T cells.

4. *Antigen sequestering.* Some self-molecules are normally hidden from the immune system because they never come into direct contact with the extracellular fluid in which the immune cells and their products circulate. Examples of such segregated antigens are the proteins of the cornea (the clear anterior portion of the eye) and thyroglobulin, a complex protein sequestered within the hormone-secreting structures of the thyroid gland.

Occasionally, the immune system fails to make the distinction between self-antigens and foreign antigens, unleashing its deadly powers against one or more of the body's own tissues. A condition in which the immune system fails to recognize and tolerate self-antigens associated with particular tissues is known as an **autoimmune disease**, of which myasthenia gravis is an example. People with this condition erroneously produce antibodies against the acetylcholine receptors on their own skeletal-muscle fibers (see p. 210).

Autoimmune diseases may arise from a number of different causes:

1. A reduction in suppressor T-cell activity or an imbalance in the ratio of suppressor to helper T cells specific for self-antigens may be responsible for the development of some autoimmune conditions.

2. Normal self-antigens may be modified by factors such as drugs, environmental chemicals, viruses, or genetic mutations so that they are no longer recognized and tolerated by the immune system.

3. Exposure of normally inaccessible self-antigens sometimes induces an immune attack against these antigens. Since the immune system is usually never exposed to hidden self-antigens, it does not "learn" to tolerate them. Inadvertent exposure of these normally inaccessible antigens to the immune system because of tissue disruption caused by injury or disease can lead to a rapid immune attack against the affected tissue, just as if these self-proteins were foreign invaders. *Hashimoto's disease,* which involves the production of antibodies against thyroglobulin and destruction of the thyroid gland's hormone-secreting capacity, is one such example.

4. Exposure of the immune system to a foreign antigen structurally almost identical to a self-antigen may induce the production of antibodies or activated T lymphocytes that not only interact with the foreign antigen but also cross-react with the closely similar body antigen. An example is the streptococcal bacteria responsible for "strep throat." The bacteria possess antigens that are structurally very similar to self-antigens in the tissue covering the heart valves of some individuals, in which case the antibodies produced against the streptococcal organisms may also bind with this heart tissue. The resultant inflammatory response is responsible for the heart-valve lesions associated with *rheumatic fever*.

The major histocompatibility complex is the code for surface-membrane–bound human leukocyte-associated antigens unique for each individual.

What is the nature of the self-antigens that the immune system learns to recognize as markers of a person's own cells? These self-antigens are plasma-membrane–bound glycoproteins (proteins with sugar attached); they are known as **human leukocyte-associated antigens** or **HLA antigens** because they were first discovered in leukocytes, but actually they are present in all the other cells as well (except erythrocytes, which lack nuclear DNA to direct HLA antigen synthesis). Each person has a group of genes, the **major histocompatibility complex** or **MHC**, which serves as a code for the synthesis of self-antigens. Although more than one hundred different HLA antigens have been identified in human tissue, each individual has a code for only four of these possible antigens. Because of the tremendous number of different combinations possible, the exact pattern of HLA antigens varies from one individual to another, except in identical twins, who have the same MHC-encoded HLA antigens.

T cells typically bind with HLA self-antigens only when they are in association with a foreign antigen, such as a viral protein, also displayed on the cell surface. In the case of cytotoxic T cells, the outcome of this binding is destruction of the infected body cell. Since T cells do not bind to HLA self-antigens in the absence of foreign antigen, normal body cells are protected from lethal immune attack.

T cells do bind with HLA antigens present on the surface of *transplanted cells* in the absence of foreign antigen. The ensuing destruction of the transplanted cells is responsible for rejection of transplanted or grafted tissues. Presumably, some of the T cells "mistake" the foreign HLA antigens of the donor cells because they resemble the combination of a conventional foreign antigen complexed with HLA self-antigens.

To minimize the rejection phenomenon, the tissues of donor and recipient are matched according to HLA antigens as closely as possible. Therapeutic procedures to suppress the immune system then follow. In the past, the primary immunosuppressive tools included radiation therapy and drugs aimed at destroying the actively multiplying lymphocyte populations, plus anti-inflammatory drugs that suppressed the growth of all lymphoid tissue. However, these measures not only suppressed the T cells that were primarily responsible for rejecting transplanted tissue but also depleted the antibody-secreting B cells. Unfortunately, the treated individual was left with little specific immune protection against bacterial and viral infections. In recent years a new therapeutic agent, *cyclosporin*, has become extremely useful in selectively depressing T cell–mediated immune activity while leaving B-cell humoral immunity essentially intact. Specifically, cyclosporin blocks interleukin 2, the lymphokine secreted by helper T cells that is required for expansion of the selected cytotoxic T-cell clone. Furthermore, a new technique under investigation holds promise of completely preventing rejection of transplanted tissues even from an unmatched donor. This technique involves the use of tailor-made antibodies that block two specific facets of the rejection process. If proven to be safe and effective, the technique will have a tremendous impact on tissue transplantation.

The major histocompatibility (*histo* means "tissue") complex was so named because these genes and the self-antigens they encode were first discerned in relation to tissue typing (similar to blood typing), which is done to obtain the most compatible matches for tissue grafting and transplantation. It is important to realize, however, that transfer of tissue from one individual to another does not normally occur in nature. The natural function of HLA antigens lies in their ability to direct the responses of T cells, not in their artificial role in rejecting transplanted tissue.

Each individual has two main classes of MHC-encoded glycoproteins (HLA antigens) that are differentially recognized by cytotoxic T and helper T cells (● Fig. 12–19). Cytotoxic T cells are able to respond to foreign antigen only in association with **class I MHC glycoproteins**, which are found on the surface of virtually all nucleated body cells. **Class II MHC glycoproteins**, which are recognized by helper T cells, are restricted to the surface of a few special types of immune cells, such as B cells, cytotoxic T cells, and macrophages. The class I and II markers serve as signposts to guide cytotoxic and helper T cells to the precise cellular locations where their immune capabilities can be most effective. The binding of suppressor T cells is less well understood.

Because cytotoxic T cells cannot dispose of free viruses, bacteria, or other antigens circulating within the host's body fluids, it would be inefficient for this subpopulation of T cells to recognize and bind with such extracellular antigens. To carry out their role of dealing with pathogens that have invaded host cells, it is appropriate that cytotoxic T cells bind only with cells of the organism's own body that have been infected by viruses; that is, with foreign antigen in association with self-antigen. Since any nucleated body cell can be invaded by viruses, essentially all cells display class I MHC glycoproteins, enabling cytotoxic T cells to attack any invaded host cell. In contrast, helper T cells can bind with foreign antigen only when it is found on the surfaces of immune cells with which it interacts—namely, those bearing class II glycoproteins. These include the macrophages, which "present" antigen to helper T cells, and B and T cells, whose activity is enhanced by helper T cells. The capabilities of helper T cells would be squandered if these cells were able to bind with body cells other than these special immune cells. Thus, spe-

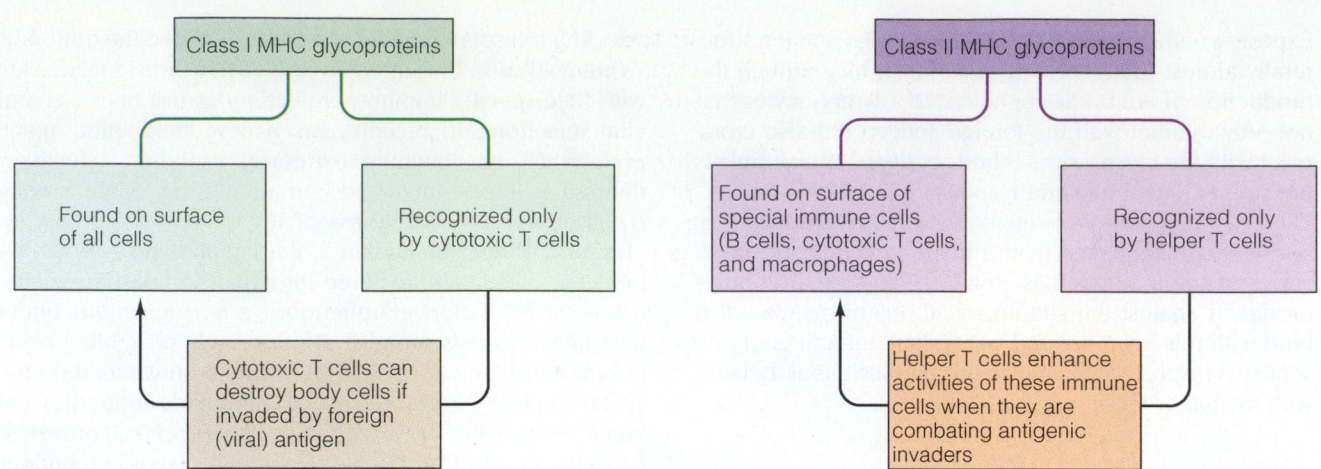

• **FIGURE 12–19   Distinctions between Class I and Class II Major Histocompatibility Complex (MHC) Glycoproteins**
Specific binding requirements for cytotoxic and helper T cells ensure that these cells bind only with the targets that they can influence. Cytotoxic T cells can recognize and bind with foreign antigen only when the antigen is in association with class I MHC glycoproteins, which are found on the surface of all body cells. This requirement is met when a virus invades a body cell, whereupon the cell is destroyed by the cytotoxic T cells. Helper T cells, which enhance the activities of other immune cells, can recognize and bind with foreign antigen only when it is in association with class II MHC glycoproteins, which are found only on the surface of these other immune cells.

cific binding requirements for the various T cells help ensure the appropriate T-cell responses.

Immune surveillance against cancer cells involves an interplay among cytotoxic T cells, natural killer cells, macrophages, and interferon.

Besides destruction of virus-infected host cells, another important function generally attributed to the T-cell system is its role in recognizing and destroying newly arisen, potentially cancerous tumor cells before they have a chance to multiply and spread, a process known as **immune surveillance**. Any normal cell may be transformed into a cancer cell if mutations occur within its genes responsible for controlling cell division and growth. Such mutations may occur by chance alone or, more frequently, by exposure to **carcinogenic** (cancer-causing) factors such as ionizing radiation, certain environmental chemicals, certain viruses, or physical irritants.

Cellular multiplication and growth are normally under strict control, but the regulatory mechanisms are largely unknown. Cell multiplication in an adult is generally restricted to the replacement of lost cells. Furthermore, cells normally respect their own place and space in the body's society of cells. If a cell that has been transformed into a tumor cell manages to escape destruction, however, it defies the normal controls on its proliferation and position. Unrestricted multiplication of a single tumor cell results in a **tumor** that consists of a clone of cells identical to the original mutated cell.

If the mass is slow growing, stays put in its original location, and does not infiltrate into the surrounding tissue, it is considered to be a **benign** tumor. In contrast, the transformed cell may multiply rapidly and form an invasive mass that lacks the altruistic behavior characteristic of normal cells. Such invasive tumors are known as **malignant** tumors, commonly referred to as **cancer**. Malignant tumor cells usually do not ad-

here well to the neighboring normal cells, with the result that often some of the cancer cells break away from the parent tumor. These "emigrant" cancer cells are transported through the blood to new territories, where they continue to proliferate, forming multiple malignant tumors. **Metastasis** is the term applied to this spread of cancer to other parts of the body.

If a malignant tumor is detected early, before it has metastasized, it can be removed surgically. Once cancer cells have dispersed and seeded multiple cancerous sites, surgical elimination of the malignancy is impossible. In this case, agents that interfere with rapidly dividing and growing cells, such as certain chemotherapeutic drugs, are employed in an attempt to destroy the malignant cells. Unfortunately, these agents are also detrimental to normal body cells, especially rapidly proliferating cells such as blood cells and the cells lining the digestive tract.

Untreated cancer is eventually fatal in most cases for several interrelated reasons. The uncontrollably growing malignant mass crowds out normal cells by vigorously competing with them for space and nutrients, yet the cancer cells are unable to take over the functions of the cells they are destroying. Cancer cells typically remain immature and do not become specialized, often resembling embryonic cells instead (• Fig. 12–20). Such dedifferentiated malignant cells lack the ability to perform the specialized functions of the normal cell type from which they mutated. Affected organs gradually become disrupted to the point that they are no longer able to perform their life-sustaining functions, and death results.

Even though many body cells undergo mutations throughout a person's lifetime, most of these mutations do not result in malignancy for three reasons:

1. Only a fraction of the mutations involve loss of control over the cell's growth and multiplication. More frequently, other facets of cellular function are altered.

2. Evidence suggests that a cell usually becomes cancerous only after an accumulation of multiple independent mu-

tations. A single mutation generally is not sufficient. This requirement could contribute at least in part to the much higher incidence of cancer in older individuals, in whom mutations have had more time to accumulate in a single cell lineage. Alternatively, a few cancers have been shown to be caused by tumor viruses, which permanently alter particular DNA sequences of the cells they invade.

3. Potentially cancerous cells that do arise are usually destroyed by the immune system early in their development. Presumably, the immune system recognizes cancer cells because they bear new and different surface antigens alongside the cell's normal self-antigens, either because of genetic mutation or invasion by a tumor virus. Immune surveillance against cancer depends on an interplay among three types of immune cells—cytotoxic T cells, natural killer cells, and macrophages—as well as interferon. Not only are these immune cells able to directly attack and destroy cancer cells, but all three of them also secrete interferon. Interferon, in turn, inhibits multiplication of cancer cells and increases the killing ability of the immune cells (•Fig. 12–21).

Because natural killer cells, unlike the cytotoxic T cells, do not require prior exposure and sensitization to a cancer cell before being able to launch a lethal attack, they are probably the single most important weapon the body has in its war against cancer. Cytotoxic T cells are believed to be especially important in defending against the few kinds of virus-induced cancer. Upon contacting a cancer cell, both of these killer cells release perforin and other toxic chemicals that destroy the targeted mutant cell (•Fig. 12–22). Macrophages, in addition to clearing

away the remains of the dead victim cell, are able to engulf and destroy cancer cells intracellularly.

The fact that cancer does sometimes occur means that cancer cells must occasionally be able to escape these immune mechanisms. Why or how immune surveillance fails to nip newly formed cancer cells in the bud remains unclear. Other facets of the immune system may even be involved. For example, **blocking antibodies** that interfere with T-cell function have been identified. Although B cells and antibodies are not believed to play a direct role in cancer defense, B cells, upon viewing a mutant cancer cell as an alien to normal self, may produce antibodies against it. These antibodies, for unknown reasons, do not activate the complement system, which could destroy the cancer cells. Instead, the antibodies are able to bind with the antigenic sites on the cancer cell, "hiding" these sites from recognition by cytotoxic T cells. The coating of a tumor cell by blocking antibodies thus protects the harmful cell from attack by deadly T cells. Other mechanisms undoubtedly are involved in cancer development to account for the fact that natural killer cells, macrophages, and interferon, which are not dependent on binding with specific antigenic sites, also occasionally fail to stop the relentless growth and proliferation of mutant cancer cells.

A regulatory loop appears to link the immune system and the nervous and endocrine systems.

From the preceding discussion, it is obvious that complex controlling factors operate within the immune system itself.

• **FIGURE 12–21   Immune Surveillance against Cancer** Anticancer interactions of cytotoxic T cells, natural killer cells, macrophages, and interferon.

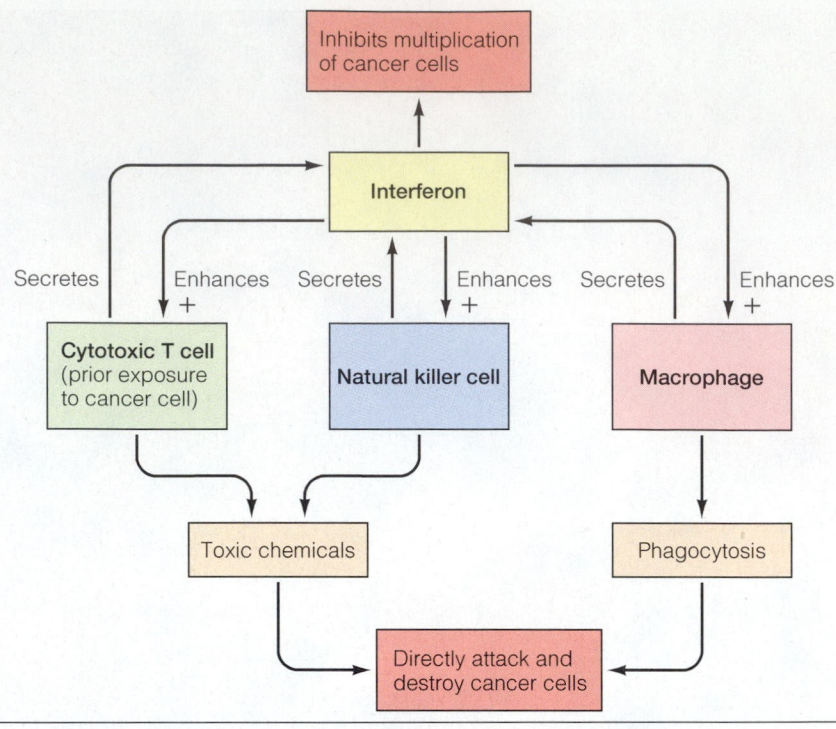

• **FIGURE 12–22   Cytotoxic T Cell Destroying Cancer Cell** Upon contacting a cancer cell with which it can specifically bind, a cytotoxic T cell releases toxic chemicals such as perforin, which destroy the cancer cell.

Until recently, the immune system was believed to function independently of other control systems in the body. Investigations along a number of lines indicate, however, that there are important links between the immune system and the body's two major control systems, the nervous and endocrine systems. Apparently, the immune system both influences and is influenced by the nervous and endocrine systems. For example, interleukin 1 can turn on the stress response by activating a sequence of nervous and endocrine events that result in the secretion of cortisol, one of the major hormones released during stress. This linkage between a mediator of the immune response and a mediator of the stress response is appropriate. Cortisol mobilizes the body's nutrient stores so that metabolic fuel is readily available to keep pace with the body's energy demands at a time when the person is sick and may not be eating enough (or, in the case of an animal, may not be able to search for food). Furthermore, cortisol mobilizes amino acids, which serve as building blocks to repair any tissue damage sustained during the encounter that triggered the immune response. In the reverse direction, lymphocytes are responsive to blood-borne signals from the nervous system and from certain endocrine glands. These important immune cells possess receptors for a wide variety of neurotransmitters, hormones, and other chemical mediators. Among the mediators that influence lymphocyte activity are endorphins and enkephalins, a group of related compounds that serve as neurotransmitters in the CNS and are secreted into the blood as part of the neuroendocrine pathway involved in the control of cortisol release. (For a discussion of the possible effects of exercise on immune defense, see the accompanying boxed feature, A Closer Look at Exercise Physiology.)

EXERCISE: A HELP OR HINDRANCE TO IMMUNE DEFENSE?

For years people who engage in moderate exercise regimens have claimed they have fewer colds when they are in good aerobic condition. In contrast, elite athletes and their coaches have often complained about the number of upper respiratory infections that the athletes seem to contract at the height of their competitive seasons. The results of recent scientific studies lend support to both of these claims. The impact of exercise on immune defense depends on the intensity of the exercise.

Animal studies have shown that high-intensity exercise after experimentally induced infection results in more severe infection. Moderate exercise performed prior to infection or to tumor implantation, on the other hand, results in less severe infection and slower tumor growth in experimental animals.

Recent studies on humans lend further support to the hypothesis that exhaustive exercise suppresses immune defense whereas moderate exercise stimulates the immune system. A survey of 2,300 runners competing in the 1987 Los Angeles Marathon indicated that those who trained more than sixty miles a week had twice the number of respiratory infections of those who trained less than twenty miles a week in the two months preceding the race. In another study, ten elite athletes were asked to run on a treadmill for three hours at the same pace they would run in competition. Blood tests after the run indicated that natural-killer-cell activity had decreased by 25% to 50%, and this decrease lasted for six hours. The runners also showed a 60% increase in the stress hormone cortisol, which is known to suppress immunity. Other studies have shown that athletes have lower resting salivary IgA levels compared with control subjects and that their respiratory mucosal immunoglobulins are decreased after prolonged exhaustive exercise. These results suggest a lower resistance to respiratory infection following high-intensity exercise. Because of these results, researchers in the field recommend that athletes keep exposure to respiratory viruses to a minimum by avoiding crowded places or anyone with a cold or flu for the first six hours after strenuous competition.

On the other hand, a study evaluating the effects of a moderate exercise program in which a group of women walked forty-five minutes a day, five days a week, for fifteen weeks found that the walkers' antibody levels and natural-killer-cell activity increased throughout the exercise program. Other studies using moderate exercise on stationary bicycles in subjects over the age of sixty-five showed increases in natural-killer-cell activity as large as those found in young people.

Unfortunately, the few studies conducted on those infected with human immunodeficiency virus (HIV, the AIDS virus) have not found an improvement in immune function with exercise. The studies have shown that HIV-positive patients can gain strength through resistance training and improve psychological well-being through exercise and that they suffer no detrimental effects from moderate exercise.

≡ 🔴 IMMUNE DISEASES

Immune deficiency diseases reduce resistance to foreign invaders.

Abnormal functioning of the immune system can lead to immune diseases in two general ways: deficiency diseases and inappropriate immune attacks. Deficiency diseases occur when the immune system fails to respond adequately to foreign invasion. The condition may be congenital (present at birth) or acquired (nonhereditary), and it may specifically involve impairment of either antibody-mediated immunity, cell-mediated immunity, or both. In a rare hereditary condition known as **severe combined immunodeficiency,** both B and T cells are lacking. Its victims have extremely limited defenses against pathogenic organisms and usually die in infancy unless maintained in a germ-free environment (that is, live in a "bubble"). Acquired (non-hereditary) immune deficiency states can arise from inadvertent destruction of lymphoid tissue during prolonged therapy with anti-inflammatory agents, such as cortisol derivatives, or from cancer therapy aimed at destroying rapidly dividing cells (which unfortunately includes lymphocytes as well as cancer cells). The most recent and tragically the most common acquired immune deficiency disease is AIDS, which, as described earlier, is caused by HIV, a virus that invades and incapacitates the critical helper T cells.

Inappropriate immune attacks against harmless environmental substances are responsible for allergies.

The other category of immune diseases involves inappropriate specific immune attacks that cause reactions harmful to the body. These include (1) *autoimmune responses,* in which the immune system turns against one of the body's own tissues; (2) *immune-complex diseases,* which involve overexuberant antibody responses that "spill over" and damage normal tissue; and (3) *allergies.* The first two conditions have been described earlier in this chapter, so we will now concentrate on allergies.

TABLE 12–9 Immediate versus Delayed Hypersensitivity Reactions

CHARACTERISTIC	IMMEDIATE HYPERSENSITIVITY REACTION	DELAYED HYPERSENSITIVITY REACTION
Time of onset of symptoms after exposure to allergen	Within 20 min	Within 1 to 3 days
Type of immune response involved	Antibody-mediated immunity against allergen	Cell-mediated immunity against allergen
Immune effectors involved	B cells, IgE antibodies, mast cells, basophils, histamine, slow-reactive substance of anaphylaxis, eosinophil chemotactic factor	T cells
Allergies commonly involved	Hay fever, asthma, hives, anaphylactic shock in extreme cases	Contact allergies such as allergies to poison ivy, cosmetics, and household cleaning agents
Treatment	Antihistamines (partially effective); adrenergic drugs to counteract effects of histamine and slow-reactive substance of anaphylaxis; anti-inflammatory agents such as cortisol derivatives	Anti-inflammatory agents such as cortisol derivatives

An **allergy** is the acquisition of an inappropriate specific immune reactivity, or **hypersensitivity**, to a normally harmless environmental substance, such as dust or pollens. The offending agent, known as an **allergen**, may itself be an antigen, or it may be a hapten that becomes antigenic only after it combines with a body protein. Subsequent reexposure of a sensitized individual to the same allergen elicits an immune attack, which may vary from a mild, annoying reaction to a severe, body-damaging reaction that may even be fatal.

Allergic responses can be classified into two different categories: immediate hypersensitivity and delayed hypersensitivity (Table 12–9). In **immediate hypersensitivity,** the allergic response appears within about twenty minutes after a sensitized individual is exposed to an allergen, whereas in **delayed hypersensitivity**, the reaction is not generally manifested until a day or so following exposure. The difference in timing is due to the different mediators involved. A particular allergen may activate either B-cell or T-cell responses. Immediate allergic reactions involve B cells and are elicited by antibody interactions with an allergen; delayed reactions involve T cells and the more slowly responding process of cell-mediated immunity against the allergen. Let us examine the causes and consequences of each of these reactions in more detail.

Immediate Hypersensitivity In immediate hypersensitivity, the antibodies involved and the events that ensue upon exposure to an allergen differ from the typical antibody-mediated response to bacteria. The most common allergens that provoke immediate hypersensitivities are pollen grains, bee stings, penicillin, certain foods, molds, dust, feathers, and animal fur. (Actually, people allergic to cats are not allergic to the fur itself. The true allergen is in the cat's saliva, which is deposited on the fur during licking. Likewise, people are not allergic to dust or feathers per se, but to tiny mites that inhabit the dust or feathers and eat the scales constantly being shed from the skin.) For unclear reasons, these allergens bind to and elicit the synthesis of IgE antibodies rather than the IgG antibodies associated with bacterial antigens. When an individual with an allergic tendency is first exposed to a particular allergen, compatible B cells synthesize IgE antibodies specific for it. More importantly, memory cells are also formed that are primed for a more powerful response on subsequent reexposure to the same allergen.

In contrast to the antibody-mediated response elicited by bacterial antigens, IgE antibodies do not freely circulate. Instead, their tail portions attach to mast cells and basophils. Binding of an appropriate allergen with the attached IgE antibodies triggers the release of several chemical mediators from the involved mast cells and basophils. A single mast cell (or basophil) may be coated with a number of different IgE antibodies, each able to bind with a different allergen. Thus, the mast cell can be triggered to release its chemical products by any one of a number of different allergens (• Fig. 12–23). These released chemicals are responsible for the reactions that characterize immediate hypersensitivity. The following are among the most important chemicals released during immediate allergic reactions:

1. **Histamine**, which brings about vasodilation and increased capillary permeability.
2. **Slow-reactive substance of anaphylaxis** (**SRS-A**), which induces prolonged and profound contraction of smooth muscle, especially of the small respiratory airways.
3. **Eosinophil chemotactic factor,** which specifically attracts eosinophils to the area. Interestingly, eosinophils release enzymes that inactivate SRS-A and may also inhibit

● **FIGURE 12–23   Role of IgE Antibodies and Mast Cells in Immediate Hypersensitivity** B-cell clones are converted into plasma cells, which secrete IgE antibodies on contact with the allergen for which they are specific. The Fc tail portion of all IgE antibodies, regardless of the specificity of their Fab arm regions, binds to receptor proteins specific for IgE tails on mast cells and basophils. Unlike B cells, each mast cell bears a variety of antibody surface receptors for binding different allergens. When an allergen combines with an IgE receptor specific for it on the surface of a mast cell, the mast cell releases histamine and other chemicals by exocytosis. These chemicals elicit the allergic response.

histamine, perhaps serving as an "off switch" to limit the allergic response.

Symptoms vary depending on the site, allergen, and mediators involved. Most frequently, the reaction is localized to the body site in which the IgE-bearing cells first come into contact with the allergen. If the reaction is limited to the upper respiratory passages after a person inhales an allergen such as ragweed pollen, the released chemicals bring about the symptoms characteristic of **hay fever**—for example, nasal congestion caused by histamine-induced localized edema and sneezing and runny nose caused by increased mucus secretion in response to local irritation. If the reaction is concentrated primarily within the bronchioles (the small respiratory airways that lead to the tiny air sacs within the lungs), **asthma** results. Contraction of the smooth muscle in the walls of the bronchioles narrows or constricts these passageways,

making breathing difficult. Localized swelling in the skin because of allergy-induced histamine release causes **hives**.

Treatment of localized immediate allergic reactions with antihistamines often offers only partial relief of the symptoms, because some of the manifestations are invoked by other chemical mediators not blocked by these drugs. For example, antihistamines are not particularly effective in treating asthma, the most serious symptoms of which are invoked by SRS-A. Adrenergic drugs (which mimic the sympathetic nervous system) are helpful through their vasoconstrictor-bronchodilator actions in countering the effects of both histamine and SRS-A. Anti-inflammatory drugs such as cortisol derivatives may be necessary to block the inflammatory response.

A life-threatening systemic reaction can occur if the allergen becomes blood-borne or if very large amounts of chemicals are released from the localized site into the circulation.

When large amounts of these chemical mediators gain access to the blood, the extremely serious systemic (involving the entire body) reaction known as **anaphylactic shock** occurs. Widespread vasodilation and a massive shift of plasma fluid into the interstitial spaces as a result of a generalized increase in capillary permeability cause severe hypotension (see p. 341) that can lead to circulatory failure. Concurrently, pronounced bronchiolar constriction can lead to respiratory failure. The victim may suffocate because of an inability to move air through the narrowed airways. Unless countermeasures, such as injection of a vasoconstrictor-bronchodilator drug, are undertaken immediately, anaphylactic shock is frequently fatal. This reaction is why even a single bee sting or a single dose of penicillin can be so dangerous in individuals sensitized to these allergens.

Delayed Hypersensitivity Some allergens invoke delayed hypersensitivity, a T cell–mediated immune response, rather than an immediate, B cell–IgE antibody response. Among these allergens are poison ivy toxin and certain chemicals to which the skin is frequently exposed, such as cosmetics and household cleaning agents. Most commonly, the response is characterized by a delayed skin eruption that reaches its peak intensity one to three days following contact with an allergen to which the T system has previously been sensitized. To illustrate, poison ivy toxin is a hapten that may bind with skin proteins with which it comes into contact. The toxin itself does not harm the skin upon contact, but it activates T cells specific for the toxin, including formation of a memory component. Upon subsequent exposure to the toxin, activated T cells diffuse into the skin within a day or two, combining with the poison ivy toxin that is present. The resultant interaction gives rise to the tissue damage and discomfort typically associated with the condition. The best relief is obtained from application of anti-inflammatory preparations, such as those containing cortisol derivatives.

More long-term relief from both immediate and delayed hypersensitivities can sometimes be obtained by **desensitization injections (allergy shots)**. This therapeutic regimen involves regular injections of small but increasing quantities of the offending allergen. Through this process, the individual (with varying degrees of success) gradually becomes less and less sensitive to natural exposure to the allergen. It seems ironic that deliberately exposing a person to a known allergen can make that person less sensitive to the allergen. The mechanism by which desensitization is accomplished is currently uncertain, but the leading theory proposes that it occurs because of the production of blocking IgG antibodies specific for the allergen injected (• Fig. 12–24). When the individual is naturally exposed to the allergen again, complementary blocking IgG antibodies bind with the allergen so that it is not available to bind with the IgE antibodies attached to mast cells and basophils. Thus, these histamine-laden cells are not stimulated to release their symptom-causing chemicals. An alternative proposal suggests that desensitization is brought about by activation of suppressor T cells, which thwart the synthesis of IgE antibodies specifically directed against the allergen.

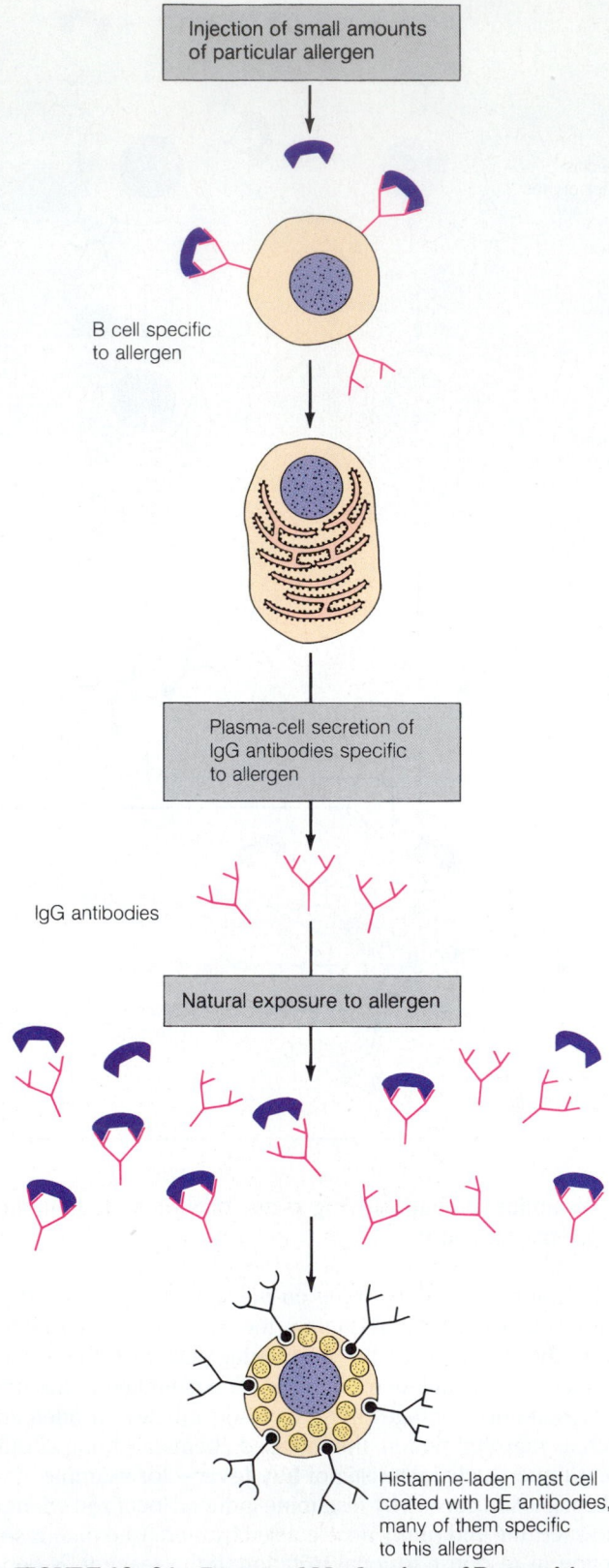

Injection of small amounts of particular allergen

B cell specific to allergen

Plasma-cell secretion of IgG antibodies specific to allergen

IgG antibodies

Natural exposure to allergen

Histamine-laden mast cell coated with IgE antibodies, many of them specific to this allergen

• **FIGURE 12–24 Proposed Mechanism of Desensitization** Researchers speculate that during desensitization, B cells specific to a particular allergen secrete IgG antibodies in response to injection of small amounts of that allergen. On subsequent natural exposure to the allergen, the IgG antibodies bind with newly entering allergens, thereby blocking the allergens' attachment to mast-cell IgE receptors and reducing symptoms. Without allergen attached, mast cells do not release histamine and other symptom-causing chemicals.

EXTERNAL DEFENSES

The body's defenses against foreign microbes are not limited to the intricate, interrelated immune mechanisms that destroy the microorganisms that have actually invaded the body. In addition to the internal immune defense system, the body is equipped with external defense mechanisms designed to prevent microbial penetration wherever body tissues are exposed to the external environment. The most obvious external defense is the **skin**, or **integument**, which covers the outside of the body.

The skin consists of an outer protective epidermis and an inner connective-tissue dermis.

The skin, which is the largest organ of the body, not only serves as a mechanical barrier between the external environ-ment and the underlying tissues but is dynamically involved in defense mechanisms and other important functions as well. The skin consists of two layers, an outer *epidermis* and an inner *dermis* (● Fig. 12–25).

The **epidermis** consists of numerous layers of epithelial cells. The inner epidermal layers are composed of cube-shaped cells that are living and rapidly dividing, whereas the cells in the outer layers are dead and flattened. The epidermis has no direct blood supply. Its cells are nourished only by diffusion of nutrients from a rich vascular network in the un-derlying dermis. The newly forming cells in the inner layers constantly push the older cells closer to the surface, farther and farther from their nutrient supply. This, coupled with the fact that the outer layers are continuously subjected to pres-sure and "wear and tear," causes these older cells to die and become flattened. Epidermal cells are tightly bound together by spot desmosomes (see p. 55), which interconnect with in-

● **FIGURE 12–25**   **Anatomy of the Skin**   The skin consists of two layers, a keratinized outer epidermis and a richly vascularized inner connective-tissue dermis. Special infoldings of the epidermis form the sweat glands, sebaceous glands, and hair follicles. The epidermis contains four types of cells: keratinocytes, melanocytes, Langerhans cells, and Granstein cells. The skin is anchored to underlying muscle or bone by the hypodermis, a loose, fat-containing layer of connective tissue.

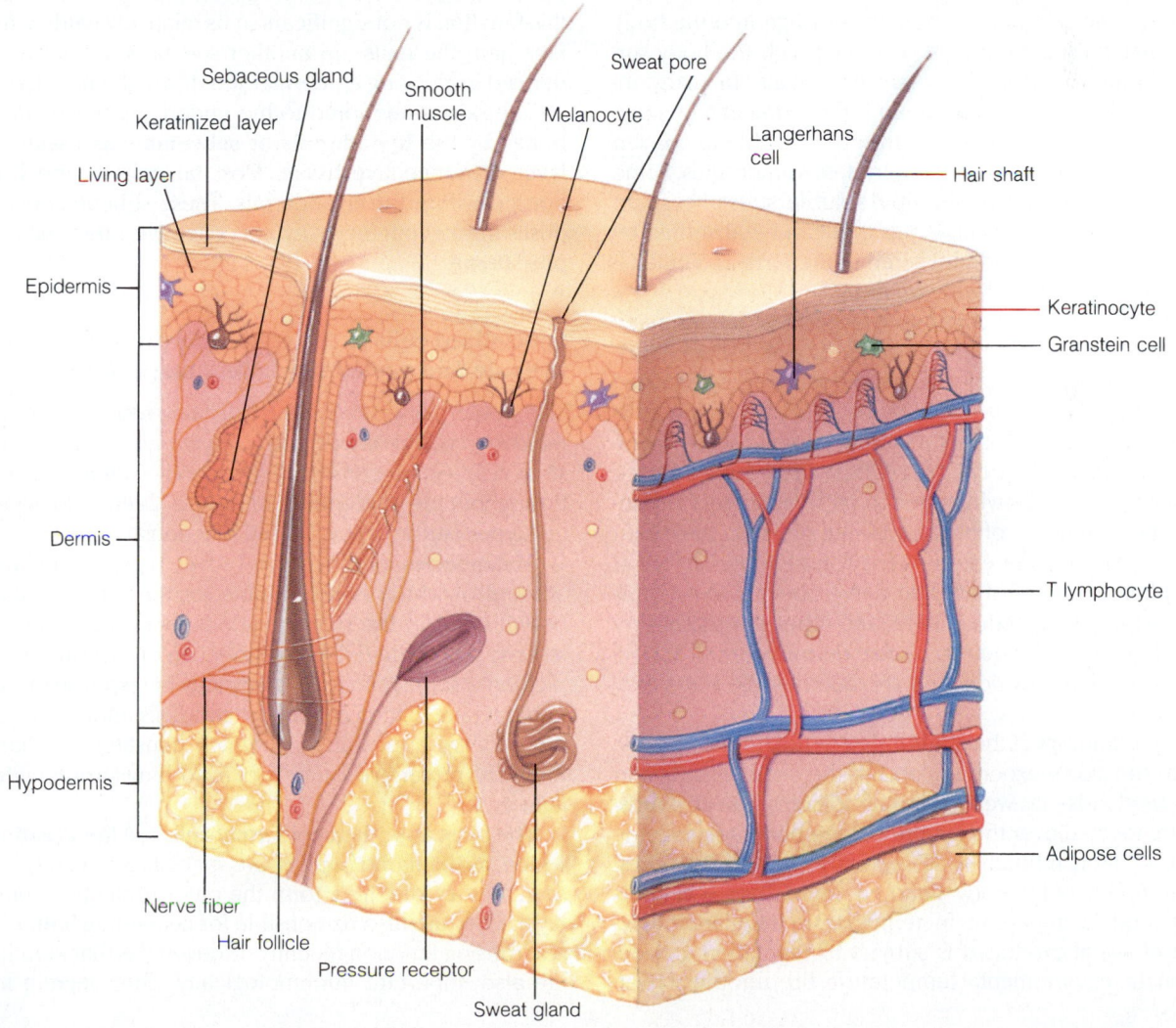

tracellular keratin filaments (see p. 41) to form a strong, cohesive covering. During maturation of a keratin-producing cell, keratin filaments progressively accumulate and cross-link with each other within the cytoplasm. As the outer cells die, only this fibrous keratin core remains, forming flattened, hardened scales that provide a tough, protective **keratinized layer**. As the scales of the outermost keratinized layer slough or flake off through abrasion, they are continuously replaced by means of cell division in the deeper epidermal layers. The rate of cell division, and consequently the thickness of the keratinized layer, varies in different regions of the body. It is thickest in the areas where the skin is subjected to the most pressure, such as the bottom of the feet.

The keratinized layer is airtight, fairly waterproof, and impervious to most substances. It serves to resist passage in both directions between the body and the external environment. For example, it minimizes loss of water and other vital constituents from the body. This protective layer's value in holding in body fluids becomes obvious in severe burns. Not only can bacterial infections occur in the unprotected underlying tissue, but even more serious are the systemic consequences of loss of body water and plasma proteins, which escape from the exposed, burned surface. The resultant circulatory disturbances can be life-threatening.

Likewise, the skin barrier impedes passage into the body of most materials that come into contact with the body surface, including bacteria and toxic chemicals. In many instances, the skin modifies compounds that come into contact with it. For example, epidermal enzymes are able to convert many potential carcinogens into harmless compounds. Some materials, however, especially lipid-soluble substances, are able to penetrate intact skin. Drugs that can be absorbed by the skin are sometimes administered in the form of a cutaneous "patch" impregnated with the drug.

The **dermis** is a connective-tissue layer that contains many elastin fibers (for stretch) and collagen fibers (for strength), as well as an abundance of blood vessels and specialized nerve endings. The dermal blood vessels not only supply both the dermis and epidermis but also play a major role in temperature regulation. The caliber of these vessels, and hence the volume of blood flowing through them, is subject to control to vary the amount of heat exchange between these skin surface vessels and the external environment. Receptors at the peripheral endings of afferent nerve fibers in the dermis detect pressure, temperature, pain, and other somatosensory input. Efferent nerve endings in the dermis control blood-vessel caliber, hair erection, and secretion by the skin's exocrine glands.

Special infoldings of the epidermis into the underlying dermis form the skin's exocrine glands—the sweat glands and sebaceous glands—as well as the hair follicles. **Sweat glands**, which are located over the majority of the body, release a dilute salt solution through small openings, the sweat pores, onto the surface of the body. Evaporation of this sweat cools the skin and is important in temperature regulation. The amount of sweat produced is subject to regulation and depends on the environmental temperature, the amount of heat-

generating muscular activity, and various emotional factors (for example, a person often sweats when nervous). A special type of sweat gland located in the axilla (armpit) and pubic region produces a protein-rich sweat that supports the growth of surface bacteria, which give rise to a characteristic odor. In contrast, most sweat as well as the secretions from the sebaceous glands contain chemicals that are generally highly toxic to bacteria.

The cells of the **sebaceous glands** produce an oily secretion known as **sebum** that is released into adjacent hair follicles. From there the oily sebum flows to the surface of the skin, oiling both the hairs and the outer keratinized layers of the skin to help waterproof them and prevent them from drying and cracking. Insufficient protection by sebum is evidenced by chapped hands or lips. The sebaceous glands are particularly active during adolescence, causing the oily skin prominent in teenagers.

Each **hair follicle** is lined by special keratin-producing cells, which secrete keratin and other proteins that form the hair shaft. Hairs increase the sensitivity of the skin's surface to tactile (touch) stimuli. In some lower species, this function is more exquisitely fine-tuned. For example, the whiskers on a cat are extremely sensitive in this regard. An even more important role of hair in lower species is heat conservation, but this function is not significant in us relatively hairless humans. Like hair, the nails are another special keratinized product derived from living epidermal structures, the nail beds.

The skin is anchored to the underlying tissue (muscle or bone) by the **hypodermis** or **subcutaneous tissue**, a loose layer of connective tissue. Most fat cells in the body are housed within the hypodermis. These subcutaneous fat deposits throughout the body are collectively referred to as **adipose tissue**.

Specialized cells in the epidermis produce keratin and melanin and participate in immune defense.

The epidermis contains four distinct resident cell types— *melanocytes, keratinocytes, Langerhans cells,* and *Granstein cells*—plus transient T lymphocytes that are scattered throughout the epidermis and dermis. Each of these resident cell types performs specialized functions.

Melanocytes produce the brown pigment **melanin**, the amount of which is responsible for the different shades of brown color in the skin of various races. In addition to hereditary determination of melanin content, the amount of this pigment can be increased transiently in response to exposure to ultraviolet light rays from the sun. This additional melanin, the outward appearance of which constitutes a "tan," performs the protective function of absorbing harmful ultraviolet light rays.

The most abundant epidermal cells are the **keratinocytes**, which, as the name implies, are specialists in keratin production. As they die, they form the outer protective keratinized layer. They are also responsible for generating hair and nails. A surprising function recently discovered is that keratinocytes are also important immunologically. They secrete interleukin 1 (a product also secreted by macrophages), which influences the maturation of T cells that tend to localize in the skin. Interestingly, the epithelial cells of the thymus have been shown to bear anatomical, molecular, and functional similarities to those of the skin. Apparently, some post-thymic steps in T-cell maturation take place in the skin under keratinocyte guidance.

The two other epidermal cell types also play a role in immunity. Both **Langerhans cells**, which migrate to the skin from the bone marrow, and **Granstein cells**, the most recently discovered and least understood type of epidermal cells, serve as antigen-presenting cells. Langerhans cells present antigen to helper T cells, thereby facilitating their responsiveness to skin-associated antigens. In contrast, it appears that Granstein cells interact with suppressor T cells, probably serving as a ''brake'' on skin-activated immune responses. It is significant that Langerhans cells are more susceptible to damage by ultraviolet radiation (as from the sun) than are Granstein cells. Loss of Langerhans cells as a result of exposure to ultraviolet radiation can detrimentally lead to a predominate suppressor signal rather than the normally dominant helper signal, leaving the skin more vulnerable to microbial invasion and cancer cells.

The various epidermal components of the immune system are collectively termed **skin-associated lymphoid tissue** or **SALT** (• Fig. 12–26). Recent research suggests that the skin probably plays an even more elaborate role in specific immune defense than described here. This is appropriate, because the skin serves as the major interface with the external environment.

In addition, the epidermis synthesizes vitamin D in the presence of sunlight. The cell type that produces vitamin D is undetermined. Vitamin D, which is derived from a precursor molecule closely related to cholesterol, promotes the absorption of Ca^{++} from the digestive tract into the blood. Dietary supplements of vitamin D are usually required because typically the skin is not exposed to sufficient sunlight to produce adequate amounts of this essential chemical.

Protective measures within body cavities that communicate with the external environment discourage pathogen invasion into the body.

The human body's defense system must guard against entry of potential pathogens not only through the outer surface of the body but also through the internal cavities that communicate directly with the external environment—namely, the digestive system, the genitourinary system, and the respiratory system. These systems employ various strategies to destroy microorganisms entering through these routes.

Saliva secreted into the mouth at the entrance of the digestive system contains an enzyme that lyses certain bacteria. Many of the surviving bacteria that are swallowed are killed by the strongly acidic gastric juice that they encounter in the stomach. Farther down the tract, the intestinal lining is endowed with gut-associated lymphoid tissue. These defensive mechanisms are not 100% effective, however. Some bacteria do manage to survive and reach the large intestine (the last portion of the digestive tract), where they continue to flourish. Surprisingly, this normal microbial population provides a natural barrier against infection within the lower intestine. These harmless resident flora competitively suppress the growth of potential pathogens that have managed to escape the antimicrobial measures of earlier parts of the digestive tract. Occasionally, orally administered antibiotic therapy against one infection within the body may actually induce another infection in the intestinal tract. By knocking out some of the normal intestinal flora, an antibiotic may permit an antibiotic-resistant pathogenic species to overgrow.

Within the genitourinary (reproductive and urinary) sys-

• **FIGURE 12–26 Skin-Associated Lymphoid Tissue (SALT)**

tem, would-be invaders encounter hostile conditions in the acidic urine and acidic vaginal secretions. The genitourinary organs also produce a sticky mucus, which, like flypaper, entraps small invading particles. Subsequently, the particles are either engulfed by phagocytes or are swept out as the organ empties (for example, they are flushed out with urine flow).

The respiratory system is likewise equipped with several important defense mechanisms against inhaled particulate matter. The respiratory system is the largest surface of the body that comes into direct contact with the increasingly polluted external environment. The surface area of the respiratory system that is exposed to the air is thirty times that of the skin. Larger airborne particles are filtered out of the inspired air by hairs at the entrance of the nasal passages. Lymphoid organs, the *tonsils* and *adenoids,* provide immunological protection against inspired pathogens near the beginning of the respiratory system. Farther down in the respiratory airways, millions of tiny hairlike projections known as cilia (see p. 37) constantly beat in an outward direction. The respiratory airways are coated with a layer of thick, sticky mucus secreted by epithelial cells within the airway lining. This mucus sheet, laden with any inspired particulate debris (such as dust) that adheres to it, is constantly moved upward to the throat by ciliary action. This moving "staircase" of mucus is known as the **mucus escalator**. The dirty mucus is either expectorated (spit out) or in most cases swallowed without the person even being aware of it; any undigestible foreign particulate matter is subsequently eliminated in the feces. Besides keeping the lungs clean, this mechanism is an important defense against bacterial infection, because many bacteria enter the body on dust particles. Also contributing to defense against respiratory infections are antibodies secreted in the mucus. In addition, an abundance of phagocytic specialists called the **alveolar macrophages** scavenge within the air sacs (alveoli) of the lungs. Further respiratory defenses include coughs and sneezes. These commonly experienced reflex mechanisms involve forceful outward expulsion of material in an attempt to remove irritants from the trachea (*coughs*) or nose (*sneezes*).

Cigarette smoking suppresses these normal respiratory defenses. The smoke from a single cigarette can paralyze the cilia for several hours, with repeated exposure eventually leading to ciliary destruction. Failure of ciliary activity to sweep out a constant stream of particulate-laden mucus enables inspired carcinogens to remain in contact with the respiratory airways for prolonged periods. Furthermore, cigarette smoke incapacitates alveolar macrophages. Particulates in cigarette smoke not only overwhelm the macrophages, but certain components of cigarette smoke also have a direct toxic effect on the macrophages, reducing their ability to engulf foreign material. In addition, noxious agents in tobacco smoke irritate the mucous linings of the respiratory tract, resulting in excess mucus production, which may partially obstruct the airways. "Smoker's cough" is an attempt to dislodge this excess stationary mucus. These and other direct toxic effects on lung tissue lead to the increased incidence of lung cancer and chronic respiratory diseases associated with cigarette smoking. Air pollutants include some of the same substances found in cigarette smoke and can similarly affect the respiratory system. We will examine the respiratory system in greater detail in the next chapter.

 ## CHAPTER IN PERSPECTIVE: FOCUS ON HOMEOSTASIS

We could not survive beyond early infancy were it not for the body's defense mechanisms. These mechanisms resist and eliminate potentially harmful foreign agents with which we continuously come into contact in our hostile external environment and also destroy abnormal cells that often arise within the body. Homeostasis can be optimally maintained, and thus life sustained, only if the cells are not physically injured or functionally disrupted by pathogenic microorganisms or are not replaced by abnormally functioning cells, such as traumatized cells or cancer cells. The immune defense system, a complex, multifaceted, interactive network of leukocytes, their secretory products, and plasma proteins, contributes indirectly to homeostasis by keeping other cells alive so that they can perform their specialized activities to maintain a stable internal environment. The immune system protects the other healthy cells from foreign agents that have gained entrance to the body, eliminates newly arisen cancer cells, and clears away dead and injured cells to pave the way for replacement with healthy new cells.

The skin contributes indirectly to homeostasis by serving as a protective barrier between the external environment and the remainder of the body cells. It helps prevent harmful foreign agents such as pathogens and toxic chemicals from entering the body and helps prevent the loss of precious internal fluids from the body. The skin also contributes directly to homeostasis by helping maintain body temperature by means of the sweat glands and adjustments in skin blood flow. The amount of heat carried to the body surface for dissipation to the external environment is determined by the volume of warmed blood flowing through the skin.

Other systems that have internal cavities in contact with the external environment, such as the digestive, genitourinary, and respiratory systems, also have defense capabilities to prevent harmful external agents from entering the body through these avenues.

 CHAPTER SUMMARY

Introduction

Foreign invaders and newly arisen mutant cells are immediately confronted with multiple interrelated defense mechanisms aimed at destroying and eliminating anything that is not part of the normal self. These mechanisms, collectively referred to as immunity, include both nonspecific and specific immune responses. Nonspecific immune responses nonselectively defend against foreign material even upon initial exposure to it. Specific immune reponses are selectively targeted against particular invaders for which the body has been specially prepared after a prior exposure. Leukocytes and their derivatives are the major effector cells of the immune system and are reinforced by a number of different plasma proteins. Leukocytes are produced in the bone marrow, then circulate transiently in the blood. They spend most of their time, however, on defense missions in the tissues. Some lymphocytes are also produced, differentiated, and perform their defense activities within lymphoid tissues strategically located at likely points of foreign infiltration.

The most common invaders are bacteria and viruses. Bacteria are self-sustaining, single-celled organisms, which produce disease by virtue of the destructive chemicals they release. Viruses are protein-coated nucleic-acid particles, which invade host cells and take over the cellular metabolic machinery for their own survival to the detriment of the host cell. In addition to defending against these microbes and mutant cells, the immune cells also clean up cellular debris, preparing the way for tissue repair.

Nonspecific Immune Responses

Nonspecific immune responses, which form a first line of defense against atypical cells (foreign, mutant, or injured cells) even upon first exposure to them, include inflammation, interferon, natural killer cells, and the complement system.

Inflammation is a nonspecific response to foreign invasion or tissue damage mediated largely by the professional phagocytes (neutrophils and monocytes-turned-macrophages) and their secretions. The phagocytic cells destroy foreign and damaged cells both by phagocytosis and by the release of lethal chemicals. Histamine-induced vasodilation and increased permeability of local vessels at the site of invasion or injury permit enhanced delivery of more phagocytic leukocytes and inactive plasma-protein precursors crucial to the inflammatory process, such as clotting factors and components of the complement system. These vascular changes are also largely responsible for the observable local manifestations of inflammation—swelling, redness, heat, and pain.

Interferon is nonspecifically released by virus-infected cells and transiently inhibits viral multiplication in other cells to which it binds. Interferon further exerts anticancer effects by slowing division and growth of tumor cells as well as by enhancing the power of killer cells.

Natural killer cells nonspecifically lyse and destroy virus-infected cells and cancer cells on first exposure to them.

Upon being activated by locally released factors or microbes themselves at the site of invasion, the complement system directly destroys the foreign invaders by lysing their membranes and also augments other aspects of the inflammatory process.

Specific Immune Responses

Following initial exposure to a microbial invader, specific components of the immune system become specially prepared to selectively attack the particular foreigner. The immune system is able not only to recognize foreign molecules as different from self-molecules—so that destructive immune reactions are not unleashed against the body itself—but also to distinguish between millions of different foreign molecules. The cells of the specific immune system, the lymphocytes, are each uniquely equipped with surface membrane receptors that are able to bind

lock-and-key fashion with only one specific complex foreign molecule, which is known as an antigen. The tremendous variation in antigen-detecting ability between different lymphocytes arises from the shuffling around of a few different gene segments, coupled with a high incidence of somatic mutation, during lymphocyte development. Those lymphocytes produced by chance that are able to attack the body's own antigen-bearing cells are eliminated or suppressed so that they are prevented from functioning. In this way, the body is able to "tolerate" (not attack) its own antigens. The major surface antigens of all nucleated cells are known as human leukocyte-associated (HLA) antigens, which are coded for by the major histocompatibility complex (MHC), a group of genes with DNA sequences unique for each individual.

There are two broad classes of specific immune responses: antibody-mediated immunity and cell-mediated immunity. In both instances, the ultimate outcome of a particular lymphocyte binding with a specific antigen is destruction of the antigen, but the effector cells, stimuli, and tactics involved are different. Plasma cells derived from B lymphocytes (B cells) are responsible for antibody-mediated immunity, whereas T lymphocytes (T cells) accomplish cell-mediated immunity. B cells develop from a lineage of lymphocytes that originally matured within the bone marrow. The T-cell lineage arises from lymphocytes that migrated from the bone marrow to the thymus to complete their maturation.

After being activated by antigen associated with a foreign invader, a lymphocyte (either a B cell or a T cell, depending on the characteristics of the antigen) rapidly proliferates, producing a clone of its own kind that can specifically wage battle against the invader. Some of the newly developed lymphocytes do not participate in the attack but become memory cells that lie in waiting, ready to launch a swifter and more forceful attack should the same foreigner ever invade the body again. B cells and T cells have different targets because their requirements for antigen recognition differ. Each B cell recognizes specific free extracellular antigen that is not associated with cell-bound self-antigens, such as that found on the surface of bacteria. Appropriately, the activated B cell differentiates into a plasma cell, which is specialized to secrete freely circulating antibodies that besiege the freely existing invading bacteria (or other foreign substance) that induced their production. Antibodies do not directly destroy the foreign material. Instead, they intensify lethal nonspecific immune mechanisms already called into play by the foreign invasion. Antibodies activate the complement system, enhance phagocytosis, and stimulate killer cells.

T cells, in contrast, have a dual binding requirement of foreign antigen in association with HLA self-antigens on the surface of one of the body's own cells. Two different types of host cells meet this requirement: (1) virally invaded host cells and (2) other immune cells with foreign antigen attached on their surface. The presence of different classes of self-antigens on the surface of these foreign antigen–bearing host cells causes three different types of T cells to differentially interact with them: (1) Cytotoxic T cells are able to bind only with virus-infected host cells, whereupon they release toxic substances that kill the infected cell. (2) Helper T cells can only bind with other T cells, B cells, and macrophages that have encountered foreign antigen. Subsequently, helper T cells enhance the immune powers of these other effector cells by secreting specific chemical mediators. (3) Suppressor T cells suppress both T and B cells that have been activated by antigen, thereby preventing the immune system from overresponding and potentially damaging normal host cells. Such differential activation of the various types of lymphocytes assures that the appropriate specific immune response ensues to dispose of the particular enemy efficiently. Moreover, B cells, the various T cells, and macrophages reinforce each other's defense strategies, primarily by releasing a number of important secretory products.

In a process known as immune surveillance, natural killer cells, cytotoxic T cells, macrophages, and the interferon that they collectively secrete normally eradicate newly arisen cancer cells before they have a chance to spread.

Immune Diseases

Occasionally, through a deficiency of B or T cells, the immune system fails to defend normally against bacterial or viral infections, respectively. In contrast, in certain instances the immune system becomes overzealous. In autoimmune disease, the immune system erroneously turns against one of the person's own tissues that it no longer recognizes and tolerates as self. With immune-complex diseases, body tissues are inadvertently destroyed as an overabundance of antigen-antibody complexes activates excessive quantities of lethal complement, which destroys surrounding normal cells as well as the antigen. Allergies occur when the immune system inappropriately launches a symptom-producing, body-damaging attack against an allergen, a normally harmless environmental antigen.

External Defenses

The body surfaces exposed to the outside environment—both the outer covering of skin and the linings of internal cavities that communicate with the external environment—serve not only as mechanical barriers to deter would-be pathogenic invaders but also play an active role in thwarting entry of bacteria and other unwanted materials.

The skin consists of two layers: an outer avascular, keratinized epidermis and an inner connective-tissue dermis. The epidermis contains four cell types: melanocytes, keratinocytes, Langerhans cells, and Granstein cells. Melanocytes produce a brown pigment, melanin, the amount of which is responsible for the varying shades of brown skin color. Melanin protects the skin by absorbing harmful ultraviolet radiation. The most abundant cells are the keratinocytes, producers of the tough keratin that forms the outer protective layer of the skin. This physical barrier discourages bacteria and other harmful environmental agents from entering the body and prevents water and other valuable body substances from escaping. Keratinocytes further serve immunologically by secreting interleukin 1, which enhances post-thymic T-cell maturation within the skin. Langerhans cells and Granstein cells also function in specific immunity by presenting antigen to helper T cells and suppressor T cells, respectively.

The dermis contains (1) blood vessels, which nourish the skin and play an important role in regulating body temperature; (2) sensory nerve endings, which provide information about the external environment; and (3) several exocrine glands and hair follicles, which are formed by specialized invaginations of the overlying epithelium. The skin's exocrine glands include sebaceous glands, which produce sebum, an oily substance that softens and waterproofs the skin, and sweat glands, which produce cooling sweat. Hair follicles produce hairs, the distribution and function of which are minimal in humans. Additionally, the skin synthesizes vitamin D in the presence of sunlight.

Besides the skin, the other main routes by which potential pathogens enter the body are (1) the digestive system, which is defended by an antimicrobial salivary enzyme, destructive acidic gastric secretions, gut-associated lymphoid tissue, and harmless colonic resident flora; (2) the genitourinary system, which is protected by destructive acidic and particle-entrapping mucus secretions; and (3) the respiratory system, whose defense depends on alveolar macrophage activity and on secretion of a sticky mucus that traps debris, which is subsequently swept out by ciliary action. Other respiratory defenses include nasal hairs, which filter out large inspired particles; reflex cough and sneeze mechanisms, which expel irritant materials from the trachea and nose, respectively; and the tonsils and adenoids, which defend immunologically.

REVIEW EXERCISES

Objective Questions (Answers on p. D–3.)

1. The complement system can only be activated by antibodies. (True or false?)

2. Specific immune responses are accomplished by neutrophils. (True or false?)

3. Damaged tissue is always replaced by scar tissue. (True or false?)

4. Active immunity against a particular disease can be acquired only by actually having the disease. (True or false?)

5. A secondary response has a more rapid onset, is more potent, and has a longer duration than a primary response. (True or false?)

6. _____ forms a doughnut-shaped complex that imbeds in a microbial surface membrane, causing osmotic lysis of the victim cell.

7. _____ is a collection of phagocytic cells, necrotic tissue, and bacteria.

8. _____ refers to the localized response to microbial invasion or tissue injury that is accompanied by swelling, heat, redness, and pain.

9. A chemical that enhances phagocytosis by serving as a link between a microbe and the phagocytic cell is known as a(n) _____.

10. _____ refer collectively to all of the chemical messengers other than antibodies secreted by lymphocytes.

11. Which of the following statements concerning leukocytes is (are) *incorrect*?
    a. Monocytes are transformed into macrophages.
    b. T lymphocytes are transformed into plasma cells that secrete antibodies.
    c. Neutrophils are highly mobile phagocytic specialists.
    d. Basophils release histamine.
    e. Lymphocytes arise in part from lymphoid tissues.

12. Using the answer code at the right, indicate whether the following characteristics of the specific immune system apply to antibody-mediated immunity or cell-mediated immunity:

____ 1. involves secretion of antibodies

____ 2. mediated by B cells

____ 3. mediated by T cells

____ 4. is accomplished by thymic-educated lymphocytes

____ 5. is triggered by the binding of specific antigens to complementary lymphocyte-bearing receptors

____ 6. involves formation of memory cells in response to initial exposure to an antigen

____ 7. is primarily aimed against virus-infected host cells

____ 8. protects primarily against bacterial invaders

____ 9. directly destroys targeted cells

____ 10. is involved in rejection of transplanted tissue

____ 11. requires binding of a lymphocyte to a free extracellular antigen

____ 12. requires dual binding of a lymphocyte with both foreign antigen and self-antigens present on the surface of a host cell

a. antibody-mediated immunity

b. cell-mediated immunity

c. both antibody-mediated and cell-mediated immunity

13. Using the answer code at the right, indicate whether the following characteristics apply to the epidermis or dermis:

____ 1. is the inner layer of skin

____ 2. has layers of epithelial cells that are dead and flattened

____ 3. has no direct blood supply

____ 4. contains sensory nerve endings

____ 5. contains keratinocytes

____ 6. contains melanocytes

____ 7. contains rapidly dividing cells

____ 8. is mostly connective tissue

a. epidermis

b. dermis

14. Match the following:

____ 1. a family of proteins that nonspecifically defend against viral infection

____ 2. a response to tissue injury in which neutrophils and macrophages play a major role

____ 3. a group of plasma proteins that, when activated, bring about destruction of foreign cells by attacking their plasma membranes

____ 4. lymphocytelike entities that spontaneously lyse tumor cells and virus-infected host cells

a. complement system

b. natural killer cells

c. interferon

d. inflammation

Essay Questions

1. Distinguish between bacteria and viruses.
2. Summarize the functions of each of the lymphoid tissues.
3. Distinguish between specific and nonspecific immune responses.
4. Compare the life history of B cells and T cells.
5. What is an antigen?
6. Describe the structure of an antibody. List and describe the five subclasses of immunoglobulins.
7. In what ways do antibodies exert their effect?
8. Describe the clonal selection theory.
9. Compare the functions of B cells and T cells. What are the roles of the three types of T cells?
10. Summarize the functions of macrophages in immune defense.
11. What mechanisms are believed to be involved in tolerance?
12. What is the importance of class I and class II MHC glycoproteins?
13. Describe the factors that contribute to immune surveillance against cancer cells.
14. Distinguish among immune deficiency disease, autoimmune disease, immune-complex disease, immediate hypersensitivity, and delayed hypersensitivity.
15. What are the immune functions of the skin?

 POINTS TO PONDER

1. Compare the defense mechanisms that come into play in response to bacterial and viral pneumonia.
2. Why does the fact that HIV, the AIDS virus, frequently mutates make it difficult to develop vaccine against this virus?
3. What impact would failure of the thymus to develop embryonically have on the immune system after birth?
4. Medical researchers are currently working on ways to "teach" the immune system to view foreign tissue as "self." What useful clinical application will the technique have?
5. When someone looks at you, are the cells of your body they are viewing dead or alive?

Respiratory System

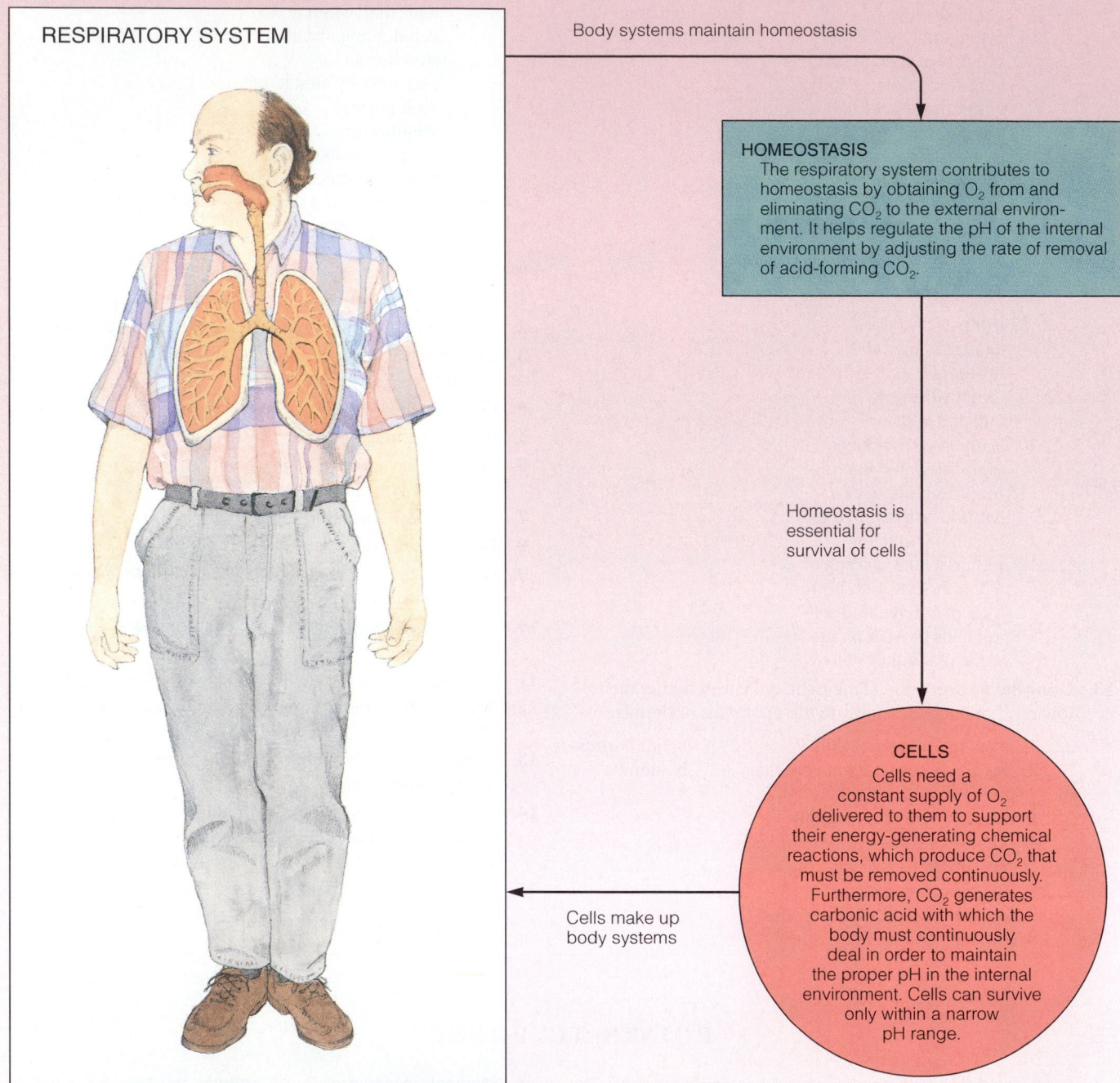

RESPIRATORY SYSTEM

Body systems maintain homeostasis

HOMEOSTASIS
The respiratory system contributes to homeostasis by obtaining O_2 from and eliminating CO_2 to the external environment. It helps regulate the pH of the internal environment by adjusting the rate of removal of acid-forming CO_2.

Homeostasis is essential for survival of cells

CELLS
Cells need a constant supply of O_2 delivered to them to support their energy-generating chemical reactions, which produce CO_2 that must be removed continuously. Furthermore, CO_2 generates carbonic acid with which the body must continuously deal in order to maintain the proper pH in the internal environment. Cells can survive only within a narrow pH range.

Cells make up body systems

Energy is essential for sustaining life-supporting cellular activities, such as protein synthesis and active transport across plasma membranes. The cells of the body need a continual supply of O_2 to support their energy-generating chemical reactions. The CO_2 produced during these reactions must be eliminated from the body at the same rate it is produced to prevent dangerous fluctuations in pH (that is, to maintain the acid-base balance), because CO_2 generates carbonic acid.

Respiration involves the sum of the processes that accomplish ongoing passive movement of O_2 from the atmosphere to the tissues to support cellular metabolism, as well as the continual passive movement of metabolically produced CO_2 from the tissues to the atmosphere. The **respiratory system** contributes to homeostasis by exchanging O_2 and CO_2 between the atmosphere and the blood. The blood transports O_2 and CO_2 between the respiratory system and tissues.

CHAPTER CONTENTS AT A GLANCE

INTRODUCTION
Internal and external respiration
Anatomical considerations
RESPIRATORY MECHANICS
Pressure considerations
Respiratory cycle
Airway resistance
Elastic behavior of the lungs
   Compliance; elastic recoil
   Alveolar surface tension; pulmonary surfactant
Work of breathing
Lung volumes and capacities
Pulmonary and alveolar ventilation
GAS EXCHANGE
Concept of partial pressure
Gas exchange across the pulmonary and systemic capillaries
GAS TRANSPORT
Oxygen transport
Carbon-dioxide transport
Abnormalities in blood gas content
  ● CONCEPTS, CHALLENGES, AND CONTROVERSIES Effects of
    Heights and Depths on the Body
CONTROL OF RESPIRATION
Respiratory control centers
Generation of respiratory rhythm
Chemical inputs affecting the magnitude of ventilation
  ● A CLOSER LOOK AT EXERCISE PHYSIOLOGY How to Find Out
    How Much Work You're Capable of Doing

INTRODUCTION

The respiratory system does not participate in all steps of respiration.

The primary function of respiration is to obtain O_2 for use by the body's cells and to eliminate the CO_2 the cells produce. Most people think of respiration as the process of breathing in and breathing out. In physiology, however, respiration has a much broader meaning. **Internal** or **cellular respiration** refers to the intracellular metabolic processes carried out within the mitochondria, which use O_2 and produce CO_2 during the derivation of energy from nutrient molecules (see p. 33). The **respiratory quotient** (**R.Q.**), the ratio of CO_2 produced to O_2 consumed, varies depending on the foodstuff consumed. When carbohydrate is being used, the R.Q. is 1; that is, for every molecule of O_2 consumed, one molecule of CO_2 is produced: $C_6H_{12}O_6 + 6O_2 \rightarrow 6CO_2 + 6H_2O + ATP$. For fat utilization, the R.Q. is 0.7; for protein, it is 0.8. On a typical American diet consisting of a mixure of these three nutrients, resting O_2 consumption averages about 250 ml/min, and CO_2 production averages about 200 ml/min, for an average R.Q. of 0.8:

$$R.Q. = \frac{CO_2 \text{ produced}}{O_2 \text{ consumed}} = \frac{200 \text{ ml/min}}{250 \text{ ml/min}} = 0.8$$

External respiration refers to the entire sequence of events involved in the exchange of O_2 and CO_2 between the external environment and the cells of the body (● Fig. 13–1). External respiration, the topic of this chapter, encompasses four steps:

1. Air is alternately moved in and out of the lungs so that exchange of air can occur between the atmosphere (external environment) and the air sacs (**alveoli**) of the lungs. This exchange is accomplished by the mechanical act of **breathing**, or **ventilation**. The rate of ventilation is regulated so that the flow of air between the atmosphere and the alveoli is adjusted according to the body's metabolic needs for O_2 uptake and CO_2 removal.

2. Oxygen and CO_2 are exchanged between air in the alveoli and blood within the pulmonary (*pulmonary* refers to "lung") capillaries by the process of diffusion.

3. Oxygen and CO_2 are transported by the blood between the lungs and tissues.

4. Exchange of O_2 and CO_2 takes place between the tissues and the blood by the process of diffusion across the systemic (tissue) capillaries.

The respiratory system does not accomplish all the steps of respiration; it is involved only with ventilation and the exchange of O_2 and CO_2 between the lungs and blood. The circulatory system carries out the remainder of the respiratory process.

The respiratory system additionally performs the following nonrespiratory functions:

- It provides a route for water loss and heat elimination. Inspired atmospheric air is humidified and warmed by the respiratory airways before it is expired. Moistening of inspired air is essential to prevent the alveolar linings from drying out. Oxygen and CO_2 cannot diffuse through dry membranes.

- It enhances venous return (see the "respiratory pump," p. 332).

- It contributes to the maintenance of normal acid-base balance by altering the amount of H^+-generating CO_2 exhaled (chapter 15).

- It enables speech, singing, and other vocalization.

- It defends against inhaled foreign matter (see p. 408).

- It removes, modifies, activates, or inactivates various materials passing through the pulmonary circulation. All blood returning to the heart from the tissues must pass through the lungs before being returned to the systemic circulation. The lungs, therefore, are uniquely situated to partially or completely remove specific materials that have been added to the blood at the tissue level before they have a chance to reach other parts of the body by means of the arterial system. For example, prostaglandins, a collection of chemical messengers released in numerous tissues to mediate particular local responses, may spill into the blood, but they are inactivated during passage through the lungs so that they cannot exert systemic effects. On the

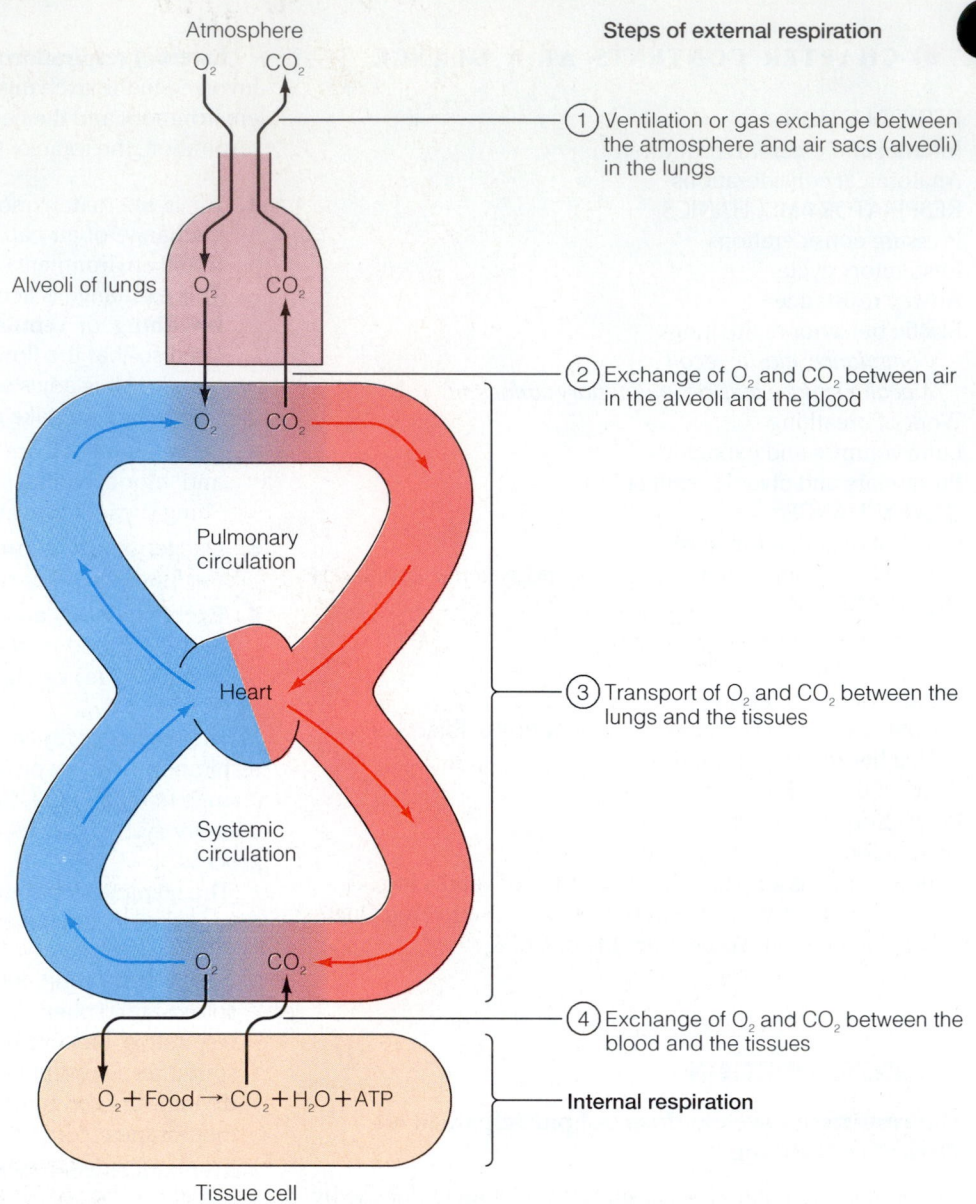

• **FIGURE 13–1   External and Internal Respiration** External respiration encompasses the steps involved in the exchange of O_2 and CO_2 between the external environment and tissue cells. Internal respiration refers to the intracellular metabolic reactions involving the utilization of O_2 to derive energy (ATP) from food, producing CO_2 as a by-product.

Atmosphere
O_2 CO_2

Alveoli of lungs O_2 CO_2

① Ventilation or gas exchange between the atmosphere and air sacs (alveoli) in the lungs

O_2 CO_2

② Exchange of O_2 and CO_2 between air in the alveoli and the blood

Pulmonary circulation

Heart

③ Transport of O_2 and CO_2 between the lungs and the tissues

Systemic circulation

O_2 CO_2

④ Exchange of O_2 and CO_2 between the blood and the tissues

O_2 + Food → CO_2 + H_2O + ATP

— Internal respiration

Tissue cell

---

other hand, the lungs activate angiotensin II, a hormone that plays an important role in regulating the concentration of Na^+ in the extracellular fluid.

■ The nose, a part of the respiratory system, serves as the organ of smell (chapter 6).

The respiratory airways conduct air between the atmosphere and alveoli.

The **respiratory system** includes the respiratory airways leading into the lungs, the lungs themselves, and the structures of the thorax (chest) involved in producing movement of air through the airways into and out of the lungs. The **respiratory airways** are tubes that carry air between the atmosphere and the alveoli, the latter being the only site where exchange of gases can take place between air and blood. The airways (• Fig. 13–2a) begin with the **nasal passages (nose)**. The na-

sal passages open into the **pharynx** (throat), which serves as a common passageway for both the respiratory and digestive systems. Two tubes lead from the pharynx—the **trachea (windpipe)**, through which air is conducted to the lungs, and the **esophagus**, the tube through which food passes to the stomach. Air normally enters the pharynx through the nose, but it can enter by the mouth as well when the nasal passages are congested; that is, you can breathe through your mouth when you have a cold. Because the pharynx serves as a common passageway for food and air, reflex mechanisms exist to close off the trachea during swallowing so that food enters the esophagus and not the airways. The esophagus remains closed except during swallowing to prevent air from entering the stomach during breathing.

Located at the entrance of the trachea is the **larynx**, or **voice box**, the anterior protrusion of which forms the "Adam's apple." The **vocal cords**, two bands of elastic tissue that lie

Nasal passages
Mouth
Pharynx
Larynx
Trachea
Cartilaginous ring
Right bronchus
Bronchiole
Left bronchus
Terminal bronchiole

(a)

Smooth muscle
Terminal bronchiole
Respiratory bronchiole
Branch of pulmonary artery
Branch of pulmonary vein
Alveolus
Pulmonary capillaries

(b)

• **FIGURE 13–2   Anatomy of the Respiratory System** (a) The respiratory airways. (b) Enlargement of the alveoli (air sacs) at the terminal end of the airways. Most alveoli are clustered in grapelike arrangements at the end of the terminal bronchioles. A few individual alveoli bud off laterally along the last portion of the airways. Because gas exchange can take place across them, these smallest of the airways are termed *respiratory bronchioles.*

across the opening of the larynx, can be stretched and positioned in different shapes by laryngeal muscles. As air is moved past the taut vocal cords, they vibrate to produce the many different sounds of speech. The lips, tongue, and soft palate modify the sounds into recognizable sound patterns. During swallowing, the vocal cords assume a function not related to speech; they are brought into tight apposition to each other to close off the entrance to the trachea.

Beyond the larynx, the trachea divides into two main branches, the right and left **bronchi**, which enter the right and left lungs, respectively. Within each lung, the bronchus continues to branch into progressively narrower, shorter, and more numerous airways, much like the branching of a tree. The smaller branches are known as **bronchioles**. Clustered at the ends of the terminal bronchioles are the alveoli, the tiny air sacs where gas exchange between air and blood takes place (Fig. 13–2b).

To permit airflow in and out of the gas-exchanging portions of the lungs, the continuum of conducting airways from the entrance through the terminal bronchioles to the alveoli

must remain open. The trachea and larger bronchi are fairly rigid, nonmuscular tubes encircled by a series of cartilaginous rings that prevent the tubes' compression. The smaller bronchioles have no cartilage to hold them open. Their walls contain smooth muscle that is innervated by the autonomic nervous system and is sensitive to certain hormones and local chemicals. These factors, by varying the degree of contraction of bronchiolar smooth muscle and hence the caliber of these small terminal airways, are able to regulate the amount of air passing between the atmosphere and each cluster of alveoli.

The gas-exchanging alveoli are small, thin-walled, inflatable air sacs encircled by a jacket of pulmonary capillaries.

The lungs are ideally structured for their function of gas exchange. Recall that according to Fick's law of diffusion (see p. 58), the shorter the distance through which diffusion must take place, the greater the rate of diffusion. Also, the greater

• **FIGURE 13–3   Alveoli and Associated Pulmonary Capillaries**   (a) A scanning electron micrograph of alveoli showing the rich capillary network surrounding them. [From *Tissues and Organs: A Text-Atlas of Scanning Electron Microscopy* by Richard G. Kessel and Randy H. Kardon. Copyright © 1979 by W. H. Freeman and Company. Reprinted with permission.] (b) A schematic representation of a detailed electron-microscope view of an alveolus and surrounding capillaries. A single layer of flattened Type I alveolar cells forms the alveolar walls. Type II alveolar cells embedded within the alveolar wall secrete pulmonary surfactant. Wandering alveolar macrophages are found within the alveolar lumen. (The size of the cells and respiratory membrane is exaggerated compared to the size of the alveolar and pulmonary capillary lumens. The diameter of an alveolus is actually about 1,500 times larger than the intervening space between air and blood.)

the surface area across which diffusion can take place, the greater the rate of diffusion.

The alveoli are clusters of thin-walled, inflatable, grape-like sacs at the terminal branches of the conducting airways (• Fig. 13–3a). The alveolar walls consist of a single layer of flattened **Type I alveolar cells** (Fig. 13–3b). The dense network of pulmonary capillaries encircling each alveolus is also only one cell-layer thick. The interstitial space between an alveolus and the surrounding capillary network forms an extremely thin barrier, with only 0.2 μm separating the air in the alveoli from the blood in the pulmonary capillaries. (A sheet of tracing paper is about fifty times thicker than this air-to-blood barrier.) The thinness of this barrier facilitates gas exchange.

Furthermore, the alveolar air-blood interface presents a tremendous surface area for exchange. There are about 300 million alveoli in the lungs, each about 300 μm (⅓ mm) in diameter. So dense are the pulmonary capillary networks that each alveolus is encircled by an almost continuous sheet of blood. The total surface area thus exposed between alveolar air and pulmonary capillary blood is about 75 square meters (about the size of a tennis court). In contrast, if the lungs consisted of a single hollow chamber of the same dimensions in-

stead of being divided into myriad alveolar units, the total surface area would be only about 1/100 square meter.

In addition to the thin, wall-forming Type I cells, the alveolar epithelium also contains **Type II alveolar cells** (Fig. 13–3b), which secrete **pulmonary surfactant**, a phospholipoprotein complex that facilitates lung expansion (to be described later). Also present within the lumen of the air sacs are the defensive alveolar macrophages.

Minute **pores of Kohn** present in the alveolar walls permit airflow between adjacent alveoli, a process known as **collateral ventilation**. These passageways are especially important in allowing fresh air to enter an alveolus whose terminal conducting airway is blocked because of disease.

The lungs occupy much of the thoracic cavity.

There are two **lungs**, each divided into several lobes and each supplied by one of the bronchi. The lung tissue itself consists of the series of highly branched airways, the alveoli, the pulmonary blood vessels, and large quantities of elastic connective tissue. The only muscle within the lungs is the smooth muscle in the walls of the arterioles and bronchioles, both of which are subject to control. There is no muscle within the alveolar walls to cause them to inflate and deflate during the breathing process. Rather, it is through changes in the dimensions of the thorax that corresponding changes in lung volume are produced.

The lungs occupy most of the volume of the **thoracic (chest) cavity**, the only other structures in the chest being the heart and associated vessels, the esophagus, the thymus, and some nerves. The outer chest wall (**thorax**) is formed by twelve pairs of curved **ribs**, which join the **sternum** (breastbone) anteriorly and the **thoracic vertebrae** (backbone) posteriorly. The rib cage provides bony protection for the lungs and heart. The **diaphragm**, which forms the floor of the thoracic cavity, is a large, dome-shaped sheet of skeletal muscle that completely separates the thoracic cavity from the abdominal cavity. It is penetrated only by the esophagus and blood vessels traversing between the thoracic and abdominal cavities. The thoracic cavity is enclosed at the neck by muscles and connective tissue. The only communication between the thorax and the atmosphere is through the respiratory airways into the alveoli. Like the lungs, the chest wall contains considerable amounts of elastic connective tissue.

A pleural sac separates each lung from the thoracic wall.

Separating each lung from the thoracic wall and other surrounding structures is a double-walled, closed sac called the pleural sac (●Fig. 13–4). The dimensions of the **pleural cavity** within the pleural sac are greatly exaggerated in the illustration to aid visualization; in reality the layers of the pleural sac are in close contact with one another. The surfaces of the pleura secrete a thin **intrapleural fluid**, which lubricates the pleural surfaces as they slide past each other during respiratory movements. **Pleurisy**, an inflammation of the pleural sac, is accompanied by painful breathing because each inflation and deflation of the lungs cause a "friction rub."

RESPIRATORY MECHANICS

Interrelationships among atmospheric, intra-alveolar, and intrapleural pressures are important in respiratory mechanics.

Air tends to move from a region of higher pressure to a region of lower pressure, that is, down a **pressure gradient**. Air flows in and out of the lungs during the act of breathing by moving down alternately reversing pressure gradients established between the alveoli and the atmosphere by cyclical respiratory-muscle activity. Three different pressure considerations are important in ventilation (●Fig. 13–5):

1. **Atmospheric (barometric) pressure** is the pressure exerted by the weight of the air in the atmosphere on objects on the earth's surface. At sea level it equals 760 mm Hg

(a)

(b)

●FIGURE 13–4 Pleural Sac
(a) Pushing a lollipop into a water-filled balloon produces a relationship analogous to that of the double-walled closed pleural sac surrounding each lung and separating it from the thoracic wall. (b) Schematic representation of the relationship of the pleural sac to the lungs and thorax. One layer of the pleural sac, the *visceral pleura*, closely adheres to the surface of the lung (*viscus* means "organ") then reflects back on itself to form another layer, the *parietal pleura*, which lines the interior surface of the thoracic wall (*paries* means "wall"). The relative size of the pleural cavity between these two layers is grossly exaggerated for the purpose of visualization.

Atmosphere
760 mm Hg

Airways (represents all airways collectively)

Atmospheric pressure (the pressure
exerted by the weight of the gas in the
atmosphere on objects on the earth's
surface—760 mm Hg at sea level)

Thoracic wall (represents entire thoracic cage)

760 mm Hg

Intra-alveolar pressure (the pressure within
the alveoli—760 mm Hg when equilibrated
with atmospheric pressure)

Pleural sac (space represents pleural cavity)

Intrapleural pressure (the pressure within
the pleural sac—the pressure exerted
outside the lungs within the thoracic
cavity, usually less than atmospheric
pressure at 756 mm Hg)

Lungs (represents all alveoli collectively)

756 mm Hg

• **FIGURE 13–5   Pressures Important in Ventilation**

(• Fig. 13–6). Atmospheric pressure diminishes with increasing altitude above sea level as the column of air above the earth's surface correspondingly decreases. Minor fluctuations in atmospheric pressure occur at any height because of changing weather conditions (that is, when barometric pressure is rising or falling).

• **FIGURE 13–6   Atmospheric Pressure** The pressure exerted on objects by the atmospheric air above the earth's surface at sea level can push a column of mercury to a height of 760 mm. Therefore, atmospheric pressure at sea level is considered to be 760 mm Hg.

760 mm

Vacuum

Pressure exerted by
atmospheric air above
the earth's surface

Sea level

Mercury (Hg)

2. **Intra-alveolar pressure**, also known as **intrapulmonary pressure**, is the pressure within the alveoli. Because the alveoli communicate with the atmosphere through the conducting airways, air quickly flows down its pressure gradient any time intra-alveolar pressure differs from atmospheric pressure; the airflow continues until the two pressures equilibrate (become equal).

3. **Intrapleural pressure** is the pressure within the pleural sac. Also known as **intrathoracic pressure**, it is the pressure exerted outside the lungs within the thoracic cavity. The intrapleural pressure is usually less than atmospheric pressure, averaging 756 mm Hg at rest. Just as blood pressure is recorded using atmospheric pressure as a reference point (that is, a systolic blood pressure of 120 mm Hg is 120 mm Hg greater than the atmospheric pressure of 760 mm Hg or in reality 880 mm Hg), 756 mm Hg is sometimes referred to as a pressure of − 4 mm Hg, although there is really no such thing as an *absolute* negative pressure. A pressure of − 4 mm Hg is just negative when compared with the normal atmospheric pressure of 760 mm Hg. To avoid confusion, we will use absolute positive values throughout our discussion of respiration.

Intrapleural pressure does not equilibrate with atmospheric or intra-alveolar pressure, because there is no direct communication between the pleural cavity and either the atmosphere or the lungs. Since the pleural sac is a closed sac with no openings, air cannot enter or leave despite any pressure gradients that might exist between it and surrounding regions.

The intrapleural fluid's cohesiveness and the transmural pressure gradient hold the lungs and thoracic wall in tight apposition, even though the lungs are smaller than the thorax.

The thoracic cavity is larger than the unstretched lungs because the thoracic wall grows more rapidly than the lungs during development. However, two forces—the intrapleural fluid's cohesiveness and the transmural pressure gradient—hold the thoracic wall and lungs in close apposition, stretching the lungs to fill the larger thoracic cavity.

The polar water molecules in the intrapleural fluid resist being pulled apart because of their attraction to each other. The resultant cohesiveness of the intrapleural fluid tends to hold the pleural surfaces together. Thus, the intrapleural fluid can be considered very loosely as a "stickiness" or "glue" between the lining of the thoracic wall and the lung. Have you ever tried to pull apart two smooth surfaces held together by a thin layer of liquid, such as two wet glass slides? If so, you know that the two surfaces act as if they are stuck together by the thin layer of water. Even though you can easily slip the slides back and forth relative to reach other (just as the intrapleural fluid facilitates movement of the lungs against the interior surface of the chest wall), you can pull the slides apart only with great difficulty, because the molecules within the intervening liquid resist being separated. This relationship is partly responsible for the fact that changes in thoracic dimension are always accompanied by corresponding changes in lung dimension; that is, when the thorax expands, the lungs, being stuck to the thoracic wall by virtue of the intrapleural fluid's cohesiveness, do likewise.

An even more important reason that the lungs follow the movements of the chest wall is the **transmural pressure gradient** that exists across the lung wall (*trans* means "across"; *mural* means "wall") (● Fig. 13–7). The intra-alveolar pressure, equilibrated with atmospheric pressure at 760 mm Hg, is greater than the intrapleural pressure of 756 mm Hg, so a greater pressure is pushing outward than is pushing inward across the lung wall. This net outward pressure differential, the transmural pressure gradient, pushes out on the lungs, stretching or distending them. Because of this pressure gradient, the lungs are always forced to expand to fill the thoracic cavity.

A similar transmural pressure gradient exists across the thoracic wall. The atmospheric pressure pushing inward on the thoracic wall is greater than the intrapleural pressure pushing outward on this same wall, so the chest wall tends to be "squeezed in" or compressed compared to what it would be in an unrestricted state. The effect of the transmural pressure gradient across the lung wall is much more pronounced, however, because the highly distensible lungs are influenced by this modest pressure differential to a much greater extent than is the more rigid thoracic wall.

Because neither the lungs nor the thoracic wall are in their natural position when they are held in apposition to each other, they constantly try to assume their own inherent dimensions. The stretched lungs have a tendency to pull inward

---

● **FIGURE 13–7 Transmural Pressure Gradient** Across the lung wall, the intra-alveolar pressure of 760 mm Hg pushes outward, whereas the intrapleural pressure of 756 mm Hg pushes inward. This 4 mm Hg difference in pressure constitutes a transmural pressure gradient that pushes out on the lungs, stretching them to fill the larger thoracic cavity. Across the thoracic wall, the atmospheric pressure of 760 mm Hg pushes inward whereas the intrapleural pressure of 756 mm Hg pushes outward. This 4 mm Hg difference in pressure constitutes a transmural pressure gradient that pushes inward and compresses the thoracic wall.

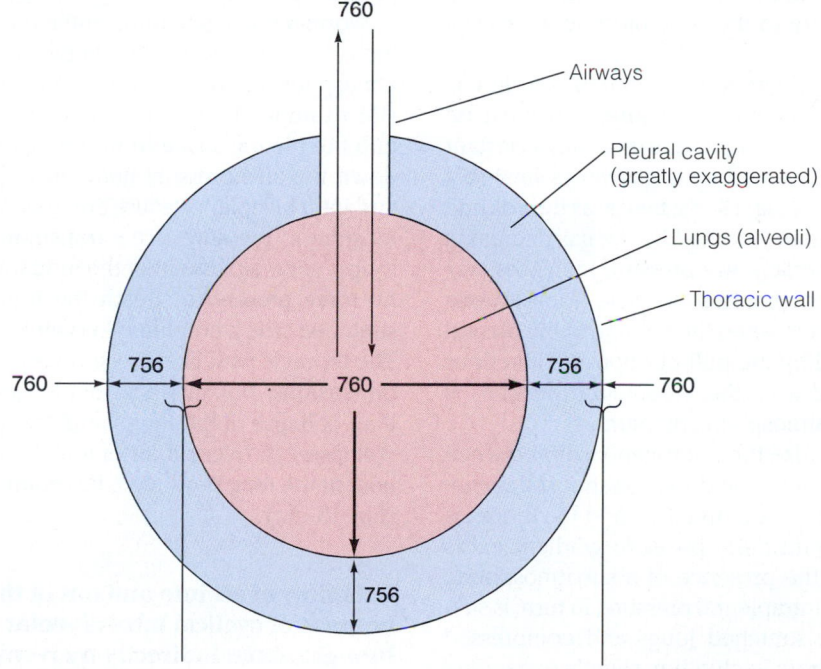

Transmural pressure gradient across lung wall = intra-alveolar pressure — intrapleural pressure

Transmural pressure gradient across thoracic wall = atmospheric pressure — intrapleural pressure

Numbers are mm Hg pressure.

• FIGURE 13-8 Boyle's Law Each container has the same number of gas molecules. Given the random motion of gas molecules, the likelihood of a gas molecule striking the interior wall of the container and exerting pressure varies inversely with the volume of the container at any constant temperature. The gas in container B exerts more pressure than the same gas in larger container C but less pressure than the same gas in smaller container A. This relationship is stated as Boyle's law: $P_1V_1 = P_2V_2$. As the volume of a gas increases, the pressure of the gas decreases proportionately; conversely, the pressure increases proportionately as the volume decreases.

A

B

Closed container with given number of gas molecules

C

Piston

Pressure gauge

Volume = 1/2
Pressure = 2

Volume = 1
Pressure = 1

Volume = 2
Pressure = 1/2

away from the thoracic wall, whereas the compressed thoracic wall tends to move outward away from the lungs. The transmural pressure gradient and intrapleural fluid's cohesiveness, however, prevent these structures from pulling away from each other except to the slightest degree. Even so, the resultant ever-so-slight expansion of the pleural cavity is sufficient to drop the pressure in this cavity by 4 mm Hg, bringing the intrapleural pressure to the subatmospheric level of 756 mm Hg.

To understand how the slight increase in intrapleural volume causes a drop in intrapleural pressure, you must be aware of **Boyle's law.** Boyle's law states that at any constant temperature, the pressure exerted by a gas varies inversely with the volume of the gas (• Fig. 13-8); that is, as the volume of a gas increases, the pressure exerted by the gas decreases proportionately, and conversely, the pressure increases proportionately as the volume decreases. Because of this inverse volume-pressure relationship, when the volume of the pleural cavity is slightly increased by the pull in opposite directions of the lungs and thoracic wall, the intrapleural pressure is slightly decreased below atmospheric pressure.

It is important to recognize the interrelationship between the transmural pressure gradient and the subatmospheric intrapleural pressure. The lungs are stretched and the thorax is compressed because a transmural pressure gradient exists across their walls due to the presence of a subatmospheric intrapleural pressure. The intrapleural pressure, in turn, is subatmospheric because the stretched lungs and compressed thorax tend to pull away from each other, slightly expanding the pleural cavity and dropping the intrapleural pressure below atmospheric pressure.

If the intrapleural pressure were ever to equilibrate with atmospheric pressure, the transmural pressure gradient would be abolished. As a result, the lungs and thorax would separate and assume their own inherent dimensions. This is exactly what happens if air is permitted to enter the pleural cavity, a condition known as **pneumothorax** ("air in the chest"). (The intrapleural fluid's cohesiveness cannot hold the lungs and thoracic wall in apposition in the absence of the transmural pressure gradient.)

Normally, air does not enter the pleural cavity because there is no communication between the cavity and either the atmosphere or alveoli. However, if the chest wall is punctured (for example, by a stab wound or a broken rib), air rushes into the pleural space from the higher atmospheric pressure down the air's pressure gradient (• Fig. 13-9a). Intrapleural and intra-alveolar pressure are now both equilibrated with atmospheric pressure, so a transmural pressure gradient no longer exists across either the lung wall or the chest wall. With no force present to stretch the lung, it collapses to its unstretched size, a condition known as **atelectasis** (Fig. 13-9b). The thoracic wall likewise springs outward to its unrestricted dimensions, but this has much less serious consequences than collapse of the lung. Similarly, pneumothorax and lung collapse can occur if air enters the pleural cavity through a hole in the lung produced, for example, by a disease process (Fig. 13-9c).

Bulk flow of air into and out of the lungs occurs because of cyclical intra-alveolar pressure changes brought about indirectly by respiratory-muscle activity.

Because air flows down a pressure gradient, the intra-alveolar pressure must be less than atmospheric pressure for air to flow into the lungs during inspiration. Similarly, the intra-

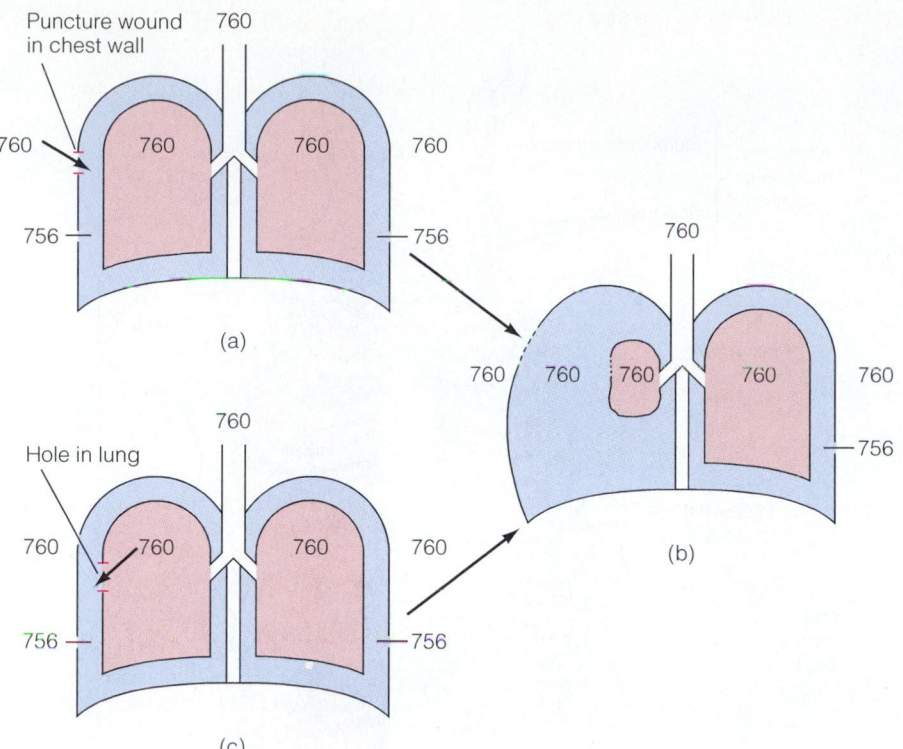

Puncture wound in chest wall

(a)

Hole in lung

(c)

(b)

Numbers are mm Hg pressure.

• **FIGURE 13–9   Pneumothorax**
(a) Traumatic pneumothorax. A puncture in the chest wall permits air to enter the pleural cavity from the atmosphere down the air's pressure gradient, abolishing the transmural pressure gradient. (b) When the transmural pressure gradient is abolished, the lung collapses to its unstretched size (atelectasis), and the chest wall springs outward. (c) Spontaneous pneumothorax. A hole in the lung wall permits air to enter the pleural cavity from the lungs down the air's pressure gradient, abolishing the transmural pressure gradient.

alveolar pressure must be greater than atmospheric pressure for air to flow out of the lungs during expiration. Intra-alveolar pressure can be changed by altering the volume of the lungs, in accordance with Boyle's law. The respiratory muscles that accomplish breathing do not act directly on the lungs to change their volume. Instead, these muscles change the volume of the thoracic cavity, causing a corresponding change in lung volume because the thoracic wall and lungs are linked together by the intrapleural fluid's cohesiveness and the transmural pressure gradient.

Let us follow the changes that occur during one respiratory cycle—that is, one breath in (**inspiration**) and out (**expiration**). Before the beginning of inspiration, the respiratory muscles are relaxed, no air is flowing, and intra-alveolar pressure is equal to atmospheric pressure. At the onset of inspiration, the **inspiratory muscles**—the *diaphragm* and *external intercostal muscles* (• Fig. 13–10 and Table 13–1)—are stimulated to contract, resulting in enlargement of the thoracic cavity. The major inspiratory muscle is the diaphragm, a sheet of skeletal muscle that forms the floor of the thoracic cavity and is innervated by the **phrenic nerve**. The relaxed diaphragm assumes a dome shape that protrudes upward into the thoracic cavity. When the diaphragm contracts upon stimulation by the phrenic nerve, it descends downward, enlarging the volume of the thoracic cavity by increasing its vertical dimension (• Fig. 13–11a). The abdominal wall, if relaxed, can be seen to bulge outward during inspiration as the descending diaphragm pushes the abdominal contents downward and forward.

Two sets of **intercostal** (*inter* means "between"; *costa* means "rib") **muscles**, lie between the ribs—the external intercostal muscles lie on top of the internal intercostal muscles. Upon contraction of the **external intercostal muscles**, whose fibers run downward and forward between adjacent ribs, the ribs are elevated upward and outward, further enlarging the thoracic cavity in both the anteroposterior (front-to-back) and lateral (side-to-side) dimensions (Fig. 13–11a). **Intercostal nerves** activate these intercostal muscles.

As the thoracic cavity enlarges, the lungs are also forced to expand to fill the larger thoracic cavity. As the lungs enlarge, the intra-alveolar pressure drops because the same number of air molecules now occupy a larger lung volume. In a typical inspiratory excursion, the intra-alveolar pressure drops 1 mm Hg to 759 mm Hg (• Fig. 13–12a). Since the intra-alveolar pressure is now less than atmospheric pressure, air flows into the lungs down the pressure gradient from higher to lower pressure. Air continues to enter the lungs until no further gradient exists—that is, until intra-alveolar pressure equals atmospheric pressure. Thus, lung expansion is not caused by movement of air into the lungs; instead, air flows into the lungs because of the fall in intra-alveolar pressure brought about by lung expansion.

During inspiration, the intrapleural pressure falls to 754 mm Hg as a result of expansion of the thorax. The resultant increase in the transmural pressure gradient during inspiration assures that the lungs are stretched to fill the expanded thoracic cavity.

● **FIGURE 13-10   Anatomy of the Respiratory Muscles**

Deeper inspirations (more air breathed in) can be accomplished by contracting the diaphragm and external intercostal muscles more forcefully and by bringing the **accessory inspiratory muscles** into play to further enlarge the thoracic cavity. Contraction of these accessory muscles, which are lo- cated in the neck (Fig. 13-10 and Table 13-1), raises the ster- num and elevates the first two ribs, enlarging the upper por- tion of the thoracic cavity. As the thoracic cavity increases even further in volume than under resting conditions, the lungs likewise expand even more, dropping the intra-alveolar

TABLE 13-1 Actions of the Respiratory Muscles		
MUSCLES	**RESULT OF MUSCLE CONTRACTION**	**TIMING OF STIMULATION TO CONTRACT**
Inspiratory Muscles		
Diaphragm	Descends downward, increasing vertical dimension of thoracic cavity	Every inspiration; primary muscle of inspiration
External intercostal muscles	Elevate ribs upward and outward, enlarging thorax in both front-to-back and side-to-side dimensions	Every inspiration; play secondary complementary role to primary action of diaphragm
Neck muscles (scalenus, sternocleidomastoid)	Raise sternum and elevate first two ribs, enlarging upper portion of thoracic cavity	Only during forceful inspiration; accessory inspiratory muscles
Expiratory Muscles		
Abdominal muscles	Increase intra-abdominal pressure, which exerts upward force on diaphragm to decrease vertical dimension of thoracic cavity	Only during active (forced) expiration
Internal intercostal muscles	Flatten thorax by pulling ribs downward and inward, decreasing front-to-back and side-to-side dimension of thoracic cavity	Only during active (forced) expiration

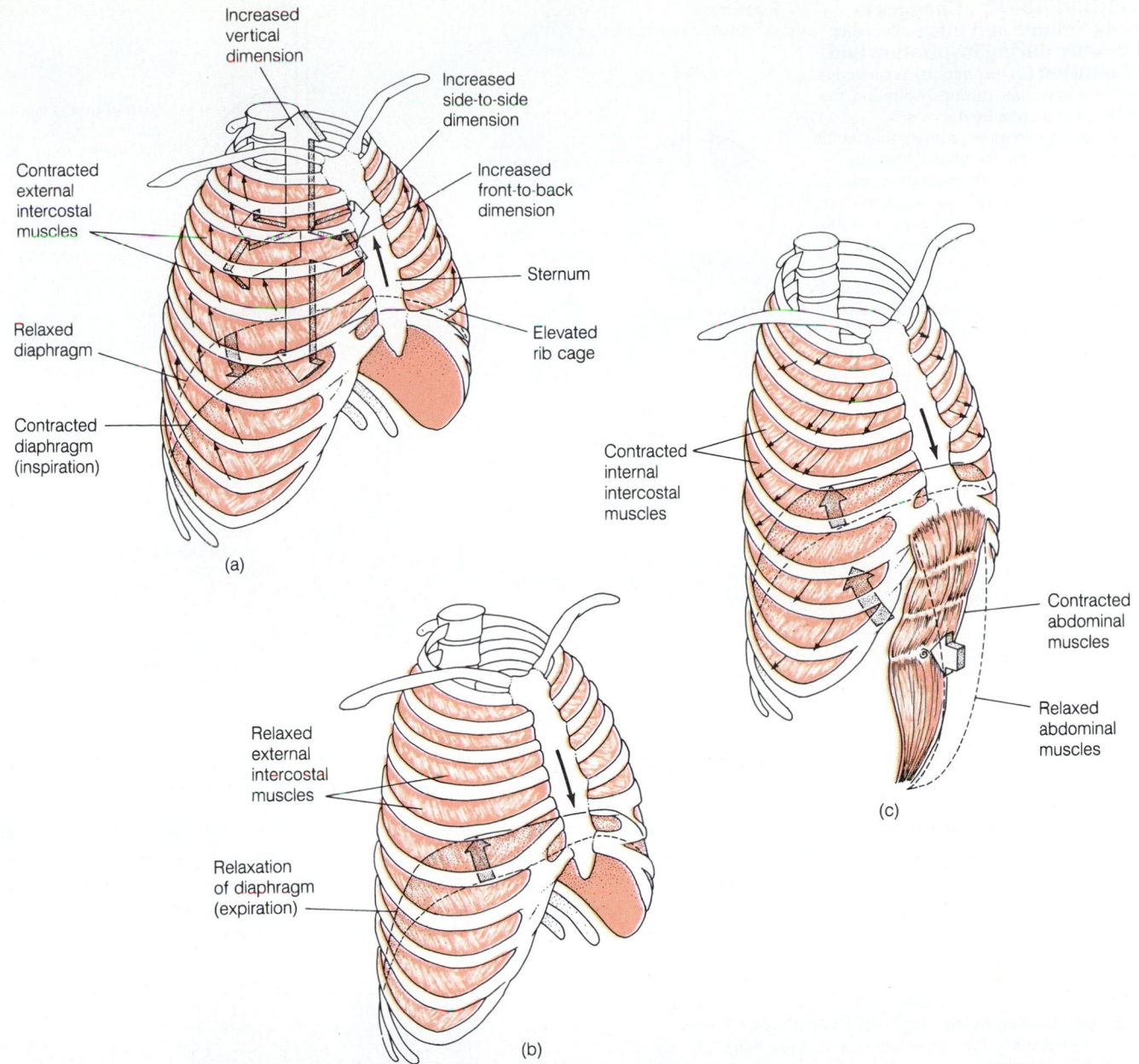

Increased vertical dimension

Increased side-to-side dimension

Increased front-to-back dimension

Contracted external intercostal muscles

Sternum

Elevated rib cage

Relaxed diaphragm

Contracted diaphragm (inspiration)

(a)

Contracted internal intercostal muscles

Contracted abdominal muscles

Relaxed abdominal muscles

(c)

Relaxed external intercostal muscles

Relaxation of diaphragm (expiration)

(b)

● **FIGURE 13–11   Respiratory Muscle Activity during Inspiration and Expiration**   (a) Inspiration, during which the diaphragm descends on contraction, increasing the vertical dimension of the thoracic cavity. Contraction of the external intercostal muscles elevates the ribs to enlarge the thoracic cavity from front-to-back and side-to-side. Note the three-dimensional arrows, which depict expansion of the thoracic cavity. The thin, solid black arrows depict movement of the sternum and ribs. The wide, stippled arrow depicts movement of the diaphragm. (b) Quiet passive expiration, during which the diaphragm relaxes, reducing the volume of the thoracic cavity from its peak inspiratory size. As the external intercostal muscles relax, the elevated rib cage falls because of the force of gravity. This also reduces the volume of the thoracic cavity. (c) Active expiration, during which contraction of the abdominal muscles increases the intra-abdominal pressure, exerting an upward force on the diaphragm. This reduces the vertical dimension of the thoracic cavity further than it is reduced during quiet passive expiration. Contraction of the internal intercostal muscles decreases the front-to-back and side-to-side dimensions by flattening the ribs.

pressure even further. Consequently, a larger inward flow of air occurs before equilibration with atmospheric pressure is achieved; that is, a deeper breath occurs.

At the end of inspiration, the inspiratory muscles relax. The diaphragm assumes its original dome-shaped position when it relaxes; the elevated rib cage falls because of gravity

when the external intercostals relax; and the chest wall and stretched lungs recoil to their preinspiratory size because of their elastic properties, much as a stretched balloon would upon release (Fig. 13–11b). As the lungs recoil and become smaller in volume, the intra-alveolar pressure rises, because the greater number of air molecules contained within the

• FIGURE 13–12   Changes in Lung Volume and Intra-alveolar Pressure during Inspiration and Expiration (a) Inspiration. As the lungs increase in volume during inspiration, the intra-alveolar pressure decreases, establishing a pressure gradient that favors the flow of air into the alveoli from the atmosphere; that is, an inspiration occurs. (b) Expiration. As the lungs recoil to their pre-inspiratory size upon relaxation of the inspiratory muscles, the intra-alveolar pressure increases, establishing a pressure gradient that favors the flow of air out of the alveoli into the atmosphere; that is, an expiration occurs.

Equilibrated; no net movement of air

760

760

Preinspiratory size of thorax

Preinspiratory size of lungs

756

Before inspiration

760

Size of thorax on contraction of inspiratory muscles

759

Size of lungs as they are stretched to fill the expanded thorax

754

During inspiration

(a)

760

Size of thorax on relaxation of inspiratory muscles

761

Size of lungs as they recoil

756

During expiration

(b)

Numbers are mm Hg pressure.

larger lung volume at the end of inspiration are now compressed into a smaller volume. In a resting expiration, the intra-alveolar pressure increases about 1 mm Hg above atmospheric level to 761 mm Hg (Fig. 13–12b). Air now leaves the lungs down its pressure gradient from high intra-alveolar pressure to lower atmospheric pressure. Outward flow of air ceases when intra-alveolar pressure becomes equal to atmospheric pressure and a pressure gradient no longer exists. • Figure 13–13 summarizes the intra-alveolar and intrapleural pressure changes that take place during one respiratory cycle.

• FIGURE 13–13   Intra-alveolar and Intrapleural Pressure Changes throughout Respiratory Cycle

■ During inspiration, intra-alveolar pressure is less than atmospheric pressure.

■ During expiration, intra-alveolar pressure is greater than atmospheric pressure.

■ At the end of both inspiration and expiration, intra-alveolar pressure is equal to atmospheric pressure, because the alveoli are in direct communication with the atmosphere and air continues to flow down its pressure gradient until the two pressures equilibrate.

■ Throughout the respiratory cycle, the intrapleural pressure is lower than the intra-alveolar pressure. Thus, a transmural pressure gradient always exists, and the lung is always stretched to some degree, even during expiration.

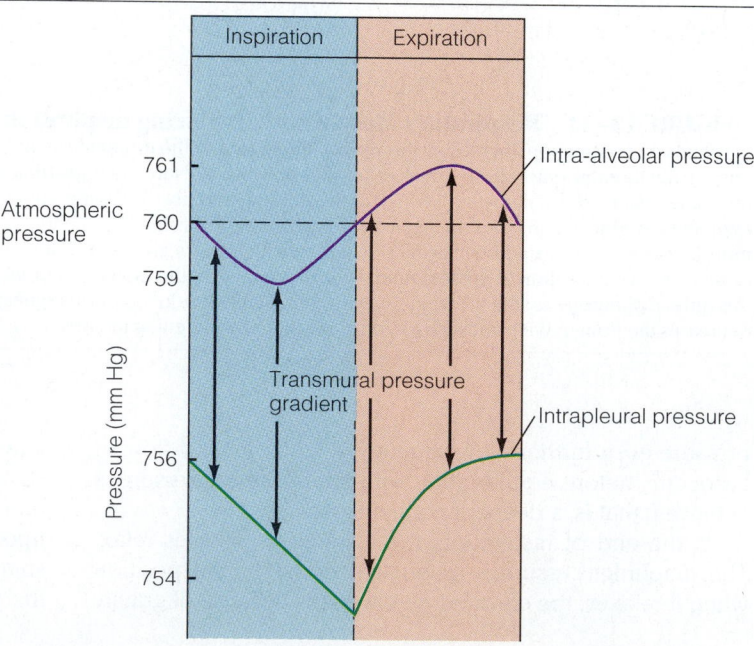

Expiration is normally a *passive* process during quiet breathing since it is accomplished by elastic recoil of the lungs on relaxation of the inspiratory muscles, with no muscular exertion or energy expenditure required. In contrast, inspiration is *always active,* because it is brought about only by contraction of inspiratory muscles at the expense of energy utilization. To empty the lungs more completely and more rapidly than is accomplished during quiet breathing, as during the deeper breaths accompanying exercise, expiration does become active. The intra-alveolar pressure must be increased even further above atmospheric pressure than can be accomplished by simple relaxation of the inspiratory muscles and elastic recoil of the lungs. To produce such a **forced** or **active expiration**, **expiratory muscles** must contract to further reduce the volume of the thoracic cavity and lungs (Figure 13–10 and Table 13–1). The most important expiratory muscles are (unbelievable as it may seem at first) the *muscles of the abdominal wall.* As the abdominal muscles contract, the resultant increase in intra-abdominal pressure exerts an upward force on the diaphragm, pushing it further up into the thoracic cavity than its relaxed position, thus decreasing the vertical dimension of the thoracic cavity even more. The other expiratory muscles are the **internal intercostal muscles**, whose contraction pulls the ribs downward and inward, flattening the chest wall and further decreasing the size of the thoracic cavity; this action is just the opposite of that of the external intercostal muscles (Fig. 13–11c).

As the active contraction of the expiratory muscles further reduces the volume of the thoracic cavity, the lungs also become further reduced in volume because they do not have to be stretched as much to fill the smaller thoracic cavity; that is, they are permitted to recoil to an even smaller volume. The intra-alveolar pressure increases further as the air in the lungs is confined within this smaller volume. The differential between intra-alveolar and atmospheric pressure is even greater now than during passive expiration, so more air leaves down the pressure gradient before equilibration is achieved. In this way, the lungs are emptied more completely during forceful, active expiration than during quiet, passive expiration.

During forceful expiration, the intrapleural pressure exceeds atmospheric pressure, but the lungs do not collapse. Because the intra-alveolar pressure is also increased correspondingly, a transmural pressure gradient still exists across the walls of the lungs, keeping them stretched to fill the thoracic cavity. For example, if the pressure within the thorax increases 10 mm Hg, the intrapleural pressure becomes 766 mm Hg and the intra-alveolar pressure becomes 770 mm Hg—still a 4 mm Hg pressure difference.

Because the diaphragm is the major inspiratory muscle and its relaxation also causes expiration, paralysis of the intercostal muscles alone does not seriously influence quiet breathing. Disruption of diaphragm activity caused by nerve or muscle disorders, however, leads to respiratory paralysis. Fortunately, the phrenic nerve arises from the spinal cord in the neck region (cervical segments 3, 4, and 5) and then descends to the diaphragm at the floor of the thorax, instead of arising from the thoracic region of the cord as might be expected. For this reason, individuals completely paralyzed below the neck as a result of traumatic severance of the spinal cord are still able to breathe, even though they have lost use of all other skeletal muscles in their trunk and limbs.

Airway resistance becomes an especially important determinant of airflow rates when the airways are narrowed by disease processes.

Thus far we have discussed airflow in and out of the lungs as a function of the magnitude of the pressure gradient between the alveoli and the atmosphere, with the pressure gradient changing through alterations in the dimensions of the thoracic cavity and subsequently of the lungs. However, just as flow of blood through the blood vessels depends not only on the pressure gradient but also on the resistance to the flow offered by the vessels, so it is with airflow:

$$F = \frac{\Delta P}{R}$$

where

F = airflow rate
ΔP = difference between atmospheric and intra-alveolar pressure (pressure gradient)
R = resistance of airways, determined by their radii

The primary determinant of resistance to airflow is the radius of the conducting airways. We ignored airway resistance in our preceding discussion of pressure-gradient–induced airflow rates because, in a healthy respiratory system, the radius of the conducting system is sufficiently large that resistance remains extremely low. Therefore, the pressure gradient between the alveoli and the atmosphere is normally the primary factor determining the airflow rate. Indeed, the airways offer such low resistance that normally only very small pressure gradients of 1–2 mm Hg need be created to achieve adequate rates of airflow in and out of the lungs. (By comparison, it would take a pressure gradient 250 times greater to move air through a smoker's pipe than through the respiratory airways at the same flow rate.)

Normally, modest adjustments in airway size can be accomplished by autonomic nervous-system regulation to suit the body's needs. Parasympathetic stimulation, which occurs in quiet, relaxed situations when the demand for airflow is not high, promotes bronchiolar smooth-muscle contraction, which increases airway resistance by producing **bronchoconstriction**. In contrast, sympathetic stimulation and to a greater extent its associated hormone, epinephrine, bring about **bronchodilation** and decreased airway resistance by promoting bronchiolar smooth-muscle relaxation (Table 13–2). Thus, during periods of sympathetic domination, when increased demands for O_2 uptake are actually or potentially placed on the body, bronchodilation occurs to assure that the pressure gradients established by respiratory-muscle activity are able to achieve maximum airflow rates with minimum resistance. Because of this bronchodilator action, epinephrine or similar drugs are useful therapeutic tools to counteract airway constriction in patients with bronchial spasms.

TABLE 13-2 Factors Affecting Airway Resistance

STATUS OF AIRWAYS	EFFECT ON RESISTANCE	FACTORS PRODUCING THE EFFECT

Bronchoconstriction	↓ radius, ↑ resistance to airflow	*Pathological factors:*
		Allergy-induced spasm of airways caused by:
		    Slow-reactive substance of anaphylaxis
		    Histamine
		Physical blockage of airways caused by:
		    Excess mucus
		    Edema of walls
		    Airway collapse
		Physiological control factors:
		Neural control: parasympathetic stimulation
		Local chemical control: ↓ CO_2 concentration

Bronchodilation	↑ radius, ↓ resistance to airflow	*Pathological factors:* none *Physiological control factors:* Neural control: sympathetic stimulation (minimal effect) Hormonal control: epinephrine Local chemical control: ↑ CO_2 concentration

Resistance becomes an extremely important impediment to airflow when airway lumens become narrowed as a result of disease. We have all transiently experienced the effect that increased airway resistance has on breathing when we have a cold. We know how difficult it is to produce an adequate airflow rate through a "stuffy nose" when the nasal passages are narrowed as a result of swelling and mucus accumulation.

More serious is **chronic obstructive pulmonary disease** (**COPD**), a group of lung diseases characterized by increased airway resistance resulting from the narrowing of the lumen of the lower airways. When airway resistance increases, a larger pressure gradient must be established to maintain even a normal airflow rate. For example, if resistance is doubled by narrowing of airway lumens, ΔP must be doubled through increased respiratory-muscle exertion to induce the same flow rate of air in and out of the lungs as a normal individual accomplishes during quiet breathing. Accordingly, patients with COPD have to work harder to breathe.

Chronic obstructive pulmonary disease encompasses three chronic (long-term) diseases: asthma, chronic bronchitis, and emphysema. In **asthma**, airway obstruction is due to (1) profound constriction of the smaller airways caused by allergy-induced spasm of the smooth muscle in the walls of these airways (see p. 402); (2) plugging of the airways by excess secretion of a very thick mucus; and (3) thickening of the walls of the airways due to inflammation and histamine-induced edema.

Chronic bronchitis is a long-term inflammatory condition of the lower respiratory airways, generally triggered by frequent exposure to irritating cigarette smoke, polluted air, or allergens. In response to the chronic irritation, the airways become narrowed because of prolonged edematous thickening of the airway linings, coupled with overproduction of a thick mucus. Despite frequent coughing associated with the chronic irritation, the plugged mucus often cannot be satisfactorily removed, especially since the ciliary mucus escalator is immobilized by the irritants (see p. 408). Pulmonary bacterial infections frequently occur, because the accumulated mucus serves as an excellent medium for bacterial growth.

Emphysema is characterized by collapse of the smaller airways and a breakdown of alveolar walls. This irreversible condition can arise in two different ways. Most commonly, emphysema results from excessive release of destructive enzymes such as *trypsin* from alveolar macrophages in response to chronic exposure to cigarette smoke or other inhaled chemical irritants. The lungs are normally protected from damage by these enzymes by α_1-*antitrypsin,* a protein that inhibits trypsin. Excessive secretion of these destructive enzymes in response to chronic irritation, however, can overwhelm the protective capability of α_1-antitrypsin so that these enzymes destroy not only foreign materials but lung tissue as well. Loss of lung tissue leads to the breakdown of alveolar walls and collapse of small airways, the characteristics of emphysema. Less frequently, emphysema arises from a genetic inability to produce α_1-antitrypsin so that the unprotected lung tissue gradually disintegrates under the influence of even small amounts of macrophage-released enzymes in the absence of chronic exposure to inhaled irritants.

When airway resistance is increased as a result of chronic obstructive lung disease of any type, expiration is more difficult to accomplish than inspiration. The smaller airways, lacking the cartilaginous rings that hold the larger airways open, are held open by the same transmural pressure gradient that distends the alveoli. Expansion of the thoracic cavity during inspiration indirectly dilates the airways even further than their expiratory dimensions, similar to alveolar expansion, so airway resistance is lower during inspiration than during expiration. In a normal individual, the airway resistance is always so low that the slight variation occurring between inspiration and expiration is not noticeable. When airway resistance has substantially increased, however, such as during an asthmatic attack, the difference between inspiration and expiration is quite noticeable. Thus, an asthmatic has more

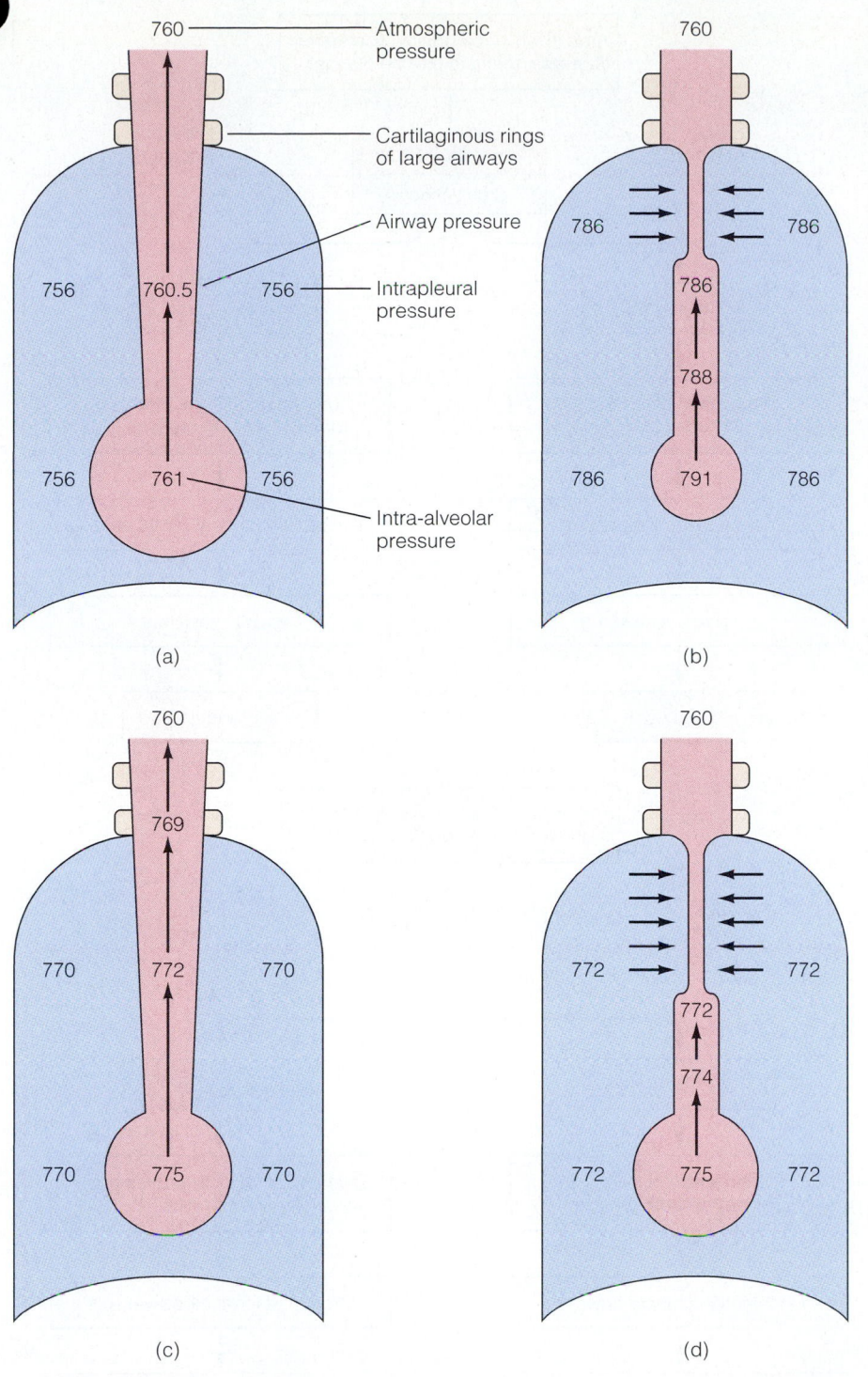

(a)

(b)

(c)

(d)

Numbers are mm Hg pressure.

• FIGURE 13–14 Airway Collapse during Forced Expiration (a) Normal quiet breathing, during which airway resistance is low so there is little frictional loss of pressure within the airways. Intrapleural pressure remains less than airway pressure throughout the length of the airways, so the airways remain open. (b) Maximal forced expiration, during which both intra-alveolar and intrapleural pressures are markedly increased. When frictional losses cause the airway pressure to fall below the surrounding elevated intrapleural pressure, the small nonrigid airways are compressed closed, blocking further expiration of air through the airway. In normal individuals, this occurs only at very low lung volumes. (c) Routine exercise. Even though intrapleural pressure is elevated during the active expiration accompanying routine vigorous activity, the airway pressure does not drop below the intrapleural pressure until the level at which the airways are held open by cartilaginous rings, so airway collapse does not occur. (d) Obstructive lung disease. Premature airway collapse occurs for two reasons: (1) the pressure drop along the airways is magnified as a result of increased airway resistance, and (2) the intrapleural pressure is higher than normal due to the loss, as in emphysema, of lung tissue that is responsible for the lung's tendency to recoil and pull away from the thoracic wall. Excessive air trapped in the alveoli behind the compressed bronchiolar segments reduces the amount of gas exchanged between the alveoli and atmosphere. Therefore, less alveolar air is "freshened" with each breath when airways collapse at higher lung volumes in patients with obstructive lung disease.

difficulty expiring than inspiring, giving rise to the characteristic "wheeze" as air is forced out through the narrowed airways.

In normal individuals, the smaller airways collapse and further outflow of air is halted only at very low lung volumes (• Fig. 13–14). Because of this airway collapse, the lungs can never be emptied completely.

Local controls act on the smooth muscle of the airways and arterioles to maximally match blood flow to airflow.

We have been talking about overall airway resistance in the lungs, but like arteriolar smooth muscle, bronchiolar smooth muscle is sensitive to local changes within its immediate en-

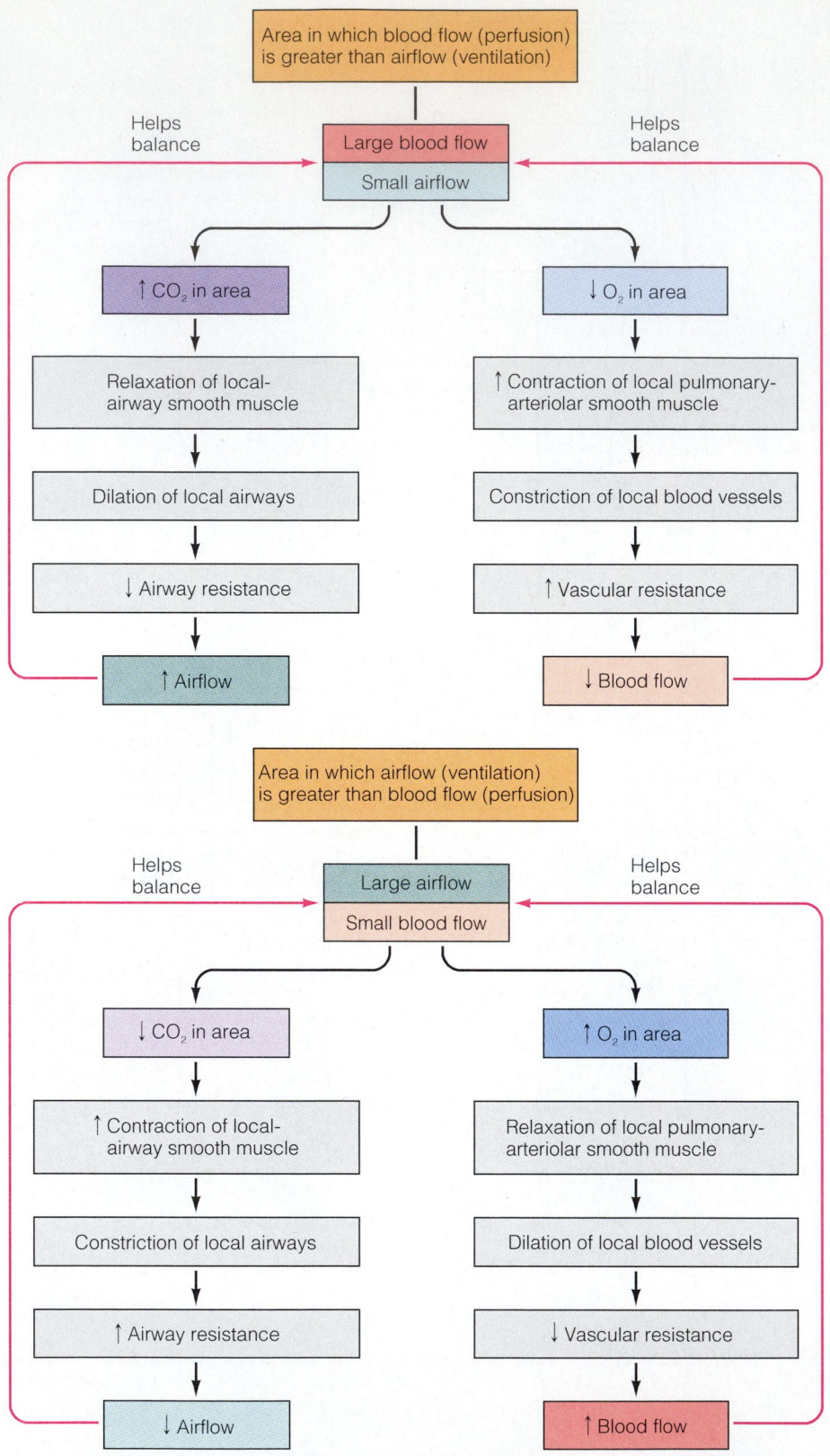

• **FIGURE 13–15   Local Controls to Match Airflow and Blood Flow to an Area of the Lung**

vironment, particularly to local CO_2 levels. If an alveolus is receiving too little airflow (ventilation) in comparison to its blood flow (perfusion), CO_2 levels will increase in the alveolus and surrounding tissue as more CO_2 is dropped off by the blood than is being exhaled into the atmosphere. This lo-

cal increase in CO_2 acts directly on the bronchiolar smooth muscle involved to induce the airway supplying the underaerated alveolus to relax. The resultant decrease in airway resistance leads to an increased airflow (for the same ΔP) to the involved alveolus, so its airflow now matches its blood supply

(● Fig. 13–15). The converse is also true. A localized decrease in CO_2 associated with an alveolus that is receiving too much air for its blood supply directly increases contractile activity of the airway smooth muscle involved, causing the airway supplying this overaerated alveolus to constrict. The result is a reduction in airflow to the overaerated alveolus.

A similar locally induced effect on pulmonary vascular smooth muscle also takes place simultaneously to maximally match blood flow to airflow. Just as in the systemic circulation, distribution of the cardiac output to different alveolar capillary networks can be controlled by adjusting the resistance to blood flow through specific pulmonary arterioles. If the blood flow is greater than the airflow to a given alveolus, the O_2 level in the alveolus and surrounding tissues will fall below normal as more O_2 than usual is extracted from the alveolus by the overabundance of blood. The local decrease in O_2 concentration causes vasoconstriction of the pulmonary arteriole supplying this particular capillary bed; the result is a reduction in blood flow to match the smaller airflow. Conversely, an increase in alveolar O_2 concentration caused by a mismatched large airflow and small blood flow brings about pulmonary vasodilation, which increases blood flow to match the larger airflow. Note that the local effect of O_2 on pulmonary arteriolar smooth muscle is appropriately just the opposite of its effect on systemic arteriolar smooth muscle (Table 13–3). In the systemic circulation, a decrease in O_2 in a tissue causes localized vasodilation to increase blood flow to the deprived area, and vice versa, which is important in matching blood supply to local metabolic needs (see p. 312).

The two mechanisms for matching airflow and blood flow function concurrently, so normally very little air or blood is wasted in the lung. Some regional differences in ventilation and perfusion exist from the top to the bottom of the lung because of gravitational effects. Nevertheless, airflow and blood flow at a particular alveolar interface are usually matched to the greatest extent possible by means of these local controls to accomplish efficient exchange of O_2 and CO_2.

Elastic behavior of the lungs is due to elastic connective-tissue fibers and alveolar surface tension.

You have learned that during the respiratory cycle, the lungs alternately expand during inspiration and recoil during expiration. What properties of the lungs enable them to behave like balloons, stretching and then snapping back to their resting position when the stretching forces are removed? Two in-

terrelated concepts are involved in pulmonary elasticity: elastic recoil and compliance.

Elastic recoil refers to how readily the lungs rebound after having been stretched. It is responsible for the lungs returning to their preinspiratory volume when the inspiratory muscles relax at the end of inspiration.

Compliance refers to how much effort is required to stretch or distend the lungs; it is analogous to how hard you have to work to blow up a balloon. (By comparison, one hundred times more distending pressure is required to inflate a child's toy balloon than to inflate the lungs.) Specifically, compliance is a measure of the magnitude of change in lung volume accomplished by a given change in the transmural pressure gradient, the force that stretches the lungs. A highly compliant lung stretches further for a given increase in the pressure difference than does a less compliant lung. Stated another way, the lower the compliance of the lungs, the larger the transmural pressure gradient that must be created during inspiration to produce normal lung expansion. In turn, a greater-than-normal transmural pressure gradient during inspiration can be achieved only by making the intrapleural pressure more subatmospheric than usual. This is accomplished by greater expansion of the thorax through more vigorous contraction of the inspiratory muscles. Therefore, the less compliant the lungs, the more work required to produce a given degree of inflation. A poorly compliant lung is referred to as a "stiff" lung, because it lacks normal stretchability. Respiratory compliance can be decreased by a number of factors, such as replacement of normal lung tissue with fibrous connective tissue as a result of breathing in asbestos fibers or similar irritants.

Pulmonary elastic behavior depends mainly on two factors: highly elastic connective tissue in the lungs and alveolar surface tension. Pulmonary connective tissue contains large quantities of elastin fibers (see p. 55). Not only do these fibers exhibit elastic properties themselves, but they are arranged into a meshwork that amplifies their elastic behavior, much like the threads in a piece of stretch-knit fabric. The entire piece of fabric (or lung) is stretchier and tends to bounce back to its original shape more than the individual threads (elastin fibers) of which the fabric is woven.

An even more important factor influencing elastic behavior of the lungs is the **alveolar surface tension** displayed by the thin liquid film that lines each alveolus. At an air-water interface, the water molecules at the surface are more strongly attracted to other surrounding water molecules than to the air above the surface. This unequal attraction produces a force known as surface tension at the surface of the liquid. Surface tension is responsible for a twofold effect. First, the liquid layer resists any force that increases its surface area; that is, it opposes expansion of the alveolus because the surface water molecules oppose being pulled apart. Accordingly, the greater the surface tension, the less compliant the lungs. Second, the liquid surface area tends to become as small as possible because the surface water molecules, being preferentially attracted to each other, try to get as close together as possible. Thus, the surface tension of the liquid lining an alveolus tends to reduce the size of the alveolus, squeezing in on the air within it (● Fig. 13–16). This property,

TABLE 13–3 Effects of Local Changes in O_2 on Pulmonary and Systemic Arterioles		
	EFFECT OF LOCAL CHANGE IN O_2	
VESSELS	**Decreased O_2**	**Increased O_2**
Pulmonary arterioles	Vasoconstriction	Vasodilation
Systemic arterioles	Vasodilation	Vasoconstriction

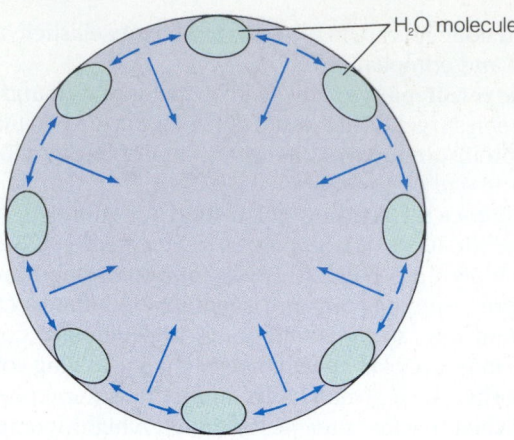

H₂O molecules

An alveolus

• **FIGURE 13–16   Alveolar Surface Tension**   The attractive forces between the water (H₂O) molecules in the liquid film that lines the alveolus are responsible for surface tension. Because of its surface tension, an alveolus (1) resists being stetched, (2) tends to be reduced in surface area or size, and (3) tends to recoil after being stretched.

along with the rebound of the stretched elastin fibers, is responsible for the lungs' elastic recoil back to their pre-inspiratory size when inspiration is over.

Pulmonary surfactant decreases surface tension and contributes to lung stability.

The cohesive forces between water molecules are so strong that if the alveoli were lined with water alone, the surface tension would be so great that the lungs would collapse; the recoil force attributable to the elastin fibers and high surface tension would exceed the opposing stretching force of the transmural pressure gradient. Furthermore, the lungs would be very poorly compliant, so exhausting muscular efforts would be required to accomplish stretching and inflation of the alveoli.

The tremendous surface tension of pure water is normally counteracted by secretion of pulmonary surfactant by the Type II alveolar cells (Fig. 13–3b). Pulmonary surfactant intersperses between the water molecules in the fluid lining the alveoli and lowers the alveolar surface tension because the cohesive force between a water molecule and an adjacent pulmonary surfactant molecule is very low. By lowering the alveolar surface tension, pulmonary surfactant provides two important benefits: (1) it increases pulmonary compliance, thus reducing the work of inflating the lungs; and (2) it reduces the lungs' tendency to recoil so that they do not collapse as readily.

Pulmonary surfactant's role in reducing the alveoli's tendency to recoil, thereby discouraging alveolar collapse, is important in helping maintain lung stability. The division of the lung into myriad tiny air sacs provides the advantage of a tremendously increased surface area for exchange of O₂ and CO₂, but it also presents the problem of maintaining the stability of all of these alveoli. Recall that the pressure generated by alveolar surface tension is directed inwards, squeezing in

on the air in the alveoli. If the alveoli are visualized as spherical bubbles, according to **LaPlace's law**, the magnitude of the inward-directed collapsing pressure is directly proportional to the surface tension and inversely proportional to the radius of the bubble:

$$P = \frac{2T}{r}$$

where

P = inward-directed collapsing pressure
T = surface tension
r = radius of bubble (alveolus)

Because the collapsing pressure is inversely proportional to the radius, the smaller the alveolus, the smaller its radius and the greater its tendency to collapse at a given surface tension. Accordingly, if two alveoli of unequal size but the same surface tension are connected by the same terminal airway, the smaller alveolus—because it generates a larger collapsing pressure—has a tendency to collapse and empty its air into the larger alveolus (• Fig. 13–17a). Small alveoli normally do not collapse and blow up larger alveoli, however, because pulmonary surfactant reduces the surface tension of small alveoli lower that of larger alveoli. Pulmonary surfactant decreases surface tension to a greater degree in small alveoli than in larger alveoli because the surfactant molecules are closer together in the smaller alveoli. The larger an alveolus, the more spread out are its surfactant molecules and the less effect they have on reducing surface tension. The surfactant-induced lower surface tension of small alveoli offsets the effect of their smaller radii in determining the inward-directed pressure. Therefore, the presence of surfactant causes the collapsing pressure of small alveoli to become comparable to that of larger alveoli and minimizes the tendency for small alveoli to collapse and empty their contents into larger alveoli (Fig. 13–17b). Pulmonary surfactant therefore helps stabilize the sizes of the alveoli and helps keep them open and available to participate in gas exchange.

A second factor that contributes to alveolar stability is the **interdependence** of neighboring alveoli. Each alveolus is surrounded by other alveoli with which it is interconnected by connective tissue. If an alveolus starts to collapse, the surrounding alveoli are stretched as their walls are pulled in the direction of the caving-in alveolus (• Fig. 13–18a). In turn, these neighboring alveoli, by recoiling in resistance to being stretched, exert expanding forces on the collapsing alveolus and thereby help keep it open (Fig. 13–18b). This phenomenon, which can be likened to a stalemated "tug of war" between adjacent alveoli, is termed *interdependence*.

The opposing forces acting on the lung (that is, the forces keeping the alveoli open and the countering forces that promote alveolar collapse) are summarized in Table 13–4.

A deficiency of pulmonary surfactant is responsible for newborn respiratory distress syndrome.

The developing fetal lungs normally do not have the ability to synthesize pulmonary surfactant until late in pregnancy. Es-

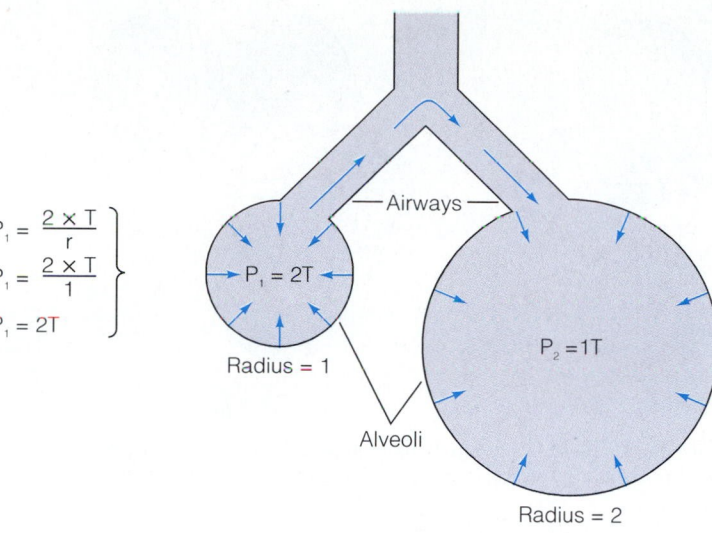

Law of LaPlace:

$$\text{Magnitude of inward-directed pressure (P) in a bubble (alveolus)} = \frac{2 \times \text{Surface tension (T)}}{\text{Radius (r) of bubble (alveolus)}}$$

$$P_1 = \frac{2 \times T}{r}$$
$$P_1 = \frac{2 \times T}{1}$$
$$P_1 = 2T$$

Airways

$P_1 = 2T$

Radius = 1

Alveoli

$P_2 = 1T$

Radius = 2

$$P_2 = \frac{2 \times T}{r}$$
$$P_2 = \frac{2 \times T}{2}$$
$$P_2 = 1T$$

$P_1 > P_2$

T = A given surface tension

(a)

$$P_1 = \frac{2 \times T}{r}$$
$$P_1 = \frac{2 \times \frac{1}{2}T}{1}$$
$$P_1 = 1T$$

Airways

$P_1 = 1T$

Alveoli

Radius = 1
Surface tension = $\frac{1}{2}T$

$P_2 = 1T$

$$P_2 = \frac{2 \times T}{r}$$
$$P_2 = \frac{2 \times T}{2}$$
$$P_2 = 1T$$

Pulmonary surfactant molecule

Radius = 2
Surface tension = T

$P_1 = P_2$

T = A given surface tension

(b)

• FIGURE 13–17   Role of Pulmonary Surfactant in Counteracting the Tendency for Small Alveoli to Collapse into Larger Alveoli (a) According to the law of LaPlace, if two alveoli of unequal size but the same surface tension are connected by the same terminal airway, the smaller alveolus—because it generates a larger inward-directed collapsing pressure—has a tendency (without pulmonary surfactant) to collapse and empty its air into the larger alveolus. (b) Pulmonary surfactant reduces the surface tension of a smaller alveolus more than that of a larger alveolus. This reduction in surface tension offsets the effect of the smaller radius in determining the inward-directed pressure. Consequently, the collapsing pressures of the small and large alveoli are comparable. Therefore, in the presence of pulmonary surfactant, a small alveolus does not collapse and empty its air into the larger alveolus.

pecially in an infant born prematurely, pulmonary surfactant may be insufficient to reduce the alveolar surface tension to manageable levels. The resultant collection of symptoms that develop are referred to as **newborn respiratory distress syndrome**. Very strenuous inspiratory efforts are required to overcome the high surface tension in an attempt to inflate the poorly compliant lungs. Adding to the dilemma, the work of breathing is further increased because the alveoli, in the ab-

sence of surfactant, tend to collapse almost completely during each expiration. It is more difficult (requires a greater transmural pressure differential) to expand a collapsed alveolus by a given volume than to increase an already partially expanded alveolus by the same volume. The situation is analogous to blowing up a new balloon. It takes more effort to blow in that first breath of air when starting to blow up a new balloon than to blow additional breaths into the already

• **FIGURE 13–18   Alveolar Interdependence**   (a) When an alveolus (in pink) in a group of interconnected alveoli starts to collapse, the surrounding alveoli (in gray) are stretched by the collapsing alveolus. (b) As neighboring alveoli recoil in resistance to being stretched, they pull outward on the collapsing alveolus. This expanding force pulls the collapsing alveolus open.

partially expanded balloon. With newborn respiratory distress syndrome, it is as though the infant must start blowing up a new balloon with every breath. Lung expansion may require transmural pressure gradients of 20 to 30 mm Hg (compared to the normal of 4 to 6 mm Hg) to overcome the tendency of surfactant-deprived alveoli to collapse.

The problem is compounded by the fact that the newborn's muscles are still weak. The respiratory distress associated with surfactant deficiency may soon lead to death as breathing efforts become exhausting or inadequate to support sufficient gas exchange.

This life-threatening condition affects 30,000 to 50,000 newborns, primarily premature infants, each year in the United States. Until the surfactant-secreting cells mature sufficiently, therapy often includes forcing air into the baby's lungs at greater-than-atmospheric, or "positive," pressure. By artificially increasing the atmospheric pressure, a sufficient pressure gradient can be established to drive air into the lungs. Recent clinical studies have also demonstrated success in treating the condition by surfactant replacement.

TABLE 13–4 Opposing Forces Acting on the Lung	
FORCES KEEPING THE ALVEOLI OPEN	**FORCES PROMOTING ALVEOLAR COLLAPSE**
Transmural pressure gradient	Elasticity of stretched pulmonary connective-tissue fibers
Pulmonary surfactant (which opposes alveolar surface tension)	Alveolar surface tension
Alveolar interdependence	

The work of breathing normally requires only about 3% of total energy expenditure.

During normal quiet breathing, the respiratory muscles must work during inspiration to expand the lungs against their elastic forces and to overcome airway resistance, whereas expiration is a passive process. Normally, the lungs are highy compliant and airway resistance is low, so only about 3% of the total energy expended by the body is used to accomplish quiet breathing. The work of breathing may be increased in four different situations:

1. *When pulmonary compliance is decreased,* more work is required to expand the lungs.
2. *When airway resistance is increased,* more work is required to achieve the greater pressure gradients necessary to overcome the resistance so that adequate airflow can occur.
3. *When elastic recoil is decreased,* passive expiration may be inadequate to expel the volume of air normally exhaled during quiet breathing. Thus, the abdominal muscles must work to aid in emptying the lungs, even when the person is at rest.
4. *When there is a need for increased ventilation,* such as during exercise, more work is required to accomplish both a greater depth of breathing (a larger volume of air moving in and out with each breath) and a faster rate of breathing (more breaths per minute).

During strenuous exercise, the amount of energy required to power pulmonary ventilation may increase up to twenty-five-fold. However, because total energy expenditure by the body is increased up to fifteen- to twenty-fold during heavy exercise, the energy used to accomplish the increased ventilation still represents only about 5% of total energy expended. In contrast, in patients with poorly compliant lungs or ob-

Total lung capacity at maximal inflation

5,700 ml

Variation in lung volume with normal, quiet breathing

Volume of lungs at end of normal inspiration (average 2,700 ml)

2,200 ml
2,700 ml

Volume of lungs at end of normal expiration (average 2,200 ml)

Difference between end-expiratory and end-inspiratory volume equals tidal volume (average 500 ml)

Minimal lung volume (residual volume) at maximal deflation

1,200 ml

• **FIGURE 13–19   Normal Range and Extremes of Lung Volume in Adult Male**

Values are average for a healthy young adult male; values for females are somewhat lower.

structive lung disease, the energy required for breathing even at rest may increase to 30% of total energy expenditure. In such cases, the individual's exercise ability is severely limited, as breathing itself becomes an exhausting exercise.

Normally, the lungs contain about 2 to 2.5 liters of air during the respiratory cycle but can be filled to over 5.5 liters or emptied to about 1 liter.

On average, in healthy young adults, the maximum amount of air that the lungs can hold is about 5.7 liters in males (4.2 liters in females). Anatomical build, age, the distensibility of the lungs, and the presence or absence of respiratory disease affect this total lung capacity. Normally, during quiet breathing, the lungs are not anywhere near maximally inflated nor are they deflated to their minimum volume. Thus, the lungs normally remain moderately inflated throughout the respiratory cycle. At the end of a normal quiet expiration, the lungs still contain about 2,200 ml of air. During each typical breath under resting conditions, about 500 ml of air are inspired and the same quantity is expired, so during quiet breathing the lung volume varies between 2,200 ml at the end of expiration to 2,700 ml at the end of inspiration (• Fig. 13–19). During maximal expiration, lung volume can be decreased to 1,200 ml in males (1,000 ml in females), but the lungs can never be completely deflated because the small airways collapse during forced expirations at low lung volumes, blocking further outflow of air (Fig. 13–14b).

An important outcome of not being able to empty the lungs completely is that, even during maximal expiratory efforts, gas exchange can still continue between blood flowing through the lungs and the remaining alveolar air. Instead of the wide fluctuations that would occur in O_2 uptake and CO_2 removal by the blood if the lungs were to completely fill and empty with each breath, the gas content of the blood leaving the lungs for delivery to the tissues normally remains remarkably constant throughout the respiratory cycle. Furthermore,

recall that it takes less effort to inflate a partially inflated alveolus than a totally collapsed one.

The changes in lung volume that occur with different respiratory efforts can be measured using a **spirometer**. Basically, a spirometer consists of an air-filled drum floating in a water-filled chamber. As the person breathes air in and out of the drum through a tube connecting the mouth to the air chamber, the drum rises and falls in the water chamber (• Fig. 13–20). This rise and fall can be recorded as a **spirogram**, which is calibrated to volume changes. The pen records inspiration as an upward deflection and expiration as a downward deflection.

• Figure 13–21 is a hypothetical example of a spirogram in a healthy young adult male. Generally, the values are lower for females. The following lung volumes and lung capacities (a lung capacity is a sum of two or more lung volumes) can be determined:

■ **Tidal volume (TV).** The volume of air entering or leaving the lungs during a single breath. Average value under resting conditions = 500 ml.

■ **Inspiratory reserve volume (IRV).** The extra volume of air that can be maximally inspired over and above the typical resting tidal volume. The IRV is accomplished by maximal contraction of the diaphragm, external intercostal muscles, and accessory inspiratory muscles. Average value = 3,000 ml.

■ **Inspiratory capacity (IC).** The maximum volume of air that can be inspired at the end of a normal quiet expiration (IC = IRV + TV). Average value = 3,500 ml.

■ **Expiratory reserve volume (ERV).** The extra volume of air that can be actively expired by maximal contraction of the expiratory muscles beyond that normally passively expired at the end of a typical resting tidal volume. Average value = 1,000 ml.

■ **Residual volume (RV).** The minimum volume of air remaining in the lungs even after a maximal expiration. Av-

● **FIGURE 13–20   A Spirometer**   A spirometer is a device that measures the volume of air breathed in and out; it consists of an air-filled drum floating in a water-filled chamber. As a person breathes air in and out of the drum through a connecting tube, the resultant rise and fall of the drum are recorded as a spirogram, which is calibrated to the magnitude of the volume change.

erage value = 1,200 ml. The residual volume cannot be measured directly with a spirometer because this volume of air does not move in and out of the lungs. It can be determined indirectly, however, through gas-dilution techniques involving inspiration of a known quantity of a harmless tracer gas such as helium.

● **FIGURE 13–21   Normal Spirogram of Healthy Young Adult Male**

TV = Tidal volume (500 ml)
IRV = Inspiratory reserve volume (3,000 ml)
IC = Inspiratory capacity (3,500 ml)
ERV = Expiratory reserve volume (1,000 ml)
RV = Residual volume (1,200 ml)
FRC = Functional residual capacity (2,200 ml)
VC = Vital capacity (4,500 ml)
TLC = Total lung capacity (5,700 ml)

■ **Functional residual capacity (FRC).** The volume of air in the lungs at the end of a normal passive expiration (FRC = ERV + RV). Average value = 2,200 ml.

■ **Vital capacity (VC).** The maximum volume of air that can be moved out during a single breath following a maximal inspiration. The subject first inspires maximally, then expires maximally (VC = IRV + TV + ERV). The VC represents the maximum volume change possible within the lungs (● Fig. 13–22). It is rarely used because the maximal muscle contractions involved become exhausting, but it is useful in ascertaining the functional capacity of the lungs. Average value = 4,500 ml.

■ **Total lung capacity (TLC).** The maximum volume of air that the lungs can hold (TLC = VC + RV). Average value = 5,700 ml.

■ **Forced expiratory volume in one second (FEV_1).** The volume of air that can be expired during the first second of expiration in a VC determination. Usually, FEV_1 is about 80% of VC; that is, normally 80% of the air that can be forcibly expired from maximally inflated lungs can be expired within 1 second. This measurement gives an indication of the maximal airflow rate that is possible from the lungs.

Measurement of the lungs' various volumes and capacities is of more than pure academic interest, because such determinations provide a useful tool to the diagnostician in various respiratory-disease states. Two general categories of respiratory dysfunction yield abnormal results during spirometry—*obstructive* and *restrictive* lung disease (● Fig. 13–23). How-

● FIGURE 13–22   X Rays of Lungs Showing Maximum Volume Change
(a) Minimum volume of the lungs at maximum expiration. (b) Maximum volume of the lungs at maximum inspiration. The difference between these two volumes is the vital capacity, which is the maximum volume of air that can be moved out during a single breath following a maximum inspiration.

(a)   (b)

ever, you should not have the impression that these are the only categories of respiratory dysfunction nor that spirometry is the only pulmonary-function test. Other conditions affecting respiratory function include (1) diseases impairing diffusion of O_2 and CO_2 across the pulmonary membranes; (2) reduced ventilation because of mechanical failure, as with neuromuscular disorders affecting the respiratory muscles, or because of depression of the respiratory control center due to alcohol, drugs, or other chemicals; (3) failure of adequate pulmonary blood flow; or (4) ventilation/perfusion abnormalities involving a poor matching of air and blood so that efficient gas exchange cannot occur. Some lung diseases are actually a complex mixture of different types of functional disturbances. To determine what abnormalities are present, the diagnostician relies on a variety of respiratory-function tests in addition to spirometry, including X-ray examination, blood-gas determinations, and tests to measure the diffusion capacity of the alveolar capillary membrane.

Alveolar ventilation is less than pulmonary ventilation because of the presence of dead space.

Various changes in volume represent only one factor in the determination of **pulmonary** or **minute ventilation**, which is

● **FIGURE 13–23   Abnormal Spirograms Associated with Obstructive and Restrictive Lung Diseases** (a) Spirogram in obstructive lung disease. Since a patient with obstructive lung disease experiences more difficulty emptying the lungs than filling them, the total lung capacity (TLC) is essentially normal, but the functional residual capacity (FRC) and the residual volume (RV) are elevated as a result of the additional air trapped in the lungs following expiration. Because the RV is increased, the vital capacity (VC) is reduced. With more air remaining in the lungs, less of the TLC is available to be used in exchanging air with the atmosphere. Another common finding is a markedly reduced FEV_1 since the airflow rate is reduced by the airway obstruction. Even though both the VC and the FEV_1 are reduced, the FEV_1 is reduced more markedly than is the VC. As a result, the FEV_1-to-VC ratio is much lower than the normal 80%; that is, much less than 80% of the reduced VC can be blown out during the first second. (b) Spirogram in restrictive lung disease. In this disease the lungs are less compliant than normal. Total lung capacity, inspiratory capacity, and VC are reduced, since the lungs cannot be expanded as normal. The percentage of the VC that can be exhaled within 1 sec is the normal 80% or an even higher percentage, because air can flow freely in the airways. Therefore, the $FEV_1/VC\%$ is particularly useful in distinguishing between obstructive and restrictive lung disease. Also, in contrast to obstructive lung disease, the RV is usually normal in restrictive lung disease.

 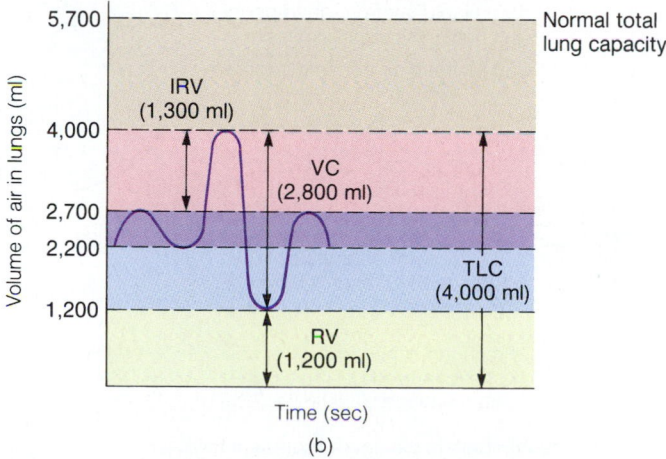

(a)   (b)

the volume of air breathed in and out in one minute. The other important factor is **respiratory rate**, which averages twelve breaths per minute.

pulmonary ventilation = tidal volume × respiratory rate
(ml/min)   (ml/breath)   (breaths/min)

At an average tidal volume of 500 ml/breath and a respiratory rate of 12 breaths per min, pulmonary ventilation is 6,000 ml

or 6 liters of air breathed in and out in one minute under resting conditions. For a brief period of time, a healthy young adult male can voluntarily increase his total pulmonary ventilation twenty-five-fold, to 150 liters/min. To increase pulmonary ventilation, both tidal volume and respiratory rate increase, but depth of breathing is increased more than frequency of breathing.

When increasing pulmonary ventilation, it is usually more

• **FIGURE 13–24   Effect of Dead-Space Volume on Exchange of Tidal Volume between Atmosphere and Alveoli** Even though 500 ml of air move in and out between the atmosphere and respiratory system and 500 ml move in and out of the alveoli with each breath, only 350 ml are actually exchanged between the atmosphere and the alveoli because of the presence of anatomic dead space (the volume of air in the respiratory airways). (a) After inspiration, before expiration. At the end of inspiration, the respiratory airways are filled with 150 ml of fresh atmospheric air from the inspiration. (b) During expiration. During the subsequent expiration, 500 ml of air are expired to the atmosphere. The first 150 ml expired are the fresh air that was retained in the airways and never used. The remaining 350 ml are "old" alveolar air that has participated in gas exchange with the blood. During the same expiration, 500 ml of gas also leave the alveoli. The first 350 ml are expired to the atmosphere; the other 150 ml of old alveolar air never reach the outside but remain in the conducting airways. (c) During inspiration. On the next inspiration, 500 ml of gas enter the alveoli. The first 150 ml to enter the alveoli are the old alveolar air that remained in the dead space during the preceding expiration. The other 350 ml entering the alveoli are fresh air inspired from the atmosphere. Simultaneously, 500 ml of air enter from the atmosphere. The first 350 ml of atmospheric air reach the alveoli; the other 150 ml remain in the conducting airways to be expired without benefit of being exchanged with the blood, as the cycle repeats itself.

advantageous to have a greater increase in tidal volume than in respiratory rate because of the presence of anatomic dead space. Not all of the inspired air gets down to the site of gas exchange in the alveoli. Part of it remains in the conducting airways, where it is not available for gas exchange. The volume of the conducting passages in an adult averages about 150 ml. This volume is considered to be **anatomic dead space** because air within these conducting airways is useless for exchange purposes. Anatomic dead space has a pronounced effect on the efficiency of pulmonary ventilation. In effect, even though 500 ml of air are moved in and out with each breath, only 350 ml are actually exchanged between the atmosphere and alveoli because of the 150 ml occupying the anatomic dead space (● Fig. 13–24).

Since the amount of atmospheric air that reaches the alveoli and is actually available for exchange with the blood is more important than the total amount breathed in and out, **alveolar ventilation**—the volume of air exchanged between the atmosphere and alveoli per minute—is more important than pulmonary ventilation. In determining alveolar ventilation, the amount of wasted air moved in and out through the anatomic dead space must be taken into account, as follows:

$$\text{alveolar ventilation} =$$
$$(\text{tidal volume} - \text{dead-space volume}) \times \text{respiratory rate}$$

With quiet breathing, alveolar ventilation is 4,200 ml/min [(500 ml/breath − 150 ml dead-space volume) × 12 breaths/min = 4,200 ml/min], whereas pulmonary ventilation is 6,000 ml/min.

To understand how important dead-space volume is in determining the magnitude of alveolar ventilation, examine the effect of various breathing patterns on alveolar ventilation in Table 13–5. If a person deliberately breathes deeply (for example, a tidal volume of 1,200 ml) and slowly (for example, a respiratory rate of 5 breaths/min), pulmonary ventilation is 6,000 ml/min, the same as during quiet breathing at rest, but alveolar ventilation increases to 5,250 ml/min compared to the resting rate of 4,200 ml/min. In contrast, if a person deliberately breathes shallowly (for example, a tidal volume of 150 ml) and rapidly (a frequency of 40 breaths/min), pulmonary ventilation would still be 6,000 ml/min; however, alveolar ventilation would be 0 ml/min. In effect, the person would only be drawing air in and out of the anatomic dead space without any atmospheric air being exchanged with the alveoli, where it could be useful. The individual could voluntarily maintain such a breathing pattern for only a few minutes before losing consciousness, at which time normal breathing would resume.

The value of reflexly bringing about a larger increase in depth of breathing than in rate of breathing when pulmonary ventilation is increased during exercise should now be apparent. It is the most efficient means of elevating alveolar ventilation. When tidal volume is increased, the entire increase goes toward elevating alveolar ventilation, while an increase in respiratory rate does not go entirely toward increasing alveolar ventilation. When respiratory rate is increased, the frequency with which air is wasted in the dead space is also increased, because a portion of *each* breath must move in and out of the dead space. As needs vary, ventilation is normally adjusted to a tidal volume and respiratory rate that meet those needs most efficiently in terms of energy cost.

We have assumed that all the atmospheric air entering the alveoli participates in exchanges of O_2 and CO_2 with pulmonary blood. However, the match between air and blood is not always perfect, because not all alveoli are equally ventilated with air and perfused with blood. Any ventilated alveoli that do not participate in gas exchange with blood because they are inadequately perfused are considered to be **alveolar dead space**. In normal persons, alveolar dead space is quite small and of little importance, but it can be increased to even lethal levels in several types of pulmonary disease.

GAS EXCHANGE

Gases move down partial pressure gradients.

The ultimate purpose of breathing is to provide a continual supply of fresh O_2 for pickup by the blood and to constantly remove CO_2 unloaded from the blood. The blood acts as a transport system for O_2 and CO_2 between the lungs and tissues, with the tissue cells extracting O_2 from the blood

BREATHING PATTERN	TIDAL VOLUME (ML/BREATH)	RESPIRATORY RATE (BREATHS/MIN)	DEAD-SPACE VOLUME (ML)	PULMONARY VENTILATION (ML/MIN)*	ALVEOLAR VENTILATION (ML/MIN)**
TABLE 13–5 **Effect of Different Breathing Patterns on Alveolar Ventilation**					
Quiet breathing at rest	500	12	150	6,000	4,200
Deep, slow breathing	1,200	5	150	6,000	5,250
Shallow, rapid breathing	150	40	150	6,000	0

*Equals tidal volume times respiratory rate.
**Equals (tidal volume minus dead-space volume) times respiratory rate.

and eliminating CO_2 into it. Gaseous exchange at both the pulmonary-capillary and tissue-capillary level involves simple passive diffusion of O_2 and CO_2 down *partial pressure gradients.* There are no active transport mechanisms for these gases. Let us see what partial pressure gradients are and how they are established.

Atmospheric air is a mixture of gases that contains about 79% nitrogen (N_2) and 21% O_2, with almost negligible percentages of CO_2, H_2O vapor, other gases, and pollutants in normally dry air. Altogether, these gases exert a total atmospheric pressure of 760 mm Hg at sea level. This total pressure is equal to the sum of the pressures that each gas in the mixture partially contributes. The pressure exerted by a particular gas is directly proportional to the percentage of that gas in the total air mixture. Every gas molecule, no matter what its size, exerts the same amount of pressure; for example, a N_2 molecule exerts the same pressure as an O_2 molecule. Since 79% of the air consists of N_2 molecules, 79% of the 760 mm Hg atmospheric pressure, or 600 mm Hg, is exerted by the N_2 molecules. Similarly, since O_2 represents 21% of the atmosphere, 21% of the 760 mm Hg atmospheric pressure, or 160 mm Hg, is exerted by O_2 (• Fig. 13-25). The individual pressure exerted independently by a particular gas within a mixture of gases is known as its **partial pressure**, designated by P_{gas}. Thus, the partial pressure of O_2 in atmospheric air, $\mathbf{P_{O_2}}$, is normally 160 mm Hg. The atmospheric partial pressure of CO_2, $\mathbf{P_{CO_2}}$, is negligible at 0.3 mm Hg.

Gases dissolved in a liquid such as blood or another body fluid are also considered to exert a partial pressure. The amount of a gas that will dissolve in the blood depends on the solubility of the gas in blood and on the partial pressure of the gas in the alveolar air to which the blood is exposed. Because the solubilities of O_2 and CO_2 in blood remain constant, the amount of O_2 and CO_2 dissolved in the pulmonary-capillary blood is directly proportional to the alveolar P_{O_2} and P_{CO_2}. The alveolar partial pressure of a particular gas can be thought of as "holding" that gas in solution in the blood.

If, as is the case with O_2, the alveolar partial pressure of a gas is *higher* than the partial pressure of that gas in the blood entering the pulmonary capillaries, the higher alveolar partial pressure drives more O_2 into the blood. Oxygen diffuses from the alveoli and dissolves in the blood until blood P_{O_2} becomes equal to alveolar P_{O_2}. Conversely, if the alveolar partial pressure of a gas is *lower* than its partial pressure in the entering blood—the situation that exists for CO_2—the lower alveolar partial pressure permits some of the CO_2 to escape from solution (that is, to no longer be dissolved) in the blood. As CO_2 comes out of solution, it diffuses into the alveoli until blood P_{CO_2} equilibrates with alveolar P_{CO_2}. Such a difference in partial pressure between pulmonary blood and alveolar air is known as a **partial pressure gradient**. A gas always diffuses down its partial pressure gradient from the area of higher to the area of lower partial pressure, similar to diffusion down a concentration gradient.

Oxygen enters and CO_2 leaves the blood in the lungs passively down partial pressure gradients.

Alveolar air is not of the same composition as inspired atmospheric air for two reasons. First, as soon as atmospheric air enters the respiratory passages, it becomes saturated with H_2O by exposure to the moist airways. Water vapor exerts a partial pressure just as any other gas does. At body temperature, the partial pressure of H_2O vapor is 47 mm Hg. Humidification of inspired air in effect "dilutes" the partial pressure of the inspired gases by 47 mm Hg, because the sum of the partial pressures must total the atmospheric pressure of

• **FIGURE 13-25 Concept of Partial Pressures** The partial pressure exerted by each gas in a mixture equals the total pressure times the fractional composition of the gas in the mixture.

Composition and partial pressures in atmospheric air

79% N_2

Partial pressure of N_2 = 600 mm Hg

Partial pressure of $N_2(P_{N_2})$ in atmospheric air = 760 mm Hg x 0.79 = 600 mm Hg

Total atmospheric pressure = 760 mm Hg

21% O_2

Partial pressure of O_2 = 160 mm Hg

Partial pressure of $O_2(P_{O_2})$ in atmospheric air = 760 mm Hg x 0.21 = 160 mm Hg

760 mm Hg. In moist air, $P_{H_2O} = 47$ mm Hg, $P_{N_2} = 563$ mm Hg, and $P_{O_2} = 150$ mm Hg.

Second, alveolar P_{O_2} is also lower than atmospheric P_{O_2} because fresh inspired air is mixed with the large volume of old air that remained in the lungs and dead space at the end of the preceding expiration (the functional residual capacity). Only about one-seventh of the total alveolar air is replaced by fresh atmospheric air with each normal breath. Thus, at the end of inspiration, less than 15% of the air in the alveoli is fresh air. As a result of humidification and the small turnover of alveolar air, the average alveolar P_{O_2} is 100 mm Hg, compared to the atmospheric P_{O_2} of 160 mm Hg.

It is logical to think that alveolar P_{O_2} would increase during inspiration with the arrival of fresh air and would decrease during expiration. Only small fluctuations of a few mm Hg occur, however, for two reasons. First, only a small proportion of the total alveolar air is exchanged with each breath. The relatively small volume of high-P_{O_2} air that is inspired is quickly mixed with the much larger volume of retained alveolar air, which has a lower P_{O_2}. Thus, the O_2 in the inspired air can only slightly elevate the total alveolar P_{O_2}. Even this potentially small elevation of P_{O_2} is diminished for another reason. Oxygen is continually moving by passive diffusion down its partial pressure gradient from the alveoli into the blood. The O_2 arriving in the alveoli in the newly inspired air simply replaces the O_2 diffusing out of the alveoli into the pulmonary capillaries. Therefore, the alveolar P_{O_2} remains relatively constant at about 100 mm Hg throughout the respiratory cycle. Because the pulmonary-blood P_{O_2} equilibrates with the alveolar P_{O_2}, the P_{O_2} of the blood likewise remains fairly constant at this same value. Accordingly, the amount of O_2 in the blood available to the tissues varies only slightly during the respiratory cycle.

A similar situation in reverse exists for CO_2. Carbon dioxide, which is continually produced by the body tissues as a metabolic waste product, is constantly added to the blood at the level of the systemic capillaries. In the pulmonary capillaries, CO_2 diffuses down its partial pressure gradient from the blood into the alveoli and is subsequently removed from the body during expiration. Like O_2, alveolar P_{CO_2} remains fairly constant throughout the respiratory cycle but at a lower value of 40 mm Hg. Ventilation constantly replenishes alveolar P_{O_2}, keeping it relatively high, and constantly removes CO_2, keeping alveolar P_{CO_2} relatively low. Thus, the appropriate partial pressure gradients between the alveoli and blood are maintained to assure that O_2 enters the blood and CO_2 leaves the blood.

The blood entering the pulmonary capillaries is systemic venous blood pumped to the lungs through the pulmonary arteries. This blood, having just returned from the body tissues, is relatively low in O_2, with a P_{O_2} of 40 mm Hg, and is relatively high in CO_2, with a P_{CO_2} of 46 mm Hg. As this blood flows through the pulmonary capillaries, it is exposed to alveolar air (● Fig. 13–26). Since the alveolar P_{O_2} at 100 mm Hg is higher than the P_{O_2} of 40 mm Hg in the blood entering the lungs, O_2 diffuses down its partial pressure gradient from the alveoli into the blood until no further gradient exists. As the blood leaves the pulmonary capillaries, it has a P_{O_2} equal

to alveolar P_{O_2} at 100 mm Hg. The partial pressure gradient for CO_2 is in the opposite direction. Blood entering the pulmonary capillaries has a P_{CO_2} of 46 mm Hg, whereas alveolar P_{CO_2} is only 40 mm Hg. Carbon dioxide diffuses from the blood into the alveoli until blood P_{CO_2} equilibrates with alveolar P_{CO_2}. Thus, the blood leaving the pulmonary capillaries has a P_{CO_2} of 40 mm Hg. As the blood passes through the lungs, it picks up O_2 and gives up CO_2 simply by diffusion down partial pressure gradients that exist between the blood and alveoli. After leaving the lungs, the blood, which now has a P_{O_2} of 100 mm Hg and a P_{CO_2} of 40 mm Hg, is returned to the heart to be subsequently pumped out to the body tissues as systemic arterial blood.

Note that blood returning to the lungs from the tissues still contains O_2 (P_{O_2} of systemic venous blood = 40 mm Hg) and that blood leaving the lungs still contains CO_2 (P_{CO_2} of systemic arterial blood = 40 mm Hg). The additional O_2 carried in the blood beyond that normally given up to the tissues represents an immediately available O_2 reserve that can be tapped by the tissue cells whenever their O_2 demands increase. The CO_2 remaining in the blood even after passage through the lungs plays an important role in the acid-base balance of the body, because CO_2 generates carbonic acid. Furthermore, arterial P_{CO_2} is important in driving respiration. This mechanism will be described later.

The amount of O_2 picked up in the lungs matches the amount extracted and used by the tissues. When the tissues metabolize more actively (for example, during exercise), more O_2 is extracted from the blood at the tissue level, reducing the systemic venous P_{O_2} even lower than 40 mm Hg—for example, to a P_{O_2} of 30 mm Hg. When this blood returns to the lungs, a larger-than-normal P_{O_2} gradient exists between the newly entering blood and alveolar air. The difference in P_{O_2} between the alveoli and blood is now 70 mm Hg (alveolar P_{O_2} of 100 mm Hg and blood P_{O_2} of 30 mm Hg), compared to the normal P_{O_2} gradient of 60 mm Hg (alveolar P_{O_2} of 100 mm Hg and blood P_{O_2} of 40 mm Hg). Therefore, more O_2 diffuses from the alveoli into the blood down the larger partial pressure gradient before blood P_{O_2} equals alveolar P_{O_2}. This additional transfer of O_2 into the blood replaces the increased amount of O_2 consumed, so O_2 uptake matches O_2 use even when O_2 consumption increases. At the same time that more O_2 is diffusing from the alveoli into the blood because of the increased partial pressure gradient, ventilation is stimulated so that O_2 enters the aveoli more rapidly from the atmosphere to replace the O_2 diffusing into the blood.

Similarly, the amount of CO_2 given up to the alveoli from the blood matches the amount of CO_2 picked up at the tissues. Once again, an increase in ventilation associated with increased activity assures that increased amounts of CO_2 delivered to the alveoli are blown off to the atmosphere.

Factors other than the partial pressure gradient influence the rate of gas transfer.

We have been discussing diffusion of O_2 and CO_2 between the blood and the alveoli as if these gases' partial pressure gradients were the sole determinants of their rates of diffusion. Re-

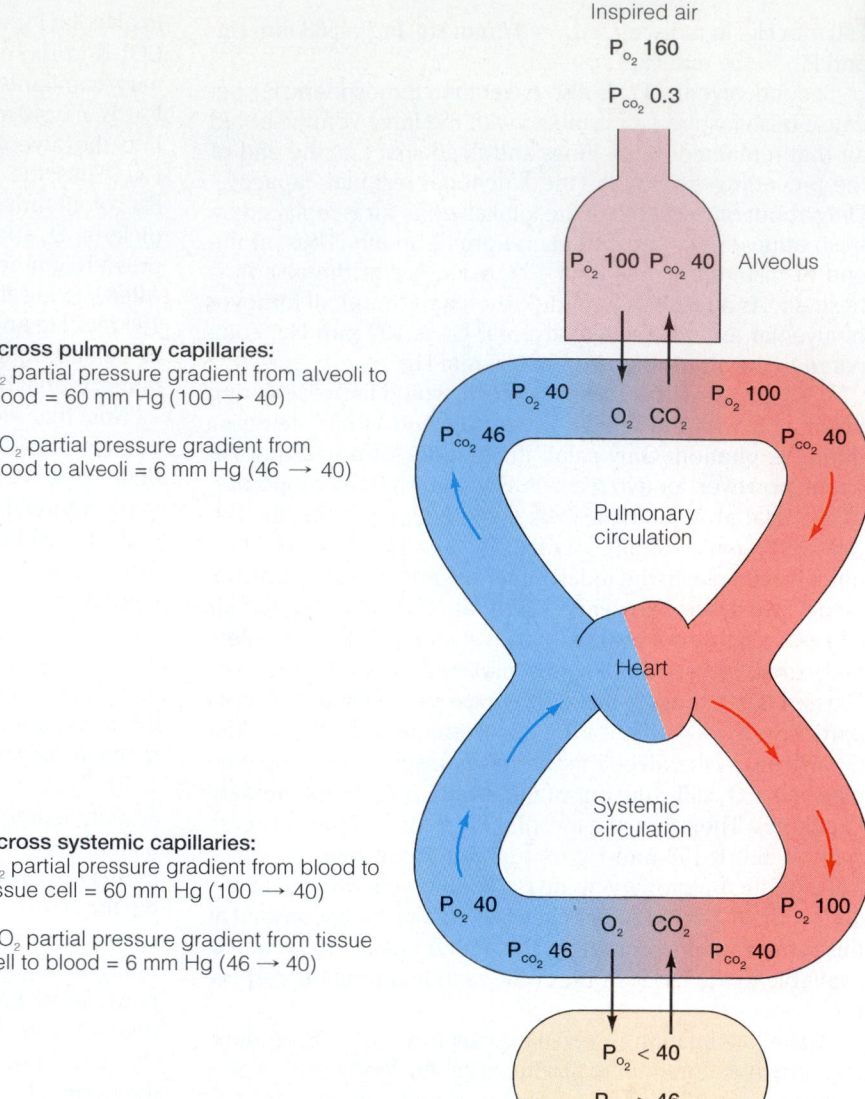

• FIGURE 13–26 Oxygen and CO₂ Exchange across Pulmonary and Systemic Capillaries Caused by Partial Pressure Gradients Alveolar P_{O_2} remains relatively high and alveolar P_{CO_2} remains relatively low because a portion of the alveolar air is exchanged for fresh atmospheric air with each breath. In contrast, the systemic venous blood entering the lungs is relatively low in O_2 and high in CO_2, having given up O_2 and picked up CO_2 at the systemic capillary level. This establishes partial pressure gradients between the alveolar air and pulmonary capillary blood that induce the passive diffusion of O_2 into the blood and CO_2 out of the blood until the blood and alveolar partial pressures become equal. The blood leaving the lungs is thus relatively high in O_2 and low in CO_2 compared to the partial pressures in the O_2-consuming, CO_2-producing tissue cells to which it is delivered. Consequently, partial pressure gradients for gas exchange at the tissue level favor the passive movement of O_2 out of the blood into the cells to support their metabolic requirements and also favor the simultaneous transfer of CO_2 into the blood. The blood then returns to the lungs to once again fill up on O_2 and dump off CO_2. The systemic arterial P_{O_2} and P_{CO_2} normally remain essentially constant, having equilibrated with the alveolar partial pressures, which remain essentially constant. In contrast, the systemic venous P_{O_2} and P_{CO_2} vary, depending on the level of metabolic activity.

Across pulmonary capillaries:
O_2 partial pressure gradient from alveoli to blood = 60 mm Hg (100 → 40)

CO_2 partial pressure gradient from blood to alveoli = 6 mm Hg (46 → 40)

Across systemic capillaries:
O_2 partial pressure gradient from blood to tissue cell = 60 mm Hg (100 → 40)

CO_2 partial pressure gradient from tissue cell to blood = 6 mm Hg (46 → 40)

Numbers are mm Hg pressure.

call that, according to Fick's law of diffusion, the rate of diffusion of a gas through a sheet of tissue also depends on the surface area and thickness of the membrane through which the gas is diffusing and on the diffusion coefficient of the particular gas (Table 13–6). Normally, changes in the rate of gas exchange are determined primarily by changes in partial pressure gradients between the blood and alveoli, because these other factors are relatively constant under resting conditions.

During exercise, however, the surface area available for exchange can be physiologically increased to enhance the rate of gas transfer. During resting conditions, some of the pulmonary capillaries are typically closed, because the normally low pressure of the pulmonary circulation is inadequate to keep all of the capillaries open. During exercise, when the

pulmonary blood pressure is raised as a result of increased cardiac output, many of the previously closed pulmonary capillaries are forced open. This increases the surface area of blood available for exchange. Furthermore, the aveolar membranes are stretched further than normal during exercise because of the larger tidal volumes (deeper breathing). Such stretching increases the aveolar surface area and decreases the thickness of the alveolar membrane. Collectively, these changes expedite gas exchange during exercise.

On the other hand, several pathological conditions can markedly reduce the pulmonary surface area and, in turn, decrease the rate of gas exchange. Most notably, surface area is reduced in emphysema because many of the alveolar walls are lost, resulting in larger but fewer chambers (• Fig. 13–27). Loss of surface area for exchange is likewise associated with

TABLE 13–6   Factors That Influence Rate of Gas Transfer across Alveolar Membrane

FACTOR	INFLUENCE ON RATE OF GAS TRANSFER ACROSS ALVEOLAR MEMBRANE	COMMENTS
Partial pressure gradients of O_2 and CO_2	Rate of transfer ↑ as partial pressure gradient ↑	Major determinant of rate of transfer
Surface area of alveolar membrane	Rate of transfer ↑ as surface area ↑	Surface area remains constant under resting conditions
		Surface area ↑ during exercise as more pulmonary capillaries open up when cardiac output increases and alveoli expand as breathing becomes deeper
		Surface area ↓ with pathological conditions such as emphysema and atelectasis
Thickness of barrier separating air and blood across alveolar membrane	Rate of transfer ↓ as thickness ↑	Thickness normally remains constant
		Thickness ↑ with pathological conditions such as pulmonary edema, pulmonary fibrosis, and pneumonia
Diffusion coefficient (solubility of gas in membrane)	Rate of transfer ↑ as diffusion coefficient ↑	Diffusion coefficient for CO_2 is 20 times that of O_2, offsetting the smaller partial pressure gradient for CO_2; therefore, approximately equal amounts of CO_2 and O_2 are transferred across the membrane

atelectic regions of the lung and also results when part of the lung tissue is surgically removed—for example, in the treatment of lung cancer.

Inadequate gas exchange can also occur when the thickness of the barrier separating the air and blood is pathologically increased. As the thickness increases, the rate of gas transfer decreases, because a gas takes longer to diffuse through the greater thickness. Thickness increases in (1) *pulmonary edema,* an excess accumulation of interstitial fluid between the alveoli and pulmonary capillaries caused by pulmonary inflammation or left-sided congestive heart failure (see p. 289), (2) *pulmonary fibrosis* involving replacement of delicate lung tissue with thick fibrous tissue in response to certain chronic irritants, and (3) *pneumonia,* which is char-acterized by inflammatory fluid accumulation within or around the alveoli. Most commonly, pneumonia is due to bacterial or viral infection of the lungs, but it may also arise from accidental aspiration (breathing in) of food, vomitus, or chemical agents.

The rate of gas transfer is directly proportional to the diffusion coefficient (D), a constant value related to the solubility of a particular gas in the lung tissues and to its molecular weight ($D \propto sol/\sqrt{mw}$). The diffusion coefficient for CO_2 is twenty times that of O_2 because CO_2 is much more soluble in body tissues than O_2 is. The rate of CO_2 diffusion across the respiratory membranes is therefore twenty times more rapid than that of O_2 for a given partial pressure gradient. This dif-ference in diffusion coefficients is normally offset by the dif-

• **FIGURE 13–27   Comparison of Normal and Emphysematous Lung Tissue** (a) A thin section of lung tissue from a normal individual. Each of the smallest clear spaces is an alveolar lumen. (b) A thin section of lung tissue from a patient with emphysema. These are photographs of the actual tissue, unmagnified. Note the loss of alveolar walls in the emphysematous lung tissue, resulting in larger but fewer alveolar chambers.

(a)   Alveolus   Bronchiole

(b)   Expanded alveolus

ference in partial pressure gradients that exist for O_2 and CO_2 across the alveolar capillary membrane. The CO_2 partial pressure gradient is 6 mm Hg (P_{CO_2} of 46 mm Hg in the blood; P_{CO_2} of 40 mm Hg in the alveoli), compared to the O_2 gradient of 60 mm Hg (P_{O_2} of 100 mm Hg in the alveoli; P_{O_2} of 40 mm Hg in the blood).

Normally, approximately equal amounts of O_2 and CO_2 are exchanged—a respiratory quotient's worth. Even though a given volume of blood spends three-fourths of a second passing through the pulmonary capillary bed, P_{O_2} and P_{CO_2} are usually both equilibrated with alveolar partial pressures by the time the blood has traversed only one-third the length of the pulmonary capillaries. This means that the lung normally has enormous diffusion reserves, a fact that becomes extremely important during heavy exercise. The time the blood spends in transit in the pulmonary capillaries is decreased as pulmonary blood flow increases with the greater cardiac output that accompanies exercise. Even when less time is available for exchange, blood P_{O_2} and P_{CO_2} are normally able to equilibrate with alveolar levels because of the lungs' diffusion reserves.

In a diseased lung in which diffusion is impeded because the surface area is decreased or the blood-air barrier is thickened, O_2 transfer is usually more seriously impaired than CO_2 transfer because of the larger CO_2 diffusion coefficient. By the time the blood reaches the end of the pulmonary capillary network, it is more likely to have equilibrated with alveolar P_{CO_2} than with alveolar P_{O_2}, because CO_2 can diffuse more rapidly through the respiratory barrier. In milder conditions, diffusion of both O_2 and CO_2 might remain adequate at rest, but during exercise, when pulmonary transit time is decreased, the blood gases, especially O_2, may not have completely equilibrated with the alveolar gases before the blood leaves the lungs.

Gas exchange across the systemic capillaries also occurs down partial pressure gradients.

Just as they do at the pulmonary capillaries, O_2 and CO_2 move between the systemic capillary blood and the tissue cells by simple passive diffusion down partial pressure gradients. Refer again to Figure 13–26. The arterial blood that reaches the systemic capillaries is essentially the same blood that left the lungs by means of the pulmonary veins, because the only two places in the entire circulatory system at which gas exchange can take place are the pulmonary capillaries and the systemic capillaries. The arterial P_{O_2} is 100 mm Hg and the arterial P_{CO_2} is 40 mm Hg, the same as alveolar P_{O_2} and P_{CO_2}.

The cells constantly consume O_2 and produce CO_2 through oxidative metabolism. Cellular P_{O_2} averages about 40 mm Hg and P_{CO_2} about 46 mm Hg, although these values are highly variable, depending on the level of cellular metabolic activity. Oxygen moves by diffusion down its partial pressure gradient from the entering systemic capillary blood (P_{O_2} = 100 mm Hg) into the adjacent cells (P_{O_2} = 40 mm Hg) until equilibrium is reached. Therefore, the P_{O_2} of venous blood leaving the systemic capillaries is equal to the tissue P_{O_2} at an average of 40 mm Hg. The reverse situation exists for

CO_2. Carbon dioxide rapidly diffuses out of the cells (P_{CO_2} = 46 mm Hg) into the entering capillary blood (P_{CO_2} = 40 mm Hg) down the partial pressure gradient created by the ongoing production of CO_2. Transfer of CO_2 continues until blood P_{CO_2} equilibrates with tissue P_{CO_2}.[1] Accordingly, the blood leaving the systemic capillaries has an average P_{CO_2} of 46 mm Hg. This systemic venous blood, which is relatively low in O_2 (P_{O_2} = 40 mm Hg) and relatively high in CO_2 (P_{CO_2} = 46 mm Hg), returns to the heart and is subsequently pumped to the lungs as the cycle repeats itself.

The more actively a tissue is metabolizing, the lower the cellular P_{O_2} falls and the higher the cellular P_{CO_2} rises. As a consequence of the larger blood-to-cell partial pressure gradients, more O_2 diffuses from the blood into the cells and more CO_2 moves in the opposite direction before blood P_{O_2} and P_{CO_2} achieve equilibrium with the surrounding cells. Thus, the amount of O_2 transferred to the cells and the amount of CO_2 carried away from the cells depend on the rate of cellular metabolism.

Note that net diffusion of O_2 occurs first between the alveoli and the blood and then between the blood and the tissues due to the O_2 partial pressure gradients created by continuous utilization of O_2 in the cells and continuous replenishment of fresh alveolar O_2 provided by alveolar ventilation. Net diffusion of CO_2 occurs in the reverse direction, first between the tissues and the blood and then between the blood and the alveoli, due to the CO_2 partial pressure gradients created by continuous production of CO_2 in the cells and the continuous removal of alveolar CO_2 through the process of alveolar ventilation (• Fig. 13–28).

⟹ GAS TRANSPORT

Most O_2 in the blood is transported bound to hemoglobin.

Oxygen picked up by the blood at the lungs must be transported to the tissues for cell use. Conversely, CO_2 produced at the cellular level must be transported to the lungs for elimination.

Oxygen is present in the blood in two forms: physically dissolved and chemically bound to hemoglobin (Table 13–7). Very little O_2 is physically dissolved in the plasma water because O_2 is poorly soluble in body fluids. The amount dissolved is directly proportional to the P_{O_2} of the blood; the higher the P_{O_2}, the more O_2 dissolved. At a normal arterial P_{O_2} of 100 mm Hg, only 3 ml of O_2 can dissolve in 1 liter of blood. Thus, only 15 ml of O_2/min can be dissolved in the normal pulmonary blood flow of 5 liters/min (the resting cardiac output). Even under resting conditions, the cells consume 250 ml of O_2/min, and this may increase up to twenty-five-fold during strenuous exercise. To deliver the O_2 required by the tissues even at rest, the cardiac output would have to be 83.3 liters/min if O_2 could only be transported in dissolved form. Obviously, there must be an additional mechanism for transporting O_2 to the tissues. This mechanism is hemoglobin (Hb). Only 1.5% of the O_2 in the blood is dissolved; the remaining

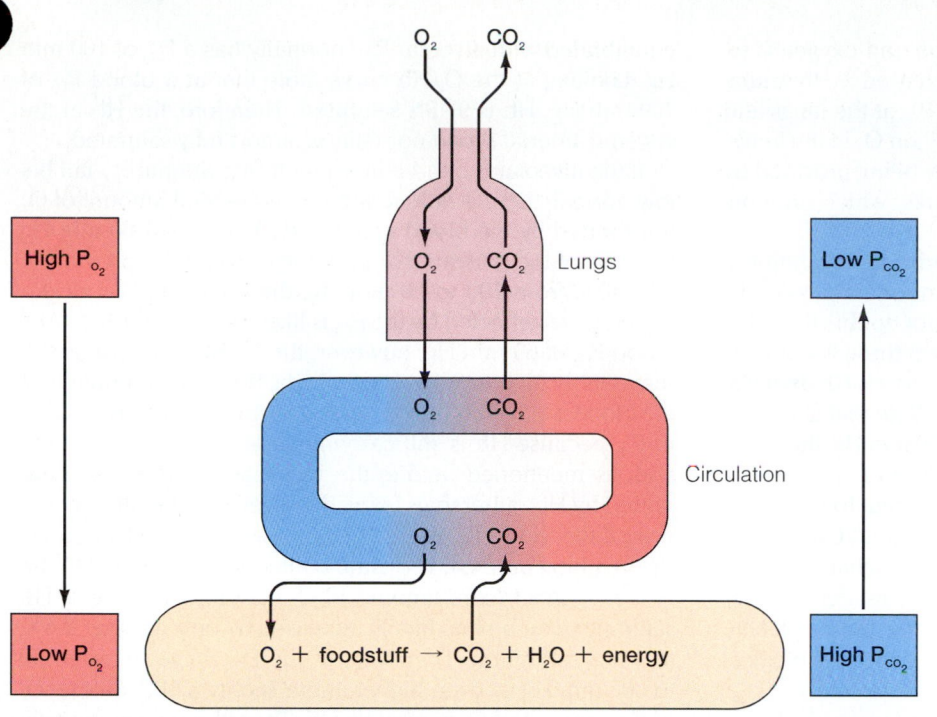

98.5% is transported in combination with Hb. *The O_2 bound to Hb does not contribute to the P_{O_2} of the blood;* thus, blood P_{O_2} is not a measure of the total O_2 content of the blood but only of the portion of O_2 that is dissolved.

Hemoglobin, an iron-bearing protein molecule contained within the red blood cells, has the ability to form a loose, easily reversible combination with O_2 (see p. 350). When not combined with O_2, Hb is referred to as **reduced hemoglobin**; when combined with O_2, it is called **oxyhemoglobin (HbO$_2$)**:

$$Hb + O_2 \quad \rightleftharpoons \quad HbO_2$$
reduced hemoglobin   oxyhemoglobin

We need to answer several important questions about the role of Hb in O_2 transport. What determines whether O_2 and Hb are combined or dissociated (separated)? Why does Hb combine with O_2 in the lungs and release O_2 at the tissues?

GAS	METHOD OF TRANSPORT IN BLOOD	PERCENTAGE CARRIED IN THIS FORM
TABLE 13–7 Methods of Gas Transport in the Blood		
O_2	Physically dissolved	1.5
	Bound to hemoglobin	98.5
CO_2	Physically dissolved	10
	Bound to hemoglobin	30
	As bicarbonate (HCO$_3^-$)	60

How can a variable amount of O_2 be released at the tissue level, depending on the level of tissue activity? How can we talk about O_2 transfer between blood and surrounding tissues in terms of O_2 partial pressure gradients when 98.5% of the O_2 is bound to Hb and thus does not contribute to the P_{O_2} of the blood at all?

The P_{O_2} is the primary factor determining the percent hemoglobin saturation.

Each of the four atoms of iron within the heme portions of a hemoglobin molecule is able to combine with an O_2 molecule, so each Hb molecule can carry up to four molecules of O_2. Hemoglobin is considered to be *fully saturated* when all of the Hb present is carrying its maximum O_2 load. The **percent hemoglobin (% Hb) saturation**, a measure of the extent to which the Hb present is combined with O_2, can vary from 0% to 100%.

The most important factor determining the % Hb saturation is the P_{O_2} of the blood, which, in turn, is related to the concentration of O_2 physically dissolved in the blood. According to the **law of mass action**, if the concentration of one of the substances involved in a reversible reaction is increased, the reaction is driven toward the opposite side. Conversely, if the concentration of one of the substances is decreased, the reaction is driven toward that side. Applying this law to the reversible reaction involving Hb and O_2 (Hb $+ O_2 \rightleftharpoons HbO_2$), when the blood P_{O_2} is increased, as it is in the pulmonary capillaries, the reaction is driven toward the right side of the equation, resulting in increased formation of HbO$_2$ (increased % Hb saturation). When the blood P_{O_2} is decreased, as it is in the systemic capillaries, the reaction is driven toward the left side of the equation and oxygen is released from Hb as HbO_2 dissociates (decreased % Hb saturation). Thus, because of the difference in P_{O_2} at the lungs and other tissues, Hb automatically "loads up" on O_2 in the lungs, where fresh supplies of O_2 are continually being provided by ventilation, and "unloads" it in the tissues, which are constantly using up O_2.

The relationship between blood P_{O_2} and % Hb saturation is not linear, however, a point that is very important physiologically. Doubling the partial pressure does not double the % Hb saturation. Rather, the relationship between these variables is depicted by an S-shaped curve known as the **O_2-Hb dissociation** (or **saturation**) **curve** (● Fig. 13–29). Note that at the upper end, between a blood P_{O_2} of 60 and 100 mm Hg, the curve flattens off, or plateaus. Within this pressure range, a rise in P_{O_2} produces only a small increase in the extent to which Hb is bound with O_2. In contrast, in the P_{O_2} range of 0 to 60 mm Hg, a small change in P_{O_2} results in a large change in the extent to which Hb is combined with O_2, as depicted by the steep lower part of the curve. Both the upper plateau and lower steep portion of the curve have physiological significance.

Significance of the Plateau Portion of the O_2-Hb Curve

The plateau portion of the curve is in the blood-P_{O_2} range that exists at the pulmonary capillaries where O_2 is being loaded onto Hb. The systemic arterial blood leaving the lungs, having equilibrated with alveolar P_{O_2}, normally has a P_{O_2} of 100 mm Hg. Looking at the O_2-Hb curve, note that at a blood P_{O_2} of 100 mm Hg, Hb is 97.5% saturated. Therefore, the Hb in the systemic arterial blood normally is almost fully saturated.

If the alveolar P_{O_2} and consequently the arterial P_{O_2} fall below normal, there is little reduction in the total amount of O_2 transported by the blood until the P_{O_2} falls below 60 mm Hg because of the plateau region of the curve. If the arterial P_{O_2} falls 40%, from 100 to 60 mm Hg, the concentration of dissolved O_2 as reflected by the P_{O_2} is likewise reduced 40%. At a blood P_{O_2} of 60 mm Hg, however, the % Hb saturation is still remarkably high at 90%. Accordingly, the total O_2 content of the blood is only slightly decreased despite the 40% reduction in P_{O_2} because Hb is still carrying an almost-full load of O_2, and, as mentioned before, the vast majority of O_2 is transported by Hb rather than being dissolved. On the other hand, even if the blood P_{O_2} is greatly increased, say, to 600 mm Hg by breathing pure O_2, very little additional O_2 is added to the blood. A small extra amount of O_2 dissolves, but the % Hb saturation can be maximally increased by only another 2½%, to 100% saturation. Therefore, in the P_{O_2} range between 60 to 600 mm Hg or even higher, there is only a 10% difference in the amount of O_2 carried by Hb. Thus, the plateau portion of the O_2-Hb curve provides a good margin of safety in O_2-carrying capacity of the blood.

Arterial P_{O_2} may be reduced because of pulmonary diseases accompanied by inadequate ventilation or defective gas exchange or by circulatory disorders that result in inadequate blood flow to the lungs. It may also fall in healthy individuals under two circumstances: (1) at high altitudes, where the total atmospheric pressure and hence the P_{O_2} of the inspired air are reduced, or (2) in O_2-deprived environments at sea level, such as would be encountered if someone were accidentally locked in a vault. Unless the arterial P_{O_2} becomes markedly reduced (falls below 60 mm Hg) in either pathological conditions or abnormal environmental circumstances, near-normal amounts of O_2 can still be carried to the tissues.

● **FIGURE 13–29 Oxygen-Hemoglobin (O_2-Hb) Dissociation (Saturation) Curve** The percent hemoglobin saturation depends on the P_{O_2} of the blood. The relationship between these two variables is depicted by an S-shaped curve with a plateau region between a blood P_{O_2} of 60 and 100 mm Hg and a steep portion between 0 and 60 mm Hg.

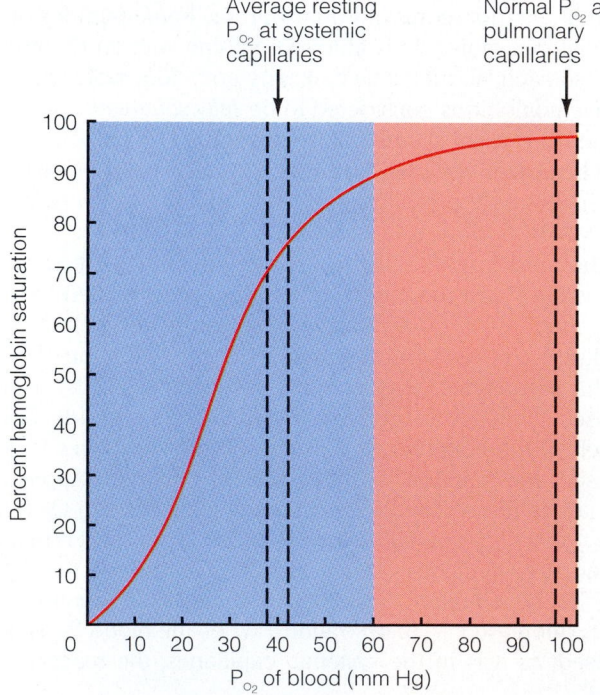

Significance of the Steep Portion of the O_2-Hb Curve

The steep portion of the curve between 0 and 60 mm Hg is in the blood-P_{O_2} range that exists at the systemic capillaries, where O_2 is being unloaded from Hb. In the systemic capillaries, the blood equilibrates with the surrounding tissue cells at an average P_{O_2} of 40 mm Hg. Note on Figure 13–29 that at a P_{O_2} of 40 mm Hg, the % Hb saturation is 75%. The blood arrives in the tissue capillaries at a P_{O_2} of 100 mm Hg with 97.5% Hb saturation. Since Hb can only be 75% saturated at the P_{O_2} of 40 mm Hg in the systemic capillaries, nearly 25% of the HbO_2 must dissociate, yielding reduced Hb and O_2. This released O_2 is free to diffuse down its partial pressure gradient from the red blood cells through the plasma and interstitial fluid into the tissue cells.

The Hb in the venous blood returning to the lungs is still normally 75% saturated. If the tissue cells are metabolizing more actively, the P_{O_2} of the systemic capillary blood falls (for example, from 40 to 20 mm Hg) because the cells are consuming O_2 more rapidly. Note on the curve that this 20 mm

Hg drop in P_{O_2} decreases the % Hb saturation from 75% to 30%; that is, about 45% more of the total HbO_2 than normal gives up its O_2 for tissue use. The normal 60 mm Hg drop in P_{O_2} from 100 to 40 mm Hg in the systemic capillaries causes about 25% of the total HbO_2 to unload its O_2. In comparison, a further drop in P_{O_2} of only 20 mm Hg results in an additional 45% of the total HbO_2 unloading its O_2 because the O_2 partial pressures in this range are operating in the steep portion of the curve. In this range, only a small drop in systemic capillary P_{O_2} can automatically make large amounts of O_2 immediately available to meet the O_2 needs of more actively metabolizing tissues. As much as 85% of the Hb may give up its O_2 to actively metabolizing cells during strenuous exercise. In addition to this more thorough withdrawal of O_2 from the blood, even more O_2 is made available to actively metabolizing cells, such as exercising muscles, by circulatory and respiratory adjustments that increase the flow rate of oxygenated blood through the active tissues.

By acting as a storage depot, hemoglobin promotes the net transfer of O_2 from the alveoli to the blood.

We still have not really clarified the role of Hb in gas exchange. Because blood P_{O_2} depends entirely on the concentration of *dissolved* O_2, we could ignore the O_2 bound to Hb in our earlier discussion of O_2 being driven from the alveoli to the blood by a P_{O_2} gradient. However, Hb does play a crucial role in permitting the transfer of large quantities of O_2 before blood P_{O_2} equilibrates with the surrounding tissues

($\bullet$ Fig. 13–30). It does so by acting as a "storage depot" for O_2, removing the O_2 from solution as soon as it enters the blood from the alveoli. Because only dissolved O_2 contributes to the P_{O_2}, the O_2 stored in Hb cannot contribute to blood P_{O_2}. When systemic venous blood enters the pulmonary capillaries, its P_{O_2} is considerably lower than the alveolar P_{O_2}, so O_2 immediately diffuses into the blood, raising the blood P_{O_2}. As soon as the P_{O_2} of the blood increases, the percentage of Hb that can bind with O_2 likewise increases, as indicated by the O_2-Hb curve. Consequently, most of the O_2 that has diffused into the blood combines with Hb and no longer contributes to blood P_{O_2}. As O_2 is removed from solution by combining with Hb, blood P_{O_2} falls to about the same level it was when the blood entered the lungs, despite the fact that the total quantity of O_2 in the blood actually has increased. Since the blood P_{O_2} is once again considerably below alveolar P_{O_2}, more O_2 diffuses from the alveoli into the blood, only to be soaked up by Hb again.

Even though we have considered this process in stepwise fashion for clarity, net diffusion of O_2 from alveoli to blood occurs continuously until Hb becomes saturated with O_2 as completely as it can be at that particular P_{O_2}. At a normal P_{O_2} of 100 mm Hg, Hb is 97.5% saturated. Thus, by soaking up O_2, Hb keeps blood P_{O_2} low and prolongs the existence of a partial pressure gradient so that a large net transfer of O_2 into the blood can take place. Not until Hb can store no more O_2 (that is, Hb is maximally saturated for that P_{O_2}) does all of the O_2 transferred into the blood remain dissolved and directly contribute to the P_{O_2}. Only at this time does the blood P_{O_2} rapidly

---

$\bullet$ **FIGURE 13–30   Hemoglobin Facilitating a Large Net Transfer of O_2 by Acting as a Storage Depot to Keep P_{O_2} Low**
(a) In the hypothetical situation in which no hemoglobin is present in the blood, the alveolar P_{O_2} and the pulmonary capillary blood P_{O_2} are at equilibrium. (b) Hemoglobin has been added to the pulmonary capillary blood. As the Hb starts to bind with O_2, it removes O_2 from solution. Since only dissolved O_2 contributes to blood P_{O_2}, the blood P_{O_2} falls below that of the alveoli, even though the same number of O_2 molecules are present in the blood as in part (a). By "soaking up" some of the dissolved O_2, Hb favors the net diffusion of more O_2 down its partial pressure gradient from the alveoli to the blood. (c) Hemoglobin is fully saturated with O_2, and the alveolar and blood P_{O_2} are at equilibrium again. The blood P_{O_2} resulting from dissolved O_2 is equal to the alveolar P_{O_2} despite the fact that the total O_2 content in the blood is much greater than in part (a) when blood P_{O_2} was equal to alveolar P_{O_2} in the absence of Hb.

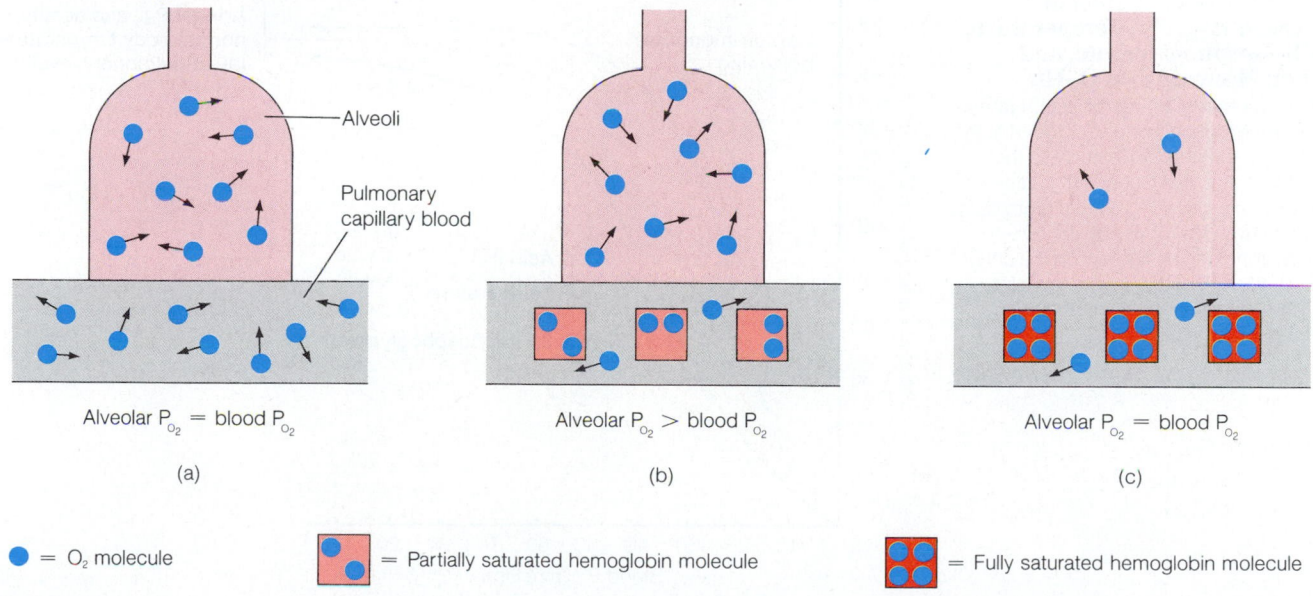

equilibrate with the alveolar P_{O_2} and bring further O_2 transfer to a halt, but this point is not reached until Hb is already loaded to the maximum extent possible. Once the blood P_{O_2} equilibrates with the alveolar P_{O_2}, no further O_2 transfer can take place, no matter how little or how much total O_2 has already been transferred.

The reverse situation occurs at the tissue level. Since the P_{O_2} of blood entering the systemic capillaries is considerably higher than the P_{O_2} of the surrounding tissue, O_2 immediately diffuses from the blood into the tissues, lowering blood P_{O_2}. When blood P_{O_2} falls, Hb is forced to unload some of its stored O_2 because the % Hb saturation is reduced. As the O_2 released from Hb dissolves in the blood, the blood P_{O_2} increases and is once again above the P_{O_2} of the surrounding tissues. This favors further movement of O_2 out of the blood, despite the fact that the total quantity of O_2 in the blood has already been reduced. Only when Hb is no longer able to release any more O_2 into solution (when Hb is unloaded to the greatest extent possible for the P_{O_2} existing at the systemic capillaries) can blood P_{O_2} become as low as in the surrounding tissue. At this time, further transfer of O_2 ceases. Hemoglobin, because it bears a large quantity of stored O_2 that can be liberated by a slight reduction in P_{O_2} at the systemic capillary level, permits the transfer of tremendously more O_2 from the blood into the cells than would be possible in its absence.

Thus, Hb plays an important role in the *total quantity* of O_2 that the blood can pick up in the lungs and drop off in the tissues. If Hb levels are reduced to one-half of normal, as in a severely anemic patient (see p. 354), the O_2-carrying capacity of the blood is reduced by 50% even though the arterial P_{O_2} is the normal 100 mm Hg with 97.5% Hb saturation. Only one-half as much Hb is available to be saturated, emphasizing once again how critical the presence of Hb is in determining the total amount of O_2 that can be picked up at the lungs and made available to the tissues.

Increased CO_2, acidity, temperature, and 2,3-diphosphoglycerate shift the O_2-Hb dissociation curve to the right.

Even though the primary factor determining the % Hb saturation is the P_{O_2} of the blood, other factors can affect the affinity, or bond strength, between Hb and O_2 and, accordingly, can shift the O_2-Hb curve (that is, change the % Hb saturation at a given P_{O_2}). These other factors are CO_2, acidity, temperature, and 2,3-diphosphoglycerate, which we will examine separately. The O_2-Hb dissociation curve with which you are already familiar (Fig. 13–29) is a typical curve at normal arterial CO_2 and acidity levels, normal body temperature, and normal 2,3-diphosphoglycerate concentration.

An increase in P_{CO_2} shifts the O_2-Hb curve to the right (Fig. 13–31). The % Hb saturation still depends on the P_{O_2}, but for any given P_{O_2}, the amount of O_2 and Hb that can be combined is reduced. This effect is important, because the P_{CO_2} of the blood increases in the systemic capillaries as CO_2 diffuses down its gradient from the cells into the blood. The presence of this additional CO_2 in the blood in effect decreases the affinity of Hb for O_2, so Hb unloads even more O_2 at the tissue level than it would if the reduction in P_{O_2} in the systemic capillaries were the only factor affecting % Hb saturation.

An increase in acidity also shifts the curve to the right. Because CO_2 generates carbonic acid (H_2CO_3), the blood becomes more acidic at the systemic capillary level as it picks up CO_2 from the tissues. The resultant reduction in Hb affinity for O_2 in the presence of increased acidity aids in releasing even more O_2 at the tissue level for a given P_{O_2}. In actively metabolizing cells, such as exercising muscles, not only is more carbonic-acid–generating CO_2 produced, but lactic acid also may be produced if the cells resort to anaerobic metabolism (see p. 236). The resultant local elevation of acid

• **FIGURE 13–31   Effect of Increased P_{CO_2}, H^+, Temperature, 2,3-Diphosphoglycerate, and Carbon Monoxide on O_2-Hb Curve** Increased P_{CO_2}, acid, temperature, and 2,3-diphosphoglycerate, as found at the tissue level, shift the O_2-Hb curve to the right. As a result, less O_2 and Hb can be combined at a given P_{O_2}, so that more O_2 is unloaded from Hb for use by the tissues. Carbon-monoxide poisoning, on the other hand, shifts the O_2-Hb curve to the left so that less O_2 is unloaded from Hb at the tissue level for a given P_{O_2}.

in the working muscles facilitates further unloading of O_2 in the very tissues that have the greatest O_2 need.

The influence of CO_2 and acid on the release of O_2 is known as the **Bohr effect**. Both CO_2 and the hydrogen-ion (H^+) component of acids are able to combine reversibly with Hb at sites other than the O_2-binding sites. The result is an alteration in the molecular structure of Hb that reduces its affinity for O_2.

In a similar manner, an elevation in temperature shifts the O_2-Hb curve to the right, resulting in more unloading of O_2 at a given P_{O_2}. An exercising muscle or other actively metabolizing cell produces heat. The resultant local elevation in temperature enhances O_2 release from Hb for use by the more active tissues.

Thus, increases in CO_2, acidity, and temperature at the tissue level, all of which are associated with increased cellular metabolism and increased O_2 consumption, enhance the effect of a drop in P_{O_2} in facilitating the release of O_2 from Hb. These effects are largely reversed at the pulmonary level, where the extra acid-forming CO_2 is blown off and the local environment is cooler. Appropriately, therefore, Hb has a higher affinity for O_2 in the pulmonary capillary environment, thus enhancing the effect of the elevation in P_{O_2} in loading O_2 onto Hb.

The preceding changes take place in the *environment* of the red blood cells, but a factor *inside* the red blood cells can also affect the degree of O_2-Hb binding: **2,3-diphosphoglycerate (DPG)**. This erythrocyte constituent, which is produced during red blood cell metabolism, can bind reversibly with Hb and reduce its affinity for O_2, just as CO_2 and H^+ do. Thus, an increased level of DPG, like the other factors, shifts the O_2-Hb curve to the right, enhancing O_2 unloading as the blood flows through the tissues. DPG production by red blood cells gradually increases whenever Hb in the arterial blood is chronically undersaturated—that is, when arterial HbO_2 is below normal. This condition may occur in individuals living at high altitudes or in those suffering from certain types of circulatory or respiratory diseases or anemia. By promoting the liberation of O_2 from Hb at the tissue level, an increased amount of DPG helps maintain O_2 availability for tissue use under circumstances associated with decreased arterial O_2 supply. However, unlike the other factors—which normally are present only at the tissue level and thus shift the O_2-Hb curve to the right only in the systemic capillaries, where the shift is advantageous in unloading O_2—DPG is present in the red blood cells throughout the circulatory system and, accordingly, shifts the curve to the right to the same degree in both the tissues and the lungs. As a result, DPG decreases the ability to load O_2 at the pulmonary level, which is the negative side of increased DPG production.

Oxygen-binding sites on hemoglobin have a much higher affinity for carbon monoxide than for O_2.

Carbon monoxide (CO) and O_2 compete for the same binding sites on Hb, but the affinity of Hb for CO is 240 times that of the bond strength between Hb and O_2. The combination of CO and Hb is known as **carboxyhemoglobin (HbCO)**. Because Hb preferentially latches onto CO, the presence of even small amounts of CO can tie up a disproportionately large share of Hb, making the latter unavailable for O_2 transport. Even though the Hb concentration and P_{O_2} are normal, the O_2 content of the blood is seriously reduced. If enough CO is present, the cells die from O_2 deprivation. Adding to the toxicity of CO, the presence of HbCO shifts the Hb-O_2 curve to the *left* (Fig. 13–31); thus, the already limited quantity of O_2-bearing Hb is unable to unload as much of its O_2 at the tissue level for a given P_{O_2}.

Fortunately, CO is not a normal constituent of inspired air. It is a poisonous gas produced during the incomplete combustion (burning) of carbon products such as automobile gasoline, coal, wood, and tobacco. Carbon monoxide is especially dangerous because it is so insidious. If CO is being produced in a closed environment so that its concentration continues to increase (for example, in a parked car with the motor running and windows closed), it can reach lethal levels without the victim ever being aware of the danger. Carbon monoxide is not detectable because it is odorless, colorless, tasteless, and nonirritating. Furthermore, for reasons to be described later, the victim has no sensation of breathlessness and makes no attempt to increase ventilation, even though the cells are O_2-starved.

The majority of CO_2 is transported in the blood as bicarbonate.

When arterial blood flows through the tissue capillaries, CO_2 diffuses down its partial pressure gradient from the tissue cells into the blood. Carbon dioxide is transported in the blood in three ways: (1) physically dissolved, (2) bound to Hb, and (3) as bicarbonate (• Fig. 13–32 and Table 13–7).

As with dissolved O_2, the amount of CO_2 *physically dissolved* in the blood depends on the P_{CO_2}. Because CO_2 is more soluble than O_2 in the blood, a greater proportion of the total CO_2 in the blood is physically dissolved compared to O_2. Even so, only 10% of the blood's total CO_2 content is carried this way at the normal systemic venous P_{CO_2} level.

Another 30% of the CO_2 combines with Hb to form **carbamino hemoglobin (HbCO$_2$)**. Carbon dioxide binds with the globin portion of Hb in contrast to O_2, which combines with the heme portions. Reduced Hb has a greater affinity for CO_2 than does HbO_2. The unloading of O_2 from Hb in the tissue capillaries therefore facilitates the picking up of CO_2 by Hb.

By far the most important means of CO_2 transport is as **bicarbonate (HCO$_3^-$)**, with 60% of the CO_2 being converted into HCO_3^- by the following chemical reaction, which takes place within the red blood cells:

$$CO_2 + H_2O \;\overset{\text{carbonic anhydrase}}{\rightleftharpoons}\; H_2CO_3 \rightleftharpoons H^+ + HCO_3^-$$

In the first step, CO_2 combines with H_2O to form **carbonic acid (H_2CO_3)**. This reaction can occur very slowly in the plasma, but it proceeds swiftly within the red blood cells because of the presence of the erythrocyte enzyme **carbonic anhydrase**, which catalyzes (speeds up) the reaction. As is characteristic

ca = Carbonic anhydrase

● **FIGURE 13–32   Carbon-Dioxide Transport in the Blood**   Carbon dioxide (CO_2) picked up at the tissue level is transported in the blood to the lungs in three ways: physically dissolved ①, bound to hemoglobin (Hb) ②, and as bicarbonate ion (HCO_3^-) ③. Hemoglobin is present only in the red blood cells, as is carbonic anhydrase, the enzyme that catalyzes the production of HCO_3^-. The H^+ generated during the production of HCO_3^- also binds to Hb. Bicarbonate moves by facilitated diffusion down its concentration gradient out of the red blood cell into the plasma, and chloride (Cl^-) moves by means of the same passive carrier into the red blood cell down the electrical gradient created by the outward diffusion of HCO_3^-.

The reactions that occur at the tissue level are reversed at the pulmonary level, where CO_2 diffuses out of the blood to enter the alveoli.

of acids, a proportion of the carbonic-acid molecules spontaneously dissociate into hydrogen ions (H^+) and bicarbonate ions (HCO_3^-). The one carbon and two oxygen atoms of the original CO_2 molecule are thus present in the blood as an integral part of HCO_3^-. This is beneficial, because HCO_3^- is more soluble in the blood than CO_2.

As this reaction proceeds, HCO_3^- and H^+ start to accumulate within the red blood cells in the systemic capillaries. The red-cell membrane has a HCO_3^--Cl^- carrier that passively facilitates the diffusion of these ions in opposite directions across the membrane. The membrane is relatively impermeable to H^+. Consequently, HCO_3^- but not H^+ diffuses down its concentration gradient out of the erythrocytes into the plasma. Since HCO_3^- is a negatively charged ion, the outflux of HCO_3^- unaccompanied by a comparable outward diffusion of positively charged ions creates an electrical gradient (see p. 59). Chloride ions (Cl^-), the dominant plasma anions, diffuse into the red blood cells down this electrical gradient to restore electric neutrality. This inward shift of Cl^- in exchange for the outflux of CO_2-generated HCO_3^- is known as the **chloride (Cl^-) shift**.

The vast majority of the accumulated H^+ within the erythrocytes following the dissociation of H_2CO_3 becomes bound

to Hb. As is the case with CO_2, reduced Hb has a greater affinity for H^+ than HbO_2 does. Therefore, the unloading of O_2 once again facilitates the pickup of CO_2-generated H^+ by Hb. Because only free dissolved H^+ contributes to the acidity of a solution, the venous blood would be considerably more acidic than the arterial blood if it were not for Hb mopping up most of the H^+ generated at the tissue level.

The fact that removal of O_2 from Hb increases the ability of Hb to pick up CO_2 and CO_2-generated H^+ is known as the **Haldane effect**. The Haldane effect and Bohr effect work in synchrony to facilitate O_2 liberation and the uptake of CO_2 and CO_2-generated H^+ at the tissue level. Increased CO_2 and H^+ cause increased O_2 release from Hb by means of the Bohr effect; increased O_2 release from Hb, in turn, causes increased CO_2 and H^+ uptake by Hb through the Haldane effect. The entire process is very efficient. Reduced Hb must be carried back to the lungs to refill on O_2 anyway. Meanwhile, after O_2 is released, Hb picks up new passengers—CO_2 and H^+—that are going in the same direction to the lungs.

The reactions that occur at the tissue level as CO_2 enters the blood from the tissues are reversed once the blood reaches the lungs and CO_2 leaves the blood to enter the alveoli (Fig. 13–32).

TABLE 13–8 Miniglossary of Clinically Important Respiratory States

Apnea   Transient cessation of breathing

Asphyxia O_2 starvation of tissues, caused by lack of O_2 in air, respiratory impairment, or inability of tissues to utilize O_2

Cyanosis   Blueness of the skin resulting from insufficiently oxygenated blood in the arteries

Dyspnea   Difficult or labored breathing

Eupnea   Normal breathing

Hypercapnia   Excess CO_2 in arterial blood

Hypernea   Increased pulmonary ventilation that matches increased metabolic demands, as in exercise

Hyperventilation   Increased pulmonary ventilation in excess of metabolic requirements, resulting in decreased P_{CO_2} and respiratory alkalosis

Hypocapnia   Below-normal CO_2 in arterial blood

Hypoventilation   Underventilation in relation to metabolic requirements, resulting in increased P_{CO_2} and respiratory acidosis

Hypoxia   Insufficient O_2 at the cellular level
   Anemic hypoxia   Reduced O_2-carrying capacity of the blood

   Circulatory hypoxia   Too little oxygenated blood delivered to the tissues; also known as stagnant hypoxia

   Histotoxic hypoxia   Cells unable to utilize O_2 available to them

   Hypoxic hypoxia   Low arterial blood P_{O_2} accompanied by inadequate Hb saturation

Respiratory arrest   Permanent cessation of breathing (unless clinically corrected)

Suffocation O_2 deprivation as a result of inability to breathe oxygenated air

Various respiratory states are characterized by abnormal blood-gas levels.

Hypoxia refers to insufficient O_2 at the cellular level. (Table 13–8 is a glossary of terms used to describe various states associated with respiratory abnormalities.) There are four general categories of hypoxia:

1. *Hypoxic hypoxia* is characterized by a low arterial blood P_{O_2} accompanied by inadequate Hb saturation. It is caused by (a) a respiratory malfunction involving inadequate gas exchange, typified by a normal alveolar P_{O_2} but a reduced arterial P_{O_2}, or (b) exposure to high altitude or to a suffocating environment where atmospheric P_{O_2} is reduced so that alveolar and arterial P_{O_2} are likewise reduced.

2. *Anemic hypoxia* refers to a reduced O_2-carrying capacity of the blood. It can be brought about by (a) a decrease in circulating red blood cells, (b) an inadequate amount of Hb within the red blood cells, or (c) CO poisoning. In all cases of anemic hypoxia, the arterial P_{O_2} is at a normal level, but the O_2 content of the arterial blood is lower than normal because of the reduction in available Hb.

3. *Circulatory hypoxia* arises when too little oxygenated blood is delivered to the tissues. Circulatory hypoxia can be restricted to a limited area as a result of a local vascular spasm or blockage. On the other hand, the body may experience circulatory hypoxia in general as a result of congestive heart failure or circulatory shock. The arterial P_{O_2} and O_2 content are typically normal, but too little oxygenated blood reaches the cells.

4. In *histotoxic hypoxia*, O_2 delivery to the tissues is normal, but the cells are unable to use the O_2 available to them. The classic example is *cyanide poisoning*. Cyanide blocks cellular enzymes essential for internal respiration.

Hyperoxia, an above-normal arterial P_{O_2}, cannot occur when a person is breathing atmospheric air at sea level. However, breathing supplemental O_2 can increase alveolar and consequently arterial P_{O_2}. Because a greater percentage of the inspired air is O_2, a greater percentage of the total pressure of the inspired air is attributable to the O_2 partial pressure, so more O_2 dissolves in the blood before arterial P_{O_2} equilibrates with alveolar P_{O_2}. Even though arterial P_{O_2} is increased, the *total* blood O_2 content is not significantly increased because Hb is nearly fully saturated at the normal arterial P_{O_2}. In certain pulmonary diseases associated with a reduced arterial P_{O_2}, however, breathing supplemental O_2 can be beneficial in establishing a larger alveoli-to-blood driving gradient, thereby improving the arterial P_{O_2}. On the other hand, a markedly elevated arterial P_{O_2}, far from being advantageous, can be dangerous. If the arterial P_{O_2} is too high, **oxygen toxicity** can occur. Even though the total O_2 content of the blood is only slightly increased, some cells can be damaged by exposure to a high P_{O_2}. In particular, brain damage and damage to the retina causing blindness are problems associated with O_2 toxicity. Therefore, O_2 therapy must be administered cautiously.

Hypercapnia refers to excess CO_2 in the arterial blood; it is caused by **hypoventilation** (ventilation inadequate to meet the metabolic needs for O_2 delivery and CO_2 removal). With most lung diseases, CO_2 accumulation in the arterial blood occurs concurrently with an O_2 deficit, because both O_2 and CO_2 exchange between the lungs and atmosphere are equally affected (● Fig. 13–33). However, when a decrease in arterial P_{O_2} is due to reduced pulmonary diffusing capacity, as in pulmonary edema or emphysema, O_2 transfer suffers more than CO_2 transfer because the diffusion coefficient for CO_2 is twenty times that of O_2. As a result, hypoxic hypoxia occurs much more readily than hypercapnia in these circumstances.

Hypocapnia, below-normal arterial P_{CO_2} levels, is brought about by hyperventilation. **Hyperventilation** occurs when a person "overbreathes," that is, when the rate of ventilation is in excess of the body's metabolic needs for CO_2 removal so that CO_2 is blown off to the atmosphere more rapidly than it is produced in the tissues and arterial P_{CO_2} falls. Hyperventilation can be triggered by anxiety states, fever, and aspirin poisoning. Alveolar P_{O_2} increases during hyperventilation as

● **FIGURE 13–33   Effects of Hypoventilation and Hyperventilation on Arterial P_{O_2} and P_{CO_2}**

more fresh O_2 is delivered to the alveoli from the atmosphere than is extracted from the alveoli by the blood for tissue consumption, and arterial P_{O_2} increases correspondingly (Fig. 13–33). However, since Hb is almost fully saturated at the normal arterial P_{O_2}, very little additional O_2 is added to the blood. Except for the small extra amount of dissolved O_2, blood O_2 content remains essentially unchanged during hyperventilation.

Increased ventilation is not synonymous with hyperventilation. Increased ventilation that matches an increased metabolic demand, such as the increased need for O_2 delivery and CO_2 elimination during exercise, is termed **hyperpnea**. During exercise, alveolar P_{O_2} and P_{CO_2} remain constant, with the increased atmospheric exchange just keeping pace with the increased O_2 consumption and CO_2 production.

The consequences of reduced O_2 availability to the tissues during hypoxia is apparent. The cells need adequate O_2 supplies to sustain their energy-generating metabolic activities. The consequences of abnormal blood CO_2 levels are less obvious. Changes in blood CO_2 concentration primarily affect acid-base balance. Hypercapnia results in an elevated production of carbonic acid. The subsequent generation of excess H^+ produces an acidic condition termed *respiratory acidosis*. Conversely, less-than-normal amounts of H^+ are generated through carbonic-acid formation in conjunction with hypocapnia. The resultant alkalotic (less acidic than normal) condition is called *respiratory alkalosis* (chapter 15). (See the accompanying boxed feature, Concepts, Challenges, and Controversies.)

CONTROL OF RESPIRATION

Respiratory centers in the brain stem establish a rhythmic breathing pattern.

Breathing, like the heartbeat, must occur in a continuous, cyclical pattern to sustain life processes. Cardiac muscle must rhythmically contract and relax to alternately empty blood from the heart and fill it again. Similarly, inspiratory muscles must rhythmically contract and relax to alternately fill the lungs with air and empty them. Both these activities are accomplished automatically without conscious effort. However, the underlying mechanisms and control of these two systems are remarkably different. Whereas the heart is able to generate its own rhythm by means of its intrinsic pacemaker activity, the respiratory muscles, being skeletal muscles, require nervous stimulation to bring about their contraction. The rhythmic pattern of breathing is established by cyclical neural activity to the respiratory muscles. In other words, the pacemaker activity that establishes the rhythmicity of breathing resides in the respiratory control centers in the brain, not in the lungs or respiratory muscles themselves. The nerve supply to the heart, not being necessary to initiate the heartbeat, only serves to modify the rate and strength of cardiac contraction. In contrast, the nerve supply to the respiratory system is absolutely essential in maintaining breathing and in reflexly adjusting the level of ventilation to match changing needs for O_2 uptake and CO_2 removal. Furthermore, unlike cardiac activity, which is not subject to voluntary control, respiratory activity can be voluntarily modified to accomplish speaking, singing, whistling, playing a wind instrument, or holding one's breath while swimming.

Neural control of respiration involves three distinct components: (1) the factors responsible for generating the alternating inspiration/expiration rhythm, (2) the factors that regulate the magnitude of ventilation (that is, the rate and depth of breathing) to match body needs, and (3) the factors that modify respiratory activity to serve other purposes. The latter modifications may be either voluntary, as in the breath control required for speech, or involuntary, as in the respiratory maneuvers involved in a cough or sneeze.

Respiratory control centers housed in the brain stem are responsible for generating the rhythmic pattern of breathing. The primary respiratory control center, the **medullary respiratory center**, consists of several aggregations of neuronal cell bodies within the medulla that provide output to the respiratory muscles. In addition, there are two other respiratory centers higher in the brain stem in the pons—the **apneustic center** and **pneumotaxic center**. These pontine centers influence the output from the medullary respiratory center (● Fig. 13–34). Exactly how these various regions interact to establish respiratory rhythmicity is unclear, but the following factors are believed to contribute.

Inspiratory and Expiratory Neurons in the Medullary Center   We rhythmically breathe in and out during quiet breathing because of alternate contraction and relaxation of the inspiratory muscles, namely, the diaphragm and external

EFFECTS OF HEIGHTS AND DEPTHS ON THE BODY

Our bodies are optimally equipped for existence at normal atmospheric pressure. Ascent into mountains high above sea level or descent into the depths of the ocean can have adverse effects on the body.

Effects of High Altitude on the Body

The atmospheric pressure progressively declines as altitude increases. At 18,000 ft above sea level, the atmospheric pressure is only 380 mm Hg—half of its normal sea level value. Since the proportion of O_2 and N_2 in the air remains the same, the P_{O_2} of inspired air at this altitude is 21% of 380 mm Hg, or 80 mm Hg, with alveolar P_{O_2} being even lower at 45 mm Hg. At any altitude above 10,000 ft, the arterial P_{O_2} falls below the safety range of the plateau region into the steep portion of the O_2-Hb curve. As a result, the % Hb saturation in the arterial blood declines precipitously with further increases in altitude.

People who rapidly ascend to altitudes of 10,000 ft or more experience symptoms of **acute mountain sickness** attributable to hypoxic hypoxia and the resultant hypocapnia-induced alkalosis. The increased ventilatory drive to obtain more O_2 causes respiratory alkalosis because acid-forming CO_2 is blown off more rapidly than it is produced. Symptoms of mountain sickness include fatigue, nausea, loss of appetite, labored breathing, rapid heart rate (triggered by hypoxia as a compensatory measure to increase circulatory delivery of available O_2 to the tissues), and nerve dysfunction characterized by poor judgment, dizziness, and incoordination.

Despite these acute responses to high altitude, millions of people live at elevations above 10,000 ft, with some villagers even residing in the Andes at altitudes greater than 16,000 ft. How do they live and function normally? They do so through the process of **acclimatization**. When a person remains at high altitude, the acute compensatory responses of increased ventilation and increased cardiac output are gradually replaced over a period of days by more slowly developing compensatory measures that permit adequate oxygenation of the tissues and restoration of normal acid-base balance. Red blood cell (rbc) production is increased, stimulated by erythropoietin in response to reduced O_2 delivery to the kidneys (see p. 351). The rise in the number of rbcs increases the O_2-carrying capacity of the blood. Hypoxia also promotes the synthesis of DPG within the rbcs, so that O_2 is unloaded from Hb more easily at the tissues. The number of capillaries within the tissues is increased, reducing the distance that O_2 must diffuse from the blood to reach the cells. Furthermore, acclimatized cells are able to use O_2 more efficiently through an increase in the number of mitochondria, the energy organelles (see p. 26). The kidneys restore the arterial pH to nearly normal by conserving acid that normally would have been lost in the urine.

These compensatory measures are not without undesirable trade-offs. For example, the greater number of circulating rbcs increases blood viscosity (makes the blood "thicker"), thereby increasing resistance to blood flow. As a result, the heart has to work harder to pump blood through the vessels (see p. 304).

Effects of Deep-Sea Diving on the Body

When a deep-sea diver descends underwater, the body is exposed to greater than atmospheric pressure. Pressure rapidly increases with sea depth as a result of the weight of the water. Pressure is already doubled by about 30 ft below sea level. The air provided by scuba equipment is delivered to the lungs at these high pressures. Recall that (1) the amount of a gas in solution is directly proportional to the partial pressure of the gas and (2) air is composed of 79% N_2. Nitrogen is poorly soluble in body tissues, but the high P_{N_2} that occurs during deep-sea diving causes more of this gas than normal to dissolve in the body tissues. The small amount of N_2 dissolved in the tissues at sea level has no known effect, but as more N_2 dissolves at greater depths, **nitrogen narcosis**, or "**raptures of the deep**," ensues. Nitrogen narcosis is believed to result from a reduction in the excitability of neurons due to the highly lipid-soluble N_2 dissolving in their lipid membranes. At 150 ft under water, divers experience a feeling of euphoria and become drowsy, similar to the effect of having a few cocktails. At lower depths, divers become weak and clumsy, and at 350 to 400 ft., they lose consciousness. *Oxygen toxicity* resulting from the high P_{O_2} is another possible detrimental effect of descending deep under water.

Another problem associated with deep-sea diving occurs during ascent. If a diver who has been submerged long enough for a significant amount of N_2 to become dissolved in the tissues suddenly ascends to the surface, the rapid reduction in P_{N_2} causes N_2 to quickly come out of solution and form bubbles of gaseous N_2 in the body. The consequences depend on the amount and location of the bubble formation. This condition is called **decompression sickness** or "**the bends**," the latter term arising because the victim often bends over in pain. Decompression sickness can be prevented by slow ascent to the surface or by gradual decompression in a decompression tank so that the excess N_2 can slowly escape through the lungs without bubble formation.

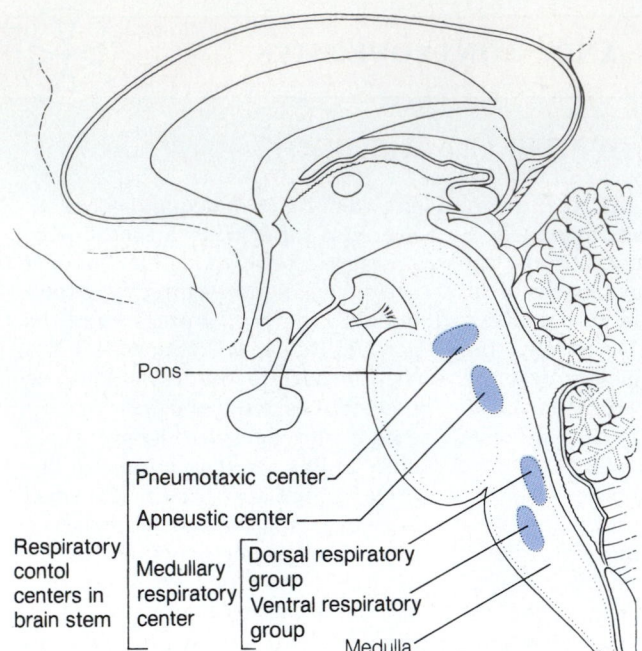

Pons

Pneumotaxic center

Apneustic center

Medullary
respiratory
center

Dorsal respiratory
group

Ventral respiratory
group

Medulla

• **FIGURE 13–34   Respiratory Control Centers in Brain Stem**

intercostal muscles, supplied respectively by the phrenic nerve and intercostal nerves. The cell bodies for the neuronal fibers composing these nerves are located in the spinal cord. Impulses originating in the medullary center terminate on these motor-neuron cell bodies (• Fig. 13–35). When these motor neurons are activated, they in turn stimulate the inspiratory muscles, leading to inspiration; when these neurons are not firing, the inspiratory muscles relax and expiration takes place.

The medullary respiratory center consists of two neuronal clusters known as the dorsal respiratory group and the ventral respiratory group (Fig. 13–34). The **dorsal respiratory group (DRG)** consists mostly of *inspiratory neurons* whose descending fibers terminate on the motor neurons that supply the

inspiratory muscles. These inspiratory neurons are believed to display pacemaker activity, repetitively undergoing self-induced action potentials similar to the SA node of the heart. When the DRG inspiratory neurons fire, inspiration takes place; when they cease firing, expiration occurs. Expiration is brought to an end as the inspiratory neurons once again reach threshold and fire. Accordingly, the DRG is generally regarded as being responsible for the basic rhythm of ventilation. The rate at which the inspiratory neurons rhythmically fire, however, is influenced by synaptic input from other areas of the brain and from elsewhere in the body. Thus, the on-off nature of the respiratory cycle is complex because of the DRG's interactions with these other regions.

The DRG has important interconnections with the **ventral respiratory group (VRG)**. The VRG is composed of *inspiratory neurons* and *expiratory neurons,* both of which remain inactive during normal quiet breathing. This region is called into play by the DRG as an "overdrive" mechanism during periods when demands for ventilation are increased. It is especially important in active expiration. No impulses are generated in the descending pathways from the expiratory neurons during quiet breathing. Only during active expiration do the expiratory neurons stimulate the motor neurons supplying the expiratory muscles (the abdominal and internal intercostal muscles). Furthermore, the VRG inspiratory neurons, when stimulated by the DRG, rev up inspiratory activity when demands for ventilation are high.

Influences from the Pneumotaxic and Apneustic Centers
The pontine centers exert "fine-tuning" influences over the medullary center to help produce normal, smooth inspirations and expirations. The pneumotaxic center sends impulses to the DRG that help "switch off" the inspiratory neurons, thereby limiting the duration of inspiration. In contrast, the apneustic center prevents the inspiratory neurons from being switched off, thus providing an extra boost to the inspiratory drive. In this check-and-balance system, the pneumotaxic center is dominant over the apneustic center, helping to bring inspiration to a halt and allowing expiration to occur normally. Without the pneumotaxic brakes, the breath-

• **FIGURE 13–35   Medullary Dorsal Respiratory Group (DRG) Control of Inspiration** Inspiration takes place when the inspiratory neurons are firing and activating the motor neurons that supply the inspiratory muscles. Expiration takes places when the inspiratory neurons cease firing, so that the motor neurons supplying the inspiratory muscles are no longer activated.

Input from other areas—
some excitatory, some inhibitory

Inspiratory neurons
in DRG
(rhythmically firing)

Medulla

Spinal cord

Phrenic nerve   +   Diaphragm

Not shown are intercostal nerves to external intercostal muscles.

TABLE 13-9 Influence of Chemical Factors on Respiration

CHEMICAL FACTOR	EFFECT ON PERIPHERAL CHEMORECEPTORS	EFFECT ON CENTRAL CHEMORECEPTORS
$\downarrow P_{O_2}$ in arterial blood	Stimulates only when arterial P_{O_2} has fallen to the point of being life-threatening (< 60 mm Hg); emergency mechanism	Directly depresses central chemoreceptors and respiratory center itself when < 60 mm Hg
$\uparrow P_{CO_2}$ in arterial blood ($\uparrow H^+$ in brain ECF)	Weakly stimulates	Strongly stimulates; is dominant control of ventilation (Levels > 70–80 mm Hg directly depress respiratory center and central chemoreceptors)
$\uparrow H^+$ in arterial blood	Stimulates; important in acid-base balance	Does not affect; cannot penetrate blood-brain barrier

ing pattern consists of prolonged inspiratory gasps abruptly interrupted by very brief expirations. This abnormal pattern of breathing is known as **apneusis**; hence, the center responsible for this type of breathing is the apneustic center. Apneusis may occur in certain types of severe brain damage.

Hering-Breuer Reflex When the tidal volume is large (greater than 1 liter), as during exercise, the **Hering-Breuer reflex** is triggered to prevent overinflation of the lungs. **Pulmonary stretch receptors** located within the smooth-muscle layer of the airways are activated by the stretching of the lungs at large tidal volumes. Action potentials from these stretch receptors travel through afferent nerve fibers to the medullary center and inhibit the inspiratory neurons. This negative feedback from the highly stretched lungs themselves helps cut inspiration short before the lungs become overinflated.

Carbon-dioxide–generated hydrogen-ion concentration in the brain extracellular fluid is normally the primary regulator of the magnitude of ventilation.

No matter how much O_2 is extracted from the blood or how much CO_2 is added to it at the tissue level, the P_{O_2} and P_{CO_2} of the systemic arterial blood leaving the lungs are held remarkably constant, indicative of the fact that arterial blood-gas content is subject to precise regulation. Arterial blood gases are maintained within the normal range exclusively by varying the magnitude of ventilation to match the body's needs for O_2 uptake and CO_2 removal. If more O_2 is extracted from the alveoli and more CO_2 dropped off by the blood because the tissues are metabolizing more actively, ventilation is increased correspondingly to bring in more fresh O_2 and blow off more CO_2.

The medullary respiratory center receives inputs that provide information about the body's needs for gas exchange. Then it responds by sending appropriate signals to the motor neurons supplying the respiratory muscles to adjust the rate and depth of ventilation to meet those needs. The two most obvious signals to increase ventilation are a decreased arterial P_{O_2} or an increased arterial P_{CO_2}. Intuitively, you would suspect that if O_2 levels in the arterial blood declined or if CO_2

accumulated, ventilation would be stimulated to obtain more O_2 or to eliminate the excess CO_2. These two factors do indeed influence the magnitude of ventilation, but not to the same degree nor through the same pathway. Also, a third chemical factor, H^+, has a notable influence on the level of respiratory activity. We will examine the role of each of these important chemical factors in the control of ventilation (Table 13–9).

Role of Decreased Arterial P_{O_2} in Regulating Ventilation Arterial P_{O_2} is monitored by **peripheral chemoreceptors** known as the **carotid bodies** and **aortic bodies**, which are located at the bifurcation of the common carotid arteries and in the arch of the aorta, respectively (● Fig. 13–36). These

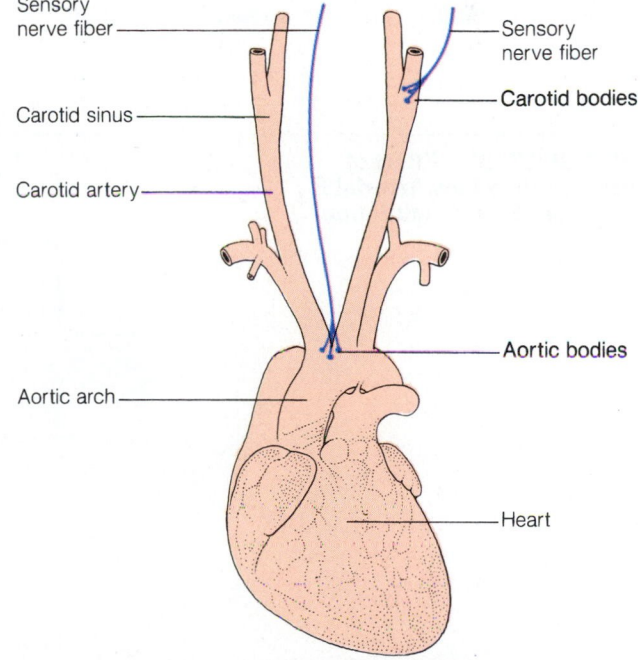

● **FIGURE 13–36   Location of Peripheral Chemoreceptors** The carotid bodies are located in the carotid sinus, and the aortic bodies are located in the aortic arch.

Sensory nerve fiber

Sensory nerve fiber

Carotid bodies

Carotid sinus

Carotid artery

Aortic bodies

Aortic arch

Heart

chemoreceptors, which respond to specific changes in the chemical content of the arterial blood that bathes them, are distinctly different from the carotid-sinus and aortic-arch baroreceptors located in the same vicinity. The latter, being important in the regulation of systemic arterial blood pressure, monitor pressure changes rather than chemical changes.

The peripheral chemoreceptors are not sensitive to modest reductions in arterial P_{O_2}. The arterial P_{O_2} must fall below 60 mm Hg (>40% reduction) before the peripheral chemoreceptors respond by sending afferent impulses to the medullary inspiratory neurons, thereby reflexly increasing ventilation. Because arterial P_{O_2} only falls below 60 mm Hg in the unusual circumstances of severe pulmonary disease or reduced atmospheric P_{O_2}, it does not play a role in the normal ongoing regulation of respiration. This fact might seem surprising at first thought, since one of the primary functions of ventilation is to provide sufficient O_2 for uptake by the blood. However, there is no need to increase ventilation until the arterial P_{O_2} falls below 60 mm Hg because of the margin of safety in % Hb saturation afforded by the plateau portion of the O_2-Hb curve. Hemoglobin is still 90% saturated at an arterial P_{O_2} of 60 mm Hg, but the % Hb saturation drops precipitously when the P_{O_2} falls below this level. Therefore, reflex stimulation of respiration by the peripheral chemoreceptors serves as an important emergency mechanism in dangerously low arterial P_{O_2} states. Indeed, this reflex mechanism is a life saver, because a low arterial P_{O_2} tends to directly depress the respiratory center, as it does all the rest of the brain (● Fig. 13–37). Except for the peripheral chemoreceptors, the level of activity in all nervous tissue becomes reduced in the face of O_2 deprivation. Were it not for stimulatory intervention of the peripheral chemoreceptors when the arterial P_{O_2} falls threateningly low, a vicious cycle ending in cessation of breathing would ensue. Direct depression of the respiratory center by the markedly low arterial P_{O_2} would further reduce ventilation, leading to an even greater fall in arterial P_{O_2}, which would even further depress the respiratory center until ventilation ceased and death occurred.

Since the peripheral chemoreceptors respond to the P_{O_2} of the blood, *not* the total O_2 content of the blood, O_2 content in the arterial blood can fall to dangerously low or even fatal levels without the peripheral chemoreceptors ever responding to reflexly stimulate respiration. Remember that only physically dissolved O_2 contributes to blood P_{O_2}. The total O_2 content in the arterial blood can be reduced in anemic states, in which in O_2-carrying Hb is reduced, or in CO poisoning, when the Hb is preferentially bound to this molecule rather than to O_2. In both cases, arterial P_{O_2} is normal so respiration is not stimulated, even though O_2 delivery to the tissues may be so reduced that the person dies from cellular O_2 deprivation.

Role of Increased Arterial P_{CO_2} in Regulating Ventilation In contrast to arterial P_{O_2}, which does not contribute to the minute-to-minute regulation of respiration, arterial P_{CO_2} is the most important input regulating the magnitude of ventilation under resting conditions. This role is appropriate, because changes in alveolar ventilation have an immediate and pronounced effect on arterial P_{CO_2}, whereas changes in ventilation have little effect on % Hb saturation and O_2 availability to the tissues until the arterial P_{O_2} falls by more than 40%. Even slight alterations from normal in arterial P_{CO_2} induce a significant reflex effect on ventilation. An increase in arterial P_{CO_2} reflexly stimulates the respiratory center, with the resultant increase in ventilation promoting elimination of the excess CO_2 to the atmosphere. Conversely, a fall in arterial P_{CO_2} reflexly reduces the respiratory drive. The subsequent decrease in ventilation allows metabolically produced CO_2 to accumulate so that P_{CO_2} can be returned to normal.

Surprisingly, given the key role of arterial P_{CO_2} in regulating respiration, there are no important receptors that monitor arterial P_{CO_2} per se. The carotid and aortic bodies are only weakly responsive to changes in arterial P_{CO_2}, so they play

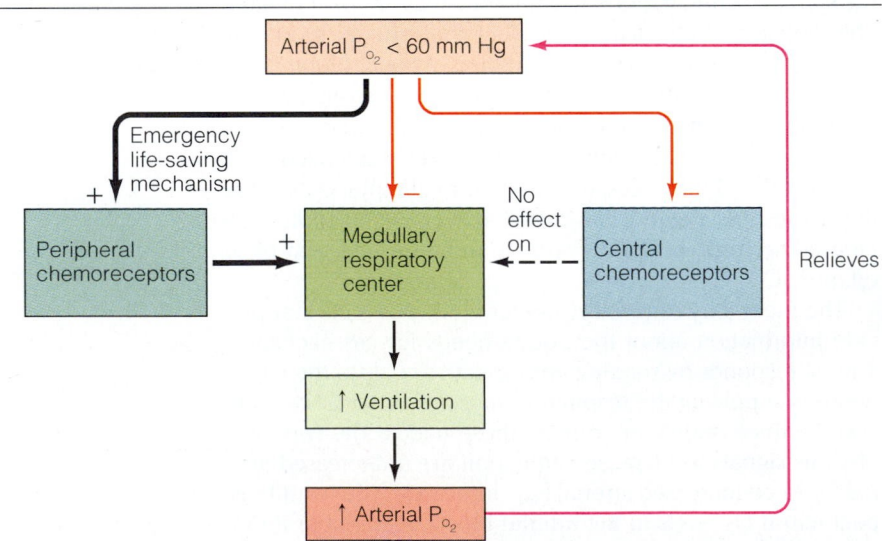

● **FIGURE 13–37   Effect of Threateningly Low Arterial P_{O_2} (< 60 mm Hg) on Ventilation**

only a minor role in reflexly stimulating ventilation in response to an elevation in arterial P_{CO_2}. More important in linking changes in arterial P_{CO_2} to compensatory adjustments in ventilation are the **central chemoreceptors**, located in the medulla in the vicinity of the respiratory center. These central chemoreceptors do not monitor CO_2 itself, however; they are sensitive to changes in CO_2-induced H^+ concentration in the brain extracellular fluid (ECF) that bathes them.

Movement of materials across the brain capillaries is restricted by the blood-brain barrier (see p. 113). Because this barrier is readily permeable to CO_2, any increase in arterial P_{CO_2} causes a similar rise in brain ECF P_{CO_2} as CO_2 diffuses down its pressure gradient from the cerebral blood vessels into the brain ECF. The increased P_{CO_2} within the brain ECF causes a corresponding increase in the concentration of H^+ according to the law of mass action as it applies to this reaction: $CO_2 + H_2O \rightleftharpoons H_2CO_3 \rightleftharpoons H^+ + HCO_3^-$. An elevation in H^+ concentration in the brain ECF directly stimulates the central chemoreceptors, which in turn increase ventilation by stimulating the respiratory center through synaptic connections (• Fig. 13–38). As the excess CO_2 is subsequently blown off, the arterial P_{CO_2} and the P_{CO_2} and H^+ concentration of the brain ECF are returned to normal. Conversely, a decline in arterial P_{CO_2} below normal is paralleled by a fall in P_{CO_2} and H^+ in the brain ECF, the result of which is a central-chemoreceptor– mediated decrease in ventilation. As CO_2 produced by cellular metabolism is consequently allowed to accumulate, arterial P_{CO_2} and P_{CO_2} and H^+ of the brain ECF are restored toward normal.

Unlike CO_2, H^+ is not readily able to permeate the blood-brain barrier, so H^+ present in the plasma cannot gain access to the central chemoreceptors. Accordingly, the central chemoreceptors are responsive only to H^+ generated within the brain ECF itself as a result of CO_2 entry. Thus, the major mechanism controlling ventilation under resting conditions is specifically aimed at regulating the brain ECF H^+ concentration, which in turn is a direct reflection of the arterial P_{CO_2}. Unless there are extenuating circumstances such as reduced availability of O_2 in the inspired air, arterial P_{O_2} is coincidentally also maintained at its normal value by the brain ECF H^+ ventilatory driving mechanism.

The powerful influence of the central chemoreceptors on the respiratory center is responsible for your inability to deliberately hold your breath for more than about a minute. While

• **FIGURE 13–38   Effect of Increased Arterial P_{CO_2} on Ventilation**

ca = Carbonic anhydrase

you hold your breath, metabolically produced CO_2 continues to accumulate in your blood and subsequently to build up the H^+ concentration in your brain ECF. Finally, the increased P_{CO_2}-H^+ stimulant to respiration becomes so powerful that central-chemoreceptor excitatory input overrides voluntary inhibitory input to respiration, so breathing resumes despite deliberate attempts to prevent it. Breathing resumes long before arterial P_{O_2} falls to the threateningly low levels that trigger the peripheral chemoreceptors. Therefore, you cannot deliberately hold your breath long enough to create a dangerously high level of CO_2 or low level of O_2 in the arterial blood.

In contrast to the normal reflex stimulatory effect of the increased P_{CO_2}-H^+ mechanism on respiratory activity, very high levels of CO_2 directly depress the entire brain, including the respiratory center, just as very low levels of O_2 do. Up to a P_{CO_2} of 70–80 mm Hg, progressively higher P_{CO_2} levels induce correspondingly more vigorous respiratory efforts in an attempt to blow off the excess CO_2. A further increase in P_{CO_2} beyond 70–80 mm Hg, however, does not further increase ventilation but actually depresses the respiratory neurons. For this reason, CO_2 must be removed and O_2 supplied in closed environments such as closed-system anesthesia machines, submarines, or space capsules. Otherwise, CO_2 could reach lethal levels, not only because of its depressant effect on respiration but also because of the resultant severe state of respiratory acidosis.

During prolonged hypoventilation caused by certain types of chronic lung disease, an elevated P_{CO_2} occurs simultaneously with a markedly reduced P_{O_2}. In most cases, the elevated P_{CO_2} (acting via the central chemoreceptors) and the reduced P_{O_2} (acting via the peripheral chemoreceptors) are *synergistic;* that is, the combined stimulatory effect on respiration exerted by these two inputs together is greater than the sum of their independent effects. However, some patients with severe chronic lung disease lose their sensitivity to an elevated arterial P_{CO_2}. In the presence of a prolonged increase in H^+ generation in the brain ECF as a result of long-standing CO_2 retention, enough HCO_3^- may cross the blood-brain barrier to buffer, or "neutralize," the excess H^+. The additional HCO_3^- combines with the excess H^+, removing it from solution so that it no longer contributes to free H^+ concentration. By raising the brain ECF HCO_3^- concentration, the brain ECF H^+ concentration is restored to normal despite the fact that arterial P_{CO_2} and brain ECF P_{CO_2} remain high. The central chemoreceptors are no longer aware of the elevated P_{CO_2} because the brain ECF H^+ is normal. Since the central chemoreceptors no longer reflexly stimulate the respiratory center in response to the elevated P_{CO_2}, the drive to eliminate CO_2 is blunted in such patients; that is, their level of ventilation is abnormally low considering their high arterial P_{CO_2}. In these patients, the hypoxic drive to ventilation becomes their primary respiratory stimulus, in contrast to normal individuals, in whom the arterial P_{CO_2} level is the dominant factor governing the magnitude of ventilation. Ironically, administering O_2 to such patients to relieve the hypoxic condition can markedly depress their drive to breathe by elevating the arterial P_{O_2} and removing the primary driving stimulus for respiration. Because of this danger, O_2 therapy must be administered cautiously in patients with long-term pulmonary diseases.

Role of Increased Arterial H^+ Concentration in Regulating Ventilation Changes in arterial H^+ concentration cannot influence the central chemoreceptors because H^+ does not readily cross the blood-brain barrier. However, the aortic and carotid body peripheral chemoreceptors are highly responsive to fluctuations in arterial H^+ concentration, in contrast to their weak sensitivity to deviations in arterial P_{CO_2} and their unresponsiveness to arterial P_{O_2} until it falls 40% below normal.

Any change in arterial P_{CO_2} brings about a corresponding change in the H^+ concentration of the blood as well as of the brain ECF. These CO_2-induced H^+ changes in the arterial blood are detected by the peripheral chemoreceptors; the result is reflex stimulation of ventilation in response to an increase in arterial H^+ concentration and depression of ventilation in association with a decrease in arterial H^+ concentration. However, these changes in ventilation mediated by the peripheral chemoreceptors are far less important than the powerful central-chemoreceptor mechanism in adjusting ventilation in response to changes in CO_2-generated H^+ concentration.

The peripheral chemoreceptors do play a major role in adjusting ventilation in response to alterations in arterial H^+ concentration unrelated to fluctuations in P_{CO_2}. In many situations, even though P_{CO_2} is normal, arterial H^+ concentration is changed by the addition or loss of non-carbonic acid from the body. For example, arterial H^+ concentration increases during diabetes mellitus because excess H^+-generating keto acids are abnormally produced and added to the blood. A rise in arterial H^+ concentration reflexly stimulates ventilation by means of the peripheral chemoreceptors. Conversely, the peripheral chemoreceptors reflexly suppress respiratory activity in response to a fall in arterial H^+ concentration resulting from nonrespiratory causes. Changes in ventilation by this mechanism are extremely important in regulating the acid-base balance of the body. By changing the magnitude of ventilation, the amount of H^+-generating CO_2 that is eliminated can be varied. The resultant adjustment in the amount of H^+ added to the blood from CO_2 can compensate for the nonrespiratory-induced abnormality in arterial H^+ concentration that first elicited the respiratory response (● Fig. 13–39).

Exercise profoundly increases ventilation, but the mechanisms involved are unclear.

Alveolar ventilation may increase up to twenty-fold during heavy exercise to keep pace with the increased demand for O_2 uptake and CO_2 output. (Table 13–10 highlights the changes in O_2- and CO_2-related variables that occur during exercise.) The cause of increased ventilation during exercise is still largely speculative. It would seem logical that changes in the "big three" chemical factors—decreased P_{O_2}, increased P_{CO_2}, and increased H^+—could account for the increase in ventilation. This does not appear to be the case, however.

● FIGURE 13–39 **Effect of Increased Arterial Non-Carbonic-Acid–Generated Hydrogen Ion (Non-CO_2-H^+) on Ventilation**

TABLE 13–10	Oxygen- and Carbon-Dioxide–Related Variables during Exercise	
O_2- OR CO_2-RELATED VARIABLE	**CHANGE**	**COMMENT**
O_2 utilization	Marked increase	Active muscles are oxidizing nutrient molecules more rapidly to meet their increased energy needs
CO_2 production	Marked increase	More actively metabolizing muscles produce more CO_2
Alveolar ventilation	Marked increase	By mechanisms not completely understood, alveolar ventilation keeps pace with or even slightly exceeds the increased metabolic demands during exercise
Arterial P_{O_2}	Normal or slight ↑	Despite marked increase in O_2 utilization and CO_2 production during exercise,
Arterial P_{CO_2}	Normal or slight ↓	alveolar ventilation keeps pace with or even slightly exceeds the stepped-up rate of O_2 consumption and CO_2 production
O_2 delivery to muscles	Marked increase	Although arterial P_{O_2} remains normal, O_2 delivery to muscles is greatly increased by the increased blood flow to exercising muscles accomplished by increased cardiac output coupled with local vasodilation of active muscles
O_2 extraction by muscles	Marked increase	Increased utilization of O_2 lowers the P_{O_2} at the tissue level, which results in more O_2 unloading from hemoglobin; this is enhanced by ↑ P_{CO_2}, ↑ H^+, and ↑ temperature
CO_2 removal from muscles	Marked increase	The increased blood flow to exercising muscles removes the excess CO_2 produced by these more actively metabolizing tissues
Arterial H^+ concentration		
Mild to moderate exercise	Normal	Since carbonic-acid–generating CO_2 is held constant in arterial blood, arterial H^+ concentration does not change
Heavy exercise	Modest increase	In heavy exercise, when muscles resort to anaerobic metabolism, lactic acid is added to the blood

- Despite the marked increase in O_2 utilization during exercise, arterial P_{O_2} does not decrease but remains normal or may actually increase slightly. This is because the increase in alveolar ventilation keeps pace with or even slightly exceeds the stepped-up rate of O_2 consumption.

- Likewise, despite the marked increase in CO_2 production during exercise, arterial P_{CO_2} does not increase but remains normal or decreases slightly. This is because the extra CO_2 is removed as rapidly or even more rapidly than it is produced as a result of the increase in ventilation.

- During mild or moderate exercise, H^+ concentration does not increase because H^+-generating CO_2 is held constant. During heavy exercise, H^+ concentration does increase somewhat because of the release of H^+-generating lactic acid into the blood as a result of anaerobic metabolism in the exercising muscles. Even so, the elevation in H^+ concentration resulting from lactic-acid formation is not sufficient to account for the large increase in ventilation accompanying exercise.

Some investigators argue that the constancy of the three chemical regulatory factors during exercise is evidence that ventilatory responses to exercise are actually being controlled by these factors—particularly by P_{CO_2}, because it is normally the dominant control during resting conditions. According to this reasoning, how else could alveolar ventilation be increased in exact proportion to CO_2 production, thereby keeping the P_{CO_2} constant? This proposal, however, cannot account for the observation that during heavy exercise, alveolar ventilation may be increased relatively more than the increase in CO_2 production, thereby actually causing a slight decline in P_{CO_2}. Also, ventilation increases abruptly at the onset of exercise (within seconds), long before changes in arterial blood gases could become important influences on the respiratory center (which requires a matter of minutes).

Researchers have suggested that a number of other factors, including the following, play a role in the ventilatory response to exercise:

1. *Reflexes originating from body movements.* Joint and muscle receptors excited during muscle contraction reflexly stimulate the respiratory center, abruptly increasing ventilation. Even passive movement of the limbs (for example, someone else alternately flexing and extending a person's knee) may increase ventilation several-fold through activation of these receptors, even though no actual exercise is occurring. Thus, the mechanical events of exercise are believed to play an important role in coordinating respiratory activity with the increased metabolic requirements of the active muscles.

2. *Increase in body temperature.* Much of the energy generated during muscle contraction is converted to heat rather than to actual mechanical work. Heat-loss mechanisms such as sweating are frequently unable to keep pace with the increased heat production that accompanies increased physical activity, so body temperature often increases sightly during exercise. Since an increase in body temperature stimulates ventilation, this exercise-related heat production undoubtedly contributes to the respira-

tory response to exercise. For the same reason, increased ventilation often accompanies a fever.

3. *Epinephrine release.* The adrenal medullary hormone epinephrine also stimulates ventilation. The level of circulating epinephrine is increased during exercise in response to the sympathetic nervous-system discharge that accompanies increased physical activity.

4. *Impulses from the cerebral cortex.* Especially at the onset of exercise, the motor areas of the cerebral cortex are believed to simultaneously stimulate the medullary respiratory neurons and activate the motor neurons of the exercising muscles. This is similar to the cardiovascular adjustments initiated by the motor cortex at the onset of exercise. In this way, the motor region of the brain calls forth increased ventilatory and circulatory responses to support the increased physical activity that it is about to orchestrate. These anticipatory adjustments are unusual in that regulatory steps are taken *before* any homeostatic factors have actually changed. In the usual case, regulatory adjustments take place *after* a factor has been altered in an attempt to restore homeostasis.

None of these factors or combinations of factors are fully satisfactory in explaining the abrupt and profound effect exercise has on ventilation, nor can they completely account for the high degree of correlation between respiratory activity and the body's needs for gas exchange during exercise. (For a discussion of how measurement of O_2 consumption during exercise can be used to determine a person's maximum work capacity, see the accompanying boxed feature, A Closer Look at Exercise Physiology.)

Ventilation can be influenced by factors unrelated to the need to supply O_2 or remove CO_2.

Respiratory rate and depth can be modified for reasons other than the need to supply O_2 or remove CO_2. Protective reflexes such as sneezing and coughing temporarily govern respiratory activity in an effort to expel irritant materials from the respiratory passages. Inhalation of particularly noxious agents frequently triggers immediate cessation of ventilation. Pain originating anywhere in the body reflexly stimulates the respiratory center (for example, one "gasps" with pain). Involuntary modification of breathing also occurs during the expression of various emotional states, such as laughing, crying, sighing, and groaning. The emotionally induced modifications are mediated through connections between the limbic system in the brain (which is responsible for emotions) and the respiratory center. In addition, the respiratory center is reflexly inhibited during swallowing, when the airways are closed to prevent food from entering the lungs.

Human beings also have considerable voluntary control over ventilation. Voluntary control of breathing is accomplished by the cerebral cortex, which does not act on the respiratory center in the brain stem but instead sends impulses directly to the motor neurons in the spinal cord that supply the respiratory muscles. We can voluntarily hyperventilate ("overbreathe") or, at the other extreme, hold our breath, but only for a brief period of time. The resulting chemical

HOW TO FIND OUT HOW MUCH WORK YOU'RE CAPABLE OF DOING

The best single predictor of a person's work capacity is the determination of the maximum volume of O_2 the person is capable of using per minute to oxidize nutrient molecules for energy production. *Maximal O_2 consumption, or max VO_2,* is measured by having the person engage in exercise, usually on a treadmill or bicycle ergometer (a stationary bicycle with variable resistance). The work load is incrementally increased until the person becomes exhausted. Expired air samples collected during the last minutes of exercise, when O_2 consumption is at a maximum because the person is working as hard as possible, are analyzed for the percentages of O_2 and CO_2 they contain. Furthermore, the volume of air expired is measured. Equations are then employed to determine the amount of O_2 consumed, taking into account the percentages of O_2 and CO_2 in the inspired air, the total volume of air expired, and the percentages of O_2 and CO_2 in the exhaled air.

Maximal O_2 consumption depends on three systems. The respiratory system is essential for ventilation and exchange of O_2 and CO_2 between the air and blood in the lungs. The circulatory system is required to deliver O_2 to the working muscles. Finally, the muscles must have the oxidative enzymes available to use the O_2 once it has been delivered.

Regular aerobic exercise can improve max VO_2 by making the heart and respiratory system more efficient, thereby delivering more O_2 to the working muscles. Exercised muscles themselves become better equipped to use O_2 once it is delivered. The number of functional capillaries increases, as do the number and size of mitochondria, which contain the oxidative enzymes.

Maximal O_2 consumption is measured in liters per minute and then converted into milliliters per kilogram of body weight per minute so that large and small people can be compared. As would be expected, athletes have the highest values for maximal O_2 consumption. The max

VO_2 for male cross-country skiers has been recorded to be as high as 94 ml O_2/kg/min. Distance runners maximally consume between 65 and 85 ml O_2/kg/min, and football players have max VO_2 values between 45 and 65 ml O_2/kg/min, depending on the position they play. Sedentary young men maximally consume between 25 and 45 ml O_2/kg/min. Female values for max VO_2 are 20% to 25% lower than for males when expressed as ml/kg/min of total body weight. The difference in max VO_2 between females and males is only 8% to 10% when expressed as ml/kg/min of lean body weight, however, because females generally have a higher percentage of body fat (the female sex hormone estrogen promotes fat deposition).

Available norms are used to classify people as being low, fair, average, good, or excellent in aerobic capacity for their age group. Exercise physiologists use max VO_2 measurements to prescribe or adjust training regimens to help people achieve their optimal level of aerobic conditioning.

changes in the arterial blood directly and reflexly influence the respiratory center, which in turn overrides the voluntary input to the motor neurons of the respiratory muscles. Other than these extreme forms of deliberately controlling ventilation, we also control our breathing to perform such voluntary acts as speaking, singing, and whistling.

During apnea, a person subconsciously "forgets to breathe," whereas during dyspnea, a person consciously feels that ventilation is inadequate.

Apnea is the transient cessation of ventilation with the expectation that breathing will resume spontaneously. The condition is called **respiratory arrest** if breathing does not resume. Because ventilation is normally decreased and the central chemoreceptors are less sensitive to the arterial P_{CO_2} drive during sleep, especially REM sleep (see p. 138), apnea is most likely to occur during this time. Victims of **sleep apnea** may stop breathing for a few seconds or up to one or two minutes as many as five hundred times a night. Mild forms of sleep apnea are not dangerous unless the sufferer has pulmonary or circulatory disease, the consequences of which can be compounded by recurrent bouts of apnea.

In exaggerated cases of sleep apnea, the victim may be unable to recover from an apneic period, and death results. This is the case in **sudden infant death syndrome (SIDS)**, or "crib death." With this tragic form of sleep apnea, an otherwise healthy two- to five-month-old infant is found dead in his or her crib for no apparent reason. The underlying cause of SIDS is the subject of intense investigation. Most evidence suggests that the baby "forgets to breathe" as a result of the immaturity of the respiratory control mechanisms, either in the brain stem or in the chemoreceptors that monitor the body's respiratory status. For example, on autopsy, more than half the victims have been shown to have poorly developed carotid bodies, the more important of the peripheral chemoreceptors.

In contrast to sleep apnea, in which the victim unconsciously stops breathing, people who have **dyspnea** have the subjective sensation that they are not getting enough air; that is, they feel "short of breath." Dyspnea is the mental anguish associated with the unsatiated desire for more adequate ventilation. It often accompanies the labored breathing characteristic of obstructive lung disease or the pulmonary edema associated with congestive heart failure. In contrast, during exercise a person can breathe very hard without experiencing the sensation of dyspnea, because such exertion is not accompanied by a sense of anxiety over the adequacy of ventilation. Surprisingly, dyspnea is not directly related to chronic elevation of arterial P_{CO_2} or reduction of P_{O_2}. The subjective feeling of air hunger may occur even when alveolar ventilation and the blood gases are normal. Some individuals experience dyspnea when they *perceive* that they are short of air even though this is not actually the case, such as when they are in a crowded elevator.

CHAPTER IN PERSPECTIVE: FOCUS ON HOMEOSTASIS

The respiratory system contributes to homeostasis by obtaining O_2 from and eliminating CO_2 to the external environ-

ment. All body cells ultimately need an adequate supply of O_2 to use in oxidizing nutrient molecules to generate ATP. Brain cells, which are especially dependent on a continual supply of O_2, die if deprived of O_2 for more than four minutes. Even cells that can resort to anaerobic ("without O_2") metabolism for energy production, such as strenuously exercising muscles, can do so only transiently by incurring an O_2 debt that ultimately must be repaid (see p. 236).

As a result of these energy-yielding metabolic reactions, large quantities of CO_2 are produced that must be eliminated from the body. Because CO_2 and H_2O form carbonic acid, adjustments in the rate of CO_2 elimination by the respiratory system are important in the regulation of acid-base balance in the internal environment. Cells can survive only within a narrow pH range.

CHAPTER SUMMARY

Introduction

Internal respiration refers to the intracellular metabolic reactions that utilize O_2 and produce CO_2 during energy-yielding oxidation of nutrient molecules. External respiration encompasses the various steps involved in the transfer of O_2 and CO_2 between the external environment and tissue cells. The respiratory and circulatory systems function together to accomplish external respiration.

The respiratory system accomplishes exchange of air between the atmosphere and the lungs through the process of ventilation. Exchange of O_2 and CO_2 between the air in the lungs and the blood in the pulmonary capillaries takes place across the extremely thin walls of the air sacs, or alveoli. Respiratory airways conduct air from the atmosphere to this gas-exchanging portion of the lungs. The lungs are housed within the closed compartment of the thorax, the volume of which can be changed by contractile activity of surrounding respiratory muscles.

Respiratory Mechanics

Ventilation, or breathing, is the process of cyclically moving air in and out of the lungs so that old alveolar air that has already participated in exchange of O_2 and CO_2 with the pulmonary capillary blood can be exchanged for fresh atmospheric air. Ventilation is mechanically accomplished by alternately shifting the direction of the pressure gradient for airflow between the atmosphere and alveoli through cyclical expansion and recoil of the lungs. Alternate contraction and relaxation of the inspiratory muscles (primarily the diaphragm) indirectly produce periodic inflation and deflation of the lungs by cyclically expanding and compressing the thoracic cavity, with the lungs passively following its movements.

Because energy is required for contraction of the inspiratory muscles, inspiration is an active process, but expiration is passive during quiet breathing because it is accomplished by elastic recoil of the lungs on relaxation of inspiratory muscles at no energy expense. For more forceful active expiration, contraction of the expiratory muscles (namely, the abdominal muscles) further

decreases the size of the thoracic cavity and lungs, which further increases the intra-alveolar-to-atmospheric pressure gradient. The larger the gradient between the alveoli and atmosphere in either direction, the larger the airflow rate, because air continues to flow until the intra-alveolar pressure equilibrates with atmospheric pressure.

Besides being directly proportional to the pressure gradient, airflow rate is also inversely proportional to airway resistance. Because airway resistance, which depends on the caliber of the conducting airways, is normally very low, airflow rate usually depends primarily on the pressure gradient established between the alveoli and the atmosphere. If airway resistance is pathologically increased by chronic obstructive pulmonary disease, the pressure gradient must be correspondingly increased by more vigorous respiratory muscle activity to maintain a normal airflow rate.

The lungs can be stretched to varying degrees during inspiration and then recoil to their pre-inspiratory size during expiration because of their elastic behavior. Pulmonary compliance refers to the distensibility of the lungs—how much they stretch in response to a given change in the transmural pressure gradient, the stretching force exerted across the lung wall. Elastic recoil refers to the phenomenon of the lungs returning to their resting position during expiration. Pulmonary elastic behavior depends on the elastic connective-tissue meshwork within the lungs and on alveolar surface tension/pulmonary surfactant interaction. Alveolar surface tension, which is due to the attractive forces between the surface water molecules in the liquid film lining each alveolus, tends to resist the alveolus being stretched upon inflation (decreases compliance) and tends to return it back to a smaller surface area during deflation (increases lung rebound). If the alveoli were lined by water alone, the surface tension would be so great that the lungs would be poorly compliant and would tend to collapse. Type II alveolar cells secrete pulmonary surfactant, a phospholipoprotein that intersperses between the water molecules and lowers the alveolar surface tension, thereby increasing the compliance of the lungs and counteracting the tendency for alveoli to collapse.

The lungs can be filled to over 5½ liters upon maximal inspiratory effort or emptied to about 1 liter upon maximal expiratory effort. Normally, however, the lungs operate at "half-full." The lung volume typically varies from about 2 to 2½ liters as an average tidal volume of 500 ml of air is moved in and out with each breath.

The amount of air moved in and out of the lungs in one minute, the pulmonary ventilation, is equal to tidal volume × respiratory rate. However, not all of the air moved in and out is available for O_2 and CO_2 exchange with the blood because part of it occupies the conducting airways, known as the anatomic dead space. Alveolar ventilation, the volume of air exchanged between the atmosphere and alveoli in one minute, is a measure of the air actually available for gas exchange with the blood. Alveolar ventilation equals (tidal volume minus the dead space volume) × respiratory rate.

Gas Exchange

Oxygen and CO_2 move across body membranes by passive diffusion down partial pressure gradients. Net diffusion of O_2 occurs first between the alveoli and the blood and then between the blood and the tissues as a result of the O_2 partial pressure gradients created by continuous utilization of O_2 in the cells and continuous replenishment of fresh alveolar O_2 provided by ventilation. Net diffusion of CO_2 occurs in the reverse direction, first between the tissues and the blood and then between the blood and the alveoli, as a result of the CO_2 partial pressure gradients created by continuous production of CO_2 in the cells and continuous removal of alveolar CO_2 through the process of ventilation.

Gas Transport

Because O_2 and CO_2 are not very soluble in the blood, they must be transported primarily by mechanisms other than simply being physically dissolved. Only 1.5% of the O_2 is physically dissolved in the blood, with 98.5% chemically bound to hemoglobin (Hb). The primary factor that determines the extent to which Hb and O_2 are combined (the % Hb saturation) is the P_{O_2} of the blood. The relationship between blood P_{O_2} and % Hb saturation is such that in the P_{O_2} range found in the pulmonary capillaries, Hb is still almost fully saturated even if the blood P_{O_2} falls as much as 40%; this provides a margin of safety by assuring near-normal O_2 delivery to the tissues despite a substantial reduction in arterial P_{O_2}. On the other hand, in the P_{O_2} range found in the systemic capillaries, large increases in Hb unloading occur in response to a small local decline in blood P_{O_2} associated with increased cellu-

lar metabolism; thus, more O_2 is provided to match the increased tissue needs.

Carbon dioxide picked up at the systemic capillaries is transported in the blood by three methods: (1) 10% is physically dissolved; (2) 30% is bound to Hb; and (3) 60% is in the form of bicarbonate (HCO_3^-). The erythrocyte enzyme carbonic anhydrase catalyzes the conversion of CO_2 to HCO_3^- according to the reaction: $CO_2 + H_2O \rightleftharpoons H_2CO_3 \rightleftharpoons H^+ + HCO_3^-$. The generated H^+ binds to Hb. These reactions are all reversed in the lungs as CO_2 is eliminated to the alveoli.

Control of Respiration

Ventilation involves two distinct aspects, both of which are subject to neural control: (1) rhythmic cycling between inspiration and expiration and (2) regulation of the magnitude of ventilation, which in turn depends on control of respiratory rate and depth of tidal volume. Respiratory rhythm is primarily established by pacemaker activity displayed by inspiratory neurons located in the respiratory control center in the medulla of the brain stem. When these inspiratory neurons autonomously fire, impulses ultimately reach the inspiratory muscles to bring about inspiration. When the inspiratory neurons cease firing, the inspiratory muscles relax and expiration takes place. If active expiration is to occur, the expiratory muscles are activated by output from the medullary expiratory neurons at this time. This basic rhythm is smoothed out by a balance of activity in the apneustic and pneumotaxic centers located higher in the brain stem in the pons. The apneustic center prolongs inspiration, whereas the more powerful pneumotaxic center limits inspiration.

Three chemical factors play a role in determining the magnitude of ventilation: the P_{CO_2}, P_{O_2}, and H^+ concentration of the arterial blood. The dominant factor in the minute-to-minute regulation of ventilation is the arterial P_{CO_2}. An increase in arterial P_{CO_2} is the most potent chemical stimulus for increasing ventilation. Changes in arterial P_{CO_2} alter ventilation primarily by bringing about corresponding changes in the brain ECF H^+ concentration, to which the central chemoreceptors are exquisitely sensitive. The peripheral chemoreceptors are responsive to an increase in arterial H^+ concentration, which likewise reflexly brings about increased ventilation. The resultant adjustment in arterial H^+-generating-CO_2 is important in maintaining the acid-base balance of the body. The peripheral chemoreceptors also reflexly stimulate the respiratory center in response to a marked reduction in arterial P_{O_2} (<60 mm Hg). This response serves as an emergency mechanism to increase respiration when the arterial P_{O_2} levels fall below the safety range provided by the plateau portion of the O_2-Hb curve.

 REVIEW EXERCISES

Objective Questions (Answers on p. D–3.)

1. Breathing is accomplished by alternate contraction and relaxation of muscles within the lung tissue. (True or false?)

2. The alveoli normally empty completely during maximal expiratory efforts. (True or false?)

3. Alveolar ventilation does not always increase when pulmonary ventilation increases. (True or false?)

4. Oxygen and CO_2 have equal diffusion coefficients. (True or false?)

5. Hemoglobin has a higher affinity for O_2 than for any other substance. (True or false?)

6. Rhythmicity of breathing is brought about by pacemaker activity displayed by the respiratory muscles. (True or false?)

7. The expiratory neurons send impulses to the motor neurons controlling the expiratory muscles during normal quiet breathing. (True or false?)

8. The three forces that tend to keep the alveoli open are _____, _____, and _____.

9. The two forces that promote alveolar collapse are _____ and _____.

10. _____ is a measure of the magnitude of change in lung volume accomplished by a given change in the transmural pressure gradient.

11. _____ refers to the phenomenon of the lungs snapping back to their resting size after having been stretched.

12. _____ is the erythrocytic enzyme responsible for catalyzing the conversion of CO_2 into HCO_3^-.

13. Which of the following reactions take(s) place at the pulmonary capillaries?
    a. $Hb + O_2 \rightarrow HbO_2$
    b. $CO_2 + H_2O \rightarrow H_2CO_3 \rightarrow H^+ + HCO_3^-$
    c. $Hb + CO_2 \rightarrow HbCO_2$
    d. $HbH \rightarrow Hb + H^+$

14. Using the answer code at the right, indicate which chemoreceptors are being described:
    ___ 1. stimulated by an arterial P_{O_2} of 80 mm Hg
    ___ 2. stimulated by an arterial P_{O_2} of 55 mm Hg
    ___ 3. directly depressed by an arterial P_{O_2} of 55 mm Hg
    ___ 4. weakly stimulated by an elevated arterial P_{CO_2}
    ___ 5. strongly stimulated by an elevated brain ECF H^+ induced by an elevated arterial P_{CO_2}
    ___ 6. stimulated by an elevated arterial H^+ concentration

    a. peripheral chemoreceptors
    b. central chemoreceptors
    c. both the peripheral and central chemoreceptors
    d. neither the peripheral nor the central chemoreceptors

15. Indicate the O_2 and CO_2 partial pressure relationships that are important in gas exchange by circling > (greater than), < (less than), or = (equal to) as appropriate in each of the following statements:
    a. P_{O_2} in blood entering the pulmonary capillaries is (>, <, or =) P_{O_2} in the alveoli.
    b. P_{CO_2} in blood entering the pumonary capillaries is (>, <, or =) P_{CO_2} in the alveoli.
    c. P_{O_2} in the alveoli is (>, <, or =) P_{O_2} in blood leaving the pulmonary capillaries.
    d. P_{CO_2} in the alveoli is (>, <, or =) P_{CO_2} in blood leaving the pulmonary capillaries.
    e. P_{O_2} in blood leaving the pulmonary capillaries is (>, <, or =) P_{O_2} in blood entering the systemic capillaries.
    f. P_{CO_2} in blood leaving the pulmonary capillaries is (>, <, or =) P_{CO_2} in blood entering the systemic capillaries.
    g. P_{O_2} in blood entering the systemic capillaries is (>, <, or =) P_{O_2} in the tissue cells.
    h. P_{CO_2} in blood entering the systemic capillaries is (>, <, or =) P_{CO_2} in the tissue cells.
    i. P_{O_2} in the tissue cells is (>, <, or approximately =) P_{O_2} in blood leaving the systemic capillaries.
    j. P_{CO_2} in the tissue cells is (>, <, or approximately =) P_{CO_2} in the blood leaving the systemic capillaries.
    k. P_{O_2} in blood leaving the systemic capillaries is (>, <, or =) P_{O_2} in blood entering the pulmonary capillaries.
    l. P_{CO_2} in blood leaving the systemic capillaries is (>, <, or =) P_{CO_2} in blood entering the pulmonary capillaries.

Essay Questions

1. Distinguish between internal and external respiration. List the steps in external respiration.

2. Describe the components of the respiratory system. What is the site of gas exchange?

3. Compare atmospheric, intra-alveolar, and intrapleural pressures.

4. Why are the lungs normally stretched even during expiration?

5. Explain why air enters the lungs during inspiration and leaves during expiration.

6. Why is inspiration normally active and expiration normally passive?

7. Why does airway resistance become an important determinant of airflow rates in chronic obstructive pulmonary disease?

8. Explain pulmonary elasticity in terms of elastic recoil and compliance. What are the source and function of pulmonary surfactant?

9. Define the various lung volumes and capacities.

10. Compare pulmonary ventilation and alveolar ventilation. What is the consequence of anatomic and alveolar dead space?

11. What determines the partial pressures of a gas in air and in blood?

12. List the methods of O_2 and CO_2 transport in the blood.

13. What is the primary factor that determines the percent hemoglobin saturation? What is the significance of the plateau and the steep portions of the O_2-Hb dissociation curve?

14. How does hemoglobin promote the net transfer of O_2 from the alveoli to the blood?

15. Explain the Bohr and Haldane effects.

16. Define the following: hypoxic hypoxia, anemic hypoxia, circulatory hypoxia, histotoxic hypoxia, hypercapnia, hypocapnia, hyperventilation, hypoventilation, hyperpnea, apnea, and dyspnea.

17. What are the locations and functions of the three respiratory control centers? Distinguish between the DRG and the VRG.

18. What factors contribute to rhythmicity of breathing?

1. Why is it important that airplane interiors are pressurized (that is, the pressure is maintained at sea level atmospheric pressure despite the fact that the atmospheric pressure surrounding the plane is substantially lower)? Explain the physiological value of using O_2 masks if the pressure in the airplane interior cannot be maintained.

2. Would hypercapnia accompany the hypoxia produced in each of the following situations? Explain why or why not.
   a. cyanide poisoning
   b. pulmonary edema
   c. restrictive lung disease
   d. high altitude
   e. severe anemia
   f. congestive heart failure
   g. obstructive lung disease

3. If a person lives one mile above sea level at Denver, Colorado, where the atmospheric pressure is 630 mm Hg, what would the P_{O_2} of the inspired air be once it is humidified in the respiratory airways before it reaches the alveoli?

4. Based on what you know about the control of respiration, explain why it is dangerous to voluntarily hyperventilate to lower the arterial P_{CO_2} before going underwater. The purpose of the hyperventilation is to stay under longer before P_{CO_2} rises above normal and drives the swimmer to surface for a breath of air.

5. If a person whose alveolar membranes are thickened by disease has an alveolar P_{O_2} of 100 mm Hg and an alveolar P_{CO_2} of 40 mm Hg, which of the following values for systemic arterial blood gases are most likely to exist?
   a. $P_{O_2} = 105$ mm Hg $\quad$ $P_{CO_2} = 35$ mm Hg
   b. $P_{O_2} = 100$ mm Hg $\quad$ $P_{CO_2} = 40$ mm Hg
   c. $P_{O_2} = 90$ mm Hg $\quad$ $P_{CO_2} = 45$ mm Hg

   If the person is administered 100% O_2, will the arterial P_{O_2} increase, decrease, or remain the same? Will the arterial P_{CO_2} increase, decrease, or remain the same?

14 Urinary System

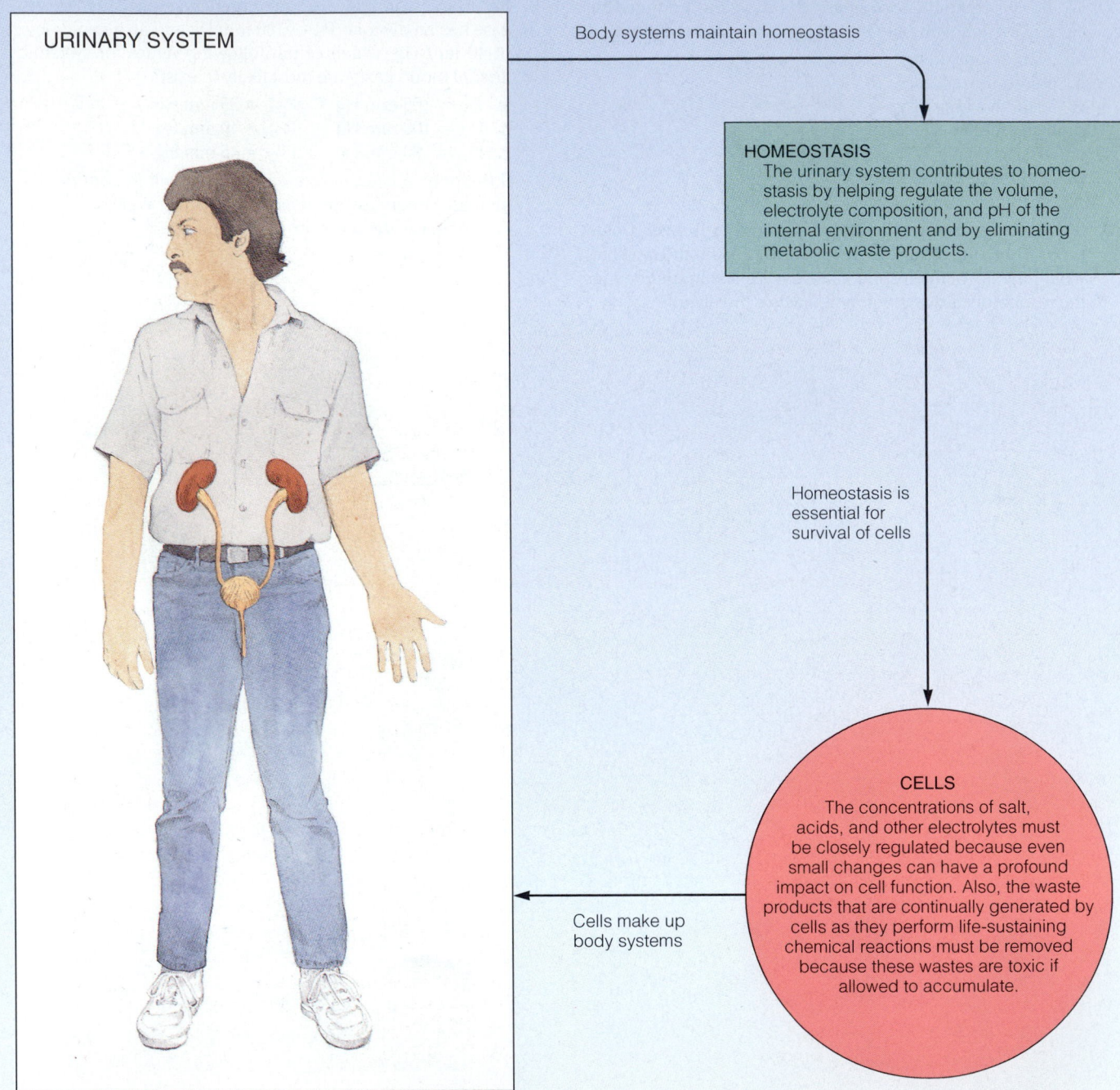

URINARY SYSTEM

Body systems maintain homeostasis

HOMEOSTASIS
The urinary system contributes to homeostasis by helping regulate the volume, electrolyte composition, and pH of the internal environment and by eliminating metabolic waste products.

Homeostasis is essential for survival of cells

CELLS
The concentrations of salt, acids, and other electrolytes must be closely regulated because even small changes can have a profound impact on cell function. Also, the waste products that are continually generated by cells as they perform life-sustaining chemical reactions must be removed because these wastes are toxic if allowed to accumulate.

Cells make up body systems

The survival and proper functioning of cells depend on the maintenance of stable concentrations of salt, acids, and other electrolytes in the internal fluid environment. Cell survival also depends on the continual removal of toxic metabolic wastes produced by the cells as they perform life-sustaining chemical reactions. The **kidneys** play a major role in maintaining homeostasis by regulating the concentration of many of the plasma constituents, especially the electrolytes and water, and by eliminating all the metabolic wastes (except CO_2, which is removed by the lungs). As plasma repeatedly filters through the kidneys, they retain constituents of value for the body and eliminate the undesirable or excess materials in the **urine**. Of special importance is the kidneys' ability to regulate the volume and osmolarity (solute concentration) of the internal fluid environment by controlling salt and water balance. Also critical is their ability to help regulate pH by controlling elimination of acid and base in the urine.

INTRODUCTION
Functions of the kidneys
Anatomical considerations
Basic renal processes
GLOMERULAR FILTRATION
Properties of the glomerular membrane
Forces responsible for glomerular filtration
Regulation of the GFR
Renal blood flow; filtration fraction
TUBULAR REABSORPTION
Transepithelial transport
Active reabsorption
   Na$^+$ reabsorption
   Secondary active reabsorption of glucose and amino acids
   Transport maximum (T_m); renal threshold
   Regulated reabsorption of $PO_4^=$ and Ca^{++}
Passive reabsorption
TUBULAR SECRETION
URINE EXCRETION AND PLASMA CLEARANCE
Urine excretion rate; plasma clearance
Excretion of urine of varying concentrations; medullary
   countercurrent system
Renal failure
● A CLOSER LOOK AT EXERCISE PHYSIOLOGY When Protein in
   the Urine Does Not Mean Kidney Disease
● CONCEPTS, CHALLENGES, AND CONTROVERSIES Dialysis:
   Cellophane Tubing or Abdominal Lining as an Artificial
   Kidney
Micturition

INTRODUCTION

The kidneys perform a variety of functions aimed at maintaining homeostasis.

There would be no animal life-forms on dry land today if it were not for the development of kidneys (or comparable organs). The simplest forms of life live in an external environment of fixed composition, the sea. Likewise, the individual cells of more complex multicellular organisms are able to function and survive only in a fluid environment of essentially constant composition similar to the sea. This salty internal fluid environment is the extracellular fluid (ECF) that bathes all the cells of the body and must be homeostatically maintained.

Simple marine organisms have no influence over the fluid environment in which they live. Exchanges between a simple life-form and its aqueous environment do not alter the external environment's composition because the volume of the organism's body is so small compared to the tremendous volume of the sea in which the organism lives. In contrast, in a complex land animal, exchanges between the cells and the ECF could notably alter the composition of this small, private

internal fluid environment were it not for mechanisms that maintain its stability. To become separated from a watery environment of fixed composition and be free to move about in a dry and ever-changing external environment, land animals have internalized their own bit of sealike water and are equipped with mechanisms to maintain its constancy.

To a large extent, terrestrial animals are able to live on dry land independent of the sea because of their kidneys, the organs that, in concert with the hormonal and neural inputs that control their function, are primarily responsible for maintaining the stability of ECF volume and electrolyte composition. By adjusting the quantity of water and various plasma constituents that are either conserved for the body or eliminated in the urine, the kidneys are able to maintain water and electrolyte balance within the very narrow range compatible with life, despite wide variations in intake and losses of these constituents through other avenues.

When there is a surplus of water or a particular electrolyte such as salt (NaCl) in the ECF, the kidneys can eliminate the excess in the urine. If there is a deficit, the kidneys cannot actually provide additional quantities of the depleted constituent, but they can limit the urinary losses of the material in short supply and thus conserve it until more of the depleted substance can be ingested. Accordingly, the kidneys can compensate more efficiently for excesses than for deficits, as is further reflected by the fact that in some instances the kidneys cannot completely halt the loss of a particular valuable substance in the urine, even though the substance may be in short supply. A prime example is the case of a H_2O deficit. Even if a person is not consuming any H_2O, the kidneys are obligated to put out about half a liter of H_2O in the urine each day to accomplish another major role as the body's "cleaners."

In addition to the kidneys' important regulatory role in maintaining fluid and electrolyte balance, they are the primary route for elimination of potentially toxic metabolic wastes and foreign compounds from the body. These wastes cannot be eliminated in solid form; they must be excreted in solution, obligating the kidneys to produce a minimum volume of around 500 ml of waste-filled urine per day. Because the H_2O eliminated in the urine is derived from the blood plasma, a person stranded without H_2O is eventually obligated to urinate himself or herself to death by depleting the plasma volume to a fatal level as H_2O is inexorably removed to accompany the wastes. Fortunately, except under such extreme circumstances, the kidneys are able to maintain stability in the internal fluid environment despite the usual variations in intake of fluids and electrolytes.

Not only are the kidneys able to adjust for wide variations in ingestion of H_2O, salt, and other electrolytes, but they also make adjustments in the urinary output of these ECF constituents to compensate for their abnormal losses through heavy sweating, vomiting, diarrhea, or hemorrhage. Thus, urine composition varies widely as the kidneys adjust for differences in intake as well as losses of various substances in an attempt to maintain the ECF within the narrow limits compatible with life.

The following are the specific functions performed by the kidneys, most of which are directed toward preserving the constancy of the internal fluid environment:

1. *Maintain H_2O balance in the body.*

2. *Regulate the quantity and concentration of most ECF ions,* including Na^+, Cl^-, K^+, HCO_3^-, Ca^{++}, Mg^{++}, $SO_4^=$, $PO_4^=$, and H^+. Even minor fluctuations in the ECF concentrations of some of these electrolytes can have profound influences. For example, changes in the ECF concentration of K^+ can potentially lead to fatal cardiac dysfunction.

3. *Maintain proper plasma volume,* thereby contributing significantly to the long-term regulation of arterial blood pressure. This function is accomplished through the kidneys' regulatory role in salt and H_2O balance.

4. *Assist in maintaining the proper acid-base balance* of the body by adjusting urinary output of H^+ and HCO_3^-.

5. *Maintain the proper osmolarity* (concentration of solutes) of body fluids, primarily through regulation of H_2O balance.

6. *Excrete (eliminate) the end products (wastes) of bodily metabolism* such as urea, uric acid, and creatinine. If allowed to accumulate, these wastes are toxic, especially to the brain.

7. *Excrete many foreign compounds* such as drugs, food additives, pesticides, and other exogenous nonnutritive materials that have gained entrance to the body.

8. *Secrete erythropoietin,* a hormone that stimulates red blood cell production (see p. 351).

9. *Secrete renin,* an enzymatic hormone that triggers a chain reaction important in the process of salt conservation by the kidneys.

10. *Convert vitamin D into its active form.*

The kidneys form the urine; the remainder of the urinary system is ductwork that carries the urine to the outside.

The **urinary system** consists of the urine-forming organs—the **kidneys**—and the structures that carry the urine from the kidneys to the outside for elimination from the body (Fig. 14–1a). The kidneys are a pair of bean-shaped organs that lie in the back of the abdominal cavity, one on each side of the vertebral column slightly above the level of the waistline. Each kidney is supplied by a **renal artery** and a **renal vein**, which, respectively, enter and leave the kidney at the medial indentation that gives this organ its beanlike form. The kidney acts on the plasma flowing through it to produce urine, conserving materials to be retained in the body and eliminating unwanted materials into the urine.

After urine is formed, it drains into a central collecting cavity, the **renal pelvis**, located at the medial inner core of each kidney (Fig. 14–1b). From there urine is channeled into the **ureter**, a smooth muscle-walled duct that exits at the me-

dial border in close proximity to the renal artery and vein. There are two ureters, one carrying urine from each kidney to the single urinary bladder.

The **urinary bladder**, which temporarily stores urine, is a hollow, distensible sac whose volume can be adjusted by varying the contractile state of the smooth muscle within its walls. Periodically, urine is emptied from the bladder to the outside through another tube, the **urethra**. The urethra in females is straight and short, passing directly from the neck of the bladder to the outside (Fig. 20–2). In males the urethra is much longer and follows a curving course from the bladder to the outside, passing through both the prostate gland and the penis (Fig. 14–1a; also see Fig. 20–1). The male urethra serves the dual function of providing both a route for elimination of urine from the bladder and a passageway for semen from the reproductive organs. The prostate gland lies below the neck of the bladder and completely encircles the urethra. **Prostatic hypertrophy** (enlargement), which often occurs during middle to older age, can partially or completely occlude the urethra, thereby impeding the flow of urine.

The parts of the urinary system beyond the kidneys merely serve as ductwork to transport urine to the outside. Once formed by the kidneys, urine is not altered in composition or volume as it moves downstream through the remainder of the urinary system.

The nephron is the functional unit of the kidney.

Each kidney is composed of about one million microscopic-sized functional units known as **nephrons**, which are bound together by connective tissue. Recall that a functional unit is the smallest unit within an organ capable of performing all of that organ's functions. Because the primary function of the kidneys is to produce urine and, in so doing, maintain constancy in the ECF composition, a nephron is the smallest unit capable of urine formation.

The arrangement of nephrons within the kidneys gives rise to two distinct regions—an outer granular-appearing region, the **renal cortex**, and an inner region of striated-appearing triangles, the **renal pyramids**, which collectively compose the **renal medulla** (Fig. 14–1b).

Knowledge of the structural arrangement of an individual nephron is essential for understanding the distinction between the cortical and medullary regions of the kidney and, more importantly, for understanding renal function. Each nephron consists of a *vascular component* and a *tubular component,* both of which are intimately related structurally and functionally (Fig. 14–2). The dominant portion of the vascular component is the **glomerulus**, a ball-like tuft of capillaries through which part of the water and solutes are filtered from the blood passing through. This filtered fluid, which is almost identical in composition to the plasma, then passes through the tubular component of the nephron, where it is modified by various transport processes that convert it into urine.

On entering the kidney, the renal artery systematically subdivides to ultimately form many small vessels known as **affer-**

Renal pyramid

Renal cortex

Renal medulla

Renal pelvis

Ureter

(b)

Renal vein

Inferior vena cava

Urinary bladder

Urethra

Renal artery

Kidney

Aorta

Ureter

(a)

● **FIGURE 14–1   The Urinary System**   (a) Location of the components of the urinary system. The pair of kidneys form the urine, which is carried by the ureters to the urinary bladder. Urine is stored in the bladder and periodically emptied to the exterior through the urethra. (b) Longitudinal section of a kidney. The kidney consists of an outer granular-appearing renal cortex and an inner striated-appearing renal medulla. The renal pelvis at the medial inner core of the kidney collects urine after it is formed.

---

ent arterioles, one of which supplies each nephron. The afferent arteriole delivers blood to the glomerular capillaries, which rejoin to form another arteriole, the **efferent arteriole**, through which blood that was not filtered into the tubular component leaves the glomerulus (● Fig. 14–3). The efferent arterioles are the only arterioles in the body that drain from capillaries. Typically, arterioles break up into capillaries that rejoin to form venules. At the glomerular capillaries, no O_2 or nutrients are extracted from the blood for use by the kidney tissues nor are waste products picked up from the surrounding tissue. Thus, arterial blood enters the glomerular capillaries through the afferent arteriole, and arterial blood leaves the glomerulus through the efferent arteriole.

The efferent arteriole quickly subdivides into a second set of capillaries, the **peritubular capillaries**, which supply the renal tissue with blood and are important in exchanges between the tubular system and blood during conversion of the filtered fluid into urine. These peritubular capillaries, as their name implies (*peri* means "around"), are intertwined around the tubular system. The peritubular capillaries rejoin to form venules that ultimately drain into the renal vein, by which blood leaves the kidney.

The tubular component of each nephron is a hollow, fluid-filled tube formed by a single layer of epithelial cells. Even though the tubule is continuous from its beginning in close proximity to the glomerulus to its ending at the renal pelvis, it is arbitrarily divided into various segments based on differences in structure and function that occur along its length

• **FIGURE 14–2   A Nephron**

Proximal tubule

Juxtaglomerular
apparatus

Efferent
arteriole

Afferent
arteriole

Distal
tubule

Collecting
duct

Bowman's
capsule

Glomerulus

Artery

Vein

Cortex

Medulla

Peritubular
capillaries

Loop of Henle

To renal
pelvis

Overview of Functions of Parts of a Nephron

Vascular component
- Afferent arteriole—carries blood to glomerulus
- Glomerulus—tuft of capillaries that filters a protein-free plasma into the tubular component
- Efferent arteriole—carries blood from glomerulus
- Peritubular capillaries—supply renal tissue; involved in exchanges with fluid in tubular lumen

Combined vascular/tubular component
- Juxtaglomerular apparatus—secretes substances involved in control of kidney function

Tubular component
- Bowman's capsule—collects glomerular filtrate
- Proximal tubule—uncontrolled reabsorption and secretion of selected substances occur here
- Loop of Henle—establishes osmotic gradient in renal medulla that is important in kidney's ability to produce urine of varying concentration
- Distal tubule—controlled reabsorption and secretion of selected substances occur here
- Collecting tubule—variable H_2O reabsorption occurs here; fluid leaving the collecting tubule is urine, which enters the renal pelvis

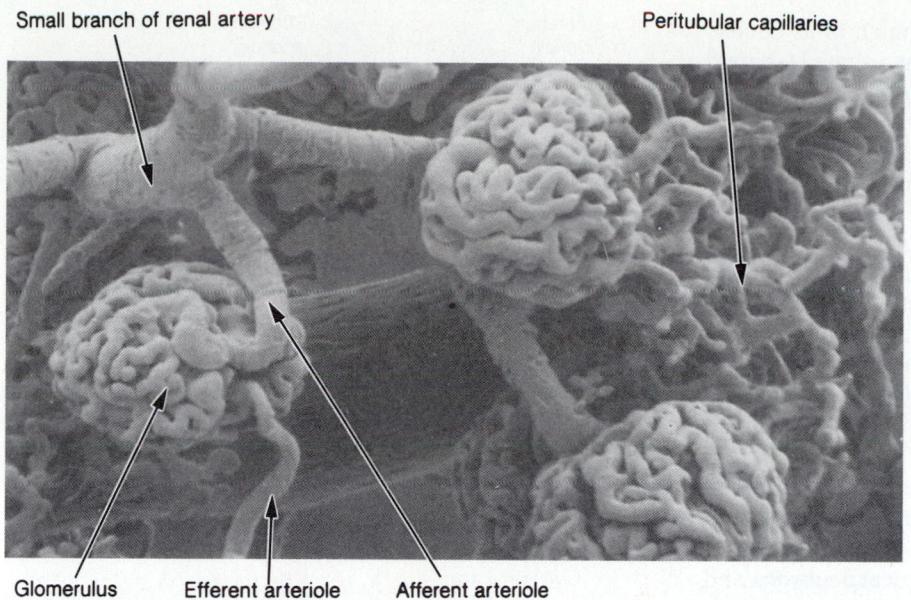

Small branch of renal artery

Peritubular capillaries

Glomerulus   Efferent arteriole   Afferent arteriole

● FIGURE 14–3   Scanning Electron Micrograph of Glomerulus and Associated Arterioles [From *Tissues and Organs: A Text-Atlas of Scanning Electron Microscopy* by Richard G. Kessel and Randy H. Kardon. Copyright © 1979 by W. H. Freeman and Company. Reprinted with permission.]

(Figs. 14–2 and ● 14–4). The tubular component begins with **Bowman's capsule**, an expanded, double-walled invagination that cups around the glomerulus to collect the fluid filtered from the glomerular capillaries. The presence of all glomeruli and associated Bowman's capsules in the cortex is responsible for the region's granular appearance.

From Bowman's capsule, the filtered fluid passes into the **proximal tubule**, which lies entirely within the cortex and is

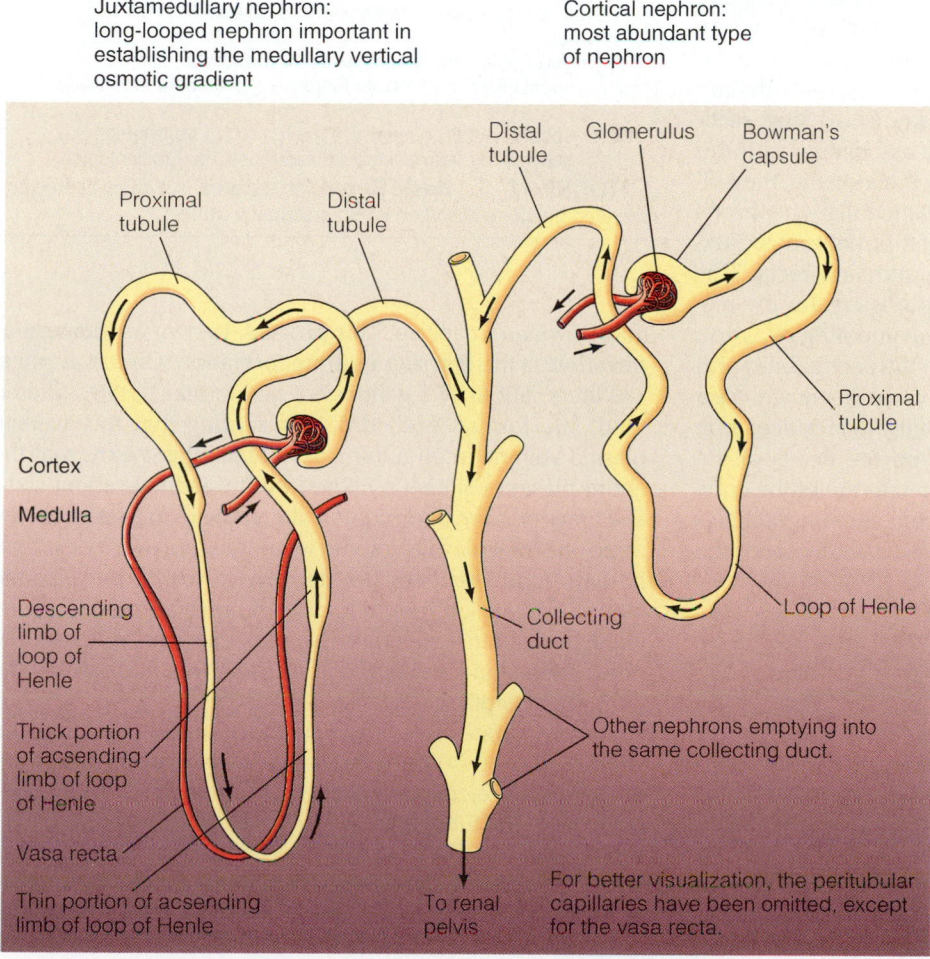

Juxtamedullary nephron: long-looped nephron important in establishing the medullary vertical osmotic gradient

Cortical nephron: most abundant type of nephron

Proximal tubule   Distal tubule

Distal tubule   Glomerulus   Bowman's capsule

Proximal tubule

Cortex

Medulla

Descending limb of loop of Henle

Thick portion of acsending limb of loop of Henle

Vasa recta

Thin portion of acsending limb of loop of Henle

Collecting duct

Other nephrons emptying into the same collecting duct.

Loop of Henle

To renal pelvis

For better visualization, the peritubular capillaries have been omitted, except for the vasa recta.

● FIGURE 14–4   Comparison of Juxtamedullary and Cortical Nephrons   The glomeruli of cortical nephrons lie in the outer cortex, whereas the glomeruli of juxtamedullary nephrons lie in the inner part of the cortex next to the medulla. The loops of Henle of cortical nephrons dip only slightly into the medulla, but the juxtamedullary nephrons have long loops of Henle that plunge deep into the medulla. The juxtamedullary nephrons' peritubular capillaries form hairpin loops known as vasa recta.

highly coiled or convoluted throughout much of its course. The next segment, the **loop of Henle**, forms a sharp U-shaped or hairpin loop that dips into the renal medulla. The *descending limb* of Henle's loop plunges from the cortex into the medulla; the *ascending limb* traverses back up into the cortex. The ascending limb returns to the glomerular region of its own nephron, where it passes through the fork formed by the afferent and efferent arterioles. Both the tubular and vascular cells at this point are specialized to form the **juxtaglomerular** (*juxta* means "next to") **apparatus**, a structure that plays an important role in regulating kidney function. Beyond the juxtaglomerular apparatus, the tubule once again becomes highly coiled to form the **distal tubule**, which also lies entirely within the cortex. The distal tubule empties into a **collecting duct** or **tubule**, with each collecting duct draining fluid from up to eight separate nephrons. Each collecting duct plunges down through the medulla to empty its fluid contents (which have now been converted into urine) into the renal pelvis.

There are two types of nephrons—**cortical nephrons** and **juxtamedullary nephrons**—distinguished by the location and length of some of their structures (Fig. 14–4). (Note the distinction between *juxtamedullary* nephrons and *juxtaglomerular* apparatus.) All nephrons originate in the cortex, but the glomeruli of cortical nephrons lie in the outer layer of the cortex, whereas the glomeruli of juxtamedullary nephrons lie in the inner layer of the cortex, adjacent to the medulla. These two nephron types differ most markedly in their loops of Henle. The hairpin loop of cortical nephrons dips only slightly into the medulla. In contrast, the loop of juxtamedullary nephrons plunges through the entire depth of the medulla. Furthermore, the peritubular capillaries of juxtamedullary nephrons form hairpin vascular loops known as **vasa recta** ("straight vessels"), which run in close association with the loops of Henle. As they course through the medulla, the collecting ducts of both cortical and juxtamedullary nephrons run parallel to the ascending and descending limbs of the juxtamedullary nephrons' loops of Henle and vasa recta. This parallel arrangement of tubules and vessels creates the medullary tissue's striated appearance. More importantly, as you will see, this arrangement, coupled with the permeability and transport characteristics of the long loops of Henle and vasa recta, plays a key role in the kidneys' ability to produce urine of varying concentrations, depending on the needs of the body. About 80% of the nephrons in humans are of the cortical type. Species with greater urine-concentrating abilities than humans, such as the desert rat, have a greater proportion of juxtamedullary nephrons.

The three basic renal processes are glomerular filtration, tubular reabsorption, and tubular secretion.

Three basic processes are involved in the formation of urine: *glomerular filtration, tubular reabsorption,* and *tubular secretion.* To aid in visualizing the relationships among these renal processes, it is useful to unwind the nephron schematically, as in ● Figure 14–5.

As blood flows through the glomerulus, filtration of protein-free plasma occurs through the glomerular capillaries

GF = Glomerular filtration, nondiscriminant filtration of a protein-free plasma from glomerulus into Bowman's capsule

TR = Tubular reabsorption, selective movement of filtered substances from tubular lumen into peritubular capillaries

TS = Tubular secretion, selective movement of nonfiltered substances from peritubular capillaries into tubular lumen

● **FIGURE 14–5 Basic Renal Processes** Anything filtered or secreted but not reabsorbed is excreted in the urine.

into Bowman's capsule. This process, known as **glomerular filtration**, is the first step in urine formation. On the average, 180 liters (about 47.5 gallons) of glomerular filtrate (filtered fluid) are formed each day. Considering that the average plasma volume in an adult is 2.75 liters, this means that the entire plasma volume is filtered by the kidneys about sixty-five times per day. If everything filtered were to pass out in the urine, the total plasma volume would be urinated in less than half an hour! This does not happen, however, because the kidney tubules and peritubular capillaries are intimately related throughout their lengths, so materials can be transferred between the fluid inside the tubules and the blood within the peritubular capillaries.

As the filtrate flows through the tubules, substances of value to the body are returned to the peritubular capillary plasma. This selective movement of substances from inside the tubule (the tubular lumen) into the blood is referred to as **tubular reabsorption**. Reabsorbed substances are not lost from the body in the urine but are carried instead by the peritubular capillaries to the venous system and then to the heart to be recirculated again. Of the 180 liters of plasma filtered per day, 178.5 liters on the average are reabsorbed, with the remaining 1.5 liters passing into the renal pelvis to be eliminated as urine. In general, substances that need to be conserved by the body are selectively reabsorbed, whereas unwanted substances that need to be eliminated remain in the urine.

The third renal process, **tubular secretion**, which refers to the selective transfer of substances from the peritubular capillary blood into the tubular lumen, provides a second route for substances to enter the renal tubules from the blood. The first means by which substances move from the plasma into the tubular lumen is by glomerular filtration. However, only about 20% of the plasma flowing through the glomerular capillaries is filtered into Bowman's capsule; the remaining 80% flows on through the efferent arteriole into the peritubular capillaries. A few substances may be discriminately transferred by tubular secretion from the plasma in the peritubular capillaries into the tubular lumen. Tubular secretion provides a mechanism for more rapidly eliminating selected substances from the plasma by extracting an additional quantity of a particular substance from the 80% of unfiltered plasma in the peritubular capillaries and adding it to the quantity of the substance already present in the tubule as a result of filtration.

Urine excretion refers to the elimination of substances from the body in the urine. It is not really a separate process but is the result of the first three processes. All plasma constituents that gain access to the tubules—that is, that are filtered or secreted—but are not reabsorbed, remain in the tubules and pass into the renal pelvis to be excreted as urine. (Do not confuse *excretion* with *secretion*.)

Glomerular filtration is largely an indiscriminate process. With the exception of blood cells and plasma proteins, all constituents within the blood—H_2O, nutrients, electrolytes, wastes, and so on—are nonselectively filtered. The highly discriminating tubular processes then go to work on the filtrate to return to the blood a fluid of the composition and volume necessary to maintain the constancy of the internal fluid environment. The unwanted filtered material is left behind in the tubular fluid to be excreted as urine. Glomerular filtration can be thought of as pushing a portion of plasma, with all of its essential components as well as those that need to be eliminated from the body, onto a tubular "conveyor belt" that terminates at the renal pelvis, which is the collecting point for urine within the kidney. All plasma constituents that enter this conveyor belt and are not subsequently returned to the plasma by the end of the line are spilled out of the kidney as urine. It is up to the tubular system to salvage by reabsorption the filtered materials that need to be preserved for the body while leaving behind substances that need to be excreted. In addition, some substances are not only filtered but are also secreted onto the tubular conveyor belt so that the amounts of these substances excreted in the urine are greater than the amounts that were filtered. For many substances, these renal processes are subject to physiological control. Thus, the kidneys handle each constituent in the plasma in a characteristic manner by a particular combination of filtration, reabsorption, and secretion.

The kidneys act only on the plasma, yet ECF consists of both plasma and interstitial fluid. The interstitial fluid is actually the true internal fluid environment of the body, because it is the only ECF that comes into direct contact with the cells. However, because of the free exchange between plasma and interstitial fluid across the capillary walls (with the exception of plasma proteins), interstitial-fluid composition reflects the composition of plasma. Thus, by performing their regulatory and excretory roles on the plasma, the kidneys maintain the proper interstitial-fluid environment for optimal cell function. Most of the remainder of this chapter will be devoted to considering how the basic renal processes are accomplished and the mechanisms by which they are carefully regulated to help maintain homeostasis.

GLOMERULAR FILTRATION

The glomerular membrane is more than one hundred times more permeable than capillaries elsewhere.

Fluid filtered from the glomerulus into Bowman's capsule must pass through the three layers that make up the **glomerular membrane** (● Fig. 14–6): (1) the wall of the glomerular capillaries, (2) an acellular gelatinous layer known as the basement membrane, and (3) the inner layer of Bowman's capsule. Collectively, these layers function as a fine molecular sieve that retains the blood cells and plasma proteins but permits H_2O and solutes of small molecular dimension to filter through. The glomerular capillary wall, which consists of a single layer of flattened endothelial cells, is perforated by many large pores, or *fenestrae*, that render it more than one hundred times more permeable to H_2O and solutes than capillaries elsewhere in the body. A *basement membrane* composed of collagen and glycoproteins is sandwiched between the glomerulus and Bowman's capsule. The collagen provides structural strength, and the glycoproteins discourage the filtration of small plasma proteins. Even though the larger plasma proteins cannot be filtered because they cannot fit through the capillary pores, the pores are actually large enough to permit passage of albumin, the smallest of plasma proteins. The glycoproteins, however, being strongly negatively charged, repel albumin and other plasma proteins because the latter are also negatively charged. Therefore, plasma proteins are almost completely excluded from the filtrate, with less than 1% of the albumin molecules escaping into Bowman's capsule. Some renal diseases characterized by excessive albumin in the urine (albuminuria) are believed to be due to disruption of the negative charges within the glomerular membrane, which causes the membrane to be more permeable to albumin even though the size of the pores remains constant.

The final layer of the glomerular membrane, which is the inner layer of Bowman's capsule, consists of **podocytes**, octopuslike cells that encircle the glomerular tuft. Each podocyte bears many elongated foot processes (*podo* means "foot") that interdigitate with foot processes of adjacent

● **FIGURE 14–6 Layers of the Glomerular Membrane** To be filtered, a substance must pass through (1) the pores between the endothelial cells of the glomerular capillary, (2) an acellular basement membrane, and (3) the filtration slits between the foot processes of the podocytes of the inner layer of Bowman's capsule.

podocytes (● Fig. 14–7), much as you slide your fingers between each other as you cup your hands around a ball. The narrow slits between adjacent foot processes, known as **filtration slits**, provide a pathway through which fluid exiting the glomerular capillaries can enter the lumen of Bowman's capsule. Thus, the route that filtered substances take across the glomerular membrane is completely extracellular—first through capillary pores, then through the acellular basement membrane, and finally through capsular filtration slits (Fig. 14–6).

● **FIGURE 14–7 Bowman's Capsule Podocytes with Foot Processes and Filtration Slits** Note the filtration slits between adjacent foot processes on this scanning electron micrograph. [From *Tissues and Organs: A Text-Atlas of Scanning Electron Microscopy* by Richard G. Kessel and Randy H. Kardon. Copyright © 1979 by W. H. Freeman and Company. Reprinted with permission.]

The glomerular-capillary blood pressure is the major force responsible for inducing glomerular filtration.

To accomplish glomerular filtration, a force must be present to drive a portion of the plasma in the glomerulus through the openings in the glomerular membrane. No active transport mechanisms or local energy expenditures are involved in moving fluid from the plasma across the glomerular membrane into Bowman's capsule. Passive physical forces similar to those acting across capillaries elsewhere are responsible for glomerular filtration. Because the glomerulus is a tuft of capillaries, the same principles of fluid dynamics that are responsible for ultrafiltration across other capillaries apply (see p. 323), except for two important differences: (1) the glomerular capillaries are much more permeable than capillaries elsewhere, so more fluid is filtered for a given filtration pressure, and (2) the balance of forces across the glomerular membrane is such that filtration occurs throughout the entire length of the capillaries. In contrast, the balance of forces in other capillaries shifts so that filtration occurs in the beginning portion of the vessel but reabsorption occurs toward the vessel's end.

Three physical forces are involved in glomerular filtration (Table 14–1): (1) glomerular-capillary blood pressure, (2) plasma-colloid osmotic pressure, and (3) Bowman's capsule hydrostatic pressure. The glomerular-capillary blood pressure is the fluid pressure exerted by the blood within the glomerular capillaries. It ultimately depends on contraction of the heart (the source of energy that produces glomerular filtration) and the resistance to blood flow offered by the afferent and efferent arterioles. The glomerular-capillary blood pressure, at an estimated average value of 55 mm Hg, is higher than capillary blood pressure elsewhere, because the diameter of the afferent arteriole is larger than that of the efferent arteriole. Since blood can more readily enter the glomerulus through the wide afferent arteriole than it can leave through the narrower efferent arteriole, glomerular-capillary blood pressure is elevated as a result of the blood damming up in the glomerular capillaries. Furthermore, because of the high resistance offered by the efferent arterioles, blood pressure does not have the same tendency to decrease along the length of the glomerular capillaries as it does along other capillaries. This elevated, nondecremental glomerular blood pressure tends to push fluid out of the glomerulus into Bow-

TABLE 14-1   Forces Involved in Glomerular Filtration

FORCE	EFFECT	MAGNITUDE (mm Hg)
Glomerular-capillary blood pressure	Favors filtration	55
Plasma-colloid osmotic pressure	Opposes filtration	30
Bowman's capsule hydrostatic pressure	Opposes filtration	15
Net filtration pressure (difference between force favoring filtration and forces opposing filtration)	Favors filtration	10

$$55 - (30 + 15) = 10$$

man's capsule along the glomerular capillaries' entire length, and it is the major force responsible for producing glomerular filtration.

Whereas glomerular-capillary blood pressure favors filtration, the two other forces acting across the glomerular membrane (plasma-colloid osmotic pressure and Bowman's capsule hydrostatic pressure) oppose filtration. Plasma-colloid osmotic pressure is caused by the unequal distribution of plasma proteins across the glomerular membrane. Because plasma proteins cannot be filtered, they are present in the glomerular capillaries but are absent in Bowman's capsule. Accordingly, the concentration of H_2O is higher in Bowman's capsule than in the glomerular capillaries. The resultant tendency for H_2O to move by osmosis down its own concentration gradient from Bowman's capsule into the glomerulus opposes glomerular filtration. This opposing osmotic force averages 30 mm Hg, which is slightly higher than across other capillaries. It is higher because considerably more H_2O is filtered out of the glomerular blood, so the concentration of plasma proteins is higher than elsewhere.

The fluid in Bowman's capsule exerts a hydrostatic (fluid) pressure that is estimated to be about 15 mm Hg. This pressure, which tends to push fluid out of Bowman's capsule, opposes the filtration of fluid from the glomerulus into Bowman's capsule.

As can be seen in Table 14–1, there is an imbalance in the forces acting across the glomerular membrane. The total force favoring filtration is attributable to the glomerular-capillary blood pressure at 55 mm Hg. The total of the two forces opposing filtration is 45 mm Hg. The net difference favoring filtration (10 mm Hg of pressure) is referred to as the **net filtration pressure**. This modest pressure is responsible for forcing large volumes of fluid from the blood through the highly permeable glomerular membrane.

The actual rate of filtration, the **glomerular filtration rate** (**GFR**), depends not only on the net filtration presssure but also on how much glomerular surface area is available for penetration and how permeable the glomerular membrane is (that is, how "holey" it is). These properties of the glomerular membrane are collectively referred to as the **filtration coefficient** (**K_f**). Accordingly:

$$GFR = K_f \times \text{net filtration pressure}$$

Normally, about 20% of the plasma that enters the glomerulus is filtered at the net filtration pressure of 10 mm Hg, producing collectively through all glomeruli 180 liters of glomerular filtrate each day for an average GFR of 125 ml/min in males and 160 liters of filtrate per day for an average GFR of 115 ml/min in females.

The most common factor resulting in a change in the GFR is an alteration in the glomerular-capillary blood pressure.

Because the net filtration pressure responsible for inducing glomerular filtration is simply due to an imbalance of opposing physical forces between the glomerular-capillary plasma and Bowman's capsule fluid, alterations in any of these physical forces can affect the GFR. We will examine the effect changes in each of these physical forces have on the GFR.

Plasma-colloid osmotic pressure and Bowman's capsule hydrostatic pressure are not subject to regulation and, under normal conditions, do not vary substantially. However, they can change pathologically and thus inadvertently affect the GFR. Because plasma-colloid osmotic pressure opposes filtration, a decrease in plasma-protein concentration, by reducing this pressure, leads to an increase in the GFR. An uncontrollable reduction in plasma-protein concentration might occur, for example, in severely burned patients who lose a large quantity of protein-rich, plasma-derived fluid through the exposed burned surface of their skin. Conversely, in situations in which the plasma-colloid osmotic pressure is elevated, such as in cases of dehydrating diarrhea, the GFR is reduced. Bowman's capsule hydrostatic pressure can become uncontrollably elevated and filtration subsequently can decrease in the presence of a urinary tract obstruction, such as a kidney stone or prostatic hypertrophy. A damming up of fluid behind the obstruction elevates capsular hydrostatic pressure.

Unlike plasma-colloid osmotic pressure and Bowman's capsule hydrostatic pressure—which may be uncontrollably altered in various disease states, thereby inadvertently altering the GFR—glomerular-capillary blood pressure can be controlled to adjust the GFR to suit the body's needs. Assuming that all other factors remain constant, the magnitude of the glomerular-capillary blood pressure depends on the rate of blood flow in each of the glomeruli, which in turn is determined largely by the magnitude of the mean systemic arterial blood pressure and the resistance offered by the afferent arterioles. GFR is controlled by two mechanisms, both of which are directed at adjusting glomerular blood flow by regulating the caliber and thus the resistance of the afferent arteriole. These are (1) autoregulation, which is aimed at preventing spontaneous changes in GFR, and (2) extrinsic sympathetic control, which is aimed at long-term regulation of arterial blood pressure.

Autoregulation of GFR Because the arterial blood pressure is the force that drives blood into the glomerulus, the glomerular-capillary blood pressure and, accordingly, the GFR would increase in direct proportion to an increase in arterial pressure if everything else remained constant (• Fig. 14–8). Similarly, a fall in arterial blood pressure would be accompanied by a decline in GFR. Such spontaneous changes in GFR are largely prevented by intrinsic regulatory mechanisms initiated by the kidneys themselves, a process known as **autoregulation** (*auto* means "self"). The kidneys can, within limits, maintain a constant blood flow in the glomerular capillaries (and thus a constant glomerular-capillary blood pressure and a stable GFR) despite changes in the driving arterial pressure. They do so primarily by altering afferent arteriolar caliber, thereby adjusting resistance to flow through these vessels. For example, if the GFR increases as a direct result of a rise in arterial pressure, the net filtration pressure and GFR can be reduced to normal by constriction of the afferent arteriole (• Fig. 14–9a), which decreases the flow of blood into the glomerulus. This local adjustment lowers the glomerular blood pressure and the GFR to normal. Conversely, when GFR falls in the presence of a decline in arterial pressure, glomerular pressure can be increased to normal by vasodilation of the afferent arteriole, which allows more

• **FIGURE 14–9   Adjustments of Afferent Arteriole to Alter GFR**   (a) Arteriolar adjustment to reduce GFR. (b) Arteriolar adjustment to increase GFR.

(a)

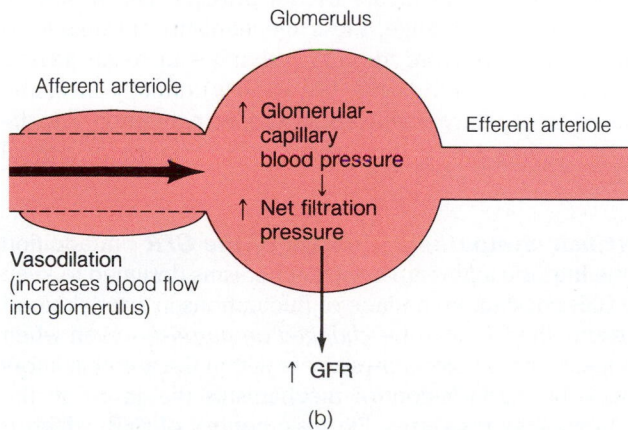

(b)

blood to enter despite the reduction in driving pressure (Fig. 14–9b). The resultant buildup of glomerular blood volume increases gomerular blood pressure, which in turn brings the GFR back up to normal.

The exact mechanisms responsible for accomplishing autoregulatory responses still elude complete understanding. Presently, two intrarenal mechanisms are thought to contribute to autoregulation: (1) a myogenic mechanism, which responds to changes in pressure within the nephron's vascular component, and (2) a tubulo-glomerular feedback mechanism, which senses changes in flow through the nephron's tubular component.

The **myogenic mechanism** is a common property of vascular smooth muscle. Arteriolar vascular smooth muscle contracts inherently in response to the stretch accompanying increased pressure within the vessel. Accordingly, the afferent arteriole automatically constricts on its own when it is stretched because of an increased arterial driving pressure. This response helps limit blood flow into the glomerulus to normal despite the elevated arterial pressure. Conversely, inherent relaxation of an unstretched afferent arteriole when pressure within the vessel is reduced increases blood flow into the glomerulus despite the fall in arterial pressure.

The **tubulo-glomerular feedback mechanism** involves the *juxtaglomerular apparatus,* which is the specialized combination of tubular and vascular cells in the region of the nephron where the tubule, after having bent back on itself, passes through the angle formed by the afferent and efferent arterioles as they join the glomerulus (Figs. 14–2 and • 14–10). Within the arteriolar walls at their point of contact with the tubule, the smooth-muscle cells are specialized to form **granular cells**, so called because they contain numerous secretory granules. Specialized tubular cells in this region are collectively known as the **macula densa**. The macula-densa cells detect changes in the rate at which fluid is flowing past them through the tubule. If the GFR is increased secondary to an elevation in arterial pressure, more fluid than normal is filtered and reaches the distal tubule (• Fig. 14–11). In response, the macula-densa cells trigger the release of vasoactive chemicals from the juxtaglomerular apparatus, which in turn cause the afferent arteriole to constrict, reducing glomerular blood flow and returning GFR to normal. The exact na-

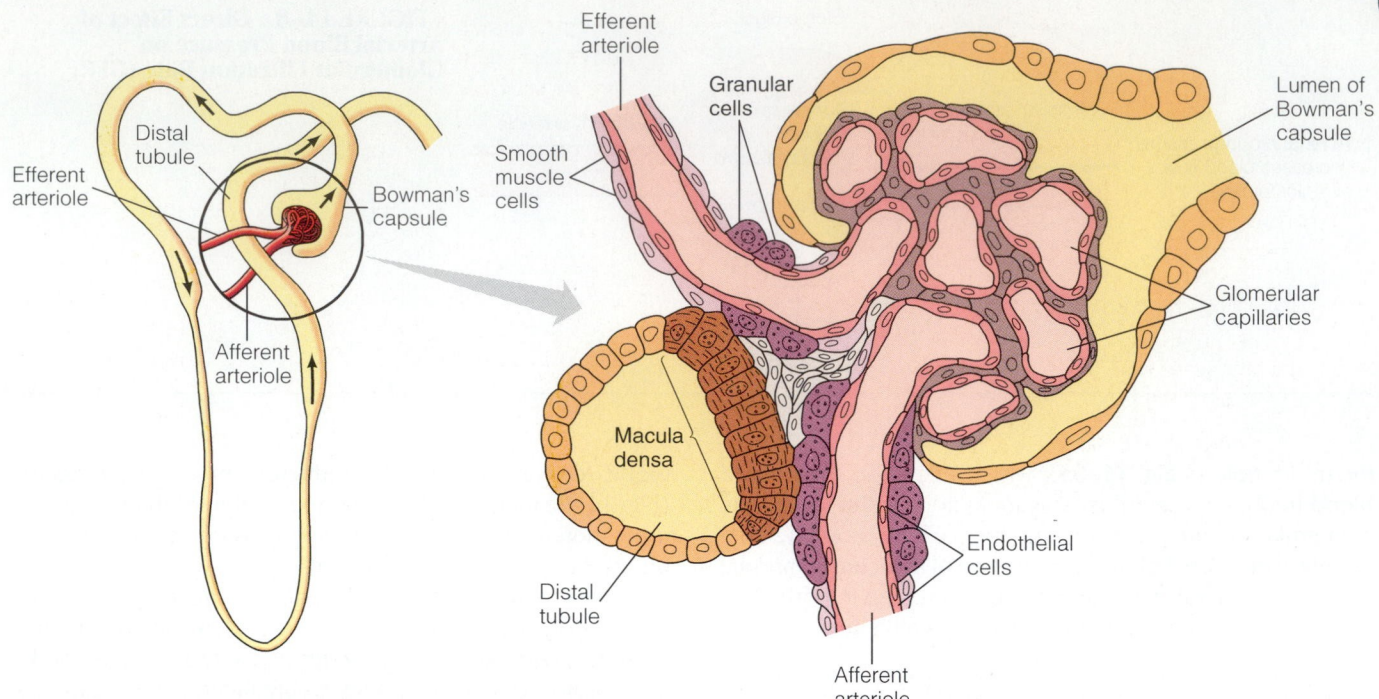

• **FIGURE 14–10** **Juxtaglomerular Apparatus** The juxtaglomerular apparatus consists of specialized vascular cells (the granular cells) and specialized tubular cells (the macula densa) at a point where the distal tubule passes through the fork formed by the afferent and efferent arterioles of the same nephron.

ture of these locally released vasoactive chemicals is currently unclear. Several chemicals have been identified, some of which are vasoconstrictors (for example, endothelin) and others vasodilators (for example, bradykinin), but their exact contributions are yet to be determined. In the opposite situation, when the macula-densa cells detect that the flow rate of fluid through the tubules is low because of a spontaneous decline in GFR accompanying a fall in arterial pressure, these cells induce afferent arteriolar vasodilation by altering the rate of secretion of the relevant vasoactive chemicals. The resultant increase in glomerular flow rate restores the GFR to normal. Thus, by means of the juxtaglomerular apparatus, the tubule of a nephron is able to monitor the rate of movement of fluid through it and adjust the GFR accordingly. This tubulo-glomerular feedback mechanism is initiated by the tubule to help each nephron regulate the rate of filtration through its own glomerulus.

The myogenic and tubulo-glomerular feedback mechanisms work in unison to autoregulate the GFR within the mean arterial blood pressure range of 80 to 180 mm Hg. Within this wide range, intrinsic autoregulatory adjustments of afferent arteriolar resistance can compensate for changes in arterial pressure, thus preventing inappropriate fluctuations in GFR, even though glomerular pressure tends to change in the same direction as arterial pressure. Normal mean arterial pressure is 93 mm Hg, so this range encompasses the transient changes in blood pressure that accompany daily activities unrelated to the need for the kidneys to regulate H_2O and salt excretion, such as the normal elevation in blood pressure accompanying exercise. Autoregulation is

important because unintentional shifts in GFR could lead to dangerous imbalances of fluid, electrolytes, and wastes. Since at least a certain portion of the filtered fluid is always excreted, the amount of fluid excreted in the urine is automatically increased as the GFR increases. If it were not for autoregulation, the GFR would increase and H_2O and solutes would be lost needlessly as a result of the rise in arterial pressure accompanying heavy exercise. On the other hand, if the GFR were too low, the kidneys would not be able to eliminate sufficient quantities of wastes, excess electrolytes, and other materials that should be excreted. Autoregulation thus greatly blunts the direct effect that changes in arterial pressure would otherwise have on GFR and subsequently on H_2O, solute, and waste excretion.

When changes in mean arterial pressure fall outside of the autoregulatory range, these mechanisms are unable to compensate. Therefore, dramatic changes in mean arterial pressure (< 80 mm Hg or > 180 mm Hg) directly cause the glomerular-capillary pressure and, accordingly, the GFR to decrease or increase in proportion to the change in arterial pressure.

Extrinsic Sympathetic Control of the GFR In addition to the intrinsic autoregulatory mechanisms designed to keep the GFR constant in the face of fluctuations in arterial blood pressure, the GFR can be *changed on purpose*—even when the mean arterial blood pressure is within the autoregulatory range—by extrinsic-control mechanisms that override the autoregulatory responses. Extrinsic control of GFR, which is mediated by sympathetic nervous system input to the afferent arterioles, is aimed at the regulation of arterial blood pressure. The parasympathetic nervous system does not exert any influence on the kidneys.

If plasma volume is decreased—for example, because of hemorrhage—the resultant fall in arterial blood pressure is detected by the arterial carotid-sinus and aortic-arch baroreceptors (see p. 334), which initiate neural reflexes to increase blood pressure toward normal. These reflex responses are coordinated by the cardiovascular control center in the brain stem and are mediated primarily through increased sympathetic activity to the heart and blood vessels. Although the resultant increase in both cardiac output and total peripheral resistance helps to increase blood pressure toward normal, the plasma volume is still reduced. In the long term, the plasma volume must be restored to normal. One of the compensations for a depleted plasma volume is a reduction in

urine output so that more fluid than normal is conserved for the body. This reduction in urine output is accomplished in part by a reduction in the GFR; if less fluid is filtered, less fluid is available to be excreted.

No new mechanism is required to decrease the GFR. It is reduced as a result of the baroreceptor reflex response to a fall in blood pressure (• Fig. 14-12). During this reflex, sympathetically induced vasoconstriction occurs in the majority of the arterioles throughout the body as a compensatory mechanism to increase total peripheral resistance. Among the arterioles that constrict in response to the baroreceptor reflex are the afferent arterioles carrying blood to the glomeruli. The afferent arterioles are innervated with sympathetic vasoconstrictor fibers to a far greater extent than the efferent arterioles. When the afferent arterioles constrict as a result of increased sympathetic activity, less blood flows into the glomeruli than normal, causing the glomerular-capillary blood pressure to fall (Fig. 14-9a). The resultant decrease in GFR in turn leads to a reduction in urine volume. In this way, some of the H_2O and salt that would have been lost in the urine is saved for the body, helping in the long term to restore the plasma volume to normal so that the short-term cardiovascular adjustments that have been made are no longer necessary. Other mechanisms described more thoroughly elsewhere, such as increased tubular reabsorption of H_2O and salt as well as increased thirst, also contribute to the long-term maintenance of blood pressure despite a loss of plasma volume by helping to restore plasma volume.

In the converse, if blood pressure is elevated (for example, because of an expansion of plasma volume following ingestion of excessive fluid), the opposite responses occur. When the baroreceptors detect a rise in blood pressure, sympathetic vasoconstrictor activity to the arterioles, including the renal afferent arterioles, is reflexly reduced, allowing afferent arteriolar vasodilation to occur. As more blood enters the glomeruli through the dilated afferent arterioles, glomerular-capillary blood pressure rises, increasing the GFR (Fig. 14-9b). As more fluid is filtered, more fluid is available to be eliminated in the urine. Contributing to the increase in urine volume is a hormonally adjusted reduction in the tubular reabsorption of H_2O and salt. By these two renal mechanisms—increased glomerular filtration and decreased tubular reabsorption of H_2O and salt—urine volume is increased and the excess fluid is eliminated from the body. A reduction in thirst and fluid intake also contributes to restoring an elevated blood pressure to normal.

The GFR can be influenced by changes in the filtration coefficient.

Thus far we have discussed changes in the GFR as a result of changes in the net filtration pressure. The rate of glomerular filtration, however, is dependent on the filtration coefficient (K_f) as well as on the net filtration pressure. For years K_f was considered to be a constant, except in disease situations in which the glomerular membrane becomes leakier than usual. Exciting new research to the contrary indicates that K_f is subject to change under physiological control. Both factors on

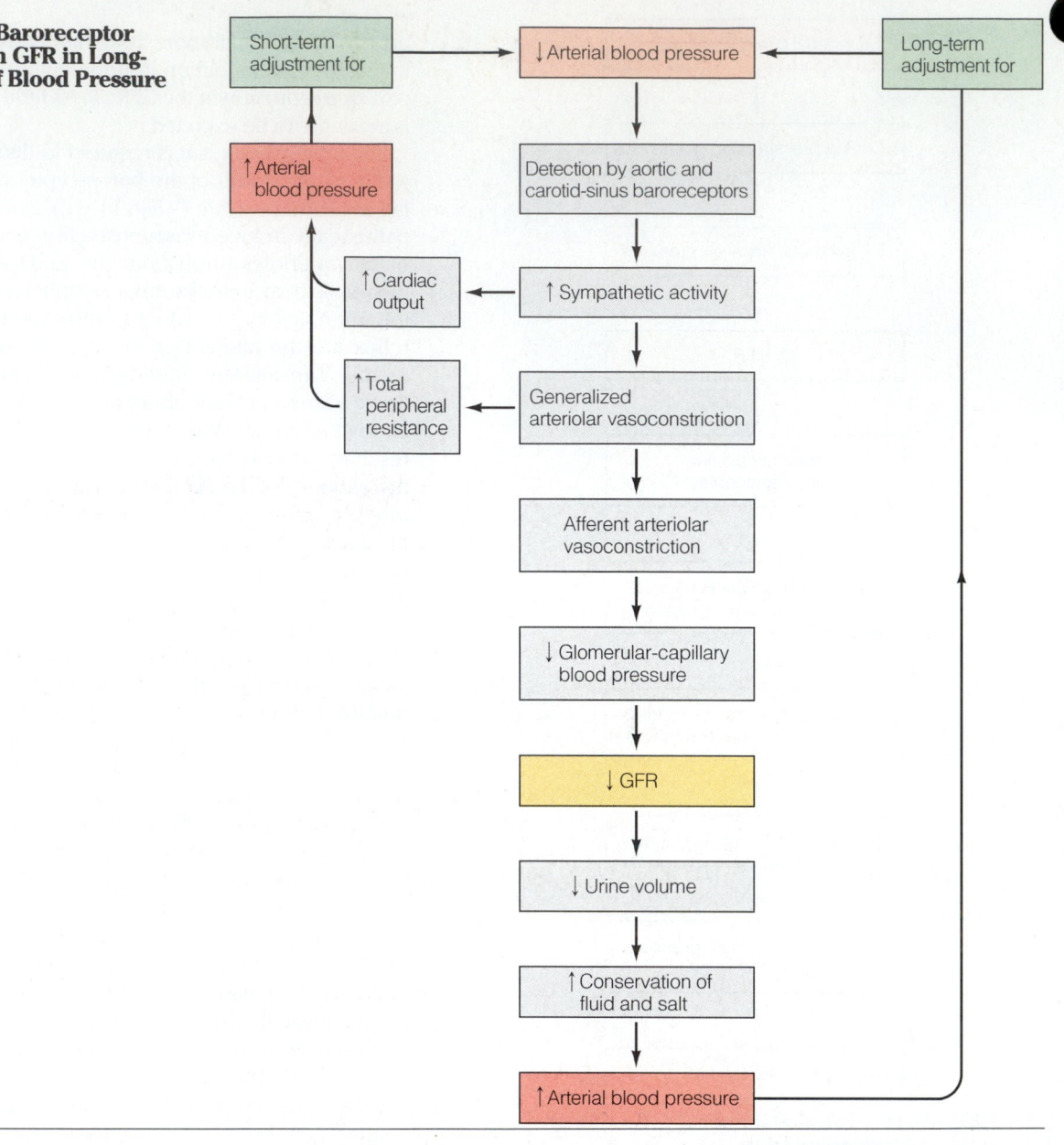

Short-term
adjustment for

↑ Arterial
blood pressure

↑ Cardiac
output

↑ Total
peripheral
resistance

↓ Arterial blood pressure

Detection by aortic and
carotid-sinus baroreceptors

↑ Sympathetic activity

Generalized
arteriolar vasoconstriction

Afferent arteriolar
vasoconstriction

↓ Glomerular-capillary
blood pressure

↓ GFR

↓ Urine volume

↑ Conservation of
fluid and salt

↑ Arterial blood pressure

Long-term
adjustment for

which K_f depends—the surface area and the permeability of the glomerular membrane—can be modified by contractile activity within the membrane.

The surface area available for filtration within the glomerulus is represented by the inner surface of the glomerular capillaries that comes into contact with blood. Each tuft of glomerular capillaries is held together by **mesangial cells**. These cells also function as phagocytes and contain contractile elements (that is, actinlike filaments). Contraction of these mesangial cells closes off a portion of the filtering capillaries, thereby reducing the surface area available for filtration within the glomerular tuft. When the net filtration pressure remains unchanged, this reduction in K_f leads to a decrease in GFR. Evidence suggests that sympathetic stimulation causes the mesangial cells to contract, thus providing a second mechanism (besides promoting afferent arteriolar vasocon-

striction) by which sympathetic activity can decrease the GFR. In addition, several hormones and local chemical mediators suspected or known to be involved in the control of other renal mechanisms, such as tubulo-glomerular feedback or tubular reabsorption, have been shown to influence mesangial-cell contractile activity.

The podocytes also possess actinlike contractile filaments, whose contraction or relaxation can, respectively, decrease or increase the number of filtration slits available in the inner membrane of Bowman's capsule by changing the shapes and proximities of the foot processes (• Fig. 14–13). The number of slits is a determinant of permeability; the more slits, the greater the permeability. Contractile activity of the podocytes, which in turn affects permeability and the K_f, appears to be under physiological control.

As the full extent of various regulatory mechanisms are be-

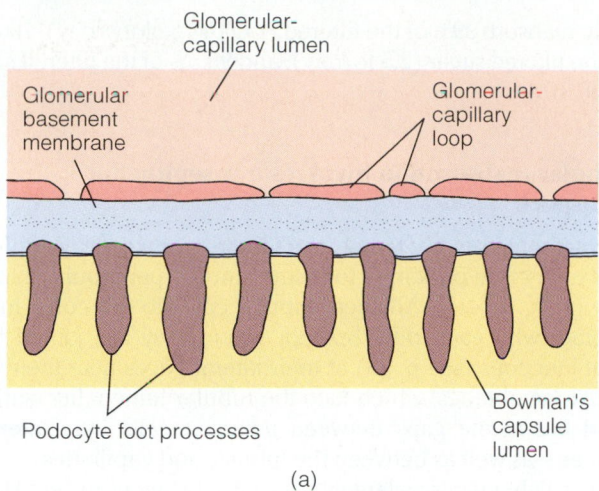

Glomerular basement membrane

Glomerular-capillary lumen

Glomerular-capillary loop

Podocyte foot processes

Bowman's capsule lumen

(a)

(b)

● **FIGURE 14–13   Decrease in Filtration Slits Caused by Podocyte Contraction** (a) Podocyte relaxation causes the bases of the foot processes to narrow, thereby increasing the number of fully open intervening filtration slits spanning a given area. (b) Podocyte contraction causes the foot processes to flatten and thus decreases the number of intervening filtration slits.

Total cardiac output (5,000 ml/min)

22.8%

Renal blood flow (1,140 ml/min)

For purpose of plasma being "adjusted" and "purified" by kidneys

77.2%

Delivered to all other tissues (3,860 ml/min)

For purpose of supplying O_2 and nutrients and removing wastes

● **FIGURE 14–14   Percentage of Cardiac Output Distributed to Kidneys**

coming understood, it is evident that control of glomerular filtration (both through adjustments in afferent arteriolar resistance and K_f) as well as control of tubular reabsorption are complexly interrelated.

The kidneys normally receive 20% to 25% of the cardiac output.

At the average net filtration pressure and K_f, 20% of the plasma that enters the kidneys is converted into glomerular filtrate. Thus, at an average GFR of 125 ml/min, the total renal plasma flow must average about 625 ml plasma/min. Because 55% of whole blood consists of plasma (that is, hematocrit = 45), the total flow of blood through the kidneys averages 1,140 ml/min. This quantity is about 22% of the total cardiac output of 5 liters (5,000 ml)/min for the kidneys, which compose less than 1% of total body weight.

The kidneys need to receive such a seemingly disproportionate share of the cardiac output because they must continuously perform their regulatory and excretory functions on the huge volumes of plasma delivered to them to maintain stability in the internal fluid environment. Most of the blood

goes to the kidneys not to supply the renal tissue but to be adjusted and purified by the kidneys. On the average, 20% to 25% of the blood pumped out by the heart each minute "goes to the cleaners" instead of serving its normal purpose of exchanging materials with the tissues (● Fig. 14–14). Only by continuously processing such a large proportion of the blood are the kidneys able to precisely regulate the volume and electrolyte composition of the internal environment and adequately eliminate the large quantities of metabolic waste products that are constantly produced.

TUBULAR REABSORPTION

Tubular reabsorption is tremendous, highly selective, and variable.

All plasma constituents except the proteins are nondiscriminantly filtered together through the glomerular capillaries. In addition to waste products and excess materials that need to be eliminated from the body, the filtered fluid also contains nutrients, electrolytes, and other substances that the body cannot afford to lose in the urine. Indeed, through the ongoing process of glomerular filtration, greater quantities of these materials are filtered per day than are even present in the entire body. It is important that the essential materials that are filtered be returned to the blood by the process of *tubular reabsorption,* the discrete transfer of substances from the tubular lumen into the peritubular capillaries.

Tubular reabsorption is a highly selective process. All constituents except plasma proteins are at the same concentration in the glomerular filtrate as in the plasma. In most cases, the quantity of each material that is reabsorbed is the amount required to maintain the proper composition and volume of the internal fluid environment. In general, the tubules have a high reabsorptive capacity for substances needed by the body and a poor or no reabsorptive capacity for substances

TABLE 14–2 Fate of Various Substances Filtered by Kidneys

SUBSTANCE	AVERAGE PERCENTAGE OF FILTERED SUBSTANCE REABSORBED	AVERAGE PERCENTAGE OF FILTERED SUBSTANCE EXCRETED
Water	99	1
Sodium	99.5	0.5
Glucose	100	0
Urea (a waste product)	50	50
Phenol (a waste product)	0	100

cally reabsorb 99% of the filtered H_2O (47 gallons/day), 100% of the filtered sugar (2.5 lb/day), and 99.5% of the filtered salt (0.36 lb/day).

Tubular reabsorption involves transepithelial transport.

Throughout its entire length, the tubule is one cell-layer thick and is in close proximity to a surrounding peritubular capillary (● Fig. 14–15). Adjacent tubular cells do not come into contact with each other except where they are joined by tight junctions (see p. 55) at their lateral edges near their luminal membranes, which face the tubular lumen. Interstitial fluid lies in the gaps between adjacent cells—the **lateral spaces**—as well as between the tubules and capillaries.

The tight junctions largely prevent substances except H_2O from moving *between* the cells, so materials must pass *through* the cells in order to leave the tubular lumen and gain entry to the blood. To be reabsorbed, a substance must traverse five distinct barriers (Fig. 14–15):

Step 1. It must leave the tubular fluid by crossing the luminal membrane of the tubular cell.

Step 2. It must pass through the cytosol from one side of the tubular cell to the other.

Step 3. It must traverse the basolateral membrane of the tubular cell to enter the interstitial fluid.

Step 4. It must diffuse through the interstitial fluid.

Step 5. It must penetrate the capillary wall to enter the blood plasma.

This entire sequence of steps is known as **transepithelial** ("across the epithelium") **transport**.

There are two types of tubular reabsorption—**passive reabsorption** and **active reabsorption**—depending on whether

of no value (Table 14–2). Accordingly, only a small percentage, if any, of filtered plasma constituents that are useful to the body are present in the urine, most having been reabsorbed and returned to the blood. Only excess amounts of essential materials such as electrolytes are excreted in the urine. For the essential plasma constituents regulated by the kidneys, absorptive capacity may vary depending on the body's needs. In contrast, a large percentage of filtered waste products are present in the urine. These wastes, which are useless or even potentially harmful to the body if allowed to accumulate, are not reabsorbed to any extent. Instead, they remain in the tubules to be eliminated in the urine. As H_2O and other valuable constituents are reabsorbed, the waste products remaining in the tubular fluid become highly concentrated.

Considering the magnitude of glomerular filtration, the extent of tubular reabsorption is tremendous: the tubules typi-

● **FIGURE 14–15 Steps of Transepithelial Transport** To be reabsorbed (move from filtrate to plasma), a substance must traverse five distinct barriers: ① across luminal cell membrane; ② through cytosol; ③ across basolateral cell membrane; ④ through interstitial fluid; and ⑤ across capillary wall.

local energy expenditure is required for transfer of a particular substance. In passive reabsorption, *all* steps in the transepithelial transport of a substance from the tubular lumen to the plasma are passive; that is, no energy is expended for the substance's net movement, which occurs down electrochemical or osmotic gradients (see p. 59). On the other hand, a substance is said to be actively reabsorbed if any one of the steps in the sequence requires energy, even if the four other steps are passive. With active reabsorption, net movement of the substance from the lumen to the plasma occurs against an electrochemical gradient. Substances that are actively reabsorbed are of particular importance to the body, such as glucose, amino acids, and other organic nutrients, as well as Na^+ and other electrolytes such as $PO_4^=$. Rather than specifically describing the reabsorptive process for each of the many filtered substances that are returned to the plasma, we will provide illustrative examples of the general mechanisms involved after first highlighting the unique and important case of Na^+ reabsorption.

An energy-dependent Na^+-K^+ ATPase transport mechanism in the basolateral membrane is essential for Na^+ reabsorption.

Sodium reabsorption is unique and complex. Eighty percent of the total energy requirement of the kidneys is used for Na^+ transport, indicative of the importance of this process. Unlike most filtered solutes, Na^+ is reabsorbed throughout the tubule, but to varying extents in different regions. Of the Na^+ filtered, 99.5% is normally reabsorbed, of which on average 67% is reabsorbed in the proximal tubule, 25% in the loop of Henle, and 8% in the distal and collecting tubules. Sodium reabsorption plays different important roles in each of these segments, as will become apparent as our discussion continues.

■ Sodium reabsorption in the proximal tubule plays a pivotal role in the reabsorption of glucose, amino acids, H_2O, Cl^-, and urea.

■ Sodium reabsorption in the loop of Henle, along with Cl^- reabsorption, plays a critical role in the kidneys' ability to produce urine of varying concentrations and volumes, depending on the body's need to conserve or eliminate H_2O.

■ Sodium reabsorption in the distal portions of the nephron is variable and subject to hormonal control, being important in the regulation of ECF volume. It is also linked in part with K^+ and H^+ secretion.

The active step in Na^+ reabsorption involves the energy-dependent Na^+-K^+ ATPase carrier located at the tubular cell's basolateral membrane (● Fig. 14–16). This carrier is the same one that is present in all cells and actively extrudes Na^+ from the cell (see p. 65). As this basolateral pump transports Na^+ out of the tubular cell into the lateral space, it keeps the intracellular Na^+ concentration low while it simultaneously builds up the concentration of Na^+ in the lateral space; that is, it moves Na^+ *against* a concentration gradient. Because the intracellular Na^+ concentration is kept low by basolateral-pump activity, a concentration gradient is established that favors the diffusion of Na^+ from its higher concentration in the tubular lumen across the luminal border through Na^+ channels into the tubular cell. Once within the cell, the Na^+ is actively extruded to the lateral space by the basolateral pump. Sodium continues to diffuse down a concentration gradient from its high concentration in the lateral space into the surrounding interstitial fluid and finally into the peritubular-capillary blood. Thus, net transport of Na^+ from the tubular lumen into the blood occurs at the expense of energy.

Aldosterone stimulates Na^+ reabsorption in the distal and collecting tubules; atrial natriuretic peptide inhibits it.

In the proximal tubule and loop of Henle, a constant percentage of the filtered Na^+ is reabsorbed regardless of the **Na^+ load** (*total amount* of Na^+ in the body fluids, *not the concentration* of Na^+ in the body fluids). The reabsorption of a small

● **FIGURE 14–16 Sodium Reabsorption** The basolateral Na^+-K^+ ATPase carrier actively transports Na^+ from the tubular cell into the interstitial fluid within the lateral space. This process establishes a concentration gradient for diffusion of Na^+ from the lumen into the tubular cell and from the lateral space into the peritubular capillary, accomplishing net transport of Na^+ from the tubular lumen into the blood at the expense of energy.

percentage of the filtered Na+ is subject to hormonal control in the distal portion of the tubule. The extent of this controlled reabsorption is inversely related to the magnitude of the Na+ load in the body. If there is too much Na+, little of this controlled Na+ is reabsorbed but is lost in the urine instead, thereby removing excess Na+ from the body. On the other hand, if Na+ is depleted, most or all of this controlled Na+ is reabsorbed, conserving for the body Na+ that otherwise would be lost in the urine. The most important and best known hormonal system involved in the regulation of Na+ is the **renin-angiotensin-aldosterone system**, which stimulates Na+ reabsorption in the distal and collecting tubules.

The Na+ load in the body is reflected by the ECF volume. Sodium and its accompanying anion Cl− account for more than 90% of the ECF's osmotic activity. Recall that osmotic pressure can be thought of loosely as a force that attracts and holds H_2O (see p. 59). When the Na+ load is above normal and the ECF's osmotic activity is therefore increased, the extra Na+ "holds" extra H_2O, expanding the ECF volume. Conversely, when the Na+ load is below normal, thereby decreasing ECF osmotic activity, less H_2O than normal can be held in the ECF, so the ECF volume is reduced. Since plasma is a component of the ECF, the most important consequence of a change in ECF volume is the corresponding change in blood pressure accompanying expansion ($\uparrow$ blood pressure) or reduction ($\downarrow$ blood pressure) of the plasma volume.

The granular cells of the juxtaglomerular apparatus (Fig. 14–10) secrete a hormone, **renin**, into the blood in response to a fall in NaCl/ECF volume/blood pressure. This function is in addition to the role the juxtaglomerular apparatus plays in autoregulation, and renin is distinct from the local vasoactive chemicals that influence glomerular blood flow. These interrelated signals for increased renin secretion are all indicative of the necessity to expand the plasma volume to increase the arterial pressure to normal on a long-term basis. Increased renin secretion, through a complex series of events, brings about increased Na+ reabsorption by the distal portion of the tubule. Chloride always passively follows Na+ down the electrical gradient established by sodium's active movement. The ultimate benefit of this salt retention is its accompanying osmotically induced H_2O retention, which helps restore the plasma volume and blood pressure.

Let us examine the mechanism by which renin secretion ultimately leads to increased Na+ reabsorption (• Fig. 14–17). Once secreted into the blood, renin acts as an enzyme to activate **angiotensinogen** into **angiotensin I**. Angiotensinogen is a plasma protein synthesized by the liver and always present in the plasma in high concentration. On passing through the lungs via the pulmonary circulation, angiotensin I is converted into **angiotensin II** by **angiotensin-converting enzyme** (**ACE**), which is abundant in the pulmonary capillaries. Angiotensin II is the primary stimulus for the secretion of the hormone **aldosterone** from the adrenal gland. The adrenal gland is an endocrine gland that produces several different hormones, each of which is secreted in response to different stimuli.

Among its actions, aldosterone increases Na+ reabsorption by the distal and collecting tubules. It does so by stimulating the synthesis of new proteins within these tubular cells. These proteins, termed **aldosterone-induced proteins**, enhance Na+ reabsorption in two ways. First, they are involved in the formation of Na+ channels in the luminal membranes of the distal and collecting tubular cells, which in turn allow greater passive inward movement of Na+ from the lumen into the cells. Second, they induce the synthesis of Na+-K+ ATPase carriers, which are inserted into the basolateral membranes of these cells. The greater passive inward flux of Na+ promotes increased active pumping of Na+ out of the cells into the lateral spaces and hence into the plasma by the basolateral Na+-K+ ATPase carriers. The net result is an increase in Na+ reabsorption. Chloride (Cl−) follows passively along the electrical gradient produced by active Na+ reabsorption.

The renin-angiotensin-aldosterone system thus promotes salt retention and a resultant H_2O retention and elevation of arterial blood pressure. Acting in negative-feedback fashion, this system alleviates the factors that triggered the initial release of renin—namely, salt depletion, plasma-volume reduction, and decreased arterial blood pressure. In addition to stimulating aldosterone secretion, angiotensin II is also a potent constrictor of arterioles, thereby directly increasing blood pressure by increasing total peripheral resistance (see p. 314). Furthermore, it stimulates thirst (increasing fluid intake) and stimulates vasopressin (a hormone that increases H_2O retention by the kidneys), both of which contribute to plasma-volume expansion and elevation of arterial pressure.

The opposite situation exists when the Na+ load, ECF and plasma volume, and arterial blood pressure are above normal. Under these circumstances, renin secretion is inhibited. Consequently, because angiotensinogen is not activated to angiotensin I and II, aldosterone secretion is not stimulated. Without aldosterone, the small aldosterone-dependent portion of Na+ reabsorption in the distal segments of the tubule does not occur. Instead, this nonreabsorbed Na+ is lost in the urine. In the absence of aldosterone, the ongoing loss of this small percentage of filtered Na+ can rapidly remove excess Na+ from the body.

Even though only about 8% of the filtered Na+ is dependent on aldosterone for reabsorption, this small loss, multiplied manyfold as the entire plasma volume is filtered through the kidneys many times per day, can lead to a sizeable loss of Na+. In the complete absence of aldosterone, 20 gm of salt may be excreted per day. With maximum aldosterone secretion, all of the filtered Na+ (and, accordingly, all of the filtered Cl−) is reabsorbed, so salt excretion in the urine is zero. The amount of aldosterone secreted, and consequently the relative amount of salt conserved versus salt excreted, usually varies between these extremes, depending on the body's needs. For example, an "average" salt consumer typically excretes about 10 gm of salt per day in the urine, a heavy salt consumer excretes more, and an individual who has lost considerable salt during heavy sweating has a lower urinary salt excretion. By varying the amount of renin and aldosterone secreted in accordance with the salt-determined fluid load in the body, the kidneys are able to finely adjust the amount of salt conserved or eliminated. In doing so, they maintain the salt load and ECF volume/arterial blood pres-

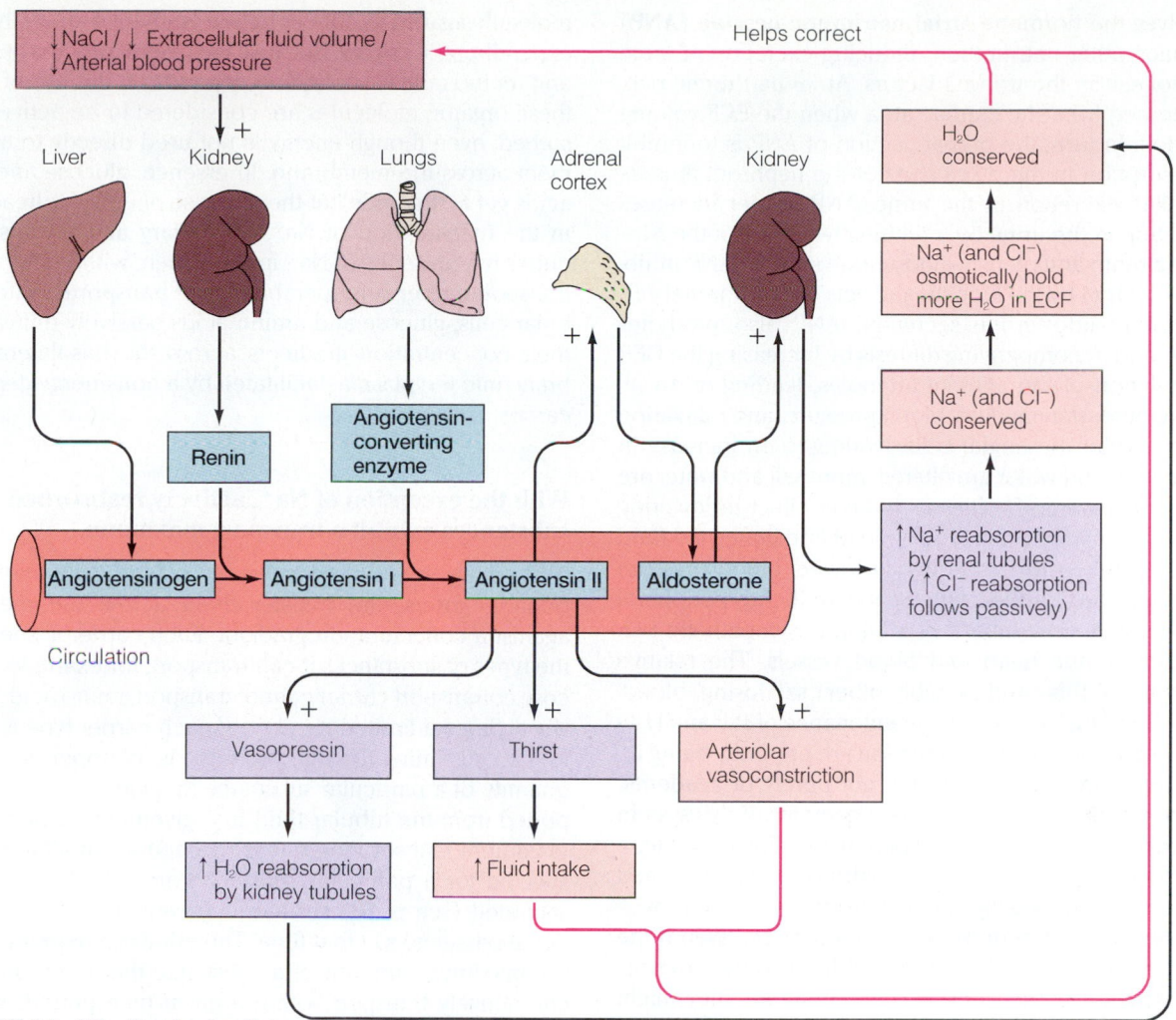

• FIGURE 14–17   Renin-Angiotensin-Aldosterone System The kidneys secrete the hormone renin in response to a reduction in NaCl/ECF volume/arterial blood pressure. Renin activates angiotensinogen, a plasma protein produced by the liver, into angiotensin I. Angiotensin I is converted into angiotensin II by angiotensin-converting enzyme produced in the lungs. Angiotensin II stimulates the adrenal cortex to secrete the hormone aldosterone, which stimulates Na^+ reabsorption by the kidneys. The resultant retention of Na^+ exerts an osmotic effect that holds more H_2O in the ECF. Together the conserved Na^+ and H_2O help correct the original stimuli that activated this renin-angiotensin-aldosterone system. Angiotensin II also exerts other effects that help rectify the original stimuli.

sure at a relatively constant level in spite of wide variations in salt consumption and abnormal losses of salt-laden fluid.

It should not be surprising that some cases of hypertension (high blood pressure) are due to abnormal increases in renin-angiotensin-aldosterone activity. This system is also responsible in part for the fluid retention and edema accompanying congestive heart failure. Because of the failing heart, the cardiac output is reduced and arterial blood pressure is low in spite of a normal or even expanded plasma volume. When a fall in blood pressure is due to a failing heart rather than a reduction in the salt/fluid load in the body, the salt- and fluid-retaining reflexes triggered by the low blood pressure are inappropriate. Sodium excretion may fall to virtually zero despite continued salt ingestion and accumulation in the body. The resultant expansion of the ECF produces edema and exacerbates the congestive heart failure as the weakened heart is unable to pump the additional plasma volume.

Because their salt-retaining mechanisms are being inap-

propriately triggered by the reduction in blood pressure, patients with congestive heart failure are placed on low-salt diets. Often, they are treated with **diuretics**, which are therapeutic agents that cause *diuresis* (increased urinary output) and thus promote loss of fluid from the body. Many of these drugs function by inhibiting tubular reabsorption of Na^+. As more Na^+ is excreted, more H_2O is also lost from the body, thus helping to remove the excess ECF. **ACE inhibitor drugs** which block the action of angiotensin-converting enzyme, are also beneficial in the treatment of congestive heart failure as well as certain cases of hypertension. By blocking the generation of angiotensin II, ACE inhibitors halt the ultimate salt- and fluid-conserving actions and arteriolar constrictor effects of the renin-angiotensin-aldosterone system.

In addition to the renin-angiotensin-aldosterone system, which is believed to exert the most powerful influence on the renal handling of Na^+, recent evidence indicates that this Na^+-retaining system is opposed by a Na^+-losing system that involves the hormone **atrial natriuretic peptide** (**ANP**) and perhaps other natriuretic ("inducing excretion of large amounts of salt in the urine") factors. Atrial natriuretic peptide is released from the cardiac atria when the ECF volume is expanded. In turn, the primary action of ANP is to inhibit Na^+ reabsorption in the distal parts of the nephron, thus increasing Na^+ excretion in the urine. ANP further increases Na^+ excretion in the urine by inhibiting two steps of the Na^+-conserving renin-angiotensin-aldosterone system. ANP inhibits renin secretion by the kidneys and acts on the adrenal cortex to inhibit aldosterone secretion. ANP also promotes natriuresis and accompanying diuresis by increasing the GFR through dilation of the afferent arterioles, leading to an increase in glomerular capillary blood pressure, and relaxation of the glomerular mesangial cells, leading to an increase in K_f. As more salt and water are filtered, more salt and water are excreted in the urine. Besides its indirect effect in lowering blood pressure by reducing the Na^+ load and hence the fluid load in the body, ANP also directly lowers blood pressure by decreasing the cardiac output and reducing peripheral vascular resistance by means of inhibiting sympathetic nervous activity to the heart and blood vessels. The relative contributions of this (and possibly other) salt-losing, blood-pressure–lowering factors in the maintenance of salt and H_2O balance and blood-pressure regulation are presently being intensely investigated. This subject is not purely of academic interest, because it is likely that derangements of this system will be found to contribute to hypertension. For example, a deficiency of a counterbalancing natriuretic system could theoretically cause long-term hypertension, since the powerful Na^+-conserving system would be unopposed. Even if the natriuretic factors are not found to contribute to the development of hypertension or congestive heart failure, they might prove to be useful in treating these circulatory disorders by reducing the Na^+ load in the body.

Glucose and amino acids are reabsorbed by Na^+-dependent secondary active transport.

Large quantities of nutritionally important organic molecules such as glucose and amino acids are filtered each day. Because these substances normally are completely reabsorbed back into the blood by energy- and Na^+-dependent mechanisms located in the proximal tubule, none of these materials are usually excreted in the urine. This rapid and thorough reabsorption early in the tubules protects against the loss of these important organic nutrients.

Even though glucose and amino acids are actively moved uphill against their concentration gradients from the tubular lumen into the blood until their concentration in the tubular fluid is virtually zero, no energy is directly used to operate the glucose or amino acid carriers. Glucose and amino acids are transferred by means of secondary active transport; a specialized cotransport carrier simultaneously transfers both Na^+ and the specific organic molecule from the lumen into the cell (see p. 66). The lumen-to-cell Na^+ concentration gradient maintained by the energy-consuming basolateral Na^+-K^+ ATPase pump drives this cotransport system and pulls the organic

molecule against its concentration gradient without the direct expenditure of energy. Because the overall process of glucose and amino acid reabsorption depends on the use of energy, these organic molecules are considered to be actively reabsorbed, even though energy is not used directly to transport them across the membrane. In essence, glucose and amino acids get a "free ride" at the expense of energy already used in the reabsorption of Na^+. Secondary active transport requires the presence of Na^+ in the lumen; without Na^+ the cotransport carrier is inoperable. Once transported into the tubular cells, glucose and amino acids passively diffuse down their concentration gradients across the basolateral membrane into the plasma, facilitated by a non-energy-dependent carrier.

With the exception of Na^+, actively reabsorbed substances exhibit a transport maximum.

All actively reabsorbed substances bind with membrane-bound carriers that transfer them across the membrane against a concentration gradient. Each carrier is specific for the types of substances it can transport; for example, the glucose cotransport carrier cannot transport amino acids, or vice versa. Since a limited number of each carrier type is present in the cells lining the tubules, there is an upper limit on the quantity of a particular substance that can be actively transported from the tubular fluid in a given period of time. The maximum reabsorption rate is reached when all the carriers specific for a particular substance are fully "occupied" or saturated (see p. 63) so that they cannot handle any additional passengers at that time. The **tubular maximum** or T_m is the maximum amount of a substance that the tubular cells can actively transport within a given time period. With the exception of Na^+, all actively reabsorbed substances display a T_m. (Sodium does not display a T_m because aldosterone promotes the synthesis of more active Na^+-K^+ ATPase carriers in the distal and collecting tubular cells as needed.) Any quantity of a substance filtered beyond its T_m fails to be reabsorbed but escapes instead into the urine.

The plasma concentrations of some but not all substances that display carrier-limited reabsorption are regulated by the kidneys. How can the kidneys regulate some actively reabsorbed substances but not others, when the renal tubules limit the quantity of each of these substances that can be reabsorbed and returned to the plasma? We will compare glucose, a substance that has a T_m but is not regulated by the kidneys, with phosphate, a T_m-limited substance that is regulated by the kidneys.

Glucose Reabsorption The normal plasma concentration of glucose is 100 mg of glucose/100 ml of plasma. Because glucose is freely filterable at the glomerulus, it passes into Bowman's capsule at the same concentration it has in the plasma. Accordingly, 100 mg of glucose are present in every 100 ml of plasma filtered. With 125 ml of plasma normally being filtered each minute (average GFR = 125 ml/min), 125 mg of glucose pass into Bowman's capsule with this filtrate every minute. The quantity of any substance filtered per minute, known as its **filtered load**, can be calculated as follows:

$$\frac{\text{filtered load}}{\text{of a substance}} = \frac{\text{plasma concentration}}{\text{of the substance}} \times \text{GFR}$$

$$\text{filtered load of glucose} = 100 \text{ mg}/100 \text{ ml} \times 125 \text{ ml/min}$$

$$= 125 \text{ mg/min}$$

At a constant GFR, the filtered load of glucose is directly proportional to the plasma glucose concentration. Doubling the plasma glucose concentration to 200 mg/100 ml doubles the filtered load of glucose to 250 mg/min, and so on (Fig. 14–18).

The T_m for glucose averages 375 mg/min; that is, the glucose carrier mechanism is capable of actively reabsorbing up to 375 mg of glucose per minute before it reaches its maximum transport capacity. At a normal plasma glucose concentration of 100 mg/100 ml, the 125 mg of glucose filtered per minute can readily be reabsorbed by the glucose carrier mechanism, because the filtered load is well below the T_m for glucose. Ordinarily, therefore, no glucose appears in the urine because all of the filtered glucose is reabsorbed. Not until the filtered load of glucose exceeds 375 mg/min is the T_m

reached. When more glucose is filtered per minute than can be reabsorbed because the T_m is exceeded, the maximum amount is reabsorbed, whereas the rest remains in the filtrate to be excreted. Accordingly, the plasma glucose concentration must be greater than 300 mg/100 ml—more than three times the normal value—before glucose starts spilling into the urine.

The plasma concentration at which the T_m of a particular substance is reached and the substance first starts appearing in the urine is known as the **renal threshold**. At the normal T_m of 375 mg/min and GFR of 125 ml/min, the renal threshold for glucose is 300 mg/100 ml. Beyond the T_m, reabsorption remains constant at its maximum rate, and any further increase in the filtered load is accompanied by a directly proportional increase in the amount of the substance excreted. For example, at a plasma glucose concentration of 400 mg/100 ml, the filtered load of glucose is 500 mg/min, 375 mg/min of which can be reabsorbed (a T_m's worth) and 125 mg/min of which is excreted in the urine. At a plasma glucose concentration of 500 mg/100 ml, the filtered load is 625 mg/min, still only 375 mg/min can be reabsorbed, and 250 mg/min spill into the urine (Fig. 14–18).

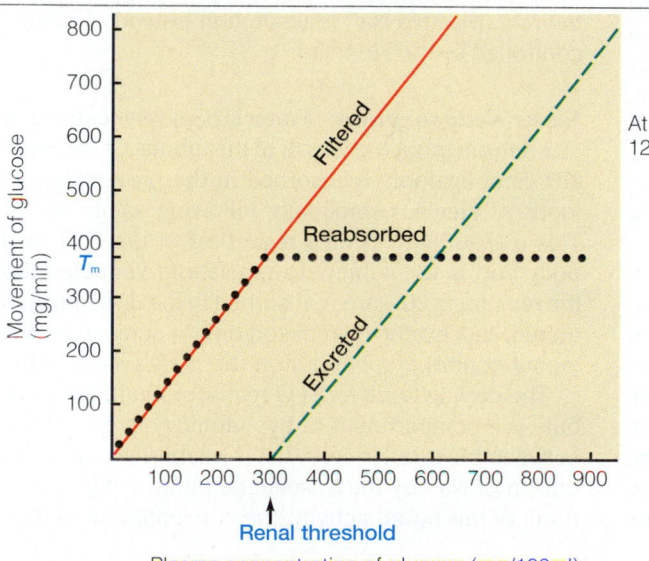

T_m = tubular maximum

FIGURE 14–18 Renal Handling of Glucose as Function of Plasma Glucose Concentration At a constant GFR, the quantity of glucose filtered/min is directly proportional to the plasma concentration of glucose. All of the filtered glucose can be reabsorbed up to the tubular maximum (T_m), the maximum amount of glucose the tubular cells can actively transport per minute. Urinary excretion of glucose does not occur until the amount of glucose filtered/min exceeds the T_m. At that point, the maximum amount of glucose is reabsorbed (a T_m's worth), and the rest remains in the filtrate to be excreted in the urine. The renal threshold is the plasma concentration at which the T_m is reached and glucose first starts appearing in the urine.

	Plasma concentration of substance × GFR = Amount of substance filtered

For example, 100 mg glucose/100 ml plasma × 125 ml plasma filtered/min = 125 mg glucose filtered/min

Plasma Concentration (mg/100ml)	GFR (ml/min)	Amount Filtered (mg/min)	T_m (mg/min)	Amount Reabsorbed (mg/min)	Amount Excreted (mg/min)
a. 80	125	100	375	100	0
b. 100	125	125	375	125	0
c. 200	125	250	375	250	0
d. 300	125	375	375	375	0
e. 400	125	500	375	375	125
f. 500	125	625	375	375	250

The plasma glucose concentration can become extremely high in diabetes mellitus, an endocrine disorder involving deficiency of insulin, a pancreatic hormone. This hormone is important in facilitating the transport of glucose into many of the body's cells. In insulin deficiency, the glucose that cannot gain entry into the cells remains in the plasma, elevating the plasma glucose concentration. Consequently, although glucose does not normally appear in the urine, it is found in the urine of persons with diabetes when the plasma glucose concentration exceeds the renal threshold, even though there has been no change in renal function.

What happens when the plasma glucose concentration falls below normal? The renal tubules, of course, reabsorb all of the filtered glucose, because the glucose reabsorptive capacity is far from being exceeded. The kidneys cannot do anything to raise a low plasma glucose level to normal. They simply return all of the filtered glucose to the plasma.

Thus, the kidneys do not influence the plasma glucose concentration over a wide range of values that varies from abnormally low levels to up to three times the normal level. Because the T_m for glucose is well above the normal filtered load, the kidneys usually conserve all of the glucose, thereby protecting against the loss of this important nutrient in the urine. The kidneys do not regulate glucose because they do not maintain glucose at some specific plasma concentration; instead, this concentration is normally regulated by endocrine and liver mechanisms, with the kidneys merely maintaining whatever plasma glucose concentration is set by these other mechanisms (except when excessively high levels overwhelm the kidney's reabsorptive capacity). The same general principle holds true for other organic plasma nutrients, such as amino acids and water-soluble vitamins.

Phosphate Reabsorption The kidneys do directly contribute to the regulation of many electrolytes, such as calcium (Ca^{++}) and phosphate ($PO_4^=$), because the renal thresholds of these inorganic ions equal their normal plasma concentrations. We will use $PO_4^=$ as an example. Our diets are generally rich in $PO_4^=$, but because the tubules can reabsorb up to the normal plasma concentration's worth of $PO_4^=$ and no more, the excess ingested $PO_4^=$ is quickly spilled into the urine, re-

storing the plasma concentration to normal. The more $PO_4^=$ is ingested beyond the body's needs, the more is excreted. In this way, the kidneys maintain the desired plasma $PO_4^=$ concentration while eliminating any excess $PO_4^=$ ingested.

Unlike the reabsorption of organic nutrients, the reabsorption of $PO_4^=$ and Ca^{++} is also subject to hormonal control. Parathyroid hormone can alter the renal thresholds for $PO_4^=$ and Ca^{++}, thus adjusting the quantity of these electrolytes conserved depending on the body's momentary needs (chapter 19).

Active Na^+ reabsorption is responsible for the passive reabsorption of Cl^-, H_2O, and urea.

Not only is the secondary active reabsorption of glucose and amino acids linked to the basolateral Na^+-K^+ pump, but the passive reabsorption of Cl^-, H_2O, and urea also depends on this active Na^+ reabsorption mechanism.

Chloride Reabsorption The negatively charged chloride ions are passively reabsorbed down the electrical gradient created by the active reabsorption of the positively charged sodium ions. The amount of Cl^- reabsorbed is determined by the rate of active Na^+ reabsorption instead of being directly controlled by the kidneys.

Water Reabsorption Water is passively reabsorbed by osmosis throughout the length of the tubule. Of the H_2O filtered, 80% is obligatorily reabsorbed in the proximal tubules and loops of Henle, osmotically following solute reabsorption. This reabsorption occurs regardless of the H_2O load in the body and is not subject to regulation. Variable amounts of the remaining 20% are reabsorbed in the distal portions of the tubule; the extent of reabsorption is subject to direct hormonal control, depending on the body's state of hydration.

The driving force for H_2O reabsorption in the proximal tubule is a compartment of hypertonicity in the lateral spaces between the tubular cells that is established by the active extrusion of Na^+ by the basolateral pump (• Fig. 14–19). As a result of this pump activity, the concentration of Na^+ rapidly

---

• **FIGURE 14–19 Water Reabsorption in Proximal Tubule** The force for H_2O reabsorption is the compartment of hypertonicity in the lateral spaces established by active extrusion of Na^+ by the basolateral pump. The dashed arrows show the direction of osmotic movement of H_2O.

diminishes in the tubular fluid and tubular cells while it simultaneously increases in the localized region within the lateral spaces. This osmotic gradient induces the passive net flow of H_2O from the lumen into the lateral spaces, either through the cells or intercellularly through "leaky" tight junctions. The accumulation of fluid in the lateral spaces results in a buildup of hydrostatic (fluid) pressure, which flushes H_2O out of the lateral spaces into the interstitial fluid and finally into the peritubular capillaries.

This return of filtered H_2O to the plasma is enhanced by the fact that the plasma-colloid osmotic pressure is greater in the peritubular capillaries than elsewhere. The concentration of plasma proteins, which is responsible for the plasma-colloid osmotic pressure, is elevated in the blood entering the peritubular capillaries because of the extensive filtration of H_2O through the glomerular capillaries upstream. The plasma proteins left behind in the glomerulus are concentrated into a smaller volume of plasma H_2O, increasing the plasma-colloid osmotic pressure of the unfiltered blood that leaves the glomerulus and enters the peritubular capillaries. This force tends to "pull" H_2O into the peritubular capillaries, simultaneous with the "push" of the hydrostatic pressure in the lateral spaces that drives H_2O toward the capillaries. By these means, 65% of the filtered H_2O—117 liters per day—is passively reabsorbed by the end of the proximal tubule. Neither the proximal tubule nor indeed any other portion of the tubule directly requires energy for this tremendous reabsorption of H_2O.

Another 15% of the filtered H_2O is obligatorily reabsorbed from the loop of Henle. Adjustable reabsorption of the remaining 20% of the filtered H_2O is accomplished in the distal and collecting tubules under control of vasopressin. The mechanisms responsible for H_2O reabsorption beyond the proximal tubule will be described in later sections.

Urea Reabsorption   In addition to Cl^- and H_2O, the passive reabsorption of urea is also indirectly linked to active Na^+ reabsorption. **Urea** is a waste product resulting from the breakdown of protein. The osmotically induced reabsorption of H_2O in the proximal tubule secondary to active Na^+ reabsorption produces a concentration gradient for urea that favors the passive reabsorption of this nitrogenous waste (• Fig. 14–20). Because of extensive reabsorption of H_2O in the proximal tubule, the original 125 ml/min of filtrate is progressively reduced, until only 44 ml/min of fluid remains in the lumen by the end of the proximal tubule (with 65% of the H_2O in the original filtrate, or 81 ml/min, having been reabsorbed). Substances that have been filtered but not reabsorbed become progressively more concentrated in the tubular fluid as H_2O is reabsorbed while they are left behind. Urea is one such substance. Urea's concentration as it is filtered at the glomerulus is identical to its concentration in the plasma entering the peritubular capillaries. The quantity of urea present within the 125 ml of filtered fluid at the beginning of the proximal tubule, however, is concentrated almost threefold in the small volume of only 44 ml left at the end of the proximal tubule. As a result, the urea concentration within the tubular fluid becomes considerably greater than the plasma urea concentration in the adjacent capillaries.

= Urea molecules

FIGURE 14–20   Passive Reabsorption of Urea at End of Proximal Tubule (a) In Bowman's capsule and at the beginning of the proximal tubule, urea is at the same concentration as in the plasma and surrounding interstitial fluid. (b) By the end of the proximal tubule, 65% of the original filtrate has been reabsorbed, leading to concentration of the filtered urea in the remaining filtrate. This establishes a concentration gradient favoring the passive reabsorption of urea.

Therefore, a concentration gradient is created for urea to passively diffuse from the tubular lumen into the peritubular-capillary plasma. Since the walls of the proximal tubules are only moderately permeable to urea, about 50% of the filtered urea is passively reabsorbed by this means.

Even though only half of the filtered urea is eliminated from the plasma with each pass through the nephrons, this removal rate is adequate. Only in impaired kidney function, when much less than half of the urea is removed, does the urea concentration in the plasma become elevated. An elevated urea level was one of the first chemical characteristics to be identified in the plasma of patients with severe renal failure. Accordingly, clinical measurement of **blood urea nitrogen (BUN)** came into use as a crude assessment of kidney function. It is now known that the most serious consequences of renal failure are not attributable to the retention of urea, which itself is not especially toxic, but rather to the accumulation of other substances that are not adequately excreted because of their failure to be properly secreted—most notably H^+ and K^+. Health professionals still often refer to renal failure as **uremia** ("urea in the blood"), indicative of the presence of excess urea in the blood, even though urea retention is not this condition's major threat.

In general, unwanted waste products are not reabsorbed.

The other filtered waste products besides urea, such as *phenol* and *creatinine,* are likewise concentrated in the tubular fluid as H_2O leaves the filtrate to enter the plasma, but they are not passively reabsorbed as urea is. Urea molecules, being the smallest of waste products, are the only wastes able to be passively reabsorbed as a result of this concentrating effect. Even though the other wastes are also concentrated in the tubular fluid, they are unable to leave the lumen down their concentration gradients to be passively reabsorbed because they are unable to permeate the tubular wall. Therefore, the waste products, failing to be reabsorbed, generally remain in the tubules and are excreted in the urine in a highly concentrated form. This excretion of metabolic wastes is not subject to physiological control. When renal function is normal, however, the excretory processes proceed at a satisfactory rate even though they are not controlled.

TUBULAR SECRETION

The most important secretory processes are those for H^+, K^+, and organic ions.

By providing a second route of entry into the tubules for selected substances, *tubular secretion* may be viewed as a supplemental mechanism that hastens the elimination of these compounds from the body. Anything that gains entry to the tubular fluid, whether by glomerular filtration or tubular secretion, and that fails to be reabsorbed is eliminated in the urine.

Tubular secretion involves transepithelial transport just as tubular reabsorption does, but now the steps are reversed. As with reabsorption, tubular secretion may be active or passive. The most important substances secreted by the tubules are hydrogen ion (H^+), potassium ion (K^+), and organic an-

ions and cations, many of which are compounds foreign to the body.

Hydrogen-Ion Secretion Renal H^+ secretion is extremely important in the regulation of the acid-base balance in the body. Hydrogen ion can be added to the filtered fluid by being secreted by the proximal, distal, and collecting tubules. The extent of H^+ secretion depends on the acidity of the body fluids. When the body fluids are too acidic, H^+ secretion increases. Conversely, H^+ secretion is reduced when the H^+ concentration in the body fluids is too low. (See chapter 15 for further detail.)

Potassium-Ion Secretion Potassium ion is an example of a substance that is selectively moved in opposite directions in different parts of the tubule; it is actively reabsorbed in the proximal tubule and actively secreted in the distal and collecting tubules. Potassium-ion reabsorption early in the tubule occurs in a constant, unregulated fashion, whereas K^+ secretion later in the tubule is variable and subject to regulation. Normally, a quantity equivalent to about 10% to 15% of the filtered K^+ is excreted in the urine. However, the filtered K^+ is almost completely reabsorbed, so most of the K^+ appearing in the urine is derived from controlled K^+ secretion rather than from filtration.

During K^+ depletion, K^+ secretion in the distal portions of the nephron is reduced to a minimum, so only the small percentage of filtered K^+ that escapes reabsorption in the proximal tubule is excreted in the urine. In this way, K^+ that normally would have been lost in the urine is conserved for the body. On the other hand, when plasma K^+ levels are elevated, K^+ secretion is adjusted so that just enough K^+ is added to the filtrate for elimination to reduce the plasma K^+ concentration to normal. Thus, K^+ secretion, not the filtration or reabsorption of K^+, is varied in a controlled fashion to regulate the rate of K^+ excretion and maintain the desired plasma K^+ concentration.

Potassium-ion secretion in the distal and collecting tubules is coupled to Na^+ reabsorption by means of the energy-dependent basolateral Na^+-K^+ pump (• Fig. 14–21). This

---

• **FIGURE 14–21 Potassium Secretion** The basolateral pump simultaneously transports Na^+ into the lateral space and K^+ into the tubular cell.

pump not only moves Na⁺ out into the lateral space but also transports K⁺ into the tubular cells. The resultant high intracellular K⁺ concentration favors the net diffusion of K⁺ from the cells into the tubular lumen. Movement across the luminal membrane occurs passively through the large number of K⁺ channels present in this barrier. By keeping the interstitial-fluid concentration of K⁺ low as it transports K⁺ into the tubular cells from the surrounding interstitial fluid, the basolateral pump encourages the passive diffusion of K⁺ out of the peritubular-capillary plasma into the interstitial fluid. Potassium exiting the plasma in this manner is subsequently pumped into the cells, from which it diffuses into the lumen. In this way, the basolateral pump actively induces the net secretion of K⁺ from the peritubular-capillary plasma into the tubular lumen.

Several factors are able to alter the rate of K⁺ secretion, the most important being the hormone aldosterone, which stimulates K⁺ secretion by the tubular cells late in the nephron simultaneous to enhancing the cell's reabsorption of Na⁺. An elevation in plasma K⁺ concentration directly stimulates the adrenal cortex to increase its output of aldosterone, which in turn promotes the secretion and ultimate urinary excretion and elimination of the excess K⁺. Conversely, a decline in plasma K⁺ concentration causes a reduction in aldosterone secretion and a corresponding decrease in aldosterone-stimulated renal K⁺ secretion.

Note that a rise in plasma K⁺ concentration directly stimulates aldosterone secretion by the adrenal cortex, whereas a fall in plasma Na⁺ concentration stimulates aldosterone secretion by means of the complex renin-angiotensin pathway. Thus, aldosterone secretion can be stimulated by two separate pathways (● Fig. 14–22). No matter what the stimulus, however, increased aldosterone secretion always promotes simultaneous Na⁺ reabsorption and K⁺ secretion. For this reason, K⁺ secretion can be inadvertently stimulated as a result of increased aldosterone activity brought about by Na⁺ depletion, ECF volume reduction, or a fall in arterial blood pressure totally unrelated to K⁺ balance. The resultant inappropriate loss of K⁺ can lead to K⁺ deficiency.

Another factor that can inadvertently alter the magnitude of K⁺ secretion is the acid-base status of the body. The basolateral pump in the distal portions of the nephron can secrete either K⁺ or H⁺ in exchange for reabsorbed Na⁺. An increased rate of secretion of either K⁺ or H⁺ is accompanied by a decreased rate of secretion of the other ion. Normally the kidneys secrete a preponderance of K⁺, but when the body fluids are too acidic and H⁺ secretion is increased as a compensatory measure, K⁺ secretion is correspondingly reduced. This reduced secretion leads to inadvertent K⁺ retention in the body fluids.

Except in the overriding circumstances of K⁺ imbalances inadvertently induced during renal compensations for Na⁺ or ECF volume deficits or acid-base imbalances, the kidneys usually exert a fine degree of control over plasma K⁺ concentration. This is extremely important, because even minor fluctuations in plasma K⁺ concentration can have detrimental consequences. Potassium ion, being the most abundant cation in the intracellular fluid, plays a key role in the membrane electrical activity of excitable tissues. Both increases

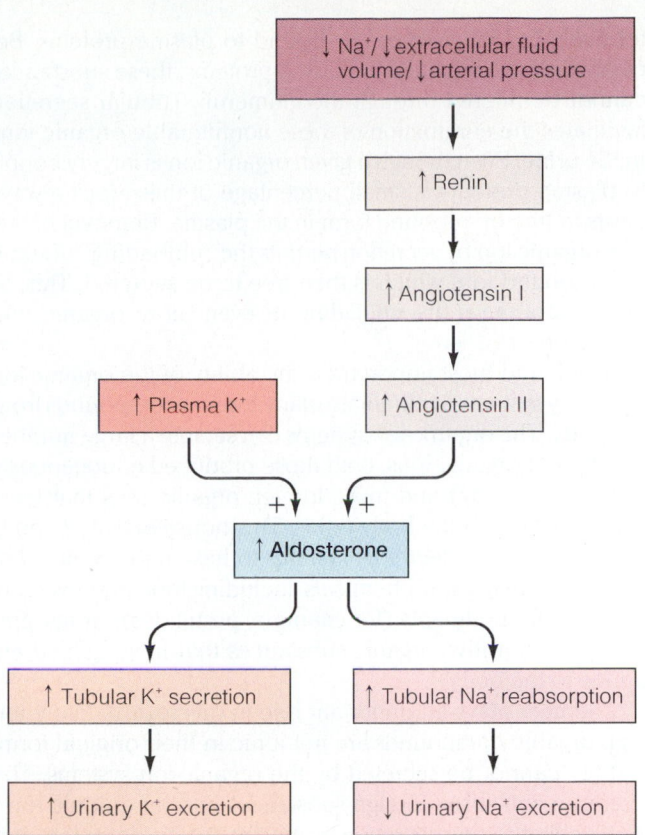

● **FIGURE 14–22** **Dual Control of Aldosterone Secretion by K⁺ and Na⁺**

or decreases in the plasma (ECF) K⁺ concentration can alter the intracellular-to-extracellular K⁺ concentration gradient, which in turn can change the resting membrane potential. A rise in ECF K⁺ concentration leads to a reduction in resting potential and a subsequent increase in excitability, especially of heart muscle. This cardiac overexcitability can lead to a rapid heart rate and even fatal cardiac arrhythmias. Conversely, a fall in ECF K⁺ concentration results in hyperpolarization of nerve and muscle cell membranes, which reduces their excitability. The manifestations of ECF K⁺ depletion are skeletal-muscle weakness, diarrhea and abdominal distention caused by smooth-muscle dysfunction, and abnormalities in cardiac rhythm and impulse conduction.

Organic Anion and Cation Secretion The proximal tubule contains two distinct types of secretory carriers, one for the secretion of organic anions and a separate system for secretion of organic cations. These systems serve several important functions. First, by adding more of a particular type of organic ion to the quantity that has already gained entry to the tubular fluid by means of glomerular filtration, these organic secretory pathways facilitate the excretion of these substances. Included among these organic ions are certain blood-borne chemical messengers such as prostaglandins that, having served their purpose, need to be rapidly removed from the blood so that their biological activity is not unduly prolonged.

Second, in some important instances, organic ions are extensively but not irreversibly bound to plasma proteins. Because they are attached to plasma proteins, these substances cannot be filtered through the glomeruli. Tubular secretion facilitates the elimination of these nonfilterable organic ions in the urine. Even though a given organic ion is largely bound to plasma proteins, a small percentage of these ions always exists in free or unbound form in the plasma. Removal of this free organic ion by secretion permits the "unloading" of some of the bound ion, which is then free to be secreted. This, in turn, encourages the unloading of even more organic ion, and so on.

Third, and most important is the ability of the organic-ion secretory systems to eliminate many foreign compounds from the body. The organic-ion systems can secrete a large number of different organic ions, both those produced endogenously (within the body) and those foreign organic ions that have gained access to the body fluids. This nonselectivity permits these organic-ion secretory systems to hasten the removal of many foreign organic chemicals, including food additives, environmental pollutants (for example, pesticides), drugs, and other non-nutritive organic substances that have gained entrance to the body.

The liver plays an important role in this regard. Many foreign organic compounds are not ionic in their original form, so they cannot be secreted by the organic-ion systems. The liver converts these foreign substances into an anionic form that facilitates their secretion by the organic-anion system and thus accelerates their elimination.

The rate of excretion of foreign organic compounds is not subject to control. Although the relatively nonselective organic-ion secretory systems enhance the removal of these substances from the body, this mechanism is not subject to physiological adjustments.

Many drugs, such as penicillin, are eliminated from the body by means of the proximal tubule organic-ion secretory systems. In order to keep the plasma concentration of these drugs at effective levels, the dosage has to be repeated on a regular, frequent basis to keep pace with the rapid removal of these compounds in the urine.

This completes our discussion of the reabsorptive and secretory processes that occur across the proximal and distal portions of the nephron. These processes are summarized in Table 14–3.

TABLE 14–3 Summary of Transport across Proximal and Distal Portions of the Nephron	
PROXIMAL TUBULE	
Reabsorption	**Secretion**
67% of filtered Na^+ actively reabsorbed, not subject to control; Cl^- follows passively	Variable H^+ secretion, depending on acid-base status of body
All filtered glucose and amino acids reabsorbed by secondary active transport; not subject to control	Organic-ion secretion; not subject to control
Variable amounts of filtered $PO_4^=$ and other electrolytes reabsorbed; subject to control	
65% of filtered H_2O osmotically reabsorbed; not subject to control	
50% of filtered urea passively reabsorbed; not subject to control	
All filtered K^+ reabsorbed; not subject to control	
DISTAL TUBULE	
Reabsorption	**Secretion**
Variable Na^+ reabsorption, controlled by aldosterone; Cl^- follows passively	Variable H^+ secretion, depending on acid-base status of body
Variable H_2O reabsorption, controlled by vasopressin	Variable K^+ secretion, controlled by aldosterone
COLLECTING DUCT	
Reabsorption	**Secretion**
Variable H_2O reabsorption, controlled by vasopressin	Variable H^+ secretion, depending on acid-base status of body

urine. Useful substances are conserved by reabsorption, so they do not appear in the urine.

A relatively small change in the quantity of filtrate reabsorbed can bring about a large change in the volume of urine formed. For example, a reduction of less than 1% in the total reabsorption rate, from 124 to 123 ml/min, increases the urinary excretion rate by 100%, from 1 to 2 ml/min.

URINE EXCRETION AND PLASMA CLEARANCE

On the average, one milliliter of urine is excreted per minute.

Typically, of the 125 ml of plasma filtered per minute, 124 ml/min are reabsorbed, so the final quantity of urine formed averages 1 ml/min. Thus, 1.5 liters of urine per day are excreted, out of the 180 liters per day filtered.

Urine contains high concentrations of various waste products plus variable amounts of the substances regulated by the kidneys, with any excess quantities having spilled into the

Plasma clearance refers to the volume of plasma cleared of a particular substance per minute.

Compared to the plasma entering the kidneys through the renal arteries, the plasma leaving the kidneys through the renal veins lacks the materials that were left behind to be eliminated in the urine. By excreting substances in the urine, the kidneys clean or "clear" the plasma flowing through them of these substances. For any substance, its **plasma clearance** is defined as the volume of plasma that is completely cleared of that substance by the kidneys per minute.[1] It does not refer to the *amount of the substance* removed but to the *volume of plasma* from which that amount was removed. Plasma clearance is actually a more useful measure than urine excretion; it is more important to know what effect urine excretion has on removing materials from the body fluids than to know the volume and composition of the discarded urine. Plasma clearance expresses the kidney's effectiveness in removing various substances from the internal fluid environment.

Plasma clearance can be calculated for any plasma constituent as follows:

$$\text{clearance rate of a substance (ml/min)} = \frac{\begin{array}{c}\text{urine concentration of the substance} \times \text{urine flow rate} \\ \text{(quantity/ml urine)} \quad \text{(ml/min)}\end{array}}{\begin{array}{c}\text{plasma concentration of the substance} \\ \text{(quantity/ml plasma)}\end{array}}$$

The plasma clearance rate for different substances varies, depending on how the kidneys handle each substance.

If a Substance Is Filtered but Not Reabsorbed or Secreted, Its Plasma Clearance Rate Equals the GFR

Assume that a plasma constituent, substance X, is freely filterable at the glomerulus but is not reabsorbed or secreted. As 125 ml/min of plasma is filtered and subsequently reabsorbed, the quantity of substance X originally contained within the 125 ml is left behind in the tubules to be excreted. Thus, 125 ml of plasma is cleared of substance X each minute. (Of the 125 ml/min of plasma filtered, 124 ml/min of the filtered fluid is returned through the process of reabsorption to the plasma minus substance X, thus clearing this 124 ml/min of substance X. In addition, the 1 ml/min of fluid lost in the urine is replaced in the long term by an equivalent volume of ingested H_2O that is already clear of substance X. Therefore, 125 ml of plasma cleared of substance X is, in effect, returned to the plasma for every 125 ml of plasma filtered/min.)

There is no endogenous chemical with the characteristics of substance X. All substances naturally present in the plasma, even wastes, are reabsorbed or secreted to some extent. However, **inulin** (do not confuse with insulin), a harmless foreign carbohydrate produced by onions and garlic, is freely filtered and not reabsorbed or secreted—an ideal substance X. Inulin can be injected and its plasma clearance determined as a clinical means of ascertaining the GFR. Since all glomerular filtrate formed is cleared of inulin, the volume of plasma cleared of inulin per minute equals the volume of plasma filtered per minute; that is, the GFR.

Although the determination of inulin plasma clearance is accurate and straightforward, it is not very convenient be-

cause inulin must be infused continuously throughout the determination to maintain a constant plasma concentration. Therefore, the plasma clearance of an endogenous substance, **creatinine**, is often used instead to give a rough estimate of the GFR. Creatinine, an end product of muscle metabolism, is produced at a relatively constant rate. It is freely filtered and not reabsorbed but is slightly secreted. Accordingly, creatinine clearance is not a completely accurate reflection of the GFR, but it does provide a close approximation and can be more readily determined than inulin clearance.

If a Substance Is Filtered and Reabsorbed but Not Secreted, Its Plasma Clearance Rate Is Always Less Than the GFR

Some or all of a reabsorbable substance that has been filtered is returned to the plasma. Because less than the filtered volume of plasma will have been cleared of the substance, the plasma clearance rate of a reabsorbable substance is always less than the GFR. For example, the plasma clearance for glucose is normally zero. All of the filtered glucose is reabsorbed along with the rest of the returning filtrate, so none of the plasma is cleared of glucose.

For a substance that is partially reabsorbed, such as urea, only part of the filtered plasma is cleared of that substance. With about 50% of the filtered urea being passively reabsorbed, only half of the filtered plasma, or 62.5 ml, is cleared of urea each minute.

If a Substance Is Filtered and Secreted but Not Reabsorbed, Its Plasma Clearance Rate Is Always Greater Than the GFR

Tubular secretion allows the kidneys to clear certain materials from the plasma more efficiently. Only 20% of the plasma entering the kidneys is filtered. The remaining 80% passes unfiltered into the peritubular capillaries. The only means by which this unfiltered plasma can be cleared of any substance during this trip through the kidneys before being returned to the general circulation is by the process of secretion. An example is H^+. Not only will the plasma that is filtered be cleared of nonreabsorbable H^+, but the plasma from which H^+ is secreted will also be cleared of H^+. For example, if the quantity of H^+ that is secreted is equivalent to the quantity of H^+ present in 25 ml of plasma, the clearance rate for H^+ will be 150 ml/min at the normal GFR of 125 ml/min. Every minute 125 ml of plasma will lose its H^+ through the process of filtration and failure of reabsorption, and 25 more ml of plasma will lose its H^+ through the process of secretion. The plasma clearance for a secreted substance is always greater than the GFR.

Just as inulin can be used clinically to determine the GFR, the plasma clearance of another foreign compound, the organic anion **para-aminohippuric acid** (**PAH**), can be used to measure renal plasma flow. Like inulin, PAH is freely filterable and nonreabsorbable. It differs, however, in that all of the PAH in the plasma that escapes filtration is secreted from the peritubular capillaries. Thus, PAH is removed from *all* of the plasma that flows through the kidneys—both from the plasma that is filtered and subsequently reabsorbed without its PAH, and from the unfiltered plasma that continues on in the peritubular capillaries and loses its PAH by means of

---

[1] Actually, plasma clearance is an artificial concept, because when a particular substance is excreted in the urine, that substance's concentration in the plasma as a whole is uniformly decreased as a result of thorough mixing in the circulatory system. However, it is useful for comparative purposes to consider clearance in effect as the volume of plasma that would have contained the total quantity of the substance that the kidneys excreted in one minute; that is, the hypothetical volume of plasma completely cleared of that substance per minute.

active secretion into the tubules. Because all of the plasma that flows through the kidneys is cleared of PAH, the plasma clearance for PAH is a reasonable estimate of the rate of plasma flow through the kidneys. Typically, renal plasma flow averages 625 ml/min, for a renal blood flow (plasma plus blood cells) of 1,140 ml/min—over 20% of the cardiac output.

Knowing PAH clearance (renal plasma flow) and inulin clearance (GFR), you can easily determine the **filtration fraction**, or that fraction of the plasma flowing through the glomeruli that is filtered into the tubules:

$$\frac{\text{filtration}}{\text{fraction}} = \frac{\text{GFR (plasma inulin clearance)}}{\text{renal plasma flow (plasma PAH clearance)}}$$

$$= \frac{125 \text{ ml/min}}{625 \text{ ml/min}} = 20\%$$

Thus, 20% of the plasma that enters the glomeruli is typically filtered.

The ability to excrete urine of varying concentrations depends on the medullary countercurrent system and vasopressin.

Having considered how the kidneys deal with a variety of solutes in the plasma, we will now concentrate on renal handling of plasma H_2O. The ECF osmolarity (solute concentration) depends on the relative amount of H_2O compared to solute. At normal fluid balance and solute concentration, the body fluids are said to be **isotonic** at an osmolarity of 300 milliosmols/liter (mosm/l). If there is too much H_2O relative to the solute load, the body fluids are **hypotonic**, which means that they are too dilute at an osmolarity less than 300 mosm/l. On the other hand, if a H_2O deficit exists relative to the solute load, the body fluids are too concentrated or are **hypertonic**, having an osmolarity greater than 300 mosm/l.

Generally speaking, the osmolarity of the ECF is uniform throughout the body. Knowing that the driving force for H_2O reabsorption throughout the entire length of the tubules is an osmotic gradient between the tubular lumen and surrounding interstitial fluid, you would expect, based on osmotic considerations, that the kidneys could not excrete urine more or less concentrated than the body fluids. Indeed, this would be the case if the interstitial fluid surrounding the tubules in the kidneys were identical in osmolarity to the remaining body fluids. Water reabsorption would proceeed only until the tubular fluid equilibrated osmotically with the interstitial fluid, and there would be no way to eliminate excess H_2O when the body fluids were hypotonic or to conserve H_2O in the presence of hypertonicity.

Fortunately, a large vertical osmotic gradient is uniquely maintained in the interstitial fluid of the medulla of each kidney. The concentration of the interstitial fluid progressively increases from the cortical boundary down through the depth of the renal medulla until it reaches a maximum of 1,200 mosm/l in humans at the junction with the renal pelvis (•Fig. 14–23). This vertical osmotic gradient remains constant regardless of the fluid balance of the body.

All values in milliosmols (mosm)/l.

• **FIGURE 14–23 Vertical Osmotic Gradient in Renal Medulla**

The presence of this gradient enables the kidneys to produce urine that ranges in concentration from 100 to 1,200 mosm/l, depending on the body's state of hydration. When the body is in ideal fluid balance, 1 ml/min of isotonic urine is formed. When the body is overhydrated (too much H_2O), the kidneys are able to produce a large volume of dilute urine (up to 25 ml/min and hypotonic at 100 mosm/l), thus eliminating the excess H_2O in the urine. Conversely, the kidneys are able to put out a small volume of concentrated urine (down to 0.3 ml/min and hypertonic at 1,200 mosm/l) when the body is dehydrated (too little H_2O), thus conserving H_2O for the body.

Unique anatomical arrangements and complex functional interactions between the various nephron components present in the renal medulla are responsible for the establishment and utilization of the vertical osmotic gradient. Recall that the hairpin loop of Henle dips only slightly into the medulla in the cortical nephrons, but in the juxtamedullary nephrons, the loop plunges through the entire depth of the medulla so that the tip of the loop lies near the renal pelvis (Fig. 14–4). Also, the vasa recta of the juxtamedullary nephrons follow the same deep hairpin loop as the long Henle's loop. Flow in both the long loops of Henle and vasa recta is considered to be countercurrent because the flow in the two closely adjacent limbs of the loop is in opposite directions. Also running through the medulla in the descending direction only on their way to the renal pelvis are the collecting tubules that serve both types of nephrons. This arrangement, coupled with the permeability and transport characteristics of these tubular segments, plays a key role in the kidneys' ability to produce urine of varying concentrations, depending on the body's needs for water conservation or elimination. Briefly, the juxtamedullary nephrons' long loops of Henle *establish the vertical osmotic gradient,* their vasa recta *prevent the dissolution of this gradient* while providing blood to the renal medulla, and the collecting tubules of all nephrons *use the gradient,* in conjunction with the hormone vasopressin, to produce urine of varying concentrations. Collectively, this entire functional organization is known as the **medullary countercurrent system**. We will examine each of its facets in greater detail.

The Long Loops of Henle of Juxtamedullary Nephrons Establish the Medullary Vertical Osmotic Gradient by Means of Countercurrent Multiplication
We will follow the filtrate through a long-looped nephron to see how this structure establishes a vertical osmotic gradient in the medulla. Immediately after the filtrate is formed, uncontrolled osmotic reabsorption of filtered H_2O occurs in the proximal tubule secondary to active Na^+ reabsorption. As a result, by the end of the proximal tubule about 65% of the filtrate has been reabsorbed, but the 35% remaining in the tubular lumen still has the same osmolarity as the body fluids. Therefore, the fluid entering the loop of Henle is still isotonic. An additional 15% of the filtered H_2O is obligatorily reabsorbed from the loop of Henle during the establishment and maintenance of the vertical osmotic gradient, with the osmolarity of the tubular fluid being altered in the process.

The following functional distinctions between the descending limb of Henle's loop (which carries fluid from the proximal tubule down into the depths of the medulla) and the ascending limb (which carries fluid up and out of the medulla into the distal tubule) are critical to the establishment of the incremental osmotic gradient in the medullary interstitial fluid.

The *descending limb:*

1. is highly permeable to H_2O; but
2. does not actively extrude Na^+ (it is the only segment of the tubule that does not do so).

The *ascending limb:*

1. actively transports NaCl out of the tubular lumen into the surrounding interstitial fluid; and
2. is always impermeable to H_2O, so salt leaves the tubular fluid without H_2O osmotically following along.

The close proximity and countercurrent flow of the two limbs allow important interactions to occur between them. Even though the flow of fluids is continuous through the loop of Henle, we will visualize what happens step by step, much like an animated movie film run so slowly that each individual frame can be viewed.

- *Initial Scene (Fig. 14–24a)* Before the vertical osmotic gradient is established, the medullary interstitial-fluid concentration is uniformly 300 mosm/l, as is the remainder of the body fluids.
- *Step 1 (Fig. 14–24b)* The active salt pump in the ascending limb is able to transport NaCl out of the lumen until the surrounding interstitial fluid is 200 mosm/l more concentrated than the tubular fluid in this limb. When the ascending-limb pump starts actively extruding salt, the medullary interstitial fluid becomes hypertonic. Water cannot follow osmotically from the ascending limb because the limb is impermeable to H_2O. However, net diffusion of H_2O does occur from the descending limb into the interstitial fluid. The tubular fluid entering the descending limb from the

proximal tubule is isotonic. Because the descending limb is highly permeable to H_2O, net diffusion of H_2O occurs by osmosis out of the descending limb into the more concentrated interstitial fluid. The passive movement of H_2O out of the descending limb continues until the osmolarities of the fluid in the descending limb and interstitial fluid become equilibrated. Thus, the tubular fluid entering the loop of Henle immediately starts to become more concentrated as it loses H_2O. At equilibrium, the osmolarity of the ascending-limb fluid is 200 mosm/l, and the osmolarities of the interstitial fluid and descending-limb fluid are equal at 400 mosm/l.

- *Step 2 (Fig. 14–24c)* If we now advance the entire column of fluid in the loop of Henle several "frames," a mass of 200 mosm/l fluid exits from the top of the ascending limb into the distal tubule, and a new mass of isotonic fluid at 300 mosm/l enters the top of the descending limb from the proximal tubule. At the bottom of the loop, a comparable mass of 400 mosm/l fluid from the descending limb moves forward around the tip into the ascending limb, placing it opposite a 400 mosm/l region in the descending limb. Note that the 200 mosm/l concentration difference has been lost at both the top and the bottom of the loop.
- *Step 3 (Fig. 14–24d)* The ascending-limb pump again transports NaCl out while H_2O passively leaves the descending limb until a 200 mosm/l difference is reestablished between the ascending limb and both the interstitial fluid and descending limb at each horizontal level. Note, however, that the concentration of tubular fluid is progressively increasing in the descending limb and progressively decreasing in the ascending limb.
- *Step 4 (Fig. 14–24e)* As the tubular fluid is advanced still farther forward, the 200 mosm/l concentration gradient is disrupted once again at all horizontal levels.
- *Step 5 (Fig. 14–24f)* Again, active extrusion of NaCl from the ascending limb, coupled with the net diffusion of H_2O out of the descending limb, reestablishes the 200 mosm/l gradient at each horizontal level.
- *Steps 6 and on (Fig. 14–24g)* As the fluid flows slightly forward again and this stepwise process continues, the fluid in the descending limb becomes progressively more hypertonic until it reaches a maximum concentration of 1,200 mosm/l at the bottom of the loop, four times the normal concentration of body fluids. Because the interstitial fluid always achieves equilibrium with the descending limb, an incremental vertical concentration gradient ranging from 300 to 1,200 mosm/l is likewise established in the medullary interstitial fluid. In contrast, the concentration of the tubular fluid progressively decreases in the ascending limb as salt is pumped out but H_2O is unable to follow. In fact, the tubular fluid even becomes hypotonic as it leaves the ascending limb to enter the distal tubule at a concentration of 100 mosm/l, one-third the normal concentration of body fluids.

Note that although a gradient of only 200 mosm/l exists between the ascending limb and surrounding fluids at each medullary horizontal level, a much larger vertical gradient ex-

(a) Initial scene.

(b) Step 1: Let the pump begin. A 200 mosm/l gradient is established at each horizontal level.

(c) Step 2: Let the fluid flow forward several "frames."

• **FIGURE 14–24** **Countercurrent Multiplication**

ists from the top to the bottom of the medulla. Even though the ascending-limb pump can generate a gradient of only 200 mosm/l, this effect is multiplied into a large vertical gradient because of the countercurrent flow within the loop. This concentrating mechanism accomplished by the loop of Henle is known as **countercurrent multiplication**.

We have artificially described countercurrent multiplication in a "stop-and-flow," stepwise fashion to facilitate understanding. It is important to realize that once the incremental medullary gradient is established, it remains constant because of the continuous flow of fluid coupled with the on-

going ascending-limb active transport activity and accompanying descending-limb passive fluxes.

If you consider only what happens to the tubular fluid as it flows through the loop of Henle, the whole process seems to be an exercise in futility. The isotonic fluid that enters the loop becomes progressively more concentrated as it flows down the descending limb, achieving a maximum concentration of 1,200 mosm/l, only to become progressively more dilute as it flows up the ascending limb, finally leaving the loop at a minimum concentration of 100 mosm/l. What is the point of concentrating the fluid fourfold and then turning around

(d) Step 3: The ascending limb pump and descending limb passive fluxes reestablish the 200 mosm/l gradient at each horizontal level.

(e) Step 4: Let the fluid flow forward several "frames" once again.

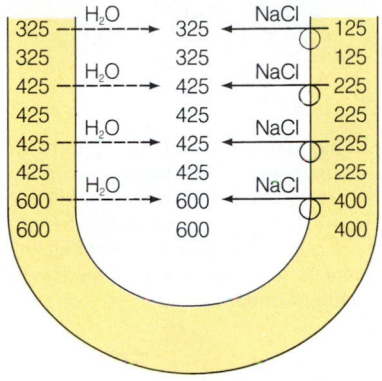

(f) Step 5: The 200 mosm/l gradient at each horizontal level is established once again.

(g) Steps 6 and on: The final vertical osmotic gradient is established and maintained by the ongoing countercurrent multiplication of the long loops of Henle.

All values in mosm/l.   ------→ = passive diffusion   ←——⊖ = active transport

and diluting it until it leaves at one-third the concentration at which it entered? Such a mechanism offers two benefits. First, it establishes a vertical osmotic gradient in the medullary interstitial fluid. This gradient, in turn, is used by the collecting ducts to concentrate the tubular fluid so that a urine *more concentrated* than normal body fluids can be excreted. Second, the fact that the fluid is hypotonic as it enters the distal portions of the tubule enables the kidneys to excrete a urine *more dilute* than normal body fluids.

Thus far we have implied that the NaCl and H_2O exchanges between the two limbs of the long loops of Henle and the interstitial fluid, driven by the ascending-limb NaCl pump, are solely responsible for the vertical osmotic gradient in the medullary interstitial fluid. Indeed, incremental trapping of salt in the medulla is the major factor contributing to medullary hypertonicity and the resulting ability of the kidneys to concentrate urine. However, accumulation of *urea* in the medullary interstitial fluid as a result of its passive recycling between the collecting tubules and the long loops of Henle also contributes to medullary hypertonicity, especially in the innermost region of the medulla. This contribution is important because the thin portion of the ascending limbs, which lies in this inner medullary region, has recently been shown to be unable to actively extrude NaCl into the interstitial fluid; only the thick portion of the ascending limb possesses active salt pumps (Fig. 14–4). However, NaCl does passively leave the thin portion of the ascending limb, driven by a concentration gradient ultimately brought about by urea recycling. We will elaborate upon the mechanism and significance of urea recycling when we discuss the role of the collecting tubules.

Countercurrent Exchange within the Vasa Recta Enables the Medulla to Be Supplied with Blood While Conserving the Medullary Vertical Osmotic Gradient

Obviously, the renal medulla must be supplied with blood to nourish the tissues in this area as well as to transport water that is reabsorbed by the loops of Henle and collecting tubules back to the general circulation. In doing so, however, it is important that circulation of blood through the medulla does not disturb the vertical gradient of hypertonicity established by the loops of Henle. Consider the situation if blood were to flow straight through from the cortex to the inner medulla and then directly into the renal vein (• Fig. 14–25a). Since capillaries are freely permeable to NaCl and H_2O, the blood would progressively pick up salt and lose H_2O through passive fluxes down concentration and osmotic gradients as it flowed through the depths of the medulla. Isotonic blood entering the medulla, upon equilibrating with each medul-

lary level, would leave the medulla very hypertonic at 1,200 mosm/l. It would be impossible to establish and maintain the medullary hypertonic gradient, because the NaCl pumped into the medullary interstitial fluid would continuously be carried away by the circulation.

This dilemma is avoided by the hairpin construction of the vasa recta, which, by looping back through the concentration gradient in reverse, allows the blood to leave the medulla and enter the renal vein essentially isotonic to incoming arterial blood (Fig. 14–25b). As blood passes down the descending limb of the vasa recta, equilibrating with the progressively increasing concentration of the surrounding interstitial fluid, it picks up salt and loses H_2O until it is very hypertonic by the bottom of the loop. Then, as blood flows up the ascending limb, salt diffuses back out into the interstitium, and H_2O reenters the vasa recta as progressively decreasing concentrations are encountered in the surrounding interstitial fluid. This passive exchange of solutes and H_2O between the two limbs of the vasa recta and the interstitial fluid is known as **countercurrent exchange**. Unlike countercurrent multiplication, it does not *establish* the concentration gradient. Rather, it *prevents the dissolution* of the gradient. Because blood enters and leaves the medulla at the same osmolarity as a result of countercurrent exchange, the medullary tissue is nourished with blood, yet the incremental gradient of hypertonicity in the medulla is preserved.

• **FIGURE 14–25   Countercurrent Exchange in the Renal Medulla**   (a) Hypothetical pattern of blood flow. If the blood supply to the renal medulla were to flow straight through from the cortex to the inner medulla, the blood would be isotonic on entering but very hypertonic on leaving, having picked up salt and lost H_2O as it equilibrated with the surrounding interstitial fluid at each incremental horizontal level. It would be impossible to maintain the vertical osmotic gradient, because the salt pumped out by the ascending limb of Henle's loop would continuously be flushed away by blood flowing through the medulla. (b) Actual pattern of blood flow. Blood equilibrates with interstitial fluid at each incremental horizontal level in both the descending limb and the ascending limb of the vasa recta, so blood is isotonic as it enters and leaves the medulla. This countercurrent exchange prevents the dissolution of the medullary osmotic gradient while providing blood to the renal medulla.

All values in mosm/l.

The Medullary Vertical Osmotic Gradient Permits Excretion of Urine of Differing Concentrations by Means of Vasopressin-Controlled, Variable H₂O Reabsorption from the Final Tubular Segments Following obligatory H_2O reabsorption from the proximal tubule (65% of the filtered H_2O) and loop of Henle (15% of the filtered H_2O), 20% of the filtered H_2O remains in the lumen to enter the distal and collecting tubules for variable reabsorption that is under hormonal control. This is still a large volume of filtered H_2O subject to regulated reabsorption; 20% × GFR(180 liters/day) = 36 liters per day are reabsorbed to varying extents, depending on the body's state of hydration. This is more than thirteen times the amount of plasma H_2O in the entire circulatory system.

The fluid leaving the loop of Henle enters the distal tubule at 100 mosm/l, so it is hypotonic to the surrounding isotonic (300 mosm/l) interstitial fluid of the renal cortex through which the distal tubule passes. The distal tubule then empties into the collecting tubule, which is bathed by progressively increasing concentrations (300 to 1,200 mosm/l) of surrounding interstitial fluid as it descends through the medulla.

In order for H_2O reabsorption to occur across a segment of the tubule, two criteria must be met: (1) an osmotic gradient must exist across the tubule, and (2) the tubular segment must be permeable to H_2O. The distal and collecting tubules are *impermeable* to H_2O except in the presence of **vasopressin**, also known as **antidiuretic hormone** (*anti* means "against"; *diuretic* means "increased urine output"),[2] which increases their permeability to H_2O. Vasopressin is produced by several specific neuronal cell bodies in the hypothalamus, a portion of the brain, then is stored in the *posterior pituitary*

gland, which is attached to the hypothalamus by a thin stalk. The hypothalamus controls the release of vasopressin from the posterior pituitary into the blood. In negative-feedback fashion, vasopressin secretion is stimulated by a H_2O deficit when H_2O must be conserved for the body and inhibited by a H_2O excess when surplus H_2O must be eliminated in the urine.

Vasopressin reaches the basolateral membrane of the tubular cells lining the distal and collecting tubules through the circulatory system, whereupon it binds with receptors specific for it (• Fig. 14–26). This binding activates the cyclic AMP second-messenger system within the tubular cells (see p. 50), which ultimately increases the permeability of the opposite luminal membrane to H_2O by increasing the number of H_2O channels in the membrane. By permitting more H_2O to permeate from the lumen, these additional channels increase H_2O reabsorption. The tubular response to vasopressin is graded; the more vasopressin present, the greater the permeability of the distal and collecting tubules to H_2O. The increase in luminal-membrane H_2O channels is not permanent, however. The channels are retrieved when vasopressin secretion decreases and cyclic AMP activity is similarly decreased. Accordingly, H_2O permeability is reduced when vasopressin secretion decreases.

When vasopressin secretion is increased in response to a H_2O deficit and the permeability of the distal and collecting tubules to H_2O is accordingly increased, the hypotonic tubular fluid entering the distal tubule is able to lose progressively more H_2O by osmosis into the interstitial fluid as the tubular fluid flows first through the isotonic cortex and then is exposed to the ever-increasing osmolarity of the medullary interstitial fluid as it plunges toward the renal pelvis (• Fig. 14–27a). As the 100 mosm/l tubular fluid is exposed to a surrounding interstitial fluid of 300 mosm/l, H_2O leaves the tubular fluid by osmosis across the now permeable tubular cells until the tubular fluid reaches a maximum concentra-

---

[2] Even though textbooks traditionally have tended to use the name *antidiuretic hormone* for this hormone, especially when discussing its actions on the kidney, investigators in the field now prefer the name *vasopressin*.

Tubular-lumen filtrate

Distal tubular cell

Peritubular-capillary plasma

Vasopressin

H₂O

H₂O

ATP

H₂O

Cyclic AMP

Increases permeability of luminal membrane to H₂O

H₂O

• **FIGURE 14–26 Mechanism of Action of Vasopressin** On binding with its receptor sites on the basolateral membrane of a distal- or collecting-tubule cell, vasopressin activates the cyclic AMP second-messenger system within the cell. Cyclic AMP in turn increases the opposite luminal membrane's permeability to H_2O by increasing the number of H_2O channels in the membrane.

• FIGURE 14–27 Excretion of Urine
of Varying Concentration Depending
on Body's Needs (a) H_2O deficit:
vasopressin present. (b) H_2O excess: no
vasopressin present.

Interstitial
fluid

Vasopressin
present

Reabsorbed H_2O
conserved for body

From ascending limb
of loop of Henle

Filtrate has concentration
of 100 mosm/l as it enters
distal and collecting tubule

Distal and collecting tubule—
permeable to H_2O

Concentration of urine may
be up to 1,200 mosm/l as it
leaves collecting duct

Small volume
of concentrated
urine

(a)

From ascending limb
of loop of Henle

Interstitial
fluid

No
vasopressin
present

No H_2O reabsorbed
in distal portion
of nephron

Filtrate has concentration
of 100 mosm/l as it enters
distal and collecting tubule

Distal and collecting tubule—
impermeable to H_2O

Concentration of urine may
be as low as 100 mosm/l as
it leaves collecting duct

Large volume
of dilute urine;
excess H_2O
eliminated

(b)

tion of 300 mosm/l. As this 300 mosm/l tubular fluid progresses farther through the tubule, it is exposed to an even higher osmolarity in the surrounding interstitial fluid. Consequently, the tubular fluid loses more H_2O by osmosis and becomes further concentrated, only to move farther forward and be exposed to an even higher interstitial fluid osmolarity and lose even more H_2O, and so on.

Under the influence of maximum levels of vasopressin, it is possible to concentrate the tubular fluid up to 1,200 mosm/l by the end of the collecting tubule. No further modification of the tubular fluid occurs beyond the collecting tubule, so what remains in the tubules at this point is urine. As a result of this extensive vasopressin-promoted reabsorption of H_2O in the late segments of the tubule, a small volume of urine concentrated up to 1,200 mosm/l can be excreted. As little as

0.3 ml of urine may be formed each minute, less than one-third the normal urine flow rate of 1 ml/min. The reabsorbed H_2O entering the medullary interstitial fluid is picked up by the peritubular capillaries and returned to the general circulation, thus being conserved for the body.

It is important to realize that although vasopressin promotes H_2O conservation by the body, it cannot completely halt urine production, even when a person is not taking in any H_2O, because a minimum volume of H_2O must be excreted with the solute wastes. Collectively, the waste products and other constituents eliminated in the urine average 600 mosm each day. Because the maximum urine concentration is 1,200 mosm/l, the minimum volume of urine that is required to excrete these wastes is 500 ml/day (600 mosm of wastes/day ÷ 1,200 mosm/l of urine = 0.5 l, or 500 ml/day, or 0.3 ml/

min). Thus, under maximal vasopressin influence, 99.8% of the 180 liters of plasma H_2O filtered per day is returned to the blood, with an obligatory H_2O loss of half a liter.

The kidneys' ability to tremendously concentrate urine to minimize H_2O loss when necessary is possible only because of the presence of the osmotic gradient in the medulla. Were it not for this gradient, the kidneys could produce a urine no more concentrated than the body fluids no matter how much vasopressin was secreted, because the only driving force for H_2O reabsorption is a concentration differential between the tubular fluid and interstitial fluid.

Conversely, when a person consumes large quantities of H_2O, the excess H_2O must be removed from the body without simultaneously losing solutes that are critical to the maintenance of homeostasis. Under these circumstances, no vasopressin is secreted, so the distal and collecting tubules remain impermeable to H_2O. The tubular fluid entering the distal tubule is hypotonic (100 mosm/l), having lost salt without an accompanying loss of H_2O in the ascending limb of Henle's loop. As this hypotonic fluid passes through the distal and collecting tubules (Fig. 14–27b), the medullary osmotic gradient is unable to exert any influence because of the late tubular segments' impermeability to H_2O. In other words, none of the H_2O remaining in the tubules can leave the lumen to be reabsorbed even though the tubular fluid is less concentrated than the surrounding interstitial fluid. Thus, in the absence of vasopressin, the 20% of the filtered fluid that reaches the distal tubule fails to be reabsorbed. Meanwhile, excretion of wastes and other urinary solutes remains constant. The net result is a large volume of dilute urine, which helps rid the body of excess H_2O. Urine osmolarity may be as low as 100 mosm/l, the same as in the fluid entering the distal tubule. Urine flow may be increased up to 25 ml/min in the absence of vasopressin, compared to the normal urine production of 1 ml/min.

Note that it would be impossible to produce urine less concentrated than the body fluids were it not for the fact that the tubular fluid is hypotonic as it enters the distal portion of the nephron. This dilution was accomplished in the ascending limb as NaCl was actively extruded but H_2O could not follow. Therefore, the loop of Henle, by simultaneously establishing the medullary osmotic gradient and diluting the tubular fluid before it enters the distal segments, plays a key role in allowing the kidneys to excrete urine that ranges in concentration from 100 to 1,200 mosm/l.

Changes in urine concentration are indicative of how much of the variably reabsorbed H_2O has been conserved for the body. Production of a large volume of dilute urine means little or none of the 20% of filtered H_2O subject to control has been returned to the plasma. In contrast, excretion of a small volume of concentrated urine implies extensive reabsorption of the controllable portion of the filtered H_2O. The extent of reabsorption varies directly with the amount of vasopressin secreted, which in turn is dependent on the body's state of hydration. Varying the amount of vasopressin secreted in proportion to the body's need for H_2O conservation enables the fine adjustments in H_2O reabsorption and excretion that are necessary to maintain proper fluid balance. Vasopressin influences H_2O permeability only in the distal and collecting tubules. It has no influence over the 80% of the filtered H_2O that is obligatorily reabsorbed without control in the proximal tubule and loop of Henle.

Water reabsorption versus excretion is only partially coupled with solute reabsorption versus excretion.

It is important to distinguish between H_2O reabsorption that mandatorily follows solute reabsorption and reabsorption of "pure" H_2O not linked to solute reabsorption.

■ *In the tubular segments that are permeable to H_2O, solute reabsorption is **always** accompanied by comparable H_2O reabsorption because of osmotic considerations.* Therefore, the total volume of H_2O reabsorbed is determined in large part by the total mass of solute reabsorbed; this is especially true of NaCl, because it is the most abundant solute in the ECF. In the proximal tubule and loop of Henle, H_2O reabsorption osmotically follows Na^+ reabsorption because these tubular segments are permeable to H_2O. In the distal and collecting tubules, Na^+ and H_2O reabsorption are not directly linked because these tubules are impermeable to H_2O in the absence of vasopressin. An aldosterone-induced increase in Na^+ reabsorption ultimately leads to increased H_2O reabsorption, however. Sodium retention in the body exerts an osmotic effect that promotes vasopressin-controlled H_2O reabsorption. Thus, the greater the salt retention within the body, the greater the H_2O retention and the larger the ECF volume. For this reason, the regulation of Na^+ reabsorption by aldosterone (followed passively by Cl^-) is the predominant factor in determining the total ECF volume.

■ *Solute excretion is **always** accompanied by comparable H_2O excretion because of osmotic considerations.* This fact is responsible for the obligatory excretion of at least a minimal volume of H_2O, even when a person is severely dehydrated.

For the same reason, when excess unreabsorbed solute is present in the tubular fluid, its presence exerts an osmotic effect to hold excessive H_2O in the lumen. This phenomenon is known as osmotic diuresis. **Diuresis** refers to increased urinary excretion, of which there are two types: water diuresis and osmotic diuresis. **Water diuresis** is increased urinary output of H_2O with little or no increase in excretion of solutes. **Osmotic diuresis**, on the other hand, involves increased excretion of both H_2O and solute caused by excess unreabsorbed solute in the tubular fluid, such as occurs in diabetes mellitus. The large quantity of unreabsorbed glucose that remains in the tubular fluid in persons with diabetes osmotically drags H_2O with it into the urine. Some diuretic drugs (drugs that promote diuresis) act by blocking specific solute reabsorption so that extra H_2O spills into the urine along with the unreabsorbed solute.

■ *A loss or gain of pure H_2O that is not accompanied by comparable solute deficit or excess in the body (that is, "free" H_2O) leads to changes in ECF osmolarity.* Such an imbalance between H_2O and solute is corrected by partially dissociating H_2O reabsorption from solute reabsorption in the distal portions of the nephron through the combined effects of osmoreceptor-controlled vasopressin secretion and the medullary osmotic gradient. Through this mechanism, free H_2O can be reabsorbed without comparable solute reabsorption to ameliorate hypertonicity of the body fluids. Conversely, to rid the body of excess pure H_2O, a large quantity of free H_2O can be excreted unaccompanied by comparable solute excretion (that is, water diuresis), thus correcting for hypotonicity of the body fluids. Water diuresis is normally a compensatory measure in response to ingestion of too much H_2O. However, excessive water diuresis accompanies alcohol ingestion. Because alcohol inhibits vasopressin secretion, the kidneys inappropriately lose too much H_2O. Typically, more fluid is lost in the urine than is consumed in the alcoholic beverage, so the body becomes dehydrated in spite of substantial fluid ingestion.

Table 14–4 summarizes how various tubular segments of the nephron handle Na^+ and H_2O and the significance of these processes.

Urea Recycling in the Renal Medulla Contributes to Medullary Hypertonicity and Helps Concentrate Urea in the Urine.

Unlike the proximal tubule, the early portion of the collecting tubule is *impermeable* to urea. Consequently, urea becomes progressively more concentrated in this segment as H_2O is reabsorbed in the presence of vasopressin, but the urea is unable to move out of the lumen down its concentration gradient. By the time the fluid reaches the last portion of the collecting tubule, its urea concentration is greater than in the surrounding interstitial fluid and the tubular fluid at the bottoms of the long Henle's loops. This concentration difference favors the diffusion of urea out of the late collecting tubule, which *is permeable* to urea, into the interstitial fluid and urea-permeable bottoms of the long loops (● Fig. 14–28).

Vasopressin increases the permeability of the late collecting tubule to urea as well as to H_2O. At first thought, this would seem to be counterproductive. Why encourage the reabsorption of urea, a waste product, just at the end of the tubule before the tubular fluid is emptied into the renal pelvis for elimination? This urea reabsorption is beneficial for two reasons: (1) reabsorbed urea contributes to medullary hypertonicity, which is important in the kidney's ability to produce hypertonic urine, and (2) urea recycling provides a mechanism for concentrating urea in the excreted fluid while economizing on the simultaneous loss of H_2O.

To elaborate, as urea diffuses out of the late collecting tubules, the concentration of urea in the inner medullary interstitial fluid and the bottoms of the long loops increases until it equilibrates with the high concentration of urea at the end of the collecting tubules. This high concentration of urea in the inner medullary interstitial fluid contributes to the medullary osmotic gradient. Furthermore, the osmotically active urea in the inner medulla pulls H_2O out of the descending limbs of the long Henle's loops as they plunge through this deep region of the medulla. Osmotic removal of H_2O from the descending limbs increases the concentration of NaCl in the tubular fluid within this segment of the tubules. As this NaCl-rich fluid starts up through the NaCl-permeable, H_2O-imper-

TABLE 14–4 Handling of Sodium and Water by Various Tubular Segments of the Nephron

| TUBULAR SEGMENT | Na+ REABSORPTION | | H2O REABSORPTION | |
	Percentage of Reabsorption in This Segment	Distinguishing Features	Percentage of Reabsorption in This Segment	Distinguishing Features
Proximal tubule	67	Active; uncontrolled; plays a pivotal role in reabsorption of glucose, amino acids, Cl⁻, H_2O, and urea	65	Passive; obligatory osmotic reabsorption following active Na^+ reabsorption
Loop of Henle	25	Active; uncontrolled; Na^+ along with Cl⁻ reabsorption from the ascending limb helps establish the medullary interstitial gradient, which is important in the kidney's ability to produce urine of varying concentrations and volumes, depending on the body's needs	15	Passive; obligatory osmotic reabsorption from descending limb as ascending limb extrudes NaCl into interstitial fluid (that is, reabsorbs NaCl)
Distal and collecting tubules	8	Active; variable and subject to aldosterone control; important in regulation of ECF volume; linked to K^+ and H^+ secretion	20	Passive; not linked to solute reabsorption; variable quantities of "free" H_2O reabsorption subject to vasopressin control; driving force is vertical osmotic gradient in medullary interstitial fluid established by long loops of Henle; important in regulation of ECF osmolarity

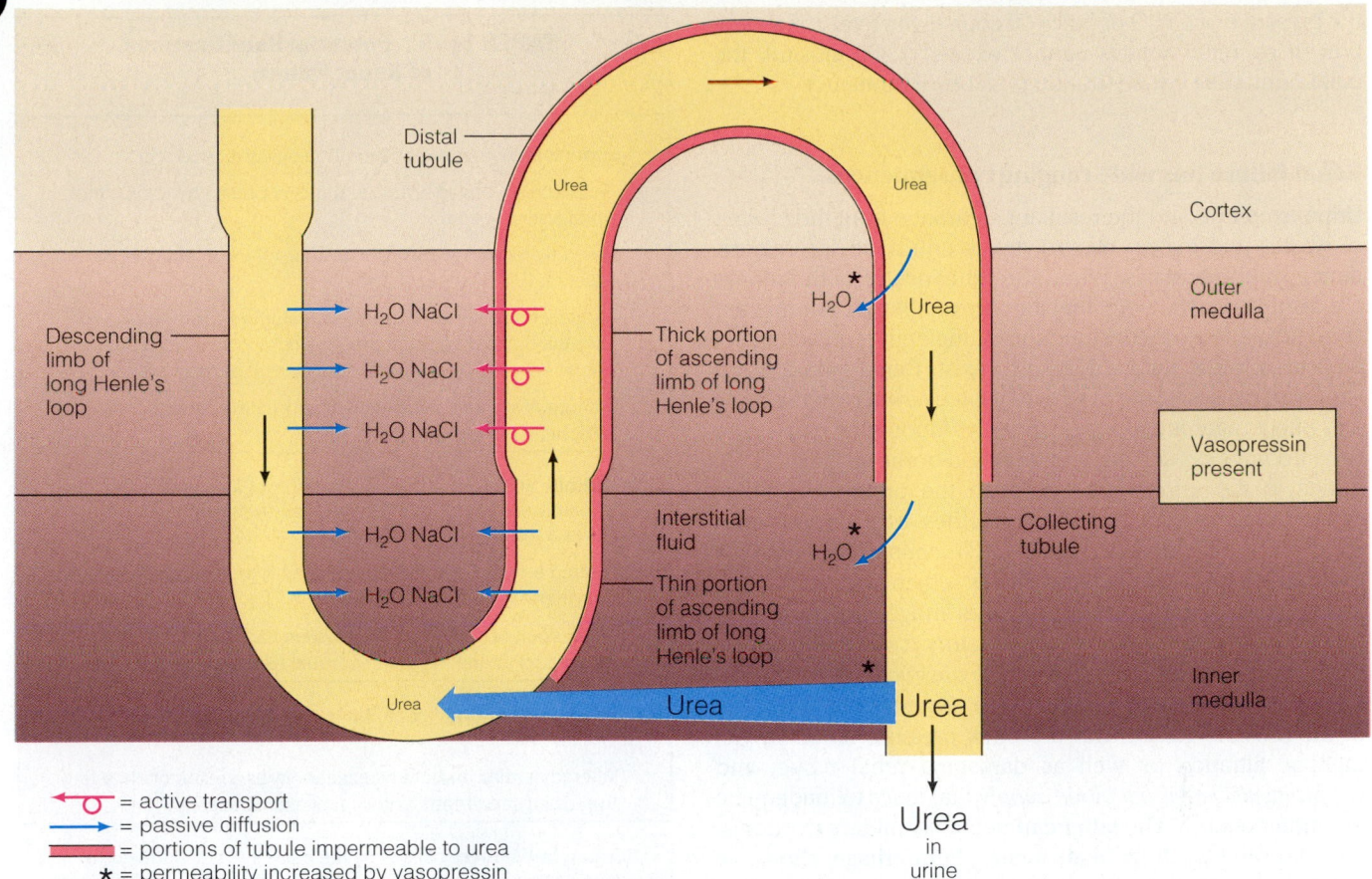

Key:
⌐σ = active transport
→ = passive diffusion
▬ = portions of tubule impermeable to urea
* = permeability increased by vasopressin

• **FIGURE 14–28 Urea Recycling in the Renal Medulla** Urea becomes more concentrated in the fluid in the early portion of the collecting tubule as H_2O is osmotically reabsorbed in the presence of vasopressin; the urea cannot move out down its concentration gradient because this segment is impermeable to urea. Urea does diffuse out of the late portion of the collecting tubule down its concentration gradient into the surrounding interstitial fluid and the nearby bottom of the long Henle's loop because these tubular segments are permeable to urea. Vasopressin increases the permeability of the late collecting tubule to urea. The entry of urea into the interstitial fluid contributes to medullary hypertonicity in the inner region of the medulla. As the tubular fluid passes through the ascending limb and the distal tubule, urea cannot leave because these segments are impermeable to it; thus, the urea cannot diffuse out even though the fluid is passing through regions with lower concentrations of urea. The tubular fluid's urea concentration increases even further as water is reabsorbed when the fluid enters the early portion of the collecting tubule once again. Thus, when vasopressin is secreted in the presence of a H_2O deficit, this urea recycling progressively concentrates urea in the tubular fluid that is excreted as urine.

meable thin portion of the ascending limbs, NaCl passively diffuses down its concentration gradient out into the interstitial fluid as the tubule passes through surrounding regions of progressively decreasing osmolarity. Thus, NaCl does leave throughout the ascending limb, but in the thin portion it leaves passively down its concentration gradient that is ultimately created by urea recycling whereas in the thick portion it is actively extruded. Passively extruded urea and NaCl both contribute to the vertical osmotic gradient in the interstitial fluid of the inner region of the medulla. Actively extruded NaCl alone is the solute responsible for the vertical osmotic gradient in the outer region of the medulla.

Meanwhile, despite the outward concentration gradient that exists for it, the urea that enters the bottom of the loop at high concentration is unable to leave as the fluid flows up through the ascending limb and distal tubule because these portions of the tubule are *impermeable* to urea. Thus, fluid bearing a high concentration of urea passes through sur-

rounding regions of lesser urea concentration in the outer medulla and cortex without losing any of its urea load. This fluid with high urea concentration then enters the early portion of the collecting tubule where the reabsorption of H_2O further increases the concentration of urea in the tubular fluid. Consequently, as this fluid flows to the end portion of the collecting tubule, even more urea diffuses out because the tubular concentration of urea is once again greater than that of the surrounding interstitial fluid. In this way, the cycle continues to repeat itself when vasopressin is secreted in response to a H_2O deficit. This recycling progressively increases the concentration of urea in both the inner medullary interstitial fluid and in the tubular fluid that is to be excreted as urine. Note that this recycling of urea does not essentially alter the total amount of urea excreted, but it does concentrate urea in the urine. The more concentrated the urea becomes, the smaller the volume of H_2O that must accompany this waste in the urine. This mechanism helps economize on the body's H_2O in the presence of a H_2O deficit. (Of course, the total concentration of excreted solutes cannot exceed 1,200 mosm/l, the maximum urine concentration possible in humans.)

Renal failure has wide-ranging consequences.

Urine excretion and the resultant clearance from the plasma of wastes and excess electrolytes is critical to the maintenance of homeostasis. When the functions of both kidneys are disrupted to the point that they are unable to perform their regulatory and excretory functions sufficiently to maintain homeostasis, **renal failure** is said to exist. Renal failure has a variety of causes, some of which begin elsewhere in the body and affect renal function secondarily. Among the causes are (1) *infectious organisms,* either blood-borne or gaining entrance to the urinary tract through the urethra; (2) *toxic agents,* such as lead, arsenic, pesticides, or even long-term exposure to high doses of aspirin; (3) *inappropriate immune responses,* such as *glomerulonephritis,* which occasionally follows streptococcal throat infections as antigen-antibody complexes leading to localized inflammatory damage are deposited in the glomeruli (see p. 384); (4) *obstruction of urine flow* due to the presence of kidney stones, tumors, or enlargement of the prostate gland, with the back pressure reducing glomerular filtration as well as damaging renal tissue; and (5) an *insufficient renal blood supply* that leads to inadequate filtration pressure. The latter can occur secondary to circulatory disorders such as heart failure, hemorrhage, shock, or narrowing and hardening of the renal arteries as a result of atherosclerosis.

Although these conditions may have different origins, almost all of them can cause some degree of nephron damage. The glomeruli and tubules may be independently affected or both may be dysfunctional. Regardless of the cause, renal failure can manifest itself either as *acute renal failure,* characterized by a sudden onset with a rapid reduction in urine formation until less than the essential minimum of around 500 ml of urine is being produced per day, or *chronic renal failure,* characterized by slow, progressive, insidious loss of renal function. A person may die from acute renal failure, or the condition may be reversible and lead to full recovery. Chronic renal failure, in contrast, is not reversible. Gradual, permanent destruction of renal tissue eventually proves to be fatal. Chronic renal failure is insidious because up to 75% of the kidney tissue can be destroyed before the loss of kidney function is even noticeable. Because of the abundant reserve of kidney function, only 25% of the kidney tissue can adequately maintain all the essential renal excretory and regulatory functions. With less than 25% of functional kidney tissue remaining, however, renal insufficiency becomes apparent. *End-stage renal failure* ensues when 90% of kidney function has been lost.

We will not sort out the different stages and symptoms associated with various renal disorders, but Table 14–5, which summarizes the potential consequences of renal failure, will give you an idea of the broad effects that kidney impairment can have. The extent of these effects should not be surprising, considering the central role the kidneys play in maintaining

---

TABLE 14–5   Potential Ramifications of Renal Failure

Uremic toxicity caused by retention of waste products

  Nausea, vomiting, diarrhea and ulcers caused by toxic effect on digestive system

  Bleeding tendency arising from toxic effect on platelet function

  Mental changes—such as reduced alertness, insomnia, and shortened attention span, progressing to convulsions and coma—caused by toxic effect on central nervous system

  Abnormal sensory and motor activity caused by toxic effect on peripheral nerves

Metabolic acidosis * caused by inability of kidneys to adequately secrete H^+ that is continually being added to body fluids as a result of metabolic activity

  Altered enzyme activity caused by action of too much acid on enzymes

  Depression of central nervous system caused by action of too much acid interfering with neuronal excitability

Potassium retention * resulting from inadequate tubular secretion of K^+

  Altered cardiac and neural excitability as a result of changing the resting membrane potential of excitable cells

Sodium imbalances caused by the inability of the kidneys to adjust Na^+ excretion to balance the changes in Na^+ consumption

  Elevated blood pressure, generalized edema, and congestive heart failure if too much Na^+ is consumed

  Hypotension and, if severe enough, circulatory shock if too little Na^+ is consumed

Phosphate and calcium imbalances arising from impaired reabsorption of these electrolytes

  Disturbances in skeletal structures caused by abnormalities in deposition of calcium phosphate crystals, which harden bone

Loss of plasma proteins as a result of increased "leakiness" of the glomerular membrane

  Edema caused by reduction in plasma-colloid osmotic pressure

Inability to vary urine concentration as a result of impairment of countercurrent system

  Hypotonicity of body fluids if too much H_2O is ingested

  Hypertonicity of body fluids if too little H_2O is ingested

Hypertension arising from combined effects of salt and fluid retention and vasoconstrictor action of excess angiotensin II

Anemia caused by inadequate erythropoietin production

Depression of immune system most likely caused by toxic levels of wastes and acids

  Increased susceptibility to infections

* Among the most life-threatening consequences of renal failure.

WHEN PROTEIN IN THE URINE DOES NOT MEAN KIDNEY DISEASE

Urinary loss of proteins usually signifies kidney disease (nephritis). However, a urinary protein loss similar to that of nephritis often occurs following exercise, but the condition is harmless, transient, and reversible. The term *athletic pseudonephritis* is used to describe this postexercise (after exercise) proteinuria (protein in the urine). Studies indicate that 70% to 80% of athletes have proteinuria after very strenuous exercise. This condition occurs in participants in both noncontact and contact sports, so it does not arise from physical trauma to the kidneys. In one study, subjects who engaged in maximal short-term running excreted more protein than when they were bicycling, rowing, or swimming at the same work intensity. The reason for this difference is unknown.

Usually, only a very small fraction of the plasma proteins that enter the glomerulus are filtered; those that are filtered are reabsorbed in the tubules, so no plasma proteins normally appear in the urine. Two basic mechanisms can cause proteinuria: (1) increased glomerular permeability with no change in tubular reabsorption or (2) impairment of tubular reabsorption. Research has shown that during mild to moderate exercise, the proteinuria that occurs results from changes in glomerular permeability, whereas during short-term exhaustive exercise, the proteinuria seems to be caused by both increased glomerular permeability and tubular dysfunction.

This reversible kidney dysfunction is believed to result from circulatory and hormonal changes that occur with exercise. Several studies have shown that renal blood flow is reduced during exercise as the renal vessels are constricted and blood is diverted to the exercising muscles. This reduction is positively correlated with the intensity of the exercise. With intense exercise, the renal blood flow may be reduced to 20% of normal. The GFR is also reduced, but not to the same extent as renal blood flow, presumably because of autoregulatory mechanisms. Some investigators propose that decreased glomerular blood flow enhances diffusion of proteins into the tubular lumen because, as the more slowly flowing blood spends more time in the glomerulus, a greater proportion of the plasma proteins have time to escape through the glomerular membrane. Hormonal changes that occur with exercise may also affect glomerular permeability. For example, renin injection is a well-recognized way to experimentally induce proteinuria. Plasma renin activity increases during strenuous exercise and may contribute to postexercise proteinuria. It is also hypothesized that maximal tubular reabsorption is reached during severe exercise, which could result in impaired protein reabsorption.

---

homeostasis. When the kidneys are unable to maintain a normal internal environment, widespread disruption of cellular activities can bring about abnormal function in other organ systems as well. By the time end-stage renal failure occurs, literally every body system has become impaired to some extent. (One of the symptoms of renal disease occurs during strenuous exercise, but it is transient and harmless. See the accompanying boxed feature, A Closer Look at Exercise Physiology.)

Since chronic renal failure is irreversible and eventually fatal, treatment is aimed at maintaining renal function by alternative methods, such as dialysis and kidney transplantation. (See the other boxed feature, Concepts, Challenges, and Controversies.)

Urine is temporarily stored in the bladder, from which it is emptied by the process of micturition.

Once urine has been formed by the kidneys, it is transmitted through the ureters to the urinary bladder. Urine does not flow through the ureters by gravitational pull alone. Peristaltic contractions of the smooth muscle within the ureteral wall propel the urine forward from the kidneys to the bladder. The ureters penetrate the wall of the bladder obliquely, coursing through the wall several centimeters before they open into the bladder cavity. This anatomical arrangement prevents backflow of urine from the bladder to the kidneys when pressure builds up in the bladder. As the bladder fills, the ureteral ends within its wall are compressed closed. Urine can still enter, however, because ureteral contractions generate sufficient pressure to overcome the resistance and push urine through the occluded ends.

The bladder wall is composed of smooth muscle lined by a special type of epithelium. It was once assumed that the bladder was an inert sac. However, both the epithelium and smooth muscle actively participate in the bladder's ability to accommodate large fluctuations in urine volume. Recently researchers learned that the epithelial lining is able to increase and decrease in surface area by the orderly process of membrane recycling as the bladder alternately fills and empties. Membrane-bound cytoplasmic vesicles are inserted by means of exocytosis into the surface membrane of the epithelial cells to expand their surface area during bladder filling; then the vesicles are withdrawn by endocytosis to shrink the surface area following emptying (see p. 24). As is characteristic of smooth muscle, bladder muscle is able to stretch tremendously without a build-up in bladder-wall tension. In addition, the highly folded bladder wall flattens out during

DIALYSIS: CELLOPHANE TUBING OR ABDOMINAL LINING AS AN ARTIFICIAL KIDNEY

Since chronic renal failure is irreversible and eventually fatal, treatment is aimed at maintaining renal function by alternative methods, such as dialysis and kidney transplantation. The process of **dialysis** bypasses the kidneys to maintain normal fluid and electrolyte balance and remove wastes artificially. In the original method of dialysis, **hemodialysis**, a patient's blood is pumped through cellophane tubing that is surrounded by a large volume of fluid similar in composition to normal plasma. Following dialysis, the blood is returned to the patient's circulatory system. Like capillaries, cellophane is highly permeable to most plasma constituents but is impermeable to plasma proteins. As blood flows through the tubing, solutes move across the cellophane down their individual concentration gradients; plasma proteins, however, remain in the blood. Urea and other wastes, which are absent in the dialysis fluid, diffuse out of the plasma into the surrounding fluid, thus cleansing the blood of these wastes. Plasma constituents that are not regulated by the kidneys and are at normal concentration, such as glucose, do not move across the cello-phane into the dialysis fluid because there is no driving force to produce their movement. (The dialysis fluid's glucose concentration is the same as normal plasma glucose concentration.) Electrolytes, such as K^+ and $PO_4^=$, that are above their normal plasma concentrations because of the inability of the diseased kidneys to eliminate excess quantities of these substances, move out of the plasma until equilibrium is achieved between the plasma and the dialysis fluid. Because the dialysis fluid's solute concentrations are maintained at normal plasma values, the solute concentration of the blood returned to the patient following dialysis is essentially normal. Hemodialysis is repeated as often as necessary to maintain the plasma composition within an acceptable level. Typically, it is done three times per week for several hours each session.

In a more recent method of dialysis, **continuous ambulatory peritoneal dialysis (CAPD)**, the peritoneal membrane (the lining of the abdominal cavity) is used as the dialysis membrane. With this method, two liters of dialysis fluid are inserted into the patient's abdominal cavity through a permanently implanted catheter. Urea, K^+, and other wastes and excess electrolytes diffuse from the plasma across the peritoneal membrane into the dialysis fluid, which is drained off and replaced several times a day. The CAPD method offers several advantages: patients can self-administer it, the patient's blood is continuously purified and adjusted, and the patient can engage in normal activities while dialysis is being accomplished. One drawback is the increased risk of peritoneal infections.

Transplantation of a healthy kidney from a donor is another option for treating chronic renal failure. A kidney is one of the few transplants that can be provided by a living donor. Because 25% of the total kidney tissue can maintain the body, both the donor and the recipient have ample renal function with only one kidney each. The biggest problem with transplantation is the possibility that the patient's immune system will reject the organ. The risk of rejection can be minimized by matching the tissue types of the donor and the recipient as closely as possible (the best donor choice is usually a close relative), coupled with the use of immunosuppressive drugs.

---

filling to increase bladder storage capacity. Since urine is continuously being formed by the kidneys, the bladder must have sufficient storage capacity to preclude the necessity for continual evacuation of the urine.

The bladder smooth muscle is richly supplied by parasympathetic fibers, stimulation of which causes bladder contraction. If the passageway through the urethra to the outside is open, bladder contraction brings about emptying of urine from the bladder. The exit from the bladder, however, is guarded by two sphincters, the *internal urethral sphincter* and the *external urethral sphincter*. A **sphincter** is a ring of muscle that, when contracted, closes off passage through an opening. The **internal urethral sphincter**—which is composed of smooth muscle and, accordingly, is under involuntary control—is not really a separate muscle but instead consists of the last portion of the bladder. Although it is not a true sphinc-ter, it performs the same function as a sphincter. When the bladder is relaxed, the anatomical arrangement of the internal urethral sphincter region closes the outlet of the bladder.

Farther down the passageway, the urethra is encircled by a layer of skeletal muscle, the **external urethral sphincter**. This sphincter is reinforced by the entire **pelvic diaphragm**, a skeletal-muscle sheet that forms the floor of the pelvis and helps support the pelvic organs. The motor neurons that supply the external sphincter and pelvic diaphragm are continuously firing at a moderate rate unless they are inhibited, thus keeping these muscles tonically contracted so that they prevent urine from escaping through the urethra. Normally, when the bladder is relaxed and filling, closure of both the internal and external urethral sphincters prevents urine from dribbling out. Furthermore, because they are skeletal muscles, the external sphincter and pelvic diaphragm are under voluntary control. They can be deliberately tightened to prevent urination from occurring even when the bladder is contracting and the internal sphincter is open.

Micturition, or **urination**, the process of bladder emptying, is governed by two mechanisms: the micturition reflex and voluntary control. The **micturition reflex** is initiated when stretch receptors within the bladder wall are stimulated (● Fig. 14–29). The bladder in an adult can accommodate up to 250 to 400 ml of urine before the tension within its walls begins to rise sufficiently to activate the stretch receptors (● Fig. 14–30). The greater the distention beyond this, the greater the extent of receptor activation. Afferent fibers from the stretch receptors carry impulses into the spinal cord and eventually, by means of interneurons, stimulate the parasympathetic supply to the bladder and inhibit the motor-neuron supply to the external sphincter. Parasympathetic stimulation of the bladder causes it to contract. No special mechanism is required to open the internal sphincter; changes in the shape of the bladder during contraction mechanically pull the internal sphincter open. Simultaneously, the external sphincter relaxes as its motor-neuron supply is inhibited. Now both sphincters are open and urine is expelled through the urethra by the force of bladder contraction. This micturition reflex, which is entirely a spinal reflex, governs bladder emptying in infants. As soon as the bladder fills sufficiently to trigger the reflex, the baby automatically wets.

● **FIGURE 14–29   Reflex and Voluntary Control of Micturition**

● **FIGURE 14–30   Pressure Changes within the Urinary Bladder as the Bladder fills with Urine**

Bladder filling, in addition to triggering the micturition reflex, also gives rise to the conscious urge to urinate. The perception of bladder fullness appears before the external sphincter reflexly relaxes, thus providing a "warning" that micturition is imminent. As a result, voluntary control of micturition, learned during toilet training in early childhood, can override the micturition reflex so that bladder emptying can take place at the person's convenience rather than at the time bladder filling first reaches the point of activating the stretch receptors. If the time is inopportune for urination when the micturition reflex is initiated, bladder emptying can be voluntarily prevented by deliberate tightening of the external sphincter and pelvic diaphragm. Voluntary excitatory impulses originating from the cerebral cortex override the reflex inhibitory input from the stretch receptors to the involved motor neurons (the relative balance of EPSPs and IPSPs), thus keeping these muscles contracted so that no urine is expelled.

Urination cannot be delayed indefinitely. As the bladder continues to fill, reflex input from the stretch receptors increases with time. Finally, reflex inhibitory input to the external-sphincter motor neuron becomes so powerful that it can no longer be overridden by voluntary excitatory input, so the sphincter relaxes and the bladder uncontrollably empties.

Micturition can also be deliberately initiated, even though the bladder is not distended, by voluntary relaxation of the external sphincter and pelvic diaphragm. Lowering of the pelvic floor allows the bladder to drop downward, which simultaneously pulls open the internal urethral sphincter and stretches the bladder wall. The subsequent activation of the stretch receptors brings about bladder contraction by means of the micturition reflex. Voluntary bladder emptying may be further assisted by contraction of the abdominal wall and respiratory diaphragm. The resultant increase in intra-abdominal pressure "squeezes down" on the bladder to facilitate its emptying.

Urinary incontinence, or the inability to prevent the discharge of urine, occurs as a result of disruption of the descending pathways in the spinal cord that mediate voluntary control of the external sphincter and pelvic diaphragm. In this case, because the components of the micturition reflex arc are still intact in the lower spinal cord, bladder emptying becomes governed by an uncontrollable spinal reflex, as it is in infants. A lesser degree of incontinence characterized by urine escaping when the bladder pressure suddenly increases transiently, such as during coughing or sneezing, can result from impairment of sphincter function. This is not uncommon in women who have borne children or in men whose sphincters have been injured during prostate surgery.

CHAPTER IN PERSPECTIVE: FOCUS ON HOMEOSTASIS

The kidneys contribute to homeostasis more extensively than any other single organ. They regulate the electrolyte composition, volume, and pH of the internal environment and eliminate all of the waste products of bodily metabolism, with the exception of respiration-removed CO_2. They accomplish these regulatory functions by eliminating in the urine substances unneeded by the body, such as metabolic wastes and excess quantities of ingested salt or water, while conserving substances useful for the body. The kidneys are able to maintain the plasma constituents they regulate within the narrow range compatible with life, despite wide variations in intake and losses of these substances through other avenues. Illustrative of the magnitude of the kidneys' task, about a quarter of the blood pumped into the systemic circulation goes to the kidneys for the purpose of being "adjusted" and "purified," with only three-quarters of the blood being used to supply all the other tissues.

Following are the specific ways in which the kidneys contribute to homeostasis:

Regulatory functions

- The kidneys regulate the quantity and concentration of most ECF electrolytes, including those important in mantaining proper neuromuscular excitability.

- They contribute to maintenance of proper pH by eliminating excess H^+ (acid) or HCO_3^- (base) in the urine.

- They help maintain proper plasma volume, which is important in the long-term regulation of arterial blood pressure, by controlling the salt balance in the body. The ECF volume, including the plasma volume, is a reflection of the total salt load in the ECF, because Na^+ and its attendant anion Cl^- are responsible for over 90% of the ECF's osmotic (water-holding) activity.

- The kidneys maintain water balance in the body, which is important in maintaining proper ECF osmolarity (concentration of solutes). This role is important in maintaining the stability of cell volume by preventing cells from swelling or shrinking as a result of water osmotically moving in or out of the cells, respectively.

Excretory functions

- The kidneys excrete the end products of metabolism in the urine. These wastes are toxic to cells if allowed to accumulate.

- They also excrete many foreign compounds that gain entrance to the body.

Hormonal functions

- The kidneys secrete erythropoietin, the hormone that stimulates red blood cell production by the bone marrow. This action contributes to homeostasis by helping to maintain the optimal O_2 content of the blood. Over 98% of O_2 in the blood is bound to hemoglobin within the red blood cells.

- They also secrete renin, the hormone that initiates the renin-angiotensin-aldosterone pathway for controlling renal-tubular Na^+ reabsorption, which is important in the long-term maintenance of plasma volume and arterial blood pressure.

Metabolic functions

- The kidneys help convert vitamin D into its active form. Vitamin D is essential for Ca^{++} absorption from the digestive tract. Calcium, in turn, exerts a wide variety of homeostatic functions.

CHAPTER SUMMARY

Introduction

The kidneys eliminate unwanted plasma constituents into the urine while conserving materials of value to the body. The urine-forming functional unit of the kidneys is the nephron, which is composed of interrelated vascular and tubular components. The vascular component consists of two capillary networks in series, the first being the glomerulus, a tuft of capillaries that filters large volumes of protein-free plasma into the tubular component. The second capillary network consists of the peritubular capillaries, which wind around the tubular component. The peritubular capillaries nourish the renal tissue and participate in exchanges between the tubular fluid and plasma. The tubular component begins with Bowman's capsule, which cups around the glomerulus

to catch the filtrate, then continues a specific tortuous course to ultimately empty into the renal pelvis. As the filtrate passes through the various regions of the tubule, the cells lining the tubules modify it, returning to the plasma only those materials necessary for maintaining the proper ECF composition and volume. What is left behind in the tubules is excreted as urine.

The kidneys perform three basic processes in carrying out their regulatory and excretory functions: (1) glomerular filtration, the nondiscriminating movement of protein-free plasma from the blood into the tubules; (2) tubular reabsorption, the selective transfer of specific constituents in the filtrate back into the blood of the peritubular capillaries; and (3) tubular secretion, the highly specific movement of selected substances from the peritubular-capillary blood into the tubular fluid. Everything that is filtered or secreted but not reabsorbed is excreted as urine.

Glomerular Filtration

Glomerular filtrate is produced as a portion of the plasma flowing through each glomerulus is passively forced under pressure through the glomerular membrane into the lumen of the underlying Bowman's capsule. The net filtration pressure that induces filtration is caused by an imbalance in the physical forces acting across the glomerular membrane. A high glomerular-capillary blood pressure favoring filtration outweighs the combined opposing forces of plasma-colloid osmotic pressure and Bowman's capsule hydrostatic pressure.

Typically, 20% to 25% of the cardiac output is delivered to the kidneys for the purpose of being acted on by renal regulatory and excretory processes. Of the plasma flowing through the kidneys, normally 20% is filtered through the glomeruli, producing an average glomerular filtration rate (GFR) of 125 ml/min. This filtrate is identical in composition to plasma except for the plasma proteins that are held back by the glomerular membrane.

The GFR can be deliberately altered by changing the glomerular-capillary blood pressure as a result of sympathetic influence on the afferent arterioles. Afferent arteriolar vasoconstriction decreases the flow of blood into the glomerulus, resulting in a reduction in glomerular blood pressure and a fall in the GFR. Conversely, afferent arteriolar vasodilation leads to increased glomerular blood flow and a rise in the GFR. Sympathetic control of the GFR is part of the baroreceptor reflex response to compensate for a change in arterial blood pressure. As the GFR is altered, the amount of fluid lost in the urine is changed correspondingly, providing a mechanism to adjust plasma volume as needed to help restore blood pressure to normal on a long-term basis.

Tubular Reabsorption

After a protein-free plasma is filtered through the glomerulus, each substance is handled discretely by the tubules, so that even though the concentrations of all constituents in the initial glomerular filtrate are identical to their concentrations in the plasma (with the exception of plasma proteins), the concentrations of different constituents are variously altered as the filtered fluid flows through the tubular system. The reabsorptive capacity of the tubular system is tremendous. Over 99% of the filtered plasma is returned to the blood through reabsorption. The major substances actively reabsorbed are Na^+ (the principal ECF cation), most other electrolytes, and organic nutrients such as glucose and amino acids. The most important passively reabsorbed substances are Cl^-, H_2O, and urea.

The pivotal event to which most reabsorptive processes are linked in some way is the active reabsorption of Na^+. An energy-dependent Na^+-K^+ ATPase carrier located in the basolateral membrane of each proximal tubular cell transports Na^+ out of the cells into the lateral spaces between adjacent cells. This transport of Na^+ induces the net reabsorption of Na^+ from the tubular lumen to the peritubular-capillary plasma, the vast majority of which takes place in the proximal tubules. The energy used to supply the Na^+-K^+ ATPase carrier is ultimately responsible for the reabsorption from the proximal tubule of Na^+, glucose, amino acids, Cl^-, H_2O, and urea. Specific cotransport carriers located at the luminal border of the proximal tubular cell are driven by the Na^+ concentration gradient to selectively transport glucose or an amino acid from the luminal fluid into the tubular cell, from which they eventually enter the plasma. Chloride is passively reabsorbed down the electrical gradient established by active Na^+ reabsorption. Water is passively reabsorbed as a result of the osmotic gradient created by active Na^+ reabsorption. Sixty-five percent of the filtered H_2O is reabsorbed from the proximal tubule in this unregulated fashion. This extensive reabsorption of H_2O increases the concentration of other substances remaining in the tubular fluid, most of which are filtered waste products. The small urea molecules are the only waste products that can passively permeate the tubular membranes. Accordingly, urea is the only waste product partially reabsorbed as a result of this concentration effect; about 50% of the filtered urea is reabsorbed. The other waste products, failing to be reabsorbed, remain in the urine in highly concentrated form.

Early in the nephron, Na^+ reabsorption occurs in constant unregulated fashion, but in the distal and collecting tubules, the reabsorption of a small percentage of the filtered Na^+ is variable and subject to control. The extent of this controlled Na^+ reabsorption depends primarily on the complex renin-angiotensin-aldosterone system. Since Na^+ and its attendant anion, Cl^-, are the major osmotically active ions in the ECF, the ECF volume is determined by the Na^+ load in the body. In turn, the plasma volume, which reflects the total ECF volume, is important in the long-term determination of arterial blood pressure. Whenever the Na^+ load/ECF volume/plasma volume/arterial blood pressure are below normal, the kidneys secrete renin, an enzymatic hormone that triggers a series of events ultimately leading to increased secretion of aldosterone from the adrenal cortex. Aldosterone increases Na^+ reabsorption from the distal portions of the tubule, thus correcting for the original reduction in Na^+/ECF volume/blood pressure.

The other electrolytes actively reabsorbed by the tubules, such as $PO_4^=$ and Ca^{++}, have their own independently functioning carrier systems. Because these carriers like the organic-nutrient cotransport carriers can become saturated, they exhibit maximal carrier-limited transport capacities, or T_m's. Once the filtered load of an actively reabsorbed substance exceeds the T_m, reabsorption proceeds at a constant maximal rate, with the additional filtered quantity of the substance being excreted in the urine.

Tubular Secretion

The kidney tubules are able to selectively add some substances to the quantity already filtered by means of the process of tubular secretion. Secretion of substances hastens their excretion in the urine. The most important secretory systems are for (1) H^+, which is important in the regulation of acid-base balance; (2) K^+, which keeps the plasma K^+ concentration at an appropriate level to maintain normal membrane excitability in muscles and nerves; and (3) organic anions and cations, which accomplishes more efficient elimination of foreign organic compounds from the body.

Urine Excretion and Plasma Clearance

Of the 125 ml/min filtered in the glomeruli, normally only 1 ml/min remains in the tubules to be excreted as urine. Only wastes and excess electrolytes not wanted by the body are left behind, dissolved in a given volume of H_2O, to be eliminated in the urine. Because the excreted material is removed or "cleared" from the plasma, the term plasma clearance refers to the volume of plasma being cleared of a particular substance each minute by means of renal activity.

The kidneys are able to excrete urine of varying volumes and concentrations to either conserve or eliminate H_2O, depending on whether the body has a H_2O deficit or excess, respectively. The kidneys are able to produce urine ranging from 0.3 ml/min at 1,200 mosm/l to 25 ml/min at 100 mosm/l by reabsorbing variable amounts of H_2O from the distal portions of the nephron. This variable reabsorption is made possible by the establishment of a vertical osmotic gradient ranging from 300 to 1,200 mosm/l in the medullary interstitial fluid by means of the loop of Henle countercurrent system and urea recycling between the collecting tubule and Henle's loops. This vertical osmotic gradient to which the hypotonic (100 mosm/l) tubular fluid is exposed as it passes through the distal portions of the nephron establishes a passive driving force for progressive reabsorption of H_2O from the tubular fluid, but the actual extent of H_2O reabsorption depends on the amount of vasopressin (antidiuretic hormone) secreted. Vasopressin increases the permeability of the distal and collecting tubules to H_2O; they are impermeable to H_2O in its absence. Vasopressin secretion increases in response to a H_2O deficit, and H_2O reabsorption increases accordingly. Vasopressin secretion is inhibited in response to a H_2O excess, thereby reducing H_2O reabsorption. In this way, adjustments in vasopressin-controlled H_2O reabsorption help correct any fluid imbalances.

Once formed, urine is propelled by peristaltic contractions through the ureters from the kidneys to the urinary bladder for temporary storage. The bladder can accommodate up to 250 to 400 ml of urine before stretch receptors within its wall initiate the micturition reflex. This reflex causes involuntary emptying of the bladder by simultaneous bladder contraction and opening of both the internal and external urethral sphincters. Micturition can transiently be voluntarily prevented until a more opportune time for bladder evacuation by deliberate tightening of the external sphincter and surrounding pelvic diaphragm.

REVIEW EXERCISES

Objective Questions (Answers on p. D–3.)

1. Part of the kidneys' energy supply is used to accomplish glomerular filtration. (True or false?)

2. Sodium reabsorption is under hormonal control throughout the length of the tubule. (True or false?)

3. Glucose and amino acids are reabsorbed by secondary active transport. (True or false?)

4. Solute excretion is always accompanied by comparable H_2O excretion. (True or false?)

5. Water excretion can occur without comparable solute excretion. (True or false?)

6. The functional unit of the kidneys is the _____.

7. _____ is the only ion actively reabsorbed in the proximal tubule and actively secreted in the distal and collecting tubule.

8. The minimum volume of obligatory H_2O loss that must accompany the excretion of wastes each day is _____ ml.

9. Indicate whether each of the following factors would (a) increase or (b) decrease the GFR, if everything else remained constant.

   ___ 1. a rise in Bowman's capsule pressure resulting from ureteral obstruction by a kidney stone

   ___ 2. a fall in plasma-protein concentration resulting from loss of these proteins from a large burned surface

   ___ 3. a dramatic fall in arterial blood pressure following severe hemorrhage (<80 mm Hg)

   ___ 4. afferent arteriolar vasoconstriction

   ___ 5. tubulo-glomerular feedback response to a reduction in tubular flow rate

   ___ 6. myogenic response of an afferent arteriole stretched as a result of an increased driving blood pressure

   ___ 7. sympathetic activity to the afferent arterioles

   ___ 8. contraction of mesangial cells

   ___ 9. contraction of podocytes

10. Which of the following filtered substances is normally *not* present in the urine at all?
    a. Na^+
    b. $PO_4^=$
    c. urea
    d. H^+
    e. glucose

11. Reabsorption of which of the following substances is *not* linked in some way to active Na^+ reabsorption?
    a. glucose
    b. $PO_4^=$
    c. H_2O
    d. urea
    e. Cl^-

In questions 12–14, indicate the proper sequence through which fluid flows as it traverses the structures in question by writing the identifying letters in the proper order in the blanks.

12. a. ureter   ___ ___ ___ ___ ___
    b. kidney
    c. urethra
    d. bladder
    e. renal pelvis

13. a. efferent arteriole   ___ ___ ___ ___ ___ ___
    b. peritubular capillaries
    c. renal artery
    d. glomerulus
    e. afferent arteriole
    f. renal vein

14. a. loop of Henle   ___ ___ ___ ___ ___ ___ ___
    b. collecting duct
    c. Bowman's capsule
    d. proximal tubule
    e. renal pelvis
    f. distal tubule
    g. glomerulus

15. Using the answer code at the right, indicate what the osmolarity of the tubular fluid is at each of the designated points in a nephron:

____ 1. Bowman's capsule
____ 2. end of proximal tubule
____ 3. tip of Henle's loop of juxtamedullary nephron
____ 4. end of Henle's loop of juxtamedullary nephron
____ 5. end of collecting duct

a. isotonic (300 mosm/l)
b. hypotonic (100 mosm/l)
c. hypertonic (1,200 mosm/l)
d. ranging from hypotonic to hypertonic (100 mosm/l to 1,200 mosm/l)

Essay Questions

1. List the functions of the kidneys.
2. Describe the anatomy of the urinary system. Describe the components of a nephron.
3. Describe the three basic renal processes; indicate how they relate to urine excretion. Distinguish between *secretion* and *excretion*.
4. Discuss the forces involved in glomerular filtration. What is the average GFR?
5. How is GFR regulated as part of the baroreceptor reflex?
6. Why do the kidneys receive a seemingly disproportionate share of the cardiac output? What percentage of renal blood flow is normally filtered?
7. List the steps in transepithelial transport.
8. Distinguish between active and passive reabsorption.
9. Describe all of the tubular transport processes that are linked to the basolateral Na^+-K^+ ATPase carrier.
10. Describe the renin-angiotensin-aldosterone system. What are the source and function of atrial natriuretic peptide?
11. To what do the terms *tubular maximum* (T_m) and *renal threshold* refer? Compare two substances that display a T_m, one that is and one that is not regulated by the kidneys.
12. What is the importance of tubular secretion? What are the most important secretory processes?
13. What is the average rate of urine formation?
14. Define plasma clearance.
15. What is responsible for the presence of a vertical osmotic gradient in the medullary interstitial fluid? Of what importance is this gradient?
16. Discuss the function of vasopressin.
17. Describe the transfer of urine to, the storage of urine in, and the emptying of urine from the bladder.

POINTS TO PONDER

1. The long-looped nephrons of animals adapted to survive with minimum water consumption, such as desert rats, have relatively much longer loops of Henle than humans have. Of what benefit would these longer loops be?
2. If the plasma concentration of substance X is 200 mg/100 ml and the GFR is 125 ml/min, the filtered load of this substance is _____.
   If the T_m for substance X is 200 mg/min, how much of the substance will be reabsorbed at a plasma concentration of 200 mg/100 ml and a GFR of 125 ml/min? _____
   How much of substance X will be excreted? _____
3. Conn's syndrome is an endocrine disorder brought about by a tumor of the adrenal cortex that secretes excessive aldosterone in uncontrolled fashion. Based on what you know about the functions of aldosterone, describe what the most prominent features of this condition would be.
4. If the urine concentration of a substance is 7.5 mg/ml of urine, its plasma concentration is 0.2 mg/ml of plasma, and the urine flow rate is 2 ml/min, what is the clearance rate of the substance? Is the substance being reabsorbed or secreted by the kidneys?
5. An accident victim suffers permanent damage of the lower spinal cord and is paralyzed from the waist down. Describe what governs bladder emptying in this individual.

15 Fluid and Acid-Base Balance

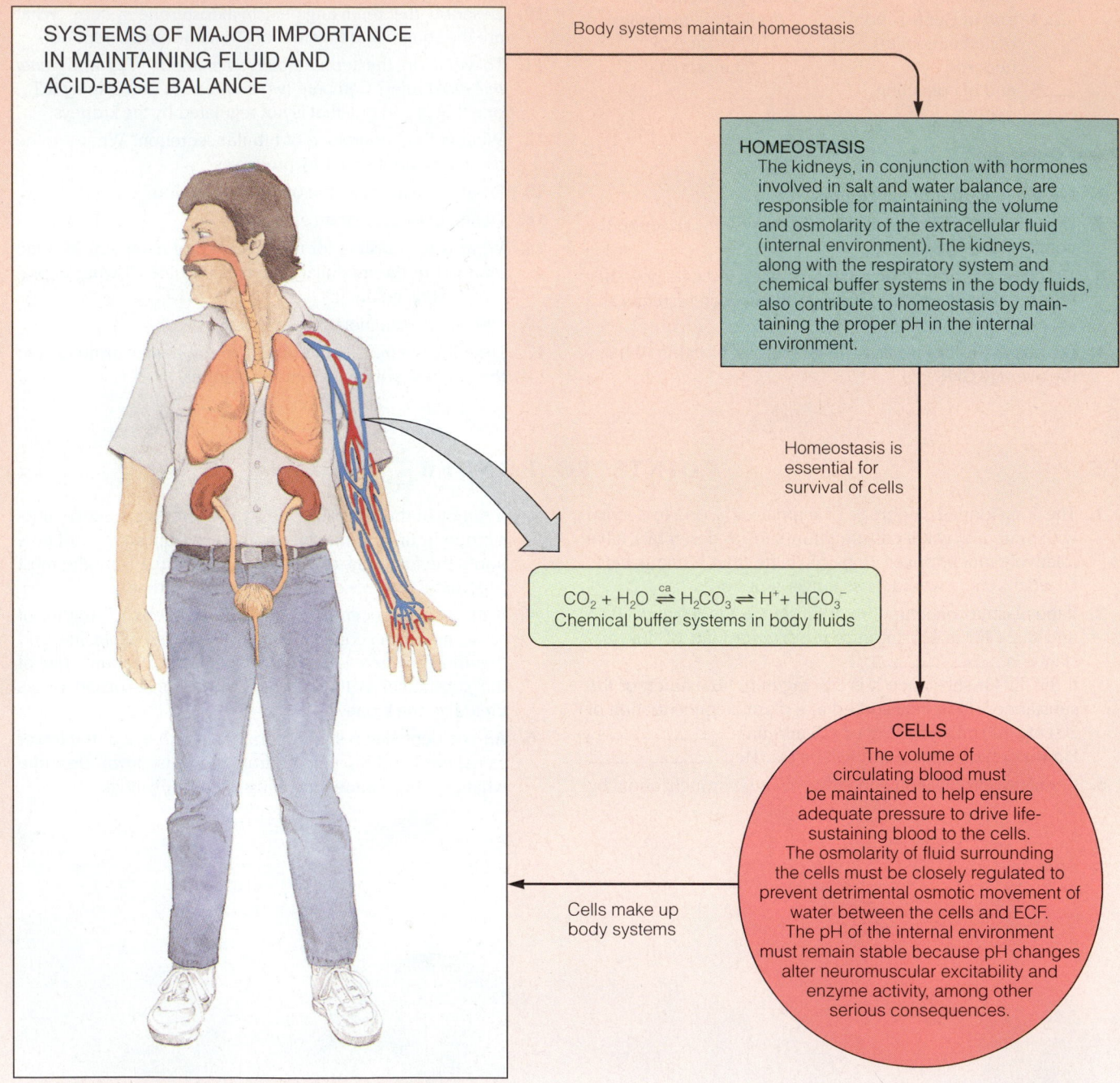

SYSTEMS OF MAJOR IMPORTANCE IN MAINTAINING FLUID AND ACID-BASE BALANCE

Body systems maintain homeostasis

HOMEOSTASIS
The kidneys, in conjunction with hormones involved in salt and water balance, are responsible for maintaining the volume and osmolarity of the extracellular fluid (internal environment). The kidneys, along with the respiratory system and chemical buffer systems in the body fluids, also contribute to homeostasis by maintaining the proper pH in the internal environment.

Homeostasis is essential for survival of cells

$$CO_2 + H_2O \overset{ca}{\rightleftharpoons} H_2CO_3 \rightleftharpoons H^+ + HCO_3^-$$
Chemical buffer systems in body fluids

CELLS
The volume of circulating blood must be maintained to help ensure adequate pressure to drive life-sustaining blood to the cells. The osmolarity of fluid surrounding the cells must be closely regulated to prevent detrimental osmotic movement of water between the cells and ECF. The pH of the internal environment must remain stable because pH changes alter neuromuscular excitability and enzyme activity, among other serious consequences.

Cells make up body systems

Homeostasis depends on maintaining a balance between the input and output of all constituents present in the internal fluid environment. Regulation of **fluid balance** involves two separate components: *control of ECF volume,* of which circulating plasma volume is a part, and *control of ECF osmolarity* (solute concentration). The kidneys control ECF volume by maintaining **salt balance** and control ECF osmolarity by maintaining **water balance**. The kidneys maintain this balance by adjusting the output of salt and water in the urine as needed to compensate for variable input and abnormal losses of these constituents.

Similarly, the kidneys contribute to the maintenance of **acid-base balance** by adjusting the urinary output of hydrogen ion (acid) and bicarbonate ion (base) as needed. Also contributing to acid-base balance are the lungs, which can adjust their rate of excretion of hydrogen-ion–generating CO_2, and the chemical buffer systems in the body fluids.

FLUID BALANCE
Balance concept
Body-fluid compartments
Factors regulated to maintain fluid balance
Control of ECF volume by regulating salt balance
● A CLOSER LOOK AT EXERCISE PHYSIOLOGY A Potentially Fatal
Clash: When Exercising Muscles and Cooling Mechanisms
Compete for an Inadequate Plasma Volume
Control of ECF osmolarity by regulating water balance
● CONCEPTS, CHALLENGES, AND CONTROVERSIES Breaching the
Blood-Brain Barrier
ACID-BASE BALANCE
Acid-base chemistry; pH
Consequences of changes in H⁺ concentration
Sources of H⁺
Three lines of defense against changes in H⁺ concentration
Chemical buffer systems
Respiratory control of pH
Renal control of pH
Acid-base imbalances

 FLUID BALANCE

Input must equal output if balance is to be maintained.

The cells of complex multicellular organisms are able to survive and function only within a very narrow range of composition of the extracellular fluid (ECF), the internal fluid environment that bathes them. The quantity of any particular substance in the ECF is considered to be a readily available internal **pool**. The amount of the substance in the pool may be increased either by transferring more in from the external environment (most commonly by ingestion) or by metabolically producing it within the body (● Fig. 15–1). Substances may be removed from the body by excretion to the outside or by being used up in a metabolic reaction. If the quantity of a substance is to remain stable within the body, its input by means of ingestion or metabolic production must be balanced by an equal output by means of excretion or metabolic consumption. This relationship, known as the **balance concept**, is extremely important in the maintenance of homeostasis. Not all input and output pathways are applicable for every body-fluid constituent. For example, salt is not synthesized or consumed by the body, so the stability of salt concentration in the body fluids depends entirely on a balance between salt ingestion and salt excretion.

The ECF pool can further be altered by transferring a particular ECF constituent into storage within the cells or bones. If the body as a whole has a surplus or deficit of a particular stored substance, the storage site can be expanded or partially depleted to maintain the ECF concentration of the substance within homeostatically prescribed limits. For example, following absorption of a meal, when more glucose is entering the plasma than is being consumed by the cells, the surfeit of glucose can be temporarily stored in muscle and liver cells in the form of glycogen. This storage depot can then be tapped between meals as necessary to maintain the plasma glucose level when no new nutrients are being added to the blood by eating. It is important to recognize, however, that internal storage capacity is limited. Although an internal exchange between the ECF and storage depot can temporarily restore the plasma concentration of a particular substance to normal, in the long run any excess or deficit of that constituent must be compensated for by appropriate adjustments in total body input or output.

Another possible internal exchange between the pool and the remainder of the body is the reversible incorporation of

● **FIGURE 15–1 Inputs to and Outputs from the Internal Pool of a Body Constituent**

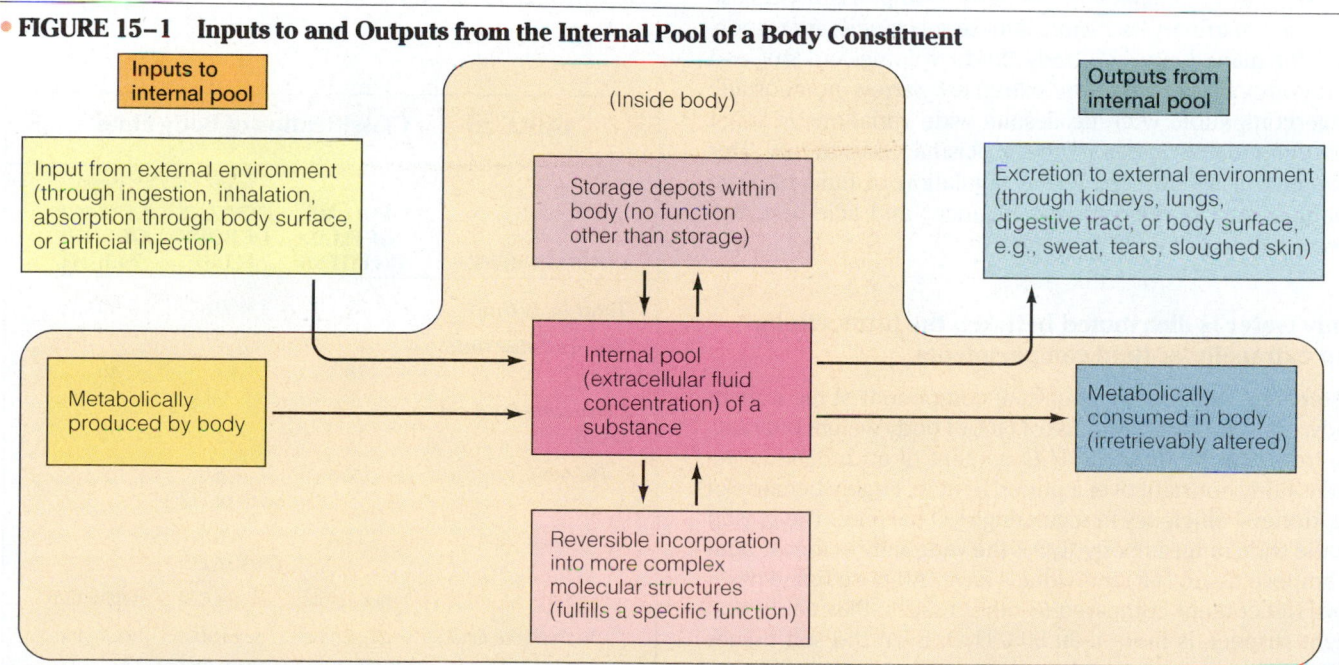

certain plasma constituents into more complex molecular structures. For example, iron is incorporated into hemoglobin within the red blood cells during their synthesis but is released intact back into the body fluids when the red cells degenerate (see p. 351). This process differs from metabolic consumption of a substance, in which the substance is irretrievably converted into another form—for example, glucose converted into CO_2 plus H_2O plus energy. It also differs from storage in that the latter serves no purpose other than storage, whereas reversible incorporation into a more complex structure serves a specific purpose.

When total body input of a particular substance equals its total body output, a **stable balance** exists. When the gains via input for a substance exceed its losses via output, a **positive balance** exists. The result is an increase in the total amount of the substance in the body. In contrast, when the losses for a substance exceed its gains, a **negative balance** exists and the total amount of the substance in the body decreases.

Changing the magnitude of any of the input or output pathways for a given substance can alter its plasma concentration. In order to maintain homeostasis, any change in input must be balanced by a corresponding change in output (for example, increased salt intake must be matched by a corresponding increase in salt output in the urine), and, conversely, increased losses must be compensated for by increased intake. Thus, maintenance of a stable balance necessitates control. However, not all input and output pathways are regulated to maintain balance. Generally, input of various plasma constituents is poorly controlled or not controlled at all. We frequently ingest salt and H_2O, for example, not because we *need* them but because we *want* them, so the intake of salt and H_2O is highly variable. Likewise, hydrogen ion (H^+) is uncontrollably generated internally and added to the body fluids. Salt, H_2O, and H^+ can also be lost to the external environment to varying degrees through the digestive tract (vomiting), skin (sweating), and elsewhere without regard for salt, H_2O, or H^+ balance in the body. Compensatory adjustments in the urinary excretion of these substances are responsible for maintaining the body fluid's volume and salt and acid composition within the extremely narrow homeostatic range compatible with life despite wide variations in input and unregulated losses of these plasma constituents. This chapter will be devoted to the regulation of fluid balance (maintenance of salt and H_2O balance) and acid-base balance (maintenance of H^+ balance).

Body water is distributed between the intracellular- and extracellular-fluid compartments.

Water is by far the most abundant component of the human body, constituting an average of 60% of body weight but ranging from 40% to 80%. The H_2O content of an individual remains fairly constant over a period of time, largely because of the kidneys' efficiency in regulating H_2O balance. The reason for the wide range in body H_2O is the variability in the amount of adipose tissue (fat) individuals have. Adipose tissue has a low H_2O content compared to other tissues. Plasma, as you might suspect, is more than 90% H_2O. Even the soft tissues such as skin, muscles, and internal organs consist of 70% to 80% H_2O. The relatively drier skeleton is only 22% H_2O. Fat, however, is the driest tissue of all, having only 10% H_2O content. Accordingly, a high body H_2O content is associated with leanness and a low body H_2O content with obesity, since a larger proportion of the body is composed of relatively dry fat in overweight individuals.

The percentage of body H_2O is also influenced by the sex and age of the individual. Women have a lower body H_2O content than men, primarily because the female sex hormone, estrogen, promotes fat deposition in the breasts, buttocks, and elsewhere. This not only gives rise to the typical female figure, but also endows women with a higher proportion of adipose tissue and, therefore, a lower body H_2O content. The percentage of body H_2O also decreases progressively with age.

Body H_2O is distributed between two major fluid compartments: the fluid within the cells, *intracellular fluid* (*ICF*), and the fluid surrounding the cells, *extracellular fluid* (*ECF*) (Table 15–1). (The terms "H_2O" and "fluid" are commonly used interchangeably. Although this usage is not entirely accurate because it ignores the solutes in the body fluids, it is acceptable when one considers the total volume of the fluids, since the major proportion of these fluids consists of H_2O.) The ICF compartment comprises about two-thirds of the total body H_2O. Even though each cell contains its own unique mixture of constituents, there are sufficient similarities between these trillions of minute fluid compartments to consider them collectively as one large fluid compartment. The remaining one-third of the body H_2O found in the ECF compartment is further subdivided into plasma and interstitial fluid. The *plasma*, which makes up about one-fifth of the ECF volume, is the fluid portion of the blood. The *interstitial fluid*, which represents the other four-fifths of the extracellular-fluid compartment, is the fluid that lies in the spaces between the cells. Interstitial fluid, sometimes also known as *tissue fluid*, constitutes the true internal environment in that it is the fluid that bathes the tissue cells.

TABLE 15–1 Classification of Body Fluid			
COMPARTMENT	VOLUME OF FLUID (IN LITERS)	PER-CENTAGE OF BODY FLUID	PER-CENTAGE OF BODY WEIGHT
Total body fluid	42	100%	60%
Intracellular fluid (ICF)	28	67	40
Extracellular fluid (ECF)	14	33	20
Plasma	2.8	6.6 (20% of ECF)	4
Interstitial fluid	11.2	26.4 (80% of ECF)	16
Lymph	negligible	negligible	negligible
Transcellular fluid	negligible	negligible	negligible

Two other minor categories are included in the ECF compartment: lymph and transcellular fluid. *Lymph* is fluid being returned from the interstitial fluid to the plasma by means of the lymphatic system, where it is filtered through lymph nodes for immune defense purposes (see p. 326). **Transcellular fluid** consists of a number of small specialized fluid volumes, all of which are secreted by specific cells into a particular body cavity to perform some specialized function. Transcellular fluid includes *cerebrospinal fluid* (surrounding, cushioning, and nourishing the brain and spinal cord); *intraocular fluid* (maintaining the shape of and nourishing the eye); *synovial fluid* (lubricating and serving as a shock absorber for the joints); *pericardial, pleural,* and *peritoneal fluids* (lubricating movements of the heart, lungs, and intestines, respectively); and the *digestive juices* (digesting ingested foods). Although these fluids are extremely important functionally, they represent an insignificant fraction of the total body H_2O. Furthermore, the transcellular compartment as a whole usually does not reflect changes in the body's fluid balance. For example, the cerebrospinal fluid does not decrease in volume when the body as a whole is experiencing a negative H_2O balance. This is not to say that these fluid volumes never change. Localized changes in a particular transcellular-fluid compartment can occur pathologically (such as too much intraocular fluid accumulating in the eyes of persons with glaucoma), but such a localized fluid disturbance does not affect the fluid balance of the body. Therefore, the trans-cellular compartment can usually be ignored when dealing with problems of fluid balance. The main exception to this generalization is when digestive juices are abnormally lost from the body during heavy vomiting or diarrhea, which can bring about a fluid imbalance.

The plasma and interstitial fluid are separated by the blood-vessel walls, whereas the ECF and ICF are separated by cellular plasma membranes.

Several barriers separate the body-fluid compartments, limiting the movement of H_2O and solutes between the various compartments to differing degrees. The two components of the ECF—plasma and interstitial fluid—are separated by the walls of the blood vessels. However, H_2O and all plasma constituents with the exception of plasma proteins are continuously and freely exchanged between the plasma and interstitial fluid by passive means across the thin, pore-lined capillary walls. Accordingly, plasma and interstitial fluid are nearly identical in composition, except that interstitial fluid lacks plasma proteins. Any change in one of these ECF compartments is quickly reflected in the other compartment because they are constantly mixing.

In contrast to the very similar composition of the vascular and interstitial-fluid compartments, the composition of the ECF differs considerably from that of the ICF ($\bullet$ Fig. 15–2). Each cell is surrounded by a highly selective plasma mem-

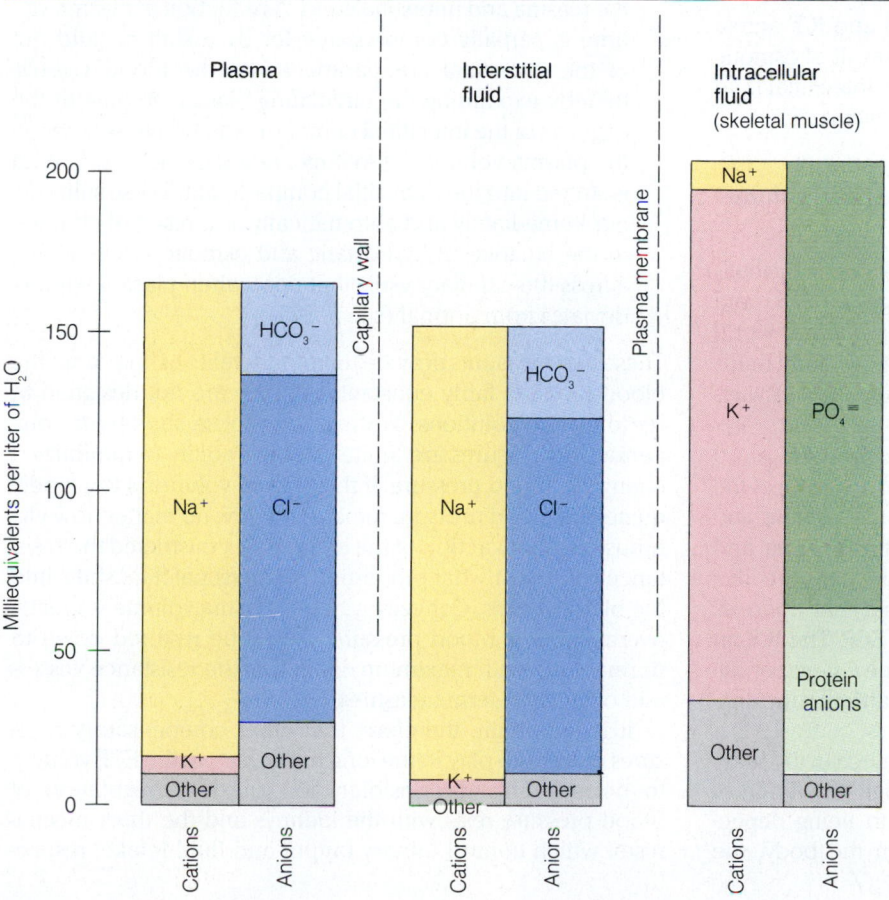

$\bullet$ **FIGURE 15–2** Ionic Composition of Major Body-Fluid Compartments brane that permits passage of certain materials while excluding others. Movement through the membrane barrier occurs by both passive and active means and may be highly discriminating. Among the major differences between the ECF and ICF are (1) the presence of cellular proteins in the ICF that are unable to permeate the enveloping membranes to leave the cells and (2) the unequal distribution of Na^+ and K^+ and their attendant anions as a result of the action of the membrane-bound Na^+-K^+ ATPase pump that is present in all cells. This pump actively transports Na^+ out of and K^+ into cells; for this reason, Na^+ is the primary ECF cation, and K^+ is primarily found in the ICF. This unequal distribution of Na^+ and K^+, coupled with differences in membrane permeability to these ions, is responsible for the electrical properties of cells, including the initiation and propagation of action potentials in excitable tissues (see chapters 3 and 4).

Except for the extremely small, electrically imbalanced portion of the total intracellular and extracellular ions that are involved in membrane potential, the majority of the ECF and ICF ions are electrically balanced. In the ECF, Na^+ is accompanied primarily by the anion Cl^- (chloride) and to a lesser extent by HCO_3^- (bicarbonate). The major intracellular anions are $PO_4^=$ (phosphate) and the negatively charged proteins trapped within the cell.

Although all cells' plasma membranes display selective permeability, all cells are freely permeable to H_2O. The movement of H_2O between the plasma and interstitial fluid across capillary walls is governed by relative imbalances between capillary blood pressure (a fluid or hydrostatic pressure) and colloid osmotic pressure (see p. 323). In contrast, the net transfer of H_2O between the interstitial fluid and ICF across the cellular plasma membranes occurs as a result of osmotic effects alone. The hydrostatic pressures of the interstitial fluid and ICF are both extremely low and fairly constant.

Fluid balance is maintained by regulating ECF volume and ECF osmolarity.

Extracellular fluid serves as an intermediary between the cells and the external environment. All exchanges of H_2O and other constituents between the ICF and the external world must occur through the ECF. Water added to the body fluids always enters the ECF compartment first, and fluid always leaves the body by way of the ECF.

Plasma is the only fluid that can be directly acted on to control its volume and composition. However, because of the free exchange across the capillary walls, if the volume and composition of the plasma are regulated, the volume and composition of the interstitial fluid bathing the cells are likewise regulated. Thus, any control mechanism that operates on the plasma in effect regulates the entire ECF. The ICF, in turn, is influenced by changes in the ECF to the extent permitted by the permeability of the membrane barriers surrounding the cells.

The factors regulated to maintain fluid balance in the body are ECF volume and ECF osmolarity. Although regulation of these two factors is closely interrelated, both being dependent on the relative NaCl and H_2O load in the body, the reasons that they are closely controlled are significantly different:

1. *Extracellular fluid volume* must be closely regulated to help *maintain blood pressure*. Maintenance of *salt balance* is of primary importance in the long-term regulation of ECF volume.

2. *Extracellular fluid osmolarity* must be closely regulated to *prevent swelling or shrinking* of the cells. Maintenance of *water balance* is of primary importance in the regulation of ECF osmolarity.

We will examine each of these factors in more detail.

Control of ECF volume is important in the long-term regulation of blood pressure.

A reduction in ECF volume, by decreasing the plasma volume, causes a fall in arterial blood pressure. Conversely, a rise in ECF volume increases the arterial blood pressure by expanding the plasma volume. Two compensatory measures come into play to transiently adjust the blood pressure until the ECF volume can be restored to normal:

1. Baroreceptor reflex mechanisms alter both cardiac output and total peripheral resistance through autonomic nervous system effects on the heart and blood vessels (see p. 334). These immediate cardiovascular responses are designed to minimize the effect a deviation in circulating volume has on blood pressure.

2. Fluid shifts occur temporarily and automatically between the plasma and interstitial fluid. A reduction in plasma volume is partially compensated for by a shift of fluid out of the interstitial compartment into the blood vessels, thereby expanding the circulating plasma volume at the expense of the interstitial compartment. Conversely, when the plasma volume is too large, much of the excess fluid is shifted into the interstitial compartment. These shifts occur immediately and automatically as a result of changes in the balance of hydrostatic and osmotic forces acting across the capillary walls that arise when plasma volume deviates from normal (see p. 325).

These two measures provide temporary relief to help keep the blood pressure fairly constant, but they are not designed to be long-term solutions. Furthermore, these short-term compensatory measures are limited in their ability to minimize a change in blood pressure. If the plasma volume is too inadequate, the blood pressure remains too low no matter how vigorous the pump action of the heart, how constricted the resistance vessels, or what proportion of interstitial fluid shifts into the blood vessels. Conversely, if the plasma volume is greatly overexpanded, blood pressure cannot be restored down to normal even with maximum dilation of the resistance vessels and other short-term measures.

It is important, therefore, that other compensatory measures come into play in the long run to restore the ECF volume to normal. This responsibility for long-term regulation of blood pressure rests with the kidneys and the thirst mechanism, which control urinary output and fluid intake, respectively. In so doing, they accomplish needed fluid exchanges between the ECF and the external environment to regulate the body's total fluid volume. Accordingly, they have an important long-term influence on arterial blood pressure.

Control of salt balance is primarily important in regulating ECF volume.

Sodium and its attendant anions account for more than 90% of the ECF's osmotic activity. Because osmotic activity can be equated with "water-holding power," the total *Na+ load* (the total quantity of NaCl, not its concentration) in the ECF determines the total amount of H_2O that will be osmotically retained in the ECF. The total mass of Na^+ salts in the ECF therefore determines the ECF's volume, and, appropriately, regulation of ECF volume depends primarily on controlling salt balance.

To maintain salt balance at a set level, salt input must equal salt output, thus preventing salt accumulation or deficit in the body. The only avenue for salt input is ingestion, which typically is well in excess of the body's need for replacement of obligatory salt losses. A half gram of salt per day is adequate to replace the small amounts of salt usually lost in the feces and sweat. In our example of a typical daily salt balance (Table 15–2), salt intake is 10.5 gm per day. (The average American salt intake is 10 to 15 gm per day, although many people are consciously reducing their salt intake.)

Since we typically consume salt in excess of our needs, it is obvious that salt intake in humans is not well controlled. Carnivores (meat eaters) and omnivores (eaters of meat and plants, like humans), which naturally get sufficient salt in fresh meat (meat contains an abundance of salt-rich ECF), normally do not manifest a physiological appetite to seek additional salt. In contrast, herbivores (plant eaters), which lack salt naturally in their diets, develop a salt hunger and will travel miles to a salt lick. Humans generally have a hedonistic rather than a regulatory appetite for salt; we consume salt because we like it rather than because we have a physiological need, except in the unusual circumstance of severe salt depletion caused by a deficiency of aldosterone, the salt-conserving hormone.

The excess ingested salt must be excreted in the urine to maintain salt balance. The three avenues for salt output are obligatory loss of salt in sweat and feces and controlled excretion of salt in the urine (Table 15–2). The total amount of sweat produced is unrelated to salt balance, being determined instead by factors that control body temperature. Aldosterone can reduce the sweat's salt content, however, thus helping to conserve salt in a hot environment. The small salt loss in the feces is not subject to control. Except when sweating heavily or during diarrhea, the body normally uncontrollably loses only about 0.5 gm of salt per day. This amount is actually the only salt that normally needs to be replaced by salt intake. Because salt consumption is typically far in excess of the meager amount needed to compensate for uncontrolled losses, the kidneys precisely excrete the excess salt in the urine to maintain salt balance. In our example, 10 gm of salt are eliminated in the urine per day so that total salt output exactly equals salt input. By regulating the rate of urinary salt excretion (that is, by regulating the rate of Na^+ excretion, with Cl^- following along), the kidneys normally keep the total Na^+ mass in the ECF constant despite any notable changes in dietary intake of salt or unusual losses through sweating, diarrhea, or other means. As a reflection of keeping the total Na^+ mass in the ECF constant, the ECF volume in turn is maintained within the narrowly prescribed limits essential for normal circulatory function.

Deviations in ECF volume accompanying changes in salt load are responsible for triggering renal compensatory responses that quickly bring the Na^+ load and ECF volume back into line. Sodium is freely filtered at the glomerulus and actively reabsorbed, but it is not secreted by the tubules, so the amount of Na^+ excreted in the urine represents the amount of Na^+ that is filtered but is not subsequently reabsorbed:

$$Na^+ \text{ excreted} = Na^+ \text{ filtered} - Na^+ \text{ reabsorbed}$$

The kidneys accordingly adjust the amount of salt excreted by controlling two processes: (1) the glomerular filtration rate (GFR) and (2) more importantly, the tubular reabsorption of Na^+.

Control of the Amount of Na^+ Filtered through Regulation of the GFR The amount of Na^+ filtered is equal to the plasma Na^+ concentration times the GFR. At any given plasma Na^+ concentration, any alteration in the GFR will correspondingly alter the amount of Na^+ filtered. Thus, control of the GFR can adjust the amount of Na^+ filtered each minute.

The GFR is deliberately changed to alter the amount of salt and fluid filtered as part of the general baroreceptor reflex response to a change in blood pressure (see Fig. 14–12, p. 477). The afferent arterioles that supply the renal glomeruli are constricted as part of the generalized vasoconstriction aimed at elevating a reduction in blood pressure. As a result of reduced blood flow into the glomeruli, GFR decreases and, accordingly, the amount of Na^+ and accompanying fluid that are filtered decreases. Consequently, excretion of salt and fluid is diminished. The conserved salt and fluid that otherwise would have been filtered and excreted help minimize the reduction in fluid volume and contribute

TABLE 15–2 Daily Salt Balance			
SALT INPUT		**SALT OUTPUT**	
Avenue	Amount (g/day)	Avenue	Amount (g/day)
Ingestion	10.5	Obligatory loss in sweat and feces	0.5
		Controlled excretion in urine	10.0
Total input	10.5	Total output	10.5

to long-term restitution of blood pressure. Conversely, an elevation in ECF volume and arterial blood pressure is reflexly countered by a baroreceptor reflex response that leads to an increase in GFR, which in turn results in enhanced salt and fluid excretion. The elimination of extra salt and fluid that otherwise would have been conserved helps relieve the expanded plasma volume.

Note that adjustments in the amount of salt filtered are accomplished as part of the general blood-pressure–regulating reflexes. Changes in Na+ load in the body are not sensed as such; instead they are monitored indirectly through the effect that Na+ ultimately has on blood pressure via its role in determining the ECF volume. Fittingly, baroreceptors that monitor fluctuations in blood pressure are responsible for bringing about adjustments in the amount of Na+ filtered and eventually excreted.

Control of the Amount of Na+ Reabsorbed through the Renin-Angiotensin-Aldosterone System The amount of Na+ reabsorbed also depends on regulatory systems that play an important role in controlling blood pressure. Although Na+ is reabsorbed throughout the tubule's length, only its reabsorption in the distal portions of the tubule is subject to control. The main factor controlling the extent of Na+ reabsorption in the distal and collecting tubule is the powerful renin-angiotensin-aldosterone system, which promotes Na+ reabsorption and thereby Na+ retention. Sodium retention, in turn, promotes osmotic retention of H_2O and the subsequent expansion of plasma volume and elevation of arterial

blood pressure. Appropriately, this Na+-conserving system is activated by a reduction in NaCl/ECF volume/arterial blood pressure (see Fig. 14–17, p. 482).

It should be apparent that control of GFR and Na+ reabsorption are highly interrelated and that both are intimately tied in with long-term regulation of ECF volume as reflected by the blood pressure. Specifically, a fall in arterial blood pressure brings about a twofold effect in the renal handling of Na+ (• Fig. 15–3): (1) a reflex reduction in the GFR to decrease the amount of Na+ filtered and (2) a hormonally adjusted increase in the amount of Na+ reabsorbed. Together these effects reduce the amount of Na+ excreted, thereby conserving for the body the Na+ and accompanying H_2O necessary to compensate for the fall in arterial pressure. (See the accompanying boxed feature, A Closer Look at Exercise Physiology.) Conversely, a rise in arterial blood pressure brings about (1) increases in the amount of Na+ filtered, and (2) a reduction in renin-aldosterone activity, which decreases salt (and fluid) reabsorption. Together these actions increase salt (and fluid) excretion, thereby eliminating the extra fluid that was expanding the plasma volume and increasing the arterial pressure.

Control of ECF osmolarity prevents changes in ICF volume.

Maintenance of fluid balance in the body depends on the regulation of both ECF volume and ECF osmolarity. Whereas regulation of ECF volume is important in the long-term con-

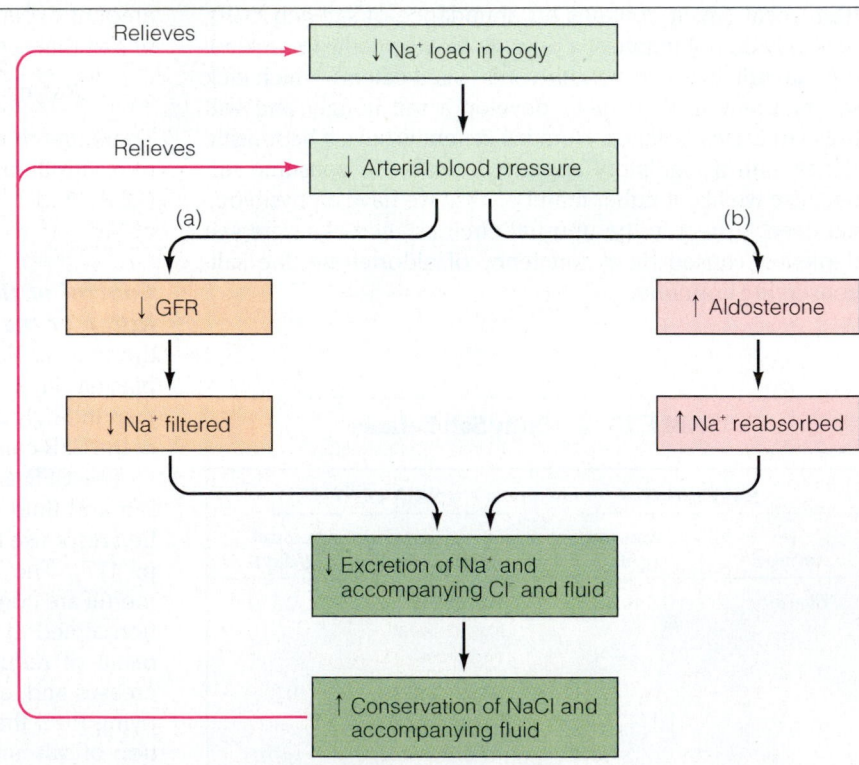

• **FIGURE 15–3   Dual Effect of Fall in Arterial Blood Pressure on Renal Handling of Na+**

(a) See Figure 14–12 for details of mechanism.
(b) See Figure 14–17 for details of mechanism.

A POTENTIALLY FATAL CLASH: WHEN EXERCISING MUSCLES AND COOLING MECHANISMS COMPETE FOR AN INADEQUATE PLASMA VOLUME

An increasing number of people of all ages are participating in walking or jogging programs to improve their level of physical fitness and decrease their risk of cardiovascular disease. For people living in environments that undergo seasonal temperature changes, exercising outdoors can become dangerous due to fluid loss during the transition from the cool days of spring to the hot, humid days of summer. If exercise intensity is not modified until the participant gradually adjusts to the hotter environmental conditions, dehydration and salt loss can indirectly lead to heat cramps, heat exhaustion, or ultimately heat stroke and death.

Acclimatization refers to the gradual adaptations the body makes in order to maintain long-term homeostasis in response to a prolonged physical change in the surrounding environment, such as a change in temperature. When a person exercises in the heat without gradually adapting to the hotter environment, the body faces a terrible dilemma. During exercise, large amounts of blood must be delivered to the muscles to supply O_2 and nutrients and to remove the wastes that accumulate from their high rate of activity. Exercising muscles also produce heat. To maintain the body temperature in the face of this extra heat, blood flow to the skin is increased so that heat from the warmed blood can be lost through the skin to the surrounding environment. If the environmental temperature is hotter than the body temperature, heat cannot be lost from the

blood to the surrounding environment despite maximal skin vasodilation. Instead, the body gains heat from its warmer surroundings, further adding to the dilemma. Because extra blood is diverted to both the muscles and the skin when a person exercises in the heat, less blood is returned to the heart, and the heart pumps less blood per beat in accordance with the Frank-Starling mechanism (see p. 286). Therefore, the heart must beat faster to deliver the same amount of blood per minute than it would in a cool environment. The increased rate of cardiac pumping further contributes to heat production.

The sweat rate also increases so that evaporative cooling can take place to help maintain the body temperature during periods of excessive heat gain. In an unacclimatized person, maximal sweat rate is about 1.5 liters per hour. Not only is water lost during sweating, but water-retaining salt is also lost. The resulting loss of plasma volume through sweating further depletes the blood supply available for muscular exercise and for cooling through skin vasodilation.

The heart has a maximum rate at which it can pump. If exercise continues at a high intensity and this maximal rate is reached, the exercising muscles win the contest for blood supply. The body responds by constricting the skin arterioles, sacrificing cooling to maintain cardiac output and blood pressure. If exercise continues, body heat continues to rise, and heat exhaustion (rapid,

weak pulse, hypotension, profuse sweating, and disorientation) or heat stroke (failure of the temperature-control center in the hypothalamus; hot, dry skin; extreme confusion or unconsciousness; and possibly death) can occur. In fact, every year people die of heat stroke in marathons run during hot, humid weather.

On the other hand, if a person exercises in the heat for two weeks at reduced, safe intensities, the body makes the following adaptations so that after acclimatization the person can do the same amount of work as was possible in a cool environment: (1) The plasma volume is increased by as much as 12%. The expansion of plasma volume provides sufficient blood to both supply the exercising muscles and direct blood to the skin for cooling. (2) The person begins sweating at a lower temperature so that the body does not get so hot before cooling begins. (3) The sweat rate increases as much as three times to 4 liters per hour with a more even distribution over the body. This increase in evaporative cooling reduces the need for cooling by skin vasodilation. (4) The sweat becomes more dilute so that less salt is lost in the sweat. The retained salt exerts an osmotic effect to hold water in the body and help maintain circulating plasma volume. These adaptations take fourteen days and will occur only if the person exercises in the heat. Being patient until these changes take place will enable the person to exercise safely throughout the summer months.

trol of blood pressure, regulation of ECF osmolarity is important in preventing changes in cell volume.

The **osmolarity** of a fluid is a measure of the concentration of the individual solute particles dissolved in it. The higher the osmolarity, the higher the concentration of solutes or, to look at it differently, the lower the concentration of H_2O. Water tends to move by osmosis down its own concentration gradi-

ent from an area of lower solute (higher H_2O) concentration to an area of higher solute (lower H_2O) concentration.

Osmosis occurs across the cellular plasma membranes only when there is a difference in concentration of non-penetrating solutes between the ECF and ICF. Solutes that can penetrate a barrier separating two fluid compartments quickly become equally distributed between the two compartments and thus do not contribute to osmotic differences. The osmotic activity across the capillary wall is due to the unequal distribution of plasma proteins. Plasma proteins are present only in the plasma (as their name implies), because they are too large to penetrate the capillary walls and enter the interstitial fluid. All other solutes are in essentially the same concentration in the plasma and interstitial fluid, so they do not contribute to any unequal distribution of H_2O that would induce osmosis across the capillary wall.

In contrast, osmotic activity across the cellular plasma membranes is directly related to any differences in ionic concentration between the ECF and ICF. Plasma proteins play no role in the osmosis of H_2O across the cell membranes because they are absent in both the interstitial fluid and the ICF that are separated by the cell membranes.

Sodium and its attendant anions, being by far the most abundant solutes in the ECF in terms of numbers of particles, account for the vast majority of the ECF's osmotic activity. In contrast, K^+ and its accompanying intracellular anions are responsible for the ICF's osmotic activity. Even though small amounts of Na^+ and K^+ passively diffuse across the plasma membrane all the time, these ions behave as if they are nonpenetrating because of Na^+-K^+ pump activity. Any Na^+ that passively diffuses down its electrochemical gradient into the cell is promptly pumped back outside, so the result is the same as if Na^+ were barred from the cells. The same holds true in reverse for K^+; it in effect remains trapped within the cells. The resulting unequal distribution of Na^+ and K^+ and their accompanying anions between the ECF and the ICF is responsible for the osmotic activity of these two fluid compartments.

Normally, the osmolarities of the ECF and ICF are the same, because the total concentration of K^+ and other effectively nonpenetrating solutes inside the cells is equal to the total concentration of Na^+ and other effectively nonpenetrating solutes in the fluid surrounding the cells. Even though the nonpenetrating solutes in the ECF and ICF differ, their concentrations are normally identical, and it is the number (not the nature) of the unequally distributed particles per volume that determines the fluid's osmolarity. Because the osmolarity of the ECF and ICF are normally equal, no net movement of H_2O usually occurs into or out of the cells. Therefore, cell volume normally remains constant, because no H_2O osmotically enters or leaves the cells.

However, any circumstance that results in a loss or gain of free H_2O (that is, loss or gain of H_2O that is not accompanied by comparable solute deficit or excess) leads to changes in ECF osmolarity. If there is a deficit of free H_2O in the ECF, the solutes become too concentrated, and the ECF osmolarity becomes abnormally high (that is, becomes *hypertonic*) (see p. 492). If there is excess free H_2O in the ECF, the solutes become too dilute, and the ECF osmolarity becomes abnormally low (that is, becomes *hypotonic*). When the ECF osmolarity changes with respect to the ICF osmolarity, osmosis takes place, with H_2O either leaving or entering the cells, depending respectively on whether the ECF is more or less concentrated than the ICF.

The osmolarity of the ECF must therefore be regulated to prevent these undesirable shifts of H_2O into or out of the cells. As far as the ECF itself is concerned, the concentration of its solutes does not really matter. However, it is crucial that ECF osmolarity be maintained within very narrow limits to prevent the cells from swelling (by osmotically gaining H_2O from the ECF) or shrinking (by osmotically losing H_2O to the ECF).

We will examine the fluid shifts that occur between the ECF and ICF when the ECF osmolarity becomes hypertonic or hypotonic relative to the ICF. Then we will consider how water balance and subsequently ECF osmolarity are normally maintained to minimize detrimental changes in cell volume.

ECF Hypertonicity Hypertonicity of the ECF, or the excessive concentration of ECF solutes, is usually associated with **dehydration**, or a negative free H_2O balance. Dehydration with accompanying hypertonicity can be brought about in three major ways: (1) insufficient H_2O intake, such as might occur during desert travel or might accompany difficulty in swallowing; (2) excessive H_2O loss, such as might occur during heavy sweating, vomiting, or diarrhea (even though both H_2O and solutes are lost during these conditions, relatively more H_2O is lost, so the remaining solutes become more concentrated); and (3) **diabetes insipidus**, a disease characterized by a deficiency of *vasopressin* (*antidiuretic hormone*), the hormone that increases the permeability of the distal and collecting tubules to H_2O and thus enhances water conservation by reducing urinary output of water. In diabetes insipidus, the kidneys cannot conserve H_2O because they are unable to reabsorb H_2O from the distal portions of the nephron in the absence of vasopressin. Such patients typically produce up to 20 liters of very dilute urine per day, compared to the normal average of 1.5 liters per day. Unless H_2O intake keeps pace with this tremendous loss of H_2O in the urine, the person quickly dehydrates. Patients complain that they spend an extraordinary amount of time day and night going to the bathroom and getting drinks. Fortunately, they can be treated with replacement vasopressin administered by nasal spray.

On rare occasions, the ECF becomes hypertonic in the absence of dehydration because of the abnormal accumulation of osmotically active solutes that do not normally contribute significantly to ECF osmotic activity. This occurs, for example, with the high blood urea levels in uremia (kidney failure).

Whenever the ECF compartment becomes hypertonic, H_2O moves out of the cells by osmosis into the more concentrated ECF until the ICF osmolarity equilibrates with the ECF. The cells shrink as H_2O leaves them. (See the accompanying boxed feature, Concepts, Challenges, and Controversies.) Of particular concern is the fact that considerable shrinking of brain neurons causes disturbances in brain function, which can be manifested as mental confusion and irrationality in moderate cases and can bring about possible delirium, convulsions, or coma in more severe hypertonic conditions. Rivaling the neural symptoms in seriousness are the circulatory disturbances that arise from a reduction in plasma volume in association with dehydration. Circulatory problems may range from a slight reduction in blood pressure to circulatory shock and death.

Other more common symptoms become apparent, even in

BREACHING THE BLOOD-BRAIN BARRIER

In the early 1980s, Neil Shay, a thirty-six-year-old cross-country truck driver, received a grim diagnosis. While making a delivery run, Shay suddenly became disoriented; he didn't know where he was or where he was going. By the time he checked into a nearby hospital, his mental fog was coupled with a searing headache. There he learned the bad news: the diagnosis was CNS lymphoma, a rare type of cancer.

Shay had three malignant tumors in his brain about the size of golf balls. None of the standard approaches for treating cancer seemed to be available in his case. The tumors were too scattered and extensive to remove surgically, and they could not be eradicated by radiation therapy. Although lymphomas elsewhere in the body frequently succumb to chemotherapy, these cancer-killing drugs could not penetrate the staunch blood-brain barrier (see p. 113), so they were unable to reach the life-threatening brain tumors. Shay was given less than a year to live.

More than a decade later, however, Shay is still alive and free of brain cancer, thanks to the ingenuity of neurosurgeon Edward Neuwelt. Using his knowledge of physiology, Neuwelt developed a pioneering method of getting chemotherapeutic drugs through the blood-brain barrier to treat the brain cancer. Applying his idea for the first time on Shay, Neuwelt threaded a catheter into one of his patient's carotid arteries, the vessels that supply the brain with blood. For thirty seconds, he pumped a concentrated solution of mannitol, a type of sugar, through the catheter, making the brain-bound blood very hypertonic. The sugar treatment was followed immediately by an infusion of potent chemotherapeutic drugs, and this time, unlike ever before, these cancer-killing agents were able to get through the blood-brain barrier.

How did the pretreatment with sugar accomplish this feat? The endothelial cells that form the walls of the brain capillaries are sealed together by tight junctions, which normally prevent substances from passing between the cells. Flooding the brain with hypertonic blood osmotically drew water from these endothelial cells. As the cells shrank, they pulled apart slightly. The chemotherapeutic drugs that followed the mannitol were able to pass through this transient breach in the blood-brain barrier and reach the cancer cells in Shay's brain. The endothelial cells quickly regained water and returned to their normal state as soon as fresh isotonic blood reached the brain, but meanwhile the life-saving drug had snuck through the barrier. The technique has been used on other brain lymphoma patients, over half of whom are alive today.

Nevertheless, this approach is not the final answer to penetrating the blood-brain barrier with chemotherapeutic and other drugs. For one thing, it is not risk-free. During the brief time the blood-brain barrier is breached, other chemicals that are usually barred from the brain can also enter. These undesirable chemicals include blood-borne hormones, such as epinephrine, that can act as neurotransmitters and set off unwanted nervous activity. Some patients undergoing this treatment have suffered seizures and developed more long-lasting neurological problems. Also, blood-borne infectious agents and toxins could enter the brain through the passages opened up for the chemotherapeutic drugs.

Other investigators are looking for alternative ways to smuggle neuropharmaceuticals across an undisturbed blood-brain barrier by capitalizing on the normal means by which substances get into the brain. As neuroscientists learn more about the brain, the possibility of developing drugs to treat a variety of brain disorders is continuing to expand. Yet these new neuropharmaceuticals are useless unless they can get into the brain. With passage between the endothelial cells blocked, substances normally get through the blood-brain barrier in one of two ways: (1) Lipid-soluble substances that must get into the brain, such as life-sustaining O_2 and regulatory steroid hormones, simply passively diffuse through the lipid plasma membranes of the brain endothelial cells. (2) Non-lipid-soluble substances that the brain needs for energy, such as glucose, and protein synthesis, such as amino acids, are transported across the brain endothelial cells by highly selective membrane-bound carriers. Twelve different specific transport systems have been identified to date.

Some foreign chemicals are already known to cross the blood-brain barrier. For example, agents such as nicotine, alcohol, and cocaine exert their effects so quickly because they are lipid soluble and pass through the blood-brain barrier with ease. Likewise, most of the drugs that are useful in treating mental illness, such as antidepressants, tranquilizers, and sedatives, are at least partially lipid soluble.

Scientists are now looking for ways to get non-lipid-soluble drugs across the barrier. These important compounds, which have largely been excluded from the brain, include antibiotics, chemotherapeutic agents, neurotransmitters, neuropeptides, nerve growth factor, and many hormones. New approaches being investigated include (1) "lipidizing" a water-soluble medication by chemically linking it to a fat group, (2) tricking one of the membrane-bound carriers to transport foreign therapeutic passengers, and (3) inducing endocytosis of drugs by brain endothelial cell membranes.

mild cases of dehydration. For example, dry skin and sunken eyeballs are indications of loss of H_2O from the underlying soft tissues, and the tongue becomes dry and parched because of suppressed salivary secretion.

ECF Hypotonicity Hypotonicity of the ECF is usually associated with **overhydration**; that is, excess free H_2O is present. When a positive free H_2O balance exists, the ECF is less concentrated (more dilute) than normal. Usually, any surplus free H_2O is promptly excreted in the urine, so hypotonicity generally does not occur. However, hypotonicity can arise in three ways: (1) Patients with renal failure who are unable to excrete a dilute urine become hypotonic when they consume relatively more H_2O than solutes. (2) Hypotonicity can occur transiently in healthy people if H_2O is rapidly ingested to such an excess that the kidneys are unable to respond quickly enough to eliminate the extra H_2O. (3) Hypotonicity can occur when excess H_2O without solute is retained in the body as a result of inappropriate secretion of vasopressin. Vasopressin is normally secreted in response to a H_2O deficit, which is relieved by increasing H_2O reabsorption in the distal portion of the nephrons. However, vasopressin secretion, and therefore hormonally controlled tubular H_2O reabsorption, can be increased in response to pain, acute infections, trauma, and other stressful situations, even when there is no H_2O deficit in the body. The increase in vasopressin secretion and resultant H_2O retention elicited in response to stress are appropriate in anticipation of potential blood loss in the stressful situation; the extra retained H_2O could minimize the effect a loss of blood volume would have on blood pressure. However, because modern-day stressful situations generally are not accompanied by blood loss, the increased vasopressin secretion is inappropriate as far as the body's fluid balance is concerned. The reabsorption and retention of too much H_2O lead to dilution of the body's solutes.

Whichever way it is brought about, excess free H_2O retention first dilutes the ECF compartment, making it hypotonic. The resultant difference in osmotic activity between the ECF and ICF induces H_2O to move by osmosis from the more dilute ECF into the cells, with the cells swelling as H_2O moves into them osmotically. Like the shrinking of cerebral neurons, pronounced swelling of brain cells also leads to brain dysfunction. Symptoms include confusion, irritability, lethargy, headache, dizziness, vomiting, drowsiness, and, in severe cases, even convulsions, coma, and death. Nonneural symptoms of overhydration include weakness caused by swelling of muscle cells and circulatory disturbances, including hypertension and edema, caused by expansion of the plasma volume.

The condition of overhydration, hypotonicity, and cellular swelling resulting from excess free H_2O retention is known as **water intoxication.** It should not be confused with the fluid retention that occurs with excess salt retention. In the latter case, the ECF is still isotonic because the increase in salt is matched by a corresponding increase in H_2O. Because the interstitial fluid is still isotonic, there is no osmotic gradient to drive the extra H_2O into the cells. The excess salt and H_2O burden is therefore confined to the ECF compartment, with

circulatory consequences being the most important concern. In the case of water intoxication, the body's H_2O content is increased relative to the salt content, and the excess H_2O becomes evenly distributed between the ECF and ICF compartments. In addition to any circulatory disturbances, symptoms caused by cellular swelling become a problem.

Isotonic Fluid Gain or Loss Let us contrast the situations of hypertonicity and hypotonicity with what happens as a result of isotonic fluid gain or loss. An example of an isotonic fluid gain is therapeutic intravenous administration of an isotonic solution, such as isotonic saline. When the isotonic fluid is injected into the ECF compartment, the ECF volume increases, but the concentration of ECF solutes remains unchanged; in other words, the ECF is still isotonic. Because the ECF's osmolarity has not changed, the ECF and ICF are still in osmotic equilibrium, so no net fluid shift occurs between the two compartments. The ECF compartment has increased in volume without causing a shift of H_2O into the cells. Thus, unless trying to correct an osmotic imbalance, intravenous fluid therapy should be isotonic to prevent fluctuations in intracellular volume and possible neural symptoms.

Similarly, in the case of an isotonic fluid loss such as occurs in hemorrhage, the loss is confined to the ECF with no corresponding loss of fluid from the ICF. Fluid does not shift out of the cells because the ECF remaining within the body is still isotonic, so there is no osmotic gradient to draw H_2O out of the cells. Of course, many other mechanisms come into play to counteract the loss of blood, but the ICF compartment is not directly affected by the loss.

Thus, when the ECF and ICF are in osmotic equilibrium, no net movement of H_2O into or out of the cells occurs regardless of whether the ECF volume increases or decreases. Shifts in H_2O between the ECF and ICF take place only when the ECF becomes more or less concentrated than the cells, and this usually arises from a loss or gain, respectively, of free H_2O.

Control of water balance by means of vasopressin and thirst is of primary importance in regulating ECF osmolarity.

Because cells, especially the brain neurons, do not function properly when they are either shrunk or swollen, it is important that ECF osmolarity be closely regulated to prevent osmotic fluid shifts between the ECF and ICF. Control of H_2O balance is crucial for regulating ECF osmolarity. Because increases in free H_2O cause the ECF to become too dilute and deficits of free H_2O cause the ECF to become too concentrated, the osmolarity of the ECF must be immediately corrected by restoring stable H_2O balance to avoid osmotically induced fluxes of H_2O into or out of the cells.

To maintain a stable H_2O balance, H_2O input must equal H_2O output. In a person's typical daily H_2O balance (Table 15–3), a little more than a liter of H_2O is added to the body by drinking liquids. Surprisingly, an amount almost equal to that is obtained from eating solid food. Recall that muscles consist of about 75% H_2O; meat is therefore 75% H_2O because it is animal muscle. Likewise, fruits and vegetables

TABLE 15-3 Daily Water Balance

WATER INPUT		WATER OUTPUT	
Avenue	**Quantity (ml/day)**	**Avenue**	**Quantity (ml/day)**
Fluid intake	1,250	Insensible loss (from lungs and non-sweating skin)	900
H_2O in food intake	1,000	Sweat	100
Metabolically produced H_2O	350	Feces	100
		Urine	1,500
Total input	2,600	Total output	2,600

consist of 60% to 90% H_2O. Therefore, people normally obtain almost as much H_2O from solid foods as from the liquids they drink. The third source of H_2O input is metabolically produced H_2O. Chemical reactions within the cells convert food and O_2 into energy, producing CO_2 and H_2O in the process. This **metabolic H_2O** produced during cellular metabolism and released into the ECF averages about 350 ml/day. The average H_2O intake from these three sources thus totals 2,600 ml/day. Another source of H_2O often employed therapeutically is intravenous infusion of fluid.

On the output side of the H_2O balance tally, a person loses close to a liter of H_2O daily without being aware of it. This so-called **insensible loss** occurs from the lungs and nonsweating skin. During the process of respiration, inspired air becomes saturated with H_2O within the airways. This H_2O is lost when the moistened air is subsequently expired. Normally, we are not aware of this H_2O loss, but it can be recognized on cold days when the H_2O vapor condenses so that we can "see our breath." The other insensible loss is the continual loss of H_2O from the skin even in the absence of sweating. Water molecules are able to diffuse through the cells of the skin and evaporate without being noticed. Fortunately, the skin is fairly waterproof because of its keratinized exterior layer, which protects against a much greater loss of H_2O by this avenue (see p. 406). When this protective surface layer is lost, such as when a person has extensive burns, fluid loss from the burned surface can increase to the point of causing serious fluid-balance problems.

Sensible loss (loss of which the person is aware) of H_2O from the skin occurs through sweating, which represents another avenue of H_2O output. At an air temperature of 68°F, an average of 100 ml of H_2O is lost daily through sweating. Loss of water from sweating can vary substantially, of course, depending on the environmental temperature and humidity and the degree of physical activity; it may range from zero up to as much as 2 liters per hour in very hot weather.

Another passageway for H_2O loss from the body is through the feces. Normally, only about 100 ml of H_2O are lost via this route each day. During the process of fecal formation in the large intestine, most of the H_2O is absorbed out of the digestive-tract lumen into the blood, thereby conserving fluid and solidifying the digestive tract's contents for elimination. Additional losses of H_2O can occur from the digestive tract through vomiting or diarrhea.

By far the most important output mechanism is urine excretion, with 1,500 ml of urine being produced daily on the average. The total H_2O output is 2,600 ml/day, the same as the volume of H_2O input in our example. This balance is not by chance. Normally, H_2O output matches H_2O intake so that the H_2O in the body remains in balance.

Of the many sources of H_2O input and output, only two can be regulated to maintain H_2O balance. On the intake side, thirst influences the amount of fluid ingested, and on the output side, the kidneys can adjust the magnitude of urine formation. Some of the other factors are regulated but not for the purpose of maintaining H_2O balance. Food intake is subject to regulation to maintain energy balance, whereas control of sweating is important in the maintenance of body temperature. Metabolic H_2O production and insensible losses are completely unregulated. Control of H_2O output in the urine is the most important mechanism in the control of H_2O balance.

Control of Water Input by Thirst **Thirst** is the subjective feeling that drives one to ingest H_2O. A **thirst center** is located in the hypothalamus in close proximity to the vasopressin-secreting cells. Thirst increases H_2O input, whereas vasopressin, by reducing urine production, decreases H_2O output.

Control of Water Output in the Urine by Vasopressin Fluctuations in ECF osmolarity caused by imbalances between H_2O input and output are quickly compensated for by adjusting the urinary excretion of H_2O without changing the usual excretion of salt; that is, H_2O reabsorption and excretion are partially dissociated from solute reabsorption and excretion, so the amount of free H_2O retained or eliminated can be varied to quickly restore ECF osmolarity to normal. Adjustments in free H_2O reabsorption and excretion are accomplished through changes in vasopressin secretion (see p. 497). Throughout most of the nephron, H_2O reabsorption is important in regulating ECF volume because salt reabsorption is accompanied by comparable H_2O reabsorption. In the distal portions of the tubule, however, variable free H_2O reabsorption can take place without comparable salt reabsorption because of the presence of a vertical osmotic gradient in the renal medulla to which the collecting tubule is exposed. Vasopressin increases the permeability of this late portion of the tubule to H_2O. Depending on the amount of vasopressin present, the amount of free H_2O reabsorbed can be adjusted as necessary to restore ECF osmolarity to normal.

Regulation of Vasopressin Secretion and Thirst Vasopressin secretion and thirst are both stimulated by a free H_2O deficit and suppressed by a free H_2O excess. Thus, appropriately, the same circumstances that call for a reduction in urinary output to conserve body H_2O also give rise to the sensation of thirst to replenish body H_2O. The predominant excitatory input for both vasopressin secretion and thirst comes from **hypothalamic osmoreceptors** located near the vasopressin-secreting cells and thirst center. These osmoreceptors monitor the osmolarity of the fluid surrounding them, which in turn reflects the concentration of the entire internal fluid environment. As the osmolarity increases (too little H_2O) and the need for H_2O conservation increases, vasopressin secretion and thirst are both stimulated (● Fig. 15–4). As a result, reabsorption of H_2O in the distal and collecting tubules is increased so that urinary output is reduced and H_2O is conserved, while H_2O intake is simultaneously encouraged. These actions restore depleted H_2O stores, thus relieving the hypertonic condition by diluting the solutes to normal concentration. On the other hand, H_2O excess, as manifested by a reduction in ECF osmolarity, brings about increased urinary output (through a decrease in vasopressin secretion) as well as suppression of thirst, which together reduce the water load in the body.

Even though the major stimulus for vasopressin secretion and thirst is an increase in ECF osmolarity, the vasopressin-secreting cells and thirst center are both influenced to a moderate extent by changes in ECF volume mediated by input from the **left atrial volume receptors**. Located in the left atrium, these volume receptors monitor the blood pressure, which is a reflection of the ECF volume. In response to a major reduction in ECF volume and arterial pressure, as during hemorrhage, the left atrial volume receptors reflexly stimulate both vasopressin secretion and thirst. The outpouring of vasopressin and increased thirst lead to decreased urine output and increased fluid intake, respectively. Furthermore, vasopressin, at the circulating levels elicited by a large decline in ECF volume and arterial pressure, exerts a potent vasoconstrictor effect on arterioles (thus giving rise to its name), in

● FIGURE 15–4   Control of
**Increased Vasopressin Secretion and
Thirst during H₂O Deficit**

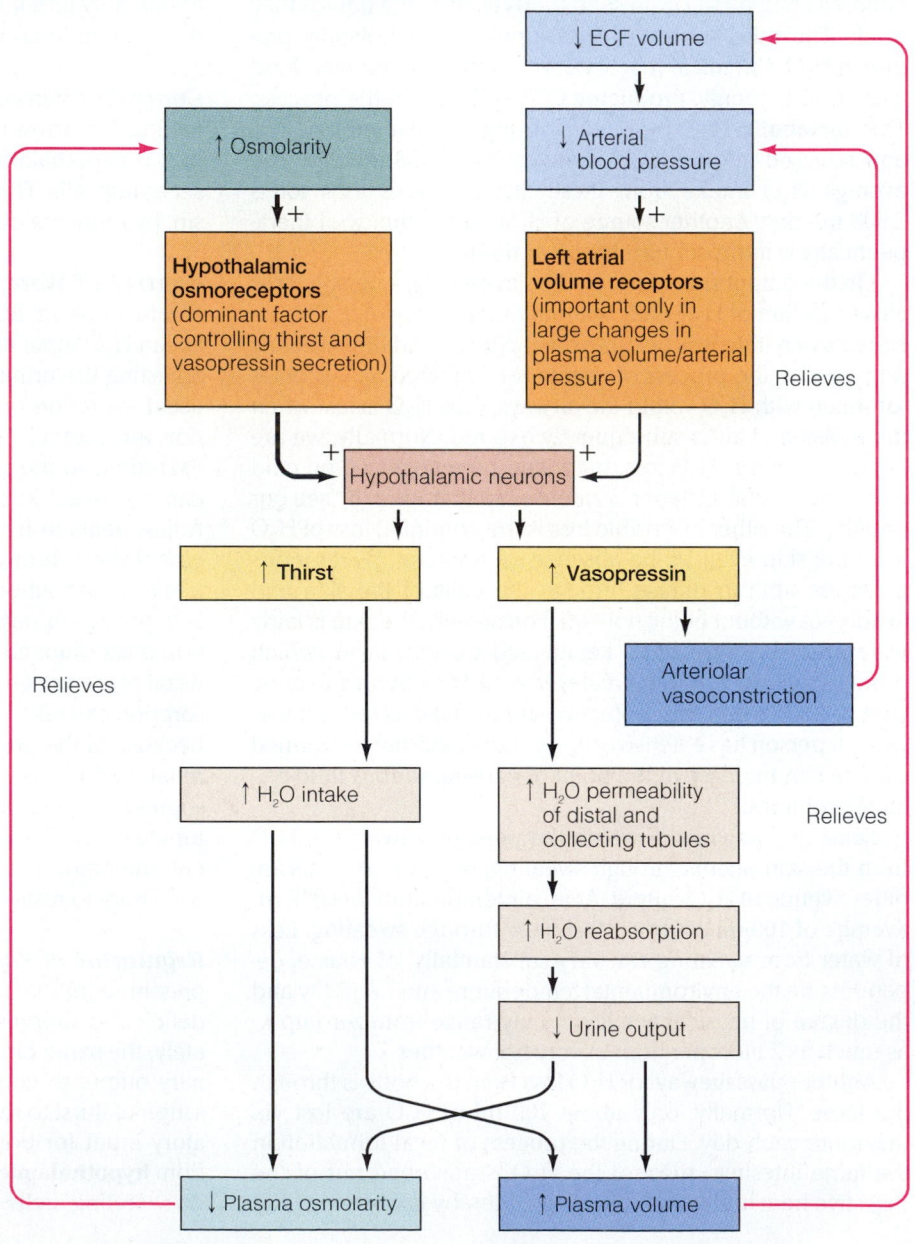

addition to having an effect on the kidney tubules. Both by helping expand the ECF and plasma volume and by increasing the total peripheral resistance, vasopressin helps relieve the low blood pressure that elicited the vasopressin secretion. Conversely, vasopressin and thirst are both inhibited when the ECF/plasma volume and arterial blood pressure are elevated. The resultant suppression of H_2O intake, coupled with the elimination of the excess ECF/plasma volume in the urine, helps to restore the blood pressure to normal.

Recall that low ECF/plasma volume and low arterial blood pressure also reflexly increase aldosterone secretion. The resultant increase in Na^+ reabsorption ultimately leads to osmotic retention of H_2O, expansion of ECF volume, and an increase in arterial blood pressure. In fact, aldosterone-controlled Na^+ reabsorption is the most important factor in regulating ECF volume, with the vasopressin and thirst mechanism playing only a supportive role. Baroreceptor input to vasopressin secretion and the thirst center becomes important only in response to fairly serious changes in ECF/plasma volume and arterial pressure. Osmoreceptor input is normally the dominant factor controlling vasopressin secretion and thirst.

Yet another stimulus for increasing both thirst and vasopressin is angiotensin II (Table 15–4). When the renin-angiotension-aldosterone mechanism is activated to conserve Na^+, angiotensin II, in addition to stimulating aldosterone secretion, acts directly on the brain to give rise to the urge to drink and concurrently stimulates vasopressin to enhance renal H_2O reabsorption. The resultant increased H_2O intake and decreased urinary output help to correct the reduction in ECF volume that triggered the renin-angiotensin-aldosterone system.

Several factors affect vasopressin secretion but not thirst. As described earlier, vasopressin is stimulated by stress-related inputs such as pain, fear, and trauma that have noth-

ing directly to do with maintaining H_2O balance. In fact, H_2O retention as a result of the inappropriate secretion of vasopressin can bring about a hypotonic H_2O imbalance. On the other hand, alcohol inhibits vasopressin secretion and can lead to ECF hypertonicity by promoting excessive free H_2O excretion.

One stimulus that promotes thirst but not vasopressin secretion is a direct effect of dryness of the mouth. Nerve endings in the mouth are directly stimulated by dryness, which causes an intense sensation of thirst that can often be relieved merely by moistening the mouth even though no H_2O is actually ingested. A dry mouth can exist when salivation is suppressed by factors unrelated to the body's H_2O content, such as nervousness, excessive smoking, or certain drugs.

Factors that affect vasopressin secretion or thirst but have nothing directly to do with the body's need for H_2O are usually short-lived. It is important to recognize that the dominant, long-standing control of vasopressin and thirst *is* directly correlated with the body's state of H_2O—namely, by the status of ECF osmolarity and, to a lesser extent, by ECF volume.

In addition, there is evidence of some kind of "oral H_2O metering," at least in animals. A thirsty animal will rapidly drink only enough H_2O to satisfy its H_2O deficit. It stops drinking before the ingested H_2O has had time to be absorbed from the digestive tract and actually return the ECF compartment to normal. Exactly what factors are involved in signaling that enough H_2O has been consumed is still uncertain. It might be a learned anticipatory response based on past experience. This mechanism seems to be less effective in humans, because we frequently drink more than is necessary to meet the needs of our bodies or, conversely, may not drink enough to make up a deficit.

In fact, even though the thirst mechanism exists to control H_2O intake, fluid consumption by humans is often influenced more by habit and sociological factors than by the need to

FACTOR	EFFECT ON VASOPRESSIN SECRETION	EFFECT ON THIRST	COMMENT
↑ *ECF osmolarity*	↑	↑	Major stimulus for vasopressin secretion and thirst
↓ *ECF volume*	↑	↑	Important only in large changes in ECF volume/arterial blood pressure
Angiotensin II	↑	↑	Part of dominant pathway for promoting compensatory salt and H_2O retention when ECF volume/arterial blood pressure are reduced
Pain, fear, trauma and other stress-related inputs	Inappropriate ↑ unrelated to body's H_2O balance	No effect	Promotes excess H_2O retention and ECF hypotonicity (resultant H_2O retention of potential value in maintaining arterial blood pressure in case of blood loss in the stressful situation)
Alcohol	Inappropriate ↓ unrelated to body's H_2O balance	No effect	Promotes excess H_2O loss and ECF hypertonicity
Dry mouth	No effect	↑	Nerve endings in mouth that ultimately give rise to sensation of thirst are directly stimulated by dryness

TABLE 15–4 Factors Controlling Vasopressin Secretion and Thirst

regulate H_2O balance. Thus, even though H_2O intake is critical in maintaining fluid balance, it is not precisely controlled in humans, who err especially on the side of excess H_2O consumption. We usually drink when we are thirsty, but we often drink even when we are not thirsty because, for example, we are on a coffee break. With H_2O intake being inadequately controlled and indeed even contributing to H_2O imbalances in the body, the primary factor involved in maintaining H_2O balance is urinary output regulated by the kidneys. Accordingly, vasopressin-controlled H_2O reabsorption is of primary importance in regulating ECF osmolarity.

Just as vasopressin secretion is the major factor controlling ECF osmolarity but plays a supportive role in controlling ECF volume, aldosterone secretion is the major factor controlling ECF volume but plays a supportive role in controlling ECF osmolarity. Researchers have recently learned that increased osmolarity accompanying H_2O deprivation acts directly on the adrenal cortex to reduce aldosterone secretion, thus decreasing renal Na^+ reabsorption and increasing urinary excretion of Na^+. It is proposed that this controlling mechanism, which is in addition to the renin-angiotensin pathway and the direct action of plasma K^+ concentration on aldosterone secretion (see Fig. 14–22, p. 489), may be important in reducing a detrimental osmotic flux of H_2O out of the cells in the face of prolonged H_2O deprivation and elevated ECF osmolarity.

ACID-BASE BALANCE

Acids liberate free hydrogen ions whereas bases accept them.

Acid-base balance actually refers to the precise regulation of **free** (that is, unbound) **hydrogen-ion (H^+)** concentration in the body fluids. A [ ] around a chemical symbol refers to the concentration of that chemical. Thus, $[H^+]$ designates H^+ concentration.

Acids are a special group of hydrogen-containing substances that *dissociate,* or separate, when in solution to liberate free H^+ and anions (negatively charged ions). Many other substances (for example, carbohydrates) also contain hydrogen, but they are not classified as acids because the hydrogen is tightly bound within their molecular structure and is never liberated as free H^+.

A strong acid has a greater tendency to dissociate in solution than a weak acid does; that is, a greater percentage of a strong acid's molecules separate into free H^+ and anions. Hydrochloric acid (HCl) is an example of a strong acid; every HCl molecule dissociates into free H^+ and Cl^- (chloride) when dissolved in H_2O. With a weaker acid such as carbonic acid (H_2CO_3), only a portion of the molecules dissociate in solution into H^+ and HCO_3^- (bicarbonate anions). The remaining H_2CO_3 molecules remain intact. Since it is only the free hydrogen ions that contribute to the acidity of a solution, H_2CO_3 is a weaker acid than HCl because H_2CO_3 does not yield as many free hydrogen ions per number of acid molecules present in solution (Fig. 15–5).

The extent of dissociation for a given acid is always constant; that is, when in solution, the same proportion of a par-

◖ = Free H^+ ◖ = Undissociated acid

▷ = Free anion

FIGURE 15–5 Comparison of a Strong and a Weak Acid (a) Five molecules of a strong acid. A strong acid such as HCl (hydrochloric acid) completely dissociates into free H^+ and anions in solution. (b) Five molecules of a weak acid. A weak acid such as H_2CO_3 (carbonic acid) only partially dissociates into free H^+ and anions in solution.

ticular acid's molecules always separate to liberate free H^+, with the other portion always remaining intact. The constant degree of dissociation for a particular acid (in this example H_2CO_3) is expressed by its **dissociation constant (K)** as follows:

$$[H^+]\,[HCO_3^-]/[H_2CO_3] = K$$

■ $[H^+]\,HCO_3^-]$ represents the concentration of ions resulting from H_2CO_3 dissociation.

■ $[H_2CO_3]$ represents the concentration of intact (undissociated) H_2CO_3.

The dissociation constant varies for different acids.

A **base** is a substance that can combine with a free H^+ and thus remove it from solution. A strong base is able to bind H^+ more readily than a weak base can.

The pH designation is used to express hydrogen-ion concentration.

The $[H^+]$ in the ECF is normally 4×10^{-8} or 0.00000004 equivalents per liter. The concept of pH has been developed to express $[H^+]$ more conveniently. Specifically, **pH** equals the logarithm (log) to the base 10 of the reciprocal of the hydrogen-ion concentration:

$$pH = \log 1/[H^+]$$

Two important points should be noted about this formula:

1. Because $[H^+]$ is in the denominator, *a high $[H^+]$ corresponds to a low pH, and a low $[H^+]$ corresponds to a high pH.* The greater the $[H^+]$, the larger the number by which 1 must be divided, and the lower the pH.

2. *Every unit change in pH actually represents a tenfold change in $[H^+]$* because of the logarithmic relationship. A log to the base 10 indicates how many times 10 must be multiplied by itself to produce a given number. For example, the log of 10 = 1, whereas the log of 100 = 2. Ten must be multiplied by itself twice to yield 100 ($10 \times 10 = 100$). Numbers less than 10 have logs less than 1. Numbers between 10 and 100 have logs between 1 and 2, and so on. Accordingly, each unit of change in pH is indicative of a tenfold change in $[H^+]$. For example, a solution with a pH of 7 has a $[H^+]$ ten times less than that of a solution with a pH of 6 (1 pH-unit difference) and one hundred times less than that of a solution with a pH of 5 (2 pH-unit difference).

The pH of pure H_2O is 7.0, which is considered to be chemically neutral. An extremely small proportion of H_2O molecules dissociate into hydrogen ions and hydroxyl (OH^-) ions. Because OH^- has the ability to bind with H^+ to once again form a H_2O molecule, it is considered to be basic. Since an equal number of acidic hydrogen ions and basic hydroxyl ions are formed, H_2O is neutral, being neither acidic nor basic. Solutions having a pH less than 7.0 contain a higher $[H^+]$ than pure H_2O and are considered to be **acidic**. Conversely, solutions having a pH value greater than 7.0 have a lower $[H^+]$ and are considered to be **basic** or **alkaline** (●Fig. 15–6a).

The pH of arterial blood is normally 7.45 and the pH of venous blood is 7.35, for an average blood pH of 7.4. The pH of venous blood is slightly lower than that of arterial blood because of H^+ generated by the formation of H_2CO_3 from CO_2 picked up at the tissue capillaries. **Acidosis** exists whenever the blood pH falls below 7.35, whereas **alkalosis** occurs when the blood pH is above 7.45 (Fig. 15–6b). Note that the reference point for determining the body's acid-base status is not the chemically neutral pH of 7.0 but the normal plasma pH of 7.4. Thus, a plasma pH of 7.2 is considered acidotic even though in chemistry a pH of 7.2 is considered basic.

Death occurs if arterial pH falls outside the range of 6.8 to 8.0 for more than a few seconds, because an arterial pH of less than 6.8 or greater than 8.0 is not compatible with life. Obviously, therefore, $[H^+]$ in the body fluids must be carefully regulated.

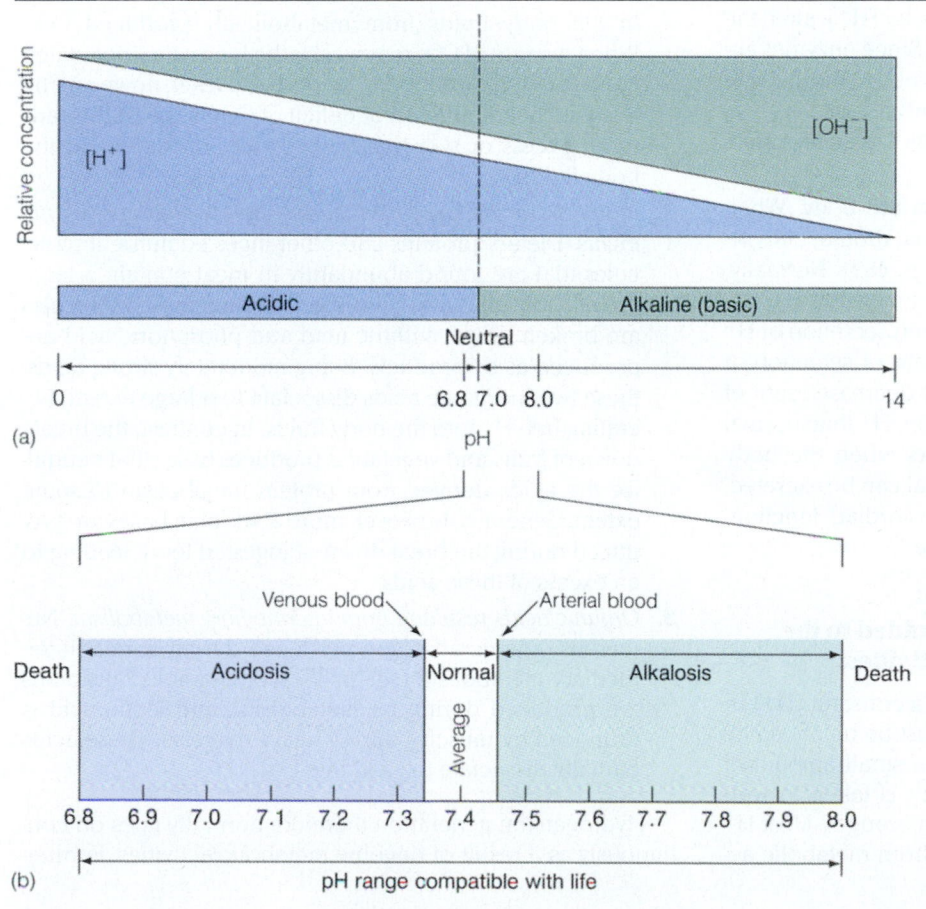

●**FIGURE 15–6 pH Considerations in Chemistry and Physiology**
(a) Relationship of pH to relative concentrations of H^+ and base under chemically neutral, acidic, and alkaline conditions. (b) Plasma pH range under normal, acidosis, and alkalosis conditions.

Fluctuations in hydrogen-ion concentration have profound effects on body chemistry.

Only a narrow pH range is compatible with life because even small changes in [H+] have dramatic effects on normal cell function. The prominent consequences of fluctuations in [H+] include the following:

- Changes in excitability of nerve and muscle cells are among the major clinical manifestations of pH abnormalities. The major clinical effect of increased [H+] (acidosis) is depression of the central nervous system. Acidotic patients become disoriented and, in more severe cases, eventually die in a state of coma. In contrast, the major clinical effect of decreased [H+] (alkalosis) is overexcitability of the nervous system, first the peripheral nervous system and later the central nervous system. Peripheral nerves become so excitable that they fire even in the absence of normal stimuli. Such overexcitability of the afferent (sensory) nerves gives rise to abnormal "pins and needles" tingling sensations. On the other hand, overexcitability of efferent (motor) nerves brings about muscle twitches and, in more pronounced cases, severe muscle spasms. Death may occur in extreme alkalosis as spasm of the respiratory muscles seriously impairs breathing. Alternatively, severely alkalotic patients may die of convulsions resulting from overexcitability of the central nervous system. In less serious situations, CNS overexcitability is manifested as extreme nervousness.

- Hydrogen-ion concentration exerts a marked influence on enzyme activity. Even slight deviations in [H+] alter the shape and activity of protein molecules. Since enzymes are proteins, a shift in the body's acid-base balance disturbs the normal pattern of metabolic activity catalyzed by these enzymes. Some cellular chemical reactions are accelerated; others are depressed.

- Changes in [H+] influence K+ levels in the body. When reabsorbing Na+ from the filtrate, the renal tubular cells secrete either K+ or H+ in exchange (see p. 489). Normally, they secrete a preponderance of K+ compared to H+. Because of the intimate relationship between secretion of H+ and K+ by the kidneys, an increased rate of secretion of one of these ions is accompanied by a decreased rate of secretion of the other. For example, if more H+ than normal is eliminated by the kidneys, as occurs when the body fluids become acidotic, less K+ than usual can be excreted. The resultant K+ retention can affect cardiac function, among other detrimental consequences.

Hydrogen ions are continually being added to the body fluids as a result of metabolic activities.

As with any other constituent, to maintain a constant [H+] in the body fluids, input of hydrogen ions must be balanced by an equal output. On the input side, only a small amount of acid capable of dissociating to release H+ is taken in with food, such as the weak citric acid found in oranges. Most H+ in the body fluids is generated internally from metabolic ac-

tivities. Normally, H+ is continually being added to the body fluids from the three following sources:

1. *Carbonic acid formation.* The major source of H+ is through H_2CO_3 formation from metabolically produced CO_2. Cellular oxidation of nutrients yields energy, with CO_2 and H_2O as end products. Catalyzed by the enzyme *carbonic anhydrase* (ca), CO_2 and H_2O form H_2CO_3, which then partially dissociates to liberate free H+ and HCO_3^-:

$$CO_2 + H_2O \overset{ca}{\rightleftharpoons} H_2CO_3 \rightleftharpoons H^+ + HCO_3^-$$

This reaction is reversible because it can proceed in either direction, depending on the concentrations of the substances involved as dictated by the *law of mass action* (see p. 443). Within the systemic capillaries, the CO_2 level in the blood increases as metabolically produced CO_2 enters from the tissues. This drives the reaction to the acid side, generating H+ as well as HCO_3^- in the process. In the lungs, the reaction is reversed: CO_2 diffuses from the blood flowing through the pulmonary capillaries into the alveoli (air sacs), from which it is expired to the atmosphere. The resultant reduction in CO_2 in the blood drives the reaction toward the CO_2 side. Hydrogen ion and HCO_3^- form H_2CO_3, which rapidly decomposes into CO_2 and H_2O once again. The CO_2 is exhaled while the hydrogen ions generated at the tissue level are incorporated into H_2O molecules.

When the respiratory system is able to keep pace with the rate of metabolism, there is no net gain or loss of H+ in the body fluids from metabolically produced CO_2. When the rate of CO_2 removal by the lungs does not match the rate of CO_2 production at the tissue level, however, the resultant accumulation or deficit of CO_2 in the body leads to an excess or shortage, respectively, of free H+ in the body fluids.

2. *Inorganic acids produced during the breakdown of nutrients.* Dietary proteins and other ingested nutrient molecules that are found abundantly in meat contain a large quantity of sulfur and phosphorus. When these molecules are broken down, sulfuric acid and phosphoric acid are produced as by-products. Being moderately strong acids, these two inorganic acids dissociate to a large extent, liberating free H+ into the body fluids. In contrast, the breakdown of fruits and vegetables produces bases that neutralize the acids derived from protein metabolism to some extent. Generally, however, more acids than bases are produced during the breakdown of ingested food, leading to an excess of these acids.

3. *Organic acids resulting from intermediary metabolism.* Numerous organic acids are produced during normal intermediary metabolism (see p. 35). For example, fatty acids are produced during fat metabolism, and lactic acid is produced by muscles during heavy exercise. These acids partially dissociate to yield free H+.

Hydrogen-ion generation therefore normally goes on continuously as a result of ongoing metabolic activities. Furthermore, in certain disease states, additional acids may be produced that further contribute to the total body pool of H⁺. For example, in diabetes mellitus, large quantities of keto acids may be produced as a result of abnormal fat metabolism. Some types of acid-producing medications may also add to the total load of H^+ that must be handled by the body. Thus, input of H^+ is unceasing, highly variable, and essentially unregulated.

The crux of H^+ balance is maintaining the normal alkalinity of the ECF (pH 7.4) despite this constant onslaught of acid. The generated free H^+ must be largely removed from solution while in the body and ultimately must be eliminated from the body so that the pH of the body fluids can remain within the narrow range compatible with life. Mechanisms must also exist to compensate rapidly for the occasional situation in which the ECF becomes too alkaline.

Three lines of defense against changes in $[H^+]$ operate to maintain the $[H^+]$ of body fluids at a nearly constant level despite unregulated input: (1) the *chemical buffer systems,* (2) the *respiratory mechanism of pH control,* and (3) the *renal mechanism of pH control.* We will look at each of these methods.

Chemical buffer systems act as the first line of defense.

A **chemical buffer system** is a mixture in a solution of two (or perhaps more) chemical compounds that minimize pH changes when either an acid or a base is added to or removed from the solution. A buffer system consists of a pair of substances involved in a reversible reaction—one substance that can yield free H^+ as the $[H^+]$ starts to fall and another that can bind with free H^+ (thus removing it from solution) when $[H^+]$ starts to rise. An important example of such a buffer system is the carbonic acid:bicarbonate (H_2CO_3:HCO_3^-) buffer pair, which is involved in the following reversible reaction:

$$H^+ + HCO_3^- \rightleftharpoons H_2CO_3$$

When a strong acid such as HCl is added to an unbuffered solution, all the dissociated H^+ remains free in the solution (●Fig. 15–7a). In contrast, when HCl is added to a solution containing the H_2CO_3:HCO_3^- buffer pair (Fig. 15–7b), the HCO_3^- immediately binds with the free H^+ to form H_2CO_3. The resultant weak H_2CO_3 dissociates only slightly compared to the marked reduction in pH that occurred when the buffer system was not present and the additional H^+ remained un-

Unbuffered NaCl solution — 3 free H⁺ present

(a)

Solution containing H₂CO₃: HCO₃⁻ buffer — 1 free H⁺ present

(b)

● **FIGURE 15–7 Action of Chemical Buffers** (a) Addition of HCl to an unbuffered solution. All of the added hydrogen ions (H⁺) remain free and contribute to the acidity of the solution. (b) Addition of HCl to a buffered solution. Bicarbonate ions (HCO₃⁻), the basic member of the buffer pair, bind with some of the added H⁺ and remove them from solution so that they do not contribute to its acidity.

TABLE 15–5 Chemical Buffers and Their Primary Roles	
BUFFER SYSTEM	**MAJOR FUNCTIONS**
Carbonic acid: bicarbonate buffer system	Primary ECF buffer against non-carbonic-acid changes
Protein buffer system	Primary ICF buffer; also buffers ECF
Hemoglobin buffer system	Primary buffer against carbonic acid changes
Phosphate buffer system	Important urinary buffer; also buffers ICF

bound. On the other hand, when the pH of the solution starts to rise because of the addition of base or loss of acid, the H^+-yielding member of the buffer pair, H_2CO_3, releases H^+ to minimize the rise in pH.

There are four buffer systems in the body: (1) the H_2CO_3:HCO_3^- buffer system, (2) the protein buffer system, (3) the hemoglobin buffer system, and (4) the phosphate buffer system. Each serves a different important role (Table 15–5).

The H_2CO_3:HCO_3^- Buffer Pair Is the Primary ECF Buffer for Non-Carbonic Acids

The H_2CO_3:HCO_3^- buffer pair is the most important buffer system in the ECF for buffering pH changes brought about by causes other than fluctuations in CO_2-generated H_2CO_3. It is a very effective ECF buffer system for two reasons. First, H_2CO_3 and HCO_3^- are abundant in the ECF, so this system is readily available to resist changes in pH. Second and more importantly, each component of this buffer pair is closely regulated. The kidneys regulate HCO_3^-, whereas the respiratory system regulates CO_2, which generates H_2CO_3. Thus, in the body the H_2CO_3:HCO_3^- buffer system includes involvement of CO_2 via the following reaction with which you are already familiar:

$$H^+ + HCO_3^- \rightleftharpoons H_2CO_3 \rightleftharpoons CO_2 + H_2O$$

When new H^+ is added to the plasma from any source other than CO_2 (for example, through lactic acid released into the ECF from exercising muscles), the preceding reaction is driven toward the right side of the equation. As the extra H^+ binds with HCO_3^-, it no longer contributes to the acidity of the body fluids, so the rise in $[H^+]$ is abated. In the converse situation, when the plasma $[H^+]$ occasionally falls below normal for some reason other than a change in CO_2 (such as the loss of plasma-derived HCl in the gastric juices during vomiting), the reaction is driven toward the left side of the equation. Dissolved CO_2 and H_2O in the plasma form H_2CO_3, which generates additional H^+ to make up for the H^+ deficit. In so doing, the H_2CO_3:HCO_3^- buffer system resists the fall in $[H^+]$.

This buffer system cannot buffer changes in pH induced by fluctuations in H_2CO_3. A buffer system cannot buffer itself. Consider, for example, the situation in which the plasma $[H^+]$ is elevated because of CO_2 retention associated with a respiratory problem. The rise in CO_2 drives the reaction to the left according to the law of mass action, resulting in an elevation

in $[H^+]$. The increase in $[H^+]$ occurs as a result of the reaction being driven to the *left* because of an increase in CO_2, so the elevated $[H^+]$ cannot drive the reaction to the *right* to buffer the increase in $[H^+]$. Only if the increase in $[H^+]$ is brought about by some mechanism other than CO_2 accumulation can this buffer system be shifted to the CO_2 side of the equation and be effective in reducing the $[H^+]$. Likewise, in the opposite situation, the H_2CO_3:HCO_3^- buffer system cannot compensate for a reduction in $[H^+]$ caused by a deficit of CO_2 by generating more H^+-yielding H_2CO_3 when the problem in the first place is a shortage of H_2CO_3-forming CO_2. Other mechanisms, to be described shortly, are available for resisting fluctuations in pH caused by changes in CO_2 levels.

The relationship between $[H^+]$ and the members of a buffer pair can be expressed according to the **Henderson-Hasselbalch equation**, which, for the H_2CO_3:HCO_3^- buffer system, is as follows:

$$pH = pK + \log [HCO_3^-]/H_2CO_3]$$

Although you do not need to know the mathematical manipulations involved, it will be helpful for you to understand how this formula is derived. Recall that the dissociation constant K for H_2CO_3 acid is

$$[H^+] [HCO_3^-]/[H_2CO_3] = K$$

and that the relationship between pH and $[H^+]$ is

$$pH = \log l/[H^+]$$

Then by solving the dissociation constant formula for $[H^+]$ (that is, $[H^+] = K \times [H_2CO_3]/[HCO_3^-]$) and replacing this value for $[H^+]$ in the pH formula, one comes up with the Henderson-Hasselbalch equation.

Practically speaking, $[H_2CO_3]$ is a direct reflection of the concentration of dissolved CO_2, henceforth referred to as $[CO_2]$, because most of the CO_2 in the plasma is converted into H_2CO_3. (The dissolved CO_2 concentration is equivalent to P_{CO_2}, as described in the chapter on respiration.) Therefore, the equation becomes

$$pH = pK + \log [HCO_3^-]/[CO_2]$$

The pK is the logarithm of l/K and, like K, always remains a constant for any given acid. For H_2CO_3, the pK is 6.1. Since the pK is always a constant, changes in pH are associated with changes in the ratio between $[HCO_3^-]$ and $[CO_2]$. Normally, the ratio between $[HCO_3^-]$ and $[CO_2]$ in the ECF is 20 to 1; that is, there is twenty times more HCO_3^- than CO_2. Plugging this ratio into our formula:

$$pH = pK + \log [HCO_3^-]/[CO_2]$$
$$= 6.1 + \log 20/1$$

The log of 20 is 1.3. Therefore, pH = 6.1 + 1.3 = 7.4 which is the normal pH of plasma.

When the ratio of $[HCO_3^-]$ to $[CO_2]$ increases above 20/1, pH increases. Accordingly, either a rise in $[HCO_3^-]$ or a fall in $[CO_2]$, both of which increase the $[HCO_3^-]/[CO_2]$ ratio if the other component remains constant, shifts the acid-base balance toward the alkaline side. In contrast, when the $[HCO_3^-]/[CO_2]$ ratio decreases below 20/1, pH decreases toward the acid side. This can occur either if the $[HCO_3^-]$ decreases or if the $[CO_2]$ increases while the other component remains constant. Because $[HCO_3^-]$ is regulated by the kidneys and $[CO_2]$ by the lungs, the pH of the plasma can be shifted up and down by kidney and lung influences. Renal and respiratory regulation of pH operate to correct changes in $[H^+]$ largely through the kidneys' and lungs' control of plasma $[HCO_3^-]$ and $[CO_2]$, respectively, to restore the ratio to normal. Accordingly,

$$pH \propto \frac{[HCO_3^-] \text{ controlled by kidney function}}{[CO_2] \text{ controlled by respiratory function}}$$

Because of this relationship, not only do both the kidneys and lungs normally participate in pH control, but renal or respiratory dysfunction can also induce acid-base imbalances by altering the $[HCO_3^-]/[CO_2]$ ratio.

The Protein Buffer System Is Primarily Important Intracellularly The most plentiful buffers of the body fluids are the proteins, including the intracellular proteins and the plasma proteins. Proteins contain both acidic and basic groups that can give up or take up H^+, thus serving as excellent buffers. Quantitatively, the protein system is most important in buffering changes in $[H^+]$ in the ICF because of the sheer abundance of the intracellular proteins. A more limited number of plasma proteins reinforces the $H_2CO_3:HCO_3^-$ system in extracellular buffering.

The Hemoglobin Buffer System Buffers Carbonic-Acid–Generated Hydrogen Ion Hemoglobin (Hb) buffers the H^+ generated from metabolically produced CO_2 in transit between the tissues and lungs. At the systemic capillary level, CO_2 continuously diffuses into the blood from the tissue cells where it is being produced. The greatest percentage of this CO_2 forms H_2CO_3, which partially dissociates into H^+ and HCO_3^-. Simultaneously, some of the oxyhemoglobin (HbO_2), releases O_2, which diffuses into the tissues. Reduced (unoxygenated) Hb has a greater affinity for H^+ than HbO_2 does. Therefore, most of the H^+ generated from CO_2 at the tissue level becomes bound to reduced Hb and no longer contributes to the acidity of the body fluids:

$$H^+ + Hb \rightleftharpoons HHb$$

At the lungs the reactions are reversed. As Hb picks up O_2 diffusing from the alveoli into the red blood cells, the affinity of Hb for H^+ is decreased, so H^+ is released. This liberated H^+ combines with HCO_3^- to yield H_2CO_3, which in turn produces CO_2, which is exhaled. Meanwhile, the hydrogen has been reincorporated into neutral H_2O molecules. Were it not for Hb, the blood would become much too acidic after picking up CO_2 at the tissues. With the tremendous buffering capacity of the Hb system, venous blood is only slightly more acidic than arterial blood despite the large volume of H^+-generating CO_2 carried in the venous blood.

The Phosphate Buffer System Is an Important Urinary Buffer The phosphate buffer system is composed of an acid phosphate salt (NaH_2PO_4) that can donate a free H^+ when the $[H^+]$ falls and a basic phosphate salt (Na_2HPO_4) that can accept a free H^+ when the $[H^+]$ rises. Basically, this buffer pair can alternately switch a H^+ for a Na^+ as demanded by the $[H^+]$:

$$Na_2HPO_4 + H^+ \rightleftharpoons NaH_2PO_4 + Na^+$$

Even though the phosphate pair is a good buffer, its concentration in the ECF is rather low, so it is not very important as an ECF buffer. Because phosphates are more abundant within the cells, this system contributes significantly to intracellular buffering, being rivaled only by the more plentiful intracellular proteins. Even more importantly, the phosphate system serves as an excellent urinary buffer. Humans normally consume more phosphate than needed. The excess phosphate filtered through the kidneys is not reabsorbed but remains in the tubular fluid to be excreted. This excreted phosphate buffers the urine as it is being formed by removing from solution the H^+ secreted into the tubular fluid. None of the other body-fluid buffer systems are present in the tubular fluid to play a role in buffering urine during its formation. Most or all of the filtered HCO_3^- and CO_2 (alias H_2CO_3) are reabsorbed, whereas Hb and plasma proteins are not even filtered.

All chemical buffer systems act immediately, within fractions of a second, to minimize changes in pH. When $[H^+]$ is altered, the involved buffer systems' reversible chemical reactions are shifted at once in favor of compensating for the change in $[H^+]$. Accordingly, the buffer systems are the *first line of defense* against changes in $[H^+]$ since they are the first mechanism to respond.

Through the mechanism of buffering, most hydrogen ions seem to "disappear" from the body fluids between the time of their generation and elimination. It must be emphasized, however, that none of the chemical buffer systems actually *eliminates* H^+ from the body. These ions are merely removed from solution by being incorporated within one of the members of the buffer pair, thus preventing the hydrogen ions from contributing to the body fluids' acidity. Because each buffer system has a limited capacity to "soak up" H^+, the H^+ that is unceasingly produced must ultimately be removed from the body. If H^+ were not eventually eliminated, soon all the body-fluid buffers would already be bound with H^+ and there would be no further buffering ability.

The respiratory and renal mechanisms of pH control actually eliminate acid from the body instead of merely suppressing it, but they respond more slowly than the chemical buffer systems. We will now turn our attention to these other defenses against changes in acid-base balance.

The respiratory system regulates hydrogen-ion concentration by controlling the rate of CO_2 removal from the plasma through adjustments in pulmonary ventilation.

The respiratory system plays an important role in acid-base balance through its ability to alter pulmonary ventilation and, consequently, to alter the excretion rate of H^+-generating CO_2. Because of this ability, it should not be surprising that

TABLE 15–6   Respiratory Adjustments to Acidosis and Alkalosis Induced by Nonrespiratory Causes

RESPIRATORY COMPENSATIONS	ACID-BASE STATUS		
	Normal (pH 7.4)	Nonrespiratory acidosis (pH 7.1)	Nonrespiratory alkalosis (pH 7.7)
Spirogram records at various pHs	$\wedge\!\wedge\!\wedge\!\wedge$	$\|\|\|\|\|\|$	$\sim\!\sim$
Respiratory rate	Normal	↑	↓
Tidal volume	Normal	↑	↓
Ventilation	Normal	↑	↓
Rate of CO_2 removal	Normal	↑	↓
Rate of carbonic-acid formation	Normal	↓	↑
Rate of H^+ generation from CO_2	Normal	↓	↑

the level of respiratory activity is governed at least in part by the arterial $[H^+]$ (Table 15–6). When arterial $[H^+]$ increases, the respiratory center in the brain stem is reflexly stimulated (see p. 456) to increase pulmonary ventilation (the rate at which gas is exchanged between the lungs and the atmosphere). As the rate and depth of breathing increase, more CO_2 than usual is blown off, so less H_2CO_3 than normal is added to the body fluids. Since CO_2 forms acid, the removal of CO_2 in essence removes acid from the body. Conversely, when arterial $[H^+]$ falls, pulmonary ventilation is reduced. As a result of slower, shallower breathing, metabolically produced CO_2 diffuses from the cells into the blood faster than it is removed from the blood by the lungs, so higher-than-usual amounts of acid-forming CO_2 accumulate in the blood, thus restoring $[H^+]$ toward normal.

The lungs are extremely important in maintaining the $[H^+]$ of the plasma. Every day they remove from the body fluids what amounts to one hundred times more H^+ derived from carbonic acid than the kidneys remove from non-carbonic-acid sources. Furthermore, the respiratory system, through its ability to regulate arterial $[CO_2]$, can adjust the amount of H^+ added to the body fluids from this source as needed to restore pH toward normal when fluctuations in $[H^+]$ from non-carbonic-acid sources occur.

Respiratory regulation acts at a moderate speed, coming into play only when the chemical buffer systems alone are unable to minimize $[H^+]$ changes. When deviations in $[H^+]$ occur, the buffer systems respond immediately, whereas adjustments in ventilation require a few minutes to be initiated. If a deviation in $[H^+]$ is not swiftly and completely corrected by the buffer systems, the respiratory system comes into action a few minutes later, thus serving as the *second line of defense* against changes in $[H^+]$.

Of course, when changes in $[H^+]$ occur because of fluctuations in $[CO_2]$ that arise from respiratory abnormalities, the respiratory mechanism cannot contribute at all to the control of pH; for example, if acidosis exists because of CO_2 accumulation caused by lung disease, the impaired lungs cannot possibly compensate for the acidosis by increasing the

rate of CO_2 removal. The buffer systems (other than the $H_2CO_3:HCO_3^-$ pair) plus renal regulation are the only mechanisms available for defending against respiratory-induced acid-base abnormalities.

The kidneys contribute powerfully to control of acid-base balance by controlling both hydrogen ion and bicarbonate concentrations in the blood.

The kidneys are the *third line of defense* against changes in $[H^+]$ in the body fluids; they require hours to days to compensate for changes in body-fluid pH, compared to the immediate responses of the buffer systems and the few-minute delay before the respiratory system responds. However, the kidneys are the most potent acid-base regulatory mechanism; not only can they vary H^+ removal, but they also can variably conserve or eliminate HCO_3^-, depending on the acid-base status of the body. For example, during renal compensation for acidosis, for each H^+ excreted in the urine, a new HCO_3^- is added to the plasma to buffer, by means of the $H_2CO_3:HCO_3^-$ system, yet another H^+ that still remains in the body fluids. By simultaneously removing acid (H^+) from and adding base (HCO_3^-) to the body fluids, the kidneys are able to restore the pH toward normal more effectively than the lungs, which can adjust only the amount of H^+-forming CO_2 in the body.

Also contributing to the kidneys' acid-base regulatory potency is their ability to return the pH almost exactly to normal. In contrast, when some nonrespiratory abnormality has altered the $[H^+]$, the respiratory system alone can return the pH to only between 50% to 75% of the way toward normal; this is because the driving force governing the compensatory ventilatory response is diminished as the pH moves toward normal. As an example, ventilation is increased in response to a rise in arterial $[H^+]$, but as the $[H^+]$ is gradually reduced because of the stepped-up removal of H^+-forming CO_2, the ventilatory response is also gradually reduced. In comparison, the kidneys continue to respond to a change in pH until compensation is essentially complete.

The kidneys control the pH of the body fluids by adjusting three interrelated factors: (1) H^+ excretion, (2) HCO_3^- excretion, and (3) ammonia (NH_3) secretion.

Hydrogen-Ion Excretion Acids are continuously being added to the body fluids as a result of metabolic activities, yet the generated H^+ must not be allowed to accumulate. Although the body's buffer systems can resist changes in pH by removing H^+ from solution, the persistent production of acidic metabolic products would eventually overwhelm the limits of this buffering capacity. The constantly generated H^+ must therefore ultimately be eliminated from the body. The lungs are able to remove only carbonic acid through the elimination of CO_2. The task of eliminating H^+ derived from sulfuric, phosphoric, lactic and other acids rests with the kidneys.

Almost all of the excreted H^+ enters the urine by means of secretion. Recall that the filtration rate of H^+ equals plasma $[H^+]$ times GFR. Since plasma $[H^+]$ is extremely low (less than in pure H_2O except during extreme acidosis, when the pH falls below 7.0), the filtration rate of H^+ is likewise extremely low. This minute amount of filtered H^+ is excreted in the urine. However, the majority of excreted H^+ gains entry into the tubular fluid by being secreted. The proximal, distal, and collecting tubules all secrete H^+. Since the kidneys normally excrete H^+, urine is usually acidic, having an average pH of 6.0. Not only do the kidneys continuously eliminate the normal amount of H^+ that is constantly being produced from non-carbonic-acid sources, but they can also alter their rate of H^+ secretion to compensate for changes in $[H^+]$ arising from abnormalities in the concentration of carbonic acid.

The magnitude of H^+ secretion depends on a direct effect of the plasma's acid-base status on the kidneys' tubular cells (● Fig. 15–8). No neural or hormonal control is involved. When the $[H^+]$ of the plasma passing through the peritubular capillaries is elevated above normal, the tubular cells respond by secreting greater-than-usual amounts of H^+ from the plasma into the tubular fluid to be excreted in the urine. Conversely, when plasma $[H^+]$ is lower than normal, the kidneys conserve H^+ by reducing its secretion and subsequent excretion in the urine. The kidneys are unable to raise plasma $[H^+]$ by reabsorbing more of the filtered H^+ because there are no reabsorptive mechanisms for H^+. The only way the kidneys can reduce H^+ excretion is by secreting less H^+.

The H^+ secretory process begins in the tubular cells with CO_2 that has come from three sources: CO_2 that has diffused into the tubular cells from either the plasma or tubular fluid or CO_2 that has been metabolically produced within the tubular cells. Under the influence of carbonic anhydrase, CO_2 and H_2O form H_2CO_3, which dissociates into H^+ and HCO_3^-. An energy-dependent carrier in the luminal membrane then transports H^+ out of the cell into the tubular lumen. In part of the nephron, this carrier transports Na^+ derived from the glomerular filtrate in the opposite direction, so H^+ secretion and Na^+ reabsorption are partially linked.

Because the chemical reactions for H^+ secretion begin with CO_2, the rate at which they proceed is influenced by $[CO_2]$. When plasma $[CO_2]$ increases, these reactions proceed more rapidly and the rate of H^+ secretion speeds up (Fig. 15–8). Conversely, the rate of H^+ secretion is reduced when plasma $[CO_2]$ falls below normal. This is especially important in renal compensations for acid-base abnormalities involving a change in H_2CO_3 caused by respiratory dysfunction. The kidneys can therefore adjust H^+ excretion to compensate for changes in both carbonic and non-carbonic acids.

Bicarbonate Excretion Before being eliminated by the kidneys, the H^+ generated from non-carbonic acids is buffered to a large extent by plasma HCO_3^-. Appropriately, therefore, renal handling of acid-base balance also involves the adjustment of HCO_3^- excretion, depending on the H^+ load in the plasma (Fig. 15–8).

The kidneys regulate plasma $[HCO_3^-]$ by two interrelated mechanisms: (1) variable reabsorption of the filtered HCO_3^- back into the plasma and (2) variable addition of new HCO_3^- to the plasma. Both of these mechanisms are inextricably linked with H^+ secretion by the kidney tubules. Every time a H^+ is secreted into the tubular fluid, a HCO_3^- is simultaneously transferred into the peritubular-capillary plasma. Whether a filtered HCO_3^- is reabsorbed or a new HCO_3^- is added to the plasma in accompaniment with H^+ secretion depends on whether filtered HCO_3^- is present in the tubular fluid to react with the secreted H^+.

Bicarbonate is freely filtered, but because the luminal membranes of the tubular cells are impermeable to the filtered HCO_3^-, it cannot diffuse into these cells. Therefore, reabsorption of HCO_3^- must occur indirectly (● Fig. 15–9). Hydrogen ion secreted into the tubular fluid combines with the filtered HCO_3^- to form H_2CO_3. Under the influence of carbonic anhydrase, which is present on the surface of the luminal membrane, H_2CO_3 decomposes into CO_2 and H_2O within the filtrate. Unlike HCO_3^-, CO_2 can easily penetrate the tubular cell membranes. Within the cells, CO_2 and H_2O, under the influence of intracellular carbonic anhydrase, form H_2CO_3, which dissociates into H^+ and HCO_3^-. Because HCO_3^- can permeate the tubular cells' basolateral membrane, it pas-

● **FIGURE 15–8 Control of Rate of Tubular H^+ Secretion**

● **FIGURE 15–9 Hydrogen-Ion Secretion Coupled with Bicarbonate Reabsorption** Since the disappearance of a HCO_3^- from the tubular fluid is coupled with the appearance of another HCO_3^- in the plasma, HCO_3^- is considered to have been "reabsorbed."

ca = Carbonic anhydrase

sively diffuses out of the cells and into the peritubular-capillary plasma. Meanwhile, the generated H^+ is actively secreted. Since the disappearance of a HCO_3^- from the tubular fluid is coupled with the appearance of another HCO_3^- in the plasma, a HCO_3^- has, in effect, been "reabsorbed." Even though the HCO_3^- entering the plasma is not the same HCO_3^- that was filtered, the net result is the same as if HCO_3^- were directly reabsorbed.

Normally, slightly more hydrogen ions are secreted into the tubular fluid than bicarbonate ions are filtered. Accordingly, all of the filtered HCO_3^- is usually reabsorbed because secreted H^+ is available in the tubular fluid to combine with it to form highly reabsorbable CO_2. The vast majority of the secreted H^+ combines with HCO_3^- and is not excreted because it is "used up" in the process of HCO_3^- reabsorption. However, the slight excess of secreted H^+ that is not matched by filtered HCO_3^- is excreted in the urine. This normal H^+

excretion rate keeps pace with the normal rate of non-carbonic-acid H^+ production.

Secretion of H^+ that is *excreted* is coupled with the *addition of new HCO_3^-* to the plasma, in contrast to the secreted H^+ that is coupled with HCO_3^- *reabsorption* and is *not excreted*, instead being incorporated into H_2O molecules. When all of the filtered HCO_3^- has been reabsorbed and additional secreted H^+ is generated by the dissociation of H_2CO_3, the HCO_3^- produced by this reaction diffuses into the plasma as a "new" HCO_3^-. It is termed "new" because its appearance in the plasma is not associated with reabsorption of filtered HCO_3^- (● Fig. 15–10). Meanwhile, the secreted H^+ combines with urinary buffers, especially basic phosphate ($HPO_4^=$), and is excreted.

When the plasma $[H^+]$ is elevated during acidosis, more H^+ is secreted than normal. At the same time, less HCO_3^- is filtered than normal because more of the plasma HCO_3^- is

● **FIGURE 15–10 Hydrogen-Ion Secretion and Excretion Coupled with Addition of New HCO_3^- to Plasma** Secreted H^+ does not combine with filtered $HPO_4^=$ and is not subsequently excreted until all of the filtered HCO_3^- has been "reabsorbed," as depicted in Figure 15–9. Once all of the filtered HCO_3^- has combined with secreted H^+, further secreted H^+ is excreted in the urine, primarily in association with urinary buffers such as basic phosphate. Excretion of H^+ is coupled with the appearance of new HCO_3^- in the plasma. The "new" HCO_3^- represents a net gain rather than being merely a replacement for filtered HCO_3^-.

ca = Carbonic anhydrase

used up in buffering the excess H^+ as HCO_3^- plus H^+ is converted to H_2CO_3. This greater-than-usual inequity between filtered HCO_3^- and secreted H^+ has two consequences. First, more of the secreted H^+ is excreted in the urine because more hydrogen ions are entering the tubular fluid at a time when fewer of them are needed to reabsorb the reduced quantities of filtered HCO_3^-. In this way, extra H^+ is eliminated from the body, making the urine more acidic than normal. Second, because the excretion of H^+ is linked with the addition of new HCO_3^- to the plasma, more HCO_3^- than usual enters the plasma that is passing through the kidneys. This additional HCO_3^- is available to buffer the excess H^+ present in the body.

In the opposite situation of alkalosis, the rate of H^+ secretion diminishes while the rate of HCO_3^- filtration increases compared to normal. When the plasma $[H^+]$ is below normal, a smaller proportion of the HCO_3^- pool is tied up buffering H^+, so the plasma $[HCO_3^-]$ is elevated above normal. As a result, the rate of HCO_3^- filtration is correspondingly increased. Not all of the filtered HCO_3^- is reabsorbed, because bicarbonate ions are in excess of secreted hydrogen ions in the tubular fluid and HCO_3^- cannot be reabsorbed without first reacting with H^+. The excess HCO_3^- is left in the tubular fluid to be excreted in the urine, thus reducing the plasma $[HCO_3^-]$ while causing the urine to become alkaline.

In short, when plasma $[H^+]$ is increased above normal during *acidosis*, renal compensation includes the following (Table 15–7):

1. Increased secretion and subsequent increased excretion of H^+ in the urine, thereby eliminating the excess H^+ and decreasing plasma $[H^+]$.
2. Reabsorption of all of the filtered HCO_3^-, plus addition of new HCO_3^- to the plasma, resulting in increased plasma $[HCO_3^-]$.

When plasma $[H^+]$ is reduced below normal during *alkalosis*, renal responses include the following:

1. Decreased secretion and subsequent reduced excretion of H^+ in the urine, resulting in conservation of H^+ and increased plasma $[H^+]$.

2. Incomplete reabsorption of filtered HCO_3^- and subsequent increased excretion of HCO_3^-, resulting in reduction of plasma $[HCO_3^-]$.

Note that to compensate for acidosis, the kidneys acidify the urine (by getting rid of extra H^+) and alkalinize the plasma (by conserving HCO_3^-) to bring the pH to normal. In the opposite case—alkalosis—the kidneys make the urine alkaline (by eliminating excess HCO_3^-) while acidifying the plasma (by conserving H^+).

Ammonia Secretion The energy-dependent H^+ carriers in the tubular cells are capable of secreting H^+ against a concentration gradient until the tubular fluid (urine) becomes 800 times more acidic than the plasma. At this point, further H^+ secretion ceases, because the gradient becomes too great for the secretory process to continue. It is impossible for the kidneys to acidify the urine beyond a gradient-limited urinary pH of 4.5. If left unbuffered as free H^+, only about 1% of the excess H^+ typically excreted daily would produce a urinary pH of this magnitude at normal urine flow rates, and elimination of the other 99% of the usually secreted H^+ load would be prevented, a situation that would be intolerable. For H^+ secretion to proceed, the majority of secreted H^+ must be buffered in the tubular fluid so that it does not exist as free H^+ and, accordingly, does not contribute to tubular acidity.

Bicarbonate cannot buffer urinary H^+ as it does in the ECF, because HCO_3^- is not excreted in the urine simultaneously with H^+. (Whichever one of these substances is in excess in the plasma is excreted in the urine.) There are, however, two important urinary buffers: (1) filtered phosphate buffers and (2) secreted **ammonia** (NH_3). Normally, secreted H^+ is first buffered by the phosphate buffer system, which is in the tubular fluid because excess ingested phosphate has been filtered but not reabsorbed. The basic member of the phosphate buffer pair binds with secreted H^+. Basic phosphate is present in the tubular fluid by virtue of dietary excess, not because of any deliberate mechanism to buffer secreted H^+. When H^+ secretion is high, the buffering capacity of urinary phosphates is exceeded, but the kidneys cannot respond by excreting more basic phosphate. Only the quantity of phos-

TABLE 15–7   Summary of Renal Responses to Acidosis and Alkalosis

ACID-BASE ABNORMALITY	H^+ SECRETION	H^+ EXCRETION	HCO_3^- REABSORPTION AND ADDITION OF NEW HCO_3^- TO PLASMA	HCO_3^- EXCRETION	pH OF URINE	COMPENSATORY CHANGE IN PLASMA pH
Acidosis	↑	↑	↑	Normal (zero; all filtered is reabsorbed)	Acidic	Alkalinization toward normal
Alkalosis	↓	↓	↓	↑	Alkaline	Acidification toward normal

phate reabsorbed, not the quantity excreted, is subject to control. As soon as all of the basic phosphate ions that are coincidentally excreted have "soaked up" H$^+$, the acidity of the tubular fluid quickly rises as more H$^+$ ions are secreted. Without additional buffering capacity from another source, H$^+$ secretion would soon be halted abruptly as the free [H$^+$] in the tubular fluid quickly rose to the critical limiting level.

When acidosis exists, the tubular cells secrete NH$_3$ into the tubular fluid once the normal urinary phosphate buffers are saturated. This NH$_3$ enables the kidneys to continue secreting additional H$^+$ ions because NH$_3$ combines with free H$^+$ in the tubular fluid to form **ammonium ion (NH$_4^+$)** as follows:

$$NH_3 + H^+ \rightarrow NH_4^+$$

The tubular membranes are not very permeable to NH$_4^+$, so the ammonium ions remain in the tubular fluid and are lost in the urine, each one taking a H$^+$ with it. Thus, NH$_3$ secretion during acidosis serves to buffer excess H$^+$ in the tubular fluid, so that large amounts of H$^+$ can be secreted into the urine before the pH falls to the limiting value of 4.5. Were it not for NH$_3$ secretion, the extent of H$^+$ secretion would be limited to whatever phosphate buffering capacity coincidentally happened to be present as a result of dietary excess.

In contrast to the phosphate buffers, which are in the tubular fluid because they have been filtered but not reabsorbed, NH$_3$ is deliberately synthesized from the amino acid *glutamine* within the tubular cells, from which it readily diffuses down its concentration gradient into the tubular fluid. The rate of NH$_3$ secretion is controlled by a direct effect on the tubular cells of the amount of excess H$^+$ to be transported in the urine. When an individual has been acidotic for more than two or three days, the rate of NH$_3$ production increases substantially. This extra NH$_3$ provides additional buffering capacity to allow H$^+$ secretion to continue after the normal phosphate buffering capacity is overwhelmed during renal compensation for acidosis.

Acid-base imbalances can arise from either respiratory dysfunction or metabolic disturbances.

Deviations from normal acid-base status are divided into four general categories, depending on the source and direction of the abnormal change in [H$^+$]. These categories are respiratory acidosis, respiratory alkalosis, metabolic acidosis, and metabolic alkalosis.

Because of the relationship between [H$^+$] and the concentrations of the members of a buffer pair, changes in [H$^+$] are reflected by changes in the ratio of [HCO$_3^-$] to [CO$_2$]. Recall that the normal ratio is 20/1. Using the Henderson-Hasselbalch equation and with pK being 6.1 and the log of 20 being 1.3, normal pH = 6.1 + 1.3 = 7.4. Determinations of [HCO$_3^-$] and [CO$_2$] provide more meaningful information about the underlying factors responsible for a particular acid-base status than do direct measurements of [H$^+$] alone. The following rules of thumb apply when examining acid-base imbalances *before any compensations take place*:

1. A change in pH that has a respiratory cause will be associated with an abnormal [CO$_2$], giving rise to a change in carbonic-acid–generated H$^+$. In contrast, a pH deviation of metabolic origin will be associated with an abnormal [HCO$_3^-$] as a result of the participation of HCO$_3^-$ in buffering abnormal amounts of H$^+$ generated from non-carbonic acids.

2. Anytime the [HCO$_3^-$]/[CO$_2$] ratio falls below 20/1, an acidosis exists. The log of any number lower than 20 is less than 1.3 and when added to the pK of 6.1 will yield an acidotic pH below 7.4. Anytime the ratio exceeds 20/1, an alkalosis exists. The log of any number greater than 20 is more than 1.3 and when added to the pK of 6.1 will yield an alkalotic pH above 7.4.

Putting these two points together:

- *Respiratory acidosis* has a ratio of less than 20/1 arising from an increase in [CO$_2$].
- *Respiratory alkalosis* has a ratio greater than 20/1 because of a decrease in [CO$_2$].
- *Metabolic acidosis* has a ratio of less than 20/1 associated with a fall in [HCO$_3^-$].
- *Metabolic alkalosis* has a ratio greater than 20/1 arising from an elevation in [HCO$_3^-$].

We will examine each of these categories separately in more detail, paying particular attention to possible causes and the compensations that occur. The "balance beam" concept, presented in ● Figure 15–11 in conjunction with the Henderson-Hasselbalch equation, wil help you better visualize the contributions of the lungs and kidneys to the causes and compensations of various acid-base disorders. The normal situation is represented in Figure 15–11a.

Respiratory Acidosis   **Respiratory acidosis** is the result of abnormal CO$_2$ retention arising from *hypoventilation* (see p. 449). As less-than-normal amounts of CO$_2$ are lost through the lungs, the resultant increase in H$_2$CO$_3$ formation and dissociation leads to an elevated [H$^+$]. Possible causes include lung disease, depression of the respiratory center by drugs or disease, nerve or muscle disorders that reduce respiratory muscle ability, or (transiently) even the simple act of holding one's breath.

In uncompensated respiratory acidosis (Fig. 15–11b), [CO$_2$] is elevated (in our example, it is doubled) while [HCO$_3^-$] is normal, so the ratio is 20/2 (10/1) and pH is reduced. Let us clarify a potentially confusing point. You might wonder why when [CO$_2$] is elevated and drives the reaction CO$_2$ + H$_2$O $\rightleftharpoons$ H$_2$CO$_3$ $\rightleftharpoons$ H$^+$ + HCO$_3^-$ to the right, we say that [H$^+$] becomes elevated but [HCO$_3^-$] remains normal when the same quantities of H$^+$ and HCO$_3^-$ are produced when CO$_2$-generated H$_2$CO$_3$ dissociates. The answer lies in the fact that normally the [HCO$_3^-$] is 600,000 times the [H$^+$]. For every one hydrogen ion and 600,000 bicarbonate ions present in the ECF, the generation of one additional H$^+$ and one HCO$_3^-$ doubles the [H$^+$] (a 100% increase) but only increases the [HCO$_3^-$] 0.00017% (from 600,000 to 600,001 ions). Therefore, an elevation in [CO$_2$] brings about a pronounced increase in [H$^+$] but [HCO$_3^-$] remains essentially normal.

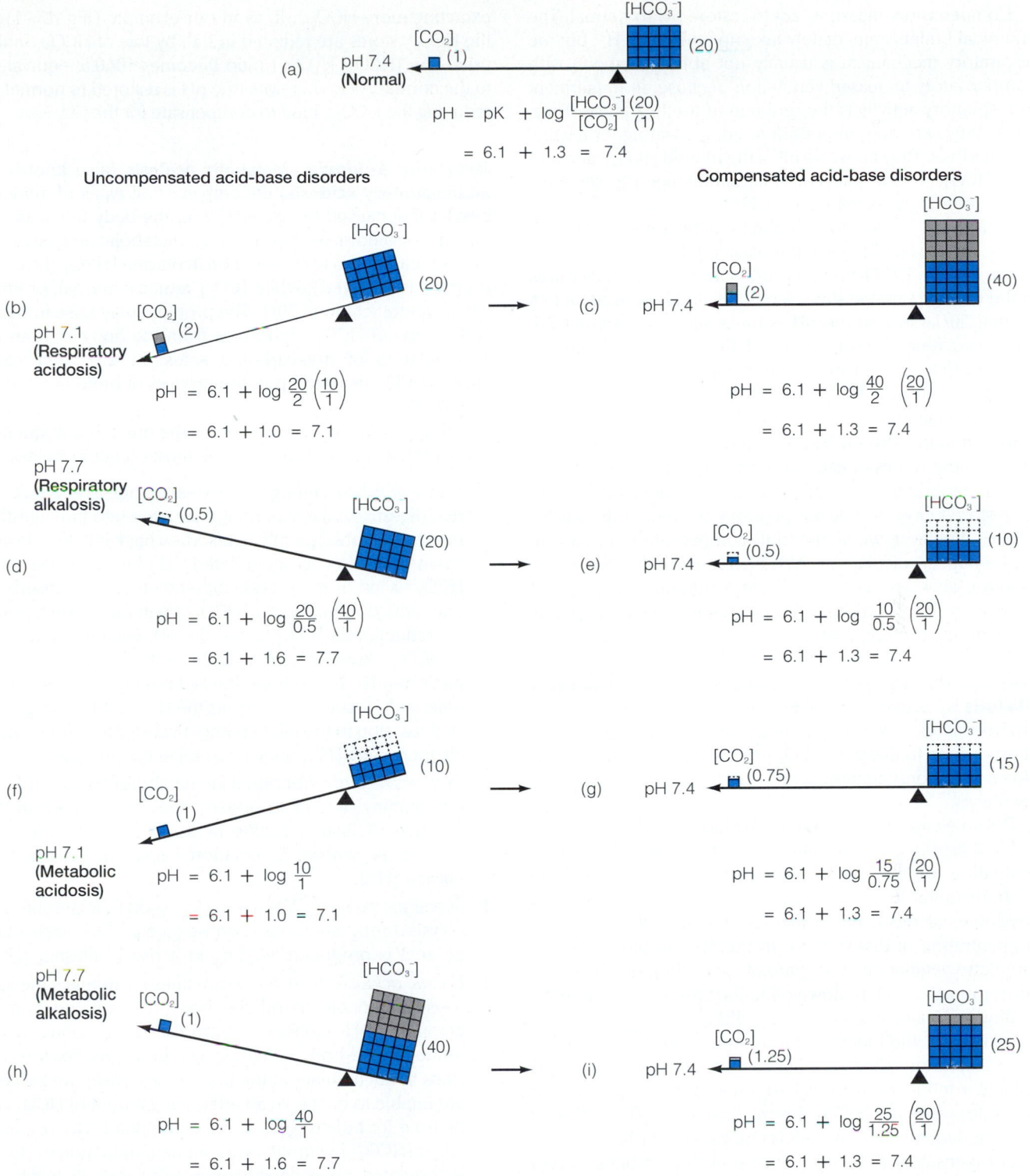

The lengths of the arms of the balance beams are not to scale.

• **FIGURE 15–11 Schematic Representation of the Relationship of HCO_3^- and CO_2 Concentrations to pH in Various Acid-Base Statuses** (a) Normal acid-base balance. The $[HCO_3^-]/[CO_2]$ ratio is 20/1. (b) Uncompensated respiratory acidosis. The $[HCO_3^-]/[CO_2]$ ratio is reduced (20/2) because CO_2 has accumulated. (c) Compensated respiratory acidosis. Compensatory retention of HCO_3^- to balance the CO_2 accumulation restores the $[HCO_3^-]/[CO_2]$ ratio to a normal equivalent (40/2). (d) Uncompensated respiratory alkalosis. The $[HCO_3^-]/[CO_2]$ ratio is increased (20/0.5) due to a reduction in CO_2. (e) Compensated respiratory alkalosis. Compensatory elimination of HCO_3^- to balance the CO_2 deficit restores the $[HCO_3^-]/[CO_2]$ ratio to a normal equivalent (10/0.5). (f) Uncompensated metabolic acidosis. The $[HCO_3^-]/[CO_2]$ ratio is reduced (10/1) due to a HCO_3^- deficit. (g) Compensated metabolic acidosis. Conservation of HCO_3^-, which partially makes up for the HCO_3^- deficit, and a compensatory reduction in CO_2 restore the $[HCO_3^-]/[CO_2]$ to a normal equivalent (15/0.75). (h) Uncompensated metabolic alkalosis. The $[HCO_3^-]/[CO_2]$ ratio is increased (40/1) due to excess HCO_3^-. (i) Compensated metabolic alkalosis. Elimination of some of the extra HCO_3^- and a compensatory increase in CO_2 restore the $[HCO_3^-]/[CO_2]$ ratio to a normal equivalent (25/1.25).

Compensatory measures act to restore pH to normal. The chemical buffers immediately take up additional H^+, but the respiratory mechanism is usually not able to respond with compensatory increased ventilation because an impairment in respiratory activity is the problem in the first place. Thus, the kidneys are most important in compensating for respiratory acidosis. They conserve all of the filtered HCO_3^- and add new HCO_3^- to the plasma while simultaneously secreting and, accordingly, excreting more H^+.

As a result, HCO_3^- stores in the body become elevated. In our example (Fig. 15–11c), the plasma $[HCO_3^-]$ is doubled, so the $[HCO_3^-]/[CO_2]$ ratio is 40/2 rather than 20/2 as it was in the uncompensated state. A ratio of 40/2 is equivalent to a normal 20/1 ratio, so the pH is once again the normal 7.4. Enhanced renal conservation of HCO_3^- has fully compensated for CO_2 accumulation, thus restoring the pH to normal, although both the $[CO_2]$ and $[HCO_3^-]$ are now distorted. Note that maintenance of normal pH depends on preserving a normal ratio between $[HCO_3^-]$ and $[CO_2]$, no matter what the absolute values of each of these buffer components are. (Compensation is never fully complete because the pH can be restored close to but not precisely to normal. In our examples, however, we assume full compensation for ease in mathematical calculations. Also bear in mind that the values used are only representative. There is actually a range of pH deviations and variations in the degree to which compensation can be accomplished.)

Respiratory Alkalosis

The primary defect in **respiratory alkalosis** is excessive loss of CO_2 from the body as a result of *hyperventilation*. When pulmonary ventilation increases out of proportion to the rate of CO_2 production, too much CO_2 is blown off. Consequently, less H_2CO_3 is formed and $[H^+]$ decreases.

Possible causes of respiratory alkalosis include fever, anxiety, and aspirin poisoning, all of which excessively stimulate ventilation without regard to the status of O_2, CO_2, or H^+ in the body fluids. Respiratory alkalosis also occurs as a result of physiological mechanisms at high altitude. When the low concentration of dissolved O_2 in the arterial blood reflexly stimulates ventilation in an attempt to obtain more O_2 for the body, too much CO_2 is blown off in the process, inadvertently leading to an alkalotic state (see p. 451).

Looking at the biochemical abnormalities in uncompensated respiratory alkalosis (Fig. 15–11d), the increase in pH reflects a reduction in $[CO_2]$ (half the normal value in our example) while the $[HCO_3^-]$ remains normal. This yields an alkalotic ratio of 20/0.5, which is comparable to 40/1.

Compensatory measures act to shift the pH back toward normal. The chemical buffer systems liberate H^+ to diminish the severity of the alkalosis. As plasma $[CO_2]$ and $[H^+]$ fall below normal because of excessive ventilation, two of the normally potent stimuli for driving ventilation are removed. This effect tends to "put brakes" on the extent to which some nonrespiratory related factor such as fever or anxiety can overdrive ventilation. Therefore, hyperventilation does not continue completely unabated. If the situation continues for a few days, the kidneys compensate by conserving H^+ and

excreting more HCO_3^-. If, as in our example (Fig. 15–11e), the HCO_3^- stores are reduced in half by loss of HCO_3^- in the urine, the $[HCO_3^-]/[CO_2]$ ratio becomes 10/.0.5, equivalent to the normal 20/1. Therefore, the pH is restored to normal by reducing the HCO_3^- load to compensate for the CO_2 loss.

Metabolic Acidosis

Metabolic acidosis (also known as **nonrespiratory acidosis**) encompasses all types of acidosis besides that caused by excess CO_2 in the body fluids. In the uncompensated state (Fig. 15–11f), metabolic acidosis is always characterized by a reduction in plasma $[HCO_3^-]$ (in our example it is halved), while $[CO_2]$ remains normal, producing an acidotic ratio of 10/1. The problem may arise from excessive loss of HCO_3^--rich fluids from the body or from an accumulation of non-carbonic acids. In the latter case, plasma HCO_3^- is used up in the process of buffering the additional H^+.

This type of acid-base disorder is the one more frequently encountered. The following are the most common causes.

1. *Severe diarrhea.* During the process of digestion, a HCO_3^--rich digestive juice is normally secreted into the digestive tract and is subsequently reabsorbed back into the plasma when digestion is completed. During diarrhea, this HCO_3^- is lost from the body rather than being reabsorbed. The reduction in plasma $[HCO_3^-]$ without a corresponding reduction in $[CO_2]$ lowers the pH. Because of the loss of HCO_3^-, less HCO_3^- is available to buffer H^+, leading to more free H^+ in the body fluids. Looking at the situation differently, loss of HCO_3^- shifts the $H^+ + HCO_3^- \rightleftharpoons CO_2 + H_2O$ reaction to the left to compensate for the HCO_3^- deficit, increasing $[H^+]$ above normal in the process.

2. *Diabetes mellitus.* Abnormal fat metabolism resulting from the inability of cells to preferentially use glucose in the absence of insulin results in the formation of excess keto acids, whose dissociation causes an increase in plasma $[H^+]$.

3. *Strenuous exercise.* When muscles resort to anaerobic glycolysis during strenuous exercise (see p. 236), excess lactic acid is produced, leading to a rise in plasma $[H^+]$.

4. *Uremic acidosis.* In severe renal failure (uremia), the kidneys are unable to rid the body of even the normal amounts of H^+ generated from the non-carbonic acids formed by the body's ongoing metabolic processes, so H^+ starts to accumulate in the body fluids. Also, the kidneys are unable to conserve an adequate amount of HCO_3^- to be used for buffering the normal acid load. The resultant fall in $[HCO_3^-]$ without a concomitant reduction in $[CO_2]$ is correlated with a decline in pH to an acidotic level.

Except in uremic acidosis, metabolic acidosis is compensated for by both respiratory and renal mechanisms as well as by chemical buffers. The buffers take up extra H^+, the lungs blow off additional H^+-generating CO_2, and the kidneys excrete more H^+ and conserve more HCO_3^-. In our example (Fig. 15–11g), these compensatory measures restore the ratio to normal by reducing $[CO_2]$ to 75% of normal and by raising $[HCO_3^-]$ halfway back toward normal (up from 50% to 75%

of the normal value). This brings the ratio to 15/0.75 (equivalent to 20/1).

Note that in compensating for metabolic acidosis, the lungs deliberately displace $[CO_2]$ from normal in an attempt to restore $[H^+]$ toward normal. Whereas in respiratory-induced acid-base disorders, an abnormal $[CO_2]$ is the *cause* of the $[H^+]$ imbalance, in metabolic acid-base disorders, $[CO_2]$ is intentionally shifted from normal as an important *compensation* for the $[H^+]$ imbalance.

When kidney disease is the cause of metabolic acidosis, complete compensation is not possible because the renal mechanism is not available for pH regulation. Recall that the respiratory system is capable of compensating only up to 75% of the way toward normal. Uremic acidosis is very serious, because the kidneys cannot help to restore the pH all the way to normal.

Metabolic alkalosis **Metabolic alkalosis** is a reduction in plasma $[H^+]$ caused by a relative deficiency of non-carbonic acids. This acid-base disturbance is associated with an increase in $[HCO_3^-]$, which, in the uncompensated state, is not accompanied by a change in $[CO_2]$. In our example (Fig. 15–11h), $[HCO_3^-]$ is doubled, producing an alkalotic ratio of 40/1. This condition arises most commonly from the following:

1. *Vomiting,* which results in abnormal loss of H^+ from the body as a result of loss of acidic gastric juices. Hydrochloric acid is secreted into the stomach lumen during the process of digestion. Bicarbonate is added to the plasma during gastric HCl secretion. This HCO_3^- is neutralized by H^+ as the gastric secretions are eventually reabsorbed back into the plasma, so normally there is no net addition of HCO_3^- to the plasma from this source. However, when this acid is lost from the body during vomiting, not only is plasma $[H^+]$ decreased, but also reabsorbed H^+ is no longer available to neutralize the extra HCO_3^- added to the plasma during gastric HCl secretion. Thus, loss of HCl in effect increases plasma $[HCO_3^-]$. (On the other hand, with "deeper" vomiting, HCO_3^- in the digestive juices secreted into the upper intestine may be lost in the vomitus, resulting in an acidosis instead of an alkalosis.)

2. *Ingestion of alkaline drugs,* such as when baking soda ($NaHCO_3$, which dissociates in solution into Na^+ and HCO_3^-) is used as a self-administered remedy for the treatment of gastric hyperacidity. By neutralizing excess acid in the stomach, HCO_3^- relieves the symptoms of stomach irritation and heartburn, but when more HCO_3^- than needed is ingested, the extra HCO_3^- is absorbed from the digestive tract and increases plasma $[HCO_3^-]$. The extra HCO_3^- binds with some of the free H^+ normally present in the plasma from non-carbonic acid sources, leading to a reduction in free $[H^+]$.

In metabolic alkalosis, the chemical buffer systems immediately liberate H^+, and ventilation is reduced so that extra H^+-generating CO_2 is retained in the body fluids. If the condition persists for several days, the kidneys conserve H^+ and excrete the excess HCO_3^- in the urine. The resultant compensatory increase in $[CO_2]$ (up 25% in our example—Fig. 15–11i) and the partial reduction in $[HCO_3^-]$ (75% of the way back down toward normal in our example) together restore the $[HCO_3^-]/[CO_2]$ ratio back to the equivalent of 20/1 at 1.25/25.

It should be obvious that an individual's acid-base status cannot be assessed on the basis of pH alone. Even though the pH is esentially normal, determinations of $[HCO_3^-]$ and $[CO_2]$ can reveal compensated acid-base disorders.

 CHAPTER IN PERSPECTIVE: FOCUS ON HOMEOSTASIS

Homeostasis depends on maintaining a balance between the input and output of all constituents present in the internal fluid environment. Regulation of fluid balance involves two separate components: control of salt balance and control of H_2O balance. Control of salt balance is primarily important in the long-term regulation of arterial blood pressure because the body's salt load affects the osmotic determination of the ECF volume, of which plasma volume is a part. An increased salt load in the ECF leads to an expansion in ECF volume, including plasma volume, which in turn causes a rise in blood pressure. Conversely, a reduction in the ECF salt load brings about a fall in blood pressure. Salt balance is maintained by constantly adjusting salt output in the urine to match unregulated, variable salt intake.

Control of H_2O balance is important in preventing detrimental changes in ECF osmolarity, which would induce detrimental osmotic shifts of H_2O between the cells and the ECF. Such shifts of H_2O into or out of the cells would cause the cells to swell or shrink, respectively. Cells, especially brain neurons, do not function normally when swollen or shrunken. Water balance is largely maintained by controlling the volume of H_2O lost in the urine to compensate for uncontrolled losses of variable volumes of H_2O from other avenues, such as through sweating or diarrhea, and for poorly regulated H_2O intake. Even though a thirst mechanism exists to control H_2O intake based on need, the amount drunk is often influenced by social custom and habit instead of thirst alone.

A balance between input and output of H^+ is also critical to maintaining the body's acid-base balance within the narrow limits compatible with life. Deviations in the internal fluid environment's pH lead to altered neuromuscular excitability, to changes in enzymatically controlled metabolic activity, and to K^+ imbalances, which can cause cardiac arrhythmias. These effects are fatal if the pH falls outside the range of 6.8 to 8.0.

Hydrogen ions are uncontrollably, continually being added to the body fluids as a result of ongoing metabolic activites, yet the ECF's pH must be kept constant at a slightly alkaline level of 7.4 for optimal body function. Like salt and H_2O balance, control of H^+ output by the kidneys is the main regulatory factor in achieving H^+ balance. Assisting the kidneys in eliminating H^+ are the lungs, which can adjust their rate of excretion of H^+-generating CO_2.

The assertion that the same input-output balance that applies to salt and H_2O also applies to H^+ homeostasis must be modified by the fact that H^+ is buffered in the body. This buffering mechanism can take up or liberate H^+, thereby transiently keeping its concentration constant within the body until its output can be brought into line with its input. Such a mechanism is not available for salt or H_2O balance.

CHAPTER SUMMARY

Fluid Balance

On average, the body fluids compose 60% of total body weight. This figure varies between individuals, depending on how much fat (a low H_2O content tissue) they possess. Two-thirds of the body H_2O is found in the intracellular fluid (ICF). The remaining one-third present in the extracellular fluid (ECF) is distributed between the plasma (20% of the ECF) and the interstitial fluid (80% of the ECF).

Because all plasma constituents are freely exchanged across the capillary walls, the plasma and interstitial fluid are nearly identical in composition, except for the lack of plasma proteins in the interstitial fluid. In contrast, the ECF and ICF have markedly different compositions because the cell membrane barriers are highly selective as to what materials are transported into or out of the cells.

The essential components of fluid balance are control of ECF volume by maintaining salt balance and control of ECF osmolarity by maintaining water balance. Because of the osmotic holding power of Na^+, the major ECF cation, a change in the body's total Na^+ content brings about a corresponding change in ECF volume, including plasma volume, which, in turn, alters the arterial blood pressure in the same direction. Appropriately, changes in ECF volume and arterial blood pressure are compensated for in the long run by Na^+-regulating mechanisms. Salt intake is not controlled in humans, but control of salt output in the urine is closely regulated. Blood-pressure–regulating mechanisms can vary the GFR, and accordingly the amount of Na^+ filtered, by adjusting the caliber of the afferent arterioles supplying the glomeruli. Simultaneously, blood-pressure–regulating mechanisms can vary the secretion of aldosterone, the hormone that promotes Na^+ reabsorption by the renal tubules. By varying Na^+ filtration and Na^+ reabsorption, the extent of Na^+ excretion in the urine can be adjusted to regulate the plasma volume and subsequently the arterial blood pressure in the long term.

Changes in ECF osmolarity are primarily detected and corrected by systems responsible for maintaining H_2O balance. The osmolarity of the ECF must be closely regulated to prevent osmotic shifts of H_2O between the ECF and ICF, because cell swelling or shrinking is deleterious, especially to brain neurons. Excess free H_2O in the ECF dilutes the ECF solutes, with the resultant ECF hypotonicity driving H_2O into the cells. An ECF free H_2O deficit, on the other hand, concentrates the ECF solutes, and consequently H_2O leaves the cells to enter the hypertonic ECF. To prevent these detrimental fluxes, regulation of free H_2O balance is accomplished largely by vasopressin and, to a lesser degree, by thirst. Changes in vasopressin secretion and thirst are both governed primarily by hypothalamic osmoreceptors, which monitor ECF osmolarity. The amount of vasopressin secreted determines the extent of free H_2O reabsorption by the distal portions of the nephrons, thereby determining the volume of urinary output. Simultaneously, the intensity of thirst controls the volume of fluid intake. However, because the volume of fluid drunk is often not directly correlated with the intensity of thirst, control of urinary output by vasopressin is the most important regulatory mechanism for maintaining H_2O balance.

Acid-Base Balance

Acids liberate free hydrogen ions (H^+) into solution; bases bind with free hydrogen ions and remove them from solution. Acid-base balance refers to the regulation of H^+ concentration ($[H^+]$) in the body fluids. To precisely maintain $[H^+]$, input of H^+ by means of metabolic production of acids within the body must continually be matched with H^+ output by urinary excretion of H^+ and respiratory removal of H^+-generating CO_2. Furthermore, between the time of its generation and elimination, H^+ must be buffered within the body to prevent marked fluctuations in $[H^+]$.

Hydrogen-ion concentration frequently is expressed in terms of pH, which is the logarithm of $1/[H^+]$. The normal pH of the plasma is 7.4, slightly alkaline compared to neutral H_2O, which has a pH of 7.0. A pH lower than normal (higher $[H^+]$ than normal) is indicative of a state of acidosis. A pH higher than normal (lower $[H^+]$ than normal) characterizes a state of alkalosis. Fluctuations in $[H^+]$ have profound effects on body chemistry, most notably, (1) changes in neuromuscular excitability, with acidosis depressing excitability, especially of the central nervous system, and alkalosis producing overexcitability of both the peripheral and the central nervous systems; (2) disruption of normal metabolic reactions by altering the structure and function of all enzymes; and (3) alterations in plasma $[K^+]$ brought about by H^+-induced changes in the rate of K^+ elimination by the kidneys.

The primary challenge in controlling acid-base balance is the maintenance of normal plasma alkalinity in the face of continual addition of H^+ to the plasma from ongoing metabolic activity. The three lines of defense for resisting changes in $[H^+]$ are (1) the chemical buffer systems, (2) respiratory control of pH, and (3) renal control of pH.

Chemical buffer systems, the first line of defense, each consist of a pair of chemicals involved in a reversible reaction, one that can liberate H^+ and the other that can bind H^+. A buffer pair acts immediately to minimize any changes in pH that occur by acting according to the law of mass action.

The respiratory system, constituting the second line of defense, normally eliminates the metabolically produced CO_2 so that H_2CO_3 does not accumulate in the body fluids. When the chemical buffers alone have been unable to immediately minimize a pH change, the respiratory system responds within a few minutes by altering its rate of CO_2 removal. An increase in $[H^+]$ arising from non-carbonic-acid sources stimulates respiration so that more H_2CO_3-forming CO_2 is blown off, compensating for the acidosis by reducing the generation of H^+ from H_2CO_3. Conversely, a fall in [H$^+$] depresses respiratory activity so that CO_2 and thus H$^+$-generating H_2CO_3 can accumulate in the body fluids to compensate for the alkalosis.

The kidneys are the third and most powerful line of defense. They require hours to days to compensate for a deviation in body-fluid pH. However, they not only eliminate the normal amount of H$^+$ produced from non-H_2CO_3 sources, but they can also alter their rate of H$^+$ removal in response to changes in both non-H_2CO_3 and H_2CO_3 acids. In contrast, the lungs can adjust only H$^+$ generated from H_2CO_3. Furthermore, the kidneys can regulate [HCO$_3^-$] in the body fluids as well. The kidneys compensate for acidosis by secreting excess H$^+$ in the urine while adding new HCO$_3^-$ to the plasma to expand the HCO$_3^-$ buffer pool. During alkalosis, the kidneys conserve H$^+$ by reducing its secretion in the urine. They also eliminate HCO$_3^-$, which is in excess because less HCO$_3^-$ than usual is tied up buffering H$^+$ when H$^+$ is in short supply.

Secreted H$^+$ that is to be excreted in the urine must be buffered in the tubular fluid to prevent the H$^+$ concentration gradient from becoming so great that it prevents further H$^+$ secretion. Normally, H$^+$ is buffered by the urinary phosphate buffer pair, which is abundant in the tubular fluid because excess dietary phosphate spills into the urine to be excreted from the body. In acidosis, when all of the phosphate buffer is already used up in buffering the extra secreted H$^+$, the kidneys secrete NH$_3$ into the tubular fluid to serve as a buffer so that H$^+$ secretion can continue.

There are four types of acid-base imbalances: respiratory acidosis, respiratory alkalosis, metabolic acidosis, and metabolic alkalosis. Respiratory acid-base disorders originate with deviations from normal [CO$_2$], whereas metabolic acid-base imbalances encompass all deviations in pH other than those caused by abnormal [CO$_2$].

REVIEW EXERCISES

Objective Questions (Answers on p. D–4.)

1. The only avenue by which materials can be exchanged between the cells and the external environment is the ECF. (True or false?)

2. Water is driven into the cells when the ECF volume is expanded by an isotonic fluid gain. (True or false?)

3. Salt balance in humans is poorly regulated because of our hedonistic salt appetite. (True or false?)

4. An unintentional increase in CO_2 is a cause of respiratory acidosis, but a deliberate increase in CO_2 is a compensation for metabolic alkalosis. (True or false?)

5. Secreted H$^+$ that is coupled with HCO$_3^-$ reabsorption is not excreted whereas secreted H$^+$ that is excreted is linked with the addition of new HCO$_3^-$ to the plasma. (True or false?)

6. The largest body-fluid compartment is the _____.

7. Specialized fluid volumes secreted by specific cells into a particular cavity within the body for a specific purpose are collectively known as _____.

8. Of the two members of the H_2CO_3:HCO$_3^-$ buffer system, _____ is regulated by the lungs whereas _____ is regulated by the kidneys.

9. Which of the following individuals would have the lowest percentage of body H_2O?
    a. a chubby baby
    b. a well-proportioned female college student
    c. a well-muscled male college student
    d. an obese, elderly woman
    e. a lean, elderly man

10. Which of the following factors does *not* increase vasopressin secretion?
    a. ECF hypertonicity
    b. alcohol
    c. stressful situations
    d. an ECF volume deficit
    e. angiotensin II

11. pH (Indicate all correct answers.)
    a. equals log 1/[H$^+$].
    b. equals pK + log [CO$_2$]/[HCO$_3^-$].
    c. is high in acidosis.
    d. falls lower as [H$^+$] increases.
    e. is normal when the [HCO$_3^-$]/[CO$_2$] ratio is 20/1.

12. Acidosis (Indicate all correct answers.)
    a. causes overexcitability of the nervous system.
    b. exists when the plasma pH falls below 7.35.
    c. occurs when the [HCO$_3^-$]/[CO$_2$] ratio exceeds 20/1.
    d. occurs when CO_2 is blown off more rapidly than it is being produced by metabolic activities.
    e. occurs when excessive HCO$_3^-$ is lost from the body such as in diarrhea.

13. The kidney tubular cells secrete NH$_3$ (Indicate all correct answers.)
    a. when the urinary pH becomes too high.
    b. when the body is in a state of alkalosis.
    c. to enable further renal secretion of H$^+$ to occur.
    d. to buffer excess filtered HCO$_3^-$
    e. when there is excess NH$_3$ in the body fluids.

14. Complete the following chart.

[HCO$_3^-$] [CO$_2$]	Uncompensated Abnormality	Possible Cause	pH
10/1	1. _____	2. _____	3. _____
20/0.5	4. _____	5. _____	6. _____
20/2	7. _____	8. _____	9. _____
40/1	10. _____	11. _____	12. _____

Essay Questions

1. Explain the balance concept.
2. Outline the distribution of body H_2O.
3. Compare the ionic composition of plasma, interstitial fluid, and intracellular fluid.

4. What factors are regulated to maintain the body's fluid balance?

5. Why is regulation of ECF volume important? How is it regulated?

6. Why is regulation of ECF osmolarity important? How is it regulated? What are the causes and consequences of ECF hypertonicity and ECF hypotonicity?

7. Outline the sources of input and output in a daily salt balance and a daily H_2O balance. Which are subject to control to maintain the body's fluid balance?

8. Distinguish between an acid and a base.

9. What is the relationship between $[H^+]$ and pH?

10. What is the normal pH of body fluids? How does this compare to the pH of H_2O? Define acidosis and alkalosis.

11. What are the consequences of fluctuations in $[H^+]$?

12. What are the body's sources of H^+?

13. Describe the three lines of defense against changes in $[H^+]$ in terms of mechanisms and speed of action.

14. List and indicate the functions of each of the body's chemical buffer systems.

15. What are the causes of the four categories of acid-base imbalances?

16. Why is uremic acidosis so serious?

POINTS TO PONDER

1. If a person loses 1,500 ml of salt-rich sweat and drinks 1,000 ml of water during the same time period, what will happen to vasopressin secretion? Why is it important to replace both the water and salt?

2. If a solute that can penetrate the plasma membrane such as dextrose (a type of sugar) is dissolved in sterile water at a concentration equal to that of normal body fluids and then is injected intravenously, what would be the impact on the body's fluid balance?

3. Explain why it is safer to treat gastric hyperacidity with antacids that are poorly absorbed from the digestive tract than with baking soda, which is a good buffer for acid but is readily absorbed.

4. Which of the following reactions would occur to buffer the acidosis accompanying severe pneumonia?
   a. $H^+ + HCO_3^- \rightarrow H_2CO_3 \rightarrow CO_2 + H_2O$
   b. $CO_2 + H_2O \rightarrow H_2CO_3 \rightarrow H^+ + HCO_3^-$
   c. $H^+ + Hb \rightarrow HHb$
   d. $HHb \rightarrow H^+ + Hb$
   e. $NaH_2PO_4 + Na^+ \rightarrow Na_2HPO_4 + H^+$

5. A person acquires salmonellosis, an intestinal bacterial infection, from improperly handled food. What impact does the resultant prolonged diarrhea have on the person's fluid and acid-base balance?

16 Digestive System

DIGESTIVE SYSTEM

Body systems maintain homeostasis

HOMEOSTASIS
The digestive system contributes to homeostasis by transferring nutrients, water, and electrolytes from the external environment to the internal environment.

Homeostasis is essential for survival of cells

CELLS
Cells need a constant supply of nutrients to support their energy-generating chemical reactions:

$$Food + O_2 \rightarrow CO_2 + H_2O + Energy.$$

Also, proper cell function depends upon maintaining the availability of water and various electrolytes.

Cells make up body systems

To maintain homeostasis, nutrient molecules used for energy production must continually be replaced by the acquisition of new, energy-rich nutrients. Similarly, water and electrolytes that are constantly lost in the urine and sweat and through other avenues must be replenished on a regular basis. The **digestive system** contributes to homeostasis by transferring nutrients, water, and electrolytes from the external environ- ment to the internal environment. The digestive system does not directly regulate the concentration of any of these con- stituents in the internal environment. It does not vary nutrient, water, or electrolyte uptake based on body needs (with few exceptions), but, rather, it optimizes conditions for digesting and absorbing what is ingested.

≡ **CHAPTER CONTENTS AT A GLANCE** ≡

INTRODUCTION
MOUTH
Chewing
Salivary secretion
PHARYNX AND ESOPHAGUS
Swallowing
STOMACH
Gastric motility
● A CLOSER LOOK AT EXERCISE PHYSIOLOGY Pregame Meal:
    What's In and What's Out?
Gastric secretion
● CONCEPTS, CHALLENGES, AND CONTROVERSIES Beaumont:
    The Beginnings of Physiology in America
Digestion in the stomach
Absorption in the stomach
PANCREATIC AND BILIARY SECRETIONS
Exocrine pancreas
Liver and the biliary system
SMALL INTESTINE
Small-intestine motility
Small-intestine secretion
Small-intestine digestion
Small-intestine absorption
LARGE INTESTINE
Large-intestine motility
Large-intestine secretion and absorption
Composition of feces; flatus
OVERVIEW OF THE GASTROINTESTINAL HORMONES

INTRODUCTION

The primary function of the **digestive system** (**alimentary system**) is to transfer nutrients (after modifying them), water, and electrolytes from the food we eat into the body's internal environment. Ingested food is essential as an energy source from which the cells can produce ATP to carry out their particular energy-dependent activities, such as active transport, contraction, synthesis, and secretion. Food is also a source of building supplies for the renewal and addition of body tissues.

Plants can capture the sun's energy and manufacture the organic molecules they need from inorganic compounds such as CO_2 and H_2O through the process of **photosynthesis**:

$$\underset{\text{(from sun)}}{\text{energy}} + CO_2 + H_2O \quad \overset{\text{plant}}{\underset{}{\xrightarrow{\text{photosynthesis}}}} \quad \underset{\text{molecules}}{\text{organic}} + O_2$$

Humans cannot harvest energy from sunlight directly, so they have to survive on second-hand energy by eating either plants or other animals that have eaten plants. In turn, humans use the organic molecules (in food) and O_2 to produce energy, with CO_2 and H_2O being formed as by-products in the process:

$$\underset{\substack{\text{molecules} \\ \text{(in food)}}}{\text{organic}} + O_2 \quad \overset{\substack{\text{human cell} \\ \text{metabolism}}}{\xrightarrow{}} \quad \underset{\substack{\text{(for use by} \\ \text{human cells)}}}{\text{energy}} + CO_2 + H_2O$$

Note that these fundamental reactions in plants and humans (as well as other animals) are the reverse of each other, thus providing a natural balance between these forms of life. (Both plants and animals perform additional chemical reactions that involve these organic molecules as well.)

The act of eating does not automatically make the preformed organic molecules in food available to the body cells as a source of fuel or as building blocks. The food first must be digested or broken down into small, simple molecules that can be absorbed from the digestive tract into the circulatory system for distribution to the cells. Normally, about 95% of the ingested food is made available for the body's use.

We will first provide an overview of the digestive system, examining the common features of the various components of the system, before we begin a detailed tour of the tract from beginning to end.

The digestive system performs four basic digestive processes.

There are four basic digestive processes: *motility, secretion, digestion*, and *absorption*.

Motility **Motility** refers to the muscular contractions that mix and move forward the contents of the digestive tract. Like vascular smooth muscle, the smooth muscle in the walls of the digestive tract maintains a constant low level of contraction known as **tone**. Tone is important in maintaining a steady pressure on the digestive-tract contents as well as in preventing the walls of the digestive tract from remaining permanently stretched following distention.

Superimposed on this ongoing tonic base are two basic types of digestive motility: propulsive movements and mixing movements. *Propulsive movements* propel or push the contents forward through the digestive tract at varying speeds, with the rate of propulsion depending on the functions accomplished by the different regions; that is, food is moved forward in a given segment at an appropriate velocity to allow that segment to "do its job." For example, transit of food through the esophagus is rapid, which is appropriate because this structure merely serves as a passageway from the mouth to the stomach. In comparison, in the small intestine, the major site of digestion and absorption, the contents are moved forward slowly, allowing sufficient time for the breakdown and absorption of food to be accomplished.

Mixing movements serve a twofold function. First, by mixing food with the digestive juices, these movements promote digestion of the food. Second, they facilitate absorption by exposing all portions of the intestinal contents to the absorbing surfaces of the digestive tract.

Movement of material through most of the digestive tract is accomplished by contraction of the smooth muscle within the walls of the digestive organs, with the exception that motility at both ends of the tract—the mouth through the early portion of the esophagus at the beginning and the external anal sphincter at the end—involves skeletal-muscle rather than smooth-muscle activity. Accordingly, the acts of chewing, swallowing, and defecation have voluntary components since skeletal muscle is under voluntary control, whereas motility accomplished by smooth muscle throughout the remainder of the tract is controlled by complex involuntary mechanisms.

Secretion A number of digestive juices are secreted into the digestive-tract lumen by exocrine glands (see p. 4) located along the route, each with its own specific secretory product or products. Each **digestive secretion** consists of water, electrolytes, and specific organic constituents that are important in the digestive process, such as enzymes, bile salts, or mucus. The secretory cells extract from the plasma large volumes of water and those raw materials necessary to produce their particular secretion (• Fig. 16–1). Secretion of all digestive juices requires energy, both for active transport of some of the raw materials into the cell (others diffuse in passively) and for synthesis of secretory products by the endoplasmic reticulum. These exocrine cells are richly endowed with mitochondria to support the extensive energy requirement necessary for secretion. The secretions are released into the digestive-tract lumen upon appropriate neural or hormonal stimulation. Normally, the digestive secretions are reabsorbed in one form or another back into the blood after their participation in digestion. Failure to do so (because of vomiting or diarrhea, for example) results in loss of this fluid that has been "borrowed" from the plasma.

Digestion **Digestion** refers to the breaking-down process whereby the structurally complex foodstuffs of the diet are converted into smaller absorbable units by the enzymes produced within the digestive system. Humans consume three different biochemical categories of energy-rich foodstuffs: *carbohydrates, proteins,* and *fats.* These large molecules are unable to cross plasma membranes intact to be absorbed from the lumen of the digestive tract into the blood or lymph. The process of digestion degrades these large food molecules into smaller nutrient molecules that can be absorbed (Table 16–1).

The simplest form of **carbohydrates** is the simple sugars or **monosaccharides** ("one sugar" molecules), such as **glucose, fructose,** and **galactose,** very few of which are normally found in the diet. Most ingested carbohydrate is in the form of **polysaccharides** ("many sugar" molecules), which consist of chains of interconnected glucose molecules. The most common polysaccharide consumed is **starch** derived from plant sources. Additionally, meat contains **glycogen,** the polysaccharide storage form of glucose in muscle. **Cellulose,** another dietary polysaccharide that is found in plant walls, cannot be digested into its constituent monosaccharides by the digestive juices secreted in humans; thus, it represents the undigested fiber or "bulk" in our diets. Besides polysaccharides, a lesser source of dietary carbohydrate is in the form of **disaccharides** ("two sugar" molecules), including **sucrose** (table sugar, which consists of one glucose and one fructose molecule) and **lactose** (milk sugar made up of one glucose and one galactose molecule).

Starch, glycogen, and disaccharides are converted through the process of digestion into their constituent monosaccharides, principally glucose with small amounts of fructose and galactose. These monosaccharides are the absorbable units for carbohydrates.

The second category of foodstuffs is **proteins,** which consist of various combinations of **amino acids** held together by peptide bonds (see p. A–10). Through the process of digestion, proteins are degraded primarily into their constituent **amino acids** as well as a few **small polypeptides** (several amino acids linked together by peptide bonds), both of which are the absorbable units for protein.

Fats represent the third category of foodstuffs. Most dietary fat is in the form of **triglycerides,** which are neutral fats, each consisting of a combination of **glycerol** with three (*tri* means "three") **fatty acid** molecules attached. During digestion, two of the fatty acid molecules are split off, leaving a **monoglyceride,** a glycerol molecule with one (*mono* means "one") fatty acid molecule attached. Thus, the end products of fat digestion are monoglycerides and free fatty acids, which are the absorbable units of fat.

• FIGURE 16–1   General Mode of Exocrine-Gland Secretion Exocrine-gland cells extract from the plasma by both active and passive means the raw materials that they need to produce their secretory product. This product is emptied into ducts, which lead, in the case of the digestive system, to the lumen of the digestive tract. Frequently, the secretion is modified as it moves through the duct by active and passive transport mechanisms within the membranes of the cells lining the duct.

Digestion is accomplished by enzymatic **hydrolysis** ("breakdown by water"). By adding H_2O at the bond site, enzymes in the digestive secretions break down the bonds that hold the small molecular subunits within the nutrient molecules together, thus setting the small molecules free. These small subunits were originally joined to form nutrient molecules by the removal of H_2O at the bond sites. Hydrolysis replaces the H_2O and frees the small absorbable units. Digestive

TABLE 16-1 Process of Digestion

enzymes are specific in the bonds they can hydrolyze. As food moves through the digestive tract, it is subjected to various enzymes, each of which breaks down the food molecules even further. In this way, large food molecules are converted to simple absorbable units in a progressive, stepwise fashion as the digestive-tract contents are propelled forward.

Absorption Digestion is completed and most absorption occurs in the small intestine. Through the process of **absorption**, the small absorbable units that result from digestion, along with water, vitamins, and electrolytes, are transferred from the digestive-tract lumen into the blood or lymph.

As we examine the digestive tract from beginning to end, we will discuss the four processes of motility, secretion, digestion, and absorption as they take place within each digestive organ (Table 16–2).

The digestive tract and accessory digestive organs make up the digestive system.

The digestive system consists of the digestive (gastrointestinal; *gastro* means "stomach") tract plus the accessory digestive organs. The **accessory digestive organs** include the *salivary glands,* the *exocrine pancreas,* and the *biliary system,* which is composed of the *liver* and *gallbladder.* These exocrine organs are located outside the wall of the digestive tract and empty their secretions through ducts into the digestive-tract lumen. They develop from outpouchings of the embryonic digestive tube and maintain their connection with the digestive tract through the ducts that are formed.

The **digestive tract** is essentially a tube about 30 feet (9 m) in length that runs through the middle of the body from the mouth to the anus. (Thirty feet is the length in a cadaver; the length is about half that in a living person because of ongoing contractions of the tract's muscular walls.) The digestive tract includes the following organs (Table 16–2): *mouth; pharynx* (throat); *esophagus; stomach; small intestine* (consisting of the *duodenum, jejunum,* and *ileum*); *large intestine* (composed of the *cecum, appendix, colon,* and *rectum*); and *anus.* It should be noted that these organs are continuous with each other and are discussed as separate entities only because of their regional modifications, which allow them to specialize in particular digestive activities.

Since the digestive tract is continuous from the mouth to the anus, the lumen of this tube, like the lumen of a straw, is continuous with the external environment. As a result, any contents within the lumen of the digestive tract are technically outside the body, just as the soda that you suck through a straw is not part of the straw. Only after a substance has been absorbed from the lumen across the intestinal wall is it considered to have become a part of the body. This fact is important because conditions essential to the digestive process can be tolerated in the digestive-tract lumen but could not be tolerated in the body proper. Consider the following examples:

■ The pH of the stomach contents falls as low as 2 as a result of the gastric secretion of hydrochloric acid (HCl), yet in the body fluids the range of pH compatible with life is 6.8 to 8.0.

■ The harsh digestive enzymes that hydrolyze food could also destroy the body's own tissues that produce them. Therefore, once they are synthesized in inactive form, these enzymes are not activated until they reach the lumen, where they actually attack the food outside the body (that is, within the lumen), thereby protecting the body tissues against self-digestion.

■ The lower portion of the intestine is inhabited by millions of living microorganisms that are normally harmless and even beneficial, yet if these same microorganisms enter the body proper (as may happen with a ruptured appendix), they may be extremely harmful or even lethal.

The wall of the digestive tract has the same general structure throughout most of its length from the esophagus to the anus, with some local variations characteristic for each region. A cross section of the digestive tube (• Fig. 16–2) reveals four major tissue layers. From the innermost layer of the tract outwards they are the *mucosa,* the *submucosa,* the *muscularis externa,* and the *serosa.*

The **mucosa** lines the luminal surface of the digestive tract. It is divided into three layers:

■ The primary component of the mucosa is a **mucous membrane**, an inner epithelial layer that serves as a protective surface as well as being modified in particular areas for secretion and absorption. The mucous membrane contains *exocrine cells* for secretion of digestive juices, *endocrine cells* for secretion of gastrointestinal hormones, and *epithelial cells* specialized for absorbing digested nutrients.

■ The **lamina propria** is a thin middle layer of connective tissue on which the epithelium rests. Small blood vessels, lymph vessels, and nerve fibers pass through the lamina propria, and it houses the gut-associated lymphoid tissue (GALT), which is important in the defense against intestinal bacteria (see p. 370).

■ The **muscularis mucosa** is a sparse outer layer of smooth muscle that lies adjacent to the submucosa.

The mucosal surface is generally not flat and smooth but is highly folded with many ridges and valleys that greatly increase the surface area available for absorption. The degree of folding varies in different areas of the digestive tract, being most extensive in the small intestine, where maximum absorption occurs, and least extensive in the esophagus, which merely serves as a transit tube. The pattern of surface folding can be modified by contraction of the muscularis mucosa. This is important primarily in exposing different areas of the absorptive surface to the luminal contents.

The **submucosa** ("under the mucosa") is a thick layer of connective tissue that provides the digestive tract with its distensibility and elasticity. It contains the larger blood and lymph vessels, both of which send branches inward to the mucosal layer and outward to the surrounding thick muscle layer. Also lying within the submucosa is a nerve network known as the *submucous plexus,* which helps control local activities of each gut region.

	DIGESTIVE ORGAN	MOTILITY
	Mouth and salivary glands	Chewing
	Pharynx and esophagus	Swallowing
	Stomach	Receptive relaxation; peristalsis
	Exocrine pancreas	Not applicable
	Liver	Not applicable
	Small intestine	Segmentation; migrating motility complex
	Large intestine	Haustrations; mass movements

Nasal passages
Mouth
Salivary glands
Pharynx
Pharyngoesophageal sphincter
Trachea
Esophagus

Gastroesophageal sphincter
Liver
Stomach
Gallbladder
Pancreas
Duodenum
Descending colon
Transverse colon
Ascending colon
Jejunum
Cecum
Ileum
Appendix
Sigmoid colon
Rectum
Anus

SECRETION	DIGESTION	ABSORPTION
Saliva -Amylase -Mucus -Lysozyme	Carbohydrate digestion begins	No foodstuffs; a few medications—for example, nitroglycerin
Mucus	None	None
Gastric juice -HCl -Pepsin -Mucus -Intrinsic factor	Carbohydrate digestion continues in body of stomach; protein digestion begins in antrum of stomach	No foodstuffs; a few lipid-soluble substances, such as alcohol and aspirin
Pancreatic digestive enzymes -Trypsin, chymotrypsin, carboxypeptidase -Amylase -Lipase Pancreatic aqueous $NaHCO_3$ secretion	These pancreatic enzymes accomplish digestion in duodenal lumen	Not applicable
Bile -Bile salts -Alkaline secretion -Bilirubin	Bile does not digest anything, but bile salts facilitate fat digestion and absorption in duodenal lumen	Not applicable
Succus entericus -Mucus -Salt (Small intestine enzymes are not secreted but function intracellularly in the brush border—disaccharidases and aminopeptidases)	In lumen, under influence of pancreatic enzymes and bile, carbohydrate and protein digestion continue and fat digestion is completely accomplished; in brush border, carbohydrate and protein digestion completed	All nutrients, most electrolytes, and water
Mucus	None	Salt and water, converting contents to feces

Surrounding the submucosa is the **muscularis externa**, the major smooth-muscle coat of the digestive tube. In most parts of the tract, it consists of two layers: an *inner circular layer* and an *outer longitudinal layer*. The fibers of the inner smooth-muscle layer (adjacent to the submucosa) run circularly around the circumference of the tube. Contraction of these circular fibers constricts or decreases the diameter of the lumen at the point of contraction. Contraction of the fibers in the outer layer, which run longitudinally along the length of the tube, accomplishes shortening of the tube. Together, contractile activity of these smooth-muscle layers produces the propulsive and mixing movements. Lying between the two muscle layers is another nerve network, the *myenteric plexus*, which, along with the submucous plexus, helps to regulate local gut activity.

The outer connective-tissue covering of the digestive tract is the **serosa**, which secretes a watery serous fluid that lubricates and prevents friction between the digestive organs and surrounding viscera. Throughout much of the tract, the serosa is continuous with the **mesentery**, which suspends the digestive organs from the inner wall of the abdominal cavity like a sling (Fig. 16–2). This attachment provides relative fixation, supporting the digestive organs in proper position, while still allowing them freedom for mixing and propulsive movements. One cause of **hernias**, or protrusions of an organ through the muscular wall of the cavity that contains them, is mesenteric tearing. Such tearing allows a portion of the digestive tube to fall free from its attachment and bulge through the abdominal wall.

Regulation of digestive function is complex and synergistic.

Digestive motility and secretion are carefully regulated to maximize digestion and absorption of the ingested food. Four factors are involved in the regulation of digestive-system function: (1) autonomous smooth-muscle function, (2) intrinsic nerve plexuses, (3) extrinsic nerves, and (4) gastrointestinal hormones.

Autonomous Smooth-Muscle Function Like self-excitable cardiac-muscle cells, some smooth-muscle cells are "pacesetter" cells that do not have a constant resting potential but rather display rhythmic, spontaneous variations in membrane potential. The prominent type of self-induced electrical activity in digestive smooth muscle is **slow-wave potentials** (see p. 251), alternatively referred to as the digestive tract's **basic electrical rhythm** (**BER**) or **pacesetter potential**. Slow waves are not action potentials and do not directly induce muscle contraction; they are rhythmic, wavelike fluctuations in membrane potential that cyclically bring the membrane closer to or farther from threshold. These slow-wave oscillations are believed to be due to cyclical variations in the rate at which the Na^+ pump transports Na^+ out of the pacesetter cell. Should these waves reach threshold at the peaks of depolarization, a volley of action potentials is triggered at each peak, resulting in repeating, rhythmical cycles of muscle contraction.

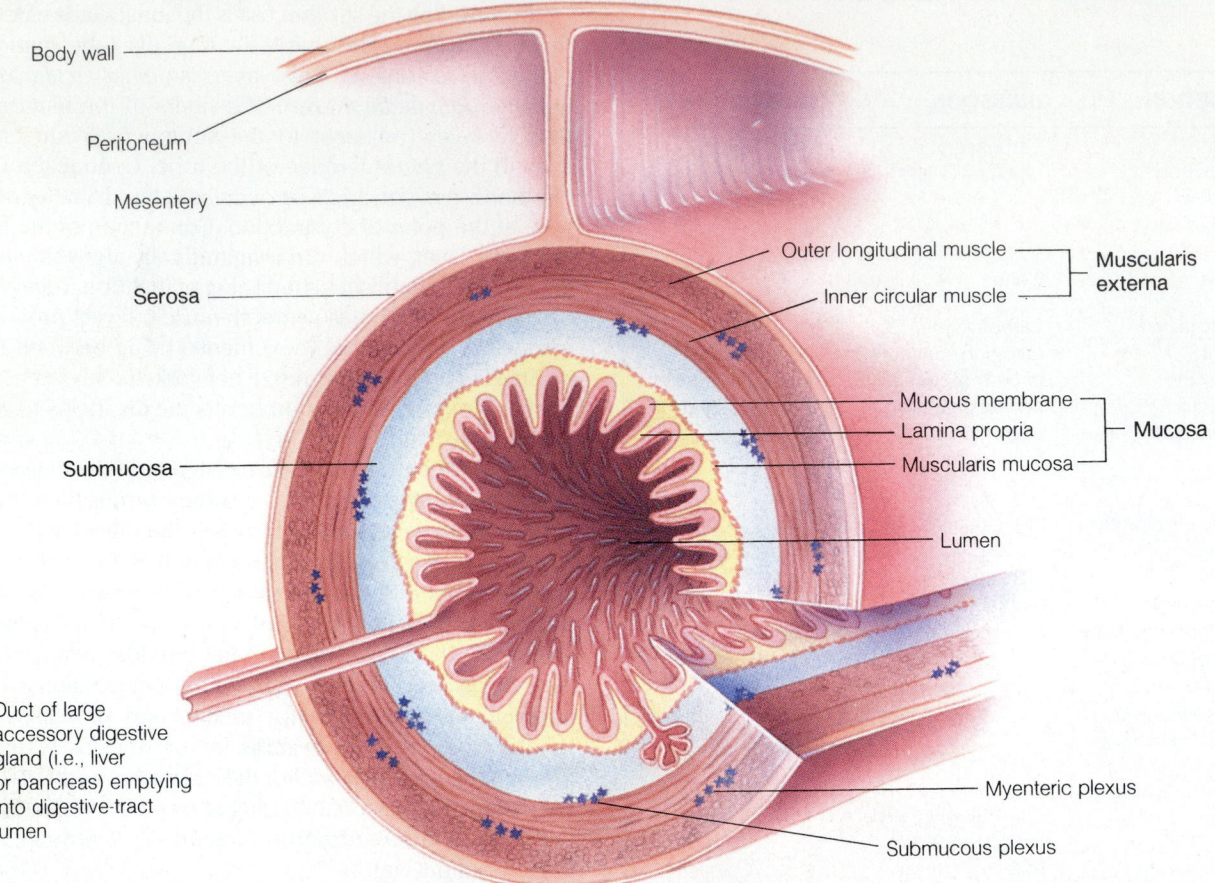

Body wall

Peritoneum

Mesentery

Serosa

Submucosa

Duct of large
accessory digestive
gland (i.e., liver
or pancreas) emptying
into digestive-tract
lumen

Outer longitudinal muscle ⎤ Muscularis
Inner circular muscle ⎦ externa

Mucous membrane ⎤
Lamina propria ⎬ Mucosa
Muscularis mucosa ⎦

Lumen

Myenteric plexus

Submucous plexus

● **FIGURE 16–2 Layers of the Digestive-Tract Wall** The digestive-tract wall consists of four major layers: from the innermost out, they
are the mucosa, submucosa, muscularis externa, and serosa.

Whether threshold is reached depends on the effect of various mechanical, nervous-system, and hormonal factors that influence the "resting" potential, or the starting point around which the slow-wave rhythm oscillates. If the starting point is nearer the threshold level, as it is when food is present in the digestive tract, the depolarizing slow-wave peak reaches threshold, so action-potential frequency and its accompanying contractile activity increase. Conversely, if the starting point is farther from threshold, as when no food is present, there is less likelihood of reaching threshold, so action-potential frequency is lowered and contractile activity is reduced.

Like cardiac muscle, sheets of smooth-muscle cells are connected by gap junctions (see p. 55), which serve as points of low electrical resistance so that electrical activity initiated in a digestive-tract pacesetter cell can spread to adjacent smooth-muscle cells. If threshold is reached and action potentials are triggered, the whole muscle sheet behaves like a functional syncytium, becoming excited and contracting as a unit. If threshold is not achieved, the oscillating electrical activity continues to sweep across the muscle without being accompanied by contractile activity.

The rate of rhythmic digestive contractile activities, such as peristalsis in the stomach, segmentation in the small intestine, and haustrations in the large intestine, depends on the inherent rate established by the involved pacesetter cells. The intensity of the contractions depends on the number of action potentials that occur when the slow-wave potential reaches threshold, which in turn depends on how long threshold is sustained. The greater the number of action potentials, the higher the cytosolic Ca^{++} concentration, the greater the cross-bridge activity, and the stronger the contraction. Thus, the level of contractility can range from low-level tone to vigorous mixing and propulsive movements by varying the cytosolic Ca^{++} concentration.

Intrinsic Nerve Plexuses   The second factor involved in the regulation of digestive-tract function is the **intrinsic nerve plexuses**. A nerve plexus is an interconnecting network of nerve cells. Two major networks of nerve fibers form the plexuses of the digestive tract: the **myenteric (Auerbach's) plexus**, which is located between the longitudinal and circular smooth-muscle layers, and the **submucous (Meissner's) plexus**, which is located in the submucosa. These two plexuses are known as intrinsic plexuses because they are located entirely within the digestive-tract wall. They run the entire length from the esophagus to the anus. Thus, unlike any other organ system, the digestive tract has its own intramural ("within wall") nervous system, which contains as many neurons as the spinal cord and endows the tract with a considerable degree of self-regulation. Together these two plexuses are often termed the **enteric** (*enteric* means "intestine") **nervous system**.

The intrinsic plexuses influence all facets of digestive-tract activity. Through innervation of the smooth-muscle cells and exocrine and endocrine cells of the digestive tract, the intrinsic plexuses directly affect digestive-tract motility, secretion of digestive juices, and secretion of gastrointestinal hormones. These intrinsic nerve networks are primarily responsible for coordinating local activity within the digestive tract. For example, if a large piece of food gets stuck in the esophagus, local contractile responses coordinated by the intrinsic plexuses are initiated to push the food forward. Intrinsic nerve activity can, in turn, be influenced by the extrinsic nerves.

Extrinsic Nerves The **extrinsic nerves** are the nerves that originate outside the digestive tract and innervate the various digestive organs—namely, nerve fibers from both branches of the autonomic nervous system. The autonomic nerves influence digestive-tract motility and secretion either by modifying ongoing activity in the intrinsic plexuses, altering the level of gastrointestinal hormone secretion, or, in some instances, acting directly on the smooth muscle and glands.

Recall that in general, the sympathetic and parasympathetic nerves supplying any given tissue exert opposing actions on that tissue. The sympathetic system, which dominates in fight-or-flight situations, tends to inhibit or slow down digestive-tract contraction and secretion. This action is appropriate considering that digestive processes are not of highest priority when the body is faced with an emergency or threat from the external environment. The parasympathetic nervous system, on the other hand, dominates in quiet, relaxed situations when general maintenance types of activities such as digestion can proceed optimally. Accordingly, the parasympathetic nerve fibers supplying the digestive tract, which arrive primarily by way of the vagus nerve, tend to increase smooth-muscle motility and promote secretion of digestive enzymes and hormones.

In addition to being called into play during generalized sympathetic or parasympathetic discharge, the autonomic nerves supplying the digestive system can be discretely activated to modify only digestive activity. One of the major purposes of specific activation of extrinsic innervation is the coordination of activity between different regions of the digestive system; for example, the act of chewing food reflexly increases not only salivary secretion but also stomach, pancreatic, and liver secretion via vagal reflexes in anticipation of the arrival of food. Another purpose of specific activation is the provision of a pathway by which factors outside the digestive system can influence digestion, as, for example, with the vagally mediated increase in digestive juices that occurs in anticipation of a meal when a person sees or smells food.

Gastrointestinal Hormones The fourth factor influencing digestive-tract activity is hormonal control. Tucked within the mucosa of certain regions of the digestive tract are endocrine-gland cells that release hormones into the blood upon appropriate stimulation. These **gastrointestinal hormones** are carried through the blood to other areas of the digestive tract, where they exert either excitatory or inhibitory influences on smooth muscle and exocrine-gland cells. In feedforward fashion (see p. 12), they also act on the endocrine cells of the pancreas to influence the secretion of pancreatic hormones, which play a key role in the uptake and storage of absorbed nutrient molecules. Gastrointestinal hormones are released primarily in response to specific local changes in the luminal contents (such as the presence of protein, fat, or acid), acting either directly on the endocrine-gland cells or indirectly through the intrinsic plexuses or extrinsic autonomic nerves. Interestingly, many of these same hormones are released from neurons in the brain, where they act as neurotransmitters and neuromodulators. During embryonic development, certain cells of the developing neural tissue migrate to the digestive system, where they become endocrine cells.

Receptor activation alters digestive activity through neural reflexes and hormonal pathways.

The wall of the digestive tract contains three different types of sensory receptors that respond to local chemical or mechanical changes in the digestive tract: (1) *chemoreceptors* sensitive to chemical components within the lumen; (2) *mechanoreceptors* (pressurereceptors) sensitive to stretch or tension within the wall; and (3) *osmoreceptors* sensitive to the osmolarity of the luminal contents. Stimulation of these receptors elicits neural reflexes or secretion of hormones, both of which alter the level of activity in the digestive system's effector cells. These effector cells include smooth-muscle cells (for modifying motility), exocrine-gland cells (for controlling secretion of digestive juices), and endocrine-gland cells (for varying secretion of gastrointestinal and pancreatic hormones) (• Fig. 16–3).

Receptor activation may elicit two types of neural reflexes—short reflexes and long reflexes. When the intrinsic nerve networks influence local motility or secretion in response to specific local stimulation, such a reflex, in which all elements of the reflex arc are located within the wall of the digestive tract itself, is known as a **short reflex**. Extrinsic autonomic nervous activity can be superimposed on the local controls to modify smooth-muscle and glandular responses, either to correlate activity between different regions of the digestive system or to modify digestive-system activity in response to external influences. Because the autonomic reflexes involve long pathways between the central nervous system and digestive system, they are known as **long reflexes**. In addition to these neural reflexes, digestive-system activity is also coordinated by the secretion of gastrointestinal hormones, which are triggered directly by local changes in the digestive tract or by short or long reflexes.

From this overview, it should be apparent that regulation of gastrointestinal function is very complex, being influenced by many synergistic, interrelated pathways designed to assure that the appropriate responses occur to digest and absorb the ingested food. Nowhere else in the body is such a level of overlapping control exercised.

We are now going to take a "tour" of the digestive tract, beginning with the mouth and ending with the anus. We will examine the four basic digestive processes of motility, secretion, digestion, and absorption at each digestive organ along the way. Table 16–2 summarizes these activities and serves as a useful reference throughout the remainder of the chapter.

MOUTH

The oral cavity is the entrance to the digestive tract.

Entry to the digestive tract is through the **mouth** or **oral cavity**. The opening is formed by the muscular **lips**, which help procure, guide, and contain the food in the mouth. The lips also serve nondigestive functions; they are important in speech (articulation of many sounds depends on a particular lip formation) and as a sensory receptor in interpersonal relationships (for example, as in kissing).

The **palate**, which forms the arched roof of the oral cavity, separates the mouth from the nasal passages. Its presence allows breathing and chewing or sucking to take place simultaneously. Embryologically, the palate is derived from projections that grow inward from the jaws on both sides and fuse in the midline of the oral cavity. Failure of these projections

to unite properly results in a cleft palate, which can interfere with sucking, eating, and speech if not surgically corrected. Toward the front of the mouth, the palate is made of bone, forming what is known as the **hard palate**. There is no bone in the portion of the palate toward the rear of the mouth; this region is called the **soft palate**. Hanging down from the soft palate in the rear of the throat is a dangling projection, the **uvula**, which plays an important role in sealing off the nasal passages during swallowing. (The uvula is the structure you elevate when you say "ahhh" so that the physician can better see your throat.)

The **tongue**, which forms the floor of the oral cavity, is composed of voluntarily controlled skeletal muscle. Movements of the tongue are not only important in guiding food within the mouth during chewing and swallowing, but they also play an important role in speech. Embedded within the tongue are the **taste buds** (see p. 190), which are dispersed on the soft palate, throat, and linings of the cheeks as well.

The **pharynx** is the cavity at the rear of the throat. It acts as a common passageway for both the digestive system (by serving as the link between the mouth and esophagus for food) and the respiratory system (by providing access between the nasal passages and trachea for air). This arrangement necessitates mechanisms (to be described shortly) to guide food and air into the proper passageways beyond the pharynx. Housed within the side walls of the pharynx are the **tonsils**, which are lymphoid organs that are part of the body's defense team.

The teeth are responsible for chewing, which breaks up food, mixes it with saliva, and stimulates digestive secretions.

The first step in the digestive process is **mastication**, or **chewing**, the motility of the mouth that involves the slicing, tearing, grinding, and mixing of ingested food by the **teeth**. The teeth are firmly embedded in and protrude from the jaw bones. The exposed portion of a tooth is covered by **enamel**, the hardest structure of the body. Enamel is formed prior to the tooth's eruption by special cells that are lost as the tooth erupts. Since it cannot be regenerated after the tooth has erupted, any defects ("cavities") that develop in the enamel must be patched by artificial "fillings," or else the surface will continue to erode into the underlying living pulp.

The upper and lower teeth normally fit together when the jaws are closed. This **occlusion** allows food to be ground and crushed between the tooth surfaces. When the teeth do not make proper contact with each other, they cannot accomplish their normal cutting and grinding action adequately. Such **malocclusion** results from abnormal positioning of the teeth and is often caused either by overcrowding of teeth too large for the available jaw space or by one jaw being displaced in relation to the other. In addition to inefficient chewing, malocclusion can cause abnormal wearing of affected tooth surfaces and dysfunction and pain of the **temporomandibular joint** (**TMJ**), where the jaw bones articulate with each other. Malocclusions can often be corrected by applying braces, which exert prolonged gentle pressure against the teeth to move them gradually to the desired position.

The purposes of chewing are (1) to grind and break food up into smaller pieces to facilitate swallowing; (2) to mix food with saliva; and (3) to stimulate the taste buds. The latter not only gives rise to the pleasurable subjective sensation of taste but also in feedforward fashion reflexly increases salivary, gastric, pancreatic, and bile secretion to prepare for the arrival of food.

The act of chewing can be voluntary, but most chewing during a meal is a rhythmic reflex brought about by activation of the skeletal muscles of the jaws, lips, cheeks, and tongue in response to the pressure of food against the oral tissues.

The teeth can exert forces much greater than those necessary to eat ordinary food. For example, the molars in an adult man can exert a crushing force of up to 200 pounds, which is sufficient to crack a hard nut, but these powerful forces are not ordinarily used. In fact, the degree of occlusion is more important than the force of the bite in determining the efficiency of chewing.

Saliva begins carbohydrate digestion but plays more important roles in oral hygiene and in facilitating speech.

Saliva, the secretion associated with the mouth, is produced by three major pairs of salivary glands—the **sublingual**, **submandibular**, and **parotid glands** (● Fig. 16–4)—that are lo-

● **FIGURE 16–4 Salivary Glands** Three major pairs of salivary glands (parotid, submandibular, and sublingual) are located in and around the oral cavity and empty into the mouth via small ducts.

cated outside the oral cavity and discharge saliva through short ducts into the mouth. In addition, minor salivary glands, the **buccal glands**, are located in the mucosa lining the cheeks.

Saliva is composed of about 99.5% H_2O and 0.5% protein and electrolytes. The most important salivary proteins—*amylase, mucus,* and *lysozyme*—contribute to the functions of saliva as follows:

1. Saliva begins digestion of carbohydrate in the mouth through action of **salivary amylase**, an enzyme that breaks polysaccharides down into disaccharides.

2. Saliva facilitates swallowing by moistening food particles, thereby holding them together, and by providing lubrication through the presence of **mucus**, which is thick and slippery.

3. Saliva exerts some antibacterial action by means of a twofold effect—first by **lysozyme**, an enzyme that lyses or destroys certain bacteria, and second by rinsing away material that may serve as a food source for bacteria.

4. Saliva serves as a solvent for molecules that stimulate the taste buds. Only molecules in solution can react with taste-bud receptors. You can demonstrate this for yourself: dry your tongue and then drop some sugar on it—you cannot taste the sugar until it is moistened. Glucose is never present in saliva and the salivary salt concentration is considerably below that in the plasma. This fact is probably important in the perception of sweet and salty tastes.

5. Saliva aids speech by facilitating movements of the lips and tongue. It is difficult to talk when the mouth feels dry.

6. Saliva plays an important role in oral hygiene by helping to keep the mouth and teeth clean. The constant flow of saliva helps to flush away food residues, shed epithelial cells, and foreign particles. Saliva's contribution in this regard is apparent to anyone who has experienced a foul taste in the mouth when salivation is suppressed for a while, such as during a fever or states of prolonged anxiety.

7. Bicarbonate buffers in the saliva neutralize acids in food as well as acids produced by bacteria in the mouth, thereby helping to prevent dental caries (cavities).

In spite of these many functions, saliva is not essential for the digestion and absorption of foods, because enzymes produced by the pancreas and small intestine can complete the digestion of food even in the absence of salivary and gastric secretion. The main problems associated with diminished salivary secretion, a condition known as **xerostomia**, are difficulty in chewing and swallowing, inarticulate speech unless frequent sips of water are taken when talking, and a rampant increase in dental caries.

The continuous low level of salivary secretion can be increased by simple and conditioned reflexes.

On the average, about 1 to 2 liters of saliva are secreted per day, ranging from a continuous spontaneous basal rate of 0.5 ml/min to a maximum flow rate of about 5 ml/min in response to a potent stimulus such as sucking on a lemon. The continuous spontaneous secretion of saliva, even in the absence of apparent stimuli, is due to constant low-level stimulation by the parasympathetic nerve endings that terminate in the salivary glands. This basal secretion is important in keeping the mouth and throat moist at all times.

In addition to this continuous, low-level secretion, salivary secretion may be enhanced by two different types of salivary reflexes (● Fig. 16–5): (1) the simple, or unconditioned, salivary reflex and (2) the acquired, or conditioned, salivary reflex. The **simple**, or **unconditioned, salivary reflex** occurs when chemoreceptors and pressurereceptors within the oral cavity respond to the presence of food. On activation, these receptors initiate impulses in afferent nerve fibers that carry the information to the **salivary center** located in the medulla of the brain stem. The salivary center, in turn, sends impulses via the extrinsic autonomic nerves to the salivary glands to promote increased salivation. Dental procedures promote salivary secretion in the absence of food in the mouth because these manipulations activate pressurereceptors in the mouth.

With the **acquired**, or **conditioned, salivary reflex**, salivation occurs without oral stimulation. Just thinking about, seeing, smelling, or hearing the preparation of pleasant food initiates salivation through this reflex. All of us have experienced such "mouth watering" in anticipation of something delicious to eat. This reflex is a learned response based on previous experience. Inputs that arise outside the mouth and are mentally associated with the pleasure of eating act through the cerebral cortex to stimulate the medullary salivary center.

The salivary center controls the degree of salivary output by means of the autonomic nerves that supply the salivary glands. Unlike the autonomic nervous system elsewhere in the body, sympathetic and parasympathetic responses in the salivary glands are not antagonistic. Both sympathetic and parasympathetic stimulation increase salivary secretion, but the quantity, characteristics, and mechanisms are different. Parasympathetic stimulation, which exerts the dominant role in salivary secretion, produces a prompt and abundant flow of watery saliva that is rich in enzymes. Sympathetic stimulation, on the other hand, produces a much smaller volume of

● **FIGURE 16–5   Control of Salivary Secretion**

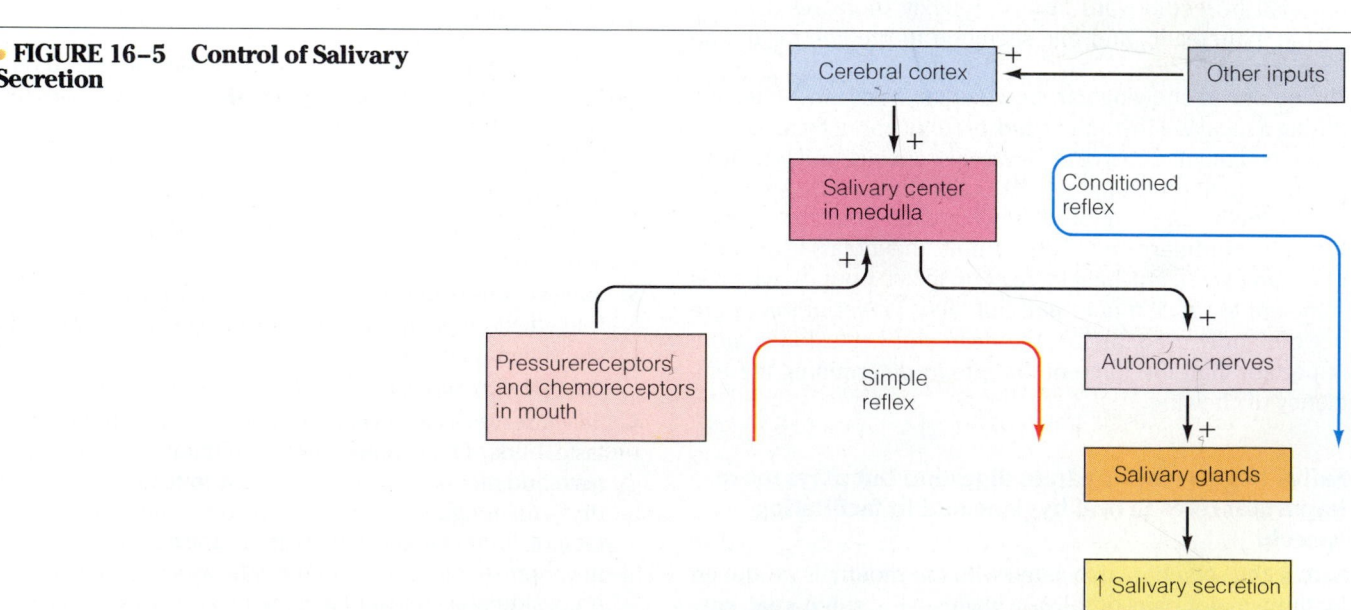

thick saliva that is rich in mucus. Because sympathetic stimulation elicits a smaller volume of saliva, the mouth feels drier than usual during circumstances when the sympathetic system is dominant, such as stress situations. Thus, people experience a dry feeling in the mouth when they are nervous about giving a speech.

Salivary secretion is the only digestive secretion entirely under neural control. All other digestive secretions are regulated by both nervous-system reflexes and hormones.

Digestion in the mouth is minimal, and no absorption of nutrients occurs.

Digestion in the mouth involves the hydrolysis of polysaccharides into disaccharides by amylase. However, most digestion by this enzyme is accomplished in the body of the stomach after the food mass and saliva have been swallowed. Acid inactivates amylase, but in the center of the food mass, where stomach acid has not yet reached, this salivary enzyme continues to function for several more hours.

No absorption of foodstuff occurs from the mouth. Importantly, some therapeutic agents can be absorbed by the oral mucosa, a prime example being a vasodilator drug, *nitroglycerin*, which is used by certain cardiac patients to relieve anginal attacks associated with myocardial ischemia (see pp. 278 and 291).

PHARYNX AND ESOPHAGUS

Swallowing is a sequentially programmed all-or-none reflex.

The motility associated with the pharynx and esophagus is **swallowing**, or **deglutition**. Most of us think of swallowing as the limited act of moving food out of the mouth into the esophagus. However, swallowing actually refers to the entire process of moving food from the mouth through the esophagus into the stomach.

Swallowing is initiated when a **bolus**, or ball of food, is voluntarily forced by the tongue to the rear of the mouth into the pharynx. The pressure of the bolus in the pharynx stimulates pharyngeal pressurereceptors, which send afferent impulses to the **swallowing center** located in the medulla. The swallowing center then reflexly activates in the appropriate sequence the muscles that are involved in swallowing. Swallowing is an example of a sequentially programmed all-or-none reflex in which multiple responses are triggered in a specific timed sequence; that is, a number of highly coordinated activities are initiated in a regular pattern over a period of time to accomplish the act of swallowing. Swallowing is initiated voluntarily, but once it is initiated, it cannot be stopped. Perhaps you have experienced this when a large piece of hard candy inadvertently slipped to the rear of your throat, triggering an unintentional swallow.

During the oropharyngeal stage of swallowing, food is directed into the esophagus and is prevented from entering the wrong passageways.

Swallowing is arbitrarily divided into two stages: the oropharyngeal stage and the esophageal stage. The **oropharyngeal stage** lasts about one second and consists of moving the bolus from the mouth through the pharynx and into the esophagus. When the bolus enters the pharynx during swallowing, it must be directed into the esophagus and prevented from entering the other openings that communicate with the pharynx. In other words, food must be prevented from reentering the mouth, from entering the nasal passages, and from entering the trachea. All of this is accomplished by the following coordinated activities (● Fig. 16–6):

- Food is prevented from reentering the mouth during swallowing by the position of the tongue against the hard palate.
- The uvula is elevated and lodges against the back of the throat, sealing off the nasal passage from the pharynx so that food does not enter the nose.
- Food is prevented from entering the trachea primarily by elevation of the larynx and tight closure of the vocal cords across the laryngeal opening, or **glottis**. The first portion of the trachea is the *larynx* or *voice box*, across which are stretched the *vocal cords* (see p. 414). During swallowing, the vocal cords serve a purpose unrelated to speech. Contraction of laryngeal muscles aligns the vocal cords in tight apposition to each other, thus sealing the glottis entrance. Also, the bolus tilts a small flap of cartilaginous tissue, the **epiglottis**, backward down over the closed glottis as further protection from food entering the respiratory airways.
- Because the respiratory passages are temporarily sealed off during swallowing, respiration is briefly inhibited so that the individual does not attempt futile respiratory efforts.
- With the larynx and trachea sealed off, pharyngeal muscles contract to force the bolus into the esophagus.

The esophagus is guarded by sphincters at both ends.

The **esophagus** is a fairly straight muscular tube that extends between the pharynx and stomach (Table 16–2). Lying for the most part in the thoracic cavity, it penetrates the diaphragm and joins the stomach in the abdominal cavity a few centimeters below the diaphragm. Occasionally, a portion of the stomach herniates through the esophageal hiatus and protrudes into the thoracic cavity, a condition known as a **hiatal hernia**.

The esophagus is guarded at both ends by sphincters. A sphincter is a ringlike muscular structure that, when closed, prevents passage through the tube it guards. The upper esophageal sphincter is the **pharyngoesophageal sphincter**, and the lower sphincter is the **gastroesophageal sphincter**.

Since the esophagus is exposed to subatmospheric intrapleural pressure (see p. 418), a pressure gradient exists between the atmosphere and the esophagus. Accordingly, if the

(a)

Nasal
passages

Hard palate

Soft palate

Uvula

Pharynx

Epiglottis

Esophagus

Trachea

Bolus

Tongue

Glottis at entrance of larynx

(b)

Swallowing center
inhibits respiratory
center in brain

Elevation of uvula
prevents food from
entering nasal passages

Position of tongue
prevents food from
reentering mouth

Epiglottis is pressed
down over closed
glottis as auxiliary
mechanism to prevent
food from entering
airways

Tight apposition of vocal cords across
glottis prevents food from entering
respiratory airways (viewed from above)

● **FIGURE 16–6   Oropharyngeal Stage of Swallowing**   (a) Position of oropharyngeal structures at rest. (b) Changes that occur during the oropharyngeal stage of swallowing to prevent the bolus of food from entering the wrong passageways.

entrance to the esophagus were not closed, air would enter the esophagus as well as the trachea with each breath. Except during a swallow, the pharyngoesophageal sphincter keeps the entrance to the esophagus closed to prevent large volumes of air from entering the esophagus and stomach during breathing. Instead, the air is directed only into the respiratory airways. Were it not for the pharyngoesophageal sphincter, the digestive tract would be subjected to large volumes of gas, which would lead to excessive **eructation** (burping). In contrast to most sphincters, passive elastic tensions in the walls of the pharyngoesophageal sphincter cause the esophagus to be closed when this sphincter is relaxed. During swallowing, this sphincter contracts, opening the sphincter and allowing the bolus to pass into the esophagus. Once the bolus has entered the esophagus, the pharyngoesophageal sphincter closes, the respiratory airways are opened, and breathing resumes. The oropharyngeal stage is complete, and about one second has passed since the swallow was first voluntarily initiated.

Peristaltic waves push the food through the esophagus.

The **esophageal stage** of the swallow now begins. The swallowing center initiates a **primary peristaltic wave** that sweeps from the beginning to the end of the esophagus, forcing the bolus ahead of it through the esophagus to the stomach. **Peristalsis** refers to ringlike contractions of the circular smooth muscle that move progressively forward with a stripping

motion, pushing the bolus ahead of the contraction (● Fig. 16–7). Thus, propulsion of food through the esophagus is an active process that does not rely on gravity. Food can be pushed to the stomach even during a headstand. The peristal-

● **FIGURE 16–7   Peristalsis in the Esophagus**   As the wave of peristaltic contraction sweeps down the esophagus, it pushes the bolus ahead of it toward the stomach.

Ringlike peristaltic
contraction sweeping
down the esophagus

tic wave takes about five to nine seconds to reach the lower end of the esophagus. Progression of the wave is controlled by the swallowing center, with innervation by means of the vagus.

Liquids, not being held up by friction against the esophageal wall, fall quickly to the lower esophageal sphincter by gravity and then must wait about five seconds until the primary peristaltic wave finally arrives before they can pass through the gastroesophageal sphincter.

If a large or sticky swallowed bolus, such as a bite of peanut butter sandwich, fails to be carried along to the stomach by the primary wave of peristalsis, the lodged bolus distends the esophagus, stimulating pressurereceptors within its walls and thus initiating a second, more forceful peristaltic wave that is mediated by the intrinsic nerve plexuses at the level of the distention. These **secondary peristaltic waves** do not involve the swallowing center, nor is the person aware of their occurrence. Distention of the esophagus also reflexly increases salivary secretion. The trapped bolus is eventually dislodged and moved forward through the combination of lubrication by the extra swallowed saliva and the forceful secondary peristaltic waves.

The gastroesophageal sphincter prevents reflux of gastric contents.

Except during swallowing, the gastroesophageal sphincter remains contracted to maintain a barrier between the stomach and esophagus, thus reducing the possibility of reflux of acidic gastric contents into the esophagus. If gastric contents do flow back into the esophagus in spite of the sphincter, the acidity of these contents irritates the esophagus, causing the esophageal discomfort known as **heartburn**. (The heart itself is not involved at all.)

The gastroesophageal sphincter relaxes reflexly as the peristaltic wave sweeps down the esophagus so that the bolus can pass into the stomach. After the bolus has entered the stomach, the gastroesophageal sphincter again contracts.

In a condition known as **achalasia**, the lower esophageal sphincter fails to relax during swallowing but instead contracts more vigorously. Food accumulates in the esophagus, which causes the esophagus to become enormously distended as the food's passage into the stomach is greatly delayed. People with this condition are prone to aspiration pneumonia because of the increased likelihood that some of the detained meal in the esophagus may spill into the pharynx and accidentally be aspirated (sucked) into the lungs. The underlying defect is apparently the result of damage to the myenteric nerve plexus in the region of the gastroesophageal sphincter.

Esophageal secretion is entirely protective.

Esophageal secretion is entirely mucus. In fact, mucus is secreted throughout the length of the digestive tract. By providing lubrication for passage of food, esophageal mucus lessens the likelihood that the esophagus will be damaged by any sharp edges in the newly entering food. Furthermore, it protects the esophageal wall from acid and enzymes in gastric juice if gastric reflux should occur.

The entire transit time in the pharynx and esophagus averages a mere six to ten seconds, too short a time for any digestion or absorption to occur in this region.

⊜ STOMACH

The stomach stores food and begins protein digestion.

The **stomach** is a J-shaped saclike chamber lying between the esophagus and small intestine. It is arbitrarily divided into three sections based on anatomical, histological, and functional distinctions (● Fig. 16–8). The **fundus** is the portion of the stomach that lies above the esophageal opening. The middle or main part of the stomach is the **body**. The smooth-muscle layers in the fundus and body are relatively thin, but the lower portion of the stomach, the **antrum**, has much heavier musculature. There are also glandular differences in the mucosa of these regions, as will be described later. The terminal portion of the stomach consists of the **pyloric sphincter**, which acts as a barrier between the stomach and the upper part of the small intestine, the duodenum.

The stomach performs several functions. The most important function is to store ingested food until it can be emptied into the small intestine at a rate appropriate for optimal digestion and absorption. It takes hours to digest and absorb a meal that was consumed in only a matter of minutes. Because the small intestine is the primary site for this digestion and absorption, it is important that the stomach store the food and

---

● **FIGURE 16–8   Anatomy of the Stomach** The stomach is divided into three sections based on structural and functional distinctions—the fundus, body, and antrum.

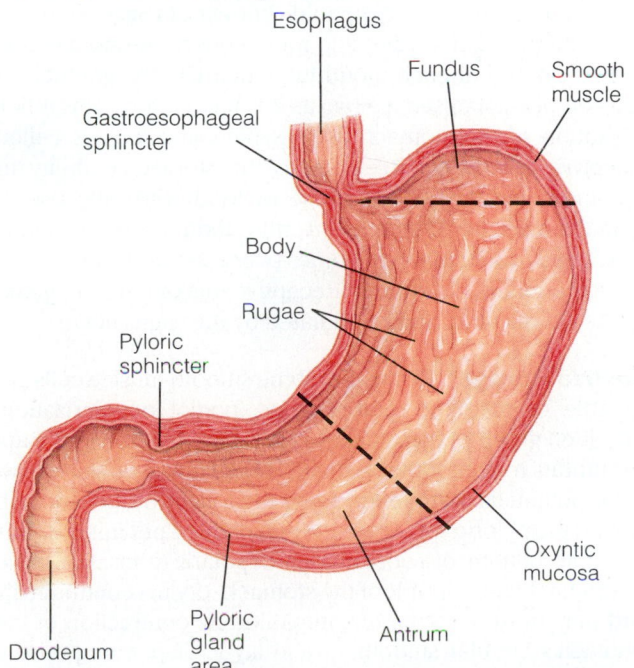

Esophagus
Fundus
Smooth muscle
Gastroesophageal sphincter
Body
Rugae
Pyloric sphincter
Oxyntic mucosa
Pyloric gland area
Antrum
Duodenum

meter it into the duodenum at a rate that does not exceed the small intestine's capacities. A second function of the stomach is to secrete hydrochloric acid (HCl) and enzymes that begin protein digestion. Finally, through the stomach's mixing movements, the ingested food is pulverized and mixed with gastric secretions to produce a thick liquid mixture known as **chyme**.

Stomach motility is complex and subject to multiple regulatory inputs.

There are four aspects of gastric motility: (1) gastric filling, (2) gastric storage, (3) gastric mixing, and (4) gastric emptying.

Gastric Filling   When empty, the stomach has a volume of about 50 ml, but it can expand to a capacity of about 1 liter (1,000 ml) during a meal. Accommodation of such a twenty-fold change in volume would create tension in the walls of the stomach and greatly increase intragastric pressure were it not for two factors: (1) the basic plasticity of the stomach smooth muscle and (2) receptive relaxation of the stomach as it fills.

Plasticity refers to the ability of smooth muscle to maintain a constant tension over a wide range of lengths, unlike skeletal and cardiac muscle, which exhibit length-tension relationships (see p. 229). Thus, when the stomach smooth-muscle fibers are stretched by gastric filling, they yield without an increase in muscle tension. Beyond a certain level of stretch, however, a stretch-activated contraction can be superimposed on this passive plastic behavior. Substantial stretch depolarizes the pacesetter cells, bringing them closer to resting potential so that the slow-wave potential is able to reach threshold and initiate contractile activity.

This basic behavior of smooth muscle is augmented by reflex relaxation of the stomach as it fills. The interior of the stomach is thrown into deep folds known as **rugae**. During a meal, the folds get smaller and flatten out as the stomach relaxes slightly with each mouthful, much like the gradual expansion of a collapsed ice bag as it is being filled. This reflex relaxation of the stomach as it is receiving food is called **receptive relaxation**; it enhances the stomach's ability to accommodate the extra volume of food with little rise in stomach pressure. Of course, if more than 1 liter of food is consumed, the stomach becomes overdistended and the person experiences discomfort. Receptive relaxation is triggered by the act of eating and is mediated by the vagus nerve.

Gastric Storage   Recall that some smooth-muscle cells are capable of rhythmic, autonomous, partial depolarization. One such group of these pacesetter cells is located in the upper fundus region of the stomach. These cells generate slow-wave potentials that sweep down the length of the stomach toward the pyloric sphincter at a rate of three per minute. This rhythmic pattern of spontaneous depolarizations, the basic electrical rhythm or BER of the stomach, occurs continuously and may or may not be accompanied by contraction of the stomach's circular smooth-muscle layer. Depending on the level of excitability in the smooth muscle, it may be brought

to threshold by this flow of current and undergo action potentials, which in turn initiate muscle contractions recognized as peristaltic waves that sweep over the stomach in pace with the BER at a rate of three per minute.

Once initiated, the peristaltic wave spreads over the fundus and body to the antrum and pyloric sphincter. Because the muscle layers are thin in the fundus and body, the peristaltic contractions in this region are weak. When the waves reach the antrum, they become much stronger and more vigorous because the muscle there is much thicker.

Since only feeble mixing movements occur in the body and fundus, food emptied into the stomach from the esopha-

---

• **FIGURE 16–9   Gastric Emptying and Mixing as a Result of Antral Peristaltic Contractions** (a) Gastric emptying. A peristaltic contraction originates in the upper fundus and sweeps down toward the pyloric sphincter, becoming more vigorous as it reaches the thick-muscled antrum. As the strong antral peristaltic contraction propels the chyme forward, a small portion of chyme is pushed through the partially open sphincter into the duodenum. The stronger the antral contraction, the more chyme is emptied with each contractile wave. (b) Gastric mixing. When the peristaltic contraction reaches the pyloric sphincter, the sphincter is tightly closed and no further emptying takes place. When chyme that was being propelled forward hits the closed sphincter, it is tossed back into the antrum. Mixing of chyme is accomplished as chyme is propelled forward and tossed back into the antrum with each peristaltic contraction.

(a)

(b)

gus is stored in the relatively quiet body without being mixed. The fundic area usually does not store food but contains only a pocket of gas. Food is gradually fed from the body into the antrum, where mixing does take place.

Gastric Mixing   The strong antral peristaltic contractions are responsible for the mixing of food with gastric secretions to produce chyme. Each antral peristaltic wave propels chyme forward toward the pyloric sphincter. Tonic contraction of the pyloric sphincter normally keeps it almost, but not completely, closed. The opening is large enough for water and other fluids to pass through with ease but too small for the thicker chyme to pass through except when a strong peristaltic antral contraction pushes it through. Even then, of the 30 ml of chyme that the antrum can hold, usually only a few milliliters of antral contents are forced into the duodenum with each peristaltic wave. Before more chyme can be squeezed out, the peristaltic wave reaches the pyloric sphincter and causes it to contract more forcefully, sealing off the exit and blocking further passage into the duodenum. The bulk of the antral chyme that was being propelled forward but failed to be pushed into the duodenum is abruptly halted at the closed sphincter and is tumbled back into the antrum, only to be propelled forward and tumbled back again as the new peristaltic wave advances (• Fig. 16–9). This tossing back and forth, called **retropulsion**, accomplishes thorough mixing of the chyme in the antrum.

Gastric Emptying   The antral peristaltic contractions, in addition to being responsible for gastric mixing, provide the driving force for gastric emptying. The amount of chyme that escapes into the duodenum with each peristaltic wave before the pyloric sphincter closes tightly depends largely on the strength of peristalsis. The intensity of antral peristalsis can vary markedly under the influence of different signals from both the stomach and the duodenum; thus, gastric emptying is regulated by both gastric and duodenal factors. By slightly depolarizing or hyperpolarizing the gastric smooth muscle, these factors influence the muscle's excitability, which in turn is a determinant of the degree of antral peristaltic activity. The greater the excitability, the more frequently the BER will generate action potentials, the greater the degree of peristaltic activity in the antrum, and the faster the rate of gastric emptying (Table 16–3).

Factors in the stomach that influence the rate of gastric emptying.   The main gastric factor that influences the strength of contraction is the amount of chyme in the stomach. Other things being equal, the stomach empties at a rate proportional to the volume of chyme in it at any given time. Distention of the stomach triggers increased gastric motility through a direct effect of stretch on the smooth muscle as well as through involvement of the intrinsic plexuses, the vagus nerve, and the stomach hormone *gastrin*. (The control and other functions of this hormone, which is secreted by special endocrine cells in the antrum, will be described later.)

Furthermore, the degree of fluidity of the chyme in the stomach influences gastric emptying. The stomach contents must be converted into a finely divided, thick liquid form before emptying. The sooner the appropriate degree of fluidity can be achieved, the more rapidly the contents are ready to be evacuated.

TABLE 16–3 Factors Regulating Gastric Motility and Emptying		
FACTORS	**MODE OF REGULATION**	**EFFECTS ON GASTRIC MOTILITY AND EMPTYING**
Within Stomach		
Volume of chyme	Distention has direct effect on gastric smooth-muscle excitability, as well as acting through intrinsic plexuses, vagus nerve, and gastrin	Increased volume stimulates motility and emptying
Degree of fluidity	Direct effect; contents must be in fluid form to be evacuated	Increased fluidity allows more rapid emptying
Within Duodenum		
Presence of fat, acid, hypertonicity, or distention	Initiates enterogastric reflex or triggers release of enterogastrones (cholecystokinin, secretin, gastric inhibitory peptide)	These factors in duodenum inhibit further gastric motility and emptying until duodenum has coped with factors already present
Outside Digestive System		
Emotion	Alters autonomic balance	Stimulates or inhibits motility and emptying
Intense pain	Increases sympathetic activity	Inhibits motility and emptying
Decreased glucose utilization in hypothalamus	Increases vagal activity	Stimulates motility; accompanied by hunger pangs

Factors in the duodenum that influence the rate of gastric emptying. In spite of these gastric influences, factors in the duodenum are of primary importance in controlling the rate of gastric emptying. The duodenum must be ready to receive the chyme and can act to delay gastric emptying by reducing peristaltic activity in the stomach until the duodenum is ready to accommodate more chyme. Even if the stomach is distended and its contents are in a liquid form, it cannot empty until the duodenum is ready to deal with the chyme.

The four most important duodenal factors that influence gastric emptying are *fat, acid, hypertonicity,* and *distention.* The presence of one or more of these stimuli in the duodenum activates appropriate duodenal receptors, thereby triggering either a neural or hormonal response that puts brakes on gastric motility and subsequently delays gastric emptying by reducing the excitability of the gastric smooth muscle. The *neural response* is mediated through both the intrinsic nerve plexuses (short reflex) and the autonomic nerves (long reflex). Collectively, these reflexes are called the **enterogastric reflex**. The *hormonal response* involves the release from the duodenal mucosa of several hormones collectively known as **enterogastrones**. These hormones are transported by the blood to the stomach, where they inhibit antral contractions to reduce gastric emptying. Three of these enterogastrones have been clearly identified: **secretin, cholecystokinin,** and **gastric inhibitory peptide**. *Secretin* was the first hormone discovered (in 1902). Because it was a secretory product that entered the blood, it was termed *secretin.* The name *cholecystokinin* derives from the fact that this same hormone is also responsible for contraction of the bile-containing gallbladder (*chole* means "bile"; *cysto* means "bladder"; and *kinin* means "contraction"). The name *gastric inhibitory peptide* is self-explanatory; it is a peptide hormone that inhibits the stomach.

Let us examine why it is important that each of these stimuli in the duodenum (fat, acid, hypertonicity, and distention) delays gastric emptying (acting through the enterogastric reflex or one of the enterogastrones).

- *Fat.* Fat is digested and absorbed more slowly than the other nutrients. Furthermore, fat digestion and absorption take place only within the small-intestine lumen. Therefore, when fat is already present in the duodenum, further gastric emptying of more fatty stomach contents into the duodenum is prevented until the small intestine has processed the fat already there. In fact, fat is the most potent stimulus for inhibition of gastric motility. This is evident when one compares the rate of emptying of a high-fat meal (after six hours some of a bacon-and-eggs meal may still be in the stomach) with that of a protein and carbohydrate meal (a meal of lean meat and potatoes may empty in three hours).
- *Acid.* Since the stomach secretes hydrochloric acid (HCl), highly acidic chyme is emptied into the duodenum, where it is neutralized by sodium bicarbonate ($NaHCO_3$) secreted into the duodenal lumen from the pancreas. Unneutralized acid irritates the duodenal mucosa and inactivates the pancreatic digestive enzymes that are secreted into the duodenal lumen. Appropriately, therefore, unneutralized acid in the duodenum inhibits further emptying of acidic gastric contents until complete neutralization can be accomplished.
- *Hypertonicity.* As molecules of protein and starch are digested in the duodenal lumen, large numbers of amino acid and glucose molecules are released. If absorption of these amino acid and glucose molecules does not keep pace with the rate at which protein and carbohydrate digestion proceeds, these large numbers of molecules remain in the chyme and increase the osmolarity of the duodenal contents. Osmolarity depends on the number of molecules present, not on their size, and one protein molecule may be split into several hundred amino acid molecules, each of which has the same osmotic activity as the original protein molecule. The same holds true for one large starch molecule, which yields many smaller but equally osmotically active glucose molecules. Since water is freely diffusible across the duodenal wall, water enters the duodenal lumen from the plasma as the duodenal osmolarity rises. Large volumes of water entering the intestine from the plasma lead to intestinal distention, and, more importantly, circulatory disturbances ensue because of the reduction in plasma volume. To prevent these effects, gastric emptying is reflexly inhibited when the osmolarity of the duodenal contents starts to rise. Thus, the amount of food entering the duodenum for further digestion into a multitude of additional osmotically active particles is reduced until absorption processes have had an opportunity to catch up.
- *Distention.* Too much chyme in the duodenum inhibits the emptying of even more gastric contents, thus allowing the distended duodenum time to cope with the excess volume of chyme it already contains before it receives an additional quantity.

Peristaltic contractions occur in the empty stomach before the next meal.

After a meal is finally emptied from the stomach, there are no more gastric factors to enhance gastric excitability, so peristaltic contractions eventually die out and the stomach remains quiet for a while. In conjunction with the sensation of hunger before the next meal, however, peristaltic contractions begin again, sweeping over the nearly empty antrum. This arousal of stomach motility appears to be mediated by increased parasympathetic activity, perhaps brought about by the hypothalamus in response to a fall in hypothalamic glucose utilization as the time for the next meal approaches. One of the leading theories of why we get hungry also involves decreased use of glucose by the hypothalamus (chapter 17). An individual may experience the sensation of hunger pangs when these peristaltic contractions are occurring, but the contractions themselves are not responsible for the sensation. Rather, both the sensation of hunger and the increased peristaltic activity are triggered simultaneously by the reduced amount of glucose being metabolized by the brain.

PREGAME MEAL: WHAT'S IN AND WHAT'S OUT?

Many coaches and athletes believe intensely in special food rituals before a competitive event. For example, a football team may always breakfast on steak before a game. Another may always include bananas in their pregame meal. Do these rituals work?

Many studies have been done to determine the effect of the pregame meal on athletic performance. Although laboratory studies have shown that substances such as caffeine improve endurance, no food substance that will greatly enhance performance has been identified. The athlete's prior training is the most important determinant of performance. Even though no particular food confers a special benefit before an athletic contest, some food choices can actually hinder the competitors. For example, a meal of steak is high in fat and could take so long to digest that it might impair the football team's performance and thus should be avoided. On the other hand, food rituals that do not impair performance, such as eating bananas, but give the athletes a morale boost or extra confidence are harmless and should be respected. People may attach special meanings to eating certain foods, and their faith in these practices can make the difference between winning and losing a game.

The greatest benefit of the pregame meal is to prevent hunger during competition. Because the stomach can take from one to four hours to empty, an athlete should eat at least three to four hours before competition begins. Excessive quantities of food should not be consumed before competition. Food that remains in the stomach during competition may cause nausea and possibly vomiting. This condition can be aggravated by nervousness, which slows digestion and delays gastric emptying by means of the sympathetic nervous system.

The best choices are foods that are high in carbohydrate and low in fat and protein. High-carbohydrate foods are recommended because they are emptied from the stomach more quickly than fat or protein are. Carbohydrates do not inhibit gastric emptying by means of cholecystokinin release, whereas fat and protein do. Fats in particular delay gastric emptying and are slowly digested. Metabolic processing of proteins yields nitrogenous wastes such as urea whose osmotic activity draws water from the body and increases urine volume, both of which are undesirable during an athletic event.

Good choices for a pregame meal include breads, pasta, rice, potatoes, gelatins, and fruit juices. Not only will these complex carbohydrates be emptied from the stomach if consumed one to four hours before a competitive event, but they also will help maintain the blood glucose level during the event.

Although it might seem logical to consume something sugary immediately before a competitive event to provide an "energy boost," beverages and foods high in sugar should be avoided because they trigger insulin release. Insulin is the hormone that enhances glucose entry into most body cells. Once the person begins exercising, insulin sensitivity increases (see p. 64), which results in a decrease in the plasma glucose level. A lowered plasma glucose level induces feelings of fatigue and an increased use of muscle glycogen stores, which can limit performance in endurance events such as the marathon. Therefore, sugar consumption just before a competition can actually impair performance instead of giving the sought-after energy boost.

Within an hour of competition, it is best for athletes to drink only plain water to ensure adequate hydration.

Emotions can influence gastric motility.

Other factors unrelated to digestion can also alter gastric motility by acting through the autonomic nerves to influence the degree of gastric smooth-muscle excitability. Even though the effect of emotions on gastric motility varies from one individual to another and is not always predictable, sadness and fear generally tend to decrease motility, whereas anger and aggression tend to increase it. In addition to emotional influences, intense pain from any part of the body tends to inhibit motility, not just in the stomach but throughout the digestive tract. This response is brought about by increased sympathetic activity and a corresponding decrease in parasympathetic activity. (For a discussion of the pregame meal before participation in an athletic event, see the accompanying boxed feature, A Closer Look at Exercise Physiology.)

The body of the stomach does not actively participate in the act of vomiting.

Vomiting, or **emesis**, the forceful expulsion of gastric contents out through the mouth, is generally perceived as being caused by abnormal gastric motility. However, vomiting is not accomplished by reverse peristalsis, as might be predicted. Actually, the stomach itself does not actively participate in the act of vomiting. The stomach, the esophagus, the gastroesophageal sphinter, and the pyloric sphincter are all relaxed during vomiting. The major force for expulsion comes, surprisingly, from contraction of the respiratory muscles—namely, the diaphragm (the major inspiratory muscle) and the abdominal muscles (the muscles of active expiration).

Vomiting begins with a deep inspiration and closure of the

glottis. The contracting diaphragm descends downward on the stomach while simultaneous contraction of the abdominal muscles compresses the abdominal cavity, increasing the intra-abdominal pressure and forcing the abdominal viscera upward. As the flaccid stomach is squeezed between the diaphragm from above and the compressed abdominal cavity from below, the gastric contents are forced into the esophagus and out through the mouth. The glottis is closed, so vomited material does not enter the respiratory airways. Also, the uvula is elevated to close off the nasal cavity.

Often when vomitus first enters the esophagus, the pharyngoesophageal sphincter remains closed so that no gastric contents enter the mouth. Distention of the esophagus by the vomitus induces secondary peristaltic waves that force the gastric contents back into the stomach. This cycle repeats itself as the contents are immediately squeezed up into the esophagus again. This is the act of *retching* or *heaves*. After a series of heaves, when the pressure becomes great enough, the person thrusts out the jaw, pulling open the pharyngoesophageal sphincter. The gastric contents are then forced through the esophagus, through the pharyngoesophageal sphincter, and out through the mouth. During this time, the duodenum contracts strongly, which may force some of the intestinal contents back into the stomach and out with the vomitus. The vomited material may therefore be stained with the yellowish bile that has entered the small intestine from the liver and gallbladder.

The vomiting cycle may be repeated several times until the stomach is emptied. Vomiting is usually preceded by profuse salivation, sweating, rapid heart rate, and the sensation of nausea, all of which are characteristic of a generalized discharge of the autonomic nervous system.

This complex act of vomiting is coordinated by a **vomiting center** in the medulla. Nausea, retching, and vomiting can be initiated by afferent input to the vomiting center from a number of receptors throughout the body. The causes of vomiting include the following:

■ Tactile (touch) stimulation of the back of the throat, which is one of the most potent stimuli. For example, sticking a finger in the back of the throat or even the presence of a tongue depressor or dental instrument in the back of the mouth is sufficient stimulation to cause gagging and even vomiting in some people.

■ Irritation or distention of the stomach and duodenum.

■ Elevated intracranial pressure, such as that caused by cerebral hemorrhage. Thus, vomiting following a head injury is considered to be a bad sign; it suggests swelling or bleeding within the cranial cavity.

■ Rotation or acceleration of the head producing dizziness, such as occurs in motion sickness.

■ Intense pain arising from a variety of organs, such as that accompanying passage of a kidney stone.

■ Chemical agents, including drugs or noxious substances that initiate vomiting (that is, **emetics**) either by acting in the upper portions of the gastrointestinal tract or by stimulating chemoreceptors in a specialized **chemoreceptor trigger zone** in the brain. Activation of this zone triggers the vomiting reflex.

■ Psychic vomiting induced by emotional factors, such as those accompanying nauseating sights and odors and even including vomiting before taking an examination or in other stressful situations.

With excessive vomiting, the body experiences large losses of secreted fluids and acids that normally would be reabsorbed. The resultant reduction in plasma volume can lead to dehydration and circulatory problems, while the loss of acid from the stomach can lead to metabolic alkalosis (see p. 537).

Vomiting is not always detrimental, however. Limited vomiting brought about by irritation of the digestive tract can provide a useful service in removing noxious material from the stomach rather than allowing it to be retained and absorbed. In fact, emetics are frequently given in the case of accidental ingestion of a poison to quickly remove the offending substance from the body.

Gastric pits are the source of gastric digestive secretions.

Each day the stomach secretes about 2 liters of gastric juice. (See the accompanying boxed feature, Concepts, Challenges, and Controversies.) The cells responsible for gastric secretion are located in the lining of the stomach, the gastric mucosa, which is divided into two distinct areas: (1) the **oxyntic mucosa**, which lines the body and fundus, and (2) the **pyloric gland area** (**PGA**), which lines the antrum (● Fig. 16–10a). The mucosal gland cells are found in **gastric pits**, which are invaginations or deep pockets in the luminal surface of the stomach (Fig. 16–10b and c). Three types of secretory cells are found in the walls of the pits in the oxyntic mucosa. The entrance or neck of the gastric pit is lined by **mucous neck cells**, which secrete a thin, watery *mucus*. (*Mucous* is the adjective; *mucus* is the noun.) The deeper portions of the pit are lined by **chief cells**, which secrete the enzyme precursor *pepsinogen*, and **parietal** (or **oxyntic**) **cells**, which secrete *HCl* and *intrinsic factor*. The parietal cells are located on the outer wall of the gastric pits and do not come into contact with the pit lumen (Fig. 16–10c). (*Parietal* means "wall," a reference to the location of these cells. *Oxyntic* means "sharp," a reference to these cells' potent HCl secretory product.) Although the parietal cells are separated from the lumen of the gastric pit by the chief cells, they transmit their HCl secretion into the lumen through fine channels, or **canaliculi**, that pass between the chief cells.

Between the gastric pits, the gastric mucosa is covered by surface epithelial cells, which secrete a thick, viscous, alkaline mucus that forms a visible layer several millimeters thick over the surface of the mucosa.

The mucous neck cells rapidly divide and serve as the parent cells of all new cells of the gastric mucosa. The daughter cells that result from cell division either migrate out of the pit to become surface epithelial cells or migrate down into the deeper parts of the pit to differentiate into chief or parietal

BEAUMONT: THE BEGINNINGS OF PHYSIOLOGY IN AMERICA

A good example of the philosophy, "When life hands you lemons, make lemonade," took place in 1822 when a hunter, Alexis St. Martin, was accidentally shot in the stomach. The wound did not heal properly; a flap formed that prevented the stomach contents from falling out, but an open passageway was left under the flap from the interior of the stomach to the outside of the abdominal wall. Such an abnormal passage leading to the surface of the body is known as a **fistula**. Rather than bemoaning his fate, St. Martin agreed to join forces with his physician, American surgeon William Beaumont, to take ad-

vantage of the situation. Beaumont was easily able to insert various foods into St. Martin's stomach and to extract them after variable periods of time to see what digestion had taken place. He was also able to collect gastric juice through the fistulous tract under various controlled conditions. This living laboratory enabled Beaumont to study the composition, functions, and regulation of gastric secretions. These experimental studies on the stomach were the beginning of our knowledge of digestive physiology. In fact, they led to Beaumont's recognition as America's "father of physiology." They demon-

strated a phenomenon that has been much more thoroughly investigated since then: digestive activities are normally regulated carefully to provide optimal conditions for the digestion and absorption of ingested food, no matter what kind or how much is eaten.

Furthermore, Beaumont observed that stomach activity varied in response to changes in emotional state unrelated to digestive processes—the basis for digestive upsets during periods of emotional stress.

• **FIGURE 16-10   The Stomach Mucosa and Its Gastric Pits**   (a) The stomach mucosa consists of the oxyntic mucosa, which lines the body and fundus, and the pyloric gland area, which lines the antrum. (b) The stomach's gland cells are located in deep invaginations, or gastric pits, in the luminal surface of the stomach. (c) These gland cells include mucous neck cells, which secrete mucus, chief cells, which secrete pepsinogen, and parietal cells, which secrete HCl and intrinsic factor. The surface epithelial cells also secrete mucus.

cells. Through this activity, the entire stomach mucosa is replaced about every three days.

The gastric pits of the PGA primarily secrete mucus and a small amount of pepsinogen; in contrast to the oxyntic mucosa no acid is secreted in this area. More importantly, endocrine cells in the PGA secrete the hormone **gastrin** into the blood. Thus, the most important gastric digestive secretions produced within the body and fundus are HCl, pepsinogen, mucus, and intrinsic factor, which are released into the gastric lumen. On the other hand, the most important product of the PGA is the hormone gastrin, which is released into the blood. We will examine each of these secretory products in more detail.

Hydrochloric Acid Secretion   The parietal cells actively secrete HCl into the lumen of the gastric pits, which in turn empty into the lumen of the stomach. The pH of the luminal contents falls as low as 2 as a result of this HCl secretion. Hydrogen ion (H^+) and chloride ion (Cl^-) are actively transported by separate pumps in the parietal cell's plasma membrane. Hydrogen ion is actively transported against a tremendous concentration gradient, with the H^+ concentration being as much as 3 to 4 million times greater in the lumen than in the blood. Because of the high energy expenditure needed to move H^+ against such a large gradient, parietal cells have an abundance of mitochondria, the energy organelles. Chloride is also actively secreted but against a much smaller concentration gradient of only 1½ times.

The secreted H^+ is not transported from the plasma but is derived instead from metabolic processes within the parietal cell (• Fig. 16–11). Whenever a H^+ is secreted, neutrality of the interior of the cell is maintained by generation of a new H^+ from carbonic acid (H_2CO_3) to replace the secreted H^+. The parietal cells contain an abundance of the enzyme carbonic anhydrase (ca). In the presence of carbonic anhydrase, H_2O readily combines with CO_2, which either has been produced within the parietal cell by metabolic processes or has diffused in from the blood. The combination of H_2O and CO_2 results in the formation of H_2CO_3, which partially dissociates to yield H^+ and HCO_3^-:

$$CO_2 + H_2O \xrightarrow{ca} H_2CO_3 \rightarrow H^+ + HCO_3^-$$

The generated H^+ replaces the one secreted. The generated HCO_3^- is moved into the plasma by the same carrier that actively transports Cl^- from the plasma into the gastric lumen, similar to the Cl^- shift that occurs in red blood cells (see p. 448). This exchange of HCO_3^- for Cl^- maintains electrical neutrality in the plasma during HCl secretion.

Although HCl does not actually digest anything and is not absolutely essential to gastrointestinal function, it does perform several functions that assist digestion. Hydrochloric acid (1) activates the enzyme precursor pepsinogen to an active enzyme, pepsin, and provides an acid medium that is optimal for pepsin activity; (2) aids in the breakdown of connective tissue and muscle fibers, thereby reducing large food particles into smaller particles; and (3) along with salivary lysozyme, kills most of the microorganisms ingested with the food, although some do escape and continue to grow and multiply in the large intestine.

Pepsinogen Secretion   The major digestive constituent of gastric secretion is **pepsinogen**, an inactive enzymatic molecule synthesized and packaged by the endoplasmic reticulum and Golgi complex of the chief cells. Pepsinogen is stored in the chief cell's cytoplasm within secretory vesicles known as **zymogen granules**, from which it is released by exocytosis (see p. 24) upon appropriate stimulation. When pep-

---

• **FIGURE 16–11   Mechanism of HCl Secretion**
The stomach's parietal cells actively secrete H^+ and Cl^- by the actions of two separate pumps. The secreted H^+ is derived from H_2CO_3 that is generated within the cell from CO_2 that is either metabolically produced in the cell or diffuses in from the plasma. The secreted Cl^- is transported into the parietal cell from the plasma. The HCO_3^- generated from H_2CO_3 dissociation is transported into the plasma in exchange for the secreted Cl^-.

● FIGURE 16–12 Pepsinogen Activation in Stomach Lumen In the lumen, hydrochloric acid (HCl) activates pepsinogen to its active form, pepsin, by cleaving off a small fragment. Once activated, pepsin autocatalytically activates more pepsinogen and begins protein digestion. Secretion of pepsinogen in the inactive form prevents it from digesting the protein structures of the cells in which it is produced. Its activation process does not commence until it reaches the lumen and comes into contact with HCl secreted by a separate cell in the gastric pit.

sinogen is secreted into the gastric lumen, HCl cleaves off a small fragment of the molecule, converting it to the active form of the enzyme, **pepsin** (● Fig. 16–12). Once formed, pepsin acts on other pepsinogen molecules to produce more pepsin. A mechanism such as this whereby an active form of an enzyme activates other molecules of the same enzyme is referred to as an **autocatalytic** ("self-activating") **process**.

Pepsin initiates protein digestion by splitting certain amino acid linkages in proteins to yield peptide fragments (small amino acid chains); it works most effectively in the acid environment provided by HCl. Because pepsin can digest protein, it must be stored and secreted in an inactive form so that it does not digest the cells in which it is formed. (The primary structural component of cells is protein.) Therefore, pepsin is maintained in the inactive form of pepsinogen until it reaches the gastric lumen, where it is activated by HCl.

Mucus Secretion The surface of the gastric mucosa is covered by a layer of mucus, which is derived from the surface

epithelial cells and mucous neck cells. This mucus serves as a protective barrier against several forms of potential injury to the gastric mucosa.

- By virtue of its lubricating properties, mucus protects the gastric mucosa against mechanical injury.
- It helps protect the stomach wall from self-digestion because pepsin is inhibited when it comes in contact with the mucus layer coating the stomach lining. (However, mucus does not affect pepsin activity in the lumen, where digestion of dietary protein proceeds without interference.)
- Being alkaline, mucus helps protect against acid injury by neutralizing HCl in the vicinity of the gastric lining.

Intrinsic Factor Secretion **Intrinsic factor**, another secretory product of the parietal cells in addition to HCl, is important in the absorption of vitamin B_{12}. Only when in combination with intrinsic factor can this vitamin be absorbed by a special transport mechanism, presumably endocytosis, in the terminal portions of the ileum. Vitamin B_{12} is essential for the normal formation of red blood cells. In the absence of intrinsic factor, vitamin B_{12} fails to be absorbed, so erythrocyte production is defective, and pernicious anemia results (see p. 354).

Occasionally, the oxyntic mucosa atrophies or degenerates. With the loss of chief and parietal cells, the stomach is unable to secrete pepsinogen, HCl, and intrinsic factor. Although pepsin and acid normally begin protein digestion in the stomach, they are not absolutely essential to protein digestion. If necessary, the pancreatic and small intestinal enzymes can completely digest proteins. The most detrimental consequence of **gastric mucosal atrophy** is the loss of intrinsic factor and the subsequent development of pernicious anemia unless vitamin B_{12} is injected therapeutically. The exact cause of gastric mucosal atrophy is uncertain, although it is suspected to be an autoimmune response (see p. 396), because many individuals with this condition have antibodies to oxyntic cells in their blood.

Gastrin Secretion Special endocrine cells, the **G cells**, located in the pyloric gland area (PGA) of the stomach secrete gastrin into the blood upon appropriate stimulation. After being carried by the blood back to the oxyntic mucosa, gastrin stimulates the parietal and chief cells, thereby promoting secretion of a highly acidic gastric juice. Gastrin is also *trophic* (growth-promoting) to the mucosa of the stomach and small intestine, thereby maintaining their secretory capabilities.

Control of gastric secretion involves three phases.

The rate of gastric secretion can be influenced by (1) factors arising before food ever reaches the stomach; (2) factors resulting from the presence of food in the stomach; and (3) factors in the duodenum after food has left the stomach. Accordingly, gastric secretion is divided into three phases—the cephalic, gastric, and intestinal phases (Table 16–4).

Cephalic Phase The cephalic phase of gastric secretion refers to the increased secretion of HCl and pepsinogen that

TABLE 16-4 Stimulation of Gastric Secretion

PHASE	STIMULI	EXCITATORY MECHANISM FOR ENHANCING GASTRIC SECRETION
Cephalic phase	Stimuli in head—seeing, smelling, tasting, chewing, swallowing food	
Gastric phase	Stimuli in stomach—protein, (peptide fragments), distention, caffeine, alcohol	
Excitatory intestinal phase	Stimuli in duodenum—digested protein products	

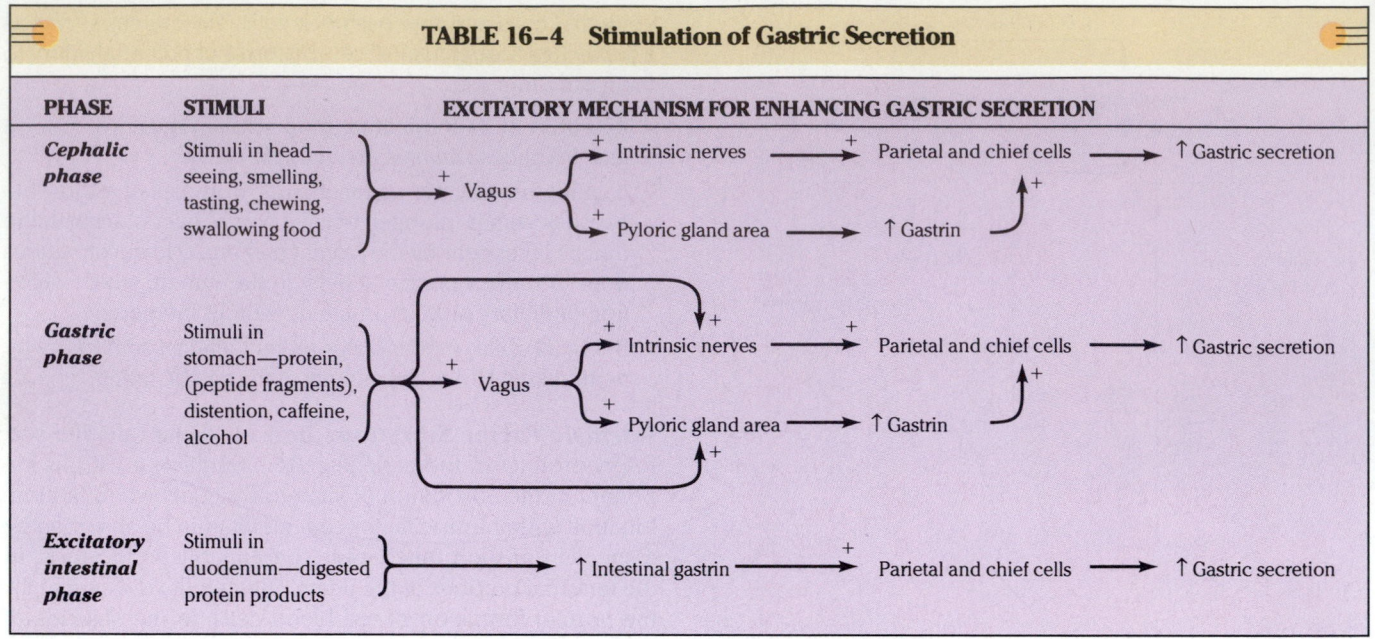

occurs in feedforward fashion in response to stimuli acting in the head (*cephalic* refers to "head") even before food reaches the stomach. Thinking about, tasting, smelling, chewing, and swallowing food increase gastric secretion by means of vagal nerve activity in two ways. First, vagal stimulation of the intrinsic plexuses promotes increased secretion of HCl and pepsinogen by the secretory cells. Second, vagal stimulation of the PGA causes the release of gastrin, which in turn further enhances secretion of HCl and pepsinogen.

Gastric Phase   The gastric phase of gastric secretion occurs when food actually reaches the stomach. Stimuli acting in the stomach—namely, protein, especially peptide fragments; distention; caffeine; or alcohol—increase gastric secretion by means of overlapping efferent pathways. For example, protein in the stomach, the most potent stimulus, initiates short local reflexes in the intrinsic nerve plexuses to stimulate the secretory cells. Furthermore, protein initiates long reflexes so that extrinsic vagal fibers to the stomach are activated. Vagal activity further enhances intrinsic nerve stimulation of the secretory cells and triggers the release of gastrin. Protein also directly stimulates the release of gastrin. Gastrin, in turn, is a powerful stimulus for further acid and pepsinogen secretion. Through these synergistic and overlapping pathways, protein induces the secretion of a highly acidic, pepsin-rich gastric juice, which continues the digestion of the protein that first initiated the process.

When the stomach is distended with protein-rich food that needs to be digested, these secretory responses are appropriate. Caffeine and, to a lesser extent, alcohol also stimulate the secretion of a highly acidic gastric juice, even when no food is present. This unnecessary acid can irritate the linings of the stomach and duodenum. For this reason, persons with ulcers or gastric hyperacidity should avoid caffeinated and alcoholic beverages.

Intestinal Phase   The third phase of gastric secretion, the intestinal phase, encompasses the factors originating in the small intestine that influence gastric secretion. The intestinal phase has both an excitatory and an inhibitory component (● Fig. 16–13).

To a limited extent, the presence of the products of protein digestion in the duodenum stimulates further gastric secretion by triggering the release of an **intestinal gastrin** that is carried by the blood to the stomach. This is the *excitatory component* of the intestinal phase of gastric secretion. It is as if the small intestine, on noting the arrival of protein fragments from the stomach, offers the stomach a "helping hand" in digesting the protein by enhancing gastric secretion.

The *inhibitory component* of the intestinal phase of gastric secretion is dominant over the excitatory component, however. The inhibitory component is important in helping to shut off the flow of gastric juices as chyme begins to be emptied into the small intestine, a topic to which we now turn our attention.

Gastric secretion gradually decreases as food empties from the stomach into the intestine.

We now know what factors turn on gastric secretion before and during a meal, but how is the flow of gastric juices shut off as chyme begins to be emptied from the stomach into the small intestine? Gastric secretion is gradually reduced by three different means as the stomach empties (Table 16–5):

■ As the meal is gradually emptied into the duodenum, the major stimulus for enhanced gastric secretion—the presence of protein in the stomach—is withdrawn.

■ After foods leave the stomach and gastric juices accumulate to such an extent that gastric pH falls very low, gastric secretion is inhibited because a high concentration of H⁺

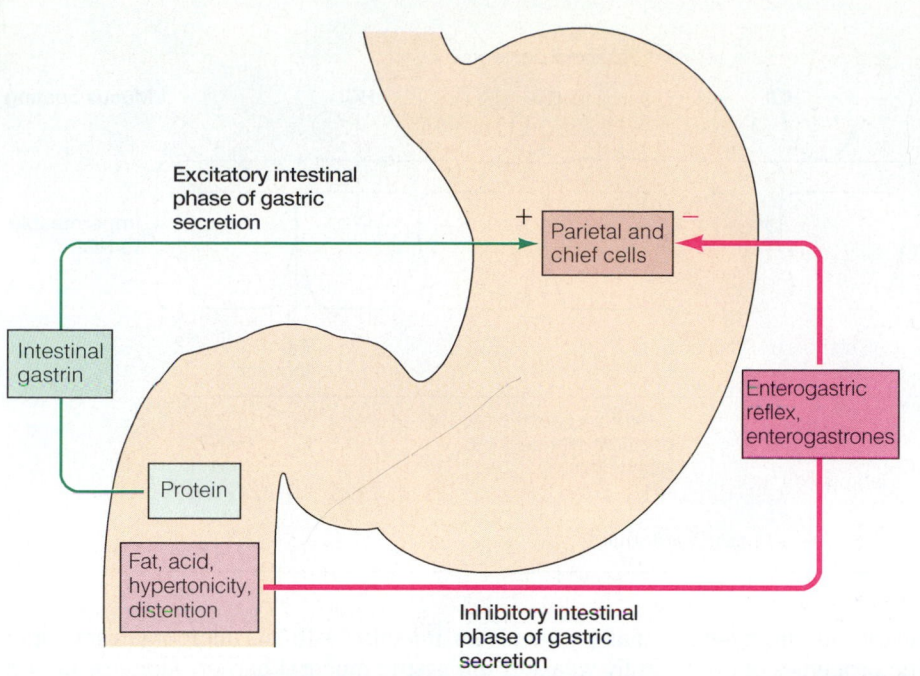

directly inhibits the PGA from releasing gastrin. As gastrin secretion declines, the most potent stimulant of gastric secretion is withdrawn.

■ The same stimuli that inhibit gastric motility (fat, acid, hypertonicity, or distention in the duodenum brought about by stomach emptying) inhibit gastric secretion as well; the enterogastric reflex and the enterogastrones suppress the gastric secretory cells while they simultaneously reduce the excitability of the gastric smooth-muscle cells. This inhibitory response is the inhibitory component of the intestinal phase of gastric secretion (Fig. 16–13).

The stomach lining is protected from gastric secretions by the gastric mucosal barrier.

How can the stomach contain strong acid contents and proteolytic enzymes without destroying itself? We already learned that mucus provides a protective coating. In addition, other barriers to mucosal acid damage are provided by the mucosal lining itself (• Fig. 16–14). First, the luminal membranes of the gastric mucosal cells are almost impermeable to H^+, so acid cannot penetrate *into* the cells and cause cellular damage. Furthermore, the lateral edges of these cells are joined together near their luminal borders by tight junctions

TABLE 16–5   Inhibition of Gastric Secretion

REGION	STIMULI	INHIBITORY MECHANISM FOR REDUCING GASTRIC SECRETION	
Body and antrum	Removal of protein and distention as stomach empties	Intrinsic nerves / Vagus / Pyloric gland area → ↓ Gastrin	→ ↓ Gastric secretion
Antrum	Accumulation of acid	Pyloric gland area → ↓ Gastrin	→ ↓ Gastric secretion
Duodenum (inhibitory intestinal phase of gastric secretion)	Fat Acid Hypertonicity Distension	Enterogastric reflex / ↑ Enterogastrones (cholecystokinin, secretin, gastric inhibitory peptide)	→ ↓ Gastric secretion

● **FIGURE 16–14   Gastric Mucosal Barrier** The gastric mucosal barrier encompasses the following factors that enable the stomach to contain acid without injuring itself: the gastric mucosal cells are joined by tight junctions that prevent HCl from penetrating between the cells ①, and the luminal membranes of these cells are impermeable to H⁺ so that HCl cannot penetrate into the cells ② . A mucous coating over the gastric mucosa offers further protection ③.

(see p. 55), so acid cannot diffuse *between* the cells from the lumen into the underlying submucosa. The properties of the gastric mucosa that enable the stomach to contain acid without injuring itself constitute the **gastric mucosal barrier**. These protective mechanisms are further enhanced by the fact that the entire stomach lining is replaced every three days. Because of rapid mucosal turnover, cells are usually replaced before they are exposed to the wear and tear of harsh gastric conditions long enough to suffer damage.

In spite of the protection provided by mucus, by the gastric mucosal barrier, and by the frequent turnover of cells, the barrier occasionally is broken so that the gastric wall is injured by its acidic and enzymatic contents. When this occurs, an erosion, or **peptic ulcer**, of the stomach wall results. (Excessive gastric reflux into the esophagus and dumping of excessive acidic gastric contents into the duodenum can lead to peptic ulcers in these locations as well.)

Until recently, the exact cause of ulcers was unknown, but in a surprising new discovery, the bacterium *Helicobacter pylori* has been pinpointed as the probable cause of up to 90%

of all peptic ulcers. Infection with this microorganism apparently weakens the gastric mucosal barrier. Alone or in conjunction with this infectious culprit, other factors are known to contribute to ulcer formation. Some chemicals can break the gastric mucosal barrier; the most important of these are ethyl alcohol and aspirin. The barrier frequently breaks in patients with preexisting debilitating conditions, such as severe injuries or infections. The persistence of stressful situations is frequently associated with ulcer formation, presumably because of excessive stimulation of gastric secretion brought about by the emotional response related to the stress.

When the gastric mucosal barrier is broken (either because it is weak or damaged or because it is overwhelmed by excessive secretion), acid and pepsin diffuse into the mucosa with serious pathophysiologic consequences (● Fig. 16–15). Acid triggers the release of *histamine,* a potent acid stimulant that is produced and stored in large amounts in the mucosa. (It is uncertain whether histamine plays a role under normal circumstances.) Released histamine stimulates secretion of more acid, which can diffuse back into the mucosa to stimu-

● **FIGURE 16–15   Ulcer Formation** When acid and pepsin are able to break through a weakened or overwhelmed gastric mucosal barrier, the acid stimulates the release of histamine that is stored in the submucosa. Histamine, in turn, stimulates the parietal cells to secrete more acid, which diffuses through the broken barrier to trigger the release of more histamine, as the vicious cycle continues. An ulcer is formed and progressively enlarges as the acid and pepsin continue to erode the gastric mucosa.

late further histamine release, triggering more acid release, and so on, thus establishing a vicious cycle. The mucosal erosion, or ulcer, continues to enlarge under the influence of increasing levels of acid and pepsin. Two of the most serious consequences of ulcers are (1) hemorrhage resulting from damage of submucosal capillaries and (2) perforation of the stomach wall due to complete erosion through the wall caused by HCl and pepsin action, resulting in the escape of potent gastric contents into the abdominal cavity.

One of the leading treatments for ulcers is an antihistamine (cimetidine) that specifically blocks H-2 receptors, the type of receptors that bind histamine released from the stomach. These receptors differ from H-1 receptors that bind the histamine involved in allergic respiratory disorders. Accordingly, traditional antihistamines used for respiratory allergies (such as hay fever and asthma) are not effective against ulcers, nor is cimetidine useful for respiratory problems. Before the discovery of the key role these unique histamine receptors play in ulcer formation, the treatment of ulcers was aimed at either neutralizing stomach acidity through the use of antacids or removing factors known to enhance gastric secretion. Accordingly, the treatment include (1) a bland diet void of caffeine and alcohol, (2) cutting the vagus nerve supply to the stomach, and (3) removal of the stomach antrum to eliminate the source of gastrin. With the discovery of the infectious component of the majority of ulcers, antibiotics have now been added to the treatment regime.

Carbohydrate digestion continues in the body of the stomach, whereas protein digestion begins in the antrum.

Two separate digestive processes take place within the stomach. Food in the body of the stomach remains as a semisolid mass, because peristaltic contractions in this region are too weak for mixing to occur. Because food is not mixed with gastric secretions in the body of the stomach, very little protein digestion occurs here. Acid and pepsin are able to attack only the surface of the food mass. Carbohydrate digestion, however, continues in the interior of the mass under the influence of salivary amylase. Even though acid inactivates salivary amylase, the unmixed interior of the food mass is free of acid.

Digestion by the gastric juice itself is accomplished in the antrum of the stomach, where the food is thoroughly mixed with HCl and pepsin, thereby initiating protein digestion.

The stomach absorbs alcohol and aspirin but no food.

No food or water is absorbed into the blood from the stomach mucosa. Carbohydrate and protein digestion have not been completed in the stomach. These large, partially digested substances are not lipid soluble, so they cannot penetrate the cell membranes. Furthermore, no special transport mechanisms are present in the stomach wall to facilitate the absorption of these nutrients. Fat digestion has not even begun in the stomach. Although it is theoretically possible for fat molecules, which are lipid soluble, to pass through the lipid portions of the cell membranes that line the stomach, this does not occur to any appreciable extent because dietary fat in the gastric lumen separates into large fat droplets that float in the chyme. This reduces the possibility that fat molecules will come into contact with the gastric mucosa. Ingested electrolytes such as Na^+ and Ca^{++} exist as non-lipid-soluble ions that can be absorbed only through use of special transport systems, which exist in the small intestine but not in the stomach. Likewise, H_2O exchange does not take place across the gastric mucosa because the stomach is impermeable to H_2O.

Even though none of the ingested food is absorbed from the stomach, two noteworthy nonnutrient substances are absorbed directly by the stomach—*ethyl alcohol* and *aspirin*. Alcohol is lipid soluble to a degree, so it can diffuse through the lipid membranes of the epithelial cells that line the stomach and enter the blood through the submucosal capillaries. Yet although alcohol can be absorbed by the gastric mucosa, it can be absorbed even more rapidly by the small intestinal mucosa, because the surface area available for absorption in the small intestine is much greater than in the stomach. Thus, alcohol absorption occurs more slowly if gastric emptying is delayed so that the alcohol remains in the stomach longer. Since fat is the most potent duodenal stimulus for inhibiting gastric motility, consumption of fat-rich foods (for example, whole milk or pizza) before or during alcohol ingestion delays gastric emptying and prevents the alcohol from producing its effects as rapidly.

Another category of substances absorbed by the gastric mucosa includes weak acids, most notably acetylsalicylic acid (aspirin). In the highly acidic environment of the stomach lumen, weak acids are almost totally unionized; that is, the H^+ and associated anion of the acid are bound together. In an unionized form, these weak acids are lipid soluble, so they can be absorbed quickly by crossing the plasma membranes of the epithelial cells that line the stomach. Most other drugs are not absorbed until they reach the small intestine, so they do not begin to take effect as quickly.

PANCREATIC AND BILIARY SECRETIONS

When gastric contents are emptied into the small intestine, they are mixed not only with juice secreted by the small-intestine mucosa but also with the secretions of the exocrine pancreas and liver that are emptied into the duodenal lumen. We will discuss the roles of each of these accessory digestive organs before we examine the contributions of the small intestine itself.

The pancreas is a mixture of exocrine and endocrine tissue.

The **pancreas** is an elongated gland that lies behind and below the stomach, above the first loop of the duodenum (● Fig. 16–16). It is a mixed gland that contains both exocrine and endocrine tissue. The predominant exocrine portion consists of grapelike clusters of secretory cells that form sacs known as **acini**, which connect to ducts that eventually empty

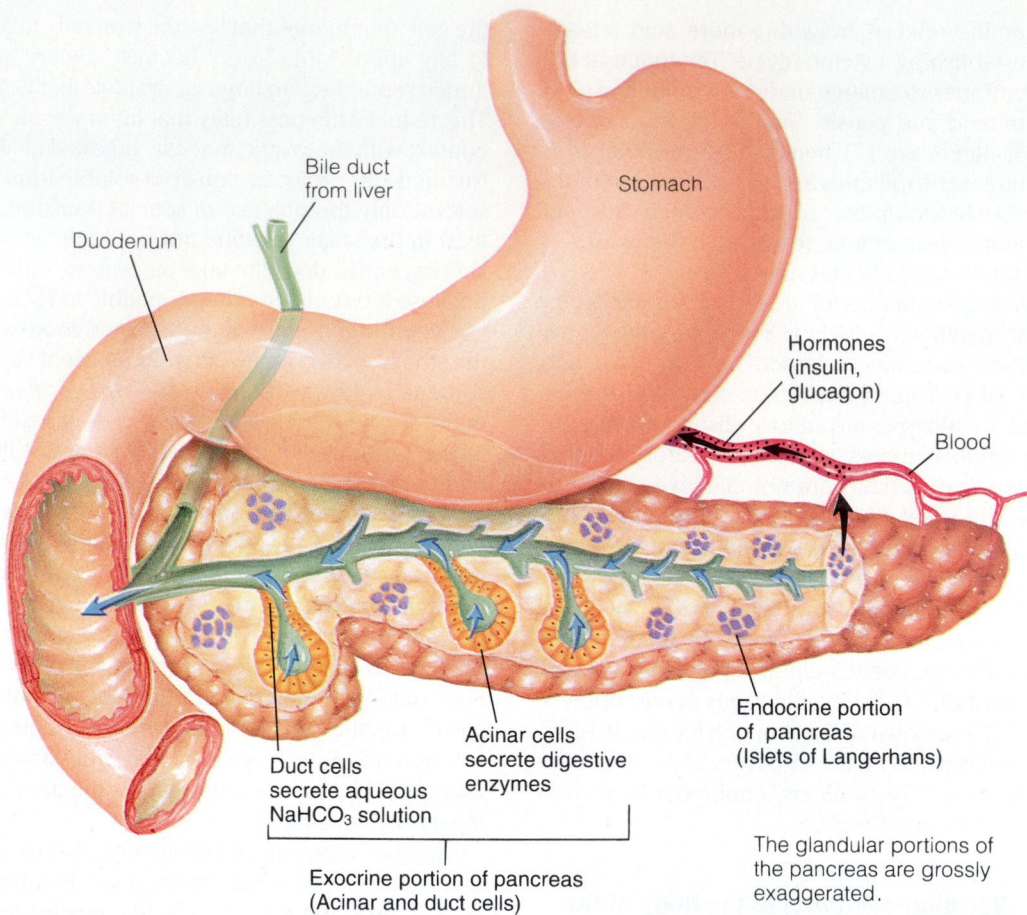

Bile duct from liver

Duodenum

Stomach

Hormones (insulin, glucagon)

Blood

Endocrine portion of pancreas (Islets of Langerhans)

Acinar cells secrete digestive enzymes

Duct cells secrete aqueous $NaHCO_3$ solution

The glandular portions of the pancreas are grossly exaggerated.

Exocrine portion of pancreas (Acinar and duct cells)

● **FIGURE 16–16   Schematic Representation of Exocrine and Endocrine Portions of the Pancreas**   The exocrine pancreas secretes into the duodenal lumen a digestive juice composed of digestive enzymes secreted by the acinar cells and an aqueous $NaHCO_3$ solution secreted by the duct cells. The endocrine pancreas secretes the hormones insulin and glucagon into the blood.

into the duodenum. The smaller endocrine portion consists of isolated islands of endocrine tissue, the **islets of Langerhans**, which are dispersed throughout the pancreas. The most important hormones secreted by the islet cells are insulin and glucagon (chapter 19). The exocrine and endocrine pancreas have nothing in common except for sharing the same location.

The exocrine pancreas secretes digestive enzymes and an aqueous alkaline fluid.

The **exocrine pancreas** secretes a pancreatic juice consisting of two components—a potent *enzymatic secretion* and an *aqueous* (watery) *alkaline secretion* that is rich in sodium bicarbonate ($NaHCO_3$). The pancreatic enzymes are actively secreted by the *acinar cells*. The aqueous $NaHCO_3$ component is actively secreted by the *duct cells* that line the early portions of the pancreatic ducts, and it then is modified as it passes down the ducts.

Like pepsinogen, pancreatic enzymes are synthesized by the endoplasmic reticulum and Golgi complex of the acinar cells, then are stored within zymogen granules and released by exocytosis as needed. The acinar cells secrete three differ-

ent types of pancreatic enzymes that are capable of digesting all three categories of foodstuffs. These pancreatic enzymes are important because they are capable of almost completely digesting food in the absence of all other digestive secretions. The three types of pancreatic enzymes are (1) **proteolytic enzymes**, which are involved in protein digestion; (2) **pancreatic amylase**, which contributes to carbohydrate digestion in a way similar to salivary amylase; and (3) **pancreatic lipase**, the only enzyme important in fat digestion.

Pancreatic Proteolytic Enzymes   The three major proteolytic enzymes secreted by the pancreas are *trypsinogen*, *chymotrypsinogen*, and *procarboxypeptidase*, each of which is secreted in an inactive form. When **trypsinogen** is secreted into the duodenal lumen, it is activated to its active enzyme form, **trypsin**, by **enterokinase**, an enzyme embedded in the luminal border of the cells that line the duodenal mucosa. Trypsin then autocatalytically activates more trypsinogen. Like pepsinogen, trypsinogen must remain inactive within the pancreas to prevent this proteolytic enzyme from digesting the cells in which it is formed. Trypsinogen remains inactive, therefore, until it reaches the duodenal lumen, where enterokinase triggers the activation process, which then proceeds autocatalytically. As further protection, the pancreatic tissue also produces a chemical known as **trypsin inhibitor**, which blocks trypsin's actions should spontaneous activation of trypsinogen inadvertently occur within the pancreas.

Chymotrypsinogen and **procarboxypeptidase**, the other pancreatic proteolytic enzymes, are converted by trypsin to their active forms, **chymotrypsin** and **carboxypeptidase**, respectively, within the duodenal lumen. Thus, once enterokinase has activated some of the trypsin, trypsin is then responsible for the remainder of the activation process.

Each of these proteolytic enzymes attacks different peptide linkages. The end products that result from this action are a mixture of amino acids and small peptide chains. Mucus secreted by the intestinal cells provides protection against digestion of the small intestine wall by the activated proteolytic enzymes.

Pancreatic Amylase Like salivary amylase, pancreatic amylase plays an important role in carbohydrate digestion by converting polysaccharides into disaccharides. Amylase is secreted in the pancreatic juice in an active form because active amylase does not present a danger to the secretory cells.

Pancreatic Lipase Pancreatic lipase is extremely important because it is the only enzyme secreted throughout the entire digestive system that can accomplish digestion of fat. Pancreatic lipase hydrolyzes dietary triglycerides into monoglycerides and free fatty acids, which are the absorbable units of fat. Like amylase, lipase is secreted in its active form because there is no risk of pancreatic self-digestion by lipase.

When pancreatic enzymes are deficient, digestion of food is incomplete. Because the pancreas is the only significant source of lipase, pancreatic enzyme deficiency results in serious maldigestion of fats. The principal clinical manifestation of pancreatic exocrine insufficiency is **steatorrhea**, or excessive undigested fat in the feces. Up to 60% to 70% of the ingested fat may be excreted in the feces. Digestion of protein and carbohydrates is impaired to a lesser degree because salivary, gastric, and small intestinal enzymes contribute to the digestion of these two foodstuffs.

Pancreatic Aqueous Alkaline Secretion Pancreatic enzymes function best in a neutral or slightly alkaline environment, yet the highly acidic gastric contents are emptied into the duodenal lumen in the vicinity of pancreatic enzyme entry into the duodenum. It is imperative that the acidic chyme be quickly neutralized in the duodenal lumen, not only to allow optimal functioning of the pancreatic enzymes but also to prevent acid damage to the duodenal mucosa. Therefore, the alkaline ($NaHCO_3$-rich) fluid secreted by the pancreas into the duodenal lumen serves the important function of neutralizing the acidic chyme as the latter is emptied into the duodenum from the stomach. This aqueous $NaHCO_3$ secretion is by far the largest component of pancreatic secretion. The volume of pancreatic secretion ranges between 1 and 2 liters per day, depending on the types and degree of stimulation.

Pancreatic secretion is hormonally regulated to maintain neutrality of the duodenal contents and to optimize digestion.

Pancreatic secretion is regulated primarily by hormonal mechanisms. A small amount of parasympathetically induced pancreatic secretion occurs during the cephalic phase of digestion, with a further token increase occurring during the gastric phase in response to gastrin. However, the predominant stimulation of pancreatic secretion occurs during the intestinal phase of digestion when chyme is in the small intestine. The release of the two major enterogastrones, secretin and cholecystokinin (CCK), in response to chyme in the duodenum plays the central role in the control of pancreatic secretion.

Of the factors that stimulate enterogastrone release, the primary stimulus specifically for secretin release is acid in the duodenum. Secretin, in turn, is carried by the blood to the pancreas, where it stimulates the duct cells to markedly increase their secretion of a $NaHCO_3$-rich aqueous fluid into the duodenum. Even though other stimuli may cause the release of secretin, it is appropriate that the most potent stimulus is acid because secretin promotes the alkaline pancreatic secretion that neutralizes the acid. This mechanism provides a control system for maintaining neutrality of the chyme in the intestine. The amount of secretin released is proportional to the amount of acid that enters the duodenum, so the amount of $NaHCO_3$ secreted parallels the duodenal acidity.

Cholecystokinin, on the other hand, is important in the regulation of pancreatic enzyme secretion. The main stimulus for release of CCK from the duodenal mucosa is the presence of fat and to a lesser extent protein products. The circulatory system transports CCK to the pancreas, where it stimulates the pancreatic acinar cells to increase enzyme secretion. Among these enzymes are lipase and the proteolytic enzymes, which appropriately bring about further digestion of the fat and protein that initiated the response and also help digest carbohydrate. In contrast to fat and protein, carbohydrate apparently does not have any direct influence on pancreatic enzyme secretion.

All three types of pancreatic enzymes are packaged together in the zymogen granules, so all the pancreatic enzymes are released together on exocytosis of the granules. Therefore, even though the *total amount* of enzymes released varies depending on the type of meal consumed (the most being secreted in response to fat), the *proportion* of enzymes released does not vary on a meal-to-meal basis. That is, a high-protein meal does not cause the release of a greater proportion of proteolytic enzymes. Evidence suggests, however, that long-term adjustments in the proportion of the types of enzymes produced may occur as an adaptive response to a prolonged change in diet. For example, a greater proportion of proteolytic enzymes are produced with a long-term switch to a high-protein diet. Cholecystokinin may play a role in pancreatic enzyme adaptation to changes in diet.

Just as gastrin is trophic to the stomach and small intestine, CCK and secretin exert trophic effects on the exocrine pancreas to maintain its integrity.

The liver performs various important functions including bile production.

Besides pancreatic juice, the other secretory product that is emptied into the duodenal lumen is **bile**. The **biliary system** includes the *liver,* the *gallbladder,* and associated ducts.

The **liver** is the largest and most important metabolic organ in the body. Its importance to the digestive system is its secretion of *bile salts,* but the liver performs a wide variety of other functions, including the following:

1. Metabolic processing of the major categories of nutrients (carbohydrates, proteins, and lipids) after their absorption from the digestive tract.
2. Detoxification or degradation of body wastes and hormones as well as drugs and other foreign compounds.
3. Synthesis of plasma proteins, including those necessary for the clotting of blood and those that transport steroid and thyroid hormones and cholesterol in the blood.
4. Storage of glycogen, fats, iron, copper, and many vitamins.
5. Activation of vitamin D, which the liver accomplishes in conjunction with the kidneys.
6. Removal of bacteria and worn-out red blood cells, thanks to its resident macrophages (see p. 358).
7. Excretion of cholesterol and bilirubin, the latter being a breakdown product derived from the destruction of worn-out red blood cells.

Given this wide range of complex functions, there is amazingly little specialization of cells within the liver. Each liver cell, or **hepatocyte** (*hepato* means "liver"; *cyte* means "cell"), appears to be able to perform the same wide variety of metabolic and secretory tasks, with the exception of the phagocytic activities carried out by the resident macrophages, which are known as **Kupffer cells**. The specialization comes from the highly developed organelles within each hepatocyte.

To carry out these wide-ranging tasks, the anatomical organization of the liver permits each hepatocyte to be in direct contact with blood from two sources: venous blood coming directly from the digestive tract and arterial blood coming from the aorta. Venous blood enters the liver by means of a unique and complex vascular connection between the digestive tract and the liver that is known as the **hepatic portal system** (● Fig. 16–17). The veins draining the digestive tract do not directly join the inferior vena cava, the large vein that returns blood to the heart. Instead, the veins from the stomach and intestine enter the hepatic portal vein, which carries the products absorbed from the digestive tract directly to the liver for processing, storage, or detoxification before they gain access to the general circulation. Within the liver, the portal vein once again breaks up into a capillary network (the liver sinusoids) to permit exchange between the blood and hepatocytes before draining into the hepatic vein, which joins the inferior vena cava. The hepatocytes also are provided with fresh arterial blood, which supplies their oxygen and delivers blood-borne metabolites for hepatic processing.

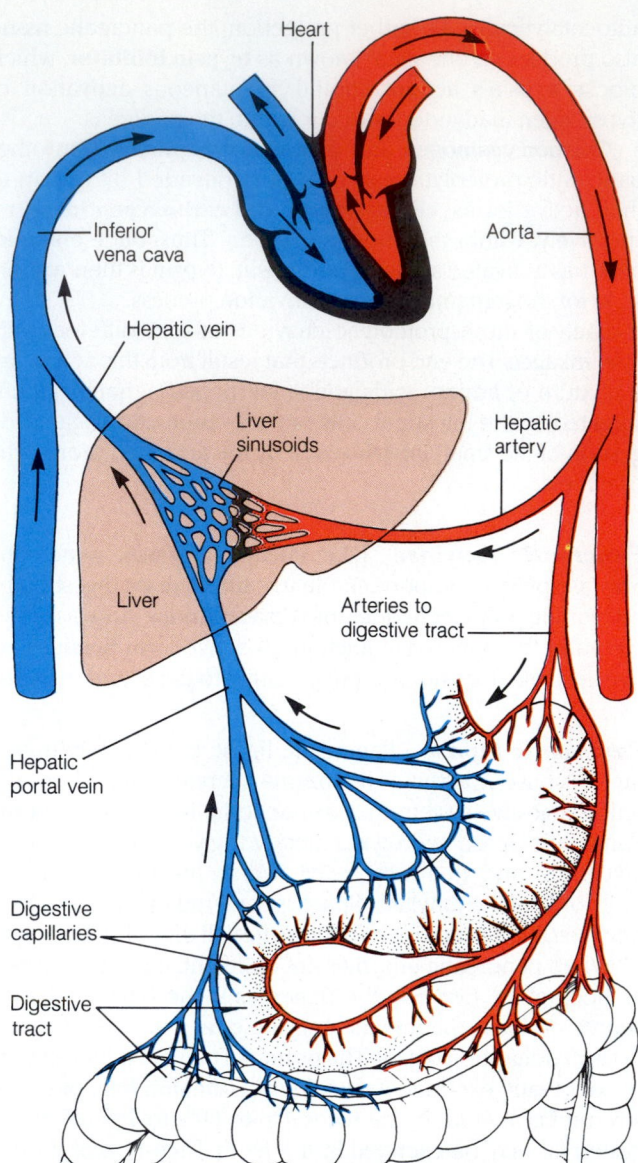

● **FIGURE 16–17   Schematic Representation of Liver Blood Flow**   The liver receives blood from two sources: (1) venous blood draining the digestive tract is carried by the hepatic portal vein to the liver for processing and storage of newly absorbed nutrients, and (2) arterial blood, which provides the liver's O_2 supply and carries blood-borne metabolites for hepatic processing, is delivered by the hepatic artery.

The liver lobules are delineated by vascular and bile channels.

The liver is organized into functional units known as **lobules**, which are hexagonal arrangements of tissue surrounding a central vein, like a six-sided angel food cake with the hole representing the central vein (● Fig. 16–18a). At the outer edge of each "slice" of the lobule are three vessels: a branch of the hepatic artery, a branch of the portal vein, and a bile duct. Blood from the branches of both the hepatic artery and the portal vein flows from the periphery of the lobule into large, expanded capillary spaces called **sinusoids**, which run

Branch of hepatic portal vein
Bile canaliculi
Central vein
Cords of hepatocytes (liver cells)
Bile duct
Branch of hepatic artery
Sinusoids
Hepatic portal vein
To hepatic duct
Hepatic artery

Branch of hepatic portal vein
Branch of hepatic artery
Bile duct
Connective tissue
Kupffer cell
Bile canaliculi
Sinusoids
Cords of hepatocytes (liver cells)
Hepatic plate
Central vein

(a)   (b)

• **FIGURE 16–18   Anatomy of the Liver** (a) Hepatic lobule. (b) Wedge of hepatic lobule.

between rows of liver cells to the central vein like spokes on a bicycle tire (Fig. 16–18b). The Kupffer cells line the sinusoids and engulf and destroy old red blood cells and bacteria that pass through in the blood. The hepatocytes are arranged between the sinusoids in plates two cell-layers thick, so that each lateral edge faces a sinusoidal pool of blood. The central veins of all of the liver lobules converge to form the hepatic vein, which carries the blood away from the liver. A thin bile-carrying channel, a **bile canaliculus**, runs between the cells within each hepatic plate. Hepatocytes continuously secrete bile into these thin channels, which carry the bile to a bile duct at the periphery of the lobule. The bile ducts from the various lobules converge to eventually form the common bile duct, which transports the bile from the liver to the duodenum. Each hepatocyte is in contact with a sinusoid on one side and a bile canaliculus on the other side.

Bile is secreted by the liver and is diverted to the gallbladder between meals.

The opening of the bile duct into the duodenum is guarded by the **sphincter of Oddi**, which prevents bile from entering the duodenum except during digestion of meals (• Fig. 16–19). When this sphincter is closed, most of the bile secreted by the liver is diverted back up into the **gallbladder**, a small saclike structure tucked beneath but not directly connected to the liver. The bile is subsequently stored and concentrated in the gallbladder between meals. After a meal, bile enters the duodenum as a result of the combined effects of gallbladder emptying and increased bile secretion by the

• **FIGURE 16–19   Enterohepatic Circulation of Bile Salts** The majority of bile salts are recycled between the liver and small intestine through the enterohepatic circulation. Bile salts secreted by the liver enter the duodenum. After participating in fat digestion and absorption, most bile salts are reabsorbed by active transport in the terminal ileum and returned through the hepatic portal vein to the liver, which resecretes them in the bile.

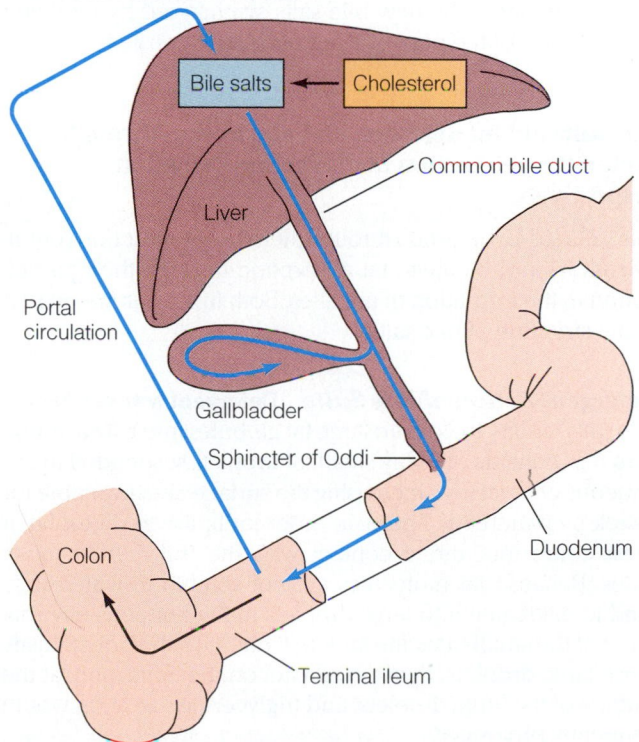

Bile salts
Cholesterol
Common bile duct
Liver
Portal circulation
Gallbladder
Sphincter of Oddi
Colon
Duodenum
Terminal ileum

liver. The amount of bile secreted per day ranges from 250 ml to 1 liter, depending on the degree of stimulation.

Bile salts are recycled through the enterohepatic circulation.

Bile consists of an *aqueous alkaline fluid* similar to the pancreatic $NaHCO_3$ secretion as well as several organic constituents, including *bile salts, cholesterol, lecithin,* and *bilirubin.* The organic constituents are derived from hepatocyte activity, whereas the water, $NaHCO_3$, and other inorganic salts are added by the duct cells. Even though bile does not contain any digestive enzymes, it is important for the digestion and absorption of fats, primarily through the activity of the bile salts.

Bile salts are derivatives of cholesterol. They are actively secreted into the bile and eventually enter the duodenum along with the other biliary constituents. Following their participation in fat digestion and absorption, most bile salts are reabsorbed back into the blood by special active transport mechanisms located in the terminal ileum, the last portion of the small intestine. From here the bile salts are returned by means of the hepatic portal system to the liver, which resecretes them into the bile. This recycling of bile salts (and some of the other biliary constituents) between the small intestine and liver is referred to as the **enterohepatic circulation** (*entero* means "intestine"; *hepatic* means "liver") (Fig. 16–19).

The total amount of bile salts in the body averages about 3 to 4 grams (g), yet 3 to 15 g of bile salts may be emptied into the duodenum in a single meal. Obviously, the bile salts must be recycled many times per day. Usually, only about 5% of the secreted bile salts escapes into the feces daily. These lost bile salts are replaced by new bile salts synthesized by the liver; thus, the size of the pool of bile salts is kept constant.

Bile salts aid fat digestion and absorption through their detergent action and micellar formation, respectively.

Bile salts aid fat digestion through their detergent action (emulsification) and facilitate fat absorption through their participation in the formation of micelles. Both functions are related to the structure of bile salts.

Detergent Action of Bile Salts
Detergent action refers to bile salts' ability to convert large fat globules into a **lipid emulsion** that consists of many small fat droplets suspended in the aqueous chyme, thus increasing the surface area available for attack by pancreatic lipase. In order for lipase to digest fat, it must come into direct contact with the triglyceride molecules. Because fat molecules are not soluble in water, they tend to aggregate into large droplets in the aqueous environment of the small-intestine lumen. If bile salts did not emulsify these large droplets, lipase could act on the lipids only at the surface of the large droplets, and triglyceride digestion would be greatly prolonged.

Bile salts exert a detergent action similar to that of the detergent you use to break up grease when you wash dishes. A bile-salt molecule contains a lipid-soluble portion (a steroid derived from cholesterol) plus a negatively charged, water-soluble portion. The lipid-soluble portion dissolves in a fat droplet, leaving the charged water-soluble portion projecting from the surface of the droplet (● Fig. 16–20a). Intestinal mixing movements break up large fat droplets into smaller ones. These small droplets would quickly recoalesce were it not for bile salts adsorbing on their surface and creating a "shell" of water-soluble negative charges on the surface of each little droplet. Because like charges repel, these negatively charged groups on the droplet surfaces cause the fat droplets to repel each other (Fig. 16–20b). This electric repulsion prevents the small droplets from recoalescing into large fat droplets, thus producing a lipid emulsion that increases the surface area available for lipase action. The increased surface area is extremely important in accomplishing rapid completion of fat digestion; without bile salts, digestion of fat would be very slow.

Micellar Formation
Bile salts—along with cholesterol and lecithin, which are also constituents of the bile—play an important role in facilitating fat absorption through micellar formation. Like bile salts, lecithin has both a lipid-soluble and a water-soluble portion, whereas cholesterol is almost totally insoluble in water. In a **micelle**, the bile salts and lecithin aggregate in small clusters with their fat-soluble portions huddled together in the middle to form a hydrophobic ("water-fearing") core while their water-soluble portions form an outer hydrophilic ("water-loving") shell (● Fig. 16–21). A micellar aggregate is about one-millionth the size of an emulsified lipid droplet. Micelles, being water soluble by virtue of their hydrophilic shells, can dissolve water-insoluble (and hence lipid-soluble) substances in their lipid-soluble cores. Micelles thus provide a handy vehicle for carrying water-insoluble substances through the watery luminal contents. The most important lipid-soluble substances thus carried are the products of fat digestion (monoglycerides and free fatty acids) as well as fat-soluble vitamins, which are all transported to their sites of absorption by means of the micelles. If they did not hitch a ride in the water-soluble micelles, these nutrients would float on the surface of the aqueous chyme (just as oil floats on top of water), never reaching the absorptive surfaces of the small intestine.

In addition, cholesterol, a highly water-insoluble substance, dissolves in the micelle's hydrophobic core. This mechanism is important in cholesterol homeostasis. The amount of cholesterol that can be carried in micellar formation depends on the relative amount of bile salts and lecithin in comparison to cholesterol. When cholesterol secretion by the liver is out of proportion to bile-salt and lecithin secretion (either too much cholesterol or too little bile salts and lecithin), the excess cholesterol in the bile precipitates into microcrystals that can aggregate into **gallstones**. One treatment of cholesterol-containing gallstones involves ingestion of bile salts to increase the bile-salt pool in an attempt to dissolve the cholesterol stones. Only about 75% of gallstones are derived

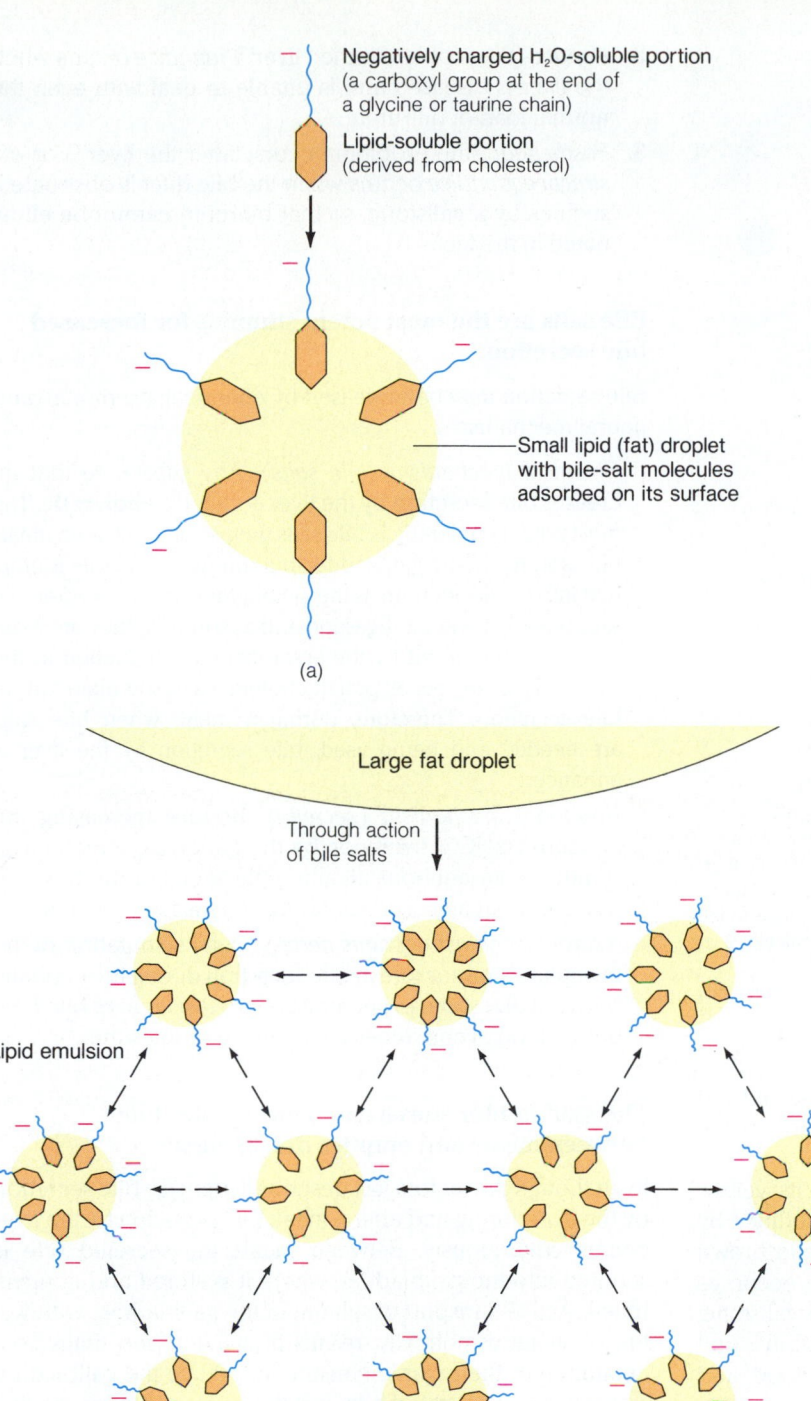

(a)

Negatively charged H₂O-soluble portion
(a carboxyl group at the end of
a glycine or taurine chain)

Lipid-soluble portion
(derived from cholesterol)

Small lipid (fat) droplet
with bile-salt molecules
adsorbed on its surface

Large fat droplet

Through action
of bile salts

Lipid emulsion

(b)

● **FIGURE 16–20 Schematic Structure and Function of Bile Salts** (a) Schematic representation of structure of bile salts and their adsorption on surface of fat droplet. A bile salt consists of a lipid-soluble portion that dissolves in the fat droplet and a negatively charged, water-soluble portion that projects from the surface of the droplet. (b) Formation of lipid emulsion through action of bile salts. Adsorption of bile salts on the surface of small fat droplets creates "shells" of negatively charged, water-soluble bile-salt components that cause the fat droplets to repel each other. This action holds the fat droplets apart and prevents them from recoalescing, thereby increasing the surface area of exposed fat available for digestion by pancreatic lipase.

from cholesterol, however. The other 25% are made up of abnormal precipitates of another bile constituent, bilirubin.

Bilirubin is a waste product excreted in the bile.

Bilirubin, the other major constituent of bile, does not play a role in digestion at all but instead is one of the few waste prod-

ucts excreted in the bile. Bilirubin is the primary bile pigment derived from the breakdown of worn-out red blood cells. The typical life span of a red blood cell in the circulatory system is 120 days. Worn-out red blood cells are removed from the blood by the macrophages that line the liver sinusoids and reside in other areas in the body. Bilirubin is the end product resulting from the degradation of the heme (iron-containing)

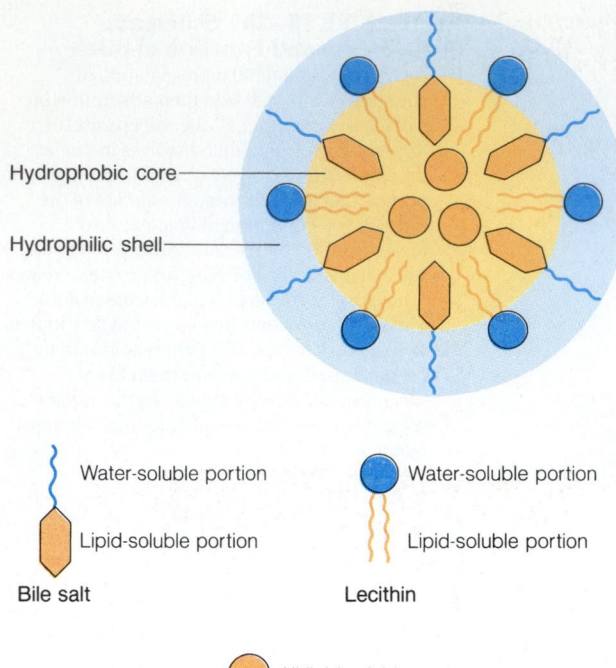

Hydrophobic core

Hydrophilic shell

Water-soluble portion

Lipid-soluble portion

Bile salt

Water-soluble portion

Lipid-soluble portion

Lecithin

All lipid-soluble

Cholesterol

● FIGURE 16–21   Schematic Representation of a Micelle Bile constituents (bile salts, lecithin, and cholesterol) aggregate to form micelles that consist of a hydrophilic (water-soluble) shell and a hydrophobic (lipid-soluble) core. Because the outer shell of a micelle is water soluble, the products of fat digestion, which are not water soluble, can be carried through the watery luminal contents to the absorptive surface of the small intestine by dissolving in the micelle's lipid-soluble core.

portion of the hemoglobin contained within these old red blood cells. This bilirubin is extracted from the blood by the hepatocytes and is actively excreted into the bile.

Bilirubin is a yellow pigment that gives bile its yellow color. Within the intestinal tract, this pigment is modified by bacterial enzymes, giving rise to the characteristic brown color of the feces. When bile secretion does not occur, as when the bile duct is completely obstructed by a gallstone, the feces are grayish white. A small amount of bilirubin is normally reabsorbed by the intestine back into the blood, and when it is eventually excreted in the urine, it is largely responsible for the urine's yellow color. The kidneys are unable to excrete bilirubin until after it has been modified during its passage through the liver and intestine.

If bilirubin is formed more rapidly than it can be excreted, it accumulates in the body and causes **jaundice**. Patients with this condition appear yellowish, with this color being seen most easily in the whites of their eyes. Jaundice can be brought about in three different ways:

1. *Prehepatic* (the problem occurs "before the liver") or *hemolytic jaundice* is due to excessive breakdown (hemolysis) of red blood cells so that the liver is presented with more bilirubin than it is capable of excreting.

2. *Hepatic* (the problem is the "liver") *jaundice* occurs when the liver is diseased and is unable to deal with even the normal load of bilirubin.

3. *Posthepatic* (the problem occurs "after the liver") or *obstructive jaundice* occurs when the bile duct is obstructed, such as by a gallstone, so that bilirubin cannot be eliminated in the feces.

Bile salts are the most potent stimulus for increased bile secretion.

Bile secretion may be increased by chemical, hormonal, and neural mechanisms.

■ *Chemical mechanism (bile salts).* Any substance that increases bile secretion by the liver is called a **choleretic.** The most potent choleretic is bile salts themselves. Between meals bile is stored in the gallbladder, but during a meal bile is emptied into the duodenum as the gallbladder contracts. After bile salts participate in fat digestion and absorption, they are reabsorbed and returned by the enterohepatic circulation to the liver, where they act as potent choleretics to stimulate further bile secretion. Therefore, during a meal, when bile salts are needed and being used, bile secretion by the liver is enhanced.

■ *Hormonal mechanism (secretin).* Besides increasing the aqueous $NaHCO_3$ secretion by the pancreas, secretin also stimulates an aqueous alkaline bile secretion by the liver ducts without any corresponding increase in bile salts.

■ *Neural mechanism (vagus nerve).* Vagal stimulation of the liver plays a minor role in bile secretion during the cephalic phase of digestion, promoting an increase in liver bile flow before food ever reaches the stomach or intestine.

The gallbladder stores and concentrates bile between meals and empties during meals.

Even though the factors just described increase bile secretion by the liver during and after a meal, bile secretion by the liver occurs continuously. Between meals the secreted bile is shunted into the gallbladder, where it is stored and concentrated. Active transport of salt out of the gallbladder, with water following osmotically, results in a five to ten times concentration of the organic constituents. Since the gallbladder stores this concentrated bile, it is the primary site for precipitation of concentrated bile constituents into gallstones. Fortunately, the gallbladder does not play an essential digestive role, so its removal as a treatment for gallstones or other gallbladder disease presents no particular problem. The bile secreted between meals is stored instead in the common bile duct, which becomes dilated.

During a meal, when chyme reaches the small intestine, the presence of food, especially fat products, in the duodenal lumen triggers the release of CCK. This hormone stimulates contraction of the gallbladder and relaxation of the sphincter of Oddi, so bile is discharged into the duodenum, where it appropriately aids in the digestion and absorption of the fat that initiated the release of CCK.

Hepatitis and cirrhosis are the most common liver disorders.

Hepatitis is an inflammatory disease of the liver that results from a variety of causes, including viral infection or exposure to toxic agents. With viral hepatitis, the viruses replicate in the nuclei of the hepatocytes, bringing about injury, inflammation, and even death of the infected cells. Toxic hepatitis occurs as a result of injury and destruction of liver cells from abusive exposure to toxic chemicals or drugs, including alcohol, carbon tetrachloride, and certain tranquilizers. Hepatitis ranges in severity from mild, reversible symptoms to acute massive liver damage, with possible imminent death resulting from acute hepatic failure.

Repeated or prolonged hepatic inflammation, usually in association with chronic alcoholism, can lead to **cirrhosis**, a condition in which damaged hepatocytes are permanently replaced by connective tissue. Liver tissue has the ability to regenerate, normally undergoing a gradual turnover of cells. If part of the hepatic tissue is destroyed, the lost tissue can be replaced by an increase in the rate of cell division. A circulating factor is apparently responsible for regulating liver-cell proliferation, although the nature and mechanism of this regulatory factor remain a mystery. There is a limit, however, to how rapidly hepatocytes can be replaced. In addition to hepatocytes, dispersed between the hepatic plates are a small number of fibroblasts (connective-tissue cells) that form a supporting framework for the liver. If the liver is repeatedly exposed to toxic substances such as alcohol so often that new hepatocytes cannot be generated rapidly enough to replace the damaged cells, the sturdier fibroblasts take advantage of the situation and overproduce. This extra connective tissue leaves little space for the hepatocytes' regrowth. Because nodules of hepatocytes that do manage to regenerate are often less functional and are frequently separated from an adequate blood supply by the thick bands of connective tissue, even the new hepatocyte growth may die. Thus, as cirrhosis develops slowly over time, active liver tissue is gradually reduced, leading eventually to chronic liver failure. Symptoms of hepatic malfunction may not appear until 70% to 80% of the liver tissue is destroyed, demonstrating that, just as with the kidneys, our bodies are endowed with a generous reserve of hepatic tissue.

SMALL INTESTINE

The **small intestine** is the site at which most digestion and absorption take place. No further digestion is accomplished after the luminal contents pass beyond the small intestine, and no further absorption of nutrients occurs, although the large intestine does absorb small amounts of salt and water. The small intestine is a tube about 21 ft (6.3 m) long with a small diameter of 1 in (2.5 cm). It lies coiled within the abdominal cavity extending between the stomach and the large intestine. It is arbitrarily divided into three segments—the **duodenum** [the first 8 in (20 cm)], the **jejunum** [8 ft (2.5 m)], and the **ileum** [12 ft (3.6 m)].

Segmentation contractions mix and slowly propel the chyme.

Segmentation, the small intestine's primary method of motility, both mixes and slowly propels the chyme. Segmentation consists of oscillating, ringlike contractions of the circular smooth muscle along the length of the small intestine; between the contracted segments are relaxed areas containing a small bolus of chyme. The contractile rings occur every few centimeters, dividing the small intestine into segments like a chain of sausages. These contractile rings do not sweep along the length of the intestine as peristaltic waves do. Rather, after a brief period of time, the contracted segments relax, and ringlike contractions appear in the previously relaxed areas (• Fig. 16–22). The new contraction forces the chyme in a previously relaxed segment to move in both directions into the now relaxed adjacent segments. A newly relaxed segment therefore receives chyme from both the contracting segment immediately ahead of it and the one immediately behind it. Shortly thereafter, the areas of contraction and relaxation alternate again. In this way, the chyme is chopped, churned, and thoroughly mixed. These contractions can be compared to squeezing a pastry tube with your hands to mix the contents. This mixing serves the dual functions of mixing the chyme with the digestive juices secreted into the small-intestine lumen and exposing all of the chyme to the absorptive surfaces of the small intestinal mucosa.

Segmentation contractions are initiated by small-intestine pacesetter cells, which produce a basic electrical rhythm (BER) similar to the gastric BER responsible for peristalsis in the stomach. If the small-intestine BER brings the circular smooth-muscle layer to threshold, segmentation contractions are induced, with the frequency of segmentation following the frequency of the BER.

The circular smooth muscle's degree of responsiveness

• **FIGURE 16–22** **Segmentation** Segmentation consists of ringlike contractions along the length of the small intestine. Within a matter of seconds, the contracted segments relax and the previously relaxed areas contract. These oscillating contractions thoroughly mix the chyme within the small-intestine lumen.

and thus the intensity of segmentation contractions can be influenced by distention of the intestine, by the hormone gastrin, and by extrinsic nerve activity. All of these factors influence the excitability of the small-intestine smooth-muscle cells by moving the "resting" potential around which the BER oscillates closer to or farther from threshold. Segmentation is slight or absent between meals but becomes very vigorous immediately after a meal. Both the duodenum and ileum start to segment simultaneously when the meal first enters the small intestine. The duodenum starts to segment primarily in response to local distention caused by the presence of chyme. Segmentation of the empty ileum, on the other hand, appears to be brought about by gastrin, which is secreted in response to the presence of chyme in the stomach, a mechanism known as the **gastroileal reflex**. Extrinsic nerves can modify the strength of these contractions. Parasympathetic stimulation enhances segmentation, whereas sympathetic stimulation depresses segmental activity.

Segmentation not only accomplishes mixing but also is the primary factor responsible for slowly moving chyme through the small intestine. How can this be when each segmental contraction propels chyme both forward and backward? The chyme progresses because the frequency of segmentation declines along the length of the small intestine. The pacesetter cells in the duodenum spontaneously depolarize at a faster rate than those farther down the tract, with the pacesetter cells in the terminal ileum exhibiting the slowest rate of spontaneous depolarization. The rate of segmentation in the duodenum is twelve contractions per minute, compared to a rate of only nine per minute in the terminal ileum. Because segmentation occurs with greater frequency in the upper part of the small intestine than in the lower part, more chyme tends to be pushed forward than is pushed backward. As a result, chyme is moved very slowly from the upper to the lower part

of the small intestine, being shuffled back and forth to accomplish thorough mixing and absorption in the process. This slow propulsive mechanism is advantageous because it allows ample time for the digestive and absorptive processes to take place. The contents usually take three to five hours to move through the small intestine.

The migrating motility complex sweeps the intestine clean between meals.

When most of the meal has been absorbed, segmentation contractions cease and are replaced between meals by the **migrating motility complex**, or "**intestinal housekeeper**." This between-meal motility consists of weak, repetitive peristaltic waves that move a short distance down the intestine before dying out. The waves start at the stomach and migrate down the intestine; that is, each new peristaltic wave is initiated at a site a little farther down the small intestine. These short peristaltic waves take about 100–150 minutes to gradually migrate from the stomach to the end of the small intestine, with each contraction "sweeping" any remnants of the preceding meal plus mucosal debris and bacteria forward toward the colon, just like a good "intestinal housekeeper." After the end of the small intestine is reached, the cycle begins again and continues to repeat itself until the next meal. It is speculated that the unconfirmed hormone **motilin** might play a role in regulating the migrating motility complex. When the next meal arrives, segmental activity is triggered again, and the migrating motility complex ceases.

The ileocecal juncture prevents contamination of the small intestine by colonic bacteria.

At the juncture between the small and large intestines, the last portion of the ileum empties into the cecum (• Fig. 16–23).

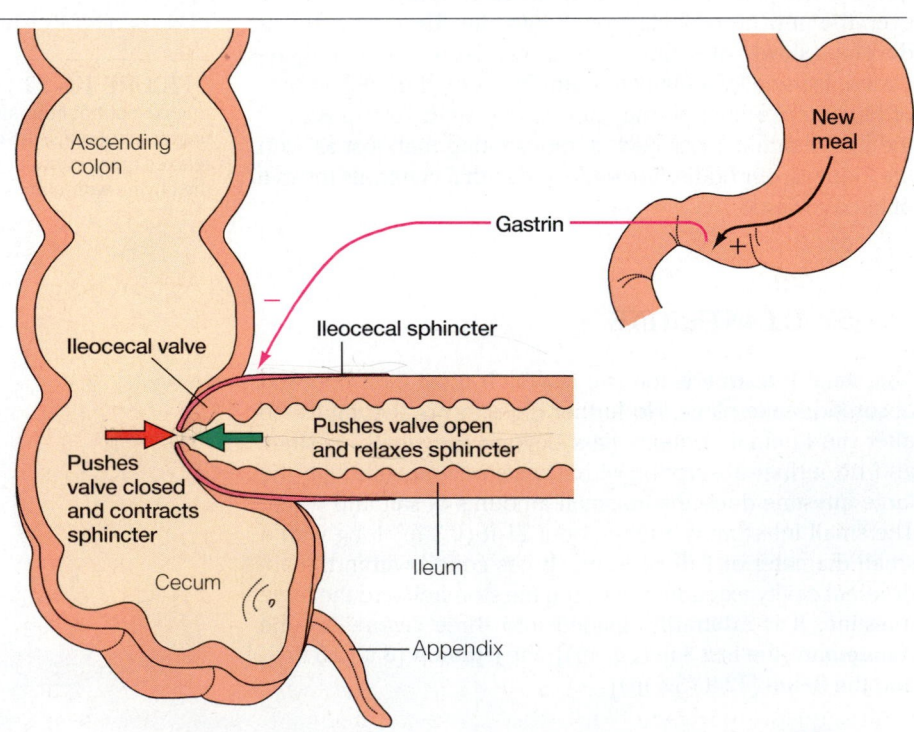

• **FIGURE 16–23 Control of Ileocecal Valve/Sphincter** The juncture between the ileum and large intestine consists of the ileocecal valve, which is surrounded by thickened smooth muscle, the ileocecal sphincter. Pressure on the cecal side pushes the valve closed and contracts the sphincter, preventing the bacteria-laden colonic contents from contaminating the nutrient-rich small intestine. The valve/sphincter opens and allows ileal contents to enter the large intestine in response to pressure on the ileal side of the valve and to the hormone gastrin secreted as a new meal enters the stomach.

Two factors contribute to this region's ability to act as a barrier between the small and large intestines. First, the anatomical arrangement is such that valvelike folds of tissue protrude from the ileum into the lumen of the cecum. When the ileal contents are pushed forward, this **ileocecal valve** is easily pushed open, but the folds of tissue are forcibly closed when the cecal contents attempt to move backward. Second, the smooth muscle within the last several centimeters of the ileal wall is thickened, forming a sphincter that is under neural and hormonal control. Most of the time this **ileocecal sphincter** remains at least mildly constricted. Pressure on the cecal side of the sphincter causes it to contract more forcibly; distention of the ileal side causes the sphincter to relax, a reaction that is mediated by the intrinsic plexuses in the area. In this way, the ileocecal sphincter prevents the bacteria-laden contents of the large intestine from contaminating the small intestine and at the same time allows the ileal contents to pass into the colon. If the colonic bacteria were to gain access to the nutrient-rich small intestine, they would multiply rapidly. Relaxation of the sphincter is enhanced through the release of gastrin at the onset of a meal, when increased gastric activity is taking place. This relaxation allows the undigested fibers and unabsorbed solutes from the preceding meal to be moved forward as the new meal enters the tract.

Small-intestine secretions do not contain any digestive enzymes.

Each day the exocrine glands located in the small-intestine mucosa secrete into the lumen about 1.5 liters of an aqueous salt and mucus solution known as the **succus entericus** (*succus* means "juice"; *entericus* means "of the intestine"). No digestive enzymes are secreted into this intestinal juice. The small intestine does synthesize digestive enzymes, but they act intracellularly within the borders of the epithelial cells that line the lumen instead of being secreted directly into the lumen.

Are any functions served by this small-intestine secretion, since no digestive enzymes are involved? The mucus in the secretion provides protection and lubrication. Furthermore, this aqueous secretion provides an abundance of H_2O to participate in the enzymatic digestion of food. Recall that digestion involves hydrolysis—bond breakage by reaction with H_2O—which proceeds most efficiently when all the reactants are in solution.

The regulation of small-intestine secretion is not clearly understood. Secretion of succus entericus does increase after a meal. The most effective stimulus for secretion appears to be local stimulation of the small-intestine mucosa by the presence of chyme.

Digestion in the small-intestine lumen is accomplished by pancreatic enzymes, whereas the small-intestine enzymes act intracellularly.

Digestion within the small-intestine lumen is accomplished by the pancreatic enzymes, with fat digestion being enhanced by bile secretion. As a result of pancreatic enzymatic activity, fats are completely reduced to their absorbable units of mon-

oglycerides and free fatty acids, proteins are broken down into small peptide fragments and some amino acids, and carbohydrates are reduced to disaccharides and some monosaccharides. Thus, fat digestion is completed within the small-intestine lumen, but carbohydrate and protein digestion have not been brought to completion.

Special actin-stiffened, hairlike projections on the luminal surface of the small intestine epithelial cells form the **brush border** (see p. 40), which contains three different categories of enzymes: (1) **enterokinase**, which activates the pancreatic enzyme trypsinogen; (2) the **disaccharidases** (**sucrase**, **maltase**, and **lactase**), which complete carbohydrate digestion by hydrolyzing the remaining disaccharides (sucrose, maltose, and lactose, respectively) into their constituent monosaccharides; and (3) the **aminopeptidases**, which hydrolyze the small peptide fragments into their amino acid components, thereby completing protein digestion. Thus, carbohydrate and protein digestion are completed intracellularly within the confines of the brush border. (Table 16–6 provides a summary of the digestive processes for the three major categories of nutrients.)

A fairly common disorder, **lactose intolerance**, involves a deficiency of lactase, the disaccharidase specific for the digestion of lactose, or milk sugar. Most children under four years of age have adequate lactase, but this may be gradually lost so that in many adults, lactase activity is diminished or absent. When milk or dairy products (except those in which lactose has already been digested by bacterial action during processing, such as some kinds of cheese and yogurt) are consumed by an individual with lactase deficiency, the undigested lactose remains in the lumen and has several related consequences. First, accumulation of undigested lactose creates an osmotic gradient that draws H_2O into the intestinal lumen. Second, bacteria residing in the large intestine possess lactose-splitting ability, so they eagerly attack the lactose as an energy source, producing large quantities of CO_2 and methane gas in the process. Distention of the intestine by both fluid and gas produces pain (cramping) and diarrhea. Depending on the extent of the lactase deficiency and the quantity of lactose ingested, symptoms can vary from mild abdominal discomfort to severe dehydrating diarrhea.

The small intestine is remarkably well adapted for its primary role in absorption.

All products of carbohydrate, protein, and fat digestion, as well as most of the ingested electrolytes, vitamins, and water, are normally absorbed by the small intestine indiscriminately. Usually, only the absorption of calcium and iron is adjusted to the body's needs. Thus, the more food that is consumed, the more that will be digested and absorbed, as those who are trying to control their weight are all too painfully aware.

Most absorption occurs in the duodenum and jejunum; very little occurs in the ileum, not because the ileum does not have absorptive capacity but because most absorption has already been accomplished before the intestinal contents reach the ileum. The small intestine has an abundant reserve absorptive capacity. In fact, about 50% of the small intestine can be removed with little interference to absorption—with

TABLE 16-6 Digestive Processes for the Three Major Categories of Nutrients

NUTRIENTS	ENZYMES FOR DIGESTING NUTRIENT	SOURCE OF ENZYMES	SITE OF ACTION OF ENZYMES	ACTION OF ENZYMES	ABSORBABLE UNITS OF NUTRIENTS
Carbohydrate	Amylase	Salivary glands	Mouth and body of stomach	Hydrolyzes polysaccharides to disaccharides	
		Exocrine pancreas	Small-intestine lumen		
	Disaccharidases (sucrase, lactase, maltase)	Small-intestine epithelial cells	Small-intestine brush border	Hydrolyzes disaccharides to monosaccharides	Monosaccharides, especially glucose
Protein	Pepsin	Stomach chief cells	Stomach antrum	Hydrolyzes protein to peptide fragments	
	Trypsin, chymotrypsin, carboxypeptidase	Exocrine pancreas	Small-intestine lumen	Attack different peptide fragments	
	Aminopeptidases	Small-intestine epithelial cells	Small-intestine brush border	Hydrolyzes peptide fragments to amino acids	Amino acids and a few small peptides
Fat	Lipase	Exocrine pancreas	Small-intestine lumen	Hydrolyzes triglycerides to fatty acids and monoglycerides	
	Bile salts (not an enzyme)	Liver	Small-intestine lumen	Emulsify large fat globules	Fatty acids and monoglycerides

one exception. If the terminal ileum is removed, vitamin B_{12} and bile salts are not properly absorbed because the specialized transport mechanisms for these two substances are located only in this region. All other substances can be absorbed throughout the length of the small intestine.

The mucous lining of the small intestine is remarkably well adapted for its special absorptive function for two reasons: (1) it has a very large surface area, and (2) the epithelial cells in this lining possess a variety of specialized transport mechanisms. The following special modifications of the small-intestine mucosa greatly increase the surface area available for absorption (• Fig. 16-24):

■ The inner surface of the small intestine is thrown into circular folds that are visible to the naked eye and increase the surface area threefold.

■ Projecting from this folded surface are microscopic finger-like projections known as **villi**, which give the lining a velvety appearance and increase the surface area by another ten times. The surface of each villus is covered by epithelial cells interspersed occasionally with mucous cells.

■ Even smaller hairlike projections known as **microvilli** (or the *brush border*) arise from the luminal surface of these epithelial cells, increasing the surface area another twenty-fold. Each epithelial cell has as many as 3,000 to 6,000 of these microvilli, which are visible only with an electron microscope. It is within the membrane of this brush border that the small-intestine enzymes perform their functions.

Altogether, the folds, villi, and microvilli provide the small intestine with a luminal surface area 600 times greater than it would have if it were a tube of the same length and diameter lined by a flat surface. In fact, if the surface area of the small intestine were spread out flat, it would cover an entire tennis court.

Malabsorption (impairment of absorption) may be caused by damage to or reduction of the surface area of the small intestine. One of the most common causes is **gluten enteropathy**. In this condition, the individual's small intestine is abnormally sensitive to gluten, a protein constituent of wheat and some other grains. Through some mechanism that is not clearly understood, exposure to gluten appears to damage the intestinal villi: the normally luxuriant array of villi is reduced, the mucosa becomes flattened, and the brush border becomes short and stubby (• Fig. 16-25). Because this loss of villi decreases the surface area available for absorption, absorption of all nutrients is impaired. The condition is treated by eliminating gluten from the diet.

Absorption across the digestive-tract wall involves transepithelial transport similar to movement of material across the kidney tubules (see p. 480). The epithelial cells that cover the surface of each villus are joined at their lateral borders by tight junctions, which limit passage of luminal contents between the cells, although the tight junctions in the small intestine are "leakier" than those in the stomach. The connective-tissue core of the villus is formed by the lamina propria. Each villus is supplied by an arteriole that breaks up

(a)

(b)

● FIGURE 16–25   Reduction in Brush Border with Gluten Enteropathy (a) Electron micrograph of the brush border of a small-intestine epithelial cell in a normal individual. (b) Electron micrograph of the short, stubby brush border of a small-intestine epithelial cell in a patient with gluten enteropathy.

● FIGURE 16–24   Small-Intestine Absorptive Surface
(a) Gross structure of the small intestine. (b) Circular folds of small-intestine mucosa, which increase the absorptive surface area threefold. (c) Microscopic fingerlike projection known as a villus. Collectively, the villi increase the surface area another tenfold. (d) Electron-microscope view of villus epithelial cell depicting the presence of microvilli on its luminal border; the microvilli increase the surface area another twentyfold. Altogether, these surface modifications increase the small-intestine absorptive surface area 600-fold.

into a capillary network within the villus (Fig. 16–24c). The capillaries rejoin to form a venule that drains away from the villus. Each villus is also provided with a terminal lymphatic vessel known as a **central lacteal**. It is into this capillary network or central lacteal that digested substances enter during the process of absorption. To be absorbed, a substance must pass completely through the epithelial cell, diffuse through the interstitial fluid within the connective-tissue core of the villus, and then cross the wall of a capillary or lymph vessel. Like renal transport, intestinal absorption may be an active or passive process, with active absorption involving energy expenditure for at least one of the steps in the transepithelial transport process.

The mucosal lining experiences rapid turnover.

Dipping down into the mucosal surface between the villi are shallow invaginations known as the **crypts of Lieberkühn** (Fig. 16–24c). Unlike the gastric pits, these intestinal crypts do not secrete digestive enzymes but function as "nurseries." The epithelial cells lining the small intestine slough off and are replaced at a rapid rate as a result of high mitotic activity in the crypts. New cells that are continually being produced at the bottom of the crypts migrate up the villi and, in the process, push off the older cells at the tips of the villi into the lumen. In this manner, more than 100 million intestinal cells are shed per minute. The entire trip from crypt to tip averages about three days, so the epithelial lining of the small intestine is replaced approximately every three days. Because of this high rate of cell division, the cells within the crypts are very sensitive to damage by radiation and anticancer drugs, both of which may inhibit cell division.

The cells undergo several changes as they migrate up the villus. The concentration of brush-border enzymes increases and the capacity for absorption improves, so the cells at the tip of the villus have the greatest digestive and absorptive capability. Just at their peak, these cells are pushed off by the newly migrating cells. Thus, the luminal contents are constantly exposed to cells that are optimally equipped to complete the digestive and absorptive functions efficiently. Furthermore, just as in the stomach, the rapid turnover of cells in the small intestine is essential because of the harsh luminal conditions. Cells exposed to the abrasive and corrosive luminal contents are easily damaged and cannot live for long, so they must be continually replaced by a fresh supply of new-born cells.

The old cells that are sloughed off into the lumen are not entirely lost to the body. These cells are digested, with the cell constituents being absorbed into the blood and reclaimed for synthesis of new cells, among other things.

Special mechanisms facilitate absorption of most nutrients.

We will now turn our attention to the mechanisms through which the specific dietary constituents are normally absorbed.

Salt and Water Absorption Sodium may be absorbed both passively and actively. When the electrochemical gradient favors movement of Na^+ from the lumen to the blood, passive diffusion of Na^+ can occur *between* the intestinal epithelial cells through the "leaky" tight junctions. Movement of Na^+ *through* the cells is energy dependent and involves at least two different carriers, similar to the process of Na^+ reabsorption across the kidney tubules (see p. 481). Sodium may passively enter the epithelial cells at their luminal surface by itself, or it may be cotransported with glucose or amino acid. It is then actively pumped out of the cell at the basolateral border into the lateral spaces between the cells where they are not joined by tight junctions. From here Na^+ eventually diffuses into the capillaries.

As with the renal tubules in the early portion of the nephron, the absorption of Cl^-, H_2O, glucose, and amino acids from the small intestine is linked to this energy-dependent Na^+ absorption. Chloride passively follows down the electrical gradient created by Na^+ absorption and can be actively absorbed as well if needed. Most H_2O absorption in the digestive tract depends on the active carrier that pumps Na^+ into the lateral spaces, resulting in a concentrated area of high osmotic pressure in that localized region between the cells,

similar to the situation in the kidneys (see p. 486). This localized high osmotic pressure induces H_2O to move from the lumen through the cell (and possibly from the lumen through the leaky tight junction) into the lateral space. Water entering the space reduces the osmotic pressure but raises the hydrostatic (fluid) pressure. As a result, H_2O is flushed out of the lateral space into the interior of the villus, where it is picked up by the capillary network. Meanwhile, more Na^+ is pumped into the lateral space to encourage more H_2O absorption.

Carbohydrate Absorption Dietary carbohydrate is presented to the small intestine for absorption mainly in the forms of the disaccharides maltose (the product of polysaccharide digestion), sucrose, and lactose (Table 16–1 and • Fig. 16–26). The disaccharidases located in the brush borders of the small-intestine cells further reduce these disaccharides into the absorbable monosaccharide units of glucose, galactose, and fructose.

Glucose and galactose are both absorbed by *secondary active transport*, in which cotransport carriers on the luminal border transport both the monosaccharide and Na^+ from the lumen into the interior of the intestinal cell. The operation of these cotransport carriers, which do not directly use energy themselves, depends on the Na^+ concentration gradient established by the energy-consuming basolateral Na^+-K^+ pump (see p. 66). Glucose (or galactose), having been concentrated in the cell by the cotransport carriers, leaves the cell down its concentration gradient to enter the blood within the villus. Fructose is absorbed into the blood solely by facilitated diffusion (passive carrier-mediated transport).

Protein Absorption Not only are ingested proteins digested and absorbed, but endogenous ("within the body") proteins that have entered the digestive-tract lumen from the three following sources are digested and absorbed as well:

1. Digestive enzymes, all of which are proteins, that have been secreted into the lumen.

2. Proteins within the cells that are pushed off from the villi into the lumen during the process of mucosal turnover.

3. Small amounts of plasma proteins that normally leak from the capillaries into the digestive-tract lumen.

About 20–40 g of endogenous protein enters the lumen each day from these three sources. This quantity can amount to more than half of the protein presented to the small intestine for digestion and absorption. All endogenous proteins must be digested and absorbed along with the dietary proteins to prevent depletion of the body's protein stores. The amino acids absorbed from both food and the endogenous protein are used primarily to synthesize new protein in the body.

The protein presented to the small intestine for absorption is primarily in the form of amino acids and a few small peptide fragments (• Fig. 16–27). Amino acids are absorbed across the intestinal cells by secondary active transport, similar to glucose and galactose absorption. Thus, glucose, galactose, and amino acids all get a "free ride" in on the energy

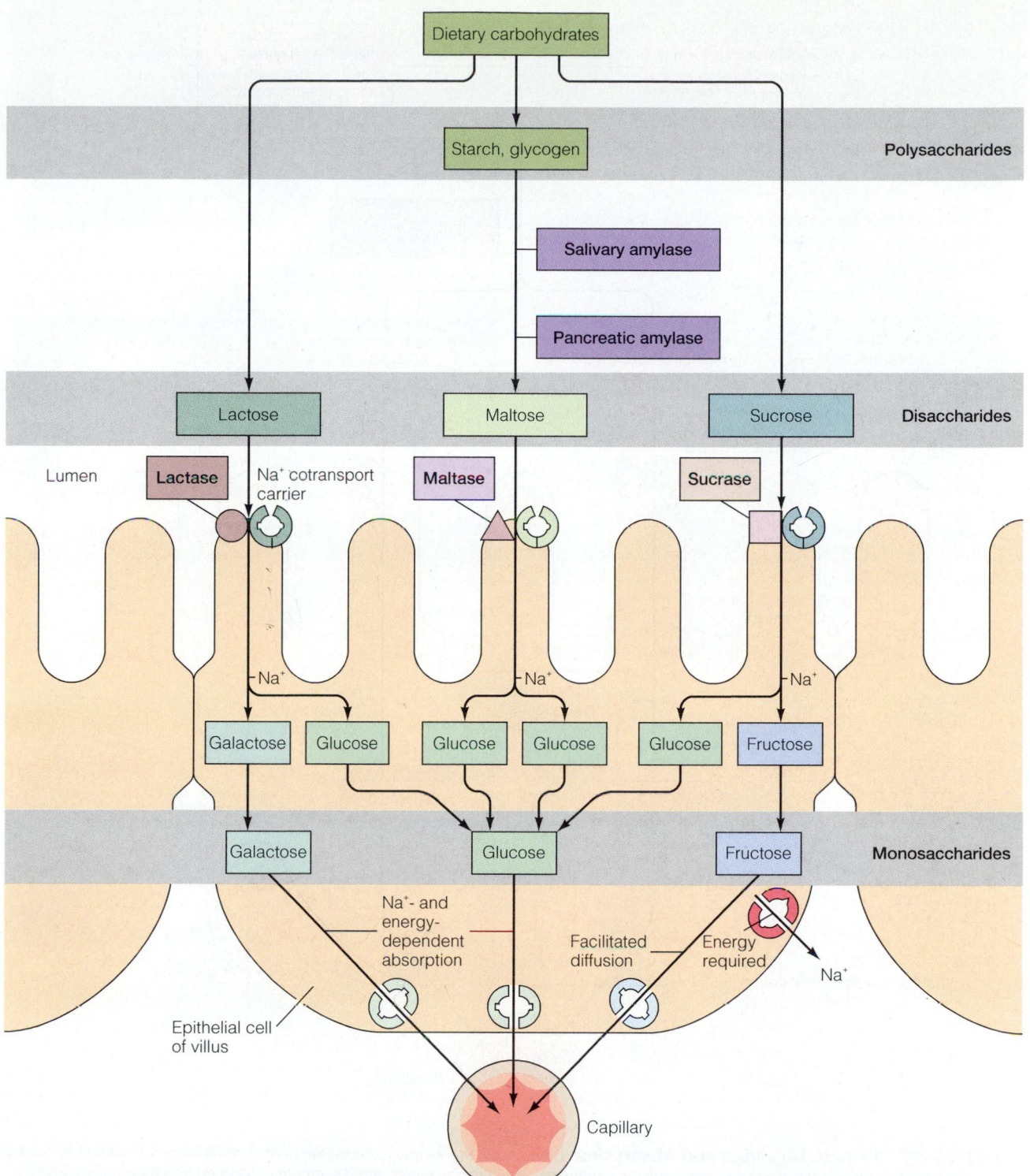

● **FIGURE 16–26 Carbohydrate Digestion and Absorption** The dietary polysaccharides starch and glycogen are converted into the disaccharide maltose through the action of salivary and pancreatic amylase. Maltose and the dietary disaccharides lactose and sucrose are converted to their respective monosaccharides by the disaccharidases (maltase, lactase, and sucrase) located in the brush borders of the small-intestine epithelial cells. The monosaccharides glucose and galactose are absorbed into the interior of the cell and eventually enter the blood by means of Na^+- and energy-dependent secondary active transport. The monosaccharide fructose is absorbed into the blood by passive facilitated diffusion.

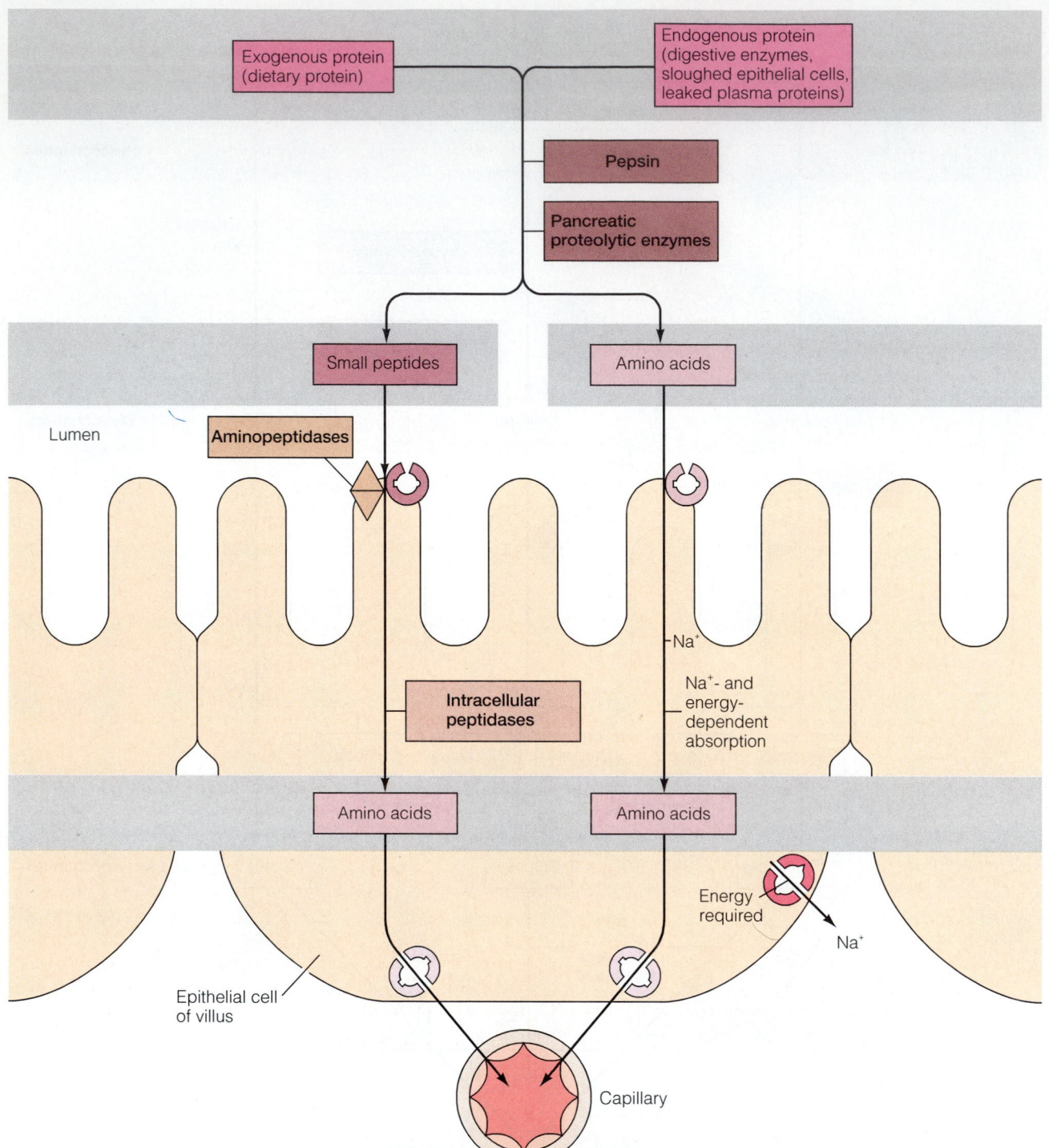

● FIGURE 16–27 Protein Digestion and Absorption Dietary and endogenous proteins are hydrolyzed to their constituent amino acids and a few small peptide fragments by gastric pepsin and the pancreatic proteolytic enzymes. Amino acids are absorbed into the small-intestine epithelial cells and eventually enter the blood by means of Na⁺- and energy-dependent secondary active transport. Various amino acids are transported by carriers specific for them. The small peptides, which are absorbed by a different type of carrier, are broken down into their amino acids by aminopeptidases in the epithelial cells' brush borders or by intracellular peptidases.

expended for Na⁺ transport. Small peptides gain entry by means of a different carrier and are broken down into their constituent amino acids by the aminopeptidases in the brush borders or by intracellular peptidases. Like monosaccharides, amino acids enter the capillary network within the villus.

Thus, absorption of the digestion end products of both carbohydrates and proteins involves special carrier-mediated transport systems that require energy expenditure and Na⁺ cotransport, and both categories of end products are absorbed into the blood.

Fat Absorption Fat absorption is quite different from carbohydrate and protein absorption because the insolubility of fat in water presents a special problem. Fat must be transferred from the watery chyme through the watery body fluids even though it is not water soluble. Therefore, fat must undergo a series of transformations to circumvent this problem during its digestion and absorption (Fig. 16–28).

When the stomach contents are emptied into the duodenum, the ingested fat is aggregated into large, oily triglyceride droplets that float in the chyme. Recall that through the bile salts' detergent action in the small intestine, the large droplets are dispersed into a lipid emulsification of small droplets, thereby exposing a much greater surface area of fat for digestion by pancreatic lipase. The products of lipase digestion (monoglycerides and free fatty acids) are also not very water soluble, so very little of these end products of fat digestion can diffuse through the aqueous chyme to reach the absorptive lining. However, biliary components facilitate absorption of these fatty end products through formation of micelles. Remember that micelles are water-soluble particles that can carry the end products of fat digestion within their lipid-soluble interiors. Once these micelles reach the luminal membranes of the epithelial cells, the monoglycerides and free fatty acids passively diffuse from the micelles through the lipid component of the epithelial cell membranes to enter the interior of these cells. As these fat products leave the micelles and are absorbed across the epithelial cell membranes, the micelles are able to pick up more monoglycerides and free fatty acids, which have been produced from digestion of further triglyceride molecules in the fat emulsion.

Bile salts continuously repeat their fat-solubilizing function down the length of the small intestine until all the fat is absorbed. Then the bile salts themselves are reabsorbed in the terminal ileum by special active transport. This is an efficient process, because relatively small amounts of bile salts can facilitate digestion and absorption of large amounts of fat, with each bile salt performing its ferrying function repeatedly before it is reabsorbed.

Once within the interior of the epithelial cells, the monoglycerides and free fatty acids are resynthesized into triglycerides. These triglycerides conglomerate into droplets and are coated with a layer of lipoprotein (synthesized by the endoplasmic reticulum of the epithelial cell), which renders the fat droplets water soluble. These large, coated fat droplets, known as **chylomicrons**, are extruded by exocytosis from the epithelial cells into the interstitial fluid within the villus. The chylomicrons subsequently enter the central lacteals rather than the capillaries because of the structural differences between these two vessels. Capillaries have a basement membrane (an outer layer of polysaccharides) that prevents chylomicrons from entering, but the lymph vessels do not have this barrier. Thus, fat can be absorbed into the lymphatics but not directly into the blood. (Fatty acids with short- or medium-length carbon chains can enter the blood, but very few of these are eaten in the normal diet.)

The actual transfer of monoglycerides and free fatty acids from the chyme across the cell membranes of the intestinal epithelial cells is a passive process, because the lipid-soluble fatty end products merely dissolve in and pass through the lipid portions of the membrane. Fat absorption is therefore said to be a passive process. However, the overall sequence of events necessary for fat absorption does require energy. For example, bile salts are actively secreted by the liver, and the resynthesis of triglycerides and formation of chylomicrons within the epithelial cells are active processes.

Vitamin Absorption Water-soluble vitamins are primarily absorbed passively with water, whereas fat-soluble vitamins are carried in the micelles and absorbed passively with the end products of fat digestion. Absorption of some of the vitamins can also be accomplished by carriers, if necessary. Vitamin B_{12} is unique in that it must be in combination with gastric intrinsic factor for absorption by special transport in the terminal ileum.

Iron and Calcium Absorption In contrast to the almost complete, unregulated absorption of other ingested electrolytes, the absorption of dietary iron and calcium may not be complete because it is subject to regulation that depends on the body's needs for these electrolytes.

Iron is essential for hemoglobin production. The normal iron intake is typically 15–20 mg/day, yet a man usually absorbs only about 0.5–1 mg/day into the blood, and a woman takes up slightly more at 1.0–1.5 mg/day (women's need for iron is greater because of the periodic loss of iron in the menstrual blood flow).

Two main steps are involved in the absorption of iron into the blood: (1) absorption of iron from the lumen into the intestinal epithelial cells and (2) absorption of iron from the epithelial cells into the blood (Fig. 16–29).

Iron is actively transported from the lumen into the epithelial cells, with women having about four times more active transport sites for iron than men. The extent to which ingested iron is taken up by the epithelial cells depends on the type of iron consumed (ferrous iron, Fe^{++}, is absorbed more easily than ferric iron, Fe^{+++}) and the presence of other substances in the lumen that can either promote or reduce iron absorption. For example, ascorbic acid (vitamin C) increases iron absorption, primarily by reducing ferric to ferrous iron. Phosphate and oxalate, on the other hand, combine with ingested iron to form insoluble iron salts that cannot be absorbed.

After active absorption into the small-intestine epithelial cells, iron has two possible fates:

1. Iron that is immediately needed for the production of red blood cells is absorbed into the blood for delivery to the bone marrow, the site of red blood cell production. Iron is transported in the blood by a plasma-protein carrier known as **transferrin**. The hormone responsible for stimulating red blood cell production, erythropoietin (see p. 351), is believed to also enhance iron absorption from the intestinal cells into the blood. The absorbed iron is then used in the synthesis of hemoglobin for the newly produced red blood cells.

2. Iron that is not immediately needed remains stored within the epithelial cells in a granular form called **ferritin**,

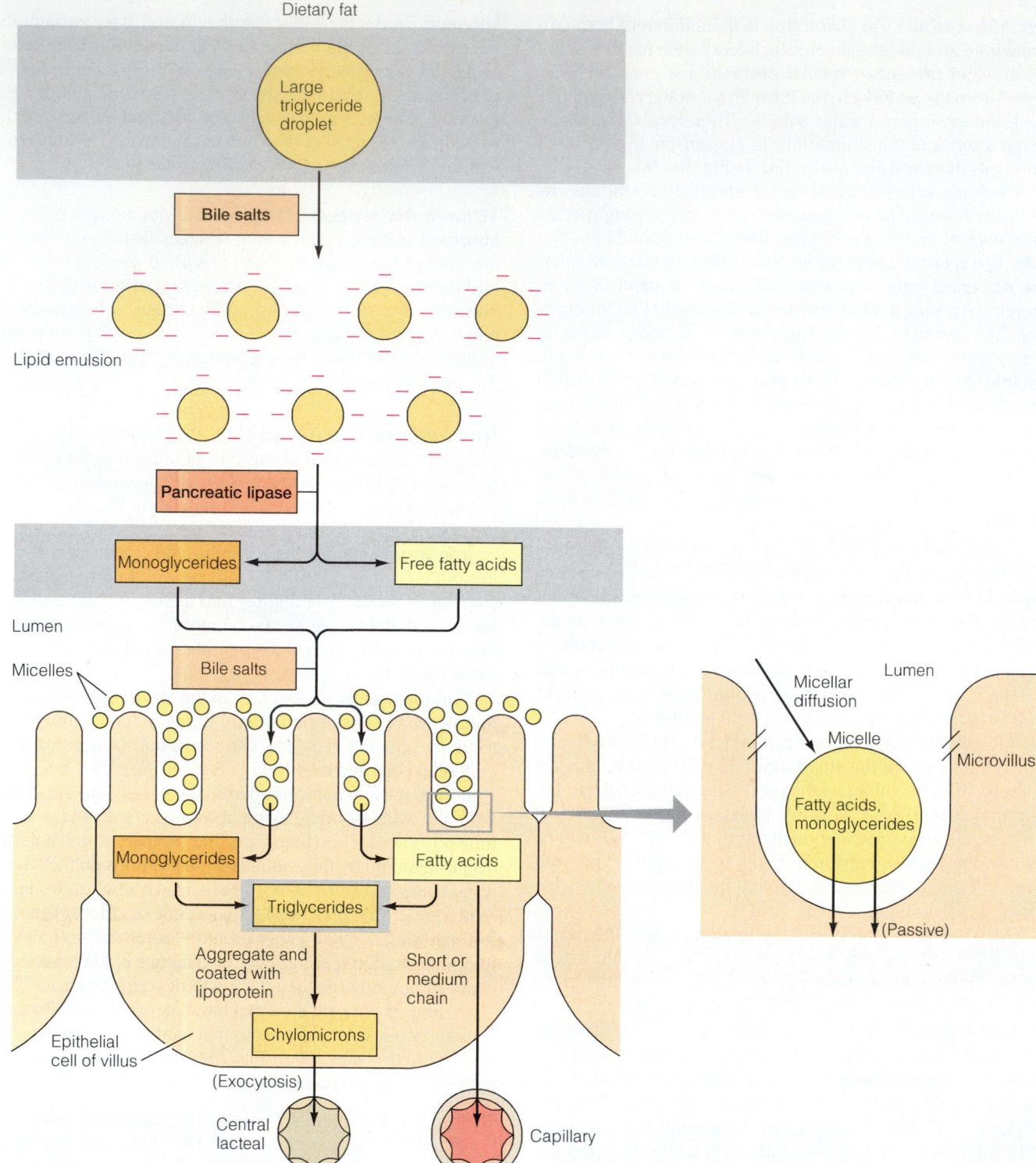

● **FIGURE 16–28 Fat Digestion and Absorption** Because fat is not soluble in water, it must undergo a series of transformations in order to be digested and absorbed. Dietary fat in the form of triglycerides is emulsified by the detergent action of bile salts. This lipid emulsion prevents the fat droplets from coalescing and thereby increases the surface area available for attack by pancreatic lipase. Lipase hydrolyzes triglycerides into monoglycerides and free fatty acids. These water-insoluble products are carried in the interior of water-soluble micelles, which are formed by bile salts and other bile constituents, to the luminal surface of the small-intestine epithelial cells. After leaving the micelles and passively diffusing through the luminal membranes, the monoglycerides and free fatty acids are resynthesized into triglycerides in the epithelial cells. These triglycerides aggregate and are coated with a layer of lipoprotein to form water-soluble chylomicrons, which are extruded through the basal membrane of the cells by exocytosis. Chylomicrons are unable to cross the basement membrane of blood capillaries, so instead they enter the lymphatic vessels, the central lacteals.

• **FIGURE 16–29 Iron Absorption**
Not all ingested iron is in a form that can be absorbed. Dietary iron that is absorbed into the small-intestine epithelial cells and is immediately needed for red blood cell production is transferred into the blood and carried to the bone marrow bound to transferrin, a plasma-protein carrier. Absorbed dietary iron that is not immediately needed is stored in the epithelial cells as ferritin, which cannot be transferred into the blood. Excess iron in the blood can be dumped into the ferritin pool. This unused iron, along with dietary iron that was not absorbed, is lost in the feces as the ferritin-containing epithelial cells are sloughed.

which cannot be absorbed into the blood. If the blood level of iron is too high, excess iron may be dumped from the blood into this unabsorbable pool of ferritin in the intestinal epithelial cells. Iron stored as ferritin is lost in the feces within three days as the epithelial cells containing these granules are sloughed off during mucosal regeneration. Large amounts of iron in the feces give them a dark, almost black color.

The amount of calcium (Ca^{++}) absorbed is also regulated. Absorption of Ca^{++} is accomplished partly by passive diffusion but mostly by active transport. Vitamin D greatly stimulates this active transport. Vitamin D can exert this effect only after it has been activated in the liver and kidneys, a process that is enhanced by parathyroid hormone. Appropriately, secretion of parathyroid hormone increases in response to a fall in Ca^{++} concentration in the blood. Normally, of the average 1,000 mg of Ca^{++} taken in daily, only about two-thirds is absorbed in the small intestine, with the rest passing out in the feces.

Most absorbed nutrients immediately pass through the liver for processing.

The venules that leave the small-intestine villi, along with those from the remainder of the digestive tract, empty into the portal vein, which carries the blood to the liver. Conse-

quently, anything absorbed into the digestive capillaries first must pass through the hepatic biochemical factory before entering the general circulation. Thus, the products of carbohydrate and protein digestion as well as the electrolytes and H_2O are channeled into the liver, where many of these products are subjected to immediate metabolic processing. Furthermore, harmful substances that may have been absorbed are detoxified by the liver before gaining access to the general circulation. After passing through the portal circulation, the venous blood from the digestive system is emptied into the vena cava and returned to the heart to be distributed throughout the body, carrying glucose and amino acids for use by the tissues.

Fat, which cannot penetrate the intestinal capillaries, is picked up by the central lacteal and enters the lymphatic system instead, thereby bypassing the hepatic portal system. Contractions of the villi, presumably accomplished by the muscularis mucosa, periodically compress the central lacteal and "milk" the lymph out of this vessel. Increased contractions of the villi are known to occur following a meal and are perhaps mediated by an unconfirmed hormone from the duodenal mucosa, **villikinin**. The lymph vessels eventually converge to form the *thoracic duct,* a large lymph vessel that empties into the venous system within the chest. By this means, fat ultimately gains access to the circulatory system. The absorbed fat is carried by the systemic circulation to the liver and to other tissues of the body. Therefore, the liver does have a chance to act on the digested fat, but not until the fat has been diluted by the blood in the general circulatory system. This dilution of fat presumably protects the liver from being inundated with more fat than it is capable of handling at one time.

Extensive absorption by the small intestine keeps pace with secretion.

The small intestine normally absorbs about 9 liters of fluid per day in the form of H_2O and solutes, including the absorbable units of nutrients, vitamins, and electrolytes. How can that be, when humans normally ingest only about 1,250 ml of fluid and consume 1,250 g of solid food (80% of which is H_2O—see p. 520) per day? Table 16–7 illustrates the tremendous daily absorptive accomplishments performed by the small intestine. Each day about 9,500 ml of H_2O and solutes enters the small intestine. Note that of this 9,500 ml, only 2,500 ml is ingested from the external environment. The remaining 7,000 ml (7 liters) of fluid consists of digestive juices that are essentially derived from plasma. Recall that plasma is the ultimate source of digestive secretions because the secretory cells extract the necessary raw materials for their secretory product from the plasma. Considering that the entire plasma volume is only about 2.75 liters, it is obvious that absorption must closely parallel secretion to prevent the plasma volume from falling sharply. Of the 9,500 ml of fluid entering the small-intestine lumen per day, about 95%, or 9,000 ml of fluid, is normally absorbed by the small intestine back into the plasma, with only 500 ml of the small-intestine contents passing on into the colon. Thus, the digestive juices are not lost from the body. After the constituents of the juices are secreted into the digestive-tract lumen and perform their function, they are returned to the plasma. The only secretory product that escapes from the body is bilirubin, a waste product that must be eliminated.

Biochemical balance among the stomach, pancreas, and small intestine is normally maintained.

Since the secreted juices are normally absorbed back into the plasma, the acid-base balance of the body is not altered by digestive processes. When secretion and absorption do not parallel each other, however, acid-base abnormalities can result. ● Figure 16–30 is a summary of the biochemical balance that normally exists among the stomach, pancreas, and small intestine. The arterial blood entering the stomach contains Cl^-, CO_2, H_2O, and Na^+, among other things. During HCl secretion, the gastric parietal cells extract Cl^-, CO_2, and H_2O from the plasma (the CO_2 and H_2O being essential for H^+ secretion) and add HCO_3^- to it (the HCO_3^- being formed in the process of generating H^+). The HCO_3^- diffuses into the plasma to replace the secreted Cl^- and to electrically balance the Na^+ in the plasma. Plasma Na^+ levels are not altered by gastric secretory processes. Because HCO_3^- is an alkaline ion, the venous blood leaving the stomach is more alkaline than the arterial blood delivered to it. The overall acid-base balance of the body is not altered, however, because the pancreatic duct cells extract a comparable amount of HCO_3^- (along with Na^+) from the plasma to neutralize the acidic gastric chyme as it is emptied into the small intestine. Within the intestinal lumen, the alkaline pancreatic $NaHCO_3$ secretion neutralizes the gastric HCl secretion, yielding NaCl and H_2CO_3. The latter molecules form Na^+ and Cl^- plus CO_2 and H_2O, respectively. All four of these constituents (Na^+, Cl^-, CO_2, and H_2O) are absorbed by the intestinal epithelium into the plasma. Note that these are exactly the same constituents present in the arterial blood entering the stomach. Thus, through these interactions, the body normally does not experience a net gain or loss of acid or base during digestion.

Diarrhea results in loss of fluid and electrolytes.

When vomiting or diarrhea occurs, these normal neutralization processes cannot take place. We have already described vomiting in the section on gastric motility. The other common digestive-tract disturbance that can lead to a loss of fluid and an acid-base imbalance is **diarrhea**. This condition is characterized by passage of a highly fluid fecal matter, often with increased frequency of defecation. Just as with vomiting, the effects of diarrhea can be either beneficial or harmful. Diarrhea is beneficial when rapid emptying of the intestine hastens the elimination of harmful material from the body. However, not only are some of the ingested materials lost, but some of the secreted materials that normally would have been reabsorbed are lost as well. Excessive loss of intestinal contents causes dehydration, loss of nutrient material, and metabolic acidosis resulting from the loss of secreted HCO_3^- (see p. 536). The abnormal fluidity of the feces in diarrhea usually occurs because the intestine is unable to absorb fluid as extensively as normal. This extra unabsorbed fluid passes out in the feces.

The following are the causes of diarrhea:

1. The most common cause of diarrhea is excessive intestinal motility, which arises either from local irritation of the

---

TABLE 16–7 Volumes Absorbed by Small and Large Intestine per Day

Volume entering small intestine per day			
	Ingested	Food eaten	1,250 g
		Fluid drunk	1,250 ml
Sources	Secreted from plasma	Saliva	1,500 ml
		Gastric juice	2,000 ml
		Pancreatic juice	1,500 ml
		Bile	500 ml
		Intestinal juice	1,500 ml
			9,500 ml
Volume absorbed by small intestine per day			9,000 ml
Volume entering colon from small intestine per day			500 ml
Volume absorbed by colon per day			350 ml
Volume of feces eliminated from colon per day			150 g

Digestive-tract lumen

Stomach parietal cell

Cl^-

$CO_2 + H_2O$

H_2CO_3

$H^+ + Cl^-$

$H^+ + HCO_3^-$

Pancreatic duct cell

$Na^+ + HCO_3^-$ ← $Na^+ + HCO_3^-$

Intestinal epithelial cell

$Na^+ + Cl^-$
and
$H^+ + HCO_3^-$

H_2CO_3

$CO_2 + H_2O$

$\begin{cases} Na^+ \\ Cl^- \end{cases}$
$\begin{cases} CO_2 \\ H_2O \end{cases}$

Blood

Cl^-

$\begin{cases} CO_2 \\ H_2O \end{cases}$

Na^+

HCO_3^-

$HCO_3^- + Na^+$

$\begin{cases} Na^+ \\ Cl^- \\ CO_2 \\ H_2O \end{cases}$

• FIGURE 16–30 Biochemical Balance among the Stomach, Pancreas, and Small Intestine When digestion and absorption proceed normally, no net loss or gain of acid or base or other chemicals from the body fluids occurs as a consequence of digestive secretions. The parietal cells of the stomach extract Cl^-, CO_2, and H_2O from and add HCO_3^- to the blood during HCl secretion. The pancreatic duct cells extract the HCO_3^- as well as Na^+ from the blood during $NaHCO_3$ secretion. Within the small-intestine lumen, pancreatic $NaHCO_3$ neutralizes gastric HCl to form NaCl and H_2CO_3, which decomposes into CO_2 and H_2O. Subsequently, the intestinal cells absorb Na^+, Cl^-, CO_2, and H_2O into the blood, thereby replacing the constituents that were extracted from the blood during gastric and pancreatic secretion.

gut wall caused by bacterial or viral infection of the intestine or from emotional stress. Rapid transit of the intestinal contents does not allow sufficient time for adequate absorption of fluid to occur.

2. Diarrhea also occurs when excess osmotically active particles, such as those found in lactase deficiency, are present in the digestive-tract lumen. These particles cause excessive fluid to enter and be retained in the lumen, thus increasing the fluidity of the feces.

3. Toxins of the bacterium *Vibrio cholerae* (the causative agent of cholera) and certain other microorganisms promote the secretion of excessive amounts of fluid by the small-intestine mucosa, resulting in profuse diarrhea.

LARGE INTESTINE

The large intestine is primarily a drying and storage organ.

The **large intestine** consists of the colon, cecum, appendix, and rectum (• Fig. 16–31). The **cecum** forms a blind-ended pouch below the junction of the small and large intestines at

the ileocecal valve. The small fingerlike projection at the bottom of the cecum is the **appendix**, a lymphoid tissue that houses lymphocytes (see p. 370). The **colon**, which makes up most of the large intestine, is not coiled like the small intestine but consists of three relatively straight portions—the *ascending colon,* the *transverse colon,* and the *descending colon.* The terminal portion of the descending colon becomes S-shaped, forming the *sigmoid colon* (*sigmoid* means "S-shaped"), then straightens out to form the **rectum** (*rectum* means "straight").

The outer longitudinal smooth-muscle layer does not completely surround the large intestine. Instead, it consists only of three separate, conspicuous, longitudinal bands of muscle, the **taeniae coli**, which run the length of the large intestine. These taeniae coli are shorter than the underlying circular smooth muscle and mucosal layers would be if the latter were stretched out flat. Because of this, the underlying layers are gathered into pouches or sacs called **haustra**, much as the material of a full skirt is gathered at the narrower waistband. The haustra are not merely passive permanent gathers, however; they actively change location as a result of contraction of the circular smooth-muscle layer.

The colon normally receives about 500 ml of chyme from

• **FIGURE 16–31 Anatomy of Large Intestine**

the small intestine each day. Since most digestion and absorption have been accomplished in the small intestine, the contents delivered to the colon consist of undigestible food residues (such as cellulose), unabsorbed biliary components, and the remaining fluid. The colon extracts more H_2O and salt from the contents. What remains to be eliminated is known as **feces**. The primary function of the large intestine is to store this fecal material before defecation. Cellulose and other indigestible substances in the diet provide bulk and help maintain regular bowel movements by contributing to the volume of the colonic contents.

Haustral contractions slowly shuffle the colonic contents back and forth while mass movements propel colonic contents long distances.

Most of the time, movements of the large intestine are slow and nonpropulsive, as is appropriate for its absorptive and storage functions. The colon's primary method of motility is **haustral contractions** initiated by the autonomous rhythmicity of colonic smooth-muscle cells. These contractions, which throw the large intestine into haustra, are similar to small intestinal segmentations but occur much less frequently.

Whereas small intestinal segmentation occurs at rates of between nine and twelve contractions per minute, there may be thirty minutes between haustral contractions. The location of the haustral sacs gradually changes as a relaxed segment that has formed a sac slowly contracts while a previously contracted area simultaneously relaxes to form a new sac. These movements are nonpropulsive; they slowly shuffle the contents in a back-and-forth mixing movement that exposes the colonic contents to the absorptive mucosa. Haustral contractions are largely controlled by locally mediated reflexes involving the intrinsic plexuses.

Because of this slow colonic movement, bacteria have time to grow and accumulate in the large intestine. In contrast, in the small intestine the contents are normally moved through too rapidly for bacterial growth to occur. Not all ingested bacteria are destroyed by salivary lysozyme and gastric HCl, so the surviving bacteria continue to thrive in the large intestine. Most colonic microorganisms are harmless in this location.

Three to four times a day, generally after meals, a marked increase in motility takes place during which large segments of the ascending and transverse colon contract simultaneously, driving the feces one-third to three-fourths of the length of the colon in a few seconds. These massive contractions, appropriately called **mass movements**, drive the colonic contents into the distal portion of the large intestine, where material is stored until defecation occurs.

When food enters the stomach, mass movements occur in the colon primarily by means of the **gastrocolic reflex**, which is mediated from the stomach to the colon by gastrin and by the extrinsic autonomic nerves. In many people, this reflex is most evident after the first meal of the day and is often followed by the urge to defecate. Thus, when a new meal enters the digestive tract, reflexes are initiated to move the existing contents farther along down the tract to make way for the incoming food. The gastroileal reflex moves the remaining small-intestine contents into the large intestine, and the gastrocolic reflex pushes the colonic contents into the rectum, triggering the defecation reflex.

Feces are eliminated by the defecation reflex.

When mass movements of the colon move fecal material into the rectum, the resultant distention of the rectum stimulates stretch receptors in the rectal wall, thus initiating the **defecation reflex**. This reflex causes the **internal anal sphincter** (which is composed of smooth muscle) to relax and the rectum and sigmoid colon to contract more vigorously. If the **external anal sphincter** (which is composed of skeletal muscle) is also relaxed, defecation occurs. Being skeletal muscle, the external anal sphincter is under voluntary control. The initial distention of the rectal wall is accompanied by the conscious urge to defecate. If circumstances are unfavorable for defecation, voluntary tightening of the external anal sphincter can prevent defecation in spite of the defecation reflex. If defecation is delayed, the distended rectal wall gradually relaxes, and the urge to defecate subsides until the next mass movement propels more feces into the rectum, once again distending the rectum and triggering the defecation reflex. During periods of nonactivity, both anal sphincters remain contracted to assure fecal continence.

When defecation does occur, it is usually assisted by voluntary straining movements that involve simultaneous contraction of the abdominal muscles and a forcible expiration against a closed glottis. This maneuver brings about a large increase in intra-abdominal pressure, which assists in elimination of the feces.

Constipation occurs when the feces become too dry.

If defecation is delayed too long, **constipation** may result. When the colonic contents are retained for longer periods of time than normal, more than the usual amount of H_2O is absorbed, causing the feces to become hard and dry. Normal variations in frequency of defecation among individuals range from after every meal to up to once a week. When the frequency is delayed beyond what is normal for a particular individual, constipation and its attendant symptoms may occur. These symptoms include abdominal discomfort, dull headache, loss of appetite sometimes accompanied by nausea, and mental depression. Contrary to popular belief, these symptoms are not caused by toxins absorbed from the retained fecal material. Although some potentially toxic substances are produced by bacterial metabolism in the colon, these substances normally pass through the portal system and are removed by the liver before they can reach the systemic circulation. Instead, the symptoms associated with constipation seem to be due to prolonged distention of the large intestine, particularly the rectum, because these sensations are promptly alleviated following relief from distention.

Possible causes for delayed defecation that might lead to constipation include (1) ignoring the urge to defecate; (2) decreased colon motility accompanying aging, emotion, or a low-bulk diet; (3) obstruction of fecal movement in the large bowel caused by a local tumor or colonic spasm; and (4) impairment of the defecation reflex, such as through injury of the nerve pathways involved.

If hardened fecal material becomes lodged in the appendix, it may obstruct normal circulation and mucus secretion in this narrow, blind-ended appendage. This blockage leads to inflammation of the appendix, or **appendicitis**. The appendix often becomes swollen and filled with pus, and the inflamed tissue may die as a result of local circulatory interference. If not surgically removed, the diseased appendix may rupture, spewing its infectious contents into the abdominal cavity.

Large-intestine secretion is protective in nature.

The large intestine does not secrete any digestive enzymes. None are needed because digestion is completed before chyme ever reaches the colon. Colonic secretion consists of an alkaline (HCO_3^-) mucus solution, whose function is to protect the large-intestine mucosa from mechanical and chemical injury. The mucus provides lubrication to facilitate passage of the feces, whereas the HCO_3^- neutralizes irritating acids produced by local bacterial fermentation. Secretion increases in response to mechanical and chemical stimulation of the colonic mucosa mediated by short reflexes and parasympathetic innervation.

No digestion takes place within the large intestine because there are no digestive enzymes. However, the colonic bacteria do digest some of the cellulose and use it for their own metabolism.

The large intestine absorbs salt and water, converting the luminal contents into feces.

Some absorption takes place within the colon but not to the same extent as in the small intestine. Because the luminal surface of the colon is fairly smooth, it has considerably less absorptive surface area than the small intestine. Furthermore, no specialized transport mechanisms are present in the colonic mucosa for absorption of glucose or amino acids, as there are in the small intestine. When excessive small-intestine motility delivers the contents to the colon before absorption of nutrients has been completed, the colon is unable to absorb these materials, and they are lost in diarrhea.

The colon normally absorbs some salt and H_2O. Sodium is

actively reabsorbed, Cl^- follows passively down the electrical gradient, and H_2O follows osmotically. Bacteria in the colon synthesize some vitamins that the colon is capable of absorbing, but this is normally not a significant contribution, except in the case of vitamin K.

Through absorption of salt and H_2O, a firm fecal mass is formed. Of the 500 ml of material entering the colon per day from the small intestine, the colon normally absorbs about 350 ml, leaving 150 g of feces to be eliminated from the body each day (Table 16–7). This fecal material normally consists of 100 g of H_2O and 50 g of solid, including undigested cellulose, bilirubin, bacteria, and small amounts of salt. Thus, contrary to popular thinking, the digestive tract is not a major excretory passageway for eliminating wastes from the body. The main waste product excreted in the feces is bilirubin. The other fecal constituents are unabsorbed food residues and bacteria, which were never actually a part of the body.

Intestinal gases are absorbed or expelled.

Occasionally, instead of fecal material passing from the anus, intestinal gas, or **flatus**, passes out. This gas is derived primarily from two sources: (1) swallowed air, with as much as 500 ml of air being swallowed during a meal, and (2) gas produced by bacterial fermentation in the colon. The presence of gas percolating through the luminal contents gives rise to gurgling sounds known as **borborygmi**. Eructation (burping) removes most of the swallowed air from the stomach, but some passes on into the intestine. Usually, very little gas is present in the small intestine, because the gas is either quickly absorbed or passes on into the colon. Most gas in the colon is due to bacterial activity, with the quantity and nature of the gas depending on the type of food eaten and the characteristics of the colonic bacteria. Some foods, such as beans, contain types of carbohydrates for which humans lack digestive enzymes. These fermentable carbohydrates enter the colon, where they are attacked by gas-producing bacteria. Much of the gas entering or forming in the large intestine is absorbed through the intestinal mucosa. The remainder is expelled through the anus.

To accomplish selective expulsion of gas when fecal material is also present in the rectum, the abdominal muscles and external anal sphincter are voluntarily contracted simultaneously. When contraction of the abdominal muscles raises the pressure sufficiently against the contracted anal sphincter, the pressure gradient forces air out at a high velocity through a slit-like anal opening that is too narrow for solid feces to escape. This passage of air at high velocity causes the edges of the anal opening to vibrate, giving rise to the characteristic low-pitched sound accompanying passage of gas.

OVERVIEW OF THE GASTROINTESTINAL HORMONES

Throughout our discussion of digestion, we have repeatedly mentioned different functions of the four accepted gastrointestinal hormones: gastrin, secretin, cholecystokinin, and gastric inhibitory peptide. We will now fit all of these functions together so that you can appreciate the overall adaptive importance of these interactions (Table 16–8).

Chyme in the stomach, especially if it contains protein, stimulates the release of gastrin, which acts on parietal and chief cells to increase secretion of HCl and pepsinogen. These two substances, in turn, are of primary importance in initiating digestion of the protein that promoted their release. Furthermore, gastrin enhances gastric motility, stimulates ileal motility, relaxes the ileocecal sphincter, and induces mass movements in the colon—functions that are all aimed at keeping the contents moving through the tract upon the arrival of a new meal. Gastrin also is trophic not only to the stomach mucosa but also to the small-intestine mucosa, helping maintain a well-developed, functionally viable digestive-tract lining. Predictably, gastrin secretion is inhibited by an accumulation of acid in the stomach and by the presence in the duodenal lumen of acid and other constituents that necessitate a delay in gastric secretion.

As the stomach empties into the duodenum, the presence of acid in the duodenum stimulates the release of secretin into the blood. Secretin performs four major interrelated functions: (1) it inhibits gastric emptying to prevent further acid from entering the duodenum until the acid that is already present is neutralized; (2) it inhibits gastric secretion to reduce the amount of acid being produced; (3) it stimulates the pancreatic duct cells to produce a large volume of aqueous $NaHCO_3$ secretion, which is emptied into the duodenum to neutralize the acid; and (4) it stimulates secretion by the liver of a $NaHCO_3$-rich bile, which likewise is emptied into the duodenum to assist in the neutralization process. Neutralization of the acidic chyme in the duodenum helps prevent damage to the duodenal walls and provides a suitable environment for the optimal functioning of the pancreatic digestive enzymes, which are inhibited by acid.

As chyme is emptied from the stomach, fat and other nutrients enter the duodenum. These nutrients, especially fat, cause the release of cholecystokinin (CCK) from the duodenal mucosa. This hormone also performs several important interrelated functions: (1) it inhibits gastric motility and secretion, thereby allowing adequate time for the nutrients already in the duodenum to be digested and absorbed; (2) it stimulates the pancreatic acinar cells to increase secretion of pancreatic enzymes, which continue the digestion of these nutrients in the duodenum (this action is especially important for fat digestion, because pancreatic lipase is the only enzyme that digests fat); and (3) it causes contraction of the gallbladder and relaxation of the sphincter of Oddi so that bile is emptied into the duodenum to aid fat digestion and absorption. Bile salts' detergent action is particularly important in enabling pancreatic lipase to perform its digestive task. Once again, the multiple effects of CCK are remarkably well adapted to dealing with the fat and other nutrients whose presence in the duodenum triggered this hormone's release.

Furthermore, it is appropriate that both secretin and CCK, which have profound stimulatory effects on the exocrine pancreas, are trophic to this tissue. Recall that CCK has also been implicated in long-term adaptive changes in the proportion

TABLE 16-8 Source, Control, and Functions of the Major Gastrointestinal Hormones

HORMONE	SOURCE	PRIMARY STIMULUS FOR SECRETION	FUNCTIONS
Gastrin	G cells of stomach's pyloric gland area	Protein in the stomach	Stimulates secretion by parietal and chief cells
			Enhances gastric motility
			Stimulates ileal motility *(Gastroilial Reflex)*
			→ Relaxes ileocecal sphincter
			Induces colonic mass movements *(Gastrocolic reflex)*
			Is trophic to stomach and small-intestine mucosa
Secretin	Endocrine cells in duodenal mucosa	Acid in duodenal lumen	Inhibits gastric emptying
			Inhibits gastric secretion
			Stimulates aqueous $NaHCO_3$ secretion by pancreatic duct cells
			Stimulates secretion of $NaHCO_3$-rich bile by liver
			Is trophic to exocrine pancreas
Cholecystokinin	Endocrine cells in duodenal mucosa	Nutrients in duodenal lumen, especially fat products and to a lesser extent protein products	Inhibits gastric emptying
			Inhibits gastric secretion
			Stimulates digestive enzyme secretion by pancreatic acinar cells
			Causes gallbladder contraction
			Causes relaxation of sphincter of Oddi
			Is trophic to exocrine pancreas
			May cause long-term adaptive changes in proportion of pancreatic enzymes
			Contributes to satiety
Gastric inhibitory peptide	Endocrine cells in duodenal mucosa	Fat, acid, hypertonicity, glucose and distention in duodenum	Inhibits gastric emptying
			Inhibits gastric secretion
			Stimulates insulin secretion by endocrine pancreas

of pancreatic enzymes produced in response to prolonged changes in diet.

Besides facilitating the digestion of ingested nutrients, CCK is an important regulator of food intake. It plays a key role in satiety, the sensation of having had enough to eat (chapter 17).

Gastric inhibitory peptide (GIP) is the most recently recognized gastrointestinal hormone. We have already mentioned its role in inhibiting gastric motility and secretion in response to fat, acid, hypertonicity and distention in the duodenum. In addition, it has been shown to stimulate insulin release by the pancreas, so it is also called **glucose-dependent insulinotrophic peptide** (once again GIP). Again, this is remarkably adaptive. As soon as the meal is absorbed, the body has to shift its metabolic gears to use and store the newly arriving nutrients. The metabolic activities of this postabsorptive phase are largely under the control of insulin (chapter 19). Stimulated by the presence of a meal in the digestive tract, GIP initiates the release of insulin in anticipation of the absorpton of the meal in a "feedforward" fashion. Insulin is especially important in promoting the uptake and stor-

age of glucose. Appropriately, glucose in the duodenum has recently been shown to increase GIP secretion.

This overview of the multiple, integrated, adaptive functions of the gastrointestinal hormones provides an excellent example of the remarkable efficiency of the human body. The list of hormones that regulate digestive function will undoubtedly grow, because a number of peptides found in the digestive tract are suspected of being hormones. Of these candidate hormones, motilin and villikinin have already been mentioned.

CHAPTER IN PERSPECTIVE: FOCUS ON HOMEOSTASIS

To maintain constancy in the internal environment, materials that are used up in the body (such as nutrients and O_2) or uncontrollably lost from the body (such as evaporative H_2O loss from the airways or salt loss in sweat) must constantly be replaced by new supplies of these materials from the external environment. All of these replacement supplies except O_2 are acquired through the digestive system. Fresh supplies of O_2 are transferred to the internal environment by the respiratory system, but all of the nutrients, H_2O, and various electrolytes needed to maintain homeostasis are acquired through the digestive system. The large, complex food that is ingested is broken down by the digestive system into small absorbable units. These small energy-rich nutrient molecules are transferred across the small-intestine epithelium into the blood for delivery to the cells to replace the nutrients constantly used for ATP production and for repair and growth of body tissues. Likewise, ingested H_2O, salt, and other electrolytes are absorbed by the intestine into the blood.

Unlike most body systems, regulation of digestive-system activities is not aimed at maintaining homeostasis. The quantity of nutrients and H_2O ingested is subject to control, but the quantity of ingested materials absorbed by the digestive tract is not subject to control, with few exceptions. The hunger mechanism governs food intake to help maintain energy balance (chapter 17), and the thirst mechanism controls H_2O in-

take to help maintain H_2O balance (chapter 15). However, we often do not heed these control mechanisms and eat and drink even when we are not hungry or thirsty. Once these materials are in the digestive tract, the digestive system does not vary its rate of nutrient, H_2O, or electrolyte uptake according to body needs (with the exception of iron and calcium); rather, it optimizes conditions for digesting and absorbing what is ingested. Truly, what you eat is what you get. The digestive system is subject to many regulatory processes, but these are not influenced by the nutritional or hydration state of the body. Instead, these control mechanisms are governed by the composition and volume of digestive-tract contents so that the rate of motility and secretion of digestive juices are optimal for digestion and absorption of the ingested food.

If excess nutrients are ingested and absorbed, the extra is placed in storage, such as in adipose tissue (fat), so that the blood level of nutrient molecules is kept at a constant level. Excess ingested H_2O and electrolytes are eliminated in the urine to homeostatically maintain the blood levels of these constituents.

CHAPTER SUMMARY

Introduction

The four basic digestive processes are motility, secretion, digestion, and absorption. Digestive activities are carefully regulated by synergistic autonomous, neural (both intrinsic and extrinsic), and hormonal mechanisms to assure that the ingested food is maximally made available to the body for energy production and as synthetic raw materials. The digestive tract consists of a continuous tube that runs from the mouth to the anus, with local modifications that reflect regional specializations for carrying out digestive functions. The lumen of the digestive tract is continuous with the external environment, so its contents are technically outside the body; this arrangement permits digestion of food without self-digestion occurring in the process.

Mouth, Pharynx, and Esophagus

Food enters the digestive system through the mouth, where it is chewed and mixed with saliva to facilitate swallowing. The salivary enzyme, amylase, begins the digestion of polysaccharides, a process that continues in the stomach after the food has been swallowed until amylase is eventually inactivated by the acidic gastric juice. More important than its minor digestive function, saliva is essential for articulate speech and plays an important role in dental health. Salivary secretion is controlled by a salivary center in the medulla, mediated by autonomic innervation of the salivary glands.

Following chewing, the tongue propels the bolus of food to the rear of the throat, which initiates the swallowing reflex. The swallowing center in the medulla coordinates a complex group of activities that result in closure of the respiratory passages and propulsion of the food through the pharynx and esophagus into the stomach.

The esophageal secretion, mucus, is protective in nature. No nutrient absorption occurs in the mouth, pharynx, or esophagus.

Stomach

The stomach, a saclike structure located between the esophagus and small intestine, stores ingested food for variable periods of time until the small intestine is ready to process it further for final absorption. The four aspects encompassing gastric motility are gastric filling, storage, mixing, and emptying. Gastric filling is facilitated by vagally mediated receptive relaxation of the stomach musculature. Gastric storage takes place in the body of the stomach, where peristaltic contractions of the thin muscular walls are too weak to mix the contents. Gastric mixing takes place in the thick-muscled antrum as a result of vigorous peristaltic contractions. Gastric emptying is influenced by factors in both the stomach and the duodenum. The volume and fluidity of chyme in the stomach tend to promote emptying of the stomach contents. The duodenal factors, which are the dominant factors controlling gastric emptying, tend to delay gastric emptying until the duodenum is ready to receive and process more chyme. The specific factors in the duodenum that delay gastric emptying by inhibiting stomach peristaltic activity are fat, acid, hypertonicity, and distention.

Carbohydrate digestion continues in the body of the stomach under the influence of the swallowed salivary amylase; protein digestion is initiated in the antrum of the stomach, where vigorous peristaltic contractions mix the food with gastric secretions, converting it to a thick liquid mixture known as chyme. Gastric secretions into the stomach lumen include (1) HCl, which activates pepsinogen, denatures protein, and kills bacteria; (2) pepsinogen, which, once activated, initiates protein digestion; (3) mucus, which provides a protective coating to supplement the gastric mucosal barrier, enabling the stomach to contain the harsh luminal contents without self-digestion; and (4) intrinsic factor, which plays a vital role in vitamin B_{12} absorption, a constituent essential for normal red blood cell production. The stomach also secretes into the blood the hormone gastrin, which plays a dominant role in regulating gastric secretion. Histamine, a potent gastric stimulant that is not normally secreted, is released into the stomach lumen with devastating effects during ulcer formation.

Both gastric motility and gastric secretion are under complex control mechanisms, involving not only gastrin but also vagal and intrinsic nerve responses and enterogastrone hormones (secretin, cholecystokinin, and gastric inhibitory peptide) secreted from the small-intestine mucosa. Regulation of the stomach is aimed at balancing the rate of gastric activity with the ability of the small intestine to handle the arrival of acidic, fat-laden contents from the stomach.

No nutrients are absorbed from the stomach.

Pancreatic and Biliary Secretions

Pancreatic exocrine secretions and bile from the liver both enter the duodenal lumen. Pancreatic secretions include (1) potent digestive enzymes from the acinar cells, which digest all three categories of foodstuff, and (2) an aqueous $NaHCO_3$ solution from the duct cells, which neutralizes the acidic contents emptied into the duodenum from the stomach. This neutralization is important to protect the duodenum from acid injury and to allow the pancreatic enzymes, which are inactivated by acid, to perform their important digestive functions. Pancreatic secretion is primarily under hormonal control, which matches the composition of the pancreatic juice with the needs in the duodenal lumen.

The liver, the body's largest and most important metabolic organ, performs many varied functions. Its contribution to digestion is the secretion of bile, which contains bile salts. Bile salts aid fat digestion through their detergent action and facilitate fat absorption through formation of water-soluble micelles that can carry the products of fat digestion to their absorptive site. Between meals, bile is stored and concentrated in the gallbladder, which is hormonally stimulated to contract and empty the bile into the duodenum during digestion of a meal. After participating in fat digestion and absorption, bile salts are reabsorbed and returned via the hepatic portal system to the liver, where they not only are resecreted but act as a potent choleretic to stimulate the secretion of even more bile. Bile also contains bilirubin, a derivative of degraded hemoglobin, which is the major excretory product in the feces.

Small Intestine

The small intestine is the main site for digestion and absorption. Segmentation, its primary motility, thoroughly mixes the food with pancreatic, biliary, and small intestinal juices to facilitate digestion; it also exposes the products of digestion to the absorptive surfaces. Between meals, the migrating motility complex sweeps the lumen clean.

The juice secreted by the small intestine does not contain any digestive enzymes. The enzymes synthesized by the small intestine act intracellularly within the brush-border membranes of the epithelial cells. These enzymes complete the digestion of carbohydrates and protein before these nutrients enter the blood. The energy-dependent process of Na^+ absorption provides the driving force for Cl^-, water, glucose, and amino acid absorption. Fat digestion is accomplished entirely in the lumen of the small intestine by pancreatic lipase. Because they are not soluble in water, the products of fat digestion must undergo a series of transformations that enable them to be passively absorbed, eventually entering the lymph. The small intestine absorbs almost everything presented to it, from ingested food to digestive secretions to sloughed epithelial cells. Only a small amount of fluid and nondigestible food residue passes on to the large intestine.

The small intestinal lining is remarkably adapted to its digestive and absorptive function. It is thrown into folds that bear a rich array of fingerlike projections, the villi, which are furnished with a multitude of even smaller hairlike protrusions, the microvilli. Altogether, these surface modifications tremendously increase the area available to house the membrane-bound enzymes and to accomplish both active and passive absorption. This impressive lining is replaced approximately every three days to assure an optimally healthy and functional presence of epithelial cells in spite of harsh luminal conditions.

Large Intestine

The colon serves primarily to concentrate and store undigested food residues and biliary waste products until they can be eliminated from the body as feces. No secretion of digestive enzymes or absorption of nutrients takes place in the colon, all nutrient digestion and absorption having been completed in the small intestine. Haustral contractions slowly shuffle the colonic contents back and forth to accomplish absorption of most of the remaining fluid and electrolytes. Mass movements occur several times a day, usually following meals, propelling the feces long distances. Movement of feces into the rectum triggers the defecation reflex, which can be voluntarily prevented by contraction of the external anal sphincter if the time is inopportune for elimination. The alkaline mucus secretion of the large intestine is primarily protective in nature.

REVIEW EXERCISES

Objective Questions (Answers on p. D–4.)

1. The extent of nutrient uptake from the digestive tract depends on the body's needs. (True or false?)

2. The stomach is relaxed during vomiting. (True or false?)

3. Acid cannot normally penetrate into or between the cells lining the stomach, which enables the stomach to contain acid without injuring itself. (True or false?)

4. Protein is continually lost from the body through digestive secretions and sloughed epithelial cells, which pass out in the feces. (True or false?)

5. Foodstuffs not absorbed by the small intestine are absorbed by the large intestine. (True or false?)

6. The endocrine pancreas secretes secretin and CCK. (True or false?)

7. A digestive reflex involving the autonomic nerves is known as a _____ reflex, whereas a reflex in which all elements of the reflex arc are located within the gut wall is known as a _____ reflex.

8. When food is broken down and mixed with gastric secretions, the resultant thick liquid mixture is known as _____.

9. The entire lining of the small intestine is replaced approximately every _____ days.

10. The two substances absorbed by specialized transport mechanisms located only in the terminal ileum are _____ and _____.

11. The most potent choleretic is _____.

12. Which of the following is *not* a function of saliva?
    a. begins digestion of carbohydrate
    b. facilitates absorption of glucose across the oral mucosa
    c. facilitates speech
    d. exerts an antibacterial effect
    e. plays an important role in oral hygiene

13. Match the following:
    ____ 1. prevents reentry of food into the mouth during swallowing
    ____ 2. triggers the swallowing reflex
    ____ 3. seals off the nasal passages during swallowing
    ____ 4. prevents air from entering the esophagus during breathing
    ____ 5. closes off the respiratory airways during swallowing
    ____ 6. prevents gastric contents from backing up into the esophagus

    a. closure of the pharyngoesophageal sphincter
    b. elevation of the uvula
    c. position of the tongue against the hard palate
    d. closure of the gastroesophageal sphincter
    e. bolus pushed to the rear of the mouth by the tongue
    f. tight apposition of the vocal cords

14. Match the substances at the right with their characteristics (answers may be used more than once):
    ____ 1. activates pepsinogen
    ____ 2. inhibits amylase
    ____ 3. essential for vitamin B_{12} absorption
    ____ 4. can act autocatalytically
    ____ 5. not normally secreted but a potent stimulant for acid secretion
    ____ 6. breaks down connective tissue and muscle fibers

    a. pepsin
    b. mucus
    c. HCl
    d. intrinsic factor
    e. histamine

____ 7. begins protein digestion
____ 8. serves as a lubricant
____ 9. kills ingested bacteria
____ 10. is alkaline
____ 11. deficient in pernicious anemia
____ 12. coats the gastric mucosa

Essay Questions

1. Describe the four basic digestive processes.

2. List the three categories of energy-rich foodstuffs and the absorbable units of each.

3. List the components of the digestive system. Describe the cross-sectional anatomy of the digestive tract.

4. What four general factors are involved in the regulation of digestive-system function? What is the role of each?

5. Describe the types of motility in each component of the digestive tract. What factors control each type of motility?

6. State the composition of the digestive juice secreted by each component of the digestive system. Describe the factors that control each digestive secretion.

7. List the enzymes involved in the digestion of each category of foodstuff. Indicate the source and control of secretion of each of these enzymes.

8. Why are some digestive enzymes secreted in inactive form? How are they activated?

9. What absorption processes take place within each component of the digestive tract? What special adaptations of the small intestine enhance its absorptive capacity?

10. Describe the absorptive mechanisms for salt, water, carbohydrate, protein, and fat.

11. What are the contributions of the accessory digestive organs? What are the nondigestive functions of the liver?

12. Summarize the functions of each of the gastrointestinal hormones.

13. What waste product is excreted in the feces?

14. How is vomiting accomplished? What are the causes and consequences of vomiting, diarrhea, and constipation?

15. Describe the process of mucosal turnover in the small intestine.

 POINTS TO PONDER

1. Why do patients who have had a large portion of their stomachs removed for treatment of stomach cancer or severe peptic ulcer disease have to eat small quantities of food frequently instead of consuming three meals a day?

2. Explain why a person whose bile duct is blocked by a gallstone experiences a painful "gallbladder attack" after eating a high-fat meal.

3. By what means would defecation be accomplished in a patient paralyzed from the waist down because of a lower spinal cord injury?

4. After bilirubin is extracted from the blood by the liver, it is conjugated (combined) with glycuronic acid by the enzyme glucuronyl transferase within the liver. Only when conjugated can bilirubin be actively excreted into the bile. For the first few days of life, the liver does not make adequate quantities of glucuronyl transferase. Explain how this transient enzyme deficiency leads to the common condition of jaundice in newborns.

5. The number of immune cells in the gut-associated lymphoid tissue is estimated to be equal to the total number of these defense cells in the rest of the body. Speculate on the adaptive significance of this extensive defense capability of the digestive system.

17 Energy Balance and Temperature Regulation

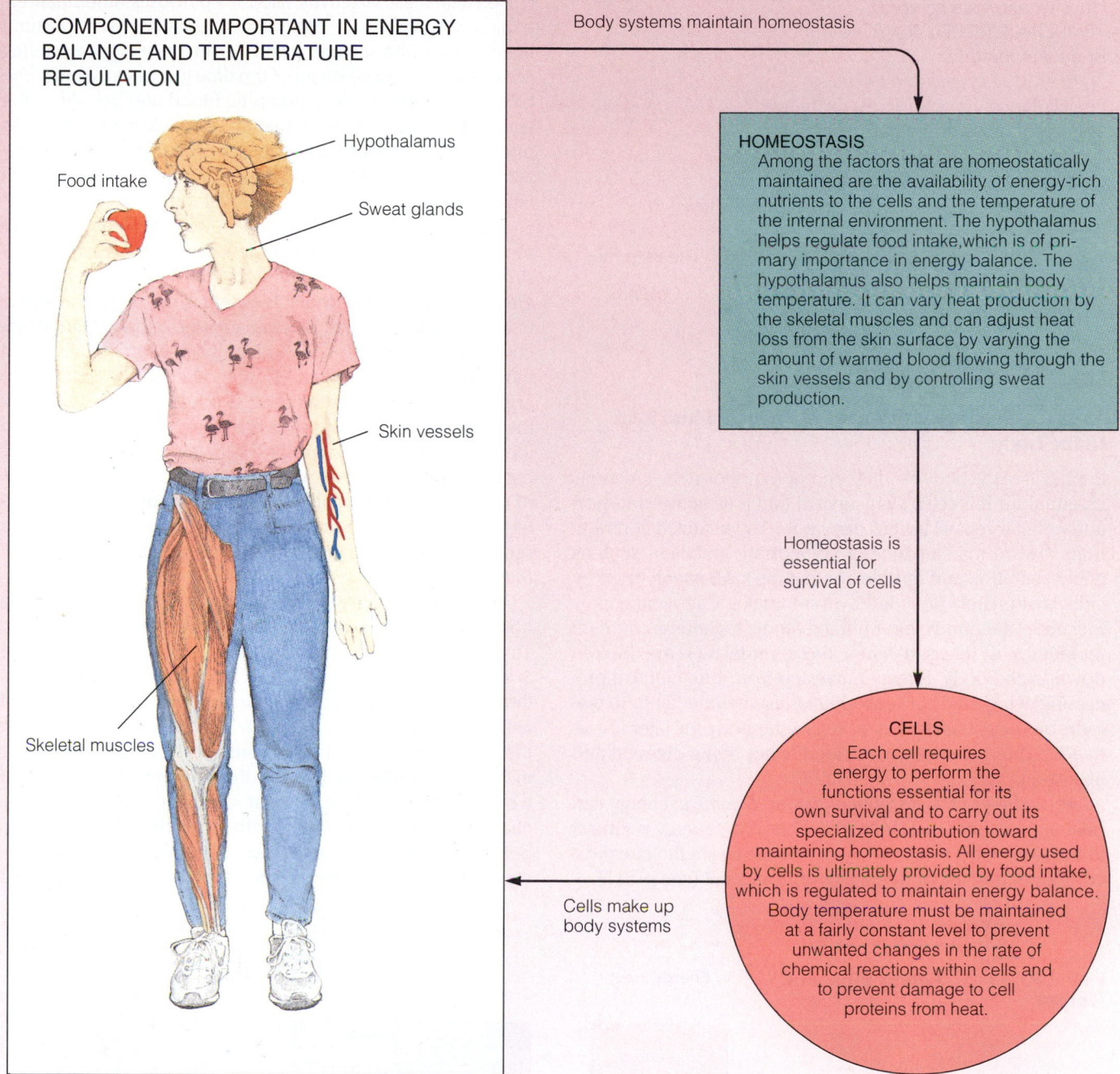

COMPONENTS IMPORTANT IN ENERGY BALANCE AND TEMPERATURE REGULATION

Food intake

Hypothalamus

Sweat glands

Skin vessels

Skeletal muscles

Body systems maintain homeostasis

HOMEOSTASIS
Among the factors that are homeostatically maintained are the availability of energy-rich nutrients to the cells and the temperature of the internal environment. The hypothalamus helps regulate food intake, which is of primary importance in energy balance. The hypothalamus also helps maintain body temperature. It can vary heat production by the skeletal muscles and can adjust heat loss from the skin surface by varying the amount of warmed blood flowing through the skin vessels and by controlling sweat production.

Homeostasis is essential for survival of cells

CELLS
Each cell requires energy to perform the functions essential for its own survival and to carry out its specialized contribution toward maintaining homeostasis. All energy used by cells is ultimately provided by food intake, which is regulated to maintain energy balance. Body temperature must be maintained at a fairly constant level to prevent unwanted changes in the rate of chemical reactions within cells and to prevent damage to cell proteins from heat.

Cells make up body systems

Food intake is essential to provide energy to power life-sustaining cell activities. The energy (caloric) value of ingested food must equal the body's total energy needs for body weight to remain constant. **Energy balance** and thus body weight are maintained by controlling food intake.

Energy expenditure by the body generates heat, which is important in **temperature regulation**. Humans are usually in external environments cooler than their bodies, so they must constantly generate heat internally to maintain their body temperatures. Also, they must have mechanisms to cool the body if it gains too much heat from heat-generating skeletal-muscle activity or from a hot external environment. Body temperature must be regulated because the rate of cellular chemical reactions is temperature dependent and overheating damages cell proteins.

The hypothalamus is the major integrating center for maintenance of both energy balance and body temperature.

CHAPTER CONTENTS AT A GLANCE

ENERGY BALANCE
Sources of energy input and output
Metabolic rate
Neutral, positive, and negative energy balance
Control of food intake
● A CLOSER LOOK AT EXERCISE PHYSIOLOGY What the Scales Don't Tell You
Obesity; anorexia nervosa
TEMPERATURE REGULATION
Body temperature
Sources of heat gain and heat loss
Physical mechanisms of heat exchange
The hypothalamus as a thermostat
Control of heat production; shivering
Control of heat loss; skin vasomotor activity
Integrated responses to cold and heat exposure
Fever
● CONCEPTS, CHALLENGES, AND CONTROVERSIES The Extremes of Heat and Cold Can Be Fatal

ENERGY BALANCE

Most food energy is ultimately converted into heat in the body.

Each cell in the body requires energy to perform the functions essential for the cell's own survival (such as active transport and cellular repair) and to carry out its specialized contributions toward maintenance of homeostatic balance (such as gland secretion and muscle contraction). All energy used by cells is ultimately provided by food intake. Chemical energy locked in the bonds that hold the atoms together in nutrient molecules is released when these molecules are broken down in the body. Energy harvested from biochemical processing of ingested nutrients either is used immediately to perform biological work or is stored in the body for later use as needed during periods when food is not being digested and absorbed.

According to the *first law of thermodynamics,* energy can neither be created nor destroyed. Therefore, energy is subject to the same kind of input-output balance as are the chemical components of the body such as H_2O and salt (see p. 511).

The energy in ingested foodstuffs constitutes energy input to the body. Energy output or expenditure falls into two categories (● Fig. 17–1): external work and internal work. **External work** refers to the energy expended when skeletal muscles are contracted to move external objects or to move the body in relation to the environment. **Internal work** constitutes all other forms of biological energy expenditure that do not accomplish mechanical work outside the body. Internal work encompasses two types of energy-dependent activities: (1) skeletal-muscle activity used for purposes other than external work, such as the contractions associated with postural maintenance and shivering, and (2) all the energy-expending activities that must go on all of the time just to sustain life. The latter include the work of pumping blood and breathing; the energy required for active transport of critical materials across plasma membranes; and the energy used during synthetic reactions essential for the maintenance, repair, and growth of cellular structures—in short, the "metabolic cost of living."

Not all energy in nutrient molecules can be harnessed to perform biological work. Energy cannot be created or destroyed, but it can be converted from one form to another. The energy in nutrient molecules that is not used to energize work is transformed into **thermal energy**, or **heat**. During biochemical processing, only about half of the energy in nutrient molecules is transferred to ATP; the other 50% of nutrient energy is immediately lost as heat. During ATP expenditure by the cells, another 25% of the energy derived from ingested food becomes heat. Since the body is not a heat engine, it cannot convert heat into work. Therefore, not more than 25% of nutrient energy is available to accomplish work, either external or internal. The remaining 75% is lost as heat during the sequential transfer of energy from nutrient molecules to ATP to cellular systems.

Furthermore, of the energy actually captured for use by the body, almost all expended energy eventually becomes heat. To exemplify, energy expended by the heart to pump blood is gradually changed into heat by friction as blood flows through the vessels. Likewise, energy used in the synthesis of cellular structural protein eventually appears as heat when that protein is degraded during the normal course of turnover of bodily constituents. Even in the performance of external work, skeletal muscles convert chemical energy into mechanical energy inefficiently, with as much as 75% of the expended energy being lost as heat. Thus, all energy that is liberated from ingested food but not directly used for movement

● **FIGURE 17–1   Energy Input and Output**

of external objects or stored in fat (adipose tissue) deposits (or, in the case of growth, as protein) eventually becomes body heat. This heat is not entirely wasted energy, however, because much of it is used to maintain body temperature.

The metabolic rate is the rate of energy use.

The rate at which energy is expended by the body during both external and internal work is known as the **metabolic rate**:

metabolic rate = energy expenditure / unit of time

Since most of the body's energy expenditure eventually appears as heat, the metabolic rate is normally expressed in terms of the rate of heat production in kilocalories per hour. The basic unit of heat energy is the **calorie**, which is the amount of heat required to raise the temperature of 1 g of H_2O 1°C. This unit is too small to be convenient when discussing the human body because of the magnitude of heat involved, so the **kilocalorie** or **Calorie**, which is equivalent to 1,000 calories, is used. When nutritionists speak of "calories" in quantifying the energy content of various foods, they are actually referring to kilocalories or Calories. Four kilocalories of heat energy are released when 1 g of glucose is oxidized or "burned," whether the oxidation takes place inside or outside the body.

The metabolic rate and consequently the amount of heat produced vary depending on the presence or absence of a variety of factors, such as exercise, food intake, shivering, and anxiety. Increased skeletal-muscle activity is the factor that can increase metabolic rate to the greatest extent. Even slight increases in muscle tone notably elevate the metabolic rate, and various levels of physical activity alter energy expenditure and heat production markedly (Table 17–1). For this rea-

son, a person's metabolic rate is determined under standardized basal conditions established to control as many of the variables that can alter metabolic rate as possible. In this way, the metabolic activity necessary to maintain the basic body functions at rest can be evaluated. Thus, the so-called **basal metabolic rate** (**BMR**) is a reflection of the body's "idling speed," or the minimal waking rate of internal energy expenditure. The BMR is measured under the following specified conditions:

1. The person should be at physical rest, having refrained from exercise for at least thirty minutes to eliminate any contribution of muscular exertion to heat production.
2. The individual should be at mental rest to minimize skeletal-muscle tone (people "tense up" when they are nervous) and to prevent a rise in epinephrine secretion. Stress elicits the secretion of the hormone epinephrine, which increases the metabolic rate.
3. The measurement should be performed at a comfortable room temperature so that the individual does not shiver. Shivering can markedly increase heat production because the purpose of these reflex oscillating skeletal-muscle contractions is to generate heat in response to cold exposure.
4. The subject should not have eaten any food within twelve hours before the BMR determination to avoid **diet-induced thermogenesis**, or the obligatory increase in metabolic rate that occurs as a consequence of food intake. This short-lived (less than twelve-hour) rise in metabolic rate is not due to digestive activities but to the increased metabolic activity associated with the processing and storage of ingested nutrients, especially by the major biochemical factory, the liver.

The rate of heat production in BMR determinations can be measured directly or indirectly. **Direct calorimetry** involves the cumbersome procedure of placing the subject in an insulated chamber with H_2O circulating through the walls. The difference in the temperature of the H_2O entering and leaving the chamber reflects the amount of heat liberated by the subject and picked up by the H_2O as it passes through the chamber. Even though this method provides a direct measurement of heat production, it is not practical because a calorimeter chamber is costly and takes up a lot of space. Therefore, a more practical method of indirectly determining the rate of heat production was developed for widespread use. With **indirect calorimetry**, the only measurement made is of the subject's O_2 uptake per unit of time, which is a simple task requiring minimal equipment. Recall that:

food + $O_2 \rightarrow CO_2 + H_2O$ + energy (mostly transformed into **heat**)

Accordingly, a direct relationship exists between the volume of O_2 used and the quantity of heat produced. This relationship also depends on the type of food being oxidized. Although carbohydrates, proteins, and fats require different amounts of O_2 for their oxidation and yield different amounts of kilocalories when oxidized, an average estimate can be made of the quantity of heat produced per liter of O_2 con-

TABLE 17–1 Rate of Energy Expenditure for 70-Kg Person during Different Types of Activity	
FORM OF ACTIVITY	**ENERGY EXPENDITURE (KCAL/HOUR)**
Sleeping	65
Awake, lying still	77
Sitting at rest	100
Standing relaxed	105
Getting dressed	118
Typewriting	140
Walking slowly on level (2.6 mi/hr)	200
Carpentry, painting a house	240
Sexual intercourse	280
Bicycling on level (5.5 mi/hr)	304
Shoveling snow, sawing wood	480
Swimming	500
Jogging (5.3 mi/hr)	570
Rowing (20 strokes/min)	828
Walking up stairs	1,100

sumed on a typical mixed American diet. This approximate value, known as the **energy equivalent of O₂**, is 4.8 kilocalories of energy liberated per liter of O_2 consumed. Using this method, the metabolic rate of a person consuming 15 l/hr of O_2 can be estimated as follows:

15	l/hr	= O_2 consumption
× 4.8	kilocalories/l	= energy equivalent of O_2
72	kilocalories/hr	= estimated metabolic rate

In this way, a simple measurement of O_2 consumption can be used to reasonably approximate heat production in the determination of metabolic rate.

Once the rate of heat production is determined under the prescribed basal conditions, it must be compared with normal values for persons of the same sex, age, height, and weight because these factors all affect the basal rate of energy expenditure. For example, a large man actually has a higher rate of heat production than a smaller man, but when expressed in terms of total surface area (which is a reflection of height and weight), the output in kilocalories per hour per square meter of surface area is normally about the same.

Thyroid hormone is the primary but not sole determinant of the rate of basal metabolism. As the level of active circulating thyroid hormone increases, the BMR increases correspondingly.

Surprisingly, the BMR is not the body's lowest metabolic rate. The rate of energy expenditure during sleep is 10% to 15% lower than the BMR, presumably because of the more complete muscle relaxation that occurs during the paradoxical stage of sleep (see p. 138).

Energy input must equal energy output to maintain a neutral energy balance.

Since energy cannot be created or destroyed, energy input must equal energy output, as represented by the following equation:

energy input = energy output

energy in	= external + internal heat ± stored
food consumed	work production energy

There are three possible states of energy balance:

- *Neutral energy balance.* If the amount of energy in food intake exactly equals the amount of energy expended by the muscles in performing external work plus the basal internal energy expenditure that eventually appears as body heat, then energy input and output are exactly in balance and body weight remains constant.
- *Positive energy balance.* If the amount of energy in food intake is greater than the amount of energy expended by means of external work and internal functioning, the extra energy taken in but not used is stored in the body, primarily as adipose tissue, so body weight increases.
- *Negative energy balance.* Conversely, if the energy derived from food intake is less than the body's immediate energy

requirements, the body must use stored energy to supply energy needs, and, accordingly, body weight decreases.

To maintain a constant body weight (with the exception of minor fluctuations caused by changes in H_2O content), energy acquired through food intake must equal energy expenditure by the body. Because the average adult maintains a fairly constant weight over long periods of time, this implies that precise homeostatic mechanisms exist to maintain a long-term balance between energy intake and energy expenditure. Theoretically, total body energy content could be maintained at a constant level by regulating the magnitude of food intake, physical activity, or internal work and heat production. Control of food intake to match changing metabolic expenditures is the major means of maintaining a neutral energy balance. The level of physical activity is principally under voluntary control. For example, there are no regulatory factors that automatically impel an obese person to exercise; in fact, in most cases overweight people are even less inclined to engage in physical activities than are their slimmer counterparts. There are mechanisms that can alter the degree of internal work and heat production, but these control measures are aimed primarily at regulating body temperature rather than total energy balance. Some studies suggest, however, that after several weeks of eating less or more than desired, small counteracting changes in metabolism may occur. For example, a compensatory increase in the body's efficiency of energy use in response to underfeeding may partially explain why some dieters become stuck at a plateau after having lost the first ten or so pounds of weight fairly easily. Similarly, a compensatory reduction in the efficiency of energy use in response to overfeeding may account in part for the difficulty experienced by very thin people who are deliberately trying to gain weight. Despite the possibility of compensatory changes in metabolism, there is no question that regulation of food intake is the most important factor in the long-term maintenance of energy balance and body weight.

Food intake is controlled primarily by the hypothalamus, but the mechanisms involved are not fully understood.

Control of food intake is primarily a function of the hypothalamus. Classically, the hypothalamus is considered to house a pair of **feeding** or **appetite centers** located in the lateral (outer) regions of the hypothalamus, one on each side, and another pair of **satiety centers** located in the ventromedial (underside middle) area. The functions of these areas have been elucidated by a series of experiments that involve either destruction or stimulation of these specific regions. Stimulation of the clusters of nerve cells designated as feeding or appetite centers makes the animal hungry, driving it to eat voraciously, whereas selective destruction of these areas suppresses eating and food-intake behavior to the point that the animal starves itself to death. In contrast, stimulation of the satiety centers signals satiety, or the feeling of having had enough to eat. Consequently, the stimulated animal refuses to eat, even if previously deprived of food. As expected, destruc-

● FIGURE 17–2   Comparison of Normal Rat with Rat Whose Satiety Center Has Been Destroyed Several months after destruction of the satiety center in the ventromedial area of the hypothalamus, the rat on the right had gained considerable weight as a result of overeating compared to its normal litter mate on the left. Rats sustaining lesions in this area also display less grooming behavior, accounting for the soiled appearance of the fat rat.

tion of this area produces the opposite effect—profound overeating and obesity because the animal never achieves a feeling of being full (● Fig. 17–2). Thus, the feeding centers tell us to eat, whereas the satiety centers tell us when we have had enough. Although it is convenient to consider these specific areas as exciting and inhibiting feeding behavior, respectively, this approach is too simplistic. Other areas of the brain as well as the liver are now known to play important roles in controlling food intake, but their contributions and interrelationships with the hypothalamus remain obscure. Undoubtedly, complex systems rather than isolated centers control feeding and satiety.

Whatever the site of final integration, a major interest is determining what input factors to these integrating centers govern feeding onset and termination as well as total energy intake. Exactly what switches feeding behavior on and off is still unclear. Even though food intake is adjusted to balance changing energy expenditures over a period of time, there are no calorie receptors per se to monitor energy input, energy output, or total body energy content. A number of proposals have been set forth, each supported by some experimental findings, but none of them alone is able to account for all observed feeding behavior. Undoubtedly, control of food in-

take does not depend on changes in a single signal but is determined by the integration of many inputs that provide information about the body's energy status. Some information is apparently used for short-term regulation of food intake, helping to control meal size and frequency. Even so, over a twenty-four-hour period the energy in ingested food rarely matches energy expenditure for that day. The correlation between total caloric intake and total energy output is excellent, however, over long periods of time. As a result, the total energy content of the body—and, consequently, body weight—remain relatively constant on a long-term basis.

The following factors are among those that have been hypothesized as contributing to the control of food intake. The relative importance of each of these factors (or possibly others) remains unclear.

The Size of Fat Stores   According to the **lipostatic** ("fat stability") **theory**, increased fat storage in adipose tissue signals satiety. The lipostatic theory is generally considered to be responsible for the long-term matching of food intake to energy expenditure so that total body energy content remains balanced and body weight remains constant. It is proposed that glycerol serves as a blood-borne signal between fat stores and the areas of the brain controlling food intake. The amount of glycerol in the blood appears to be an excellent indicator of the total amount of triglyceride fat stored in the adipose tissue. One problem with the lipostatic theory is that overweight people who have an abundance of adipose tissue still get hungry. Perhaps the percentage of filling of each adipose cell is important as well. Accordingly, persons with a greater number of adipose cells may be overweight by normal standards, yet they may still be hungry because their adipose cells are not "full."

The Extent of Gastrointestinal Distention   In addition to the mechanisms involved in the long-term control of body weight, other factors are believed to play a role in controlling the timing and size of meals. Early proposals suggested that cues of emptiness or fullness of the digestive tract signaled hunger or satiety, respectively. For example, stimulation of gastric stretch receptors has been shown to suppress food intake. However, neural input arising from stomach distention plays a more important role in controlling the rate of gastric emptying than in signaling satiety (see p. 557). Researchers now believe that internal blood-borne signals reflecting the depletion or availability of energy-producing substances are more important than stomach volume in controlling the initiation and cessation of eating.

The Extent of Glucose Utilization   The **glucostatic** ("glucose stability") **theory** proposes that satiety is signaled by increased glucose utilization, such as occurs during a meal when more glucose is available for use because it is being absorbed from the digestive tract. Conversely, after absorption of a meal is complete, and no new glucose is entering the blood, the resultant reduction in the cells' glucose use arouses the sensation of hunger.

Related to this theory is a proposal by some researchers that an increased level of insulin in the blood signals satiety. Insulin, a hormone secreted by the pancreas in response to a rise in the concentration of glucose and other nutrients in the blood following a meal, stimulates cellular uptake, utilization, and storage of glucose and other nutrients. Thus, the increase in insulin secretion that accompanies nutrient abundance and promotes increased glucose utilization would be an appropriate satiety signal.

The Intensity of Cell Power Production A recent proposal, the **ischymetric** (*ischys* means "power"; *metric* means "measure") **theory**, suggests that the signal for short-term control of food intake is not a deficit or surplus of any particular major nutrient such as glucose but is linked instead to the magnitude of cellular power (ATP) production. Changes in the availability of one or all of the nutrients to a cell may result in decreases or increases in the rate of ATP/ADP turnover, which in turn could be transduced into some sort of blood-borne or neural signal of low power (hunger) or high power (satiety).

The Level of Cholecystokinin Secretion Evidence has accumulated that **cholecystokinin (CCK)**, one of the gastro-intestinal hormones released from the duodenal mucosa during digestion of a meal, may be an important satiety signal. This theory could explain why we stop eating before the ingested food is actually digested, absorbed, and made available to meet the body's energy needs. We feel satisfied, even though the body's energy stores are still low, when adequate food to replenish the stores is in the digestive tract.

Cholecystokinin is secreted in response to the presence of nutrients in the small intestine. Through multiple effects on the digestive system, CCK facilitates the digestion and absorption of these nutrients (see p. 590). It is appropriate that this blood-borne signal, whose rate of secretion is correlated with the amount of nutrients ingested, also contributes to the sense of being filled after a meal has been consumed but before it has actually been digested and absorbed.

Influences of Neurotransmitters Various neuronal circuits within the brain are apparently involved in controlling feeding behavior. Exposure of the brain to various neurotransmitters has been shown to induce different patterns of food intake in experimental animals. For example norepinephrine and neuropeptide Y increase carbohydrate consumption whereas dopamine and serotonin suppress carbohydrate consumption. How the release of these neurotransmitters is linked to the control of food intake is unclear.

Psychosocial Influences Thus far we have described possible involuntary signals that might automatically occur to control food intake. However, as with water intake, people's eating habits are also shaped by psychological and social factors. We often do not eat merely because we are hungry or stop because we are full. Frequently, we eat out of habit (eating three meals a day on schedule no matter what our status on the hunger-satiety continuum) or because of social cus-

tom (food often plays a prime role in entertainment, leisure, and business activities).

Furthermore, the amount of pleasure derived from eating can reinforce feeding behavior. Eating foods with an enjoyable taste, smell, and texture can increase appetite and food intake. This has been demonstrated in an experiment in which rats were offered their choice of highly palatable human foods. They overate by as much as 70% to 80% and became obese. When the rats returned to eating their regular monotonous but nutritionally balanced rat chow, their obesity was rapidly reversed as their food intake was controlled once again by physiological drives rather than by hedonistic urges for the tastier offerings.

Stress, anxiety, depression, and boredom have also been shown to alter eating behavior in ways that are unrelated to energy needs in both experimental animals and humans. Thus, any comprehensive explanation of how food intake is controlled must take into account these voluntary consummatory acts that can reinforce or override the internal signals governing eating behavior.

Obesity occurs when more calories are consumed than are burned up.

Obesity is defined as excessive fat content in the adipose-tissue stores. The arbitrary boundary for obesity is generally considered to be greater than 20% overweight compared to normal standards. Obesity occurs when, over a period of time, more kilocalories are ingested in food than are used to support the body's energy needs, with the excessive energy being stored as triglycerides in adipose tissue. The causes of obesity are many and obscure. Some factors that may be involved include (1) emotional disturbances in which overeating replaces other gratifications, (2) development of an excessive number of fat cells as a result of overfeeding, (3) certain endocrine disorders such as hypothyroidism, (4) disturbances of the satiety-appetite regulatory centers in the hypothalamus, (5) hereditary tendencies, (6) the palatability of available food, and (7) lack of physical exercise.

Numerous studies have shown that, on the average, fat people do not eat any more than thin people. One possible explanation is that overweight persons do not overeat but "underexercise." Studies have shown that very low levels of physical activity are not accompanied by comparable reductions in food intake. Another explanation is that excess energy input occurs only during the time that obesity is actually developing. Some investigators suggest that the hypothalamic centers determining long-term satiety are "set at a higher level" in obese persons. Thus, overweight people do tend to maintain their weight, but at a higher set point than normal. Once obesity has developed, all that is required to maintain the condition is that energy input equals energy output.

Other studies further complicate the weight-control issue by suggesting that even though the total caloric content of food remains the same, the frequency of eating may be important in the distribution of the nutrients. In rat and monkey studies, nibblers had relatively more body protein and less body fat than animals given the same food in several large meals each day. Further studies suggest that, even though total caloric input for the day remains unchanged, foods consumed before bedtime are more likely to be put into storage. In this regard, bedtime snacks are more fattening than they would be if they were eaten earlier in the day.

The type of food ingested may also be important. Fewer calories are used to convert ingested fat into stored fat than to convert ingested carbohydrate or protein into stored fat. This difference occurs because fewer energy-consuming biochemical steps are needed to process and store dietary fat than to process and store dietary carbohydrate or protein. Therefore, if the same number of calories of fat and carbohydrate or protein are ingested, more calories are "left over" after the caloric price for processing and storing the fat has been paid than remain after the carbohydrate or protein has been processed. Accordingly, calorie for calorie, fat is more fattening than carbohydrate or protein.

Thus, our knowledge about the causes and control of obesity is still rather limited, as evidenced by the number of people who are constantly trying to stabilize their weight at a more desirable level. This is important from more than an aesthetic viewpoint. It is known that obesity, especially of the android type, can predispose an individual to illness and premature death from a multitude of disease. (See the accompanying boxed feature, A Closer Look at Exercise Physiology.)

Persons suffering from anorexia nervosa have a pathological fear of gaining weight.

The converse of obesity is generalized nutritional deficiency. The obvious causes for reduction of food intake below energy needs are lack of availability of food, interference with the swallowing or digestive mechanism, and impairment of appetite. Chronic diseases such as renal failure, cancer, and tuberculosis are commonly accompanied by a lack of appetite, which contributes to the subsequent weight loss characteristic of these "wasting" diseases. The mechanisms for this appetite suppression are obscure. Researchers have identified a protein called **cachetin** (*cachexia* refers to a chronic, wasting condition), which is released by the immune system and is believed to contribute to the development of emaciation in many patients with chronic, debilitating diseases or cancer.

Another poorly understood disorder in which lack of appetite is a prominent feature is **anorexia nervosa**. Patients with this disorder, most commonly adolescent girls and young women, have a morbid fear of becoming fat. As a result of having a distorted body image, they tend to visualize themselves as being much heavier than they actually are. Because they have an aversion to food, they eat very little and consequently lose considerable weight, perhaps even starving themselves to death. Other characteristics of the condition include altered secretion of many hormones, absence of menstrual periods, and low body temperature. It is unclear whether these symptoms occur secondarily as a result of general malnutrition or arise independently of the eating disturbance as a part of a primary hypothalamic malfunction. Many investigators think the underlying problem may be psycho-

logical rather than biological. Some experts suspect that anorexics may suffer from addiction to endogenous opiates, self-produced morphine-like substances (see p. 158) that are thought to be released during prolonged starvation.

TEMPERATURE REGULATION

Internal core temperature is homeostatically maintained at 100°F.

Humans are usually in environments cooler than their bodies, so they must constantly generate heat internally to maintain their body temperatures. Heat production ultimately depends on the oxidation of metabolic fuel derived from food.

Because cellular function is sensitive to fluctuations in internal temperature, humans homeostatically maintain body temperature at a level that is optimal for cellular metabolism to proceed in a stable fashion. Even moderate elevations of body temperature begin to cause nerve malfunction and irreversible protein denaturation. Most people suffer convulsions when the internal body temperature reaches about 106°F (41°C); 110°F (43.3°C) is considered to be the upper limit compatible with life. On the other hand, most of the body's tissues can transiently withstand substantial cooling. This characteristic is useful during cardiac surgery when the heart must be stopped. The cooled tissues need less nourishment than they do at normal body temperature because of their pronounced reduction in metabolic activity. The lower O_2 need of cooled tissues also accounts for the occasional survival of drowning victims who have been submerged in icy water considerably longer than one can normally survive without O_2.

Normal body temperature is traditionally considered to be 98.6°F (37°C). Yet there is no one "normal" body temperature because the temperature varies from organ to organ. From a thermoregulatory viewpoint, the body may conveniently be viewed as a *central core* surrounded by an *outer shell*. The temperature within the inner core, which consists of the abdominal and thoracic organs, the central nervous system, and the skeletal muscles, generally remains fairly constant. It is this internal **core temperature** that is considered to be the body temperature and that is subject to precise regulation to maintain its homeostatic constancy. The core tissues function best at a relatively constant temperature of around 100°F. The skin and subcutaneous fat constitute the outer shell. In contrast to the constant high temperature in the core, the temperature within the shell is generally cooler and may vary substantially. For example, skin temperature may fluctuate between 68° and 104°F without damage. In fact, as you will see, the temperature of the skin is deliberately varied as a control measure to help maintain the core's thermal constancy.

We are accustomed to thinking in terms of oral or rectal temperature because these are easy sites for monitoring body temperature. The average resting oral temperature is 98.6°F, with a normal range of 97 to 99°F. Rectal temperatures average about 1°F higher at 99.7°F (37.6°C), ranging from 97 to

WHAT THE SCALES DON'T TELL YOU

Body composition refers to the percentage of body weight that is composed of lean tissue and adipose tissue. The assessment of body composition is an important component in evaluating a person's health status. The age-height-weight tables used by insurance companies can be misleading for determining healthy body weight. Many athletes, for example, would be considered overweight by these charts. A football player may be 6'5" tall, weight 300 lb, but have only 12% body fat. This player's extra weight is muscle, not fat, and therefore is not a detriment to his health. A sedentary person, on the other hand, may be normal on the height-weight charts but have 30% body fat. This person should maintain body weight while increasing muscle mass and decreasing fat. Ideally, men should have 15% fat or less and women should have 20% fat or less.

The most accurate method for assessing body composition is underwater weighing. This technique is based on the fact that lean tissue is denser than water and fat tissue is less dense than water. (You can readily demonstrate this for yourself by dropping a piece of lean meat and a piece of fat into a glass of water; the lean meat will sink and the fat will float.) The most common method of underwater weighing requires the person to expel all the air from his or her lungs and then completely submerge in a tank of water while sitting in a swing that is attached to a scale. The results are used to determine body density using equations that take into consideration the density of water, the difference between the person's weight in air and underwater, and the residual volume of air remaining in the lungs. Because of the difference in density between lean and fat tissue, people who have more fat have a lower density and weigh relatively less underwater than in air compared to their lean counterparts. Body composition is then determined by means of an equation that correlates percentage fat with body density.

Another common way to assess body composition is skinfold thickness. Because approximately half of the body's total fat content is located just beneath the skin, total body fat can be estimated from measurement of skinfold thickness taken at various sites on the body. Skinfold thickness is determined by pinching up a fold of skin at one of the designated sites and measuring its thickness by means of a caliper, a hinged instrument that fits over the fold and is calibrated to measure thickness. Mathematical equations specific for the person's age and sex can be used to predict the percentage of fat from the skinfold-thickness scores. A major criticism of skinfold assessments is that accuracy depends on the investigator's skill.

There are different ways to be fat, and one way is more dangerous than the other. Obese patients can be classified into two categories—*android*, a male-type of adipose tissue distribution, and *gynoid*, a female-type distribution—based on the anatomical distribution of adipose tissue measured as the ratio of waist circumference to hip circumference. Android obesity is characterized by abdominal fat distribution (people shaped as "apples"), whereas gynoid obesity is characterized by fat distribution in the hips and thighs (people shaped as "pears"). Both sexes can display either android or gynoid obesity.

Android obesity is associated with a number of disorders, including insulin resistance, Type II (adult-onset) diabetes mellitus, excess blood lipid levels, high blood pressure, coronary heart disease, and stroke. Gynoid obesity is not associated with high risk for these diseases. Because android obesity is associated with increased risk for disease, it is most important for apple-shaped overweight individuals to reduce their fat stores.

Exercise physiologists often assess body composition as an aid in prescribing and evaluating exercise programs. Exercise generally reduces the percentage of body fat and, by increasing muscle mass, increases the percentage of lean tissue.

Research on the success of weight-reduction programs indicates that it is very difficult for people to lose weight, but when weight loss occurs, it is from the area of increased stores. Recent interesting research has shed some light on the problems of obesity. Studies indicate that the resting metabolic rate may be an inherited trait. In one study, at three months of age, babies of obese parents showed 20% less energy expenditure than babies of lean parents. Also, diet-induced thermogenesis (DIT), which is the increase in metabolism that follows consumption of carbohydrate or protein, has been shown to be lower in those who have been obese since childhood. In other words, people who have been obese since childhood are very efficient at storing the excess calories they ingest. Lean people may metabolize more of the calories they ingest, with the energy being given off as heat.

Even after obese people reduce their weight to normal levels, their DIT remains lower than someone of the same weight who has always been lean. This would be an admirable physiologic trait in times of food deprivation, but in times of food abundance, low DIT and a low metabolic rate can predispose an individual to obesity. A person with these traits has to eat less than his or her lean counterpart to maintain normal weight.

Because very-low-calorie diets are difficult to maintain, an alternative to severely cutting caloric intake to lose weight is to increase energy expenditure through physical exercise. An aerobic-exercise program further helps reduce the risk of the disorders associated with android obesity and helps to reduce fat stores.

100°F. Neither of these measurements is an absolute indication of the internal core temperature, which averages about 100°F. Even though the core temperature is held relatively constant, several factors cause it to vary slightly:

1. Most people's core temperature normally varies about 1.8°F (1°C) during the day, with the lowest level occurring early in the morning before rising (6 to 7 A.M.) and the highest point occurring in late afternoon (5 to 7 P.M.). This variation is due to an innate biological rhythm or "biological clock."

2. Women also experience a monthly rhythm in core temperature in connection with their menstrual cycle. The core temperature averages 0.9°F (0.5°C) higher during the last half of the cycle from the time of ovulation to menstruation. This mild sustained elevation in temperature was once thought to be caused by the increased secretion of progesterone, one of the ovarian hormones, during this period, but this is no longer believed to be the case. The actual cause is still undetermined.

3. The core temperature increases during exercise because of the tremendous increase in heat production by the contracting muscles. During hard exercise, the core temperature may increase to as much as 104°F (40°C). This temperature would be considered to be a fever in a resting person, but it is normal during strenuous exercise.

4. Because the temperature-regulating mechanisms are not 100% effective, the core temperature may vary slightly with exposure to extremes of temperature. For example, the core temperature may fall several degrees in cold weather or rise a degree or so in hot weather.

Thus, the core temperature can vary at the extremes between about 96° to 104°F but usually deviates less than a few degrees. This relative constancy is made possible by multiple thermoregulatory mechanisms coordinated by the hypothalamus.

Heat gain must balance heat loss to maintain a stable core temperature.

The core temperature is a reflection of the body's total heat content. To maintain a constant total heat content and thus a stable core temperature, heat input to the body must balance heat output (● Fig. 17–3). *Heat input* occurs by way of heat gain from the external environment and internal heat production, the latter being the most important source of heat for the body. Recall that most of the body's energy expenditure ultimately appears as heat. This heat is important in the maintenance of core temperature. In fact, usually more heat is generated than is required to maintain the body temperature at a normal level, so the excess heat must be eliminated from the body. *Heat output* occurs by way of heat loss from exposed body surfaces to the external environment.

Balance between heat input and output is frequently disturbed by (1) changes in internal heat production for purposes unrelated to regulation of body temperature, most notably by exercise, which markedly increases heat production,

● **FIGURE 17–3 Heat Input and Output**

and (2) changes in the external environmental temperature that influence the degree of heat gain or heat loss that occurs between the body and its surroundings. To maintain body temperature within narrow limits in spite of changes in metabolic heat production and changes in environmental temperature, compensatory adjustments must take place in heat-loss and heat-gain mechanisms.

If the core temperature starts to fall, heat production is increased and heat loss is minimized so that normal temperature can be restored. Conversely, if the temperature starts to rise above normal, it can be corrected by increasing heat loss while simultaneously reducing heat production.

We will now elaborate on the means by which heat gains and losses can be adjusted to maintain body temperature, starting with a discussion of the methods of heat exchange between the body and its surroundings.

Heat exchange between the body and the environment takes place by radiation, conduction, convection, and evaporation.

All heat loss or heat gain between the body and the external environment must take place between the body surface and its surroundings. The same physical laws of nature that govern heat transfer between inanimate objects also control the transfer of heat between the body surface and the environment. The temperature of an object may be thought of as a measure of the concentration of heat within the object. Accordingly, heat always moves down its concentration gradient; that is, down a **thermal gradient** from a warmer to a cooler region.

The body uses four mechanisms of heat transfer: radiation, conduction, convection, and evaporation. **Radiation** is the emission of heat energy from the surface of a warm body in the form of **electromagnetic waves** or heat waves, which travel through space (● Fig. 17–4a). When radiant energy strikes an object and is absorbed, the energy of the wave motion is transformed into heat within the object. The human body both emits (source of heat loss) and absorbs (source of heat gain) radiant energy. Whether the body loses or gains heat by radiation depends on the difference in temperature between the skin surface and the surfaces of various other

Direction of arrows denotes direction of heat transfer.

(a)

(b)

Snowball

Heating pad

Convection current

(c)

Liquid converted to gaseous vapor

(d)

• **FIGURE 17–4   Mechanisms of Heat Transfer** (a) Radiation—the transfer of heat energy from a warmer object to a cooler object in the form of electromagnetic waves ("heat waves"), which travel through space. (b) Conduction—the transfer of heat from a warmer object to a cooler object that is in direct contact with the warmer one. The heat is transferred through the movement of thermal energy from molecule to adjacent molecule. (c) Convection—the transfer of heat energy by air currents. Cool air warmed by the body through conduction rises and is replaced by more cool air. This process is enhanced by the forced movement of air across the body surface. (d) Evaporation—the heat of vaporization required to convert a liquid such as sweat into a gaseous vapor is absorbed from the skin.

objects in the body's environment. Because net transfer of heat by radiation is always from warmer objects to cooler ones, the body gains heat by radiation from objects warmer than the skin surface, such as the sun, a radiator, or burning logs. On the other hand, the body loses heat by radiation to objects in its environment whose surfaces are cooler than the surface of the skin, such as building walls, furniture, or trees. On the average, humans lose close to half of their heat energy through radiation.

Conduction is the transfer of heat between objects of differing temperatures that are in *direct contact* with each other (Fig. 17–4b). Heat moves down its thermal gradient from the warmer to the cooler object by being transferred from mole-

cule to molecule. All molecules are constantly in vibratory motion, with warmer molecules moving faster than cooler ones. When molecules of differing heat content touch each other, the faster moving, warmer molecule agitates the cooler molecule into more rapid motion, thereby warming up the cooler molecule. During this process, the original warmer molecule loses some of its thermal energy as it slows down and cools off a bit. Therefore, given enough time, the temperature of the two touching objects eventually equalizes.

The rate of heat transfer by conduction depends on the *temperature difference* between the touching objects and the *thermal conductivity* of the substances involved (that is, how easily heat is conducted by the molecules of the substances).

Heat can be lost or gained by conduction when the skin is in contact with a good conductor. When you hold a snowball, for example, heat moves by conduction from your hand to the snowball so that your hand becomes cold. Conversely, when you apply a heating pad to a body part, the part is warmed up as heat is transferred directly from the pad to the body.

Similarly, you either lose or gain heat by conduction to the layer of air in direct contact with your body. The direction of heat transfer depends on whether the air is cooler or warmer, respectively, than your skin. Only a small percentage of total heat exchange between the skin and environment takes place by conduction alone, however, because air is not a very good conductor of heat. (This is why swimming-pool water at 80°F feels cooler than air at the same temperature; heat is conducted more rapidly from the body surface into the water, which is a good conductor, than into the air, which is a poor conductor.)

Convection refers to the transfer of heat energy by *air (or H_2O) currents*. As the body loses heat by conduction to the surrounding cooler air, the air in immediate contact with the skin is warmed. Because warm air is lighter (less dense) than cool air, the warmed air rises while cooler air moves in next to the skin to replace the vacating warm air. The process is then repeated (Fig. 17–4c). These air movements, known as convection currents, help carry heat away from the body. If it were not for convection currents, no further heat could be dissipated from the skin by conduction once the temperature of the layer of air immediately around the body equilibrated with skin temperature.

The combined conduction-convection process of dissipating heat from the body is enhanced by forced movement of air across the body surface, either by external air movements, such as those caused by the wind or a fan, or by movement of the body through the air, such as during bicycle riding. Because forced air movement sweeps away the air warmed by conduction and replaces it with cooler air more rapidly, a greater total amount of heat can be carried away from the body over a given time period. Thus, wind makes us feel cooler on hot days, and windy days in the winter are more chilling than calm days at the same cold temperature. For this reason, weather forecasters have developed the concept of *wind chill factor*.

Evaporation is the final method of heat transfer used by the body. When water evaporates from the skin surface, the heat required to transform water from a liquid to a gaseous state is absorbed from the skin, thereby cooling the body (Fig. 17–4d). Evaporative heat loss makes you feel cooler when your bathing suit is wet than when it is dry. Evaporative heat loss occurs continually from the linings of the respiratory airways and from the surface of the skin. Heat is continuously lost through the H_2O vapor in the expired air as a result of the air's humidification during its passage through the respiratory system. Similarly, because the skin is not completely waterproof, H_2O molecules constantly diffuse through the skin and evaporate. This ongoing evaporation from the skin is completely unrelated to the sweat glands. These passive evaporative heat-loss processes are not subject to physiological control and go on even in very cold weather, when the problem is one of conserving body heat.

Sweating, on the other hand, is an active evaporative heat-loss process under sympathetic nervous control. The rate of evaporative heat loss can be deliberately adjusted by means of sweating, which is an important homeostatic mechanism to eliminate excess heat as needed. In fact, when the environmental temperature exceeds the skin temperature, sweating is the *only* avenue for heat loss, because the body is gaining heat by radiation and conduction under these circumstances.

Sweat is a dilute salt solution that is actively extruded to the surface of the skin by sweat glands dispersed all over the body. Sweat must be evaporated from the skin for heat loss to occur. If sweat merely drips from the surface of the skin or is wiped away, no heat loss is accomplished. The most important factor determining the extent of evaporation of sweat is the *relative humidity* of the surrounding air (the percentage of H_2O vapor actually present in the air compared to the greatest amount that the air can possibly hold at that temperature; for example, a relative humidity of 70% means that the air contains 70% of the H_2O vapor it is capable of holding). When the relative humidity is high, the air is already almost fully saturated with H_2O, so it has limited ability to take up additional moisture from the skin. Thus, little evaporative heat loss can occur on hot, humid days. The sweat glands continue to secrete, but the sweat simply remains on the skin or drips off, instead of evaporating and producing a cooling effect. As a measure of the discomfort associated with combined heat and high humidity, meteorologists have devised the *temperature-humidity index*.

The hypothalamus integrates a multitude of thermosensory inputs from both the core and the surface of the body.

The hypothalamus serves as the body's thermostat. The home thermostat keeps track of the temperature in a room and triggers a heating mechanism (the furnace) or a cooling mechanism (the air conditioner) as necessary to maintain the room temperature at the indicated setting. Similarly, the hypothalamus, as the body's thermoregulatory integrating center, receives afferent information about the temperature in various regions of the body and initiates extremely complex, coordinated adjustments in heat-gain and heat-loss mechanisms as necessary to correct any deviations in core temperature from the "normal setting." The hypothalamic thermostat is far more sensitive than your home thermostat. The hypothalamus is able to respond to changes in blood temperature as small as 0.01°C. The hypothalamus's degree of response to deviations in body temperature is finely matched so that precisely enough heat is lost or generated as need be to restore the temperature to normal.

To make the appropriate adjustments in the delicate balance between the heat-loss mechanisms and the opposing heat-producing and heat-conserving mechanisms, the hypothalamus must be continuously apprised of both the skin temperature and the core temperature by means of specialized temperature-sensitive receptors called **thermoreceptors**. *Peripheral thermoreceptors* monitor skin temperature throughout the body and transmit information about changes in surface temperature to the hypothalamus. The core temperature is monitored by *central thermoreceptors*, which are located in the hypothalamus itself as well as elsewhere in the central nervous system and the abdominal organs.

Two centers for temperature regulation have been identified in the hypothalamus. The *posterior region* is activated by cold and subsequently triggers reflexes that mediate heat production and heat conservation. The *anterior region,* which is activated by warmth, initiates reflexes that mediate heat loss. Let us examine the means by which the hypothalamus fulfills its thermoregulatory functions (● Fig. 17–5).

Shivering is the primary involuntary means of increasing heat production.

The body can gain heat as a result of internal heat production generated by metabolic activity or from the external environ-ment if the latter is warmer than body temperature. Because body temperature usually is higher than environmental temperature, metabolic heat production is the primary source of body heat. In a resting person, most body heat is produced by the thoracic and abdominal organs as a result of ongoing, cost-of-living metabolic activities. Above and beyond this basal level, the rate of metabolic heat production can be variably increased primarily by changes in skeletal-muscle activity or to a lesser extent by certain hormonal actions. Thus, changes in skeletal-muscle activity constitute the major way heat gain is controlled for temperature regulation.

In response to a fall in core temperature caused by exposure to cold, the hypothalamus takes advantage of the fact that increased skeletal-muscle activity generates more heat. Acting through descending pathways that terminate on the motor neurons controlling the body's skeletal muscles, the hypothalamus first gradually increases skeletal-muscle tone. (Muscle tone refers to the constant level of tension within the muscles.) Soon, shivering begins. **Shivering** consists of rhythmic, oscillating skeletal-muscle contractions that occur at a

● **FIGURE 17–5   Major Thermoregulatory Pathways**

rapid rate of ten to twenty per second. This mechanism is very effective in increasing heat production; all of the energy liberated during these muscle tremors is converted to heat because no external work is accomplished. Within a matter of seconds to minutes, internal heat production may increase two- to fivefold as a result of shivering.

Frequently, reflex changes in skeletal-muscle activity are augmented by increased voluntary, heat-producing actions such as bouncing up and down or hand clapping. Such behavioral responses appear to share neural systems in common with the involuntary physiological responses. The hypothalamus and the limbic system of which it is a part (see p. 128) are extensively involved with controlling motivated behavior. Therefore, one should not think of the hypothalamic homeostatic control systems as operating only at the subconscious level.

Although reflex and voluntary changes in muscle activity are the major means of increasing the rate of heat production, **nonshivering (chemical) thermogenesis** also plays a role in thermoregulation. Chronic cold exposure in most experimental animals brings about an increase in metabolic heat production that is independent of muscle contraction, appearing instead to involve changes in heat-generating chemical activity. In humans nonshivering thermogenesis is most important in newborns, because they lack the ability to shiver. Nonshivering thermogenesis is mediated by the hormones epinephrine and thyroid hormone, both of which increase heat production by stimulating fat metabolism. Newborns have deposits of a special type of adipose tissue known as **brown fat**, which is especially capable of converting chemical energy into heat. The role of nonshivering thermogenesis in adults remains controversial.

In the opposite situation—an elevation in core temperature caused by heat exposure—two mechanisms are employed to reduce heat-producing skeletal-muscle activity: muscle tone is reflexly reduced, and voluntary movement is curtailed. When the air becomes very warm, people often complain about it being "too hot even to move." These responses are not as effective at reducing heat production during heat exposure as are the muscular responses that increase heat production during cold exposure for two reasons. First, because muscle tone is normally quite low, the capacity to reduce it further is limited. Second, the elevated body temperature tends to increase the rate of metabolic heat production because the temperature has a direct effect on the rate of chemical reactions.

The magnitude of heat loss can be adjusted by varying the flow of blood through the skin.

Heat-loss mechanisms are also subject to control, again largely by the hypothalamus. When we are hot, we want to increase heat loss to the environment; when we are cold, we want to decrease heat loss. The amount of heat lost to the environment by radiation and conduction-convection is largely determined by the temperature gradient between the skin and the external environment. The body's central core is a heat-generating chamber in which the temperature must be maintained at approximately 100°F. Surrounding the core is an insulating shell through which heat exchanges between the body and external environment take place. In an attempt to maintain a constant core temperature, the insulative capacity and temperature of the shell can be adjusted to vary the temperature gradient between the skin and external environment, thereby influencing the extent of heat loss.

The insulative capacity of the shell can be varied by controlling the amount of blood flowing through the skin. Blood flow to the skin serves two functions. First, it provides a nutritive blood supply to the skin. Second, as blood is pumped to the skin from the heart, it has been heated in the central core and carries this heat to the skin. Most of the flow of blood through the skin is for purposes of temperature regulation; at normal room temperature, twenty to thirty times more blood flows through the skin than is needed to meet the skin's nutritional needs.

In the process of thermoregulation, skin blood flow can vary tremendously, from 400 up to 2,500 ml/min. The more blood reaching the skin from the warm core, the closer the skin's temperature is to the core temperature. The skin's blood vessels diminish the effectiveness of the skin as an insulator by carrying heat to the surface, where it can be lost from the body by radiation and conduction-convection. Accordingly, vasodilation of the skin vessels, which permits increased flow of heated blood through the skin, increases heat loss or, if the environmental temperature is above the core temperature, reduces heat gain. Conversely, vasoconstriction of the skin vessels, which reduces blood flow through the skin, decreases heat loss by keeping the warm blood in the central core, where it is insulated from the external environment. Cold, relatively bloodless skin provides excellent insulation between the core and the environment. However, the skin is not a perfect insulator, even with maximum vasoconstriction. Despite minimal blood flow to the skin, some heat can still be transferred by conduction from the deeper organs to the skin surface and then can be lost from the skin to the environment.

These skin vasomotor responses are coordinated by the hypothalamus by means of sympathetic nervous system output. Increased sympathetic activity to the skin vessels produces heat-conserving vasoconstriction in response to cold exposure, whereas decreased sympathetic activity produces heat-losing vasodilation of the skin vessels in response to heat exposure.

The hypothalamus simultaneously coordinates heat-production mechanisms and heat-loss and heat-conservation mechanisms to regulate core temperature homeostatically.

Let us now pull together the coordinated adjustments in heat production as well as heat loss and heat conservation in response to exposure to either a cold or a hot environment (Fig. 17–5 and Table 17–2). In response to cold exposure, the posterior region of the hypothalamus directs increased heat production such as by shivering, while simultaneously decreasing heat loss (that is, conserving heat) by skin vasoconstriction and other measures.

Because there is a limit to the body's ability to reduce skin

TABLE 17-2 Coordinated Adjustments in Response to Cold or Heat Exposure

IN RESPONSE TO COLD EXPOSURE (COORDINATED BY POSTERIOR HYPOTHALAMUS)		IN RESPONSE TO HEAT EXPOSURE (COORDINATED BY ANTERIOR HYPOTHALAMUS)	
Increased Heat Production	**Decreased Heat Loss (heat conservation)**	**Decreased Heat Production**	**Increased Heat Loss**
Increased muscle tone	Skin vasoconstriction	Decreased muscle tone	Skin vasodilation
Shivering	Postural changes to reduce exposed surface area (hunching shoulders, etc.)*	Decreased voluntary exercise*	Sweating
Increased voluntary exercise*			Cool clothing*
Nonshivering thermogenesis	Warm clothing*		

*Behavioral adaptations.

temperature through vasoconstriction, even maximum vasoconstriction is not sufficient to prevent excessive heat loss when the external temperature falls too low. Accordingly, other measures must be instituted to further reduce heat loss. In animals with dense fur or feathers, the hypothalamus, acting through the sympathetic nervous system, brings about contraction of the tiny muscles at the base of the hair or feather shafts to lift the hair or feathers off the skin surface. This puffing up traps a layer of poorly conductive air between the skin surface and environment, thus increasing the insulating barrier between the core and the cold air and reducing heat loss. Even though the hair-shaft muscles contract in humans in response to cold exposure, this heat-retention mechanism is ineffective because of the low density and fine texture of most human body hair. The result instead is useless goosebumps.

After maximum skin vasoconstriction has been achieved as a result of exposure to cold, further heat dissipation in humans can be prevented only by behavioral adaptations, such as postural changes that reduce as much as possible the exposed surface area from which heat can escape. These postural changes include maneuvers such as hunching over, clasping the arms in front of the chest, or curling up in a ball. Putting on warmer clothing further insulates the body from too much heat loss. Clothing entraps layers of poorly conductive air between the skin surface and the environment, thereby diminishing loss of heat by conduction from the skin to the cold external air and curtailing the flow of convection currents.

Under the opposite circumstance—heat exposure—the anterior part of the hypothalamus reduces heat production by decreasing skeletal-muscle activity and promotes increased heat loss by inducing skin vasodilation. When even maximal skin vasodilation is inadequate to rid the body of excess heat, sweating is brought into play to accomplish further heat loss through evaporation. In fact, if the air temperature rises above the temperature of maximally vasodilated skin, the temperature gradient reverses itself so that heat is gained from the environment. Sweating is the only means of heat loss under these conditions.

Humans sweat all over their bodies, but many species either lack or have limited distribution of sweat glands; for example, dogs have sweat glands only on their paw pads. These species must employ alternative methods, such as panting, to enhance heat loss during heat exposure. The hypothalamus induces a shallow, rapid breathing pattern that permits large volumes of air to move over the hot, moist tongue and respiratory airways without leading to imbalances in the blood concentration of O_2 and CO_2. The resultant increase in evaporative heat loss from the respiratory tract provides a cooling effect.

In addition to sweating, humans employ voluntary measures, such as using fans, wetting the body, drinking cold beverages, and wearing cool clothing, to further enhance heat loss. Contrary to popular belief, wearing light-colored, loose clothing is cooler than being nude. Naked skin absorbs almost all of the radiant energy that strikes it, whereas light-colored clothing reflects almost all of the radiant energy that falls on it. Thus, if light-colored clothing is loose and thin enough to permit convection currents and evaporative heat loss to occur, wearing it is actually cooler than going without any clothes at all.

Skin vasomotor activity is highly effective in controlling heat loss in environmental temperatures between the upper sixties and mid-eighties. This range, within which core temperature can be kept constant by vasomotor responses without calling supplementary heat-production or heat-loss mechanisms into play, is called the **thermoneutral zone**. When external air temperature falls below the lower limits of the ability of skin vasoconstriction to reduce heat loss further, the major burden of maintaining core temperature is borne by increased heat production, especially shivering. At the other extreme, when external air temperature exceeds the upper limits of the ability of skin vasodilation to increase heat loss further, sweating becomes the dominant factor in maintaining core temperature.

During a fever, the hypothalamic thermostat is "reset" at an elevated temperature.

Fever refers to an elevation in body temperature as a result of infection or inflammation. In response to microbial invasion, certain white blood cells release a chemical known as **endogenous pyrogen**, which among its many infection-fighting effects (see p. 374) acts on the hypothalamic thermoregulatory center to raise the setting of the thermostat (• Fig. 17–6).

• FIGURE 17–6 Fever Production

The hypothalamus now maintains the temperature at the new set level instead of maintaining normal body temperature. If, for example, endogenous pyrogen raises the set point to 102°F (as recorded orally), the hypothalamus senses that the normal prefever temperature of 98.6°F is too cold, and it initiates the cold-response mechanisms to elevate the temperature to 102°F. Shivering is initiated to rapidly increase heat production, while skin vasoconstriction is brought about to rapidly reduce heat loss, both of which drive the temperature upward. These events account for the sudden cold chills often experienced at the onset of a fever. Because the person feels cold, he or she may put on more blankets as a voluntary mechanism that helps elevate body temperature by conserving body heat. Once the new temperature is achieved, body temperature is regulated as normal in response to cold and heat but at a higher setting. Thus, fever production in response to an infection is a deliberate outcome and is not due to a breakdown of thermoregulatory mechanisms. Although the physiological significance of a fever is still unclear, many medical experts believe that a rise in body temperature has a beneficial role in fighting infection. A fever augments the inflammatory response and may interfere with bacterial multiplication (see p. 375).

Endogenous pyrogen raises the set point of the hypothalamic thermostat during fever production by triggering the local release of *prostaglandins*, which are local chemical mediators that act directly on the hypothalamus. Aspirin reduces a fever by inhibiting the synthesis of prostaglandins. Aspirin does not lower the temperature in a nonfebrile person, because prostaglandins are not present in the hypothalamus in appreciable quantities in the absence of endogenous pyrogen.

The exact molecular cause of a fever "breaking" naturally is unknown, although it presumably results from a reduction in pyrogen release or decreased prostaglandin synthesis. When the hypothalamic set point is restored to normal, the temperature at 102°F (in this example) is too high. The heat-response mechanisms are instituted to cool down the body. Skin vasodilation occurs and sweating commences. The person feels hot and throws off extra covers. The gearing up of these heat-loss mechanisms by the hypothalamus reduces the temperature to normal.

Hyperthermia can occur unrelated to infection.

Hyperthermia denotes any elevation in body temperature above the normally accepted range. The term *fever* is usually reserved for an elevation in temperature caused by resetting of the hypothalamic set point by the release of endogenous pyrogen during infection or inflammation; hyperthermia refers to all other imbalances between heat gain and heat loss that increase body temperature. Hyperthermia has a variety of causes, some of which are normal and harmless, others pathological and fatal.

The most common cause of hyperthermia is sustained exercise. As a physical consequence of the tremendous heat load generated by exercising muscles, body temperature rises during the initial stage of exercise because heat gain exceeds heat loss (• Fig. 17–7). The elevation in core temperature reflexly triggers heat-loss mechanisms (skin vasodilation and sweating), which eliminate the discrepancy between heat production and heat loss. As soon as the heat-loss mechanisms are stepped up sufficiently to equalize heat production, the core temperature stabilizes at a level slightly above the set point despite continued heat-producing exercise. Thus, during sustained exercise, body temperature initially rises, then

• FIGURE 17–7 Hyperthermia in Sustained Exercise
(a) The rate of heat production initially exceeds the rate of heat loss at the onset of exercise, so that the core temperature rises. (b) When heat-loss mechanisms are reflexly increased sufficiently to equalize the elevated heat production, the core temperature stabilizes slightly above the resting set point for the duration of the exercise.

THE EXTREMES OF HEAT AND COLD CAN BE FATAL

Heat exhaustion refers to a state of collapse, usually manifested by fainting, that is caused by reduced blood pressure brought about as a result of overtaxing the heat-loss mechanisms. Extensive sweating reduces cardiac output by depleting the plasma volume, and pronounced skin vasodilation causes a drop in total peripheral resistance. Since blood pressure is determined by cardiac output times total peripheral resistance, blood pressure falls, an insufficient amount of blood is delivered to the brain, and fainting takes place. Thus, heat exhaustion is a consequence of overactivity of the heat-loss mechanisms rather than a breakdown of these mechanisms. Because the heat-loss mechanisms have been very active, body temperature is only mildly elevated in heat exhaustion. By forcing the cessation of activity when the heat-loss mechanisms are no longer able to cope with heat gain through exercise or a hot environment, heat exhaustion serves as a safety valve to help prevent the more serious consequences of heat stroke.

Heat stroke is an extremely dangerous situation that arises from the complete breakdown of the hypothalamic thermoregulatory systems. Heat exhaustion may progress into heat stroke if the heat-loss mechanisms continue to be overtaxed. Heat stroke is more likely to occur upon overexertion during a prolonged exposure to a hot, humid environment. The elderly, in whom thermoregulatory responses are generally slower and less efficient, are particularly vulnerable to heat stroke during prolonged, stifling heat waves. So too are individuals who are taking certain common tranquilizers, because these drugs interfere with the hypothalamic thermoregulatory centers' neurotransmitter activity.

The most striking feature of heat stroke is a lack of compensatory heat-loss measures, such as sweating, in the face of a rapidly rising body temperature. No sweating occurs despite a markedly elevated body temperature because the hypothalamic thermoregulatory control centers are not functioning properly and cannot initiate heat-loss mechanisms. During the development of heat stroke, body temperature starts to climb as the heat-loss mechanisms are eventually overwhelmed by prolonged, excessive heat gain. Once the core temperature reaches the point at which the hypothalamic temperature-control centers are damaged by the heat, the body temperature rapidly rises even higher because of the complete shutdown of heat-loss mechanisms. Furthermore, as the body temperature increases, the rate of metabolism increases correspondingly because higher temperatures speed up the rate of all chemical reactions; the result is even greater heat production. This positive-feedback state sends the temperature spiraling upward. Heat stroke is a very dangerous situation that is rapidly fatal if untreated. Even with treatment to halt and reverse the rampant rise in body temperature, there is a still a high rate of mortality; the rate of permanent disability in survivors is also high because of irreversible damage caused by the high internal heat.

At the other extreme, the body can be harmed by cold exposure in two ways: frostbite and generalized hypothermia. **Frostbite** involves excessive cooling of a particular part of the body to the point where tissue in that area is damaged. If exposed tissues actually freeze, tissue damage occurs as a result of disruption of the cells by formation of ice crystals or by lack of liquid water. **Hypothermia**, a fall in body temperature, occurs when generalized cooling of the body exceeds the ability of the normal heat-producing and heat-conserving regulatory mechanisms to match the excessive heat loss. As hypothermia sets in, the rate of all metabolic processes slows down because of the declining temperature. Higher cerebral functions are the first to be affected by body cooling, leading to loss of judgment, apathy, disorientation, and tiredness, all of which diminish the cold victim's ability to initiate voluntary mechanisms to reverse the falling body temperature. As body temperature continues to plummet, depression of the respiratory center occurs, reducing the ventilatory drive so that breathing becomes slow and weak. Activity of the cardiovascular system also is gradually reduced. The heart is slowed and cardiac output decreased. Disturbances of cardiac rhythm occur, eventually leading to ventricular fibrillation and death.

---

is maintained at the higher level as long as the exercise continues.

A completely different way in which hyperthermia can be brought about is by excessive heat production in connection with abnormally high circulating levels of epinephrine or thyroid hormone that result from dysfunctions of the adrenal medulla or thyroid gland, respectively. Both of these hormones elevate the core temperature by increasing the overall rate of metabolic activity and heat production.

Another way in which hyperthermia occurs is malfunction of the hypothalamic control centers. Certain brain lesions, for example, destroy the normal regulatory capacity of the hypothalamic thermostat. When the thermoregulatory mechanisms are not functional, lethal hyperthermia may occur very rapidly. Normal metabolism produces enough heat to kill a person in less than five hours if the heat-loss mechanisms are completely shut down. In addition to brain lesions, a breakdown in the function of hypothalamic thermoregulation can also occur upon exposure to severe, prolonged heat stress. Similarly, the body can be harmed by extreme cold exposure. (See the accompanying boxed feature, Concepts, Challenges, and Controversies.)

CHAPTER IN PERSPECTIVE: FOCUS ON HOMEOSTASIS

Because energy can be neither created nor destroyed, input must equal output in the case of both the body's total energy balance and its heat-energy balance in order for body weight and body temperature, respectively, to remain constant. If total energy input exceeds total energy output, the extra energy is stored in the body, and body weight increases. Similarly, if the input of heat energy exceeds its output, body temperature increases. Conversely, if output exceeds input, body weight decreases or body temperature falls. The hypothalamus is the major integrating center for the maintenance of both a constant total energy balance (and thus a constant body weight) and a constant heat-energy balance (and thus a constant body temperature).

Body temperature, which is one of the homeostatically regulated factors of the internal environment, must be maintained within narrow limits because the structure and reactivity of the chemicals that compose the body are temperature sensitive. Deviations in body temperature outside a limited range result in protein denaturation and death of the individual if the temperature rises too high, or metabolic slowing and death if the temperature falls too low.

Body weight, in contrast, varies widely among individuals. Only the extremes of imbalances between total energy input and output become incompatible with life. For example, in the face of insufficient energy input in the form of ingested food during prolonged starvation, the body resorts to breaking down muscle protein to meet its needs for energy expenditure once the adipose stores are depleted. Body weight dwindles due to this self-cannibalistic mechanism until death finally occurs as a result of loss of heart muscle, among other things. At the other extreme, when the food energy consumed greatly exceeds the energy expended, the extra energy input is stored as adipose tissue, and body weight increases. The resultant gross obesity can also lead to heart failure. Not only must the heart work harder to pump blood to the excess adipose tissue, but obesity also predisposes the individual to atherosclerosis and heart attacks (see p. 293).

CHAPTER SUMMARY

Energy Balance

Energy input to the body in the form of food energy must equal energy output, because energy cannot be created or destroyed. Energy output or expenditure includes (1) external work performed by skeletal muscles to accomplish movement of an external object or movement of the body through the external environment and (2) internal work, which consists of all other energy-dependent activities that do not accomplish external work, including active transport, smooth- and cardiac-muscle contraction, glandular secretion, and protein synthesis. Only about 25% of the chemical energy in food is harnessed to do biological work. The rest is immediately converted to heat. Furthermore, all of the energy expended to accomplish internal work is eventually converted into heat, and 75% of the energy expended by working skeletal muscles is lost as heat. Therefore, most of the energy in food ultimately appears as body heat. The metabolic rate, which is energy expenditure per unit of time, is measured in kilocalories of heat produced/hour.

For a neutral energy balance, the energy in ingested food must equal energy expended in performing work. If more food is consumed than energy expended, the extra energy is stored in the body, primarily as adipose tissue, so body weight increases. On the other hand, if more energy is burned than is available in the food, body energy stores are used to support energy expenditure, so body weight decreases. Usually, body weight remains fairly constant over a prolonged period of time (except during growth) because food intake is adjusted to match energy expenditure on a long-term basis. Food intake is controlled primarily by the hypothalamus by means of complex, poorly understood regulatory mechanisms in which hunger and satiety are important components.

Temperature Regulation

The body can be thought of as a heat-generating core (internal organs, CNS, and skeletal muscles) surrounded by a shell of variable insulating capacity (the skin). The skin exchanges heat energy with the external environment, with the direction and amount of heat transfer depending on the environmental temperature and the momentary insulating capacity of the shell. The four physical means by which heat is exchanged between the body and external environment are (1) radiation (net movement of heat energy via electromagnetic waves); (2) conduction (exchange of heat energy by direct contact); (3) convection (transfer of heat energy by means of air currents); and (4) evaporation (extraction of heat energy from the body by the heat-requiring conversion of liquid H_2O to H_2O vapor). Because heat energy moves from warmer to cooler objects, radiation, conduction, and convection can be channels for either heat loss or heat gain, depending on whether surrounding objects are cooler or warmer, respectively, than the body surface. Normally, they are avenues for heat loss, along with evaporation resulting from sweating.

To prevent serious cellular malfunction, the core temperature must be held constant at about 100°F (equivalent to an oral temperature of 98.6°F) by continuously balancing heat gain and heat loss despite changes in environmental temperature and variation in internal heat production. This thermoregulatory balance is controlled by the hypothalamus. The hypothalamus is apprised of the skin temperature by peripheral thermoreceptors and of the core temperature by central thermoreceptors, the most important of which are located in the hypothalamus itself. The primary means of heat gain is heat production by metabolic activity, the biggest contributor being skeletal-muscle contraction. Heat loss is adjusted by sweating and by controlling to the greatest extent possible the temperature gradient between the skin and surrounding environment. The latter is accomplished by regulating the caliber of the skin's blood vessels. Vasoconstriction of the skin vessels reduces the flow of warmed blood through the skin so that skin temperature falls. The layer of cool skin between the core and the environment increases the insulating barrier between the warm core and the external air. Conversely, skin vasodilation brings more warmed blood through the skin so that skin temperature approaches the core temperature, thus reducing the insulative capacity of the skin.

Upon exposure to cool surroundings, the core temperature starts to fall as heat loss increases due to the larger-than-normal skin-to-air temperature gradient. The hypothalamus responds to reduce the heat loss by inducing skin vasoconstriction while si-

multaneously increasing heat production through heat-generating shivering. Conversely, in response to a rise in core temperature (resulting either from excessive internal heat production accompanying exercise or from excessive heat gain upon exposure to a hot environment), the hypothalamus triggers heat-loss mechanisms, such as skin vasodilation and sweating, while simultaneously decreasing heat production, such as by reducing muscle tone. In both cold and heat responses, voluntary behavioral actions also contribute importantly to maintenance of thermal homeostasis.

A fever occurs when endogenous pyrogen released from white blood cells in response to infection raises the hypothalamic set point. An elevated core temperature develops as the hypothalamus initiates cold-response mechanisms to raise the core temperature to the new set point.

REVIEW EXERCISES

Objective Questions (Answers on p. D–4.)

1. If more food energy is consumed than is expended, the excess energy is lost as heat. (True or false?)

2. All of the energy within nutrient molecules can be harnessed to perform biological work. (True or false?)

3. Each liter of O_2 contains 4.8 kilocalories of heat energy. (True or false?)

4. A body temperature greater than 98.6°F is always indicative of a fever. (True or false?)

5. Core temperature is relatively constant but skin temperature can vary markedly. (True or false?)

6. Sweat that drips off the body has no cooling effect. (True or false?)

7. Production of "goosebumps" in response to cold exposure has no value in regulating body temperature. (True or false?)

8. The posterior region of the hypothalamus triggers shivering and skin vasoconstriction. (True or false?)

9. The primary means of involuntarily increasing heat production is _____.

10. Increased heat production independent of muscle contraction is known as _____.

11. The only means of heat loss when the environmental temperature exceeds the core temperature is _____.

12. Which of the following statements concerning heat exchange between the body and the external environment is *incorrect*?

 a. Heat gain is primarily by means of internal heat production.

 b. Radiation serves as a means of heat gain but not of heat loss.

 c. Heat energy always moves down its concentration gradient from warmer to cooler objects.

 d. The temperature gradient between the skin and the external air is subject to control.

 e. Very little heat is lost from the body by conduction alone.

13. Which of the following statements concerning fever production is *incorrect*?

 a. Endogenous pyrogen is released by white blood cells in response to microbial invasion.

 b. The hypothalamic set point is elevated.

 c. The hypothalamus initiates cold-response mechanisms to increase the body temperature.

 d. Prostaglandins appear to mediate the effect.

 e. The hypothalamus is not effective in regulating body temperature during a fever.

14. Using the answer code at the right, indicate which mechanism of heat transfer is being described:

 ___ 1. sitting on a cold metal chair
 ___ 2. sunbathing on the beach
 ___ 3. a gentle breeze
 ___ 4. sitting in front of a fireplace
 ___ 5. sweating
 ___ 6. riding in a car with the windows open
 ___ 7. lying on an electric blanket
 ___ 8. sitting in a wet bathing suit
 ___ 9. fanning yourself
 ___ 10. immersion in cool water

 a. radiation
 b. conduction
 c. convection
 d. evaporation

Essay Questions

1. Differentiate between external and internal work.

2. Define metabolic rate and basal metabolic rate. Explain the process of indirect calorimetry.

3. Describe the three states of energy balance.

4. By what means is energy balance primarily maintained?

5. List the sources of heat input and output for the body.

6. What are the two hypothalamic centers for temperature regulation?

7. Discuss the compensatory measures that occur in response to a fall in core temperature as a result of cold exposure and in response to a rise in core temperature as a result of heat exposure.

POINTS TO PONDER

1. Explain how drugs that selectively inhibit CCK increase feeding behavior in experimental animals.

2. What advice would you give an overweight friend who asks for your help in designing a safe, sensible, inexpensive program for losing weight?

3. Why is it dangerous to engage in heavy exercise on a hot, humid day?

4. Describe the avenues for heat loss in a person soaking in a hot bath.

5. Humans, other mammals, and birds are *thermoregulators;* they are able to maintain a remarkably constant, rather high internal body temperature despite the body's exposure to a wide range of environmental temperatures. To maintain thermal homeostasis, thermoregulators physiologically manipulate mechanisms within their bodies to adjust heat production, heat conservation, and heat loss. In contrast, with few exceptions reptiles, amphibians, fish, and invertebrates are *thermoconformers;* their body temperatures conform to the temperature of their surroundings. Thus, their body temperatures vary capriciously with changes in the environmental temperature. Even though thermoconformers produce heat, these species cannot physiologically regulate internal heat production, nor can they control heat exchange with their environment to maintain a constant body temperature when the temperature in their surroundings rises or falls. Knowing this, do you think snakes and other thermoconformers run a fever when they have a systemic infection? Why or why not?

18 Principles of Endocrinology: Central Endocrine Glands

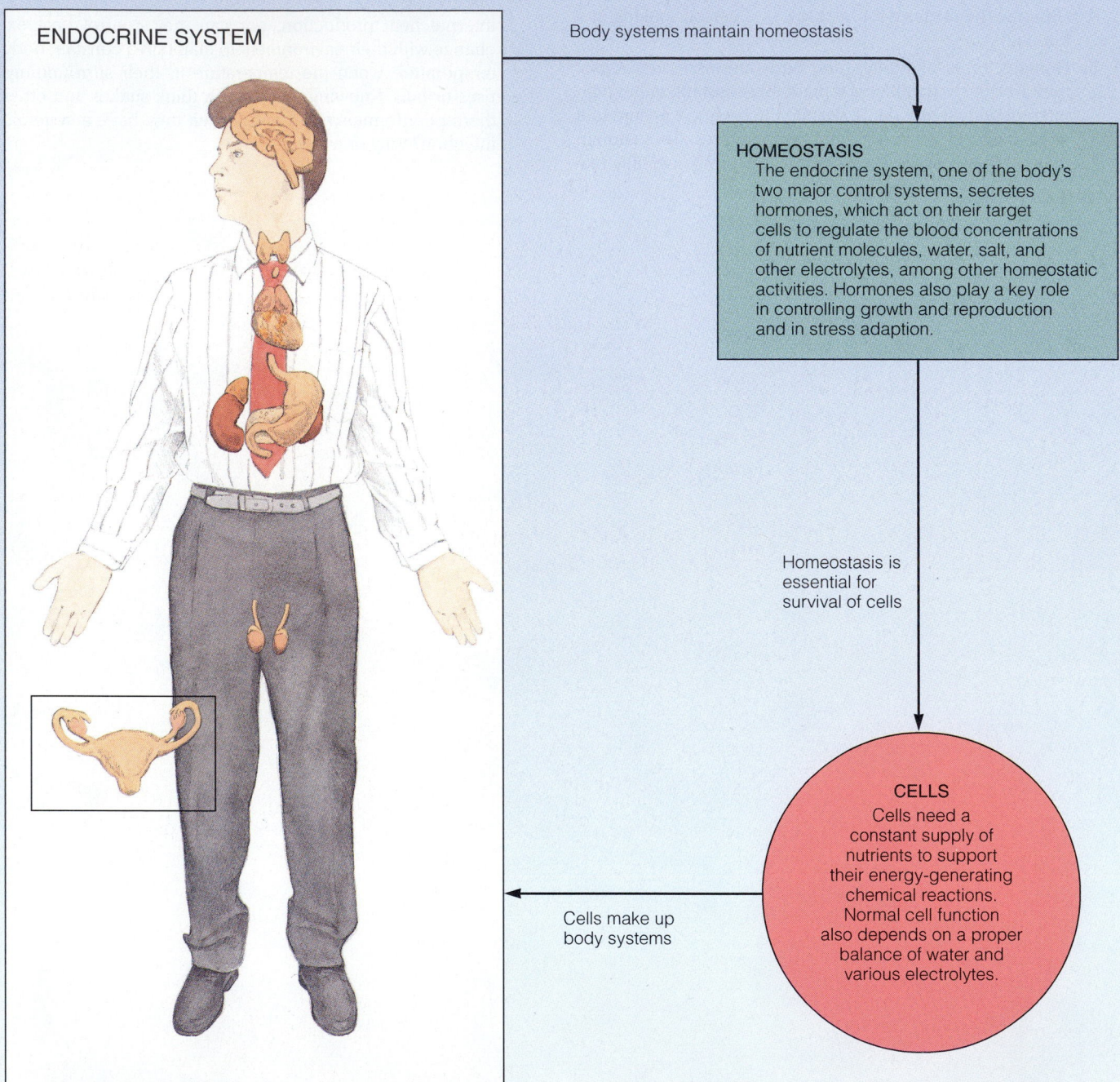

ENDOCRINE SYSTEM

Body systems maintain homeostasis

HOMEOSTASIS
The endocrine system, one of the body's two major control systems, secretes hormones, which act on their target cells to regulate the blood concentrations of nutrient molecules, water, salt, and other electrolytes, among other homeostatic activities. Hormones also play a key role in controlling growth and reproduction and in stress adaption.

Homeostasis is essential for survival of cells

CELLS
Cells need a constant supply of nutrients to support their energy-generating chemical reactions. Normal cell function also depends on a proper balance of water and various electrolytes.

Cells make up body systems

The **endocrine system** generally regulates activities that require duration rather than speed. Endocrine glands release **hormones,** blood-borne chemical messengers which act on target cells that are usually located a long distance from the endocrine gland. Most of the target cells' activities under the control of hormones are directed toward maintaining homeostasis. The central endocrine glands include the hypothalamus and pituitary gland. The hypothalamus, a part of the brain, and the posterior pituitary gland act as a unit to release hormones essential for maintaining water balance and for giving birth and breast-feeding. The hypothalamus also secretes regulatory hormones that control the hormonal output of the anterior pituitary gland. The anterior pituitary gland secretes six hormones, which in turn control the hormonal output of several peripheral endocrine glands. One anterior pituitary hormone, growth hormone, promotes growth and influences the blood concentrations of glucose and amino acids.

INTRODUCTION
GENERAL PRINCIPLES OF ENDOCRINOLOGY
Categories of hormones: peptides, amines, and steroids
Factors affecting plasma concentration of hormones
● CONCEPTS, CHALLENGES, AND CONTROVERSIES Tinkering with
  Our Biological Clocks
Types of endocrine disorders
HYPOTHALAMUS AND PITUITARY
Hypothalamus–posterior pituitary relationship
● A CLOSER LOOK AT EXERCISE PHYSIOLOGY The Endocrine
  Response to the Challenge of Combined Heat and
  Marching Feet
Hypothalamus–anterior pituitary relationship
ENDOCRINE CONTROL OF GROWTH
Factors influencing growth
Functions and control of growth hormone
● CONCEPTS, CHALLENGES, AND CONTROVERSIES Growth and
  Youth in a Bottle?
Role of other hormones in growth

INTRODUCTION

Communication is critical for the survival of the society of cells that collectively compose the body. The ability of cells to communicate with each other is essential for coordination of their diverse activities to maintain homeostasis as well as to control growth and development of the body as a whole. The **endocrine system**, which secretes blood-borne chemical messengers known as hormones, specializes in intercellular ("between cell") communications.

There are three types of cell-to-cell communication (● Fig. 18–1):

1. The most intimate means of intercellular communication is through gap junctions, which are minute tunnels that bridge the cytoplasm of neighboring cells in some types of tissue (see p. 55). Through these specialized anatomical arrangements, small molecules and ions are directly exchanged between interacting cells without ever entering the extracellular fluid (ECF). Gap junctions are especially important in permitting electrical signals to spread from one cell to the next in cardiac and smooth muscle.

● **FIGURE 18–1 Types of Intercellular Communication** (a) Gap junctions. (b) Transient direct linkup of cells. (c) Paracrine secretion. (d) Neurotransmitter secretion. (e) Hormonal secretion. (f) Neurohormone secretion. Paracrines, neurotransmitters, hormones, and neurohormones are all intercellular chemical messengers that accomplish indirect communication between cells.

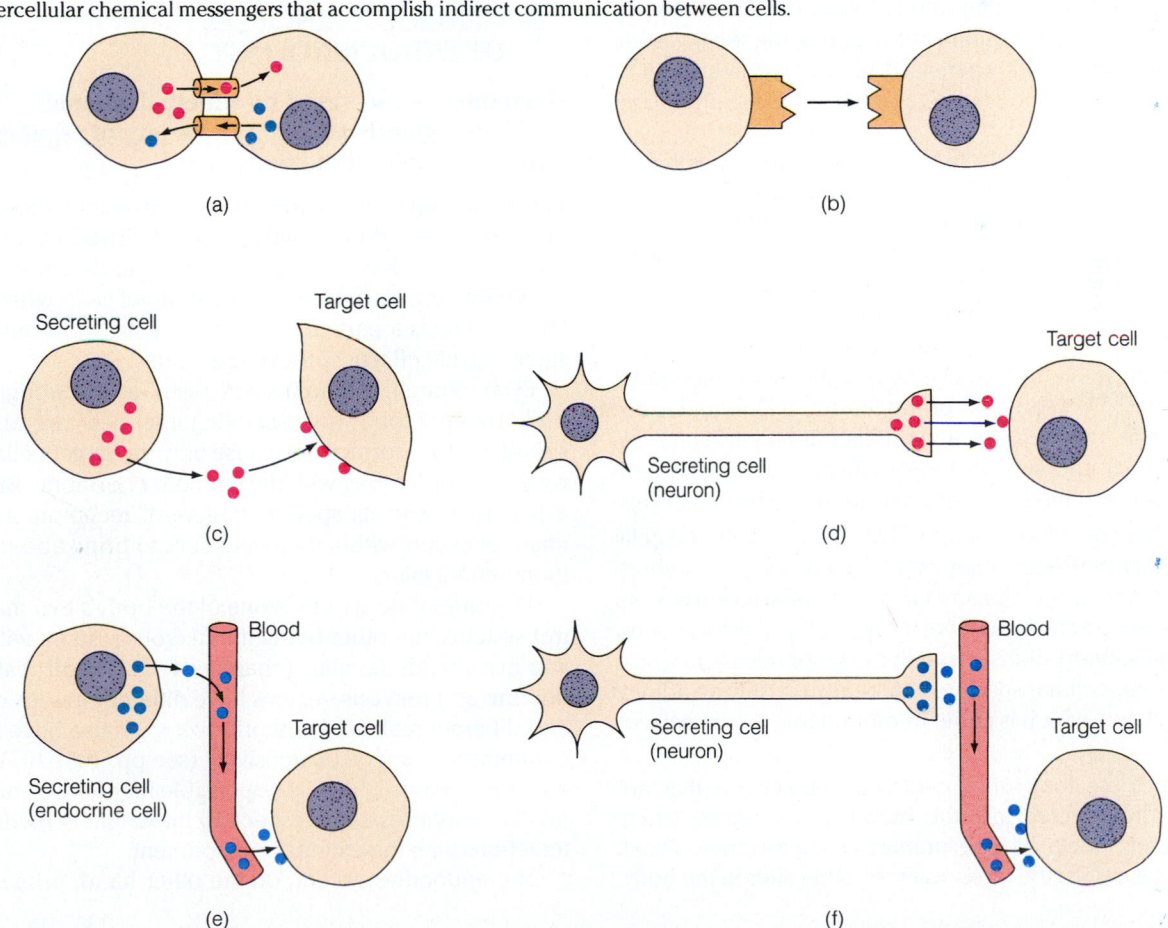

2. The presence of signaling molecules on the surface membrane of some cells permits them to directly link up and interact with certain other cells in a specialized way. This is the means by which the phagocytes of the body's defense system specifically recognize and selectively destroy only undesirable cells, such as microbial invaders, while leaving the body's own cells alone.

3. The most common means by which cells communicate with each other is through intercellular chemical messengers, of which there are four types: paracrines, neurotransmitters, hormones, and neurohormones. In each case, a specific chemical messenger is synthesized by specialized cells to serve a designated purpose. Upon being released into the ECF by appropriate stimulation, these signaling agents act on other particular cells, the messenger's **target cells**, in a prescribed manner. These four types of chemical messengers differ in their source and the distance and means by which they get to their site of action.

■ **Paracrines** are local chemical messengers whose effect is exerted only on neighboring cells in the immediate environment of their site of secretion. Since paracrines are distributed by simple diffusion, their action is restricted to short distances. They do not gain entry to the blood in any significant quantity because they are rapidly inactivated by locally existing enzymes. One example of a paracrine is histamine, which is released from mast cells during an inflammatory response within an invaded or injured tissue (see p. 371). Among other things, histamine acts on the arteriolar muscle in the vicinity to produce the local vasodilation and subsequent increased blood flow necessary to bring additional blood-borne combat supplies into the affected area.

Paracrines must be distinguished from chemicals that influence neighboring cells after being nonspecifically released during the course of cellular activity. For example, an increased local concentration of CO_2 in an exercising muscle is among the factors that promote local vasodilation by acting directly on the arteriolar smooth muscle in the vicinity (see p. 312). The resultant increased blood flow helps to meet the more active tissue's increased metabolic demands. However, CO_2 is produced by all cells and is not specifically released to accomplish this particular response, so it and similar nonspecifically released chemicals such as H^+ and K^+ are not considered to be paracrines.

■ Nerve cells (neurons) communicate directly with the cells they innervate (their target cells) by releasing very short-range chemical messengers known as **neurotransmitters** in response to action potentials (see p. 91). Like paracrines, neurotransmitters diffuse from their site of release across a narrow extracellular space to act locally on only an adjoining target cell, which is either another neuron, a muscle, or a gland.

■ **Hormones** are long-range chemical messengers that are specifically secreted into the blood by endocrine (ductless) glands (see p. 5) in response to an appropriate signal. The blood carries the messengers to other sites in the body, where they exert their effects on their target cells some distance away from their site of release.

■ **Neurohormones** are hormones released into the blood specifically by neurosecretory neurons. Like ordinary neurons, **neurosecretory neurons** possess axons and dendrites and can respond to and conduct action potentials. Instead of directly innervating target cells, however, a neurosecretory neuron releases its chemical messenger, a neurohormone, into the blood upon appropriate stimulation. The neurohormone is then distributed through the blood to the target cells. Thus, like endocrine cells, neurosecretory neurons release blood-borne chemical messengers, whereas ordinary neurons secrete short-range neurotransmitters into a confined space.

Neurotransmitters and neurohormones must be distinguished from neuromodulators (see p. 98). The first two chemical messengers are *secreted* by neurons. The latter messenger *acts* on neurons at nonsynaptic sites to bring about long-term biochemical changes in the nerve cell. A neuromodulator may be from any source, such as a locally acting neural secretion or a hormone acting at long distance.

The major topic of this chapter will be the endocrine system and its blood-borne hormonal and neurohormonal messengers. The general term "hormone" will tacitly include neurohormones.

GENERAL PRINCIPLES OF ENDOCRINOLOGY

Hormones are secreted by widely dispersed endocrine glands and exert a variety of regulatory effects throughout the body.

Endocrinology is the study of the homeostatic chemical adjustments and other activities accomplished by hormones, the secretions of the body's endocrine glands. Once secreted, a hormone travels in the blood to its target cells, where it regulates or directs a particular function. Some hormones have a single target-cell type; others have many.

Even though hormones are distributed throughout the body by the blood, only specific target cells are able to respond to each hormone because only the target cells possess receptors for binding with the particular hormone. Binding of a hormone with its specific target-cell receptors initiates a chain of events within the target cells to bring about the hormone's final effect.

The endocrine system is one of the body's two major control systems, the other being the nervous system, with which you are already familiar (chapters 4–7). Recall that the endocrine and nervous systems have different modes of action and different realms of authority, yet they also have much in common and interact extensively (see pp. 105–107). In general, the nervous system is responsible for coordinating rapid, precise responses and is especially important in mediating interactions with the external environment.

The endocrine system, on the other hand, primarily controls activities that require duration rather than speed, including the following:

1. Regulating organic metabolism and H_2O and electrolyte balance, which are important collectively in maintaining a constant internal environment.
2. Inducing adaptive changes to help the body cope with stressful situations.
3. Promoting smooth, sequential growth and development.
4. Controlling reproduction.
5. Regulating red blood cell production.
6. Along with the autonomic nervous system, controlling and integrating both circulation and the digestion and absorption of food.

The endocrine system is composed of endocrine glands that are scattered throughout the body (● Fig. 18–2). Even though the endocrine glands for the most part are not connected anatomically, they constitute a system in a functional sense. They all accomplish their functions by secreting hormones, and many functional interactions take place among the various endocrine glands.

In fact, the sole function of some hormones is regulating the production and secretion of another hormone. A hormone that has as its primary function the regulation of hormone secretion by another endocrine gland is classified functionally as a **tropic hormone**. For example, the only function of thyroid-stimulating hormone (TSH) from the anterior pituitary gland is the regulation of thyroid hormone secretion by the thyroid gland. A **nontropic hormone**, on the other hand, primarily exerts its effects on nonendocrine target tissues. Thyroid hormone, which increases the rate of O_2 consumption and the metabolic activity of almost every cell in the body, is an example of a nontropic hormone.

● **FIGURE 18–2   Endocrine System**

Solely endocrine function

Mixed function

Complete function uncertain

Pineal

Hypothalamus

Pituitary

Parathyroid

Thyroid

Thymus

Heart

Stomach

Adrenal gland

Pancreas

Duodenum

Kidney

Skin

Ovaries in female

Placenta in pregnant female

Testes in male

The following points add to the complexity of the endocrine system:

- A single endocrine gland may produce multiple hormones. The anterior pituitary, for example, secretes six different hormones; each is under different control mechanisms and has different functions, some being tropic and others having nontropic effects.

- A single hormone may be secreted by more than one endocrine gland. For example, the hypothalamus and pancreas both secrete the hormone somatostatin.

- Frequently, a single hormone has more than one type of target cell and therefore can induce more than one type of effect. As an example, vasopressin promotes H_2O reabsorption by the kidney tubules as well as vasoconstriction of arterioles throughout the body.

- A single target cell may be influenced by more than one hormone. Some cells contain an array of receptors for responding in different ways to different hormones. To illustrate, insulin promotes the conversion of glucose into glycogen within liver cells by stimulating one particular hepatic enzyme, whereas another hormone, glucagon, enhances the degradation of glycogen into glucose within liver cells by activating yet another hepatic enzyme.

- The same chemical messenger may be either a hormone or a neurotransmitter, depending on its source and mode of delivery to the target cell. Norepinephrine, which is secreted as a hormone by the adrenal medulla and released as a neurotransmitter from sympathetic postganglionic nerve fibers, is a prime example.

- Some organs are exclusively endocrine in function (they specialize in hormonal secretion alone, the anterior pituitary and thyroid glands being examples), whereas other organs of the endocrine system perform nonendocrine functions in addition to secreting hormones. For example, the testes produce sperm and also secrete the male sex hormone testosterone. Other examples of mixed organs are the ovaries, digestive tract, pancreas, kidneys, and even the brain. In each case except the brain, mixed function occurs because the organ houses nonendocrine tissue plus isolated clusters of endocrine cells that migrated to the organ during embryonic development. The endocrine function of the brain derives from the presence of neurosecretory neurons. There are no endocrine cells as such in the brain.

Because hormones are widely dispersed through the circulatory system, those that have multiple target-cell types are able to coordinate the activities of various tissues toward a common goal. Tissues such as muscle, liver, and fat all respond to hormones in a fashion that ensures that their contributions to overall intermediary metabolism are acting in concert.

Endocrine systems also provide temporal (time) coordination of function. This is particularly apparent in the endocrine control of reproductive cycles, such as the menstrual cycle, in which normal function requires highly specific patterns of change in the secretion of various hormones.

This has been a brief overview of the general functions of the endocrine system. Table 18–1 summarizes the most important specific functions of the major hormones. Some of these hormones have been introduced elsewhere and will not be discussed further; these are the gastrointestinal hormones (chapter 16), the renal hormones (erythropoietin in chapter 11 and renin in chapter 14), atrial natriuretic peptide from the heart (chapter 15), and thymosin (chapter 12). The remainder of the hormones will be described in greater detail in this and the next two chapters.

As extensive as Table 18–1 appears to be, it leaves out a variety of "candidate" or potential hormones that have not fully qualified as hormones, either because their characteristics do not quite fit the classic definition of a hormone or because they have been discovered so recently that their hormonal status has not yet been conclusively documented. The table also excludes the hormones secreted by the effector cells of the defense system (white blood cells and macrophages) and a variety of recently revealed and poorly understood growth factors that promote growth of specific tissues, such as *epidermal growth factor* and *nerve growth factor*. In addition, the table is undoubtedly incomplete in ways as yet unknown. First, it is more than likely that additional hormones will be discovered; the list of hormones still continues to grow as a result of the probing techniques of endocrinologists. Second, an exciting finding is that hormones with well-established functions sometimes exert other, quite unrelated functions as well. As an example, researchers determined that antidiuretic hormone, which controls H_2O retention by the kidneys during the formation of urine, also exerts vasoconstrictor effects on arteriolar smooth muscle, giving rise to its more commonly used alternative name of vasopressin. More recently, vasopressin has also been implicated as playing roles in fever, learning, memory, and behavior.

Hormones are chemically classified into three categories: peptides, amines, and steroids.

Hormones are not all similar chemically, but instead fall into three distinct classes according to their biochemical structure (Table 18–2): (1) peptides and proteins, (2) amines, and (3) steroids. The first two categories are both amino acid derivatives. The **peptide** and **protein hormones** consist of specific amino acids arranged in a chain of varying length; the shorter chains are peptides and the longer ones are categorized as proteins. For convenience, we will refer to this entire category as peptides. The majority of hormones fall into this class, including those secreted by the hypothalamus, anterior pituitary, posterior pituitary, pancreas, parathyroid gland, gastrointestinal tract, kidneys, liver, thyroid C cells, and heart. The **amines** are derived from the amino acid *tyrosine* and include the hormones secreted by the thyroid gland and adrenal medulla. The adrenomedullary hormones are specifically known as *catecholamines*. The **steroids**, which include the hormones secreted by the adrenal cortex, gonads, and most placental hormones, are neutral lipids derived from cholesterol.

TABLE 18–1 **Summary of Major Hormones**

ENDOCRINE GLAND	HORMONES	TARGET CELLS	MAJOR FUNCTIONS OF HORMONES
Hypothalamus	Releasing and inhibiting hormones (TRH, CRH, GnRH, GHRH, GHIH, PRH, PIH)	Anterior pituitary	Controls release of anterior pituitary hormones
Posterior pituitary (hormones stored in)	Vasopressin (antidiuretic hormone)	Kidney tubule	Increases H_2O reabsorption
		Arterioles	Produces vasoconstriction
	Oxytocin	Uterus	Increases contractility
		Mammary glands (breasts)	Causes milk ejection
Anterior pituitary	Thyroid-stimulating hormone (TSH)	Thyroid follicular cells	Stimulates T_3 and T_4 secretion
	Adrenocorticotropic hormone (ACTH)	Zona fasciculata and zona reticularis of adrenal cortex	Stimulates cortisol secretion
	Growth hormone	Bone; soft tissues	Essential but not solely responsible for growth; stimulates growth of bones and soft tissues; metabolic effects include protein anabolism, fat mobilization, and glucose conservation
		Liver	Stimulates somatomedin secretion
	Follicle-stimulating hormone (FSH)	Females: ovarian follicles	Promotes follicular growth and development; stimulates estrogen secretion
		Males: seminiferous tubules in testes	Stimulates sperm production
	Luteinizing hormone (LH) (interstitial cell-stimulating hormone—ICSH)	Females: ovarian follicle and corpus luteum	Stimulates ovulation, corpus luteum development, and estrogen and progesterone secretion
		Males: interstitial cells of Leydig in testes	Stimulates testosterone secretion
	Prolactin	Females: mammary glands	Promotes breast development; stimulates milk secretion
		Males	Uncertain
Thyroid gland follicular cells	Tetraiodothyronine (T_4 or thyroxine); triiodothyronine (T_3)	Most cells	Increases metabolic rate; essential for normal growth and nerve development
Thyroid gland C cells	Calcitonin	Bone	Decreases plasma calcium concentration
Adrenal cortex			
Zona glomerulosa	Aldosterone (mineralocorticoid)	Kidney tubules	Increases Na^+ reabsorption and K^+ secretion
Zona fasiculata and zona reticularis	Cortisol (glucocorticoid)	Most cells	Increases blood glucose at the expense of protein and fat stores; contributes to stress adaptation
	Androgens (dehydroepiandrosterone)	Females: bone and brain	Responsible for pubertal growth spurt and sex drive in females
Adrenal medulla	Epinephrine and norepinephrine	Sympathetic receptor sites throughout the body	Reinforces sympathetic nervous system; contributes to stress adaptation and blood-pressure regulation

(continued)

TABLE 18–1   Summary of Major Hormones (continued)

ENDOCRINE GLAND	HORMONES	TARGET CELLS	MAJOR FUNCTIONS OF HORMONES
Endocrine pancreas (islets of Langerhans)	Insulin (β cells)	Most cells	Promotes cellular uptake, utilization, and storage of absorbed nutrients
	Glucagon (α cells)	Most cells	Important for maintenance of nutrient levels in blood during postabsorptive state
	Somatostatin (D cells)	Digestive system	Inhibits digestion and absorption of nutrients
		Pancreatic islet cells	Inhibits secretion of all pancreatic hormones
Parathyroid gland	Parathyroid hormone (PTH)	Bone, kidneys, intestine	Increases plasma calcium concentration; decreases plasma phosphate concentration; stimulates vitamin D activation
Gonads			
Female: ovaries	Estrogen (estradiol)	Female sex organs; body as a whole	Promotes follicular development; responsible for development of secondary sex characteristics; stimulates uterine and breast growth
		Bone	Promotes closure of epiphyseal plate
	Progesterone	Uterus	Prepares for pregnancy
Male: testes	Testosterone	Male sex organs; body as a whole	Stimulates sperm production; responsible for development of secondary sex characteristics; promotes sex drive
		Bone	Enhances pubertal growth spurt; promotes closure of epiphyseal plate
Testes and ovaries	Inhibin	Anterior pituitary	Inhibits secretion of follicle-stimulating hormone
Pineal gland	Melatonin	Anterior pituitary; reproductive organs	Believed to inhibit gonadotropins; initiation of puberty possibly caused by reduction in melatonin secretion
Placenta	Estrogen (estriol); progesterone	Female sex organs	Help maintain pregnancy; prepare breasts for lactation
	Chorionic gonadotropin	Ovarian corpus luteum	Maintains corpus luteum of pregnancy
Kidneys	Renin (→ angiotensin)	Zona glomerulosa of adrenal cortex (acted on by angiotensin, which is activated by renin)	Stimulates aldosterone secretion
	Erythropoietin	Bone marrow	Stimulates erythrocyte production
Stomach	Gastrin	Digestive-tract exocrine glands and smooth muscles; pancreas; liver; gallbladder	Control of motility and secretion to facilitate digestive and absorptive processes
Duodenum	Secretin; cholecystokinin; gastric inhibitory peptide		
Liver	Somatomedins	Bone; soft tissues	Promotes growth
Skin	Vitamin D	Intestine	Increases absorption of ingested calcium and phosphate
Thymus	Thymosin	T lymphocytes	Enhances T lymphocyte proliferation and function
Heart	Atrial natriuretic peptide	Kidney tubules	Inhibits Na⁺ reabsorption

TABLE 18–2   Chemical Classification of Hormones

PROPERTIES	PEPTIDES	AMINES		STEROIDS
		Catecholamines	Thyroid Hormone	
Structure	Chains of specific amino acids, for example:	Tyrosine derivative, for example:	Iodinated tyrosine derivative, for example:	Cholesterol derivative, for example:
	Cys¹—s—s—Cys⁶—Pro⁷—Arg⁸—Gly⁹NH₂ Tyr² Asn⁵ Phe³———Gln⁴ (vasopressin)	(epinephrine)	(thyroxine, T₄)	(cortisol)
Solubility	Hydrophilic (lipophobic)	Hydrophilic (lipophobic)	Lipophilic (hydrophobic)	Lipophilic (hydrophobic)
Synthesis	In rough endoplasmic reticulum; packaged in Golgi complex	In cytosol	In colloid, an inland extracellular site	Stepwise modification of cholesterol molecule in various intracellular compartments
Storage	Large amounts in secretory granules	In chromaffin granules	In colloid	Not stored; cholesterol precursor stored in lipid droplets
Secretion	Exocytosis of granules	Exocytosis of granules	Endocytosis of colloid	Simple diffusion
Transport in blood	As free hormone	Half bound to plasma proteins	Mostly bound to plasma proteins	Mostly bound to plasma proteins
Receptor site	Surface of target cell	Surface of target cell	Inside target cell	Inside target cell
Mechanism of action	Channel changes or activation of second messenger system to alter activity of preexisting proteins that produce the effect	Activation of second messenger system to alter activity of preexisting proteins that produce the effect	Activation of specific genes to produce new proteins that produce the effect	Activation of specific genes to produce new proteins that produce the effect
Hormones of this type	All hormones from the hypothalamus, anterior pituitary, posterior pituitary, pancreas, parathyroid gland, gastrointestinal tract, kidneys, liver, thyroid C cells, heart	Only hormones from the adrenal medulla	Only hormones from the thyroid follicular cells	Hormones from the adrenal cortex and gonads plus most placental hormones (vitamin D is steroidlike)

Minor differences in chemical structure between hormones within each category often result in profound differences in biological response. For example, note the subtle difference (● Fig. 18–3) between testosterone, the male sex hormone responsible for inducing the development of masculine characteristics, and estradiol, the predominant form of estrogen, which is the feminizing female sex hormone.

The structural classification of hormones is of more than biochemical interest. The means by which a hormone is synthesized, stored, and secreted; the way it is transported in the blood; and the mechanism by which it exerts its effects at the target cell all depend on its chemical properties, the most notable being its solubility. The following differences in the solubility of the various types of hormones are critical to their function (Table 18–2):

1. All peptides and catecholamines are hydrophilic (water-loving) and lipophobic (lipid-fearing); that is, they are highly H₂O soluble and have low lipid solubility.

2. All steroid and thyroid hormones are lipophilic (lipid-loving) and hydrophobic (water-fearing); that is, they have high lipid solubility and are poorly soluble in H₂O.

OH ... OH

O ... HO

Testosterone,
a masculinizing
hormone

Estradiol,
a feminizing
hormone

• **FIGURE 18-3 Comparison of Testosterone and Estradiol**

The mechanisms of hormone synthesis, storage, and secretion vary according to the class of hormone.

Because of their chemical differences, the means by which the various classes of hormones are synthesized, stored, and secreted differ as follows.

Peptide Hormones Peptide hormones are synthesized by the same method used for the manufacture of any protein that is to be exported (see p. 19). Because they are destined to be released from the endocrine cell, the synthesized hormones must be segregated from intracellular proteins by being sequestered in a membrane-bounded compartment until they are secreted. Briefly, the synthesis of peptide hormones requires the following steps:

1. Large precursor proteins, or **preprohormones**, are synthesized by ribosomes on the rough endoplasmic reticulum. They then migrate to the Golgi complex in membrane-bounded vesicles that pinch off from the smooth endoplasmic reticulum.

2. During their journey through the endoplasmic reticulum and Golgi complex, the large preprohormone precursor molecules are pruned first to **prohormones** and finally to **active hormones**. The peptide "scraps" that are left over as a large preprohormone molecule is cleaved to form the classic hormone are often stored and cosecreted along with the hormone. This raises the possibility that these other peptides may also exert biological effects that differ from the traditional hormonal product; that is, the cell may actually be secreting multiple hormones, but the functions of the other peptide products are for the most part unknown. A known example is the cleavage of the large precursor molecule *pro-opiomelanocortin* into three active products: *adrenocorticotropic hormone (ACTH), melanocyte-stimulating hormone (MSH),* and β-*endorphin.*

3. The Golgi complex concentrates the finished hormones, then packages them into secretory vesicles that are pinched off and stored in the cytoplasm until an appropriate signal triggers their secretion. By storing peptide hormones in a readily releasable form, the gland can respond rapidly to any demands for increased secretion without the necessity of first increasing hormone synthesis.

4. Upon appropriate stimulation, the secretory vesicles fuse with the plasma membrane and release their contents to the outside by the process of exocytosis (see p. 24). Such secretion usually does not go on continuously; it is triggered only by specific stimuli. The secreted hormone is subsequently picked up by the blood for distribution.

Steroid Hormones The following steps are performed by all *steroidogenic* (steroid-producing) cells to produce and release their hormonal product:

1. Cholesterol is the common precursor for all steroid hormones. Although steroidogenic cells synthesize some cholesterol on their own, most of this raw material is derived from the low-density lipoproteins (LDL) that have been internalized into the cell and degraded by lysosomal enzymes to liberate free cholesterol (see p. 293). The uptake and degradation of LDL are regulated so that more cholesterol is made available to steroidogenic cells at times of increased need for steroid hormones. Furthermore, unused cholesterol may be chemically modified and stored in large amounts as lipid droplets within steroidogenic cells. Conversion of this major storage form of cholesterol into free cholesterol for use in steroid-hormone production is also subject to control. Thus, the provision of free cholesterol for use by steroidogenic cells can be closely coordinated with the body's overall need for the hormone product.

2. Synthesis of the various steroid hormones from cholesterol requires a series of enzymatic reactions that modify the basic cholesterol molecule, for example, by varying the type and position of side groups attached to the cholesterol framework or the degree of saturation within the steroid ring (• Fig. 18-4). Each of the conversions from cholesterol to a specific steroid hormone requires the assistance of a number of enzymes that are limited to certain steroidogenic organs. Accordingly, each steroidogenic organ is able to produce only the steroid hormone or hormones for which it has a complete set of appropriate enzymes. For example, a key enzyme necessary for the production of cortisol is found only in the adrenal cortex, so no other steroidogenic organ is able to produce this hormone. Each of the enzymes necessary for the conversion of cholesterol into a steroid hormone is localized in a specific intracellular compartment, such as the mitochondria or endoplasmic reticulum. The steroid molecule is therefore shuttled back and forth by unknown means between different compartments within the steroidogenic cell for step-by-step modification until the final secretory product is formed.

3. Unlike peptide hormones, steroid hormones are not stored after their formation. Once formed, the lipid-soluble steroid hormones immediately diffuse through the steroidogenic cell's lipid plasma membrane to enter the blood. Only the hormone precursor cholesterol is stored in significant quantities within steroidogenic cells.

~~~ = Intermediates not biologically active in humans

● **FIGURE 18–4 Steroidogenic Pathways for Major Steroid Hormones** All steroid hormones are produced through a series of enzymatic reactions that modify cholesterol molecules, such as by varying the side groups attached to them. Each steroidogenic organ can produce only those steroid hormones for which it has a complete set of the enzymes needed to appropriately modify cholesterol. For example, the testes have the enzymes necessary to convert cholesterol into testosterone (male sex hormone) whereas the ovaries possess the enzymes needed to yield progesterone and the various estrogens (female sex hormones).

Accordingly, the rate of steroid-hormone secretion is controlled entirely by the rate of hormone synthesis. In contrast, peptide-hormone secretion is controlled primarily by regulating the release of presynthesized, stored hormone.

4. Following their secretion into the blood, some steroid hormones undergo further interconversions within the blood or other organs, where they are converted into more potent or different hormones.

Amines The amines—thyroid hormone and adrenomedullary catecholamines—have unique synthetic and secretory pathways that will be thoroughly described when each of these hormones is specifically addressed. However, the amines have the following features in common:

1. They are derived from the naturally occurring amino acid tyrosine.
2. None of the enzymes directly involved in the synthesis of either of these hormone types are located in organelle compartments within the secretory cells.
3. Both types of amines are stored until they are secreted.

Water-soluble hormones are transported dissolved in the plasma, whereas lipid-soluble hormones are largely transported bound to plasma proteins.

All hormones are carried by the blood, but they are not all transported in the same manner. The hydrophilic (water-soluble) peptide hormones are transported simply dissolved in the plasma. However, lipophilic (lipid-soluble) steroids and thyroid hormone, which are poorly soluble in water, cannot dissolve in the aqueous plasma in sufficient quantities to account for their known plasma concentrations. Instead, the majority of the lipophilic hormones circulate in the blood to their target cells reversibly bound to plasma proteins. Some are bound to specific plasma proteins that are designed to carry only one type of hormone, whereas other plasma proteins, such as albumin, nondiscriminately pick up any "hitchhiking" hormone.

In the case of some hormones, 1% or less of the total hormone remains unbound. This figure is noteworthy, because only the small, unbound, freely dissolved fraction of a lipophilic hormone is biologically active (that is, free to cross capillary walls and bind with target-cell receptors to exert an effect). Once a hormone has interacted with a target cell, it is rapidly inactivated or removed so that it is no longer available to interact with another target cell. Since the carrier-bound hormone is in dynamic equilibrium with the free hormone pool, the bound form of steroid and thyroid hormones provides a large reserve of these lipophilic hormones that can be called on to replenish the active free pool. It is the magnitude of the small free effective pool rather than the total plasma concentration of a particular hormone that is monitored and adjusted to maintain normal endocrine function.

Catecholamines are unusual in that only about 50% of these hydrophilic hormones circulate as free hormone, whereas the other 50% are loosely bound to the plasma protein albumin. Because catecholamines are water soluble, the importance of this protein binding is unclear.

The chemical properties of a hormone dictate not only the means by which it is transported in the blood but also the means by which it can be artificially introduced into the blood for therapeutic purposes. Because the digestive system does not secrete enzymes that can digest steroid and thyroid hormones, these hormones, such as the sex steroids contained in birth-control pills, can be absorbed intact from the digestive tract into the blood when taken orally. None of the other types of hormones can be taken orally, because they would be attacked and converted into inactive fragments by protein-digesting enzymes. Therefore, these hormones must be administered when necessary by nonoral routes; for example, insulin deficiency (diabetes mellitus) is treated by daily injections of insulin.

Hormones generally produce their effect by altering intracellular protein activity.

Hormones must bind with target-cell receptors specific for them in order to induce their effect. However, the location of the receptors within the target cell and the mechanism by which the binding of the hormone with the receptors induces a response vary depending on the hormone's solubility characteristics. Hormones can be grouped into two categories based on the location of their receptors (Table 18–2):

1. The hydrophilic peptides and catecholamines, which are poorly soluble in lipid, are not able to pass through the lipid membrane barriers of their target cells. Instead they bind with specific receptors located on the *outer plasma-membrane surface* of the target cell.
2. The lipophilic steroids and thyroid hormone easily pass through the surface membrane to bind with specific receptors located *inside* the target cell.

Each interaction between a particular hormone and a target-cell receptor produces a highly characteristic target-cell response that differs for different hormones and differs between different target cells influenced by the same hormone. For example, one of the adrenomedullary catecholamines, epinephrine, through its ubiquitous distribution and target-cell specialization, simultaneously produces such diverse effects as contraction of vascular smooth muscle, relaxation of respiratory airway smooth muscle, and breakdown of glycogen (stored glucose) in the liver.

Even though hormones elicit a wide variety of biological

● **FIGURE 18–5   Three General Means by Which Hormones Elicit Biological Responses**

responses, all hormones ultimately influence their target cells by altering the cells' protein activity. Hormones exert an effect on their target cells' proteins through three general means: (● Fig. 18–5 and Table 18–2):

1. A few hydrophilic hormones, upon binding with a target cell's surface receptors, bring about changes in the cell's permeability (either opening or closing channels to one or more ions) by *altering the conformation (shape) of adjacent channel-forming proteins already present in the membrane.*
2. Most surface-binding hydrophilic hormones function by *activating second messenger systems* within the target cell. This activation directly *alters the activity of preexisting intracellular proteins, usually enzymes,* to produce the desired effect.
3. All lipophilic hormones function by *activating specific genes in the target cell to cause the formation of new intracellular proteins,* which in turn produce the desired effect.

Even though all hormones ultimately influence their target cells by altering intracellular protein activity, the onset of action varies depending primarily on the mode employed. Hormones that act through a second messenger system to alter a preexisting enzyme's activity elicit full action within a few minutes. In contrast, hormonal responses that require the synthesis of new protein may take up to several hours before any action is initiated. Let us examine each of the two major post-receptor (after the hormone binds with the receptor) events in more detail.

Postreceptor Events: Hydrophilic Hormones Because second messenger systems in general, and cyclic AMP (cAMP) in particular, play such a central role in hydrophilic hormone activity, it is worthwhile to briefly review the steps (see Fig. 3–5, p. 52):

1. Binding of the extracellular first messenger, the hydrophilic hormone, to its surface-membrane receptor activates the membrane-bound enzyme adenylate cyclase, which is located on the cytoplasmic side of the plasma membrane.

2. Activated adenylate cyclase converts intracellular ATP to cAMP, the intracellular second messenger.

3. Cyclic AMP triggers a preprogrammed series of biochemical steps that bring about a change in the shape and function of specific preexisting enzymatic proteins in the cell.

4. These altered enzymatic proteins are responsible for bringing about a change in cell activity. The resultant change is the target cell's ultimate physiological response to the hormone.

Once the hormone is removed, cAMP is converted to inactive cAMP by a specific chemical within the cytoplasm, and the intracellular message is "erased."

The nature of the preexisting enzymatic proteins whose activity is ultimately modified by a second messenger varies in different target cells. Thus, various target cells respond very differently to the universal mechanism of hormonally induced changes in their cAMP levels. Cyclic AMP can "turn on" (or "turn off") different cellular events depending on the different kinds of enzyme activity that are ultimately modified in the different target cells.

Most but not all hydrophilic hormones use cAMP as their second messenger. A few are known to use intracellular Ca^{++} as the second messenger; for others, the second messenger is still unknown.

Postreceptor Events: Lipophilic Hormones All lipophilic hormones (steroids and thyroid hormone) produce their effects in their target cells by enhancing the synthesis of new enzymatic or structural proteins. Their effects result from stimulation of the cell's genes as follows (• Fig. 18–6):

1. Free lipophilic hormone (hormone not bound with its plasma-protein carrier) diffuses through the plasma membrane of the target cell and binds with its specific receptor within the nucleus.

2. Each receptor has a specific region for binding with its hormone and another region for binding with DNA. Once the hormone is bound to the receptor, the hormone-receptor complex binds with DNA at a specific attachment site on DNA known as the **hormone response element** (**HRE**). Different steroid hormones and thyroid hormone, once bound with their respective receptors, attach at different HREs on DNA. For example, the estrogen-receptor complex binds at DNA's estrogen-response element.

3. Binding of the hormone-receptor complex with DNA ultimately "turns on" specific genes within the target cell.

4. The activated genes direct the synthesis of new cell protein by producing complementary messenger RNA, which enters the plasma and binds to a ribosome, the "workbench" that mediates the assembly of new proteins (p. B–7).

5. The new protein produces the target cell's ultimate physiological response to the hormone.

Plasma membrane

Target cell

Nucleus

HRE   DNA

H   R

mRNA

mRNA

H

Physiological response ← Synthesis of new protein

• FIGURE 18–6 Postreceptor Events for Lipophilic Hormones A lipophilic hormone diffuses through the plasma and nuclear membranes of its target cell and binds with a nuclear receptor specific for it. The hormone-receptor complex in turn binds with the hormone response element, a segment of DNA specific for the hormone-receptor complex. DNA binding activates specific genes, which produce complementary messenger RNA. Messenger RNA leaves the nucleus and directs the synthesis of new proteins, which accomplish the target cell's ultimate physiological response to the hormone.

H = Free lipophilic hormone
R = Lipophilic hormone receptor
HRE = Hormone response element
mRNA = Messenger RNA

As a result of this mechanism, different genes are activated by different lipophilic hormones, resulting in different biological effects.

Despite the differences in the molecular events induced by hydrophilic and lipophilic hormones, the two hormone types have several characteristics in common that have important implications for hormonal actions:

■ First, the actions of hormones are greatly amplified at the target cell. Hormones are greatly diluted by the blood and thus must exert their effect at incredibly low concentrations—as low as 1 picogram (10^{-12} gram; 1 millionth of a millionth of a gram) per ml—as opposed to the much higher localized concentration of neurotransmitter at the target cell during neural communication. Interaction of one hormonal molecule with its receptor can result in the formation of many active protein products that ultimately carry out the physiological effect. For example, one peptide hormone results in the production of numerous cAMP messengers, each in turn activating many latent enzymes (see p. 54). Similarly, one steroid-hormone–activated gene induces the formation of many messenger RNA molecules, each of which is used to make many enzymes.

■ A second common feature of hormone action is that hormones regulate the rates of existing reactions instead of initiating new reactions. Enzymes under hormonal control usually show some activity even in the absence of the hormone.

■ Finally, hormone action is relatively slow and prolonged. In general, lipophilic hormones, which act by increasing protein synthesis, take longer to produce an effect than hydrophilic hormones, which need only to activate a preformed enzyme. Once an enzyme is activated, it no longer depends on the presence of the hormone. As a result, a hormone's effect usually lasts for some time after its withdrawal.

The effective plasma concentration of a hormone is normally regulated by changes in its rate of secretion.

The primary function of hormones is the regulation of various homeostatic activities. Because hormones' effects are proportional to their concentrations in the blood, it follows that these concentrations must be subject to control according to homeostatic need. The plasma concentration of free, biologically active hormone—and thus the hormone's availability to its receptors—depends on several factors (● Fig. 18–7): (1) the hormone's rate of secretion into the blood by the endocrine gland, (2) its rate of removal from the blood by metabolic inactivation and excretion in the urine, (3) for lipophilic hormones, its extent of binding to plasma proteins, and (4) for a few hormones, its rate of metabolic activation.

Normally, the effective plasma concentration of a hormone is regulated by appropriate adjustments in the rate of its secretion. Endocrine glands do not secrete their hormones at a constant rate; the secretion rates of all hormones vary subject to control, often by a combination of several complex mechanisms. The regulatory system for each hormone will be considered in detail in subsequent sections. As you proceed through the discussion of the endocrine system, note the fol-

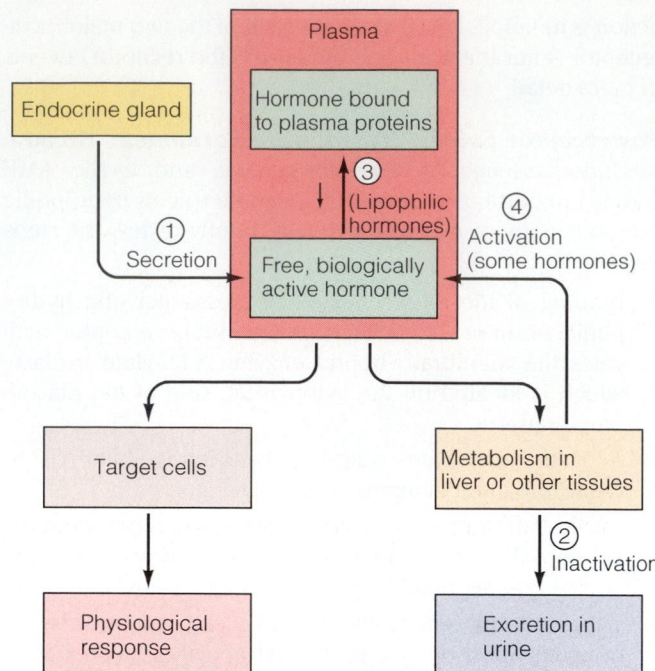

● **FIGURE 18–7   Factors Affecting the Plasma Concentration of Free, Biologically Active Hormone** The plasma concentration of free, biologically active hormone, which can interact with its target cells to produce a physiological response, depends on the hormone's rate of secretion by the endocrine gland ①, its rate of metabolic inactivation and excretion ②, its extent of binding to plasma proteins (for lipophilic hormones) ③, and its rate of metabolic activation (for a few hormones) ④.

lowing general mechanisms of controlling secretion—they are common to many different hormones.

Negative-Feedback Control   Negative feedback is a prominent feature of hormonal control systems (see p. 10). Stated simply, *negative feedback exists when the output of a system opposes a change in input*. Negative feedback maintains the plasma concentration of a hormone at a given level, similar to the way in which a home heating system maintains the room temperature at a given set point. Control of hormonal secretion provides some classic physiological examples of

● **FIGURE 18–8   Negative-Feedback Control**

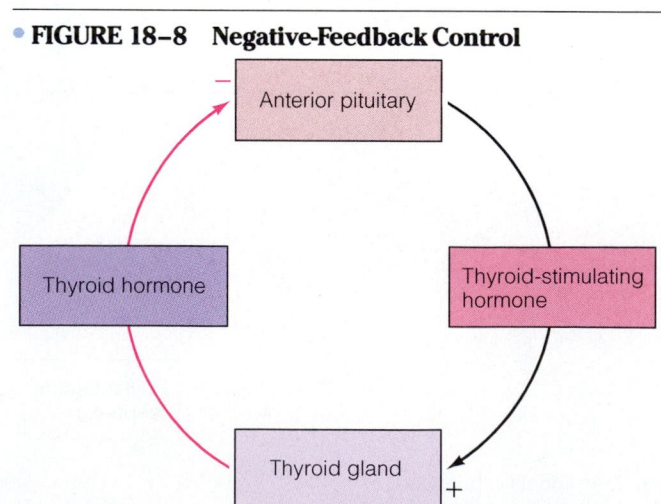

negative feedback. For example (● Fig. 18–8), the anterior pituitary secretes thyroid-stimulating hormone (TSH), which stimulates the thyroid to secrete thyroid hormone. Thyroid hormone in turn inhibits further secretion of TSH by the anterior pituitary. Negative feedback assures that once thyroid-gland secretion has been "turned on" by TSH, it will not continue unabated but instead will be "turned off" when the appropriate level of free circulating thyroid hormone has been achieved. Thus, the effect of a particular hormone's actions can bring about inhibition of its own secretion. The feedback loops often become quite complex.

Neuroendocrine Reflexes Many endocrine control systems involve **neuroendocrine reflexes**, which include neural as well as hormonal components. The purpose of such reflexes is to produce a sudden increase in hormone secretion (that is, "turn up the thermostat setting") in response to a specific stimulus, frequently a stimulus external to the body. The nervous system can influence hormonal secretion by four different means, each of which will be elaborated upon when the specific hormones involved are discussed (● Fig. 18–9). In some instances, neural input to the endocrine gland is the only factor regulating secretion of the hormone. For example,

● **FIGURE 18–9   Pathways by Which Neural Input Influences Hormonal Secretion** The nervous system influences hormonal secretion by four different means: (1) The hypothalamus actually produces the hormones that are stored in the posterior pituitary and released from this endocrine gland in response to hypothalamic stimulation. (2) The release of hormones produced by the anterior pituitary is controlled in part by hypothalamic regulatory hormones that reach the anterior pituitary by a special vascular portal system. The anterior pituitary hormones in turn control the release of a number of other hormones from other endocrine glands. (3) The release of hormones produced by the adrenal medulla, which consists of modified sympathetic postganglionic neurons, is entirely under control of sympathetic preganglionic neurons. (4) Postganglionic autonomic neurons are among many regulatory inputs to the endocrine cells of the pancreas. Gastrointestinal hormones are influenced by the autonomic nervous system in a similar way.

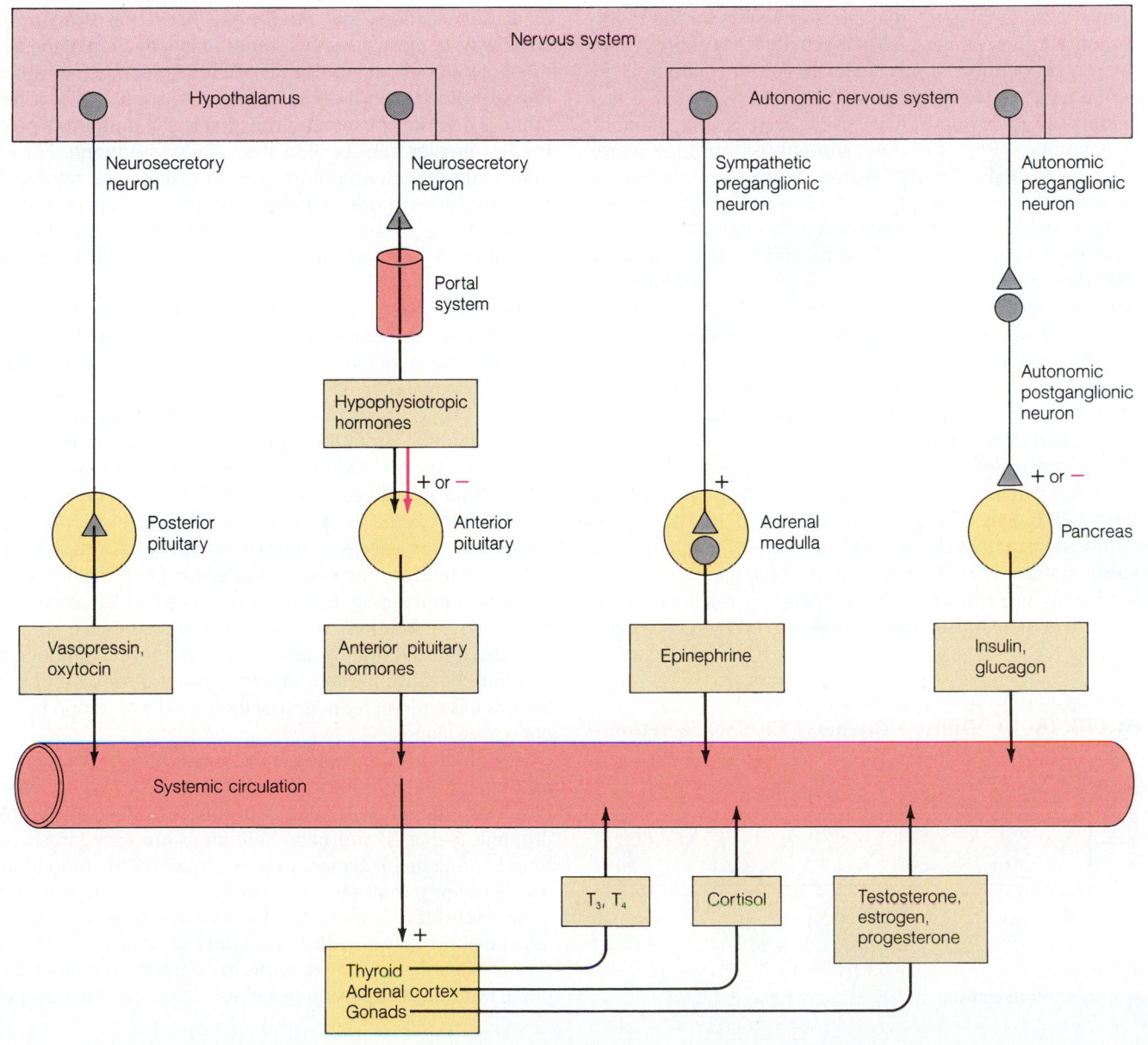

secretion of epinephrine by the adrenal medulla is solely under the control of the sympathetic nervous system. Some endocrine control systems, on the other hand, include both feedback control, which maintains a constant basal level of the hormone, and neuroendocrine reflexes, which cause sudden bursts in secretion in response to a sudden increased need for the hormone, such as increased secretion of cortisol by the adrenal cortex during a stress response.

Diurnal or Circadian Rhythms Although hormone secretion rates are usually regulated by some form of negative feedback, this does not imply that they are always maintained at a constant level. Instead, the secretion rates of all hormones rhythmically fluctuate up and down as a function of time. The most common endocrine rhythm is the **diurnal** ("day-night"), or **circadian** ("around a day"), **rhythm**, which is characterized by repetitive oscillations in hormone levels that are very regular and have a frequency of one cycle every twenty-four hours. This rhythmicity appears to be caused by endogenous oscillators similar to the self-paced respiratory neurons in the brain stem that are responsible for the rhythmic motions of breathing. Unlike the rhythmicity of breathing, however, endocrine rhythms are locked on or "entrained" to external cues, called **zeitgebers** ("time givers"), such as the light-dark cycle or the activity cycle; that is, the inherent twenty-four-hour cycles of peak and ebb of hormone secretion are set to "march in step" with cycles of light/activity and dark/inactivity. For example, cortisol secretion rises during the night, reaching its peak secretion in the morning before a person rises, then falls throughout the day to its lowest level at bedtime (• Fig. 18–10). Most investigators believe that inherent hormonal rhythmicity and the entrainment of zeitgebers are not accomplished by the endocrine organs themselves, but instead are the result of the central nervous system changing the set point of these organs. Negative-feedback control mechanisms operate to maintain whatever set point is established for that time of day.

Some endocrine cycles operate on time scales other than a circadian rhythm—some much shorter than a day and some much longer. A well-known example of the latter is the monthly menstrual cycle. Periodicity is not unique to the endocrine system. Humans have similar biological clocks for other bodily functions such as temperature regulation (see p. 603). The mechanism and purpose served by these rhyth-mic fluctuations in biological activity are unclear, but the effect of upsetting them is well known by people who experience jet lag when their inherent rhythm is out of step with external cues. (See the accompanying boxed feature, Concepts, Challenges, and Controversies.)

The effective plasma concentration of a hormone can be influenced by the hormone's transport, metabolism, and excretion.

Even though the effective plasma concentration of a hormone is normally regulated by appropriate adjustments in the rate of secretion, alterations in the transport, metabolism, or excretion of a hormone can also influence the size of its effective pool, sometimes inappropriately (Fig. 18–7). For example, since the liver synthesizes plasma proteins, liver disease may result in abnormal endocrine activity because of a change in the balance between free and bound pools of certain hormones.

Eventually, all hormones are metabolized by enzyme-mediated reactions that modify the hormonal structure in some way. In most cases this results in inactivation of the hormone. Hormone metabolism is not always a mechanism for removal of used hormones, however. In some cases, a hormone is *activated* by metabolism; that is, the hormone's product has greater activity than the original hormone. For example, after the thyroid hormone thyroxine is secreted, it is converted to a more powerful hormone by enzymatic removal of one of the iodine atoms it contains. Usually, the rate of such hormone activation is itself under hormonal control.

The liver is the most common site for metabolic hormonal inactivation (or activation, if that be the case), but some hormones are also metabolized in the kidneys, blood, or target cells. Hormonal inactivation and excretion are not subject to control. The primary means of eliminating hormones and their metabolites from the blood is by urinary excretion. Because the liver and kidneys are important in the removal of hormones from the blood by means of metabolic inactivation and urinary excretion, patients with liver or kidney disease may suffer from excess activity of certain hormones solely as a result of reduced hormone elimination. On the other hand, when liver and kidney function are normal, measurement of urinary concentrations of hormones and their metabolites provides a useful noninvasive means of assessing endocrine function, because the rate of excretion of these products in the urine is a direct reflection of their rate of secretion by the endocrine glands.

The amount of time after a hormone is secreted before it is inactivated and the means by which this is accomplished differ for different classes of hormones. In general, the hydrophilic peptides and catecholamines are easy targets for blood and tissue enzymes, so they remain in the blood only briefly (a few minutes to a few hours) before being enzymatically inactivated. In the case of some peptide hormones, the target cell actually engulfs the bound hormone by endocytosis and degrades it intracellularly. In contrast, binding of lipophilic hormones to plasma proteins renders them less vulner-

• **FIGURE 18–10**   **Diurnal Rhythm of Cortisol Secretion**

Light ☐   Dark ☐

Plasma cortisol concentration

N   M   N   M   N   M   N   M   N   M

Time of day

N is noon; M is midnight.

TINKERING WITH OUR BIOLOGICAL CLOCKS

Research shows that the hectic pace of modern life, stress, noise, pollution, and the irregular work schedules many workers follow can upset internal rhythms, illustrating how a healthy external environment affects our own internal environment—and our health.

Dr. Richard Restak, a neurologist and author, notes that the "usual rhythms of wakefulness and sleep . . . seem to exert a stabilizing effect on our physical and psychological health." The greatest disrupter of our natural circadian rhythms, says Restak, is the variable work schedule, surprisingly common in the United States and other industrialized countries. Today, one out of every four working men and one out of every six working women has a variable work schedule—shifting frequently between day and night work. In many industries, workers are on the job day and night to make optimal use of equipment and buildings. As a result, more restaurants and stores are open twenty-four hours a day and more health care workers must be on duty at night to care for accident victims.

To spread the burden, many companies that maintain shifts round the clock alter their workers' schedules. One week, employees work the day shift. The next week, they're shifted to the "graveyard shift" from 12:00 midnight to 8:00 A.M. The next week, they work the night shift from 4:00 P.M. to 12:00 midnight. Many shift workers report that they feel tired most of the time and have trouble staying awake at the job. Their work performance suffers because of fatigue and general malaise. When workers arrive home for bed, they're exhausted but can't sleep, because they're trying to doze off at a time when the body is trying to wake them up. Unfortunately, weekly changes in schedule never permit workers' internal alarm clocks to fully adjust. Most people, in fact, require four to fourteen days to adjust to a new schedule.

Workers on alternating shifts suffer more ulcers, insomnia, irritability, depression, and tension than workers on unchanging shifts. Their lives are never the same. To make matters worse, tired, irritable workers whose judgment is impaired by fatigue pose a threat to society as a whole. Consider an example.

At 4:00 A.M. in the control room of the Three Mile Island Nuclear Reactor in Pennsylvania, three operators made the first mistake in a series of errors that led to the worst nuclear accident in U.S. history. The operators did not notice warning lights and failed to observe that a crucial valve had remained open. When the morning shift operators entered the control room the next day, they quickly discovered the problems, but it was too late. Pipes in the system had burst, sending radioactive steam and water into the air and into two buildings. John Gofman and Arthur Tamplin, two radiation health experts, estimate that the radiation released from the accident will cause at least 300 and possibly as many as 900 additional fatal cases of cancer in the residents living near the troubled reactor, although other "experts" (especially in the nuclear industry) contest these projections, saying the accident will have an unnoticeable effect. Whatever the outcome, the 1979 accident at Three Mile Island will cost several billion dollars to clean up.

Late in April 1986, another nuclear power plant went amok. This accident, in the Soviet Union, was far more severe. In the early hours of the morning, two engineers were testing the reactor. They deactivated key safety systems in violation of standard operational protocol. This single error in judgment (possibly due to fatigue) led to the largest and most costly nuclear accident in the history of the world. Steam built up inside the reactor and blew the roof off the containment building. A thick cloud of radiation rose skyward and then spread throughout Europe and the world. While workers battled to cover the molten radioactive core that spewed radiation into the sky, the whole world watched in horror.

The Chernobyl disaster, like the accident at Three Mile Island, may have been the result of workers operating at a time unsuitable for clear thinking. One has to wonder how many plane crashes, auto accidents, and acts of medical malpractice can be traced to judgment errors resulting from our insistence on working against the inherent body rhythms.

Thanks to studies of biological rhythms, researchers are finding ways to reset the biological clocks, which could help lessen the misery and suffering of shift workers and could increase the performance of the graveyard shift workers. For instance, one simple measure is to put shift workers on three-week cycles to give their clocks time to adjust. And instead of shifting workers from daytime to graveyard shifts, transfer them forward, rather than backward (for example, from a daytime to a nighttime shift). It's a much easier adjustment. Bright lights can also be used to reset the biological clock as well. It's a small price to pay for a healthy work force and a safer society.

able to metabolic inactivation and prevents them from escaping into the urine. Therefore, lipophilic hormones are removed from the plasma much more slowly. They may persist in the blood for hours (steroids) or up to a week (thyroid hormone). In general, lipophilic hormones undergo a series of reactions that reduces their biological activity and enhances their H_2O solubility so that they can be freed of their plasma-protein carriers and be eliminated in the urine.

Endocrine disorders are attributable to hormonal excess, hormonal deficiency, or decreased responsiveness of the target cells.

From the preceding discussion, it should be apparent that abnormalities in a hormone's effective plasma concentration can arise from a variety of factors (Table 18–3). Endocrine disorders most commonly result from abnormal plasma concentrations of a hormone due to inappropriate rates of secretion; that is, too little hormone secreted (**hyposecretion**) or too much hormone secreted (**hypersecretion**). Occasionally, endocrine dysfunction arises because of abnormal target-cell responsiveness to the hormone, even though the plasma concentration of the hormone is normal.

Hyposecretion   If an endocrine organ is secreting too little of its hormone because of an abnormality within that organ, the condition is referred to as *primary hyposecretion*. If, on the other hand, the endocrine organ is normal but is secreting too little hormone because of a deficiency of its tropic hormone, the condition is known as *secondary hyposecretion*. The following are among the many different factors (each listed with an example) that may be responsible for hormone deficiency: (1) genetic (inborn absence of an enzyme that catalyzes synthesis of the hormone); (2) dietary (lack of iodine, which is necessary for synthesis of thyroid hormone); (3) chemical or toxic (certain insecticide residues may destroy the adrenal cortex); (4) immunologic (autoimmune antibodies may cause self-destruction of the person's own thyroid tissue); (5) other disease processes (cancer or tuberculosis may coincidentally destroy endocrine glands); (6) *iatrogenic* (physician-induced, such as surgical removal of a thyroid tumor); and (7) *idiopathic* (meaning the cause is not known).

The most common method of treating hormone hyposecretion is administration of a hormone that is the same as (or similar to, such as from another species) the one that is deficient or missing. Such replacement therapy is straightforward

in theory, but the hormone's source and means of administration involve some practical problems. The sources of hormone preparation for clinical use include (1) endocrine tissues from domestic livestock; (2) placental tissue and urine of pregnant women; (3) laboratory synthesis of hormones; and (4) "hormone factories," which consist of bacteria into which genes coding for the production of human hormones have been introduced. The method of choice for a given hormone is determined largely by its structural complexity and degree of species specificity.

Hypersecretion   Like hyposecretion, hypersecretion by a particular endocrine organ is designated as primary or secondary depending on whether the defect lies in that organ or is due to excessive stimulation from the outside. Hypersecretion may be caused by (1) tumors that ignore the normal regulatory input and continuously secrete excess hormone and (2) immunologic factors, such as excessive stimulation of the thyroid gland by an abnormal antibody that mimics the action of TSH, the thyroid tropic hormone. Excessive levels of a particular hormone may also arise from substance abuse, such as the outlawed practice among athletes of using certain steroids that increase muscle mass by promoting protein synthesis in muscle cells (see p. 239).

There are several ways of treating hormonal hypersecretion. If a tumor is the culprit, it may be surgically removed or destroyed with radiation treatment. In some instances, hypersecretion can be limited by drugs that block hormone synthesis or inhibit hormone secretion. Sometimes the condition may be treated by giving drugs that inhibit the action of the hormone without actually reducing the excess hormone secretion.

Abnormal Target-Cell Responsiveness   Endocrine dysfunction can also occur as a result of inadequate responsiveness of the target cells to the hormone, even though the effective plasma concentration of a hormone is normal. This inadequate responsiveness may be caused, for example, by an inborn lack of receptors for the hormone, such as is seen in *testicular feminization syndrome*. In this condition, receptors for testosterone, a masculinizing hormone produced by the male testes, are not produced because of a specific genetic defect. Although adequate testosterone is available, masculinization does not take place, just as if no testosterone were present. Abnormal responsiveness may also occur if the target cells for a particular hormone lack an enzyme essential to carrying out the response.

The responsiveness of a target cell to its hormone can be varied by regulating the number of its hormone-specific receptors.

In contrast to endocrine dysfunction caused by *unintentional* receptor abnormalities, the target-cell receptors for a particular hormone can be *deliberately altered* subject to physiologic control. A target cell's response to a hormone is correlated with the number of the cell's receptors occupied by molecules of that hormone, which in turn depends on the number

---

TABLE 18–3 Means by Which Endocrine Disorders Can Arise	
TOO LITTLE HORMONE ACTIVITY	**TOO MUCH HORMONE ACTIVITY**
Too little hormone secreted by endocrine gland (hyposecretion)*	Too much hormone secreted by endocrine gland (hypersecretion)*
Increased removal of hormone from blood	Reduced plasma-protein binding of hormone (too much free, biologically active hormone)
Abnormal tissue responsiveness to hormone	Decreased removal of hormone from blood
Lack of target-cell receptors	Decreased inactivation
Lack of enzyme essential to target-cell response	Decreased excretion

*Most common causes of endocrine dysfunction.

of receptors present in the target cell for that hormone as well as on the plasma concentration of the hormone. Thus, the response of a target cell to a given plasma concentration can be fine-tuned up or down by varying the number of receptors available for hormone binding.

As an illustration, when the plasma concentration of insulin is chronically elevated, the total number of target-cell receptors for insulin is reduced as a direct result of the effect that an elevated level of insulin has on the insulin receptors. This phenomenon, known as **down regulation**, constitutes an important locally acting negative-feedback mechanism that prevents the target cells from overreacting to the high concentration of insulin; that is, the target cells are *desensitized* to insulin, helping to blunt the effect of insulin hypersecretion. Down regulation of insulin is accomplished by the following mechanism. The binding of insulin to its surface receptors induces endocytosis of the hormone-receptor complex, which is subsequently attacked by intracellular lysosomal enzymes. This internalization serves a twofold purpose: it provides a pathway for degradation of the hormone, and it also plays a role in regulating the number of receptors available for binding on the target cell's surface. At high plasma insulin concentrations, the number of surface receptors for insulin is gradually reduced as a result of the accelerated rate of receptor internalization and degradation brought about by increased hormonal binding. The rate of synthesis of new receptors within the endoplasmic reticulum and their insertion in the plasma membrane do not keep pace with their rate of destruction. Over time, this self-induced loss of target-cell receptors for insulin results in a reduction in the target cell's sensitivity to the elevated hormone concentration.

A given hormone's effects are influenced not only by the concentration of the hormone itself but also by the concentration of other hormones that interact with it. Because hormones are widely distributed through the blood, target cells may be exposed simultaneously to many different hormones, giving rise to numerous complex hormonal interactions on target cells. Hormones frequently alter the receptors for other kinds of hormones as part of their normal physiological activity. A hormone can influence the activity of another hormone at a given target cell in one of three ways: permissiveness, synergism, and antagonism. With **permissiveness**, one hormone must be present in adequate amounts for the full exertion of another hormone's effect. In essence, the first hormone, by enhancing a target cell's responsiveness to another hormone, "permits" this other hormone to exert its full effect. For example, thyroid hormone increases the number of receptors for epinephrine in epinephrine's target cells, thereby increasing the effectiveness of epinephrine. Epinephrine is only marginally effective in the absence of thyroid hormone. **Synergism** occurs when the actions of several hormones are complementary and their combined effect is greater than the sum of their separate effects. An example is the synergistic action of several hormones (follicle-stimulating hormone and testosterone) required to maintain the normal rate of sperm production. Synergism probably results from each hormone's influence on the number or affinity of receptors for the other hormone.

Antagonism occurs when one hormone causes the loss of another hormone's receptors, reducing the effectiveness of the second hormone. To illustrate, progesterone (a hormone that is secreted during pregnancy and *decreases* contractions of the uterus) inhibits uterine responsiveness to estrogen (another hormone secreted during pregnancy that *increases* uterine contractions). Progesterone, by causing loss of estrogen receptors on uterine smooth muscle, prevents estrogen from exerting its excitatory effects during pregnancy and thus keeps the uterus a quiet (noncontracting) environment suitable for the developing fetus.

HYPOTHALAMUS AND PITUITARY

The pituitary gland consists of anterior and posterior lobes.

The **pituitary gland**, or **hypophysis**, is a small endocrine gland located in a bony cavity at the base of the brain just below the hypothalamus (● Fig. 18–11). If you point one finger between your eyes and another finger toward one of your ears, the imaginary point where these lines would intersect is about the location of your pituitary. The pituitary is connected to the hypothalamus by a thin stalk, the **infundibulum**, which contains nerve fibers and small blood vessels.

The pituitary has two anatomically and functionally distinct lobes, the **posterior pituitary** and the **anterior pituitary**. The posterior pituitary, being derived embryonically from an outgrowth of the brain, is composed of nervous tissue and thus is also termed the **neurohypophysis**. The anterior pituitary, in contrast, consists of glandular epithelial tissue derived embryonically from an outpouching that buds off from the roof of the mouth. Accordingly, the anterior pituitary is also known as the **adenohypophysis** (*adeno* means "glandular"). The anterior and posterior pituitary have nothing more in common than their location. The posterior pituitary is connected to the hypothalamus by a neural pathway, whereas the anterior pituitary is connected to the hypothalamus by a vascular link.

In the adenohypophysis in some species, a third, well-defined intermediate lobe is also included, but in humans this lobe is rudimentary. In lower vertebrates, the intermediate lobe secretes several **melanocyte-stimulating hormones** or **MSHs**, which regulate skin coloration by controlling the dispersion of granules containing the pigment **melanin**. By causing variable skin darkening in certain amphibia, reptiles, and fishes, MSHs play a vital role in the camouflage of these species. In humans, the small amount of MSH secreted is generally attributed to the anterior pituitary. The function of this MSH, if any, is still unclear. It is not involved in the differences in the amount of melanin deposited in the skin of various races, nor is it associated with the process of tanning. However, excessive MSH activity does cause darkening of the skin. Some evidence implicates MSH in humans in the totally different role of influencing the excitability of the nervous system, perhaps playing a role in improving memory and learning.

Hypothalamus

Bone

Anterior lobe
of pituitary

Posterior lobe
of pituitary

(a)

Hypothalamus

Optic chiasm

Infundibulum

Anterior
pituitary

Posterior
pituitary

(b)

• **FIGURE 18–11   Anatomy of the Pituitary Gland**   (a) Relation of the pituitary gland to the hypothalamus and to the remainder of the brain. (b) Schematic enlargement of the pituitary gland and its connection to the hypothalamus.

The hypothalamus and posterior pituitary form a neurosecretory system that secretes vasopressin and oxytocin.

The release of hormones from both the posterior and anterior pituitary is directly controlled by the hypothalamus, but the nature of the relationship is entirely different. The hypothala-mus and posterior pituitary form a neuroendocrine system that consists of a population of neurosecretory neurons whose cell bodies lie in two well-defined clusters in the hy-pothalamus (the **supraoptic** and **paraventricular nuclei**) and whose axons pass down through the connecting stalk to ter-minate on capillaries in the posterior pituitary (• Fig. 18–12). The posterior pituitary consists of these neuronal terminals

• **FIGURE 18–12   Relationship of the Hypothalamus and Posterior Pituitary**   The paraventricular and supraoptic nuclei both contain neurons that produce vasopressin and oxytocin. The hormone, either vasopressin or oxytocin depending on the neuron, is synthesized in the neuronal cell body in the hypothalamus and travels down the axon to be stored in the neuronal terminals within the posterior pituitary. The stored hormone is released into the systemic blood upon excitation of the neuron.

Paraventricular nucleus

Supraoptic nucleus

Hypothalamus

Hypothalamic-posterior pituitary stalk

Anterior pituitary

Posterior pituitary

Systemic arterial inflow

Systemic venous outflow

• = Vasopressin   • = Oxytocin

THE ENDOCRINE RESPONSE TO THE CHALLENGE OF COMBINED HEAT AND MARCHING FEET

When one exercises in a hot environment, maintenance of plasma volume becomes a critical homeostatic concern. Exercise in the heat results in the loss of large amounts of fluid through sweating. Simultaneously, blood is needed for shunting to the skin for cooling and for increased blood flow to nourish the working muscles. There must also be adequate venous return to maintain cardiac output. The hypothalamus–posterior pituitary neurosecretory system responds to these multiple, conflicting needs for fluid by releasing water-conserving vasopressin in an attempt to reduce urinary fluid loss to preserve plasma volume.

Studies have generally shown that exercise in heat stimulates vasopressin release, which results in decreased urinary fluid loss. In one study conducted during an eighteen-mile road march in heat, the participants' average urine output dropped to 134 ml (the normal urine output during the same time period would be about twice that much), while sweat loss averaged 4 liters. Overhydration prior to exercise appears to decrease the intensity of this response, suggesting that increased vasopressin release is related to plasma osmolarity. If fluid loss is not adequately replaced, plasma osmolarity increases. When the hypothalamic osmoreceptors detect this hypertonic condition, they promote increased secretion of vasopressin from the posterior pituitary. Some investigators believe, however, that increased vasopressin release results from other factors, such as changes in blood pressure or in renal blood flow. Regardless of the mechanism, vasopressin release is an important physiologic response to exercise in heat.

plus glial-like supporting cells called **pituicytes.** Functionally as well as anatomically, the posterior pituitary is simply an extension of the hypothalamus. The posterior pituitary does not actually produce any hormones. It simply stores and, upon appropriate stimulation, releases into the blood two small peptide hormones (actually, neurohormones), *vasopressin* and *oxytocin,* which are synthesized by the neuronal cell bodies in the hypothalamus. Both of these hydrophilic peptides are made in both the supraoptic and the paraventricular nuclei, but a single neuron is capable of producing only one of these hormones. The synthesized hormones are packaged in secretory granules that are transported down the cytoplasm of the axon (see p. 36) to be stored in the neuronal terminals within the posterior pituitary. Each terminal stores either vasopressin or oxytocin but not both. Thus, these hormones can be released independently as needed. Upon stimulatory input to the hypothalamus, either vasopressin or oxytocin is released into the blood from the posterior pituitary by exocytosis of the appropriate secretory granules. This hormonal release is triggered in response to action potentials that originate in the hypothalamic cell body and sweep down the axon to the neuronal terminal in the posterior pituitary. As in any other neuron, action potentials are generated in these neurosecretory neurons in response to synaptic input to their cell bodies.

The actions of vasopressin and oxytocin are briefly summarized here to make our endocrine story complete. They are described more thoroughly elsewhere—vasopressin in chapters 14 and 15 and oxytocin in chapter 20.

Vasopressin Vasopressin (antidiuretic hormone; ADH) has two major effects that correspond to its two names: (1) it enhances the retention of H_2O by the kidneys (an antidiuretic effect), and (2) it causes contraction of arteriolar smooth muscle (a vessel pressor effect). The first effect is of greater physiological importance. Under normal conditions, vasopressin is the primary endocrine factor that regulates urinary H_2O loss and overall H_2O balance. In contrast, typical levels of vasopressin play only a minor role in regulating blood pressure by means of the hormone's pressor effect.

The major control for hypothalamic-induced release of vasopressin from the posterior pituitary is input from hypothalamic osmoreceptors, which increase vasopressin secretion in response to a rise in plasma osmolarity. A less powerful input from the left atrial volume receptors increases vasopressin secretion in reponse to a fall in ECF volume and arterial blood pressure (see p. 522). The osmoreceptor and volume receptor mechanisms both function as negative-feedback systems to resist changes in ECF osmolarity and volume, respectively, by adjusting the H_2O load in the body through variable vasopressin activity. (See the accompanying boxed feature, A Closer Look at Exercise Physiology.)

Oxytocin Oxytocin stimulates contraction of the uterine smooth muscle to aid in expulsion of the baby during childbirth, and it promotes ejection of milk from the mammary glands (breasts) during breast-feeding. Appropriately, oxytocin secretion is increased by reflexes that originate within the birth canal during childbirth and by reflexes that are triggered when the infant suckles the breast.

The anterior pituitary secretes six established hormones, many of which are tropic to other endocrine glands.

Unlike the posterior pituitary, which releases hormones that are synthesized by the hypothalamus, the anterior pituitary

• FIGURE 18–13 Functions of Anterior Pituitary Hormones Five different endocrine cell types produce the six anterior pituitary hormones—TSH, ACTH, growth hormone, LH and FSH (produced by the same cell type), and prolactin—which exert a wide range of effects throughout the body.

itself synthesizes the hormones that it releases into the blood. Different cell populations within the anterior pituitary produce and secrete six established peptide hormones. The actions of each of these hormones will be dealt with in detail in subsequent sections. For now, a brief statement of their primary effects will be given to provide a rationale for their nomenclature (• Fig. 18–13):

1. **Growth hormone (GH, somatotropin)**, the primary hormone responsible for regulating overall body growth, is also important in intermediary metabolism.

2. **Thyroid-stimulating hormone (TSH, thyrotropin)** stimulates secretion of thyroid hormone and growth of the thyroid gland.

3. **Adrenocorticotropic hormone** (**ACTH, adrenocorticotropin**) stimulates cortisol secretion by the adrenal cortex and promotes growth of the adrenal cortex.

4. **Follicle-stimulating hormone** (**FSH**) has different functions in males and females. In females it stimulates growth and development of ovarian follicles, within which the ova, or eggs, develop. Furthermore, FSH promotes secretion of the hormone estrogen by the ovaries. In males FSH is required for sperm production.

5. **Luteinizing hormone** (**LH**) also functions differently in females and males. In females LH is responsible for ovulation, luteinization (that is, the formation of a postovulatory hormone-secreting corpus luteum in the ovary), and regulation of ovarian secretion of the female sex hormones, estrogen and progesterone. In males the same hormone stimulates the interstitial cells of Leydig in the testes to secrete the male sex hormone, testosterone, giving rise to its alternate name of **interstitial cell-stimulating hormone** (**ICSH**).

6. **Prolactin** (**PRL**) enhances breast development and milk production in females. Its function in males is uncertain, although evidence indicates that it may induce the production of testicular LH receptors.

Because they each regulate the secretion of another specific endocrine gland, TSH, ACTH, FSH, and LH are all tropic hormones. More specifically, FSH and LH are collectively referred to as **gonadotropins**, because they control secretion of the sex hormones by the gonads (ovaries and testes). Since growth hormone has recently been shown to exert its growth-promoting effects indirectly by stimulating the release of the liver hormones, the somatomedins, it too is sometimes categorized as a tropic hormone. Among the anterior pituitary hormones, prolactin is the only one that does not stimulate secretion of another hormone. Of the tropic hormones, FSH, LH, and growth hormone differ from TSH and ACTH in that the former exert nontropic functions in addition to stimulating secretion of other hormones.

Hypothalamic releasing and inhibiting hormones are delivered to the anterior pituitary by the hypothalamic-hypophyseal portal system to control anterior pituitary hormone secretion.

None of the anterior pituitary hormones are secreted at a constant rate. Even though each of these hormones has a unique control system, there are some common regulatory patterns. The two most important factors that regulate anterior pituitary hormone secretion are (1) hypothalamic hormones and (2) feedback by target-organ hormones.

Because the anterior pituitary secretes hormones that control the secretion of various other hormones, it has long had the undeserved title of "master gland." It is now known that the release of each of the anterior pituitary hormones is largely controlled by still other hormones produced by the hypothalamus and that secretion of these regulatory neurohormones, in turn, is controlled by a variety of neural and hormonal inputs to the hypothalamic neurosecretory cells (• Fig. 18–14).

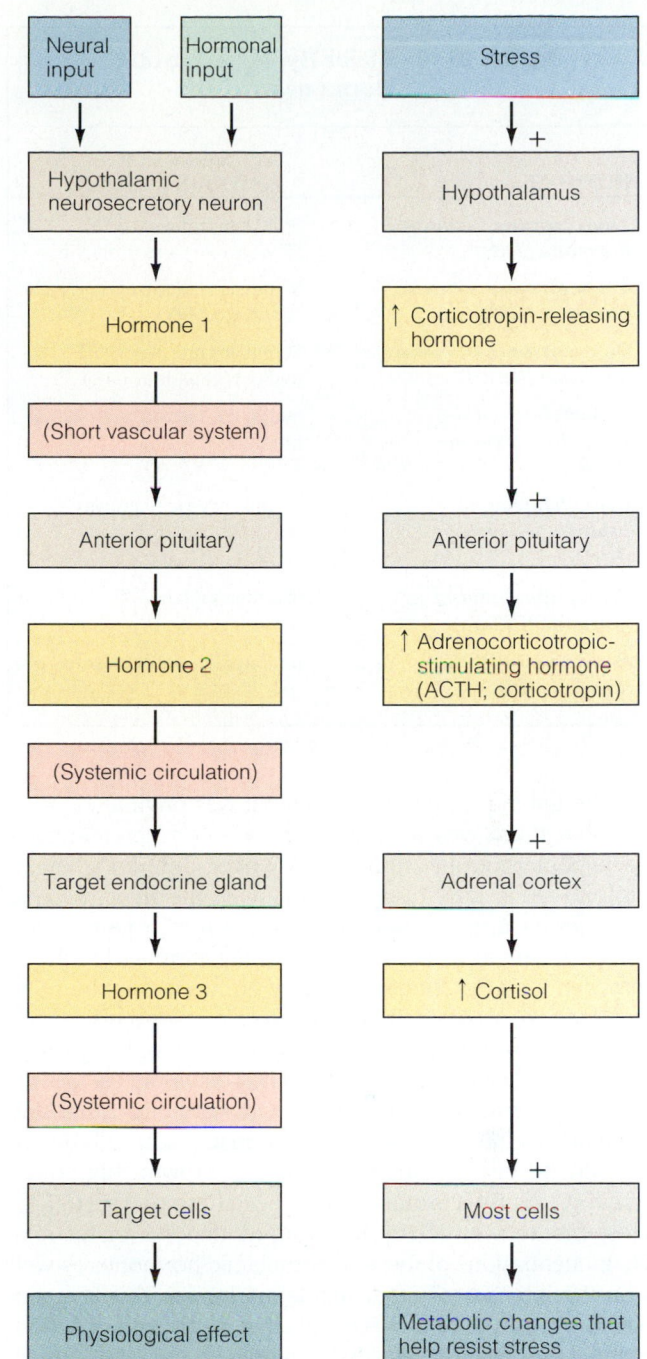

• FIGURE 18–14   Hierarchical Chain of Command in Endocrine Control

The secretion of each of the anterior pituitary hormones is stimulated or inhibited by one or more of the seven generally accepted hypothalamic **hypophysiotropic** (*hypophysis* means "pituitary"; *tropic* means "regulating") **hormones**, which are listed in Table 18–4. Depending on their actions, these small peptides are called *releasing hormones* or *inhibiting hormones*. In each case, the primary action of the hormone is apparent from its name. For example, **thyrotropin-releasing hormone** (**TRH**) stimulates the release of TSH (alias thyrotropin) from the anterior pituitary, whereas **prolactin-inhibiting hormone** (**PIH**) inhibits the release of prolactin

TABLE 18–4	Major Hypophysiotropic Hormones
HORMONE	**EFFECT ON ANTERIOR PITUITARY**
Thyrotropin-releasing hormone (TRH)	Stimulates release of TSH (thyrotropin) and prolactin
Corticotropin-releasing hormone (CRH)	Stimulates release of ACTH (corticotropin)
Gonadotropin-releasing hormone (GnRH)	Stimulates release of FSH and LH (gonadotropins)
Growth-hormone– releasing hormone (GHRH)	Stimulates release of growth hormone
Growth-hormone– inhibiting hormone (GHIH)	Inhibits release of growth hormone and TSH
Prolactin-releasing hormone (PRH)	Stimulates release of prolactin
Prolactin-inhibiting hormone (PIH)	Inhibits release of prolactin

from the anterior pituitary. Although it was originally speculated that there was a neat one-to-one correspondence— one hypophysiotropic hormone for each anterior pituitary hormone—it is now clear that many of the hypothalamic hormones have more than one effect, with their names indicating only the function that was initially attributed to them. Moreover, a single anterior pituitary hormone may be regulated by two or more hypophysiotropic hormones, which may even exert opposing effects. For example, **growth-hormone–releasing hormone** (**GHRH**) stimulates growth hormone secretion, whereas **growth-hormone–inhibiting hormone** (**GHIH**), also known as **somatostatin**, inhibits it. The output of the anterior pituitary growth hormone– secreting cells (that is, the rate of growth hormone secretion) in response to two such opposing inputs depends on the relative concentrations of these hypothalamic hormones as well as on the intensity of other regulatory inputs. This is analogous to the output of a nerve cell (that is, the rate of action-potential propagation) being dependent on the relative magnitude of excitatory and inhibitory synaptic inputs (EPSPs and IPSPs) to it.

Chemical messengers that are identical in structure to the hypothalamic releasing and inhibiting hormones and to vasopressin are produced in many areas of the brain outside the hypothalamus. Instead of being released into the blood, these messengers act locally as neurotransmitters and neuromodulators in these extrahypothalamic sites. They are thought to modulate a variety of functions that range from motor activity (TRH) to libido (GnRH) to learning (vasopressin). Thus, they provide a further example of the multiplicity of use of chemical messengers.

The hypothalamic regulatory hormones reach the anterior pituitary by means of a unique vascular link. In contrast to the direct neural connection between the hypothalamus and posterior pituitary, the anatomical and functional link between the hypothalamus and anterior pituitary is an unusual capillary-to-capillary connection, the **hypothalamo-hypophyseal portal system**. A portal system is a vascular arrangement in which venous blood flows directly from one capillary bed through a connecting vessel to another capillary bed without passing through the systemic circulation. The largest and best known portal system is the hepatic portal system, which drains intestinal venous blood directly into the liver for immediate processing of absorbed nutrients (see p. 570). Although much smaller, the hypothalamo-hypophyseal portal system is no less important, because it provides a critical link between the brain and much of the endocrine system. It begins in the base of the hypothalamus with a group of capillaries that recombine into small portal vessels, which pass down through the connecting stalk into the anterior pituitary. Here they branch to form most of the anterior pituitary capillaries, which in turn drain into the systemic venous system (• Fig. 18–15).

Note that almost all of the blood supply to the anterior pituitary must first pass through the hypothalamus. Since materials can be exchanged between the blood and surrounding tissue only at the capillary level, the hypothalamo-hypophyseal portal system provides a route where releasing and inhibiting hormones can be picked up at the hypothalamus and delivered immediately and directly to the anterior pituitary at relatively high concentrations, completely bypassing the systemic circulation. The hypophysiotropic hormones would be considerably more diluted if they had to travel to the anterior pituitary by means of the usual systemic circulatory route. This type of chemical messenger distribution is intermediate between a classic endocrine system, in which the hormone is released into the general circulation, and a neuronal system, in which the neurotransmitter is released into a confined space (that is, a synapse).

The axons of the neurosecretory neurons that produce the hypothalamic regulatory hormones terminate on the capillaries at the origin of the portal system. These hypothalamic neurons secrete their hormones in the same way as the hypothalamic neurons that produce vasopressin and oxytocin. The hormone is synthesized in the cell body and then transported to the axon terminal. It is stored there until its release into an adjacent capillary when, upon appropriate stimulation, an action potential is generated in the neuron. The major difference is that the hypophysiotropic hormones are released into the portal vessels, which deliver them to the anterior pituitary, where they control the release of anterior pituitary hormones into the systemic circulation. In contrast, the hypothalamic hormones stored in the posterior pituitary are themselves released into the systemic circulation (• Fig. 18–16).

Knowing that the secretion of anterior pituitary hormones is largely controlled by the hypothalamic releasing and inhibiting hormones, the next logical question is: What regulates the secretion of these hypophysiotropic hormones? Like other neurons, the neurons secreting these regulatory hormones receive abundant input of information (both neural and hormonal and both excitatory and inhibitory) that they must

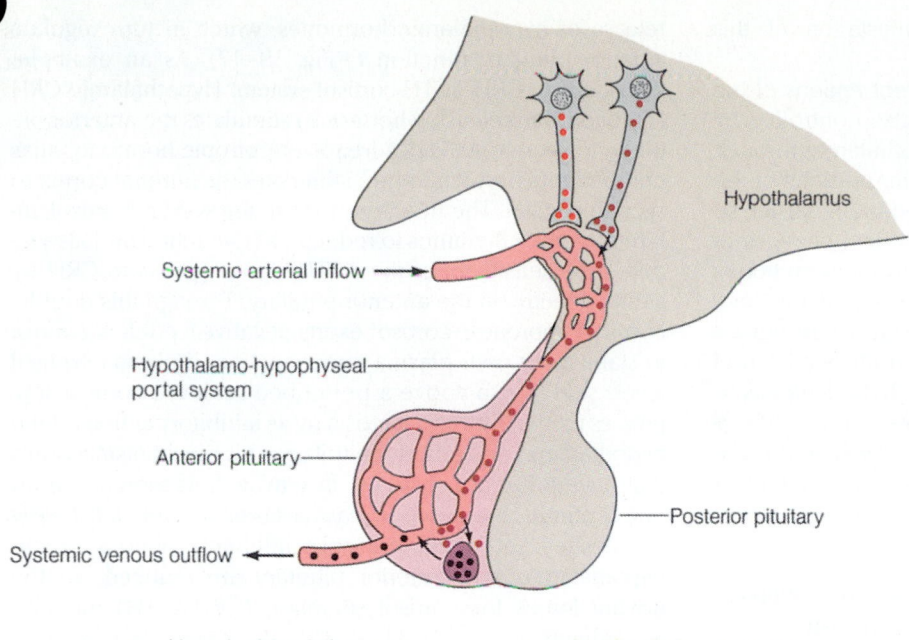

Hypothalamus

Systemic arterial inflow

Hypothalamo-hypophyseal portal system

Anterior pituitary

Systemic venous outflow

Posterior pituitary

● = Hypophysiotropic hormone   • = Anterior pituitary hormone

integrate. Studies are still in progress to unravel the complex neural input from many diverse areas of the brain to the hypophysiotropic secretory neurons. Some of these inputs carry information about a variety of environmental conditions. One example is the marked increase in the secretion of corticotropin-releasing hormone (CRH) in response to stress-

ful situations (Fig. 18–14). There are also numerous neural connections between the hypothalamus and the portions of the brain that are concerned with emotions (the limbic system—see p. 127). Thus, secretion of hypophysiotropic hormones is greatly influenced by emotions. The menstrual irregularities sometimes experienced by women who are

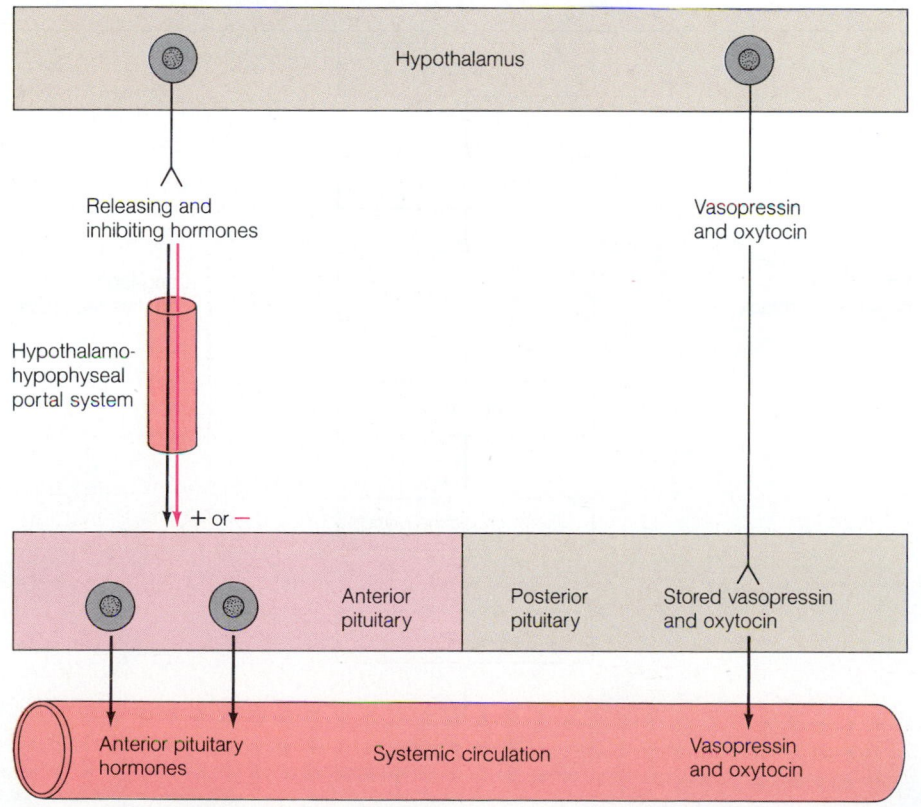

Hypothalamus

Releasing and inhibiting hormones

Vasopressin and oxytocin

Hypothalamo-hypophyseal portal system

+ or −

Anterior pituitary

Posterior pituitary

Stored vasopressin and oxytocin

Anterior pituitary hormones

Systemic circulation

Vasopressin and oxytocin

emotionally upset are a common manifestation of this relationship.

In addition to being regulated by different regions of the brain, the hypophysiotropic neurons are also controlled by various chemical inputs that reach the hypothalamus through the blood. Unlike other regions of the brain, portions of the hypothalamus are not guarded by the blood-brain barrier, so the hypothalamus can easily monitor chemical changes in the blood. In some instances, hypophysiotropic secretion is influenced by the immediate metabolic state of the individual. For example, one or more of the metabolic responses produced by growth hormone (such as an elevated blood glucose level) negatively feed back to the hypothalamus to control growth hormone secretion. The more common blood-borne factors that influence hypothalamic neurosecretion are the negative-feedback effects of either anterior pituitary or target-organ hormones, to which we now turn our attention.

In general, feedback by target-organ hormones strives to maintain relatively constant rates of anterior pituitary hormone secretion.

In most cases, hypophysiotropic hormones initiate a three-hormone sequence: (1) hypothalamic releasing hormone, (2) anterior pituitary tropic hormone, and (3) peripheral target-organ hormone (Fig. 18–14). With one exception, in addition to producing its physiological effects, the target-organ hormone also acts to suppress secretion of the tropic hormone that is driving it. This suppression, designated **long-loop negative feedback**, is accomplished by the target-organ hormone acting either directly on the pituitary itself or on the release of hypothalamic hormones, which in turn regulate anterior pituitary function (• Fig. 18–17). As an example, consider the CRH-ACTH-cortisol system. Hypothalamic CRH (corticotropin-releasing hormone) stimulates the anterior pituitary to secrete ACTH (adrenocorticotropic hormone, alias corticotropin), which in turn stimulates the adrenal cortex to secrete cortisol. The final hormone in the system, cortisol, inhibits the hypothalamus to reduce CRH secretion and also reduces the sensitivity of the ACTH-secreting cells to CRH by acting directly on the anterior pituitary. Through this double-barreled approach, cortisol exerts negative-feedback control to stabilize its own plasma concentration. If plasma cortisol levels start to rise above a prescribed set level, cortisol suppresses its own further secretion by its inhibitory actions at the hypothalamus and anterior pituitary. This mechanism assures that once a hormonal system is activated, its secretion does not continue unabated. If plasma cortisol levels fall below the desired set point, cortisol's inhibitory actions at the hypothalamus and anterior pituitary are reduced, so the driving forces for cortisol secretion (CRH-ACTH) increase accordingly.

Similarly, the other target-organ hormones act by means of long-loop negative feedback. The goal of such feedback is to maintain relatively constant levels of the target-organ hormone. Remember, however, that the diurnal rhythms are superimposed on this type of stabilizing negative-feedback regulation. Furthermore, other controlling inputs may break through the negative-feedback control to alter hormone secretion (that is, change the set level) at times of special need.

The one exception to the negative-feedback relationship just described is the long-loop positive-feedback effect of es-

• **FIGURE 18–17 Negative Feedback in Hypothalamic–Anterior Pituitary Control Systems**

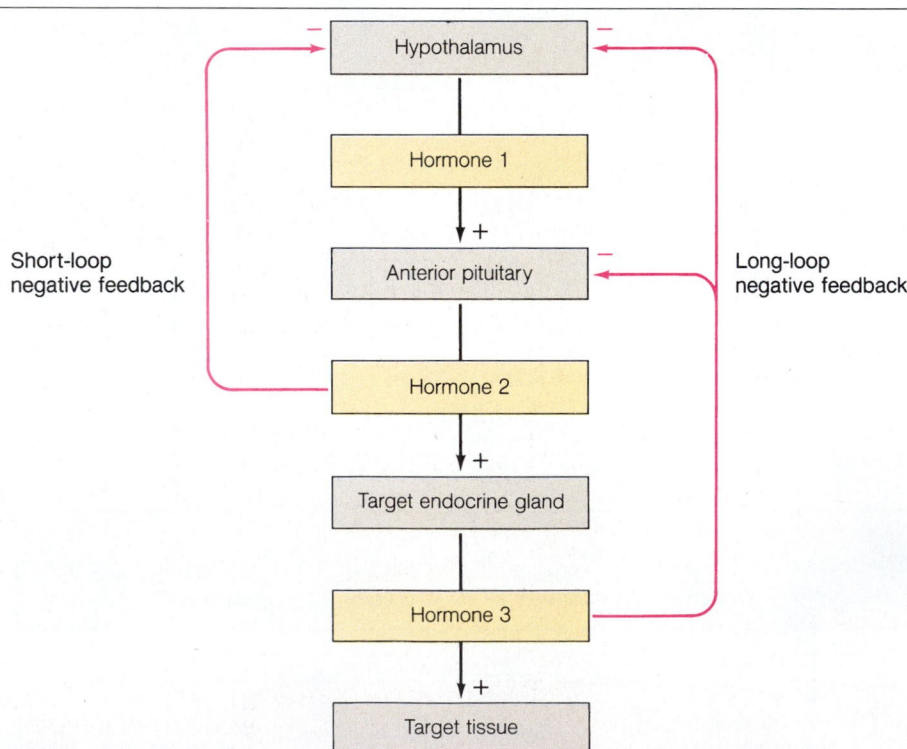

trogen on LH secretion. This causes an extremely dramatic rise in LH secretion that is responsible for triggering ovulation. This relationship will be discussed further in chapter 20.

Besides the physiologically more important long-loop feedback mechanism just described, evidence exists for **short-loop negative feedback**, which refers to a pituitary hormone's action at the hypothalamus to inhibit the release of its stimulatory neurohormone (Fig. 18–17). For example, prolactin, which is the only anterior pituitary hormone that does not participate in a three-hormone sequence, is believed to act directly on the hypothalamus to influence secretion of hypophysiotropic hormones that control prolactin secretion. It is possible that anterior pituitary tropic hormones employ similar short-loop feedback on their respective hypophysiotropic neurons. The recent discovery that a substantial portion of the blood flowing from the anterior pituitary may actually go back to the hypothalamus before entering the general circulation suggests that this may be the case. Such retrograde flow of blood makes it plausible that pituitary hormones may reach the brain in sufficiently high concentrations to influence releasing-hormone secretion. The actual existence and importance of such mechanisms in humans remain uncertain.

In addition, other hormones *outside* a particular sequence may also exert important influences, either stimulatory or inhibitory, on the secretion of hypothalamic or anterior pituitary hormones within a given sequence. For example, even though estrogen is not in the direct chain of command for prolactin secretion, this sex steroid notably enhances prolactin secretion by the anterior pituitary. This is but one example of the common phenomenon in the endocrine system that one seemingly unrelated hormone can have pronounced effects on the secretion or actions of another hormone.

ENDOCRINE CONTROL OF GROWTH

The detailed functions and control of all of the anterior pituitary hormones except growth hormone are discussed elsewhere in conjunction with the peripheral target tissues that they influence; for example, thyroid-stimulating hormone is covered in the next chapter with the discussion of thyroid gland. Accordingly, growth hormone is the only anterior pituitary hormone that will be elaborated on at this time.

Growth depends on growth hormone but is influenced by other factors as well.

In growing children, continuous net protein synthesis occurs under the influence of growth hormone as the body steadily gets larger. Weight gain alone is not synonymous with growth, because weight gain may occur as a result of retention of excess H_2O or fat without true structural growth of tissues. Growth requires net synthesis of proteins and includes lengthening of the long bones (the bones of the extremities) as well as increases in the size and number of cells in the soft tissues throughout the body.

Although, as the name implies, growth hormone is absolutely essential for growth, it alone is not wholly responsible for determining the rate and final magnitude of growth in a given individual. An individual's maximum growth capacity is *genetically determined*. Attainment of this full growth potential further depends on:

- *An adequate diet,* including sufficient total protein and ample essential amino acids to accomplish the protein synthesis necessary for growth. Malnourished children never achieve their full growth potential. The growth-stunting effects of inadequate nutrition are most profound when they occur in infancy. In severe cases, the child may be locked in to irreversible stunting of body growth and brain development. About 70% of the total growth of the brain occurs in the first two years of life. On the other hand, one cannot exceed his or her genetically determined maximum by eating more than an adequate diet. The excess food intake produces obesity instead of growth.

- *Freedom from chronic disease and stressful environmental conditions.* Stunting of growth under adverse circumstances is due in large part to the prolonged stress-induced secretion of cortisol from the adrenal cortex. Cortisol exerts several potent antigrowth effects, such as promoting protein breakdown, inhibiting growth in the long bones, and blocking the secretion of growth hormone. Even though sickly or stressed children do not grow well, if the condition is rectified before adult size is achieved, they can rapidly catch up to their normal growth curve through a remarkable spurt in growth.

- *A normal milieu of growth-influencing hormones.* In addition to the absolutely essential growth hormone, other hormones, including thyroid hormone, insulin, and the sex hormones, play secondary roles in the promotion of growth.

The rate of growth is not continuous nor are the factors responsible for promoting growth the same throughout the growth period. *Fetal growth* seems to be largely independent of hormonal control, with the size at birth being determined principally by genetic and environmental factors. Hormonal factors begin to play an important role in the regulation of growth after birth. Genetic and nutritional factors also strongly affect growth during this period.

Children display two periods of rapid growth—a *postnatal growth spurt* during their first two years of life and a *pubertal growth spurt* during adolescence (Fig. 18–18). From the age of two until puberty, the *rate* of linear growth progressively declines, even though the child is still growing. Before puberty there is little sexual difference in height or weight. During puberty, a marked acceleration in linear growth takes place because of the lengthening of the long bones. Puberty begins at about age eleven in girls and thirteen in boys and lasts for several years in both sexes. The mechanisms responsible for the pubertal growth spurt are not clearly understood. Apparently, both genetic and hormonal factors are involved. Some evidence indicates that growth hormone secretion is elevated during puberty and thus may contribute to the ac-

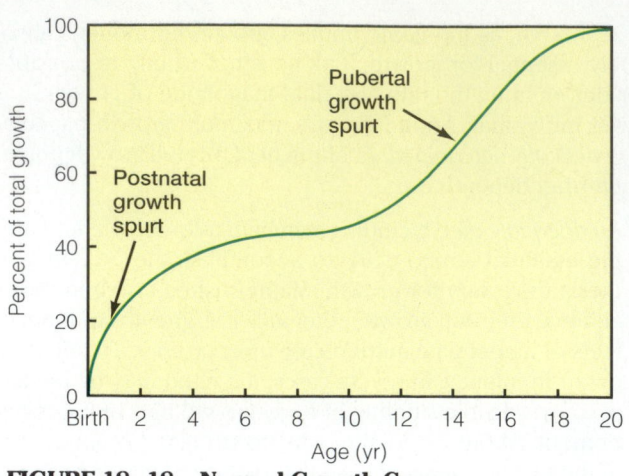

● FIGURE 18–18   Normal Growth Curve

celeration of growth during this time. Furthermore, **andro-gens** ("male" sex hormones), whose secretion increases dramatically at puberty, also contribute to the pubertal growth spurt by promoting protein synthesis and bone growth. The potent androgen from the male testes, testosterone, is of greatest importance in promoting a sharp increase in height in adolescent boys, whereas the less potent adrenal androgens from the adrenal gland, which also manifest a sizable increase in secretion during adolescence, are most likely important in the female pubertal growth spurt. Although estrogen secretion by the ovaries also begins during puberty, it is unclear what role this "female" sex hormone may play in the pubertal growth spurt in girls. There is no doubt that testosterone and estrogen both ultimately act on bone to halt its further growth so that full adult height is attained by the end of adolescence.

Growth hormone is essential for growth, but it also exerts metabolic effects not related to growth.

Growth hormone is the most abundant hormone produced by the anterior pituitary, even in adults in whom growth has already ceased. The continued high secretion of growth hormone beyond the growing period implies that this hormone must have important influences other than on growth. Its growth-promoting effects are fairly well understood. Its metabolic actions not related to growth are known, but their physiological role remains somewhat nebulous.

Metabolic Actions Unrelated to Growth   Growth hormone increases fatty acid levels in the blood by enhancing the breakdown of triglyceride fat stored in adipose tissue, and it increases blood glucose levels by decreasing glucose uptake by muscles. Muscles use the mobilized fatty acids instead of glucose as a metabolic fuel. Thus, the overall metabolic effect of growth hormone is to mobilize fat stores as a major energy source while conserving glucose for glucose-dependent tissues such as the brain. The brain can use only glucose as its metabolic fuel, yet nervous tissue cannot store glycogen (stored glucose) to any extent. This metabolic pattern is suitable for maintaining the body during prolonged

fasting or other situations when the body's energy needs exceed available glucose stores.

Growth-Promoting Actions on Soft Tissues   When tissues are responsive to its growth-promoting effects, growth hormone stimulates growth of both soft tissues and the skeleton. Growth of soft tissues is accomplished by (1) increasing the number of cells (**hyperplasia**) by stimulating cell division and (2) increasing the size of cells (**hypertrophy**) by favoring synthesis of proteins, the main structural component of cells.

Growth hormone stimulates almost all aspects of protein synthesis while it simultaneously inhibits protein degradation. It promotes the uptake of amino acids (the raw materials for protein synthesis) by cells and decreases blood amino acid levels in the process. Furthermore, it stimulates the cellular machinery responsible for accomplishing protein synthesis according to the cell's genetic code.

Growth-Promoting Actions on Bones   Growth of the long bones resulting in increased height is the most dramatic effect of growth hormone. **Bone** is a living tissue. Being a form of connective tissue, it consists of cells and an extracellular organic matrix that is produced by the cells. The bone cells that produce the organic matrix are known as **osteoblasts** ("bone-formers"). The organic matrix is composed of collagen fibers in a mucopolysaccharide-rich semisolid gel referred to as *ground substance*. This matrix has a rubbery consistency and is responsible for the tensile strength of bone (the resilience of bone to breakage when tension is applied). Bone is made hard by precipitation of calcium phosphate crystals within the matrix. These inorganic crystals provide the bone with compressional strength (the ability of bone to hold its shape when squeezed or compressed). If bones were composed entirely of inorganic crystals, they would be brittle, like pieces of chalk. Bones have structural strength approaching that of reinforced concrete, yet they are not brittle and are much lighter in weight as a result of the structural blending of an organic scaffolding hardened by inorganic crystals. **Carti-lage** is similar to bone, except that living cartilage is not calcified.

A long bone basically consists of a fairly uniform cylindrical shaft, the **diaphysis**, with a flared articulating knob at either end, an **epiphysis**. In a growing bone, the diaphysis is separated at each end from the epiphysis by a layer of cartilage known as the **epiphyseal plate** (● Fig. 18–19a). The central cavity of the bone is filled with bone marrow, which is the site of blood-cell production (chapter 11).

Growth in *thickness* of bone is achieved by the addition of new bone on top of the already existing bone on the outer surface. This growth occurs through activity of osteoblasts within the **periosteum**, a connective-tissue sheath that covers the outer bone surface. As new bone is being deposited by osteoblast activity on the external surface, other cells within the bone, the **osteoclasts** ("bone-breakers"), dissolve the bony tissue on the inner surface adjacent to the marrow cavity. In this way, the marrow cavity is enlarged to keep pace with the increase in the circumference of the bone shaft.

Growth in *length* of long bones is accomplished by a dif-

Articular cartilage

Bone of epiphysis

Epiphyseal plate

Bone of diaphysis

Marrow cavity

(a)

Cartilage

Calcified cartilage

Bone

Bone of epiphysis

Bone of epiphysis

Resting chondrocytes

Chondrocytes undergoing cell division

Causes thickening of epiphyseal plate

Older chondrocytes enlarging

Calcification of extracellular matrix (entrapped chondrocytes die)

Dead chondrocytes cleared away by osteoclasts

Epiphyseal plate

Diaphysis

Osteoblasts swarming up from diaphysis and depositing bone over persisting remnants of disintegrating cartilage

(b)

● **FIGURE 18–19   Anatomy and Growth of Long Bones**   (a) Anatomy of long bones. (b) Two sections of the same epiphyseal plate at different times, depicting the growth in length of long bones.

ferent mechanism than growth in thickness. Bones grow in length as a result of proliferation of the cartilage cells in the epiphyseal plates (Fig. 18–19b). During growth, new cartilage cells (**chondrocytes**) are produced through cell division on the outer edge of the plate adjacent to the epiphysis. As new chondrocytes are being formed on the epiphyseal border, the older cartilage cells toward the diaphyseal border are enlarging. This combination of proliferation of new cartilage cells and hypertrophy of maturing chondrocytes causes the epiphyseal plate to temporarily increase in width. This thickening of the intervening cartilaginous plate causes the bony epiphysis to be pushed farther away from the diaphysis. Soon the matrix surrounding the oldest hypertrophied cartilage be-

comes calcified. Since cartilage lacks its own capillary network, the survival of cartilage cells depends on diffusion of nutrients and O_2 through the ground substance, a process that is prevented by the deposition of calcium salts. As a result, the old nutrient-deprived cartilage cells on the diaphyseal border die. As osteoclasts clear away the dead chondrocytes and the calcified matrix that imprisoned them, the area is invaded by osteoblasts, which swarm upward from the diaphysis, trailing their capillary supply with them. These new tenants lay down bone around the persisting remnants of disintegrating cartilage until the inner region of cartilage on the diaphyseal side of the plate is entirely replaced by bone. When this **ossification** (bone-forming) process is completed, the bone on the diaphyseal side has increased in length, and the epiphyseal plate has been returned to its original thickness. The cartilage that has been replaced by bone on the diaphyseal end of the plate is equivalent in thickness to the new cartilaginous growth on the epiphyseal end of the plate. Thus, bone growth is made possible by the growth and death of cartilage, which acts like a "spacer" to push the epiphysis farther out while it provides a framework for future formation of bone on the end of the diaphysis.

As the extracellular matrix produced by an osteoblast becomes calcified, the osteoblast like its chondrocyte predecessor becomes entombed by the matrix that it has deposited around itself. Unlike chondrocytes, however, osteoblasts trapped within a calcified matrix do not die because they are supplied by nutrients transported to them through small canals that the osteoblasts themselves form by sending out cytoplasmic processes around which the bony matrix is deposited. Thus, within the final bony product, a network of permeating tunnels radiates from each entrapped osteoblast, serving as a lifeline system for nutrient delivery and waste removal. The entrapped osteoblasts, which are now called **osteocytes**, retire from active bone-forming duty, since their imprisonment prevents them from laying down any new bone. However, they are involved in the hormonally regulated exchange of calcium between bone and the blood. This exchange is under the control of parathyroid hormone (to be discussed in the next chapter), not growth hormone.

Growth hormone promotes growth of bone in both thickness and length. It stimulates the proliferation of epiphyseal cartilage, thereby making space for more bone formation, and also stimulates osteoblast activity. Growth hormone is able to promote lengthening of long bones as long as the epiphyseal plate remains cartilaginous or is "open." At the end of adolescence, under the influence of the sex hormones, these plates completely ossify or "close" so that the bones can grow no further in length despite the presence of growth hormone. Thus, after the plates are closed, the individual does not grow any taller.

Growth hormone exerts its growth-promoting effects indirectly by stimulating somatomedins.

The growth-producing actions of growth hormone (increased cell division, enhanced protein synthesis, and bone growth) are not accomplished directly by its effects on the target cells. These effects are directly brought about by peptide mediators known as **somatomedins**, whose synthesis in turn is induced by growth hormone. These peptides are also referred to as **insulin-like growth factors** (**IGF**) because they are structurally and functionally similar to insulin. Two somatomedins—IGF I and IGF II—have been identified.

The major site of somatomedin production is the liver, which releases this peptide product into the blood. However, somatomedin production has also been demonstrated in a variety of other tissues. It has been proposed that somatomedins that are produced locally in target tissues may act through paracrine means for at least some of the growth-hormone–induced effects. Such a mechanism could account for the fact that blood levels of growth hormone are no higher, and indeed circulating somatomedin levels are lower, during the first several years of life compared to adult values, even though growth is quite rapid during the postnatal period. Local production of somatomedins in target tissues may possibly be more important than delivery of blood-borne somatomedins during this time.

To complicate the situation further, the production and activity of somatomedins are also controlled by a number of factors other than growth hormone, including age and nutritional status. As a result, changes in circulating somatomedin levels do not always coincide with changes in growth hormone secretion. For example, fasting decreases IGF I levels even though it increases growth hormone secretion. Furthermore, the two different somatomedins are not always secreted in parallel. A dramatic increase in IGF I levels accompanies the moderate increase in growth hormone at puberty, which may, of course, be an important factor in the pubertal growth spurt. In contrast, IGF II concentrations do not appear to increase during puberty. These few examples serve to illustrate the confusing state of knowledge regarding the role and control of these important growth hormone mediators.

Growth hormone secretion is regulated by two hypophysiotropic hormones.

Two antagonistic regulatory hormones from the hypothalamus are involved in the control of growth hormone secretion: growth-hormone–releasing hormone (GHRH), which is stimulatory, and growth-hormone–inhibiting hormone (GHIH or somatostatin), which is inhibitory (● Fig. 18–20). (Note the distinction between *somatotropin*, alias growth hormone; *somatomedin*, the liver hormone that directly mediates the effects of growth hormone; and *somatostatin*, which inhibits growth hormone secretion.) Any factor that increases growth hormone secretion could theoretically do so either by stimulating GHRH release or inhibiting GHIH release. It is not known which of these pathways is used in each specific case.

As with the other hypothalamus–anterior-pituitary axes, negative-feedback loops participate in the regulation of growth hormone secretion. Both growth hormone and the somatomedins inhibit pituitary secretion of growth hormone, presumably by stimulating GHIH release from the hypothalamus. The somatomedins may also exert direct effects on the anterior pituitary to inhibit the effects of GHRH on growth hormone release.

Interacting with or overriding the basic negative-feedback control system are a number of factors that influence growth hormone secretion. Growth hormone secretion displays a well-characterized diurnal rhythm. Through most of the day, growth hormone levels tend to be low and fairly constant. Approximately one hour after the onset of deep sleep, however, growth hormone secretion markedly increases up to five times the daytime value, then rapidly drops over the next several hours.

Superimposed on this diurnal undulation in growth hormone secretion are further bursts in secretion that occur in response to exercise, stress, and hypoglycemia (low blood

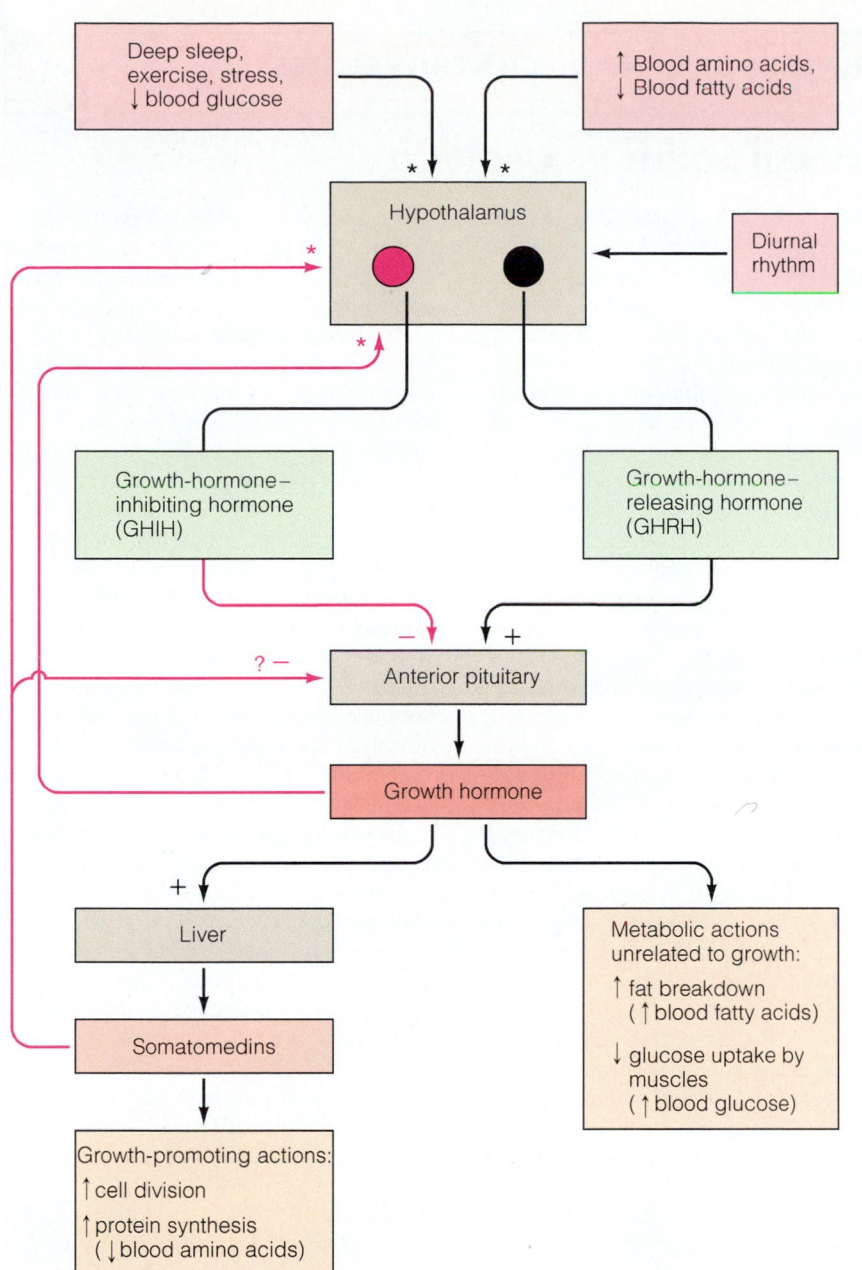

● FIGURE 18–20   Control of Growth Hormone Secretion

Deep sleep, exercise, stress, ↓ blood glucose

↑ Blood amino acids, ↓ Blood fatty acids

*

*

Hypothalamus

*

*

Diurnal rhythm

Growth-hormone–inhibiting hormone (GHIH)

Growth-hormone–releasing hormone (GHRH)

−

+

? −

Anterior pituitary

Growth hormone

+

Liver

Somatomedins

Growth-promoting actions:
↑ cell division
↑ protein synthesis
(↓ blood amino acids)
↑ bone growth

Metabolic actions unrelated to growth:
↑ fat breakdown
(↑ blood fatty acids)
↓ glucose uptake by muscles
(↑ blood glucose)

*These factors all increase growth hormone secretion, but it is unclear whether they do so by stimulating GHRH or inhibiting GHIH, or both.

*These factors inhibit growth hormone secretion in negative-feedback fashion, but it is unclear whether they do so by stimulating GHIH or inhibiting GHRH or inhibiting the anterior pituitary itself.

glucose), the major stimuli for increased secretion. The benefit of increased growth hormone secretion during these situations when energy demands outstrip the body's glucose reserves is presumably the conservation of glucose for the brain and the provision of fatty acids as an alternative energy source for muscle.

Because growth hormone uses up fat stores and promotes synthesis of body proteins, it encourages a change in body composition away from adipose deposition toward an increase in muscle protein. Accordingly, the increase in growth hormone secretion that accompanies exercise may at least in part mediate exercise's effects in reducing the percentage of body fat while increasing the lean body mass.

An abundance of blood amino acids, such as after a high protein meal, also enhances the secretion of growth hormone, which in turn promotes the use of these amino acids

GROWTH AND YOUTH IN A BOTTLE?

Unlike most hormones, growth hormone's structure and activity differ among species. As a result, growth hormone from animal sources is ineffective in the treatment of growth hormone deficiency in humans. In the past, the only source of human growth hormone was pituitary glands from human cadavers. This supply was never adequate, and it has been removed from the market because of fear of viral contamination. Recently, however, the supply of human growth hormone has become unlimited through the technique of genetic engineering. The gene that directs the synthesis of growth hormone in humans has been introduced into bacteria, converting the bacteria into human-growth-hormone–synthesizing "factories."

Even though genetic engineering has succeeded in providing an adequate supply of growth hormone, the solution has given rise to problems for the medical community regarding the circumstances under which synthetic growth hormone treatment is appropriate. Synthetic human growth hormone has currently been approved by the Food and Drug Administration only for use in growth-hormone–deficient children.

Growth hormone therapy is also being considered for the treatment of other disorders including (1) promoting faster healing of the skin in severely burned patients; (2) counteracting the wasting, or loss of body mass, that accompanies chronic, debilitating diseases; and (3) treating obesity. In preliminary studies using growth hormone in obese patients, the subjects did not lose weight, but their bodies became leaner (more muscle, less fat) during the treatment.

Another group for whom replacement growth hormone therapy may be beneficial is the elderly. Scientists suspect that in many people growth hormone production may start to dwindle after age forty. This decline may contribute to some of the characteristic signs of aging, such as decreased muscle mass (growth hormone promotes synthesis of proteins, including muscle protein), increased fat deposition (growth hormone promotes leanness by mobilizing fat stores for use as an energy source), and thinner, sagging skin (growth hormone promotes proliferation of skin cells).

A recent study suggests that some of these consequences of aging may be partially reversed or counteracted through the use of synthetic growth hormone in people over age sixty who have a measurable growth hormone deficit. Elderly growth hormone–deficient experimental subjects who were provided with supplemental growth hormone demonstrated an increase in muscle mass, a reduction in fatty tissue, and a thickening of the skin. Even though these early studies are exciting, scientists caution that synthetic growth hormone should not be viewed as a potential cure for old age. It is too early to tell whether the changes will last, and whether the experimental treatment will prove to be safe. Also, researchers advise that the drug would not be expected to produce similar results in elderly people who do not have a growth hormone deficiency. Investigators hope that synthetic growth hormone will find future use in strengthening muscle and bone sufficiently in the many elderly who do have growth hormone deficits to help reduce the incidence of bone-breaking falls that often lead to disability and loss of independence.

An ethical dilemma is whether the drug should be made legally available for others who have normal growth hormone levels but desire the product's growth-promoting actions for cosmetic or athletic reasons, such as normally growing teenagers who wish to attain even greater stature. The drug is already being used illegally by some athletes and bodybuilders.

Hesitancy in making the drug more readily available stems in part from the fact that synthetic growth hormone is a double-edged hormone. Although it has positive effects such as promoting growth and muscle mass, it also has negative effects, such as possibly causing diabetes or high blood pressure when used for an extended time or in large doses. Furthermore, a recent British study revealed that supplemental growth hormone therapy in children who do not lack the hormone causes redistribution of the body's fat and protein. The investigators compared two groups of otherwise healthy six- to eight-year-olds who were among the shortest for their age. One group consisted of children who were receiving growth hormone, the other of children who were not. At the end of six months, the children taking the synthetic hormone had outpaced the untreated group in growth by more than 1.5 inches per year. However, the untreated children added both muscle and fat as they grew, whereas the treated children became unusually muscular and lost up to 76% of their body fat. The loss of fat became especially obvious in their faces and limbs, giving them a rawboned, gangly appearance. It is unclear what long-term effects—either deleterious or desirable—these dramatic changes in body composition might have. Scientists also express concern that these readily observable physical changes may be accompanied by more subtle abnormalities in organs and cells. Thus, the debate about whether to make growth hormone available to normal but short children is likely to continue until further investigation demonstrates if it is safe and appropriate.

for protein synthesis. Growth hormone release is also stimulated by a decline in the level of blood fatty acids. Because of the fat-mobilizing actions of growth hormone, such regulation contributes to the maintenance of fairly constant blood fatty acid levels.

Note that the known regulatory inputs for growth hormone secretion are aimed at adjusting the levels of glucose, amino acids, and fatty acids in the blood. There are no known growth-related signals that influence growth hormone secretion. The whole issue of what really controls growth is complicated by the fact that growth hormone levels during early childhood, a period of quite rapid linear growth, are similar to those found in normal adults. As mentioned earlier, the poorly understood control of somatomedin activity may be important in this regard. Another related question is, why aren't tissues still responsive to growth hormone's growth-promoting effects in adulthood? We know that we do not grow any taller after adolescence because the epiphyseal plates have closed, but why don't the soft tissues continue to grow through hypertrophy and hyperplasia under the influence of growth hormone? One speculation is that the levels of growth hormone may only be high enough to produce its growth-promoting effects during the bursts in secretion that occur in deep sleep. It is interesting to note that time spent in deep sleep is greatest in infancy and gradually declines with age. Still, even as we age, we spend some time in deep sleep, yet we do not gradually get larger. Further research will be needed to unravel some of these mysteries.

Abnormal growth hormone secretion results in aberrant growth patterns.

Diseases related to both deficiencies and excesses of growth hormone can occur. The effects on the pattern of growth are much more pronounced than the metabolic consequences.

Growth Hormone Deficiency Growth hormone deficiency may be caused by a pituitary defect (lack of growth hormone) or occur secondarily to hypothalamic dysfunctions (lack of GHRH). Hyposecretion of growth hormone in a child results in **dwarfism**. The predominant feature is short stature caused by retarded skeletal growth. Less obvious characteristics include poorly developed musculature (reduction in muscle protein synthesis) and excess subcutaneous fat (less fat mobilization).

In addition, growth may be thwarted because the tissues fail to respond normally to growth hormone. **Laron dwarfism** is an example of this condition. Its symptoms resemble those of severe growth hormone deficiency even though blood levels of growth hormone are actually high. In some instances, growth hormone levels are adequate and target-cell responsiveness is normal, but somatomedins are deficient. African pygmies are an interesting example. Their short stature is attributable to a genetic deficit of the most potent of the somatomedins.

The onset of growth hormone deficiency in adulthood after growth is already complete produces relatively few symptoms. Growth-hormone–deficient adults tend to have reduced muscle strength (less muscle protein) as well as decreased bone density (less osteoblast activity during ongoing bone remodeling). (See the accompanying boxed feature, Concepts, Challenges, and Controversies.)

Growth Hormone Excess Hypersecretion of growth hormone is most often caused by a tumor of the growth-hormone–producing cells of the anterior pituitary. The symptoms that result depend on the age of the individual when the abnormal secretion begins. If overproduction of growth hormone begins in childhood before the epiphyseal plates are closed, the principal manifestation of the disorder is a rapid growth in height without distortion of body proportions. Appropriately, this condition is known as **gigantism**. If not treated by removal of the tumor or by drugs that block the effect of growth hormone, the individual may reach a height of eight feet or more. All of the soft tissues grow correspondingly, so the body is still well proportioned.

If growth hormone hypersecretion occurs after adolescence when the epiphyseal plates have already closed, further growth in height is prevented. Under the influence of excess growth hormone, however, the bones become thicker and the soft tissues, especially connective tissue and skin, proliferate. This disproportionate growth pattern produces a disfiguring condition known as **acromegaly** (*acro* means "extremity"; *megaly* means "large"). Bone thickening is most obvious in the extremities and face. A marked coarsening of the features to an almost apelike appearance gradually develops as the jaws and cheekbones become more prominent because of the thickening of the facial bones and the skin (• Fig. 18–21). The hands and feet enlarge, and the fingers and toes become greatly thickened. Peripheral nerve disor-

• **FIGURE 18–21 Patient with Acromegaly** Note the prominent cheekbones and jaw caused by thickening of the facial bones and skin.

ders often occur as a result of entrapment of nerves by overgrowth of connective tissue or bone or both.

Visual disturbances may accompany either gigantism or acromegaly. This is not due to excessive growth hormone per se but rather to the fact that a growth-hormone–secreting tumor causes enlargement of the anterior pituitary gland, which lies in close proximity to the optic chiasm, the point at which the nerve fibers passing from the eyes cross over on their way to the visual cortex of the brain (Fig. 18–11b).

Other hormones besides growth hormone are essential for normal growth.

Several other hormones in addition to growth hormone contribute in special ways to overall growth.

▪ *Thyroid hormone* is essential for growth but is not itself directly responsible for producing growth-promoting effects. It plays a permissive role in promoting skeletal growth; the actions of growth hormone are fully manifested only in the presence of adequate amounts of thyroid hormone. As a result, growth is severely stunted in hypothyroid children, but hypersecretion of thyroid hormone does not cause excessive growth.

▪ *Insulin* is believed to be an important growth-promoting factor, as is evidenced by a high correlation between insulin levels and growth. Growth failure often accompanies insulin deficiency, and hyperinsulinism is frequently associated with excessive growth. Since insulin promotes protein synthesis, its growth-promoting effects should not be surprising. However, insulin's growth-promoting effects may also arise from a different mechanism other than its direct effect on protein synthesis. Insulin is structurally similar to the somatomedins and may interact with the somatomedin (IGF 1) receptor, which is very similar to the insulin receptor.

▪ *Androgens,* which are believed to play an important role in the pubertal growth spurt, are powerful stimulants of protein synthesis in many organs. Androgens stimulate linear growth, promote weight gain, and increase muscle mass. The most potent androgen, testicular testosterone, is responsible for the development of the heavier musculature in men as compared to women. These androgenic growth-promoting effects depend on the presence of growth hormone. Androgens have virtually no effect on body growth in the absence of growth hormone, but in its presence, they synergistically enhance linear growth. Although androgens stimulate growth, they ultimately stop further growth by promoting closure of the epiphyseal plates.

▪ *Estrogens,* like androgens, ultimately terminate linear growth by stimulating complete conversion of the epiphyseal plates to bone. However, the effects of estrogens on growth prior to bone maturation are not well understood. Some studies suggest that large doses of estrogen may even inhibit further body growth.

Several factors contribute to the average height differences between men and women. First, since puberty occurs about two years earlier in girls than in boys, on the average boys have two more years of prepubertal growth than girls. As a result, boys are usually several inches taller than girls at the start of their respective growth spurts. Second, boys experience a greater androgen-induced growth spurt than girls before their respective gonadal steroids seal their long bones from further growth; this results in greater heights in men than in women on the average. Third, recent evidence suggests that androgens "imprint" the brains of males during development, giving rise to a "masculine" secretory pattern of growth hormone characterized by higher cyclical peaks, which are speculated to contribute to the greater height of males.

In addition to these hormones that exert overall effects on body growth, a number of poorly understood peptide *growth factors* have been identified that stimulate mitotic activity of specific tissues (for example, epidermal growth factor).

 ## CHAPTER IN PERSPECTIVE: FOCUS ON HOMEOSTASIS

The endocrine system is one of the body's two major control systems, the other being the nervous system. Through its relatively slowly acting hormonal messengers, the endocrine system generally regulates activities that require duration rather than speed. Most of these activities are directed toward maintaining homeostasis, as exemplified by the following:

▪ Hormones help maintain the proper concentration of nutrients in the internal environment by directing chemical reactions involved in the cellular uptake, storage, and release of these molecules. Furthermore, the rate at which these nutrients are metabolized is controlled in large part by the endocrine system.

▪ The control of H_2O balance, which is responsible for maintaining ECF osmolarity and proper cell volume, is largely accomplished by hormonal regulation of H_2O reabsorption by the kidneys during urine formation.

▪ Salt balance, which is important in the maintenance of ECF volume and arterial blood pressure, is achieved by hormonally controlled adjustments in salt reabsorption by the kidneys during urine formation. Likewise, hormones act on various target cells to maintain the plasma concentration of calcium and other electrolytes.

▪ The endocrine system orchestrates a wide range of adjustments that help the body maintain homeostasis in response to stressful situations.

▪ The endocrine and nervous systems work in concert to control the circulatory and digestive systems, which in turn carry out important homeostatic activities.

Unrelated to homeostasis, hormones direct the growing process and control most aspects of the reproductive system.

General Principles of Endocrinology

Hormones are long-distance chemical messengers secreted by the ductless endocrine glands into the blood, which transports them to specific target sites where they regulate or direct a particular function by altering protein activity within the target cells. Even though hormones are able to reach all tissues via the blood, they exert their effects only at their target cells because these cells alone have unique receptors for binding the hormone. The endocrine system is especially important in regulating fuel metabolism, H_2O and electrolyte balance, growth, and reproduction.

Hormones are grouped into three categories—based on differences in their mode of synthesis, storage, secretion, transport in the blood, and interaction with target cells—peptides, steroids, and amines, the latter including thyroid hormone and adrenomedullary catecholamines. Peptides and catecholamines are hydrophilic; steroid and thyroid hormones are lipophilic. Hydrophilic hormones are synthesized and packaged for export by the endoplasmic reticulum/Golgi complex route, stored in secretory vesicles, and released by exocytosis on appropriate stimulation. They dissolve freely in the blood for transport to their target cells where they bind with surface-membrane receptors. Upon binding, a hydrophilic hormone triggers a chain of intracellular events by means of a second messenger system that ultimately alters preexisting cellular proteins, usually enzymes, which exert the effect leading to the target cell's response to the hormone.

Steroids are synthesized by modifications of stored cholesterol by means of enzymes specific for each steroidogenic tissue. Steroids are not stored in the endocrine cells. Being lipophilic, they diffuse out through the lipid membrane barrier as soon as they are synthesized. Control of steroids is directed at their synthesis. Lipophilic steroids and thyroid hormone are both transported in the blood largely bound to carrier plasma proteins, with only free, unbound hormone being biologically active. Lipophilic hormones readily enter through the lipid membrane barriers of their target cells and bind with nuclear receptors. Hormonal binding activates the synthesis of new intracellular proteins that carry out the hormone's effect on the target cell.

The effective plasma concentration of each hormone is normally controlled by regulated changes in the rate of hormone secretion. Secretory output of endocrine cells is influenced by one or more of three different types of direct regulatory inputs: (1) neural input, which increases hormone secretion in response to a specific need and also governs diurnal variations in secretion; (2) input from another hormone, which involves either stimulatory input from a tropic hormone or inhibitory input from a target cell hormone in negative-feedback fashion; and (3) changes in the plasma concentration of an organic nutrient or an electrolyte being regulated by the hormone, also acting in negative-feedback fashion. In all cases, the regulated secretory output is designed to accomplish particular adjustments needed to maintain homeostasis, promote growth, or control reproduction.

Endocrine dysfunction arises when too much or too little of any particular hormone is secreted. Symptoms are generally referable to exaggerated or insufficient target-cell responses normally controlled by the hormone.

Hypothalamus and Pituitary

The pituitary gland consists of two distinct lobes, the posterior pituitary and the anterior pituitary. The hypothalamus, a portion of the brain, secretes nine peptide hormones; two are stored in the posterior pituitary, and seven are carried through a special vascular link to the anterior pituitary, where they either stimulate or inhibit the release of particular anterior pituitary hormones.

The posterior pituitary is essentially a neural extension of the hypothalamus. Two small peptide hormones, vasopressin and oxytocin, are synthesized within the cell bodies of neurosecretory neurons located in the hypothalamus, from which they pass down the axon to be stored in nerve terminals within the posterior pituitary. These hormones are independently released from the posterior pituitary into the blood in response to action potentials originating in the hypothalamus.

The anterior pituitary secretes six different peptide hormones that it produces itself. With the exception of prolactin, which stimulates milk secretion, the five other anterior pituitary hormones stimulate and maintain other endocrine tissues; that is, they are tropic. The sole function of thyroid-stimulating hormone (TSH) is to stimulate secretion of thyroid hormone, and that of adrenocorticotropic hormone (ACTH) is to stimulate secretion of cortisol by the adrenal cortex. The gonadotropic hormones—follicle-stimulating hormone (FSH) and luteinizing hormone (LH)—stimulate production of gametes (eggs and sperm) as well as secretion of sex hormones. Growth hormone stimulates growth and exerts metabolic effects as well.

The anterior pituitary releases its hormones into the blood at the bidding of releasing and inhibiting hormones from the hypothalamus. The hypothalamus, in turn, is influenced by a variety of neural and hormonal controlling inputs. Both the hypothalamus and the anterior pituitary are inhibited in negative-feedback fashion by the product of the target organ in the hypothalamus/anterior pituitary/target-organ axis.

Endocrine Control of Growth

Growth hormone promotes growth indirectly by stimulating the liver's production of somatomedins, which act directly on bone and soft tissues to cause growth. The growth hormone/somatomedin pathway causes growth by stimulating protein synthesis, cell division, and lengthening and thickening of bones. Growth hormone also directly exerts metabolic effects unrelated to growth on the liver, adipose tissue, and muscle, such as conservation of carbohydrates and mobilization of fat stores.

Growth hormone secretion by the anterior pituitary is regulated in negative-feedback fashion by two hypothalamic hormones, growth-hormone–releasing hormone and growth-hormone–inhibiting hormone. Growth hormone levels are not highly correlated with periods of rapid growth. The primary signals for increased growth hormone secretion are related to metabolic needs rather than growth, namely, deep sleep, stress, exercise, and low blood glucose levels.

Objective Questions (Answers on p. D–4.)

1. One endocrine gland may secrete more than one hormone. (True or false?)

2. One hormone may influence more than one type of target cell. (True or false?)

3. All endocrine glands are exclusively endocrine in function. (True or false?)

4. A single target cell may be influenced by more than one hormone. (True or false?)

5. Each steroidogenic organ has all of the enzymes necessary to produce any steroid hormone. (True or false?)

6. Hyposecretion or hypersecretion of a specific hormone can occur even though its endocrine gland is perfectly normal. (True or false?)

7. Inhibition of the anterior pituitary by a target-organ hormone is known as short-loop negative feedback. (True or false?)

8. Growth hormone levels in the blood are no higher during the early childhood growing years than during adulthood. (True or false?)

9. A hormone that has as its primary function the regulation of another endocrine gland is classified functionally as a _____ hormone.

10. Self-induced reduction in the number of receptors for a specific hormone is known as _____.

11. Activity within the cartilaginous layer of bone known as the _____ is responsible for linear growth of long bones.

12. Indicate the relationships among the hormones in the hypothalamic/anterior pituitary/adrenal cortex system by using the following answer code to identify which hormone belongs in each blank:
   a. cortisol
   b. ACTH
   c. CRH
   (1) _____ from the hypothalamus stimulates the secretion of (2) _____ from the anterior pituitary. (3) _____ in turn stimulates the secretion of (4) _____ from the adrenal cortex. In negative-feedback fashion, (5) _____ inhibits secretion of (6) _____ and furthermore reduces the sensitivity of the anterior pituitary to (7) _____.

13. Match the following:
   ___ 1. hormone
   ___ 2. neurohormone
   ___ 3. neurotransmitter
   ___ 4. paracrine

   a. local chemical mediator secreted by a variety of cells
   b. local chemical mediator secreted by neurons
   c. long-distance chemical mediator secreted by endocrine glands
   d. long-distance chemical mediator secreted by neurons

Essay Questions

1. List the overall functions of the endocrine system.

2. Compare the three categories of hormones in terms of chemical structure; mechanisms of synthesis, storage, and secretion; transport in the blood; and interaction with target cells.

3. By what means is the plasma concentration of a hormone normally regulated?

4. List and briefly state the functions of the posterior pituitary hormones.

5. List and briefly state the functions of the anterior pituitary hormones.

6. Compare the relationship between the hypothalamus and posterior pituitary with the relationship between the hypothalamus and anterior pituitary. Describe the role of the hypothalamo-hypophyseal portal system and the hypothalamic releasing and inhibiting hormones.

7. Describe the actions of growth hormone that are unrelated to growth. What are growth hormone's growth-promoting actions? What is the role of somatomedins?

8. Discuss the control of growth hormone secretion.

POINTS TO PONDER

1. A new supervisor at a local hospital decides to rotate the nursing staff to a different shift every week so that one group of employees is not always "stuck" on an undesirable shift. From a physiological viewpoint, do you think this proposal is advisable?

2. Would you expect the concentration of hypothalamic releasing and inhibiting hormones in a systemic venous blood sample to be higher, lower, or the same as the concentration of these hormones in a sample of hypothalamo-hypophyseal portal blood?

3. A patient displays symptoms of excess cortisol secretion. What factors could be measured in a blood sample to determine whether the condition is caused by a defect at the hypothalamic/anterior pituitary level or the adrenal cortex level?

4. Gigantism due to a pituitary tumor is usually treated by surgical removal of the pituitary gland. What hormonal replacement therapy do you think would have to be instituted following this procedure?

5. A black market for growth hormone abuse already exists among weight lifters and other athletes. What actions of growth hormone would induce a full-grown athlete to take supplemental doses of this hormone? What are the potential detrimental side effects?

19 Peripheral Endocrine Glands

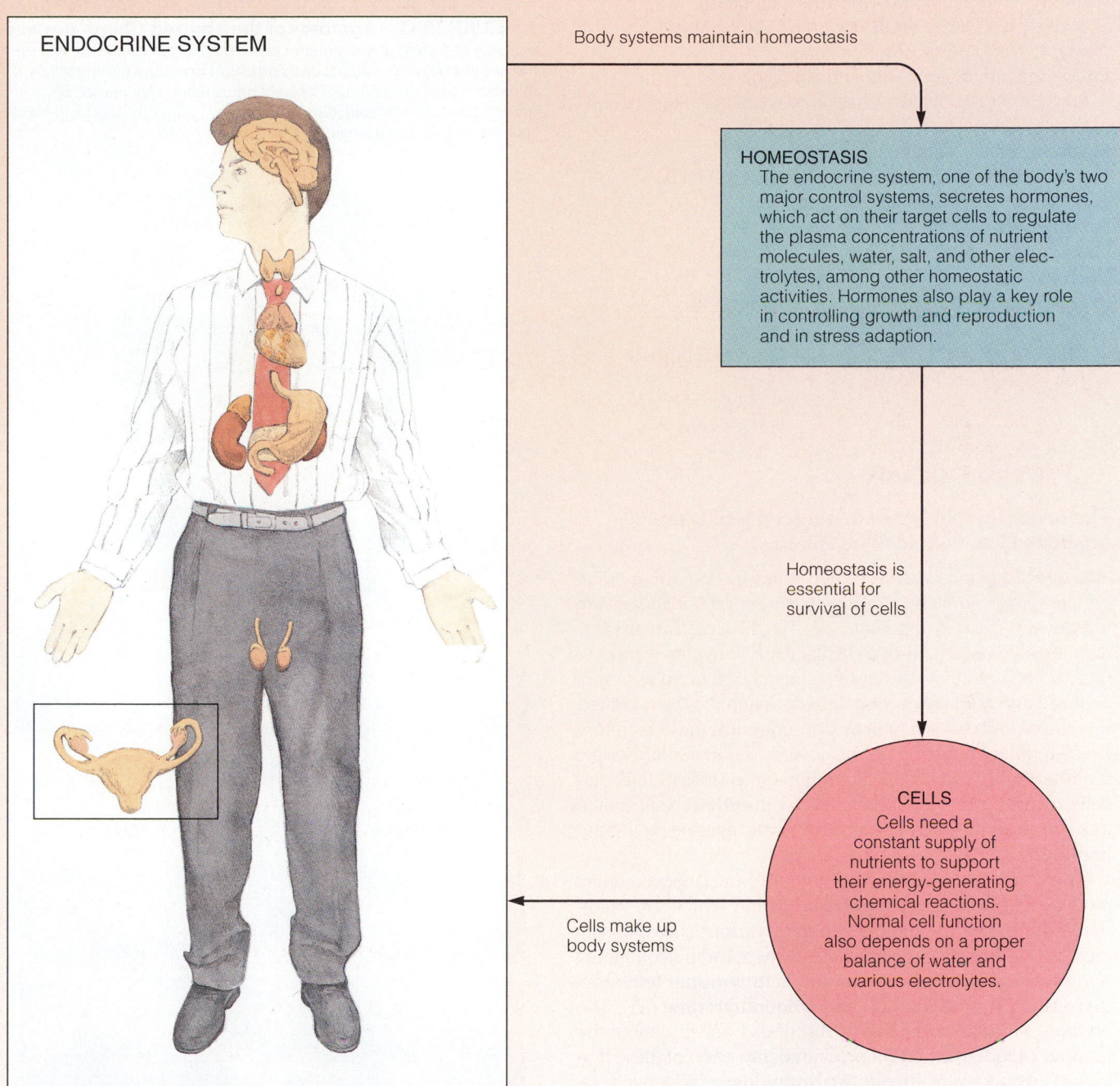

ENDOCRINE SYSTEM

Body systems maintain homeostasis

HOMEOSTASIS
The endocrine system, one of the body's two major control systems, secretes hormones, which act on their target cells to regulate the plasma concentrations of nutrient molecules, water, salt, and other electrolytes, among other homeostatic activities. Hormones also play a key role in controlling growth and reproduction and in stress adaption.

Homeostasis is essential for survival of cells

CELLS
Cells need a constant supply of nutrients to support their energy-generating chemical reactions. Normal cell function also depends on a proper balance of water and various electrolytes.

Cells make up body systems

The endocrine system, by means of the blood-borne hormones it secretes, generally regulates activities that require duration rather than speed. The peripheral endocrine glands include the **thryoid gland**, which controls the body's basal metabolic rate; the **adrenal glands**, which secrete hormones important in the metabolism of nutrient molecules, adaptation to stress, and maintenance of salt balance; the **endocrine pancreas**, which secretes hormones important in the metabolism of nutrient molecules; and the **parathryoid glands**, which secrete a hormone important in Ca^{++} metabolism.

CHAPTER CONTENTS AT A GLANCE

THYROID GLAND
Anatomy of the thyroid gland
Thyroid hormone
ADRENAL GLANDS
Adrenocortical steroids
Adrenomedullary catecholamines
Hormonal and neural roles in stress adaptation
ENDOCRINE CONTROL OF FUEL METABOLISM
Absorptive and postabsorptive states
Endocrine pancreas: insulin and glucagon
● CONCEPTS, CHALLENGES, AND CONTROVERSIES Diabetics and Insulin: Some Have It and Some Don't
Metabolic effects of other hormones
ENDOCRINE CONTROL OF CALCIUM METABOLISM
Importance of free plasma calcium
Parathyroid hormone
Calcitonin
Vitamin D
Relationship of phosphate metabolism to calcium metabolism
● A CLOSER LOOK AT EXERCISE PHYSIOLOGY Osteoporosis: The Bane of Brittle Bones

THYROID GLAND

The major thyroid hormone secretory cells are organized into colloid-filled spheres.

The **thyroid gland** consists of two lobes of endocrine tissue joined in the middle by a narrow portion of the gland, giving it a bow-tie–shaped appearance (● Fig. 19–1a). The gland is even located in the appropriate place for a bow tie, lying over the trachea just below the larynx. The major thyroid secretory cells are arranged into hollow spheres, each of which forms a functional unit called a **follicle**. Consequently, these secretory cells are often referred to as *follicular cells*. On a microscopic section (Fig. 19–1b), the follicles appear as rings of follicular cells enclosing an inner lumen filled with **colloid**, a substance that serves as an "inland" extracellular storage site for thyroid hormones.

The chief constituent of the colloid is a large, complex molecule known as **thyroglobulin**, within which are incorporated the thyroid hormones in their various stages of synthesis. The follicular cells produce two iodine-containing hormones derived from the amino acid tyrosine: **tetraiodothyronine** (T_4 or **thyroxine**) and **triiodothyronine** (T_3). The prefixes *tetra* and *tri* and the subscripts 4 and 3 denote the number of iodine atoms incorporated into each of these hormones, respectively. These two hormones, collectively referred to as **thyroid hormone**, are important regulators of overall basal metabolic rate.

Interspersed in the interstitial spaces between the follicles is another secretory cell type, the **C cells**, so called because they secrete the peptide hormone **calcitonin**, which plays a role in calcium metabolism. Calcitonin is not related in any way to the two other major thyroid hormones. We will restrict our discussion of thyroid hormones to the secretions of the follicular cells, deferring coverage of calcitonin until a later section dealing with endocrine regulation of calcium balance.

● **FIGURE 19–1   Anatomy of the Thyroid Gland** (a) Gross anatomy of thyroid gland, anterior view. The thyroid gland lies over the trachea just below the larynx and consists of two lobes connected by a thin strip called the isthmus. (b) Light-microscope appearance of thyroid gland. The thyroid gland is composed primarily of colloid-filled spheres enclosed by a single layer of follicular cells.

Thyroid gland

Right lobe   Trachea   Isthmus   Left lobe

(a)

Follicular cell   Colloid

(b)

All of the steps of thyroid hormone synthesis occur on the large thyroglobulin molecule, which subsequently stores the hormones.

The basic ingredients for thyroid hormone synthesis are tyrosine and iodine, both of which must be taken up from the blood by the follicular cells. Tyrosine, an amino acid, is synthesized in sufficient amounts by the body, so it is not an essential dietary requirement. The iodine needed for thyroid hormone synthesis, on the other hand, must be obtained from dietary intake. The synthesis, storage, and secretion of thyroid hormone involve the following steps:

1. All steps of thyroid hormone synthesis take place on the thyroglobulin molecules within the colloid. Thyroglobulin itself is produced by the endoplasmic reticulum/Golgi complex of the thyroid follicular cells. Tyrosine becomes incorporated in the much larger thyroglobulin molecules as the latter are being produced. Once produced, tyrosine-containing thyroglobulin is exported from the follicular cells into the colloid by exocytosis (Step 1 in Figure 19–2).

2. The thyroid captures iodine from the blood and transfers it into the colloid by means of a very active "iodine pump" or "iodine-trapping mechanism"—the powerful, energy-requiring carrier proteins located in the outer membranes of the follicular cells (Step 2). Almost all of the iodine in the body is moved against its concentration gradient to become trapped in the thyroid for the purpose of thyroid hormone synthesis. Iodine serves no other purpose in the body.

3. Within the colloid, iodine is quickly attached to a tyrosine within the thyroglobulin molecule. Attachment of one iodine to tyrosine yields **monoiodotyrosine (MIT)** (Step 3a). Attachment of two iodines to tyrosine yields **diiodotyrosine (DIT)** (Step 3b).

4. Next, a coupling process occurs between the iodinated tyrosine molecules to form the thyroid hormones. Coupling of two DITs (each bearing two iodine atoms) yields **tetraiodothyronine (T_4 or thyroxine)**, the four-iodine form of thyroid hormone (Step 4a). Coupling of one MIT (with one iodine) and one DIT (with two iodines) yields **triiodothyronine** or **T_3** (with three iodines) (Step 4b). Coupling does not occur between two MIT molecules.

Because these reactions occur within the thyroglobulin molecule, all the products remain attached to this large protein. Thyroid hormones remain stored in this form in the colloid until they are split off and secreted. It is estimated that

---

FIGURE 19–2 Synthesis, Storage, and Secretion of Thyroid Hormone Tyrosine-containing TGB produced within the thyroid follicular cells is transported into the colloid by exocytosis ①. Iodine is actively transported from the blood into the colloid by the follicular cells ②. Attachment of one iodine to tyrosine within the TGB molecule yields MIT ③a; attachment of two iodines yields DIT ③b. Coupling of two DITs yields T_4 ④a; coupling of one MIT and one DIT yields T_3 ④b. On appropriate stimulation, the thyroid follicular cells engulf a portion of TGB-containing colloid by phagocytosis ⑤. Lysosomes attack the engulfed vesicle and split the iodinated products from TGB ⑥. T_3 and T_4 diffuse into the blood ⑦a. MIT and DIT are deiodinated, and the freed iodine is recycled for synthesis of more hormone ⑦b.

TGB = Thyroglobulin
I = Iodine
MIT = Monoiodotyrosine

DIT = Diiodotyrosine
T_3 = Triiodothyronine
T_4 = Tetraiodothyronine (thyroxine)

* Organelles not drawn to scale. Endoplasmic reticulum/Golgi complex are proportionally too small.

sufficient thyroid hormone is normally stored in the colloid to supply the body's needs for several months.

The follicular cells phagocytize thyroglobulin-laden colloid to accomplish thyroid hormone secretion.

The release of the thyroid hormones into the systemic circulation requires a rather complex process for two reasons. First, before their release, T_4 and T_3 are still bound within the thyroglobulin molecule. Second, these hormones are stored at an inland extracellular site, the follicular lumen; before they can enter the blood vessels that course through the interstitial spaces, they must be transported completely across the follicular cells. The process of thyroid hormone secretion essentially involves the follicular cells "biting off" a piece of colloid, breaking the thyroglobulin molecule down into its component parts, and "spitting out" the freed T_4 and T_3 into the blood. Upon appropriate stimulus for thyroid hormone secretion, the follicular cells internalize a portion of the thyroglobulin-hormone complex by phagocytizing a piece of colloid (Step 5 of Fig. 19–2). Within the cells, the membrane-bounded droplets of colloid coalesce with lysosomes, whose enzymes split off the biologically active thyroid hormones, T_4 and T_3, as well as the inactive iodotyrosines, MIT and DIT (Step 6). The thyroid hormones, being very lipophilic, pass freely through the outer membranes of the follicular cells and into the blood (Step 7a). The MIT and DIT are of no endocrine value. The follicular cells contain an enzyme that swiftly removes the iodine from MIT and DIT, allowing the freed iodine to be recycled for synthesis of more hormone (Step 7b). This highly specific enzyme will remove iodine only from the worthless MIT and DIT, not the valuable T_4 or T_3.

Most of the secreted T_4 is converted into T_3 outside the thyroid.

About 90% of the secretory product released from the thyroid gland is in the form of T_4, yet T_3 is about four times more potent in its biological activity. However, most of the secreted T_4 is converted into T_3, or *activated*, by being stripped of one of its iodines in the liver and kidneys. About 80% of the circulating T_3 is derived from secreted T_4 that has been peripherally stripped. Therefore, T_3 is the major biologically active form of thyroid hormone at the cellular level, even though the thyroid secretes mostly T_4.

For the most part, both T_4 and T_3 are transported bound to specific plasma proteins.

Once released into the blood, the highly lipophilic thyroid hormone molecules very quickly bind with several plasma proteins. Less than 1% of the T_3 and less than 0.1% of the T_4 remain in the unbound (free) form. This is remarkable considering that only the free portion of the total thyroid hormone pool has access to the target-cell receptors and thus is able to exert an effect.

Three different plasma proteins are important in thyroid hormone binding: *thyroxine-binding globulin* selectively binds only thyroid hormones—55% of the circulating T_4 and 65% of the T_3—even though its name specifies only "thyroxine" (T_4); *albumin* nonselectively binds many lipophilic hormones, including 10% of the T_4 and 35% of the T_3; and *thyroxine-binding prealbumin* binds the remaining 35% of the T_4.

Thyroid hormone is the primary determinant of the body's overall metabolic rate and is also important for bodily growth and normal development and function of the nervous system.

Virtually every tissue in the body is affected either directly or indirectly by thyroid hormone. The effects of T_3 and T_4 can be grouped into several overlapping categories.

Effect on Metabolic Rate Thyroid hormone increases the body's overall basal metabolic rate or "idling speed" (see p. 597). This hormone is the most important regulator of the body's rate of O_2 consumption and energy expenditure under resting conditions.

Compared to other hormones, the action of thyroid hormone is "sluggish." Only after a delay of several hours is the metabolic response to thyroid hormone detectable, and the maximal response is not evident for several days. The duration of the response is also quite long, partially because thyroid hormone is not rapidly degraded but also because the response continues to be expressed for days or even weeks after the plasma thyroid hormone concentrations have returned to normal.

Calorigenic Effect Closely related to thyroid hormone's overall metabolic effect is its **calorigenic** (heat-producing) effect. Increased metabolic activity results in increased heat production.

Effect on Intermediary Metabolism In addition to increasing the general metabolic rate, thyroid hormone modulates the rates of many specific reactions involved in fuel metabolism. The effects of thyroid hormone on the metabolic fuels are multifaceted; not only can it influence both the synthesis and degradation of carbohydrate, fat, and protein, but small or large amounts of the hormone may induce opposite effects. For example, the conversion of glucose to glycogen, the storage form of glucose, is facilitated by small amounts of thyroid hormone, but the reverse—the breakdown of glycogen into glucose—occurs with large amounts of the hormone. Similarly, adequate amounts of thyroid hormone are essential for the protein synthesis needed for normal bodily growth, yet protein degradation effects predominate at high doses. In general, at abnormally high plasma levels of thyroid hormone, such as in thyroid hypersecretion, the overall effect is to favor consumption rather than storage of fuel, as manifested by depletion of liver glycogen stores, depletion of fat stores, and muscle wasting resulting from protein degradation. (The main structural component of cells is protein. Muscle cells are particularly rich in structural protein because they are packed full of contractile elements made of protein—that is, the actin and myosin filaments.)

Sympathomimetic Effect Any action similar to one produced by the sympathetic nervous system is known as a **sympathomimetic** ("sympathetic-mimicking") **effect**. Thyroid hormone increases target-cell responsiveness to catecholamines (epinephrine and norepinephrine), the chemical messengers used by the sympathetic nervous system and its hormonal reinforcements from the adrenal medulla. Thyroid hormone presumably accomplishes this permissive action by causing a proliferation of specific catecholamine target-cell receptors (see p. 630). Because of this action, many of the effects observed when thyroid hormone secretion is elevated are similar to those that accompany activation of the sympathetic nervous system (a sympathomimetic effect).

Effect on the Cardiovascular System Through its effect of increasing responsiveness of the heart to circulating catecholamines, thyroid hormone increases heart rate and force of contraction, thus increasing cardiac output. In addition, in response to the heat load generated by the calorigenic effect of thyroid hormone, peripheral vasodilation occurs to carry the extra heat to the body surface for elimination to the environment (see p. 607).

Effect on Growth and the Nervous System Thyroid hormone is essential for normal growth. The growth-promoting effect of thyroid hormone seems to be secondary to its effects on growth hormone. Thyroid hormone not only stimulates growth hormone secretion but also promotes the effects of growth hormone (or somatomedins) on the synthesis of new structural proteins and on skeletal growth. Thyroid-deficient children have stunted growth that is reversible with thyroid replacement therapy. Unlike excess growth hormone, however, excess thyroid hormone does not result in excessive growth.

Thyroid hormone plays a crucial role in the normal development of the nervous system, especially the CNS, an effect impeded in children with thyroid deficiency from birth. Thyroid hormone is also essential for normal CNS activity in adults. Abnormal thyroid hormone levels are associated with behavioral changes. Furthermore, the conduction velocity of peripheral nerves varies directly with the availability of thyroid hormone.

Thyroid hormone is regulated by the hypothalamus-pituitary-thyroid axis.

Thyroid-stimulating hormone (TSH), the thyroid tropic hormone from the anterior pituitary, is the most important physiological regulator of thyroid hormone secretion (Fig. 19–3). Almost every step of thyroid hormone synthesis and release is stimulated by TSH.

In addition to enhancing thyroid hormone secretion, TSH is responsible for maintaining the structural integrity of the thyroid gland. In the absence of TSH, the thyroid atrophies (decreases in size) and secretes its hormones at a very

● **FIGURE 19–3   Regulation of Thyroid Hormone Secretion**

low rate. Conversely, it undergoes hypertrophy (increase in the size of each follicular cell) and hyperplasia (increase in the number of follicular cells) in response to excess TSH stimulation.

Thyroid hormone, in negative-feedback fashion, "turns off" TSH secretion, whereas the hypothalamic **thyrotropin-releasing hormone** (TRH), in tropic fashion, "turns on" TSH secretion by the anterior pituitary. In the hypothalamus-pituitary-thyroid axis, inhibition is exerted primarily at the level of the anterior pituitary. Like the other negative-feedback loops, the one between thyroid hormone and TSH tends to maintain stability of thyroid hormone output.

Day-to-day regulation of free thyroid hormone levels is apparently accomplished by negative feedback between the thyroid and anterior pituitary whereas long-range adjustments are mediated by the hypothalamus. Unlike most of the other hormonal systems, the hormones in the thyroid axis in an adult normally do not undergo sudden, wide swings in secretion. The relatively steady rate of thyroid hormone secretion is in keeping with the sluggish, long-lasting responses that this hormone induces; there would be no adaptive value in suddenly increasing or decreasing plasma thyroid hormone levels. The only known factor that increases TRH (and, accordingly, TSH and thyroid hormone) secretion is exposure to cold in infants; this is a highly adaptive mechanism in newborn infants. The dramatic increase in heat-producing thyroid hormone secretion is thought to contribute to the maintenance of body temperature in the face of the abrupt drop in surrounding temperature at birth, as the infant goes from the mother's warm body to the cooler environmental air. A similar TSH response to cold exposure does not occur in adults, although it makes sense physiologically and does occur in many types of experimental animals.

Various types of stress are known to inhibit TSH and thyroid hormone secretion, presumably through neural influences on the hypothalamus, although the adaptive importance of this inhibition is unclear.

Abnormalities of thyroid function include both hypothyroidism and hyperthyroidism.

Normal thyroid function is referred to as **euthyroidism**. Abnormalities of thyroid function are among the most common of all endocrine disorders. They fall into two major categories—*hypothyroidism* and *hyperthyroidism*—reflecting deficient and excess thyroid hormone secretion, respectively. A number of specific causes can give rise to each of these conditions (Table 19–1). Whatever the cause, the consequences of too little or too much thyroid hormone secretion are largely predictable, based on a knowledge of the functions of thyroid hormone.

Hypothyroidism   **Hypothyroidism** can result (1) from primary failure of the thyroid gland itself; (2) secondary to a deficiency of TRH, TSH or both; or (3) from an inadequate dietary supply of iodine. The symptoms of hypothyroidism are largely referable to a reduction in overall metabolic activity. Among other things, a patient with hypothyroidism has a reduced basal metabolic rate; displays poor tolerance of cold (lack of calorigenic effect); has a tendency to gain excessive weight (not burning fuels at normal rate); is easily fatigued (lower energy production); has a slow, weak pulse (caused

by a reduction in the rate and strength of cardiac contraction and a lowered cardiac output); and exhibits slow reflexes and slow mentation (because of the effect on the nervous system). The latter is characterized by diminished alertness, slow speech, and poor memory.

Another notable characteristic is an edematous condition caused by infiltration of the skin with complex, water-retaining carbohydrate molecules, presumably as a result of altered metabolism. The resultant puffy appearance, primarily of the face, hands, and feet, is known as **myxedema**. In fact, the term *myxedema* is often used as a synonym for hypothyroidism in an adult because of the prominence of this symptom.

If an individual has hypothyroidism from birth, a condition known as **cretinism** develops. Because adequate levels of thyroid hormone are essential for normal growth and CNS development, cretinism is characterized by dwarfism and mental retardation as well as other general symptoms of thyroid deficiency. The mental retardation is preventable if replacement therapy is started promptly, but it is not reversible once it has developed for a few months after birth, even with later treatment with thyroid hormone. For this reason, an important public health measure is the recent establishment of neonatal screening programs in which congenital hypothyroidism can be detected from a single drop of a newborn infant's blood.

Treatment of hypothyroidism, with one exception, consists of replacement therapy through the administration of exogenous thyroid hormone. The exception is hypothyroidism caused by iodine deficiency, in which the remedy is adequate dietary iodine.

Hyperthyroidism   The most common cause of **hyperthyroidism** is **Grave's disease**, an autoimmune disease in which the body erroneously produces **thyroid-stimulating immunoglobulin** (**TSI**), an antibody whose target is the TSH receptors on the thyroid cells. Thyroid-stimulating immunoglobulin stimulates both secretion and growth of the thyroid in a manner similar to TSH. Unlike TSH, however, TSI is not subject to negative-feedback inhibition by thyroid hormone, so thyroid secretion and growth continue unchecked (•Fig. 19–4). Less frequently, hyperthyroidism occurs secondary to excess TRH or TSH or in association with a hypersecreting thyroid tumor.

TABLE 19–1 Types of Thyroid Dysfunctions			
THYROID DYSFUNCTION	**CAUSE**	**PLASMA CONCENTRATIONS OF RELEVANT HORMONES**	**GOITER PRESENT?**
Hypothyroidism	Primary failure of thyroid gland	↓ T_3 and T_4, ↑ TSH	Yes
	Secondary to hypothalamic or anterior pituitary failure	↓ T_3 and T_4, ↓ TRH and/or ↓ TSH	No
	Lack of dietary iodine	↓ T_3 and T_4, ↑ TSH	Yes
Hyperthyroidism	Abnormal presence of thyroid-stimulating immunoglobulin (TSI) (Grave's disease)	↑ T_3 and T_4, ↓ TSH	Yes
	Secondary to excess hypothalamic or anterior pituitary secretion	↑ T_3 and T_4, ↑ TRH and/or ↑ TSH	Yes
	Hypersecreting thyroid tumor	↑ T_3 and T_4, ↓ TSH	No

• **FIGURE 19–4   Role of Thyroid-Stimulating Immunoglobulin in Grave's Disease** Thyroid-stimulating immunoglobulin, an antibody erroneously produced in the autoimmune condition of Grave's disease, binds with the TSH receptors on the thyroid gland and continuously stimulates thyroid hormone secretion outside the normal negative-feedback control system.

• **FIGURE 19–5   Patient Displaying Exophthalmos** Abnormal fluid retention behind the eyeballs causes them to bulge forward.

As expected, the hyperthyroid patient has an elevated basal metabolic rate. The resultant increase in heat production leads to excessive perspiration and poor tolerance of heat. In spite of the increased appetite and food intake that occur in response to the increased metabolic demands, body weight typically falls because the body is burning fuel at an abnormally rapid rate. Net degradation of endogenous carbohydrate, fat, and protein stores occurs. The resultant loss of skeletal-muscle protein results in weakness. Various cardiovascular abnormalities are associated with hyperthyroidism, caused both by the direct effects of thyroid hormone and by its interactions with catecholamines. Heart rate and strength of contraction may increase so much that the individual has palpitations (an unpleasant awareness of the heart's activity). In severe cases, the heart may fail to meet the body's metabolic demands in spite of increased cardiac output. Nervous-system involvement is manifested by an excessive degree of mental alertness to the point where the patient is irritable, tense, anxious, and excessively emotional.

A prominent feature of Grave's disease but not of the other types of hyperthyroidism is **exophthalmos** (bulging eyes) (• Fig. 19–5). Complex, water-retaining carbohydrates are deposited behind the eyes for reasons still being debated. The resultant fluid retention behind the eyes pushes the eyeballs forward so that they bulge from their bony orbit. The eyeballs may bulge so far that the lids cannot completely close, in which case the eyes become dry, irritated, and prone to corneal ulceration. Even after correction of the hyperthyroid condition, these troublesome eye symptoms may persist.

Three general methods of treatment are available for suppressing excess thyroid hormone secretion: surgical removal of a portion of the oversecreting thyroid gland; administration of radioactive iodine, which, after being concentrated in the thyroid gland by the iodine pump, selectively destroys thyroid glandular tissue; and use of antithyroid drugs that specifically interfere with thyroid hormone synthesis.

A goiter may or may not accompany either hypothyroidism or hyperthyroidism.

A **goiter** refers to an enlarged thyroid gland. Because of the location of the thyroid over the trachea, a goiter is readily palpable and usually highly visible (• Fig. 19–6). A goiter will occur whenever there is excessive stimulation of the thyroid gland by either TSH or TSI (Table 19–1). Note from the table that a goiter may accompany hypothyroidism or hyperthyroidism, but it need not be present in either condition. Knowing the hypothalamus-pituitary-thyroid axis and feedback control, we can predict which types of hypo- and hyperthyroidism will be accompanied by a goiter:

■ Hypothyroidism secondary to hypothalamic or anterior pituitary failure will not be accompanied by a goiter because

• **FIGURE 19–6   Patient with Goiter**

the thyroid gland is not being adequately stimulated, let alone excessively stimulated.

- With hypothyroidism caused by thyroid gland failure or lack of iodine, a goiter does develop because the circulating level of thyroid hormone is so low that there is little negative-feedback inhibition on the anterior pituitary, and TSH secretion is therefore elevated. Recall that TSH acts on the thyroid to increase the size and number of follicular cells and to increase their rate of secretion. If the thyroid cells are incapable of secreting hormone because of a lack of a critical enzyme or lack of iodine, no amount of TSH will be able to induce these cells to secrete T_3 and T_4. However, TSH can still promote hypertrophy and hyperplasia of the thyroid with a consequent paradoxical enlargement of the gland (that is, a goiter), even though the gland is still underproducing.

Similarly, a goiter may or may not accompany hyperthyroidism:

- Excessive TSH secretion resulting from a hypothalamic or anterior pituitary defect would obviously be accompanied by a goiter because of overstimulation of thyroid growth. Since the thyroid gland in this circumstance is also capable of responding to excess TSH with increased hormone secretion, hyperthyroidism is present with this goiter.

- In Grave's disease, a hypersecreting goiter occurs because TSI promotes growth of the thyroid as well as enhancing secretion of thyroid hormone. Because the high levels of circulating T_3 and T_4 inhibit the anterior pituitary, TSH secretion itself is low. In all other cases when a goiter is present, TSH levels are elevated and are directly responsible for excessive growth of the thyroid.

- Hyperthyroidism resulting from overactivity of the thyroid in the absence of overstimulation, such as that caused by an uncontrolled thyroid tumor, is not accompanied by a goiter. The spontaneous secretion of excessive amounts of T_3 and T_4 inhibits TSH, so there is no stimulatory input to promote growth of the thyroid.

ADRENAL GLANDS

Each adrenal gland consists of an outer, steroid-secreting adrenal cortex and an inner, catecholamine-secreting adrenal medulla.

There are two **adrenal glands**, one embedded above each kidney in a capsule of fat (*adrenal* means "next to the kidney") (Fig. 19–7a). Each adrenal is actually composed of two endocrine organs, one surrounding the other (Fig. 19–7b). The inner portion, the **adrenal medulla**, secretes

• **FIGURE 19-7   Anatomy of the Adrenal Glands**   (a) Location and structure of the adrenal glands. (b) Layers of the adrenal cortex.

(a)

(b)

catecholamines; the outer layers composing the **adrenal cortex** secrete a variety of steroid hormones. Derived embryologically from distinct structures, the adrenal medulla and cortex secrete hormones belonging to different chemical categories, whose functions, mechanisms of action, and regulation are entirely different.

The adrenal cortex secretes mineralocorticoids, glucocorticoids, and sex hormones.

About 80% of the adrenal gland is composed of the cortex, which consists of three different layers or zones: the **zona glomerulosa**, the outermost layer; the **zona fasciculata**, the middle and largest portion; and the **zona reticularis**, the innermost zone (Fig. 19–7b). The adrenal cortex produces a number of different **adrenocortical hormones**, all of which are steroids derived from the common precursor molecule, cholesterol. Slight variations in structure confer different functional capabilities on the various adrenocortical hormones. On the basis of their primary actions, the adrenal steroids can be divided into three categories: (1) **mineralocorticoids**, mainly *aldosterone*, which influence mineral (electrolyte) balance; (2) **glucocorticoids**, primarily *cortisol*, which play a major role in glucose metabolism as well as in protein and lipid metabolism; and (3) **sex hormones** identical or similar to those produced by the gonads (testes in males, ovaries in females).

The production of adrenocortical and other steroid hormones requires a series of steps in which the cholesterol molecule undergoes various enzymatic modifications. The different functional types of adrenal steroids are produced in anatomically distinct portions of the adrenal cortex. This zonation is the result of differential distribution of the enzymes that are required to catalyze the different biosynthetic pathways leading to the formation of each of these steroids. Of the two major adrenocortical hormones, aldosterone is produced exclusively in the zona glomerulosa. Cortisol synthesis is limited to the two inner layers of the cortex, with the zona fasciculata being the major source of this glucocorticoid. No other steroidogenic tissues have the capability of producing either mineralocorticoids or glucocorticoids.

In contrast, the adrenal sex hormones, also produced by the inner cortical zones, are produced in far greater abundance in the gonads. In both sexes, the zona fasciculata and zona reticularis produce both *androgens,* or "male" sex hormones, and *estrogens,* or "female" sex hormones. Because the enzymes required for the production of these sex hormones are found in very low concentrations in the adrenocortical cells, androgens and estrogens are normally produced in very small quantities from this source. The testes are the main site of production for androgens (of which testosterone is the most powerful and most abundant) and the ovaries for estrogens. Accordingly, males have a preponderance of circulating androgens, whereas in females, estrogens are predominant. However, no hormones are unique to either males or females (except those from the placenta during pregnancy) because small amounts of the sex hormone of the opposite sex are produced by the adrenal cortex in both sexes.

Under normal circumstances, the adrenal androgens and estrogens are not sufficiently abundant or powerful to induce masculinizing or feminizing effects, respectively. The only adrenal sex hormone that has any biological importance is the androgen **dehydroepiandrosterone** (**DHEA**). The testes' primary androgen product is the potent testosterone, but the most abundant adrenal androgen is the much weaker DHEA. Adrenal DHEA is overpowered by testicular testosterone in males but is of physiological significance in females, who otherwise lack androgens. This adrenal androgen is responsible for androgen-dependent processes in the female such as growth of pubic and axillary (armpit) hair, enhancement of the pubertal growth spurt, and development and maintenance of the female sex drive.

Being lipophilic, the adrenocortical hormones are all carried in the blood extensively bound to plasma proteins. About 60% of circulating aldosterone is protein bound, primarily to nonspecific albumin. Cortisol is 98% bound, mostly to a plasma protein specific for it called **corticosteroid-binding globulin** (**transcortin**). Likewise, 98% of DHEA is bound, in this case exclusively to albumin.

Mineralocorticoids' major effects are on electrolyte balance and blood-pressure homeostasis.

The actions and regulation of the primary adrenocortical mineralocorticoid, **aldosterone**, are described thoroughly elsewhere (chapters 14 and 15). To highlight aldosterone activity, its principal site of action is on the distal tubules of the kidney, where it promotes Na^+ retention and enhances K^+ elimination during the formation of urine. The promotion of Na^+ retention by aldosterone secondarily induces osmotic retention of H_2O, thereby causing expansion of the ECF volume, which is important in long-term regulation of blood pressure.

Mineralocorticoids are *essential for life*. Without aldosterone, a person rapidly dies from circulatory shock because of the marked fall in plasma volume caused by excessive losses of H_2O-holding Na^+. With most other hormonal deficiencies, death is not imminent, even though a chronic hormonal deficiency may eventually lead to a premature death.

Aldosterone secretion is increased by (1) activation of the renin-angiotensin-aldosterone system by factors related to a reduction in Na^+ and a fall in blood pressure and (2) direct stimulation of the adrenal cortex by a rise in plasma K^+ concentration (see Fig. 14–22, p. 489). In addition to its effect on aldosterone secretion, angiotensin promotes growth of the zona glomerulosa, in a manner similar to the effect of TSH on the thyroid. The adrenal tropic hormone ACTH primarily affects the inner cortical zones and has only a weak effect in stimulating aldosterone secretion. Thus, unlike cortisol regulation, the regulation of aldosterone secretion is largely independent of anterior pituitary control.

Glucocorticoids exert metabolic effects and have an important role in adaptation to stress.

Cortisol, the primary glucocorticoid, plays an important role in carbohydrate, protein, and fat metabolism; exhibits significant permissive actions for other hormonal activities; and helps people resist stress.

Metabolic Effects   The overall effect of cortisol's metabolic actions is to increase the concentration of blood glucose at the expense of protein and fat stores. Specifically, cortisol performs the following functions:

- It stimulates hepatic **gluconeogenesis** (*gluco* means "glucose"; *neo* means "new"; *genesis* means "production"), which refers to the conversion of noncarbohydrate sources (namely, amino acids) into carbohydrate within the liver. Between meals or during periods of fasting, when no new nutrients are being absorbed into the blood for utilization and storage, the glycogen (stored glucose) in the liver tends to become depleted as it is broken down to release glucose into the blood. Gluconeogenesis is an important factor in replenishing hepatic glycogen stores and thus in maintaining normal blood glucose levels between meals. This is essential because the brain can use only glucose as its metabolic fuel, yet nervous tissue cannot store glycogen to any extent. The concentration of glucose in the blood must therefore be maintained at an appropriate level to adequately supply the glucose-dependent brain with nutrients.

- It inhibits glucose uptake and use by many tissues, but not the brain, thus sparing glucose for use by the brain, which absolutely requires it as a metabolic fuel.

- It stimulates protein degradation in many tissues, especially muscle. By breaking down a portion of muscle proteins into their constituent amino acids, cortisol increases the concentration of blood amino acids. These mobilized amino acids are available for use in gluconeogenesis or wherever else they are needed, such as for repair of damaged tissue or synthesis of new cellular structures.

- It facilitates lipolysis (*lysis* means "breakdown"), the breakdown of lipid (fat) stores in adipose tissue, thus releasing free fatty acids into the blood. The mobilized fatty acids are available as an alternative metabolic fuel for tissues that can use this energy source in lieu of glucose, thereby conserving glucose for the brain.

Permissive Actions   Cortisol is extremely important for its permissiveness. For example, cortisol must be present in adequate amounts to permit the catecholamines to induce vasoconstriction. A person lacking cortisol, if untreated, may go into circulatory shock in a stressful situation that demands immediate widespread vasoconstriction.

Role in Adaptation to Stress   Cortisol plays a key role in adaptation to stress. **Stress** refers to the generalized, nonspecific response of the body to any factor that overwhelms, or threatens to overwhelm, the body's compensatory abilities to maintain homeostasis. Contrary to popular usage, the agent inducing the response is correctly called a *stressor*, whereas *stress* refers to the state induced by the stressor. The following types of noxious stimuli illustrate the range of factors that can induce a stress response: *physical* (trauma, surgery, intense

heat or cold); *chemical* (reduced O_2 supply, acid-base imbalance); *physiological* (heavy exercise, hemorrhagic shock, pain); *psychological* or *emotional* (anxiety, fear, sorrow); and *social* (personal conflicts, changes in life-style). Stress of any kind is one of the major stimuli for increased cortisol secretion.

Although cortisol's precise role in adapting to stress is not known, a speculative but plausible explanation might be as follows. A primitive human or an animal wounded or faced with a life-threatening situation must forgo eating. A cortisol-induced shift away from protein and fat stores in favor of expanded carbohydrate stores and increased availability of blood glucose would help protect the brain from malnutrition during the imposed fasting period. Also, the amino acids liberated by protein degradation would provide a readily available supply of building blocks for tissue repair should physical injury occur. Thus, an increased pool of glucose, amino acids, and fatty acids is available for use as needed.

Other Physiological Actions   Cortisol is also known to alter mood and behavior. The underlying mechanisms are unclear.

Anti-Inflammatory and Immunosuppressive Effects
When cortisol or synthetic cortisol-like compounds are administered to yield higher than physiological concentrations of glucocorticoids (that is, *pharmacological levels*), not only are all of the metabolic effects increased in magnitude, but several important new actions not evidenced at normal physiological levels are seen. The most noteworthy of glucocorticoids' pharmacological effects are anti-inflammatory and immunosuppressive effects. Synthetic glucocorticoids have been developed that maximize the anti-inflammatory and immunosuppressive effects of these steroids while minimizing the metabolic effects.

Administration of large amounts of glucocorticoid inhibits almost every step of the inflammatory response, making these steroids effective drugs in treating conditions in which the inflammatory response itself has become a destructive process, such as rheumatoid arthritis. It is important to recognize that glucocorticoids used in this manner do not affect the underlying disease process; they merely suppress the body's responses to the disease. Because glucocorticoids also exert multiple inhibitory effects on the overall immune process, such as "knocking out of commission" the white blood cells responsible for antibody production and destruction of foreign cells, these agents have also proved to be useful in the management of various allergic disorders and in the prevention of organ transplant rejections.

When these steroids are employed therapeutically, they should be used only when warranted and then only sparingly for several important reasons. First, because they suppress the normal inflammatory and immune responses that form the backbone of the body's defense system, a glucocorticoid-treated individual has limited ability to resist infections. Second, in addition to the anti-inflammatory and immunosuppressive effects readily exhibited at pharmacological levels, other less desirable effects may also be observed with prolonged exposure to supraphysiological concentrations of glucocorticoids. These effects include development of gastric ulcers, high blood pressure, atherosclerosis, and menstrual irregularities. Third, high levels of exogenous glucocorticoids act in negative-feedback fashion to suppress the hypothalamus-pituitary axis that drives normal glucocorticoid secretion and maintains the integrity of the adrenal cortex. Prolonged suppression of this axis can lead to irreversible atrophy of the cortisol-secreting cells of the adrenal gland and thus to permanent inability of the body to produce its own cortisol.

Cortisol secretion is directly regulated by ACTH.

Cortisol secretion by the adrenal cortex is regulated by a long-loop negative-feedback system involving the hypothalamus and anterior pituitary (● Fig. 19–8). Adrenocorticotropic hormone (ACTH) from the anterior pituitary stimulates the adrenal cortex to secrete cortisol. ACTH is derived from a large

● **FIGURE 19–8   Control of Cortisol Secretion**

precursor molecule, **pro-opiomelanocortin**, produced within the endoplasmic reticulum of the anterior pituitary's ACTH-secreting cells. Prior to secretion, this large precursor is pruned into ACTH and several other biologically active peptides, namely, *melanocyte-stimulating hormone (MSH)* (see p. 631) and a morphine-like substance, β-*endorphin* (see p. 158). The possible significance of these multiple secretory products from a single precursor molecule will be addressed later.

Being tropic to the zona fasciculata and zona reticularis, ACTH stimulates both the growth and secretory output of these two inner layers of the cortex. In the absence of adequate amounts of ACTH, these layers shrink considerably, and cortisol secretion is drastically reduced. Recall that angiotensin, not ACTH, is responsible for maintaining the size of the zona glomerulosa. Like the actions of TSH on the thyroid gland, ACTH enhances many steps in the synthesis of cortisol.

The ACTH-producing cells, in turn, secrete only at the command of corticotropin-releasing hormone (CRH) from the hypothalamus. The feedback control loop is completed by cortisol's inhibitory actions on CRH and ACTH secretion by the hypothalamus and anterior pituitary, respectively.

In addition to controlling cortisol secretion, ACTH (not the pituitary gonadotropic hormones) controls adrenal androgen secretion. In general, cortisol and DHEA output by the adrenal cortex parallel each other. However, adrenal androgens feed back outside of the hypothalamus-pituitary-adrenal cortex loop. Instead of inhibiting CRH, DHEA inhibits gonadotropin-releasing hormone, just as testicular androgens do. Furthermore, sometimes adrenal androgen and cortisol output diverge from each other—for example, at the time of puberty, adrenal androgen secretion undergoes a marked surge, but cortisol secretion does not change. This enhanced secretion initiates the development of androgen-dependent processes in females. In males the same thing is accomplished primarily by testicular androgen secretion, which is also aroused at puberty. The nature of the pubertal inputs to the adrenals and gonads is still unresolved.

The negative-feedback system for cortisol maintains a relatively constant average level of cortisol secretion that is punctuated by alternating bursts of modest secretion separated by "silent" periods of little or no secretion. The amount of cortisol secreted with each burst does not vary much from one secretory episode to the next. However, the total amount of cortisol secreted during a given period of time can be changed by varying the *frequency* of secretory bursts. Reminiscent of summation in skeletal-muscle contraction, one burst "adds on" to the previous burst, increasing the plasma concentration of cortisol beyond that achieved by a single burst. Thus, the mean plasma concentration of cortisol can be elevated by stimulating the secretory bursts to occur more frequently (● Fig. 19–9).

Superimposed on the basic negative-feedback control system are two additional factors that influence plasma cortisol concentrations by varying the frequency of secretory bursts: these are *diurnal rhythm* and *stress*. Recall that there is a characteristic diurnal rhythm in plasma cortisol concentration, with the highest level occurring in the morning and the lowest

● **FIGURE 19–9 Changes in Plasma Cortisol Concentration Caused by Changes in the Frequency of Bursts of Hormone Secretion**

Plasma hormone levels

Bursts of hormone secretion

0   Time (hours)   24

level at night (see Fig. 18–10, p. 628). This diurnal rhythm, which is intrinsic to the hypothalamus-pituitary control system, is related primarily to the sleep-wake cycle. The peak and low levels are reversed in a person who works at night and sleeps during the day. Such time-dependent variations in secretion are of more than academic interest for several reasons. First, it is important clinically to know at what time of day a blood sample was taken when interpreting the significance of a particular value. Second, the linking of cortisol secretion to day-night activity patterns raises serious questions about the common practice of swing shifts at work (that is, constantly switching day and night shifts among employees). Third, because cortisol helps a person resist stress, increasing attention is being given to the time of day various surgical procedures are performed.

The other major factor that is independent of, and in fact can override, the stabilizing negative-feedback control is stress. Dramatic increases in cortisol secretion, mediated by the central nervous system through enhanced activity of the CRH-ACTH system, occur in response to all kinds of stressful situations. The magnitude of the increase in plasma cortisol concentration is generally proportional to the intensity of the stressful stimulation; a greater increase in cortisol levels is evoked in response to severe stress than to mild stress.

The adrenal gland may secrete too much or too little of any one of its hormones.

Although uncommon, there are a number of different disorders of adrenocortical function. Excessive secretion may occur with any of the three categories of adrenocortical hormones. Accordingly, three main patterns of symptoms resulting from hyperadrenalism can be distinguished, depending on which hormone type is in excess: aldosterone hypersecretion, cortisol hypersecretion, and adrenal androgen hypersecretion (Table 19–2).

Aldosterone Hypersecretion Excess mineralocorticoid secretion may be caused by (1) a hypersecreting adrenal tumor made up of aldosterone-secreting cells (**primary hyperaldosteronism** or **Conn's syndrome**) or (2) inappropriately high activity of the renin-angiotensin system (**secondary hyperaldosteronism**). The latter may be produced by any number of conditions that cause a chronic reduction in arterial

blood flow to the kidneys, thereby excessively activating the renin-angiotensin-aldosterone sytem. An example is atherosclerotic narrowing of the renal arteries.

The symptoms of both primary and secondary hyperaldosteronism are related to the exaggerated effects of aldosterone—namely, excessive Na⁺ retention (**hypernatremia**) and K⁺ depletion (**hypokalemia**). Also, high blood pressure (hypertension) is generally present, at least partially because of excessive Na⁺ and fluid retention.

Cortisol Hypersecretion Excessive cortisol secretion (**Cushing's syndrome**) can be caused by (1) overstimulation of the adrenal cortex by excessive amounts of CRH and/or ACTH; (2) adrenal tumors that uncontrollably secrete cortisol independent of ACTH; or (3) ACTH-secreting tumors located in places other than the pituitary, most commonly in the lung. Whatever the cause, the prominent characteristics of this syndrome are related to the exaggerated effects of glucocorticoid, with the main symptoms being reflections of excessive gluconeogenesis. When too many amino acids are converted into glucose, the body suffers from combined glucose excess (high blood glucose) and protein shortage. Because the resultant hyperglycemia and glucosuria (glucose in the urine) mimic diabetes mellitus, the condition is sometimes referred to as adrenal diabetes. For reasons that are unclear, some of the extra glucose is deposited as body fat in locations characteristic for this disease, typically in the abdomen, face, and above the shoulder blades. This abnormal fat distribution in the latter two locations are descriptively called a "moon-face" and "buffalo hump." The appendages, in contrast, remain thin.

Besides the effects attributable to excessive glucose production, other effects arise from the widespread mobilization of amino acids from body proteins for use as glucose precursors. Loss of muscle protein leads to muscle weakness and fatigue. The protein-poor, thin skin of the abdomen becomes overstretched by the excessive underlying fat deposits. As a result, the subdermal tissues rupture, causing the formation of irregular, reddish-purple linear streaks. The weakness of blood vessel walls caused by depletion of structural protein results in an excessive tendency to bruise and to form ecchymoses (small patches of subcutaneous bleeding). Formation of collagen, a major structural protein of connective tissue, is depressed. This interferes with scar formation, so wounds

TABLE 19-2 Major Adrenocortical Abnormalities

ABNORMALITY	CONDITION	CAUSE	SYMPTOMS
Excess aldosterone	Conn's syndrome (primary hyperaldosteronism)	Hypersecreting tumor of zona glomerulosa	Hypernatremia; hypokalemia; hypertension
	Secondary hyperaldosteronism	Inappropriately high activity of renin-angiotensin system	
Excess cortisol	Cushing's syndrome	Excess CRH and/or ACTH caused by hypothalmic or anterior pituitary disease; hypersecreting tumor of inner layers of adrenal cortex; ACTH-secreting tumor in lung	Glucose excess; protein shortage; abnormal fat distribution
Excess androgen	Adrenogenital syndrome	Lack of enzyme in cortisol pathway	Inappropriate masculinization in all but adult males
Deficient cortisol and aldosterone	Addison's disease (primary adrenocortical insufficiency)	Destruction or idiopathic atrophy of adrenal cortex	Related to cortisol deficiency: poor response to stress; hypoglycemia; lack of permissiveness for many metabolic activities
Deficient cortisol	Secondary adrenocortical insufficiency	Insufficient ACTH caused by hypothalamic or anterior pituitary failure	Related to aldosterone deficiency: hyperkalemia; hyponatremia; hypotension (if severe enough, fatal)

heal poorly. Furthermore, loss of the proteinaceous collagen framework of bone leads to weakness of the skeleton, so fractures may result from little or no apparent injury.

Adrenal Androgen Hypersecretion Excess adrenal androgen secretion, a masculinizing condition, is more common than the extremely rare feminizing condition of excess adrenal estrogen secretion. Either condition is referred to as **adrenogenital syndrome**, emphasizing the pronounced effects excessive adrenal sex hormones have on the genitalia and associated sexual characteristics.

The symptoms that result from excess androgen secretion depend on the sex of the individual and the age when the hyperactivity first begins. Because androgens exert masculinizing effects, a woman with this disease tends to develop a male pattern of body hair, a condition referred to as **hirsutism**. She usually also acquires other male secondary sexual characteristics such as deepening of the voice and more muscular arms and legs. The breasts become smaller, and menstruation may cease as a result of androgen suppression of the woman's hypothalamus-pituitary-ovarian pathway for her own female sex hormone secretion.

Female infants born with adrenogenital syndrome manifest male-type external genitalia because excessive androgen secretion occurs early enough during fetal life to induce development of their genitalia along male lines, similar to the development of males under the influence of testicular androgen. The clitoris, which is the female homologue of the male penis, enlarges under androgen influence and takes on a pen-

ile-like appearance so that in some cases it is difficult at first to determine the child's sex. This condition is known as **female pseudohermaphroiditism**. (A true hermaphrodite has the gonads of both sexes.)

Excessive adrenal androgen secretion in prepubertal boys causes them to prematurely develop male secondary sexual characteristics—for example, deep voice, beard, enlarged penis, and sex drive. This condition is referred to as **precocious pseudopuberty** to differentiate it from true puberty, which occurs as a result of increased testicular activity. In precocious pseudopuberty, the androgen secretion from the adrenal cortex is not accompanied by sperm production or any other gonadal activity because the testes are still in their nonfunctional prepubertal state.

Overactivity of adrenal androgens in adult males has no apparent effect because of the already existing male sex characteristics.

The adrenogenital syndrome is most commonly caused by an inherited enzymatic defect in the cortisol steroidogenic pathway. The pathway for synthesis of androgens branches from the normal biosynthetic pathway for cortisol (see Fig. 18–4). When an enzyme specifically essential for synthesis of cortisol is deficient, the result is decreased secretion of cortisol. The decline in cortisol secretion removes the negative-feedback effect on the hypothalmus and anterior pituitary so that levels of CRH and ACTH increase considerably (● Fig. 19–10). The defective adrenal cortex is incapable of responding to this increased ACTH secretion with cortisol output and instead shunts more of its cholesterol precursor into

• **FIGURE 19–10 Hormonal Interrelationships in Adrenogenital Syndrome** The adrenocortical cells that are supposed to produce cortisol produce androgens instead because of a deficiency of a specific enzyme essential for cortisol synthesis. Because no cortisol is secreted to act in negative-feedback fashion, CRH and ACTH levels are elevated. The adrenal cortex responds to the increased ACTH by further increasing androgen secretion. The excess androgen produces virilization and inhibits the gonadotropin pathway, with the result that the gonads stop producing sex hormones and gametes.

- - - - -→ = Normal pathway that does not occur
ACTH = Adrenocorticotropic hormone
GnRH = Gonadotropin-releasing hormone

FSH = Follicle-stimulating hormone
LH = Luteinizing hormone
CRH = Corticotropin-releasing hormone

the androgen pathway. The result is excess DHEA production. This excess androgen does not inhibit ACTH but, rather, inhibits the gonadotropins. Because gamete production is not stimulated in the absence of gonadotropins, individuals with adrenogential syndrome are sterile. Of course, the victims also exhibit symptoms of cortisol deficiency.

The symptoms of adrenal virilization, sterility, and cortisol deficiency are all reversed by glucocorticoid therapy. Administration of exogenous glucocorticoid replaces the cortisol deficit, and, more dramatically, also inhibits the hypothalamus and pituitary so that ACTH secretion is suppressed. Once ACTH secretion is reduced, the profound stimulation of the adrenal cortex ceases, and androgen secretion declines markedly. Removal of the large quantities of adrenal androgens from the circulation allows the masculinizing characteristics to gradually recede and normal gonadotropin secretion to resume. Without understanding how these hormonal systems are related, it would be very difficult to comprehend how glucocorticoid administration could dramatically reverse symptoms of masculinization and sterility.

Adrenocortical Insufficiency If one adrenal gland is nonfunctional or removed, the other healthy organ can take over the function of both through hypertrophy and hyperplasia. Therefore, both glands must be affected before adrenocortical insufficiency occurs.

In **primary adrenocortical insufficiency**, also known as **Addison's disease**, all layers of the adrenal cortex are undersecreting. This condition is most commonly due to idiopathic atrophy of the gland. Although unproven, the most probable cause is autoimmune destruction of the gland by erroneous production of adrenal-cortex–attacking antibodies. **Secondary adrenocortical insufficiency** may occur because of a pituitary or hypothalamic abnormality, resulting in insufficient ACTH secretion. In Addison's disease, both cortisol and aldosterone are deficient, whereas in the secondary form of the condition, only cortisol is deficient, because aldosterone secretion does not depend on ACTH stimulation.

Symptoms associated with cortisol deficiency are as would be expected—poor response to stress, hypoglycemia (low blood glucose) caused by reduced gluconeogenic activity, and lack of permissive action for many metabolic activities. With the primary form of the disease, hyperpigmentation (darkening of the skin) resulting from excessive secretion of ACTH is also seen. Because the pituitary is normal, the decline in cortisol secretion brings about an uninhibited elevation in ACTH output. Recall that both ACTH and melanocyte-stimulating hormone (MSH) are produced from the same large precursor molecule, pro-opiomelanocortin. As a result, MSH activity also becomes elevated when the blood levels of ACTH are very high.

The symptoms associated with aldosterone deficiency in

Addison's disease are the most threatening. If severe enough, the condition is fatal because aldosterone is essential for life. However, the loss of adrenal function may develop slowly and insidiously so that aldosterone secretion may be subnormal but not totally lacking. Patients with aldosterone deficiency display K^+ retention (**hyperkalemia**) caused by reduced K^+ loss in the urine and Na^+ depletion (**hyponatremia**) caused by excessive urinary loss of Na^+. The former results in disturbances in cardiac rhythm. The latter results in a fall in ECF volume, including a reduction in circulating blood volume, which in turn leads to low blood pressure (hypotension).

The catecholamine-secreting adrenal medulla is a modified sympathetic postganglionic neuron.

The adrenal medulla is actually a modified part of the sympathetic nervous system. A sympathetic pathway consists of two neurons in sequence—a *preganglionic neuron* originating in the CNS, whose axonal fiber terminates on a second peripherally located *postganglionic neuron,* which in turn terminates on the effector organ (see p. 198). The neurotransmitter released by sympathetic postganglionic fibers is norepinephrine, which interacts locally with the innervated organ by binding with specific target receptors known as *adrenergic receptors.*

The adrenal medulla is composed of modified postganglionic sympathetic neurons. Unlike ordinary postganglionic sympathetic neurons, those in the adrenal medulla do not possess axonal fibers that terminate on effector organs. Instead, the ganglionic cell bodies within the adrenal medulla release their chemical transmitter directly into the circulation upon stimulation by the preganglionic fiber (see Fig. 7–2, p. 202). In this case, the transmitter qualifies as a hormone instead of a neurotransmitter. Like sympathetic fibers, the adrenal medulla does release norepinephrine, but its most abundant secretory output is a similar chemical messenger known as **epinephrine**. Both epinephrine and norepinephrine belong to the chemical class of catecholamines, which are derived from the amino acid tyrosine. Epinephrine is the same as norepinephrine except that it has a methyl group added to it.

Catecholamine synthesis is accomplished almost entirely within the cytosol of the adrenomedullary secretory cells, with only one step taking place within hormonal storage granules. Once produced, epinephrine and norepinephrine are stored in **chromaffin granules**, which are similar to the transmitter storage vesicles found in sympathetic nerve endings. Accordingly, adrenomedullary tissue is often called *chromaffin tissue.* The chromaffin granules have a very efficient, active transport system for catecholamine uptake; as a result, the concentration of epinephrine in the chromaffin granules is at least 25,000 times greater than that in the cytosol. Segregation of catecholamines in chromaffin granules protects them from being destroyed by cytosolic enzymes during storage. Furthermore, packaging of these hormones into granules is a prerequisite for their export from the adrenomedullary cells. Catecholamines are secreted into the circulation by exocytosis of chromaffin granules; their release is analogous to the release mechanism for secretory vesicles that contain stored peptide hormones or the release of norepinephrine at sympathetic postganglionic terminals.

Of the total adrenomedullary catecholamine output, epineprine accounts for 80% and norepinephrine for 20%. Whereas epinephrine is produced exclusively by the adrenal medulla, the bulk of norepineprhine in the body is produced by sympathetic postganglionic fibers. Adrenomedullary norepinephrine is generally secreted in quantities too small to exert significant effects on target cells. Therefore, for practical purposes, we can assume norepinephrine effects are predominantly mediated directly by the sympathetic nervous system and epinephrine effects are brought about exclusively by the adrenal medulla.

Epinephrine and norepinephrine have differing affinities for three distinctive receptor types: α, β_1, and β_2 adrenergic receptors (see p. 203). Some target cells have only α receptors, some have only β_2 receptors, and some have both α and β_2 receptors; β_1 receptors are found almost exclusively in the heart. In general, the responses elicited by activation of α and β_1 receptors are excitatory in nature, whereas the responses to stimulation of β_2 receptors are inhibitory.

Neither catecholamine is totally specific for particular receptor types, but each has markedly different affinities. Norepinephrine binds predominantly with α and β_1 receptors located in the vicinity of postganglionic sympathetic-fiber terminals. Hormonal epinephrine, which is able to reach all α and β_1 receptors by means of its circulatory distribution, interacts with these receptors with approximately the same potency as neurotransmitter norepinephrine. In addition, ephinephrine activates β_2 receptors, over which the sympathetic system exerts little influence. Epinephrine is at least ten times more potent than norepinephrine at β_2 receptors. In fact, many of the essentially epinephrine-exclusive β_2 receptors are located at tissues not even supplied by the sympathetic system but reached by epinephrine through the blood. An example is skeletal muscle, where epinephrine exerts metabolic effects such as promoting the breakdown of stored glycogen.

Epinephrine and norcpinephrine exert similar effects in many tissues, with epinephrine generally reinforcing sympathetic activity. For example, epinephrine and norepinephrine both accelerate the heart rate by binding with cardiac β_1 receptors, and both induce generalized arteriolar vasoconstriction of the skin, digestive tract, and kidneys through their mutual α-receptor activation. Yet there are some important differences in response that can be explained on the basis of differential activation of different receptor types. As an example, epinephrine, through its exclusive β_2-receptor activation, brings about vasodilation of the blood vessels that supply skeletal muscles and the heart. This effect is in addition to its generalized vasoconstrictor effect mediated by α-receptor stimulation. Also, epinephrine is able to exert some unique effects, such as its metabolic effects, because it reaches places not supplied by sympathetic fibers.

It is important to realize, however, that epinephrine functions only at the bidding of the sympathetic nervous system, which is solely responsible for stimulating its secretion from the adrenal medulla. Epinephrine secretion always accompanies a generalized sympathetic discharge, so sympathetic activity indirectly exerts control over the actions performed by epinephrine. By having the more versatile circulating epinephrine at its call, the sympathetic system has a means of reinforcing its own neurotransmitter effects plus a way of exerting additional actions on tissues that it does not directly innervate.

Epinephrine reinforces the sympathetic nervous system and exerts additional metabolic effects as well.

Adrenomedullary hormones are not essential for life, but virtually all organs in the body are affected by these catecholamines. They play important roles in stress responses, regulation of arterial blood pressure, and control of fuel metabolism. Collectively, the sympathetic nervous system and adrenomedullary epinephrine mobilize the body's resources to support peak physical exertion in the face of impending danger. The sympathetic and epinephrine actions constitute a fight-or-flight response that prepares the individual to combat an enemy or flee from danger. The following sections discuss epinephrine's major effects, which it accomplishes in collaboration with the sympathetic transmitter norepinephrine or performs alone to complement direct sympathetic response.

Effects on Organ Systems The sympathetic system and epinephrine both exert widespread effects on organ systems that are ideally suited for fight-or-flight responses (see p. 200). Under the influence of epinephrine and the sympathetic system, the rate and strength of cardiac contraction increase, resulting in increased cardiac output, and their generalized vasoconstrictor effects increase total peripheral resistance. Together these effects cause an increase in arterial blood pressure, thus assuring an appropriate driving pressure to force blood to the organs that are most vital for meeting the emergency. Meanwhile, vasodilation of coronary and skeletal-muscle blood vessels induced by epinephrine and local metabolic factors shifts blood to the heart and skeletal muscles from other vasoconstricted regions of the body.

Because of their profound influence on the heart and vasculature, the adrenomedullary catecholamines and sympathetic system also play an important role in the ongoing maintenance of arterial blood pressure.

Epinephrine (but not norepinephrine) dilates the respiratory airways to reduce the resistance encountered in moving air in and out of the lungs. Epinephrine also reduces digestive activity and inhibits bladder emptying, both activites that can be "put on hold" during a fight-or-flight situation.

Metabolic Effects Epinephrine exerts some important metabolic effects, even at blood hormone concentrations lower than those required for eliciting the cardiovascular responses. In general, epinephrine prompts the mobilization of stored carbohydrate and fat to provide immediately available energy for use as needed to fuel muscular work. Specifically, epinephrine increases the blood glucose level by several different mechanisms. First, it stimulates both hepatic (liver) gluconeogenesis and **glycogenolysis**, the latter referring to the breakdown of stored glycogen into glucose, which is released into the blood. Epinephrine also stimulates glycogenolysis in skeletal muscles. Because of the difference in enzyme content between liver and muscle, however, muscle glycogen cannot be converted directly to glucose. Instead, lactic acid is released into the blood as a result of the breakdown of muscle glycogen. This lactic acid is removed from the blood by the liver and converted into glucose, so epinephrine's actions on skeletal muscle indirectly contribute to an increase in blood glucose levels. Epinephrine and the sympathetic system may further add to this hyperglycemic effect by inhibiting the secretion of insulin, the pancreatic hormone primarily responsible for removal of glucose from the blood, and by stimulating glucagon, another pancreatic hormone that promotes hepatic glycogenolysis and gluconeogenesis. In addition to increasing blood glucose levels, epinephrine also increases the blood fatty acids level by promoting lipolysis.

Epinephrine's metabolic effects are appropriate for fight-or-flight situations. The elevated levels of glucose and fatty acids provide additional fuel to power the muscular movement required by the situation and also assure adequate nourishment for the brain during the crisis when no new nutrients are being consumed. Muscles can use fatty acids for energy production, but the brain cannot.

Because of its other widespread actions, epinephrine also increases the overall metabolic rate. Under the influence of epinephrine, many tissues metabolize at a faster rate. For example, the work of the heart and respiratory muscles is increased, and the pace of liver metabolism is stepped up. Thus, epinephrine as well as thyroid hormone can increase the metabolic rate.

Other Effects Epinephrine affects the central nervous system to promote a state of arousal and increased CNS alertness. This permits "quick thinking" to help cope with the impending emergency. Many drugs that are used as stimulants or sedatives probably exert their effects by altering catecholamine levels in the CNS.

Both epinephrine and norepinephrine cause sweating, which helps the body rid itself of extra heat generated by increased muscular activity. Also, epinephrine acts on smooth muscles within the eyes to dilate the pupil and flatten the lens. These actions adjust the eyes for more encompassing vision so that the whole threatening scene can be quickly viewed.

Sympathetic stimulation of the adrenal medulla is solely responsible for epinephrine release.

Catecholamine secretion by the adrenal medulla is controlled entirely by sympathetic input to the gland. When the sympathetic system is activated, it simultaneously triggers a surge of adrenomedullary catecholamine release, flooding the circulation with up to three hundred times the normal concentration of epinephrine. Although a number of different factors have been shown to influence adrenal catecholamine secretion, they all act by increasing preganglionic sympathetic impulses to the adrenal medulla. Among the major factors that stimulate increased adrenomedullary output are a variety of stressful conditions such as physical or psychological trauma, hemorrhage, illness, exercise, hypoxia (low arterial O_2), cold exposure, and hypoglycemia (low blood glucose). The amount of epinephrine released depends on the type and intensity of the stressful stimulus.

Adrenomedullary dysfunction is very rare.

Adrenomedullary hyposecretion is not a recognized clinical entity. No adverse effects have been attributed to a deficiency of epinephrine, presumably not because a deficiency never occurs but because the majority of epinephrine's functions can be duplicated by activation of the sympathetic nervous system alone.

The only catecholamine disorder is a **pheochromocytoma**, a rarely occurring catecholamine-secreting tumor. These tumors are usually, but not always, located in the adrenal medulla. Pheochromocytomas may contain up to twenty times more catecholamines per gram of tissue than normal medullary tissue, with release of these hormones not being subject to neural control. The symptoms of this condition are directly attributable to the actions of excessive amounts of catecholamines, the most common of which are high blood pressure, rapid heart rate, palpitations, excessive sweating, and high blood glucose.

The stress response is a generalized, nonspecific pattern of neural and hormonal reactions to any situation that threatens homeostasis.

Since both components of the adrenal gland play an extensive role in responding to stress, this is an appropriate place to pull together the various major factors involved in the stress response. Recall that a variety of noxious physical, chemical, physiological, and psychosocial stimuli that threaten to overwhelm the body's compensatory ability to maintain homeostasis can elicit a stress response. Different stressors may produce some specific responses characteristic of that stressor; for example, the body's specific response to cold exposure is shivering and skin vasoconstriction, whereas the specific response to bacterial invasion includes increased phagocytic activity and antibody production. In addition to their specific response, however, all stressors also produce a similar nonspecific, generalized response regardless of the type of stressor (• Fig. 19–11).

Dr. Hans Selye was the first to recognize this commonality of responses to noxious stimuli in what he called the **general adaptation syndrome**. When a stressor is recognized, both nervous and hormonal responses are called into play to bring about defensive measures to cope with the emergency. The result is a state of intense readiness and mobilization of biochemical resources.

To appreciate the value of the multifaceted stress response, imagine a primitive cave dweller who has just seen a

• **FIGURE 19–11   Action of Stressor on Body**

large wild beast lurking in the shadows. The major neural response to such a stressful stimulus is generalized activation of the sympathetic nervous system. The resultant increase in cardiac output and ventilation as well as the diversion of blood from vasoconstricted regions of suppressed activity, such as the digestive tract and kidneys, to the more active vasodilated skeletal muscles and heart prepare the body for a fight-or-flight response. Simultaneously, the sympathetic system calls forth hormonal reinforcements in the form of a massive outpouring of epinephrine from the adrenal medulla. Epinephrine strengthens sympathetic responses and reaches places not innervated by the sympathetic system to perform additional functions, such as mobilizing carbohydrate and fat stores.

Besides epinephrine, a number of other hormones are involved in the overall stress response (Table 19–3). The predominant hormonal response is activation of the CRH-ACTH-cortisol system. Recall that cortisol's role in helping the body cope with stress is presumed to be related to its metabolic effects. Cortisol breaks down fat and protein stores while expanding carbohydrate stores and increasing the availability of blood glucose. A logical assumption is that the increased pool of glucose, amino acids, and fatty acids is available for use as needed, such as to sustain nourishment to the brain and provide building blocks for repair of damaged tissues.

Besides the effects of cortisol in the hypothalamus-pituitary-adrenal cortex axis, there is evidence that ACTH may play a role in resisting stress. Since ACTH is one of several peptides that facilitate learning and behavior, it is possible that an increase in ACTH during psychosocial stress helps the body cope more readily with similar stressors in the future by facilitating the learning of appropriate behavioral responses. Furthermore, ACTH is not released alone from its anterior pituitary storage vesicles. Pruning of the large pro-opionmelanocortin precursor molecule yields not only ACTH but also morphine-like β-endorphin and similar compounds. These compounds are cosecreted with ACTH upon stimulation by CRH during stress. It has been hypothesized that β-endorphin, as a potent endogenous opiate (see p. 158), might exert a role in mediating analgesia (reduction of pain perception) should physical injury be inflicted during stress. It is further speculated that these cosecreted peptides have possible roles in learning, mood alterations, and appetite suppression, among other things. The precise contributions of these ACTH-related compounds during stress is unclear, but their role is a subject of considerable interest.

TABLE 19-3 Major Hormonal Changes during the Stress Response

HORMONE	CHANGE	PURPOSE SERVED
Epinephrine	↑	Reinforces sympathetic nervous system to prepare the body for "fight or flight"
		Mobilizes carbohydrate and fat energy stores; increases blood glucose and blood fatty acids
CRH-ACTH-cortisol	↑	Mobilizes energy stores and metabolic building blocks for use as needed; increases blood glucose, blood amino acids, and blood fatty acids
		ACTH facilitates learning and behavior
Glucagon	↑	Act in concert to increase blood glucose and blood fatty acids
Insulin	↓	
Renin-angiotensin-aldosterone	↑	Conserve salt and H_2O to expand plasma volume; help sustain blood pressure when acute loss of plasma volume occurs
Vasopressin	↑	Vasopressin and angiotensin II cause arteriolar vasoconstriction to increase blood pressure
		Vasopressin facilitates learning

Other hormonal responses besides those of cortisol contribute to the overall metabolic response to stress. The sympathetic nervous system and the epinephrine secreted at its bidding both inhibit insulin and stimulate glucagon. These hormonal changes act in concert to elevate the blood levels of glucose and fatty acids. Epinephrine and glucagon, whose blood levels are elevated during stress, promote hepatic glycogenolysis and (along with cortisol) hepatic gluconeogenesis. However, insulin, whose secretion is suppressed during stress, opposes the breakdown of liver glycogen stores. All of these effects contribute to increasing the concentration of blood glucose. The primary stimulus for insulin secretion is a rise in blood glucose; in turn, a primary effect of insulin is to lower the blood glucose. If it were not for the deliberate inhibition of insulin during the stress response, the hyperglycemia caused by stress would stimulate secretion of glucose-lowering insulin. As a result, the elevation in blood glucose could not be sustained. Stress-related hormonal responses also promote a release of fatty acids from fat stores, because lipolysis is favored by epinephrine, glucagon, and cortisol but opposed by insulin.

In addition to the hormonal changes that mobilize energy stores during stress, other hormones are simultaneously called into play to sustain blood volume and blood pressure during the emergency. The sympathetic system and epinephrine have major responsibilities in acting directly on the heart and blood vessels to improve circulatory function. In addition, the renin-angiotensin-aldosterone system is activated as a consequence of a sympathetically induced reduction of blood supply to the kidneys (see p. 482). Vasopressin secretion is also increased during stressful situations. Collectively, these hormones expand the plasma volume by promoting retention of salt and H_2O. Presumably, the enlarged plasma volume serves as a protective measure to help sustain blood pressure should acute loss of plasma fluid occur through hemorrhage or heavy sweating during the impending period of danger. Vasopressin and angiotensin also have direct vasopressor effects, which would be of benefit in maintaining an adequate arterial pressure in the face of acute blood loss. Vasopressin is further believed to facilitate learning, which has implications for future adaptation to stress.

The multifaceted stress response is coordinated by the hypothalamus.

All of the individual responses to stress just described are either directly or indirectly influenced by the hypothalamus (● Fig. 19–12). The hypothalamus receives input concerning physical and emotional stressors from virtually all areas of the brain and from many receptors throughout the body. In response, the hypothalamus directly activates the sympathetic nervous system, secretes CRH to stimulate ACTH and cortisol release, and triggers the release of vasopressin. Sympathetic stimulation, in turn, brings about the secretion of epinephrine, with which it has a conjoined effect on the pancreatic secretion of insulin and glucagon. Furthermore, vasoconstriction of the renal afferent arterioles by the catecholamines indirectly triggers the secretion of renin by reducing the flow of oxygenated blood through the kidneys. Renin, in turn, sets in motion the renin-angiotensin-aldosterone mechanism. In this way, the hypothalamus integrates the responses of both the sympathetic nervous system and endocrine system during stress.

Activation of the stress response by chronic psychosocial stressors may be harmful.

Acceleration of cardiovascular and respiratory activity, retention of salt and H_2O, and mobilization of metabolic fuels and building blocks can be of benefit in response to a physical stressor, such as an athletic competition. Most of the stressors in our everyday lives are psychosocial in nature however, yet they induce these same magnified responses. Stressors such as anxiety about an exam, conflicts with loved ones, or impatience while sitting in a traffic jam can elicit a stress response. Although the rapid mobilization of body resources is appropriate in the face of real or threatened physical injury, it is generally inappropriate in response to nonphysical stress. If no extra energy is demanded, no tissue damaged, and no blood lost, body stores are being broken down and fluid retained needlessly, probably to the detriment of the emotionally stressed individual. In fact, there is strong circumstantial evidence for a link between chronic exposure to psychosocial stressors and the development of pathological conditions

● **FIGURE 19–12** **Integration of Stress Response by Hypothalamus**

such as atherosclerosis and high blood pressure, although no definitive cause-and-effect relationship has been ascertained. As a result of "unused" stress responses, could hypertension result from too much sympathetic vasoconstriction? Too much salt and H_2O retention? Too much vasopressin and angiotensin pressor activity? A combination of these? Other factors? Recall that these same diseases can develop with prolonged exposure to pharmacological levels of glucocorticoids. Could long-standing lesser elevations of cortisol, such as might occur in the face of continual psychosocial stressors, do the same thing, only more slowly? Considerable work remains to be done to evaluate the contributions the stressors in our everyday life-styles make toward disease production.

ENDOCRINE CONTROL OF FUEL METABOLISM

All three classes of nutrient molecules can be used to provide cellular energy and, to a large extent, can be interconverted.

We have just discussed the metabolic changes that are elicited during the stress response. Now we will concentrate on the metabolic patterns that occur in the absence of stress, including the hormonal factors that govern this normal metabolism.

Metabolism refers to all of the chemical reactions that occur within the cells of the body. Those reactions involving the

TABLE 19–4 Summary of Reactions in Fuel Metabolism

METABOLIC PROCESS	REACTION	CONSEQUENCE
Glycogenesis	Glucose → Glycogen	↓ Blood glucose
Glycogenolysis	Glycogen → Glucose	↑ Blood glucose
Gluconeogenesis	Amino acids → Glucose	↑ Blood glucose
Protein synthesis	Amino acids → Protein	↓ Blood amino acids
Protein degradation	Protein → Amino acids	↑ Blood amino acids
Fat synthesis (lipogenesis or triglyceride synthesis)	Fatty acids and glycerol → Triglycerides	↓ Blood fatty acids
Fat breakdown (lipolysis or triglyceride degradation)	Triglycerides → Fatty acids and glycerol	↑ Blood fatty acids

degradation, synthesis and transformation of the three classes of energy-rich organic molecules—protein, carbohydrate, and fat—are collectively known as **intermediary metabolism** or **fuel metabolism** (Table 19–4).

During the process of digestion, large nutrient molecules (**macromolecules**) are broken down into their smaller absorbable subunits as follows: proteins are converted into amino acids, complex carbohydrates into monosaccharides (mainly glucose), and triglycerides (dietary fats) into monoglycerides and free fatty acids. These absorbable units are transferred from the digestive-tract lumen into the blood, either directly or by way of the lymph (chapter 16).

These organic molecules are constantly exchanged between the blood and the cells of the body (● Fig. 19–13). The chemical reactions in which the organic molecules participate within the cells are categorized into two metabolic processes: anabolism and catabolism. **Anabolism** refers to the buildup or synthesis of larger organic macromolecules from the small organic molecular subunits. Anabolic reactions generally require the input of energy in the form of ATP. These reactions result in either (1) the manufacture of materials needed by the cell, such as cellular structural proteins or secretory product, or (2) storage, as in glycogen (the storage form of glucose) or fat reservoirs, of excess ingested nutrients that are not immediately needed for energy production or as cellular building blocks. **Catabolism**, on the other hand, refers to the breakdown or degradation of large, energy-rich organic molecules within cells. Catabolism encompasses two levels of breakdown: (1) hydrolysis (see p. 544) of large cellular organic macromolecules into their smaller subunits, similar to the process of digestion except that the reactions take place within the cells of the body instead of within the digestive-tract lumen (for example, release of glucose by the catabolism of stored glycogen), and (2) oxidation of the smaller subunits, such as glucose, to yield energy for ATP production (see p. 33). Indeed, even the structural components of cells represent stored energy, albeit an "expensive" energy source, because they contain energy-rich proteins that can be cannibalized if necessary to yield energy. As an alternative to energy production, the smaller, multipotential organic subunits derived from intracellular hydrolysis may be released into the blood. These mobilized glucose, fatty

acid, and amino acid molecules can then be used as needed for energy production or cellular synthesis elsewhere in the body.

In an adult, the rates of anabolism and catabolism are generally in balance, so the adult body remains in a dynamic steady state and appears unchanged even though the organic molecules that determine its structure and function are continuously being turned over. During growth, anabolism exceeds catabolism.

In addition to being able to resynthesize catabolized organic molecules back into the same type of molecules, many cells of the body, especially liver cells, have the ability to convert most types of small organic molecules into other types—as in, for example, the transformation of amino acids into glucose or fatty acids. Because of these interconversions, adequate nourishment can be provided by a wide range of molecules present in different types of foods. There are limits, however. **Essential nutrients** such as the essential amino acids and vitamins cannot be formed in the body by conversion from another type of organic molecule.

The major fate of both ingested carbohydrates and fats is catabolism to yield energy. Amino acids are predominantly used for protein synthesis, but can be used to supply energy after being converted to carbohydrate or fat. Thus, all three categories of foodstuff can be used as fuel, and excesses of any foodstuff can be deposited as stored fuel.

The brain must be continuously supplied with glucose, even between meals when no new nutrients are being taken up from the digestive tract.

At a superficial level, fuel metabolism appears relatively simple: the amount of nutrients in the diet must be sufficient to meet the body's needs for energy production and cellular synthesis. This simple relationship is complicated, however, by two important considerations. First, dietary fuel intake is intermittent, not continuous. As a result, excess energy must be absorbed during meals and stored for use during fasting periods between meals, when dietary sources of metabolic fuel are not available (Table 19–5). Excess circulating glucose is stored as *glycogen,* a large molecule consisting of interconnected glucose molecules, in the liver and muscle. Be-

FIGURE 19–13 Summary of Major Pathways Involving Organic Nutrient Molecules

cause glycogen is a relatively small energy reservoir, less than a day's energy needs can be stored in this form. Once the liver and muscle glycogen stores are "filled up," additional glucose is transformed into fatty acids and glycerol, which are used to synthesize *triglycerides* (glycerol with three fatty acids attached), primarily in adipose tissue (fat) and to a lesser extent in muscle. Excess circulating fatty acids derived from dietary intake also become incorporated into triglycerides. Ex-

cess circulating amino acids not needed for protein synthesis are not stored as extra protein but are converted to glucose and fatty acids, which ultimately end up being stored as triglycerides. Thus, the major site of energy storage for excess nutrients of all three classes is adipose tissue. Normally, sufficient triglyceride is stored to provide energy for about two months, more so in an overweight individual. Consequently, during any prolonged period of fasting, the fatty acids re-

TABLE 19-5   Stored Metabolic Fuel in Body

METABOLIC FUEL	CIRCULATING FORM	STORAGE FORM	MAJOR STORAGE SITE	PERCENTAGE OF TOTAL BODY ENERGY CONTENT (AND CALORIES*)	RESERVOIR CAPACITY	ROLE
Carbohydrate	Glucose	Glycogen	Liver, muscle	1% (1,500 calories)	Less than a day's worth of energy	First energy source; essential for brain
Fat	Free fatty acids	Triglycerides	Adipose tissue	77% (143,000 calories)	About 2 months' worth of energy	Primary energy reservoir; energy source during a fast
Protein	Amino acids	Body proteins	Muscle	22% (41,000 calories)	Death results long before capacity is utilized because of structural and functional impairment	Source of glucose for brain during a fast; last resort to meet other energy needs

*Actually refers to kilocalories; see p. 597.

leased from triglyceride catabolism serve as the primary source of energy for most tissues. The catabolism of stored triglycerides frees glycerol as well as fatty acids, but quantitatively speaking, the fatty acids are far more important. Catabolism of stored fat yields 90% fatty acids and 10% glycerol by weight. Glycerol can be converted to glucose by the liver and contributes in a small way to maintaining blood glucose during a fast.

As a third energy reservoir, a substantial amount of energy is stored as *structural protein,* primarily in muscle, the most abundant protein mass in the body. Protein is not the first choice to tap as an energy source, however, because it serves other essential functions; in contrast, the glycogen and triglyceride reservoirs serve solely as energy depots.

The second factor complicating fuel metabolism is that the brain normally depends on the delivery of adequate blood glucose as its sole source of energy. Consequently, it is essential that the blood glucose concentration be maintained above a critical level. The blood glucose concentration is typically 100 mg glucose/100 ml plasma and is normally maintained within the narrow limits of 70–110 mg/100 ml. Liver glycogen is an important reservoir for maintaining blood glucose levels during a short fast. However, liver glycogen is depleted relatively rapidly, so during a longer fast, other mechanisms must be used to assure that the energy requirements of the glucose-dependent brain are met. First, when new dietary glucose is not entering the blood, tissues not obligated to use glucose shift their metabolic gears to burn fatty acids instead, thus sparing glucose for the brain. Fatty acids are made available by catabolism of triglyceride stores as an alternative energy source for non-glucose-dependent tissues. Second, amino acids can be converted to glucose by gluconeogenesis, whereas fatty acids cannot. Thus, once glycogen stores are depleted despite glucose sparing, new glucose supplies for the brain are provided by the catabolism of body proteins and conversion of the freed amino acids into glucose.

Metabolic fuels are stored during the absorptive state and are mobilized during the postabsorptive state.

From the preceding discussion, it should be obvious that the disposition of organic molecules depends on the body's metabolic state. There are two functional metabolic states related to eating and fasting cycles—the absorptive state and the postabsorptive state, respectively (Table 19–6). Following a

TABLE 19-6   Comparison of Absorptive and Postabsorptive States

METABOLIC FUEL	ABSORPTIVE STATE	POST-ABSORPTIVE STATE
Carbohydrate	Glucose providing major energy source	Glycogen degradation and depletion
	Glycogen synthesis and storage	Glucose sparing to conserve glucose for brain
	Excess converted to and stored as triglyceride fat	Production of new glucose through gluconeogenesis
Fat	Triglyceride synthesis and storage	Triglyceride catabolism
		Fatty acids providing major energy source for non-glucose-dependent tissues
Protein	Protein synthesis	Protein catabolism
	Excess converted to and stored as triglyceride fat	Amino acids used for gluconeogenesis

meal, ingested nutrients are being absorbed and entering the blood during the **absorptive state** or **fed state**. During this time, glucose is plentiful and serves as the major energy source. Very little of the absorbed fat and amino acids is used for energy during the absorptive state because most cells prefer to use glucose when it is available. Extra nutrients not immediately used for energy or structural repairs are channeled into storage as glycogen or triglycerides.

The average meal is completely absorbed in about four hours. Therefore, on a typical three-meals-a-day diet, no nutrients are being absorbed from the digestive tract during late morning, late afternoon, and throughout the night. These times constitute the **postabsorptive state** or **fasting state**. During this state, endogenous energy stores are mobilized to provide energy, while gluconeogenesis and glucose sparing are used to maintain the blood glucose at an adequate level to nourish the brain. The synthesis of protein and fat is curtailed. Instead, stores of these organic molecules are catabolized for glucose formation and energy production, respectively. Carbohydrate synthesis does occur through gluconeogenesis, but the utilization of glucose for energy is greatly reduced.

Note that the blood concentration of nutrients does not fluctuate markedly between the absorptive and postabsorptive states. During the absorptive state, the glut of absorbed nutrients is swiftly removed from the blood and placed into storage; during the postabsorptive state, these stores are catabolized to maintain the blood concentrations at levels necessary to sustain tissue energy demands.

As already described, during these alternating metabolic states, various tissues play different roles. The liver plays the primary role in maintenance of normal blood glucose levels. It stores glycogen when excess glucose is available, releases glucose into the blood when needed, and is the principal site for metabolic interconversions such as gluconeogenesis. Adipose tissue serves as the primary energy storage site and is important in the regulation of fatty acid levels in the blood. Muscle is the primary site of amino acid storage and is the major energy user. The brain normally can use only glucose as an energy source, yet it does not store glycogen, making it mandatory that blood glucose levels be maintained.

Several other organic intermediates play a lesser role as energy sources—namely, glycerol, lactic acid, and ketone bodies. As mentioned earlier, *glycerol* derived from triglyceride hydrolysis (it is the backbone to which the fatty acid chains are attached) can be converted to glucose by the liver. Similarly, *lactic acid,* which is produced by the incomplete catabolism of glucose via glycolysis in muscle (• Fig. 19–14; see also p. 236), can also be converted to glucose in the liver. **Ketone bodies** are a group of compounds produced by the liver during glucose sparing. Unlike other tissues, when the liver uses fatty acids as an energy source, it oxidizes them only to acetyl CoA, which it is unable to process through the citric-acid cycle for further energy extraction. Thus, the liver does not degrade fatty acids all the way to CO_2 and H_2O for maximum energy release. Instead, it partially extracts the available energy and converts the remaining energy-bearing acetyl CoA molecules into ketone bodies, which it releases into the blood. Ketone bodies serve as an alternative energy source for tissues capable of oxidizing them further by means of the citric-acid cycle.

During long-term starvation, the brain starts using ketones instead of glucose as a major energy source. Since death resulting from starvation is usually due to protein wasting rather than hypoglycemia (low blood glucose), prolonged survival without any caloric intake requires that gluconeogenesis be kept to a minimum while the energy needs of the brain are not compromised. A sizable portion of cell protein can be catabolized without serious cellular malfunction, but a point is finally reached at which a cannabilized cell is no longer able to function adequately. To ward off the fatal point of failure as long as possible during prolonged starvation, the brain starts using ketones as a major energy source, correspondingly decreasing its use of glucose. Use by the brain of this fatty acid "table scrap" left over from the liver's "meal" limits the necessity of mobilizing body proteins for glucose production to nourish the brain. Both of the major metabolic adaptations to prolonged starvation—a decrease in protein catabolism and use of ketones by the brain—are attributable to the high levels of ketones in the blood at the time. Utilization of ketones by the brain occurs only when the blood ketone level is high. The high blood levels of ketones also directly inhibit protein degradation in muscle. Thus, ketones are responsible for sparing body proteins while satisfying the brain's energy needs.

The pancreatic hormones, insulin and glucagon, are most important in regulating fuel metabolism.

How does the body "know" when to shift its metabolic gears from one of net anabolism and nutrient storage to one of net catabolism and glucose sparing? The flow of organic nutrients along metabolic pathways is influenced by a variety of hormones, including insulin, glucagon, epinephrine, cortisol, and growth hormone. Under most circumstances, the pancreatic hormones, insulin and glucagon, are the dominant hormonal regulators that shift the metabolic pathways back and forth from net anabolism to net catabolism and glucose sparing, depending on whether the body is in a state of feasting or fasting, respectively.

The **pancreas** is an organ composed of both exocrine and endocrine tissues. The exocrine portion of the pancreas secretes a watery alkaline solution and digestive enzymes through the pancreatic duct into the digestive-tract lumen. Scattered throughout the pancreas between the exocrine cells are clusters, or "islands," of endocrine cells known as the **islets of Langerhans** (see Fig. 16–16, p. 568). The most abundant pancreatic endocrine-cell type is the β (**beta**) **cell**, the site of *insulin* synthesis and secretion. Next most important are the α (**alpha**) **cells**, which produce *glucagon*. The **D** (**delta**) **cells** are the pancreatic site of *somatostatin* synthesis, whereas the least common islet cells, the **PP cells**, secrete *pancreatic polypeptide.*

The most important pancreatic hormones in the regulation of fuel metabolism are insulin and glucagon. Accordingly, we will pay the most attention to these two pancreatic hormones, giving only a brief highlight now of the other two endocrine

(number C) = Number of carbon atoms present in molecule
n = Number of 2-carbon fatty acids present in a fatty acid chain

● **FIGURE 19–14 Major Pathways in Energy Production by, Storage of, and Interconversions between Organic Nutrient Molecules**

pancreatic secretory products. **Somatostatin** is also produced by the hypothalamus, where it serves to inhibit the secretion of growth hormone and TSH. Furthermore, somatostatin is produced by cells lining the digestive tract, where it is suspected of acting locally as a paracrine to inhibit most digestive processes. Pancreatic somatostatin also exerts a variety of inhibitory effects on the digestive system, the overall effect of which is to inhibit digestion of nutrients and to decrease nutrient absorption. Somatostatin is released from the pancreatic D cells in direct response to an increase in blood glucose and blood amino acids during absorption of a meal. By exerting its inhibitory effects, pancreatic somatostatin acts in negative-feedback fashion to put the brakes on the rate at which the meal is being digested and absorbed, thereby preventing excessive plasma levels of nutrients. In addition, pancreatic somatostatin may play an important role in the local regulation of pancreatic hormone secretion. The secretion of insulin, glucagon, pancreatic polypeptide, and somatostatin itself

is decreased by the local presence of somatostatin, but the physiological importance of such paracrine function has not been determined.

Little is known about **pancreatic polypeptide**. Its effects seem to be concerned primarily with inhibiting gastrointestinal function. Like somatostatin, pancreatic polypeptide appears to have no direct effects on carbohydrate, protein, or fat metabolism.

Insulin lowers blood glucose, amino acid, and fatty acid levels and promotes anabolism of these small nutrient molecules.

Insulin has important effects on carbohydrate, fat, and protein metabolism. It lowers the blood levels of glucose, fatty acids, and amino acids and promotes their storage. As these nutrient molecules enter the blood during the absorptive state, insulin promotes their cellular uptake and conversion into glycogen, triglycerides, and protein, respectively. Insulin exerts its many effects either by altering transport of specific blood-borne nutrients into cells or by altering the activity of the enzymes involved in specific metabolic pathways.

Actions on Carbohydrates The maintenance of blood glucose homeostasis is a particularly important function of the pancreas. Circulating glucose concentrations are determined by the balance that exists among the following processes (• Fig. 19–15): glucose absorption from the digestive tract; transport of glucose into cells; cellular (primarily hepatic) glucose production; and (abnormally) urinary excretion of glucose. Insulin exerts a fourfold effect to lower blood glucose levels and promote carbohydrate storage as follows:

1. It facilitates glucose transport into most cells. Glucose molecules cannot readily penetrate most cell membranes in the absence of insulin. Most tissues, therefore, are highly dependent on insulin for uptake of glucose from the blood and for its subsequent use. Insulin enhances the carrier-mediated mechanism for facilitated diffusion of glucose into these insulin-dependent cells by the phenomenon of **transporter recruitment**. Glucose gains entry to cells only by means of plasma-membrane carriers known as **glucose transporters**. Insulin-dependent cells maintain an intracellular pool of extra glucose transporters. These transporters are inserted into the plasma membrane in response to increased insulin secretion, thus increasing the transport of glucose into the cell. When insulin secretion decreases, the extra transporters are retrieved from the membrane and returned to the intracellular pool.

   Several tissues are not dependent on insulin for their glucose uptake—namely, the brain, working muscles, and the liver. The brain, which requires a constant supply of glucose for its minute-to-minute energy needs, is freely permeable to glucose at all times. For reasons that are unclear, skeletal-muscle cells are not dependent on insulin

for their glucose uptake during exercise, even though they are dependent at rest. This fact is important in the management of diabetes mellitus (insulin deficiency), as will be described later. The liver is also not dependent on insulin for glucose uptake; however, insulin does enhance the metabolism of glucose by the liver by stimulating the first step in glucose metabolism, the phosphorylation of glucose to form glucose-6-phosphate. The phosphorylation of glucose as it enters the cell keeps the intracellular concentration of "plain" glucose low so that a gradient favoring the facilitated diffusion of glucose into the cell is maintained.

2. Insulin stimulates **glycogenesis**, the production of glycogen from glucose, in both skeletal muscle and the liver.

3. Insulin inhibits **glycogenolysis**, the breakdown of glycogen into glucose. By inhibiting the degradation of glycogen, it likewise favors carbohydrate storage and decreases glucose output by the liver.

4. Insulin further decreases hepatic glucose output by inhibiting gluconeogenesis, the conversion of amino acids into glucose in the liver. Insulin accomplishes this in two ways: by decreasing the amount of amino acids in the blood that are available to the liver for gluconeogenesis, and by inhibiting the hepatic enzymes required for the conversion of amino acids into glucose.

Thus, insulin decreases the concentration of blood glucose by promoting the cells' uptake of glucose from the blood for utilization and storage, while simultaneously blocking the two mechanisms by which the liver releases glucose into the blood (glycogenolysis and gluconeogenesis). Insulin is the only hormone capable of lowering the blood glucose level.

Actions on Fat Insulin exerts multiple effects to lower blood fatty acids and promote triglyceride storage as follows:

1. It increases the transport of glucose into adipose-tissue cells, as it does for most cells of the body. Glucose serves

---

• **FIGURE 19–15 Factors Affecting Blood Glucose Concentration**

Factors that increase blood glucose

Factors that decrease blood glucose

Glucose absorption from digestive tract

Hepatic glucose production:
—Through glycogenolysis of stored glycogen
—Through gluconeogenesis

Blood glucose

Transport of glucose into cells:
—For utilization for energy production
—For storage
• as glycogen through glycogenesis
• as triglycerides

Urinary excretion of glucose (occurs only abnormally, when blood glucose level becomes so high it exceeds the reabsorptive capacity of kidney tubules during urine formation)

☐ = Factors subject to hormonal control to maintain blood glucose level as a precursor for the formation of fatty acids and glycerol, which are the raw materials for triglyceride synthesis.

2. It activates enzymes that catalyze the production of fatty acids from glucose derivatives.

3. It promotes the entry of fatty acids from the blood into adipose-tissue cells.

4. It inhibits lipolysis (fat breakdown), thus reducing the release of fatty acids from adipose tissue into the blood.

Collectively, these actions favor removal of glucose and fatty acids from the blood and promote their storage as triglycerides.

Actions on Protein Insulin lowers blood amino acid levels and enhances protein synthesis as follows:

1. It promotes the active transport of amino acids from the blood into muscles and other tissues. This effect decreases the circulating amino acid level and provides the building blocks for protein synthesis within the cells.

2. It increases the rate of amino acid incorporation into protein by stimulating the cells' protein-synthesizing machinery.

3. It inhibits protein degradation.

The collective result of these actions is a protein anabolic effect. For this reason, insulin is essential for normal growth.

In short, insulin stimulates biosynthetic pathways that lead to increased glucose utilization, increased carbohydrate and fat storage, and increased protein synthesis. In so doing, this hormone lowers the blood glucose, fatty acid, and amino acid levels. This metabolic pattern is characteristic of the absorptive state. Indeed, insulin secretion rises during this state

and is responsible for shifting metabolic pathways to net anabolism.

When insulin secretion is low, the opposite effects occur. The rate of glucose entry into cells is reduced, and net catabolism rather than net synthesis of glycogen, triglycerides, and protein occurs. This pattern is reminiscent of the postabsorptive state; indeed, insulin secretion is reduced during the postabsorptive state. However, the other major pancreatic hormone, glucagon, also plays an important role in shifting from absorptive to postabsorptive metabolic patterns, as will be described later.

The primary stimulus for increased insulin secretion is an increase in blood glucose concentration.

The primary control of insulin secretion is a direct negative-feedback system between the pancreatic β cells and the concentration of glucose in the blood flowing to them. An elevated blood glucose level, such as occurs during absorption of a meal, directly stimulates synthesis and release of insulin by the β cells. The increased insulin, in turn, reduces the blood glucose to normal while it promotes utilization and storage of this nutrient. Conversely, a fall in blood glucose below normal, such as occurs during fasting, directly inhibits insulin secretion. Lowering the rate of insulin secretion shifts metabolism from the absorptive to the postabsorptive pattern. Thus, this simple negative-feedback system is able to maintain a relatively constant supply of glucose to the tissues without requiring the participation of nerves or other hormones.

In addition to plasma glucose concentration, the following inputs are involved in the regulation of insulin secretion, (Fig. 19–16).

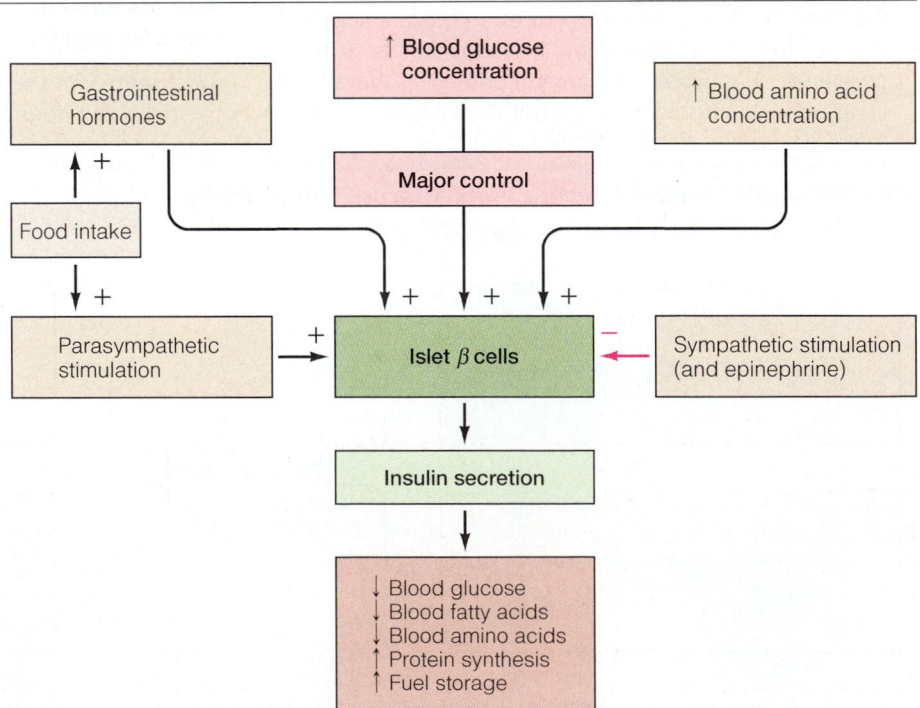

• **FIGURE 19–16 Factors Controlling Insulin Secretion**

- An elevated plasma amino acid level, such as occurs following ingestion of a high-protein meal, directly stimulates the β cells to increase insulin secretion. In negative-feedback fashion, the increased insulin enhances the entry of these amino acids into the cells, lowering the blood amino acid level while promoting protein synthesis.

- The major gastrointestinal hormones secreted by the digestive tract in response to the presence of food, especially gastric inhibitory peptide (see p. 591), stimulate pancreatic insulin secretion in addition to having direct regulatory effects on the digestive system. Through this control, insulin secretion is increased in "feedforward" or anticipatory fashion even before nutrient absorption increases the concentration of glucose and amino acids in the blood.

- The autonomic nervous system also directly influences insulin secretion. The islets are richly innervated by both parasympathetic (vagal) and sympathetic nerve fibers. The increase in parasympathetic activity that occurs in response to food in the digestive tract stimulates insulin release. This, too, is a feedforward response in anticipation of nutrient absorption. In contrast, sympathetic stimulation and the concurrent increase in epinephrine both inhibit insulin secretion. The reduction in insulin allows the blood glucose level to increase, an appropriate response to the circumstances under which generalized sympathetic activation occurs—namely, stress (fight or flight) and exercise. In both of these situations, extra fuel is needed for increased muscle activity.

The symptoms of diabetes mellitus are characteristic of an exaggerated postabsorptive state.

Diabetes mellitus is by far the most common of all endocrine disorders. The acute symptoms of diabetes mellitus are attributable to inadequate insulin action. Because insulin is the only hormone capable of lowering blood glucose levels, one of the most prominent features of diabetes mellitus is elevated blood glucose levels, or **hyperglycemia**. *Diabetes* literally means "syphon" or "running through," a reference to the large urine volume accompanying this condition. A large urine volume occurs both in diabetes mellitus due to insulin insufficiency and in diabetes insipidus due to vasopressin deficiency. *Mellitus* means "sweet"; *insipidus* means "tasteless." The urine of patients with diabetes mellitus acquires its sweetness from excess blood glucose that spills into the urine, whereas the urine of patients with diabetes insipidus contains no sugar, so it is tasteless. (Aren't you glad you were not a health professional in the time when these two conditions were distinguished on the basis of the taste of the urine?)

There are two major variants of diabetes mellitus, differing in the capacity for pancreatic insulin secretion: *Type I diabetes,* characterized by a lack of insulin secretion, and *Type II diabetes,* characterized by normal or even increased insulin secretion but reduced sensitivity of insulin's target cells to its presence. (See the accompanying boxed feature, Concepts, Challenges, and Controversies.)

The acute consequences of diabetes mellitus can be grouped according to the effects of inadequate insulin action on carbohydrate, fat, and protein metabolism (● Fig. 19–17, p. 678). The figure looks overwhelming, but the numbers correspond to the numbers in the following discussion to help you work your way through this complex disease step-by-step.

Since the postabsorptive metabolic pattern is induced by low insulin activity, the changes that occur in diabetes mellitus are an exaggeration of this state, with the exception of hyperglycemia. In the usual fasting state, the blood glucose level is slightly below normal. Hyperglycemia, the hallmark of diabetes mellitus, arises from reduced glucose uptake by cells, coupled with increased output of glucose from the liver (① on Fig. 19–17). Hepatic output of glucose increases as the glucose-producing processes of glycogenolysis and gluconeogenesis proceed unchecked in the absence of insulin. Because most of the body's cells are unable to use glucose without the assistance of insulin, an ironic extracellular glucose excess occurs coincident with an intracellular glucose deficiency—"starvation in the midst of plenty." Even though the non-insulin-dependent brain is adequately nourished during diabetes mellitus, further consequences of the disease lead to brain dysfunction, as you will see shortly.

When the blood glucose rises to the level when the amount of glucose filtered exceeds the tubular cells' capacity for reabsorption, glucose appears in the urine (*glucosuria*) ②. Glucose in the urine exerts an osmotic effect that draws H_2O with it, producing an osmotic diuresis characterized by *polyuria* (frequent urination) ③. The excess fluid lost from the body leads to dehydration ④, which in turn can ultimately lead to peripheral circulatory failure because of the marked reduction in blood volume ⑤. Circulatory failure, if uncorrected, can lead to death because of low cerebral blood flow ⑥ or secondary renal failure due to inadequate filtration pressure ⑦. Furthermore, cells lose water as the body becomes dehydrated as a result of an osmotic shift of water from the cells into the hypertonic extracellular fluid ⑧. Brain cells are especially sensitive to shrinking so that nervous-system malfunction ensues ⑨ (see p. 518). Another characteristic symptom of diabetes mellitus is *polydipsia* (excessive thirst) ⑩, which is actually a compensatory mechanism that takes place to counteract the dehydration.

The story is still not complete. In the face of intracellular glucose deficiency, appetite is stimulated, leading to *polyphagia* (excessive food intake) ⑪. In spite of the increased food intake, however, progressive weight loss occurs as a result of the effects of insulin deficiency on fat and protein metabolism. Triglyceride synthesis decreases while lipolysis increases, resulting in large-scale mobilization of fatty acids from triglyceride stores ⑫. The increased blood fatty acids are used to a large extent by the cells as an alternative energy source. Increased liver utilization of fatty acids results in the release of excessive ketone bodies into the blood, causing *ketosis* ⑬. Since the ketone bodies include several different acids such as acetoacetic acid that result from the incomplete breakdown of fat during hepatic energy production, this developing ketosis leads to progressive metabolic acidosis ⑭. Acidosis depresses the brain and, if severe enough, can lead to diabetic coma and death ⑮.

DIABETICS AND INSULIN: SOME HAVE IT AND SOME DON'T

There are two distinct types of diabetes mellitus (see the accompanying table). **Type I (insulin-dependent** or **juvenile-onset) diabetes mellitus,** which accounts for 10% to 20% of all cases of diabetes, is characterized by a lack of insulin secretion. In **Type II (non-insulin-dependent** or **maturity-onset) diabetes mellitus,** insulin secretion may be normal or even increased, but insulin's target cells are less sensitive than normal to this hormone. Although either type can first be manifested at any age, Type I has a greater prevalence in children, whereas the onset of Type II more generally occurs in adulthood, giving rise to the age-related designations of the two conditions. Genetic as well as environmental factors appear to be important in the development of both types of diabetes mellitus.

Type I diabetes is an autoimmune process involving the erroneous, selective destruction of the pancreatic β cells. Although self-produced antibodies against β cells have been identified in Type I diabetics, strong evidence implicates activated T lymphocytes as the major β-cell-killing agents. T lymphocytes are a type of white blood cell that normally attacks virus-invaded cells and cancer cells (see p. 392). In Type I diabetes, T lymphocytes are apparently activated and erroneously attack the pancreatic β cells. Because Type I diabetics suffer a total or near-total lack of insulin secretion by their pancreatic β cells, they require exogenous insulin for survival. This dependence is the basis for the al-

Comparison of Type I and Type II Diabetes Mellitus

CHARACTERISTIC	TYPE I DIABETES	TYPE II DIABETES
Level of insulin secretion	None or almost none	May be normal or exceed normal
Typical age of onset	Childhood	Adulthood
Percentage of diabetics	10–20%	80–90%
Basic defect	Destruction of β cells	Reduced sensitivity of insulin's target cells
Associated with obesity?	No	Usually
Genetic and environmental factors important in precipitating overt disease?	Yes	Yes
Speed of development of symptoms	Rapid	Slow
Development of ketosis	Common if untreated	Rare
Treatment	Insulin injections; dietary management	Dietary control and weight reduction; occasionally oral hypoglycemic drugs

A compensatory measure for metabolic acidosis is increased ventilation to blow off extra acid-forming CO_2 [16]. Exhalation of one of the ketone bodies, acetone, causes a "fruity" breath odor. Sometimes, because of this odor, a patient collapsed in a diabetic coma is unfortunately mistaken by passersby for a "wino" passed out in a state of drunkenness. (This situation illustrates the merits of medical-alert identification tags.) Persons with Type I diabetes are much more prone to develop ketosis than are Type II diabetics.

The effects of a lack of insulin on protein metabolism result in a net shift toward protein catabolism. The net breakdown of muscle proteins leads to wasting and weakness of skeletal muscles [17] and, in child diabetics, a reduction in overall growth. Reduced amino acid uptake coupled with increased protein degradation results in excess amino acids in the blood [18]. The increased circulating amino acids can be used for additional gluconeogenesis, which further aggravates the hyperglycemia [19].

As you can readily appreciate from this overview, diabetes mellitus is a complicated disease that can lead to disturbances in carbohydrate, fat, and protein metabolism and in fluid and acid-base balance. It can also have repercussions on the circulatory system, kidneys, respiratory system, and nervous system.

In addition to these potential consequences of untreated diabetes, which can be explained on the basis of insulin's short-term metabolic effects, numerous long-range complications of this disease frequently occur that are not as easy to explain. These chronic complications, which account for the shorter life expectancy of diabetics, primarily involve degenerative disorders of the vasculature and nervous system. Cardiovascular lesions are the most common cause of premature death in diabetics. Heart disease and strokes occur with greater incidence than in nondiabetics. Because vascular lesions often develop in the kidneys and retinas of the eyes, kidney failure is a very common long-term complication of diabetes, and diabetes is the second leading cause of blindness in the United States. Extremities may become gangrenous as a result of impaired delivery of blood to these tissues, which leads to the necessity of amputating toes or even whole

DIABETICS AND INSULIN: SOME HAVE IT AND SOME DON'T (continued)

ternative name insulin-dependent diabetes mellitus for this form of the disease.

Type II diabetics, on the other hand, do secrete insulin in varying amounts. In fact, insulin levels may be normal or even exceed those in nondiabetics. The basic problem in Type II diabetes is not lack of insulin but reduced sensitivity of insulin's target cells to its presence, usually because of down-regulation (see p. 631) of insulin receptors in association with obesity. Chronic overeating by an obese person results in the secretion of increased amounts of insulin to maintain the blood glucose at normal levels by putting the excess nutrients in storage. In response to chronic hyperinsulinemia, the number of insulin receptors gradually becomes reduced over time. The resultant decrease in sensitivity to insulin in obese but otherwise normal individuals is overcome by secretion of additional insulin. In this way, the excess nutrients are stored despite the decreased availability of insulin receptors, so blood glucose homeostasis is maintained. In obese diabetes-prone individuals,

however, the sustained overtasking of the pancreas by chronic excessive nutrient intake eventually exceeds the reserve secretory capacity of the genetically weak β cells. Even though insulin secretion may be normal or somewhat elevated, symptoms of insulin insufficiency develop because the amount of insulin is still inadequate to prevent significant hyperglycemia in the presence of excess nutrient absorption.

The symptoms in Type II diabetes are usually slower in onset and less severe than in Type I diabetes. Whereas Type I diabetics are permanently insulin dependent, dietary control and weight reduction may be all that is necessary to completely reverse the symptoms in Type II diabetics. Therefore, Type II diabetes is alternatively known as non-insulin-dependent diabetes. As the magnitude of insulin secretion decreases in connection with reduced caloric intake, the number of insulin receptors gradually returns to normal and so too does target-tissue responsiveness to insulin. Exercise is also useful in the management of both types of diabetes, because working muscles

are not insulin dependent. Exercising muscles take up and use some of the excess glucose in the blood, thus reducing the overall need for insulin.

Occasionally, Type II diabetics are administered drugs that remove excess nutrients from the blood by driving the weakened β cells to secrete more insulin than they do on their own. These therapeutic agents are referred to as *oral hypoglycemic drugs* since they can be taken orally to bring about a reduction in blood glucose. In contrast, insulin must be administered by injection because this peptide hormone would be digested by proteolytic enzymes if it were to be swallowed. Although oral hypoglycemics alleviate the symptoms of diabetes, their use is controversial because they overwork already weakened β cells; the end result may be the complete "burnout" of these cells so that they are no longer able to produce insulin at all. In such a case, the previously non-insulin-dependent patient would have to be placed on insulin therapy for life. Weight reduction and exercise are safer and more permanent solutions.

---

limbs. In addition to the problems related to the circulatory system, degenerative lesions in nerves lead to multiple neuropathies that result in dysfunction of the brain, spinal cord, and peripheral nerves.

The mechanisms responsible for the development of these long-range vascular and neural degenerative complications are largely unknown. There is considerable disagreement about whether managing the diabetic patient's condition so that blood glucose levels are continuously kept within reasonable limits diminishes the incidence of these chronic abnormalities.

Treatment of diabetes depends on insulin injections (for Type I), dietary management, weight control, and exercise. Because exercising muscles take up and use some of the excess glucose in the blood, the overall need for insulin is reduced. Current research on several fronts may dramatically change the approach to diabetic therapy. New treatments may include such innovations as β cell transplants and implanted, glucose-sensitive insulin-releasing devices. Recent advances in understanding the underlying molecular defects

in diabetes have even made investigators hopeful that safe therapies may be developed within this decade to prevent new cases of diabetes.

Insulin excess causes brain-starving hypoglycemia.

Let us now look at the opposite of diabetes mellitus, insulin excess, which is characterized by **hypoglycemia** (low blood glucose) and can occur in two different ways. First, insulin excess can occur in a diabetic patient when too much insulin has been injected for the person's caloric intake and exercise level, resulting in so-called **insulin shock**. Second, an abnormally high blood insulin level may occur in a nondiabetic individual with a β cell tumor or in whom the β cells are overresponsive to glucose. The true incidence of the latter condition, so-called **reactive hypoglycemia**, is a subject of intense controversy, because laboratory measurements to confirm the presence of low blood glucose during the time of symptoms have not been performed in the majority of people who have been diagnosed as having the condition. In mild

• **FIGURE 19–17   Acute Effects of Diabetes Mellitus** The acute consequences of diabetes mellitus can be grouped according to the effects of inadequate insulin action on carbohydrate, fat, and protein metabolism. These effects ultimately cause death through a variety of pathways. See pages 675–676 for an explanation of the circled numbers.

cases, the symptoms of hypoglycemia, such as tremor, fatigue, sleepiness, and inability to concentrate, are nonspecific. Because these symptoms could also be attributable to emotional problems or other factors, a definitive diagnosis based on symptoms alone is impossible to make.

The consequences of insulin excess are primarily manifestations of the effects of hypoglycemia on the brain. Recall that the brain relies on a continuous supply of blood glucose for its nourishment and that glucose uptake by the brain does not depend on insulin. With insulin excess, more glucose than necessary is driven into the other insulin-dependent cells of the body. The result is a lowering of the blood glucose level so that not enough glucose is left in the blood to be delivered to the brain. The brain literally starves in the presence of hypoglycemia. The symptoms, therefore, are primarily referable to depressed brain function, which, if severe enough, may rapidly progress to unconsciousness and death. Persons with overresponsive β cells usually do not become sufficiently hypoglycemic to manifest these more serious consequences, but they do show milder symptoms of depressed CNS activity.

The treatment of hypoglycemia depends on the cause. In the case of a diabetic with insulin overdose, something sugary should be taken at the first indication of a hypoglycemic attack. Prompt treatment of severe hypoglycemia is imperative if brain damage is to be prevented. Note that a diabetic can lose consciousness and die from either diabetic ketoacidotic coma caused by prolonged insulin deficiency or acute hypoglycemia caused by insulin shock. Fortunately, the other accompanying signs and symptoms differ sufficiently between the conditions to enable medical caretakers to administer appropriate therapy, either insulin or glucose. For example, ketoacidotic coma is accompanied by deep, labored breathing (a compensatory measure for the metabolic acidosis) and fruity breath (due to exhalation of ketone bodies), whereas insulin shock is not.

Ironically, even though reactive hypoglycemia is characterized by a low blood glucose level, persons with this disorder are treated by limiting their intake of sugar and other glucose-yielding carbohydrates to prevent their β cells from overresponding to a high glucose intake. With low carbohydrate intake, the blood glucose does not rise as much during the absorptive state. Since blood glucose elevation is the primary regulator of insulin secretion, the β cells are not stimulated as much with a low-carbohydrate meal as with a typical meal. Accordingly, reactive hypoglycemia is less likely to occur. Giving a symptomatic individual with reactive hypoglycemia something sugary, as is done with a hypoglycemic diabetic, temporarily alleviates the symptoms, because the blood glucose level is transiently restored to normal so that the brain's energy needs are once again satisfied. As soon as the extra glucose triggers further insulin release, however, the situation is merely aggravated.

Glucagon in general opposes the actions of insulin.

Even though insulin plays a central role in controlling the metabolic adjustments between the absorptive and postabsorptive states, the secretory product of the pancreatic islet α cells, **glucagon**, is also very important. Many physiologists view the insulin-secreting β cells and the glucagon-secreting α cells as a coupled endocrine system whose combined secretory output is a major factor in the regulation of fuel metabolism.

Glucagon affects many of the same metabolic processes that are influenced by insulin, but in most cases glucagon's actions are opposite to those of insulin. The major site of action of glucagon is the liver, where it exerts a variety of effects on carbohydrate, fat, and protein metabolism.

Actions on Carbohydrate The overall effects of glucagon on carbohydrate metabolism result in an increase in hepatic glucose production and release and thus an increase in blood glucose levels. Glucagon exerts its hyperglycemic effects by decreasing glycogen synthesis, promoting glycogenolysis, and stimulating gluconeogenesis.

Actions on Fat Glucagon also antagonizes the actions of insulin with regard to fat metabolism by promoting fat breakdown and inhibiting triglyceride synthesis. Glucagon enhances hepatic ketone production (**ketogenesis**) by promoting the conversion of fatty acids to ketone bodies. Thus, the blood levels of fatty acids and ketones increase under glucagon's influence.

Actions on Protein Glucagon inhibits hepatic protein synthesis and promotes degradation of hepatic protein. Stimulation of gluconeogenesis further contributes to glucagon's catabolic effect on hepatic protein metabolism. Although glucagon promotes protein catabolism in the liver, it does not have any significant effect on blood amino acid levels because it does not affect muscle protein, the major protein store in the body.

Glucagon secretion is increased during the postabsorptive state.

Considering the catabolic effects of glucagon on the body's energy stores, you would be correct in assuming that glucagon secretion is increased during the postabsorptive state and decreased during the absorptive state, just the opposite of insulin secretion. In fact, insulin is sometimes referred to as a "hormone of feasting" and glucagon as a "hormone of fasting." Insulin tends to put nutrients in storage when their blood levels are high, such as following a meal, whereas glucagon promotes catabolism of nutrient stores between meals to keep up the blood nutrient levels, especially blood glucose.

Like insulin secretion, the major factor regulating glucagon secretion is a direct effect of the blood glucose concentration on the endocrine pancreas. In this case, the pancreatic α cells increase glucagon secretion in response to a fall in blood glucose. The hyperglycemic actions of this hormone tend to restore the blood glucose level to normal. Conversely, an increase in blood glucose concentration, such as occurs after a meal, inhibits glucagon secretion, which likewise tends to restore the blood glucose level to normal. Thus, there is a direct negative-feedback relationship between blood glucose concentration and the α cells' rate of secretion, but it is in the opposite direction of the effect of blood glucose on the β cells; in other words, an elevated blood glucose level inhibits glucagon secretion but stimulates insulin secretion, whereas a fall in blood glucose level leads to increased glucagon secretion and decreased insulin secretion (• Fig. 19–18). Since glucagon raises blood glucose and insulin decreases blood glucose, the changes in secretion of these pancreatic hormones in response to deviations in blood glucose work together homeostatically to restore blood glucose levels to normal. Similarly, a fall in blood fatty acid concentration directly stimulates glucagon output and inhibits insulin output by the pancreas, both of which are negative-feedback control mechanisms to restore the blood fatty acid level to normal.

The opposite effects exerted by the blood concentrations of glucose and fatty acids on the pancreatic α and β cells are appropriate for regulating the circulating levels of these nutrient molecules, because the actions of insulin and glucagon

• **FIGURE 19–18   Complementary Interactions of Glucagon and Insulin**

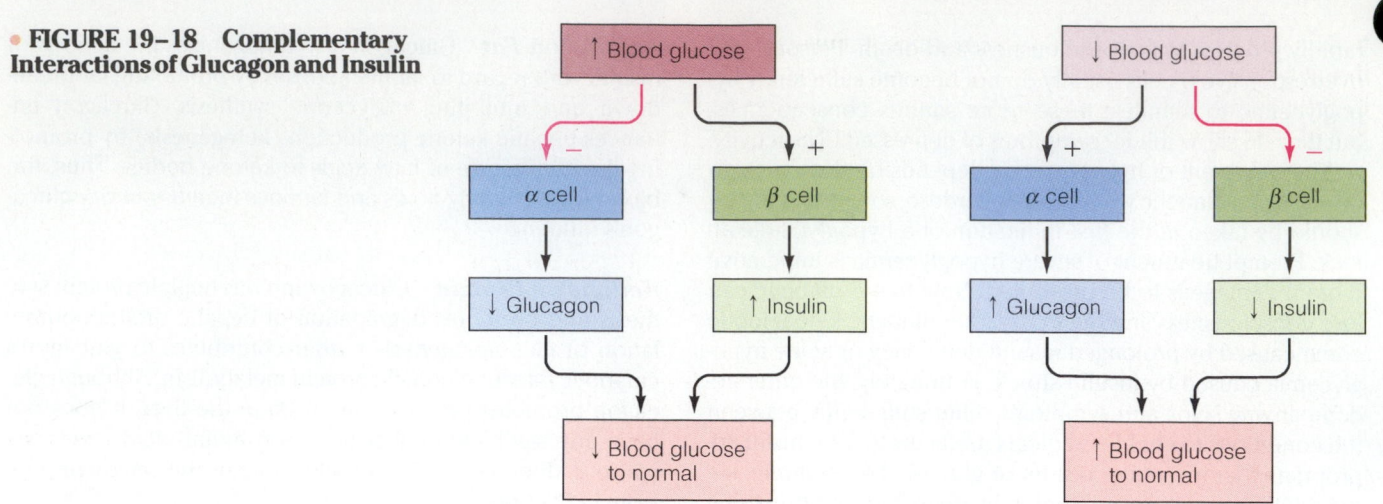

on carbohydrate and fat metabolism oppose one another. The effect of blood amino acid concentration on the secretion of these two hormones is a different story. A rise in blood amino acid concentration stimulates *both* glucagon and insulin secretion. Why this seeming paradox, since glucagon does not exert any effect on blood amino acid concentration? The identical effect of high blood amino acid levels on both glucagon and insulin secretion makes sense if you consider the concomitant effects these two hormones have on blood glucose levels (• Fig. 19–19). If during absorption of a pro-

• **FIGURE 19–19   Counteracting Actions of Glucagon and Insulin on Blood Glucose during Absorption of High-Protein Meal**

High-protein meal (little carbohydrate)

↑ Blood amino acid concentration

β cells (+) → ↑ Insulin

↑ Insulin → Promotes cellular uptake and assimilation of amino acids

↑ Insulin → ↑ Glucose uptake by cells

↑ Insulin → ↓ Hepatic glucose output

↑ Glucose uptake by cells + ↓ Hepatic glucose output → Hypoglycemia

α cells (+) → ↑ Glucagon

↑ Glucagon → ↑ Hepatic glucose output

↑ Hepatic glucose output → Hyperglycemia

(Effects counteract each other)

Hypoglycemia + Hyperglycemia → Blood glucose remains normal

tein-rich meal the rise in blood amino acids stimulated only insulin secretion, hypoglycemia might result. Because little carbohydrate is available for absorption following consumption of a high-protein meal, the amino acid–induced increase in insulin secretion would drive too much glucose into the cells, causing a sudden, inappropriate drop in the blood glucose level. However, the simultaneous increase in glucagon secretion elicited by elevated blood amino acid levels increases hepatic glucose production. Since the hyperglycemic effects of glucagon counteract the hypoglycemic actions of insulin, the net result is maintenance of normal blood glucose levels (and prevention of hypoglycemic starvation of the brain) during absorption of a meal that is high in protein but low in carbohydrates.

Glucagon excess can aggravate the hyperglycemia of diabetes mellitus.

No known clinical abnormalities are attributable to glucagon deficiency or excess per se. However, diabetes mellitus is frequently accompanied by excess glucagon secretion, because insulin is required for glucose to gain entry into the α cells, where it can exert control over glucagon secretion. As a result, diabetics frequently have a high rate of glucagon secretion concurrent with their insulin insufficiency because the elevated blood glucose is not able to inhibit glucagon secretion as it normally would. Since glucagon is a hormone that raises blood glucose, its excess intensifies the hyperglycemia of diabetes mellitus. For this reason, some insulin-dependent diabetics respond best to a combination of insulin and somatostatin therapy. By inhibiting glucagon secretion, somatostatin indirectly helps to achieve better reduction of the elevated blood glucose concentration than can be accomplished by insulin therapy alone.

Epinephrine, cortisol, growth hormone, and thyroid hormone also exert direct metabolic effects.

The pancreatic hormones are the most important regulators of normal fuel metabolism. However, several other hormones exert direct metabolic effects even though control of their secretion is keyed to factors other than transitions in metabolism between feasting and fasting states (Table 19–7).

The stress hormones, epinephrine and cortisol, both increase blood levels of glucose and fatty acids. In addition, cortisol mobilizes amino acids by promoting protein catabolism. Neither of these hormones plays important roles in the regulation of fuel metabolism under resting conditions, but they are important for the metabolic responses to stress. Cortisol also appears to contribute to the maintenance of blood glucose concentration during long-term starvation.

Growth hormone has protein anabolic effects in muscle. In fact, this is one of its growth-promoting features. Although growth hormone can elevate the blood levels of glucose and fatty acids, it is normally of little importance to the overall regulation of fuel metabolism. Deep sleep, stress, exercise, and severe hypoglycemia stimulate growth hormone secretion, possibly to provide fatty acids as an energy source and

spare glucose for the brain under these circumstances. Growth hormone, like cortisol, appears to contribute to the maintenance of blood glucose concentrations during starvation.

The effects of thyroid hormone on intermediary metabolism are decidedly mixed. It stimulates both anabolic and catabolic actions as well as the overall metabolic rate. Catabolic effects tend to dominate at higher thyroid hormone levels. However, changes in thyroid hormone secretion are usually not important for fuel homeostasis for two reasons. First, control of thyroid hormone secretion is not directed toward maintenance of nutrient levels in the blood. Second, the onset of thyroid hormone action is too slow to have any significant effect on the rapid adjustments required to maintain normal blood levels of nutrients. During starvation, less T_4 is converted to the more potent T_3. This shift in the pattern of circulating thyroid hormone is probably in large part responsible for the reduction in basal metabolic rate that usually occurs with prolonged fasting.

Note that, with the exception of the anabolic effects of growth hormone on protein metabolism, all of the metabolic actions of these other hormones are opposite to those of insulin. Insulin alone is able to reduce blood glucose and blood fatty acid levels, whereas glucagon, epinephrine, cortisol, and growth hormone all increase blood levels of these nutrients. Because of this, these other hormones are considered to be *insulin antagonists*. Thus, the main reason diabetes mellitus has such devastating metabolic consequences is that no other control mechanism is available to pick up the slack to promote anabolism when insulin activity is insufficient, so the catabolic reactions promoted by other hormones are allowed to proceed unchecked. The only exception is protein anabolism stimulated by growth hormone.

ENDOCRINE CONTROL OF CALCIUM METABOLISM

Plasma calcium must be closely regulated to prevent changes in neuromuscular excitability.

Besides regulating the concentration of organic nutrient molecules in the blood by manipulation of anabolic and catabolic pathways, the endocrine system also regulates the plasma concentration of a number of inorganic electrolytes. As you already know, aldosterone controls Na^+ and K^+ concentrations in the ECF. Three other hormones—*parathyroid hormone, calcitonin,* and *vitamin D*—control calcium (Ca^{++}) and phosphate ($PO_4^=$) metabolism. These hormonal agents concern themselves with regulation of plasma Ca^{++}, and in the process, plasma $PO_4^=$ is also maintained. Plasma Ca^{++} concentration is one of the most tightly controlled variables in the body. The need for the precise regulation of plasma Ca^{++} stems from its critical influence on so many body activities.

About 99% of the Ca^{++} in the body is in crystalline form within the skeleton and teeth. Of the remaining 1%, about 0.9% is found intracellularly within the soft tissues; less than

TABLE 19–7 Summary of Hormonal Control of Fuel Metabolism

HORMONE	MAJOR METABOLIC EFFECTS				CONTROL OF SECRETION	
	Effect on Blood Glucose	Effect on Blood Fatty Acids	Effect on Blood Amino Acids	Effect on Muscle Protein	Major Stimuli for Secretion	Primary Role in Metabolism
Insulin	↓ +Glucose uptake +Glycogenesis −Glycogenolysis −Gluconeogenesis	↓ +Triglyceride synthesis −Lipolysis	↓ +Amino acid uptake	↑ +Protein synthesis −Protein degradation	↑ Blood glucose ↑ Blood amino acids	Primary regulator of absorptive and postabsorptive cycles
Glucagon	↑ +Glycogenolysis +Gluconeogenesis −Glycogenesis	↑ +Lipolysis −Triglyceride synthesis	No effect	No effect	↓ Blood glucose; ↑ Blood amino acids	Regulation of absorptive and postabsorptive cycles in concert with insulin; protection against hypoglycemia
Epinephrine	↑ +Glycogenolysis +Gluconeogenesis −Insulin secretion +Glucagon secretion	↑ +Lipolysis	No effect	No effect	Sympathetic stimulation during stress and exercise	Provision of energy for emergencies and exercise
Cortisol	↑ +Gluconeogenesis −Glucose uptake by tissues other than brain; glucose sparing	↑ +Lipolysis	↑ +Protein degradation	↓ +Protein degradation	Stress	Mobilization of metabolic fuels and building blocks during adaptation to stress
Growth hormone	↑ −Glucose uptake by muscles; glucose sparing	↑ +Lipolysis	↓ +Amino acid uptake	↑ +Protein synthesis −Protein degradation +Synthesis of DNA and RNA	Deep sleep Stress Exercise Hypoglycemia	Promotion of growth; normally little role in metabolism; mobilization of fuels plus glucose sparing in extenuating circumstances

0.1% is present in the ECF. Approximately half of the plasma Ca^{++} is either bound to plasma proteins and therefore is restricted to the plasma or is complexed with $PO_4^=$ and is not free to participate in chemical reactions. The other half of the plasma Ca^{++} is freely diffusible and can readily pass into the interstitial fluid and interact with the cells. The free Ca^{++} in the plasma and interstitial fluid is considered to be a single pool. Only this free Ca^{++} is biologically active and subject to regulation; it constitutes less than one-thousandth of the total Ca^{++} in the body.

This small, freely diffusible fraction of ECF Ca^{++} plays a vital role in a number of essential activities, including the following:

1. *Neuromuscular excitability.* Even minor variations in the concentration of free ECF Ca^{++} can have a profound and immediate impact on the sensitivity of excitable tissues. A fall in free Ca^{++} results in overexcitability of nerves and muscles; conversely, a rise in free Ca^{++} depresses neuro-

muscular excitability. These effects result from the influence of Ca^{++} on membrane permeability to Na^+. A decrease in free Ca^{++} increases Na^+ permeability, with the resultant influx of Na^+ moving the resting potential closer to threshold. Consequently, in the presence of **hypocalcemia** (low blood Ca^{++}), excitable tissues may be brought to threshold by normally ineffective physiologic stimuli so that skeletal muscles discharge and contract (go into spasm) "spontaneously" (in the absence of normal stimulation). If severe enough, spastic contraction of the respiratory muscles results in death by asphyxiation. **Hypercalcemia** (elevated blood Ca^{++}), on the other hand, is also life-threatening because it causes cardiac arrhythmias accompanied by generalized depression of neuromuscular excitability.

2. *Excitation-contraction coupling in cardiac and smooth muscle.* Entry of ECF Ca^{++} into cardiac- and smooth-muscle cells, resulting from increased Ca^{++} permeability

in response to an action potential, triggers the contractile mechanism. Calcium is also necessary for excitation-contraction coupling in skeletal-muscle fibers, but in this case the Ca^{++} is released from intracellular Ca^{++} stores in response to an action potential. Some of the increase in cytosolic Ca^{++} in cardiac-muscle cells is also derived from internal stores.

Note that a *rise in cytosolic Ca^{++}* within a muscle cell causes contraction, whereas an *increase in free ECF Ca^{++}* decreases neuromuscular excitability and reduces the likelihood of contraction occurring. Unless this point is kept in mind, it is difficult to understand why low plasma Ca^{++} levels induce muscle hyperactivity when Ca^{++} is necessary to switch on the contractile apparatus. We are talking about two different Ca^{++} pools, which exert different effects.

3. *Stimulus-secretion coupling.* The entry of Ca^{++} into secretory cells, which results from increased permeability to Ca^{++} in response to appropriate stimulation, triggers the release of the secretory product by exocytosis. This process is important for the secretion of neurotransmitters by nerve cells and for peptide and catecholamine hormone secretion by endocrine cells.

4. *Maintenance of tight junctions between cells.* Calcium forms part of the intercellular cement that holds particular cells tightly together.

5. *Clotting of blood.* Calcium serves as a cofactor in several steps of the cascade of reactions that lead to clot formation.

Because of the profound effects of deviations in free Ca^{++}, especially on neuromuscular excitability, the plasma concentration of this electrolyte is regulated with extraordinary precision. In addition, intracellular Ca^{++} (not the free ECF Ca^{++}) serves as a second messenger in many cells and is involved in cell motility and cilia action. Finally, the Ca^{++} in bone and teeth is essential for the structural and functional integrity of these tissues.

Control of calcium metabolism includes regulation of both calcium homeostasis and calcium balance.

Maintenance of the proper plasma concentration of free Ca^{++} differs from regulation of Na^+ and K^+ in two important regards. Sodium and K^+ homeostasis is maintained primarily by regulating the urinary excretion of these electrolytes so that controlled output matches uncontrolled input. In the case of Ca^{++}, in contrast, not all of the ingested Ca^{++} is absorbed from the digestive tract, with the extent of absorption being hormonally controlled depending on the Ca^{++} status of the body. In addition, bone serves as a large Ca^{++} reservoir that can be drawn on to maintain the free plasma Ca^{++} concentration within the narrow limits compatible with life should dietary intake become too low. Exchange of Ca^{++} between the ECF and bone is also subject to control. Similar inhouse stores are not available for Na^+ and K^+.

Regulation of Ca^{++} metabolism depends on hormonal control of exchanges between the ECF and three other compartments: bone, kidneys, and intestine. Accordingly, control of Ca^{++} metabolism encompasses two aspects. First, regulation of **calcium homeostasis** involves the immediate adjustments required to maintain a constant free plasma Ca^{++} concentration on a minute-to-minute basis. This is largely accomplished by rapid exchanges between the bone and ECF and to a lesser extent by modifications in urinary excretion of Ca^{++}. Second, regulation of **calcium balance** involves the more slowly responding adjustments in intestinal Ca^{++} absorption as well as adjustments in urinary Ca^{++} excretion required to maintain a constant total amount of Ca^{++} in the body. Control of Ca^{++} balance ensures that Ca^{++} intake is equivalent to Ca^{++} excretion over the long term (weeks to months).

Parathyroid hormone (PTH), the principal regulator of Ca^{++} metabolism, acts directly or indirectly on all three of these effector sites. It is the primary hormone responsible for maintenance of Ca^{++} homeostasis and is essential for maintenance of Ca^{++} balance, although vitamin D also contributes in important ways to Ca^{++} balance. The third Ca^{++}-influencing hormone, calcitonin, is not essential for maintenance of either Ca^{++} homeostasis or balance. It serves a backup function during the rare times of extreme hypercalcemia. We will examine the specific effects of each of these hormonal systems in more detail.

Parathyroid hormone raises free plasma calcium levels by its effects on bone, kidneys, and intestine.

Parathyroid hormone (**PTH**) is a peptide hormone secreted by the **parathyroid glands**, four small glands located on the back surface of the thyroid gland, one in each corner. Like aldosterone, PTH *is essential for life.* The overall effect of PTH is to increase the Ca^{++} concentration of plasma (and, accordingly, of the entire ECF), thereby preventing hypocalcemia. In the complete absence of PTH, death ensues within a few days, usually because of asphyxiation caused by hypocalcemic spasm of respiratory muscles. By its actions on bone, kidneys, and intestine, PTH raises the plasma Ca^{++} level when it starts to fall so that hypocalcemia and its effects are normally avoided. This hormone also acts to lower plasma $PO_4^=$ concentration.

Actions on Bone Recall that 99% of the Ca^{++} in the body is found in the skeleton. (See Table 19–8 for other functions of the skeleton.) Bone is a living tissue that is composed prin-

TABLE 19–8   Functions of the Skeleton
Support
Protection of vital internal organs
Assistance in body movement by giving attachment to muscles and providing leverage
Manufacture of blood cells (bone marrow)
Storage depot for Ca^{++} and $PO_4^=$, which can be exchanged with plasma to maintain plasma concentrations of these electrolytes

cipally of an organic extracellular matrix impregnated with **hydroxyapatite crystals**, which consist primarily of precipitated $Ca_3(PO_4)_2$ (calcium phosphate) salts (see p. 640). Normally, $Ca_3(PO_4)_2$ salts are in solution in the ECF, but the conditions within the bone are suitable for these salts to precipitate (crystallize) around the collagen fibers in the matrix. By mobilizing some of these Ca^{++} stores in the bone, PTH raises the plasma Ca^{++} concentration when it starts to fall.

In spite of the apparent inanimate nature of bone, bone constituents are continually being turned over. **Bone deposition** (formation) and **bone resorption** (removal) normally go on concurrently so that bone is constantly being remodeled, much as people remodel buildings by tearing down walls and replacing them. Bone remodeling serves two purposes: (1) it keeps the skeleton appropriately "engineered" for maximum effectiveness in its mechanical uses, and (2) it helps to maintain the plasma Ca^{++} level.

Mechanical factors are responsible for adjusting the strength of bone in response to the demands placed on it. The greater the physical stress and compression to which a bone is subjected, the greater the rate of bone deposition. For example, the bones of athletes are more massive and stronger than those of sedentary individuals. On the other hand, loss of bone mass occurs in response to removal of mechanical stress, as in persons who undergo prolonged bed confinement or those in space flight. Early astronauts lost up to 20% of their bone mass during their time in orbit. Therapeutic exercises can limit or prevent such loss of bone.

The relative rates of bone resorption and deposition are also influenced by hormones. During the childhood years,

• **FIGURE 19–20 Organization of Bone into Osteons** (a) An osteon, the structural unit of most bone, consists of concentric lamellae (layers of osteocytes entombed by the bone they have deposited around themselves) surrounding a central canal. A small blood-vessel branch traverses the central canal. (b) A magnification of the lamellae within an osteon. A network of small canals, the canaliculi, interconnect the entombed osteocytes with each other and with the central canal. Long cytoplasmic processes extend from osteocyte to osteocyte within the canaliculi.

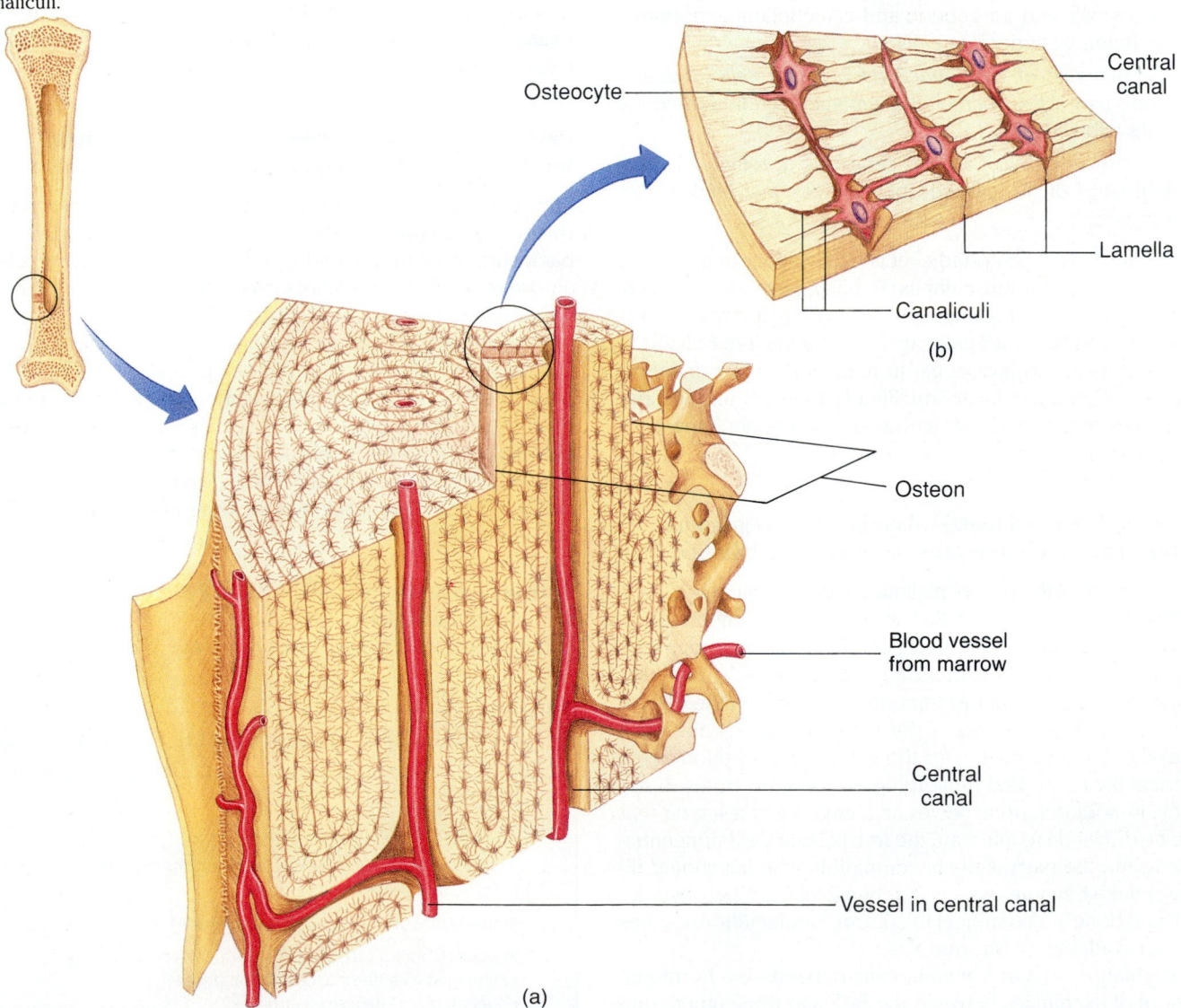

growth hormone promotes deposition of bone to accomplish skeletal growth (see p. 642). Throughout life, parathyroid hormone uses bone as a "bank" from which it withdraws Ca^{++} as needed to maintain the plasma Ca^{++} level.

Recall that three types of bone cells are present in bone (see p. 640). The *osteoblasts* secrete the extracellular organic matrix within which the calcium phosphate crystals precipitate. The *osteocytes* are the retired osteoblasts imprisoned within the bony wall that they have deposited around themselves. The *osteoclasts* resorb bone in their vicinity by releasing acids that dissolve the calcium phosphate crystals and enzymes that break down the organic matrix.

Parathyroid hormone has two major effects on bone that raise plasma Ca^{++} concentration. First, it induces a fast Ca^{++} efflux into the plasma from the small *labile pool* of Ca^{++} in the bone fluid. Second, by stimulating bone dissolution, it promotes a slow transfer into the plasma of both Ca^{++} and $PO_4^=$ from the *stable pool* of bone minerals in the bone itself. As a result, ongoing bone remodeling is tipped in favor of bone resorption over bone deposition.

Most bone is organized into **osteon** units, each of which consists of a **central canal** surrounded by concentrically arranged **lamellae**. Lamellae are layers of osteocytes entombed within the bone that they have deposited around themselves (• Fig. 19–20). The osteons typically run parallel to the long axis of the bone. Blood vessels penetrate the bone either from the outer surface or the marrow cavity and run through the central canals. Osteoblasts are present along the outer surface of the bone and along the inner surfaces lining the central canals. The surface osteoblasts and entombed osteocytes are connected by an extensive network of small, fluid-containing canals, the **canaliculi**, which allow substances to be exchanged between trapped osteocytes and the circulation. These small canals also contain long, filmy cytoplasmic extensions of osteocytes and osteoblasts that are connected to each other, much as if these cells were "holding hands." The interconnecting cell network, which is called the **osteocytic-osteoblastic bone membrane**, separates the mineralized bone itself from the plasma within the central canals (• Fig. 19–21a).

The small labile pool of Ca^{++} is present in the bone fluid that lies between this bone membrane and the adjacent bone both within the canaliculi and along the surface of the central canal. Parathyroid hormone's earliest effect is to activate membrane-bound Ca^{++} pumps to promote movement of Ca^{++}, without the accompaniment of $PO_4^=$, from the bone

(a)

(b)

• **FIGURE 19–21 Relationship of Mineralized Bone, Bone Cells, Bone Fluid, and the Blood**
(a) Schematic representation of osteocytic-osteoblastic bone membrane. The entombed osteocytes and surface osteoblasts are interconnected by long cytoplasmic processes that extend from these cells and connect to each other within the canaliculi. This interconnecting cell network, the osteocytic-osteoblastic bone membrane, separates the mineralized bone from the plasma in the central canal. Bone fluid lies between the membrane and the mineralized bone.
(b) Schematic representation of fast and slow exchange of Ca^{++} between the bone and the plasma. In a fast exchange, Ca^{++} is moved from the labile pool in the bone fluid into the plasma by means of PTH-activated Ca^{++} pumps in the osteocytic-osteoblastic bone membrane. In a slow exchange, Ca^{++} is moved from the stable pool in the mineralized bone into the plasma by means of PTH-induced dissolution of the bone.

fluid, across the osteocytic-osteoblastic bone membrane into the plasma. This transfer process is called **osteolysis**. Movement of Ca^{++} out of the labile pool across the bone membrane accounts for the fast exchange between the bone and the plasma (Fig. 19–21b). Because of the large surface area of the osteocytic-osteoblastic membrane, small movements of Ca^{++} across individual cells are amplified into large Ca^{++} fluxes between the bone and the plasma. After Ca^{++} is pumped out, the bone fluid is believed to be replenished with Ca^{++} from the partially mineralized bone along the adjacent bone surface. Thus, the fast exchange of Ca^{++} does not involve the resorption of completely mineralized bone, and bone mass is not decreased. Through this means, PTH draws Ca^{++} out the "quick-cash branch" of the bone bank and rapidly increases the plasma Ca^{++} level without actually entering the bank (that is, without breaking down mineralized bone itself). Under normal conditions, this exchange is much more important for the maintenance of plasma Ca^{++} concentration than is the slow exchange.

Under conditions of chronic hypocalcemia, such as might occur with dietary Ca^{++} deficiency, PTH influences the slow exchange of Ca^{++} between the bone itself and the ECF by promoting actual localized dissolution of bone. It does so by stimulating osteoclasts to gobble up bone, increasing the formation of more osteoclasts, and transiently inhibiting the bone-forming activity of the osteoblasts. Bone contains such a great abundance of Ca^{++} in comparison to the plasma (more than one-thousand times as much) that even when PTH promotes increased bone resorption, there are no immediate discernible effects on the skeleton because such a minute amount of bone is affected. Yet the negligible amount of Ca^{++} "borrowed" from the bone bank can be lifesaving in terms of restoring the free plasma Ca^{++} level to normal. The borrowed Ca^{++} is then redeposited in the bone at another time when Ca^{++} supplies are more abundant. Meanwhile, the plasma Ca^{++} level has been maintained, without sacrificing the integrity of the bone. However, prolonged excess PTH secretion over months or years is eventually evidenced by the formation of cavities throughout the skeleton that are filled with very large, overstuffed osteoclasts.

When PTH promotes dissolution of the calcium phosphate crystals in bone to harvest their Ca^{++} content, both Ca^{++} and $PO_4^=$ are released into the plasma. An elevation in plasma $PO_4^=$ is undesirable, but PTH deals with this dilemma by its actions on the kidneys.

Actions on Kidneys Parathyroid hormone stimulates Ca^{++} conservation and promotes $PO_4^=$ elimination by the kidneys during the formation of urine. Under the influence of PTH, the kidneys are able to reabsorb more of the filtered Ca^{++}, so less Ca^{++} escapes into the urine. This effect increases the plasma Ca^{++} level and decreases urinary Ca^{++} losses. (It would be futile to dissolve bone to obtain more Ca^{++} only to lose it in the urine.)

At the same time that it stimulates renal Ca^{++} reabsorption, PTH also increases urinary $PO_4^=$ excretion by decreasing $PO_4^=$ reabsorption. As a result, PTH causes a fall in plasma $PO_4^=$ levels at the same time it increases Ca^{++} concentrations.

This PTH-induced removal of extra $PO_4^=$ from the body fluids is essential for preventing reprecipitation of the Ca^{++} freed from the bone. Because of the solubility characteristics of calcium phosphate salt, the product of the plasma concentration of Ca^{++} times the plasma concentration of $PO_4^=$ must remain an approximately constant value. Therefore, an inverse relationship exists between the plasma concentrations of Ca^{++} and $PO_4^=$; for example, when the plasma $PO_4^=$ level rises, some of the plasma Ca^{++} is forced back into the bone through hydroxyapatite crystal formation, thus reducing the plasma Ca^{++} level and keeping the calcium phosphate product constant. This inverse relationship occurs because the free ions in the ECF are in equilibrium with the bone crystals.

Recall that both Ca^{++} and $PO_4^=$ are released from the bone when PTH promotes bone dissolution. Because PTH is secreted only when plasma Ca^{++} falls below normal, the released Ca^{++} is needed to restore plasma Ca^{++} to normal, yet the released $PO_4^=$ tends to increase plasma $PO_4^=$ levels above normal. If the plasma $PO_4^=$ level were allowed to rise above normal, some of the plasma Ca^{++} would have to be redeposited back in the bone along with the $PO_4^=$ in order to keep the calcium phosphate product constant. This redeposition of Ca^{++} would lower plasma Ca^{++}, just the opposite of the desired effect. Therefore, PTH acts on the kidneys to decrease the reabsorption of $PO_4^=$ by the renal tubules. This increases the urinary excretion of $PO_4^=$ and lowers its plasma concentration, even though extra $PO_4^=$ is being released from the bone into the blood. Such action prevents the self-defeating redeposition of released Ca^{++} back into the bone.

The third important action of PTH on the kidneys (besides increasing Ca^{++} reabsorption and decreasing $PO_4^=$ reabsorption) is to enhance the activation of vitamin D by the kidneys.

Action on the Intestine Although PTH has no direct effect on the intestine, it indirectly increases both Ca^{++} and $PO_4^=$ reabsorption from the small intestine by means of its role in vitamin D activation. This vitamin, in turn, directly increases intestinal absorption of Ca^{++} and $PO_4^=$.

The primary regulator of PTH secretion is the plasma concentration of free calcium.

All the effects of PTH are aimed at raising the plasma Ca^{++} levels. Appropriately, PTH secretion is increased in response to a fall in plasma Ca^{++} concentration and decreased by a rise in plasma Ca^{++} levels. The secretory cells of the parathyroid glands are directly and exquisitely sensitive to changes in free plasma Ca^{++}. Since PTH regulates plasma Ca^{++} concentration, this relationship forms a simple negative-feedback loop for controlling PTH secretion without involving any nervous or other hormonal intervention (• Fig. 19–22).

Calcitonin lowers the plasma calcium concentration but is not important in the normal control of calcium metabolism.

Calcitonin, the hormone produced by the C cells of the thyroid gland, also exerts an influence on plasma Ca^{++} levels.

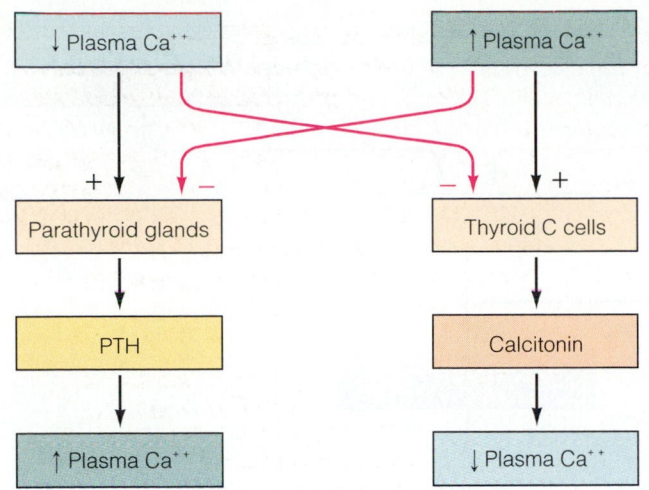

● **FIGURE 19–22   Negative-Feedback Loops Controlling Parathyroid Hormone (PTH) and Calcitonin Secretion**

Like PTH, calcitonin has two effects on bone, but in this case, both effects *decrease* plasma Ca^{++} levels. First, on a short-term basis, calcitonin decreases Ca^{++} movement from the bone fluid into the plasma. Second, on a long-term basis, calcitonin decreases bone resorption by inhibiting the activity of osteoclasts. The suppression of bone resorption results in decreased plasma $PO_4^{=}$ levels as well as a reduced plasma Ca^{++} concentration. The hypocalcemic and hypophosphatemic effects of calcitonin are due entirely to this hormone's actions on bone. It has no effect on the kidneys or intestine.

As with PTH, the primary regulator of calcitonin release is the free plasma Ca^{++} concentration, but in contrast to its effect on PTH release, an increase in plasma Ca^{++} stimulates calcitonin secretion and a fall in plasma Ca^{++} inhibits calcitonin secretion (Fig. 19–22). Since calcitonin reduces plasma Ca^{++} levels, this system constitutes a second simple negative-feedback control over plasma Ca^{++} concentration, one that is opposed to the PTH system.

Most evidence suggests, however, that calcitonin plays little or no role in the normal control of Ca^{++} or $PO_4^{=}$ metabolism. Although calcitonin will protect against hypercalcemia, this condition rarely occurs under normal circumstances. Moreover, neither thyroid removal nor calcitonin-secreting tumors alter circulating levels of Ca^{++} or $PO_4^{=}$, implying that this hormone is not normally essential to the maintenance of Ca^{++} or $PO_4^{=}$ homeostasis. Calcitonin may play a role in protecting skeletal integrity when there is a large Ca^{++} demand, such as during pregnancy or breast-feeding.

Vitamin D is actually a hormone that increases calcium absorption in the intestine.

The final factor involved in the regulation of Ca^{++} metabolism is **cholecalciferol**, or **vitamin D**, a steroidlike compound that is essential for Ca^{++} absorption in the intestine. Strictly speaking, vitamin D should be considered a hormone because it can be produced in the skin from a precursor related to cholesterol (7-dehydrocholesterol) on exposure to sunlight. It is

subsequently released into the blood to act at a distant target site, the intestine. The skin, therefore, is actually an endocrine gland and vitamin D a hormone. Traditionally, however, this chemical messenger has been considered a vitamin for two reasons. First, it was originally discovered and isolated from a dietary source and tagged as a vitamin. Second, even though the skin would be an adequate source of vitamin D if it were exposed to sufficient sunlight, indoor dwelling and clothing in response to cold weather and social customs preclude significant exposure of the skin to sunlight most of the time. At least part of the essential vitamin D must therefore be derived from dietary sources.

Regardless of its source, vitamin D is biologically inactive when it first enters the blood from either the skin or the digestive tract. It must be activated by two sequential biochemical alterations that involve the addition of two hydroxyl ($-OH$) groups (● Fig. 19–23). The first of these reactions occurs in the liver and the second in the kidneys. The end result is production of the active form of vitamin D, 1,25-$(OH)_2$-vitamin D_3. The kidney enzymes that are involved in the second step of vitamin D activation are stimulated by PTH in response to a fall in plasma Ca^{++}. To a lesser extent, a fall in plasma $PO_4^{=}$ also enhances the activation process.

The most dramatic and biologically important effect of activated vitamin D is to increase Ca^{++} absorption in the intestine. Unlike most dietary constituents, dietary Ca^{++} is not indiscriminately absorbed by the digestive system. In fact, the majority of ingested Ca^{++} is typically not absorbed but is lost instead in the feces. When needed, more dietary Ca^{++} is absorbed into the plasma under the influence of vitamin D. Independently of its effects on Ca^{++} transport, the active form of vitamin D also increases intestinal $PO_4^{=}$ absorption. Furthermore, vitamin D increases the responsiveness of bone to PTH. Thus, vitamin D and PTH have a closely interdependent relationship (● Fig. 19–24).

Parathyroid hormone is principally responsible for controlling Ca^{++} homeostasis, because the actions of vitamin D are too sluggish for it to contribute substantially to the minute-to-minute regulation of plasma Ca^{++} concentration. However, both PTH and vitamin D are essential to Ca^{++} balance, the process that ensures that, over the long term, Ca^{++} input into the body is equivalent to Ca^{++} output. When dietary Ca^{++} intake is reduced, the resultant transient fall in plasma Ca^{++} level stimulates PTH secretion. The increased PTH has two effects that are important for the maintenance of Ca^{++} balance: (1) it stimulates Ca^{++} reabsorption by the kidneys, thereby decreasing Ca^{++} output, and (2) it activates vitamin D, which increases the efficiency of uptake of ingested Ca^{++}. Since PTH also promotes bone resorption, a substantial loss of bone minerals occurs if Ca^{++} intake is reduced for a prolonged period, even though bone is not directly involved in maintaining Ca^{++} input and output in balance.

Phosphate metabolism is controlled by the same mechanisms that regulate calcium metabolism.

Plasma $PO_4^{=}$ concentration is not as tightly controlled as plasma Ca^{++} concentration. Phosphate is regulated directly

• FIGURE 19–23 Activation of Vitamin D

Precursor in skin (7-dehydrocholesterol)

Dietary vitamin D

Sunlight

Vitamin D$_3$

Hydroxyl group (OH)

Liver enzymes

25-OH D$_3$

Hydroxyl group

PTH $\overset{+}{\leftarrow}$ ↓ Plasma Ca^{++}

+

Kidney enzymes

+

↓ Plasma PO$_4^=$

1, 25-(OH)$_2$ D$_3$ (active vitamin D)

Promotes intestinal absorption of Ca^{++} and PO$_4^=$

by vitamin D and indirectly by the plasma Ca^{++}-PTH feedback loop. To illustrate, a fall in plasma PO$_4^=$ concentration (**hypophosphatemia**) exerts a twofold effect to help restore the circulating PO$_4^=$ level to normal (• Fig. 19–25). First, because of the inverse relationship between the PO$_4^=$ and Ca^{++} concentrations in the plasma, a fall in plasma PO$_4^=$ causes an increase in plasma Ca^{++}, which directly suppresses PTH secretion. In the presence of reduced PTH, PO$_4^=$ reabsorption by the kidneys increases, returning plasma PO$_4^=$ concentration toward normal. Second, a fall in plasma PO$_4^=$ also results in increased activation of vitamin D, which then promotes PO$_4^=$ absorption in the intestine. This further helps to alleviate the initial hypophosphatemia.

Note that Ca^{++} balance is not compromised by these changes. Although the increase in activated vitamin D stimulates Ca^{++} absorption, the concurrent fall in PTH produces a compensatory increase in urinary Ca^{++} excretion because less of the filtered Ca^{++} is reabsorbed.

Disorders in calcium metabolism may arise from abnormal levels of parathyroid hormone or vitamin D.

The primary disorders that affect Ca^{++} metabolism are too much or too little PTH or a deficiency of vitamin D.

PTH Hypersecretion   Excess PTH secretion, or **hyperparathyroidism**, which is usually due to a hypersecreting tumor in one of the parathyroid glands, is characterized by hypercalcemia and hypophosphatemia. The affected individual can be asymptomatic or symptoms can be severe, depending on the magnitude of the problem. The following are among the possible consequences:

■ Hypercalcemia reduces the excitability of muscle and nervous tissue, leading to muscle weakness and neurological disorders, including decreased alertness, poor memory, and depression. Cardiac disturbances may also occur.

■ Excessive mobilization of Ca^{++} and PO$_4^=$ from skeletal stores leads to thinning of bone, which may result in skeletal deformities and increased incidence of fractures.

- An increased incidence of formation of Ca⁺⁺-containing kidney stones occurs because of the excess quantity of Ca⁺⁺ being filtered through the kidneys. These stones may impair renal function. Passage of the stones through the ureters causes extreme pain. Because of these potential multiple consequences, hyperparathyroidism has been called a disease of "bones, stones, and abdominal groans."

- To further account for the "abdominal groans," hypercalcemia can cause digestive disorders such as peptic ulcers, nausea, and constipation.

PTH Hyposecretion   Because of the parathyroid glands' close anatomical relation to the thyroid, the most common cause of deficient PTH secretion, or **hypoparathyroidism**, used to be inadvertent removal of the parathyroid glands (before their existence was known) during surgical removal of the thyroid gland (for treatment of thyroid disease). If all of the parathyroid tissue was removed, these patients, of course, died, because PTH is essential for life. In fact, physicians were puzzled why some patients died soon after thyroid removal even though there were no apparent surgical complications. Now that the location and importance of the parathyroid

glands have been discovered, surgeons are careful to leave parathyroid tissue during thyroid removal. Rarely, hyposecretion of the parathyroid glands occurs as a result of failure of the parathyroid tissue.

Hypoparathyroidism leads to hypocalcemia and hyperphosphatemia. The symptoms are primarily referable to increased neuromuscular excitability caused by the reduction in the level of free plasma Ca⁺⁺. In the complete absence of parathyroid hormone, death is imminent because of hypocalcemic spasm of respiratory muscles. With a relative deficiency rather than a complete absence of PTH, milder symptoms of increased neuromuscular excitability are evident. Muscle cramps and twitches derive from spontaneous activity in the motor nerves, while tingling and pins-and-needles sensations result from spontaneous activity in the sensory nerves. Mental changes include irritability and paranoia.

Vitamin D Deficiency   The major consequence associated with vitamin D deficiency is impaired intestinal absorption of Ca⁺⁺. In the face of reduced Ca⁺⁺ uptake, PTH maintains the plasma Ca⁺⁺ level at the expense of the bones. As a result, the bone matrix is not properly mineralized because Ca⁺⁺ salts

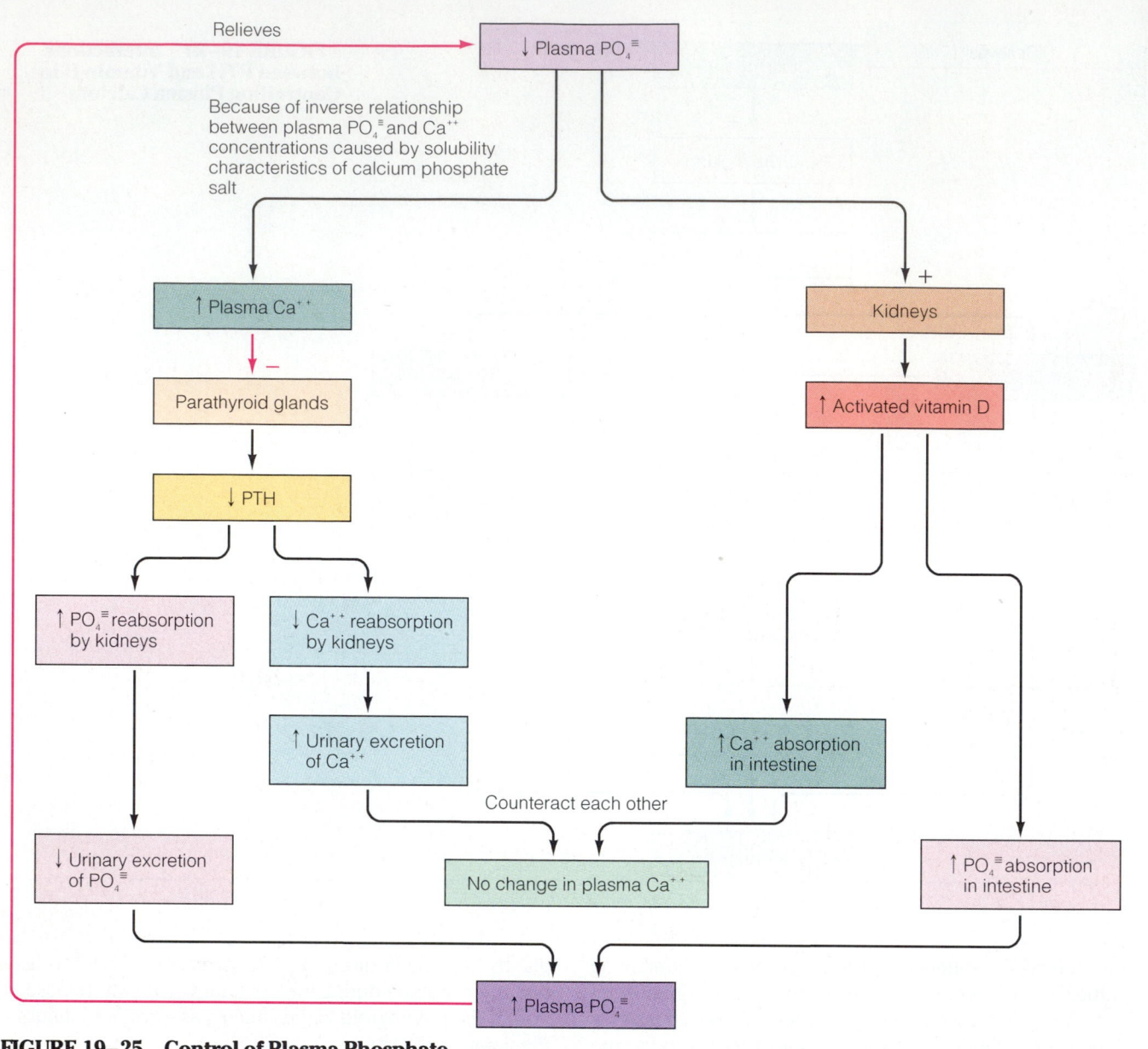

Relieves

↓ Plasma PO₄≡

Because of inverse relationship between plasma PO₄≡ and Ca⁺⁺ concentrations caused by solubility characteristics of calcium phosphate salt

↑ Plasma Ca⁺⁺

Kidneys +

Parathyroid glands −

↑ Activated vitamin D

↓ PTH

↑ PO₄≡ reabsorption by kidneys

↓ Ca⁺⁺ reabsorption by kidneys

↑ Urinary excretion of Ca⁺⁺

↑ Ca⁺⁺ absorption in intestine

Counteract each other

↓ Urinary excretion of PO₄≡

No change in plasma Ca⁺⁺

↑ PO₄≡ absorption in intestine

↑ Plasma PO₄≡

• **FIGURE 19–25 Control of Plasma Phosphate**

are not available for deposition. The demineralized bones become soft and deformed, bowing to the pressures of weight bearing, especially in children. This condition is known as **rickets** in children and **osteomalacia** in adults.

Osteoporosis The rates of bone formation and resorption are about equal throughout most of adult life, so total bone mass usually remains fairly constant during this period. Exceptions include minor hormonally induced fluctuations in bone mass to maintain Ca⁺⁺ homeostasis or mechanically induced adjustments in bone mass in response to changes in compressional weight bearing (for example, taking up exercise or being confined to bed). Such is not the case, however, during the first twenty and last twenty years of an average life

span. During the first two decades of life, when growth is occurring, bone deposition exceeds bone resorption under the influence of growth hormone (see p. 642). In contrast, by fifty to sixty years of age, bone resorption often exceeds bone formation. The result is a reduction in bone mass known as **osteoporosis**. This condition is characterized by a diminished laying down of organic matrix as a result of reduced osteoblast activity rather than abnormal bone calcification. The underlying cause of osteoporosis is uncertain. Plasma Ca⁺⁺ and PO₄≡ levels are normal, as are PTH and vitamin D concentrations. Osteoporosis occurs with greatest frequency in postmenopausal women, suggesting that estrogen withdrawal plays a role. (See the accompanying boxed feature, A Closer Look at Exercise Physiology.)

OSTEOPOROSIS: THE BANE OF BRITTLE BONES

Osteoporosis, a decrease in bone density resulting from reduced deposition of the bone's organic matrix, is a major health problem in the United States. It is responsible for the greater incidence of bone fractures among women over the age of fifty than among the population at large. Because bone mass is reduced, the bones are more susceptible to fracture in response to a fall, blow, or lifting action that normally would not strain stronger bones. Osteoporosis is the underlying cause of approximately 1.2 million fractures each year, of which 530,000 are vertebral fractures and 227,000 are hip fractures. The cost of rehabilitation is in excess of $6 billion per year. The cost in pain and suffering is not measurable. One-half of all American women have spinal pain and deformity by age seventy-five.

There appear to be two types of osteoporosis, caused by different mechanisms. Type I osteoporosis affects women soon after menopause and is characterized by vertebral crush fractures or fractures of the arm just above the wrist. It is hypothesized that these fractures occur as a result of the reduction in bone density that accompanies the estrogen deficiency of menopause. Type II osteoporosis occurs in men as well as women, although it affects females twice as often as males. It is

characterized by hip fractures as well as fractures at other sites. Because Type II osteoporosis occurs later in life, the decreased ability to absorb Ca^{++} associated with advancing age may play a key role in the development of the condition, although estrogen deficiency probably also contributes, accounting for the higher incidence in women.

Estrogen replacement therapy, Ca^{++} supplementation, and a regular weight-bearing exercise program are among the therapeutic approaches used to minimize or reverse bone loss. Because treatment of osteoporosis is difficult and often less than satisfactory, prevention is by far the best approach to managing this disease. Development of strong bones to begin with before menopause through a good Ca^{++}-rich diet and adequate exercise appears to be the best preventive measure. A large reservoir of bone at midlife may delay the clinical manifestations of osteoporosis in later life. Continued physical activity throughout life appears to retard or prevent bone loss, even in the elderly.

It is well documented that osteoporosis can result from disuse—that is, from reduced mechanical loading of the skeleton. Space travel has clearly shown that lack of gravity results in a decrease in bone density. Studies of athletes, on the other

hand, demonstrate that physical activity increases bone density. Within groups of athletes, bone density correlates directly with the load that the bone must bear. If one looks at athletes' femurs (thigh bones), the greatest bone density is found in weight lifters, followed in order by throwers, runners, soccer players, and finally swimmers. In fact, the bone density of swimmers does not differ from that of nonathletic controls. Swimming does not place any strain on bones. The bone density in the playing arm of male tennis players has been found to be as much as 35% greater than in their other arm; female tennis players have been found to have 28% greater density in their playing arm than in their other arm. One study found that very mild activity in nursing-home patients, whose average age was eighty-two years, not only slowed bone loss but even resulted in bone buildup over a thirty-six-month period. Thus, exercise is a good defense against osteoporosis.

The exact mechanism responsible for an increase in bone mass as a result of exercise is unknown. According to one proposal, exercise places strain on bone, which causes changes in electrical potential that induce bone formation.

CHAPTER IN PERSPECTIVE: FOCUS ON HOMEOSTASIS

A number of peripherally located endocrine glands play key roles in maintaining homeostasis, primarily by means of their regulatory influences over the rate of various metabolic reactions and over electrolyte balance. These endocrine glands all secrete hormones in response to specific stimuli. The hormones, in turn, exert effects that act in negative-feedback fashion to resist the change that induced their secretion, thus maintaining stability in the internal environment. The specific

contributions of the peripheral endocrine glands to homeostasis include the following:

■ Two closely related hormones secreted by the thyroid gland, tetraiodothyronine (T_4) and triiodothyronine (T_3), increase the overall metabolic rate. Not only does this action influence the rate at which nutrient molecules and O_2 within the internal environment are used by the cells, but it also produces heat, which is a contributing factor to the control of body temperature. T_4 and T_3 also stimulate a number of specific metabolic reactions, most of which influence the plasma concentrations of nutrient molecules.

- The adrenal cortex secretes three classes of hormones. Aldosterone, the primary mineralocorticoid, is essential for Na$^+$ and K$^+$ balance. Because of Na$^+$'s osmotic effect, Na$^+$ balance is critical to maintaining the proper ECF volume and arterial blood pressure. This action is essential for life. Without aldosterone's Na$^+$- and H$_2$O-conserving effect, so much plasma volume would be lost in the urine that death would quickly ensue. Maintenance of K$^+$ balance is essential for homeostasis because changes in extracellular K$^+$ have a profound impact on neuromuscular excitability, thus jeopardizing normal functioning of the heart, among other detrimental effects.

- Cortisol, the primary glucocorticoid secreted by the adrenal cortex, increases the plasma concentrations of glucose, fatty acids, and amino acids above normal. Although these actions disrupt the maintenance of stable concentrations of these molecules in the internal environment, they contribute to homeostasis indirectly by making the molecules readily available as energy sources or building blocks for tissue repair to help the body adapt to stressful situations.

- The sex hormones secreted by the adrenal cortex do not contribute to homeostasis.

- The major hormone secreted by the adrenal medulla, epinephrine, generally reinforces activities of the sympathetic nervous system. It contributes to homeostasis directly by its role in blood-pressure regulation. Epinephrine also contributes to homeostasis indirectly by helping prepare the body for peak physical responsiveness in fight-or-flight situations. This includes increasing the plasma concentrations of glucose and fatty acids above normal to provide additional energy sources to support increased physical activity.

- The two major hormones secreted by the endocrine pancreas, insulin and glucagon, are important in shifting metabolic pathways between the absorptive and postabsorptive states to maintain the appropriate plasma levels of nutrient molecules.

- Parathyroid hormone from the parathyroid glands is critical to the maintenance of the plasma concentration of Ca^{++}. PTH is essential for life because of Ca^{++}'s effect on neuromuscular excitability. In the absence of PTH, death rapidly occurs due to asphyxiation resulting from pronounced spasms of the respiratory muscles.

CHAPTER SUMMARY

Thyroid Gland

The thyroid gland contains two types of endocrine secretory cells: (1) follicular cells, which produce the iodine-containing hormones, T$_4$ (thyroxine or tetraiodothyronine) and T$_3$ (triiodothyronine), collectively known as thyroid hormone, and (2) C cells, which synthesize a Ca^{++}-regulating hormone, calcitonin.

Thyroid hormone is the primary determinant of the overall metabolic rate of the body. By accelerating the metabolic rate of most tissues, it increases heat production. Thyroid hormone also enhances the actions of the chemical mediators of the sympathetic nervous system. Through this and other means, thyroid hormone indirectly increases cardiac output. Finally, thyroid hormone is essential for normal growth as well as the development and function of the nervous system.

Thyroid hormone secretion is regulated by a negative-feedback system between hypothalamic TRH, anterior pituitary TSH, and thyroid gland T$_3$ and T$_4$. The feedback loop maintains thyroid hormone levels relatively constant. Cold exposure in newborn infants is the only input to the hypothalamus known to be effective in increasing TRH and thereby thyroid hormone secretion.

Adrenal Glands

Each of the pair of adrenal glands consists of two separate endocrine organs—an outer steroid-secreting adrenal cortex and an inner catecholamine-secreting adrenal medulla. The adrenal cortex secretes three different categories of steroid hormones: mineralocorticoids (primarily aldosterone), glucocorticoids (primarily cortisol), and adrenal sex hormones (primarily the weak androgen, dehydroepiandrosterone).

Aldosterone regulates Na$^+$ and K$^+$ balance and is important for blood-pressure homeostasis, which is accomplished second-arily as a result of the osmotic effect of Na$^+$ in maintaining the plasma volume, a lifesaving effect. Control of aldosterone secretion is related to Na$^+$ and K$^+$ balance and blood-pressure regulation and is not influenced by ACTH.

Cortisol helps regulate fuel metabolism and is important in stress adaptation. It increases the blood levels of glucose, amino acids, and fatty acids and spares glucose for use by the glucose-dependent brain. The mobilized organic molecules are available for use as needed for energy or for repair of injured tissues. Cortisol secretion is regulated by a negative-feedback loop involving hypothalamic CRH and pituitary ACTH. The most potent stimulus for increasing activity of the CRH/ACTH/cortisol axis is stress.

Dehydroepiandrosterone is responsible in females for the sex drive and growth of pubertal hair.

The adrenal medulla is composed of modified sympathetic postganglionic neurons, which secrete the catecholamine epinephrine into the blood in response to sympathetic stimulation. For the most part, epinephrine reinforces the sympathetic system in its general systemic "fight-or-flight" responses and in its maintenance of arterial blood pressure. Epinephrine also exerts important metabolic effects, namely, increasing blood glucose and blood fatty acids. The primary stimulus for increased adrenomedullary secretion is activation of the sympathetic system by stress.

Endocrine Control of Fuel Metabolism

Intermediary or fuel metabolism refers collectively to the synthesis (anabolism), breakdown (catabolism), and transformations of the three classes of energy-rich organic nutrients—carbohydrate, fat, and protein—within the body. Glucose and fatty acids derived respectively from carbohydrates and fats are primarily used as metabolic fuels, whereas amino acids derived from proteins are primarily used for the synthesis of structural and enzymatic proteins.

During the absorptive state following a meal, the excess absorbed nutrients not immediately needed for energy production or protein synthesis are stored to a limited extent as glycogen in the liver and muscle but mostly as triglycerides in adipose tissue. During the postabsorptive state between meals when no new nutrients are entering the blood, the glycogen and triglyceride stores are catabolized to release nutrient molecules into the blood. If necessary, body proteins are degraded to release amino acids for conversion into glucose. It is essential to maintain the blood glucose concentration above a critical level even during the postabsorptive state, because the brain depends on blood-delivered glucose as its energy source. Tissues not dependent on glucose switch to fatty acids as their metabolic fuel, sparing glucose for the brain.

These shifts in metabolic pathways between the absorptive and postabsorptive state are hormonally controlled. The most important hormone in this regard is insulin. Insulin is secreted by the β cells of the islets of Langerhans, the endocrine portion of the pancreas. The other major pancreatic hormone, glucagon, is secreted by the α cells of the islets. Insulin is an anabolic hormone; it promotes the cellular uptake of glucose, fatty acids, and amino acids and enhances their conversion into glycogen, triglycerides, and proteins, respectively. In doing so, it lowers the blood concentrations of these small organic molecules. Insulin secretion is increased during the absorptive state, primarily by a direct effect of an elevated blood glucose on the β cells, and is largely responsible for directing the organic traffic into cells during this state.

Glucagon mobilizes the energy-rich molecules from their stores during the postabsorptive state. Glucagon, which is secreted in response to a direct effect of a fall in blood glucose on the pancreatic α cells, in general opposes the actions of insulin.

Endocrine Control of Calcium Metabolism

Changes in the concentration of free, diffusible plasma Ca^{++}, the biologically active form of this ion, produce profound and life-threatening effects, most notably on neuromuscular excit-

ability. Hypercalcemia reduces excitability whereas hypocalcemia brings about overexcitability of nerves and muscles. If it is severe enough, fatal spastic contractions of respiratory muscles can occur.

Three hormones regulate the plasma concentration of Ca^{++} (and concurrently regulate $PO_4^=$)—parathyroid hormone (PTH), calcitonin, and vitamin D. PTH, whose secretion is directly increased by a fall in plasma Ca^{++} concentration, acts on bone, kidneys, and the intestine to raise the plasma Ca^{++} concentration. In so doing, it is essential for life by preventing the fatal consequences of hypocalcemia. The specific effects of PTH on bone are to promote Ca^{++} movement from the bone fluid into the plasma in the short term and to promote localized dissolution of bone by enhancing activity of the osteoclasts (bone-dissolving cells) in the long term. Dissolution of the calcium phosphate bone crystals releases $PO_4^=$ as well as Ca^{++} into the plasma. Parathyroid hormone acts on the kidneys to enhance the reabsorption of filtered Ca^{++}, thereby reducing the urinary excretion of Ca^{++} and increasing its plasma concentration. Simultaneously, PTH reduces renal $PO_4^=$ reabsorption, in this way increasing $PO_4^=$ excretion and lowering plasma $PO_4^=$ levels. This is important because a rise in plasma $PO_4^=$ would force the deposition of some of the plasma Ca^{++} back into the bone. Furthermore, PTH facilitates the activation of vitamin D, which in turn stimulates Ca^{++} and $PO_4^=$ absorption from the intestine.

Vitamin D can be synthesized from a cholesterol derivative in the skin when exposed to sunlight, but frequently this endogenous source is inadequate, so vitamin D must be supplemented by dietary intake. From either source, vitamin D must be activated first by the liver and then by the kidneys (the site of PTH regulation of vitamin D activation) before it can exert its effect on the intestine.

Calcitonin, a hormone produced by the C cells of the thyroid gland, is the third factor that regulates Ca^{++}. In negative-feedback fashion, calcitonin is secreted in response to an increase in plasma Ca^{++} concentration and acts to lower plasma Ca^{++} levels by inhibiting activity of bone osteoclasts. Calcitonin is unimportant except during the rare condition of hypercalcemia.

REVIEW EXERCISES

Objective Questions (Answers on p. D–4.)

1. The response to thyroid hormone is detectable within a few minutes after its secretion. (True or false?)

2. "Male" sex hormones are produced in both males and females by the adrenal cortex. (True or false?)

3. Adrenal androgen hypersecretion is usually due to a deficit of an enzyme crucial to cortisol synthesis. (True or false?)

4. Excess glucose and amino acids as well as fatty acids can be stored as triglycerides. (True or false?)

5. Insulin is the only hormone that can lower blood glucose levels. (True or false?)

6. The most life-threatening consequence of hypocalcemia is its impact on reducing blood clotting. (True or false?)

7. All ingested Ca^{++} is indiscriminately absorbed in the intestine. (True or false?)

8. The $Ca_3(PO_4)_2$ bone crystals form a labile pool from which Ca^{++} can rapidly be extracted under the influence of PTH. (True or false?)

9. The lumen of the thyroid follicle is filled with _____, the chief constituent of which is a large, complex glycoprotein known as _____.

10. The common large precursor molecule that yields ACTH, MSH, and β-endorphin is known as _____.

11. _____ refers to the conversion of glucose into glycogen. _____ refers to the conversion of glycogen into glucose. _____ refers to the conversion of amino acids into glucose.

12. The three major tissues that are not dependent on insulin for their glucose uptake are _____, _____, and _____.

13. The three compartments with which ECF Ca^{++} is exchanged are _____, _____, and _____.

14. Which of the following hormones does *not* exert a direct metabolic effect?
  a. epinephrine
  b. growth hormone
  c. aldosterone
  d. cortisol
  e. thyroid hormone

15. Which of the following are characteristic of the postabsorptive state?
  a. glycogenolysis
  b. gluconeogenesis
  c. lipolysis
  d. glycogenesis
  e. protein synthesis
  f. triglyceride synthesis
  g. protein degradation
  h. increased insulin secretion
  i. increased glucagon secretion
  j. glucose sparing

16. Indicate the primary circulating form and storage form of each of the three classes of organic nutrients:

	PRIMARY CIRCULAT-ING FORM	PRIMARY STORAGE FORM
Carbohydrate	1. _____	2. _____
Fat	3. _____	4. _____
Protein	5. _____	6. _____

Essay Questions

1. Describe the steps of thyroid hormone synthesis.

2. What are the effects of T$_3$ and T$_4$? Which is the more potent? What is the source of most circulating T$_3$?

3. Describe the regulation of thyroid hormone.

4. Discuss the causes and symptoms of both hypothyroidism and hyperthyroidism. When does a goiter occur?

5. What hormones are secreted by the adrenal cortex? What are the functions and control of each of these hormones?

6. Discuss the causes and symptoms of each type of adrenocortical dysfunction.

7. What is the relationship of the adrenal medulla to the sympathetic nervous system? What are the functions of epinephrine? How is epinephrine release controlled?

8. Define stress. Describe the neural and hormonal responses to a stressor.

9. Define fuel metabolism, anabolism, and catabolism.

10. Distinguish between the absorptive and postabsorptive states with regard to the dispensation of nutrient molecules.

11. Name the two major cell types of the islets of Langerhans, and indicate the hormonal product of each.

12. Compare the functions and control of insulin secretion with those of glucagon secretion.

13. What are the consequences of diabetes mellitus? Distinguish between Type I and Type II diabetes mellitus.

14. Why must plasma Ca^{++} be closely regulated?

15. Discuss the contributions of parathyroid hormone, calcitonin, and vitamin D to Ca^{++} metabolism. Describe the source and control of each of these hormones.

16. Discuss the major disorders in Ca^{++} metabolism.

 POINTS TO PONDER

1. Iodine is naturally present in salt water and is abundant in soil along the coastal regions. Fish and shellfish living in the ocean and plants grown in coastal soil take up iodine from their environment. Fresh water does not contain iodine, and the soil becomes more iron poor the farther inland. Knowing this, explain why the midwestern United States was once known as an *endemic goiter belt* because of the high incidence of goiter in this region. Why is this region no longer an endemic goiter belt even though the soil is still iodine poor?

2. One type of endocrine disorder is dubbed "diabetes of bearded ladies." Based on this description and your knowledge of the endocrine system, what underlying defect do you think causes this condition?

3. Why would an infection tend to increase the blood glucose level of a diabetic individual?

4. Tapping the facial nerve at the angle of the jaw in a patient with moderate hyposecretion of a particular hormone elicits a characteristic grimace on that side of the face. What endocrine abnormality could give rise to this so-called *Chvostek's sign?*

5. The only drugs currently approved by the Food and Drug Administration for treatment of advanced osteoporosis are estrogen and calcitonin. Explain the physiological rationale for the use of these agents. Both of these drugs have serious drawbacks, however. Estrogen supplementation has been linked with an increased risk of breast cancer. Calcitonin is very expensive and must be injected, both deterrents to patient compliance. Investigators are currently evaluating a new oral drug, *etidronate*, for treatment of osteoporosis. Explain why this nonhormonal drug, which suppresses osteoclast activity, might be a useful osteoporosis therapy.

REPRODUCTIVE SYSTEM

Body systems maintain homeostasis

HOMEOSTASIS
The reproductive system does not contribute to homeostasis but is essential for perpetuation of the species.

Homeostasis is essential for survival of cells

CELLS

Cells make up body systems

Normal functioning of the **reproductive system** is not aimed toward homeostasis and is not necessary for survival of an individual, but it is essential for survival of the species. Only through reproduction can the complex genetic blueprint of each species survive beyond the lives of individual members of the species.

≡ **CHAPTER CONTENTS AT A GLANCE** ≡

INTRODUCTION
Anatomy of the reproductive system
Sex determination and differentiation
MALE REPRODUCTIVE PHYSIOLOGY
Testicular function
 Scrotal location
 Testosterone secretion
 Spermatogenesis
Puberty
Male reproductive tract
Accessory sex glands
SEXUAL INTERCOURSE BETWEEN MALES AND FEMALES
Male sex act
Female sex act
FEMALE REPRODUCTIVE PHYSIOLOGY
Ovarian function
 Oogenesis
 Follicular and luteal phases
Menstrual cycle
● A CLOSER LOOK AT EXERCISE PHYSIOLOGY Menstrual
 Irregularities: When Cyclists and Other Female Athletes Do
 Not Cycle
Puberty; menopause
Fertilization
● CONCEPTS, CHALLENGES, AND CONTROVERSIES Reproductive-
 Related Technology: Taking the Making of Life into Our
 Own Hands
Implantation; placentation
Gestation
● A CLOSER LOOK AT EXERCISE PHYSIOLOGY Exercise and
 Pregnancy: A Compatible or Contraindicated
 Combination?
Parturition
Lactation

≡ ● **INTRODUCTION**

The reproductive system is not essential for survival of the individual, but it is necessary for survival of the species and has a profound impact on a person's life.

The central theme of this book has been the physiological processes aimed at maintaining homeostasis to ensure survival of the individual. We are now going to disembark from this theme to discuss the reproductive system, which primarily serves the purpose of perpetuating the species. Normal functioning of the reproductive system is not aimed toward homeostasis and is not necessary for survival of an individual, but it is essential for survival of the species. Only through reproduction can the complex genetic blueprint of each species survive beyond the lives of individual members of the species.

Even though the reproductive system does not contribute to homeostasis and is not essential for survival of an individual, it still plays an important role in a person's life. For example, the manner in which people relate as sexual beings contributes in significant ways to psychosocial behavior and has important influences on how people view themselves and how they interact with others. Reproductive function also has a profound effect on society. The universal organization of societies into family units provides a stable environment that is conducive for perpetuating our species. On the other hand, the population explosion and its resultant drain on dwindling resources have recently led to worldwide concern with the means by which reproduction can be limited.

Reproductive capability depends on an intricate relationship among the hypothalamus, anterior pituitary, reproductive organs, and target cells of the sex hormones. In addition to these basic biological processes, sexual behavior and attitudes are deeply influenced by emotional factors and the sociocultural mores of the society in which the individual lives. We will concentrate on the basic sexual and reproductive functions that are under nervous and hormonal control and will not examine the psychological and social ramifications of sexual behavior.

The reproductive system includes the gonads and reproductive tract.

Reproduction depends on the union of male and female **gametes** (**reproductive** or **germ cells**), each with a half set of chromosomes, to form a new individual with a full, unique set of chromosomes. Unlike the other body systems, which are essentially identical in the two sexes, the reproductive systems of males and females are remarkably different, befitting their different roles in the reproductive process. The male and female **reproductive systems** are designed to enable union of genetic material from the two sexual partners, and the female system is equipped to house and nourish the offspring to the developmental point at which it can survive independently in the external environment.

The **primary reproductive organs**, or **gonads**, consist of a pair of **testes** in the male and a pair of **ovaries** in the female. In both sexes, the mature gonads perform the dual function of (1) producing gametes (**gametogenesis**), that is, **spermatozoa** (**sperm**) in the male and **ova** (**eggs**) in the female, and (2) secreting sex hormones, specifically, **testosterone** in males and **estrogen** and **progesterone** in females.

In addition to the gonads, the reproductive system in each sex includes a **reproductive tract** encompassing a system of ducts that are specialized to transport or house the gametes after they are produced, plus **accessory sex glands** that empty their supportive secretions into these passageways. In females, the **breasts** are also considered to be accessory reproductive organs. The externally visible portions of the reproductive system are known as **external genitalia**.

The **secondary sexual characteristics** are the many external characteristics that are not directly involved in reproduction but that distinguish males and females, such as body configuration and hair distribution. In humans, for example, males have broader shoulders whereas females have curvier hips, and males have beards while females do not. Testosterone in the male and estrogen in the female are responsible for the development and maintenance of these characteristics. In some species, the secondary sexual characteristics are of great importance in courting and mating behavior; for example, the rooster's headdressing attracts the female's attention, and the stag's antlers are useful to ward off other males. In humans, the differentiating marks between males and females do serve to attract the opposite sex, but attraction is also strongly influenced by the complexities of human society and cultural behavior.

The essential reproductive functions of the male are (1) production of sperm (*spermatogenesis*) and (2) delivery of sperm to the female. The sperm-producing organs, the testes, are suspended outside the abdominal cavity in a skin-covered sac, the **scrotum**, which lies within the angle between the legs. The male reproductive system is designed to deliver sperm to the female reproductive tract in a liquid vehicle, **semen**, which is conducive to sperm viability. The major male accessory sex glands, whose secretions provide the bulk of the semen, are the *seminal vesicles, prostate gland,* and *bulbourethral glands* (• Fig. 20–1). The **penis** is the organ used to deposit semen in the female. Sperm exit the testes through the *epididymis, ductus (vas) deferens, ejaculatory duct,* and *urethra,* the latter being a canal that runs the length of the penis.

The female's role in reproduction is more complicated than the male's. The essential female reproductive functions include (1) the production of ova (*oogenesis*); (2) reception of sperm; (3) transport of the sperm and ovum to a common site for union (*fertilization* or *conception*); (4) maintenance of the developing fetus until it can survive in the outside world (*gestation* or *pregnancy*), including formation of the *placenta,* the organ of exchange between mother and fetus; (5) giving birth to the baby (*parturition*); and (6) nourishing the infant after birth by milk production (*lactation*). The product of fertilization is known as an **embryo** during the first two months of intrauterine development. Beyond this time, it is recognizable as human and is known as a **fetus** during the remainder of gestation.

The ovaries and female reproductive tract lie within the pelvic cavity (• Fig. 20–2a and b). The female reproductive tract consists of two **oviducts** (**uterine** or **Fallopian tubes**), which pick up ova upon ovulation and serve as the site for fertilization; the thick-walled hollow **uterus**, which is primarily responsible for maintaining the fetus during its development and expelling it at the end of pregnancy; and the **vagina**, a muscular, expansible tube connecting the uterus to the external environment. The lowest portion of the uterus, the **cervix**, projects into the vagina and contains a single small opening, the **cervical canal**. Sperm are deposited in the vagina by the penis during sexual intercourse. The cervical canal serves as a pathway for sperm through the uterus to the site of fertilization in the oviduct, and when greatly dilated during parturition, it serves as the passageway for delivery of the baby from the uterus.

The **vaginal opening** is located in the **perineal region** between the urethral opening anteriorly and the anal opening posteriorly (Fig. 20–2c). It is partially covered by a thin mucous membrane, the **hymen**, which can be physically disrupted in a variety of ways, including by the first sexual intercourse. The vaginal and urethral openings are surrounded laterally by two pairs of skin folds, the **labia minora** and **labia majora**. The smaller labia minora are located medially to the more prominent labia majora. The **clitoris**, a small erotic structure composed of tissue identical to the penis, lies at the anterior end of the folds of the labia minora. The female external genitalia are collectively referred to as the **vulva**.

• **FIGURE 20–1 Anatomy of Male Reproductive Organs**

Fimbriae
Ovary
Oviduct

Uterus

Urinary bladder
Pubic bone

Urethra

Clitoris

Labium minora

Vertebral column

Cervix

Rectum

Vagina

Anus

Labium majora

(a)

Oviduct

Ovarian vessels

Endometrium

Myometrium

Cervical canal

Cervix

Vagina

Fimbriae

Ovary

Uterus

Clitoris

Opening
of urethra

Hymen

Labium
minora

Labium
majora

Vaginal
opening

Perineum

Anus

(c)

(b)

● FIGURE 20–2   Anatomy of Female Reproductive Organs (a) Sagittal section. (b) Posterior view. (c) Perineal view.

Reproductive cells each contain a half set of chromosomes.

The DNA molecules that carry the cell's genetic code are not randomly crammed into the nucleus but are precisely organized into **chromosomes** (see p. B–3). Each chromosome consists of a different DNA molecule that contains a unique set of genes. **Somatic** (body) **cells** contain forty-six chromosomes (the **diploid number**), which can be sorted into twenty-three pairs on the basis of various distinguishing features. Chromosomes composing a matched pair are termed **homologous chromosomes**, one member of each pair having been derived from the individual's maternal parent and the other member from the paternal parent. Gametes (that is, sperm and eggs) contain only one member of each homologous pair for a total of twenty-three chromosomes (the **haploid number**).

Gametogenesis is accomplished by meiosis.

Most cells in the human body have the ability to reproduce themselves, a process important in growth, replacement, and repair of tissues. Cell division involves two components: division of the nucleus and division of the cytoplasm. Nuclear division in somatic cells is accomplished by **mitosis**. In mitosis, the chromosomes replicate (make duplicate copies of themselves), then the identical chromosomes are separated so that a complete set of genetic information (that is, a diploid number of chromosomes) is distributed to each of the two new daughter cells. Nuclear division in the specialized case of gametes is accomplished by **meiosis**, in which only a half set of genetic information (that is, a haploid number of chromosomes) is distributed to each of four new daughter cells (see Fig. B-11).

During meiosis, a specialized diploid germ cell undergoes one chromosome replication followed by two nuclear divisions. In the first meiotic division, the replicated chromosomes do not separate into two individual, identical chromosomes but remain joined together. The doubled chromosomes sort themselves into homologous pairs, and the pairs separate so that each of two daughter cells receives a half set of doubled chromosomes. During the second meiotic division, the doubled chromosomes within each of the two daughter cells separate and are distributed into two cells, yielding four daughter cells, each containing a half set of chromosomes, a single member of each pair. During this process, the maternally and paternally derived chromosomes of each homologous pair are distributed to the daughter cells in random assortments containing one member of each chromosome pair without regard for its original derivation. That is, not all of the mother-derived chromosomes go to one daughter cell and the father-derived chromosomes to the other cell. More than 8 million (2^{23}) different mixtures of the twenty-three paternal and maternal chromosomes are possible. This genetic mixing provides novel new combinations of chromosomes.

Thus, sperm and ova each have a unique haploid number of chromosomes. When fertilization takes place, a sperm and ovum fuse to form the start of a new individual with forty-six chromosomes, one member of each chromosomal pair having been inherited from the mother and the other member from the father (• Fig. 20-3).

The sex of an individual is determined by the combination of sex chromosomes.

Whether individuals are destined to be males or females is a genetic phenomenon determined by the sex chromosomes they possess. As the twenty-three chromosome pairs are separated during meiosis, each sperm or ovum receives only one member of each chromosome pair. Twenty-two of the chromosome pairs are **autosomal chromosomes** that code for gen-

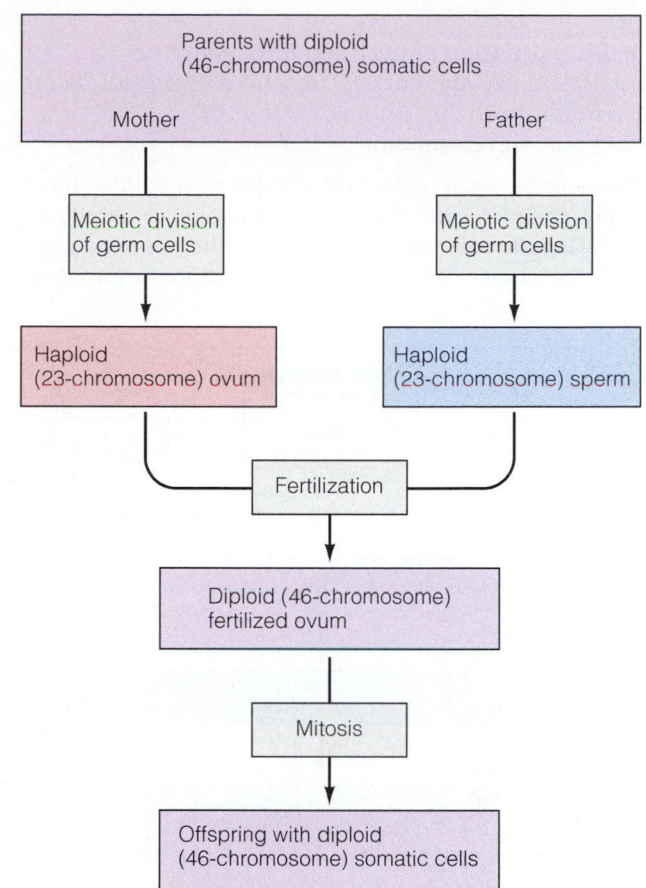

• **FIGURE 20-3 Chromosomal Distribution in Sexual Reproduction**

eral human characteristics as well as for specific traits such as eye color. The remaining pair of chromosomes are the **sex chromosomes**, of which there are two genetically different types—a larger **X chromosome** and a smaller **Y chromosome**. **Sex determination** depends on the combination of sex chromosomes: **genetic males** have both an X and a Y sex chromosome; **genetic females** have two X sex chromosomes. Thus, the genetic difference responsible for all of the anatomical and functional distinctions between males and females is the single Y chromosome. Males have it; females do not.

As a result of meiosis during gametogenesis, all chromosome pairs are separated so that each daughter cell contains only one member of each pair, including the sex chromosome pair. When the XY sex chromosome pair separates during sperm formation, half the sperm receive an X chromosome and the other half a Y chromosome. In contrast, during oogenesis, every ovum receives an X chromosome because separation of the XX sex chromosome pair yields only X chromosomes. During fertilization, combination of an X-bearing sperm with an X-bearing ovum produces a genetic female, XX, whereas union of a Y-bearing sperm with an X-bearing ovum results in a genetic male, XY. Thus, genetic sex is determined at the time of conception and depends on which type of sex chromosome is contained within the fertilizing sperm.

Sex differentiation along male or female lines depends on the presence or absence of masculinizing determinants during critical periods of embryonic development.

Differences between males and females exist at three levels: genetic, gonadal, and phenotypic (anatomical) sex (● Fig. 20–4). **Genetic sex**, which depends on the combination of sex chromosomes at the time of conception, in turn deter- mines **gonadal sex**, that is, whether testes or ovaries develop. The presence or absence of a Y chromosome determines gonadal differentiation. For the first month and a half of gestation, all embryos have the potential to differentiate along either male or female lines, because the developing repro- ductive tissues of both sexes are identical and indifferent. Go- nadal specificity appears during the seventh week of intra- uterine life when the indifferent gonadal tissue of a genetic male begins to differentiate into testes under the influence of

● **FIGURE 20–4   Sex Differentiation**

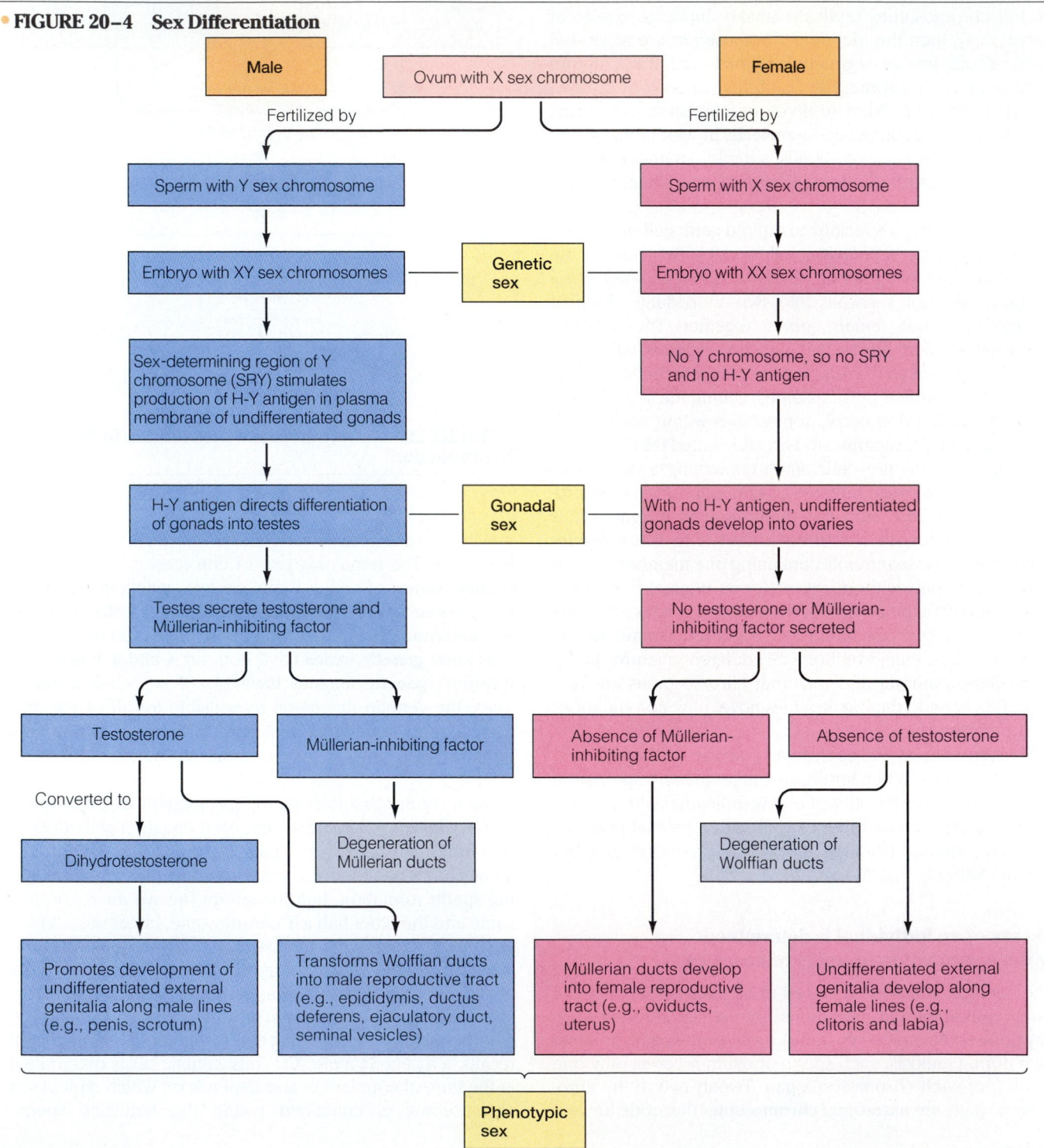

the **sex-determining region** of the Y chromosome (**SRY**), the single gene that is responsible for sex determination. This gene triggers a chain of reactions that lead to physical development of a male. The sex-determining region of the Y chromosome "masculinizes" the gonads (induces their development into testes) by stimulating production of **H-Y antigen** by primitive gonadal cells. H-Y antigen, which is a specific plasma-membrane protein found only in males, directs differentiation of the gonads into testes. Because genetic females lack the SRY gene and consequently do not produce H-Y antigen, their gonadal cells never receive a signal for testicular formation, so the undifferentiated gonadal tissue starts developing during the ninth week into ovaries instead.

Phenotypic sex, the apparent anatomic sex of an individual, depends on the genetically determined gonadal sex. **Sexual differentiation** refers to the embryonic development of the external genitalia and reproductive tract along either male or female lines. As with the undifferentiated gonads, embryos of both sexes have the potential to develop either male or female reproductive tracts and external genitalia. Differen-

tiation into a male-type reproductive system is induced by **androgens**, which are masculinizing hormones secreted by the developing testes. Testosterone is the most potent androgen. The absence of these testicular hormones in female fetuses results in the development of a female-type reproductive system. By ten to twelve weeks of gestation, the sexes can easily be distinguished by the anatomical appearance of the external genitalia.

Male and female external genitalia develop from the same embryonic tissue. In both sexes, the undifferentiated external genitals consist of a *genital tubercle,* paired *urethral folds* surrounding a urethral groove, and, more laterally, *genital (labioscrotal) swellings* (● Fig. 20–5). The **genital tubercle** gives rise to exquisitely sensitive erotic tissue—in males the **glans penis** (the cap at the distal end of the penis) and in females the clitoris. The major distinctions between the glans penis and clitoris are the smaller size of the clitoris and penetration of the glans penis by the urethral opening. The urethra is the tube through which urine is transported from the bladder to the outside and also serves in males as a passageway for exit

● **FIGURE 20–5** **Sexual Differentiation of External Genitalia** (a) Undifferentiated stage (7 wk). (b) Male development. (c) Female development.

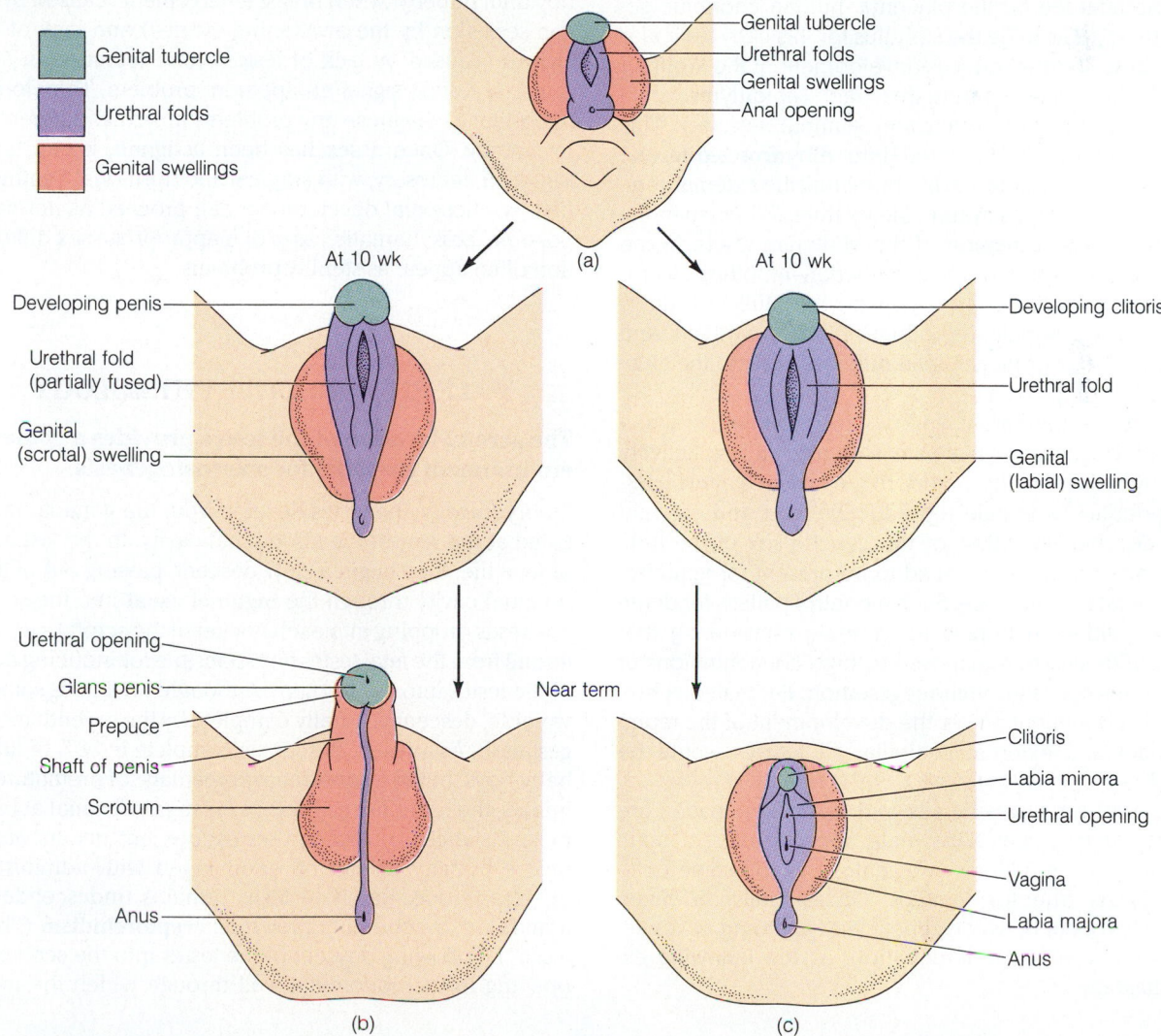

of semen through the penis to the outside. In males the **urethral folds** fuse around the urethral groove to form the penis, which encircles the urethra. The **genital swellings** similarly fuse to form the scrotum and **prepuce**, a fold of skin that extends over the end of the penis and more or less completely covers the glans penis. In females, the urethral folds and genital swellings do not fuse at midline but develop instead into the labia minora and labia majora, respectively. The urethral groove remains open, providing access to the interior through the urethral opening and vaginal orifice.

Although the male and female external genitalia develop from the same undifferentiated embryonic tissue, this is not the case with the reproductive tracts. Two primitive duct systems—the Wolffian ducts and the Müllerian ducts—develop in all embryos. In males, the reproductive tract develops from the **Wolffian ducts** and the Müllerian ducts degenerate, whereas in females, the **Müllerian ducts** differentiate into the reproductive tract and the Wolffian ducts regress. Since both duct systems are present before sexual differentiation occurs, the early embryo has the potential to develop either a male or female reproductive tract. Development of the reproductive tract along male or female lines is determined by the presence or absence of two hormones secreted by the fetal testes—testosterone and Müllerian-inhibiting factor (Fig. 20–4). A hormone released by the placenta, human chorionic gonadotropin, appears to be the stimulus for this early testicular secretion. Testosterone induces development of the Wolffian ducts into the male reproductive tract (epididymis, ductus deferens, ejaculatory duct, and seminal vesicles). This hormone, after being converted into **dihydrotestosterone (DHT)**, is also responsible for differentiating the external genitalia into the penis and scrotum. Meanwhile, Müllerian-inhibiting factor causes regression of the Müllerian ducts. In the absence of testosterone and Müllerian-inhibiting factor in females, the Wolffian ducts regress, the Müllerian ducts develop into the female reproductive tract (oviducts and uterus), and the external genitalia differentiate into the clitoris and labia.

Note that the indifferent embryonic reproductive tissue passively develops into a female structure unless actively acted on by masculinizing factors. In the absence of male testicular hormones, a female reproductive tract and external genitalia develop regardless of the genetic sex of the individual. Ovaries do not even need to be present for feminization of the fetal genital tissue. Such a control pattern for determining sex differentiation is appropriate considering that fetuses of both sexes are exposed to high concentrations of female sex hormones throughout gestation. If female sex hormones exerted influence over the development of the reproductive tract and external genitalia, all fetuses would be feminized.

In the usual case, genetic sex and sex differentiation are compatible; that is, a genetic male appears to be a male anatomically and functions as a male, and the same compatibility holds true for females. Occasionally, however, discrepancies occur between genetic and anatomic sexes because of errors in sex differentiation, as the following examples illustrate:

- If the testes in a genetic male fail to properly differentiate and secrete hormones, the result is the development of an apparent anatomic female in a genetic male, who, of course, will be sterile.

- Because testosterone acts on the Wolffian ducts to convert them into a male reproductive tract but the testosterone derivative DHT is responsible for masculinization of the external genitalia, a genetic deficiency of the enzyme that converts testosterone into DHT results in a genetic male with testes and a male reproductive tract but with female external genitalia.

- The adrenal gland normally secretes a weak androgen, *dehydroepiandrosterone,* in insufficient quantities to masculinize females. However, pathologically excessive secretion of this hormone in a genetically female fetus during critical developmental stages imposes differentiation of the reproductive tract and genitalia along male lines (see adrenogenital syndrome, p. 661).

Sometimes these discrepancies between genetic sex and apparent sex are not recognized until puberty, when the discovery produces a psychologically traumatic gender identity crisis. For example, a masculinized genetic female with ovaries but with male-type external genitalia may be reared as a boy until puberty, when breast enlargement (caused by estrogen secretion by the awakening ovaries) and lack of beard growth (caused by lack of testosterone secretion in the absence of testes) signal an apparent problem. Therefore, it is important to diagnose any problems in sexual differentiation in infancy. Once a sex has been assigned, it can be reinforced, if necessary, with surgical and hormonal treatment so that psychosexual development can proceed as normally as possible. Less dramatic cases of inappropriate sex differentiation often appear as sterility problems.

≡● MALE REPRODUCTIVE PHYSIOLOGY

The scrotal location of the testes provides a cooler environment essential for spermatogenesis.

Embryonically, the testes develop from the gonadal ridge located at the rear of the abdominal cavity. In the last months of fetal life, they begin a slow descent, passing out of the abdominal cavity through the **inguinal canal** into the scrotum, one testis dropping into each pocket of the scrotal sac. Testosterone from the fetal testes is responsible for inducing descent of the testes into the scrotum. Although the time is somewhat variable, descent is usually complete by the seventh month of gestation. As a result, descent is complete in 98% of full-term baby boys, but in a substantial percentage of premature male infants, the testes are still within the inguinal canal at birth. In most instances of retained testes, descent occurs naturally before puberty or can be encouraged with administration of testosterone. Rarely, a testis remains undescended into adulthood, a condition known as **cryptorchidism** ("hidden testis"). Following descent of the testes into the scrotum, the opening in the abdominal wall through which the inguinal canal passes closes snugly around the sperm-carrying duct and blood vessels that traverse between each testis and the abdominal cavity. Incomplete closure or rupture of this opening permits abdominal viscera to slip through, resulting in an **inguinal hernia**.

The temperature within the scrotum averages several degrees Celsius less than normal body (core) temperature. Descent of the testes into this cooler environment is essential, because spermatogenesis is temperature sensitive and cannot occur at normal body temperature. Therefore, a cryptorchid is unable to produce viable sperm.

The position of the scrotum in relation to the abdominal cavity can be varied by a spinal reflex mechanism that plays an important role in regulating testicular temperature. Reflex contraction of scrotal muscles upon exposure to a cold environment raises the scrotal sac to bring the testes closer to the warmer abdomen. Conversely, relaxation of the muscles upon exposure to heat permits the scrotal sac to become more pendulous, moving the testes farther from the warm core of the body.

The testicular Leydig cells secrete masculinizing testosterone.

The testes perform the dual function of producing sperm and secreting testosterone. About 80% of the testicular mass consists of highly coiled **seminiferous tubules**, within which spermatogenesis takes place. The endocrine cells that produce testosterone—the **Leydig cells** or **interstitial cells**—are located in the connective tissue (interstitial tissue) between the seminiferous tubules (see Fig. 20–6b). Thus, the portions of the testes that produce sperm and secrete testosterone are structurally and functionally distinct.

Testosterone is a steroid hormone derived from a cholesterol precursor molecule, as are the female sex hormones, estrogen and progesterone. The Leydig cells contain a high concentration of the enzymes required to direct cholesterol through the testosterone-yielding pathway (see p. 623). Once produced, some of the testosterone is secreted into the blood, where it is transported, primarily bound to plasma proteins, to its target sites of action. A substantial portion of the newly synthesized testosterone goes into the lumen of the seminiferous tubules, where it plays an important role in sperm production.

Most but not all of testosterone's actions are ultimately directed toward ensuring delivery of sperm to the female. The effects of testosterone can be grouped into five categories: (1) effects on the reproductive system before birth; (2) effects on sex-specific tissues after birth; (3) other reproduction-related effects; (4) effects on secondary sexual characteristics; and (5) nonreproductive actions (Table 20–1).

Effects on the Reproductive System before Birth Before birth, testosterone secretion by the fetal testes is responsible for masculinizing the reproductive tract and external genitalia and for promoting descent of the testes into the scrotum, as already described. After birth, testosterone secre-

TABLE 20–1   Effects of Testosterone
EFFECTS BEFORE BIRTH
Maculinizes reproductive tract and external genitalia
Promotes descent of testes into scrotum
EFFECTS ON SEX-SPECIFIC TISSUES
Promotes growth and maturation of reproductive system at puberty
Essential for spermatogenesis
Maintains reproductive tract throughout adulthood
OTHER REPRODUCTIVE EFFECTS
Development of sex drive at puberty
Control of gonadotropin hormone secretion
EFFECTS ON SECONDARY SEXUAL CHARACTERISTICS
Induces male pattern of hair growth (e.g., beard)
Causes voice to deepen because of thickening of vocal cords
Promotes muscle growth responsible for male body configuration
NONREPRODUCTIVE ACTIONS
Exerts protein anabolic effect
Promotes bone growth at puberty and then closure of epiphyses
May induce aggressive behavior

tion ceases, and the testes and remainder of the reproductive system remain small and nonfunctional until puberty.

Effects on Sex-Specific Tissues after Birth **Puberty** refers to the period of arousal and maturation of the previously nonfunctional reproductive system, culminating in attainment of sexual maturity and the ability to reproduce. Its onset usually occurs sometime between the ages of ten and fourteen; on the average it begins about two years earlier in females than in males. Usually lasting three to five years, puberty encompasses a complex sequence of endocrine, physical, and behavioral events. **Adolescence** is a broader concept that refers to the entire transition period between childhood and adulthood, not just sexual maturation.

At puberty, the Leydig cells start secreting testosterone once again, and spermatogenesis is initiated in the seminiferous tubules for the first time. Testosterone is responsible for growth and maturation of the entire male reproductive system. Under the influence of the pubertal surge in testosterone secretion, the testes enlarge and become capable of spermatogenesis, the accessory sex glands enlarge and become secretory, and the penis and scrotum enlarge. Ongoing testosterone secretion is essential for spermatogenesis and for maintaining a mature male reproductive tract throughout adulthood. Following **castration** (surgical removal of the testes) or testicular failure caused by disease, the sex organs regress in size and function.

Other Reproduction-Related Effects Testosterone is responsible for development of sexual libido at puberty and helps to maintain the sex drive in the adult male. Stimulation of this behavior by testosterone is important for facilitating delivery of sperm to females. In humans, libido is also influenced by many interacting social and emotional factors. Once libido has developed, testosterone is no longer absolutely required for its maintenance. Castrated males often remain sexually active but at a reduced level.

In another reproduction-related function, testosterone participates in the normal negative-feedback control of gonadotropin hormone secretion by the anterior pituitary, a topic that will be covered more thoroughly later.

Effects on Secondary Sexual Characteristics All male secondary sexual characteristics depend on testosterone for their development and maintenance. These nonreproductive male characteristics induced by testosterone include (1) the male pattern of hair growth (for example, beard and chest hair and, in genetically predisposed men, baldness); (2) a deep voice caused by enlargement of the larynx and thickening of the vocal cords; (3) thick skin; and (4) the male body configuration (for example, broad shoulders, heavy arm and leg musculature) as a result of protein deposition. A male castrated before puberty (a **eunuch**) does not mature sexually nor does he develop secondary sexual characteristics.

Nonreproductive Actions Testosterone exerts several important effects not related to reproduction. It has a general protein anabolic (synthesis) effect and promotes bone growth, thus contributing to the more muscular physique of males and to the pubertal growth spurt. Ironically, testosterone not only stimulates bone growth but eventually prevents further growth by sealing the growing ends of the long bones (that is, ossifying or "closing" the epiphyseal plates—see p. 642). Testosterone also stimulates oil secretion by the sebaceous glands. This effect is most striking during the adolescent surge of testosterone secretion, predisposing the young man to develop acne.

In animals, testosterone induces aggressive behavior, but whether it influences human behavior other than in the area of sexual behavior is an unresolved issue. Even though some athletes and bodybuilders who take testosterone-like anabolic androgenic steroids to increase muscle mass have been observed to display more aggressive behavior (see p. 239), it is unclear to what extent general behavioral differences between the sexes are hormonally induced or are a result of social conditioning.

Once initiated at puberty, testosterone secretion and spermatogenesis occur continuously throughout the male's life. Testicular efficiency gradually declines after forty-five to fifty years of age, however, even though men in their seventies and beyond may continue to enjoy an active sex life and some even father a child at this late age. The gradual diminution in circulating testosterone levels and in sperm production is not caused by a decrease in stimulation but probably arises instead from degenerative changes associated with aging that occur in the small testicular blood vessels. This gradual decline is often termed "male menopause," although it is not deliberately programmed as is female menopause.

Spermatogenesis yields an abundance of highly specialized, mobile sperm.

About 800 feet (250 meters) of sperm-producing seminiferous tubules are packed within the testes (• Fig. 20–6a). Two functionally important cell types are present in these tubules: *germ cells,* most of which are in various stages of sperm development, and *Sertoli cells,* which provide crucial support for spermatogenesis (Fig. 20–6b and c). **Spermatogenesis** is a complex process by which relatively undifferentiated primordial germ cells, the **spermatogonia** (each of which contains a diploid complement of forty-six chromosomes), proliferate and are converted into extremely specialized, motile spermatozoa (sperm), each bearing a randomly distributed haploid set of twenty-three chromosomes.

Microscopic examination of a seminiferous tubule reveals layers of germ cells in an anatomical progression of sperm development, starting with the least differentiated in the outer layer and moving inward through various stages of division to the lumen, where the highly differentiated sperm are ready for exit from the testes (Fig. 20–6b and c). Spermatogenesis takes sixty-four days for development from a spermatogonium to a mature sperm. Up to several hundred million sperm may reach maturity daily. Spermatogenesis encompasses three major stages: *mitotic proliferation, meiosis,* and *packaging* (• Fig. 20–7).

Mitotic Proliferation Spermatogonia located in the outermost layer of the tubule continuously divide mitotically, with all new cells bearing the full complement of forty-six chromosomes that are identical to those of the parent cell. Such proliferation provides a continual supply of new germ cells. Following mitotic division of a spermatogonium, one of the daughter cells remains at the outer edge of the tubule as an undifferentiated spermatogonium, thus maintaining the germ cell line. The other daughter cell starts moving toward the lumen while undergoing the various steps required to form sperm, which will be released into the lumen. In humans, the sperm-forming daughter cell divides mitotically twice more to form four identical **primary spermatocytes**. After the last mitotic division, the primary spermatocytes enter a resting phase during which the chromosomes are duplicated and the doubled strands remain together in preparation for the first meiotic division.

Meiosis During meiosis, each primary spermatocyte (with forty-six doubled chromosomes) forms two **secondary spermatocytes** (each with twenty-three doubled chromosomes) during the first meiotic division, finally yielding four **spermatids** (each with twenty-three single chromosomes) as a result of the second meiotic division.

No further division takes place beyond this stage of spermatogenesis. Each spermatid is remodeled into a single spermatozoon. Because each sperm-producing spermatogo-

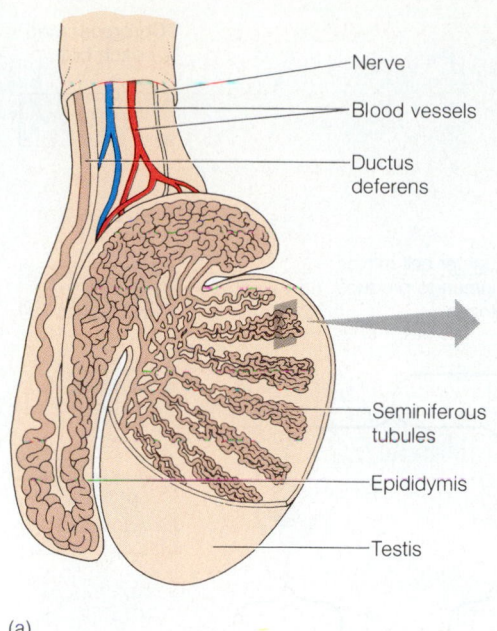

(a)

Nerve

Blood vessels

Ductus deferens

Seminiferous tubules

Epididymis

Testis

Spermatogonium

Spermatozoon

Cytoplasm of Sertoli cell

Interstitial tissue

(b)

Leydig cell

Lumen of seminiferous tubule

Varying stages of sperm development

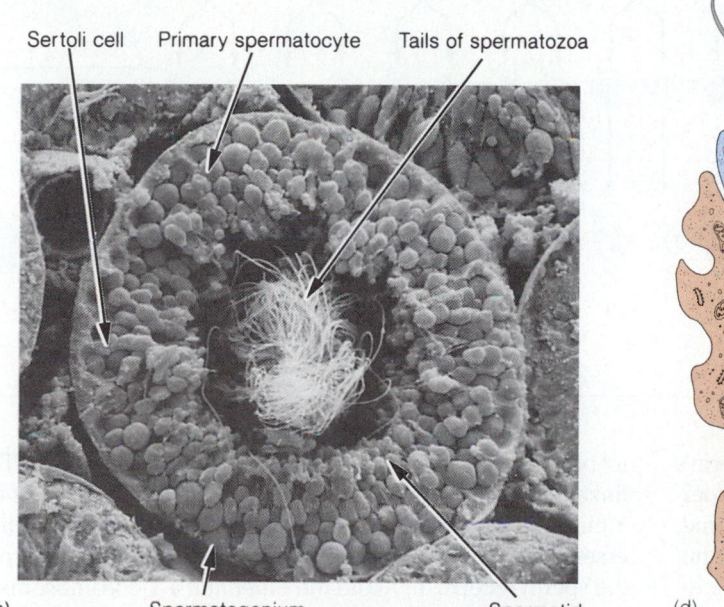

Sertoli cell   Primary spermatocyte   Tails of spermatozoa

(c)

Spermatogonium   Spermatid

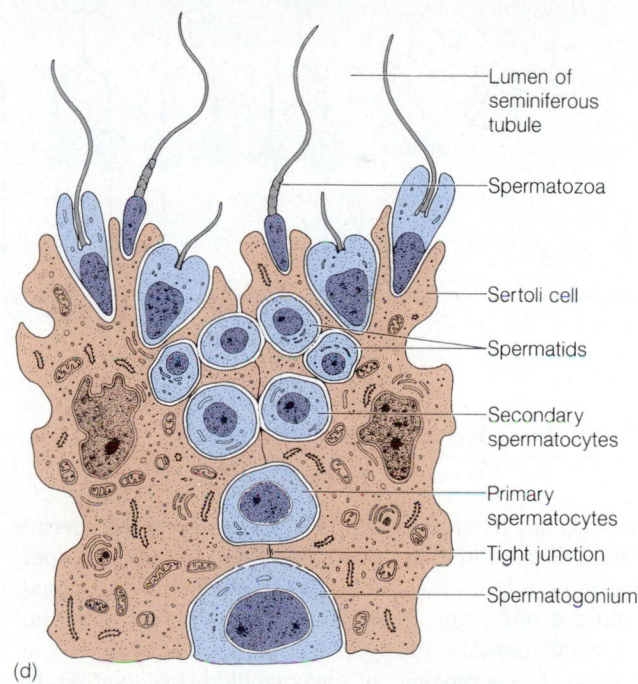

Lumen of seminiferous tubule

Spermatozoa

Sertoli cell

Spermatids

Secondary spermatocytes

Primary spermatocytes

Tight junction

Spermatogonium

(d)

FIGURE 20–6   **Testicular Anatomy Depicting Site of Spermatogenesis** (a) Longitudinal section of testis showing location and arrangement of the seminiferous tubules, the sperm-producing portion of the testis. (b) Light micrograph of cross section of seminiferous tubule. The undifferentiated germ cells (the spermatogonia) lie in the periphery of the tubule, and the differentiated spermatozoa are in the lumen, with the various stages of sperm development in between. (c) Scanning electron micrograph of cross section of seminiferous tubule. [From *Tissues and Organs: A Text-Atlas of Scanning Electron Microscopy* by Richard G. Kessel and Randy H. Kardon. Copyright © 1979 by W. H. Freeman and Company. Reprinted with permission.] (d) Relationship of Sertoli cells to developing sperm cells.

nium mitotically produces four primary spermatocytes and each primary spermatocyte meiotically yields four spermatids (spermatozoa-to-be), the spermatogenic sequence in humans can theoretically produce sixteen spermatozoa each time a spermatogonium initiates this process. Usually, however, some cells are lost at various stages, so the efficiency of productivity is rarely this high.

Packaging Even after meiosis, spermatids still resemble undifferentiated spermatogonia structurally, except for their half complement of chromosomes. Production of extremely specialized, mobile spermatozoa from spermatids requires extensive remodeling, or packaging, of cellular elements, a process known as **spermiogenesis**. Sperm are essentially "stripped-down" cells in which most of the cytosol and the

Stages

Chromosomes
in each cell

| Spermatogonium | 46 |

Mitotic
proliferation — Spermatogonia — One daughter cell remains at outer edge of seminiferous tubule to maintain germ cell line — One daughter cell moves toward lumen to produce spermatozoa — 46

Primary spermatocytes — 46 (doubled)

Meiosis — Secondary spermatocytes — 23 (doubled)

Spermatids — 23

Packaging — Spermatozoa — 23

• **FIGURE 20–7 Spermatogenesis**

organelles not needed for the task of delivering the sperm's genetic information to an ovum have been extruded. A **spermatozoon** has four parts (• Fig. 20–8): a head, an acrosome, a midpiece, and a tail. The **head** consists primarily of the nucleus, which contains the sperm's complement of genetic information. The **acrosome**, an enzyme-filled vesicle at the tip of the head, is used as an "enzymatic drill" for penetrating the ovum. The acrosome is formed by aggregation of vesicles produced by the endoplasmic reticulum/Golgi complex before these organelles are discarded. Mobility for the spermatozoon is provided by growth of a long, whiplike **tail** out of one of the centrioles. Movement of the tail, which occurs as a result of relative sliding of its constituent microtubules (see p. 37), is powered by energy generated by the mitochondria concentrated within the **midpiece** of the sperm.

Until sperm maturation is complete, the developing germ cells arising from a single primary spermatocyte remain joined by cytoplasmic bridges. These connections, which result from incomplete cytoplasmic division, permit cytoplasm

to be exchanged between the four developing sperm. This linkage is important, because the X chromosome but not the Y chromosome contains genes that code for cellular products essential for development of the sperm. During meiosis, half the sperm receive an X and the other half a Y chromosome. If it were not for the sharing of cytoplasm so that all the haploid cells are provided with the products coded for by X chromosomes until sperm development is complete, the Y-bearing, male-producing sperm would not be able to develop and survive.

Throughout their development, sperm remain intimately associated with Sertoli cells.

In addition to the spermatogonia and developing sperm cells, the seminiferous tubules also house the **Sertoli cells**. The **Sertoli cells** form a ring that extends from the outer basement membrane to the lumen of the tubule. Each Sertoli cell spans the entire distance from the outer membrane to the fluid-filled

(a)

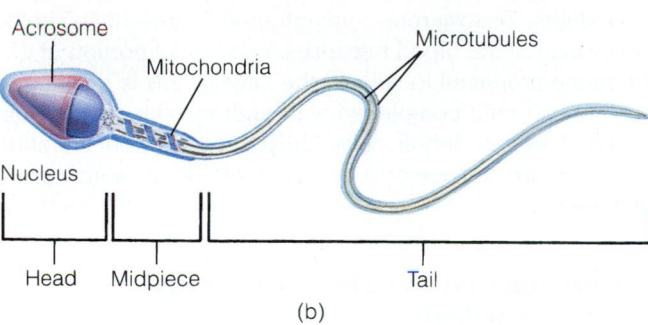

Acrosome

Mitochondria

Microtubules

Nucleus

Head | Midpiece | Tail

(b)

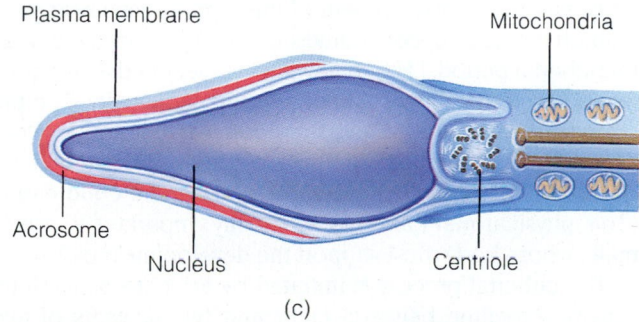

Plasma membrane

Mitochondria

Acrosome

Nucleus | Centriole

(c)

● **FIGURE 20–8 Anatomy of a Spermatozoon** (a) A phase-contrast photomicrograph of human spermatozoa. (b) Schematic representation of spermatozoon in "frontal" view. (c) Longitudinal section of head portion of spermatozoon in "side" view.

lumen (Fig. 20–6d). Adjacent Sertoli cells are joined by tight junctions (see p. 55) at a point slightly beneath the outer membrane. Spermatogonia are tucked between the Sertoli cells at the outer perimeter of the tubule in the spaces between the basement membrane and the tight junctions.

During spermatogenesis, developing sperm cells arising from spermatogonial mitotic activity pass through the tight junctions, which transiently separate to make a path for them, then migrate toward the lumen in intimate association with the adjacent Sertoli cells. The cytoplasm of the Sertoli cells envelops the migrating germ cells, which remain buried within these cytoplasmic recesses throughout their development.

The supportive Sertoli cells perform the following functions essential for spermatogenesis:

1. The tight junctions between adjacent Sertoli cells form a **blood-testes barrier**. Because this barrier prevents blood-borne substances from passing between the cells to gain entry to the tubular lumen, only selected molecules that are able to pass through the Sertoli cells reach the luminal fluid. As a result, the composition of the intratubular fluid varies considerably from that of the blood. The unique composition of this fluid that bathes the germ cells is believed to be critical for the later stages of sperm development. The blood-testes barrier also prevents the antibody-producing cells in the extracellular fluid from reaching the tubular sperm factory, thus preventing the formation of antibodies against the highly differentiated spermatozoa.

2. Since the secluded developing sperm cells do not have direct access to blood-borne nutrients, the Sertoli cells provide nourishment for them.

3. The Sertoli cells have an important phagocytic function. They engulf the cytoplasm extruded from the spermatids during their remodeling and destroy defective germ cells that fail to successfully complete all stages of spermatogenesis.

4. The Sertoli cells secrete into the lumen **seminiferous tubule fluid**, which "flushes" the released sperm from the tubule into the epididymis for storage and further processing.

5. An important component of this Sertoli secretion is **androgen-binding protein**. As the name implies, this protein binds androgens (that is, testosterone), thus maintaining a very high level of this hormone within the lumen of the seminiferous tubules. This high local concentration of testosterone is essential for sustaining sperm production. Androgen-binding protein is necessary to retain testosterone within the lumen because this steroid hormone is lipid soluble and could easily diffuse across the plasma membranes and leave the lumen.

6. The Sertoli cells are the site of action for control of spermatogenesis by both testosterone and follicle-stimulating hormone (FSH). The Sertoli cells themselves release another hormone, *inhibin,* which acts in negative-feedback fashion to regulate FSH secretion.

The two anterior pituitary gonadotropic hormones, LH and FSH, control testosterone secretion and spermatogenesis.

The testes are controlled by the two gonadotropic hormones secreted by the anterior pituitary, **luteinizing hormone (LH)** and **follicle-stimulating hormone (FSH)**, which are named for their functions in females (see p. 635). These hormones act on separate components of the testes (● Fig. 20–9). Luteinizing hormone acts on the Leydig cells to regulate tes-

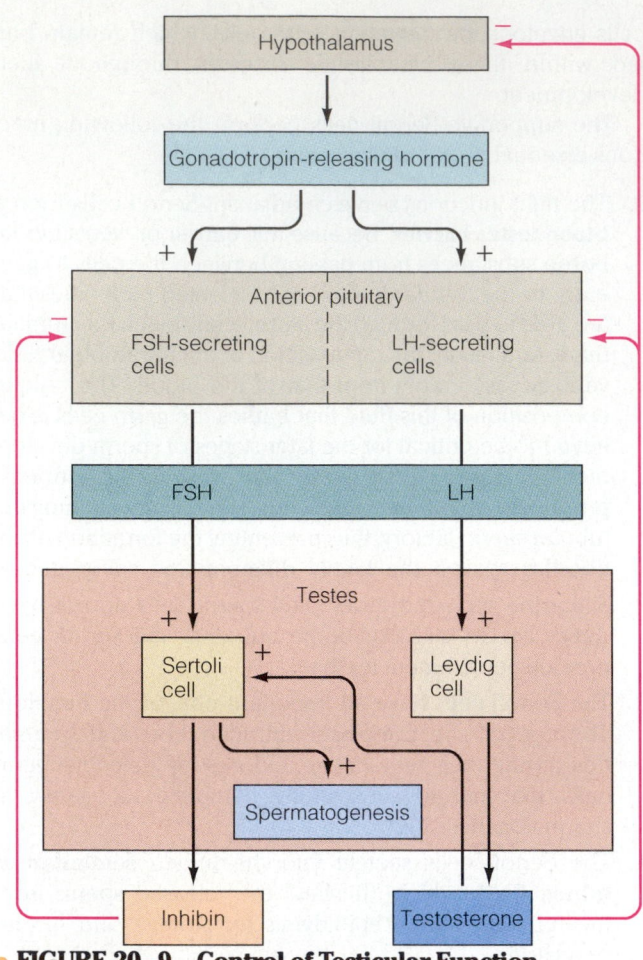

• **FIGURE 20–9   Control of Testicular Function**

acts in negative-feedback fashion to inhibit LH secretion in two ways. The predominant negative-feedback effect of testosterone is to decrease the episodes of GnRH release by acting on the hypothalamus, thus indirectly decreasing both LH and FSH release by the anterior pituitary. Secondly, testosterone acts directly on the anterior pituitary to reduce the responsiveness of the LH-secretory cells to GnRH. The latter action explains why testosterone exerts a greater inhibitory effect on LH secretion than on FSH secretion.

The testicular inhibitory signal specifically directed at controlling FSH secretion is the peptide hormone **inhibin**, which is secreted by the Sertoli cells. Inhibin is believed to act directly on the anterior pituitary to inhibit FSH secretion. This feedback inhibition of FSH by a Sertoli cell product is appropriate, because FSH stimulates spermatogenesis by acting on the Sertoli cells.

Both testosterone and FSH play critical roles in controlling spermatogenesis, each exerting its effect by acting on the Sertoli cells. Testosterone is essential for both mitosis and meiosis of the germ cells, whereas FSH is required for spermatid remodeling. Testosterone concentration is much higher in the testes than in the blood because a substantial portion of this hormone produced locally by the Leydig cells is retained in the luminal fluid complexed with androgen-binding protein secreted by the Sertoli cells. Only this high concentration of testicular testosterone is adequate to sustain sperm production.

Gonadotropin-releasing hormone activity increases at puberty.

Even though the fetal testes secrete testosterone, which directs masculine development of the reproductive system, after birth the testes become quiescent until puberty. During the prepubertal period, LH and FSH are not secreted at adequate levels to stimulate any significant testicular activity. The prepubertal delay in the onset of reproductive capability allows time for the individual to mature physically (though not necessarily psychologically) enough to handle childrearing. (This physical maturation is especially important in the female, whose body must support the developing fetus.)

The pubertal process is initiated by an increase in GnRH activity sometime between eight and twelve years of age. Early in puberty, GnRH secretory pulses occur only at night, causing brief nocturnal increases in LH secretion and, accordingly, testosterone secretion. The duration of episodic GnRH secretion gradually increases as puberty progresses until the adult pattern of GnRH, FSH, LH, and testosterone secretion is established. Under the influence of the rising levels of testosterone during puberty, the physical changes that encompass the secondary sexual characteristics and reproductive maturation become evident.

The low level of GnRH activity during the prepubertal period appears to be due to active inhibition of GnRH release by both hormonal and neural mechanisms. Before puberty, the hypothalamus is extremely sensitive to the negative-feedback actions of testosterone, so the very small amounts of testosterone produced by the prepubertal testes are able to in-

tosterone secretion, accounting for its alternative name in males—*interstitial-cell–stimulating hormone* (*ICSH*). Follicle-stimulating hormone acts on the seminiferous tubules, specifically the Sertoli cells, to enhance spermatogenesis. (There is no alternative name for FSH in males.) Secretion of both LH and FSH from the anterior pituitary is stimulated, in turn, by a single hypothalamic hormone, **gonadotropin-releasing hormone (GnRH)** (see p. 636).

Once every two to three hours, GnRH is released from the hypothalamus in secretory bursts, with no secretion occurring in between. Since GnRH stimulates the gonadotropic hormone–secretory cells in the anterior pituitary, this pulsatile pattern of hypothalamic secretion results in similar episodic bursts in LH and FSH secretion.

Even though GnRH stimulates both LH and FSH secretion, the blood concentrations of these two gonadotropic hormones do not always parallel each other for two reasons. First, LH is removed from the blood more rapidly between the secretory bursts than is the more slowly metabolized FSH, so the pulsatile variations in blood levels of LH are much more pronounced than are those of FSH. Second, two other regulatory factors besides GnRH—testosterone and inhibin—differentially influence the secretory rate of LH and FSH. Testosterone, the product of LH stimulation of the Leydig cells, hibit GnRH release. At puberty, the hypothalamus becomes less sensitive to feedback inhibition by testosterone. Because the low levels of testosterone no longer suppress the hypothalamus, GnRH and gonadotropic hormone levels rise. Also, during the prepubertal period, the hypothalamic GnRH cells are subjected to direct neural inhibition, which is similarly removed at puberty.

The factors responsible for removing these inhibitory mechanisms, thus initiating puberty in humans, remain a mystery. Three proposals, none of which is fully satisfactory, have been put forth. One possibility is a preprogrammed, age-related reduction in inhibitory activity. A second suggestion links attainment of a critical body weight or percentage of body fat to increased GnRH release. The leading proposal focuses on a potential role for the hormone **melatonin**, which is secreted by the **pineal gland** within the brain (see Fig. 18–2). Melatonin, whose secretion decreases during exposure to the light and increases during exposure to the dark, has an antigonadotropic effect in many species. Light striking the eyes inhibits the nerve pathways that are responsible for stimulating melatonin secretion. In many seasonally breeding species, the overall decrease in melatonin secretion in connection with longer days and shorter nights initiates the mating season. Some researchers suggest that an observed reduction in the overall rate of melatonin secretion at puberty in humans—particularly during the night, when the peaks in GnRH secretion first occur—is the trigger for the onset of puberty.

The pineal gland is also believed to have extremely widespread influences besides its possible link with the reproductive system. It may synchronize a gamut of functions (for example, temperature regulation, anterior pituitary hormone metabolism, fat deposition, and immune responses) in relation to the seasons of the year.

The ducts of the reproductive tract store and concentrate sperm and increase their motility and fertility as well.

The remainder of the male reproductive system (besides the testes) is designed for delivery of sperm to the female reproductive tract. Essentially, it consists of (1) a tortuous pathway of tubes that transport sperm from the testes to the outside of the body; (2) several glands, which contribute secretions that are important to the viability and motility of the sperm; and (3) the penis, which is designed to penetrate and deposit the sperm within the vagina of the female.

A comma-shaped **epididymis** is loosely attached to the rear surface of each testis (Figs. 20–1 and 20–6a). After sperm are produced in the seminiferous tubules, they are swept into the epididymis as a result of the pressure created by the continual secretion of tubular fluid by the Sertoli cells. The epididymal ducts from each testis converge to form a large, thick-walled, muscular duct called the **ductus (vas) deferens**. The ductus deferens from each testis passes up out of the scrotal sac and runs back through the inguinal canal into the abdominal cavity, where it eventually empties into the urethra at the neck of the bladder (Fig. 20–1). The urethra carries

TABLE 20–2	Functions of Components of Male Reproductive System

TESTES

Produce sperm

Secrete testosterone

EPIDIDYMIS AND DUCTUS DEFERENS

Serve as sperm exit route from testes

Serve as site for maturation of sperm for motility and fertility

Concentrate and store sperm

SEMINAL VESICLES

Supply fructose to nourish ejaculated sperm

Secrete prostaglandins that stimulate motility within male and female reproductive tracts to help transport sperm

Provide bulk of semen

Provide precursors for clotting of semen

PROSTATE GLAND

Secretes alkaline fluid that neutralizes acidic vaginal secretions

Triggers clotting of semen to keep sperm in vagina during penis withdrawal

BULBOURETHRAL GLANDS

Secrete mucus for lubrication

sperm out of the penis during ejaculation, the forceful expulsion of semen from the body.

These ducts perform several important functions (Table 20–2). First, the epididymis and ductus deferens serve as the sperm's exit route from the testes. As they leave the testes, the sperm are incapable of either movement or fertilization. They gain both capabilities during their passage through the epididymis. This maturational process is stimulated by the testosterone retained within the tubular fluid bound to androgen-binding protein. Sperm's capacity to fertilize is enhanced even further by exposure to secretions of the female reproductive tract. This enhancement of sperm's capacity in the male and female reproductive tracts is known as **capacitation**. The epididymis also concentrates the sperm a hundredfold by absorbing most of the fluid that enters from the seminiferous tubules. The maturing sperm are slowly moved through the epididymis into the ductus deferens by rhythmic contractions of the smooth muscle in the walls of these tubes. The ductus deferens serves as an important site for sperm storage. Because the tightly packed sperm are relatively inactive and their metabolic needs are accordingly low, they can be stored in the ductus deferens for many days, even though they have no nutrient blood supply and are nourished only by simple sugars present in the tubular secretions.

In a **vasectomy**, a common sterilization procedure in males, a small segment of each ductus deferens (alias vas deferens, hence the term *vasectomy*) is surgically removed after it passes from the testis but before it enters the inguinal canal, thus blocking the exit of sperm from the testes. The sperm that build up behind the tied-off testicular end of the severed ductus are removed by phagocytosis. Although this procedure blocks sperm exit, it does not interfere with testosterone activity because the Leydig cells secrete testosterone into the blood, not through the ductus deferens. Thus, there should be no diminution of testosterone-dependent masculinity or libido following a vasectomy.

The accessory sex glands contribute the bulk of the semen.

Several accessory sex glands—the seminal vesicles and prostate—empty their secretions into the duct system before it joins the urethra (Fig. 20–1). A pair of saclike *seminal vesicles* empty into the last portion of the two ductus deferens, one on each side. The short segment of duct that passes beyond the entry point of the seminal vesicle to join the urethra constitutes the *ejaculatory duct*. The *prostate* is a large single gland that completely surrounds the ejaculatory ducts and urethra. In a significant number of men, prostatic hypertrophy (enlargement) occurs in middle to older age. Difficulty in urination is often encountered as the enlarging prostate impinges on the portion of the urethra that passes through the prostate. Another pair of accessory sex glands, the *bulbourethral glands,* drain into the urethra after it has passed through the prostate just before it enters the penis. Numerous mucussecreting glands are also located along the length of the urethra.

During ejaculation, the accessory sex glands contribute secretions that provide support for the continuing viability of the sperm inside the female reproductive tract. These secretions constitute the bulk of the **semen**, which consists of a mixture of accessory sex gland secretions, sperm, and mucus. Sperm make up only a small percentage of the total ejaculated fluid.

Although the accessory sex gland secretions are not absolutely essential for fertilization, they do make contributions that greatly facilitate the fertilization process (Table 20–2):

- The **seminal vesicles** (1) supply fructose, which serves as the primary energy source for ejaculated sperm; (2) secrete *prostaglandins,* which are believed to stimulate contractions of the smooth muscle in both the male and female reproductive tracts, thereby helping to transport sperm from their storage site in the male to the site of fertilization in the female oviduct; (3) provide more than half the semen, which helps to wash the sperm into the urethra and also dilutes the thick mass of sperm, thus enabling them to develop motility; and (4) secrete fibrinogen, a precursor of fibrin, which forms the meshwork of a clot (see p. 360).
- The **prostate gland** (1) secretes an alkaline fluid that neutralizes the acidic vaginal secretions, an important function because sperm are more viable in a slightly alkaline environment, and (2) provides clotting enzymes and fibrinolysin. The prostatic clotting enzymes act on fibrinogen from

the seminal vesicles to produce fibrin, which "clots" the semen, thus helping to keep the ejaculated sperm in the female reproductive tract during withdrawal of the penis. Shortly thereafter, the seminal clot is broken down by fibrinolysin, a fibrin-degrading enzyme from the prostate, thus releasing motile sperm within the female tract.

- During sexual arousal, the **bulbourethral glands** secrete a mucuslike substance that provides lubrication for sexual intercourse.

Prostaglandins are ubiquitous, locally acting chemical messengers.

Although **prostaglandins** were first identified in the semen and were believed to be of prostate-gland origin (hence their name, even though they are actually secreted into the semen by the seminal vesicles), their production and actions are by no means limited to the reproductive system. These twenty-carbon fatty acid derivatives are among the most ubiquitous chemical messengers in the body. They are produced in virtually all tissues from arachidonic acid, a fatty acid constituent of the phospholipids within the plasma membrane. Prostaglandins (and other closely related arachidonic-acid derivatives that are often included for convenience in the category of prostaglandins, namely, *prostacyclins, thromboxanes,* and *leukotrienes*) are among the most biologically active compounds known. Upon appropriate stimulation, arachidonic acid is split from the plasma membrane by a membrane-bound enzyme and then is converted into the appropriate prostaglandin, which acts locally within or near its site of production. After prostaglandins act, they are rapidly inactivated by local enzymes before they gain access to the blood, or if they do reach the circulatory system, they are swiftly degraded on their first pass through the lungs so that they are not dispersed through the systemic arterial system.

Prostaglandins are designated as belonging to one of three groups—PGA, PGE, or PGF—according to structural variations in the five-carbon ring that they contain at one end (• Fig. 20–10). Within each group, prostaglandins are further identified by the number of double bonds present in the two side chains that project from the ring structure (for example, PGE_1 has one double bond and PGE_2 has two double bonds).

• **FIGURE 20–10   Structure and Nomenclature of Prostaglandins**

Letter designation (PGA, PGE, PGF) denotes structural variations in the five-carbon ring

Number designation (e.g., PGE_1, PGE_2) denotes number of double bonds present in the two side chains

TABLE 20-3 Known or Suspected Actions of Prostaglandins

BODY SYSTEM ACTIVITY	ACTIONS OF PROSTAGLANDINS
Reproductive system	Promote sperm transport by action on smooth muscle in male and female reproductive tracts
	Important in menstruation
	Play role in ovulation
	Contribute to preparation of maternal portion of placenta
	Contribute to parturition
Respiratory system	Some promote bronchodilation, others bronchoconstriction
Urinary system	Increase renal blood flow
	Increase excretion of water and salt
Digestive system	Inhibit HCl secretion by stomach
	Stimulate intestinal motility
Nervous system	Influence neurotransmitter release and action
	Act at hypothalamic "thermostat" to increase body temperature
Endocrine system	Enhance cortisol secretion
	Influence tissue responsiveness to hormones in many instances
Circulatory system	Influence platelet aggregation
Fat metabolism	Inhibit fat breakdown
Defense system	Promote many aspects of inflammation, including fever and development of pain

Prostaglandins exert a bewildering variety of effects. Not only are slight variations in prostaglandin structure accompanied by profound differences in biological action, but the same prostaglandin molecule may even exert opposite effects in different tissues. Besides enhancing sperm transport in semen, these abundant chemical messengers are known or suspected to exert other actions in the female reproductive system and in the respiratory, urinary, digestive, nervous, and endocrine systems, in addition to having effects on platelet aggregation, fat metabolism, and inflammation (Table 20-3).

As prostaglandins' various actions are better understood, new ways of manipulating them therapeutically are becoming available. A classic example is the use of aspirin, which blocks the conversion of arachidonic acid into prostaglandins, for fever reduction and pain relief. Prostaglandin action is also therapeutically inhibited in the treatment of premenstrual symptoms and menstrual cramping. Furthermore, specific prostaglandins have been medically administered in such diverse situations as inducing labor, treating asthma, and treating gastric ulcers.

SEXUAL INTERCOURSE BETWEEN MALES AND FEMALES

Ultimately, union of male and female gametes in humans requires delivery of semen into the female vagina through the **sex act**, also known as **sexual intercourse**, **coitus**, or **copulation**.

The male sex act is characterized by erection and ejaculation.

The *male sex act* involves two components: (1) **erection**, or hardening of the normally flaccid penis to permit its entry into the vagina, and (2) **ejaculation**, or forceful expulsion of semen into the urethra and out of the penis (Table 20-4). In addition to these strictly reproductive-related components, the **sexual response cycle** also encompasses broader physiological responses that can be divided into four phases:

1. The *excitement phase,* which includes erection accompanied by testicular vasocongestion (engorgement with blood) and heightened sexual awareness.

2. The *plateau phase,* which is characterized by intensification of these responses, plus more generalized body responses, such as steadily increasing heart rate, blood pressure, respiratory rate, and muscle tension.

3. The *orgasmic phase,* which includes ejaculation as well as other responses that culminate the mounting sexual excitement and are collectively experienced as an intense physical pleasure.

4. The *resolution phase,* which returns the genitalia and body systems to their prearousal state.

The human sexual response is a multicomponent experience that, in addition to these physiological phenomena, also encompasses emotional, psychological, and sociological factors. We will examine only the physiological aspects of sex.

Erection is not caused by contraction of skeletal muscles within the penis, as might be expected, but by engorgement of the penis with blood. The penis consists almost entirely of **erectile tissue** made up of three columns of spongelike vascular spaces extending the length of the organ (Fig. 20-1). In the absence of sexual excitation, the erectile tissues contain little blood, because the arterioles that supply these vascular chambers are constricted. As a result, the penis remains small and flaccid. During sexual arousal, these arterioles reflexly dilate and the erectile tissue fills with blood, causing the penis to enlarge both in length and width and to become more rigid. A larger buildup of blood and further enhancement of erection is achieved by a reduction in venous outflow. The veins that drain the erectile tissue are compressed as a result of engorgement and expansion of the vascular spaces brought about by increased arterial inflow. These local vascular responses—reflexly increased arterial inflow and mechanically reduced venous outflow—transform the penis into a hardened, elongated organ capable of penetrating the vagina.

TABLE 20–4   Components of the Male Sex Act

COMPONENT OF MALE SEX ACT	DEFINITION	ACCOMPLISHED BY
Erection	Hardening of the normally flaccid penis to permit its entry into the vagina	Engorgement of the penis erectile tissue with blood as a result of marked parasympathetically induced vasodilation of penile arterioles and mechanical compression of veins
Ejaculation		
Emission phase	Emptying of sperm and accessory sex gland secretions (semen) into the urethra	Sympathetically induced contraction of the smooth muscle in the walls of the ducts and accessory sex glands
Expulsion phase	Forceful expulsion of semen from the penis	Motor-neuron–induced contraction of the skeletal muscles at the base of the penis

The erection reflex is a spinal reflex triggered by stimulation of highly sensitive mechanoreceptors located in the glans penis, which caps the tip of the penis. Tactile stimulation of the glans reflexly triggers increased parasympathetic and decreased sympathetic activity to the penile arterioles; the result is vasodilation of these arterioles and an ensuing erection (• Fig. 20–11). This response is the major instance of direct parasympathetic control over blood-vessel caliber in the body. Recent evidence suggests that parasympathetic stimulation brings about relaxation of penile arteriolar smooth muscle by means of nitric oxide (alias endothelial-derived relaxing factor), which is known to cause arteriolar vasodilation in response to local tissue changes elsewhere in the body (see p. 312). Arterioles are typically supplied only by sympathetic

nerves, with increased sympathetic activity producing vasoconstriction and decreased sympathetic activity resulting in vasodilation. Concurrent parasympathetic stimulation and sympathetic inhibition of penile arterioles accomplish vasodilation more swiftly and in greater magnitude than is possible in other arterioles supplied only by sympathetic nerves. Through this efficient means of rapidly increasing blood flow into the penis, complete erection can be accomplished as quickly as five to ten seconds. At the same time, parasympathetic impulses promote secretion of lubricating mucus from the bulbourethral glands and the urethral glands in preparation for coitus.

This basic spinal reflex can be either facilitated or inhibited by higher brain centers through descending path-

• **FIGURE 20–11   Erection Reflex**

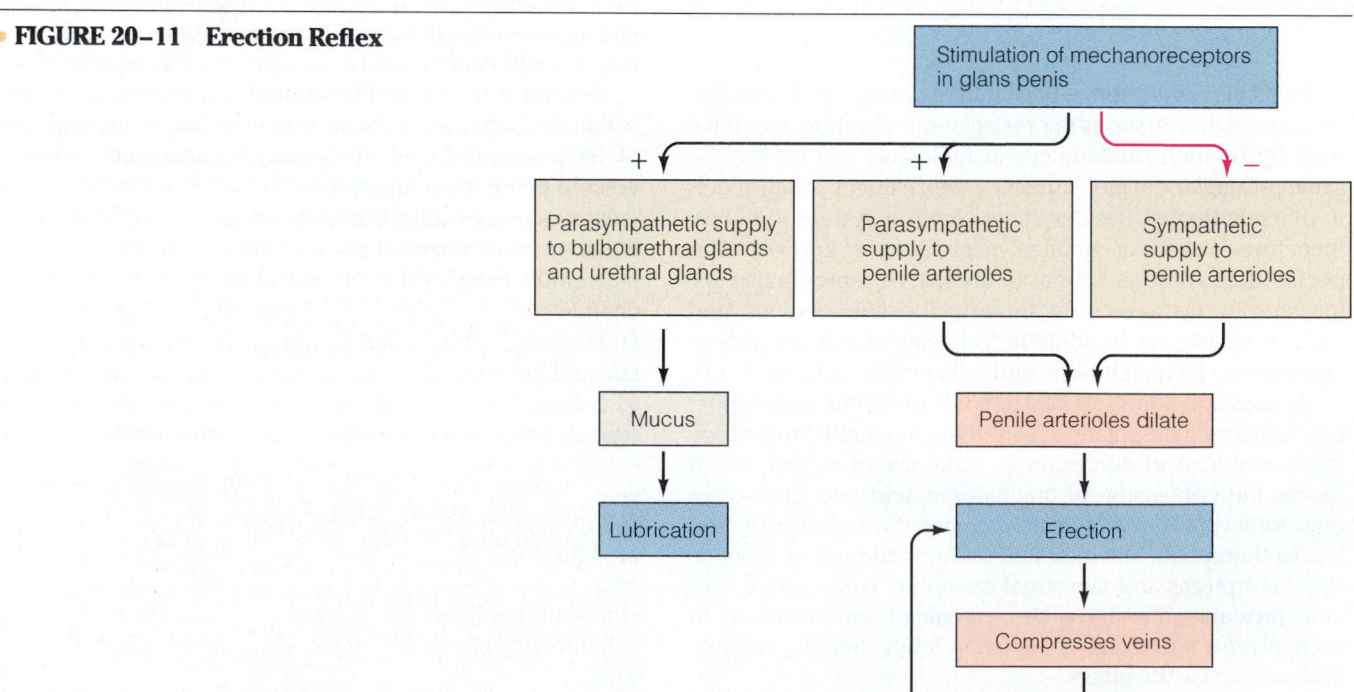

ways that also terminate on the autonomic nerves supplying the penile arterioles. As an example of facilitation, psychic stimuli, such as viewing something sexually exciting, can induce an erection in the complete absence of tactile stimulation of the penis. On the other hand, failure to achieve an erection in spite of appropriate stimulation (**impotence**) may occur as a result of inhibition of the erection reflex by higher brain centers. In fact, impotence is more commonly caused by psychological influences than by actual physical limitations. A man who becomes overly anxious about his ability to perform the sex act may well be on his way to failure. Physical causes of impotence include nerve damage, certain medications that interfere with autonomic function, and problems with blood flow through the penis.

The second component of the male sex act is *ejaculation*. Like erection, ejaculation is accomplished by a spinal reflex. The same types of tactile and psychic stimuli that induce erection cause ejaculation when the level of excitation intensifies to a critical peak. The overall ejaculatory response occurs in two phases: emission and expulsion. First, sympathetic impulses cause sequential contraction of smooth muscles in the prostate, reproductive ducts, and seminal vesicles. This contractile activity delivers prostatic fluid, then sperm, and finally seminal vesicle fluid (collectively, *semen*) into the urethra. This phase of the ejaculatory reflex is known as **emission**. During this time, the sphincter at the neck of the bladder is tightly closed to prevent semen from entering the bladder and urine from being expelled along with the ejaculate through the urethra. Second, the filling of the urethra with semen triggers nerve impulses that activate a series of skeletal muscles at the base of the penis. Rhythmic contractions of these muscles occur at 0.8-second intervals and increase the pressure within the penis, forcibly expelling the semen through the urethra to the exterior. This is the **expulsion** phase of ejaculation.

The rhythmic contractions that occur during semen expulsion are accompanied by involuntary rhythmic throbbing of pelvic muscles and peak intensity of the overall body responses that were climbing during the earlier phases. Heavy breathing, a heart rate of up to 180 beats per minute, marked generalized skeletal-muscle contraction, and heightened emotions are characteristic. These pelvic and overall systemic responses that culminate the sex act are associated with an intense pleasure characterized by a feeling of release and complete gratification, an experience known as **orgasm**.

During the resolution phase following orgasm, sympathetic vasoconstrictor impulses slow the inflow of blood into the penis so that the erection subsides. A deep relaxation ensues, often accompanied by a feeling of fatigue. Muscle tone returns to normal while the cardiovascular and respiratory systems return to their prearousal level of activity. Once ejaculation has occurred, a temporary refractory period of variable duration ensues before sexual stimulation can trigger another erection. Males are therefore unable to experience multiple orgasms within a matter of minutes as females sometimes do.

The volume and sperm content of the ejaculate depend on the length of time between ejaculations. The average volume of semen is 3 ml, ranging from 2.5 to 6 ml, the higher volumes following periods of continence. An average human ejaculate contains about 300–400 million sperm (120 million/ml). Both quantity and quality of the sperm are important determinants of fertility. A man is considered to be clinically infertile if his sperm concentration falls below 20 million/ml of semen. Even though only one spermatozoon actually fertilizes the ovum, large numbers of accompanying sperm are needed to provide sufficient acrosomal enzymes to break down the barriers surrounding the ovum until the victorious sperm penetrates into the ovum's cytoplasm. The quality of sperm also must be taken into account when assessing the fertility potential of a semen sample. The presence of substantial numbers of sperm with abnormal motility or structure, such as sperm with distorted tails, reduces the chances of fertilization.

The female sexual cycle parallels that of males in many ways.

Both sexes experience the same four phases of the sexual cycle—excitement, plateau, orgasm, and resolution. Furthermore, the physiological mechanisms responsible for orgasm are fundamentally the same in males and females.

The excitement phase in females can be initiated by either physical or psychological stimuli. Tactile stimulation of the clitoris and surrounding perineal area is an especially powerful sexual stimulus. These stimuli trigger spinal reflexes that bring about parasympathetically induced vasodilation of arterioles throughout the vagina and external genitalia. The resultant inflow of blood becomes evident as swelling of the labia and erection of the clitoris. The latter—like its male homologue, the penis—is composed largely of erectile tissue. Vasocongestion of the vaginal capillaries forces fluid out of the vessels into the vaginal lumen. This fluid, which is the first positive indication of sexual arousal, serves as the primary lubricant for intercourse. Additional lubrication is provided by the mucus secretions from the male and by mucus released during sexual arousal from glands located at the outer opening of the vagina. Also during the excitement phase in the female, the nipples become erect and the breasts enlarge as a result of vasocongestion. In addition, the majority of women show a *sex flush* during this time, which is caused by increased blood flow through the skin.

During the plateau phase, the changes initiated during the excitement phase intensify, while systemic responses similar to those in the male (such as increased heart rate, blood pressure, respiratory rate, and muscle tension) occur. Further vasocongestion of the lower third of the vagina during this time reduces its inner capacity so that it tightens around the thrusting penis, thus heightening tactile sensation for the male. Simultaneously, the uterus raises upward, lifting the cervix and enlarging the upper two-thirds of the vagina. This ballooning or **tenting effect** creates a space for ejaculate deposition.

If erotic stimulation continues, the sexual response culminates in orgasm as sympathetic impulses trigger rhythmic contractions of the pelvic musculature at 0.8-second intervals, the same rate as in males. The contractions occur most intensely in the engorged lower third of the vaginal canal; hence this region is often termed the **orgasmic platform**. Systemic responses identical to those of the male orgasm also occur. In fact, the orgasmic experience in females parallels that of males with two exceptions. First, there is no female counterpart to ejaculation. Second, females do not become refractory following an orgasm, so they can respond immediately to continued erotic stimulation and achieve multiple orgasms. If stimulation continues, the sexual intensity only diminishes to the plateau level following orgasm and can quickly be brought to a peak again. Women have been known to achieve as many as twelve successive orgasms in this manner.

During resolution, pelvic vasocongestion and the systemic manifestations gradually subside. As with males, this is a time of great physical relaxation for females.

⇛ FEMALE REPRODUCTIVE PHYSIOLOGY

Complex cycling characterizes female reproductive physiology.

Female reproductive physiology is much more complex than male reproductive physiology. Unlike the continuous sperm production and essentially constant testosterone secretion characteristic of the male, release of ova is intermittent, and secretion of female sex hormones displays wide cyclical swings. The tissues influenced by these sex hormones also undergo cyclical changes, the most obvious of which is the monthly menstrual cycle. During each cycle, the female reproductive tract is prepared for the fertilization and implantation of an ovum released from the ovary at ovulation. If fertilization does not occur, the cycle repeats itself. If fertilization does occur, the cycles are interrupted while the female system adapts to nurture and protect the newly conceived human being until it has developed into an individual capable of living outside the maternal environment. Furthermore, the female continues her reproductive duties after birth by producing milk (lactation) for the baby's nourishment. Thus, the female reproductive system is characterized by complex cycles that are interrupted only by more complex changes should pregnancy ensue.

The ovaries, as the primary female reproductive organs, perform the dual function of producing ova (oogenesis) and secreting the female sex hormones, estrogen and progesterone. These hormones act together to promote fertilization of the ovum and to prepare the female reproductive system for pregnancy. Estrogen in the female is responsible for many functions similar to those carried out by testosterone in the male, such as maturation and maintenance of the entire female reproductive system and establishment of female secondary sexual characteristics. In general, the actions of estrogen are important to preconception events. Estrogen is essential for ova maturation and release, development of physical characteristics that are sexually attractive to males, and transport of sperm from the vagina to the site of fertilization in the oviduct. Furthermore, estrogen contributes to breast development in anticipation of lactation. The other

ovarian steroid, progesterone, is important in preparing a suitable environment for nourishing a developing embryo/fetus and for contributing to the breasts' ability to produce milk.

As in males, reproductive capability begins at puberty in females, but unlike males, who have reproductive potential throughout life, female reproductive potential ceases during middle age at **menopause**.

Chromosome division in oogenesis parallels that in spermatogenesis, but there are major qualitative and quantitative sexual differences in gametogenesis.

Oogenesis contrasts sharply with spermatogenesis in several important aspects, even though the identical steps of chromosome replication and division take place during gamete production in both sexes. The undifferentiated primordial germ cells in the fetal ovaries, the **oogonia** (comparable to the spermatogonia), divide mitotically to give rise to six to seven million oogonia by the fifth month of gestation, at which time mitotic proliferation ceases. During the last part of fetal life, the oogonia begin the early steps of the first meiotic division but do not complete it. Known now as **primary oocytes**, they contain forty-six replicated chromosomes, which are gathered into homologous pairs but do not separate. The primary oocytes remain in this state of meiotic arrest for years until they are prepared for ovulation.

Before birth, each primary oocyte is surrounded by a single layer of **granulosa cells** to form a **primary follicle**. Oocytes that fail to be incorporated into follicles degenerate, and at birth only about two million primary follicles remain, each containing a single primary oocyte capable of producing a single ovum. No new oocytes or follicles appear after birth; the follicles already present in the ovaries at birth serve as a reservoir from which all ova throughout the reproductive life of a female must arise. Of these follicles, only about four hundred will mature and release ova. The pool of primary follicles present at birth gives rise to an ongoing trickle of developing follicles. A follicle is destined for one of two fates once it starts to develop: it will reach maturity and ovulate, or it will degenerate to form scar tissue, a process known as **atresia**. Until puberty, all of the follicles that start to develop undergo atresia in the early stages without ever ovulating. Even for the first few years after puberty, many of the cycles are **anovulatory** (that is, no ovum is released). Of the initial pool of follicles, 99.98% never ovulate but instead undergo atresia at some stage in development. By menopause, which occurs on average in a woman's early fifties, few, if any, primary follicles remain, having either already ovulated or become atretic. From this point on, the woman's reproductive capacity ceases.

This limited gamete potential, which is already determined at birth in females, is in sharp contrast to the continual process of spermatogenesis in males, who have the potential to produce several hundred million sperm in a single day. Furthermore, there is considerable chromosome wastage in oogenesis compared with spermatogenesis.

The primary oocyte within a primary follicle is still a diploid cell that contains forty-six chromosomes. From puberty until menopause, a portion of the resting pool of follicles starts developing into **secondary** (or **antral**) **follicles** on a cyclical basis. The number of developing follicles at any time is roughly proportional to the size of the pool, but the mechanisms that determine which follicles in the pool will develop during a given cycle are unknown. Development of a secondary follicle is characterized by growth of the primary oocyte and by expansion and differentiation of the surrounding cell layers. Oocyte enlargement is due to a buildup of cytoplasmic materials that will be needed by the early embryo. Just before ovulation, the primary oocyte, whose nucleus has been in meiotic arrest for years, completes its first meiotic division. This division yields two daughter cells, each receiving a set of twenty-three doubled chromosomes, analogous to the formation of secondary spermatocytes (● Fig. 20–12). However, almost all of the cytoplasm remains with one of the daughter cells, now called the **secondary oocyte**, which is destined to become the ovum. The chromosomes of the other daughter cell together with a small share of cytoplasm form the **first polar body**. In this way, the ovum-to-be loses half of its chromosomes to form a haploid gamete but retains all of its nutrient-rich cytoplasm.

It is actually the secondary oocyte and not the mature ovum that is ovulated and fertilized, but common usage refers to the developing female gamete as an ovum even in its primary and secondary oocyte stages. Sperm entry into the secondary oocyte is needed to trigger the second meiotic division. Oocytes that are not fertilized never complete this final division. During this division, a half set of chromosomes along with a thin layer of cytoplasm is extruded as the **second polar body**. The other half set of twenty-three unpaired chromosomes remains behind in what is now the **mature ovum**. These twenty-three maternal chromosomes unite with the twenty-three paternal chromosomes of the penetrating sperm to complete fertilization. If the first polar body has not already degenerated, it too undergoes the second meiotic division at the same time the fertilized secondary oocyte is dividing its chromosomes.

Thus, the steps involved in chromosome distribution during oogenesis parallel those of spermatogenesis except that the cytoplasmic distribution and time span for completion sharply differ. Just as four haploid spermatids are produced by each primary spermatocyte, four haploid daughter cells are produced by each primary oocyte. In spermatogenesis, each daughter cell develops into a highly specialized, motile spermatozoon unencumbered by unessential cytoplasm and

● **FIGURE 20–12   Oogenesis**

organelles, its only destiny being to supply half of the genes for a new individual. In oogenesis, however, of the four daughter cells, only the one destined to become the ovum receives cytoplasm. This uneven distribution of cytoplasm is important, because the ovum, in addition to providing half the genes, provides all of the cytoplasmic components needed to support early development of the fertilized ovum. The large, relatively undifferentiated ovum contains numerous nutrients, organelles, and structural and enzymatic proteins. The three other cytoplasm-scarce daughter cells, the polar bodies, rapidly disintegrate, their chromosomes being deliberately wasted.

Note also the considerable difference in time required for completion of spermatogenesis and oogenesis. It takes about two months for a spermatogonium to develop into fully remodeled spermatozoa. In contrast, development of an oogonium (developed before birth) to a mature ovum requires anywhere from eleven years (beginning of ovulatory cycles at onset of puberty) to fifty years (end of ovulation at onset of menopause). There is considerable speculation that the older age of ova released by women in their late thirties and forties accounts for the higher incidence of genetic abnormalities, such as Down's syndrome, in children born to women in this age range.

The ovarian cycle consists of alternating follicular and luteal phases.

After the onset of puberty, the ovary constantly alternates between two phases—the **follicular phase**, which is dominated by the presence of *maturing follicles*, and the **luteal phase**, which is characterized by the presence of the *corpus luteum* (to be described shortly). This cycle is normally interrupted only by pregnancy and is finally terminated by menopause. The average ovarian cycle lasts twenty-eight days, but this varies among women and among cycles in any particular

• **FIGURE 20–13   Follicular Development**   (a) Stages in follicular development from primary follicle through mature follicle. (b) Scanning electron micrograph of developing secondary follicle. (c) Ovary (actual size) showing presence of mature follicle.

woman. The follicle operates in the first half of the cycle to produce a mature egg ready for ovulation at midcycle. The corpus luteum takes over during the second half of the cycle to prepare the female reproductive tract for pregnancy if fertilization of the released egg occurs.

At any given time throughout the cycle, a portion of the primary follicles are starting to develop. However, only those that do so during the follicular phase, when the hormonal milieu is right to promote their maturation, continue beyond the early stages of development. The others, lacking hormonal support, undergo atresia. During follicular development, as the primary oocyte is synthesizing and storing materials for future use if fertilized, important changes are taking place in the cells surrounding the reactivated oocyte in preparation for the egg's release from the ovary (● Fig. 20–13). First, the single layer of *granulosa cells* in a primary follicle proliferates to form several layers that surround the oocyte. These granulosa cells secrete a thick, gel-like material that covers the oocyte and separates it from the surrounding granulosa cells. This intervening membrane is known as the **zona pellucida**. At the same time, specialized ovarian connective-tissue cells at the edge of the growing follicle proliferate and differentiate to form an outer layer of **thecal cells**. The thecal and granulosa cells, collectively known as **follicular cells**, function as a unit to secrete estrogen. Of the three physiologically important estrogens—estradiol, estrone, and estriol—**estradiol** is the principal ovarian estrogen.

The hormonal environment that exists during the follicular phase promotes enlargement and development of the follicular cells' secretory capacity, converting the primary follicle into a secondary, or antral, follicle capable of estrogen secretion. This stage of follicular development is characterized by the formation of a fluid-filled **antrum** in the midst of the granulosa cells. The follicular fluid originates partially from transudation (passage through capillary pores) of plasma and partially from follicular cell secretions. As the follicular cells start producing estrogen, some of this hormone is secreted into the blood for distribution throughout the body. However, a portion of the estrogen collects in the hormone-rich antral fluid.

The oocyte has reached full size by the time the antrum begins to form. The shift to an antral follicle initiates a period of rapid follicular growth. During this time, the follicle increases in size from a diameter of less than 1 mm to 12–16 mm shortly before ovulation. Part of the follicular growth is due to continued proliferation of the granulosa and thecal cells, but most is due to a dramatic expansion of the antrum. As the follicle grows, estrogen is produced in increasing quantities. One of the follicles usually grows more rapidly than the others, developing into a **mature** (**preovulatory**, **tertiary**, or **Graafian**) **follicle** within about fourteen days after the onset of follicular development. The antrum occupies most of the space in a mature follicle. The oocyte, surrounded by the zona pellucida and a single layer of granulosa cells, is displaced asymmetrically at one side of the growing follicle in a little mound that protrudes into the antrum.

The greatly expanded mature follicle bulges on the ovarian surface, creating a thin area that ruptures to release the oocyte at **ovulation**. Rupture of the follicle is facilitated by the release from the follicular cells of enzymes that digest the connective tissue in the follicular wall. The bulging wall is thus weakened so that it balloons out even further to the point that it can no longer contain the rapidly expanding follicular contents.

Just before ovulation, the oocyte completes its first meiotic division. The ovum (secondary oocyte), still surrounded by its tightly adhering zona pellucida and granulosa cells (now called the **corona radiata**), is swept out of the ruptured follicle into the abdominal cavity by the leaking antral fluid (● Fig. 20– 14). The released ovum is quickly drawn into the oviduct, where fertilization may or may not take place.

The other developing follicles that failed to reach maturation and ovulate undergo degeneration, never to be reactivated. Occasionally, two (or perhaps more) follicles reach maturation and ovulate at about the same time. If both are fertilized, **fraternal twins** result. Because fraternal twins arise from separate ova fertilized by separate sperm, they share no more in common than any other two siblings except for the same birth date. **Identical twins**, on the other hand, develop from a single fertilized ovum that completely divides into two

● **FIGURE 20–14   Ovulation and Development and Degeneration of Corpus Luteum**   (a) Rupture of mature follicle and release of ovum (secondary oocyte) at ovulation. (b) Formation of corpus luteum from old follicular cells following ovulation. (c) Degeneration of corpus luteum if released ovum is not fertilized.

separate, genetically identical embryos at a very early stage in development.

Rupture of the follicle at ovulation signals the end of the follicular phase and ushers in the luteal phase. The ruptured follicle that is left behind in the ovary following release of the ovum undergoes a rapid change. The granulosa and thecal cells remaining in the remnant follicle first collapse into the emptied antral space that has been partially filled by clotted blood. These old follicular cells soon undergo a dramatic

structural transformation to form the **corpus luteum**, in a process called **luteinization** (Fig. 20–14). The follicular-turned-luteal cells hypertrophy and are converted into very active steroidogenic (steroid-hormone–producing) tissue. Abundant storage of cholesterol, the steroid precursor molecule, in lipid droplets within the corpus luteum gives this tissue a yellowish appearance, hence its name (*corpus* means "body"; *luteum* means "yellow"). The corpus luteum becomes highly vascularized as blood vessels from the thecal region invade

● **FIGURE 20–15   Correlation between Hormonal Levels and Cyclical Ovarian and Uterine Changes** During the follicular phase (the first half of the ovarian cycle), the ovarian follicle secretes estrogen under the influence of FSH, LH, and estrogen itself. The low but rising levels of estrogen (1) inhibit FSH secretion, which declines during the last part of the follicular phase, and (2) incompletely suppress tonic LH secretion, which continues to rise throughout the follicular phase. When the follicular output of estrogen reaches its peak, the high levels of estrogen trigger a surge in LH secretion at midcycle. This LH surge brings about ovulation of the mature follicle. Estrogen secretion plummets when the follicle meets its demise at ovulation.

The old follicular cells are transformed into the corpus luteum, which secretes progesterone as well as estrogen during the luteal phase (the last half of the ovarian cycle). Progesterone strongly inhibits both FSH and LH, which continue to decrease throughout the luteal phase. The corpus luteum degenerates in about two weeks if the released ovum has not been fertilized and implanted in the uterus. Progesterone and estrogen levels sharply decrease when the corpus luteum degenerates, removing the inhibitory influences on FSH and LH. As these anterior pituitary hormone levels start to rise again upon the withdrawal of inhibition, they begin to stimulate the development of a new batch of follicles as a new follicular phase is ushered in.

Concurrent uterine phases reflect the influences of the ovarian hormones on the uterus. Early in the follicular phase, the highly vascularized, nutrient-rich endometrial lining is sloughed off (the uterine menstrual phase). This sloughing results from the withdrawal of estrogen and progesterone when the old corpus luteum degenerated at the end of the preceding luteal phase. Late in the follicular phase, the rising levels of estrogen cause the endometrium to thicken (the uterine proliferative phase). After ovulation, progesterone from the corpus luteum brings about vascular and secretory changes in the estrogen-primed endometrium to produce a suitable environment for implantation (the uterine secretory, or progestational, phase). When the corpus luteum degenerates, a new ovarian follicular phase and uterine menstrual phase begin.

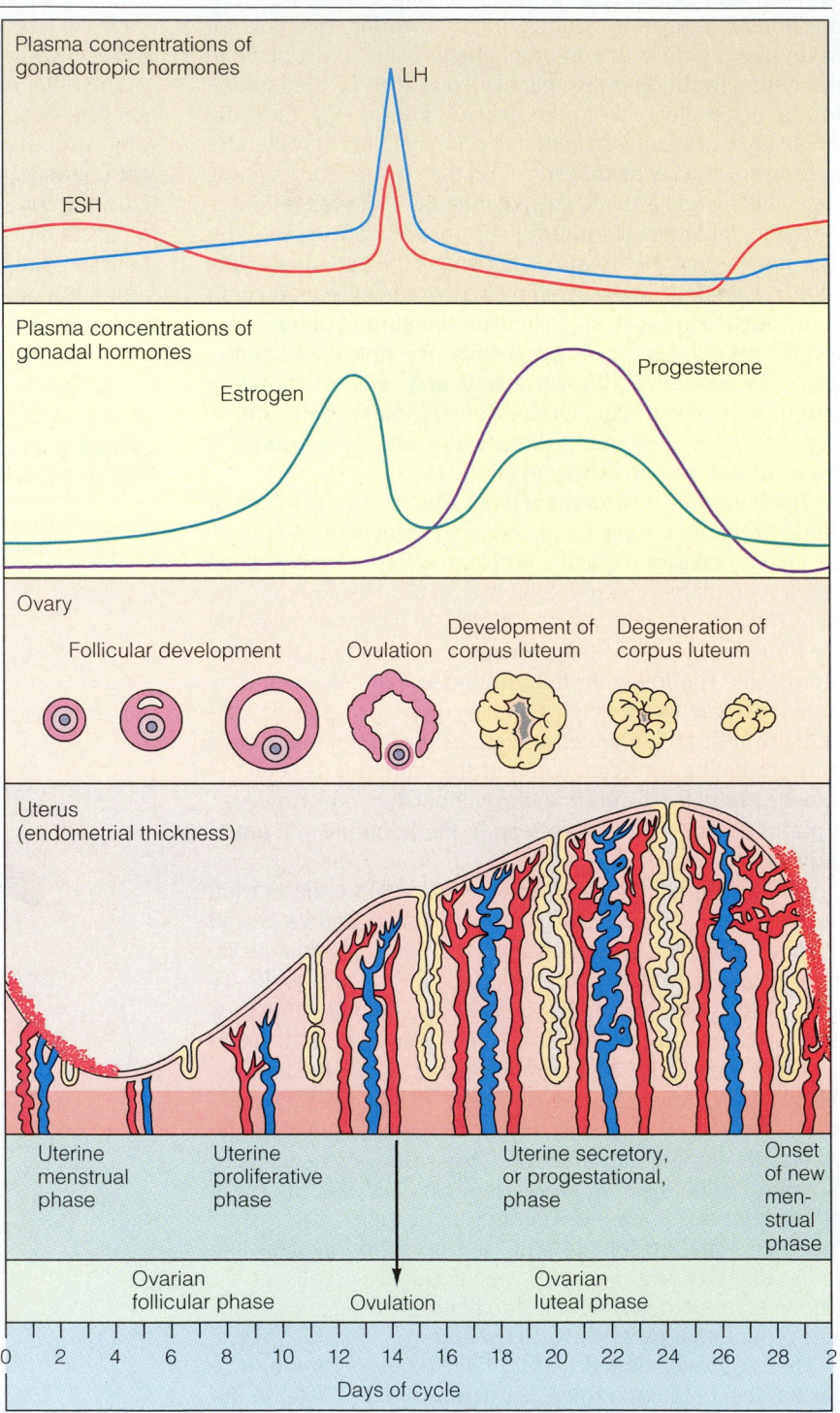

the luteinizing granulosa. These changes are appropriate for the corpus luteum's function, which is to secrete abundant quantities of progesterone along with lesser amounts of estrogen into the blood. Estrogen secretion in the follicular phase followed by progesterone secretion in the luteal phase is essential for preparing the uterus to be a suitable site for implantation of a fertilized ovum. The corpus luteum becomes fully functional within four days after ovulation, but it continues to increase in size for another four or five days. If the released ovum is not fertilized and does not implant, the corpus luteum degenerates within fourteen days after its formation. The luteal cells degenerate and are phagocytized, the vascular supply is withdrawn, and connective tissue rapidly fills in to form a fibrous tissue mass known as the **corpus albicans** ("white body"). The luteal phase is now over, and one ovarian cycle is complete. A new wave of follicular development, which begins when degeneration of the old corpus luteum is completed, signals the onset of a new follicular phase.

If fertilization and implantation do take place, the corpus luteum continues to grow and produce increasing quantities of progesterone and estrogen instead of degenerating. Now called the **corpus luteum of pregnancy**, this ovarian structure persists until the end of pregnancy. It provides the hormones essential for the maintenance of pregnancy until the developing placenta is able to take over this crucial function.

The ovarian cycle is regulated by complex hormonal interactions among the hypothalamus, anterior pituitary, and ovarian endocrine units.

The ovary has two related endocrine units—the estrogen-secreting follicle during the first half of the cycle and the corpus luteum, which secretes both progesterone and estrogen, during the last half of the cycle. These units are sequentially triggered by complex cyclical hormonal relations among the hypothalamus, anterior pituitary, and these two ovarian endocrine units.

As in the male, gonadal function in the female is directly controlled by the anterior pituitary gonadotropic hormones, follicle-stimulating hormone (FSH) and luteinizing hormone (LH). These hormones, in turn, are regulated by pulsatile episodes of hypothalamic gonadotropin-releasing hormone (GnRH) and feedback actions of gonadal hormones. Differing from the male, however, control of the female gonads is complicated by the cyclic nature of ovarian function. For example, the effects of FSH and LH on the ovaries depend on the stage of the ovarian cycle. Also in contrast to the male, FSH is not strictly responsible for gametogenesis nor is LH solely responsible for gonadal hormone secretion. We will consider control of follicular function, ovulation, and the corpus luteum separately, using • Figure 20–15 as a means of integrating the various concurrent and sequential activities that take place throughout the cycle.

Control of Follicular Function   The factors that initiate follicular development are poorly understood. The early stages of preantral follicular growth and oocyte maturation do not require gonadotropic stimulation. Hormonal support is required, however, for antrum formation, further follicular development, and estrogen secretion. Estrogen, FSH, and LH are all needed. Antrum formation is induced by FSH. Both FSH and estrogen stimulate proliferation of the granulosa cells. Both LH and FSH are required for synthesis and secretion of estrogen by the follicle, but these hormones act on different cells and at different steps in the estrogen production pathway (• Fig. 20–16). Both granulosa and thecal cells participate in estrogen production. The conversion of cholesterol into estrogen requires a number of sequential steps, the last of which is conversion of androgens into estrogens (see Fig. 18–4, p. 623). Thecal cells readily produce androgens but have limited capacity to convert them into estrogens. Granulosa cells, on the other hand, are readily able to convert androgens into estrogens but are unable to produce androgens in the first place. Luteinizing hormone acts on the thecal cells to stimulate androgen production, whereas FSH acts on the granulosa cells to promote the conversion of thecal andro-

• **FIGURE 20–16   Production of Estrogen by Ovarian Follicle** Stimulated by LH, thecal cells convert cholesterol into androgen, which diffuses into the granulosa cells. The granulosa cells, stimulated by FSH, in turn convert this androgen into estrogen. Part of the estrogen remains in the ovarian follicle and contributes to antral formation, while the other part is secreted into the blood.

gens (which diffuse into the granulosa cells from the thecal cells) into estrogens. Because low basal levels of FSH are sufficient to promote this final conversion to estrogen, the rate of estrogen secretion by the follicle primarily depends on the circulating level of LH, which continues to rise during the follicular phase. Furthermore, as the follicle continues to grow, more estrogen is produced simply because more estrogen-producing follicular cells are present.

Part of the estrogen produced by the growing follicle is secreted into the blood and is responsible for the steadily increasing plasma estrogen levels during the follicular phase. The remainder of the estrogen remains within the follicle, contributing to the antral fluid and stimulating further granulosa cell proliferation.

The secreted estrogen, in addition to acting on sex-specific tissues such as the uterus, inhibits the hypothalamus and anterior pituitary in negative-feedback fashion (● Fig. 20–17). The low but rising levels of estrogen characterizing the follicular phase act directly on the hypothalamus to inhibit GnRH secretion, thus suppressing GnRH-prompted release of FSH and LH from the anterior pituitary. However, estrogen's primary effect is directly on the pituitary itself. Estrogen reduces

the sensitivity to GnRH of the cells that produce gonadotropic hormones, especially the FSH-producing cells.

This differential sensitivity of FSH- and LH-producing cells induced by estrogen is at least in part responsible for the fact that the plasma FSH level, unlike the plasma LH concentration, declines during the follicular phase as the estrogen level rises (Fig. 20–15). Another contributing factor to the fall in FSH during the follicular phase is secretion of inhibin by the follicular cells. Inhibin preferentially inhibits FSH secretion by acting at the anterior pituitary, just as it does in the male. The decline in FSH secretion brings about atresia of all but the single most mature of the developing follicles.

In contrast to FSH, LH secretion continues to rise slowly during the follicular phase despite inhibition of GnRH (and thus indirectly, LH) secretion. This seeming paradox is due to the fact that estrogen alone cannot completely suppress **tonic** (low-level, ongoing) **LH secretion**; both estrogen and progesterone are required to completely inhibit tonic LH secretion. Since progesterone does not appear until the luteal phase of the cycle, the basal level of circulating LH slowly increases during the follicular phase under incomplete inhibition by estrogen alone.

Control of Ovulation Ovulation and subsequent luteinization of the ruptured follicle are triggered by an abrupt, massive increase in LH secretion. This **LH surge** brings about four major changes in the follicle:

1. It halts estrogen synthesis by the follicular cells.

2. It reinitiates meiosis in the oocyte of the developing follicle, apparently by blocking release of an *oocyte maturation-inhibiting substance* produced by the granulosa cells. This substance is believed to be responsible for arresting meiosis in the primary oocytes once they are wrapped within granulosa cells in the fetal ovary.

3. It triggers production of specific, locally acting prostaglandins, which induce ovulation by promoting vascular changes that cause rapid swelling of the follicle while inducing enzymatic digestion of the follicular wall. Together these actions lead to rupture of the weakened wall that covers the bulging follicle.

4. It causes differentiation of follicular cells into luteal cells. Because the LH surge triggers both ovulation and luteinization, formation of the corpus luteum automatically follows ovulation. Thus, the midcycle burst in LH secretion is a dramatic point in the cycle; it terminates the follicular phase and initiates the luteal phase.

The two different modes of LH secretion—the tonic secretion of LH responsible for promoting ovarian hormone secretion and the LH surge that causes ovulation—not only occur at different times and produce different effects on the ovaries but also are controlled by different mechanisms. Tonic LH secretion is partially suppressed by the inhibitory action of the low, rising levels of estrogen during the follicular phase and is completely suppressed by the increasing levels of progesterone during the luteal phase. Since tonic LH secretion stimulates both estrogen and progesterone secretion, this is a typical negative-feedback control system.

● **FIGURE 20–17 Feedback Control of FSH and Tonic LH Secretion during Follicular Phase**

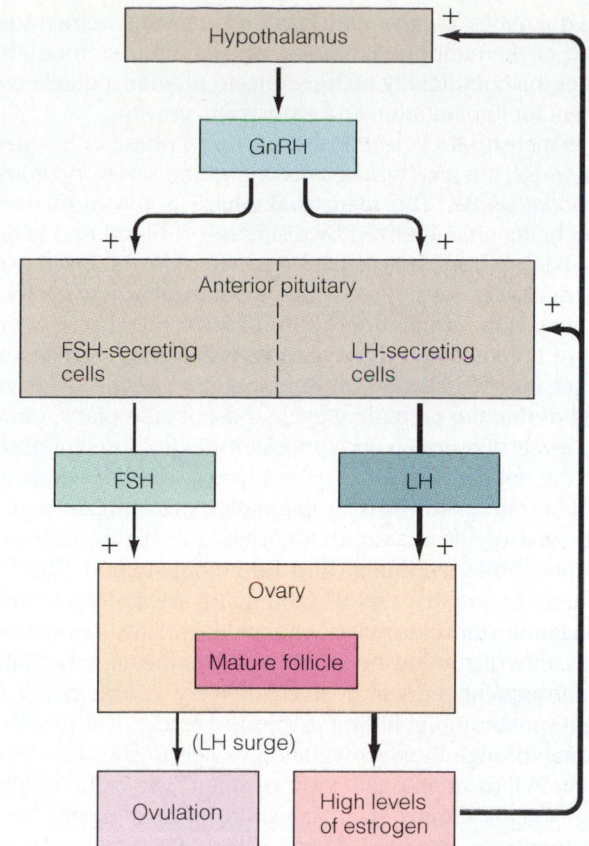

opment of the corpus luteum, LH stimulates ongoing steroid hormone secretion by this ovarian structure. Under the influence of LH, the corpus luteum secretes both progesterone and estrogen, with progesterone being its most abundant hormonal product. The plasma progesterone level increases for the first time during the luteal phase. No progesterone is secreted during the follicular phase (except for a slight output of progesterone by the about-to-rupture follicle under the influence of the LH surge). Therefore, the follicular phase is dominated by estrogen and the luteal phase by progesterone (Fig. 20–15).

A transitory drop in the level of circulating estrogen occurs at midcycle as the estrogen-secreting follicle meets its demise at ovulation. The estrogen level climbs again during the luteal phase because of the corpus luteum's activity, although it does not reach the same peak as during the follicular phase. What keeps the moderately high estrogen level during the luteal phase from triggering another LH surge? Progesterone. Even though a high level of estrogen stimulates LH secretion, progesterone, which dominates the luteal phase, powerfully inhibits LH secretion as well as FSH secretion (● Fig. 20–19). Inhibition of FSH and LH by progesterone prevents new follicular maturation and ovulation during the luteal phase. Under progesterone's influence, the reproductive system is gearing up to support the just-released ovum, should it be fertilized, instead of preparing other ova for release. No inhibin is secreted by the luteal cells.

The corpus luteum functions for two weeks, then degenerates if fertilization does not occur. The mechanisms respon-

In contrast, the LH surge is triggered by a *positive-feedback* effect. Whereas the low, rising levels of estrogen early in the follicular phase *inhibit* LH secretion, the high level of estrogen that occurs during peak estrogen secretion late in the follicular phase (Fig. 20–15) *stimulates* LH secretion and initiates the LH surge (● Fig. 20–18). Thus, LH enhances estrogen production by the follicle, and the resultant peak estrogen concentration stimulates LH secretion. The high plasma concentration of estrogen acts directly on the hypothalamus to increase the frequency of GnRH pulses, thereby increasing both LH and FSH secretion. It also acts directly on the anterior pituitary to specifically increase the sensitivity of LH-secreting cells to GnRH. The latter effect accounts for the much greater surge in LH secretion compared to FSH secretion at midcycle. There is no known role for the modest midcycle surge in FSH that accompanies the pronounced and pivotal LH surge. Because only a mature, preovulatory follicle, not follicles in earlier stages of development, is capable of secreting sufficiently high levels of estrogen to trigger the LH surge, ovulation is not induced until a follicle has reached the proper size and degree of maturation. In a way, then, the follicle lets the hypothalamus know when it is ready to be stimulated to ovulate. The LH surge lasts for only one to two days at midcycle, just before ovulation.

Control of the Corpus Luteum   Luteinizing hormone "maintains" the corpus luteum; that is, after triggering devel-

● **FIGURE 20–19   Feedback Control during Luteal Phase**

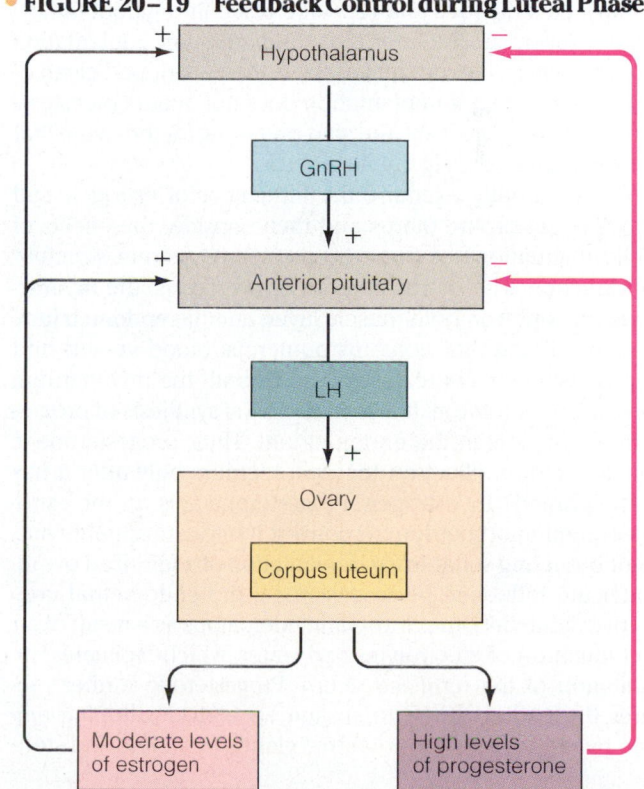

sible for degeneration of the corpus luteum are not fully understood. The declining level of circulating LH, driven down by inhibitory actions of progesterone, undoubtedly contributes to the corpus luteum's downfall. Prostaglandins and estrogen released by the luteal cells themselves may play a role. Demise of the corpus luteum terminates the luteal phase and sets the stage for a new follicular phase. As the corpus luteum degenerates, plasma progesterone and estrogen levels fall rapidly because these hormones are no longer being produced (Fig. 20–15). Withdrawal of the inhibitory effects of these hormones on the hypothalamus allows FSH and tonic LH secretion to modestly increase once again. Under the influence of these gonadotropic hormones, another batch of primary follicles are induced to mature as a new follicular phase begins.

The uterine changes that occur during the menstrual cycle reflect hormonal changes during the ovarian cycle.

The fluctuations in circulating levels of estrogen and progesterone that occur during the ovarian cycle induce profound changes in the uterus, giving rise to the **menstrual cycle**, or **uterine cycle**. Because it reflects hormonal changes that occur during the ovarian cycle, the menstrual cycle averages twenty-eight days, as does the ovarian cycle, although there is considerable variation from this mean even in normal adults. This variability is primarily a reflection of differing lengths of the follicular phase; the duration of the luteal phase is fairly constant. The outward manifestation of the cyclical changes that occur in the uterus is the menstrual bleeding that takes place once during each menstrual cycle (that is, once a month). Less obvious changes take place throughout the cycle, however, as the uterus is prepared for implantation should a released ovum be fertilized, then is stripped clean of its prepared lining if implantation does not occur (menstruation), only to repair itself and start preparing for the ovum that will be released during the next cycle.

We will briefly examine the influences of estrogen and progesterone on the uterus and then consider the effects of cyclic fluctuations of these hormones on uterine structure and function. The uterus consists of two layers: the **myometrium**, the outer smooth-muscle layer; and the **endometrium**, the inner lining that contains numerous blood vessels and glands. Estrogen stimulates growth of both the myometrium and the endometrium. It also induces the synthesis of progesterone receptors in the endometrium. Thus, progesterone is able to exert an effect on the endometrium only after it has been "primed" by estrogen. Progesterone acts on the estrogen-primed endometrium to convert it into a hospitable and nutritious lining suitable for implantation of a fertilized ovum. Under the influence of progesterone, the endometrial connective tissue becomes loose and edematous as a result of an accumulation of electrolytes and water, which facilitates implantation of the fertilized ovum. Progesterone further prepares the endometrium to sustain an early-developing embryo by inducing the endometrial glands to secrete and store large quantities of glycogen and by causing tremendous growth of the endometrial blood vessels. Progesterone also reduces the contractility of the uterus to provide a quiet environment for implantation and embryonic growth.

The menstrual cycle consists of three phases: the *menstrual phase*, the *proliferative phase,* and the *secretory* or *progestational phase*. The **menstrual phase** is the most overt phase, being characterized by discharge of blood and endometrial debris from the vagina. By convention, the first day of menstruation is considered to be the start of a new cycle. It coincides with termination of the ovarian luteal phase and onset of the follicular phase. As the corpus luteum degenerates because fertilization and implantation of the ovum released during the preceding cycle did not take place, circulating levels of estrogen and progesterone drop precipitously. Since the net effect of estrogen and progesterone is preparation of the endometrium for implantation of a fertilized ovum, withdrawal of these steroids deprives the highly vascular, nutrient-rich uterine lining of its hormonal support. The fall in ovarian hormone levels also stimulates release of a uterine prostaglandin that causes vasoconstriction of the endometrial vessels, thus disrupting the blood supply to the endometrium. The subsequent reduction in O_2 delivery causes death of the endometrium, including its blood vessels. The resulting bleeding through the disintegrating vessels flushes the dying endometrial tissue into the uterine lumen. The entire uterine lining sloughs during each menstrual period except for a deep, thin layer of epithelial cells and glands from which the endometrium will regenerate. The same local uterine prostaglandin also stimulates mild rhythmic contractions of the uterine myometrium. These contractions help expel the blood and endometrial debris from the uterine cavity out through the vagina as **menstrual flow**. Excessive uterine contractions caused by the overproduction of prostaglandin are responsible for the menstrual cramps (**dysmenorrhea**) experienced by some women.

The average blood loss during a single menstrual period is 50 to 150 ml. Blood that seeps slowly through the degenerating endometrium clots within the uterine cavity, then is acted on by fibrinolysin, a fibrin dissolver that breaks down the fibrin that forms the meshwork of the clot. Therefore, blood in the menstrual flow usually does not clot because it has already clotted and the clot has been dissolved before it passes out of the vagina. When blood flows rapidly through the leaking vessels, however, it may not be exposed to sufficient fibrinolysin, so blood clots may appear when the menstrual flow is most profuse. In addition to the blood and endometrial debris, large numbers of leukocytes are found in the menstrual flow. These white blood cells play an important defense role in helping the raw endometrium resist infection.

Menstruation typically lasts for about five to seven days after degeneration of the corpus luteum, coinciding in time with the early portion of the ovarian follicular phase (Fig. 20–15). Withdrawal of estrogen and progesterone upon degeneration of the corpus luteum leads simultaneously to sloughing of the endometrium (menstruation) and development of new follicles in the ovary under the influence of rising gonadotropic hormone levels. The drop in gonadal hormone secretion removes inhibitory influences from the hypothalamus and anterior pituitary, so FSH and LH secretion rise and a new follicular phase begins. After five to seven days under the influence of FSH and LH, the newly growing follicles are secreting sufficient quantities of estrogen to induce repair and growth of the endometrium.

Thus, menstrual flow ceases, and the **proliferative phase** of the uterine cycle begins concurrent with the last portion of the ovarian follicular phase as the endometrium starts to repair itself and proliferate under the influence of estrogen from the newly growing follicles. When the menstrual flow ceases, a thin endometrial layer less than 1 mm thick remains. Estrogen stimulates proliferation of epithelial cells, glands, and blood vessels in the endometrium, increasing this lining to a thickness of 3 to 5 mm. The estrogen-dominant proliferative phase lasts from the end of menstruation to ovulation. Peak estrogen levels trigger the LH surge responsible for ovulation.

Following ovulation, when a new corpus luteum is formed, the uterus enters the **secretory**, or **progestational**, **phase**, which coincides in time with the ovarian luteal phase. The corpus luteum secretes large amounts of progesterone and estrogen. Progesterone acts on the thickened, estrogen-primed endometrium to convert it to a richly vascularized, glycogen-filled tissue. This period is called either the secretory phase, because the endometrial glands are actively secreting glycogen, or the progestational ("before pregnancy") phase, in reference to the development of a lush endometrial lining capable of supporting an early embryo. If fertilization and implantation do not occur, the corpus luteum degenerates and a new follicular phase and menstruation phase begin once again. (For the effects of exercise, see the accompanying boxed feature, A Closer Look at Exercise Physiology.)

Fluctuating estrogen and progesterone levels produce cyclical changes in cervical mucus.

Hormonally induced changes also take place in the cervix during the ovarian cycle. Under the influence of estrogen during the follicular phase, the mucus secreted by the cervix is abundant, clear, and thin. This change, which is most pronounced when estrogen is at its peak and ovulation is approaching, facilitates passage of sperm through the cervical canal. After ovulation, under the influence of progesterone from the corpus luteum, the mucus becomes thick and sticky, essentially forming a plug across the cervical opening. This plug constitutes an important defense mechanism by preventing entry from the vagina into the uterus of bacteria that might threaten a pregnancy should conception have occurred. Sperm also cannot penetrate this thick mucus barrier.

Pubertal changes in females are similar to those in males, but menopausal changes are unique to females.

Regular menstrual cycles are absent in both young and aging females, but for different reasons.

Pubertal Changes The female reproductive system does not become active until puberty. Unlike the fetal testes, the fetal ovaries do not need to be functional, because feminization of the female reproductive system automatically takes place in the absence of fetal testosterone secretion without the presence of female sex hormones. The female reproductive system remains quiescent from birth until puberty, which occurs at about eleven years of age, because hypothalamic GnRH is actively suppressed by mechanisms that are similar to those in the prepubertal male. As in the male, removal of these inhibitory influences by mechanisms that are not clearly understood is responsible for the onset of puberty.

The resultant secretion of estrogen by the activated ovaries induces growth and maturation of the female reproductive tract as well as development of the female secondary sexual characteristics. Estrogen's prominent action in the latter regard is to promote fat deposition in strategic locations such as the breasts, buttocks, and thighs, giving rise to the typical curvaceous female figure. Enlargement of the breasts at puberty is due primarily to fat deposition in the breast tissue and not to functional development of the mammary glands. Three other pubertal changes in females—growth of axillary and pubic hair, the pubertal growth spurt, and development of libido—are attributable to a spurt in adrenal androgen secretion at puberty, not to estrogen. The pubertal rise in estrogen does close the epiphyseal plates, however, halting further growth in height, similar to the effect of testosterone in males.

Menopausal Changes The cessation of a woman's menstrual cycles at menopause sometime between the age of forty-five and fifty-five years is attributable to the limited supply of ovarian follicles present at birth. Once this reservoir is depleted, ovarian cycles, and hence menstrual cycles, cease. Thus, the termination of reproductive potential in a middle-aged woman is "preprogrammed" at her own birth. Evolutionarily, menopause may have developed as a mechanism to prevent pregnancy in women beyond the time that they could likely rear a child before their own death.

Menopause is preceded by a period of progressive ovarian failure characterized by increasingly irregular cycles, dwindling estrogen levels, and a host of physical and emotional changes. This entire period of transition from sexual maturity to cessation of reproductive capability is known as the **climacteric**. The absence of ovarian estrogen is responsible for the physical postmenopausal changes that occur, such as vaginal dryness, which can cause discomfort during sex, and gradual atrophy of the genital organs. However, postmenopausal women still have a sex drive because of their adrenal androgens. It is unclear whether the emotional symptoms associated with waning ovarian function, such as depression and irritability, are caused by estrogen withdrawal or are psychological reactions to the implications of menopause.

Males do not experience a similar complete gonadal failure for two reasons. First, their germ cell supply is unlimited because of continuing mitotic activity of the spermatogonia. Second, gonadal hormone secretion in males is not inextricably dependent on gametogenesis, as it is in females. If fe-

MENSTRUAL IRREGULARITIES: WHEN CYCLISTS AND OTHER FEMALE ATHLETES DO NOT CYCLE

Since the 1970s, as women in growing numbers have participated in a variety of sports requiring vigorous training regimens, researchers have become increasingly aware that many women experience changes in their menstrual cycles as a result of athletic participation. These changes are referred to as *athletic menstrual cycle irregularity* (*AMI*). The menstrual cycle dysfunction can vary in severity from amenorrhea (cessation of menstrual periods) to oligomenorrhea (cycles at irregular or infrequent intervals) to cycles that are normal in length but are anovulatory (no ovulation) or that have a short or inadequate luteal phase.

In early research studies using surveys and questionnaires to determine the prevalence of the problem, the frequency of these sport-related disorders varied from 2% to 51%. In contrast, the rate of occurrence of menstrual cycle dysfunction in females of reproductive age in the general population is 2% to 5%. A major problem of determining the frequency of menstrual cycle irregularity by survey is the questionable accuracy of recall of menstrual periods. Furthermore, without blood tests to determine hormone levels throughout the cycle, a woman would not know whether she was anovulatory or had a shortened luteal phase. Studies in which hormone levels have been determined throughout the menstrual cycle have demonstrated that seemingly normal cycles in athletes frequently have a short luteal phase (less than ten days long with low progesterone levels).

In a study conducted to determine whether strenuous exercise spanning two menstrual cycles would induce menstrual disorders, twenty-eight initially untrained college women with documented ovulation and luteal adequacy served as subjects. The women participated in an eight-week exercise program in which they initially ran four miles per day and progressed to ten miles per day by the fifth week. They were expected to participate daily in three and one-half hours of moderate-intensity sports. Only four women had normal menstrual cycles during the training. Abnormalities that occurred as a result of training included abnormal bleeding, delayed menstrual periods, abnormal luteal function, and loss of LH surge. All women returned to normal cycles within six months after training. The results of this study suggest that the frequency of AMI with strenuous exercise may be much greater than indicated by questionnaire alone. In other studies using low-intensity exercise regimens, AMI was much less frequent.

The mechanisms of AMI are unknown at present, although studies have implicated rapid weight loss, decreased percentage of body fat, dietary insufficiencies, prior menstrual dysfunction, stress, age at onset of training, and the intensity of training as factors that play a role. Epidemiologists have shown that if girls participate in vigorous sports before **menarche** (the first menstrual period), menarche is delayed. On the average, athletes have their first period when they are about three years older than their nonathletic counterparts. Furthermore, females who participate in sports before menarche seem to have a higher frequency of AMI throughout their athletic careers than those who begin to train after menarche. Hormonal changes that have been found in female athletes include (1) severely depressed FSH levels, (2) elevated LH levels, (3) low progesterone during the luteal phase, (4) low estrogen levels in the follicular phase, and (5) an FSH/LH environment totally unbalanced as compared to age-matched nonathletic women. The preponderance of evidence indicates that cycles return to normal once vigorous training is stopped.

The major problem associated with athletic amenorrhea is a reduction in bone-mineral density. Studies have shown that the mineral density in the vertebrae of the lower spine of those with athletic amenorrhea is lower than in athletes with normal menstrual cycles and lower than in age-matched nonathletes. However, amenorrheic runners have higher bone-mineral density than amenorrheic nonathletes, presumably because the mechanical stimulus of exercise helps to retard bone loss. Studies have shown that amenorrheic athletes are at higher risk for stress-related fractures than athletes with normal menstrual cycles. One study, for example, found stress fractures in six of eleven amenorrheic runners but in only one of six runners with normal menstrual cycles. The mechanism for bone loss is probably the same as is found in postmenopausal osteoporosis—lack of estrogen (see p. 690). The problem is serious enough that an amenorrheic athlete should discuss the possibility of estrogen replacement therapy with her physician.

There may be some positive benefits of athletes' menstrual dysfunction. A recent epidemiologic study to determine if the long-term reproductive and general health of women who had been college athletes differed from that of college nonathletes showed that former athletes had less than half the lifetime occurrence rate of cancers of the reproductive system and half the breast cancer occurrence compared to nonathletes. Because these are hormone-sensitive cancers, the delayed menarche and lower estrogen levels found in women athletes may play a key role in decreasing the risk of cancer of the reproductive system and breast.

male sex hormones were produced by separate tissues unrelated to those responsible for gametogenesis, as they are in the male, cessation of estrogen and progesterone secretion would not automatically accompany termination of oogenesis.

The oviduct is the site of fertilization.

You have now learned about the events that take place if fertilization does not occur. Because the primary function of the reproductive system is, of course, reproduction, we will next turn our attention to the sequence of events that ensue when this function is accomplished. (See the accompanying boxed feature, Concepts, Challenges, and Controversies.)

Fertilization, the union of male and female gametes, normally occurs in the **ampulla**, the upper third of the oviduct (Fig. 20–2). Thus, both the ovum and the sperm must be transported from their gonadal site of production to the ampulla (● Fig. 20–20).

Ovum Transport to the Oviduct At ovulation the ovum is released into the abdominal cavity, but it is quickly picked up by the oviduct. The dilated end of the oviduct cups around the ovary and contains **fimbriae**, fingerlike projections that contract in a sweeping motion to guide the released ovum into the oviduct. Furthermore, the fimbriae are lined by cilia, which are fine hairlike projections that beat in waves toward the interior of the oviduct, further assuring the ovum's passage into the oviduct (see p. 37). Within the oviduct, the ovum is rapidly propelled by peristaltic contractions and ciliary action to the ampulla.

Conception can take place during a very limited time span in each cycle (the **fertile period**). If not fertilized, the ovum begins to disintegrate within twenty-four hours and is subsequently phagocytized by cells that line the reproductive tract. Fertilization must therefore occur within twenty-four hours after ovulation, when the ovum is still viable. Sperm can survive about two days in the female reproductive tract, so sperm deposited within forty-eight hours before to twenty-four hours after ovulation may be able to fertilize the released ovum, although these times are subject to considerable variation.

Occasionally, an ovum fails to be transported into the oviduct and remains instead in the peritoneal cavity. Rarely, such an ovum gets fertilized, resulting in an **ectopic (abdominal) pregnancy**, in which the fertilized egg implants in the rich vascular supply to the digestive organs rather than in its usual site in the uterus. If this unusual pregnancy proceeds to term, the baby must be delivered surgically, because the normal vaginal exit is not available to it.

Sperm Transport to the Oviduct Once sperm are deposited in the vagina at ejaculation, they must travel through the cervical canal, through the uterus, and then up to the egg in the upper third of the oviduct (Fig. 20–20). The first sperm arrive in the oviduct within a half hour after ejaculation. Even though sperm are mobile by means of whiplike contractions of their tails, thirty minutes is much too soon for a sperm's own mobility to transport it to the site of fertilization. To accomplish this formidable journey, sperm need the help of the female reproductive tract. The first hurdle is passage through the cervical canal. The cervical mucus is too thick to permit sperm penetration throughout most of the cycle because of high progesterone or low estrogen levels. Only when estrogen levels are high, as occurs in the presence of a mature follicle about to ovulate, does the cervical mucus become thin and watery enough to permit sperm to penetrate. Sperm migrate up the cervical canal under their own power. The canal re-

● **FIGURE 20–20   Sperm and Ovum Transport to Site of Fertilization**

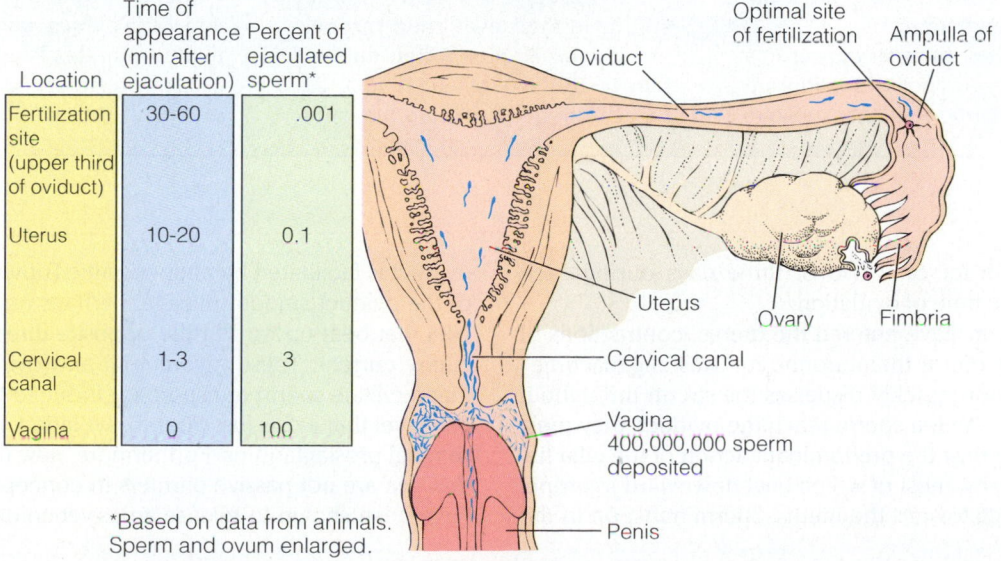

Location	Time of appearance (min after ejaculation)	Percent of ejaculated sperm*
Fertilization site (upper third of oviduct)	30-60	.001
Uterus	10-20	0.1
Cervical canal	1-3	3
Vagina	0	100

*Based on data from animals. Sperm and ovum enlarged.

REPRODUCTIVE-RELATED TECHNOLOGY: TAKING THE MAKING OF LIFE INTO OUR OWN HANDS

Implicit in our ever-increasing understanding of the mechanisms underlying physiological functions is our growing ability to correct or compensate for defective functions or to manipulate various functions to our advantage. Most advances in the area of biomedical technology have been welcomed as new ways to save, prolong, or enhance human lives. However, one area of new technological control over biological processes has aroused considerable moral, ethical, and legal controversy—namely, our growing power to determine the very existence of new individuals. Reproductive-related technological advances have opened up many new avenues for those desiring to control their potential progeny through artificial means. A brief listing of capabilities already available or under development will indicate how extensive our control over the future generation is:

- A variety of methods of *contraception* are available to avoid pregnancy by preventing the sperm and egg from joining or, once united, to prevent the fertilized egg from implanting in the uterus where it could develop into another human being.
- An unwanted pregnancy can be terminated by several different methods of *abortion* that remove the developing individual from its supportive uterine environment.
- An unborn child can be unobtrusively "viewed" in its mother's womb through *ultrasound* techniques. Furthermore, a sample of its cells can be extracted from the uterus through *amniocentesis* or *placental biopsy* and analyzed to determine the presence or absence of numerous genetic disorders. If defects are discovered, the parents have to make the difficult choice of terminating the pregnancy or knowingly bringing a child into the world who will be physically or mentally impaired.
- Sperm can be frozen and stored in *sperm banks* where a woman can "shop" for a "father" with particular traits for her future child; the woman can then undergo *artificial insemination* with the chosen sperm and have a normal pregnancy and birth without active participation or even knowledge of the male.
- Infertile women who fail to ovulate (do not release eggs) can be given *"fertility pills,"* which are hormones that promote ovulation.
- Infertile women who have a mechanical blockage in the pathway where the egg and sperm unite can still have their own child through *in vitro fertilization,* often dubbed the "test-tube baby" technique. After hormonally promoting multiple ovulation, the physician collects the eggs through a small incision in the woman's abdomen, then incubates them with sperm donated by the father-to-be. If fertilization takes place in this test-tube environment, the fertilized egg is placed in the mother's uterus, where it develops as would a naturally fertilized egg.
- In vitro fertilization often results in fertilization of more than one egg. Viable unused *embryos can be frozen* for future use.
- Because many externally fertilized eggs fail to implant when inserted in the uterus, several variations on in vitro fertilization have been developed. With *gamete intra-Fallopian transfer (GIFT),* unfertilized eggs and sperm are mixed and inserted into the oviduct (Fallopian tube). Thus, this procedure bypasses a mechanical blockage in the upper oviduct while permitting fertilization and implantation to proceed naturally. With *zygote intra-Fallopian transfer (ZIFT),* a zygote resulting from in vitro fertilization is placed in the oviduct rather than in the uterus so that it can leisurely descend before implanting as zygotes normally do. Using GIFT and ZIFT techniques, infertility clinics have reported implantation rates two to three times those of ordinary in vitro fertilization procedures.

mains penetrable for only two or three days during each cycle, around the time of ovulation.

Once the sperm have entered the uterus, contractions of the myometrium churn them around in "washing-machine" fashion. This action quickly disperses the sperm throughout the uterine cavity. When sperm reach the oviduct, they must travel upward against the predominant action of the cilia lining this passageway, most of which beat downward to propel the nonmotile egg toward the uterus. Sperm transport in the oviduct is facilitated by antiperistaltic (upward) contractions of the oviduct smooth muscle, perhaps assisted by tracts of cilia that beat upward in the opposite direction of the main ciliary current. These myometrial and oviduct contractions that facilitate sperm transport are induced by the high estrogen level that exists just prior to ovulation, perhaps aided by seminal prostaglandins. Furthermore, new research indicates that ova are not passive partners in conception. Mature eggs have been shown to release an as-yet-unidentified chemical

REPRODUCTIVE-RELATED TECHNOLOGY: TAKING THE MAKING OF LIFE INTO OUR OWN HANDS (continued)

- The sperm of some men are "lazy" or lack the necessary acrosomal enzymes to penetrate the formidable barriers surrounding an egg. Working under powerful magnification and using a thin needle, an infertility specialist can overcome this problem by inserting a single spermatozoon directly into an egg, a technique known as *micro-injection*.

- Another technique under development to facilitate sperm penetration of an egg is *zona drilling*. In this procedure, a tiny hole is drilled in the zona pellucida, the outer membrane surrounding an unfertilized egg. When the prepared egg is mixed with sperm, the hole serves as a ready-made passageway for entry of sperm. This new technology offers hope for couples with sperm insufficiencies or zona abnormalities.

- Although sperm and embryos can be successfully frozen and stored, unfertilized eggs have not been able to withstand similar processing. However, researchers have accomplished a major step toward eventual success in *egg freezing*. When this technique is perfected, a young woman facing the loss of ovarian function because of surgery or chemotherapy for cancer could opt to collect and store eggs for future use before the ovary-destroying procedure is

performed. A more controversial use of the egg-freezing technique might be its employment by women who wish to postpone childrearing for professional or other reasons. A women's fertility declines sharply and the risk of having children with genetic abnormalities such as Down's syndrome rises after age forty. Evidence indicates that the age of the eggs, not the reproductive tract, is responsible for the decline in fertility and rise in fetal abnormalities. Accordingly, if egg freezing were available, a women could collect healthy eggs during her most fertile years and put them on ice for later use.

- It is technically possible, although legally questionable, for a couple unable to have children because of the woman's inability to bear a child to "rent" another woman's uterus to carry their child. The *surrogate mother* can be artificially inseminated with the father's sperm, or an in vitro fertilized egg from the couple can be introduced into the "rented" uterus.

- In vitro fertilization techniques and *genetic engineering* make possible the screening of early embryos for potential genetic defects before implantation. Many genetic abnormalities, such as sickle-cell anemia and cystic fibrosis, can be detected in the genetic code from

the moment of fertilization. Because an early embryo can tolerate the loss of a cell without impairing normal development, it is possible to remove a single cell from a sixteen-cell embryo and analyze it for genetic defects before implanting it in the female reproductive tract.

While these new technologies are being hailed as modern miracles, they are also making us tread on new moral, ethical, and legal ground. What rights do the partners who created a frozen embryo have, for example, or what rights does the frozen embryo itself have? Because legislation has not kept pace with technological capabilities, many ambiguities are being decided through precedent-setting lawsuits. For example, a woman and her ex-husband recently engaged in a highly controversial legal battle over whether she could attempt to become pregnant after their divorce using frozen embryos that the couple had created while still married. The court ruled that she could not use the embryos against his will. In another case, the court ruled that frozen embryos could not inherit the estate left by a wealthy, childless couple who died in a plane crash, leaving the embryos as their only direct "heirs."

---

that attracts sperm and causes them to propel themselves toward the waiting female gamete.

Even around ovulation time when sperm can penetrate the cervical canal, of the several hundred million sperm deposited in a single ejaculate, only a few thousand make it to the oviduct (Fig. 20–20). The fact that only a very small percentage of the deposited sperm ever reach their destination is one reason why sperm concentration must be so high (>20 million/ml of semen) for a man to be fertile. The other rea-

son is that the acrosomal enzymes of many sperm are needed to break down the barriers surrounding the ovum (● Fig. 20–21).

Fertilization　The tail of the sperm is used to maneuver for final penetration of the ovum. To fertilize an ovum, a sperm must first pass through the corona radiata and zona pellucida surrounding it. The acrosomal enzymes, which are exposed as the acrosomal membrane disrupts on contact with the co-

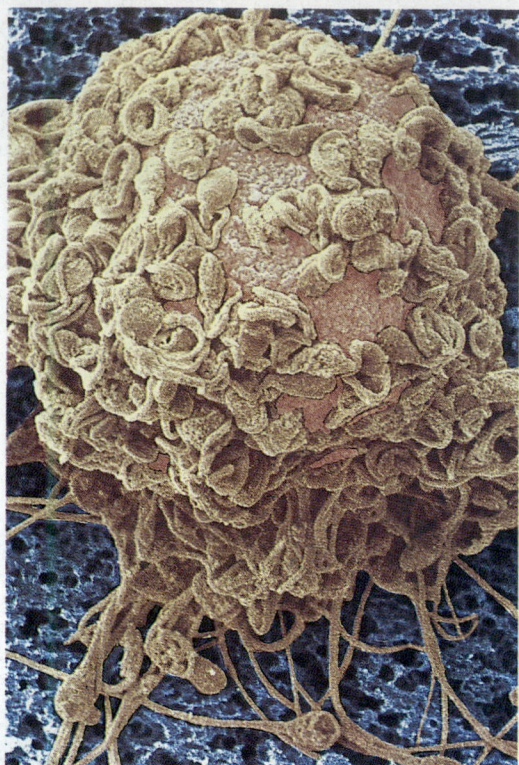

● FIGURE 20–21   Scanning Electron Micrograph of Sperm Amassed at Surface of Ovum

rona radiata, enable the sperm to tunnel a path through these protective barriers (● Fig. 20–22). Sperm are able to penetrate the zona pellucida only after binding with specific receptor sites on the surface of this layer. (Only sperm of the same species are able to bind to these receptors and pass through.) The first sperm to reach the ovum itself fuses with the plasma membrane of the ovum (actually a secondary oocyte), triggering a chemical change in the ovum's surrounding membrane that makes this outer layer impenetrable to the entry of any more sperm. This phenomenon is known as **block to polyspermy** ("many sperm").

The head of the fused sperm is gradually pulled into the ovum's cytoplasm by a growing cone that engulfs it. The sperm's tail is frequently lost in this process, but it is the head that carries the crucial genetic information anyway. Penetration of the sperm into the cytoplasm triggers the final meiotic division of the secondary oocyte. Within an hour, the sperm and egg nuclei fuse. In addition to contributing its half of the chromosomes to the fertilized ovum, now called a **zygote**, the victorious sperm also activates ovum enzymes that are essential for the early embryonic developmental program.

The blastocyst implants in the endometrium through the action of its trophoblastic enzymes.

During the first three to four days following fertilization, the zygote remains within the ampulla, because a constriction between the ampulla and the remainder of the oviduct canal prevents further movement of the zygote toward the uterus. The zygote is not idle during this time, however. It is rapidly undergoing a number of mitotic cell divisions to form a solid ball of cells called the **morula** (● Fig. 20–23). Meanwhile, the rising levels of progesterone from the newly developing corpus luteum that formed after ovulation stimulate release of glycogen from the endometrium into the reproductive-tract lumen for use as energy by the early embryo. The nutrients stored in the cytoplasm of the ovum can sustain the product of conception for less than a day. The concentration of secreted nutrients increases more rapidly in the small confines of the ampulla than in the uterine lumen. About three to four days after ovulation, progesterone is being produced in sufficient quantities to relax the oviduct constriction, thus permitting the morula to be rapidly propelled into the uterus by oviductal peristaltic contractions and ciliary activity. The temporary delay before the developing embryo passes into the uterus allows sufficient nutrients to accumulate in the uterine lumen to support the embryo until implantation can take place. If the morula arrives prematurely, it dies.

When the morula descends to the uterus, it floats freely within the uterine cavity for another three to four days, living on the endometrial secretions and continuing to divide. During the first six to seven days after ovulation, while the developing embryo is in transit in the oviduct and floating in the uterine lumen, the uterine lining is simultaneously being prepared for implantation under the influence of luteal-phase progesterone. During this time, the uterus is in its secretory or progestational phase, storing up glycogen and becoming richly vascularized.

Occasionally, the morula fails to descend into the uterus and continues instead to develop and implant in the lining of the oviduct. This leads to a **tubal pregnancy**, which must be terminated. Such a pregnancy can never succeed, because the oviduct cannot expand as the uterus does to accommodate the growing embryo. The first warning of a tubal pregnancy is pain caused by stretching of the oviduct by the growing embryo. If not removed, the enlarging embryo will rupture the oviduct, causing possibly lethal hemorrhage.

In the usual case, by the time the endometrium is suitable for implantation (about a week after ovulation), the morula has descended to the uterus and continued to proliferate and differentiate into a **blastocyst** capable of implantation (Fig. 20–23). Thus, the week's delay after fertilization and before implantation takes place allows time for both the endometrium and the developing embryo to be prepared for implantation.

A **blastocyst** is a single-layered sphere of cells encircling a fluid-filled cavity, with a dense mass of cells grouped together at one side (● Fig. 20–24). This dense mass, called the **inner cell mass**, is destined to become the fetus itself. The remainder of the blastocyst will never be incorporated into the fetus but will serve a supportive role during intrauterine life. The thin outermost layer, the **trophoblast**, is responsible for accomplishing implantation, after which it develops into the fetal portion of the placenta. The fluid-filled cavity, the **blastocoele**, will become the **amniotic sac**, which surrounds and cushions the fetus throughout gestation.

Corona radiata

Cytoplasm of ovum

Spermatozoa

Nucleus of ovum undergoing second meiotic division

First polar body

Zona pellucida

Plasma membrane of ovum

Enzyme-filled acrosome

Spermatozoon head bearing sperm's nucleus

Path tunneled through barriers surrounding ovum by acrosomal enzymes exposed upon disruption of acrosomal membrane

Spermatozoon that has accomplished fertilization

(a)

(b)

● **FIGURE 20–22   Process of Fertilization**   (a) Schematic representation of sperm tunneling the barriers surrounding the ovum. (b) Scanning electron micrograph of a spermatozoan in which the acrosomal membrane has been disrupted and the acrosomal enzymes (in red) are exposed.

When the blastocyst is ready to implant, its surface becomes sticky. By this time the endometrium is ready to accept the early embryo. The blastocyst adheres to the uterine lining on its inner-cell-mass side (● Fig. 20–25a). **Implantation** begins when the trophoblastic cells overlying the inner cell mass release protein-digesting enzymes upon contact with the endometrium. These enzymes digest pathways between the endometrial cells, thus permitting fingerlike cords of trophoblastic cells to penetrate into the depths of the endometrium, where they continue to digest uterine cells (Fig. 20–25b). Through its cannabilistic actions, the trophoblast performs the dual functions of (1) accomplishing implantation as it carves out a hole in the endometrium for the blastocyst and (2) making metabolic fuel and raw materials available for the developing embryo as the advancing trophoblastic projections break down the nutrient-rich endometrial tissue.

Stimulated by the invading trophoblast, the endometrial tissue at the site of contact undergoes dramatic changes that enhance its ability to support the implanting embryo. In response to a chemical messenger released by the blastocyst, the underlying endometrial cells secrete prostaglandins, which act to locally increase vascularization, produce edema, and enhance nutrient storage. The endometrial tissue so modified at the implantation site is called the **decidua**. It is into this super-rich decidual tissue that the blastocyst becomes embedded. After the blastocyst burrows into the decidua by means of trophoblastic activity, a layer of endometrial cells covers over the surface of the hole, completely burying the blastocyst within the uterine lining (Fig. 20–25c). The trophoblastic layer continues to digest the surrounding decidual cells, providing energy for the embryo until the placenta develops.

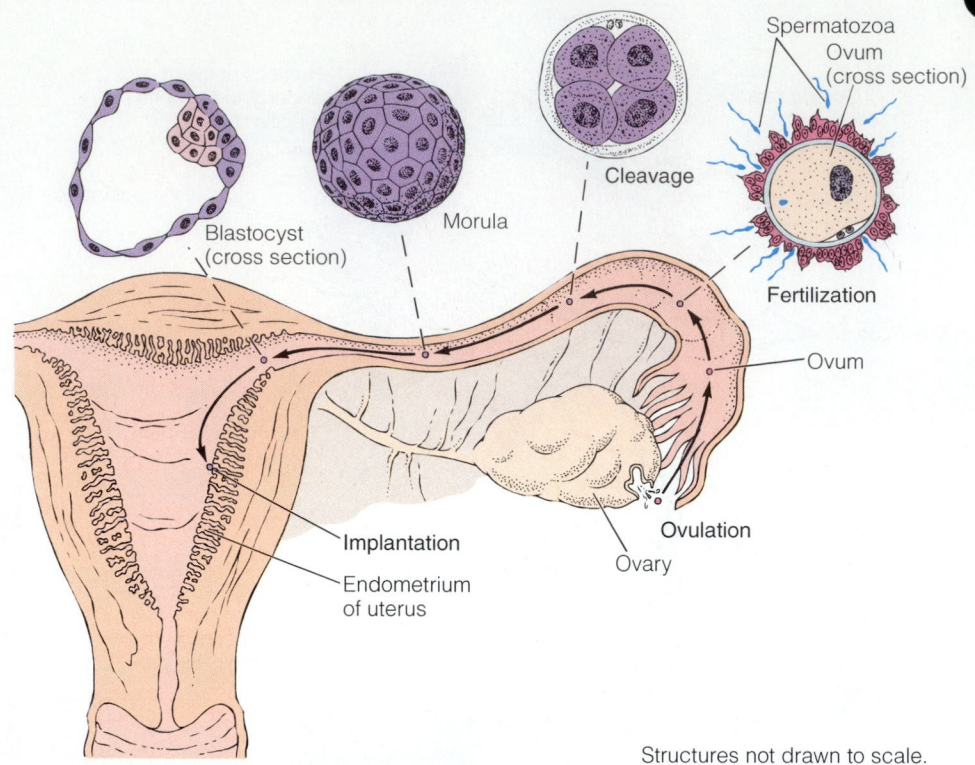

● **FIGURE 20–23 Early Stages of Development from Fertilization to Implantation** Note that the fertilized ovum progressively divides and differentiates into a blastocyst as it moves from the site of fertilization in the upper oviduct to the site of implantation in the uterus.

Spermatozoa
Ovum (cross section)
Cleavage
Morula
Blastocyst (cross section)
Fertilization
Ovum
Implantation
Endometrium of uterus
Ovulation
Ovary

Structures not drawn to scale.

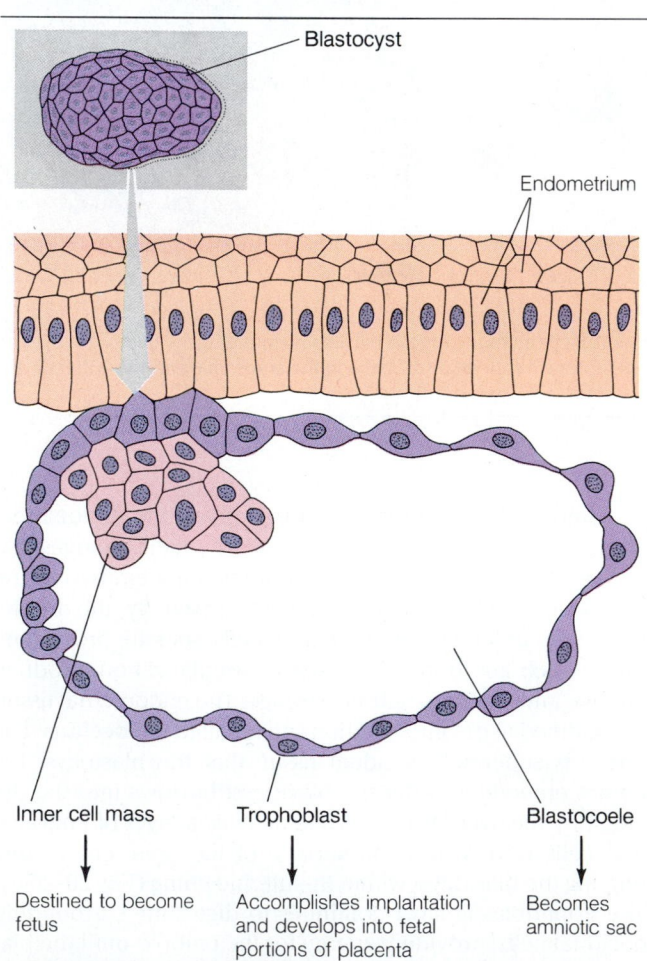

Blastocyst
Endometrium
Inner cell mass
Trophoblast
Blastocoele

Inner cell mass	Trophoblast	Blastocoele
Destined to become fetus	Accomplishes implantation and develops into fetal portions of placenta	Becomes amniotic sac

● **FIGURE 20–24 Blastocyst Adhering to Endometrial Lining and About to Implant**

The placenta is the organ of exchange between maternal and fetal blood.

The glycogen stores in the endometrium are only sufficient to nourish the embryo during its first few weeks. To sustain the growing embryo/fetus for the duration of its intrauterine life, the **placenta**, a specialized organ of exchange between the maternal and fetal blood, is rapidly developed (● Fig. 20–26). The placenta is derived from both trophoblastic and decidual tissue.

By day twelve, the embryo is completely embedded in the decidua. By this time the trophoblastic layer is two cell-layers thick and is called the **chorion**. As the chorion continues to release enzymes and expand, it forms an extensive network of cavities within the decidua. As decidual capillary walls are eroded by the expanding chorion, these cavities fill with maternal blood, which is prevented from clotting by an anticoagulant produced by the chorion. Fingerlike projections of chorionic tissue extend into the pools of maternal blood. Soon the developing embryo sends out capillaries into these chorionic projections to form **placental villi**. Some villi extend completely across the blood-filled spaces to anchor the fetal portion of the placenta to the endometrial tissue, but most simply project into the pool of maternal blood.

Each placental villus contains embryonic (later fetal) capillaries surrounded by a thin layer of chorionic tissue, which separates the embryonic/fetal blood from the maternal blood in the intervillus spaces. There is no actual mingling of maternal and fetal blood, but the barrier between them is extremely thin. To visualize this relationship, think of your hands (the fetal capillary blood vessels) in rubber gloves (the chorionic tissue) immersed in water (the pool of maternal blood). Only the rubber gloves separate your hands from the water. In the

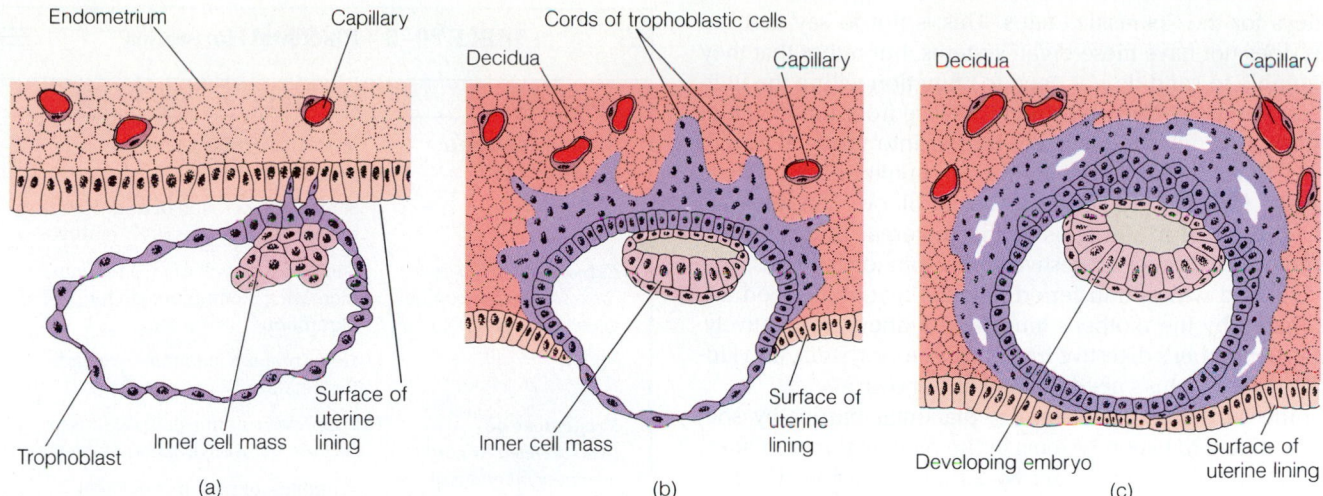

Endometrium   Capillary   Cords of trophoblastic cells

                              Decidua   Capillary   Decidua   Capillary

                    Surface of   Surface of
                    uterine   uterine
Trophoblast   lining   lining
         Inner cell mass   Inner cell mass   Developing embryo   Surface of
                                                                                                  uterine lining
         (a)   (b)   (c)

• **FIGURE 20–25   Implantation of the Blastocyst** (a) Free-floating blastocyst adheres to the endometrial lining. (b) Cords of trophoblastic cells tunnel into the endometrium, carving out a hole for the blastocyst. The boundaries between the cells in the advancing trophoblastic tissue disintegrate. (c) When implantation is finished, the blastocyst is completely buried in the endometrium.

same way, only the thin chorionic tissue (plus the capillary wall of the fetal vessels) separates the fetal and maternal blood. It is across this extremely thin barrier that all materials are exchanged between these two bloodstreams. This entire system of interlocking maternal (decidual) and fetal (chorionic) structures makes up the placenta.

Even though not fully developed, the placenta is well established and operational by five weeks after implantation. By this time, the heart of the developing embryo is pumping blood into the placental villi as well as to the embryonic tissues. Throughout gestation, fetal blood continuously traverses between the placental villi and the circulatory system of the fetus by means of the **umbilical artery** and **umbilical vein**, which are wrapped within the **umbilical cord**, a lifeline between the fetus and placenta (Fig. 20–26). The maternal blood within the placenta is continuously replaced as fresh blood enters through the uterine arterioles, percolates through the intervillus spaces, where it exchanges substances with fetal blood in the surrounding villi, then exits through the uterine vein.

During intrauterine life, the placenta performs the functions of the digestive system, the respiratory system, and the

• **FIGURE 20–26   Placentation** Schematic representation of interlocking maternal and fetal structures that form the placenta.

Pool of maternal blood

Placental villus

Uterine decidual tissue

Maternal arteriole

Maternal venule

Fetal vessels

Chorionic tissue

Chorion   Amniotic sac   Umbilical cord   Placenta

kidneys for the "parasitic" fetus. This is not to say that the fetus does not have these organ systems, but rather that they are unable to (and do not need to) function within the uterine environment. Nutrients and O_2 diffuse from the maternal blood across the thin placental barrier into the fetal blood, while CO_2 and other metabolic wastes simultaneously diffuse from the fetal blood into the maternal blood. The nutrients and O_2 brought to the fetus in the maternal blood are acquired by the mother's digestive and respiratory systems, and the CO_2 and wastes transferred into the maternal blood are eliminated by the mother's lungs and kidneys, respectively. Thus, the mother's digestive tract, respiratory system, and kidneys serve the fetus's needs as well as her own.

Some substances traverse the placental barrier by special mediated-transport systems in the placental membranes, whereas others move across by simple diffusion. Unfortunately, many drugs, environmental pollutants, other chemical agents, and microorganisms in the mother's bloodstream are also able to cross the placental barrier, and some of them may be harmful to the developing fetus. Individuals born limbless as a result of exposure to thalidomide, a tranquilizer taken by pregnant women before this drug's devastating effects on the growing fetus were known, serve as a grim reminder of this fact. Similarly, newborns who have become "addicted" during gestation by their mother's use of an abusive drug such as heroin suffer withdrawal symptoms after birth. Even more common chemical agents such as aspirin, alcohol, and agents in cigarette smoke can reach the fetus and have adverse effects. Likewise, fetuses can acquire AIDS before birth if their mothers are infected with the virus. A pregnant woman should therefore be very cautious about potentially harmful exposures from any source.

In addition to serving as an organ of exchange, the placenta in some way acts as a protective barrier to prevent the embryo from being immunologically rejected by the mother. The embryo is actually a "foreigner" because it is half derived from genetically different paternal chromosomes. Furthermore, the placenta becomes a temporary endocrine organ during pregnancy, a topic to which we now turn our attention.

Hormones secreted by the placenta play a critical role in the maintenance of pregnancy.

The placenta has the remarkable capacity to secrete a number of peptide and steroid hormones essential for the maintenance of pregnancy. The most important are *human chorionic gonadotropin, estrogen,* and *progesterone* (Table 20–5). Serving as the major endocrine organ of pregnancy, the placenta is unique among endocrine tissues in two regards. First, it is a transient tissue. Second, secretion of its hormones is not subject to extrinsic control, in contrast to the stringent, often complex mechanisms that regulate the secretion of other hormones. Instead, the type and rate of placental hormone secretion depends primarily on the stage of pregnancy.

One of the first events following implantation is secretion by the developing chorion of **human chorionic gonadotropin** (**hCG**), a peptide hormone that prolongs the life span of the

TABLE 20–5 Placental Hormones	
HORMONE	FUNCTION
Human chorionic gonadotropin	Maintains corpus luteum of pregnancy
	Stimulates secretion of testosterone by developing testes in XY embryos
Estrogen (also secreted by corpus luteum of pregnancy)	Stimulates growth of myometrium, increasing uterine strength for parturition
	Helps prepare mammmary glands for lactation
Progesterone (also secreted by corpus luteum of pregnancy)	Suppresses uterine contractions to provide quiet environment for fetus
	Promotes formation of cervical mucus plug to prevent uterine contamination
	Helps prepare mammary glands for lactation
Human chorionic somatomammotropin	Helps prepare mammary glands for lactation
	Believed to reduce maternal utilization of glucose so that greater quantities of glucose may be shunted to the fetus
Relaxin (also secreted by corpus luteum of pregnancy)	Softens cervix in preparation for cervical dilation at parturition
	Loosens connective tissue between pelvic bones in preparation for parturition

corpus luteum. Recall that during the ovarian cycle, the corpus luteum degenerates and the highly prepared, luteal-dependent uterine lining sloughs if fertilization and implantation do not occur. When fertilization does occur, the implanted blastocyst saves itself from being flushed out in menstrual flow by producing hCG. This hormone, which is functionally similar to LH, stimulates and maintains the corpus luteum so that it does not degenerate. Now called the **corpus luteum of pregnancy**, this ovarian endocrine unit grows even larger and produces increasingly greater amounts of estrogen and progesterone for an additional ten weeks until the placenta takes over secretion of these steroid hormones. As a result of the persistence of circulating estrogen and progesterone, the thick, pulpy endometrial tissue is maintained instead of sloughing. Accordingly, menstruation ceases during pregnancy.

Stimulation by hCG is necessary to maintain the corpus luteum of pregnancy because LH, which maintains the corpus luteum during the normal luteal phase of the uterine cycle, is suppressed through feedback inhibition by the high levels of progesterone. Suppression of the anterior pituitary hormones by the high progesterone levels also precludes further follicular maturation and ovulation for the duration of the pregnancy.

• **FIGURE 20–27   Secretion Rates of Placental Hormones**

The maintenance of a normal pregnancy depends on high concentrations of progesterone and estrogen. Thus, production of hCG is critical during the first trimester to maintain ovarian output of these hormones. In a male fetus, hCG also stimulates the precursor Leydig cells in the fetal testes to secrete testosterone, which masculinizes the developing reproductive tract.

The secretion rate of hCG increases rapidly during early pregnancy to save the corpus luteum from demise. Peak secretion of hCG occurs about sixty days after the end of the last menstrual period (• Fig. 20–27). By the tenth week of pregnancy, hCG output declines to a low rate of secretion that is maintained for the duration of gestation. The fall in hCG occurs at a time when the corpus luteum is no longer needed for its steroidal hormone output because the placenta has begun to secrete substantial quantities of estrogen and progesterone. The corpus luteum of pregnancy partially regresses as

hCG secretion dwindles, but it is not converted into scar tissue until after delivery of the baby.

Human chorionic gonadotropin is eliminated from the body in the urine, where its detection forms the basis of pregnancy diagnosis tests. This hormone can be detected in the urine as early as the first month of pregnancy, about two weeks after the first missed menstrual period. Since this is before the growing embryo can be detected by physical examination, the test permits early confirmation of pregnancy.

A frequent early clinical sign of pregnancy is **morning sickness**, a daily bout of nausea and vomiting that often occurs in the morning but can take place at any time of day. Since this condition usually appears shortly after implantation and coincides with the time of peak hCG production, it is speculated that this early placental hormone may trigger the symptoms, perhaps by acting on the chemoreceptor trigger zone in the vomiting center (see p. 560).

A logical question is why the developing placenta does not start producing estrogen and progesterone in the first place instead of secreting hCG, which in turn stimulates the corpus luteum to secrete these two critical hormones. The answer is that, for different reasons, the placenta is unable to produce sufficient quantities of either estrogen or progesterone in the first trimester of pregnancy. In the case of estrogen, the placenta does not have all of the enzymes needed for complete synthesis of this hormone. Estrogen synthesis requires a complex interaction between the placenta and the fetus (• Fig. 20–28). The placenta is able to convert the androgen hormone produced by the fetal adrenal cortex, dehydroepiandrosterone (DHEA), into estrogen. The placenta is unable to produce estrogen until the fetus has developed to the point that its adrenal cortex is secreting DHEA into the blood. The placenta extracts DHEA from the fetal blood and converts it into estrogen, which it then secretes into the maternal blood. Because both fetal and placental tissues participate in this biosynthetic pathway, estrogen is said to be produced by the **fetoplacental unit**. The primary estrogen synthesized by this unit is **estriol**, in contrast to the main estrogen product of the ovaries, estradiol. Consequently, measurement of estriol levels in the maternal urine can be used clinically to assess the viability of the fetus.

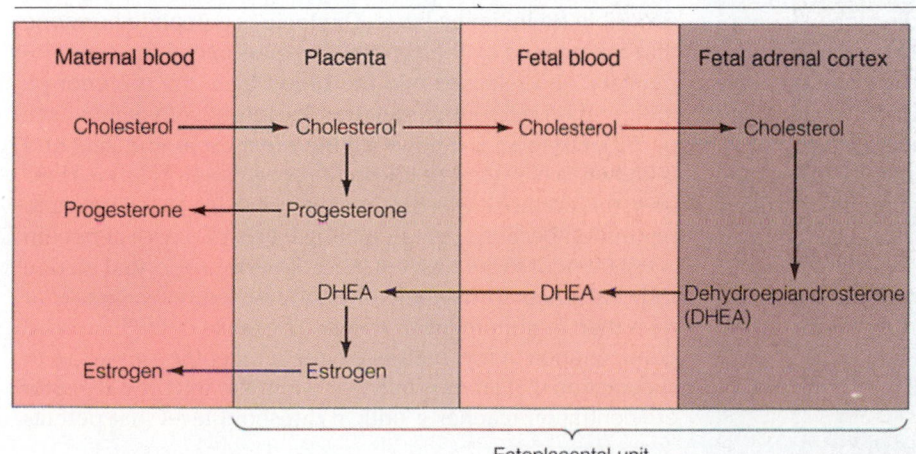

• **FIGURE 20–28   Secretion of Estrogen and Progesterone by the Placenta**   The placenta secretes increasing quantities of progesterone and estrogen into the maternal blood after the first trimester. The placenta itself has the ability to convert cholesterol into progesterone but lacks some of the enzymes necessary to convert cholesterol into estrogen. However, the placenta can convert DHEA derived from cholesterol in the fetal adrenal cortex into estrogen when DHEA reaches the placenta by means of the fetal blood.

In the case of progesterone, the placenta can synthesize this hormone soon after implantation. Even though the early placenta possesses the enzymes necessary to convert cholesterol extracted from the maternal blood into progesterone, it does not produce much of this hormone because the amount of progesterone produced is proportional to placental weight. The placenta is simply too small in the first ten weeks of pregnancy to produce sufficient quantities of progesterone to maintain the endometrial tissue. The notable increase in circulating progesterone in the last seven months of gestation reflects placental growth during this period.

Estrogen stimulates growth of the myometrium, which increases in size throughout pregnancy. The stronger uterine musculature is needed to expel the fetus during labor. Estriol also promotes development of the ducts within the mammary glands through which milk will be ejected during lactation. Progesterone performs various roles throughout pregnancy. Its primary function is to prevent miscarriage by suppressing contractions of the uterine myometrium. Progesterone also promotes formation of a mucus plug in the cervical canal to prevent vaginal contaminants from reaching the uterus. Finally, placental progesterone stimulates the development of milk glands in the breasts in preparation for lactation.

Maternal body systems respond to the increased demands of gestation.

The period of **gestation** (**pregnancy**) is about thirty-eight weeks from conception (forty weeks from the end of the last menstrual period). During gestation, the embryo/fetus continues to grow and develop to the point of being able to leave its maternal life-support system. Meanwhile, a number of physical changes take place within the mother to accommodate the demands of the pregnancy. The most obvious change is uterine enlargement. The uterus expands and increases in weight more than twenty times, exclusive of its contents. The breasts enlarge and develop the capability of producing milk. Body systems other than the reproductive system also make needed adjustments. The volume of blood increases by 30%, and the cardiovascular system responds to the increasing demands of the growing placental mass. The weight gain experienced during pregnancy is due only in part to the weight of the fetus. The remainder is primarily caused by the increased weight of the uterus, including the placenta, and the increased blood volume. Respiratory activity is increased by about 20% to handle the additional fetal requirements for O_2 utilization and CO_2 removal. Urinary output increases, and the kidneys excrete the additional wastes from the fetus. The increased metabolic demands of the growing fetus result in increased nutritional requirements for the mother. In general, the fetus takes what it needs from the mother, even if this leaves the mother with a nutritional deficit. For example, if the mother does not consume sufficient Ca^{++} in her diet, Ca^{++} will be mobilized from the maternal bones to assure adequate calcification of the fetal bones. (See the accompanying boxed feature, A Closer Look at Exercise Physiology.)

Parturition is accomplished by a positive-feedback cycle.

Parturition (**labor**, **delivery**, or **birth**) requires (1) dilation of the cervical canal to accommodate passage of the fetus from the uterus through the vagina and to the outside and (2) contractions of the uterine myometrium that are sufficiently strong to expel the fetus. During the first two trimesters of gestation, the uterus remains relatively quiet because of the inhibitory effect of the high levels of progesterone on the uterine musculature. During the last trimester, the uterus becomes progressively more excitable so that mild contractions (**Braxton-Hicks contractions**) are experienced with increasing strength and frequency. Sometimes these contractions become regular enough to be mistaken as the onset of labor, a phenomenon called "false labor."

Several other events take place near the end of pregnancy in preparation for parturition. The cervix begins to soften (or "ripen") as a result of the dissociation of its connective-tissue fibers. This cervical softening is believed to be caused by **relaxin**, a peptide hormone produced by the corpus luteum of pregnancy and by the placenta. Relaxin also "relaxes" the birth canal by loosening the connective tissue between the pelvic bones. Meanwhile, the fetus shifts downward (the baby "drops") and is normally oriented so that the head is in contact with the cervix in preparation for exiting through the birth canal. In a **breech birth**, any part of the body other than the head approaches the birth canal first.

Rhythmic, coordinated contractions, usually painless at first, begin at the onset of true labor. As labor progresses, the contractions occur with increasing frequency and intensity and are accompanied by increasing discomfort. These strong, rhythmic contractions force the fetus against the cervix, resulting in dilation of the cervix. Then, after having dilated the cervix sufficiently for passage of the fetus, these contractions force the fetus out through the birth canal.

The exact factors responsible for triggering this change in uterine contractility and thus initiating parturition are not clearly established, although endocrine factors are believed to be the most important. According to the leading proposal, a dramatic, progressive increase in the concentration of myometrial receptors for oxytocin, probably induced by the increasing levels of estrogen during pregnancy, is ultimately responsible for initiating labor. **Oxytocin** is a peptide hormone that is produced by the hypothalamus, stored in the posterior pituitary, and released into the blood from the posterior pituitary upon nervous stimulation by the hypothalamus (see p. 633). Oxytocin is a powerful uterine muscle stimulant and is known to play the key role in the progression of labor. However, this hormone was discounted as serving as the trigger to parturition because the circulating levels of oxytocin remain constant prior to the onset of labor. The discovery that uterine responsiveness to oxytocin is one hundred times greater at term than in nonpregnant women (because of the increased concentration of myometrial oxytocin receptors) gives rise to speculation that labor is initiated when the oxytocin-receptor concentration reaches a critical threshold level that permits

EXERCISE AND PREGNANCY: A COMPATIBLE OR CONTRAINDICATED COMBINATION?

Researchers have established that a regular exercise program is beneficial for optimal health, and a large number of young women participate in both competitive and noncompetitive exercise. Many of these women do not want to interrupt their exercise programs during pregnancy.

The questions most often asked about exercise during pregnancy are (1) How does exercise influence the course of pregnancy and the growing fetus, and (2) How does pregnancy alter the capacity to exercise?

One of the most serious concerns about exercise during pregnancy involves the intensity of the exercise and the resultant hyperthermia and redistribution of blood flow. Hyperthermia is not uncommon during high-intensity exercise (see p. 609), and body temperatures of 40°C have been recorded in long-distance runners. Pregnancy might compound this problem of elevated body temperature because the ratio of surface area to body mass becomes smaller due to the presence of a growing fetus. Thus, it might be more difficult to dissipate body heat. The fetus may also be an additional source of heat. However, reports indicate that peak core body temperatures reached during exercise decrease more rapidly in late pregnancy than in the nonpregnant state. These reports also indicate that the temperatures reached during exercise decrease more rapidly as pregnancy progresses, suggesting that the dissipation of body heat during pregnancy is somehow enhanced.

During exercise, blood flow is redistributed to the working muscles and to the skin (for cooling) at the expense of the uterine area. It is still unknown whether uterine blood flow during exercise is maintained at a level that is sufficient for the fetus. However, noninvasive ultrasound investigations found that intervillus blood flow in healthy pregnant women was not decreased by moderate exercise.

Several reports have indicated that a woman who engages in moderate exercise on a regular basis will have a better course of pregnancy than a sedentary woman. There is no evidence that moderate exercise results in harm to the fetus. The exercise should involve large muscle groups, be aerobic, and not require delicate balance. Swimming is suggested as an ideal exercise. The cool water keeps the internal temperature down, and the buoyancy provided by the water tends to offset gravity, preventing joint injury. This type of exercise also reduces the chance of the pregnant uterus being in a position to compress the descending aorta and disrupt blood flow to the uterus, as might be the case in supine exercise or stationary bicycling. Exercise in water may also decrease swelling in the lower legs.

It is recommended that exercise during pregnancy be moderate, with exercising heart rates kept below 160 beats per minute. New strenuous exercise programs should not be started during the first or third trimesters, but a gradual program may be started during the second trimester.

A proper exercise program during pregnancy may result in improved aerobic and muscular fitness, easier labor and recovery from labor, and an enhanced feeling of psychological well-being. Exercise generally can be resumed about eight weeks after delivery or as advised by a physician.

---

the onset of strong, coordinated contractions in response to ordinary levels of circulating oxytocin.

Furthermore, one new study suggests that the uterus itself synthesizes oxytocin. Investigators have noted a steady increase throughout gestation in activity of a gene in the uterus of pregnant rats that codes for oxytocin synthesis. This oxytocin-coding activity of the uterine gene peaks at 150 times normal levels just before the onset of labor. Therefore, the local level of oxytocin in the uterus is considerably higher than the blood concentration of oxytocin at the onset of labor. What triggers oxytocin production in the uterus and what role this local oxytocin plays in the onset of labor are presently unclear.

The idea that maternal factors determine the onset of labor has recently been challenged by the finding that in sheep at least the trigger for parturition comes not from the mother but from the fetus. When the paraventricular nucleus in a fetal sheep's hypothalamus reaches a critical level of maturation, it signals the anterior pituitary to increase ACTH secretion (see p. 659). ACTH, in turn, causes increased cortisol secretion by the adrenal cortex. Upon reaching the placenta by means of the fetal blood, fetal cortisol promotes the conversion of progesterone into estrogen, which is secreted into the maternal blood. Since progesterone inhibits uterine contractility and estrogen enhances it, this fetal-induced shift in circulating maternal hormones brings about the uterine contractions associated with the onset of labor. Whether this mechanism is also operable in humans remains to be determined. In fact, this and the oxytocin-receptor mechanism may both contribute to the onset of parturition in complementary fashion.

Once triggered by whatever mechanism, myometrial contractions progressively increase in frequency, strength, and duration throughout labor until the uterine contents are ex-

• FIGURE 20-29 Positive
Hormonal Feedback during
Parturition

Triggers onset
or parturition
(leading proposal)

Responsible
for progression
of parturition

pelled. At the beginning of labor, contractions lasting thirty seconds or less occur about every twenty-five to thirty minutes; by the end, they last sixty to ninety seconds and occur every two to three minutes.

Myometrial contractions incessantly increase as labor progresses because a positive-feedback cycle involving oxytocin and prostaglandin ensues (• Fig. 20–29). Each uterine contraction begins at the top of the uterus and sweeps downward, forcing the fetus toward the cervix. Pressure of the fetus against the cervix accomplishes two things. First, the fetal head pushing against the softened cervix acts as a wedge to dilate the cervical canal. Second, cervical stretch stimulates the release of oxytocin through a neuroendocrine reflex. Stimulation of receptors in the cervix in response to fetal pressure produces a neural signal that travels up the spinal cord to the hypothalamus, which in turn triggers oxytocin release from the posterior pituitary. This additional oxytocin promotes more powerful uterine contractions. As a result, the fetus is pushed more forcefully against the cervix, stimulating the release of even more oxytocin, and so on. This cycle is reinforced as oxytocin stimulates prostaglandin production by the decidua. As a powerful myometrial stimulant, prostaglandin further enhances uterine contractions. Oxytocin secretion, prostaglandin production, and uterine contractions continue to increase in positive-feedback fashion throughout labor until the pressure on the cervix is relieved by delivery.

Labor is divided into three stages: (1) cervical dilation, (2) delivery of the baby, and (3) delivery of the placenta (• Fig. 20–30). At the onset of labor or sometime during the first stage, the membranes surrounding the amniotic sac or "bag of waters" rupture. As the amniotic fluid escapes out of the vagina, it helps to lubricate the birth canal. During the first stage, the cervix is forced to dilate to accommodate the diameter of the baby's head, usually to a maximum of 10 cm. This stage is the longest, lasting from several hours to as long as twenty-four hours in a first pregnancy. If another part of the fetus's body other than the head is oriented against the cervix, it is generally less effective than the head as a wedge. The head has the largest diameter of the baby's body. If the baby approaches the birth canal feet first, the cervix may not be dilated sufficiently by the feet to permit passage of the head. Without medical intervention in such a case, the baby's head would remain stuck behind the too-narrow cervical opening.

The second stage of labor, the actual birth of the baby, begins once cervical dilation is complete. When the infant begins to move through the cervix and vagina, stretch receptors in the vagina activate a neural reflex that triggers contractions of the abdominal wall in synchrony with the uterine contractions. These abdominal contractions greatly increase the force pushing the baby through the birth canal. The mother can help deliver the infant by voluntarily contracting the abdominal muscles at this time in unison with each uterine contraction (that is, "push" with each "labor pain"). Stage two is usually much shorter than the first stage and lasts thirty to ninety minutes. The infant is still attached to the placenta by the umbilical cord at birth. The cord is tied and severed, with the stump shriveling up in a few days to form the **umbilicus** (**navel**).

Shortly after delivery of the baby, a second series of uterine contractions causes the placenta to separate from the myo-

Placenta   Urinary   Pubic
           bladder   bone

Urethra

Vagina

Cervix

Rectum

(a)

Partially dilated cervix

(b)   (c)   Placenta   Uterus   Umbilical cord

                                                                        (d)

● **FIGURE 20–30   Stages of Labor**   (a) Position of the fetus near the end of pregnancy. (b) First stage of labor: cervical dilation. (c) Second stage of labor: delivery of the baby. (d) Third stage of labor: delivery of the placenta.

metrium and be expelled through the vagina. Delivery of the placenta or **afterbirth** constitutes the third stage of labor, which is typically the shortest stage, being completed within fifteen to thirty minutes after the baby is born. After the placenta is expelled, continued contractions of the myometrium constrict the uterine blood vessels supplying the site of placental attachment to prevent hemorrhage.

After delivery, the uterus shrinks to its pregestational size, a process known as **involution**, which takes four to six weeks to complete. During involution, the remaining endometrial tissue that was not expelled with the placenta gradually disintegrates and sloughs off, producing a vaginal discharge called **lochia** that continues for three to six weeks following parturition. After this period, the endometrium is restored to its nonpregnant state.

Involution occurs largely because of the precipitous fall in circulating estrogen and progesterone when the placental source of these steroids is lost at delivery. The process is facilitated in mothers who breast-feed their infants because of the oxytocin released in response to suckling. In addition to playing an important role in lactation, this periodic nursing-induced postpartum release of oxytocin promotes myometrial contractions that help maintain uterine muscular tone, thus enhancing involution. Involution is usually complete in

about four weeks in nursing mothers but takes about six weeks in those who do not breast-feed.

Lactation requires multiple hormonal inputs.

Milk (or its equivalent) is essential for survival of the newborn. Accordingly, during gestation the **mammary glands**, or **breasts**, are prepared for **lactation (milk production)**.

The breasts in nonpregnant females are composed mostly of adipose tissue and a rudimentary duct system. The size of the breasts is determined by the amount of adipose tissue, which has nothing to do with the ability to produce milk. Under the hormonal environment present during pregnancy, the mammary glands develop the internal glandular structure and function necessary for milk production. A breast capable of lactating consists of a network of progressively smaller ducts that branch out from the nipple and terminate in lobules (● Fig. 20–31a). Each *lobule* is made up of a cluster of saclike epithelial-lined *alveoli* that constitute the milk-producing glands. Milk is synthesized by the epithelial cells, then secreted into the alveolar lumen, which is drained by a milk-collecting duct that transports the milk to the surface of the nipple (Fig. 20–31b).

During pregnancy, the high concentration of estrogen pro-

(a)

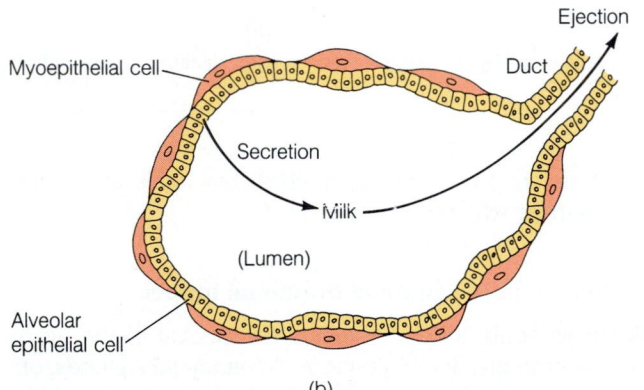

(b)

● **FIGURE 20–31   Mammary Gland Anatomy**   (a) Internal structure of the mammary gland, lateral view. (b) Schematic representation of the microscopic structure of an alveolus within the mammary gland. The alveolar epithelial cells secrete milk into the lumen. Contraction of the surrounding myoepithelial cells ejects the secreted milk out through the duct.

motes extensive duct development, whereas the high level of progesterone stimulates abundant alveolar-lobular formation. Elevated concentrations of **prolactin** (an anterior pituitary hormone stimulated by the rising levels of estrogen) and **human chorionic somatomammotropin** (a peptide hormone produced by the placenta) also contribute to mammary gland development by inducing the synthesis of enzymes needed for milk production.

Most of these changes in the breasts occur during the first

half of gestation, so the mammary glands are fully capable of producing milk by the middle of pregnancy. However, milk secretion does not occur until parturition. The high estrogen and progesterone concentrations during the last half of pregnancy prevent lactation by blocking prolactin's stimulatory action on milk secretion. Prolactin is the primary stimulant of milk secretion. Thus, even though the high levels of placental steroids induce the development of the milk-producing machinery in the breasts, they prevent these glands from becoming operational until the baby is born and milk is needed. The abrupt decline in estrogen and progesterone that occurs with loss of the placenta at parturition initiates lactation. (The functions of estrogen and progesterone during gestation and lactation as well as throughout the reproductive life of females are summarized in Table 20–6.)

Once milk production begins after delivery, two hormones are critical for maintaining lactation: (1) *prolactin,* which acts on the alveolar epithelium to promote secretion of milk, and (2) *oxytocin,* which produces **milk ejection**. The latter refers to the forced expulsion of milk from the lumen of the alveoli out through the ducts. Release of both of these hormones is stimulated by a neuroendocrine reflex triggered by suckling (● Fig. 20–32). Milk cannot be directly sucked out of the alveolar lumen by the infant. Instead, milk must be actively squeezed out of the alveoli into the ducts and hence toward the nipple by contraction of specialized **myoepithelial cells** (musclelike epithelial cells) that surround each alveolus (Fig. 20–31b). Suckling of the breast by the infant stimulates sensory nerve endings in the nipple, resulting in initiation of action potentials that travel up the spinal cord to the hypothalamus. Thus activated, the hypothalamus triggers a burst of oxytocin release from the posterior pituitary. Oxytocin, in turn, stimulates contraction of the myoepithelial cells in the breasts to bring about milk ejection or "milk letdown." Milk letdown continues only as long as the infant continues to nurse. In this way, the milk-ejection reflex assures that milk exits the breasts only when and in the amount needed by the baby. Even though the alveoli may be full of milk, the milk cannot be released without oxytocin. The reflex can become conditioned to stimuli other than suckling, however. For example, the infant's cry can trigger milk letdown, thus causing a spurt of milk to leak from the nipples. On the other hand, psychological stress, acting through the hypothalamus, can easily inhibit milk ejection. For this reason, a positive attitude toward breast-feeding as well as a relaxed environment are essential for successful breast-feeding.

Suckling not only triggers oxytocin release but also stimulates prolactin secretion. Prolactin output by the anterior pituitary is controlled by two hypothalamic secretions: **prolactin-inhibiting hormone** (**PIH**) and **prolactin-releasing hormone** (**PRH**). Throughout most of the female's life, PIH is the dominant influence, so prolactin is normally not secreted. During lactation, a burst in prolactin secretion occurs each time the infant suckles. Afferent impulses initiated in the nipple upon suckling are carried by the spinal cord to the hypothalamus. This reflex ultimately leads to prolactin release by the anterior pituitary, although it is unclear whether this outcome is accomplished by inhibition of PIH or stimulation of PRH secretion or both. Prolactin then acts on the alveolar epithelium to promote secretion of milk to replenish the milk lost from the alveoli by milk ejection (Fig. 20–32).

Concurrent stimulation by suckling of both milk ejection and milk production assures that the rate of milk synthesis keeps pace with the baby's needs for milk. The more the infant nurses, the more milk is removed by letdown and the more milk is produced.

In addition to prolactin, which is the most important factor controlling synthesis of milk, at least four other hormones are essential for their permissive role in milk production:

these are cortisol, insulin, parathyroid hormone, and growth hormone.

Milk is composed of water, triglyceride fat, the carbohydrate lactose (milk sugar), a number of proteins, vitamins, and the minerals calcium and phosphate. The composition of milk differs during the early part of lactation. **Colostrum**, the milk produced for the first five days postpartum, contains lower concentrations of fat and lactose but higher concentrations of proteins, most notably lactoferrin and immunoglobulins. These proteins provide early protection against infection until the newborn's own defense system develops. *Lactoferrin* has bactericidal activities, and *immunoglobulins* are antibodies. Furthermore, recent evidence suggests that breast milk may actually stimulate development of the newborn's own immune capabilities. Infants who are bottle-fed on a formula made from cow's milk do not have this protective advantage provided by colostrum and, accordingly, have a higher incidence of colds and other infections than breast-fed babies. Also, the digestive system of a newborn is better equipped to handle human milk than cow milk–derived formula, so bottle-fed babies tend to have more digestive upsets.

Breast-feeding is also advantageous for the mother. Oxytocin release triggered by nursing hastens uterine involution. In addition, suckling suppresses the menstrual cycle by inhibit-

TABLE 20–6 Actions of Estrogen and Progesterone

ESTROGEN

Effects on Sex-Specific Tissues

Essential for egg maturation and release

Stimulates growth and maintenance of entire female reproductive tract

Stimulates granulosa-cell proliferation, which leads to follicle maturation

Thins cervical mucus to permit sperm penetration

Enhances transport of sperm to oviduct by stimulating upward contractions of uterus and oviduct

Stimulates growth of endometrium and myometrium

Induces synthesis of progesterone receptors in endometrium and, during gestation, of myometrial receptors for oxytocin

Other Reproductive Effects

Promotes development of secondary sex characteristics

Controls GnRH and gonadotropin secretion

   Low levels inhibit secretion

   High levels responsible for triggering LH surge

Stimulates duct development in breasts during gestation

Inhibits milk-secreting actions of prolactin during gestation

Nonreproductive Effects

Promotes fat deposition

Closes epiphyseal plates

Reduces blood cholesterol (incidence of atherosclerosis lower in females until after menopause)

Exerts vascular effects (deficiency produces "hot flashes" at menopause)

PROGESTERONE

Prepares suitable environment for nourishment of developing embryo/fetus

Promotes formation of thick mucus plug in cervical canal

Inhibits hypothalamic GnRH and gonadotropin secretion

Stimulates alveolar development in breasts during gestation

Inhibits milk-secreting actions of prolactin during gestation

Inhibits uterine contractions during gestation

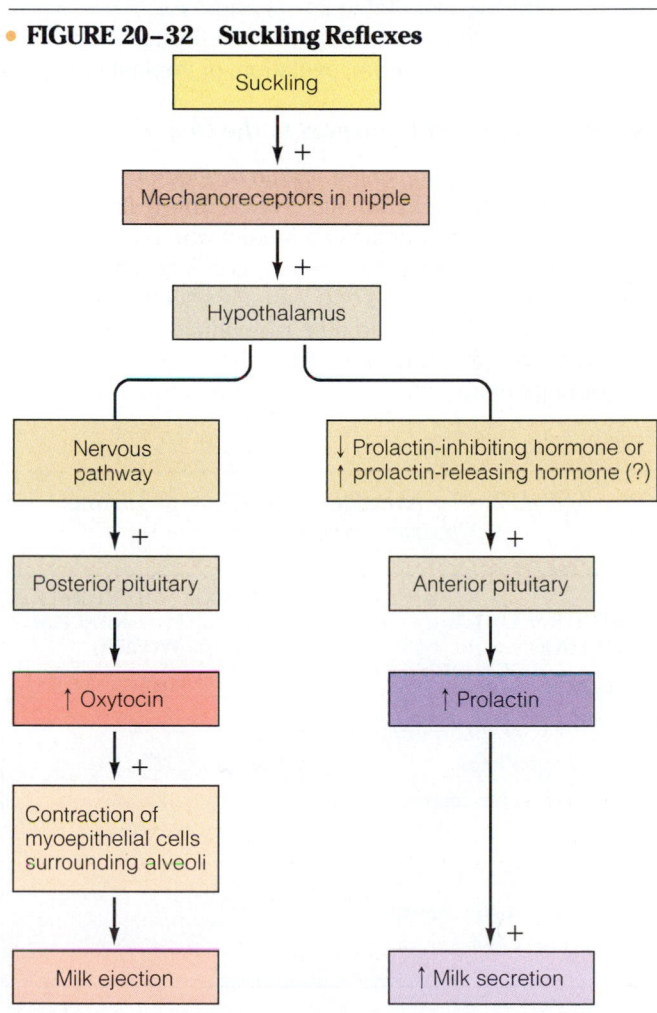

• **FIGURE 20–32 Suckling Reflexes**

ing LH and FSH secretion, probably through inhibition of GnRH. Lactation, therefore, prevents ovulation and serves as a means of preventing pregnancy (although it is not 100% effective as a means of contraception), thus permitting all the mother's resources to be directed toward the newborn instead of being shared with a new embryo.

When the infant is weaned, two mechanisms contribute to the cessation of milk production. First, without suckling, prolactin secretion is not stimulated, thus removing the primary stimulus for continued milk synthesis and secretion. Also because of the lack of suckling, milk letdown does not occur in the absence of oxytocin release. Since milk production does not immediately shut down, milk accumulates in the alveoli, causing engorgement of the breasts. The resultant pressure buildup acts directly on the alveolar epithelial cells to suppress further milk production. Cessation of lactation at weaning therefore results from a lack of suckling-induced stimulation of both prolactin and oxytocin secretion.

Contraceptive techniques act by blocking sperm transport, ovulation, or implantation.

Couples wishing to engage in sexual intercourse but avoid pregnancy have a number of methods of **contraception** ("against conception") available. These methods, which range in effectiveness (Table 20–7) and ease of use, act by blocking one of three major steps in the reproductive process: sperm transport to the ovum, ovulation, or implantation.

Blockage of Sperm Transport to the Ovum

■ *Natural contraception* or the *rhythm method* of birth control relies on abstinence from intercourse during the woman's fertile period. The woman can predict when ovulation is to occur based on keeping careful records of her menstrual cycles. This technique is only partially effective because of variability in cycles. The time of ovulation can be determined more precisely by recording body temperature each morning before getting up. Body temperature rises slightly about a day after ovulation has taken place. The temperature-rhythm method is not useful in determining when it is safe to engage in intercourse before ovulation, but it can be helpful in determining when it is safe to resume sex after ovulation. This technique is also employed by couples as a method of fertility. The rise in body temperature occurs during the time period that the ovulated ovum may still be viable and capable of being fertilized.

One group of scientists is working on developing a rapid, sensitive, at-home test using a few drops of urine to detect (1) a rise in estrogen, a "red-light" signal that ovulation will occur in about four days and intercourse should be avoided (if contraception is the goal) and (2) a rise in progesterone, a "green-light" signal that ovulation is past and sexual activity can be resumed.

■ *Coitus interruptus* involves withdrawal of the penis from the vagina before ejaculation occurs. This method is only moderately effective, however, because timing is difficult and some sperm may pass out of the urethra prior to ejaculation.

■ *Chemical contraceptives,* such as spermicidal ("sperm-killing") jellies, foams, creams, and suppositories, when inserted into the vagina are toxic to sperm for about an hour following application.

■ *Barrier methods* mechanically prevent sperm transport to the oviduct. For males, the condom is a thin, strong rubber or latex sheath placed over the erect penis prior to ejaculation to prevent sperm from entering the vagina. For females, the diaphragm is a flexible rubber dome that is inserted through the vagina and positioned over the cervix to block sperm entry into the cervical canal. It is held in position by lodging snugly against the vaginal wall. The diaphragm must be fitted by a trained professional and must be left in place for at least six hours but no longer than twenty-four hours after intercourse. Barrier methods are often used in conjunction with spermicidal agents for increased effectiveness. The cervical cap is a recently developed alternative to the diaphragm. Smaller than a diaphragm, the cervical cap, which is coated with a film of spermicide, cups over the cervix and is held in place by suction.

The contraceptive sponge, a nonprescription polyurethane device, once placed in the vagina next to the cervix, acts as a physical deterrent to sperm transport and also releases a spermicidal agent for up to twenty-four hours. The sponge has been implicated as the underlying cause of toxic shock syndrome in a small number of this product's users.

A condomlike contraceptive for females is the latest barrier method developed. It is an open-ended polyurethane (clear plastic) cylindrical pouch with a flexible ring at both ends. The ring at the closed end of the device is inserted into the vagina and fits over the cervix, similar to a diaphragm. The ring at the open end of the pouch is positioned outside of the vagina over the external genitalia.

■ *Sterilization,* which involves surgical disruption of either the ductus deferens (vasectomy) in men or the oviduct

TABLE 20–7 Average Failure Rate of Various Contraceptive Techniques	
CONTRACEPTIVE METHOD	**AVERAGE FAILURE RATE (ANNUAL PREGNANCIES/ 100 WOMEN)**
None	90
Natural (rhythm) methods	20–30
Coitus interruptus	23
Chemical contraceptives	20
Barrier methods	10–15
Oral contraceptives	2–2.5
Implanted contraceptives	1
Intrauterine device	4

(tubal ligation) in women, is considered to be a permanent method of preventing sperm and ovum from uniting.

Prevention of Ovulation

■ *Oral contraceptives* or *birth-control pills* prevent ovulation by suppressing gonadotropin secretion. These pills, which contain synthetic estrogen-like and progesterone-like steroids, are taken for three weeks, either in combination or in sequence, and then are withdrawn for one week. These steroids, just as the natural steroids produced during the ovarian cycle, inhibit GnRH and thus FSH and LH secretion. As a result, follicle maturation and ovulation do not take place so conception is impossible. The endometrium responds to the exogenous steroids by thickening and developing secretory capacity, just as it would to the natural hormones. When these synthetic steroids are withdrawn after three weeks, the endometrial lining sloughs and menstruation occurs, as it normally would upon degeneration of the corpus luteum. In addition to blocking ovulation, oral contraceptives also prevent pregnancy by increasing the viscosity of cervical mucus, which makes sperm penetration more difficult, and by decreasing muscular contractions in the female reproductive tract, which reduces sperm transport to the oviduct.

Oral contraceptives are available only by prescription. They have been shown to increase the risk of intravascular clotting, especially in women who also smoke tobacco.

■ A new approach to contraception is a *long-acting subcutaneous* ("under the skin") *implantation* of synthetic progesterone, which acts similarly to oral contraceptives by blocking ovulation and thickening the cervical mucus to prevent sperm transport. Unlike oral contraceptives, however, which must be taken on a regular basis, this new contraceptive, once implanted, is effective for five years. Six matchstick-sized capsules containing the steroid are inserted under the skin in the inner arm above the elbow. The capsules slowly release the synthetic progesterone at a nearly steady rate for five years, thus sustaining their contraceptive effect for a prolonged period.

Blockage of Implantation

■ Blockage of implantation is most commonly accomplished by insertion of a small *intrauterine device (IUD)* into the uterus by a physician. The IUD's mechanism of action is not completely understood, although most evidence suggests that the presence of this foreign object in the uterus induces a local inflammatory response that prevents implantation of a fertilized ovum. Although the IUD is a convenient birth-control method because it does not require ongoing attention by the user, it is no longer as popular as it once was because of reported complications associated with its use, the most serious of which are pelvic inflammatory disease, permanent infertility, and uterine perforation.

■ Implantation can also be blocked by the so-called *morning-after pill,* which is a different type of oral contraceptive than the usual birth-control pill. The high-estrogen-content morning-after pill is taken during the early luteal phase within a few days after conception may have occurred. It prevents implantation by inducing premature degeneration of the corpus luteum so that the developing endometrium's hormonal support is withdrawn. Because of side effects, such as nausea and vomiting, and because of the increased risks of cardiovascular disease associated with high doses of estrogen, this contraceptive method is not used on a routine basis. It is beneficial for one-time usage in special circumstances, however, such as for rape victims who may have conceived.

Future Possibilities

■ A future birth-control technique currently being explored is development of a vaccine that induces the formation of antibodies against human chorionic gonadotropin so that this essential corpus-luteum–supporting hormone is not effective should pregnancy occur.

■ Other investigators are working on a vaccine against the protein in a sperm's head that normally binds to the receptor sites on the zona pellucida surrounding the egg.

■ Still another possibility under investigation is manipulation of the anterior-pituitary secretion of FSH and LH by GnRH-like drugs.

When contraceptive practices fail or are not used and an unwanted pregnancy results, women sometimes turn to abortion to terminate the pregnancy. Although surgical removal of an embryo/fetus is legal in the United States, the practice of abortion is fraught with emotional, ethical, and political controversy.

Adding to the controversy is the discovery by a French pharmaceutical in 1980 of an "abortion pill," RU 486, which terminates an early pregnancy by chemical interference rather than by standard surgical procedures. RU 486 (named after Roussel-Uclaf, the company that developed the drug) is a progesterone antagonist. It binds tightly with the progesterone receptors on the target cells but does not evoke progesterone's usual effects and prevents progesterone from binding and acting. Deprived of progesterone activity, the highly developed endometrial tissue sloughs off, carrying the implanted embryo with it. RU 486 is typically given in conjunction with a prostaglandin that induces uterine contractions to help expel the endometrium and embryo. As of yet, the makers of RU 486 are unwilling to market the drug in the United States, largely because of the heated controversy surrounding the abortion issue.

The end is a new beginning.

Reproduction is an appropriate way to end our discussion of physiology from cells to systems. The single cell resulting from the union of male and female gametes divides mitotically and differentiates into a multicellular individual made up of a number of different organ systems that interact cooperatively to maintain homeostasis (that is, stability in the internal environment). All of the life-supporting homeostatic processes introduced throughout this book begin all over again at the start of a new life.

 ## CHAPTER IN PERSPECTIVE: FOCUS ON HOMEOSTASIS

The reproductive system is unique in that it is not essential for homeostasis or for survival of the individual, but it is essential for sustaining the thread of life from generation to generation. Reproduction depends on the union of male and female gametes (reproductive cells), each with a half set of chromosomes, to form a new individual with a full, unique set of chromosomes. Unlike the other body systems, which are essentially indentical in the two sexes, the reproductive systems of males and females are remarkably different, befitting their different roles in the reproductive process.

The male system is designed to continuously produce huge numbers of motile spermatozoa that are delivered to the female during the sex act. Male gametes must be produced in abundance for two reasons: (1) only a small percentage of them survive the hazardous journey through the female reproductive tract to the site of fertilization, and (2) the cooperative effort of many spermatozoa is required to break down the barriers surrounding the female gamete (ovum or egg) to enable one spermatozoon to penetrate and unite with the ovum.

The female reproductive system undergoes complex changes on a monthly cyclical basis. During the first half of the cycle, a single nonmotile ovum is prepared for release. During the second half, the reproductive system is geared toward preparing a suitable environment for supporting the ovum if fertilization (union with a spermatozoon) occurs. If fertilization does not occur, the prepared supportive environment within the uterus sloughs off, and the cycle starts over again as a new ovum is prepared for release. If fertilization occurs, the female reproductive system adjusts to support growth and development of the new individual until it can survive on its own on the outside.

There are three important parallels in the male and female reproductive systems, even though they differ considerably in structure and function. First, the same set of undifferentiated reproductive tissues in the embryo can develop into either a male or female system, depending on the presence or absence, respectively, of male-determining factors. Second, the same hormones—namely, hypothalamic GnRH and anterior pituitary FSH and LH—control reproductive function in both sexes. In both cases, gonadal steroids and inhibin act in negative-feedback fashion to control hypothalamic and anterior pituitary output. Third, the same events take place in the developing gamete's nucleus during sperm formation and egg formation, although males produce millions of sperm in one day, whereas females produce only about four hundred ova in a lifetime.

 # CHAPTER SUMMARY

Introduction

Both sexes produce gametes (reproductive cells), sperm in males and ova (eggs) in females, each of which bears one member of each of the twenty-three pairs of chromosomes present in human cells. Union of a sperm and an ovum at fertilization results in the beginning of a new individual with twenty-three complete pairs of chromosomes, half from the father and half from the mother.

The reproductive system is anatomically and functionally distinct in males and females. Males produce sperm and deliver them into the female. Females produce ova, accept sperm delivery, and provide a suitable environment for supporting development of a fertilized ovum until the new individual can survive on its own in the external world. In both sexes, the reproductive system consists of (1) a pair of gonads, testes in males and ovaries in females, which are the primary reproductive organs that produce the gametes and secrete sex hormones, and (2) a reproductive tract composed of a system of ducts and associated glands that respectively provide a passageway and supportive secretions for the gametes. The externally visible portions of the reproductive system constitute the external genitalia.

Sex determination is a genetic phenomenon dependent on the combination of sex chromosomes at the time of fertilization, an XY combination being a genetic male and an XX combination a genetic female. Sex differentiation refers to the embryonic development of the gonads, reproductive tract, and external geni-

talia along male or female lines, which gives rise to the apparent anatomic sex of the individual. In the presence of masculinizing factors, a male reproductive system develops; in their absence, a female system develops.

Male Reproductive Physiology

Spermatogenesis (sperm production) occurs in the highly coiled seminiferous tubules within the testes. Leydig cells located in the interstitial spaces between these tubules secrete the male sex hormone testosterone into the blood. Testosterone is secreted before birth to masculinize the developing reproductive system, then its secretion ceases until puberty, at which time it begins once again and continues throughout life. Testosterone is responsible for maturation and maintenance of the entire male reproductive tract, for development of secondary sexual characteristics, and for stimulating libido.

The testes are regulated by the anterior pituitary hormones, luteinizing hormone (LH) and follicle-stimulating hormone (FSH). These gonadotropic hormones, in turn, are under control of hypothalamic gonadotropin-releasing hormone (GnRH). Testosterone secretion is regulated by LH stimulation of the Leydig cells and, in negative-feedback fashion, testosterone inhibits gonadotropin secretion. Spermatogenesis requires both testosterone and FSH. Testosterone stimulates the mitotic and meiotic divisions required to transform the undifferentiated diploid germ cells, the spermatogonia, into undifferentiated haploid spermatids. The remodeling of spermatids into highly specialized motile spermatozoa is stimulated by FSH. A spermatozoon consists only of a DNA-packed head bearing an enzyme-filled acrosome at its tip for penetrating the ovum, a midpiece containing the metabolic machinery for energy production, and a whiplike motile tail. Also present in the seminiferous tubules are Sertoli cells, which protect, nurse, and enhance the germ cells throughout their development. Sertoli cells also secrete inhibin, a hormone that inhibits FSH secretion, thus completing the negative-feedback loop.

The still immature sperm are flushed out of the seminiferous tubules into the epididymis by fluid secreted by the Sertoli cells. The epididymis and ductus deferens store and concentrate the sperm and increase their motility and fertility prior to ejaculation. During ejaculation, the sperm are mixed with secretions released by the accessory glands, which contribute the bulk of the semen. The seminal vesicles supply fructose for energy and prostaglandins, which promote smooth-muscle motility in both the male and female reproductive tracts to enhance sperm transport. The prostate gland contributes an alkaline fluid for neutralizing the acidic vaginal secretions. The bulbourethral glands release lubricating mucus.

Sexual Intercourse between Males and Females

The male sex act consists of erection and ejaculation, which are part of a much broader systemic, emotional response that typifies the male sexual response cycle. Erection is a hardening of the normally flaccid penis that enables it to penetrate the female vagina. Erection is accomplished by marked vasocongestion of the penis brought about by reflexly induced vasodilation of the arterioles supplying the penile erectile tissue. When sexual excitation reaches a critical peak, ejaculation, which consists of two stages, occurs: (1) emission, the emptying of semen (sperm and accessory sex gland secretions) into the urethra, and (2) expulsion of semen from the penis. The latter is accompanied by a set of characteristic systemic responses and intense pleasure referred to as orgasm.

Females experience a sexual cycle similar to males, with both having excitation, plateau, orgasmic, and resolution phases. The major differences are that women do not ejaculate and they are capable of multiple orgasms. During the female sexual response, the outer third of the vagina constricts to grip the penis while the inner two-thirds expands to create space for sperm deposition.

Female Reproductive Physiology

In the nonpregnant state, female reproductive function is controlled by a complex, cyclical negative-feedback control system between the hypothalamus (GnRH), anterior pituitary (FSH and LH), and ovaries (estrogen, progesterone, and inhibin). During pregnancy, placental hormones become the main controlling factors.

The ovaries perform the dual and interrelated functions of oogenesis (producing ova) and secreting estrogen and progesterone. Two related ovarian endocrine units sequentially accomplish these functions: the follicle and the corpus luteum. Oogenesis and estrogen secretion take place within an ovarian follicle during the first half of each reproductive cycle (the follicular phase). At approximately midcycle, the maturing follicle releases a single ovum (ovulation). The empty follicle is then converted into a corpus luteum, which produces progesterone as well as estrogen during the last half of the cycle (the luteal phase). This endocrine unit is responsible for preparing the uterus as a suitable site for implantation should the released ovum be fertilized. If fertilization and implantation do not occur, the corpus luteum degenerates. The consequent withdrawal of hormonal support for the highly developed uterine lining causes it to disintegrate and slough, producing menstrual flow. Simultaneously, a new follicular phase is initiated. Menstruation ceases and the uterine lining (endometrium) repairs itself under the influence of rising estrogen levels from the newly maturing follicle.

If fertilization does take place, it occurs in the oviduct as the released egg and sperm deposited in the vagina are both transported to this site. The fertilized ovum begins to divide mitotically. Within a week it grows and differentiates into a blastocyst capable of implantation. Meanwhile, the endometrium has become richly vascularized and stocked with stored glycogen under the influence of luteal-phase progesterone. It is into this especially prepared lining that the blastocyst implants by means of enzymes released by the blastocyst's outer layer. These enzymes digest the nutrient-rich endometrial tissue, accomplishing the dual function of carving out a hole in the endometrium for implantation of the blastocyst while at the same time releasing nutrients from the endometrial cells for use by the developing embryo.

Following implantation, an interlocking combination of fetal and maternal tissues, the placenta, develops. The placenta is the organ of exchange between the maternal and fetal blood and also acts as a transient, complex endocrine organ that secretes a number of hormones essential for pregnancy. Human chorionic gonadotropin, estrogen, and progesterone are the most important of these hormones.

At parturition, rhythmic contractions of increasing strength, duration, and frequency accomplish the three stages of labor: dilation of the cervix, birth of the baby, and delivery of the placenta (afterbirth). Once the contractions are initiated at the onset of labor, a positive-feedback cycle is established that progressively increases their force. As contractions push the fetus against the cervix, secretion of oxytocin, a powerful uterine muscle stimulant, is reflexly increased. The extra oxytocin causes stronger contractions, giving rise to even more oxytocin release, and so on. This positive-feedback cycle progressively intensifies until cervical dilation and delivery are accomplished.

During gestation, the breasts are specially prepared for lactation. The elevated levels of placental estrogen and progesterone, respectively, promote development of the ducts and alveoli in the mammary glands. Prolactin stimulates the synthesis of enzymes essential for milk production by the alveolar epithelial cells. However, the high gestational level of estrogen and progesterone prevents prolactin from promoting milk production. Withdrawal of the placental steroids at parturition initiates lactation. Lactation is sustained by suckling, which triggers the release of oxytocin and prolactin. Oxytocin causes milk ejection by stimulating the myoepithelial cells surrounding the alveoli to squeeze the secreted milk out through the ducts. Prolactin stimulates the production of more milk to replace the milk ejected as the baby nurses.

Objective Questions (Answers on p. D–5.)

1. It is possible for a genetic male to have the anatomical appearance of a female. (True or false?)
2. Testosterone secretion essentially ceases from birth until puberty. (True or false?)
3. The pineal gland secretes more melatonin during the light than during the dark. (True or false?)
4. Females do not experience erection. (True or false?)
5. Most of the lubrication for sexual intercourse is provided by the female. (True or false?)
6. If a follicle fails to reach maturity during one ovarian cycle, it can finish maturing during the next cycle. (True or false?)
7. Low but rising levels of estrogen inhibit tonic LH secretion whereas high levels of estrogen stimulate the LH surge. (True or false?)
8. Spermatogenesis takes place within the _____ of the testes, stimulated by the hormones _____ and _____.
9. During estrogen production by the follicle, the _____ cells under the influence of the hormone _____ produce androgens, and the _____ cells under the influence of the hormone _____ convert these androgens into estrogens.
10. The source of estrogen and progesterone during the first ten weeks of gestation is the _____.
11. Detection of _____ in the urine is the basis of pregnancy diagnosis tests.
12. Which of the following statements concerning chromosomal distribution is *incorrect*?
    a. All human somatic cells contain twenty-three chromosomal pairs for a total diploid number of forty-six chromosomes.
    b. Each gamete contains twenty-three chromosomes, one member of each chromosomal pair.
    c. During a meiotic reduction division, the members of the chromosome pairs regroup themselves into the original combinations derived from the individual's mother and father for separation into haploid gametes.
    d. Sex determination depends on the combination of sex chromosomes, an XY combination being a genetic male, XX a genetic female.
    e. The sex chromosome content of the fertilizing sperm determines the sex of the offspring.
13. When the corpus luteum degenerates,
    a. circulating levels of estrogen and progesterone rapidly decline.
    b. FSH and LH secretion start to rise as the inhibitory effects of the gonadal steroids are withdrawn.
    c. the endometrium sloughs.
    d. Both a and b are correct.
    e. All of the above are correct.

14. Match the following:
    ___ 1. secrete(s) prostaglandins
    ___ 2. increase(s) motility and fertility of sperm
    ___ 3. secrete(s) an alkaline fluid
    ___ 4. provide(s) fructose
    ___ 5. storage site for sperm
    ___ 6. concentrate(s) the sperm a hundredfold
    ___ 7. secrete(s) fibrinogen
    ___ 8. provide(s) clotting enzymes
    ___ 9. contain(s) erectile tissue

    a. epididymis and ductus deferens
    b. prostate gland
    c. seminal vesicles
    d. bulbourethral glands
    e. penis

15. Using the answer code at the right, indicate when each event takes place during the ovarian cycle:
    ___ 1. development of antral follicles
    ___ 2. secretion of estrogen
    ___ 3. secretion of progesterone
    ___ 4. menstruation
    ___ 5. repair and proliferation of the endometrium
    ___ 6. increased vascularization and glycogen storage in the endometrium

    a. occurs during the follicular phase
    b. occurs during the luteal phase
    c. occurs during both the follicular and luteal phases

Essay Questions

1. What are the primary reproductive organs, the gametes, the sex hormones, the reproductive tract, the accessory sex glands, the external genitalia, and the secondary sexual characteristics in males and females?
2. List the essential reproductive functions of the male and of the female.
3. Discuss the differences between males and females with regard to genetic, gonadal, and phenotypic sex.
4. What parts of the male and female reproductive systems develop from each of the following: genital tubercle, urethral folds, genital swellings, Wolffian ducts, and Müllerian ducts?
5. Discuss the source and functions of testosterone.
6. Describe the three major stages of spermatogenesis. Discuss the functions of each part of a spermatozoon. What are the roles of Sertoli cells?
7. Discuss the control of testicular function.
8. Compare the sex act in males and females.
9. Compare oogenesis with spermatogenesis.
10. Describe the events of the follicular and luteal phases of the ovarian cycle. Correlate the phases of the uterine cycle with the ovarian cycle.
11. How are the ovum and spermatozoa transported to the site of fertilization? Describe the process of fertilization.

12. Describe the process of implantation and placenta formation.

13. What are the functions of the placenta? What hormones does the placenta secrete?

14. What is the role of human chorionic gonadotropin?

15. What is the leading proposal for the mechanism that initiates parturition? What are the stages of labor? What is the role of oxytocin?

16. Describe the hormonal factors that play a role in lactation.

17. Summarize the actions of estrogen and progesterone.

POINTS TO PONDER

1. The hypothalamus releases GnRH in pulsatile bursts once every two to three hours, with no secretion occurring in between. A promising line of research for a new method of contraception involves administration of GnRH-like drugs. In what way could such drugs act as contraceptives when GnRH is the hypothalamic hormone that triggers the chain of events leading to ovulation? (Hint: The anterior pituitary is "programmed" to respond only to the normal pulsatile pattern of GnRH.)

2. Occasionally, testicular tumors composed of interstitial cells of Leydig may secrete up to one hundred times the normal amount of testosterone. When such a tumor develops in young children, they grow up much shorter than their genetic potential. Explain why. What other symptoms would be present?

3. What type of sexual dysfunction might arise in men taking drugs that inhibit sympathetic nervous system activity as part of the treatment for high blood pressure?

4. Explain the physiological basis for administering a posterior pituitary extract to induce or facilitate labor.

5. The symptoms of menopause are sometimes treated with supplemental estrogen and progesterone. Why wouldn't treatment with GnRH or FSH and LH also be effective?

A Review of Chemical Principles

 APPENDIX A CONTENTS AT A GLANCE

ATOMS, ELEMENTS, COMPOUNDS, AND MOLECULES
CHEMICAL BONDS
CHEMICAL REACTIONS
FORMULAS, EQUATIONS, AND THE MOLE
SOLUTIONS
SUSPENSIONS AND COLLOIDS
INORGANIC AND ORGANIC CHEMICALS
ACIDS, BASES, AND SALTS
FUNCTIONAL GROUPS OF ORGANIC MOLECULES
CARBOHYDRATES
LIPIDS
PROTEINS
NUCLEIC ACIDS
HIGH-ENERGY BIOMOLECULES

The chemical nature of all matter makes an understanding of chemistry helpful in many areas of study, including human physiology. This appendix contains a brief discussion of some basic chemical concepts that you are encouraged to examine as necessary while you study the material in the textbook.

ATOMS, ELEMENTS, COMPOUNDS, AND MOLECULES

All matter is made up of tiny particles called **atoms**. These particles are too small to be seen individually, even with the most powerful electron microscopes available today. However, the work of generations of scientists has led to an understanding of many characteristics of atoms that help us explain and understand the behavior of matter.

Even though they are tiny, atoms are composed of three types of smaller particles. **Protons** and **neutrons** are particles of nearly identical mass and make up the nucleus of an atom. Protons carry a positive charge, whereas neutrons have no charge. **Electrons**, the third type of particle found in atoms, move rapidly around the central nucleus (Fig. A–1). Electrons have a much smaller mass than protons and neutrons and are negatively charged. The charge of a proton exactly matches that of an electron, but it is opposite in sign. In all atoms, the number of protons in the nucleus is equal to the number of electrons moving around the nucleus, so their charges balance and the atoms are neutral.

A pure substance that contains only one type of atom is called an **element**. A pure sample of the element carbon contains only carbon atoms, even though the atoms might be arranged in a form called diamond or in a form called graphite.

Pure substances called **compounds** contain more than one type of atom. Pure water, for example, is a compound that contains atoms of hydrogen and atoms of oxygen in a 2:1 ratio, regardless of whether the water is in the form of liquid, solid (ice), or vapor (steam). A **molecule** is the smallest unit of a pure substance that has the properties of that substance and is capable of a stable, independent existence. For example, a molecule of water consists of two atoms of hydrogen and one atom of oxygen, held together by chemical bonds.

Exactly what are we talking about when we refer to a "type" of atom? That is, what makes carbon, hydrogen, and oxygen atoms different? The answer is the number of protons in the nucleus. Regardless of where they are found, all hydrogen atoms have one proton in the nucleus, all carbon atoms have six, and all oxygen atoms have eight. Of course, these numbers also represent the number of electrons moving

 FIGURE A–1   Electrons Move Rapidly Around a Massive Nucleus

TABLE A–1   Characteristics of Selected Elements

NAME	SYMBOL	NUMBER OF PROTONS	ATOMIC NUMBER	ATOMIC WEIGHT (amu)
Hydrogen	H	1	1	1.01
Carbon	C	6	6	12.01
Nitrogen	N	7	7	14.01
Oxygen	O	8	8	16.00
Sodium	Na	11	11	22.99
Magnesium	Mg	12	12	24.31
Phosphorus	P	15	15	30.97
Sulfur	S	16	16	32.06
Chlorine	Cl	17	17	35.45
Potassium	K	19	19	39.10
Calcium	Ca	20	20	40.08

around each nucleus, because the number of electrons and protons in an atom are equal. The number of protons in the nuclei of the atoms of an element is called the **atomic number** of the element.

As expected, tiny atoms have tiny masses. For example, the actual mass of a hydrogen atom is 1.67×10^{-24} g, that of a carbon atom is 1.99×10^{-23} g, and that of an oxygen atom is 2.66×10^{-23} g. These very small numbers are inconvenient to work with in calculations, so a system of relative masses has been developed. These relative masses simply compare the actual masses of the atoms to each other. Suppose the actual masses of two people were determined to be 45.50 kg and 113.75 kg. Their relative masses are determined by dividing each mass by the smaller mass of the two: 45.50/45.50 = 1.00 and 113.75/45.50 = 2.50. Thus, the relative masses of the two people are 1.00 and 2.50; these numbers simply express the fact that the mass of the heavier person is 2.50 times that of the other person. The relative masses of atoms are called **atomic masses** or **atomic weights** and are given in atomic mass units (amu). In this system, hydrogen atoms, the least massive of all atoms, have an atomic weight of 1.01 amu. The atomic weight of carbon atoms is 12.01 amu, and that of oxygen atoms is 16.00 amu. Thus, oxygen atoms have a mass about 16 times that of hydrogen atoms. Table A–1 gives the atomic weights and some other characteristics of the elements that are most important physiologically.

≡◉ CHEMICAL BONDS

Since all matter is made up of particles called atoms, we must conclude that the atoms somehow are held together to form matter. The forces holding atoms together are called **chemical bonds**. Not all chemical bonds are formed in the same way, but all of them involve the electrons of atoms. It is now understood that the electrons of each atom have energies and other characteristics that allow them to be classified into groupings called **shells**. In general, electrons will belong to the lowest energy shell possible, but the specific shells of an atom have maximum capacities that cannot be exceeded. For example, the first or lowest energy shell of every atom can contain a maximum of only two electrons, while the second or next highest energy shell can contain a maximum of eight electrons. Different atoms have different numbers of electrons in the various shells. Hydrogen atoms have only one electron, so it is in the first shell. Helium atoms have two electrons, which are both in the first shell and fill it. Carbon atoms have six electrons, two in the first shell and four in the second shell, while the eight electrons of oxygen are arranged with two in the first shell and six in the second shell.

Having filled electron shells provides an energy benefit. That is, the average energy per electron is lower for the second shell when the shell holds the maximum number of eight electrons instead of any number from one to seven. This energy benefit leads to a general statement about the electronic behavior of atoms: *atoms tend to undergo processes that result in a filled outermost electron shell.* Thus, it is the electrons of the outer or higher energy shell that determine the bonding characteristics of an atom.

Consider sodium atoms (Na) and chlorine atoms (Cl). Sodium atoms have eleven electrons, two in the first shell, eight in the second shell, and one in the third shell. Chlorine atoms have seventeen electrons, two in the first shell, eight in the second shell and seven in the third shell. Because eight electrons are required to fill the second and third shells, sodium atoms have one electron more than is needed to provide a filled second shell, while chlorine atoms have one less electron than is needed to fill the third shell. Each sodium atom can lose an electron to a chlorine atom, leaving each sodium with ten electrons, eight of which are in the second shell, which is now the outer shell occupied by electrons. By accepting one electron, each chlorine atom has a total of eighteen electrons, with eight of them in the third or outer shell.

As a result of giving up and accepting electrons, the sodium atoms and chlorine atoms have achieved filled outer shells, but now each atom is unbalanced electrically. While each sodium now has ten electrons, it still has eleven protons in the nucleus and a net electrical charge of +1. Similarly, each chlorine now has eighteen electrons, but only seventeen protons. Thus, each chlorine has a −1 charge. Charged atoms such as these are called **ions**. Positively charged ions are called **cations**, whereas negatively charged ions are called **anions**. Since opposite charges attract, sodium ions (Na^+) and charged chlorine atoms, now called *chloride* ions (Cl^-), are attracted toward each other. It is this attraction that bonds the ions together in the compound **sodium chloride, NaCl**, which is common table salt. A sample of sodium chloride actually contains sodium and chloride ions in a three-dimensional geometric arrangement called a crystal lattice. The ions of opposite charge occupy alternate sites within the lattice (● Fig. A−2).

It is not energetically favorable for an atom to give up or accept more than three electrons. In spite of this, carbon atoms, which have four electrons in their outer shell, are known to form compounds. This observation led scientists to propose another bonding mechanism for atoms. Atoms that would have to lose or gain four or more electrons to achieve outer-shell stability usually bond by *sharing* electrons. Thus, a carbon atom can share its four outer electrons with the four electrons of four hydrogen atoms, as shown in Equation A−1; the outer-shell electrons are shown as dots around the symbol of each atom (the resulting compound is methane [CH_4], which is a gas made up of individual CH_4 molecules):

shared
electron pairs

$$\cdot \overset{\displaystyle \cdot}{\underset{\displaystyle \cdot}{C}} \cdot + 4 \cdot H \rightarrow H \overset{\displaystyle \cdot \cdot}{\underset{\displaystyle \cdot \cdot}{:C:}} H$$

shared
electron pairs   Eq. A−1

Each electron that is shared by two atoms is counted toward the number of electrons needed to fill the outer shell of each atom. Thus, each carbon atom shares four pairs or eight electrons and so has eight in its outer shell. Each hydrogen shares one pair or two electrons and so has a filled outer shell. (Remember, hydrogen atoms need only two electrons to complete their outer shell, which is the first shell.) Atoms that share a pair of electrons are both attracted toward the shared pair. This mutual attraction bonds the atoms together in what is called a **covalent bond**.

Covalent bonds also form between some identical atoms. For example, two hydrogen atoms can complete their outer shells by sharing one electron pair made from the single electrons of each atom, as shown in Equation A−2:

$$H\cdot + \cdot H \rightarrow H:H \qquad \text{Eq. A−2}$$

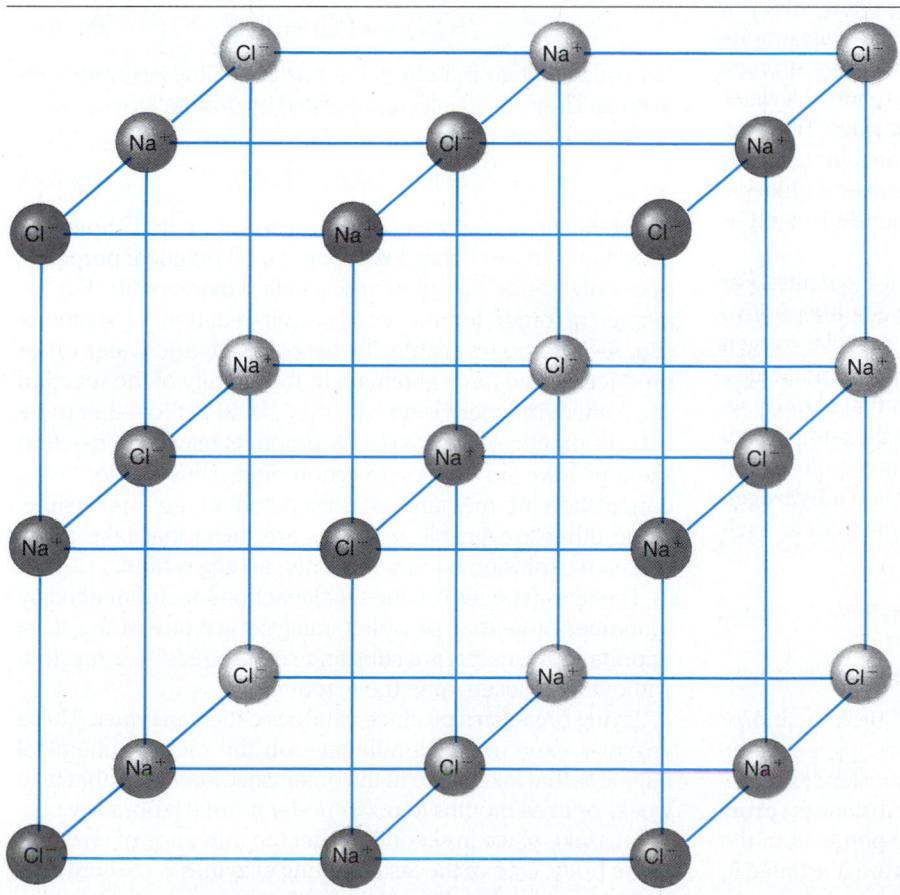

● **FIGURE A−2   Crystal Lattice for Sodium Chloride (Table Salt)**

Thus, hydrogen gas consists of individual H_2 molecules. Several other nonmetallic elements also exist as molecules because covalent bonds form between identical atoms; examples include chlorine (Cl_2), oxygen (O_2), and nitrogen (N_2).

One of the most familiar covalently bonded substances is water (H_2O). Equation A–3 represents the formation of its covalent bonds:

$$H\cdot + \cdot \overset{\cdot\cdot}{\underset{\cdot\cdot}{O}} : \;\rightarrow\; H : \overset{\cdot\cdot}{\underset{H}{O}} : \qquad\qquad \text{Eq. A–3}$$

The water molecule is sometimes represented as

$$H - \underset{\underset{H}{|}}{O},$$

where the nonshared electron pairs are not shown and the covalent bonds or shared pairs are represented by dashes. The water molecule is a good example of a **polar molecule**; that is, a molecule in which the electrons are not distributed uniformly. Polar molecules result because shared electrons are unequally attracted by the atoms that share them. When the atoms sharing an electron pair are identical, the electrons are attracted equally by both atoms and so are shared equally; such molecules are **nonpolar**. Examples of molecules containing equally shared electrons are H_2, O_2, and N_2. When the sharing atoms are not identical, the shared pair of electrons is pulled closer to one atom than the other. The side of the molecule to which the electrons are pulled is electrically negative compared to the other side. In water molecules, the oxygen atom pulls shared electrons more strongly than do the hydrogen atoms. Thus, the oxygen side of a water molecule is more negative than the hydrogen sides. The electron distribution is not uniform, and the water molecule is polar. Atoms of oxygen, nitrogen, phosphorus and chlorine strongly attract electrons when they are bonded to other atoms.

Polar molecules are attracted to other polar molecules. For example, in water there is an attraction between the positive hydrogen ends of some molecules and the negative oxygen ends of others. Hydrogen is not a part of all polar molecules, but when it is covalently bonded to an atom that strongly attracts electrons to form a covalent molecule, the attraction of the positive (hydrogen) end of the polar molecule to the negative end of another polar molecule is called a **hydrogen bond**. Thus, the polar attractions of water molecules to each other is an example of hydrogen bonding.

CHEMICAL REACTIONS

Processes in which chemical bonds are broken and/or formed are called **chemical reactions**. Reactions are represented by equations in which the reacting substances (**reactants**) are written on the left, the produced substances (**products**) are written on the right, and an arrow points from the reactants to the products. These conventions are illustrated in Equation A–4:

$$\underset{\text{reactants}}{A + B} \;\rightarrow\; \underset{\text{products}}{C + D} \qquad \text{Eq. A–4}$$

Let's look at a specific example, the combustion of methane gas, CH_4, as shown in Equation A–5:

$$CH_4 + 2O_2 \rightarrow CO_2 + 2H_2O \qquad \text{Eq. A–5}$$

According to this equation, one molecule of methane gas reacts with two molecules of oxygen gas to produce one molecule of carbon dioxide gas and two molecules of water vapor. Coefficients such as the 2s to the left of the O_2 and H_2O are used so that the total number of each type of atom is the same on the left and right sides of the equation. Thus, the reaction above has one carbon atom on the left (in one CH_4) and one on the right (in one CO_2). Four hydrogen atoms are on the left (in one CH_4) and four on the right (in two H_2Os). There are four oxygen atoms on the left (in two O_2s) and four on the right (in one CO_2 and two H_2Os). Equations in which the atoms of each type appear in equal numbers on both sides are called **balanced equations**.

Under appropriate conditions, the products of a reaction can be changed back to the reactants. For example, carbon dioxide gas dissolves in and reacts with water to form carbonic acid, H_2CO_3:

$$CO_2 + H_2O \rightarrow H_2CO_3 \qquad \text{Eq. A–6}$$

Carbonic acid is not very stable, however, and as soon as some is formed, part of it decomposes to give carbon dioxide and water:

$$H_2CO_3 \rightarrow CO_2 + H_2O \qquad \text{Eq. A–7}$$

Reactions that go in both directions are called **reversible reactions**. They are usually represented by double arrows pointing in both directions:

$$CO_2 + H_2O \rightleftharpoons H_2CO_3 \qquad \text{Eq. A–8}$$

Theoretically, every reaction is reversible. Often, however, conditions are such that a reaction, for all practical purposes, goes only in one direction and is called **irreversible**. For example, in order for the combustion reaction of methane (Eq. A–5) to be reversible, the gaseous CO_2 and water vapor products would have to remain in the vicinity of the reaction site (otherwise, there is no way they could get together to react). In practice, these gaseous products leave the reaction site and have no chance to recombine. Thus, the ordinary combustion of methane is considered to be irreversible. Some other irreversible reactions are those that take place when an explosion occurs and when an egg is fried.

The rates (speeds) of chemical reactions are influenced by a number of factors, of which catalysts are one of the most important. A **catalyst** is a substance that speeds up a reaction without being used up in the reaction.

Living organisms produce catalysts called **enzymes**. These enzymes exert amazing influence on the rates of chemical reactions that take place in the organisms. Reactions that take weeks or even months to occur under normal laboratory conditions take place in seconds under the influence of enzymes in the body. One of the fastest-acting enzymes is **carbonic anhydrase**, which catalyzes the reaction between carbon dioxide and water to form carbonic acid. This reaction is important in the transport of carbon dioxide from tissue cells, where it is produced metabolically, to the lungs, where it is excreted. The equation for the reaction was shown in Equation A–6. Each molecule of carbonic anhydrase catalyzes the conversion of 36 million CO_2 molecules per minute. Enzymes are important in essentially every chemical reaction that takes place in living organisms.

FORMULAS, EQUATIONS, AND THE MOLE

We have already discussed the concept of atomic weights. The idea of comparing masses by using relative values is also useful in discussing molecules. Since molecules are made up of atoms, the relative mass of a molecule is simply the sum of the relative masses (atomic weights) of the atoms found in the molecule. The relative masses of molecules are called **molecular masses** or **molecular weights**. The molecular weight of water, H_2O, is thus the sum of the atomic weights of two hydrogen atoms and one oxygen atom, or 1.01 amu + 1.01 amu + 16.00 amu = 18.02 amu.

Not all compounds exist in the form of molecules. Ionically bonded substances such as sodium chloride consist of three-dimensional arrangements of sodium ions (Na^+) and chloride ions (Cl^-) in a 1:1 ratio. The formulas for ionic compounds reflect only the ratio of the ions in the compound and should not be interpreted in terms of molecules. Thus, the formula for sodium chloride, NaCl, indicates that the ions combine in a 1:1 ratio. It is convenient to apply the concept of relative masses to ionic compounds even though they do not exist as molecules. The **formula weight** for such compounds is defined as the sum of the atomic weights of the atoms found in the formula. Thus, the formula weight of magnesium chloride, an ionic compound with the formula $MgCl_2$, is equal to the sum of the atomic weights of one magnesium atom and two chlorine atoms, or 24.31 amu + 35.45 amu + 35.45 amu = 95.21 amu.

As we have seen, chemical reactions can be represented by equations and discussed in terms of numbers of molecules, atoms, and ions reacting with each other. However, it is much more convenient to describe reactions in terms of amounts of reactants and products that can readily be measured in units such as grams. Using the mole concept makes such descriptions possible. A **mole** (abbreviated mol) of a pure element or compound is the amount of material contained in a sample of the pure substance that has a mass in grams equal to the substance's atomic weight (for elements) or the molecular weight or formula weight (for compounds). Thus, one mole of potassium, K, would be a sample of the element with a mass of 39.10 g. Similarly, a mole of water, H_2O, would have a mass of 18.02 g, and a mole of sodium chloride, NaCl, would be a sample with a mass of 58.44 g.

The fact that atomic weights, molecular weights, and formula weights are relative masses leads to a fundamental characteristic of moles. One mole of hydrogen atoms has a mass of 1.01 g, and one mole of oxygen atoms has a mass of 16.00 g. The ratio of atomic weights for the two elements is 16.00/1.01,

the same as the ratio of the masses of one mole of each element: 16.00 g/1.01 g. Therefore, one mole of hydrogen contains exactly the same number of hydrogen atoms as the number of oxygen atoms in one mole of oxygen. Thus, it is possible and sometimes useful to think of a mole as a specific number of particles. This number, called **Avogadro's number**, is equal to 6.02×10^{23}.

SOLUTIONS

Most chemical reactions in the body take place between reactants that have dissolved to form solutions. **Solutions** are homogeneous mixtures containing a relatively large amount of one substance called the **solvent** and smaller amounts of one or more substances called **solutes**. Salt water, for example, contains mostly water, which is thus the solvent, and a smaller amount of salt, which is the solute. Water is the solvent in most solutions found in the human body.

When ionic solutes are dissolved in water to form solutions, the resulting solution will conduct electricity. This is not true for most covalently bonded solutes. For example, a saltwater solution conducts electricity, but a sugar-water solution does not. When salt dissolves in water, the solid lattice of Na^+ and Cl^- is broken down, and the individual ions are separated and distributed uniformly throughout the solution. These mobile, charged ions conduct electricity through the solution. Solutes that form ions in solution and conduct electricity are called **electrolytes**. Some very polar covalent molecules also behave this way. When sugar dissolves, however, individual covalently bonded sugar molecules leave the solid and become uniformly distributed throughout the solution. These uncharged molecules cannot conduct a current. Solutes that do not form conductive solutions are called **nonelectrolytes**.

The amount of solute dissolved in a specific amount of solution can vary. For example, a saltwater solution might contain 1 g of salt in 100 milliliters (ml) of solution, or it could contain 10 g of salt in 100 ml of solution. Both solutions are saltwater solutions, but they have different concentrations of solute. The **concentration** of a solution indicates the relationship between the amount of solute and the amount of solvent or the amount of solution. Concentrations can be given in a number of different units.

Concentrations given in terms of **molarity** (abbreviated M) give the number of moles of solute in exactly one liter of solution. Thus, a half molar (0.50 M) solution of NaCl would contain one-half mole or 29.22 g of NaCl in each liter of solution.

Concentrations given in terms of **molality** (abbreviated m) give the number of moles of solute dissolved in 1,000 g (1.00 kg) of solvent. Thus, a half molal (0.50 m) solution of NaCl would contain NaCl and water in the ratio of 29.22 g NaCl to each 1,000 g (or 1,000 ml) of water.

Note the difference between molarity and molality. Molarity gives the amount of solute in a specific amount of solution, whereas molality gives the amount of solute dissolved in a specific amount of solvent.

When the solute is an electrolyte, it is sometimes useful to express the concentration of the solution in a unit that gives information about the amount of ionic charge in the solution. This is done by expressing concentration in terms of **normality** (abbreviated N). The normality of a solution gives the number of equivalents of solute in exactly one liter of solution. An **equivalent** of an electrolyte is the amount that produces one mole of positive (or negative) charges when it dissolves. The number of equivalents of an electrolyte can be calculated by multiplying the number of moles of electrolyte by the total number of positive charges produced when one formula unit of the electrolyte dissolves. Consider sodium chloride (NaCl) and calcium chloride ($CaCl_2$) as examples. The ionization reactions for one formula unit of each solute are:

$$NaCl \rightarrow Na^+ + Cl^- \hspace{2cm} \text{Eq. A--9}$$
$$CaCl_2 \rightarrow Ca^{++} + 2Cl^- \hspace{1.5cm} \text{Eq. A--10}$$

Thus, one mole of NaCl produces one mole of positive charges (Na^+) and so contains one equivalent:

$$(1 \text{ mole NaCl})(1) = 1 \text{ equivalent,}$$

where the number 1 used to multiply the 1 mole of NaCl came from the +1 charge on Na^+.

One mole of $CaCl_2$ produces one mole of Ca^{++}, which is two moles of positive charge. Thus, one mole of $CaCl_2$ contains two equivalents:

$$(1 \text{ mole } CaCl_2)(2) = 2 \text{ equivalents,}$$

where the number 2 used in the multiplication came from the +2 charge on Ca^{++}.

If two solutions were made such that one contained one mole of NaCl per liter and the other contained one mole of $CaCl_2$ per liter, the NaCl solution would contain one equivalent of solute per liter and would be one normal (1 N). The $CaCl_2$ solution would contain two equivalents of solute per liter and would be two normal (2 N).

Another expression of concentration frequently used in physiology is **osmolarity** (abbreviated Osm), which indicates the total *number* of solute particles in a liter of solution instead of the relative weights of the specific solutes. The osmolarity of a solution is the product of M and n, where n is the number of moles of solute particles obtained when 1 mole of solute dissolves. Because nonelectrolytes such as glucose do not dissociate in solution, n = 1 and the osmolarity (n times M) is equal to the molarity of the solution. For electrolyte solutions, the osmolarity exceeds the molarity by a factor equal to the number of ions produced upon dissociation of each molecule in solution. For example, because a NaCl molecule dissociates into two ions, Na^+ and Cl^-, the osmolarity of a 1 M solution of NaCl is $2 \times 1 \text{ M} = 2 \text{ Osm}$.

SUSPENSIONS AND COLLOIDS

Suspensions, like solutions, consist of two or more components, with much more of one component than the others. In solutions, the component present in largest amount is called the solvent. In suspensions, it is called the **dispersing medium**. In solutions, the components present in smaller amounts are called solutes; in suspensions, they are called **dispersed phases**. An important difference between solutions and suspensions is the size of the particles dissolved or dispersed. In solutions, solute particles are ions or small molecules. Dispersed-phase particles are much larger than ions or small molecules. When the dispersed-phase particles are no more than about a hundred times the size of the largest solution solute particles, the suspension is called a **colloid** or **colloidal dispersion**. The dispersed-phase particles of colloids generally do not settle out. They are small enough to be kept in suspension by the constant buffeting they receive from the motion of dispersing-medium molecules. All dispersed-phase particles of colloids carry electrical charges of the same sign. Thus, they repel each other and, despite their collisions with other dispersed-phase particles, do not get together to form larger particles that would settle. When dispersed-phase particles are larger than those in colloids, the dispersed phase will settle out. Such mixtures are usually called suspensions.

The relatively large size of dispersed-phase particles in colloids or suspensions causes them to scatter light. As a result, colloids and suspensions appear cloudy. Solutions are clear because the dissolved solute particles are too small to scatter light. In the body, the dispersing medium of most colloids is water, and the colloids are liquids. However, colloids can occur in all three states, depending on the state of the dispersing medium.

INORGANIC AND ORGANIC CHEMICALS

Chemicals are commonly classified into two categories: inorganic and organic. The original criterion used for this classification was the origin of the chemicals. Those that came from living or once-living sources were *organic,* and those that came from other sources were *inorganic*. Today, the basis for classification is the element carbon. **Organic** chemicals are generally those that contain carbon. All others are classified as **inorganic**. A few carbon-containing chemicals are also classified as inorganic; the most common are pure carbon in the form of diamond and graphite, carbon dioxide (CO_2), carbon monoxide (CO), carbonates such as limestone ($CaCO_3$), bicarbonates such as baking soda ($NaHCO_3$), and cyanides such as sodium cyanide (NaCN).

The unique ability of carbon atoms to bond to each other and form networks of carbon atoms results in an interesting fact. Even though organic chemicals are only required to contain one specific element, carbon, millions of these compounds have been identified. Some were isolated from natural plant or animal sources, and many have been synthesized in laboratories. Inorganic chemicals include all the other 108 elements and their compounds; the number of known inorganic chemicals is estimated to be about 250,000.

Another result of carbon's ability to bond to itself is the large size of some organic molecules. Molecules classified as organic range in size from methane, CH_4, a small, simple molecule with one carbon atom, to molecules such as DNA that contain as many as a million carbon atoms. Large molecules such as this are often called **macromolecules**. Macromolecules include many naturally occurring molecules such as DNA and many molecules that are synthetically produced. In this latter category are numerous substances that are widely used today, including synthetic textiles (e.g., nylon, dacron, and orlon) and plastics (e.g., lucite, plexiglass, and teflon). Another general name given to these materials is **polymer**, which means "many units," reflecting the fact that polymeric macromolecules are made by bonding together a large number of smaller molecules.

ACIDS, BASES, AND SALTS

Acids, bases, and salts are among the most common and important compounds studied in chemistry. Both inorganic and organic compounds are known that fit into these three categories. Until late in the nineteenth century, these substances were classified on the basis of such properties as taste or by the color changes induced in certain dyes. Acids taste sour, bases bitter, and salts salty. Litmus, a dye, is red in the presence of acids and blue in the presence of bases. These and other observations led to the correct conclusions that acids and bases are chemical opposites and that salts are produced when acids and bases react with each other. Today, acids and bases are defined in more precise ways.

In 1887, Swedish chemist Svante Arrhenius proposed a theory defining acids and bases. He said that an *acid* is any substance that will dissociate or break apart when dissolved in water and in the process release a hydrogen ion, H^+. Similarly, *bases* are substances that dissociate when dissolved in water and in the process release a hydroxide ion, OH^-. Hydrogen chloride (HCl) and sodium hydroxide (NaOH) are examples of Arrhenius acids and bases; their dissociations in water are represented in Equations A–11 and A–12, respectively:

$$HCl \rightarrow H^+ + Cl^- \qquad \text{Eq. A–11}$$
$$NaOH \rightarrow Na^+ + OH^- \qquad \text{Eq. A–12}$$

Note that the hydrogen ion is a bare proton, the nucleus of a hydrogen atom. Also note that both HCl and NaOH would behave as electrolytes.

Arrhenius did not know that free hydrogen ions cannot exist in water. It is now believed that they covalently bond to water molecules to form hydronium ions, as shown in Equation A–13:

$$H^+ + :\overset{..}{O} - H \rightarrow \left[H - \overset{..}{O} - H \right]^+ \qquad \text{Eq. A–13}$$

In 1923, Johannes Bronsted in Denmark and Thomas Lowry in England proposed an acid-base theory that took this behavior into account. They defined an **acid** as any hydrogen-containing substance that donates a proton (hydrogen ion) to another substance, and a **base** as any substance that accepts a proton. According to these definitions, the acidic behavior of HCl given in Equation A–11 is written as shown in Equation A–14:

$$HCl + H_2O \rightleftharpoons H_3O^+ + Cl^- \qquad \text{Eq. A–14}$$

Note that this reaction is shown to be reversible, and the hydronium ion is represented as H_3O^+.

In Equation A–14, the HCl acts as an acid in the forward (left-to-right) reaction, while water acts as a base. In the reverse reaction (right-to-left), the hydronium ion gives up a proton and thus is an acid, while the chloride ion, Cl^-, accepts the proton and so is a base. It is still a common practice to use equations such as A–11 to simplify the representation of the dissociation of an acid, even though it is recognized that equations like A–14 are more correct.

At room temperature, **inorganic salts** are crystalline solids that contain the positive ion (cation) of an Arrhenius base such as NaOH and the negative ion (anion) of an acid such as HCl. Salts can be produced by mixing solutions of appropriate acids and bases, allowing a neutralization reaction to occur. In **neutralization reactions**, the acid and base react to form a salt and water. Most salts that form are water soluble and can be recovered by evaporating the water. Equations A–15 and A–16 are neutralization reactions:

$$HCl + NaOH \rightarrow NaCl + H_2O \qquad \text{Eq. A–15}$$
$$H_2SO_4 + Cu(OH)_2 \rightarrow CuSO_4 + 2H_2O \qquad \text{Eq. A–16}$$

When acids or bases are used as solutes in solutions, the concentrations can be expressed as normalities just as they were earlier for salts. An equivalent of acid is the amount that gives up one mole of H^+ in solution. Thus, one mole of HCl is also one equivalent, but one mole of H_2SO_4 is two equivalents. Bases are described in a similar way, but an equivalent is the amount of base that gives one mole of OH^-.

FUNCTIONAL GROUPS OF ORGANIC MOLECULES

The study of organic compounds is simplified by the concept of functional groups. All organic compounds can be classified according to the functional group or groups they contain. **Functional groups** are specific combinations of atoms that generally react in the same way, regardless of the number of carbon atoms in the molecule to which they are attached. For example, all *aldehydes* contain a functional group that contains one carbon atom, one oxygen atom, and one hydrogen atom covalently bonded in a specific way

$$\overset{O}{\underset{}{\overset{\|}{(-C-H)}}}.$$

The carbon atom in an aldehyde group forms a single covalent bond with the hydrogen atom and a **double bond** (a bond in which two covalent bonds are formed between the same atoms, designated by a double line between the atoms) with the oxygen atom. The aldehyde group is attached to the rest of the molecule by a single covalent bond extending to the left of the carbon atom. Most reactions of aldehydes involve this group, so most aldehyde reactions are the same regardless of the size and nature of the rest of the molecule to which the aldehyde group is attached. Reactions of physiological importance often occur between two functional groups or between one functional group and a small molecule such as water.

CARBOHYDRATES

Carbohydrates are organic compounds of tremendous biological and commercial importance. They are widely distributed in nature and include such familiar substances as starch, table sugar, and cellulose. Carbohydrates have four important functions in living organisms: they provide energy, they supply carbon atoms for the synthesis of cell components, they serve as a stored form of chemical energy, and they form part of the structural elements of some cells.

Carbohydrates contain carbon, hydrogen, and oxygen. They acquired their name because most of them contain these three elements in an atomic ratio of one carbon to two hydrogens to one oxygen. This ratio suggests that the general formula is CH_2O and that the compounds are simply carbon hydrates or carbohydrates. It is now known that they are not hydrates of carbon, but the name persists. All carbohydrates have a large number of functional groups per molecule. The most common functional groups in carbohydrates are *alcohol, ketone* and *aldehyde:*

$$(- OH), (- \overset{\overset{\text{O}}{\|}}{C} -), (- \overset{\overset{\text{O}}{\|}}{C} - H),$$
alcohol   ketone   aldehyde

or functional groups formed by reactions between pairs of these three.

The simplest carbohydrates are simple sugars, also called **monosaccharides**. As their name indicates, they consist of single (*mono* means "one") units called saccharides. The molecular structure of *glucose,* an important monosaccharide, is shown in • Figure A–3a. In solution, most glucose molecules assume the ring form shown in Figure A–3b. Other common monosaccharides are *fructose, galactose,* and *ribose.*

• **FIGURE A–3   Forms of Glucose**   (a) Chain. (b) Ring.

(a)

(b)

• **FIGURE A–4   A Simplified Representation of Glycogen**

Disaccharides (*di* means "two") are sugars formed by a reaction between two monosaccharide molecules. Some common examples of disaccharides are *sucrose* (common table sugar) and *lactose* (milk sugar). Sucrose molecules are formed from one glucose and one fructose molecule. Lactose molecules each contain one galactose and one glucose unit.

The large number of functional groups on carbohydrate molecules makes it possible for large numbers of simple carbohydrate molecules to bond together and form long chains and branched networks. These substances are called **polysaccharides** (*poly* means "many"), a name that indicates they contain many saccharide units. Three common polysaccharides that are made up entirely of glucose units are glycogen, starch, and cellulose.

Glycogen is a storage carbohydrate found in animals. It is a highly branched polysaccharide that averages a branch every eight to twelve glucose units. The structure of glycogen is represented in • Figure A–4, where each circle represents one glucose unit.

Starch, a storage carbohydrate of plants, consists of two fractions, amylose and amylopectin. Amylose consists of long, essentially unbranched chains of glucose units. Amylopectin is a highly branched network of glucose units averaging twenty-four to thirty glucose units per branch. Thus, it is less highly branched than glycogen. *Cellulose,* a structural carbohydrate of plants, exists in the form of long, unbranched chains of glucose units.

The bonding between the glucose units of cellulose is slightly different than the bonding between the glucose units of glycogen and starch. Humans have digestive enzymes that catalyze the breaking (hydrolysis) of the glucose-to-glucose bonds in starch but lack the necessary enzymes to hydrolyze cellulose glucose-to-glucose bonds. Thus, starch is a food for humans, but cellulose is not.

LIPIDS

The group of compounds called lipids is made up of substances with widely different compositions and molecular structures. Unlike carbohydrates, which are classified on the

$$CH_2-OH$$
$$CH-OH$$
$$CH_2-OH$$
Glycerol

$$HO-\overset{\overset{\displaystyle O}{\|}}{C}-(CH_2)_{14}CH_3$$
Fatty Acid
(saturated)

$$HO-\overset{\overset{\displaystyle O}{\|}}{C}-(CH_2)_7CH=CH(CH_2)_7CH_3$$
Fatty Acid
(unsaturated)

$$CH_2-O-\overset{\overset{\displaystyle O}{\|}}{C}-(CH_2)_7CH=CH(CH_2)_7CH_3$$
$$CH-O-\overset{\overset{\displaystyle O}{\|}}{C}-(CH_2)_{14}CH_3$$
$$CH_2-O-\overset{\overset{\displaystyle O}{\|}}{C}-(CH_2)_{16}CH_3$$
Triglyceride

• **FIGURE A–5   Triglyceride Components and Structure**

basis of their molecular structure, substances are classified as lipids on the basis of their solubility. **Lipids** are compounds that are insoluble in water but soluble in nonpolar solvents. Thus, lipids are the waxy, greasy, or oily compounds found in plants and animals. Lipids repel water, a useful characteristic of the protective wax coatings found on some plants. Fats and oils are energy-rich and have relatively low densities. These properties account for the use of fats and oils as stored energy in plants and animals. Still other lipids occur as structural components, especially in cellular membranes.

Simple lipids contain just two types of components, fatty acids and alcohols. **Fatty acid molecules** consist of a hydrocarbon chain with a *carboxylic acid* functional group (—COOH) on the end. The hydrocarbon chain can be of variable length, but natural fatty acids always contain an even number of carbon atoms. The hydrocarbon chain can also contain one or more double bonds between carbon atoms. Fatty acids with no double bonds are called **saturated fatty acids**, whereas those with double bonds are called **unsaturated fatty acids**. The more double bonds present, the higher the degree of unsaturation. The most common alcohol found in simple lipids is **glycerol** (glycerin), a three-carbon alcohol that has three alcohol functional groups (—OH).

Simple lipids called fats and oils are formed by a reaction between the carboxylic acid group of three fatty acids and the three alcohol groups of glycerol. The resulting lipid is called

a **triglyceride** or **triacylglycerol**. Such lipids are classified as fats or oils on the basis of their melting points. *Fats* are solids at room temperature, whereas *oils* are liquids. Their melting points depend on the degree of unsaturation of the fatty acids of the triglyceride. The melting point goes down with increasing degree of unsaturation. Thus, oils contain more unsaturated fatty acids than do fats. Examples of the components of fats and oils and a typical triglyceride molecule are shown in • Figure A–5.

When triglycerides form, a molecule of water is released as each fatty acid reacts with glycerol. Adipose tissue in the body contains triglycerides. When the body uses adipose tissue as an energy source, the triglycerides react with water to release free fatty acids into the blood. The fatty acids can be used as an immediate energy source by many organs. In the liver, free fatty acids are converted into compounds called **ketone bodies**. Two of the ketone bodies are acids and one is the ketone called acetone.

Complex lipids contain more than two types of components. The different complex lipids usually contain three or more of the following components: glycerol, fatty acids, phosphoric acid, an alcohol other than glycerol, and a carbohydrate. Those that contain phosphoric acid are called **phospholipids**. • Figure A–6 contains representations of a few complex lipids; it emphasizes the components but does not give details of the molecular structures.

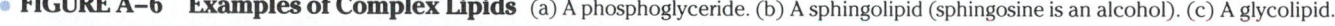

• **FIGURE A–6   Examples of Complex Lipids**   (a) A phosphoglyceride. (b) A sphingolipid (sphingosine is an alcohol). (c) A glycolipid.

(a)   (b)   (c)

A Review of Chemical Principles   **A–9**

(a)

(b)

● **FIGURE A–7   The Steroid Ring System** (a) Detailed. (b) Simplified.

Steroids are lipids that have a unique structural feature consisting of a fused carbon ring system containing three six-membered rings and a single five-membered ring (● Fig. A–7). Different steroids possess this characteristic ring struc-

● **FIGURE A–8   Examples of Steroidal Compounds**

Cholesterol

Cortisol

ture but have different functional groups and carbon chains attached.

Cholesterol, a steroidal alcohol, is the most abundant steroid in the human body. It is a component of cell membranes and is used by the body to produce other important steroids that include bile salts, male and female sex hormones, and adrenocortical hormones. The structures of cholesterol and cortisol, an important adrenocortical hormone, are given in ● Figure A–8.

PROTEINS

The name **protein** is derived from the Greek word *proteios*, which means "of first importance." It is certainly an appropriate term for these very important biological compounds. Proteins are indispensable components of all living things, where they play crucial roles in all biological processes.

Proteins are macromolecules made up of subunits called **amino acids**. Hundreds of different amino acids, both natural and synthetic, are known, but only twenty are commonly found in natural proteins. Each amino acid molecule has three important parts: an amino functional group ($-NH_2$), a carboxylic acid functional group ($-COOH$), and a characteristic side chain or R group. These components are shown in ● Figure A–9.

Amino acids form long chains as a result of reactions between the amino group of one amino acid and the carboxylic acid group of another amino acid. This reaction is illustrated in Equation A–17, in which the carboxylic acid group is shown in an expanded form for clarity:

Eq. A–17

Notice that after the two molecules react, the ends of the product still have an amino group and a carboxylic acid group that can react to extend the chain length. The covalent bond formed in the reaction is called a **peptide bond**.

On a molecular scale, proteins are immense molecules.

● **FIGURE A–9   The General Structure of Amino Acids**

Side chain (different for each amino acid)

Thr—Lys—Pro—Thr—Tyr—Phe—Phe—Gly—Arg— · · · · · ·

● **FIGURE A–10 A Portion of the Primary Protein Structure of Human Insulin**

Their size can be illustrated by comparing a glucose molecule to a molecule of hemoglobin, a protein. Glucose has a molecular weight of 180 amu and a molecular formula of $C_6H_{12}O_6$. Hemoglobin, a relatively small protein, has a molecular weight of 65,000 amu and a molecular formula of $C_{2952}H_{4664}O_{832}N_{812}S_8Fe_4$. The many atoms in a protein are not arranged in a random way. In fact, proteins have a high degree of structural organization that plays an important role in their behavior in the body.

The first level of protein structure is called the **primary structure**. It is simply the order in which amino acids are bonded together to form the protein chain. Amino acids are frequently represented by three-letter abbreviations such as Gly for glycine and Arg for arginine. When this practice is followed, the primary structure of a protein can be represented as shown in ● Figure A–10, which shows part of the primary structure of human insulin.

The second level of protein structure, called the **secondary structure**, results when hydrogen bonding occurs between the amino hydrogen of one amino acid in the primary chain and the carboxyl oxygen

$$\underset{(-\,C\,-)}{\overset{\overset{O}{\|}}{}}$$

of another amino acid in the same or another chain. When the hydrogen bonding occurs between amino acids in the same chain, the chain assumes a coiled, helical shape called the alpha (α) helix, which is by far the most common secondary structure found in natural proteins (● Fig. A–11).

The third level of structure in proteins is the **tertiary structure**. It results when functional groups of the side chains of amino acids in the protein chain react with each other. Several different types of interactions are possible, as shown in ● Figure A–12. Tertiary structures can be visualized by letting

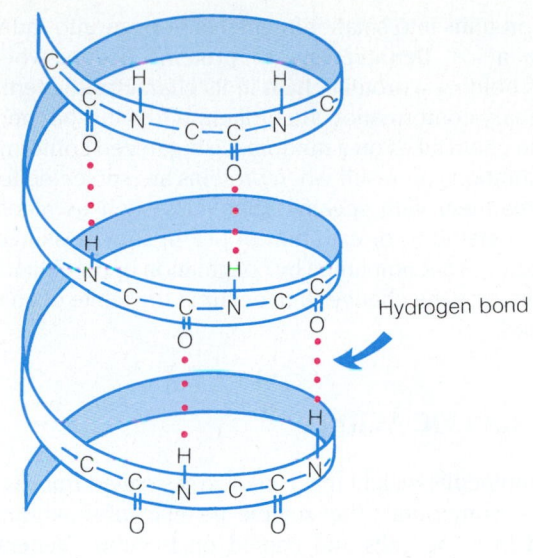

Hydrogen bond

α-helix

● **FIGURE A–11   Secondary Structures of Proteins**

a length of wire represent the chain of amino acids in the primary structure of a protein. Next, imagine the wire is wound around a pencil to form a helix, which represents the secondary structure. The pencil is removed, and the helical structure is now folded back on itself or carefully wadded into a ball. Such folded or spherical structures represent the tertiary structure of a protein.

One of the important functions of proteins is to serve as enzymes that catalyze the many essential chemical reactions of the body. In addition to catalyzing reactions, proteins can undergo reactions themselves. Two of the most important are hydrolysis and denaturation. Notice that according to Equation A–17, the formation of peptide bonds releases water molecules. Under appropriate conditions, it is possible to reverse such reactions by adding water to the peptide bonds and breaking them. **Hydrolysis** reactions of this type convert

● **FIGURE A–12   Side Chain Interactions Leading to Tertiary Protein Structure**

large proteins into smaller fragments or even into individual amino acids. **Denaturation** of proteins occurs when the bonds holding a protein chain in its characteristic tertiary or secondary conformation are broken. When this happens, the protein chain takes on a random, disorganized conformation. Denaturation can result when proteins are subjected to heating, treatment with specific chemicals such as alcohol or heavy metal ions, or extremes of pH. In some instances, denaturation is accompanied by coagulation or precipitation, as illustrated by the changes that occur in the white of an egg as it is fried.

⊜ NUCLEIC ACIDS

High-molecular-weight macromolecules called **nucleic acids** are the compounds that enable genetic information to be stored in living cells and passed on to future generations. These important biomolecules are classified into two categories: **ribonucleic acids** (**RNA**) and **deoxyribonucleic acids** (**DNA**). Deoxyribonucleic acids are found primarily in the nuclei of cells, and ribonucleic acids are found primarily in the cytoplasm that surrounds cell nuclei.

Both types of nucleic acid are made up of units called **nucleotides**, which, in turn, are composed of three simpler components. Each nucleotide contains an organic base, a sugar, and phosphoric acid. The three components are chemically bonded together with the sugar molecule lying between the base and the phosphoric acid. In RNA, the sugar is ribose, whereas in DNA it is deoxyribose. When nucleotides bond together to form nucleic acid chains, the bonding is between the phosphoric acid of one nucleotide and the sugar of another. Thus, the resulting nucleic acids consist of chains of alternating phosphoric acid and sugar molecules, with a base molecule extending out of the chain from each sugar molecule (see Fig. B–1, p. B–2).

The chains of nucleic acid assume structural features somewhat like those found in proteins. DNA occurs in the form of two chains that mutually coil around one another to form the well-known double helix. Some RNA occurs in essentially straight chains, while in other types the chain forms specific loops or helices.

⊜ HIGH-ENERGY BIOMOLECULES

Certain molecules in the body store energy that is released during the metabolism of foods and make it available to the parts of the cells where it is needed to do specific cellular work. The primary substance that performs this function is **adenosine triphosphate**, or **ATP**. The structure of this molecule is shown in ● Figure A–13. Energy is stored in the phosphate bonds of this molecule and is released to the cells when a phosphate bond reacts with water to form adenosine diphosphate (ADP) and inorganic phosphate (P_i) (Eq. A–18).

$$ATP \rightarrow ADP + P_i + \text{energy for use by cell} \qquad \text{Eq. A–18}$$

More energy is released when the second phosphate reacts with water in the same way and ADP is converted to adenosine monophosphate (AMP).

Under the influence of an enzyme, AMP can be converted to a cyclic form called **cyclic AMP** or **cAMP**, which affects the activities of a number of enzymes that are involved in important reactions in the body.

● **FIGURE A–13   The Structure of ATP**

B Storage, Replication, and Expression of Genetic Information

 APPENDIX B CONTENTS AT A GLANCE

DEOXYRIBONUCLEIC ACID (DNA) AND CHROMOSOMES
DNA structure; genes
Chromosomes and chromatin
PROTEIN SYNTHESIS
Complementary base pairing
Types of RNA
Steps of protein synthesis
CELL DIVISION
Mitosis
Meiosis
MUTATIONS

 ## DEOXYRIBONUCLEIC ACID (DNA) AND CHROMOSOMES

The nucleus perpetuates the genetic blueprint and serves as the control center of the cell.

The nucleus of the cell houses **deoxyribonucleic acid** (**DNA**), the genetic blueprint that is unique for each individual. This genetic material serves two essential functions. First, DNA contains "instructions" for assembling the structural and enzymatic proteins of the cell. Cellular enzymes, in turn, control the formation of other cellular structures and also determine the functional activity of the cell by regulating the rate at which metabolic reactions proceed. The nucleus serves as the cell's control center by directly or indirectly controlling almost all cell activities through the role its DNA plays in governing protein synthesis. Since cells make up the body, the DNA code determines the structure and function of the body as a whole. The DNA an organism possesses not only dictates whether the organism is a human, a toad, or a pea but also determines the unique physical and functional characteristics of that individual, all of which ultimately depend on the proteins produced under DNA control. Second, by replicating (making copies of itself), DNA perpetuates the genetic blueprint within all new cells formed within the body and is responsible for passing on genetic information from parents to children. We will first examine the coding mechanism used by DNA and then turn our attention to the means by which DNA replicates itself and controls protein synthesis.

Deoxyribonucleic acid is a double helix composed of nucleotides arranged in a particular sequence unique for each individual.

Deoxyribonucleic acid is a huge molecule, composed in humans of millions of nucleotides arranged into two long, paired strands that spiral around each other to form a double helix. Each **nucleotide** has three components: (1) a *nitrogenous base,* a ring-shaped organic molecule containing nitrogen; (2) a five-carbon ring-shaped sugar molecule, which in the case of DNA is *deoxyribose;* and (3) a phosphate group. Nucleotides are joined end to end by linkages between the sugar of one nucleotide and the phosphate group of the adjacent nucleotide to form a long polynucleotide ("many nucleotide") strand with a sugar-phosphate backbone and bases projecting out one side (● Fig. B–1). There are four different bases in DNA—the double-ringed bases **adenine** (**A**) and **guanine** (**G**) and the single-ringed bases **cytosine** (**C**) and **thymine** (**T**). The two polynucleotide strands within a DNA molecule are wrapped around each other and oriented so that their bases all project to the interior of the helix. The strands are held together by weak hydrogen bonds (see p. A–4) formed between the bases of adjoining strands. Base pairing is highly specific: adenine only pairs with thymine and guanine only pairs with cytosine (● Fig. B–2).

The composition of the repetitive sugar-phosphate backbones that form the "sides" of the DNA "ladder" is identical for every molecule of DNA, but the sequence of the linked bases that form the "rungs" varies among different DNA molecules. The particular sequence of bases in a DNA molecule serves as "instructions" or a "code" that dictates the assembly of amino acids into a given order for the synthesis of specific **polypeptides** (chains of amino acids linked by peptide bonds; see p. A–10). A **gene** is a stretch of DNA that codes for the synthesis of a particular polypeptide. Polypeptides, in turn, are folded into a three-dimensional configuration to form a functional protein. Not all portions of a DNA molecule code for structural or enzymatic proteins. Some stretches of DNA code for proteins that regulate genes. Other segments appear to be important in organizing and packaging DNA

Thymine

Base

Adenine

Cytosine

Guanine

Phosphate

Nucleotide

Sugar

= Sugar-phosphate backbone of polynucleotide strand

● **FIGURE B–1   Polynucleotide Strand**   Sugar-phosphate bonds link adjacent nucleotides together to form a polynucleotide strand with bases projecting to one side. The sugar-phosphate backbone is identical in all polynucleotides, but the sequence of the bases varies.

A = Adenine	G = Guanine		= Sugar-phosphate backbone
T = Thymine	C = Cytosine		

● FIGURE B–2 Complementary Base Pairing in DNA (a) Two polynucleotide strands held together by weak hydrogen bonds formed between the bases of adjoining strands—adenine always paired with thymine and guanine always paired with cytosine. (b) Arrangement of the two bonded polynucleotide strands of a DNA molecule into a double helix.

within the nucleus. Still other regions are "nonsense" base sequences that have no apparent significance.

Deoxyribonucleic acid is precisely packaged within the nucleus.

The DNA molecules within each human cell, if lined up end to end, would extend more than 2 m (2,000,000 μm), yet these molecules are packed into a nucleus that is only 5 μm in diameter. These molecules are not randomly crammed into the nucleus but are precisely organized into **chromosomes**. Each chromosome consists of a different DNA molecule and contains a unique set of genes. **Somatic** (body) **cells** contain forty-six chromosomes (the **diploid number**), which can be sorted into twenty-three pairs on the basis of various distinguishing features. Chromosomes composing a matched pair are termed **homologous chromosomes**, one member of

each pair having been derived from the individual's maternal parent and the other member from the paternal parent. **Germ (reproductive) cells** (that is, sperm and eggs) contain only one member of each homologous pair for a total of twenty-three chromosomes (the **haploid number**). Union of a sperm and an egg results in a new diploid cell with forty-six chromosomes, consisting of a set of twenty-three chromosomes from the mother and another set of twenty-three from the father.

The packaging and compression of DNA molecules into discrete chromosomal units are accomplished in part by nuclear proteins associated with DNA. Two classes of proteins—histone and nonhistone proteins—bind with DNA. **Histones** form bead-shaped bodies that play a key role in packaging DNA into its chromosomal structure. The **nonhistones** are believed to be important in gene regulation. The complex formed between the DNA and its associated proteins

● FIGURE B–3   Levels of Organization of DNA   (a) Double helix of DNA molecule. (b) DNA molecule wound around histone proteins, forming a "beads-on-a-string" structure. (c) Further folding and supercoiling of DNA-histone complex. (d) Rodlike chromosomes, the most condensed form of DNA, which are visible in the cell's nucleus during cell division.

is known as **chromatin**. The long threads of DNA within a chromosome are wound round histones at regular intervals, thus compressing a given DNA molecule to about one-sixth its fully extended length. This "beads-on-a-string" structure is further folded and supercoiled into higher and higher levels of organization to further condense DNA into rodlike chromosomes that are readily visible by means of a light microscope during cell division (● Fig. B–3). When the cell is not dividing, the chromosomes partially "unravel" or decondense to a less compact form of chromatin that is indistinct under a light microscope but appears as thin strands and clumps with an electron microscope. The decondensed form of DNA is its working form; that is, it is the form used as a template for protein assembly.

≡● PROTEIN SYNTHESIS

Complementary base pairing serves as the foundation for both DNA replication and the initial step of protein synthesis.

During replication, the two DNA strands "unzip" as the weak bonds between the paired bases are enzymatically broken. New nucleotides present within the nucleus pair with the exposed bases from each strand (● Fig. B–4). New adenine-bearing nucleotides pair with exposed thymine-bearing nucleotides in an old strand, and new guanine-bearing nucleotides pair with exposed cytosine-bearing nucleotides in an old strand. This complementary base pairing is initiated at one end of the old strands and proceeds in an orderly fashion

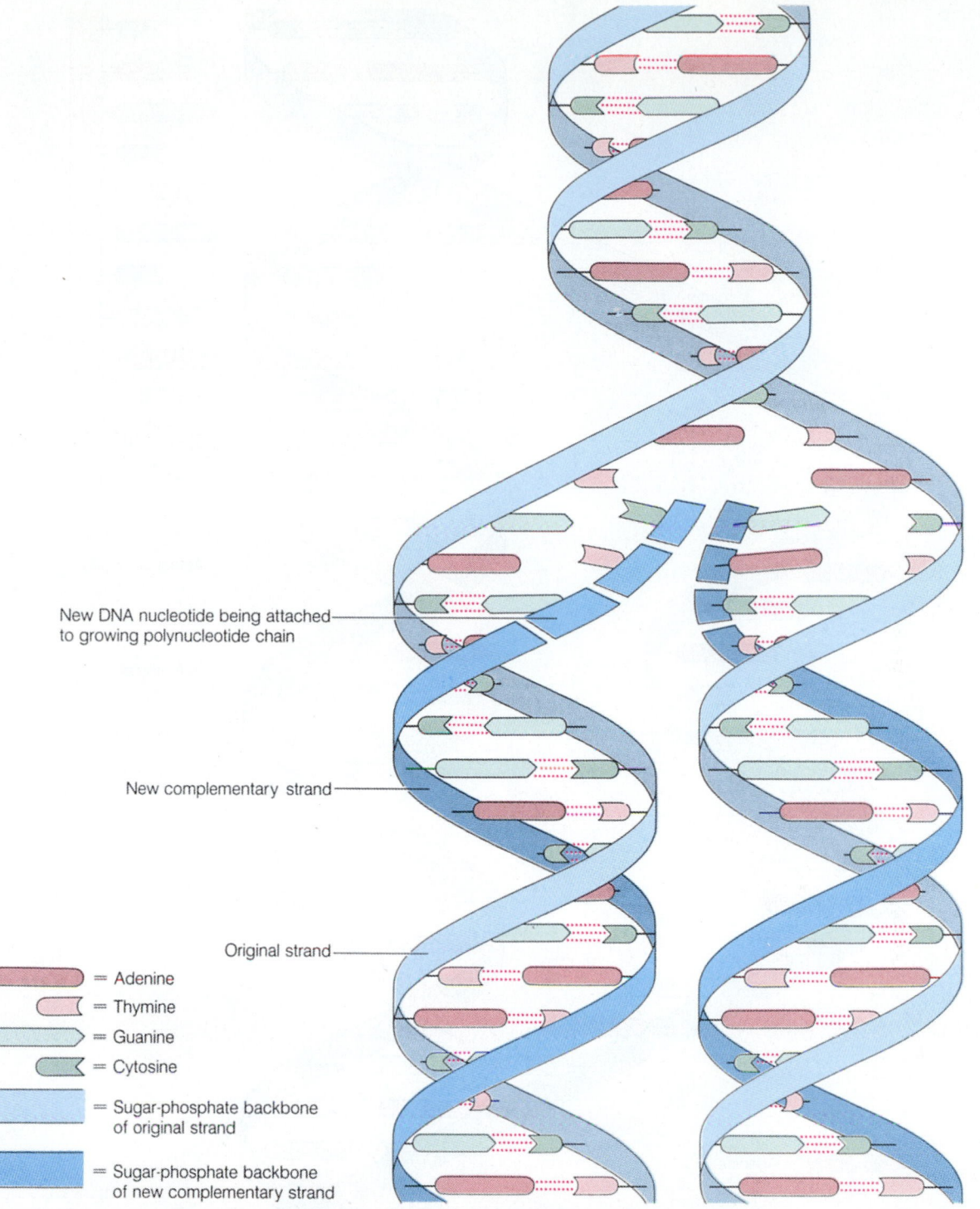

New DNA nucleotide being attached to growing polynucleotide chain

New complementary strand

Original strand

= Adenine
= Thymine
= Guanine
= Cytosine
= Sugar-phosphate backbone of original strand
= Sugar-phosphate backbone of new complementary strand

● **FIGURE B–4 Complementary Base Pairing during DNA Replication** During DNA replication, the DNA molecule is unzipped, and each old strand directs the formation of a new strand; the result is two identical double-helix DNA molecules.

to the other end. The new nucleotides attracted to and thus aligned in a prescribed order by the old nucleotides are sequentially joined by sugar-phosphate linkages to form two new strands that are complementary to each of the old strands. This replication process results in two complete double-stranded DNA molecules, one strand within each molecule having come from the original DNA molecule and one strand having been newly formed by complementary base pairing. These two DNA molecules are both identical to the original DNA molecule, with the "missing" strand in each of the original separated strands having been produced as a result of the imposed pattern of base pairing. This replication process, which occurs only during cell division, is essential for assuring the perpetuation of the genetic code in both of

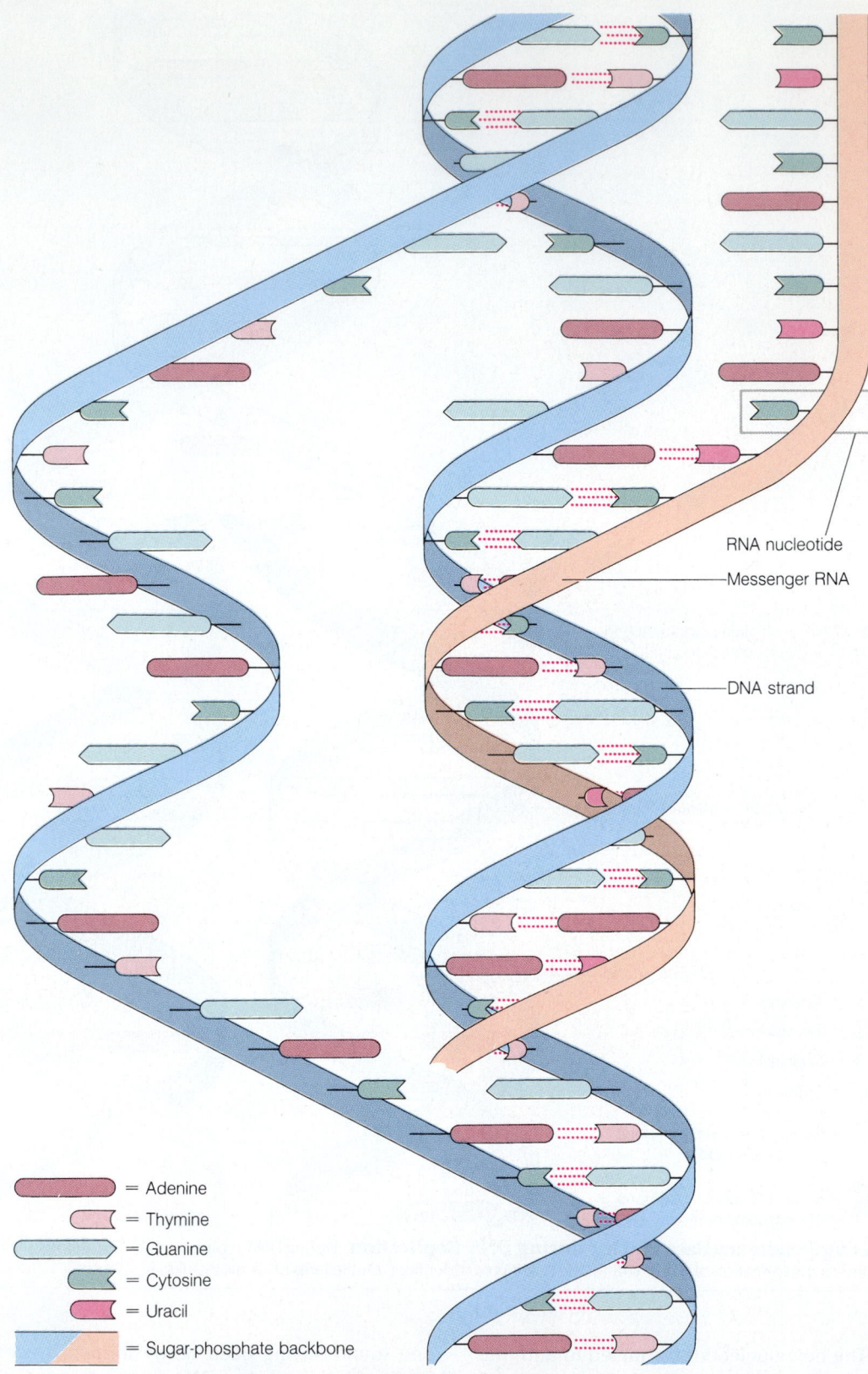

RNA nucleotide

Messenger RNA

DNA strand

= Adenine
= Thymine
= Guanine
= Cytosine
= Uracil
= Sugar-phosphate backbone

• **FIGURE B–5 Complementary Base Pairing during DNA Transcription** During DNA transcription, a messenger RNA molecule is formed as RNA nucleotides are assembled by complementary base pairing at a given segment of one strand of an unzipped DNA molecule (that is, a gene).

the new daughter cells. The duplicate copies of DNA are separated and evenly distributed to the two halves of the cell before it divides.

At other times when DNA is not replicating in preparation for cell division, it serves as a blueprint for dictating cellular protein synthesis. How is this accomplished when DNA is sequestered within the nucleus and protein synthesis is carried out by ribosomes within the cytoplasm? Several types of another nucleic acid, **ribonucleic acid** (**RNA**), serve as the "go-between." Ribonucleic acid differs structurally from DNA in three regards: (1) the five-carbon sugar in RNA is *ribose* instead of deoxyribose, the only difference between them being the presence in ribose of a single oxygen atom that is absent in deoxyribose; (2) RNA contains the closely related base **uracil** instead of thymine, with the three other bases being the same as in DNA; and (3) RNA is single-stranded and not self-replicating. All RNA molecules are produced in the nucleus using DNA as a template or mold, then exit the nucleus through openings in the nuclear membrane known as **nuclear pores**. These pores are large enough for passage of RNA molecules but preclude passage of the much larger DNA molecules.

The DNA instructions for assembling a particular protein coded in the base sequence of a given gene are "transcribed" into a molecule of **messenger RNA** (**mRNA**). The segment of the DNA molecule to be copied uncoils, and the base pairs separate to expose the particular sequence of bases in the gene. In any given gene, only one of the DNA strands is used as a template for transcribing RNA, with the copied strand varying for different genes along the same DNA molecule. The beginning and end of a gene within a DNA strand are designated by particular base sequences that serve as "start" and "stop" signals. **Transcription** is accomplished by complementary base pairing of free RNA nucleotides with their DNA counterparts in the exposed gene (• Fig. B–5). The same pairing rules apply except that uracil, the RNA nucleotide substitute for thymine, pairs with adenine in the exposed DNA nucleotides. As soon as the RNA nucleotides pair with their DNA counterparts, sugar-phosphate bonds are formed to join the nucleotides together into a single-stranded RNA molecule that is released from DNA once transcription is complete. The original conformation of DNA is then restored. The RNA strand is much shorter than a DNA strand, because only a one-gene segment of DNA is transcribed into a single RNA molecule. The length of the finished RNA transcript varies, depending on the size of the gene. Within its nucleotide base sequence, this RNA transcript contains instructions for assembling a particular protein. Note that the message is coded in a base sequence that is *complementary to, not identical to,* the original DNA code.

Messenger RNA delivers the final coded message to the ribosomes for **translation** into a particular amino acid sequence to form a given protein. Thus, genetic information flows from DNA (which can replicate itself) through RNA to protein. This is accomplished first by *transcription* of the DNA code into a complementary RNA code, followed by *translation* of the RNA code into a specific protein (• Fig. B–6). The

• **FIGURE B–6   Flow of Genetic Information from DNA through RNA to Protein by Transcription and Translation**

structural and functional characteristics of the cell as determined by its protein composition can be varied, subject to control, depending on which genes are "switched on" to produce mRNA.

Free nucleotides present in the nucleus cannot be randomly joined together to form either DNA or RNA strands, because the enzymes required to link together the sugar and phosphate components of nucleotides are active only when bound to DNA. This assures that DNA, mRNA, and protein assembly occur only according to genetic plan.

Three forms of RNA participate in protein synthesis.

Besides messenger RNA, two other forms of RNA are required for translation of the genetic message into cellular protein: these are ribosomal RNA and transfer RNA. Messenger RNA carries the coded message from nuclear DNA to a cytoplasmic ribosome, where it directs the synthesis of a particular protein. **Ribosomal RNA** (**rRNA**) is an essential component of ribosomes, the "workbenches" for protein synthesis. Ribosomal RNA "reads" the base-sequence code of mRNA and translates it into the appropriate amino acid sequence during protein synthesis. **Transfer RNA** (**tRNA**) transfers the appropriate amino acids in the cytosol to their designated site in the amino acid sequence of the protein under construction.

Twenty different amino acids are used to construct proteins, yet only four different nucleotide bases are used to code for these twenty amino acids. In the "genetic dictionary," each different amino acid is specified by a **triplet code** that consists of a specific sequence of three bases in the DNA nucleotide chain. For example, the DNA sequence ACA (adenine, cytosine, adenine) specifies the amino acid cysteine, whereas the sequence ATA specifies the amino acid tyrosine. Each DNA triplet code is transcribed into mRNA as a complementary code word, or **codon**, consisting of a sequenced order of the three bases that pair with the DNA triplet. For example, the DNA triplet code ATA is transcribed as UAU (uracil, adenine, uracil) in mRNA.

Sixty-four different DNA triplet combinations (and, accordingly, sixty-four different mRNA codon combinations) are possible using the four different nucleotide bases (4^3). Of these possible combinations, sixty-one code for a specific amino acid and the remaining three serve as "stop signals." A stop signal acts as a "period" at the end of a "sentence" that specifies the amino acid sequence in a particular protein; that is, ribosomal RNA releases the finished polypeptide product when it reaches a stop codon. Because sixty-one triplet codes each specify a particular amino acid and there are twenty different amino acids, a given amino acid may be specified by more than one base-triplet combination. For example, tyrosine is specified by the DNA sequence ATG as well as by ATA. In addition, one DNA triplet code, TAC (mRNA codon sequence AUG) functions as a "start signal" in addition to specifying the amino acid methionine. This code marks the place on mRNA where translation is to begin so that the message is started at the correct end and thus reads in the right direction. Interestingly, the same genetic dictionary is used universally; a given three-base code stands for the same amino acid in all living things, including microorganisms, plants, and animals.

●**FIGURE B–7 Ribosomal Assembly and Protein Translation** (a) On binding with a messenger RNA (mRNA) molecule, the small ribosomal subunit joins with the large subunit to form a functional ribosome. A transfer RNA (tRNA), charged with its specific amino acid passenger, binds to mRNA by means of complementary base pairing between the tRNA anticodon and the first mRNA codon positioned in the first ribosomal binding site. (b) Another tRNA molecule attaches to the next codon on mRNA positioned in the second ribosomal binding site. (c) The amino acid from the first tRNA is linked to the amino acid on the second tRNA. The first tRNA detaches. (d) The mRNA molecule shifts forward one codon (a distance of a three-base sequence). Another charged tRNA moves in to attach with the next codon on mRNA, which has now moved into the second ribosomal binding site. The amino acids from the tRNA in the first ribosomal site are linked with the amino acid in the second site. This process continues, with the polypeptide chain continuing to grow, until a stop codon is reached and the polypeptide chain is released.

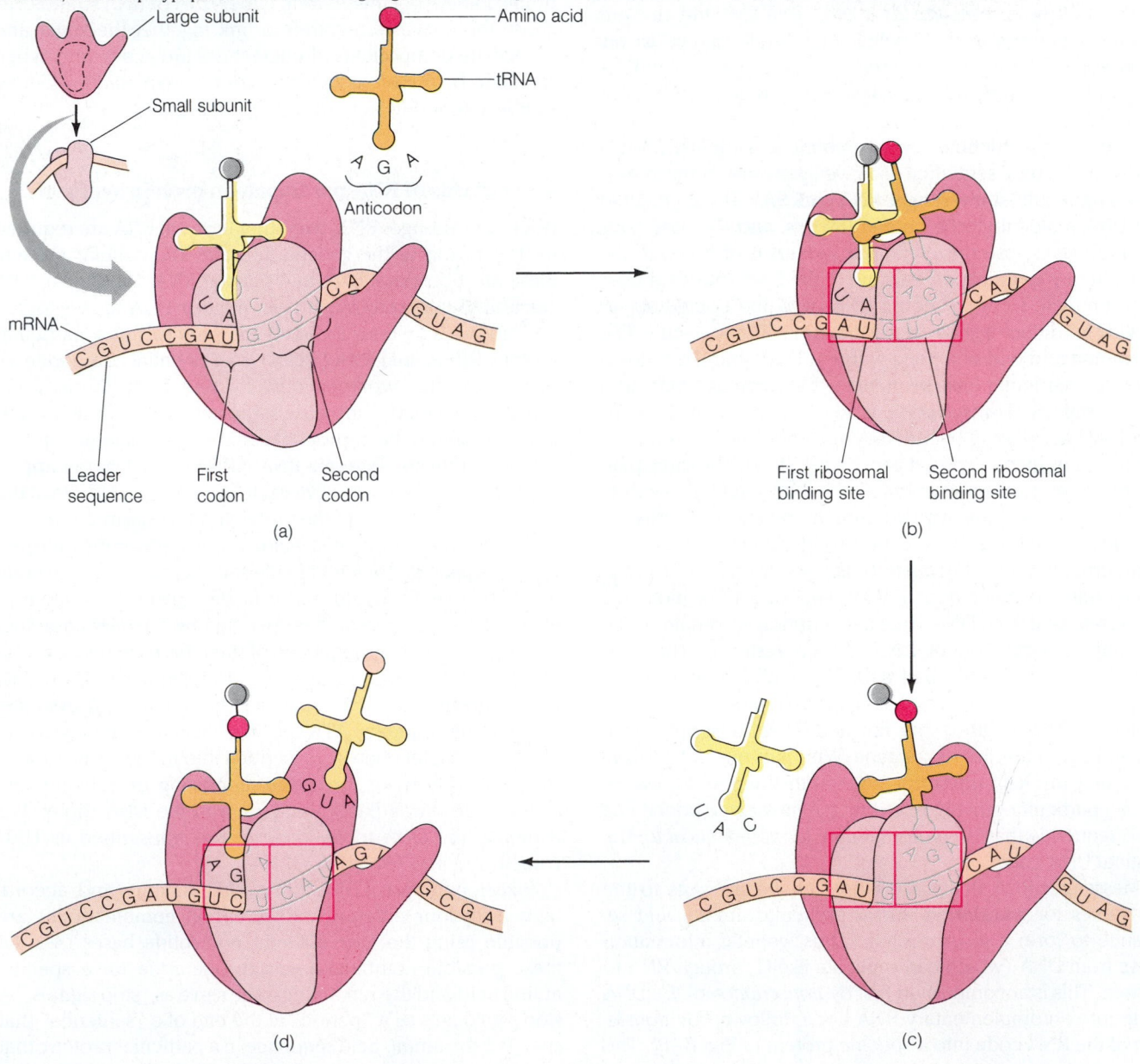

The three steps of protein synthesis are initiation, elongation, and termination.

A ribosome brings together all components that participate in protein synthesis—mRNA, tRNA, and amino acids—and provides the enzymes and energy required for linking the amino acids together. The nature of the protein synthesized by a given ribosome is determined by the mRNA message that is being translated. Each mRNA serves as a code for only one particular polypeptide.

A ribosome is an rRNA-protein structure organized into two subunits of unequal size. Only when a protein is being synthesized are these subunits brought together (•Fig. B–7a). During assembly of a ribosome, an mRNA molecule attaches to the smaller of the ribosomal subunits by means of a *leader sequence*, a section of mRNA that precedes the start codon. The small subunit with mRNA attached then binds to a large subunit to form a complete, functional ribosome. When the two subunits unite, a groove is formed that accommodates the mRNA molecule as it is being translated.

Free amino acids in the cytoplasm are not able to "recognize" and bind directly with their specific codons in mRNA. Transfer RNA is required to bring the appropriate amino acid to its proper codon. Even though tRNA is single-stranded, as are all RNA molecules, it is folded back onto itself into a T shape with looped ends (•Fig. B–8). The open-ended stem portion recognizes and binds to a specific amino acid. There are at least twenty different varieties of tRNA, each able to bind with only one of the twenty different kinds of amino acids. A tRNA is said to be "charged" when it is carrying its passenger amino acid. The loop end of a tRNA opposite the amino acid binding site contains a sequence of three exposed bases, known as the **anticodon**, which is complementary to the mRNA codon that specifies the amino acid being carried. Through complementary base pairing, a tRNA can bind with mRNA and insert its amino acid into the protein under construction only at the site designated by the codon for the amino acid. For example, the tRNA molecule that binds with tyrosine bears the anticodon AUA, which can pair only with the mRNA codon UAU, which specifies tyrosine. This dual binding function of tRNA molecules assures that the correct amino acids are delivered to mRNA for assembly in the order specified by the genetic code. Transfer RNA can only bind with mRNA at a ribosome, so protein assembly does not occur except in the confines of a ribosome.

Protein synthesis is initiated when a charged tRNA molecule bearing the anticodon specific for the start codon binds at this site on mRNA. A second charged tRNA bearing the anticodon specific for the next codon in the mRNA sequence then occupies the site next to the first tRNA (Fig. B–7b). At any given time, a ribosome can accommodate only two tRNA molecules bound to adjacent codons. Through enzymatic ac-

• FIGURE B–8 Structure of tRNA Molecule (a) The open end attaches to free amino acids. (b) The anticodon loop attaches to a complementary mRNA codon.

tion, a peptide bond is formed between the two amino acids that are linked to the stems of the adjacent tRNA molecules (Fig. B-7c). The linkage is subsequently broken between the first tRNA and its amino acid passenger, leaving the second tRNA with a chain of two amino acids. The uncharged tRNA molecule (that is, the one minus its amino acid passenger) is released from mRNA. The ribosome then moves along the mRNA molecule by precisely three bases, a distance of one codon, so that the tRNA bearing the chain of two amino acids is moved into the number one ribosomal site for tRNA. Then, an incoming charged tRNA with a complementary anticodon for the third codon in the mRNA sequence occupies the number two ribosomal site that was vacated by the second tRNA (Fig. B-7d). The chain of two amino acids subsequently binds with and is transferred to the third tRNA to form a chain of three amino acids. Through repetition of this process, amino acids are subsequently added one at a time to a growing polypeptide chain in the order designated by the mRNA codon sequence as the ribosomal translation machinery moves stepwise along the mRNA molecule one codon at a time. This process is rapid. Up to ten to fifteen amino acids can be added per second. Elongation of the polypeptide chain continues until the ribosome reaches a stop codon in the mRNA molecule, at which time the polypeptide is released. The polypeptide is then folded and modified into a full-fledged protein. The ribosomal subunits dissociate and are free to reassemble into another ribosome for translation of other mRNA molecules.

Protein synthesis is energetically expensive. Attachment of each new amino acid to the growing polypeptide chain requires a total investment of splitting four high-energy phosphate bonds—two to charge tRNA with its amino acid, one to bind tRNA to the ribosomal-mRNA complex, and one to move the ribosome forward one codon.

A number of copies of a given protein can be produced from a single mRNA molecule before the latter is chemically degraded. As one ribosome moves forward along the mRNA molecule, a new ribosome attaches at the starting point on mRNA and also starts translating the message. Attachment of many ribosomes to a single mRNA molecule results in a polyribosome. Multiple copies of the identical protein are produced as each ribosome moves along and translates the same message (Fig. B-9). The released proteins are used within the cytosol, except for the few that move into the nucleus through the nuclear pores.

In contrast to the cytosolic polyribosomes, recall that ribosomes directed to bind with the rough endoplasmic reticulum (ER) feed their growing polypeptide chains into the ER lumen. The resultant proteins are subsequently packaged for export out of the cell or for replacement of membrane components within the cell.

Control of gene activity and protein transcription are incompletely understood.

Since each somatic cell in the body has the identical DNA blueprint, you might assume that they would all produce the same proteins. This is not the case, however, because different cell types are able to transcribe different sets of genes and thus synthesize different sets of structural and enzymatic proteins. For example, only red blood cells are able to synthesize hemoglobin, even though all body cells carry the DNA instructions for hemoglobin synthesis. Only about 7% of the DNA sequences in a typical cell are ever transcribed into mRNA for ultimate expression as specific proteins.

Control of gene expression is believed to involve gene regulatory proteins that activate ("switch on") or repress ("switch off") the genes that code for specific proteins within a given cell. Various DNA segments that do not code for structural and enzymatic proteins code for synthesis of these regulatory proteins. The molecular mechanisms by which these regulatory genes, in turn, are controlled in human cells are only beginning to be understood. In some instances, regulatory proteins are controlled by **gene-signaling factors** that bring about differential gene activity among various cells to accomplish specialized tasks. The largest group of known gene-signaling factors in humans is the hormones. Some hormones exert their homeostatic effect by selectively altering the transcription rate of the genes that code for enzymes that are in turn responsible for catalyzing the reaction(s) regulated by the hormone. For example, the hormone cortisol promotes the breakdown of fat stores by stimulating synthesis of the enzyme that catalyzes the conversion of stored fat into its component fatty acids. In other cases, gene action appears to be time-specific; that is, certain genes are expressed only at a certain developmental stage in the individual. This is especially important during embryonic development.

 FIGURE B-9 A Polyribosome A polyribosome is formed by numerous ribosomes simultaneously translating mRNA.

Ribosome   Protein

mRNA

CELL DIVISION

Mitosis is essential for cell reproduction.

Most cells in the human body have the ability to reproduce themselves, a process important in growth, replacement, and repair of tissues. The rate at which cells divide is highly variable. Cells within the deeper layers of the intestinal lining divide every few days to replace cells that are continually sloughed off the surface of the lining into the lumen of the digestive tract. In this way, the entire intestinal lining is replaced about every three days. At the other extreme are nerve cells, which permanently lose the ability to divide beyond a certain period of growth and development. Consequently, when nerve cells are lost through trauma or disease, they cannot be replaced. In between these two extremes are cells that divide infrequently except when needed to replace damaged or destroyed tissue. The factors that control the rate of cell division remain obscure.

Recall that cell division involves two components: nuclear division and cytoplasmic division (cytokinesis). Nuclear division in somatic cells is accomplished by mitosis, in which a complete set of genetic information (that is, a diploid number of chromosomes) is distributed to each of two new daughter cells.

A cell capable of dividing alternates between periods of mitosis and nondivision. The interval of time between cell division is known as **interphase**. Since mitosis takes less than an hour to complete, the vast majority of cells in the body at any given time are in interphase.

Replication of DNA and growth of the cell take place during interphase in preparation for mitosis. Although mitosis is a continuous process, it displays four distinct phases: **prophase**, **metaphase**, **anaphase**, and **telophase** (● Fig. B–10a).

■ Prophase:
1. Chromatin condenses and becomes microscopically visible as chromosomes. The condensed duplicate strands of DNA, known as **sister chromatids**, remain joined together within the chromosome at a point called the **centromere** (● Fig. B–11).
2. The centriole pair divides, and the daughter centrioles move to opposite ends of the cell, where they assemble between them a mitotic spindle made up of microtubules.
3. The membrane surrounding the nucleus starts to break down.

■ Metaphase:
1. The nuclear membrane completely disappears.
2. The forty-six chromosomes, each consisting of a pair of sister chromatids, align themselves at the midline, or equator, of the cell. Each chromosome becomes attached to the spindle by means of several spindle fibers that extend from the centriole to the centromere of the chromosome.

■ Anaphase:
1. The centromeres split, converting each pair of sister chromatids into two identical chromosomes, which separate and move toward opposite poles of the spindle. The spindle fibers are responsible for pulling the chromosomes toward the poles by an unknown mechanism.
2. At the end of anaphase, an identical set of forty-six chromosomes is present at each of the poles, for a transient total of ninety-two chromosomes in the soon-to-be-divided cell.

■ Telophase:
1. The cytoplasm divides through formation and gradual tightening of an actin contractile ring at the midline of the cell, thus forming two separate daughter cells, each with a full diploid set of chromosomes.
2. The spindle fibers disintegrate.
3. The chromosomes uncoil to their decondensed chromatin form.
4. A nuclear membrane reforms in each new cell.

Cell division is complete with the end of telophase. Each of the new cells now enters interphase.

Meiosis is essential for formation of reproductive cells.

Nuclear division in the specialized case of germ cells is accomplished by **meiosis**, in which only a half set of genetic information (that is, a haploid number of chromosomes) is distributed to each daughter cell. Meiosis differs from mitosis in several important regards (Fig. B–10b). Specialized diploid germ cells undergo one chromosome replication followed by two nuclear divisions to produce four haploid germ cells.

■ Meiosis I:
1. During prophase of the first meiotic division, the members of each homologous pair of chromosomes line up side by side to form a **tetrad**, which is a group of four sister chromatids with two identical chromatids within each member of the pair.
2. The process of crossing over occurs during this period, when the maternal and paternal copy of each chromosome are paired. **Crossing over** involves a physical exchange of chromosome material between nonsister chromatids within a tetrad (● Fig. B–12). This process yields new chromosome combinations, thus contributing to genetic diversity.
3. During metaphase, the twenty-three tetrads line up at the equator.
4. At anaphase, homologous chromosomes, each consisting of a pair of sister chromatids joined at the centromere, separate and move toward opposite poles. Maternally and paternally derived chromosomes migrate to opposite poles in random assortments of one member of each chromosome pair without regard for its original derivation. This genetic mixing provides novel new combinations of chromosomes.
5. During the first telophase, the cell divides into two cells. Each cell contains twenty-three chromosomes consisting of two sister chromatids.

■ Meiosis II:
1. Following a brief interphase in which no further replication occurs, the twenty-three unpaired chromosomes line up at the equator, the centromeres split, and the sister chromatids separate for the first time into independent chromosomes that move to opposite poles.

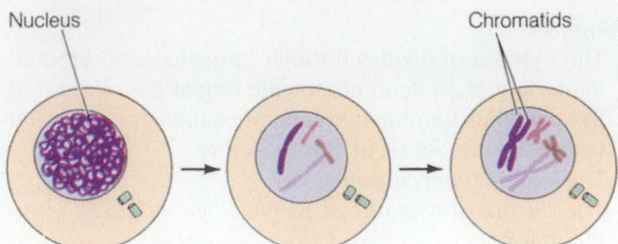

Nucleus   Chromatids

Interphase   DNA replication

(a)

Paired
homologous
chromosomes

Tetrad

Interphase   DNA replication   Prophase I   Metaphase I   Anaphase I

(b)

• **FIGURE B–10   A Comparison of Events in Mitosis and Meiosis** (a) Mitosis. (b) Meiosis.

2. During cytokinesis, each of the daughter cells derived from the first meiotic division forms two new daughter cells. The end result is four daughter cells, each containing a haploid set of chromosomes.

Union of a haploid sperm and haploid egg results in a zygote (fertilized egg) that contains the diploid number of chromosomes. Development of a new multicellular individual from the zygote is accomplished by mitosis and cell differentiation. Since DNA is normally faithfully replicated in its entirety during each mitotic division, all cells in the body possess an identical aggregate of DNA molecules. Structural and functional variations between different cell types result from differential gene expression.

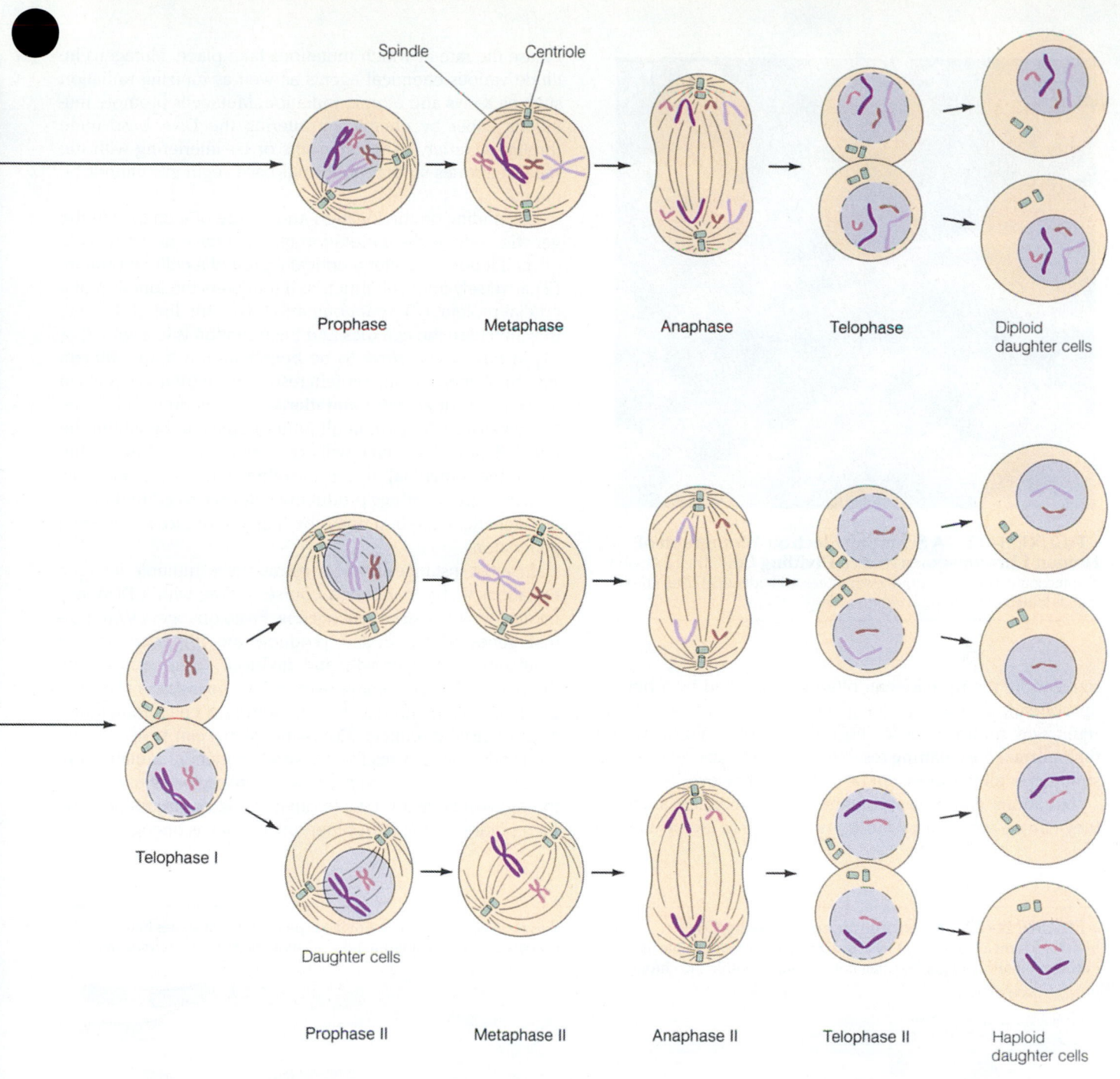

Spindle Centriole

Prophase   Metaphase   Anaphase   Telophase   Diploid daughter cells

Telophase I

Daughter cells

Prophase II   Metaphase II   Anaphase II   Telophase II   Haploid daughter cells

≡ MUTATIONS

Mutations can be harmless, deleterious, fatal, or beneficial.

It is estimated that in the vicinity of 10^{16} cell divisions take place in the body during the course of a person's lifetime to accomplish growth, repair, and normal cell turnover. Because more than 3 billion nucleotides must be replicated during each cell division, it is no wonder that "copying errors" occasionally occur. Any change in the DNA sequence is known as a **point** (or **gene**) **mutation**. A point mutation arises when a base is inadvertently substituted, added, or deleted during the replication process.

When a base is inserted in the wrong position during DNA

• **FIGURE B–11   A Scanning Electron Micrograph of Human Chromosomes from a Dividing Cell** The replicated chromosomes appear as double structures, with identical sister chromatids joined at a common centromere.

replication, the mistake can often be corrected by a built-in "proofreading" system. Repair enzymes remove the newly replicated strand back to the defective segment, at which time normal base pairing resumes to resynthesize a corrected strand. Not all mistakes can be corrected, however.

Mutations can arise spontaneously by chance alone or they can be induced by **mutagens**, which are factors that in-

crease the rate at which mutations take place. Mutagens include various chemical agents as well as ionizing radiation such as X-rays and atomic radiation. Mutagens promote mutations either by chemically altering the DNA base code through a variety of mechanisms or by interfering with the repair enzymes so that abnormal base segments cannot be cut out.

Depending on the location and nature of a change in the genetic code, a given mutation may (1) have no noticeable effect if it does not alter a critical region of a cellular protein; (2) adversely alter cell function if it impairs the function of a crucial protein; (3) be incompatible with the life of the cell, in which case the cell dies and the mutation is lost with it; or (4) in rare cases, prove to be beneficial if a more efficient structural or enzymatic protein results. If a mutation occurs in a body cell (a **somatic mutation**), the outcome will be reflected as an alteration in all future copies of the cell in the affected individual, but it will not be perpetuated beyond the life of the individual. If, on the other hand, a mutation occurs in a sperm- or egg-producing cell (**germ-cell mutation**), the genetic alteration may be passed on to succeeding generations.

In most instances, cancer results from multiple somatic mutations that occur over a course of time within DNA segments known as **proto-oncogenes**. Proto-oncogenes are normal genes whose coded products are important in the regulation of cell growth and division. These genes have the potential of becoming overzealous **oncogenes** ("cancer genes"), which induce the uncontrolled cell proliferation characteristic of cancer. Proto-oncogenes can become cancer producing as a result of several sequential mutations in the gene itself or by changes in adjacent regions that regulate the proto-oncogenes. Less frequently, tumor viruses become incorporated in the DNA blueprint and act as oncogenes.

• **FIGURE B–12   Phenomenon of Crossing Over** (a) During prophase I of meiosis, each homologous pair of chromosomes lines up side by side to form a tetrad. (b) Physical exchange of chromosome material occurs between nonsister chromatids. (c) As a result of this crossing over, new combinations of genetic material are formed within the chromosomes.

Centromere

(a)

(b)

(c)

C Control, Functions, and Interactions of Immune Response Effectors

IMMUNE RESPONSE EFFECTORS	DESCRIPTION	CONTROLLING MECHANISM(S)	FUNCTION(S)
Acute-phase proteins[a]	Collection of proteins from the liver	Stimulated by LEM	Exert multitude of wide-ranging effects associated with inflammation, tissue repair, and immune cell activity
Angry macrophages[b]	Large tissue-bound phagocytic cells	Stimulated by MIF	Display superpower phagocytic ability
Antibodies[a]	Special plasma proteins secreted by plasma cells; same as gamma globulins or immuno-globulins	Production and release stimulated by presence of "matching" antigen, which might be any large, complex molecule that triggers an immune attack against itself	Amplify nonspecific immune responses against the antigen that stimulated their production In some instances, physically hinder antigen by neutralization or agglutination
Autoantibodies[a]	Antibody proteins against self-antigens (own body proteins)	Breakdown of mechanisms for tolerance of self-antigens	Attack one or more of body's own tissues
B lymphocytes (B cells) (see also Plasma cells)[b]	White blood cells	Activated by macrophage-presented antigen; stimulated by helper T cells, IL-1, interferon, B-cell growth factor, and helper T cells	Converted into plasma cells, which produce antibodies Secrete lymphokines
Basophils[b]	White blood cells	Binding of allergen to IgE antibodies coating the basophils	Release histamine; involved in immediate allergic reactions
B-cell growth factor[a]	Lymphokine secreted by helper T cells	Release stimulated by macrophage-presented antigen and IL-1	Enhances antibody secretion by activated B-cell clone
Blocking antibodies[a]	Antibody proteins	Production induced by cancer cells	Protect harmful cancer cells from attack by cytotoxic T cells
Complement system[a]	Group of inactive plasma proteins that can be sequentially activated	Activated by antibody (classical pathway) and by carbohydrate chains on surface of foreign invaders (alternate pathway); several steps stimulated by kinins	Destroys foreign cells by punching holes in their plasma membranes Facilitates every step of inflammatory process (e.g., various components serve as opsonins, stimulate secretion of histamine, or activate kinins)
Cortisol[a]	Steroid hormone secreted by adrenal cortex	Stimulated by stress	Anti-inflammatory action at pharmacological levels and possibly at physiological levels, the latter of which may prevent stress-activated immune responses from overshooting
Chemotaxins[a]	Local chemical mediators	Released from a variety of sources at site of inflammation	Attract phagocytes to site of inflammation
Cytotoxic T cells[b]	Form of T lymphocytes	Stimulated by interferon, IL-1, and helper T cells; activated by macrophage-presented antigen	Destroy virally infected cells or cancer cells Secrete interferon Mediate delayed hypersensitivity

[a] Chemical mediator.
[b] Effector cell.

IMMUNE RESPONSE EFFECTORS	DESCRIPTION	CONTROLLING MECHANISM(S)	FUNCTION(S)
Endogenous pyrogen (EP)[a]	Chemical mediator released by monocytes/macrophages; identical to LEM and IL-1	Released in presence of microbes	Causes development of fever
Eosinophil chemotactic factor[a]	Chemical mediator in allergic responses	Released by mast cells or basophils activated by allergen	Attracts eosinophils to site of allergen contact
Eosinophils[b]	White blood cells	Stimulated by eosinophil chemotactic factor and presence of internal parasites	Involved in allergic manifestations Destroy parasitic worms
Fibrinogen (inactive); fibrin (active)[a]	Final plasma protein in clotting cascade	Activated by tissue thromboplastin in injured area and chemical mediators released from microbe-stimulated phagocytes	Forms interstitial fluid clots to wall off injured or invaded area
Fibroblasts[b]	Connective-tissue cells	Start dividing in area vacated by lost tissue cells	Form scar tissue
Gamma globulin (see Antibodies)[a]			
Helper T cells[b]	Form of T lymphocyte	Activated by macrophage-presented antigen; stimulated by IL-1	Enhance activity of B cells, cytotoxic T cells, and suppressor T cells through secretion of B-cell growth factor and T-cell growth factor (IL-2) Activate macrophages
Histamine[a]	Chemical mediator released from mast cells and basophils	Stimulated by chemical mediators released from microbe-activated phagocytes, complement system, and allergens	Important in inflammation Important in immediate allergic reactions Acts on microcirculation to cause localized vasodilation and increased capillary permeability
Immuno-globulin (see Antibodies)[a]			
Interferon[a]	Family of proteins	Released from cells invaded by virus	Protects noninvaded cells from viral invasion Increases killing ability of cytotoxic T cells, NK cells, and phagocytes Stimulates production of antibodies Slows cell division and suppresses tumor growth; exerts anticancer effect
Interleukin 1 (IL-1)[a]	Chemical mediator released by macrophages; identical to EP and LEM	Released in presence of microbes	Enhances proliferation and differentiation of both B and T lymphocytes
Interleukin 2 (IL-2)[a]	Lymphokine secreted by helper T cells; identical to T-cell growth factor	Released by presence of macrophage-presented antigen; stimulated by IL-1	Augments activity of all T cells
Kallikrein[a]	Chemical mediator released from neutrophils	Released in presence of microbes	Activates kinins
Killer (K) cells[b]	Lymphocytelike cells	Stimulated by antibody on surface of target cell	Lyse plasma membrane of target cell

[a] Chemical mediator.
[b] Effector cell.

IMMUNE RESPONSE EFFECTORS	DESCRIPTION	CONTROLLING MECHANISM(S)	FUNCTION(S)
Kininogen (inactive); kinin (active)[a]	Plasma protein	Stimulated by kallikrein released from neutrophils and by complement component	Stimulates complement system Enhances vasodilation and increases capillary permeability in inflammation Activates nearby pain receptors Acts as powerful chemotaxin
Lactoferrin[a]	Chemical mediator secreted by neutrophils	Stimulated by presence of microbes	Binds iron so it is unavailable for essential use by invading bacteria
Leukocyte endogenous mediator (LEM)[a]	Chemical mediator released by monocytes/ macrophages; identical to EP and IL-1	Released in presence of microbes	Decreases plasma iron Stimulates granulopoiesis Increases release of acute-phase proteins
Lymphocytes[b]	White blood cells—two types: B and T lymphocytes	Stimulated by products specific for each type of cell (*see* specific cells)	Responsible for specific immune defense
Lymphokines[a]	Chemical mediators, other than antibodies, secreted by lymphocytes (including B- and T-cell growth factor and MIF)	Released by presence of macrophage-presented antigen	Enhance all activity of B and T cells Act as chemotaxins Retain macrophages in inflamed area
Macrophage migration inhibition factor (MIF)[a]	Lymphokine secreted by helper T cells	Released by presence of macrophage-presented antigen	Prevents macrophages from migrating out of inflamed area Increases phagocytic power of macrophages (activates angry macrophages)
Macrophages[b]	Large, tissue-bound phagocytic specialists	Stimulated by presence of microbes; enhanced by interferon; activated by helper T cells; stimulated by antibodies on surface of foreign material; attracted by chemotaxins; linked to target cell by opsonin	Important in nonspecific defense; first line of defense Phagocytize foreign material and cellular debris Secrete chemical mediators that exert a variety of effects (*see* Phagocytic secretions) Clear way for tissue repair by removing debris Secrete interferon Process and present antigen to B and T cells, thereby assisting in antibody formation and T-cell sensitization
Memory cells[b]	B and T cells produced during intial infection but remaining dormant until subsequent invasion by same antigen	Stimulated by subsequent invasion of same antigen	Expand activated clone Launch swifter, more powerful attack on subsequent exposure
Monocytes[b]	White blood cells	Various factors stimulate their derivatives, the macrophages (*see* Macrophages)	Converted into large, tissue-bound macrophages after leaving the blood
Natural killer (NK) cells[b]	Lymphocytelike cells	Stimulated by interferon	Nonspecifically lyse virus-infected cells and tumor cells on first exposure to them Secrete interferon
Neutrophils[b]	White blood cells	Stimulated by presence of microbes, by antibodies on surface of foreign material, and by circulating antibodies; attracted by chemotaxins; linked to target cell by opsonins	Important in nonspecific defense Important in inflammatory process; highly mobile phagocytic specialists Release chemicals that exert a variety of effects (*see* Phagocytic secretions)

[a] Chemical mediator.
[b] Effector cell.

IMMUNE RESPONSE EFFECTORS	DESCRIPTION	CONTROLLING MECHANISM(S)	FUNCTION(S)
Opsonins[a]	Antibodies and one of activated proteins of complement system	Produced in response to factors that stimulate antibody secretion (*see* Antibodies) and complement activation (*see* Complement system)	Enhance phagocytosis by linking foreign cell to phagocytic cell
Phagocytic secretions (EP/LEM/IL-1 and others)[a]	Chemical mediators released from neutrophils or macrophages	Released in presence of microbes	Directly kill microbes extracellularly
			Trigger clotting and anticlotting systems
			Activate kinins
			Induce development of fever
			Decrease plasma iron
			Stimulate granulopoiesis
			Stimulate secretion of acute-phase proteins
			Enhance proliferation and differentiation of both B and T cells
Plasma cells[b]	Activated form of B lymphocyte	Specific B cells that match invading antigen proliferate and are converted into antibody-secreting plasma cells; stimulated by B-cell growth factor, helper T cells, and interferon	Secrete customized antibodies specific to invading antigen
Plasminogen (inactive); plasmin (active)[a]	Final plasma protein in anticlotting cascade	Activated by chemical mediators released from microbe-stimulated phagocytes	Slowly dissolves interstitial fluid clot formed during inflammatory response
Prostaglandin[a]	Fatty acid–derived local chemical mediator	Released in hypothalamus by EP	Raises set point of hypothalamic thermostat to produce fever
Slow reactive substance of anaphylaxis (SRS-A)[a]	Chemical mediator released from mast cells and basophils in allergic reactions	Released by binding of allergen to IgE antibodies coating mast cells and basophils	Induces profound, prolonged constriction of small airways
			Increases capillary permeability
Suppressor T cells[b]	Form of T lymphocyte	Stimulated by helper T cells	Limit immune reaction in check-and-balance fashion
T-cell growth factor[a]	Lymphokine secreted by helper T cells	Release stimulated by macrophage-presented antigen and IL-1	Augments activity of all T cells
Thymosin[a]	Collection of hormones produced by thymus	Unknown	Enhances proliferation of T-cell colonies
			Enhances immune capabilities of existing T cells
T lymphocytes (T cells) (see Cytotoxic T cells, Helper T cells, and Suppressor T cells)[b]	White blood cells	Activated by macrophage-presented antigen; stimulated by interferon, IL-1, and helper T cells	Responsible for cell-mediated immunity (kills cells through nonphagocytic means)
			Secrete lymphokines

[a] Chemical mediator.
[b] Effector cell.

D Answers to End-of-Chapter Objective Questions

CHAPTER 1 Homeostasis: The Foundation of Physiology

(Questions on p. 13.)

1. e
2. b
3. c
4. T
5. F (Unicellular organisms and cells in a multicellular organism all perform similar basic functions essential for the cell's survival.)
6. T
7. muscle tissue, nervous tissue, epithelial tissue, connective tissue
8. secretion
9. exocrine, endocrine, hormones
10. intrinsic, extrinsic
11. 1.d, 2.g, 3.a, 4.e, 5.b, 6.j, 7.h, 8.i, 9.c, 10.f

CHAPTER 2 Cellular Physiology

(Questions on p. 43.)

1. plasma membrane
2. deoxyribonucleic acid (DNA), nucleus
3. organelles, cytosol, cytoskeleton
4. endoplasmic reticulum, Golgi complex
5. oxidative, hydrogen peroxide
6. adenosine triphosphate (ATP)
7. F (Human cells cannot be seen by the unaided eye; they are about ten times smaller than the smallest point visible to the naked eye.)
8. F (Lysosomal enzymes are less potent in the cytosol than in the acidic environment within the lysosome. Furthermore, cells can usually tolerate the limited damage accompanying the inadvertent rupture of a few lysosomes because most cell parts are renewable.)
9. T
10. 1.b, 2.a, 3.b
11. 1.b, 2.c, 3.c, 4.a, 5.b, 6.c, 7.a, 8.c

CHAPTER 3 Plasma Membrane and Membrane Potential

(Questions on p. 76.)

1. T
2. F (The hydrophobic regions correspond to the light space; the hydrophilic regions correspond to the two dark layers.)
3. T

4. F (Only 20% of the membrane potential is *directly* generated by the Na^+–K^+ pump. The other 80% is caused by the passive diffusion of K^+ and Na^+ down concentration gradients. The Na^+–K^+ pump *indirectly* contributes to membrane potential by maintaining these concentration gradients.)
5. opening or closing specific membrane channels, activating an intracellular second messenger
6. pinocytosis, phagocytosis, endocytosis
7. negative, positive
8. 1.b, 2.a, 3.b, 4.a, 5.c, 6.b, 7.a, 8.b
9. 1.a, 2.a, 3.b, 4.a, 5.b, 6.a, 7.b
10. 1.c, 2.b, 3.a., 4.a, 5.c, 6.b, 7.c, 8.a, 9.b

CHAPTER 4 Neuronal Physiology

(Questions on p. 102.)

1. T
2. F (The membrane is returned to resting potential after it reaches the peak of an action potential as a result of K^+ efflux. The Na^+–K^+ pump gradually restores the original concentration gradients following an action potential by pumping out the Na^+ and pumping in the K^+ that left during the action potential.)
3. F (Even though K^+ leaves the cell during an action potential, there is still much more K^+ inside than outside the cell following the action potential because only an extremely small percentage of the total intracellular K^+ leaves.)
4. F (Synapses operate in one direction only; the presynaptic neuron influences the postsynaptic neuron, but the postsynaptic neuron cannot influence the presynaptic neuron.)
5. refractory period
6. axon hillock
7. synapse
8. temporal summation
9. spatial summation
10. convergence, divergence
11. 1.b, 2.a, 3.a, 4.b, 5.b, 6.a

CHAPTER 5 Central Nervous System

(Questions on p. 148.)

1. F (The major function of CSF is to serve as a shock-absorbing fluid to prevent brain damage during sudden, jarring movements of the head.)

2. F (The brain cannot produce ATP under anaerobic conditions.)
3. F (Damage to the left hemisphere brings about paralysis and loss of sensation on the right side of the body because of fiber crossover.)
4. T
5. F (The right hemisphere specializes in verbal and analytical skills whereas the left side excels in artistic and musical ability.)
6. F (The brain has considerable plasticity; the organizational pattern of sensory and motor regions of the cortex is subject to constant subtle modifications based on use-dependent competition for cortical space.)
7. habituation
8. consolidation
9. dorsal, ventral
10. receptor, afferent pathway, integrating center, efferent pathway, effector
11. 1.a, 2.c, 3.a and b, 4.b, 5.a, 6.c, 7.c
12. 1.c, 2.e, 3.a, 4.b, 5.f, 6.d
13. 1.d, 2.c, 3.f, 4.e, 5.a, 6.b

CHAPTER 6 Peripheral Nervous System: Afferent Division; Special Senses

(Questions on p. 195.)
1. transduction
2. adequate stimulus
3. T
4. F (Only afferent information that reaches conscious levels of the brain is considered to be sensory information.)
5. T
6. T
7. T
8. F (Displacement of the round window just dissipates pressure. Displacement of the basilar membrane and the resultant activation of the organ of Corti's hair cells generate neural impulses that are perceived as sound sensations.)
9. T
10. F (Each taste receptor responds in varying degrees to all four primary tastes.)
11. F (Adaptation to odors involves some sort of CNS adaptation, not olfactory receptor adaptation.)
12. 1.a, 2.b, 3.c, 4.c, 5.c
13. 1.f, 2.h, 3.l, 4.d, 5.i, 6.e, 7.b, 8.j, 9.a, 10.g, 11.c, 12.k

CHAPTER 7 Peripheral Nervous System: Efferent Division

(Questions on p. 211.)
1. T
2. F (Action potentials are transmitted on a one-to-one basis at neuromuscular junctions, but one action potential in a presynaptic neuron does not bring about one action potential in a postsynaptic neuron at a synapse. A postsynaptic neuron is brought to threshold and has an action potential only upon summation of EPSPs.)

3. c
4. c
5. sympathetic, parasympathetic
6. adrenal medulla
7. 1.a, 2.b, 3.a, 4.b, 5.a, 6.a, 7.b
8. 1.b, 2.b, 3.a, 4.a, 5.b, 6.b, 7.a

CHAPTER 8 Muscle Physiology

(Questions on p. 256.)
1. F (The contractile response lasts about one hundred times longer than the action potential.)
2. F (The velocity of shortening depends on the magnitude of the load as well as on the ATPase activity of its fibers.)
3. F (When a muscle is maximally stretched, the thin filaments are completely pulled out from between the thick filaments so that no cross-bridge activity and consequently no contraction can occur. Maximal tension is achieved when the muscle is at its optimal length and the thin filaments optimally overlap the thick filaments.)
4. T
5. F (Slow wave potentials initiate action potentials only when the automatic depolarizing swing in potential reaches threshold. Whether threshold is reached depends on the starting point of the membrane potential at the onset of its depolarizing swing.)
6. T
7. concentric, eccentric
8. alpha, gamma
9. denervation atrophy, disuse atrophy
10. b
11. a, b, e
12. 1.f, 2.d, 3.c, 4.e, 5.b, 6.g, 7.a
13. 1.a, 2.a, 3.a, 4.b, 5.b, 6.b

CHAPTER 9 Cardiac Physiology

(Questions on p. 297.)
1. intercalated discs, desmosomes, gap junctions
2. bradycardia, tachycardia
3. adenosine
4. F (The left ventricle is a stronger pump because it pumps the same quantity of blood as the right ventricle does but must pump into the high-pressure, high-resistance systemic circulation whereas the right ventricle pumps into the low-pressure, low-resistance pulmonary circulation.)
5. F (The heart is located approximately midline within the thoracic cavity.)
6. F (The only point of electrical contact between the atria and ventricles is the AV node.)
7. T
8. d
9. d
10. e
11. less than, greater than, less than, greater than, less than
12. AV, systole, semilunar, diastole
13. 1.e, 2.a, 3.d, 4.b, 5.f, 6.c, 7.g
14. 1.c, 2.a, 3.b

CHAPTER 10 Blood Vessels and Blood Pressure

(Questions on p. 345.)

1. T
2. F (Blood flow is continuous through the capillaries. The driving force for continued blood flow to the tissues during cardiac diastole is provided by the elastic recoil of the arterial walls that had been stretched by blood during cardiac systole.)
3. T
4. T
5. F (Capillaries vary in pore size.)
6. T
7. a, c, d, e, f
8. 1.b, 2.a, 3.b, 4.a, 5.a, 6.a, 7.b, 8.a, 9.b, 10.a, 11.b, 12.a, 13.a
9. 1.a, 2.a, 3.b, 4.a, 5.b, 6.a

CHAPTER 11 Blood

(Questions on p. 366.)

1. T
2. F (Hemoglobin normally carries CO_2 and H^+ from the tissues to the lungs. It also has a high affinity for carbon monoxide.)
3. T
4. T
5. F (White blood cells are in the blood only while in transit from their site of production and storage in the bone marrow or lymphoid organs to their site of action in the tissues.)
6. T
7. lymphocytes
8. liver
9. d
10. a
11. 1.c, 2.f, 3.b, 4.a, 5.g, 6.d, 7.h, 8.e, 9.f
12. 1.e, 2.c, 3.b, 4.d, 5.g, 6.f, 7.a, 8.h

CHAPTER 12 Body Defenses

(Questions on p. 410.)

1. F (The complement system can also be activated by the presence of any foreign invader.)
2. F (Specific immune responses are accomplished by lymphocytes.)
3. F (In tissues capable of regeneration, lost cells are replaced by cell division of surrounding, healthy, organ-specific cells so that perfect repair is accomplished.)
4. F (Active immunity can also be acquired through vaccination.)
5. T
6. membrane-attack complex
7. pus
8. inflammation
9. opsonin
10. lymphokines
11. b
12. 1.a, 2.a, 3.b, 4.b, 5.c, 6.c, 7.b, 8.a, 9.b, 10.b, 11.a, 12.b

13. 1.b, 2.a, 3.a, 4.b, 5.a, 6.a, 7.a, 8.b
14. 1.c, 2.d, 3.a, 4.b

CHAPTER 13 Respiratory System

(Questions on p. 461.)

1. F (The respiratory muscles directly change the size of the thoracic cavity, not the lungs. There are no muscles in the lungs, except for the smooth muscle in the walls of the airways and blood vessels.)
2. F (A volume of air, known as the residual volume, remains in the lungs after a maximal expiration because air is trapped in the alveoli as a result of compression of the small nonrigid airways by the high intrapleural pressure accompanying the forceful expiration.)
3. T
4. F (The diffusion coefficient for CO_2 is twenty times that of O_2.)
5. F (Hemoglobin has a much higher affinity for CO than for O_2.)
6. F (Rhythmicity of breathing is brought about by pacemaker activity of the inspiratory neurons, not by the respiratory muscles themselves.)
7. F (Impulses from the expiratory neurons to the expiratory muscles occur only during active expiration, not during normal quiet breathing.)
8. transmural pressure gradient, pulmonary surfactant action, alveolar interdependence
9. pulmonary elasticity, alveolar surface tension
10. compliance
11. elastic recoil
12. carbonic anhydrase
13. a
14. 1.d, 2.a, 3.b, 4.a, 5.b, 6.a
15. a.<, b.>, c.=, d.=, e.=, f.=, g.>, h.<, i. approximately =, j. approximately =, k.=, l.=

CHAPTER 14 Urinary System

(Questions on p. 508.)

1. F (Glomerular filtration is a passive process, requiring no energy expenditure by the kidneys.)
2. F (Sodium reabsorption is not under hormonal control in the proximal tubule and loop of Henle. Aldosterone increases sodium reabsorption in the distal and collecting tubules.)
3. T
4. T
5. T
6. nephron
7. potassium
8. 500
9. 1.b, 2.a, 3.b, 4.b, 5.a, 6.b, 7.b, 8.b, 9.b
10. e
11. b
12. b, e, a, d, c
13. c, e, d, a, b, f

14. g, c, d, a, f, b, e
15. 1.a, 2.a, 3.c, 4.b, 5.d

CHAPTER 15 Fluid and Acid-Base Balance

(Questions on p. 539.)
 1. T
 2. F (With an isotonic fluid gain, there is no change in ECF osmolarity to induce a fluid shift between the ECF and cells, so the fluid gain is confined to the ECF compartment.)
 3. F (Salt balance is carefully regulated. Despite wide variations in salt intake, a stable salt balance is maintained by controlling the urinary excretion of salt.)
 4. T
 5. T
 6. intracellular fluid
 7. transcellular fluid
 8. $[H_2CO_3]$, $[HCO_3^-]$
 9. d
 10. b
 11. a, d, e
 12. b, e
 13. c
 14. 1. metabolic acidosis, 2. diabetes mellitus, 3. pH = 7.1, 4. respiratory alkalosis, 5. anxiety, 6. pH = 7.7, 7. respiratory acidosis, 8. pneumonia, 9. pH = 7.1, 10. metabolic alkalosis, 11. vomiting, 12. pH = 7.7

CHAPTER 16 Digestive System

(Questions on p. 593.)
 1. F (The digestive system does not vary nutrient, water, or electrolyte uptake based on body needs, but instead it absorbs all digestible food that is ingested.)
 2. T
 3. T
 4. F (The protein components of digestive secretions and sloughed epithelial cells are digested and absorbed.)
 5. F (The large intestine does not have the ability to absorb nutrients.)
 6. F (The major hormones secreted by the endocrine pancreas are insulin and glucagon. Secretin and CCK, which are secreted by the small intestine, act on the exocrine pancreas.)
 7. long, short
 8. chyme
 9. three
 10. vitamin B_{12}, bile salts
 11. bile salts
 12. b
 13. 1.c, 2.e, 3.b, 4.a, 5.f, 6.d
 14. 1.c, 2.c, 3.d, 4.a, 5.e, 6.c, 7.a, 8.b, 9.c, 10.b, 11.d, 12.b

CHAPTER 17 Energy Balance and Temperature Regulation

(Questions on p. 612.)
 1. F (The excess energy is stored in the body, primarily as adipose tissue.)

 2. F (Only about 25% of the chemical energy in nutrient molecules is harnessed to do biological work.)
 3. F (Oxygen does not contain heat energy. For each liter of O_2 consumed in oxidizing nutrient molecules, 4.8 kilocalories of heat are liberated from the food on the average.)
 4. F (Body temperature normally varies several degrees Fahrenheit.)
 5. T
 6. T
 7. T
 8. T
 9. shivering
 10. nonshivering thermogenesis
 11. sweating
 12. b
 13. e
 14. 1.b, 2.a, 3.c, 4.a, 5.d, 6.c, 7.b, 8.d, 9.c, 10.b

CHAPTER 18 Principles of Endocrinology; Central Endocrine Organs

(Questions on p. 648.)
 1. T
 2. T
 3. F (Some endocrine glands exert nonendocrine effects in addition to secreting hormones.)
 4. T
 5. F (Each steroidogenic organ has a limited set of enzymes for producing only a given type or types of steroid hormones.)
 6. T
 7. F (Inhibition of the anterior pituitary and hypothalamus by a target-organ hormone is known as long-loop negative feedback. Short-loop negative feedback involves inhibition of the hypothalamus by an anterior pituitary hormone.)
 8. T
 9. tropic
 10. down regulation
 11. epiphyseal plate
 12. 1.c, 2.b, 3.b, 4.a, 5.a, 6.c, 7.c
 13. 1.c, 2.d, 3.b, 4.a

CHAPTER 19 Peripheral Endocrine Organs

(Questions on p. 693.)
 1. F (Only after a delay of several hours is the metabolic response to thyroid hormone first detectable, and the maximal response is not evident for several days.)
 2. T
 3. T
 4. T
 5. T
 6. F (The most life-threatening consequence of hypocalcemia is overexcitability of nerves and muscles.)
 7. F (Absorption of ingested Ca^{++} from the intestine is controlled by vitamin D.)

8. F (The labile pool of Ca^{++} is the bone fluid. The stable pool of Ca^{++} is the bone crystals.)
9. colloid, thyroglobulin
10. pro-opiomelanocortin
11. glycogenesis, glycogenolysis, gluconeogenesis
12. brain, working muscles, liver
13. bone, kidneys, digestive tract
14. c
15. a, b, c, g, i, j
16. 1. glucose, 2. glycogen, 3. free fatty acids, 4. triglycerides, 5. amino acids, 6. body proteins

CHAPTER 20 Reproductive System

(Questions on p. 744.)
1. T
2. T
3. F (Melatonin secretion decreases during exposure to light.)
4. F (The clitoris becomes erect during sexual arousal.)
5. T
6. F (A follicle that fails to reach maturity undergoes atresia; that is, it degenerates and forms scar tissue.)
7. T
8. seminiferous tubules, FSH, testosterone
9. thecal, LH, granulosa, FSH
10. corpus luteum of pregnancy
11. human chorionic gonadotropin
12. c
13. e
14. 1.c, 2.a, 3.b, 4.c, 5.a, 6.a, 7.c, 8.b, 9.e
15. 1.a, 2.c, 3.b, 4.a, 5.a, 6.b

E Supplemental Reading List

The following list includes more advanced textbooks and relevant articles in lay science journals. The list is not meant to be comprehensive but is merely a starting point for those interested in pursuing specific topics in more depth. Publications in specialty science journals are not included.

GENERAL PHYSIOLOGY REFERENCES

Berne, R. M., and M. N. Levy, eds. *Physiology*. 2d ed. St. Louis: C. V. Mosby, 1988.

Guyton, A. C. *Textbook of Medical Physiology*. 8th ed. Philadelphia: W. B. Saunders, 1991.

Johnson, L. R., ed. *Essential Medical Physiology*. New York: Raven Press, 1991.

Mountcastle, V. B. ed. *Medical Physiology*. 14th ed. 2 vols. St. Louis: C. V. Mosby, 1979–1980.

Patton, H. D., A. F. Fuchs, B. Hille, A. M. Scher, and R. Steiner, eds. *Textbook of Physiology*. 21st ed. Philadelphia: W. B. Saunders, 1989.

Schmidt, R. F., and G. Thews, eds. *Human Physiology*. 2d ed. Berlin: Springer-Verlag, 1989.

Vick, R. L. *Contemporary Medical Physiology*. Menlo Park, Calif.: Addison-Wesley, 1984.

EXERCISE PHYSIOLOGY REFERENCES

Brooks, G. A., and T. D. Fahey. *Exercise Physiology: Human Bioenergetics and Its Applications*. New York: John Wiley & Sons, 1985.

DeVries, H. A. *Physiology of Exercise*. 4th ed. Dubuque, Iowa: W. C. Brown, 1986.

Komi, P. V., ed. *Strength and Power in Sport*. London: Blackwell Scientific Publications, 1992.

Lamb, D. R. *Physiology of Exercise: Responses and Adaptations*. 2d ed. New York: Macmillan, 1984.

McArdle, W. D., F. D. Katch, and V. L. Katch. *Exercise Physiology: Energy, Nutrition, and Human Performance*. 3d ed. Philadelphia: Lea & Febiger, 1991.

Noble, B. J. *Physiology of Exercise and Sport*. St. Louis: Times Mirror, Mosby College Publications, 1986.

Rosato, F. D. *Fitness and Wellness*. 2d ed. St. Paul: West Publishing Co., 1990.

Ryan, A. J., and F. L. Allman, eds. *Sports Medicine*. 2d ed. San Diego: Academic Press, 1989.

Shephard, R. J., and P. D. Aistrand, eds. *Endurance in Sport*. London: Blackwell Scientific Publications, 1992.

Teitz, C. C., ed. *Scientific Foundations of Sports Medicine*. Philadelphia: Decker, 1989.

PRINCIPLES OF HOMEOSTASIS AND REGULATION (CHAPTER 1)

Adolph, E. F., ed. *The Development of Homeostasis*. New York: Academic Press, 1960.

———. *Origins of Physiological Regulations*. New York: Academic Press, 1968.

Jones, R. W. *Principles of Biological Regulation: An Introduction to Feedback Systems*. New York: Academic Press, 1973.

Langley, L. L., ed. *Homeostasis: Origin of the Concept*. Stroudsburg, Pa.: Dowden, Hutchison, & Ross, 1973.

CELLULAR PHYSIOLOGY (CHAPTERS 2 AND 3)

Alberts, B., D. Bray, J. Lewis, M. Raff, K. Roberts, and J. Watson. *Molecular Biology of the Cell*. 2d ed. New York: Garland Publishing, 1989.

Allen, R. D. "The Microtubule as an Intracellular Engine." *Scientific American* 256 (February 1987).

Berridge, M. J. "The Molecular Basis of Communication within the Cell." *Scientific American* 253 (October 1985).

Bretscher, M. S. "The Molecules of the Cell Membrane." *Scientific American* 253 (October 1985).

Byrne, J. H., and S. G. Schultz. *An Introduction to Membrane Transport and Bioelectricity*. New York: Raven Press, 1988.

Carafoli, E., and J. T. Penniston. "The Calcium Signal." *Scientific American* 253 (November 1985).

"CF Gene Therapy on Horizon." *Science News* 141 (January 11, 1992).

Chien, S., ed. *Molecular Biology in Physiology*. New York: Raven Press, 1989.

Darnell, J., H. Lodish, and D. Baltimore. *Molecular Cell Biology*. 2d ed. New York: Scientific American Books, 1990.

"Defect in Cystic Fibrosis Protein Found." *Science News* 141 (April 4, 1992).

Ezzell, C. "Sticky Situations: Picking Apart the Molecules That Glue Cells Together." *Science News* 141 (June 13, 1992).

Hanson, B. "Message in a Barrel." *Discover* 13 (June 1992).

"How Cells Absorb Glucose." *Scientific American* 266 (January 1992).

Linder, M. E., and A. G. Gilman. "G Proteins." *Scientific American* 267 (July 1992).

Matthews, G. G. *Cellular Physiology of Nerve and Muscle.* London: Blackwell Scientific Publications, 1986.

The Molecules of Life: Readings from Scientific American. New York: Scientific American Books, 1986.

Murray, M. "Life on the Move." *Discover* 12 (March 1991).

Neher, E., and B. Sakmann. "The Patch Clamp Technique." *Scientific American* 266 (March 1992).

"New Perspective on Cystic Fibrosis." *Science News* 138 (November 24, 1990).

Rasmussen, H. "The Cycling of Calcium as an Intracellular Messenger." *Scientific American* 261 (October 1989).

Rothman, J. E. "The Compartmental Organization of the Golgi Apparatus." *Scientific American* 253 (September 1985).

Satir, P. "Mechanism of Ciliary Movement—What's New?" *News in Physiological Science* 4 (August 1989).

Weber, K., and M. Osborn. "The Molecules of the Cell Matrix." *Scientific American* 253 (October 1985).

NERVOUS SYSTEM (CHAPTERS 4, 5, 6, AND 7)

Abelsen, P. H., E. Butz, and S. H. Snyder, eds. *Neuroscience.* Washington, D.C.: American Association for the Advancement of Science, 1985.

"Adult Neurons: Not Too Old to Divide." *Science News* 141 (April 4, 1992).

Alkon, D. L. "Memory Storage and Neural Systems." *Scientific American* 261 (July 1989).

Aoki, C., and P. Siekevitz. "Plasticity in Brain Development." *Scientific American* 259 (December 1988).

Appenzeller, O. *The Autonomic Nervous System.* 3d ed. New York: Elsevier, 1982.

Bloom, F. E., and A. Lazerson. *Brain, Mind and Behavior.* New York: Scientific American Books, 1988.

Borg, E., and S. A. Counter. "The Middle-Ear Muscles." *Scientific American* 261 (August 1989).

Bradford, H. F. *Chemical Neurobiology: An Introduction to Neurochemistry.* New York: Scientific American Books, 1986.

Chollar, S. "The Nerves of Some People." *Discover* 12 (February 1991).

Ciriello, J., F. R. Calaresu, L. P. Renaud, and C. Polosa. *Organization of the Autonomic Nervous System—Central and Peripheral Mechanisms.* New York: Alan R. Liss, 1987.

Damasio, A. R., and H. Damasio. "Brain and Language." *Scientific American* 267 (September 1992).

Darian-Smith, I., ed. *Sensory Processes.* New York: Oxford University Press, 1984.

Dowling, J. E. *Neurons and Networks: An Introduction to Neuroscience.* Boston: The Belknap Press of Harvard University Press, 1992.

Ezzell, C. "Alzheimer's Alchemy." *Science News* 141 (March 7, 1992).

———. "Memories Might Be Made of This." *Science News* 139 (May 25, 1991).

Fackelmann, K. A. "Myelin on the Mend." *Science News* 137 (April 7, 1990).

Fischbach, G. D. "Mind and Brain." *Scientific American* 267 (September 1992).

Fitzgerald, M. J. T. *Neuroanatomy Basic and Applied.* London: Bailliere Tindall, 1985.

Freeman, W. J. "The Physiology of Perception." *Scientific American* 264 (February 1991).

Goldman-Rakic, P. S. "Working Memory and the Mind." *Scientific American* 267 (September 1992).

Goldstein, G. W., and A. L. Betz. "The Blood-Brain Barrier." *Scientific American* 255 (September 1986).

"Hints of Another Signaling System in Brain." *Science News* 137 (January 27, 1990).

"How the Nose Knows." *Discover* 13 (January 1992).

Jahn, A. F., and J. R. Santos-Sacchi, eds. *Physiology of the Ear.* New York: Raven Press, 1988.

Kandel, E. R., and R. D. Hawkins. "The Biological Basis of Learning and Individuality." *Scientific American* 267 (September 1992).

Kandel, E. R., and J. H. Schwartz. *Principles of Neural Science.* 3d ed. New York: Elsevier, 1991.

Katzman, R., and T. Saitoh. "Advances in Alzheimer's Disease." *The FASEB Journal* 5 (March 1991).

Kimelberg, H. K., and M. D. Norenberg. "Astrocytes." *Scientific American* 260 (April 1989).

Krut, Z. L., and C. J. Pycock. *Neurotransmitters and Drugs.* 2d ed. Baltimore: University Park Press, 1983.

Kuffler, S. W., J. G. Nicholls, and A. R. Martin. *From Neuron to Brain.* 3d ed. Sunderland, Mass.: Sinauer Associates, 1992.

"Left Brain May Serve as Language Director." *Science News* 141 (March 7, 1992).

Levitan, J. B., and L. K. Kaczmarek. *The Neuron: Cell and Molecular Biology.* New York: Oxford University Press, 1991.

Livingstone, M. "Art, Illusion, and the Visual System." *Scientific American* 258 (January 1988).

Llinás, R. R., ed. *The Biology of the Brain: From Neurons to Networks.* New York: Scientific American Books, 1989.

———. *The Workings of the Brain: Development, Memory and Perception.* New York: Scientific American Books, 1990.

Lundberg, J. M., J. Pernow, and J. S. Lacroix. "Neuropeptide Y: Sympathetic Cotransmitter and Modulator?" *News in Physiological Science* 4 (February 1989).

McKean, K. "Pain." *Discover* 7 (October 1986).

Masland, R. H. "The Functional Architecture of the Retina." *Scientific American* 255 (December 1986).

Melzack, R. "Phantom Limbs." *Scientific American* 266 (April 1992).

"Memory in a Neuron." *Scientific American* 260 (January 1989).

Miller, S. K. "Picking Up Parkinson's Pieces." *Discover* 12 (May 1991).

"Molecular Custodians Sweep Away Odorants." *Science News* 136 (December 9, 1989).

Montgomery, G. "Molecules of Memory." *Discover* 10 (December 1989).

Müller, E. E., and G. Nistico. *Brain Messengers and the Pituitary*. San Diego: Academic Press, 1989.

Nathans, J. "The Genes for Color Vision." *Scientific American* 260 (February 1989).

Ottoson, D. *Physiology of the Nervous System*. New York: Oxford University Press, 1983.

"Parkinson's Progress." *Discover* 13 (March 1992).

Petit, T. L., and G. O. Ivy, eds. *Neural Plasticity: A Lifespan Approach*. New York: Alan R. Liss, 1987.

Pickles, J. O. *An Introduction to the Physiology of Hearing*. San Diego: Academic Press, 1988.

Plum, F., ed. *Higher Functions of the Brain*. New York: Oxford University Press, 1987.

Radgetsky, P. "The Brainiest Cells Alive." *Discover* 12 (April 1991).

Risau, W. "Differentiation of Blood-Brain Barrier Endothelium." *News in Physiological Science* 4 (August 1989).

Rock, I., ed. *The Perceptual World*. New York: Scientific American Books, 1990.

Schnapf, J. L., and D. A. Baylor. "How Photoreceptor Cells Respond to Light." *Scientific American* 256 (April 1987).

Selkoe, D. J. "Amyloid Protein and Alzheimer's Disease." *Scientific American* 265 (November 1991).

Shatz, C. J. "The Developing Brain." *Scientific American* 267 (September 1992).

Shepherd, G. M. *Neurobiology*. New York: Oxford University Press, 1988.

Springer, S. P., and G. Deutsch. *Left Brain, Right Brain*. New York: Scientific American Books, 1985.

Stryer, L. "The Molecules of Visual Excitation." *Scientific American* 251 (July 1987).

Stux, G. *Basics of Acupuncture*. Berlin: Springer-Verlag, 1991.

Wang, M., and A. Freeman. *Neural Function*. Boston: Little, Brown, 1987.

Weiss, R. "Shadows of Thoughts Revealed." *Science News* 138 (November 10, 1990).

Zeki, S. "The Visual Image in Mind and Brain." *Scientific American* 267 (September 1992).

Zivin, J. A., and D. W. Choi. "Stroke Therapy." *Scientific American* 265 (July 1991).

MUSCLE (CHAPTER 8)

Bourne, G. H., ed. *The Structure and Function of Muscle*. 2d ed. 4 vols. New York: Academic Press, 1972–1974.

Bower, B. "Pumped Up and Strung Out." *Science News* 140 (July 13, 1991).

Entman, M. L., and W. B. VanWinkle. *Sarcoplasmic Reticulum in Muscle Physiology*. Boca Raton, Fla.: CRC Press, 1986.

Franklin, D. "Stuck on Steroids." *In Health* 4 (May/June 1990).

Gaesser, G. A., and G. A. Brooks. "Metabolic Bases of Excess Post-Exercise Oxygen Consumption: A Review." *Medicine and Science in Sports and Exercise* 16 (1984).

"Gene Fix for Muscular Dystrophy Defect." *Science News* 140 (September 21, 1991).

Huxley, A. *Reflections on Muscle*. Princeton, N.J.: Princeton University Press, 1980.

Jones, D. A., and J. M. Round. *Skeletal Muscle in Health and Disease: A Textbook of Muscle Physiology*. Manchester, England: Manchester University Press, 1990.

Junge, D. *Nerve and Muscle Excitation*. 3d ed. Sunderland, Mass.: Sinauer Associates, 1991.

Keynes, R. D., and D. J. Aidley. *Nerve and Muscle*. Cambridge: Cambridge University Press, 1981.

Komi, P. V., ed. *Strength and Power in Sport*. London: Blackwell Scientific Publications, 1992.

"Muscular Dystrophy: New Focus on Myoblasts. *Science News* 141 (March 14, 1992).

Peachy, L. D. *Skeletal Muscle*. New York: Oxford University Press, 1983.

Rios, E., and G. Pizarro. "Voltage Sensors and Calcium Channels of Excitation-Contraction Coupling." *News in Physiological Science* 3 (December 1988).

Shephard, R. J., and P. O. Aistrand, eds. *Endurance in Sport*. London: Blackwell Scientific Publications, 1992.

Squire, J. M. *Molecular Mechanisms in Muscular Contraction*. Boca Raton, Fla.: CRC Press, 1990.

CIRCULATORY SYSTEM (CHAPTERS 9, 10, AND 11)

Berne, R. M. *Cardiovascular Physiology*. 6th ed. St. Louis: C. V. Mosby, 1992.

"A Better Red." *Scientific American* 266 (February 1992).

Bohr, D. F., A. P. Somlyo, and H. V. Sparks, Jr., eds. *Vascular Smooth Muscle*. New York: Oxford University Press, 1980.

Brown, M. S. and J. Goldstein. "How LDL Receptors Influence Cholesterol and Atherosclerosis." *Scientific American* 251 (November 1984).

Daniel, T. O., and H. E. Ives. "Endothelial Control of Vascular Function." *News in Physiological Science* 4 (August 1989).

Folkow, B. "Salt and Hypertension." *News in Physiological Science* 5 (October 1990).

Fozzard, H. A., E. Haber, R. B. Jennings, A. M. Katz, H. E. Morgan, eds. *The Heart and Cardiovascular System, Scientific Foundations*. 2d ed. 2 vols. New York: Raven Press, 1991.

Garfein, O. B., ed. *Current Concepts in Cardiovascular Physiology*. San Diego: Academic Press, 1990.

Goerke, J., and A. H. Mines. *Cardiovascular Physiology*. New York: Raven Press, 1988.

Golde, D. W. "The Stem Cell." *Scientific American* 265 (December 1991).

Hirst, G. D. S., F. R. Edwards, N. J. Bramich, and M. F. Klemm. "Neural Control of Cardiac Pacemaker Potentials." *News in Physiological Science* 6 (August 1991).

Honig, C. R. *Modern Cardiovascular Physiology*. 2d ed. Boston: Little, Brown, 1988.

Katz, A. M. *Physiology of the Heart*. 2d ed. New York: Raven Press, 1992.

Kuo, L., M. J. Davis, and W. M. Chilian. "Endothelial Modulation of Arteriolar Tone." *News in Physiological Science* 7 (February 1992).

Lawn, R. M. "Lipoprotein(a) in Heart Disease." *Scientific American* 266 (June 1992).

Little, R. C. *Physiology of the Heart and Circulation.* 4th ed. Chicago: Year Book Medical Publishers, 1989.

Miller, V. M. "Interactions between Neural and Endothelial Mechanisms in Control of Vascular Tone." *News in Physiological Science* 6 (April 1991).

Mohrman, D. E., and L. J. Heller. *Cardiovascular Physiology.* 3d ed. New York: McGraw-Hill, 1991.

Renkin, E. M., and C. C. Michel, eds. *Microcirculation.* New York: Oxford University Press, 1984.

Robinson, T. F., S. M. Factor, and E. H. Sonnenblick. "The Heart as a Suction Pump." *Scientific American* 254 (June 1986).

Shepherd, J. T., and P. M. Vanhoutte. *The Human Cardiovascular System: Facts and Concepts.* New York: Raven Press, 1979.

Smith, J. J. *Circulatory Physiology: The Essentials.* 3d ed. Baltimore: Williams & Wilkins, 1990.

Smith, J. J., and J. P. Kampine. *Circulatory Physiology.* Baltimore: Williams & Wilkins, 1984.

Snyder, S. H., and D. S. Bredt. "Biological Roles of Nitric Oxide." *Scientific American* 266 (May 1992).

Sparks, H. V. *Essentials of Cardiovascular Physiology.* Minneapolis: University of Minnesota Press, 1987.

Stray-Gundersen, J. "Unethical Alterations of Oxygen-Carrying Capacity in Endurance Athletes." *News in Physiological Science* 3 (December 1988).

Vanhoutte, P. M., and T. F. Luscher, eds. *Endothelium-Derived Vasoactive Factors.* Boca Raton, Fla.: CRC Press, 1990.

IMMUNE DEFENSE (CHAPTER 12)

Barrett, J. T. *Textbook of Immunology.* 5th ed. St. Louis: C. V. Mosby, 1988.

Cohen, I. R. "The Self, the World, and Autoimmunity." *Scientific American* 258 (April 1988).

Edelson, R. L., and J. M. Fink. "The Immunologic Function of Skin." *Scientific American* 252 (June 1985).

Greaves, M. F., J. J. T. Owen, and M. C. Raff. "T and B Lymphocytes: Origins, Properties, and Roles in Immune Responses." *Excerpta Medica.* Amsterdam: 1973.

Johnson, H. M., J. K. Russell, and C. H. Pontzer. "Superantigens in Human Disease." *Scientific American* 266 (April 1992).

Marrack, P., and J. Kappler. "The T Cell and Its Receptor." *Scientific American* 254 (February 1986).

Milstein, C. "From Antibody Structure to Immunological Diversification of Immune Response." *Science* 231 (March 1986).

"Overview: Tolerating Self." *Scientific American* 263 (September 1990).

Rennie, J. "The Body Against Itself." *Scientific American* 263 (December 1990).

Roitt, I. M., J. Brostoff, and D. Male. *Immunology.* 2d ed. St. Louis: C. V. Mosby, 1989.

Smith, K. A. "Interleukin 2." *Scientific American* 262 (March 1990).

Tonegawa, S. "The Molecules of the Immune System." *Scientific American* 253 (October 1985).

von Boehmer, H., and P. Kisielow. "How the Immune System Learns about Self." *Scientific American* 265 (October 1991).

Weissmann, G. "Aspirin." *Scientific American* 264 (January 1991).

RESPIRATORY SYSTEM (CHAPTER 13)

Cherniak, N. S., and J. G. Widdicombe, eds. *Control of Breathing.* New York: Oxford University Press, 1986.

Crystal, R. G., J. B. West, P. J. Barnes, N.S. Cherniak, and E. R. Weibel, eds. *The Lung, Scientific Foundations.* 2 vols. New York: Raven Press, 1991.

Farhi, L. E., and S. M. Tenney, eds. *Gas Exchange.* New York: Oxford University Press, 1987.

Macklem, P. T., and Mead, J., eds. *Mechanics of Breathing.* New York: Oxford University Press, 1986.

Malthew, O. P., ed. *Respiratory Function of the Upper Airway.* New York: Dekker, 1988.

Mines, A. H. *Respiratory Physiology.* 2d ed. New York: Raven Press, 1986.

Taylor, A. E., K. Rehder, R. E. Hyatt, and J. C. Parker. *Clinical Respiratory Physiology.* Philadelphia: W. B. Saunders, 1989.

West, J. B. *Respiratory Physiology: The Essentials.* 4th ed. Baltimore: Williams & Wilkins, 1990.

URINARY SYSTEM AND ACID-BASE BALANCE (CHAPTERS 14 AND 15)

Brenner, B. M., and F. C. Rector. *The Kidney.* Philadelphia: W. B. Saunders, 1991.

Cantin, M., and J. Genest. "The Heart as an Endocrine Gland." *Scientific American* 254 (February 1986).

Carmines, P. K., and J. T. Fleming. "Control of the Renal Microvasculature by Vasoactive Peptides." *The FASEB Journal* 4 (December 1990).

Cowley, A. W., J. F. Liard, and D. A. Ausiello, eds. *Vasopressin.* New York: Raven Press, 1988.

Davenport, H. W. *The ABC of Acid-Base Chemistry.* 7th ed. Chicago: University of Chicago Press, 1978.

Gottschalk, C. W., R. W. Berliner, and G. H. Giebisch. *Renal Physiology: People and Ideas.* New York: Oxford University Press, 1987.

Grossman, S. P. *Thirst and Sodium Appetite.* San Diego: Academic Press, 1990.

Haperin, M. L., and M. B. Goldstein. *Fluid, Electrolyte and Acid-Base Emergencies.* Philadelphia: W. B. Saunders, 1988.

Klalr, S. *The Kidney and Body Fluids in Health and Disease.* New York: Plenum Medical Book Co., 1984.

Koushanpour, E., and W. Kriz. *Renal Physiology: Principles, Structure, and Function.* 2d ed. New York: Springer-Verlag, 1986.

Marsh, D. J. *Renal Physiology.* New York: Raven Press, 1983.

"New Insights into the Mechanisms of Action of Aldosterone." *News in Physiological Science* 5 (June 1990).

Robinson, J. R. *Reflections on Renal Function*. 2d ed. London: Blackwell Scientific Publications, 1988.

Rose, B. D. *Clinical Physiology of Acid-Base and Electrolyte Disorders*. 3d ed. New York: McGraw-Hill, 1989.

Schneider, E. G. "In Water Deprivation, Osmolarity Becomes an Important Determinant of Aldosterone Secretion." *News in Physiological Science* 5 (October 1990).

Seldin, D. W., and G. Giebisch, eds. *The Kidney: Physiology and Pathophysiology*. 2d ed. New York: Raven Press, 1992.

————. *The Regulation of Acid-Base Balance*. New York: Raven Press, 1989.

————. *The Regulation of Potassium Balance*. New York: Raven Press, 1989.

————. *The Regulation of Sodium and Chloride Balance*. New York: Raven Press, 1989.

Sullivan, L. P., and J. J. Grantham. *Physiology of the Kidney*. 2d ed. Philadelphia: Lea & Febiger, 1982.

Valtin, H. *Renal Function: Mechanisms Preserving Fluid and Solute Balance in Health*. 2d ed. Boston: Little, Brown, 1983.

Vander, A. J. *Renal Physiology*. 4th ed. New York: McGraw-Hill, 1991.

Windhager, E. E., ed. *Renal Physiology*. New York: Oxford University Press, 1992.

DIGESTIVE SYSTEM (CHAPTER 16)

Cooke, H. J. "Role of the 'Little Brain' in the Gut in Water and Electrolyte Homeostasis." *The FASEB Journal* 3 (February 1989).

Davenport, H. W. *A Digest of Digestion*. 2d ed. Chicago: Year Book Medical Publishers, 1978.

————. *Physiology of the Digestive Tract*. 5th ed. Chicago: Year Book Medical Publishers, 1982.

Field, M., and R. A. Frizzell, eds. *Intestinal Absorption and Secretion*. New York: Oxford University Press, 1991.

Forte, J. G., and S. G. Schlutz, eds. *Salivary, Gastric, and Hepatobiliary Secretion*. New York: Oxford University Press, 1989.

Granger, D. N., J. A. Barrowman, and D. R. Kvietys. *Clinical Gastrointestinal Physiology*. Philadelphia: W. B. Saunders, 1985.

Johnson, L. R., J. Christensen, M. I. Grossman, E. D. Christensen, and S. G. Schultz, eds. *Physiology of the Gastrointestinal Tract*. 2d ed. 2 vols. New York: Raven Press, 1987.

Lewis, L. D., and J. A. Williams. "Cholecystokinin: A Key Integrator of Nutrient Assimilation." *News in Physiological Science* 5 (August 1990).

Makhlouf, G. M., and S. G. Schultz, eds. *Neural and Endocrine Biology*. Digestion series. New York: Oxford University Press, 1989.

Sernka, T., and E. Jacobson. *Gastrointestinal Physiology: The Essentials*. Baltimore: Williams & Wilkins, 1983.

Wood, J. D. "Communication between Minibrain in Gut and Enteric Immune System. *News in Physiological Science* 6 (April 1991).

Wood, J.D., and S. G. Schultz, eds. *Motility and Circulation*.
Digestion series, Parts 1 and 2. New York: Oxford University Press, 1989.

ENERGY BALANCE AND TEMPERATURE REGULATION (CHAPTER 17)

Bligh, J. *Temperature Regulation in Mammals and Other Vertebrates*. New York: Elsevier, 1973.

Bloch, G. B. "Fat Master." *Discover* 11 (April 1990).

Hamilton, E. M. N., E. N. Whitney, and F. S. Sizer. *Nutrition: Concepts and Controversies*. 5th ed. St. Paul: West Publishing Co., 1991.

Harris, R. B. S. "Role of Set-Point Theory in Regulation of Body Weight." *The FASEB Journal* 4 (December 1990).

Lytle, L. D. "Control of Eating Behavior." *Nutrition and the Brain*. Edited by R. J. Wurtman and J. J. Wurtman. New York: Raven Press, 1977.

Shitzer, A., and R. Eberhart, eds. *Heat Transfer in Medicine and Biology*. New York: Plenum Press, 1985.

Weiss, P. L. "Fat and Fiction." *Science News* 138 (September 1, 1990).

"Why Do We Stop Eating?" *News in Physiological Science* 5 (April 1990).

ENDOCRINE AND REPRODUCTIVE SYSTEMS (CHAPTERS 18, 19, AND 20)

Atkinson, M. A., and N. K. Maclaren. "What Causes Diabetes?" *Scientific American* 263 (July 1990).

Bolander, F. F. *Molecular Endocrinology*. San Diego: Academic Press, 1989.

Burger, H., and D. DeKretser. *Comprehensive Endocrinology: The Testis*. New York: Raven Press, 1981.

Carmichael, S. W., and H. Winkler. "The Adrenal Chromaffin Cell." *Scientific American* 253 (August 1985).

Crapo, L. *Hormones: Messengers of Life*. New York: Scientific American Press, 1985.

"Deciding to Be Born." *Discover* 13 (May 1992).

"Eggs Not Silent Partners in Conception." *Science News* 139 (April 6, 1991).

Elmer-Dewitt, P. "Making Babies." *Time*. September 30, 1991.

Evans, R. M. "The Steroid and Thyroid Hormone Receptor Superfamily." *Science* 240 (May 13, 1988).

Fackelmann, K. A. *Sex Protection: Balancing the Equation."* *Science News* 141 (March 14, 1992).

————. "Zona Blasters." *Science News* 138 (December 15, 1990).

"Fetus Tells Mother: It's Time for Labor." *Science News* 140 (September 21, 1991).

Goodman, H. M. *Basic Medical Endocrinology*. New York: Raven Press, 1988.

Griffin, J. E., and S. R. Ojeda. *Textbook of Endocrine Physiology*. New York: Oxford University Press, 1988.

Hedge, G. A., H. D. Colby, and R. L. Goodman. *Clinical Endocrine Physiology*. Philadelphia: W. B. Saunders, 1987.

"Heightened Concern over Growth Hormone." *Science News* 138 (December 8, 1990).

"Hormone May Restore Muscle in Elderly." *Science News* 138 (July 14, 1990).

Illnerová, H. "Mammalian Circadian Clock and Its Resetting." *News in Physiological Science* 6 (June 1991).

Knobil, E., and J. D. Neill, eds. *The Physiology of Reproduction.* 2 vols. New York: Raven Press, 1988.

Lamb, J. C., IV, ed. *Physiology and Toxicology of Male Reproduction.* San Diego: Academic Press, 1988.

Lienhard, G. E., J. W. Slot, D. E. James, and M. M. Mueckler. "How Cells Absorb Glucose." *Scientific American* 266 (January 1992).

"Maleness Gene May Be a Master Gene Switch." *Science News* 140 (August 3, 1991).

Martin, C.R. *Endocrine Physiology.* New York: Oxford University Press, 1985.

Müller, E. E., and G. Nisticò. *Brain Messengers and the Pituitary.* San Diego: Academic Press, 1989.

"New Clues to Diabetes' Cause and Treatment." *Science News* 140 (December 21 and 28, 1991).

Norman, A. W., and G. Litwack. *Hormones.* San Diego: Academic Press, 1987.

Orci, L., J. Vassali, and A. Perrelet. "The Insulin Factory." *Scientific American* 259 (September 1988).

Serra, G. *Comprehensive Endocrinology: The Ovary.* New York: Raven Press, 1983.

Snyder, S. H. "The Molecular Basis of Communication between Cells." *Scientific American* 253 (October 1985).

Stroh, M. "The Root of Impotence: Does Nitric Oxide Hold the Key?" *Science News* 142 (July 4, 1992).

Tepperman, J. *Metabolic and Endocrine Physiology.* 5th ed. Chicago: Year Book Medical Publishers, 1987.

Ulmann, A., G. Teutsch, and D. Philibert. "RU 486." *Scientific American* 262 (June 1990).

Villanúa, M. A., C. Agrasal, and A. I. Esquifino. "New Perspectives in the Research of Pineal Gland." *News in Physiological Science* 5 (February 1990).

Wilson, J. D., and D. W. Foster, eds. *Williams' Textbook of Endocrinology.* 8th ed. Philadelphia: W. B. Saunders, 1992.

GLOSSARY

A band one of the dark bands that alternate with light (I) bands to create a striated appearance in a skeletal- or cardiac-muscle fiber when these fibers are viewed with a light microscope

absorptive state the metabolic state following a meal when nutrients are being absorbed and stored

accessory digestive organs exocrine organs located outside the wall of the digestive tract that empty their secretions through ducts into the digestive-tract lumen

accessory sex glands glands that empty their secretions into the reproductive tract

accommodation the ability to adjust the strength of the lens in the eye so that both near and far sources can be focused on the retina

acetylcholine (ACh) the neurotransmitter released from all autonomic preganglionic fibers, parasympathetic postganglionic fibers, and motor neurons

acetylcholinesterase (AChE) an enzyme present in the motor end-plate membrane of a skeletal-muscle fiber that inactivates acetylcholine

ACh see *acetylcholine*

AChE see *acetylcholinesterase*

acid a hydrogen-containing substance that yields a free hydrogen ion and anion on dissociation

acidosis blood pH of less than 7.35

ACTH see *adrenocorticotropic hormone*

actin the contractile protein that forms the backbone of the thin filaments in muscle fibers

active expiration emptying of the lungs by contracting the expiratory muscles; more complete emptying than when at rest; also called *forced expiration*

active force a force that requires expenditure of cellular energy (ATP) in the transport of a substance across the plasma membrane

active reabsorption when any one of the five steps in the transepithelial transport of a substance reabsorbed across the kidney tubules requires energy expenditure

active transport active carrier-mediated transport involving transport of a substance against its concentration gradient across the plasma membrane

acuity discriminative ability; the ability to discern between two different points of stimulation

acute myocardial infarction death of heart-muscle cells caused by disruption of their blood supply; a heart attack

adaptation a reduction in receptor potential in spite of sustained stimulation of the same magnitude

adenosine diphosphate (ADP) the two-phosphate product formed from the splitting of ATP to yield energy for the cell's use

adenosine triphosphate (ATP) the body's common energy "currency," which consists of an adenosine with three phosphate groups attached; splitting of the high-energy, terminal phosphate bond provides energy to power cellular activities

ADH see *vasopressin*

adipose tissue the tissue specialized for storage of triglyceride fat; found under the skin in the hypodermis

ADP see *adenosine diphosphate*

adrenal cortex the outer portion of the adrenal gland; secretes three classes of steroid hormones: glucocorticoids, mineralocorticoids, and sex hormones

adrenal medulla the inner portion of the adrenal gland; an endocrine gland, which is a modified sympathetic ganglion that secretes the hormones epinephrine and norepinephrine into the blood in response to sympathetic stimulation

adrenergic fibers nerve fibers that release norepinephrine as their neurotransmitter

adrenocorticotropic hormone (ACTH) an anterior pituitary hormone that stimulates cortisol secretion by and promotes growth of the adrenal cortex

aerobic a condition in which oxygen is available

aerobic exercise exercise that can be supported by ATP formation accomplished by oxidative phosphorylation because adequate O_2 is available to support the muscle's modest energy demands; also called *endurance-type exercise*

afferent arteriole the vessel that carries blood into the glomerulus of the kidney's nephron

afferent division the portion of the peripheral nervous system that carries information from the periphery to the central nervous system

afferent neuron the class of neuron that possesses a sensory receptor at its peripheral ending and carries information to the central nervous system

after hyperpolarization a slight, transient hyperpolarization that sometimes occurs at the end of an action potential

agranulocytes leukocytes that do not contain granules, including lymphocytes and monocytes

albumin the smallest and most abundant of the plasma proteins; binds and transports many water-insoluble substances in the blood; contributes extensively to plasma-colloid osmotic pressure

aldosterone the adrenocortical hormone that stimulates Na⁺ reabsorption by the distal and collecting tubules of the kidney's nephron during urine formation

alkalosis blood pH of greater than 7.45

allergy acquisition of an inappropriate specific immune reactivity to a normally harmless environmental substance

all-or-none law an excitable membrane either responds to a stimulus with a maximal action potential that spreads nondecrementally throughout the membrane, or it does not respond with an action potential at all

alpha cells the type of endocrine pancreatic cell that secretes the hormone glucagon

alpha motor neuron the type of motor neuron that innervates ordinary skeletal-muscle fibers

alveolar surface tension the surface tension of the fluid lining the alveoli in the lungs; see *surface tension*

alveolar ventilation the volume of air exchanged between the atmosphere and alveoli per minute; equals (tidal volume minus dead-space volume) times respiratory rate

alveoli the air sacs across which O_2 and CO_2 are exchanged between the blood and air in the lungs

amines hormones derived from the amino acid tyrosine; include thyroid hormone and catecholamines

amoeboid movement white blood cells' "crawling" movement, which is similar to the means by which amoebae move

anabolism the buildup or synthesis of larger organic molecules from small organic molecular subunits

anaerobic a condition in which oxygen is not present

anaerobic exercise high-intensity exercise that can be supported by ATP formation accomplished by anaerobic glycolysis for brief periods of time when O_2 delivery to a muscle is inadequate to support oxidative phosphorylation

analgesic pain-relieving

anatomy the study of body structure

androgen a masculinizing "male" sex hormone; includes testosterone from the testes and dehydroepiandrosterone from the adrenal cortex

anemia a reduction below normal in O_2-carrying capacity of the blood

anions negatively charged ions that have gained one or more electrons in their outer shell

ANP see *atrial natriuretic peptide*

antagonism actions opposing each other; with hormones, when one hormone causes the loss of another hormone's receptors, reducing the effectiveness of the second hormone

anterior pituitary the glandular portion of the pituitary that synthesizes, stores, and secretes six different hormones, including growth hormone, TSH, ACTH, FSH, LH, and prolactin

antibody an immunoglobulin produced by a specific activated B lymphocyte (i.e., plasma cell) against a particular antigen; binds with the specific antigen against which it is produced and promotes the destruction of the antigenic invader by augmenting nonspecific immune responses already initiated against the antigen

antibody-mediated immunity a specific immune response accomplished by antibody production by B cells

antidiuretic hormone see *vasopressin*

antigen a large, complex molecule that triggers a specific immune response against itself when it gains entry into the body

antrum (of ovary) the fluid-filled cavity formed within a developing ovarian follicle

antrum (of stomach) the lower portion of the stomach

aorta the large vessel that carries blood from the left ventricle

aortic valve a one-way valve that permits the flow of blood from the left ventricle into the aorta during ventricular emptying but prevents the backflow of blood from the aorta into the left ventricle during ventricular relaxation

appetite centers see *feeding centers*

aqueous humor the clear watery fluid in the anterior chamber of the eye; provides nourishment for the cornea and lens

arterioles highly muscular, high-resistance vessels, the caliber of which can be changed subject to control to determine the distribution of cardiac output

artery a vessel that carries blood away from the heart

ascending tract a bundle of nerve fibers of similar function that travels up the spinal cord to transmit signals derived from afferent input to the brain

asthma an obstructive pulmonary disease characterized by profound constriction of the smaller airways caused by allergy-induced spasm of the smooth muscle in the walls of these airways

astrocyte a type of glial cell in the brain; major functions include holding the neurons together in proper spatial relationship and inducing the brain capillaries to form tight junctions important in the blood-brain barrier

atherosclerosis a progressive, degenerative arterial disease that leads to gradual blockage of affected vessels, thereby reducing blood flow through them

atmospheric pressure the pressure exerted by the weight of the air in the atmosphere on objects on the earth's surface; equals 760 mm Hg at sea level

ATP see *adenosine triphosphate*

ATPase possesses ATP-splitting ability

ATP synthetase the enzyme within the mitochondrial inner membrane that phosphorylates ADP to ATP

atrial natriuretic peptide (ANP) a peptide hormone released from the cardiac atria that promotes urinary loss of Na⁺

atrioventricular (AV) node a small bundle of specialized cardiac cells located at the junction of the atria and ventricles that serves as the only site of electrical contact between the atria and ventricles

atrioventricular (AV) valve a one-way valve that permits the flow of blood from the atrium to the ventricle during filling of the heart but prevents the backflow of blood from the ventricle to the atrium during emptying of the heart

atrium (atria, plural) the upper chamber of the heart that receives blood from the veins and transfers it to the ventricle

atrophy decrease in mass of an organ

autoimmune disease characterized by erroneous production of antibodies against one of the body's own tissues

autonomic nervous system the portion of the efferent division of the peripheral nervous system that innervates smooth and cardiac muscle and exocrine glands; composed of two subdivisions, the sympathetic nervous system and the parasympathetic nervous system

autorhythmicity the ability of an excitable cell to rhythmically initiate its own action potentials

AV nodal delay the delay in impulse transmission between the atria and ventricles at the AV node to allow sufficient time for the atria to become completely depolarized and contract, emptying their contents into the ventricles, before ventricular depolarization and contraction occur

AV valve see *atrioventricular valve*

axon a single, elongated tubular extension of a neuron that conducts action potentials away from the cell body; also known as a *nerve fiber*

axon hillock the first portion of a neuronal axon plus the region of the cell body from which the axon leaves; the site of action-potential initiation in most neurons

axon terminals the branched endings of a neuronal axon, which release neurotransmitter that influences target cells in close association with the axon terminals

baroreceptor reflex an autonomically mediated reflex response that influences the heart and blood vessels to oppose a change in mean arterial blood pressure

baroreceptors receptors located within the circulatory system that monitor blood pressure

basal metabolic rate the minimal waking rate of internal energy expenditure; the body's "idling speed"

basal nuclei several masses of gray matter located deep within the white matter of the cerebrum of the brain; play an important inhibitory role in motor control

base a substance that can combine with a free hydrogen ion and remove it from solution

basic electrical rhythm (BER) self-induced electrical activity of the digestive-tract smooth muscle

basilar membrane the membrane that forms the floor of the middle compartment of the cochlea and bears the organ of Corti, the sense organ for hearing

basophils white blood cells that synthesize, store, and release histamine, which is important in allergic responses, and heparin, which hastens the removal of fat particles from the blood

BER see *basic electrical rhythm*

beta cells the endocrine pancreatic cells that secrete the hormone insulin

bicarbonate (HCO_3^-) the anion resulting from dissociation of carbonic acid, H_2CO_3

bile salts cholesterol derivatives secreted in the bile that facilitate fat digestion through their detergent action and facilitate fat absorption through their micellar formation

biliary system the bile-producing system consisting of the liver, gall bladder, and associated ducts

bilirubin a bile pigment, which is a waste product derived from the degradation of hemoglobin during the breakdown of old red blood cells

bipolar cells the nerve cells in the middle layer of the retina; synapse with the photoreceptors of the eye

blastocyst the developmental stage of the fertilized ovum by the time it is ready to implant; consists of a single-layered sphere of cells encircling a fluid-filled cavity

blood-brain barrier special structural and functional features of the brain capillaries that limit access of materials from the blood into the brain tissue

B lymphocytes (B cells) white blood cells that produce antibodies against specific targets to which they have been exposed

body of the stomach the main or middle part of the stomach

body systems a collection of organs that perform related functions and interact to accomplish a common activity that is essential for survival of the whole body—for example, the digestive system

bone marrow the soft, highly cellular tissue that fills the internal cavities of bones and is the source of most blood cells

Bowman's capsule the beginning of the tubular component of the kidney's nephron that cups around the glomerulus and collects the glomerular filtrate as it is formed

Boyle's law at any constant temperature, the pressure exerted by a gas varies inversely with the volume of the gas

brain the most anterior, most highly developed portion of the central nervous system

brain stem the portion of the brain that is continuous with the spinal cord and serves as an integrating link between the spinal cord and higher brain levels and controls many life-sustaining processes, such as breathing, circulation, and digestion

bronchioles the small, branching airways within the lungs

bronchoconstriction narrowing of the respiratory airways

bronchodilation widening of the respiratory airways

brush border the collection of microvilli projecting from the luminal border of epithelial cells lining the digestive tract and kidney tubules

buffer see *chemical buffer system*

bulbourethral glands male accessory sex glands that secrete mucus for lubrication

bulk flow movement in bulk of a protein-free plasma across the capillary walls between the blood and surrounding interstitial fluid; encompasses ultrafiltration and reabsorption

bundle of His a tract of specialized cardiac cells that rapidly transmits an action potential down the interventricular septum of the heart

calcitonin a hormone secreted by the thyroid C cells that lowers plasma Ca^{++} levels

calcium balance maintenance of a constant total amount of Ca^{++} in the body; accomplished by slowly responding adjustments in intestinal Ca^{++} absorption and in urinary Ca^{++} excretion

calcium homeostasis maintenance of a constant free plasma Ca^{++} concentration; accomplished by rapid exchanges of Ca^{++} between the bone and ECF and to a lesser extent by modifications in urinary Ca^{++} excretion

calmodulin an intracellular Ca^{++}-binding protein that, upon activation by Ca^{++}, induces a change in structure and function of another intracellular protein; especially important in smooth-muscle excitation-contraction coupling

CAMs see *cell adhesion molecules*

capillaries the thin-walled, pore-lined, smallest blood vessels, across which exchange between the blood and surrounding tissues takes place

carbonic anhydrase the enzyme that catalyzes the conversion of CO_2 and H_2O into carbonic acid, H_2CO_3

cardiac cycle one period of systole and diastole

cardiac muscle the specialized muscle found only in the heart

cardiac output (CO) the volume of blood pumped by each ventricle each minute; equals stroke volume times heart rate

cardiovascular control center the integrating center located in the medulla of the brain stem that controls mean arterial blood pressure

carrier-mediated transport transport of a substance across the plasma membrane facilitated by a carrier molecule

carrier molecule membrane proteins that, by undergoing reversible changes in shape so that specific binding sites are alternately exposed at either side of the membrane, are able to bind with and transfer particular substances unable to cross the plasma membrane on their own

cascade a series of sequential reactions that culminates in a final product, such as a clot

catabolism the breakdown or degradation of large, energy-rich molecules within cells

catalase an antioxidant enzyme found in peroxisomes that decomposes potent hydrogen peroxide into harmless H_2O and O_2.

catecholamines the chemical classification of the adrenomedullary hormones

cations positively charged ions that have lost one or more electrons from their outer shell

C cells the thyroid cells that secrete calcitonin

cell the smallest unit capable of carrying out the processes associated with life; the basic unit of both structure and function of living organisms

cell adhesion molecules (CAMs) proteins that protrude from the surface of the plasma membrane and form loops or other appendages that the cells use to grip ahold of each other and the surrounding connective-tissue fibers

cell body the portion of a neuron that houses the nucleus and organelles

cell-mediated immunity a specific immune response accomplished by activated T lymphocytes, which directly attack unwanted cells

center a functional collection of cell bodies within the central nervous system

central chemoreceptors receptors located in the medulla near the respiratory center that respond to changes in the H^+ concentration of the extracellular fluid resulting from changes in arterial Pco_2 and adjust respiration accordingly

central lacteal the terminal lymphatic vessel that supplies each of the small-intestinal villi

central nervous system (CNS) consists of the brain and spinal cord

central sulcus a deep infolding of the brain surface that runs roughly down the middle of the lateral surface of each cerebral hemisphere and separates the parietal and frontal lobes

centrioles a pair of short cylindrical structures within a cell that form the mitotic spindle during cell division

cerebellum the portion of the brain attached at the rear portion of the brain stem and concerned with maintaining proper position of the body in space and subconscious coordination of motor activity

cerebral cortex the outer shell of gray matter in the cerebrum; site of initiation of all voluntary motor output and final perceptual processing of all sensory input as well as integration of most higher neural activity

cerebral hemispheres the cerebrum's two halves, which are connected by a thick band of neuronal axons

cerebrospinal fluid a special cushioning fluid that is produced by, surrounds, and flows through the central nervous system

cerebrum the division of the brain that consists of the basal nuclei and cerebral cortex

channels small water-filled pathways through the plasma membrane; formed by membrane proteins that span the membrane and provide highly selective passage for small water-soluble substances such as ions

chemical bonds the forces holding atoms together

chemical buffer system a mixture in a solution of two or more chemical compounds that minimize pH changes when either an acid or a base is added to or removed from the solution

chemical mediator a chemical that is secreted by a cell that influences an activity outside the cell

chemical messenger–gated channels channels that open or close in response to the binding of a specific messenger with a membrane receptor site that is in close association with the channel

chemotaxin a chemical released at an inflammatory site that attracts phagocytes to the area

chief cells the stomach cells that secrete pepsinogen

cholecystokinin (CCK) a hormone released from the duodenal mucosa primarily in response to the presence of fat; inhibits gastric motility and secretion, stimulates pancreatic enzyme secretion, and stimulates gallbladder contraction

cholesterol a type of fat molecule that serves as a precursor for steroid hormones and bile salts and is a stabilizing component of the plasma membrane

cholinergic fibers nerve fibers that release acetylcholine as their neurotransmitter

chronic obstructive pulmonary disease a group of lung diseases characterized by increased airway resistance resulting from narrowing of the lumen of the lower airways

chyme a thick liquid mixture of food and digestive juices

cilia motile, hairlike protrusions from the surface of cells lining the respiratory airways and oviducts

ciliary body the portion of the eye that produces aqueous humor and contains the ciliary muscle

ciliary muscle a circular ring of smooth muscle within the eye whose contraction increases the strength of the lens to accommodate for near vision

circulatory shock when mean arterial blood pressure falls so low that adequate blood flow to the tissues can no longer be maintained

citric-acid cycle a cyclical series of biochemical reactions that involves the further processing of intermediate breakdown products of nutrient molecules, resulting in the generation of carbon dioxide and the preparation of hydrogen carrier molecules for entry into the high energy–yielding electron-transport chain

CNS see *central nervous system*

cochlea the snail-shaped portion of the inner ear that houses the receptors for sound

collecting tubule the last portion of tubule in the kidney's nephron that empties into the renal pelvis

colloid the thyroglobulin-containing substance enclosed within the thyroid follicles

complement system a collection of plasma proteins that are activated in cascade fashion on exposure to invading microorganisms, ultimately producing a membrane-attack complex that destroys the invaders

compliance the distensibility of a hollow, elastic structure, such as a blood vessel or the lungs; a measure of how easily the structure can be stretched

concave a curved surface that converges light rays

concentration gradient a difference in the concentration of a particular substance between two adjacent areas

concentric contraction an isotonic muscle contraction during which the muscle shortens

conduction transfer of heat between objects of differing temperatures that are in direct contact with each other

conduction by local current flow the means by which an action potential is propagated throughout a nonmyelinated nerve fiber; local current flow between an active and adjacent inactive area brings the inactive area to threshold, triggering an action potential in a previously inactive area

cones the eye's photoreceptors used for color vision in the light

congestive heart failure the inability of the cardiac output to keep pace with the body's needs for blood delivery, with blood damming up in the veins behind the failing heart

connective tissue serves to connect, support, and anchor various body parts; distinguished by relatively few cells dispersed within an abundance of extracellular material

contractile component the sarcomere-containing myofibrils within a muscle fiber that are capable of shortening on excitation

contractile proteins myosin and actin, whose interaction brings about shortening (contraction) of a muscle fiber

controlled variable some factor that can vary but is controlled to reduce the amount of variability and keep the factor at a relatively steady state

convection transfer of heat energy by air or water currents

convergence the converging of many presynaptic terminals from thousands of other neurons on a single neuronal cell body and its dendrites so that activity in the single neuron is influenced by the activity in many other neurons

convex a curved surface that diverges light rays

core temperature the temperature within the inner core of the body (abdominal and thoracic organs, central nervous system, and skeletal muscles) that is homeostatically maintained at about 100°F

cornea the clear, anteriormost outer layer of the eye through which light rays pass to the interior of the eye

coronary artery disease atherosclerotic plaque formation and narrowing of the coronary arteries that supply the heart muscle

coronary circulation the blood vessels that supply the heart muscle

corpus luteum the ovarian structure that develops from a ruptured follicle following ovulation

cortisol the adrenocortical hormone that plays an important role in carbohydrate, protein, and fat metabolism and helps the body resist stress

cranial nerves the twelve pairs of peripheral nerves that arise from the brain stem

cross bridges the myosin molecules' globular heads that protrude from a thick filament within a muscle fiber and interact with the actin molecules in the thin filaments to bring about shortening of the muscle fiber during contraction

cyclic adenosine monophosphate (cyclic AMP or cAMP) an intracellular second messenger derived from adenosine-triphosphate (ATP)

cyclic AMP see *cyclic adenosine monophosphate*

cytoplasm the portion of the cell interior not occupied by the nucleus

cytoskeleton a complex intracellular protein network that acts as the "bone and muscle" of the cell

cytosol the semiliquid portion of the cytoplasm not occupied by organelles

cytotoxic T cells the population of T cells that destroy host cells bearing foreign antigen, such as body cells invaded by viruses or cancer cells

dead-space volume the volume of air that occupies the respiratory airways as air is moved in and out and thus is not available to participate in the exchange of O_2 and CO_2 between the alveoli and atmosphere

dehydration a water deficit in the body

dehydroepiandrosterone (DHEA) the androgen (masculinizing hormone) secreted by the adrenal cortex in both sexes

dendrites projections from the surface of a neuron's cell body that carry signals toward the cell body

deoxyribonucleic acid (DNA) the cell's genetic material, which is found within the nucleus and provides codes for protein synthesis and serves as a blueprint for cell replication

depolarization a reduction in membrane potential from resting potential; movement of the potential from resting toward 0 mV

dermis the connective-tissue layer that lies under the epidermis in the skin; contains the skin's blood vessels and nerves

descending tract a bundle of nerve fibers of similar function that travels down the spinal cord to relay messages from the brain to efferent neurons

desmosome an adhering junction between two adjacent but nontouching cells formed by the extension of filaments between the cells' plasma membranes; most abundant in tissues that are subject to considerable stretching

DHEA see *dehydroepiandrosterone*

diabetes insipidus an endocrine disorder characterized by a deficiency of vasopressin

diabetes mellitus an endocrine disorder characterized by inadequate insulin action

diaphragm a dome-shaped sheet of skeletal muscle that forms the floor of the thoracic cavity; the major inspiratory muscle

diastole the period of cardiac relaxation and filling

diencephalon the division of the brain that consists of the thalamus and hypothalamus

diet-induced thermogenesis (DIT) the obligatory increase in metabolic rate that occurs as a result of the processing and storage of ingested nutrients

diffusion random collisions and intermingling of molecules as a result of their continuous thermally induced random motion

digestion the breaking-down process whereby the structurally complex foodstuffs of the diet are converted into smaller absorbable units by the enzymes produced within the digestive system

diploid number a complete set of forty-six chromosomes (twenty-three pairs) as found in all somatic cells

distal tubule a highly convoluted tubule that extends between the loop of Henle and collecting duct in the kidney's nephron

DIT see *diet-induced thermogenesis*

diurnal rhythm repetitive oscillations in hormone levels that are very regular and have a frequency of one cycle every twenty-four hours, usually linked to the light-dark cycle

divergence the diverging or branching of a neuron's axon terminals so that activity in this single neuron influences the many other cells with which its terminals synapse

DNA see *deoxyribonucleic acid*

dorsal root ganglion a cluster of afferent neuronal cell bodies located adjacent to the spinal cord

down regulation a reduction in the number of receptors for (and thereby the target cells' sensitivity to) a particular hormone as a direct result of the effect that an elevated level of the hormone has on its own receptors

eccentric contraction an isotonic contraction during which the muscle lengthens as it resists being stretched by an external force

ECG see *electrocardiogram*

edema swelling of tissues as a result of excess interstitial fluid

EDV see *end-diastolic volume*

EEG see *electroencephalogram*

effector organs the muscles or glands that are innervated by the nervous system and carry out the nervous system's orders to bring about a desired effect, such as a particular movement or secretion

efferent division the portion of the peripheral nervous system that carries instructions from the central nervous system to effector organs

efferent neuron the class of neuron that carries information from the central nervous system to the effector organs

efflux movement out of the cell

elastic recoil rebound of the lungs after having been stretched

electrical gradient a difference in charge between two adjacent areas

electrocardiogram (ECG) the graphic record of the electrical activity that reaches the surface of the body as a result of cardiac depolarization and repolarization

electrochemical gradient the simultaneous existence of an electrical gradient and concentration (chemical) gradient for a particular ion

electroencephalogram (EEG) a graphic record of the collective postsynaptic potential activity in the cell bodies and dendrites located in the cortical layers under a recording electrode

electrolytes solutes that form ions in solution and conduct electricity

embolus a freely floating clot

emphysema a pulmonary disease characterized by a collapse of the smaller airways and a breakdown of alveolar walls

end-diastolic volume (EDV) the volume of blood in the ventricle at the end of diastole when filling is complete

endocytosis internalization of extracellular material within a cell as a result of the plasma membrane forming a pouch that contains the extracellular material, then sealing at the surface of the pouch to form a small, intracellular membrane-bound vesicle with the contents of the pouch trapped inside

endogenous opiates endorphins and enkephalins, which bind with opiate receptors and are important in the body's natural analgesic system

endogenous pyrogen a chemical released from macrophages during inflammation that acts by means of local prostaglandins to raise the set point of the hypothalamic thermostat to produce a fever

endometrium the lining of the uterus

endoplasmic reticulum an organelle consisting of a continuous membranous network of fluid-filled tubules and flattened sacs, partially studded with ribosomes; synthesizes proteins and lipids for formation of new cell membrane and other cell components and manufactures products for secretion

endothelial-derived relaxing factor (EDRF) a local chemical mediator released from the endothelial cells lining an arteriole that diffuses locally to cause relaxation of the arteriolar smooth muscle in the vicinity

endothelium the thin, single-celled layer of epithelial cells that lines the entire circulatory system

end-plate potential (EPP) the graded potential generated at the motor end plate of a skeletal-muscle fiber in response to binding with acetylcholine released from the motor neuron terminal

end-systolic volume (ESV) the volume of blood in the ventricle at the end of systole when emptying is complete

endurance-type exercise see *aerobic exercise*

enterogastrones hormones secreted by the duodenal mucosa that inhibit gastric motility and secretion; include secretin, cholecystokinin, and gastric inhibitory peptide

enterohepatic circulation the recycling of bile salts and other bile constituents between the small intestine and liver by means of the hepatic portal vein

enzyme a special protein molecule that speeds up a particular chemical reaction in the body

eosinophils white blood cells that are important in allergic responses and in combating internal parasite infestations

epidermis the outer layer of the skin, consisting of numerous layers of epithelial cells, with the outermost layers being dead and flattened

epinephrine the primary hormone secreted by the adrenal medulla; important in preparing the body for "fight-or-flight" responses and in regulation of arterial blood pressure

epiphyseal plate a layer of cartilage that separates the diaphysis (shaft) of a long bone from the epiphysis (flared end); the site where bones grow in length before the cartilage ossifies (turns into bone)

epithelial tissue a functional grouping of cells specialized in the exchange of materials between the cell and its environment; lines and covers various body surfaces and cavities and forms secretory glands

EPSP see *excitatory postsynaptic potential*

equilibrium potential the potential that exists when the concentration gradient and opposing electrical gradient that exist for a given ion exactly counterbalance each other so that there is no net movement of the ion

erythrocytes red blood cells, which are bags of hemoglobin enclosed in plasma membrane that transport O_2 and to a lesser extent CO_2 and H^+ in the blood

erythropoiesis erythrocyte production by the bone marrow

erythropoietin the hormone released from the kidney in response to a reduction in O_2 delivery to the kidneys; stimulates the bone marrow to increase erythrocyte production

esophagus a straight muscular tube that extends between the pharynx and stomach

estrogen feminizing "female" sex hormone

ESV see *end-systolic volume*

excitable tissue tissue capable of producing electrical signals when excited; includes nervous and muscle tissue

excitation-contraction coupling the series of events linking muscle excitation (the presence of an action potential) to muscle contraction (filament sliding and sarcomere shortening)

excitatory postsynaptic potential (EPSP) a small depolarization of the postsynaptic membrane in response to neurotransmitter binding, thereby bringing the membrane closer to threshold

excitatory synapse a synapse in which the postsynaptic neuron's response to neurotransmitter release is a small depolarization of the postsynaptic membrane, bringing the membrane closer to threshold

exocytosis fusion of a membrane-bound intracellular vesicle with the plasma membrane, followed by the opening of the vesicle and the emptying of its contents to the outside

expiration a breath out

expiratory muscles the skeletal muscles whose contraction reduces the size of the thoracic cavity and allows the lungs to recoil to a smaller size, bringing about movement of air from the lungs to the atmosphere

external intercostal muscles inspiratory muscles whose contraction elevates the ribs, thereby enlarging the thoracic cavity

external work energy expended by contracting skeletal muscles to move external objects or to move the body in relation to the environment

extracellular fluid (ECF) all of the body's fluid found outside the cells; consists of interstitial fluid and plasma

extracellular matrix an intricate meshwork of fibrous proteins embedded in a watery, gel-like substance; secreted by local cells

extrinsic controls regulatory mechanisms initiated outside an organ that alter the activity of the organ; accomplished by the nervous and endocrine systems

extrinsic nerves the nerves that originate outside the digestive tract and innervate the various digestive organs

facilitated diffusion passive carrier-mediated transport involving transport of a substance down its concentration gradient across the plasma membrane

fatigue inability to maintain muscle tension at a given level despite sustained stimulation

feedforward mechanism a response designed to prevent an anticipated change in a controlled variable

feeding (appetite) centers neuronal clusters in the lateral regions of the hypothalamus that drive the individual to eat

fibrinogen a large soluble plasma protein that is converted into an insoluble, threadlike molecule that forms the meshwork of a clot during blood coagulation

Fick's law of diffusion the rate of net diffusion of a substance across a membrane is directly proportional to the substance's concentration gradient, the membrane's permeability to the substance, and the surface area of the membrane, and inversely proportional to the substance's molecular weight and the diffusion distance

fight-or-flight response the changes in activity of the various organs innervated by the autonomic nervous sytem in response to sympathetic stimulation, which collectively prepare the body for strenuous physical activity in the face of an emergency or stressful situation, such as a physical threat from the outside environment

fire when an excitable cell undergoes an action potential

first messenger an extracellular messenger, such as a hormone, that binds with a surface membrane receptor and activates an intracellular second messenger to carry out the desired cellular response

flagellum the single, long, whiplike appendage that serves as the tail of a spermatozoon

follicle (of ovary) a developing ovum and the surrounding specialized cells

follicle-stimulating hormone (FSH) an anterior pituitary hormone that stimulates ovarian follicular development and estrogen secretion in females and stimulates sperm production in males

follicular cells (of ovary) collectively, the granulosa and thecal cells

follicular cells (of thyroid gland) the cells that form the walls of the colloid-filled follicles in the thyroid gland; secrete thyroid hormone

follicular phase the phase of the ovarian cycle dominated by the presence of maturing follicles prior to ovulation

forebrain the division of the brain that consists of the diencephalon and cerebrum

Frank-Starling law of the heart intrinsic control of the heart, such that increased venous return resulting in increased end-diastolic volume leads to an increased strength of contraction and increased stroke volume; i.e., the heart normally pumps out all of the blood returned to it

frontal lobes the lobes of the cerebral cortex that lie at the top of the brain in front of the central sulcus and are responsible for voluntary motor output, speaking ability, and elaboration of thought

FSH see *follicle-stimulating hormone*

fuel metabolism see *intermediary metabolism*

functional syncytium a group of smooth- or cardiac-muscle cells that are interconnected by gap junctions and function electrically and mechanically as a single unit

functional unit the smallest component of an organ that can perform all the functions of the organ

gametes reproductive or germ cells, each containing a haploid set of chromosomes; sperm and ova

gamma motor neuron the type of motor neuron that innervates the fibers of a muscle-spindle receptor

ganglion a collection of neuronal cell bodies located outside the central nervous system

ganglion cells the nerve cells in the outermost layer of the retina; their axons form the optic nerve

gap junction a communicating junction formed between adjacent cells by small connecting tunnels that permit passage of charge-carrying ions between the cells so that electrical activity in one cell is spread to the adjacent cell

gastrin a hormone secreted by the pyloric gland area of the stomach that stimulates the parietal and chief cells to secrete a highly acidic gastric juice

gestation pregnancy

glomerular filtration filtration of a protein-free plasma from the glomerular capillaries into the tubular component of the kidney's nephron as the first step in urine formation

glomerular filtration rate (GFR) the rate at which glomerular filtrate is formed

glomerulus a ball-like tuft of capillaries in the kidney's nephron that filters water and solute from the blood as the first step in urine formation

glucagon the pancreatic hormone that raises blood glucose and blood fatty acid levels

glucocorticoids the adrenocortical hormones that are important in intermediary metabolism and in helping the body resist stress; primarily cortisol

gluconeogenesis the conversion of amino acids into glucose

glycogen the storage form of glucose in the liver and muscle

glycogenesis the conversion of glucose into glycogen

glycogenolysis the conversion of glycogen into glucose

glycolysis a biochemical process that takes place in the cell's cytosol and involves the breakdown of glucose into two pyruvic-acid molecules

GnRH see *gonadotropin-releasing hormone*

Golgi complex an organelle consisting of sets of stacked, flattened membranous sacs; processes raw materials transported to it from the endoplasmic reticulum into finished products, sorts the finished products, and directs them to their final destination

gonadotropin-releasing hormone (GnRH) the hypothalamic hormone that stimulates the release of FSH and LH from the anterior pituitary

gonadotropins FSH and LH; hormones that are tropic to the gonads

gonads the primary reproductive organs that produce the gametes and secrete the sex hormones; testes and ovaries

G protein a membrane-bound intermediary, which, when activated upon binding of an extracellular first messenger to a surface receptor, activates the enzyme adenylate cyclase on the intracellular side of the membrane in the cAMP second-messenger system

gradation of contraction variable magnitudes of tension produced in a single whole muscle

graded potential a local change in membrane potential that occurs in varying grades of magnitude; serves as a short-distance signal in excitable tissues

granulocytes leukocytes that contain granules, including neutrophils, eosinophils, and basophils

granulosa cells the layer of cells immediately surrounding a developing oocyte within an ovarian follicle

gray matter the portion of the central nervous system composed primarily of densely packaged neuronal cell bodies and dendrites

growth hormone an anterior pituitary hormone that is primarily responsible for regulating overall body growth and is also important in intermediary metabolism

H⁺ see *hydrogen ion*

haploid number the number of chromosomes found in gametes; a half set of chromosomes, one member of each pair for a total of twenty-three chromosomes

hapten a low-molecular-weight organic substance that is not antigenic by itself but can become antigenic if it attaches to body proteins

Hb see *hemoglobin*

hCG see *human chorionic gonadotropin*

heart failure an inability of the cardiac output to keep pace with the body's demands for supplies and removal of wastes

helper T cells the population of T cells that enhances the activity of other immune-response effector cells

hematocrit the percentage of blood volume occupied by erythrocytes as they are packed down in a centrifuged blood sample

hemoglobin a large iron-bearing protein molecule found within erythrocytes that binds with and transports most of the O_2 in the blood; also carries some of the CO_2 and H^+ in the blood

hemolysis rupture of red blood cells

hemostasis the stopping of bleeding from an injured vessel

heparin a chemical released from basophils and mast cells that hastens removal of fat particles from the blood following a fatty meal and is used as an anticoagulant in blood samples

hepatic portal system a complex vascular connection between the digestive tract and liver such that venous blood from the digestive system drains into the liver for processing of absorbed nutrients before being returned to the heart

hippocampus the elongated, medial portion of the temporal lobe that is a part of the limbic system and is especially crucial for forming long-term memories

histamine a chemical released from mast cells or basophils that brings about vasodilation and increased capillary permeability; important in allergic responses

homeostasis maintenance by the highly coordinated, regulated actions of the body systems of relatively stable chemical and physical conditions in the internal fluid environment that bathes the body's cells

hormone a long-distance chemical mediator that is secreted by an endocrine gland into the blood, which transports it to its target cells

hormone response element (HRE) the specific attachment site on DNA for a given steroid hormone and its nuclear receptor

host cell a body cell infected by a virus

HRE see *hormone response element*

human chorionic gonadotropin (hCG) a hormone secreted by the developing placenta that stimulates and maintains the corpus luteum of pregnancy

hydrogen ion (H$^+$) the cationic portion of a dissociated acid

hydrolysis the digestion of a nutrient molecule by the addition of water at a bond site

hydrostatic pressure the pressure exerted by fluid on the walls that contain it

hypercalcemia high blood Ca^{++} levels

hyperglycemia elevated blood glucose concentration

hyperplasia an increase in the number of cells

hyperpolarization an increase in membrane potential from resting potential; the potential becomes even more negative than at resting potential

hypersecretion too much of a particular hormone secreted

hypertension sustained above-normal mean arterial blood pressure

hypertonic having an osmolarity greater than normal body fluids; more concentrated than normal

hypertrophy increase in the size of an organ as a result of an increase in the size of its cells

hyperventilation overbreathing; when the rate of ventilation is in excess of the body's metabolic needs for CO$_2$ removal

hypocalcemia low blood Ca^{++} levels

hypophysiotropic hormones hormones secreted by the hypothalamus that regulate the secretion of anterior pituitary hormones; see also *releasing hormone* and *inhibiting hormone*

hyposecretion too little of a particular hormone secreted

hypotension sustained below-normal mean arterial blood pressure

hypothalamo-hypophyseal portal system the vascular connection between the hypothalamus and anterior pituitary gland used for the pickup and delivery of hypophysiotropic hormones

hypothalamus the brain region located beneath the thalamus that is concerned with regulating many aspects of the internal fluid environment such as water and salt balance and food intake; serves as an important link between the autonomic nervous system and endocrine system

hypotonic having an osmolarity less than normal body fluids; more dilute than normal

hypoventilation underbreathing; ventilation inadequate to meet the metabolic needs for O$_2$ delivery and CO$_2$ removal

hypoxia insufficient O$_2$ at the cellular level

I band one of the light bands that alternate with dark (A) bands to create a striated appearance in a skeletal- or cardiac-muscle fiber when these fibers are viewed with a light microscope

IGF see *insulin-like growth factor*

immune surveillance recognition and destruction of newly arisen cancer cells by the immune system

immunity the body's ability to resist or eliminate potentially harmful foreign materials or abnormal cells

immunoglobulins antibodies; gamma globulins

impermeable passage of a particular substance through the plasma membrane is prohibited

implantation the burrowing of a blastocyst into the endometrial lining

inflammation an innate, nonspecific series of highly interrelated events, especially involving neutrophils, macrophages, and local vascular changes, that are set into motion in response to foreign invasion or tissue damage

influx movement into the cell

inhibin a hormone secreted by the Sertoli cells of the testes or the ovarian follicles that inhibits FSH secretion

inhibiting hormone a hypothalamic hormone that inhibits the secretion of a particular anterior pituitary hormone

inhibitory postsynaptic potential (IPSP) a small hyperpolarization of the postsynaptic membrane in response to neurotransmitter binding, thereby moving the membrane farther from threshold

inhibitory synapse a synapse in which the postsynaptic neuron's response to neurotransmitter release is a small hyperpolarization of the postsynaptic membrane, moving the membrane farther from threshold

inorganic refers to substances that do not contain carbon; from nonliving sources

inspiration a breath in

inspiratory muscles the skeletal muscles whose contraction enlarges the thoracic cavity, bringing about lung expansion and movement of air into the lungs from the atmosphere

insulin the pancreatic hormone that lowers blood levels of glucose, fatty acids, and amino acids and promotes their storage

insulin-like growth factor (IGF) synonymous with somatomedins

integrating center a region that determines efferent output based on processing of afferent input

integument the skin and underlying connective tissue

intercostal muscles the muscles that lie between the ribs; see also *external intercostal muscles* and *internal intercostal muscles*

interferon a chemical released from virus-invaded cells that provides nonspecific resistance to viral infections by transiently interfering with replication of the same or unrelated viruses in other host cells

interleukin 1 a multipurpose chemical mediator released from macrophages that enhances B-cell activity

interleukin 2 a chemical mediator secreted by helper T cells that augments the activity of all T cells

intermediary metabolism the collective set of intracellular chemical reactions that involve the degradation, synthesis, and transformation of small nutrient molecules; also known as fuel metabolism

intermediate filaments threadlike cytoskeletal elements that play a structural role in parts of the cells subject to mechanical stress

internal environment the body's aqueous extracellular environment, which consists of the plasma and interstitial fluid and which must be homeostatically maintained in order for the cells to make life-sustaining exchanges with it

internal intercostal muscles expiratory muscles whose contraction pulls the ribs downward and inward, thereby reducing the size of the thoracic cavity

internal respiration the intracellular metabolic processes carried out within the mitochondria that use O_2 and produce CO_2 during the derivation of energy from nutrient molecules

internal work all forms of biological energy expenditure that do not accomplish mechanical work outside of the body

interneuron the class of neuron that lies entirely within the central nervous system and is important for integration of peripheral responses to peripheral information as well as for the abstract phenomena associated with the "mind"

interstitial fluid the portion of the extracellular fluid that surrounds and bathes all of the body's cells

intra-alveolar pressure the pressure within the alveoli

intracellular fluid (ICF) the fluid collectively contained within all of the body's cells

intrapleural pressure the pressure within the pleural sac

intrinsic controls local control mechanisms that are inherent to an organ

intrinsic factor a special substance secreted by the parietal cells of the stomach that must be combined with vitamin B_{12} in order for this vitamin to be absorbed by the intestine; deficiency produces pernicious anemia

intrinsic nerve plexuses interconnecting networks of nerve fibers within the digestive tract wall

ion an atom that has gained or lost one or more of its electrons so that it is not electrically balanced

IPSP see *inhibitory postsynaptic potential*

iris a pigmented smooth muscle that forms the colored portion of the eye and controls pupillary size

islets of Langerhans the endocrine portion of the pancreas that secretes the hormones insulin and glucagon into the blood

isometric contraction a muscle contraction in which the development of tension occurs at constant muscle length

isotonic having an osmolarity equal to normal body fluids

isotonic contraction a muscle contraction in which muscle tension remains constant as the muscle fiber changes length

juxtaglomerular apparatus a cluster of specialized vascular and tubular cells at a point where the ascending limb of the loop of Henle passes through the fork formed by the afferent and efferent arterioles of the same nephron in the kidney

killer cells destroy a target cell that has been coated with antibodies by lysing its membrane

lactation milk production by the mammary glands

lactic acid an end product produced from pyruvic acid during anaerobic glycolysis

larynx the "voice box" at the entrance of the trachea; contains the vocal cords

lateral inhibition the phenomenon in which the most strongly activated signal pathway originating from the center of a stimulus area inhibits the less excited pathways from the fringe areas by means of lateral inhibitory connections within sensory pathways

lateral sacs the expanded saclike regions of a muscle fiber's sarcoplasmic reticulum; store and release calcium, which plays a key role in triggering muscle contraction

law of mass action if the concentration of one of the substances involved in a reversible reaction is increased, the reaction is driven toward the opposite side, and if the concentration of one of the substances is decreased, the reaction is driven toward that side

left ventricle the heart chamber that pumps blood into the systemic circulation

length-tension relationship the relationship between the length of a muscle fiber at the onset of contraction and the force the fiber can achieve on a subsequent tetanic contraction

lens a transparent, biconvex structure of the eye that refracts (bends) light rays; its strength can be adjusted to accommodate for vision at different distances

leukocyte endogenous mediator (LEM) a chemical mediator secreted by macrophages that is identical to endogenous pyrogen and exerts a wide array of effects associated with inflammation

leukocytes white blood cells, which are the immune system's mobile defense units

Leydig cells the interstitial cells of the testes that secrete testosterone

LH see *luteinizing hormone*

LH surge the burst in LH secretion that occurs at midcycle of the ovarian cycle and triggers ovulation

limbic system a functionally interconnected ring of forebrain structures that surrounds the brain stem and is concerned with emotions, basic survival and sociosexual behavioral patterns, motivation, and learning

lipid emulsion a suspension of small fat droplets held apart as a result of adsorption of bile salts on their surface

loop of Henle a hairpin loop that extends between the proximal and distal tubule of the kidney's nephron

lumen the interior space of a hollow organ or tube

luteal phase the phase of the ovarian cycle dominated by the presence of a corpus luteum

luteinization the formation of a postovulatory corpus luteum in the ovary

luteinizing hormone (LH) an anterior pituitary hormone that stimulates ovulation, luteinization, and secretion of estrogen and progesterone in females and stimulates testosterone secretion in males

lymph interstitial fluid that is picked up by the lymphatic vessels and returned to the venous system, meanwhile passing through the lymph nodes for defense purposes

lymphocytes white blood cells that provide immune defense against targets for which they are specifically programmed

lymphoid organs organs that produce and store lymphocytes, such as lymph nodes and tonsils

lymphokines all chemicals other than antibodies that are secreted by lymphocytes

lysosomes organelles consisting of membrane-bound sacs containing powerful hydrolytic enzymes that destroy unwanted material within the cell, such as internalized foreign material or cellular debris

macrophages large, tissue-bound phagocytes

mast cells cells located within connective tissue that synthesize, store, and release histamine, as during allergic responses

mature follicle an ovarian follicle that is ready to ovulate

mean arterial blood pressure the average pressure responsible for driving blood forward through the arteries into the tissues throughout the cardiac cycle; equals cardiac output times total peripheral resistance

mechanically gated channels channels that open or close in response to stretching or other mechanical deformation

mechanistic approach explanation of body functions in terms of mechanisms of action, that is, the ''how'' of events that occur in the body

mechanoreceptor a sensory receptor sensitive to mechanical energy, such as stretching or bending

medullary respiratory center several aggregations of neuronal cell bodies within the medulla that provide output to the respiratory muscles and receive input important for regulating the magnitude of ventilation

meiosis cell division in which the chromosomes replicate followed by two nuclear divisions so that only a half set of chromosomes is distributed to each of four new daughter cells

melanocyte-stimulating hormones (MSHs) several closely related hormones produced by the anterior pituitary in humans and by the intermediate lobe of the pituitary in lower vertebrates; regulate skin coloration by controlling the dispersion of melanin granules in lower vertebrates; presumably involved with memory and learning in humans

membrane-attack complex a collection of the five final activated components of the complement system that aggregate to form a porelike channel in the plasma membrane of an invading microorganism, with the resultant leakage leading to destruction of the invader

membrane potential a separation of charges across the membrane; a slight excess of negative charges lined up along the inside of the plasma membrane and separated from a slight excess of positive charges on the outside

memory cells B or T cells that are newly produced in response to a microbial invader; they do not participate in the current immune response against the invader but instead remain dormant, ready to launch a swift, powerful attack should the same microorganism invade again in the future

menstrual cycle the cyclical changes in the uterus in accompaniment with the hormonal changes in the ovarian cycle

menstrual phase the phase of the menstrual cycle characterized by sloughing of endometrial debris and blood out through the vagina

metabolic acidosis acidosis resulting from any cause other than excess accumulation of carbonic acid in the body

metabolic alkalosis alkalosis caused by a relative deficiency of non-carbonic acid

metabolic rate energy expenditure/unit of time

micelle a water-soluble aggregation of bile salts, lecithin, and cholesterol that has a hydrophilic shell and a hydrophobic core; carries the water-insoluble products of fat digestion to their site of absorption

microfilaments cytoskeletal elements made of actin molecules (as well as of myosin molecules in muscle cells); play a major role in various cellular contractile systems and serve as a mechanical stiffener for microvilli

microtrabecular lattice cytoskeletal element consisting of a meshwork of exceedingly fine interlinked filaments that suspends and functionally links larger cytoskeletal elements and various organelles

microtubules cytoskeletal elements made of tubulin molecules arranged into long, slender, unbranched tubes that help maintain asymmetrical cell shapes and coordinate complex cell movements

microvilli actin-stiffened, nonmotile, hairlike projections from the luminal surface of epithelial cells lining the digestive tract and kidney tubules; tremendously increase the surface area of the cell exposed to the lumen

micturition the process of bladder emptying; urination

milk ejection the squeezing out of milk produced and stored in the alveoli of the breasts by means of contraction of the myoepithelial cells that surround each alveolus

millivolts the unit used to measure membrane potential

mineralocorticoids the adrenocortical hormones that are important in Na^+ and K^+ balance; primarily aldosterone

mitochondria the energy organelles, which contain the enzymes for oxidative phosphorylation

mitosis cell division in which the chromosomes replicate before nuclear division so that each of the two daughter cells receives a full set of chromosomes

molecule a chemical substance formed by linking atoms together; the smallest unit of a given chemical substance

monocytes white blood cells that emigrate from the blood, enlarge, and become macrophages, large tissue phagocytes

monosaccharides simple sugars, such as glucose; the absorbable unit of digested carbohydrates

motor activity movement of the body accomplished by contraction of skeletal muscles

motor end plate the specialized portion of a skeletal-muscle fiber that lies immediately underneath the terminal button of the motor neuron and possesses receptor sites for binding acetylcholine released from the terminal button

motor neurons the neurons that innervate skeletal muscle; their axons constitute the somatic nervous system

motor unit one motor neuron plus all of the muscle fibers it innervates

motor-unit recruitment the progressive activation of a muscle fiber's motor units to accomplish increasing gradations of contractile strength

mucosa the innermost layer of the digestive tract that lines the lumen

multipotential stem cells precursor cells that reside in the bone marrow and continuously divide and differentiate to give rise to each of the types of blood cells

multiunit smooth muscle a smooth-muscle mass that consists of multiple discrete units that function independently of each other and that must be separately stimulated by autonomic nerves to contract

muscarinic receptor type of cholinergic receptor found at the effector organs of all parasympathetic postganglionic fibers

muscle fiber a single muscle cell, which is relatively long and cylindrical shaped

muscle tension the force produced during muscle contraction by shortening of the sarcomeres, resulting in stretching and tightening of the muscle's elastic connective tissue and tendon, which transmits the tension to the bone to which the muscle is attached

muscle tissue a functional grouping of cells specialized for contraction and force generation

myelin an insulative lipid covering that surrounds myelinated nerve fibers at regular intervals along the axon's length; each patch of myelin is formed by a separate myelin-forming cell that wraps itself jelly-roll fashion around the neuronal axon

myelinated fibers neuronal axons covered at regular intervals with insulative myelin

myocardial ischemia inadequate blood supply to the heart tissue

myocardium the cardiac muscle within the heart wall

myofibril a specialized intracellular structure of muscle cells that contains the contractile apparatus

myometrium the smooth-muscle layer of the uterus

myosin the contractile protein that forms the thick filaments in muscle fibers

Na^+-K^+ pump a carrier that actively transports Na^+ out of the cell and K^+ into the cell

natural killer cells naturally occurring, lymphocytelike cells that nonspecifically destroy virus-infected cells and cancer cells by directly lysing their membranes on first exposure to them

negative balance when the losses for a substance exceed its gains so that the total amount of the substance in the body decreases

negative feedback a regulatory mechanism in which a change in a controlled variable triggers a response that opposes the change, thus maintaining a relatively steady set point for the regulated factor

nephron the functional unit of the kidney; consisting of an interrelated vascular and tubular component, it is the smallest unit that can form urine

nerve a bundle of peripheral neuronal axons, some afferent and some efferent, which are enclosed by a connective-tissue covering and follow the same pathway

nervous system one of the two major control systems of the body; in general, coordinates rapid activities of the body, especially those involving interactions with the external environment

nervous tissue a functional grouping of cells specialized for initiation and transmission of electrical signals

net diffusion the difference between two opposing movements

net filtration pressure the net difference in the hydrostatic and osmotic forces acting across the glomerular membrane that favors the filtration of a protein-free plasma into Bowman's capsule

neuroendocrine refers to an interaction between the nervous and endocrine systems

neurohormones hormones released into the blood by neurosecretory neurons

neuromodulators chemical messengers that bind to neuronal receptors at nonsynaptic sites (that is, not at the subsynaptic membrane) and bring about long-term changes that subtly depress or enhance synaptic effectiveness

neuromuscular junction the juncture between a motor neuron and a skeletal-muscle fiber

neuron a nerve cell, typically consisting of a cell body, dendrites, and an axon; specialized to initiate, propagate, and transmit electrical signals

neuropeptides large, slowly acting peptide molecules released from axon terminals along with classical neurotransmitters; most neuropeptides function as neuromodulators

neurotransmitter the chemical messenger that is released from the axon terminal of a neuron in response to an action potential and influences another neuron or an effector with which the neuron is anatomically linked

neutrophils white blood cells that are phagocytic specialists and are important in inflammatory responses and defense against bacterial invasion

nicotinic receptor type of cholinergic receptor found at all autonomic ganglia and the motor end plates of skeletal-muscle fibers

nitric oxide a recently identified local chemical mediator released from endothelial cells and other tissues; exerts a wide array of effects ranging from causing local arteriolar vasodilation to acting as a toxic agent against foreign invaders to serving as a unique type of neurotransmitter

nociceptor a pain receptor that is sensitive to tissue damage

nonspecific immune responses inherent defense responses that nonselectively defend against foreign or abnormal material, even upon initial exposure to it; see also *inflammation, interferon, natural killer cells,* and *complement system*

nontropic hormone a hormone that exerts its effects on nonendocrine target tissues

norepinephrine the neurotransmitter released from sympathetic postganglionic fibers

nucleus (of brain) a functional aggregation of neuronal cell bodies within the brain

nucleus (of cells) a distinct spherical or oval structure, which is usually located near the center of a cell and which contains the cell's genetic material, deoxyribonucleic acid (DNA)

occipital lobes the lobes of the cerebral cortex that are located posteriorly and are responsible for initially processing visual input

O_2-Hb dissociation curve a graphic depiction of the relationship between arterial P_{O_2} and percent hemoglobin saturation

oligodendrocytes the myelin-forming cells of the central nervous system

oogenesis egg production

opsonin body-produced chemicals that link bacteria to macrophages, thereby making the bacteria more susceptible to phagocytosis

optic nerve the bundle of nerve fibers that leaves the retina relaying information about visual input

optimal length the length before the onset of contraction of a muscle fiber at which maximal force can be developed upon a subsequent tetanic contraction

organelles distinct, highly organized, membrane-bound intracellular compartments, each containing a specific set of chemicals for carrying out a particular cellular function

organic refers to substances that contain carbon; from living or once-living sources

organ of Corti the sense organ of hearing within the inner ear that contains hair cells whose hairs are bent in response to sound waves, setting up action potentials in the auditory nerve

organs a distinct structural unit composed of two or more types of primary tissue organized to perform a particular function or functions—for example, the stomach

osmolarity a measure of the concentration of solute molecules in a solution

osmosis movement of water across a membrane down its own concentration gradient toward the area of higher solute concentration

osteoblasts bone cells that produce the organic matrix of bone

osteoclasts bone cells that resorb bone in their vicinity

otolith organs sense organs in the inner ear that provide information about rotational changes in head movement; include the utricle and saccule

oval window the membrane-covered opening that separates the air-filled middle ear from the upper compartment of the fluid-filled cochlea in the inner ear

overhydration excess water in the body

ovulation release of an ovum from a mature ovarian follicle

oxidative phorphorylation the entire sequence of mitochondrial biochemical reactions that efficiently extracts energy from the nutrients in food and transforms it into ATP, producing CO_2 and H_2O in the process

oxyhemoglobin hemoglobin combined with O_2

oxyntic mucosa the mucosa that lines the body and fundus of the stomach; contains gastric pits lined by mucous neck cells, parietal cells, and chief cells

oxytocin a hypothalamic hormone that is stored in the posterior pituitary and stimulates uterine contraction and milk ejection

pacemaker activity self-excitable activity of an excitable cell in which its membrane potential gradually depolarizes to threshold on its own

Pacinian corpuscle a rapidly adapting skin receptor that detects pressure and vibration

pancreas a mixed gland composed of an exocrine portion, which secretes digestive enzymes and an aqueous alkaline secretion through a duct into the duodenal lumen, and an endocrine portion, which secretes the hormones insulin and glucagon into the blood

paracrine a local chemical messenger whose effect is exerted only on neighboring cells in the immediate vicinity of its site of secretion

parasympathetic nervous system the subdivision of the autonomic nervous system that dominates in quiet, relaxed situations and promotes body maintenance activities such as digestion and emptying of the urinary bladder

parathyroid glands four small glands located on the posterior surface of the thyroid gland that secrete parathyroid hormone

parathyroid hormone (PTH) a hormone that raises plasma Ca^{++} levels

parietal cells the stomach cells that secrete hydrochloric acid and intrinsic factor

parietal lobes the lobes of the cerebral cortex that lie at the top of the brain behind the central sulcus and contain the somatosensory cortex

partial pressure the individual pressure exerted independently by a particular gas within a mixture of gases

partial pressure gradient a difference in the partial pressure of a gas between two regions that promotes the movement of the gas from the region of higher partial pressure to the region of lower partial pressure

parturition delivery of a baby

passive expiration expiration accomplished during quiet breathing as a result of elastic recoil of the lungs on relaxation of the inspiratory muscles, with no energy expenditure required

passive force a force that does not require expenditure of cellular energy to accomplish transport of a substance across the plasma membrane

passive reabsorption when none of the steps in the transepithelial transport of a substance reabsorbed across the kidney tubules requires energy expenditure

pathogens disease-causing microorganisms, such as bacteria or viruses

pathophysiology abnormal functioning of the body associated with disease

pepsin; pepsinogen an enzyme secreted in inactive form by the stomach that, once activated, begins protein digestion

peptide hormones hormones that consist of a chain of specific amino acids of varying length

percent hemoglobin saturation a measure of the extent to which the hemoglobin present is combined with O_2

perception the conscious interpretation of the external world as created by the brain from a pattern of nerve impulses delivered to it from sensory receptors

peripheral chemoreceptors the carotid and aortic bodies, which respond to changes in arterial P_{O_2}, P_{CO_2}, and H^+ and adjust respiration accordingly

peripheral nervous system consists of nerve fibers that carry information between the central nervous system and other parts of the body

peristalsis ringlike contractions of the circular smooth muscle of a tubular organ that move progressively forward with a stripping motion, pushing the contents of the organ ahead of the contraction

peritubular capillaries capillaries that intertwine around the tubules of the kidney's nephron; supply the renal tissue and participate in exchanges between the tubular fluid and blood during the formation of urine

permeable passage of a particular substance through the plasma membrane is permitted

permissiveness when one hormone must be present in adequate amounts for the full exertion of another hormone's effect

pernicious anemia the anemia produced as a result of intrinsic factor deficiency

peroxisomes organelles consisting of membrane-bound sacs that contain powerful oxidative enzymes that detoxify various wastes produced within the cell or foreign compounds that have entered the cell

pH equals the logarithm to the base 10 of the reciprocal of the hydrogen-ion concentration (pH = log 1/[H$^+$])

phagocytosis a type of endocytosis in which large multimolecular, solid particles are engulfed by a cell

pharynx the back of the throat, which serves as a common passageway for the digestive and respiratory systems

phosphorylation addition of a phosphate group to a molecule

photoreceptor a sensory receptor responsive to light

phototransduction the mechanism of converting light stimuli into electrical activity by the rods and cones of the eye

phrenic nerve the nerve that supplies the diaphragm

physiology the study of body functions

pinocytosis the type of endocytosis in which the cell internalizes fluid

pitch the tone of a sound, determined by the frequency of vibrations (i.e., whether a sound is a C or G note)

pituitary gland a small endocrine gland connected by a stalk to the hypothalamus; consists of the anterior pituitary and posterior pituitary

placenta the organ of exchange between the maternal and fetal blood; also secretes hormones that support the pregnancy

plaque a deposit of cholesterol and other lipids, perhaps calcified, in thickened, abnormal smooth-muscle cells within blood vessels as a result of atherosclerosis

plasma the liquid portion of the blood

plasma cell an antibody-producing derivative of an activated B lymphocyte

plasma clearance the volume of plasma that is completely cleared of a given substance per minute

plasma-colloid osmotic pressure the force caused by the unequal distribution of plasma proteins between the blood and surrounding fluid that encourages fluid movement into the capillaries

plasma membrane a protein-studded lipid bilayer that encloses each cell, separating it from the extracellular fluid

plasma proteins the proteins that remain within the plasma, where they perform a number of important functions; include albumins, globulins, and fibrinogen

plasticity the ability of portions of the brain to assume new responsibilities in response to the demands placed on it

platelets specialized cell fragments in the blood that participate in hemostasis by forming a plug at a vessel defect, among other things

pleural sac a double-walled, closed sac that separates each lung from the thoracic wall

polycythemia excess circulating erythrocytes

polysaccharides complex carbohydrates consisting of chains of interconnected glucose molecules

positive balance when the gains via input for a substance exceed its losses via output so that the total amount of the substance in the body increases

positive feedback a regulatory mechanism in which the input and the output in a control system continue to enhance each other so that the controlled variable is progressively moved farther from a steady state

postabsorptive state the metabolic state after a meal is absorbed during which endogenous energy stores must be mobilized and glucose must be spared for the glucose-dependent brain

posterior pituitary the neural portion of the pituitary that stores and, on hypothalamic stimulation, releases into the blood two hormones produced by the hypothalamus, vasopressin and oxytocin

postganglionic fiber the second neuron in the two-neuron autonomic nerve pathway; originates in an autonomic ganglion and terminates on an effector organ

postsynaptic neuron the neuron that conducts its action potentials away from a synapse

preganglionic fiber the first neuron in the two-neuron autonomic nerve pathway; originates in the central nervous system and terminates on an autonomic ganglion

pressure gradient a difference in pressure between two regions that drives the movement of blood or air from the region of higher pressure to the region of lower pressure

presynaptic facilitation enhanced release of neurotransmitter from a presynaptic axon terminal as a result of excitation of another neuron that terminates on the axon terminal

presynaptic inhibition a reduction in the release of neurotransmitter from a presynaptic axon terminal as a result of excitation of another neuron that terminates on the axon terminal

presynaptic neuron the neuron that conducts its action potentials toward a synapse

primary active transport a carrier-mediated transport system in which energy is directly required to operate the carrier and move the transported substance against its concentration gradient

primary follicle a primary oocyte surrounded by a single layer of granulosa cells in the ovary

primary motor cortex the portion of the cerebral cortex that lies anterior to the central sulcus and is responsible for voluntary motor output

progestational phase see *secretory phase*

prolactin an anterior pituitary hormone that stimulates breast development and milk production

proliferative phase the phase of the menstrual cycle during which the endometrium is repairing itself and thickening following menstruation; lasts from the end of the menstrual phase until ovulation

pro-opiomelanocortin a large precursor molecule produced by the anterior pituitary that is cleaved into adrenocorticotropic hormone, melanocyte-stimulating hormone, and endorphin

proprioception awareness of body position

prostaglandins local chemical mediators that are derived from a component of the plasma membrane, arachidonic acid

prostate gland a male accessory sex gland that secretes an alkaline fluid, which neutralizes acidic vaginal secretions

protein kinase an enzyme that phosphorylates and thereby induces a change in the shape and function of a particular intracellular protein

proteolytic enzymes enzymes that digest protein

proximal tubule a highly convoluted tubule that extends between Bowman's capsule and the loop of Henle in the kidney's nephron

PTH see *parathyroid hormone*

pulmonary artery the large vessel that carries blood from the right ventricle to the lungs

pulmonary circulation the closed loop of blood vessels carrying blood between the heart and lungs

pulmonary surfactant a phospholipoprotein complex secreted by the Type II alveolar cells that intersperses between the water molecules that line the alveoli, thereby lowering the surface tension within the lungs

pulmonary valve a one-way valve that permits the flow of blood from the right ventricle into the pulmonary artery during ventricular emptying but prevents the backflow of blood from the pulmonary artery into the right ventricle during ventricular relaxation

pulmonary veins the large vessels that carry blood from the lungs to the heart

pulmonary ventilation the volume of air breathed in and out in one minute; equals tidal volume times respiratory rate

pupil an adjustable round opening in the center of the iris through which light passes to the interior portions of the eye

Purkinje fibers small terminal fibers that extend from the bundle of His and rapidly transmit an action potential throughout the ventricular myocardium

pyloric gland area the specialized region of the mucosa in the antrum of the stomach that secretes gastrin

pyloric sphincter the juncture between the stomach and duodenum

radiation emission of heat energy from the surface of a warm body in the form of electromagnetic waves

reabsorption the net movement of interstitial fluid into the capillary

receptor see *sensory receptor* or *receptor site*

receptor potential the graded potential change that occurs in a sensory receptor in response to a stimulus; generates action potentials in the afferent neuron fiber

receptor site membrane protein that binds with a specific extracellular chemical messenger, thereby bringing about a series of membrane and intracellular events that alter the activity of the particular cell

reduced hemoglobin hemoglobin that is not combined with O_2

reflex any response that occurs automatically without conscious effort; the components of a reflex arc include a receptor, afferent pathway, integrating center, efferent pathway, and effector

refraction bending of a light ray

refractory period the time period when a recently activated patch of membrane is refractory (unresponsive) to further stimulation; prevents the action potential from spreading backward into the area through which it has just passed, thereby assuring the unidirectional propagation of the action potential away from the initial site of activation

regulatory proteins troponin and tropomyosin, which play a role in regulating muscle contraction by either physically covering or exposing the sites of interaction between the contractile proteins

regulatory T cells helper and suppressor T cells

releasing hormone a hypothalamic hormone that stimulates the secretion of a particular anterior pituitary hormone

renal cortex an outer granular-appearing region of the kidney

renal medulla an inner striated-appearing region of the kidney

renal threshold the plasma concentration at which the T_m of a particular substance is reached and the substance first starts appearing in the urine

renin an enzymatic hormone released from the kidneys in response to a decrease in Na^+/ECF volume/arterial blood pressure; activates angiotensinogen

renin-angiotensin-aldosterone system the salt-conserving system triggered by the release of renin from the kidneys, which activates angiotensin, which stimulates aldosterone secretion, which stimulates Na^+ reabsorption by the kidney tubules during the formation of urine

repolarization return of membrane potential to resting potential following a depolarization

reproductive tract the system of ducts that are specialized to transport or house the gametes after they are produced

residual volume the minimum volume of air remaining in the lungs even after a maximal expiration

respiration the sum of processes that accomplish ongoing passive movement of O_2 from the atmosphere to the tissues, as well as the continual passive movement of metabolically produced CO_2 from the tissues to the atmosphere

respiratory acidosis acidosis resulting from abnormal retention of CO_2 arising from hypoventilation

respiratory airways the system of tubes that conducts air between the atmosphere and the alveoli of the lungs

respiratory alkalosis alkalosis caused by excessive loss of CO_2 from the body as a result of hyperventilation

respiratory rate breaths per minute

resistance hindrance of flow of blood or air through a passageway (blood vessel or respiratory airway, respectively)

resting membrane potential the membrane potential that exists when an excitable cell is not displaying an electrical signal

reticular activating system (RAS) ascending fibers that originate in the reticular formation and carry signals upward to arouse and activate the cerebral cortex

reticular formation a network of interconnected neurons that runs throughout the brain stem and initially receives and integrates all synaptic input to the brain

retina the innermost layer in the posterior region of the eye that contains the eye's photoreceptors, the rods and cones

ribosomes special ribonucleic acid (RNA)-protein complexes that synthesize proteins under the direction of nuclear DNA

right atrium the heart chamber that receives venous blood from the systemic circulation

right ventricle the heart chamber that pumps blood into the pulmonary circulation

rods the eye's photoreceptors used for night vision

round window the membrane-covered opening that separates the lower chamber of the cochlea in the inner ear from the middle ear

salivary amylase an enzyme produced by the salivary glands that begins carbohydrate digestion in the mouth and continues in the body of the stomach after the food and saliva have been swallowed

saltatory conduction the means by which an action potential is propagated throughout a myelinated fiber, with the impulse jumping over the myelinated regions from one node of Ranvier to the next node

SA node see *sinoatrial node*

sarcomere the functional unit of skeletal muscle; the area between two Z lines within a myofibril

sarcoplasmic reticulum a fine meshwork of interconnected tubules that surrounds a muscle fiber's myofibrils; contains expanded lateral sacs, which store calcium that is released into the cytosol in response to a local action potential

satiety centers neuronal clusters in the ventromedial regions of the hypothalamus that inhibit feeding behavior

saturation when all of the binding sites on a carrier molecule are occupied

Schwann cells the myelin-forming cells of the peripheral nervous system

secondary active transport a transport mechanism in which a carrier molecule for glucose or an amino acid is driven by a Na^+ concentration gradient established by the energy-dependent Na^+ pump to transfer the glucose or amino acid uphill without directly expending energy to operate the carrier

secondary follicle a developing ovarian follicle that is secreting estrogen and forming an antrum

secondary sexual characteristics the many external characteristics that are not directly involved in reproduction but that distinguish males and females

second messenger an intracellular chemical that is activated by binding of an extracellular first messenger to a surface receptor site and that triggers a preprogrammed series of biochemical events, which result in altered activity of intracellular proteins to control a particular cellular activity

secretin a hormone released from the duodenal mucosa primarily in response to the presence of acid; inhibits gastric motility and secretion and stimulates secretion of a $NaHCO_3$ solution from the pancreas

secretion release, on appropriate stimulation, of substances that have been produced by a cell to the cell's exterior

secretory phase the phase of the menstrual cycle characterized by the development of a lush endometrial lining capable of supporting a fertilized ovum; also known as the *progestational phase*

secretory vesicles membrane-bound sacs containing proteins that have been synthesized and processed by the endoplasmic reticulum/Golgi complex of the cell and that will be released to the cell's exterior by exocytosis on appropriate stimulation

segmentation the small intestine's primary motility; consists of oscillating, ringlike contractions of the circular smooth muscle along the small intestine's length

self-antigens antigens that are characteristic of a person's own cells

semen a mixture of accessory sex gland secretions and sperm

semicircular canals sense organ in the inner ear that detects rotational or angular acceleration or deceleration of the head

semilunar valves the aortic and pulmonary valves

seminal vesicles male accessory sex glands that supply fructose to ejaculated sperm and secrete prostaglandins

seminiferous tubules the highly coiled tubules within the testes that produce spermatozoa

sensory afferent a pathway coming into the central nervous system that carries information that reaches the level of consciousness

sensory input includes somatic sensation and special senses

sensory receptor an afferent neuron's peripheral ending, which is specialized to respond to a particular stimulus in its environment

septum the muscular partition that separates the right and left halves of the heart

series-elastic component the noncontractile portions of a skeletal-muscle fiber, including the connective tissue and sarcoplasmic reticulum

Sertoli cells cells located in the seminiferous tubules that support spermatozoa during their development

single-unit smooth muscle the most abundant type of smooth muscle; made up of muscle fibers that are interconnected by gap junctions so that they become excited and contract as a unit; also known as *visceral smooth muscle*

sinoatrial (SA) node a small specialized autorhythmic region in the right atrial wall of the heart that has the fastest rate of spontaneous depolarizations and serves as the normal pacemaker of the heart

skeletal muscle striated muscle, which is attached to the skeleton and is responsible for movement of the bones in purposeful relation to one another; innervated by the somatic nervous system and under voluntary control

slow-wave potentials self-excitable activity of an excitable cell in which its membrane potential undergoes gradually alternating depolarizing and hyperpolarizing swings

smooth muscle involuntary muscle innervated by the autonomic nervous system and found in the walls of hollow organs and tubes

somatic cells body cells, as contrasted with reproductive cells

somatic nervous system the portion of the efferent division of the peripheral nervous system that innervates skeletal muscles; consists of the axonal fibers of the alpha motor neurons

somatic sensation sensory information arising from the body surface, including somesthetic sensation and proprioception

somatomedins hormones secreted by the liver or other tissues in response to growth hormone that act directly on the target cells to promote growth

somatosensory cortex the region of the parietal lobe immediately behind the central sulcus; the site of initial processing of somesthetic and proprioceptive input

somesthetic sensations awareness of sensory input such as touch, pressure, temperature, and pain from the body's surface

sound waves traveling vibrations of air that consist of regions of high pressure caused by compression of air molecules alternating with regions of low pressure caused by rarefaction of the molecules

spatial summation the summing of several postsynaptic potentials arising from the simultaneous activation of several excitatory (or several inhibitory) synapses

special senses vision, hearing, taste, and smell

specific immune responses responses that are selectively targeted against particular foreign material to which the body has previously been exposed; see also *antibody-mediated immunity* and *cell-mediated immunity*

specificity ability of carrier molecules to transport only specific substances across the plasma membrane

spermatogenesis sperm production

sphincter a voluntarily controlled ring of skeletal muscle that controls passage of contents through an opening into or out of a hollow organ or tube

spinal reflex a reflex that is integrated by the spinal cord

spleen a lymphoid organ located in the upper left part of the abdomen that stores lymphocytes and platelets and destroys old red blood cells

state of equilibrium no net change in a system is occurring

steady state synonymous with *state of equilibrium*

steroids hormones derived from cholesterol

stimulus a detectable physical or chemical change in the environment of a sensory receptor

stress the generalized, nonspecific response of the body to any factor that overwhelms, or threatens to overwhelm, the body's compensatory abilities to maintain homeostasis

stretch reflex a monosynaptic reflex in which an afferent neuron originating at a stretch-detecting receptor in a skeletal muscle terminates directly on the efferent neuron supplying the same muscle to cause it to contract and counteract the stretch

stroke volume (SV) the volume of blood pumped out of each ventricle with each contraction or beat of the heart

subcortical regions the brain regions that lie under the cerebral cortex, including the basal nuclei, thalamus, and hypothalamus

submucosa the connective-tissue layer of the digestive tract that lies under the mucosa and contains the larger blood and lymph vessels and a nerve network

substance P the neurotransmitter released from pain fibers

subsynaptic membrane the portion of the postsynaptic cell membrane that lies immediately underneath a synapse and contains receptor sites for the synapse's neurotransmitter

summation (of twitches) the addition of two or more muscle twitches as a result of rapidly repetitive stimulation, resulting in greater tension in the fiber than is produced by a single action potential

suppressor T cells the population of T cells that suppresses the activity of the other T cells, serving to limit immune responses in check-and-balance fashion

surface tension the force at the liquid surface of an air-water interface resulting from the greater attraction of water molecules to the surrounding water molecules than to the air above the surface; a force that tends to decrease the area of a liquid surface and resists stretching of the surface

sympathetic nervous system the subdivision of the autonomic nervous system that dominates in "fight-or-flight" emergency or stressful situations and prepares the body for strenuous physical activity

synapse the specialized junction between two neurons where an action potential in the presynaptic neuron influences the membrane potential of the postsynaptic neuron by means of the release of a chemical messenger that diffuses across the small cleft that separates the two neurons

synergism when several actions are complementary so that their combined effect is greater than the sum of their separate effects

systemic circulation the closed loop of blood vessels carrying blood between the heart and body systems

systole the period of cardiac contraction and emptying

T₃ see *triiodothyronine*

T₄ see *thyroxine*

tactile refers to touch

target-cell receptors receptors located on a target cell that are specific for a particular chemical mediator

target cells the cells that a particular extracellular chemical messenger such as a hormone or a neurotransmitter influences

teleological approach explanation of body functions in terms of their particular purpose in fulfilling a bodily need, that is, the "why" of body processes

temporal lobes the lobes of the cerebral cortex that are located laterally and are responsible for initially processing auditory input

temporal summation the summing of several postsynaptic potentials occurring very close together in time because of successive firing of a single presynaptic neuron

tension see *muscle tension*

terminal button a motor neuron's enlarged knoblike ending that terminates near a skeletal muscle fiber and releases acetylcholine in response to an action potential in the neuron

testosterone the male sex hormone secreted by the Leydig cells of the testes

tetanus a smooth, maximal muscle contraction that occurs when the fiber is stimulated so rapidly that it does not have a chance to relax at all between stimuli

thalamus the brain region that serves as a synaptic integrating center for preliminary processing of all sensory input on its way to the cerebral cortex

thecal cells the outer layer of specialized ovarian connective-tissue cells in a maturing follicle

thermoreceptor a sensory receptor sensitive to specific chemicals

thick filaments specialized cytoskeletal structures within skeletal muscle that are made up of myosin molecules and interact with the thin filaments to accomplish shortening of the fiber during muscle contraction

thin filaments specialized cytoskeletal structures within skeletal muscle that are made up of actin, tropomyosin, and troponin molecules and interact with the thick filaments to accomplish shortening of the fiber during muscle contraction

thoracic cavity the chest cavity

threshold potential the critical potential that must be reached before an action potential is initiated in an excitable cell

thromboembolism a condition characterized by the presence of thrombi and emboli in the circulatory system

thrombus an abnormal clot attached to the inner lining of a blood vessel

thymus a lymphoid organ located midline in the chest cavity that processes T lymphocytes and produces the hormone thymosin, which maintains the T-cell lineage

thyroglobulin a large, complex molecule on which all steps of thyroid hormone synthesis and storage take place

thyroid gland a bilobed endocrine gland that lies over the trachea and secretes three hormones—thyroxine and triiodothyronine, which regulate overall basal metabolic rate, and calcitonin, which contributes to control of calcium balance

thyroid hormone collectively, the hormones secreted by the thyroid follicular cells, namely, thyroxine and triiodothyronine

thyroid-stimulating hormone (TSH) an anterior pituitary hormone that stimulates secretion of thyroid hormone and promotes growth of the thyroid gland

thyroxine the most abundant hormone secreted by the thyroid gland; important in the regulation of overall metabolic rate; also known as *tetraiodothyronine* or T_4

tidal volume the volume of air entering or leaving the lungs during a single breath

tight junction an impermeable junction between two adjacent epithelial cells formed by the sealing together of the cells' lateral edges near their luminal borders; prevents passage of substances between the cells

tissue (1) a functional aggregation of cells of a single specialized type, such as nerve cells forming nervous tissues; (2) the aggregate of various cellular and extracellular components that make up a particular organ, such as lung tissue

T lymphocytes (T cells) white blood cells that accomplish cell-mediated immune responses against targets to which they have been previously exposed; see also *cytotoxic T cells, helper T cells,* and *suppressor T cells*

T_m see transport maximum

tone the ongoing baseline of activity in a given system or structure, as in muscle tone, sympathetic tone, or vascular tone

tonic LH secretion the low-level, ongoing secretion of LH that occurs throughout most of the ovarian cycle

total peripheral resistance the resistance offered by all of the peripheral blood vessels, with arteriolar resistance contributing most extensively

trachea the "windpipe"; the conducting airway that extends from the pharynx and branches into two bronchi, one of which enters each lung

tract a bundle of nerve fibers (axons of long interneurons) with a similar function within the spinal cord

transduction conversion of stimuli into action potentials by sensory receptors

transepithelial transport the entire sequence of steps involved in the transfer of a substance across the epithelium between either the renal tubular lumen or digestive-tract lumen and the blood

transmural pressure gradient the pressure difference across the lung wall (intra-alveolar pressure is greater than intrapleural pressure) that stretches the lungs to fill the thoracic cavity, which is larger than the unstretched lungs

transporter recruitment the phenomenon of inserting additional transporters (carriers) for a particular substance into the plasma membrane, thereby increasing membrane permeability to the substance, in response to an appropriate stimulus

transport maximum (T_m) the maximum rate of a substance's carrier-mediated transport across the membrane when the carrier is saturated; known as *tubular maximum* in the kidney tubules

transverse tubule (T tubule) a perpendicular infolding of the surface membrane of a muscle fiber; rapidly spreads surface electric activity into the central portions of the muscle fiber

triglycerides neutral fats composed of one glycerol molecule with three fatty acid molecules attached

triiodothyronine (T_3) the most potent hormone secreted by the thyroid follicular cells; important in the regulation of overall metabolic rate

trophoblast the outer layer of cells in a blastocyst that is responsible for accomplishing implantation and developing the fetal portion of the placenta

tropic hormone a hormone that stimulates the secretion of another hormone

tropomyosin one of the regulatory proteins found in the thin filaments of muscle fibers

troponin one of the regulatory proteins found in the thin filaments of muscle fibers

TSH see *thyroid-stimulating hormone*

T tubule see *transverse tubule*

tubular maximum (T_m) the maximum amount of a substance that the renal tubular cells can actively transport within a given time period

tubular reabsorption the selective transfer of substances from the tubular fluid into the peritubular capillaries during the formation of urine

tubular secretion the selective transfer of substances from the peritubular capillaries into the tubular lumen during the formation of urine

twitch a brief, weak contraction that occurs in response to a single action potential in a muscle fiber

tympanic membrane the eardrum, which is stretched across the entrance to the middle ear and vibrates when struck by sound waves that are funneled down the external ear canal

Type I alveolar cells the single layer of flattened epithelial cells that forms the wall of the alveoli within the lungs

Type II alveolar cells the cells within the alveolar walls that secrete pulmonary surfactant

ultrafiltration the net movement of a protein-free plasma out of the capillary into the surrounding interstitial fluid

ureter a duct that transmits urine from the kidney to the bladder

urethra a tube that carries urine from the bladder to outside the body

urine excretion the elimination of substances from the body in the urine; anything filtered or secreted and not reabsorbed is excreted

vagus nerve the tenth cranial nerve, which serves as the major parasympathetic nerve

vasoconstriction the narrowing of a blood vessel lumen as a result of contraction of the vascular circular smooth muscle

vasodilation the enlargement of a blood vessel lumen as a result of relaxation of the vascular circular smooth muscle

vasopressin a hormone secreted by the hypothalamus, then stored and released from the posterior pituitary; increases the permeability of the distal and collecting tubules of the kidney to water; also known as *antidiuretic hormone (ADH)*

vaults a recently discovered organelle shaped like octagonal barrels; believed to serve as transporters for messenger RNA from the nucleus to sites of protein synthesis; may be important in cellular movement

vein a vessel that carries blood toward the heart

velocity speed

vena cava the large veins that empty blood into the right atrium

venous return the volume of blood returned to each atrium per minute from the veins

ventilation the mechanical act of moving air in and out of the lungs; breathing

ventricle the lower chamber of the heart that pumps blood into the arteries

vesicle a small, intracellular, fluid-filled, membrane-bound sac

vestibular apparatus the component of the inner ear that provides information essential for the sense of equilibrium and for coordinating head movements with eye and postural movements; consists of the semicircular canals, utricle, and saccule

villus (villi, plural) microscopic fingerlike projections from the inner surface of the small intestine

virulence the disease-producing power of a pathogen

visceral afferent a pathway coming into the central nervous system that carries subconscious information derived from the internal viscera

visceral smooth muscle see *single-unit smooth muscle*

viscosity the friction developed between molecules of a fluid as they slide over each other during flow of the fluid; the greater the viscosity, the greater the resistance to flow

vital capacity the maximum volume of air that can be moved out during a single breath following a maximal inspiration

voltage-gated channels channels that open or close in response to changes in membrane potential

white matter the portion of the central nervous system composed of myelinated nerve fibers

Z line a flattened disc-like cytoskeletal protein that connects the thin filaments of two adjoining sarcomeres together

zona fasciculata the middle and largest layer of the adrenal cortex; major source of cortisol

zona glomerulosa the outermost layer of the adrenal cortex; sole source of aldosterone

zona reticularis the innermost layer of the adrenal cortex; produces cortisol, along with the zona fasciculata

INDEX

A ⁻ (*See* intracellular protein anions)
A bands, 216, 218–19
Abdominal muscles
  role
    in respiration, 422, 425
    in vomiting, 559
Abdominal pregnancy, 725
Abortion, 726, 741
ABO system of blood groups, 389
Absolute refractory period, 90
Absorption
  across capillaries. *See* Reabsorption
  by the digestive system, 545
    of carbohydrates, 580, 581f
    of fat, 572
    in the large intestine, 589–90
    of protein, 580–82
    in the small intestine, 577–79, 580–87
  across renal tubules. *See* Tubular
    reabsorption
Absorptive state, 671
Accessory digestive organs, 545–49
Accessory inspiratory muscles, 422–23
Accessory sex glands, 696
  male, 710
Acclimatization
  to heat during exercise, 517
  to high altitudes, 451
Accommodation in vision, 166–69
  in hyperopic eyes, 168f
  mechanism of, 167f
  in myopic eyes, 168f
ACE inhibitor drugs, 483
Acetic acid, derivation from pyruvic acid,
  28
Acetylcholine (ACh), 106, 206–7
  and contraction of skeletal muscle,
    215
  deficiency in Alzheimer's disease,
    132–33
  effect on heart, 284
  and neuromuscular junction, 205, 222
  neurotransmitter in the autonomic ner-
    vous system, 198, 199
Acetylcholinesterase (AChE), 207
  effect of organophosphates on, 209
Acetyl coA. *See* Acetyl coenzyme A
Acetyl coenzyme A, and the citric-acid
  cycle, 27f

ACh. *See* Acetylcholine
Achalasia, 555
AChE. *See* Acetylcholinesterase
Acid-base balance, 466, 510, 524–37
  abnormalities of, 534–37
  in the digestive system, 586
  regulation of, 488–89, 524–37
Acidity of blood, 447
Acidosis, 525
  in diabetes mellitus, 675
  effects of
    on neuromuscular excitability, 526
    on potassium levels, 526
  renal response to, 532–33
  respiratory, 450, 534–36, 536f
  respiratory adjustments to, 530
  See also Metabolic (nonrespiratory) aci-
    dosis; respiratory acidosis
Acids, 525
  definition of, 524, A7
  effects on arteriolar radius, 312
  effects on enterogastrone secretion, 558
  sources in the body, 526
  See also Hydrogen ion; pH
Acinar cells
  pancreatic, 567–68, 590
Acquired immune deficiency syndrome
  (AIDS), 395
  transfer to fetus, 732
Acquired reflexes
  salivary, 552
  See Conditioned reflexes
Acromegaly, 645–46, 645f
Acrosomal enzymes, and fertility, 727
Acrosome, 706
ACTH. *See* Adrenocorticotropic hormone
Actin, 39, 39f, 217–18, 217f, 223f
  in amoeboid movement, 39
  in the contractile ring of cytokinesis, 39f
  helix of, 218
  in microfilaments, 39
  microvilli stiffened by, ear, 188
  in muscle contractile systems, 39, 220f
  in platelets, 358
  See also Thin filaments
Actin binding site, 216, 216f
Actin filaments
  in skeletal muscle, 217–18, 218f
  in smooth muscle, 247

Actin-myosin contractile apparatus in
  capillaries, 319
Action potentials, 79, 80–83, 101–2, 108,
  144
  all-or-none law applied to, 91
  in cardiac muscle pacemaker cells,
    268–69, 268f
  in cardiac muscle contractile cells, 272,
    272f
  changes in membrane potential during,
    81f
  comparison with graded potentials, 79,
    86
  duration of
    in cardiac muscle, 268–74
    in nerve cells, 81
    in skeletal muscle, 225, 225f
  effect of immediate early genes (IEGs)
    on, 135
  effect of parasympathetic stimulation on,
    283
  effect of sympathetic nervous system
    stimulation, 285
  frequency of, 90
  initiation of
    at the axon hillock in postsynaptic
      neurons, 96
    beyond the receptor in afferent neu-
      rons, 152–53, 153f
  and permeability changes, 83f
  plateau, contractile cardiac-muscle cells,
    272–73
  propagation of
    by conduction by local current flow,
      84–86, 85f
    by saltatory conduction, 86–88, 88f
  refractory period
    in the heart, 274, 274f
    in nerves, 89–91
    in skeletal muscles, 274
  in single-unit smooth muscle fibers, 248
  threshold potential for, 81
  velocity of propagation, 88
Activated T cells, 393
Active area of an excitable membrane, 79
  during an action potential, 84–85
Active (forced) expiration, 425
Active forces, 57
Active hyperemia, 312

Active immunity, 387–88
    versus passive immunity, 389
Active reabsorption across the epithelium, 481–82
Active transport, 64–67, 67f
    ATP use in, 65f
    primary, 66
    secondary, 66
    See also Passive transport; Transport
Acuity, 155–57
    and structure of the eye, 173
Acupuncture, 158
    physiological basis of, 160
Acute mountain sickness, 451
Acute myocardial infarction, outcomes of, 293
Acute-phase proteins, 375, C1
Acute renal failure. *See* Renal failure
Adaptation, 153–54
    of the eye to light, 173–74
    versus habituation, 154
    to odors, 193
    role of cortisol in, 658
Adaptation syndrome, general, 665
Addiction to anabolic steroids, 239
Addison's disease, 662
A-delta fibers, 158
Adenine (A), B1
Adenohypophysis, 631
    See also Anterior pituitary
Adenoids, 370
Adenosine
    in ATP, 27
    role in cardiac vasodilation, 290, 312
Adenosine diphosphate (ADP)
    and energy production, 27
    formation during a power stroke, with energy release, 224
    and platelet aggregation, 358–59
    See also Adenosine triphosphate
Adenosine monophosphate (AMP), A12
Adenosine triphosphate (ATP), 12, 27–34
    and anabolism, 668
    anaerobic formation of, 236
    energy transfer during muscle contraction and relaxation, 234–36, 236f
    heat release on processing of, 596
    and ischymetric theory, 600
    in membrane transport, 64–65, 65f
    in muscle contraction, 223, 233–34
    recharge-expenditure cycle of, 33
    synthesis
        under aerobic conditions, 32–33
        under anaerobic conditions, 32–33
        in the citric-acid cycle, 28
        in the electron transport chain, 30–32, 32f
        by oxidative phosphorylation, 32–33
        via the chemiosmotic mechanism, 31–32
Adenylate cyclase, 51
    and hydrophilic hormones, 625
Adequate stimulus, 152
ADH. *See* Vasopressin

Adhesions, cell-to-cell, 54–56
    See also Desmosome; Extracellular matrix; Gap junctions; Tight junctions
Adipose tissue, 35
    effects of growth hormone on, 643
    effects of insulin on, 674
    fat inclusions in, 35
    and fuel metabolism, 669
    triglycerides in, A9
    See also Fat
Adolescence, 703
ADP. *See* Adenosine diphosphate
Adrenal androgens
    and adrenogenital syndrome, 661–62, 662f, 702
    control of, 659
    functions of in females, 723
Adrenal cortex, 657
    abnormalities in secretion of, 660–63
    and ACTH, 659–60
    aldosterone production in, 338
    anatomy of, 656f
    and control of ECF volume and osmolarity, 524
    control of secretion of, 657, 659–60
    cortisol production in, 622
    hormones, major functions, and target cells, 619
    and stress, 658, 660, 665
Adrenal diabetes, 660
Adrenal glands, 202, 649, 656–67, 656f
    See also Adrenal cortex; Adrenal medulla
Adrenal medulla, 202, 656–57, 663, 664–65
    abnormalities of, 665
    hormones, major functions, and target cells of hormones, 619
    sympathetic nervous system and, 663–64
    See also Epinephrine; Norepinephrine
Adrenergic drugs, and treatment of allergies, 403
Adrenergic fibers, 200
Adrenocortical hormones, 657
    See also Adrenal androgens; Aldosterone; Cortisol
Adrenocortical insufficiency (Addison's disease), primary, 662–63
Adrenocorticotropic hormone (ACTH), 622, 635
    control of cortisol secretion, 659–60, 659f
    in hypothalamic-anterior pituitary control systems, 638
Adrenocorticotropic hormone (ACTH or adrenocorticotropin), and stress, 665
Adrenogenital syndrome, 661–62, 662f, 702
Aerobic conditions, 32–33
    energy yield, compared with anaerobic conditions, 33f
Aerobic exercise, 34
    benefits of, 34
    definition of, 34, 234

induction of long-term adaptive changes in muscle by, 238
    in management of Type II diabetes mellitus, 64
    and risk of hypertension, 339
Aerobic metabolism in the heart, 267
Afferent arterioles, 466–67
    effect on glomerular-capillary blood pressure, 473
    factors influencing in regulation of GFR, 474, 475f
Afferent division, peripheral nervous system, 107, 151–210
    See also Afferent neurons; Afferent pathway; Receptors; Somatosensory sensation; Special sensation
Afferent neurons, 107–8
    characteristics and functions of, 206
    and control of motor unit output, 240
    firing rate, and mean arterial pressure, 335f
    reactions in habituation and sensitization, 131
Afferent pathway
    as part of reflex arc, 142, 144–45
Afferent peripheral nerves, 146
Afterbirth, 736–37
After hyperpolarization, 81, 82
Afterload, 288
Age/aging
    and body water, 511
    and cardiac reserve, 288
    and changes in vision, 167–68
    and decline in muscle mass and strength, 240
    effect on the autonomic nervous system, 204
    and the effects of growth hormone, 642
    growth hormone therapy for, 644
    and the male reproductive system, 704
    and osteoporosis, 691
    and prostatic hypertrophy, 710
    and the reproductive system, 714, 723
    and thymus activity, 381
Agglutination, 383–84, 384f
    in antibody reaction, 389
Aggregation of platelets, 358–59
Agonists, 203
AIDS
    and the need for blood substitutes, 353
    virus causing, 395–96
    See Acquired immune deficiency syndrome
Air
    composition
        in the atmosphere, 438
        in lung alveoli, 438–39
    partial pressures of, 438, 438f
Airflow rates, 425–27
    matching of
        with blood flow, 429
Air pollution, effect on the respiratory system, 408

Airway resistance
and the sympathetic nervous system, 201
See also Resistance, of airways
Airways
cilia in, 37–38, 37f
See also Respiratory airways
Albumins, 349
concentration and function in plasma, 349
filtration through the glomerular membrane, 471
loss in kidney disease, 503
and thyroid hormone transport, 652
Alcohol
absorption by digestive tract, 567
in brain, 519
effect on gastric secretions, 564
effect on vasopressin action, 523
and placental transfer, 732
and toxic hepatitis and cirrhosis, 575
Aldehydes, A7–8
Aldosterone
and control of ECF volume and osmolarity, 524
control of secretion of
by potassium, 489, 489f
by renin-angiotensin-aldosterone system, 482–83, 483f, 657
deficiency of, and salt balance, 515
effect of
on potassium secretion, 489, 657
on salt in sweat, 515
on sodium reabsorption, 481–84
functions of, 482, 657
and homeostasis in the endocrine system, 692
hypersecretion of, 660
levels of, in Conn's syndrome, 338
mechanism of action, 482
synthesis of, 657
and water retention, 521
Aldosterone-induced protein, 482
Alimentary system. *See* Digestive system
Alkaline solutions, 525
secretion of
by the exocrine pancreas, 568, 569
by the large intestine, 589
by the liver, 572
by the prostate gland, 710
Alkalosis, 525, 526
effect of on neuromuscular excitability, 526
renal response to, 533
respiratory, 450
respiratory response to, 530
Allergen, 402
Allergy, 369, 401–4
and basophils, 357, 402
and eosinophils, 357, 402
shots for, 404
All-or-none law, applied to nerve fibers, 91
Alpha 1-antitrypsin (α1-antitrypsin), 426
Alpha (α) cells, 671, 679
Alpha (α) globulin, 349

Alpha motor neurons
distinction from gamma motor neurons, 243
inhibition of, 246
See also Motor neurons
Alpha (α) receptors, 203, 663
Alpha rhythm, 124f
Alternate complement pathway, 379–80
Altitude, and low oxygen levels in blood, 447
Aluminum, concentration in degenerating neurons, 133
Alveolar dead space. *See* Dead space
Alveolar macrophages, 408
effect of cigarette smoking on, 408
Alveolar surface tension, and elastic behavior of the lungs, 429–30
Alveolar ventilation, 437
effect of exercise on, 456–57
See also Ventilation
Alveoli, (in lungs), 412, 415–16, 415f
air composition in, 438–39
gas exchange across, 438
interdependence of, 430
stability of, 431f
surface tension of, 429–30, 430f
Alveoli (in breasts), 737
Alzheimer, Alois, 132
Alzheimer's disease, 132–33
damage to the hippocampus in, 131
Amblyopia (lazy eye), 105
Amines
as hormones, 618, 623
Amino acids, 543, A10
active transport of
across intestine, 580
across kidney tubules, 484
as absorbable units of protein from digestive tract, 580
digestion of proteins to, 580
effects of
on glucagon secretion, 680
on insulin secretion, 675
effects on
of cortisol, 658
of growth hormone, 643, 645
of insulin, 674, 680
and gluconeogenesis, 658
metabolism of, 667–71, 669f
structure of (general), A10f
tubular reabsorption of, 484
Aminopeptidases, 544f, 577–78
in brush borders, 582
digestion by, 577
Ammonia secretion by renal tubules, 533
Ammonium ion, 534
Amnesia, 131
Amniocentesis, 726
Amniotic sac, 728, 735
Amoeba, 5, 6
Amoeboid movement, 39, 39f
Amplification of signals in a second messenger system, 53–54, 54f

Amplitude of sound waves, 178
Ampulla
of the ear, semicircular canal, 188
of the oviduct, 722
Amylase, 544f
pancreatic, 568, 569
salivary, 551, 567
Amyloid precursor protein (APP), 133
Amylopectin, A8
Amylose, A8
Anabolic androgenic steroids, 238
Anabolic effect of testosterone, 704
Anabolism, 668
Anaerobic conditions, 32–33
Anaerobic exercise, 34, 236
Analgesia
by aspirin-like drugs, 158
built-in system for, 158
proposed pathway for, 159f
Anaphase, B11
Anaphylactic shock, 404
Anatomic dead space, 437
and tidal volume, 437
See Dead space
Anatomy, definition of, 2
See also Anatomy or structure of specific organs
Andrenergic receptors in arteriolar smooth muscle, 315
Androgen, 701
effects of
on development at puberty in females, 723
on growth, 640, 646
production
by the adrenal cortex, 657
by the testes, 657
See also Adrenal androgens; Dehydroepiandrosterone; Testosterone
Androgen-binding protein, 707
Androgens, estrogen production using, 719–20
Android obesity, 602
Anemia, 354–55
and oxygen-carrying capacity of the blood, 446
and oxygen levels in blood, 447
as result of kidney failure, 502
Anemic hypoxia, 449
Angina pectoris, 290–91
Angiotensin-converting enzyme (ACE), 482
Angiotensin I, 482
See also Renin-angiotensin-aldosterone system
Angiotensin II
and ACE inhibitors, 483
effect on arterioles, 315–16
and thirst, 482
and vasopressin secretion, 482
See also Renin-angiotensin-aldosterone system
Angiotensinogen, 349, 482
Angry macrophages, 395

Anions
  definition of, A3
  movement along an electrical gradient,
    59, 59f
Anorexia nervosa, 601
Anovulatory cycles, 714
ANP, 484
Antagonism, and hormone targets, 631
Antagonistic muscles, 144
Antagonists, 203
Anterior pituitary (adenohypophysis), 631,
    633–39, 639f
  ACTH secretion by, 659
  and adrenal cortex secretion, 659
  and female reproduction, 719–22
  gonadotropic hormones of
    and female reproduction, 719
    and male reproduction, 707–8
  hormones, major functions, and target
    cells of hormones, 619
  link with the hypothalamus, 637f
  regulation of secretion, 635–39, 635f,
    637f
  response to estrogen concentration, 721
  See also Specific hormones
Anterior pituitary gland, enlargement of,
    and visual disturbances, 646
Anterograde amnesia, 131
Anthrax bacteria, 388
Antibiotics, destruction of intestinal flora
    by, 408
Antibodies, 358, C1
  and active versus passive immunity,
    387–88
  anti-A and anti-B, 389
  clonal selection theory of production of,
    385, 385f
  functions of, 383, 384f
  mechanism of action, 382–83
  responses to microbes, 384f
  structure of, 382f
  subclasses of, 382
Antibody-mediated (humoral) immunity,
    380
  See also B Lymphocytes
Anticipatory response. See Feedforward
Anticoagulants, 364
Anticodons in tRNA, B9, 9f
Antidiuretic hormone (ADH), 497–99
  See also Vasopressin
Antigen
  blood-type, of erythrocytes (red blood
    cells), 389
  definition of, 381
  macrophage presenting of to lympho-
    cytes, 391, 390f
  protection against by antibodies, 381–83
  stimulation of B lymphocytes for anti-
    body production, 385–87
  and T-cell activation, 392, 397–98
  See also Allergen; Self-antigens
Antigen-binding fragments (Fab)(variable
    region) of antibody, 383

Antigen sequestering as a tolerance mech-
    anism, 397
Antihistamines
  and treatment of allergies, 403
  and treatment of ulcers, 567
Anti-inflammatory drugs, use in transplan-
    tation, 397
Anti-inflammatory effects of cortisol, 658
Antiserum (antitoxin), 388
Antral follicle, 715
Antral peristaltic contractions, 557
Antrum (of stomach), 555
  pyloric gland area of, 560
  as site of gastric mixing, 555, 555f
Antrum (of ovarian follicle), 715f
  changes in, during follicular develop-
    ment, 717
  formation of, 719
Anxiety, alkalosis associated with, 536
Aorta, 263, 305
Aortic arch baroreceptor, 334
Aortic bodies
  arterial oxygen monitoring by, 453–54
  response to arterial hydrogen ion con-
    centration, 456
  See also Peripheral chemoreceptors
Aortic valve
  anatomy of, 264
  opening and closing during cardiac
    cycle, 279f, 280
  relation to heart sounds, 281
Aphasias, 123
Aplastic anemia, 354–55
Aplysia, habituation and sensitization in,
    134f
Aplysia, experiments on to study memory,
    131
Apnea, 449, 459–60
Apneusis, 453
Apneustic centers, 450, 452–53
Apolipoprotein B-100 (apo-B), 295
Appendicitis, 589
Appendix, 370, 587
Appetite centers, 598–99
Appetite loss, at high altitudes, 451
Aqueous humor, 162
Arachidonic acid
  precursor for prostaglandins, 360,
    710–11
Arachnoid mater, 112f, 113
Arachnoid villi, 112f, 113
Arginine, A11
Aristotle, 302
Arousal system in sleep-wake cycle, 139
Arrhenius, Svante, A7
Arrhythmia, detection on an ECG, 276
Arsenic
  and aplastic anemia, 354
  and kidney disease, 502
Arterial baroreceptors. See also
    Baroreceptors
Arterial blood pressure, 307f, 332–43
  as afterload of cardiac contraction, 288

  baroreceptor regulation of, 336f
  determinants of, 334f
  deviations from normal. See Hypoten-
    sion; Hypertension
  diastolic, 306, 308
  direct effect on GFR, 474, 475f, 476
  effects of sympathetic nervous system
    and epinephrine on, 664
  long-term regulation of, 514–15
  mean, 299
  measurement of, 308
  regulation by control of ECF volume
    by means of salt balance, 516,
    516f
  short-term regulation of, 516
  sympathetic nervous system effect on,
    201
  systolic, 306, 308
Arterial branches, 305
Arteries, 299, 301–2, 305–10
  See also Arterial blood pressure; Blood
    vessels
Arteriolar resistance
  effect on distribution of cardiac output,
    308–10
  extrinsic controls of, 314–16, 482
  local controls of, 310–14
  sympathetic nervous system and, 201,
    425
Arterioles, 299, 302, 305, 308–16, 469f
  branching of capillaries from, 321f
  penile, 712
  vasoconstriction of, 521–22
  See also Arteriolar resistance; Blood
    vessels
Arthritis, 377
  rheumatoid, 385
Artificial insemination, 726
Artificial pacemaker, 271
Ascending colon, 587
Ascending limb of loop of Henle, 493–94
Ascending tracts, 139–40
  spinal cord, 142f, 143f
Ascorbic acid (vitamin C), and iron ab-
    sorption, 583
Asphyxia, 449
Aspirin
  absorption by stomach, 567
  effect on gastric mucosal barrier, 566
  and fever reduction, 609
  interaction with prostaglandins, 711
  for pain relief, 158
  and placental transfer, 732
  and renal failure, 502
Aspirin poisoning, alkalosis associated
    with, 536
Association areas of the cerebral cortex,
    123
Asthma, 403, 426
  treatment of, 377
Astigmatism, 166
Astrocytes, 109, 114f
  functions of, 111, 111f

Asynchronous recruitment of motor units, 227
Atelectasis, 420
Atheromas, 291
Atherosclerosis, 290–93
    and coronary artery disease, 291
    and hypertension, 338
    and kidney failure, 502
    and stress, 667
    and thromboembolism, 292
Atherosclerotic plaque, 292
Athletic menstrual cycle irregularity (AMI), 724
Athletic pseudonephritis, 503
Atmospheric pressure, 417–18, 418f
    and partial pressures, 438
Atomic masses, definition of, A2
Atomic weights, definition of, A2
Atoms, definition of, A1–2
ATP. See adenosine triphosphate
ATPase, 216
ATPase activity
    of different skeletal muscle fiber types, 234
    at myosin cross bridge, 223
    in a Na$^+$-K$^+$ pump, 64–65
ATP synthetase, 31–32, 31f
Atresia, 714
Atria, 261
    See also Atrial contraction
Atrial contraction
    effect of parasympathetic stimulation on, 284
    effect of sympathetic stimulation on, 285
    initiation of, 280
    in relation to ventricular activity, 271, 280
Atrial excitation, 271–72
Atrial fibrillation, detection on an ECG, 277
Atrial flutter, detection on an ECG, 276–77
Atrial natriuretic peptide (ANP), 481–84
Atrioventricular (AV) node, 269
Atrioventricular (AV) valves, 264
Atrioventricular bundle. See Bundle of His
Atrophy, 240
Atropine, 203
Auditory cortex, primary, 119f, 186
Auditory nerve, 184
Auditory tube. See Eustachian tube
Auerbach's plexus, 548–549
Autoantibodies, C1
Autocatalysis
    in formation of pepsin, 562
    in formation of trypsin, 568
Autoimmune diseases, 369, 396
    Addison's disease, 662
    Grave's disease, 654
    multiple sclerosis, 88–89
    myasthenia gravis, 210, 396
Autoimmune response, gastric atrophy, 563
Autonomic nerve pathway
    efferent, 108, 109f

Autonomic nervous system, 107, 198–204, 204f
    control of the ciliary muscle by, 167
    effects on heart, 283–85, 285f
    efferent neurons of, 206
    homeostatic activities of, 210
    innervation of smooth muscle, 248, 251–52
    innervation of the heart, 296
    and insulin secretion, 675
    muscles innervated by, 214
    regulation of airway size, 425
    versus the somatic nervous system, 205
    See also Parasympathetic nervous system; Sympathetic nervous system
Autoregulation of the GFR, 474–76
Autorhythmic cells
    in cardiac muscle, 267, 267f
    depolarization rates, 271f
Autorhythmicity of heart, 268
Autosomal chromosomes, 698–99
Autotransfusion, 342
AV-nodal delay, 272
AV node. See Atrioventricular (AV) node
Avogadro's number, A5
AV valves, 265f
Axon, 83, 84f
Axonal transport, 36–37, 36f
Axon hillock, 83–84, 84f, 102
    initiation of action potential at, 96
Axon terminal, 83, 84f
    of alpha motor neurons, 205–6
    of presynaptic neuron, 91, 92f
    structures terminating on, 91

Backward failure, of the heart, 289
Bacteria
    description, 369
    intestinal, 408, 577, 588, 590
    and neutrophilia, 357
    nonspecific immune responses to, 372f
    specific immune responses to, 383
    summary of defenses against, 391
    See also Intestinal bacteria
Baking soda. See Sodium bicarbonate
Balance
    of calories. See Energy balance
    of fluid. See Fluid balance
    of heat. See Temperature regulation
    of hydrogen ion. See Acid-base balance
    of organic nutrients. See Energy balance
    See also Posture and balance
Balance concept, 511
Balanced chemical equations, A4
Balance, sense of. See Posture and balance
Banding of striated muscle, 216
    changes during contraction, 220f
Barbiturates, effect on smooth ER in liver, 23
Barometric pressure, 417–18, 418f
Baroreceptor reflex, 334–37, 336f, 337f
    and control of GFR, 477, 478f

    and control of salt balance, 516
    and ECF volume, 514
Baroreceptors
    adaptation of in hypertension, 339
    aortic-arch, 335f
    and blood-pressure regulation, 333–34
    carotid-sinus, 335f
    and salt balance, 515
Barrier
    blood-brain, 112, 113–15
    blood-testes, 707
Barrier methods of birth control, 740
Barriers to infection. See External defenses; Nonspecific immune defenses; Specific immune defenses
Basal bodies, 37, 38
Basal ganglia. See Basal nuclei
Basal metabolic rate (BMR), 597–98
    in hyperthyroidism, 655
Basal nuclei (basal ganglia), 117, 125–26, 242
    role of
        in control of muscle tone, 136
Basement membrane, 471
Base pairing
    in DNA, B3f
    See also Complementary base pairing
Bases
    definition of, A7
    in DNA, B1
    in acid-base systems, 524
Basic (alkaline) solutions, 525
Basic digestive processes, 542
Basic electrical rhythm (BER)
    of the colon, 548 588
    definition, 547
    of small intestine, 548, 575–76
    of the stomach, 548, 556
Basic reflex, 142
    withdrawal reflex as, 144
Basic renal processes, 470–71, 470f
Basilar membrane, 179, 182f, 184
    pitch discrimination by, 184–87
Basolateral border, 55, 56f
Basolateral membrane
    ATPase carrier at, 481, 481f
    permeation by bicarbonate ion, 531–32
    sodium reabsorption by in the renal tubules, 481, 481f
Basolateral pump, 486, 486f
    active potassium ion (K$^+$) secretion by, 488f, 489
    active sodium reabsorption by, 481, 481f
Basophils, 357, 370, C1
B-cell growth factor, 391, C1
B cells. See B lymphocytes
Beaumont, William, 561
Bee sting, allergic reaction to, 404
Behavioral patterns
    effect of anabolic steroids on, 239
    limbic system control of, 128
    in paradoxical sleep, 138–39
    in slow-wave sleep, 138

Behavioral reactions in the experience of
    pain, 161
Bends, in deep-sea diving, 451
Benign tumor, 398
BER. *See* Basic electrical rhythm
Beta amyloid in neuritic plaque, 132
Beta (β) cells, 671, 674, 679–80
    excess insulin secretion by, 679
    and oral hypoglycemic drugs, 677
Beta (β)-endorphins, 622, 665
    ACTH precursor, 659
Beta (β) globulin, 349
Beta (β, β_1, β_2) receptors, 203
Beta rhythm, 124
Bicarbonate excretion, 531–33
Bicarbonate ion (HCO_3^-), 350
    formation of, 447–48
    influence on brain ECF H^+, 456
    as part of carbonic acid-bicarbonate
            buffer system, 527–29
    regulation by kidneys, 530
    relation to transport of CO_2 in blood,
            447–48, 448f
    secretion of by pancreas, 558
Bicarbonate reabsorption, and hydrogen
        ion changes, 532f
Biceps muscle, 144–45, 233–34
Bicuspid valve. *See* Left atrioventricular
        (AV) valve
Bicycle ergometer, 278, 459
Bile, 570
    composition of, 572
    control of secretion of, 574
    excretion of cholesterol into, 294
    storage in gallbladder, 571, 574
Bile canaliculus, 571, 571f
Bile pigments, 574
Bile salts, 571f
    as choloretics, 574
    effect of on gastric mucosal barrier, 566
    effect on of saturated fatty acids, 294
    structure and functions of, 573f, 583
    synthesis from cholesterol, 572
Biliary system, 570–75
Bilirubin, 573–74
Binding sites
    of carriers, in facilitated diffusion, 64
    and competition in carrier-mediated
            transport, 63–64
Binocular vision, 176
Biofeedback, 200
Biological clocks, 629
    and body temperature, 603
    and hormone production, 628
Biomolecules, A12
Bipolar neurons in retina, 169, 171
Birth. *See* Parturition
Birth control. *See* Contraception
Birth control pills. *See* Oral contraceptives
Black widow spider venom, 209
Bladder (urinary), 503–6
Blastocoele, 728
Blastocyst
    implantation of, 728–30, 730f, 731f

Blindness
    in diabetes, 676
    in glaucoma, 162
Blind spot, 169, 170f
Blocking antibodies, 399, C1
Block to polyspermy, 728
Blood, 299–343, 347–65
    carbon dioxide transport by, 447
    circulation of oxygenated and deoxy-
            genated, 261f
    as connective tissue, 5
    distribution of. *See* Cardiac output
    functions of, 259
    hormone transport by, 623–24
    oxygen transport by, 442–43
    packed cell volume, 348
    reconditioning of, 321
    summary of constituents and functions,
            348
    types of cells of, 356
    See also Erythrocytes; Leukocytes;
            Plasma; Platelets
Blood-brain barrier, 112, 113–15, 519
    and partial pressure of carbon dioxide,
            455
Blood cells
    origins of, 351, 356, 357f, 358
    summary of functions of, 348
    regulation of production of, 351, 357
    typical blood cell count, 357
See also Erythrocytes; Leukocytes; Platelets
Blood coagulation (clotting), 360–64
Blood-colloid osmotic pressure. *See* Col-
            loid osmotic pressure
Blood doping, 352
Blood flow
    determinants of through a particular tis-
            sue, 309–14
    to reconditioning versus nonrecondi-
            tioning organs, 300–301
    related to pressure and resistance, 302
    to various organs at rest and during exer-
            cise, 310
    velocity of, 319
Blood glucose. *See* Glucose
Blood groups, ABO system, 389
Blood pressure
    and cardiac output, 484
    regulation of, and salt balance, 515
    See also Arterial blood pressure
Blood reservoir, veins as, 328
Blood substitute research, 353
Blood-testes barrier, 707
Blood types, 388–89
Blood urea nitrogen (BUN), 487–88
Blood vessels, 299–343
    effect of hypertension on, 339
    function of, 259
    of skin, role in temperature regulation,
            607
    summary of features of vascular compo-
            nents, 305
Blood-vessel walls, separation of plasma
            and interstitial fluid by, 512–14

Blood viscosity, 304
B lymphocytes (B cells), 358, 370, 381–92,
        C1
    and allergies, 402, 403f
    and B-cell growth factor, 391
    and clonal selection theory, 385, 385f
    and immediate hypersensitivity, 402,
            403f
    antibody production by, 381–82, 385
    interactions of
        with helper T cells and macrophages,
                390f, 391
    maturation of, 380–81, 381f
    and memory cells, 385f, 386
    and plasma cells, 381–82, 382f, 385–86,
            385f
    versus T lymphocytes, 392
BMR. *See* Basal metabolic rate
Body (of stomach), 555, 555f
    oxyntic glandular mucosa of, 560
    as site of gastric storage, 556
Body (as whole)
    composition of, 512
    overview, 10
    systems in, 5
    systems in, role in homeostasis, 6
    temperature of, 601–11
    weight gains during pregnancy, 734
    weight of, 598
    whole, 5
Body fluid compartments, 512
    See also Extracellular fluid; Intracellular
            fluid
Body fluids, classification of, 512
Body temperature
    and fever, 608–609
    and hyperthermia, 609–11
    and hypothermia, 610
    normal range of, 601–603
    regulation of, 605–608
Body water. *See* Water
Bohr effect, 447, 448
Bolus, swallowing of, 553
Bonding
    in cellulose, A8
    in organic chemicals, A6–7
Bone, 5
    abnormalities of, 690
    as calcium store, 683
    deposition of, 684–85
    effects on
        of estrogen, 646
        of growth hormone, 642
        of parathyroid hormone, 683
        of testosterone, 646, 704
    growth of, 640–42
    labile pool of, 685–86
    osteocytic-osteoblastic membrane of,
            685
    stable pool of, 685
    structure of, 640–41, 641f, 684f, 685, 685f
Bone fluid, 685, 685f
Bone marrow, 351, 640
    and B-cell maturation, 380, 381f

blood cell production in, 351, 357f
destruction of, and anemia, 354–55
erythrocyte production by, 351
and leukemia, 358
leukocyte production by, 356–357
as lymphoid tissue, 370
platelet production by, 358
summary of functions of, 371
Bone resorption, 684–85
suppression of, 687
Borborygmi, 590
Botulism, 209–10
Bowman's capsule, 469
hydrostatic pressure of, 473
podocytes of, 471
podocytes with foot processes, 472f
See Glomerular filtration
Boyle's law, 420, 420f
Bradycardia, detection on an ECG, 276
Bradykinin, 158
Brain, 107
components of, 117
need of
for glucose, 658, 668–70
for oxygen, 301
atherosclerosis leading to damage to, 291
and blood-brain barrier, 112, 113–15
blood flow to, 315
chemoreceptor trigger zone, 560
control of hypophysiotropic neurons, 637–38
damage, causes and results of, 115–16
effect on
of dehydration, 518
of diabetes mellitus, 675, 677
of extracellular fluid hypertonicity, 518
of extracellular fluid hypotonicity, 520
of insulin excess, 678–79
of overhydration, 520
exchanges with cerebrospinal fluid, 114f
frontal section of, 125
hydrogen-ion concentration in the extracellular fluid of, 453
ketone utilization by during starvation, 671
mapping of. See Maps, somatotopic
observations of human, 110
plasticity of, 116, 122
protection of, 109
secretion of hormones by, 619, 632–33, 632f, 635–36, 635f, 637f
sensitivity to water balance, 537
tumors, 112f
use-dependent competition for cortical space, 122
water imbalance and functioning of, 518, 520
See also Specific brain components
Brain cells, 3
Brain death, legal determination of, 124–25
Brain stem, 117, 136–39

cardiovascular control center of, 285, 315, 335
respiratory centers in, 450–53
responses to arterial hydrogen ion changes, 530
role of
in hearing, 186
in the multineuronal system of motor control, 242
Braxton-Hicks contractions, 734
Breasts, 737
as accessory reproductive organs, 696
Breathing. See Respiratory mechanics; Ventilation
Breech birth, 734
Broca's area, 122–23
Bronchi, 415, 415f
Bronchioles, 415, 415f
constriction of in asthma, 403
Bronchoconstriction, 427
Bronchodilation, 427
Brown fat, 607
Brush border, 40
in gluten enteropathy, 579f
as site of small intestine enzyme activity, 577
See also Microvilli
Buccal glands, 551
"Buffalo hump," 660
Buffering, 526–29, 526f
of blood by chemical buffer systems, 527–29
and homeostasis, 538
in saliva, 552
Buffy coat, 348
Bulbourethral glands, 697, 710
Bulk flow across capillaries, 321–25, 323f
Bulk flow of air in and out of lungs, 420–25
BUN, 487–88
Bundle of His, 269
Burns
consequences of skin loss in, 406
and edema, 327
and the GFR, 474

C5–C9. See Membrane-attack complex
Cachectin, 601
Caffeine
effect on gastric secretions, 564
and exercise, 559
Calcitonin, 650, 686–87
Calcium
absorption by the digestive tract, 583, 585
balance, 683
and cyclic AMP, 53
cytosolic, 50
disorders of, 688–89
distribution of in body, 681–82
endocrine control of, 681–91
homeostasis, 683
movement in cells, 50–51
in neurotransmitter release, 206

permeability to, heart contractile cells, 285
in plasma, 682
regulation of, 681–87
renal handling of, 486
response to beta amyloid in cells, 133
role of
in blood coagulation, 361, 683
in cardiac autorhythmic cells, 269
in cardiac muscle action potential changes, 273
in chemotaxis, 373
in contraction of smooth muscle and skeletal muscle, 252
in declarative memory formation, 133–34
in digestive tract motility, 548
in excitation-contraction coupling, 223f
in exocytosis, 68
in habituation, 132
in muscular dystrophy, 240–41
in neuromuscular excitability, 682
in sensitization, 132
in skeletal muscle contraction, 218–25, 251
in smooth muscle activity, 247, 682
in smooth muscle contraction, 251, 682
in stimulation of atrial contractile cells, 283
in stimulus-secretion coupling, 683
in turning cross bridges on, 219f
salts of, and bone, 640, 681, 684
as a second messenger, 52–53, 53f, 75
summary of functions of, 682
Calcium balance, 683
and renal failure, 502
Calcium carbonate crystals, of otoliths, 188
Calcium homeostasis, 683
Calcium ion channels, foot proteins as, 222–23
Calmodulin, 52–53
role in smooth muscle activity, 247
Calorie, 597
Calorigenic effect, 652
Calorimetry, direct and indirect, 597
cAMP. See Cyclic adenosine monophosphate
cAMP-dependent protein kinases, 52
CAMs. See cell adhesion molecules
Canaliculi
bile, 571, 571f
in bone, 685
for hydrochloric acid secretion, 560
Cancer, 398–99
cells of, compared with normal cells, 399f
defenses against, 379, 398–99
destruction of endocrine glands in, 630
genetic mutations and, B14
lung, 441
Capacitance vessels, veins as, 328
Capacitation, 709

Capillaries, 299, 302, 305, 317–28, 343
  and the blood-brain barrier, 113
  brain, 114f
  bulk flow across, 321–25
  control of blood flow through, 319–321, 322f
  diffusion across, 320
  effect of histamine on, 371–72
  functions of, 317
  of the glomerulus, 473
  lymphatic. *See* Lymphatic capillaries
  permeability of, 319
  pulmonary. *See* Pulmonary capillaries
  systemic. *See* Systemic capillaries
Capillary bed, 321f
Capillary blood pressure, 323
  and colloid osmotic pressure, 514
  effect on of gravity during standing, 330, 331f
Capillary wall
  bulk flow across, 323f
  electron micrograph, 317f
Carbamino hemoglobin, 447
Carbohydrates, A8
  dietary, 543
  digestion and absorption by digestive system, 567, 577, 578, 580, 581f
  effect on
    of cortisol, 658
    of diabetes mellitus, 675, 678f
    of epinephrine, 664
    of glucagon, 679, 681
    of growth hormone, 640
    of insulin, 673
    of thyroid hormone, 652
  fuel metabolism, 668–71, 669f
  membrane, 47–48, 50
  respiratory quotient of, 413
  storage of, 670
  structure of, A8
Carbon dioxide (CO_2)
  and acid-base balance, 450, 526, 528, 529–30
  arterial partial pressure of, 454–55
  and carbamino hemoglobin, 447–48
  control of ventilation by, 453
  diffusion coefficient of, 441
  effect of
    on airway resistance, 426, 427–28
    on systemic arterioles, 312
  exchange of
    across placenta, 732
    between alveoli and blood, 413, 438–39
    between blood and tissues, 413, 441–42
  formation of carbonic acid by, 450, 526, 528, 529–30
  and homeostasis in the respiratory system, 460
  inhibition of ventilation by, 453, 455f
  net diffusion gradients for between tissues and lungs, 442, 443f

partial pressure of, and exercise, 457
production of
  in the citric-acid cycle, 28, 33
  regulation of by respiratory system, 453, 454–55, 455f
  relation to hydrogen ion concentration in the blood, 526
  and respiratory quotient, 413
  solubility of, 441
  summary of partial pressures and exchanges of, 440f
  transport in blood, 448f
Carbonic acid
  and carbonic acid-bicarbonate buffer system, 528–29
  dissociation constant of, 524
  relation to CO_2 production, 450, 526, 528
Carbonic anhydrase, A4–5
  in erythrocytes, 350
  role in HCl secretion by stomach, 562
Carbon monoxide (CO)
  binding to hemoglobin, 350
  effect on the O_2Hb curve, 446f
Carbon tetrachloride, and liver disease, 575
Carboxyhemoglobin ($HbCO_2$), 447
Carboxypeptidase, 544f, 569
Carcinogens, 398
Cardiac arrhythmias
  and potassium ion (K^+) concentration, 489
  in response to abnormal ion concentrations, heart, 274
Cardiac atria, atrial natriuretic peptide release from, 484
Cardiac contractility, sympathetic nervous system and, 201
Cardiac cycle, 278–82, 282f
Cardiac excitation, 272, 272f
Cardiac muscle, 4, 213, 214, 254
  action potential
    in autorhythmic (pacemaker) cells, 268–69
    in contractile cells, 273
  anatomy and functional distinctions of, 267–68, 267f
  characteristics of, 248–49
  comparison with skeletal and smooth muscle, 246
  excitation-contraction coupling compared with smooth muscle, 682–83
  gradation of, 287
  myopathies, 278
  summary of characteristics, 246–54
Cardiac muscle fibers, 267
Cardiac output, 283–89
  and arterial blood pressure, 332–33, 334f, 335f
  control of, 288f
  determinants of, 283, 288f
  distribution of, 300–301, 300f
  effect of ANP on, 484

effects on
  of exercise, 311f
  of gravity, 330
  percentage distributed to kidneys, 479, 479f
Cardiac physiology, 258–96
Cardiac reserve, 283
Cardiac suction, effect on venous return, 332
Cardiac tamponade, 268
Cardiogenic shock, 340
Cardiovascular control center, 285, 315, 335
Cardiovascular hypertension, 338
Cardiovascular system
  anabolic steroid damage to, 239
  basic organization of, 301f
  See Circulatory system
Caries, 551
Carotid bodies, 453–54
  effect on
    of arterial H^+, 456, 457f
    of P_{O_2}, 453, 454–55, 455f
    of P_{CO_2}, 453–54, 454f
  in sudden infant death syndrome, 459
  See also Peripheral chemoreceptors
Carotid-sinus baroreceptor, 334
  See also Baroreceptors
Carrier-mediated transport, 62–66, 62f, 582
Carrier molecules/carriers, 48
Cartilage, role in bone growth, 640
Cascade
  cAMP and gAMP, 53–54
  clotting, 361
Castration, 703
Catabolism, 668
Catalase, 26
Catalysts, A4
Cataract, 167
Catecholamines, 128–29, 618, 623, 624, 663–64
  effect of thyroid hormone on, 653
  as hormones, 618, 663
  receptors for, 624
  as sympathetic postganglionic neurotransmitter, 200
Cations
  definition of, A3
  movement along an electrical gradient, 59, 59f
Cauda equina, 139
C cells, 650
  thyroid gland, 686–87
CCK. *See* Cholecystokinin
Cecum, 587
Cell adhesion molecules (CAMs), 49–50
Cell body, 83, 84f
Cell culture, brain cells, 3
Cell division, B10–13
  effect on of growth hormone, 640
  of olfactory neurons, 191
  role of microtubules in, 38, 38f
  See also Meiosis; mitosis

Cell-mediated immune response, 358, 380, 392–401
  See also T lymphocytes
Cell membranes. *See* Plasma membrane
Cell organelles. *See* Organelles
Cell(s), 2–3
  basic cell functions, 2
  composition of, 16
  contributions to homeostasis, 6
  definition of, 2
  electrical properties of. *See* Membrane potential
  exchanges with surrounding environment, 5–6
  interactive groups of, 12
  in multicellular organisms, 2
  normal and cancerous, 399f
  organization within body, 4
  specialized functions of, 3–4
  structure of, 16–17, 20f
  summary of cell structures and functions, 18
  total number in body, 2
  unicellular organisms as, 2
  See also Cytosol; Organelles; Plasma membrane
Cells of Leydig, regulation of secretions by LH, 635
Cell-to-cell adhesions, 54–56
Cell-to-cell communication. *See* Communication
Cellular plasma membranes
  osmotic activity across, 218
  separation of plasma and interstitial fluid by, 512–14
Cellular respiration, 412
Cellulose, 543, A8
Center in brain, definition of, 141
Central axon, 108
Central chemoreceptors, 453–56
Central fatigue, 236
Central lacteal, 579, 585–86
Central nervous system (CNS), 107, 146
  and development of sports skills, 144
  effect on
    of changes in hydrogen ion concentration, 526
    of epinephrine, 664
    of thyroid hormone, 653
    uses of afferent information by, 152
  See also Brain; spinal cord
Central sulcus, 119
Central thermoreceptors, 606
Centrioles, 18, 20f, 38
  identity with basal bodies, 38
  mitotic spindle, 38
Cephalic phase
  of bile secretion, 574
  of gastric secretion, 563–64
  of pancreatic secretion, 569
Cerebellum, 117, 135–36, 135f
  role in memory, 131, 136

role in motor activity, 121, 242
  symptoms of cerebellar disease, 136
Cerebral cortex, 115–25
  functional areas of, 119f
  layers of, 118
  linkages among regions of, 124f
  lobes of, 118, 118f
  plasticity of, 116, 122
  role in
    control of ventilation, 458
    emotions and behavior, 128
    language ability, 122
    learning and memory, 131
    motor activity, 120
    perception of sound and vision, 119
    somatosensory perception, 119
Cerebral cortex-hypothalamic pathway, 337
Cerebral dominance, 124
Cerebral hemispheres, 118
  specialization of, 123–24
Cerebrocerebellum, 136
Cerebrospinal fluid (CSF), 112, 113
  extraction by means of spinal tap, 112f
  as transcellular fluid, 513
Cerebrovascular accident (stroke), 115
Cerebrum, 117, 118
Cerumen, 181
Cervical canal, 697
  sperm transport through, 725–26
Cervical mucus
  cyclical changes in, 723, 725
  and oral contraceptives, 741
Cervix, 697
  during parturition, 734
C fibers, 158
Channel regulation, and chemical messengers, 51f
Channels, 48, 48f
  calcium ion (Ca^{++})
    changes during simple learning, 132–33
    role in hormone secretion, 68
    gating mechanism of, 81–82
  in plasma membrane, 48, 48f
  regulation as a postreceptor event, 50–51, 51f
  responses to chemical messengers, 75
  sodium, voltage-gated, 81
Charge, unequal distribution across a membrane, 70
Chemical bonds
  definition of, A2–4
  double, in organic compounds, A7–8
  peptide bonds, A10
Chemical buffer systems, 527–29, 527f
Chemical classification of hormones, 621
Chemical contraceptives, 740
Chemical elements. *See* Elements
Chemical energy from food, 27–28
Chemical gradient, 57, 57f
Chemical mechanism for stimulus of bile secretion, 574

Chemical messenger-gated channels, 81
  of photoreceptors, 170
  at subsynaptic membrane, 91
Chemical messengers, 50
  comparison of nervous and endocrine system messengers, 105
  in the immune system, 393
  prostaglandins as, 711
  types of, 615–18
Chemical principles, A1–13
Chemical reactions, A4–5
Chemical senses (taste and smell), 190–93
Chemical thermogenesis. *See* Nonshivering (chemical) thermogenesis
Chemiosmotic mechanism, and adenosine triphosphate (ATP), 31–32
Chemoreceptors, 152
  central, 453–56
  peripheral
    effect on arterial blood pressure, 337
    and respiration, 453
  for taste and smell, 190–93
Chemoreceptor trigger zone, 560
Chemotaxins, 373
  complement as, 380
  induction of phagocyte migration by, 374
  produced by T cells, 395
Chernobyl disaster, 629
Chewing, 549, 550
Chief cells, 560, 561f
Chloride channels in cystic fibrosis, 49
Chloride ion (Cl^-)
  distribution of in body, 514
  equilibrium potential of, 74
  permeability of
    at inhibitory synapse, 92–93
    in resting membrane, 74
  related to sodium load and ECF volume, 481
  renal handling of, 486
  role of in generation of IPSP, 92
Chloride ion shift, 448
  in the gastric lumen, 562
Cholecalciferol (vitamin D), 687–88, 688f
Cholecystokinin (CCK)
  characteristics of, 558
  effects of
    on gallbladder contraction, 574
    on gastric motility, 558
    on gastric secretions, 565
    on pancreatic digestive enzyme secretion, 569
    on sphincter of Oddi, 574
  functions of, 98
  overview of actions, 590–91
  and satiety, 600
Cholera, 54, 77, 587
Choleretic substances, 574
Cholesterol, A10
  and atherosclerosis, 293–95
  bile salt derivatives of, 572
  dietary manipulation of, 293

Cholesterol, (continued)
  effect on of saturated and polyunsaturated fatty acits, 294f, 295
  in gallstones, 572
  and high-density lipoprotein (HDL), 293–95
  hormone synthesis from, 657
  and the liver, 293–95, 294f
  and low-density lipoprotein (LDL), 293–95
  metabolism of, 293–95
  in micelles, 572–73
  in plasma membranes, 46
  as steroid hormone precursor, 622
  structure of, A10
  uptake by cells, 294f
Cholinergic fibers, 199
Cholinergic receptors, 203
Chondrocytes, 641
Chordae tendineae, 264
Chorion, 730
Chorionic gonadotropin, 732
Choroid, 162
Choroid plexus, 113
Chromaffin granules, 663
Chromatin, B4
  in prophase, B11
Chromosomes, 696, B3–4, B4f
  distribution during sexual reproduction, 699f
  division during oogenesis, 714–16
  number of, 698, B3
  scanning electron micrograph of, B14f
  sex, 698–701
  and spermatogenesis, 704
  See also Autosomes; Genes; X-chromosomes; Y-chromosomes
Chronic bronchitis, 426
Chronic disease, and growth, 639
Chronic lung disease, hypoventilation in, 456
Chronic obstructive pulmonary disease (COPD), 426
Chronic renal failure, 502
  See Renal failure
Chylomicrons, 583
Chyme, 556, 556f
  gastrin release stimulated by, 590
  production of, 557
Chymotrypsin, 544f, 569
Chymotrypsinogen, 569
Cigarette smoking
  and atherosclerosis, 295
  effect of, on respiratory system, 408
  and high-density lipoproteins, 295
  and placental transfer, 732
Cilia, 19, 37, f
  in airways, 37–38, 37f, 408
  effect of cigarette smoking on, 408
  in oviducts, 37, 725, 726
Ciliary body, 162
Ciliary muscle, 167
Cimetidine, 567
Circadian rhythm, 628

Circular (constrictor) muscle of iris, 162–63
Circular smooth muscle in digestive tract. See Muscularis externa
Circulatory hypoxia, 449
Circulatory shock, 340–41, 341f, 343
  as result of aldosterone deficiency, 657
Circulatory system, 258, 259, 299
  blood volume, distribution of, 329f
  components of, 259
  function of, 5, 7, 259
  history of knowledge of, 302–3
  homeostasis interaction with the endocrine system, 646
  homeostasis maintenance by, 295
  See also Blood; Blood vessels; Heart
Cirrhosis, 575
Cisternae, 23
Citric-acid cycle (Krebs cycle), 27, 33
Cl⁻. See Chloride
Class I and II MHC glycoproteins, 397–98, 398f
Classical complement pathway, 379–80
Classification of body fluids, 512
Clathrin, 24
Clearance, plasma, 490–506
Climacteric, 723
Clitoris, 697, 701
  erection of, 713
Clonal anergy, and tolerance, 397
Clonal deletion as a tolerance mechanism, 397
Clonal selection theory, 385, 385f
Clone, 385, 385f
Clostridium botulinum toxin, 209–10
Clot dissolution, 363
Clot retraction, 363
Clotting, 363f, 683
  of blood, 360–64, 364f
  of blood in varicose veins, 332
  final step in hemostasis, 359
  globulin participation in, 349
  of interstitial fluid in inflammation, 373
  risk of, and birth-control pills, 741
  role of heparin in, 357
  of semen, 710
  See also Blood coagulation
Clotting cascade, 361
Clotting factors, 365
CNS. See Central nervous system
Coactivation, 244–45
Coagulation (clotting) of blood, 360–64
Coated vesicles, 23–24, 24f
Cocaine, 99
  in brain, 519
  effect on the parasympathetic nervous system, 202
Cochlea, 178f, 182, 182f, 184
Cochlear duct, 179, 182–83
Cochlear nerve. See Auditory nerve
Coding of sensory information, 155–56
Codons in mRNA, B7
Coiling in DNA, B4f
Coitus, 711–14

Cold
  adverse effects of, 610
  effect of
    on arteriolar radius and blood flow, 313
    on thyroid hormone in infants, 654
  response of body to, 607–608
  See also Cooling
Collagen, 55, 406, 660–61
  in basement membrane, 471
  response of factor XII to, 361
Collateral circulation in the heart, 292
Collateral ganglia, 199, 199f
Collaterals of axons, 83, 84f
Collateral ventilation, 416
Collecting duct or tubule, 470
  control of
    by aldosterone, 482
    by vasopressin, 497–98
  and reabsorption
    of sodium, 481
    of water, 486–87
  and secretion of
    hydrogen ion, 488, 531, 532f
    potassium ion, 488–89, 488f
  summary of transport across, 490
Colloid of thyroid gland, 650
Colloid osmotic pressure, 323
  and bulk flow across capillaries, 323–24, 323f
  and capillary blood pressure, 514
  effect on glomerular filtration, 473
  of interstitial fluid, 324
  of plasma, 323, 473
Colloids
  definition of, A6, A6
Colon, 587
Color blindness/defectiveness, 174
Color vision, 174
Colostrum, 739
  IgA in, 388
Coma
  in acidosis, 526
  in diabetes mellitus, 675, 676, 679
  state of consciousness in, 137
Communication
  between cells, 615–16, 615f
  in the endocrine and nervous systems, 106
  role of the afferent division of the peripheral nervous system in, 194
Competition for carrier binding sites, 63–64
Complementary base pairing
  in DNA, B3f
  in DNA transcription, B6f
  in protein synthesis, B4–7
Complement C5–C9. See Membrane attack complex (MAC)
Complement system, 371, 379–80, C1
  activation of, 383, 384f
Complete heart block, 271
  detection on an ECG, 277, 277f
  See Heart block

Complex cells in the visual system, 177
Complex lipids, A9, 9f
Compliance of lungs, 429
Compliance of vessels, 304–5
Computerized axial tomography (CAT)
    scans, 110
Concave lens, 164, 165f
Concentration
    definition of, 59, 59f
    gradients in cells, 73
    of ions in resting nerve cells, 71
Concentration gradients, 57–58, 57f
    and exchange across capillary walls,
        322f
Concentration of a solution, definition of,
    A5
Concentric contraction, 231
Conception. See Fertilization
Conditioned reflexes, 142
Conducting system, heart, 269, 269f
Conduction, as method of heat transfer,
    604–5
Conduction by local current flow
    in muscle, 207
    in nerve, 84–86, 84f
Conduction, of action potentials. See
    Action potentials
Conductive deafness, 186–87
Cones, 162, 169
    color-selective, 170
    comparison with rods, 173
    sensitivity to different wavelengths
        (colors), 171f
Congestive heart failure, 289, 329
    as a complication of hypertension,
        338
    edema with, 328
    pulmonary edema associated with,
        459–60
    role of salt-retaining, fluid-retaining
        reflexes in, 483
Connective tissue, 5, 55
Connexons, 56
Conn's syndrome, 660
    and hypertension, 338
Conscious awareness, 128
Consciousness, 137–39
Consolidation, 129
Constant (Fc) region of an antibody, 383
Constipation, 589
Continuous ambulatory peritoneal dialysis
    (CAPD), 504
Contraception, 726–27, 740–42
Contraceptive sponge, 740
Contractile cells in cardiac muscle, 267
Contractile proteins, 217
Contractile systems in nonmuscle cells, 39
    in capillaries, 319–20
    and the GFR, 478–79
    podocytes, 479f
Contractility of the heart, 267, 286–88, 288f,
    295–96
    extrinsic control of, 286–88
    in heart failure, 289

intrinsic control of, 285–86
    in response to drugs, 274
Contraction time
    in skeletal muscle, 225
    in smooth muscle, 253
Control of adrenal androgens, 659
Controlled variable, 11
Control system, 11
Convection, 605
Convergence of light rays, 164
Convergence of neurons, 99–101, 101f
Convex lens, 164, 165f
Cooling, effect of on body, 601, 610
COPD. See Chronic obstructive pulmonary
    disease
Copulation, 711–14
Core temperature, 601, 603
Cornea, 162, 396
Cornified layer of skin. See Keratinized
    layer of skin
Corona radiata, 717, 727
Coronary artery disease, 291–93
Coronary blood flow, 290–95, 295f
    in coronary artery disease, 291
    effect of epinephrine on, 664
Coronary circulation. See Coronary blood
    flow
Coronary heart disease, and android obe-
    sity, 602
Corpus albicans, 719
Corpus callosum, 118
Corpus luteum, 717, 717f
    control of, 721–22
    formation of, 718–19
    of pregnancy, 719, 732
Cortex. See Adrenal cortex; Cerebral
    cortex; Motor cortex; Olfac-
    tory cortex; Premotor cortex; Re-
    nal cortex; Visual cortex
Cortical gustatory area, 191
Cortical lobes, 118–20, 118f
Cortical nephron, 470
Corticospinal (pyramidal) system, 242
Corticospinal tract, 140
Corticosteroid-binding globulin, 657
Corticotropin. See Adrenocorticotropic
    hormone
Corticotropin-releasing hormone, 659
    in hypothalamic-anterior pituitary con-
        trol systems, 638
    and stress, 665
Corticotropin-releasing hormone (CRH),
    release in response to stress, 637
Cortisol, C1
    anti-inflammatory action of, 378, 658–
        359
    control in hypothalamic-anterior pitu-
        itary control systems, 638
    diurnal control of secretion of, 628, 628f
    effect of
        on carbohydrate metabolism, 658
        on fuel metabolism, 658
        on the immune system, 400
        on growth, 639

on protein metabolism, 658
        pharmacological, 658
    and homeostasis in the endocrine sys-
        tem, 692
    link with immune system, 400
    metabolic actions, summary of, 682
    and pain transmission, 160
    and permissiveness, 657–58
    and regulation of fuel metabolism, 681
    stimulation of production of, by ACTH,
        635
    and stress, 658, 665
    synthesis of, 622
Cortisol hypersecretion, 660–61
Cortisone and other cortisol-like drugs. See
    Glucocorticoids
Cotransport. See Secondary active
    transport
Cotransport carriers in kidneys and intes-
    tines, 66
Cough, 408, 458
Countercurrent exchange, 496, 496f
Countercurrent multiplication, 493–96,
    496f
Covalent bonds, definition of, A3–4
Cranial nerves, 136–37
Craniosacral parasympathetic division of
    autonomic nervous system, 199f
Cranium, 112
Creatine kinase, 234
Creatine phosphate, 234
Creatinine
    filtering as waste, 488
    for plasma clearance measurements,
        491
Cretinism, 654
CRH. See Corticotropin-releasing hormone
Cristae, mitochondrial, 26, 26f
Critical developmental period of brain,
    105
Cross-bridge cycle, 221f, 224f, 254
Cross bridges, 216, 273
    changes during muscle contraction,
        218–20, 220f
    of smooth muscle, 253, 548
    special sites of, 223–24
Crossed extensor reflex, 145–46, 146f
Crossing over, in meiosis, B11, B14f
Cryptorchidism, 702–3
Crypts of Lieberkühn, 579–80
Crystal lattice for sodium chloride (table
    salt), A3
CSF. See Cerebrospinal fluid
Cupula, 188
Curare, 209–10
Current
    decremental spread in graded poten-
        tials, 80
    flow in propagation of action potentials,
        79–80
Cushing's syndrome (adrenal diabetes),
    660
Cyanide poisoning, 449
Cyanosis, 449

Cyclic adenosine monophosphate (cyclic AMP), A12
and hydrophilic hormones, 625
as a second messenger, 75
triggering by serotonin, 132–33
in short-term memory, 133
Cyclic adenosine monophosphate (cyclic AMP; cAMP)
in a second messenger pathway, 50–51, 53–54
and vasopressin, 497, 497f
Cyclical activity in female reproductive physiology, 714
Cyclic AMP. See Cyclic adenosine monophosphate (Cyclic AMP)
Cyclic GMP. See Cyclic guanosine monophosphate
Cyclic guanosine monophosphate (cyclic GMP)
and postreceptor events, 53
role in vision, 170
Cyclosporin, 397
Cysteinuria, 63
Cystic fibrosis, 49
detection in the embryo, 727
Cystic fibrosis transmembrane conductance regulator (CFTR), 49
Cytokinesis, 38, B11
actin in, 39f
in meiosis II, B12
Cytoplasm, 17, 18
Cytoplasmic bridges, among germ cells, 706
Cytosine (C), B1
Cytoskeleton, 17, 18–19, 35–42, 43
elements of, 35
specialized elements of in skeletal muscle, 216
Cytosol, 17, 18, 20f, 34–35, 43
glycolysis in, 27–28
Cytosolic calcium, 683
Cytosolic enzymes in breakdown of APP, 133
Cytosolic polyribosomes, B10
Cytotoxic T cells, 392, C1
and cancer defense, 399, 400f
and defense against virus-invaded host cells, 393–394, 393f, 394f
enhancement by interferon, 379
function of, 393–94, 393f
and HLA self-antigens, 397
interactions with
other T cells, 392
interferon, 379

Dark adaptation, 173
D (δ) cells, 671
Dead space, 435–37
alveolar, 437
Deafness, 186–87
Decibels (dB), 178–79
Decidua, 729

Declarative memory, 131, 133
Decompression sickness in deep-sea diving, 451
Decremental current flow, 80
Deep-sea diving, effects on the body, 451
Defecation, 589
Defense mechanisms of body, 368–408
Deglutition. See Swallowing
Dehydration, 518
Dehydroepiandrosterone (DHEA), 657, 702, 733
Delayed hypersensitivity, 402, 404
Delivery of a baby. See Parturition
Denaturation of proteins, A12
Dendrites, 83, 84f
in synapses, 91
Denervation atrophy of skeletal muscle, 240
Dense bodies in smooth muscle, 247
Dental caries, 551
Deoxyribonucleic acid (DNA), 18, A7, A12, B1–4
in chromosomes, 698
interaction with lipophilic hormones, 625
nuclear, and B-cell development, 386
organization of, B2f
replication of, B4–B6, B4f
role in protein synthesis, B4–B7
transcription of, B7
Deoxyribose in DNA, B1
Dephosphorylation in active transport, 64
Depolarization, 80
and action potentials, 80, 101
of heart muscle, 271–72
and positive feedback to sodium ion channels, 82
rate change, in response to sympathetic system stimulation, 285
response of ion channels to, 82
in the SA node, 283
Depression, and catecholamine deficiency in, 129
Depth perception, 176
Dermatome, 141
Dermis, 406
Descending colon, 587
Descending tracts, 140
spinal cord, 142f, 143f
Desensitization injections (allergy shots), 404, 404f
Desmosome, 55, 55f, 405–6
of intercalated discs, 267
Detergent action of bile salts, 572
Development, from fertilization to implantation, 730f
DHEA. See Dehydroepiandrosterone
DHT. See Dihydrotestosterone
Diabetes insipidus, 518, 675
Diabetes mellitus, 63, 499, 675–79, 675f
and acid production, 527
acute effects of, 678f
and android obesity, 602

as a cause of metabolic acidosis, 536
and exercise, 64
and glucagon excess, 681
glucose levels in, 486
side-effect of growth hormone therapy, 644
Dialysis, kidney, 504
Diapedesis, 373
Diaphragm, contraceptive, 740
Diaphragm, respiratory, 417, 421
major role in inspiration, 425
role in vomiting, 559–60
Diaphysis, 640
Diarrhea
causes and consequences of, 586–87
and fluid imbalance, 513, 515
and the GFR, 474
and metabolic acidosis, 536
Diastole
and cardiac filling, 278
and coronary circulation, 290
Diastolic murmur, 282
Diastolic pressure, 306
measurement of, 308
Dicrotic notch, 280
Diencephalon, 117
Dietary fiber. See Cellulose
Diet-induced thermogenesis, 597, 602
Diffusion, 57, 57f
across capillary walls, 320, 321
coefficient of, 441–42
Fick's law of, 415
of water (osmosis), 59–62
Digestion, 543, 545f
in the mouth, 553
by pancreatic enzymes, 568–69
in small intestine, 577
in the stomach, 567
summary of, 578
Digestive enzymes
in digestive tract, summary, 544f
as endogenous proteins, 580
in lysosomes, 24–25
Digestive juices, 513
magnitude of, 586
secretion of, 5
Digestive secretion, 543
Digestive system, 5, 541–92
biochemical balance among stomach, pancreas, and small intestine, 586, 587f
components of, 545–47, 547f
and control of food intake, 599
defense mechanisms of, 407
digestion and absorption of nutrients by, summary, 544, 546–47, 578
effect of prostaglandins on, 711
functions of, 5, 7
hormones secreted by. See Gastrointestinal hormones
overview of activities, 546–47
regulation of, general concepts, 547–50, 550f

role in energy conversion from food to ATP, 27
sympathetic nervous system and, 201
Digestive tract, 5
  components of, 545
  endocrine function of, 549, 590–91
  layers of, 545–47, 548f
  wall of (cross section), 548f
Digitalis, 289
  effect on heart contractility, 274
Dihydrotestosterone (DHT), 702
Diiodotyrosine (DIT), 651
2,3-Diphosphoglycerate (DPG), 447
Diploid number, chromosomes, 698, B3
Diplopia, 176
Direct calorimetry, 597
Disaccharidases, 577–78
Disaccharides, 543, A8
Discrimination test for receptive fields, 156
Discriminative ability, 155–56, 156f
  of ears, 184–87
  of eyes, 173
Digestive system, homeostasis controlled by the endocrine system, 646
Dispersed phase, definition of, A6
Dispersing medium of a colloid, definition of, A6
Dissociation constant (K), 524
  related to Henderson-Hasselbalch equation, 528
Dissociation of acids and bases, A7
Distal tubule, 470
  control of
    by aldosterone, 482
    by vasopressin, 497–98
  and reabsorption
    of sodium, 481
    of water, 486
  and secretion
    of hydrogen ion, 488, 531, 532f
    of potassium, 488–89, 488f
  summary of transport across, 490
Distensibility. See Compliance
Distention
  and gastric emptying, 558
  gastrointestinal, and satiation, 599
Disuse atrophy of skeletal muscle, 240
DIT. See Diiodotyrosine
Diuresis, 499
Diuretics, 483
Diurnal rhythm, 628, 628f
  and cortisol control, 659
  in secretion of growth hormone, 642–43
Divergence of light rays, 163, 164f
Divergence of neurons, 99–101, 101f
DNA. See Deoxyribonucleic acid
Dopamine
  deficiency in Parkinson's disease, 99, 126
  excess in schizophrenia, 128–29
  levels of in MPP+ poisoning, 100
  as PIH, 635–36

Dorsal (posterior) horn, spinal cord, 141
Dorsal respiratory group (DRG), 452, 452f
Dorsal root, spinal cord, 141
Dorsal root ganglion, 141
Double bonds, A7–8
  and saturation of fatty acids, A9
Double helix in DNA, B4f
Down regulation, 631
  of insulin receptors
    in Type II diabetes mellitus, 677
Down's syndrome
  and maternal age, 715, 727
DPG, production at high altitudes, 451
Dreaming, 139
Driving force for water reabsorption, 486
Drugs
  and dry mouth, 521
  for immunosuppression, 397
Dry mouth, 521
Dual innervation by the autonomic nervous system, 200–202
Dual pump function of the heart, and homeostasis, 295
Duchenne muscular dystrophy, 241
Duct of an accessory digestive gland, 548f
Ductless glands (endocrine glands), 5
  See also Endocrine glands
Ductus arteriosus, 262
Ductus (vas) deferens, 697, 709
Duodenal mucosa, CCK release from, 569
Duodenum, 557, 575
  emptying of the stomach into, 590
  hormones, major functions, and target cells of hormones, 620
  See also Small intestine
Dural sinuses, 113
Dura mater, 112–13, 112f
Dwarfism, 645
Dynein, 37
Dynorphin, role in natural analgesia, 158
Dyslexia, 123
Dysmenorrhea, 722
Dyspnea, 449, 459–60
Dystrophin, deficiency in muscular dystrophy, 241

Ear, 177–90
  functions of major components, 179
  role in
    equilibrium and coordination, 187–90
    hearing 178–86
  structure of, 178f, 182f
  summary of hearing function, 186
  See also External ear; Inner ear; Middle ear
Ear bones, 178f, 182f
Ear canal, 178f, 181
Ear drum. See Tympanic membrane
Eating, control of, 598
Eccentric contraction, 231
ECF. See Extracellular fluid

ECG. See Electrocardiogram
Ectopic (abdominal) pregnancy, 725
Ectopic foci, 271
  in response to abnormal ion concentrations, heart muscle, 274
Ectoplasm (plasma gel), 39
Edema, 327–28
  in congestive heart failure, 483
  and gravitational effects, 330
  in inflammatory response, 372
  pulmonary
    effects on gas exchange, 441
EDV. See End-diastolic volume
EEG. See Electroencephalogram
Effector, 11
Effector organs, 107
  as part of reflex arc, 144
Efferent arterioles, 467
  effect on glomerular-capillary blood pressure, 473
Efferent autonomic nerve pathways, 109f
Efferent axons, 108
Efferent division of peripheral nervous system, 107, 198–210
Efferent motor neurons, reactions in habituation and sensitization, 131
Efferent nerves, and alkalosis, 526
Efferent neurons, 107–8
  characteristics and functions of, 206
Efferent pathway, 144–45
Egg freezing, and fertility, 727
Einthoven, William, 274
Ejaculation, 710, 711
Ejaculatory duct, 697, 710
E K+. See Equilibrium potential for K+
Elastic recoil, 429–30
Elastin, 5, 55
  in arteries, 305, 306f
  in lungs, 429–30
  in veins, 328
Electrical defibrillation, 271
Electrical gradient, movement along, 59, 59f
Electrical signaling system, 146
Electric potential. See Membrane potential
Electrocardiogram (ECG), 274–78
  wave forms, from lead II, 277f
Electrochemical gradient, 59
Electroencephalogram (EEG), 124–25, 124f
  patterns during sleep, 138
Electrolytes
  definition of, A5
  effects of mineralocorticoids on, 657
  in plasma, 365
Electromagnetic spectrum, 163, 164f
Electromagnetic ("heat") waves 603–4
Electrons
  definition of, A1–2, A1–2, 1f
Electron-transport chain, 27, 30–31f
Elements, characteristics of selected, A2
Elephantiasis, 328, 328f
Embolus, 291, 364

Embryo, 697
  early stages of, 728
  nourishment of, 728–29
  relation to placenta, 730–31
  twisting of the heart axis, 266f
Emesis. See Vomiting
Emetics, 560
Emission, 713
Emotion
  and gastric motility, 559
  physical effects of, 458
  role of the limbic system in, 127–28
Emotional reactions in the experience of
    pain, 161
Emphysema, 426–27
  effects of
    on airway collapse, 426–27
    on airway resistance, 426
    on alveolar walls, 440, 441f
    on gas exchanges, 440, 441f
Emulsification of fat, 572–73, 573f
E Na$^+$. See Equilibrium potential for Na$^+$
Enamel, 551
Encephalitis, 117
End-diastolic volume (EDV), 280
  relationship with stroke volume, 286,
    287f
Endocardium, 267
Endocrine control, 635f
Endocrine glands, 4, 5, 616
  disorders of, 630
  location of, 617f
  temporal coordination by, 618
  summary of, 619–20
Endocrine hypertension, 338
Endocrine pancreas, 568f, 649, 671–81
  hormones, major functions, and target
    cells of hormones, 620, 671–81
  See also Insulin; Glucagon
Endocrine system
  characteristics of hormones and target
    cells, 105, 616–31
  comparison with nervous system, 105–9,
    614–46, 646f
  components of, 10
  control of calcium metabolism, 681–87
  control of fuel metabolism, 667–81
  control of growth, 639–46
  definition of, 105
  effects of prostaglandins on, 711
  general functions of, 617
  links to the immune system, 399–401
  specificity of, 106
Endocrinology, definition, 616
Endocytosis, 24, 25f, 67–68
Endogenous opiates, 158
Endogenous pyrogen (EP), 374–75, 608–9,
    C2
  See also Interleukin 1 (IL-1); Leukocyte
    endogenous mediator (LEM)
Endolymph, 183
  of the cochlea, 183
  of the vestibular apparatus, 188

Endometrium, uterine, 722
  changes in
    during gestation and implantation,
    729
    during menstrual cycle, 722
Endoplasmic reticulum (ER), 18, 18–19,
    18, 20f, 41f, 43
  agranular. See Smooth ER
  binding with ribosomes, B10
  formation of the acrosome by, 706
  granular. See Rough ER
  lipoprotein synthesis by, 583
  pancreatic enzyme synthesis by, 568
  rough, 19–21, 20f
    and production of thyroglobulin, 651
  structure and functions, 19–21, 20f, 22f
  smooth
    structure and functions, 19, 20f,
    22–23, 22f
Endoplasm (plasma sol), 39
Endorphin hypothesis, 160
Endorphins, 158
  β-endorphin, 622, 665
    ACTH precursor, 659
  leukocyte activity mediation by, 400
  See also Enkephalins
Endothelial cells, 305
  and blood pressure regulation, 338
  capillary membrane regulation by,
    318
  functions of, 312
  in lymphatic capillaries, 325
Endothelial-derived relaxing factor
    (EDRF), 312
Endothelin, 313
Endothelium, endocardium, 267
End-plate potential (EPP), 207
End-stage renal failure, 502
End-systolic volume (ESV), 280
Energy
  from oxidative phosphorylation, 32–33f
  for protein synthesis, B10
  See also Adenosine triphosphate (ATP)
Energy balance, 595–601
Energy equivalent of O_2, 598
Energy expenditure (output), 596, 596f
  categories of, 596
  requirement for, general, 33, 596
  the work of breathing, 432–33
Energy yield
  of citric acid cycle, 27, 28
  of electron-transport chain, 27, 33
  of glycolysis, 27
  of nutrients
    under aerobic conditions, 33f
    under anaerobic conditions, 33f
  of oxidative phosphorylation
  See also Adenosine triphosphate; Energy
    balance
Enkephalins
  as endogenous opiates, 158
  leukocyte activity mediation by, 401
Enteric nervous system, 549

Enterogastric reflex, 558
Enterogastrones, 558, 569
  See also Cholecystokinin; Gastric inhibi-
    tory peptide; Secretin
Enterohepatic circulation, 572
Enterokinase, 568, 577
Environment, relationship of pollutants to
    Parkinson's disease, 100
Enzymes, A4–5, A4
  clearing of odorant molecules by, 193
  digestive, A8
  hydrolytic, 43
  influence of changes in hydrogen ion
    concentration on, 526
  membrane-bound, 49
  oxidative, 26
  as proteins, A11
  responses to pH changes, 526
  role in protein synthesis, B9–10
Eosinophil chemotactic factor, 402–3,
    C2
Eosinophilia, 357
Eosinophils, 356, 357, 369, C2
EP. See Endogenous pyrogen
Ependymal cells, 109
  functions of, 111, 111f
Epicardium, 267
Epidermal growth factor (EGF), 3, 618
Epidermis, 405
Epididymis, 697, 709
Epiglottis, 553
Epilepsy, 124
Epinephrine
  and adrenergic receptor specificity, 202,
    664–65
  control of secretion of, 628
  effect of
    on adipose tissue, 664
    on airways, 425
    on arterial blood pressure, 315, 664
    on arterioles, 315
    in brain, 519
    on digestive activity, 549
    on eyes, 163
    in exercise, 237
    on glucagon secretion, 664
    in heart contractility changes, 287
    on heart rate, 285, 664
    on heat production, 607
    on insulin secretion, 664
    on metabolic rate, 664
    on sweat glands, 605
    on ventilation, 458
  and heart contractility, 296
  and homeostasis in the endocrine sys-
    tem, 692
  metabolic actions of, summary, 682
  and nonshivering thermogenesis, 607
  and permissive action of thyroid hor-
    mone, 653
  in platelet plugs, 360
  receptors for, 202f, 624
  and regulation of fuel metabolism, 681

secretion of
    by the adrenal medulla, 663
    in pheochromocytoma, 338
and stress, 664–65
and thyroid hormone, 631
summary of metabolic actions of
Epiphyseal plate, 640
Epiphysis, 640
Epithelial cells
    of the gastric mucosa, 560, 561f
    of the mucosal membrane, 545
Epithelial tissue, 4
    as barriers between body fluids and external environment, 4
    tight junctions in, 55
    of urinary bladder, 503
Epithelial transport, absorption across the digestive-tract wall, 578
Epithelium, active reabsorption across, 481–82
EPP. See End-plate potential
EPSP. See Excitatory postsynaptic potential
Epstein-Barr virus, 358
Equations, chemical, balancing of, A4
Equilibrium
    osmotic, 60
    and resting potential, nerve cell, 73
    state of, 57, 57f
Equilibrium potential
    for K^+ (E_{K^+}), 71f, 72
    for Na^+ (E_{Na^+}), 72, 72f
Equivalents, chemical, definition of, A6
Erasistratus, 302–3
Erectile tissue, 711
Erection, 711–12, 712f
Eructation, 554
Erythroblastosis fetalis (hemolytic disease of the newborn), 391
Erythrocytes (red blood cells), 349–55, 364–65
    and blood doping, 352
    blood-type antigens of, 389
    and blood viscosity, 304
    characteristics of, 347
    disorders of, 354–55
    glycolysis in, 34
    and hematocrit, 348, 354f
    production of, 351
    red blood cell count, 349
Erythropoiesis, 351–54, 352f, 354f
Erythropoietin, 466
    anemia caused by lack of, and renal failure, 502
    deficiency in renal anemia, 355
    production at high altitudes, 451
    synthetic, 352
Esophageal stage of swallowing, 554–55
Esophagus, 414, 553–55
Essential hypertension. See Hypertension
Essential nutrients, 668
Estradiol, 717
    structural comparison with testosterone, 622f

Estriol, 733
Estrogen, 696, 714
    actions of, summary, 739
    and atherosclerotic coronary artery disease, 295
    and body water, 511
    deficiency of, and osteoporosis, 691
    effect of
        on breasts, 737–38
        on cervical mucus, 725
        on epiphyseal plate, 646
        on female reproductive tract, 738
        on FSH secretion, 728
        on growth, 646
        on LH secretion, 720–21, 720f, 721f
        on oxytocin receptors, 734
        on prolactin, 738
        on secondary sexual characteristics, 723
        on the uterus, 722
    and high-density lipoproteins, 295
    during menopause, 723
    negative feedback by, 720
    in ovarian control, 719–22
    positive feedback by, 721
    during pregnancy, 734
    production of, by an ovarian follicle, 719
    promotion of secretion by FSH, 635
    regulation of secretion by LH, 635
    role in follicular development, 719
    secretion of
        by the adrenal cortex, 657
        by the corpus luteum, 719, 721
        in the male, 657
        by the ovarian follicle, 717, 719
        in pregnancy, 732–33, 733f
Estrogen replacement therapy, and osteoporosis, 691
Estrone, 717
ESV. See End-systolic volume
Eunuch, 704
Eupnea, 449
Eustachian tube, 181
Euthyroidism, 654
Evaporation, 605
Evolution of mitochondria, proposed, 26
Excitable tissues, 79
Excitation-contraction coupling
    calcium release in, 222f, 223f
    in cardiac muscle, 273, 273f, 682–83
    in skeletal muscle, 220, 222
    in smooth muscle, 248–50, 251f, 682–83
Excitatory postsynaptic potential (EPSP), 92
    comparison to end-plate potential, 207
Excitatory synapse, 92
Excretory functions of the kidneys, 506
Exercise
    and blood doping, 352
    and atherosclerosis, 295
    cardiovascular responses in, 337–38
    control of ventilation during, 456–58

    effect of
        on body temperature, 609, 609f
        on breathing patterns, 437
        on development of collateral circulation in the heart, 292
        on fat content in body, 602
        on glucose uptake by muscles during exercise, 64, 677
        on growth hormone secretion, 642–43
        on immune defense, 401
        on kidney function, 503
        on management of diabetes mellitus, 64
        on menstrual cycles, 724
        on metabolic rate, 597
        on muscle mass, 238
        on oxygen release from hemoglobin, 445
        on plasma glucose levels, 559
        on plasma insulin levels, 64
        on pulmonary surface area, 440
        on time for gas exchange in lungs, 440
        on work of breathing, 432
    and energy expenditure, 598
    and heat, 633
    heat production by, and oxygen levels in blood, 447
    and high-density lipoproteins, 295
    and hypertension, 340
    hyperthermia in, 610, 610f
    and maximal oxygen consumption, 459
    and metabolic acidosis, 536
    and metabolic rate, 597
    muscle adaptation to, 238
    and obesity, 602
    and osteoporosis, 691
    overheating during, 517
    P_{O_2}, P_{CO_2}, and H^+ during, 456
    pregame meal, 559
    during pregnancy, 735
    "runner's high," 158
    and stress tests, 278
Exercise physiology, 7
Exocrine cells of the mucosal membrane, 545
Exocrine glands, 4–5, 5f
    defect in cystic fibrosis, 49
Exocrine pancreas, 568–69, 568f
Exocrine secretion, 543f
Exocytosis, 24, 24f, 67–68, 91
Exophthalmos, 655, 655f
Expiration, 421
    active, 425
    effect of increased airway resistance (obstructive lung disease) on, 426–27, 427f
    lung volume change in, 435f
    related to lung volume, 425
Expiratory muscles, 425
Expiratory neurons, 450–51
Expiratory reserve volume (ERV), 433
Expulsion phase of ejaculation, 713
Extension of joint, 230, 232f

External anal sphincter, 589
External auditory meatus (ear canal), 179
External cardiac compression, 260
External defenses, 405–8
External ear, 181
  components of, 179
External environment, 5
  aquatic compared to dry land, 465
External eye muscles, 177
External genitalia, 696
  differentiation of, 701, 701f
External intercostal muscles, 421
External respiration, 413–14, 414f
External urethral sphincter, 504
External work, 596, 596f
Extracellular fluid (ECF)
  brain, hydrogen ion concentration of,
    453–56
  components of, 6f, 512
  control of osmolarity, 514, 520
  exchanges with intracellular fluid across
    capillary wall, 317, 321, 512–513
  hypertonicity of, 518
  hypotonicity of, 520
  internal environment of the body, 5
  ionic composition of, 71, 514
  osmolarity. See Extracellular fluid, con-
    trol of osmolarity
  pools of particular constituents, 511
  volume of
    related to arterial blood pressure,
      514–15
    related to sodium load, 482–83
Extracellular matrix, 54–55
Extrafusal fibers, 243
Extrapyramidal system. See Multineuronal
  system
Extrasystole, 271, 277f
Extrinsic control, 10–11
  of the arteriolar radius and resistance,
    310–314, 316f
  of contractility of the heart, 286–87
  of stroke volume, 285, 285f
Extrinsic innervation of the digestive system, 549
Extrinsic nerves, 549
  mediation of the gastrocolic reflex, 589
  role in segmentation contractions, 576
Extrinsic pathway for clot formation,
  361–62, 362f
Extrinsic sympathetic control of the glomerular filtration rate (GFR),
  476–77
Eye, 161–77
  abnormalities of, 166, 167–69, 168f
  accommodation of, 166–69
  control of light entering, 162–63
  dark and light adaptation, 173
  functions of major components of, 161
  movement of by external eye muscles,
    177
  phototransduction by, 170–73, 172f
  protective mechanisms of, 177

receptor cells in, 170–73, 171f
  refractive ability of, 164–66
  structure of, 161–62, 162f
  sympathetic nervous system and, 201
Eyelashes, 177
Eyelids, 177
Eye muscles, external, 177

Fab region of antibody, 383
Facilitated diffusion, 62–66, 62f
  definition of, 64
  of glucose, 66
Factor VIII, and hemophilia, 364
Factor X, 361
Factor XII (Hageman factor), 361, 363f
Factor XIII, fibrin stabilization by, 361
FAD; FADH2, 28
Fainting, 339–40
Fallopian tubes, 697
  See also Oviducts
Farsightedness. See Hyperopia
Fast-glycolytic fibers, 237
  switch to, in response to weight training,
    238
Fasting state. See Postabsorptive (fasting)
  state
Fast-oxidative fibers, 237
Fast pain pathway, 158
Fat
  body stores of, 599
    adipose tissue, 512
    and body water, 511
    brown, 607
    effect of growth hormone on, 643
  determination of percentage of in body,
    602
  sympathetic nervous system effect on,
    201
  nutrient, 543
    absorption of, 585–86
    and control of gastric emptying, 558
    digestion and absorption of, 569, 572,
      578, 583, 584f
    effect on
      of cortisol, 658
      of diabetes mellitus, 675
      of epinephrine, 664
      of glucagon, 679
      of growth hormone, 640
      of insulin, 673–74
    and fuel metabolism, 670
    metabolic pathways for, 668–71, 669f
    and nonshivering thermogenesis, 607
    and release of CCK from the duodenal
      mucosa, 569
    transport in the lymphatic system,
      326–27
    water content of, 512
  See also Adipose tissue
Fatigue, 236
  at high altitudes, 451
Fat inclusions in adipose tissue, 35
Fat synthesis, and fuel metabolism, 668

Fatty acids, 543, 544f, A9
  as absorbable units of dietary triglycer-
    ide fat, 543, 544f, 583, 584f
  and cholesterol metabolism, 295
  effect on
    of cortisol, 658
    of diabetes mellitus, 675
    of epinephrine, 664
    of glucagon, 679–81
    of insulin, 674
  and fuel metabolism, 670
Fc region of antibody, 382f, 383
Feces, 588
  color of related to bilirubin, 574
  and defecation reflex, 589
  sodium in, 515
  water losses in, 521
Feedback
  in anterior pituitary hormone secretion,
    638–39
  in the ovarian cycle, 718f
  tubuloglomerular, 476, 477f
  See also Negative feedback; Positive
    feedback
Feedforward
  control by gastrointestinal hormones,
    549
  in insulin secretion, 591, 675
  mechanisms for, 12
  See also Anticipatory response
Feeding (appetite) centers, 598–99
Female pseudohermaphroditism, 661
Female reproductive physiology, 698f,
  714
  age-related changes, 723–24
  control of, 719–22, 718f
  cyclical activity of, 714, 719–23, 718f
  essential reproductive functions, 697
  fertilization, 725–28
  gestation, 734
  implantation, 729, 731f
  lactation, 737–40
  menstrual cycle, 722–23
  oogenesis, 714–16, 715f
  ovarian cycle, 716–22
  ovarian functions, 714
  parturition, 734–37, 737f
  placentation, 730–32, 731f
  secondary sexual characteristics,
    696–97, 723
  sexual cycle, 713–14
  uterine cycle, 722–23
  See also Estrogen; Progesterone
Females, adrenal androgens in, 723
Female sex drive, and DHEA, 657
Fenestrae in the glomerular capillary wall,
  471
Ferritin, 583–84
Fertile period, 725
Fertility, 713
Fertility pills, 726
Fertilization, 697, 725–28, 729f
Fetoplacental unit, 733

Fetus, 697
    circulation of, 262–63
FEV$_1$, 434
Fever, 374–75, 608–9, 609f
    alkalosis associated with, 536
Fiber, dietary. *See* Cellulose; Psyllium seed
        husks
Fiber types, skeletal muscle, 237–38
Fibrillation, 271
Fibrin, C2
    in inflammation, 373
    role in blood coagulation, 360
Fibrinogen, C2
    in blood coagulation, 349, 360
    in semen, 710
Fibrinolysin, in menstrual flow, 722
Fibroblasts, 39, 55, 363, C2
    liver, 574
Fibronectin, 55
Fibrous skeleton
    of the heart, 265, 266f
Fick's law of diffusion, 58, 58, 415, 440
Fight-or-flight response, 200–201, 337
    and epinephrine, 664
Filaments, muscle. *See* Thick filaments;
        Thin filaments
Filariasis, 328
Filtered load, 485
Filtration
    across capillary walls. *See* Ultrafiltration
    across glomerular membranes. *See* Glo-
        merular filtration
Filtration coefficient (Kf), 474, 477–79
Filtration fraction, 492
Filtration slits, 473, 473f, 478, 479f
Fimbriae, 725
Final common pathway, 2
Firing of action potential, 81, 85–86, 85f
First heart sound, 281
First law of thermodynamics, 596
First line of defense against changes in hy-
        drogen ion concentration, 529
First messenger, 50, 75
First-order sensory neuron, 154–55
First polar body, 715
Fistula, 561
Flaccid paralysis, 243
Flagellum, 19, 37, 38f
Flatus, 590
Flavine adenine dinucleotide (FAD), 28
Flexion of joint, 230, 232f
Flow. *See* Airflow; Blood flow; Laminar
        flow; Turbulent flow
Flow rate
    relationship to pressure and resistance,
        302
    versus resistance, 310
    versus velocity of flow, 317–18
Fluid balance, 510
    control of ECF osmolarity, 514, 520–21
    control of ECF volume, 515
Fluidity of plasma membrane, 46–47
Fluid mosaic model, 47, f

Fluid pressure. *See* Hydrostatic pressure
Flux. *See* Net diffusion
Focus, eye's maintenance of, 166f
Folic acid, and nutritional anemia, 355
Follicle
    hair, 406
    ovarian (*See* Ovarian follicles)
    thyroid, 650, 650f
Follicle-stimulating hormone (FSH), 635
    in females, 719
    in males, 707
    role in spermatogenesis, 707
Follicular cells, 717
    ovarian, 717
Follicular phase, 715–19
    control of, 719–20, 720f
Food
    energy from, 596
    intake of, 598–99
Foot proteins in the sarcoplasmic reticu-
        lum-lateral sac gap, 222–23
Foramen ovale, 262
Force
    muscle tension and, 233
Forced expiration, 425
Forced expiratory volume in one second
        (FEV1), 434
Foreign chemicals, excretion by kidneys,
        466, 490
Forgetting, 130
Formula, chemical, A5
Formula weight, A5
Forward failure, of the heart, 289
Fossa ovalis, 263
Fovea, 169
    representation in visual cortex, 176
Frank-Starling curve, 287f
    in heart failure, 288, 289f
Frank-Starling law of the heart, 286, 286f
Fraternal twins, 717
Free hydrogen ion (H$^+$). *See* Acid-base
        balance; Hydrogen ion
Free water, 499–500, 520
Frequency of sound, 178
Frequency code, 153
Frontal lobes, 118, 118f, 120
Frostbite, 610
Fructose, 543, 544, 580, A8
    as an energy source for sperm, 710
FSH (*See* Follicle-stimulating hormone)
Fuel metabolism, 18, 667–81
    hormonal control of, 682
Functional groups of organic molecules,
        A7–8
Functional hyperemia. *See* Active
        hyperemia
Functional murmur, 281
Functional residual capacity (FRC), 434,
        439
Functional syncytium, 248, 253f
    atrial and ventricular, 267
    in the digestive tract, 548
    of heart muscle, 287

Functional unit, 216
    of kidney, 466
    of skeletal muscle, 216
Fundamental cellular activities, 2
Fundus, stomach, 555, 555f, 556

GABA. *See* Gamma-aminobutyric acid
Galactose, 543, 544, 580, A8
Galen, 302–3
Gallbladder, 571, 571f, 574, 590
Gallstones, 572–73, 574
GALT. *See* Gut-associated lymphoid tissue
Gamete intra-Fallopian transfer (GIFT), 726
Gametes, 696
    chromosomes of, 698–99
    See also Ova; Sperm
Gametogenesis, 696, 699
    See also Oogenesis; Spermatogenesis
Gamma-aminobutyric acid (GABA), 99
    astrocytes' use of, 109
Gamma (γ) globulin, 349, 382, C2
    See also Antibodies; Immunoglobulins
Gamma motor neurons, 243, 244f
Ganglion, 141
Ganglion cells
    optic nerve, 171
    in the retina, 169, 173
        on-center and off-center, 175, 175f
Gap junctions, 55–56, 56f, 548, 615
    of intercalated discs, 267
    between single-unit smooth muscle
        fibers, 248
Gas exchange, 437–42
    across pulmonary capillaries, 439
    across systemic capillaries, 440, 440f
    factors affecting, 439–42
Gas, intestinal, 590
Gas transport in blood, 442–50
Gastric emptying, 557–58
Gastric filling, 556
Gastric inhibitory peptide (glucose-
        dependent insulinotropic
        peptide), 558, 590, 591, 675
Gastric mixing, 557
Gastric mucosal atrophy, 563
Gastric mucosal barrier, 565–67, 566f
Gastric phase of gastric secretion, 564
Gastric pits, 560–64, 564f
Gastric secretions, 563–64
    See also Stomach, secretions
Gastric storage, 556
Gastrin
    effect of
        on colonic motility, 589
        on gastric secretion, 563–65
        on gastric motility, 557
        on ileocecal sphincter, 577, 576f
        on small intestinal motility, 576
        on mass movements, 589
    intestinal, 564
    overview of actions, 590
    source of, 562
Gastrocolic reflex, 589

Gastroesophageal sphincter, 553, 555f
Gastroileal reflex, 576
Gastrointestinal hormones, overview, 549, 590–91
   See also Cholecystokinin; Gastric inhibi-tory peptide; Secretin
Gastrointestinal system. *See* Digestive system
Gates of membrane channels, 81
G cells, 563
Gelation in amoeboid movement, 39
Gene mutation, B13
General adaptation syndrome, 665
Generator potential, 152–53, 153f
Genes
   activation by lipophilic hormones, 624
   antibody, 386–87
   coding for muscular dystrophy, 241
   definition of, B1–2
   for direction of erythropoietin synthesis, 353
   expression of, B10
   for smell receptors, 193
Gene-signaling factors, B10
Gene transcription B7
Genetic abnormalities
   association with Alzheimer's disease, 133
   cystic fibrosis, 49, 727
   leading to emphysema, 426
   and maternal age, 715, 727
   sickle-cell anemia, 727
   See also Muscular dystrophy
Genetic code, B14
Genetic controls of endocrine organs, 630
Genetic dictionary, B7
Genetic engineering, 727
   growth hormone synthesis, 644
   See also Recombinant DNA technology
Genetic female, 699
Genetic information, B1–14
   transcription and translation of, B7f
Genetic male, 699
Genetic sex, 699–700
Genital swelling, 702
Genital tubercle, 701
Germ cells
   chromosomes of, B3
   mutation of, B14
   in the seminiferous tubule, 704
Gestation (pregnancy), 697, 734
GFR. *See* Glomerular filtration rate
GH. *See* Growth hormone
GHIH. *See* Growth hormone-inhibiting hormone (GHIH)
GHRH. *See* Growth hormone-releasing hormone (GHRH)
GIP. *See* Gastric inhibitory peptide (glucose-dependent insulino-tropic peptide)
Gland secretion, 4
Glans penis, 701, 712
Glaucoma, 162, 513
Glial cells, 109

functions of, 111
reproduction of, 111
Gliomas, 111
Globin portion of hemoglobin, 350
Globulins, 349
Glomerular-capillary blood pressure, effect of on glomerular filtration, 473
Glomerular filtration, 471–79
   description of, 470, 470f
   filtered load, 485
   filtration fraction, 492
   forces responsible for, 473–74
   See also Glomerular filtration rate
Glomerular filtration rate (GFR)
   control of
      autoregulation, 474–76
      extrinsic, 476
      in maintenance of salt balance, 515–16
      as part of baroreceptor reflex, 477, 478f, 515
   direct effect of arterial blood pressure on, 474
   factors affecting, 474–77
   in long-term regulation of arterial blood pressure, 514–15
   magnitude of, 474
   measurement of by inulin clearance, 491
   and plasma clearance rates, 491–92
   and salt balance, 515
Glomerular membrane, 471–72, 472f
Glomerulus, 466, 469f
Glottis, 553
Glucagon
   and blood glucose levels, 680f
   control of secretion of, 679–81, 680f
   effect of
      on carbohydrate, fat, and protein, 681
   effect on
      of epinephrine, 664
      of somatostatin, 671
   excess secretion of, 681
   and homeostasis in the endocrine sys-tem, 692
   interactions with insulin, 680f
   and postabsorptive state, 679–81
   role in diabetes mellitus, 681
   role in fuel metabolism, 682
   source of, 671–72
   and stress, 666
Glucocorticoids, 657–658
   as anti-inflammatory drugs, 376–78
   for allergy treatment, 403
   in therapy for adrenal virilization, 662
   See also Cortisol
Gluconeogenesis
   definition of, 658, 668
   effect on
      of cortisol, 658
      of epinephrine, 664
      of glucagon, 679
      of insulin, 673
   excessive, 660

hepatic, 666
inhibition by insulin, 673
during diabetes mellitus, 675
during starvation, 671
Glucose, 543
   as absorbable unit of carbohydrate digestion, 543, 544, 580, 581f
   concentration in blood plasma, 484
      factors affecting, 673, 673f
   dependency of brain on, 115, 658
   effect of
      on food intake, 599–600
      on glucagon secretion, 679
      on growth hormone secretion, 642–43, 643f
      on insulin secretion, 674–75
   effect on
      of cortisol, 666
      of diabetes mellitus, 675
      of exercise, 64
      of glucagon, 679
      of growth hormone, 640
      of thyroid hormone, 652
   facilitated diffusion of, 66
   and gluconeogenesis, 658
   and glycogenesis, 673
   and glycogenolysis, 673
   homeostasis of, 673
   interactions with insulin and glucagon, 680f
   metabolism of, 668
   renal handling of, 484–86, 486f
   sparing of for brain, 640, 670, 671, 681
   storage as glycogen, 35
   structure of, A8f
   transport into exercising muscle, 64
   uptake during exercise, 64
Glucose-dependent insulinotropic peptide, 591
Glucose transporters, 673
Glucostatic theory, 599–699
Glucosuria, 660, 675
Glutamate
   astrocytes' use of, 109
   toxic release of, after cell destruction, 115
Gluten enteropathy, 578, 579f
Glycerol, 543
   as component of triglycerides, 543
   effect on food intake, 599
   and formation of glucose, 671
   role in fat metabolism, 670
   and saturation of fatty acids, A9
   structure of, A9
Glycine, 99, A11
Glycogen, A8, 8f
   effects on
      of cortisol, 658
      of epinephrine, 664
      of glucagon, 679
      of insulin, 673
      of thyroid hormone, 652
   glucose storage as, 35, 511, 668–69

in liver, 35
  release stimulated by progesterone, 728
  in skeletal muscle, 35
  storage in inclusions, 35
  storage of nutrients as, 236
  structure of, A8
  sympathetic nervous system effect on stores of, 201
  in the uterine endometrium, 723
Glycogenesis, 668
  insulin stimulation of, 673
Glycogenolysis, 664, 668
  inhibition by insulin, 673
Glycolipid, structure of, A9f
Glycolysis, 27–28, 28f
  importance in red blood cells, 34, 350
  metabolism using, 236
  role in skeletal muscle, 234
Glycolytic enzymes, 350
Glycoproteins, 47
  in basement membrane, 471
  HLA antigens as, 397
GnRH. See Gonadotropin-releasing hormone
Goiter, 655–56, 655f
Golgi apparatus. See Golgi complex
Golgi complex, 18, 23–24, 24f, 43, 568, 622, 651
Golgi tendon organs, 97, 243–46
Gonadal sex, 700–701
Gonadotropic hormones. See Gonadotropins
Gonadotropin-releasing hormone (GnRH), 708–9, 708f, 719
Gonadotropins, 635
  See also Chorionic gonadotropin; Follicle-stimulating hormone; Luteinizing hormone
Gonads
  hormones of, major functions, and target cells of hormones, 620
  See also Ovaries; Testes
Goosebumps, 608
G protein, 51
GPSP. See Grand postsynaptic potential
Graafian follicle, 717
  See also Mature follicle
Gradation of contraction
  in cardiac muscle, 267
  in heart muscle, 287
  in smooth muscle, 251–53
Graded exercise test, 278
Graded potentials, 79–80, 79f, 80f, 101–2
  comparison with action potentials, 86
  See also End-plate potential; Excitatory postsynaptic potential; Grand postsynaptic potential; Inhibitory postsynaptic potential; Pacemaker activity; Receptor potential; Slow-wave potential
Grand postsynaptic potential (GPSP), 94–95, 95f
Granstein cells, 407

Granular cells, 475
Granulocyte colony-stimulating factor, 357
Granulocytes, polymorphonuclear, 356
Granuloma, 374
Granulopoiesis, 375
Granulosa cells
  changes in, during follicular development, 717
  estrogen production by, 719
  in the primary follicle, 714
  See also Follicular cells, ovarian
Grave's disease, 654–55, 655f
Gravity
  effect on circulation, 330, 331f
  effect on muscle, 209
  and orthostatic hypotension, 339
Gray matter
  in brain, 118
  periaqueductal, in analgesic system, 158
  in spinal cord, 139, 141, 142f
Ground substance in bone matrix, 640
Growth, 639–46
  effect on
    of androgens, 646, 657, 704, 723
    of cortisol, 639
    of estrogen, 646, 723
    of growth hormone, 640, 642, 644, 645
    of insulin, 646
    of somatomedins, 642
    of thyroid hormone, 646, 653
  factors dependent on, 639
  rate of, 639–40, 640f
Growth hormone (GH), 634
  abnormalities of, 645–46
  actions of, 640–42
  control of secretion of, 642–43, 645, 643f
  differences in males and females, 646
  effect of
    on amino acids, 640
    on bone, 642
    on fat and carbohydrate, 640
    on growth, 640, 642
    on protein, 640
  and fasting, 681
  metabolic effects, summary, 682
  relation to somatomedins, 642
  source
    in body 634, 640
    for clinical use, 644
Growth-hormone-inhibiting hormone (GHIH), 636, 642–43
  See also Somatostatin
Growth-hormone-releasing hormone (GHRH), 636, 642–43
GTP. See Guanosine triphosphate
Guanine (G), B1
Guanosine diphosphate (GDP), 28
Guanosine triphosphate (GTP), 28
Gustation, 190
  See also Taste
Gut-associated lymphoid tissue (GALT), 370, 407
Gynoid obesity, 602

H^+. See Hydrogen ion
Habituation, 131
  in Aplysia, 134f
  calcium ion channel activity in, 132
  comparison with adaptation, 154
Hageman factor, 361
Hair
  distribution, as a secondary sexual characteristic, 704
  role in temperature regulation, 608
  in tactile discrimination, 406
Hair cells, 184
  loss of because of loud noise, 639, 639f
  microvilli (brush border) in, 40
  of the organ of Corti, 182–84, 183f, 186, 186f
  and rotation of the head, 188, 188f
  of semicircular canals, 187f, 188
  of utricle, 188–89, 189f
Hair follicle, 406
Haldane effect, 448
Haploid number, chromosomes, 3, 699
Haptens, 381
  poison ivy toxin, 404
Hardening of the arteries, 291–93
Hard palate, 550
Harvey, William, 302–3
Hashimoto's disease, 396
Haustra, 587
Haustrations in the large intestine, 548, 588–89
Hayfever, 403
Hb. See Hemoglobin
$HbCO_2$. See Carbamino hemoglobin
HbO_2. See Oxyhemoglobin
hCG. See Human chorionic gonadotropin
HCO_3^-
  See Bicarbonate ion
H_2CO_3. See Carbonic acid
HDL. See High-density lipoproteins
Head of a sperm, 706
Headaches, 116–17
Head-righting reflexes, 144
Hearing, 161
  definition of, 178
  threshold of, 181
  See also Ear
Hearing aids, 187
Heart, 259
  anatomy of, 260–78
  blood flow to, 290–95
  cardiac cycle, 278–81, 279f
  cardiac output, 283–89, 288f
  conducting system of, 269f, 272
  coronary circulation, 290
  and diabetes mellitus, 676–77
  disorders of, 288–90
  as a dual pump, 261–64
  effect on
    of epinephrine, 285, 663
    of hypertension, 339
    of potassium abnormalities, 489–90
    of shock, 343

Heart, (continued)
    of the sympathetic nervous system, 200–201
    of thyroid hormones, 653
  electrical activity of, 268–74
  and electrocardiogram, 274–78, 276f, 277f
  embryonic development of, 265, 266f
  external compression as a life-saving measure, 260
  of fetus, 262–63
  fibrous skeleton of, 264, 266f
  function of, 259, 261–62
  hormones, major functions, and target cells of, 620
  layers of, 267
  mechanical activity of, 278–81
  as a muscle. *See* Cardiac muscle
  normal sounds of, 281
  oxygen needs of, 290–91
  pacemaker activity of, 269
  refractory period of, 274
  responses to exercise, 7, 377–38
  role in
    of calcium, 273
    of potassium, 273
    of sodium, 273
sequence of excitation in, 271–72
  syncytial nature of, 267
  valves of, 264–65, 265f, 282
  See also entries for Cardiac; Coronary
Heart attack, 278
  risk of, and exercise, 34
  See also Acute myocardial infarction
Heartbeat. *See* Heart rate
Heart block
  complete, 271
  detection on an ECG, 277, 277f
Heartburn, 555
Heart failure. *See* Congestive heart failure
Heart murmurs, 281
Heart rate
  changes in during sleep, 138–39
  control of, 283–85
  as determinant of cardiac output, 283
  effect on
    of epinephrine, 285, 663
    of parasympathetic stimulation, 284
    of sympathetic stimulation, 285
    of thyroid hormone, 653
  relation to ventricular filling, 281, 281f
  SA node as pacemaker, 269
  sympathetic nervous system effect on, 201
Heart sounds, 281
Heat
  balance in body, 603, 603f. *See also* Temperature regulation
  definition of, 596
  effect on arteriolar radius and blood flow, 313
  environmental
    and exercise, 517, 633
    response of body to, 608, 610

input, 603
loss of (output), 603–5
production of
  by nonmuscle sources, 606–7
  by nonshivering thermogenesis, 607
  by shivering, 606–7
  by skeletal muscle, 233
  by thyroid hormone, 652
production by contracting muscles, 254
transfer mechanisms, 604f
Heat exhaustion, 517, 610
Heat gain mechanisms, 606–7
Heat loss mechanisms, 607–8
Heat stroke, 517, 610
Heaves, 560
Helicotrema, 183f, 184
Heliobacter pylori, 566
Helper T cells, 392, 394–95, C2, C4
  in AIDS, 395, 396f
  and class II MHC glycoproteins, 397
  functions of, 392, 394–95
  interactions of
    with B cells, 391
    with B cells and macrophages, 390f
    with B cells, macrophages and T cells, 397–98, 398f
  with Langerhans cells of skin, 407
  interleukin 2 secretion by, 394
  lymphokine secretion by, 393
Hematocrit, 348
  of males versus females, 349f
  normal, and under disease conditions, 354f
Heme groups in hemoglobin, 350
Hemiplegia, 243
Hemodialysis, 504
Hemoglobin, 347, 350, 350f, A11
  buffering reactions of, 529
  forms of, 443
  and iron, 583
  oxygen-hemoglobin dissociation (saturation) curve, 442–43, 446–47
  percent saturation of, 443–44, 444f
  reduced, 443
  as storage depot for oxygen, 445–46, 445f
  structure of, 350, 350f
  transport by of
    carbon dioxide, 447
    carbon monoxide, 447
    hydrogen ion, 448
    oxygen, 350, 442–47
Hemolysis, 355
  of erythrocytes attacked by antibodies, 389
Hemolytic anemia, 355
Hemolytic disease of the newborn, 391
Hemolytic jaundice, 574
Hemophilia, 364
Hemopoiesis, 357f
Hemopoietic marrow, 351
Hemorrhage
  consequences and compensations of, 342f

isotonic fluid loss in, 520
and kidney failure, 502
spontaneous, in hypertension, 338
Hemorrhagic anemia, 355
Hemostasis, 347, 359
Henderson-Hasselbalch equation, 528, 534
Henle's loop. *See* Loop of Henle
Heparin, 357
  anticoagulant, 364
Hepatic jaundice, 574
Hepatic portal system, 570, 570f, 571f
Hepatitis, 575
Hepatocytes, 570
  replacement of, 575
Hering-Breuer reflex, 453
Hernia, 547
  hiatal, 553
Heroin, 100
  placental transfer of, 732
Herpes virus, 394
Hiatal hernia, 553
High altitude
  effect of on arterial P_{O_2}, 444
  effects on the body, 451
High blood pressure
  and android obesity, 602
  side-effect of growth hormone therapy, 644
  and stress, 667
  See also Hypertension
High-density lipoproteins, reduction in, by anabolic steroid use, 239
High-density lipoproteins (HDL), 293
  and cholesterol metabolism, 295
Higher sensory areas, 120
Hippocampus, 131
  long-term potentiation in, 133
Hirsutism, 661
Histamine, C2
  and allergy, 402
  from basophils, 357
  edema induced by, 426
  effect on arteriolar radius, 313
  effect on capillaries, 319
  and IgE antibodies, 402, 403f
  and immediate hypersensitivity, 402
  and inflammation, 372
  as a paracrine, 616
  receptors for, 567
  release from mast cells, 374, 380
  release in the gastric system, 565–66
  and ulcerogenesis, 566–67, 566f
  and vascular changes, 371, 372, 616
Histocompatibility antigens. *See* Human leukocyte-associated antigens
Histone proteins, B3–4
  in DNA, B4f
Histotoxic hypoxia, 449
Hives, 403
HLA antigen. *See* Human leukocyte-associated antigens
Homeostasis, 5–6, 45
  and acid-base balance, 586
  and the afferent division, peripheral nervous system, 193–94

of blood glucose, 673, 673f
blood's contribution to, 364
blood vessels' roles in, 343
in body weight, 597
of calcium ion (Ca^{++}), 687
of cholesterol, 293, 572
in core temperature regulation, 607–8
defense systems and, 408
and the digestive system, 591–92
effects of mineralocorticoids on, 657
and the efferent division, peripheral nervous system, 210
endocrine system role in, 646–47, 691–92
and exercise, 7
and fluid balance, 537–38
and heart functions, 295–96
and heat-energy balance, 611
hypothalamus control of, 126–27
maintenance by body systems, 7, 8–9, 9f
maintenance by the hypothalamus, 115
maintenance by the kidneys, 465–66, 506
maintenance through membrane transport, 74–75
and motivation, 128
and the nervous system, 146–47
and neuronal physiology, 101
and renal failure, 502–3
and reproductive system, 742
and the respiratory system, 460
roles of thrombin in, 361f
skeletal muscles participation in, 254
and solute concentrations in blood, 321
and temperature regulation, 601
and the ventilatory response to exercise, 458
Homologous chromosomes, 698, B3
Hormone response element (HRE), interaction with lipophilic hormones, 625
Hormones, 614
amplification of actions, 626
antagonism of, 631
anterior pituitary, 634f
binding to target-cell receptors, 106
chemical classes of, 618
as chemical messengers, 105
control of cyclical ovarian and uterine changes, 718f
control of secretion of, 626–28
definition, 5, 616
down regulation of, 631
excretion of, 628
factors influencing effective plasma concentration of, 626
gastrointestinal, 549
general functions of, 617
and growth, 639
hypersecretion of, 630
hyposecretion of, 630
leukocyte receptors for, 400
mechanisms of actions, 624
mechanisms of synthesis, storage, and secretion of, 622–23

metabolic activation and inactivation of, 628–29
and neural input, 627, 627f
as neuromodulators, 98, 107
permissiveness of, 631
placental secretion of, 732–34
protein, 67
exocytosis of, 68
receptors for, 625
control of, 630–31
summary, major hormones, 619–21
synergism of, 631
target-tissue responsiveness to, 630–31
target-tissue specificity for, 106
transport in blood, 364
See also Specific hormones
Host cell, 369
destruction by cytotoxic T cells of virally-invaded, 393, 397
Hot environment. See Heat
Human chorionic gonadotropin (hCG), 732
release by the placenta, 702
vaccine against as a birth control method, 741
Human chorionic somatomammotropin, 738
placental secretion of, 732
Human immunodeficiency virus (HIV), 395, 396f
Human leukocyte-associated antigens (HLA antigens), 397
Humoral immunity. See Antibody-mediated (humoral) immunity; B lymphocytes
Hunger, 558–59
Hyaline membrane disease. See Newborn respiratory distress syndrome
H-Y antigen, 701
Hydrocephalus, 113
Hydrochloric acid
secretion in the stomach, 556, 562, 562f
as a strong acid, 524
Hydrogen bonds, A4
Hydrogen molecules, covalent bonds of, A3–4
Hydrogen ion pump, 65
Hydrogen ions (H$^+$)
active transport of, 65, 562, 562f
binding with hydrogen carrier molecules, 28
buffering of, 527–29
concentration in arterial blood, effect on ventilation, 456, 457f
concentration in brain extracellular fluid, effect on ventilation, 455
control of
by chemical buffers, 527–29
by kidneys, 530–34
by respiratory system, 529–530
effect of
on enzyme activity, 526
on neuromuscular excitability, 526
on potassium levels, 489, 526

in the electron-transport chain, 31–32
and homeostasis, 537
relation to carbon dioxide concentrations, 526, 529–30
release during energy-yielding reactions, 28
sources of, in body, 526–27
transfer in energy-yielding reactions, 28
tubular secretion of, 488
See also Acid; pH
Hydrogen peroxide (H$_2$O$_2$), 26
Hydrolysis, 544–46, 546f, A11
enzymatic, 43, 544–45, 544f
Hydrophilic compounds, 46
Hydrophilic hormones, peptides and catecholamines, 621
Hydrophilic peptides, receptors for, 624
Hydrophobic compounds, 46
Hydrostatic pressure, 60–61, 61f
in Bowman's capsule, 474
in interstitial fluid, 324
See also Bulk flow; Glomerular filtration
Hydroxyapatite crystals, 684
Hymen, 697
Hypercalcemia, 682, 688
Hypercapnia, 449
Hypercomplex cells in the visual system, 177
Hyperemia
active, 312
reactive, 313
Hyperglycemia, 660, 675
caused by stress, 666
and glucagon excess, 681
Hyperkalemia, 663
Hypernatremia, 660–61
Hypernea, 449
Hyperopia, 168, 168f
Hyperoxia, 449
Hyperparathyroidism, 689
Hyperpigmentation, and cortisol deficiency, 662
Hyperplasia, 640
in skeletal muscle, 238
Hyperpnea, 450
Hyperpolarization, 80
and action potential, 80
of photoreceptors, 170–71
of a postsynaptic neuron, 92–93, 99
Hypersecretion of the endocrine glands, 630
Hypersensitivity, 402
Hypertension, 338–39
and abnormalities in renin-angiotensin-aldosterone system, 483
adaptation of baroreceptors in, 339
causes of, 338
complications of, 339
and exercise, 34, 339
and hyperaldosteronism, 660
as result of kidney failure, 502
treatment of, 339
Hyperthermia, 609–11, 609f
and exercise during pregnancy, 735

Hyperthyroidism, 654–55
Hypertonicity
  definition of, 492
  as the driving force for water reabsorption, 486–87
  of duodenal contents, effect on gastric emptying, 558
  of extracellular fluid, 518
Hypertrophy
  of the heart, 267–68, 288
  skeletal muscle, 238
Hyperventilation, 449–50, 450f
  alkalosis associated with, 536
Hypnosis, pain suppression by, 158
Hypocalcemia, 682, 686
Hypocapnia, 449
Hypocapnia-induced alkalosis, 451
Hypodermis (subcutaneous tissue). *See* Adipose tissue
Hypoglycemia, 677–79
  and cortisol deficiency, 662
  as a stimulus for growth hormone secretion, 642–43
Hypokalemia, 660
Hyponatremia, 663
Hypoparathyroidism, 689
Hypophosphatemia, 688
Hypophysiotropic hormones, 635–36
  regulation of growth hormone secretion by, 642–45
Hypophysiotropic secretory neurons, 637
Hyposecretion of endocrine glands, 630
Hypotension, 338, 339
  See also Circulatory shock
Hypothalamic inhibiting hormones, 635
  See also specific hormones
Hypothalamic osmoreceptors,
  and thirst, 521–22
  and vasopressin secretion, 521–22, 522f
Hypothalamic releasing hormones, 635
  See also specific hormones
Hypothalamo-hypophyseal portal system, 636
Hypothalamus, 115, 117, 125
  and anorexia nervosa, 601
  control of autonomic nervous system output, 203
  control of food intake, 598–99
  control of vasopressin release by, 497
  as endocrine gland, 632–33, 635–36
  in fever production, 608
  functions of, summary, 126–27
  hormones, major functions, and target cells of hormones, 619
  and limbic system, 127
  link with the anterior pituitary, 637f
  link with the posterior pituitary, 632f, 637f
  location of, 127f
  osmoreceptors in, 521
  and the pituitary, 631–39
  response to estrogen concentration, 781

role of
  in arousal of motility in empty stomach, 558
  in cortisol secretion, 659
  in emotion, 126, 127
  in the female reproductive function, 719–22
  in lactation, 738
  in male reproductive function, 708, 708f
  in parturition, 734–35
  in puberty, 708–9
  in stress, 666–67, 667f
  in temperature regulation, 605–9
  summary of functions of, 126
Hypothalamus-pituitary-adrenal cortex axis, 659, 665
Hypothalamus-pituitary portal vessels. *See* Hypothalamo-hypophyseal portal system
Hypothalamus-pituitary-thyroid axis, role in thyroid hormone secretion, 653
Hypothermia, 610
Hypothyroidism, 654
  and goiter, 655–56
Hypotonicity of extracellular fluid, causes and consequences, 518
Hypotonic solutions, 492
Hypoventilation, 449
  in chronic lung disease, 456
Hypovolemic shock, 340
  See also Circulatory shock
Hypoxia, 449
  circulatory, 449
Hypoxic hypoxia, 449
  at high altitudes, 451
H zone, 216

Iatrogenic damage to the endocrine glands, 630
I bands, 216, 218–19
ICF. *See* Intracellular fluid
ICSH. *See* Interstitial cell-stimulating hormone
Identical twins, 717–18
Idiopathic hypertension, 338–39
Idiopathic hyposecretion of the endocrine glands, 630
IgA
  in colostrum, 388
  response to exercise, 401
IgD, 382
IgE
  antibodies to, and mast cells, 403f
  association with allergies, 402–3
IgG, 382
  blocking antibodies, role in desensitization, 404
  as an opsonin, 383
  passive immunity conferred by, 388
IgM, 382, 385
IL-1. *See* Interleukin 1 (IL-1)

IL-2. *See* Interleukin 2 (IL-2)
Ileocecal valve/sphincter, 576–77, 576f
Ileum, 575
  See also Small intestine; Terminal ileum
Imaging techniques for study of the human brain, 110
Immediate early genes (IEGs), 135
Immediate hypersensitivity, 402
  role of IgE and mast cells in, 403f
Immune-complex disease, 383–85, 502
Immune defense system, 368, 369–404, 408
  components and functions of, 8–9, 369
  effect on
    of exercise, 401
    of kidney failure, 502
    of prostaglandins, 71
  nonspecific immune responses, 371–80
  role of skin, 406–7
specific immune responses, 380–404
  summary of effector cells and chemical mediators, C1–C4
Immune disease, 401–4
  See also Allergy; Autoimmune disease; Autoimmune response; Immune-complex disease; Immune responses
Immune response effectors, C1–4
Immune responses
  nonspecific, 370
  specific, 370
Immune surveillance against cancer, 369, 398–99, 400f
Immunity, 355–56, 369
  active, 387–88
  active versus passive, 387–88
  long-term, 387, 387f
  natural, 388–89
  See also Immune defense system
Immunoglobulins (Ig), 349, 382, C2
  in colostrum, 739
  See also Antibodies
Immunosuppressive effects of cortisol, 658
Impermeability, definition of, 57
Implantation, 728–30, 731f
  prevention by morning-after pill, 741
Impotence, 713
Inactive area of an excitable membrane, 79
Inclusions, 18, 35, 35f
Incus, 179, 181
Indirect calorimetry, 597
Inertia, and perception of motion, 188
Infection
  resistance to bacterial, 371–75, 379, 391
  resistance to viral, 378, 393–94, 395
Infectious agents, pathways for movement in the nervous system, 37
Infectious mononucleosis, 358
Inferior vena cava. *See* Vena cava
Inflammation, 371–78, 378f
Infundibulum, 631
Inguinal canal, 702
Inguinal hernia, 703

Inhibin
  in females, 720
  in males, 707
  role in spermatogenesis, 708, 708f
Inhibiting hormones, 635–36
Inhibitory postsynaptic potential (IPSP),
        93
Inhibitory synapse, 91–93
Inner cell mass, 728
Inner circular smooth muscle layer (of
        digestive tract), 547
Inner ear (cochlea)
  structure of, 178f, 179
  transduction of sound into action poten-
        tials by, 184–86, 185f
  transmission of sound waves in, 182–83
Inner ear (vestibular apparatus), compo-
        nents of, 179
Inner sheath, 37
Innervated blood vessels, 202
Inorganic acids, generated from catabo-
        lism of dietary proteins, 526
Inorganic chemicals, A6–7
  carbon-containing, A6–7
Inorganic salts, A7
Insensible loss, 521
Insertion of muscle, 230
Inspiration, 421–24
  effect on
        of newborn respiratory distress syn-
        drome on, 430–32
        of restrictive lung disease, 434–35
  related to pressure changes and lung
        volumes, 421–23
Inspiratory capacity (IC), 433
Inspiratory muscles, 421, 425
Inspiratory neurons, 450–51
  pacemaker activity of, 452
Inspiratory reserve volume (IRV), 433
Insufficient valve (heart), 282
Insulin, 486, 671–72
  and absorptive state, 674, 679
  control of secretion of, 674f, 679–81
  deficiency of, 675
  and diabetes mellitus, 675–77
  down regulation of, 631
        and Type II diabetes mellitus, 677
  effect of
        on amino acids, 680
        on carbohydrates, 673
        on carrier-mediated transport of glu-
        cose, 63
        on fat, 673–74
        on fuel metabolism, 672–73
        on glucagon, 679
        on growth, 646
        on liver, 673
        on proteins, 674
  effect on
        of blood glucose, 674
        of epinephrine, 664
        of exercise, 64
        of gastric inhibitory peptide, 591
  of parasympathetic nervous system,
        675
  excess secretion of, 677–79
  and exercise, 559
  feedforward release of, 591
  and homeostasis in the endocrine sys-
        tem, 692
  hypersecretion, and modification of tar-
        get cells, 631
  metabolic effects, summary of, 682
  and postabsorptive state, 674, 679
  role in fuel metabolism, 682
  and satiety, 600
  source of, 671
  and stress, 666
  structure of, 22f
  synthesis of, 22f
Insulin antagonists, 681
Insulin-dependent diabetes, 677
Insulin-like growth factors, 642
  See also Somatomedins
Insulin resistance, and android obesity, 602
Insulin shock, 677, 679
Integration by postsynaptic neurons, 94–96
Integrating center, 142
Integrator, 11
Integument. See Skin
Integumentary system, 9
Intensity of light, 163
Intensity of sound, 178, 180f
Intention tremor, 136
Interatrial pathway, 271
Intercalated discs, 267, 267f
Intercellular chemical messengers. See
        Hormones; Neuromodulation/
        neuromodulators; Neurotransmit-
        ters; Paracrines
Intercellular communication, 615–16,
        615f
Intercostal muscles, 421
Intercostal nerves, 421
Interdependence of alveoli, 430, 432f
Interferon, 378–79, 378f, C2
  and cancer defense, 400f
  functions of, 377
  and viral defense, 371
Interleukin 1 (IL-1), 375, C2
  interaction with interleukin 2 (IL-2),
        394–95
  secretion by keratinocytes, 406–7
  See also Endogenous pyrogen; Leuko-
        cyte endogenous mediator
Interleukin 2 (IL-2), 394–95, C2
  blockage by cyclosporin, 397
Intermediary metabolism. See Fuel
        metabolism
Intermediate filaments, 40–41
Internal anal sphincter, 589
Internal (cellular) respiration, 413, 414f
Internal environment, 5
  afferent information about, 194
  and kidneys, 465
  urinary system, 471
Internal genitalia. See Reproductive tract
Internal intercostal muscles, 425
Internal pool. See Pool
Internal urethral sphincter, 504
Internal work, 596, 596f
Interneurons, 108–9
  characteristics and functions of, 206
Internodal pathway, 271–72
Interphase, B11
Interstitial cells in testes. See Leydig cells
Interstitial cell-stimulating hormone
        (ICSH), 635, 708
Interstitial fluid, 5, 6, 322f
  of the brain, 113
  colloid-osmotic pressure in, 324
  composition of, 513f
  hydrostatic pressure in, 324
  as intermediary between blood and
        cells, 321, 322f
  percentage of extracellular fluid, 512
  percentage of total body water, 512
  shifts between capillaries in mainte-
        nance of plasma volume and
        blood pressure, 324–25
  as true internal environment, 471,
        513–14
Intestinal bacteria. See Bacteria, intestinal
Intestinal gas, 590
Intestinal gastrin, 564
Intestinal housekeeper, 576
Intestinal phase of gastric secretion, 564,
        565f
Intestinal segmentation, 588
Intra-alveolar pressure, 418, 418f
  changes during the respiratory cycle,
        424
  and transmural pressure gradient, 419
Intracellular enzymes, small intestine, 577
Intracellular fluid (ICF)
  buffers of, 529
  exchanges with extracellular fluid,
        513–14
  ionic composition of, 71, 513f
  percentage of total body water, 512
Intracellular protein anions (A⁻), 71
Intrafusal fibers, 243
Intramural nervous system
  of the digestive tract, 548
  See also Intrinsic nerve plexuses
Intraocular fluid, 512
  See also Aqueous humor; Vitreous
        humor
Intrapleural fluid, 417, 418–20
Intrapleural pressure, 418, 418f
  changes during the respiratory cycle,
        424, 425
  and transmural pressure gradient, 419,
        419f
Intrapulmonary pressure. See Intra-alveolar
        pressure
Intrathoracic pressure. See Intrapleural
        pressure
Intrauterine device (IUD), 741

Intrinsic control, 10
  of contractile activity in arteriolar
      smooth muscle and arteriolar
      radii, 309–10, 316f
  of stroke volume, 285–86, 285f, 286f
Intrinsic factor, 563
  deficiency in anemia, 354
Intrinsic nerve plexuses, 548–49
Intrinsic pathway
  for clot formation, 361, 362f
Inulin, 491
In vitro fertilization, 726
Involuntary branch of nervous system. See
      also Autonomic nervous system
Involuntary muscle, 214
Involution of the uterus, 737
  role of lysosomes, 25
Iodine pump, 651
Ion channels. See Channels
Ion concentration gradient in secondary
      active transport, 66
Ions, 59
  changes in concentration during an
      action potential, 83f
  definition of, A3
  movement through cell junctions, 55–56
IPSP. See Inhibitory postsynaptic potential
Iris, 162–63
Iron
  absorption from the digestive tract, 583,
      585f
  bacterial needs for, 374, 375
  and bilirubin, 573–74
  in hemoglobin, 350
Iron-deficiency anemia, 354
Irreversible reactions, A4
Irreversible shock, 343
Ischemic hypoxia. See Circulatory hypoxia
Ischymetric theory, 600
Islets of Langerhans, 568, 671
Isometric contraction, 230–32
Isotonic contraction, 230–32
Isotonic fluid gain or loss, 520
Isotonic solution, 492
Isovolumetric ventricular contraction, 280
Isovolumetric ventricular relaxation, 280

Jaundice, 574
Jejunum, 575
  See also Small intestine
Jenner, Edward, 388
Joint proprioceptors, 155
Joints, 233
Juvenile-onset diabetes, 677
Juxtaglomerular apparatus, 470
  anatomy of, 475–76, 476f
  role of
      in autoregulation, 475–76
      in renin secretion, 482
Juxtamedullary nephrons, 468f, 470
  role in establishing medullary osmotic
      gradient, 492

K_f. See Filtration coefficient
Kallikrein, C2
K cells. (See killer cells)
K (dissociation constant), 524
K^+ equilibrium potential, 71–72, 71f
Keratin, 41, 406
Keratinized layer of skin, 406
Keratinocytes, 406–7
Keto acids
  in diabetes mellitus, 527
  See also Ketone bodies
Ketogenesis, effect of glucagon on, 679
Ketone bodies, A9
  in diabetes mellitus, 675
  as an energy source, 671
  source of, 671
Ketosis, 676
Kidney disease
  and edema, 327
  and hormone excretion, 628
  and renal failure, consequences of,
      502–3
  and transfusion reaction, 389
  See also Renal failure
Kidney dysfunction, from a transfusion
      reaction, 389
Kidneys, 464–506
  ability to produce urine of varying con-
      centrations, 492–99
  activation of vitamin D by, 687
  active transport of glucose in, 66
  ammonia secretion by, 533–34
  anatomy of, 466–470, 467f, 468f, 469f
  basic renal processes, 470–71, 470f
  blood flow to, 466–67, 491–92
  and carbonic acid-bicarbonate buffer
      system, 527, 529
  carrier saturation in the membranes of,
      63
  and control of acid-base balance, 530
  and control of erythrocyte production,
      351
  and diabetes mellitus, 486
  disease of. See Kidney disease; Renal
      failure
  effect on
      of aldosterone, 481–82
      of parathyroid hormone, 686
      of vasopressin, 497–99, 497f, 498f
  and erythropoiesis, 351
  excretion of foreign compounds, 490
  functions of, 466
  and glomerular filtration, 470, 471–79
  handling by
      of chloride, 486
      of foreign compounds, 480
      of glucose and amino acids, 484–86
      of organic ions, 489–90
      of urea, 487
      of wastes, 488
  and homeostasis, 465–66
  hormones, major functions, and target
      cells of hormones, 620

  and hypertension, 338
  and medullary countercurrent system,
      492–96
  and medullary vertical osmotic gradient,
      492–99
  percentage of cardiac output received
      by, 479, 492
  regulation by
      of acid-base balance, 530
      of bicarbonate ion, 530, 531–33
      of calcium, 686
      of extracellular fluid (plasma) vol-
          ume, 514–16
      of hydrogen ion, 488, 530, 531, 533
      of osmolarity, 514, 516–17, 521
      of phosphate, 486
      of potassium, 488–89
      of sodium (salt), 481–84
      of water, 486–87
  response of, to congestive heart failure,
      288
  role in acclimatization to high altitudes,
      451
  stones, 474, 502, 689
  structure of, 466
  transplantation of, 503
  and tubular reabsorption, 470, 479–88
  and tubular secretion, 470, 488–90
  and waste excretion, 471, 490
Killer (K) cells, C2
  mechanism of action, 394f
  stimulation by, 383
  stimulation of, 384f
Kilocalorie (Calorie), 597
Kinesin, 36–37
Kinin, 374, C3
  activation in inflammation, 380
  relationship with neutrophils in inflam-
      matory response, 375f
Kininogen, C3
Kinocilium, 187f, 188
Knee-jerk reflex. See Patellar tendon reflex
Krebs cycle. See Citric-acid cycle
Kupffer cells, 570

Labeled lines, 155
Labia majora, 697, 702
Labia minora, 697, 702
Labor, 734–37
  stages of, 737f
Lacrimal glands, 177
Lactase, 544f, 577
Lactation (milk production), 697, 714,
      737–40,
Lacteal. See Central lacteal
Lactic acid
  and angina pectoris, 291
  effect of oxygen and carbon dioxide in
      blood, 446–47
  as energy source, 671
  formation during heavy exercise, 458
  formation of, by glycolysis, 236
  and metabolic acidosis, 536

and muscle soreness, 236
and recovery from oxygen debt, 237
reduction in blood doping, 352
as source of hydrogen ion, 526
Lactoferrin, 374, C3
in colostrum, 739
Lactose, 543, 577, 580, A8
Lactose intolerance, 577
Lamina propria, 545
in villi, 578–79
Laminar flow, 281, 308
Langerhans cells, 407
Language ability, components of, 122–23
LaPlace's law, 430
Large intestine, 587–90
structure of, 588f
Laron dwarfism, 645
Larynx (voice box), 414–15, 553
effect of testosterone on, 704
Latch phenomenon, 254
Latent pacemaker, 269, 271
Latent period, between stimulation and
contraction of muscle, 225
Lateral geniculate nucleus, 176, 176f
Lateral horn, spinal cord, 141
Lateral inhibition, 156, 157f
Lateral sacs of sarcoplasmic reticulum,
221–22
Lateral spaces in kidney tubules, 481
and sodium reabsorption, 481
and water reabsorption, 486–87
Lateral spaces in the small intestine, and
water reabsorption, 580
Lateral spinothalamic tract, 140–41
Law of mass action, 443–44, 455, 526
Laws
All or none law, 91
Boyle's law, 420, 420f
Fick's law of diffusion, 58, 58f, 415, 440
first law of thermodynamics, 596
Frank-Starling law of the heart, 286
LaPlace's law, 430
law of conservation of energy. See First
law of thermodynamics
law of mass action, 443–44, 455, 526
law of specific nerve energies, 152
Poiseuille's law, 305
LDL. See Low-density lipoproteins
Lead, and renal failure, 502
Lead (for an ECG), 275, 276f
Leader sequence in mRNA, 21, 21f
roles in protein synthesis, B9
Leaky valve, 282
Lean versus fat tissue, effect of growth hormone on, 645
Learning
definition of, 129
effect of MSH on (proposed), 631
habituation and sensitization as, 131–32
and stress, 666
Left atrial volume receptors
control of thirst, 522–23
control of vasopressin secretion, 522–23

relation to control of arterial blood pressure, 337
Left atrioventricular (AV) valve
anatomy of, 264
opening and closing of during cardiac
cycle, 279f, 280
relation to heart sounds, 287
Left cerebral hemisphere, 124
LEM. See Leukocyte endogenous mediator
Length-tension relationship
in cardiac muscle, 254, 285–86
in skeletal muscle, 229–230, 230f
in smooth muscle, 253
Lens, 162, 164, 167
Leukemia, 358
Leukocyte endogenous mediator (LEM),
375, C3
See also Endogenous pyrogen; Inter-
leukin 1
Leukocytes, 347, 355–58, 365, 369–70
control of, 356–57
functions of, 357–58, 369–70
and leukemia, 358
in menstrual flow, 722
production of, 356, 357f
role in inflammation, 373–74
types of, 356, 369–70
See also B lymphocytes; Basophils; Eo-
sinophils; Monocytes; Neutrophils;
T lymphocytes
Leukotrienes, 711
Lever action of muscles, 223–34
Levodopa (L-dopa), 99
Leydig cells (interstitial cells), 703, 707–8
LH. See Luteinizing hormone
LH surge, 720–21, 721f
Libido in males, 704
Life-saving measures, cardiopulmonary
resuscitation, 260–61
Ligamentum arteriosum, 263
Light, reaction of the eye to, 163
Light adaptation, 173–74
Light ray, 163
refraction of, 164–65
Limbic association cortex, 123
Limbic system , 127f
role of
in control of behavioral patterns, 128
in emotions, 127–28
in memory, 131
in sense of smell, 193
and ventilation, 458
Lipase, 544f, 572
pancreatic, 568, 569, 583
Lipid bilayer, 46–47, 47f, 48, 48f
as a diffusion barrier, 47–50
Lipid emulsion in chyme, 572
Lipids, A8–10
in myelin, 86
Lipid solubility, membrane permeability
determination by, 56–59
Lipolysis, 668
Lipophilic hormones, 625–26, 625f

removal from plasma, 629
steroids and thyroid hormones, 621
Lipophilic steroids, receptors for, 624
Lipoproteins, 293
and fat absorption, 583
Lipostatic theory, 599
Lips, 550
Liver, 570–75
activation of vitamin D by, 687
adverse effects of anabolic androgenic
steroids on, 239
anatomy of, 571f
angiotensin synthesis in, 482
and bile secretion, 571, 574
and bilirubin removal from blood,
573–74
cirrhosis, 575
detoxification and degradation actions
of, 589
disease of, and edema, 327
effect on
of cortisol, 658
of epinephrine, 664
of glucagon, 679
of insulin, 673
of secretion, 574
of vagus stimulation, 574
as endocrine gland, 642
formation of ketone bodies in, A9
functions of, summary, 570
glycogen synthesis in, 670
and hepatic portal system, 570
hepatitis, 575
hormones, major functions, and target
cells of hormones, 620
macrophages of, 570
metabolism by, 670
response to shock, 343
role of
in cholesterol metabolism, 294–95,
294f
in excretion of foreign compounds,
490
role in fuel metabolism, 671
secretion of somatomedins by, 642
storage of glycogen by, 35, 35f, 668–69
synthesis by
of acute phase proteins, 375
of angiotensinogen, 482
of bile salts, 571
of cholesterol 494–95, 494f
of clotting factors,
of ketone bodies, 671
of plasma protein, 349
l_o (See optimal length (l_o) of muscle)
Load, 230
Load-velocity relation of skeletal muscle,
232–33, 232f
Lobes of cerebral cortex, 118–20, 118f
Lobules, liver, 570–71
Local current flow, 102
conduction by, 80f
in graded potentials, 79

Localization of sound, 181
Localization of stimulus. *See* Coding of sensory information
Lochia, 737
Lockjaw, 99
Long bones, anatomy and growth of, 641f
Longitudinal smooth muscle of digestive tract. *See* Muscularis externa
Long-loop negative feedback in anterior pituitary hormone secretion, 638
Long-loop positive feedback in anterior pituitary hormone secretion, 638–39
Long reflex, 549
Long-term memory, 129, 134–35
    comparison with short-term memory, 130
Long-term potentiation (LTP), 133
Loop of Henle, 470
    and reabsorption
        of chloride, 495
        of sodium, 481, 493
        of water, 486, 493
    role in establishing medullary vertical osmotic gradient, 496–99
    and urea recycling, 500–502, 501f
Loudness, 178
Loudness discrimination, 185–86
Low-density lipoproteins (LDL)
    and cholesterol metabolism, 293, 295
Lower esophageal sphincter. *See* Gastro-esophageal sphincter
Lumen, 4
Luminal border, 55
Lung cancer, 441
Lung compliance, 429–30
Lung mechanics. *See* Respiratory mechanics
Lungs
    atelectasis of, 420
    elastic behavior of, 429–30
    elastin in, 5
    and emphysema, 426–27
    nonrespiratory functions of, 413
    and pneumothorax, 420
    pressure changes within during the respiratory cycle, 420–25, 425f
    removal of blood-borne prostaglandins, 413, 711
    size of in relation to size of thoracic cavity, 418–20
    stability of alveoli, 430
    structure of, 416–17
    and transmural pressure gradient, 419
    volumes and capacities of, 433–35, 433f
Luteal phase, 718–19
Luteinization, 718
Luteinizing hormone (LH), 635, 639
    in females, 719
        surge, 720, 721, 721f
        tonic secretion of, 720, 720f
    in males, 707–8, 708f

Lymph, 325
    absorption of fat into, 583, 584f
    as a component of extracellular fluid compartment, 512, 513
Lymphatic capillaries, 325, 326f
Lymphatic system, 325–27, 327f
Lymph nodes, 326, 370
Lymphocytes, 349, 356, C3
    number of, 356, 381
    origin of, 358
    overview of functions, 370
    response to antigens processed by macrophages, 391
    See also B lymphocytes; T lymphocytes
Lymphoid tissues, 356, 370, 370f
    functions of, 371
    gut-associated, 407
Lymphokines, 393, C3
Lymphoma, CNS, 519
Lymph vessels
    blockage of, and edema, 328
Lysis of infected cells by cytotoxic T cells, 393, 393f
Lysosomes, 18, 20f, 24–26, 26f, 43
    role in breakdown of APP, 133
    role in endocytosis, 68
Lysozyme, 551

MAC. *See* Membrane-attack complex
Macromolecules, 668, A7
    proteins as, A10–12
Macrophage migration-inhibition factor, 395, C3
Macrophages, C1, C3
    of alveoli. *See* Alveolar macrophages
    and cancer defense, 399, 400f
    and clot dissolution, 363
    derivation from monocytes, 358
    in destruction of erythrocytes, 573
    in inflammation, 371
    interactions
        with B cells and T cells, 390f, 397–98
        with interferon, 379
    lifespan of, 374
    of liver. *See* Kupffer cells
    processing of antigens, 391
    production of NO by, 312–13
    secretion of interleukin 1 by, 391
    summary of actions and interactions, 376–77f
    See also Phagocytes
Macula densa, 475
Macula lutea, 169–70, Ff
Magnesium ion (Mg^{++}), role in muscle contraction, 223
Magnetic resonance imaging (MRI), 110
Major histocompatibility complex (MHC), 397
Malabsorption, 578–79
Malaria, 355
Male reproductive physiology, 702–11
    accessory sex glands, 696, 710
    control of, 707–9

essential male reproductive functions, 697, 697f
    secondary sexual characteristics, 696–97
    sex act, 711–14
    and sperm production, 704–8
    structures involved, 696–97
    testicular function, 703–9
    and testosterone secretion, 696, 703–4
Male reproductive system, components of, summary, 709
Male sterility, and dynein arm or inner sheath defects, 37–38
Malignant tumor, 398
    See also Cancer
Malleus, 179, 181
Malocclusion, 551
Malpighi, Marcello, 303
Maltase, 544f, 577
Maltose, 577, 580
Mammary glands, 737
    anatomy of, 738
    See also Breasts
Mannitol, 519
Maps
    somatotopic
        of the primary motor cortex, 121, 121f
        individuality of, 122
        of the somatosensory cortex, 120, 120f
    technology for obtaining for brain, 110, 110f
    tonotopic, 186
    See also Motor homunculus; Sensory homunculus
Margination, 373
Mass action, law of, 443–44, 455, 526
Mass movements, 589
Mast cells
    and allergies, 402
    and IgE antibodies, 403f
    and immediate hypersensitivity reactions, 402
    similarity to basophils, 357
Mastication, 550
    See also Chewing
Matrix
    of bone, 642
    extracellular, 54–55
    of a mitochondrion, 26, 26f
Mature follicle, 716, 716f, 717
Mature ovum, 715
Maturity-onset diabetes, 677
Maximal oxygen consumption (max VO_2), 459
Maximal tubular capacity. *See* Tubular maximum
Maximal work load, 459
Maximum expiration, 435f
Maximum inspiration, 435f
Max VO_2. *See* Maximal oxygen consumption
Mean arterial pressure, 308
    See also Arterial blood pressure
Mechanically gated channels, 81–82

Mechanical nociceptor, 157–58
Mechanistic approach, 2
Mechanoreceptors, 152
Medial geniculate nucleus, 186
Mediated transport, 62–66, 62f
    active transport, 64–66
    facilitated diffusion, 62–64, 62f
Medulla
    adrenal. *See* Adrenal medulla
    brain, 136–37
        cardiovascular control centers in, 335
        and control of autonomic function,
            203
        respiratory centers in, 450, 452, 452f
        swallowing center in, 553
        vomiting center in, 560
    kidney. *See* Renal medulla
Medullary countercurrent system, 493
    impairment in kidney failure, 502
Medullary respiratory center, 450
Megakaryocytes, 358
Meiosis, 699, 706, 720, B11–13, B12f
    in the production of oocytes, 715
    of the secondary oocyte, 728
    of sperm, 704–5
    stages of, B13f
Meissner's plexus. *See* Submucous
        (Meissner's) plexus
Melanin, 406, 631
Melanocytes, 406
Melanocyte-stimulating hormone, 622, 631
    and pro-opiomelanocortin, 659, 662
Melatonin, and male development at
        puberty, 709
Membrane-attack complex (MAC), 379,
        380f
Membrane-bound enzymes, 49
Membrane carbohydrates, 47–48, 50
Membrane junctions, 55, 55f
Membrane lipids, 46
Membrane permeability, 57
Membrane potential, 45, 45f, 68–102, 70f,
        514
    action potentials, 80–83
    changes in, during an action potential,
        81f
    definition of, 68
    end-plate potential, 207
    equilibrium potential, 71, 72
    excitatory postsynaptic potential, 92, 93f
    factors responsible for establishing and
        maintaining, 70–74
    graded potentials, 79–80, 80f, 86
    grand postsynaptic potential, 94–95
    inhibitory postsynaptic potential, 92, 93f
    measurement of, 68–69
    pacemaker activity, 268–69
    receptor potential, 152–53
    resting, 72–73, 71f, 73f
    role of Na$^+$-K$^+$ pump in, 71, 83
    slow-wave potential, 548
    specialized use of by nerve and muscle
        cells, 74, 79

threshold potential, 81
    types of changes in, 80f
    units of measure of (mV), 68–69
Membrane proteins, 47
Membranes
    diffusion through, 58f
    fluidity of, 46–47
    permeability of, 56–59
        relationship to surface area and thick-
            ness, 58
    protein structure of, 48
    structure of, 75
    transport across. *See* Membrane trans-
        port; Plasma membrane
Membrane transport, 56–68, 75–76
    active, 64
    carrier-mediated, 62–66, 62f
    by osmosis, 59
    summary of mechanisms of, 69
Memory, 129–35
    effect of MSH on (proposed), 631
Memory cells, 387f, C3
    in the immune system, 386
    T cells, 393
Memory consolidation, 131
Memory storage, 130f
Memory trace, 129, 131
Menarche, 724
Meniere's disease, 190
Meninges, 112, 112f
Meningiomas, 111, 112
Meningitis, 117
Menopause, 714, 723–24
Menstrual cramps, 722
Menstrual cycle, 628, 714
    and body temperature, 603
    irregularities of in athletes, 724
    phases of, 722–23
    uterine changes during, 722–23
Menstrual flow, 722–23
Menstrual phase, menstruation, 722
Meperidine (MPPP), a synthetic heroin, 100
Mesangial cells, 478
Mesentery, 547
Messenger-gated channels, 91
Messenger RNA (mRNA), B7
    induction by hormones, 626
Messengers, chemical. *See* Chemical
        messengers
Metabolic fuel storage, 670–71
Metabolic functions of the kidneys, 506
Metabolic (nonrespiratory) acidosis, 535f,
        536
    as a consequence of kidney failure, 502
    in diabetes mellitus, 676
Metabolic (nonrespiratory) alkalosis, 535f,
        537
    as a consequence of vomiting, 560
Metabolic rate, 597
    effects on
        of epinephrine, 664
        of thyroid hormone, 652, 691
Metabolic water, 521

Metabolism
    cellular, 2
    compensatory changes in to maintain
        energy balance, 598
    definition of, 667–68
    effect of epinephrine on, 664
    of hormones, 628
    in skeletal muscle, 234–41
    water formation in the electron-transport
        chain, 30–31, 33, 33f
    See also Intermediary metabolism; Meta-
        bolic rate
Metaphase, B11
Metarterioles, 319, 321f
Metastasis, 398
Methane, A6
    combustion of, A4
Methyl mercaptan, human ability to detect
        by odor, 193
MHC glycoproteins, 397–98, 398f
Micelles, 572–73, 573f, 574f
    and fat absorption, 583
    formation by bile salts, 572
Microbes. *See* Bacteria; Pathogens; Viruses
Microcirculation, 302
Microfilaments, 38–40, 41f
    actin in, 39
Microglia, 109, 111
    functions of, 111, 111f
Microorganisms. *See* Bacteria; Pathogens;
        Viruses
Microtrabecular lattice, 41–42, 41f
Microtubules, 35–38, 38f, 41f
Microvilli (brush border), 19, 39–40, 40f
    in hair cells of inner ear, 40
    in small intestine, 40, 40f, 578, 579f
    in taste buds, 190
Micturition (urination), 505, 505f
Midbrain, 136
Middle ear, 178f, 181–82, 182f
    components of, 179
Middle ear bones. *See* Ossicles
Midpiece of a sperm, 706
Migrating motility complex (intestinal
        housekeeper), 576
Milk, composition of, 739
Milk ejection, 738–39
Milk production. *See* Lactation
Milk sugar. *See* Lactose
Millivolt (mV), 68–69
Mineralocorticoids, 657
    See also Aldosterone
Minute ventilation, 435–36
MIT. *See* Monoiodotyrosine
Mitochondria, 18, 20f, 18, 26–27, 26f, 41f, 43
    of cardiac-muscle cells, 267
    citric-acid cycle reactions in, 27–28
    in digestive exocrine glands, 543
    electron-transport chain in, 29–32, 33
    energy yield compared with burning,
        32–33f
    evolution of, proposed mechanism, 26
    functions of, 26–27

Mitochondria, (continued)
  in oxidative muscle fibers, 237
  oxidative phosphorylation pathway in,
      32f
  production at high altitudes, 451
  in skeletal muscle, 216
  in spermatozoa, 706
  structure of, 26–27, 26f
Mitosis, 38, B10–11, B12f
  stages of, B13f
Mitotic proliferation of sperm, 704
Mitotic spindle, formation of, 38
Mitral valve, 264
  See also Left (AV) atrioventricular valve
Mixing movements. *See* Motility
M line, 216
Modalities of stimuli, 152
Molality, definition of, A5
Molarity, definition of, A5
Molecular masses, A5
Molecular weight, A5
  and rate of diffusion, 58
Mole (mol), A5
Monoacylcerols. *See* Monoglycerides
Monocytes, 356, 358, 370, 373, C3
  See also Macrophages
Monoglycerides (monoacylglycerols), 543,
      544f, 583
Monoiodotyrosine (MIT), 651
Mononuclear agranulocytes, 356
Monosaccharides, 543, A8
  as absorbable units of carbohydrate
      digestion, 544f
Monosynaptic reflex, 145
  See also Stretch reflex
"Moon-face," 660
Morning-after pill, 741
Morning sickness, 733
Morphine, 158
  analgesic properties of, 158
  pain relief by
    comparison with acupuncture and
        placebos, 160
Morula, 728
Motilin, 576
Motility (digestive tract), 542–43
  control of, general features, 547–49
  of pharynx and esophagus, 554–55, 554f
  of large intestine, 587
  mixing movements, 542
  propulsive movements, 542
  of small intestine, 575–76
  of stomach, 556–58, 556f
  summary of, 546–47
Motion sickness, 190
Motivation, definition of, 128
Motor activity, 121–22
  control of, 240–46, 242f
    relationship among the thalamus, basal
        nuclei, and cortex in, 126f
    role of the cerebellum in, 136
    spinal cord role in coordination of,
        143–44
Motor cortex. *See* Premotor cortex; Pri-

mary motor cortex; Supplemen-
    tary motor area
Motor end-plate, 206
Motor homunculus, 120
Motor neurons, 106, 107
  of external urethral sphincter, 505
  innervation of skeletal muscles by, 97
  and motor units, 226
  and neuromuscular junction, 205
  output controlled by, 240
  and voluntary control over ventilation,
      458–59
  See also Alpha motor neurons; Gamma
      motor neurons
Motor program for voluntary tasks, 121
Motor unit, 226, 227f
Motor-unit recruitment, 226–227
Mountain sickness, acute, 451
Mouth, 550–53
Movement in digestive tract. *See* Motility
mRNA. *See* Messenger RNA
MSH. *See* melanocyte-stimulating hormone
Mucosa, 545
  turnover rate
    in the small intestine, 579–80
    in the stomach, 566
Mucous membrane, 545
Mucous neck cells (in the stomach), 560
Mucus
  cervical, 723
  in the digestive tract, 551, 560, 563
  esophageal, 555
  in the genitourinary system, 408
  in the olfactory mucosa, 192
  in the reproductive tract, 710, 714
  in the respiratory airways, 408, 426
Mucus escalator, 408
Müllerian ducts, 702
Müllerian inhibiting substance, 702
Multicellular organisms, 2
Multineuronal system, 242
Multiple sclerosis, 88–89
Multiunit smooth muscle, 247–48
  See also Smooth muscle
Murmur, heart, 281
Muscarinic receptors, 203
Muscle, 213–54
  functions of, 213, 214
  glucose utilization by, 64
  storage of glycogen by, 35
  types of, 3, 248–49
  See also Cardiac muscle; Muscle fiber;
      Skeletal muscle; Smooth muscle
Muscle, cardiac. *See* Cardiac muscle
Muscle cells
  as excitable tissues, 79
  specialization of, 101
Muscle contractile systems, actin in, 39
Muscle fatigue
  in skeletal muscle
    causes of, 236
    comparison of in different fiber types,
        237–38
  in smooth muscle, 254

Muscle fibers, 205–6, 215, 234–36
  adaptation to demands placed on them,
      238
  cardiac muscle, 267
  and motor unit, 226
  relation to motor neuron at neuromuscu-
      lar junction, 205–7, 207f, 208f
  skeletal muscle, 215–16, 215f
  smooth muscle, 248, 250f
  types of, 237–38
Muscle protein, 652–53
  effect on
    of cortisol, 660
    of diabetes mellitus, 676
    of growth hormone, 645
  and fuel metabolism, 670
Muscle, skeletal. *See* Skeletal muscle
Muscle, smooth. *See* Smooth muscle
Muscle spindles, 155, 243–46, 246f
  function of, 245f
Muscle stretch receptors, 155
  See also Golgi tendon organs; Muscle
      spindles
Muscle tension
  determinants of whole muscle tension,
      226
  effect on
    of frequency of stimulation, 227–29,
        228f
    of motor-unit recruitment, 226–27
    of muscle length at onset of contrac-
        tion, 229–30
  relation to cross-bridge activity, 230
  relation to load in isotonic and isometric
      contractions, 231–32
  relation to series-elastic component, 230
  See also Contraction; Muscle spindles;
      Muscle tone; Skeletal muscle
Muscle tissue, 4
  percentage of body comprised of, 214
Muscle tone
  effect of on metabolic rate and heat pro-
      duction, 606
  inhibition of by basal nuclei, 125
  in paradoxical sleep, 138
  regulation of by spinocerebellum, 136
  in slow-wave sleep, 138
Muscular dystrophy, 240, 241
Muscularis externa, 547
Muscularis mucosa, 545
Muscular system, 7–8, 8f
  See also Cardiac muscle; Skeletal mus-
      cle; Smooth muscle
Mutagens, B14
Mutations, B13–14
mV, 68–69
Myasthenia gravis, 54, 209–10, 396
Myelin, 86
  formation from oligodendrocytes, 86,
      111
  formation from Schwann cells, 86
  role in action potential propagation,
      86–88
  role in multiple sclerosis, 88–89

Myelinated fibers, 86, 87f
Myenteric (Auerbach's) plexus, 548–549
Myoblasts, 240
    transplantation of, 241
Myocardial damage, 292f
Myocardial infarction, 278
    acute, outcomes of, 293
Myocardial ischemia, 278
Myocardial toxic factor, 343
Myocardium, 267
Myoepithelial cells, 738
Myofibrils, 215f, 216, 216f
Myogenic activity
    of arterioles, 313
    in cardiac muscle, 254, 267
    in single-unit smooth muscle, 250
Myogenic mechanism of renal autoregula-
        tion, 475
Myoglobin, 235–36
    in heart muscle, 267
    in oxidative muscle fibers, 237
Myometrium, 722
    role in parturition, 735–36
    role in sperm transport, 726
Myopia, 168, 168f
Myosin, 39
    activation of, in skeletal muscle, 219–
        25
    activation of, in smooth muscle, 251
    in platelets, 358
    structure of, 217–18
    See also Thick filaments
Myosin ATPase site, 216, 216f
Myosin cross-bridge, 223f
Myosin filaments
    in smooth muscle, 247, 251f
Myosin kinase, role in smooth muscle
        activity, 247
Myxedema, 654
Müllerian ducts, 702
Müllerian inhibiting substance, 702

NaCl. See Salt
NAD/NADH+, 28
Na+ equilibrium potential, 72, 72f
Nails, 406
Na+-K+ ATPase carriers, synthesis of, 482
Na+-K+ ATPase pump (Na+-K+ pump),
        65–66, 66f
    effect on membrane potential, 71
    functions of, 65
    and ion distribution in a resting nerve
        cell, 73
    mechanism of, 65–66
    role of
        in concentration gradient mainte-
            nance by, 83
        in maintenance of a resting mem-
            brane potential by, 74, 74f
        in potassium secretion by kidneys,
            488–89
        in secondary active transport, 484
        in sodium reabsorption by kidneys,
            481, 481f

Narcolepsy, 139
Nasal passages (nose), 414
National Institute of Aging, 132
National Institute of Drug Abuse, 239
Natriuretic hormone. See Atrial natriuretic
        peptide
Natural contraception, 740
Natural immunity, 388–91
Natural killer (NK) cells, 371, 379, C3
    and cancer defense, 399, 400f
    and exercise, 401
Nausea, at high altitudes, 451
Nearsightedness. See Myopia
Necrosis in acute myocardial infarction,
        278
Negative balance
    of energy, 598
    general definition, 512
    of water. See Dehydration
Negative feedback, 10, 11, 11f
    and blood glucose, 679
    in control of cortisol, 659
    in control of gonadotropin secretion by
        the anterior pituitary, 704
    in control of parathyroid hormone and
        calcitonin secretion, 687
    in control of thyroid hormone, 653–54
    definition of, 10, 11, 11f
    and regulation of homeostasis by the
        endocrine system, 691
    in hormone systems, 626–27, 626f
    in hypothalamic-anterior pituitary con-
        trol systems, 638f
    in insulin secretion, 631
    in pancreatic β cells, 674
    in regulation of growth hormone, 642
    in the renin-angiotensin-aldosterone sys-
        tem, 482
    in spermatogenesis, 707
    by suppressor T cells, 395
    by testosterone, role in spermatogenesis,
        708
    in tonic LH secretion, 720
    in vasopressin release, 497
Neostigmine in treatment for myasthenia
        gravis, 210
Nephrons
    anatomy of, 466–70, 468f, 470f
    juxtamedullary versus cortical, 469f
    summary of transport functions of, 490
    See also Glomerular filtration; Medullary
        countercurrent system; Tubular
        reabsorption; Tubular secretion;
        Urine excretion
Nernst equation, and equilibrium mem-
        brane potential, 72
Nerve
    definition of, 143
    See also Cranial nerves; Spinal nerves
Nerve cells
    as excitable tissues, 79
    relation to muscle fiber at neuromuscu-
        lar junction, 206
    resting, permeability of, 71

resting potential of, 81
role of
    in homeostasis, 101
Nerve deafness (See Sensorineural
        deafness)
Nerve fibers, 83
    bundles of, in spinal cord tracts, 139–
        40
Nerve growth factor, 618
Nerve plexus. See Intrinsic nerve plexus
Nervousness, and dry mouth, 521
Nervous system
    classes of neurons, 107–9, 109f
    comparison with endocrine system,
        105–9
    components and functions of, 8f
    development of, 117
    of the digestive tract, 548
    divisions of, 107, 108f
    effect of MSH on (proposed), 631
    effect on
        of androgens, 646
        of diabetes mellitus, 676–77
        of insulin excess, 677–78
        of prostaglandins, 711
        of thyroid hormone, 653
    links to endocrine system, 107, 627
    links to the immune system, 399–401
    metabolism during starvation, 671
    organization of, 107–9, 109f
    role of sleep during development of,
        139
    somatic, 107, 204–5
    specificity of, 106
    See also Central nervous system; Periph-
        eral nervous system
Nervous tissue, 4
Net diffusion, 57
Net filtration pressure
    across capillary walls, 324, 325f
    across glomerular membrane, 474
Neural input, and hormonal secretion, 627,
        627f
Neuritic plaques (senile plaques), 132
Neuroendocrine reflexes, 627–28
Neuroendocrinology, 107
Neurofibrillary tangles, 132
Neurofilaments, 41
Neurogenic hypertension, 338
Neurogenic activity of muscles, 247–48
Neurogenic shock, 341
Neuroglia, 109, 114
Neurohormones, 616
Neurohypophysis. See also Posterior
        pituitary
Neuromodulation/neuromodulators,
        96–98, 98–99, 616
Neuromuscular fatigue, 236
Neuromuscular junction, 205–10
    comparison with synapse, 208
    events at, 206–7, 208f
    role of acetylcholinesterase, 207
    vulnerability of, 209–10
Neuronal integration, 102

Neurons
  alpha motor neurons, 204–5
  brain, 114f
  changes with memory, 133–34
  classes of, 107–9, 206
  definition of, 83
  developmental association with endo-
      crine cells, 549
  infection by viruses, 394
  and neuromodulators, 98
  relation to nerves, 141
  structure of, 83, 84f
  structures terminating on, 91
  See also Afferent neurons; Interneurons;
      Motor neurons; Postganglionic fi-
      ber; Preganglionic fiber; Synapse
Neuropeptides, 94, 96–98, 102
  neuropeptide Y, 204
Neurosecretory neurons, 616, 636
Neurotransmitters, 91, 93–94, 94, 96–98,
      98, 616
  of the autonomic nervous system,
      198–200
  as chemical messengers, 105
  classical, 94, 94, 97, 102
  comparison with hormones, 106–7
  function at synapses, 91–93
  general description of, 91, 616
  leukocyte receptors for, 400
  at neuromuscular junction, 206
  nitric oxide as, 313
  removal from synapse, mechanism of, 94
  and satiety, 600
  See also Acetylcholine; Norepinephrine
Neutral energy balance, 598
Neutralization in immune defense, 383,
      384f
Neutralization reactions, A7
Neutrons, definition of, A1–2
Neutrophilia, 357
Neutrophils, 357, 369, C3
  relationship with the kinin system, 375
  See also Phagocytes
Neuwelt, Edward, 519
Newborn respiratory distress syndrome,
      430–32
Nicotinamide adenine dinucleotide
      (NAD), 28
Nicotine in brain, 519
Nicotinic receptors, 203
Night blindness, 174
Nipple, 737–38
Nitric oxide (NO)
  and blood pressure regulation, 338
  and erection, 712
  and maintenance of arteriole pressure,
      312–13
  metabolism of nitroglycerin to, 292
  as a retrograde messenger in memory
      formation, 134
Nitrogen, 438
Nitrogen narcosis, 451
Nitroglycerin as a vasodilator, 292
NK cells. *See* Natural killer (NK) cells

Nociceptors, 152, 157–58
Nodes of Ranvier, 86, 87f
Nonelectrolytes, definition of, A5
Nonhistone proteins, B3–4
Noninsulin-dependent diabetes, 677
Nonpolar molecule, 46, 57, A4
Nonrespiratory acidosis. *See* Metabolic
      (nonrespiratory) acidosis
Nonrespiratory alkalosis. *See* Metabolic
      (nonrespiratory) alkalosis
Nonshivering (chemical) thermogenesis,
      607
Nonspecific immune responses, 371
  to bacterial invasion, summary of, 391
  enhancement of by antibodies, 384f
  role of
      complement system, 379–80
      inflammation, 371–75
      interferon, 378–79
      natural killer cells, 379
Nontropic hormones, 617
Noradrenaline. *See* Norepinephrine
Norepinephrine
  adrenergic receptor specificity of, 202f
  as an adrenomedullary hormone, 663
  effect of
      on arterioles, 315
      in heart contractility changes, 287
      on heart rate, 285
  neurotransmitter in the autonomic ner-
      vous system, 198
  neurotransmitter in the sympathetic ner-
      vous system, 200
  and pain transmission, 160
  possible role in depression, 128–29
  receptors for, 203
  release of, and calcium ion changes in
      aging, 288
  release of, in initial response to heart
      failure, 288
  secretion by adrenal medulla, 663
  secretion in pheochromocytoma, 338
  See also Sympathetic nervous system
Normality of a solution, definition of, A6
Nuclear division of cells, B11
Nuclear pores, 34, B7
Nuclei in brain, definition, 125
Nuclei, multiple, in muscle cells, 216
Nucleic acids, A12
Nucleotides, A12
  in DNA, B1–2, B1
Nucleus of cells, 17, 18, 20f
  See also Deoxyribonucleic acid
Nursing. *See* Lactation
Nutrient molecules, major energy produc-
      tion pathways, 672f
Nutrients, homeostasis controlled by the
      endocrine system, 646
Nutritional anemia, 354
Nutritional requirement, during pregnancy,
      734
Nutritional status, and the effects of growth
      hormone, 642
Nystagmus, 136

Oat bran, and cholesterol levels, 295
Obesity, 600–601
  in diabetes mellitus, 677
  growth hormone therapy for, 644
Obstructive jaundice, 574
Obstructive lung disease, 434–35
  See also Chronic obstructive pulmonary
      disease
Occipital lobes, 118, 118f, 119
Occlusion, 551
Off-center ganglion cell, 175, 175f
Off response, 153–54
O_2-Hb dissociation (saturation) curve, 444,
      444f
Olfaction, 190, 192–93
Olfactory bulb, 192–93, 192f
Olfactory cortex. *See* Primary olfactory
      cortex
Olfactory mucosa, 192, 192f
Olfactory nerve, 192
Olfactory pathways, 193, 193f
Oligodendrocytes, 86, 87f, 109
  functions of, 111, 111f
  myelination by, 89
On-center ganglion cell, 175, 175f
Oncogenes, B14
Oncotic pressure, 323–24
  See Colloid osmotic pressure
Oocyte. *See* Oogenesis; Primary oocyte;
      Secondary oocyte
Oocyte maturation-inhibiting substance,
      720
Oogenesis, 697, 715f
  chromosome division in, 714–16, 716f
  comparison with spermatogenesis, 714,
      715–16
  relation to ovarian cycle, 716–17
Oogonia, 714
Opiate receptors, 158
Opsin, 170
Opsonins, 374, 374f, C4
  antibodies as, 384
  complement as, 380
  IgG as, 383
Optic chiasm, 176, 176f
Optic disc, 169
Optic nerve, 169, 176f
Optic radiations, 176f
Optic tract, 176–77f
Optimal length (l_o) of muscle, 229
Oral cavity, 550–53
Oral contraceptives, 741
Oral hypoglycemic drugs, 677
Origin, of muscle, 230
Organelles, 17, 18–34, 43
  See also Endoplasmic reticulum; Golgi
      complex; Lysosomes; Mitochon-
      dria; Peroxisomes; Vaults
Organic acids, generated during metabo-
      lism, 526
Organic chemicals, 6–7
  functional groups of, A7–8
Organic ions, secretion by kidneys, 489–90
Organic metabolism. *See* Fuel metabolism

Organ of Corti, 179, 182–84, 182f, 183f, 186, 186f
Organophosphates, inhibition of acetylcholinesterase, 209–10
Organs, 5
Organ systems. See Body, systems in
Orgasm, 713
Orgasmic platform, 714
Oropharyngeal stage of swallowing, 553, 554f
Orthostatic hypotension, 339–40
Osmolarity (Osm), 464
    definition of, 517, A6
    ECF, 537
    effect of on arteriolar radius, 312
    and gastric emptying, 558
    need for regulating, 514
    renal regulation of, 466, 492, 492f
Osmoreceptors, 152
    digestive tract, 549
    hypothalamic, 521–22
Osmosis, 59–62, 62f
    and bulk flow across capillaries, 323
    to maintain ICF-ECF water balance, 518
    and water exchange between extracellular and intracellular fluids, 514, 515
    and water reabsorption in kidney tubules, 486–87
Osmotic diuresis, 499
    and diabetes mellitus, 675
Osmotic effects, net transfer of water across cellular plasma membranes, 514
Osmotic force due to plasma proteins, across glomerular membrane, 474
Osmotic gradient, 349
    and concentration of urine, 499
Osmotic pressure, 60–61, 61f
Ossicles, 181–82
Ossification in bone matrix, 641–42
Osteoblasts, 640, 685
    effect on
        of growth hormone, 642
        of parathyroid hormone, 686
Osteoclasts, 640
    effect on
        of parathyroid hormone, 686
Osteocytes, 642, 685
Osteocytic-osteoblastic bone membrane, 685, 685f
Osteolysis, 685
Osteomalacia, 690
Osteons, 684f, 685
Osteoporosis, 690–91
Otoconia. See Otoliths
Otolith organs, 187, 187f, 188
Otoliths, 188
Outer longitudinal smooth muscle layer (of digestive tract), 547
Oval window of the ear, 179, 181–82, 184
Ovarian follicles
    control of, 719–20, 720f
    conversion to corpus luteum, 718–19, 171f

development of, 716
    estrogen production by, 719, 719f
    and menopause, 723
Ovarian hormones. See Estrogen; Progesterone
Ovaries, 696
    control of, 719–22
    cyclical activity of, 718f, 720–22
    functions of, 696, 714
    hormones produced by, 719–22
    and menopause, 723
    and oogenesis, 714–16, 714f
    and puberty, 723
Overeating, effects of on body, 610
Overhydration, 520
Oviducts, 697
    cilia in, 37
    and fertilization, 725–28, 725f
    and transport of fertilized ovum, 728
Ovulation, 717
    and body temperature, 741
    control of, 720–21
    inhibited by nursing, 740
    and oral contraceptives, 741
Ovum, 696, 715, 715f
    development of, 714–16
    fertilization of, 727–28
    transport to the oviduct, 725, 725f
Oxidative enzymes, 26
Oxidative phosphorylation, 32, 43
    comparison with burning, 32, 32f
    in skeletal muscle, 234–36
Oxidative reactions
    within mitochondria, 27–28
    within peroxisomes, 26
Oxygen
    abnormalities of, 449
    consumption in cells, 2
    content of in blood, 442–43
    control of ventilation by, 453–54, 454f
    diffusion coefficient of, 441
    effect of
        on hemoglobin's affinity for CO_2, 448
        on pulmonary arterioles, 429
        on red blood cell production, 351, 352f
        on systemic arterioles, 312, 429
    energy equivalent of, 598
    exchange of
        across the placenta, 732
        between alveoli and blood, 413, 438–39
        between blood and tissues, 442
    heart's utilization of, 267
    need for
        brain, 115
        cooled tissues, 601
        heart, 290–91, 290f
    net diffusion gradients for between lungs and tissues, 442, 443f
    partial pressure, 438–39
        effects on percent hemoglobin saturation, 443, 444–45
        and exercise, 457

and gradient across pulmonary capillaries, 439
    and gradient across systemic capillaries, 442
    influence of hemoglobin on, 444–45
    and respiratory quotient, 413
    solubility of, 441, 442
    transport by erythrocytes, 349–50
    transport in blood, 442–43
    use of, in oxidative phosphorylation, 32–33, 32f, 33f
Oxygen debt, 236–37
Oxygen toxicity, 449
    in deep-sea diving, 451
Oxyhemoglobin (HbO_2), 443, 444
Oxyntic cells (parietal cells), 560
Oxyntic mucosa, 555f, 560
    atrophy of, 563
Oxytocin, 633
    control of, 734–36, 739
    functions of, 633, 734–35, 763f, 739
    and milk ejection, 739
    and parturition, 734–35
    positive feedback in release of, 12
    and relation of hypothalamus and posterior pituitary, 632–33, 632f
P_{CO_2}. See Carbon dioxide, partial pressure of; Partial pressure gradient
P K^+, 82
    See also Potassium, membrane permeability to
P Na^+, 82
    See also Sodium, membrane permeability to
P_{O_2}. See Oxygen, partial pressure of; Partial pressure gradient
Pacemaker of the heart, 269–71, 271f
Pacemaker (pacesetter) activity, 249–50, 253
    artificial pacemaker, 271
    in cardiac muscle, 254
    of heart autorhythmic cells, 268, 268f
    of inspiratory neurons, 452
    potential of digestive tract smooth muscle, 547–48
    of small intestine, 575
    in the stomach, 556
Pacinian corpuscle, 154, 154f
Packaging
    of DNA, B3–4
    of sperm, 705–6
    See also Spermiogenesis
Packed cell volume, 348
    See Hematocrit
PAH. See Para-aminohippuric acid
Pain, 157–61
    effect of, on ventilation, 458
    of inflammation, 374
    phantom, 155
    relief of, studies of acupuncture, 160
Pain pathway, 159f
Pain receptors, 152
Palate, 550
Palpitations in hyperthyroidism, 655

Pancreas
  defect in cystic fibrosis, 49
  digestive secretions produced by,
      567–71
  hormones produced by, 671–79
  secretion of bicarbonate by, 558
  See also Endocrine pancreas; Exocrine
      pancreas; Glucagon; Insulin; Pan-
      creatic enzymes
Pancreatic aqueous sodium bicarbonate
    secretion, 558
Pancreatic enzymes
  amylase, 568, 569
  chymotrysinogen, 568–69
  lipase, 568, 569, 572, 577, 583
  procarboxypeptidase, 568–69
  trypsinogen, 568–69
Pancreatic polypeptide, 671–72
Pancreatic somatostatin, 671
Pantothenic acid, 28
Papillary muscles, 264
Para-aminohippuric acid (PAH), 491–92
Paracrines, 616
Paradoxical sleep (REM sleep), 138
  requirement for, 139
  versus slow-wave sleep, 138
Paralysis, 243
Paraplegia, 243
Paraquat, potential nerve damage from,
    100
Parasympathetic dominance, 200
Parasympathetic nervous system
  comparison with sympathetic nervous
      system, 198
  effect of
    on airway resistance, 425
    on the arterial blood pressure, 336f
    on bile secretion, 574
    on blood vessels supplying penis and
        clitoris, 315, 712, 713
    on the digestive system (vagus nerve),
        549
    on esophageal peristalsis, 555
    on the eye, 163
    on gastric secretion, 564
    on the heart, 283–84
    on insulin secretion, 675
    on salivary secretion, 552
    on the urinary bladder, 505
  general effects of, 200
  neurotransmitters of, 198
  role of
    in airway regulation, 425
    in control of near vision, 167
    in erection, 712
  vagus nerve of, 137
Parasympathetic tone, 200
Parathyroid glands, 5, 649, 685
  hormones, major functions, and target
      cells of hormones, 620
Parathyroid hormone (PTH), 683–86,
    686–687
  and homeostasis in the endocrine sys-
      tem, 692

Paraventricular nucleus, 632, 632f
Parietal lobes, 118, 118f, 119–20
Parietal (oxyntic) cells, 560
Parietal pleura, 417f
Parietal-temporal-occipital association cor-
    tex, 123
Parkinson's disease, 99, 100, 126
Parotid glands, 551
Partial pressure, 438, 438f
Partial pressure gradients, 438, 440f
  across pulmonary capillaries, 439
  across systemic capillaries, 440f
Particle size, and membrane permeability,
    57
Parturition (labor, delivery, birth), 697,
    734–37, 737f
  feedback among hormones during, 736f
Passive forces, 57
Passive immunity, 387–88
  versus active immunity, 389
  cell-mediated, 393
Passive reabsorption, 481–82
Passive transport
  diffusion down concentration gradient,
      57–58
  movement along electrical gradient, 59
  osmosis, 59–62
Pasteur, Louis, 388
Patellar tendon reflex, 244, 246f
Pathogens, leukocyte attack on, 356
Pathophysiology, 12
Pelvic diaphragm, 504–5
Penicillin
  allergic reaction to, 404
  as a hapten, 381
  urinary excretion of, 490
Penis, 697
  development of, 702
  erection and ejaculation of, 711
Pepsin, 544f, 563
Pepsinogen, 560, 562–63, 563f
Peptic ulcer, 565
Peptide bonds, A10–11
Peptide hormones, 618, 622
  definition of, 618
  endocytosis of, 628
  hormones included in this class, 618
  inactivation of, 628
  mechanism of action, 625
  mechanism of synthesis, storage, and
      secretion, 622
  receptor location, 625
  postreceptor events of, 625
  transport of, 623–24
  summary of, 621
Peptides, digestion and absorption of, 582
Percent hemoglobin (% Hb) saturation,
    443–44, 444f
Perception, 151, 152
Perforin, 393, 394, 400f
Periaqueductal gray matter, 158
Pericardial fluid, 268, 513
Pericardial sac, 268, 268f
Pericarditis, 268

Perilymph
  of cochlea, 183, 183f, 184
  of vestibular apparatus, 188
Perineal region, 697
Periosteum, 640
Peripheral axon (afferent fiber), 108
Peripheral chemoreceptors, 453–54, 453f
  See also Aortic bodies; Carotid bodies
Peripheral nervous system (PNS), 107, 141
  See also Afferent division; Autonomic
      nervous system; Efferent division;
      Somatic nervous system
Peripheral thermoreceptors, 606
Peristalsis
  due to slow-wave potentials, 548
  esophageal, 554, 554f
  in small intestine, 576
  in the stomach, 556, 556f, 558–59
  in ureters, 503
Peritoneal fluid, 513
Peritubular capillaries, 467
  See also Tubular reabsorption; Tubular
      secretion
Permeability
  alteration by hormones, 624
  of capillaries, active regulation of, 319
  of capillary walls in edema, 327–28
  changes during an action potential, 83f
  definition of, 57
  and Fick's law, 58
  of membranes, 70–74
    in resting nerve cells, 71, 73
    at the threshold potential, 82
Permissiveness
  of cortisol, 658
  in hormone secretion, 631
  of thyroid hormone on catecholamine
      actions, 653
Pernicious anemia, 354, 563
Peroxisomes, 18, 20f, 26, 43
Peyer's patches, 370
PF3, 361
PGA. *See* Pyloric gland area
PGA/PGE/PGF. *See* Prostaglandins
pH
  in acidosis and alkalosis, 525, 525f
  buffering in blood, 365
  of cells, and carbon dioxide levels, 460
  constant, in blood, 349
  definition of, 524–25
  effect of buffering on, 528
  and Henderson-Hasselbalch equation,
      538
  and hydrogen ion concentration,
      524–25
  of normal body fluids, 525f
  of pure water, 525
  range compatible with life, 525
  regulation of, 527–34
  of the stomach, 545
  See also Hydrogen ions (H^+)
Phagocytes
  in brain, 111
  in inflammation, 373

life span of, 374
mesangial cells as, 478
microglia as, 111
role in inflammation, 376f
secretions of, 374–75
See also Macrophages; Neutrophils
Phagocytic cells in lymph nodes, 326
Phagocytic secretions, C4
Phagocytosis, 68
enhancement
by antibodies, 383, 384f
by complement, 380
by opsonins, 374, 374f
in inflammation, 373–74
of peripheral axons, damaged, 89
role in thyroid hormone secretion, 652
of an unfertilized ovum, 725
Phantom pain, 155
Pharyngoesophageal sphincter, 553–54
Pharynx
connection to middle ear, 181
and the digestive system, 550, 553
and the respiratory system, 414
Phasic receptors, 153–54, 154f
Phenol, filtering as waste, 488
Phenotypic sex, 701
Pheochromocytoma, 338, 665
Phosphate
distribution of in body fluid compartments, 514
effect of on vitamin D activation, 687, 688f
effect on
of calcitonin, 687
of parathyroid hormone, 685, 686, 687–88
regulation of, 690f
by kidney, 687–88
by vitamin D, 687
relation to calcium, 681, 687–88
renal handling of, 486
as a urinary buffer, 533–34
Phosphate buffer system as a urinary buffer, 529
Phosphate imbalances, and renal failure, 502
Phospholipase C, 52
Phospholipids, 46–47, 47f, A9
in a lipid bilayer, 47f
platelet factor PF3, 361
Phosphorylation, 52
in active transport mechanisms, 64–65
of myosin in smooth muscle, 247
Photons, 163
Photopigments, 170, 173
Photoreceptors, 152, 169, 170–73, 173f
Photosynthesis, 542
Phototransduction, 170–73
and action potential, visual pathway, 172f
Phrenic nerve, 421, 425
Physiological dead space. *See* Dead space
Physiology, definition of, 2
Pia mater, 112f, 113

PIH. *See* Prolactin-inhibiting hormone
Pineal gland
hormones, major functions, and target cells of hormones, 620
and male development at puberty, 709
Pinna, 179, 181
Pinocytosis, 68
Pitch discrimination, 184
Pitch of sound waves, 178, 180f
Pituicytes, 633
Pituitary gland (hypophysis), 631–38
anatomy of, 632f
anterior, 633–39, 639f
and the hypothalamus, 631–39
posterior, 632–33, 632f
See also Anterior pituitary; Posterior pituitary
Placebo effect, 160
Placenta, 697, 730–34
as afterbirth, 736–37
and foreign chemicals, 731–32, 731f
hormones produced by, 620, 732–34
Placental biopsy, 726
Placental hormones
rate of secretion, 733f
summary, 732
Placental villi, 730
Plaque (in atherosclerosis), 291, 292
Plasma, 5, 348–49, 513f
and hematocrit, 348
percentage of body water, 513
similarities to interstitial fluid, 513
as source of raw materials for digestive secretions, 586
summary of constituents' functions, 347
water in, 511
Plasma cells, 381–82, 381f, 385–86, C4
Plasma clearance, 490–506
definition of, 490–91
Plasma-colloid osmotic pressure, 323–24, 473
in edema, 327
Plasma membrane, 17, 20f, 18, 41f, 45–68
appearance, 46
composition and functions, 46–50
definition of, 45
membrane permeability, 57
membrane potential, 68–74
membrane transport, 56–68
See also Membrane
Plasma osmolarity, and heat, 633
Plasma proteins, 349, 518
binding
of peptide hormones to, 623–24
of thyroid hormone to, 652
in blood coagulation, 360–64
and blood viscosity, 304
buffering role of, 349
and bulk flow across capillaries, 323–24, 327
and glomerular filtration, 474
and hormone transport in the blood, 623
in inflammatory response, 371–73
loss of in kidney disease, 502

synthesis in response to hemorrhage, 343
Plasma sol, 39
Plasma volume
effect on
of dehydration, 518
of diarrhea, 586–87
of heat stroke, 610
of vomiting, 560
and exercise in heat, 517, 633
maintenance by the kidneys, 466
regulation of
by control of glomerular filtration rate, 477
by factors controlling ECF volume, 514–16
related to
extracellular fluid volume, 514–16
sodium load, 481–82, 515
and sweating, 517
See also Extracellular fluid, volume of
Plasmin, C4
fibrinolysis of clots by, 363–64
Plasminogen, 363–64
Plasticity
of the brain, 116, 122
of smooth muscle, stomach, 556
Platelet aggregation, 359–360, 359f, 363f
Platelet factor (PF3), 361
Platelet plug, 358, 358f
Platelets, 347, 359–60, 365
aggregation of in hemostasis, 359–60
origin and structure of, 358
role in blood coagulation, 361, 363f
Pleural cavity, 417
Pleural fluid, 513
Pleural sac, 417, 417f
Pleurisy, 417
Pluripotential stem cells, 351
Pneumonia, 441
Pneumotaxic center, 450, 453f
Pneumothorax, 420, 421f
PNS. *See* Peripheral nervous system
Podocytes, 471–72, 472f
role in regulation of filtration coefficient, 478
Point mutation, B13
Poiseuille's law, 305
Poison ivy, 377, 381, 404
Poisonous spiders, 209
Polar body, 715–16
Polarization, 80
Polar molecules, 46, A4
lipid solubility of, 57
Poliomyelitis, 205
Polycythemia, 355
Polycythemia vera, 355
Polydipsia, 675
Polymers, definition of, A7
Polymodal nociceptor, 157–58
Polymorphonuclear granulocytes. *See also* Basophils; Eosinophils; Neutrophils
Polynucleotide, B2f

Polypeptides
and protein digestion, 543
Polyphagia, 675
Polyribosomes, 35, B10, 10f
Polysaccharides, A8
Polyspermy, block of, 728
Polysynaptic reflex, 145
Polyunsaturated fatty acids. *See* Unsaturated fatty acids
Polyuria, 675
Pons, 136
Pool of particular ECF constituent, 511
Population code, 153
Pores
between alveoli, 416
capillary, 319, 321
in glomerular membrane, 471–72, 472f
nuclear, 34, B7
in plasma membranes. *See* Channels
Pores of Kohn, 416
Portal system. *See* Hepatic portal system; Hypothalamo-hypophyseal portal system
Portal vein, transport of blood from the digestive tract to the liver, 585
Positive balance, 512
Positive energy balance, 598
Positive feedback, 11–12
among hormones during parturition, 736f
in kallikrein production, 374
in the LH surge, 721, 721f
in opening of ion channels, 82, 82f
Positive water balance. *See* Overhydration
Positron emission tomography (PET), 110
Postabsorptive (fasting) state, 671
glucagon control of, 679–81
insulin control of, 674, 679
Posterior parietal cortex, 121–22
Posterior pituitary (neurohypophysis), 497, 631–33
hormones of, major functions, and target cells of hormones, 619, 631–33
link with the hypothalamus, 632f, 637f
See also Oxytocin; Vasopressin
Postganglionic autonomic fiber, 198, 251–52
Postganglionic nerve terminals, 253f
Postganglionic neuron, 663
Posthepatic jaundice, 574
Postnatal growth spurt, 639
Postreceptor events, 50–54, 54f, 75, 625
cyclic AMP second messenger system, 52f
of hydrophilic hormones, 625
of lipophilic hormones, 625
Postsynaptic neuron, 91–94, 94f
Postural hypotension, 339
Posture and balance, roles of
basal nuclei, 125
cerebellum, 136
crossed extensor reflex, 145–46

lower brain regions and spinal cord, 137, 145
multineuronal pathway, 243
muscle and joint proprioceptors, 155
vestibular information, 190
Potassium
action in nerve and muscle cells, 50
active transport of, 65–66
and aldosterone, 489, 489f, 657
in the brain, 111
in cardiac pacemaker activity, 269
distribution of in body fluid compartments, 514
effect on heart of abnormal concentration of, 489
equilibrium potential of, 71f, 72
influence of changes in hydrogen ion concentration on, 526
membrane permeability to
at inhibitory synapse, 92–93
at motor end plate, 207
at peak of action potential, 82
at resting potential, 72–73, 73f
at threshold potential, 82
related to Na$^+$ reabsorption, 488–89
renal regulation of, 488–89, 488f
retention, as consequence of kidney failure, 502
role of
in action potentials, 81–82
in excitation of muscle
in generation of end-plate potential, 207
in generation of an EPSP, 92
in generation of an IPSP, 92–93
in local control of arteriolar radius, 312
in postsynaptic potential generation, 92–93
voltage-gated channels for, 82
See also Na$^+$-K$^+$ pump
Potential. *See* Membrane potential
Power stroke, 220, 221f
PP cells, 671
Practice, role in memory, 130–31
Precapillary sphincters, 319–20, 324
complementary action with arterioles, 322f
Precipitation of antibody-antigen complexes, 383
Precocious pseudopuberty, 661
Prefrontal association cortex, 123
Pregame meal, 559
Preganglionic fiber, 198
Preganglionic neuron, 663
Preganglionic sympathetic impulses, 665
Pregnancy
abdominal, 725
changes in body systems during, 734
tubal, 728
See also Gestation; Parturition; Placenta
Pre-hepatic jaundice, 574
Preload, 286

Premature beat, 271
Premotor cortex, 121
Preprohormones, synthesis in the rough endoplasmic reticulum, 622
Prepuce, 702
Presbyopia, 168
Pressure. *See* Arterial blood pressure; Atmospheric pressure; Hydrostatic pressure; Intra-alveolar pressure; Intrapleural pressure; Osmotic pressure; Partial pressure
Pressure autoregulation, 313–14, 314f
Pressure gradient
in blood vessels, 302–3, 304f
during expiration, 425
during inspiration, 421, 425–26
between lungs and atmosphere, 417, 419–20
Pressure reservoir, arteries as, 306f
Presynaptic facilitation, 98–99
Presynaptic inhibition, 98–99, 99f
Presynaptic inputs to motor neurons, 97
Presynaptic knob, 91
Presynaptic neuron, 91
PRF. *See* Prolactin-releasing factor
Primary active transport, 66
Primary adrenocortical insufficiency (Addison's disease), 662–63
Primary auditory cortex, 186
Primary (annulospiral) endings, 243, 244f
Primary follicle, 714–15, 715f
Primary hyperaldosteronism (Conn's syndrome), 660
Primary hypertension, 338–39
Primary hyposecretion of endocrine organs, 630
Primary immune response, 386, 386f
Primary motor cortex, 120
influence on motor neurons, 240
Primary odors, 193
Primary olfactory cortex, 193
Primary oocytes, 714–15, 714f
Primary peristaltic wave, 554–55
Primary reproductive organs (gonads), 696
See also Ovaries; Testes
Primary sensory cortex. *See* Somatosensory cortex
Primary spermatocytes, 704
Primary structure of proteins, A11
Primary tastes, 191
Primary tissue types, 5
Primary visual area. *See also* Visual cortex
PRL. *See* Prolactin
Procarboxypeptidase, 569
Procedural memory, 131
Progestational phase, of the menstrual cycle, 722, 723
Progesterone, 696, 714
actions of, summary, 739
antagonism of, as birth control method, 741
effect of
on breasts, 738

on cervical mucus, 723
on estrogen receptors, 631
on prolactin, 738
on the uterus, 722, 728
negative feedback by, 719, 720
regulation of secretion by LH, 635
secretion of
by the corpus luteum, 721
by the placenta, 732–33, 733f
synthetic, and oral contraceptives, 741
Prohormones, 622
Projection of sensory input, 155
Prolactin (PRL), 635
control of secretion of, 639
function in females, 635, 738–39
function in males, 635
and lactation, 738, 739
Prolactin-inhibiting hormone (PIH),
635–36, 738
Prolactin-releasing factor (PRF), 738
Proliferative phase, of the menstrual cycle,
722, 723
Pro-opiomelanocortin, 662, 665
ACTH precursor, 659
cleavage of, 622
Propagation of action potentials. See Action potentials, propagation of
Prophase, B11, B14
Proprioception, 119, 151, 155
in muscle movement, 242
role in sports skills, 155
Propulsive movements. See Motility
Prostacyclin, 710
and platelet aggregation, 360
Prostaglandins, 710–11, 710f, C4
actions of, summary, 711
and active hyperemia, 312
and contractions during labor, 736
and corpus luteum degeneration, 722
effect of
on arteriolar radius, 312
on oviducts,726
on uterus, 726, 736
and fever, 374–75, 609
and menstrual flow, 722
and ovulation, 720
and pain, 158
and parturition, 736
removal by lungs, 413
role in blastocyst implantation, 729
source of, 710
structure and nomenclature of, 710d11,
710f
Prostate gland, 697, 710
hypertrophy of, influence on urinary
system, 474, 502
Prostatic hypertrophy, 466
Protein buffer system, 529
Protein channels in plasma membrane.
See Channels
Protein hormones, 67, 618
exocytosis of, 68
Protein kinase, 53–54

Proteins, 543, A10–12
absorption in the small intestine, 580,
582
acute phase, 375
as antigens, 381
as buffers, 529
channel, sensitivity to charge, 82
conformation, A10-A11
digestion and absorption of by digestive
system, 543, 556, 562–63, 568–69,
577, 578, 580–82, 582f
dietary, 543
effect of on gastric secretion, 564
effect on
of cortisol, 658
of diabetes mellitus, 675
of glucagon, 679
of growth hormone, 640, 643
of insulin, 674
of thyroid hormone, 652–53
and fuel metabolism, 670
membrane, 47–48
metabolic pathways for, 668
muscle, effects of testosterone, 704
negatively charged intracellular, 71
precursor, 349
processing and sorting of, in the Golgi
complex, 23
production of in rough endoplasmic
reticulum, 19–22
respiratory quotient of, 413
secretion of, 24
as source of glucose. See
Gluconeogenesis
structure of, A10-A11
synthesis of. See Protein synthesis
translation by RNA, B8f
Protein synthesis, B4–10
effect on
of thyroid hormone, 652
and fuel metabolism, 668
of new cellular components, 21
role of
in long-term memory, 135
ribosomes, 21, B7-B8, B8f
the rough endoplasmic reticulum,
19–21
transfer RNA, B7
of secretory products, 22
steps of, B7
and transcription of DNA genetic information, B5f, B7
and translation of messenger RNA, B7,
B8f
Proteinuria
in exercise, 503
in kidney disease, 503
Proteolytic enzymes, pancreatic, 568–69
Prothrombin, 361
Protons
definition of, A1–2
hydrogen ion as, A7
Proto-oncogenes, 14

Proximal tubule, 469
and reabsorption
of chloride, 486
of glucose and amino acids, 484–86
of phosphate, 486
of potassium, 488
of sodium, 481
of urea, 487f
of water, 486–87, 486f
and secretion of
hydrogen ion, 488
organic ions, 489–90
potassium, 488–89
PR segment, 275
Pseudonephritis, athletic, 503
Pseudopod, 39
Psychological fatigue, 236
Psychosocial influences on satiety, 600
Psychosocial stressors, 666–67
Psyllium-seed husks, effects on cholesterol,
295
PTH. See Parathyroid hormone
Pubertal growth spurt, and DHEA, 657
Puberty, 708–9
female, 723
growth spurt of, 639
male, physical changes during, 703
timing difference in males and females,
646
Pulmonary arterioles, effect of oxygen on,
429
Pulmonary artery, 262
Pulmonary capillaries
angiotensin-converting enzyme in, 482
gas exchange across, 439–40
surrounding the alveoli, 415
Pulmonary circulation, 259–60, 259f
blood flow through, 262–63
gas exchange across capillaries, 439–40f
and lymphatic system, 327f
pressures in, 263
resistance of, 263
Pulmonary compliance. See Compliance
of lungs
Pulmonary edema. See Edema, pulmonary
Pulmonary fibrosis, 441
Pulmonary stretch receptors, 453
Pulmonary surfactant, 416, 430–32
Pulmonary valve, 264
anatomy of, 265f
relation to heart sounds, 281
Pulmonary vascular resistance, 281
in fetus, 262
Pulmonary veins, 262
Pulmonary ventilation, 435–37
See also Ventilation
Pulse deficit, detection on an ECG, 277
Pulse pressure, 308
Pump
active transport mechanisms as, 65
the heart as, 261
Pumping action, coordination of, in the
heart, 271

"Punishment" center, 128
Pupils, 162
  control of the size of, 162–63, 162f
Purkinje fibers/system, 269, 271
Pus, 374
P wave, 275
Pygmies, lack of somatomedins in, 645
Pyloric gland area (PGA), 560
  G cells, 563
  See also Gastrin
Pyloric sphincter, 555–56, 556f
Pyramidal cells, 118
Pyramidal system, 242
Pyruvic acid, 236
  acetic acid derived from, 28
  conversion to glucose, 237
  conversion to lactic acid under anaero-
      bic conditions, 236
  formation during glycolysis, 27–28
  and O$_2$ debt, 236–37

QRS complex or wave, 275
Quadriceps femoris, 156, 244
Quadriplegia, 243
Quality of sound. See Timbre

Rabies virus, 388
Radial (dilator) muscle of iris, 162, 163
Radiation
  electromagnetic, effect on white cell
      production, 358
  heat transfer by, 603–4
Radiation exposure
  aplastic anemia caused by, 355
  immunosuppression caused by, 397
Radius, relation to resistance, 304, 304f
  in airways, 425
  in blood vessels, 303–4, 304f
  See also Resistance
Rapid eye movement (REM) sleep, 138,
      139
RAS. See Reticular activating system
Rates of reaction, alteration by hormones,
      626
Reabsorption (across capillaries), 323, 325f
  See also Tubular reabsorption
Reactive hyperemia, 313
Reactive hypoglycemia, 677–78
Readiness potential in activation of the
      motor cortex, 120
Receptive field, 155, 156f
  of a cone ganglion cell, 175
  two-point threshold of discrimination
      test for, 156
Receptive relaxation, 556
Receptor cells, hair cells of the ear, 184
Receptor-operated channel. See Chemical
      messenger-gate channel
Receptor potential, 152–53, 153f
  of olfactory receptors, 192, 192f
  of taste buds, 191

Receptors (in afferent neuron), 108,
      152–57
  activation of
  adaptation of, 153–54
  coding of information from. See Coding
  generator potential, 152–53
  as part of reflex arc, 142
  receptor potential, 152–53
  sensitivity of, 152
  as transducers, 152–53
  types of, categorized by adequate stimu-
      lus, 152
  types of, categorized by speed of adapta-
      tion, 153–54, 154f
  See also Specific receptors by name
Receptor sites, in plasma membrane,
      48–49
  down regulation of, 631
  for hormones, 624–25, 630–31
  for epinephrine and norepinephrine, 663
  regulation of, 54
  and postreceptor events, 50–54
Reciprocal innervation, 144
Recombinant DNA technology, production
      of interferon, 379
Reconditioning organs, 321
Recruitment of motor units, 226–227
Rectum, 587
Red blood cell count, 349
Red blood cell ghost preparation, 61–62
Red blood cells
  breakdown of, 573–74
  membrane fluidity and functions of, 47
  production of. See Erythropoiesis
  See also Erythrocytes
Red fibers, 237
Red marrow, 351
Reduced hemoglobin (Hb), 443
  See also Hemoglobin
Referred pain, 141
Reflex arc, 142, 145
Reflexes
  definition of, 142
  See also Specific reflexes by name
Refraction, 163–66, 166f
  effect on visual field of each eye, 175
Refractory period, 89–91, 91f
  absolute, 90
  for erection, 713
  of heart muscle, 274, 274f
  relative, 90
  of skeletal muscle, 274
Regeneration of cut peripheral axon, 89,
      89f
Regeneration tube, 89, 89f
Regulatory functions of the kidneys, 506
Regulatory proteins, 218
Regulatory T cells, 392
Regurgitation in the heart, 282
Rehabilitation costs, and osteoporosis, 691
Rehearsal, role in memory, 130–31
Rejection, of transplanted tissue, 504
Relative humidity, 605

Relative refractory period, 90
Relaxation time for muscle, 225
Relaxin, 734
  placental secretion of, 732
Releasing hormones, 635–36
REM. See Rapid eye movement (REM)
Remembering, 130
REM sleep. See Paradoxical sleep
Renal anemia, 355
Renal cortex, 466
Renal failure, 502–3
  and diabetes mellitus, 676–77
  and ECF hypotonicity, 520
  and hypertension, 338
Renal function. See Kidneys
Renal hypertension, 338
Renal medulla, 466
  countercurrent exchange within, 496f
  recycling of urea in, 500–506, 506f
  vertical osmotic gradient in, 492, 492f
Renal pelvis, 466
Renal plasma flow, measurement of by
      PAH clearance 491–92
Renal pyramids, 466
Renal threshold, 485, 485f
Renin, 482
Renin-angiotensin-aldosterone system, 482,
      483f, 657
  and control of ECF volume and osmolar-
      ity, 524
  and control of sodium reabsorption by
      kidneys, 482–83, 483f, 657
  effect of on mineralocorticoids, 657
  and hypertension, 343
  and long-term regulation of blood pres-
      sure, 516
  and stress, 666
Replication of DNA, B1
Repolarization, 80
  atrial and ventricular, 280–81
  of heart muscle, 273
  phase of an action potential, 82
Reproduction of cells, 2–3
Reproductive system, 695–742
  adverse effects of anabolic androgenic
      steroid abuse on, 239
  anatomy of, 696–97, 697f, 698f
  changes with aging, 715
  components and functions of, 8f, 10
  defense mechanisms in, 407–8
  differentiation along female or male
      lines, 700–2, 700f
  prostaglandin actions on, 710–11
  sex determination and differentiation,
      699–702
  sexual intercourse, 711–14
  See also Female reproductive physi-
      ology; Male reproductive
      physiology
Reproductive technology, 726–27
Reproductive tract, 696
  development along male or female lines,
      702

Residual bodies, 25
Residual volume (RV), 433–34
Resistance
    of airways, 425–27
    to flow in blood vessels, 303–4
    versus flow rate, 310
    relation to flow, 425
    See also Arteriolar resistance; Total
        peripheral resistance
Respiration
    control of, 450–60
    definitions of, 413
    effect on venous return, 332
    inhibition of during swallowing, 553
    and insensible water loss, 521
    steps of, 413, 414f
    See also Respiratory mechanics; Respira-
        tory system; Ventilation
Respiratory acidosis, 450, 534–36, 536f
Respiratory airways, 414
    collapse of in forced expiration, 426–27,
        427f
    effect on
        of carbon dioxide, 427–28
        of epinephrine, 664
        of slow reactive substance of anaphy-
            laxis, 402
    resistance of, 426
Respiratory alkalosis, 450, 535f, 536
Respiratory arrest, 449, 459
Respiratory bronchioles, 415f
Respiratory centers, 450–53, 453f
    effect of body cooling on, 610
Respiratory chain. *See* Electron-transport
    chain
Respiratory distress syndrome of newborn,
    430–32
Respiratory mechanics, 417–37
    alveolar ventilation, 437
    airway resistance, 425–27
    control of ventilation, 450–58
    elastic behavior of the lungs, 429–30
    lung volumes and capacities, 433–35
    muscle activity during breathing,
        421–23, 422f, 423f, 425
    pressure relationships involved in venti-
        lation, 421–25, 424f
    role of pulmonary surfactant, 430, 431f
    spirometry, 433, 434f
    work of breathing, 432–33
Respiratory mucosal immunoglobulins,
    401
Respiratory muscles, 422, 422f
    paralysis of, from poisoning, 209–10
    role in vomiting, 560
    summary of activities of, 422
Respiratory paralysis, 209–10, 425
Respiratory pump, 332, 333f
Respiratory quotient (R.Q.), 413
Respiratory rate, 436
    changes in during paradoxical sleep, 138
Respiratory system, 412–60
    abnormalities of, 434–35

alveolar ventilation, 437
and carbonic acid-bicarbonate buffer
    system, 527, 529–30
components of, 7, 415f
control of, 450–58
defense mechanisms in, 408
effect on
    of acidosis and alkalosis, 530
    of cigarette smoking, 408
    of curare, 209–10
    of organophosphates, 209–10
    of prostaglandins, 711
factors affecting airway resistance,
    425–29
functions of, 5, 413
gas exchange in, 437–42
lung volumes and capacities, 433–35
muscles of, summary, 422
nonrespiratory functions of, 413
opposing forces acting on lungs, sum-
    mary, 432
pressure changes throughout respiratory
    cycle, summary, 424
respiratory mechanics, 417–37
role in acid-base balance, 529–30
ventilation, 435–37
Restak, Richard, 629
Resting membrane potential, 72–73, 73f
    status of K^+ and Na^+ channels during,
        82
    See also Membrane potential
Resting tremors, 126
Restrictive lung disease, 434–35, 435f
Retching, 560
Reticular activating system (RAS), 137,
    137f, 152
Reticular formation, 137
    role of
        in the analgesic system, 158
        in the multineuronal system of motor
            control, 242
        in the sleep-wake cycle, 137–39
Reticulocytes, 351
Retina, 162, 170f
    damage to
        in glaucoma, 162
        in hyperoxia, 449
        in hypertension, 339
    layers of, 169f
    phototransduction by, 170–73
    preliminary processing of visual informa-
        tion by, 174–75
Retinene, 170, 174
Retrograde amnesia, 131
Retropulsion, 557
Reversible reactions, A4
Reversible shock, 343
"Reward" center, 128
Rheumatic fever, 282, 397
Rheumatoid arthritis, glucocorticoid
    therapy for, 658
Rh factor, 390
Rh-negative, 390

Rhodopsin, 170, 174
Rh-positive, 390
Rhythm, beta, 124
Rhythm method of birth control, 740
Ribonucleic acid (RNA), A12, B7
Ribophorins, 21, 21f
Ribose, A8
Ribosomal RNA (rRNA), B7
Ribosomes, 18, 20f, 41, 41f
    assembly of, B8f
    cytosolic, 18, 21, 21f, 34
    in RER, 19, 20f, 21, 21f
    role of
        in peptide hormone synthesis, 622
        in protein synthesis, 21, 34–35, B9
    translation of RNA instructions, B7
Ribs, movements during breathing, 423f
Rickets, 690
Right atrioventricular (AV) valve, 264
Right cerebral hemisphere, 124
Rigor complex, 224f
Rigor mortis, 225
Risks of heart disease, 295
RNA. *See* Ribonucleic acid
Rods, 162, 169, 173
    comparison with cones, 173
Rough endoplasmic reticulum (rough ER),
    19–21, 20f
    binding with ribosomes, B10
    in plasma cells, 381
    preprohormone synthesis in, 622
Round window of the ear, 179, 182f, 184
R.Q., 413
rRNA. *See* Ribosomal RNA
RU 486, 741
Rugae, of the stomach, 555f, 556
"Runner's high," 158

Saccule, 179, 187, 187f, 188, 189
Salbutamol, 203
Salicylates, 376–78
    See also Aspirin
Saliva, 407, 551–53
Salivary amylase. *See* Amylase, salivary
Salivary center, 552
Salivary glands, 551–53, 551f
    innervation of, 202
Salivary secretion, 551–53, 553f
Salt
    appetite for, 515
    balance, 465–66
    and hypertension, 339, 340
    imbalances caused by kidney failure,
        502
    regulation of, 482–83, 515–516
    structure of, A3
    See also Sodium; Sodium chloride
SALT, acronym. *See* Skin-associated lym-
    phoid tissue
Saltatory conduction, 86, 102
Salt balance, 510
    homeostasis controlled by the endocrine
        system, 646

Salt load, regulation of, 482–83
SA node. *See* Sinoatrial (SA) node
Sarcomere, 215f, 216
    during contraction, 219, 220f
Sarcoplasmic reticulum, 221–22
Satiety centers, 598–99, 598f
Satiety signals, 599–600
Saturated fatty acids, A9
    effect of, on cholesterol, 295
Saturation in carrier-mediated transport, 63
Scala media, 182–83, 183f
Scala tympani, 179, 183, 183f
Scala vestibuli, 179, 183, 183f
Scar tissue, 363, 375
Schizophrenia, 129
Schwann cells, 86, 87f
    formation of regeneration tubes by, 89
Sclera, 162
Scrotum, 697, 702–3
Seasonal cycles, due to pineal gland control, 709
Sebaceous glands, 406, 704
Sebum, 406
Secondary active transport, 66, 67f
    in kidney tubules, 484
    in small intestine, 580
Secondary adrenocortical insufficiency, 662
Secondary (antral) follicles, 715, 715f
Secondary (flower-spray) endings, 243, 244f
Secondary hyperaldosteronism, 660
Secondary hypertension, 338
Secondary hyposecretion of endocrine organs, 630
Secondary immune response, 386, 386f
Secondary oocyte, 715
    fertilization of, 728
Secondary peristaltic wave, 555
Secondary sexual characteristics, 696–97
    male, 704
Secondary spermatocytes, 704
Secondary structure of proteins, A11, 11f
Second heart sound, 281
Second line of defense against changes in hydrogen ion concentration, 530
Second messenger systems
    activation by hydrophilic hormones, 624
    cyclic AMP, 132–33
    role in declarative memory formation, 133–34
    mechanism of action, 50–54
    neurotransmitters utilizing, 94
    See also Calcium; Cyclic adenosine monophosphate
Second-order sensory neuron, 154–55
Second polar body, 715
Secretin
    characteristics of, 558
    effect of
        on bile secretion, 574
        on gastric motility, 558
            on gastric secretion, 565
            on pancreatic aqueous NaHCO$_3$ secretion, 569
    overview of, 590–91
Secretion
    exocrine versus endocrine, 4–5
    mechanism of, 24
Secretion (of digestive juices), 543, 543f
    of biliary system, control of, general features, 548–49
    of esophagus, 555
    of exocrine pancreas, 568–69
    of large intestine, 589
    of liver, 571–72
    of salivary glands, 551–53
    of small intestine, 577
    of stomach, 560–67
    summary of, 546–47, 586
Secretion, tubular, 488–90
    See also Tubular secretion
Secretory phase of the menstrual cycle, 722
Secretory products, protein synthesis of, 22
Secretory vesicles, 18, 22–24, 24f
    transport of, 36–37
Seeing. *See* Eye; Vision
Segmentation
    in the small intestine, 548, 575–76, 575f
Selective permeability, 57, 514
Selective transport, between the ECF and the ICF, 75
Selegiline, pharmaceutical for treating Parkinson's disease, 100
Self-antigens
    as human leukocyte-associated antigens, 397
    as major histocompatibility complex-coded antigens, 397
    and T-cell activation, 392
    in tolerance and autoimmune disease, 396
Self-excitable smooth muscle, 249
Self-generated electrical activity in smooth muscle, 252f
Selye, Hans, 665
Semen, 697, 710, 713
Semicircular canals, 179, 187, 187f
Semilunar valves, 264
    See Aortic valve; Pulmonary valve
Seminal vesicles, 697, 710
Seminiferous tubules, 703
    fluid of, 707
    and Sertoli cells, 707
    and spermatogenesis, 704
Senile dementia, 132
Sensation perceived related to stimulus modality, 152
    See also Proprioception; Somatic sensation; Somesthetic sensation; Special sensation
Sensible loss of water, 521
Sensitivity
    of diagnostic tests, stress test, 278
    of the eye, 173–74
Sensitization, 131
    in *Aplysia*, 134f
    calcium ion channel activity in, 132–33
Sensitized T cells. *See* T lymphocytes, activated
Sensor, 11
Sensorineural deafness, 186–87
Sensory afferent, 154
Sensory coding. *See* Coding of sensory information
Sensory deprivation, and long-term memory, 134–35
Sensory homunculus, 119
    related to receptive field size, 156
Sensory information/input, 151, 154
Sensory modalities. *See* Modalities
Sensory neurons, receptive field of, 155
Sensory receptor. *See* Receptors, in afferent neuron
Septic (septicemic) shock, 340–41, 364
    See also Circulatory shock
Septum, 262
Series-elastic component, 230, 231f
Serosa, 547
Serotonin, 128–29
    and pain transmission, 160
    participation in sensitization and habituation, 132–33
    in platelet plugs, 360
Sertoli cells, 704, 706–7
    role in spermatogenesis, 708, 708f
Serum, 363
Serum sickness, 388
Set point, 11
Severe combined immunodeficiency, 401
Sex, and body water, 511
Sex act, 711–14
Sex chromosomes, 699–701
Sex determination, 699
Sex flush, 713
Sex hormones, 696
    from the adrenal cortex, 657
    See also Androgen; Estrogen; Progesterone; Testosterone
Sexual differentiation, 700–702, 710f
Sexual intercourse, 711–14
    lubrication for, 713
Sexual response cycle
    in females, 713–14
    in males, 711
Shells, electron, definition of, A2
Shivering, 606–7
Shock. *See* Neurogenic shock; Circulatory shock
Short-loop negative feedback in the hypothalamus-pituitary system, 639
Short reflex, 549
Short-term memory, 129
    comparison with long-term memory, 130
    mechanisms of, 131–34
Sickle-cell anemia, 355
    detection in the embryo, 727
SIDS. *See* Sudden-infant-death syndrome

Smooth muscle, (continued)
   of urinary bladder wall, 503–4
   vascular
      in arteries, 305
      in arterioles, 308, 309f
      in veins, 328
Sneeze, 408, 458
Sniffing, in odor perception, 192
Sodium
   absorption by digestive tract, 580
   absorption in the large intestine, 589–90
   active transport
      across the plasma membrane, 65–66
      across renal tubules, 481
      in the intestine, 580
   amount filtered, 515–16
   and atrial natriuretic peptide, 484
   balance of, 515–16, 516f
   concentration in extracellular fluid, 514
   control of urinary excretion of
      by control of filtration of, 477, 515–16
      by control of tubular reabsorption of, 481–84, 516
   cotransport with glucose, 66
   distribution of in body fluid compartments, 71
   equilibrium potential of, 72, 72f
   in generation of resting membrane potential, 72–73, 73f
   handling by loop of Henle, 493
   imbalances caused by renal failure, 502
   membrane permeability to
      at an activated receptor, 153
      at an excitatory synapse, 92
      at a motor end plate, 207
      at peak of action potential, 82
      at resting potential, 71, 73
      at threshold potential, 81
   relation to ECF volume/ plasma volume/ arterial blood pressure, 515–16
   renal handling of, 481–84, 481f
   and renin-angiotensin-aldosterone system, 482, 483f, 657
   reabsorption of, in the urinary system, 481–84, 481f
   retention of in congestive heart failure, 289
   role of
      in an action potential, 81–82
      in EPSP, 91–92
      in heart action potential changes, 273
      in muscle excitation, 207
      in nerve and muscle cells, 50
      in photoreceptors, 170
      in secondary active transport, 66
      in slow-wave potentials, 249
      in tubular reabsorption of
         amino acids, 484
         chloride, 486
         glucose, 484
         urea, 487, 487f
         water, 486, 486f
   See also Na+-K+ pump

Sodium bicarbonate
   secretion by the exocrine pancreas, 558
   secretion by liver, 590
   self-medication with, as cause of metabolic alkalosis, 537
Sodium chloride, A3
   role in establishing kidney's medullary vertical osmotic gradient, 493–95
   See also Salt; Sodium
Sodium ion (Na+) channels, 482
Sodium load, 481–82, 515
Sodium-potassium pump. See Na+-K+ ATPase pump
Soft palate, 550
Solutes
   concentration of, and osmosis, 59–60
   definition of, A5
   nonpenetrating, and water movement across a membrane, 60–61, 61f
Solutions, A5–6
Solvents, definition of, A5
Somatic cells, 698
   chromosomes of, B3
Somatic mutation, B14
   as the source of diverse B cells, 386–87
Somatic nervous system, 107, 204–5
   versus the autonomic nervous system, 205
   efferent neurons of, 206
   muscles innervated by, 214
Somatic sensation, 151, 154
Somatomedins, 642, 653
   deficiency of, and dwarfism, 645
   regulation of growth hormone secretion by, 642
Somatosensory cortex, 119–20, 120f
   ascending pain pathways to, 158
Somatosensory pathways, labeling of, 154–55
Somatostatin, 636
   from pancreas, 671–72
   regulation of growth hormone secretion by, 642
   in therapy for diabetes mellitus, 681
Somatotopic map, of brain. See Maps, somatopic
Somatotropin, 634
   See also Growth hormone
Somesthetic sensations, 119, 151
Sound localization, 181
Sound transduction, 184f
Sound waves, 178, 179, 180f
   effect on tympanic membrane, 182
Space travel
   effect on bone, 684
   effect on muscles, 209
Spastic paralysis, 243
Spatial summation, 95, 96, 102
Special sensation, 154, 161
Specific immune responses, 380–81
   to bacterial invasion, summary, 391
   B lymphocytes and antibody-mediated immunity, 381–92

general concepts, 380–81
   in immune diseases, 401–4
   T lymphocytes and cell-mediated immunity, 392–99
   to viral invasion, summary, 395
   See also Autoimmune diseases; Immune disease; B lymphocytes; T lymphocytes
Specificity
   of carrier-mediated transport, 63
   of stress test, 278
   of the nervous and endocrine systems, 106
Specific nerve energies, law of, 152
Speech, 552
   See also Language ability
Speech impediments, 123
Spermatids, 704
Spermatogenesis, 704–5, 705f
   comparison with oogenesis, 714
   control of, 708, 708f
   role of the anterior pituitary, 707–8, 708f
   role of the Sertoli cells, 705f, 706–7
   sensitivity to temperature, 703
   steps of, 706f
   and vasectomy, 709–10
Spermatogonia, 704, 707
Spermatozoa (sperm), 696 706, 707f
   anatomy of, 706, 707f
   capacitation of, 709
   and cervical mucus, 723
   concentration in ejaculate, 713
   electron micrograph of, 728
   and fertilization, 727–28
   flagella of, 37
   numbers produced per day, 704
   production of, 704–6, 705f, 706f
   relationship with Sertoli cells, 706–7
   and semen, 697, 710, 713
   transport
      in males, 710
      to the oviduct in females, 725–26, 725f
Sperm banks, 726
Spermicidal jellies, foams, 740
Spermiogenesis, 696, 705–6
Sperm production, promotion of, by FSH, 635
Sphincter of Oddi, 571, 571f, 590
Sphincters, 232
   anal, 589
   urinary, 504
Sphingolipid, structure of, A9f
Sphygmomanometer, 306–8, 306f
Spider, poisonous, 209
Spike, action potential, 81
Spinal cord, 139–46
   cross-sectional anatomy of, 141f
   organization of gray matter of, 141, 142f
   origin of spinal nerves from, 139, 140f, 141, 141f
   tract in white matter of, 139–41, 142f, 143f
Spinal nerves, 139, 140f

Sight. *See* Eye; Vision
Sigmoid colon, 587
Signal-recognition protein, 21
Simple cells in the visual system, 177
Simple reflexes, 142
   See also Basic reflexes
Single-unit (visceral) smooth muscle, 248
   See also Smooth muscle
Sinoatrial (SA) node, 269
  autonomic influences on, 283–84, 284f
Sinusoids, liver, 570–71, 571f
Sister chromatids, B11
Skeletal muscle, 4, 213, 214, 215–246
  in the anal sphincters, 589
  ATP requirements of, 234
  atrophy of, 209
  blood flow to, 310–12
  comparison with cardiac and smooth
    muscle, 248–49
  concentric contraction of, 231–32
  contraction
    duration of, 225
    gradation of, 226–30
    mechanism of, 218–25
  control of, 241–46
  cross-bridge cycle of, 218–25
  differences in male and female, 704
  eccentric contraction of, 231–32
  in the digestive tract, 543
  effect of venous return, 330, 330f, 332f
  effect on
    of epinephrine, 663, 664
    of exercise, 238–39
    of hypocalcemia and hypercalcemia,
     682
    of insulin, 674
    of testosterone, 704
  efficiency of, 233–34
  embryonic development of, 216
  energy metabolism of, 234–36
  excitation of, 205–6
  excitation-contraction coupling of,
    220–25
  fatigue of, 236
  fiber types of, 237–38
  glycogen in, 35, 670
  Golgi tendon organ of, 97, 243–46
  gradation of, 226–30
  heat production by, 606–7
  hypertrophy and hyperplasia, 238
  innervation of skeletal muscle cells by,
    207f
  isotonic and isometric contractions of,
    230
  lactic acid production by, 236
  length-tension relationship of, 229, 230f
  levels of organization of, 215f
  lever action of, 233–34
  mechanics of, 233–34
  metabolism and fiber types, 234–41
  molecular basis of contraction, 218–34
  motor units of, 226–227, 227f
  muscle spindle of, 243–45, 244f

  muscular dystrophy in, 240
  paralysis of, 243
  receptors for monitoring length and ten-
    sion of, 243
  relaxation of, 225
  repair of, 241
  role of calcium ion in, 219–25, 249
  sliding-filament mechanism of, 219
  stretch reflex of, 244, 245f
  structure of, 215–18
  summary of characteristics, 248–49
  in temperature regulation, 607
  tension
    determinants of whole muscle, 226
    relation to load in isotonic and iso-
     metric contractions, 231
  tetanus, 228
  transmission of action potentials to, 88
  twitch, 226
  velocity of shortening related to load,
    232–33
  and work, 596
Skeletal-muscle arterioles, sympathetic ner-
    vous system effect on, 201
Skeletal-muscle pump
  effect on venous return, 330, 330f, 332f
Skeleton
  components of, 7
  functions of, 683
  water in, 511
Skin arterioles, hypothalamic control over,
    338
Skin-associated lymphoid tissue (SALT),
    407, 407f
Skin (integument), 4, 405–7, 408
  blood flow to, 607
  as an endocrine gland, 687
  hormones of, major functions, and target
    cells of hormones, 620
  immune defense activities of, 406–7,
    407f
  and insensible loss of water, 521
  intermediate filaments in, 41
  structure and functions of, 405f
  and sweat, 521
  in temperature regulation, 601, 608
  and vitamin D synthesis, 687
Skin proprioceptors, 155
Skin thermoreceptors, 606
Skull, 112
Sleep, 137–39
  EEG for observation of stages of, 124
  and growth hormone secretion, 642
  and metabolic rate, 598
  need for, 139
  neural mechanism of, 139
  summary comparing slow-wave and
    paradoxical sleep, 138
  types of, 137–38
Sleep apnea, 459
Sleep attacks, 139
Sleep-wake cycle, 137, 139
  and cortisol control, 660

Sliding-filament mechanism, 219
  in cardiac muscle, 254
  and length-tension relationship, 229–30
  in skeletal muscle, 219
Slow-oxidative fibers, 237
Slow pain pathway, 158
Slow-reactive substance of anaphylaxis
    (SRS-A), 402
Slow-wave potentials
  digestive tract, 547
  in smooth muscle, 249, 252f
Slow-wave sleep, 137–38
  versus paradoxical sleep, 138
Small intestine, 575–87
  absorption by, 577–86
  and diarrhea, 586–87
  entry into of
    bile, 571–72
    exocrine pancreatic secretions,
     567–68
    gastric contents, 556f, 557–58
  intracellular enzymes, 577
  malabsorption, 578
  motility, 575–77
  secretion, 577
  structure of, 579f
Smallpox vaccination, 388
Smell, sense of, 161, 192–93
Smoking
  and dry mouth, 521
  and risks of taking birth-control pills, 741
  "smoker's cough," 408
Smooth endoplasmic reticulum (smooth
    ER), 19, 20f, 22–23, 22f
Smooth muscle, 4, 213, 214, 246–54
  in arterioles, 308–9
  autonomic nervous system effects on,
    251–52, 253f
  categories of, 247–48
  cells of, 247, 250
  characteristics of, 248–49
  comparison with cardiac and skeletal
    muscle, 246, 248–49
  cross-bridge activity of, 247, 254
  in the digestive tract, 542–43
  duration of contraction, 253
  energy consumption by, 254
  excitation-contraction coupling of, 247
  factors that can modify, 251–52
  gradation of contraction of, 250–53
  latch phenomenon of, 254
  length-tension relationship of, 252–53
  mechanism of contraction, 247
  multiunit, 247
  pacemaker activity of,
  relaxation of, 247
  role of
    calcium, 247, 251, 251f, 252f
    calmodulin, 247
  self-excitation of, 247–48
  single-unit, 248
  slow-wave potentials of, 249
  structure of, 247, 251f

Spinal reflex, erection as, 712
Spinal reflexes, 142–46
  See also Specific reflexes by name
Spinal tap, 139
Spindle, muscle. See Muscle spindle
Spinocerebellum, 135–36, 136f
Spinothalamic tract, lateral, 140–41
Spirogram, 433, 434f
  in obstructive or restrictive lung disease,
    435
Spirometer, 433, 434f
Spleen, red blood cell destruction in, 351
Spontaneous depolarization, 249, 252
Sports skills, control of, and practice, 144
Spot desmosomes, 55, 55f
SRS-A. See Slow-reactive substance of
    anaphylaxis
St. Martin, Alexis, 561
Stable balance, 512
Stagnant hypoxia. See Circulatory hypoxia
Stapes, 179, 181
Starch, 543, A8
Starling's law of the capillaries. See Bulk
    flow
Starling's law of the heart. See Frank-
    Starling law of the heart
Starvation (prolonged fasting)
  brain nourishment during, 671
  role in
    of cortisol and growth hormone, 681
State of equilibrium, 57, 57f
States of consciousness, 137
Steady state, 11, 57, 57f
  of a resting cell, 73
Steatorrhea, 569
Stellate cells, 118
Stem cells
  leukocyte origin in, 356
  megakaryocytes derived from, 358
Stenotic valve, 282
Stereocilia, 187f, 188
Sterility, 702
Sterilization, for birth control, 740–41
Sternum, 260
Steroid abuse, 239
Steroid anti-inflammatory drugs. See
    Glucocorticoids
Steroid hormones, 622–23
  biosynthetic pathways, 623f
  cholesterol as a precursor in synthesis of,
    293
  definition, 618
  hormones included in this class, 618
  inactivation of, 629
  mechanism of action of, 624, 625–26
  postreceptor events for, 625–26, 625f
  summary of characteristics, 621
  synthesis of, 622–23
  transport of in blood, 623–24
Steroidogenic pathways, 623f
Steroids, A10, 10f
  as hormones, 618
STH. See Growth hormone

Stimulus, 142
  effect of, on receptor potential, 152
  intensity coding of, 153, 154
  localization of, 154, 156–57
Stimulus-secretion coupling, 683
Stomach, 5, 555–67
  absorption by, 567
  anatomy of, 555f
  digestion by, 567
  functions of, 555
  gastric mucosa and gastric pits, 561f
  gastric mucosal barrier, 565–67
  hormones, major functions, and target
    cells of, 620
  hydrochloric acid secretion by, 562, 562f
  motility of, 556–60
  secretion of, 560–65
  ulcers, 566–67
  vomiting, 559
Storage
  for oxygen, in Hb complexes, 445–46f
  See also Glycogen
Streptococci, 373, 397, 502
Stress, 665–66, 665f
  and cortisol control, 659
  CRH release in response to, 637
  effect of
    on cortisol secretion, 658
    on growth hormone secretion, 642–43
    on the immune system, 400
    on insulin secretion, 666, 675
    on the sympathetic nervous system,
      200–201
  emotional, and fainting, 339–40
  and food intake, 600
  fuel homeostasis in, 665
  and general adaptation syndrome, 665
  and growth, 639
  major hormonal responses to, 666
  and pain perception, 158–59
  responses to, summary, 666
  and ulcer formation, 566
Stressor, 658
Stress relaxation response, 253
Stress test, 278
Stretch receptors in lungs, 453
Stretch receptors in muscle, 243–46
Stretch reflex, 145, 244
Striated muscles. See Cardiac muscle;
    Skeletal muscle
Stroke, 291
  and android obesity, 602
Stroke volume
  control of, 285–86
  definition of, 280
  as determinant of cardiac output, 283
  effect on
    of arterial blood pressure, 288
    of sympathetic stimulation, 286–88
    of venous return, 285
  extrinsic control of, 286–88
  factors controlling, summary, 285
  intrinsic control of, 285–86

  sympathetic stimulation of, 287, f
Stroking in muscle contraction, 219–20
Strong acid, 524
  hydrochloric acid, 524
  versus weak acid, 524f
Strychnine poisoning, 99
ST segment, 275
Subarachnoid space, 113
Subcortical regions of the brain, 125–35
Subcutaneous implantation of progester-
    one, 741
Subcutaneous tissue. See Adipose tissue
Subjective feelings, and brain function, 110
Sublingual salivary glands, 551
Submandibular salivary glands, 551
Submucosa, 545
Submucous (Meissner's) plexus, 545, 548
  See also Intrinsic nerve plexuses
Substance abuse, hypersecretion of the en-
    docrine glands caused by, 630
Substance P, 158, 159f
Substantia nigra, destruction by MPP$^+$, 100
Subsynaptic membrane, 91
Succus entericus, 577
Suckling reflexes, 738–39, 739f
Sucrase, 544f, 577
Sucrose, 580, A8
Sudden-infant-death syndrome (SIDS), 459
Suffocation, 449
Sugar. See Glucose
Summation
  spatial, 96
  temporal, 95
  of twitches, 228
Superior vena cava. See Vena cava
Supplementary motor area, 120–21
Suppressor T cells, 392, C4
  in autoimmune disease, 396
  interactions with
    B and T cells, 395
    Granstein cells in skin, 407
  role in tolerance, 396
Supraoptic nucleus, anatomy of, 632, 632f
Surface tension
  of alveoli, 429–30, 430f, 431f
Surfactant. See Pulmonary surfactant
Surrogate mother, 697, 727
Suspensions, definition of, A6
Suspensory ligaments, 167
SV. See Stroke volume
Swallowing, 553
Swallowing center, 553
Sweat glands, 5, 406
  innervation of, 202, 605
  sympathetic nervous system effect on, 201
Sweating, 521
  effect of epinephrine and norepineph-
    rine on, 664
  and extracellular fluid hypertonicity, 515
  and sodium loss, 517
  and temperature control, 517
  and temperature regulation, 605
  and water loss, 521

Sympathetic dominance, 200
Sympathetic ganglion chain, 199, 199f
Sympathetic nerve fibers, control of the
    ciliary muscle, 167
Sympathetic nervous system, 198–204
    comparison with parasympathetic ner-
        vous system, 198
    dominance of, 200
    effect of
        on afferent arterioles of kidneys,
            476–77
        on airway resistance, 201
        on the arterial blood pressure, 201,
            336f
        on arteriolar resistance, 201, 315, 425
        on cardiac conducting system, 285
        on cardiac contractility, 201, 287
        on the digestive system, 201, 549
        on the eye, 163, 201
        on fat stores, 201
        on the female reproductive tract dur-
            ing orgasm, 713–14
        on glomerular filtration rate, 476–77
        on glycogen stores, 201
        on hair follicles, 608
        on the heart rate, 201, 283, 284–85
        on insulin secretion, 675
        on the male reproductive tract during
            ejaculation, 712
        on salivary glands, 552
        on skeletal muscle arterioles, 201
        on stroke volume of the heart, 205,
            287
        on sweat glands, 201
        on the urinary bladder, 201
        on various organs, 201
        on venous return, 330
    general effects of, 200–201
    neurotransmitters of, 200
    relationship to the adrenal medulla,
        663–64
    response to heart failure, 288
    and stress, 665
    summary of specific effects of, 201
Sympathetic tone, 200
Sympathetic trunk, 199
Sympathomimetic effect, of thyroid hor-
    mone, 653
Synapse, 91
    anatomy of, 92–93f
    comparison with neuromuscular junc-
        tion, 208
    drugs and diseases affecting, 99
    excitatory, 92
    general mechanism of action, 91, 93f
    inhibitory, 91–92
    neuromodulation of, 98
    presynaptic inhibition and facilitation of,
        98
Synapse transmission, drugs and diseases
    affecting, 99
Synaptic cleft, 91
    removal of neurotransmitters from, 94

Synaptic delay, 93
Synaptic knob, 91
Synaptic pathways, 102
Synaptic vesicles, 91
Synergism in hormone secretion, 631
Synovial fluid, 513
Synthetic glucocorticoids, 658
Systemic arterial blood pressure. See Arte-
    rial blood pressure
Systemic capillaries, gas exchange across,
    440, 440f, 442, 526
    See also Capillaries
Systemic circulation, 262, 263
    blood flow through, 259f, 260
    gas exchange across capillaries, 440
    lymphatic system, 327f
    pressures in, 309f
    resistance of, 263. See also total periph-
        eral resistance
Systole
    and cardiac emptying, 278
    and coronary circulation, 290
Systolic murmur, 282
Systolic pressure, 306
    measurement of, 308

T3. See Triiodothyronine
T4. See Tetraiodothyronine
T_m. See Transport maximum
Tachycardia, 277f
    detection on an ECG, 276
Tactile discrimination, 156
Taeniae coli, 587
Tail of a sperm, 706
Target cells/tissues (general)
    definition of, 105, 616
    endocrine dysfunction due to inade-
        quate responsiveness of, 630
    receptors of for various classes of hor-
        mones, 625
    receptors of for various classes of neuro-
        transmitters, 203
    responses of, abnormal, 630
Taste buds, 190, 191f, 550, 551
Taste pore, 190
Taste, sense of, 161, 190–92, 192f, 551
Tay-Sachs disease, 26
T-cell growth factor (interleukin 2 or IL-2),
    394, C4
T cells. See T lymphocytes (T cells)
TDF (Testicular determining factor). See
    H-Y antigen
Tears, 177
Tectorial membrane, 179, 183f, 184
Teeth, 551
Teleological approach, 2
Telophase, B11
Temperature, body
    and cold exposure, 606–8
    core, 601
    disorders of, 610
    effect of
        on arterial blood pressure, 338

on hemoglobin saturation, 446–47,
    446f
on protein shape and function, 601
on rate of chemical reactions, 601
on ventilatory response to exercise,
    458
effect on
    of biological clock, 603
    of exercise, 609
    of extremes in environmental tem-
        perature, 610
    of menstrual cycle, 603
and fever production, 608–9
and heat balance, 603
and heat exposure, 607–8
and heat transfer mechanisms, 603–5,
    604f
normal, 601–3
regulation of, 601–11
and shivering, 606–7
summary of thermoregulatory pathways,
    606f
and sweating, 605
Temperature-humidity index, 605
Temporal lobes, 118, 118f, 119
    role in auditory processing, 119
    role in memory consolidation, 131
Temporal summation, 95, 102
Temporomandibular joint, 551
Tension, muscle
    versus length of fibers, 252
    See Muscle tension
Tenting effect, 714
Terminal bronchiole, 415f
Terminal button, 206
Terminal ganglia, 199, 199f
Terminal ileum, 578
    recycling of bile salts from, 572
    vitamin B_{12} absorption from, 563, 583
Tertiary follicle. See Mature follicle
Tertiary structure of proteins, A11, 11f
Testes, 696, 703–9
    age-related changes, 705f
    control of testicular function, 707–9,
        708f
    descent of, 702
    development of in male embryos, 700
    hormone produced by, 696, 703
    H-Y antigen as determinant of develop-
        ment, 701
    spermatogenesis by, 704–7
    structure of, 703, 705f
    testosterone secretion by, 696, 703
Testicular feminization syndrome, 630
Testosterone, 239, 696–97, 703–4
    and closure of epiphyseal plate, 642,
        646, 704
    conversion to dihydrotestosterone, 702
    defect in target cells for, 630
    effects of
        before birth, 701, 702
        on erythropoiesis, 354
        on growth, 646

during puberty and adulthood, 703–4
  on secondary sexual characteristics, 704
  on sex-drive, 704
  on sex-specific tissues, 703
  on spermatogenesis, 708
  and muscle fiber, 238
  negative-feedback control of LH and FSH by, 708
  and pubertal growth spurt, 646, 703
  and sperm production, 708
  synergism with FSH, 708
  structural comparison with estradiol, 622f
"Test tube baby," 726
Tetanus (disease), 99
Tetanus (of muscle), 228
  prevention of, in cardiac muscle, 274
Tetanus toxin, 99
Tetrad, formation in meiosis, B11
Tetraiodothyronine (T₄ or thyroxine), 650, 651
  See also Thyroid hormone
Thalamus, 125f, 126
  connection with the motor cortex, 242
  relation with basal nuclei, and cerebral cortex in motor control, 126
  role of
    in auditory perception, 186
    in conscious experience, 126, 137
    in pain perception, 158
    in preliminary sensory processing, 126
    in visual information processing, 176–77
  sensory recognition by, 119
Thecal cells, 717, 719–20
Thermal conductivity, 604
Thermal energy. See Heat
Thermal gradient, 603
Thermal motion of molecules, 57
Thermal nociceptor, 157–58
Thermoneutral zone, 608
Thermoreceptors, 152
  skin, 606
Thermoregulation. See Temperature, regulation of
Thermoregulatory pathways, 606f
Thick filaments
  and length-tension relationship, 229
  and sliding-filament mechanism of contraction, 219
  structure of
    in cardiac muscle, 254
    in skeletal muscle, 216–18, 217f
    in smooth muscle, 248
Thin filaments
and length-tension relationship, 229
  and sliding-filament mechanism of contraction, 219
  structure of
    in cardiac muscle, 254
    in skeletal muscle, 216–18, 218f
    in smooth muscle, 248

Third line of defense against changes in hydrogen-ion concentration, 530
Third-order sensory neuron, 155
Thirst, 482, 521
Thirst center, 521
Thoracic cavity and walls, 417
  size compared to lung volume, 418–19
Thoracic duct, and fat absorption, 585–86
Thoracic vertebrae, 417
Thoracolumbar sympathetic division of autonomic nervous system, 199f
Thorax. See Thoracic cavity and walls
Three Mile Island accident, 629
Threshold potential, 81, 91
  permeability at, 82
Throat. See Pharynx
Thrombin, 360
  roles in hemostasis, 361f
Thrombocytes. See Platelets
Thrombocytopenia purpura, 364
Thromboembolism, 291, 291f, 364
Thromboplastin, tissue. See Tissue thromboplastin
Thromboxane A₂, 360
  in platelet plugs, 360
Thrombus, 291, 364
Thucydides, 388
Thymic education, 393
Thymic hormone. See Thymosin
Thymine (T), B1
Thymosin, 381, C4
Thymus, 370, 380–81, 417
  hormones, major functions, and target cells of hormones, 620
Thyroglobulin, 650–52
  tolerance of, 396
Thyroid gland, 649
  abnormalities of, 654–56
  anatomy of, 650f
  failure of, and goiter, 656
  and homeostasis in the endocrine system, 691
  hormones, major functions, and target cells of hormones, 619
Thyroid hormone, 623, 650–55, 652f
  abnormalities in secretion, 654
  activation of, 652
  control of secretion of, 653f
  duration of response to, 652, 653
  effect of
    on the cardiovascular system, 653
    on growth, 646, 652
    on heat production, 607, 652
    on intermediary metabolism, 652
    on the metabolic rate, 598, 652
    on the nervous system, 653
    during starvation, 681
  effect on
    of cold, 654
    of stress, 654
  functions of, 652–53
  and hyperthyroidism and hypothyroidism, 654–55

  modification of target cells by, 631
  and nonshivering thermogenesis, 607
  as a nontropic hormone, 617
  permissive action on epinephrine, 653
  receptors for, 624
  summary of characteristics, 621
  synthesis, storage and secretion of, 651–52, 651f
  transport of in blood, 652
Thyroid-stimulating hormone (TSH or thyrotropin), 634
  functions of, 653–54
  in goiter development, 656
  negative feedback in secretion of, 627
  as a tropic hormone, 617
Thyroid-stimulating immunoglobulin (TSI), 654–55
Thyrotropin. See Thyroid-stimulating hormone (TSH or thyrotropin)
Thyrotropin-releasing hormone (TRH), 635, 653–54
Thyroxine (T4), activation of, 628
  See Tetraiodothyronine; Thyroid hormone
Thyroxine-binding globulin, 652
Thyroxine-binding prealbumin, 652
Tidal volume (TV), 433
  and anatomic dead space, 437
  effect of dead-space volume on, 436f
  and the Hering-Breuer reflex, 453
Tight junctions, 55, 56f
  and absorption across the digestive-tract wall, 578
  in brain capillaries, 113
  in gastric mucosal barrier, 565–66
  in kidney tubules, 481
  maintenance of, 683
  between Sertoli cells, 707
Timbre of sound, 180f, 181
Tissue fluid. See Interstitial fluid
Tissue plasminogen activator (tPA), 364
Tissue repair, 375
Tissues, 4–5
Tissue thromboplastin, 362, 373
Tissue typing, 397
T lymphocytes (p3T; cells), 358, 370, 392–401, C4
  activated, 393
  versus B lymphocytes (B cells), 392
  cytotoxic, enhancement by interferon, 379
and delayed hypersensitivity reactions, 404
  functions of, 392–95
  influence of skin keratinocytes on, 407
  interactions with B cells and macrophages, 390f, 391, 392
  and lymphokine secretion, 393
  memory cells, 393
  origin and maturation of, 380–81, 381f
  subpopulations of, 392
  See also Cytotoxic T cells; Helper T cells; Suppressor T cells
Tm. See Tubular maximum

TMJ. *See* Temporomandibular joint
Tolerance in the immune system, 395–97
Tone
 of blood vessels. *See* Vascular tone
 of the digestive tract muscles, 542
 of muscles. *See* Muscle tone
 parasympathetic, 200
 of a sound (pitch), 178
 sympathetic, 200
Tongue, 191, 550
Tonic luteinizing hormone secretion, 720
Tonic receptors, 153, 154f
Tonsils, 370, 550–51
Total body fluid, 512
Total lung capacity (TLC), 434
Total peripheral resistance, 314, 316f
 control of in regulation of arterial blood
  pressure, 338
 effect of epinephrine on, 664
 in polycythemia, 355
 See also Arteriolar resistance
Touch discrimination. *See* Tactile
  discrimination
Toxic chemicals, aplastic anemia caused
  by, 354
Toxic substances, effect on endocrine
  organs, 630
TP interval, 275
TPR. *See* Total peripheral resistance
Trachea (windpipe), 414
Tracts in central nervous system, 139–40
Tranquilizers, adverse effects of, 575
Transcellular fluid, 512, 513
Transcortin, 657
Transcription of DNA instructions, 7
Transduction, 152
Transepithelial transport, 480–81, 480f
Transferrin, 583
Transfer RNA (tRNA), B7, B9, B9f
Transfusion reaction, 389–91
Transmural pressure gradient, 419–20, 419f
 and compliance, 429
Transplantation
 of kidneys, 503, 504
 rejection of, 397
Transplants of brain tissue, 100
Transport
 by albumins and globulins in blood,
  349
 of gases in blood, 442–50
 of hormones, 628–29
 membrane. *See* Membrane transport
 role of capillaries, 343
 selective, blood-brain, 114f
Transporter recruitment for blood glucose,
  673
Transport maximum (T_m), 63
 in active reabsorption, 484–86
Transport mechanisms, selective, 62–68
Transport vesicles, 22, 22f
Transverse colon, 587
Transverse tubule (T tubule), 220–22, 222f
Treadmill, motorized, 278, 459

TRH. *See* Thyrotropin-releasing hormone
Triacylglycerols. *See* Triglycerides
Tricarboxylic acid cycle. *See* Citric-acid
  cycle
Triceps muscle, 144–45
Tricuspid valve, 264
 See also Right atrioventricular (AV)
  valve
Triglycerides (triacylglycerols), 543, 544f,
  A9, A9f
 digestion and absorption of, 569, 572,
  583, 584f
 effect on
  of cortisol, 658
  of diabetes mellitus, 675
  of glucagon, 679
  of growth hormone, 640
  of insulin, 674
 as nutrient storage, 669
 synthesis of, 668
 See also Fat, body stores of; Fat, nutrient
Triiodothyronine (T_3), 650, 651–52
 See also Thyroid hormone
Trilaminar structure of plasma membrane,
  46, 46f, 47
Triplet code for amino acids, B7–8
tRNA. *See* Transfer RNA
Trophoblast, 728–29
Tropic hormones, 617, 635
Tropomyosin, 218, 218f, 219f, 223f, 273
 in cardiac muscle, 254
Troponin, 218, 218f, 219f, 223f, 274
 in cardiac muscle, 254
Trypsin, 544f, 568–69
 role in emphysema, 426
Trypsin inhibitor, 569
Trypsinogen, 568–69, 577
TSH. *See* Thyroid-stimulating hormone
TSI. *See* Thyroid-stimulating
  immunoglobulin
T tubules, 220–22
 role in calcium ion movement, heart,
  273
Tubal ligation, 740–41
Tubal pregnancy, 728
Tuberculosis, 373–74
 destruction of endocrine glands in, 630
Tubular component of nephrons, 466,
  467–69, 469f
 See also Tubular reabsorption; Tubular
  secretion
Tubular maximum (T_m), 484–86
Tubular reabsorption, 470–71, 470f,
  479–88
 active, 480
 of amino acids, 484
 of chloride, 486
 of glucose, 484–85
 passive, 480
 of phosphate, 486
 of potassium, 488
 and salt balance, 515
 of sodium, 481–84, 481f

 summary of, 490
 and transepithelial transport, 480–81,
  480f
 and tubular maximum (T_m), 484
 of urea, 487–88
 of water, 486–87, 486f
Tubular secretion, 470f, 471, 488–90
 of foreign compounds, 490
 of hydrogen ion, 488
 of organic ions, 489–90
 of potassium, 488–89, 488f
 summary of, 490
Tubulin, 35–36, 36f, 37
Tubuloglomerular feedback mechanism of
  autoregulation, 476, 477f
Tumors, 398
 hypersecretion of the endocrine glands
  caused by, 630
Turbulent flow
 in arteries during blood pressure measurement, 308
 and heart murmurs, 281–82, 282f
T wave, 275
Twitch, 226
 summation of, 228, 228f
Tympanic membrane (eardrum), 178f, 179,
  181–82, 183f
Type I alveolar cells, 416, 416f
Type I diabetes mellitus, 675, 676
Type I fibers, 237
Type II alveolar cells, 416, 416f
Type II diabetes mellitus, 675, 676
Type IIa fibers, 237
Type IIb fibers, 237
Tyrosine, 618
 in thyroglobulin, 651

Ulcers, 565–66, 566f
Ultrafiltration, 323, 324
Ultrasound for observation of a fetus, 726
Umbilical artery, 731
Umbilical cord, 731, 736
Umbilical vein, 731
Umbilicus (navel), 736
Unconditioned salivary reflex, 552
Unicellular organisms, 2
Universal blood donors, 389
Universal blood recipients, 389–90
Unsaturated fatty acids, A9
 effect on cholesterol, 295
Unstriated (smooth) muscle, 214
Upper esophageal sphincter. *See* Pharyngo-
  esophageal sphincter
Upright posture and cardiovascular system,
  339
Uracil in RNA, 7
Urea, 487
 accumulation in the medullary intersti-
  tial fluid, 495–96
 recycling of, in the renal medulla,
  500–506, 506f
 renal handling of, 487–88, 487f

Uremia, 487–88
    as cause of ECF hypertonicity, 518
Uremic acidosis, 536
Uremic toxicity, 502
Ureter, 466, 503
Urethra, 466
    as part of male reproductive system, 697
    as part of urinary system, 466
Urethral fold, 702
Urinary bladder, 466, 505
    sympathetic nervous system effect on, 201
Urinary buffer
    ammonia as, 533–34
    phosphate system as, 529, 533–34
Urinary excretion of hormones, 628
Urinary incontinence, 505–6
Urinary system, 8f, 464–506
    components of, 7, 8f, 467f
    defense mechanisms of, 407–8
    functions of, 466
    See also Kidneys; Micturition
Urination. See Micturition
Urine, 464
    concentration of, 499
Urine excretion
    magnitude of, 490–506
    and micturition (urination) reflex, 505
    relation to basic renal processes, 471
Use-dependent competition for cortical space, 122
Uterine cycle. See Menstrual cycle
Uterine tubes. See Oviducts
Uterus, 697
    changes in
        during implantation and pregnancy, 739–40
        during the menstrual cycle, 722–23
    contraction of, 735–36
    role in parturition, 734–37
    sperm transport in, 726
Utricle, 179, 187, 187f, 188–89, 189f
Uvula, 550, 553

Vaccination, 386, 388
    against hCG for birth control, 741
Vagina, 697
Vagus nerve, 137
    effect on the digestive system, 549
    effect on the heart, 283, 285
    role in stimulation of bile secretion, 574
    See also Parasympathetic nervous system
Valves
    heart, 264–66, 266f
        opening and closing during cardiac cycle, 280
        relation to heart sounds, 281
    heart, defective, 282
    lymphatic, 325, 326f
    mechanism of action. 264, 264f
    venous, 331–32

Varicose veins, 332
Varicosities in autonomic nerve fibers, 200
Vasa recta
    anatomy of, 470
    role in countercurrent exchange, 492, 496
Vascular component of nephron, 466, 468f
    See also Glomerular filtration
Vascular resistance. See Arteriolar resis-
    tance; Blood flow; Blood pres-
    sure; Resistance, to flow in blood
    vessels
Vascular spasm
    in coronary artery disease, 290
    in hemostasis, 358
Vascular tone, 309
Vasculature. See Blood vessels
Vas deferens. See Ductus (vas) deferens
Vasectomy, 709–10, 740
Vasoconstriction, 309, 309f
Vasodilation, 309, 309f
    to increase oxygen delivery to cardiac cells, 290
    in inflammation, 371
    for treatment of angina pectoris, 292
Vasogenic shock, 340–41
Vasomotor control of body temperature, 607–8
Vasopressin (antidiuretic hormone or ADH), 75, 633
    abnormalities in secretion of, 518
    control of secretion of, 521–23, 522f
    and diabetes insipidus, 518
    effect of
        on arteriolar radius and arterial blood pressure, 315–16, 521–22, 521f
        on water reabsorption in distal and collecting tubules, 497–99, 497f, 521–22
    effect on
        of alcohol, 523
        of stress, 523, 666
    and hypothalamus-posterior pituitary re-
        lationship, 632–33
    inappropriate secretion of, 520, 523
    mechanism of action on kidney tubules, 497–99f
    and medullary vertical osmotic gradient, 497–99, 498f
    role in urine concentration, 492
    secretion of
        and heat, 633
        and stress, 666
        and thirst, 521–24, 522f
Vaults, 34, 43
Veins, 262, 299, 302, 305, 328–32
    anatomy of, 328
    capacity of, 328
    effect on of gravity, 330, 331f
    factors affecting venous return, 328–32, 329f
    functions of, 328
    pressure in, 329

valves of, 331, 333f
    varicose, 332
    See also Blood vessels
Velocity of blood flow, 319
Vena cava, 262, 305
Venom of poisonous snakes, 388
Venous capacity, 328–29
Venous pressure
    and capillary pressure, 331, 331f
    and edema, 328
    effect of gravity on, 330, 331f
Venous return, 329, 329f
Venous valves, function of, 333f
Ventilation, 413
    alveolar
        effect of breathing patterns on, 437
    changes in lung volume during, 435f
    control of, 450–60
        by apneustic center, 450, 452–53
        by arterial hydrogen-ion concentra-
            tion, 456, 457f
        by arterial P_{O_2}, 453–54, 454f
        by brain ECF hydrogen-ion concentra-
            tion, 455–56
        by arterial P_{CO_2}, 454–55, 455f
        by cerebral cortex, 458
        by chemical factors, summary of, 453
        by epinephrine, 458
        during exercise, 456–58
        by Hering-Breur reflex, 453
        by joint and muscle receptors, 458
        by medullary respiratory center, 450–51, 453
        by pneumotaxic center, 450, 452–53
        by pulmonary stretch receptors, 453
        by protective reflexes, 458
        by temperature, 458
        voluntary, 458–59
    effect on
        of gravity, 429
        of dead space, 436f, 437
        of diabetes mellitus, 675
        of high altitude, 451
    matching of with blood flow to lungs, 427–29, 428f
    measurement of. See Spirometry
    mechanics of. See Respiratory mechanics
    pulmonary, 435–37
    See also Hyperventilation;
        Hypoventilation
Ventral (anterior) horn, spinal cord, 141
Ventral respiratory group (VRG), 452
Ventral root, 141
    spinal cord, 141
Ventricles
    brain, 112f
    heart, anatomy of, 262
    heart, emptying of, 280–81
    heart, excitation of, 268–74
    heart, filling of, 281, 281f
Ventricular activity, atrial contraction and, 271

Ventricular diastole, 280, 305–6
Ventricular excitation, 272
Ventricular fibrillation, detection on an ECG, 277, 277f
Ventricular systole, 289, 305–6
Venules, 302
Verapamil, 274
Vertebral column, 112
Vertigo, 190
Very-low-density lipoproteins (VLDL), 293
Vesicular transport, 67–68
    across capillaries, 319
    in varying bladder volume, 503
    See also Endocytosis; Exocytosis
Vestibular apparatus, 178, 187–90, 187f
    processing and utilization of vestibular information, 190, 190f
    saccule, 189
    semicircular canals, 188, 188f
    utricle, 189, 189f
Vestibular membrane, 182f, 183f, 184
Vestibular nerve, 188
Vestibular nuclei, input and output of, 190f
Vestibulocerebellum, 135
Vestibulocochlear nerve, 188
Vibrio cholerae, 587
Villi, 578–79
Villikinin, 585
Viral hepatitis, transmission by blood transfusion, 353
Virulence, 369
Viruses
    as a cause of tumors, 399
    description of, 369
    invasion of host cells by, 395
    prevention of reproduction by interferon, 378
    summary of defenses against, 395
    viral hepatitis, 575
Visceral afferent information, 154, 200
Visceral organs, innervation by autonomic nervous system, 200–202
Visceral pleura, 417f
Visceral smooth muscle, 248
Viscosity of blood, 303–4, 349
    at high altitudes, 451
    in polycythemia, 355
Visible light, 163
Vision, 161
    binocular, 176
    color, 170, 174
    daytime, 173
    nighttime, 173
    See also Eye
Visual cortex, 176–77
    color vision center in, 174
Visual field, 174, 177f
Visual pathway, 174–77, 177f
Vital capacity (VC), 434
Vitamin A
    derivatives of, in photopigments, 170
    and night blindness, 174
Vitamin B, derivative in acetyl CoA, 28

Vitamin B complex
    Vitamin B_{12}, 563
    Vitamin B_{12}, and anemia, 354
Vitamin D, 585, 687
    activation of, in the kidneys, 466, 686
    deficiency of, 689–90
    functions of, 687
    interactions with parathyroid hormone, 687
Vitamin K
    and clotting, 364
    production by intestinal bacteria, 590
Vitamins, absorption of by small intestine, 583
Vitreous humor, 162
VLDL. See Very low density lipoproteins
VO_2 max. (See maximal O_2 consumption)
Vocal cords, 414–15, 553
Voice box. See Larynx
Voltage-gated channels, 81–82, 101–2
    for calcium, 91, 206
    for potassium, 82
    for sodium, 81f, 96
Voluntary branch of the nervous system, 200
    See also Somatic nervous system
Voluntary movement, 146–47
Voluntary muscle, 214
Vomiting, 559–60
    alkalosis associated with, 537
    and fluid imbalance, 513
Vomiting center, 560
VRG. See Ventral respiratory group
Vulva, 697

Walking, 136
Wastes, excretion of, 466, 480, 488
Water
    absorption of
        in the large intestine, 589–90
        in the small intestine, 580
    balance of, 520–24
    concentration of, defined, 59–62
    content of in various tissues, 512
    control of, 520–24
    covalent bonds of, A4
    diffusion of (osmosis), 59
    distribution of in body fluid compartments, 512
    driving force for absorption across epithelia, 486–87
    formation by oxidative phosphorylation, 30–31
    intoxication, 520
    production in the electron-transport chain, 30–31, 33f
    reabsorption in kidneys, 486–87, 486f, 497–98
    renal handling of
        by the distal tubule and collecting duct, 499
        by the loop of Henle, 493

    by the proximal tubule, 486
    relation to solute handling, 499–500
    retention of, in congestive heart failure, 288–89
    sources of input and output, 521
    and thirst, 523
    urinary excretion of, 497–500
    and vasopressin, 497–99, 521–22
Water balance, 465–66, 510, 521
Water diuresis, 499–500
Water intoxication, 520
Wavelength of light, 163, 163f
    selective response to, in the eye, 174
Weak acid, 524
    aspirin, 567
    versus strong acid, 524f
Weight. See Body weight
Weight training, 97
Wernicke's area, 122–23
White blood cell count, 356
White blood cells, 67–68, 355–58
White fibers, 238
White matter
    in the brain, 118
    in the spinal cord, 139–40, 143f
Whole body (total body), 5
Withdrawal of the penis as birth control method, 740
Withdrawal reflex, 144–45, 145f
    and the crossed extensor reflex, 146f
Wolffian ducts, 702
Work, 233
Working memory, 129–30
Work schedule, stress from rotating, 629

X chromosomes, 699, 706
    defect for muscular dystrophy on, 241
Xerostomia, 552

Y chromosomes, 699, 706
    sex-determining region of (SRY), 700

Zeitgebers in cyclic functions, 628
Z line, 216, 218
Zona drilling, and fertility, 727
Zona fasciculata, 657, 659
Zona glomerulosa, 657
    and angiotensin, 659
Zona pellucida, 717, 727–28
Zona reticularis, 657, 659
Zygote, 728
    formation of, B12
Zygote intra-Fallopian transfer (ZIFT), 726
Zymogen granules
    in the pancreas, 569
    in the stomach, 562–63

CREDITS

Physiology (Philadelphia: W.B. Saunders Company, 1987), Figure 1–13, p. 28. 18–21 Lester V. Bergman and Associates, Inc. 19–1b Elizabeth R. Walker, Associate Professor, and Dennis O. Overman, Associate Professor, Department of Anatomy, School of Medicine, West Virginia University. 19–5 Lester V. Bergman and Associates, Inc. 19–6 Lester V. Bergman and Associates, Inc. 19–9 Adapted with permission from George A. Hedge, Howard D. Colby, and Robert L. Goodman, Clinical Endocrine Physiology (Philadelphia: W.B. Saunders Company, 1987), Figure 4–4, p. 80. 19–20 Modified and redrawn with permission from Alexander P. Spence and Elliott B. Mason, Human Anatomy and Physiology, 3d ed. (Menlo Park, CA: The Benjamin/Cummings Publishing Company, 1987), Figure 6–3, p. 123. 20–6b Elizabeth R. Walker, Associate Professor, and Dennis O. Overman, Associate Professor, Department of Anatomy, School of Medicine, West Virginia University. 20–6c From Tissues and Organs: A Text-Altas of Scanning Electron Microscopy by Richard G. Kessel and Randy H. Kardon. Copyright ©1979 by W.H. Freeman and Company. Reprinted with permission. 20–8a David M. Phillips/ Visuals Unlimited. 20–13b Dr. P. Bagavandoss, Developmental and Reproductive Biology, University of Michigan Medical School. 20–21 David Scharf. 20–22b Photo Lennart Nilsson, A Child is Born, Dell Publishing Company. Table 20–7 All data (with the exception of chemical and implanted contraceptives) for women 20–29 years old taken from Carr, B.R., and Griffin, J.E., in Williams' Textbook of Endocrinology, 7th ed. J.D. Wilson and D.W. Foster, editors (Philadelphia: W.B. Saunders Company, 1985). Figure A–1 Reprinted by permission from page 61 of Chemistry: An Introduction by Michael R. Slabaugh and Spencer L. Seager. Copyright © 1988 by West Publishing Company. All rights reserved. Figure A–2 Reprinted by permission from page 113 of Chemistry: An Introduction by Michael R. Slabaugh and Spencer L. Seager. Copyright © 1988 by West Publishing Company. All rights reserved. Figure A–3 Adapted by permission from page 520 of Chemistry For Today: General, Organic, and Biochemistry by Spencer L. Seager and Michael R. Slabaugh. Copyright © 1987 by West Publishing Company. All rights reserved. Figure A–4 Adapted by permission from page 532 of Chemistry For Today: General, Organic, and Biochemistry by Spencer L. Seager and Michael R. Slabaugh. Copyright © 1987 by West Publishing Company. All rights reserved. Figure A–5 Adapted by permission from pages 547 and 548 of Chemistry For Today: General, Organic, and Biochemistry by Spencer L. Seager and Michael R. Slabaugh. Copyright © 1987 by West Publishing Company. All rights reserved. Figure A–6 Adapted by permission from pages 553 and 555 of Chemistry For Today: General, Organic, and Biochemistry by Spencer L. Seager and Michael R. Slabaugh. Copyright © 1987 by West Publishing Company. All rights reserved. Figure A–8 Adapted by permission from pages 560 and 563 of Chemistry For Today: General, Organic, and Biochemistry by Spencer L. Seager and Michael R. Slabaugh. Copyright © 1987 by West Publishing Company. All rights reserved. Figure A–9 Adapted by permission from page 574 of Chemistry For Today: General, Organic, and Biochemistry by Spencer L. Seager and Michael R. Slabaugh. Copyright © 1987 by West Publishing Company. All rights reserved. Figure A–10 Adapted by permission from page 589 of Chemistry For Today: General, Organic, and Biochemistry by Spencer L. Seager and Michael R. Slabaugh. Copyright © 1987 by West Publishing Company. All rights reserved. Figure A–11 Adapted by permission from page 591 of Chemistry For Today: General, Organic, and Biochemistry by Spencer L. Seager and Michael R. Slabaugh. Copyright © 1987 by West Publishing Company. All rights reserved. Figure A–12 Adapted by permission from page 595 of Chemistry For Today: General, Organic, and Biochemistry by Spencer L. Seager and Michael R. Slabaugh. Copyright © 1987 by West Publishing Company. All rights reserved.

Figure A–13 Adapted by permission from page 687 of Chemistry For Today: General, Organic, and Biochemistry by Spencer L. Seager and Michael R. Slabaugh. Copyright © 1987 by West Publishing Company. All rights reserved. B–11 From Christine J. Harrison et al.: "Cytogenics," Cell Genetics 35: 21–27 (1983), Figure 3B. Reprinted with permission from Dr. Christine J. Harrison and S. Karger AG, Basel.

ILLUSTRATION CREDITS

Wayne Clark
4–9a; 4–11a,b; 4–18 top; 11–3; 13–10; 13–11

Original layout

4–16a; 4–19 top; 4–20; 10–18a; 10–19; 12–2; 13–3b; 14–2; 14–13; 20–13a,c; 20–14; 20–25

Darwen and Vally Hennings
5–11; 5–14; 5–16; 5–17; 5–27; 5–28; 6–11; 6–35; 6–40a–c; 9–1; 9–3a,c; 10–2; 10–26a; 13–34; 13–36; 16–1; 16–7; 16–17; 16–22; 18–11

Original layout

Table 5–3

Carlyn Iverson
5–22; 5–24; 5–26b,c; 9–6b; 16–8; 16–10

Reflective conversion

Table 5–3; 5–7; 5–8; 5–9; 5–10; 5–20; 10–18a; 10–25; 16–12; 16–16; 20–8b,c

Sandra McMahon
2–26; 6–9; 6–44; 8–2; 8–10; 8–24; 9–5; 12–25; 16–2; 16–18; 16–24; 16–31; 18–19; 20–1; 20–2; 20–31a

Elizabeth Morales-Denney
2–1; 2–2a; 2–5; 2–6a; 2–10a; 3–3

Precision Graphics
8–21; 8–28; 10–17; 12–7; 12–17 bottom; 14–6; 14–10; 14–28; 16–9; 19–21a

Electronic conversion

2–17; 2–18; 2–22; 3–6; 4–16a,c; 6–36; 8–4; 8–5; 8–6; 8–7; 8–16; 10–18b; 10–19; 12–17 top; 14–2; 14–4; Table 14–1; Table 16–1; 16–6; 19–21b; B–1

Publication Services
4–1; 8–1; 8–11b; 9–16; 10–3; 11–8; 12–4; 13–17

Electronic conversion

1–3; 1–5; 2–12; 2–13; 2–14; 2–15; 3–4; 3–5; 3–20; 3–25; 3–26; 3–27; p. 77; 4–2; 4–3; 4–4; 4–5; 4–7; 4–8; 4–15; 4–17; 4–19; 5–1; 5–2; 5–5; 5–6; 5–12; 5–15; 5–18; 5–19; 5–21; 6–3; 6–4; 6–7; 6–8; 6–12; 6–25; 6–26; 6–30; 6–31; 6–34; 6–38; 6–43; 6–45; 6–47; 7–1; 7–2; 8–11a; 8–12; 8–13; 8–15; 8–17; 8–20; 8–22; 8–23; 8–29; 8–31; 9–4; 9–10; 9–12; 9–15; 9–17; 9–19; 9–20; 9–21; 9–22; 9–23; 9–24; 9–25; 9–26; 9–27; 9–28; 9–29; 9–31; 10–1; 10–4; 10–7; 10–8b; 10–10b–d; 10–13; 10–14; 10–16; 10–20; 10–23; 10–24; 10–29; 10–35; 10–37; 10–38; 10–40; 10–41; 11–4; 11–9; 11–11; 11–12; 11–13; 11–14; 12–3; 12–5; 12–8; 12–10; 12–11; 12–13; 12–14; 12–15; 12–19; 12–21; 12–26; 13–1; 13–6; 13–7; 13–9; 13–12; 13–13; 13–14; 13–15; 13–21; 13–24; 13–26; 13–28; 13–29; 13–31; 13–32; 13–33; 13–35; 13–37; 13–38; 13–39; 14–11; 14–12; 14–13; 14–17; 14–18; 14–20; 14–22; 14–23; 14–29; 14–30; 15–1; 15–3; 15–4; 15–8; 15–9; 15–10; 15–11; 16–3; 16–5; 16–11; 16–13; 16–14; 16–15; 16–19; 16–23; 16–26; 16–27; 16–28; 16–29; 16–30; 17–1; 17–3; 17–5; 17–7; 18–5; 18–6; 18–7; 18–8; 18–10; 18–13; 18–14; 18–18; 18–20; 19–2; 19–3; 19–4; 19–8; 19–9; 19–10; 19–11; 19–12; 19–13; 19–14; 19–15; 19–16; 19–17; 19–18; 19–19; 19–22; 19–23; 19–24;

19–25; 20–3; 20–4; 20–9; 20–11; 20–15; 20–16; 20–17; 20–18; 20–19; 20–27; 20–29; 20–32

Rolin Graphics

4–18 bottom; Concepts, Challenges, and Controversies box art p. 262; 9–30; 10–9; 13–23; 13–25; 19–20

Original layout

3–20; p. 77; 4–3; 4–4; 4–5; 4–8; 4–15; 4–17; 4–19 bottom; 6–4; 6–12; 6–25; 6–33a; 6–34; 6–45; 8–13; 8–15; 8–16; 8–17; 8–20; 8–26; 8–31; 9–10; 9–12; 9–15; 9–17; 9–19; 9–20 top; 9–21; 9–23a; 9–25; 9–27; 9–29; 10–7; 10–8b; 10–13; 10–24; 10–37; 12–13; 13–13; 13–21; 13–29; 13–31; 13–33; 14–18; 14–30; 15–2; 15–4; 17–7; 18–10; 18–18; 19–9; 20–27

Electronic conversion

2–7; 2–8; 3–13; 3–18; 4–10; 4–12; 4–13; 4–14; 4–20; 5–25; 5–26a; 5–29; 6–5; 6–10; 6–19a,c,d; 6–20; 6–21; 6–24; 6–28; 6–37; 6–41; 6–42; 7–4; 8–8; 8–9; 8–14; 8–25; 8–26; 8–32; 8–33; 9–6a; 9–9; 9–11; 9–14; 9–34; 9–35; 9–36; 10–21; 13–3b; 20–20; 20–22a; 20–23

Reflective conversion

9–7; 12–2; 20–30

John and Judy Waller

1–1; 1–2; 2–3; 2–4; 2–9a,b; 2–11; 2–20a; 2–21; 2–23; 2–24; 3–2; 3–7; 3–8; 3–9; 3–10; 3–11; 3–12; 3–14; 3–15; 3–16; 3–17; 3–19; 3–21; 3–22; 3–23; 3–24; 3–28; 4–6; 4–11d; 5–13; 5–23; 6–6; 6–14; 6–15a; 6–16; 6–17; 6–18; 6–29; 6–33b,c; 8–30; 9–8 art; 9–13; 10–6; 10–11; 10–12; 10–22; 10–26b; 10–31b; 10–32; 10–33; 10–39; 11–1; 11–5; 12–6; 12–12; 12–16; 12–23; 12–24; 13–4; 13–5; 13–8; 13–16; 13–18; 13–19; 13–30; 14–5; 14–8; 14–9; 14–14; 14–15; 14–16; 14–19; 14–21; 14–24; 14–25; 14–26; 14–27; 15–5; 15–6; 15–7; 16–20; 16–21; 17–6; 18–1; 18–3; 18–4; 18–9; 18–12; 18–15; 18–16; 18–17; 20–5 (final rendering); 20–6a,d (final rendering); 20–7; 20–8b,c (original final rendering); 20–10; 20–12; 20–13a,c (final rendering); 20–14 (final rendering); 20–20 (original final rendering); 20–22a (original final rendering); 20–23 (original final rendering); 20–24; 20–25 (final rendering); 20–28; 20–31b; B–2; B–3; B–4; B–5; B–6; B–7; B–8; B–9; B–10; B–12

Original layout

1–3; 1–5; 2–7; 2–8; 2–12; 2–13; 2–14; 2–15; 2–17; 2–18; 2–22; 3–4; 3–5; 3–6; 3–13; 3–18; 3–25; 3–26; 3–27; 4–2; 4–7; 4–10; 4–12; 4–13; 4–14; 4–16c; 5–1; 5–2; 5–5; 5–12; 5–15; 5–18; 5–19; 5–21; 6–3; 6–7; 6–8; 6–26; 6–28; 6–38; 6–43; 7–1; 7–2; 7–4; 8–4; 8–5; 8–6; 8–7; 8–8; 8–9; 8–11a; 8–12; 8–14; 8–22; 8–23; 8–25; 8–29; 8–32; 8–33; 9–3b; 9–4; 9–14; Table 9–2; 9–22; 9–23b; 9–24; 9–26; 9–28; 9–31; 9–35; 10–1; 10–4; 10–10b-d; 10–14; 10–16; 10–18b; 10–20; 10–21; 10–23; 10–25b; 10–29; 10–35; 10–38; 10–40; 10–41; 11–4; 11–9; 11–11; 11–12; 11–13; 11–14; 12–3; 12–5; 12–8; 12–10; 12–11; 12–14; 12–15; 12–17 top; 12–19; 12–21; 12–26; 13–1; 13–6; 13–7; 13–9; 13–12; 13–14; 13–15; 13–24; 13–26; 13–28; 13–32; 13–35; 13–37; 13–38; 13–39; 14–4; Table 14–1; 14–11; 14–12; 14–17; 14–20; 14–22; 14–23; 14–29; 15–1; 15–3; 15–8; 15–9; 15–10; 15–11; Table 16–1; 16–3; 16–5; 16–11; 16–12; 16–13; 16–14; 16–15; 16–19; 16–26; 16–27; 16–28; 16–29; 16–30; 17–1; 17–3; 17–5; 18–5; 18–6; 18–7; 18–8; 18–13; 18–14; 18–20; 19–2; 19–3; 19–4; 19–8; 19–10; 19–11; 19–12; 19–13; 19–14; 19–15; 19–16; 19–17; 19–18; 19–19; 19–21b; 19–22; 19–23; 19–24; 19–25; 20–3; 20–4; 20–9; 20–11; 20–15; 20–16; 20–17; 20–18; 20–19; 20–29; 20–32; B–1

Cyndie C.H.-Wooley

Front inside cover endsheet model; Chapter 1 model; 1–4 model; Chapter 2 model; Chapter 3 model; Chapter 4 model; Chapter 5 model; 5–4; Chapter 6 model; 6–32; 6–46; Chapter 7 model; Chapter 8 model; 8–18; 8–19; Chapter 9 model; 9–2; 9–18; Chapter 10 model; 10–8a,c; 10–30; 10–31a; 10–34; 10–36; Chapter 11 model; Chapter 12 model; 12–1; Chapter 13 model; 13–2; 13–20; Chapter 14 model; 14–1; Chapter 15 model; Chapter 16 model; Table 16–2; 16–4; Chapter 17 model; 17–4; Chapter 18 model; 18–2; Chapter 19 model; 19–1a; 19–7; Chapter 20 model; 20–5; 20–26

Original layout

6–41; 6–42; 6–47; 8–26; 16–6; 20–5; 20–6a,d; 20–8b,c; 20–20; 20–22a; 20–23; 20–30